2010 IEEE International Reliability Physics Symposium

(IRPS 2010)

Anaheim, California, USA
2-6 May 2010

Pages 1-598

IEEE Catalog Number:	CFP10RPS-PRT
ISBN:	978-1-4244-5430-3

Copyright © 2010 by the Institute of Electrical and Electronic Engineers, Inc
All Rights Reserved

Copyright and Reprint Permissions: Abstracting is permitted with credit to the source. Libraries are permitted to photocopy beyond the limit of U.S. copyright law for private use of patrons those articles in this volume that carry a code at the bottom of the first page, provided the per-copy fee indicated in the code is paid through Copyright Clearance Center, 222 Rosewood Drive, Danvers, MA 01923.

For other copying, reprint or republication permission, write to IEEE Copyrights Manager, IEEE Service Center, 445 Hoes Lane, Piscataway, NJ 08854. All rights reserved.

***This publication is a representation of what appears in the IEEE Digital Libraries. Some format issues inherent in the e-media version may also appear in this print version.**

IEEE Catalog Number:	CFP10RPS-PRT
ISBN 13:	978-1-4244-5430-0
Library of Congress No.:	1541-7026
ISSN:	82-640313

Additional Copies of This Publication Are Available From:

Curran Associates, Inc
57 Morehouse Lane
Red Hook, NY 12571 USA
Phone: (845) 758-0400
Fax: (845) 758-2633
E-mail: curran@proceedings.com
Web: www.proceedings.com

TABLE OF CONTENTS

ESREF BEST PAPER

ESREF A Review on the Reliability of GaN-Based Laser Diodes...1
  Nicola Trivellin, Matteo Meneghini, Enrico Zanoni, Kenji Orita, Masaaki Yuri, Gaudenzio Meneghesso

SESSION 2A: TRANSISTORS: BTI, HOT CARRIER

2A.1 The Statistical Analysis of Individual Defects Constituting NBTI and Its Implications for Modeling DC- and AC-Stress 7
  Hans Reisinger, Tibor Grasser, Wolfgang Gustin, Christian Schlünder

2A.2 The Time Dependent Defect Spectroscopy (TDDS) for the Characterization of the Bias Temperature Instability 16
  T. Grasser, H. Reisinger, P.-J. Wagner, F. Schanovsky, W. Goes, B. Kaczer

2A.3 Origin of NBTI Variability in Deeply Scaled pFETs 26
  B. Kaczer, T. Grasser, J. Roussel, J. Franco, R. Degraeve, L.-A. Ragnarsson, E. Simoen, G. Groeseneken, H. Reisinger

2A.4 Two Independent Components Modeling for Negative Bias Temperature Instability 33
  Vincent Huard

2A.5 Recovery-Free Electron Spin Resonance Observations of NBTI Degradation....................................43
  J. T. Ryan, P. M. Lenahan, T. Grasser, H. Enichlmair

SESSION 2B: TRANSISTORS BTI, HOT CARRIERS

2B.1 PBTI Relaxation Dynamics After AC Vs. DC Stress in High-k/Metal Gate Stacks 50
  K. Zhao, J. H. Stathis, A. Kerber, E. Cartier

2B.2 Off State Incorporation into the 3 Energy Mode Device Lifetime Modeling for Advanced 40nm CMOS Node 55
  A. Bravaix, C. Guérin, D. Goguenheim, V. Huard, D. Roy, C. Besset, S. Renard, Y. Mamy Randriamihaja, E. Vincent

2B.3 Mobility Enhancement Due to Charge Trapping & Defect Generation: Physics of Self-Compensated BTI 65
  Ahmad Ehteshamul Islam, Muhammad Ashraful Alam

2B.4 Understanding Noise Measurements in MOSFETs: The Role of Traps Structural Relaxation 73
  D. Veksler, G. Bersuker, S. Rumyantsev, M. Shur, H. Park, C. Young, K. Y. Lim, W. Taylor, R. Jammy

SESSION 2C: NANOELECTRONICS

2C.1 Reliability Study of Bilayer Graphene - Material for Future Transistor and Interconnect 80
  Tianhua Yu, Eun-Kyu Lee, Benjamin Briggs, Bhaskar Nagabhirava, Bin Yu

2C.2 Failure Analysis of Resistive Switching Devices 84
  An Chen

2C.3 Charging and Discharging Characteristics of Metal Nanocrystals in Degraded Dielectric Stacks....89
  Z. Z. Lwin, K. L. Pey, Y. N. Chen, P. K. Singh, S. Mahapatra

2C.4 Characterization of Gate-All-Around Si-NWFET, Including R_{sd}, Cylindrical Coordinate Based 1/f Noise and Hot Carrier Effects...94
  Rock-Hyun Baek, Hyun-Sik Choi, Hyun Chul Sagong, Sang-Hyun Lee, Gil-Bok Choi, Seung Hyun Song, Chan-Hoon Park, Jeong-Soo Lee, Yoon-Ha Jeong, Chang-Ki Baek, Dae Mann Kim, Yun Young Yeoh, Kyoung Hwan Yeo, Dong-Won Kim, Kinam Kim

2C.5 Thermal Disturbance and Its Impact on Reliability of Phase-Change Memory Studied by the Micro-Thermal Stage 99
  SangBum Kim, Byoungil Lee, Mehdi Asheghi, G. A. M. Hurkx, John Reifenberg, Kenneth Goodson, H.-S. Philip Wong

SESSION 2D: RELIABILITY IN MANUFACTURING

2D.1 Mature Processability and Manufacturability by Characterizing V_T and V_{MIN} Behaviors Induced by NBTI and AHTOL Test 104
Jongwoo Park, Sungmok Ha, Sunme Lim, Jae-Yoon Yoo, Junkyun Park, Kidan Bae, Gunrae Kim, Min Kim, Yongshik Kim

2D.2 Validating Foundry Technologies for Extended Mission Profiles 111
K. van Dijk, P. A. J. Volf, C. Detcheverry, A. Yau, P. Ngan, Z. Liang, F. G. Kuper

2D.3 Non-Destructive Current-Ramp Dielectric Breakdown (IRDB) for Fast BEOL Reliability Monitoring 117
Kok-Yong Yiang, Rick Francis, Amit Marathe, Oliver Aubel

2D.4 Modeling of Cu IMD-TDDB Caused by Extrinsic Defects 120
T. Ouchi, K. Makabe, M. Ogasawara, E. Murakami, N. Yoshioka

2D.5 Product Failures: Power-Law or Exponential Voltage Dependence? 125
Amr Haggag, Keith Forbes, Gary Anderson, Dave Burnett, Peter Abramowitz, Mohamed Moosa

SESSION 2E: COMPOUND SEMICONDUCTORS

2E.1 Reliability Status of GaN Transistors and MMICs in Europe 129
M. Dammann, M. Cäsar, H. Konstanzer, P. Waltereit, R. Quay, W. Bronner, R. Kiefer, S. Müller, M. Mikulla, P. J. van der Wel, T. Rödle, F. Bourgeois, K. Riepe

2E.2 Effect of Trapping on the Critical Voltage for Degradation in GaN High Electron Mobility Transistors 134
Sefa Demirtas, Jesús A. del Alamo

2E.3 Reliability Assessment in Different HTO Test Conditions of AlGaN/GaN HEMTs 139
N. Malbert, N. Labat, A. Curutchet, C. Sury, V. Hoel, J.-C. de Jaeger, N. Defrance, Y. Douvry, C. Dua, M. Oualli, M. Piazza, C. Bru-Chevallier, J.-M. Bluet, W. Chikhaoui

2E.4 High Temperature On- and Off-State Stress of GaN-On-Si HEMTs with In-Situ Si_3N_4 Cap Layer 146
Denis Marcon, Farid Medjdoub, Domenica Visalli, Marleen Van Hove, Joff Derluyn, Jo Das, Stefan Degroote, Maarten Leys, Kai Cheng, Stefaan Decoutere, Robert Mertens, Marianne Germain, Gustaaf Borghs

2E.5 Identification of Electronic Traps in AlGaN/GaN HEMTs Using UV Light-Assisted Trapping Analysis 152
M. Tapajna, R. J. T. Simms, M. Faqir, M. Kuball, Y. Pei, U. K. Mishra

SESSION 2F: HIGH VOLTAGE DEVICES

2F.1 Reliability of SiC Power Devices and Its Influence on Their Commercialization – Review, Status, and Remaining Issues 156
Michael Treu, Roland Rupp, Gerald Sölkner

2F.2 Effects of Negative Differential Resistance in High Power Devices and Some Relations to DMOS Structures 162
Roman Baburske, Josef Lutz, Birk Heinze

2F.3 Investigation of Monotonous Increase in Saturation-Region Drain Current During Hot Carrier Stress in N-Type Lateral Diffused MOSFET with STI 170
Yu-Hui Huang, J. R. Shih, Y. H. Lee, Sunnys Hsieh, C. C. Liu, Kenneth Wu, H. L. Chou

2F.4 Modeling the Lifetime of a Lateral DMOS Transistor in Repetitive Clamping Mode 175
E. Riedlberger, R. Keller, H. Reisinger, W. Gustin, A. Spitzer, M. Stecher, C. Jungemann

2F.5 Low-Side Driver's Failure Mechanism in a Class-D Amplifier Under Short Circuit Test and a Robust Driver Device 182
Jian-Hsing Lee, J. R. Shih, Tong-Chern Ong, Kenneth Wu

SESSION 3A: SOFT ERRORS

3A.1 On the Radiation-Induced Soft Error Performance of Hardened Sequential Elements in Advanced Bulk CMOS Technologies 188
N. Seifert, V. Ambrose, B. Gill, Q. Shi, R. Allmon, C. Recchia, S. Mukherjee, N. Nassif, J. Krause, J. Pickholtz, A. Balasubramanian

3A.2 Effect of Multiple-Transistor Charge Collection on SET Pulse Widths..198
J. R. Ahlbin, M. J. Gadlage, N. M. Atkinson, B. L. Bhuva, A. F. Witulski, W. T. Holman, L. W. Massengill, P. H. Eaton, B. Narasimham

3A.3 LEAP: Layout Design through Error-Aware Transistor Positioning for Soft-Error Resilient Sequential Cell Design..203
Hsiao-Heng Kelin Lee, Klas Lilja, Mounaim Bounasser, Prasanthi Relangi, Ivan R. Linscott, Umran S. Inan, Subhasish Mitra

3A.4 Alpha-Particle-Induced Soft Errors and Multiple Cell Upsets in 65-nm 10T Subthreshold SRAM..213
Hiroshi Fuketa, Masanori Hashimoto, Yukio Mitsuyama, Takao Onoye

3A.5 SEILA: Soft Error Immune Latch for Mitigating Multi-Node-SEU and Local-Clock-SET 218
Taiki Uemura, Yoshiharu Tosaka, Hideya Matsuyama, Ken Shono, Chihiro J. Uchibori, Keiji Takahisa, Mitsuhiro Fukuda, Kichiji Hatanaka

SESSION 3B: MICRO-ELECTRONIC SYSTEMS AND MEMS

3B.1 Evolving MEMS Qualification Requirements 224
Andrew Olney

3B.2 EFM Study of Injected Charges in the Silicon Nitride of an Electrostatic Actuated MEMS 231
Antoine Nowodzinski, Didier Bloch, Adam Koszewski, Thibaut Toussaint

3B.3 A Novel Low Cost Failure Analysis Technique for Dielectric Charging Phenomenon in Electrostatically Actuated MEMS Devices..237
U. Zaghloul, F. Coccetti, G. J. Papaioannou, P. Pons, R. Plana

3B.4 Accelerated Testing of RF-MEMS Contact Degradation through Radiation Sources....................246
A. Tazzoli, M. Barbato, V. Giliberto, G. Monaco, S. Gerardin, P. Nicolosi, A. Paccagnella, G. Meneghesso

SESSION 3C: FAILURE ANALYSIS

3C.1 X-Ray Computed Tomography for Non-Destructive Failure Analysis in Microelectronics............252
Mario Pacheco, Deepak Goyal

3C.2 Impact on Device Performance and Monitoring of a Low Dose of Tungsten Contamination by Dark Current Spectroscopy...259
F. Domengie, J. L. Regolini, D. Bauza, P. Morin

3C.3 Electromigration of NiSi Poly Gated Electrical Fuse and Its Resistance Behaviors Induced by High Temperature 265
Han-Byul Kang, Jongwoo Park, Gun-Rae Kim, Hyun-Woo Park, Woon-Hak Lee, Joo-Byoung Yoon

3C.4 Determination of the Local Electric Field Strength by Energy Dispersive Photon Emission Microscopy 271
T. Geinzer, R. Heiderhoff, J. C. H. Phang, L. J. Balk

3C.5 Explosion Phenomenon of High Resistance Via During TEM Sample Preparation Using FIB 277
Pan Liu, Irene Tee, Soo Sien Seah, Chi Wen Soo, Ye Chen, Zhi Qiang Mo

SESSION 3D: PROCESS AND INTEGRATION RELIABILITY

3D.1 Gate Oxide Effect on Wafer Level Reliability of Next Generation DRAM Transistors 282
Yu Gyun Shin, Kab-Jin Nam, Heedon Hwang, Jeong Hee Han, Sangjin Hyun, Siyoung Choi, Joo-Tae Moon

3D.2 Reliability Characterization of 32nm High-K and Metal-Gate Logic Transistor Technology 287
Sangwoo Pae, Ashwin Ashok, Jingyoo Choi, Tahir Ghani, Jun He, Seok-hee Lee, Karen Lemay, Mark Liu, Ryan Lu, Paul Packan, Chris Parker, Richard Purser, Anthony St. Amour, Bruce Woolery

3D.3 Reliability Studies on a 45NM Low Power System-on-Chip (SoC) Dual Gate Oxide High-K / Metal Gate (DG HK+MG) Technology..293
C. Prasad, P. Bai, S. Gannavaram, W. Hafez, J. Hicks, C.-H. Jan, J. Lin, M. Jones, K. Komeyli, R. Kotlyar, K. Mistry, I. Post, C. Tsai

3D.4 Re-Consideration of Influence of Silicon Wafer Surface Orientation on Gate Oxide Reliability from TDDB Statistics Point of View...299
Yuichiro Mitani, Akira Toriumi

SESSION 3E: PHOTOVOLTAIC DEVICES

3E.1 Photovoltaic Module Reliability Studies at the Florida Solar Energy Center................306
Neelkanth G. Dhere, Shirish A. Pethe, Ashwani Kaul

3E.2 Intrinsic Reliability of Amorphous Silicon Thin Film Solar Cells................312
M. A. Alam, S. Dongaonkar, Y. Karthik, S. Mahapatra, D. Wang, M. Frei

3E.3 Correlations of Capacitance-Voltage Hysteresis with Thin-Film CdTe Solar Cell Performance During Accelerated Lifetime Testing................318
David S. Albin, Joseph A. del Cueto

3E.4 Solar Cell Interface Stability Probed by Charge Extraction................323
R. L. Graham, C. E. France, S. A. Carter, G. B. Alers

SESSION 3F: THIN FILM DEVICES

3F.1 Reliability Aspects of Organic Light Emitting Diodes................327
Thomas Riedl, Thomas Winkler, Hans Schmidt, Jens Meyer, Daniel Schneidenbach, Hans-Hermann Johannes, Wolfgang Kowalsky, Thomas Weimann, Peter Hinze

3F.2 Light, Bias, and Temperature Effects on Organic TFTs................334
N. Wrachien, A. Cester, N. Bellaio, A. Pinato, M. Meneghini, A. Tazzoli, G. Meneghesso, K. Myny, S. Smout, J. Genoe

3F.3 Reliability of (100) and (110) Oriented Single-Grain Si TFTs without Seed Substrate................342
Tao Chen, Ryoichi Ishihara, C. I. M. Beenakker

3F.4 Non-Uniform Threshold Voltage and Non-Saturating Drain Current in Amorphous-Si TFT After Saturation-Mode Bias Temperature Stress................347
C. R. Wie, Z. Tang

SESSION 4A: DEVICE DIELECTRIC BREAKDOWN

4A.1 New Insight into the TDDB and Post Breakdown Reliability of Novel High-k Gate Dielectric Stacks................354
K. L. Pey, N. Raghavan, X. Li, W. H. Liu, K. Shubhakar, X. Wu, M. Bosman

4A.2 High-k Gate Stack Breakdown Statistics Modeled by Correlated Interfacial Layer and High-k Breakdown Path................364
G. Ribes, P. Mora, F. Monsieur, M. Rafik, F. Guarin, G. Yang, D. Roy, W. L. Chang, J. Stathis

4A.3 Impact of Charge Trapping on the Voltage Acceleration of TDDB in Metal Gate/High-k n-Channel MOSFETs................369
A. Kerber, A. Vayshenker, D. Lipp, T. Nigam, E. Cartier

4A.4 Mechanism of High-k Dielectric-Induced Breakdown of the Interfacial SiO$_2$ Layer................373
G. Bersuker, D. Heh, C. D. Young, L. Morassi, A. Padovani, L. Larcher, K. S. Yew, Y. C. Ong, D. S. Ang, K. L. Pey, W. Taylor

4A.5 Characterization of Millisecond-Anneal-Induced Defects in SiON and SiON/Si Interface by Gate Current Fluctuation Measurement................379
Tsunehisa Sakoda, Keita Nishigaya, Tomohiro Kubo, Mitsuaki Hori, Hiroshi Minakata, Yuko Kobayashi, Hiroko Mori, Katsuji Ono, Katsuto Tanahashi, Naoyoshi Tamura, Toshifumi Mori, Yoshiharu Tosaka, Hideya Matsuyama, Chioko Kaneta, Koichi Hashimoto, Masataka Kase, Yasuo Nara

4A.6 Gate Dielectric Reliability in the Sub Threshold Regime................385
Paul E. Nicollian, Cathy A. Chancellor, Anand T. Krishnan

SESSION 4B: SOFT ERRORS

4B.1 Soft Error Assessments for Servers................391
K. Paul Muller, Pia N. Sanda

4B.2 Contribution of Low-Energy (< 10 MeV) Neutrons to Upset Rate in a 65 nm SRAM................395
Brian D. Sierawski, Kevin M. Warren, Robert A. Reed, Robert A. Weller, Marcus M. Mendenhall, Ronald D. Schrimpf, Robert C. Baumann, Vivian Zhu

4B.3 Scaling Trends of Neutron Effects in MLC NAND Flash Memories................400
S. Gerardin, M. Bagatin, A. Paccagnella, G. Cellere, A. Visconti, S. Beltrami, C. Andreani, G. Gorini, C. D. Frost

4B.4 SEE Test and Modeling Results on 45nm SRAMs with Different Well Strategies................407
Gilles Gasiot, Slawosz Uznanski, Philippe Roche

4B.5 Fidelity of Energy Spectra at Neutron Facilities for Single-Event Effects Testing 411
S. P. Platt, A. V. Prokofiev, Xiao Xiao Cai

SESSION 4C: PROCESS AND INTEGRATION RELIABILITY

4C.1 Mobile and Stable Hydrogen Species in the Interface Layer Between Poly Silicon and Gate Oxynitride 417
Ziyuan Liu, Shuu Ito, Shoichi Hiroshima, Shin Koyama, Mariko Makabe, Markus Wilde, Katsuyuki Fukutani
4C.2 Novel TDDB Mechanism for p-FET Accelerated by Hydrogen from HfSiON Film 424
Izumi Hirano, Koichi Kato, Yasushi Nakasaki, Shigeto Fukatsu, Yuichiro Mitani, Masakazu Goto, Seiji Inumiya, Katsuyuki Sekine, Motoyuki Sato

SESSION 4C: ASSEMBLY AND PACKAGING

4C.3 Buckling, Wrinkling and Debonding in Thin Film Systems 430
S. Goyal, K. Srinivasan, G. Subbarayan, T. Siegmund
4C.4 Reliability of Microelectronics Packaging in the Era of EnergyWise and Borderless Networks 440
Li Li, Jie Xue
4C.5 Comparison of Two Calibration Methods for a Package Stress Measurement Testchip 446
Christian Djelassi, Helmut Köck, Michael Glavanovics
4C.6 Reliability Evaluation Methodology for Chip-Package Interaction - Die Corner Edge Failure Mechanism 453
Melida Chin, Amit Marathe

SESSION 4D: ESD AND LATCHUP

4D.1 A Failure Levels Study of Non-Snapback ESD Devices for Automotive Applications 458
Yiqun Cao, Ulrich Glaser, Stephan Frei, Matthias Stecher
4D.2 Understanding Transient Latchup Hazards and the Impact of Guard Rings 466
Farzan Farbiz, Elyse Rosenbaum
4D.3 A Novel TCAD-Based Methodology to Minimize the Impact of Parasitic Structures on ESD Performance 474
Nicholas Olson, Gianluca Boselli, Akram Salman, Elyse Rosenbaum
4D.4 On the Differences Between 3D Filamentation and Failure of N & P Type Drain Extended MOS Devices Under ESD Condition 480
Mayank Shrivastava, S. Bychikhin, D. Pogany, Jens Schneider, M. Shojaei Baghini, Harald Gossner, Erich Gornik, V. Ramgopal Rao
4D.5 Predictive Simulation of CDM Events to Study Effects of Package, Substrate Resistivity and Placement of ESD Protection Circuits on Reliability of Integrated Circuits 485
Vrashank Shukla, Nathan Jack, Elyse Rosenbaum

SESSION 4E: PHOTOVOLTAIC DEVICES

4E.1 Thin-Film Photovoltaics: What are the Reliability Issues and Where Do They Occur? 494
James R. Sites
4E.3 Thermoreflectance Imaging of Defects in Thin-Film Solar Cells 499
D. Kendig, G. B. Alers, A. Shakouri
4E.4 SEAM and EBIC Studies of Morphological and Electrical Defects in Polycrystalline Silicon Solar Cells 503
L. Meng, D. Nagalingam, C. S. Bhatia, A. G. Street, J. C. H. Phang
4E.5 Photovoltaic (PV) Cells Characterization Using Advanced Optical Tools 508
Franco Stellari, Steven E. Steen, Kathryn C. Fisher, Xiaoyan Shao

SESSION 4F: COMPOUND SEMICONDUCTORS

4F.1 Reliability Assessment of State-of-the-Art GaN HEMT by Means of Cellular Monte Carlo Simulation 516
Fabio Alessio Marino, Diego Guerra, Stephen M. Goodnick, Marco Saraniti

4F.2 A Study of the Failure of GaN-Based LEDs Submitted to Reverse-Bias Stress and ESD Events 522
Matteo Meneghini, Augusto Tazzoli, Enrico Ranzato, Nicola Trivellin, Gaudenzio Meneghesso, Enrico Zanoni, Maura Pavesi, Manfredo Manfredi, Rainer Butendeich, Ulrich Zehnder, Berthold Hahn

4F.3 Temperature Assessment of AlGaN/GaN HEMTs: A Comparative Study by Raman, Electrical and IR Thermography 528
N. Killat, M. Kuball, T.-M. Chou, U. Chowdhury, J. Jimenez

4F.4 Analysis of Interface-Trap Effects in Inversion-Type InGaAs/ZrO$_2$ MOSFETs 532
L. Morassi, G. Verzellesi, A. Padovani, L. Larcher, P. Pavan, D. Veksler, Injo Ok, G. Bersuker

4F.5 Degradation of III-V Inversion-Type Enhancement-Mode MOSFETs 536
N. Wrachien, A. Cester, E. Zanoni, G. Meneghesso, Y. Q. Wu, P. D. Ye

SESSION 5A: INTERCONNECT AND BEOL DIELECTRICS

5A.1 E- and Square Root E-Model Too Conservative to Describe Low Field Time Dependent Dielectric Breakdown 543
K. Croes, Zs. Tökei

5A.2 Study of Leakage Mechanism and Trap Density in Porous Low-k Materials 549
Gianni Giai Gischia, Kristof Croes, Guido Groeseneken, Zsolt Tökei, Valery Afanas'ev, Larry Zhao

5A.3 New Voltage Ramp Dielectric Breakdown Methodology Based on Square Root E Model for Cu/Low-k Interconnect Reliability 556
Mingte Lin, James W. Liang, K. C. Su

5A.4 A Simple Electrical Method for Etch Bias and Process Reliability Determination 562
Kok-Yong Yiang, Melida Chin, Amit Marathe, Oliver Aubel

5A.5 Comprehensive Investigations of CoWP Metal-Cap Impacts on Low-k TDDB for 32nm Technology Application 566
F. Chen, M. Shinosky, B. Li, C. Christiansen, T. Lee, J. Aitken, D. Badami, E. Huang, G. Bonilla, T.-M. Ko, T. Kane, Y. Wang, M. Zaitz, L. Nicholson, M. Angyal, C. Truong, X. Chen, G. Yang, S. B. Law, T. J. Tang, S. Petitdidier, G. Ribes, M. Oh, C. Child, H. Sawada, A. Kolics, O. Rigoutat, N. Gilbert

SESSION 5B: INTERCONNECT AND BEOL DIELECTRICS

5B.1 EM and SM Induced Degradation Dynamics in Copper Interconnects Studied Using Electron Microscopy and X-Ray Microscopy 574
Ehrenfried Zschech, René Hübner, Oliver Aubel, Paul S. Ho

5B.2 Effects of Cap Layer and Grain Structure on Electromigration Reliability of Cu/Low-k Interconnects for 45 nm Technology Node 581
L. Zhang, J. P. Zhou, J. Im, P. S. Ho, O. Aubel, C. Hennesthal, E. Zschech

5B.3 Study of Stress Migration and Electromigration Interaction in Copper/Low-k Interconnects 586
A. Heryanto, K. L. Pey, Y. K. Lim, W. Liu, J. Wei, N. Raghavan, J. B. Tan, D. K. Sohn

5B.4 Electromigration and Stress-Induced-Voiding in Dual Damascene Cu/Low-k Interconnects: A Complex Balance Between Vacancy and Stress Gradients 591
K. Croes, C. J. Wilson, M. Lofrano, B. Vereecke, G. P. Beyer, Zs. Tokei

SESSION 5C: MEMORY

5C.1 The New Scaling Limitation of the Floating Gate Cell in NAND Flash Memory 599
Yong Seok Kim, Dong Jun Lee, Chi Kyoung Lee, Hyun Ki Choi, Seong Soo Kim, Jai Hyuk Song, Du Heon Song, Jeong-Hyuk Choi, Kang-Deog Suh, Chilhee Chung

5C.2 Investigation of the Threshold Voltage Instability After Distributed Cycling in Nanoscale NAND Flash Memory Arrays 604
Christian Monzio Compagnoni, Carmine Miccoli, Riccardo Mottadelli, Silvia Beltrami, Michele Ghidotti, Andrea L. Lacaita, Alessandro S. Spinelli, Angelo Visconti

5C.3 Optimal Cell Design for Enhancing Reliability Characteristics for Sub 30 nm NAND Flash Memory 611
Eun Suk Cho, Hyun Jung Kim, Byoung Taek Kim, Jai Hyuk Song, Du Heon Song, Jeong-Hyuk Choi, Kang-Deog Suh, Chilhee Chung

5C.4 Impact of the Current Density Increase on Reliability in Scaled BJT-Selected PCM for High-Density Applications 615
An. Redaelli, A. Pirovano, I. Tortorelli, F. Ottogalli, A. Ghetti, L. Laurin, A. Benvenuti

SESSION 5D: MEMORY

5D.1 Trade-Off Between Data Retention and Reset in NiO RRAMs 620
D. Ielmini, F. Nardi, C. Cagli, A. L. Lacaita

5D.2 Chip-Level Reliability Study of Barrier Engineered (BE) Floating Gate (FG) Flash Memory Devices 627
Hang-Ting Lue, JiFong Pan, C. S. Chang, Szu-Yu Wang, Y. F. Chang, Y. C. Lee, M. H. Liaw, Y. J. Chen, K. F. Chen, Chester Lo, I. J. Huang, T. T. Han, M. S. Chen, W. P. Lu, T. Yang, K. C. Chen, Kuang-Yeu Hsieh, Chih-Yuan Lu

5D.3 Source/Drain Dopant Concentration Induced Reliability Issues in Charge Trapping NAND Flash Cells 634
Yin-Jen Chen, Lit Ho Chong, Shang-Wei Lin, Teng-Hao Yeh, Kuan-Fu Chen, Jyun-Siang Huang, Cheng-Hsien Cheng, Shaw-Hung Ku, Nian-Kai Zous, I-Jen Huang, Tzung-Ting Han, Tzu-Hsuan Hsu, Hang-Ting Lue, Ming-Shiang Chen, Wen-Pin Lu, Kuang-Chao Chen, Chih-Yuan Lu

5D.4 Performance and Reliability Optimizations of BE-SONOS NAND Flash Using SiON Bandgap-Tuning Tunneling Barrier 639
Jeng-Hwa Liao, Jung-Yu Hsieh, Hang-Ting Lue, Ling-Wu Yang, Tahone Yang, Kuang-Chao Chen, Chih Yuan Lu

SESSION 5E: CIRCUIT RELIABILITY

5E.1 Flexible Electronics: What Can It Do? What Should It Do? 644
Sameer M. Venugopal, David R. Allee, Manuel Quevedo-Lopez, Bruce Gnade, Eric Forsythe, David Morton

5E.2 On the Bias Dependence of Time Exponent in NBTI and CHC Effects 650
Jyothi B. Velamala, Vijay Reddy, Rui Zheng, Srikanth Krishnan, Yu Cao

5E.3 Managing SRAM Reliability from Bitcell to Library Level 655
Vincent Huard, Remy Chevallier, Chittoor Parthasarathy, Anand Mishra, Natalia Ruiz-Amador, Flore Persin, Vincent Robert, Alejandro Chimeno, Emmanuel Pion, Nicolas Planes, David Ney, Florian Cacho, Neeraj Kaapor, Vishal Kulshrestra, Sanjeev Chopra, Nicolas Vialle

5E.4 Prediction of NBTI Degradation for Circuit Under AC Operation 665
Y. S. Tsai, N. K. Jha, Y.-H. Lee, R. Ranjan, Wayne Wang, J. R. Shih, M. J. Chen, J. H. Lee, K. Wu

5E.5 An Extensive and Improved Circuit Simulation Methodology for NBTI Recovery 670
Haldun Kufluoglu, V. Reddy, A. Marshall, J. Krick, T. Ragheb, C. Cirba, A. Krishnan, C. Chancellor

SESSION 5F: CIRCUIT RELIABILITY

5F.1 Adaptive Sensing and Design for Reliability 676
P. Singh, D. Sylvester, D. Blaauw

5F.2 Statistical Evaluation of Dynamic Junction Leakage Current Fluctuation Using a Simple Arrayed Capacitors Circuit 683
Kenichi Abe, Takafumi Fujisawa, Hiroyoshi Suzuki, Shunichi Watabe, Rihito Kuroda, Shigetoshi Sugawa, Akinobu Teramoto, Tadahiro Ohmi

5F.3 Scaling Reliability and Modeling of Ferroelectric Capacitors 689
Antonio G. Acosta, John Rodriguez, Borna Obradovic, Scott Summerfelt, Tamer San, Keith Green, Ted Moise, Srikanth Krishnan

5F.4 Measurement of Neutron-Induced Single Event Transient Pulse Width Narrower Than 100ps 694
Hideyuki Nakamura, Katsuhiko Tanaka, Taiki Uemura, Kan Takeuchi, Toshikazu Fukuda, Shigetaka Kumashiro

SESSION 6A: INTERCONNECT AND BEOL DIELECTRICS

6A.1 New Electromigration Validation: Via Node Vector Method 698
Young-Joon Park, Palkesh Jain, Srikanth Krishnan

6A.2 Electromigration Mechanisms in Cu Nano-Wires 705
M. H. Lin, S. C. Lee, A. S. Oates

6A.3 Degradation and Failure Analysis of Copper and Tungsten Contacts Under High Fluence Stress 712
Thomas Kauerauf, Geni Butera, Kristof Croes, Steven Demuynck, Christopher J. Wilson, Philippe Roussel, Chris Drijbooms, Hugo Bender, Melina Lofrano, Bart Vandevelde, Zsolt Tokei, Guido Groeseneken

6A.4 Effective Thermal Characteristics to Suppress Joule Heating Impacts on Electromigration in Cu/Low-k Interconnects 717
S. Yokogawa, H. Tsuchiya, Y. Kakuhara

6A.5 Assessing the Degradation Mechanisms and Current Limitation Design Rules of SiCr-Based Thin-Film Resistors in Integrated Circuits...724
Yuan Li, David Donnet, Andrzej Grzegorczyk, Jan Cavelaars, Fred Kuper

SESSION 6C: MEMORY

6C.1 Role of Holes and Electrons During Erase of TANOS Memories: Evidences for Dipole Formation and Its Impact on Reliability 731
Luca Vandelli, Andrea Padovani, Luca Larcher, Antonio Arreghini, Geert Van den bosch, Malgorzata Jurczak, Jan Van Houdt, Vincenzo Della Marca, Paolo Pavan

6C.2 Reset Current Distributions in Phase Change Memories...738
A. Calderoni, M. Ferro, D. Ventrice, D. Ielmini, P. Fantini

6C.3 Random Telegraph Signal Noise in Phase Change Memory Devices 743
Davide Fugazza, Daniele Ielmini, Simone Lavizzari, Andrea L. Lacaita

6C.4 Reliability of Ferroelectric Random Access Memory Embedded within 130nm CMOS 750
J. Rodriguez, K. Remack, J. Gertas, L. Wang, C. Zhou, K. Boku, J. Rodriguez-Latorre, K. R. Udayakumar, S. Summerfelt, T. Moise, D. Kim, J. Groat, J. Eliason, M. Depner, F. Chu

SESSION 6E: EXTREME ENVIRONMENTS

6E.1 Electronic Failures in Spacecraft Environments...759
Douglas J. Sheldon

6E.2 Single Event Transient Pulse Width Measurements in a 65-nm Bulk CMOS Technology at Elevated Temperatures 763
M. J. Gadlage, J. R. Ahlbin, B. L. Bhuva, L. W. Massengill, R. D. Schrimpf

6E.3 Practicality of Evaluating Soft Errors in Commercial Sub-90 nm CMOS for Space Applications..................768
Jonathan A. Pellish, Kenneth A. LaBel

POSTER PRESENTATIONS

BD - DEVICE DIELECTRIC BREAKDOWN POSTERS

BD.1 Analysis of the Breakdown Spots Spatial Distribution in Large Area MOS Structures...................775
E. Miranda, E. O'Connor, P. K. Hurley

BD.2 New Statistical Model to Decode the Reliability and Weibull Slope of High-k and Interfacial Layer in a Dual Layer Dielectric Stack...778
N. Raghavan, K. L. Pey, W. H. Liu, X. Li

BD.3 Role of Interface Layer in Stress-Induced Leakage Current in High-k/Metal-Gate Dielectric Stacks...787
W. L. Chang, J. H. Stathis, E. Cartier

BD.4 A Compact Analytic Model for the Breakdown Distribution of Gate Stack Dielectrics 792
Santi Tous, Ernest Y. Wu, Jordi Suñe

BD.5 Time Dependent Dielectric Breakdown and Stress Induced Leakage Current Characteristics of 8Å EOT HfO_2 N-MOSFETs..799
Robert O'Connor, Greg Hughes, Thomas Kaureauf, Lars-Åke Ragnarsson

BD.6 Frequency-Dependent Charge-Pumping: The Depth Question Revisited 804
F. Zhang, K. P. Cheung, J. P. Campbell, J. Suehle

CD - COMPOUND SEMICONDUCTORS POSTERS

CD.1 Progressive Schottky Junction Reaction Induced Degradation in Pt-Sunken Gate InP HEMT MMICs for High Reliability Applications 807
Y. C. Chou, D. L. Leung, M. Biedenbender, D. Buttari, D. C. Eng, R. S. Tsai, C. H. Lin, L. S. Lee, X. B. Mei, M. Wojtowicz, M. E. Barsky, R. Lai, A. K. Oki, T. R. Block

CD.2 Electrical Stress Induced Degradation in InAs – AlSb HEMTs...813
S. DasGupta, R. A. Reed, R. D. Schrimpf, D. M. Fleetwood, X. Shen, S. T. Pantelides, J. Bergman, B. Brar

CD.3 InAlAs/InGaAs MHEMT Degradation During DC and Thermal Stressing 818
E. A. Douglas, K. H. Chen, C. Y. Chang, L. C. Leu, C. F. Lo, B. H. Chu, F. Ren, S. J. Pearton

CR - CIRCUIT RELIABILITY

CR.1 A Built-In Aging Detection and Compensation Technique for Improving Reliability of Nanoscale CMOS Designs 822
Hamed F. Dadgour, Kaustav Banerjee

CR.2 A Test Concept for Circuit Level Aging Demonstrated by a Differential Amplifier 826
Florian R. Chouard, Christoph Werner, Doris Schmitt-Landsiedel, Michael Fulde

CR.3 The Hot Carrier Degradation Rate Under AC Stress 830
Guido T. Sasse, Jaap Bisschop

EL - ESD AND LATCHUP POSTERS

EL.1 ESD Protection for High-Speed Receiver Circuits 835
Nathan Jack, Elyse Rosenbaum

EL.2 On the Failure Mechanism and Current Instabilities in RESURF Type DeNMOS Device Under ESD Conditions 841
Mayank Shrivastava, Jens Schneider, Maryam Shojaei Baghini, Harald Gossner, V. Ramgopal Rao

EL.3 Characterization of High-k/Metal Gate Stack Breakdown in the Time Scale of ESD Events 846
Yang Yang, James Di Sarro, Robert J. Gauthier, Kiran Chatty, Junjun Li, Rahul Mishra, Souvick Mitra, Dimitris E. Ioannou

EL.5 Robust High Current ESD Performance of Nano-Meter Scale DeNMOS by Source Ballasting 853
Amitabh Chatterjee, Forrest Brewer, Harald Gossner, Sameer Pendharkar, Charvaka Duvvury

EL.6 A Bending N-Well Ballast Layout to Improve ESD Robustness in Fully-Silicided CMOS Technology 857
Yong-Ru Wen, Ming-Dou Ker, Wen-Yi Chen

FA - FAILURE ANALYSIS POSTERS

FA.1 Isolating Marginally Defective Gate Using Photoperturbation Induced via a C-AFM Laser Beam 861
Hung Sung Lin, Mong Sheng Wu

FA.2 A Novel Sample Preparation Technique for Visualizing Invisible Defects Embedded in Poly Gate 865
Hung Sung Lin

FA.3 A Case Study of High Temperature Pass Analysis Using Thermal Laser Stimulation Technique 870
Hung Sung Lin, Mong Sheng Wu

FA.4 High Spatial and Temporal Resolution Thermal Imaging for LSI Circuits with Phase Microscopy 874
Tomonori Nakamura, Hidenao Iwai, Toyohiko Yamauchi, Hirotoshi Terada, Hitoshi Iida

FA.5 Reliability of Electronic Equipment Exposed to Chlorine Dioxide Used for Biological Decontamination 879
G. E. Derkits, M. L. Mandich, W. D. Reents, J. P. Franey, C. Xu, D. Fleming, R. Kopf, S. Ryan

HV - HIGH VOLTAGE DEVICES POSTERS

HV.1 Analysis of HCS in STI-Based LDMOS Transistors 881
Susanna Reggiani, Stefano Poli, Elena Gnani, Antonio Gnudi, Giorgio Baccarani, Marie Denison, Sameer Pendharkar, Rick Wise, Sridhar Seetharaman

IC - INTERCONNECT AND BEOL DIELECTRICS POSTERS

IC.1 Lifetime Extrapolation for Electromigration Tests at Wafer Level with a Dedicated Device 887
C. Chappaz, P. Nakkala

IC.2 Ultra-Low-k Dielectric Degradation Before Breakdown 890
T. Breuer, U. Kerst, C. Boit, E. Langer, H. Ruelke

IC.3 Analysis of the Impact of Linewidth Variation on Low-K Dielectric Breakdown 895
Muhammad M. Bashir, Linda Milor

IC.4 Effect of Pre-Existing Void in Sub-30nm Cu Interconnect Reliability 903
Zungsun Choi, Matsuda Tsukasa, Jong Myeong Lee, Gil-Heyun Choi, Siyoung Choi, Joo-Tae Moon

IC.5 Study of Upstream Electromigration Bimodality and Its Improvement in Cu Low-k Interconnects 906
 W. Liu, Y. K. Lim, F. Zhang, H. Liu, Y. H. Zhao, A. Y. Du, B. C. Zhang, J. B. Tan, D. K. Sohn, L. C. Hsia

IC.6 Modeling of Stress Evolution of Electroplated Cu Films During Self-Annealing 911
 Rui Huang, Werner Robl, Thomas Detzel, Hajdin Ceric

IC.7 The TDDB Failure Mode and Its Engineering Study for 45nm and Beyond in Porous Low k Dielectrics Direct Polish Scheme 918
 Chia-Lin Hsu, Kuan-Ting Lu, Wen-Chin Lin, Jeh-Chieh Lin, Chih-Hsien Chen, Teng-Chun Tsai, Climbing Huang, J. Y. Wu, Dung-Ching Perng

IC.8 Resistance Trace Modeling and Electromigration Immortality Criterion Based on Void Growth Saturation 922
 P. Lamontagne, D. Ney, L. Doyen, E. Petitprez, Y. Wouters

IC.9 Practical Considerations of Process Corner Evaluation for Deep-Sub Micron Technology Nodes Using the Example of Its Impact on Electromigration 926
 Oliver Aubel, Thomas Hoffmann

IC.10 Investigating the Electro-Thermal Origin of Breakdown in Low-K/Cu Dielectrics Under Short Duration over Stressed Pulsed Regime 932
 Amitabh Chatterjee, Forrest Brewer, S. C. Lee, A. S. Oates

IC.11 Study of Electric Field-Based Lifetime Projection Method in IMD TDDB 938
 W. Zhang, X. Zeng, W. Liu, Y. K. Lim, J. F. Liu, E. C. Chua

IC.12 On the Physical Interpretation of the Impact Damage Model in TDDB of Low-k Dielectrics 943
 J. R. Lloyd

IC.13 Reliability and Performance Limiting Defects in Low-k Dielectrics for Use as Interlayer Dielectrics 947
 B. C. Bittel, P. M. Lenahan, S. King

MY - MEMORY POSTERS

MY.1 A High-Endurance (>100K) BE-SONOS NAND Flash with a Robust Nitrided Tunnel Oxide/Si Interface 951
 Szu-Yu Wang, Hang-Ting Lue, Tzu-Hsuan Hsu, Pei-Ying Du, Sheng-Chih Lai, Yi-Hsuan Hsiao, Shih-Ping Hong, Ming-Tsung Wu, Fang-Hao Hsu, Nan-Tzu Lian, Jung-Yu Hsieh, Ling-Wu Yang, Tahone Yang, Kuang-Chao Chen, Kuang-Yeu Hsieh, Chih-Yuan Lu

MY.2 Transition of Erase Mechanism for MONOS Memory Depending on SiN Composition and Its Impact on Cycling Degradation 956
 Shosuke Fujii, Jun Fujiki, Naoki Yasuda, Ryota Fujitsuka, Katsuyuki Sekine

MY.3 Use of Random Telegraph Signal as Internal Probe to Study Program/Erase Charge Lateral Spread in a SONOS Flash Memory 960
 Y. L. Chou, J. P. Chiu, H. C. Ma, Tahui Wang, Y. P. Chao, K. C. Chen, C. Y. Lu

MY.4 Bias Temperature Instability of Binary Oxide Based ReRAM 964
 Z. Fang, H. Y. Yu, W. J. Liu, K. L. Pey, X. Li, L. Wu, Z. R. Wang, Patrick G. Q. Lo, B. Gao, J. F. Kang

MY.5 Reliability Constraints for TANOS Memories Due to Alumina Trapping and Leakage. 966
 Salvatore M. Amoroso, Aurelio Mauri, Nadia Galbiati, Claudia Scozzari, Evelyne Mascellino, Elisa Camozzi, Armando Rangoni, Tecla Ghilardi, Alessandro Grossi, Paolo Tessariol, Christian Monzio Compagnoni, Alessandro Maconi, Andrea L. Lacaita, Alessandro S. Spinelli, Gabriella Ghidini

MY.6 Variability Effects on the V_T Distribution of Nanoscale NAND Flash Memories 970
 Alessio Spessot, Alessandro Calderoni, Paolo Fantini, Alessandro S. Spinelli, Christian Monzio Compagnoni, Fabrizio Farina, Andrea L. Lacaita, Andrea Marmiroli

MY.7 NAND Flash Reliability Degradation Induced by HCI in Boosted Channel Potential 975
 Milim Park, Sukkwang Park, Seokwon Cho, Dong-Kyu Lee, YeonJoo Jeong, Chonga Hong, Ho Seok Lee, Myoung Kwan Cho, Kun-Ok Ahn, Yohwan Koh

NA - NANOELECTRONICS POSTERS

NA.1 Interface-Trap Modeling for Silicon-Nanowire MOSFETs 977
 Zuhui Chen, Xing Zhou, Guojun Zhu, Shihuan Lin

NA.2 Applicability of Dual Layer Metal Nanocrystal Flash Memory for NAND 2 or 3-Bit/Cell Operation: Understanding the Anomalous Breakdown and Optimization of P/E Conditions 981
 Pawan Singh, C. Sandhya, Kshitij Auluck, Gaurav Bisht, M. Sivatheja, Ralf Hofmann, Gautam Mukhopadhyay, Souvik Mahapatra

PI - PROCESS AND INTEGRATION RELIABILITY POSTERS

PI.1 Pattern-Independent, Fine-Morphology Ni-Pt Silicide Formation by Partial Conversion with Low Metal-Consumption Ratio 988
Takuya Futase, Takeshi Kamino, Naoto Hashikawa, Yutaka Inaba, Tetsuo Fujiwara, Hirohiko Yamamoto, Hisanori Tanimoto

PI.2 Disconnection of NiSi Shared Contact and Its Correction Using NH_3 Soak Treatment in Ti/TiN Barrier Metallization 995
Takuya Futase, Kota Funayama, Naoto Hashikawa, Hiroshi Tobimatsu, Hirohiko Yamamoto, Hisanori Tanimoto

PI.3 Analysis of Statistical Variation in NBTI Degradation of HfO_2/SiO_2 FETs 1001
H. Yoshimoto, D. Hisamoto, Y. Shimamoto, R. Tsuchiya, I. Yanagi, T. Arigane, K. Torii, K. Funayama, T. Hashimoto, H. Makiyama, K. Horita, T. Iwamatsu, K. Shiga, M. Mizutani, M. Inoue, T. Kaneoka

RM - RELIABILITY IN MANUFACTURING POSTERS

RM.1 Method of Deciding Burn-In Stress Voltage in Conceptual Design Phase 1004
Jae Yong Seo, Noh Seok Park, Hyung-Jin Park, Hong Sik Park, Woo Sup Kim, Se Young Lim, Hyun Kim, Nam Hyun Cha, Ju Seong Kang, Byung Se So

RM.2 Device-Level Reliability Simulation for High Temperature Applications of a Modular CMOS Foundry Process 1006
Markus Ackermann

RM.3 Accurate Projection of V_{ccmin} by Modeling "Dual Slope" in FinFET Based SRAM, and Impact of Long Term Reliability on End of Life V_{ccmin} 1008
H. Park, S. C. Song, S. H. Woo, M. H. Abu-Rahma, L. Ge, M. G. Kang, B. M. Han, J. Wang, R. Choi, J. W. Yang, S. O. Jung, G. Yeap

SE - SOFT ERRORS POSTERS

SE.1 System-Level Analysis of Soft Error Rates and Mitigation Trade-Off Explorations 1014
Zhe Ma, Francky Catthoor, Frank Vermunt, Teun Hendriks

SE.2 Soft Errors from Neutron and Proton-Induced Multiple-Node Events 1019
Ethan H. Cannon

SE.3 Effects of Multi-Node Charge Collection in Flip-Flop Designs at Advanced Technology Nodes 1026
Vijay B. Sheshadri, Bharat L. Bhuva, Robert A. Reed, Robert A. Weller, Marcus H. Mendenhall, Ron D. Schrimpf, Kevin M. Warren, Brian D. Seirawski, Shi-Jie Wen, Ricky Wong

SE.4 Analysis of Soft Error Rates in Combinational and Sequential Logic and Implications of Hardening for Advanced Technologies 1031
N. N. Mahatme, I. Chatterjee, B. L. Bhuva, J. Ahlbin, L. W. Massengill, R. Shuler

SE.5 Thermal Neutron Soft Error Rate for SRAMs in the 90nm-45nm Technology Range 1036
ShiJie Wen, Richard Wong, Michael Romain, Nelson Tam

TF - THIN FILM DEVICES POSTERS

TF.1 Evaluation of Self-Heating and Hot Carrier Degradation of Poly-Si Thin-Film Transistors Using Charge Pumping Technique 1040
Xiaowei Lu, Mingxiang Wang, Kai Sun, Lei Lu

XT - TRANSISTORS: BTI, HOT CARRIER POSTERS

XT.1 PBTI Response to Interfacial Layer Thickness Variation in Hf-Based HKMG nFETs 1044
D. P. Ioannou, E. Cartier, Y. Wang, S. Mittl

XT.2 HCI and NBTI Including the Effect of Back-Biasing in Thin-BOX FD-SOI CMOSFETs 1049
T. Ishigaki, R. Tsuchiya, Y. Morita, H. Yoshimoto, N. Sugii, S. Kimura

XT.3 The Understanding of Strain-Induced Device Degradation in Advanced MOSFETs with Process-Induced Strain Technology of 65nm Node and Beyond 1053
M. H. Lin, E. R. Hsieh, Steve S. Chung, C. H. Tsai, P. W. Liu, Y. H. Lin, C. T. Tsai, G. H. Ma

XT.4 Effect of Strain on Negative Bias Temperature Instability of Germanium p-Channel Field-Effect Transistor with High-k Gate Dielectric 1055
Bin Liu, Phyllis Shi Ya Lim, Yee-Chia Yeo

XT.5 A Robust Ultrafast Switching Methodology for Device Parameter Characterization of Bias-Temperature Instability 1058
Y. Z. Hu, D. S. Ang, Z. Q. Teo, G. A. Du

XT.6 Impact of Hydrogen on Recoverable and Permanent Damage Following Negative Bias Temperature Stress 1063
T. Aichinger, S. Puchner, M. Nelhiebel, T. Grasser, H. Hutter

XT.7 A Multi-Probe Correlated Bulk Defect Characterization Scheme for Ultra-Thin High-k Dielectric 1069
M. Masuduzzaman, A. E. Islam, M. A. Alam

XT.8 Dependence of the Negative Bias Temperature Instability on the Gate Oxide Thickness 1073
Gregor Pobegen, Thomas Aichinger, Michael Nelhiebel, Tibor Grasser

XT.9 Interpretation of PBTI/TDDB Predicted Lifetime Based on Trap Characterization by TSCIS in V_{th}-Adjusted Transistors 1078
S. Sahhaf, R. Degraeve, V. Srividya, M. Cho, T. Kauerauf, G. Groeseneken

XT.10 Improvements of NBTI Reliability in SiGe p-FETs 1082
J. Franco, B. Kaczer, M. Cho, G. Eneman, G. Groeseneken, T. Grasser

XT.11 A Model for NBTI in Nitrided Oxide MOSFETs without Hydrogen or Diffusion 1086
P. M. Lenahan

XT.12 A Generalized, I_B-Independent, Physical HCI Lifetime Projection Methodology Based on Universality of Hot-Carrier Degradation 1091
Dhanoop Varghese, Muhammad Ashraful Alam, Bonnie Weir

XT.13 Positive and Negative Bias Temperature Instability on Sub-Nanometer EOT High-k MOSFETs 1095
Moonju Cho, Marc Aoulaiche, Robin Degraeve, Ben Kaczer, Jacopo Franco, Thomas Kauerauf, Philippe Roussel, Lars Å. Ragnarsson, Joshua Tseng, Thomas Y. Hoffmann, Guido Groeseneken

XT.14 Hot-Carrier Degradation in Undoped-Body ETSOI FETs and SOI FINFETs 1099
Miaomiao Wang, Pranita Kulkarni, Kangguo Cheng, Ali Khakifirooz, V. S. Basker, Hemanth Jagannathan, Chun-Chen Yeh, Vamsi Paruchuri, Bruce Doris, Huiming Bu, Chung-Hsun Lin, James H. Stathis, Kingsuk Maitra, Philip J. Oldiges

XT.15 NBTI Lifetime Prediction in SiON p-MOSFETs by H/H2 Reaction-Diffusion(RD) and Dispersive Hole Trapping Model 1105
S. Deora, V. D. Maheta, S. Mahapatra

XT.16 Product NBTI Distribution and Voltage Dependence - Impact of Relaxation and Droops 1115
Amr Haggag, Ning Liu, Peter Abramowitz, Mohamed Moosa, Gary Anderson, David Burnett, Sanjay Parihar, Glenn Abeln, Jack Higman

XT.17 Analysis of the Relationship Between Random Telegraph Signal and Negative Bias Temperature Instability 1117
Yasumasa Tsukamoto, Seng Oon Toh, Changhwan Shin, Andrew Mairena, Tsu-Jae King Liu, Borivoje Nikolic

XT.18 Energy Resolved Spin Dependent Trap Assisted Tunneling Investigation of SILC Related Defects 1122
J. T. Ryan, P. M. Lenahan, A. T. Krishnan, S. Krishnan

Author Index

PREFACE

On behalf of the IRPS 2010 Management Committee and the IRPS Board of Directors, it is my pleasure to present the 48th edition of the International Reliability Physics technical proceedings. Within its pages are the manuscripts detailing the technical presentations and posters that are the heart of the 2010 symposium. This volume would not be possible without the efforts of many individuals, especially the authors whose work was selected for its outstanding technical quality.

These proceedings and the symposium are the result of the dedication and tireless efforts of many individuals who form the IRPS "family". I wish to express my deepest thanks to my management committee who have volunteered many hours to make this year's symposium a reality and especially the consultants who continually show the utmost professionalism and competence every year in coordinating the logistics of the IRPS including registration, arrangements, and publication.

This year the International Reliability Physics Symposium is pleased to have two keynote speakers. These keynotes speakers will address issues related to a new IRPS technical focus on the reliability of renewable energy systems and devices. I am delighted to introduce our first 2010 keynote speaker, Mr. Clark Gellings, Fellow at the Electric Power Research Institute (EPRI) and IEEE Fellow. Mr. Gellings will describe the efforts currently underway to evolve a "Smart Grid" which will make extensive use of digital devices in power generation, power delivery and in our homes and offices and in the appliances and devices that use electricity.

Our second 2010 keynote speaker, Dr. Harald Schone, manages JPL's Electronics Parts Engineering Office which establishes reliability and quality standards for all JPL missions with a special emphasis on enabling technologies that push traditional concepts of reliability. Dr. Schone's keynote address entitled "Trading Margin with Knowledge" outlines Jet Propulsion Lab's constant push to fly missions which endure the severe environments of outer planet research to better understand the genesis of our planets and ultimately find life in our solar system.

This year's symposium will feature our traditionally strong technical program including a new emphasis on the reliability challenges of implementing renewable energy sources and "Smart Grid" technology. The technical program will consist of more than 103 contributed talks and an additional 22 invited talks, as well as a poster reception featuring 77 posters. The symposium will also include a large range of stimulating tutorials over two days, a Reliability Year-In-Review Seminar, informal workshops on topics of interest and an equipment demonstration program where hands-on evaluations are welcomed. There are lots of opportunities to be involved in increasing your understanding of this technically important and ever changing field.

Our signature "Virtual IRPS" will also be available shortly after the symposium. This DVD provides the video, audio, and presentation material for all of the platform presentations given at all of the technical sessions. We hope that you have an enriching and valuable experience at IRPS this year and look forward to your participation and attendance at future symposiums

Tom Moore
2010 IRPS General Chair

Keynote Address

Using a Smart Grid to Evolve a Reliable Power System

Clark Gellings
Electric Power Research Institute

The Nation's Power Delivery Infrastructure is under stress, aging and barely able to meet society's needs. Despite having spawned the digital age - the electric infrastructure is still an analog system. In fact, the electric power industry is the only industry in the western world that has not modernized itself through the use of sensors, communications and computational ability. In particular, today's power system is incapable of efficiently meeting our needs for highly reliable, digital - grade power supplied from low carbon emitting generating sources. However, a number of efforts are now underway to evolve a "Smart Grid". The Smart Grid will require extensive use of digital devices in power generation, power delivery and in our homes and offices and in the appliances and devices that use electricity.

Clark W. Gellings holds the position of Fellow at the Electric Power Research Institute (EPRI) and is responsible for technology strategy in areas concerning energy efficiency, demand response, renewable energy resources, and other clean technologies.

Gellings joined EPRI in 1982 progressing through a series of technical management and executive positions including seven vice president positions. He was also chief executive officer of several EPRI subsidiaries. Prior to joining EPRI, he spent 14 years with Public Service Electric & Gas Company.

Gellings received distinguished awards from a number of organizations, including The Illuminating Engineering Society, the Association of Energy Services Professionals, and the South African Institute of Electrical Engineers. He is a 2003 recipient of CIGRE's (International Council on Large Electric Systems) Attwood Award for notable contributions. He is the 2010 recipient of EnergyBiz Magazine's KITE (Knowledge, Innovation, Technology, Excellence) Lifetime Achievement Award. He is currently a Member of the Board of the University of Minnesota's Technology Leadership Institute and several National Academy of Sciences Committees.

Gellings has a Bachelor of Science in electrical engineering from Newark College of Engineering in New Jersey, a Master of Science degree in mechanical engineering from New Jersey Institute of Technology, and a Master of Management Science from the Wesley J. Howe School of Technology Management at Stevens Institute of Technology.

Gellings is a registered Professional Engineer, a Fellow in the Institute of Electrical and Electronics Engineers, a Fellow in the Illuminating Engineering Society, and a Distinguished Member and President of the U.S. National Committee of CIGRE.

Keynote Address

Trading Margin with Knowledge

Harald Schöne,
JPL Electronic Parts Engineering

Dr. Schöne's keynote address entitled "Trading Margin with Knowledge" outlines Jet Propulsion Lab's constant push to fly missions which endure the severe environments of outer planet research. To better understand the genesis of our planets and ultimately find life in our solar system, JPL is descending to the hot pressure cooker of Venus, looking under many a rock on Mars and penetrates the icy surfaces of Io and Europa. None of these missions are possible unless we are adopting new approaches that eliminate the risk of using new technologies. Dr. Schone outlines JPL's approach to technology insertion and qualification for a number of key components.

Since 2005, Harald Schöne has been managing JPL's Electronics Parts Engineering Office which establishes and reviews the EEE part radiation, reliability and quality standards for all JPL missions. A special emphasis is placed on bringing enabling technologies to missions that push traditional concepts of reliability. He has a 20 year carrier in space technology program management, research and development. During his tenure at the Air Force Research Lab, he was part of the Office of the Secretary of Defense team to define the direction of Space Electronics Technology Development for the next 20 years. Harald Schöne has received his Ph.D. ('90) in Atomic Physics from the Max-Plank Institute in Heidelberg, his Masters in Nuclear Physics ('86) and his Bachelors in Physics and Chemistry in ('84).

MARK YOUR CALENDAR AND START PLANNING TO ATTEND,

AND POSSIBLY

CONTRIBUTE A PAPER TO THE FOLLOWING SYMPOSIUM:

2011 IEEE INTERNATIONAL

RELIABILITY PHYSICS SYMPOSIUM

April 10 - 14, 2011

Hyatt Regency Monterey Hotel & Spa on Del Monte Golf Course • Monterey, California

2009 INTERNATIONAL RELIABILITY PHYSICS SYMPOSIUM
SYMPOSIUM OFFICERS

GENERAL CHAIR .... T.M. Moore, Omniprobe

VICE-GENERAL CHAIR.. J.H. Stathis, IBM Research

SECRETARY ....G. Meneghesso, University of Padua

FINANCE..C. Henderson, Semitracks

SYMPOSIUM COMMITTEE CHAIRS

TECHNICAL PROGRAM ....E. Ogawa, Broadcom

PUBLICITY .... P. Chaparal, Nat'l Semiconductor

REGISTRATION .... R. Kaplar, Sandia National Labs

ARRANGEMENTS .... S. Krishnan, Texas Instruments

AUDIO VISUAL...G. La Rosa, IBM

PUBLICATIONS .... E. Rosenbaum, Univ. of Illinois

TUTORIAL ....M. Porter, Medtronics

EQUIPMENT DEMOS ....A. Haggag, Freescale

WORKSHOP...Y. Chen, NASA

CONSULTANTS .... Widerkehr and Associates

CONSULTANTS ....D.F. Barber, Scien-Tech Associates, Inc.

BOARD OF DIRECTORS

R.C. Lacoe, Chair
The Aerospace Corp.

E. I. Cole Jr., Sandia National Labs	N.R. Mielke Intel	J. H. Stathis IBM
L.A. Kasprzak Siemens	T. M. Moore Omniprobe	J.S. Suehle NIST
J.W. McPherson Texas Instruments	H.A. Schafft NIST	W.R. Tonti IBM

2010 IRPS OFFICERS AND COMMITTEES

GENERAL CHAIR
Tom Moore
Omniprobe, Inc.

VICE GENERAL CHAIR

Jim Stathis
IBM

FINANCE

Chris Henderson, *Chair*
Semitracks

PUBLICATIONS

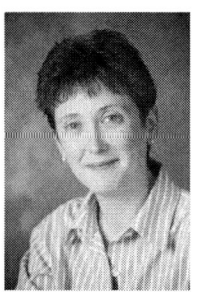

Elyse Rosenbaum, *Chair*
University of Illinois

ARRANGEMENTS

Srikanth Krishnan, Chair
Texas Instruments

AUDIO VISUAL

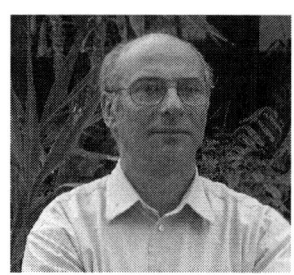

Giuseppe La Rosa, Chair
IBM

EQUIPMENT DEMOS

Amr Haggag, *Chair*
Freescale

PUBLICITY

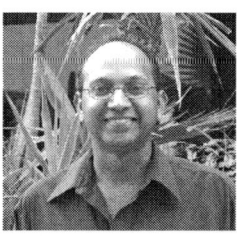

Prasad Chaparala, *Chair*
National Semi.

REGISTRATION

Robert Kaplar, *Chair*
Sandia National Labs

SECRETARY

Gaudenzio Meneghesso,
Chair
University of Padua

TUTORIALS

Mark Porter, *Chair*
Medtronics

WORKSHOP

Yuan Chen, *Chair*
NASA
Jason Ryan, Vice Chair
Penn State Univ.

TECHNICAL PROGRAM COMMITTEE

Ennis Ogawa, *Chair*
Broadcom

ASSEMBLY & PACKAGING

Jeffery Coffin, Chair
  IBM
Charlie Zhai
  Nvidia
Richard Blish
  Spansion
Rajen Dias
  Intel
Steve Groothius
  SimuTech Group
Lakshmi Ramanathan
  Freescale
Sidharth Srivastava
  Microsoft
Ahmer Syed
  Amkor
Jeffrey Suhling
  Auburn University
Jianmin Qu
  Northwestern University

DEVICE DIELECTRIC BREAKDOWN

Sangwoo Pae, *Chair*
  Intel
Tanya Nigam, Vice Chair
  GLOBALFOUDRIES
Gennadi Bersuker
  SEMATECH
Barry Linder
  IBM
Enrique Miranda
  Universitat Autonoma
  de Barcelona
Yuuichiro Mitani
  Toshiba
Frederic Monsieur
  IBM
Motoyuki Sato
  Selete
J. R. Shih
  TSMC
Kin-Leong Pey
  National Taiwan University
Thomas Pompl
  Infineon

COMPOUND SEMICONDUCTORS

Gaudenzio Meneghesso, *Chair*
  Univ. of Padua
Jose Jimenez, Vice Chair
  TriQuint Semiconductor
Brian Skromme
  ASU
Michael Dammann
  Fraunhofer Inst.
Martin Kuball
  University of Bristol
Tetsuzo Ueda
  Panasonic Corporation
Aris Christou
  University of Maryland
Denis Marcon
  IMEC
Kikkawa Toshi
  Fujitsu
Giovanni Verzellesi
  Univ. of Modena and
  Reggio Emilia
Rama Vetury
  RFMD

CIRCUIT RELIABILITY

Yu Cao, *Chair*
  ASU
Xiaojun Li, Vice Chair
  Intel
Kanak Agarwal
  IBM
Aditya Bansal
  IBM
Vincent Huard
  STMicroelectronics

Vijay Reddy
  Texas Instruments
Chris Kim
  Univ. of Minnesota
Subhasish Mitra
  Stanford University

ESD AND LATCHUP

Kai Esmark, *Chair*
  Infineon

Junjun Li, *Vice Chair*
  IBM
Melanie Etherton
  Freescale
Jonathan Brodsky
  Texas Instruments
Howard Tang
  UMC
Vladislav Vashchencko
  NSC
Hans van Zwol
  NXP Semiconductors

EXTREME ENVIRONMENTS

Bert Vermeire, *Chair*
  ASU
Patrick McCluskey, *Vice Chair*
  University of Maryland

Hugh Barnaby
  ASU
Nadim Haddad
  BAE Systems
Skip Parks
  Univ. of Arizona
Ron Pease
  RLP Research
Esko Mikkola
Douglas Sheldon
  NASA

FAILURE ANALYSIS

Ed Cole, *Chair*
  Sandia National Labs
John Guravage
  Mediatek Wireless
Chris Henderson
  Semitracks
Manny Ma
  Micron
Philippe Perdu
  CNES
Christine Toh
  Cambridge Silicon
  Radio
Ludwig Balk
  Univ. Wuppertal
Jacob Phang
  NUS

HIGH VOLTAGE DEVICES

Mathias Stecher, *Chair*
  Infineon
Sameer Pendharka, *Vice Chair*
  Texas Instruments

Young Chung
  Freescale
Merlyne DeSouza
  Sheffield University
Jong Mun Park
  Austriamicrosystems
Ayman Shibib
  Int'l Rectifiers
Paul van der Wel
  NXP Semiconductors

INTERCONNECT & BEOL DIELECTRICS

Armin Fischer, *Chair*
  Infineon
Oliver Aubel, *Vice Chair*
  GLOBALFOUNDARIES
Jim Lloyd, *Vice Chair*
  Univ. Albany
Martin Gall
  Freescale
Y. K. Lim
  Chartered Semi.
A.S. Oates
  TSMC
Y.J. Park
  Texas Instruments
Wen Wu
  Novellus
Walter Yao
  AMD
Kristof Croes
  IMEC
Christine Hau-Riege
  Qualcomm
Baozhen Li
  IBM
Mingte Lin
  UMC
Guillaume Ribes
  STMicroelectronics

MEMORY

Hang-Ting Lue, *Chair*
  Macronix
Angelo Visconti, *Vice Chair*
  Nymonyx
Su Jin Ahn
  Samsung Electronics
Hanmant Belgal
  Intel
Steve Chung
  NCTU
Horacio Gasquet
  Freescale
Todd Marquart
  Micron
Christian Monzio Compagnoni
  Politecnico di Milano

Kiyomi Naruke
  Toshiba
Keum Hwan Noh
  Hynix

MICRO-ELECTRONIC SYSTEMS & MEMS

David Grosjean, *Chair*
  Analog Devices
Cora Salm, *Vice Chair*
  University of Twente
Osamu Tabata
  Univ. of Kyoto
Toshiyuki Tsuchiya
  Univ. of Kyoto
Martin Van der Heide
  Kionix
Paul van der Wel
  NXP

NANOELECTRONICS

Jeff Peterson, *Chair*
Intel

Curt Richter, *Vice Chair*
  NIST
An Chen
  AMD
Luigi Colombo
  Texas Instruments
Alain Diebold
  Univ. of Albany
Neeraj Kumar Jha
  TSMC
Yue Kuo
  Texas A&M
James Lu
  Georgia Tech

PROCESS AND INTEGRATION RELIABILITY

Terrence Hook, *Chair*
  IBM

Chris Connor, *Vice Chair*
  Intel
Koji Eriguchi
  Univ. Kyoto
Salvatore Lombardo
  CNR-IMM
Nobuyuki Mise
  Hitachi
Homi Mogul
  Qualcomm
Luigi Pantisano
  IMEC

PHOTOVOLTAIC DEVICES

David Albin, *Chair*
  NREL

Glenn Alers, *Vice Chair*
  UCSC

RELIABILITY IN MANUFACTURING

Sriram Kalpat, *Chair*
  Qualcomm
Lin Cong, *Vice Chair*
  Nvidia
Werner Kanert
  Infineon
Paul Ngan
  NXP
Gautam Verma
  Altera
James Walls
  Freescale
ShiJie Wen
  Cisco
Robert Wu
  Broadcom

SOFT ERRORS

Norbert Seifert, *Chair*
  Intel
Jeff Wilkinson, *Vice Chair*
  Medtronic
Rob Baumann
  Texas Instruments
Ethan Cannon
  Boeing
Mike Dion
  Rockwell Collins

Eishi Ibe
  Hitachi
Hajime Kobayashi
  Sony
Helmut Puchner
  Cypress Semiconductor
Philippe Roche
  STMicroelectronics
Yoshiharu Tosaka
  Fujitsu
Kevin Warren
  Vanderbilt University
ShiJie Wen
  Cicso
Georg Georgakos
  Infineon
Balaji Narasimham
  Broadcom
Charles Recchia
  Intel
Kenneth Rodbell
  IBM
Charles Slayman
  SunMicrosystems

THIN FILM DEVICES

Gaudenzio Meneghesso, *Chair*
  Univ. of Padua
Andrea Cester, *Vice Chair*
  Univ. Padua
Jan Genoe
  IMEC
Ryoichi Ishihara
  TU Delft
Yue Kuo
  Texas A&M

TRANSISTOR

Jason Campbell, *Chair*
  NIST
Chadwin Young, *Vice Chair*
  SEMATECH
Ben Kaczer
  IMEC
Felice Crupi
  University of Calabria
Tibor Grasser
  TU Vienna
Amr Haggag
  Freescale
Giuseppe La Rosa
  IBM
Ming-Fu Li
  Fudan University
Souvik Mahapatra
  IIT Bombay
David Esseni
  Univ. of Udine
Chetan Prasad
  Intel
Hans Reisinger
  Infineon

BOARD OF DIRECTORS

Ron Lacoe, Chair
The Aerospace Corp.

E. I. Cole, Jr.
Sandia National Labs

L. A. Kasprzak
Siemens

J.W. McPherson
Texas Instruments

N.R. Mielke
Intel

T. M. Moore
Omniprobe

H.A. Schafft
NIST

J. H. Stathis
IBM

J. S. Suehle
NIST

W.R. Tonti
IBM

CONSULTANTS

SITE SELECTION/ARRANGEMENTS/
EQUIPMENT DEMONSTRATIONS/AUDIO-
VISUAL/SIGNS

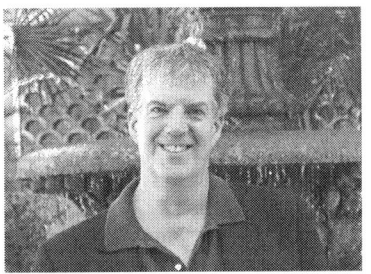

David Barber
Scien-Tech Assoc., Banner Elk, NC

PUBLICATIONS/REGISTRATION /
TECHNICAL PROGRAM SUPPORT/WEB
PAGE

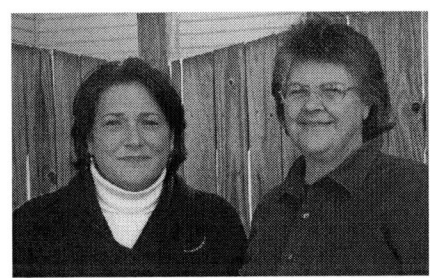

Phyllis Mahoney and Wendy Walker
Widerkehr and Associates
Montgomery Village, MD

HISTORIAN

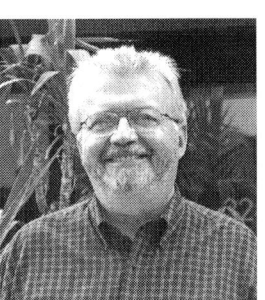

Bernie Petruchka

2009 IRPS Paper Awards to be Recognized at the 2010 IRPS

Best Paper Award

To: R.G. Filippi, P.-C. Wang, A. Brendler, P.S. McLaughlin, J. Poulin B. Redder,
J.R. Lloyd and J.J. Demarest, IBM

For the paper entitled, "The Effect of a Threshold Failure Time and Bimodal Behavior on the Electromigration Lifetime of Copper Interconnects"

Outstanding Paper Award

To: J.P. Campbell, J. Qin, K.P. Cheung, L.C. Yu, J.S. Suehle, A. Oates and K. Sheng,
NIST, University of Maryland, Rutgers University and TSMC

For the paper entitled, "Random Telegraph Noise in Highly Scaled nMOSFETs"

Best Poster Award

To: S. Malobabic, D. Ellis, J. J. Liou, J.A. Salcedo, J-J Jajjar and Y. Zhou,
University of Central Florida, Analog Devices

For the poster entitled, "Very Fast Transient Simulation and Measurement Methodology for ESD Technology Development"

2008 PAPER AWARDS

BEST PAPER AWARD

AN ENERGY-LEVEL PERSPECTIVE OF BIAS TEMPERATURE INSTABILITY

TIBOR GRASSER, WOLFGANG GOES AND BEN KACZER

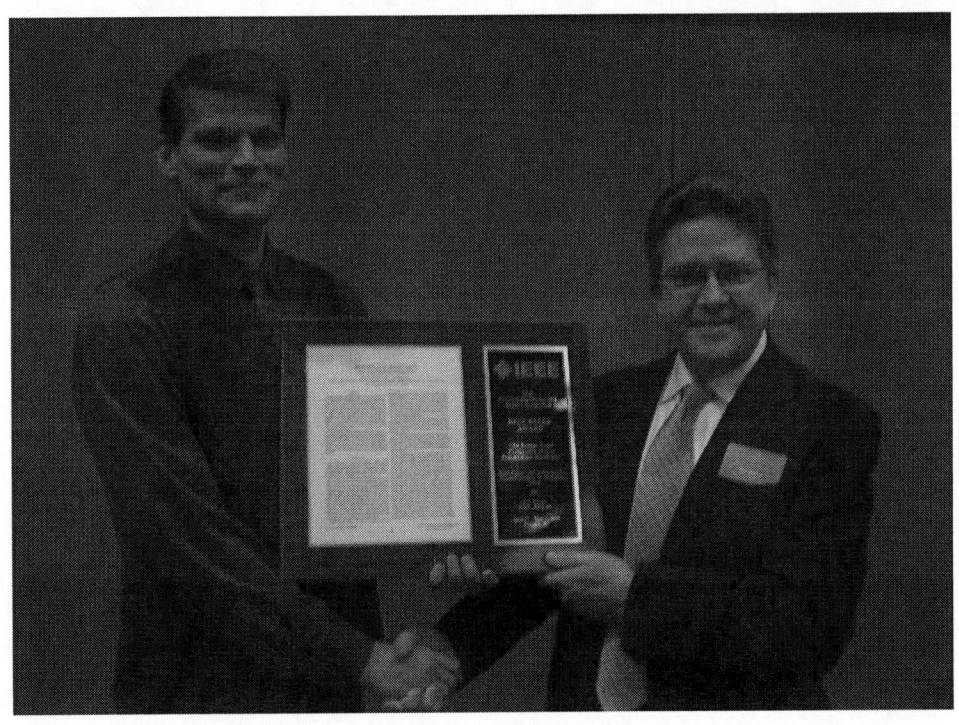

Tibor Grasser (left) accepting the Award from Tom Moore, 2008 Technical Program Chair

2008 PAPER AWARDS

BEST OUTSTANDING PAPER

PHYSICAL MODELING OF SINGLE-TRAP RTS
STATISTICAL DISTRIBUTION IN FLASH MEMORIES

ANDREA GHETTI, MAURO BONANOMI, ANGELO VISCONTI,
C. MONZIO COMPAGNONI, ALESSANDRO S. SPINELLI, AND ANDREA L. LACAITA

Angelo Visconti, Alessandro Spinelli and Christian Monzio Compagnoni (left) accepting the Award
from Tom Moore, 2008 Technical Program Chair

2008 PAPER AWARDS

BEST STUDENT PAPER

MODELING OF MAJORITY AND MINORITY CARRIER TRIGGERED EXTERNAL LATCHUP

FARZAN FARBIZ AND ELYSE ROSENBAUM

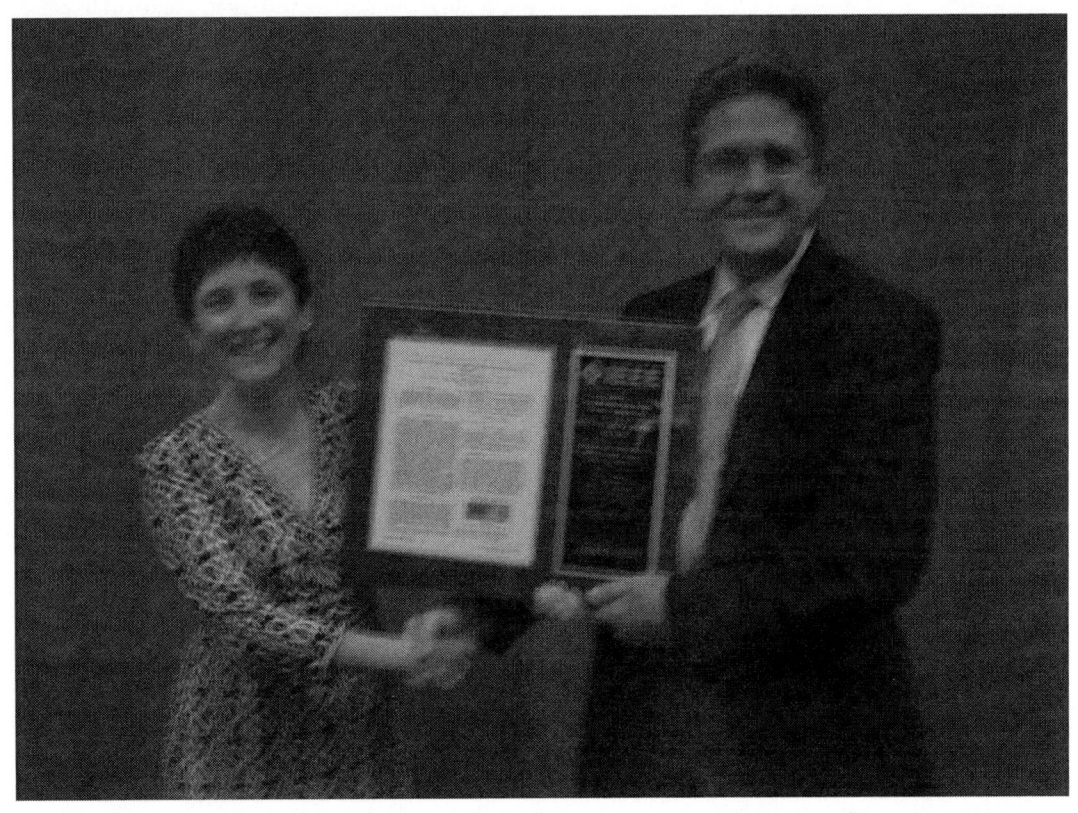

Elyse Rosenbaum (left) accepting the Award from Tom Moore, 2008 Technical Program Chair

2010 IRPS Tutorial Program

Mark Porter, Medtronics

[111] BASIC CONCEPTS AND METHODS FOR RELIABILITY DATA ANALYSIS — W. Meeker — Iowa State University—Reliability assurance processes in manufacturing industries require data-driven information for making product-design decisions. Life tests, accelerated life tests, and accelerated degradation tests are commonly used to collect reliability data. Data from products in the field provide another important source of useful reliability information. These reliability studies typically yield data that are censored and/or truncated, require the use of less familiar distributions like the Weibull, the lognormal, and the gamma, and call for inferences that involve extrapolation.

This tutorial will present and discuss the analyses of several different life data analysis applications in the area of product reliability and materials evaluation. The analyses illustrate the use of a mix of proven traditional techniques, enhanced and brought up to date with modern computer-based methodology. Methods used in the analyses include probability plotting, maximum likelihood estimation, analysis of data with multiple failure modes, acceleration models, accelerated life testing, accelerated degradation testing, and the analysis of recurrence data from repairable systems.

This tutorial will focus primarily on graphical presentation of the results of the analyses and will describe the applications, data, concepts, statistical methods, and interpretation of the results.

[112] EMERGING TRENDS IN FA FOR 32NM AND BELOW — M. Bruce – Independent Consultant —This tutorial describes emerging trends in FA for 32nm technology nodes and below. The state of tools and techniques will be described for design debug and identifying defects at a level required to keep pace with scaling. The semiconductor industry faces enormous challenges as defects move from being physically viewable to atomic level perturbations that cause circuits to malfunction. A holistic approach will be presented from design debug to manufacturing defect analysis.

[113] PARASITIC AND RELIABILITY ISSUES OF GAN-BASED HIGH ELECTRON MOBILITY TRANSISTORS — M. Damman and G. Meneghesso — Fraunhofer - Institut für Angewandte Festkörperphysik and University of Padova — Gallium Nitride represents an almost ideal material for the fabrication of high power microwave devices and circuits: its high energy gap (3.4 eV vs. 1.4 eV for GaAs) is reflected into a very high breakdown field (3500 kV/cm); piezoelectric and spontaneous polarization effects within AlGaN/GaN result in 2D gas densities above 10^{13} cm^{-2}, 5 times higher than for GaAs-based HEMTs, without requiring doping of the barrier layer. Saturation and overshoot velocity are around 3×10^7 cm/s, with relatively good electron mobility values (1200 cm^2/V.s). Epitaxial structures can be grown on silicon carbide with limited lattice mismatch, thus exploiting the excellent thermal conductivity and semi-insulating properties of this material, suitable for RF and microwave device operation. GaN HEMTs are therefore extremely promising for power electronics applications from power conditioning to microwave amplifiers and transmitters. Front-end applications are also interesting, due to the intrinsic robustness and survivability of high breakdown voltage GaN HEMTs, coupled with reasonable noise figures. Satellite communications, high performance radars and commercial ground base stations currently represent target system applications. Within these applications, it is expected that GaN-based transistors will replace the so far used Si-transistors in the near future mainly due to their larger bandwidth capabilities, their higher operating voltages as well as their higher linear efficiencies.

In the last years, GaN HEMTs have been subject to various optimization processes, starting from the material properties, to the control of surface and buffer properties aimed at reducing transient phenomena, gate-lag effects and the "current collapse" problems. A better control of short-channel effects, gate current, and degradation phenomena at high electric fields, together with the development of suitable structures for the management of the electric field (using T-shaped and Γ-shaped gates and field-plates) have lead to the progressive increase of the operating drain voltage from 12 V to 24 V and 48 V.
In this tutorial recent reliability data under DC and RF operation and the current understanding of the major degradation mechanisms will be presented. In particular, failure modes and mechanisms of GaN HEMTs, identified within the framework of various accelerated tests, including step-stress short tests (<150 hours) and life tests with a duration exceeding 3000 hours, will be presented. The proposed methodology includes a detailed characterization of the main parasitic effects in GaN HEMT devices (gate leakage current, current collapse, kink effect, etc.) by means of DC and pulsed electrical measurements. The observed failure modes are subsequently analyzed by two-dimensional device simulations in order to validate hypotheses on physical failure mechanisms. Electroluminescence microscopy and spectroscopy is adopted to evaluate hot carrier effects in GaN HEMTs, and as a powerful failure analysis tool.

[121] ADVANCED GATE STACK RELIABILITY: CORRELATING STRUCTURAL AND ELECTRICAL CHARACTERISTICS — G. Bersuker – SEMATECH — Relentless device scaling challenges the traditional "empirical" approach to device characterization, which becomes too costly and time consuming. Introduction of new materials and complex multi-component gate stack structures shift emphasis to instabilities caused by process-related defects rather than stress-generated ones, which are usually associated with the time dependency of device characteristics. This points to a growing need to identify the nature of defects affecting electrical characteristics of

devices, specifically their reliability, that would allow developing physics-based predictive degradation models, as well as provide helpful feedback to the fabrication process optimization efforts. The process development can be greatly accelerated when reliability characteristics can be correlated to the electrically-active atomic structural features of the dielectric stacks. In this tutorial, we focus on analysis and interpretation of electrical data provided by a variety of measurement techniques in an effort to understand the gate stack material characteristics contributing to instability of electrical parameters.

[122] NBTI: CONFUSION, FRUSTRATION, AND...PROMISE? — J. Campbell — NIST —The negative-bias temperature instability (NBTI) is a reliability problem that, in the last ten years, has risen from relative obscurity to become the most important reliability problem in advanced pMOSFET devices. Even though a significant effort has been spent trying to eliminate NBTI signatures (negative threshold voltage shift and transconductance degradation after inversion gate stress at elevated temperatures), the issue still persists. NBTI's elusiveness is due to the fact that NBTI-induced degradation relaxes very quickly after the conclusion of stress. This makes NBTI characterizations quite tedious and clouds the fundamental understanding of the degradation/relaxation mechanism. In high-$\square$ gate stacks, the situation is complicated further by inherent fast charge trapping issues and the emergence of an additional positive bias temperature instability (PBTI) component. While this depiction of NBTI may seem hopeless, there have been renewed efforts to uncover the fundamental physical mechanisms that govern the process. This tutorial summarizes noteworthy NBTI experimental observations/techniques and discusses how these observations might be leading towards a fundamental understanding of NBTI. A general aim is to examine how these observations validate/invalidate the current understanding of NBTI.

[123] NBTI MEASUREMENT: HOW TRICKY COULD IT GET? — D.S. Ang — Nanyang Technological University — Just when we thought we had completely understood the long-standing negative bias-temperature instability (NBTI) phenomenon, observation of non-negligible recovery effect has forced us to rethink the manner by which measurement of transistor degradation should be carried out. This tutorial will provide an overview of the numerous measurement methodologies proposed in recent years to address the challenge posed by NBTI recovery and discuss their relative merits and demerits. It will also introduce the ultrafast switching (UFS) method and show how this method could provide more comprehensive information regarding transistor degradation – threshold voltage shift, mobility degradation and charge pumping current – using a single experimental set-up. During the development of this technique, an important insight into the behavior of the drain current versus gate voltage curve in the high gate voltage regime was derived and will also be discussed. In short, we will show that the severe "bending" of the I_d-V_g curve in the high gate voltage regime, previously thought to be a consequence of mobility attenuation by surface roughness

scattering, is actually the result of interface degradation due to rapid charge trapping. This finding raises an important concern over NBTI characterization in the high gate voltage regime (e.g. the "on-the-fly" technique) using simplified I_d-V_g models. We will wrap up the talk with a discussion on some of the results obtained using the UFS method that may potentially offer a better understanding of the elusive NBTI phenomenon.

[124] CMOS CHANNEL HOT CARRIER QUALIFICATION FROM PHYSICS TO END OF LIFE PROJECTIONS — G. LaRosa and S. Rauch – IBM —This tutorial covers the state of the art understanding on the physics of channel hot carrier and currently experienced challenges with CMOS scaling. Its implications to a rigorous technology qualification methodology more closely related to circuit/product operations are also provided. It is shown how to apply DC CHC models, developed from accelerated stress conditions, to actual circuit waveforms. The use of idealized waveforms to produce simplified AC projection equations to be used in design manuals is explained. We cover recent proposals for the correct treatment of PMOSFET AC hot carrier, and the distinction between hot carrier and NBTI in PMOSFETs. These ideas apply to high-$\square$ technologies which exhibit NMOSFET PBTI as well. Finally, an overview of the industry adopted qualification targets to be used for CHC End of Life projections is given. As an example, the commonly used "T0.1" or similar lifetime adopted as a Foundry target is described and proven not to be indicative of actual product lifetime or failure rate and not providing guidance to ensure reliable system design and product test strategy.

[131] MECHANISMS, MODELING, MEASUREMENT AND MITIGATION OF SOFT ERRORS — C. Slayman, K. Warren, and J. Wilkinson – Independent Consultant, ISDE, Medtronic — Soft errors continue to be a key reliability concern for electronics. Unlike hard failures, a soft error's transient nature and varying manifestations makes it difficult to develop appropriate strategies for test and characterization. Without reliable data on the rate and modes of occurrence it is not possible to know if a system can meet its reliability objectives or to design appropriate mitigations. The tutorial addresses these questions for engineers working on systems for terrestrial applications. The tutorial is divided into 3 sessions of 1.5 hours each.

Physics of Soft Errors: The first session solidifies the participant's background in soft errors by describing the nuclear reactions, charge deposition and circuit responses that create soft errors. The session will cover the history of soft errors, their manifestations, and relevant test standards. Results from nuclear modeling will be introduced to illustrate key concepts.

Testing and Simulation: Session 2 takes the participants through the details of the JESD89A soft error testing standard including real-time testing and accelerated test methods for terrestrial cosmic rays, alpha particles and thermal neutrons. Descriptions of modeling to supplement test methods are presented. Specialty test methods such as ion microbeam and heavy ion testing will also be briefly described.

Mitigation and Case Studies: In the last session a variety of techniques are discussed for mitigating the effect of soft errors. The subjects will range from IC process adjustments, through circuits, and up to system level recovery. Case studies will also be reported from the literature and the authors' experiences.

[141] PHYSICS-BASED MATHEMATICAL MODELING OF BATTERIES — P. Gomadam — Medtronic Energy and Component Center —Batteries have current and potential applications as power sources to a very wide variety of electrical and electronic devices – from small handheld devices (e.g. watches and cell-phones) to those demanding very high energy and power (e.g. electric vehicles and grid storage systems). It is, therefore, critical to understand the working of batteries so as to be able to optimally design and reliably manufacture them for a given application. Building batteries and testing them in the lab is the most common approach to understanding performance, but it is highly resource-consuming and has limited applicability. Another approach is to understand the battery from the fundamental physical and chemical processes occurring during operation, and to use them in combination with mathematics, to develop a theoretical, predictive model of the battery. In addition to offering great insight, this approach drastically minimizes the expensive building-and-testing required, as well as significantly widens applicability. Further, for applications with very high reliability requirements, it is critical to have performance accurately characterized ahead of time and, thus, predictive mathematical models form necessary and valuable tools. Furthermore, many applications involve batteries operating together with mechanical components or as part of sophisticated electronic circuitry, where understanding the interaction between the battery and the other components of the system are important. Using the mathematical model of the battery, with the models of the other components, offers a most efficient and reliable approach to understanding the system.

In this tutorial, we present the development of a physics-based mathematical model that predicts the performance of a lithium-based non-rechargeable battery, relevant for a class of implantable medical devices. The mathematical model is a set of coupled ordinary differential and algebraic equations governing the electrochemical reaction kinetics of the battery, and requiring a numerical solution to predict battery performance. We present data obtained from individual testing of the main components of the battery electrodes, and show how they were compared with the model to obtain important thermodynamic and kinetic parameters. The model predictions, made using these parameters, are compared to the performance of manufactured batteries over a wide range of design and operating conditions. We present the comparisons and show that the model accurately predicts battery performance. Finally, we provide an overview of how battery models are typically used as part of implantable medical electronics product development.

[132] PROCESS INTEGRATION FOR COPPER INTERCONNECTS IN LOW-K DIELECTRICS — J. Gambino — IBM — Interconnect processes using copper wiring and low-□ dielectrics are reviewed for advanced technology nodes. First, the structure and properties of low-□ materials and barrier layers are described. Porous dielectrics are generally required to achieve □ < 2.5, but these materials are mechanically weak and chemically reactive, making integration very challenging.

Next, integration issues are described including patterning, cleans, metallization, chemical mechanical polishing (CMP), and packaging. The patterning process is becoming increasingly difficult because of the smaller dimensions, the need for reduced line width variation, and the use of dielectrics that contain carbon. Hardmasks are increasingly being used to solve problems such as resist erosion and resist strip damage of the low-□ material. Metallization is more difficult because the liner and Cu seed thicknesses must be drastically reduced, to avoid increasing the line resistance and allow void-free Cu plating to occur. The physical vapor deposition (PVD) techniques that have been used in previous technology generations are preferred for process simplicity. However, other deposition methods, such as atomic layer deposition (ALD) for the liner and direct plating to replace the seed layer, may be required as interconnect dimensions are reduced. CMP is increasingly challenging because of the low mechanical strength of the low-□ dielectrics and the need for reduced dishing and dielectric erosion. New planarization methods are being used to address these problems. Packaging is increasingly difficult because of the low mechanical strength of the dielectrics. Dicing, wirebonding, molding, and underfill processes must be optimized when packaging chips that have low-□ dielectrics in the stack. The effect of each of these interconnect processes on reliability will be discussed.

[142] FLASH MEMORY RELIABILITY: A COMBINED CHARACTERIZATION AND MODELING APPROACH — L. Larcher and P. Pavan — University of Modena — This course will present the basic concepts of the Flash memory reliability, employing dedicated physical models to understand mechanisms behind the observed reliability phenomena. This perspective provides both a simple explanation for experimental characterization results, as well as extraction guidelines for device reliability improvements.
The course will address the following points:
- Basic fundamentals of Flash memory reliability
- NOR and NAND Flash reliability requirements
- Key role of tunnel oxide: charge trapping and defect generation effects on endurance, disturbs and retention
- Reliability issues in nano-Flash devices (RTN, etc.)
- Reliability issues related to high-k material introduction (band-gap engineered tunneling barriers, charge trapped memory devices MANOS)

[143] SILICON SYSTEM DESIGN FOR HIGH AVAILABILITY APPLICATIONS — A. Silburt — Cisco Systems — This tutorial presents a methodology for developing a specification for soft error performance of an

integrated hardware/software system that must achieve highly reliable operation. The methodology enables tradeoffs between reliability and cost to be made during the early silicon design and SW architecture phase. An accelerated measurement technique using neutron beam irradiation is also described that ties the final system performance to the reliability model and specification. The methodology is illustrated for the design of a line card for an internet core router.

[144] RELIABILITY CHANLLENGES IN THE PHOTOVOLTAIC INDUSTRY — A. Terao — SunPower— Photovoltaic (PV) modules are essentially semiconductor devices, most of them made on slices of silicon. However, some of their characteristics make then very different than other electronic devices. They are optoelectronic devices handling high currents, encapsulated in large low-cost packages, stringed in series to create high DC voltages.

When it comes to reliability, these differences have important consequences in terms of failure analysis techniques, physics of failure, accelerated life test, statistical analysis, etc. The requirements are very high too: more than twenty-five-year life expectancy in uncontrolled outdoor stress conditions.

This tutorial will present the challenges faced by this young PV industry growing at an exponential rate. It will focus on the reliability of the main components of photovoltaic systems, namely solar cells and PV modules but will also briefly cover entire systems, from residential rooftops to utility-scale power plants.

[211] THEORETICAL ASPECTS OF RELIABILITY STATISTICS AND DATA ANALYSIS — K. Croes, P. Roussel, and G. Groeseneken – IMEC — The statistical aspects of different stages in reliability research will be theoretically discussed. Four such stages are considered: a) the design of the test structures, b) the choice of test conditions, c) the actual performance of the test and d) the final data analysis.

- The influence of test structure and choice of failure criterion on the final result of a reliability study will be discussed.
- The theory about planning reliability experiments will be covered. The question of how to obtain more reliable estimates with a given amount of test samples and test time will be addressed.
- The specifications of a test system (resolution, accuracy, gradients, etc.) will be linked to the final interpretation of the data.
- The analysis of the final data will be the bigger part of the tutorial. Here, we will introduce the different failure types, failure time distributions, the least-squares and maximum likelihood fitting method together with their specific confidence interval calculation techniques. Besides this, two more advanced models will be discussed: bimodal distributions and models for HBD distributions including SBD.

[221] MEMS TEST, YIELD, AND RELIABILITY — I. De Wolf – IMEC —This course will discuss MEMS reliability from a very broad perspective, starting with metrology tests addressing planarity and stress gradients; yield testing; functional testing using electrical, optical or mechanical actuation or detection schemes; and reliability testing.

Possible ways to tackle reliability, through standard testing or through a FMEA/physics-of-failure driven methodology will be presented.

A short overview and discussion of the large number of failure mechanisms that can occur in MEMS and MEMS packages will be given.

The main focus of the course is on characterization and test methods that can be used to study the reliability of MEMS and of wafer-level MEMS packages. In many cases existing techniques can be used or adapted for MEMS testing and reliability assessment through simple modifications. Several examples of this will be given. In some cases dedicated instrumentation has to be used.

MEMS reliability is often linked to materials related reliability issues. It requires information on material properties. Techniques and test structures that can be used to gather this information will be discussed.

It will be shown that several reliability issues in MEMS can be solved by a proper material choice and design, by packaging, or by applying alternative actuation schemes.

[222] RELIABILITY CHALLENGES OF 3D INTEGRATION — I. De Wolf – IMEC —3D integration is a technology that allows for the vertical stacking and connecting of layers of basic electronic components, such as integrated circuits (IC). This technology promises better performance and smaller and cheaper systems, linking various designs and applications (logic, memory, analog, passives, sensors, etc) together in 3D.

However, as is the case for each new technology, this might bring unknown reliability issues. 3D technology in general goes together with several non-standard processing steps such as the fabrication and filling of high aspect-ratio through silicon vias (TSVs), the thinning of chips or wafers, and the stacking and interconnection of these chips. Not only each of these processing steps, but also the total 3D stack brings new reliability issues.

In this course various yield and reliability related challenges encountered in the 3D-SIC (stacked IC) process which is under development in IMEC will be discussed. This include problems related to

- TSV formation: stress and stability of the Cu in the TSV, stress induced in the silicon and possible impact on active devices, barrier issues, possible effects of TSVs on the BEOL, effect of TSV on chip strength, etc.

- Chip thinning: thinning-induced damage in the Si or the devices, mechanical stress, potential thermo-mechanical effects, wafer strength, etc.
- Bonding: SnCu micro-bump reliability, IMC growth, Cu-Cu bonding issues, bonding-induced stresses and damage, particles, co-planarity issues, etc.
- Thermal issues in stacked ICs.

In addition some failure analysis challenges will be discussed.

[231] DESIGN IN RELIABILITY IN ADVANCED CMOS — V. Huard – STMicroelectronics — The continuous scaling of CMOS technologies down to sub-micron range inevitably yields to reliability challenges such as Negative Bias Temperature Instability (NBTI), Time-Dependent Dielectric Breakdown (TDDB) and Hot Carrier Injection (HCI). All these effects contribute to degrading transistor temporal performance though impacting overall product performance. Improvements in reliability were made historically by the means of process improvements which is proving to be more and more difficult in advanced CMOS nodes. Supporting product reliability in advanced nodes will require understanding, simulating and mitigating reliability aspects during both the process development stage as well as during the design stage.

Design for Reliability in advanced CMOS requires a set of predictive modeling and simulation tools called Design-In Reliability. This set should integrate leading edge reliability mechanisms, diagnose their impact on product operations, and allow evaluating performance/reliability trade-offs.

This tutorial will present both basic and advanced topics on reliability modeling and simulation tools, including: underlying reliability physics of leading reliability mechanisms, transistor-level reliability modeling approach within the framework of compact models, some ageing effects in full-custom designs (SRAM, RFCMOS, Analog) and simulation approach for hierarchical reliability analysis.

[232] ADVANCED ESD DESIGN AND QUALIFICATION ISSUES — C. Duvvury — Texas Instruments — Component level IC ESD has been constantly challenging for silicon technologies at every turn of the technology node scaling and with the development of higher speed circuits. These include the effects from new transistor structures to the recent trend for SoC. This tutorial will first review the most common protection design techniques used by the industry today and then describe the severe impact coming from technology scaling on the design capability. Equally important to the component ESD is the issue of system level protection at the board level. The tutorial will also address this important topic and give an insight into the strategies for coupling both types of protection design. Finally with the approach of 32nm and 22nm technologies there is now a new paradigm shift in ESD target levels for realistic and practical qualification. This roadmap for ESD will be reviewed.

[241] FAILURE ANALYSIS OF PHOTOVOLTAIC MODULES AND CELLS — G. Alers – USCS — The efficiency of a photovoltaic module can degrade through many mechanisms. Isolating these mechanisms and making quantitative measurements of degradation rates is the challenge of failure analysis. Failure mechanisms for modules and cells will be quite different. The normal procedure for understanding the failure of a module is (1) measuring the electrical performance including light and dark IV curves, (2) visual inspection, and (3) thermal imaging. Based on the results of these initial screening tests additional tests may be performed. This tutorial will begin with an introduction to the basics of how to analyze IV curves at the module and cell level. Thermal imaging techniques will then be reviewed including far-infrared, near infrared, lock-in thermography and thermal reflectance imaging. Electroluminescence and photoluminescence are additional evaluation tools that are now commonly used for fully packaged modules. One of the most versatile techniques is Laser Beam Induced Current (LBIC) which has been used extensively at the cell level to map out efficiency. At the cell level, most of the failure analysis tools used for the microelectronics industry are also applicable to photovoltaic materials. A list of common failure modes for both modules and cells will be presented with their characteristic signatures.

[242] BUILDING IN RELIABILITY AND QUALIFICATIONS OF PV ELECTRONICS PRODUCTS — P. Chaparala – National Semiconductor Corp. — In recent years, there has been a significant increase in deployment of distributed Balance of System (BOS) products such as power optimizers and micro-inverters in the photovoltaic industry. Market for smart solar panels, where IC electronics are integrated into panels for communication, security and power optimization is expected to grow significantly in the near future. Warranties of this new generation of power electronics products are expected to match with solar panel warranties of more than 20-years in outdoor environments. Lack of industry-standard reliability qualification tests and acceleration models makes it quite challenging to build and validate long-term reliability into PV-electronics products cost-effectively. In this tutorial, various PV-industry qualification standards, models and test methods to qualify PV-electronics will be reviewed. Also, building-in reliability methods at various product development phases, including concept, design, prototype and manufacturing of power electronic system products will be presented.

2010 IRPS RELIABILITY YEAR IN REVIEW SEMINAR

James H. Stathis, IBM

GATE DIELECTRICS - Jason Campbell - The gate dielectrics year-in-review includes a comprehensive examination of the past year's reports which detail gate stack reliability issues and the corresponding physical mechanisms which limit the performance and lifetimes of advanced devices. The intent of this presentation is to critically examine and publicize the most important advancements in understanding and mitigating various instabilities (BTI, TDDB, Noise, etc...) in a variety of technologically relevant (oxide, oxynitride, and high-k) gate stacks. In addition, this review includes a short discussion regarding reliability issues involved with the adoption of alternative channel materials and the instabilities associated with resistive/phase-change memories.

INTERCONNECT RELIABILITY - Fen Chen - This review will provide an overview of 2009-2010 publications dealing with interconnect reliability mechanisms including breakdown of BEOL dielectrics, electromigartion, and stress migration for an IRPS audience. The literatures of recent trends in BEOL interconnect process development such as doped Cu seeds, metal-cap, and airgap for 45nm and beyond, and in reliability analysis methodologies for addressing interconnect reliability challenges also will be reviewed.

FAILURE ANALYSIS - Philippe Perdu - "More Moore" and "More than Moore" trends have triggered incredible challenges for microelectronic failure analysis. It mostly concerns Fault Isolation, Chip access, Defect Localization, Sample Preparation and Physical analysis. In addition to a general overview through key papers, a more specific focus on defect localization will show the advanced solutions developed to find the defect.

NANO-CMOS - Shinichi Takagi - The Nano-electronics session in YIR will focus on the recent progress of nano-scaled MOSFETs. After brief summary of advanced CMOS platform and the critical issues, three directions of future nano FETs, ultrathin body/3D CMOS, new channel material CMOS and emerging ultralow voltage FET concepts will be reviewed.

A review on the reliability of GaN-based laser diodes

Nicola Trivellin, Matteo Meneghini, Enrico Zanoni
Department of Information Engineering
University of Padova
Padova, Italy
+39 049 827 7625, nicola.trivellin@dei.unipd.it

Kenji Orita, Masaaki Yuri,
Tsuyoshi Tanaka, Daisuke Ueda
Panasonic Corp.
Takatsuki City, Osaka, Japan

Gaudenzio Meneghesso
Department of Information Engineering
University of Padova
Padova, Italy

Abstract— **University of Padova in collaboration with Panasonic Corp. has developed in the recent years an in depth reliability analysis of Blu-Ray InGaN Laser Diodes (LD) submitted to CW stress at different driving conditions. The reliability analysis has been focused towards a) the identification of the effects of current, temperature and optical field and b) the identification of the physical mechanism related to degradation. Results show that LD devices exhibit a gradual threshold current increase, while slope efficiency is almost not affected by the ageing treatment. Degradation rate is found to depend on stress temperature and on current level, while it does not significantly depend on the optical field in the cavity. Within this paper we demonstrate that: (i) the degradation rate shows a linear correlation with stress current level; (ii) the Ith increase is correlated to the decrease in non-radiative lifetime (τ_{nr}); (iii) stress temperature acts as an accelerating factor for LD degradation; (iv) pure thermal storage does not significantly degrade LDs characteristics.**

Keywords: Laser diode, Gallium Nitride, reliability, non raditive lifetime, degradation.

I. INTRODUCTION

Since the first introduction of Compact Disc (CD) systems in 1982, optical data storage has become more and more fundamental for mass storage applications. Over the last decade, the evolution of this technology led to a rapid increase in data storage capacity, from CDs' 650MB to 4.7 GB/layer of DVDs and reaching 25 GB/layer in 2006 with Blu-Ray Disc (BD). These achievements have been reached thanks to the intensive research carried out on the LD sources. In particular, Blu-Ray technology requires the use of low-wavelength laser diodes, emitting at 405 nm: this kind of devices can be grown using gallium nitride, a wide bandgap material that has recently gained attention due to its excellent optoelectronic properties. Since the first demonstration of pulsed LD operation by Nakamura in 1996 [1], InGaN LD technology has been subjected to significant improvement in order to increase both output power, needed for high speed data writing, and reliability, required for mass market applications. In particular over the last years, intensive research has been carried out to study the causes of device failure and limited lifetime. Some

authors [2] have indicated impurity diffusion through threading dislocations as a possible cause for the threshold current increase detected during stress. However only few works have analyzed the dependence of the degradation kinetics on the operating conditions like current, temperature and optical power. With this work we have extensively analyzed the effects of different operating conditions on device optical and electrical characteristics in order to understand the various mechanisms responsible for device degradation. Through a technique proposed by Van Opdorp [3], we have investigated the effect of stress on the non-radiative recombination rates, in order to confirm that the threshold current increase is strongly correlated to the degradation of the radiative efficiency of the active region. Finally, with this work we demonstrate that InGaN LD degradation is an electro-thermally activated mechanism, possibly correlated to the increase in defect density during operation.

II. EXPERIMENTAL DETAILS

The reliability studies were carried out on InGaN/GaN LDs. Devices are grown on GaN substrate by use of Metal Organic Vapor Deposition (MOCVD). Devices structure consist in an AlGaN cladding layer, an n-GaN guiding layer, a triple InGaN/GaN quantum-well structure, a p-AlGaN electron blocking layer, and a p-GaN contact layer. Devices active region is composed by three InGaN quantum wells (QW), emitting at 3.06 eV. High density carrier injection is obtained with superficial ohmic stripe contact, the optical confinement is achieved by gain guiding, thus generating an optical cavity of 1.5 x 600 µm. Average threshold current density and slope efficiency of the LDs were measured to be 3.2 kA/cm^2 (about 29 mA) and 1.6 W/A for the analyzed samples.

Several devices has been subjected to electrothermal stresses, with a maximum of 6 devices per each stress condition. By means of temperature-controlled (TEC) fixtures, controlled by an ILX Lightwave LDC-3916 instrument, LDs were submitted to current stress at different case temperature T_c. The use of active TEC system guarantees minimal self-heating. Using the method described by Xi et al. [4] we have

978-1-4244-5430-3/10 $26.00 © 2010 IEEE

calculated the junction temperature Tj during stress. This method is based on two different phases: thermal mapping and self heating transient. The result from the thermal mapping is an I-V-T map of the device, it is obtained with ultrashort pulsed (~4us, to avoid self heating during measurement) electrical characterization of the device at different fixed temperature. Temperature has been controlled by means of TEC peltier cells. In the second phase to determine the final temperature during self heating, the LD voltage is monitored while a constant current bias is applied to the device at a fixed case temperature. It is therefore possible to determine operating temperature from the thermal map. Results showed that for T_c equal to 30°C, 50°C and 70°C the corresponding T_j during 70 mA stress is 44°C, 63°C, and 80°C respectively. By means of this method, it is also possible to evaluate the thermal resistance: we have calculated that the junction-to-fixture thermal resistance of the samples is about 52 K/W.

High sensitivity subthreshold optical measurement and electrical characterization have been performed by means of an HP 4155A Semiconductor Parameter Analyzer and a large area photodiode. In order to better understand the degradation process, we have defined a technique for the evaluation of the non-radiative recombination lifetime (τ_{nr}) of the carriers in the active region. This technique is based on the method proposed by Van Opdorp [3], and allows to correlate the degradation of the lasing properties of the laser diodes to the worsening of the radiative properties of active region. By this technique it is possible to extrapolate τ_{nr} from subthreshold optical measurements. A brief description of this technique is reported in the following.

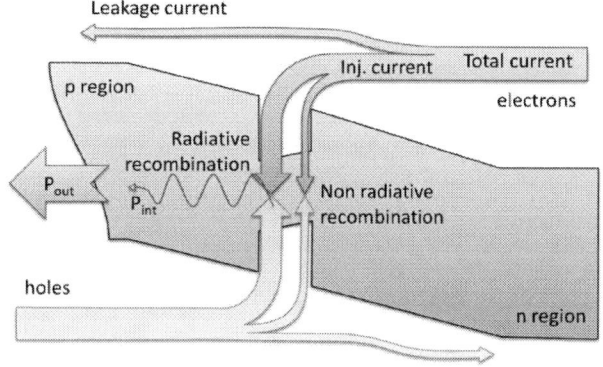

Fig.1: Schematic view of the active region and total current splitted in leakage, radiative and non-radiative current.

As presented if Fig. 1 the current flowing through the device can be expressed as the sum of the current leaking through shunt paths (I_{leak}) and a third order polynomial accounting for radiative, non-radiative and Auger recombination. Assuming the electron (n) and hole (p) densities in the QWs to be equal, from [5] we obtain that the non-radiative term is expressed as An, the radiative term as Bn^2 and the Auger as Cn^3.

$$I = qV(An + Bn^2 + Cn^3) + I_{leak} \qquad (1)$$

It is worth noticing that in the following we will consider the Auger term negligible, because C is in the order of 10^{30} cm^6/s and the analysis is focused in the low current density polarization region. Considering the external efficiency (η_{ext}), the injection efficiency (η_{inj}), the volume of the cavity (V), the energy of the emitted photon (hν), the output emitted power (P_{out}), the extraction efficiency (E) and the radiative emission coefficient (B), we can obtain the following relation:

$$\frac{1}{\eta_{ext}} = \frac{1}{E\eta_{inj}} + \frac{1}{\tau_{nr}\eta_{inj}} \sqrt{\frac{V}{EB}} \sqrt{\frac{1}{P_{out}/(h\nu)}} \qquad (2)$$

It is therefore possible to extrapolate the non radiative lifetime τ_{nr} from the slope of a plot of $1/\eta_{ext}$ as a function of P_{out} in the subthreshold region. .

III. RESULTS

When the devices are submitted to constant current stress, a gradual degradation is detectable. The degradation mostly consists in a threshold current (I_{th}) increase, while slope efficiency (SE) is almost not affected by ageing treatment. Fig.1 shows the comparison between a fresh device and the same device subjected to a stress test carried out at a constant current of 100mA with a case temperature of 75°C (resulting in a junction temperature of 83°C); the ageing duration is 2000h. Since the reliability of the LD is mostly related to increase in Ith, we have considered this term as a degradation monitor. Applying the method introduced by Van Opdorp [3] we have plotted the variation in the $1/\eta_{ext}$ plot, Fig. 3, before and after the degradation (with the same condition as Fig. 2). A clear increase in slope in the linear region is detected, indicating a decrease in non radiative lifetime during degradation. To completely understand the ageing behavior of the LD devices we have extrapolated the dependence of the I_{th} as a function of current, temperature and optical power. As a consequence, using the method already described, non radiative lifetime has been extrapolated and compared with the degradation kinetics of the LDs.

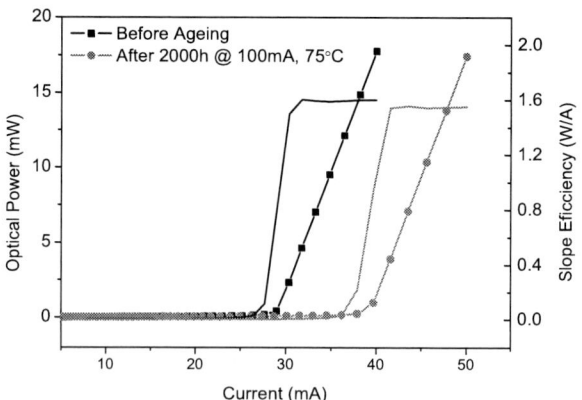

Fig.2: Effects of stress on optical characteristics of a LD, a sensible increase of threshold current is detectable, but almost no variation in slope efficiency, all measurement are performed at 25°C.

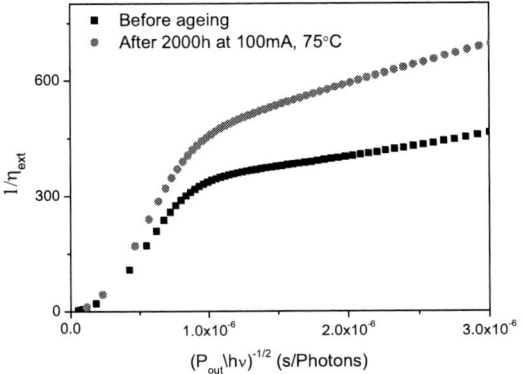

Fig.3: A clear increase of slope is noticeable in the $1/\eta_{ext}$ plot as a consequence of stress; this imply a decrease in non radiative lifetime, measurement performed at 25°C.

A. Effects of driving current

The results of stress tests carried out for different LD devices at various current levels (in the range 40 - 100mA, corresponding to 4.4 and 11.1 kA/cm^2 respectively) with a constant case temperature of 75°C, are summarized in Fig. 4 and 5. Previous studies demonstrated that the I_{th} increase has a square root dependence on stress time [6-7]. This behavior is confirmed by Fig. 4: experimental data can be fitted accordingly to:

$$\Delta I_{th} = I_0 + D\sqrt{t} \qquad (3)$$

Where I_0 is close to zero, while D is a prefactor that defines the degradation rate.

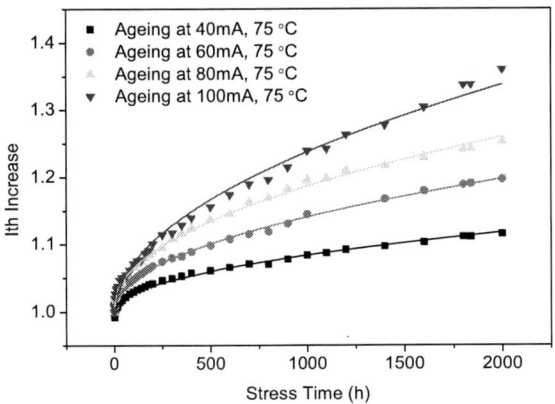

Fig.4: The kinetics of threshold current increase for different driving currents, the solid lines represent square root of time fitting of the data, measurement performed at 25°C.

Results show that the degradation kinetics are strongly determined by the stress current level: to completely understand the role of stress current in determining the Ith increase, we have plotted the degradation parameter D as a function of stress current (Fig. 5). Results demonstrate that an increase in stress current induces a linear increase in the degradation rate.

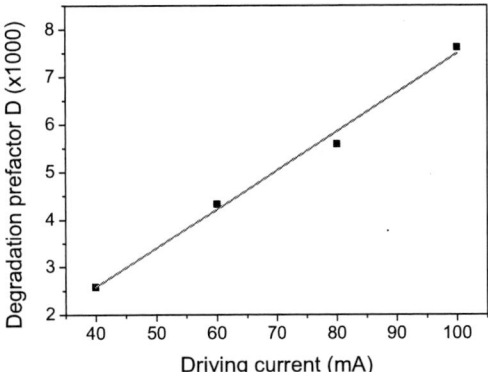

Fig.5: Plot of prefactor D, obtained by fitting after 2000h stress test, as a function of driving current.

In Fig. 6 we report the effects of current-induced degradation on the electrical characteristics of the devices. Stress determines an increase in LD reverse current and low-voltage forward current. The magnitude of the reverse-current increase was different for each analyzed LD, indicating a dependence on sample quality. This effect can be attributed to an increase of the parasitic leakage paths due to the generation of defects, and/or to the generation of parasitic conduction paths at the active region sidewalls.

In Fig. 7 we report the subthreshold EL spectrum at 20mA of a device before and after 2000h of current stress at 80mA, 70°C.

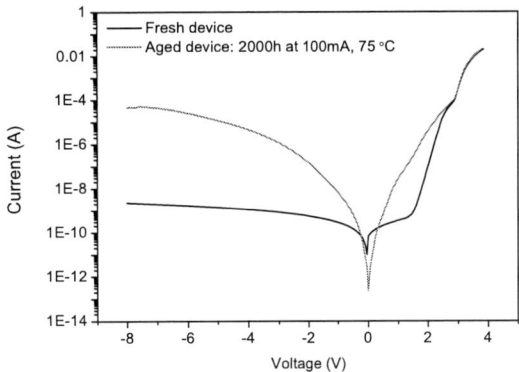

Fig.6: Comparison of electrical characteristics of an untreated LD and a device subjected to 100mA, 75°C stress for 2000h.

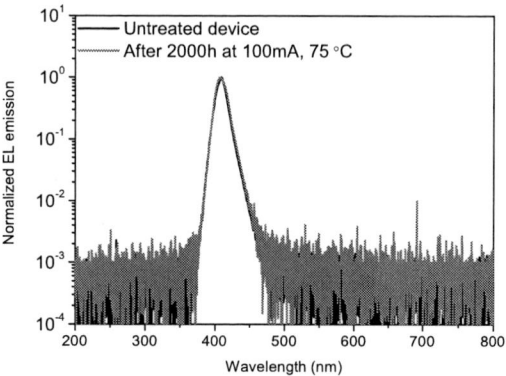

Fig.7: Effects of stress on EL output from LD, measurement performed in subthreshold region, 20mA at 25°C.

B. Temperature effects

To further understand LD degradation process we have submitted the devices to constant optical power stress at different temperatures (ranging from 50 to 80°C). Results show that an increase in the operating temperature determines a faster degradation (Fig. 8). With this data it has been possible to extrapolate the activation energy for the temperature, this value is 250 meV. This result indicate a quite low value implying that an increase in the operating temperature determines only a limited increase of the degradation rate. Up to now we have only showed effects of temperature during electrical stress. To clarify the role of temperature, it is necessary to understand if LD degradation is also present in high temperature storage stress tests. Pure thermal reliability tests (at 180°C) have been performed, whose results are presented in Fig.8.

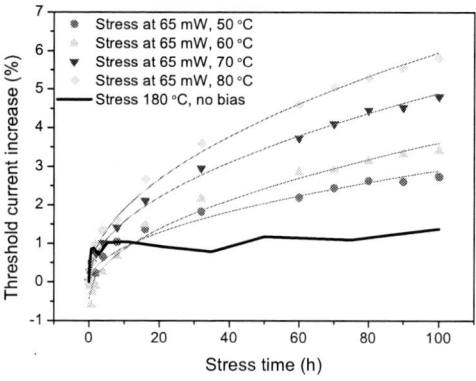

Fig.8: The kinetics of threshold current increase for different operating temperature, the solid red lines represent square root of time fitting of the data, the black solid line represent the kinetic of a pure thermal stress carried out at 180°C, measurement performed at 25°C.

During high temperature stress tests (no bias applied to the junction) only a limited degradation is observable, with a kinetic extinguishing after few hours. Since temperature alone does not induce a strong degradation, electrical and/or optical stress must be considered as the main cause for devices degradation. These last results confirm that temperature only acts as an accelerating factor for LDs degradation.

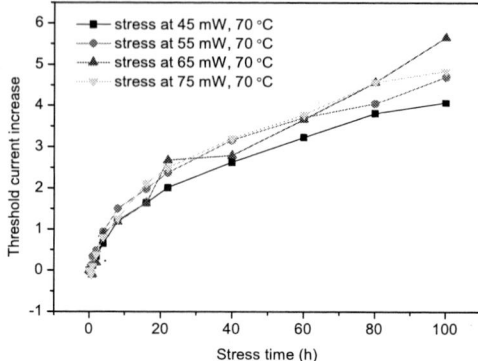

Fig.9: The kinetics of threshold current increase for different output optical power, measurement performed at 25°C.

C. Optical field effects

Optical field is generally a factor not considered for GaN LED devices, because of the low optical power density which is involved in these structures and the incoherency of the light emitted. LD devices have a structure much more complicated that could be influenced by the high optical field and power density.

Stress carried out at different optical filed are presented in Fig. 9, with an output optical power of 45 to 75mW, and a temperature of 70°C. Results have shown that the optical field has a limited interaction with the device degradation being the variation of degradation between measurement errors. This result imply that non-radiative lifetime and defects density are modified by the optical field; also indicate a good stability of the device, in particular DBR mirror structures to optical field.

D. Non radiative lifetime variations

By means of the technique described above we have extrapolated the non-radiative recombination rate A (proportional to $1/\tau_{nr}$) during device degradation. As a result, we have found that stress induces the increase in the non-radiative recombination rate, $(A=1/\tau_{nr})$, well correlated with the Ith increase. It is important to remark that these two parameters (Ith and τ_{nr}) have been evaluated with different and uncorrelated measurement procedures. A summary of the 300 h stress test carried out for various LDs at different current levels (from 40 to 100mA) is plotted in Fig. 10. Results show a linear correlation between A and Ith variation, thus indicating that degradation is correlated to the increase in non-radiative recombination rate. This work thus indicates that the LD degradation is a phenomena related to the decrease in carrier lifetime. To support our thesis we report here the variation in electrical characteristic of one LD during the stress tests (Fig. 5). The visible increase in the reverse bias current indicates an increase in generation/recombination factors inside the active region.

978-1-4244-5430-3/10 $26.00 © 2010 IEEE

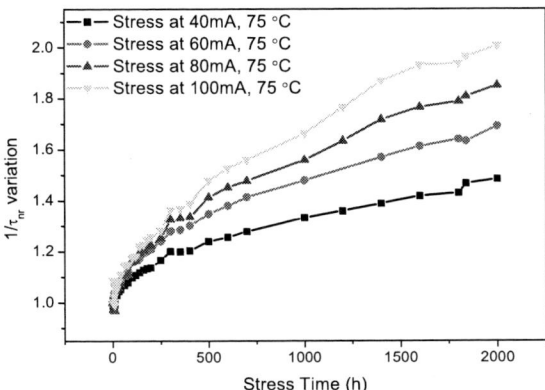

Fig.10: Plot of variation in non radiative lifetime during device degradation.

To understand if the variation in carriers lifetime is related to non-radiative or intra-band radiative recombination we have reported here the results of normalized EL spectrum measured before and after the degradation (Fig.6). Results show that no new peaks nor emission shift are generated during the LD lifetime, confirming that the recombination lifetime decrease is due to solely non-radiative paths.

E. Role of different current levels

Fig. 10 shows the correlation between $A = 1/\tau_{nr}$ and Ith where each dot represent this correlation for a different ageing time. The linear behavior indicate that a strict correlation between non radiative lifetime and threshold current is present also for different driving currents. Moreover the plots have the same slope for different current levels, indicating that the degradation mechanism is the same, but amplified by the increased current level. This result further confirms that the degradation is an electro-thermally activated phenomena.

IV. CONCLUSION

In conclusion with this work we have analyzed the degradation of 405 nm Blu-Ray LDs. We have demonstrated that the degradation kinetics are strongly determined by the stress current level. Moreover thermal Activation Energy has been extrapolated to be equal to 250 meV [8]. This value indicates that the temperature does act as an accelerating factor for device degradation: however, temperature has a limited effect on LD reliability, since the activation energy value is quite low compared to previous reports [9, 10]. With this reliability study we wanted to investigate the physical mechanism responsible for degradation. By means of optical measurement it has been possible to extrapolate the non-radiative recombination rate of the carriers. By this technique we have demonstrated that Ith increase is strongly related to τ_{nr} decrease, thus indicating that point defects propagation or generation in the active region occurs during device degradation. Defect density increase in the active region can also enhance the recombination and reverse bias current. The gradual increase in LD reverse current, observed in electrical measurement, supports this hypothesis. The EL spectrum does not vary significantly before/after stress: this fact confirms that no trap-assisted radiative recombination is generated during stress.

In conclusion our reliability study demonstrates that gradual degradation of Blu-Ray InGaN LDs is an electro-thermally activated process caused by the increase in non-radiative recombination rate that reduces active region efficiency. Therefore Blu-Ray LDs reliability may be ameliorated with an optimized growth procedure, focused on dislocation and point defect reduction: this approach will reduce the diffusion of impurities from the bulk to the active region.

ACKNOWLEDGMENT

This work has been co-funded by Panasonic Corp.

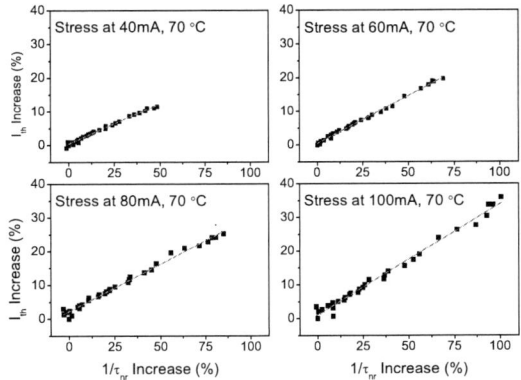

Fig.10: The four plots represent the correlation between threshold current increase and non radiative lifetime increase during LD ageing for different driving currents.

REFERENCES

[1] S. Nakamura et al. InGaN-based multi-quantum-well structure laser diodes Jpn. J. Appl. Phys. 1996; 35, L74-76.

[2] O. H. Nam, K. H. Ha, J. S. Kwak, S. N. Lee, K. K. Choi, T. H. Chang, et al. Characteristics of GaN-based laser diodes for post-DVD applications phys. stat. sol. (a) 2004; 201, No. 12 2717-20.

[3] C. van Opdorp, Method for determining effective nonradiative lifetime and leakage losses in double-heterostructure lasers, J. Appl. Phys. 1981; Vol. 52, pp. 3827.

[4] Y. Xi and E. F. Schubert, Junction-temperature measurement in GaN ultraviolet light emitting diodes using diode forward voltage method, Appl. Phys. Lett. 2004; vol. 85, pp. 2163-5.

[5] R. Olshansky, J. Manning and W. Powazinik, Measurement of Radiative and Nonradiative Recombination Rates in InGaAsP and AlGaAs Light Sources IEEE Journal of Quantum Electronics, 1984; Vol. Qe-20, No. 8.

[6] L. Marona, P. Wisniewski, P. Prystawko, I. Grzegory, T. Suski, S. Porowski, et al. , Degradation mechanisms in InGaN laser diodes grown on bulk GaN crystals, Appl. Phys. Lett. 2006; vol. 88, pp. 201111-3.

[7] S. Tomiya, T. Hino, S. Goto, M. Takeya, and M. Ikeda, Dislocation related issues in the degradation of GaN-based laser diodes, IEEE

Journal of Selected Topics in Quantum Electronics 2004; vol. 10, no. 6, pp. 1277-86.

[8] M. Meneghini, G. Meneghesso, N. Trivellin, E. Zanoni, K. Orita, M. Yuri, and D. Ueda, Extensive analysis of the degradation of Blu-Ray laser diodes, IEEE Electron Device Letters 2008; vol. 9, no. 6, pp. 578-81.

[9] S. Nakamura, InGaN-based laser diodes with an estimated lifetime of longer than 10,000 hours, Proc. SPIE 1998vol. 3283 (2), pp. 2-13.

[10] M. Kneissl, D. P. Bour, L. Romano, C. G. Van de Walle, J. E. Northrup, W. S. Wong, et al., Performance and degradation of continuous-wave InGaN multiple-quantum-well laser diodes on epitaxially laterally overgrown GaN substrates, Appl. Phys. Lett. Vol. 77, 2000 No. 13, pp. 1931-3.

The statistical analysis of individual defects constituting NBTI and its implications for modeling DC- and AC-stress

Hans Reisinger, Tibor Grasser * , Wolfgang Gustin and Christian Schlünder
Infineon Technologies AG
München, Germany
Email: Hans.Reisinger@infineon.com

* Institute for Microelectronics, TU Wien
Wien, Austria

Abstract — **The physical origin of the Negative Bias Temperature Instability (NBTI) is still under debate. In this work we analyze the single defects constituting NBTI. We introduce a new measurement technique stimulating a charging of these defects. By employing a statistical analysis of many stochastic stimulation processes of the same defect we are able to determine the electric field and the temperature dependence of these defects with great precision. Based on our experiments we present and verify a new, physics-based, quantitative model allowing a precise prediction of NBTI degradation and recovery. This model takes the stress history into account and also provides a prediction for degradation due to AC-NBTI and an understanding of the special features seen in conjunction with AC-NBTI.**

Keywords-NBTI; recovery; random telegraph noise; AC-stress;

INTRODUCTION

30 years of research [1] still did not bring about any consensus regarding the physical origin of NBTI. So far only ESR experiments [2] have provided some insight into the microscopic processes. The majority of electrical experiments has been done on large pFETs and thus averages over many defects (see table 1). We will show that this averaging obscures the properties of the individual defects and leads to misinterpretations, for example of the thermal activation. In this work we analyze the properties of the single, discrete events constituting NBTI degradation and recovery in small FETs. We have developed analysis techniques based on those known from the Random Telegraph Signal (RTS) studies of the 1980s [3, 4]. RTS is caused by capture and emission of charge in oxide traps. To our knowledge RTS studies have been done for nMOSFETs only [4], the FETs kept in a quasi steady state. In contrast, our method uses multiple high-gate-field charging pulses to excite the defect state to a charged state. This technique is similar to the deep level transient spectroscopy (DLTS) [6], but it is essential to vary the length of the charging pulses, because of the wide distribution of capture time constants. This is why we term the technique time dependent defect spectroscopy (TDDS). Stimulated by recent work on small pFETs [5, 7] we employ for the first time a highly accurate extraction of NBTI-relevant defect parameters. Capture and emission time constants τ_c and τ_e, (corresponding to stress and recovery) have been seen before [5, 8]. However, to extract temperature- and field-dependencies of those τ_cs and τ_es, and thus to gain insight in physical processes, a highly accurate determination of averaged time constants is required. Thus we extract averaged values for τ_c and τ_e from repetitive measurements. Typically 100 or more excitations and measurements are done for each τ_c and τ_e determination for a

given defect and condition. Main parameters are the gate voltage, the length of the excitation pulse and the temperature. The purpose of the present work is to give an overview over the implications regarding our new understanding of the NBTI phenomena and a modeling of NBTI which includes recovery and AC-stress. A detailed description of a mathematical model which can be used in a degradation simulator and the corresponding parameter extraction as well as a description of our AC-stress experiments (effect of frequency, measuring delay, etc.) both are beyond the scope of this paper and will be presented in subsequent publications.

We will begin with a detailed description of our new measurement technique in section I followed by a short discussion of our sample properties and geometry dependent parameters in section II. Section III will discuss the basic properties of our physical model. Since the exact nature of the involved defects cannot be determined from an electrical characterization the model will remain phenomenological. A real microscopic, physical defect model as well as more and more detailed extractions of the defect properties are found in [9, 10]. The statistical analysis and defect parameter extraction will be demonstrated in section IV. Our mathematical model in section V, based on the defect analysis, will be used to discuss the missing thermal activation for NBTI short-term stress in VI. From section VII on we switch back to data from large FETs, treating practical modeling and model verification. AC-NBTI features will be shown and explained in section VIII. We conclude with a short discussion of permanent degradation in IX followed by the conclusions section X.

I. EXPERIMENTS

Our defect spectroscopy experiments were done in the following way: A stress pulse was applied to the gate in order to occupy (i.e. to

FET-name and dimension (in µm)	wide		narrow		minimal	
	W	L	W	L	W	L
	10	0.1	0.2	0.12	0.11	0.1
number of carriers in channel at Vg=VT-200mV	15000		370		170	
number #Nit at a density DNit=1E11/cm^2	1000		24		11	
ΔDNit causing a ΔVT=50mV (in cm^{-2})	4.9×10^{11}		4.9×10^{11}		4.9×10^{11}	
makes a number Δ#Nit	4900		120		50	
ΔVT caused by a single trapped carrier (at interface)	0.01mV		0.43mV		<u>1.0 mV</u>	

Table 1 Geometries of the used pFETs and some useful numbers. Oxide thickness is 2.2nm. Measured specific capacitance in inversion is 1.3µF/cm^2. Real measured step-heights for the minimal FET (cmp. underlined value) range from 0.2mV to 5mV. DNit denotes the charged interface-states density.

978-1-4244-5430-3/10 $26.00 © 2010 IEEE

charge positively) individual defects. After the end of the charging pulse the threshold voltage *VT* was continuously recorded for a time of typically 1000s. In order to determine different capture time constants the stress pulses ranged from 200ns to 100s. *VT* was directly recorded using our fast feedback loop [11]. In contrast to a direct measurement of the steps in the drain current *ID* with a fixed gate voltage *Vgate* our method is somewhat simpler because it saves the effort of recalculating the *VT*-step from the *ID*-step (which varies with *ID*). Capture and emission are stochastic processes. Thus each pulse/readout was done repetitively (32 to 256 times) to allow a statistical analysis. A rough determination of individual capture and emission times (within a factor 2, like in Fig. 3, for example) could be done by recording just a few traces. However, if precise parameter dependencies (on *Vgate*, or T, cmp. Fig. 8) are required a statistical analysis of **many** traces has to be done. To ensure the same pre-pulse condition for each pulse/readout sequence full recovery had to be awaited after each stress pulse (cmp. Fig. 5). A measuring sequence for a given gate voltage/temperature, just varying the stress pulse width from µs to 10s, takes a time of several days. It is worth noting that all information, also about capture, is gathered during recovery. Capture events occur in the stress period and in principle could be sensed from *ID*-steps occurring during stress. However, the total number in the channel of a minimal FET during recovery is 200 (cmp. table 1). At stress, in strong inversion this number is higher by a factor of >10. Thus a single carrier has an effect in *ID* smaller by more than a factor 10, and is below the noise level in general.

II. SAMPLES

We use production-quality pFETs with a 2.2nm nitrided oxide. Such oxides around 2nm have been the subject of many NBTI-investigations of different research groups (see references in [11]). To anticipate criticism claiming our samples would exhibit an extraordinarily high defect level or defects of a special kind Fig.1 shows an inter-technology comparison of a couple of oxides. This comparison is done by a normalization of the degradation *ΔVT* resulting in the surface density of charged defects (see right hand scale of Fig. 1). It includes a series of "clean", non-nitrided oxide with different thicknesses as a reference and different samples published in the literature with a moderate nitridation level of about 6%. Fig. 1 shows that our samples' defect density is at the same level or better than samples studied in other NBTI work.

We have chosen the geometries to be small enough to conveniently resolve the effect from a single carrier in the channel

(typical *ΔVT*=1mV, see table 1) and large enough to have at least a handful of active defects (≈20 average step-height defects for *ΔVT*=20mV). An overview of geometry-dependent sample properties is given in table 1. The data in table 1 were calculated using the simple charge-sheet model and thus give only average step heights in *VT*. Real individual step heights are long known to be larger or smaller than this average by factors of 10 [4]. The origin and size of step height variations in FETs of a similar geometry as the ones used here is explained in [15]. The step height of each individual defect is its "fingerprint", allowing to distinguish between the defects. In order to increase this difference in the individual step heights all our measurements were done in saturation. During the measurements, i.e. the recovery traces (cmp. Fig. 3) the drain current *ID* was kept constant while recording the gate voltage *VT*. A typical level for *ID* was 1µA**W/L*. For our "minimal FETs" the precision we achieved is sufficient to determine minimum *ΔVT* step heights of 0.2mV. This value is well below the average expected step height of 1mV (see table 1). Based on the analysis of our single defects we will draw conclusions about the NBTI phenomena observed in large FETs. So it has to be made sure that in fact these single defects are responsible for NBTI and **do not** just represent a special kind of defects playing a minor role for "real" NBTI. For this purpose Fig. 2 shows the comparison of a wide FET with a narrow FET. In order to average over the statistical variation of degradation in the small FETs 25 of these small FETs have been measured sequentially and summed up. An example of recovery traces, for a single small FET and the sum over 25 small FETs is shown in Fig. 3. All these traces are showing the clear, distinct steps in *VT* from single defects. The

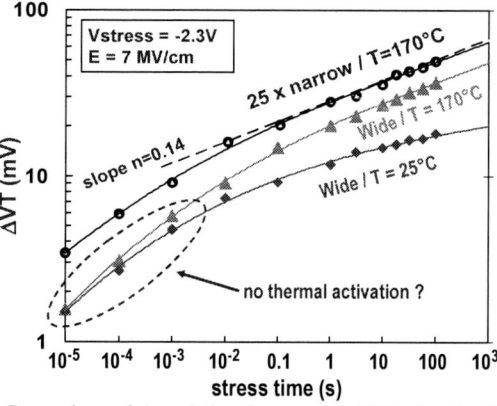

Figure 2. Comparison of degradation for a narrow (0.2µm) and a 10µm wide FET at 170°C. The narrow FET data is from summing up 25 single 0.2µm wide FET data. For the wide FET the thermal activation, and the "missing" thermal activation for µs to ms times are shown.

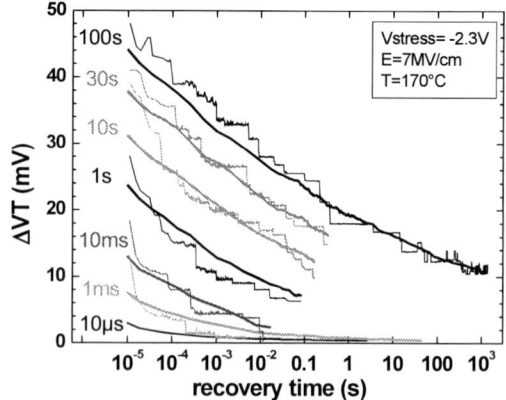

Figure 3. Comparison of the recovery traces for a single narrow (traces with steps) and the average over 25 different 0.2µm wide FETs (smooth traces). Labels denote the stress time preceding the recovery trace. The recovery behavior of the wide devices (see Fig. 2 in ref.[17]) is identical to the narrow FETs.

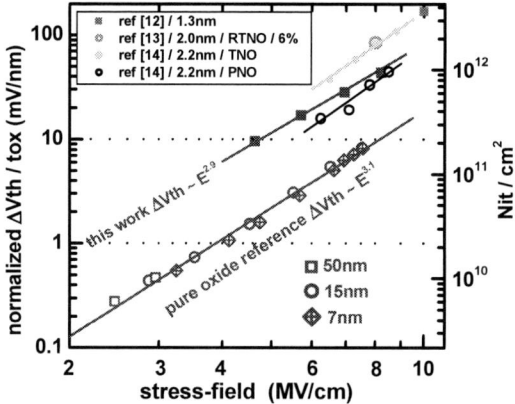

Figure.1. Comparison of different technologies from different sources. Plotted is the normalized *ΔVT* after a stress time of 10ks at 175°C and the corresponding surface charge density (right hand scale). Data from [12, 13, 14] have been extrapolated to 10ks. Oxide parameters are given in legend, as far as available. The pure oxide reference contains thicknesses from 7nm to 50nm.

978-1-4244-5430-3/10 $26.00 © 2010 IEEE

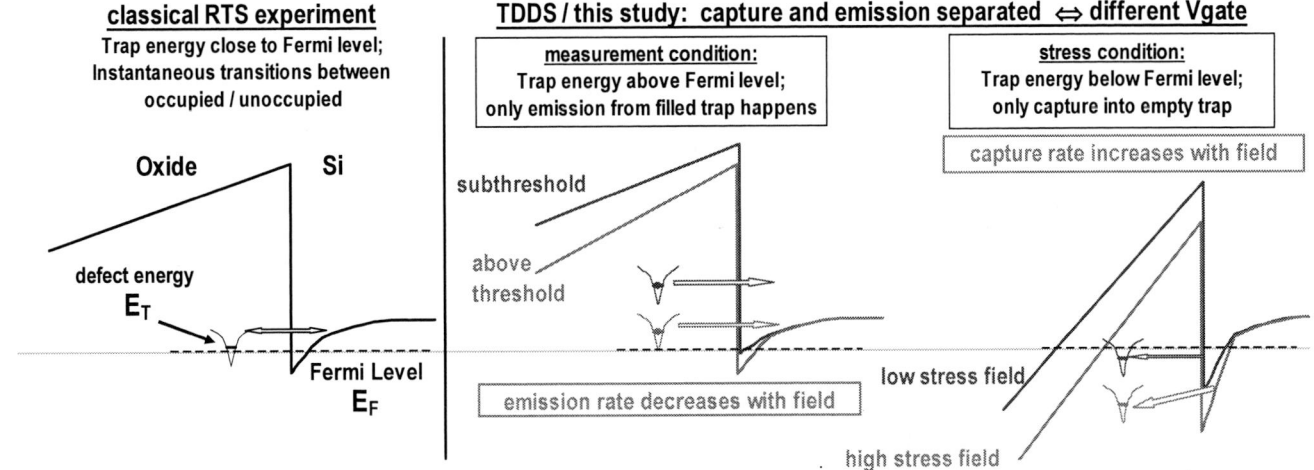

Figure 4.   Schematic band-diagrams, showing the inversion layer and defect levels in the oxide in the vicinity of the substrate. The diagram is meant to be polarity independent, showing the conduction band in an nMOSFET **or** the valence band of a pMOSFET when flipped upside down. Increasing potential energy corresponds to the upward direction. States above / below E_F are empty / occupied, respectively. Each of the 3 diagrams contains the same **single trap** shown for **two different fields** (red/blue) for the TDDS-side. Arrows just indicate the charge carrier transitions between an initial and a final state. These arrows are **not supposed** to indicate that the charge transfer process is due to elastic or inelastic tunneling. In contrast the charge transfer is thermally activated for all cases, including RTS. RTS and TDDS do not differ with respect to sample type, nature of traps or physical processes involved. The main difference for TDDS is the stimulation of capture at a field higher than the field for emission, and the way of data analysis. **Capture rate** increases with increasing field, due to an increasing transition probability and an increasing density of initial states available in the inversion layer. **Emission rate** increases with decreasing field, due to an increasing density of final, empty states. Note that emission to the gate instead to the inversion layer is possible but neglected for simplicity.

agreement of the NBTI degradation and recovery in wide and narrow FETs is quite satisfactory. Degradation in the narrow FETs is 30% higher than in the wide FET, presumably due to edge/stress effects (STI). Such a geometry dependence of NBTI has been reported previously [16].

III. PHENOMENOLOGICAL MODEL

For a basic understanding of our results it is useful and necessary to reconsider the experiments studying random telegraph signals or noise (RTS or RTN), especially the model by Kirton/Uren [4]. This model has been successful to explain the basic phenomena seen in the RTS experiments of nFETS, done mainly in the 1980ies. This model also explains the main phenomena seen in our experiments, and thus of NBTI in general. A major issue of the model is its failure to properly predict the bias dependence of capture and emission, see [10] for a discussion. The essential assumptions of the model are that RTN signals are generated by capture and emission of charges in oxide defects. Capture and emission involve a structural relaxation of the oxide matrix surrounding the defect. It is this structural relaxation which comprises the potential barrier ruling the transition of charge. Thus a transition requires, like in a chemical reaction, an overcoming of a barrier by **heavy** particles. As a consequence the transitions are thermally activated. Since the energy level of the charged defects is field dependent the reaction rate for both directions, capture and emission of charge, is gate voltage dependent. The experiments in this work and also low temperature experiments [20] have clearly shown that tunneling of light particles (electrons or holes), which is nearly temperature independent, does not contribute to NBTI. At this point we want to mention that Shockley-Read-Hall (SRH) recombination also has been tried (unsuccessfully) to explain the NBTI temperature dependence. However, the SRH model has been developed for interaction of semiconductor bulk traps with carriers which are not separated spatially from these traps. It never has been intended to explain the interaction of charge carriers in the silicon space charge layer with oxide defects and thus totally fails to explain even qualitatively any of the phenomena corresponding to oxide traps. A model sketching

the band diagrams corresponding to capture and emission is given in Fig. 4. We want to stress that Fig. 4 is simplified and meant only to illustrate the basic properties of capture and emission of charge. It does not show the energy barriers corresponding to structural relaxation in any way. A model which treats the microscopic physical processes is presented in [10].

IV. STATISTICAL ANALYSIS OF AN INDIVIDUAL DEFECT

Before we come to the statistical analysis of single defects some introducing and clarifying remarks are in order:

(1) As indicated in Fig. 4 all defects have an equilibrium occupation probability of 100% (E_T below E_F) during stress and 0% (E_T above E_F) during recovery. This is a good approximation as long as E_T is not within an energetic distance of <50mV to the Fermi level E_F. This is actually happening for some defects (see RTN signal in Fig. 5), but is unlikely, and will be neglected for the sake of simplicity.

(2) When switching from stress (defect 100% occupied) to recovery it takes a while (i.e. the individual, *Vgate* and T-dependent emission time constant τ_e) to reach the new 0% equilibrium occupation level.

(3) The same is true vice versa when switching from recovery to stress.

(4) Capture and emission are stochastic processes. Like for radioactive decay only probabilities for capture and emission during a certain time can be given.

(5) A given defect always is either occupied or un-occupied. Only its average occupancy (averaged over many capture or emission processes) can be given by eq. 1 and 2.

(6) RTS is a static experiment (see Fig. 4, left side) and analyses the mark-space ratio for a defect having - on the average - its equilibrium occupancy, which is determined by the Fermi-Dirac distribution function. Thus RTS is able to determine (without assumptions or model) a defect energy for a given *Vgate*.

(7) For our TDDS experiment capture and emission are happening at different *Vgate* values. So - in contrast to RTS - the defect energy **cannot** be determined directly from an experiment at a single field. A determination of the defect energy requires the measurement of the field dependence of capture and emission.

978-1-4244-5430-3/10 $26.00 © 2010 IEEE

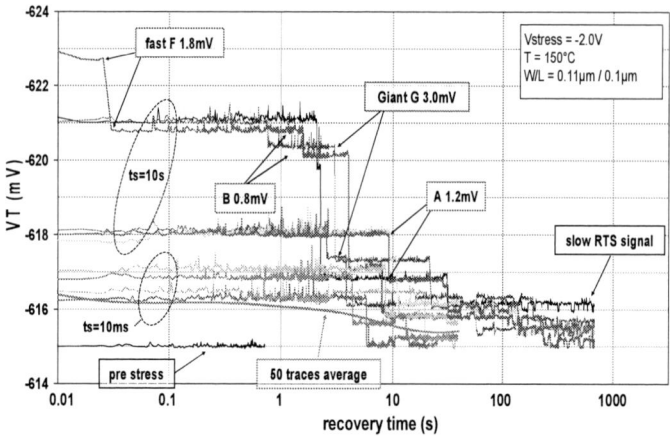

Figure 5. An example of steps from emission of positive charge from defects during recovery, all from the same device. Stress pulses are applied repetitively (typ. 100 times). Some traces, 4 after 10ms stress-pulses (reddish color) and 6 after 10s stress (bluish) were chosen as examples. Individual defects can be distinguished by their step heights, allowing to name the traps, A, F, G, etc. The smooth trace is an average over 50 traces showing the exp(-t/τ_e) (eq. 2) behavior.

(8) Major drawback of the RTS method is that the energy of a defect causing an RTS signal has to be in the vicinity of the Fermi-level. This only happens per chance, and thus allows an analysis of only a fraction of the defects. In contrast our TDDS allows the analysis of **all** the defects contributing to NBTI, provided they are within our experimental window for τ_c and τ_e. This window is from μs to >1000s, which is much wider than for RTN as shown in Fig. 9 of ref. [10].

An example for the responses of roughly 10 defects to stress pulses is given in Fig. 5. Let us first examine the recovery traces with stress time ts=10ms. As seen in Fig. 5 for ts=10ms only a single defect with a emission time around 10s happens to be active. During the stress the oxide field pulls its energy level E_T far below E_F (see Fig. 4). Please note that the valence band in Fig. 4 is flipped upside down in order to have rising energy up and occupied levels below E_F. For the given stress parameters the defect's capture time constant is <10ms, thus its occupation probability is nearly unity after stress. After the stress has been released (corresponding to recovery time=0 in Fig.5) E_T is above E_F again, the defect's equilibrium occupancy

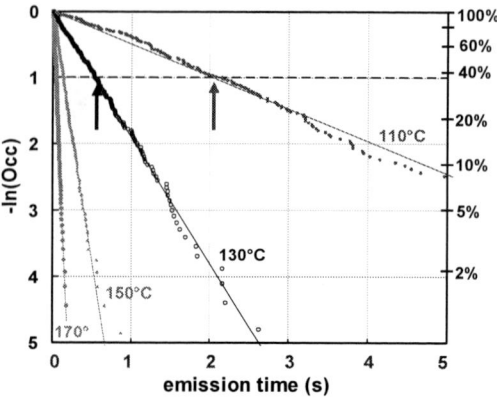

Figure 7. Example for the statistical analysis of emission times of a single defect at 4 temperatures. Plotted is the decreasing occupancy of a defect, starting with 100% just after capture, i.e. at zero emission time. Arrows mark the time constants τ_e. The straight line behavior ensures that only a single defect is contained in each curve. Each dot corresponds to an emission event in a recovery trace like in Fig. 5.

level ≈0). It will be discharged by emission at a random time around a characteristic emission time τ_e. After applying stress, capture of (positive) charge will happen in the same stochastic manner around a characteristic capture time τ_c. Averaged either over **many** capture or emission events of the same defect under the same condition (as the 50-traces average in Fig. 5 shows, or all our single-defect data) **or** a single stress pulse or recovery trace averaged over **many** equivalent defects (in a large FET) the average occupancy or occupation probability P during stress time ts and recovery time tr will be described by the following equations:

(1) $$P(ts) = P(ts = 0) \times [1 - \exp(ts / \tau_c)]$$

describes the increase in occupancy due to capture during stress, and

(2) $$P(tr) = P(tr = 0) \times \exp(tr / \tau_e)$$

describes the decay of occupancy due to emission during recovery. For the sake of simplicity the Fermi distribution function is neglected and assumed to be 1 during stress and 0 during recovery. Eq. (1) and (2) gives average occupancies, any given defect always is either occupied or empty.

An example for the statistical determination of **capture** time constants according to eq. (1) is shown in Fig. 6 and an example for the statistical determination of **emission** time constants according to eq. (2) is shown in Fig. 7. In Fig. 7 the temperature dependence of τ_e of a given defect is investigated. Each emission time of this defect, extracted from a recovery trace, is plotted as a dot, in the standard way, allowing to extract τ_e corresponding to the slope of a straight line. The fit to the straight line also is a proof that the defects behave in the expected statistical way (eq. 2) and that **only one** defect is contained in the distribution. Mixing another defect into the distribution, by chance with a similar step-height, hardly to distinguish, would lead to a deviation from the straight line. In order to minimize the statistical error in the determination of τ_c the stress pulse width ts has to be chosen to have a similar value as τ_c. As seen in Fig. 6 ts increasing in steps of a factor 10 delivers a satisfactory accuracy for all τ_cs. Fig. 8 shows extracted τ_cs and τ_es for a couple of defects as an Arrhenius plot. Our statistical analysis allows a very precise determination of activation energies E_A. E_As are higher than the ones extracted from conventional experiments, in agreement with the findings from ultra-fast temperature changes [18]. We will discuss these E_As again in section VI . It should be noted, that the thermal activation of the single defects has **exactly** an Arrhenius

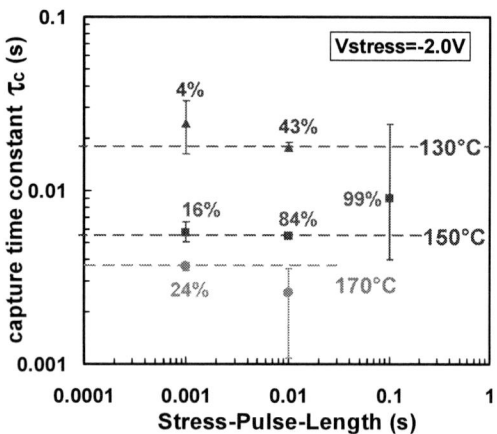

Figure 6. Example for the statistical analysis of capture time constants from repetitive capture attempts (by doing stress pulses of 1ms, 10ms and 100ms) for a single defect at 3 temperatures. Labels denote the capture probability, equal to the ratio r=number capture events / number of attempts. note that r is the quantity which is directly measured. As seen the statistical errors (see error bars) in the determined τ_c can be minimized, or at least kept small by choosing the stress pulse length suitable to have the capture probability between 15% and 90%.

978-1-4244-5430-3/10 $26.00 © 2010 IEEE

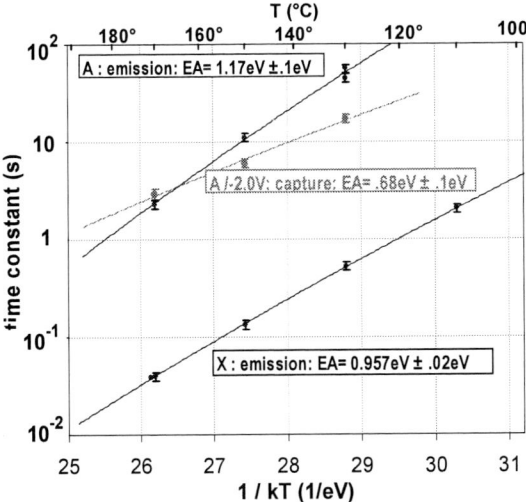

Figure 8. Arrhenius plot for defect A as well as capture and emission for trap X. E_A values are higher than in conventional experiments. Note the considerable accuracy in determining τ values. Each point is extracted from 256 emission events of the given defect.

behavior, in contrast to a conventional determination of thermal activation done on wide FETs [19].

This work shows only one example for parameter extraction, namely the T-dependence determination in Figs. 6, 7, 8. In an analogous way also the *Vgate* dependence of τ_cs and τ_es can be determined. These results are presented in [10].

V. NEW NBTI MODEL

With the findings so far we can be sure that the model of distributed time constants as proposed in [7] correctly describes the recoverable part of NBTI. Some considerations regarding permanent NBTI will be discussed in section IX. The model is equivalent to eq. (1) and (2), its equivalent circuit is shown in Fig. 9. Each defect is represented by a capacitor C which can be charged or discharged from a signal line applying stress- or recovery-voltage. The charge or voltage on C is equivalent to the individual ΔVT caused by the defect. The complete model for a FET thus consists of just one RC-element for each defect. The value of C for each defect will be roughly the same and corresponds to the value of ΔVT produced by this defect. The values for the charging resistor R_c and the discharging resistor R_e determine the capture and emission time constants R_cC and R_eC and are widely distributed. All the time constants are a function of *Vgate* and temperature. We will employ this NBTI model in the following sections.

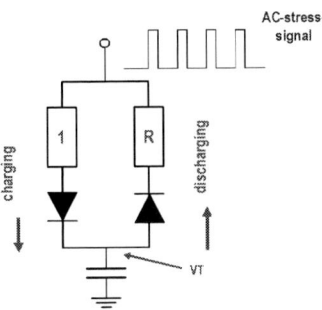

Figure 9. A single RC-element representing one given defect with a capture and emission time constants τ_c and τ_e having a ratio τ_e / τ_c =R. The RC-element is an equivalent to the equations (1) and (2). The voltage VT on C corresponds to the ΔVT produced by this defect. A real FET contains a large number of these defects (see Fig. 13), the individual VTs have to be summed up. The signal applied to the upper terminal indicates charging of the RC-element by an AC stress.

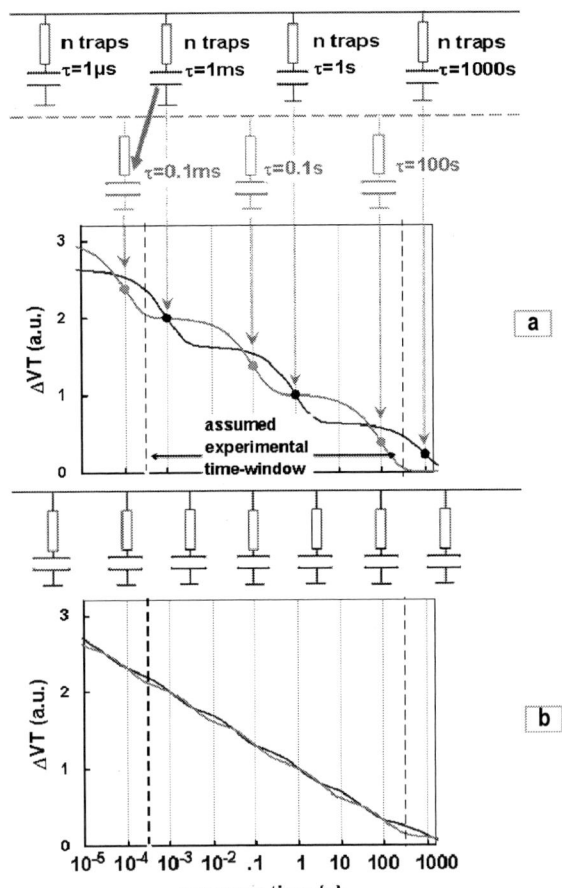

Figure 10. Simplified simulation of NBTI using the equivalent circuit from the top of the Figure. The τ's are aligned with the X-axis. When the τ's are widely separated (only one in 3 decades in Fig. a) distinct bumps are produced. A change of τ (by increased temperature, indicated by the red arrow) shifts the distinct bumps in the recovery curves. With densely, equally distributed RC's, ranging from ns to 1E6s (Fig. b) it is obvious that a shift left/right of this ladder has **no effect** in the experimental time window.

VI. THE MISSING-THERMAL-ACTIVATION RIDDLE

In numerous studies by different research groups (see references in [11]), including our own [11] an apparently T-independent short-term degradation has been found. Fig. 2 shows a further example for this behavior. The standard explanation for this behavior has been that - apart from "real" NBTI - there is a special class of oxide defects which is charged and discharged by elastic quantum mechanical tunneling (QM) in and out of traps. The barrier ruling this QM-tunneling is in the order of the Si/SiO$_2$ valence band offset. Thus the barrier is 4eV>>kT which means that the effect is temperature independent. Typical contributions from this assumed QM-tunneling to the total ΔVT have been found to be in the regime of 10 to 20% [11]. In strong contrast, **all** the capture/emission events we analyzed, for <1µs up to ks, were strongly thermally activated (cmp. Figs. 6, 7, 8 and Fig. 8 in ref. [10]). In total we analyzed the thermal activation of roughly 100 defects and since this seemingly non-thermally-activated contribution is actually existing in our samples (see Fig. 2), we would by no means have overlooked such a class of not-thermally-activated defects. The solution of the riddle is given in Fig. 10: A simplified equivalent circuit describing charging and discharging of the defects is shown on top of Fig. 10. Only a few defects with time constants separated by 3 decades in times are taken into account. For the sake of simplicity τ_cs and τ_es are assumed to be equal, which is not correct but is good enough for the purpose. The time constants of the RC-elements are aligned with the

978-1-4244-5430-3/10 $26.00 © 2010 IEEE

time axis in Fig. 10. Now let us consider an increase of temperature by 100°C: For an E_A=1eV this means that both τ_c and τ_e of any given defect (i.e. any of the drawn RC-elements) are accelerated by a factor of roughly 1000. Thus each of the RC-elements in Fig. 10 (each representing an individual defect or a class of defects) is getting faster, that means is shifted left by 3 decades. So, given that **all C-values are equal -** meaning the same number of traps in each class - the RC ladder just has shifted, but its effect has not changed: Were the 1ms traps were at low temperature, at the higher temperature the 1s-traps take over their "task" and so on. The apparent degradation behavior **did not change** due to the increase in temperature, the apparent activation energy is zero. Thus the strongly (a factor of 1000 for a change of 100°C) activated behavior of the individual defects is not seen in the large FETs due to the mixing of all the defects. The above postulation that all the C-values must be equal is equivalent to a uniform, flat distribution of time constants in log(t). This flat distribution is in turn equivalent to a degradation curve ΔVT vs ts which is linear in log(ts). This is in fact a standard observation in NBTI. To summarize, we think all conclusions about non-thermally-activated behavior for short times in large FETs were wrong. A convincing further proof for this assumption is - except our experimental results - the low temperature experiment in [20] showing that all threshold shift vanishes below T=150K.

VII. PRACTICAL MODELING

A distribution map of defect properties based on few defects is shown in Fig. 11. Clearly this map does not contain a number of defects large enough to be representative for a "real", say 0.5μm-wide FET. A 0.5μm wide FET would have filled ≈250 defects at a ΔVT=50mV (cmp. table 1). To fill this map with >250 measured defects in order to get a modeling-relevant spectral-density-map would be a task taking at least months. However, with our new understanding of trap properties, we now have the justification to switch back to a wide FET (with >4000 defects) and can extract a full defect density map just from measuring full-recovery traces from a single FET. This justification is mainly based on the fact that the number of defects and their properties do not change during stress (see also points 2, 3, 4, 5 in the conclusions).

The principles of extracting such a map, of gathering all the information needed, are outlined in Figs. 12 and 13. The "difference"-curve actually is a recovery curve containing **all** defects having a capture time constant **between** 10ms and 100ms. All these defects then are separated in one-decade-wide bins of different emission time constants as shown in Fig. 13. The filling level of the bins is given simply in mV-units, corresponding to the ΔVT generated by each bin. The density levels from Fig. 12 (denoted as labels in Fig. 12 and 13) contain the information to fill just one slice in Fig. 13. The other slices of Fig. 13 are filled in an analogous way. When all the squares in the defect density map Fig. 13 are filled this map comprises a complete set of parameters allowing a complete and straightforward modeling of NBTI for a given stress voltage and temperature. It includes recovery and the response of the sample to AC-stress or to any arbitrary sequence of stress-recovery cycles. An example for a calculated degradation curve and a recovery-curve in comparison to the experimental curves is given in Fig. 14. As seen the fit is perfect. This is not a surprise, however. Our model and the way of parameter extraction actually very much resembles a Fourier transformation: A set of data in the time-domain, (example Fig. 12) is just converted into a set of coefficients in the frequency domain (example Fig. 13), i.e. the set of "amplitudes" in the 2-dimensional spectral plot. Fig. 14 is quasi a re-conversion into the time-domain. If done correctly this re-conversion, like a Fourier transformation exactly reproduces the experimental data as demonstrated in Fig. 14. To prove the claim that the model is able to predict recovery and

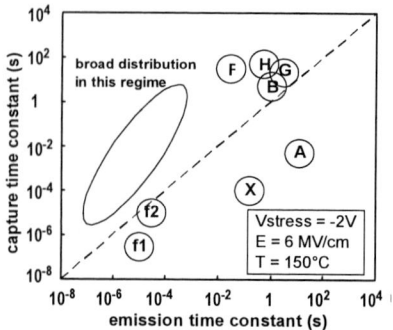

Figure 11. Correlation of τ_c/τ_e of a couple of selected defects from the statistical analysis of single defects (cmp. Fig. 5). τ_e may be larger or smaller than τ_c by orders of magnitude.

degradation for any arbitrary stress signal another model verification is shown in Fig. 15. Note that the stress condition is different from the example in Figs. 12, 13 and 14. For the model verification a series of stress/recovery sequences, with increasing times, has been applied to a sample and then simulated. As seen the quality of fit is very good. Unfortunately the log-time axis obscures all the μs to 1s time effects which are occurring for the long times also. To re-gain some of this information the experimental points have been chosen to be equidistant on a log-time scale.

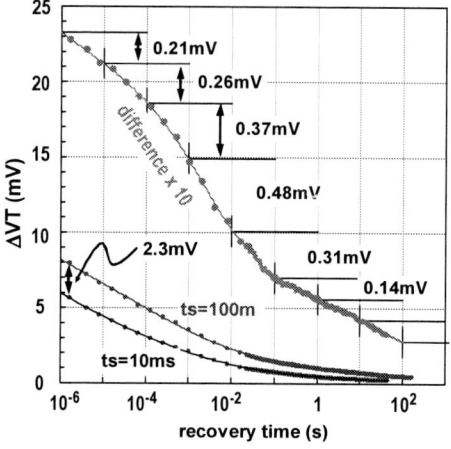

Figure 12. Example for extraction of spectral trap densities from recovery traces: The difference between ts=10ms and ts=100ms traces is due to the class of defects with 10ms<τ_c<100ms. So this class can be separated and then divided into "bins" of different τ_e s (i.e. recovery times). Experimental parameters given in Fig. 14.

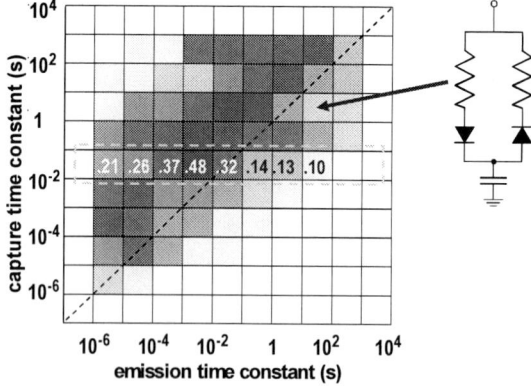

Figure 13. Complete spectral density map. The τ_c-slice analyzed in Fig. 12 is marked by a frame and labeled with the numbers from Fig. 12. Spectral density is simply given in ΔVT-units (=mV). Experimental parameters given in Fig. 14. Each square corresponds to a class of defects with given τ_c and τ_e as indicated by the RC-element.

978-1-4244-5430-3/10 $26.00 © 2010 IEEE

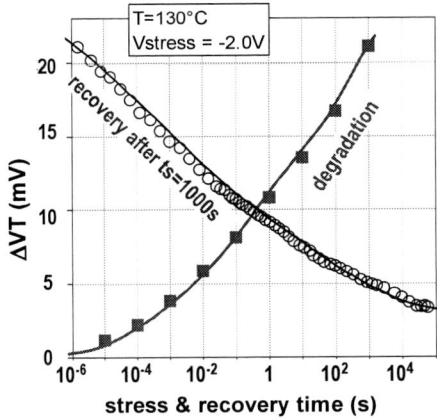

Figure 14. Using the parameter-table given in Fig. 13 a simulation of degradation to ts=1000s and subsequent recovery is done. Dots are experimental points; lines are calculated values.

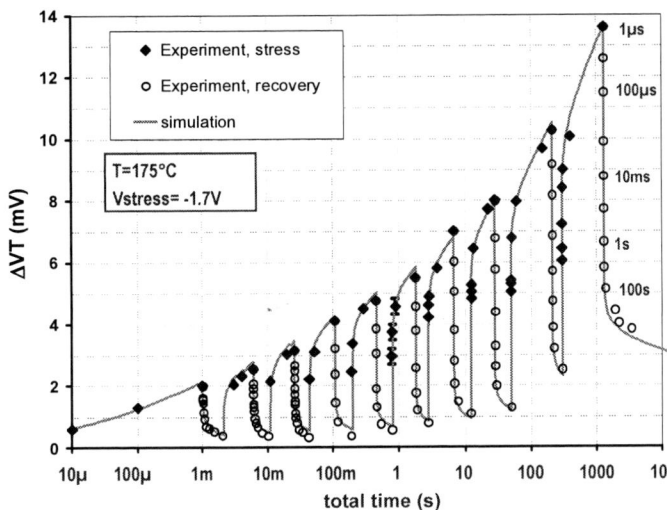

Figure 15. Example for model verification. Vstress is close to the maximum operating voltage of 1.6V. A stress/recovery sequence with increasing times was chosen. Agreement between experiment and model is in the sub-mV regime. Obviously all the experimentally observed time constants from μs to >10^3s are correctly captured by the simulation. Readouts during the stress phases (filled dots) have been measured with a measuring delay of 1μs and a total interruption of stress of 10μs. Measuring delay as well as this interruption are also considered in the simulation. The dents in the simulation, seen between 2ms and 5ms, are due to this interruption. Labels at the rightmost recovery branch denote the recovery time for each point, valid for each of the recovery branches.

A more detailed discussion of a fully automated extraction of the model parameters, of the hi-quality noise-free measurements needed, and of the required approximations needed for a fast and effective calculation of degradation, for example in a simulator like RelXpert cannot, due to limited space, be presented in this paper. These topics as well as gate voltage and temperature dependence will be discussed in a separate publication.

VIII. AC-STRESS

Existing studies about AC-NBTI, e.g. [21] are showing a relation of ΔVT vs. duty cycle having the typical S-like shape. The reason for this shape could not been explained yet. Fig. 16 shows measured AC-stress data in comparison with a very simple simulation. The simulation perfectly fits the experimental data, proving that the AC-behavior can be fully understood by our model. The fit-curve in

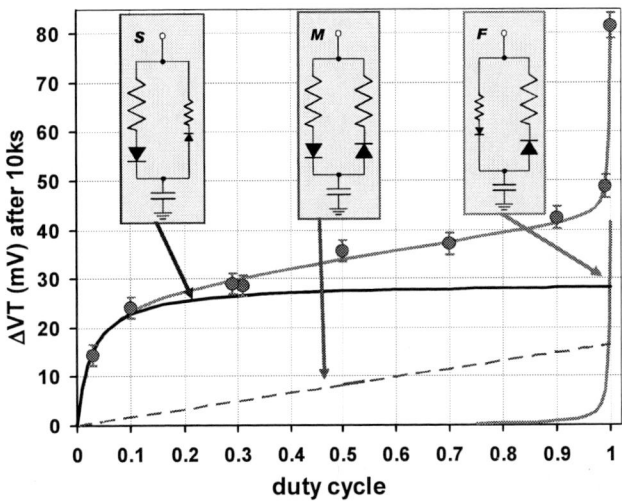

Figure 16. Degradation under AC stress at -2.7V at 175°C / 100kHz as a function of duty cycle. The full line trough the data is a best fit. It is composed as a sum from 3 contributions (the 2 full and the dashed lines), corresponding to 3 types of equivalent circuits S, M, F (=defects) as described in the text. The magnitudes of the 3 contributions are roughly equal.

Fig. 16 is "assembled" from a sum of 3 contributions, corresponding to 3 classes of defects. The 3 classes differ by the ratio between capture- and emission time constant. Class "S" has an emission time constant much slower than capture $\tau_e \gg \tau_c$. Looking at the equivalent circuit it is clear that this is the behavior of an electronic peak detector. The C in the peak detector is almost fully charged as soon as the duty cycle D is just above zero. Class "F" has capture much slower than emission, $\tau_c \gg \tau_e$. It is the opposite of the peak detector. D has to be very close to 100% to get a little charge into C, depending on the value of the ratio τ_e/τ_c. Finally class "M" has an emission time equal to capture time. It gives the linear response seen in Fig. 16 and is responsible for the linear center part of the experimental curve. The assumption that there are just 3 classes of defects is unrealistic. However, the shape of the curve in Fig. 16 is mainly determined by the contributions by the "extreme" classes "F" and "S". Thus the consideration of just 3 classes gives a perfect fit. It is noteworthy that mainly the ratio τ_c / τ_e determines the shape of the curve, while the absolute values are not important, as long as these time constants are longer than the AC-period. This explains a good deal of the observed independence of AC-degradation on frequency. Given that the measuring delay is much shorter than the stress-time (by more than 5 orders of magnitude) there also is no influence of the measuring delay. In our experiment the ΔVT measurement is always done after the completion of the stress-phase of an AC stress cycle. Measurement and stress signal are synchronized. Like for DC-stress we define the short (2μs) recovery phase before VT is determined as the measurement delay.

AC-stress measurements are easy to do, and will provide a lot of information about the sample. An experimental setup with a short measuring delay should be used, and the measurement-timing should be synchronized with the AC-signal. In this work we could only briefly sketch our experimental AC-results, the modeling and the derivations done. It is clear that the calculation of the response of an RC-element with e.g. a 1000s time constant, for a 10 year AC stress, with a total of 10^{15} AC-periods will need a couple of appropriate approximations in the calculation. We will provide a more detailed study of AC-NBTI, including data for more technologies in a subsequent publication.

IX. PERMANENT DEGRADATION

This study exclusively deals with the recoverable part of NBTI. Our mathematical model (represented by the spectral distribution in Fig. 13), however, is general and may include a permanent part without any changes. In our "single-defect" experiments, though the total stress-times applied to individual samples have been >10000s, there has been no evidence for a permanent degradation. In long stress experiments, reaching a high level of degradation (above 10^{12}cm^{-2} cmp. Fig. 1), we actually find experimental evidence for permanent degradation, and believe that it is really existing. So far we found it impossible to clearly distinguish a defect with a long emission time >10^5s, which appears in a distribution like the one of Fig. 13, from a defect which might appear in a distinct τ_e-column located at τ_e>>10^5s. Permanent defects may be different from the recoverable ones, and may have a different physical origin. An experimental proof for permanent defects, let alone a determination of their field- and temperature-dependencies so far just failed because it would require >10^7s-times, while most experimenters are limited to times somewhat above 10^5s.. Several studies have claimed to separate recoverable from permanent, or fast from slow components in NBTI-degradation [22]. We think that any distinction between fast and slow is still arbitrary and the distinction recoverable / permanent not yet justified by experimental data. There might also be the possibility that a potentially permanent part of NBTI is due to a damage of the gate oxide from highly-energetic, bond-breaking carriers like in time dependent dielectric breakdown tests. This would mean that additional oxide defects - the permanent ones - are generated at high field only. This effect would be different from NBTI. Due to a higher field acceleration these high field effects, causing TDDB, would not contribute to degradation at use voltage. We do not claim that this is true or proven, but it is a hypothesis which is worth to be considered.

X. CONCLUSIONS

Our new defect spectroscopy technique has been able to resolve some open questions in NBTI. We draw the following conclusions. All these conclusions are based on the experimental observations only and do not require further assumptions or models.

(1) The recoverable part of NBTI is completely driven by pre-existing defects. For a moderate degradation below a $5*10^{11}$cm^{-2} charge defect density no generation of defects has been observed.

(2) Each defect is characterized by a capture and an emission time constant τ_c and τ_e, both a function of temperature an electric field.

(3) τ_c and τ_e have a wide distribution and are not correlated (see Figs. 11, 13).

(4) Recovery of a given defect is only a matter of defect properties, absolutely independent of the previous stress-time and stress-voltage. This fact rules out that any diffusion is involved in degradation or recovery [23].

(5) Defects act uncorrelated with each other. This is plausible, assuming a density of 10^{12}cm^{-2} their typical distance is 10nm which is more than the screening length in the inversion layer [24]. Exceptions will be shown in ref. [10]

(6) All processes, also short term 1µs to 1ms, are thermally activated with activation energy typically 1eV. As explained in section VI the seemingly T-independent degradation at short times is an artifact.

(7) The new understanding of NBTI leads to a straightforward modeling, especially of AC-NBTI, based on a simple equivalent circuit like in Fig. 9 and a set of empiric parameters.

(8) If a defect density map like the one in Fig. 13 describes degradation and recovery it is clear that the concept of "universal recovery" cannot be correct but at the most an approximation.

REFERENCES

[1] J. H. Stathis and S. Zafar, "The Negative Bias Temperature Instability in MOS Devices: A Review", MR Vol. 46, no.2-4, 2006, p. 270

[2] J. P. Campbell, P. M. Lenahan, A. T. Krishnan, and S. Krishnan, "Identification of the atomic-scale defects involved in the negative bias temperature instability in plasma-nitrided p-channel metal-oxide-silicon field-effect transistors", J. Appl. Phys., Vol. 103, no. 4, 2008, pp. 044505-1-11

[3] K. S. Ralls , W. J. Skocpol, L. D. Jackel, R. E. Howard, L. A. Fetter, R. W. Epworth, and D. M. Tennant, "Discrete Resistance Switching in Submicrometer Silicon Inversion Layers: Individual Interface Traps and Low-Frequency (1/f?) Noise", Phys. Rev. Lett. 52, pp.228-231, 1984

[4] M. J. Kirton and M. J. Uren, "Noise in solid-state microstructures: A new perspective on individual defects, interface states and low frequency (1/f) noise", Adv. Phys., vol. 38, pp. 367-468, 1989

[5] V. Huard, C.R. Parthasarathy, and M. Denais, "Single-Hole Detrapping Events in pMOSFETs NBTI Degradation, 2005 IIRW final report, p. 5

[6] A. Karwath and M. Schulz, "Deep level transient spectroscopy on single, isolated interface traps in field-effect transistors", Appl. Phys. Lett. 52, p. 634, 1988

[7] B. Kaczer, T. Grasser, J. Martin-Martinez, E. Simoen, M. Aoulaiche, Ph. J. Roussel, G. Groeseneken, "NBTI from the perspective of defect states with widely distributed time scales", Proc. IRPS 2009, pp. 55-60.

[8] H.C. Ma, J.P. Chiu, C.J. Tang, T. Wang and C.S. Chang, "Investigation of Post-NBT Stress Current Instability Modes in HfSiON Gate Dielectric pMOSFETs", Proc. IRPS 2009, pp. 51-54

[9] T. Grasser, H. Reisinger, W. Goes, Th. Aichinger, Ph. Hehenberger, P.-J. Wagner, M. Nelhiebel, J. Franco, and B. Kaczer, "Switching Oxide Traps as the Missing Link Between Negative Bias Temperature Instability and Random Telegraph Noise", IEDM technical digest 2009 , pp. 729-732

[10] T. Grasser, H. Reisinger, P.-J. Wagner, F. Schanovsky, W. Goes, B. Kaczer, "The Time Dependent Defect Spectroscopy (TDDS) for the Characterization of the Bias Temperature Instability", IRPS 2010 (in press)

[11] H. Reisinger, O. Blank, W. Heinrigs, W. Gustin, and C. Schlünder, "A comparison of very fast to very slow components in degradation and recovery due to NBTI and bulk hole trapping to existing physical models", IEEE TDMR, Vol. 7, No. 1, p. 119 (2007)

[12] C. Shen, M.-F. Li, C.E. Foo, T, Yang, D.M. Huang, A. Yap, G.S. Samudra, Y.-C. Yeo, "Characterization and Physical Origin of Fast Vth Transient in NBTI of pMOSFETs with SiON Dielectric", IEDM 2006, p.333

[13] V. Maheta; C. Olsen, K. Ahmed, and S. Mahapatra, "The Impact of Nitrogen Engineering in Silicon Oxynitride Gate Dielectric on Negative-Bias Temperature Instability of p-MOSFETs: A Study by Ultrafast On-The-Fly IDLIN Technique", IEEE TED, Vol.55, No. 7, p. 1630

[14] S. Mahapatra, K. Ahmed, D. Varghese, A. E. Islam, G. Gupta, L. Madhav, D. Saha and M. A. Alam, "On the Physical Mechanism of NBTI in Silicon Oxynitride p-MOSFETs: Can Differences in Insulator Processing Conditions Resolve the Interface Trap Generation versus Hole Trapping Controversy?, IRPS 2007, p. 1

[15] B. Kaczer, Ph. J. Roussel, J. Franco, R. Degraeve, L.-A. Ragnarsson, E. Simoen, G. Groeseneken, T. Grasser, H. Reisinger, "Origin of NBTI Variability in Deeply Scaled pFETs", IRPS 2010 (in press)

[16] G. Math, C. Benard, J. Ogier, D. Goguenheim, "Geometry effects on the NBTI degradation of PMOS transistors", 2008 IIRW final report, pp. 60-63

[17] H. Reisinger, T. Grasser and C. Schlünder, "A study of NBTI by the statistical analysis of the properties of individual defects in pMOSFETs", 2009 IIRW final report, pp. 60-63

[18] T. Aichinger, M. Nelhiebel, T. Grasser, "Unambiguous Identification of the NBTI Recovery Mechanism using Ultra-Fast Temperature Changes", Proc. IRPS 2009, pp. 2-7

[19] B. Kaczer, V. Arkhipov, R. Degraeve, N. Collaert, G. Groeseneken, M. Goodwin, "Disorder-controlled-kinetics model for negative bias temperature instability and its experimental verification", IRPS 2005, pp. 381-387

[20] R.G. Southwick III, W.B. Knowlton, B. Kaczer, and T. Grasser, "On the thermal activation of negative bias temperature instability", 2009 IIRW final report, p. 36

[21] R. Fernadez, B. Kaczer, A. Nackeaerts, R. Rodriguez, M. Nafria, G. Groeseneken, "AC NBTI studied in the 1 Hz - 2 GHz range on dedicated on-chip CMOS circuits", IEDM technical digest 2006, pp.337-340

[22] V. Huard, C. Parthasarathy, N. Rallet, C. Guerin, M. Mammase, D. Barge, C. Ouvrard, "New characterization and modeling approach for NBTI degradation from transistor to product level", IEDM 2007, p. 797

[23] K. O. Jeppson and C. M. Svensson, "Negative bias stress of MOS devices at high electric fields and degradation of MNOS devices", J. App. Phys. Vol. 48, No.5, 1977, pp. 2004-14

[24] T. Ando, A. B. Fowler and F. Stern, "Electronic properties of two dimensional systems", Rev. Mod. Phys. Vol. 54, pp.437-672, 1982

The Time Dependent Defect Spectroscopy (TDDS) for the Characterization of the Bias Temperature Instability

T. Grasser[*], H. Reisinger[•], P.-J. Wagner[*], F. Schanovsky[*], W. Goes[*], B. Kaczer[○]

[*]Institute for Microelectronics, TU Wien, Austria   [•] Infineon, Munich, Germany   [○] IMEC, Leuven, Belgium

Abstract—We introduce a new method to analyze the statistical properties of the defects responsible for the ubiquitous recovery behavior following negative bias temperature stress, which we term time dependent defect spectroscopy (TDDS). The TDDS relies on small-area metal-oxide-semiconductor field effect transistors (MOSFETs) where recovery proceeds in discrete steps. Contrary to techniques for the analysis of random telegraph noise (RTN), which only allow to monitor the defect behavior in a rather narrow window, the TDDS can be used to study the capture and emission times of the defects over an extremely wide range. We demonstrate that the recoverable component of NBTI is due to thermally activated hole capture and emission in individual defects with a very wide distribution of time constants, consistent with nonradiative multiphonon theory previously applied to the analysis of RTN. The defects responsible for this process show a number of peculiar features similar to anomalous RTN previously observed in nMOS transistors. A quantitative model is suggested which can explain the bias as well as the temperature dependence of the characteristic time constants. Furthermore, it is shown how the new model naturally explains the various abnormalities observed.

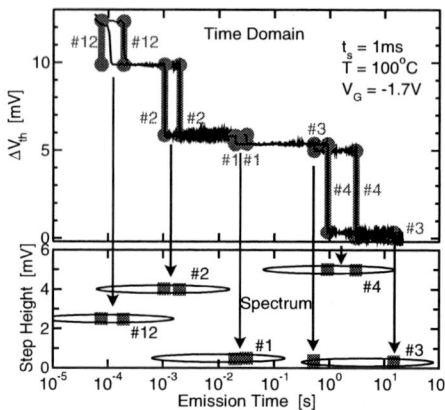

Fig. 1: Two typical threshold voltage recovery traces of a previously stressed small-area pMOSFET. The measured data are given by the (slightly noisy) thin black lines in the top part of the figure. The thick blue and red lines together with the symbols mark the automatically extracted emission times and step heights which are unambiguous fingerprints of each defect and build the spectral map (bottom).

I. INTRODUCTION

The lifetime of metal-oxide-semiconductor field effect transistors (MOSFETs) is limited by unavoidable degradation during operation. One of the most critical degradation mechanisms in p-channel MOS-FETs is the negative bias temperature instability (NBTI). As the name implies, it is observed when a pMOSFET is subjected to negative bias at the gate with the other terminals grounded. The degradation is considerably accelerated at elevated temperatures. Although the detrimental impact of negative bias temperature stress (NBTS) has been known for more than 40 years and is of highest relevance to the semiconductor industry, the detailed physics behind the phenomenon are still highly controversial [1]. The most popularized explanation of this phenomenon invokes a reaction-diffusion (RD) driven depassivation of hydrogen-passivated interface states [2, 3], initially obscured by elastic (temperature-independent) hole trapping in the oxide [4].

The first clues to the limitations of the RD theory were found when the recovery of the degradation was studied in greater detail [5–8]: most importantly, it was found that the degradation consists of a recoverable component on top of a nearly permanent part [7]. Also, recovery starts on time scales faster than accessible by experiment (below microseconds), continues longer than is usually measured (up to weeks), and is strongly sensitive to the applied gate bias. This is inconsistent with a diffusive process involving neutral hydrogen species but is intuitively compatible with a hole trapping mechanism with widely distributed time constants [9, 10].

In a first attempt to explain experimental data, simple elastic hole trapping models were used [7] which explain the wide distribution of time constants by assuming that hole traps are spatially distributed into the depth of the oxide. Due to the necessary tunneling transition from the channel, traps deeper in the oxide have larger time constants [11]. However, for the ultra-thin oxides employed in modern devices having thicknesses of 1-2 nm, the maximum time constants explainable with such a model are in the order of milliseconds [11]. Yet, experimentally observed recovery continues unabated for weeks and months, seemingly linear on a semi-log scale [5, 8, 12].

Another phenomenon where the assumption of a wide distribution of time constants is commonly invoked is the $1/f$ noise [13–15]. In small-area devices only a handful of defects are present and their time constants can be experimentally assessed from the dispersion of random telegraph noise (RTN) properties [15]. These time constants are temperature activated and can be very sensitive to the applied gate bias. In particular, models originally developed for RTN and $1/f$ noise qualitatively capture many features typically associated with NBTI [10]. As such it has been speculated that the defects responsible for RTN are also the cause of NBTI [9, 10]. Like $1/f$ noise, NBTI is typically studied on large-area devices, where a large ensemble of defects acts simultaneously. Consequently, the individual defect properties are averaged out and unambiguous identification of the physics is difficult. In the following, we use small-area devices, which allow us to clearly and unambiguously identify and study the individual defects constituting the macroscopically observable behavior. In order to guarantee that the discrete steps observed in the recovery traces of small-area devices [9, 16] are truly the main ingredient of the log-like recovery ubiquitous in large area devices [9, 17], we have compared the average of 25 small-area device recovery traces to a single large-area device of comparable effective width, with favorable outcome [18].

II. THE TIME DEPENDENT DEFECT SPECTROSCOPY

We proceed by introducing an intuitive analysis method, which we term *time dependent defect spectroscopy* (TDDS). It uses small-area devices, where recovery after stress proceeds in discrete steps occurring at stochastic times [9, 16, 18–20]. For details of the measurement procedure see [18]. We conclude that just like in RTN, each discrete step is due to the emission of a single trapped carrier [16]. As such, the TDDS is similar in spirit to the successful deep-level transient spectroscopy (DLTS) technique [21], which has also been applied to small devices [22]. However, rather than assuming that after application of a charging pulse for a certain amount of time

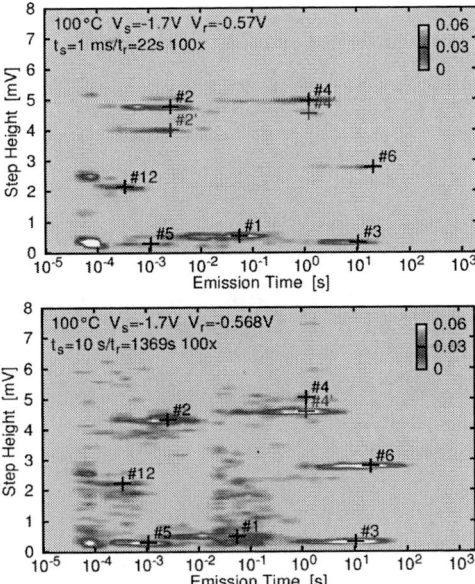

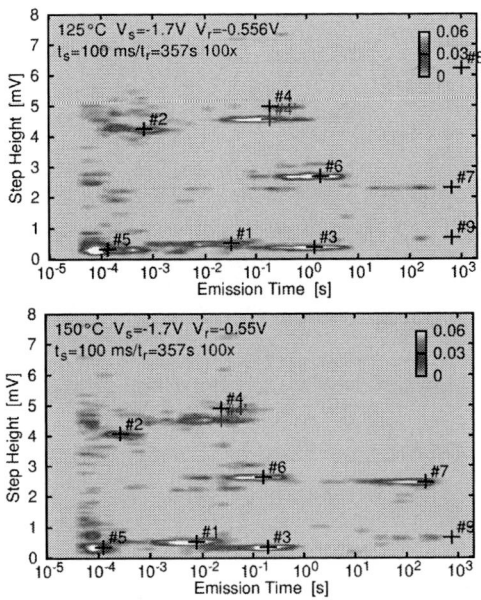

Fig. 2: Spectral maps at two stress times, $t_s = 1\,$ms (top) and $t_s = 10\,$s (bottom). With increasing stress time, the number of defects in the map increases as defects with $\bar{\tau}_c \lesssim t_s$ have a significant probability of being charged after stress. Note the increased noise level (hits outside clusters) although the same number of defects constitute the map. The width of each cluster is given by the exponential distribution of τ_e and the extracted $(d, \bar{\tau}_e)$ are marked by '+'.

Fig. 3: Temperature dependence of the spectral maps after $t_s = 100\,$ms. With increasing temperature, the emission times decrease. As also the capture time constant decreases, the clusters associated with each defect appear after a shorter stress time at higher temperatures. Defects #2 and #5 partially move outside our experimental window, while #7, #8, and #9 enter. Nonetheless, the parameters of #7, #8, and #9 cannot be reliably extracted at these temperatures.

all defects are fully charged [22], we exploit the fact that the capture time constants are widely distributed, an issue also observed in DLTS literature as a non-saturating behavior in the DLTS spectra [23]. It is particularly this time-dependence of the TDDS spectral maps that allows for a detailed assessment of the physical phenomenon.

For the TDDS, a number of stress/recovery experiments has to be performed for a reliable characterization of the stochastic process. For this, the device is repeatedly stressed for the same amount of time and the subsequent recovery trace is recorded. The statistical properties of the discrete steps in the recovery traces are then analyzed by collecting the heights d and emission times τ_e of each emission event, see Fig. 1. The collected pairs (τ_e, d) are then binned into a 2D histogram, see Fig. 1 (bottom).

As has been demonstrated in RTN literature [15, 24], the step height observed in the threshold voltage shifts does not depend on the microscopic physics of the defect but rather on the interaction between the lateral position of the defect and the random potential inside the channel [24] and thus allows for the identification of individual defects. Step heights larger than the value obtained from the charge-sheet approximation, $qt_{ox}/(\varepsilon WL)$ with t_{ox} being the oxide thickness, ε the oxide permittivity, W and L the device width and length, indicate a defect in a so-called current percolation path [25] and is commonly referred to as a 'giant' step in RTN literature [15, 24]. For the device used in this study, the charge sheet approximation gives about 0.9 mV, while experimentally step heights of more than 6 mV have been observed in our samples. For even higher values and details on the distribution of the step heights see [26].

In order to gather sufficiently accurate statistics, the stress and recovery cycle is repeated $N = 100$ times and the entries in the 2D histogram are normalized by N to obtain the spectral map after the stress time t_s. The whole procedure is then repeated M times with increasing stress time, where we used $t_s \in \{1\,\mu s, 10\,\mu s, \ldots, 10\,s\}$. The obtained set of M spectral maps associated with a particular stress and recovery condition is then used to thoroughly analyze the defects.

A. Example Spectral Maps

Two examples for such spectral maps obtained from a production quality pMOSFET with a 2.2 nm thick plasma-nitrided oxide [8, 18] stressed at $-1.7\,$V and $100\,$°C are shown in Fig. 2. Clearly, marked clusters of (τ_e, d) pairs evolve, the intensity of which typically follows $P_c = 1 - \exp(-t_s/\bar{\tau}_c(V_{stress}))$, with $\bar{\tau}_c(V_{stress})$ as the average capture time at the stress bias $V_G = V_{stress}$. This expression for P_c is valid as long as $\bar{\tau}_c(V_{stress}) \ll \bar{\tau}_e(V_{stress})$, which is mostly the case. Should $\bar{\tau}_c(V_{stress}) \sim \bar{\tau}_e(V_{stress})$ the defect produces RTN making the extraction more problematic as $P_c(t_s \to \infty) = \bar{\tau}_e/(\bar{\tau}_e + \bar{\tau}_c) < 1$, requiring the extraction of both $\bar{\tau}_c(V_{stress})$ and $\bar{\tau}_e(V_{stress})$. The position on the emission time axis remains fixed with stress time while the step height is subject to slight variations, depending on the most dominant current conduction path [24, 25] which can change with the occurrence of additional charged defects. Each cluster corresponds to an individual defect and is labeled accordingly in the maps.

When the procedure is repeated at higher temperatures, the same clusters are observed but clearly have shorter emission times, see Fig. 3, confirming that we are dealing with a thermally activated emission process. Also, all clusters appear already after shorter stress times, demonstrating the thermally activated nature of charge capture. For example, after a 100 ms stress the cluster belonging to defect #7 is already fully developed at 150 °C while it only gradually begins to form at 125 °C. Recall that the spectral maps show the emission times at the recovery voltage V_{relax}, while P_c is controlled by $\bar{\tau}_c$ and $\bar{\tau}_e$ at V_{stress}. This is in contrast to RTN measurements where the capture and emission times are always recorded at the same bias condition.

B. Theoretical Considerations

For the analysis of the spectral maps we need to understand what theory predicts. In order to capture the stochastic nature of the trapping and detrapping process, the model equations, which are given by coupled partial differential equation systems, are solved

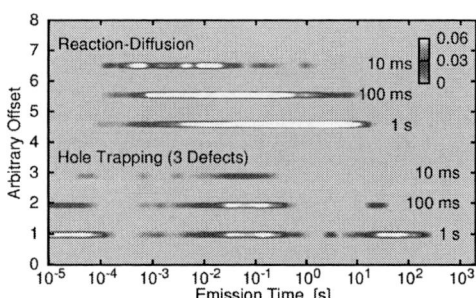

Fig. 4: Theoretical spectral maps obtained from the stochastic solution of the reaction-diffusion model and a simple hole-trapping model for three stress times. The top part of the figure shows the solution of the RD model which produces a single moving cluster, which has a maximum at $\tau_e \sim t_s$ and increases in intensity, in stark contradiction to the experimental data. The lower part shows a solution of a hole-trapping model, which perfectly resembles the experimentally observed behavior.

using the Stochastic Simulation Algorithm [20, 27]. As can be seen in Fig. 4 (top), the conventional RD model predicts the occurrence of relatively wide clusters in the spectral map where 80% of the emission events occur in about 3.8 decades around the average emission time [20] compared to the 1.3 decades observed experimentally. Furthermore, the cluster resulting from a diffusion-limited process would move to larger emission times for increasing stress time, while our experiments only reveal clusters fixed with stress time. And finally, this RD cluster should increase in magnitude as a function of $t_s^{1/6}$ [1], while the maximum amplitude of the charge trapping clusters is obtained from the case $P_c = 1$, corresponding to a 100% capture probability. As a consequence, the RD cluster would begin to dominate the spectral maps at a very early time. In addition, the initial charge trapping step would lead to temperature independent clusters in the spectral maps while not a single temperature-independent cluster has been observed. In short, not a single prediction of the stochastic RD model is consistent with our experiments.

In contrast, a simple hole capture and emission model produces narrow clusters with fixed emissions times for a particular recovery condition (cf. Fig. 4, bottom), in excellent agreement with the experimental data. While this feature can be reproduced by any first-order reaction model, the bias and temperature dependencies of the capture and emission time constants is crucial. As a reference we recall the frequently used model based on a simple nonradiative multiphonon (NMP) process, as used for instance by Kirton and Uren [15]

$$\bar{\tau}_c = \tau_0\, e^{\beta \Delta E_B}\, \frac{N_V}{p}, \tag{1}$$

$$\bar{\tau}_e = \bar{\tau}_c\, e^{\beta E_{TF}} \approx \tau_0\, e^{\beta(\Delta E_B + \Delta E_T)}\, e^{xF/V_T}, \tag{2}$$

with $\tau_0^{-1} = N_V v_{th} \sigma_0 e^{-x/x_0}$ and $E_{TF} = E_T - E_F$. Assuming to first-order that the oxide is charge free, the oxide trap level depends on the applied bias via $E_T = E_{T_0} - q\varphi_s + qxF$, which gives $E_{TF} = E_{VF} + \Delta E_T + qxF$, with $\Delta E_T = E_{T_0} - E_{V_0}$. F is the modulus of the oxide field, p the surface hole concentration, ΔE_B the (bias-independent) NMP barrier, $\beta^{-1} = k_B T$, $V_T = k_B T/q$, x the distance of the defect into the oxide, v_{th} the thermal hole velocity, φ_s the surface potential, x_0 from a simple WKB approximation for a large and thin barrier of height ϕ as $x_0 = \hbar/(2\sqrt{2m\phi})$, and σ_0 the capture cross section. Note that in this model $\bar{\tau}_c$ only weakly depends on bias as $1/p$, while depending on the trap depth x, a rather strong exponential dependence on F is predicted for $\bar{\tau}_e$. However, around V_{th}, F depends only weakly on V_G, and so does $\bar{\tau}_e$.

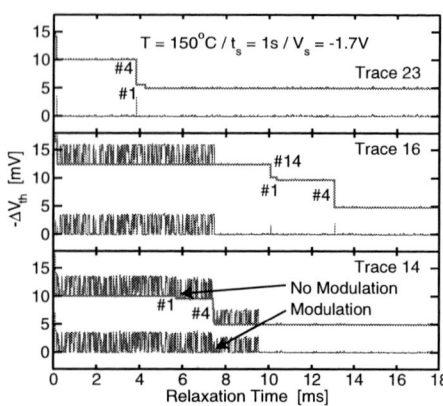

Fig. 5: In addition to the discrete recovery steps, the traces can contain temporary RTN, which disappears after about the average slow emission time, $\bar{\tau}_e^s$, of the defect, see the three selected traces above. The extracted step heights (brown) are subtracted from the raw data (red) to yield the noise trace (blue). Just like with any other defect, the tRTN step height can be modulated by a change in charge in another defect, in the above example defect #4. By contrast, the emission event associated with #1 does not modulate the step height. Although the tRTN defect has a step-height and average (slow) emission time similar to #4, it is a different defect as emission events of #4 are completely uncorrelated (middle and bottom). Also, the step height of the tRTN defect has a different bias dependence.

Since such a charge trapping mechanism is a Poisson process, just like charge capture and emission leading to RTN [15], the capture and emission times are exponentially distributed. Consequently, the probability of τ_e falling into bin i is given by $P_e = P(\tau_i \leq \tau_e < \tau_{i+1}) = \exp(-\tau_i/\bar{\tau}_e) - \exp(-\tau_{i+1}/\bar{\tau}_e)$. However, in order for the defect to be discharged, it has to be previously charged after the stress time t_s, the probability of which is given by P_c. In total, the probability of having an event in bin i is given by $P = P_c P_e$. By fitting the set of spectral maps to $P(\tau_c, \tau_e)$, the characteristic parameters $\bar{\tau}_e$ and $\bar{\tau}_c$ as well as the step height d can be extracted for each cluster.

C. Peculiarities

Although a first-order reaction can describe many clusters with very good accuracy, deviations from this behavior occur quite regularly. The most commonly observed feature is the splitting of the cluster into two peaks due to the electrostatic interaction when another trapped charge in the same percolation path modulates the step height of the defect. More complex deviations, like different emission times for the two peaks, have also been observed but will be discussed in a separate publication.

While the spectral maps are dominated by defects which discharge in the expected manner, RTN appeared on a few traces in our particular device after a stress time of about 1 s. Quite intriguingly, the RTN disappeared after a certain time, for instance after about a few milliseconds at 150 °C. We consequently suggest to term this phenomenon *temporary RTN*, or tRTN. The occurrence of this RTN was found to follow a stochastic process, similar to normal charge capture, that is, with increasing stress time and increasing T the number of traces showing tRTN increased. Also, the tRTN capture and emission times showed the same temperature and bias dependence as known from normal RTN.

Consequently, four characteristic time constants are required for such a defect: the *slow capture time constant* $\bar{\tau}_c^s$ initiating the tRTN process, the *fast capture and emission time constants* $\bar{\tau}_c^f$ and $\bar{\tau}_e^f$ determining the on and off times of the tRTN process, and the *slow emission time constant* $\bar{\tau}_e^s$ eventually terminating the tRTN signal.

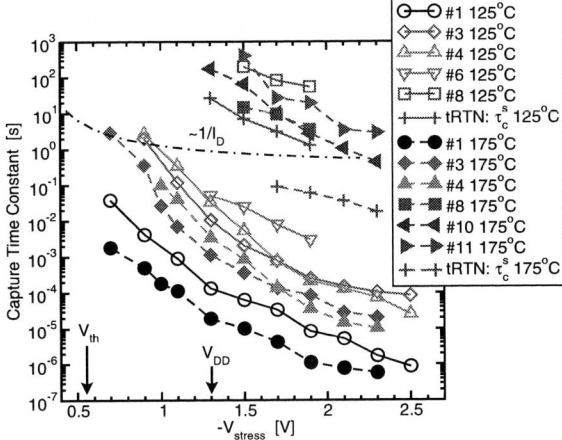

Fig. 6: Voltage and temperature dependence of the capture time constant for 7 defects. Defect #6 was visible during the initial experiments only (taken at 125 °C) and then disappeared permanently, #10 and #11 were outside our experimental window at 125 °C. A strong field/voltage dependence slightly different from exponential is observed for all defects. Also shown is the $1/I_D$ tendency expected from the standard SRH-like models which is considerably weaker.

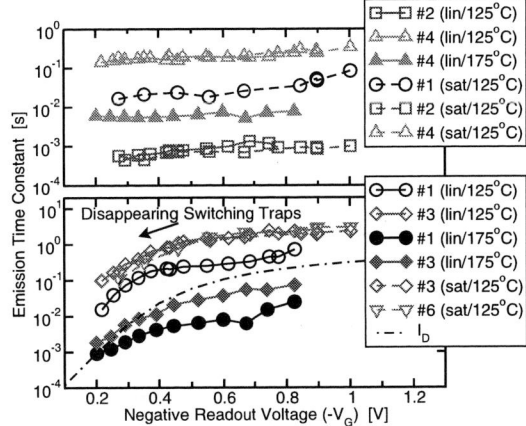

Fig. 7: Gate voltage dependence of the emission time constants at 125 °C and 175 °C, measured in the linear and the saturation regime. Defects #1/sat, #2, and #4 (top) may be also described by a standard model such as (2) because they show only a weak (if at all) field dependence at lower voltages. In contrast to that, $\bar{\tau}_e$ of defects #1/lin, #3, and #6 (bottom) shows a pronounced V_G dependence at low bias. This behavior seems to coincide with the interfacial hole concentration, which is roughly proportional to I_D. Such a behavior cannot be explained with a standard model.

An example for the occurrence of tRTN is shown in Fig. 5. Following the construction of the discrete approximation of ΔV_{th}, the discrete signal is subtracted from the original trace to yield the *noise trace*. As can be seen in Fig. 5, tRTN is visible from the onset of recovery and disappears on average after $\bar{\tau}_e^s \sim 8\,\text{ms}$. Just like the step height of 'normal' defects, the step height or tRTN can be modulated by the charge state of other defects, in this case by the emission of charge from #4 but not by #1. The time constants of the tRTN defect show a very strong bias dependence, that is, tRTN is only visible in a narrow window. Also, tRTN shows a marked temperature dependence in all four characteristic time constants.

Also interesting is the observation that particularly heavier stresses resulted in defects *disappearing* (observed regularly for #6, #7, and #8) for a certain period. For example, #7 was inactive for more than a month during measurements at 125 °C after having been stressed a 100 times at −1.7 V/175 °C for 10 s. Contrary to that, defect #1 was present during the whole duration of the study and only disappeared once for two hours after having been stressed at −1.7 V/75 °C for 10 ms. A similar behavior was previously observed in RTN and was explained by the defect having an additional metastable state [28].

We particularly note that defects seem to disappear from the map without any impact on the threshold voltage shift, meaning there is no permanent degradation directly linked to their disappearance.

III. EXTRACTION OF DEFECT PARAMETERS

To study the bias and temperature dependence of the basic defect parameters we recorded 80+ sets of spectral maps (with 5 to 8 maps per set) by continuously measuring on a single device over a period of three months and for a few weeks after a three months break. Quite intriguingly, the spectral maps remained basically the same for the whole duration. Although the individual traces show an increasing offset with increasingly heavier stress, probably due to a permanent degradation component, the spectral maps remained basically unaffected by this offset.

Within our range of voltages, temperatures, stress and relaxation times, 13 defects could be identified on this device and the bias and voltage dependence of their parameters extracted, at least to a certain degree. Defects that stayed within our experimental window and did not disappear could be studied over the whole operational regime of the transistors, starting from slightly above threshold up to close to oxide breakdown.

As an example, the extracted capture time constants are shown in Fig. 6, which are clearly temperature activated with an activation energy of about 0.6 eV and depend only approximately in an exponential manner on the stress bias. As has been reported for RTN [15], this is incompatible with the $1/p \sim 1/I_D$ dependence of the SRH-like model (2). In fact, for all defects which could be traced over a wider window, $\log(\bar{\tau}_c)$ also showed a marked non-linear field dependence which can be fit by a quadratic relation $-c_1 F + c_2 F^2$. It is particularly worthwhile to point out that this non-linearity is different for all defects and must as such be related to the details of the defect properties as in this regime both F and p depend linearly on V_G. Also note that were these defects studied only within the typical RTN measurement window, this non-linearity would have gone unnoticed.

From a reliability point of view this non-linearity in $\log(\bar{\tau}_c)$ is of fundamental importance as it connects the degradation accumulated in an accelerated stress regime to the degradation level expected under operating conditions. The observed non-linearity in the individual defects is also consistent with the observed non-linearity in macroscopic experiments, where a $\Delta V_{th} \sim \log(\bar{\tau}_c)$ behavior could be expected from averaging a large number of defects [10]. Accurate back-extrapolation to operating conditions is thus only possible when this non-linearity is properly accounted for.

Also of particular practical interest is the bias dependence of $\bar{\tau}_e$, as a strong bias dependence would allow one to efficiently remove the charges from the oxide by applying a small positive bias. In fact, a strong bias dependence of the macroscopically measured ΔV_{th} has been previously observed [7, 12, 29], albeit of a more complicated form than previously anticipated. It is worthwhile to recall that this issue was one of the first problems noted regarding reaction-diffusion theory [12], as the assumed diffusion of a neutral species is difficult to reconcile with any bias dependence.

Fig. 7 shows $\bar{\tau}_e(V_G)$ for five selected defects as measured in the linear and saturation regime. Defects #2 and #4 are insensitive to changes in V_G and have the same $\bar{\tau}_e$ in both regimes. On the

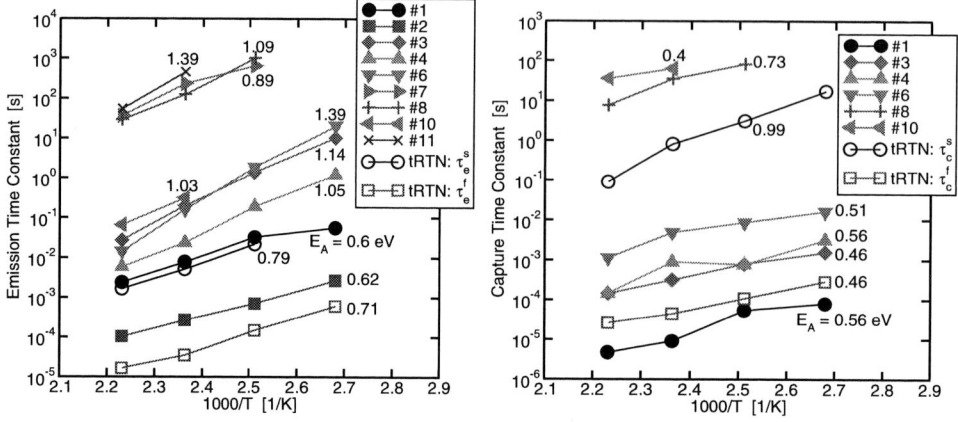

Fig. 8: Left: Arrhenius plot of the emission time constant $\bar{\tau}_e$ for a recovery bias of $-0.55\,\text{V}$. The approximate activation energies are given for each defect where a wide spread is observed. **Right:** Arrhenius plot of the capture time constant $\bar{\tau}_c$ for a stress bias of $-1.7\,\text{V}$. While the extraction of $\bar{\tau}_c$ is less reliable compared to the extraction of $\bar{\tau}_e$, the spread in activation energies seems to be smaller.

other hand, defects #3 and #6 demonstrate a very strong sensitivity when $|V_G|$ moves below $|V_{th}|$. Interestingly, this regime coincides with a large drop in the interfacial hole concentration. As with the bias dependence of $\bar{\tau}_c$, this is incompatible with a SRH-like model which predicts only a weak bias dependence in this regime. We have previously associated such a behavior with a switching trap [10] which differs from a 'normal' trap by the fact that, once created, its charge state can be controlled by the Fermi-level in the substrate prior to annealing. The most interesting case is thus defect #1, which in the saturation regime only shows a weak bias dependence but demonstrates switching trap behavior in the linear regime. The simplest conclusion to draw from this is that in fact both types of behavior are special cases of one and the same general defect type.

The temperature dependencies of $\bar{\tau}_e$ and $\bar{\tau}_c$ are shown in Fig. 8, demonstrating a wide spread in the activation energies, with the spread in $\bar{\tau}_e$ appearing larger than that observed in $\bar{\tau}_c$. As also observed in Fig. 6, the defect responsible for tRTN appears in no way different compared to the other defects, leading us to conclude that tRTN is just another special form of the general defect type.

A. The TDDS Measurement Window

Typical RTN measurements require the capture and emission times to be about equal and within the measurement window. Furthermore, the measurement becomes difficult to deconvolute when there are contributions from more than a single defect. The situation is fundamentally different and considerably improved for the TDDS:

- The basic requirement is only that the time constants fall within the experimental window. Typical time windows would be from $10\,\mu\text{s}$ up to 100 or even $1\,\text{ks}$ (covering about 8 decades in time), where the upper bounds are mostly "limited by the experimenters' patience" [30]. Occasionally, a defect may have $\bar{\tau}_c$ outside the experimental window while $\bar{\tau}_e$ can be fully characterized. In our particular device this is the case for defects #2, #5, and #12, depending on the temperature and bias: Due to the temperature dependence of the time constants, adjustment of the temperature allows to shift defects into the window. Similar considerations hold for $\bar{\tau}_c$, which can be brought into the window by appropriately selecting the stress voltage and temperature.

- Even when more than a dozen defects contribute to the spectral map, each of them can be analyzed provided that their capture times are separated by about a decade and the step height by

more than about $0.5\,\text{mV}$ (for the current device). The latter can often be enforced by suitably adjusting the drain bias during recovery, which may significantly impact the step height. For example, defects #1 and #3 have a step height of about $0.5\,\text{mV}$ at $V_D = -1.2\,\text{V}$. At $|V_G| < |V_{th}|$, the much stronger bias sensitivity of #3 results in a near overlap of the respective clusters on the spectral map. In contrast, at $V_D = -0.1\,\text{V}$ the step heights are $0.3\,\text{mV}$ and $1.5\,\text{mV}$ and no overlap occurs.

- The gate bias dependence of the capture time constant can be measured over nearly the whole operation range of the transistor, from slightly above threshold (in order to provide a sufficient probability of charging the fastest defects) up to oxide breakdown.

- The gate bias dependence of the emission time constant is somewhat restricted since at low $|V_G|$ the minimum time constant obtainable with our equipment can increase significantly due to the small drain current. On the other hand, at high $|V_G|$, about $1\,\text{V}$ in the present example technology, the drain current becomes too large to allow for a sufficiently accurate extraction of ΔV_{th}.

The strength of the TDDS becomes apparent from Fig. 9, where the capture and emission times of defect #1 are shown as a function of $|V_G|$. Using the conventional RTN analysis method, this defect could have only been monitored in the rather narrow window where $\bar{\tau}_c \sim \bar{\tau}_e$, not to mention the tremendous difficulties caused by the interference of other defects. Interestingly, when measured in the saturation regime $\bar{\tau}_e(V_G)$ is practically bias independent for $|V_G| < |V_{th}|$, while in the linear regime the switching trap behavior becomes apparent.

IV. MODEL

As has been noted previously, a number of features in $\bar{\tau}_c$ and $\bar{\tau}_e$ have been revealed by the TDDS which are inconsistent with standard theories:

- A non-linear bias dependence of $\log(\bar{\tau}_c)$, which is different for every defect. Here we remark that standard NMP theories predict a linear field dependence [31] for the strong electron-phonon coupling case, so the reason for this discrepancy must be investigated.

- A strong bias dependence of $\bar{\tau}_e$ for $|V_G| < |V_{th}|$ in some defects, depending on the drain (and probably substrate) bias. Such a strong bias dependence is likely due to the existence of a metastable state [20].

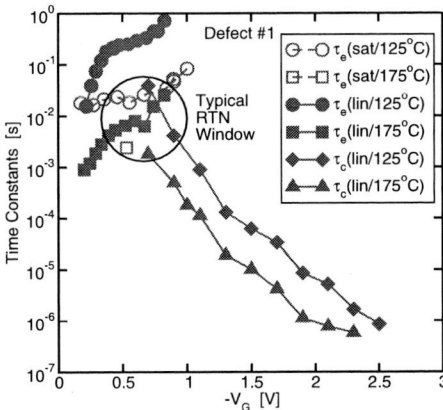

Fig. 9: The TDDS allows the measurement of the capture and emission times over an extremely wide range. Shown are also the emission times measured in the linear ($V_D = -0.1\,\text{V}$) and saturation regimes ($V_D = -1.2\,\text{V}$). For comparison a typical RTN measurement window is shown which requires $\bar{\tau}_e \sim \bar{\tau}_c$. When measured in the linear regime, the defect acts like a switching trap, while in the saturation regime this feature is not visible.

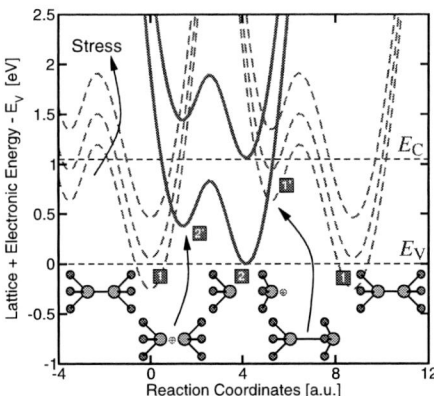

Fig. 10: Configuration coordinate diagram showing the adiabatic total energy potentials corresponding to the four defect configurations. The energy is given relative to the valence band edge E_V. In state 1 and 1' the electron is at the defect site, that is, the electronic contribution to the total energy shifts the potential up with increasing NBTS (more negative V_G). In state 2' and 2, the electron is either (approximately) at the valence band or conduction band edge. The relative position of these potentials with respect to E_V does not change with bias.

- Occurrence of transient RTN, which is also likely due to a metastable state, similar to what has been suggested for anomalous RTN [28].

A particularly intriguing observation is that a defect can behave as a switching trap or not, depending on the drain bias. Considering that the capture time constants of all defects appear similar enough, it is thus most tempting to assume that all defects are basically of the same type and that the differences observed in $\bar{\tau}_e$ are merely due to small configurational differences. It should thus be obvious that a model that can describe such a wealth of information has to be more complicated than standard models and has to include metastable defect states.

In the following we will extend our previously suggested semiempirical model [10] which is based on the Harry-Diamonds-Lab switching trap model [32] for the E' center. As the exact microscopic nature of the defect responsible for recoverable NBTI has not been unanimously identified, it is worth highlighting that the E' center serves mainly as an inspiriation for the model but that the model can be applied to any defect which has the required metastable states. The configuration coordinate (CC) diagram of the extended version is shown in Fig. 10 and considers four different configurations of the defect, as have for instance been used by Poindexter for the E' center [33]. Two of the states are electrically neutral (1 and 1') while two of them correspond to the singly positively charged state (2' and 2). In each charge state the defect is represented by a double well, with the first of the two states being the equilibrium state and the other a secondary (metastable) minimum. Transitions involving charge exchange with the substrate are assumed to occur predominantly between 1 and 2' as well as 2 and 1'. On the other hand, transitions between 1 and 1' as well as between 2' and 2 are assumed to be purely thermally activated. In our previous model [10] the metastable state 1' had already been introduced in order to capture the bias dependence observed in the macroscopic measurements, in particular the switching trap behavior. While the fundamental assumptions employed in the old model appear to be basically correct, the rate equations where formulated in an *ad hoc* manner and the metastable state of defect 2 (2') was neglected. However, as will be shown below, to fully capture the non-linearity in the experimental data, state 2' has to be considered and the rate equations have to

be derived more rigorously using nonradiative multiphonon (NMP) theory [34–36].

We assume that the time dynamics of the defect can be described by a simple stochastic process $X(t)$, a homogeneous continuous-time jump Markov process [37], see Fig. 11 (top). The defect can be in one of its four states i, described by the probabilities $p_i(t) = P\{X(t) = i\}$, with $i = 1, 1', 2, 2'$. Naturally, $\sum p_i(t) = 1$ at all times. Following conventional Markov process theory, the transition probabilities from state i to state j can be written as

$$P\{X(t+h) = j \,|\, X(t) = i; i \neq j\} = k_{ij}h + O(h), \qquad (3)$$

with h being an infinitesimally small time step, $\lim_{h\to 0} O(h)/h = 0$, and k_{ij} as the transition rate. Together with the probability that the defect stays in state i

$$P\{X(t+h) = i \,|\, X(t) = i\} = 1 - h \sum_{j, i \neq j} k_{ij} + O(h), \qquad (4)$$

it is straight forward to derive the partial differential equation (the master equation) controling the time evolution of the probabilities p_i

$$\frac{\partial p_i(t)}{\partial t} = \sum_j \Big(p_j(t)k_{ji} - p_i(t)k_{ij} \Big). \qquad (5)$$

We neglect the transitions which cause a change in both the charge state and a considerable change in the defect configuration at the same time, that is, $k_{12} = k_{21} = k_{2'1'} = k_{1'2'} = 0$. Being a linear equation in $p_i(t)$, the expectation values of the probabilities, $E\{p_i(t)\} = f_i(t)$, are controlled by the same equation as (5), with f_i replacing p_i, which is then the conventional reaction rate equation. In other words, the stochastic description and the reaction rate formalism are 'consistent in the mean' for the linear problem at hand [38]. The advantage of using the stochastic description (5) lies in the fact that it also allows the stochastic simulation of RTN and tRTN as well as a simple calculation of the probability density function (p.d.f.) describing the stochastic capture and emission processes [39], see Fig. 11. With the p.d.f. at hand, the effective capture and emission times as well as their variances can be calculated analytically and compared to experimental data.

978-1-4244-5430-3/10 $26.00 © 2010 IEEE

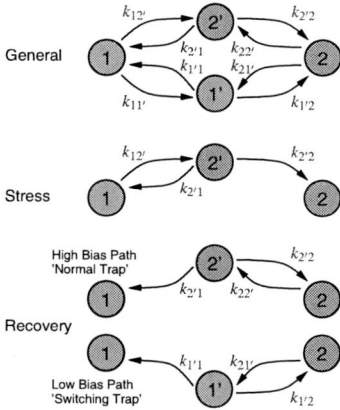

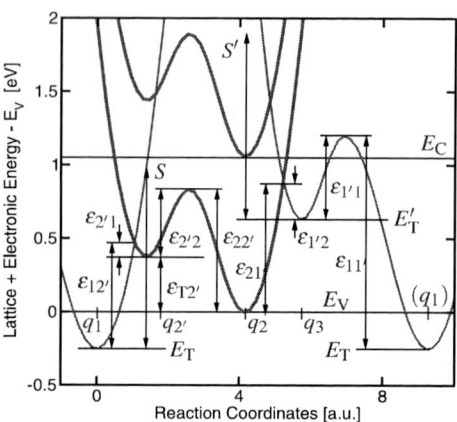

Fig. 11: State transition rate diagram for the general case (top), stress case (middle), and the two important pathways for recovery (bottom). During stress and recovery the Markov chain is described by a Birth-Death process since we are only interested in the first passage time from the initial to the final state, the expectation value of which gives the time constant.

The details of the derivation of the transition rates k_{ij} will be published elsewhere and we restrict ourselves to summarizing the main assumptions together with the results. Most assumptions employed in the model are standard and frequently used in NMP theory with a few important exceptions:

- We assume that the electron-phonon system can be dealt with in the Born-Oppenheimer approximation and describe the defect energy using adiabatic potentials.
- As the model is based on reported properties of the E' center, which has been suggested to undergo severe structural relaxation upon charge capture, we restrict our derivation to the strong electron-phonon coupling case.
- Since our prime concern is charge trapping at higher temperatures, such as observed during NBTS, the final equations can be considerably simplified (semiclassical approximation).
- To be somewhat more general than the frequently used theories, we consider both linear and quadratic electron-phonon coupling modes [40].
- In order to highlight the important features of the model, the transition rates are only given for the case of charge exchange with the substrate valence band. Charge exchange with the conduction band, which is important in accumulation (PBTI), as well as charge exchange with the gate can be dealt with analogously.
- Quite frequently one finds that the NMP equations derived for bulk traps are used for oxide defects, that is, the NMP process merely introduces a bias-independent thermal barrier ΔE_B [15, 41]. Our previously suggested model [10, 20] suffered from the same problem and in order to properly account for the experimentally observed bias dependence we empirically introduced a field assisted acceleration term, $\exp(-F^2/F_c^2)$, valid for the bulk case only [42, 43]. In fact, however, in oxide traps the relative position of the parabolas changes with bias [31, 36, 44], quite naturally introducing the required strong bias dependence into ΔE_B, in addition to some other features.

Under the above assumptions the relations for the transition rates involving charge transfer are

$$k_{12'} = \sigma v_{\text{th}} p e^{-\beta \varepsilon_{12'}}, \qquad k_{2'1} = \sigma v_{\text{th}} p e^{-\beta \varepsilon_{12'}} e^{-\beta(E_{\text{TF}} - \varepsilon_{\text{T2'}})}, \quad (6)$$

$$k_{1'2} = \sigma v_{\text{th}} p e^{-\beta \varepsilon_{1'2}}, \qquad k_{21'} = \sigma v_{\text{th}} p e^{-\beta \varepsilon_{1'2}} e^{-\beta E_{\text{T'F}}}. \quad (7)$$

Fig. 12: Definition of the symbols used in the model. Note that the reaction coordinate describing the transition $1 \leftrightarrow 2'$ is different from the one describing $2 \leftrightarrow 1'$, which is why the potential describing states 1 and 1' is plotted twice, once to the left and once to the right of state 2. In this simple model only the expansions around q_1 and $q_{2'}$ determine $1 \leftrightarrow 2'$ while the expansions around q_2 and q_3 determine $2 \leftrightarrow 1'$.

For simplicity we assume all capture cross sections to be equal and given by a simple WKB approximation for a large and thin barrier of height ϕ as $\sigma = \sigma_0 \exp(-x/x_0)$, with $x_0 = \hbar/(2\sqrt{2m\phi})$ and v_{th} the thermal velocity.

For the calculation of the barriers $\varepsilon_{12'}$ and $\varepsilon_{1'2}$ the adiabatic potentials in each state i are expanded quadratically around the minima q_i, which in general results in different vibrational frequencies ω_i. Note that for linear electron-phonon coupling all frequencies are the same, $\omega_i = \omega$ [40]. Then, by setting $\omega_i = R_i \omega_j$ the rates can be approximately given as

$$\varepsilon_{ij} \approx \frac{S_i \hbar \omega_i}{(1 + R_i)^2} + \frac{R_i}{1 + R_i} \varepsilon_{\text{T}i} + \frac{R_i}{4 S_i \hbar \omega_i} \varepsilon_{\text{T}i}^2, \quad (8)$$

where S_i is the Huang-Rhys factor characterizing the number of phonons required for the optical transition. The trap depth $\varepsilon_{\text{T}i}$ is either $\varepsilon_{\text{T}1} = E_V - E_T + \varepsilon_{\text{T}2'}$ or $\varepsilon_{\text{T}1'} = E_V - E_T'$. We remark that the above equation is exact for linear electron-phonon coupling ($R_i = 1$), and gives the familiar result $\varepsilon_{ij} = (\varepsilon_{\text{T}i} + S_i \hbar \omega)^2/(4 S_i \hbar \omega)$.

For simplicity, we assume the transition between states 1 and 1' as well as 2' and 2 to be bias independent but to occur along different reaction coordinates. Consequently, we do not calculate the barriers via intersections of the parabolas but consider them as explicit parameters. Obviously, $\varepsilon_{22'} = \varepsilon_{\text{T}2'} + \varepsilon_{2'2}$ and $\varepsilon_{11'} = \varepsilon_{1'1} + E_T' - E_T$ and we use $k_{mn} = v_m \exp(-\beta \varepsilon_{mn})$ where $v_m \sim 10^{13} \, \text{s}^{-1}$.

Naturally, the prediction of the model depends heavily on an accurate description of the defect potentials which are described by the parameters $S\hbar\omega = S_1 \hbar \omega_1$, $S'\hbar\omega' = S_{1'} \hbar \omega_{1'}$, $\Delta E_T = E_{T_0} - E_{V_0}$, $\Delta E_T' = E_{T_0}' - E_{V_0}$, $\varepsilon_{22'}$, $\varepsilon_{1'1}$, $R = R_1$, and $R' = R_{1'}$. In addition, the trap depth x as well as the capture cross section ($\sim 10^{-15} \, \text{cm}^{-2}$) and the attempt frequency ($\sim 10^{13} \, \text{s}^{-1}$) are required. While it appears that the model has more parameters than one might wish it to have, one has to recall that this is a consequence of the multitude of features it can explain, including the full bias and temperature dependence. Should one aim at a crude, first-order approximation, one could neglect the metastable states (no tRTN, no switching trap behavior, linear bias dependence of $\log(\bar{\tau}_c)$) to obtain a rough model of the defect with the three parameters $S\hbar\omega$, ΔE_T, x only. Care has to be taken, though, that the physical meaning of these effective parameters may then be questionable.

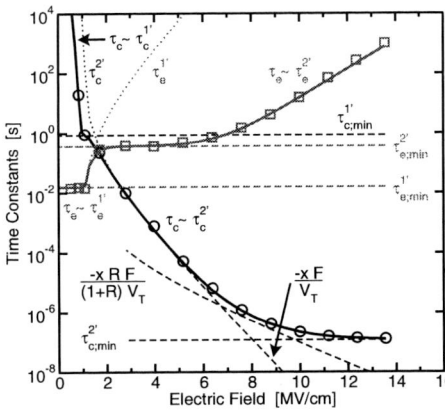

Fig. 13: Bias dependence of the effective capture and emission times according to the model. The symbols are obtained from a Monte Carlo simulation of the stochastic process, while the dashed lines give the individual contributions to the effective time constants (solid lines).

V. APPROXIMATE SOLUTIONS

Although the solution of the master equation (5) is in principle straight forward to obtain, it does not provide significant insight into the behavior of the defect. In particular, depending on the defect configuration various complicated transition patterns are possible, most notably patterns which would be recognized as RTN and anomalous RTN. However, during both stress and recovery the rates become highly asymmetric, strongly favoring a transition to 2 during stress and back to 1 during recovery, see Fig. 11. The most likely path during stress is from 1 to 2, while during recovery the defect may either recover via 2' or 1', the latter becoming particularly important at low $|V_G|$. Thus, in the following approximate solutions for these two important limiting cases will be given, while the details of the derivation will be published elsewhere. We only consider the dominant charge exchange with the valence band in the substrate. Also, the second-order term in ε_{Ti} defined by (8) will be neglected and the oxide will be assumed to be charge-free, in order to provide a simple relationship between the electrostatic defect level and the (constant) field in the oxide. Finally, Boltzmann statistics will be assumed which is valid for $\bar{\tau}_e$ at low $|V_G|$ but rather crude for $\bar{\tau}_c$ at high $|V_G|$. Nonetheless, the qualitative features of the model are not affected by either of these approximations.

The capture time constant is calculated from the expectation value of the first passage time from state 1 to state 2, see Fig. 11. The transition may proceed either via state 2', which is the dominant case at high bias, or via state 1' at lower biases (not shown in Fig. 11). For the first case we obtain $\bar{\tau}_c^{2'}$, while the latter case is described by $\bar{\tau}_c^{1'}$, which, in total, results in

$$1/\bar{\tau}_c \approx 1/\bar{\tau}_c^{2'} + 1/\bar{\tau}_c^{1'}. \quad (9)$$

Conversely, the emission time constant is calculated from the expectation value of the first passage time from state 2 to state 1 which may also proceed either via state 2' or via state 1'. For the first case we obtain $\bar{\tau}_e^{2'}$, while the latter case is described by $\bar{\tau}_e^{1'}$, which, again, gives

$$1/\bar{\tau}_e \approx 1/\bar{\tau}_e^{2'} + 1/\bar{\tau}_e^{1'}. \quad (10)$$

It is worth emphasizing that the capture and emission times described in such a way are not exponentially distributed. Nonetheless, it can be shown that under most bias conditions the time constants are

dominated by a single transition, in which case the p.d.f. becomes nearly exponential, thereby justifying the concept of an effective first-order process with effective capture and emission times (9) and (10).

The individual contributions to the time constants are

$$\bar{\tau}_c^{2'} = \bar{\tau}_{c;\min}^{2'}\left(1 + \frac{N_1}{p}\exp\left(-\frac{xF}{V_T}\right)\right) + \tau_0\frac{N_2}{p}\exp\left(-\frac{xR}{1+R}\frac{F}{V_T}\right), \quad (11)$$

$$\bar{\tau}_c^{1'} = \bar{\tau}_{c;\min}^{1'} + \tau_0\frac{N_3}{p}\exp\left(-\frac{xR'}{1+R'}\frac{F}{V_T}\right), \quad (12)$$

$$\bar{\tau}_e^{2'} = \bar{\tau}_{e;\min}^{2'} + \tau_{2'}\exp\left(\frac{x}{1+R}\frac{F}{V_T}\right), \quad (13)$$

$$\bar{\tau}_e^{1'} = \bar{\tau}_{e;\min}^{1'}\left(1 + e^{\beta E_{T'F}}\right) + \tau_{1'}\exp\left(\frac{x}{1+R'}\frac{F}{V_T}\right). \quad (14)$$

with the same τ_0 as used in (2) and the temperature-dependent but field-independent auxiliary quantities

$$N_1 = N_V\exp(\beta(\varepsilon_{T2'} - \Delta E_T)),$$

$$N_2 = N_V\exp\left(\beta\left(\frac{S\hbar\omega}{(1+R)^2} - \frac{R(\Delta E_T - \varepsilon_{T2'})}{1+R}\right)\right),$$

$$N_3 = N_V\exp\left(\beta\left(\frac{S'\hbar\omega'}{(1+R')^2} - \frac{R'\Delta E_T'}{1+R'}\right)\right)(1 + \exp(\beta(\Delta E_T' - \Delta E_T))),$$

$$\tau_{2'} = \tau_0\exp\left(\beta\left(\frac{S\hbar\omega}{(1+R)^2} + \frac{\Delta E_T - \varepsilon_{T2'}}{1+R}\right)\right)(1 + \exp(\beta\varepsilon_{T2'})),$$

$$\tau_{1'} = \tau_0\exp\left(\beta\left(\frac{S'\hbar\omega'}{(1+R')^2} - \frac{\Delta E_T'}{1+R'}\right)\right).$$

Interestingly, since each time constant contains a contribution from a purely thermal transition, the minimum value is bounded by $\bar{\tau}_{c;\min}^{2'} = 1/k_{2'2}$, $\bar{\tau}_{c;\min}^{1'} = 1/k_{13}$, $\bar{\tau}_{e;\min}^{2'} = 1/k_{22'}$, and $\bar{\tau}_{e;\min}^{1'} = 1/k_{31}$. An evaluation of the analytic capture and emission time models against a Monte Carlo simulation of the full model is given in Fig. 13 for a switching trap, where around $V_G = V_{th}$ the dominant pathway changes from via 2' to 1'.

A. Normal Kinetics

Under 'normal' kinetics we understand the case where the impact of the metastable states is not directly obvious, that is, no switching behavior can be observed (no transition to 1'). This is the case when $\Delta E_T'$ is too large to give a significant occupancy of state 1'. In this case, the time constants are

$$\bar{\tau}_c = \bar{\tau}_{c;\min}^{2'}\left(1 + \frac{N_1}{p}\exp\left(-\frac{xF}{V_T}\right)\right) + \tau_0\frac{N_2}{p}\exp\left(-\frac{xR}{1+R}\frac{F}{V_T}\right), \quad (15)$$

$$\bar{\tau}_e = \bar{\tau}_{e;\min}^{2'} + \tau_{2'}\exp\left(\frac{x}{1+R}\frac{F}{V_T}\right). \quad (16)$$

Both time constants consist now of two terms, where the first one denotes the impact of the relaxation barrier $\varepsilon_{2'2}$. For capture, the field dependence of the two terms is different, resulting in a non-linear overall exponential field dependence and eventual saturation at $\bar{\tau}_c = \bar{\tau}_{c;\min}^{2'}$ for high fields. For example, for $R = 1$, the argument of the exponential field term is initially $-xF/V_T$ and gradually reduces to $-xF/2V_T$ with increasing field. In addition, a weaker bias dependence is introduced by the same $1/p$ dependence of both factors. Note that this $1/p$ dependence is the only field dependence of $\bar{\tau}_c$ in the standard SRH-like model (2).

For emission, the term due to the relaxation barrier is bias independent and dominates for small fields. This is similar to (2) where irrespective of the exponential F dependence only a weak field dependence is obtained, as F depends weakly on V_G below V_{th}. For $R = 1$ one obtains $xF/2V_T$ for the argument of the exponential field term at high fields. In standard SRH-like models the argument is twice as large, namely xF/V_T. Note how NMP theory with a linear

978-1-4244-5430-3/10 $26.00 © 2010 IEEE

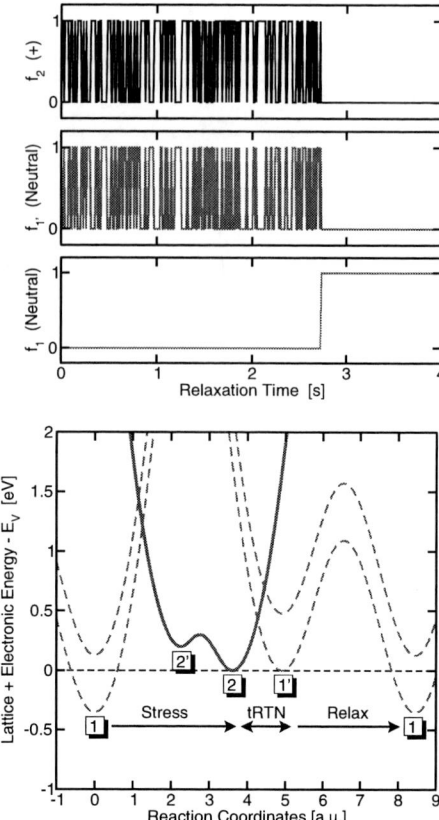

Fig. 14: Stochastic simulation (top) of a defect configuration (bottom) that leads to tRTN. During stress, state 2 is occupied, during the initial recovery phase the defect switches between states 2 and 1', while eventually the defect anneals. Such a behavior can only be observed in a very narrow window of V_G where the minima of states 2 and 1' are close to each other.

coupling term only ($R = 1$) introduces a symmetric field dependence in both $\bar{\tau}_c$ and $\bar{\tau}_e$ which becomes asymmetric for $R \neq 1$. Nonetheless, for medium values of F the ratio $\bar{\tau}_c/\bar{\tau}_e$ is controlled by $-xF/V_T$, just as in the SRH-like model (2).

B. Switching Trap Kinetics

Under the condition that the metastable state 1' is moved close to E_V and the barrier separating 2 and 1' is low enough, a transition $2 \rightarrow 1'$ can occur and the defect may even switch back and forth between states 1' and 2, see Fig. 11 (bottom). For normal switching traps, however, these transitions are too fast to be directly observable by the measurement equipment which only records the average value.

In the switching trap configuration, the impact of the metastable state 1' becomes evident in $\bar{\tau}_e$, see Fig. 13. At low enough bias, annealing of the defect back to state 1 will now occur via state 1'. Note that although even during stress the pathway $1 \rightarrow 1' \rightarrow 2$ is theoretically possible, we have so far not observed a defect compatible with such a configuration.

Introducing $f_n = (1 + k_{1'2}/k_{2'1})^{-1} = (1 + \exp(\beta E_{T'F}))^{-1}$, which is the probability that the trap level E_T' is occupied by an electron, that is, neutral, we can express the emission time constant for $\bar{\tau}_e^{1'} \lesssim \bar{\tau}_e^{2'}$ as

$$\bar{\tau}_e = \frac{\bar{\tau}_{e;min}^{1'}}{f_n} + \tau_{1'} \exp\left(\frac{x}{1+R'}\frac{F}{V_T}\right). \quad (17)$$

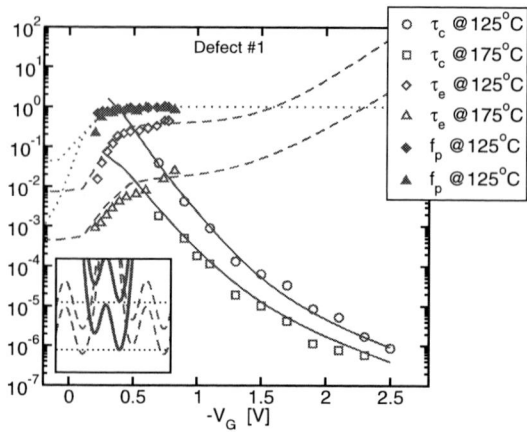

Fig. 15: Simulated capture and emission time constants for defect #1 compared to the experimental values. The CC diagram shown in the inset is similar to that of the tRTN case shown in Fig. 14 with the difference that the barrier between states 2 and 1' is rather small. The experimental occupation probability f_p is given by the filled symbols.

This is a remarkable result: when $E_F > E_T'$, the defect is neutral ($f_n = 1$) and the emission time is given by the bias independent value $\bar{\tau}_{e;min}^{1'}$. As soon as the defect level E_T' moves above the Fermi level, the probability f_n will decrease, thereby strongly increasing $\bar{\tau}_e$. This strong bias dependence explains the typical switching trap characteristics around the threshold voltage observed experimentally in Fig. 7. For large $|V_G|$, on the other hand, $\bar{\tau}_e^{1'}$ will become larger than $\bar{\tau}_e^{2'}$ and the pathway $2 \rightarrow 2' \rightarrow 1$ dominates the emission time.

An interesting special configuration of the switching trap leads to tRTN, namely when the transitions between states 1' and 2 are slow enough to fall within the experimental window, as discussed in Section VI.

VI. QUALITATIVE MODEL BEHAVIOR

A stochastic simulation of a defect configuration leading to tRTN is shown in Fig. 14. During stress, state 2 becomes occupied ($p_2 = 1$), while during the initial recovery phase the defect switches back and forth between states 2 and 1' ($p_2 + p_{1'} = 1$), visible as tRTN. Eventually, the defect anneals by a transition from 1' to 1 ($p_1 = 1$). Such a behavior can only be observed in a very narrow window of V_G where the minima of the states 2 and 1' are close to each other. Furthermore, the barrier between states 2 and 1' must be large enough to cause capture and emission times within the experimental window. For transitions faster than experimentally observable, the measurement equipment will record the averaged signal $E\{P\{X(t) = 2 | X(t) = 2 \vee 1'\}\} = f_p = 1 - f_n$ and the defect will appear as a switching trap.

VII. QUANTITATIVE MODEL BEHAVIOR

The simulated capture and emission time constants calibrated to the experimental data available for defect #1 are shown in Fig. 15. As can be seen from the CC diagram shown in the inset, defect #1 is similar to the schematic tRTN case shown in Fig. 14 with the difference that the barrier between states 2 and 1' is rather small. As a consequence, the fluctuations between states 2 and 1' are too fast and cannot be resolved by the measurement equipment and the defect appears like a switching trap.

In contrast, Fig. 16 shows the model calibrated to the data of defect #4. Here the metastable state 1' is energetically too high and has thus no apparent effect on the measured capture and emission times.

978-1-4244-5430-3/10 $26.00 © 2010 IEEE

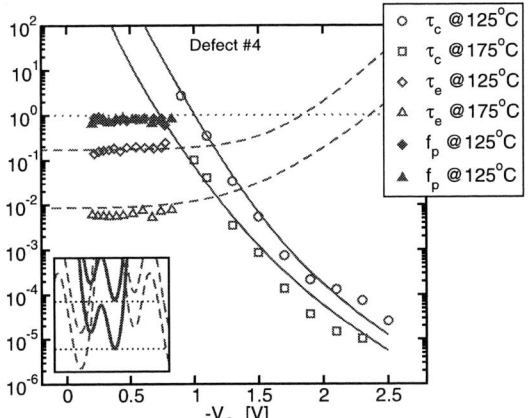

Fig. 16: Similar to Fig. 15, but now for defect #4. In contrast to defect #1, E'_T is too high to be significantly populated and no switching trap behavior can be observed. This can be clearly seen in the CC diagram shown in the inset. As a consequence, the emission time constant is insensitive to the gate bias.

As can be seen from these examples, the suggested model can naturally explain all experimentally observed features. In particular the initially surprising occurrence of tRTN, switching traps, as well as the non-linearities in $\log(\bar{\tau}_c)$ and $\log(\bar{\tau}_e)$ have been explained consistently within a single model.

VIII. Conclusions

We have suggested the powerful time dependent defect spectroscopy (TDDS) for the analysis of defects leading to NBTI and RTN. Application of this new method in a large number of experiments leads us to the following conclusions: *(i)* NBTI recovery is due to discharging of individual defects with a wide distribution of time scales. There is no diffusion involved. *(ii)* Both capture and emission time constants are temperature activated, consistent with nonradiative multiphonon theory. Furthermore, the time constants are uncorrelated. No signs of a temperature-independent elastic tunneling process could be found. *(iii)* The capture time constants show a very strong field dependence. Similarly, the bias dependence of the emission time constant around V_{th} may be either weak or strong, depending on the configuration of the defect. In addition, the visibility of this switching trap behavior my depend on the bias conditions. *(iv)* The total number of defect precursors is pre-existing and no signs of newly created defects which create emission events within our experimental window could be found in our short to medium term stress experiments. *(v)* The existence of metastable states becomes obvious due to disappearing defects, transient RTN, and a bias dependencies stronger than expected in a simpler model. *(vi)* Finally, the defects responsible for the recoverable component of NBTI are identical to those causing RTN. Conceptually, the difference is as follows: in an RTN experiment only a limited number of defects having capture and emission time constants within the experimental window are visible. Due to the strong bias dependence of the capture time constant, many more defects contribute to NBTI. These defects can be visualized in a TDDS setup. Consequently, NBTI can be considered the non-equilibrium response of these defects while RTN is a consequence of their quasi-stationary behavior.

We have finally suggested a comprehensive defect model which can describe the standard and switching trap defect behavior as well as all the anomalies observed, such as disappearing defects and transient RTN. Although the model is based on reported properties

of the E' center, in particular the switching behavior, we do at the moment not have sufficient evidence to claim that the NBTI recoverable component is solely (or at all) due to E' centers, particularly in nitrided and high-k oxides. Nonetheless, as the model uses empirical potentials which are adjusted to fit the experimental data, it is perfectly general and can be applied to a broad range of defects having metastable states. The model can accurately describe the bias dependence of the capture time constant which thus allows back-extrapolation from stress to operating conditions. Furthermore, the model accurately describes the bias dependence of the emission time constant around V_{th}, which is important for our understanding of the functioning of a transistor in a circuit where the gate bias is not just switched between static stress and recovery voltages.

IX. Acknowledgments

The research leading to these results has received funding from the European Community's Seventh Framework Programme under grant agreement n°216436 (project ATHENIS). We also gratefully acknowledge stimulating discussions with Th. Aichinger, Ph. Hehenberger, Ph. Roussel, M. Uren and M. Kirton.

References

[1] T. Grasser et al., in *Defects in Microelectronic Materials and Devices*, edited by D. Fleetwood et al. (Taylor and Francis/CRC Press, 2008), pp. 1–30.
[2] K. Jeppson et al., JAP **48**, 2004 (1977).
[3] M. Alam et al., MR **47**, 853 (2007).
[4] S. Mahapatra et al., T-ED **56**, 236 (2009).
[5] S. Rangan et al., *IEDM* (2003), pp. 341–344.
[6] B. Kaczer et al., APL **86**, 1 (2005).
[7] V. Huard et al., MR **46**, 1 (2006).
[8] H. Reisinger et al., T-DMR **7**, 119 (2007).
[9] B. Kaczer et al., *IRPS* (2009), pp. 55–60.
[10] T. Grasser et al., *IRPS* (2009), pp. 33–44.
[11] T. Tewksbury, Ph.D. Thesis, MIT, 1992.
[12] B. Kaczer et al., *IRPS* (2005), pp. 381–387.
[13] A. McWhorter, Sem.Surf.Phys 207 (1957).
[14] M. Weissman, Rev.Mod.Phys **60**, 537 (1988).
[15] M. Kirton et al., Adv.Phys. **38**, 367 (1989).
[16] V. Huard et al., *IIRW* (2005), pp. 5–9.
[17] B. Kaczer et al., *IRPS* (2008), pp. 20–27.
[18] H. Reisinger et al., *IRPS* (2010), (in print).
[19] H. Reisinger et al., *IIRW* (2009), pp. 30–35.
[20] T. Grasser et al., *IEDM* (2009), pp. 729–732.
[21] D. Lang, JAP **45**, 3023 (1974).
[22] A. Karwath et al., APL **52**, 634 (1988).
[23] D. Vuillaume et al., PRB **34**, 1171 (1986).
[24] A. Asenov et al., T-ED **50**, 839 (2003).
[25] H. Mueller et al., JAP **79**, 4178 (1996).
[26] B. Kaczer et al., *IRPS* (2010), (in print).
[27] D. Gillespie, J.Comp.Phys. **22**, 403 (1976).
[28] M. Uren et al., PRB **37**, 8346 (1988).
[29] T. Grasser et al., *IEDM* (2007), pp. 801–804.
[30] K. Ralls et al., PRL **52**, 228 (1984).
[31] A. Avellán et al., JAP **94**, (2003).
[32] A. Lelis et al., T-NS **41**, 1835 (1994).
[33] E. Poindexter et al., J.Electrochem.Soc. **142**, 2508 (1995).
[34] H. Huang et al., Proc.R.Soc.A **204**, 406 (1950).
[35] C. Henry et al., PRB **15**, 989 (1977).
[36] W. Fowler et al., PRB **41**, 8313 (1990).
[37] D. Gillespie, *Markov Processes* (AP, 1992).
[38] D. McQuarrie, J.Appl.Prob. **4**, 413 (1967).
[39] H. Ibe, *Markov Processes for Stochastic Modeling* (AP, 2009).
[40] C. Kelley, PRB **20**, 5084 (1979).
[41] P. Restle et al., IBM J.Res.Dev. **34**, 227 (1990).
[42] S. Makram-Ebeid et al., PRB **25**, 6406 (1982).
[43] S. Ganichev et al., Phys.SS **39**, 1703 (1997).
[44] A. Palma et al., PRB **56**, 9565 (1997).

Origin of NBTI Variability in Deeply Scaled pFETs

B. Kaczer, T. Grasser[1], Ph. J. Roussel, J. Franco[2], R. Degraeve, L.-A. Ragnarsson, E. Simoen,
G. Groeseneken[2], H. Reisinger[3]

IMEC, Kapeldreef 75, B-3001 Leuven, Belgium, phone: +32-16-281-557, email: ben.kaczer@imec.be
[1]Christian Doppler Laboratory for TCAD, Institute for Microelectronics, TU Wien, A-1040 Vienna, Austria
[2]also ESAT, KU Leuven, Leuven, Belgium
[3]Infineon Technologies AG, D-81730 Munich, Germany

Abstract—*The similarity between Random Telegraph Noise and Negative Bias Temperature Instability (NBTI) relaxation is further demonstrated by the observation of exponentially-distributed threshold voltage shifts corresponding to single-carrier discharges in NBTI transients in deeply scaled pFETs. A SPICE-based simplified channel percolation model is devised to confirm this behavior. The overall device-to-device ΔV_{th} distribution following NBTI stress is argued to be a convolution of exponential distributions of uncorrelated individual charged defects Poisson-distributed in number. An analytical description of the total NBTI threshold voltage shift distribution is derived, allowing, among other things, linking its first two moments with the average number of defects per device.*

Keywords: *pFET, Negative Bias Temperature Instability, Random Telegraph Noise, variability, Random Dopant Fluctuations*

I. INTRODUCTION

The large, micrometer-sized FET devices of the past CMOS technologies were considered identical in terms of electrical performance. Similarly, the application of a given stress resulted in an identical parameter shift in all devices. With the gradual downscaling of the FET devices, the oxide dielectric was the first to reach nanometer dimensions, thus introducing the first stochastically distributed reliability mechanism—the time dependent dielectric breakdown [1]. With the shrinking of *lateral* device dimensions to atomic levels, variation between devices appeared due to effects such as random dopant fluctuation and line edge roughness. Similarly, application of a fixed stress in such devices results in a distribution of the parameter shift. Understanding these distributions will be crucial for correctly predicting the reliability of future deeply downscaled technologies. The purpose of this paper is to further illuminate the causes of the variation of NBTI in deep submicron devices, already discussed in Refs. [2, 3].

Charging and discharging of individual oxide defects has been readily observable in sub-micron FETs in the form of random telegraph noise (RTN). Recently, threshold voltage steps due to individual defects were also observed in NBTI relaxation transients [3-6]. We have already argued in [4] that both effects are in fact but the two facets of the same mechanism, with RTN being the channel/gate dielectrics system in the state of dynamic equilibrium, while NBTI relaxation corresponding to the perturbed system returning to this equilibrium (Fig. 1) [7]. Here we will show that, identically to RTN amplitude distribution, the *individual* NBTI relaxation steps are exponentially distributed in amplitude [8]. The exponential distribution will be confirmed with a simplified percolation stochastic model. Combined with the assumption of the Poisson-distributed number of trapped gate oxide charges, an *analytical* description of the *total* NBTI threshold voltage shift distribution is then derived. This allows, among other things, linking its first two moments with the average number of defects per device.

Finally, we will argue that NBTI in future downscaled devices will be treated as a stochastic ensemble of individual defects, Poisson-distributed in number per device, with each defect described by its impact on the channel conduction discussed here and its capture and emission times [5, 9].

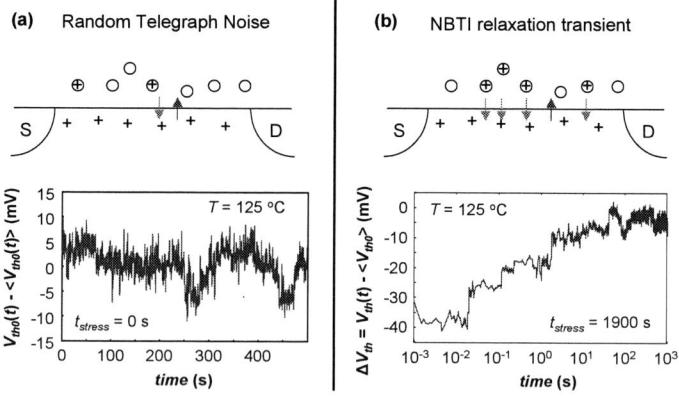

Figure 1. (a) At constant bias conditions, oxide defects are charged by channel carriers and subsequently discharged back into the channel with a wide range of time constants controlled by a nonradiative multiphonon emission process [9]. The system is in *dynamic* equilibrium, manifested by low-frequency noise or Random Telegraph Noise (RTN) in small devices. (b) Following the perturbation by NBTI stress, excess charged oxide defects gradually discharge and the system is returning to the dynamic equilibrium of (a), resulting in long NBTI transients.

This work is part of IMEC's Industrial Affiliation Program, funded by IMEC's core partners: Intel, Texas Instruments, Micron, Infineon, NXP, ST, Panasonic, TSMC, Samsung, and Elpida.

II. EXPERIMENTAL

pFETs with nominal gate length L = 70 nm (metallurgic length $L \sim 35$ nm), width W = 90 nm, and HfO_2 dielectrics with EOT = 0.8 nm were used in this study. Lanthanum has been incorporated in the oxide to boost the complementary nFET performance [10], but this was deemed to have no effect on the pFET NBTI [11] and especially on the NBTI variation discussed here.

All measurements were done at T = 125 °C. To compensate for the variability of these aggressively scaled devices, the initial threshold voltage V_{th0} of each DUT was first automatically determined using a fixed I_S criterion. The DUT was then stressed at $V_G = V_{th0} - 1.2V$ using the extended Measure-Stress-Measure (eMSM) sequence [12]. Specifically, the source current I_S in the linear regime (V_D = -0.1 V) was recorded during a series of 7 stress and relaxation phases. The relaxation phase I_S measured at $V_G \sim V_{th0}$ was converted to $V_{th}(t_{stress}, t_{relax})$ using the corresponding initial I_S-V_G curve. In our aggressively-scaled devices we estimate the V_{th} measurement accuracy to be ~ 1 mV. This is related to the relatively small current I_S at low V_G and V_D and to the conversion of I_S into V_{th} given the sub-threshold slope of our devices. Subsequently, ΔV_{th} was calculated as $\Delta V_{th} = V_{th}(t_{stress}, t_{relax}) - V_{th0}$. Note that because of the large RTN in the aggressively scaled devices, V_{th0} itself is not a fixed value, but rather fluctuates with time by up to ~ 10 mV, as shown in Fig. 1a. Consequently, it is possible to obtain small *positive* NBTI ΔV_{th} in a fraction of our devices, particularly for shorter stress times.

The resulting total threshold voltage shifts were found to be uncorrelated with the initial V_{th0} (Fig. 2), which confirmed that the NBTI mechanism is decoupled from sources of the V_{th0} variation and can be studied separately.

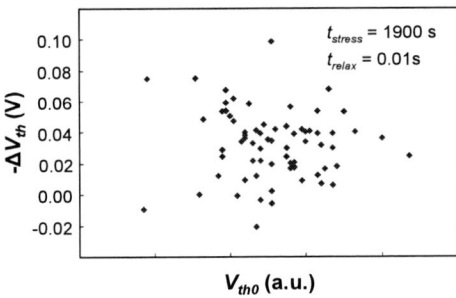

Figure 2. The lack of any correlation between the pFET initial threshold voltage V_{th0} and NBTI ΔV_{th} allows us to study the NBTI mechanisms independently of the pFET variation.

III. RESULTS

A. Observation of single discharge events

Fig. 3a shows a typical result of the MSM stress measurement. As already reported previously [3,4,6], clear steps caused by single discharge events are visible in the NBTI relaxation transients. However, the average step height is significantly larger than those reported earlier. A *single discharging event in many devices routinely exceeded 15 mV, and in several devices exceeded 30 mV*, the NBTI lifetime

criterion presently used by some groups. For comparison, ΔV_{th} of less than 2 mV would be expected based on a simple charge sheet approximation. As will be shown below, the large observed step height amplitude is due to the aggressively scaled dimensions of the pFETs used.

The individual down-steps were detected in relaxation traces together with the corresponding relaxation times in all measured pFETs [9]. The step-detecting algorithm was designed to work automatically even in the presence of RTN in the traces. In order to ensure this, a detection resolution of 1 mV was used. Consequently, no steps smaller than 1 mV were detected.

An example of the result of this extraction is given in Fig. 3b. As already discussed previously [4], we did not observe any obvious correlation between the step height and trap emission time—all step heights appeared to be equally likely at all measured relaxation times in our 72 high-k pFETs. We have previously argued that this property is required in order to observe the long "featureless" $\log(t_{relax})$-like relaxation tails [4]. This, however, also implies that defects exist with emission times faster than our measurement setup and we are therefore analyzing only the *visible*, but representative subset of all defects.

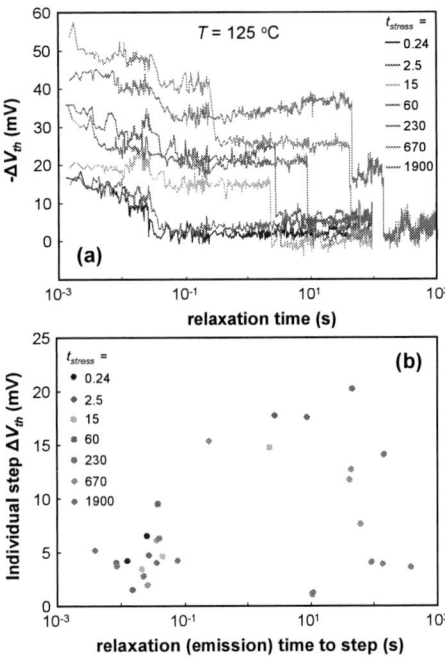

Figure 3. (a) A typical result of the eMSM sequence obtained on a single device: 7 NBTI relaxation transients following stress for the indicated times. Steps of varying heights due to single discharge events are clearly visible. (b) Step heights and the corresponding relaxation (emission) times for individual defects extracted from (a) [9].

B. Single discharge ΔV_{th} distribution

A histogram of the step heights from transients following the longest stress time is constructed from all 72 pFET devices in Fig. 4a. The figure shows that *the distribution of NBTI relaxation step heights is exponential,* with their probability distribution function (PDF) being

978-1-4244-5430-3/10 $26.00 © 2010 IEEE

$$f_1(\Delta V_{th}, \eta) = \frac{1}{\eta} e^{-\frac{\Delta V_{th}}{\eta}}, \qquad (1)$$

where the scaling factor η is the mean ΔV_{th} value for a *single* charge. The cumulative distribution function (CDF) corresponding to Eq. (1) is then

$$F_1(\Delta V_{th}, \eta) = 1 - e^{-\frac{\Delta V_{th}}{\eta}} \qquad (2)$$

and the variance of this distribution is $\sigma^2 = \eta^2$. We note that the exponential distribution has been repeatedly reported for RTN amplitudes [13-15]. This similarity further strengthens the link shown in Fig. 1. We moreover note that the large range of possible ΔV_{th}'s gives each defect its *individual signature*, which e.g. allows tracing its properties under various stress conditions [5,6].

The maximum-likelihood fit to individual ΔV_{th}'s following the longest stress, shown in the *cumulative* plot in Fig. 4b, confirms the exponential distribution and allows extracting the average ΔV_{th} shift per single discharge $\eta = 4.75 \pm 0.3$ mV in our devices. An exponential distribution is also observed for shorter stress times, including the shortest t_{stress} shown in Fig. 4a, but given the amount of collected data we only *assume* it has the same η. We note that η varying with the stress time could indicate e.g. charging of defects at varying depths. This is not expected for the NBTI mechanism, which has been repeatedly shown to be occurring at or very close to the Si/SiO$_2$ interface [11]. This also agrees with the non-correlation between single ΔV_{th}'s and emission times discussed above.

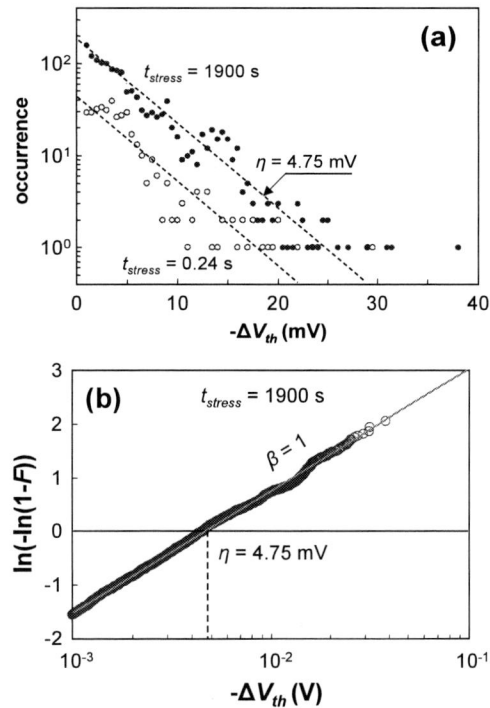

Figure 4. (a) Histograms of NBTI transient step heights for 72 devices show a clear exponential distribution. (b) CDF of the $t_{stress} = 1900$ s data in (a) shown in Weibull plot confirms exponential distribution (i.e., Weibull with shape factor $\beta = 1$).

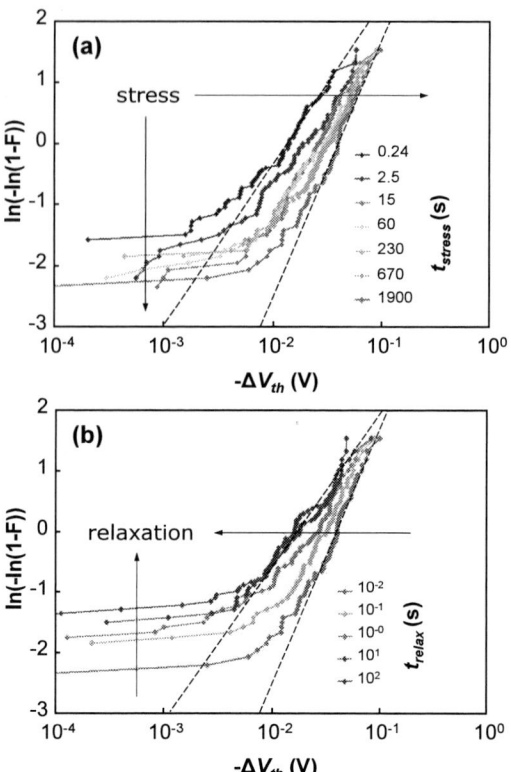

Figure 5. (a) Cumulative distribution of the total ΔV_{th} for 72 pFETs following stress at indicated stress times. With increasing t_{stress} the mean $<\Delta V_{th}>$ increases, while the fraction of devices with negligible ΔV_{th} decreases. (b) Cumulative distribution of ΔV_{th} for the same devices during relaxation following the longest stress of 1900 s. The opposite trends are observed.

C. Total ΔV_{th} distribution

Fig. 5a shows the distribution of the *total* ΔV_{th} of 72 pFETs for increasing stress times, corresponding to an increasing number of charged defects. Such total ΔV_{th} distributions are typically reported for a particular technology [3]. In Fig. 5a we note that the mean and maximum ΔV_{th}'s are increasing with stress, its *relative* deviation is decreasing (the distribution is getting *relatively* tighter). Perhaps surprisingly, Fig. 5a also demonstrates that a fraction of devices exists with negligible ΔV_{th} even after the longest stress. Overall, this fraction decreases with increasing stress time, i.e., with increasing mean ΔV_{th}. The opposite trends are observed in Fig. 5b when the devices are left to relax after the longest stress.

IV. DISCUSSION

As we will now show, all the trends as well as the total ΔV_{th} distribution itself, can be fully analytically described if i) the number of defects per device is assumed to follow a Poisson distribution, while ii) the impact of each individual defect on ΔV_{th} is exponentially distributed.

A. Single discharge ΔV_{th} distribution

The exponential distribution of single-charge ΔV_{th} can be understood if non-uniformities in the pFET channel due to random dopant fluctuations (RDF) are considered [13-15]. The threshold voltage of such a device corresponds to the formation of a conduction (percolation) path in the random dopant potential between Source and Drain (Fig. 6a). To zeroth order, depending on the position of the NBTI-stress-generated oxide charge, the conduction path could be either unaffected or obstructed by the new charged defect. In the latter case, the drop in the current has to be compensated by an increase of the gate voltage, resulting in the observed ΔV_{th}.

A.1. Simplified channel percolation model

An accurate reproduction of this process is typically done through computation-intensive physics-based device simulations with RDF, line edge roughness, and other realistic effects [15,16]. Here we show that the essence of the mechanism can be qualitatively captured in a very simplified channel percolation model. We emphasize that in contrast to the all-encompassing device simulations, our aim is to keep the model as simple as possible and to include only the bare minimum of assumptions. We find it very instructional that the minimalist model correctly reproduces most of the common observations.

In our simplified model (Fig. 6b), a mesh of "elementary" FETs with random V_{th}'s, representing variations in the local potential, is set up to represent the channel of our pFETs. For the sake of simplicity, a uniform distribution of the random "elementary" V_{th}'s is used and short-channel effects are not considered. A script is used to generate 400 instances of the randomized mesh, to call SPICE to solve them, and to extract the V_{th0} of the simulated pFET. As is typically experimentally observed, the resulting V_{th0}'s are normally distributed (Fig. 7a) and their variance scales reciprocally with the FET area (Fig. 7b).

A number n of "charged defects" is then inserted, each represented by an additional V_{th} shift of one random "elementary" FET in the netlist. A fixed value of V_{th} shift for these "single oxide charges" is assumed, representing the fact that NBTI charges are occurring very close to (i.e., at a fixed distance from) the substrate interface. Subsequently, a new V_{th} is calculated, resulting in ΔV_{th} for each pFET instance. For a single additional charged defect, the simplified model shows the ΔV_{th} distribution to be Weibull-distributed with $\beta = 0.8$ for a range of dimensions of the channel mesh (Fig. 8). This confirms that the above described process can be responsible for the observed exponential distribution of step heights (cf. Fig. 4b).

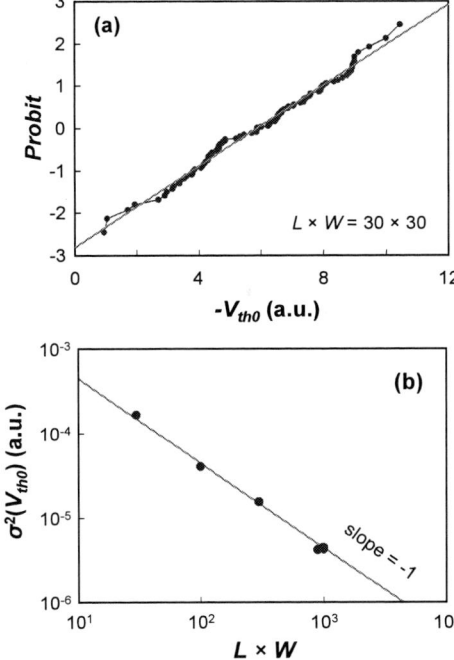

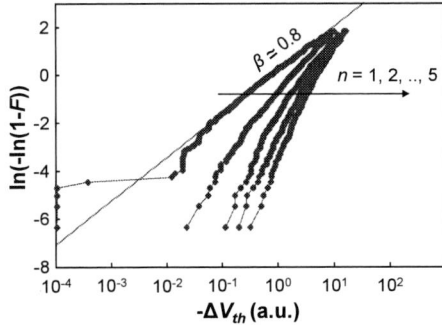

Figure 7. The simple percolation model correctly reproduces (a) the normal distribution of initial threshold voltages V_{th0} and (b) the variance σ^2 of V_{th0} scaling reciprocally with the $L \times W$ (Pelgrom's rule) [18].

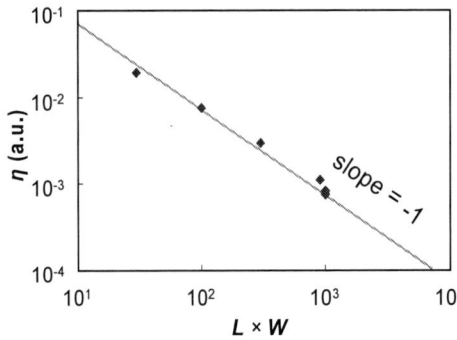

Figure 8. Cumulative ΔV_{th} distributions generated for an increasing number of gate oxide defects n by the simplified channel percolation model. For a single charged defect ($n = 1$), the model well reproduces the observed exponential distribution in Fig. 4b.

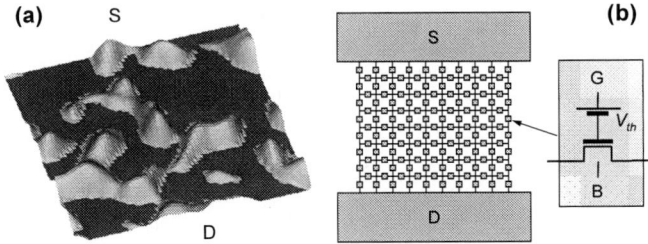

Figure 6. (a) An illustration of a percolation path in a random potential (from [17]) such as that between FET source and drain. (b) A mesh of "elementary" FETs with random V_{th}'s (voltage source in series with gate) representing (a) can be readily solved with SPICE.

Figure 9. The step heights due to individual discharge events will increase with both decreasing L and W, explaining why large single-charge steps can be expected in very small devices.

The simplified model predicts that η scales inversely with both W and L (Fig. 9), i.e., the smaller the device the larger the steps, thus explaining the large observed value of η. This can be understood as the impact of a single charge relatively increasing as the device becomes smaller. We note here that a more thorough discussion of the dependence of η on W, L, EOT, and channel doping in the framework of RTN is given in Ref. [15], which infers $\eta \sim L^{-1/2}$ for short devices.

The simplified model also predicts the *total* ΔV_{th} distribution for the number of defects $n > 1$ (Fig. 8). Since the subsequent charge lateral locations are uncorrelated, the overall ΔV_{th} distribution can be readily expressed as a convolution of individual exponential distributions (Eq. 1), and the PDF and the CDF are respectively described by

$$f_n(\Delta V_{th}, \eta) = \frac{e^{-\frac{\Delta V_{th}}{\eta}}}{(n-1)!} \frac{\Delta V_{th}^{\ n-1}}{\eta^n} \qquad (3)$$

and

$$F_n(\Delta V_{th}, \eta) = 1 - \frac{\Gamma(n, \Delta V_{th}/\eta)}{(n-1)!}. \qquad (4)$$

The fraction in Eq. 4 can be also recognized as the regularized gamma function. CDF in Eq. 4 well describes the result in Fig. 8 for $\beta = 1$.

B. Total ΔV_{th} distribution

An actual population of stressed devices will consist of devices with a *different* number n of visible oxide defects in each device. That number will be Poisson distributed [2-4,16]. The *total* ΔV_{th} distribution can be therefore obtained by summing distributions F_n weighted by the Poisson probability

$$P_N(n) = \frac{e^{-N} N^n}{n!}. \qquad (5)$$

In Eq. 5, N is the mean number of defects in the FET gate oxide and is related to the oxide trap (surface) density N_{ot} as $N = W L N_{ot}$ (note that N is not an integer).

This reasoning then results in the *total* ΔV_{th} CDF given by

$$F_N(\Delta V_{th}, \eta) = \sum_{n=1}^{\infty} \frac{e^{-N} N^n}{n!} F_n(\Delta V_{th}, \eta). \qquad (6)$$

The corresponding PDF is

$$f_N(\Delta V_{th}, \eta) = e^{-N}\left[\delta(\Delta V_{th}) + N \frac{e^{-\frac{\Delta V_{th}}{\eta}}}{\eta}\ {}_0\widetilde{\mathcal{F}}_1(2; N\frac{\Delta V_{th}}{\eta})\right], \quad (7)$$

where the hypergeometric function ${}_0\widetilde{\mathcal{F}}_1(2; x)$ can be also written in terms of the modified Bessel function I_1 as ${}_0F_1(2; x) = x^{-1/2} I_1(2x^{1/2})$. The Dirac $\delta(\Delta V_{th})$ term represents the fraction of devices with 0 V shift [14], which decreases with increasing N.

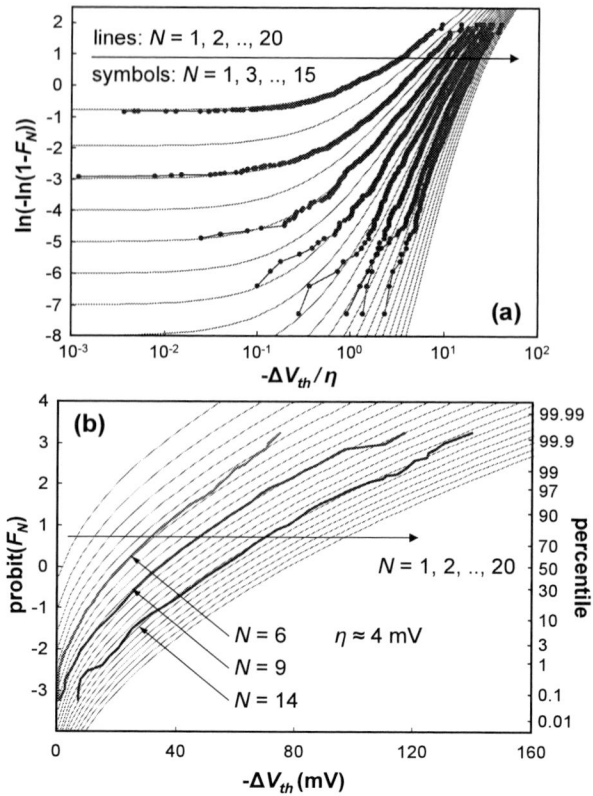

Figure 10. Eq. 6 for several values of the average number of defects N (lines). (a) Weibull plot emphasizes the fraction of devices with $\Delta V_{th} \sim 0$ V, cf. Fig. 5. Monte Carlo calculation with 1000 samples is included for comparison (symbols). (b) Eq. 6 in a probit plot rescaled to fit experimental distributions from Fig. 10 of Ref. [3], with the corresponding values of N and η readily extracted.

The CDF of Eq. 6 is plotted in Fig. 10 for several values of N. We can see that it has the same properties as the distributions obtained on the limited population in Fig. 5; specifically, the fraction of devices showing negligible ΔV_{th} varies with $\langle \Delta V_{th} \rangle$. For completeness, the CDF is also compared with a simple, 1000-sample Monte Carlo calculation assuming i) exponentially distributed individual ΔV_{th} steps ii) Poisson-distributed in number (Fig.10a). Fig. 10b then shows Eq. 6 in a probit plot, such as that used in Refs. [2,3,16]. For comparison, measured total ΔV_{th} distributions from Ref. [3] are excellently fitted by Eq. 6, which further supports our assumptions above.

C. Implications

The advantages of describing the total ΔV_{th} distribution in terms of Eqs. 6 and 7 are their relative simplicity and tangibility of the variables, while the analytical description allows further statistical treatment. The mean of the above-derived distribution is

$$\langle \Delta V_{th} \rangle = N\eta, \qquad (8)$$

i.e., it should be independent of FET gate area $L \times W$ provided N and η are respectively directly and inversely proportional to $L \times W$. The variance of the distribution is then

$$\sigma_{\Delta Vth}^2 = 2N\eta^2, \tag{9}$$

i.e., it increases with decreasing gate area. The relative deviation $\sigma_{\Delta Vth}/<V_{th}> = (2N)^{-1/2}$ is therefore decreasing with increasing N, as observed in Fig. 5.

With the value of η extracted from the single discharge step height histogram in Fig. 2 earlier, we can use Eq. 8 to convert $<\Delta V_{th}>$ to the average number of trapped defects N. In our devices, N increases from 2.6 (t_{stress} = 0.24 s, Fig. 1) to 6.9 (t_{stress} = 1900 s), and then decreases to 3.4 (t_{relax} = 10 s). These values correspond to effective trap densities of $1\text{-}2\times10^{11}$ cm^{-2}, typically observed for NBTI in large devices.

We acknowledge that obtaining the value of η as in Fig. 2 could be rather laborious. Eqs. 8 and 9, however, allow us expressing both N and η in terms of $<\Delta V_{th}>$ and $\sigma_{\Delta Vth}^2$ as

$$N = \frac{2\langle \Delta V_{th} \rangle^2}{\sigma_{\Delta Vth}^2} \tag{10}$$

and

$$\eta = \frac{\sigma_{\Delta Vth}^2}{2\langle \Delta V_{th} \rangle}. \tag{11}$$

This means that both N and η can be extracted from the first two moments of a measured total NBTI distribution, *without having to characterize individual step heights*. This way we *independently* obtain N increasing from 1.9 to 4.6 with stress and η varying between 7 and 9 mV. For the limited population of devices measured, these values are very close to those obtained directly by counting individual ΔV_{th} step heights in Fig. 4. Note that other effects potentially increasing $\sigma_{\Delta Vth}$, such as the variance of the initial V_{th0} (Fig. 1a) and FET variability (line-edge roughness, work-function fluctuations, etc.) have not been considered here.

D. Reliability projection of future downscaled devices

Finally, we review the described concepts to illustrate the paradigm shift in understanding NBTI reliability in aggressively scaled devices. Fig. 11a illustrates the reliability projection in large devices of the past. Since all devices are expected to behave identically, there is no device-to-device variation. Measuring and extrapolating one stress condition per device is therefore sufficient. Additionally, there is also the application-independent hard failure criterion.

In contrast to that, future failure criteria will be application-dependent and will depend specifically on the circuit immediately surrounding the device in question. Each deca-nanometer device will be described by the Poisson-distributed total number of defects, each of which is characterized by i) voltage and temperature dependent capture time τ_c and ii) emission time τ_e, and iii) its impact on the FET current (e.g. via ΔV_{th}). This is schematically illustrated in Fig. 11b. The PDFs of all three parameters are known: i) and ii) appear to be uniform on the log scale (at least within the measured $\sim10^{-6}-$

10^6 s range so far) [12]. In this description, "permanent" defects can be seen as those with $\tau_e \sim \infty$. Significant progress in describing the voltage and temperature dependence of τ_c and τ_e using nonradiative multiphonon theory has been made [9]. The exponential PDF of iii) has been justified here. Fig. 11c then schematically summarizes the progress of NBTI in the three hypothetical devices shown in Fig. 11b. For simplicity, only continuous stress is illustrated, although the specified parameters for each device (Fig. 11b) allow evaluating the degradation following an arbitrary waveform [5]. The origin of the NBTI variability at a given time or a given ΔV_{th} is apparent.

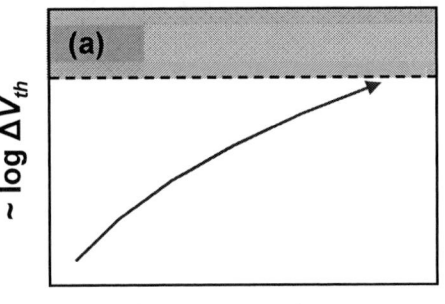

Large devices

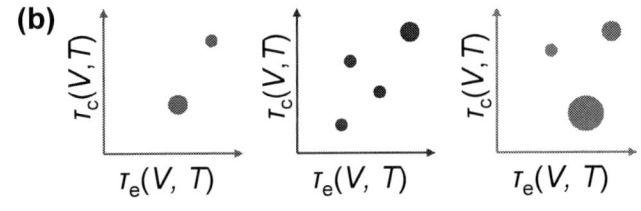

Deeply scaled devices

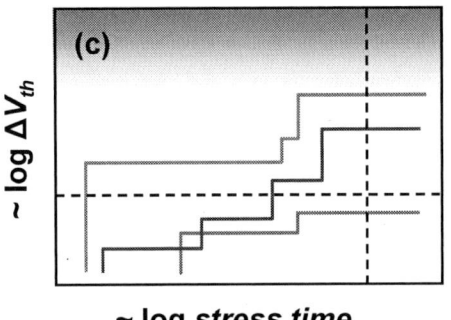

Figure 11. (a) All large devices behave identically upon stress and are expected to fail when reaching the projected "hard" degradation criterion. (b) NBTI degradation in deeply scaled devices can be described in terms of the total number of defects in each device, their (voltage and temperature dependent) capture and emission times, and their impact on the device (demarked by the size of the point). (c) Schematic showing the progress of stress in the three devices described in (b). The origin of the NBTI variability is apparent.

V. CONCLUSIONS

The correspondence between RTN and NBTI relaxation was further strengthened by our observation of exponentially distributed step heights in NBTI transients in deeply scaled pFETs. A simplified channel percolation model was devised to illustrate and to confirm this behavior. The overall ΔV_{th} distribution was argued to be a convolution of exponential distributions of uncorrelated individual charged defects. The analytical description derived for this distribution should prove useful for both reliability data analysis and simulations of deeply-scaled CMOS circuitry. The proposed picture allows us to predict that the reliability of future deca-nanometer devices will be treated as a stochastic ensemble of the Poisson-distributed total number of defects, each characterized by capture and emission times and its impact on the FET current.

REFERENCES

[1] R. Degraeve, G. Groeseneken, R. Bellens, M. Depas, and H. E. Maes, "A consistent model for the thickness dependence of intrinsic breakdown in ultra-thin oxides," *Int. Electron Devices Meeting Tech. Dig.*, pp. 863 – 866, 1995.

[2] S. E. Rauch, "Review and Reexamination of Reliability Effects Related to NBTI Statistical Variations", *IEEE T. Dev. Mat. Rel.* 7, p. 524, 2007.

[3] V. Huard C. Parthasarathy, C. Guerin, T. Valentin, E. Pion, M. Mammasse, N. Planes, L. Camus, "NBTI degradation: From Transistor to SRAM Arrays," *Proc. Int. Rel. Phys. Symp.*, p. 289, 2008.

[4] B. Kaczer, T. Grasser, J. Martin-Martinez, E. Simoen, M. Aoulaiche, Ph. J. Roussel, and G. Groeseneken, "NBTI from the Perspective of Defect States with Widely Distributed Times," *Proc. Int. Rel. Phys. Symp.*, p. 55, 2009.

[5] H. Reisinger, T. Grasser, W. Gustin, and C. Schlünder, "The Statistical Analysis of Individual Defects Constituting NBTI and its Implications for Modeling DC- and AC-Stress," presented at *Int. Reliab. Phys. Symp.*, 2010.

[6] T. Grasser, H. Reisinger, W. Goes, Th. Aichinger, Ph. Hehenberger, P.-J. Wagner, M. Nelhiebel, J. Franco, and B. Kaczer, "Switching Oxide Traps as the Missing Link Between Negative Bias Temperature Instability and Random Telegraph Noise, " *Int. Electron Devices Meeting Tech. Dig.*, pp. 729-732, 2009.

[7] A. Karwath and M. Schulz, "Deep level transient spectroscopy on single, isolated interface traps in field-effect transistors", *Appl. Phys. Lett.* **52**, pp. 634-636, 1988.

[8] B. Kaczer, Ph. J. Roussel, T. Grasser, and G. Groeseneken, "Statistics of multiple trapped charges in the gate oxide of deeply-scaled MOSFET devices—application to NBTI," accepted to *Electron Dev. Lett.*

[9] T. Grasser, H. Reisinger, P.-J. Wagner, F. Schanovsky, W. Goes, and B. Kaczer, "The Time Dependent Defect Spectroscopy (TDDS) for the Characterization of the Bias Temperature Instability", presented at *Int. Reliab. Phys. Symp.*, 2010.

[10] L.-Å. Ragnarsson, Z. Li, J. Tseng, T. Schram, E. Rohr, M. J. Cho, T. Kauerauf, T. Conard, Y. Okuno, B. Parvais, P. Absil, S. Biesemans, and T. Y. Hoffmann, "Ultra Low-EOT (5 Å) Gate-First and Gate-Last High Performance CMOS Achieved by Gate-Electrode Optimization", *Int. Electron Devices Meeting Tech. Dig.*, pp. 663-666, 2009.

[11] B. Kaczer, A. Veloso, Ph. J. Roussel, T. Grasser, and G. Groeseneken, "Investigation of Bias-Temperature Instability in work-function-tuned high-k/metal-gate stacks", *J. Vac. Sci. Technol. B* **27**, pp. 459-462, 2009.

[12] B. Kaczer, T. Grasser, Ph. J. Rousse, J. Martin-Martinez, R. O'Connor, B. J. O'Sullivan, and G. Groeseneken, "Ubiquitous Relaxation in BTI stressing—New Evaluation and Insights", *Proc. Int. Reliab. Phys. Symp.*, p. 20, 2008.

[13] A. Asenov, R. Balasubramaniam, A. R. Brown, and J. H. Davies, "RTS Amplitudes in Decananometer MOSFETs: 3-D Simulation Study," *IEEE T. Electron Dev.*, vol. 50, no. 3, p. 839, 2003.

[14] K. Takeuchi, T. Nagumo, S. Yokogawa, K. Imai, and Y. Hayashi, "Single-Charge-Based Modeling of Transistor Fluctuations Based on Statistical Measurement of RTN Amplitude," *Symp. VLSI Tech.*, p. 54, 2009.

[15] A. Ghetti, C. M. Compagnoni, A. S. Spinelli, and A. Visconti, "Comprehensive Analysis of Random Telegraph Noise Instability and Its Scaling in Deca–Nanometer Flash Memories," *IEEE T. Electron Dev.*, vol. 56, no. 8, pp. 1746-1752, 2009.

[16] M. F. Bukhori, S. Roy, A. Asenov, "Simulation of statistical aspects of reliability in nano CMOS," presented at *Int. Integ. Rel. Workshop*, 2009.

[17] http://www.ibiblio.org/e-notes/Perc/contour.htm .

[18] M. J. M. Pelgrom, A. C. J. Duinmaijer, and A, P. G. Welbers, "Matching Properties of MOS Transistors", IEEE J. Solid-State Circ. **24**, pp. 1433-1440, 1989.

Two independent components modeling for Negative Bias Temperature Instability

Vincent Huard

STMicroelectronics – 850 rue Jean Monnet 38926 Crolles, FRANCE
phone: + 33(0)438922907 ; fax : + 33(0)438923227 ; email : vincent.huard@st.com

Abstract—Based on vast experimental dataset obtained from different technologies (pure or nitrided SiO2 and HK), we suggest that Negative Bias Temperature Instability is made of two independent components, presenting different voltage and temperature acceleration factors as well as process dependences. The recoverable part, subject to fast transient effects, is shown to obey field-assisted LRME hole trapping/detrapping processes. The permanent part is shown to be made of an equal number of interface traps and positive fixed charges, as resulting from hydrogen transfer to oxygen bridge. This hydrogen transfer was shown for the firs time to be reversible allowing in-depth analysis of the microscopic mechanisms at play.

I. INTRODUCTION

Following pioneering works [1,2], it is now widely accepted that Negative Bias Temperature Instability (NBTI) is made of a quickly recoverable component (D_R) on top of a slowly recovering or even permanent component (D_P) [3,4]. In spite of the large dataset collected worldwide, there are still a lot of discussions concerning the microscopic mechanisms lying behind the overall macroscopic behavior. Generally, a consensus starts to be achieved on the contribution of hole trapping in the recoverable part while the permanent component is explained by the generation of interface states. First believed to be independent components, a recent approach using two tightly coupled components [5] allows to elegantly reproducing most of the experimental observations. This approach is based on the fact that the degradation presents a very broad scalability for various stress and recovery times, for various stress voltages and temperatures and for various gate stacks nature. In this paper, we will make use of our extensive dataset to demonstrate that the observed broad scalability is only apparent whenever one component dominates over another one. We will show that in a more general case where both contributions are of equal importance, there is no longer scalability. This fact, coupled to several process dependence evidences, will be explained as the two components are not tightly coupled but actually made of a permanent part D_P related to interface traps creation by Si-H bond breaking and a recoverable part D_R related to hole trapping/detrapping in oxide defects. The theoretical description of microscopic mechanisms will be further discussed.

II. EXPERIMENTAL OBSERVATIONS

For this study, we had gathered a large dataset of NBTI degradation on pMOS made of pure oxide (PO) SiO2, thermally nitrided oxide (TNO), plasma nitrided oxide (PNO) and high-k HfSiON stacks (HGK). Dielectric physical thicknesses are ranging from 1.3nm to 12nm. Electrical degradation is monitored by either using simple DC On-The-Fly (OTF) approach or in combination with relaxation sequences using a wide range of stress voltages (-0.6V to -2.2V) and stress temperatures (25°C to 200°C). The recorded OTF degradation in I_{Dlin} was converted to ΔV_{th} using the procedure described in [6]. This procedure allows intrinsically to get rid of any contamination by mobility variations. In addition, we had checked using ultra fast measurements setup (measuring down to 1us) that our experimental OTF setup allows capturing fast enough the first OTF readpoint, thus avoiding any misleading interpretations.

As discussed and observed experimentally previously [1,4,7-9], the overall NBTI degradation is made of two macroscopic components. For short stress times (t_s < 10s, depending on the gate stacks and stress conditions), the degradation is dominated by a recoverable component (D_R) which is well described by a logarithmic time dependence (cf. figure 1 for typical example for two different gate stacks). The recovery phases can be described by the coexistence of the remaining part of the recoverable component D_R as well as a

roughly permanent contribution D_P. The recovery dynamics is strongly influenced by the gate nature stacks, as shown in figure 2. The permanent part amplitude is solely impacted by the stress time and voltage [5,7].

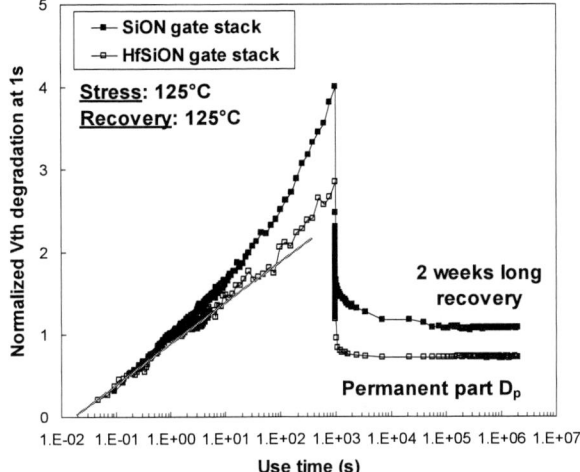

Fig. 1: NBTI degradation/recovery scaled at 1s for 125C for both 1.7nm SiON PNO (filled symbols) and 2.3nm HfSiON (open symbols) gate stacks for 7MV/cm vertical oxide field. Initial degradation phase (for ts<10s) is well reproduced by logarithmic time dependence (red line). Long recovery phases (up to two weeks, at 125C for one day, then 25C) had been conducted to evidence without ambiguity the permanent part D_P contribution (in opposition with the fast recoverable part D_R).

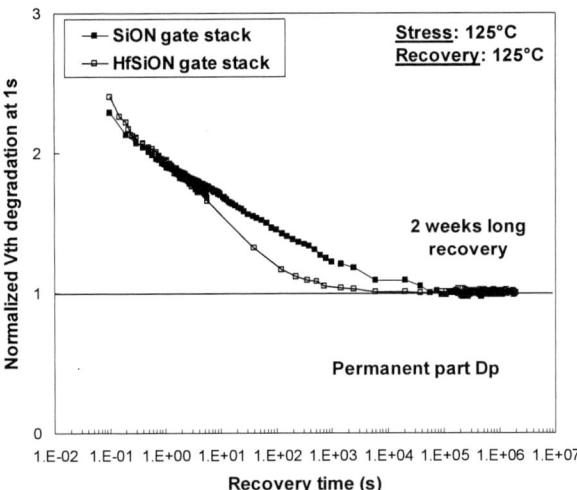

Fig. 2: NBTI recovery scaled on the permanent part for 125C for both 1.7nm SiON PNO and 2.3nm HfSiON gate stacks following 1000s stress at 7MV/cm vertical oxide field. Long recovery phases (up to two weeks, at 125C for one day, then 25C) had been conducted to evidence without ambiguity the permanent part D_P contribution. It is also worth noticing the difference of recovery dynamics for different gate stacks.

The main question left today to solve out is whether or not these two components (D_P and D_R) are elements of the same microscopic mechanism, as suggested by [5], and so they are two tightly coupled components or quite differently they are strictly independent compo- nents, as suggested by [7]. This fundamental question can only be answered based on key experimental observations that we will brief- ly summarize before developing our model.

A. Degradation scalability

The assumption of two tightly coupled components is based on the apparent broad scalability of the NBTI degradation in both stress and recovery phases for various stress voltages and temperatures [4,10]. Similarly to previous studies, scalability is also observed in our experimental dataset (cf. figure 3 for temperature scalability, red symbols). Nevertheless, this apparent scalability can only be ob- served in our dataset under specific stress conditions where one component (here D_R for instance) dominates over the other one. As far as it can be guessed from the figures of [4,10], in spite of their large dataset, only situations where component D_R is dominant over D_P one are considered thus giving the apparent perspective of scala- bility. Nevertheless, it is difficult to conclude on scalability (and tight coupling) when one component dominates over another one.

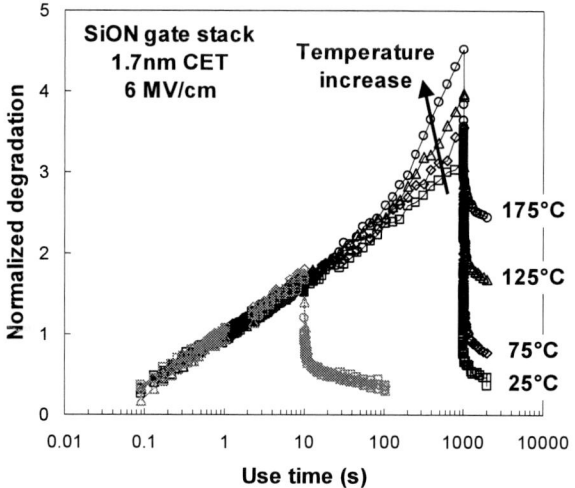

Fig. 3: NBTI stress/recovery phases for 1.7nm SiON PNO gate stacks for various stress temperatures ranging from 25°C to 175°C, scaled for 1 second of stress. For short stress times (red symbols), the D_R component dominates over the D_P one and an apparent temperature scalability is observed. For longer stress times, the D_P component starts getting importance and the tem- perature scalability cannot be longer observed.

On the contrary, for conditions where both components are equivalent in importance, temperature and voltage scalability is no longer respected (figure 3 for temperature scalability (black symbols) and figure 4 for voltage scalability). This conclusion was reached for a broad range of gate stacks including SiO2, SiON, HfSiON gate stacks nature with electrical thicknesses ranging from 1.3nm to 6.5nm. Both components have different temperature and voltage acceleration factors which imply that two independent microscopic mechanisms are at play in the overall degradation without any tight coupling.

Before discussing the physics of the microscopic mechanisms ly- ing behind the macroscopic behavior described just above, it is worth emphasizing some specific process dependences which also provide strong evidences that the overall NBTI degradation results from the co-existence of two independent microscopic mechanisms.

B. Process dependence

Though an advanced CMOS process route allows a large number of parameters to play with, only process splits which are known to have significant impact on NBTI degradation will be presented here to enlighten specific behaviors which would help to separate contri- butions of the different microscopic mechanisms.

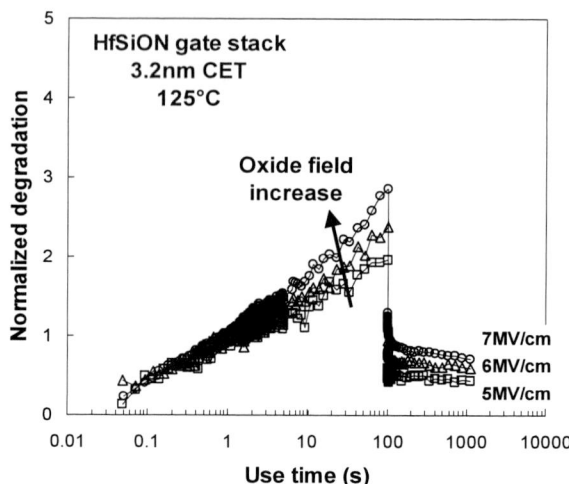

Fig. 4: NBTI stress/recovery phases for 3.2nm HfSiON gate stacks for various vertical oxide fields, scaled for 100 msecs of stress. For short stress times, the voltage scalability is observed but for longer stress times, while the D_P component starts getting impor- tance, the voltage scalability cannot be longer observed.

Hydrogen species

Among all impurities, hydrogen is the most widely found in CMOS process. In a process viewpoint, hydrogen is an important impurity in due its ability to react with a wide variety of lattice im- perfections. Such reactions affect the electronic properties of the material by removal of electronic states from the band gap. The pas- sivation of dangling bonds present at the interface is an important step in improvement of transistor electrical parameters. Hydrogen has a stable isotope variant, the deuterium. Deuterium is heavier die the additional neutron in its nucleus. Due to energetically similar valence band orbitals, deuterium and hydrogen atoms present similar binding energies on Si atoms. Nevertheless, SiD bonds are shown to be more resistant to hot-carrier stress than SiH bonds, the so-called isotope effect [11,12]. This isotope effect is explained by the heavier mass of deuterium, making it more difficult to extract from interface, compared to the hydrogen atom, similarly to the well-known elec- tronic surface desorption as described by Menzel and Gomer [13]. Similarly to hot-carrier stress, deuterium was shown to reduce NBTI degradation [14]. In our experiments, deuterium atoms were intro- duced in place of hydrogen atoms by a deuterium-based forming gas anneal. Incorporation of deuterium atoms at the SiO2/substrate inter- face was confirmed by SIMS profiles. Charge-pumping experiments on fresh devices showed that the number of passivated interface traps was similar between hydrogen- and deuterium-based samples.

Though deuterium splits was realized on several gate stacks, the main conclusions can be summarized in figure 5. While the recover- able part D_R remains identical independently of the anneal nature, the permanent part D_P is reduced with deuterium anneal. These results show that only the permanent part is related to any hydrogen species displacement at or close to the interface. The degradation reduction is proportional to the square root of the deuterium/hydrogen mass ratio ($\sim \sqrt{2}$), as expected for a diffusion phenomenon [15].

Nitrogen

NBTI degradation is known since the late sixties, but it really be- came a serious problem for CMOS technologies since the incorpora- tion of nitrogen was needed both to keep the gate leakage current at a reasonable level and to avoid boron atoms diffusion. Many research- ers already reported that increasing the nitrogen content into the oxide yields to a higher NBTI degradation [7,16]. Although it is known that holes and hydrogen species are involved into the degra- dation, their exact interaction with the incorporated nitrogen atoms remains mostly unsolved. In this part, the role of incorporated nitrogen atoms in NBTI degradation would be discussed.

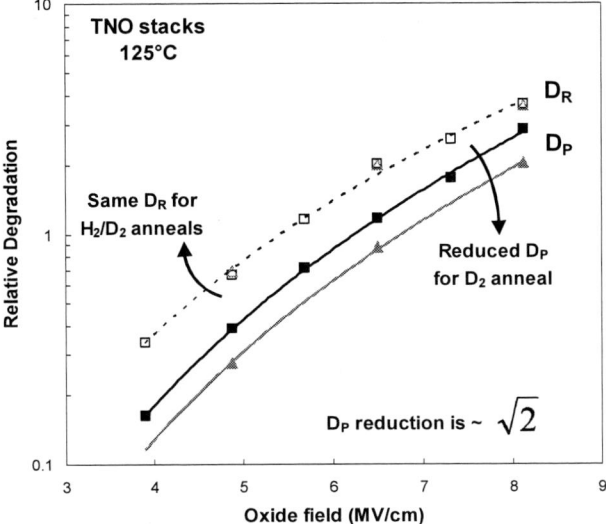

Fig. 5: Final anneals in H_2 (squares) or D_2 (triangles) atmosphere impact only the permanent part D_P (filled symbols), with a reduction equals to $\sqrt{2}$, while the recoverable part D_R (open symbols) remains unchanged.

The first step to understand the role of nitrogen incorporation is to compare NBTI degradation on pure oxide (i.e. no nitrogen) gate stacks versus nitrided ones. Figure 6 shows that the degradation increase is solely related to recoverable part D_R enhancement while the permanent part D_P is not impacted by nitrogen incorporation.

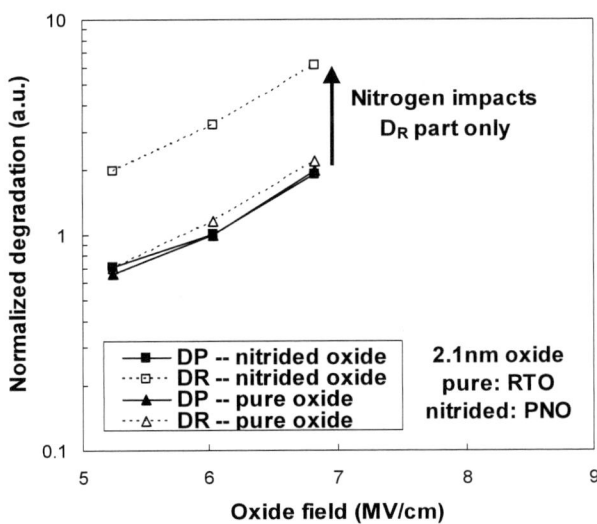

Fig. 6: Nitrogen incorporation into gate stack only impacts the recoverable part D_R while the permanent part D_P remains unchanged. Squares symbols relate to nitrided gate stack (PNO) and triangles symbols relate to pure gate stack (RTO).

The positive role of nitrogen incorporation on gate leakage current and boron diffusion can be reached for a given nitrogen total content without specific spatial localization in the oxide. On the contrary, the nitrogen-related traps have to set close to the Si/SiO_2 interface to have an influence on electrical parameters such as the threshold voltage or the drive current. Different nitridation processes such as rapid thermal nitridation (RTN) or the decoupled plasma nitridation (DPN) are known by SIMS profiles to create nitrogen-rich layer close to the substrate interface for the RTN and close to the

poly interface for the DPN. Figure 7 shows that in the second case, the NBTI degradation is strongly decreased because even if the nitrogen-related traps still exist their relative influence on electrical parameters is less important since they are located far from this interface. On another hand, whatever the nitrogen atoms profile within the dielectric the permanent part D_P remains unchanged, only related to the vertical oxide field.

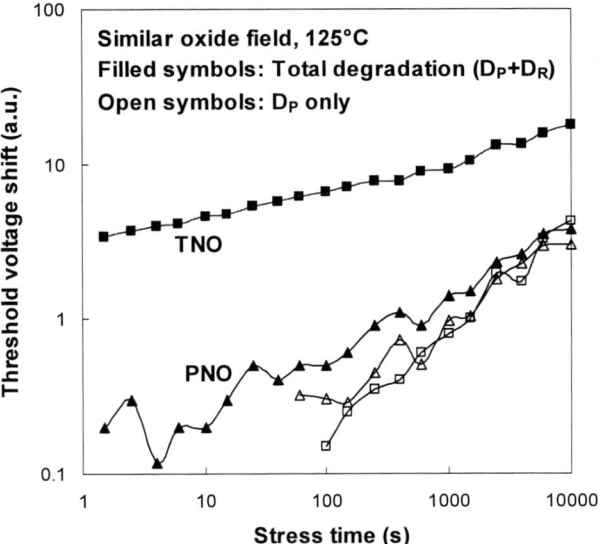

Fig. 7: Gate stacks with similar nitrogen content and similar electrical thicknesses can have different total NBTI degradation (filled symbols). The difference comes from the different nitridation process resulting in different nitrogen dose at interface. On the contrary, for similar oxide fields, the permanent part D_P is the same independently of the nitrogen incorporation process.

This apparent independence of the permanent part D_P as a function of the nitrogen presence (either process-related incorporation and nitrogen dose) has been observed to demonstrate some "universal" behavior on a very broad gate stacks datasets (cf. figure 8).

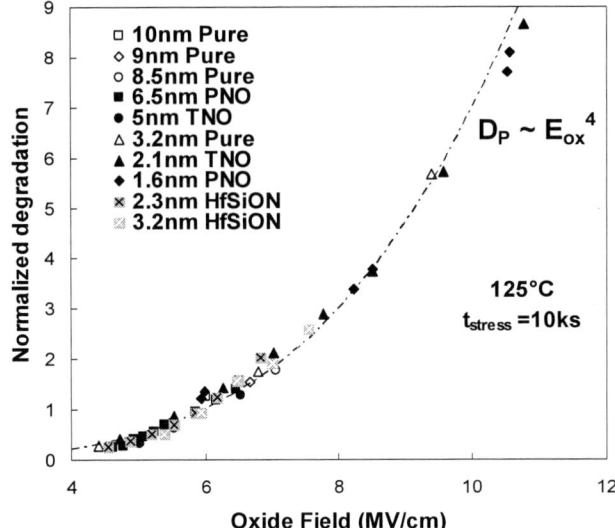

Fig. 8: The independence of the permanent part D_P generation with the nitrogen content and gate stack process is universally observed on different gate stacks: pure $SiO2$, SiON (thermally- or plasma-nitrided) and HfSiON gate stacks. All D_P degradations were normalized to 6 MV/cm vertical oxide field. Dotted line represents the E_{ox}^4 dependence which fits correctly the overall broad dataset.

978-1-4244-5430-3/10 $26.00 © 2010 IEEE   35

More interestingly, similar conclusions can be reached while analyzing large Design-Of-Experiments (DOE) process split experiments (cf. figure 9). Though large nitrogen dose splits had been experienced, the permanent part D_P remains unchanged by the presence of nitrogen atoms (similarly to figure 8). The oxide field dependence (represented here by the electrical thickness variations for a given gate voltage VG) is close to the already reported E_{ox}^4 dependence [17].

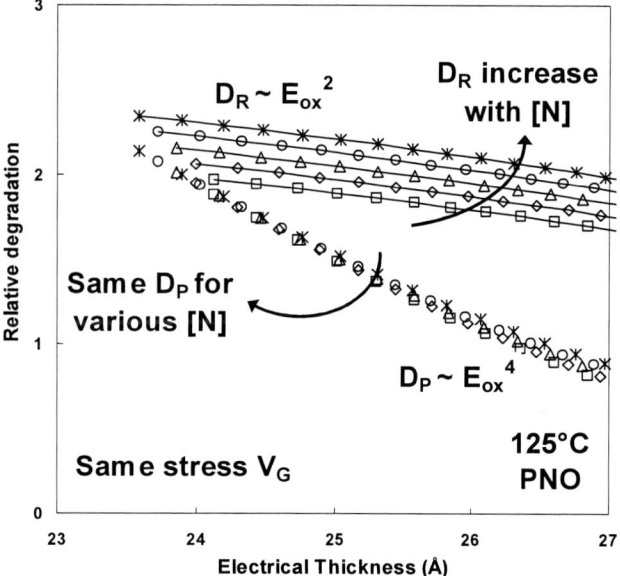

Fig. 9: Based on large process split DOE experiments, the independence of the permanent part D_P generation with the nitrogen content/process is confirmed inline with previous findings of figure 8. Electrical thickness variations also confirm an E_{ox}^4 dependence for D_P. On another hand, D_R shows strong nitrogen dose dependence, as well as an E_{ox}^2 dependence, quite different than D_P.

On another hand, the recoverable part D_R shows strong dependence to the presence of nitrogen atoms in the dielectric, D_R increasing with nitrogen dose. The recoverable part D_R also differs from the permanent one D_P with a smaller oxide field dependence, about E_{ox}^2 dependence. These results are inline with the absence of voltage scalability presented in figure 4, where D_P presents larger oxide field dependence than D_R.

All the results can be further summarized in the table below. By reading carefully these results, it is possible to conclude that the two components (D_P and D_R) **are not** elements of the same microscopic mechanism, as suggested by [5], and so they **are not** two tightly coupled components but on the contrary they are strictly **independent** components, as suggested by [7].

Dependence	Permanent part	Recoverable part
Oxide field	E_{ox}^4	E_{ox}^2
Temperature	0.15-0.25 eV	T^2 (~0.08eV)
Hydrogen species	YES	NO
Nitrogen dose	NO	YES

After demonstrating the existence of two independent components in NBTI degradation based on different acceleration factors and process dependences, we will discuss the physics of microscopic mechanisms lying behind the macroscopic behavior.

III. RECOVERABLE PART

Though the relationship between hole trapping and recoverable part makes consensus [5,7,8], the exact microscopic mechanism remains not very well known. In order to achieve in-depth understanding of the microscopic mechanisms, it is possible to study emission and capture times for single defects using very small transistors as first shown by [18,19].

The technique, first developed on flash EEPROM memories [20] is based on the subthreshold drain current which allows studying the recovery with increased accuracy. After a stress without interruption, the gate voltage is decreased down to the subthreshold part (typically around 0.1-0.2V) and the subthreshold drain current is used as a monitor of the recovery (i.e. drain voltage, usually small, is kept similar between stress and recovery). Unlike GIDL current, subthreshold current flows through the entire channel and potentially can be used to detect charge exchange between the oxide and the substrate all along the channel. Since subthreshold current exhibits an exponential dependence on oxide charge, a small change of oxide charge can result in a significant variation of the subthreshold current. The threshold voltage shift will be modelled as:

$$\Delta V_{th}(t_r) = \frac{1}{S}\ln(\frac{I_d(t_r)}{I_d(0)}) \quad \text{eqn. 1}$$

where t_r is the recovery time and S is the subthreshold slope (decade/mV), as obtained through I_d-V_g characteristics of devices. Due to the subthreshold slope value and the accuracy of our probing system, we have estimated by noise analysis approach on fresh devices that V_{th} shift will be known with a standard deviation (~accuracy) of about 50µV. It is worth noticing that the fresh device subthreshold current is stable, proving that the measurement procedure does not introduce additional trapping/detrapping due to the small gate voltage used [19].

A. Recovery for small gate area devices

For large gate area pMOS devices, subthreshold current Id-time recovery trace shows after NBTI stress a recovery with logarithmic time dependence, in agreement with OTF results [2,21]. Under similar NBTI stress, an ultra-small gate area pMOS device will only get few holes trapped. The discrete nature of the detrapping events can be observed on recovery traces as shown in figure 10.

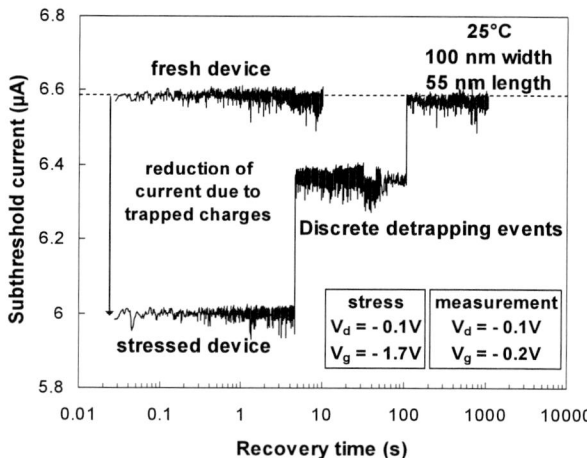

Fig. 10: Subthreshold current evolution with time for a fresh and NBTI stressed ultra small gate area 1.9nm-thick pMOS device demonstrating the discrete nature of the detrapping events in presence of few defects (two defects in this case).

Several recovery traces of single defects are obtained by repeatedly stressing and recovering the same device, while ensuring complete recovery in between two successive sequences. Most of the single defects experiments were led at 25°C by applying low-to-medium vertical oxide fields, so that the permanent part generation is negligible. Due to the stochastic nature of the trapping/detrapping phenomenon, every single recovery traces (and corresponding emission times) are different (cf. figure 11) even though the exact same defect is experimentally characterized. Only by averaging a large number of

978-1-4244-5430-3/10 $26.00 © 2010 IEEE

recovery traces, the expected poissonian behavior $\sim A.Occ(t_s,V_{gs}).\exp(-t_r/\tau_e)$ is retrieved (figure 12, red symbols)[22], with t_s the stress time, A the trap height, Occ the trap occupancy after the stress, V_{gs} the stress gate voltage, τ_e the characteristic emission time.

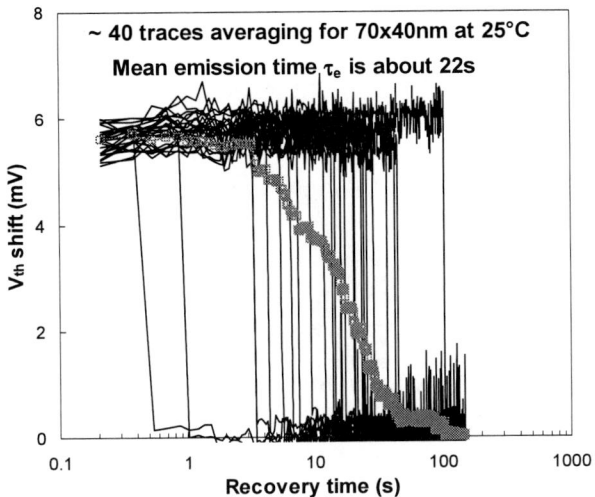

Fig. 11: V_{th} recovery traces (~40, black lines) for a single defect at 25°C following a 10s stress at 7MV/cm. Only when recovery traces are averaged (red symbols), the poissonian behaviour $\sim A\exp(-t_r/\tau_e)$ is retrieved, with a mean emission time τ_e of 22s in this case.

The trap occupancy after the stress is a direct monitor of the mean capture time. It can be easily understand by considering that the trap occupancy only reaches unity when the stress time t_s, during which the trap might be filled, is longer than trap characteristic capture time τ_c. The mean capture time τ_c can be experimentally monitored by characterizing the trap occupancy dependence to the stress time (cf. figure 12).

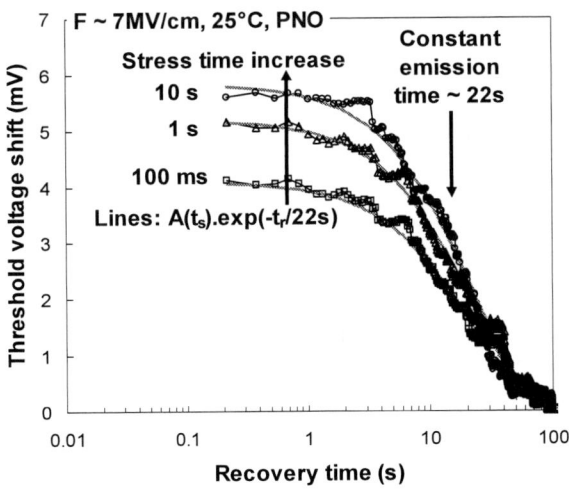

Fig. 12: Averaged recovery traces for a single giant defect for different stress times. Trap occupancy is changed up to unity for t_s=10s allowing the characterization of the mean capture time for this defect. It is worth noticing that independently of the stress time, the emission time remains constant (~22s in this case).

On devices where several defects are present, it is possible to evidence that the mean capture and emission times are uncorrelated. Figure 13 presents the case of one device with two single defects. Interestingly, the defect with the longest capture time (trap 2) (visible only after 100ms stress) presents a short emission time (~0.65s) while the first defect (trap) has a capture time close to 1ms but

presents large emission time of about 11s. This situation can only be explained in the case of perfectly uncorrelated defects.

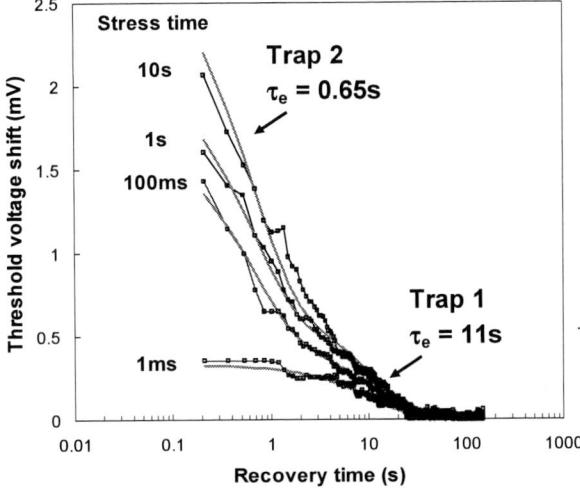

Fig. 13: Averaged recovery traces for two single defects for different stress times ranging from 1ms to 10s. Trap 1 occupancy is already close to 1 with 1ms of stress. In spite of short capture time, its characteristic emission time is about 11s (so 4 decades longer). On another hand, Trap 2 is characterized by a long capture time (>10s), while its emission time is about 0.65s. These two defects are a good example to demonstrate that capture and emission times are uncorrelated.

By studying the mean capture characteristic times for several single defects, it is observed that they all present oxide field dependence stronger than the expected linear dependence for either direct tunneling approach [23] or lattice-relaxation multi-phonon emission (LRME) process [24] (cf. figure 14). On another hand, a field-assisted LRME process was proposed some time ago to explain the high-field deep trap transients [25,26]. This approach was recently used to explain the recoverable part of the NBTI degradation for large area devices [5]. The field-assisted LRME process is expected to present a capture time oxide field dependence as $\exp(-F_{ox}^2/F_c^2)$ (cf. figure 16). From our single defects experiments, we had been able to extract a dataset of 30-40 single defects to be analyzed. Figure 15 shows that all the single defects we had monitored in our experiments follows a $\exp(-F_{ox}^2/F_c^2)$ oxide field dependence, assuming a mean critical field F_c value of 3 MV/cm (figure 14).

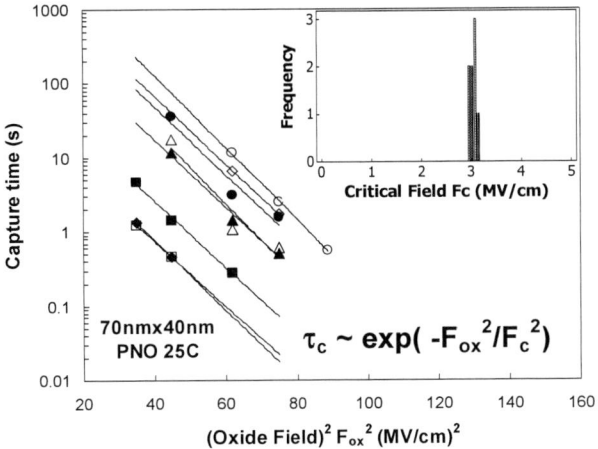

Fig. 14: Capture times for several single defects present similar oxide field dependence which is well explained in MP-FAT approach by a 3MV/cm critical field (F_c histogram in inset).

$$\tau_c = \tau_0 e^{\frac{\Delta E_B}{kT}} \frac{N_V}{p} e^{-\frac{E_{ox}^2}{E_c^2}} \qquad \tau_e = \frac{e^{\frac{\Delta E_B}{kT}}}{\upsilon}\left(1+\frac{p}{N_V}e^{\frac{\Delta E_T'}{kT}}\right)$$

$$\tau_0^{-1} = N_V v_p^{th} \sigma_p e^{-\frac{x}{xp}}$$

Fig. 15: Capture and emission times for the field-assisted LRME model [5,24,25] with ΔE_B the LRME barrier, p the hole surface density, $\Delta E_T'$ the energy level of the defect after lattice relaxation (switching trap energy level) with respect to the valence band edge, E_c the LRME critical field and $\upsilon=10^{13}$ s^{-1} the phonon frequency.

Based on the capture time experimental dataset (shown in figure 14) and the corresponding LRME equations (figure 15)[22], it is possible to extract the related LRME barrier ΔE_B for every single defect. For that purpose, the trap oxide depth is defined by its electrical influence at interface (i.e. the trap step height) and the hole surface density is calculated using self-consistent approach and considering the impact of quantization in the inversion layer. Interestingly, the LRME barrier ΔE_B presents a trend to decrease with defect distance into the oxide (cf. figure 16). In order to understand this surprising feature, the impact of 10000 defects has been simulated in the oxide with random LRME barrier, oxide position and trap energy level. Self-consistent approach has been used to calculate channel properties. Other parameters used during simulations are summarized in figure 18. By considering only the defects showing properties identical to the experimental capture times window ranging from 50ms to 50s, it is possible to reproduce the overall feature observed experimentally with good accuracy (fig. 16).

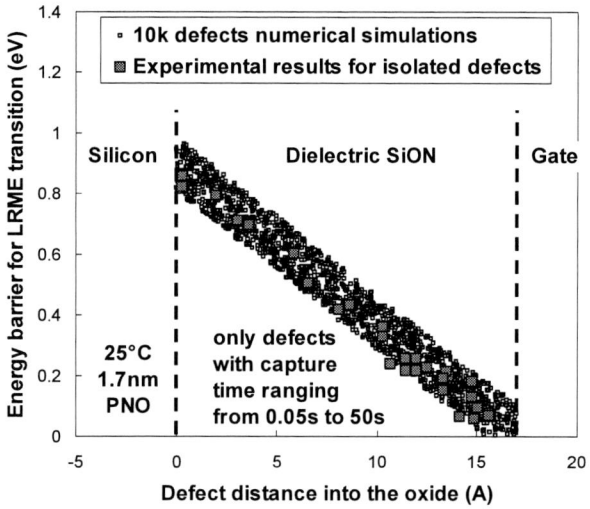

Fig. 16: Experimental LRME energy barriers from single defects (red filled squares) as extracted from capture time oxide field dependence. LRME barrier presents a trend to decrease with defect distance into the oxide. This particular configuration can be understood by numerical simulation results of 10000 random defects (random wrt LRME barrier, oxide position and trap energy)(open squares). This trend is only related to the fact that experimentally only capture times ranging from 50ms to 50s were investigated.

Emission times were also determined experimentally for the same single defects. Considering the LRME barrier extracted from capture times, it is possible to experimentally assess the energy level of the defect after lattice relaxation by studying the emission times (cf. figure 15). Interestingly, the trap level after lattice relaxation presents a trend to increase with respect to defect distance into the oxide. This is also explained by numerical simulations with the same assumptions than for capture times (cf. figure 17). Nevertheless, it is impor-

tant to notice that the defect levels after lattice relaxation lie within silicon bandgap, from silicon valence band edge up to 0.9eV above.

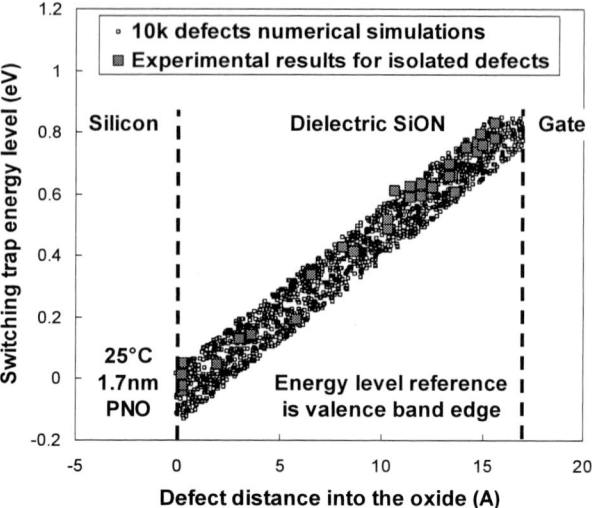

Fig. 17: Experimental switching trap energy level after lattice relaxation from single defects (red filled squares) as extracted from emission times. Trap energy level presents a trend to increase with defect distance into the oxide. This particular configuration can be understood by the numerical simulation results of 10000 random defects (random with respect to LRME barrier, oxide position and trap energy level) (black open squares). This trend is only related to the fact that experimentally only capture times ranging from 50ms to 50s were investigated.

All these experiments led on single defects show that field-assisted LRME process is the driving force lying behind hole trapping and detrapping phenomenon, also called recoverable part D_R.

C. Macroscopic behavior modeling

Based on these inputs on microscopic mechanisms, we have performed both experimental measurements and numerical simulations on large area pMOSFET transistors. Due to the dielectric amorphous nature, the field-assisted LRME model parameters can spread over a broad range. Traps positions distribution within the oxide was defined accordingly to the experimental one observed at single defects level. Defect levels and LRME barrier are assumed to be homogeneously distributed. 10000 defects were generated with individual set of parameters. Oxide field and temperature dependences (cf. figures 19-20) at macroscopic level are well reproduced. The parameter values assumed in the simulations are listed in figure 18. Both the initial logarithmic time behaviour as well as the quadratic oxide field and temperature dependence are reproduced.

$$\sigma_p = 1.2\,10^{-14} cm^{-2}, F_c = 3MV/cm$$

$$x_p = 0.5\,\overset{\circ}{A}, v_p^{th} = 10^5\, m.s^{-1}$$

Fig. 18: Parameter values assumed in the simulations.

D. Physical nature of hole trap defects

In parallel to the discussion concerning the microscopic description of the capture/emission rates, it is also relevant to point out the physical nature of the defects involved in the trapping/detrapping phenomenon. Though the presence of nitrogen atoms into the dielectric is well known to enhance hole trapping during NBTI stress phase, the recent studies discussing field-assisted LRME processes only consider oxygen vacancy defects (mostly present in the sub-stochiometric interfacial layer) as the main contributor [5].

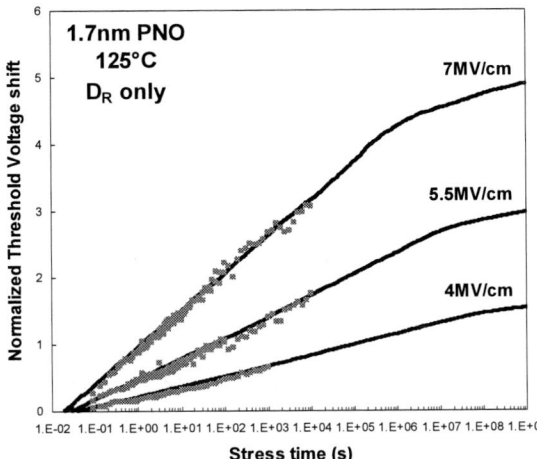

Fig. 19: Experimental threshold voltage shifts oxide field dependence related to trapping/detrapping component (D_R) (red symbols) are well reproduced assuming experimental trap distribution as observed from single defect experiments.

In this context, it is not possible to explain experimental results showing strong nitrogen content impact on NBTI degradation such as in figures 6 and 7. To better understand the intimate link between NBTI and nitrogen content, highly accurate SIMS profiles were realized on gate stacks similar to our single defects experiments (described above). Figure 21 compares the trap position distribution (histogram bars) within the oxide (as observed from single defects experiments) with the nitrogen [N] dose SIMS profile (square symbols).

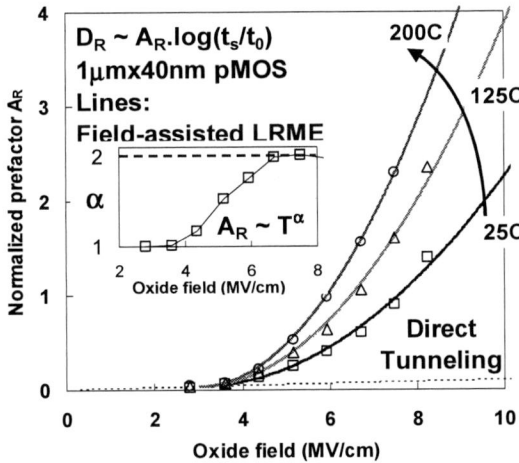

Fig. 20: Experimental threshold voltage shifts oxide field and temperature dependence related to trapping/detrapping component (D_R) (red symbols) are well reproduced assuming experimental trap distribution as observed from single defect experiments.

For the first time, we have experimentally evidenced at single trap level the strong correlation between the trap numbers and the nitrogen content for almost all the oxide depth. This correlation can only be explained by the fact that hole traps are nitrogen-related centers (like K centers) [27,28]. Nevertheless, it is also important to notice that more traps are found in the interfacial layer than expected from nitrogen content which might be explained by the contribution of oxygen vacancy centers, also called $E_\gamma{'}$ centers (oxygen-vacancy centers)[26 and references therein] as recently suggested [5].

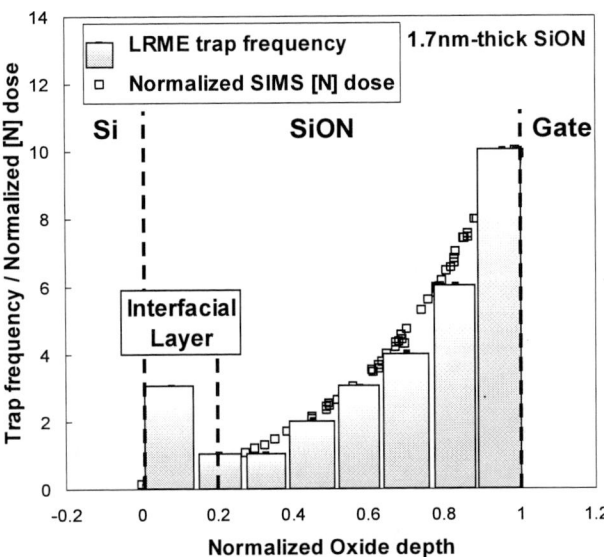

Fig. 21: Experimental trap position distribution (histogram bars) as a function of oxide depth and normalized SIMS nitrogen dose [N] (square symbols). It is worth noticing the strong correlation between nitrogen content and trap numbers in almost all the oxide depth, at the exception of the interfacial layer where more traps are observed.

This correlation between nitrogen atoms presence and hole trapping/detrapping processes is further demonstrated at macroscopic level using large devices. Figure 22 evidences that the correlation exists between amplitude factor A_R of the recoverable part and the nitrogen content. Large DOE experiments have been used to construct this dataset in addition to non-nitrided (pure) oxide. It is worth noticing that A_R increases exponentially with nitrogen dose, independently of oxide field.

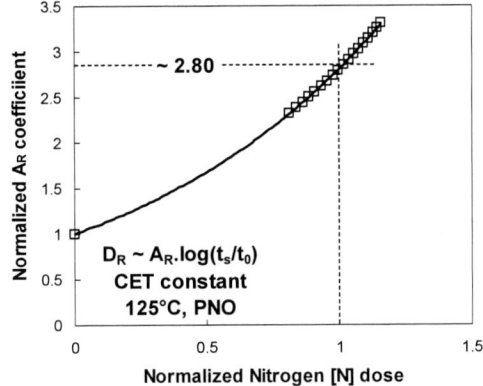

Fig. 22: NBTI recoverable part prefactor A_R is intimately linked to nitrogen [N] dose. For control 1.7nm PNO stacks, the ratio is about 2.8 compared to pure oxide stacks. This is true whatever the oxide field as shown in figure 6.

Another way to demonstrate the intimate link between nitrogen atoms presence and trapping/detrapping processes is based on the Low Frequency Noise (LFN) measurements, used as a monitor of carrier exchange between channel and oxide traps. LFN measurements were averaged over 20 large devices various nitrogen doses. Strong correlation is evidenced between LFN amplitude and the nitrogen dose, in a very similar way than recoverable part amplitude A_R, confirming the intimate link between nitrogen and hole trapping (figure 23). All these experiments confirm the fact that most of hole trap precursors are related to the presence of nitrogen atoms and not solely to oxygen vacancies as previously claimed [5].

Finally, we would conclude our discussion on the recoverable by showing evidences that the nitrogen-based defects related to hole trapping are pre-existent to the stress (i.e. introduced by process choices). Though evidences were already provided by our group [7], a new approach is here discussed. Since LFN measurements are known to be related to oxide trap density and are now known to be correlated to nitrogen dose, we have characterized LFN on large devices before and after strong and long NBTI stress.

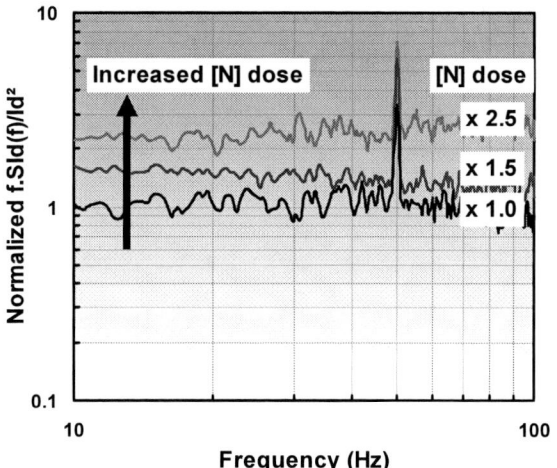

Fig. 23: Low Frequency Noise (LFN) averaged on 20 samples and monitored at 25°C shows strong correlation with nitrogen [N] dose for 1.7nm PNO stacks. It is worth noticing that, since LFN characterizes also trapping/detrapping processes within the oxide, and it shows similar behaviour than NBTI recoverable part.

In spite of the large contribution of permanent part (up to 3 times the recoverable part), we haven't been able to evidence any oxide trap buildup by LFN measurements (inline with previous results [7]), (cf. figure 24). This set of experiments further demonstrates one one hand that defects involved in hole trapping are pre-existent to the NBTI stress and on another hand that both recoverable and permanent parts are not two tightly coupled mechanisms.

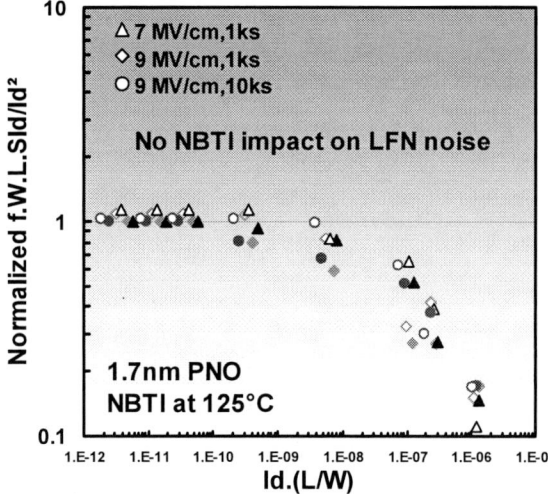

Fig. 24: Low Frequency Noise is monitored and averaged on 20 samples prior (filled symbols) and after (open symbols) NBTI stress. In spite of large permanent part increase, LFN remains unchanged demonstrating that no oxide defects are generated, trapping phenomenon only occurring in pre-existing defects.

The absence of the generation of new trapping defects during NBTI stress was also demonstrated previously [7,30] by using transient measurements of V_{th} shifts [29].

In conclusion, the so-called recoverable part is well explained by field-assisted LRME hole traping/detrapping processes either at microscopic level (as confirmed by single defects study) or at macroscopic level. In addition, these trapping processes were shown to occur on pre-existent defects which are either oxygen vacancies or nitrogen-related centers as introduced by nitridation.

IV. PERMANENT PART

In parallel to the trapping/detrapping processes at play during NBTI degradation, another component can be identified after long-term recovery phases (cf. figure 1 and 2) as a permanent degradation. This permanent component D_P is not affected by fast transient phenomena, like the recoverable part D_R, but depends only on the stress conditions. Many authors have already related the permanent part to the generation of interface traps. These traps are made of unpaired electrons in dangling orbital at interface which can be occupied by zero, one or two electrons, which would make the same defect positively charged (donor-like), neutral or negatively charged (acceptor-like), depending on the Fermi level at the interface [30-31]. In general, high frequency (>500kHz) Charge Pumping (CP) experiments allow monitoring interface traps densities N_{it}. Moreover, CP measurements introduce uncontrolled recovery with fast transient effects.

In our study, we have turned this drawback into an advantage by considering long-term recovery phases. Figure 25 shows that though N_{it} level (as monitored by CP) remains unchanged by one week recovery, V_{th} shifts recover down to the constant N_{it} level after one week. It demonstrates for the first time without ambiguity that the interface traps do not get passivated in this timeframe and are the main contributor of the permanent part.

Fig. 25: V_{th} shifts and interface traps density N_{it} (by CP) measurements show different behaviours during NBTI recovery. N_{it} level remains identical while V_{th} shifts present large transient effects to finally recover down to N_{it} level.

Several microscopic models have been already proposed to explain why SiH bonds are broken at the interface in spite of the absence of energetic carriers [33][34]. Some models imply the generation of positive fixed charges due to the binding of the interface-released hydrogenated species within the oxide. Capacitance-voltage (C-V) curves demonstrate an asymmetric shift (cf. inset in figure 26), with large V_{th} shift, smaller midgap voltage (V_{mg}) shift and no flat-band voltage (V_{fb}) shift.

These results are consistent with previous results, which were explained by an increase of donor-like interface traps [35]. Though, this explanation denies the amphoteric nature of the interface traps.

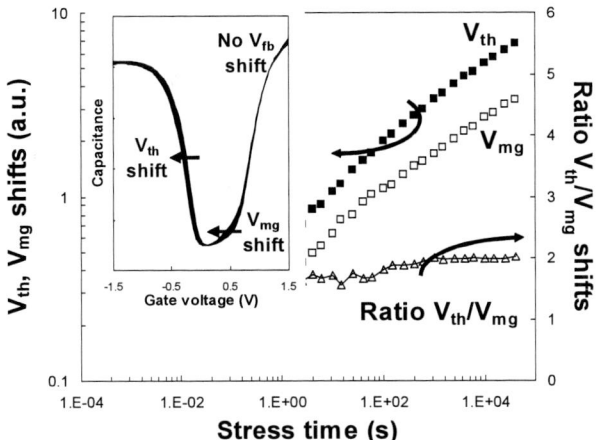

Fig. 26: V_{th} (filled squares) and V_{mg} (midgap voltage) (open squares) shifts for 2nm-thick nitrided oxide and their ratio (triangles), which presents a constant value about 2 for about 5 decades of stress times. Inset shows the Capacitance-Voltage curves the shifts are extracted from, showing that there is no observable V_{fb} shift.

For such defects, total charge at V_{mg} is null. The presence of V_{mg} shift has to be related to the creation of positive fixed charges during the stress [36]. The 2:1 correlation between V_{th} and V_{mg} shifts can only be explained by a similar number of interface traps than fixed charges. This is confirmed by the absence of V_{fb} shift, where positive fixed charges and acceptor interface traps are neutralizing each other. The 1:1 ratio between positive fixed charges and interface traps favors microscopic models where interface traps creation is related to the transfer of H atoms into the oxide, preferentially to a precursor site where a hole is trapped [37]. This hydrogen transfer to a precursor site is also experimentally evidenced by the process dependence related to hydrogen species (cf. figure 5).

Following [38], we claim that interface traps are related to neutral hydrogen atom transfer from interface to oxygen-silicon bridge. Such transfer is favored by hole trapping and yields to positive fixed charge formation in a 1:1 ratio with interface traps (dangling bonds) (cf. figure 27). This theoretical approach predicts forward energy barrier E_2-E_1 about 1.5eV (also confirmed by [45]) and a reverse energy barrier E_3-E_2 about 3eV.

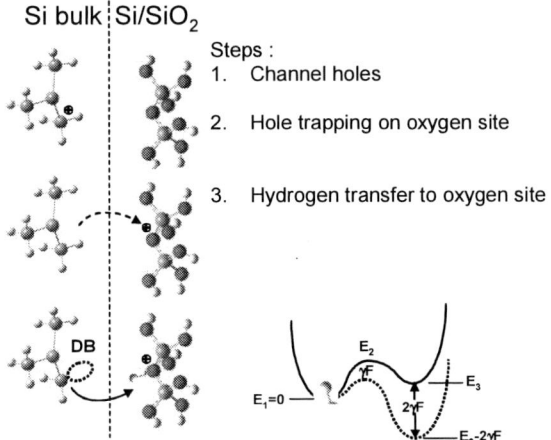

Fig. 27: Permanent part microscopic mechanism description: the transfer of neutral hydrogen on neighbouring oxygen atoms is favoured by preliminary hole trapping on the oxygen site [37]. Once transferred, the hydrogen atom stabilizes the defect into a positive fixed charge while creating a related inter-face traps (dangling bond DB) in a 1:1 ratio.

A. Forward interface traps reaction rate

Based on Jeppson work [39] and related RD models, the SiH bonds dissociation is reaction-limited at short times [7,41]:

$$N_{it}(t) = (1-e^{-k_F N_o t}) \approx k_F N_o t \quad \text{eqn. 2}$$

where N_o is the number of SiH bonds left to be broken and k_F is the oxide field dependent forward dissociation rate, which is assumed constant for a given oxide field (i.e. all bonds are identical). At longer times, the diffusion of hydrogenated species controls the trap generation process, yielding $N_{it}(t) \sim M.t^{0.25}$.

Determining N_o by CP measurements, we have evidenced that the forward dissociation rate k_F is not constant (cf. figure 28) but distributed as Fermi derivative distribution $g(E,\sigma)$ (with σ about 0.1eV) [6-7], consistently with [42-43]. This experimental observation is in contradiction with RD model where k_F is constant. From k_F values spread, dissociation energies are shown to follow Fermi derivative distribution $g(E,\sigma)$ (with σ about 0.1eV) [6-7], consistently with [42-43], resulting in a global defect rate [24]:

$$\frac{\Delta N_{it}}{N_{it\max}}(t) = \int_0^\infty g(E_d,\sigma)(1-e^{-k_F(E_d)N_o(k_F)t})dE_d \propto \frac{1}{1+(\frac{t}{\tau})^{-\alpha}} \quad \text{eqn.4}$$

assuming $\tau = \tau_o \exp(\frac{E_d(E_{ox})}{kT})$ and $\alpha = \frac{kT}{\sigma}$ for $\tau_{min} < t < \tau$, T being the bond temperature and τ_{min} the time constant of the weakest defect.

The median dissociation energy for a given oxide field is shown to follow a linear relationship with oxide field (cf. figure 29). The median dissociation energy in absence of oxide field is thus estimated to be equal to 1.5eV, in close agreement with the theoretical estimations [38,45] for the migration barrier of hydrogenated species within the oxide. The predicted linear dependence of the time exponent with temperature was confirmed experimentally by many authors [6-7,44], providing a second experimental determination for the distribution spread σ of about 0.1 eV.

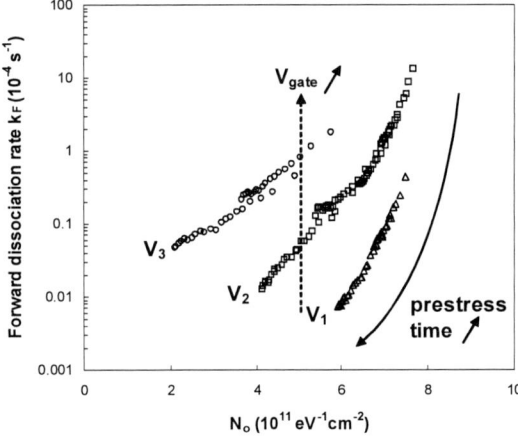

Fig. 28: Forward dissociation rate k_F as a function of N_o, i.e. the total number of SiH bonds left for three different stress voltages V_i. k_F is distributed over a large range of values.

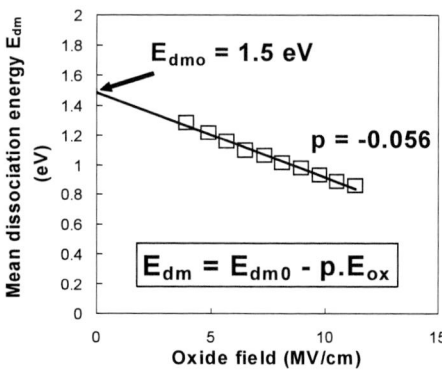

Fig. 29: Linear dependence of median dissociation energies with oxide field showing linear dependence and zero-field mean dissociation energy about 1.5eV inline with theoretical descriptions.

B. Reverse interface traps reaction rate

Though the permanent component appears stable (i.e. no recovery) in conventional timeframe, our double well model predicts the possibility for the hydrogen atom to transfer back to passivate interface dangling bonds. Long thermal anneals (up to 7 months) in absence of gate voltage have been conducted to observe for the first time without ambiguity the interface traps repassivation (cf. figure 30). It is worth noticing that the relative interface traps recovery is independent of the gate dielectric nature (pure vs nitrided) and dielectric thickness (here ranging from 1.8nm to 6.5nm).

From these interface traps recovery experiments with long thermal anneals and accordingly to our proposed model (fig. 27), we extracted an estimate for the energy barrier for the hydrogen-bound state on silicon-oxygen bridge, with $(E_2-E_3)=2.3eV$.

In conclusion, interface traps creation is due to hydrogen transfer to oxygen bridges in the dielectric in presence of holes. This process is reaction-limited and is activated back by long thermal anneals.

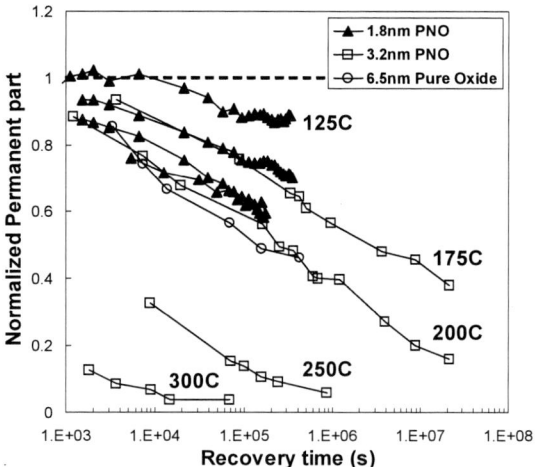

Fig. 30: Permanent part can be reduced to zero by interface traps repassivation under long thermal anneals in absence of gate voltage. Thick oxides data are obtained from V_{th} and CP measurements (empty symbols)[47] and thin oxides (filled symbols) are obtained from V_{th} only (original to this study).

CONCLUSIONS

This study suggests that the Negative Bias Temperature Instability degradation is a combination of two independent components, which both present different voltage, temperature and process dependences. The permanent component is made in equal proportion of interface traps and positive fixed charges. The reaction-limited hydrogen transfer is favored by hole trapping on oxygen bridges in the dielectric close to the interface. For the first time, it was shown that this component can suffer total recovery after long thermal anneals.

The recoverable component is made of hole trapping/detrapping processes to pre-existent defects in the dielectric. These defects are shown without ambiguity to be related to both oxygen vacancies (as in [5]) and nitrogen-related centers. The field-related LRME model allows explaining the various dependences either at microscopic or macroscopic levels. Finally, this in-depth approach to NBTI physical modeling over a large range of process technologies opens the way to reliability predictive modeling.

ACKNOWLEDGEMENTS

The author is in debt with many people who supported this research work since many years from now. Special thanks are addressed to C.R. Parthasarathy, M. Denais, and F. Monsieur for discussions on NBTI, to P. Mora and D. Barge for providing samples and finally to C. Leyris for LFN measurements.

In the loving memory of Louis REY who, though not directly implied in this research work, always provided strong support and advices to the author.

REFERENCES

[1] V. Huard et al., IEEE IRPS (2004)
[2] S. Rangan, IEEE IEDM (2003)
[3] C. Shen et al., IEEE IEDM (2006)
[4] T. Grasser et al., IEEE IRPS (2008)
[5] T. Grasser et al., IEEE IRPS (2009)
[6] V. Huard et al., Microelectr. Reliab. (2006)
[7] V. Huard et al., IEEE IEDM (2007)
[8] H. Reisinger et al., IEEE IRPS (2006)
[9] J.F. Zhang et al., IEEE IEDM (2007)
[10] T. Grasser, IEEE IRPS Tutorial (2008)
[11] J.W. Lyding, et al., Appl. Phys. Lett., 68, p.2526 (1996)
[12] E. Li et al., IEEE IRPS (1999).
[13] D. Menzel and R. Gomer, J. Chem. Phys., 41, p.331 (1964)
[14] N.Kimizuka et al., IEEE VLSI (2000)
[15] S. Zafar, IEEE VLSI (2004)
[16] N.Kimizuka et al., IEEE VLSI (1999)
[17] A. Haggag et al., IEEE IRPS (2007)
[18] C.T. Chan et al., IEEE VLSI (2005)
[19] V.Huard et al., IEEE IIRW (2005)
[20] T. Wang et al. , IEEE Trans. Elec. Dev. (1998)
[21] M. Denais et al., IEEE IEDM (2004)
[22] T. Grasser elal., IEEE IEDM (2009)
[23] A. McWhorter, Semi. Surf. Phys. (1957)
[24] M. Kirton et al., Adv.Phys. (1989)
[25] S. Makram-Ebeid and M. Lannoo, Phys. Rev B (1982)
[26] S. Ganichev et al., Phys. SS (1997)
[27] E. H. Poindexter el al., J. Electro. Soc. (1995)
[28] D.T. Trick et al., J. Appl. Phys. (1998)
[29] Y. Nissan-Cohen et al., J. Appl. Phys. (1986)
[30] C.R. Parthasarathy et al., IEEE IEDM (2003)
[31] E.H.Poindexter et al., Prog.Surf. Sci. (1983)
[32] D.K. Schroder et al., J. Appl.Phys. (2003)
[33] C.E. Blat et al., J. Appl. Phys. (1991).
[34] K. Kushida-Abdelghafar et al., Appl. Phys. Lett. (2002)
[35] V. Reddy et al. , IEEE Micr. Rel. (2005)
[36] S. Tsujikawa et al., IEEE IRPS (2003).
[37] J. Ushio et al., Appl. Phys. Lett. (2002)
[38] T. Mazuimi et al., SISPAD (2007)
[39] K.O. Jeppson and C.M. Svensson, J. Appl. Phys. (1977)
[40] S. Ogawa and N. Shiono, Phys. Rev. B (1995)
[41] H. Kufluoglu and M.A. Alam, IEEE IEDM (2004)
[42] A. Stesmans, Phys. Rev. B (2000)
[43] R. Devine et al. , Apl. Phys. Lett. (1997)
[44] B. Kaczer et al., IEEE IRPS (2005)
[45] S.T. Pantelides et al., IEEE Trans. Nucl. Sci. (2000)
[46] S. Chakravarthi et al., IEEE IRPS (2004)
[47] C. Benard et al., IEEE IIRW (2008)

Recovery-Free Electron Spin Resonance Observations of NBTI Degradation

J.T. Ryan and P.M. Lenahan
The Pennsylvania State University
212 EES Building
University Park, PA 19490, USA
phone: 814-863-4630, jtr16@psu.edu

T. Grasser
Institute for Microelectronics, Technical University of Vienna
Gusshausstrasse 27-29/E360
1040 Vienna, Austria

H. Enichlmair
austriamicrosystems AG
Schloss Premstaetten, A-8141
Unterpremstaetten, Austria

Abstract— We have developed an approach to perform "on the fly" electron spin resonance (OTF-ESR) measurements of negative bias temperature instability (NBTI) defect generation. This OTF-ESR approach allows for an atomic-scale identification of the defects involved in NBTI free of any recovery contamination. We demonstrate that, during NBTI stressing at elevated temperature and modest negative oxide bias, positively charged oxygen vacancy sites (E' centers) are generated. Upon removal of the NBTI stressing conditions, the E' center density quickly recovers to that of its pre-stress values. When similar measurements are made with zero oxide bias at elevated temperature or negative oxide bias at room temperature, the E' defect density does not change. These observations strongly indicate that NBTI is triggered by inversion layer hole capture at an E' precursor site which then leads to the depassivation of nearby interface states.

Keywords: electron spin resonance, negative bias temperature instability, E' centers, on the fly

I. INTRODUCTION AND BACKGROUND

The negative bias temperature instability (NBTI) is perhaps the most important reliability problem facing modern CMOS technology [1-3]. Traditionally, NBTI has been explained in terms of the reaction-diffusion model [1-3]. Although the reaction-diffusion model makes physical sense, many variations of this general idea exist in the literature and none of them are able to fully describe the phenomenon [1-3]. A fundamental and complete understanding of the physical processes involved in NBTI is not yet available.

In the general reaction-diffusion model, a reaction takes place during a negative bias temperature stress (NBTS) which depassivates a silicon-hydrogen bond at the Si/SiO_2 interface [1-3]. A Si/SiO_2 interface state (apparently a P_b center in pure SiO_2 devices [4, 5]) is created and the hydrogenic species diffuses into the gate stack and into the gate poly silicon; this process leads to the observed shift in threshold voltage and degradation of drive current [1-3]. The observation of NBTI recovery [6-9], a process in which much of the NBTI damage disappears, is explained as the reversal of this process. When the NBTS is removed, some of the hydrogenic species diffuse back to the interface and repassivate the interface states [1-3].

It should be noted that NBTI recovery has only recently been studied in detail [6-9] and is perhaps the most challenging aspect of NBTI. Recent studies of NBTI recovery [6-9] call into question the validity of conclusions drawn in earlier NBTI studies in which recovery was not accounted for. Additionally, recovery cannot be fully explained by the reaction-diffusion model. As noted by Grasser *et al.*, the reaction-diffusion model predicts a universal recovery phenomena nearly independent of the hydrogenic species involved [1]. Additionally, in contrast to some experimental studies [10-13], Grasser *et al.* [1] note that the reaction-diffusion model fails to predict recovery which depends on gate bias, temperature and process conditions.

Recent conventional electron spin resonance (ESR) observations of Fujieda *et al.* [4] on simple Si/SiO_2 capacitors and electrically-detected magnetic resonance (EDMR) observations of Campbell *et al.* [5, 14, 15] on fully processed transistors indicate that in pure SiO_2 structures, NBTI is dominated by Si/SiO_2 interface states (P_b centers). When subject to very severe NBTS conditions, Campbell *et al.* [5, 14,

---

Work at Penn State supported by Texas Instruments Custom Funding through the Semiconductor Research Corporation. Part of this work has received funding from the European Community's Seventh Framework Programme under Grant Agreement No. 216436 (project ATHENIS).

978-1-4244-5430-3/10 $26.00 © 2010 IEEE

15] also observed E' center generation. Although their experimental observations are somewhat tenuous, Campbell *et al.* suggest that E' centers could trigger the NBTI process via an E' center/P_b center hydrogen exchange [5, 14, 15]. This general idea, that NBTI is caused by an E'/P_b center hydrogen exchange triggered by hole capture at an E' site, has been expressed by Lenahan [16, 17] who provides simple thermodynamics based arguments to this effect. He also noted [16, 17] that the correlation between Si/SiO$_2$ P_b centers and E' oxide defects frequently observed in ESR studies of variously stressed Si/SiO$_2$ device structures may be relevant to NBTI degradation. Conley *et al.* had previously proposed a somewhat similar model to explain how oxide trapped holes trigger interface trap generation in MOS radiation damage [18, 19]. The experimental results of Conley *et al.* [20, 21] clearly demonstrate that multiple E'/P_b center reactions are thermodynamically and kinetically possible. Fig. 1 provides schematic drawings of the two types of interface states commonly found in (100) Si/SiO$_2$, the P_{b0} (top) and P_{b1} (bottom) centers [22]. Both P_{b0} and P_{b1} centers are silicon dangling bond defects in which the central silicon atom is back-bonded to three other silicon atoms. Both defects are located precisely at the Si/SiO$_2$ interface. Fig. 2 illustrates schematic drawings of two E' centers commonly found in pure SiO$_2$: a neutral oxygen vacancy (top) and a positively charged oxygen vacancy (bottom) [22]. E' centers are silicon dangling bond defects in which the central silicon atom is back-bonded to three oxygen atoms.

following thermal oxidation of SiO$_2$; Si/SiO$_2$ P_b interface states are hydrogen passivated. Lenahan [16, 17] argues that if the oxide stress generates a large number of E' oxide defects (Fig. 3b), and if the bonding energies of the Si-H bond at the hydrogen passivated P_b center and a Si-H bond at a hydrogen passivated E' center are about the same, the illustration of Fig. 3b is thermodynamically unstable. That is, completely hydrogen passivated P_b centers in the vicinity of completely unpassivated E' centers is thermodynamically unstable.

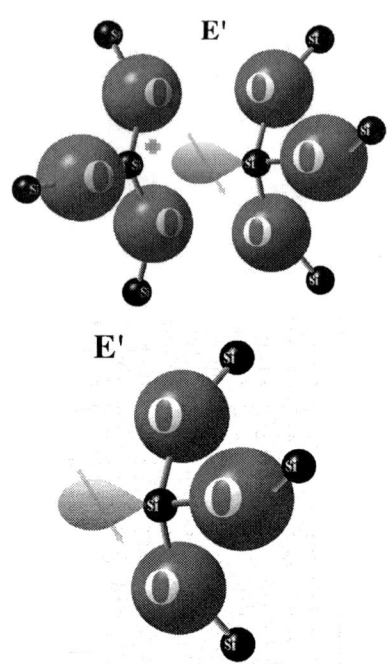

Figure 2. Schematic drawings of two E' centers commonly found in pure SiO$_2$; a neutral oxygen vacancy (top) and a positively charged oxygen vacancy (bottom). E' centers dominate charge trapping in pure SiO$_2$ dielectrics. They are silicon dangling bond defects in which the central silicon atom is back-bonded to oxygen atoms.

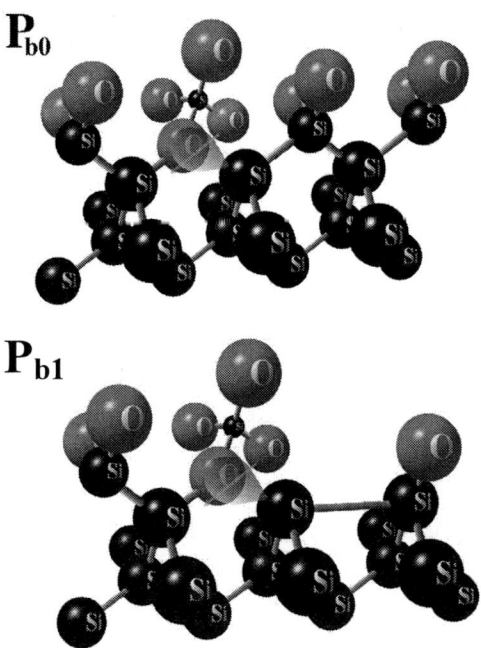

Figure 1. Schematic drawings of P_{b0} (top) and P_{b1} (bottom) Si/SiO$_2$ interface states. P_{b0} and P_{b1} defects dominate interface trapping in (100) Si/SiO$_2$. Both are silicon dangling bond defects in which the central silicon atom is back-bonded to three other silicon atoms. The main differences between them are their dangling bond axis of symmetry and electronic density of states [23-27].

Fig. 3a schematically illustrates a Si/SiO$_2$ interface before (Fig. 3a) and after (Fig. 3b) oxide stressing. The interface of Fig. 3a is typical of oxides which received a forming gas anneal

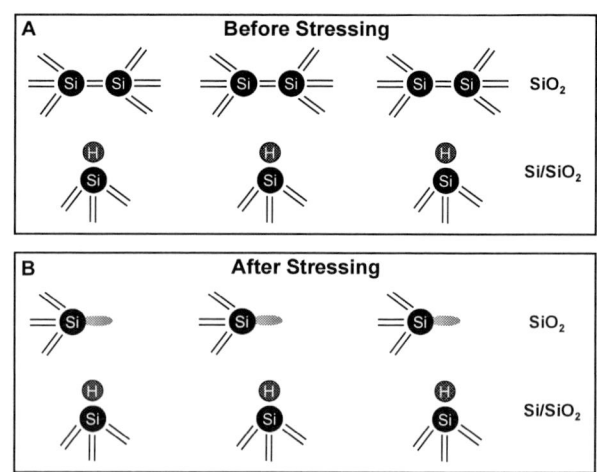

Figure 3. Two simplified illustrations of the Si/SiO$_2$ interface. (a) A perfect interface prior to stressing in which all P_b center precursors are hydrogen passivated. (b) After oxide stressing which created large numbers of E' centers in the oxide. Note that the post-stressing illustration (b) is thermodynamically unstable.

978-1-4244-5430-3/10 $26.00 © 2010 IEEE   44

The Gibbs free energy of the system:

$$G = H - TS, \qquad (1)$$

where H is enthalpy (sum of energy plus pressure times volume), T is absolute temperature, and S is entropy, would be lowered if some hydrogen from the passivated P_b centers transfers to some of the unpassivated E' centers in the oxide [16, 17]. This hydrogen transfer would cost little energetically and thus little enthalpy since the two bond energies are roughly equal [16, 17].

However, the hydrogen transfer would greatly increase the entropy of the system; since the entropy of the system is defined as

$$S = k \ln(\Omega), \qquad (2)$$

where k is Boltzmann's constant and Ω is the number of microscopic configurations responsible for the macroscopic system, the configurational entropy would increase from $k \ln(1)$ for the case of Fig. 3b to $k \ln(M)$, a large increase, if one hydrogen was removed from any of the M hydrogen passivated P_b center sites. [16, 17]. The removal of a second hydrogen would lead to a configurational entropy of $k \ln[(M)(M-1)/2]$ and so on for additional removal of hydrogen. Additionally, the configurational entropy due to the unpassivated E' centers would increase from $k \ln(1)$ to $k \ln(N)$, a large increase, if one hydrogen were added to any of the N unpassivated E' center sites [16, 17]. This is schematically illustrated in Fig. 4.

Thus, these simple statistical thermodynamics arguments [16, 17] indicate that the transfer of hydrogen from passivated interface states to unpassivated oxide defects is thermodynamically favored and provides a very plausible explanation for the triggering mechanisms of NBTI.

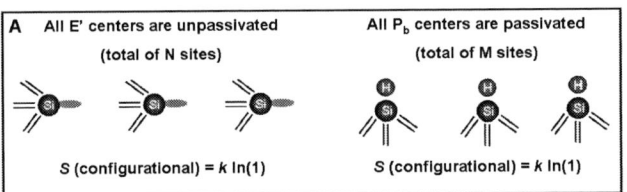

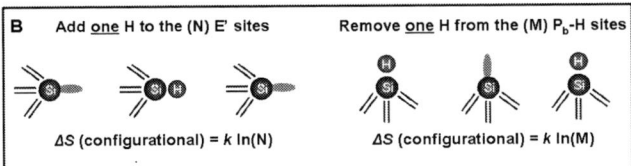

Figure 4. (a) A schematic illustration of an oxide with N unpassivated E' center sites and M hydrogen passivated P_b center sites. In both cases, the configurational entropy, S, is given by $k \ln(1)$. (b) A schematic illustration of the effect of transferring one hydrogen from a passivated P_b center site to an unpassivated E' center site. In both cases, there is a large increase in the configurational entropy.

Quite recently, Grasser *et al.* [28] have developed a comprehensive quantitative two-stage model for NBTI in pure SiO_2 devices based on these ideas [14-17]. In the Grasser *et al.* model, NBTI is triggered by inversion layer hole capture at an E' center precursor site (a neutral oxygen vacancy). In the

model proposed by Grasser *et al.*, the NBTI degradation process is initiated (stage one) when inversion layer hole capture occurs at E' precursor sites (neutral oxygen vacancy) [28]. The hole capture leads to positively charged E' centers (paramagnetic defects observable with ESR) in the oxide, similar to the schematic drawing provided in Fig. 3b. In stage one, the system is in a recoverable state where the positively charged E' center can very quickly emit a hole leading to full recovery back to the precursor state.

Grasser *et al.* argue that this recoverable charge trapping state is responsible for NBTI recovery in which much of the damage very quickly heals once the NBTI stress is removed [28]. However, if the NBTI stress is maintained, the system can also proceed to stage two of the model (permanent degradation). Following the arguments [16, 17] discussed above, Grasser *et al.* note that the large number of unpassivated E' centers in the vicinity of hydrogen passivated P_b interface states in stage one is thermodynamically unstable. The oxide silicon dangling bonds (E' centers) created in the stage one process triggers the creation of P_b centers through the P_b/E' hydrogen exchange process discussed above [28]. This leads to a poorly recoverable P_b interface state and the transferred hydrogen essentially "locks in" the positive charge on the E' center site (rendering it diamagnetic and unobservable with ESR).

The positively charged E' center and the newly created P_b interface state in the Grasser *et al.* model is consistent with much of the NBTI literature [1-3] suggesting the degradation is due to interface state generation and/or oxide charge build up. The interface state generation aspect of the model is consistent with the magnetic resonance observations of Campbell *et al.* [5, 14, 15] and Fujieda *et al.* [4] who indicate that NBTI is dominated by interface state generation (P_b centers). Additionally, the model is consistent with the somewhat tenuous arguments of Campbell *et al.* [5, 14, 15] who observed E' centers following very harsh NBTI stress and suggest that E' centers may play an important role in NBTI degradation.

The comprehensive quantitative model of Grasser *et al.* [28] explains NBTI degradation over a wide range of bias voltage and stress temperature, the observed asymmetry between stress and recovery, and the strong sensitivity to bias and temperature during recovery. Perhaps more importantly, the model attributes recovery to charge trapping/detrapping via a non-radiative multi-phonon process [29], which accurately predicts both the very fast and slow recovery phenomena [6-9, 30], due to a very broad distribution of time constants. Since the reaction-diffusion model relies on diffusion of hydrogenic species (a much slower process than tunneling) to account for recovery, it predicts a much slower recovery which is incompatible with experimental observations [6-9]. Additionally, the model predicts that paramagnetic E' centers will be present during stress and will very quickly recover upon removal of stress [28]. However, prior to this work the existence and role of these E' centers had not yet been conclusively demonstrated.

As mentioned previously, Campbell *et al.* were only able to report somewhat tenuous E' experimental observations in NBTI stressed devices [14, 15]. This is so for two reasons.

First, and most importantly, the EDMR technique of spin dependent recombination (SDR) used [31, 32] does not permit observations at significant negative bias; the stress biasing conditions must be altered so that electron and hole quasi Fermi levels are split more or less symmetrically about the intrinsic Fermi level at the Si/SiO$_2$ interface (that is, the stress biasing must be changed in order to make the measurement and thus introduces recovery). Secondly, SDR is only marginally adequate for E' center detection because only those E' centers very close to the interface can contribute to SDR. Conventional ESR does permit E' center detection at any gate bias if the center is positively charged, as should be the case under stress conditions. In this work, we utilize a newly developed on-the-fly ESR technique to detect positively charged E' centers during NBTI stressing of MOS structures. The newly developed technique permits a recovery free glimpse into the dynamics of NBTI and our results provide insight into the E'/P$_b$ center NBTI triggering mechanisms discussed above.

II. EXPERIMENTAL DETAILS

The samples used in this study are simple Si/SiO$_2$ blanket wafers with 49.5nm thermally grown SiO$_2$ oxides. The samples received a forming gas anneal following thermal oxidation. ESR measurements were performed before, during, and after the samples were subjected to a NBTS of V_G = -25V (<5MV/cm oxide electric field) at 100°C. OTF-ESR measurements were performed by first applying negative gate bias via corona ions and then loading the biased samples into a heated quartz dewar situated inside the ESR microwave resonance cavity. The gate bias was monitored before and after stress with a Kelvin probe. ESR measurements were made on a commercially available Bruker Instruments X-band spectrometer with a TE$_{104}$ microwave cavity and a calibrated weak pitch spin standard. Some measurements also utilized a calibrated SiO$_2$ E' standard [33].

III. RESULTS AND DISCUSSION

Fig. 5 illustrates pre-stress ESR spectra for the forming gas annealed sample (bottom trace) and a nearly identical sample which did not receive a forming gas anneal (top trace) at identical spectrometer gain. The spectrometer settings used were chosen to permit the observation of both P$_b$ and E' defects and are not optimized for either defect. The sample which did not receive the forming gas (top trace) displays three spectra with g = 2.0063 (P$_{b0}$ Si/SiO$_2$ interface states), g = 2.0036 (P$_{b1}$ Si/SiO$_2$ interface states), and g = 2.0006 (E' oxide defects). The sample which did receive the forming gas anneal (bottom trace) displays a much weaker signal with g = 2.0069 which is consistent with a low density of Si/SiO$_2$ P$_{b0}$ centers. The second integral of the ESR signal is proportional to the number of defects present and, as expected, the forming gas annealed sample has far fewer defects present. Since Si/SiO$_2$ samples which did not receive forming gas anneals are not technologically relevant, only results taken on the sample which did receive the forming gas anneal are shown in the remainder of this chapter.

Fig. 6 illustrates three ESR traces taken at room temperature for the sample with forming gas. The top trace

was taken on the as-processed sample, the middle trace was taken with the sample biased with -25V at room temperature (bias maintained for several hours during measurement), and the bottom trace taken after removing the negative bias. The room temperature corona bias of -25V (middle trace) does not result in an increase of interface states (P$_b$ centers) or oxide defects (E' centers). It does, of course, suppress the P$_{b0}$ signal because these defects are interface traps and can respond to the substrate silicon Fermi level [23]. (The negative bias renders most of the P$_{b0}$ centers positive and diamagnetic and thus ESR inactive.) Fig. 7 illustrates three ESR traces taken at zero volts bias for the sample. The top trace was taken on the as-processed sample at room temperature, the middle trace was taken with the sample at elevated temperature (100°C), and the bottom trace taken after returning the sample to room temperature. The elevated temperature at zero volts bias (middle trace) does not result in an increase of interface states or oxide defects.

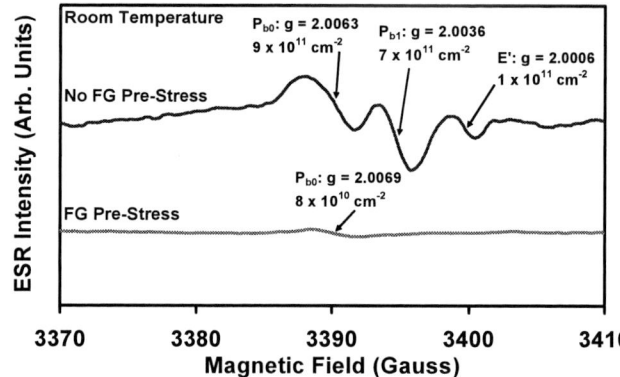

Figure 5. Comparison of pre-stress ESR spectra plotted with identical spectrometer gain for the sample treated with forming gas (bottom trace) and a nearly identical sample which was not treated with forming gas following oxidation (bottom trace). As expected, the sample which received the forming gas anneal is of much higher quality. (The overlapping P$_{b0}$ and P$_{b1}$ spectra results in a slightly lower P$_{b0}$ g value in the non forming gas annealed sample.)

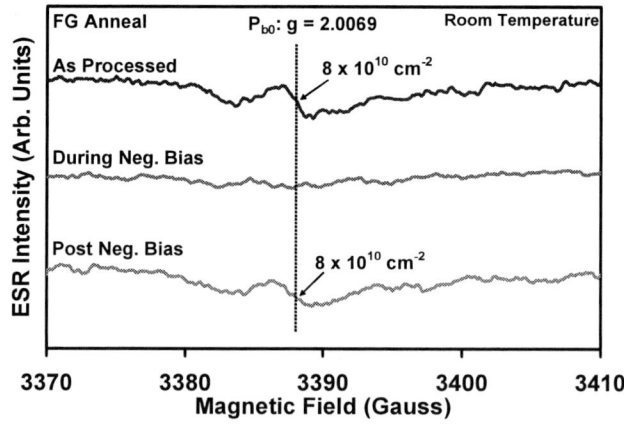

Figure 6. Room temperature ESR traces taken on the sample which received a forming gas anneal as-processed (top trace), with -25V bias (middle trace), and after removal of negative bias (bottom trace). Note that the negative bias alone does not generate additional P$_b$ interface states or E' defects within the resolution of the measurement.

Fig. 8 illustrates three OTF-ESR traces taken before (top trace), during (middle trace) and after (bottom trace) NBTI stressing (V_G = -25V and T = 100°C) on the FG annealed sample. As mentioned previously, in the as-processed case (top trace) we observe a weak spectrum consistent with Si/SiO_2 P_{b0} centers. During NBTI stress (middle trace), we observe the clear generation of Si/SiO_2 P_{b1} centers (g = 2.0034) and SiO_2 E' centers (2.0006). Upon removal of the stress, the g = 2.0006 E' center signal completely recovers while some of the P_{b1} centers remain. This result clearly demonstrates that positively charged E' centers are generated during NBTI stress and very quickly recover upon removal of the stress; that is, positively charged oxygen vacancy sites are generated during stress and very quickly recover. Furthermore, in contrast to Fig. 6 (negative bias only) and Fig.7 (elevated temperature only), the positively charged E' centers are only present when both elevated temperature and negative bias are applied. The fast recovery phenomenon observed also explains the absence of E' center spectra in previous SDR NBTI studies at moderate stress conditions [4, 5, 14, 15]; since SDR is a relatively slow measurement and since the stress biasing conditions must be altered, a large amount of E' center recovery is introduced.

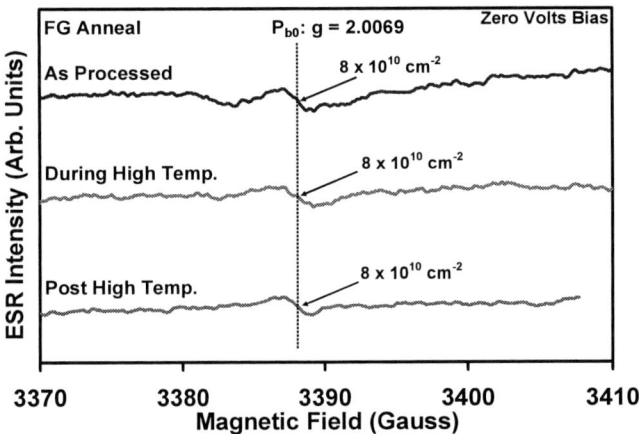

Figure 7. Three ESR traces for the sample which received the forming gas anneal as-processed (top trace), with zero volts bias at 100°C (middle trace), and after cooling the sample back to room temperature (bottom trace). Note that the elevated temperature alone does not generate additional Pb interface states or E' defects.

As mentioned previously, the spectrometer settings used in Figs. 5-8 were chosen to permit the observation of both Si/SiO_2 P_b centers and SiO_2 E' centers and are not optimized for either defect; the E' center density is underrepresented in these traces. (There is a significant difference in E' and P_b spin lattice relaxation times which leads to this under-representation [34].) In an attempt to further demonstrate that E' centers (positively charged oxygen vacancy sites) are present during NBTI stressing, Fig. 9 shows three OTF-ESR traces taken on the forming gas annealed sample before (top trace), during (middle trace) and after NBTI stressing (bottom trace). In this figure, the spectrometer settings are optimized for the observation of E' centers. When NBTS stressing is applied (middle trace), a clear signal with $g_{\parallel}$=2.0016 and $g_{\perp}$ = 2.0006 appears which is characteristic of an E' center. Upon removal of the NBTI

stress (bottom trace), the E' signal completely recovers. Fig. 10 further demonstrates the identification of this signal as due to an E' center by comparing the during NBTI stress spectra of Fig. 9 (top trace) with that of a commercially available E' standard (bottom trace) [33]. Note the close correspondence between the g values and the line shapes which are characteristic to this type of defect.

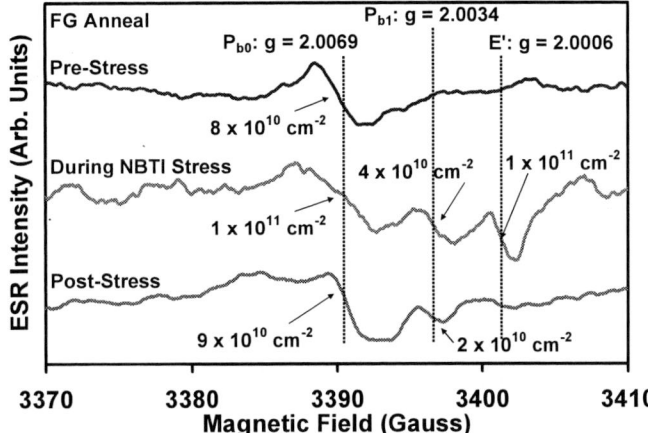

Figure 8. Three ESR traces for the sample which received the forming gas anneal. Note the clear generation of an E' signal during NBTI stress (middle trace), as well as P_{b1} center generation, and the nearly complete recovery of the E' defects post-stress (bottom trace).

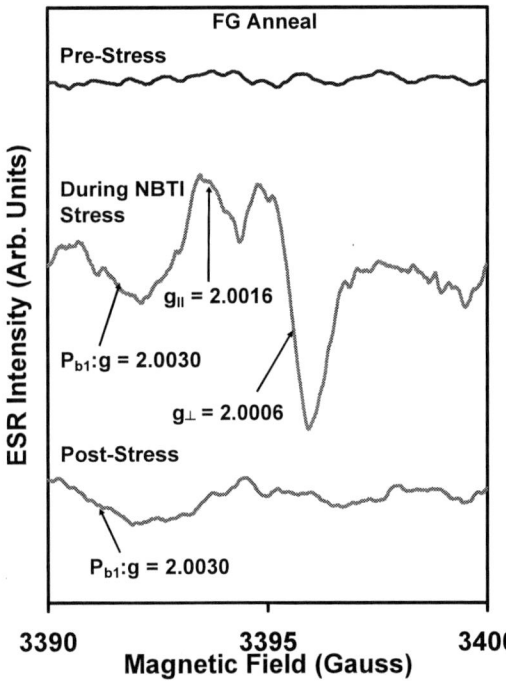

Figure 9. Three ESR traces taken on the sample which received the forming gas anneal. In these traces, the spectrometer settings are optimized to observe E' centers. Note the clear generation of an E' spectrum ($g_{\parallel}$=2.0016 and $g_{\perp}$ = 2.0006) during stress (middle trace) and its subsequent recovery post-stress (bottom trace).

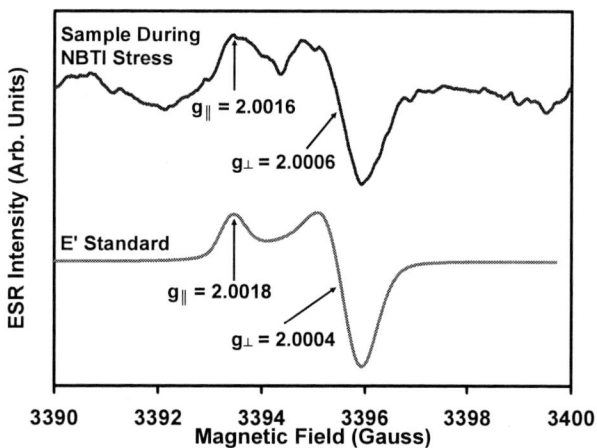

Figure 10. Comparison of the forming gas annealed sample during NBTI stress from Fig. 9 (top trace) and a commercially available E' standard (bottom trace). The standard sample signal-to-noise is much higher because the standard has orders of magnitude more E' centers. Note the close correspondence between the g values and line shapes. The gain of the sample trace is approximately 10,000 times larger that used for the E' standard; all other spectrometer settings are identical. (Note that the precision of g is ± 0.0002.)

IV. CONCLUSIONS

In summary, we present results which demonstrate that E' centers are generated in Si/SiO_2 MOS structures when subjected to modest negative oxide bias at elevated temperature. We further demonstrate that these E' centers recover once the stressing conditions are removed. Only the combination of negative oxide bias and elevated temperature results in E' center generation. These results are consistent with and strongly support the suggestions of Campbell et al. [14, 15] and Lenahan [16] as well as the more recent comprehensive two-stage model proposed by Grasser et al. [28] in which NBTI is triggered by the tunneling of electrons from a neutral E' center precursor to unoccupied valence band states. Our results also clearly support (and also provide an explanation for) previously reported results indicating that the very rapid portion of recovery involves the annihilation of oxide trapped holes.

REFERENCES

[1] T. Grasser, W. Goes, and B. Kaczer, "Toward engineering modeling of negative bias temperature instability," in Defects in Microelectronic Materials and Devices, D. M. Fleetwood, S. T. Pantelides, and R. D. Schrimpf, Eds. Boca Raton, London, New York: CRC Press, 2009, pp. 399-436.

[2] G. LaRosa, "Negative bias temperature instabilities in pMOSFET devices," in Reliability Wearout Mechanisms in Advanced CMOS Technologies, A. W. Strong, E. Y. Wu, R. P. Vollertsen, J. Sune, G. LaRosa, S. E. Rauch, and R. P. Sullivan, Eds. Hoboken: Wiley, 2009, pp. 331-439.

[3] D. K. Schroder, "Negative bias temperature instability: What do we understand?," Microelectronics Reliability, pp. 841-852, 2007

[4] S. Fujieda, Y. Miura, M. Saitoh, E. Hasegawa, S. Koyama, and K. Ando, "Interface defects responsible for NBTI in plasma-nitrided SiON/Si(100) systems," Applied Physics Letters, pp. 3677-3679, 2003

[5] J. P. Campbell, P. M. Lenahan, C. J. Cochrane, A. T. Krishnan, and S. Krishnan, "Atomic-scale defects involved in NBTI," IEEE Transactions on Device and Materials Reliability, pp. 540-557, 2007

[6] G. Chen, M. F. Li, C. H. Ang, J. Z. Zheng, and D. L. Kwong, "Dynamic nbti of p-MOS transistors and its impact on MOSFET scaling," IEEE Electron Device Letters, pp. 734-736, 2002

[7] M. Ershov, S. Saxena, H. Karbasi, S. Winters, S. Minehane, J. Babcock, R. Lindley, P. Clifton, M. Redford, and A. Shibkov, "Dynamic recovery of negative bias temperature instability in p-type metal-oxide-semiconductor field-effect transistors," Applied Physics Letters, pp. 1647-1649, 2003

[8] S. Tsujikawa, T. Mine, K. Watanabe, Y. Shimamoto, R. Tsuchiya, K. Ohnishi, T. Onai, J. Yugami, and S. i. Kimura, "Negative bias temperature instability of pMOSFETs with ultra-thin SiON gate dielectrics," IEEE Int. Reliab. Phys. Symp., pp. 183-188, 2003

[9] W. Abadeer and W. Ellis, "Behavior of NBTI under ac dynamic circuit conditions," IEEE Int. Reliab. Phys. Symp., pp. 17-22, 2003

[10] M. Ershov, R. Lindley, S. Saxena, A. Shibkov, S. Minehane, J. Babcock, S. Winters, H. Karbasi, T. Yamashita, P. Clifton, and M. Redford, "Transient effects and characterization methodology of negative bias temperature instability in pMOS transistors," IEEE Int. Reliab. Phys. Symp., pp. 606-607, 2003

[11] B. Kaczer, V. Arkbipov, R. Degraeve, N. Collaert, G. Groeseneken, and M. Goodwin, "Disorder-controlled-kinetics model for negative bias temperature instability and its experimental verification," IEEE Int. Reliab. Phys. Symp., pp. 381-387, 2005

[12] H. Reisinger, O. Blank, W. Heinrigs, A. Muhlhoff, W. Gustin, and C. Schlunder, "Analysis of NBTI degradation- and recovery-behavior based on ultra fast VT-measurements," IEEE Int. Reliab. Phys. Symp., pp. 448-453, 2006

[13] C. Shen, M. F. Li, C. E. Foo, T. Yang, D. M. Huang, A. Yap, G. S. Samudra, and Y. C. Yeo, "Characterization and physical origin of fast vth transient in NBTI of pMOSFETs with SiON dielectric," IEEE IEDM Technical Digest, pp. 333-336, 2006

[14] J. P. Campbell, P. M. Lenahan, A. T. Krishnan, and S. Krishnan, "Observations of NBTI-induced atomic-scale defects," IEEE Transactions on Device and Materials Reliability, pp. 117-122, 2006

[15] J. P. Campbell, P. M. Lenahan, A. T. Krishnan, and S. Krishnan, "Direct observation of the structure of defect centers involved in the negative bias temperature instability," Applied Physics Letters, pp. 204106, 2005

[16] P. M. Lenahan, "Atomic scale defects involved in MOS reliability problems," Microelectronic Engineering, pp. 173-181, 2003

[17] P. M. Lenahan, "Deep level defects involved in MOS device instabilities," Microelectronics Reliability, pp. 890-898, 2007

[18] J. F. Conley and P. M. Lenahan, "Molecular-hydrogen, E' center hole traps, and radiation-induced interface traps in MOS devices," IEEE Transactions on Nuclear Science, pp. 1335-1340, 1993

[19] J. F. Conley, P. M. Lenahan, B. D. Wallace, and P. Cole, "Quantitative model of radiation induced charge trapping in SiO_2," IEEE Transactions on Nuclear Science, pp. 1804-1809, 1997

[20] J. F. Conley, P. M. Lenahan, A. J. Lelis, and T. R. Oldham, "Electron spin resonance evidence that e'(gamma) centers can behave as switching oxide traps," IEEE Transactions on Nuclear Science, pp. 1744-1749, 1995

[21] J. F. Conley, P. M. Lenahan, A. J. Lelis, and T. R. Oldham, "Electron spin resonance evidence for the structure of a switching oxide trap: Long term structural change at silicon dangling bond sites in SiO_2," Applied Physics Letters, pp. 2179-2181, 1995

[22] P. M. Lenahan and J. F. Conley, "What can electron paramagnetic resonance tell us about the Si/SiO_2 system?," Journal of Vacuum Science and Technology B, pp. 2134-2153, 1998

[23] J. P. Campbell and P. M. Lenahan, "Density of states of P_{b1} Si/SiO_2 interface trap centers," Applied Physics Letters, pp. 1945-1947, 2002

[24] J. W. Gabrys, P. M. Lenahan, and W. Weber, "High-resolution spin-dependent recombination study of hot-carrier damage in short-channel mosfets - si^{29} hyperfine spectra," Microelectronic Engineering, pp. 273-276, 1993

[25] T. D. Mishima and P. M. Lenahan, "A spin-dependent recombination study of radiation-induced P_{b1} centers at the (001) Si/SiO_2 interface," IEEE Transactions on Nuclear Science, pp. 2249-2255, 2000

[26] E. H. Poindexter, G. J. Gerardi, M. E. Rueckel, P. J. Caplan, N. M. Johnson, and D. K. Biegelsen, "Electronic traps and P_b centers at the Si/SiO_2 interface - band gap energy distribution," Journal of Applied Physics, pp. 2844-2849, 1984

[27] K. L. Brower, "Structural features at the Si/SiO_2 interface," Z. Phys. Chem., pp. 177-189, 1987

[28] T. Grasser, B. Kaczer, W. Goes, T. Aichinger, P. Hehenberger, and M. Nelhiebel, "A two-stage model for negative bias temperature instability," IEEE Int. Reliab. Phys. Symp., pp. 33-44, 2009

[29] T. Grasser, H. Reisinger, P. J. Wagner, F. Schanovsky, W. Goes, and B. Kaczer, "The time dependent defect spectroscopy (TDDS) for the characterization of the bias temperature instability," IEEE Int. Reliab. Phys. Symp., pp. 2010

[30] B. Kaczer, T. Grasser, J. Martin-Martinez, E. Simoen, M. Aoulaiche, P. J. Roussel, and G. Groeseneken, "NBTI from the perspective of defect states with widely distributed time scales," IEEE Int. Reliab. Phys. Symp., pp. 55-60, 2009

[31] D. Kaplan, I. Solomon, and N. F. Mott, "Explanation of large spin-dependent recombination effect in semiconductors," Journal De Physique Lettres, pp. L51-L54, 1978

[32] D. J. Lepine, "Spin-dependent recombination on silicon surface," Physical Review B, pp. 436-441, 1972

[33] "E' standard available from wilmad lab glass."

[34] P. M. Lenahan and P. V. Dressendorfer, "Hole traps and trivalent silicon centers in metal-oxide silicon devices," Journal of Applied Physics, pp. 3495-3499, 1984

PBTI Relaxation Dynamics After AC vs. DC Stress in High-k/Metal Gate Stacks

K. Zhao [1], J. H. Stathis [1], A. Kerber [2] and E. Cartier [1]
[1] IBM Research Division, [2] GLOBALFOUNDRIES
T. J. Watson Research Center, 1101 Kitchawan Rd, P.O. BOX 218, Yorktown Heights, NY, 10598, USA
phone: 914-945-1305; fax: 914-945-2141; e-mail: kzhao@us.ibm.com

Abstract

A detailed study on PBTI relaxation after AC and DC stress in high-k nFETs is reported. First, V_t shift during AC and DC stress are examined, showing that the PBTI time evolution depends on the stress mode due to the relaxation effect. Then, comparison of relaxation after different stress types reveals large difference in the relaxation behavior at short times, whereas AC and DC relaxation are observed to merge at longer times. The "time-to-merge" rapidly increases with stress time and it strongly depends on the duty cycle. From a series of "Stress-Relax-Stress" measurement, we also demonstrate that the charge trapping and de-trapping process are highly correlated through "trap level". A simple model from a trap distribution point of view is proposed to rationalize the above observations. These observations provide new insight into the trapping dynamics during PBTI.

Introduction

With the introduction of high-k/metal gate, PBTI has become one of the main reliability issues for advanced nFETs. It is generally accepted that the electron trapping in the bulk of the high-k layer, e.g., at oxygen vacancies, is the root cause for the V_t instability associated with PBTI [1-3]. In previous publications, it was suggested that the charge trapping and de-trapping characteristics during and after stress exhibit widely distributed time scales, which are strongly related to the trap distributions in energy and/or in space [4-7]. In this paper, we report on a systematic study of the PBTI degradation and relaxation subject to AC and DC stress. Measurements of long term relaxation after AC stress and the comparison to the relaxation after DC stress are shown to be a powerful new approach to investigate the trapping dynamics in MG/HK stacks. The results have practical implications for PBTI lifetime estimation for CMOS circuits.

Experiments

The samples comprised nFETs with Hafnium based high-k dielectric layer on a SiO_2 interlayer (IL) with TiN/poly-Si electrodes. EOT of the gate dielectric stacks is ~1.2 nm. Several devices with different interfacial layer process conditions were compared, yielding very similar results to those shown here, indicating a general underlying mechanism. Typical waveforms for DC stress and AC stress with 50% duty cycle are illustrated in Fig. 1. During AC stress, the gate bias is alternating between V_{stress} and zero. The net stress time is defined as $Ts = N \times C \times T$, where N is the total number of the stress cycles, C is duty cycle and T is the period. In all cases, AC stress is compared to DC stress with the same net stress time T_s. To monitor the PBTI relaxation, a fresh device is stressed for T_s, then the drain current I_d is measured at sense bias set approximately equal to the initial V_t. Finally, the V_t shift is calculated from the I_d degradation by comparing the I_d value with a reference I_d-V_g curve measured before

the device was stressed. For both stress and relaxation measurements, the first I_d point is measured ~300 μs after the end of stress.

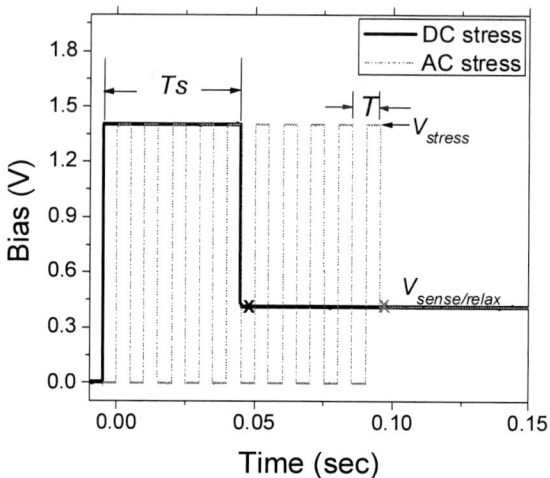

FIGURE 1, TYPICAL WAVEFORMS USED TO STUDY THE PBTI RELAXATION SUBJECT TO AC AND DC STRESS. RELAXATION IS MEASURED ~300 μSEC AFTER THE STRESS IS REMOVED. FOR AC STRESS, THE NET STRESS TIME IS DEFINED AS $T_s = N \times C \times T$.

Results and Discussion

A. Comparison of PBTI: AC stress vs. DC stress

Figure 2 shows the PBTI time evolution of two identically processed devices under DC and AC (100Hz) stress. We minimized the measurement points to reduce the interruption from sense measurement. As can be seen, in both cases, V_t shift follows a power law time dependence. However, at each stress time, AC stress always produces smaller V_t shift comparing to DC stress. It is noticed that the difference (both relative and absolute) of V_t shift between AC stress and DC stress increases over stress time, giving a smaller power law exponent for the AC stress case (n=0.122 for AC stress versus n=0.132 for DC stress). This suggests conventional DC stress measurement will underestimate the projected lifetime as devices are subject to switching in a circuit. The duty cycle dependence of the V_t-shift is shown in Fig 3. For these AC data, the stress time per cycle was kept constant at $C \times T = 5m\sec$. Above observations illustrate that PBTI can be largely affected by the stress mode (DC, AC, frequency, duty cycle...) and the root cause for this is the fact that for PBTI, relaxation occurs immediately after the removal of stress bias [8]. A clear understanding of the relaxation dynamics is essential for

978-1-4244-5430-3/10 $26.00 © 2010 IEEE

developing robust models to predict PBTI behavior subject to complex stress conditions.

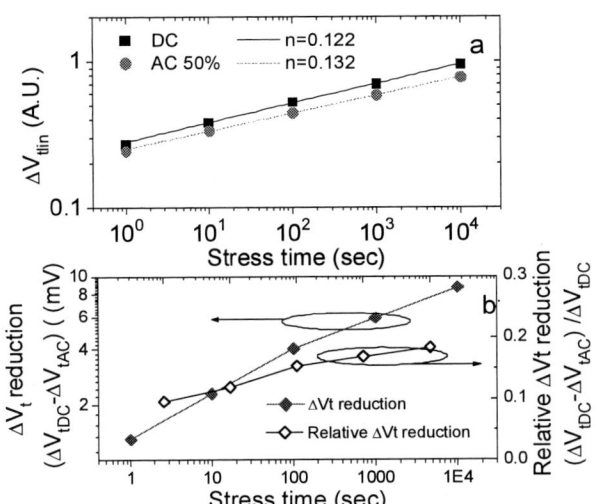

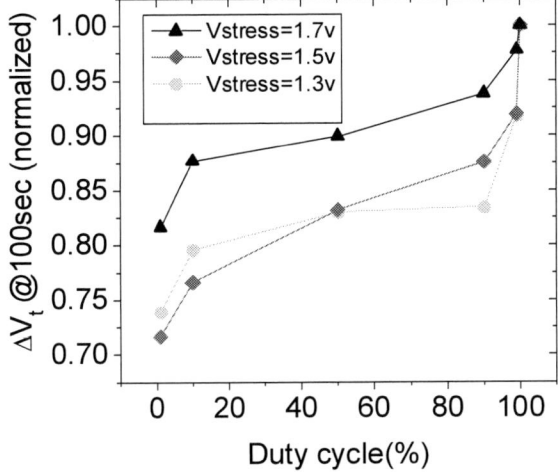

FIGURE 2, COMPARISON OF V_t SHIFTS DURING AC AND DC STRESS. A) V_t SHIFT SUBJECT TO AC STRESS SHOWS SHALLOWER SLOPE. B) BOTH THE ABSOLUTE AND THE RELATIVE DIFFERENCE BETWEEN AC AND DC STRESS GROWS OVER STRESS TIME.

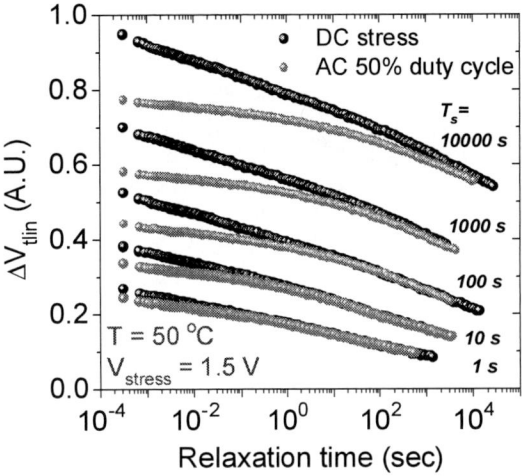

FIGURE 3, DEPENDENCE OF THE V_t SHIFT ON THE AC STRESS DUTY CYCLE AT DIFFERENT STRESS VOLTAGES. NOTICE THE ABSOLUTE V_t SHIFT INCREASES WITH STRESS VOLTAGE. THE OVERALL TREND WITH DUTY CYCLE REMAINS THE SAME.

B. Comparison of PBTI relaxation: AC vs. DC

In Figs. 4-8 the V_t-relaxations after AC stress and DC stress are compared in detail. Fig. 4 compares the relaxation traces measured for identically processed nFETs after being stressed for different stress times Ts, under AC (100Hz) and DC stress. V_{stress} and V_{relax} were kept the same for all measurements. As can be seen, for the same stress time T_s, the first V_t-shift value is smaller for AC stress case. More interestingly, it is noticed that right after the removal of

stress, for AC stress case, the relaxation curves are much shallower than those after DC stress. However, at longer relaxation times, the relaxations after the two stress types gradually merge together with remarkable consistence. This observation is the key finding that motivates our investigation in this work. It clearly indicates that for the same total stress time Ts, AC stress results in less trapped charges than DC stress. The traps responsible for this difference are mostly shallow traps with short characteristic emission times. Another implication to the practical PBTI measurement is that, the initial relaxation is much reduced (shallower slope) under AC stress, thus the need for ultra-fast measurements is less stringent in the AC case. Similar observation can also be found in Fig. 5, where relaxation traces measured after AC stress with different duty cycles show difference at beginning but merge to each other at long relaxation time.

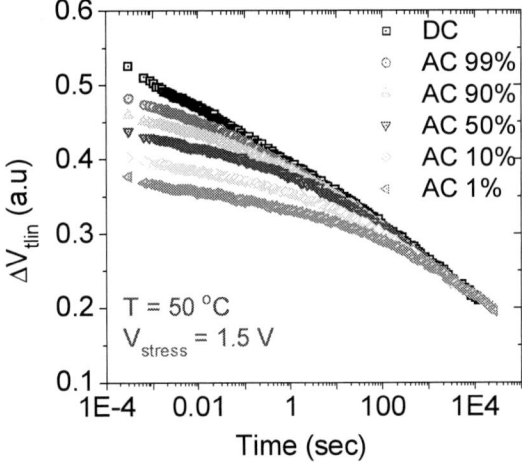

FIGURE 4, RELAXATION TRACES MEASURED AFTER AC AND DC STRESS WITH STRESS TIME CHANGING FROM 1SEC TO 10000SEC. NOTICE THAT THE AC RELAXATIONS ALWAYS START WITH LOWER V_t SHIFT AND SHALLOWER SLOPE, BUT THEY GRADUALLY MERGES WITH THE CORRESPONDING DC RELAXATION AT LONGER RELAXATION TIMES.

FIGURE 5, COMPARISON OF RELAXATION TRACES AFTER DC STRESS AND AC STRESS WITH DIFFERENT DUTY CYCLES. ALL THE RELAXATIONS MERGE INTO EACH OTHER AT LONGER RELAXATION TIMES.

To quantify the comparison, we take the ratio between AC relaxation and DC relaxation as shown in Fig. 6(a). As the two relaxation curves gradually merge into each other, the ratio approaches 1. We define "time to merge" as the time when this ratio is equal to 98%. In Fig. 6(b), it shows that "time to merge" strongly depends on the total stress time. The longer the stress time is, the longer it takes for the AC relaxation to merge with the DC relaxation, indicating more traps at deeper levels are affected at longer stress times. Similarly in Fig. 7, it shows that the "time to merge" increases when duty cycle becomes smaller, indicating for the AC stress with smaller duty cycle (longer relaxation time for each AC stress cycle), traps at deeper levels can be de-trapped during the relaxation period of each AC stress cycle. These observations lead to a physical picture that the relaxation dynamics for DC and AC cases are governed by the distribution of the trap occupancy.

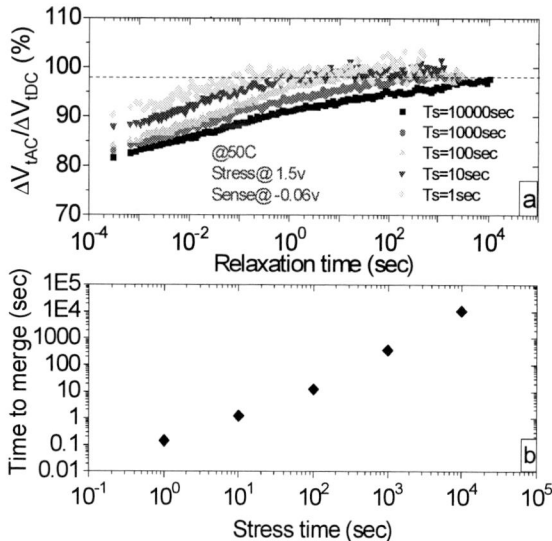

FIGURE 6, A) ILLUSTRATES THE PROCEDURE USED TO QUANTIFY WHEN THE AC RELAXATION MERGES WITH THE DC RELAXATION. B) SHOWS STRONG DEPENDENCE OF "TIME TO MERGE" ON THE TOTAL STRESS TIME Ts.

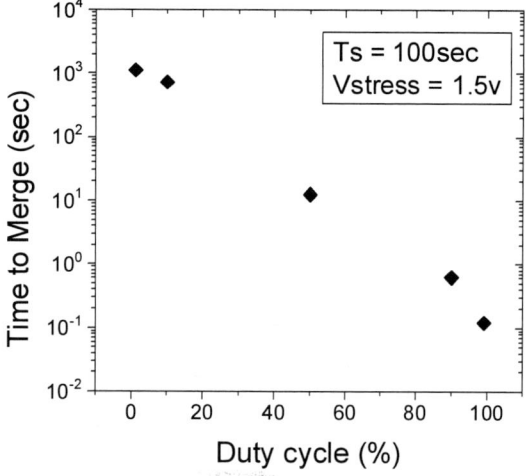

FIGURE 7, DEPENDENCE OF "TIME TO MERGE" ON THE AC STRESS DUTY CYCLE. Ts. SMALLER DUTY CYCLE RESULTS IN LONGER "TIME TO MERGE".

C. "Stress-Relax-Stress" measurement

To further demonstrate how the trapping/de-trapping events occur during AC stress, we conduct a series of experiments, where we monitor V_t-shift during a "Stress-Relax-Stress" sequence, which in some way can represent a single cycle of AC stress.

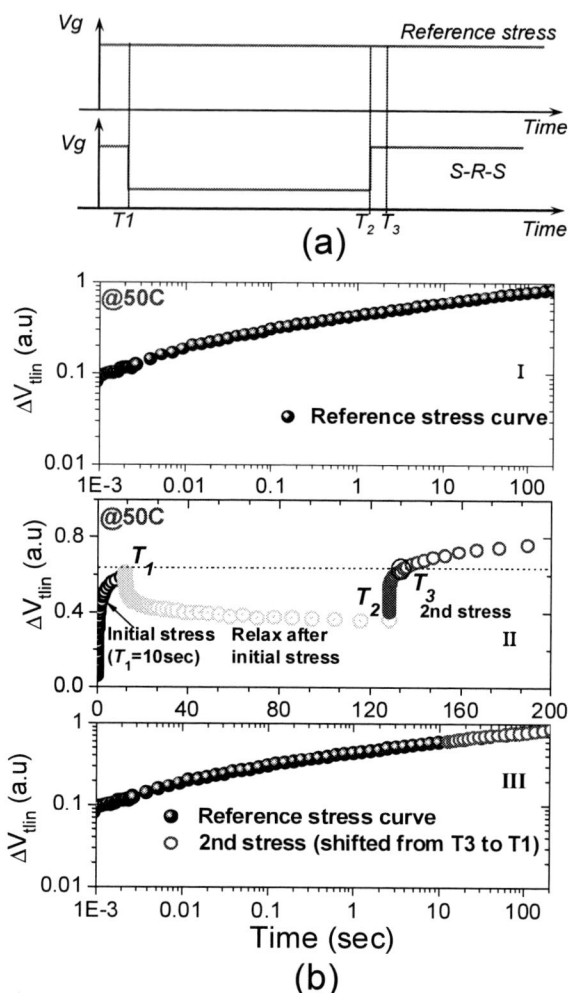

FIGURE 8, A) SCHEMATIC DRAWING OF THE TYPICAL WAVEFORMS USED IN THE "S-R-S" EXPERIMENT. B) PLOT I: MEASURED REFERENCE VT SHIFT CURVE. PLOT II: VT SHIFT MEASURED DURING A "STRESS-RELAX-STRESS" SEQUENCE. PLOT III: COMPARISON OF THE 2ND STRESS CURVE (THE PART AFTER T3) WITH THE REFERENCE VT SHIFT CURVE.

Two identically processed devices were measured in this experiment. For the first device, a reference V_t-shift curve was measured under a constant stress bias V_{stress}. Then, for the second device, V_t-shift was monitored while the device was first stressed for T1 sec at V_{stress}, then relaxed to T2 and then re-stressed at Vstress again. Typical waveforms used in the experiment are schematically shown in Fig. 8 (a). In Fig. 8 (b), the measured reference V_t-shift curve is shown as plot I and the V_t-shift measured from the second device is shown as plot II. As can be seen, for the second device, during T1, the V_t-shift evolves the same way as the reference curve. During the relaxation phase, the V_t-shift decreases and reaches to a certain value at T2. Then, during the 2nd stress phase, V_t-shift first grows back at

978-1-4244-5430-3/10 $26.00 © 2010 IEEE   52

T3 to the original value before relaxation, i.e., $\Delta V_{tlin}(T_3) = \Delta V_{tlin}(T_1)$ and then continues to grow. In plot III, we shift the curve in plot II from T3 to T1 and compare the resulting curve (the part after T3) with the reference curve. As can be seen, the 2nd stress curve (the part after T3) follows a very similar path as the reference curve (the part after T1), indicating the traps released during the relaxation phase (from T1 to T2) were most refilled during the 2nd stress, resulting similar trap occupancy at T3 as that at T1. *This further implies that traps that are first released during a relaxation phase have higher probability to be refilled during the subsequent stress phase.*

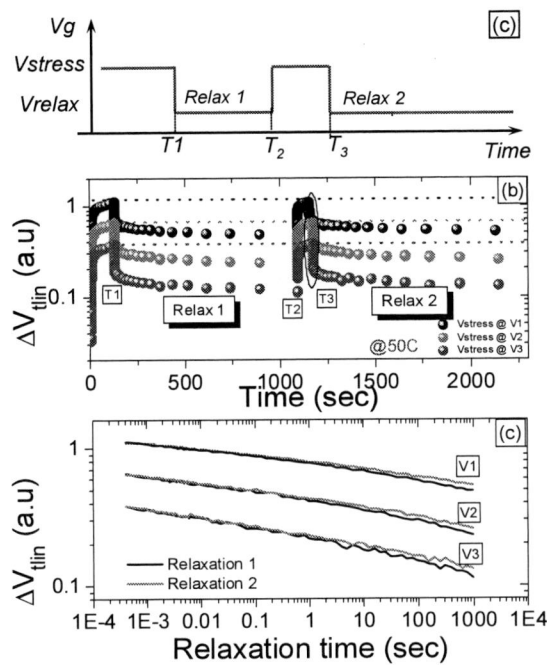

FIGURE 9, A) SCHEMATIC DRAWING OF THE WAVE-FORMS USED IN THE "S-R-S-R" EXPERIMENT. B) MEASURED V_T-SHIFT CURVES FOR THREE DEVICES UNDER DIFFERENT STRESS BIASES. NOTICE THAT V_T-SHIFT AT T3 IS THE SAME AS THAT AT T1. C) COMPARISON OF V_T- SHIFT RELAXATION AT T1 AND T3, SHOWING SLIGHTLY BUT CONSISTENTLY SHALLOWER SLOPE FOR "RELAXATION 2".

However, the trap occupancy at T3 can not be identical to that at T1 since that would be contradictory to the observation in which at beginning, AC relaxation always shows shallower slope than DC relaxation. To identify the subtle difference in trap occupancy between T1 and T3, we conduct the experiment shown in Fig. 9. The experimental waveform is schematically shown in Fig. 9(a). In this experiment, three devices were tested at different stress conditions. Each device first went through a Stress-Relax-Stress sequence. Here, T3 was chosen so that the V_t-shift was restored back to its value at T1, i.e., $\Delta V_{tlin}(T_3) = \Delta V_{tlin}(T_1)$. Then, once V_t-shift was restored at T3, the devices were immediately relaxed again. Fig. 9(b) shows the V_t-shift measured during S-R-S-R sequence at three different stress conditions. Notice that for each device, V_t-shift level at T3 was the same as that at T1. Fig. 9(c) shows the comparison between "relaxation 1" and "relaxation 2". It is found that for all three devices tested, "relaxation 2" is quite similar as "relaxation 1". However, at longer relaxation time, "relaxation 2" deviates from "relaxation 1" and shows a slightly but consistently shallower slope. This indicates that at T3,

although V_t-shift value is the same as that at T1, the trap occupancy distribution, however, is indeed altered, resulting less shallow traps and more deep traps.

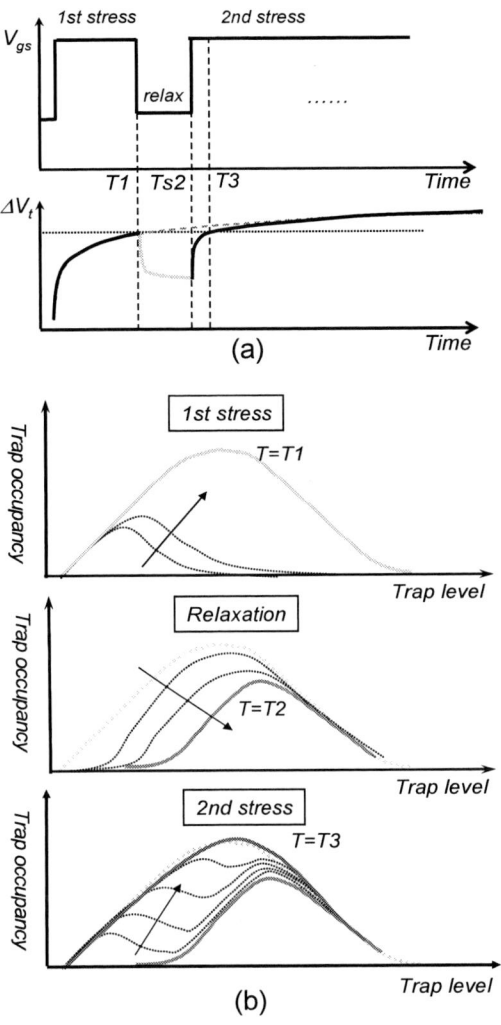

FIGURE 10, A) SCHEMATIC DRAWING OF THE TYPICAL WAVEFORMS AND THE TYPICAL V_T- SHIFT MEASURED IN THE "S-R-S" EXPERIMENT. B) SCHEMATIC ILLUSTRATION OF THE TRAP OCCUPANCY EVOLUTION DURING TRAPPING (STRESS) AND DE-TRAPPING (RELAXATION). NOTICE THAT AT THE END OF THE S-R-S SEQUENCE, TRAP OCCUPANCY IS SLIGHTLY ALTERED, RESULTING LESS SHALLOW TRAPS AND MORE DEEP TRAPS.

From the above observations, the following can be concluded:

(a) Giving the same Vt shift value at T1 and T3, i.e. , $\Delta V_{tlin}(T_3) = \Delta V_{tlin}(T_1)$, the total number of occupied traps at T3 is equal to that at T1.

(b) Similar stress and relaxation dynamics indicate the traps occupancy distributions at T1 and T3 are similar.

(c) Subtle difference in the relaxation dynamics (slightly shallower slope) suggests the trap occupancy distribution is although similar but indeed altered during the first relaxation and the re-stress process, resulting less shallow traps and more deep traps.

In Fig. 10, a physical picture is schematically drawn from the above observations and summaries the key conclusions.

D. "Shallow Trap Modulation Model" for AC stress

All the above experimental observations clearly suggest that the PBTI stress and relaxation dynamics are strongly correlated and the physical quantity that links the two processes is the "trap level", which strongly determines the characteristic charge trapping/de-trapping time during stress and relaxation. With this understanding, we can rationalize the AC relaxation with a simple model which involves the modulation of the trap occupancy during AC stress, as illustrated in Fig. 11.

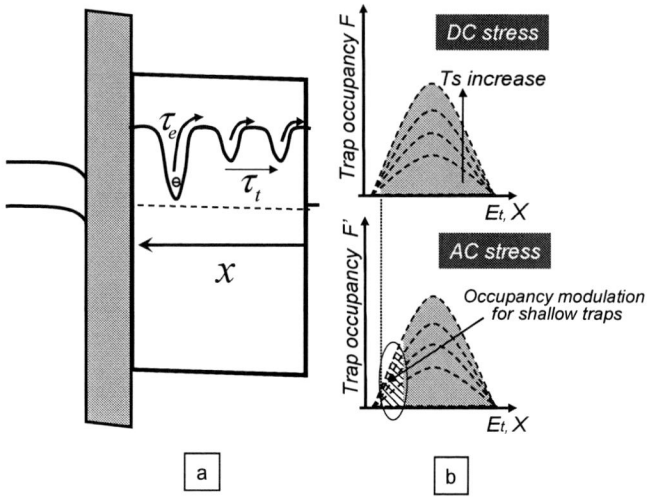

FIGURE 11, a) SCHEMATIC ILLUSTRATION OF THE DE-TRAPPING PROCESS IN HIGH-K DIELECTRIC LAYER. b) SCHEMATIC COMPARISON OF THE TRAP FILLING PROCESS SUBJECT TO DC AND AC STRESS. NOTICE THAT FOR AC STRESS CASE, OCCUPANCY MODULA-TION AFFECTS MOSTLY THE SHALLOW TRAPS.

In this model, we assume the electron de-trapping time $\tau = \tau_e + \tau_t$, where τ_e is the characteristic time for an electron to be emitted out of a trap state and τ_t is the time for the emitted electron to transport from the trap location x to the contact. The emission time, τ_e, is mostly dependent on the trap energy, $\tau_e \propto e^{E/kT}$ and the transport time, τ_t, is dependent on the trap location, $\tau_t \propto e^{x/\lambda}$. If this applies, the energetically shallower traps as well as those located close to the contacts tend to be de-trapped faster. Note that it is not determined at this point which process is the dominant one. Thus, we include both trap state energy as well as the trap location in the model. During DC stress, filling of trap states follows a distribution function $F(E, x, T_s)$. While during AC stress, filling of trap states follows a different distribution function $F'(E, x, Ts)$. For the same amount of stress time, Ts, the difference between F and F' is caused by the additional de-trapping events occurring at the short relaxation intervals during AC stress. Since the relaxation interval is rather short during AC stress (~ms), it mostly modulate the occupancy of energetically shallow traps and traps close to the contacts, while the occupancy of deep

and/or distant traps is hardly affected. In other words, for each stress time, AC stress populates less shallow traps, while deep states are expected to have a similar occupancy for both stress conditions. Therefore, at the beginning, AC relaxation starts at a lower V_t shift value and it shows a smaller initial relaxation rate. Then, as relaxation proceeds, when deeper traps become more and more dominant, the AC relaxation merges with the DC relaxation. Similarly, this can also explain the increase in the "time-to-merge" for longer stress times and also for smaller duty cycles.

CONCLUSIONS

The PBTI relaxation after AC is shown to differ strongly from the relaxation after DC stress at short times, while the relaxation becomes independent of the stress type beyond a critical "time-to-merge". Experimental evidence shows that the trapping and de-trapping process are strongly correlated and the electron trapping/de-trapping characteristic times are strongly tied to the depth of the trap state. During AC stress, occupancy of shallow traps is strongly modulated by the short relaxation intervals, while the deeper traps are hardly affected.

ACKNOWLEDGMENT

Author thanks Dr. Sufi Zafar, Dr. Miaomiao Wang from IBM and Dr. Frederic Monsieur from ST Microelectronics for their insightful discussions. This work was performed by the Research Alliance Teams at various IBM Research and Development Facilities.

REFERENCES

[1]   A. Kerber, et al, G. H. Author, "Origin of the threshold voltage instability in SiO2/HfO2 dual layer gate dielectrics" IEEE Electron Device Letters, 2003, pp. 87-89.

[2]   J. L. Gavartin, A. L. Shluger, "Modeling HfO₂/SiO₂/Si interface " Applied Physics Letters, 2007, pp. 2412-2415.

[3]   E. Cartier, B. P. Linder, V. Narayanan, V. K. Paruchuri, "Fundamental understanding and optimization of PBTI in nFETs with SiO2/HfO2 gate stack," IEEE International Electron Device Meeting, 2006, pp. 1-4.

[4]   R. Biswas, Y. P. Li, B. C. Pan, "Enhanced stability of deuterium in silicon" Applied Physics Letters, 1998, 72, pp. 3500-3503.

[5]   B. Kaczer, T. Grasser, P. J. Roussel, J. M. Martinez, R. O. Connor, B. J. O'Sullivan, G. Groeseneken, "Ubiquitous relaxation in BTI stressing – new evaluation and insights" IRPS, 2008, pp. 20-27.

[6]   T. Aichinger, M. Nelhiebel, T. Grasser, "Unambiguous identification of the NBTI recovery mechanism using ultra-fast temperature changes," IRPS, 2009, pp. 2-7.

[7]   D. Ielmini, M. Manigrasso, F. Gattel, G. Valentini, "A unified model for permanent and recoverable NBTI based on hole trapping and structure relaxation," IRPS, 2009, pp. 26-32.

[8]   S. Ramey, C. Prasad, M. Agostinelli, S. Pae, S. Walstra, S. Gupta, J. Hicks, "Frequency and recovery effects in High-k BTI degradation" IRPS, 2009, pp. 1023-1027.

978-1-4244-5430-3/10 $26.00 © 2010 IEEE

Off State Incorporation into the 3 energy mode Device Lifetime Modeling for advanced 40nm CMOS node

A. Bravaix, C. Guérin[*], D. Goguenheim

ISEN-IM2NP, UMR CNRS 6242, Maison des Technologies, place G. Pompidou, 83000 Toulon, France
phone: (+33) 4.94.03.89.92, alain.bravaix@isen.fr

V. Huard, D. Roy, C. Besset, S. Renard, Y. Mamy Randriamihaja, E. Vincent

STMicroelectronics, Crolles 2 alliance, 850 rue Jean Monnet, 38926 Crolles, France
* now at Roth and Rau Switzerland AG, 23 rue de la Maladière, 2000 Neuchâtel, Switzerland
phone: (+33) 43.892.2907, vincent.huard@st.com, phone (+41) 32.718.3394, chloe.guerin@roth-rau.ch

Abstract: Hot-Carrier degradation is analyzed with 3 mode lifetime modeling extended to the cases of PMOSFETs and Off state modes in last CMOS nodes. Damage worsens in subthreshold region with positive temperature activation due to interface traps generation in the gate-drain overlap (GDO) and localized charge trapping into the spacer oxide. Care has been done on the distinct impact of the measuring bias and stressing conditions in Sub-V_T regime. The latter can be much more degraded than On-state parameters showing the amphoteric nature of Si-H bonds breaking rates throughout the channel-GDO. Off-mode damage has been included in the 3 mode energy device lifetime giving a useful modeling for any AC waveforms suitable for digital to analog operations.

Keywords: Off Mode, Gate-Induced Drain Leakage, Cold Carriers, Band to Band Tunneling, Hot Carriers, Interface traps, Oxide traps, Multi Vibrational Excitation, High Temperature.

1. INTRODUCTION

The demand for Low-Power (LP) circuits in actual IC industry has driven the market for portable electronic devices and mobile multimedia by the continuous scaling of device geometry and strong limitation of power consumption [1-3]. The mastering of standby (active) power consumption is straightly dependent on leakage currents which may have distinct origins during Off (On) state operation in MOSFET structure for digital [4] and analog [5] as well as memory application during idle state in SRAM cell [6, 7] and data retention time in DRAM cell [8]. The need for highly reliable circuits with strong performance requirements has concentrated more attention during last decades due to the importance of Negative-Bias Temperature Instability (NBTI) [9-11] and Channel Hot-Carriers (CHC) [12-14]. However, these mechanisms may interact causing enhanced or slow down of the overall damage process when bias is reduced or put to 0V. As a consequence, the examinations of standby mode, Idle state (sleepy-awaked) and Off mode take on great interests not only facing power management [3-8, 15] but also for device reliability issues [16-21] regarding resilience phases and Non-Conducting Hot-Carrier (NCHC) reliability.

CHC degradation has recently led to a renewed interest [16, 17] as it still represents a significant proportion of device to circuit aging effects [18] in contrast to what was expected with the power supply voltage (V_{DD}) reduction in most advanced CMOS nodes. The main defect related to MOSFET originates from the depassivation of $Si_3 \equiv SiH$ leaving silicon Dangling Bonds (DB) ($Si_3 \equiv Si\bullet$) as Pb_o, Pb_1, centers [22]. The growth/recovery of this defect generation (N_{it}) has been successfully described by the Reaction-Diffusion (RD) model for NBTI [23, 9], consistent with time power-law dependence (slope n) based on the forward/back diffusion of H-species in the bulk of the oxide. Signature of the defects are attributed to the distinct n values of the generation rate and used to distinguish NBTI and CHC damage [24] where the contribution of broken $\equiv Si-O$ bonds (n = 0.7-0.8) is considered in addition to broken $\equiv Si-H$ bonds. It has been further shown that broken $\equiv Si-O$ bonds are involved in the degradation of Drain-Extended NMOS (DeNMOS) in the Off-mode [25], leading to a universal dependence.

This paper extends the CHC to Cold Carrier (CCC) damage modeling related to the permanent N_{it} based on a 3 mode energy-driven [14, 16] to the cases of Off mode characterized by the subthreshold regime ($V_{GS} < V_T$), zero volt operating ($V_{GS} = 0$) condition and to NCHC, *e.g.* when $V_{GS} < 0$ in NMOSFET ($V_{GS} > 0$ in PMOS). The first section studies the Off mode specificity in 40nm CMOS node as a function of stressing V_{GS} and V_{DS}. Care has been done to distinguish the measuring bias conditions (VG_m, VD_m) from the impacts of On/Off stressing damage in both N- and P- MOSFETs. Section II extends our general CHC to CCC modeling to the permanent degradation in PMOSFETs (section II.A). Section II.B incorporates the distinct Off-mode degradation to the former CHC-CCC modeling. This is distinguished in two modes, *i.e.* under subthreshold regime ($0 < V_{GS} < V_T$) where the carrier energy acquisition dominates the mechanism until the depletion mode ($V_{GS} \leq 0$), where diode configuration leads to unique $I_{DS} = -I_{SUB}$ dependence. This enables to determine the overall device lifetime evaluation for all voltage conditions, consistent with digital and analog functioning.

978-1-4244-5430-3/10 $26.00 © 2010 IEEE

I. OFF MODE DEGRADATION

Transistor degradation in Off-mode was historically investigated in previous works by studying the CHC effects in long-channel CMOS nodes (V_{DD}=5V-3.3V) for digital application in NMOSFETs [26, 27] and PMOSFETs [28, 29]. This mode needed careful analysis due to the detrimental effects of Gate-Induced Drain Leakage (GIDL) currents [30-34] focusing on the charge loss issues for memory applications as EEPROM, refresh time in DRAM and power consumption. GIDL currents have been first explained by Band-to-Band Tunneling (BBT) which occurs between the surface channel and the drain extension till the Gate-Drain Overlap (GDO) region. GIDL is consequently observed in saturation mode (VD_m= V_{DD}) in **Fig.1a** with a net increase in ID_{BBT} related to the high field in the GDO length with $V_{DG} = V_D - V_G$ through a small L_G dependence. BBT was described by a one dimensional (1D) vertical electric field dependence [30, 31] which has been improved by a quasi 2D model including the indirect phonon-assisted tunneling [32] to account for the lateral electric field (E_{Lat}) and doping profile variation for different GDO structures with Lightly-Doped Drain (LDD). This is justified in short channel devices as doping increases both in the channel as in the drain from LDD to Medium Doped Drain (MDD) levels leading to a larger influence of E_{Lat} and vertical fields F_{ox} on BBT respectively. Then, the drain architecture has a net effect on GIDL current and the spread of the defects region [32-35] which may extends from the spacer till the GDO. Some authors further analyzed GIDL by the effect of Band-Trap-Band mechanism [35] where doubled doped drain FET exhibits a larger extension of ΔN_{it}. As will be shown in section I.B, GIDL issue is strongly dependent on V_{DG} condition whether CHC *vs.* CCC population is involved or not in the degradation process.

A. Off mode characteristics in ultra-thin gate-oxide

The main difference between thick and thin gate-oxide is the larger magnitude in Off-currents (I_{Off}) coming from leakage currents in ultra-thin T_{ox} which may have distinct origins. I_{Off} importance (**Fig.1**) can be attributed partly to the ability to switch to Off-state from V_{TS} (in saturation) by the subthreshold slope (S) effect which is contained in the ID_{SubVT} curve, the Drain-Induced Barrier Lowering effect (DIBL) which reduces $V_{TS}(V_{DD})$ and by the GIDL component (I_{BBT}). In MOSFETs with T_{ox}= 1.7nm, the direct tunneling gate current (IG_{DT}) occurs with increasing V_{GS} (and V_{DG}) (**Fig.1**) similarly to the drain-bulk junction leakage (I_{DBJn}) with V_{DB}. Therefore, the main I_{Off} contributions can be distinguished by the different components of leakage currents with:

$$I_{Off} = ID_{SubVT} + I_{GIDL} + I_{SUB} + IG_{DT} \quad (1)$$

Hence, turning to depletion mode in NMOSFETs ($VG_m < 0$) BBT component is the dominant term with $ID_{BBT} \approx I_{SUB}$ as IG_{DT} remains very small, *e.g.* at 6% of ID_{BBT} at VG_m= -0.55V. Increasing E_{Lat} under CHC mode, we have at high V_{DG}:

$$ID_{BBT} = I_{SUB} + IG_{DT} = IB_{II} + I_{DBJn} + IG_{DT} \quad (2)$$

where IB_{II} represents the Impact Ionization (II) current induced first by energetic carriers injected (vertically) from the gate by direct tunneling $IG_{DT} = IG_{EVB}$ in the GDO region, and second, by ID_{BBT} component where electrons are swept (laterally) by the drain field in the same region. However the IB_{II} contribution from the gate IG_{EVB} **Fig.1** is unlikely to be dominant in the GIDL current (as IG_{HVB} from the drain) as the

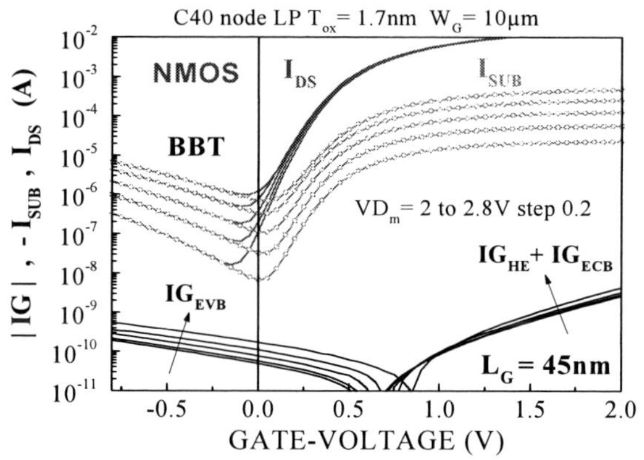

FIG. 1: I_{DS}, I_{SUB} and I_G measurements as a function of VG_m in saturation mode in core NMOSFETs under CHC voltage conditions with VD_m= 2 to 2.8V in L_G= 45nm NMOSFET.

ratio IG_{DT}/I_{SUB} at VG_m= -0.8V is 5.10^{-4} and 8.10^{-5} for VD_m= 2 and 2.8V respectively. In contrast the gate current in On-state becomes much larger in strong saturation as it is composed of the thermionic injected current and the ECB gate tunneling current [36]. The IG curves are relatively independent of VD_m in saturation ($VG_m > 0$) indicating that CHC population is overwhelmed by the ECB tunneling component [37, 16]. As observed in **Fig.1b**, the ID_{BBT} increases with V_{DG} (so with negative VG_m) under CHC bias conditions more markedly than IG_{EVB}. However in our case $Fs \cong (V_{DG} -1.12)/3T_{ox}$ = 2.5-2.8 MV/cm [30] which is already a quite high electric field. At high field **Fig.1b** confirms that BBT becomes the dominant contribution in GIDL current in agreement with results obtained in longer CMOS nodes [30-32] and in high-voltage DeNMOS [25] which characterizes the NCHC regime.

Off-mode characteristics in PMOSFET show **Fig.2** smaller On/Off I_{SUB} and I_G magnitudes in comparison to NMOS which arises from the reduced E_{Lat}. The Off-mode injection mechanisms for holes in PMOS are related to IG_{HVB} coming from the P+ poly Si gate in the GDO region, the ID_{BBT} arising laterally from the channel holes which may trigger the II phenomenon deep into the drain. In order to account for the slightly smaller doping profile in PMOS devices, we have characterized the Off-mode at larger $|VD_m|$ condition (0.4V) which allows compensating the smaller II rate and lateral field condition than in NMOS. These VD_m conditions make clearly appear **Fig.2** twice the CHC contribution in gate-current in On-state; *i.e.* by the Hot Electron (HE) injections observed in IG_e at low $|VG_m|$, and hole injection in the IG_h portion at high

978-1-4244-5430-3/10 $26.00 © 2010 IEEE

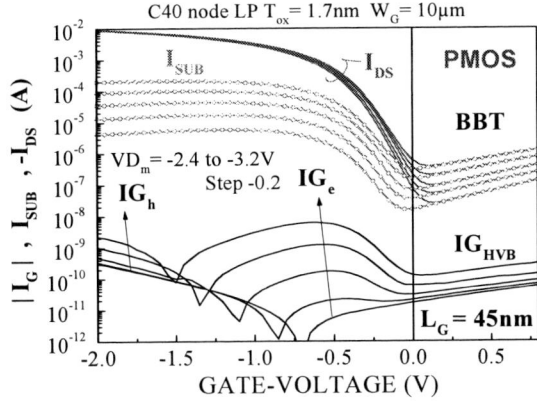

FIG. 2: I_{DS}, I_{SUB} and I_G measurements in saturation mode as a function of VG_m in core PMOSFETs under CHC mode with VD_m= -2.4 to -3.2V (L_G= 45nm) showing IG_e, IG_h components.

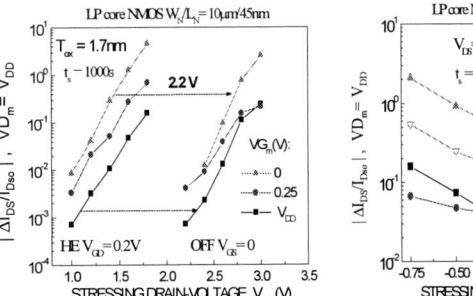

FIG. 3: $|\Delta I_{DS}/I_{Dso}|$ measured in saturation mode FWD at t_s= 1000s (a) Comparison of the V_{DS} dependence between On-mode HE stressing (V_{GD}= 0.2V) with Off-mode (V_{GS}= 0) in L_G= 45nm for various sensing VG_m from V_{DD}, Sub-VT to Off-mode. (b) V_{GS} dependence (V_{DS}= 2.8V) with same bias under sub-V_T measurements in NMOSFETs L_G = 40nm (T_{ox}= 1.7nm). I_{SUB} − V_{GS} dependence is added to the right plot.

$|VG_m|$, both curves in correlation to the maximum F_{ox} (E_{Lat}) variation This points out that large V_{DS} put the regime close to drain avalanche which may generate Hot Hole (HH) injections with negative V_{GS} that may take larger efficiency in the damage process [37]. Therefore, the equality (2) remains valid for PMOSFET (**Fig.2**) in the range $0.2V < V_{GS} \leq 0.8V$, where the increase in GIDL current satisfies $ID_{BBT} = I_{SUB} + IG_{DT}$.

B. CHC to Off mode damage and NCHC in NMOSFETs

One important point of the previous subsection is that the acceleration voltage for Off-mode damage requires much higher stressing fields than under CHC damage which is now related to high V_{GS} conditions [14, 16]. This is observed in core NMOSFETs **Fig.3a** where a difference of 2.2V is necessary to obtain similar degradation in saturated I_{DS}. The stressing V_{GS} dependence in Off-mode **Fig.3b** follows roughly the channel current dependence (**Fig.3a**) whereas the degradation is minimum at VG_m = -0.25V and increases with the substrate current in On-mode and Subthreshold regime. The stressing V_{GS} dependence in Off-mode **Fig.3b** follows roughly the channel current dependence (**Fig.1a**) whereas the degradation is minimum at VG_m = -0.25V and increases with the substrate current in On-mode and Sub-V_T regime. Off-mode at VG_m= 0 exhibit a turnaround that needs to be clarified with the help of sub-threshold characteristics. **Figs.4a,b** show DC stressing in NMOSFETs for different V_{GS} corresponding to to Sub-V_T regime (V_{GS}= 0.25V), Off mode (0V) and NCHC (-0.25V). Comparison is done with On mode stressing under worst-case HE condition (V_{GS}/V_{DS}= 1.8/2 V). In contrast to HE condition (**Fig.3a**), V_{DG} needs to be large enough to observe degradation due to the smaller amount of CCC in weak inversion (Sub-V_T) and tunneling carriers from the junction ($V_{GS} \leq 0$). Turning from NCHC till Sub-V_T stressing, a pivot point is observed in the exponential I_{DS} - V_{GS} curves due to the combination of ΔV_T, slope reduction (1/S) defined by S = $(dLog(I_{DS})/dVG_m)^{-1}$= 2.302 $(kT/e)(dVG_m/d\psi_s)$, and GIDL rise with V_{GS}. This first shows that S variation exhibits

a pivot between the charging of donor type N_{it} to acceptor type when the device is biased from depletion to weak inversion where minority carriers (electrons) from the channel equals the hole concentration at the surface (n_s=n_i), *i.e.* when $VG_m(\psi_s) \approx \phi_p$, with ϕ_p the Fermi potential $e\phi_p$=E_i - E_{Fp}. Therefore the pivot is related to the discharge of interface traps. Because Off-mode stressing involves tunneling carriers localized at the drain this evidences amphoteric defects (Pb_o centers) which are positively charged by holes under BBT mode in depletion and negatively charged when $VG_m(\psi_s) > \phi_p$ according to the change of charge state with ψ_s.

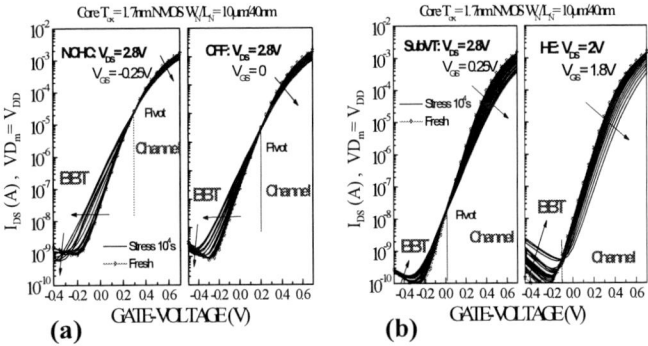

FIG. 4: Sub-threshold characteristics in NMOSFETs (a) stressed under NCHC (V_{GS}=-0.25V) and Off mode (0V); (b) under Sub-V_T mode (left) till worst-case On-mode HE (right) condition (V_{GS}/V_{DS}= 1.8V/2V).

When On-mode stressing is performed (**Fig.4b**), GIDL increase becomes larger in Off-mode as a much larger ΔN_{it} is obtained (negatively charged) which extends from GDO region towards the channel. This is known as giving a cumulative effect from the large ΔV_T and mobility reduction above the channel [38]. Therefore, Off-mode damage shows very different magnitude of degradation depending on the bias measurement (VG_m) in **Fig.4a,b** and stressing voltage (V_{GS}, V_{DS}) due to the three regions in series between spacer, GDO and channel. The change in pivot point location with V_{GS}, VG_m (**Fig.4a,b**) explains why a turnaround may happen in damage effect **Fig.3b** between stressing bias $V_{GS}=0$ and weak inversion ($V_{GS}=0.25V$) in the sensing range $VG_m=0$ to $0.25V$.

In ultra-thin T_{ox} BBT electron damage leads to a larger effect **Figs.5a,b** which increases from weak inversion (Sub-V_T) to depletion mode in contrast to thicker T_{ox} (3.1nm) which exhibits smaller impact in Off-stressing damage. In both stressed NMOS nodes we emphasize that the worst-case Off-mode damage corresponds to On-stressing case carried out under HE damage (**Fig.4b**) according to the larger degraded channel to GDO region which induces a net (1/S) reduction enlarged with ΔV_T in agreement with the high density of ΔN_{it}. Because electron trapping is unlikely above the channel in NMOS device with $T_{ox}=1.7nm$ [16], the highly localized donor-type ΔN_{it} above the MDD region increases the GIDL current from depletion to $VG_m(\psi_s)=\phi_p$. Moreover, the strongest degradation effect is seen in deep depletion under HE stress (**Figs.5a,b**) as the GDO damage is maximum until the sidewall oxide spacer till weak inversion measurements ($VG_m=0.25V$). However, localized electron trapping in the GDO region, *i.e.*

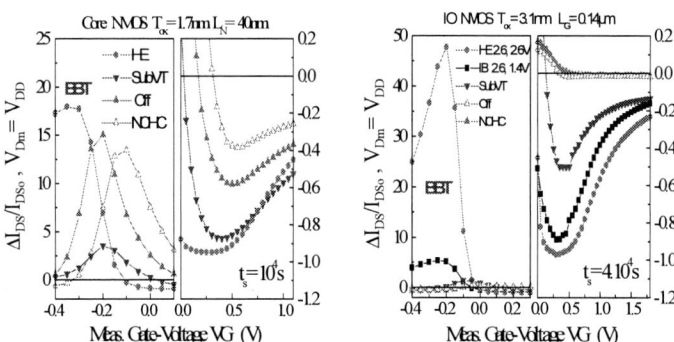

FIG. 5: I_{DS} - VG_m degradation measured in saturation mode forward ($VD_m=V_{DD}$) in NMOSFETs between On-mode (HE, IB cases), Sub-VT ($V_{GS}=0.25V$), Off-mode ($V_{GS}-0$) and NCHC ($V_{GS}=-0.25V$) **(a)** in core $L_G=40nm$ with HE stress at $V_{DS}/V_{GS}=2/1.8$ V and $V_{DS}=2.8V$ for the other Off-state stress conditions **(b)** in IO $L_G=0.14\mu m$ with HE, IB conditions ($V_{DS}=2.6V$) and $V_{DS}=3.4V$ for Sub-VT, Off and NCHC modes.

out of the channel and into the thick spacer oxide, might lead to a similar I_{DS} reduction in On-mode [38] which is typically induced by HE stress. In that case we would not expect to observe a current increase from the pivot to weak depletion (**Fig.4a**), which reduces GIDL current in strong depletion for $V_{GS}=0$ and $-0.25V$ stresses suggesting possible hole trapping into the spacer as found in previous technologies [39].

The comparison to IO NMOSFETs with thicker T_{ox} **Fig.5b** gives arguments for a larger impact of trapped electrons in the GDO region [33] which is observed (**Fig.4b,5b**) by a larger $\Delta V_T>0$ in addition to S variation under HE stress. NCHC stressing (**Fig.5b**) shows a strongly reduced damage both in depletion and above the channel ($VG_m > V_T$). The weak inversion sensing bias shows the larger impact of defects on I_{DS} (**Figs.5a,b** right) independently of the type of On/Off stressing which is directly linked to the high electric field in saturation ($VD_m = V_{DD}$) typically encountered in analog applications [5]. Regarding the importance of Off/On mode

degradation for device lifetime evaluation (section II.B), it is necessary to compare the time-dependence for each mechanism. As the measuring condition VG_m, VD_m has a great impact on the magnitude of the degraded (I_{DS}) curves, we limit the following study to the saturated drain current degradation measured at $VD_m= V_{DD}$ as a function of VG_m sensing from strong saturation to depletion modes. **Fig. 6** shows that I_{DS} degradation follows a classical time-power law in the first time decades according to the generation of interface traps due to broken Si-H bonds from the Sub-V_T mode till $V_{GS}= 0V$ stress. A net saturation effect is observed due to the change in series-resistance effect with GDO damage, smaller in saturation than in linear mode, and to the modification of the stress levels with time. The NCHC stress mode (**Fig.6**) exhibits a larger slope (n= 0.64) value which shows a distinct mechanism in depletion mode stressing where BBT dominates at the drain.

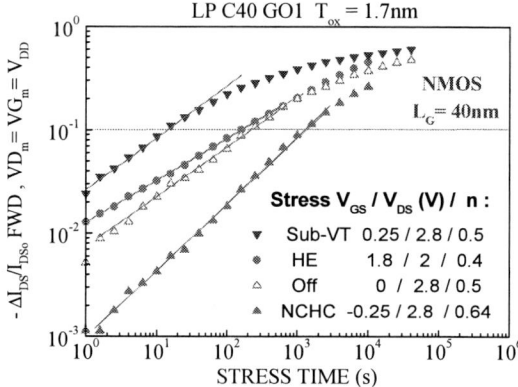

FIG. 6: Time-dependence of the drive current measured in saturation (FWD) in NMOSFETs. DC stressing were carried out under worst-case On-mode (HE), Sub-V_T ($V_{GS}=0.25V$), Off-mode (0V) and NCHC ($V_{G3}=-0.25V$). Kinetics follows a time power-law with exponent n between 0.5 and 0.64.

Much larger degradation rates (n=0.8) have been previously found in LDeMOS [25] which was modeled by the broken of $\equiv$Si-O- bonds leading to the universal RD description. In these high voltage devices saturation effects were also observed in the kinetics and attributed to the distribution in bond energy [40, 41] rather than to the spatial dispersion of degradation [25]. This is indeed explained in the laterally drain extended structure (with long channel) due to the highly localized peak lateral field at the gate-edge giving no large impact of the Off-mode stressing on the internal electric field. This was measured by the small $I_{SUB}(VG_m)$ reduction related to on-state II current. Changing the sensing bias to the Off-state ($VG_m= 0$) that for the same stressing conditions in **Fig.7** makes a great difference in the time dependences. This arises from the competing effect previously described between the On-damage above the channel and the Off-damage from the MDD overlap length till the spacer. This also results from the change in the injection mechanism with V_{GS} and the effect of the growing defect region on the electric field. The larger saturation effect is observed under Sub-V_T stress ($V_{GS}=0.25V$)

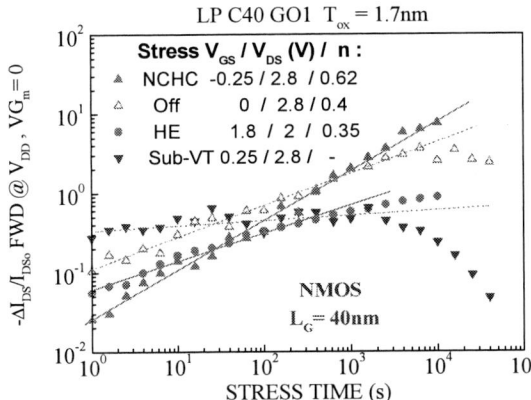

FIG. 7: Time-dependence of saturated current measured at $VD_m = V_{DD}$, $VG_m = 0V$ (FWD) in NMOSFETs. Same DC stressing conditions than in Fig. 8 where a net saturation effect is seen under Sub-V_T (V_{GS}=0.25V) stress and Off-mode (0V) while NCHC (V_{GS}= -0.25V) and HE stresses follow a time power-law with n=0.62 and 0.35 respectively,

where few electrons are accelerated and BBT begins to increase (**Fig.4b** left) leading to ΔN_{it} charged by holes which extend from the drain. With V_{GS} reduction to 0V and -0.25V more holes are injected from the GDO in a shorter region under the spacer which shows a larger time-power-law at short time for both On-mode sensing at $VG_m=V_{DD}$ and 0V. In contrast, HE stress condition shows much larger channel extension consistent with acceptor type N_{it}. In core devices (L_G= 40nm), $I_{SUB}(t_s)/I_{SUB}(0)$ during stressing at V_{GS}= 0 shows a continuous increase in agreement with the electric field increase. This suggests that hole trapping may occur into the spacer region which may lead to I_{DS} current increase (**Fig.4a** right) in the same manner than localized donor N_{it} on the Sub-V_T current. This latter effect can be a consequence of II triggered into the drain by cold carriers through the BBT injection mode as $-I_{SUB}(t_s)$ is increased with V_{DS} at V_{GS}=0. This is found under high E_s condition in agreement with results in thicker T_{ox} **[26]**.

The strong n variation from the 0.5 power-law dependence **Fig.7** partly originates from the exponential dependence of I_{DS} with VG_m. According to our measurement condition in saturation ($VD_m = V_{DD}$, $VG_m = 0$) and introducing the subthreshold slope S, V_T can be extracted assuming a constant level of $I_{DS}= ID_{o,VT}$, giving a simplified Sub-I_{DS} relationship for non degraded device:

$$I_{Off,o} = ID_{o,VT} \exp(-\frac{V_{To}}{S_o} Ln10) \quad (3)$$

as the degraded drain current originates from the shift in threshold voltage $V_{TF} = V_{To} + \Delta V_T$ and slope reduction (S^{-1}) with $S_F= S_o + \Delta S$, (3) for degraded device satisfies:

$$I_{Off, F} = ID_{o,VT} \exp(-\frac{V_{To}+\Delta V_T}{S_o + \Delta S} Ln10) \quad (4)$$

Giving a useful relationship after simplification with:

$$Log(\frac{I_{Off,F}}{I_{Off,o}}) = \frac{\Delta S V_{To} - S_o \Delta V_T}{S_o(S_o + \Delta S)}$$

$$Log(1+\frac{\Delta I_{Off}}{I_{Off,o}}) \propto \frac{\Delta I_{Off}}{I_{Off,o}} \approx 2.3 \frac{\Delta S V_{To} - S_o \Delta V_T}{S_o^2} \quad (5)$$

Where here the I_{Off} rising is considered with the Sub-V_T current increase in **Figs.4a**, *e.g.* posing $I_{Off,F} = I_{Off,o}+\Delta I_{Off}$ for stressing at V_{GS}= -0.25V and 0V, that can also be a smaller variation for V_{GS}=0.25V (**Figs.4b**) or else with an I_{Off} reduction under channel HE condition. Relation (5) points out that the Sub-I_{DS} transistor parameter does not degrade linearly with ΔV_T but intrinsically with ΔS which is first affected by ΔN_{it} and the extension of the damaged regions. This is in contrast to On-mode sensing parameters as found by the direct correlation above V_T under CHC **[16]** and NBTI **[42]**.

C. CHC to Off mode damage and NCHC in PMOSFETs

The hot-carrier degradation in PMOSFETs has taken more attention these last years **[37, 42, 43]** due to the increasing importance of II holes and temperature effects with respect to electrons in the damage mechanisms. **Fig.8** shows that the Off-mode degradation in PMOSFET results mainly in $|I_{DSP}|$ increase from depletion mode (V_{GS}=+0.5V) and Off-mode (0V) with no pivot. The Sub-V_T stress (V_{GS}=-0.25V) exhibits same behavior but with a larger channel damage observed by the large ΔV_T in strong inversion with stress time (arrow in

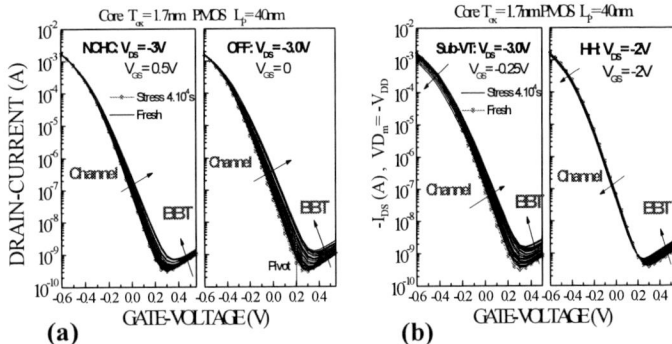

FIG. 8: Sub-threshold characteristics in PMOSFETs **(a)** stressed under NCHC (V_{GS}= 0.5V) and Off mode (0V); **(b)** under Sub-V_T mode at V_{GS}= -0.25V (left) till worst-case On-mode HH stressing (right) at V_{GS}/V_{DS}= -2V/-2V in L_G = 40nm.

Fig.8b left). Only CHC damage at $V_{GS}= V_{DS}$= -2V exhibits a current reduction with smaller Sub-I_{DS} degradation compared to other bias but with still a GIDL increase in Off-state condition. The implication of channel holes needs to increase further the surface electric field which evidences the main contribution of BBT holes to the drain (electrons to the well). The I_{Off} increase and S reduction characterized in saturation mode (**Fig.8a,b**) are first consistent with acceptor-like ΔN_{it} localized in the GDO region while sensing above V_{TP} always

978-1-4244-5430-3/10 $26.00 © 2010 IEEE

shows the generation of ΔN_{it} positively charged ($\Delta V_{TP} < 0$) leading to $|I_{DSP}|$ and g_m reductions. Hence, in dual manner to

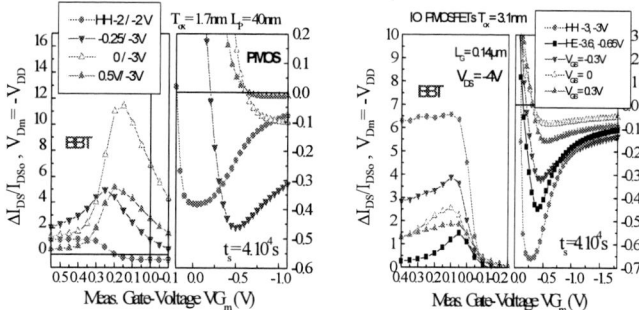

FIG. 9: I_{DS} - VG_m degradation in PMOSFETs measured in saturation mode forward ($VD_m = -V_{DD}$) between On-mode (HH, HE cases), Sub-VT ($V_{GS} = -0.25, -0.3V$), Off-mode ($V_{GS} = 0$) and NCHC ($V_{GS} = 0.3, 0.5V$) **(a)** in core $L_G = 40nm$ with HH stress at $V_{DS}/V_{GS} = -2/-2$ V and $V_{DS} = -3V$ for the other Off-state stress conditions **(b)** in IO $L_G = 0.14\mu m$ under HH, HE conditions ($V_{DS} = -3.6V$) and $V_{DS} = -4V$ for all Off-mode stressing.

electrons in NMOS, the BBT holes in PMOS **Fig.9a** trigger II into the drain and generate ΔN_{it} where interface states within the band gap are charged by tunneling electrons as found in longer device [34]. Charge-Pumping (CP) results using the differential technique [44] have evidenced electron trapping from the GDO region in addition to ΔN_{it} showing a larger ΔN_{it} rate under Sub-V_T stress compared to the smaller ΔN_{it} rate under BBT (not shown here). The degradation effect in weak inversion for $V_{TP} < VG_m < 0$ is also maximum in thin and thick T_{ox} **Fig.9b** according to the maximum ΔS and ΔV_T impact with (5).

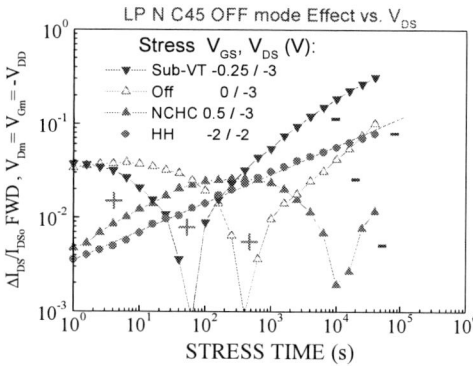

FIG. 10: Time-dependence of saturated ΔI_{DS} (-V_{DD}, -V_{DD}) in PMOSFETs for the Sub-V_T ($V_{GS} = -0.25V$), Off-mode ($0V$), NCHC ($V_{GS} = 0.5V$) stresses showing turnaround in kinetics while HH stressing (-2V, -2V) gives n= 0.3.

The previous arguments are supported by the observation in **Fig.10** that the competing mechanisms induce turnaround effects in the time dependence of the saturated current with first a positive $\Delta I_{DS}/I_{DSo}$ (localized N_{ot}-, N_{it}^{Acc}) before coming negative at longer stress time (channel spread N_{it}^{Don}).

Changing the sensing bias to $VG_m = 0V$ **Fig.11** makes again a great difference in the time-dependence with the On-mode result (**Fig.10**). As the large $|I_{DS}|$ increase in the Sub-V_T

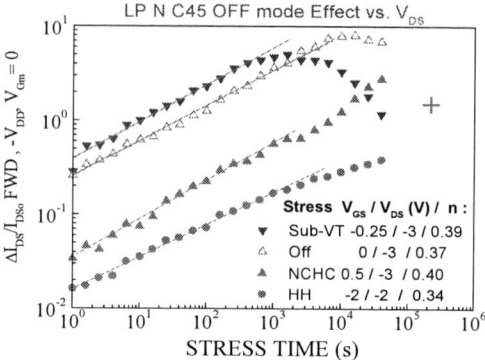

FIG. 11: Time-dependence of saturated ΔI_{DS} (FWD) in PMOSFETs measured at $VG_m = 0$ ($VD_m = -V_{DD}$) for the Sub-V_T ($V_{GS} = -0.25V$), Off-mode ($0V$), NCHC ($V_{GS} = 0.5V$) and HH stressing (-2V, -2V).

region dominates without a pivot, kinetics follows a time power-law with exponents n = 0.34-0.40 with a large saturation effect for Sub-V_T stress ($V_{GS} = -0.25V$). This is due to the strong ΔV_T at long stress time (channel part) which induces a net ΔS decrease in the opposite direction at the onset of strong inversion (**Fig.8b** left) which stands out the net saturation effect of the Sub-V_T stress for the degraded Off current measured at $VG_m = 0$ in PMOSFETs.

D. Temperature acceleration effects in Off-mode

As one consequence of scaling is the large temperature increase in core blocs, it is mandatory to examine its effect on the Off-mode damage in comparison to CHC damage. We have shown the reverse temperature effect in ultra-thin gate-oxide MOSFETs by the net I_{SUB} increase at High Temperature (HT) in NMOSFETs under CHC [17, 18]. This arises from the spread of the electron energy distribution function with EES and temperature which modifies the thermal tail ($\propto 1/nkT$) particularly at low V_{DS} [13, 14]. **Fig.12** shows that electron

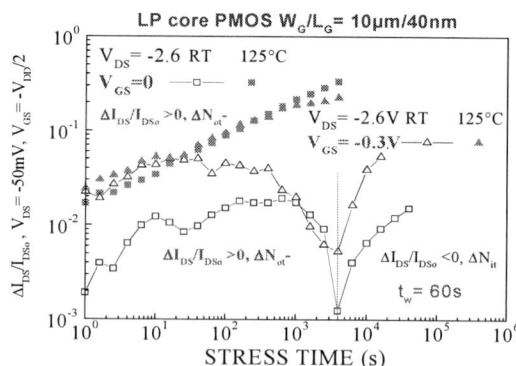

FIG. 12: Off state degradation in core $L_G = 40nm$ PMOSFETs for the two bias conditions of Fig.17 showing the trapping/detrapping effect $\Delta N_{ot}^{-/o}$ at RT which let ΔN_{it}, while the electron trapping remains dominant into the sidewall spacer at HT. This induces a series resistance decrease at HT although the use of a waiting time ($t_w = 60s$) before measurements.

978-1-4244-5430-3/10 $26.00 © 2010 IEEE

trapping in PMOS occurs into the spacer oxide at HT with a net temperature activation (V_{GS}= 0 and -0.3V) leading to a series resistance decrease. For same stressing at Room Temperature (RT) a competing trapping/detrapping mechanism occurs observed by turnarounds in the linear I_{DS} leaving ΔN_{it} taking the lead at RT and long stress duration. This behavior was observed since 130nm node (T_{ox}= 2nm) in correlation to the smaller trapping efficiency in thin T_{ox} [37]. NMOSFET with L_G= 40 nm shows **Fig.13** a similar positive temperature activation for stressing at V_{GS}= 0 consistent with the generation of ΔN_{it} with an apparent small activation energy E_a= 50meV. This supports the main contribution of BBT which is a phonon-assisted mechanism according to section I.A-B. Moreover no enhanced damage is seen at HT under Sub-V_T stressing (V_{GS}= 0.3V) due to the small II current and phonon scattering. This behavior is also observed in long channel device L_G= 0.14μm (E_a = 60 meV) using larger drain field condition (insert of **Fig.13**) according to the predominance of ΔN_{it} [18].

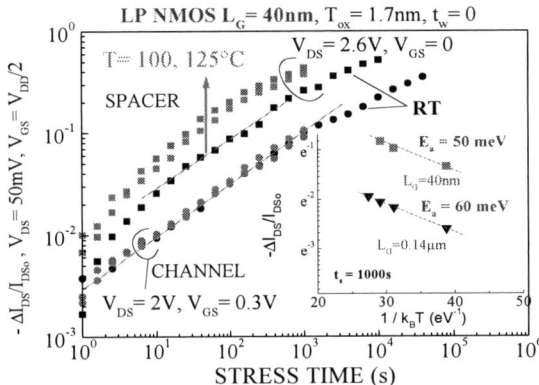

FIG. 13: : I_{DS} reduction in core NMOSFETs in off state showing no N_{it} activation above the channel (V_{GS}= 0.3V) as no MVE mechanism occurs due to low V_{GS} while BBT electrons from the junction at higher field condition induce ΔN_{it} with an apparent activation of 50-60meV.

The detailed results of this section have shown that Off-mode degradation can be interpreted as a composite picture **Fig.14** where Si-H bonds are generated from the spacer and extend to the channel with an amphoteric nature. This can be distinguished **Figs.14a,b** by a net electron trapping localized into the thick spacer oxide in PMOSFETs (hole trapping in NMOS) which may extend through the GDO length in addition to interface traps in this region. These localized N_{it} are charged by tunneling carriers (at medium E_s)and generated by higher energetic carriers (at large E_s) in series to the channel N_{it} which are charged from weak to strong inversion modes (electrons in NMOS *vs.* holes in PMOS), *i.e.* with an opposite charge state than ΔN_{it} of the drain side. Then, V_{GS} and the surface electric field ($\propto V_{DG}$) govern the efficiency of the injection mechanism between (cold) tunneling carriers localized into the drain which lead to the charging of interface traps, and quasi-ballistic high energetic carriers which generate larger N_{it} and may reach deeper traps spreading the distribution towards the channel.

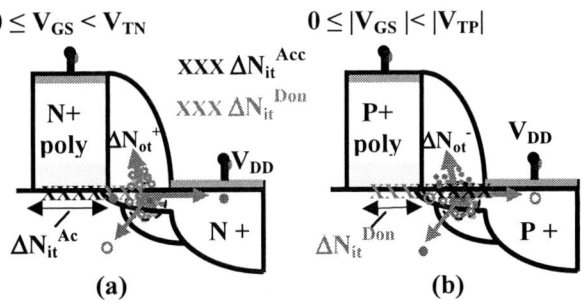

FIG. 14: Schematic illustrations of the distinct degraded regions under Off-mode Stressing, **(a)** in NMOSFETs, **(b)** in PMOSFETs with the simplified view of the series connection of the channel part (ΔN_{it}-/+) and the GDO to spacer region.

II. ENERGY DRIVEN LIFETIME MODELING

We have developed a three-mode lifetime modeling [14, 16, 17] based on the classical field (voltage) accelerating model at high energy (called mode 1), *i.e.* based on the Lucky Electron Model (LEM) [21], a new energy-driven scheme where interactions between carriers (mode 2) play an additional role in the acquisition of electron energy and distribution function [19, 20], and a third degradation mechanism at low carrier energy [14, 16] (mode 3), specially devoted to Channel Cold Carriers (CCC). This latter mode has related ΔN_{it} to the physical dependence of Multi Vibrational Excitation (MVE) of Si-H bonds [16, 17]. The transfer to an energy driven scheme is done by introducing scattering rates $S_{ii}(E)$ for the II mechanism and $S_{it}(E)$ for ΔN_{it} [14, 19, 20] where $S_{it} \propto (E - \varphi_{it})^{Pit}$ with value $p_{it} \cong 11$ and $\varphi_{it} = 1.5$eV. This relates bond breaking rate for mode 1 with α= 1 and for mode 2 with α= 2 :

$$R_b \propto \frac{1}{\tau} \cdot \propto S_{it}\left(E_{domit}\right) I_{ds}^{\alpha} \qquad (6)$$

A. Channel Cold Carrier (CCC) in NMOS and PMOSFETs

The observation of the strong deviation in lifetime dependence makes clearly appear a threshold in the channel current to enter into the low energy range that we call mode 3. This is correlated to the high I_{DS} range that has been previously characterized in NMOSFETs [14, 16] by the worsening of damage with the vanishing of the L_G and V_{DS} dependence with $V_{GS} \geq V_{DS}$. CCC is distinguished from modes 1-2 by this direct correlation to channel current under MVE degradation showing **Fig.15a** the demarcation in core NMOSFETs for various stressing voltages. The bond breaking rate under MVE has been simplified as [16, 17]:

$$R_{bMVE} \propto \frac{1}{\tau_{MVE}} \propto \left[S_{MVE} \cdot \left(\frac{I_{ds}}{W}\right) \right]^{N_{ph}} \exp\left(\frac{-E_{emi}}{k_B T}\right) \qquad (7)$$

with S_{MVE} slightly dependent on the electron energy for NMOSFETs in the form $S_{MVE} \propto (eV_{DS} - \hbar\omega)^{0.5}$ giving a typical power-law with $1/\tau_{MVE} \propto V_{DS}^{NPh/2} (I_{DS}/W)^{NPh}$.

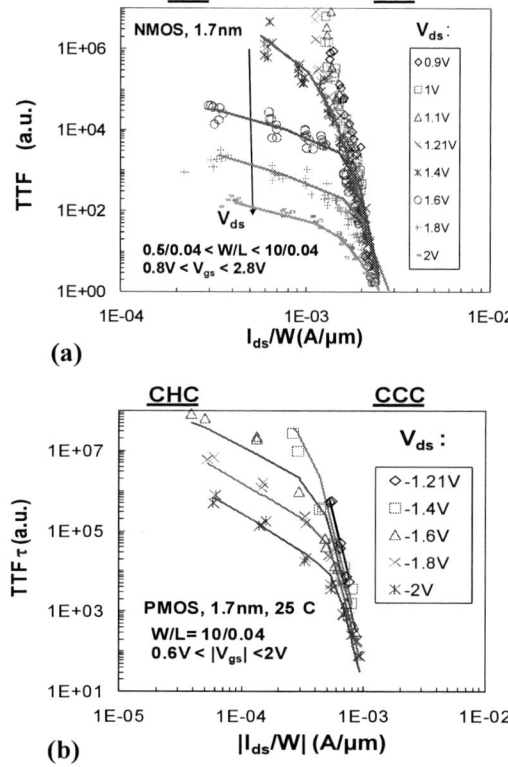

(a)

(b)

FIG. 15: **(a)** Two energy ranges in the device lifetime for permanent damage (ΔID_{sat} FWD) with a strong V_{DS} dependence at moderate I_{ds} (mode 1, 2) and a single dependence at higher I_{ds} under mode 3 **(a)** in NMOSFETs, **(b)** in PMOSFETs with waiting time t_w= 60s before sensing, in devices with L_G= 40nm, T_{ox}= 1.7nm, lines represent the complete modeling according to formulations (6) and (7).

This CCC lifetime dependence is also observed in **Fig.15b** under channel hole current in PMOSFETs where the reduction in ΔID_{sat} FWD is intentionally measured with delay using a waiting time (t_w= 60s). This guaranty to stay mostly sensitive to interface traps with no effect of hole trapping/detrapping [18, 42, 43]. Lifetime plots give slope values -10 and -17.7 for PMOS and NMOS respectively. This is in agreement with vibrational excitation of the Si-H bond with bending mode ($\hbar\omega$=80meV) rather than stretching mode [17]. In the case of PMOSFET the desorption process is modeled by vibrational heating of hydrogen caused by inelastic scattering of tunneling holes with the Si-H 5σ hole resonance [45]. This implies with (7) that the number of phonon levels reached under hole excitation is almost half the one under electron excitation ($N_{Ph,h} \approx N_{Ph,n}/2$). This has been found by STM results showing power-law dependences on current and bias where electrons tunneling from the substrate may excite the Si-H 5σ hole resonance which upon de-excitation can transfer energy to the H atom [45]. Our simplification for S_{MVE} is based on the cross section determined under the electron-phonon scattering [17] which differs in the case of hole-phonon scattering in addition to the fact that the density of states is different for holes. This explains the difference in the power-law exponent and smaller current threshold of the MVE mechanism for holes.

Furthermore, as device lifetime plotted as a function of the drain current (I_{DS}/W) exhibits a weak V_{DS} dependence, it was necessary to include this effect by a simplified form for S_{MVE} which allowed us to take into account more accurately the current dependence for mode 3 [26, 27]. This is performed **Fig.16** by comparing the product $\tau_{MVE}.(V_{DS} - \hbar\omega)^{(E_B/2\hbar\omega)}$ as a function of I_{DS}/W between NMOS and PMOSFETs.

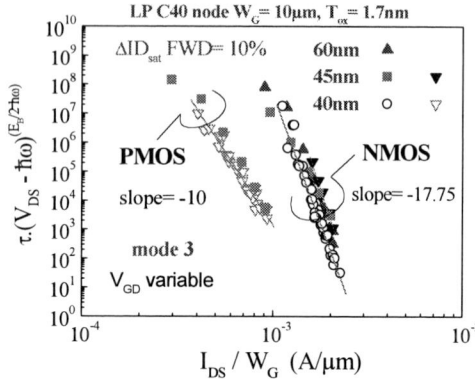

FIG. 16 : Extraction of the mode 3 parts from Figs.24a,b for ΔID_{sat} FWD and various high $V_{GD} \geq 0$ stressing voltages showing the current threshold I_{th} where I_{th} (PMOS) $\approx$ I_{th} (NMOS) /2 under MVE mode.

The extracted current exponent in NMOS is 17-20 which corresponds to E_B= 1.4-1.5eV while it reduces to 10-12 in PMOS according to our theoretical assumption that the mechanisms of H desorption is mainly correlated to bending vibrational modes. However, this energy modeling is a simplified approach due to the fact that E_B, separating the ground state to the transport state of the Si-H bond, is considered here constant while it rather follows a broad distribution of activation energies [11, 37, 40, 41].

B. Off-mode incorporation into the lifetime modeling

Following the results of section I-II.A, we include here Off-mode damage in the full CHC to CCC modeling. The former is a very different mechanism than MVE as V_{GS} is low in Sub-V_T regime ($V_{TP} < V_{GS} < V_{TN}$) and involves a small amount of carriers. However, there are similarities with the energy driven approach as (cold) BBT carriers require a high field condition to significantly degrade the drain. Hence, one can consider that increasing E_s, BBT carriers participate to the damage whether they reach the dominant energy under the local maximum electric field and may induce carrier multiplication in the junction [31, 32]. Off-mode stressing incorporation is first done by extending the usual CHC stressing condition **Fig.17** to the bias where device operates in Sub-V_T regime, and Off-mode. Comparing Sub-V_T and Off mode bias ranges in **Figs.5,9**, it is not easy to choose the right lifetime monitor as one can find very different magnitude of parameter degradation between the Off-, On- operation. This issue is accentuated by the fact that time dependence can follow very different accelerating law and time exponents from the 0.5 value depending on VG_m, VD_m. For the sake of simplicity On-mode parameters are chosen here in order to

978-1-4244-5430-3/10 $26.00 © 2010 IEEE   62

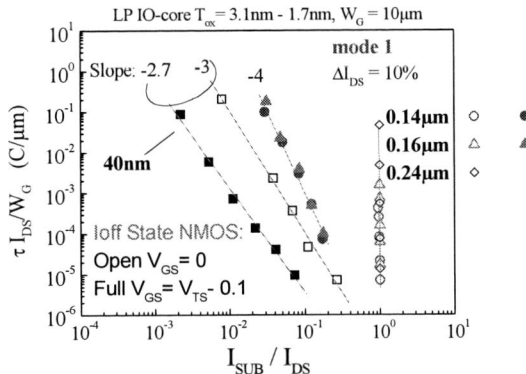

FIG. 17: Lifetime plots for NMOSFETs in off state stressing match the ΔN_{it} dependence in core devices while IO devices with $V_{GS}= 0$ exhibit infinite slope ($I_{SUB}= -I_{DS}$) due to predominance of the spacer-junction damaged injecting region.

determine the Off-mode stress impact on speed performance and switch. Lifetime plots in Off-mode stress **Fig.17** with on-mode sensing $\Delta I_{DS}(V_{DD}, V_{DD})$ follow the high energy dependence described by mode 1 according to combined effect of spatial spread of ΔN_{it} and accelerating field in core 40nm NMOS. In contrast, long channel IO devices stressed at $V_{GS}= 0$ show infinite slope as $I_{SUB}= -I_{DS}$ (diode configuration) at high field (section I.A **Figs.1-2**). This has been explained by the main contribution of the degraded sidewall spacer as a series resistance variation in the drain region. When stressing is performed in Sub-V_T regime in long channel case, the spatial spread takes the lead again which is shown by the energy dependence in the lifetime plot in similar way as shorter devices. The change of V_{GS} from Sub-V_T towards strong depletion ($V_{GS}= -0.75$) confirms **Fig.18** the net increase

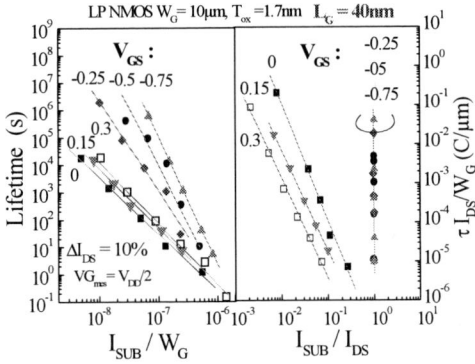

FIG. 18: Device lifetime in Off state with $V_{GS}= 0.3V$ to $-0.75V$ in $L_G= 40nm$ NMOSFETs plotted as a function of I_B/W_G (left) or I_B/I_{DS} (right) according to the mode 1-2 modeling.

in device lifetime as a function of I_{SUB} due to the vanishing of CHC damage above the channel with respect to the damage in the spacer region as depicted in **Fig.14**. Same data plotted as a function of mode 1 (**Fig.18** right) shows that high energy carriers increase with V_{DG} with a lifetime exponent $m_3= (\phi_{it} /\phi_{i,e})= 2.7$ to 3 consistent with the LEM picture under Sub-V_T stressing when the spatial spread is involved. Turning to NCHC regime with $V_{GS}< 0$ shows the same effect than in

Fig.17 in thicker T_{ox} and longer channel, *i.e.* an infinite slope as I_{SUB} tends towards ($-I_{DS}$). This is explained by a pure stressing diode configuration in this mode which is strongly dependent on E_s at the drain where BBT carriers may trigger the II process at high V_{DG}, according to the results of Section I. Because NCHC degradation follows the gate-voltage dependence of the substrate current in correlation to the appearance of the BBT mechanism, it is consequently more adapted to directly expressed device lifetime as $\tau - I_{SUB}/W_G$.

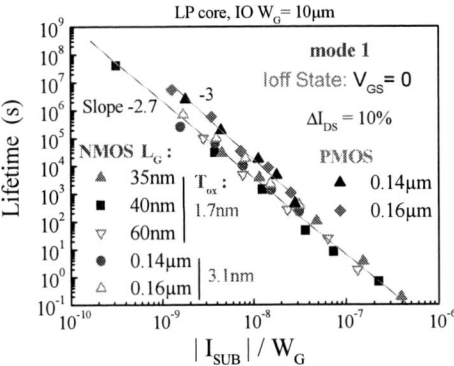

FIG. 19: Off-mode device lifetime τ vs. I_{SUB}/W_G in both IO and core devices in NMOSFETs and only in IO devices in PMOSFETs showing the I_{SUB} universal power-law dependence of ionization rate from the drain junction.

The lifetime plot indeed exhibits universal power-law dependence in **Fig.19** for NMOS and PMOS IO devices. The results validate the LEM approach (mode 1) as the Off-mode degradation is directly dependent on high energy carriers with power-law exponent -2.7 and -3 in accordance to results of **Fig.17**. This way is suitable only in IO PMOS ($T_{ox}= 3.1nm$) as core PMOSFETs with ultra-thin gate-oxide present competing damage mechanisms (section I.B-C) which complicate the use of accelerating techniques. With the principle of QS transfer, inverter waveform can be used in **Fig.20** to include the Off-mode phases as it represents a general form of digital operation. This is simply done first for $V_{In} < V_{TN}$ in NMOS and $V_{Out} > V_{DD}+ V_{TP}$ for PMOSFETs, and then, during half of the period where NMOS and PMOS alternatively switches to the Off-state, *i.e.* when $V_{In}= 0$, $V_{Out}=V_{DD}$ and $V_{In}=V_{DD}$, $V_{Out}=0$ for NMOS and PMOS.

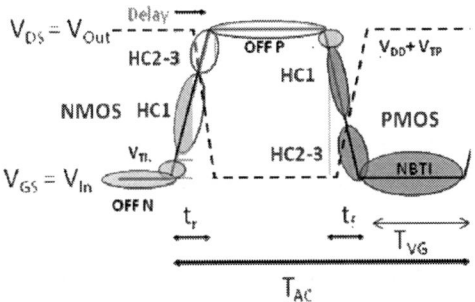

FIG. 20: Inverter-like pulse waveform used for the DC-AC transfer of each accelerating degradation rates in the three mode energy modeling devoted to core and IO CMOS nodes. The Off-mode phases are included both in NMOS and PMOS for the use of QS device lifetime extrapolation.

With the QS scheme, the results of previous sections have evidenced the high V_{DG} conditions required to observe Off-mode degradation in contrast to CHC damage which needs smaller V_{DS} and much reduced V_{GD} in MOSFETs [14, 16, 18]. Hence Off-mode damage can be included in the full modeling for digital operation. This is performed by completing the 3 mode On-state lifetime (τ_{ON}) developed here for NMOS and PMOS by the Off-state contributions (τ_{OFF}) in two phases $1/\tau_{OFF} = 1/\tau_{VG=0} + 1/\tau_{SubVT}$ useful for any pulse waveform shapes. Device lifetime extraction (τ) is then deduced for all stressing conditions by assuming that all degradation modes compete in parallel leading to a general device lifetime modeling expressed by:

$$R_{it} = \frac{1}{\tau} = \sum_i \frac{1}{\tau_{ON,i}} + \sum_j \frac{1}{\tau_{OFF,J}} \qquad (8)$$

$$\frac{1}{\tau_{ON}} = C_1 \cdot \left(\frac{I_{ds}}{W}\right)^{a_1} \cdot \left(\frac{I_{bs}}{I_{ds}}\right)^{m_1} + C_2 \cdot \left(\frac{I_{ds}}{W}\right)^{a_2} \cdot \left(\frac{I_{bs}}{I_{ds}}\right)^{m_1} + C_3 \cdot V_{ds}^{a_3/2} \cdot \left(\frac{I_{ds}}{W}\right)^{a_3} \cdot \exp\left(\frac{-E_{emi}}{k_B T}\right)$$

$$\frac{1}{\tau_{OFF}} = C_4 \left(\frac{I_{Sub}}{W_G}\right)^{m_2} + C_5 \cdot \left(\frac{I_{ds}}{W_G}\right) \cdot \left(\frac{I_{Sub}}{I_{ds}}\right)^{m_3} \qquad (9)$$

where C_1 (SVE), C_2 (EES) and C_3 (MVE), C4(Off-state), C5(Sub-VT) are all the subsequent constants determined in their respective dominant mode. This modeling has been validated for various T_{ox}= 5nm, 3.2nm and 1.7nm, under a large set of voltages conditions (V_{GS}, V_{DS}), temperatures and device geometries in both NMOSFET and IO PMOSFET.

CONCLUSION

Off-mode damage in 40nm nodes is related to BBT mechanisms whereas V_{GS} and the surface electric field ($\propto V_{DG}$) govern the injected carrier efficiency between (cold) tunneling carriers into the drain which lead to the charging of interface traps (low to medium E_s), and high energetic carriers (high E_s) which generate ΔN_{it} and trapped charges in the spacer oxide. We have shown *(1)* the strong effect of sensing bias (VG$_m$, VD$_m$) on the magnitude of I_{DS} degradation from Sub-V_T regime to Off- and depletion modes, *(2)* the amphoteric nature of defects consistent with broken Si-H bonds charged by the tunneling holes (electrons) in NMOS (PMOS) in the drain region in series with ΔN_{it} from the channel, *(3)* The small trapping of localized holes (electrons) in NMOS (PMOS) detected in the GDO region by I-V and CP.

The Off-damage contribution has been included in the three-mode device lifetime modeling developed in an energy driven framework. The modeling has been extended to PMOSFET which shows smaller threshold (I_{DS}) current for MVE triggering with smaller power-law exponent -10. The acceleration of degradation in Sub-V_T regime corresponds to high energetic carriers modeling with the LEM through usual E_{max} lifetime monitoring giving power-law exponents -2.7(-3). Off-mode damage (V_{GS}=0) related to smaller amount of BBT carriers has been characterized by the direct I_{SUB} (= -I_{DS})

accelerating power-law. This enables to directly include the Off-mode contributions in the general lifetime modeling suitable for quasi-static DC-AC transfers and applied to digital and analog operations.

REFERENCES

[1] Y. Taur; C.H. Wann, D.J. Frank, in IEDM Tech. Dig., 1998, p. 789.
[2] D.J. Frank *et al.*, *Proc. of* the IEEE, 2001, Vol. 89 (3), p. 259.
[3] T. Skotnicki, *et. al.*, Trans. Electron Dev., 2008, Vol. 55, N°1, p. 96.
[4] K. Roy, S. Mukhopadhyay, H. M. Meimand, *Proc. of* the IEEE, 2003, Vol. 91, N° 2, p. 305.
[5] R. Thewes *et al.*, in IEDM Tech. Dig., 1999, p. 81.
[6] S. P. Mohanty, V. Mukherjee, R. Velagapudi, *Proc. of* 14th IEEE Int.Workshop on Logic and Synthesis (IWLS), 2003, p. 248.
[7] V. Mukherjee, S. P. Mohanty, E. Kougianos, Proc. of the IEEE Int. Conf. on Computer Design (ICCD), 2005, p. 431.
[8] A. Weber *et al.*,, Solid State Elec., 2006, Vol. 50, p. 613.
[9] S. Mahapatra, P. B. Kumar, M. A. Alam, Trans. Electron Dev., 2004, Vol. 51, N° 9, p. 1371.
[10] V. Huard, M. Denais, C. Parthasarathy, Microelec. Reliab., 2006, Vol. 46, No. 1, p. 1.
[11] T. Grasser, *Tutorial Proc. of* Int. Reliab. Phys. Symp. (IRPS), 2008, Topic 113, p. 113.
[12] J.E. Chung *et al.*, Trans. Elec. Dev., 1991, Vol. 38, N°6, p. 1362.
[13] S.E. Rauch, G. La Rosa, F.J. Guarin, Trans. on Device and Materials Reliab., 2001, Vol. 1, N°2, p. 113.
[14] C. Guérin, V. Huard, A. Bravaix, Trans. on Device and Materials Reliab., 2007, Vol. 7, N°2, p. 225.
[15] M. Miyazaki *et al.*, IEEE J. Solid-State Cir., 2002, Vol. 37, N°2, p. 210.
[16] A. Bravaix *et al.*, *Proc. of* Int. Reliab. Phys. Symp. (IRPS), 2009, p. 531.
[17] C. Guérin, V. Huard, A. Bravaix, J. Appl. Phys., 2009, Vol. 105, p. 114513.
[18] V. Huard, *Tutorial Proc. of* Int. Reliab. Phys. Symp. (IRPS), 2009, Topic 121, p. 113.
[19] S.E. Rauch, G. La Rosa, IEEE Trans. on Device and Material Reliab., 2005, Vol. 5, N° 4, p. 701.
[20] G. La Rosa, S.E. Rauch, Microelec. Reliab., 2007, Vol. 47, p. 552.
[21] C. Hu *et al.*, Trans. Electron Dev., 1985, Vol. 48, N°4, p. 375.
[22] E. H. Poindexter, P. J. Caplan, B. E. Deal, R. R. Razouk, J. Appl. Phys., 1981, Vol. 52, p. 879.
[23] M. A. Alam, in *IEDM Tech. Dig.*, 2003, p. 345.
[24] S. Mahapatra, D. Saha, D. Varghese, P. B. Kumar, Trans. Electron Dev., 2006, Vol. 53, N°7, p. 1583.
[25] D. Varghese *et al.*, Trans. Electron Dev., 2007, Vol. 54, N°10, p. 2669.
[26] M. Orlowski, S.W. Sun, P. Blakey, R. Subrahmanyan, Electron Device Lett., 1990, Vol. 11, N°12, p. 593.
[27] B.S. Doyle, K.R. Mistry, Trans. Electron Dev., 1992, Vol. 39, N° 7, p. 1774.
[28] H. Fang, P. Fang, J-T. Yue, Trans. Electron Dev., 1994, Vol. 15, N° 11, p. 463.
[29] M. Rodder, Trans. Electron Dev., 1990, Vol. 11, N° 8, p. 346.
[30] J. Chen *et al.*, Electron Device Lett., 1987, Vol. 8, N° 11, p. 515.
[31] Y. Igura, H. Matsuoka, E. Takeda, Trans. Electron Dev., 1989, Vol. 10, N°5, p. 227.
[32] S. A. Parke *et al.*, Trans. Electron Dev., 1992, Vol. 39, N°7, p. 1694.
[33] G.Q. Lo *et al.*, Trans. Electron Dev., 1991, Vol. 12, N°1, p. 5.
[34] G.Q. Lo *et al.*, Trans. Electron Dev., 1991, Vol. 12, N°12, p. 710.
[35] T. Wang *et al.*, Electron Device Lett., 1995, Vol. 16, N°12, p. 566.
[36] Y. Shi *et al.*, Trans. Electron Dev., 1998, Vol. 45, N°11, p. 2355.
[37] A. Bravaix *et al.*, Microelec. Reliab., 2004, Vol. 44, N°1, p. 65.
[38] D. Vuillaume *et al.*, Trans. Electron Dev., 1993, Vol. 40, N°4, p. 773.
[39] B. Doyle *et al.*, Trans. Electron Dev., 1990, Vol. 37, N° 8, p. 1869.
[40] K. Hess *et al.*, *in* IEDM Tech. Dig., 2000, p. 10.
[41] A. Haggag *et al.*, Int. Reliab. Phys. Symp. (IRPS), 2001, p. 271.
[42] C.R. Parthasarathy *et al.*, Trans. on Dev. and Materials Reliab., 2007, Vol. 7, N°1, p. 130.
[43] V. Huard *et al.*, in IEDM Tech. Dig., 2007, p. 797.
[44] W. Chen *et al.*, Trans. Electron Dev., 1993, Vol. 40, N°1, p. 187.
[45] K. Stokbro *et al.*, Phys. Rev. Lett., 1998, Vol. 80, N°12, p. 2618.

Mobility Enhancement due to Charge Trapping & Defect Generation: Physics of Self-Compensated BTI

Ahmad Ehteshamul Islam and *Muhammad Ashraful Alam*

Department of Electrical and Computer Engineering, Purdue University
West Lafayette, Indiana, USA
Phone: 765-532-6514; Fax: 765-494-2706; Email: aeislam@ieee.org, alam@purdue.edu

Abstract—**Threshold voltage V_T of a transistor degrades with time both due to the formation of defects at the oxide/Si interface, as well as charge trapping into bulk defects – a phenomenon commonly known as Bias Temperature Instability (BTI). However, we have shown earlier that with appropriate mobility vs. vertical effective electric field characteristics, transistor's drivability (*i.e.*, drain current) can be made far less sensitive to the NBTI-induced threshold voltage degradation ΔV_T, than previously presumed. Higher steepness of the mobility-field characteristics results in an increase in mobility due to interface defects, which can self-compensate the effect of ΔV_T on drain current. In this paper, for the first time we analyze the *additional effect of PBTI-induced ΔV_T in NMOS transistor* parameters and show that mobility at constant gate voltage always increases with PBTI, irrespective of the mobility-field steepness. Therefore, self-compensation for PBTI is even more pronounced compared to NBTI. Next, we demonstrate the *consequence of self-compensation* via an intuitive analysis in simple digital circuits and show that lifetime of digital ICs increases dramatically once we incorporate the effect of self-compensation by using appropriate sign for mobility variation at constant gate voltage. This might in turn reduce the requirement of different circuit level optimization techniques, currently employed to manage transistor variabilities. Finally, we establish the *importance of flatter transfer characteristics* for self-compensation, which can be obtained through advanced CMOS technologies.**

Keywords- Bias Temperature Instability, degradation, threshold voltage, effective mobility, drain current, transconductance, digital circuit, inverter, SRAM cell, self-compensation, strain, lifetime.

I. INTRODUCTION

Time-dependent degradation of transistor parameters, commonly known as Bias Temperature Instability or BTI, is one of the major reliability concerns in current CMOS technologies [1]. BTI for a PMOS transistor originates from interface defect or N_{IT} generation and hole trapping into oxide defects N_{HT} [2] (see Fig. 1a) and is enhanced under the influence of negative gate bias and high temperature, and is termed as Negative BTI or NBTI. On the other hand, BTI for a NMOS transistor originates from electron trapping into oxide defects N_{ET} [3] (see Fig. 1b), and is known as Positive BTI or PBTI. PBTI occurs due to the application of positive gate bias, and increases at high temperature [4] [5]. The impact of BTI on transistor level is generally observed as a change in threshold voltage V_T (specifically, NBTI decreases V_T and PBTI increases V_T), and in the reduction of transistor drivability or drain

current I_D. Even though the physical origin of such V_T shift due to BTI degradation is still debated [2] [6], its impact on digital circuits is well-known [7]-[11], *i.e.*, change in BTI-induced V_T shift (*i.e.*, ΔV_T) increases the gate delay (for example, rise and fall delay of a single-stage inverter increases due to NBTI and PBTI, respectively). The impact of such delay degradation in digital circuit has been routinely observed in ring oscillator measurements [5] [12]-[15]. Since device designers often feel that process modifications (*e.g.*, choosing plasma nitridation over thermal nitridation for reducing NBTI [16]) cannot adequately address the BTI problem without unacceptable loss of device performance, a number of approaches have been undertaken to alleviate its effect at the circuit level. For example, a standard option is to operate the transistors with an extra guard band voltage, over and above the required voltage for nominal operation [15] [16] so that the circuit remains functional despite the BTI degradation, including other process-related variation. Similarly, there are proposals for using adaptive body bias [17]-[19] or adaptive power supply (*i.e.*, circuit sleeping) [20] for minimizing the impact of degradation. Moreover, algorithms involving area-resizing of transistors along the critical path of a digital circuit have also been proposed for optimizing circuit-level degradation [9]. All these circuit level optimization techniques, when applied to an operational IC, are pretty often coupled with a degradation monitor (also known as 'Silicon Odometer' [21]) to identify the IC's degradation level at different stages of operation.

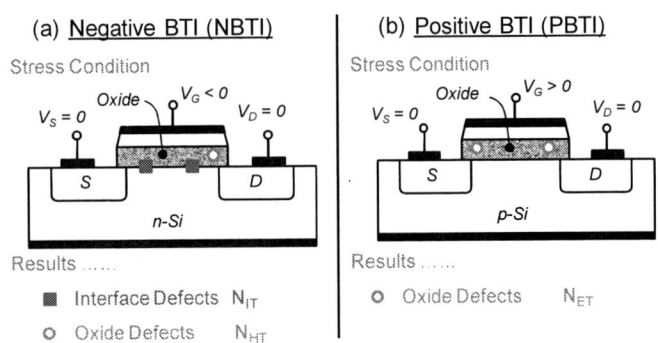

Figure 1. Stress conditions and the defects under consideration for (a) NBTI and (b) PBTI stress.

In addition to the circuit level solutions so far adopted to optimize the BTI issues in CMOS architecture, there are additional process-related options that could potentially

978-1-4244-5430-3/10 $26.00 © 2010 IEEE

alleviate BTI concerns. As we have discussed in [22], the effect of N_{IT} generation on the drain current (I_D) of a transistor can be self-compensated by choosing appropriate mobility vs. effective vertical electric field (μ_{eff}-E_{eff}) characteristics. Higher steepness of the mobility-field characteristics results in an *increase of mobility due to N_{IT} generation* (contrary to the classical observation of mobility decrease due to defects [25][26]), which can self-compensate the effect of ΔV_T on I_D. And under certain favorable conditions, it might even be possible to have negligible change in I_D (*i.e.*, $\Delta I_D \sim 0$).

In section II of this paper, we first systematically study the variation of transistor parameters (*e.g.*, threshold voltage V_T, sub-threshold slope SS, effective mobility μ_{eff}, transconductance g_m, *etc.*) at different levels of NBTI and PBTI stress. *In section III*, we use the variation of the above transistor parameters to study the impact of BTI on drain current I_D. In this process, we identify (for the first time) the presence of self-compensation for PBTI-induced N_{ET} in high-κ NMOS transistors. Then *in section IV*, we perform an intuitive circuit analysis to identify the consequences of self-compensation in digital circuits and memories. Next *in section V*, we establish the significance of having flatter transfer (I_D-V_G) characteristics for achieving self-compensation and show its relation with steeper μ_{eff}-E_{eff} characteristics. Later *in section VI*, we show that, in addition to strain (which is used as a parameter for achieving self-compensation in [22]), reduced temperature can also increase the μ_{eff}-E_{eff} steepness and thus enable one to obtain better self-compensation. *Finally*, we conclude by discussing the promise of obtaining self-compensation in advanced substrates involving strained-Si, as well as III-V materials.

II. IMPACT OF DEFECTS ON V_T, SS, G_M, AND μ_{EFF}

A. *Effect of NBTI-induced Interface Defects*

It is well-known in NBTI literature that the generation of N_{IT} increases $|V_T|$ (*i.e.*, decreases V_T), SS (Fig. 2a), and decreases maximum transconductance $g_{m,max}$ (Fig. 2b). Since $I_{D,lin} = (W/L)\mu_{eff}C_{di}(V_G-V_T)V_{DS}$ and $\mu_{eff} = \mu_0/(1+\theta_1(V_G-V_T))$ [23], one can write –

$$g_{m,max} \equiv \max\left(\frac{\partial I_{D,lin}}{\partial V_G}\right) \sim \frac{W}{L}\mu_0 C_{di}V_{DS} \quad (1)$$

where, W is channel width, L is channel length, C_{di} is the dielectric capacitance, μ_0 is the field-independent mobility parameter, and θ_1 is a signature of mobility-field steepness. According to (1), a decrease in $g_{m,max}$ due to N_{IT} (see Fig. 2b) indicates a decrease in μ_0 or mobility at constant effective vertical field ($\mu_{eff}@E_{eff}$). Such variation in $\mu_{eff}@E_{eff}$ before and after N_{IT} generation can be explained if we remember the well-known fact that an increase in N_{IT} always reduces mobility through Coulomb scattering [24]-[27] (Fig. 3a). As a result, when measured at constant E_{eff}, μ_{eff} always decreases due to N_{IT} generation, as shown by the A → B transitions in Fig. 3b.

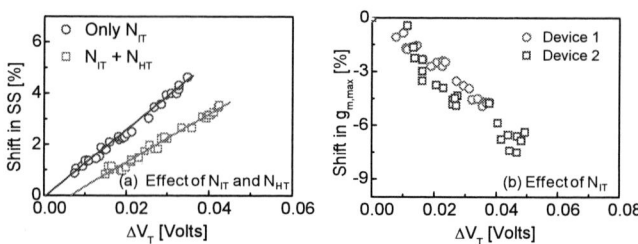

Figure 2. NBTI stress generates N_{IT} and results hole trapping N_{HT} into oxide defects. (a) N_{IT} (and associated ΔV_T) causes a linear increase in SS. However, similar to electron trapping N_{ET} into oxide defects (Fig. 4a), N_{HT} does not affect SS – thus causes only a lateral shift in the ΔSS vs. ΔV_T plot. (b) N_{IT} introduces Coulomb scattering, hence reduces $\mu_{eff}@E_{eff}$ or $g_{m,max}$. See (1) for the relation between $\mu_{eff}@E_{eff}$ and $g_{m,max}$.

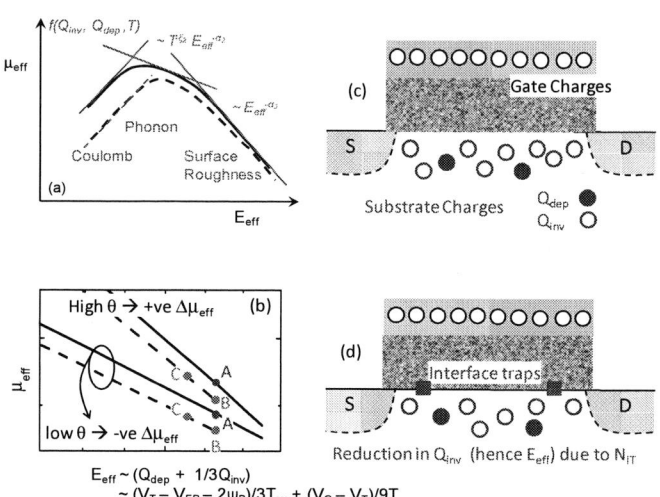

Figure 3. (a) A schematic representation of the effect of N_{IT} generation on μ_{eff}-E_{eff} plot. N_{IT} induces extra coulomb scattering and thus reduces μ_{eff}, especially at low E_{eff}. (b) μ_{eff} vs. E_{eff} in both pre- and post-N_{IT} generation conditions for different steepness of μ_{eff} vs. E_{eff} characteristics. Although μ_{eff} decreases at constant operating E_{eff} (A → B); at constant V_G, E_{eff} decreases with degradation (B → C) and thus gives rise to increase/decrease in μ_{eff} (A → C), depending on the steepness of the μ_{eff} vs. E_{eff} plot. Physics of E_{eff} reduction: (c-d) At constant V_G, there should be equal amount of gate and substrate charges in order to maintain the electrostatics. As a result, effect of N_{IT} generation, when measured at constant V_G, should consider a corresponding reduction in substrate charge (c→d). Since in an inverted MOS, Q_{dep} remains invariant of N_{IT} generation, $E_{eff} \sim Q_{dep} + \eta Q_{inv}$ reduces with N_{IT}.

However, one has to note that digital circuit operation is dictated by the change of mobility at constant gate voltage ($\mu_{eff}@V_G$), but not $\mu_{eff}@E_{eff}$. So knowledge of $\mu_{eff}@E_{eff}$ can not be sufficient to perform SPICE analysis for N_{IT} generation. Here, one should consider how E_{eff} changes during N_{IT} generation, where E_{eff} is defined as [23] [26]:

$$E_{eff} = \frac{1}{\varepsilon_{Si}}\left(Q_{dep}+\eta Q_{inv}\right) \sim \frac{V_T-V_{FB}-2\psi_B}{3T_{ox}}+\eta\frac{\left(V_G-V_T\right)}{3T_{ox}} \quad (2)$$

In (2), Q_{dep} and Q_{inv} are the depletion and inversion charges within the substrate respectively, ε_{Si} is the dielectric constant of silicon, ψ_B is the difference between the substrate Fermi level and the intrinsic Fermi level of silicon, and $\eta = 1/2$ for NMOS and 1/3 for PMOS respectively. Since the PMOS transistor is biased in inversion during NBTI stress at constant V_G, N_{IT} generation reduces $Q_{inv} \sim C_G(V_G\text{-}V_T)$ (see Fig. 3c→d) and hence E_{eff}. As a result of this E_{eff} reduction at constant V_G, there should always be some increase in μ_{eff}, as shown by the B → C transitions in Fig. 3b.

Since $\mu_{eff}@V_G$ will depend on both A→B and B→C transitions, depending on the steepness of the μ_{eff} vs. E_{eff} relationship (Fig. 3b), for transistors with larger steepness B→C transition will dominate over the A→B transition, with an overall increase in $\mu_{eff}@V_G$. Conversely, for transistors with smaller steepness, A→B transition dominates the B→C transition with an overall decrease in $\mu_{eff}@V_G$. Therefore, the steepness of μ_{eff} vs. E_{eff} relationship is an important factor in identifying the sign and magnitude of $\Delta\mu_{eff}@V_G$ due to N_{IT}. We have experimentally demonstrated these considerations in [22] by tuning μ_{eff} vs. E_{eff} steepness with strain.

B. Effect of trapping into Oxide Defects during BTI

Trapping into bulk oxide defects (*i.e.*, hole trapping N_{HT} for NBTI stress and electron trapping N_{ET} for PBTI stress) increase $|V_T|$. However, since the oxide defects are located away from the channel, charge trapping into these defects has relatively small (and often negligible) impact on *SS*, $g_{m,max}$, and $\mu_{eff}@E_{eff}$. Fig. 4 supports this conclusion by showing PBTI-related N_{ET} has no impact *SS* and $g_{m,max}$. Likewise, NBTI-related N_{HT} appears to have no effect on *SS* [24] for certain class of PMOS transistors, having dominance of N_{HT} over N_{IT} at short stress time [2][16][28]. Therefore, in these transistors, initial part of ΔV_T shows negligible *SS* variation (see the red squares of Fig. 2a). However, at higher level of ΔV_T, N_{IT} generation dominates over N_{HT} during NBTI stress and *SS* begins to show a linear increase with ΔV_T for the remaining stress durations.

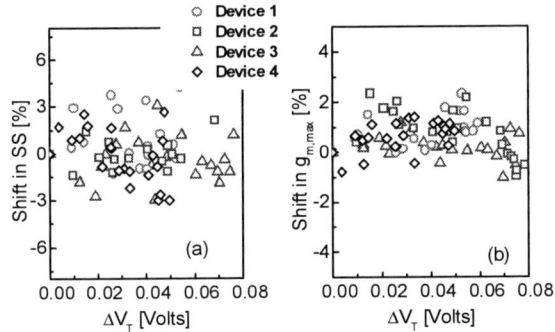

Figure 4. PBTI-induced N_{ET} on high-κ NMOS transistors has negligible impact on (a) sub-threshold slope $SS \equiv \partial(log_{10}I_D)/\partial V_G$ and (b) maximum transconductance $g_{m,max}$ (thus $\mu_{eff}@E_{eff}$ according to (1)). The result is also consistent with [5]. Noise in Fig. 4a depends on the range of V_G used in *SS* determination.

Now that we know *SS* and $g_{m,max}$ are relatively insensitive to trapping in oxide defects, let us study how μ_{eff} is affected by such trapping. Since $\mu_{eff}@E_{eff}$ or $g_{m,max}$ remains unchanged due to N_{ET} (Fig. 4b) and E_{eff} always reduces due to N_{ET} at constant V_G, $\Delta\mu_{eff}@V_G$ for N_{ET} (or, in general, due to oxide defects) will **always** be positive (see Fig. 5c). Therefore, we obtain comparable values of positive $\Delta\mu_{eff}@V_G$ for the high-κ NMOS transistor (studied in this paper) and the highly strained PMOS transistor of [22] (compare Fig. 5c with Fig. 5a), even though

μ_{eff}-E_{eff} steepness for the high-κ NMOS transistor is smaller (see Fig. 5b) than that of highly strained PMOS transistor of [22]. This is because the presence of N_{IT} causes negative $\Delta\mu_{eff}@E_{eff}$, which dictates that μ_{eff} - E_{eff} steepness be enhanced for N_{IT} to achieve overall positive $\Delta\mu_{eff}@V_G$. On the other hand, $\Delta\mu_{eff}@E_{eff}$ is negligible for N_{ET} (Fig. 4b), thus N_{ET}-related $\Delta\mu_{eff}@V_G$ is always positive, irrespective of μ_{eff}-E_{eff} steepness.

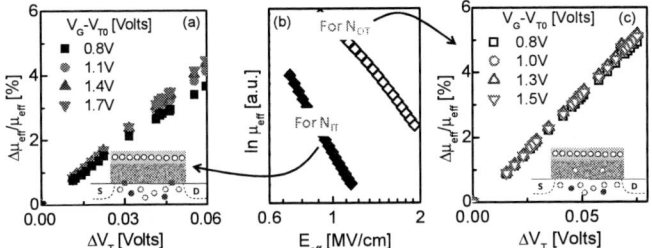

Figure 5. (a) $\Delta\mu_{eff}$ at different $(V_G\text{-}V_{T0})$ for the highly strained PMOS transistor in [22]. (b) Comparison of θ between the high-κ NMOS transistor under study and the highly strained PMOS transistor of [22]. (b) Although NMOS transistor has smaller θ, presence of negligible $\Delta\mu_{eff}@E_{eff}$ and E_{eff} reduction at constant V_G stress provide positive $\Delta\mu_{eff}@V_G$ due to N_{ET}.

C. Summary: Impact of BTI on transistor parameters

Even though $|V_T|$ always increases with BTI, the change in *SS*, $g_{m,max}$ and μ_{eff} depend on the type of transistor and the type of defects under study. In general, N_{IT} increases *SS* and reduces $g_{m,max}$, while oxide defects (N_{ET} or N_{HT}) has negligible impact on those two parameters. More importantly, μ_{eff} does not always reduce due to BTI, as presumed in almost all the circuit analysis, so far reported in literature. Depending on the type of transistor (specifically, depending on its mobility-field characteristics) and the type of defects (interface/oxide), μ_{eff} can increase due to BTI. This observation has important consequences on the drivability (and hence circuit performance) of modern CMOS transistors, as discussed next.

III. IMPACT OF DEFECTS ON DRAIN CURRENT AND ITS SELF-COMPENSATION

In the last section, we have discussed the effect of BTI-induced interface/oxide defect on transistor parameters. However, the performance of logic/memory circuits (*i.e.*, charging/ discharging of load capacitance C_L by an inverter, static noise margin or SNM of an SRAM cell) is dictated essentially by changes in transistor drivability or I_D. In this section, we explain the effect of defect on I_D and show that the variation of I_D (*i.e.*, ΔI_D) in BTI-degraded circuits differs significantly from the traditional view of this problem.

A. Variation of I_D in linear regime ($I_{D,lin}$)

Since $I_{D,lin}$ depends on both inversion charge $Q_{inv} \sim C_G(V_G\text{-}V_T)$ and μ_{eff}, $\Delta I_{D,lin}$ depends on both ΔQ_{inv} and $\Delta\mu_{eff}$. Therefore, we can write –

$$\left.\frac{\Delta I_{D,lin}}{I_{D,lin0}}\right|_{V_G} = \left.\frac{\Delta\mu_{eff}}{\mu_{eff0}}\right|_{V_G} + \frac{\Delta Q_{inv}}{Q_{inv0}} = \left.\frac{\Delta\mu_{eff}}{\mu_{eff0}}\right|_{V_G} - \frac{\Delta V_{T,lin}}{V_G - V_{T,lin0}}. \quad (3)$$

Here, the subscript 0 indicates pre-BTI stress values. Since, second term in the right-hand side of (3) is always negative,

$\Delta I_{D,lin}$ would have been negative as well (*i.e.*, transistor will have reduced drivability), **unless** $\Delta I_{D,lin}$ is compensated by the positive contribution from $\Delta \mu_{eff} @ V_G$. In conventional transistor with small μ_{eff}-E_{eff} steepness, the contribution from $\Delta \mu_{eff}$ is relatively small, *i.e.*, $\Delta \mu_{eff} @ V_G \sim \Delta \mu_{eff} @ E_{eff}$ – such that $\Delta I_{D,lin}$ is always negative for classical transistors. However, as discussed in section II-A and also in [22], for uniaxially strained transistors with high μ_{eff}-E_{eff} steepness, $\Delta \mu_{eff} @ V_G$ can be positive during N_{IT} generation. Indeed, for a highly strained transistor, with high positive $\Delta \mu_{eff} @ V_G$, as shown in Fig. 5a, $\Delta I_{D,lin}$ becomes negligible and total self-compensation may be achieved at $V_G - V_{T0} \geq 1.4V$ (Fig. 6a). At lower $V_G - V_{T0}$, the second term of (3) is more negative, and hence compensation is not complete.

Similarly, self-compensation of $I_{D,lin}$ is also present for N_{ET} during PBTI stress. Since $\Delta \mu_{eff} @ V_G$ is always positive for oxide defects (see discussions in section II-B), $\Delta I_{D,lin}$ can indeed be totally compensated at higher $(V_G - V_{T0})$ and partially compensated at lower $(V_G - V_{T0})$ through mobility improvement (see Fig. 6b).

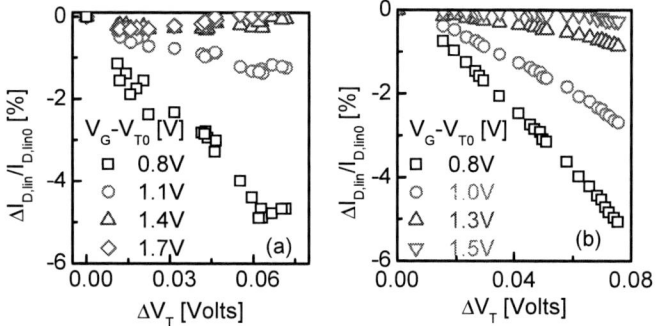

Figure 6. (a) $\Delta I_{D,lin}$ at different V_G-V_{T0} for the highly strained PMOS transistor, having $\Delta \mu_{eff} @ V_G$ values of Fig. 5a. (b) $\Delta I_{D,lin}$ at different V_G-V_{T0} for the high-κ NMOS transistor, having $\Delta \mu_{eff} @ V_G$ values of Fig. 5c.

B. Variation of I_D in saturation regime ($I_{D,sat}$)

Since analysis of drain current in saturation regime is more complicated than its analysis in linear regime, we offer a simple estimate of the relevance of self-compensation in the saturation regime by using the scattering theory of Lundstrom *et al.* [29]. According to this scattering theory for linear regime of source-to-drain transport, source-injected carriers are continuously scattered over the entire channel length (L) and hence a change in mobility in all parts of the channel will affect the current transport. On the other hand, in the saturation regime, electric field along the channel is very high. As a result, for calculating current transport, we only need to consider scattering of source-injected carriers over a characteristics distance (ℓ) away from the top of the energy barrier, located near the source end. Carriers that have crossed ℓ indeed face scattering, but these scattered carriers can not overcome the thermal barrier to reach the source side and always gets collected at the drain end. Hence, a change in mobility or $\Delta \mu_{eff}$ will have reduced impact on $I_{D,sat}$ (having effective scattering length of ℓ), compared to $I_{D,lin}$ (having scattering length of L). Thus, the overall change in ΔI_D due to

BTI degradation in linear and saturation regions can be approximated as,

$$\frac{\Delta I_{D,lin(sat)}}{I_{D,lin(sat)0}}\Bigg|_{V_G} = \left(1 - B_{lin(sat)}\right)\frac{\Delta \mu_{eff}}{\mu_{eff0}}\Bigg|_{V_G} - \frac{\Delta V_{T,lin(sat)}}{V_G - V_{T,lin(sat)0}}. \quad (4)$$

Now, the ballistic coefficients in the linear regime (B_{lin}) is ~ 0, in the saturation regime (B_{sat}) ranges from $0 << B_{sat} < 1$ [29][30]. Therefore, one expects self-compensation to be more effective for $\Delta I_{D,lin}$ compared to $\Delta I_{D,sat}$. For example, in Fig. 6, we have observed total compensation in linear regime between the two terms of the right-hand side of (4) at higher $(V_G$-$V_{T0})$. However, for the same transistors at similar $(V_G$-$V_{T0})$, we observe reduced self-compensation in the saturation regime (see Fig. 7). Nevertheless, in both cases of Fig. 7, the improvement is still significant and relevant for reduction in design margins.

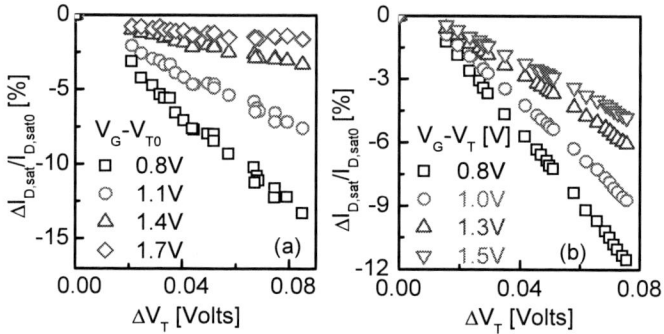

Figure 7. (a) $\Delta I_{D,sat}$ at different V_G-V_{T0} for the highly strained PMOS transistor, having $\Delta I_{D,lin}$ values of Fig. 6a. (b) $\Delta I_{D,sat}$ at different V_G-V_{T0} for the high-κ NMOS transistor, having $\Delta I_{D,lin}$ values of Fig. 6b.

C. Variation of OFF-state leakage (I_{OFF})

Finally, we discuss the presence of self-compensation on the variation of I_{OFF} in the sub-threshold region $(V_G < V_T)$.[1] Since I_{OFF} for a transistor is dominated by thermionic emission (when band-to-band tunneling is negligible)[2], we can write $|I_{OFF}| \sim exp\ [-q|V_T|/mk_BT]$ [23] and its differentiation in the sub-threshold regime results –

$$\frac{\Delta I_{OFF}}{I_{OFF}} = \frac{1}{mk_BT}\left(-|\Delta V_T| + \frac{\Delta m}{m}|V_T|\right) \quad (5)$$

where, $m = 1 + (C_D + C_{IT})/C_{di}$ is the body-effect co-efficient, C_D is the depletion-layer capacitance, and C_{IT} is the interface trap capacitance. Thus (5) suggests that a partial compensation between ΔV_T and Δm (which obviously has different physics compared to the self-compensation between $\Delta \mu_{eff}$ and ΔV_T, discussed so far) will only be possible for $|V_T|/m \sim 1$. However, modern transistors has $m > 1$ and $|V_T| \sim 0.3$V [31].

---

[1]   In addition to the digital circuit operation considered here, I_{OFF} plays the major role in I_{DDQ}-based BTI degradation monitoring, reported in [12].

[2]   If I_{OFF} is dominated by band-to-band tunneling (as the case for transistors having low bandgap Ge substrates [32]), any change in V_T will directly reflect in a corresponding change in I_{OFF}. Hence, self-compensation for I_{OFF} is also unexpected in such short-channel transistors.

978-1-4244-5430-3/10 $26.00 © 2010 IEEE

Therefore, self-compensation of I_{OFF} is impossible in CMOS transistors.

IV. SELF-COMPENSATION IN DIGITAL CIRCUITS

So far we have explored the implication of self-compensation on individual transistor parameters and demonstrated that (relative) self-compensation can be routinely achieved under various conditions in modern CMOS transistors. However, we have shown in section III that self-compensation in not equally effective at all bias conditions. Indeed, during circuit operation, a transistor goes through a variety of bias conditions, therefore the obvious question is: 'Without perfect self-compensation at all bias conditions, will there be any significant effect of this phenomena in actual circuit operation?'. In this section, we answer this question by studying the consequence of self-compensation in simple logic and memory circuits. We use measured I-V characteristics before and after NBTI degradation to obtain performance response of two basic circuits (i.e., single-stage inverter, SRAM) and show how self-compensation improves the circuit performance significantly.

A. Effect of BTI on single-stage inverter

First, consider the effect of NBTI self-compensation in an inverter driving a constant load $C_L \sim 6fF$ (Fig. 8a). Since C_L is typically dominated by a comparatively large interconnect capacitance (over transistor's intrinsic capacitance), delay in other digital circuits (e.g., ring oscillator, combinatorial gates, etc.) would follow similar trends. For the inverter under study, only the PMOS transistor is subjected to NBTI-induced $\Delta V_T \sim 70mV$ due to N_{IT} generation. On the other hand, the NMOS transistor has matched I-V characteristics (with PMOS) in pre-NBTI stress condition. Later, for a particular input voltage V_{IN} (with frequency =1GHz and rise/fall time =1ps), we calculate the output voltage V_{OUT} transient using –

$$\frac{dV_{OUT}}{dt} = \frac{I_P(V_{IN}, V_{OUT}) - I_N(V_{IN}, V_{OUT})}{C_L}, \quad (6)$$

where, I_P and I_N are currents through the PMOS and NMOS transistors of the inverter, respectively, as shown in Fig. 8a. After simulating the V_{OUT} transients, we compare V_{IN} with V_{OUT} and estimate the rise (τ_R) and fall (τ_F) delay of the inverter (see Fig. 8b for a schematic). Since we only consider NBTI degradation, the fall delay remains (approximately) unaffected due to stress. On the other hand, the rise delay for a self-compensated transistor (having $\Delta I_{D,lin} \sim 0$ for $|V_G - V_{T0}| > 1.1V$) increases by ~5% (i.e., $\Delta \tau_R \sim 5\%$), which is approximately half of the rise delay degradation for a non-compensated transistor (having $\Delta \tau_R \sim 11\%$ for similar NBTI-induced ΔV_T). Thus, considering that delay degradation due to NBTI for a circuit has similar time-exponent ($n \sim 1/6$) as an individual transistor, as observed in [8], self-compensation may provide ~64 times increase in NBTI lifetime (Fig. 8c) – possibly eliminating the NBTI degradation in CMOS reliability concern!

Similarly, for the case of PBTI degradation in an NMOS transistor, we expect that $\Delta \tau_F$ for a self-compensated transistor to be much smaller than that for a non-compensated transistor.

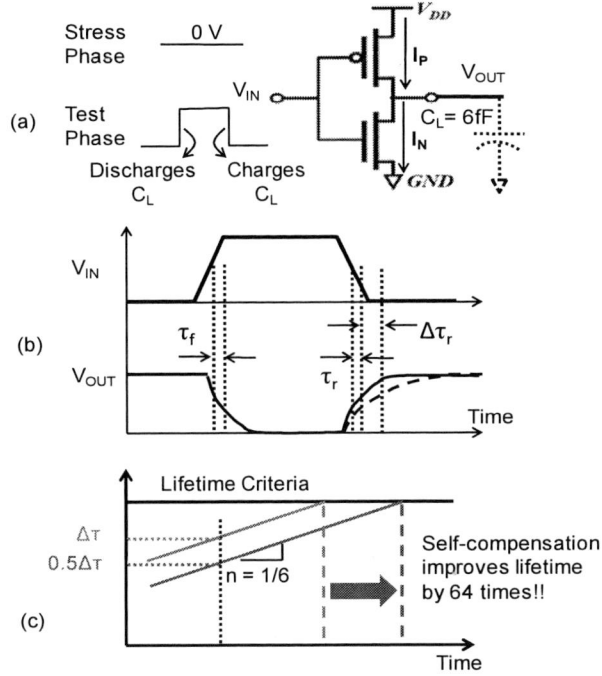

Figure 8. (a) Configuration of the inverter used in section IV-A. In stress phase (V_{IN} =0), there is NBTI degradation in PMOS transistor. (b) The impact of NBTI degradation (i.e., change in rise τ_R and fall delay τ_F) is monitored by applying a test pulse as V_{IN} and calculating V_{OUT} using (6). (c) Self-compensation reduces the rise delay degradation by 50% – thus can increase NBTI lifetime by ~64 times (assuming time exponent for delay degradation, $n \sim 1/6$, as obtained in [9]).

B. Effect of BTI on SRAM performance

Next, we use a conventional 6T SRAM cell (Fig. 9a) for explaining the effect of BTI degradation on SRAM cells. One common feature of NBTI degradation in SRAM cell is the change in static noise margin (SNM, defined in Fig. 9b) [8] [11] [33] [34]. For example, for a particular bit storage, if V_L = '0' and V_R = '1', then transistor PR will be under NBTI degradation and transistor NL will be under PBTI degradation. As a result, $|V_T|$ for PR (due to NBTI) will increase and, therefore, shift the PR-NR of Fig. 9b towards left, thus reduces SNM_1 and increases SNM_2. Similarly, increase in V_T for NL (due to PBTI) shifts up the PL-NL of Fig. 9b, thus also reduces SNM_1 and increases SNM_2. Thus, the storage condition of V_L = '0' and V_R = '1', always reduces SNM_1 and increases SNM_2; whereas, the opposite happens for the storage condition of V_L = '1' and V_R = '0'.[3]

Now to understand the effect of self-compensation on NBTI degradation, we compare the SNM variation for the two cases: in one case we use a self-compensated transistor for PR having $\Delta I_{D,lin}$ and $\Delta I_{D,sat}$ of Fig. 6a and Fig. 7a respectively, and in the second case we use a non-compensated transistor for PR. We observe that self-compensation reduces SNM_1 degradation from 20mV (or 2.75% for the non-compensated

---

[3] Actual signal probabilities of V_L and V_R in practical memory operation are far more complicated and should be subjected to a statistical evaluation [8].

transistor) to 7.2mV (or 2% for the compensated transistor) and increases SNM$_2$ improvement from 19.5mV (or 2.68% for the non-compensated transistor) to 23.7mV (or 6.64% for the compensated transistor). Similar improvement in SNM$_1$ degradation is also expected for a self-compensated transistor, subjected to PBTI stress.

However, SNM is not the only performance criteria for the operation of an SRAM cell. Read delay of an SRAM cell (studied using BL$_B$ = BL = '1') depends on the strength of NMOS excess transistors and thus it is unaffected by NBTI [33], but should be affected by PBTI. On the other hand, NBTI (PBTI) weakens the PMOS (NMOS) compared to NMOS (PMOS) and thus should cause a decrease (increase) in write delay. Similar to the improvement in delay degradation for logic circuits (see section IV-A), self-compensation is again expected to reduce the read and write delay variations.

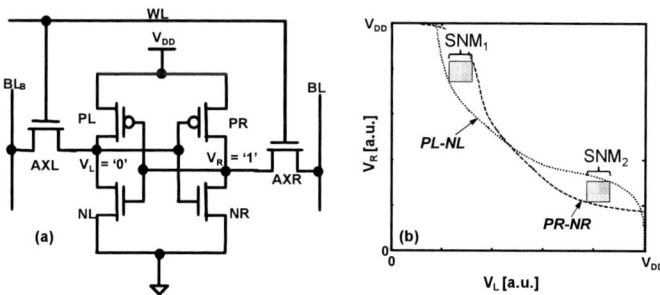

Figure 9. (a) Schematic of a 6T-SRAM Cell. PL and PR are PMOS transistors within the storage framework; whereas NL and NR are corresponding NMOS transistors. AXL and AXR are NMOS transistors used for accessing (reading or writing) stored data. (b) Static Noise Margin (SNM) calculation using voltage-transfer characteristics of back-to-back inverters. This figure is re-drawn, based on [8].

C. Large-scale SPICE analysis

So far we have analyzed the effect of self-compensation on circuit performance using the measured I-V characteristics of a transistor before and after BTI degradation and then calculating the V_{IN}-V_{OUT} relationship through (6). However, for large scale circuit simulation, a detailed SPICE simulation would be necessary. To perform that, one should consider the effect of E_{eff} reduction, which we have experimentally observed under BTI degradation. Such E_{eff} reduction ultimately provides the opportunity of having positive $\Delta\mu_{eff}@V_G$ in a transistor having certain μ_{eff}-E_{eff} steepness, which is needed for obtaining partial or full self-compensation. These issues related to detailed SPICE simulation will be discussed in a future article.

D. Summary:Imapct of self-compensation in circuits

Though classical circuit-level BTI analysis [8]-[12][33]-[35] ignores the effect of self-compensation (i.e., $\Delta\mu_{eff}@V_G$ is always considered to be negative), we have shown that self-compensation should be present in advanced CMOS technology [39], which (i) have significant amount of strain (~1 GPa or more) within the channel, (ii) use high-κ dielectric within the gate stack, (iii) have recent concerns with PBTI, in addition to NBTI, and (iv) may need to use novel orientation, as well as, III-V as substrate material [38]. Simple circuit simulations that are performed in this section indeed show the improvement in circuit-level BTI degradation through self-

compensation. However, a detailed SPICE simulation will indeed to helpful to draw a more general conclusion in this matter.

V. SELF-COMPENSATION AND SMALL G$_{M,ON}$

In the preceding analysis, we have shown the importance of μ_{eff} - E_{eff} steepness to compensate the effect of ΔV_T on ΔI_D, thus reducing the BTI concern in integrated circuits. Indeed, the requirement of steep μ_{eff} - E_{eff} characteristics for self-compensation is easy to explain graphically: Since $I_D \sim Q_{inv}\mu_{eff}$, transistor's I_D-V_G characteristics depends on the functional dependencies of Q_{inv} and μ_{eff} on V_G. In classical transistors, μ_{eff} decreases weakly with V_G [26]. As a result, dependence of I_D on V_G in ON state (i.e., $g_{m,ON}$) is dominated by the Q_{inv} - V_G ($Q_{inv} \sim V_G - V_T$ increases linearly with V_G) relationship. This results larger $g_{m,ON}$ in classical transistors and makes them susceptible to ΔV_T (Fig. 10b). Such effect of ΔV_T (resultant ΔI_D) in classical transistor, for example, can be optimized by using an extra guard-band voltage (V_{DD}-V_1) during IC design, so that even after maximum ΔV_T, I_D remains above $I_{D,min}$ - thus ensuring nominal performance (e.g., circuit delay, clock frequency, etc.) for a technology.

On the other hand, in a self-compensated transistor, higher negative steepness of the μ_{eff} - E_{eff} relationship matches the positive steepness of the Q_{inv} -V_G relationship – making the I_D-V_G relationship at ON state relatively flat (Fig. 10c). Such flat I_D-V_G characteristic during ON state (i.e., small $g_{m,ON}$) makes the I_D of a transistor immune to ΔV_T. Moreover, self-compensation would possibly reduce the need for guard-band voltage (V_{DD}-V_1) and thereby allow the transistors to operate at smaller V_{DD}. Hence, it is desirable to design transistors not only with higher μ_{eff} (for better performance), but also with higher μ_{eff} - E_{eff} steepness (for better reliability).

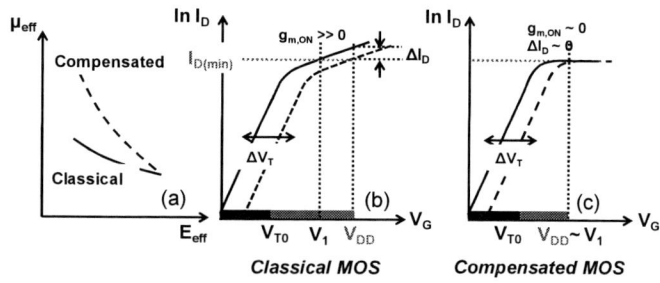

Figure 10. (a) μ_{eff}-E_{eff} and (b-c) transfer characteristics of classical and ΔV_T-compensated MOS transistors. Classical transistor handles ΔV_T using an extra guard-band voltage (V_{DD}-V_1), which can be reduced through self-compensating ΔV_T using small $g_{m,ON}$.

VI. OBTAINING SELF-COMPENSATION

One can achieve self-compensation by increasing the steepness of μ_{eff} - E_{eff} characteristics through uniaxial strain [22]. Uniaxial strain reduces inter-valley phonon scattering by introducing valley-splitting [36], and therefore makes surface roughness scattering (having larger μ-E steepness [26] compared to the phonon component) the dominant component in the μ_{eff} - E_{eff} characteristics. However, one might presume that the presence of self-compensation is a property of strained technology and may not be applicable in general. To refute

978-1-4244-5430-3/10 $26.00 © 2010 IEEE

such presumption, we study self-compensation in one particular transistor at different temperature. Since a transistor operated at low temperature has reduced phonon scattering (like the case for higher uniaxial strain), one can have higher μ_{eff} - E_{eff} steepness at lower temperature. We experimentally observe such an increase in the μ_{eff} - E_{eff} steepness with reduction in temperature (see Fig. 11a).[4] And consequently, higher μ_{eff} - E_{eff} steepness gets reflected in the reduced $\Delta I_{D,lin}$, even at $|V_G - V_{T0}| \sim 0.8V$ for the highly strained PMOS transistor of Fig. 6a.

Therefore, higher μ_{eff} - E_{eff} steepness or small $g_{m,ON}$ is indeed required for designing a self-compensated transistor. Encouragingly, the recent reports of I_D-V_G characteristics in Intel's 45-nm strained MOS technology [5] and III-V transistors [38] indicate the presence of higher μ_{eff} - E_{eff} steepness or small $g_{m,ON}$. Thus advanced substrates involving strained-Si and III-V show the possibility of self-compensating ΔV_T with reduced use of guard-band or any complicated circuit-level optimization.

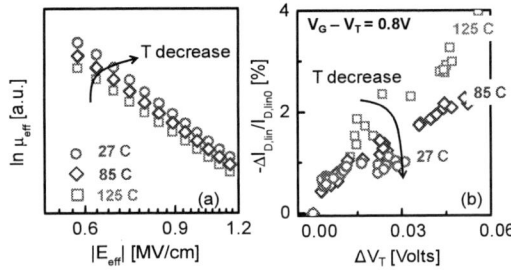

Figure 11. (a) Reduction in temnperature increases the μ_{eff} - E_{eff} steepness for the highly-strained PMOS transistor of Fig. 6a. (b) The transistor, operated at reduced temperature, is less sensitive to N_{IT}-induced $\Delta I_{D,lin}$ at $|V_G - V_{T0}| = 0.8V$.

VII. CONCLUSION

Therefore, we illustrate the existence of self-compensation both for interface (originally proposed in [22]) and oxide defects. We establish that the concept of self-compensation is mainly related to the μ_{eff}-E_{eff} steepness and is more pronounced for ΔV_T related to oxide defects compared to that for interface defect. We also perform simple circuit analysis and show that the lifetime of digital circuit increases dramatically, once we incorporate the effect of self-compensation. Hence, the presence of self-compensation might obviate (or at least substantially reduce) the need for guard-band voltage [15], expensive design algorithms [9] or external circuits [17]-[19] for mitigating the effect from BTI.

MOS technology has so far evolved to maximize transistor's performance by introducing new concepts/materials. Here, we speculate that the new substrate materials (*e.g.*, strained-Si, III-V) currently being studied for performance improvement, can also provide better resilience against the effects of BTI. Increase in negative steepness of μ_{eff} - E_{eff} relationship or smaller $g_{m,ON}$ (*i.e.*, flatter I_D-V_G) has been identified as the requirement for such self-compensation, which

---

[4]   Similar increase in steepness is already shown for both PMOS and NMOS transistors in [40].

is can be easily achieved in strained/III-V transistors. The use of such degradation-free technology would not only simplify IC design, but also help reduce the supply voltage (and power dissipation) by significantly curtailing the guard-band necessary to account for ΔV_T in classical transistors.

ACKNOWLEDGEMENTS

We gratefully acknowledge Sematech for the high-κ NMOS transistors and Dr. Khaled Ahmed (Applied Materials) for the strained transistors used in this study. We further acknowledge Network for Computational Nanotechnology (NCN) for computational resources, Birck Nanotechnology Center (BNC) for experimental facilities, and Taiwan Semiconductor Manufacturing Corp. (TSMC), Applied Materials (AMAT), and Nanoelectronic Research Initiative (NRI) for financial support during this work. In addition, we thank Charles Augustine (Purdue University), Prof. Kaushik Roy (Purdue University), Prof. Mark S. Lundstrom (Purdue University) and Prof. Kevin Cao (Arizona State University) for insightful discussions on related issues.

REFERENCES

[1]   K. Bernstein, D. J. Frank, A. E. Gattiker, et al., "High-performance CMOS variability in the 65-nm regime and beyond," IBM Journal of Research and Development, vol. 50, pp. 433-449, Jul-Sep 2006.

[2]   A. E. Islam, H. Kufluoglu, D. Varghese, et al., "Recent Issues in Negative Bias Temperature Instability: Initial Degradation, Field-Dependence of Interface Trap Generation, Hole Trapping Effects, and Relaxation," *IEEE Transactions on Electron Devices*, vol. 54, pp. 2143-2154, 2007.

[3]   G. Ribes, J. Mitard, M. Denais, *et al.*, "Review on high-k dielectrics reliability issues," *IEEE Transactions on Device and Materials Reliability*, vol. 5, pp. 5-19, 2005.

[4]   D. Heh, C. D. Young, and G. Bersuker, "Experimental evidence of the fast and slow charge trapping/detrapping processes in high-k dielectrics subjected to PBTI stress," *IEEE Electron Device Letters*, vol. 29, pp. 180-182, 2008.

[5]   S. Pae, M. Agostinelli, M. Brazie, *et al.*, "BTI reliability of 45 nm high-k plus metal-gate process technology," *International Reliability Physics Symposium Proceedings*, pp. 352-357, 2008.

[6]   T. Grasser, B. Kaczer, W. Goes, T. Aichinger, P. Hehenberger, and M. Nelhiebel, "A Two-Stage Model for Negative Bias Temperature Instability," *International Reliability Physics Symposium Proceedings*, pp. 33-44, 2009.

[7]   B. C. Paul, K. Kang, H. Kufluoglu, *et al.*, "Impact of NBTI on the temporal performance degradation of digital circuits," *IEEE Electron Device Letters*, vol. 26, pp. 560-562, 2005.

[8]   K. Kang, H. Kufluoglu, K. Roy, *et al.*, "Impact of Negative Bias Temperature Instability in Nano-Scale SRAM Array: Modeling and Analysis," *IEEE Transactions on Computer-Aided Design of Integrated Circuits and Systems*, vol. 26, pp. 1770-1781, 2007.

[9]   B. C. Paul, K. H. Kang, H. Kufluoglu, *et al.*, "Negative bias temperature instability: Estimation and design for improved reliability of nanoscale circuits," *IEEE Transactions on Computer-Aided Design of Integrated Circuits and Systems*, vol. 26, pp. 743-751, 2007.

[10]   W. P. Wang, V. Reddy, A. T. Krishnan, *et al.*, "Compact Modeling and simulation of circuit reliability for 65-nm CMOS technology," *IEEE Transactions on Device and Materials Reliability*, vol. 7, pp. 509-517, 2007.

[11]   S. V. Kumar, C. H. Kim, and S. S. Sapatnekar, "Impact of NBTI on SRAM read stability and design for reliability," *Proceedings of International Symposium on Quality Electronic Design*, pp. 210-215, 2006.

[12]   K. Kang, K. Kim, A. E. Islam, *et al.*, "Characterization and estimation of circuit reliability degradation under NBTI using on-line I-DDQ

978-1-4244-5430-3/10 $26.00 © 2010 IEEE

measurement," *IEEE Design Automation Conference*, pp. 358-363, 2007.

[13] J. J. Kim, R. Rao, S. Mukhopadhyay, *et al.*, "Ring oscillator circuit structures for measurement of isolated NBTI/PBTI effects," *Proceedings of IEEE International Conference on Integrated Circuit Design and Technology*, pp. 163-166, 2008.

[14] V. Reddy, A. T. Krishnan, A. Marshall, *et al.*, "Impact of negative bias temperature instability on digital circuit reliability," in *IEEE International Reliability Physics Symposium*, pp. 248-254, 2002.

[15] Y. H. Lee, W. McMahon, N. Mielke, *et al.*, "Managing bias-temperature instability for product reliability," *Proceedings of International Symposium on VLSI Technology*, pp. 52-53, 2007.

[16] S. Mahapatra, K. Ahmed, D. Varghese, A. E. Islam, G. Gupta, L. Madhav, D. Saha, and M. A. Alam, "On the Physical Mechanism of NBTI in Silicon Oxynitride p-MOSFETs: Can Differences in Insulator Processing Conditions Resolve the Interface Trap Generation versus Hole Trapping Controversy?," *International Reliability Physics Symposium Proceedings*, pp. 1-9, 2007.

[17] S. M. Martin, K. Flautner, T. Mudge, *et al.*, "Combined dynamic voltage scaling and adaptive body biasing for lower power microprocessors under dynamic workloads," *IEEE/ACM International Conference on CAD*, pp. 721-725, 2002.

[18] J. W. Tschanz, J. T. Kao, S. G. Narendra, *et al.*, "Adaptive body bias for reducing impacts of die-to-die and within-die parameter variations on microprocessor frequency and leakage," *IEEE Journal of Solid-State Circuits*, vol. 37, pp. 1396-1402, 2002.

[19] K. Kang, K. Kim, and K. Roy, "Variation resilient low-power circuit design methodology using on-chip phase locked loop," *ACM/IEEE Design Automation Conference*, pp. 934-939, 2007.

[20] M. Agarwal, B. C. Paul, M. Zhang, *et al.*, "Circuit failure prediction and its application to transistor aging," *Proceedings of IEEE VLSI Test Symposium*, pp. 277-284, 2007.

[21] M. Alam, "Reliability- and process-variation aware design of integrated circuits," *Microelectronics Reliability*, vol. 48, pp. 1114-1122, 2008.

[22] A. E. Islam and M. A. Alam, "On the possibility of degradation-free field effect transistors," *Applied Physics Letters*, vol. 92, p. 173504, 2008.

[23] Y. Taur and T. Ning, *Fundamentals of Modern VLSI Devices*: Cambridge University Press, 1998.

[24] A. E. Islam, V. D. Maheta, H. Das, *et al.*, "Mobility Degradation Due to Interface Traps in Plasma Oxinitride PMOS Devices," in *IEEE International Reliability Physics Symposium*, pp. 87-96, 2008.

[25] A. T. Krishnan, V. Reddy, S. Chakravarthi, *et al.*, "NBTI impact on Transistor and Circuit: Models, Mechanisms and Scaling Effects," in *International Electron Devices Meeting (IEDM) Technical Digest*, pp. 349-352, 2003.

[26] S. Takagi, A. Toriumi, M. Iwase, *et al.*, "On the Universality of Inversion Layer Mobility in Si Mosfets .1. Effects of Substrate Impurity Concentration," *IEEE Transactions on Electron Devices*, vol. 41, pp. 2357-2362, 1994.

[27] S. Villa, A. L. Lacaita, L. M. Perron, *et al.*, "Physically-based model of the effective mobility in heavily-doped n-MOSFET's," *IEEE Transactions on Electron Devices*, vol. 45, pp. 110-115, 1998.

[28] A. E. Islam, S. Mahapatra, S. Deora, et al., "On The Differences Between Ultra-fast NBTI Experiments and Reaction-Diffusion Theory", *International Electron Devices Meeting (IEDM)*, pp. 733-736, 2009.

[29] M. S. Lundstrom, "On the mobility versus drain current relation for a nanoscale MOSFET," *IEEE Electron Device Letters*, vol. 22, pp. 293-295, 2001.

[30] F. Assad, Z. Ren, D. Vasileska et al., "On the performance limits for Si MOSFET's: A theoretical study," *IEEE Trans. Electron Devices*, vol. 47, pp. 232-240, 2000.

[31] International Technology Roadmap for Semiconductors, 2009 Edition.

[32] T. Krishnamohan, D. Kim, C. D. Nguyen, C. Jungemann, Y. Nishi, and K. C. Saraswat, "High-mobility low band-to-band-tunneling strained germanium double-gate heterostructure FETs: Simulations," *IEEE Trans. Electron Devices*, vol. 53, no. 5, pp. 1000–1009, 2006.

[33] S. Bhardwaj, W. P. Wang, R. Vattikonda, *et al.*, "Predictive modeling of the NBTI effect for reliable design," *Proceedings of the IEEE Custom Integrated Circuits Conference*, pp. 189-192, 2006.

[34] J. C. Lin, A. S. Oates, and C. H. Yu, "Time dependent V-ccmin degradation of SRAM fabricated with high-k gate dielectrics," *Proceedings of International Reliability Physics Symposium*, pp. 439-444, 2007.

[35] K. K. Saluja, S. Vijayakumar, W. Sootkaneung, *et al.*, "NBTI Degradation: A Problem or a Scare?," *International Conference on VLSI Design*, pp. 137-142, 2008.

[36] S. E. Thompson, G. Y. Sun, Y. S. Choi, *et al.*, "Uniaxial-process-induced strained-Si: Extending the CMOS roadmap," *Ieee Transactions on Electron Devices*, vol. 53, pp. 1010-1020, 2006.

[37] R. K. Kirschman, *Low-temperature electronics*. NY: IEEE, 1986.

[38] C. L. Hinkle, A. M. Sonnet, R. A. Chapman, *et al.*, "Extraction of the Effective Mobility of $In_{0.53}Ga_{0.47}As$ MOSFETs," *IEEE Electron Device Letters*, vol. 30, pp. 316-318, 2009.

[39] A. E. Islam and M. A. Alam, "Self-Optimizing Defect Generation for Advanced CMOS Substrates", *International Integrated Reliability Workshop (IIRW)*, pp. 97-101, 2009.

[40] B. H. Cheng and J. Woo, "A temperature-dependent MOSFET inversion layer carrier mobility model for device and circuit simulation," *IEEE Trans. Electron Devices*, vol. 44, pp. 343-345, 1997.

Understanding noise measurements in MOSFETs: the role of traps structural relaxation

D. Veksler, G. Bersuker, S. Rumyantsev[1], M. Shur[1], H. Park, C. Young, K. Y. Lim, W. Taylor, R. Jammy

SEMATECH, Austin, TX 78741 and Albany, NY 12203, USA
phone: 518-649-1141; e-mail: Dmitry.veksler@SEMATECH.org
[1]ECSE, Rensselaer Polytechnic Institute, Troy, NY, USA

Abstract— The presented theoretical analysis of random telegraph signal (RTS) and 1/f noise data provides consistent interpretation of the measurement results allowing trap characteristics to be extracted and the atomic structure of oxide traps to be identified. We emphasize the critical role of the lattice structural relaxation associated with charge trapping/detrapping, which represents one of the major factors controlling electron capture/emission times.

Keywords - Electrical noise, random telegraph signal, MOSFET characterization, configurational relaxation of traps

I. INTRODUCTION

Random telegraph signal (RTS) noise, which results from fluctuations of the drain (gate) current between two or more discrete values, becomes more pronounced as device feature sizes are scaled, strongly affecting device performance. Analysis of RTS data can provide information on the properties of individual electron traps, such as their capture/emission times and their energy and spatial positions. This information, which can be extracted based on the ratio of trap capture and emission times [1-4], could be extremely valuable for improving the fabrication process if it would impart greater understanding of the defect.

A series of publications (see [4-6] and references therein) has noted that the atomic structure of traps and changes to it caused by electron capture might significantly affect the trapping dynamics. Consequently, the conventionally used elastic trapping model [7], in which the time of an injected electron capture by a bulk defect is controlled by the electron tunneling from/to the substrate, might be inadequate, as demonstrated in [8,9]. It points to a need to use a more comprehensive RTS analysis to correctly identify essential defect characteristics.

The concept of multi-phonon-assisted (non-elastic) tunneling [4-6], which takes into account the electron-phonon interaction, describes the rearrangement of the lattice atoms caused by the electrostatic coupling between the atom nuclei and the trapping electron. Such a lattice rearrangement, which requires the atoms to leave their positions of equilibrium, is associated with the system overcoming an energy barrier, which retards the electron trapping process.

Using this multi-phonon tunneling process analysis, the characteristics of the defects contributing to RTS in high-k/metal gate (HKMG) and SiON transistors and in the tunnel oxide of the charge trapping memory devices (TANOS) were obtained. Along with the trap energy and location, we extracted the trap relaxation energy, which is found to be an important identification marker of the defect. A comparison of the extracted values with those calculated for certain types of the oxygen vacancies in SiO_2 allows the traps contributing to the RTS to be identified.

II. MODEL DESCRIPTION

The capture time within the conventional Shockley-Read-Hall (SRH) approximation is

$$\tau_c = (v_t \times n \times \sigma)^{-1}, \qquad (1)$$

where n is the bulk density of the inversion charge carriers, v_t is their thermal velocity, and the capture cross-section σ is calculated by taking into account both (i) tunneling to (from) the trap, D and (ii) lattice relaxation, R [10], associated with the electron capture (emission):

$$\sigma = \sigma_0 \times D \times R, \qquad (2)$$

where

$$D = \exp(-(x_T / \lambda)), \quad R = e^{-(2\bar{n}+1)S} \sqrt{\left(\frac{\bar{n}+1}{\bar{n}}\right)^p} I_p\left(2 S \sqrt{\bar{n}(\bar{n}+1)}\right) \quad (3)$$

Here σ_0 is the electronic capture cross-section, $S = E_{relax}/\hbar\omega$, $p = (E_{CSi} - E_T)/\hbar\omega$, $\bar{n} = 1/(e^{\hbar\omega/kT} - 1)$, T is the ambient temperature, E_{CSi} is the Si conduction band edge, E_F is the Fermi level in the semiconductor (we assume that the trap does not exchange carriers with the gate metal), ω is the characteristic phonon frequency in the dielectric, and λ is the characteristic tunneling length for the electrons.

The model parameters are trap position, x_T; change of the total energy of the electronic subsystem after trapping, E_0; and trap relaxation energy, E_{relax}, which is a measure of the displacements of the lattice atoms around the defect caused by the trapped charge (Fig. 1). Eq. (3) is obtained by taking into account the electron-phonon (lattice vibrations) coupling in the Schrödinger equation (see Refs. [10-12]). Thus, a description of the trapping process requires the full system energy to be consideration.

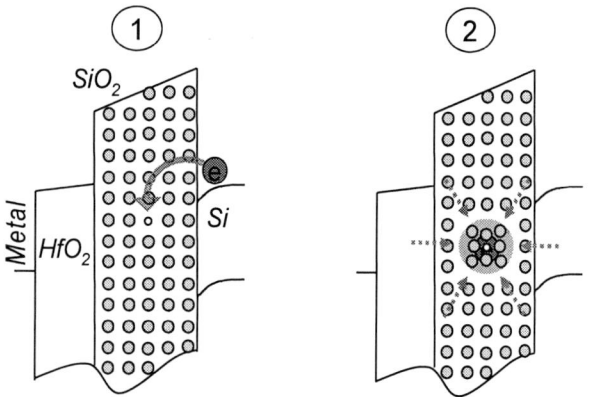

Figure 1. Schematic representation of the electron trapping process: geometric relaxation of the trap, caused by the trapped charge, involves displacements of the lattice atoms around the defect, switching its configuration.

The adiabatic potentials (Fig. 2a) for the total system include lattice vibrations (phonons) along the generalized coordinate, Q, before (electron resides in the channel) and after (i.e., the electron is in the trap and the atomic structure is distorted to accommodate the additional charge) trapping.

A ratio of the emission-to-capture times can be found from a conventionally used detailed balance equation:

$$\tau_e / \tau_c = \exp(-((E_T + e \cdot F_{ox} \cdot x_T) - E_F)/k \cdot T) \qquad (4)$$

The ratio τ_e/τ_c by itself is frequently used for extracting the trap position and energy [1-4]. However, separate evaluation of emission and capture times using Eqs. (1-4) allows the trap relaxation energy to be extracted, which is a critical identifier of the defect structure.

For a strong electron-phonon coupling ($S >> p$) and high temperature ($kT >> \hbar\omega$), Eq. (2) can be approximated by

$$\sigma = \sigma_0 \times \exp\left(-\left(\frac{x_T}{\lambda}\right)\right) \times \exp\left(-\left(\frac{E_B}{k \cdot T}\right)\right) \qquad (5)$$

Note that, as follows from Eq. (5), an electron capture must overcome an energy barrier.

The barrier height can be evaluated from crossing two parabolas corresponding to the initial and final adiabatic potentials of the system, as illustrated in Fig. 1a:

$$E_B = (E_{relax} - E_0)^2 / 4E_{relax}. \qquad (6)$$

Thus, the barrier height depends on the trap energy and on the gate bias [12, 13]. Even when there is no offset between the initial and final (after the electron trapping) energies of the system (the trap level is aligned with the Si conduction band $E_0 = 0$ in Fig. 2), the trap structural relaxation forms a barrier, which reduces the probabilities of capture and emission. In

other words, the trap structural relaxation leads to the electron capture and emission becoming thermally activated processes.

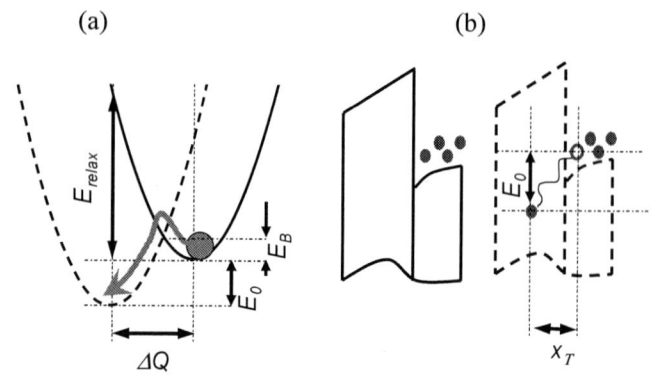

Figure 2. Total energy (a) and single electron (b) diagrams (Q – the generalized coordinate of the atomic displacements in the system). The "Initial" (solid line) and "Final" (dashed line) states correspond to the system with an empty and filled trap, respectively. E_{relax} is the energy corresponding to displacements of the lattice atoms, ΔQ, caused by electron trapping. E_0 is the energy lost by electronic subsystem after a trapping event. The diagram (a) indicates that electron trapping is associated with the system transitioning over the barrier E_B.

III. EXPERIMENT

A. RTS measurements and characteristic times extraction

The experimental data on temperature-dependent two-level RTS noise (caused by a single trap) in a small (W/L = 0.3μm/0.1μm) HiK MOSFET were collected in the temperature range of 300K-345K. The gate stack of the device was composed of 1 nm SiO$_2$/3 nm HfO$_2$/TaN. A 200 mm wafer was placed on a hotplate and biased using the battery-powered biasing box. Drain current was amplified with the low noise current preamplifier SR570SR. 40 GS/s oscilloscope with large memory (Lecroy WP740Zi) was used to record the current traces (10 Mpts/ch). Only a few MOSFETs out of over 50 devices have demonstrated a two-level RTS signal. A typical $I_d(t)$ trace (zoomed-in to show the details) is shown in Fig. 3a. The low current state corresponds to a filled trap. The high current state corresponds to an empty trap. The average capture, τ_C, and emission, τ_E, times were extracted by fitting the distributions of time periods in the low current state (for emission time) and high current state (for capture time) with the $\alpha \cdot \text{Exp}(-t_{hi,lo}/\tau_{C,E})$ dependence. (See Fig. 3b.)

Time resolution of the measurement setup was limited by the preamplifier bandwidth and became a very important factor as the characteristic capture and emission time decreased significantly with increasing temperature. Figure 4 demonstrates how the extraction error can be introduced into the interpretation of the experimental data. The dashed line in Fig.4 represents the threshold used at each moment to determine whether the current is high (when the measured value of the current is above the threshold) or low (when the value is below the threshold). As one can see (Fig. 4c), depending on the chosen threshold current value, the switching event between high-low states and back may be missed.

978-1-4244-5430-3/10 $26.00 © 2010 IEEE

current value being equal to $(3I_{high}+I_{low})/4$ and $(I_{high}+3I_{low})/4$, respectively.

Figure 5 shows the dependencies of extracted capture and emission times vs. the gate bias. The borders of the shaded areas correspond to the times extracted at two different values of the threshold current, mentioned above. Note that at 345 K and V_{GS}=0.43 V, the capture time becomes comparable to the time resolution limit of the experimental setup leading to a greater than order of magnitude uncertainty in the extracted average emission time. Further in the text, we compare the average values for capture and emission times with the theoretical values.

B. Low frequency noise measurments

The low frequency noise in TANOS devices (gate stack: 4.5 nm SiO_2/7 nm SiN/12 nm Al_2O_3; W/L = 0.3 μm/0.1 μm) was measured using the standard noise measurement setup. A SR570 spectrum analyzer was used to measure the noise spectral power density. The "quiet" battery-powered biasing box was employed to maximize the sensitivity of the system. The samples and the biasing box were placed in the shielded metallic chest. Fresh TANOS devices and devices that underwent 1000 cycle program/erase stress were studied.

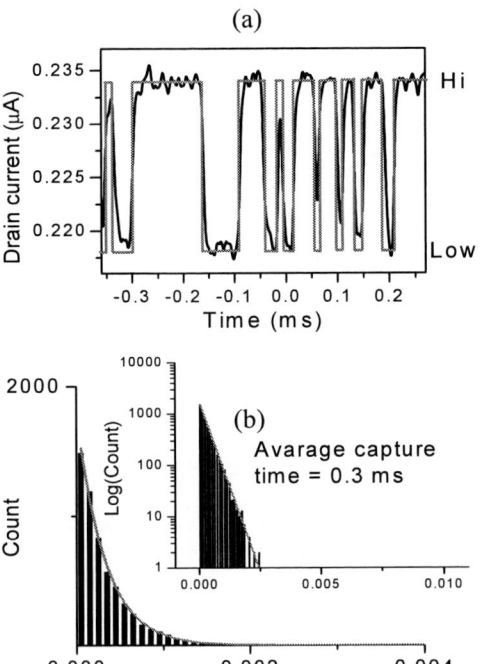

Figure 3. (a) Typical two-level RTS signal in HiK devices at T=330 K, VGS=0.13 V. Red line is drawn to guide the eye. Time resolution is estimated as 5 μs. (b) The average capture time is extracted by fitting capture time distribution by the exponent. Inset shows exponential fit of capture time distribution in the log scale.

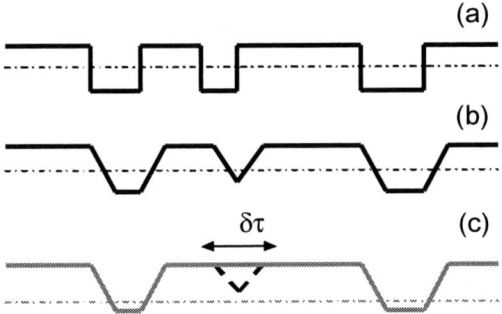

Figure 4. Effect of the setup bandwidth on the extracted parameter values. (a) Schematic of the actual trace of RTS signal. (b) Schematic of a measured RTS trace with the effect of the amplifier/circuitry delay. (c) An opportunity for overestimating the capture time (time in high current state) when emission time is getting close to the setup resolution limit and the high-low-high switching event is missed.

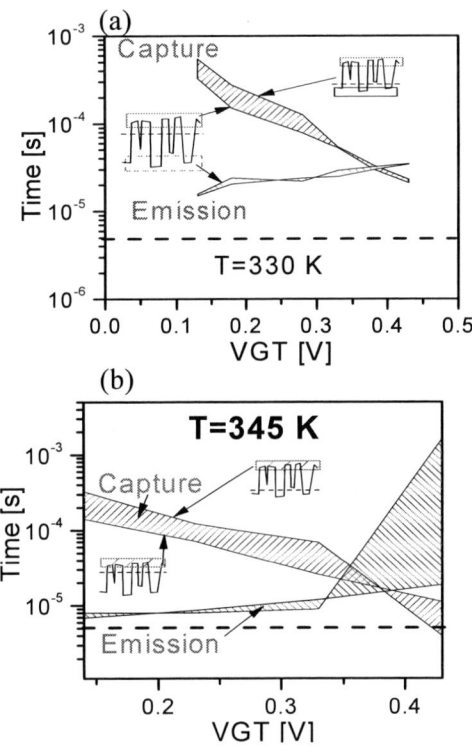

Figure 5. Extracted capture and emission time dependencies vs. gate overdrive voltage at 330 K and 345 K. Dashed line shows the time resolution limit of the measurement setup. The borders of the shaded areas correspond to the extraction procedure performed with the threshold values $(3I_{high}+I_{low})/4$ and $(I_{high}+3I_{low})/4$, respectively. Shaded areas show the possible values of the capture and emission times.

If the average emission time (time in the low current state) is comparable to the time resolution limit of the experimental setup, the extracted capture time can be significantly overestimated. The average emission time still can be obtained using the tail of the Poisson distribution of time periods spent in the low state. The same is true for the opposite case: emission time can be overestimated, when capture time diminishes and gets closer to the resolution limits. Therefore, to ensure the reliable extraction of the trap parameters, we performed the extraction procedure twice, with the threshold

The characteristic times for individual traps were extracted by fitting the experimental spectra (Fig. 6) with the sum of the minimum necessary number of Lorentzians. It is well known that spectral power density of the current noise of a single trap located in the oxide has a Lorentzian shape [1]:

$$\frac{S_I}{I^2} \propto \frac{\tau}{1+(\omega\tau)^2} \qquad (6)$$

The corner frequency of the spectrum, $f_C = 1/\tau$, is determined simultaneously by both characteristic capture and emission times of the trap:

$$1/\tau = 1/\tau_e + 1/\tau_c = (\tau_e + \tau_c)/\tau_e\tau_c \qquad (7)$$

The trap causing the peak at f ~ 30 kHz (Fig. 6) was a pre-existing one (since it was observed in the spectra of both a fresh device and one that experienced stress). The other trap causing peak at ~ 1 kHz appears to be the result of program/erase stress.

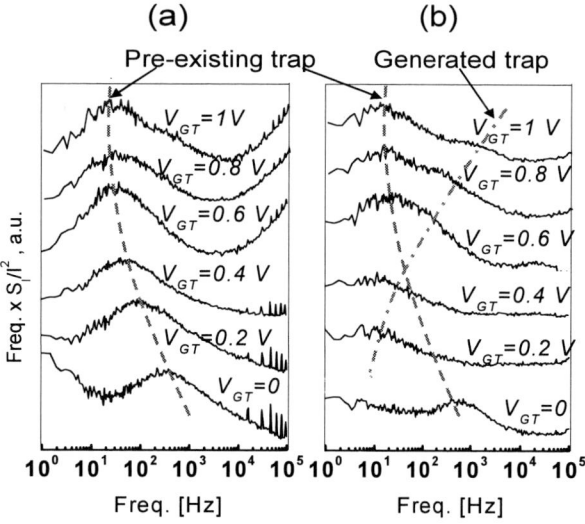

Figure 6. Normalized noise spectral density measured in TANOS #2 device , (TANOS#1 data are not shown) (a) before and (b) after stress. EOT = 13 nm. Pre-existing (measured on fresh devices) and generated (measured after program/erase operation) traps are shown by broken and dotted lines, respectively.

IV. RESULTS AND DISCUSSION

A. Thermal activation of capture/emission processes and trap relaxation

Figure 7 shows the experimental dependencies for capturing the emission time of a "slow" trap in the gate stack of two HiK devices [14] and a relatively "fast" trap (this work) at room temperature. The dependencies were reproduced theoretically with Eqs. (1-4).

In both cases, the traps were found to reside in the interfacial SiO_2 layer 0.63 nm and 0.35 nm away from the substrate, respectively. One needs to assume significant (E_{relax} = 1.86 eV and E_{relax} = 1.71 eV) trap relaxation to reproduce the experimental data. See Table I for the parameters used.

Characteristic times for Trap #2 at elevated temperatures extracted from the RTS data at 330 K and 345 K were reproduced assuming exactly the same parameter values (see Fig. 8).

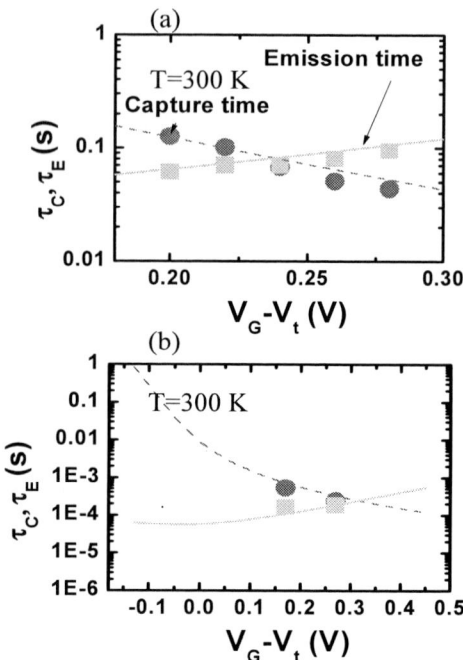

Figure 7. Extracted (symbols) and calculated (lines) capture (circles) and emission (squares) times for 1 nm SiO_2/3 nm HfO_2/TaN nFET. (a) Trap #1 - "slow" trap. (b) Trap #2 - "fast" trap (different device). T= 300 K.

This indicates that the height of the barriers for thermal activation of capture and emission is determined by the trap structural relaxation and can be characterized with a single value of the relaxation energy across the whole temperature and gate voltage ranges of the experimental data.

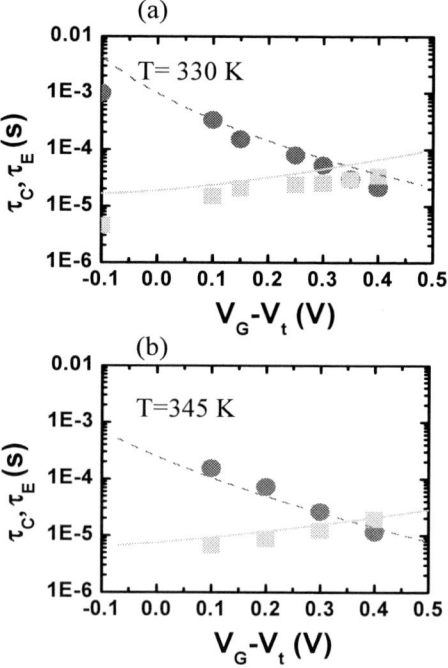

Figure 8. Extracted (dots) and calculated (curves) dependencies of the capture (circles) and emission (squares) times of Trap #2 (See Fig. 7b for T=300 K) vs. gate bias at (a) 330 K and (b) 345 K.

The extracted E_{relax} value was compared to those obtained by *ab initio* calculations of the various possible defect configurations in amorphous SiO_2 [15]. The calculations show that electron trapping by neutral oxygen vacancies' V^0 center is accompanied by a large structural relaxation (1.5 ÷ 2 eV) that closely matches the extracted E_{relax} value.

This allows us to identify the traps observed in high-k MOSFETs as neutral and negatively charged (upon capturing an electron) oxygen vacancies.

B. Role of the interfacial traps

Further development of the model was necessary to describe the sub-threshold RTS data measured on 1.4 nm SiON/poly-Si devices [8,9]. In the sub-threshold regime, the density of the carriers in the channel is very low, and one must take into account the electrons supplied by the interfacial states, as illustrated in Fig. 9. The total capture (emission) time is then calculated as

$$\tau_c = (1/\tau_{C-1} + 1/\tau_{C-2})^{-1}, \quad \tau_e = (1/\tau_{e-1} + 1/\tau_{e-2})^{-1} \quad (8)$$

with the capture/emission time for the electron from/to the interfacial state calculated as

$$\frac{1}{\tau_{c,e-2}} = a \frac{\omega}{2\pi} \frac{1}{1 + e^{\pm \frac{E_t - E_F}{kT}}} \times$$
$$\times \exp\left(-\left(\frac{x_T}{\lambda}\right)\right) \times \exp\left(-\left(\frac{E_{relax}}{4k \cdot T}\right)\right) \quad (9)$$

where a is the parameter having a value on the order of unity and reflecting the capture probability when the motion of the lattice atoms, oscillating at the phonon frequency, bring the electron captured at the interfacial trap or at the bulk trap above the trapping barrier. The Fermi-Dirac distribution function in Eq. (9) takes into account the probability of the interfacial state being filled/empty. It is assumed that the characteristic time to fill the interfacial states is less than the resolution of the experimental setup.

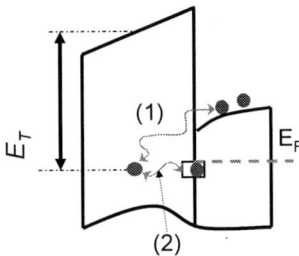

Figure 9. Schematic of the electron trapping process contributing to RTS in the regimes above the threshold (process 1) and subthreshold (process 2 - from the interfacial states).

The trap position relative to the Si interface and extracted value of the relaxation energy (2.07 eV) in the thin SiON dielectric are similar to those of the 1nm interfacial SiO_2 layer in a high-k stack.

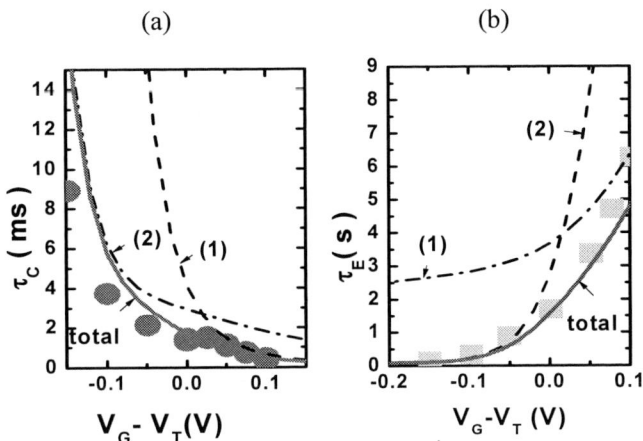

Figure 10. Experimental (symbols) and calculated (lines) (a) emission times, and (b) capture times vs. gate bias, V_g, in the SiON/poli-Si nFET [7]. Calculated dependencies for both processes (1) and (2) are also shown. "total" designates the combined contribution of both processes (See Eq.9). See Table 1 for the parameter values.

C. Pre-existing and stress- induced traps in MOSFETs with thick SiO_2

In contrast, only defects with low relaxation energies (0.13÷0.44 eV) were identified in TANOS FETs with a thick tunnel oxide of 4.5 nm. In these devices, the characteristic times obtained from the low frequency noise spectra of Fig. 6 were reproduced theoretically using Eqs. (1-3,7), see Fig. 11. Both pre-existing and stress-generated (by the program/erase operation) traps were found deep in the SiO_2 dielectric, exhibiting similar characteristics.

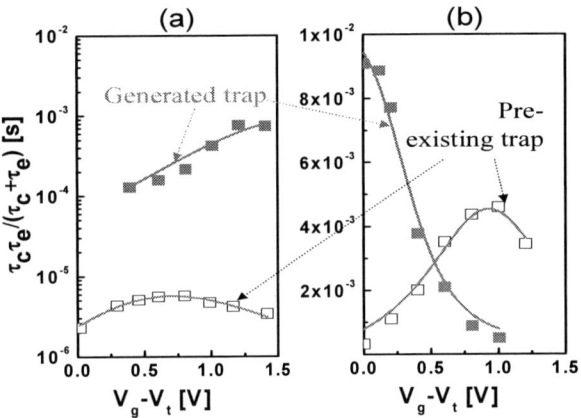

Figure 11. Analysis of 1/f noise data in TANOS devices. Open and filled symbols correspond to pre-existing and program/erase stress-generated characteristic times $(1/\tau_c + 1/\tau_e)^{-1}$ extracted from 1/f noise measurements for (a) TANOS#1 and (b) TANOS#2. Lines are the theoretical fit.

The trap positions relative to the Si/SiO_2 interface and extracted values of the trap energy are shown in Fig. 12. The other parameters are in Table I.

978-1-4244-5430-3/10 $26.00 © 2010 IEEE

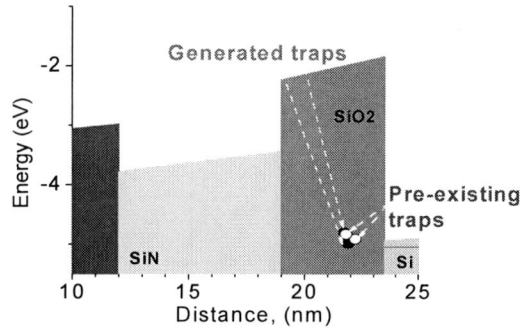

Figure 12. Schematic of the TANOS band diagram with the calculated energy/depth positions of the pre-existing (open dots) and generated by program/erase operation (filled dots) traps.

D.   What can be learnt about the trap structure

Table I summarizes the parameters extracted for all the experimental data considered in this paper. We attempted to use relaxation energy as well as the trap energy as trap-identifying parameters. We also tried to understand why we observe traps, located at different distances from the Si channel, in the thin interfacial layers of the HiK MOSFET and SiON MOSFET vs. those in the thick (4.5 nm) tunneling SiO_2 layer of TANOS FETs.

Traps in the high-k devices and in the SiON MOFET were found to have a rather high relaxation energy. Significant structural relaxation of the dielectric lattice around the trap upon capturing an electron can control the capture and emission speed by itself, extending the characteristic times far beyond the tunneling time value. Thus, the description used here explains the mismatch between the calculated times for the electron to tunnel to/from traps and the much longer measured capture/emission times, which was pointed out in [8,9].

As mentioned above, these traps might be identified as neutral oxygen vacancies, V^0, which are converted to negative vacancies, V^-, when an electron is captured. It is also known that the interfacial SiO_2 layer in high-k MOSFETs is rich with oxygen vacancies, as the HfO_2 scavenges oxygen from the underlying layers.

In contrast, low E_{relax} values for the stress-generated traps deep in the SiO_2 of TANOS devices are in agreement with the relaxation energies of the SiO_2 defects contributing to the trap-assisted tunneling, which were extracted from simulations of SILC data [16].

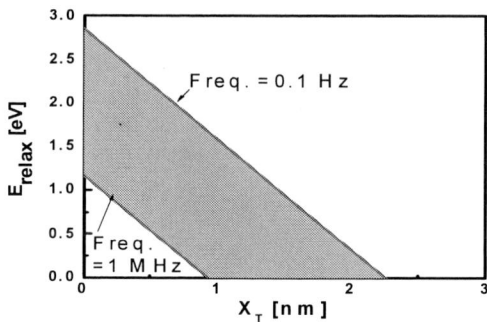

Figure 13. Correlation diagram of the trap relaxation energy, E_{relax}, vs. its distance from the substrate, x_T, for the traps that can contribute to 1/f noise measurements within the frequency range of 0.1 Hz-1 MHz. Estimations assumed $n = 10^{18}$ cm^{-3}, $E_0 = 0$ and $\sigma_0 = 10^{-14}$ cm^2. Due to a limited frequency range, traps with high relaxation energies are detectable only when close to the substrate.

Note that the practical available frequency range of the noise measurement setup limits the observable spatial and relaxation energy windows: the defects with high relaxation energies can be observed only when located closer to the interface with the substrate (to keep the characteristic times below the upper limit determined by the measurement frequency), while the defects with low relaxation energies can be observed if they are deeper in the oxide (Fig. 13).

CONCLUSION

We report a self-consistent analysis of the charge exchange process between the semiconductor and defects (electron traps) in the dielectric to identify the defect responsible for RTS. The theoretical description employed reproduces experimental data with a single set of the defect characteristics across the entire range of temperatures and gate biases in a variety of the gate stacks (high-k/metal, SiON/poly) and devices (transistors, TANOS memory) used in the experiments.

We emphasize that in high-k MOSFETs the observed critical RTS characteristics—the trapping/detrapping times—are mostly controlled by the displacements of the lattice atoms caused by electron trapping. This allows the relaxation energy associated with this lattice re-arrangement, which represents a fundamental feature of the defect's atomic structure and can be used as a defect identifying characteristic, to be extracted. The experimental values obtained for the relaxation energies allowed us to speculate that the defects observed by the drain current RTS in the high-k/metal and SiO_2/poly-Si gate stacks are oxygen vacancies in the SiO_2 layer close to the Si substrate.

TABLE I.   PARAMETERS EXTRACTED USING THE TRAPS RELAXATION MODEL

Sample	x_T nm	E_T^* eV	σ_0 cm^2	E_{relax} eV	Trap type
HKMG nMOSFET #1 (Trap # 1)	0.63	3.02	1.3 10^{-14}	1.86	V-
HKMG nMOSFET #2 (Trap # 2)	0.35	3.03	2 10^{-14}	1.71	V-
SiON MOSFET [7]	0.4	3.3	1.5 10^{-14}	2.07	V-
TANOS#1 Trap 1	1.3	2.96	1.5 10^{-14}	0.13	
TANOS#1 Trap 2 (induced)	1.76	2.81	1.5 10^{-14}	0.06	
TANOS#2 Trap 3	1.7	2.84	1.5 10^{-14}	0.36	
TANOS#2 Trap 4 (induced)	1.6	3.0	1.5 10^{-14}	0.44	

*)E_T is the absolute value of the trap in respect to the bottom of the oxide conduction band.

V. ACKNOWLEDGMENT S

The authors would like to thank Dr. J. P. Campbell and Dr. K. P. Cheung of NIST for providing experimental data for the SiON MOSFET and for useful discussions.

REFERENCES

[1] C. Leyris et al., Microel. Reliability 47, p. 41, (2007).

[2] C. M. Chang et al., IEDM, 2008, p. 787.

[3] S. Lee et al., IEDM, 2009, 32.2.

[4] M. J. Kirton and M. J. Uren. Adv. Phys. 38, p. 367 (1989).

[5] M. J. Kirton, M. J. Uren, Appl. Phys. Lett. 48, p. 1270 (1986).

[6] M. B. Weissman, Rev. Mod. Phys. 60, p. 537 (1988).

[7] A. L. McWhorter, in Semic. Surf. Phys, p. 207 (1957).

[8] J. P. Campbell et al., Proc. IEEE IRPS, 2009, pp. 382–388.

[9] J. P. Campbell et al., Proc. IEEE IIRW, 2008, p. 105.

[10] C. H. Henry and D. V. Lang, Phys. Rev. B, 15, p. 989, (1977).

[11] Yu. E. Perlin, Sov. Phys-Usp. (Usp Phys Nauk) v. 6, p. 542 (1964).

[12] W. B. Fowler et al., Phys. Rev. B, v 41, p. 8313 (1990).

[13] A. Avellán, D. Schroeder, and W. Krautschneider J. Appl. Phys. 94, p. 703 (2003).

[14] D. Veksler et al., Proc. IEEE IIRW, 2009, LN-2.

[15] Anna Kimmel et al., ECS Trans. 19 (2), p. 3 (2009).

[16] G. Bersuker et al., IEDM, 2008, pp. 791-794.

Reliability Study of Bilayer Graphene - Material for Future Transistor and Interconnect

Tianhua Yu, Eun-Kyu Lee, Benjamin Briggs, Bhaskar Nagabhirava, and *Bin Yu

College of Nanoscale Science and Engineering
University at Albany, State University of New York
Albany, New York, 12203, USA
TEL: 1-518-956-7492, *E-mail: byu@uamail.albany.edu

Abstract— Graphene is considered to be promising candidate for future transistor and interconnects material in integrated circuits because of its high intrinsic mobility and current-carrying capacity outperforming Cu. Particularly, bilayer graphene (BLG) systems offer controllable and wide band gap tunability without the need for nontrivial atomically precise nanoribbon patterning, which is indispensable to band gap engineering of monolayer graphene. Hence, novel devices consisting of BLG as both transistors and interconnects in combination with well-established Cu interconnects is conceivable. In this frame, this study has aimed to address reliability limiting factors of BLG/Cu contacts and current-carrying capacity.

Keywords- Bilayer graphene, contact, interconnect, current annealing, breakdown.

I. INTRODUCTION

Scaling and reliability issues in state-of-the-art copper interconnect have driven researchers to seek alternative materials for next-generation on-chip interconnects technology [1] As the feature size of Cu interconnects is scaling down into nanometer range, resistivity increases with decreasing line width under size effect due to electron scattering at the grain boundaries and interfaces [2]. This increased resistivity leads to more Joule heating and hence limit the reliability of Cu interconnects. Electromigration (EM) is diffusion-controlled mass transport, driven by electron current flow in metal lines. It can be a serious reliability threat when the dimension of Cu interconnects approaches nanoscale range [3]. The maximum current limit in copper was reported to be approximately 10^6 A/cm^2 [4]. Due to the high thermal conductivity, there have been a number of reported studies on breakdown mechanism of carbon-based interconnects, such as multiwall carbon nanotube (MW-CNT) [5], carbon nanofiber (CNF) [6], and graphene nanoribbon (GNR) [7]. The current-carrying capacity of MW-CNT, CNF and GNR are found to be on the order of 10^8 A/cm^2, 10^6 A/cm^2, and 10^8 A/cm^2, respectively.

Graphene, single or a few layers of graphite, has been considered as an emerging material candidate for high-performance transistors and interconnects [1][8] because of its superior intrinsic carrier mobility (200,000cm^2/V-s, 200 times higher than Si) [9], excellent thermal conductivity (4800W/m-K, 10 times higher than Cu) [10], large current-carrying capacity (superior to copper in nanometer-sized wires [11]), and reduced vulnerability to electromigration. Because graphene can be patterned using the conventional microelectronics processes, the transition from copper needn't integrate new manufacturing technique into circuit fabrication.

Among the available material configurations of graphene systems, AB-stacked bilayer graphene (BLG) offers unique electrical flexibility with tunable band-gap up to a few hundred meV by applying a perpendicular electric field [12]. Hence, engineering design employing BLG as a new material platform for both electronic devices and local interconnects in combination with the well-established Cu interconnects is conceivable. However, little experimental evidence has been ever reported on the electrically induced breakdown behavior of bilayer graphene. In this work, we will investigate the reliability limiting factors of BLG-to-copper contacts and the current-carrying capacity in BLG/Cu hybrid system.

II. SAMPLE FABRICATION

Aiming at potential on-chip applications with hybrid design of BLG-based devices/local interconnects and state-of-the-art copper interconnects, we explore the breakdown behavior in BLG with buried Cu structure. Cross-section view of the test structure fabrication process is shown in Fig. 1.

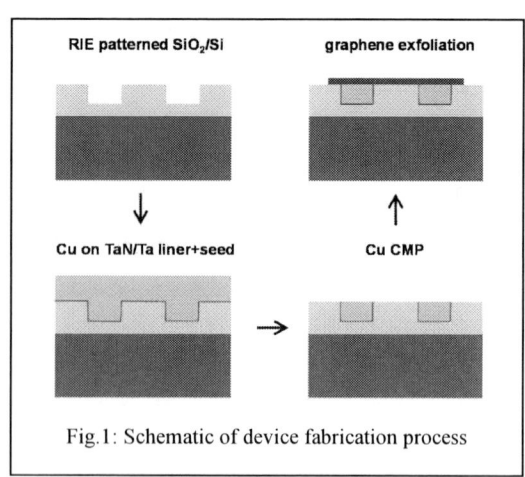

Fig.1: Schematic of device fabrication process

978-1-4244-5430-3/10 $26.00 © 2010 IEEE

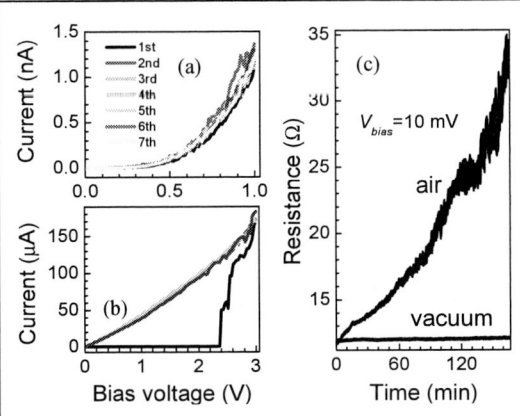

Fig.2: Electrical current-induced thermal annealing effect on BLG-Cu contact: (a) with low voltage (0~1V) sweep, and (b) with high voltage (0~3V) sweep, and (c) post-annealing time-dependent resistance behavior under different ambient conditions

Due to optical contrast requirement to visually identify the atomically-thin graphene flakes, 100±5 nm thick SiO_2 is thermally grown on p-type Si substrate (industry-standard 300mm wafer). Firstly, trenches in SiO_2 with 1 μm wide and 60 nm deep are fabricated by optical lithography and reactive ion etching (RIE). Next, Ta/TaN seeding layers are deposited using physical vapor deposition, followed by Cu electroplating to fill the trenches. Finally, chemical mechanical planarization (CMP) process is used to ensure the formation of an ultra-flat metal contact surface for graphene. Bilayer graphene flakes are obtained from high-purity graphite (Kish Graphite, Toshiba Ceramics) by mechanical exfoliation. Graphene thickness is visually identified under high-resolution optical microscope and further confirmed by unique 2D band spectral signature in Raman scattering measurements [13]. All electrical measurements are performed with standard lock-in methods. Agilent B1500A semiconductor parameter analyzer is used for I-V measurements with DC voltage sweep. All measurements in this work are done at room temperature.

III. EXPERIMENT AND RESULTS

Thermal annealing has been reported to help improving the conduction in graphene system by lowering the contact resistance [14]. In this work, we perform "local" annealing via DC current induced heating on BLG-to-copper contacts of different sizes on the same test sample. Electrical annealing is also expected to avoid possible contamination on graphene surface during furnace annealing process. Before any annealing is performed, the measured contact resistance is almost infinite due to extremely high contact resistance.

Firstly, we perform current annealing with repeated voltage sweeping from 0V to 1V for several cycles. The measured resistance is still very high on the order of GΩ (in Fig. 2(a)). However, when the voltage sweep range is increased from 0V to 3V, samples exhibit nearly linear I-V characteristics. The resistance, subsequently reading under low voltage bias (5 mV), is found to be on the order of kΩs (in Fig. 2(b)) and proportional to the BLG aspect ratio (not shown in the figure), indicating insignificant BLG-to-Cu resistance. The contact

resistance is found to be unchanged after several repeated high-bias cycling. The contact resistance is extracted to be ~280Ω·um² via measuring a set of testing structures with different length and width of graphene layer. When BLG is kept in vacuum after high-bias current cycles, total resistance remains constant. However, when BLG is exposed to air after annealing, continuous degradation was observed shown in Fig. 2(c): nearly 3 times higher total resistance (from 12kΩ to 35kΩ) after 150 minutes. Since the resistance change due to adsorption of gas molecule on graphene surface is smaller than 1kΩ [15], this post-annealing time-dependency behavior in ambient condition is mainly due to contact degradation. When high-bias current annealing is conducted again, total resistance drops down to the initial value. While the reason of BLG/Cu contact degradation in air is still under investigation, all the rest electrical measurements in this study are performed in vacuum (~ 10^{-6} Torr).

In order to determine the BLG current-carrying capacity, we perform I-V tests by increasing voltage bias until graphene breakdown occurs. As current density increasing, there is a threshold current at which breakdown occurs, resulting in an abrupt drop in current. By measuring over 20 samples, we find there are two failure modes depending on the stressing current density and the contact size. When large BLG sample (width > 5μm) is tested (Fig. 3(a)), a physical break is clearly seen in the BLG flake (Fig. 3(b)). Before graphene burnt down, the I-V curve is initially linear, and starts to saturate as voltage increasing due to self-heating. Similar non-linearity in I-V curve was observed in CNTs, which indicates the self-heating effects at high-bias. The BLG breakdown occurs when the current increases to about 4mA, and abrupt decrease to 0. If we assume the thickness of BLG is ~0.7nm, current-carrying capacity will be on the order of 10^8 A/cm², comparable to the breakdown current in GNR and CNT [5][7].

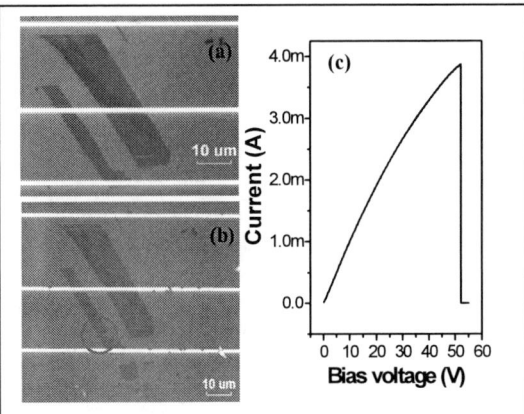

Fig.3: Typical breakdown behavior in BLG devices (a) before breakdown, (b) after breakdown and (c) high bias I-V charaterstics showing breakdown

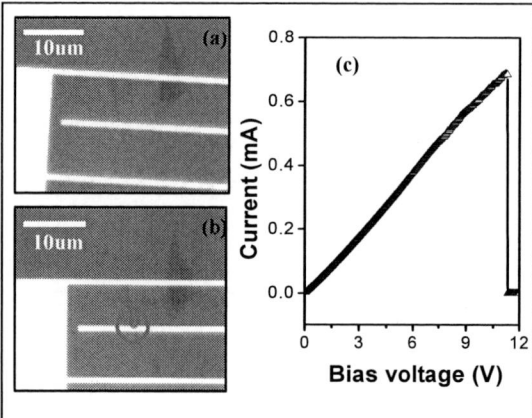

Fig.4: Typical breakdown behavior in BLG devices (a) before breakdown, (b) after breakdown and (c) high bias I-V charaterstics showing breakdown

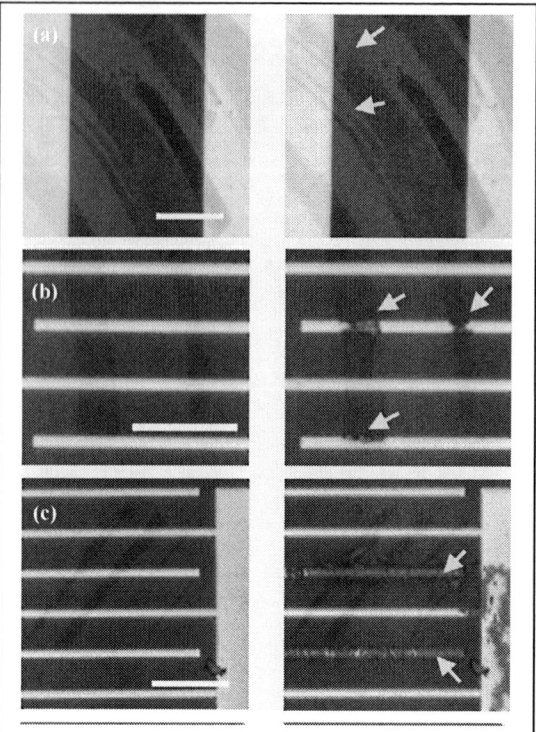

Fig.5: Different types of failure sites developed after breakdown in BLG devices of various geometries: arrows indicate (a) cuts and(b,c) contact damages; white scale bars in the images are for 10 um

For BLG samples with scaled contact (width 0.5~1 μm), breakdown originates from contact damage instead of physical cut on graphene and the failure sites are located in metal fingers where Cu voids are observed in Fig.4(b). Fig. 4(c) is the I-V curve under high-bias sweep, showing breakdown occurs at a lower current (~0.7mA) without noticeable current non-linearity. The corresponding current density is about 10^7 A/cm^2, beyond typical value at which electromigration occurs in Cu. In this case, we may conclude that breakdown happens first on copper with electromigration failure due to scaled contact area. However, we didn't find any contact damage in Fig. 3(b) and the breakdown current density is about 10^8 A/cm^2 on graphene, which is much higher than copper EM limit. The two results seem to be inconsistency. Therefore, we need to perform more measurements to confirm our conclusion.

Fig.5 shows the different types of failure after breakdown a set of parallel BLGs across Cu contacts: (a) BLG cuts and (b) (c) Cu/BLG contact damage. In this work, we use the repeated current stress method to study the breakdown one by one until the total resistance increase to infinite: when a visible drop in current is found during the measurement, the device testing is stopped at this point and the voltage ramp will be then repeated from 0V. Due to the more than one graphene across the contact, the maximum breakdown current up to 20 mA is much higher than single graphene. We also found that contact failure happened at the cathode end (Fig.5(b) and Fig.5(c)), which is consistent with the electromigration damage of Cu interconnect due to the tensile stress caused by mass transport [16]. Under high current density, mass of copper will accumulate on the anode side, causes local compression and eventually the mass is squeezed out of the surface to form protrusions. While on the cathode side, mass depletion causes tension and vacancy accumulation. Voids nucleating under tensile stress will grow and coalesce until avoid forms which leads to an electrical failure [17]. If we study the case in fig. 3, although the current density of graphene is on the order of 10^8 A/cm^2, we didn't find the voids or failure on copper contact due to the large contact area on cathode side.

Figures 6(a) and 6(b) compare the *I-V* characteristics with different types of breakdown failures, contact damage and BLG cuts, as shown in Figure 5(c) and 5(a), respectively. Both *I-V* characteristics exhibit similar behaviors, i.e., abrupt decrease in current and breakdown under current stress with about the same order of magnitude. Measurements on multiple samples with single BLG flake on two Cu fingers 25 μm apart were also performed in order to determine the dependence of breakdown current on sample resistance. In these samples, most of the samples were observed to be broke down due to contact damages. Fig. 6(c) shows the graphene breakdown current as function of graphene width/length ratio (i.e. reciprocal of resistance). BLGs with higher resistance are more prone to failure. It is found that breakdown current is proportional to the aspect ratio, indicating Joule heating related breakdown mechanism [7]. For CNF, a relation of breakdown current $I_{BR} \propto \rho^{-0.5}$ was proposed [6]. The linear relation ($I_{BR} \propto \rho^{-1}$) extracted from BLG samples suggests a faster breakdown with increasing resistance as compared with CNF. This indicates that factors that cause high resistance also results in a reduction in breakdown threshold, e.g. in-plane or edge defects. In addition to BLG samples, breakdown tests are performed on trilayer graphene (TLG) samples. Similar linear correlation between aspect ratio and breakdown current is observed (Fig. 6(c)). Compared with BLG, TLG exhibits improved breakdown property due to the addition of one more graphene monolayer.

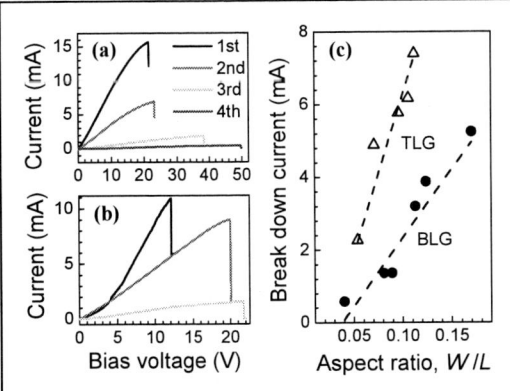

Fig.6: I-V characteristics showing sample breakdown originating from (a) cuts and (b) contact damages. (c) measured aspect ratio vs. breakdown current for BLG and Trilayer graphene (TLG) samples; dashed lines are linear fitting curve.

IV. SUMMARY

From the experimental results obtained, BLGs are found to display an impressive current-carrying capacity on the order of $10^8 A/cm^2$. DC current-induced annealing reduces BLG/Cu contact resistance but contact contamination can happen directly from room ambient. The BLG/Cu contact damage plays dominant role in breakdown where small-area contacts are made. The observed linear dependence of BLG breakdown current on the size of contact area (and hence resistance) suggests Joule-heating as the breakdown mechanisms. Similar measurements were observed in trilayer graphene with slightly improved breakdown characteristics.

REFERENCES

[1] J. Brooks, "Characterization of graphene-based interconnects," in NNIN REU Research Accomplishments, 2008, pp.128-129.

[2] Y. Hou and C. Tan, "Size effect in Cu nano-interconnects and its implication on electromigration", in 2nd IEEE International Nanoelectronics Conference, 2008, pp.610-613.

[3] N. Michael, C. Kim, P. Gillespie, and R. Augur, "Electromigration failure in ultra-fine copper interconnects", J. Electron. Mater. vol.32, no.10, pp. 988-993, 2003.

[4] P. Wang and R. G. Filippi, "Electromigration threshold in copper interconnects," Appl. Phys. Lett., vol. 78, pp. 3578-3581, Jun. 2001.

[5] P. G. Collins, M. Hersam, M. Arnold, R. Martle, and Ph. Avouris, "Current saturation and electrical breakdown in multiwalled carbon nanotubes," Phys. Rev. Lett., vol. 86, no. 14, pp. 3128-3131, Apr. 2001.

[6] M. Suzuki, Y. Ominami, Q. Ngo, and C. Y. Yang, "Current-induced breakdown of carbon nanofibers," J. Appl. Phys., vol. 101, no.11, pp. 114307, June 2007.

[7] R. Murali, Y. Yang, K. Brenner, T. Beck, and J. D. Meindl, "Breakdown current density of graphene nanoribbons," Appl. Phys. Lett., vol. 94, no. 24, pp. 3114-3117, June 2009.

[8] Yuji Awano, "Graphene for VLSI: FET and Interconnect Applications," IEDM Tech. Dig., pp. 233-236, Dec. 2009.

[9] X. Du, I. Skachko, A. Barker, and E. Y. Andrei, "Approaching ballistic transport in suspended graphene," Nat. Nano., vol. 3, no. 8, pp. 491-495, Aug. 2008.

[10] A. A. Balandin, S. Ghosh, W. Bao, I. Calizo, D. Teweldebrhan, F. Miao, and C. N. Lau, "Superior thermal conductivity of single-layer graphene," Nano Lett., vol. 8, no. 3, pp. 902-907, Feb. 2008.

[11] A. Naeemi and J. D. Meindl, "Conductance modeling for grapheme nanoribbon (GNR) interconnects," IEEE Electron Dev. Lett., vol. 28, no. 5, pp. 428-431, May 2007.

[12] Y. Zhang, T. Tang, C. Girit, Z. Hao, M. C. Martin, A. Zettl, M. F. Crommie, Y. R. Shen, and F. Wang, "Direct observation of a widely tunable bandgap in bilayer graphene," Nature, vol. 459, pp.820-823, June 2009.

[13] A. C. Ferrari, J. C. Meyer, V. Scardaci, C. Casiraghi, M. Lazzeri, F. Mauri, S. Piscanec, D. Jiang, K. S. Novoselov, S. Roth, and A. K. Geim, "Raman spectrum of graphene and graphene Layers," Phys. Rev. Lett., vol. 97, no. 18, pp.187401 Oct. 2006.

[14] K. I. Bolotin, K. J. Sikes, Z. Jiang, M. Klima, G. Fudenberg, J. Hone, P. Kim and H.L. Storme, "Ultrahigh electron mobility in suspended graphene," Solid State Comm., vol. 146, no. 9, pp. 351-355, 2008.

[15] T. Wehling, K. Novoselov, S. Morozov, E. Vdovin, M. Katsnelson, A.Geim, and A. Lichtenstein, "Molecular Doping of Graphene", Nano Lett., vol.8, no.1 pp.173-177, November 2008.

[16] P. Ho, K Lee, S. Yoon, X. Lu and E. Ogawa, "Effect of low k dielectrics on electromigration reliability for Cu interconnects," Mater. Sci. Semicond. Process. vol. 7, pp.157–163, August 2004.

[17] C. Basaran, and M. Lin, "Damage mechanics of electromigration induced failure", Mech. Mater. vol.40, pp.66–79, 2008.

Failure Analysis of Resistive Switching Devices

An Chen

Strategic Technology Group,
GLOBALFOUNDRIES,
Sunnyvale, CA 94085, USA
Phone: 1-408-462-4039, Email: an.chen@globalfoundries.com

Abstract— **Cycling failures for resistive switching devices are discussed based on array statistics measured on Cu_2O metal-insulator-metal (MIM) devices. Four types of failures can be identified under rigorous testing conditions. The rate of these failures can be reduced by optimizing operation methods, which has significant impact on cycling endurance and yield. Failures related to data loss may have their origin in material stability.**

Keywords – resistive switching, switching failure, cycling endurance, retention

I. INTRODUCTION

Resistive switching devices based on metal oxides have attracted great attention owing to promising characteristics for applications in next-generation non-volatile memories and programmable logic. These devices can be repeatedly switched between a high-resistance state (OFF state) and a low-resistance state (ON state) by voltage or current. Both states have been shown to be nonvolatile and the switching can occur at relatively low voltage with the speed as fast as several ns. These resistive switching devices are usually made in a metal-insulator-metal (MIM) structure. Their simple two-terminal structures and highly scalable sizes may enable extremely high device density in novel architectures (e.g., cross-bar array). They are also compatible with CMOS processing and can be built in via locations in CMOS architectures. Various metal oxides have demonstrated resistive switching behaviors; however, the exact switching mechanisms are not clear and the reported switching characteristics vary in a wide range [1-13].

Fig. 1 plots published cycling endurance of some resistive switching metal oxides. The data collectively show a general trend of lower cycling endurance at higher ON/OFF ratio. Large variation of cycling endurance exists even on the same target materials, as shown in the figure. This is partially because properties of these metal oxides are very sensitive to compositions and processing conditions. The difference in operation methods and testing criteria also contributes to the variation in the measured cycling endurance [1-12].

Endurance and reliability are key parameters of these resistive switching devices and may eventually determine their application space. Although many papers have published interesting resistive switching characteristics of various metal oxides, reliability and failure reasons of these devices have rarely been discussed. This paper intends to provide a phenomenological analysis of the failure types and propose some methods for yield improvement, using Cu_2O-based resistive switching device as an example.

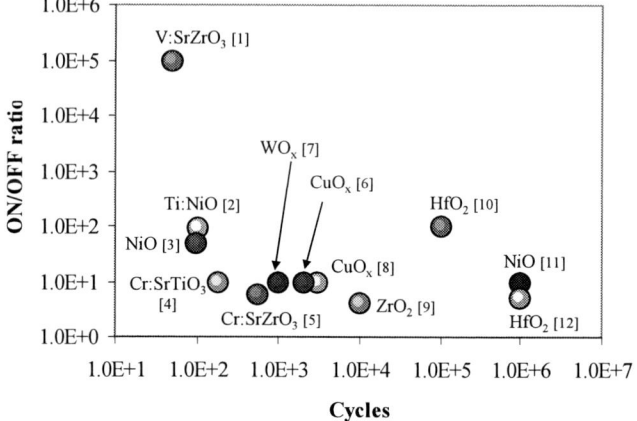

Fig. 1 Cycling endurance and ON/OFF ratios reported in literature on some resistive switching devices based on various metal oxides.

II. EXPERIMENTS

A. *Cu_2O-based Resistive Switching Devices*

Cu_2O MIM devices are integrated in standard 0.18 µm CMOS architecture as shown in Fig. 2(a), and 64 kbit arrays are fabricated [6]. The Cu via with diameter of 0.18 µm defines bottom electrode. 120Å thick Cu_2O films are grown by thermal oxidation of the Cu via as shown in Fig. 2(b).

Each MIM device is connected to a serial transistor that provides both selection and current-limit functions, as shown in Fig. 2(c). Drain voltage V_d is applied across MIM (and the transistor) for switching, and gate voltage V_g imposes the current limit during switching.

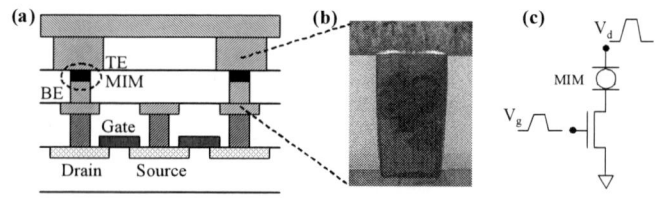

Fig. 2 (a) Cross-section view of Cu_2O-based MIM resistive switching devices integrated in a CMOS architecture; (b) cross-section TEM image of Cu_2O MIM; (c) schematics of 1T-1R configuration and voltage pulses applied for switching.

978-1-4244-5430-3/10 $26.00 © 2010 IEEE

Typical DC switching current-voltage (I-V) characteristics is shown in Fig. 3. Switching from OFF to ON state is called "set" and ON to OFF "reset". A unipolar switching behavior (i.e., set and reset with the same voltage polarity) is observed with Cu$_2$O MIM devices. In AC operation, these devices can be switched with voltage pulses as short as 50 ns.

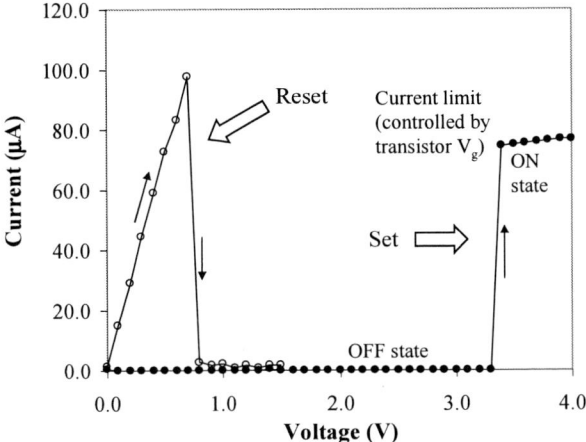

Fig. 3 Typical switching I-V curve of Cu$_2$O MIM devices.

The 1T-1R structure enables two operation methods to switch Cu$_2$O MIM devices, namely "ramp-V$_d$" and "ramp-V$_g$" methods respectively. Table I explains the two methods and summarizes their effects. Ramp-V$_g$ method switches devices by changing the current limit (with fixed voltage across MIM devices), which is found to be more effective than ramp-V$_d$ method to set and reset devices to predefined ON- and OFF-state resistance targets.

TABLE I.   RAMP-V$_D$ AND RAMP-V$_G$ OPERATIONS

	Operation	Effects
Ramp-V$_d$	Fix V$_g$ and ramp up V$_d$ until devices are set/reset or maximum voltage is reached.	• Wider R$_{on}$ distribution • Shallower reset depth
Ramp-V$_g$	Fix V$_d$ and ramp up V$_g$ until devices are set/reset or maximum voltage is reached.	• Narrower R$_{on}$ distribution • Deeper reset depth

B. Cycling Testing Algorithm

Fig. 4 illustrates the voltage pulse sequence in one cycle during the cycling test. In each cycle, before and after the set and reset operations, device states are verified by reading their resistance to ensure that the ON and OFF states are not only reached but also maintained. This double-verification between set and reset operations reveals failure types related to data loss, which may not be exposed in less rigorous testing conditions. The delay between the set/reset/read operations is on the order of ms in the memory array tester.

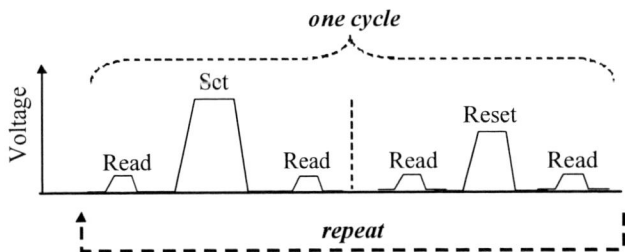

Fig. 4 The operation sequence in one cycle, with read verification before and after both set and reset, to explore all possible failure reasons.

III.   ANALYSIS OF FAILURE TYPES

Four types of failures have been observed in experiments as illustrated in Fig. 5: set failure, reset failure, and two types of data loss.

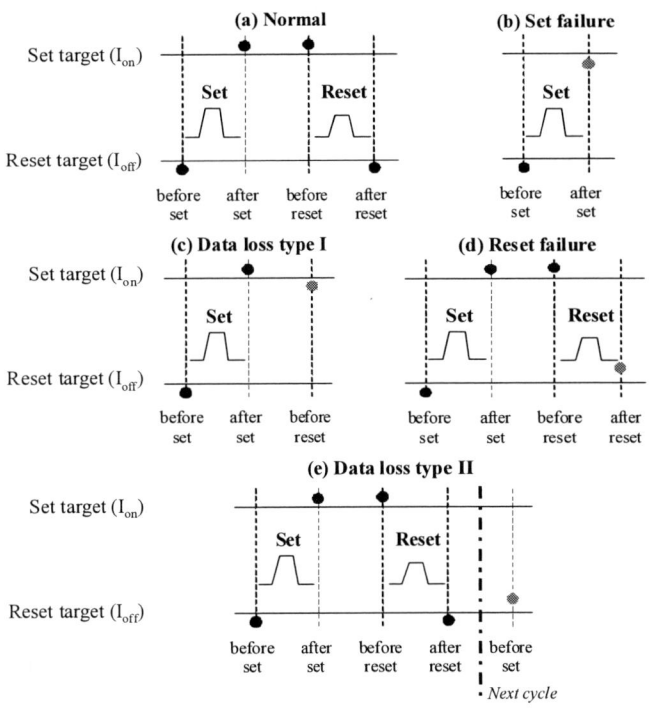

Fig. 5 Illustrations of normal switching behavior (a) and four types of switching failures (b) – (e); the dots represent read current levels.

A. Failures related to switching

Set failure (Fig. 5(b)) and reset failure (Fig. 5(d)) are the failures to reach the defined ON-state and OFF-state targets after maximum number of voltage pulses have been applied. It has been observed that ramp-V$_g$ method provides better switching yield for both set and reset operations. This is because the device resistance is affected more by the current limit than by the applied voltage during switching [14]. Fig. 6 shows the distribution of ON-state read current after devices are set by these two methods. Ramp-V$_d$ method results in a tail of high-resistance devices that fail to reach the set criterion. With ramp-V$_g$ method, ON-state resistance distribution is narrower and 100% set yield can be achieved consistently.

978-1-4244-5430-3/10 $26.00 © 2010 IEEE   85

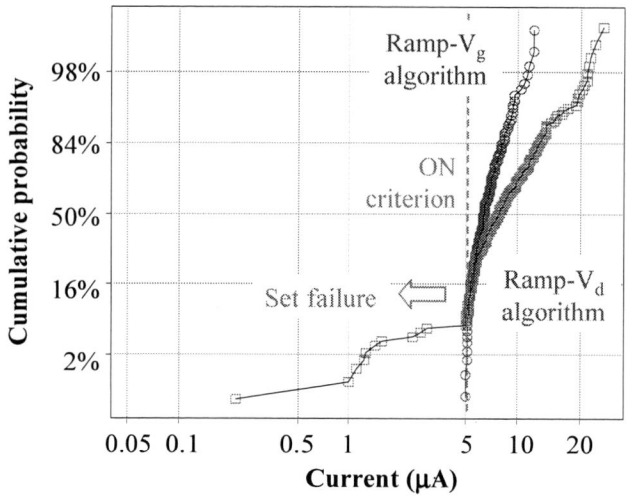

Fig. 6 Distribution of ON-state read current after set operation with ramp-V_d and ramp-V_g methods.

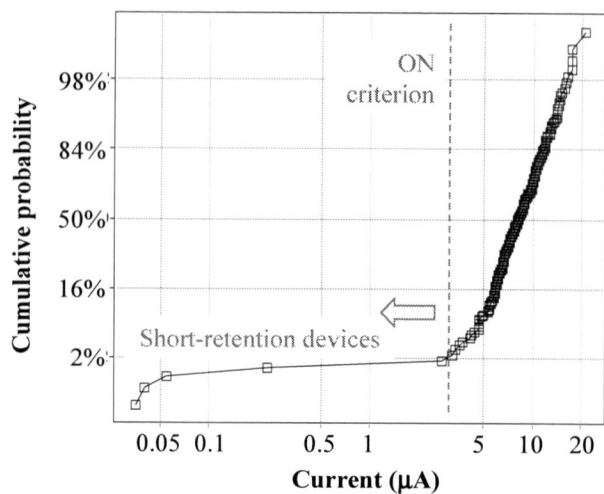

Fig. 8 Distribution of ON-state current read again after a successful set operation, showing type-I data loss.

Reset failure has the highest failure rate among the four failure types and is the most important failure reason for cycling endurance. Fig. 7 shows the reset ratios (defined as the ratio of the read current after reset divided by the read current before reset) for different reset voltages. The error bars around each data point show 25% - 75% data distribution. Low voltage (e.g., ≤ 0.8 V) is insufficient for reset; therefore, most device resistance is unchanged, resulting in the reset ratio close to 1. If the voltage is too high (e.g., ≥ 1.8 V), devices tend to be set further into lower resistance state instead of staying in the OFF state, as shown by the reset ratios above 1. This is due to the high electric field that causes further set switching or material damage. Most devices can be reset with voltage in the medium range, as shown by the low value (<10⁻²) and narrow distribution of reset ratio between 1.2 V and 1.4 V. Under optimum voltages, nearly 100% reset yield can be achieved.

the set operation. Although these devices were all successfully set above the ON criterion (≥ 3 μA at 0.3 V) and verified by the first read, a very small number of devices already fell below this criterion in the second read. It has been found that retention (τ) depends exponentially on an activation energy E, i.e., $\tau = \tau_0 \cdot \exp(E/kT)$ [15]. Statistics measured from Cu_2O memory array show that devices with E ≈ 0.9 eV have the retention of ~ 10 days. Retention shortens exponentially with decreasing activation energy; therefore, devices with small activation energy may lose ON-state data within the testing time. They may also be easily disturbed by voltage spikes during array testing. The existence of large retention variation indicates a wide range of activation energy, which is also confirmed by temperature-accelerated retention tests [15]. Type-I data loss failure may be reduced by material engineering to improve ON-state retention.

In comparison, type-II data loss (the loss of OFF-state retention) is less intuitive because the OFF-state is generally more stable. Fig. 9 shows the bake test of the ON and OFF

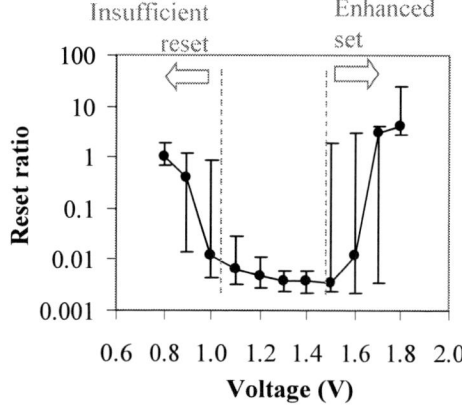

Fig. 7 Reset ratio *vs.* reset voltage; error bars shows 25%-75% data distribution.

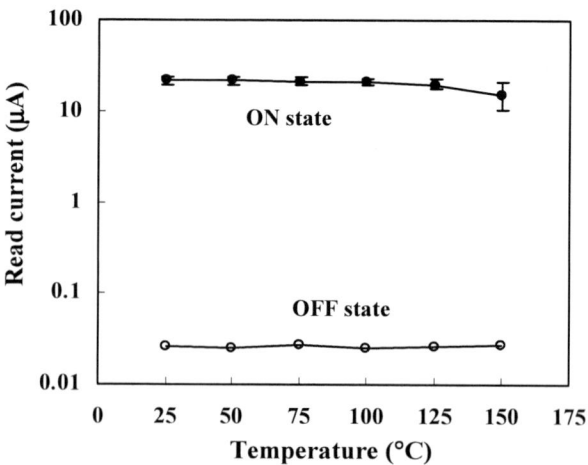

Fig. 9 Read current (at 0.3 V) of ON and OFF state measured at temperature from 25°C up to 150°C.

B. Failures related to data loss

Type-I data loss is the loss of ON-state retention. Fig. 8 shows a distribution of ON-state current read again right after

states up to 150°. While ON-state read current starts to decrease above 100°C, OFF-state read current is almost unchanged. The type-II data loss may be caused by the following reasons: (1) read disturb, (2) shallow reset, and (3) material instability. Various experiments have been performed to systematically study each failure reason and search for solutions. Some OFF state devices display resistance variation under continuous reading and the variation range can be reduced by reading at lower voltage, which provides evidence of the read disturb cause. Read disturb may be minimized by lowering read voltage and designing array structures for better isolation. It is also found that devices forced to reset to higher resistance levels (deep reset) exhibit less type-II data loss than devices reset to lower resistance levels (shallow reset), confirming the shallow reset cause. This cause can be minimized by forced deep reset. It is observed that the ramp-V_g method can reset devices to deeper levels (higher resistance) than the ramp-V_d method, which helps to reduce the type-II data loss. Although read disturb and shallow reset are relatively easy to identify and can be reduced by optimizing operation conditions, intrinsic material instability cannot be excluded as a failure cause in these tests. Instability issues have to be addressed by material improvement. Table II summarizes these tests and solutions.

TABLE II.   CAUSES AND SOLUTIONS FOR TYPE-II DATA LOSS FAILURE

Causes	Evidence	Solutions
Read disturb	Continuous reading of reset devices shows resistance variation	Reduce read voltage and minimize equipment disturb
Shallow reset	Shallow reset devices have more type II data loss than deep reset devices	Forced deep reset
Material instability	Different materials and processes show different rates of type II data loss	Material and structure engineering

Although the rate of both data loss failures is low, they could still cause severe cycling yield degradation. However, most data loss failures only cause small resistance change around the ON- and OFF-state criteria. Therefore, an operation margin can be designed around the set and reset criteria to tolerate small range of variation. Devices are set (reset) to certain resistance target, but read with the ON-state (OFF-state) resistance criterion higher (lower) than the set (reset) target. Cycling yield is significantly improved with the operation margin.

C. Cycling Endurance

Reliability is a major challenge for resistive switching devices. The failure rate in each cycle has to be sufficiently low to achieve acceptable cycling endurance and final yield for practical applications. If devices fail with an average failure rate of r every cycle, n-cycle yield is $Y(n) = (1-r)^n$. With 1% failure rate every cycle, 100-cycle yield is only 36.7%. This is assuming failed devices are excluded in the future cycles. However, the failure types discussed above are not necessarily permanent, especially the set and reset failures. Most devices that fail to set or reset only slightly miss the targets, and only a small percentage of devices show signs of irreversible damage. Therefore, it may be possible to improve cycling yield with some intelligent design of "repair" rather than discarding all failure devices. This is beyond the scope of this paper.

An interesting observation is that the measured cycling yield fits well with a two-stage model: $Y(n) = (1-r_1)^k \cdot (1-r_2)^{n-k}$, where $Y(n)$ is n-cycle yield. The failure rate r_1 in the first $(1, k)$ cycling stage is higher than the failure rate r_2 in the second $(k+1, n)$ cycling stage. This indicates that devices surviving the earlier cycles are usually more robust and have lower failure rate in the following cycles. This cycling behavior can be understood from the fact that the data loss failures have their origins in material property and would be exposed as soon as cycling starts. Therefore, devices failing for data loss could be identified and excluded from the test in early cycles. Devices surviving early cycles are expected to have less data loss failures and are therefore more robust. This two-stage cycling behavior may be utilized to design a screening process to eliminate devices with higher failure rate; and consequently more robust devices could be selected as final products.

IV.   DISCUSSION AND SUMMARY

Table III summarizes the four failure types with their impact on cycling yield, as well as solutions to minimize these failures. Set failure can be almost eliminated with optimized switching conditions and 100% set yield has been consistently achieved. Reset failure is the most important cause of cycling yield degradation. Although nearly 100% reset yield is achievable in one cycle, reset failures still exist during cycling and may gradually degrade final yield. Data loss failures have lower rate than reset failure but are not negligible.

TABLE III.   FAILURE TYPES AND SOLUTIONS

Failure Types	Solutions	Impact on cycling
Set failure	• Optimize operation voltage • Use ramp-V_g method	100% set yield can be achieved
Reset failure	• Optimize operation voltage • Use ramp-V_g method	Most important for cycling yield
Type I data loss	• Material/structure engineering to improve ON-state retention • Adopt set margin to tolerate small variation	Low rate
Type II data loss	• Material/structure engineering to improve stability • Forced deeper reset • Adopt reset margin to tolerate small variation	Low rate

Set and reset switching yield can be improved by using ramp-V_g method instead of ramp-V_d method. The ramp-V_g method is made possible because of the serial transistor.

However, the use of three-terminal planar transistors as the selection devices constrains the highest achievable density. Proposals of two-terminal selection devices may help to achieve truly functional cross-bar architectures, but they may not enable the switching method based on I_{limit} control.

In summary, the failure mechanisms of resistive switching devices are analyzed in this paper. A rigorous testing condition is designed to expose all the failure types. Reset switching failure represents the most important failure reason, followed by two types of data loss. Cycling endurance can be improved by designing the operation methods appropriately to minimize each failure type. The upper limit of cycling endurance may be imposed by intrinsic material properties and array structures, which has to be optimized by material engineering and structural design. The two-stage cycling behavior observed in these devices can be utilized in a screening process to exclude devices with high failure rate from final products. Analysis in this paper is based on statistics measured by array characterization, and can be applied on other resistive switching materials. Clear understanding of the switching mechanism is important for the optimization of resistive switching devices.

REFERENCES

[1] C.C. Lin, C.Y. Lin M.H. Lin C.H. Lin, and T.Y. Tseng, "Voltage-polarity-independent and high-speed resistive switching properties of V-doped SrZrO3 thin films," IEEE Trans. Electron. Dev., vol. 54, no. 12, pp. 3146-3151, December 2007.

[2] K. Tsunoda, et al, "Low power and high speed switching of Ti-doped NiO ReRAM under the unipolar voltage source of less than 3 V," IEDM Tech. Digest, pp. 767-770, December 2007.

[3] U. Russo, D. Ielmini, C. Cagli, and A.L. Lacaita, "Filament Conduction and Reset Mechanism in NiO-Based Resistive-Switching Memory

(RRAM) Devices," IEEE Trans. Electron. Dev., vol. 56, no. 2, pp. 186-192, Feburary 2009.

[4] D. Lee, et al, "Resistance switching characteristics of metal oxide and schottky junction for nonvolatile memory applications," NVMTS, pp. 89-93, November 2006.

[5] A. Beck, J. G. Bednorz, Ch. Gerber, C. Rossel, and D. Widmer, "Reproducible switching effect in thin oxide films for memory applications," Appl. Phys. Lett., vol. 77, pp. 139-141, July 2000.

[6] A. Chen, et al, "Non-volatile resistive switching for advanced memory applications," IEDM Tech. Digest, pp. 764-767, December 2005.

[7] C.H. Ho, et al, "A highly reliable self-Aligned graded oxide WOx resistance memory: conduction mechanisms and reliability ," Symposium VLSI Tech., pp. 228-229, June 2007.

[8] H. B. Lv, et al, "Resistive memory switching of CuxO films for a nonvolatile memory application," IEEE Electron Dev. Lett., vol. 29, no. 4, pp. 309-311, April 2008.

[9] C.Y. Lin, et al, "Effect of top electrode material on resistive switching properties of ZrO2 film memory devices," IEEE Electron Dev. Lett., vol. 28, no. 5, pp. 366-368, May 2007.

[10] H.Y. Lee, et al, "HfO2 bipolar resistive memory device with robust endurance using AlCu as electrode," Symposium VLSI TSA, pp. 21-23 April 2008.

[11] I.G. Baek, et al, "Highly scalable non-volatile resistive memory using simple binary oxide driven by asymmetric unipolar voltage pulses," IEDM Tech. Digest, pp. 587-590, December 2004.

[12] Y.S. Chen, et al, "Forming-free HfO2 bipolar RRAM device with improved endurance and high speed operation," Symposium VLSI TSA, pp. 37-38, April 2009.

[13] S. Q. Liu, N. J. Wu, and A. Ignatiev, "Electric-pulse-induced reversible resistance change effect in magnetoresistive films," Appl. Phys. Lett., vol. 76, pp. 2749-2751, May 2000.

[14] A. Chen, S. Haddad, Y.C. Wu, T.N. Fang, S. Kaza, and Z. Lan, "Erasing characteristics of Cu2O metal-insulator-metal resistive switching memory," Appl. Phys. Lett., vol. 92, pp. 013503-1-3, January 2008.

[15] A. Chen, S. Haddad, and Y.C. Wu, "A Temperature-accelerated method to evaluate data retention of resistive switching non-volatile memory," IEEE Electron Dev. Lett., vol. 29, pp. 38-40, January 2008.

Charging and Discharging Characteristics of Metal Nanocrystals in Degraded Dielectric Stacks

Z. Z. Lwin, K. L. Pey[#], Y. N. Chen

Microelectronics Center, School of Electrical and Electronic Engineering
Nanyang Technological University
Singapore - 639798
[#]Ph: +65-67906371, E-mail: eklpey@ntu.edu.sg

P. K. Singh, S. Mahapatra
Indian Institute of Technology Bombay
Mumbai - 400076, India

Abstract— The conduction mechanisms of dielectric breakdown (BD) in MOS capacitor structure with nanocrystals (NCs) embedded in bi-layer gate stacks (SiO_2/Al_2O_3) are studied systematically. Using a unique stressing methodology of inducing a BD path in one dielectric layer, the charging and discharging phenomenon of the metal NCs and leakage mechanism in the degraded gate stacks are found to be strongly dependent on the lateral charge tunneling/hopping among the NCs. It is found that the localized BD not only affects charge holding capability of the affected NCs, but also provides a leakage path for the charges stored in the surrounding NCs. Thus, the discharging of NCs via the BD path is not a localized phenomenon.

Keywords- Metal nanocrystal; Dielectric breakdown; Discharging

I. INTRODUCTION & EXPERIMENT

Nanocrystal (NC) memories are promising due to their large storage capability, long retention time and scalability. However, the cumulative effect of the charge trapping/detrapping process in the dielectric during the program/erase cycling operation reduces the endurance [1]. Many research studies have been carried out on the dielectric breakdown (BD) and retention characteristics of metal and semiconductor NCs [1-3]. However, there is a lack of systematic study on the influence of a one-layer dielectric BD on the charging and discharging characteristics of NCs. In this paper, we investigate this charging and discharging behavior of metal NCs embedded in a degraded dielectric stack.

The samples used in this study are Au-NCs embedded in a bi-layer stack comprising SiO_2 (40Å) as tunnel oxide and Al_2O_3 (60Å) as blocking oxide on an n-Si substrate with an As doping of 3×10^{19} cm^{-3}. Fig. 1(a) shows a schematic of NCs embedded bi-layer gate stacks. From the planar view TEM micrograph shown in Fig. 1(b), the extracted density of NC is about 3×10^{12} cm^{-2}. The average size of the NCs is around 3 nm, as displayed in Fig. 1(c). Constant voltage stressing (CVS) with $V_{g\text{-stress}}$ = 4.5~5.5V, which is much lower than the BD voltage of 7V, in substrate injection mode was employed to induce BD in the gate stack.

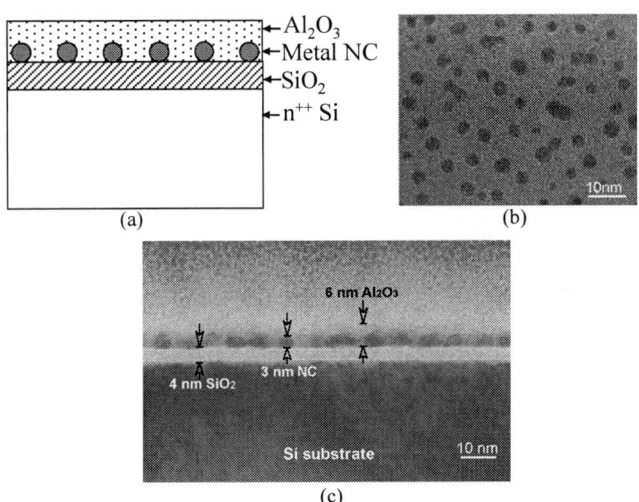

Figure 1. **(a)** Schematic of the NC-embedded gate stack used in our study. **(b)** Planar view TEM micrograph of Au-NCs. **(c)** Cross-sectional view TEM micrograph of a Ru-NC sample which has exactly the same structure as the Au-NC samples.

II. ONE-LAYER BREAKDOWN & I_G BLOCKADE

In order to achieve sequential BD of the individual dielectric layers in the NC-embedded SiO_2/high-κ gate stacks, current limited CVS is used. The stressing was automatically halted when the gate leakage current, I_g, reached a current compliance, I_{gl}, and subsequently, I_g-V_g measurements were performed. This cycle was repeated by setting a higher I_{gl} without changing the $V_{g\text{-stress}}$. Fig. 2(a) shows the evolution of I_g using a current limited CVS with $V_{g\text{-stress}}$=5.5 V in the substrate injection mode at room temperature. I_{gl} was initially set to 100 nA. After I_g reached 100 nA (solid line), I_{gl} was increased to 500 nA (open circles) and the stressing was continued, while $V_{g\text{-stress}}$ remained unchanged at 5.5 V. The similar procedure was repeated for I_{gl} = 1 µA, 10 µA and 0.1 A. During the electrical stress test, we performed I_g-V_g measurement on the fresh sample, after one-layer BD and after two-layer BD, respectively, as illustrated in Fig. 2(b).

978-1-4244-5430-3/10 $26.00 © 2010 IEEE

Comparing with the fresh I_g, I_g after an one-layer BD increases by about 2-3 orders of magnitude, while I_g after a two-layer BD increases by more than 7 orders of magnitude. Hence, a low I_{gl} of 100-200 nA (typically ~2 times of the pre-BD stress induced I_g) was used to induce one-layer BD first and post-BD I_g-V_g measurements were performed to differentiate between one- and two-layer BD, respectively.

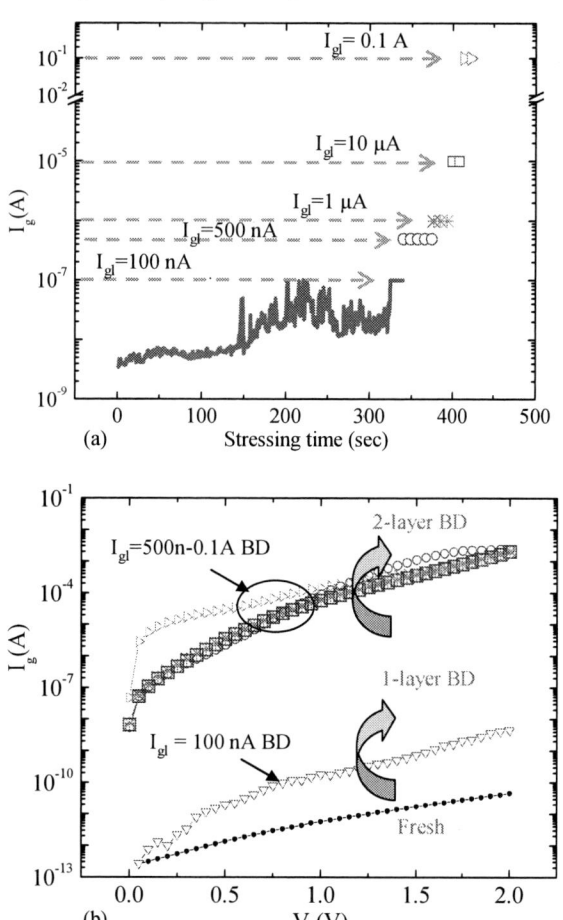

Figure 2. **(a)** Evolution of I_g using CVS, $V_{g\text{-stress}} = 5.5V$, in substrate injection mode with I_{gl} ranging from 100nA to 100mA. **(b)** I_g-V_g at different stages of progressive BD.

As shown in Figs. 3 and 4, two different types of I_g evolution were observed under CVS. In Fig. 3(a), after an instantaneous, huge jump (i.e., the occurrence of a dielectric BD event), I_g evolved into a high conduction state. The post-BD I_g-V_g shown in Fig. 3(b) suggests that a 2-layer dielectric BD occurred. Fig. 4 illustrates a new 3-stage I_g evolution, which is different from the usual I_g evolution during a progressive breakdown as depicted in Fig. 3. In Stage 1, I_g fluctuation was observed (black solid circles) when I_{gl} was set to 50 nA. After I_{gl} was changed to 200 nA, I_g experienced a sudden, abrupt increase (blue solid triangles), denoted as Stage 2. Then, the experiment was halted and the post-BD I_g-V_g measurement was carried out, as shown in Fig. 4(b). It is noted that I_g increased by 2 orders of magnitude as compared to the pre-stress I_g, indicating the rapid jump in I_g in Stage 2 is due to a one-layer BD. Then the stressing was continued by setting to

higher I_{gl} (300 nA and 500 nA) and I_g increased with I_{gl}. However, when the stressing was continued with $I_{gl} > 500$ nA, a subsequent abrupt drop of I_g by ~2 orders of magnitude occurred and after which, I_g remained at a very low current level (green open circles) in Stage 3 for a prolonged time. We name this phenomenon in Stage 3 as "I_g blockade". Similar behaviors were observed in other Au (Fig. 5) and Ir (Fig. 6) NC-embedded gate stacks.

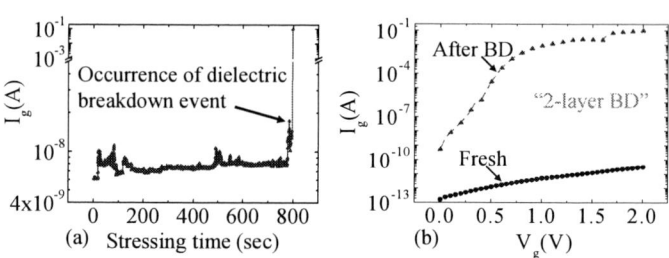

Figure 3. **(a)** Evolution of I_g using a CVS with $V_{g\text{-stress}} = 4.5V$ in substrate injection mode. **(b)** I_g-V_g before stress and after a 2-layer BD, respectively.

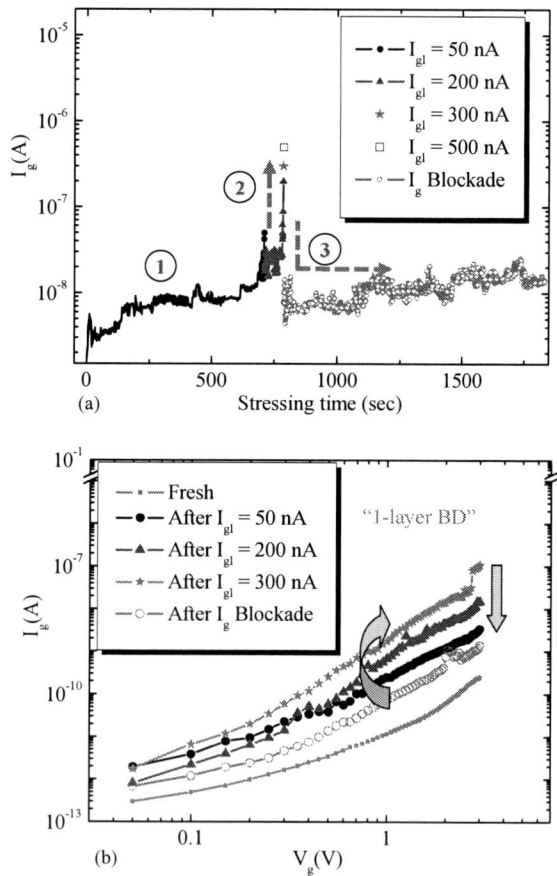

Figure 4. A new 3-stage I_g evolution under CVS in an Au-NC sample denoted as Sample 1. **(a)** I_g under 5V CVS (substrate injection) with I_{gl} ranging from 50 nA to 500 nA. Stage 1 - I_g fluctuations, Stage 2 - rapid jump in I_g due to a one-layer BD, and Stage 3 - I_g blockade are shown. **(b)** I_g-V_g trends at different stages of progressive BD and after I_g blockade, respectively. The arrows show I_g evolution during progressive BD and after I_g blockade (i.e., significant decrease in I_g).

978-1-4244-5430-3/10 $26.00 © 2010 IEEE

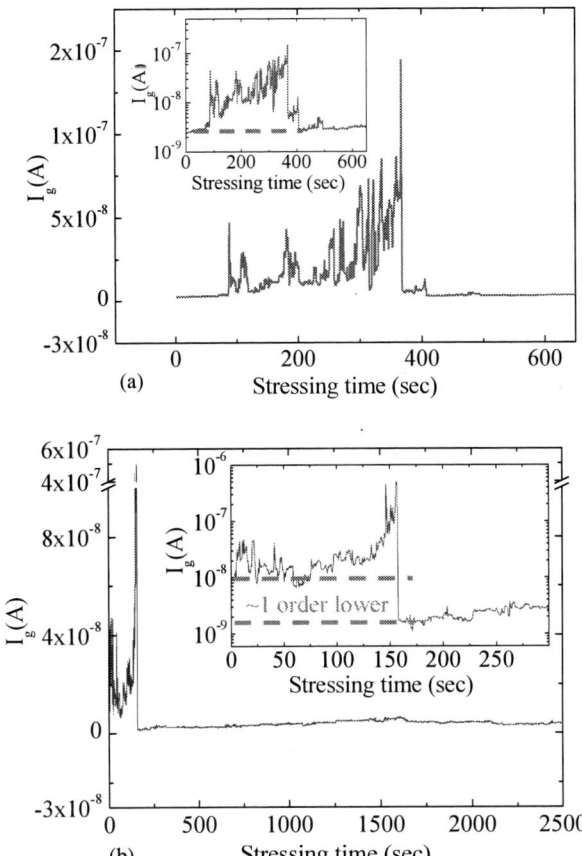

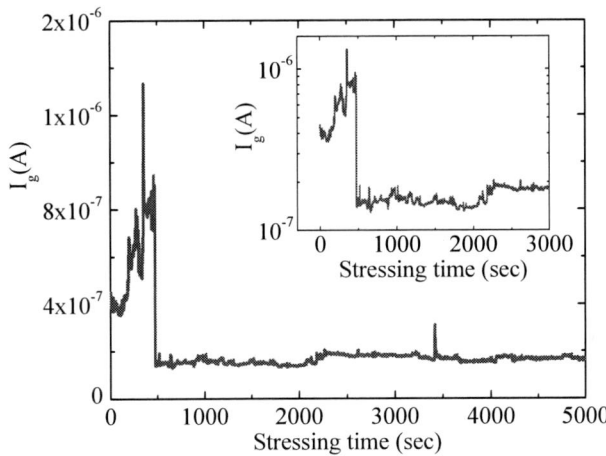

Figure 5. 3-stage I_g evolution under CVS in Au NC-embedded samples. **(a)** Sample 2 with $I_{g\text{-max}}$ = 200 nA under CVS of 4.8V. **(b)** Sample 3 with $I_{g\text{-max}}$ = 500 nA under $V_{g\text{-stress}}$=5.5 V. The insets show the same data in a logarithmic ordinate scale and the horizontal lines are guides to eyes.

As shown in Fig. 7, we propose a lateral charging model via the percolation (i.e., breakdown) path formed in one layer of the SiO_2/Al_2O_3 stack to explain the I_g blockade phenomenon. The I_g fluctuation observed in Stage 1 in Figs. 4-6 is closely related to charging and discharging of traps during the CVS stressing. When one layer is broken down in Stage 2, a localized electric field enhancement occurs at the BD path/NC interface as shown in Fig. 8, resulting in a lateral charging to the near-by NCs via the BD path such that all energy states in affected NCs were occupied by the injected electrons from the substrate injection stress until no more electron injection into the NCs is possible. After the NCs are charged through the BD path, I_g decreases due to the increased electrostatic repulsive potential among the charged NCs and the Coulomb Blockade effect in these charged NCs.

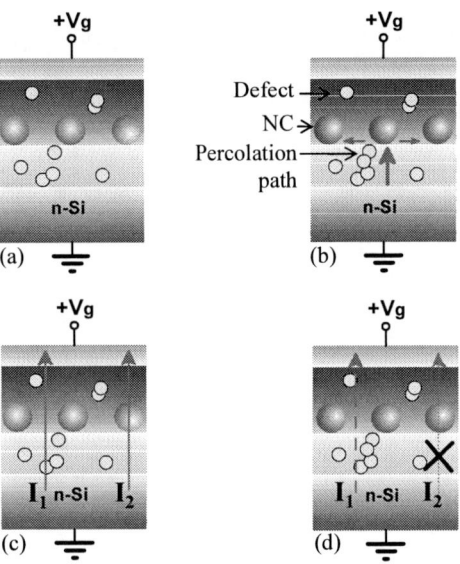

Figure 7. Schematics of **(a)** defect generation in dielectric film during CVS, **(b)** lateral charging through a percolation path, **(c)** leakage components before breakdown, $I_g = I_1 + I_2$ where I_1 is the direct/FN/trap assisted tunneling (TAT) current, and I_2 is the resonant tunneling current through NCs, and **(d)** after one-layer BD and lateral charging occurred, I_1 decreases due to the increased electrostatic potential between the charged NCs. I_2 is eliminated by the Coulomb blockade effect. In this illustration, we assume the BD path to be in the SiO_2 layer.

Figure 6. 3-stage I_g evolution in Ir-NC embedded gate stacks for CVS = 2.6V. The inset shows the same data in a logarithmic ordinate scale.

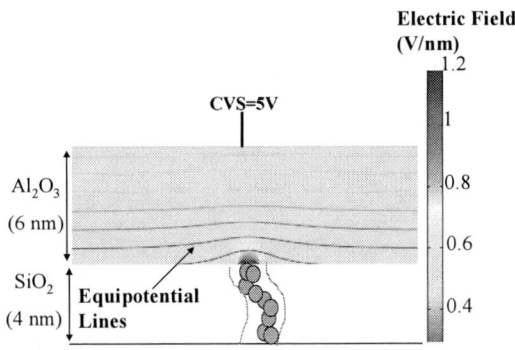

Figure 8. Simulation of localized electric-field enhancement at the BD path and SiO_2/high-κ interface. For simplicity, we ignored the presence of NCs.

978-1-4244-5430-3/10 $26.00 © 2010 IEEE

Fig. 9 shows the schematic energy band diagrams of the proposed lateral charging model when SiO_2 is assumed to have broken down. For simplicity, the energy band diagram of a charged NC (NC1) with an uncharged neighboring NC (NC2) is considered. With a positive bias applied to the gate electrode, electrons in the substrate can tunnel through the tunneling oxide and charge NC1. After a one-layer BD, electrons can conduct through the BD path and charge NC1 easily. The charging process will be continued when another electron has energy larger than the charging energy [4],

$$E_c = e^2/2C, \qquad (1)$$

where $C = 4\pi\varepsilon_0\varepsilon_r R$ is the self-capacitance of the charged NC with radius R. After NC1 has been charged, the electrostatic potential of NC1 increases, and then the lateral charging to the near-by NC2 could occur. Nevertheless, after NCs are completely charged under the applied constant voltage, further charging or tunneling through the tunneling oxide is prohibited by the energy barrier E_c, resulting in an I_g blockade. When the bias is removed, the charges stored in NCs could discharge through the BD path gradually as illustrated in Fig. 9(b).

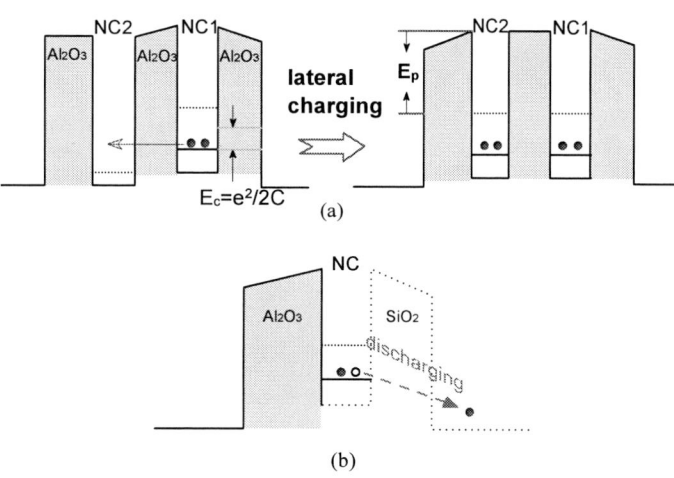

Figure 9. Schematic energy band diagram of **(a)** a lateral charging process between a charged NC (NC1) and an uncharged neighboring NC (NC2) and **(b)** a discharging process through the BD path. The BD path is assumed to be in the SiO_2 layer.

III. DISCHARGING OF NCS IN DEGRADED DIELECTRIC

The I_g blockade phenomenon allows us to study charging and discharging characteristics of NCs and their lateral interaction via the BD path. The relaxation current, I_{relax}, due to charge trapping/detrapping and dielectric polarization/relaxation [5-7], was measured after the removal of a 5 sec, +3V (V_{on}) stress on the sample of interest (e.g. after the occurrence of the I_g blockade). It is noted that I_{relax} follows the Curie-von Schweidler law [8] and the slope of I_{relax} is independent of the V_{on} magnitude, as illustrated in Fig. 10. In the NC-embedded bi-layer gate stacks, I_{relax} is contributed mainly by the NCs and the high-κ layer since I_{relax} of SiO_2 is negligible [9]. Fig. 11 shows I_{relax} for Samples 1-3 before stressing and after the occurrence of an I_g blockade. We suggest that in the fresh sample before CVS, I_{relax} is mainly contributed by the discharging of the as-deposited shallow traps

and polarization/relaxation in the high-κ layer. However, after the I_g blockade, a larger and flatter (i.e., longer decay time) I_{relax} was observed suggesting that the discharging of the electrons from NCs in the vicinity of the BD path has taken place. It is believed that I_{relax} discharge takes place through the BD path as well.

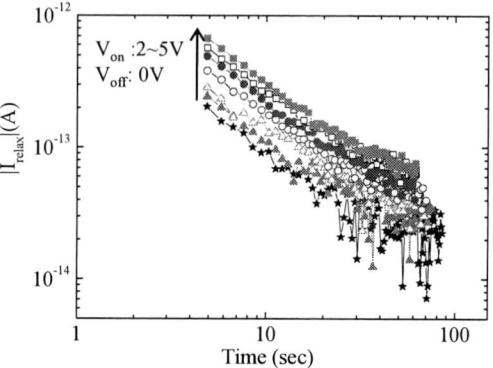

Figure 10. I_{relax} at different V_{on} (from 2 to 5V with step of 0.5V) of an unstressed fresh device.

As shown in Fig. 11, the slope of $\log(I_{relax})$-$\log(t)$ curves after the occurrence of the I_g blockade in Samples 1, 2 and 3 are 0.25, 0.60 and 0.83, respectively; smaller slopes indicate longer discharging time (t_d). The transient discharging time constant (τ) is defined as the time taken for I_{relax} to decay to 37% (or $1/e$) of its initial value. As shown in Fig. 11(c), it will take $t_d = 5\tau$ (for constant slope) for NCs to reach a completely discharged state. In the fresh samples before CVS, t_d is ~60 sec for all three samples. On the other hand, the estimated t_d for Samples 1, 2 and 3 at Stage 3 are 280, 120 and 100 sec, respectively. Since the gate is at zero-bias during the I_{relax} measurement, the charge density and electric field will decrease with more electrons tunneling out of the NCs [3]. Hence the decay slope would decrease with time. Therefore, t_d extrapolated here is the minimum discharging time. The difference in t_d after the I_g blockade among the 3 different samples is attributed to the difference in the BD hardness of the percolation path. Comparing Samples 2 and 3 under the same stressing conditions (see Table. 1), the I_g blockade in Sample 2 happened after I_g reached $I_{g\text{-}max}=200$ nA, while in Sample 3, it occurred after $I_{g\text{-}max}=500$ nA was reached, indicating that Sample 3 has experienced a harder BD than that of Sample 2. As a result, I_g of Stage 3 in Sample 3 is one order of magnitude lower than that of the fresh sample, while that in Sample 2 is similar to its pre-TDDB stress induced I_g (refer to the insets in Fig. 5). Thus, it is believed that sample with high breakdown hardness has more lateral distribution of defects in the dielectric[10]; hence more NCs can be charged and discharged more easily due to a more conductive path.

TABLE I.   STRESSING CONDITION OF AU-NC SAMPLES

Sample[a]	Stressing condition	$I_{g\text{-}max}$ before I_g blockade
1 (Fig. 4)	Current limited CVS	500 nA
2 (Fig. 5(a))	CVS	200 nA
3 (Fig. 5(b))	CVS	500 nA

[a] Au-NC in SiO_2/Al_2O_3 dielectric stack

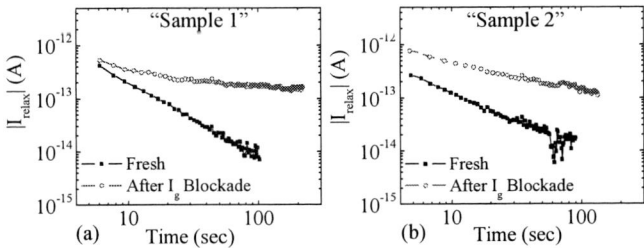

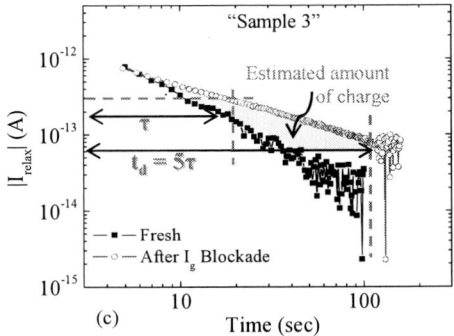

Figure 11. I_{relax} before stress and after I_g blockade in (a) Sample 1 (Fig. 4) (b) Sample 2 (Fig. 5(a)) and (c) Sample 3 (Fig. 5(b)).

The number of discharged electrons during the I_g blockade period can be estimated from Figs. 11(b) and (c) and is found to be $4\sim8\times10^7$ in Samples 2 and 3. Since the electronic energy levels are discrete in nanoparticles because of the electron wave function confinement, the average electronic energy level spacing of successive quantum levels, δ, for a 3 nm Au NC estimated from Kubo gap equation [11] is

$$\delta = 4E_f / 3n \qquad (2)$$

where E_f is the Fermi energy of bulk material and n is the total number of valance electrons in the nanocrystal. With $\delta = 0.3$ meV and a calculated potential well depth, $E_p = 4.1$ eV (see Fig. 9(a)) for 3nm Au NC in between SiO_2 and Al_2O_3, we estimate that each Au NC can store a maximum of 27320 electrons. This corresponds to ~2928 NCs affected by the BD path in Sample 2 and 3. However, as depicted in Fig. 12, the number of NCs just above (or below) the BD spot is only $11\sim50$ NCs since the localized BD occurs in one-layer dielectric could have radial distribution of $30\sim50$ nm [12]. It implies that the discharge diameter is much larger (up to ~380 nm for a given NC density of 3×10^{12} cm^{-2}) than the BD spot. Therefore, a very localized BD path will not just affect the charge-holding capability of the NCs situated directly above (or below) the BD spot. Rather, additional charges stored in the surrounding NCs will also tend to gradually leak via the BD path, suggesting that the discharging of NCs via the BD path is not a localized phenomenon.

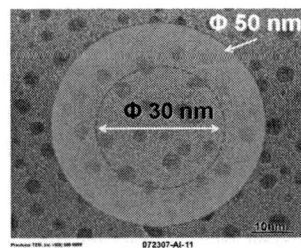

Figure 12. Radial distribution of a percolation path. The diameter of the BD path in SiO2/SiON gate dielectrics is estimated to be 30 nm – 50 nm at the early stage of a BD [12]. All dimensions are shown to scale.

IV. SUMMARY

We have found that charging and discharging of metal NCs embedded in Al_2O_3/SiO_2 gate stacks can take place easily via a BD path in the dielectrics. The results show that discharging of the stored electrons in the gate stacks is not a localized phenomenon but instead the leakage can occur from a large amount of NCs near the neighborhood of the BD path as well.

ACKNOWLEDGMENT

ZZ Lwin would like to acknowledge the Singapore EDB-GLOBALFOUNDRIES scholarship for funding her PhD study.

REFERENCES

[1] P. K. Singh, G. Bisht, M. Sivatheja, C. Sandhya, G. Mukhopadhyay, S. Mahapatra, et al., "Reliability of single and dual Layer Pt nanocrystal devices for NAND flash applications: A 2-region model for endurance defect generation," in *Reliability Physics Symposium, 2009 IEEE International*, 2009, pp. 301-306.

[2] P. K. Singh, G. Bisht, R. Hofmann, K. Singh, N. Krishna, and S. Mahapatra, "Metal Nanocrystal Memory With Pt Single- and Dual-Layer NC With Low-Leakage Al2O3 Blocking Dielectric," *Electron Device Letters, IEEE*, vol. 29, pp. 1389-1391, 2008.

[3] W. Guan, S. Long, M. Liu, Q. Liu, Y. Hu, et al., "Modeling of retention characteristics for metal and semiconductor nanocrystal memories," *Solid-State Electronics*, vol. 51, pp. 806-811, 2007.

[4] P. Beecher, A. J. Quinn, E. V. Shevchenko, H. Weller, and G. Redmond, "Insulator-to-Metal Transition in Nanocrystal Assemblies Driven by in Situ Mild Thermal Annealing," *Nano Letters*, vol. 4, pp. 1289-1293, 2004.

[5] Y. Chia-Han, K. Yue, W. Rui, L. Chen-Han, and K. Way, "Failure analysis of nanocrystals embedded high-k dielectrics for nonvolatile memories," in *Reliability Physics Symposium, 2008. IRPS 2008. IEEE International*, 2008, pp. 46-49.

[6] R. Wan, J. Yan, Y. Kuo, and J. Lu, "Dielectric Breakdown and Charge Trapping of Ultrathin ZrHfO/SiON High-k Gate Stacks," *Jpn. J. Appl. Phys.*, vol. 47, pp. 1639-1641, 2008.

[7] X. Zhen, L. Pantisano, A. Kerber, R. Degraeve, E. Cartier, S. De Gendt, et al., "A study of relaxation current in high-κ dielectric stacks," *Electron Devices, IEEE Transactions on*, vol. 51, pp. 402-408, 2004.

[8] K. J. Andrew, , "Dielectric relaxation in solids," *Journal of Physics D: Applied Physics*, vol. 32, p. R57, 1999.

[9] L. Wen, K. Yue, and K. Way, "Dielectric relaxation and breakdown detection of doped tantalum oxide high-k thin films," *Device and Materials Reliability, IEEE Transactions on*, vol. 4, pp. 488-494, 2004.

[10] X. Li, C. H. Tung, K. L. Pey, and V. L. Lo, "The physical origin of random telegraph noise after dielectric breakdown," *Applied Physics Letters*, vol. 94, p. 132904, 2009.

[11] C. N. R. Rao, G. U. Kulkarni, P. J. Thomas, and P. E. Peter, "Size-Dependent Chemistry: Properties of Nanocrystals," *Chemistry - A European Journal*, vol. 8, pp. 28-35, 2002.

[12] X. Li, C. H. Tung, and K. L. Pey, "The radial distribution of defects in a percolation path," *Applied Physics Letters*, vol. 93, p. 262902, 2008.

Characterization of Gate-All-Around Si-NWFET, including R_{sd}, Cylindrical Coordinate Based 1/f Noise and Hot Carrier Effects

Rock-Hyun Baek[†], Hyun-Sik Choi[†], Hyun Chul Sagong[†], Sang-Hyun Lee[†], Gil-Bok Choi[†],
Seung Hyun Song[†], Chan-Hoon Park[†], Jeong-Soo Lee[†‡], and Yoon-Ha Jeong[†‡]
[†]Dept. of Electronic and Electrical Engineering, Pohang University of Science and Technology (POSTECH)
[‡]National Center for Nanomaterials Technology (NCNT)
Pohang 790-784, Republic of Korea
Phone: +82-54-279-2897, E-mail: rock8201@postech.ac.kr

Chang-Ki Baek, Dae Mann Kim
School of Computational Sciences, Korea Institute for Advanced Study (KIAS)
Seoul 130-722, Republic of Korea

Yun Young Yeoh, Kyoung Hwan Yeo, Dong-Won Kim, Kinam Kim
Semiconductor R&D Center, Samsung Electronics Co., Ltd.
Yongin 449-711, Gyeonggi, Republic of Korea

Abstract—**In this paper, we introduce the cylindrical coordinate based flicker noise model for Silicon NanoWire Field Effect Transistor (Si-NWFET) with Gate-All-Around (GAA) structure. For the accurate extraction of the volume trap density, N_t, with 1/f noise modeling, the parameters which represent the intrinsic channel properties are determined by rejecting the series resistance R_{sd} effect. Due to the random distribution of traps in Si-NWFETs, the 1/f noise data are obtained by averaging the drain current power spectral density, S_{id}, for several devices. By using the proposed 1/f model, the extracted volume trap density is compared for three different oxide processes (ISSG/RTO/GNOx) and verified by hot carrier stress test.**

Keywords-flicker noise; 1/f; cylindrical coordinate; Gate-All-Around (GAA); R_{sd}; series resistance; MOSFET; twin silicon nanowire; TSNWFET

I. INTRODUCTION

The Si-NWFET is generally viewed as a promising next generation device, exhibiting a high on/off current ratio, excellent gate controllability through Gate-All-Around structure, and immunity from the short channel effects [1]-[4]. However, the series resistance R_{sd} is a significant parameter resulting from 1-D or 3-D contact configurations between the channel and Source/Drain [5]. Moreover, because of the nano scale three dimensional process and radial shaped oxide interface more traps are formed therein than conventional planar MOSFETs. Flicker noise becomes dominant as the channel length is scaled down and is a suitable criterion to analyze the oxide trap density of Si-NWFETs.

In this study, R_{sd} is extracted unambiguously with the use of the Y-function technique [6]. Next, by rejecting the parasitic R_{sd}, the intrinsic channel properties are extracted and 1/f model for Gate-All-Around structure is described accurately using

carrier number fluctuation and mobility fluctuation due to oxide traps. Additionally the hot carrier stress effects are examined.

II. DEVICE FABRICATION AND MEASUREMENT

All of the Gate-All-Around Twin Silicon NanoWireFETs (TSNWFETs) were fabricated in Samsung Electronics Co., with a common 80 nm channel width (and/or the circumference) but different channel lengths. To reduce the Source/Drain overlap and physical gate length, photoresist trimming was used in gate definition. There are three different oxide processes: in-situ steam generated oxide (ISSG) with T_{ox}=3.3 nm, rapid thermal oxide (RTO) with T_{ox}=1.5 nm and clean oxide annealed in N_2 atmospheres, namely nitride gated oxide (GNOx) with T_{ox}=2.8 nm. The process steps and flows are the same as those originally reported in [1]-[2]. For the linear operation of the device to extract series resistance R_{sd}, V_{gs} was swept from 0 V to 1.8 V at V_{ds}=0.05 V, V_{bs}= 0 V by using the Keithley 4200. Current power spectral density S_{id} is measured by using the ProPlus BTA9812A/B at V_{gs}-V_T=0.5 V and V_{ds}=0.1 V.

III. 1/F NOISE MODELING FOR SI-NWFET

A. 1/f noise characteristics of Si-NWFET

Figure 1 shows the dispersion and/or the non-uniformity of power spectral density, S_{id}. These data were obtained from 13 Si-NWFET samples with the same W/L=80/86 nm. S_{id} exhibits much larger dispersion, amounting to as much as 2~3 orders of magnitude. Since S_{id} is caused by random dopant fluctuation [7] S_{id} dispersion again points to substantial variations in oxide surrounding the channel, e.g. cross-sectional areas, interface roughness, etc. Consequently, extracting the volume trap

978-1-4244-5430-3/10 $26.00 © 2010 IEEE

density N_t by using the single device is meaningless and inaccurate. However, the average S_{id} value shows the 1/f trend as clearly shown in Fig. 1 [8]. It implies Si-NWFET shows an arbitrary spatial distribution in the individual case, but their averaging value is analogous to uniform spatial distribution. It is similar with the summation of Lorentzian noise makes 1/f noise shape. For this reason, all the R_{sd} extraction and 1/f modeling in this paper are performed with average value of I_{ds} and S_{id}.

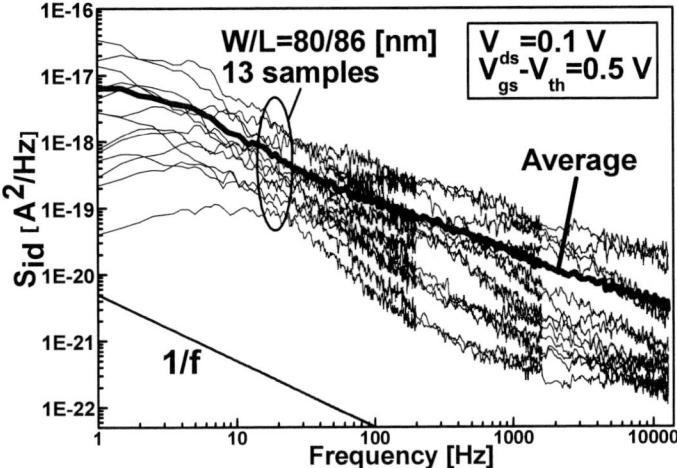

Figure 1.   Noise dispersion of Si-NWFET with W/L=80/86 nm.

B. 1/f noise model for Si-NWFET with GAA structure

Unlike planar MOSFETs, a Si-NWFET has a cylindrically shaped oxide. Therefore, the conventional 1/f model [9] developed for planar structures must be modified and transcribed into cylindrical coordinates, r, ϕ, z, as shown in Fig. 2. Here, T_{ox} is the oxide thickness, R is the channel radius, ΔZ is the differential channel length, and $2\pi R\Delta Z$ is the area element of the channel boundary or the oxide interface.

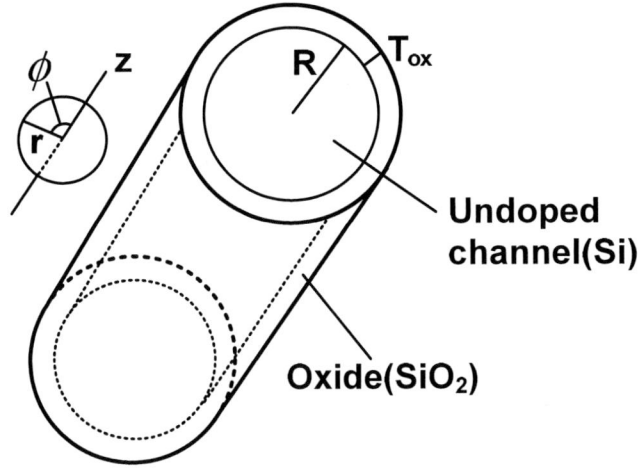

Figure 2.   Cylindrical coordinate system for describing Si-NWFET.

In the Si-NWFET, the current power spectral density $S_{\Delta Id}(z,f)$ for the area element $2\pi R\Delta Z$ is given by

$$S_{\Delta Id}(z,f) = \left[\frac{I_{ds}}{\Delta N}(1 \pm \alpha\mu N)\right]^2 S_{\Delta Nt}(z,f) \qquad (1)$$

where α is the scattering coefficient, μ is the carrier mobility, N is the number of channel carriers per unit area $\Delta W\Delta Z$, $\Delta N = N2\pi R\Delta Z$, I_{ds} is the drain current and $S_{\Delta Nt}$ is the power spectral density of the mean-square fluctuation in the number of occupied traps over the area $2\pi R\Delta Z$.

For specifying $S_{\Delta Nt}(z,f)$ in $S_{\Delta Id}(z,f)$, the classical number fluctuation [10] has to be modified to account for cylindrical device geometry and is re-expressed as

$$S_{\Delta Nt}(z,f) =$$
$$\int_{E_v}^{E_c}\int_0^{2\pi}\int_R^{R+T_{ox}} 4N_t(E,r,\phi,z)\Delta z \cdot f_t(1-f_t) \qquad (2)$$
$$\cdot \frac{\tau(E,r,\phi,z)}{1+\omega^2\tau(E,r,\phi,z)^2} dr(rd\phi)dE$$

where E_c is the conduction band, E_v is the valence band, $N_t(E, r, \phi, z)$ is the trap distribution in space and energy level, $f_t = [1+\exp(E-E_{fn})/kT]^{-1}$ is the trap occupancy function, $\omega = 2\pi f$ is the angular frequency, and $\tau(E, r, \phi, z)$ is the trapping time constant. In general, the oxide traps are presumed to be uniformly distributed in space with the largest noise contribution arising from traps near the quasi Fermi level E_{fn}, $N_t(E, r, \phi, z) = N_t(E) \approx N_t(E_{fn})$ [9]. Moreover, the probability of an electron penetrating into oxide decreases exponentially with the distance from the interface [9]. In Si-NWFETs, the trapping time constant is described as $\tau = \tau_s\exp(\gamma(r-R))$ because the trapping carriers penetrate the Si-SiO$_2$ interface from all directions. In this case, τ_s is the time constant at the interface, typically 10^{-10} s [11], and γ is the attenuation coefficient of the electron wave function in the oxide, typically 10^8 cm^{-1}. Thus, upon inserting (2) into (1) and performing the integrations with respect to trap energy levels and the channel length, one finds the total power spectral density of drain current $S_{Id}(f)$ throughout the entire channel, (4) through (3).

$$S_{\Delta Nt}(z,f) = N_t(E_{fn})\frac{kT2\pi R\Delta z}{\gamma f}\left(1-\frac{\ln 2\pi f\tau_s}{\gamma R}\right). \quad (3)$$

$$S_{Id}(f) = \frac{1}{L^2}\int_0^L S_{\Delta Id}(z,f)\Delta z\, dz$$
$$= \frac{kTI_{ds}^2}{\gamma f^n(2\pi R)L}(\frac{1}{N}+\alpha\mu)^2 N_t(E_{fn}) \qquad (4)$$
$$\cdot \left(1-\frac{\ln 2\pi f\tau_s}{\gamma R}\right).$$

In here, L is the channel length and n is the frequency exponent.

C. Series resistance R_{sd} extraction

To apply the 1/f noise model, devices parameters (R_{sd}, μ, V_{th}) used in (4) must be extracted. The accurate extraction of R_{sd}, together with other device parameters, μ, V_{th} [6] makes it possible to analyze and model 1/f noise measured in devices fabricated with three different oxides (ISSG/RTO/GNOx).

We have extracted R_{sd} in Si-NWFET unambiguously, using the Y-function technique, viz $Y = I_{ds}/g_m^{1/2}$, in conjunction with I_{ds} and g_m data [6]. In short channel planar MOSFETs, the mobility degradation due to surface roughness scattering strongly affects both g_m and I_{ds}, as illustrated in Fig. 3. In contrast, g_m and I_{ds} in Si-NWFET are fairly insensitive to the mobility degradation, as shown in Fig. 4. Rather, the Y-function therein exhibits the linear behavior in strong inversion compared to planar FET (Fig. 5), enabling thereby accurate R_{sd} extraction. This points to volume inversion in Si-NWFETs with the inverted carriers being relatively free of surface scattering.

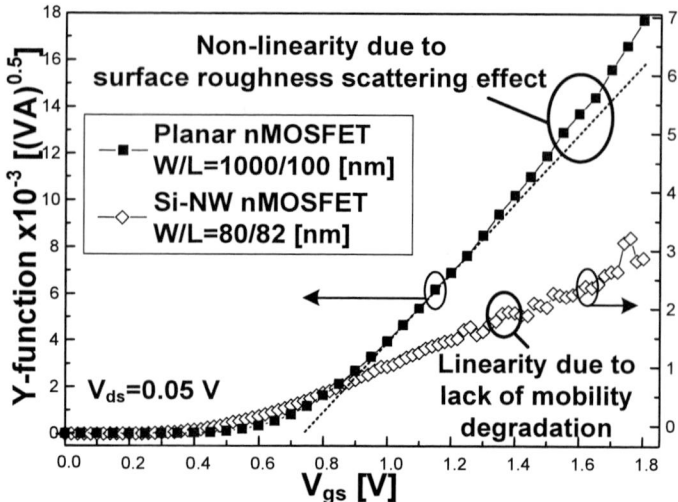

Figure 5. Comparison of the Y-function($Y = I_{ds}/g_m^{1/2}$) for a planar nMOSFET (T_{ox}=1.5 nm) and an n-type Si-NWFET (ISSG, T_{ox}=3.3 nm).

Fig. 6 summarizes the normalized R_T, R_{ch}, and R_{sd} measured from Si-NWFET with W/L=80/82 nm. A more detailed description of the extraction process can be found in [6].

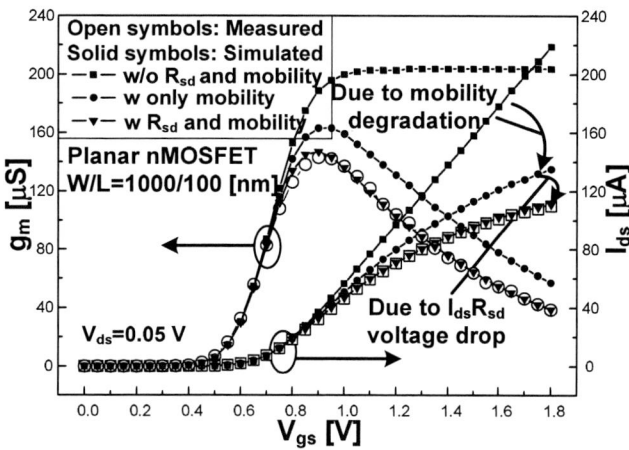

Figure 3. Measured data and simulated curves for a planar MOSFET (T_{ox}=1.5 nm) with W/L=1000/100 nm. Compared with Si-NWFET (in Fig. 4), the mobility degradation effect is pronounced.

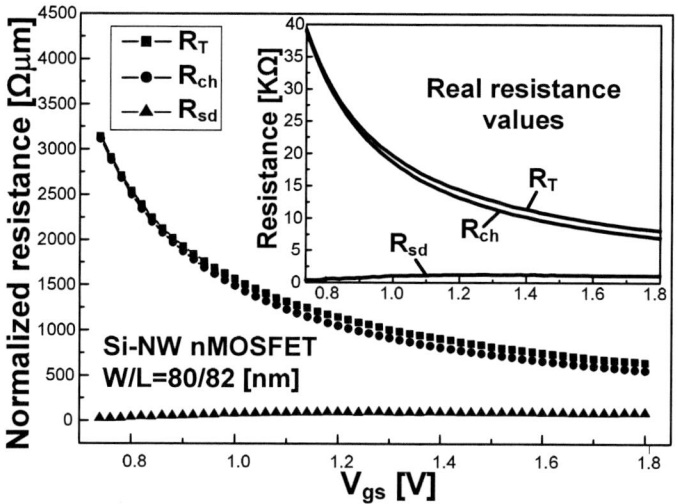

Figure 6. Normalized total (R_T), channel (R_{ch}), and series (R_{sd}) resistance of Si-NWFET with W/L=80/82 nm. Inset shows the real values.

IV. RESULTS AND DISCUSSIONS

A. Volume trap density (N_t) extraction

By using (4) and extracted device parameters (R_{sd}, μ, V_{th}) in Table.1, the averaged S_{id} data are modeled as a function of the channel length. Here, scattering coefficient α ($\approx 10^{-15}$) derived from the same TSNWFET [12] was used and the carrier density N was calculated by averaging N_s and N_d in source and drain regions, respectively using [13], which reported the charge based equations for GAA MOSFETs.

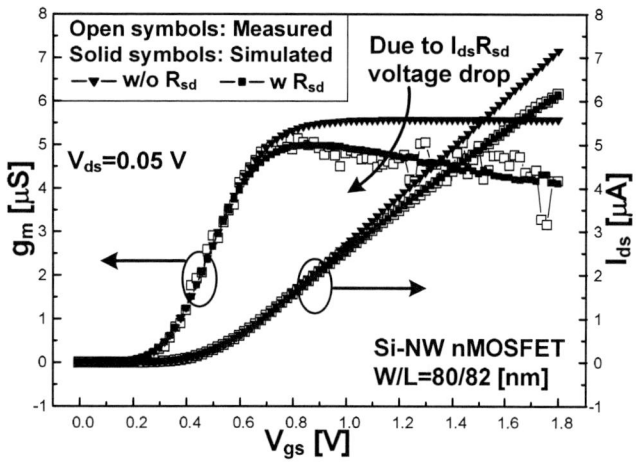

Figure 4. Measured data and simulated curves for the Si-NWFET (ISSG, T_{ox}=3.3 nm) with W/L=80/82 nm. The R_{sd} effect is clearly exhibited.

TABLE 1. THE EXTRACTED PARAMETERS OF 3 DIFFERENT OXIDE PROCESS

Oxide types	W/L [nm]	μ [cm^2/Vs]	V_{th} [V]	R_{sd} [KΩ]
ISSG oxide (T_{ox}=3.3 nm)	80/86	78.9	0.55	1.26
	80/132	115.9	0.59	1.82
	80/164	119.9	0.58	1.87
RTO (T_{ox}=1.5 nm)	80/86	41.8	0.41	1.63
	80/132	61.0	0.38	1.52
	80/164	73.4	0.40	1.69
GNOx (T_{ox}=2.8 nm)	80/86	48.7	0.44	1.11
	80/132	80.2	0.46	1.63
	80/164	86.0	0.43	1.91

Figure 7 shows the 1/f noise data obtained from Si-NWFETs with an ISSG oxide with different channel lengths and theoretical fits to the data based on cylindrical the flicker noise model. Because the parasitic effect of R_{sd} has been eliminated, parameters reflecting the intrinsic channel properties are used for fitting the 1/f data. As demonstrated in the figure, the cylindrical 1/f noise model fits the measured data rather well.

In Fig. 8, the volume trap density $N_t(E_{fn})$ extracted from devices with three different oxide processes are presented. Regardless of channel length, the extracted $N_t(E_{fn})$ values are nearly the same in each oxide type consistent with the flicker noise model. Devices with RTO show the largest trap density, while those with ISSG oxide show the lowest trap level. This indicates that optimizing the oxide process is crucial in Si-NWFET for reducing relatively large trap density, compared with other devices.

B. Hot Carrier Stress Effect

To verify the accuracy of extracted volume trap density N_t, we executed another reliability experiment, hot carrier (HC) stress. As we apply 1000s HC stress and 1000s relaxation, pre-existing traps formed during fabrication and stress induced traps can be obtained qualitatively.

Figure 9 shows the V_{th} shift due to hot carrier stress. The shift is more pronounced for shorter channel lengths as reported in [14]. In the case of ISSG and GNOx (rectangular and triangle symbols), no appreciable recovery was observed in relaxation time; it thus appears that ΔV_{th} is caused by permanent oxide damage. However, the recovery process is apparent for the case of RTO (empty circle symbols). For the similar vertical E-field, another stress condition ($V_{gs}=V_{th}+0.85$ V, $V_{ds}=V_{th}+1.7$ V) was also applied to RTO. Ths recovery is caused by the pre-existed trap density of RTO: largest among the oxide processes considered. Therefore, Fig. 9 experimentally verifies the results shown in Fig. 8.

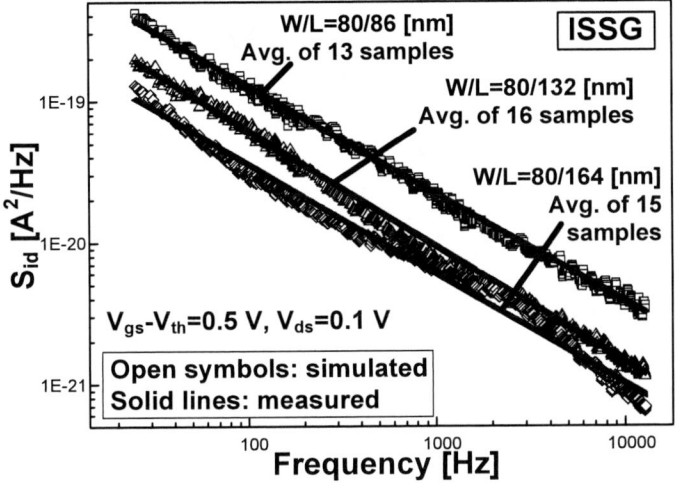

Figure 7. The 1/f noise data and theoretical fits in the linear and strong inversion region V_{gs}-V_{th}=0.5 V, V_{ds}=0.1 V, from Si-NWFETs for ISSG with different channel lengths.

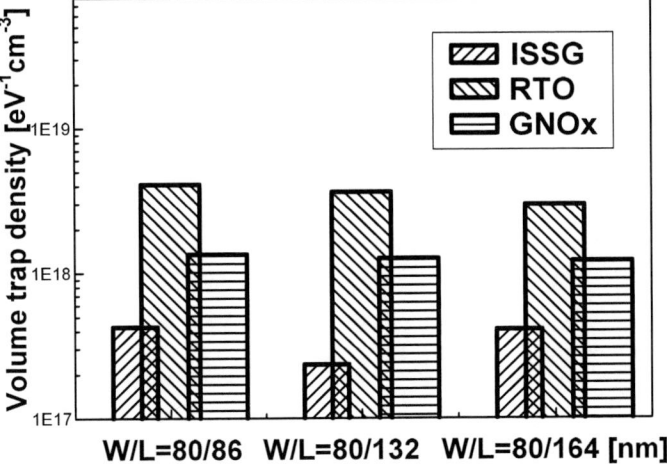

Figure 8. The extracted volume trap densities for the three different oxides (ISSG/RTO/GNOx) and channel lengths.

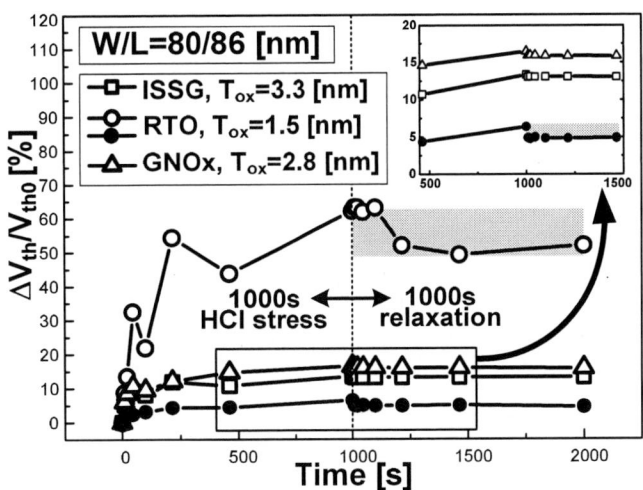

Figure 9. V_{th} shift due to HCI for different oxides and the same W/L=80/86 nm. Only RTO shows the recovery in both HC stress condition: $V_{gs}=V_{ds}=V_{th}+1.7$ V (Open symbols) and $V_{gs}=V_{th}+0.85$ V, $V_{ds}=V_{th}+1.7$ V (Solid symbols).

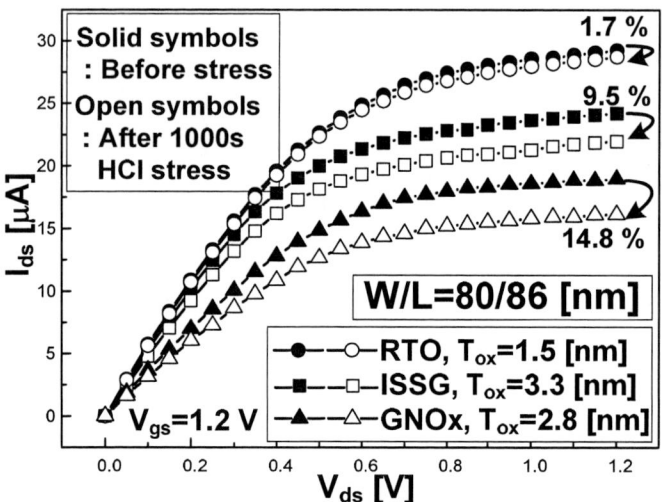

Figure 10. I_{ds}-V_{ds} curve before and after 1000s HC stress with similar vertical E-field condition ($V_{gs}=V_{ds}=V_{th}+1.7$ V for ISSG and GNOx, $V_{gs}=V_{th}+0.85$ V, $V_{ds}=V_{th}+1.7$ V for RTO).

Figure 10 shows I_{ds} degradation after 1000s hot carrier stress which gives an almost the same oxide electric field. As mentioned in [14], because direct tunneling is more dominant than oxide trap generation induced by hot carriers for thin oxide below 2 nm of thickness, RTO was degraded only 1.7%, which is less than the others. Though RTO is more reliable to defects generated by HCI, it shows an obvious recovery characteristic as shown in inset of Fig. 9. It means RTO originally has more pre-existed traps than the others. This fact is another evidence supporting the Fig. 8 and consistent with result of Fig. 9.

V. CONCLUSIONS

The accurate extraction of R_{sd} made it possible to characterize the 1/f noise in terms of the intrinsic properties of the channel for TSNWFETs. The cylindrical coordinate based flicker noise model for Si-NWFETs with GAA structure was successfully developed and applied to three different oxide processes. The trap density was lowest for ISSG, while highest for RTO. These findings were correlated with hot carrier stress effects.

ACKNOWLEDGMENT

This research was supported by the World Class University (WCU) program through the Korea Science and Engineering Foundation funded by the Ministry of Education, Science and Technology (Project No. R31-2008-000-10100-0). Also, this work was supported by the National Research Foundation of Korea by the Korean Government (2009-0089200) and partially supported by the BK21 program, the National Center for Nanomaterials Technology (NCNT) in Korea, and the "System IC 2010" project of Korea Ministry of Knowledge Economy. The authors thank Samsung Electronics Co., Ltd. for device fabrication.

REFERENCES

[1] S. D. Suk, S.-Y. Lee, S.-M, Kim, E.-J. Yoon, M.-S. Kim, M. Li, C. W. Oh, K. H. Yeo, S. H. Kim, D.-S. Shin, K.-H. Lee, H. S. Park, J. N. Han, C. J. Park, J.-B. Park, D.-W. Kim, D. Park, and B.-I. Ryu, "High performance 5 nm radius Twin Silicon Nanowire MOSFET (TSNWFET) : fabrication on bulk Si wafer, characteristics, and reliability," in *Proc. IEDM Tech. Dig.,* 2005, pp. 717-720.

[2] K. H. Yeo, S. D. Suk, M. Li, Y.-Y. Yeoh, K. H. Cho, K.-H. Hong, S. K. Yun, M. S. Lee, N. M. Cho, K. H. Lee, D. H. Hwang, B. K. Park, D.-W. Kim, D. Park, and B.-I. Ryu, "Gate-All-Around (GAA) Twin Silicon Nanowire MOSFET (TSNWFET) with 15 nm length gate and 4 nm radius nanowires," in *Proc. IEDM Tech. Dig.,* 2006, pp. 1-4.

[3] S. D. Suk, M. Li, Y. Y. Yeoh, K. H. Yeo, K. H. Cho, I. K. Ku, H. Cho, W. J. Jang, D.-W. Kim, D. Park, and W.-S. Lee, "Investigation of nanowire size dependency on TSNWFET," in *Proc. IEDM Tech. Dig.,* 2007, pp. 891-894.

[4] S. D. Suk, K. H. Yeo, K. H. Cho, M. Li, Y. Y. Yeoh, K.-H. Hong, S.-H. Kim, Y.-H. Koh, S. G. Jung, W. J. Jang, D.-W. Kim, D. Park, and R.-I. Ryu, "Gate-All-Around Twin Silicon Nanowire SONOS Memory," in *Proc. VLSI Symp. Tech. Dig.,* 2007, pp. 142-143.

[5] J. Hu, Y. Liu, C. Z. Ning, R. Dutton, and S.-M. Kang, "Fringing field effects on electrical resistivity of semiconductor nanowire-metal contacts," *Appl. Phys. Lett.,* vol. 92, pp. 083503, 2008.

[6] R. H. Baek, C. K. Baek, S. W. Jung, Y. Y. Yeoh, D.-W. Kim, J.-S. Lee, Dae M. Kim and Y. H. Jeong, "Characteristics of the series resistance extracted from Si-nanowire FETs using the Y-function technique," accepted for publication in the *IEEE Trans. Nanotechnol.,* 2009. This article is available in the *IEEE Xplorer.*

[7] A. Asenov, "Random dopant induced threshold voltage lowering and fluctuations in sub-0.1 mm MOSFET's: A 3-D "Atomistic" simulation study," *IEEE Trans. Electron Devices,* vol. 45, no. 12, pp. 2505-2513, Dec. 1998.

[8] J. Zhuge, R. Wang, R. Huang, Y, Tian, L. Zhang, D.-W. Kim, D. Park, and Y. Wang, "Investigation of low-frequency noise in Silicon Nanowire MOSFETs," *IEEE Electron Device Lett.,* vol. 30, no. 1, pp. 57-60, Jan. 2009.

[9] K. K. Hung, P. K. Ko, C. Hu, and Y. C. Cheng, "A unified model for the flicker noise in Metal-Oxide Semiconductor Field-Effect Transistors," *IEEE Trans. Electron Devices,* vol. 37, no. 3, pp. 654-665, Mar. 1990.

[10] S. Christensson, I. Lundström and C. Svensson, "Low frequency noise in MOS transistors-I THEORY," *Solid-State Electron.,* vol. 11, pp. 797-812, 1968.

[11] R. Jayaraman, and C. G. Sodini, "A 1/f noise technique to extract the oxide trap density near the conduction band edge of silicon," *IEEE Trans. Electron Devices,* vol. 36, no. 9, pp. 1773-1782, Sep. 1989.

[12] S. Yang, K. H. Yeo, D.-W. Kim, K.-I. Seo, D. Park, G. Y. Jin, K. S. Oh, and H. Shin, "Random telegraph noise in N-type and P-type silicon nanowire transistors," in *Proc. IEDM Tech. Dig.,* 2008, pp. 765-768.

[13] B. Iñíguez, D. Jiménez, J. Roig, H. A. Hamid, L. F. Marsal, and J. Pallarès, "Explicit continuous model for long-channel undoped surrounding gate MOSFETs," *IEEE Trans. Electron Devices,* vol. 52, no. 8, pp. 1868-1873, Aug. 2005.

[14] Y. Y. Yeoh, S. D. Suk, K. H. Yeo, D.-W. Kim, G. Y. Jin, and K. S. Oh, "Investigation on hot carrier reliability of Gate-All-Around Twin Si Nanowire Field Effect Transistor," in *Proc. IEEE Int. Reliability Phys. Symp.,* 2009, pp. 400-404.

978-1-4244-5430-3/10 $26.00 © 2010 IEEE

Thermal Disturbance and its Impact on Reliability of Phase-Change Memory Studied by the Micro-Thermal Stage

SangBum Kim[+], Byoungil Lee[+], Mehdi Asheghi[#], G. A. M. Hurkx[*], John Reifenberg[#], Kenneth Goodson[#], and H.-S. Philip Wong[+]

[+] Department of Electrical Engineering, Stanford University, Stanford, CA, U.S.A.
Tel.: 1-650-723-9484, Fax: 1-650-723-4659, E-mail: kimsangb@stanford.edu

[#] Department of Mechanical Engineering, Stanford University, Stanford, CA, U.S.A.

[*] NXP_TSMC Research Center, Leuven, Belgium.

Abstract— **In this paper, we study thermal disturbance and its impact on reliability using a novel measurement structure – the micro-thermal stage (MTS). The small thermal time constant of the MTS extends the time-scale of temperature dependence measurement to ~100 μs. The reliability of phase-change memory (PCM) is evaluated in terms of data retention and variation of the high resistance (RESET) state resistance (R_{RESET}) and the threshold switching voltage (V_{th}). We experimentally show how the impact of thermal disturbances on retention is accumulated and its dependence on the electric field. The thermal disturbance effect on R_{RESET} variation changes with time and it is the largest for the shortest time delay after RESET programming. Thermal disturbance can cause at least 25 and 100% variation for R_{RESET} and V_{th} respectively in the given thermal disturbance scenario. We propose an effective method to exploit thermal disturbance to make multi-bit operation more robust.**

Keywords-reliability; thermal disturbance; program-disturb; thermal crosstalk; proximity disturbance; Phase-change memory;

I. INTRODUCTION

Phase-change memory (PCM) is one of the most mature candidates for a next generation non-volatile memory [1], [2]. As a candidate for future memory devices, scalability is the key issue. Aggressive scaling scenarios on PCM and migration to a $4F^2$ cross-point array structure result in faster scaling for cell-to-cell distances which leads to a larger temperature rise in adjacent cells [3], [4]. Therefore, understanding the impact of thermal disturbance (TD) on PCM reliability is important.

The main concern for TD has been its impact on retention for high resistance (RESET) state [3], [4]. However, this can be a too simplistic picture because many characteristics of the RESET state are also dependent on temperature such as RESET resistance (R_{RESET}) [5], threshold switching voltage (V_{th}) [6], and their drift behaviors [7]-[11]. Therefore, TD results in larger variation for R_{RESET} and V_{th}, which is one of main concerns for memory reliability and makes multi-bit realization more difficult. TD effect is not static because the RESET state continuously changes after RESET programming due to the drift behavior. Therefore, TD can have different impact on R_{RESET} and V_{th} variation depending on the time when it affects the memory cell (delay time).

In this paper, we experimentally evaluate the impact of TD on retention, R_{RESET}, and V_{th} variation using a PCM cell integrated with an external heater and their impact on reliability of PCM is discussed.

II. MICRO-THERMAL STAGE

When a selected cell in the PCM array is programmed, the adjacent cells are thermally disturbed due to the heat diffusion. The programming time is typically less than a few hundred nanoseconds and thus TD can be treated as a short heat pulse. To experimentally emulate short time-scale thermal fluctuations as TDs, we implement the micro-thermal stage (MTS). The MTS is designed to enable precise long-range temperature control in a microsecond time-scale. Fig. 1 shows the MTS. The MTS heater is integrated directly on top of the lateral PCM cell to control the local temperature in-situ during the measurement. The total thermal mass of the MTS is much less compared to the conventional thermal stages where the whole sample and chuck are heated. Therefore, the MTS lowers thermal time constant to a few microseconds and we extend the time-scale of temperature dependence measurement

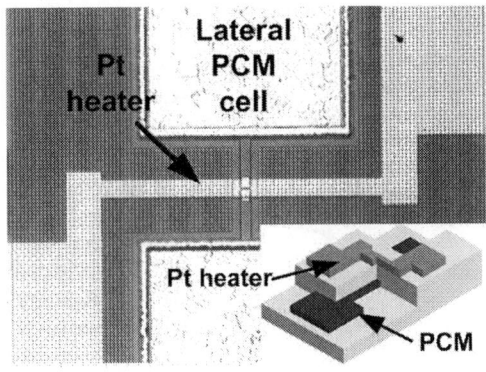

Figure 1. Microscope image of the micro-thermal stage (MTS). A Pt MTS heater is integrated on top of the lateral phase-change memory (PCM) cell which was reported elsewhere [12]. The inset figure shows the MTS heater overlapped region over PCM programming region.

from ~100 μs to ~10 s with current MTS design. Pt is chosen as a MTS heater material because the temperature of the heater can be calibrated using the temperature coefficient of resistance and its resistivity is appropriate for effective heating for the geometry of the heater. The electrical pulse to the MTS heater generates Joule heat and controls the temperature of the programmed region of the memory cell.

III.   EXPERIMENTAL RESULTS

A.   Crystallization Time

We first investigate the impact of TD on retention. Retention failure occurs when the phase-change material in the RESET state is crystallized into the low resistance (SET) state and the crystallization process can be accelerated by TD. Therefore, retention characteristic can be characterized by the temperature dependence of crystallization time (t_{crys}). t_{crys} at various temperatures has been determined by monitoring current through the PCM cell with a small bias while being heated by the MTS heater in time-scale as short as hundreds of microseconds and detecting the first sudden increase in current. In measurement of current through the PCM cell, multiple steps are observed (Fig. 2). This observation suggests that multiple crystalline paths are formed during crystallization as schematically depicted in Fig. 3. Fig. 4 shows that t_{crys} follows an Arrhenius behavior with activation energy of 1.84 eV over a large temperature range between 240 and 130 °C with t_{crys} between 100 μs and 10 s across 5 orders in time-scale. We compare t_{crys} with and without TDs (Fig. 4). A 75 μs long heating pulse at 240 °C is applied 100 μs after RESET programming as a TD by the MTS and t_{crys} is measured afterwards. The measured t_{crys} with TD shows a constant shift in log-scale time, which means that the ratio between t_{crys} with and without TD is constant throughout large temperature range between 130 °C and 240 °C. The 75 μs long heating pulse at 240 °C corresponds to ~40 % of t_{crys} at 240 °C, which corresponds to the ratio of difference in t_{crys} with and without TD (the inset figure in Fig. 4). When the heating pulse width is reduced to 25 μs, the ratio is reduced accordingly. In most of retention measurements [3], [4], it has been assumed that TDs at different temperatures and for different durations can be simply added. Result in Fig. 4 experimentally verifies this assumption for the first time. Equation (1) summarizes this relationship between retention and TD.

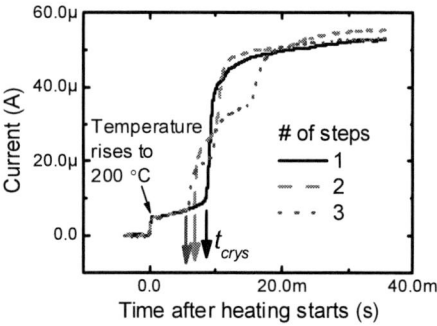

Figure 2.   Number of crystallization steps varies between 1 to several. Crystallization behaviors are represented by the current through PCM cells. Reading voltage is 0.1V. Cells are annealed at 200 °C for crystallization.

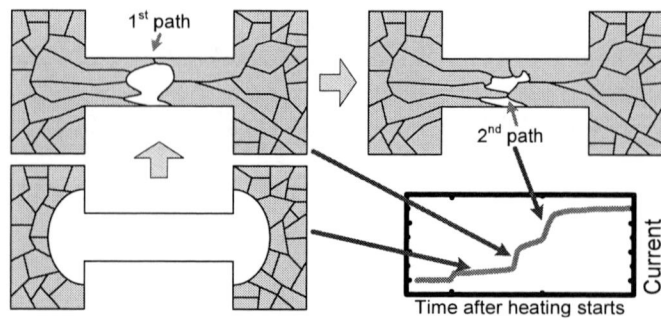

Figure 3.   Schematic crystallization process for growth dominated material. Non-uniformity in growth speed can result in multiple conduction paths.

$$r = \int v_{crys}(T(t))dt \qquad (1)$$

where $T(t)$ is the temperature as a function of time, $v_{crys}(T)$ is the crystallization speed as a function of temperature, and r is the degree of the crystallization between '0' (no crystallization) and '1' (the formation of the crystalline conductive path). Once $t_{crys}(T)$ is experimentally found, $v_{crys}(T)$ is given by the following equation.

$$v_{crys}(T) = 1/t_{crys}(T). \qquad (2)$$

In a cross-point array with diode selection devices, unselected cells can be reverse-biased while being thermally disturbed by the adjacent selected cell. If the electric field has an effect on the crystallization process, this can result in different t_{crys} and retention behavior. Therefore we have measured t_{crys} dependence on electric field by applying different amount of read bias on the PCM cell. Fig. 5 shows that larger read bias results in shorter t_{crys}. In other words, larger read bias shifts the curve in Fig. 5 to the right by temperature difference less than 30 °C. This small difference in temperature can be attributed to the Joule heating in the phase-change material by the read bias. The Joule heating caused by the read bias is usually insignificant because RESET state is highly resistive. However, the Joule heating in the phase-change material becomes significant at higher temperatures because amorphous phase-change material becomes much more conductive as temperature increases [5]. Therefore, the actual temperature of the programmed region is higher than the temperature set by the MTS heater due to Joule heat generated in phase-change material, which explains the temperature shift in Fig. 5. The direct dependence of t_{crys} on the electric field itself has not been observed for read biases between 0.1 and 0.4 V.

B.   RESET Resistance and Threshold Switching Voltage Drift

TD can cause changes in R_{RESET} and V_{th} in two different ways. First, if TD changes the temperature of the cell when R_{RESET} and V_{th} are read, the measurement results will change even when amorphous region has the same trap density.

978-1-4244-5430-3/10 $26.00 © 2010 IEEE

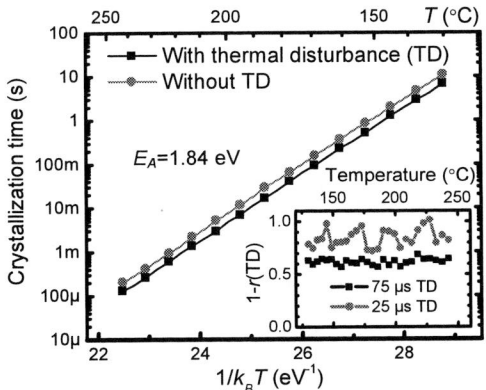

Figure 4. Crystallization time (t_{crys}) with and without thermal disturbance (TD). For t_{crys} with TD, cells are disturbed at 240 °C for 75 µs before annealing for crystallization. Constant difference in log-scale crystallization time suggests that their ratio is constant. The inset figure shows that given TD results in a fixed amount of crystallization between 130 and 240 °C, i.e. for at least 5 orders in time. 'r' is the degree of crystallization between '0' (no crystallization) and '1' (the formation of the first crystalline conductive path).

Second, TD can cause difference in trap densities because the drift behavior is expedited at high temperatures causing faster trap decay. Difference in trap densities will result in different R_{RESET} and V_{th} even though they are read at the same temperature.

First, we investigate how much effect TD has on R_{RESET} drift and how it depends on the delay time between the RESET programming event and the TD event. Electrical pulses applied to the PCM cell and the MTS heater are shown in Fig. 6. Though not intuitive, the well-known structural relaxation (SR) model [10], [13] suggests that 1) the same amount of annealing (temperature and duration) has the largest impact on instantaneous R_{RESET} if the delay time for annealing ('t_B' in Fig. 6) is the shortest, 2) the accelerated drift caused by annealing results in slowing down of the drift right afterwards, and 3) final value of R_{RESET} would not depend on the delay time for annealing ('t_B' in Fig. 6) if the amount of annealing is the same. All these implicit inferences of the SR model are verified experimentally for the first time in the following experiments. PCM cells in the RESET state are annealed with the MTS heater at 60 °C for 600 µs and the delay time for annealing ('t_B' in Fig. 6) is varied between 300 µs and 110 ms. Cell resistances

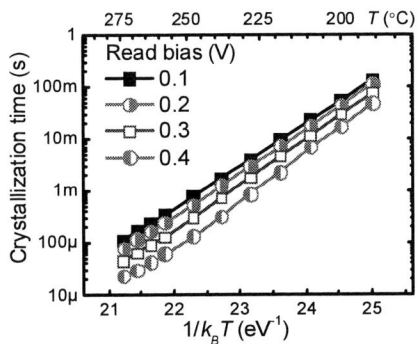

Figure 5. Crystallization time dependence on read bias between 0.1 and 0.4 V. Curves are shifted to right as read bias increases. Temperature difference caused by Joule heating in phase-change material is responsible for differences.

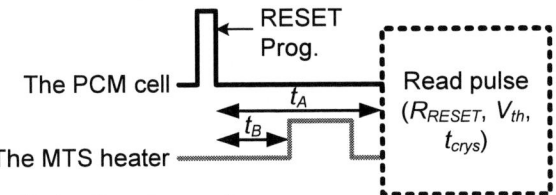

t_A : Delay time for read
t_B : Delay time for annealing (or thermal disturbance)

Figure 6. Pulse input profiles for the phase-change memory cell (top) and micro-thermal stage heater (bottom). t_A: The delay time between the RESET programming event and the read event. t_B: The delay time between the RESET programming event and the TD (or annealing) event. Read pulse shapes are determined by properties to be measured and reading temperature (T_R).

are measured at room temperature. Fig. 7(a) shows that TD at the shortest delay time for annealing (300 µs, black squares) has the largest effect resulting in ~25% increase for instantaneous R_{RESET} right after annealing. On the other hand, TD at 1.2 ms causes smaller increase and TD at 11 ms and 110 ms do not cause any meaningful R_{RESET} increase. Therefore, TD at earlier delays after RESET programming should be prevented to reduce variation in R_{RESET}. In addition, when the drift coefficient is significantly increased by annealing, the drift coefficient right afterwards gets lower than normal as can be seen from data 'C' in Fig. 7(b). This is because traps with certain activation energy already decayed earlier during

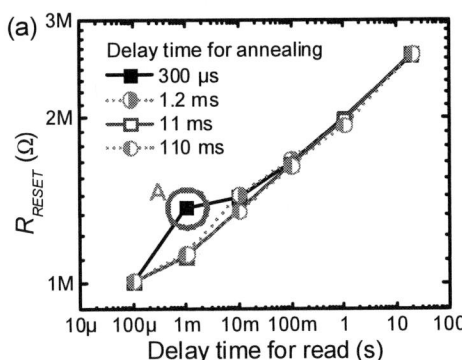

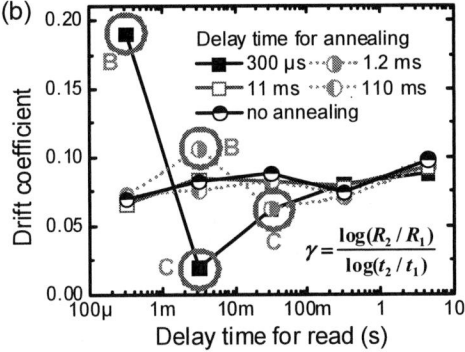

Figure. 7 (a) RESET resistance (R_{RESET}) and (b) drift coefficient (γ) for the same amount of annealing (60 °C for 600 µs) with different delay times for annealing. Resistances are measured at room temperature. Each data point is averaged over 150 measurements. Annealing at the shortest delay time (300 µs) increases R_{RESET} by 25 % instantaneously ('A'), while annealing at larger delays shows much less increase for R_{RESET}. 'B': Instantaneous drift coefficient increases more if the delay time for annealing is shorter. 'C': Drift coefficient is reduced right after annealing.

annealing and there is less number of traps remaining that would have decayed right after annealing. *However, the drift coefficient is recovered long time after annealing because traps that decay at that time-scale have large activation energy and the impact of short TD on those traps can be ignored due to their large activation energy.* Finally, R_{RESET} measured at the same delay time for read ('t_A' in Fig. 6) is the same if the amount of annealing (temperature and duration) is the same regardless of their differences in delay time for annealing ('t_B' in Fig. 6), as can be seen from Fig. 7(a), e.g. two samples with 300 µs and 1.2 ms of delay time for annealing merge into the same R_{RESET} at 10 ms. This is because each trap has its own decay time constant which depends only on temperature regardless of the delay time.

We experimentally show that measured V_{th} and its drift are affected by TD as well. Fig. 8 shows the effect of annealing on V_{th} drift. V_{th} is measured at room temperature by a triangular pulse that ramps up. When the voltage amplitude ramps and reaches at V_{th} of the PCM cell, the current through the PCM cell suddenly increases. The measurement results clearly show that V_{th} drifts in time and it drifts faster at higher annealing temperatures. V_{th} dependence on measurement temperature is shown in Fig. 9. PCM cells are not annealed but the temperature is raised by the MTS heater right before the measurement. Results clearly show that measured V_{th} decreases as measurement temperature increases even though they have the same trap density. Reduced V_{th} by TD can be problematic for reliable writing operation of the PCM array. When the selected cell is being programmed, adjacent cells are thermally disturbed while they are in reverse bias. If the amount of reverse bias is larger than the reduced V_{th}, adjacent cells in the RESET state can be switched to the partial SET state. Combining these effects of TD on V_{th}, current measurement data show that TD at 65 °C alone could lead to ~100 % variation (from 1.4 V to 0.7 V) in V_{th}.

C. Multi-bit Operation

Random TD can be harmful to robust operation of multi-bit memory cells by causing larger R_{RESET} and V_{th} variations [14]. However, if we deliberately use the large effect of annealing in early delays on drift, we can make multi-bit operation more reliable by suppressing the amount of drift. Fig. 10 shows that 600 µs long annealing at 300 µs delay and 60 °C provides 25 % larger room in resistance for middle resistance levels at 1 ms delay. The temperature and duration of annealing can be optimized for the largest resistance margin with minimum energy consumption and the smallest reduction in retention time.

IV. CONCLUSION

Using a novel measurement structure called the micro-thermal stage (MTS), we study thermal disturbance (TD) effect on not only retention but also RESET resistance (R_{RESET}) and threshold switching voltage (V_{th}) variations. We experimentally verify how to add up the impacts of TD on retention. We show that TD can cause at least 25% variation in R_{RESET} and 100% in V_{th}. Experimental results support details of the structural relaxation model with distributed activation energy. We

propose a scheme to deliberately use this TD effect to make larger resistance margin for multi-level operation.

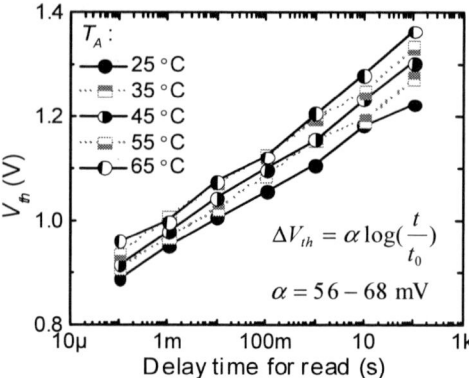

Figure 8. Threshold switching voltage (V_{th}) drift dependence on annealing temperature (T_A). Annealing at the elevated temperature makes V_{th} drift faster resulting in higher V_{th} at the given time after RESET programming. Cells are programmed and V_{th} is measured at room temperature. Annealing started at 1 µs after RESET programming and the PCM cell is brought back to room temperature before V_{th} measurement. The inset equation describes the V_{th} drift behavior where α is the V_{th} drift coefficient which increases as T_A increases.

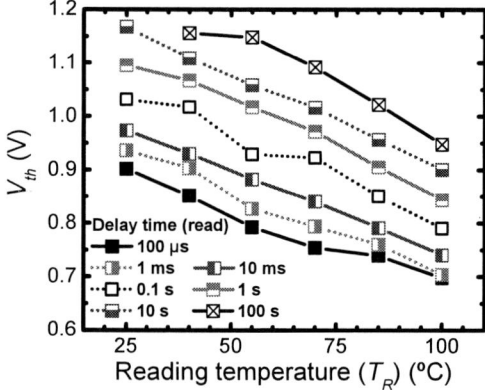

Figure 9. Threshold switching voltage (V_{th}) dependence on reading temperature (T_R) for various delay time for read. V_{th} decreases as T_R increases due to increased conductivity at higher T_R. At the given T_R, V_{th} increases as time elapses after RESET programming. Cells are programmed and remains at the room temperature. Then, the temperature is raised to T_R right before V_{th} reading.

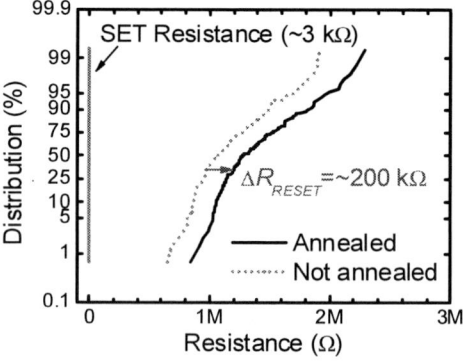

Figure 10. Larger resistance margin is achieved by annealing the cell. The RESET resistance is measured at 1ms after programming. The additional margin is ~25% larger by annealing the cell at 60 °C between 300 and 900 µs delay (for 600 µs). The widened margin can be utilized for robust multi-level operation.

ACKNOWLEDGMENT

This work is supported in part by NXP, the National Science Foundation, the Global Research Collaboration (GRC) of the Semiconductor Research Corporation (SRC), the member companies of the Stanford Non-Volatile Memory Technology Research Initiative (NMTRI), and the MSD Focus Center, one of six research centers funded under the Focus Center Research Program (FCRP), a Semiconductor Research Corporation entity. S. Kim is additionally supported by the Samsung Fellowship.

REFERENCES

[1] J. H. Oh et al., "Full integration of highly manufacturable 512Mb PRAM based on 90nm technology," in *IEDM Tech. Dig.*, pp. 49-52, 2006.

[2] G. Servalli, "A 45nm generation phase change memory technology," in *IEDM Tech. Dig.*, pp. 5.7.1-5.7.4, 2009.

[3] A. Pirovano, A. L. Lacaita, A. Benvenuti, F. Pellizzer, S. Hudgens, and R. Bez, "Scaling analysis of phase-change memory technology," in *IEDM Tech. Dig.*, pp. 699-702, 2003.

[4] U. Russo, D. Ielmini, A. Redaelli, and A. L. Lacaita, "Modeling of programming and read performance in phase-change memories-Part II: Program disturb and mixed-scaling approach," *IEEE Trans. Electron Devices*, vol. 55, no. 2, pp. 515-522, 2008.

[5] D. Ielmini, and Y. Zhang, "Analytical model for subthreshold conduction and threshold switching in chalcogenide-based memory devices," *J. Appl. Phys.*, vol. 102, p. 054517, 2007.

[6] D. Ielmini, "Threshold switching mechanism by high-field energy gain in the hopping transport of chalcogenide glasses," *Phys. Rev. B*, vol. 78, p. 035308, 2008.

[7] A. Pirovano, A. L. Lacaita, F. Pellizzer, S. A. Kostylev, A. Benvenuti, and R. Bez, "Low-field amorphous state resistance and threshold voltage drift in chalcogenide materials," *IEEE Trans. Electron Devices*, vol. 51, no. 5, pp. 714-719, 2004.

[8] I. V. Karpov, M. Mitra, D. Kau, G. Spadini, Y. A. Kryukov, and V. G. Karpov, "Fundamental drift of parameters in chalcogenide phase change memory," *J. Appl. Phys.*, vol. 102, p. 124503, 2007.

[9] D. Ielmini, D. Sharma, S. Lavizzari, and A. L. Lacaita, "Reliability impact of chalcogenide-Structure relaxation in phase-change memory (PCM) cells-Part I: Experimental study," *IEEE Trans. Electron Devices*, vol. 56, no. 5, pp. 1070-1077, 2009.

[10] S. Lavizzari, D. Ielmini, D. Sharma, and A. L. Lacaita, "Reliability impact of chalcogenide-Structure relaxation in phase-change memory (PCM) cells-Part II: Physics-based modelling," *IEEE Trans. Electron Devices*, vol. 56, no. 5, pp. 1078-1085, 2009.

[11] D. Ielmini and M. Boniardi, "Common signature of many-body thermal excitation in structural relaxation and crystallization of chalcogenide glasses," *Appl. Phys. Lett.*, vol. 94, p. 091906, 2009.

[12] D. Tio Castro et al., "Evidence of the thermo-electric thomson effect and influence on the program conditions and cell optimization in phase-change memory cells," in *IEDM Tech Dig.*, pp. 35-38, 2007.

[13] D. Ielmini, S. Lavizzari, D. Sharma, and A. L. Lacaita, "Temperature acceleration of structural relaxation in amorphous $Ge_2Sb_2Te_5$," *Appl. Phys. Lett.*, vol. 92, p. 193511, 2008.

[14] D.-H. Kang et al., "Two-bit cell operation in diode-switch phase change memory cells with 90nm technology," in *VLSI Symp. Tech. Dig.*, pp. 98-99, 2008.

MATURE PROCESSABILITY AND MANUFACTURABILITY BY CHARACTERIZING V_T AND V_{MIN} BEHAVIORS INDUCED BY NBTI AND AHTOL TEST

Jongwoo Park*[1], Sungmok Ha[1], Sunme Lim[2], Jae-Yoon Yoo[2], Junkyun Park[1], Kidan Bae[1], Gunrae Kim[1], Min Kim[1] and Yongshik Kim[2]

Technology Reliability[1]/Technology Development[2], System LSI division, Samsung Electronics

San #24 Nongseo-Dong Giheung-Gu, Yongin-City, Gyeonggi-Do, Korea 446-711

jongwoo.s.park@samsung.com (email); 82-31-209-1344 (phone); 82-31-209-4312 (fax)

ABSTRACT

A systematical reliability assessment for technology process that is essential for technology feasibility and qualification is presented by addressing physical and electrical characterization and reliability evaluation. By varying the duty cycle of enhanced pulsed radio frequency (eprf) technique used for the gate oxynitridation, the effects of nitrogen concentration and profile at SiO_2/Si interface on V_T and V_{min} shift of thin oxide pMOSFET (~20A) and SRAM, which result from negative bias temperature stability (NBTI) and accelerated high temperature operating life (AHTOL) stress test, are meticulously investigated. Using secondary ion mass spectrometry (SIMS) and high resolution Rutherford back scattering (H-RBS), nitrogen concentration and profile at the interface are carefully characterized. It is found that pMOSFET device processed with 10% of eprf provides ~2× longer NBTI lifetimes than with 20% of eprf due to lower nitrogen concentration at the interface. Furthermore, V_{min} shift of SRAM with 10% of eprf, which is caused by AHTOL test conditioned at 140°C with 1.4× V_{dd}, is ~3~4× less than with 20% of eprf. In fact, a nano-probing technique elucidates that V_{min} shift is mainly attributed to the mismatch of V_T between pull-up (PU) transistors in SRAM induced by NBTI stemmed from AHTOL test. It is also empirically shown that V_{min} shift behavior is in good agreement with the read margin rather than the write. Accordingly, a stabilized V_{min} drift behavior consistently adheres to the write margin. Hence, the optimization of interfacial nitrogen concentration results in less pMOSFET NBTI degradation so as to efficiently suppress V_{min} shift of SRAM. Besides, increasing PU transistor size that decreases the γ value (the ratio of I_{on} current of PG to PU) can also reduce V_{min} shift during AHTOL test. Finally, mature processability and manufacturability are attained by characterizing V_T and V_{min} behaviors for pMOSFET and SRAM from the front-end-of-line (FEOL) process optimization and SRAM bit-cell design aspect.

INTRODUCTION

For technology process reliability qualification below 90nm nodes, technical hurdles arisen from reliability perspectives in the earlier stage of development are V_T and V_{min} degradation induced by negative bias temperature instability (NBTI) and accelerated highly temperature operating life (AHTOL) test in transistor and product level with SRAM as a technology vehicle. Tremendous demands for higher drive current and better device performance push gate oxide thickness toward its material limit without any compromise in gate oxide integrity as well as device performance. Hence, nitrogen implantation process during oxide growth and integration becomes crucial and is an inevitable process step particularly for pMOSTFET to protect SiO_2/Si interface from boron diffusion, which is typically used as P^+ implantation. It is well known that nitrogen profile at SiO_2 interface significantly impacts on the gate oxide integrity and influences on lifetime projection at normal operating conditions [1-3]. In fact, throughout literature, there are many studies regarding these issues from

reliability perspectives coupled with HCI, NBTI and TDDB [4-5]. It is well known that controversy still exists in regards to the effects of nitrogen on MOSFET device performance and reliability [6-8]. The minimum operating voltage (V_{min}) of SRAM is prone to increase due to V_T variations as technology process is scaled down, which hinders a wide range power scaling of memory rich system on chip necessary for mobile application. Indeed, the product standby currents and regulator designs depends on transistor reliability. Since SRAM cell maintains stable data even under parasitic noise and device mismatch in SRAM cell, the understanding of impact of NBTI degradation on SRAM bit-cell stability is crucial. Nevertheless, a few papers [9-11] are available to anticipate the behaviors of V_{min} shift of SRAM induced by AHTOL test in conjunction with physical phenomena of pMOSFET V_T NBTI degradation from the front-end-of line (FEOL) process and bit cell design perspectives.

In this paper, a systematical reliability assessment for technology feasibility and qualification that focuses on pMOSFET V_T NBTI and SRAM V_{min} shift is presented by addressing physical and electrical characterization and reliability assessment. Enhanced pulsed radio frequency (eprf) technique is adopted for the gate oxide nitridation, and the effects of nitrogen concentration and profile at SiO_2/Si interface on V_T and V_{min} shift of thin oxide pMOSFET (~20A) and SRAM induced by NBTI and AHTOL stress test are painstakingly investigated. In consequence, the framework that can control fundamental phenomena of V_T and V_{min} shift driven by NBTI and AHTOL test are presented from the FEOL process and bit cell design aspects.

EXPERIMENTAL

The pMOSFET devices with a dielectric thickness of ~20A and n-type poly Si as gate material are prepared. For the gate oxide nitridation, the enhanced pulsed radio frequency (eprf) technique is chosen for better uniformity of nitrogen at the interface compared with rapid thermal oxynitride (RTO) deposition. By varying the duty cycle of eprf, pMOSFET devices having different nitrogen concentrations at SiO_2/Si interface are processed through the standard front-end-of-line (FEOL) process. Then, they are subjected to NBTI test conditioned at 140C with various V_G stress conditions. ΔV_T shift induced by NBTI degradation is measured by a fast measurement [12]. Using secondary ion mass spectrometry (SIMS) and high resolution Rutherford back scattering (H-RBS), physical analyses of nitrogen concentration and profile at the interface are carefully characterized.

In addition, circuit level reliability assessment of ΔV_{min} shift resulting from AHTOL test is investigated using 6T SRAM cell with 10 and 20% of eprf. At regular time bases of AHTOL stress test conditioned at 140°C with 1.4× V_{dd}, ΔV_{min} is measured at room and hot temperatures with respect to deterioration of the read and write performance of SRAM. In particular, for SRAM suffering from lager V_{min} shift related failure, the nano-probing technique is adopted for measuring changes in V_T and I_{dsat} of pull-up (PU), pull-down (PD) and pass gate (PG) transistors in 6T SRAM.

978-1-4244-5430-3/10 $26.00 © 2010 IEEE

RESULTS AND DISCUSSION

Since the gate oxynitridation that prevents boron penetration and reduces the gate leakage is crucial for dielectric integrity in the earlier stage of the FEOL process development, the early reliability assessment (ERA) aims at the diagnosis of weakest process and understanding of intrinsic degradation mechanism based on fast turnaround results prior to technology process qualification. No doubt, such closed loop activity cannot be underestimated to refine the baseline for a straightforward technology process reliability qualification that can leverage mature process so as to facilitate volume production.

TABLE I contains the ERA and technology process qualification divided into two parts of L1 and L2 qualification. L1 rather focuses on intrinsic reliability adhered to the FEOL and back-end-of-line (BEOL) process. Whereas, package and technology related reliability assessments that include the precondition followed by environmental test, AHTOL and EFR test fall into L2 category. Between L1 and L2 qualification, other tests, *e.g.*, efuse, soft error rate (SER) and chip-package-interaction (CPI), are typically conducted and closed. After a full qualification is granted upon the completion of L1 and L2, the ongoing reliability monitoring (ORM) is launched at regular time bases to monitor process and secure mature manufacturability. Note that TABEL I is given to explain the qualification procedure for technology process and product qualification for the logic and foundry.

TABLE I. TECHNOLOGY PROCESS QUALIFICATION METRICS FOR INTRINSIC PURE PROCESS QUALIFICATION FOR L1 AND RELEVANT FUNCTIONAL TEST BY USING TECHNOLOGY VEHICLE FOR L2 QUALIFICATION.

Early Reliability Assessment for Technology Feasibility: HCI, NBTI, TDDB, EM, IMD-TDDB, HTOL etc.			
Technology Process Qualification			
L1		L2	
FEOL	BEOL	Package related	Technology related
HCI	EM	Precond.	EFR/HTOL
BTI	SM	HTS	
Intrinsic/Extrinsic	TC	TC	
TDDB	TDDB	HAST/THB	

Figure 1 (a-b) shows TOF-SIM characterization of the nitrogen profile and distribution at SiO₂/Si interface fabricated by thermal and eprf oxynitridation on Si substrate with and without poly Si in the sample preparation. As shown, the nitrogen profile and distribution inherently rely on the gate dielectric nitridation scheme. In Fig. 1(a), both thermal and eprf oxynitridation result in a symmetrical nitrogen profile at the interface. However, the appearance of nitrogen distribution is different, narrow for the eprf and broad for thermal nitridation. In case for the eprf, the nitrogen peak exists toward poly-Si, while the thermal oxynitridation yields the nitrogen peak near SiO₂/Si interface. It is reported that the nitrogen formation near SiO₂/Si interface results in enhanced NBTI degradation with respect to increase of interface states and hole trapping [13, 14]. In Fig. 1(b), the nitrogen peak with the eprf is rarely shifted after sputtering in comparison to thermal oxynitridation. In contrast, the nitrogen distribution of thermal oxynitridation is far moved after sputtering. This discrepancy implies that the eprf enables the nitrogen to be uniformly distributed at SiO₂/Si interface. In consequence, the eprf provides a better nitrogen profile and uniformity at the interface than thermal oxynitridation. These results are well agreed with [2, 15]. It is found that NBTI degradation increases with increasing nitrogen concentration in the gate dielectric and is aggravated particularly for thermal oxynitridation, when compared to plasma nitridation

with similar nitrogen dose [16]. It is also found that NBTI degradation depends on the interfacial nitrogen concentration rather than total nitrogen dose [17]. Accordingly, the eprf oxynitridation is adopted as the baseline for further process integration and development.

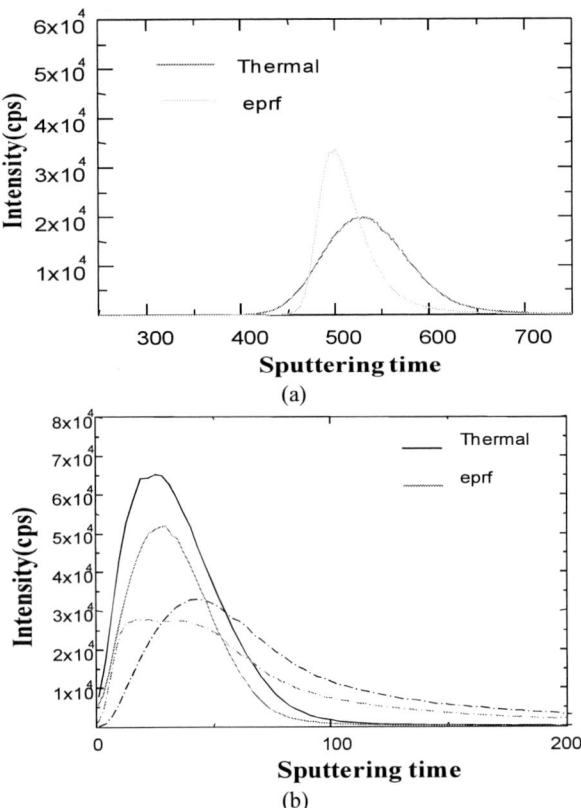

Fig. 1. TOF-SIMS characteriztion for the nitrogen profile at SiO₂/Si interface fabricated by thermal and eprf oxynitridation: (a) nitrogen profile and (b) uniformity; from the samples prepared with and without poly-Si on oxynitridation deposited on Si substrate. Note that solid and dotted line represent the nitrogen distribution before and after sputtering in Fig. 1(b).

Figure 2 reveals the nitrogen profile and concentration at SiO₂/Si interface fabricated by 10% and 20% of eprf characterized by SIMS and H-RBS, respectively. Since SIMS can underestimate nitrogen concentration due to a limited resolution, H-RBS that provides a quantitative depth resolution is used as an alternative technique. As shown, the interfacial nitrogen concentration increases as the duty cycle of eprf increases. On the comparative bases, the appearance of nitrogen profile is physically similar, except the nitrogen concentration at the interface. In turn, SIMS results are well agreed with H-RBS, which is summarized in TABLE II. The interfacial nitrogen concentration achieved from 10 and 20% of eprf is 6.4 and 8.6%, respectively.

TABLE II. NITROGEN CONCENTRATION AT SiO₂/Si INTERFACE MEASURED BY H-RBS ANALYSIS.

Gate oxide nitridation	N (%)	O (%)	Si (%)
1 RTO +eprf 10%	6.4	48.2	43.3
2 RTO +eprf 20%	8.6	50.8	42.8

RTO stands for the rapid thermal oxidation.

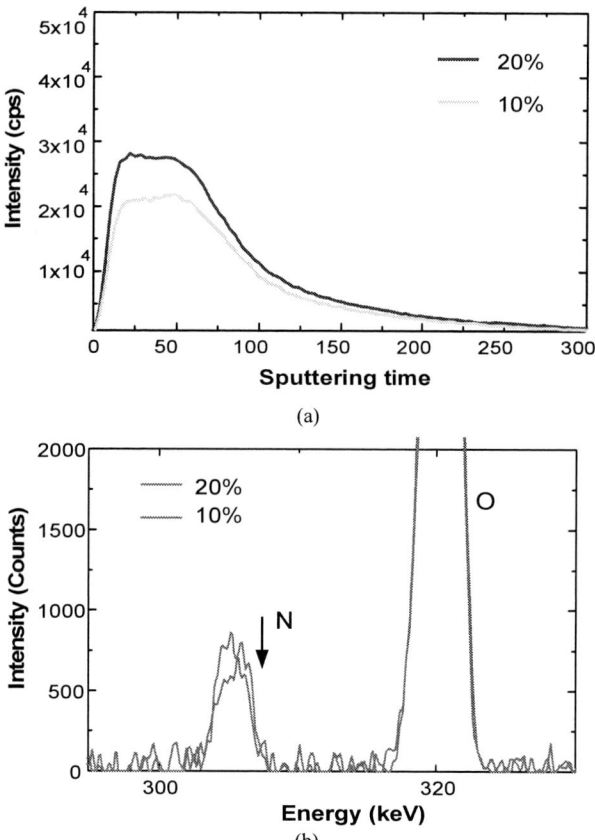

Fig. 2. Nitrogen profile and concentration at SiO_2/Si interface fabricated by 10% and 20% of eprf; (a) TOF-SIMS and (b) H-RBS analysis.

To investigate nitrogen enhanced NBTI degradation, pMOSFET devices fabricated with 10, 20 and 30% of eprf oxynitridation are subjected to NBTI stress test at 140°C with different V_G stress conditions. Note that ΔV_{TFS} (V_T forward saturate, from now defined as ΔV_T) was measured by a fast measurement [12] to minimize the recovery during measurement.

In Fig. 3(a), time exponent factor, n, for NBTI degradation established from different nitrogen concentrations is parallel to each other and, in turn, it is ~0.16. Such constant n implies that charge trapping and interface states responsible for NBTI induced ΔV_T degradation are independent of interfacial nitrogen concentration. Suggested that the origin of defect generation would be identical. Identical n from a wide range of nitrogen concentration asserts that the diffusion is governed by Arrehenius model, and its corresponding Ea is ~0.10eV [18]. Since NBTI degradation model for ΔV_T is empirically determined and expressed as f {temperature, length, width, V_G, time}, the model compatibility (actual vs. model degradation) can be established as ΔV_T data measured is compared to the calculated ΔV_T by using the degradation model parameters extracted. As shown in Fig. 3(b), it is apparent that ΔV_T increases with increasing the interfacial nitrogen concentration. Note that the dependency of length, width and V_G are carefully estimated during NBTI stress test. However, they are not included herein due to relatively weak dependency. The lifetime can be calculated when t in NBTI degradation model is rearrange then solved to ΔV_T. It is shown in Fig. 3(c) that NBTI lifetime significantly depresses with increasing the duty cycles of eprf, an order of magnitude gap in lifetime projection at the normal operation. In other words, the interfacial nitrogen concentration and profile play an important role in nitrogen enhanced NBTI degradation. It is reported that nitrogen is more effective hydrogen

oriented hole trapping center than oxygen and, hence, enhances NBTI [19]. Although an empirical value of time exponent n of 0.16 and activation energy of 0.10eV cohesively adhere to the R-D model with molecular hydrogen diffusion, the interpretation of nitrogen enhanced NBTI degradation linked to effect of hydrogen and its recovery phenomena entangled with interface states and hole trapping is still a subject of intensive debate [20, 21]. In spite, this is beyond the scope of this paper.

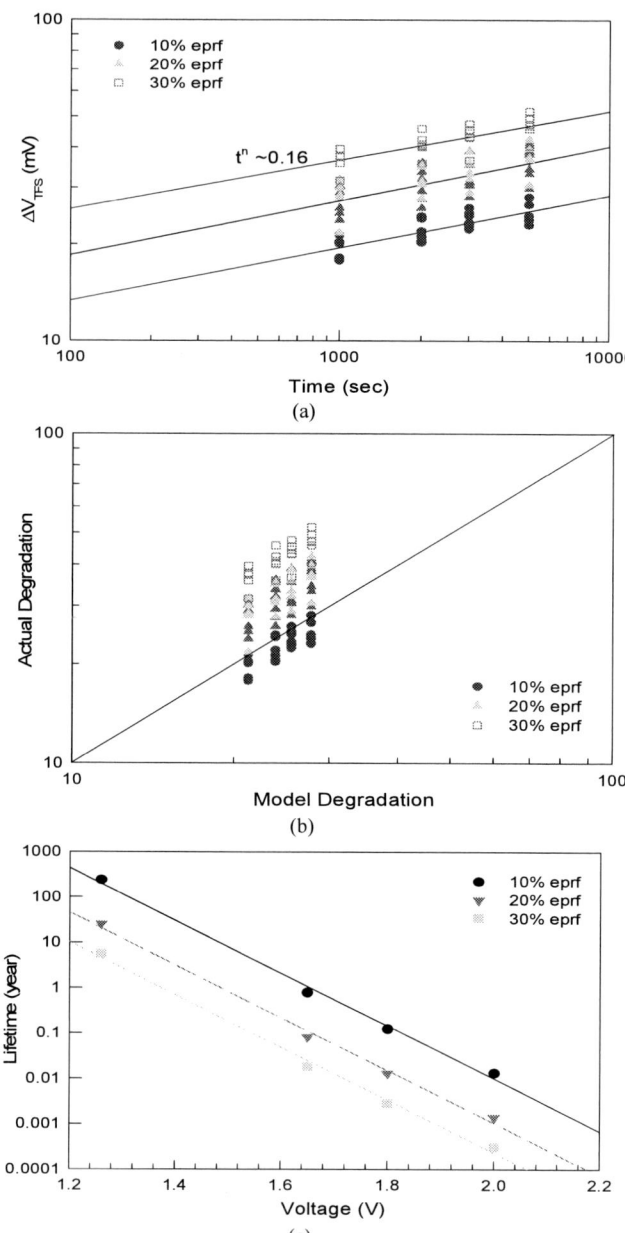

Fig. 3. NBTI degradation of pFET as a function of the nitrogen concentration by varying the duty cycle of eprf: (a) ΔV_T shift induced by NBTI stressed at 140C with V_G= -1.7V, -1.8V and -2.0V, (b) model compatibility (actual vs. model degradation) and (c) lifetime projection at the normal operation extrapolated from stress conditions.

During process development and design rule selection phase, the understanding of impact of NBTI degradation on SRAM bit-cell stability is crucial because SRAM cell enables to maintain stable data even under parasitic noise and device mismatch in SRAM cell. Moreover, product standby currents and regulator designs are highly rely on transistor reliability. Such that physical understanding of these major concerns are empirically attempted to be explored in regards to V_{min} behaviors and changes in the read margin (RM) and writer margin (WM) at room and hot temperatures throughout AHTOL test. Note that the RM and WM can be defined as the minimum voltage for reading and writing memory.

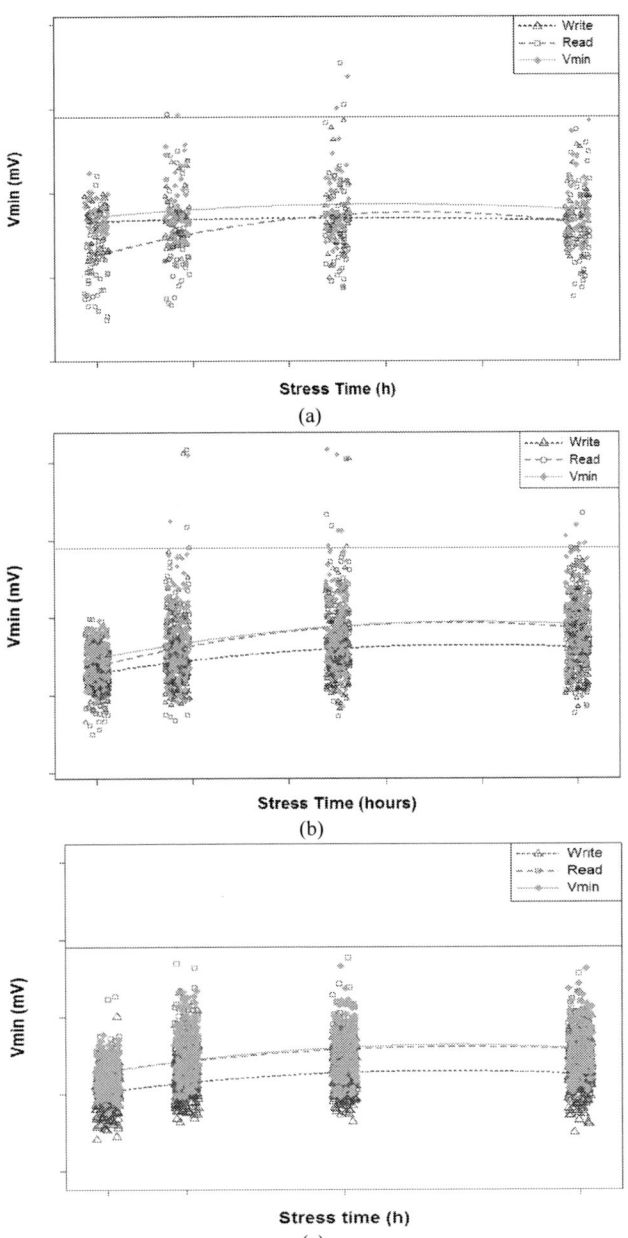

Fig. 4. V_{min} behaviors of SRAM fabricated with (a) 10% of eprf (b) 20% of eprf and (c) bit-cell optimization (PU size increase) as a function AHTOL test. Note that V_{min} shift, the RM and WM of SRAM are measured at hot temperatures (worst case) represented by a solid green, long dash red and dotted blue line, respectively.

We characterize V_{min} behaviors with respect to changes in the RM and WM measured at room (not shown) and hot temperatures (worst case) during AHTOL test. Figure 4(a-c) shows V_{min} behaviors of SRAM fabricated with (a) 10%, (b) 20% of eprf and (3) bit-cell optimization by increasing PU transistor, as a function

AHTOL test. It is found that V_{min} shift behaviors (solid green line) are in good agreement with the RM (long dash red line) rather than the WM (dotted blue line). This implies that V_{min} behavior is rather determined by the RM than the WM. In case for 10% of eprf, which is the WM dominant at T0 before stress test, the RM is rapidly degraded then later saturated after 500h of stressing. Whereas, the WM is rarely changed during AHTOL test [see Fig. 4(a)]. Obviously, only the RM degrades with increasing stressing. Remind that V_{min} of SRAM is determined by the relatively weaker margin between the RM and WM. Accordingly, V_{min} behaviors of 20% of eprf, which is the RM dominant at T0, are in good agreement with the RM. With increasing stress, the WM is also slightly deteriorated [see Fig. 4(b)]. As such, more V_{min} degradation occurs during AHTOL test if SRAM is the RM dominant at T0. Again, V_{min} shift is governed by the RM degradation rather than the WM. In Fig. 4(c), changing a bit cell design, increase of PU transistor, results in the similar V_{min} trends as 20% of eprf. However, there is no low V_{min} shift failure even after 1000h of AHTOL test. This indicates that a bit cell design can efficiently suppress V_{min} shift. It is worth of noting that none of SRAM modules is failed due to process induced hard failure. Note that the horizontal line in Fig. 4 represents the criterion for low $V_{cc\,min}$ failure.

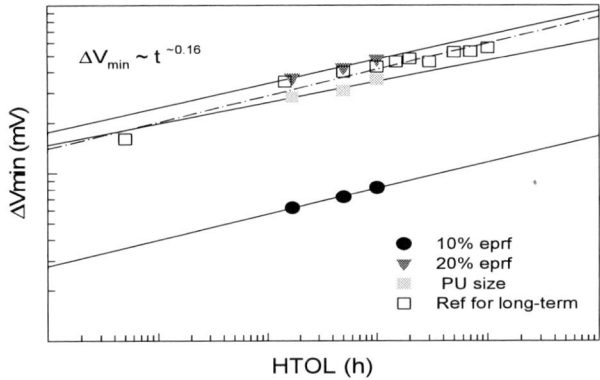

Fig. 5. The average of ΔV_{min} of SRAM fabricated with 10% and 20% of eprf and increasing PU size as a function of AHTOL test.

Figure 5 shows the average of ΔV_{min} of SRAM fabricated with 10% and 20% of eprf and increasing PU size, respectively. It is shown that ΔV_{min} with 10% of eprf is ~3~4× less than with 20% of eprf. Even though ΔV_{min} behaviors of increasing PU size are similar to 20% of eprf, which is dominated by the RM, the magnitude of V_{min} shift is ~2× less than V_{min} of 20% of eprf. Such V_{min} degradation is ascribed to the RM dominant at T0. Comparatively, 10% of eprf provides steadfast V_{min} shift with increasing stressing. It is interesting to note that time exponent of n that represents the average of ΔV_{min} degradation as a function of AHTOL test is well matched with NBTI degradation of pMOSFET previously explained by the R-D model [15] with molecular hydrogen diffusion [see Fig. 3(a)]. Note that, in Fig. 5, the opened square represents the reference exposed to long-term stress test in order to establish n of ΔV_{min} shift with a statistical certainty. Based on the coincidence, one might argue that ΔV_{min} is entirely attributed to pMOSFET V_T shift caused by NBTI. However, there is lack of correlation between initial and post V_{min} value due primarily to random NBTI degradation on SRAM bit cell. It is, therefore, not feasible to precisely anticipate the final V_{min} shift from the initial V_{min} prior to the HTOL test, although pMOSFET V_T shift induced by NBTI is qualitatively established. As such, optimizing interfacial nitrogen concentration and increasing PU transistor that can minimize V_T and V_{min} shift induced by NBTI and AHTOL test are inevitable to avoid fundamental degradation from the FEOL process and bit cell design perspectives. In addition, voltage screening prior to stress test is a benefit to eliminate either initially weakest cell or process related hard failure.

Even though we assert an assumption that ΔV_{min} solely results from pMOS V_T NBTI during AHTOL test, SRAM reliability assessment at normal operations is still uncertain due to lack of universal information associated with the voltage exponent (γ) and activation energy (Ea). In turn, establishing these parameters become more complicated with respect to V_{min} failures that includes soft and process induced hard failure representing single or multi-bit fail. Although these parameters are of important for ensuring reliability prediction at normal operations, a legacy type of empirical manner with different voltages and temperatures often fails because of randomness in V_{min} behavior and small number of samples failed under the given stress conditions. Hence, the way suggested below is an attempt to evaluate Ea and γ is based on physics based and technology relevant model in a manner of a fact that V_{min} shift of circuit is dictated by pMOS V_T shift induced by NBTI.

Since NBTI degradation model can be represented by:

$$\Delta V_T \propto A_0 \exp \cdot (-\frac{Ea1}{kT}) V^\alpha \cdot t^n \qquad (1)$$

Regarding the nominal and accelerated operations, Eqn. (1) can be convert to

$$\Delta V_T \propto A_0 \exp \cdot (-\frac{Ea1}{kT_{acc}}) V_{acc}{}^\alpha \cdot t_{acc}{}^n \qquad (2)$$

$$\Delta V_T \propto A_0 \exp \cdot (-\frac{Ea1}{kT_{nom}}) V_{nom}{}^\alpha \cdot t_{nom}{}^n \qquad (3)$$

Then solving Eqns. (2) and (3) yield

$$\left(\frac{t_{nom}}{t_{acc}}\right)^n = \exp\left[\frac{Ea1}{n \cdot k}\left(\frac{1}{T_{nom}} - \frac{1}{T_{acc}}\right)\right] \cdot \exp\left[\frac{\alpha}{n} \cdot \ln\left(\frac{V_{acc}}{V_{nom}}\right)\right] \qquad (4)$$

By using $\ln(x) = 2 [(x-1)/(x+1)]$, Eqn. (4) can be developed and simplified as

$$\frac{t_{nom}}{t_{acc}} = \exp\left[\frac{Ea1}{n \cdot k}\left(\frac{1}{T_{nom}} - \frac{1}{T_{acc}}\right)\right] \cdot \exp\left[\frac{2\alpha}{n(V_{acc}+V_{nom})} \cdot (V_{acc} + V_{nom})\right] \qquad (5)$$

And, it is equivalent to A_F

$$A_F = T_{AF} \cdot V_{AFF} = \exp\left[\frac{Ea2}{k}\left(\frac{1}{T_{nom}} - \frac{1}{T_{acc}}\right)\right] \cdot \exp[\gamma \cdot (V_{acc} - V_{nom})] \qquad (6)$$

Ea2 and γ for AHTOL test are defined as

$$Ea2 = \frac{Ea1}{n} \quad \text{and} \quad \gamma = \frac{2\alpha}{n(V_{acc}+V_{nom})} \qquad (7)$$

Hence, Ea2 and γ responsible for V_{min} shift controlled by NBTI during AHTOL are 0.7eV and ~16, respectively. It is worth of noting that Ea of 0.7 is widely chosen for product qualification in many industries. Indeed, this value represents the gate oxide related defect generation [22]. It is found that γ is rapidly evolved from 4 to 16, since technology node enters into deep submicron technology nodes. However, if *n* and α in Eqn. (1) is large associated with the recovery phenomena and larger voltage acceleration from higher V_G stress, Ea2 and γ fall into out of the logical range, leading to erroneous lifetime projections. Upon the assumption of Ea2=0.7 and γ=4, 100h of HTOL stress test at 140T_j with 1.4×V_{dd} exceeds >10 years lifetime requirement in normal operation. Such that both parameters can be adopted for monitoring technology process from a conservative point of view.

Although Eqn (7) provides a useful framework in the case for product reliability assessment with unknown Ea and γ from AHTOL test, a quantitative manner that precisely predict V_{min} behaviors estimated from NBTI induced V_T shift is the area ripe for the future study.

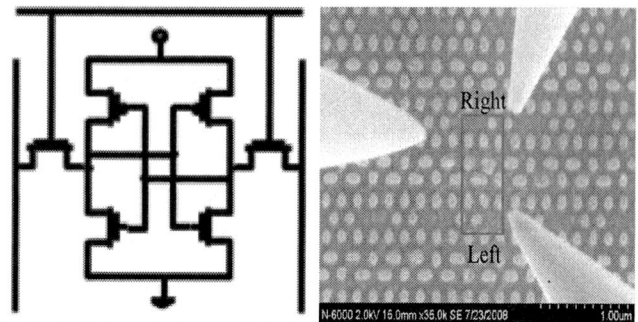

Fig. 6. (a) Schematic layout of 6T-SRAM and (b) illustration of the nano-probing to measure V_{tsat} and I_{dsat} from PU, PD and PG transistors.

In an effort of failure analysis, the nano-probing is conducted on the SRAMs suffered from larger V_{min} shift after AHTOL stress test. Figure 6 (a) shows the schematic layout of 6T SRAM cell consisting of a cross-over inverter pair [pull-up (PU) and pull-down (PD)] and two access transistors [pass gate (PG)] that connect to the inverter and the bit-line. In order to increase SRAM cell density, a small FET device is used that enables to capture process induced variation. The electron micrograph of a nano-probing technique to measure V_{tsat} (mV) and I_{dsat} (μA) from PU, PD, and PG transistor is shown in Fig. 6(b).

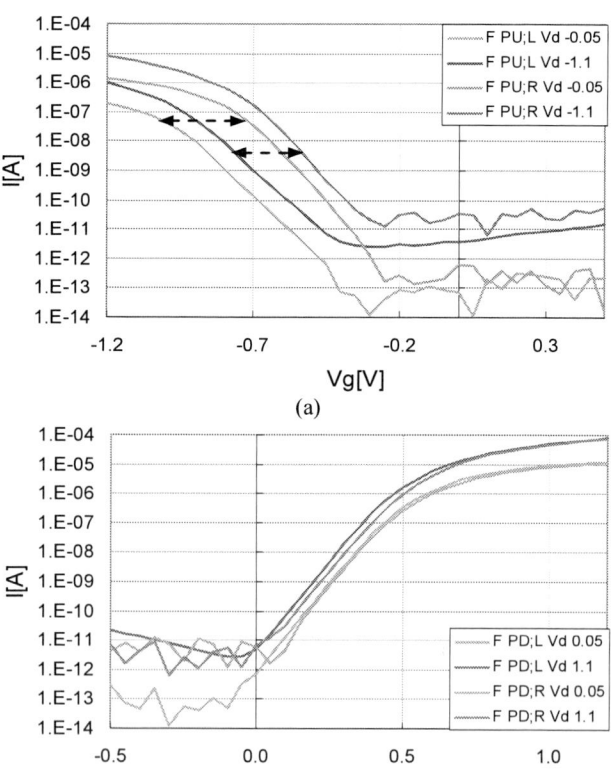

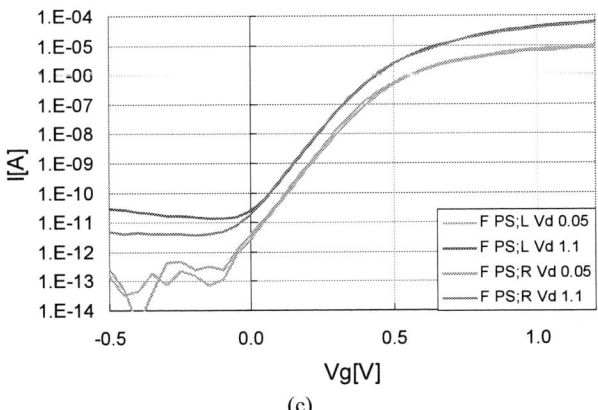

(c)

Fig. 7. Result of the nano-probing on: (a) PU, (b) PD and (c) PG in SRAM (with 20% of eprf) stressed at AHTOL test conditioned at 140°C with 1.4×Vdd. Larger V_T mismatch between L (left) and R (right) PU transistor due to NBTI degradation results in the read disturbance failure.

Figure 7 (a-c) reveals the results from the nano-probing on PU, PD and PG transistors in 6T SRAM that collapse V_{min} margin during AHTOL test conditioned at 140°C with 1.4×V_{dd}. As potential degradation mechanisms, NBTI in PU and PBTI in PD are taken into accounted. It is, however, found that nMOSFET V_T shift induced PBTI is far less than 10% of NBTI degradation. As such, the effect of PBTI on PD is eliminated from a practically reliability concern. Due to a relatively short switching time, PBTI on PG device is negligible. In fact, Fig. 7 supports that the premises on PD and PG degradation are legitimate. In Fig. 7(a), V_T mismatch between PU transistors (left and right) in SRAM (marked by an arrow) is differentiate from PD and PG [see Fig 7(b) and (c)]. Since NBTI degradation is stemmed from AHTOL test, larger V_T mismatch between PU transistors results in larger V_{min} shift and then leads the read disturbance failure. This indicates that the degradation of PU transistor enables bit cell to switch to the RM dominated so that the WM decreases. In consequence, V_{min} drift becomes worse with increasing stress. It is found that SRAM V_{min} stability depends on PU transistor immunity against NBTI. As results, bit cell design, in which V_{min} behaviors are dominated by the WM, is a key design feature to ensure a stabilized V_{min} behavior.

TABLE III contains ΔV_T (mV) and I_{dsat} (%) from the nano-proving measurement on the bit cell showing larger V_{min} shift after AHTOL stress test. It is found that ΔV_T mismatch between L and R PU transistors is well consistent with Fig. 7(a). Such discrepancy results in high read current (I_{read}) and triggers the read disturbance failure after AHTOL test.

TABLE III. V_T AND IDSAT COMPARISON BETWEEN GOOD AND FAILED SRAM BIT CELL AFTER 500H OF AHTOL STRESS TEST MEASURED BY NANO-PROBING. AS SHOWN, A FAILED DEVICE SHOWS LARGER V_T AND IDSAT MISMATCH SPECIALLY FOR PU BETWEEN THE LEFT (L) AND RIGHT (R) IN 6T SRAM.

20% eprf	Transistor	ΔV_T (mV) mismatch	I_{dsat} mismatch (%)
Good	PU (L vs. R)	30	8.53
	PG (L vs. R)	30	4.66
	PD (L vs. R)	70	9.78
Fail	PU (L vs. R)	105	26.79
	PG (L vs. R)	70	0.57
	PD (L vs. R)	50	13.41

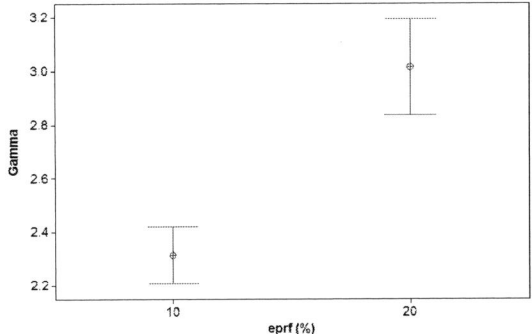

Fig. 8. γ (the ratio V_T of PG to PU) vs. nitrogen concentration controlled by the duty cycle of eprf. As shown, 10% of eprf yields lower γ than 20%, resulting in less ΔV_T and ΔV_{min} shift from NBTI and AHTOL test.

As shown in Fig. 8, the γ value (the ratio of V_T of PG to PU) of 10% of eprf is much lower than that of 20% of eprf. This can be an indicative of NBTI induced V_T shift of pMOSFET. Recall that 10% of eprf results in less V_T and V_{min} and the WM are little changed. Whereas, the larger V_T and V_{min} deterioration occur from 20% of eprf during NBTI and AHTOL test. This is because of PU V_T degradation induced by NBTI stemmed from HTOL test, which is well consistent with the results summarized in TABLE III.

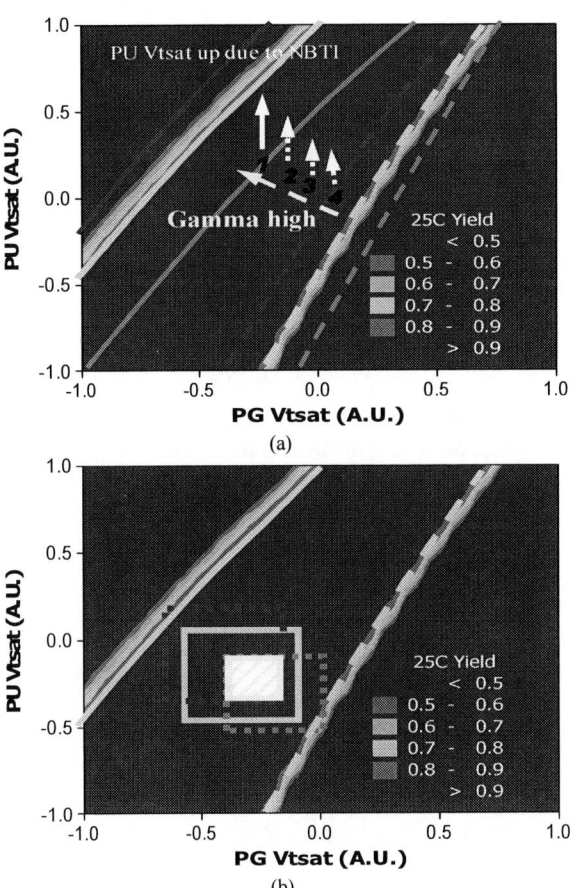

Fig. 9. Simulated Milky-Way plot of SRAM: (a) the RM and WM associated with V_T of PU and PG at the process variation corners and (b) a sweet SRAM yield zone at three temperatures. Note that the solid and dotted line indicate the read and write limit. Three temperatures are drawn with blue for -40°C, green for 25°C and magenta for 125°C.

978-1-4244-5430-3/10 $26.00 © 2010 IEEE

Figure 9 is so called Milky-Way plot [23] of memory cell, which illustrates the dependency of the RM and WM operation margin of PU and PG V_T and a sweet SRAM yield zone at three temperatures (blue at -40°C; green at 25°C and magenta at 125°C). This simulated plot is constructed with respect to 6σ V_T variation with respect to wafer-to-wafer and lot-to-lot variation in order to explore a sweet V_{min} based yield zone within the electrical module spec. The vertical and horizontal axes represent V_T for pMOS (PU) and nMOS (PG) in memory cell. In addition, the read (disturbance) and write limit are drawn by the solid and dotted line, respectively. As such, the RM and WM are subjected to be failed out of these lines. Vice versa, both RM and WM properly function between these lines. Hence, the RM becomes worst as PG and PU V_T moves to the left and upward, FS process corner (left hand upper corner). There is no WM at SF corner (right hand lower corner). Suggested that SRAM yield at V_{dd} can be severely compromised, if bit cell design is improperly taken from the area where the RM and WM fail. Even if bit cell is designed with higher PU V_T and lower PG V_T within the RM and WM pass region, they are easily vulnerable against NBTI during stress due to the higher γ, as mentioned previously. Remind that the WM is proportional to γ. It is worth of noting that the RM is inverse proportional to WM.

In Fig. 9(a), the RM and WM pass region shift to the WM fail area and become narrow with increasing temperature. Since NBTI results in PU V_T degradation, SRAM bit cell with the higher γ is more susceptible against NBTI and, in turn, leads to the read disturbance fail. Hence, in order to amplify both RM and WM with respect to NBTI, small γ taken from either process optimization or increasing PU size is inevitable. This implies that the RM and WM line are subjected to be changed based on bit cell design so as to influence the RM and WM pass area. In Fig. 9(b), with increasing temperatures, a sweet spot for SRAM yield (a square), in which both RM and WM properly function with NBTI immunity, becomes slime. Note that the overlaid area implies the highest SRAM yield from three temperatures. To secure mature processability and manufaturability leveraged by lessening process variations, Fig. 9 demonstrates how to explore a sweet spot for SRAM yield at V_{dd} with respect to the RM and WM. In consequence, a care must be taken for bit cell design in the earlier stage of development to achieve yield as well as reliability.

CONCLUSIONS

A systematical reliability assessment that is necessary for technology feasibility and qualification is presented by addressing V_T and V_{min} shift of thin oxide pMOSFET (~20A) and SRAM induced by NBTI and AHTOL test. It is manifested that physical characterization and understanding of nitrogen concentration and profile at the SiO_2/Si interface by using SIMS and H-RBS are a crucial prerequisite particularly for NBTI induced pMOSFET V_T shift in the FEOL process. It is found that lower interfacial nitrogen concentration achieved by 10% of eprf gives a visible benefit of NBTI reliability compared to 20% of eprf. Such profit can further expand to circuit level reliability in regard to SRAM performance represented by V_{min} shift, which is attributed to the mismatch of V_T between PU transistors in 6T SRAM dominated by NBTI during AHTOL test. It is empirically shown that V_{min} shift behavior is in good agreement with the RM rather than the WM. In addition, V_{min} stability is related to the γ value and, in fact, is proportional to the WM. It is also found that a stabilized V_{min} drift is obtained by increasing PU size to enhance the WM. Although pMOSFET NBTI V_T shift and SRAM V_{min} instability are critical reliability concerns in advance technology process development, such fundamental phenomena can be effectively controlled. Finally, base on pMOSFET NBTI V_T and SRAM V_{min} characterization, mature processability and manufaturability are attained from the FEOL process optimization and SRAM bit cell design aspect.

REFERENCE

[1] T. Sujikawa et al, IEEE 41th Int. Rel. Phy. Symp., 2003, pp. 183-8.

[2] S. S. Tan et al, Microelectronics Reliability, 45, 2005, pp. 19-30.

[3] D. Varghese et al, IEEE tans. Electronic Devices, Vol. 54, No. 7, 2007, pp. 1672-80.

[4] T. Kuroi et al, IEEE VLSI Symp., 1996, pp. 210.

[5] N. Kimizuka et al, IEEE VLSI Symp., 1994, pp. 105.

[6] Y. Okada et al, IEEE VLSI Symp., 1999, pp. 117.

[7] N. Kimizuka et al, IEEE VLSI Symp., 2000, pp. 92.

[8] P. A. Kraus et al, IEEE VLSI Symp., 2003, pp. 143.

[9] V. Reddy et al, IEEE 40th Int. Rel., Phy. Symp., 2002, pp. 248-54.

[10] K. Mueller et al, IEEE 42th Int. Rel. Phy. Symp., 2004, pp. 426-9.

[11] G. La Rosa et al, IEEE 44th Int. Rel., Phy. Symp., 2006, p .274-82.

[12] M. Denais et al, IEDM 2004, pp. 109-112.

[13] K. Kushida-Abdelghafar et al, Applied physics Lett., Vol. 81, No. 23, 2002, pp. 4362-64.

[14] J. H. Stathis et al, Microelectronics Reliability, 46, 2006. pp. 270-86.

[15] S. Mahapatra et al, IEEE trans. Device Materials. Reliability, Vol. 8, N. 1, 2008, pp. 35-45.

[16] Y. He et al, Solid State Electronics, 49, 2005. pp. 57-61.

[17] T. B. Hook et al, Microelectronics Reliability, 45, 2005. pp. 47-56.

[18] M. Alam et al, Microelectronics Reliability, 47, 2007. pp. 853-62.

[19] J. B. Yang et al, J. Electro. Chem. Soc., 154, 12, 2007, pp. G255-61.

[20] V. Huard et al, Microelectronics Reliability, 46, 2006, pp. 1-23.

[21] A. E. Islam et al, IEEE tans. Electronic Devices, Vol. 54, No. 9, 2007, pp. 2143-54.

[22] JEP 122C, Failure mechanism and models for semiconductor devices.

[23] Y. Morita et al, VLSI Circuits Digest Symp., 2006. pp. 13.

Validating Foundry Technologies for Extended Mission Profiles

K. van Dijk[1], P.A.J. Volf[1], C. Detcheverry[1], A. Yau[2], P. Ngan[3], Z. Liang[1], F.G. Kuper[1,4]

[1]NXP Semiconductors, Nijmegen, The Netherlands
[2]TSMC, Taiwan, ROC
[3]NXP Semiconductors, San Jose, USA
[4]University of Twente, The Netherlands

Primary Contact: Kitty van Dijk, phone: +31-24-3532742; e-mail: kitty.van.dijk@nxp.com
Second Contact: Paul Volf, phone: +31-24-3536818; e-mail: paul.volf@nxp.com

Abstract— **This paper presents a process qualification and characterization strategy that can extend the foundry process reliability potential to meet specific automotive mission profile requirements. In this case study, data and analyses are provided that lead to sufficient confidence for pushing the allowed mission profile envelope of a process towards more aggressive (automotive) applications.**

Keywords; qualification strategy, HTOL, ppm level, test screens, guard band, SRAM Vddmin

I. INTRODUCTION

A number of semiconductor companies are moving towards asset-light company configurations that retain a strong portfolio in specialty processes and utilize external foundries for the generic advanced CMOS processes [1,2]. Foundry processes are generally qualified according to standard conditions. For these processes, there is a gap between what a foundry demonstrates at process qualification and what is for instance required for automotive mission profiles (e.g. extended lifetime temperature range). There are also additional requirements for validation of robustness, proof of low ppm customer return capability, and wider test temperature ranges. Another complication is that Automotive Business typically utilizes mature technology nodes, while the foundry development team has already started focusing on newer generations.

To bridge this gap, a qualification strategy, utilizing dedicated test-chips, was developed and exploited for released 65nm and 90nm technology nodes. In this paper, the approach to proof process capability and to enable product capability is explained, and results are shown.

II. QUALIFICATION SET-UP

This automotive robustness validation addresses two requirements: first to show process capability to cover the automotive mission profile envelope and lifetime requirements, and second to provide the means to actually reach that goal on products in production. Therefore, in addition to the normal product qualification tests originally performed by the foundry at process release, extended qualification tests were performed on internal validation chips. This qualification setup includes the following features: test chip contents, characterization condition ranges, extended reliability stress conditions (time, temperature, and voltage), and process corner conditions.

The validation chips contain all relevant IP-blocks, such as standard logic, SRAM, ROM, and ring oscillators (ringos). A complete electrical characterization is executed before and after life tests, covering functional tests, Iddq, and speed searches, over the extended voltage range from 0.75*Vdd to 1.15*Vdd, in addition to Vddmin and data retention (DRET) searches on all relevant IPs. Read-points are performed at –40C, 25C, and +125C (3T testing). The aggressive mission profile with longer use time at higher temperatures requires longer equivalent high temperature operating life (HTOL) stress time, e.g. 3000 hours instead of 1000 hours. It is a good practice to qualify automotive products beyond the mission profile to prove sufficient lifetime margin (a step towards robustness validation [3,4]). Therefore, the HTOL was extended to 4000 hours, using 140 dies with a dynamic stress, with an active clock (HTOL), and 60 devices with a static stress, not clocked, to mimic stand-by conditions (SHTOL). Additionally, accelerated HTOL stresses, using voltage, (AHTOL), with 40 dies, or temperature (150HTOL), with 20 dies, were performed to validate the foundry's qualification for intrinsic effects. Before starting the HTOL, a split was made in voltage stress conditions to establish a safe voltage stress condition. Furthermore, samples from a process corner lot were used (Typical, Slow, Fast, 130, 65, 65 dies respectively). The resulting set-up is summarized in Table 1.

TABLE I.   SUMMARY OF THE DOE SETUP ON V-SCREEN AND HTOL

Process split	Voltage Screen setting			HTOL setting		
	V-stress	*Voltage*	*Duration*	*Oven Temperature*	*Vdd*	*Read-points [hrs]*
Typical	Dynamic stress	1.7*Vdd	1000 ms	T=125C (HTOL)	Vdd+10%	48,168,500,1000,2000,4000
Slow	Dynamic stress	2*Vdd	1000 ms	T=125C (SHTOL)	Vdd+10%	48,168,500,1000,2000,4000
Fast	Static stress	2*Vdd	100 ms	T=125C (AHTOL)	Vdd+40%	48,168,500,1000,2000
	No stress			T=150C (150HTOL)	Vdd+10%	48,168,500,1000

The data analyses consists of (1) a "normal" pass/fail check, (2) an intrinsic parametric shift study, and (3) a parametric outlier check, by means of a regression analysis between t=0 and end-read-point, with a 4 sigma limit on the residuals. The intrinsic shift analysis links observed product parameter shift to Wafer Level Reliability (WLR) results and models [5], such that process robustness can be validated, process capability proven and lifetime guard bands obtained [2,6]. From the detailed outlier detection analysis, on the data after lifetime, additional tests and settings for specific screens are derived, thus enabling low ppm failure rate for products.

III. PROVING PROCESS CAPABILITY

For investigating the process capability, key product properties (speed on ringos, Taa (address access time), and gate delay, Iddq, Vddmin, and DRET) were studied over the entire temperature range on process corners at t=0, and over the lifetime for different HTOL conditions.

Robustness validation dictates within-spec-behavior over the entire process window and mission profile. Thus all IP and all process shifts have to be assessed on corner silicon. Differences in shifts between corners have indeed been observed. Slow corner products, measured at low temperature, give the highest degradation for speed over lifetime. But even at t=0, the conclusions of a corner validation can be far reaching. For instance, a proprietary SRAM bit-cell layout, approved by the foundry, was found to be too sensitive to the process lithography variation in the default slow corner. The specific usage requirements for the bit-cell, necessitated slightly modified poly dimensions inside the cell. In combination with the foundries' standard SRAM cell OPC (Optical Proximity Correction), the applied slow corner lithography settings, and the process variation, this caused a low level of poly-poly shorts as shown in Fig. 1. Together with the foundry a new slow corner setting was defined which was more representative of the actual process variation. An additional FEM (Focus Energy Matrix) wafer was processed to prove sufficient process window around this new setting. Since this was observed in an early stage, during IP qualification, it will support defining litho process settings for products requiring a non-standard process target.

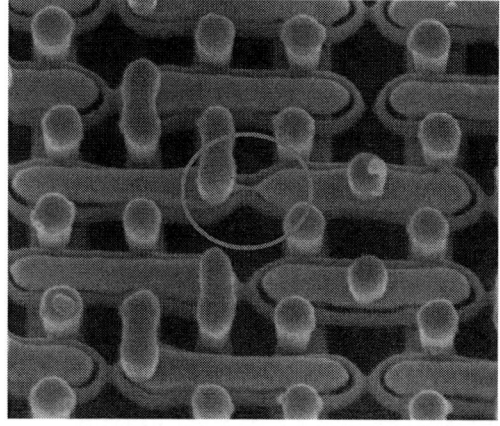

Figure 1. Random poly-poly short in SRAM array

Extensive shift analyses over lifetime have been performed to understand the different acceleration conditions of the HTOL stress. Those analyses help understand acceleration condition effects on lifetime behavior. With this understanding, HTOL conditions can be defined to fit the product mission profiles while avoiding extremely long HTOL experiments. The extensive shift analyses also deliver safe guard band settings for production tests to ensure low ppm levels in the field.

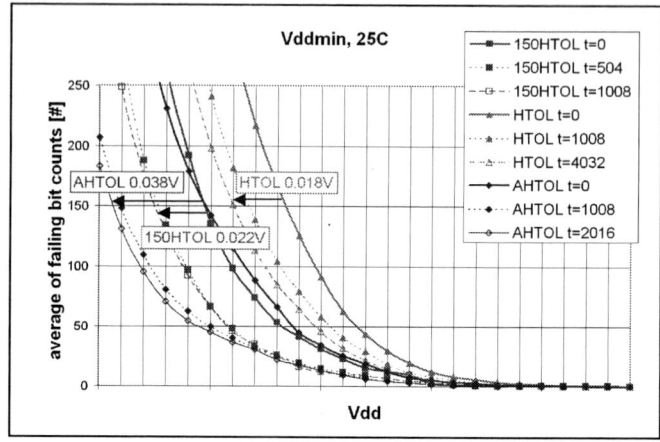

Figure 2. SRAM Vddmin shift for different HTOL conditions

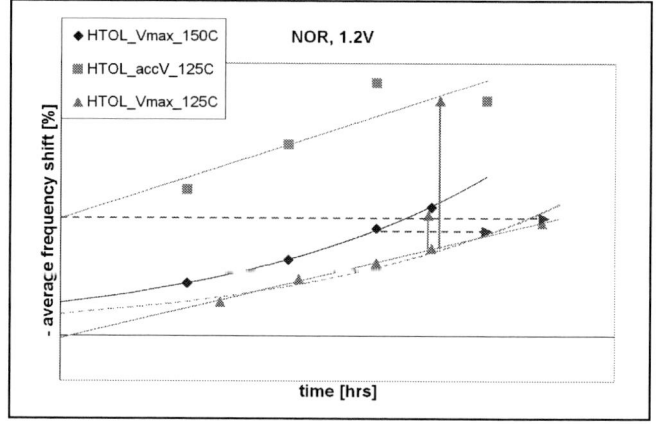

Figure 3. % Lifetime shift of a NOR ringo vs time; arrows indicate general direction, solid vertical arrows: acceleration in shift, dashed horizontal arrows: acceleration in time

The impact of voltage and temperature acceleration has been investigated for various parameters (speed, leakage, Vddmin, DRET). At the end read point, voltage accelerated HTOL has degraded all parameters about twice as fast as standard HTOL. For temperature accelerated HTOL this degradation was 10 to 20% higher than for standard HTOL. This is shown in Fig. 2 for the Vddmin shift of the SRAM at room temperature. A similar shift behavior, a factor 2 difference between voltage acceleration and normal HTOL, and 10 to 20% between temperature acceleration and normal HTOL, is seen for the decrease in ringo frequency over lifetime, see Fig. 3, solid vertical arrows.

The performance shift has to be assessed over time in order to determine the acceleration factor, see the dashed horizontal arrows in Fig. 3. This performance shift was assessed using different ringos: an inverter (representative of clock circuits), a NOR (PMOST limited), a NAND (NMOST limited) and a complex Boolean cell (representing multi-stage cells). All ringos show a decrease in frequency over lifetime for all DOE conditions: HTOL conditions, process corners, read-point temperatures, and core voltages. It is expected that NBTI will be the dominating reliability risk for low frequency, up to a few hundred MHz, digital circuits. From the data this is supported by the fact that the highest relative shift was found for NOR ringos, as NOR ringos are the most sensitive to NBTI. Table 2 shows an overview of acceleration factors for 65nm and 90nm technologies for different intrinsic reliability mechanisms.

TABLE II.   OVERVIEW OF VOLTAGE AND TEMPERATURE ACCELERATION FACTORS FOR 65NM AND 90NM TECHNOLOGIES

AFV (voltage acceleration)		V2	V1	**AFV**
exp[γ(V2-V1)]	NBTI	1.68	1.32	~ 30 - 80
	HCI-pmost*	1.68	1.32	~ E+08
	HCI-nmost*	1.68	1.32	~ E+08
(V2/V1)^(n)	TDDB-nmost	1.68	1.32	~ E+05
	TDDB-pmost	1.68	1.32	~ E+05
	Electromigration	1.68	1.32	~ 1 - 2
	Stress migration	N/A		
	TDDB dielectric	N/A		
AFT (temperature acceleration)		T1 (°C)	T2 (°C)	**AFT**
exp[(Ea/k)*((1/T1)-(1/T2))]	NBTI	125	150	~ 3 - 5
	HCI-pmost*	125	150	~ 1 - 3
	HCI-nmost*	125	150	~ 1 - 3
	TDDB-nmost	125	150	~ 2 - 4
	TDDB-pmost	125	150	~ 2 - 4
	Electromigration	125	150	~ 4 - 6
	Stress migration	125	150	~ 4 - 6
	TDDB dielectric	125	150	~ 3 - 5

* worst case stress condition

TDDB and HCI typically have very high voltage acceleration factors but here will have a negligible impact on parameter shift, within the measurement accuracy. The reason is that for TDDB, the stressed voltage is far below the breakdown voltage, even for the voltage accelerated HTOL setting. For HCI, the worst-case condition is present only during a very small fraction of the time, due to the relatively low frequency of the ringos (which are in line with the typical product speeds).

The experimental ringo data, see for example Fig. 3, was used to estimate the acceleration to provide supporting data for our automotive release. The data was fitted using a power law [6] or logarithmically, subsequently the acceleration was calculated over a range of shift (along the y-axis of Fig. 3). A temperature acceleration factor of approximately 3 to 4 was found to match all the data fairly well. The voltage acceleration factor ranged between 10 and 5000, due to the noise of the measurements and the fits, with a mean between 50 and 60. This is approximately the acceleration factor of NBTI, as found from Wafer Level Reliability measurements.

The measurement data support NBTI to be the dominating aging mechanism, i.e. the NOR ringo shifts most, and the subsequent acceleration calculation support the WLR acceleration factor for NBTI as provided by the Foundry. Therefore these WLR factors can be safely used to determine proper HTOL and Burn-In settings. Use of these factors allow long mission profile times to be covered with acceptable time-to-market HTOL durations. Also for Burn-In, trading time with

voltage acceleration, based on these factors, allows to better balance Burn-In effectiveness against consuming product lifetime margin. Furthermore, these computed intrinsic acceleration factors are higher than the foundry qualification and monitoring acceleration factors which are based on extrinsic defects, showing the foundry's conservative attitude towards product reliability. Consequently, the foundry's monitoring covers longer lifetimes with respect to intrinsic defects than is used in their calculations.

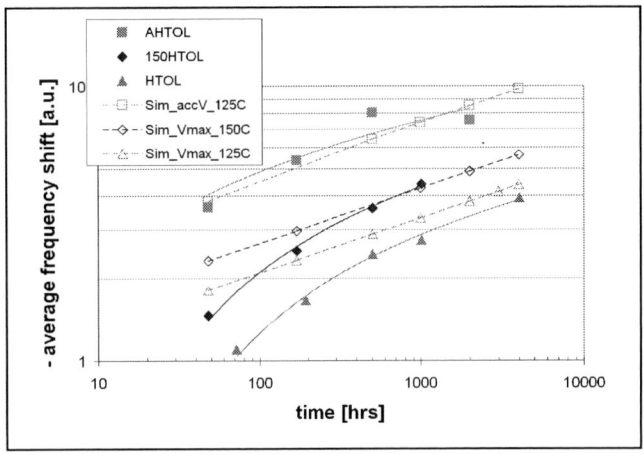

Figure 4.   % Lifetime shift of an inverter ringo for HTOL, voltage and temperature accelerated HTOL conditions; silicon (closed symbols) compared to simulation (open symbols)

An important goal of the intrinsic shift analysis over HTOL is to set product parameter guard bands for t=0 hour test to safeguard low ppm over lifetime. The different acceleration factors (V, T) complicate a direct translation of observed shifts to guard bands. If such direct application of the HTOL shifts results in only a relatively small yield loss, these can be used "as-is". A better approach is to apply reliability models to recalculate the HTOL accelerated shifts into realistic product use cases [7]. Once validated, such model is a valuable tool for analogue IP development. For this study, the different HTOL acceleration conditions were simulated for the ringos using Presto [8] (an in-house reliability simulator). Only NBTI degradation, as modeled directly from the WLR data of the process release, was taken into account. Fig. 4 shows experimental and simulation results comparing the different HTOL conditions (HTOL, voltage and temperature accelerated HTOL). Although the curvatures of the lines do not align, which could be attributed to the absence of a complete NBTI model including recovery and the accuracy of the measurements for very small shifts (at 48 and 168 hours), the difference between voltage and temperature acceleration compared to normal conditions is well predicted by simulation. For these dynamic conditions, models based on WLR data, measured in a conventional way, thus based on data where the NBTI is partly recovered, can be used. For the static condition there is no recovery during HTOL (only some recovery during the read-points). Therefore, the models to simulate static behavior should be based on WLR data that captures both the permanent and the recoverable damage, as in "on-the-fly" WLR measurements. Fig. 5 shows that when conventional

978-1-4244-5430-3/10 $26.00 © 2010 IEEE

measurements are taken to model dynamic behavior, and "on-the-fly" measurements are taken to model static behavior, the trend of the simulation lines fits indeed the trend of the experimental data. Cooperation with the Foundry is ongoing to obtain better NBTI models, including recovery, and HCI models, but in essence it is shown that the lifetime shifts can be extrapolated to use conditions, using the simulation results, and can thus be used for speed reliability guard band settings.

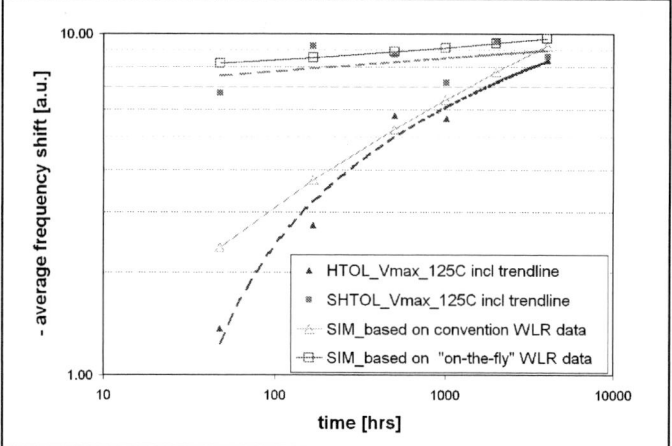

Figure 5. Trend in shift of a NAND ringo for dynamic and static lifetime behaviour; silicon (closed symbols) compared to simulation (open symbols)

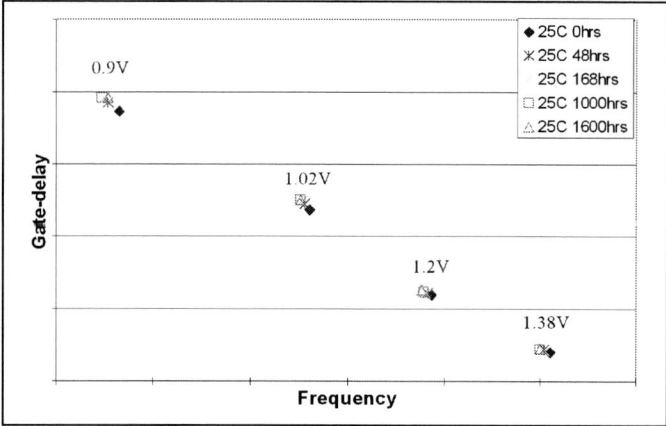

Figure 6. % Lifetime shift of a NOR ringo vs % lifetime shift of the Gate-delay test, at different Vdd

The speed degradation over lifetime on ringos is found consistent with the increase in memory Taa (address access time) and gate delay on on-board DSP cores. Fig. 6 plots the shift of a NOR ringo versus the gate-delay measurement shift, at different Vdd points. This validates the use of ringos for speed degradation assessment. Such strong correlation confirms the suitability of ringos to evaluate product speed. Therefore system integrity self test (SIST) modules, containing ringos, are placed within products [9], and used in all advanced technology nodes.

For understanding the shifts of the SRAM Vddmin over lifetime, one should take into account that the SRAM Vddmin behavior is determined by two underlying mechanisms: a read limited and a write limited Vddmin. Literature [5] and

simulations predict that these two mechanisms behave differently over lifetime: the write limitation improves over lifetime, while read limitation degrades over lifetime. Our own tests, where the write and read limited contributions were explicitly separated from the overall SRAM Vddmin, show that there is a difference in temperature behavior. The SRAM is write-limited at –40C and 25C, whereas it is read limited at 125C. Note that this crossover temperature between write and read limited Vddmin behavior depends on process corner and SRAM compiler. Fig. 7 shows the resulting lifetime behavior on silicon at –40C and 125C. This confirms the importance to separate the read limited and write limited tests and to apply 3T testing in order to fully understand the SRAM behavior at process qualification. This understanding, combined with the separation of the tests, is a necessity for future product test optimization.

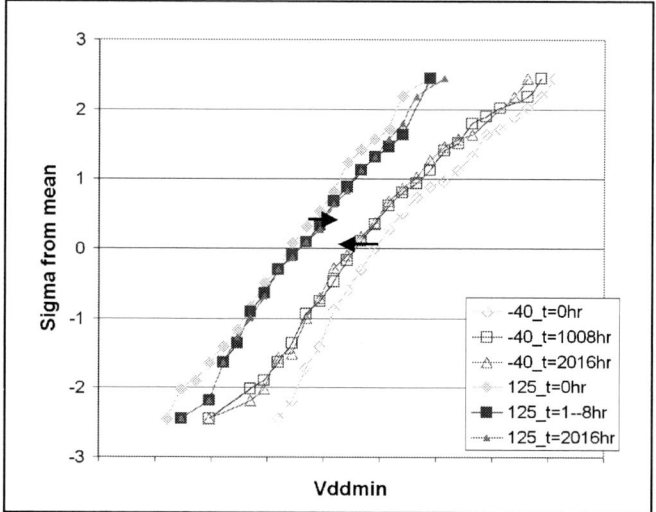

Figure 7. Vddmin shift of the SRAM read-points at –40C (open) compared to read-points at 125C (closed)

The extended test-chip-based analysis demonstrates an intrinsic process capability to cover automotive mission profiles, provided that products are produced with strict guard bands and with attention paid to temperature dependent properties. The only process related impact was that a new slow corner setting had to be defined for the 65nm node.

IV. ENABLING PRODUCT CAPABILITY

To ensure the capability to reach the automotive quality at product level two steps were taken. The known Voltage screen (V-screen) was explored and to investigate the need for additional screens, all parametric outliers from the qualification were studied.

Automotive applications require effective screens at wafer level to achieve single digit ppm levels. V-screen has been proven to be effective [10] in reducing ppm levels at the 65nm, 90nm and earlier nodes. However, the applied stress voltage determines the ability to screen out latent defects. The foundry conservatively advises a 1.4*Vdd V-screen setting where a higher setting is preferred to achieve low ppm levels for automotive products. The foundry WLR results and models give the maximum voltage stressing for 1sec at 25C, to

978-1-4244-5430-3/10 $26.00 © 2010 IEEE

consume less than 1% of the lifetime, to be 2.2*Vdd, due to HCI for dynamic or TDDB for static V-stress. A practical safe V-stress setting is between 1.4*Vdd and 2.2*Vdd . This higher V-stress setting needs to be validated for safety, its effect on lifetime evaluated for hard fails and parametric drifts. Our study applied both 1.7*Vdd and 2.0*Vdd V-stress settings. The study showed, as expected, no stress induced functional fails or significant parametric shifts after full lifetime stress according to the automotive mission profile. Data shows that a dynamic V-screen setting of 1.7*Vdd for 1000ms is safe for digital circuitries within this mission profile.

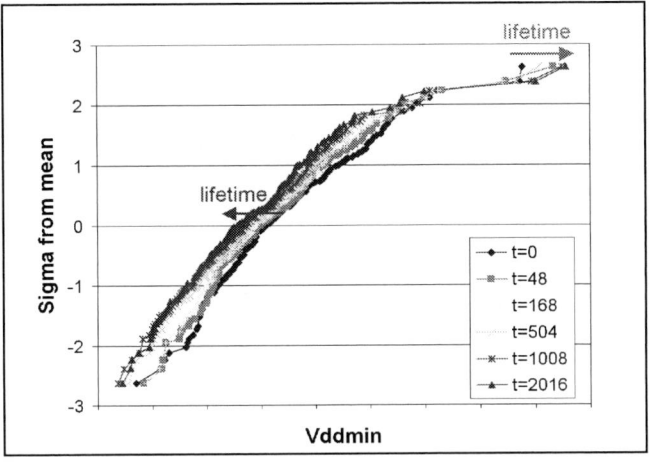

Figure 8.   Tailing on the vddmin of an SRAM. The tail shifts to higher Vddmin while the bulk shifts to lower Vddmin

During the 65nm node extended mission profile qualification, the SRAM Vddmin test was found to be the only test that showed extrinsic shift behavior over lifetime. Fig. 8 shows that these outliers already appeared at t=0 hours in the tail of the Vddmin distribution. These Vddmin outliers shift over lifetime to higher Vddmin, while at this temperature all other samples shift towards lower Vddmin. This result shows the need of a large sample size and a parametric shift analysis in addition to pass/fail information for lifetime evaluation. The Vddmin outliers, potential reliability fails, emphasize the importance of guard band settings, so that these are screened out at t=0 hours.

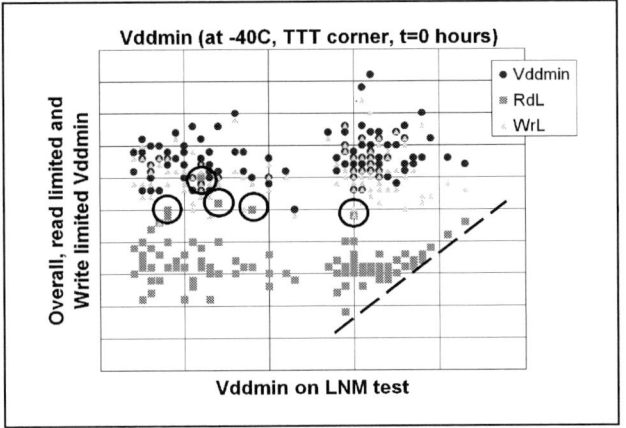

Figure 9. Correlation between the LNM test and the Vddmin tests

All our qualification test programs (for IP qualification, extended mission profile qualification and process qualification) include an extensive SRAM Vddmin test suite to detect even small extrinsic deviations in lifetime behavior. The test suite consists of a read limited Vddmin test, a low noise margin test (LNM, [10]) to detect cells with weak Static Noise Margin (SNM), a write limited Vddmin test, and a DRET test. The 90nm node Vddmin investigations at t=0 hours and over lifetime have been of great value during process development. For example, a "read disturb" failure mode, related to bits with a low noise margin, was found in this node. The separation of read limited (RdL), write limited (WrL) and LNM Vddmin bitmaps successfully localized gates with deviating Vt's that occurred only once per million, supporting a fast solution in time for the process release. Analyses at t=0 hours revealed that extrinsic Vddmin bits at hot were mostly caused by LNM related and read limited bits, whereas at cold the extrinsic bits seemed to be mostly write limited. Lifetime analyses show that there were more fails at hot than at cold. The fails at hot are well known from literature [5,10], as LNM fails that shift toward higher Vddmin over lifetime. A fail at cold was caused by a write limited bit. Typically, the write operation improves over lifetime, shifting towards lower Vddmin, due to degradation of the PMOST (the pull-up transistor) by NBTI. But, if the (absolute) PMOST Vt is already extremely high at t=0 hours, a further increase in Vt will actually make the write operation more difficult, leading to an increased Vddmin.

The foundry proposes a screening strategy based on the LNM test. The t=0 hours data in Fig. 9 suggest that at lower temperature the extrinsic Vddmin behavior can only be partially explained by SNM deviations. Samples with cells with a poor SNM, meaning a high LNM Vddmin (the points furthest to the right in Fig. 9), show an elevated read limited Vddmin but not necessarily a higher overall Vddmin. But at t=0 hours there are also read limited fails (circled RdL points in Fig. 9) that do not have a poor SNM, indicating a second limiting read mechanism. Also the observed write limited fail over lifetime indicates that SNM might not be a sufficiently complete model for extrinsic lifetime behavior and thus a screen based on the LNM test alone might be insufficient. Additional screens, based on read and write limited Vddmin behavior, may be necessary to reach low ppm.

Next to the bit-cell stability (SNM), also the memory compiler will influence Vddmin outlier behavior. Note that not all SRAM compilers are placed on the vehicles used for extended mission profile qualifications. The Vddmin behavior over lifetime of the memory compilers, that are not on these vehicles, can be validated by comparing their Vddmin behavior at t=0 hours to the behavior of the qualified compilers at t=0 hours. For this, the SRAM robustness for Vddmin behavior is expressed in the number of Vddmin outlier bits as a ratio to the total number of bits tested. This ratio is determined for all underlying tests (read limited, write limited, LNM). Vddmin outlier behavior on compilers with deviating ratios is investigated because it will impact yield and product ppm rate.

V. CONCLUDING REMARKS

To bridge the gap between foundry process qualification and requirements from customers with an extended mission profile, for instance for automotive, a qualification strategy was developed. This paper illustrates that such an extended qualification strategy proves process capability and enables product capability. It results in: (1) sharper process corner definition, (2) sufficient supportive data for the used acceleration factors, (3) proof of process capability for extended mission profiles, (4) guard band settings for production test programs, and (5) relevant screens and additional test suites that facilitate the release of low ppm automotive products.

It is concluded that an extended process qualification and characterization strategy can validate release of a standard foundry process supporting safe launch of future automotive products.

ACKNOWLEDGMENTS

Qualification of a standard CMOS process for products with an extended mission profile is an extensive project, requiring the support by many teams; reliability teams at TSMC and NXP, process development teams at TSMC and NXP and the NXP quality and reliability centers in Nijmegen and Taiwan for executing the lifetime experiments. Special thanks to Arjan Mels, Patrick van de Steeg, P.J. Huang, Remco de Haar, Rene Wientjes, and Xiao-Mei Zhang from NXP, and J.M. Huang from TSMC. The support of the Dutch government for the knowledge workers project Resilience for Automotive is greatly appreciated.

REFERENCES

[1] S.Y. Pai, J.K. Jerry Lee, and NG Kenny, "Reliability framework in a Fabless-Foundry environment," *IRPS 2009*, pp 229-235

[2] G. Verma, K. Wu, B. Euzent, and C. Huang, "Advanced process & product reliability development in fabless environment," *IRPS 2009*, pp 236-243

[3] ZVEI Handbook for Robustness Validation of Semiconductor Devices in Automotive Applications, April 2007

[4] SAE Standard J1879, Handbook for Robustness Validation of Semiconductor Devices in Automotive Applications, October 2007

[5] V. Huard, C. Parthasarathy, C. Guerin, T. Valentin, E. Pion, M. Mammasse, N. Planes, and L. Camus, "NBTI degradation: From transistor to SRAM arrays," *IRPS 2008*, pp. 289-300

[6] V. Reddy, J. Carulli, A. Krishnan, W. Bosch, and B. Burgess, "Impact of negative bias temperature instability on product parametric drift," *ITC 2004*, pp. 148-155

[7] T. Nigam, B. Parameshwaran, and G. Krause, "Accurate product lifetime predictions based on device-level measurements," *IRPS 2009*, pp. 634-639

[8] M. Kole, "Circuit realiability simulation based on Verilog-A," *IEEE International Behavioral Modeling and Simulation Workshop 2007*, pp. 58-63

[9] V. Petrescu, M. Pelgrom, H. Veendrick, P. Pavithran, and J. Wieling, "A Signal-Integrity Self-Test Concept for Debugging Nanometer CMOS ICs," *ISSCC 2006*, pp. 544-545

[10] A. Wang, C.H. Wu, R.Y. Shiue, H.M. Huang, and K. Wu, "New screen methodology for ultra thin gate oxide technology," *IRPS 2004*, pp. 659-660

Non-Destructive Current-Ramp Dielectric Breakdown (IRDB) for Fast BEOL Reliability Monitoring

Kok-Yong Yiang, Rick Francis, and Amit Marathe

Technology and Reliability Development
GLOBALFOUNDRIES, Inc.
1050 E Arques Ave, MS 143, Sunnyvale CA 94085, USA
Tel: +1-408-462-4135; E-mail: kok-yong.yiang@globalfoundries.com

Oliver Aubel

Quality and Reliability Engineering
GLOBALFOUNDRIES Dresden Module One LLC & Co. KG
Wilschdorfer Landstrasse 101, 01109 Dresden, Germany

Abstract—A fast current-ramp dielectric breakdown (IRDB) method is developed for critical dielectric reliability assessment. Unlike the conventional voltage-ramp method, the non-destructive nature of IRDB still allows wafers to be shipped for revenue after reliability monitoring. This is an important criterion for reduced-cost, high-volume manufacturing.

Keywords- Dielectric Breakdown, VRDB, IRDB

I. INTRODUCTION

The general health of BEOL dielectrics is conventionally monitored on scribeline interdigitated comb/serpentine structures, using a voltage-ramp method to detect uncharacteristically low breakdown strengths. Ironically, an upper voltage limit has to be imposed to prevent catastrophic dielectric breakdown, which can cause severe burn spots. This is because if even a single breakdown event occurs, the entire wafer has to be scrapped, due to the possibility of burn spots/debris spilling over the scribelines and contaminating the chip. Fig. 1 exemplifies some typical post-breakdown burning of Cu/SiCOH via comb structures, and the potential for the chip to be contaminated by spillover debris.

The need to impose an upper voltage limit presents a conflict between reliability and high-volume manufacturing. Setting too low a voltage limit only provides yield/defectivity statistics but not reliability metrics, since no breakdown voltage (V_{bd}) can be measured. On the other hand, setting too high a voltage limit can drastically increase the risk of dielectric breakdown, which leads to wafer wastage and, consequently, a huge burden on manufacturing.

In this paper, we will present a current-ramp method that achieves both the reliability and manufacturing objectives (i.e., the actual V_{bd} values from various scribeline positions of the wafer can be obtained, without ever incurring a single burn spot).

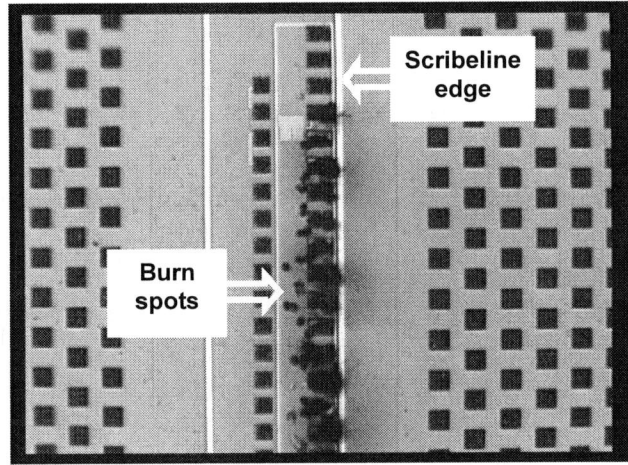

Figure 1. Burn marks and debris spillages from scribeline Cu/SiCOH via comb structures after voltage-ramp dielectric breakdown (VRDB).

II. EXPERIMENTAL

The procedure for a conventional voltage-ramp dielectric breakdown (VRDB) method is well documented [1-3]. Dielectric breakdown during voltage-ramp occurs suddenly and catastrophically without any prior warning, as exemplified by the current-voltage (I-V) graph in Fig. 2; by then, a burn mark would have been created. The burning is the aftermath of a thermal runaway condition at breakdown, whereby the sudden discharge of a high (uncontrolled) leakage current, coupled with a steadily held bias voltage, provides an immense amount of dissipated power at a localized breakdown spot. To control the extent of burning (i.e., dissipated power), it is imperative to control the amount of current flowing through the localized breakdown spot. A current-ramp method is therefore a suitable candidate for this purpose.

978-1-4244-5430-3/10 $26.00 © 2010 IEEE

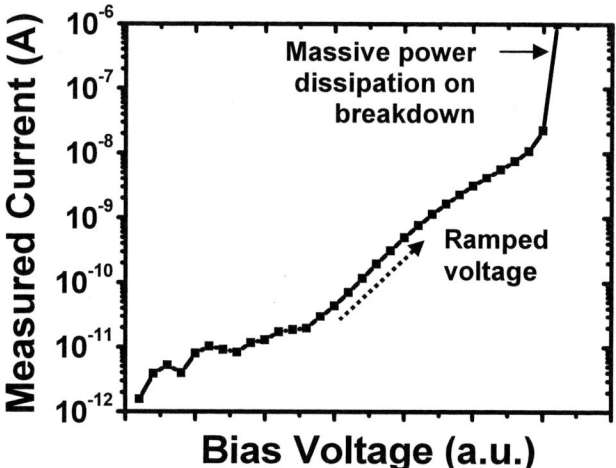

Figure 2. Typical VRDB leakage trace for an interdigitated via comb structure. On dielectric breakdown, uncontrolled leakage current, coupled with a steadily held bias voltage, creates massive power dissipation, which leads to burning.

Fig. 3 shows a typical leakage trace for an interdigitated via comb structure, using the current-ramp dielectric breakdown (IRDB) method. The structure is at minimum designed via/metal pitch and fabricated on 45 nm wafer technology. The bias current was increased in logarithmic intervals from a low start current (5pA) at a fixed ramp rate (0.1 current decade/sec), and the voltage was measured at each ramp step.

During the ramp, the measured voltage increased monotonically, before it began to drop just prior to catastrophic dielectric breakdown. The drop in voltage indicated the onset of breakdown, whereby the dielectric was beginning to weaken; if the current-ramp were to be stopped at this instant (to prevent excessive power dissipation), the highest measured voltage can then be taken as the V_{bd} of the dielectric.

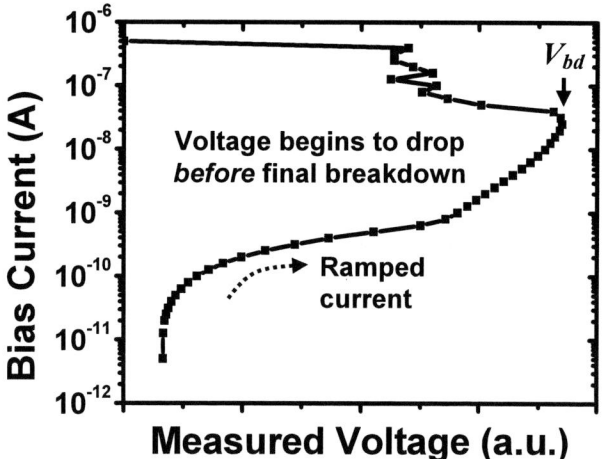

Figure 3. Typical IRDB current-voltage (I-V) trace for an interdigitated via comb structure. Voltage is measured as current is increased in logarithmic intervals. Highest measured voltage is the breakdown voltage (V_{bd}).

Fig. 4 shows the collective IRDB traces of 16 samples on the same wafer using the above IRDB criterion (i.e., the ramped current was stopped when the measured voltage just began to drop). The effective V_{bd} from the IRDB measurements were equivalent to the V_{bd} obtained using the conventional VRDB method, as shown in Fig. 5. This demonstrates the robustness and accuracy of IRDB in measuring V_{bd}, based on the aforementioned criterion.

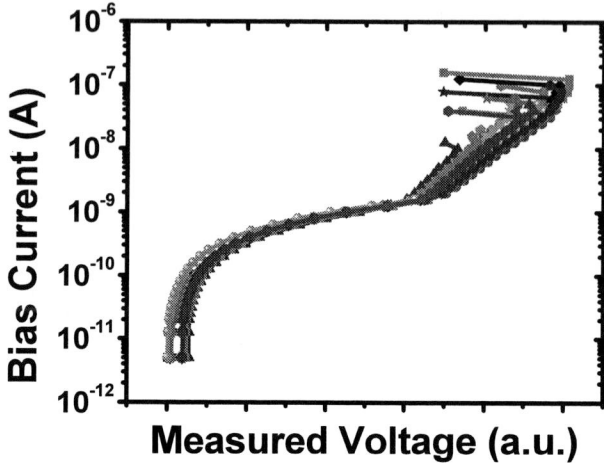

Figure 4. Collective IRDB traces of 16 samples. To prevent burning, the ramped current is stopped as soon as voltage begins to drop.

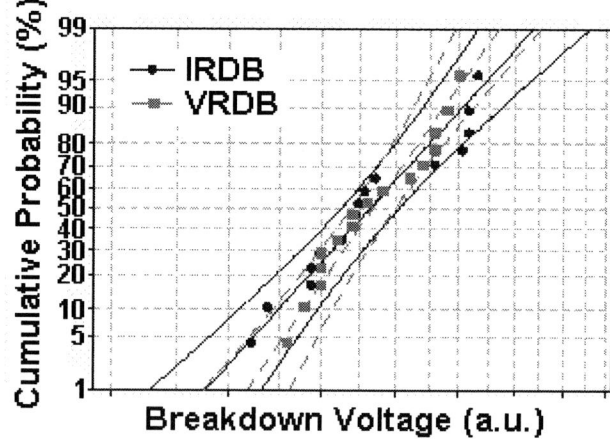

Figure 5. Breakdown voltages (V_{bd}) obtained from IRDB are consistent with those using conventional VRDB.

High-magnification optical microscopy showed no visible burn marks on all IRDB samples. In the most severe case, only small, almost indiscernible bulges at localized failure sites were created, as shown in Fig. 6. The wafer passivation remained intact and non-blistered, and the bulges were mechanically confined within the scribelines by the seal ring. This demonstrates that, by controlling the stress current, the IRDB method can successfully limit the amount of dissipated power (heat) during a breakdown event and thereby prevent catastrophic burning.

978-1-4244-5430-3/10 $26.00 © 2010 IEEE

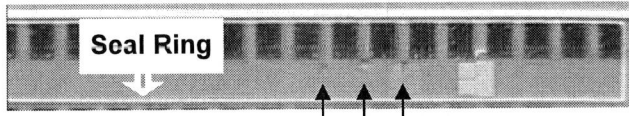

Figure 6. At the most severe case, only small bulges are created after IRDB. Wafer passivation remains intact (i.e., not blistered), no burn marks are created, and the bulges (indicated by black arrows) are mechanically confined within the scribelines by the seal ring.

This is further exemplified by Fig. 7 and Table I, which show that the severity of the localized bulge closely correlated with the amount of dissipated power, P (= current $\times V_{bd}$), at the onset of dielectric breakdown. This shows that controlling the current is vitally important in preventing catastrophic burning—a crucial factor which the conventional VRDB method is clearly deficient.

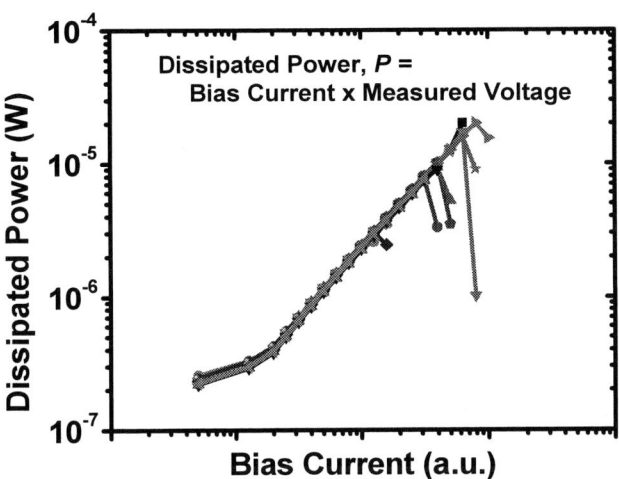

Figure 7. Plot of dissipated power, P, during IRDB. Amount of dissipated power is controlled at every step up to onset of dielectric breakdown.

TABLE I. SEVERITY OF LOCALIZED BULGE CLOSELY CORRELATES WITH THE DISSIPATED POWER, P (= CURRENT $\times V_{BD}$), AT THE ONSET OF DIELECTRIC BREAKDOWN.

Sample	Power (W)*	Bulge?
1	5.51E-07	No
2	2.34E-06	No
3	3.02E-06	No
4	3.02E-06	No
5	3.74E-06	No
6	3.89E-06	No
7	6.04E-06	No
8	7.49E-06	No
9	7.56E-06	No
10	9.33E-06	Very minor
11	1.01E-05	Yes
12	1.02E-05	Extremely minor
13	1.26E-05	Very minor
14	1.57E-05	Very minor
15	1.71E-05	Yes
16	2.01E-05	Minor

Increasing Power

III. SUMMARY AND CONCLUSION

We have demonstrated an IRDB method that successfully meets both reliability and manufacturing objectives by yielding breakdown voltage (V_{bd}) information without creating any burn spots on the wafer. This enables wafers to be shipped for revenue even after monitoring. IRDB also has the following compelling advantages over the conventional VRDB method:

a. Dies exhibiting abnormally low V_{bd} may survive burn-in, but present long-term time-dependent dielectric breakdown (TDDB) risks in the field. Using IRDB, these weak dies can be blacklisted individually, while healthy dies with high V_{bd} can still be harvested for production. This has a direct, positive impact on overall wafer yield.

b. Current-ramp on a logarithmic scale is much faster than the conventional voltage-ramp on a linear scale; IRDB is therefore even more compatible with high-volume manufacturing than VRDB.

c. Since V_{bd} can be obtained by IRDB to give a good reliability indicator of dielectric health, there is less need for full-scale TDDB monitoring, which is resource-intensive and time-consuming, and requires production wafers to be diced and therefore wasted.

d. The impending use of porous ultra-low-k dielectrics, which have comparatively lower glass transition temperature (T_g) and mechanical strength, will exacerbate the degree of burning and debris spillage using the conventional VRDB method. The non-catastrophic nature of the IRDB method therefore provides a highly-compatible reliability monitoring solution as we scale towards future technology nodes.

IV. REFERENCES

[1] Jinyoung Kim, Ennis T. Ogawa, and Joe W. McPherson, "Time Dependent Dielectric Breakdown Characteristics of Low-k Dielectric (SiOC) Over a Wide Range of Test Areas and Electric Fields," *Proceedings of IRPS*, pp. 399-404 (2007).

[2] F. Chen *et al.*, "Investigation of CVD SiCOH Low-k Time-dependent Dielectric Breakdown at 65nm Node Technology," *Proceedings of IRPS*, pp. 501-507 (2005).

[3] JESD35-A, "Procedure for the Wafer-Level Testing of Thin Dielectrics," *JEDEC Solid State Technology Association*, April 2001.

Modeling of Cu IMD-TDDB caused by extrinsic defects

T. Ouchi, K. Makabe*, M. Ogasawara*, E. Murakami*, and N. Yoshioka
Yield Management / Process & Device Analysis Engineering Development dept.
Renesas Technology Corp.
751, Horiguchi, Hitachinaka-shi, Ibaraki-ken, Japan
Phone: +81-29-270-2986; e-mail: ouchi.tomohiko@renesas.com

Abstract—In this paper, a prediction method for degradation failure ratio (B-mode) of Cu IMD-TDDB is studied. In this study, B-mode failure is assumed to be caused by dielectric thinning due to random defects (thinning model). Its probability is calculated by Critical Area Analysis (CAA). Prediction is in good agreement with measurement results. This method can be applied to estimate reliability of various LSI products.

Keywords; degradation failure, thinning model, Critical Area Analysis, random defects, TDDB, Reliability, LSI

I. INTRODUCTION

Recently, it is becoming increasingly difficult to ensure reliability of LSI because of miniaturization and new materials introduction. In addition, the reliability requirements are severe, in particular, for automotive applications. Therefore, prediction and designing methods for reliability are becoming more important.

Cu IMD (inter-metal dielectrics) – TDDB (time dependent dielectric breakdown) is one of the typical failure modes, which has been discussed intensively [1]-[3]. Yokogawa et al. showed that the degradation failure (B-mode) of Cu IMD-TDDB is related to the initial (A-mode) defect density experimentally [4]. These defects are mainly originated from the surface condition of the dielectric film. However, after optimizing surface treatment for Cu process, the main cause of defects must be random particles.

In this work, we attempt to predict the B-mode failure ratio of Cu IMD-TDDB by assuming the main cause is random defect induced dielectric (Thinning model) [5]. The IMD-TDDB failure ratio is predicted from initial defect density and kill ratio that calculated by Critical Area Analysis (CAA) [6].

We confirmed good agreement between the prediction and actual failure ratio. This result means that B-mode is mainly caused by extrinsic defects.

II. EXPERIMENTAL

In this model, soft-short critical area and defect density are required to derive the degradation failure ratio of IMD-TDDB. "Soft-short" means that the defects don't immediately cause short failure but the dielectrics are very thin. Soft-short critical area is calculated with CAA [6]. Failure ratio of initial breakdown is used to derive defect density.

CAA gives the cumulative failure ratio due to defects that exist below a certain defect-metal spacing. We need to convert defect-metal spacing to breakdown voltage in order to compare prediction with ramp voltage test results. Furthermore, failure ratio as a function of time is needed to design reliability for LSI. For these purposes, we use the oxide thinning. In this model, defects are modeled as effective oxide thinning spot. We also use single-via test structure [7] to determine the breakdown voltage and the lifetime at the oxide thinning spot experimentally.

III. RESULTS AND DISCUSSION

A. Measurement results of Cu IMD-TDDB

Figure 1 shows the results of breakdown voltage measurement of large-scale comb test structure. The horizontal axis denotes breakdown voltage and the vertical axis denotes cumulative failure ratio. There are some samples between intrinsic failure (C-mode) and initial failure (A-mode). We call this early failure B-mode. Traditionally, intrinsic failure ratio is used to design LSI reliability. However, it is important to consider B-mode failure to design for reliability accurately.

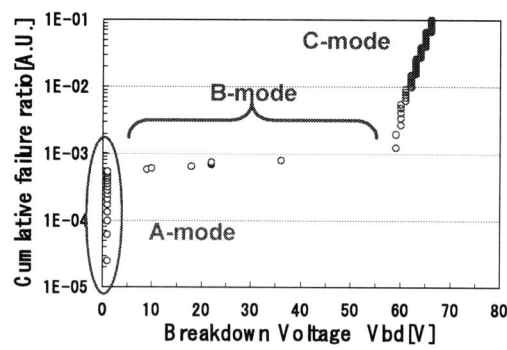

B. IMD-TDDB caused by extrinsic random defects

Among failure samples, dielectric thinning was found as the cause of B-mode. Figure 2 shows one of the SEM images at the failure point. Physical failure analysis found extrinsic defect at this point. Therefore, our IMD-TDDB model is based on extrinsic defects.

Figure 3 shows our model of B-mode failure. Some defects do not cause short, but defect-metal spacing is very small. We assume that B-mode failure is caused by these soft-short defects.

C. Simulation results and discussion

To predict failure ratio by random defects, CAA is a well-known method. Figure 4 shows usual CAA calculation where, only hard-short defects are treated. To treat soft-short events, CAA method is extended [6]. Figure 5 shows soft-short critical area and procedure of deriving B-mode failure ratio (FB).

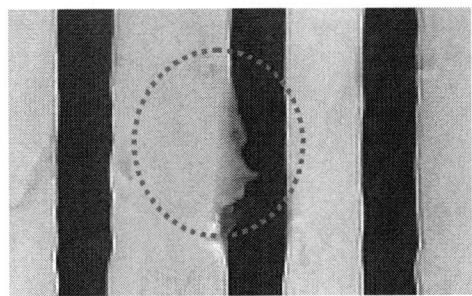

Fig.2 SEM image of B-mode failure point
Dielectrics thinning are found at the failure point.

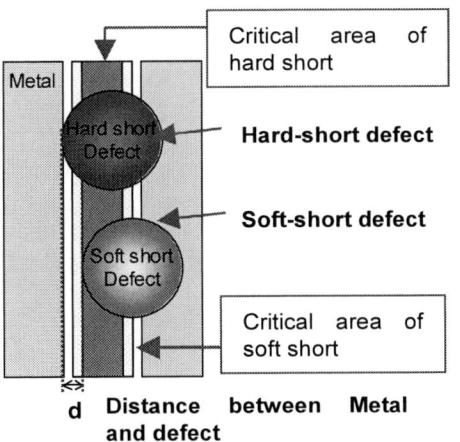

Fig.3. Schematic of defect types.
Hard-short defects cause A-mode failure. B-mode failure is assumed to be caused by soft-short defects. Dielectric thinning causes decrease in breakdown voltage.

Defects having the distance to metal less than d are treated as failure defects. The probability of failure (POF) curve is shifted to the left by the distance, d. The soft-short critical area is obtained by subtracting hard-short critical area from the area calculated from this shifted POF. B-mode failure ratio is obtained from hard-short defect density and soft-short critical area. We use A-mode defect density as hard-short defect density.

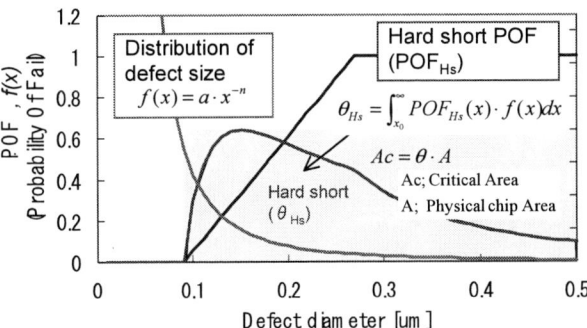

Fig.4 Description of usual CAA
To obtain kill ratio (θ), POF and defect distribution function are required. POF is determined by physical layout pattern only. Power function is used as a defect distribution function empirically.
Ac defined as $\theta*A$ is critical area with A being the chip area. Ac is the virtual area in which existing defect results in failure.

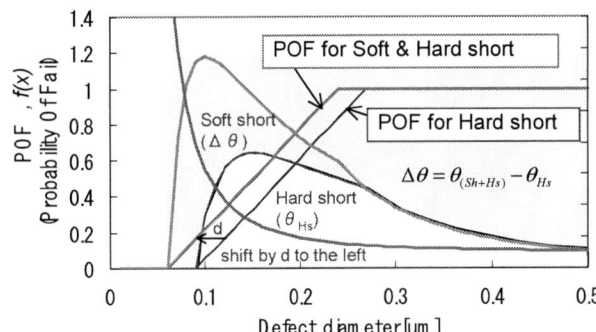

Fig.5 Description of soft-short kill ratio and failure ratio derivation procedure.
Soft-short kill ratio ($\Delta\theta$) is obtained by subtracting Hard-short kill ratio from total kill ratio. Total kill ratio is calculated from shifted POF.
To derive B-mode failure using CAA, defect density(D_0) of hard-short(A-mode) and soft-short, critical area($=\Delta\theta*A$), and chip area(A) are required.

$$F_B(d) = 1 - \exp(-D_0 \cdot \Delta\theta(d) \cdot A)$$

CAA gives cumulative failure ratio as a function of defect-metal distance d. To compare this result to actual data, distance must be converted to breakdown voltage. Single-via measurement enables this conversion (See Appendix). Figure 6 shows simple illustration of single-via test structure. As shown in Ref [7], breakdown voltage is determined by via-metal distance.

Figure 7 shows the prediction and the actual data of breakdown voltage. Horizontal axis is breakdown voltage, and vertical is cumulative failure ratio. It shows good agreement between prediction and actual data. Therefore, we conclude that B-mode failure is mainly caused by random defects.

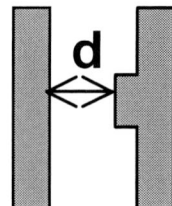

$$V_{bd} = d \cdot E_{bd}$$

V_{bd}; Breakdown voltage

d; metal - via distance

E_{bd}; Breakdown electric field

Fig.6 Schematic of Single-via test structure

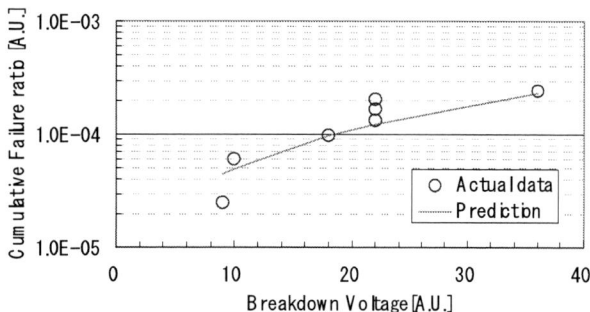

Fig.7 Comparison of experimental result and prediction of B-mode breakdown failure.

D. LSI reliability design

CAA enables prediction of failure ratio for any design pattern. However, reliability designers are interested in lifetime instead of breakdown voltage. Figure 8 shows the TDDB lifetime as a function of the electric field. By using these data, distances can be converted to the lifetime (See Appendix). Here, the electric field is calculated from the defect-metal distance and circuit's operating voltage.

Figure 9 shows B-mode failure ratio at an arbitrary operation time. Even for actual products, we can use the same procedure. As a result, we can discuss reliability of any products.

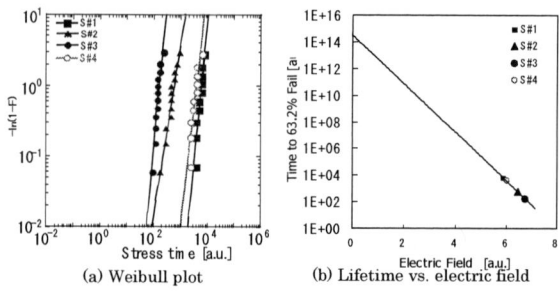

(a) Weibull plot (b) Lifetime vs. electric field

Fig.8 TDDB test results for single-via test structure
S#1:const voltage, space 1x,
S#2: const voltage, space 0.91x,
S#3: const voltage, space 0.87x,
S#4:constant field stress

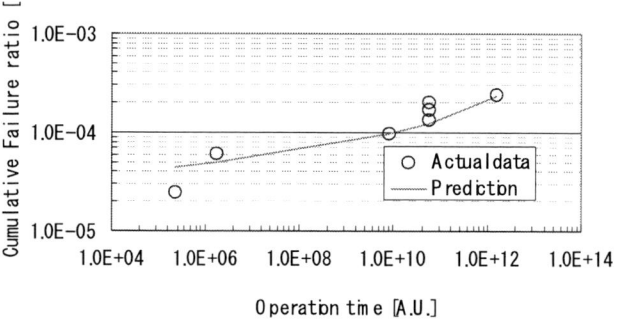

Fig.9 Prediction of B-mode failure ratio under the operation

B-mode failure ratio at any operation time can be calculated by CAA. This method can be applied to actual products easily.

Figure 10 shows application for DSP module designed according as 130nm design rules. Predictions of B-mode failure at an arbitrary voltage are shown. 70% and 50% shrink data are also shown. With miniaturization, B-mode is increasing more than A-mode.

Figure 11 shows reliability improvement by wire spreading. Wire spreading is the operations to increase inter-metal spacing. As a result, both yield and reliability are improved. Improvement rate of reliability by wire spreading is increases with miniaturization.

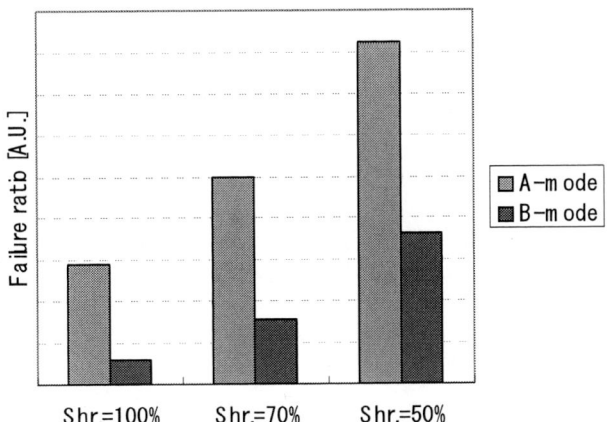

Fig.10 Predicted Failure ratios. B-mode failure ratio is calculated at an arbitrary breakdown voltage. B-mode becomes important with miniaturization.

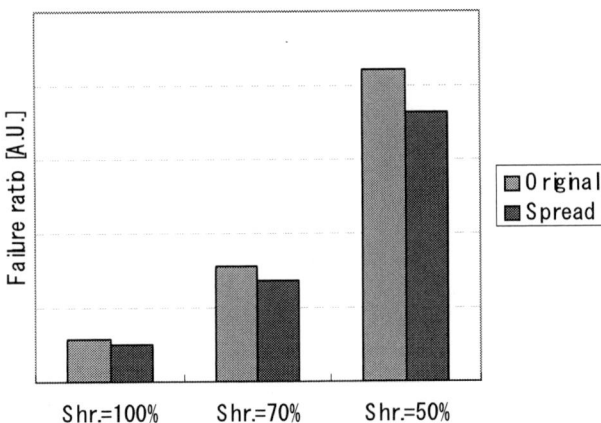

Fig.11 Improvement of B-mode failure ratio by wire spreading
Efficiency of wire spread increases with miniaturization.

IV. CONCLUSION

We modeled B-mode of Cu IMD-TDDB based on dielectric thinning due to random defects. The prediction shows good agreement with the actual data. This indicates that the main cause of B-mode is random defects.

In addition, we converted distance to operation time. As a result, we can apply this method to the reliability design of actual products easily.

Appendix

In this appendix, we derive the relation between the TDDB failure distribution function and the distance distribution function calculated using the CAA model. Consider a structure with an extrinsic defect. The extrinsic defect is modeled as a thinning spot. The intrinsic failure and the extrinsic failure is competing, so the whole failure distribution (F) is calculated by using

$$1 - F(t_0, s) = (1 - F_i(t_0)) \cdot (1 - F_e(t_0, s)) \qquad (A1)$$

Where F_i is the intrinsic failure distribution, F_e is the extrinsic failure distribution function, s is the distance at thinning spot and t_0 is a particular operation time.

If we consider the early failure, then (A1) is approximated as follows:

$$F(t_0, s) = F_e(t_0, s) \qquad (A2) \quad .$$

In general, there is a distribution of distance at thinning spots. If the distance density function is p(s), then the expected failure is

$$F(t_0) = \int_0^{s_{design}} p(s) \cdot F_e(t_0, s) ds \qquad (A3) .$$

Next, we use a single-via test structure to model the thinning spots. The schematic layout of the single-via test structure is shown in Fig. A. The 1-via is intentionally misaligned from the metal line to simulate the thinning spots. The lifetime variation of this single-via structure is small as shown in Fig.8. Therefore, we approximate the F_e with step function as follows:

$$F_e(t_0, s) = \begin{cases} 0 & t_0 < t_{63.2}(s) \\ 1 & t_0 > t_{63.2}(s) \end{cases} \qquad (A4)$$

$$t_{63.2}(s) = K \cdot 10^{-a \cdot V_{op}/s} \qquad (A5)$$

Where $t_{63.2}(s)$ is a time to 63.2% fail, K and a are a constant number, Vop is an operation voltage.

Inserting (A4) to (A3), the result is

$$F(t_0) = \int_0^{s_0} p(s) ds \qquad (A6),$$

Where $\quad s_0 = -a \cdot V_{op} / \log_{10}\left(\dfrac{t_0}{K}\right) \qquad (A7) .$

(A6) means

$$\Pr\{T_{bd} < t_0\} = \Pr\{\text{distance at thinning spot } s < s_0\}.$$

Using the same procedure, we can get following relation:

$$\Pr\{V_{bd} < v_{bd0}\} = \Pr\{\text{distance at thinning spot } s < s_0\}$$

Where $\quad s_0 = v_{bd0} / E_{bd} \qquad (A8) .$

Therefore, we can use the classical thinning model itself [5]. All parameters (K, a, E_{bd}) are determined from the single-via structure experimentally.

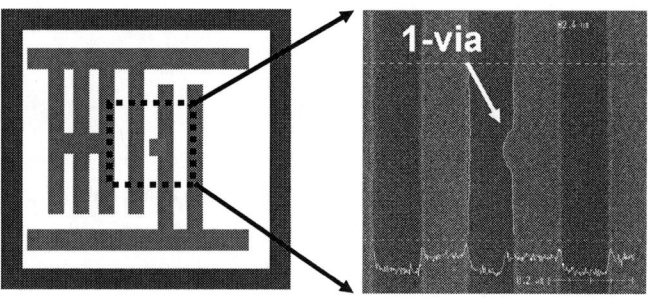

Fig. A. Schematic layout of single-via test structure (After Ref.7)

ACKNOWLEDGEMENTS

The authors would like to thank T. Furusawa and S. Miyata for helpful discussion, and F. Arakawa, Y. Kato, and K. Yoshino for physical analysis.

REFERENCES

[1] J. Noguchi, et al., IEEE ED-48(n), p.1340(2001)
[2] K. Makabe et al., Proc. ADMETA 2004, p. 44 (2004)
[3] A. Oates, IRPS 2009, Interconnects Year in Review (2009)
[4] S. Yokogawa et al, IRPS 2008, p144 (2008)
[5] J. C. Lee et al, IEEE trans. vol. ED-35, 12 p.2268 (1988)
[6] G. A. Allan et al, IEEE trans. Vol.11-1, p146 (1998)
[7] T. Kamoshima et al., Proc. IITC 2009, p. 185 (2009)

Product Failures: Power-Law or Exponential Voltage Dependence?

Amr Haggag, Keith Forbes, Gary Anderson, Dave Burnett, Peter Abramowitz, Mohamed Moosa

Freescale Semiconductor, Austin, Texas

amr.haggag@freescale.com

Abstract: The product failures voltage acceleration has traditionally been modelled with exponential voltage dependence. However with voltage scaling, the voltage acceleration parameter (VAP) in an exponential model has increased as V^{-1} – as expected for dielectric breakdown in either back-end or front-end. This suggests an exponential model is probably quite conservative and a power-law model may be more appropriate for 90nm and beyond. Even if an exponential model continues to be used, this understanding can help assess the amount of conservatism built in such a model.

Keywords: product; extrinsic; powerlaw; exponential; TDDB; voltage acceleration.

Introduction

Voltage scaling has played a pivotal role in enabling semiconductor products to be offered with increasing levels of functionality while maintaining cost, power and performance constraints. At the same time, new product definition and design cycles ranging from six months to two years mandate a requirement to provide realistic projections of product reliability as a function of target operating conditions. This paper shares some experience in this arena, identifying some critical mechanisms that affect product reliability and how to realistically project product reliability from proper interpretation of early life failure rate (ELFR) and high temp operating lifetest (HTOL) voltage stress data.

Voltage Acceleration Parameter Model

A. Fail Modes Separated:

Analysis of ELFR and HTOL failures is complicated by the introduction of failure modes with various voltage and temperature dependencies. ELFR and HTOL tests are performed according to JEDEC standards. JEDEC standards use an exponential voltage acceleration parameter for product extrinsic failures (JESD74) and long term stressing in technology certification (JP001). The voltage acceleration parameter should be measured for each technology to be confirmed appropriate for each device.

It is recognized that different failure mechanisms will have different voltage and temperature acceleration. For example, gate dielectric breakdown is best

described with a power-law voltage acceleration model (JP001). This power-law voltage dependence has been demonstrated in [1]. Also intra-layer dielectric breakdown in the front-end or back-end has been demonstrated to follow an exponential sqrt model due to Poole-Frenkel conduction in these dielectrics [2,3,4]. Hence the voltage acceleration parameter (VAP) in an exponential model is not constant and follows:

$$VAP \sim -\frac{d\ln(TTF)}{dV}$$

$$VAP(metal, via) \sim 0$$

$$VAP((intra-layer-dielectrics) \sim -\frac{d\ln(Ae^{-B\sqrt{V}})}{dV} \sim \frac{1}{2}\frac{B}{\sqrt{V}}$$

$$VAP(gate\_dielectrics) \sim -\frac{d\ln(CV^{-N})}{dV} \sim \frac{N}{V}$$

(1)

B. Fail Modes Combined:

Defects tend to cluster as has been evident in the semiconductor industry for decades, which simply implies that defects are more likely to be found in groups rather than by themselves. Due to the tendency of defects to cluster the failure distribution follows the well known negative binomial [5]:

$$F \sim 1 - (1 + AD/\alpha)^{-\alpha} \quad \textbf{(2)}$$

where the defect density D is a time-dependent $D \sim Dmax*(t/tmax)^{\beta}$. For large AD/$\alpha$ (more clustering) this converges to **Pareto** and for small AD/α (less clustering) this converges to **Weibull**. In the case of heavy clustering we have the **Pareto** distribution:

$$F \sim 1 - (t/tc)^{-\alpha\beta} \quad \textbf{(3)}$$

Combining fail modes, the fail distribution becomes

$$F_{all} \sim 1 - \Pi_i(1 - F_i) = 1 - \Pi_i(t/tc_i)^{-\alpha\beta_i}$$
$$\sim 1 - (t/tc_{all})^{-\alpha\beta_{all}}$$

where

$$\beta_{all} = \sum_i \beta_i$$

$$tc_{all} = \Pi_i(tc_i)^{\beta_i/\beta_{all}}$$

$$VAP_{all} = d\ln(tc_{all})/dV = \sum_i \frac{\beta_i}{\beta_{all}}VAP_i$$

(4)

Combining the VAP expressions for metal/via, intra-layer dielectric and gate dielectrics fails, we get (assuming time-dependence exponent β~0.4-0.6 for dielectric defect generation is typically double that for metal/via defect generation β~0.2-0.3):

$$VAP_{all} = \frac{1}{5}(0) + \frac{2}{5}(\frac{1}{2}\frac{B}{\sqrt{V}}) + \frac{2}{5}(\frac{N}{V}) \sim \frac{17}{V} \qquad (5)$$

Here we have used for extrinsic gate dielectric fails the same power-law (N~38-43) as intrinsic gate dielectric fails [1,6]. Similarly for extrinsic intra-layer dielectric fails we have used the same sqrt voltage dependence (B~6-9V^-1/2) as intrinsic intra-layer dielectric fails [2,3,4]. This gives a VAP for all fail modes of ~17/V in an exponential model.

It is therefore expected that as voltage scales, the VAP in each technology will increase with 17/V. This suggests that a power-law 17 voltage dependence as opposed to an exponential voltage dependence may be more appropriate to describe product fails. Figure 1 summarizes the voltage acceleration parameter trend for IBM/TI and FSL [7,8,9] observed as mid stress voltage V is scaled with Vdd. Typically V ~1.4x Vdd.

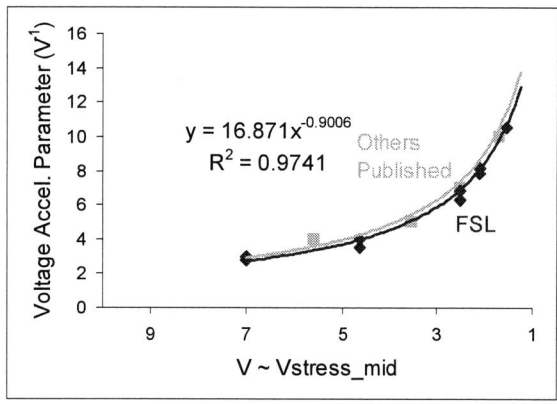

Figure 1 – Product VAP in an exponential model follows almost 17/V equivalent to a power-law 17 model. This implies that for 1V technology, whereas VAP~12 makes sense from 1.5V stress to 1.3V stress acceleration i.e. V ~ 1.4V, VAP~15 is more realistic from 1.3V stress to 1V use acceleration i.e. V ~ 1.15V.

Voltage Acceleration Parameter Discussion

Dielectric failures are becoming prevalent in recent CMOS technologies and hence are expected to drive the voltage dependence of product failures. These dielectric failures are typically a combination of gate dielectric failures and/or intra-layer dielectric failures. Below is an example showing dependence of product

failure rate on gate stack integration for six month worth of development lots.

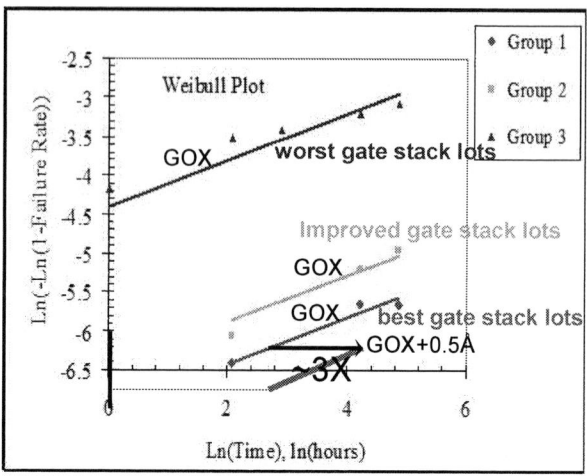

Figure 2 – HTOL fallout vs time for sixth months worth of lots grouped in terms of HTOL fail level.

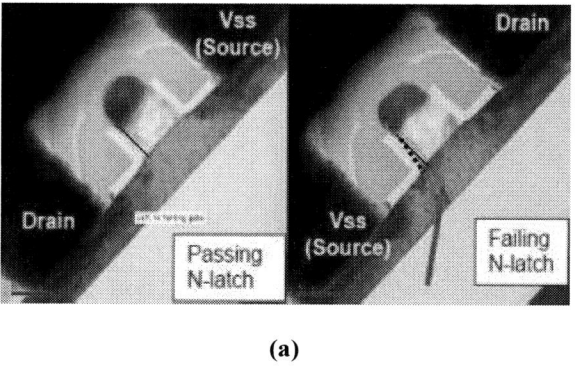

(a)

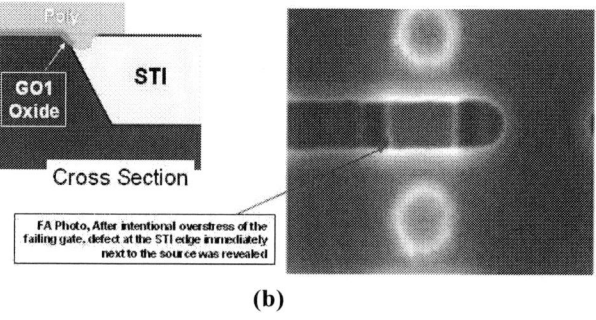

(b)

Figure 3 – HTOL fails FA revealed gate oxide leakage in the NMOS related to poly foots and STI edge thinning.

The failure rate was strongly correlated to the gate stack integration (STI corner rounding and poly profile). As the gate stack integration improved, the fallout improved. Then when the gate oxide was thickened by 0.5Å (i.e. 1.78X reduction in Jg0), the fallout was reduced by a further 3X consistent with gate dielectric thickness scaling dependence. This is illustrated in Fig. 2. Furthermore, failure analysis of HTOL fails in Fig. 3 revealed gate oxide breaks (leakage) in NMOS with the failing NMOS showing a slight poly foot and STI thinning compared to passing NMOS [10, 11].

However, it is important to note that gate dielectric breakdown is not the only dominant mode in advanced technologies, intra-layer dielectric breakdown may also occur causing product failures. The intra-layer dielectric breakdown may either be contact failures or low-k failures. For this reason the voltage acceleration that is measured will be a weighted average of both gate dielectric breakdown and intra-layer dielectric breakdown per Equation (5). Based on that equation a power-law 17 is expected to be observed for large sample voltage acceleration experiments (~3000 units) as extracted from the failure plots below.

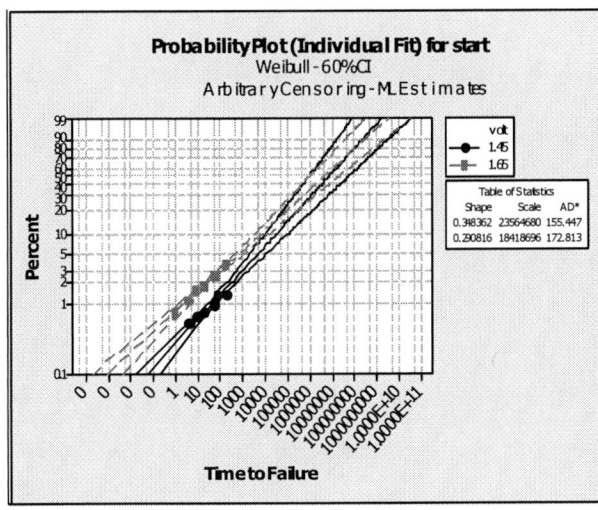

Figure 4 – HTOL fails versus voltages from which a power-law 17 voltage dependence was extracted.

Exponential vs Power-Law Risk

One can use the power-law model to assess conservatism when an exponential model is still used for product failures. It is important to note that under stress conditions, one cannot see the difference between exponential and power-law, and the difference becomes important when extrapolating to use conditions as shown in Figure 1. Nevertheless, given the product voltage acceleration is expected to be driven by dielectric breakdown (either intra-layer-dielectrics in the back end / front end or gate dielectrics

in the front end), the VAP is expected to increase with voltage scaling.

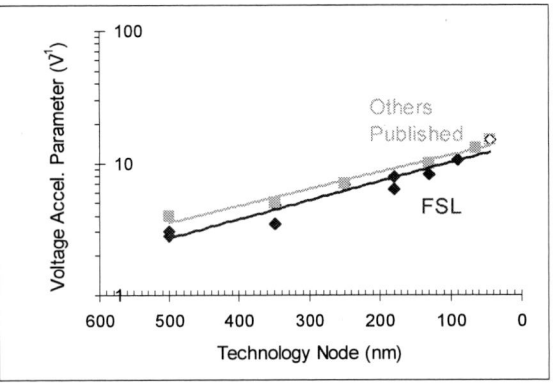

(a)

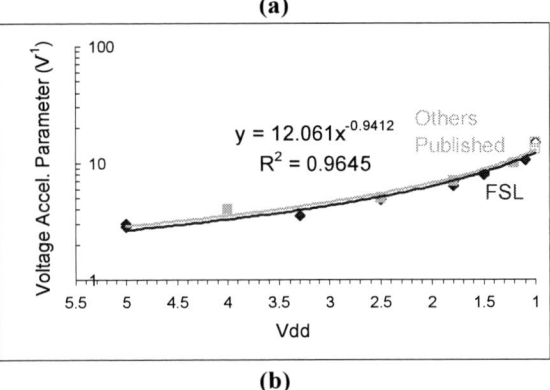

(b)

Figure 5 – Product VAP in an exponential model plotted vs (a) technology dimension (b) technology Vdd. It clearly shows more universality vs technology Vdd and follows a dependence VAP~12/Vdd consistent with VAP~17/V (V~1.4xVdd) and the dielectric breakdown explanation.

We plotted the voltage acceleration parameter from an exponential model [7,8,9] vs technology dimension as in Fig. 5(a) and vs technology Vdd as in Fig. 5(b). It clearly shows more universality vs technology Vdd and follows a dependence VAP~12/Vdd consistent with VAP~17/V (V~1.4x Vdd) and the dielectric breakdown explanation.

Conclusion

In the 90nm and beyond nodes, dielectric breakdown (gate or intra-layer) will continue to play an important role in product failures and given its non-exponential dependence, it is not unreasonable to expect that product failures voltage acceleration is moving towards a power-law dependence as opposed to the classical exponential dependence. Even if an exponential model continues to be used, this understanding can help assess the amount of conservatism built with the exponential VAP chosen for product failure rate predictions.

References

[1] P.E. Nicollian et. al., p. 197-208, IRPS 2007.
[2] Kok-Yong Yiang et. al., p. 476-480, IRPS 2009.
[3] D. Gajewski et. al., p. 53-55, P2ID 2002.
[4] F. Chen et. al., p. 132-137, IRPS 2008.
[5]T.S.Barnett et. al., p.110, IEEE Design & Test of Computers 2006
[6] E. Y. Wu et. al. p. 26 – 30, IRPS 2006.
[7]J.M. Carulli et. al., p.118, IEEE Design & Test of Computers 2006
[8]IBM, Q&R ASIC App Note, SA14-2280-03, rev3 1999 and PPC 750GX DD1.1, rev1, Qual Report, 2004.
[9]M.F. Zakaria et. al., p.88, IEEE Design & Test of Computers 2006
[10] A. Haggag et. al., p. 541-544, IRPS 2006.
[11] Y.H. Lee et. al., p. xx - xx, IEDM 2007.

Reliability Status of GaN Transistors and MMICs in Europe

Invited Paper

M. Dammann, M. Cäsar, H. Konstanzer, P. Waltereit, R. Quay, W. Bronner, R. Kiefer, S. Müller, M. Mikulla,
Fraunhofer Institute for Applied Solid State Physics (IAF)
Tullastr. 72, 79108 Freiburg, Germany
Email: michael.dammann@iaf.fraunhofer.de, phone: ++49 761 5159 517; fax: ++49 761 515971517

P. J. van der Wel, T. Rödle
NXP Semiconductors
Gerstweg 2, 6534 AE, Nijmegen, The Netherlands

F. Bourgeois, K. Riepe
United Monolithic Semiconductors
Wilhelm-Runge-Strasse 11, 89081 Ulm, Germany

Abstract—Recent DC- and RF-reliability results of European GaN HEMTs for high frequency power and MMIC applications between 2 and 18 GHz will be presented. The DC-stress test experiments have been performed at high current and high voltage settings in order to test the devices in the different regimes during large signal operation. GaN HEMTs and one stage MMICs have also been tested under RF-operation conditions and the correlation to DC-stress tests has been investigated.

Keywords: GaN HEMT, GaN MMIC, reliability, DC-stress test, RF-stress test.

I. INTRODUCTION

The final objective of current European activities is to develop a source for reliable high end GaN transistors and MMICs [1]. Different large-scale projects in Europe such as Korrigan [2] focus on the manufacturing technology and reliability improvement of GaN transistors. The aim of ESA's new GaN Reliability Enhancement and Technology Transfer Initiative (GREAT2) is to enable the manufacture of high reliability, space compatible, microwave transistors and integrated circuits [3]. These projects include reliability testing activities and assessment of failure modes [4]. The difference in test methods and conditions makes it difficult to compare reliability data of different European GaN players. Table 1 lists the different operating life test conditions used at different European teams. Although the table is not exhaustive it shows that more teams rely on voltage or temperature accelerated DC-stress test than on the more complicated RF-stress tests. The published reliability looks promising, i.e. based on a failure criterion of 10%-Idss degradation an extrapolated life time of

more than 10,000 h at a channel temperature of 90°C at a drain voltage of 50 V for 0.25 µm devices with Mo/Au gate from Alcatel-Thales has been determined under DC-operation conditions [7]. In addition, an extrapolated life time of more than 20 years has been found for 0.5 µm devices from IAF at a channel temperature of 90°C and a drain voltage of 50V under DC-operation conditions [10]. At higher channel temperatures ranging from 140°C to 200°C excellent device reliability and stability of the gate leakage current under DC- and RF-operation has been achieved by improving the gate processing technology [5].

In joint collaboration projects Fraunhofer IAF and United Monolithic Semiconductors have developed a GaN50 process with a gate length of 0.5 µm optimized for power applications in the L- and S-band and a GaN25 process with a gate length of 0.25 µm for MMIC applications in the X-band and the Ku-band. NXP Semiconductors plans the introduction of GaN RF power transistors for next generation mobile communication systems using the GaN50 technology in the near future. In this paper recent DC- and RF-reliability results of the GaN25 and GaN50 process will be presented. The reliability of the GaN25 process was investigated using 0.48 mm wide HEMTs and one stage MMICs, whereas the GaN50 technology was tested using HEMTs with larger gate periphery of 2.4 mm and 7.2 mm, respectively.

978-1-4244-5430-3/10 $26.00 © 2010 IEEE

TABLE 1. OPERATING LIFE TEST CONDITIONS OF EUROPEAN TEAMS

stress test	test conditions	team	Ref
RF life time test	T_{ch}=160°C, V_d=50 V, 2 GHz, P_{out}=2.5 W/mm	IAF	[5]
RF step stress test	3 db compression, Increase of V_d by 5 V every 30 min, V_{dmax}=30 V	Uni Modena	[6]
High temperature operating (Idq)	T_{ch}=140°C and 200°C, V_d=50 V, P_{dc}=2.5 W/mm or	IAF	[5]
	T_{ch}=150°C and 175°C, V_d=25 V, P_{dc}=10 W/mm	Uni Bordeaux	[7]
	T_{ch}=204, 232, 260°C, V_d=25 V, P_{dc}=6 W/mm	III-V Lab	[8]
HTRB: High temperature reverse bias	T_{ch}=130°C, V_d = 46 V, V_g = -6 V	III-V Lab	[8]
	T_{ch}=175°C, V_d = 70 V, V_g = -5 V	Uni Bordeaux	[7]
HTRB step stress test	increase of V_d by 5 V every 30 min, V_{dmax}=120 V	FBH	[9]

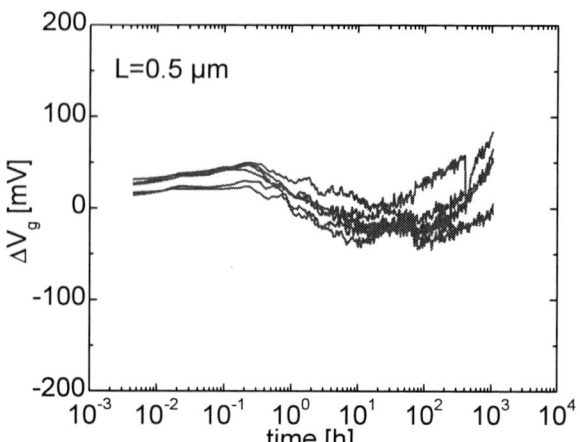

Fig. 1: Shift of gate voltage during high constant current DC-stress test (Idss) of 7.2 mm wide GaN HEMT.

Stress conditions: V_d=11V, T_{ch}=220°C, I_d=325 mA/mm

II. TECHNOLOGY

A. GaN50 Power Process

The GaN50 process is a 0.5 µm gate length technology optimized for GaN HEMTs in power applications, intended for instance in future telecommunication systems. The processing technology has previously been described by Waltereit et al. [11]. After processing, the HEMT devices with a gate periphery of 7.2 mm have been soldered in a standard package using AuSn. The thermal resistance of the package is 5.7 K/W and has been determined using infrared microscopy. The devices have not been exposed to any burn-in before DC- and RF-stress test.

B. GaN25 MMIC Process

The GaN25 process is a 0.25 µm gate length technology optimized for MMIC applications between 6 and 18 GHz. The AlGaN/GaN heterostructures are grown on semi-insulating SiC substrates by MOCVD with sheet resistance non-uniformities better than 2%. The growth procedure is optimized for both a highly insulating buffer as well as low trap densities. Processing is performed in microstrip transmission line technology consisting of frontside processing, substrate thinning to 100 µm, and backside processing including front-to-back substrate via holes. The MMICs feature thin film resistors, high-voltage capacitors and inductors for impedance matching to a 50 Ω environment. Device fabrication is performed using standard processing techniques involving electron-beam and optical lithography: stepper alignment for frontside and contact mask alignment for backside device definition. Special attention is paid towards performance and reliability optimization which is mainly achieved by a combination of epitaxial growth optimization as

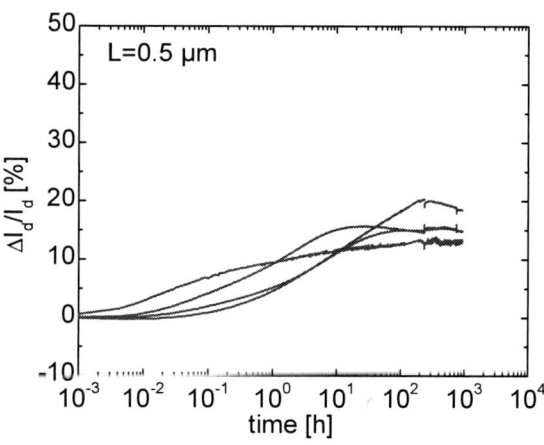

Fig. 2: Drain current during constant voltage DC-stress test (Idq) of 7.2 mm wide GaN HEMT.

Stress conditions: V_d=50 V, T_{ch}=155°C, I_d=50 mA/mm

well as modifications in the passivation and gate modules. The process technology exhibits good uniformity across a single wafer as well as high reproducibility from wafer to wafer [12]. After processing, 0.48 mm wide HEMT devices and single stage MMICs with a gate width of 1 mm have been diced and attached in a package using an epoxy resin with good thermal conductivity (3 W/mK). The thermal resistance of the package and the channel temperature has been determined using Raman spectroscopy measurements [13]. The devices have not been exposed to any burn-in before DC-and RF-stress test.

978-1-4244-5430-3/10 $26.00 © 2010 IEEE

III. RELIABILITY OF GAN50 PROCESS

A. DC-Stress Test

During large signal RF-operation the GaN HEMT is driven in different bias regimes and must be stable under high-current and high-voltages extremes. In order to test the long term stability in the different regimes, devices with a gate length of 0.5 μm and a gate periphery of 7.2 mm were stressed under three DC-operation conditions listed in Table 2. The channel temperature was determined using infrared microscopy. During the high-current low-voltage stress test (Idss stress test) the gate voltage is regulated to keep the drain current at a constant level. After temperature stabilization of around 0.5 h, the gate voltage initially decreases before it starts to increase as shown in Fig. 1. The overall change of the gate voltage at a channel temperature of 220°C after 3000h is less than 100 mV. A similar finding was observed during the high-voltage low-current stress test (Idq stress test) shown in Fig. 2. Under the fixed gate voltage setting of this test the drain current first increases before it starts to decrease. Under constant current operation this behaviour would correspond to an initial decrease followed by an increase of the gate voltage as observed during the Idss stress test. The drift of some electrical parameters of all three DC-stress tests was measured by intermediate measurements and are listed in Table 3. Except of a strong decrease of the gate and drain leakage current and a slight negative threshold voltage shift all other electrical parameters show little or no degradation after 1000 h of stress time. This is also the case for the high temperature reverse bias stress test (HTRB stress test) where the devices were stressed at a drain voltage of 100 V under off-state condition at a channel temperature of 175°C.

TABLE 2. DC STRESS TEST CONDITIONS

Stress test	V_d [V]	I_d [mA/mm]	T_{ch} [°C]	remark
IDSS stress test	11	325	220	I_d=const
Idq stress test	50	50	155	V_g=const
HTRB stress test	100	0	175	V_g=-7 V

TABLE 3. CHANGE OF ELECTRICAL PARAMETERS DURING STRESS

Parameter	Value at t0	ΔIdq 1000h (%)	ΔIdss 1000h (%)	ΔHTRB 1000h (%)
Vt	-1.46 V	+58 to -94 mV	-10 to -68 mV	-10 to -90 mV
Id leak	24 μA/mm	-92	-57	-83
Ig leak	25 μA/mm	-84	-43	-83
Idss	245 mA/mm	5	1	2
Gm max [mS/mm - %]	213 mS/mm	0	-2	-1
Rs	2.43 Ωmm	0	-1	0
Rd	4.21 Ωmm	3	2	4

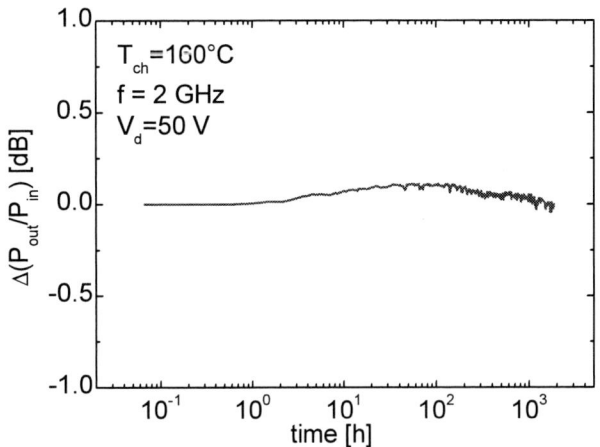

Fig. 3: The change of gain during RF-stress test at 2 GHz and V_d=50 V. The on-package RF-output power during stress was 7 W.

DUT: GaN HEMT, W=2.4 mm, L=0.5 μm.

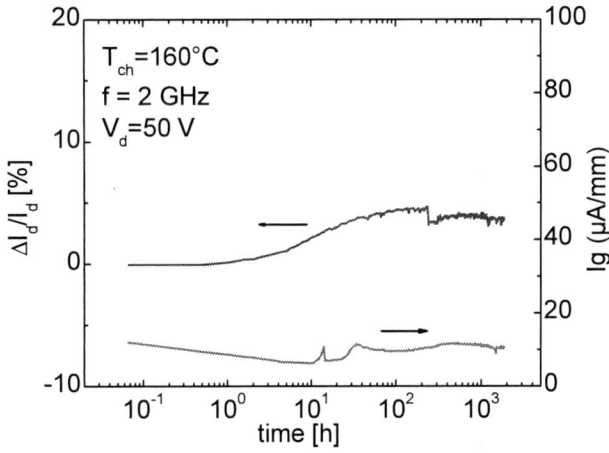

Fig. 4: The change of drain current and the absolute gate current during RF-stress test shown in Fig. 3.

DUT: GaN HEMT, W=2.4 mm, L=0.5 μm.

B. RF-Stress Test

The long term stability of a 2.4 mm power FET has been measured at T_{ch}=160°C, 2 GHz and a package level output power density of 2.5 W/mm using an external matching circuit. As shown in Fig. 3, the degradation of the gain is less than 0.3 dB after 2000 h of stress time. The drain current during RF-stress test shown in Fig. 4 first increases and then decreases as observed during the DC-stress. The gate current stays lower than 20 μA/mm for the whole duration of the test.

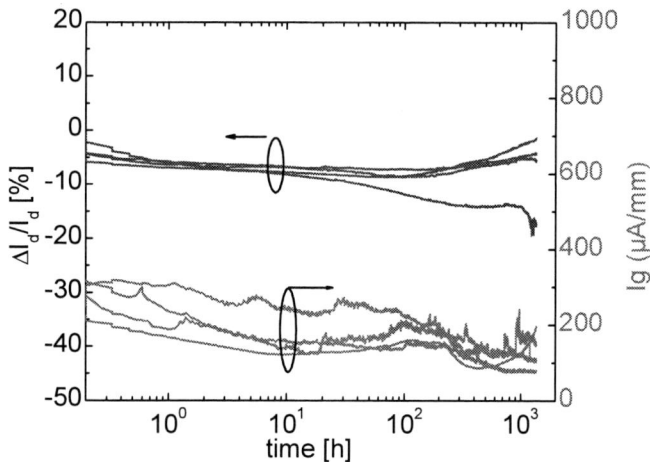

Fig. 5: Drain current degradation during constant voltage DC-stress test (Idq settings) of GaN HEMTs.

Stress conditions: V_d=30V, T_{ch}=260°C, I_d=150 mA/mm

DUT: GaN HEMT, W=0.48 mm, L=0.25 µm.

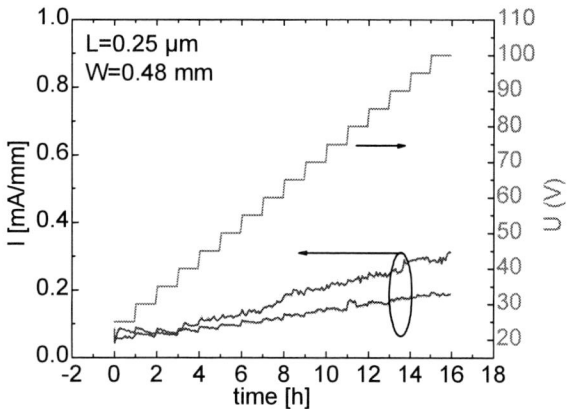

Fig. 6: Gate and drain current densities during drain voltage step-stress-test of two devices at T_{ch}=150°C.

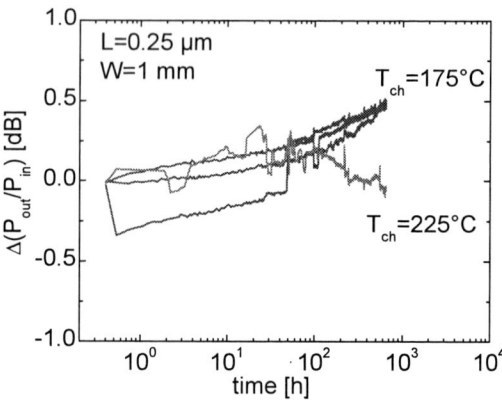

Fig. 7: The change of gain during RF-stress test of one stage MMIC at 10 GHz and V_d=30 V.

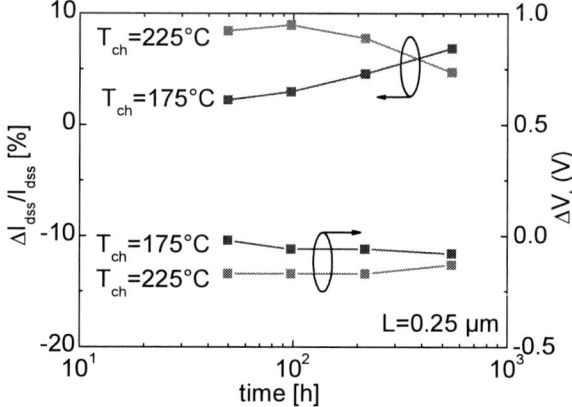

Fig. 8: The relative change of Idss measured at V_d=10 V and V_g=0V and the shift of threshold voltage during RF-stress test shown in Fig. 7.

IV. RELIABILITY OF GAN25 PROCESS

A. DC-Stress Test of GaN HEMT

The long term stability of the GaN25 process was investigated using an Idq stress test at a drain voltage of 30V and a drain current of 150 mA/mm at a channel temperature of 260°C as shown in Fig. 5. The drain current first decreases and then tends to increase. One device reaches almost the failure criterion of 20% drain current degradation after 1500h of stress time, whereas the other three devices degrade less than 10%. The gate current tends to decrease during the stress and is lower than 0.4 mA/mm during the whole test.

The drain voltage robustness of the GaN25 process was tested using the high temperature reverse bias step stress test. During this test the drain voltage is increased by 5 V every hour under off-state conditions at a channel temperature of 150°C. As shown in Fig. 6, the gate and drain current densities are always lower than 0.3 mA/mm and do not increase during the plateau phase even at the highest stress voltage of 100 V.

B. RF-Stress Test of one stage MMIC

The reliability under RF-operation of the GaN25 technology was tested using one-stage MMICs having a gate periphery of 1 mm. Fig. 7 shows the result from a RF-test at 10 GHz with a supply voltage of 30 V at a channel temperature of 175 and 225°C as determined by Raman spectroscopy [12]. At channel temperature of 175°C the gain increases by about 0.5 dB after 900 h of stress. At 225°C channel temperature the gain first increases and then starts to decrease. During the RF-stress test the MMICs were periodically cooled down and some electrical parameters were measured automatically on the test bench. Fig. 8 shows the result of the Idss and the threshold voltage intermediate measurements. At T_{ch}=175°C a negative threshold voltage shift of about 100 mV and an Idss increase by about 10% were observed. At 225°C the threshold voltage shows the same shift, but stabilizes after 50 h of stress and the Idss value first increases and then starts to decrease.

V. Discussion

A. GaN50 Process

Very good long term stability under DC- and RF-operation at a drain voltage of 50V and a channel temperatures above 150°C has been achieved. There is a good correlation between the drain current degradation during DC-stress and gain degradation during RF-stress. Both first increase and then decrease indicating the involvement of two degradation mechanisms. Since the two degradation mechanisms have opposite drift behaviour it is rather difficult to extract life time values from the DC- and RF- stress tests results. The root cause of these two degradation mechanisms is unclear and is still under investigation. The first one leads to a negative shift of the threshold voltage and thus to a drain current increase. This shift is probably not caused by the charging of traps since it cannot be recovered by UV light illumination or temperature storage. Possibly the negative threshold voltage shift is caused by the generation of positively charged traps below the gate contact. The second degradation mechanism leads to a decrease of the drain current. The latter mechanism can be partly recovered by temperature storage and is therefore partly caused by the charging of slow traps with capture or emission time constants in the range of hours [10].

B. GaN25 Process

As for the GaN50 technology the gain first increases and then decreases during RF-stress testing of the MMICs at 10 GHz. Therefore we assume that for the GaN25 and GaN50 technology the same two degradation mechanisms are involved. The life time of GaN HEMTs fabricated using the GaN25 process exceeds 1000 h at a channel temperature of 260°C and a drain voltage of 30 V. As shown by the intermediate measurements the initial increase of drain current and output power are caused by a negative threshold voltage shift. It was also found that during RF-stress the gain degradation correlates well to the Idss degradation. A good correlation between Idss degradation during DC-stress and gain degradation during RF-stress was also observed by Singhal et al. [14]. The HTRB step stress test shows no indication of the onset of irreversible gate current increase which may be caused by the inverse piezoelectric field effect at high drain voltages [15].

Conclusion

The reliability of GaN HEMTs and MMICs fabricated using GaN25 and GaN50 processes have been investigated by DC- and RF-life time tests. Both process technologies show very promising long term stability under DC and RF-operation. GaN HEMT transistors fabricated using the GaN50 process. The DC-life time of a GaN HEMTs fabricated using the GaN25 process exceeds 1000 h at a channel temperature of 260°C and a drain voltage of 30 V.

Acknowledgements

The authors would like to thank J. Pomeroy for the Raman spectroscopy measurements. This work was financially supported by the German Ministry of Education and Research (BMBF), the German Ministry of Defence (BMVg) and the European Space Agency (GREAT2 project, contract 21499/08/NL/PA).

References

[1] H. Blanck, J. Splettstöser, D. Floriot, "GaN technology for RF electronics: Development Status in Europe, " in Technical Digest of IEEE CSIC Symposium, 2009, pp. 22-25.

[2] G. Gauthier and F. Reptin, "KORRIGAN: Development of GaN HEMT Technology in Europe," in Proc. CS Mantech Conf., 2006, pp. 49–51.

[3] http://www.great2-project.com

[4] G. Meneghesso, G. Verzellesi, F. Danesin, F. Rampazzo, F. Zanon, A. Tazzoli, M. Meneghini, E. Zanoni, "Reliability of GaN High-Electron-Mobility Transistors: State of the Art and Perspectives," IEEE Transactions On Device And Materials Reliability, 2008, pp. 332-343.

[5] M. Dammann, M. Cäsar, P. Waltereit, W. Bronner, H. Konstanzer, R. Quay, S. Müller, M. Mikulla, O. Ambacher, P. van der Wel, T. Rödle, R. Behtash, F. Bourgeois, K. Riepe, „Reliability of AlGaN/GaN HEMTs under DC- and RF-operation," in ROCS Workshop Greensboro, NC, USA 11 Oct 2009, pp 19-32.

[6] A. Chini, V. Di Lecce, M. Esposto, G. Meneghesso, E. Zanoni, "RF Degradation of GaN HEMTs and its correlation with DC stress and I-DLTS measurements," in Proceedings of the 4th European Microwave Integrated Circuits Conference, 2009, pp. 132-135.

[7] N. Malbert, N. Labat, A. Curutchet, C. Sury, V. Hoel, J.-C. de Jaeger, N. Defrance, Y. Douvry, C. Dua, M. Oualli, C. Bru-Chevallier, J.-M. Bluet, W. Chikhaoui, "Characterisation and modelling of parasitic effects and failure mechanisms in AlGaN/GaN HEMTs," Microelectronics Reliability, 2009, 49, pp. 1216-1221.

[8] A. Sozza, C. Dua, E. Morvan, B. Grimber, S.L. Delage, "A 3000 hours DC Life Test on AlGaN/GaN HEMT for RF and microwave applications," Microelectronics Reliability, 2005, 45, pp. 1617-1621.

[9] P. Ivo, A. Glowacki, R. Pazirandeh, E. Bahat-Treidel, R. Lossy, J. Würfl, C. Boit, G. Tränkle, "Influence of GaN cap on robustness of AlGaN/GaN HEMTs," in Reliability Physics Symposium, 2009, pp. 71-76.

[10] M. Dammann, W. Pletschen, P. Waltereit,W. Bronner, R. Quay, S. Müller, M. Mikulla, O Ambacher, P.J. van der Wel, S. Murad, T. Rödle, R Behtash, F Bourgeois, K. Riepe, M. Fagerlind, E.Ö. Sveinbjörnsson, "Reliability and Degradation Mechanism of AlGaN/GaN HEMTs for Next Generation Mobile Communication Systems," Microelectronics Reliability, 2009, 49, pp. 474-477.

[11] P. Waltereit, W. Bronner, R. Quay, M. Dammann, R. Kiefer, S. Müller, M. Musser, J. Kühn, F. van Raay, M. Seelmann, M. Mikulla, O. Ambacher, F. van Rijs, T. Rödle, and K. Riepe, "GaN HEMT and MMIC development at Fraunhofer IAF: performance and reliability," Phys. Status Solidi, 2009, A 206, pp. 1215–1220.

[12] P. Waltereit, W. Bronner, R. Kiefer, R. Quay, J. Kühn, F. van Raay, M. Dammann, S. Müller, M. Mikulla, O. Ambacher, "High efficiency and low leakage AlGaN/GaN HEMTs for a robust, reproducible and reliable X-band MMIC space technology", Mantech 2010, accepted for publication.

[13] M. Kuball, J. M. Hayes, M. J. Uren, T. Martin, J. C. H. Birbeck, R. S. Balmer, B. T. Hughes, "Measurement of Temperature in Active High-Power AlGaN/GaN HFETs Using Raman Spectroscopy," IEEE Electron Device Letters, 2002, pp. 7-9.

[14] S. Singhal, T. Li, A. Chaudhari, A.W. Hanson, R. Therrien, J.W. Johnson, W. Nagy, J. Marquart, P. Rajagopal, J.C. Roberts, E.L. Piner, I.C. Kizilyalli, K.J. Linthicum, "Reliability of large periphery GaN-on-Si HFETs," Microelectronics Reliability, 2006, 46, pp. 1247–1253.

[15] Joh et al., J. A. del Alamo, " Critical Voltage for Electrical Degradation of GaN High-Electron Mobility Transistors," IEEE Electron Device Letters, 2008, pp. 287-289.

Effect of Trapping on the Critical Voltage for Degradation in GaN High Electron Mobility Transistors

Sefa Demirtas, Jesús A. del Alamo

Microsystems Technology Laboratories
Massachusetts Institute of Technology
Cambridge, MA 02139
TEL: +1 ‐ 617 ‐ 253 ‐ 1620, FAX: +1 ‐ 617 ‐ 258 ‐ 7393, E ‐ mail: sefa@mit.edu

Abstract— We have performed $V_{DS} = 0$ V and OFF–state step–stress experiments on GaN–on–Si and GaN–on–SiC high electron mobility transistors under UV illumination and in the dark. We have found that for both stress conditions, UV illumination decreases the critical voltage for the onset of degradation in gate current in GaN–on–Si HEMTs in a pronounced way, but no such decrease is observed on SiC. This difference is attributed to UV–induced electron detrapping, which results in an increase in the electric field and, through the inverse piezoelectric effect, in the mechanical stress in the AlGaN barrier of the device. Due to the large number of traps in GaN–on–Si, this effect is clearer and more prominent than in GaN–on–SiC, which contains fewer traps in the fresh state.

Keywords– GaN HEMTs, critical voltage, degradation, UV illumination, detrapping

I.   INTRODUCTION

GaN high electron mobility transistors (HEMTs) are promising devices for high power and high frequency applications. Si is an attractive substrate for GaN HEMTs because of its lower cost, availability in large diameters and sophisticated technology base. However, the larger lattice and thermal mismatch between GaN and Si as compared to the more commonly used substrate, SiC, results in more defects and certain kinds of dislocations [1–2]. Although it has been recently reported that no misfit dislocations are observed and that screw dislocations are present at a density lower than 10^7 cm^{-2} in GaN on Si, the density of edge and mixed dislocations are on the order of $1–5 \times 10^9$ cm^{-2} whereas it is in the 10^8 cm^{-2} range for GaN on SiC [1–3]. The concern is that these defects may compromise the reliability of GaN HEMTs on Si.

We have previously shown that GaN–on–Si HEMTs are affected by a similar degradation mechanism as GaN–on–SiC HEMTs, namely a sharp increase of the gate current at a certain voltage that has been termed the *critical voltage* [4–5]. We have attributed this degradation to defect generation due to excessive mechanical stress introduced by the inverse piezoelectric effect at high voltage [4–5]. In GaN–on–Si HEMTs we have observed relatively high critical voltages with a rather wide range of values even for devices in the same reticle. Separately, we detected significant electron trapping associated with pre–existing traps in virgin devices, much more than in typical GaN–on–SiC HEMTs [6]. In this work, we report an intriguing finding: the high number of traps that are present in GaN–on–Si HEMTs seems to be partially responsible for the high critical voltages for the onset of electric field–induced degradation of the gate current. We postulate that electron trapping during electrical stress leads to lower electric fields and increased robustness against gate current degradation. This finding should be of wide applicability regardless of the characteristics of the substrate.

II.   EXPERIMENTAL

We studied experimental Al$_x$Ga$_{1-x}$N/GaN HEMTs on Si with a 17.5 nm thick $x = 0.26$ AlGaN barrier. The gate width is 2×25 µm and the gate–source, gate–drain separation and the gate length are 1, 3 and 0.5 µm, respectively. The devices include a source-connected field plate [7–8]. These devices are representative of the capabilities of the GaN–on–Si technology. They exhibit $P_{out} = 3.9$ W/mm and *PAE* = 62% at 2.14 GHz under $V_{DS} = 28$ V. They also show excellent electrical reliability: an average lifetime in excess of 10^7 hours with $E_a = 1.7 – 2.0$ eV has been reported [9–11].

The Al$_x$Ga$_{1-x}$N/GaN HEMTs on SiC substrate used in this study have a 16 nm AlGaN barrier with $x = 0.28$ [12]. The gate width is also 2×25 µm, however the drain and source are located symmetrically at a distance of 2 µm from the gate. Since the peak electric field that is responsible for the onset of inverse piezoelectric degradation is encountered at the vicinity of the gate [5], the difference in gate–source and gate–drain spacing is acceptable when comparing the reliability of both technologies. GaN–on–SiC HEMTs have an integrated field plate [12–13], which plays a similar role to the source field plate in GaN–on–Si HEMTs. These experimental devices exhibit $P_{out} = 8$ W/mm and *PAE* = 62% at 10 GHz under $V_{DS} = 40$ V.

We performed $V_{DS} = 0$ V step–stress experiments where we grounded the drain and source and stepped the gate voltage starting from –5 V to –80 V by decreasing 1 V every 30 seconds. Important DC figures of merit were continuously measured by a benign characterization suite. In this stress scheme, both the drain and source sides of the gate are stressed with increasing electric field over time. This is a very harsh stress bias condition when compared with the normal operation of the device. We use this stress because it reveals important aspects of the physics of degradation in GaN HEMTs in an accelerated manner [4–5].

978-1-4244-5430-3/10 $26.00 © 2010 IEEE

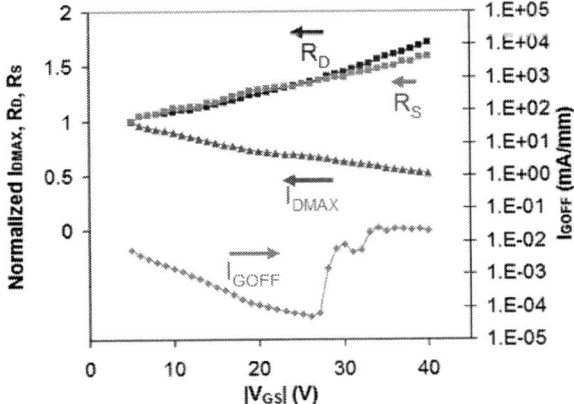

Fig. 1. Normalized I_{DMAX}, R_S, R_D (left axis) and I_{GOFF} (right axis) as a function of $|V_{GS}|$ in a $V_{DS} = 0$ V step–stress experiment on a GaN–on–Si HEMT. V_{GS} is stepped from -5 V by -1 V steps every 30 seconds.

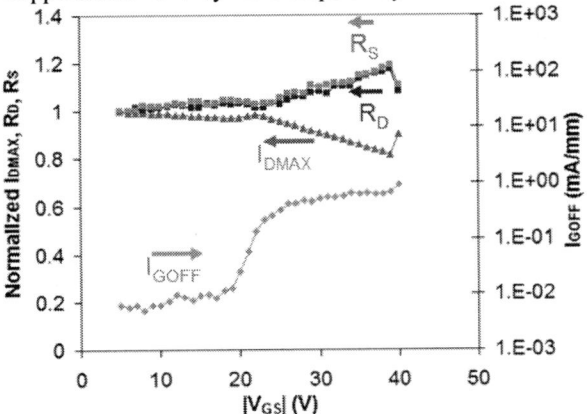

Fig. 2. Normalized I_{DMAX}, R_S, R_D (left axis) and I_{GOFF} (right axis) as a function of $|V_{GS}|$ in a $V_{DS} = 0$ V step–stress experiment on a GaN–on–SiC HEMT. V_{GS} is stepped from -5 V by -1 V steps every 30 seconds.

We also performed OFF–state step–stress experiments, which are more representative of the normal use conditions for these HEMTs. In this approach, V_{GS} is kept constant at -5 V while V_{DS} is stepped in 1 V steps from 5 V to 80 V every 30 seconds. Since the voltage across the gate and source is constant throughout the experiment, it is the drain side of the device that is exposed to high electrical stress. This difference is also why $V_{DS} = 0$ V stress is harsher than OFF–state stress [4]. On the other hand, $V_{DS} = 0$ V and OFF–state step–stress experiments are similar in that the channel is always OFF and there is negligible drain current and negligible self heating taking place in the device. This condition helps to isolate the effects of current and temperature from that of electric field, as the field is found to be the main cause of degradation in GaN HEMTs [4–5].

III. RESULTS

Figs. 1 and 2 show the results of typical $V_{DS} = 0$ V step–stress experiments in a GaN–on–Si HEMT and a GaN–on–SiC

This work was funded by DARPA (WBGS program through ARL) and Office of Naval Research (MURI)

HEMT, respectively. The graphs show the maximum current, I_{DMAX} (defined at $V_{DS} = 5$ V and $V_{GS} = 2$ V); source resistance, R_S and drain resistance, R_D on the left axis (all normalized to their initial values) and the OFF–state gate current, I_{GOFF} (defined at $V_{DS} = 0.1$ V and $V_{GS} = -5$ V) on the right axis as a function of $|V_{GS}|=|V_{GD}|$. These experiments stop at -40 V.

In both devices, at a certain voltage, there is a sharp rise in I_{GOFF} that is irreversible. This voltage is what we term the critical voltage, V_{CRIT}. The increase in I_{GOFF} is attributed to the creation of defects in the AlGaN barrier layer by excessive mechanical stress introduced by the inverse piezoelectric effect [4–5]. In our experiments, we see this pattern of degradation in both types of devices also under OFF–state step–stress conditions although the degradation is somewhat larger in the $V_{DS} = 0$ V condition due to its harsher nature (not shown).

While the distinct critical voltage behavior of the degradation of I_{GOFF} in GaN–on–Si HEMTs is similar to that of GaN–on–SiC devices, there are also marked differences between the patterns of degradation in both devices. For stress voltages below the critical voltage, I_{GOFF} and I_{DMAX} are observed to decrease starting from the beginning of the stress, and R_D and R_S increase in GaN–on–Si HEMTs (Fig. 1). In GaN–on–SiC HEMTs these figures of merit remain largely unchanged below V_{CRIT} (Fig. 2). Along with other evidence [6], this unique behavior of GaN–on–Si HEMTs derives from the prominent electron trapping that these devices suffer from in their virgin state. Another important observation from GaN–on–Si HEMTs that is relevant here is shown in Fig. 3. For these devices, V_{CRIT} varies significantly from device to device even when they are nominally identical and at close proximity on the wafer [6].

In order to investigate the role of trapping on the critical voltage, we performed identical stress experiments, except for UV illumination, on pairs of neighboring identical devices. One of these experiments was performed in the dark whereas in the other, its neighbor was illuminated by 365 nm UV light. The intensity of the UV light source at a distance of 4 cm from the device was measured to be 1.56 mW/cm^2, which corresponds to 2.9×10^{15} photons/cm^2–s. Separately we verified that 365 nm UV light greatly enhances electron detrapping [6, 14]. Fig. 4 shows a typical result in GaN–on–Si HEMTs under $V_{DS} = 0$ V

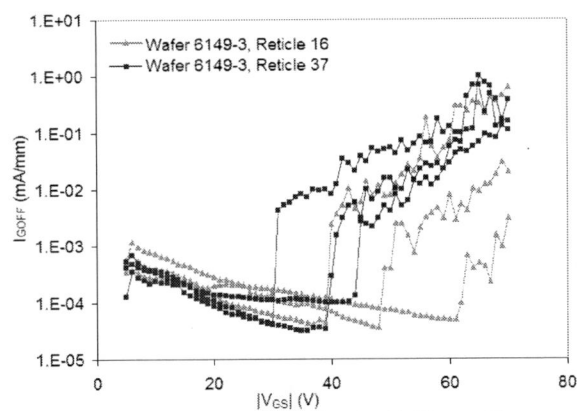

Fig. 3. I_{GOFF} vs. stress $|V_{GS}|$ for six GaN–on–Si devices in typical $V_{DS} = 0$ V step–stress experiments. The distribution of V_{CRIT} is very broad even for devices in close proximity. Very high V_{CRIT} values are also observed.

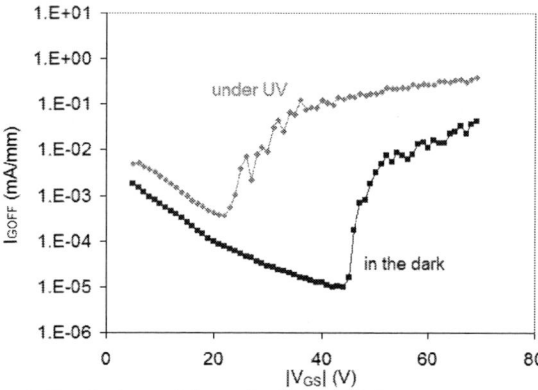

Fig. 4. Evolution of I_{GOFF} in $V_{DS} = 0$ V step–stress experiments on two neighboring GaN–on–Si devices, where one of them was stressed under UV illumination (365 nm) and the other in the dark. Under UV illumination V_{CRIT} is significantly smaller than in the dark.

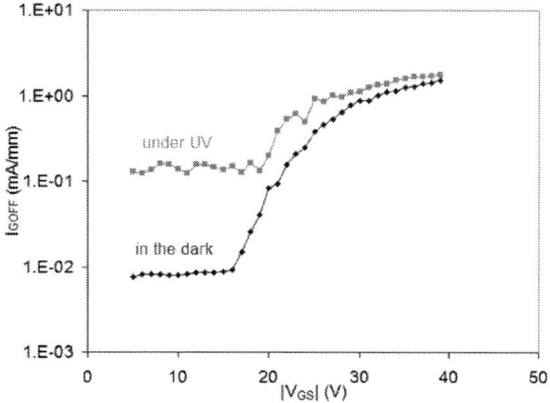

Fig. 5. Evolution of I_{GOFF} in $V_{DS} = 0$ V step–stress experiments on two neighboring GaN–on–SiC devices, where one of them was stressed under UV illumination (365 nm) and the other in the dark. No significant effect of UV on V_{CRIT} is observed.

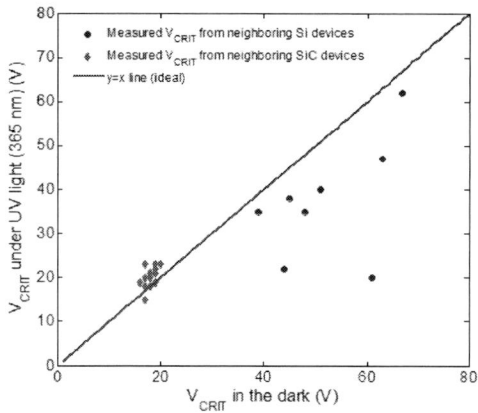

Fig. 6. Comparison of V_{CRIT} in $V_{DS} = 0$ V step–stress experiments for closest–neighbor pairs of devices. One device from each pair is stressed under UV and the other in the dark. UV illumination generally results in a lower value of V_{CRIT} for GaN–on–Si devices (8 pairs shown). V_{CRIT} values in the dark and under illumination are negligibly different for GaN–on–SiC devices (15 pairs shown).

step–stress conditions. Under illumination, V_{CRIT} is significantly smaller than in the dark. In contrast with this observation, in a similar experiment, V_{CRIT} in typical GaN–on–SiC HEMTs did not change (Fig. 5). UV light also produces an increase in I_{DMAX} and a reduction in R_D and R_S relative to the dark, which proves that UV light reaches close to the intrinsic device and successfully enhances detrapping of electrons during stress.

Fig. 6 shows the results obtained on several such pairs of $V_{DS} = 0$ V step–stress experiments. Each point in the figure corresponds to a pair of neighboring identical devices. The abscissa is the critical voltage when a device was stressed in the dark and the ordinate is the critical voltage when its neighbor was stressed under UV illumination. On average, V_{CRIT} in GaN–on–Si HEMTs is 29% smaller under UV illumination while it is about the same as in the dark in GaN HEMTs on SiC. The spread in V_{CRIT} values in the dark and under light for GaN on Si HEMTs in Fig. 6 is rather broad reflecting the relatively wide distribution of V_{CRIT} values that was shown in Fig. 3 even among devices spaced by short distances.

In order to further explore this shift in V_{CRIT} under UV illumination in GaN–on–Si HEMTs, similar experiments were performed under different stress bias conditions and with different wavelengths. Fig. 7 shows OFF–state step–stress experiments performed on two neighboring devices on Si in the dark and under UV light illumination. In this stress condition, a high electric field appears on the drain side of the gate, which is not covered by a field plate and is ensured to be exposed to UV illumination. An average reduction of V_{CRIT} of 17% was observed as a result of UV illumination (Fig. 8).

The effects of 254 nm UV as well as visible microscope light were also investigated although results are not shown here. The intensity at 4 cm from the devices was measured to be 1.62 mW/cm^2 for 254 nm UV light, which corresponds to 2.1×10^{15} photons/cm^2–s. We found that 254 nm UV light was less effective in detrapping electrons due to the different absorption rate by the AlGaN barrier and/or less number of incident photons, hence it had a smaller effect on V_{CRIT} (only 11% decrease in V_{CRIT} under $V_{DS} = 0$ V step–stress). Visible microscope light resulted in no shift in the average V_{CRIT}, which is consistent with the observation that it induces negligible detrapping.

IV. DISCUSSION

Our experiments suggest that native traps (those that exist in the virgin device) play a role in determining the critical voltage for high–voltage gate current degradation of GaN HEMTs. We know from earlier experiments that the application of high voltage results in electron trapping in GaN HEMTs [15]. This trapping reduces the sheet carrier concentration in the high field region of the channel and, in consequence, the peak electric field at the gate edge. This manifests itself, among other ways, in a reduction in the gate current as the voltage increases in electrical stress experiments as can be seen in Figs. 1 and 4. A similar reduction in gate leakage current has also been observed by other authors on GaN–capped devices on SiC substrates after ON and OFF state

978-1-4244-5430-3/10 $26.00 © 2010 IEEE

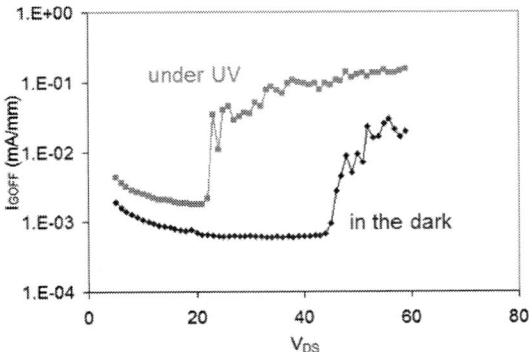

Fig. 7. OFF state step–stress experiments on two neighboring GaN–on–Si devices, where one of them was stressed under UV illumination (365 nm) and the other in the dark. Similar to $V_{DS} = 0$ V step–stress, V_{CRIT} is significantly smaller under UV illumination than in the dark.

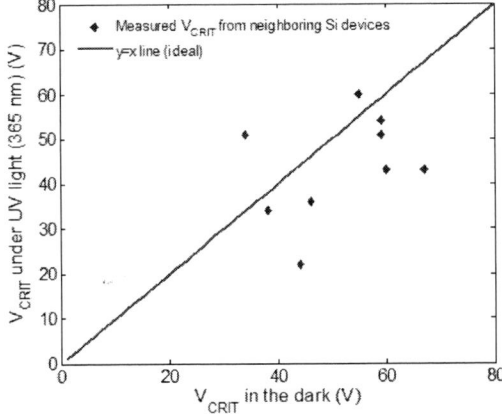

Fig. 8. Comparison of V_{CRIT} under OFF–state step–stress experiments for closest–neighbor pairs of devices on Si when one device from each pair is stressed under UV and the other in the dark. UV illumination in OFF–state stress generally results in a lower reduction in value of V_{CRIT} than $V_{DS} = 0$ V stress for GaN–on–Si devices (9 pairs shown).

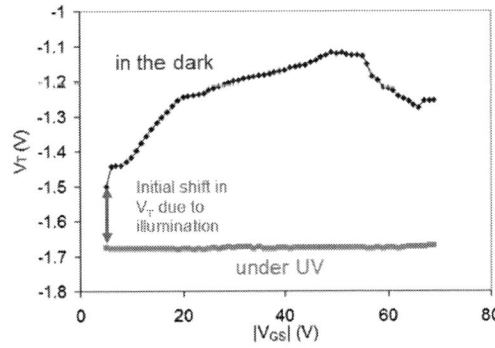

Fig. 8. Evolution of V_T in the $V_{DS} = 0$ V step–stress experiments shown in Fig. 4. Under UV illumination, V_T stays rather constant as opposed to shifting in the positive direction due to trapping in the dark. The initial shift in V_T observed for the neighboring device in the dark can be considered as a relative measure of UV–induced detrapping from the native traps in the fresh device.

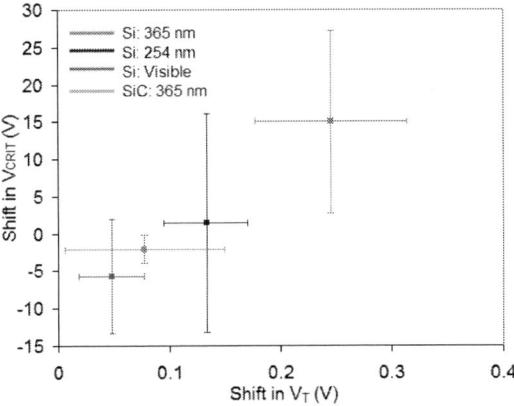

Fig. 7. Correlation between the average initial shift in V_T under several illumination schemes and the resulting shift in V_{CRIT} obtained from $V_{DS} = 0$ V step–stress experiments on pairs of neighboring devices. The error bars represent one standard deviation on each side around the mean value. Under conditions in which V_T shifts in a substantial way as a result of light illumination, V_{CRIT} is also seen to be reduced.

stress experiments [16]. The application of UV illumination causes electron detrapping and increases the electric field. This is seen in the increase in the gate leakage current observed at the beginning of the experiments in Figs. 4 and 7, which is not of a photovoltaic nature [17]. Through the inverse piezoelectric effect, the increase in the electric field should result in an increase in the elastic energy density stored in the AlGaN barrier, which will therefore reach the critical value for defect formation at a lower critical voltage [18].

A correlation between trap density in the virgin device and the shift in V_{CRIT} can be obtained by examining the shift in the threshold voltage induced by UV light. In our hypothesis, UV illumination results in electron detrapping. This should produce a negative shift in V_T. The larger the shift, the higher the concentration of traps in the device. Fig. 9 shows the evolution of V_T for the $V_{DS} = 0$ V step–stress experiments illustrated in Fig. 4. In the dark, V_T shifts positive as traps get filled with

electrons. Under UV, electron trapping is largely suppressed and V_T remains unchanged throughout the experiment. The amount of V_T change at the beginning of the experiment can be considered as a qualitative measure of the number of traps in the virgin device. Hence, the correlation between trapping and change in V_{CRIT} can be verified by comparing the initial shift in V_T as a result of illumination and the corresponding decrease in V_{CRIT}. Fig. 10 displays the average initial shift in V_T on the horizontal axis and the corresponding average shift in V_{CRIT} on the vertical axis with error bars. This figure shows that they both correlate and that the correlation for different types of illumination follows a general trend that is also consistent. The shift in V_{CRIT} and V_T are both minor in GaN–on–SiC HEMTs regardless of the illumination scheme, which is consistent with the fact that they have few native traps.

V. CONCLUSIONS

In conclusion, we have studied the degradation of GaN–on–Si and GaN–on–SiC HEMTs under UV illumination and in the dark. We have observed a pronounced decrease in V_{CRIT} for GaN–on–Si HEMTs under UV light in both $V_{DS} = 0$ V and OFF–state step–stress experiments. We have proposed a mechanism where UV–induced electron detrapping from native traps increases the sheet carrier concentration in the channel, which results in an increase in the peak electric field and in the elastic energy stored in the AlGaN barrier of the HEMT through the inverse piezoelectric effect. This decreases the critical voltage for defect formation. A similar shift in V_{CRIT} was not observed in GaN–on–SiC HEMTs where trapping in the virgin devices is minimal. Our work suggests that the existence of an increased number of traps in GaN–on–Si HEMTs might enhance the robustness of the devices to high voltage degradation of the gate current.

ACKNOWLEDGMENT

This work was funded by the DARPA WBGS program (Mark Rosker, program manager) under ARL contract #W911QX–05–C–0087 (Alfred Hung, COTR) and by the Office of Naval Research Grant #N00014–08–1–0655 (Paul Maki and Harry Dietrich, Program Managers). We also acknowledge collaboration with Nitronex Corporation and TriQuint Semiconductor and discussions with Dr. Jungwoo Joh.

REFERENCES

[1] P. Rajagopal et al., "Large–area, device quality GaN on Si using a novel transition layer scheme," Mater. Res. Soc. Sympos. Proc. vol. 743, L1. 2 (2002).

[2] J. C. Roberts, J. W. Cook, Jr., P. Rajagopal, E. L. Piner, and K. Linthicum, "AlGaN transition layers on Si (111) substrates – observations of microstructure and impact on material quality," Mater. Res. Soc. Sympos. Proc. vol. 1068, C06–03 (2008).

[3] R. Quay, *Gallium Nitride Electronics*, illustrated ed. Berlin: Springer, 2008.

[4] J. Joh and J. A. del Alamo, "Critical voltage for electrical degradation of GaN high–electron mobility transistors," IEEE Electron Device Lett., vol. 29, no.4, pp. 287–289 (2008).

[5] J.A. del Alamo and J. Joh, "GaN HEMT reliability," Microelectronics Reliability, vol. 49, Issues 9–11, 20th Europ. Sympos. Reliab. Electron Devices, Fail. Phys. and Analysis, pp. 1200–1206 (2009).

[6] S. Demirtas and J. A. del Alamo, "Critical voltage for electrical reliability of GaN high electron mobility transistors on Si substrate," Reliab. of Compound Semicond. Workshop, pp. 53–56 (2009).

[7] A. W. Hanson et al., "Development of GaN transistor process for linear power applications," Comp. Sem. MANTECH, pp. 107–110 (2004).

[8] J. W. Johnson et al., "Material, process and device development of GaN–based HFETs on silicon substrates," Electrochem. Soc. Proc., vol. 2004–06, pp. 405–419 (2004).

[9] E.L. Piner et al., "Device degradation phenomena in GaN HFET technology: status, mechanisms, and opportunities," Internat. Electron Devices Mtg., pp.1–4 (2006).

[10] S. Singhal., et al., "Reliability of large periphery GaN–on–Si HFETs," Reliab. of Compound Semicond. Workshop, pp. 135–149 (2005).

[11] S. Singhal et al., "Qualification and reliability of a GaN process platform," Comp. Sem. MANTECH, pp. 83–86 (2007).

[12] U. Chowdhury et al., "TEM observation of crack– and pit–shaped defects in electrically degraded GaN HEMTs," IEEE Electron Device Lett., vol. 29, pp. 1098–1100 (2008).

[13] D. Fanning et al., "High voltage GaAs pHEMT technology for S–band high power amplifiers," Comp. Sem. MANTECH, pp. 173–176 (2007).

[14] G. Koley, V. Tilak, L. F. Eastman and M. G. Spencer, "Slow transients observed in AlGaN/GaN HFETs: Effects of SiN_x passivation and UV illumination," IEEE Trans. Electron Devices, vol. 50, no.4, pp. 886–893 (2003).

[15] J. Joh and J. A. del Alamo, "Impact of electrical degradation on trapping characteristics of GaN high electron mobility transistors," Internat. Electron Devices Mtg., pp.1–4 (2008).

[16] G. Meneghesso, F. Rampazzo, P. Kordos, G. Verzellesi and E. Zanoni, "Current collapse and high–electric–field reliability of unpassivated GaN/AlGaN/GaN HEMTs," IEEE Trans. Electron Devices, vol. 53, no. 12, pp. 2932–2941 (2006).

[17] J. Joh, L. Xia and J. A. del Alamo, "Gate current degradation mechanisms of GaN high electron mobility transistors," Internat. Electron Devices Mtg., pp. 385–388 (2007).

[18] J. Joh and J. A. del Alamo, "A model for the critical voltage for electrical degradation of GaN high electron mobility transistors," Reliab. of Compound Semicond. Workshop, pp. 3–6 (2009).

Reliability Assessment in Different HTO Test Conditions of AlGaN/GaN HEMT's

N. Malbert, N. Labat, A. Curutchet, C. Sury

IMS Laboratory-University of Bordeaux
33405 Talence cedex, France
phone: +33 0(5) 4000 2859; fax: +33 0(5) 5637 1545; e-mail: nathalie.malbert@ims-bordeaux.fr

V. Hoel, J.-C. de Jaeger, N. Defrance , Y. Douvry

Institut d'Electronique, de Microélectronique et de Nanotechnologie
IEMN – University of Lille
Villeneuve d'Ascq, France

C. Dua, M. Oualli, M. Piazza

Alcatel-Thales III-V Lab, c/o Alcatel CIT
Marcoussis, France

C. Bru-Chevallier, J.-M. Bluet, W. Chikhaoui

Institut des Nanotechnologies de Lyon
INL – University of Lyon
Villeurbanne, France

Abstract — This study reports on a reliability investigation of AlGaN/GaN HEMTs submitted to life tests in High Temperature Operating (HTO) conditions at 150°C, 175°C, 275°C and 320°C. These life tests showed two different degradation steps of the drain current. One is occurring in the first tens of hour of the life test and characterized by a decrease of the drain current. The evolution of the electrical characteristics during ageing does not depend on the bias conditions but rather on the channel temperature. This degradation mode is characterized by a high activation energy of 1.2eV. The small changes of electrical characteristics observed during the life tests results from a combination of trap-related effects before stabilization sets in.

The second failure mechanism observed during the HTO tests at 275°C and 320°C results in a higher drain current decrease. Moreover, no stabilization of the parameter drifts was observed before the end of the tests. Pulsed I-V measurements show a large evolution of gate lag and drain lag rates after HTOT275 and HTOT320 life tests. LF 1/f noise after the life tests at high temperature drastically increased by more than two orders of magnitude while it hardly changed after 2000 hours of life test at 150°C and 175°C. It results that the temperature is considered as an acceleration factor of the degradation affecting the conduction channel. TEM observations revealed similar damages in the gate finger cross-section of aged devices after the HTOT275 and HTOT320 life tests. These results could point out a degradation mechanism associated with the inverse piezoelectric effect.

Keywords-AlGaN/GaN HEMT – Reliability – HTO Life test – LF noise- DLTS-

I. INTRODUCTION

GaN based high electron mobility transistors (HEMTs) have emerged as very attractive candidates for high temperature, high-voltage, and high-power operation at microwave as well as lower frequencies [1, 2].

For this reason, GaN-HEMT reliability has been addressed by an increasing number of works. Physical mechanisms likely to explain degradation of electrical parameters are related to the Schottky or ohmic contacts stability and the influence of deep levels and surface and/or interface traps [3, 4, 5]. To gain more insight into possible physical origins of GaN HEMT failure mechanisms, research efforts continue to focus on the implementation of various life tests in on and off-state DC and RF stress conditions. The interpretation of results includes the possibility that enhancement of the electric-field induced strain and its related relaxation becomes relevant [6].

The motivation for this work is the in depth understanding of the physical mechanisms that affect the carrier transport properties in AlGaN/GaN HEMTs and then penalise their performances and their reliability. As these devices suffer from the effect of traps and surface states partially coupled with material spontaneous polarization and piezoelectric effects, the methodology is based on cross-characterisation analysis.

978-1-4244-5430-3/10 $26.00 © 2010 IEEE

II. Technology Under Test

The HEMT under test were fabricated by MOCVD on SiC substrates. The epitaxial structure is composed of a buffer layer, an undoped 1.7 μm GaN layer, and an undoped 22 nm $Al_{0.24}Ga_{0.76}N$ barrier layer. The sheet resistance $R_\square$ is 527±33 Ω/ and the bi-dimensional sheet carrier density N_S is about $1.05 \times 10^{13} cm^{-2}$ [7]. The ohmic contacts are Ti/Al/Ni/Au evaporated stacks annealed at 900°C for 30 seconds under nitrogen atmosphere. Mushroom-shaped Mo/Au 0.25 μm gates were fabricated using electron-beam lithography. The HEMTs have a 2x75 μm-finger or a 8x75 μm-finger gate topology. The drain-source distance is 3.25 μm while the gate-source one is 1μm. The devices have been passivated with SiO_2/SiN layers of 50/100 nm thickness using plasma enhanced chemical vapor deposition. After passivation opening, the thick interconnection Ti/Pt/Au metallization is evaporated.

Specific test structures such as Transmission Line Method (TLM), Gated Transmission Line Method (GTLM) and FATFET ones with gate length of 100 μm have been used for complementary electrical characterisation and analysis of the origin of the suspected mechanism.

III. Deep Level Identification Before Ageing

Trapping effects were studied by DC pulsed measurements using 100ns pulses with 0.3% duty cycle. Pulsed $I_{DS}(V_{DS})$ characteristics were determined for different quiescent bias points chosen to simultaneously eliminate the thermal effect and to reveal the gate and drain lag effects. At V_{DS}=10V, a decrease of the drain current of 7.3% and 14.7% corresponds respectively to the gate lag effect and drain lag effect. These values are similar to those currently obtained for recent technologies as reported in [8, 9, 10].

The traps located in the devices under test have been characterized by capacitance DLTS (Figure 1) on FATFETs [11]. Six trap activation energies have been identified at 0.12 eV, 0.15 eV, 0.21 eV, 0.42 eV, 0.49 eV and 0.94 eV between 100 K and 450 K. Some of them are already known and their location is either at the surface (0.15 eV and 0.21 eV) [12], at the heterointerface (0.42 eV and 0.49 eV) [13] or in the buffer layer (0.94 eV) [14].

Low frequency noise measurements were also performed on virgin devices. A generation-recombination contribution was detected with a cut-off frequency of 40Hz at room temperature. The evolution of the cut-off frequency with temperature varying from 300K to 400K was analyzed in an Arrhenius. Activation energy was found to be 0.59 eV. This trap is probably the same as the one detected by the DLTS at 0.49eV.

IV. High Temperature Operating Tests

A. Life test conditions

The purpose of the life tests is to compare degradations related to electrical and temperature stresses. Channel temperatures T_{ch} were estimated from physical simulations and 3D thermal modeling. The potential profile between gate and drain was simulated using the commercial software

---

Identify applicable sponsor/s here. *(sponsors)*

ATLAS/Blaze from Silvaco. ANSYS-pro software allowed the thermal modeling by finite elements. Hypotheses and procedures used for these simulations are described more in detail in [15, 16].

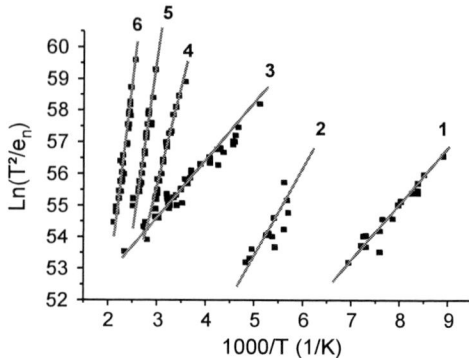

Figure 1. Arrhenius plot idenfying six traps from DLTS characterization (1 : 0.15 eV, 2 : 0.21 eV, 3 : 0.12 eV, 4 : 0.42 eV, 5 : 0.49 eV, 6 : 0.94 eV)

A set of 14 packaged 2x75μm devices was submitted to DC stress for 2000 to 3500 cumulated hours. Three types of stress conditions were applied to this set of devices, defining mainly semi-on state operation at different channel temperatures. Six devices were submitted to each of the following HTO tests (HTOT) and two devices to the I_{DQ} ageing test. The stress conditions are denoted hereafter with the label attributed to each test:

- HTOT175: T_{ch}=175°C, V_{DS}=25V, I_{DS}=417mA/mm

- HTOT150: T_{ch}=150°C, V_{DS}=25V, I_{DS}=417mA/mm

- I_{DQ}90: T_{ch} =90°C, V_{DS}=50V, V_{GS} so that I_{DS}=50mA/mm

Seven other 8x75μm packaged devices were submitted to HTO tests at higher temperatures. The test conditions were set at:

- HTOT275: T_{ch} =275°C, V_{DS}=25V, I_{DS}=417mA/mm,

- HTOT320: T_{ch} =320°C, V_{DS}=25V, I_{DS}=417mA/mm

During the HTO tests, the dissipated power was kept constant by automatic control of V_{GS}. These tests are considered as the most severe because of possible hot electrons effects [3]. Drain and gate currents and voltages were monitored during the test and read-out measurements were regularly performed in situ to study the degradation.

B. Parameter variation during the HTO tests

Figure 2 shows the evolution of $I_{DSS}@V_{GS}$=0V as a function of ageing time in the different HTO tests and the I_{DQ} one. All values correspond to the average variation of I_{DSS} except for the curve HTO 175. Only one transistor was aged for 3500 hours at 175°C. This evolution is presented in figure 2 since the values are in good agreement with those of the other transistors aged in the same conditions.

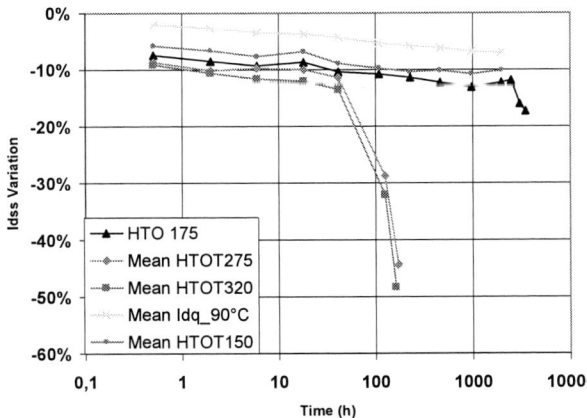

Figure 2. I_{DSS} variation observed during the different HTO tests

TABLE I. PARAMETER VARIATION OBSERVED AFTER THE DIFFERENT HTO TESTS AND THE I_{DQ} ONE AT 2000 HOURS

Test	HTO 175°C		HTO 150°C	Idq
Gate periphery	2x75μm	8x75μm	2x75μm	8x75μm
I_{DS} @ V_{DS}=8V & V_{GS}=0V	-13%	-13%	-10%	-7%
I_{DS} @ V_{DS}=5V & V_{GS}=0V	-17%	-18%	-14%	-11%
I_{DS} @ V_{DS}=2V & V_{GS}=0V	-19%	-20%	-17%	-16%
G_{mMAX} @ V_{DS}=8V	-5%	-5%	-4%	-5%
G_m @ V_{GS}=0V & V_{DS}=8V	-20%	-25%	-31%	-25%
Shift of V_{th}	0.15V	0.2V	0.1V	0V
R_{on}	+16%	+23%	+18%	+10%

These life tests showed two different degradation steps of the drain current. One is similar to all life test conditions and characterized by a decrease of the drain saturation current mainly occurring in the first hours of the life tests. The average I_{DSS} degradations in all life test conditions (I_{DQ} test, HTOT150, HTOT175 (for 2x75μm and 8x75μm devices), HTOT275 and HTOT320) are similar for the first 40 hours of test. The slopes of the I_{DSS} variation with time appear roughly parallel until a drastic I_{DSS} decrease is observed at a time threshold that could be temperature dependent. Indeed, in figure 2, this drastic I_{DSS} decrease is not reached within the stress time at 150°C, occurs after approximately 2500 hours in the HTO test at 175°C and around 42 hours at 275°C and 320°C. It looks as though the higher the temperature the shorter the time was.

Further assumptions are not reliable, since the time determined for the HTOT175 relies only on one sample and the time for HTOT275 corresponds to a series of devices with scattered electrical parameters.

1) First degradation mechanism
The detailed analysis of the mechanism responsible for the first degradation, has been reported in [17] for the life tests HTO 150°C, HTO 175°C and I_{DQ} 90°C performed during 2000 hours lifetime test. These life tests have mainly induced a shift of the pinch-off voltage in the range of 0.1-0.2 V while the Schottky contact remained stable. The maximal transconductance has weakly changed while its value at V_{GS}=0V and V_{DS}=8V has decreased at least by 20%. LF drain current noise measured at 2000 hours of life tests has demonstrated that there is no creation of new traps after the life tests. The weak evolution of the 1/f drain current noise has confirmed that there is no degradation in the channel. Moreover, pulsed I-V measurements show a weak evolution of the gate lag and drain lag rates.

Complementary ageing tests performed on TLM test vehicles have demonstrated the stability of the ohmic contact resistance up to 300°C and a drastic increase of 156% after 2000 hours at 350°C [18]. Consequently, the weak evolution of the electrical characteristics related to the first mechanism cannot be explained by an increase of the ohmic contact resistance but is more related to a combination of trap-related effects before stabilization sets in.

The activation energy associated to the first degradation mechanism common to all life-tests was calculated for a 20% I_{DSS} decrease considered as the failure criterion. The extracted value is within a range of E_A=0.8 – 1.2eV (figure 3). The uncertainty on the estimated value of E_A is a direct consequence of the limited number of collected data and especially the non homogeneous behavior of the 3 transistors aged at 275°C. This activation energy of about 1.2 eV has already been reported for a similar rather slow degradation of I_{DSS} [19, 20, 21].

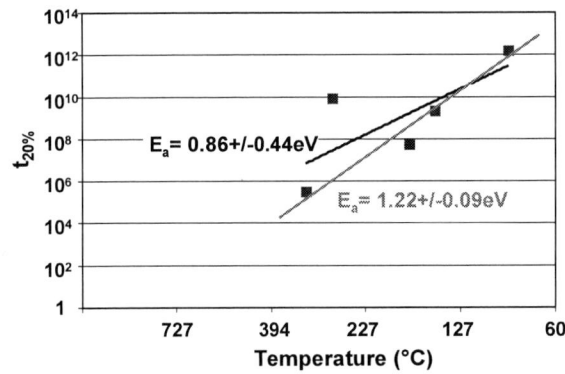

Figure 3. Activation energy of the first failure mechanism with a -20% of I_{DSS} decrease as the failure criterion

2) Second degradation mechanism
In contrast to the first one, the second failure mechanism observed during the HTO tests at 275°C and 320°C has induced a high drain current drop associated with a high increase of the on-state drain-source resistance Ron in the ohmic regime as plotted in figure 4. Moreover, there has been no stabilization of the parameter drifts observed before the end of the tests.

The variation of the DC parameters is reported in Table II after only 126 hours of ageing tests. The shift of the pinch-off voltage is in the range of 0.2-0.3 V while the Schottky gate characteristic presents rather small change. The maximal transconductance has significantly changed and its value at V_{GS}=0V and V_{DS}=8V has decreased at least by 30%. These results are consistent with the assumption of a different origin for this second failure mechanism.

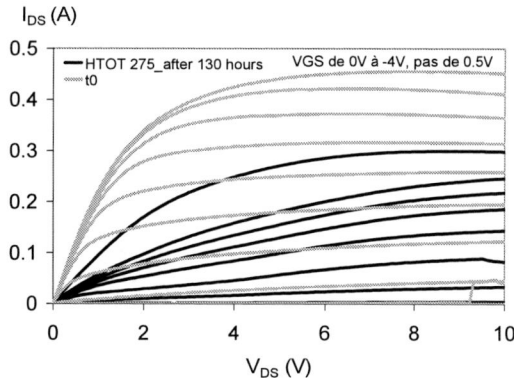

Figure 4. I_{DS}-V_{DS} characteristics before (grey lines) and after (black lines) the HTOT320 life test

TABLE II. PARAMETER VARIATION OBSERVED AFTER OCCURRENCE OF THE SECOND FAILURE MECHANISM DURING HTO TESTS

Test	HTO 175°C	HTO 275°C	HTO 320°C
Gate periphery	8x75µm	8x75µm	8x75µm
Cumulated time	3500 hrs	126 hrs	126 hrs
I_{DS} @ V_{DS}=8V & V_{GS}=0V	-17%	-29%	-32%
I_{DS} @ V_{DS}=5V & V_{GS}=0V	-22%	-34%	-38%
I_{DS} @ V_{DS}=2V & V_{GS}=0V	-24%	-40%	-45%
G_{mMAX} @ V_{DS}=8V	-7%	-14%	-26%
G_m @ V_{GS}=0V & V_{DS}=8V	-28%	-32%	-34%
Shift of V_{th}	+0.2V	+0.3V	+0.3V
R_{on}	+23%	+72%	+81%

V. ELECTRICAL ANALYSIS AFTER AGEING

To get deeper insight on the physical origin of the second failure mechanism, complementary electrical analysis has been performed by DC pulsed characterization and low frequency drain noise measurements.

A. DC-pulsed characteristics

The figure 5 shows the evolution of the drain current determined at V_{DS}=8V and V_{GS}=0V, before ageing and after the different HTO tests (150°C, 175°C, 275°C and 320°C) measured in pulsed conditions for different quiescent points. These different considered quiescent points are: V_{DS0} = 0V and V_{GS0} = 0V, V_{DS0} = 0V and V_{GS0} = -5V, V_{DS0} = 15V and V_{GS0} = -5 V. A linear decrease of the drain current after 3500 hrs for HTOT150 and HTOT175 ageing tests is observed. The degradation of the device performance is quite similar in both cases. As shown in figures 6 and 7, this result is consistent with the drain current evolution I_{DS} observed after 1000 and 2000 hours in pulsed conditions.

A clear decrease of I_{DS} is also observed in figure 5 after the HTOT275 and HTOT320 life tests (at 226 hours). This phenomenon is linked to the second degradation mechanism revealed in figure 3.

Figure 8 presents the gate-lag and drain-lag drops measured in pulsed conditions for different quiescent points after the different HTO tests. The gate-lag drop is the difference in percentage between I_{DS} (0V, 0V) and I_{DS} (-5V, 0V) and the drain lag drop is between I_{DS} (-5V, 0V) and I_{DS} (-5V, 15V). Regarding the gate-lag and the drain-lag lag determined in open channel conditions (V_{DS}=8V and V_{GS}=0V), the same comment can be made as previously. The evolution of the gate-lag and the drain-lag is quite similar for HTOT150 and HTOT175 tests. A significant increase of the drain lag is noted for ageing tests performed at higher temperature such as the HTOT275 and HTOT320 ones. The signature of the second mechanism is also remarkable on the lag effects, in particular for HTOT275. The evolution between HTOT275 and HTOT320 must be analyzed with precaution because of the scattering of DC characteristics observed before the ageing tests for devices which were submitted to the HTOT275 life test.

To illustrate these results, an estimation of the expected output power P_0 was determined from the drain-lag characteristics measurement. Indeed, under this quiescent bias point the $I_{DS}(V_{DS})$ characteristics correspond to dynamic conditions associated with an operating bias in class A. In this case, an estimated power decrease of around 25% is observed between the virgin devices and aged ones which have been submitted to the HTOT150 and HTOT175 life tests. But, a drastic decrease of 75% is estimated between the virgin devices and aged ones which have been submitted to the HTOT275 and HTOT320 life tests. Finally, for HTO ageing tests performed at temperature lower than 175°C, the time introduced linear and small degradation. When the temperature is equal or higher than 275°C, a drastic failure mechanism occurs or the first one is amplified and the drop of the current appeared earlier.

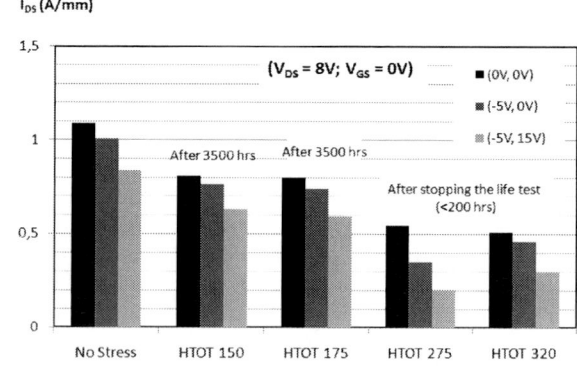

Figure 5. Evolution of the drain current (Vds=8V, Vgs=0V) measured in pulsed conditions for different quiescent points after the different HTO tests

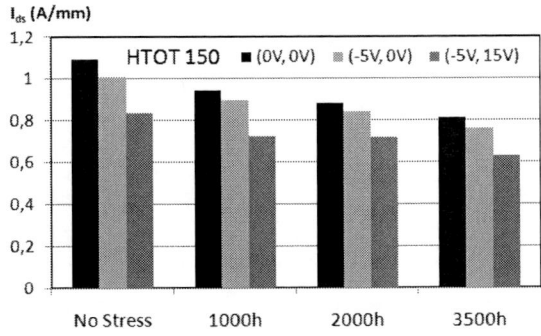

Figure 6. Evolution of the drain current I_{ds} (mA/mm) after 1000, 2000 and 3500 hrs for HTOT150 tests measured in pulsed conditions for different quiescent points

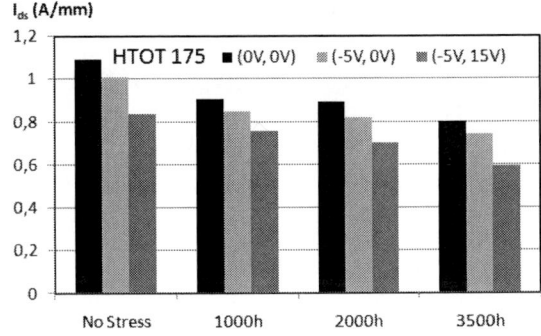

Figure 7. Evolution of the drain current I_{DS} (mA/mm) after 1000, 2000 and 3500 hrs for HTOT175 tests measured in pulsed conditions for different quiescent points

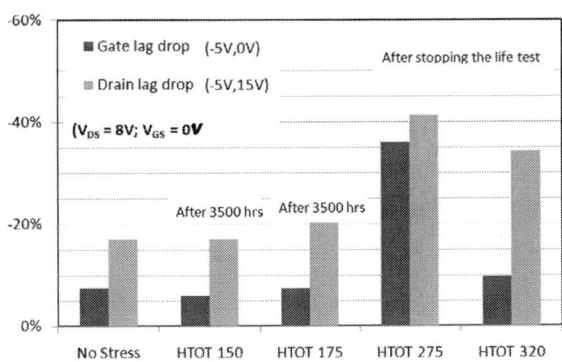

Figure 8. Gate-lag and drain-lag drops measured in pulsed conditions for different quiescent points after the different HTO tests [gate lag: drop between I_{DS} (0V, 0V) and I_{DS} (-5V, 0V)] [drain lag: drop between I_{DS} (-5V, 0V) and I_{DS} (-5V, 15V)]

B. Low frequency noise analysis

Low frequency (LF) noise measurements have been performed to study the evolution of the drain current excess noise, especially the 1/f noise contribution, which is related to drain-source conductivity fluctuations. It will be analysed with respect to the drain source resistance drop in static regime.

As virgin devices as well as aged transistors issued from HTOT175 test have a 2x75µm gate periphery while the other aged samples have a 8x75µm gate periphery, the 1/f noise level has been normalized taking into account the ratio between the two gate widths.

Figure 9 shows a comparison of the drain current noise spectra of virgin devices and aged ones after 226 hours of HTOT320 life test. The LF drain current noise level is roughly higher in aged devices than in virgin ones.

The LF noise spectra of virgin devices exhibit a G-R noise contribution, with a cut-off frequency in the range of 40 Hz at 300K; it is related to a deep trap with activation energy of 0.59eV [22]. In the LF noise spectra of the aged devices, the cut-off frequency of the G-R noise component remains in the same range. Therefore, the signature of the preeminent traps has not changed during the life test and no new trap (with a different activation energy) has been created.

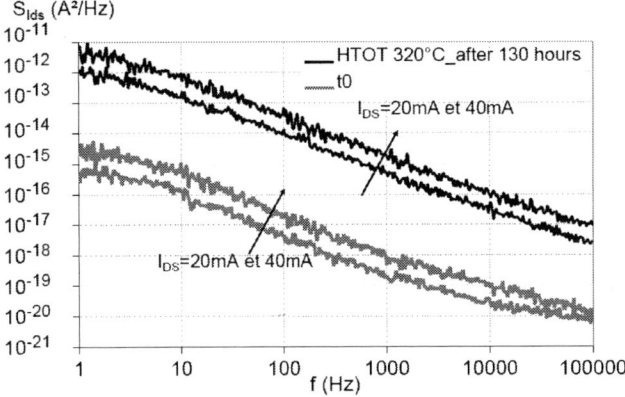

Figure 9. Comparison of the drain current noise spectra measured at V_{GS}=0V for I_{DS}=20mA and 40mA before (grey spectra) and after (black spectra) the HTOT320 test

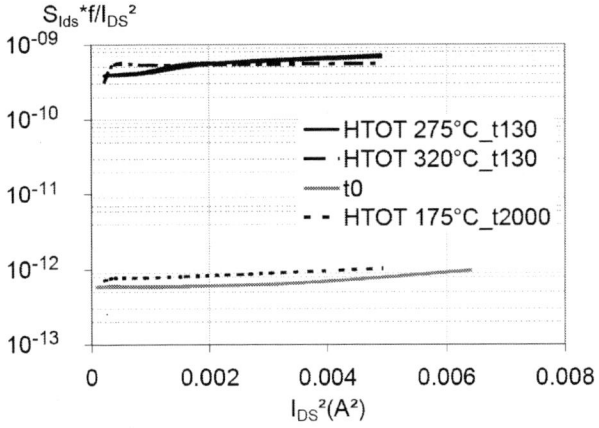

Figure 10. Drain Current noise normalized to the squared drain current before (grey line) the life tests, after 2000 hours of HTOT175 test (dashed line), and after 226 hours of HTOT320 test (plain black line) and after 288 hours of HTOT275 test (dashed and dotted black line)

In figure 10, the 1/f component of the drain current noise normalized to the squared drain current has been plotted to exhibit its evolution after the life test. As expected from the static regime results for devices submitted to the HTOT275 and HTOT320 tests, the drastic increase of the drain-source resistance has induced a drastic increase of the 1/f noise level by more than two orders of magnitude.

Regarding these results related to the second mechanism, the temperature is considered as an acceleration factor of the degradation affecting the conduction channel.

VI.   TEM ANALYSIS OF FAILED DEVICES

TEM lamellas were prepared from 4 transistors; one unstressed transistor used as reference and 3 others chosen among the transistors submitted to the HTO tests at 175°C, 275°C and 320°C respectively. Only one lamella was prepared for each transistor. As mentioned by Park et al. [23] one drawback of TEM analysis is the extremely localized observation area and as a result it hardly representative of the overall features of the devices. On the other hand, TEM observations of the lamellas have revealed similar crystal damages in the gate finger cross-section of only transistors submitted to the HTO tests (figure 11). It seems reasonable to associate these defects with the observed current decrease. These defects are located at the gate edge on the drain side and are deep enough to contact the GaN layer.

The SiO$_2$ passivation layer on the surface also appears damaged close to the defect. All transistors studied by TEM reached the second stage of drain current decrease but do not present the same level of current reduction. However, it is difficult to clearly correlate the current degradation due to the second failure mechanism with the size of the defect.

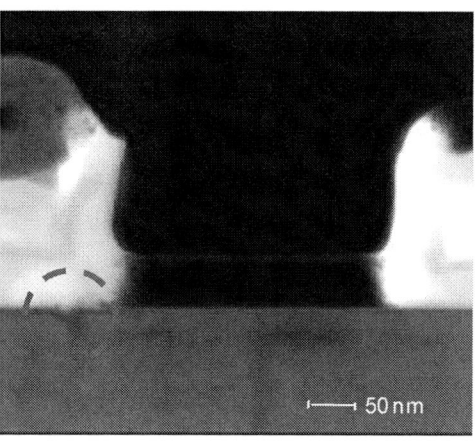

Figure 11. TEM micrograph of one gate out of the eight ones of one transistor aged during 473hrs at 275°C showing a defect on the drain side of the gate

Nevertheless, similar defects have already been reported [23, 24]. In these works it has been demonstrated that in AlGaN/GaN HEMTs submitted to high bias voltage, large electric field appears at the drain side of the gate edge. This can result in very large mechanical stress which adds to the lattice mismatch tensile strain. If the elastic energy exceeds a critical value, crystallographic defects are formed. This mechanism is known as inverse piezoelectric effect.

Complementary analysis of aged transistors having not reached the second stage of drain current degradation are on-going in order to investigate if this defect is only associated to the second degradation mechanism.

VII.   CONCLUSION

This study reports on a reliability investigation of AlGaN/GaN HEMTs submitted to life tests in High Temperature Operating (HTO) conditions at 150°C, 175°C, 275°C and 320°C. These life tests showed two different degradation steps of the drain current.

The first mechanism is occurring in the first hours of the life test and characterized by a decrease of the drain current. The average I$_{DSS}$ degradations in all test conditions appear similar until 40 hours of test. The evolution of the electrical characteristics after ageing does not depend on the bias conditions but rather more on the channel temperature. This degradation mode is characterized by high activation energy of 1.2 eV. The associated evolution of electrical characteristics is weak and cannot be obviously explained by a single physical mechanism. It results rather more from a combination of trap-related effects before stabilization.

On the contrary to the first degradation mechanism, the second failure mechanism observed during the HTO tests at 275°C and 320°C induced a higher drain current drop associated. This second mechanism occured after about 50 hours of HTOT275 and HTOT320 life tests and after 2000 hours of HTOT175 test. Moreover, no stabilization of the parameter drifts was observed before the end of the tests.

To gain more insight about possible physical origins of those failure mechanisms, the implementation of cross-characterisation analysis has been performed. Pulsed I-V measurements showed a large evolution of gate lag and drain lag rates after high temperature life tests at 275°C and 320°C. Low frequency 1/f noise drastically increased by more than two orders of magnitude after the life tests at high temperature while it did not change after 2000 hours of life test at 150°C and 175°C. We conclude that the temperature is considered as an acceleration factor of the degradation affecting the conduction channel. TEM observations revealed similar crystal damages in the gate finger cross-section of only aged devices after the life tests at 175°C, 275°C and 320°C. As these defects were found deep enough to contact the GaN layer and located at the gate edge on the drain side, it is possible to link them with the large degradation of the drain current observed in. Similar defects have already been reported and these results suggest that this degradation mechanism could be associated with the inverse piezoelectric effect.

In this work, two degradation mechanisms were demonstrated in AlGaN/GaN HEMTs submitted to high temperature operating tests. One is likely related to trapping effects and a second one associated to inverse piezoelectric effect. Both mechanisms lead to electrical parameters degradation, the latter being more detrimental.

ACKNOWLEDGMENT

This work is supported by the CARDYNAL ANR research project which partners are co-authors of this paper. The authors would also like to thank the European Defense Agency for their financial participation to the fabrication of devices.

REFERENCES

[1] A.Chini, D. Buttari, R.Coffie, S.Heikman, S.Keller, U.K. Mishra, "12W/mm power density AlGaN/GaN HEMTs on sapphire substrate", Electronics Letters, Vol. 40, N° 1, January 2004.

[2] Y. F.Wu, A. Saxler, M.Moore, P. Smith, S. Sheppard, P. M. Chavarkar, T. Wisleder, U. K.Mishra and P. Parikh, "30-W/mm GaN HEMTs by field plate optimization", IEEE Electron Device Lett, Vol. 25, pp. 117–119, March 2004.

[3] A. Sozza, C. Dua, E. Morvan, M. A. Di Forte-Poisson, S. Delage, F. Rampazzo, A. Tazzoli, F. Danesin, G. Meneghesso, E. Zanoni, A. Curutchet, N. Malbert, N. Labat, B. Grimbert and J. C. De Jaeger, "Evidence of traps creation in GaN/AlGaN/GaN HEMTs after a 3000 hour on-state and off-state hot-electron stress", in IEDM Tech. Digest, 2005, pp. 601-604.

[4] G. Meneghesso, F. Rampazzo, P. Kordoš, G. Verzellesi, and E. Zanoni, "Current collapse and high-electric-field reliability of unpassivated GaN/AlGaN/GaN HEMTs", IEEE Trans. Electron. Devices, vol. 53, pp. 2932-2941, 2006.

[5] M. Faqir, G. Verzellesi, F. Fantini, F. Danesin, F. Rampazzo, G. Meneghesso, E. Zanoni, A. Cavallini, A. Castaldini, N. Labat, A. Touboul, C. Dua, " Characterization and analysis of trap-related effects in AlGaAs-GaN HEMTs», Microelectronics Reliability, Vol. 47, pp 1639-1642, 2007

[6] J. Joh and J. del Alamo, "Mechanisms for electrical degradation of GaN high electron mobility transistors", in IEDM Tech. Digest, 2006, pp. 415-418.

[7] M-A. Di Forte Poisson, M. Magis, M. Tordjman, R. Aubry, M. Peschang, S.L. Delage, J. Di Persio, B. Grimbert, V. Hoel, E. Delos, D. Ducatteau, and C. Gaquiere, "LP-MOCVD Growth of GaAlN/GaN Heterostructures on Silicon Carbide: Application to HEMT's Devices," Journal of crystal growth, vol. 272, n° 1-4, pp. 305-311, 2004.

[8] J-C. De Jaeger., V. Hoel, N. Defrance, Y. Douvry, C. Gaquiere, M-A. Poisson., J. Thorpe, H. Lahreche, R. Langer, "Microwave power performance on AlGaN/GaN HEMTs on composite substrate" Proceedings of the 4th European Microwave Integrated Circuits Conference, EuMIC 2009, Rome, Italy, september 28-29, 2009, 144-147

[9] N. Defrance, J. Thorpe, Y. Douvry, V. Hoel, J. C. De Jaeger, C. Gaquiere, Xiao Tang, M. A. Di Forte-Poisson, R. Langer, M. Rousseau, and H. Lahreche, "AlGaN/GaN HEMT High Power Densities on SiC/SiO₂/poly-SiC Substrates," IEEE Electron Device Lett, vol. 30, Issue 6, pp. 596–598, June 2009.

[10] D. Ducatteau, A. Minko, V. Hoël, E. Morvan, E. Delos, B. Grimbert, H. Lahreche, P. Bove, C. Gaquière, J. C. De Jaeger, and S. Delage, "Output power density of 5.1/mm at 18 GHz with an AlGaN/GaN HEMT on Si substrate," IEEE Electron Device Lett, vol. 27, Issue 1, pp.7-9, January 2006.

[11] W. Chikhaoui, J.M. Bluet, P. Girard, G. Bremond, C. Bru-Chevallier , C. Dua, R. Aubry, "Deep levels investigation of AlGaN/GaN heterostructure transistors," Physica B vol. 404, pp 4877–4879, 2009.

[12] T. Mizutani, A. Kawano, S. Kishimoto, and K. Maezawa, H. Ueno, T. Ueda, and T. Tanaka "AlGaN/GaN MIS-HEMTs with HfO₂ gate insulator," Phys. Stat. Sol., vol. 4, n° 7, pp. 2700–2703, 2007.

[13] A. Y. Polyakov, N. B. Smirnov, A. V. Govorkov, J. Kim, F. Ren, G. T. Thaler, M. E. Overberg, R. Frazier R, C. R. Abernathy, S. J. Pearton, C. M. Lee, J. I. Chyi, R. G. Wilson and J. M. Zavada, "Comparison of the electrical and luminescent properties of p-layer-up and n-layer-up GaN/InGaN light emitting diodes and the effects of Mn doping of the upper n-layer," Solid-State Electronics, vol. 47, Issue 6, pp. 981-987, June 2003.

[14] Z.-Q. Fang, D. C. Look, D. H. Kim, and I. Adesida, "Traps in AlGaN/GaN/SiC heterostructures studied by deep level transient spectroscopy," Appl. Phys. Lett., vol. 87, Issue 18,. October 2005.

[15] R. Aubry, J. C. Jacquet, J. Weaver, O. Durand, P. Dobson, G. Mills, M.-A. Di Forte-Poisson, S. Cassette, S. L. Delage, "SThM Temperature Mapping and Nonlinear Thermal Resistance Evolution With Bias on AlGaN/GaN HEMT Devices," IEEE Trans. Electron. Devices, vol. 54, Issue 3, pp. 385–390, March 2007.

[16] J.-C. Jacquet, R. Aubry, H. Gérard, E. Delos, N. Rolland, Y. Cordier, Bussutil, M. Rousseau, and S. L. Delage, "Analytical transport model of AlGaN/GaN HEMT based on electrical and thermal measurement," in Proc. 12th GAAS Symp., Amsterdam, The Netherlands, Oct. 11–15, 2004, pp. 235-238,

[17] N. Malbert, N. Labat, A. Curutchet, C. Sury, V. Hoel, J.-C. De Jaeguer, N. Defrance, Y. Douvry, C. Dua, M. Oualli, C. Bru-Chevallier, J.-M. Bluet and W. Chikhaoui, " Characterisation and modelling of parasitic effects and failure mechanisms in AlGaN/GaN HEMTs," Microelectronics Reliability, vol. 49, pp. 1216-1221, 2009.

[18] M. Piazza, C. Dua, M. Oualli, E. Morvan, D. Carisetti and F. Wyczisk, "Degradation of Ti/Al/Ni/Au as ohmic contact metal for GaN HEMTs," Microelectronics Reliability, vol. 49, pp. 1222-1225, 2009.

[19] S. Singhal, J.C. Roberts, P. Rajagopal, T. Li, A.W. Hanson, R. Therrien, J.W. Johnson, I.C. Kizilyalli and K.J. Linthicum, "GaN-On-Si failure mechanisms and reliability improvements," in Proc. IRPS 2006, San Jose, CA, USA, pp. 95-98.

[20] J. L. Jimenez and U. Chowdhury, "X-band GaN FET Reliability," in Proc. IRPS 2008, Phoenix, AZ, USA, pp. 429-435.

[21] E. Zanoni, G. Meneghesso, M. Meneghini, A. Tazzoli, N. Ronchi, A. Stocco, F. Zanon, A. Chini, G. Verzellesi, A. Cetronio, C. Lanzieri and M. Peroni, "Long-term stability of Gallium Nitride High Electron Mobility Transistors: a reliability physics approach," Microwave Integrated Circuits Conference, EuMIC 2009, Rome, Italy, pp. 212-217, September 2009.

[22] C. Sury, A. Curutchet, N. Malbert and N. Labat, "Low Frequency Noise Evolution of AlGaN/GaN HEMT after 2000 hours of HTRB and HTO life tests," in Proc. AIP for ICNF 2009, Pisa, Italy, pp. 625-628.

[23] S.Y. Park, C. Floresca, U. Chowdhury, J. L. Jimenez, C. Lee, E. Beam, P. Saunier, T. Balistreri and M. J. Kim, "Physical degradation of GaN HEMT devices under high drain bias reliability testing," Microelectron Reliab (2009), doi:10.1016/j.microrel.2009.02.01.

[24] J.A. del Alamo and J. Joh, "GaN HEMT Reliability", Microelectronics Reliability, vol. 49, pp. 1200-1206, 2009.

High Temperature On- and Off-state Stress of GaN-on-Si HEMTs with In-situ Si₃N₄ Cap Layer

Denis Marcon, Farid Medjdoub, Domenica Visalli, Marleen Van Hove, Joff Derluyn, Jo Das, Stefan Degroote, Maarten Leys, Kai Cheng, Stefaan Decoutere, Robert Mertens, Marianne Germain, and Gustaaf Borghs

IMEC
Leuven, Belgium
+32 16 28 8404, Denis.Marcon@imec.be

Denis Marcon, Domenica Visalli, Robert Mertens and Gustaaf Borghs are also with
K.U. Leuven
Leuven, Belgium

Abstract—In this work the stability of Gallium Nitride based high electron mobility transistors grown on 4-in Si substrate (GaN-on-Si HEMTs) were tested both in off-state at high drain voltage (200 V) and in on-state at large gate voltage (+2 V) with low drain bias (5 V). In each stress experiment the ambient temperature was fixed at 200°C. Remarkably, despite the considerably large drain voltage used in the off-state stress on only 5 µm gate-drain spaced transistors, negligible signs of degradation were observed after more than 350 hours of testing. Similar results were obtained after the on-state stress. In fact, only small degradation signs were reported in spite of the large gate current and high junction temperature the devices have to withstand during the on-state stress. These results show the robustness of these devices to operate under high electric field conditions, high temperature and to withstand also a large gate current for considerable time.

Keywords: failure mechanisms, GaN-HEMT, reliability

I. INTRODUCTION

Nowadays, Gallium Nitride High Electron Mobility Transistors grown on Si substrate (GaN-on-Si HEMTs) attract a lot of attention since they combine high voltage and high frequency capabilities [1] - [2] with the opportunity of processing large-diameter wafers, up to 200 mm [3]. This results in a high performance low cost technology.

In contrast to the large amount of publications reporting impressive performance of GaN-based devices, the number of reliability studies is relatively limited [4] - [11]. The reliability analysis of power switching devices should consider that during the normal switching operation the transistor is continuously switched from an off-state at a high drain voltage to an on-state in a low impedance. In the first case, the transistor has to withstand a high electric field peak under the gate edge at the drain side which might cause stability problems such as a considerable increase of the leakage currents [4] - [11]. In the second case, the device degradation might be induced by the high current density in the channel and through the Schottky gate (for large gate bias). Consequently, despite the intrinsic high voltage and current capabilities, reliability issues might limit the operating bias.

In this work the stability of 5 µm gate-drain spaced GaN-on-Si HEMTs was tested in both on- and off-state conditions with the base plate temperature set at 200°C. For the off-state condition the drain voltage was set to 200 V. The on-state condition was defined at a drain voltage of 5 V while +2 V was applied to the gate.

In spite of the harsh conditions the transistors have to withstand during the on- and off-state stress only small signs of degradation were noticed. As will be discussed in the subsequent sections, these achievements result from the simultaneous optimization of epilayer quality and gate technology as well as the used of the in-situ Si₃N₄ cap.

II. THERMAL STABILITY ENHANCEMENT BY MEANS OF IN-SITU Si₃N₄ CAP LAYER

In this section is reported a preliminary experiment performed in order to elucidate the effect of two different cap layers in the thermal stability of the GaN/AlGaN heterostructure.

The two dimensional electron gas (2DEG) of GaN-based HEMTs is generally obtained by the growth of a thin AlGaN layer on top of a GaN layer [12]. The piezoelectric polarization of the AlGaN barrier layer, which is induced by the lattice mismatch with the GaN channel layer, has a crucial role in defining the 2DEG properties [12].

978-1-4244-5430-3/10 $26.00 © 2010 IEEE

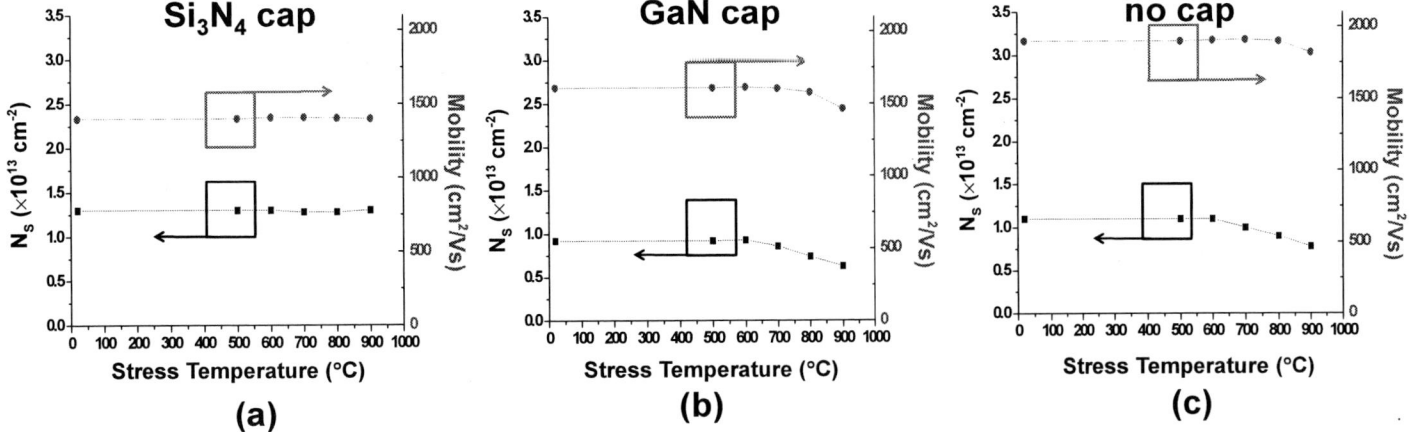

Figure 1. 2DEG concentration (black dots) and mobility (blue dots) measured at room temperature and during the temperature step-test for a GaN/Al$_{0.3}$Ga$_{0.7}$N epilayer capped with (a) Si$_3$N$_4$, (b) GaN and (c) not capped. The temperature was stepped from 500°C to 900°C every 30 min with a step of 100°C.

The strain of the AlGaN layer might be changed by modifying the Al content [12] or by applying an external electric field [11]. It is reported that excessive intrinsically or extrinsically induced strain of the AlGaN layer results in crystallographic defects causing a degradation of the 2DEG properties [7], [8], [10], [11], [13]. For these reasons the AlGaN layer is considered as one of the main weak point of the GaN-based technology.

Recently, it has been shown that the use of GaN cap layer on top of the AlGaN barrier layer might enhance the robustness of the entire structure [14].

In this section the thermal stability of three structures nominally identical except for the cap layer has been compared. These structures consist of a GaN buffer layer epitaxially grown by Metal Organic Chemical Vapor Deposition (MOCVD) on Si substrate followed by a 22 nm of Al$_{0.3}$Ga$_{0.7}$N layer. Finally the structure called A was capped by 3 nm Si$_3$N$_4$ layer, the structure B was terminated with 2 nm GaN layer and the structure C was left uncapped as reference. Both cap layers were deposited in-situ in the MOCVD chamber.

The stability of the structures was evaluated by means of a temperature step-stress experiment in vacuum. The temperature was ramped up from 500°C to 900°C every 30 min using a step of 100°C. After each step the oven was cooled down in order to perform room temperature Hall measurements necessary to identify irreversible material degradation such as a drop of the carrier concentrations.

As can be observed in Fig. 1, only the structure A (capped with Si$_3$N$_4$) preserved the material properties up to 900°C (Fig. 1a). For the other two structures (capped with GaN and uncapped) the electron concentrations decrease at 700°C (Fig. 1b and 1c). This degradation is attributed to the strain relaxation of the AlGaN barrier layer resulting in the reduction of the 2DEG concentration. In a reported experiment, an uncapped epi-structure consisting of an unstrained In$_{0.17}$Al$_{0.83}$N barrier layer subjected to similar thermal tests has shown stability above 900°C [15], which indicates the key role of strain in the material degradation observed.

This experiment indicates that the Si$_3$N$_4$ is effective in strengthening the AlGaN barrier layer by preventing the relaxation phenomena [16]. This could be a key in the device reliability enhancement under harsh conditions (high electric field/ high junction temperature).

III. DEVICE DESCRIPTION

For the on- and off-state stress experiment the epitaxial layer was also grown by MOCVD on 4-in Si (111) substrate. The layer stack consists of a 2-µm-thick AlGaN buffer layer, 150 nm GaN channel layer and a 25 nm Al$_{0.35}$Ga$_{0.65}$N barrier layer followed by a 50-nm-thick in-situ Si$_3$N$_4$. More details on the epitaxial growth can be found elsewhere [17]. The 50 nm Si$_3$N$_4$ cap layer allows defining the gate field plate directly on top of the thick in-situ Si$_3$N$_4$ layer and thus avoiding the deposition of an ex-situ dielectric layer. Consequently, interface states between in- and ex-situ passivation can be avoided near the sensitive AlGaN surface.

Device isolation was obtained by mesa etching using Cl$_2$-based chemistry. The ohmic contacts were fabricated after selectively removing the in-situ Si$_3$N$_4$ layer by ICP using SF$_6$ gas. The ohmic contact metallization was formed by depositing Ti/Al/Mo/Au followed by rapid thermal annealing at 850°C. The same etching recipe was also used to define the embedded gate in the in-situ dielectric layer. The gate metallization was Ni/Au (20 nm/200 nm). Finally, the devices were passivated with 200-nm-thick PECVD Si$_3$N$_4$ layer.

The device geometries used for this study were: gate length L$_G$ = 1.5 µm with an overhang of 1 µm, gate-source spacing L$_{GS}$ = 1.5 µm, gate-drain spacing L$_{GD}$ = 5 µm and device width W = 0.2 mm.

The saturated drain current (I$_{DSS}$) measured at V$_{DS}$ = 5 V and V$_{GS}$ = 0 V was 580 mA/mm with a R$_{ON}$ as low as 6 Ω·mm.

IV. RELIABILITY TEST DESCRIPTION

Several devices were subjected to off-state stress (V$_{GS}$ = -7 V) at very high drain voltage (V$_{DS}$ = 200 V) at a chuck temperature (T$_{CHUCK}$) of 200°C. In such a test the transistor under stress has to withstand a very high electric field

978-1-4244-5430-3/10 $26.00 © 2010 IEEE

condition. Typical reported degradation signs are a sudden increase of the gate and drain leakage currents, a drop of the I_{DSS} and an increase of the trapping phenomena [4] - [11]. For this reason, an extensive DC and pulsed characterization was performed at room temperature before and after stress. Additionally, the gate, drain and source leakage currents were monitored during the stress whereas the I_{DSS} was extracted every 10 hours without cooling down the system. The failure criterion was defined as a 10% drop of the I_{DSS} or an increase of the drain/gate leakage current above 1 mA/mm.

Another batch of devices was tested under on-state stress condition. In this case the gate voltage was fixed at +2 V whereas the drain was biased at 5 V. Also in this experiment the chuck temperature was fixed at 200°C. Under such conditions the junction temperature ($T_{Junction}$), estimated by means of simulations [18] which were calibrated by Raman spectroscopy [19], was 280°C. Also in this case, an extensive device characterization was performed at room temperature before and after stress. As previously, the failure criterion was defined as a 10% drop of I_{DSS}.

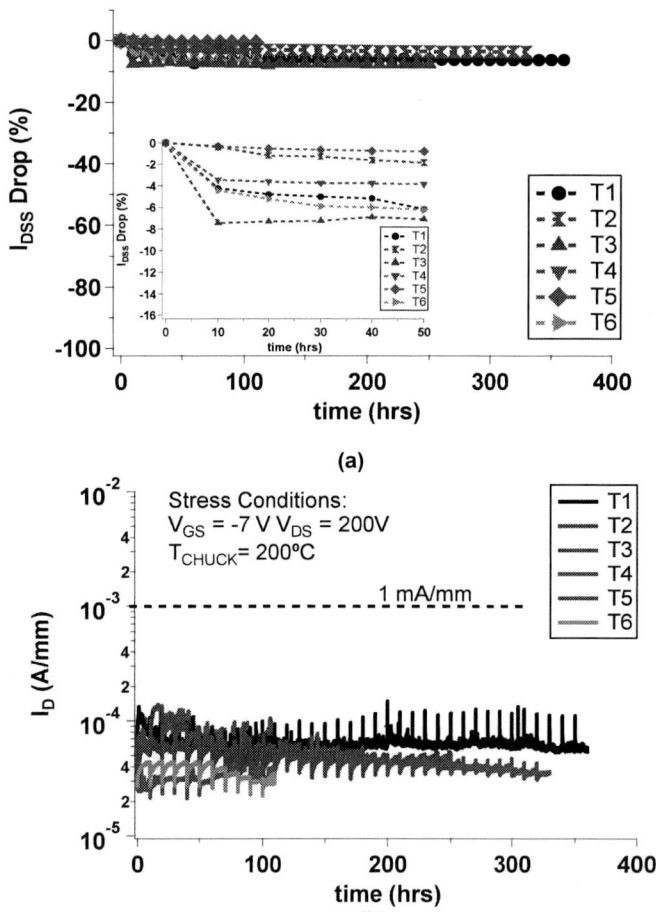

(a)

(b)

Figure 2. (a) Saturated drain current (I_{DSS}) variations measured at regular intervals and (b) drain current (I_D) monitored during the off-state stress for at a set of 6 devices indicated as T1-T6. The stress conditions are indicated in (b). In the inset of (a) is shown the I_{DSS} evolution during the first 50 hours of off-state stress.

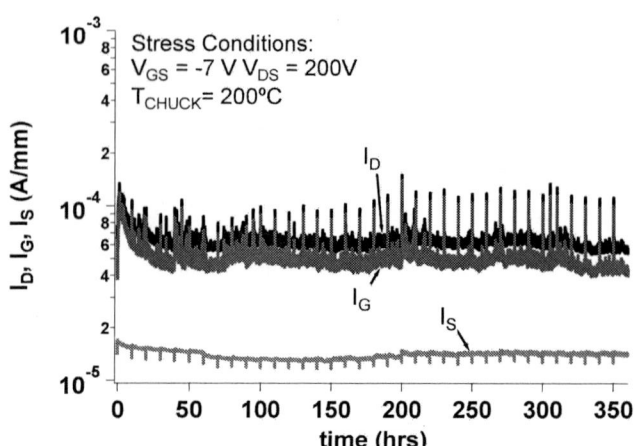

Figure 3. Drain (black), gate (dark grey) and source (light grey) currents measured during the 360 hours of the off-state stress for the transistors labeled as T1 in Fig. 2a and 2b.

V. OFF-STATE STRESS

Fig. 2a shows the evolution of I_{DSS} at 200°C and in Fig. 2b the drain leakage current during the stress is plotted. For the whole set of devices both parameters remained relatively stable and far from the failure levels defined above. The initial I_{DSS} variation (inset Fig. 2a) was attributed to temporary trapping effects due to the severe bias condition used; in fact, it was observed only in measurements performed immediately after the stress at the stress temperature.

The subsequent analysis will be focused on the transistor stressed for the longest time (360 hours). This transistor is indicated in Fig. 2a and 2b as T1.

The evolution of the gate, drain and source current monitored during the stress is shown in Fig. 3. As expected most of the current flows between the gate and the drain contact. Additionally, it is possible to notice that the drain current is larger than the gate current due to the contribution of the source current. This implies that a conduction path might exist between the source and the drain, located most probably at the III-N/Si interface at the substrate level [20]. Nevertheless, no abrupt changes of the leakage current during the stress were observed.

By comparing the DC characteristic before and after the stress (Fig. 4a, 4b and 4c), it is possible to notice only negligible differences. In particular important parameters such as the threshold voltage and R_{ON} are unaffected by the stress. The only minor changes were reported on the reverse gate diode characteristics (Fig. 4a) and it was attributed to the stabilization of the gate Schottky contact (i.e. burn-in).

The comparison of pulsed I-V measurements performed before and after the stress is useful to detect the formation of stress-induced traps [9]. In Fig. 5, the DC output characteristics are compared with the dynamic characteristics obtained by pulsing both gate and drain from the quiescent bias point $V_{GS} = -7$ V and $V_{DS} = 50$ V before and after stress. Clearly, in spite of the harsh stress conditions trapping effects on the dynamic R_{ON} or I_{DSS} were not noticeable.

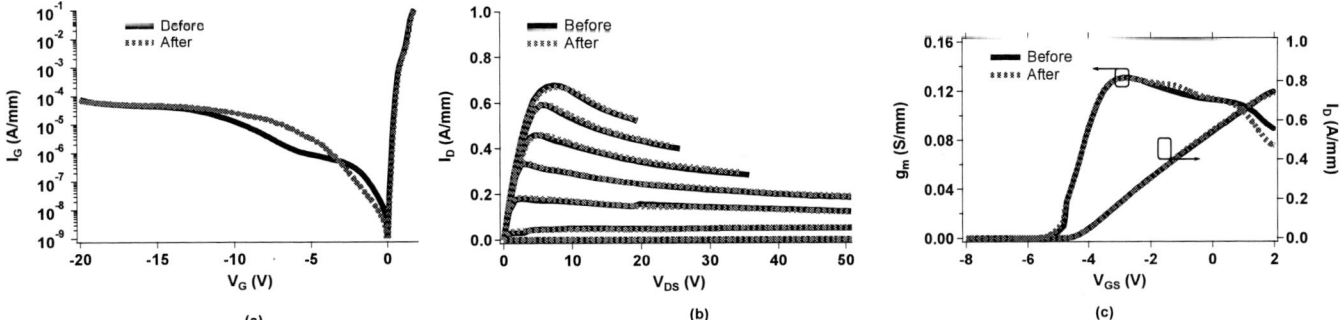

Figure 4. Comparison of the characteristic taken at room temperature before (continuous black lines) and after (dotted grey lines) 360 hours of high temperature off-state stress at $V_{DS} = 200$ V (device T1 in Fig. 2a and 2b); (a) I_G-V_G gate diode curve, (b) I_D-V_{DS} characteristic measured by sweeping the V_{GS} from 1V in 1V steps and (c) I_D-V_{GS} and g_m-V_{GS} characteristic measured at $V_{DS} = 10$ V.

The reported robustness of these devices under high electric field conditions has been obtained through the meticulous optimization process which involved both material and process technology. The high material quality is reflected in the small DC-to-RF current dispersion phenomena present in these devices (Fig. 5). Moreover, as it was discussed in the section II, the in-situ Si_3N_4 deposition technique enhances the thermal stability of the AlGaN layer avoiding the formations of crystallographic defects within the considered stress conditions.

Concerning the process technology, it has been observed that the gate process step has a considerable impact on the final device yield and reliability. Additionally, it has also been observed that the gate edge definition has a key role in the overall device robustness since it is the location where the electric field is concentrated [21]. For this reason, it is important to obtain a rounded gate edge that avoids local electric field crowding [22] and spread the electric field at the gate edge towards the drain using T-shape gates [21], [22].

VI. ON-STATE STRESS

In the case of the on-state stress, the devices have to withstand a large current through the gate and a high channel temperature.

The evolution of the I_{DSS} measured at $T_{CHUCK} = 200°C$ at regular intervals during the on-state stress is shown in Fig. 6a whereas the gate current constantly monitored during the

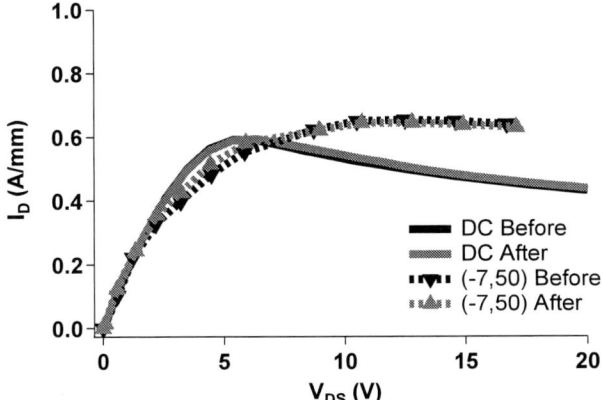

Figure 5. DC (continuous lines) and pulsed (dotted lines) I_D-V_{DS} characteristics at $V_{GS} = 0$ V measured before (black) and after (grey) 360 hours of high temperature off-state stress at $V_{DS} = 200$ V (device T1 in Fig. 2a and 2b). The pulsed I-V characteristic were obtained by pulsing from the quiescent bias point ($V_{GS} = -7$ V, $V_{DS} = 50$ V) with a pulsed width of 400 ns and a period of 1 ms.

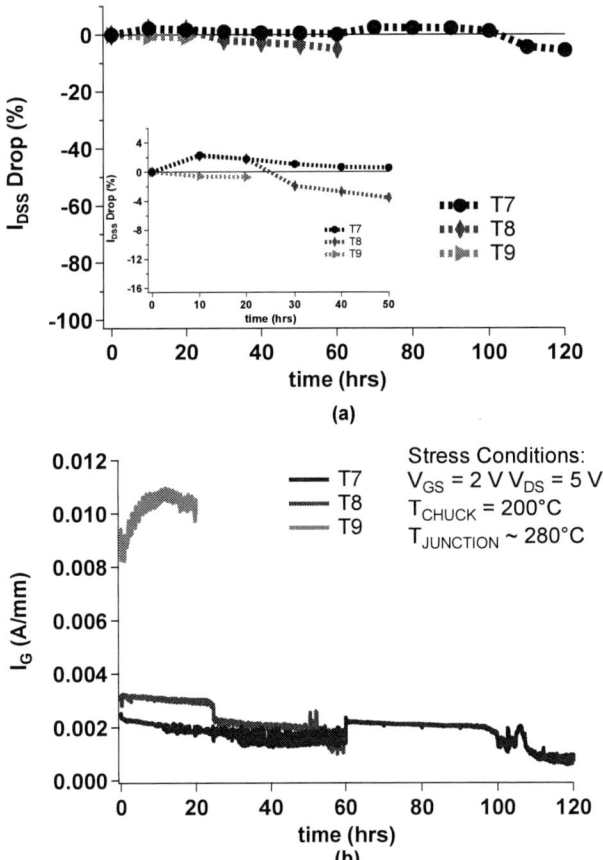

Figure 6. (a) Saturated drain current (I_{DSS}) variations measured at regular intervals and (b) gate current (I_G) monitored during the on-state stress for at a set of 3 devices indicated as T7-T9. The stress conditions are indicated in (b). In the inset of (a) is shown the I_{DSS} evolution during the first 50 hours of on state stress.

978-1-4244-5430-3/10 $26.00 © 2010 IEEE   149

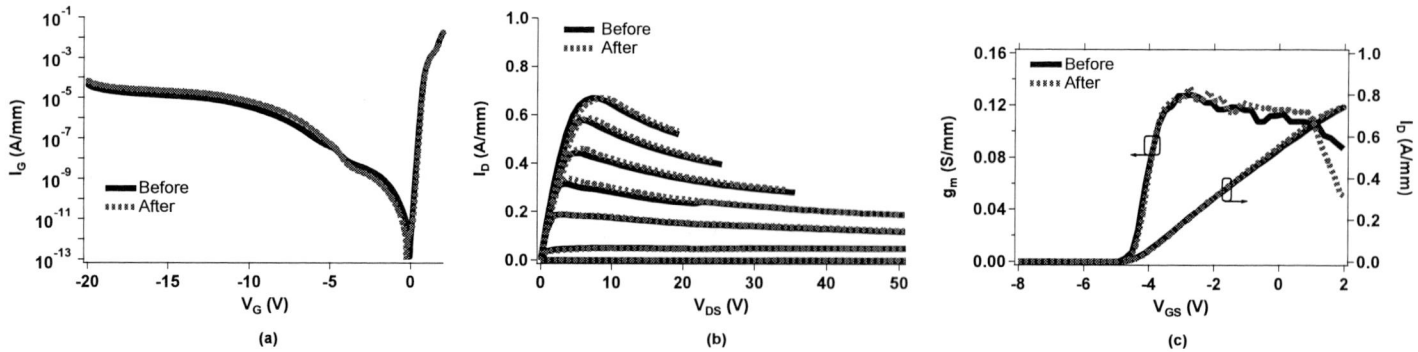

Figure 7. Comparison of the characteristic taken at room temperature before (continuous black lines) and after (dotted grey lines) 120 hours of high temperature on-state stress (device T7 in Fig. 6a and 6b); (a) I_G-V_G gate diode characteristic (b) I_D-V_{DS} characteristic measured by sweeping the V_{GS} from 1V in 1V steps and (c) I_D-V_{GS} and g_m-V_{GS} characteristic measured at V_{DS} = 10 V.

stress is shown Fig. 6b. As possible to observe, the gate current in all the devices was above 1 mA/mm.

Compared to the results obtained from off-state stress, in this case the evolution of I_{DSS} behaves differently as it shows a gradual drop over the time (Fig. 6a and its inset). In spite of this, there are only minor degradation signs noticeable when comparing the characteristics taken at room temperature before and after 120 hours of on-state stress (Fig. 7). The changes to be pointed out are an increase of the R_{ON} from 6 $\Omega \cdot$mm to 7 $\Omega \cdot$mm (Fig. 7b and Fig. 8) and a drop of the trans-conductance (g_m) at gate voltages beyond 1V (Fig. 7c). In fact, important parameters such as the threshold voltage and the maximum of the g_m were unchanged (Fig. 7b and 7c). Furthermore, the gate characteristic showed almost no changes (Fig. 7a).

As it has been speculated in [4], the drop of the g_m only for large gate voltage is attributed to the formation of defects over a wide damaged region extending laterally from the gate edge toward the drain side. This might also explain the increase of the R_{ON}. Additionally, such a damaged region might dramatically enhance the DC-to-RF current dispersion phenomena [4] but in this case the dynamic behavior characterized by means of pulsed measurements has shown no

changes (Fig. 8). In contrast with the explanation given, this indicates that no additional noticeable traps were induced by the stress.

The reasons behind the discrepancy between the small degradation observed in the DC behavior and the absence of degradation in the dynamic behavior are still under investigation.

VII. CONCLUSION

In this work, 5 µm gate-drain spaced GaN-on-Si HEMTs were electrically stressed. In order to accelerate the degradation process, the temperature was set to 200°C. Part of the devices were stressed under high electric field condition (off-state and drain biased at 200 V). Remarkably, the DC and the pulsed characterization performed before and after stressed revealed almost no changes after more than 350 hours.

The second batch of devices was stressed in on-state condition at a large gate voltage (+2 V). Also in this case the temperature was set to 200°C resulting in an estimated channel temperature of 280°C. The characterization performed at room temperature before and after stress has shown only small degradation signs which are still under investigation. Nevertheless, no changes were observed in the dynamic parameters investigated by means of pulsed measurement.

These performances result from the combination of several process optimizations mainly focused on the quality of the in in-situ Si_3N_4 cap layer, on the embedded gate technology and on the epilayer quality.

REFERENCES

[1] N. Ikeda, S. Kaya, J. Li, Y. Sato, S. Kato, and S. Yoshida, "High power AlGaN/GaN HFET with a high breakdown voltage of over 1.8 kV on 4 inch Si substrates and the suppression of current collapse", *Proc. Int. Symp. Power Semiconductor Devices and IC's*, 2008, p. 287.

[2] S. Hoshi, M. Itoh, T. Marui, H. Okita, Y. Morino, I. Tamai, F. Toda, S. Seki, and T. Egawa "12.88 W/mm GaN High Electron Mobility Transistor on Silicon Substrate for High Voltage Operation", *Appl. Phys. Express*, vol. 2, pp. 061001, 2009.

[3] A.R. Boyd, S. Degroote, M. Leys, F. Schulte, O. Rockenfeller, M. Luenenbuerger, M. Germain, J. Kaeppeler, and M. Heuken, "Growth of GaN/AlGaN on 200 mm diameter silicon (111) wafers by MOCVD", *Phys. Status Solidi C*, vol. 6, pp. S1045, 2009.

[4] G. Meneghesso, G. Verzellesi, F. Danesin, F. Rampazzo, F. Zanon, A. Tazzoli, M. Meneghini, and E. Zanoni, "Reliability of GaN High-Electron-Mobility Transistors: State of the Art and Perspectives", *IEEE*

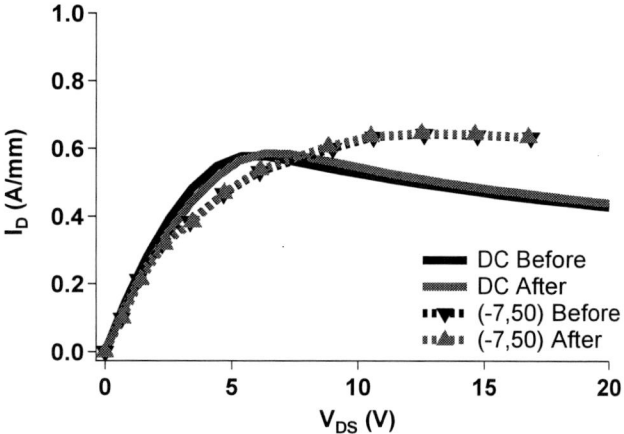

Figure 8. DC (continuous lines) and pulsed (dotted lines) I_D-V_{DS} characteristics at V_{GS} = 0 V measured before (black) and after (grey) 120 hours of high temperature on-state stress (device T7 in Fig. 6a and 6b). The pulsed I-V characteristic were obtained by pulsing from the quiescent bias point (V_{GS} = - 7 V, V_{DS} = 50 V) with a pulsed width of 400 ns and a period of 1 ms.

Trans. Device & Mater. Reliab., vol. 8, no. 2, pp.332, 2008 and references therein.

[5] Y. Inoue, S Masuda, M. Kanamura, T. Ohki, K. Makiyama, N. Okamoto, K. Imanishi, T. Kikkawa, N. Hara, H. Shigematsu, and K. Joshin, "Degradation-mode analysis for highly reliable GaN HEMT," *Proc. IEEE MTT-S*, 2007, p. 639.

[6] D. Marcon A. Lorenz, J. Derluyn, J. Das, F. Medjdoub, K. Cheng, S. Degroote, M. Leys, R. Mertens, M. Germain, and G. Borghs, "GaN-on-Si HEMT stress under high electric field condition", *Phys. Status Solidi C,* vol 6, pp. S1024, 2009.

[7] J.L. Jimenez, U. Chowdhury,"X-Band GaN FET reliability," *Proc. IEEE Int. Rel. Phys. Symp*, 2008, pp.429-435.

[8] U. Chowdhury, J. L. Jimenez, C. Lee, E. Beam, P. Saunier, T. Balistreri, S. Park, T. Lee, J. Wang, M.J. Kim, J. Joh, and J. A. del Alamo, "TEM Observation of Crack- and Pit-Shaped Defects in Electrically Degraded GaN HEMTs", *IEEE Electron Device Lett.*, vol. 29, no. 10, pp. 1098, 2008.

[9] E. Zanoni, F. Danesin, M. Meneghini, A. Cetronio, C. Lanzieri, M. Peroni, and G. Meneghesso, "Localized Damage in AlGaN/GaN HEMTs Induced by Reverse-Bias Testing," *IEEE Electron Device Lett.*, vol.30, no.5, pp.427-429, 2009.

[10] J. Joh and J. A. del Alamo "Impact of Electrical Degradation on Trapping Characteristics of GaN High Electron Mobility Transistors", *IEDM Tech. Dig.*,2008, pp. 461-464.

[11] J. Jungwoo, Feng Gao, T. Palacios, J.A. del Alamo, "A model for the critical voltage for electrical degradation of GaN high electron mobility transistors", *Reliability of Compound Semiconductors*, 2009,pp.3-6.

[12] O. Ambacher, J. Smart, J. R. Shealy, N. G. Weimann, K. Chu, M. Murphy,W. J. Schaff, L. F. Eastman R. Dimitrov, L. Wittmer, M. Stutzmann, W. Rieger, and J. Hilsenbeck, "Two-dimensional electron gases induced by spontaneous and piezoelectric polarization charges in N- and Ga-face AlGaN/GaN heterostructures," *J. Appl. Phys.*, vol. 85, pp. 3222-3233, 1999.

[13] D.W. Gotthold, S.P. Guo, R. Birkhahn, B. Albert, D. Florescu, B. Peres, "Time dependent degradation of AlGaN/GaN heterostructures grown on silicon carbide", *J. Electron. Mater.*,vol. 33, no. 5, pp.408–411, 2004.

[14] P. Ivo, A. Glowacki, R. Pazirandeh, E. Bahat-Treidel, R. Lossy, J. Wurfl, C. Boit, G. Trankle,"Influence of GaN cap on robustness of AlGaN/GaN HEMTs," *Proc. IEEE Int. Rel. Phys. Symp*, 2009, pp.71-75.

[15] F. Medjdoub, J.-F. Carlin, M. Gonschorek, E. Feltin, M.A. Py, D. Ducatteau, C. Gaquiere, N. Grandjean, E. Kohn, "Can InAlN/GaN be an alternative to high power / high temperature AlGaN/GaN devices?," in *Proc. IEDM Tech. Dig*, 2006, pp. 927-930.

[16] J. Derluyn, S. Boeykens, K. Cheng, R. Vandersmissen, J. Das, W. Ruythooren, S. Degroote, M. Leys, M. Germain, and G. Borghs "Improvement of AlGaN/GaN high electron mobility transistor structures by in situ deposition of a Si₃N₄ surface layer", *J. Appl. Phys.*, vol. 98, pp. 054501, 2005.

[17] K. Cheng, M. Leys, J. Derluyn, K. Balachander, S. Degroote, M.Germain, and G. Borghs., "AlGaN-based heterostructures grown on 4 inch Si(111) by MOVPE", *Phys. Status Solidi C*, vol. 5, pp. 1600, 2008.

[18] J. Das, H. Oprins, H. Ji, A. Sarua, W. Ruythooren, J. Derluyn, M. Kuball, M. Germain, and G. Borghs, "Improved Thermal Performance of AlGaN/GaN HEMTs by an Optimized Flip-Chip Design" *IEEE Trans. on Electron Devices*, vol. 53, no. 11, pp. 2696-2702, 2006

[19] M. Kuball, J. M. Hayes, M. J. Uren, T. Martin, J. C. H. Birbeck,R. S. Balmer, and B. T. Hughes, "Measurement of temperature in active high-Power AlGaN/GaN HFETs using Raman spectroscopy," *IEEE Electron Device Lett.*, vol. 23, no. 1, pp. 7–9, 2002.

[20] D. Visalli, M. Van Hove, J. Derluyn, K. Cheng, S. Degroote, M. Leys, M. Germain, G. Borghs, "AlGaN/GaN/AlGaN Double Heterostructures on Silicon Substrates for High Breakdown Voltage Field-Effect Transistors with low On-Resistance," *Phys. Stat. Sol. (c)* Vol. 6, pp. S988-S991, 2009.

[21] Y. S. Karmalkar, M. S. Shur, G. Simin, M. A. Khan, "Field-Plate Engineering for HFETs", *Electron Device Lett*, vol 52, no. 12, pp. 2534-2540, 2005.

[22] J.W. Johnson, J. Gao, K. Lucht, J. Williamson, C. Strautin, J. Riddle, R. Therrien, P. Rajagopal, J.C. Roberts, A. Vescan, J.D. Brown, A. Hanson,

S. Singhal, R. Borges, E.L. Piner, and K.J. Linthicum "Material, Process, and Device Development of GaN-based HFETs on Silicon Substrates", *Electrochemical Society Proceedings*, 2004, pp. 405-419.

Identification of Electronic Traps in AlGaN/GaN HEMTs using UV light-assisted Trapping Analysis

M. Ťapajna, R. J. T. Simms, M. Faqir, M. Kuball

H. H. Wills Physics Laboratory
University of Bristol
Tyndall Avenue, Bristol, BS1 8TL, UK
phone: (+44) –(117)- 3318-109, milan.tapajna@bristol.ac.uk

Y. Pei, U. K. Mishra

Department of Electrical and Computer Engineering
University of Santa Barbara California
Santa Barbara, CA 93106, USA

Abstract—**UV light-assisted trapping analysis in conjunction with electroluminescence studies was employed to identify the location of traps generated in AlGaN/GaN HEMTs submitted to on-state stress. Our results indicate that UV light-assisted trapping is closely related to traps in the access region close to the gate edges. An increase in the dominant electronic trap density spatially located within the AlGaN layer underneath the gate and in the access region close to the drain side of the gate edge was found to be the most pronounced degradation mechanism for the stress conditions investigated. This trap level was found to be located 0.5 eV below the AlGaN conduction band.**

Keywords-AlGaN/GaN HEMTs; reliability; UV light-assisted trapping; electroluminescence; trap location;

I. INTRODUCTION

In order to utilize the full potential of the AlGaN/GaN material system for high electron mobility transistors (HEMTs) for communication and radar applications, it is necessary to address reliability issues, i.e., the degradation of performance during HEMT operation [1]-[6]. Detailed knowledge of the physics behind the degradation processes is essential to enable future reliability improvements, however, this is still lacking mostly due to the difficulty in identifying the nature of the electronic traps generated during stressing and their location within the device. In the present paper, UV light-assisted trapping analysis in conjunction with an electroluminescence study is developed and employed to identify the location of traps generated in AlGaN/GaN HEMT submitted to on-state stress. It is demonstrated that UV light-assisted trapping analysis represents powerful tool for trap identification in AlGaN/GaN HEMTs.

II. EXPERIMENTAL DETAILS

AlGaN/GaN HEMTs were grown on SiC substrates using metal-organic chemical vapor deposition (MOCVD), 25 nm $Al_{0.26}Ga_{0.74}N$ on top of a 2 µm semi-insulating GaN buffer,

with 0.7 nm AlN layer in between. For device fabrication, standard Ti/Al/Ni/Au source/drain ohmic metallization was used and annealed at 870 °C for 30 s, while I-shaped Ni/Au was used for the gate contact. Devices were passivated using 160 nm SiN_x deposited by plasma-enhanced CVD. HEMTs with source-gate spacing, gate-drain spacing, gate length, and gate width of 0.5, 2, 0.7, and 300 µm, respectively, were subjected to on-state stressing. On-state stress with V_{gs}=0 V and V_{ds}=30 V for 40 hours was used (channel temperature ~300°C).

Trapping analysis similar to [7] was employed to determine relative trap densities and activation energies. After applying a filling pulse with V_{gs}=-10 V and V_{ds}=0 V, HEMT is stepped into linear regime with V_{gs}=1 V and V_{ds}=0.5 V and drain current (I_d) is monitored in the logarithmic scale from 10^{-3} to 10^3 s. The key new addition presented in this work is that trapping characteristics were measured after UV light exposure using spectrally filtered UV light ranging from λ=290 to 380 nm ($\Delta\lambda$~1.14 nm), i.e., with photon energies higher than AlGaN E_g (4.1 eV), higher than the GaN E_g (E_g=3.4 eV) but below the AlGaN E_g, and below the GaN band-gap, respectively. The HEMT is first illuminated by UV light for

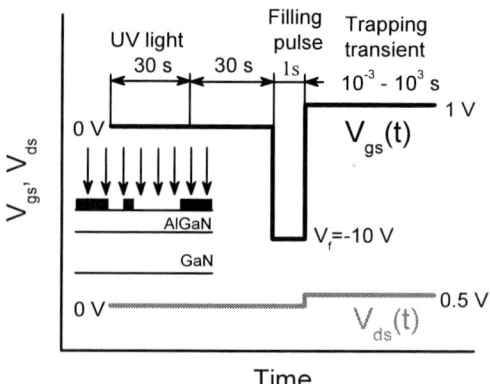

Figure 1. Illustration of UV light-assisted trapping analysis voltage waveform and device illumination (inset).

Funding from ONR and ONR Global through (N00014-08-1-1091) DRIFT program (monitored by Paul Maki) is gratefully acknowledged.

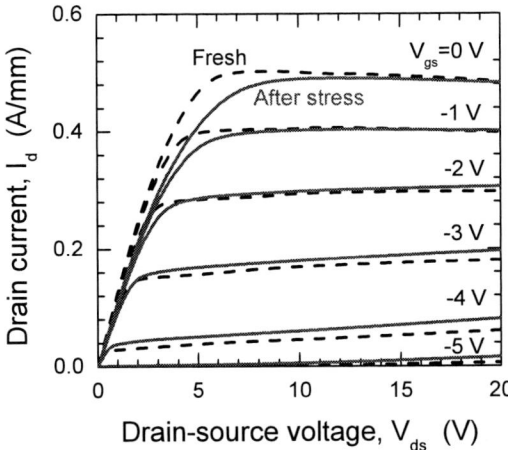

Figure 2. Output characteristics measured on AlGaN/GaN HEMT before (dashed lines) and after (solid lines) on-state stress.

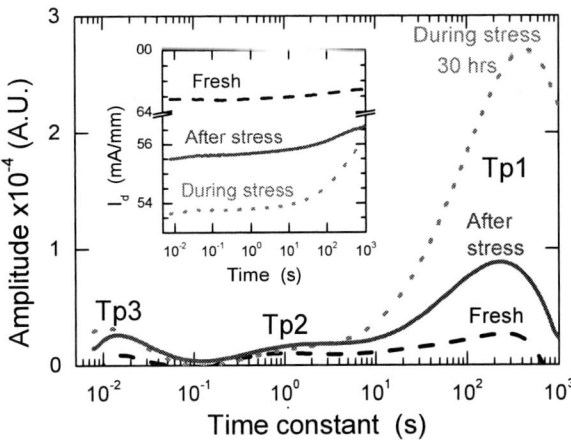

Figure 4. Time constant analyses (without UV light exposure) measured before (dashed line), during (30 hrs; dotted line) and 1-day after (solid line) on-state stress. Inset shows corresponding drain current transients.

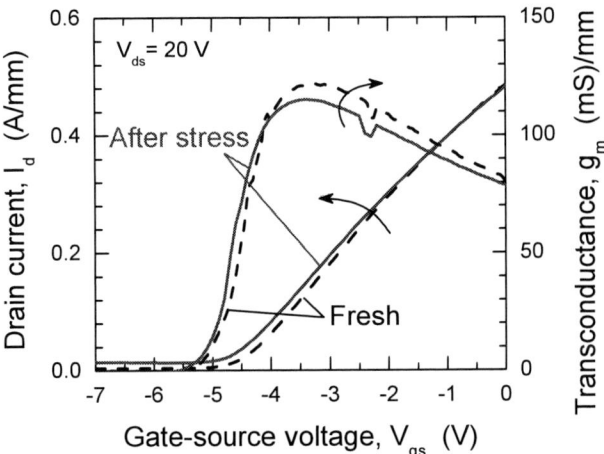

Figure 3. Transfer characteristics and transconductance as a function of V_{gs} measured before (dashed lines) and after (solid lines) on-state stress.

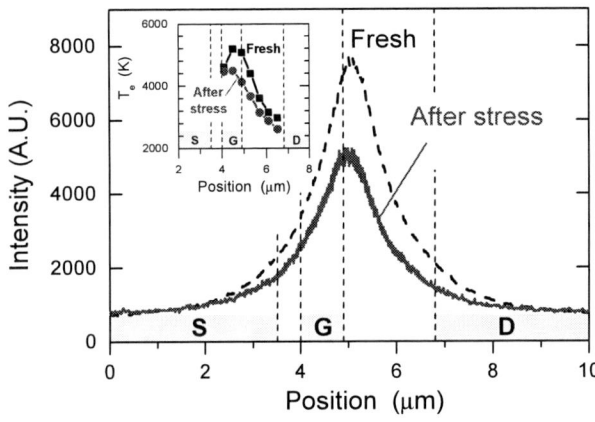

Figure 5. EL intensity across AlGaN/GaN HEMT before and after 40 hour on-state stress measured at V_{ds}=30 V and V_{gs}=0. EL measurement was preformed in the center of the device finger in the distance of 75 µm from the mesa etch. The inset shows hot electron temperature before and after stress.

30 s followed by 30 s for reaching an equilibrium (Fig. 1). Changes in the trap occupation after the UV illumination are measured. Electroluminescence (EL) spatial imaging was used to determine the information on electric field strength distribution and hot electron temperature distribution within the devices, to complement the electrical trap analysis.

III. RESULTS AND DISCUSSION

On-state stress resulted in a drain current decrease mostly in the linear regime (up to ~15%) and a slight permanent negative threshold voltage (V_{th}) shift from -4.5 V to -4.7 V, as shown in Figure 2 and 3 for a representative device. Gate leakage current (I_g) increased from 3.1±0.5 to 15±2 mA/mm (measured at V_{gs}=-7 V and V_{ds}=20 V on three devices stressed). Such changes in the electrical characteristics can be attributed to the generation of traps in the device. Figure 4 shows the time constant analyses derived from I_d transients (inset) measurements

without UV light illumination before, during and one-day after on-state stress. Following an analysis similar to our previous work [8], a detailed analysis of Figure 4 reveals that the fresh devices studied here exhibit two temperature dependent traps denoted in the following as Tp1 and Tp2 with activation energies of 0.45 and 0.12 eV respectively, and one temperature independent trap (Tp3). Time constant analysis illustrates that on-stressing results in particular in a filling of Tp1 trap states during the stress, but also a small permanent increase in Tp1 trap density, as measurements performed one day after on-state stress still showed an increase Tp1 signal. This is in contrast to the result of off-state stress [8], where the density of all three traps significantly increased. This different signature of on- versus off-state stress may be due to the differences in leakage current, electric field distribution, temperature, and converse piezo-electric stresses between both stress conditions.

978-1-4244-5430-3/10 $26.00 © 2010 IEEE 153

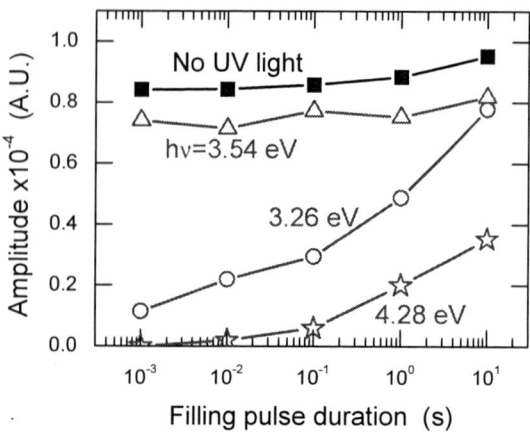

Figure 6. Time constant analysis and drain current transients without (solid squares) and after (open symbols) UV light exposure.

Figure 7. Tp1 peak amplitude as a function of filling pulse duration without (solid squares) and after UV light exposure (open symbols) with different photon energies.

EL distribution across the device channel, from source to drain contact, before and after stress is shown in the Figure 5. Somewhat larger changes in EL intensity are apparent on the drain side of the gate edge extending towards the drain (reduction by about ~30%) while a smaller EL intensity decrease was observed in the source-gate access region (~15%). As EL intensity originates from intraband optical transition of hot-electrons [9], [10], it is related to the electric field strength. This indicates that traps generation, most likely of Tp1, is somewhat more pronounced at the drain side of the gate. Similar conclusions can be drawn from an electron temperature (T_e) profile (inset of Fig. 5) determined from the high energy tail of the EL spectrum, with the most pronounced changes taking place on the drain side of the gate contact.

While a standard electrical trap analysis coupled with EL analysis identifies the lateral location of traps in a device affected by stress, it does not give access to information whether these are surface, bulk AlGaN or bulk GaN traps. To identify in which device layer the traps are located, a UV light-assisted trapping analysis was developed and employed. Figure 6 shows time constant analysis performed following the procedure described in Fig. 1 using UV light illumination with photon energy of 3.26, 3.54, and 4.28 eV. Only Tp1 is affected by the illumination used, showing its amplitude slightly decreasing after $\hbar\omega$=3.26 eV illumination, increasing with $\hbar\omega$=3.54 eV, and a significant decrease after $\hbar\omega$=4.28 eV illumination, while Tp2 and Tp3 did not show a change large enough to be observable. In the following, we restrict the discussion to the dominant Tp1 trap in the devices.

The strong dependence of the Tp1 amplitude on UV light suggests that besides traps located in AlGaN/GaN underneath the gate, transient characteristics are also sensitive to traps located at the access regions close to the gate edges as these parts are directly exposed to the UV light. Filling of these traps can be governed via the surface leakage current and electron tunneling. Moreover, the dependence of the Tp1 amplitude on different photon energy illuminations strongly suggests that Tp1 is located either in the AlGaN barrier layer or at its surface. From temperature dependent measurements, we

determined an activation energy for the Tp1 of about 0.45 eV [8]. UV light with $\hbar\omega$=3.26 eV results in electron emission from the Tp1 leading to partial trap ionization in the access regions. The electrical pulse to fill Tp1 then becomes less efficient, decreasing the Tp1 peak amplitude as observed experimentally. Consistently, an increase in the absolute value of the drain current was observed (inset of Fig. 6). For UV light with $\hbar\omega$=3.54 eV photon energy, both electron emission from the Tp1 and transitions of electrons from the AlGaN valance band to the trap state are possible. We speculate that electron capture dominates over the emission process, then leading to subsequent Tp1 amplitude increase, and absolute drain current decrease compared to 3.26 eV in agreement with our observation. In contrast, UV light with $\hbar\omega$=4.28 eV introduces an additional direct electron-hole pair generation in the AlGaN by band-to-band transitions. However, significant absolute drain current increase (Fig. 6 inset) suggests electron emission from the traps to be a dominant process over their population that leads to strong trap ionization. As a result, electrical filling becomes less efficient in populating Tp1 trap, decreasing its peak amplitude as shown in Figure 6.

To confirm this interpretation, Figure 7 shows the Tp1 amplitude as a function of filling pulse duration (t_f) without UV light exposure and after exposure to UV light with $\hbar\omega$=3.26, 3.54 and 4.28 eV. Without UV light exposure, Tp1 amplitude is almost unaffected by filling pulse duration indicating fast filling of the traps. After UV light with $\hbar\omega$=3.26 eV, Tp1 amplitude drops abruptly when t_f=10^{-3} s compared to no UV light exposure and then it increases with t_f increasing. On the contrary, after UV light exposure with $\hbar\omega$=3.54 eV prior to the transient measurement, changes in the t_f do not affect significantly the Tp1 amplitude, while there is again a strong dependence on filling pulse length after 4.28 eV illumination. This difference is consistent with the aforementioned hypothesis that Tp1 is a bulk or surface AlGaN trap. UV light with $\hbar\omega$=3.26 eV results only in emission of electrons from Tp1 and longer filling pulse is then necessary to fill these traps again. As illustrated in Figure 8, exposure of the device by the

978-1-4244-5430-3/10 $26.00 © 2010 IEEE   154

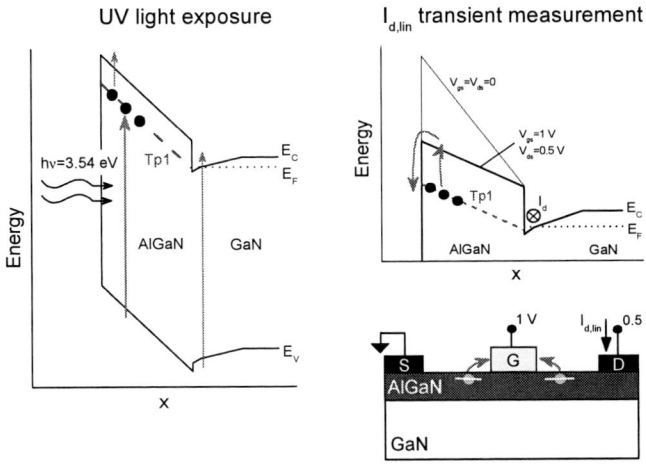

Figure 8. Schematic illustration of excitation processes in band diagram of AlGaN/GaN HEMT access region after exposure of UV light with photon energy of 3.54 eV (left) and subsequent detrapping mechanism illustrated in band diagram (upper right) and HEMT (lower right).

UV light with $\hbar\omega=3.54$ populates the levels in AlGaN so that the length of the filling pulse during subsequent I_d transient measurement has a small effect on Tp1 amplitude. However, UV light with photon energy $\hbar\omega=4.28$ eV depopulates all of the traps that can be understood as a result of more light being absorbed in AlGaN layer. Consequently, longer filling pulse duration is necessary to fill Tp1.

Degradation in the AlGaN/GaN HEMTs investigated here under on-state stress is consistent with the generation of donor-like Tp1 traps in the AlGaN layer underneath the gate as well as in the access region close to the drain side of the gate contact. Traps generated underneath the gate can cause a negative shift of V_{th} (Fig. 3) while those generated at the drain side of the gate edge can be responsible for electric field decreasing in this region decreasing the EL intensity (Fig. 5). Moreover, traps located close to the drain side of the gate edge would also be consistent with slight decrease in the transconductance at low V_{gs} while g_m remains almost unchanged at high V_{gs} voltages, suggesting negligible trap generation close to the drain (Fig. 3). Similar trap evolution during on-state stress to that reported for off-state stress [8] indicates a similar trap generation mechanism to off-state stress, namely impurity diffusion into AlGaN close to the drain side of the gate driven by increased leakage current and high electric field, however other contributing mechanisms are also possible. Although 10-hour-long on-state stress reported in [8] did not produce any significant device degradation, prolonged stressing time studied here caused more pronounced

degradation consistent with progressive, diffusion driven process. However, since increased filling of the traps during on-state stress was observed (Fig. 4), contributions of hot-electron degradation may also play a role.

IV. CONCLUSION

UV light-assisted trapping analysis together with EL measurements was used to determine trap location generated in AlGaN/GaN HEMTs under on-state stress. The results suggest that traps are generated either in the AlGaN bulk or its surface spatially located underneath the gate and in the access region close to the drain side of the gate. In agreement with our previous data, degradation mechanisms in the studied AlGaN/GaN HEMTs are consistent with impurity diffusion into AlGaN at the drain side of the gate driven by increased leakage current and high electric field, however, hot-electrons may play an additional role.

REFERENCES

[1] S. Singhal, J. C. Roberts, P. Rajagopal, T. Li, A. W. Hanson, R. Therrien, J. W. Johnson, I. C. Kizilyalli, and K. J. Linthicum, "GaN-on-Si failure mechanism and reliability improvements," IEEE Int. Reliab. Phys. Symp., 2006, pp. 95-98.

[2] T. Ohki, M. Kanamura, N. Okamoto, K. Imanishi, K. Makiyama, K. Joshin, T. Kikkawa, and N. Hara, "Effect of gate edge silicidation on gate leakage current in AlGaN/GaN HEMTs," Int. Conf. Compound Semiconductor MANTECH Tech. Dig., 2008, pp. 249-252.

[3] J. L Jimenez and U. Chowdhury, "X-band GaN FET reliability," IEEE Int. Reliab. Phys. Symp. Dig., 2008, pp. 429-435.

[4] J. Joh and J. A. del Alamo, "Critical voltage for electrical degradation of GaN high-electron mobility transistors," IEEE Electron Device Lett., vol. 29, no. 4, pp. 287-289, Apr. 2008.

[5] G. Meneghesso, G. Verzellesi, F. Danesin, F. Rampazzo, F. Zanon, A. Tazzoli, M. Meneghini, and E. Zanoni, "Reliability of GaN high-electron-mobility transistors: State of the art and perspectives," IEEE Trans. Device and Materials Rel., vol. 8, no. 2, pp. 332-343, June 2008.

[6] R. J. Trew, D. S. Green, and J. B. Shealy, "AlGaN/GaN HFET reliability," IEEE Microwave Mag., vol. 10, no. 4, pp. 116–127, June 2009.

[7] J. Joh and J. del Alamo, "Impact of electrical degradation on trapping characteristics of GaN high mobility transistors," IEDM Tech. Dig., 2008, pp. 1-4.

[8] M. Ťapajna, R. J. T. Simms, Y. Pei, U. K. Mishra, and M. Kuball, EDL, "Integrated optical and electrical analysis: Identifying location and properties of traps in AlGaN/GaN HEMTs during electrical stress," submitted to IEEE Electron Dev. Lett.

[9] N. Shigekawa, K. Shiojima, and T. Suemitsu, "Optical study of high-biased AlGaN/GaN high-electron-mobility transistors," J. Appl. Phys., vol. 92, no. 1, pp. 531-535, July 2002.

[10] J. W. Pomeroy, M. Kuball, M. J. Uren, K. P. Hilton, R. S. Balmer, and T. Martin, "Insight into electroluminescent emission from AlGaN/GaN field effect transistors using micro-Raman thermal analysis," Appl. Phys. Lett., vol. 88, no. 2, p. 023507, Jan. 2006.

Reliability of SiC Power Devices and its Influence on their Commercialization – Review, Status, and Remaining Issues

Michael Treu
Technology Development SiC and High Voltage MOS Devices
Infineon Technologies Austria AG
Villach, Austria
phone: (43) - (5) - 1777-6076, michael.treu@infineon.com

Roland Rupp, Gerald Sölkner
Technology Development SiC and High Voltage MOS Devices
Infineon Technologies AG
Neubiberg, Germany

Abstract—**The following paper will give an overview about the main reliability aspects of silicon carbide power devices. After a brief review of the key device concepts it covers reliability topics of bipolar devices, Schottky diodes, metal oxide semiconductor field effect devices, and junction field effect devices. Special attention is paid to the influence of the different reliability topics on the commercialization of the different device types. It will be shown that for some device types the reliability is at a very high level being not hampering commercialization (e.g. Schottky diodes or junction field effect devices) whereas other devices concepts still need improvement until commercialization (e.g. bipolar devices).**

silicon carbide; bipolar degradation; MOS; gate oxide integrity; packaging; Schottky diode

I. INTRODUCTION

Silicon carbide (SiC) is a very attractive base material for power semiconductor devices. Some of the striking physical properties are a wide band gap of about 3 eV, an about ten times higher critical electrical field, and an about three to five times higher thermal conductivity compared to silicon (depending on doping and temperature). The 1st device type which has been commercially available is the SiC Schottky diode, which has been introduced in 2001 by Infineon Technologies followed by the introduction of diodes by CREE and lately ST Microelectronics. The application domains of SiC Schottky Diodes are first of all boost diodes in switch mode power supplies running with high switching frequency and/or highest efficiency. Although research activities started in the late 1980s these are the only SiC based commercial power devices for mass applications so far. Most important for this are the device costs compared to silicon devices. Nevertheless massive cost downs in the last years brought SiC diode prices in competitive ranges and attractive costs can be expected in the future also for switches. The 2nd important reason are reliability concerns for some device types, whereas such device types are well established in the silicon world.

These are for example bipolar or metal oxide semiconductor (MOS) devices. However first announcements about commercial metal oxide semiconductor field effect transistors (MOSFETs) have been made by CREE together with Powerex [1].

The main reason for the reliability issues is the high amount of crystal defects in SiC wafers compared to silicon wafers. The most prominent and killing defect is the so called micro pipe, which is a hollow core dislocation with a diameter in the nm or μm range and lengths of several centimeters through a crystal. Being a defect with densities of hundreds per cm^2 in the 1990s this defect is in the mean time less important since the densities have been reduced to only a few per cm^2 or even a few per wafer. Furthermore micro pipes normally lead to reduced yield but not to reliability issues. The amount of screw, edge, and basal plane dislocations is still in the range of 10^2 cm^{-2} to 10^4 cm^{-2} [2].

The following sections will describe the functionality and some design aspects of the main power device types followed by a review of the reliability topics of each type.

II. BIPOLAR DEVICES

Bipolar SiC devices target applications where blocking capabilities of at least some kV are required. These are mainly industrial applications like motor drives, large power supplies, or inverters for renewable energies where reliability requirements are traditionally very high since the equipment lifetime can reach quite some decades.

Today bipolar devices in SiC suffer from the so called forward voltage degradation (in the case of diodes) or current gain degradation (in the case of bipolar junction transistors, BJTs). It means that the forward voltage drop of a pn diode increases within minutes or hours during regular current operation. In the case of BJTs this leads to a significant reduction of the current gain. This phenomenon was reported to the SiC community in 1999 at the international conference for SiC and related devices by the group of ABB without being documented in the proceedings. At first some people in the community have been very skeptical if this is really a severe issue. But being back in their labs most developers had to admit that they simply did not stress the devices at high enough current density for a longer time and that in fact it was more or less an epidemic issue at that point of time.

Shortly later in the year 2000, first publications by the same group revealed that the root cause are basal plane dislocations (BPDs) which expand fast into triangular shaped stacking faults during the presence of electron hole recombination [3], [4], [5]. Figure 1 show a schematic cross section of a pn diode and the degradation mechanism. Since SiC wafers used for homo epitaxy have a 4° or 8° off-orientation of the crystallographic c-axis against the surface, these stacking faults start at the substrate epitaxy interface and grow inclined to the surface throughout the epitaxy and – according to the off-orientation – cover a large device area. The stacking faults in turn decrease the minority carrier life time and form a barrier for the current flow, both leading to increased voltage drop.

Skowronski et al. [6], Stahlbush et al. [7], and Ha et al. [8] made detailed analysis of the mechanisms of stacking fault generation, the nature and orientation of the involved BPDs which laid the ground for upcoming work proposing first solutions to overcome the issue. It was observed that about 90% of the BPDs being present in the substrate convert into threading edge dislocations TEDs during homoepitaxial growth. These kinds of dislocations do not lead to any V_f degradation. For Si-faced 4H-SiC the conversion rate depends on the off-orientation angle, substrate preparation and epitaxy growth parameters. To enhance the conversion rate it was proposed in 2005 by Sumakeris et al. to prepare the surface either by etching the surface in molten KOH or a combination of lithography and dry etching [9]. The KOH etching produces etch pits at every dislocation and generates a surface nearly matching the basal plane enabling a high conversion probability into TEDs. This method improved the yield of V_f drift-free 20A 10kV devices dramatically equivalent to a BPD density of $\sim 10/cm^2$ [10]. Unfortunately the blocking yield dropped nearly to zero unless a sophisticated polishing process was used to cope with the surface roughness after KOH etching. The lithographic version of the approach resulted in a higher blocking yield but was less effective in eliminating the V_f drift. Overall the method was an approach giving very valuable input how to address the issue but did not finally

enable a high overall yield suitable for commercial production of SiC pn diodes.

Recently Kallinger et al. published very promising results for 4° off-oriented 4H SiC substrates [11]. This work showed that it is possible to decrease the amount of BPD below $1/cm^2$ by choosing a proper combination of process parameters without any surface structuring. Reference samples with KOH etched surface confirmed also a high BPD to TED conversion rate but the BPD density stayed about three times higher. A further optimization and proof of the stability of this approach might be a very good starting point for the development of commercial SiC pn diodes in the next years. In case, that a save 100% elimination a BPDs will not be possible, it will be crucial, to develop a fast and reliable screening test, to sort out devices with degradation probability already on wafer level. Only such a screening will allow sufficient reliability on module level, where typically many of single pn diode chips will be assembled per unit

Figure 2 shows the schematic cross section of a vertical power BJT. In the case of BJTs also the surface has to considered as an additional source for device degradation. On the one hand the interface states act as recombination centers and affect the current gain. Additionally it is reported that stacking faults start growing at the surface near the emitter in the presence of electron hole recombination and degrade the initial current gain of ~30 by 50% after a 16 hours stress with ~450 A/cm^2 [12] even without the presence of initial BPDs. A proper processing of the surface passivation layers can reduce the degradation. Domeij et al. claim that they improved the processing and reached current gains of about 20 to 35 being stable even after 660 hours of base current stress [13]. It remains open if high collector current stress causes any degradation in these devices and if they can commercially compete with their faster switching competitors MOSFET or junction FET.

III. Schottky Diodes

Figure 3 shows the cross section of a pure Schottky diode and a merged pn Schottky diode. The important design parameters for a Schottky diode are the choice of the Schottky metal leading to a certain Schottky barrier height, the drift layer

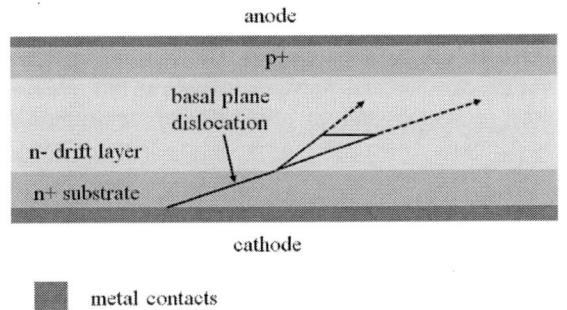

Figure 1. a) Schematic cross section of a vertical pn diode with an examplary schematic of a basal plane dislocation growing stacking fault.

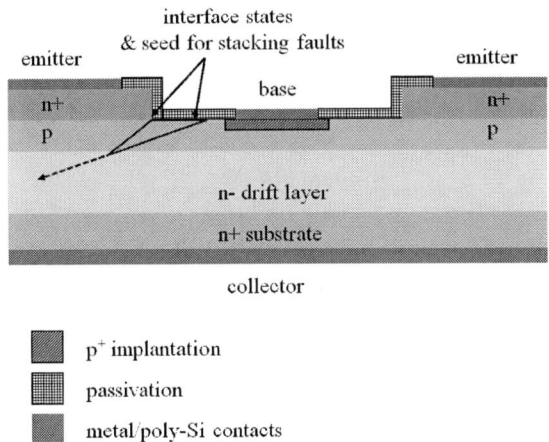

Figure 2. a) Schematic cross section of a bipolar junction transistor (BJT) with an examplary schematic of a growing stacking fault.

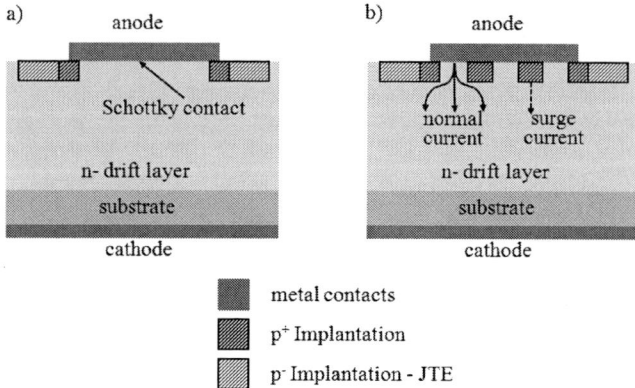

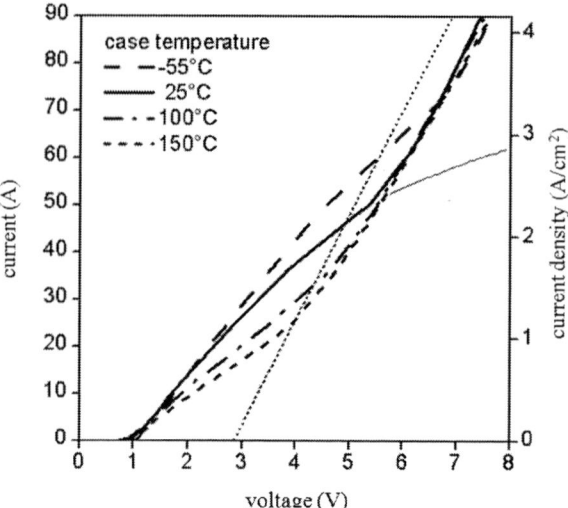

Figure 4. Forward characteristic of a merged pn Schottky diode as shown in Figure 3 b) with the case temperature as a parameter. The thin solid line indicates the behavior of pure Schottky diode and the thin dashed line indicates the substrate resitivity shifted by the pn threshold voltage of 2.8V.

Figure 3. a) Schematic cross section of a Schottky diode including a junction termination extension (JTE) and b) the same device with additional merged pn junctions a so called junction barrier Schottky diode (JBS).

thickness, and it's doping. These parameters influence the electric field at the Schottky interface and thus the stress level at this interface. For a merged pn Schottky diode the situation is slightly different. The p+ areas in the active area can reduce the electric field at the Schottky interface by a shielding effect which can be influenced by the distance between the p+ areas. This is an important degree of freedom to reduce the electrical field stress level at the rectifying interface in certain limits independent of the drift layer doping. The question is now how the electrics field stress at the interface alters the reliability of the device. In general there are no mechanisms known describing the degradation or life time of a Schottky contact under electrical stress; it is assumed and a matter of long term experience that Schottky interfaces do not degrade under electrical stress alone. Nevertheless the electric field causes a certain reverse current leading to heat dissipation. This effect gets important if the Schottky barrier is reduced locally e.g. by crystal defects leading to local temperature increase which can degrade the Schottky interface and thus lead to device failure over time. Rupp et al. examined this effect at the example of an epitaxially overgrown micropipe [14]. In this work the experimentally observed degradation was explained with numerical simulation of the temperature distribution at and around an assumed localized leakage path. The conclusion was that it is possible to define a critical leakage current for a localized defect which is higher than the average current of all devices. If the leakage current is below this critical current no degradation will occur. Thus it is possible to sort out all critical devices by defining a proper leakage current limit without losing disproportional yield.

Another reliability topic is that most applications using SiC Schottky diodes require a certain surge current capability of the device which is significantly higher than the normal operation current. If this requirement is not fulfilled device failures in the field may occur during line cycle drop outs or similar irregularities at the mains. This requirement cannot be fully addressed with a structure as shown in Figure 3 a) since this purely unipolar device has a positive temperature coefficient for high currents. In the case of the junction barrier Schottky (JBS) diode shown in Figure 3 b) the pn junctions takes over the current leading to a behavior shown in Figure 4. With this kind of structure a surge current capability of more than

4000 A/cm^2 for a 10 ms pulse or several ten kA/cm^2 for a 10 μs pulse can be reached compared to about 2000 A/cm^2 for a 10 ms pulse in the case of a standard Schottky diode [15][16].

On the other hand it has to be considered that the device acts as bipolar devices during these surge current conditions and that bipolar degradation may occur. Brosselard et al. investigated 1.2 kV and 3.5 kV JBS diodes under normal and surge current operation. It appeared that the 3.5 kV devices showed bipolar degradation whereas the 1.2 kV devices were stable [17]. The bipolar degradation of the 3.5 kV devices increased also resulted in an increase in forward voltage in unipolar current operation. This proves that the stacking faults form a barrier to the majority carriers or at least impede carrier mobility. The difference between the two devices was attributed to a higher local current density near the edge termination in the case of the 3.5 kV device resulting in stacking fault generation. It can also be argued that the thinner epitaxy of the 1.2 kV device did allow the stacking faults to cover a significant area of the device. This conclusion is supported by work of Holz et al. proofing that no bipolar degradation is occurring in 600 V JBS diodes during a pulse test applying more than 4000 A/cm^2 [18].

Besides such SiC specific failure mechanisms of course a "usual" thermo mechanical degradation due to extensive temperature cycling and related wear out mechanisms will occur in SiC based devices in a similar manner as in standard Si devices. However, in this case standard reliability tests like temperature cycling and power cycling can be easily used to guarantee reliability and to predict wear out time for given mission profiles.

Infineon's 2nd SiC diode generation thinQ!2G™ which is delivered to the field since the beginning of 2006 shows a failure rate of less than 0.15 ppm. This proves also that the pn junctions of the junction termination extension of these devices are stable under reverse bias. Thus it can be stated that the SiC Schottky diode technology is very reliable and that there is no

978-1-4244-5430-3/10 $26.00 © 2010 IEEE

interference of any reliability topics with the commercial use of such devices.

IV. MOS Devices

Metal oxide semiconductor field effect transistors (MOSFETs) have of course been of interest since the very beginning of SiC device development as this device is the most successful device concept in the Si world. Forming a native oxide SiC is a promising candidate for the realization of this concept. Figure 5 shows the cross section of a fully implanted MOSFET. Unlike for MOSFETs on Si the gate oxide is grown after the definition of the p+ and n+ implantation due to the high temperature of about 1700°C needed for post implantation annealing. Additionally a significant amount of implantation defects remains even after this annealing [19]. This results in three areas differing in the oxide growth and stress conditions. In area 1 the oxide grows on SiC implanted with a high nitrogen dose. In area 2 the oxide grows on epitaxial SiC with a comparably low number of crystal defects. In area 3 the oxide grows on aluminum doped SiC. Special processing sequences can also define this p area by epitaxy rather than ion implantation. In the area 1 and 3 the electrical field stress level in the oxide is defined by the choice of the oxide thickness and the gate voltage. Both parameters can only be chosen in a limited range due to common targets for the threshold voltage and the transconductance. Area 2 is also special compared to MOSFETs on Si since the electrical fields in SiC devices are designed to be about 10 times higher than in Si devices. Due to the difference of the dielectric constant of SiC and SiO$_2$ the electric field in the oxide is about 3 times higher than in the SiC. Designing the device with a critical electrical field in SiC of about 2 MV/cm next to the oxide would lead to an unacceptably high field of about 6 MV/cm in the oxide. Nevertheless by decreasing the distance d the electrical field stress in the oxide can be reduced to levels of about 2 to 3 MV/cm commonly used for Si devices.

Commonly used oxidation schemes known from Si technology in the 1990s lead to a very high density of interface states, a poor inversion channel mobility of ≤ 1 Vs/cm^2 (on 4H-SiC) and a large stress induced drift of the threshold voltage

(V_{th}) for n-channel devices on 4H-SiC [20]. Fortunately already at that time it was possible to give hints that at least the intrinsic time dependent dielectric breakdown of oxides on SiC can reach the same level like for oxides on Si [21][22]. To overcome the channel mobility issue it was suggested to use nitrogen assisted oxidation to remove carbon residues from the SiC-SiO$_2$ interface to increase the carrier mobility. In 1999 it was published that the use of NO or N$_2$O during oxidation leads to a removal of carbon clusters and nitrogen incorporation at the SiC-SiO$_2$ interface [23][24]. In the following years it was shown by numerous groups that nitric oxides increase the channel mobility [25] and can also improve dielectric quality [26]. Nitric oxides are still the dielectrics of choice and time dependent dielectric breakdown (TDDB) measurements of such oxides clearly showed that the intrinsic oxide quality can fulfill automotive and industrial quality requirements [27]. Recently promising results have been also reported concerning the threshold voltage stability. Grieb et al. showed that deposited and NO annealed samples show a threshold voltage shift of less than 0.2 V even after 83 h stress at 2 MV/cm and 130°C. What is still an open issue is the relatively high amount of TDDB extrinsics of 10 cm^{-2} to 30 cm^{-2} or even more. Even if the devices are about a factor 10 smaller than Si devices this results in a defect density per chip which at least 10 times higher than for Si MOSFETs. Nakamura et al. [29] claim that those extrinsic defects can be screened before shipment. Nevertheless in the case of Si technologies wafers with such high defect densities are scrapped and it has still to be proven that a reliably screening test can be done.

V. Cosmic Radiation

Cosmic radiation constantly showers our earth with highly energetic particles, mainly protons and nuclei of 10^9-10^{17} eV. Via collisions with air molecules secondary particles with energies up to 1 GeV reach sea level [30]. For Si devices it has been shown in the 1990s that cosmic radiation can cause avalanche carrier multiplication which may lead to a so called single event burnout (SEB) [31]. Being relevant at ground level this effect gains even more importance for power electronics being used at high altitudes e.g. in airplanes, since the particle density in 13 km altitude is about 300 times higher than at sea level.

The much higher breakdown electric field, wider band gap and higher melting temperature compared to Si may lead to the assumption that SiC devices have superior radiation hardness than Si devices. On the other hand, SiC devices are designed to operate at correspondingly higher internal electric field strengths. Therefore, massive radiation-induced carrier multiplication which is the basic mechanism behind SEB of power diodes could occur in SiC devices as well as in devices based on Si. As a matter of fact, SEB of SiC Schottky diodes irradiated by 63 MeV protons has been observed and can be explained by the model of massive carrier multiplication, which is well established for Si devices [32].

Figure 6 shows a failure rate comparison of 600 V SiC devices and a 600 V Si pin diode as a reference. The failure rates are shown in arbitrary units since the exact acceleration factor of the used proton beam compared to the natural cosmic

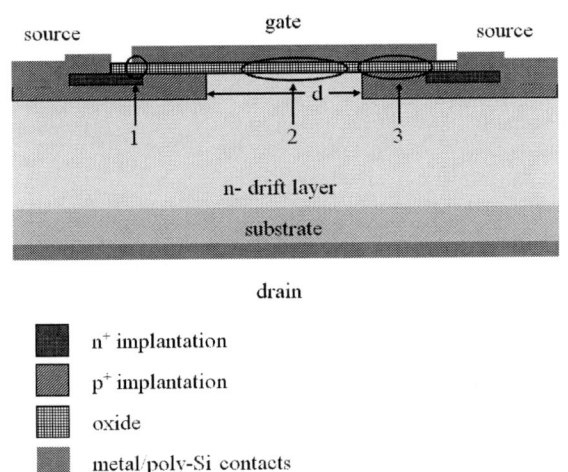

Figure 5. Schematic cross section of a MOSFET with three regions with different oxide growth and stress conditions and the distance d between the p+ body.

978-1-4244-5430-3/10 $26.00 © 2010 IEEE

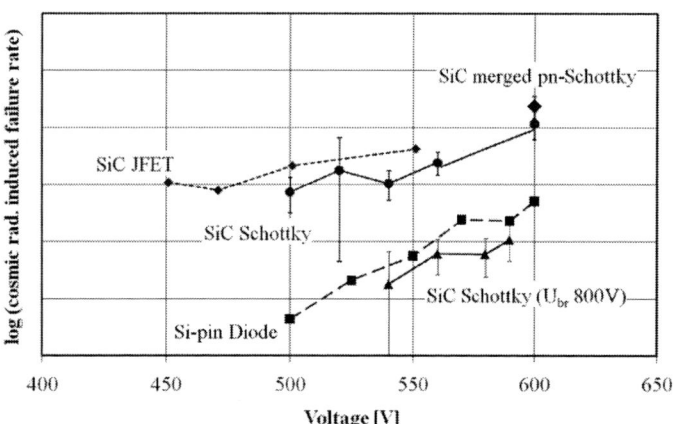

Figure 6. Voltage-dependent failure rates of 600 V SiC devices. Data of a 800 V SiC Schottky diode and of a standard 600 V silicon diode are given as a comparison.

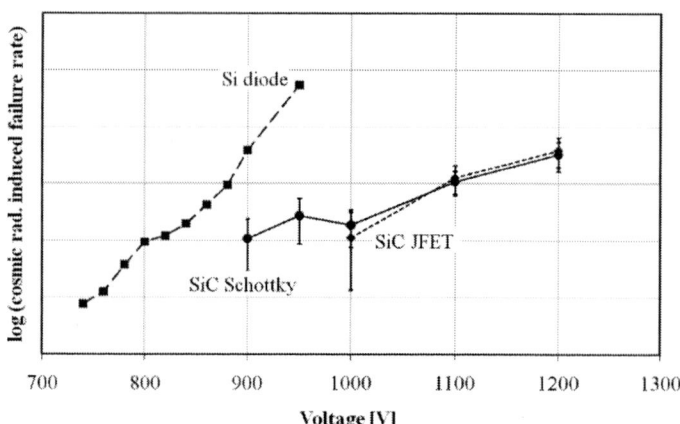

Figure 7. Voltage-dependent failure rates of 1200 V SiC devices. Data of a standard 1200 V silicon diode are given as a comparison.

radiation is still under investigation. As mentioned before the real total failure of Infineon's SiC Schottky diodes is below 0.15 ppm. This means that also cosmic radiation failures are not an issue in the 600 V class. The graph shows that SiC devices indeed do not have higher radiation hardness by default. For the same rated blocking voltage the failure rates of the SiC devices are about 1 to 2 orders of magnitude higher. Due to the absence of a Schottky barrier the SiC JFET allows the design in of higher electrical fields compared to the diodes without sacrificing leakage current limits. This might be the reason for slightly higher failure rates compared to the diodes. The comparison with the 800 V Schottky diode shows that the failure rate can be easily decreased by two orders of magnitude if needed for high radiation and high reliability applications.

Figure 7 show the equivalent results for 1200 V devices. In this case SiC devices indeed show benefits compared to their Si counterparts. Especially if the applied voltage approaches the rated blocking voltage, SiC devices show a dramatically lower failure rate than Si devices. The reason for this is not yet understood. Nevertheless this is a very nice coincidence since in a lot of high voltage applications need a dramatically larger device area compared to 600 V applications to achieve the needed current ratings. This means that at high voltage SiC devices could differentiate against their Si counterparts in the future when large area devices become cheaper and attractive in high volume applications.

VI. CONCLUSION

Today the high reliability of SiC Schottky diodes up to 1200 V blocking voltage is clearly supported by the field experience based on commercial mass production. This proves that Schottky contacts as well as pn junctions (edge termination) can be designed properly and effectively tested by wafer level test. Furthermore no degradation is observed during unipolar current conduction. Due to this positive experience it is likely that the invention of SiC switches for the mass market maybe lead by concepts which are also based on unipolar operation and pure use of pn junctions. A prominent example for such a device is a junction field effect transistor (JFET). Nevertheless the mass market calls for a switch which is normally-off. Although normally-off JFETs can be designed,

this kind of dimensioning is not easy to be produced stable in high volume. Thus mainly MOSFETs will have to provide the normally-off behavior in future. For this concept extrinsic gate oxide breakdown and threshold voltage stability are the demanding tasks. Some work shows that there seems no major blocking point to tackle these issues in the future.

Bipolar device operation today is only under control for blocking voltages slightly above 1 kV and helps to improve surge current capability of Schottky diodes or provides body diode operation for unipolar switches in the same blocking voltage range. Purely bipolar devices for higher blocking voltages seem to have still some way to go until commercialization due to the forward voltage degradation of thick epitaxial layers. This is issue is accompanied by the demand of relatively large chip sizes to support high current ratings. Thus very low defect densities have to be achieved to generate cost competitive devices. Nevertheless it was possible to show that the cosmic radiation hardness of SiC devices for high blocking voltages is better compared to their Si counterparts. This is a promising perspective, since cosmic radiation failures are a prominent failure mode for high voltage Si devices.

REFERENCES

[1] http://www.pwrx.com/TechnicalDocument.aspx?id=966

[2] Paper ISCRM 2010 status crystal defecte

[3] J.P. Bergmann, H. Lendenmann, P.A. Nilsson, and P. Skytt, "Crystal defects as source of anomalous forward voltage increase of 4H-SiC diodes", Mat. Sci. For., vols 353-356, pp.299-302, 2001

[4] H. Lendenmann, F. Dahlquist, J.P. Bergmann, H. Bleichner and C. Hallin, "High-power SiC Diodes: characteristics, reliability and relation to metarial defects", Mat. Sci. For., vols 389-393, pp. 1259-1264, 2002

[5] H. Lendenmann, J.P. Bergmann, F. Dahlquist, and C. Hallin, "Degradation in SiC bipolar devices: source and consequence of electrically active dislocations in SiC", Mat. Sci. For., vols 433-436, pp.901-906, 2003

[6] M. Skowronski, J.Q. Liu, W.M. Vetter, M. Dudley, C. Hallin, and H. Lendenmann, "Recombination-enhanced defect motion in forward-biased 4H-SiC pn diodes" J. Appl. Phys., vol. 92, no. 8, pp. 4699-4704, 2002

[7] R.E. Stahlbush, M.E. Twigg, J.J. Sumakeris, K.G. Irvine, and P.A. Losec, "Mechanisms of stacking fault growth in SiC pin diodes", Mat. Res. Soc. Symp. Proc., vol. 815, pp. 103-113, 2004

[8] S. Ha, P. Mieszkowski, M. Skowronsky, and L. Rowland, "Dislocation conversion in 4H silicon carbide epitaxy", J. Cryst. Growth, vol. 244, no. 3-4., pp. 257- 266, 2002

[9] J.J. Sumakeris, J.P Bergman, M.K. Das, C. Hallin, B.A. Hull, E. Janzén, H. Lendenmann, M.J. O'Loughlin, M.J. Paisley, S. Ha, M. Skowronski, J.W. Palmour, and C.H Carter, "Techniques for minimizing the basal plane dislocation density in SiC epilayers to reduce V_f drift in SiC bipolar power devices", Mat. Scie. For., vols. 527-529, pp. 141-146, 2006

[10] M.K. Das, J.J. Sumakeris, B.A. Hull, and J. Richmond, "Evolution of drift-free, high power 4H-SiC PiN diodes", Mat. Scie. For., vols. 527-529, pp. 1329-1334, 2006

[11] B. Kallinger, B. Thomas, and J. Friedrich, " Influence of substrate preparation and epitaxial growth paramters on the dislocation densities in 4H-SiC epitaxial layers", Mat. Scie. For., vols. 600-603, pp. 143-146, 2009

[12] A. Agarwal, S. Krishnaswami, J. Richmond, C. Capell, S.-H. Rye, J. Palmour, B. Geil, D. Katsis, C. Scoozie, and R. Stahlbush, "Influence of basal plane dislocation induced stacking faults on the current gain in SiC BJTs", Mat. Sci. For., vols. 527-529, p.1409-1412, 2006

[13] M. Domenji, C. Zaring, A.O. Konstantinov, M. Nawaz, J.-O. Svedberg, K. Gumaelius, I. Keri, A. Lindgren, B. Hammarlund, M. Östling, M. Reimark, "2.2 kV SiC BJTs with low VCESAT, fast swicthing and short-circuit capabiliy", Mat. Scie. For., vols. 645-648, pp. 1033-1036, to be published

[14] R. Rupp, M. Treu, P. Türkes, H. Beermann, T. Scherg, H. Preis, and H. Cerva, "Influence of overgrown micropipes in the active area of SiC Schottky diodes on long term reliability", Mat. Sci. For., vols. 483- 485, p. 925-8, 2004

[15] M. Treu, R. Rupp, C.S. Tai, P. Blaschitz, J. Hilsenbeck, H. Brunner, D. Peters, R. Elpelt, and T. Reinmann, "A surge current stable and avalanche rugged SiC merged pn Schottky diode blocking 600V especially suited for PFC applications", Mat. Scie. For., vols. 527-529, pp. 1155-1158, 2006

[16] B. Heinze, J. Lutz, M. Nuemeister, R. Rupp, and M. Holz, "Surge current ruggedness of silicon carbide Schottky and merged-pin-Schottky diodes", Proc. 20th Int. Symp. Power Sem. Dev. IC's, pp. 245-248, 2008

[17] P. Brosselard, N. Camara, X. Jordà, M. Vellvehi, E. Bano, J. Millan, and P. Godignon, "Reliability aspects of high voltage 4H-SiC JBS doides", Mat. Scie. For., vols. 600-603, pp. 935-603, 2007

[18] M. Holz, J. Hilsenbeck, and R. Rupp, "Reliability aspects of SiC Schottky diodes", Phys. Stat. Solidi, vol. 206, no. 10, pp. 2295-2307, 2009

[19] R. Stief, "Dotierung von 4H-SiC durch Ionenimplantation", Erlanger Berichte Mikroelektronik, Band 2000,1, Shaker Verlag, 2000

[20] Private communication R. Schörner

[21] M. Treu, E.P Burte, R. Schörner, P. Friedrichs, and H. Ryssel, "Reliability of metal-oxide-semiconductor capacitors on nitrogen implanted 4H-silicon carbide", J. Appl. Phys., vol. 84, no. 5, pp. 2943-2948, 1998

[22] M. Treu, R. Schörner, P. Friedrichs, R. Rupp, A. Wiedenhofer, D. Stephani, and H. Ryssel, "Reliability and degradation of metal-oxide-semiconductors on 4H- and 6H-silicon carbide", Mat. Scie. For., vols. 338-342, pp. 1089-1092, 1999

[23] P. Tanner, S. Dimitrijev, H.F. Li, D. Sweatman, K.E. Prince, andH.B. Harrison, "SIMS analysis of nitrided oxides grown on 4H-SiC", J. Electr. Mat., vol. 28, no.2 , pp. 109-111, 1999

[24] H.F. Li, S. Dimitrijev, D. Sweatman, H.B. Harrison, P.Tanner, and B. Feil, "Investigations of nitric oxide and Ar annealed SiO2/SiC interfaces ba X-ray photoelectron spectroscopy", J. Appl. Phys., vol. 86, no8, pp 4316-4321, 1999

[25] M.K. Das, B.A. Hull, S. Krishnaswami, F. Husna, S. Haney, A. LElis, C.J. Scozzie, and J.D. Scofield, " Improved 4H-SiC MOS Interfaces produced via two independent processes: metal enhanced oxidation and 1300°C NO Anneal", Mat. Scie. For., vols. 527-529, pp. 967-970, 2006

[26] T. Kimoto, H. Kawano, M. Noborio, J. Suda, and H. Matsunami, "Improved dielectric and interface properties of $H-SiC MOS structures processed by oxide depoistion and N2O annealing", Mat. Scie. For., vols. 527-529, pp. 987-900, 2006

[27] M. Treu, R. Rupp, P. Blaschitz, K. Rüschenschmidt, Th. Sekinger, P. Friedrichs, R. Elpelt, and D. Peters, "Strategic considerations for unipolar SiC switch options: JFET vs. MOSFET", 42nd Ind. Appl. Conf., pp. 324-330, 2007

[28] M. Grieb, M. Noborio, D. Peters, A.J. Bauer, P. Friedrichs, T. Kimoto, and H. Ryssel, "Comparison of the threshold-voltage stability of SiC MOSFETs with thermally grown and deposited gate oxides", Mat. Scie. For., vols. 645-648, pp. 681-684, to be published

[29] T. Nakamura, "SiC Schoky diodes and MOSFETs for automotive applications", Mat. Scie. For., vols. 645-648, to be published

[30] J.F. Ziegler, "Terrestrial cosmic ray intensities", IBM J. Res. Dev., vol. 42, no.1, pp. 117-139, 1998

[31] H. Kabza, H.-J. Schulze, Y. Gerstenmaier, P. Voss, J. Wilhelmi, W. Schmid, F. Pfirsch, K. Platzöder, Proc. 6th Int. Symp. Power Semic. Dev. ICs, Davos, pp. 9-12, 1994

[32] G. Sölkner, W. Kaindl, M. Treu, and D. Peters, "Reliability of SiC power devices against cosmic radiatio-induced failure", Mat. Scie. For., vols. 556-557, pp. 851-856, 2007

Effects of Negative Differential Resistance in High Power Devices and some Relations to DMOS Structures

Roman Baburske, Josef Lutz, Birk Heinze
Chemnitz University of Technology
Chair of Power Electronics and EMC
Chemnitz, Germany
Email: Roman.Baburske@etit.tu-chemnitz.de
Josef.Lutz@etit.tu-chemnitz.de
Birk.Heinze@etit.tu-chemnitz.de

Abstract—**Due to the feedback of free carriers to the electric field, branches with negative differential resistance (NDR) occur in high power devices [1]. NDR usually leads to current filaments. These current filaments might be non-destructive if they move. Most critical effects are found if impact ionization occurs at the n^--n^+-junction with a positive electrical feedback to avalanche or dynamic avalanche at the pn-junction. Against dangerous double-sided dynamic avalanche, countermeasures have been found in high power devices. These structures suppress or limit the formation of electric fields at the n^--n^+-junction. They lead to devices with dramatically increased ruggedness [1]. Even if ESD and DMOS devices operate at lower voltage, the very high current density levels lead to a density of free carriers above the specific doping of the layer, and free carriers dominate the shape of the electric field. Similar effects of negative differential resistance and moving filaments have been found in ESD protection devices. Snap-back effects have been found in DMOS devices. Possible relations are discussed.**

I. Introduction

The characteristic of DMOS structures has typically a branch with negative differential resistance (NDR) at high current densities, e.g. Fig. 1. Such a NDR-branch indicate the

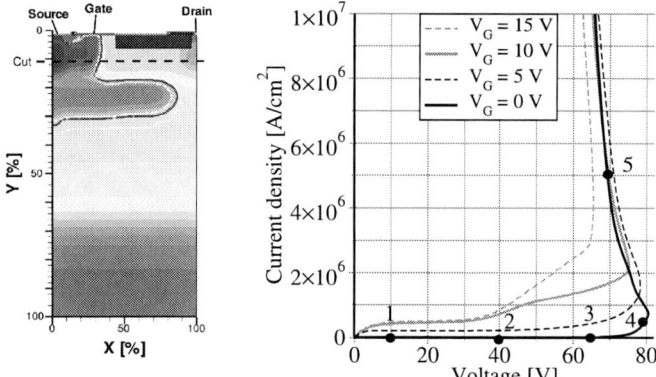

Fig. 1. **Left:** Doping profil of the simulated DMOS. **Right:** Simulated output characteristic of the DMOS.

ability of the device to switch into a state with inhomogeneous current distributions. If current filaments with a very high

current density appear, the device can be destroyed due to strong local heating. In DMOS devices, ESD events are critical, because they stress the device with a current pulse of high magnitude.

The reason of NDR-branches are a modification of the electric field distribution due to avalanche-generated free charge carriers, Fig. 2, lines 3-4. At significant high current

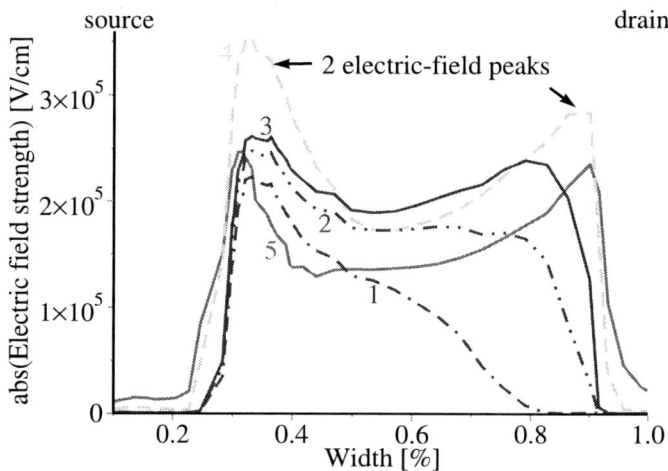

Fig. 2. Electric field strength and space charge at different current densities (marked in Fig. 1) at zero gate voltage.

densities, the DMOS has two avalanche regions, at the p^+-n^- junction and the n^--n^+ junction. Therefore, the electric field distribution is characterized by two peaks and a hammock-like shape.

Similar NDR effects can be observed in high voltage bipolar devices, Fig. 3. They are induced by avalanche breakdown, second breakdown [3], [4] or latching. They also indicate the ability to build up current filaments. This occurs during hard transient stress, for example during dynamic avalanche of high power diodes. A clear understanding, of what happens inside the device is of great importance to find countermeasures preventing the device destruction.

The aim of this paper is to explain NDR-effects and filament

978-1-4244-5430-3/10 $26.00 © 2010 IEEE

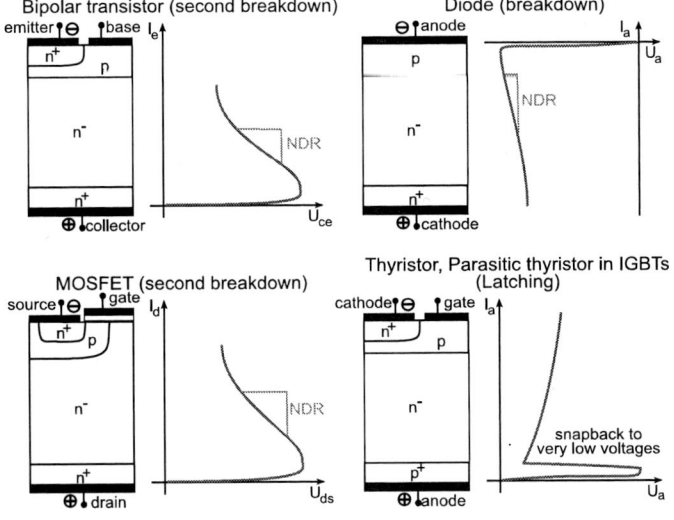

Fig. 3.   S-shape Quasistationary characteristics of power devices [2].

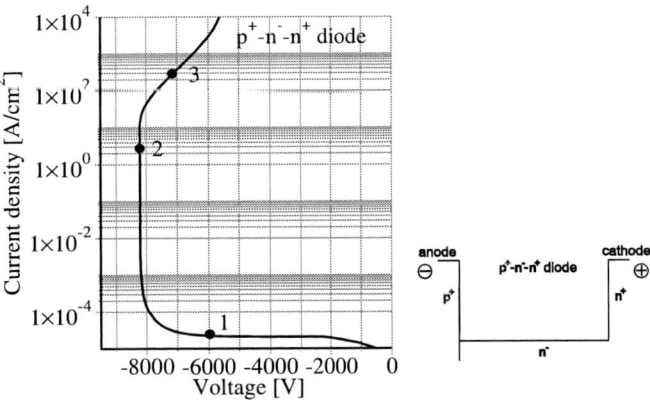

Fig. 4.   **Left:** Reverse Characteristic of a p$^+$-n$^-$-n$^+$ diode. **Right:** Schematic doping profil of a p$^+$-n$^-$-n$^+$ diode.

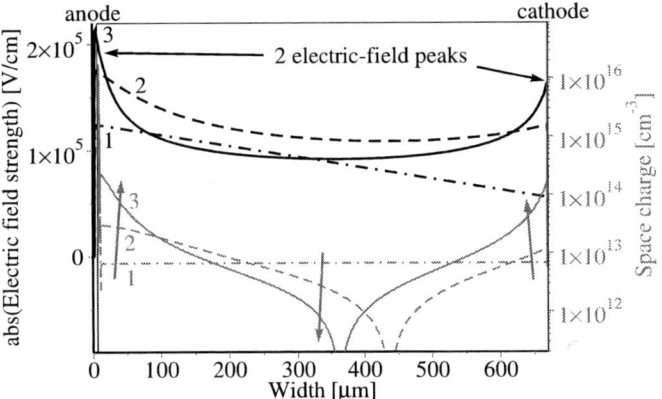

Fig. 5.   Electric field strength and space charge (absolute value) at different current densities (marked in Fig. 4) during the reverse state of the p$^+$-n$^-$-n$^+$ diode.

behavior in power devices on the base of a p$^+$-n$^-$-n$^+$ diode and to compare the results with effects in DMOS structures. In Section II, the appearance of NDR-branches in reverse characteristics is presented and countermeasures are analyzed. Section III and IV deal with dynamic avalanche which occurs during reverse recovery of power devices and can lead to filamentation and finally to the device destruction. Finally, Section V compares DMOS devices with power devices.

II. REVERSE CHARACTERISTIC OF POWER DIODES

At very high blocking voltages, avalanche generation occurs inside a semiconductor device. The avalanche generation rate

$$G_{ava} = \alpha_n n v_n + \alpha_p p v_p \qquad (1)$$

depends on ionization rates (α_n, α_p), the densities of the free charges (n, p) and the carrier velocities (v_n, v_p). The ionization rates increase with the electric field strength and the carrier velocities are of the size of the saturation velocities due to the high electric field strengths necessary for avalanche generation. For the simulated p$^+$-n$^-$-n$^+$ diode reverse characteristic, Fig. 4 avalanche generation leads to a strong increase of the current density at above 8000 V. Above a certain level of current density (10 A/cm^2) the voltage decreases with increasing current density. This implies a branch of having negative differential resistance in the reverse characteristic, which is caused by the space charge effect of carriers [5]. This can be shown by considering the shape of the electric field and the space charge at different state of the reverse characteristic, Fig. 5. At the state 1, the diode conducts only a very small reverse current density. The free carrier density is very small and the gradient of the electric field

$$\frac{dE}{dx} = \frac{q}{\varepsilon} N_{eff} = \frac{q}{\varepsilon} (N_D + p - n) \qquad (2)$$

is mainly determined by the doping density in the low doped region (N_D). The situation changes during avalanche. The generated holes move towards the anode and increase the space charge (N_{eff}) in front of the p$^+$-n$^-$-junction. Whereas, the electrons moving towards cathode decrease the space charge in the middle of the low doped region. If the electron density is high enough a second electric field peak occurs at the n$^-$-n$^+$-junction. The result is an Egawa-type electric field [5] with two peaks and decreased values of the electric field strength in the middle. The area under the electric field curve decreases for increasing current densities resulting in the NDR-branch. To prevent a diode destruction caused by reaching the NDR-region in reverse state, the NDR-branch has to be shifted to high reverse current densities. This can be done by increasing N_D, leading to higher electron densities necessary for compensation. However, this decreases the blocking capability for a fixed chip thickness and worsens the cosmic ray ruggedness. Another opportunity is to implement buffer structures in front of the n$^+$-region [6] and the p$^+$-region [7], [8]. Fig. 6 shows that an cathode-side n-buffer layer leads to a postponement of the NDR-branch to higher current densities. The additional branch with positive differential resistance is connected to penetrations of the space charge region into the buffer layer, Fig. 7. The NDR-branch starts if the higher doping of the buffer layer is compensated.

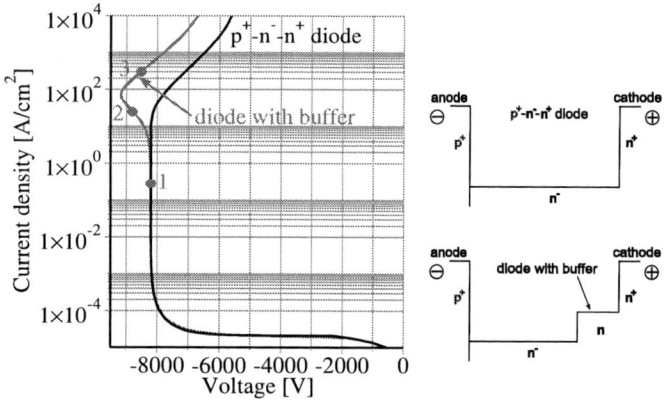

Fig. 6. **Left:** Reverse Characteristic of a p^+-n^--n^+ diode and a diode with boxbuffer. **Right:** Schematic doping profiles of a p^+-n^--n^+ diode and a diode with boxbuffer.

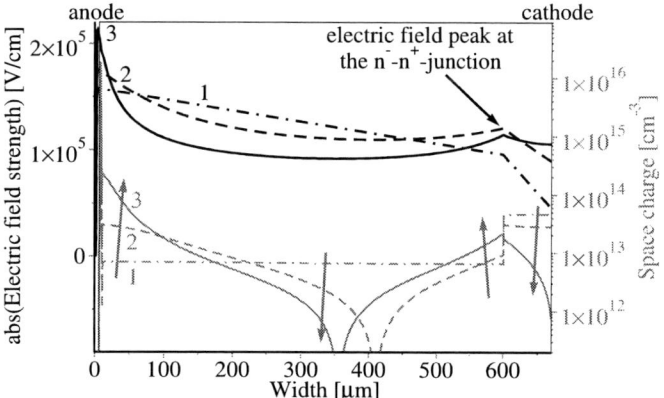

Fig. 7. Electric field strength and space charge (absolute value) at different current densities (marked in Fig. 6) during the reverse state of the diode with a boxbuffer.

III. DYNAMIC AVALANCHE IN POWER DIODES

In bipolar devices, the low doped region is flooded by a charge carrier plasma in the forward state, Fig. 8, state t_0. During the reverse recovery process, the plasma is ex-

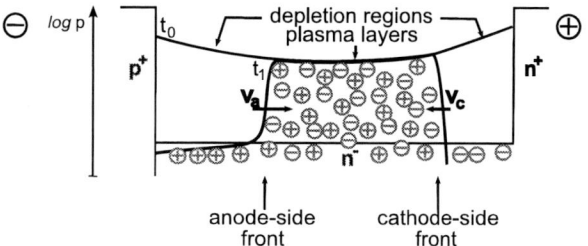

Fig. 8. Extraction of the charge carrier plasma during reverse-recovery period of a p^+-n^--n^+ diode.

tracted. Depletion regions arise surrounding a plasma layer which shrinks during the recovery process (t_1). The holes are extracted through the anode-side depletion layer and the electrons are extracted through the cathode-side depletion layer. Fig. 9, left shows the expansion of the anode-side depletion layer at for increasing reverse current density for a simulated reverse recovery process. If there is no avalanche in the anode-side depletion layer, the electron density can be

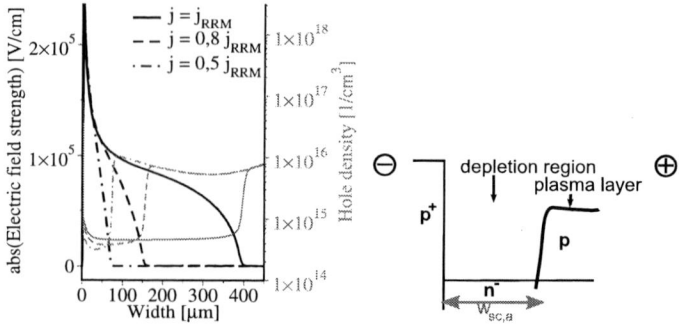

Fig. 9. **Left:** Electric field strength and hole density in the anode-side depletion region during reverse recovery of a diode with a blocking capability of 8 kV. **Right:** Model to simulate the characteristic of the anode-side depletion region.

assumed as zero. According to Eq. 2, the gradient of the electric field strength is proportional to $N_D + p$ and, thus, significantly higher as in the static reverse state. This leads to dynamic avalanche, although the voltage across the depletion layer is well below the static breakdown voltage of the diode. Similar to the static state, the electric field shape significantly differs from the expected triangular shape due to avalanche generated electrons which lower the dE/dx in the middle of the depletion region. The question is, if the deflection of the electric field curve could be high enough that the voltage area under the field curve decreases with increasing current density. To answer that, the characteristic of the depletion layer can be simulated by replacing the plasma layer with an hole-injecting p-region, Fig. 9, right. The characteristics, Fig. 10, left, show a NDR branch for high current densities [9]. Starting from

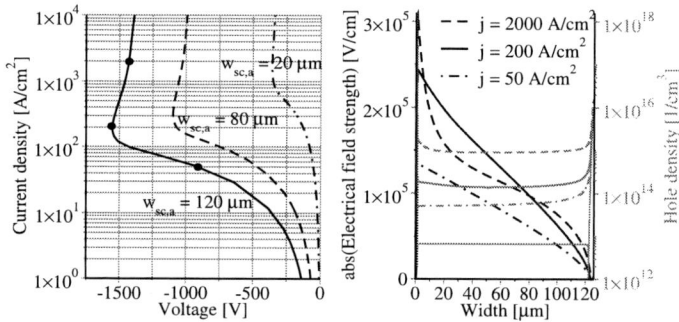

Fig. 10. **Left:** Characteristics of an anode-side depletion region of a given structure ($T = 400$ K). **Right:** Electrical field strength in the 120 μm depletion region at different current densities.

low current densities ($w_{sc,a} = 120$ μm), the voltage increases with increasing current density, because the hole density and the gradient of the electric field increase. With peak values of the electric field strengths above 2×10^5 V/cm, avalanche begins. However, a high deflection of the electric field shape appears not before the electron density reaches the level of N_D [10], Fig. 10, right. This low avalanche is to be called dynamic avalanche of the first degree [11]. If the current density is increased more, the dynamic avalanche increases, resulting in a high deflection of the electric field curve. The characteristic achieves the bistable point which is followed by

978-1-4244-5430-3/10 $26.00 © 2010 IEEE

an NDR-branch. The depletion layer is under conditions of dynamic avalanche of the second degree [11]. The higher the width of the depletion layer the lower the current density at the bistable point, because the hole dominating areas and the electron dominating areas are expanded for higher depletion regions increasing the deflection of the electric field curve. Similar characteristics can be calculated for the cathode-side depletion region by replacing the plasma layer by an electron injecting n-region.

If the diode attains a bistable point of a depletion region during reverse recovery, the current density becomes inhomogeneous in lateral direction. Areas with higher current densities arise, which are called filaments or current tubes. Starting from the bistable point the operating point divides into parts with higher current density and parts with lower current density. The width of the depletion regions remains for a moment constant. Therefore, both working points are on the same characteristic. The voltage over the depletion layer decreases. That is the reason, why the voltage curve of a reverse-recovery process reveals the onset of filaments typically with little voltage decreases [9], [12], Fig. 11. After the onset of

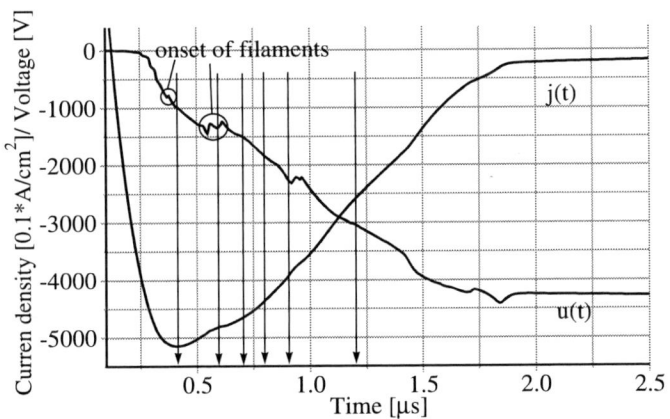

Fig. 11. Voltage and current of the p$^+$-n$^-$-n$^+$ diode with a blocking capability of 8 kV during reverse recovery process [13].

filaments, the depletion layer width expands with different velocities in the area of the filament and outside the filament. Hence, the working points are on different characteristics.

IV. DESTRUCTION MECHANISM OF POWER DIODES DURING REVERSE RECOVERY

Initially the appearance of filaments has been seen as the reason for the diode destruction due to strong local heating caused by high current densities in the area of filaments [9], [14]. However, diode reverse recovery measurements show a rugged behavior also if strong dynamic avalanche occurs in the anode-side plasma layer [15], [16]. Further analysis of the filament behavior shows that anode-side filaments move laterally, because of a strong plasma extraction in the vicinity of the filament and the resulting higher expansion of the anode-side depletion region, Fig. 12 (next page). As the voltage drop across the depletion layer is at every point equal, the peaks of the electric field strengths decrease in the vicinity

of the filament and the filament extinguish at that position and a new filament appears at a adjacent position where the width of the depletion region is lower [9], Fig. 12 (next page). The filament as heat source moves preventing strong local heating and consequently diode destruction. However, what provokes the destruction of the diode? It is conspicuous in Fig. 12 that the cathode-side filament behaves different, it stands fixed at $x = 400\ \mu m$. The width of the depletion region shrinks in the vicinity of the cathode-side filament due to the velocity saturation of the electrons and holes in the high field region [13], [17]. Therefore, the peak of the electric field remains in the center of the filament and the filament can only thermally move. The thermal movement depends strongly on the temperature gradient and is generally significantly slower [18]. Furthermore, the fixation of the cathode-side filament inhibits the onset of multiple filaments. In the anode-side depletion region multiple filaments appear, if a filament extinguish, the reverse current still increases and there are multiple position with conditions for strong dynamic avalanche.

Fig. 13 shows a destructive reverse-recovery simulation of a p$^+$-n$^-$-n$^+$ diode. At 0.7 μs, one cathode-side filament and

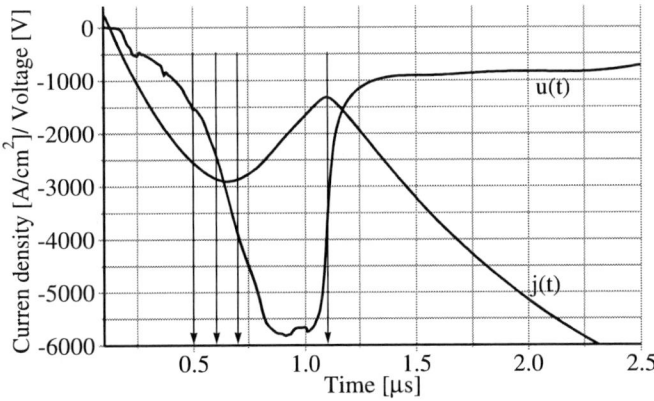

Fig. 13. Voltage and current of the p$^+$-n$^-$-n$^+$ diode with a blocking capability of 8 kV during a destructive reverse recovery process.

multiple anode-side filaments exist, Fig. 14. Almost the whole

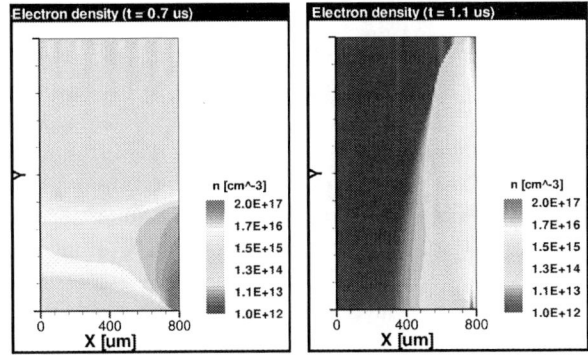

Fig. 14. Electron density during reverse recovery at certain points in time marked in Fig. 13 (top: anode, bottom: cathode)

reverse current is flowing through one single cathode-side filaments. The charge-carrier density becomes higher than the

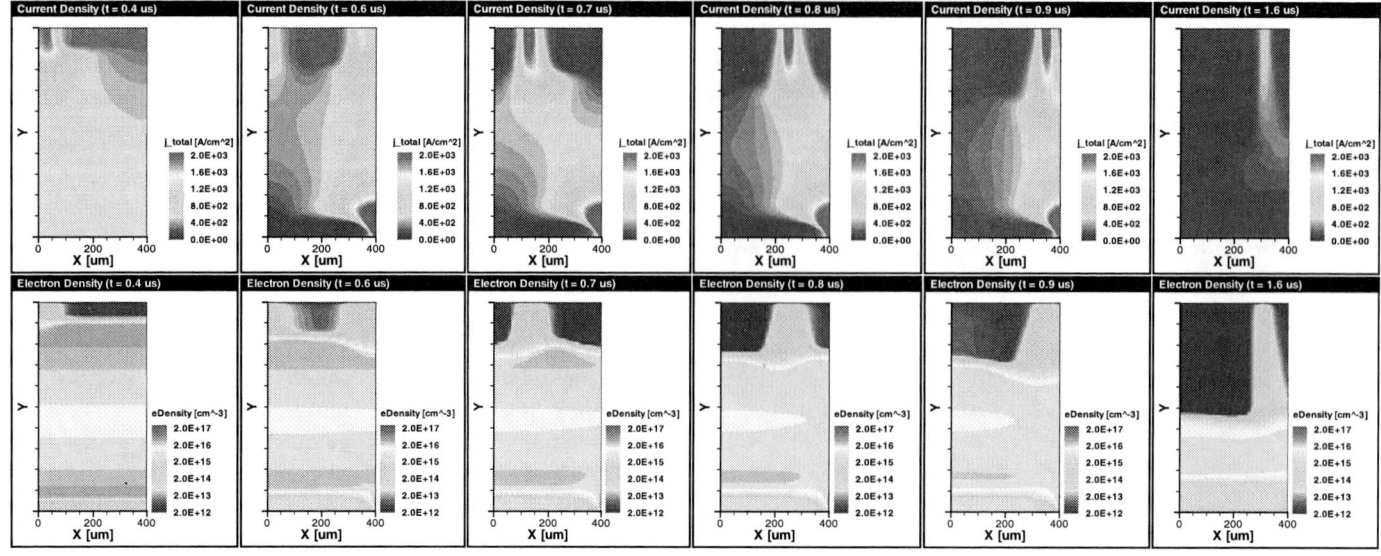

Fig. 12. Current density and electron density during reverse recovery at certain points in time marked in Fig. 11 (top: anode, bottom: cathode)

doping density and compensates the n⁺-region, Fig. 15. A

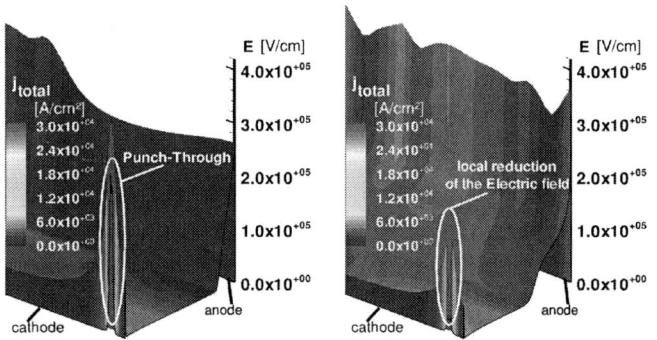

Fig. 15. Electric field strength (profile) and current density (color range) in the area of the cathode-side filament during punch-through; **Left:** $t = 0.5\ \mu$s, **Right:** $t = 0.6\ \mu$s.

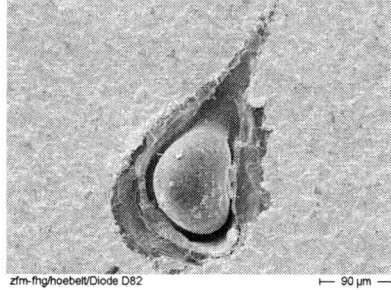

Fig. 16. SEM picture of a 6-kV diode after failure. The melted silicon is solidified to a bead.

punch-through appears, that leads together with the heating of the filament area to a carrier injection from the metal-semiconductor contact. Additionally charge carriers are generated by Auger generation due to high temperature and high carrier density. After plasma layer exhaustion, Fig. 14, right, a anode-side filament appears across from the cathode-side filament. Both filaments build one high conducting filament connecting anode and cathode. At the beginning of the reverse recovery, the emerging filaments are avalanche-generated. Therefore, they have a negative temperature feedback, because avalanche generation decreases with increasing temperature. However, the filaments migrate into thermal-generated filaments which exists due to injection from the metal-semiconductor contact and Auger generation, both with a positive temperature feedback. The thermal-generated filament leads to thermal runaway, which can be identified by a strong increase in the diode current and a break-in of the diode voltage, Fig. 13. A typical failure picture shows a point-shaped hole, Fig. 16. Although the final destruction is caused

by thermal runaway, we call the whole destruction mechanism electrothermal runaway, because it can be only understood by considering the dynamic electrothermal processes of the filaments.

The explained destruction mechanism can be prevented by suppressing the formation of cathode-side filaments. That can be effectively done by implementing the concept of Controlled Injection of Backside Holes (CIBH) [16]. The CIBH diode has p-doped islands and a n-buffer in front of the n⁺-region, Fig. 17, left. There is an additional avalanche region between the p-islands and the n⁺-region during reverse recovery. This avalanche region injects holes into the middle region. The additional hole current density inhibits the formation of a cathode-side depletion layer [17]. Figs. 17, right and 18 compare the reverse recovery of a p⁺-n⁻-n⁺ diode and a diode with an implemented CIBH structure. The voltage and current curves are quiet similar. The CIBH diode has a slightly higher reverse recovery charge due to the additional injected holes. That implies a slightly higher reverse recovery time. At 1.0 μs, the p⁺-n⁻-n⁺ diode shows a cathode-side depletion region and a cathode-side filament, whereas the charge carrier plasma is still connected to the cathode-side emitter in the CIBH diode. The set up of a cathode-side filament is suppressed by the

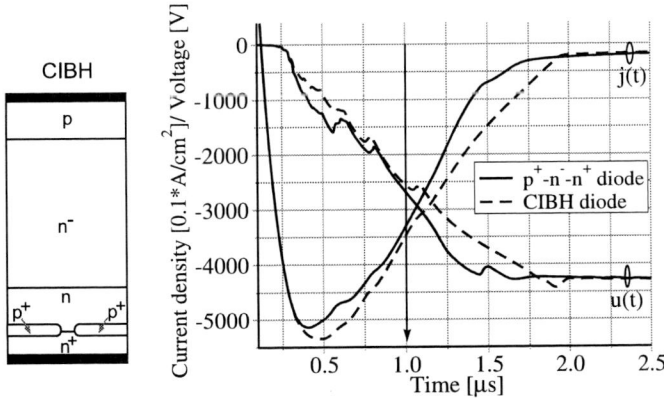

Fig. 17. **Left:** Schematic of the CIBH diode (top: anode, bottom: cathode). **Right:** Voltage and current of the p^+-n^--n^+ diode and a CIBH diode both with a blocking capability of 8 kV during reverse recovery process [13].

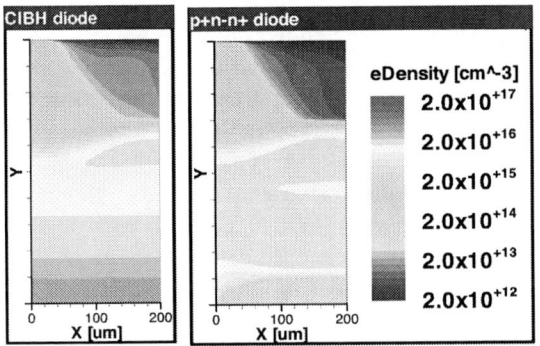

Fig. 18. Electron density during turn-off of the CIBH-diode and the p^+-n^--n^+ diode (Fig. 17) at $t = 1.0\ \mu s$ (top: anode, bottom: cathode).

CIBH concept. To see that from a different point of view, multiple cathode-side filaments are generated by the CIBH structure, which are homogeneously distributed over the active area. The diode current is no longer flowing through one single filament, but divides into several parts. This counteracts strong local heating and destruction.

V. COMPARISON

To compare the reasons for NDR effects between high voltage power devices and DMOS transistors, p^+-n^--n^+ structures with the same doping profile of the emitter regions ($N_{peak} = 1 \times 10^{18}$ cm^{-3}, $d_{junction} = 1\ \mu$m) have been analyzed. However, the n^--region has been designed for different blocking capabilities of 60 V and 5000 V and a trapezoidal shape of the electric field during breakdown with the avalanche-parameters of Fulop [19]. The results for the 5-kV structure, Fig. 19, are in consistency with Figs. 4 to 6. There is an additional small branch with a positive differential resistance at about 10 A/cm^2. This behavior is caused by the high penetration depth of the emitter-regions, which are acting like a buffer structures. The NDR-branch is due to the hammock-like shape of the electric field as described above. At about 1×10^6 A/cm^2, there begins again a branch with a positive differential resistance due to the punch-through of the electric field at the contacts. The structure behaves like a resistor. This

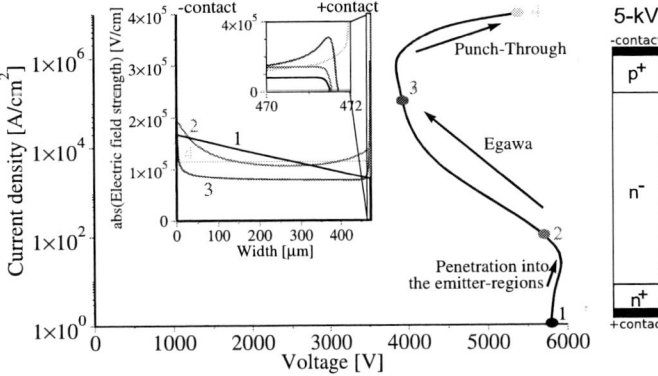

Fig. 19. **Left:** Static characterisitic of a p^+-n^--n^+ structure with a blocking capability of around 5 kV. **Inset:** Electric field strengths at certain current densities. **Right:** Schematic of the simulated structure.

second branch with positive differential resistance plays no role for high voltage structures since the devices are destroyed far below such high current densities.

The doping density of the 60-V structure is with 6×10^{15} cm^{-3} much more higher compared to the 5-kV structure ($N_D = 1.2 \times 10^{13}$ cm^{-3}) leading to a higher gradient of the electric field at breakdown. The branch with the positive differential resistance is relatively large compared to the 5-kV structure, Fig. 20. The NDR-branch due to the Egawa-effect

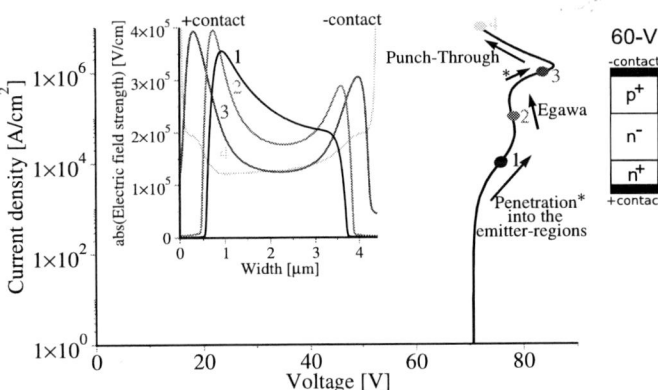

Fig. 20. **Left:** Static characterisitic of a p^+-n^--n^+ structure with a blocking capability of around 60 V. **Inset:** Electric field strengths at certain current densities. **Right:** Schematic of the simulated structure.

appears at higher current densities, because higher electron and hole densities are necessary to modify the electric field shape in the higher doped n^--region. At around 1×10^6 A/cm^2, the characteristic has a second branch with positive differential resistance, because the penetration of the electric field into the emitter regions dominates again due to very high densities of free charge carriers. With increasing current densities, the electric field reaches the metal-contacts. This leads contrary to the 5-kV structure to an NDR-branch. The electric field strengths decreases near to the metal-contacts, but increases in the n^- region due to resistive behavior. In the 60-V structure with a low width of 4.4 μm, the first effect is dominating.

978-1-4244-5430-3/10 $26.00 © 2010 IEEE

DMOS devices exhibit a parasitic transistor structure. To investigate the influence of the parasitic transistor, an additional n^+-region ($N_{peak} = 1 \times 10^{20}$ cm^{-3}, $d_{junction} = 0.1$ μm) was implemented into the 60-V structure and connected to the -contact, Fig. 21. The characteristic shows an additional

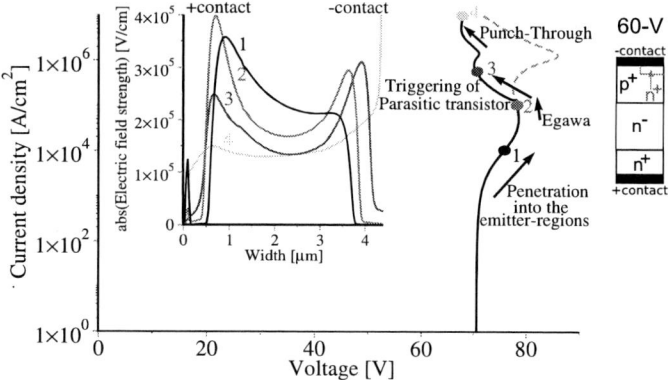

Fig. 21. **Left:** Static characterisitic of a p^+-n^--n^+ structure with an additional n^+-region and a blocking capability of around 60 V. **Inset:** Electric field strengths at certain current densities (cut at $x = 0.75$ μm). **Right:** Schematic of the simulated structure.

NDR-branch starting at about 1×10^5 A/cm^2. At lower current densities, the current is completely flowing through the p^+-doped area, Fig. 22. However, at a current density something

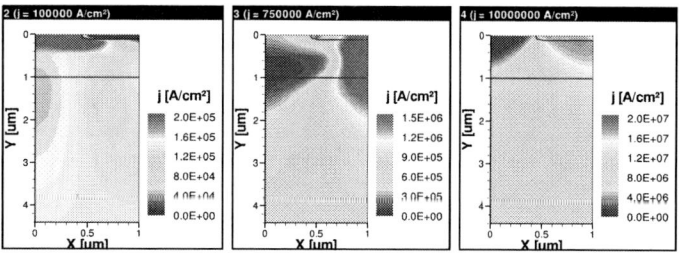

Fig. 22. Current density distribution at certain points marked in Fig. 21 (top: -contact, bottom: +contact)

higher than 1×10^5 A/cm^2, the parasitic transistor is triggered leading to an decrease of the electric-field strengths.

Comparing the NDR effects in power diodes with the NDR effects in the LDMOS structures, the following similarities were found:

- The initial increase in the current density is due to avalanche generation.
- The distribution of the electric field strengths disposes two peaks, one at the p^+-n^- junction and one at the $n^- - n^+$ junction. Between the two peak, the distribution has a hammock-like shape.
- An additional branch with positive differential resistance can be achieved by using buffer structures or high penetration depths of the doping profiles of the emitter regions [20].

However, there are also differences between both devices:

- The current density at which the NDR-branch occurs is significantly higher in DMOS devices.

- This is related to the higher doping of the n^--region (up to three orders of magnitude).
- The width of the n^--region of the LDMOS is much smaller.
- There are multiple reasons for NDR-branches in DMOS structures: Egawa-effect, triggering of parasitic transistor, punch-through of the electric field. Modern DMOS structures have a lower sheet resistance of the p^+ region and a lower extension of the source region. This measures can avoid the triggering of the parasitic transistor.

Both, power diodes and DMOS structures, have in common that NDR-branches in their static characteristics indicates a possible onset of filaments (=current tubes) during transient strain. Like in power diodes, filaments can thermally move due to negative temperature coefficient of avalanche generation in DMOS devices [21], [22]. Thermal filaments can not jump to other positions. They move continuously with decreasing temperature gradients. This could be the reason for the lower ESD ruggedness of DMOS transistors with stripe cells [20]. If filaments are locally fixed, they cause strong local heating and a device destruction. A typical failure shows a point-shaped melting on the silicon chip. However, the most critical case in continuous operation of power diodes is the reverse recovery process under hard switching conditions. Under this conditions, filaments move due to a stronger plasma extraction in the vicinity of the filament. This mechanism leads to higher lateral velocities of the filaments than thermal movement. During ESD events in DMOS structures, there is no plasma-layer. The behavior is mainly determined by the static characteristic.

VI. CONCLUSION

This paper deals with NDR effects in power diodes and gives some relations to DMOS structures. The NDR-branch in the reverse characteristic of power diodes is determined by the modification of the electric-field distribution to a hammock-like shape caused by the high density of free charges. Beside this effect, NDR-branches caused by triggering of the parasitic transistor and a punch-through at the metal-contact occur in DMOS structures. For high-voltage power diodes and DMOS devices have in common, that the static characteristic can be improved by implementing buffer structures which leads to additional branches with positive differential resistance. Contrary to the situation in high-voltage power diodes during reverse recovery, filaments can only thermally move in DMOS devices during ESD events. For that reason, filaments are only able to move continuously, but not to jump to other positions.

REFERENCES

[1] J. Lutz, R. Baburske, M. Chen, B. Heinze, M. Domeij, H. P. Felsl, and H.-J. Schulze, "The nn$^+$-junction as the key to improved ruggedness and soft recovery of power diodes," *IEEE Trans. Electron Devices*, vol. 56, no. 11, pp. 1–2, Nov. 2009.

[2] D. Silber, "Second breakdown, latching and thermal runaway - an overview," in *ECPE workshop "Power Semiconductor Robustness" Proceedings*, 24-25 June 2009.

[3] J. Kirk, C.T., "A theory of transistor cutoff frequency (ft) falloff at high current densities," *Electron Devices, IRE Transactions on*, vol. 9, no. 2, pp. 164–174, March 1962.

[4] P. Hower and V. Krishna Reddi, "Avalanche injection and second breakdown in transistors," *Electron Devices, IEEE Transactions on*, vol. 17, no. 4, pp. 320–335, Apr 1970.

[5] H. Egawa, "Avalanche characteristics and failure mechanism of high voltage diodes," *IEEE Trans. Electron Devices*, vol. 13, no. 11, pp. 754–758, Nov. 1966.

[6] B. Heinze, H. P. Felsl, A. Mauder, H.-J. Schulze, and J. Lutz, "Influence of buffer structures on static and dynamic ruggedness of high voltage fwds," in *Proc. ISPSD*, Santa Barbara, May 2005, pp. 215–218.

[7] J. Vobecky and P. Hazdra, "Radiation-enhanced diffusion of palladium for a local lifetime control in power devices," *IEEE Trans. Electron Devices*, vol. 54, no. 6, pp. 1521–1526, Jun. 2007.

[8] M. Chen, J. Lutz, H. P. Felsl, and H. J. Schulze, "Analysis of a p^+-p^--n^--n^+ diode structure," in *Proc. ISPSD*, Orlando, Florida, May 2008, pp. 153–156.

[9] J. Oetjen, R. Jungblut, U. Kuhlmann, J. Arkenau, and R. Sittig, "Current filamentation in bipolar power devices during dynamic avalanche breakdown," *Solid-State Electronics*, vol. 44, no. 1, pp. 117–123, 2000.

[10] H. Schlangenotto, J. Serafin, F. Sawitzki, and H. Mauder, "Improved recovery of fast power diodes with self-adjusting p emitter efficiency," *IEE Electron Device Lett.*, vol. 10, no. 7, pp. 322–324, Jul. 1989.

[11] J. Lutz and M. Domeij, "Dynamic avalanche and reliability of high voltage diodes," *Microelectronic Reliability*, vol. 43, pp. 529–536, 2003.

[12] F. J. Niedernostheide, F. Falck, H.-J. Schulze, and U. Kellner-Werdehausen, "Current-density patterns induced by avalanche injection phenomena in high voltage diodes during turn-off," *Ann. Phys. (Leipzig)*, vol. 13, no. 7-8, pp. 414–422, 2004.

[13] R. Baburske, B. Heinze, F.-J. Niedernostheide, J. Lutz, and D. Silber, "On the formation of stationary destructive cathode-side filaments in p^+-n^--n^+ diodes," in *Proc. ISPSD*, Barcelona, Spain, Jun. 2009, pp. 41–44.

[14] A. Porst, "Ultimate limits of an IGBT (MCT) for high voltage applications in conjunction with a diode," in *Power Semiconductor Devices and ICs, 1994. ISPSD '94. Proceedings of the 6th International Symposium on*, May-3 Jun 1994, pp. 163–170.

[15] M. Domeij, B. Breitholtz, M. Oestling, and J. Lutz, "Stable dynamic avalanche in si power diodes," *Applied Physics Letters*, vol. 74, no. 21, pp. 3170–3172, May. 1999.

[16] M. Chen, J. Lutz, M. Domeij, H. P. Felsl, and H. J. Schulze, "A novel diode structure with controlled injection of backside holes (CIBH)," in *Proc. ISPSD*, Naples, June 2006, pp. 9–12.

[17] R. Baburske, B. Heinze, J. Lutz, and F.-J. Niedernostheide, "Charge-carrier plasma dynamics during the reverse-recovery period of p^+-n^--n^+ diodes," *IEEE Trans. Electron Devices*, vol. 55, no. 8, pp. 2164–2172, Aug. 2008.

[18] F. J. Niedernostheide, F. Falck, H.-J. Schulze, and U. Kellner-Werdehausen, "Influence of joule heating on current filaments induced by avalanche injection," *IEE Proc.-Circuits Devices Syst.*, vol. 153, no. 1, pp. 3–10, 2006.

[19] W. Fulop, "Calculation of avalanche breakdown voltages of silicon p-n junctions," *Solid-State Electron.*, vol. 10, no. 1, pp. 39–43, Jan. 1967.

[20] A. Podgaynaya, D. Pogany, E. Gornik, and M. Stecher, "Improvement of the electrical safe operating area of a DMOS transistor during ESD events," in *Reliability Physics Symposium, 2009 IEEE International*, April 2009, pp. 437–442.

[21] D. Pogany, S. Bychikhin, E. Gornik, M. Denison, N. Jensen, G. Groos, and M. Stecher, "Moving current filaments in ESD protection devices and their relation to electrical characteristics," in *Reliability Physics Symposium Proceedings, 2003. 41st Annual. 2003 IEEE International*, March-4 April 2003, pp. 241–248.

[22] M. Denison, M. Blaho, P. Rodin, V. Dubec, D. Pogany, D. Silber, E. Gornik, and M. Stecher, "Moving current filaments in integrated DMOS transistors under short-duration current stress," *Electron Devices, IEEE Transactions on*, vol. 51, no. 10, pp. 1695 – 1703, Oct. 2004.

Investigation of Monotonous Increase in Saturation-Region Drain Current during Hot Carrier Stress in N-type Lateral Diffused MOSFET with STI

Yu-Hui Huang, J.R. Shih, Y.H. Lee, Sunnys Hsieh, C.C. Liu, Kenneth Wu, and H.L. Chou

High Voltage Reliability Section, Mainstream Technology Quality & Reliability Department
Taiwan Semiconductor Manufacturing Company, No. 121 Park Ave. 3, Hsinchu Science Park, Taiwan 300
Tel: 886-3-5636688-7022079, FAX: 886-3-5633211 e-mail: yhhuangx@tsmc.com

Abstract—Monotonous increase of saturation drain current Id_{sat} but linear-region drain current Id_{lin} reduction during hot carrier injection (HCI) stress is observed in N-type Lateral Diffused MOSFET. But the phenomenon of Id_{sat} increase is contrary to what we typically observed during HCI stress. The increase of Id_{sat} has been attributed to the increase of saturation substrate current Ib_{sat} after HCI stress. TCAD simulations showed that the lateral electric field increases under the high gate bias when a significant amount of electron trapping occurs along the STI corner in the drift region. The trapped electrons will change the distribution of localized electric potential and will result in the substrate current Ib increase. It is also observed that the 1st Ib peak at lower Vgs degrades, consistent with the reduction of drain and source current, due to HCI induced electron trapping. In another word, the electron trapping has two competing effects - one is with current degradation at lower Vgs and the other is with the electric field enhancement that causes the Id_{sat} to increase at higher Vgs.

Introduction

Lateral Diffused MOS (LDMOS) transistors have been widely used in many high voltage applications such as LCD panels, LED, and power management. To sustain high voltage, a lightly doped region called drift region and shallow trench isolation (STI) are utilized. Because of the high voltage operation, hot-carrier-injection (HCI) due to impact-ionization will be the cause of device performance degradation in LDMOS [1]-[4]. Anomalous hot-carrier-induced increase of drain saturation current in NLDMOS without STI has been reported in the literature [5]. The authors attributed this phenomenon to significant Si/SiO_2 interface N_{it} formation in the drift region, thus enhanced the Kirk effect. In this study, a similar increase in saturation drain current Id_{sat} measured at Vgs/Vds=20V/20V rather than degradation is also observed when device is stressed under the condition of maximum substrate (Ib) current . TCAD simulation was performed to understand the physical root cause that was responsible for the increase of Id_{sat}. Instead of N_{it} formation in the N⁻ drift region as reported by Chen, *et al* [5], our study indicated that N_{it} formation along STI bottom corner and followed by electrons filled-in could also enhance the Ib_{sat} and Id_{sat} increase.

Experimental

The N-type high-voltage LDMOS discussed in this article is fabricated using an industry standard 0.25um technology. The STI and HV N-Well drift region are introduced in both drain and source regions for sustaining high voltage and symmetrical utilization, as illustrated in Fig. 1. The width and length are 20um and 1.5um, respectively. The gate oxide thickness is around 630A, and the operational voltage for Vgs and Vds are both at 20V.

During hot carrier injection, the NLDMOS is stressed under the maximum Ib condition (i.e., Vgstr = 5V and Vdstr

= 1.1*Vcc = 22V) at room temperature 25°C. Device parameters, Vt, transconductance Gm, Id_{lin} and Id_{sat}, were monitored during the stress. TCAD simulation tools, TSUPREM4 and MEDICI, are used to simulate the changes of electric field, equi-potential and the impact ionization rate before and after HCI stress.

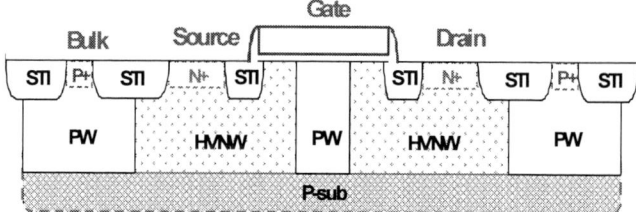

Fig. 1. The device scheme of high voltage nMOS transistor (HVNMOS) that consists of a lightly doped HVNW and a shallow trench isolation (STI) region on both drain and source sides with operation voltage of Vgs/Vds=20/20V. The width and length are 20um and 1.5um, respectively, and the gate oxide thickness is around 630A.

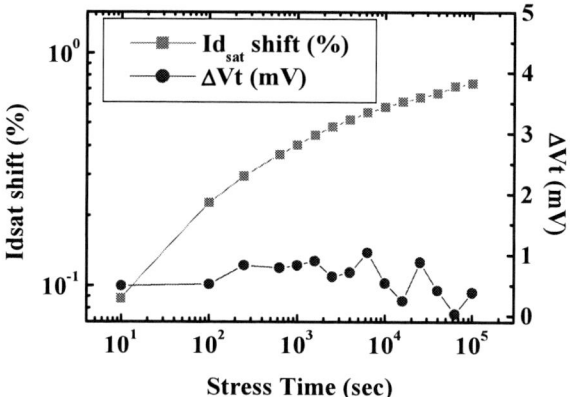

Fig. 2. The drain saturation current Id_{sat} at Vgs/Vds=20V/20V monotonously increases during HCI stress (@ Vgs/Vds=5V/22V). However, the threshold voltage Vt almost keeps unchanged during HCI stress, indicating no or little damage in the channel during HCI stress.

Results

Fig. 2 shows the device Id_{sat} and Vt shift during HCI stress. Vt remains unchanged (ΔVt <1mV) for stress up to 100K seconds, indicating little or no damage in the channel region. The Id_{sat}, on the other hand, shows a noticeable improvement during the HCI stress, which could be the result of oxide electron trapping induced effective channel shortening. Detailed investigation will be discussed in the next section.

Since both the Id_{lin} and the transconductance Gm degrade with stress time (Fig. 3), it indicates that the HCI-induced damage is located in the drift region. Because the device acts as a resister at linear condition, its drift region with lighter concentration dominates the Id_{lin} behavior than channel. It is also observed that the maximum Ib current, measured at Vgs/Vds = 5V/20V, degrades as a function of the stress time

978-1-4244-5430-3/10 $26.00 © 2010 IEEE

(Fig. 4). This is consistent with the reduction in drain and the source currents, as also shown in Fig. 4. On the other hand, the Ib current shows a different trend when biased at high Vgs. Fig. 5 shows that the 1st substrate current Ib peak (at Vgs ~ 5V) degrades under the HCI stress, while Ib at Vgs>15V significantly increases which might be due to an enhanced Kirk-effect after HCI stress or other mechanisms.

The Id$_{sat}$ improvement measured at Vgs/Vds = 20/20V during HCI stress is somewhat unusual (Figs. 2 and 6). This finding is quite different from what we observed before, where both Id$_{lin}$ and Id$_{sat}$ degraded at the same time. It implies that another mechanism responsible for current increase is competing with current degradation caused by hot carrier stress. To further investigate the mechanism responsible for the Id$_{sat}$ increase, the source current Is$_{sat}$ and the substrate current Ib$_{sat}$ in the saturation region Vgs/Vds=20V/20V are also characterized. Fig. 6 shows that the source current Is$_{sat}$ slightly degrades while the Id$_{sat}$ monotonously increases during HCI stress. There exists the opposite behavior between Is$_{sat}$ and Id$_{sat}$ but the same behavior between Ib$_{sat}$ and Id$_{sat}$. It seems that the abnormal increase of Id$_{sat}$ during HCI stress is caused by the increase of Ib$_{sat}$. Comparing the pre- and post-stressed Ib-Vg curves in Fig. 5, the Ib at Vgs=5V is reduced after HCI stress, but it is significantly increased when Vgs is larger than 15V. Comparing the current difference between Id$_{sat}$ and Is$_{sat}$ ($\Delta I = Id_{sat} - Is_{sat}$) with Ib$_{sat}$, we can find the amount of Ib$_{sat}$ increase nearly equals to the increase of ΔI ($\Delta I = Id_{sat} - Is_{sat}$).

DISCUSSION

To investigate the abnormal Id$_{sat}$ improvement, TCAD simulation is performed. Fig. 7 shows the TCAD simulation of the impact ionization rate, electric injection current, and the e-field under HCI stress. Results show a noticeable impact-ionization and the electron injection current in the drift region near the STI corner, which is in agreement with the experiment data of Id$_{lin}$ and Gm degradation as expected. Furthermore, there exists secondary impact ionization near the channel region, smaller by 1000 times (illustrated in orange color), which can account for the 1st Ib peak reduction as well as drain current degradation at lower Vgs.

The Ib$_{sat}$ increase during HCI stress implies that the lateral electric field may be also increased. Because there exists high field crowding along the STI corner during HCI stress, it is postulated that the charges trapped along the STI bottom corner may play an important role on the Id$_{sat}$ increase. To identify which charge type dominates the Id$_{sat}$ increase, TCAD simulations with positive and negative charges at a density level of 10^{12}/cm2 placed along STI corner are performed to study their effects on the device reliability characteristics.

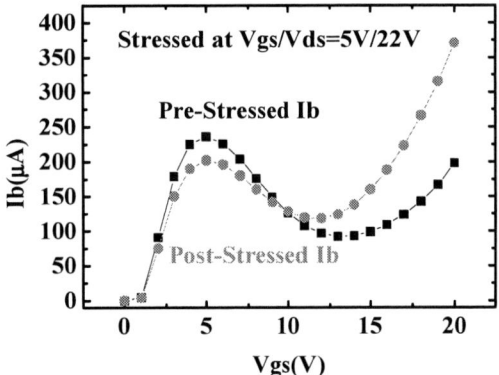

Fig. 5. The substrate current at Vgs/Vds=5V/20V (the maximum Ib condition before stress) is degraded after HCI stress because both of the drain current and the source current degraded at the same time, while the substrate current Ib is significantly increased when Vgs is larger than 15V.

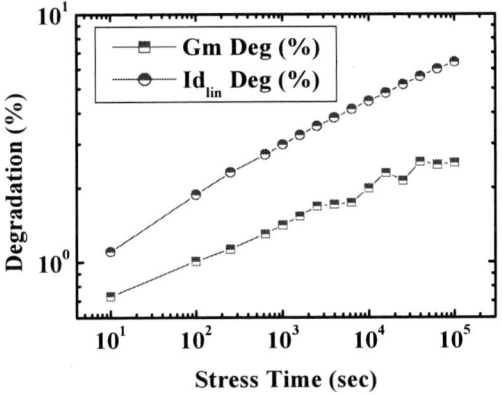

Fig. 3. The degradations of the linear drain current Id$_{lin}$ and the maximum Gm at Vgs/Vds=5V/0.1V indicate that the damage sites exist in the drift region during HCI stress.

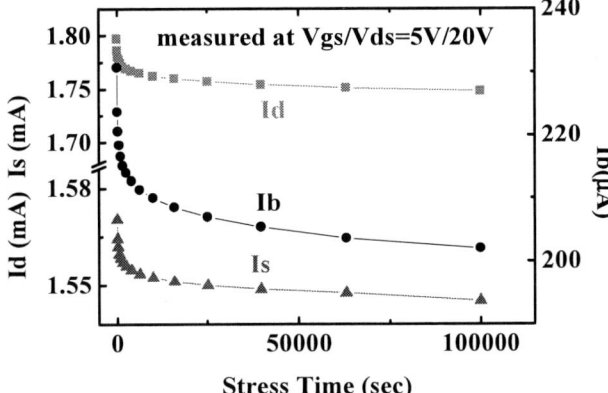

Fig. 4. The substrate current Ib measured at the maximum Ib condition Vgs/Vds=5V/20V degrades during HCI stress (@ Vgs/Vds=5V/22V) because the impact ionization significantly degrades both drain and source current measured at Vgs/Vds=5V/20V.

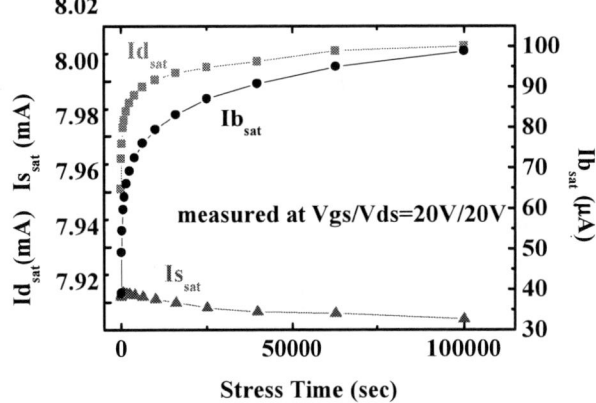

Fig. 6. The saturation drain current Id$_{sat}$ at Vgs/Vds=20V/20V increases during stress, while the source current Is$_{sat}$ slightly degrades. It indicates the increase of Id$_{sat}$ is caused by the increase of substrate current Ib$_{sat}$ through the enhanced impact-ionization during HCI stress.

978-1-4244-5430-3/10 $26.00 © 2010 IEEE

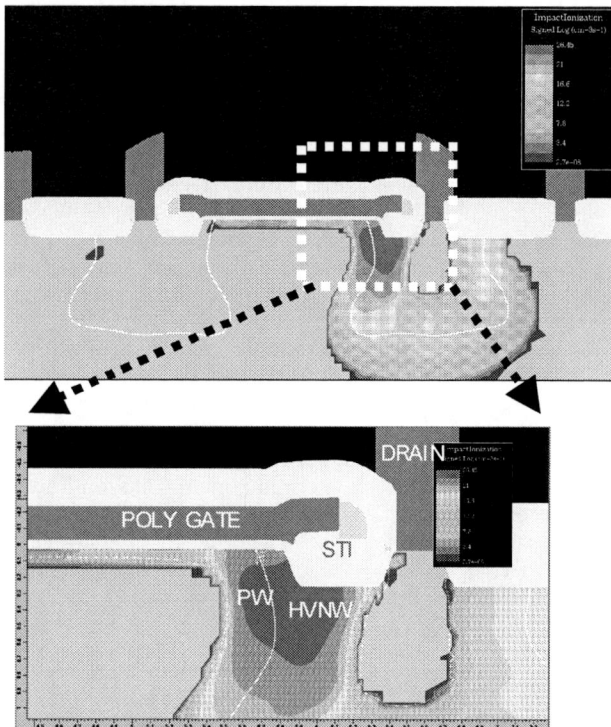

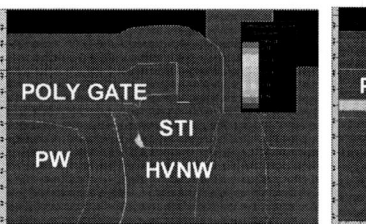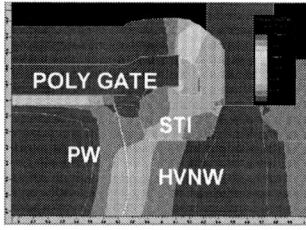

(a) Impact-ionization contour at stress condition Vgs/Vds=5V/22V

(b) Electron injection current contour and electric field magnitude at stress condition Vgs/Vds=5V/22V

Fig. 7. TCAD simulation showed that (a) the impact ionization and (b) the electron injection current are primarily located near the STI corner under the HCI stress condition at Vgs/Vds=5V/22V which is the peak-Ib condition.

Fig. 8 shows the equi-potential distribution with and without the charge placement. Compared to the no trapped charge case, the negative charges will push the equi-potential lines toward the right side and will result in a more crowded electric potential distribution near the STI bottom corner (Fig. 8b). As a result, a larger lateral electric field builds up (denoted by the red circles in Fig. 9) in the STI bottom area and causes larger impact ionization with Ib_{sat} increase (denoted by the red circles in Figs. 10 and 11) compared to the no charge trapping case. But if the positive charges are placed at the STI corner, as shown in Fig. 8c, the equi-potential lines will move toward the left side and will result in a more sparse electric potential distribution near the STI bottom and a smaller localized lateral electric field (denoted by the black squares in Fig. 9), leading to a smaller impact ionization and Ib_{sat} reduction (denoted by the black squares in Figs. 10 and 11).

Besides the e-field effect in the STI area, one may argue that similar behavior of the electric potential re-distribution could also happen along the gate Si/SiO$_2$ interface, as reported by Chen, et al [5]. TCAD simulations of charges placed along the surface of the drift region are also performed to clarify it. Results show that the electric potential distribution is almost unchanged no matter if negative charges or positive charges are trapped along the surface of the drift region, as illustrated in Fig. 12. Simulation also indicated that trapped charge placed along the surface of the drift region does not affect the e-field, impact ionization, and hole current distribution compared to the no trapping case, as illustrated in Figs. 13-15. It is to be noted that the difference between this study and the work by Chen [5] primarily comes from the gate overlap difference. When the gate fully covers the drift region and extends to STI, high gate bias confines the electric potential contours along the surface of the drift region, the contribution of charge trapped along the surface of drift region is therefore negligible. However, gate didn't fully cover the drift region in the work of SY Chen, and therefore would lead to the apparent electric field enhancement. Thus, the Ib_{sat} growth was also observed in that work. But it was proposed in that work the imaged positive charge in drift-region due to the HCI induced negative N_{it} would further reduce the concentration of drift-region, leading to the enhanced Kirk-effect that makes Ib current grow.

To have a clear comparison of the charge trapping effect on the electric field at higher Vgs by different trapping polarities, TCAD simulation results are summarized in Table-1. These results demonstrate that only the trapped electrons along the STI corner can enhance the localized electric field effect, which results in the more severe impact ionization and the increase of Ib_{sat} during HCI stress and further enhance the Id_{sat}.

CONCLUSION

HCI induced NLDMOS Id_{sat} increase is observed and analyzed. It is concluded that the impact ionization induced electron trapping in the drift region has two competing effects. The trapped electrons near the channel region will degrade the current measured at the maximum-Ib condition or lower Vgs. The electrons trapped along STI corner will cause the relocation of the electric potential and is responsible for the monotonous Id_{sat} increase, which is contributed by the Ib_{sat} increase, especially at high Vgs region.

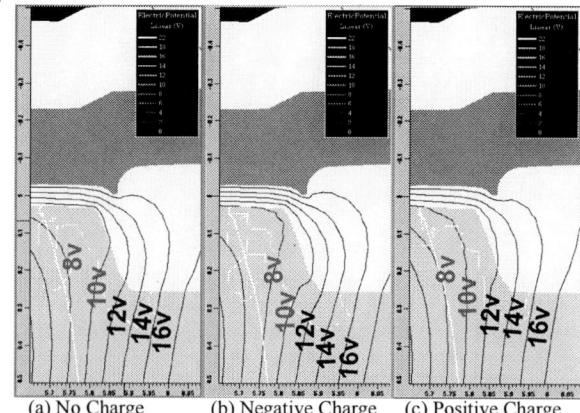

(a) No Charge   (b) Negative Charge   (c) Positive Charge

Fig. 8. The electric potential distribution is apparently changed after charges are placed along the STI corner where the impact ionization occurs. (a) No charge is placed. (b) The negative charge trapping moves the equi-potential lines toward the right, resulting in a more dense potential distribution near the STI corner. (c) The positive charge trapping results in a more sparse potential distribution near the STI corner.

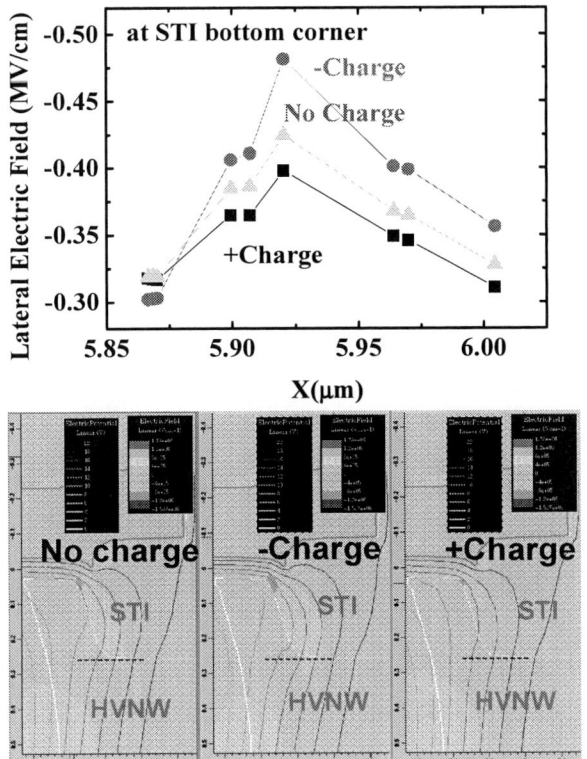

Fig. 9. The lateral electric field comparison at the cutline sketched in the TCAD figure shows E (-charge) > E (no charge) > E (+charge) near the STI corner after charges are placed along the STI corner.

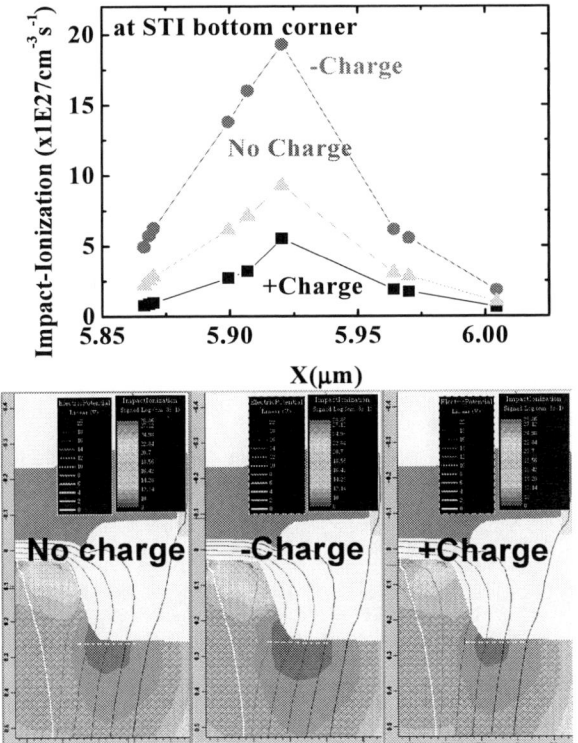

Fig. 10. The impact-ionization rate comparison at the cutline sketched in the TCAD figure shows II (-charge) > II (no charge) > II (+charge) near the STI corner after charges are placed along the STI corner.

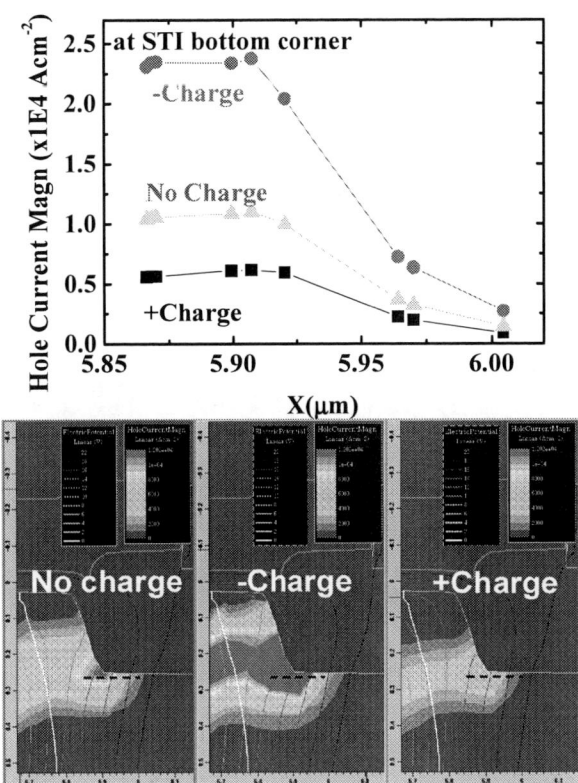

Fig. 11. The Hole-Current magnitude comparison at the cutline sketched in the TCAD figure shows Hole (-charge) > Hole (no Charge) > Hole (+charge) near the STI corner after charges are placed along the STI corner.

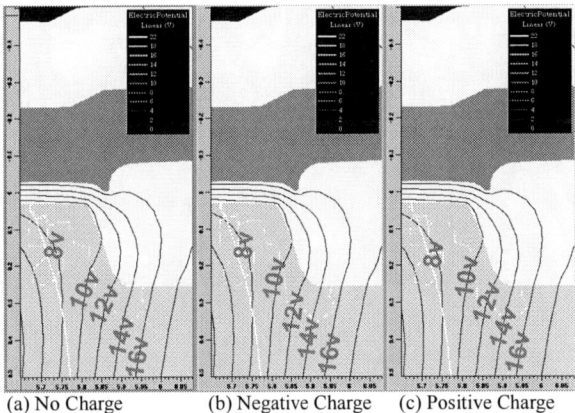

(a) No Charge   (b) Negative Charge   (c) Positive Charge

Fig.12. No change in the electric potential distribution is observed no matter negative charges or positive charges are trapped along the surface of the drift region. The reason is that the electric field along the surface of the drift region is relatively large. Therefore the contribution of charges trapped there can be neglected.

978-1-4244-5430-3/10 $26.00 © 2010 IEEE   173

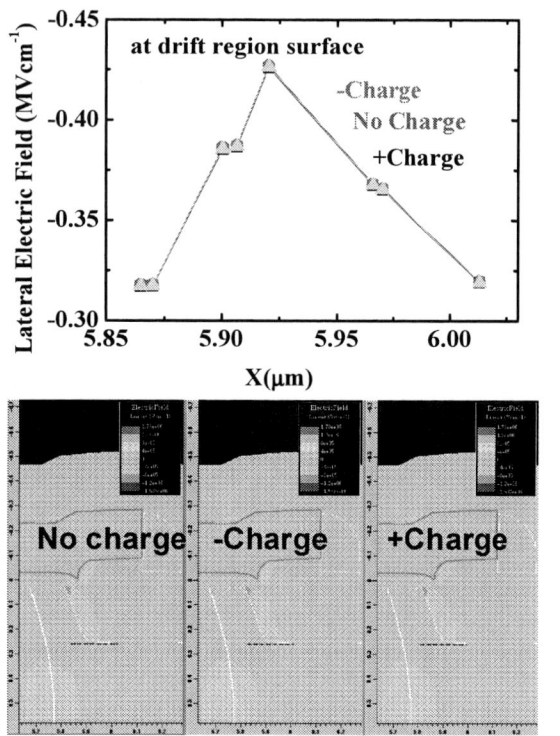

Fig. 13. The lateral electric field comparison at the cutline sketched in the TCAD figure shows no difference between negative and positive charges placed along the surface of the drift-region.

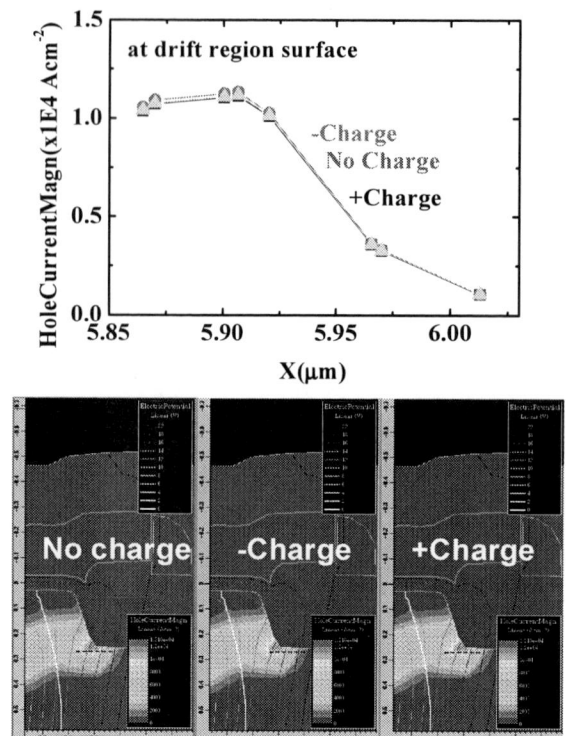

Fig. 15. The Hole-Current magnitude at the cutline sketched in the TCAD figure with negative and positive charges placed along the surface of the drift-region.

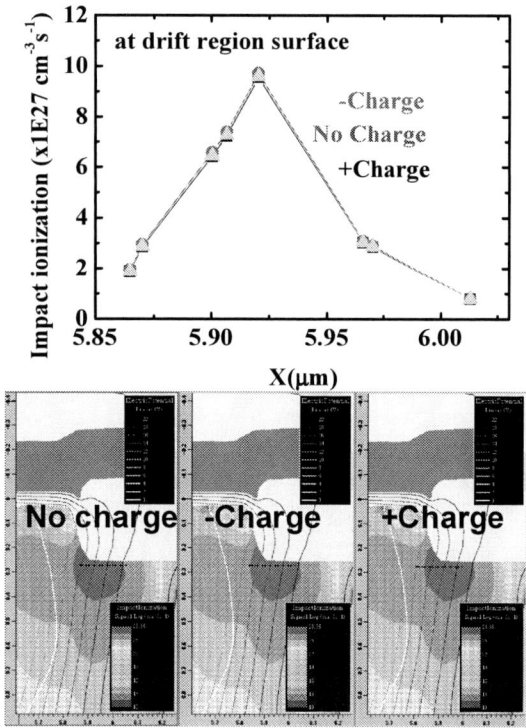

Fig. 14. The Impact ionization rate comparison at the cutline sketched in the TCAD figure with negative and positive charges placed along the surface of the drift-region.

Table-1 TCAD Simulation Summary

Trapped Region	Along the STI corner		The surface of the drift region	
Charge Polarity	(-)	(+)	(-)	(+)
Effect	Enhance the localized lateral electric field	Retard the localized lateral electric field	None	

REFERENCE

[1] P. Moens, G. Bosch, and G. Groeseneken, "Hot-Carrier Degradation Phenomena in Lateral and Vertical DMOS Transistors." IEEE Trans. on Electron Devices, Vol. 51, no. 4, p. 623, 2004.

[2] J. Chen, K. Wu, K. Lin, Y. Su, and S. Hsu, "Hot-Carrier Reliability in Submicrometer 40V LDMOS Transistors with thick Gate Oxide." IEEE IRPS, p.560, 2005.

[3] C. Cheng, J. Lin, T. Wang, T. Hsieh, J. Tzeng, Y. Jong, R. Liou, S. Pan, and S. Hsu, "Physics and Characterization of Various Hot-Carrier Degradation Modes in LDMOS by Using a Three-Region Charge-Pumping Techniquel." ." IEEE Trans. on Device and Material Reliability, Vol. 6, no. 3, p. 358, 2006.

[4] K. Wu, J. Chen, Y. Su, J. Lee, Y. Lin, S. Hsu, and J. Shih, "Anomalous Reduction of Hot-Carrier-Induced ON-Resistance Degradation in n-Type DEMOS Transistors." IEEE Trans. on Device and Material Reliability, Vol. 6, no. 3, p.371, 2006.

[5] S. Chen, J. Chen, J. Lee, K. Wu, C. M. Liu, and S. L. Hsu, "Anomalous Hot-Carrier-Induced Increase in Saturation-Region Drain Current in n-Type Lateral Diffused Metal–Oxide–Semiconductor Transistors." IEEE Trans. on Electron Devices, Vol. 55, no. 5, p. 1137, 2008.

Modeling the Lifetime of a Lateral DMOS Transistor in Repetitive Clamping Mode

E. Riedlberger, R. Keller, H. Reisinger, W. Gustin, A. Spitzer, M. Stecher

Infineon Technologies AG
Munich, Germany
+49-89-234-51769, eva.riedlberger@infineon.com

C. Jungemann

Institute for Microelectronics and Circuit Theory
Bundeswehr University
Neubiberg, Germany

Abstract — **Hot carrier degradation is critical for LDMOS transistors especially in applications where inductive loads are repetitively switched. In this work, a model for predicting the hot carrier degradation of an LDMOS in dynamic operation conditions is developed and verified for a device driving an inductive load in repetitive clamping mode. Device simulations are performed using the hydrodynamic model. Based on these simulations the physical mechanism of hot carrier degradation is investigated. The results are verified experimentally by photon-emission microscopy. Monte-Carlo simulation delivers profound insight into the spatial and energy distribution of the carriers impinging on the Si/SiO$_2$-interface.**

LDMOS; inductive load; model; switching; clamping; hot carrier degradation; hot carrier stress; inductance;

INTRODUCTION

Lateral double-diffused Metal-Oxide-Semiconductor (LDMOS) transistors are attractive for medium-voltage smart-power applications due to their compatibility with complementary metal-oxide-semiconductor (CMOS) process flow [1]. Typical applications are in automotive electronics, display drivers, digital audio and power management [2]. Because integrated LDMOS transistors are widely used in safety-critical hardware, their reliability is of prime importance. In general, four possible failure mechanisms of power devices operating in typical application conditions are distinguished: Electric and electro-thermal failures respectively can occur by triggering the parasitic bipolar transistor at small time scales (<1µs). For long stress times from seconds to years there is failure of the device due to the shift of electrical parameters (e.g. the On-resistance R_{on}) out of their specifications which finally leads to circuit malfunction [2]. This is due to the generation of interface traps and/or the trapping of carriers in the oxide. Both damage types are the result of the injection of accelerated high energy "hot" carriers into the oxide [3]. In some applications also bias temperature instability effects may play a role [2]. LDMOS transistors are especially vulnerable to hot carrier degradation because of high electric fields and current densities in proximity to the Si/SiO$_2$-interface [4].

State-of-the-art qualification techniques provide lifetime predictions for a transistor in DC-mode. However, devices integrated in typical application environments are customarily driven in active-cycling mode, e.g. when switching an inductive load. In that case conventional lifetime extraction techniques are not capable to model the degradation of the device as a function of the specific signal waveform. Instead, a worst case approximation is made. This in turn leads to a systematic underestimation of the lifetime of the device in dynamic application mode. However, with progressing shrinking of the devices the lifetime margin decreases because the electric fields increase and thus hot carrier degradation becomes accelerated. As a consequence, more appropriate qualification techniques need to be developed to make lifetime predictions based on the particular dynamic operating conditions.

Within this work, a model for predicting the lifetime of a device in arbitrary dynamic application conditions is presented. It is verified for a clamped n-channel LDMOS transistor that drives an inductive load in repetitive switching mode. Additionally, a comprehensive understanding of the underlying hot carrier degradation mechanism is developed by different experimental and simulation approaches.

In the first section, the device under test is introduced. An overview of the dynamic stress conditions that an LDMOS experiences when switching an inductive load in repetitive clamping mode is given in section II. In section III the parameters of the V_D- and V_G-dependency of DC-degradation are extracted from experimental data in accordance to [5]. Experimental and simulation results on the hot carrier degradation mechanism are presented in section IV. Finally in section V, a model for predicting the device degradation in dynamic application conditions is developed. It is verified for a clamped LDMOS transistor that drives an inductive load in repetitive switching mode. A summary is given in the last section.

978-1-4244-5430-3/10 $26.00 © 2010 IEEE

I. DEVICE DESCRIPTION

The devices used in this study are n-channel lateral 60V double-diffused MOS transistors (Fig. 1) which have been processed in a Smart Power Technology. The lowly-doped n-type drift region was realized by a LOCOS (local oxidation of silicon) process and features the typical edges called "bird's beaks". It is overlapped by the gate-poly. The devices own a double RESURF (reduced surface field) layer resulting from the implantation of a deep body. The threshold voltage is V_{Th}=1.5V. All measurements were performed at room temperature. A small gate width (W=28(36) µm) minimizes self-heating. The On-resistance R_{on} turned out to be the most critical parameter with regard to hot carrier degradation. It was measured at V_D=0.25V and V_G=5V. Transistors with both connected and separate source- and body-contacts have been fabricated.

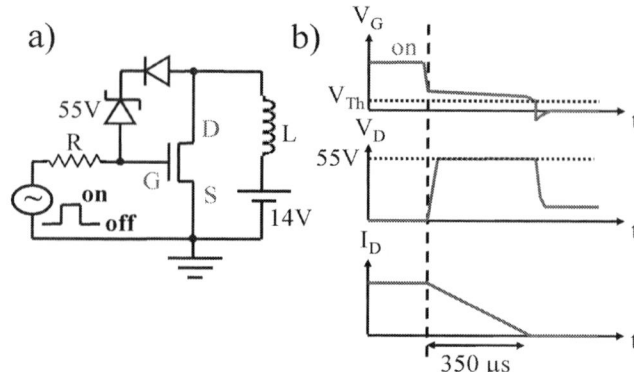

Figure 2   a) Clamped LDMOS embedded in a circuit with an inductive load b) Switch-off behavior of V_G, V_D and I_D

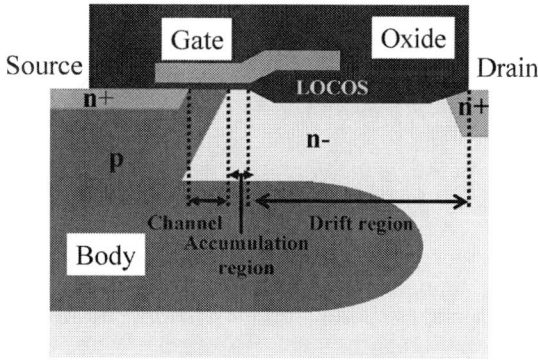

Figure 1   Schematics of the investigated device. A double RESURF-layer was realized by the implantation of a deep body.

II. DYNAMIC STRESS CONDITIONS

In Fig. 2a, a schematics of the LDMOS embedded in a circuit is shown. The setup corresponds to a typical automotive application environment. While the source is at ground-potential, the drain of the transistor is series-connected to an inductor and a DC voltage source (V_{Supply}=14V). The gate is connected highly resistive to a pulse generator. Zener-diodes between gate and drain constitute the so-called clamping structure and limit the drain voltage to V_D=55V. When the gate voltage V_G is below the threshold voltage V_{Th}, the transistor is off and V_{Supply} drops almost exclusively across the device. Due to the absence of current flow and a noteworthy voltage no degradation is observed. When the gate voltage is increased above V_{Th}, the transistor becomes low-resistive. As a consequence V_{Supply} drops predominately across the load. Due to the very low drain voltage no noticeable hot carrier degradation occurs. However, when the transistor is switched off, the induction voltage imposed by the inductor causes a voltage overshoot at the drain which is limited to V_D=55V by the clamping structure. Fig. 2b shows the evolution of the gate voltage V_G, the drain voltage V_D and the drain current I_D with time during switch-off. V_D increases rapidly until the clamping voltage is reached and then remains almost constant during switch-off. The drain current I_D experiences a linear decrease. V_G drops to a value slightly above V_{Th} and then is slowly reduced below threshold while the magnetic energy of the

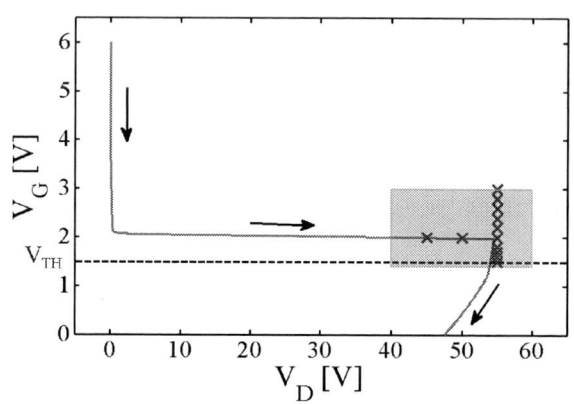

Figure 3   Typical switching path of an LDMOS during switch-off. The most critical operation points with regard to hot carrier degradation are schematized by a blue rectangle ("critical operation area"). The crosses indicate the operation points at which DC-stress experiments have been carried out.

inductor is dissipated. Fig. 3 shows the switching path during switch-off as given by circuit simulation. The highlighted blue rectangle defines a critical operation area for the present application with regard to hot carrier degradation. DC-degradation measurements have been performed at the operation points indicated as crosses in Fig. 3.

III. EXTRACTION OF THE DC MODEL PARAMETERS

For modeling the degradation of an LDMOS in the aforementioned application environment, a detailed analysis of the DC hot carrier degradation in the critical operation area is required. The On-Resistance R_{on} is the most critical parameter with regard to hot carrier stress. No threshold voltage shift is observed. The V_D- and V_G-dependency of R_{on}-degradation in DC-mode is depicted in Fig. 4 and Fig. 5 as symbols. Three measurements were performed per stress condition. The degradation curves exhibit saturation for enhanced stress times which is typical for drain extended transistors [6]. No initial decrease of R_{on} is observed (t_{min}=17s). The V_D-dependency of R_{on}-degradation is shown in Fig. 4. V_G is kept constant at

V_G=2V. ΔR_{on} increases monotonously with increasing V_D. The degradation of R_{on} at V_D=const=55V and different V_G is depicted in Fig. 5. Coming from threshold voltage (V_{Th}=1.5V), the shift of R_{on} increases and reaches its maximum at around V_G=2.2V. For higher V_G the degradation decreases again. An appropriate method for modeling the DC-degradation is the Dreesen-Model [6]. In contrast to a simple time-power-law it takes into account the self-limiting behavior of the degradation of an LDMOS and is, slightly adapted, written as follows:

$$\frac{\Delta R_{on}}{R_{on}^{\;0}} = \frac{C_1(V_G) \cdot (\xi \cdot t)^{n(V_G)}}{1 + C_2(V_G) \cdot (\xi \cdot t)^{n(V_G)}} \quad (1)$$

with

$$\xi(V_G, V_D) = \frac{I_{Source}(V_G, V_D)}{W} \cdot \left(\frac{I_{Body}(V_G, V_D)}{I_{Source}(V_G, V_D)}\right)^{\beta} \quad (2)$$

$\beta := \dfrac{\varphi_{it,e}}{\varphi_{ii}}$ for electrons and $\beta := \dfrac{\varphi_{it,h} \cdot \lambda_e}{\varphi_{ii} \cdot \lambda_h} + 1$ for holes [6,7,8].

φ_{ii} is the minimum energy that a hot electron must possess in order to create an impact ionization (φ_{ii}=1.3eV) [7]. $\varphi_{it,e}$ and $\varphi_{it,h}$ are the critical energies that electrons and holes respectively must have for generating damage at the interface. In a simple approach those are the energies necessary to overcome the band-offsets at the Si/SiO$_2$-interface ($\varphi_{it,e}$=3.3eV [9] for electrons and $\varphi_{it,h}$=4.5eV [9] for holes). λ_e and λ_h are the corresponding mean free paths in silicon and are assumed to be to each other as λ_h/λ_e=0.724 [10]. Therefore we expect β to take a value close to β=2.5 in the case of an electron induced degradation mechanism and a value around β=5.8 if the degradation is mainly driven by hot hole injection. W denotes the gate width of the transistor. $C_1(V_G)$, $C_2(V_G)$, $n(V_G)$ and β are fitting parameters. Assuming that the degradation mechanism is not a function of V_D, they are independent on V_D [5]. $\xi(V_G, V_D)$ denotes the so-called acceleration factor [6]. It represents the rate of hot carriers possessing enough energy to overcome the energy barrier at the Si/SiO$_2$-interface. For simplicity ξ is assumed to be constant, though it undergoes some degradation with increasing stress time. The system would be over-determined if the individual degradation curves were modeled separately each with four fitting parameters. Therefore in a first step the exponent β is extracted by fitting the degradation curves in Fig. 4 together. This is allowed as C_1, C_2 and n are assumed to be independent on V_D. The resulting curve fits are depicted as lines. We obtain β=2.8. This is close to the expected value for an electron-induced degradation

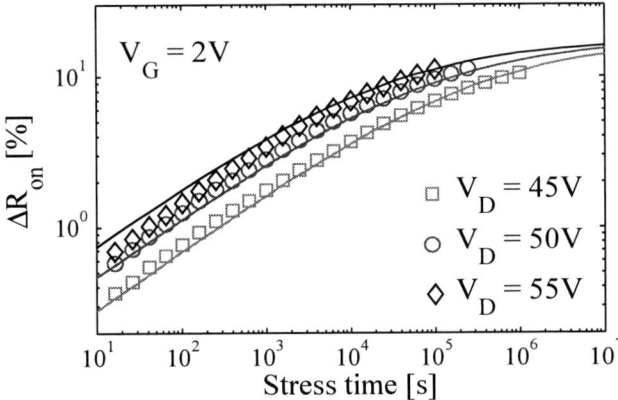

Figure 4   Symbols: Experimental DC stress measurements at V_G=2V, Lines: Curve Fit

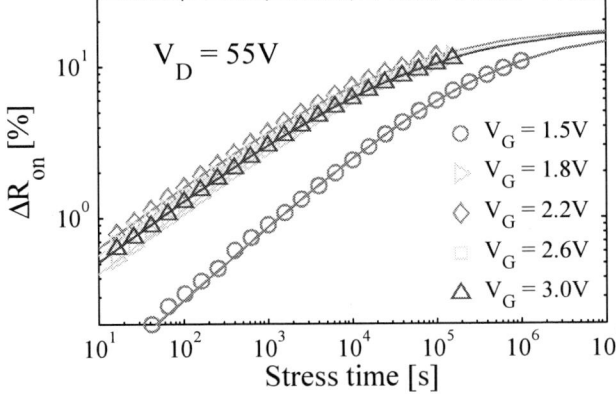

Figure 5   Symbols: Experimental DC stress measurement at V_D=55V Lines: Curve Fit

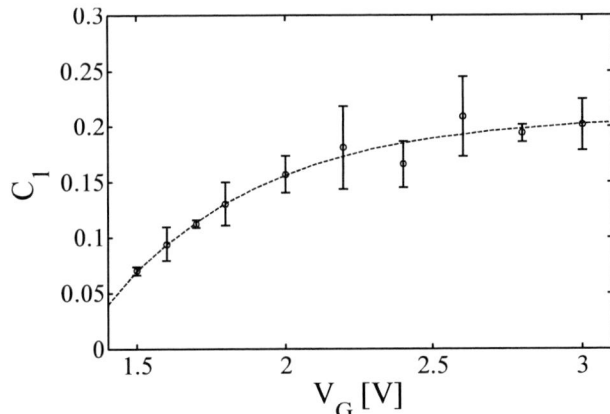

Figure 6 V_G-dependency of the fitting parameter C_1

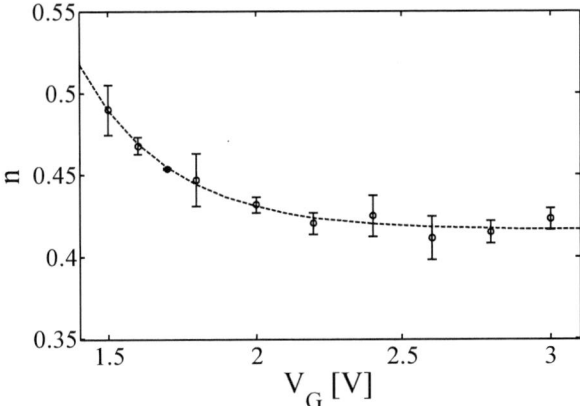

Figure 7 V_G-dependency of the fitting parameter n

mechanism and corresponds to values reported in literature (e.g. β=2.7 in [5]). In a next step β is introduced into equation (1) and the V_G-dependency of the model parameters C_1, C_2 and n is extracted by fitting the degradation data for each stress experiment separately (lines in Fig. 5): The model parameters C_1 and n are depicted in Fig. 6 and Fig. 7 as a function of V_G. The curve progression is in agreement to previous work [5]. The saturation level C_1/C_2 of R_{on} is constant (C_1/C_2=18%) throughout the investigated V_G-range. Since the V_G-dependency of the fitting parameters requires further explanation, the mechanism of hot carrier injection is investigated in the following.

IV. PHYSICAL ANALYSIS

In order to gain a detailed understanding of the hot carrier degradation within the critical operation area, two-dimensional process and device (TCAD) simulations had been carried out. Because of high electric fields inside the device the energy transport of the carriers needs to be taken into account. Hence the simulations were performed using the hydrodynamic approach within the Sentaurus Device Simulator [11]. T-Suprem4 [11] was used for the process simulation. Fig. 8

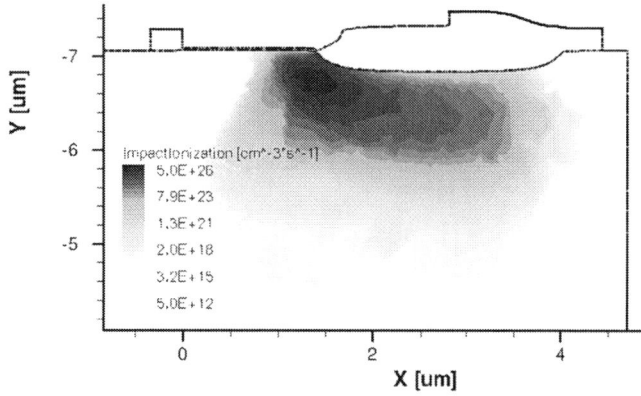

Figure 8   Device simulation (Hydrodynamic approach) at V_D=55V, V_G=2.2V: The spatial distribution of impact ionization has a maximum in proximity to the channel-side bird's beak.

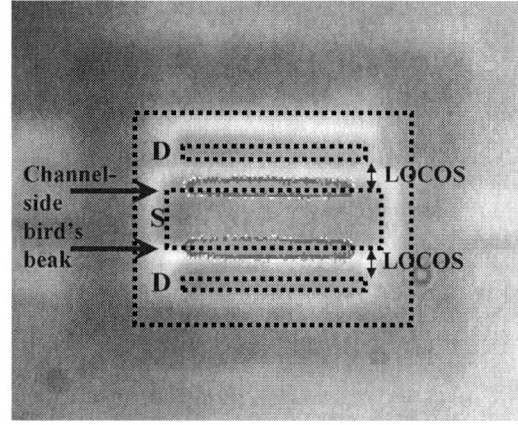

Figure 9   Top view on the device: Superposition of layout and CCD-image yields that the position of maximum impact ionization is located in proximity to the channel-side bird's beak. (S: Source, D: Drain)

depicts the spatial distribution of the avalanche generation rate at the worst stress condition. The position of maximum impact ionization is located in proximity to the channel-side bird's beak. Its position does not change significantly within the critical operation area. As the magnitude of avalanche generation is strongly coupled to the density of hot carriers [7], degradation of the Si/SiO$_2$-interface is expected to occur predominately at the channel-side bird's beak and/or in the accumulation region. The position of maximum impact ionization as given by simulation is verified experimentally by back-side photon-emission microscopy [12]: A CCD-camera detects radiation in the near infra-red and the visible range. It is thus sensitive to photons originating from electron-hole pair recombination and radiant scattering of accelerated carriers. As both effects go along with impact ionization, the spatial intensity distribution of the emitted light is correlated to the impact ionization distribution inside the device. The position of maximum impact ionization is received by overlaying the intensity distribution with the device geometry. The result is shown in Fig. 9 and agrees well with device simulation. Furthermore, device simulation shows that the electric field in the accumulation region and at the channel-side bird's beak accelerates holes towards the interface. This result suggests that the degradation might be dominated by the injection of hot holes. However, the value for β (β=2.8) that we gain from fitting of the degradation data is indicative for injection of hot electrons.

Hydrodynamic simulations do not model correctly especially hot electrons. Full-band Monte-Carlo simulation [13] is performed to cope with this problem. It aims at revealing the spatial and energy distribution of electrons and holes impinging on the oxide interface in the accumulation region and at the channel-side bird's beak. The small inset in Fig. 10 shows the electron and hole hot spot respectively at the channel-side bird's beak which had been approximated by a step-wise function for Monte-Carlo simulation. Electrons and holes are accelerated in opposite directions by the strong electric field in proximity to the bird's beak. As a consequence the position of maximum impingement of hot holes is located

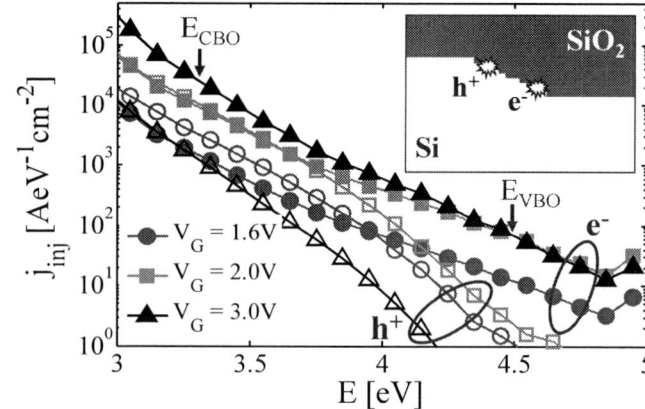

Figure 10   Device simulation (Monte-Carlo): Energy distribution of the electron (filled symbols) and hole (open symbols) impingement current density at their respective hot spots for different gate voltages (V_D=55V). Inset: The positions of maximum impingement for electrons and holes respectively are located at the channel-side bird's beak.

978-1-4244-5430-3/10 $26.00 © 2010 IEEE

in the upper part of the bird's beak whereas the position of maximum impingement of hot electrons is shifted further down the LOCOS. Fig. 10 shows the energy distribution of electron and hole impingement current density at their respective hot spots. The rate of electrons with energies that exceed the Si-SiO$_2$ conduction band offset (E$_{CBO}$=3.3eV [9]) is several orders of magnitude higher than the rate of holes possessing enough energy to overcome the valence band offset (E$_{VBO}$=4.5eV [9]). Nevertheless, for decreasing V$_G$ we observe that the rate of hot holes increases with regard to the rate of hot electrons. This is in good correlation to the experimental results by P. Moens et al. [2] who could show that hot hole injection is most serious at small gate voltages. In conclusion, Monte-Carlo simulation as the most physics-based approach suggests that degradation of the device is predominately caused by the injection of hot electrons in the lower part of the bird's beak. Furthermore the position of the electron injection spot does not change for V$_{Th}$≤V$_G$≤3.0V within the given exactness.

In the next step, the dependency of R$_{on}$ on the position of the interface damage and on different damage types is investigated. Therefore a uniform distribution of interface traps and fixed charges with density N=5·10^{11}cm^{-2} and width W=0.1μm is introduced at different locations at the Si/SiO$_2$-interface. The effect on R$_{on}$ is evaluated by hydrodynamic device simulations. The results are depicted in Fig. 11. X' denotes the distance of the interface traps and fixed charges from the end of the channel along the Si/SiO$_2$-interface. Fixed negative trapped charges Q$_{e-}$ increase R$_{on}$ because they repel electrons from the interface and thus increase the resistance of the current path along the interface which is in accumulation while measuring R$_{on}$. Positive trapped charges Q$_{h+}$ on the other hand increase the conductance [5]. Since R$_{on}$ increases with increasing stress time, hot hole injection is not supposed to be the dominant degradation mechanism. Charges in the accumulation region have only small effect on R$_{on}$ because of the strong gate coupling. When the charges are shifted down the LOCOS, the impact on R$_{on}$ is strongly increased because the gate coupling decreases due to the enhanced oxide thickness. It reaches its maximum in the lower part of the bird's beak. Further shift of the position of damage decreases the

impact on R$_{on}$ again because the fraction of current flowing next to the interface decreases with increasing oxide thickness [5]. In addition, acceptor- and donor-type interface traps (N$_{it}$^{A}, N$_{it}$^{D}) were introduced and in a first approach positioned in the middle of the band gap. Acceptor-type interface traps carry the charge of one electron when occupied whereas donor-type interface traps then carry the charge of one hole. The MOS-capacitance in the accumulation region and at the channel-side bird's beak is in accumulation while measuring R$_{on}$. The simulation shows that the occupancy of acceptor-type interface traps is very high within wide parts of the band gap. As a consequence they are supposed to have a major impact on R$_{on}$. In conclusion, the degradation of R$_{on}$ is assumed to be dominated by the generation of acceptor-type interface traps and/or negative oxide trapped charges. Donor-type interface traps may be generated during stress, but are considered to be irrelevant for R$_{on}$-degradation as they are only rarely occupied within wide parts of the band gap. This is in agreement to reports that give evidence of an amphoteric nature of interface traps [14, 15]. Due to the elevated position of the Fermi level during R$_{on}$-measurement it is not possible to distinguish between negative oxide trapped charges and interface traps. However, the interface of the accumulation region and at the

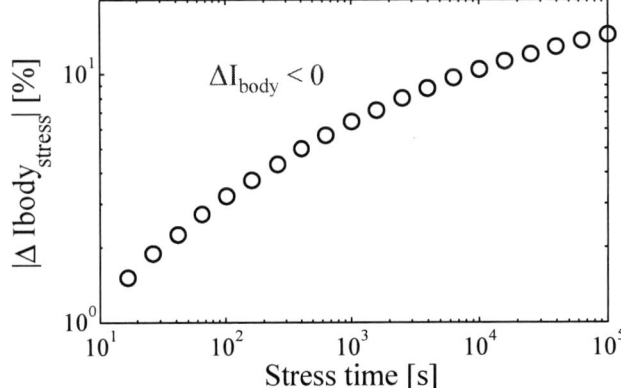

Figure 12 Experimental degradation of I$_{Body}$ during stress (V$_D$=55V, V$_G$=2V). I$_{Body}$ decreases with increasing stress time.

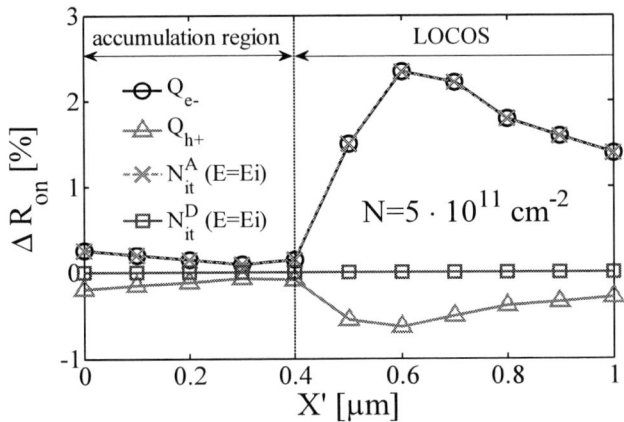

Figure 11 Device Simulation (Hydrodynamic approach): The influence of different damage types (N=5·10^{11}cm^{-2}, W=0.1μm) on R$_{on}$ is investigated as a function of the position X' along the Si/SiO$_2$-interface.

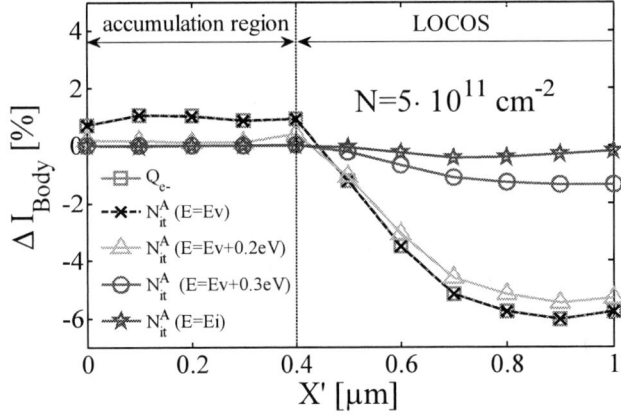

Figure 13 Device Simulation (Hydrodynamic approach): The influence of the energy level of acceptor-type interface traps (N=5·10^{11}cm^{-2}, W=0.1μm) on I$_{Body}$ during stress is investigated as a function of the position X' along the Si/SiO$_2$-interface

channel-side bird's beak is close to the onset of inversion at V_D=55V and V_G=2V, which is a typical stress condition. The body-current I_{Body} experiences distinct degradation during stress and is an excellent indicator for a more detailed investigation of the degradation mechanism [3]. In Fig. 12 the experimental degradation of I_{Body} during DC-stress is depicted at V_D=55V and V_G=2V. I_{Body} decreases monotonously with increasing stress time. Fig. 13 shows the influence of fixed negative trapped charges and acceptor-type interface traps on the degradation of I_{Body} as given by simulation. The largest shift is given for fixed negative charges and for acceptor-type interface traps with energy equal to the valence band edge. Increase of the energy level of the interface traps decreases the occupancy and thus reduces the I_{Body}-shift. Monte-Carlo simulation suggests that interface damage is predominately located in the LOCOS region. Then negative oxide trapped charges and acceptor-type interface traps near the valence band edge lead to a distinct decrease of I_{Body}. Donor-type interface traps have an opposite effect on ΔI_{Body}, but are uncritical as long as their occupancy is below the occupancy of the acceptor-type interface traps.

In summary, we have shown by hydrodynamic device simulation that the hot spot is situated in proximity to the channel-side bird's beak. Experimental verification is provided by back-side photon-emission microscopy. Monte-Carlo simulation suggests that the degradation is dominated by the injection of hot electrons in the lower part of the channel-side bird's beak. No shift of the electron injection spot is observed. Simulation of the impact of different damage types on I_{Body} during stress and on R_{on} indicates that the interface damage is due to fixed negative oxide trapped charges and/or acceptor-type interface traps near the valence band edge at the channel-side bird's beak. We do not find evidence for a change of the degradation mechanism for $V_{Th} \leq V_G \leq 3.0V$. This is in good agreement to the constant saturation level C_1/C_2 of degradation. Nevertheless, though the V_G-dependency of C_1 and n in Fig. 6 and Fig. 7 is not fully understood yet, a function can be extracted that defines C_1, C_2 and n for the whole range $V_{Th} \leq V_G \leq 3.0V$.

V. MODELING THE DEGRADATION IN DYNAMIC CONDITIONS

In the following, a model for predicting the degradation of a device in arbitrary switching conditions is developed. Since the linearization of the DC-model is not sufficiently accurate, a non-linear time-dependent system needs to be solved, requiring a numerical approach. In the following, the basic concepts of the model are sketched: The upper picture in Fig. 14 shows an arbitrary simple switching cycle of an LDMOS passing through the stress conditions $V_{G,1}$, $V_{G,2}$ and $V_{G,3}$ repetitively. The lower picture schematizes the concept of adding up the different contributions of damage. The system remains in each stress condition for the time interval Δt. During the initial time interval in stress condition $V_{G,1}$ the device reaches damage level D_1 (e.g. ΔR_{on}). Provided that the injection spot does not move during the different stress stages, the degradation of the device at a certain stress stage is explicitly dependent on the pre-existing damage. Thus the degradation of the device during the second time interval follows the DC-stress curve $V_{G,2}$ with

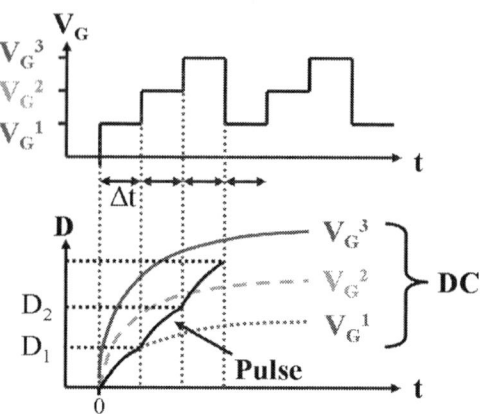

Figure 14 Method for predicting the degradation in pulsed operation mode from DC-degradation data

the starting point corresponding to the damage level D_1. With this method, the contributions of all time intervals are summed up until the end of stress is reached. This method is valid as long as the signal waveform is quasi-stationary, i.e. the switching speed is low compared to intrinsic charging effects. For sufficiently small time intervals, arbitrary signal waveforms may be modeled by linearization. We define the R_{on}-shift as

$$D := \frac{\Delta R_{on}}{R_{on}^0} . \tag{3}$$

In DC-mode, for V_G=const. and V_D=const., the degradation per time $\partial D / \partial t$ can be written as

$$\frac{\partial D}{\partial t} = \frac{C_1 \cdot \xi^n \cdot n \cdot t^{n-1}}{(1 + C_2 \cdot (\xi \cdot t)^n)^2} . \tag{4}$$

For calculation of the degradation in arbitrary dynamic mode, $\partial D / \partial t$ is not a unique function of t anymore. A better variable is however the damage level D. Therefore the variable t in equation (1) is substituted by transforming equation (1) to

$$t = \left(\frac{D}{C_1 \cdot \xi^n - D \cdot C_2 \cdot \xi^n} \right)^{\frac{1}{n}} . \tag{5}$$

$\partial D / \partial t$ then looks like

$$\frac{\partial D}{\partial t} = \frac{C_1 \cdot \xi^n \cdot n \cdot \left(\dfrac{D}{C_1 \cdot \xi^n - D \cdot C_2 \cdot \xi^n} \right)^{\frac{n-1}{n}}}{\left(1 + C_2 \cdot \xi^n \cdot \left(\dfrac{D}{C_1 \cdot \xi^n - D \cdot C_2 \cdot \xi^n} \right) \right)^2} . \tag{6}$$

As $C_1(V_G)$, $C_2(V_G)$, $n(V_G)$ and $\xi(V_G, V_D)$ can be complicated functions of t, equation (6) cannot be solved analytically. Instead, a numerical approach is used for calculating the damage level D. For very small t we can write

$$D(t + \Delta t) = \frac{\partial D}{\partial t}(D(t), V_G(t), V_D(t)) \cdot \Delta t + D(t) \tag{7}$$

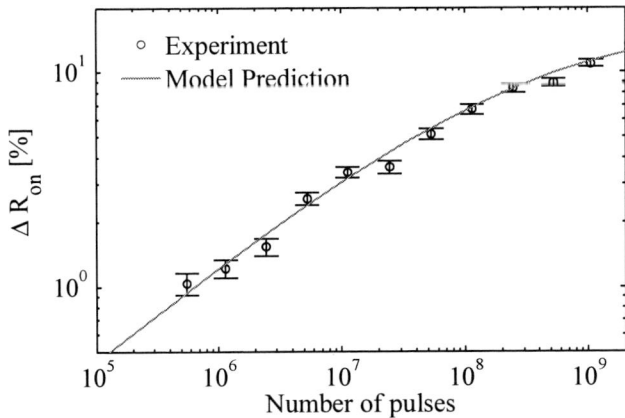

Figure 15 Symbols: Experimental degradation of R_{on} in repetitive clamping mode. Line: Model prediction

For the calculation of the degradation of the device in an arbitrary switching mode, the signal waveforms $V_G(t)$ and $V_D(t)$ must be known. Additionally, β, $C_1(V_G)$, $C_2(V_G)$ and $n(V_G)$ must have been extracted before as well as the source- and body-current for all operation points. Then the degradation of the device can be evaluated. The red line in Fig. 15 gives the calculated degradation of R_{on} as a function of the number of pulses for a clamped LDMOS in a stress condition that corresponds to the repetitive switching of an inductive load. As the scaling of the dynamic test setup with a real inductance is difficult to be realized technically, the coil was substituted by a pulsed current-source impressing a triangular current pulse through the clamped device. By this method the voltage- and current propagation corresponding to a real clamped inductive load during switch-off could be reproduced appropriately. The symbols in Fig. 15 represent the experimental R_{on}-degradation of an LDMOS in repetitive clamping mode. Good correlation is found between the model prediction and the experimental degradation. Due to the small transistor size and limitation of V_G to low values during clamping the dissipated power does not exceed 20mW. The solution of the heat-conduction equation [16] gives an estimate for the upper limit $\Delta T=5°C$ for self-heating during repetitive clamping and at the corresponding DC-stress conditions. The total influence of temperature fluctuations on the degradation is estimated to have a value smaller than one percent of R_{on}.

VI. CONCLUSION

The DC hot carrier degradation mechanism of an n-type LDMOS was investigated by both simulation and experimental investigations. Our findings indicate that the degradation is dominated by electron injection which leads to the formation of acceptor-type interface traps and/or negative oxide trapped charges. A model was developed that predicts the degradation for operation in arbitrary dynamic application mode as long as the signal waveform is quasi-stationary. The model prediction for degradation of an LDMOS in repetitive clamping mode

correlates very well to the measurement and thus is shown to be a valuable method to be integrated into circuit design.

ACKNOWLEDGEMENT

The authors would like to thank S. Aresu, H. Nielen, G. Schindler, R. Kraus, U. Brunner, S. Müller, T. Grasser and S. Decker for fruitful discussions and technical support. This work was financed by MEDEA+ in the framework of the 2T204 ELIAS project.

REFERENCES

[1] W. Kanert, "Reliability Challenges for Power Devices under Active Cycling", Proc. of the Int. Reliability Physics Symp., 2009, pp. 409-415

[2] P. Moens, and G. Van den bosch, "Characterization of Total Safe Operating Area of Lateral DMOS Transistors", IEEE Transactions on Device and Materials Reliability, Vol. 6, 2006, pp. 349-357

[3] P. Heremans, R. Bellens, G. Groeseneken, A. v. Schwerin, H. E. Maes, M. Brox, and W. Weber, "The Mechanisms of Hot-Carrier Degradation", in Hot Carrier Design Considerations for MOS Devices and Circuits, C. T. Wang, Eds. New York, Van Nostrand Reinhold, 1992

[4] P. Moens, G. Van den bosch, C. De Keukeleire, R. Degraeve, M. Tack, and G. Groeseneken, "Hot Hole Degradation Effects in Lateral nDMOS Transistors", IEEE Transactions on Electron Devices, Vol. 51, 2004, pp. 1704-1710

[5] P. Moens, J. Mertens, F. Bauwens, P. Joris, W. De Ceunick, and M. Tack, "A Comprehensive Model for Hot Carrier Degradation in LDMOS Transistors", Proc. of the Int. Reliability Physics Symp., 2007, pp.492-497

[6] R. Dreesen, K. Croes, J. Manca, W. De Ceunick, L. De Schepper, A. Pergoot, and G. Groeseneken, "Modelling hot-carrier degradation of LDD NMOSFETs by using a high-resolution measurement technique", Microelectronics Reliability, Vol. 39, 1999, pp. 785-790

[7] C. Hu, S. C. Tam, F.-C. Hsu, P.-K. Ko, T-Y. Chan, and K. W. Terrill, "Hot-Electron-Induced MOSFET degradation – Model, Monitor, and Improvement", IEEE Journal of Solid-State Circuits, Vol. 20, 1985, pp. 295-305

[8] R. Bellens, P. Heremans, G. Groeseneken, and H. E. Maes, "A new procedure for lifetime prediction of n-channel MOS-transistors using the charge pumping technique.", Proc. of the International Reliability Physics Symp., 1988, pp. 8-14

[9] E. Bersch, S. Rangan, R. A. Bartynski, E. Garfunkel, and E. Vescovo, "Band offsets of ultrathin high-κ oxide films with Si", Physical Review B 78, 085114, 2008

[10] J. J. Tzou, C. C. Yao, R. Cheung, and H. W. K. Chan, "Hot-Carrier-Induced Degradation in P-Channel LDD MOSFET's", IEEE Electron Device Letters, Vol. 7, 1986, pp. 5-7

[11] SDevice and T-SUPREM4, distributed by Synopsys Inc., http://www.synopsys.com

[12] C. Boit, "New physical techniques for IC functional analysis of on-chip devices and interconnects", Applied Surface Sience 252, 2005, pp.18-23

[13] C. Jungemann, and B. Meinerzhagen, "Hierarchical Device Simulation: The Monte-Carlo Perspective", Springer, Wien, 2003

[14] L.-Å. Ragnarsson, and P. Lundgren, "Electrical characterization of P_b centers in (100)Si-SiO$_2$ structures: The influence of surface potential on passivation during post metallization anneal", Journal of Applied Physics, Vol. 88, 2000, pp. 938-942

[15] E. Poindexter, "MOS interface states: overview and physiochemical perspective", Semiconductor Science and Technology, Vol. 4, 1989, pp. 961-969

[16] G. Chen, "Nanoscale Heat Transfer and Nanostructured Thermoelectrics", IEEE Transactions on Components and Packaging Technologies, Vol. 29, 2006, pp. 238-246

978-1-4244-5430-3/10 $26.00 © 2010 IEEE

LOW-SIDE DRIVER'S FAILURE MECHANISM IN A CLASS-D AMPLIFIER UNDER SHORT CIRCUIT TEST AND A ROBUST DRIVER DEVICE

Jian-Hsing Lee, J.R. Shih, Tong-Chern Ong and Kenneth Wu

Technology Quality and Reliability Division, Taiwan Semiconductor Manufacturing Company, jhlee@tsmc.com

ABSTRACT

The failure mechanism in a class-D audio amplifier under short-circuit test is analyzed. The damage, always in the low-side driver, is due to high current induced thermal run-away, which occurs during the shutdown after the over-current is detected. However, this high current doesn't come from the over-current itself since the current is limited to below that the transistor in the class-D amplifier can sustain. Instead, the damage is caused by the displacement current when there is a large voltage change at the output of the class-D amplifier. Although the shutdown circuit is designed to prevent the high current flowing through the transistors of the class-D amplifier, it cannot prevent the current coming from the class-D amplifier itself. To eliminate the damage, the output transistors should be designed robust enough to against the low-pass filter induced large voltage swing.

INTRODUCTION

Two damage modes have been reported when an amplifier IC is connected to an inductive load. The first one is the unclamped inductive switching [1] caused by turning off the driver rapidly. The second one is the diode recovery stress [2] caused by the rapid commutation of the current of the parasitic diode in the output transistor from forward to reverse bias. In addition to the above two damage modes, another damage mode during the short-circuit test is observed when inductor is connected to a capacitor in a full-bridge class-D amplifier. During the shutdown mode of the short-circuit test, the inductor pulls down the voltage of the low-side driver to below -0.7V to keep the current flowing. After the stored energy of the inductor is dissipated completely, the junction capacitor of the high-side driver pulls up the voltage of the low-side driver to equalize the potentials of two class-D amplifier's output nodes. This can induce a large current flowing through the low-side driver to damage the transistor even the induced voltage is still bellow the avalanche breakdown voltage of the transistor.

EXPERIMENT, FAILURE MECHANISM AND SIMULATION

The full-bridge class-D audio amplifier discussed in this article for short-circuit test consists of the conventional double-diffusion devices with high-voltage (HV) 18V CMOS process. Fig. 1 shows the asymmetry layout structure and cross-section of the low side driver in a full-bridge class-D audio amplifier. The source is composed of a high-dose N+ implant. While the drain is composed of the high-dose N+ implant and a low-dose NDD implant. The high-dose N+ implant is enclosed by the low-dose NDD implant and is away from the poly gate to form a reduced surface field (RESURF) region [3] to sustain the high voltage.

In order to investigate the root cause of the class-D amplifier failure during the short-circuit test, a 500MHz and 4G samples/sec digital oscilloscope was used to capture the voltage and current waveforms of the key nodes in the class-D amplifier. The voltage is measured by the voltage-probe, and the current is measured by a 1mA-to-5mA current-probe (Tek CT-1). In addition, commercial simulators (TSUPREM-4 and MEDICI) were used to simulate the transient behavior of the device and study the failure mechanism. So, a robust device for low-side driver of the class-D amplifier can be developed. TSUPREM4 is used for modeling device fabrication, and MEDICI is used for electrical behavior prediction. The apparatus used to evaluate the high current characteristics of the device for simulating the electrical behaviors of the output transistor of the class-D amplifier during the short-circuit test is a high power pulse generator (Agilent HP 8114A).

(a)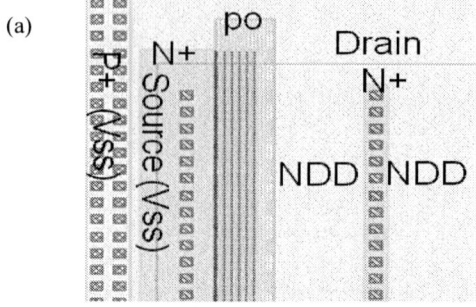

(b).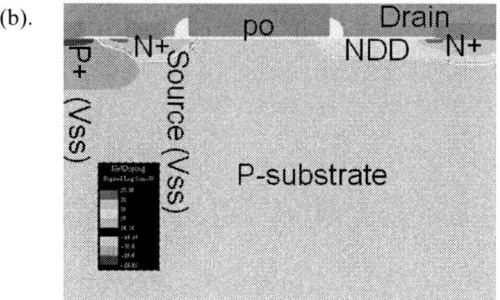

Fig. 1 (a) Layout (b) cross-section for 18V NDD NMOS

A. Short-Circuit Test

Fig. 2 shows the schematic diagram of a full-bridge audio class-D amplifier. A loudspeaker and two LC low-pass filters are in series with the full-bridge amplifier [4]. The full-bridge amplifier is used to drive the LC low-pass filter to amplify and reproduce the audio signal. Pulse width modulation (PWM) is used to transfer the analog audio signal into the on -or off states of output devices. There is a shutdown circuit to shut off the amplifier when the over-current detection circuit detects current out of limitation [4], [5]. This current is caused by the accidental short between two output drivers in the class-D amplifier.

At the beginning of the short-circuit test, the Out_1 is at high voltage state when PMOS P1 turns on and NMOS N1 turns off. Out_2 is at low voltage state when PMOS P2 turns off and NMOS N2 turns on (Fig. 3a). This will form a loop composed of the high side driver P1 of the Out_1, inductors and the low side driver N2 of the Out_2 when the switch (S in Fig. 2) is closed for short-circuit test. Owing to the impedances of the components in this loop, Out_2L's voltage is pulled up to ~8V in an instant as shown Fig. 4a. The short-circuit current (I_L in Fig. 3a) ramps up with a slope $dI_L/dt = Vcc/(2 \times L1)$ as shown in Fig. 4b. When the current level reaches the over-current limit ~4.5A in Fig. 4b, the shutdown circuit will be turned on to shut off all the transistors in the class-D amplifier. In the meantime, it can be observed that the output voltage of the shutdown circuit has changed from 3.3V to 0V as shown in Fig. 4a. Owing to the inductors, the short-circuit current (I_L in Fig. 3b) cannot be changed immediately. Its slope is changed from positive to negative value (Fig. 4b). Then, the short-circuit current gradually ramps down to generate the induced electromotive forces across the two inductors, which pulls down driver Out_1's voltage to below 0V (~-1.2V Fig. 4a) and pulls up driver Out_2's voltage to above Vcc (~15.7V Fig. 4a). So, the diode D_{N1} of the low-side drivers and the diode D_{P2} of the high-side drivers are forward biased to keep the short-circuit current flowing until the energy stored in the inductors is dissipated completely. It can also be found that the voltage of the Out_1 is pulled up from negative to positive and the voltage of the Out_2 is

978-1-4244-5430-3/10 $26.00 © 2010 IEEE   182

pulled down immediately. Finally, the voltages of the two output nodes fall to the same voltage as the Out_2L. Since the two output nodes (Out_1 and Out_2) of the class-D amplifier are connected to Out_2L via an inductor, the three nodes will have same potential when the short current decreases to zero.

From the failure analysis (FA) result of Fig. 5, we can find that only the low-side driver N1 of the driver Out_1 is damaged after the short-circuit test. However, we cannot find any damage at the high-side driver P1 of the driver Out_1 and at the drivers P2 and N2 of the driver Out_2. Apparently, the catastrophic damage was caused by thermal run-away induced by high power generation. It is suspected that the current level might reach a level beyond what the device can sustain. However, the device dimension of the class-D amplifier is designed to be wide enough to sustain 5A current at least. Before going into the shutdown mode, the short-circuit current as shown in Fig. 4b is still too small to damage the two turned-on transistors P1 and N2 in Fig. 3a. During the shutdown mode, diodes D_{P1} and D_{N2} can also sink the short-circuit current without inducing any injury before the current decreases to zero. So, the damage should not be caused by the short-circuit current as shown in Fig. 4b. From the voltage waveforms of the two output nodes of the class-D amplifier as shown in Fig. 4a, it can be observed that there are some discontinuous points at the shutdown mode. One discontinuous point is at the beginning transient of the shutdown mode, where Out_2's voltage switches from 0V to high voltage (~15V), while Out_1's voltage switches from high voltage (~13V) to negative voltage (~ -1.5V). At this transient, the parasitic diodes of the Out_1 and Out_2 are forward biased. If the damage can be found at the device N1 of the Out_1, similar damage should also be observed at the device P2 of the Out_2. Another one discontinuous point is at the transient period of the short current decreasing to zero. Although the voltages of the Out_2 and Out_1 all fall to the same voltage (~8V), Out_1's voltage is switched from negative voltage to positive voltage, while the voltage of the Out_2 is always biased at positive level. Apparently, the electrical behaviors of the two outputs during this transient should be different. The device N2 is always biased at revered mode. For the device N1, however, it is a bi-mode stress, which includes the forward mode and reverse mode.

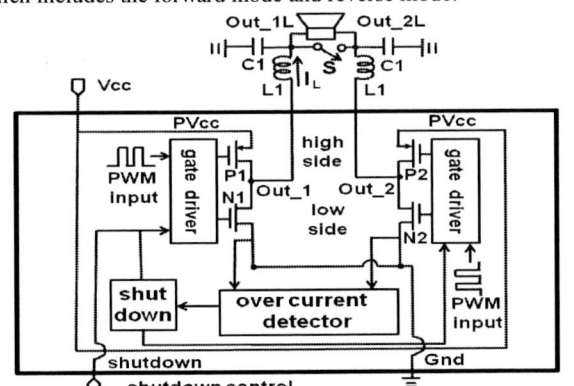

Fig. 2 Functional schematic diagram of a class-D audio amplifier

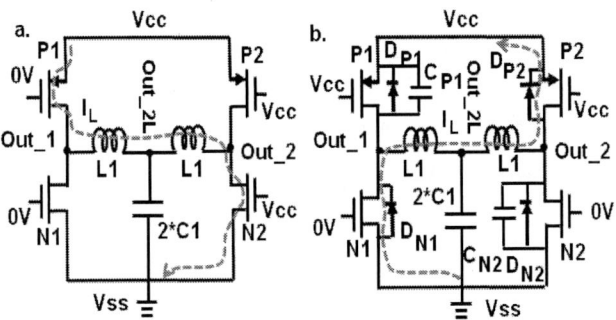

Fig. 3 Current path a. before shutdown, b. after shutdown

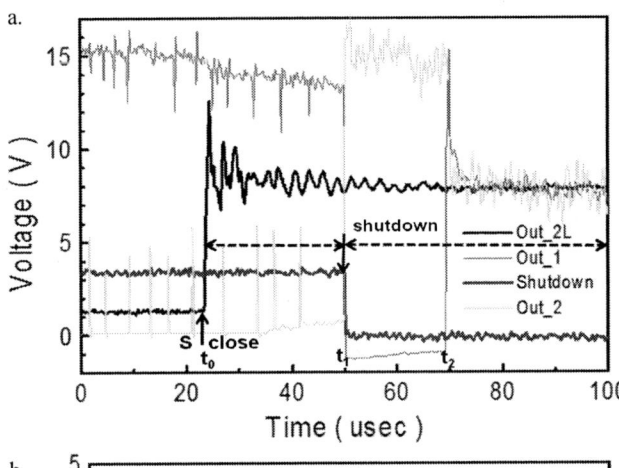

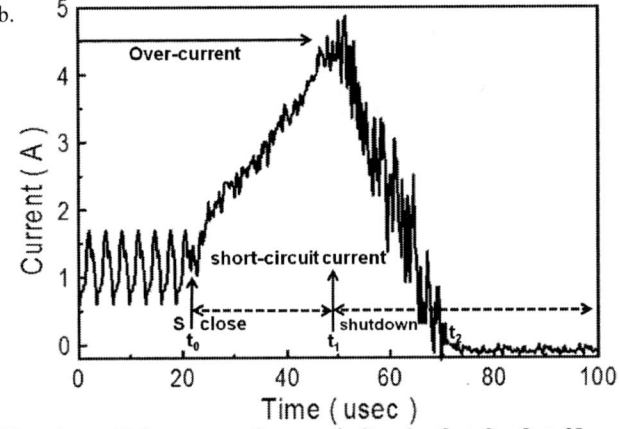

Fig. 4 a. Voltage waveforms of Out_1, Out_2, Out_2L and shutdown , b. short-circuit current waveform I_L in Fig. 2 when S closes.

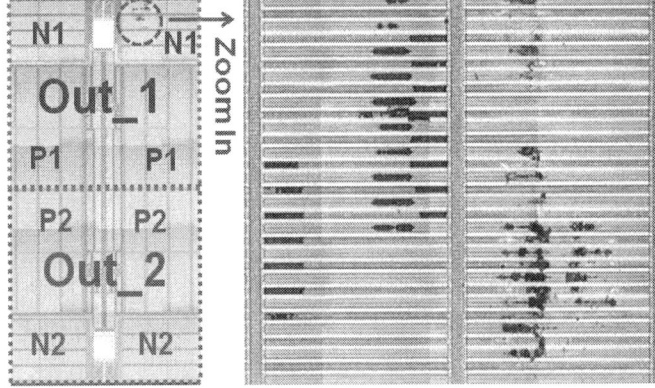

Fig. 5 Chip's failure site locates at the low side driver N1 of the Out_1 after short circuit test.

B. Failure Mechanism

In order to identify the failure mechanism, the repeated pulses with different base level are applied to device's drain with source, gate and P-substrate (Fig. 6) grounded to simulate the stress on the two low-side drivers (N1 and N2 in Fig. 2) during the short-circuit test. For the pulse with zero base level $V_L=0V$, the current cannot flow through the device except the on-off transients of the pulse if the pulse level V_H is below transistor's avalanche breakdown voltage. During the positive cycle, the voltage can be kept at a constant until the avalanche breakdown (V_H 19.5V) occurs, since the device is

978-1-4244-5430-3/10 $26.00 © 2010 IEEE

always at the off state, as shown in Fig. 7a. After further increasing the voltage level, an apparent snapback phenomenon can be observed since the parasitic npn bipolar transistor is turned on. But, it can also be observed that the turn-on voltage (V_{trig}) of the device under the pulse with V_H 19.5V is decreased from 19.5V to 6V at the beginning of the second cycle of the pulse. It is because that the device had been damaged after the first cycle. In other words, the pulse with a zero base level will not damage the transistor if the V_H is below 19.5V. This is why the transistor N2 in Fig. 2 is not damaged by the short-circuit test.

Fig. 7b shows that the currents can be observed at the positive cycles of the pulse with the negative base level since the device turns on even the voltage level (V_H=15V or V_H=16V) is below the avalanche breakdown voltage. This indicates that the negative base level of the pulse can reduce the turn-on threshold voltage (V_{trig}) of the device. In addition, the abnormal current increase and voltage decrease also can be observed at the second cycle of the pulse with V_H=16V due to damage. This implies that the negative base level of the pulse also can decrease the immunity of the device to against the voltage swing during the short-circuit test. Fig. 8 shows the simulated voltage waveform to reproduce the stress condition of the low side driver Out_1 in the class-D amplifier during the shutdown mode for investigating the electrical behavior of the device under the pulse with negative base level. Fig. 9 shows the electron concentration profiles of the device under a pulse for each time interval in Fig. 8. Before the pulse (T_0 in Fig. 8), the electron concentration is quite low as shown in Fig. 9a. As the device biased at negative voltage (T_1 in Fig. 8), the drain will inject a lot of electrons into the P-well since the pn junction is forwarded. It can observe that the electrons flood the whole P-well as shown in Fig. 9b. When the pulse switches from negative to positive level, the electrons generated by the forward bias (D_{N1} in Fig. 6) cannot be annihilated instantly and are still stored in the P-well. Fig. 9c shows that there are still existed huge electrons in the P-well region when the pulse switches to zero voltage. Unlike the case with zero base level, the electron concentration in Fig. 9d is already higher than the required concentration to turn on the npn bipolar transistor even though the applied voltage is below the avalanche breakdown voltage. So, the parasitic npn bipolar transistor can be turned on more easily than under the pulse condition with zero base level. As the bipolar transistor turns on, the power often results in device damage due to thermal run-away. From Fig. 7b, the damage threshold of the device for the pulse with negative base level is only 16V, which is smaller than the avalanche breakdown voltage of the device. This is consistent with the result that the transistor N1 of the amplifier in Fig. 2 can be damaged after the short-circuit test.

Although the voltage at Out_1 is pulled from negative to positive, the short-circuit current is decreased to zero at the transient t_2 in Fig. 4. So, the damages at transistor N1 of the class-D amplifier in Fig. 5 should be not caused by the power product of the short-circuit current and voltage across the Out_1 at the transient t_2. Based on the above discussion, the equivalent circuit of the class-D amplifier can be depicted as in Fig. 10. Since the voltage of the shutdown circuit is still kept at 0V at the transient t2, all HV MOS transistors will be turned off. But the parasitic diodes (D_{N1} and D_{P2}) are forward biased before transient t_2 in Fig. 4. So, the low-side driver N1 and high-side driver P2 behave as the npn bipolar transistor and pnp bipolar transistor, respectively. In the meantime, the drain junctions of high-side driver P1 and low-side driver N2 behave as the capacitors since the drain junctions are reverse biased. As the short-circuit current decreases to zero, the induced electromotive force across the two inductors disappears and it results in quick voltage changes at Out_1 and Out_2 to equalize the potentials of Out_1, Out_2 and Out_2L. This will induce the large displacement currents $Cd\overline{V}/dt$ (I_{C1} and I_{C2} in Fig. 10) to flow through the turned-on transistors since all the transistors of the class-D amplifier are with huge dimension. Due to the larger voltage change, the displacement current flowing through the low-side driver N1 will be much larger than the current flowing through the high-side driver P2. The larger the current flows through

the device, the more the power is dissipated on the device. This is why the damage only occurs at the low-side driver N1, but at the high-side driver P2, although the diodes of both transistors are forward biased before the transient t_2.

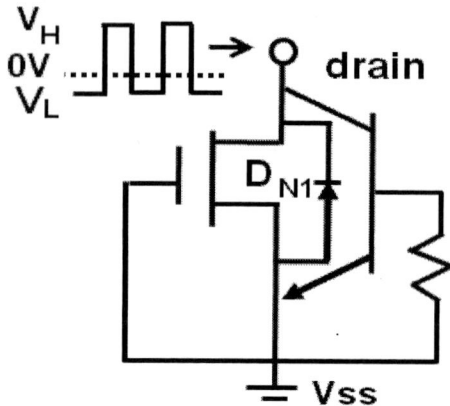

Fig. 6 Device under a repeated pulse with different base level

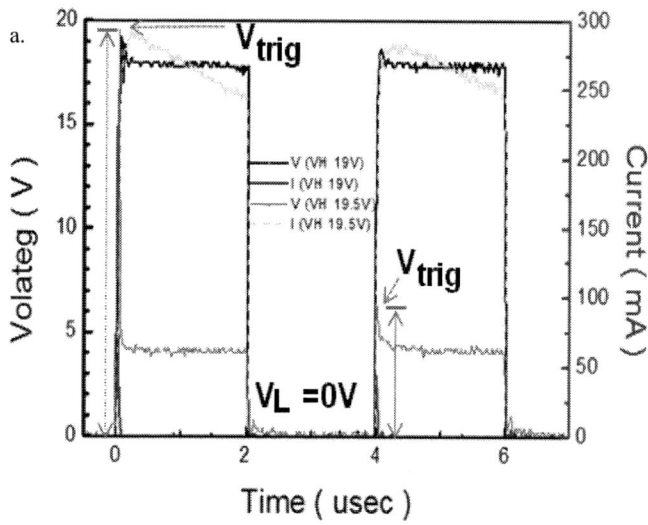

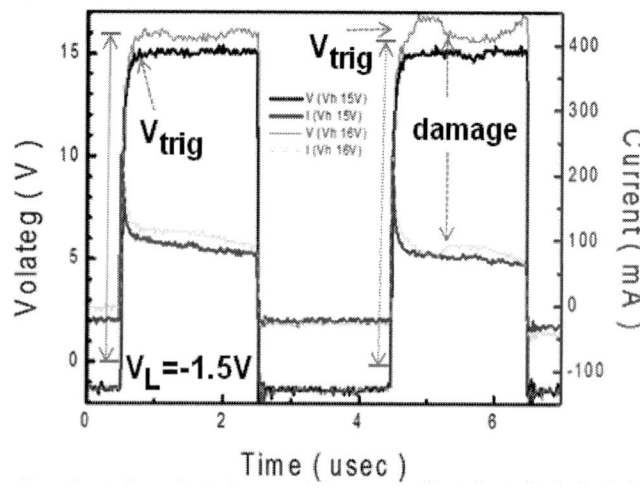

Fig. 7 Drain voltage and current waveforms of the transistor under pulse with a. zero base level, b. negative base level.

978-1-4244-5430-3/10 $26.00 © 2010 IEEE

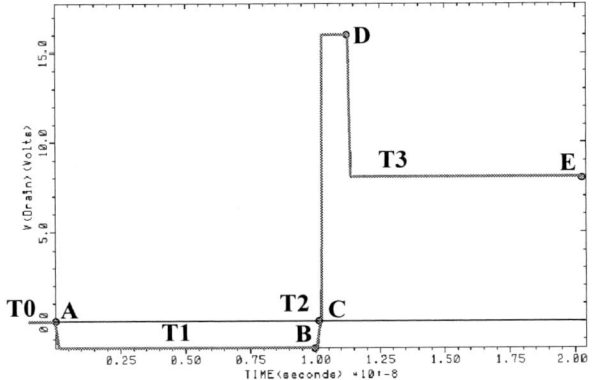

Fig. 8 The simulated voltage waveform of the transistor under a negative base level pulse

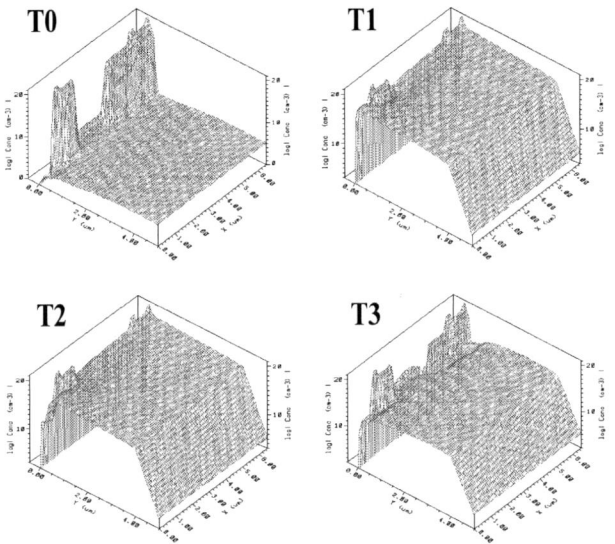

Fig. 9 Electron concentration distributions at (a) T_0, (b) T_1, (c) T_2, and (d) T_3 in Fig. 8

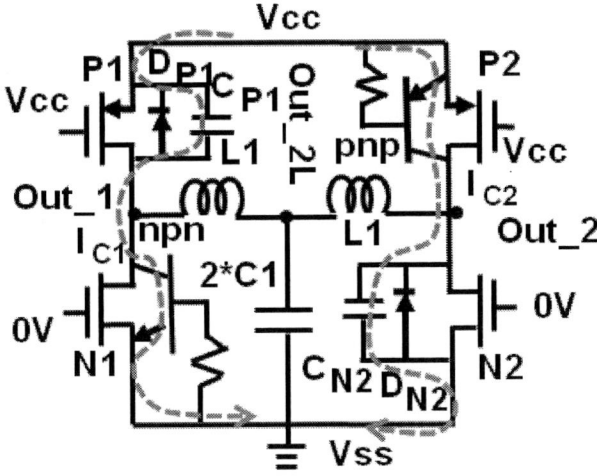

Fig. 10 Equivalent circuit of the class-D amplifier under short-circuit test at the transient t_2 in Fig. 4.

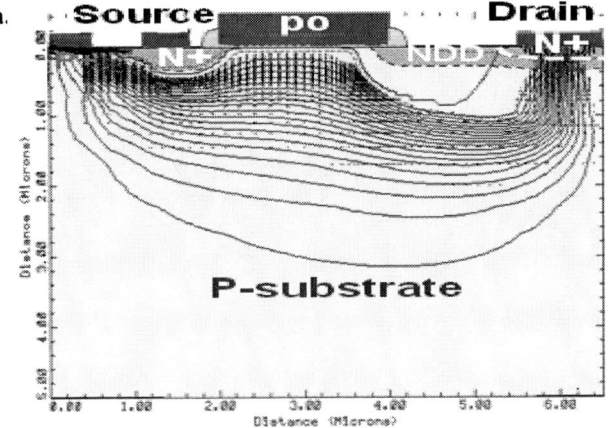

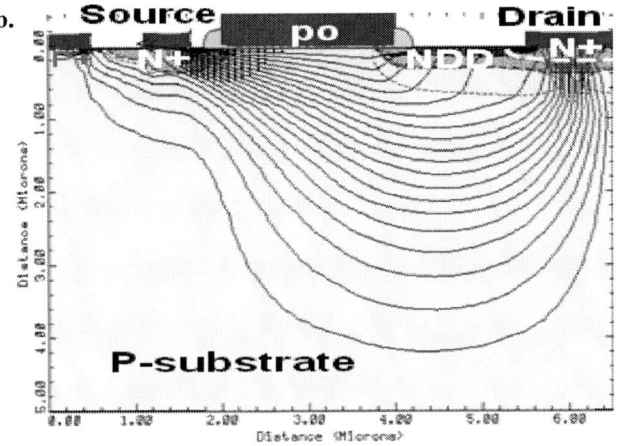

Fig. 11 The simulated current flow lines of conventional HV transistor a. biased at V_{trig}, b. after npn bipolar turns on.

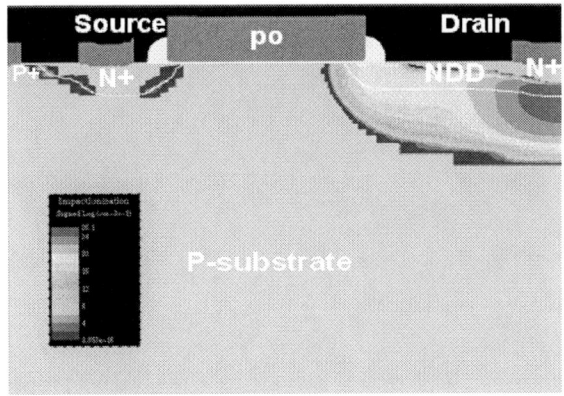

Fig. 12 The simulated impact-ionization rate of conventional transistor biased at V_{trig}

C. A Robust HV Device For Short-Circuit Test

Fig. 11 shows the simulated current flow of the conventional HV transistor biased at V_{trig} and after npn bipolar transistor turns on. This transistor has at long lightly doped NDD region with a shallower

978-1-4244-5430-3/10 $26.00 © 2010 IEEE 185

junction depth than the junction at N+ drain region as shown in Fig. 1b. The long NDD is used to form the reduced surface field (RESURF) region [3] to sustain the high voltage. It can be found that most of the currents does not flow through the NDD region, but are crowded underneath the N+ region. During the normal operation, the current will flow laterally through the channel and NDD to the N+ region. However, this conventional structure has the low field and high resistivity region in the lateral direction, but high field and low resistivity below the N+ region. The impact-ionization rate at the NDD region below the N+ implant is apparently higher than that any other NDD region as shown in Fig. 12. As the device is biased at V_{trig}, it will cause the device to go into the avalanche breakdown at the junction below the N+ implant and result in the current crowding there (Fig. 11a). Even the parasitic npn bipolar transistor is turned on, the current is still difficult to flow through this NDD region and still crowded at the region below the N+ implant. It can find that only a small amount of current flows laterally through the NDD region. Most of the currents still flow vertically through the NDD region b as shown in Fig. 11b.

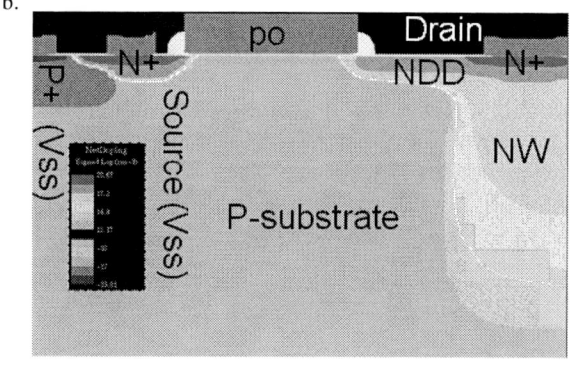

Fig. 13 (a) Layout, (b) Cross-section and profile of the proposed HV NMOS structure with NW strap

In order to improve the uniformity of electrical field distribution in the transistor, the N-well is added to the NDD region below the N+ region as shown in Fig. 13. Unlike the conventional transistor in Fig. 1, the maximum electrical field of the conventional transistor is located at the poly edge. Fig. 14 shows the impact-ionization rate at the poly edge is apparently higher than that at other regions when the device is biased at V_{trig}. As a result, the breakdown location can be moved to the junction nearby the poly edge, resulting in the current crowding there (Fig. 15a). So, the current flow is changed from vertical direction (Fig. 11a) to lateral direction (Fig. 15a). Since the current needs to flow a high resistance region, the Vt1 of the new HV NMOS (Fig. 7) is increased significantly by compared with the Vt1 of the conventional HV NMOS (Fig. 16). As the npn bipolar

transistor is turned on, the currents flow more uniformly along the NDD/PW and NW/PW junctions as shown in Fig. 15b. Fig. 16 shows the failure threshold of this new HV device is increased to 43V and 39V for the pulses with zero base level and negative base level, respectively. From Fig. 4a, the stress condition of the transistor during the short-circuit test is apparently much lower than the failure threshold of the modified HV transistor. Thus, the proposed transistor can survive well during the short-circuit test.

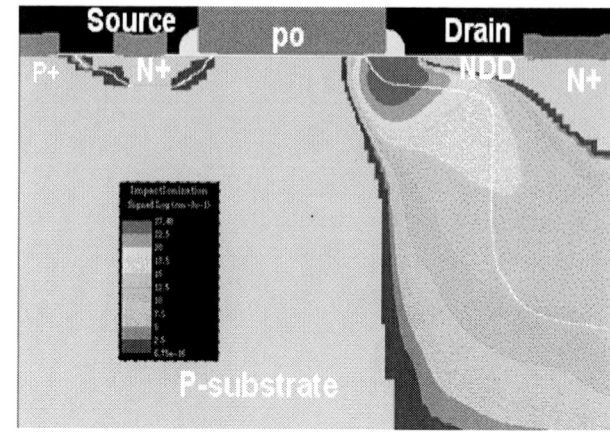

Fig. 14 The simulated impact-ionization rate of the modified HV device biased at V_{trig}.

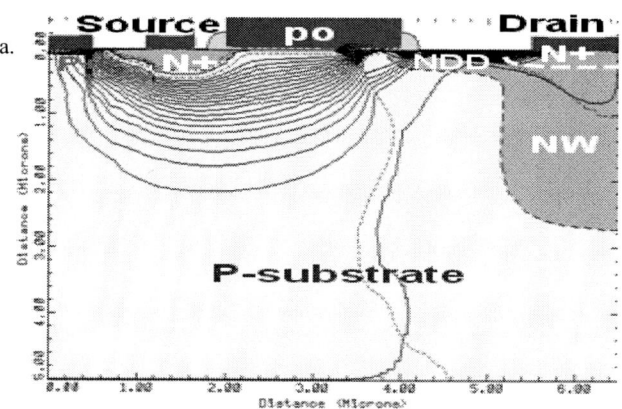

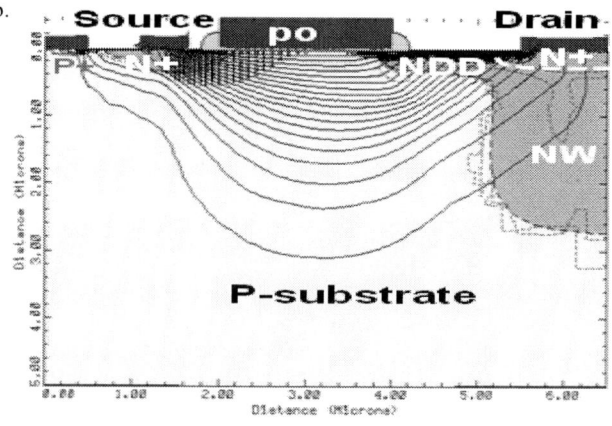

Fig. 15 The simulated current flow-lines of the modified transistor (a) biased at V_{trig}, (b) after npn bipolar transistor turns on

CONCLUSIONS

2237, 2005

If a class-D amplifier needs to be qualified by short-circuit test, the transistor not only needs to be designed with enough large dimension to sustain the short-circuit current, it also needs to be designed with a robust structure to prevent the thermal run-away due to large and short transient current. To prevent such kind of damage, a modified device structure with N-well implant strap beneath the N+ and NDD regions is proposed. With this robust driver, the system designer does not need to use discrete Schottky diode to eliminate this kind of damage.

a.

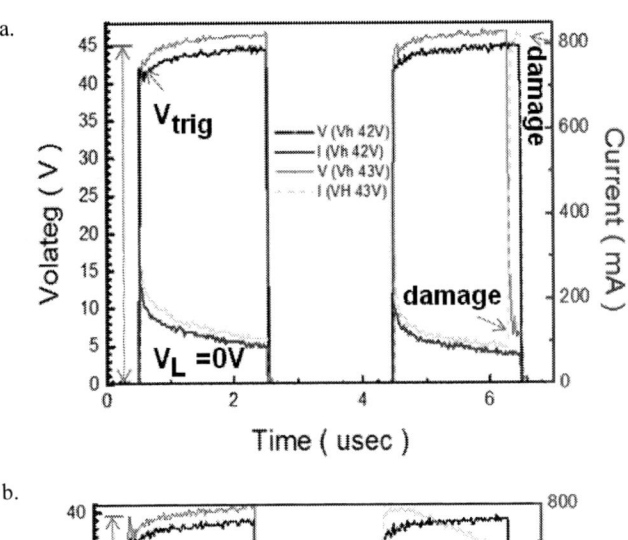

b.

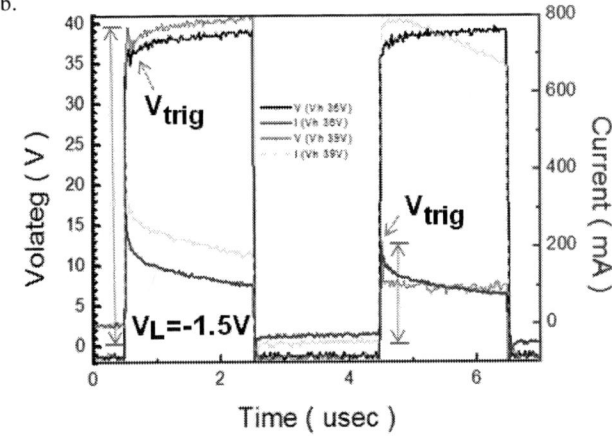

Fig. 16 Drain voltage and current waveforms of the transistor in Fig. 13 under (a) zero base level pulses, (b) negative base level pulses

REFERENCES

[1] Rainer Constapel, M. S. Shekar, and Richard K. William, "Unclampedl Inductive Switching of Integrated Quasi-Vertical DMOSFETs," ISPSD, p. 219, 1996.

[2] John P. Phipps and Kim Gauen, " New Insights Affect Power MOSFET Ruggedness ," APEC, p. 290, 1988.

[3] Adriaan W. Ludikhuize, "A Review of RESURF Technology," ISPSD, p. 11, 2000.

[4] TI technique text "TPA3007D1".

[5] Marco Berkhout, "Integrated Overcurrent Protection System for Class-D Audio Power Amplifiers," IEEE J. solid-state cir cuits, p.

On The Radiation-Induced Soft Error Performance of Hardened Sequential Elements in Advanced Bulk CMOS Technologies

N. Seifert [1], V. Ambrose [2], B. Gill [1], Q. Shi [1], R. Allmon [2], C. Recchia [1], S. Mukherjee [2], N. Nassif [2], J. Krause [2], J. Pickholtz [2], A. Balasubramanian [1]

[1] Technology Manufacturing Group, Intel Corporation, [2] Intel Architecture Group, Intel Corporation

Norbert Seifert, Hillsboro, OR, 971-214-1700; email: Norbert.Seifert@intel.com

Vinod Ambrose, Santa Clara, CA, phone: 408-765-0487; e-mail: Vinod.Ambrose@intel.com

ABSTRACT

Test chips built in a 32nm bulk CMOS technology consisting of hardened and non-hardened sequential elements have been exposed to neutrons, protons, alpha-particles and heavy ions. The radiation robustness of two types of circuit-level soft error mitigation techniques has been tested: 1) SEUT (Single Event Upset Tolerant), an interlocked, redundant state technique, and 2) a novel hardening technique referred to as RCC (Reinforcing Charge Collection). This work summarizes the measured soft error rate benefits and design tradeoffs involved in the implemented hardening techniques.

Neutron; Alpha particle; neutron; proton; heavy ion; space; terrestrial; single event effects; SEE; soft errors; SE; hardened;mitigation

INTRODUCTION

Ionizing radiation is known to cause noise bursts in silicon (Si) substrates of modern integrated circuits (ICs) [1-3]. If the amount of charge collected at reverse-biased junctions is larger than the so-called critical charge (Qcrit), an upset occurs [4]. Due to the relatively low flux rates in the radiation environments of interest in this work, faults are induced by single particles and all radiation induced phenomena are referred to as single event effects (SEE). In memory type cells radiation-induced faults are called single event upsets (SEU). SEUs are stable in time until the upset devices are re-written. An entirely different class of radiation-induced faults is formed by single event transients (SETs). SETs occur in static combinational logic where the node voltage is always restored in the case of a particle strike. Radiation induced glitches per se do not result in errors on the chip or system level until the glitch is captured by a receiving storage element [5]. SETs in clock networks are discussed in reference [6].

More than 95% of all upsets at sea-level are either due to a) high-energy neutrons, or b) alpha particles emitted from radioactive isotopes located within ~50µm of the active Si surface [3, 7]. In contrast, soft error upsets in a space environment mainly result from a) protons trapped in belts by earth's magnetosphere in the case of low earth orbits, and b) heavy ions in geosynchronous orbits. For a detailed description of the different radiation environments, please see reference [8].

Neutrons and high energy protons, in contrast to alpha particles or heavy ions, do not directly ionize Si but generate electron hole pairs via secondary ions that are created in nuclear reactions [2, 9]. There are two classes of nuclear collisions: elastic and inelastic scattering. In most elastic events, only a small amount of energy is transferred onto the target nucleus, which recoils but does not change its intrinsic energy state. In case of an inelastic event, secondary protons, neutrons, and pions are produced, and an excited intermediate nucleus is formed. This nucleus subsequently de-excites by the emission of other secondary particles, and it is finally transformed into a stable and lighter residual nucleus. The secondary fragments from the second reaction stage consist of protons, neutrons, light ions, and heavy residual nuclei. The heavy recoiling nuclei typically deposit a large amount of charge in a small volume, whereas the secondary light fragments deposit charge over path lengths that are large compared to typical device dimensions. Low-energy secondary protons deposit appreciable ionization energies (per unit track length) in the Si substrate. For modern technologies, characterized by low Qcrit values, these low energy protons might be a significant contributor to device upset rates through direct ionization rather than nuclear reactions [10].

If the radiation event deposits sufficient charge, more than a single device or bit may be affected, creating a so-called multi-cell upsets (MCU) as opposed to a single bit upsets (SBU) [11]. Technology scaling is known to increase the fraction of MCU clusters dramatically, with important implications for future memory architectures in systems utilizing error correction codes (ECC). Recent studies have demonstrated the increased sensitivity of memory type devices in the presence of MCU and charge sharing[1] for various radiation environments [11, 12, 13].

Most published SER trend data are for terrestrial environments. However, there is no reason why similar trends should not be expected for space applications. With process scaling, most authors agree that the total SER/bit is decreasing for SRAM devices [11, 14, 15, 16]. Because most SRAM arrays are nowadays protected, SRAM trends are becoming of lesser importance. This is in contrast to random logic, which is much more difficult and expensive to protect [17, 18]. No industry-wide agreement seems to exist for logic devices [3, 19, 20]. However, to the best of our knowledge, logic SER on the chip-level is expected to increase per generation if no additional mitigation techniques are implemented. MCU rates show an exponential increase with process scaling [11].

In a recent publication it was speculated that the return on investment of some popular design mitigation techniques that rely on separation in space, such as interleaving in memory arrays, or hardened devices utilizing local redundancy, might suffer greatly with continued process scaling [21]. For older technologies conventional radiation-hardened-by-design (RHBD) approaches such as Dual Interlocked Cell (DICE), Built-in Soft Error Resilience (BISER), Single Event Upset Tolerant (SEUT) or Triple Modular Redundancy designs (TMR) provide excellent protection against SEU [22, 23, 24, 25]. With technology scaling, charge collection at multiple nodes due to a single particle strike is becoming more probable. One of the key objectives of our work is to measure and characterize the SER benefit of mitigation techniques that rely on hardening by redundancy. To quantify the impact of scaling, the authors designed and implemented hardened devices with different minimum node separation design

---

[1] The term charge sharing is somewhat misleading. It denotes collection of charge by two or more nodes (within the same or different cells) due to the one particle strike (i.e., one single event).

rules on two test chips. The implemented and tested technique is SEUT [24]. However, the learning gained by this study is expected to be applicable to any circuit-level mitigation scheme that relies on separation in space of redundant state nodes. Experimental SER results for various radiation environments are presented. Another key objective of this work is to introduce and characterize the soft error sensitivity of a novel circuit level mitigation technique called RCC (Reinforcing Charge Collection). RCC promises very low power and area overheads at sufficiently low upset rates and most importantly is expected to show better technology scaling properties than redundancy based techniques.

TEST CHIP DESIGNS

Two different soft error (SE) test chips have been designed and built in a 32nm CMOS bulk technology. Each test chip (TC) contains thousands of instantiations of several flavors of sequential elements each, all chained together in a shift register fashion [24]. All relevant elements are briefly summarized in Table 1.

TABLE 1. SUMMARY OF TEST CHIPS AND IMPLEMENTED SEQUENTIAL DESIGNS.

Test Chip	Sequential Design	Description
TC1	Ref. Latch	Standard library reference latch
	SEUT800	SEUT latch with sensitive diffusion separation of about 800nm
	SEUT150	SEUT latch with sensitive diffusion separation of about 150nm
	RCC2	RCC latch with sensitive diffusion separation of 5 poly widths
	RCC1	RCC latch with one poly width diffusion separation
TC2	SEUT800	SEUT latch with sensitive diffusion separation of about 800nm
	SEUT600	SEUT latch with sensitive diffusion separation of about 600nm
	SEUT400	SEUT latch with sensitive diffusion separation of 400nm
	SEUT150	SEUT latch with sensitive diffusion separation of about 150nm

In the following sections, upset modes and key properties of each investigated design style (SEUT and RCC) will be explained.

Local Redundancy Hardened Designs

Many redundancy based hardened designs have been published and tested [22, 23, 24, 27, 28, 29, 30]. Formal design and analysis techniques have also been developed for SEU immune circuits [31, 32]. Most designs are single error correcting circuits except BISER [23]—which is an error blocking design. Single error correcting designs, such as SEUT [24, 30], recover upsets to a correct state after radiation induced pulse is removed without waiting for the next clock signal. Error blocking designs use two redundant memory elements

and add error blocking logic at the output to block error propagation. BISER reuses scan circuits as redundant memory to reduce area and power penalties and a C-element at the output to block error propagation. SEUT and BISER devices as well as other redundancy based hardened designs such as TMR are expected to show very similar radiation properties and dependencies on circuit parameters such as critical state node separation, critical charge and diffusion areas [21].

Redundancy based hardened designs discussed in this work can only recover strikes when charge is collected by one node (with the exception of clock node strikes, see below). It is very important to separate "critical nodes" in space to minimize the amount of charge collected at those sensitive nodes. Therefore, SER reduction is strongly dependent on critical node spacing. The farther critical nodes are separated, the higher the SER reduction (see below). The flipside of larger spacing is larger cell areas. Hence, trade-offs must be made to balance SER reduction and cell area growth. Technology scaling reduces spacing by about ~0.7x each generation. This imposes a big challenge on hardened circuit design for current and future generation designs.

A second major upset mechanism of hardened sequential elements is clock node strikes [16]. Please note that this upset mechanism does not involve charge collection at more than one node and therefore can be a significant SER contributor. In principle one can distinguish two modes of clock node upsets [6]: a) Radiation-induced race which reflects a false opening of the receiving sequential and data racing through it. For non-critical paths this mode is the dominating clock node upset mechanism. b) Radiation-induced clock jitter, where the clock edge is shifted by the particle strike such that for critical paths, data will not be latched correctly. Due to the very slow clock speeds applied in our experiments, our shift register test chips were not sensitive to radiation-induced jitter.

Finally, for non error blocking schemes and strikes that yield transient glitches (SETs) only, pulses could propagate to downstream logic and could potentially be latched by downstream sequential elements—similar to other forms of noise in combinational logic. This soft error contributor of hardened sequential elements is difficult to quantify (by simulation or measurements), but nevertheless might become an important SER contributor in future technologies. However, our test chips are not sensitive to this upset mode due to the implemented design and test methodology.

The core design of all tested local redundancy hardened cells is the storage element shown in Figure 1, which replaces the classical cross-coupled inverter non-hardened memory element. The fully interrupted SEUT circuit features redundant data signals (d0 and d2) to reduce the overall cell SER[2]. During a normal write operation, clock is high and input passgates are on. The transistors controlled by clock inputs to SEUT are off. Data is written into SEUT inputs d0 and d2 which controls qp2, qn2, qp5 and qn5 setting states d1 and d3 correctly.

---

[2] Another option to reduce SER would be to protect clock nodes by implementing redundant clocks.

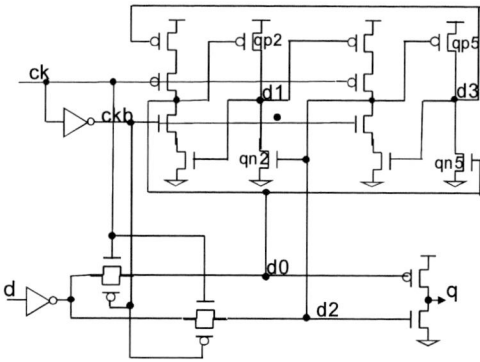

FIGURE 1. SEUT STORAGE CELL [24, 30]

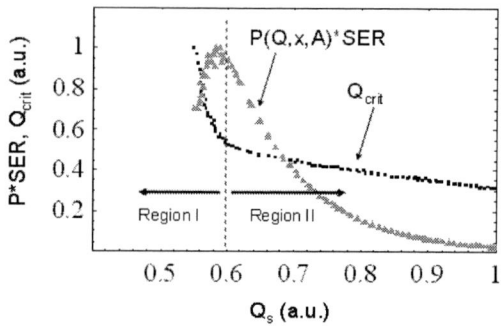

FIGURE 2. SEUT QCRIT AND P(Q,X,A) VS QS [21]

Qcrit simulation results and a SER model of SEUT (and other local redundancy hardened designs) have been introduced and discussed in reference [21]. In the following, key findings are briefly summarized (see Figure 2).

The key innovation that differentiates the hardened circuit-level simulation methodology from traditional ones is that not one but two (or more) current sources are attached to the nodes that are simultaneously collecting charge [21]. One has to differentiate between charge collection at the primary and secondary nodes[3]: Qcrit of the primary node is simulated by iterating the collected charge at the primary node as a function of charge collected at the secondary node Qs until the circuit fails. Two distinct regions can be observed. At low Qs values Qcrit initially decreases steeply with increasing Qs. Therefore a Qs threshold exists below which the primary node cannot be upset, independent on how much charge is collected on the secondary node. In other words, a minimum amount of charge needs to be collected at the electrically coupled secondary node, or the device cannot be upset by charge sharing. The soft error rate under charge sharing conditions SER_{CS} then equals the integral over the product of the SBU soft error rate $SER(Qcrit_p(Qs(x))$ at the primary node p with $P(Qs(x)|Qcrit(Qs(x)))$ [21, 33]

$$SER_{CS}(p,s,x) \propto \int_{0}^{\infty} \frac{P(Q_s(x) \mid Qcrit_P(Q_s(x))) *}{SER(Qcrit_p(Q_s(x))dQ_s} \tag{1}$$

$P(Qs(x)|Qcrit(Qs(x)))dQs^4$ denotes the conditional probability that charge Q in the interval [Qs, Qs+dQs] is collected at the secondary node, given that Qcrit or more is collected at the primary node due to the same single event. The probability $P(Qs(x)|Qcrit(Qs(x)))$ decreases steeply with node (actually diffusion) separation (x) which can be determined from layout [21]. The integrand in equation (1) P*SER contributes significantly only in a small Qs range due to the Qs dependence of P and the Qcrit(Qs) dependence of SER (see Figure 2) [21].

---

[3] Amusan calls the two charge collecting nodes active and passive [13].

[4] Old terminology is P(Q,x,A). Diffusion area dependence is dropped here for better readability.

Reinforcing Charge Collection (RCC) Design

RCC is best explained using a static storage element, typically consisting of a pair of cross-coupled inverters. In each inverter, the OFF device's diffusion (referred to as victim diffusion) is vulnerable to collecting ionizing-particle-induced charge that can disrupt the stored state. The ON device's diffusion[5] (referred here as reinforcing diffusion), on the other hand, collects charge that reinforces the stored state (in the case of the RCC design at least). If the charge generated by a particle strike can be collected in both the victim and reinforcing diffusions (charge sharing), the critical charge (Qcrit) needed to upset the stored state can be increased, thus reducing SER.

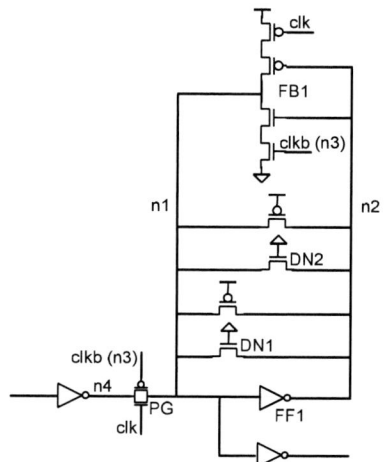

FIGURE 3. RCC SCHEMATIC

The victim diffusion is fully reverse biased making it an efficient collector of the particle-induced charge. The reinforcing diffusion initially has no externally applied reverse bias, but has only the built-in potential, making it a weak collector. However, even this weak collection serves to increase Qcrit [34]. Furthermore, once the victim diffusion begins collecting charge the electric field across the victim diffusion's depletion region quickly collapses [25]. Simultaneously, since the reverse bias across the reinforcing diffusion increases, its depletion region widens and the charge collection efficiency increas-

---

[5] I.e., ON during normal circuit operation when the particle strike occurs. In general charge collection that occurs simultaneously can be reinforcing or weakening. Both, victim and reinforcing diffusions can be primary or secondary nodes.

es. In modern day circuits, these field fluctuations occur in pico-second time scales. Therefore, charge collection in victim and reinforcing nodes is a highly dynamic process that requires mixed-mode device simulations to correctly model. Kawakami et al. [34] have published such simulations, and have shown the increase in Qcrit theorized here. However, they did not do these simulations in the context of radiation hardening. In this work, cells were specifically designed to maximize charge sharing, and Si test structures were used to show its effectiveness.

In order for charge sharing to occur, the victim and reinforcing nodes need to be physically close to each other (see equation 2 below). Typical layout of these cross-coupled structures might already have these nodes physically close to each other. However, test structure measurements show that if these nodes are within a minimum design rule dimension of each other, the closer proximity leads to a dramatic SER reduction. In this work, "dummy" gates (OFF transistors) have been used to bring the diffusions of the same type within one poly dimension. This greatly increases the probability that charge generated by a particle will be collected by both diffusions, thus reducing SER. Introduction of the dummy gates does cost leakage power. In addition, storage node capacitance and area will increase in most cases, since there is less opportunity for diffusion sharing in layout. This costs dynamic power. The impact of increased capacitance and area on SER can be simulated [36], and is shown later in Tables 3 and 5. Figure 3 is a RCC latch schematic that shows two pairs of dummy devices that are OFF, and whose sole purpose is to minimize victim-to-reinforcing diffusion separation. One pair allows charge sharing between victim and reinforcing diffusions of the cross-coupled inverters. The other pair allows charge sharing between the input pass gate diffusion and its complement node. The layout stick diagram for such an arrangement is shown in Figure 4. Only the N diffusions are shown; P diffusions have a similar arrangement.

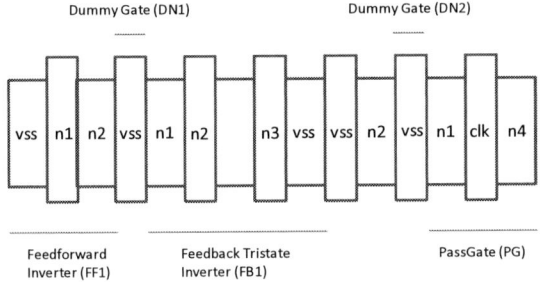

FIGURE 4. RCC LAYOUT DIAGRAM FOR N DIFFUSIONS

A test chip with the latch shown in Figure 3 was built in a 32nm technology (TC1; Table 1). The separation between victim and reinforcing nodes is one poly width for this cell. A cell with no dummy gates was also placed on the test chip. The minimum separation between victim and reinforcing nodes is increased in this version due to standard layout techniques that incorporate a shared power diffusion between critical diffusions (RCC2 in Table 1). The control latch has the same size devices as both RCC flavors implemented on TC1.

The SER of non-redundancy hardened devices under charge sharing conditions (CS), for nodes p (primary) and s (secondary) is given by [33]

$$SER_{CS}(p,s) = SER(Qcrit_p(0)) + \Delta SER_{CS}$$

$$\Delta SER_{CS}(p,s,x) =$$

$$\propto \int_0^{\infty} \left[P(Q_s(x) \mid Qcrit_p(Q_s(x))) * \begin{pmatrix} SER(Qcrit_p(Q_s(x)) - \\ SER(Qcrit_p(0)) \end{pmatrix} \right] dQ_s \quad (2)$$

where SER denotes the nominal soft error rate when charge sharing is ignored and ΔSER_{CS} the correction term in the presence of charge sharing. Symbols have the same meanings as in the previous section. Qcrit (0) denotes the critical charge of the primary node if only the primary node collects charge. It is important to realize that ΔSER_{CS} can be positive or negative depending on the node distances (x), states involved and diffusion types. $P(Q_s(x) \mid Qcrit(Q_s(x)))$ is a steeply decreasing function with increasing $Q_s(x)$ and hence with increasing x. In the case of redundancy hardened devices the main design objective is to reduce SER by maximizing the separation (x) of nodes that increase Qcrit if simultaneous charge collection occurs[6]. In contrast, the key mitigation concept behind RCC devices is to minimize distances of diffusions of nodes that reinforce the stored state if charge is collected simultaneously.

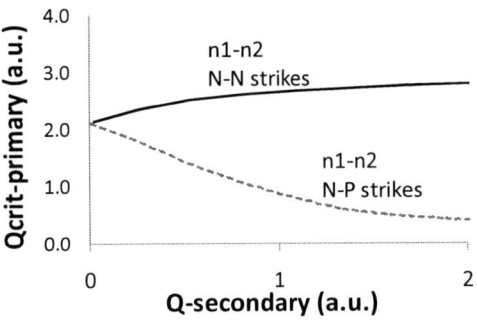

FIGURE 5. SIMULATED CRITICAL CHARGES OF A NON-HARDENED LATCH IN THE PRESENCE OF CHARGE SHARING BETWEEN NODES N1 AND N2.

There are, however, nodes that need to be kept separated even in the RCC design. For instance, N-N strikes (denoting NMOS strikes in the case of charge collecting nodes n1 and n2) result in an increase in critical charges and a correspondingly decrease in SER (solid line in Figure 5). N-P strikes in the same inverter can also decrease SER. In contrast, N-P strikes (denoting strikes where charge is collected in the OFF NMOS on one side and the OFF PMOS on the other side) results in an increase in SER (dashed line in Figure 5). For N-N strikes, one might wonder how it is possible that NMOS diffusions on both sides of the cross-coupled inverter collect charge simultaneously. One of the NMOS must be ON, and the junction therefore is not reversed biased. As explained earlier, the assumption behind equation (2) is that the radiation induced SET propagates turning off the NMOS on the other side allowing charge to be collected there efficiently during this time period. Logic in modern technologies is sufficiently fast and SETs sufficiently wide such that both NMOS on either side of the cross-coupled devices are reversed biased and charge is temporarily collected at nearly the same time. However, both devices are not off for equal lengths in time and equation (2) still needs to be adjusted and fitted to experimental or simulation data.

---

[6] The RCC concept can be combined with local redundancy schemes such as the one implemented in SEUT to achieve even better SER performance levels.

EXPERIMENTAL SETUP

Logic test vehicles built in a 32nm high-k + metal gate process [35] have been exposed to neutron, proton, alpha-particle and heavy ion radiation. The sensitivity of the test structures to alpha-particle radiation was studied by placing Thorium-232 foils on the wire bonded test chips described above. Neutron SER data were collected at the Los Alamos Laboratory Weapons Neutrons Research (WNR) facility, New Mexico (see Figure 6 for typical setup). Proton and heavy-ion irradiation experiments were conducted at the Indiana University Cyclotron facility (IUCF) and Texas A&M University facility, respectively. The investigated proton energy range was 27 – 200MeV. Parts were exposed to heavy-ion beams with linear energy transfer (LET) values ranging from 2.8 to 71 MeV/(mg/cm^2).

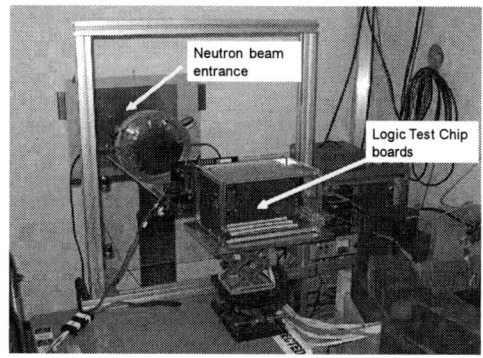

FIGURE 6. TYPICAL EXPERIMENTAL SETUP DURING ACCELERATED NEUTRON SEU TESTING IN THE ICE HOUSE AT WNR [21].

Each of the test chip designs is based on shift register topology and contains 3 inputs: Din, ClkA, ClkB and one output Dout. Each shift register consists of two identical latches: master and slave. The clock input of all of the master latches on a chip is connected to ClkA and similarly, all of the slave latches clock input is connected to ClkB[7]. In order to write a desired test pattern on a chip such as all ones, all zeros, checkerboard, etc, the Din signal is set to appropriate value (i.e. either 0 or 1) and a pulse on ClkA is applied followed by non-overlapping pulse on ClkB. Thus, writing a checkerboard test pattern (10101010...) to shift registers will result in writing 110011001100....sequence in latches of the chip. It is not possible to read both master and slave latches during one experimental run in this design. To read out the data from all the slave latches, a pulse on ClkA is applied followed by non-overlapping pulse on ClkB and data is captured from Dout output pin. Similarly, to read data from all master latches, a pulse on ClkB is applied followed by a non-overlapping pulse on ClkA.

Our test chips can be operated in two testing modes. In mode 1 devices are sensitive to charge sharing induced upsets only and in mode 2 they are sensitive to clock node strikes as well as to upsets induced by charge sharing. The test condition for charge sharing experiments is to have same polarity data stored in the latch as well as at the input pin of the latch. A required test condition for clock nodes strikes is that the data stored inside the latch should be of opposite polarity of the data at the input pin of the latch. The clock node sensitive SER testing is done by writing a checkerboard pattern,

---

[7] Two non overlapping clocks to eliminate the risk of race

stopping the clock during the pre-defined exposure time, and after stopping the beam shifting out the content stored in master latches. The charge sharing testing is done by writing checkerboard pattern, waiting for exposure time, and reading out data from slave latches.

DESIGN TRADEOFFS

The SEUT based hardened cells are subjected to the same strict timing, area and power constraints as non-hardened designs. The design goal is to minimize timing arc changes, while minimizing the increase of area and power compared to the non-hardened cells. We have studied a wide range of SEUT drive strengths and the results of this study are summarized below. Please note that Table 2 only lists overheads for devices implemented in TC1 and TC2.

More than 95% setup and clock to out timing arcs of SEUT based cells meet the design target with average of 13% setup time degradation and 3.6% clock to out delay degradation. Only about 3.6% of the arcs are above the targeted margin. A substantial increase in routing and a corresponding increase in gate and diffusion capacitance make it prohibitively expensive to further power up the cell to bring the arcs below the allowed threshold.

The active power is estimated to increase from 40% to 150% depending on drive strength and averages at about 100%. Power consumption could be lower if the timing requirement is more relaxed. In general, smaller drive strength cells incur a higher percentage power increase. The main reason is that devices in small non-hardened cells are already close to minimum device size. Thus, device sizes are almost double in hardened cells to create redundant paths. In larger cells, devices could simply be split between two redundant paths. For the similar reasons, clock power increase is slightly higher than data power increase in terms of percentage.

The overall area overhead ranges from 50% to 180% with averages at about 100%. In general, an increase in overhead can be observed with decreasing drive strengths. As mentioned earlier and demonstrated experimentally in the next section, sufficient diffusion separation is crucial to achieving low SEU rates. However, increasing critical diffusion spacing implies cell area growth. All critical diffusions are identified through formal analysis of SEUT circuit operation and verified with circuit simulations. Shown in Table 2 are area, power and timing overheads of a typical small drive strength SEUT latch implemented in TC1 and TC2 as compared to the non-hardened control latch of the same drive strength. Also shown are the overheads for an RCC1 latch as compared to the TC1 reference latch of same drive strength.

TABLE 2 AREA, POWER AND TIMING OVERHEAD RESULTS: HARDENED COMPARED TO NON-HARDENED CELLS:

Device	Area	Active Power	Delay	Setup
Small drive strength SEUT latch	+120%	+114%	+4.9%	+11%
RCC1	+10%	+28%	+6.4%	+19%

The RCC layout technique results in modest area increases. Large cells, such as flip-flops with scan, see smaller area increases than small cells, such as a 1-read/1-write 8T register file cell. Modest power increases are also incurred due to increased diffusion capacit-

978-1-4244-5430-3/10 $26.00 © 2010 IEEE

ance and interconnect capacitance. Similarly, the timing penalties are also relatively small. The power and timing estimates shown in Table 2 are based on layout extracted capacitances.

EXPERIMENTAL RESULTS AND DISCUSSION

In this section measured relative upset rates of tested SEUT and RCC devices with respect to the TC1 non-hardened reference design are presented and discussed for terrestrial and space radiation environments. SEUT and RCC SER ratios were computed by dividing the measured upset rates or cross sections with corresponding ones measured for the TC1 reference latch at the same conditions (voltage, particle energy, etc), i.e.

$$\text{SER - Ratio}_{\text{SEUT/RCC}} = \frac{\text{SER}_{\text{SEUT/RCC}}}{\text{SER}_{\text{referencelatch}}}$$

(3)

Unless explicitly mentioned, data were collected under charge sharing conditions (not sensitive to clock node strikes) and with beams at normal incidence to the chips.

Cosmic Ray Testing Results

Despite the fact that in our typical experimental setup tens of test chips each with tens of thousands of hardened and non-hardened devices are daisy chained together, only very few upsets are detected in a typical run that can last several days at a white neutron beam facility such as WNR[8]. High-energy proton facilities (such as IUCF) are much more accessible than white neutron facilities and usually offer much higher particle beam fluxes. A good correlation between white neutron beam and high-energy proton SER results is therefore of great importance for accurate hardened device characterization and SER modeling purposes. The authors of this publication have tested several hardened test chips over the last few years and exposed them to high-energy proton and white neutron beams and in most cases observed a very good correlation. In particular for all SEUT devices reported in this work the correlation was excellent (all within error bars[9]) as illustrated in Figure 7 which compares high-energy proton (198MeV) and neutron (WNR) results for TC1 SEUT devices as a function of critical node separation.

---

[8] Strictly speaking, a white beam contains all energies at equal intensity. The WNR beam spectrum does not but matches that of atmospheric neutrons at sea-level. For the purposes of this paper, we continue to refer to the WNR beam as "white"

[9] All error bars in this publication denote 90% confidence levels assuming Poisson statistics

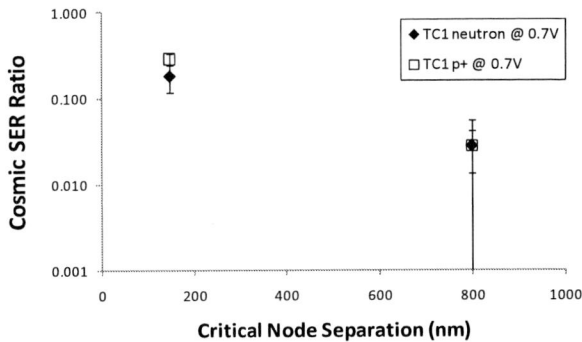

FIGURE 7. RELATIVE SER PERFORMANCE OF SEUT DEVICES AS A FUNCTION OF NODE SEPARATION UNDER 198MEV·P+ AND WNR NEUTRON BEAM IRRADIATION

The fact that relative upset rates under high-energy proton and white neutron beams correlate is not surprising. We have reported in previous studies that for instance multi-cell upset (MCU) probabilities and trends of 45nm SRAMs agree as well [11]. At low proton energies MCU probabilities [11] and upset rates of SEUT devices (Figure 8) start deviating from high-energy proton and consequently from WNR neutron beam results. The physical interpretation is that, on average, the charge cloud generated by secondary particles formed in nuclear proton or neutron target nuclei (mainly Si) reactions is larger at higher incident proton or neutron energies. It also indicates that the WNR neutron beam upset cross sections for MCU and upsetting hardened SEUT devices is dominated by high-energy neutrons.

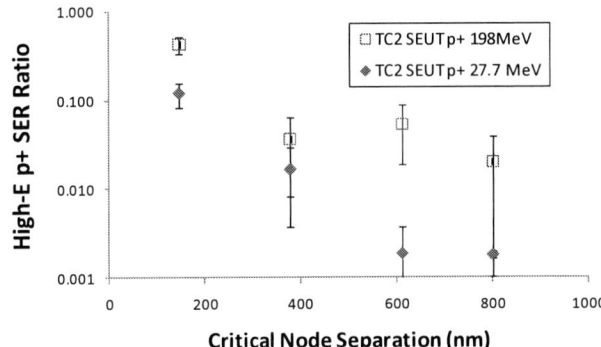

FIGURE 8. SER PROTON E-DEPENDENCE ON NODE SEPARATION FOR TC2 SEUT STRUCTURES AT 1V.

Data depicted in Figures 7-9 underline the exponential dependence of the relative SER performance of SEUT devices on critical node separation. As discussed in the design tradeoffs section, SEUT devices with relaxed node separation requirements have somewhat better power and area performance numbers and so a balance between reliability performance and cost can be struck.

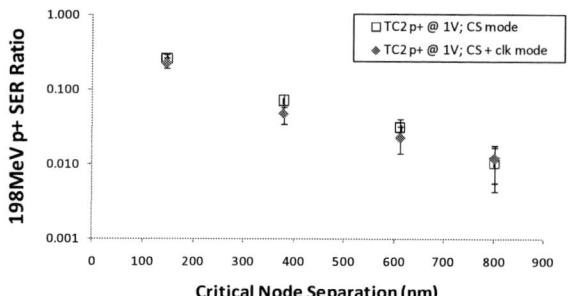

FIGURE 9. RELATIVE P+ SER PERFORMANCE OF SEUT DEVICES IMPLEMENTED ON TC2. SER RESULTS ARE SHOWN FOR CS ONLY AND CS + CLOCK NODE UPSET MODES.

All tested SEUT devices tested in this work show a low susceptibility to clock node upsets when compared to results quoted for SEUT 45nm designs reported in reference [21] (Figure 9). Even for SEUT800 devices, upset rates measured in charge sharing and charge sharing plus clock node SER testing modes are within error bars. One explanation might be that the N to P separations of diffusions located on the same clock buffers are significantly smaller in devices built in the 32nm technology than in the previously reported ones that were built in a 45nm process. Charge collected on N and P devices in the same inverter along data- or clock paths have a similar impact on SER as the layout placement of the same diffusion type (N-N or P-P) of cross-coupled inverters in memory type cell. This is consistent with data collected on 32nm test chips not discussed in this work [38].

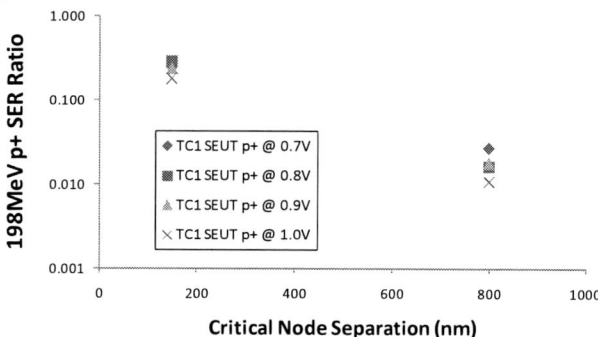

FIGURE 10. VOLTAGE DEPENDENCE OF RELATIVE SER PERFORMANCE OF SEUT DEVICES UNDER HIGH-ENERGY P+ IRRADIATION

SER benefits diminish with reduced power supply voltages for SEUT (Figure 10)[10] and RCC (Table 3) devices. Less charge (lower Qcrit at lower voltages) has a higher probability to be collected over larger distances [19] and therefore larger node separations of state weakening diffusions would be needed in both cases, SEUT and RCC, for iso-performance.

High energy proton-induced upset rates and neutron beam results also correlate for RCC type devices (Table 3). However, in the case of RCC1 the measured SER benefit was consistently higher for neutron beam testing. The authors of this work speculate that with only

---

[10] Error bars have been omitted in figure 10 to improve readability. Error bars are of the same order of magnitude as shown in Figure 9.

one poly width separation between state reinforcing diffusions, even low energy reaction products generated by low-energy neutrons deposit sufficient charge to increase Qcrit and reduce the SER susceptibility of RCC1 devices (upper line in Figure 5). In contrast, only high-energy protons or neutrons generate charge clouds large enough to result in sufficient simultaneous charge collection at state weakening diffusions that lower Qcrit[11]. In the implemented RCC designs these diffusions have separations of >300nm.

In the results summarized in Tables 3 and 5 the control latch has the same size devices as those implemented in RCC devices. However, storage node capacitances and diffusion areas are different due to differences in layout, and the addition of the dummy devices in RCC1. The SER reduction for just the capacitance and area changes is derived through SPICE simulations, and shown in the "no RCC" column. The SPICE simulation methodology has been described in detail before [36], and is not covered here. Low and high voltage measurements, and 90% CI are shown for measured data.

TABLE 3. NEUTRON AND PROTON SER RATIO RESULTS FOR RCC TYPE DEVICES

Device	Vcc [V]	Neutron measured SER reduction	Neutron simulated SER reduction (no RCC)	Proton measured SER reduction	
				27 MeV	198 MeV
RCC1	0.7	3.8x ± 30%	1.2x	2.0x ± 10%	2.5x ± 10%
RCC2	0.7	1.1x ± 30%	1.0x	1.1x ± 10%	1.3x ± 10%
RCC1	1.0	NA		NA	3.2x ± 10%
RCC2	1.0	NA		NA	1.3x ± 10%

Heavy-Ion Testing Results

Figure 11 depicts relative SER results for TC1 SEUT and RCC2 devices as a function of LET. Error bars have been omitted for better readability (except for RCC2 devices). Unfortunately no RCC1 results are available at the time of writing of this paper. At high LETs, both RCC2 and SEUT upset rates are either equal or even worse than that of the non-hardened reference latch. We expect RCC1 devices to show a somewhat better performance than RCC2 ones, but the overall benefit in heavy ion dominated orbits (such as geosynchronous ones; see below) is expected to be worse than in terrestrial radiation environments. Clock node strikes again do not seem to contribute significantly to the overall soft error rate but at the highest LET values. RCC2 performance remains poor down to the lowest LET values investigated (2.8 MeV/(mg/cm²)).

---

[11] Such as N-P devices located on opposite sides of the RCC cross-coupled devices (lower line in Figure 5)

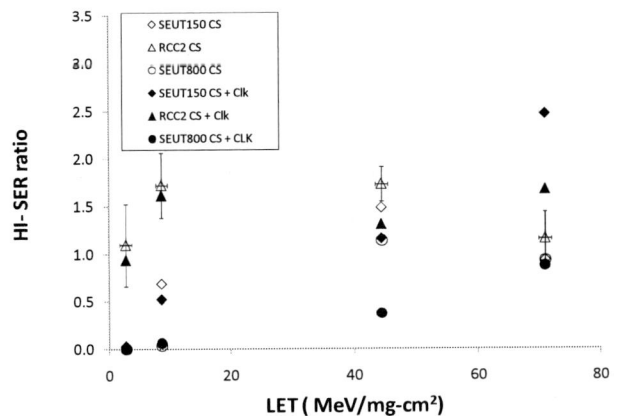

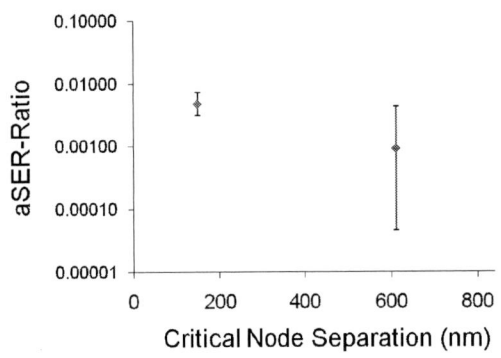

FIGURE 12. ALPHA-PARTICLE INDUCED RELATIVE SER FOR TC2 SEUT DEVICES AT 0.7V.

FIGURE 11. HEAVY-ION SEUT AND RCC2 RESULTS (TC1) FOR CS ONLY AND CS + CLK NODE TESTING MODES.

SER performance in geosynchronous orbits (100 mil of Aluminum shielding and solar quiet mode) has been estimated using Creme96 and results are summarized in Table 4 [37]. Please note that error bars are expected to be large[12]. Actual performance in orbit is likely worse than what is shown in Table 4, since only normal incidence data have been collected, and work by Amusan et al show that those yield optimistic results [13]. Nevertheless, performance of the only tested RCC device (RCC2) is disappointing. It is speculated that even at low to moderate LET values sufficient charge is collected at distances that cover the separation of sensitive state weakening nodes. MCU data reported in reference [11] indicate average charge cloud dimensions in excess of 1μm at high LETs which is well beyond the separation of N-P diffusions (Qcrit impact see Figure 5) in the tested RCC flavor.

TABLE 4. APPROXIMATE RELATIVE UPSET RATES IN GEOSYNCHRONOUS ORBITS (CREME96). CS + CLK NODE UPSET DATA APPLIED.

Device	SER -Ratios
SEUT150	0.64
SEUT800	0.04
RCC2	2.14

Alpha-particle Testing Results

SEUT (Figure 12) as well as RCC1 results (Table 5) demonstrate that the alpha-particle SER contribution can be neglected for typical, modern ambient alpha-particle radiation environments and fluxes. Even for SEUT150 a reduction of the order of ~200x can be expected, versus ~2-3x for high-energy proton or neutron irradiation. For SEUT devices with node separations > 600nm no upsets have been observed even after weeks of continuous testing under accelerated alpha-particle flux conditions.

---

[12] We have not tried to estimate 90% confidence level values of the mean for all tested devices due to the complex convolution of raw data error bars with uncertainties in non-linear fitting parameters (Weibull parameters in Creme96).

Alpha particle results for RCC1 and RCC2 are summarized in Table 5. The SER reduction relative to the non-RCC control latch on the same test chip is shown. The smaller separation between victim and reinforcing nodes in RCC1 results in a larger reduction compared to RCC2. The SER improvements purely due to the RCC effect are relatively modest at lower voltages. Comparing the last two columns in Table 5 for RCC1 at 0.7V yields an approximately 2x SER reduction due to the RCC effect alone. However, in real world applications, the full 10.1x SER reduction will be seen, and this includes the impact of increased capacitance and area due to the addition of dummy devices in RCC1.

TABLE 5. ALPHA-PARTICLE SER RATIO RESULTS FOR RCC TYPE DEVICES

	Voltage [V]	Measured SER reduction	Simulated SER reduction due to capacitance, area changes only (no RCC)
RCC1	0.7	10.1x ± 10%	5.0x
RCC2	0.7	2.4x ± 10%	1.5x
RCC1	1.0	57.0x	8.0x
RCC2	1.0	4.0x	3.0x

Technology Scaling Impact

The measured SEU data discussed in the previous section demonstrates that of the order of 30x SER reductions for SEUT devices and about 3x for RCC devices built in an advanced 32nm technology can be achieved with typical overheads of the order of 100% and 10%, respectively. We would like to emphasize that on the chip-level RCC might be more efficient than SEUT, despite the ~10x advantage in SER for SEUT devices. This is best illustrated by a simple example. Let's assume the RCC SER reduction is 3x at a 20% area cost (a rather conservative assumption), whereas it is 30x for a 100% area overhead in the case of SEUT. The design goal shall be a chip-level SER reduction of y FIT. How many RCC and SEUT latches (N_{RCC} and N_{SEUT}) are needed to achieve this goal and what are the involved area overheads?

978-1-4244-5430-3/10 $26.00 © 2010 IEEE

$$y = N_{RCC} * (x - x/3) = N_{SEUT} * (x - x/30)$$

$$\Rightarrow \frac{N_{RCC}}{N_{SEUT}} = 1.45 \tag{4}$$

where x denotes non-hardend device FIT/cell. Only ~50% more RCC devices (relative to SEUT devices) are needed to achieve a chip-level SER reduction of y FIT. However, the assumed area overhead of RCC devices is 5x lower than for SEUT devices. Therefore, the overall chip-level area overhead is about 3x lower for RCC devices in the above example.

RCC fills the void where using SEUT would be over-kill. As long as there are other sources besides sequential elements contributing to the product SER[13] there is limited ROI in making all sequential elements SEUT. A better approach that leads to lower overheads but comparable SER would be to protect only highly vulnerable sequential elements with SEUT, and convert others to RCC as needed

Preserving the same absolute node separation and therefore SER benefit will be increasingly difficult and costly for SEUT type devices as we continue to scale. This will translate into more hardened devices needed to achieve the same level of SER reduction in future semiconductor technologies at a similar area overhead cost. Since the SER benefit diminishes exponentially with node separation, a corresponding larger number of hardened devices that rely on some form of local redundancy[14] would be needed. For instance, assuming simple scaling of our 32nm SEUT800 devices, the device level SER is projected to increase by roughly 3x the next few technology generations under neutron or proton radiation[15].

For RCC type designs we also expect the SER benefit at constant cost to diminish somewhat with scaling. However, the rate at what this will occur is expected to be slower than for SEUT type devices. The main reason for the expected slower rate of diminishing returns for RCC devices is based on the fact that the separation of both, reinforcing and state weakening diffusions will decrease with scaling. As long as the SER contribution due to secondary victim nodes is relatively small, technology scaling should to first order not impact the radiation robustness of this design technique (neglecting voltage scaling). In the extreme case when the impact of state weakening secondary victim nodes can be completely neglected, an improvement in SER is even expected with scaling (again, ignoring Vcc scaling).

Both hardened mitigation techniques show diminishing returns as power supply voltages are scaled, further reducing the ROI of circuit-level mitigation techniques discussed in this work.

The authors of this work do not see radiation hardened sequential elements as mutually exclusive with other SER mitigation methods, but rather as complementing in logic structures where other techniques such as parity or residue checking are not practical [18]. Placing hardened sequential elements offer a high degree of flexibility and if done correctly can yield good SER performance with little chip-level area and power tradeoffs. Although not addressed in this work, it is important to remember that an intelligent placement of hardened devices requires a solid architectural vulnerability factor (AVF) analysis of the structures of interest [18]. It would not make much sense to protect devices with very low AVF values.

---

[13] Such as combinational logic
[14] SEUT, BISER, DICE, circuit-level TMR, etc
[15] Assuming 0.7x scaling of distances per generation and no scaling in supply voltage (the latter assumption is very optimistic)

CONCLUSIONS

A novel circuit hardening technique called RCC (Reinforcing charge collection) is introduced. RCC exploits the fact that charge collection at state reinforcing nodes will increase Qcrit and hence reduce SER with respect to non-hardened devices of similar performance and design targets.

RCC devices and devices designed in a conventional local redundancy technique (SEUT) have been implemented on two test chips built in a 32nm bulk CMOS process. The radiation robustness of both mitigation schemes was tested for several different radiation environments. Our results indicate that even for hardened devices manufactured in such an advanced technology, SER reduction levels of the order of 30x for SEUT devices and 3x for RCC type devices can be expected for terrestrial applications. Overheads of the implemented designs are of the order of 100% for SEUT devices and 20% or less for RCC devices.

We show that RCC devices despite their modest SER reduction of ~3x, can be a more efficient mitigation technique on the chip-level. In space environments, SEUT devices are expected to fare much better than RCC based designs, however.

For SEUT devices diminishing returns are expected as we continue to scale our technologies. A somewhat better scaling performance is predicted for RCC devices. The authors believe that in the short term redundancy hardened devices will continue to play an important role for protecting few important non-arrayed architectural state elements (with high architectural vulnerability [18]), whereas RCC is a better option for protecting large numbers of logic memory elements in random logic. However, with technology and voltage scaling, both mitigation techniques are projected to eventually become inadequate and too costly.

ACKNOWLEDGEMENTS

The authors would like to thank David Parkhouse, Venkatesh Govindarajulu, Scott Lorion and Marcus Sherwin for their efforts in Test Chip design and characterization and Steven Uffner for his help with the experimental setups.

REFERENCES

[1] T.C. May and M.H. Woods, "A New Physical Mechanism for Soft Errors in Dynamic Memories", proceedings of the International Reliability Physics Symposium (IRPS), pp. 33-40, 1978

[2] J. F. Ziegler and W. A. Lanford, "Effect of Cosmic Rays on Computer Memories", Science, Vol. 206, No. 4420, pp. 776-788, 1979

[3] R.C. Baumann, "Radiation-induced soft errors in advanced semiconductor technologies", IEEE Transactions on Device and Materials Reliability, Volume 5, Issue 3, pp. 305 – 316, 2005

[4] L.B. Freeman, "Critical charge calculations for a bipolar SRAM array", IBM J. Res. Develop. Vol.40, No.1, pp119-129, 1996.

[5] H.T Nguyen, Y. Yagil, N. Seifert, M. Reitsma, "Chip-level soft error estimation method", IEEE Transactions on Device and Materials Reliability, Volume 5, No. 3, pp. 365 – 381, Sept. 2005

[6] N. Seifert, P. Shipley, M.D. Pant, V. Ambrose, B. Gill, "Radiation Induced Clock Jitter and Race", International Physics Reliability Symposium (IRPS, San Jose, CA), April 2005, pp.21

[7] J.F. Ziegler, "Terrestrial cosmic rays", IBM Journal of Research and Development, volume 40, No. 1, 1996, pp. 19-3

[8] G. C. Messenger, M.S. Ash, "Single Event Phenomena", Chapman & Hall (New York 1997).

[9] F. Wrobel, J. M. Palau, M. C. Calvet, O. Bersillon, H. Duarte, " Incidence of multi-particle events on soft error rates caused by n-Si nuclear reactions," IEEE Trans. Nucl. Sci., vol. 47, pp. 2580 – 2585 , Dec. 2000

[10] K. Rodbell, D. Heidel, H. Tang, M. Gordon, P. Oldiges, and C. Murray, "Low-energy proton-induced single-event-upsets in 65 nm node, silicon on-insulator, latches and memory cells," IEEE Trans. Nucl. Sci., vol. 54, No. 6, pp. 2474–2479, Dec. 2007.

[11] N. Seifert, B. Gill, K. Foley, P. Relangi, "Multi-Cell Upset Probabilities of 45nm High-k + Metal Gate SRAM Devices in Terrestrial and Space Environments", Proceedings of the International Reliability Physics Symposium (IRPS), pp. 181-186, 2008

[12] M. H. Quinn, K. Morgan, P. Graham, J. Krone, M. Caffrey, "Static Proton and Heavy Ion Testing of the Xilinx Virtex-5 Device", IEEE Radiation Effects Data Workshop, pp. 177-184, 2007

[13] O. A. Amusan, L.W. Massengill, M.P. Baze, B.L. Bhuva, A.F. Witulski, J.D. Black, A. Balasubramanian, M.C. Casey, D.A. Black, J.R. Ahlbin, R.A. Reed, M.W. McCurdy, "Mitigation Techniques for Single-Event-Induced Charge Sharing in a 90-nm Bulk CMOS Process", IEEE Transactions on device and Materials Reliability, Vol. 9, No. 2, pp.311-317, 2009

[14] E. H. Cannon, D. D. Reinhardt, M. S. Gordon, and P. S. Makowenskyj, "SRAM SER in 90, 130, and 180nm bulk and SOI technologies", Proc. Int'l Reliability Physics Symp. (IRPS), pp. 300-304, 2004.

[15] P. Shivakumar, M. Kistler, S.W. Keckler, D. Burger, and L. Alvisi, "Modeling the effect of technology trends on the soft error rate of combinational logic," in Proc. IEEE Dependable Systems and Networks Conf., pp. 389 – 398, June 2002

[16] M. J. Gadlage, P. H. Eaton, J. M. Benedetto, and T. L. Turflinger, "Comparison of Heavy Ion and Proton Induced Combinatorial and Sequential Logic Error Rates in a Deep Submicron Process", IEEE Transactions on Nuclear Science, Vol. 52, No.6, pp. 2120-2124, 2005

[17] M. Nicolaidis, "Design for soft error mitigation", IEEE Trans. On Device and Materials Reliability, Vol,5, Issue 3, pp. 405-418, 2005.

[18] S. Mukherjee, "Architecture Design for Soft Errors", (Boston, Morgan-Kaufmann), 2008.

[19] N. Seifert, P. Slankard, M. Kirsch, B. Narasimham, V. Zia, C. Brookreson, A. Vo, S. Mitra, B. Gill, J. Maiz, "Radiation-Induced Soft Error Rates of Advanced CMOS Bulk Devices", Proceedings of the IEEE International Physics Symposium, pp. 217-225, 2006

[20] T. Heijmen, P. Roche, G. Gasiot, Keith R. Forbes, and D. Giot, "A Comprehensive Study on the Soft-Error Rate of Flip-Flops From 90-nm Production Libraries", IEEE Transactions on Device and Materials Reliability, Vol. 7, Issue 1, pp. 84 – 96, 2007

[21] N. Seifert, B. Gill, V. Zia, M. Zhang, V. Ambrose, "On the Scalability of Redundancy based SER Mitigation Schemes", Proceedings of IEEE International Conference on Integrated Circuit Design and Technology (ICICDT), pp. 1-9, 2007

[22] T. Calin, M. Nicolaidis and R. Velazco, "Upset Hardened Memory Design for Submicron CMOS Technology", IEEE Trans Nuclear Science, vol. 43, pp.2874-2878, Dec. 1996.

[23] S. Mitra, N. Seifert, M. Zhang, Q. Shi, K.S. Kim, "Robust system design with built-in soft-error resilience", Computer, Volume 38, Issue 2, pp.43 – 52, 2005.

[24] P. Hazucha, T. Karnik, S. Walstra, B. Bloechel, J. Tschanz, J. Maiz, K. Soumyanath, G. Dermer, S. Narendra, V. De, S. Borkar, "Measurements and analysis of SER tolerant latch in a 90 nm dual-Vt CMOS process", Proceedings of IEEE Custom Integrated Circuits Conference, pp. 617-620, 2003.

[25] D.G. Mavis, P.H. Eaton, "SEU and SET Modeling and Mitigation in Deep Submicron Technologies", proceedings of the IEEE International Reliability Physics Symposium, pp. 293-305, 2007

[26] M. N. Liu and S. Whitaker, "Low Power SEU Immune CMOS Memory Circuits", IEEE Trans Nuclear Science, vol. 39, pp.1679-1684, Dec. 1992

[27] R. Velazco, D. Bessot, S. Duzellier, R. Ecoffet and R. Koga, "Two CMOS Memory Cells Suitable for the Design of SEU-Tolerant VLSI Circuits", IEEE Trans Nuclear Science, vol. 41, pp.2229-2234, Dec. 1994.

[28] M. J. Berry, "Radiation Resistant SRAM Memory Cell", Oct. 1992, U.S. Patent Number 5,157,635.

[29] J. G. Dooley, "SEU-Immune Latch for Gate Array, Standard Cell, and Other ASIC Applications", May 1994. U.S. Patent Number 5,311,070.

[30] P. Hazucha, T. Karnik, S. Walstra, B. A. Bloechel, J. W. Tschanz, J. Maiz, K. Soumyanath, G. E. Dermer, S. Narendra, V. De and S. Borkar, "Measurements and Analysis of SER-Tolerant Latch in a 90-nm Dual-Vt CMOS Process", IEEE Journal of Solid-State Circuits, Vol. 39, No. 9, Sept., 2004.

[31] Q. Shi, "Design of SEU Immune Circuits", PhD dissertation, University of New Mexico, Dec., 2000.

[32] Q. Shi and G. Maki, "New Design Techniques for SEU Immune Circuits", 9th NASA Symposium on VLSI Design, 2000.

[33] N. Seifert, "Radiation-induced Soft Errors: A Chip-level Modeling Perspective", to be published in Foundations and Trends in Electronic Design Automation (2010).

[34] Y. Kawakami, M. Hane, H. Nakamura, T. Yamada, and K. Kumagai, "Investigation of Soft Error Rate Including Multi-Bit Upsets in Advanced SRAM Using Neutron Irradiation Test and 3D Mixed-Mode Device Simulation", IEEE International Technical Digest Electron Devices Meeting (IEDM), pp. 38.4.1, 2004

[35] P. Packan et al., "High Performance 32nm Logic Technology Featuring 2nd Generation High-k + Metal Gate Transistors", IEEE International Technical Digest Electron Devices Meeting (IEDM), pp659-662, 2009

[36] S.V. Walstra and Changhong Dai, "Circuit-level modeling of soft errors in integrated circuits", IEEE Transactions on Device and Materials Reliability, Volume 5, Issue 3, pp. 358 – 364, 2005

[37] Creme96 code: https://creme96.nrl.navy.com; Crème stands for Cosmic Ray Effects on Micro Electronics

[38] N. Seifert et al., unpublished

Effect of Multiple-Transistor Charge Collection on SET Pulse Widths

J. R. Ahlbin, M. J. Gadlage, N. M. Atkinson, B. L. Bhuva, A. F. Witulski, W. T. Holman, L. W. Massengill

Dept. of Elec. Eng. and Comp. Sci.
Vanderbilt University
Nashville, TN USA
phone: (01) - (615) - 343-6705, e-mail: jon.ahlbin@vanderbilt.edu

P. H. Eaton
Micro-RDC
Albuquerque, NM USA

B. Narasimham
Broadcom Corp.
Irvine, CA USA

Abstract— **New heavy-ion data from a 130 nm bulk CMOS process shows a counterproductive result in using a common single-event charge collection mitigation technique. Guard bands can reduce single-event pulse widths for normal strikes, but increase them for angled strikes. Calibrated 3D TCAD mixed-mode modeling has identified a multiple-transistor charge collection mechanism that explains the experimental data, namely that angled strikes result in charge collection in the normally ON device that increases the restoring current on the struck device.**

Keywords-soft error, single event, single event transient, SET, SER, pulse width, charge sharing, radiation environment

I. INTRODUCTION

As technology scales, the close proximity of transistors along with lower transistor currents and nodal capacitances has resulted in increased vulnerability of circuits to soft errors [1]. The International Technology Roadmap for Semiconductors (ITRS) has identified soft errors as a key reliability concern for future fabrication processes [2]. One of the major factors that affect the soft error vulnerability of a circuit are single-event transients (SETs) which can be caused by an energetic particle strike on an Integrated Circuit (IC). As the pulse width of SETs increases, the soft error failures in time (FIT) rate also increases because the probability to latch the SET as a soft error has a direct correlation with pulse width. Previously, designers have used guard bands, which

This work was supported in part by the DTRA Radiation-Hardened Microelectronics Program and Cisco Systems.

J. R. Ahlbin, M. J. Gadlage, N.M Atkinson, B. L. Bhuva, A. F. Witulski, W. T. Holman, and L. W. Massengill are with Vanderbilt University, Nashville TN 37235 USA (e-mail: jon.ahlbin@vanderbilt.edu).

P. H. Eaton is with Microelectronics Research and Development Corp., Albuquerque, NM 87110 USA.

B. Narasimham is with Broadcom Corp., Irvine CA 92617 USA.

are well contacts that encircle the individual transistors, to reduce the amount of charge collected by each transistor. For an individual transistor this yields lower charge collection at each transistor, resulting in shorter SET pulses, and a lower soft error rate.

For advanced technologies, the decrease in minimum transistor-to-transistor spacing has resulted in multiple transistors, or nodes, being susceptible to the charge deposited from a single particle strike compared to older processes where only one transistor was affected [3]. When guard bands are used around each individual transistor, the designers expect the charge collection by each transistor to decrease, resulting in shorter SET pulses. However, recent work has shown that multiple transistors that collect charge and that are electrically related may affect the operation of the overall circuit [4]. Current conclusions made about the effectiveness of guard bands to reduce the soft error rate of a circuit do not take into account the electrical relationship among multiple transistors collecting charge [5]. In this work when guard bands are used, interactions between transistors in an inverter cell are shown to lengthen SET pulse widths, resulting in a higher than expected soft error rate for the circuit. A multiple-transistor charge collection mechanism is proposed to help explain the experimental results.

DYNAMICS OF TRANSISTOR CHARGE COLLECTION

The temporal charge collection efficiencies at a node struck by an ion determine the pulse characteristics – including the width of the resulting voltage SET. These temporal mechanisms include classical drift, diffusion, and parasitic amplification at the struck node [6]. Additionally, in scaled technologies there exist other mechanisms such as the removal of charge by substrate/well contacts and the collection effects of neighboring transistors (e.g. charge sharing) [7]. Charge sharing – the envelopment of proximal transistors or nodes by the charge cloud generated by the single-event - has been shown to be prevalent at the 130 nm technology node, with increasing impact on single-event response of a circuit as technology scales towards smaller device geometries [3, 7].

For digital CMOS circuits, the conduction state of the transistor is an important factor in determining the amount of charge collected at the drain node. Within an inverter, for example, where either the pMOS or the nMOS transistor is OFF, any charge collected by the OFF transistor will result in a perturbation in the output voltage. On the other hand, charge collected by an ON transistor will strengthen the logic voltage at the output by effectively increasing the ON transistor current. For older technologies, where individual transistors were far apart, only one transistor collected charge after a particle strike. While in advanced technologies where transistors are close together, both the pMOS and nMOS transistors within an inverter may collect charge. This multiple transistor charge collection now causes the voltage of the inverter output to be affected by two competing mechanisms. The first mechanism causes the voltage perturbation and the second one strengthens the pre-strike output voltage. Hence the resultant SET pulse width is determined not only by charge collection mechanisms for each transistor, but also by the electrical interaction between the two transistors. In the following sections, the effects of charge collection by multiple transistors with and without guard bands within a logic gate are studied experimentally and through 3D mixed-mode TCAD simulations.

II. EXPERIMENTAL SET MEASUREMENT RESULTS

The test circuit was fabricated in an IBM 130 nm bulk CMOS technology. The test chip design was based on the autonomous pulse-width measurement technique described in [8] and contained two variants of otherwise identical target inverter chains – with and without guard bands around individual transistors as shown in Fig. 1 (after [5]). It was assumed the presence of guard bands would reduce the total charge collected by each transistor for shorter SET pulses, and lower soft error rates [9]. The inverter transistors were designed to be drive-current matched with pMOS W/L of 720 nm x 120 nm and nMOS W/L of 240 nm x 120 nm. These target chains were both laid out in a serpentine arrangement of 4 rows of 25 inverters each. Other than the use of guard bands, the two layouts were identical both in transistor sizing and spacing.

The circuits were tested at the Lawrence Berkley National Laboratory with 30 MeV-cm^2/mg Krypton at a 60° angle of incidence. The test fixture was arranged so that particle penetration was from a South-to-North direction as shown in Fig. 1. For the layout used in this design, this will result in the particle depositing charge through both pMOS and nMOS transistors within a logic gate.

A comparison of SET pulse widths for the transistors with guard bands and without guard bands is shown in Fig. 2. In the figure, SET cross-section is defined as the number of SETs measured divided by a fluence of 5 x 10^7 particles/cm^2. The pulse width distribution for the guard band variant skews towards the longer, ~1ns type SET pulse widths, while the pulse width distribution for the non-guard band variant skews toward the shorter, ~300 ps type SET pulse widths. In particular, the differences are noticeable between the pulse width ranges of 240 ps to 480 ps and 720 ps to 960 ps.

The main reason for this behavior is charge-sharing between multiple transistors within a logic gate. This result is in contrast to Narasimham et al. who tested the same circuit design but at a normal incidence with Xe [5]. The authors found that the guard-band mitigation scheme was effective at reducing the number of SETs with pulse widths over 960 ps. The difference in data sets is due to charge collection by either pMOS or nMOS transistors, but not both, because the incident ion is at normal incidence in [5]. When only one transistor collects charge, the presence of guard-bands reduces the charge collected by the hit transistor, resulting in shorter SET pulses. In summary, when both the transistors within an inverter collect charge, the presence of guard-bands actually increases the SET pulse width as observed in our data. The mechanism responsible for this behavior is discussed below with the help of 3D mixed-mode TCAD simulations.

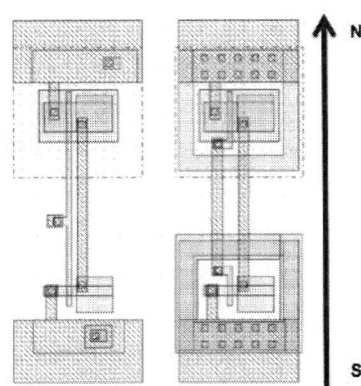

Fig. 1. Layout of 130 nm inverter without/with guard bands and arrow showing the path of an ion strike in the south-to-north direction (after [5]).

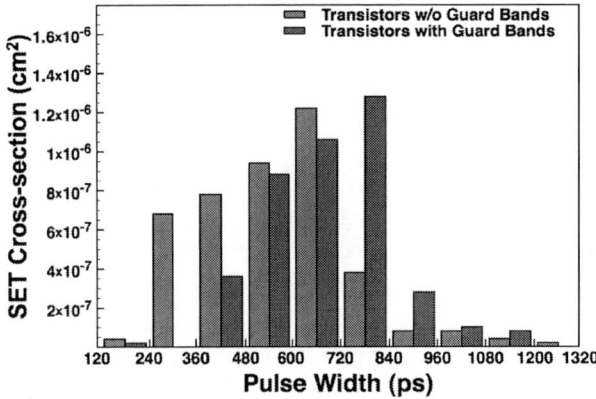

Fig. 2. Experimental results at 60° and 30 MeV-cm^2/mg Krypton ion with 130 nm pulse width measurement circuits.

III. CHARGE COLLECTION BY PMOS AND NMOS TRANSISTORS AFTER A PARTICLE STRIKE

Three-dimensional mixed-mode technology computer aided design (TCAD) finite-element device simulations, with appropriately calibrated transistor models, are an excellent tool to model charge transport mechanisms. Mixed-mode simulations allow specific transistors in a circuit to be modeled fully in 3D TCAD while the rest of the circuit can be implemented using compact models. Such an approach is very efficient and effective in understanding various mechanisms involved in single-event effects. 3D TCAD simulations were performed for the 130 nm CMOS technology node where both pMOS and nMOS transistors of a single inverter and an adjacent pMOS were modeled with/without guard bands as shown in Fig. 3. The TCAD models were calibrated to the 130 nm IBM CMOS8RF compact models in the process design kit. For each mixed-mode TCAD simulation, the 3D TCAD model of the inverter was the 2nd inverter in a nine inverter long chain with an additional pMOS device from the 3rd inverter. The simulated ion strike had a fixed LET of 30 MeV-cm^2/mg and struck the center of the drain of the nMOS transistor at a 60° angle from South-to-North direction as shown in Fig. 4. This allowed the incident particle to hit the nMOS transistor drain and pass under the pMOS transistor, which is in contrast to a normal incident ion strike as shown in Fig. 5. The physical models used included: Fermi-Dirac statistics, SRH recombination and Auger recombination, and the Philips mobility model. The incident heavy ions were modeled using a Gaussian radial profile with a characteristic 1/e radius of 50 nm and a Gaussian temporal profile with a characteristic decay time of 2 ps.

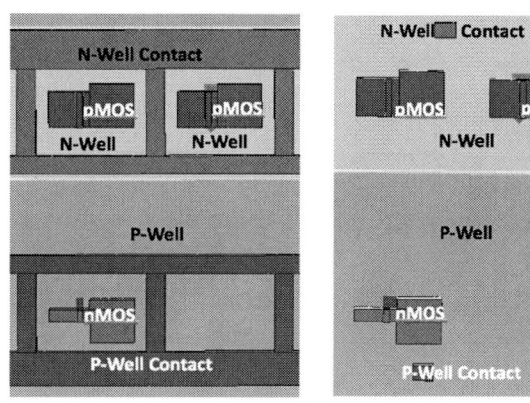

Fig. 3. TCAD models of 130 nm inverter and adjacent pMOS transistor with/without guard bands.

Within an inverter, if the pMOS transistor is ON and nMOS transistor is OFF, the output voltage is HIGH. When the nMOS transistor collects a sufficient amount of electrons after a particle strike to be upset, the output voltage will change to LOW. The pMOS transistor at this point provides current to restore the output node voltage. It is the amount of current provided by the pMOS transistor in combination with

the temporal profile of the charge collection and the inherent parasitics of the circuit design that determines the pulse width of the SET [10]. If the pMOS transistor also collects charge after a hit, the effective current for restoring the output node back to its original state will increase accordingly, resulting in a shorter SET pulse. For the TCAD simulations, the angular incidence of the ion strike ensures that both the pMOS and nMOS will collect charge, resulting in a shorter SET pulse width.

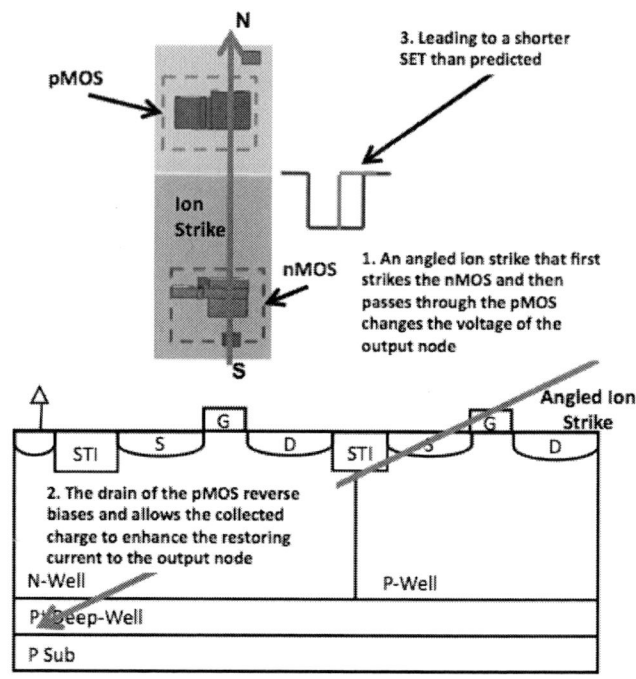

Fig. 4. Two-dimensional diagram illustrating the multiple transistor charge collection mechanism that leads to a shorter than predicted SET pulse width.

Fig. 4 shows the steps involved in the charge collection process by both the nMOS and pMOS transistors and the subsequent reduction in the resultant SET. First, the ion strikes the nMOS transistor while it is in the OFF state. The nMOS transistor drain collects charge and perturbs the output node voltage of the inverter. The switch in output node voltage causes the drain of the pMOS to reverse-bias forcing any transistor collected charge deposited in the n-well by the angled ion strike to go to the output node. This collected charge is seen as an additional current that enhances the restoring power of the pMOS transistor and helps shorten the SET.

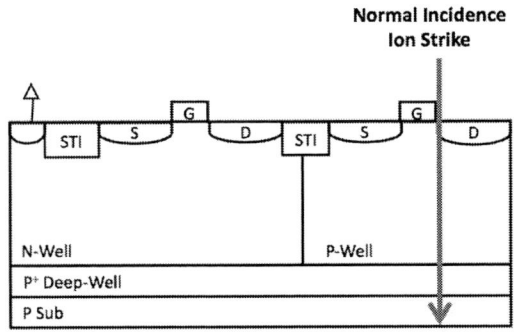

Fig. 5. Two-dimensional diagram illustrating a normal incidence ion strike.

Using the capabilities of 3D mixed-mode TCAD simulations by monitoring the current and voltage of each transistor within the model, Fig. 6 shows the drain current plot of the pMOS transistors with and without guard bands. The additional spike in current for the without-guard bands condition is caused by the charge collected by the pMOS transistor drain after the strike. This additional charge enhances the restoring current at the output node of the inverter causing the voltage of the output node to quickly return to its original state. The resulting output node voltage of the struck nMOS is shown in Fig. 7. A similar spike in current for with-guard-bands condition is not observed because the guard bands serve as a charge sink and help to maintain the substrate/well potential.

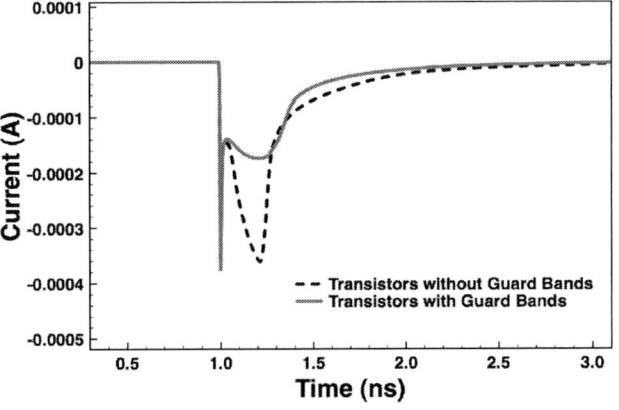

Fig. 6. Plot showing SET drain current pulses from pMOS transistors with and without guard bands for an ion strike of 30 MeV-cm^2/mg at a 60° angle perpendicular to the n-well.

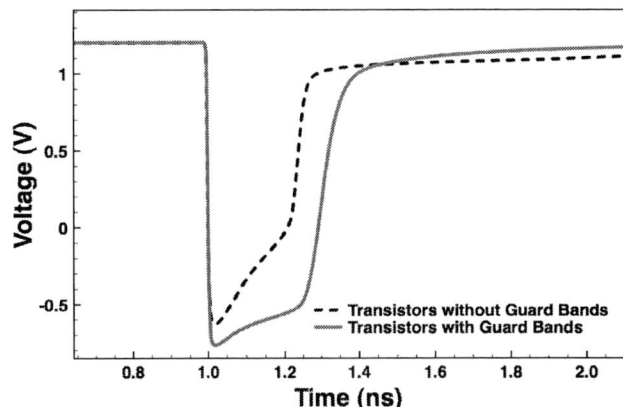

Fig. 7. Plot showing SET pulse widths from struck nMOS transistors with and without guard bands for an ion strike of 30 MeV-cm^2/mg at a 60° angle perpendicular to the n-well.

IV. CONCLUSIONS

A single-event transient mechanism has been found that explains the observed difference in SET pulse width between transistors with and without guard bands. In contrast to previous results at normal incidence, experimental data has shown angled particle strikes causing SET pulse widths to be longer for transistors with guard bands than transistors without guard bands. 3D TCAD simulations of digital SET pulse widths show that angled particle strikes can cause multiple-transistors in an inverter cell to collect charge. Multiple transistor charge collection leads to enhanced restoring current in the inverter cell creating a shorter SET pulse than if only one transistor collected charge. Care should be taken during tests and simulations to include angular incidence strikes in order to assure that SET pulse width modulation via charge sharing is properly addressed. Since the actual environment of exposure may occur over a large solid angle, analyses without this important mechanism may predict incorrect soft error rates.

ACKNOWLEDGMENT

The computational portion of this work was conducted through Vanderbilt University's Advanced Computing Center for Research and Education (ACCRE).

REFERENCES

[1] R. C. Baumann, "Single event effects in advanced CMOS Technology," in *Proc. IEEE Nuclear and Space Radiation Effects Conf. Short Course Text*, 2005.

[2] International Technology Roadmap for Semiconductors, 2007 Edition[Online]. Available http://www.itrs.net/Links/2007ITRS/Home2007.htm

[3] O. A. Amusan, A. F. Witulski, L. W. Massengill, B. L. Bhuva, P. R. Fleming, M. L. Alles, A. L. Sternberg, J. D. Black, and R. D. Schrimpf, "Charge Collection and Charge Sharing in a 130 nm CMOS Technology," IEEE Trans. Nucl. Sci., vol. 53, no. 6, pp. 3253-3258, Dec. 2006.

[4] J. R. Ahlbin, L. W. Massengill, B. L. Bhuva, B. Narasimham, M. J. Gadlage, and P. H. Eaton, "Single-Event Transient Pulse Quenching in Advanced CMOS Logic Circuits", *IEEE Trans. On. Nucl. Sci.*, vol. 56, no. 6, pp. 3050-3056, Dec. 2009.

[5] B. Narasimham, B. L. Bhuva, R. D. Schrimpf, L. W. Massengill, M. J. Gadlage, O. A. Amusan, W. T. Holman, A. F. Witulski, W. H. Robinson, J. D. Black, J. M. Benedetto, and P. H. Eaton, "Effects of Guard Bands and Well Contacts in Mitigating Long SETs in Advanced CMOS Processes," *IEEE Trans. Nucl. Sci.*, vol. 55, no.3 , pp. 1708-1713, June 2008.

[6] L. W. Massengill, "SEU modeling and prediction techniques," *IEEE NSREC Short Course*, pp. III-1–III-93, 1993.

[7] O.A. Amusan, L. W. Massengill, B. L. Bhuva, S. DasGupta, A. F. Witulski, and J. R. Ahlbin, "Design Techniques to Reduce SET Pulse Widths in Deep Submicron Combinational Logic," *IEEE Trans. Nucl. Sci.*, vol. 54, no. 6, pp. 2060-2064, Dec. 2007.

[8] B. Narasimham, V. Ramachandran, B. L. Bhuva, R. D. Schrimpf, A. F. Witulski, W. T. Holman, L. W. Massengill, J. D. Black, W. H. Robinson, D. McMorrow, "On-chip characterization of single event transient pulse widths", *IEEE Trans. on Dev. and Mat. Rel.,* vol. 6, p. 542-549, 2006.

[9] J. D. Black, A. L. Sternberg, M. L. Alles, A. F. Witulski, B. L. Bhuva, L. W. Massengill, J. M. Benedetto, M. P. Baze, J. L. Wert, and M. G. Hubert, "HBD layout isolation techniques for multiple node charge in mitigation," *IEEE Trans. Nucl. Sci.*, vol. 52, no. 6, pp. 2536–2541, Dec. 2005.

[10] S. DasGupta, A. F. Witulski, B. L. Bhuva, M. L. Alles, R. A. Reed, O. A. Amusan, J. R. Ahlbin, R. D. Schrimpf, & L. W. Massengill, "Effect of Well and Substrate Potential Modulation on Single Event Pulse Shape in Deep Submicron CMOS," Nuclear Science, IEEE Transactions on , vol.54, no.6, pp.2407-2412, Dec. 2007.

LEAP: Layout Design through Error-Aware Transistor Positioning for Soft-Error Resilient Sequential Cell Design

Hsiao-Heng Kelin Lee[1], Klas Lilja[3], Mounaim Bounasser[3], Prasanthi Relangi[1], Ivan R. Linscott[1], Umran S. Inan[1], Subhasish Mitra[1,2]

[1] Department of Electrical Engineering, [2] Department of Computer Science, Stanford University, [3] Robust Chip Inc.

Abstract— This paper presents a new layout design principle called LEAP which is an acronym for **L**ayout Design through **E**rror-**A**ware Transistor **P**ositioning. This principle extends beyond traditional layout techniques, such as node separation, and significantly improves the soft error resilience of digital circuits with negligible performance cost. In this study, we applied the LEAP technique to the Dual Interlocked Storage Cell (DICE) and designed a new sequential element called LEAP-DICE. This element retains the circuit topology and transistor sizing of DICE but has a new layout based on the LEAP principle. Radiation experiments using an 180nm CMOS test chip demonstrate that our LEAP-DICE flip-flop encounters 5X fewer errors on average compared to our reference DICE flip-flop, and 2,000X fewer errors on average compared to a conventional D-flip-flop. Our LEAP-DICE flip-flop imposes negligible power and delay costs and 40% flip-flop-level area costs compared to our reference DICE flip-flop.

Index Terms—LEAP, soft error, single event upset (SEU), single event multiple upset (SEMU), multiple bit upset (MBU) latch, flip-flop, dual interlocked storage cell (DICE), layout, proton irradiation.

I. INTRODUCTION

Radiation-induced soft errors are a major concern for robust systems, especially those targeting enterprise applications [Karnik 04, Baumann 05, Meaney 05, Mitra 05, Seifert 07, Gill 09]. Unlike large SRAM arrays that can be protected using error-correcting codes and bit interleaving, soft error protection of *sequential elements*, i.e., latches and flip-flops, is challenging. Traditional techniques for designing soft-error-resilient sequential elements generally address single errors caused by single event upsets [Calin 96, Mavis 02, Mitra 05, Shuler 05]. However, with technology scaling, the charge deposited by a particle strike can be simultaneously collected and shared by multiple circuit nodes in the same well [Olson 05, Seifert 07, Amusan 08]. In addition to charge sharing, the particle strike direction also determines how much charge is deposited, as well as charge distribution among multiple nodes [Amusan 07, Baze 08]. When the incoming particle is a proton or neutron, interactions between the incident particle and the silicon nucleus can produce secondary particles generating multiple ionization tracks [Koga 96]. Hence, soft error resilience techniques for sequential elements must focus on *Single Event Multiple Upsets* or *SEMUs*, also referred to as *Multiple-Bit Upsets* or *MBUs*, where a single energetic particle strike can result in upsets in multiple circuit nodes [Dodd 03].

The major contributions of this paper are:

1. We present a new soft-error-resilient layout design of a sequential element, called *LEAP-DICE*, based on the popular *Dual Interlocked Storage Cell* or *DICE* circuit design [Calin 96]. The LEAP-DICE has been designed based on a new layout technique called **Layout Design through Error-Aware Transistor Positioning or LEAP** (patent pending [IP 08]). LEAP-DICE preserves the circuit design (including transistor sizing) of the original DICE, but uses a special layout design to improve the robustness of the resulting sequential element by reducing its susceptibility to SEMUs. Note that LEAP-DICE maintains immunity to single errors similar to the original DICE design.

2. Experimental results obtained from 200 MeV proton irradiation on 180nm CMOS test chips demonstrate that our LEAP-DICE flip-flop encounters 5X fewer errors on average compared to our reference DICE flip-flop (for static testing and for dynamic testing at 50 kHz). Moreover, our LEAP-DICE flip-flop encounters 2,000X fewer errors on average compared to a traditional D flip-flop.

3. Our LEAP-DICE flip-flop does not rely on traditional node separation to achieve robustness to SEMUs [Seifert 06, Seifert 07]. It introduces negligible power and delay costs compared to our reference DICE flip-flop. However, the intra-cell routing in LEAP-DICE is more complex, leading to a 40% area overhead at the cell level compared to the DICE flip-flop.

The rest of this paper is organized as follows. In Section II, we describe the LEAP principle. Section III applies the LEAP design principle to a DICE design. Section IV describes our test chip implementation. Section V presents the experimental setup for single event testing and experimental results, followed by conclusions in Section VI.

II. LEAP PRINCIPLE

LEAP (Layout Design through **E**rror-**A**ware Transistor **P**ositioning) is based on:

1. An analysis of the circuit response to a single event for each individual drain contact node in the layout.

2. Placement of each drain contact node in the layout based on the above analysis, such that multiple drain contact nodes act together to cancel (fully or partially) the overall effect of the single event on the circuit.

To explain the LEAP principle, we first discuss the effects of energetic particle strikes, or *Single Event Transients* (*SETs*), in CMOS technology. When an energetic particle strikes in the vicinity of a MOS transistor, electrons and holes are injected into the silicon. The injected charge is transported by drift and diffusion, causing reverse bias currents in all pn-junctions reached by the charge, and eventually removed by charge collection or recombination. For an NMOS transistor, a net negative charge will be collected at the source and drain contacts [Dodd 03], resulting in a positive current pulse (into the silicon), whereas for a PMOS transistor, net positive charge collected at the source or drain contacts results in a negative current pulse.

For a single inverter, when an energetic particle hits the drain contact node of the PMOS transistor, the positive charge collected at this node raises the inverter output voltage. If the inverter output is initially LOW (i.e., logic 0) and enough charge is collected, then the logic value of the inverter output can change (Fig. 2.1a). Once the injected excess charge is swept out or recombined, the node voltage is restored by the "on" NMOS transistor. Likewise, when the NMOS transistor of an inverter is hit while the inverter output is HIGH (i.e., logic 1), the output voltage is temporarily lowered (Fig. 2.1b).

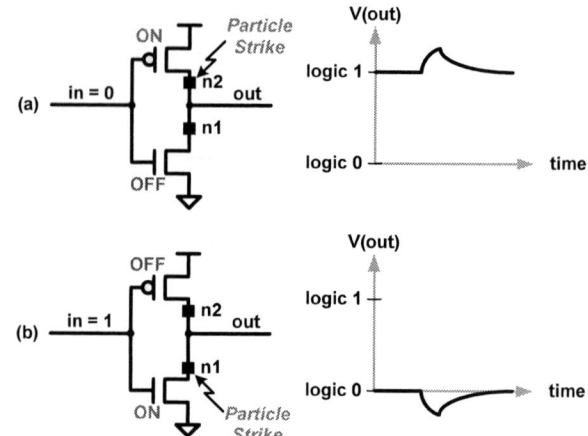

Figure 2.2. Effect of an energetic particle strike on the drain contact node of an "on" (a) PMOS transistor, (b) NMOS transistor in an inverter.

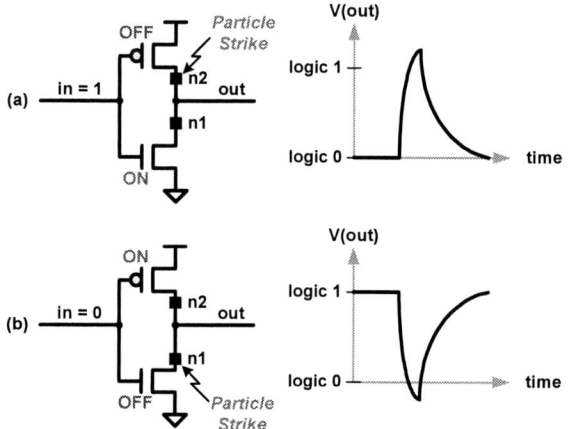

Figure 2.1. Effect of an energetic particle strike on the drain contact node of an "off" (a) PMOS transistor, (b) NMOS transistor in an inverter.

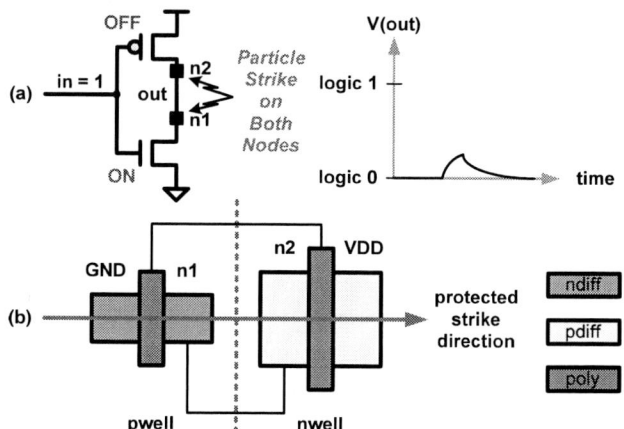

Figure 2.3. LEAP principle for an inverter through transistor alignment. (a) Reduced charge collection when a particle hits both NMOS and PMOS drain contact nodes of an inverter simultaneously. (b) Horizontal transistor alignment to reduce charge collection.

When the inverter output is HIGH, a single event particle strike on the drain contact node of the PMOS transistor does not change the logic value of that circuit node, but rather drives the inverter output voltage higher than the supply voltage until the "on" PMOS transistor removes the excess charge (Fig. 2.2a). In the same way, if the inverter output is LOW, a particle strike on the drain contact node of the NMOS transistor pulls the voltage lower than the ground voltage, and reinforces the output state of the inverter (Fig. 2.2b).

As an initial illustration of how LEAP utilizes the different effects of charge collection on multiple nodes to reduce single event sensitivity, consider the case where the drain contact nodes of the PMOS and NMOS transistors in an inverter are simultaneously hit by a particle strike. In the inverter example shown in Fig. 2.3, the positive charge collected by the PMOS transistor is offset by the negative charge collected by the NMOS transistor, resulting in lower total charge collection at the output node. The extent of the charge reduction depends on the relative sizes of both drain contact nodes as well as the exact strike direction hitting both nodes.

The inverter example in Fig. 2.3 shows how single event charge collection in multiple drain contact nodes, sharing the same circuit node, can be used to reduce the effect of the single event transient on the circuit. The LEAP technique also considers interactions between multiple circuit nodes. To illustrate this, consider a pair of cross-coupled inverters (Fig. 2.4a). The state of this circuit is the state of the latch formed by the two inverters: "state0" (A=0, B=1) and "state1" (A=1, B=0). A single event affecting drain node n1 will pull circuit node A LOW and turn on transistor M4, driving circuit node B HIGH and pushing the latch state toward "state0". Conversely, a single event affecting drain contact node n3 will pull circuit node B LOW and turn on transistor M2, driving circuit node A HIGH and pushing the latch state toward "state1". If an energetic particle simultaneously strikes both drain contact nodes n1 and n3, charge collection on drain contact node n3 reduces the effect of the charge collection at node n1 (for any initial state of the latch). The result will be a higher LET upset threshold for a single event affecting both n1 and n3, than for

978-1-4244-5430-3/10 $26.00 © 2010 IEEE   204

a single event affecting only n1. The increase in LET upset threshold can be quantified using simulations capable of accurately modeling the charge collection (discussed in Section III). Fig. 2.4b shows a layout of a cross-coupled inverter which utilizes charge cancellation for SEMU resilience along the horizontal direction.

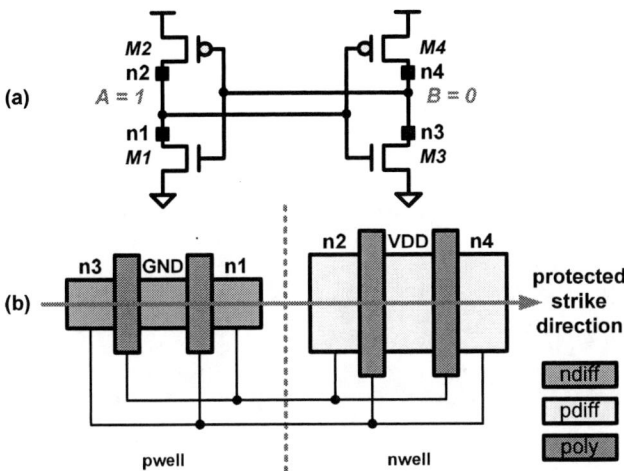

Figure 2.4. LEAP principle for a cross-coupled inverter pair. (a) Circuit schematic. (b) Transistor alignment to reduce charge collection in the horizontal direction.

LEAP places the drain contact nodes in the layout to take advantage of the opposing single event effects discussed above. Since LEAP does not depend on physical separation of sensitive circuit nodes, which may require large area overheads, LEAP-based designs can be more compact. For example, one study in 90nm bulk CMOS showed that the separation technique may provide a 10X reduction in charge collection at the expense of increased node separation from 0.14μm to 2μm [Amusan 08].

III. LEAP-DICE DESIGN

The *Dual Interlocked Storage Cell* or *DICE* is a storage element that relies on dual redundancy of internal circuit nodes to achieve soft-error resilience [Calin 96]. A DICE storage element is immune to single event upsets affecting single circuit nodes, but is vulnerable to single event multiple upsets affecting multiple circuit nodes [Baze 08]. The basic 8-transistor DICE storage element (without access or clocking transistors) is shown in Fig. 3.1.

When the drain contact node of an "off" transistor in the DICE design, e.g., drain contact node n1 in Fig. 3.1, is hit by a particle, the circuit node connected to this drain contact node, e.g., circuit node A, can temporarily switch its logic state. This can turn off a previously "on" transistor, e.g., M7, and turn on a previously "off" transistor, e.g., M4, causing the following behaviors in the remaining circuit nodes:

1. Circuit node B is now driven by both "on" transistors M3 and M4. This results in voltage contention at circuit node B.

2. The change in the voltage of circuit node B reduces the current drive of M6. As a result, node C is now weakly driven by M6.

3. Circuit node D is now left floating (M7 and M8 are "off").

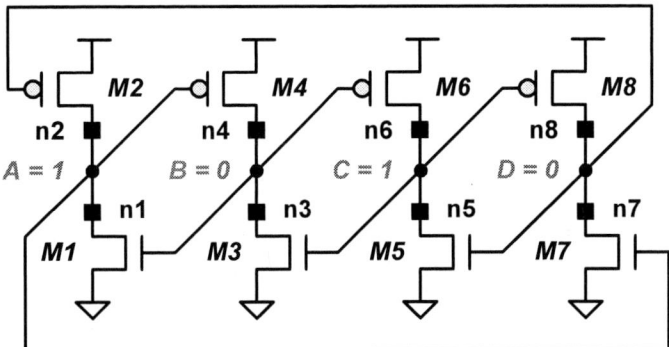

Figure 3.1. 8-transistor Dual Interlocked Storage Cell (DICE) [Calin 96]. Transistor drain contact nodes are labeled using square dots. Internal circuit nodes connecting the drain terminals are labeled using round dots.

If no other circuit node in the DICE storage element is affected by the single event, circuit node A will eventually recover its original state, and the storage element will continue to produce correct outputs. However, if an additional drain contact node of an "off" transistor is hit by the same single event, an upset may be induced in the storage element.

To protect the DICE storage element from such undesirable situations, we apply the LEAP principle discussed in Section II. For any two transistors T1 and T2 that can be simultaneously "off", we place the drain contact node of another transistor T3 between the drain contact nodes of T1 and T2 such that:

1. The drain contact nodes of T1, T2 and T3 lie along the horizontal direction.

2. T3 is "on" when T1 and T2 are "off.

3. The drain contact node of T3 is directly connected to the drain contact node of either T1 or T2.

Whenever both "off" transistors are struck by an energetic particle simultaneously, the inserted drain contact node of the "on" transistor is also hit and its collected charge offsets the charge collected at the drains of the "off" transistors, thus reduces the overall charge collected at the shared circuit node between the "on" transistor and one of the "off" transistors.

Fig. 3.2 shows the LEAP-DICE layout implementation for the DICE storage element which satisfies the above conditions.

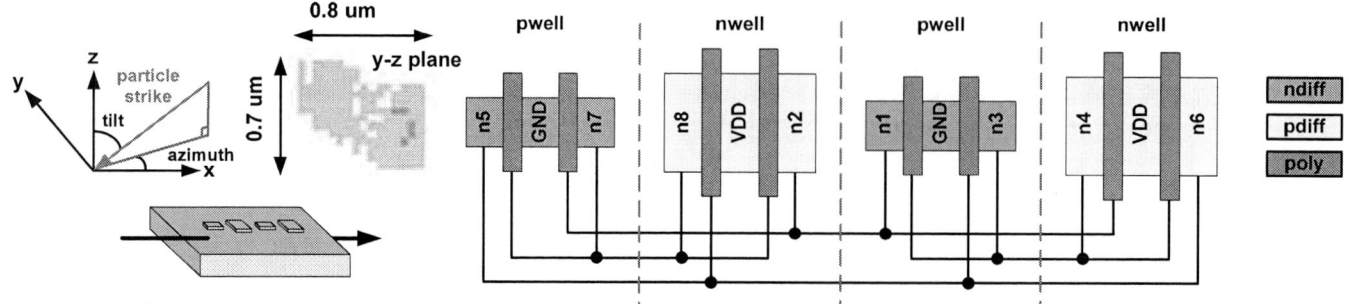

Figure 3.2. LEAP-DICE storage cell layout with projected cross-section region on y-z plane for 85° tilt and 180° azimuth.

An additional benefit of this layout is that, since all transistors are aligned horizontally, the number of possible particle tracks hitting multiple circuit nodes in LEAP-DICE is confined to a narrow incident angle range around the horizontal direction, further reducing its susceptibility to SEMUs. The analysis of SEMUs on more than two drain contact nodes becomes quite complex, and we only focus our discussion in this paper on reducing SEMUs related to simultaneous hits on the drain contact nodes of two "off" transistors.

Accurate single event simulations can provide effective quantitative assessment of LEAP for a specific circuit and layout. We have used the tool Accuro [Lilja 09, RCI 10] which accurately simulates single event charge distribution and charge collection while fully accounting for layout, substrate, and circuit details. This single event simulation tool provides accuracy similar to a full 3D technology CAD (TCAD) device simulation, and is fast enough to run a very large number of single event experiments as required for cross-section and LET threshold prediction (100,000+ single event simulations).

To evaluate the LEAP-DICE storage cell layout and compare it to the DICE storage cell layout, we implemented two layouts in 90nm CMOS technology for the basic DICE circuit (Fig. 3.1): the LEAP-DICE layout (Fig. 3.2) and DICE layout (Fig. 3.5). Minimum device widths were used for the NMOS devices with a P/N width ratio of 2.4. Note that both layouts use exactly the same circuit and device sizing.

For selected incoming angles (defined using "tilt" and "azimuth" in Fig. 3.2) of the single event generating particle, Accuro scans the entire layout area and finds the LET upset threshold at every scan point in the layout. A "snapshot" from one such simulation is shown in Fig. 3.3. Initial charge generation, governed by the LET of the particle, is injected into the 3D structure along the trajectory of the particle, and the charge transport and collection at the drain contact nodes are simulated. At the end of the simulation, the voltage on the DICE output node is monitored to determine if the latch is upset. Fig. 3.4 shows the DICE output voltage (circuit node n4 in Fig. 3.1) for different LET values for a particle strike at a particular angle of incidence.

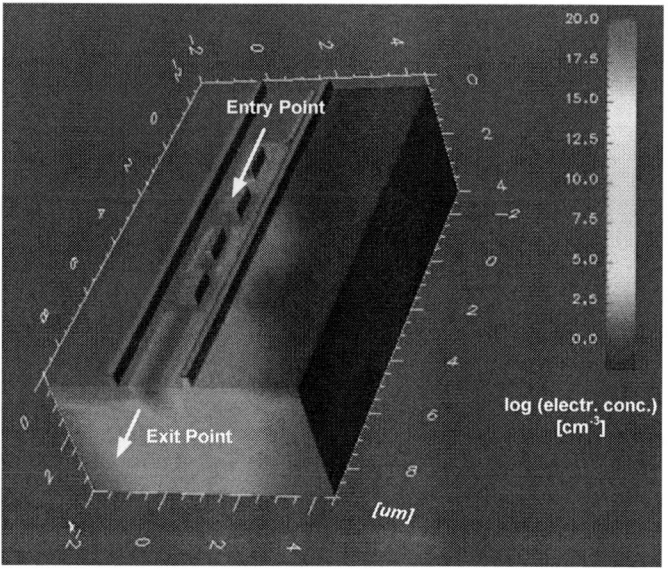

Figure 3.3. Simulation "snapshot" for the LEAP-DICE latch structure used in the simulation, with color coded electron concentration during a single event with azimuth 180°, tilt 85°.

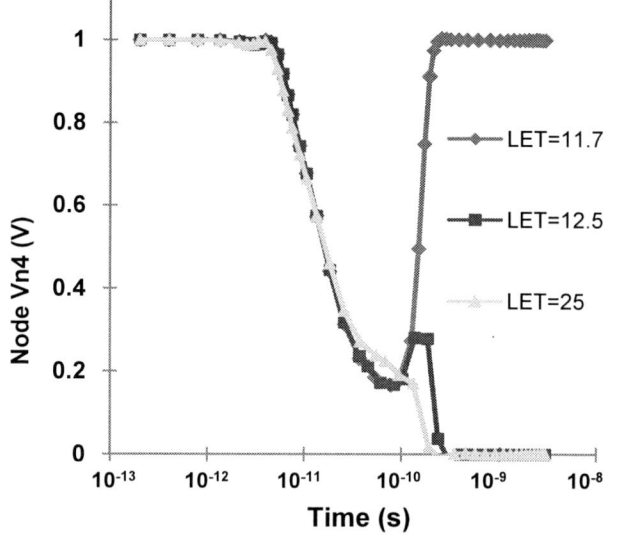

Figure 3.4. Simulated DICE output voltage for single events with different LET in the LEAP-DICE. Angle of incidence tilt = 85°, azimuth = 180°, position marked in Fig. 3.3.

978-1-4244-5430-3/10 $26.00 © 2010 IEEE

As discussed earlier, the LEAP-DICE storage cell (Fig. 3.3) can only be upset when the angle of incidence is in a narrow cone around tilt = 90° and azimuth = 0° or 180° due to the horizontal alignment of the transistors. For a conventional layout of the DICE storage cell (Fig. 3.5), the latch can be upset for multiple azimuth angles around 90° tilt. We determined one sensitive direction to be around an azimuth angle of 305°, corresponding to strikes directions hitting the following pair of drain contact nodes: n2-n3, n4-n5 and n6-n7. We also show simulation results for another sensitive direction with an azimuth angle of 220°, corresponding to strike directions hitting drain contact nodes n1 and n8. Fig. 3.2 and 3.5 show the cross-section regions where both LEAP-DICE and DICE layouts are upset at the given LET and angle. The cross-section region is the area in the plane perpendicular to the particle trajectory where an incident particle strike can upset DICE or LEAP-DICE. For improved visibility, this cross-section region is first projected to the side of the simulation structure (x-z, and y-z planes), then rotated to the plane of the layout and centered.

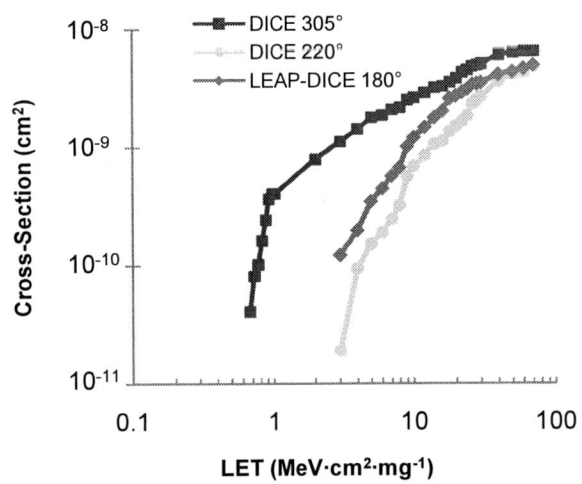

Figure 3.6. Cross-section of DICE (85° tilt / 305° azimuth and 85° tilt / 220° azimuth) and LEAP-DICE (85° tilt / 180° azimuth) as a function of LET.

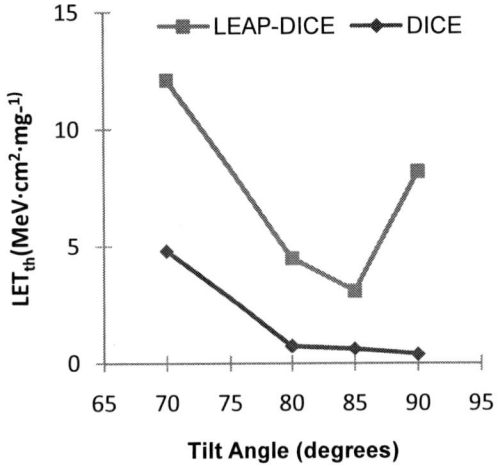

Figure 3.7. LET upset threshold as a function of tilt angle for LEAP-DICE at an azimuth angle of 180°, and for DICE at an azimuth angle of 305°.

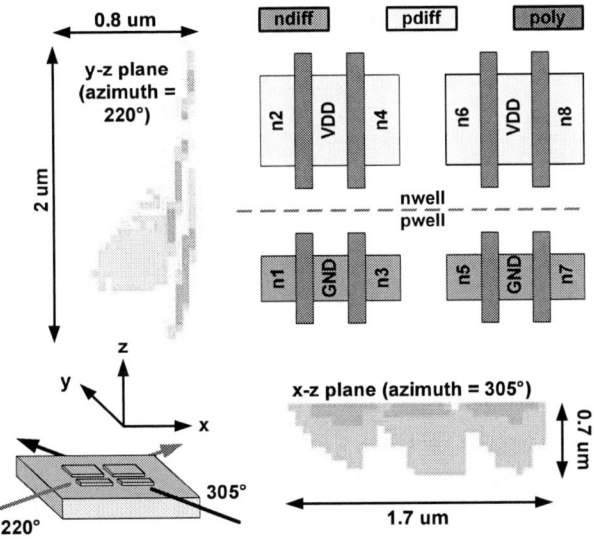

Figure 3.5. Layout configuration with projected cross-section regions for DICE at LET = 3/30 MeV·cm^2·mg^{-1}: blue/green region on x-z plane for particle strike direction with 85° tilt / 305° azimuth, orange/green region on y-z plane for particle strike direction with 85° tilt / 220° azimuth.

Figure 3.6 shows the cross-sections (area in the circuit vulnerable to upsets) of DICE and LEAP-DICE as a function of LET for the selected directions in each layout. The lowest LET upset threshold for LEAP-DICE is almost an order of magnitude larger than the lowest LET upset threshold for DICE (at 305° azimuth). Note that, even at an angle quite far from the worst case (220° azimuth), the LET upset threshold of the DICE is still as low as the upset threshold of the LEAP-DICE. The LEAP-DICE implementation has its lowest LET upset threshold around 85° tilt, whereas the DICE has the minimum at 90° tilt. The simulated LET upset threshold as a function of tilt angle, at the worst case azimuth angle, is shown in Fig. 3.7.

While the simulations presented here are relatively limited in scope, they show the effectiveness of the LEAP principle and quantify the reduction in LET upset threshold and cross-section for LEAP-DICE compared to DICE.

IV. TEST CHIP IMPLEMENTATION

To evaluate the effectiveness of LEAP-DICE, we implemented three flip-flop designs on a test chip in 180nm CMOS technology. Each flip-flop consists of one minimum-sized input inverter, one master latch, one slave latch and one minimum-sized output inverter (Fig. 4.1). For the 180nm technology used in this paper, a minimum-sized inverter consists of the smallest-width NMOS transistor with a P/N width ratio is 2.4. The master and slave latches have separate clock inputs, MCLK and SCLK. The three flip-flop designs are described next.

978-1-4244-5430-3/10 $26.00 © 2010 IEEE   207

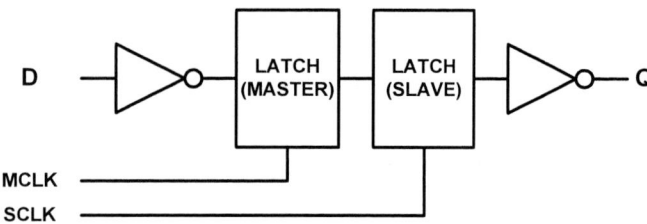

Figure 4.1. Master-slave flip-flop implementation.

1. We implemented a master-slave flip-flop named *"BASIC FF"* using conventional D latches (*"BASIC D-Latch"*) based on tri-state inverters (Fig. 4.2). To make the flip-flop radiation-hardened, each active NMOS or PMOS region is enclosed by a guard ring to reduce overall charge collection [Clein 99, Black 05, Olson 07, Narasimham 08]. The layout of "BASIC FF" is shown in Fig. 4.3, with a close-up of the "BASIC D-Latch" layout shown in Fig. 4.4.

2. We implemented a master-slave flip-flop called *"DICE FF"* based on the *"DICE Latch"* shown in Fig. 4.5. The layout of "DICE FF" is shown in Fig. 4.6, with a close-up view of the "DICE Latch" layout in Fig. 4.7.

3. We implemented a master-slave flip-flop called *"LEAP-DICE FF"* based on the LEAP principle. This flip-flop uses latches named *"LEAP-DICE Latch"* with the same circuit schematic as "DICE Latch" (Fig. 4.5), but uses the layout design principles described Section III. The layout for "LEAP-DICE FF" is shown in Fig. 4.8, with a close-up of the "LEAP-DICE Latch" in Fig. 4.9. The transistor drain contact nodes from Fig. 3.2, viz., n5, n7, n8, n2, n1, n3, n4 and n6, are aligned along the horizontal direction in Fig. 4.9.

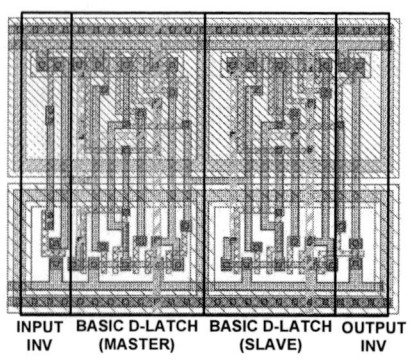

Figure 4.3. Layout of "BASIC FF".

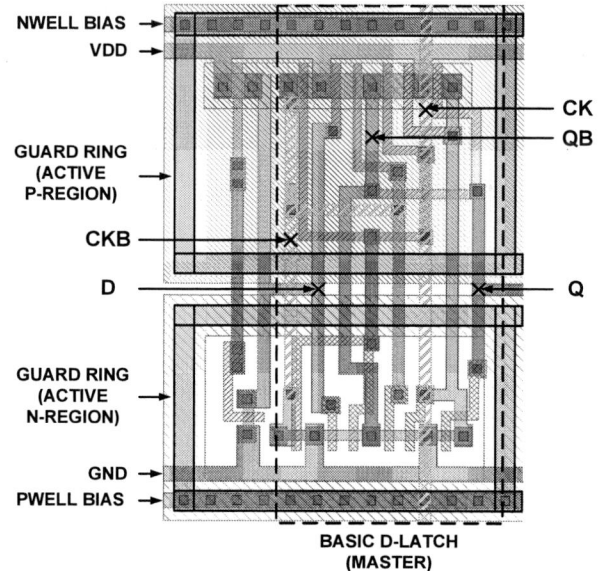

Figure 4.4. Close-up of "BASIC D Latch" layout.

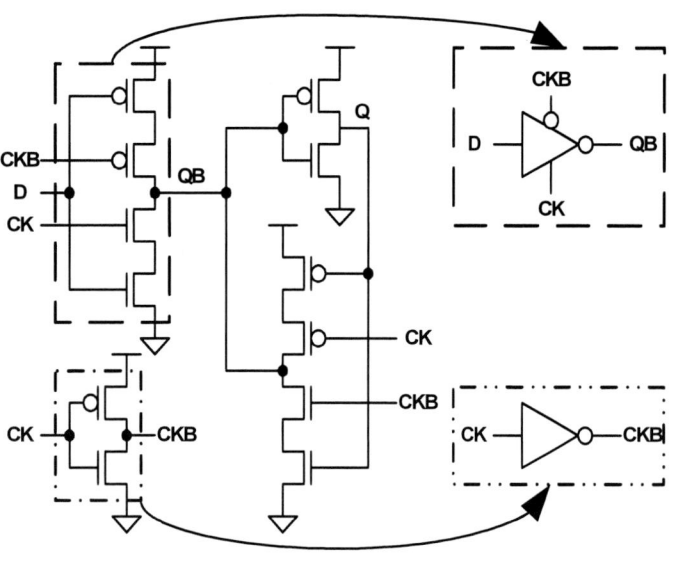

Figure 4.2. Schematic of "BASIC D-Latch". All NMOS transistors have an aspect ratio of 0.28μm/0.18μm, all PMOS transistors have an aspect ratio of 0.68μm/0.18μm.

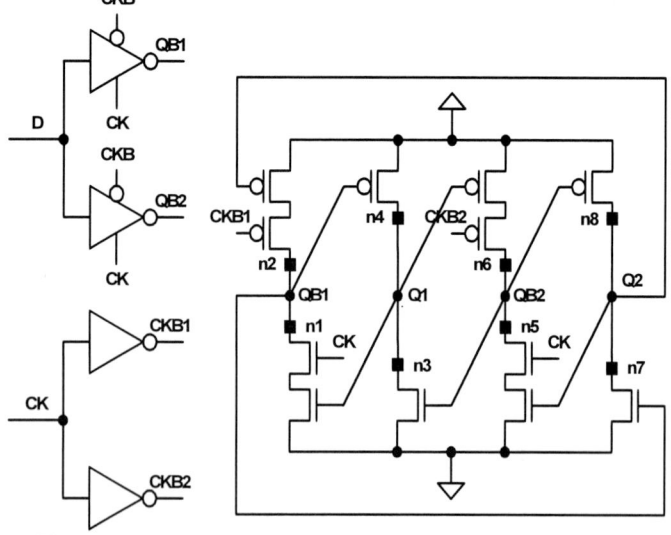

Figure 4.5. Schematic of "DICE Latch". Same transistor sizing as in Fig. 4.2.

978-1-4244-5430-3/10 $26.00 © 2010 IEEE   208

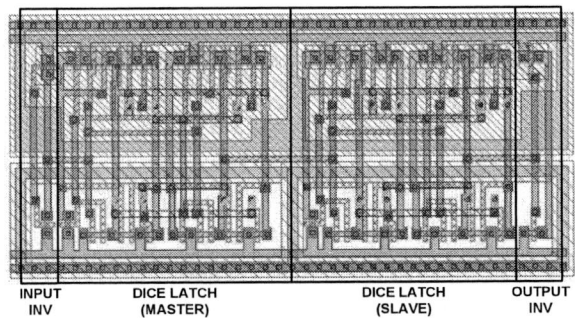

Figure 4.6. Layout of "DICE FF".

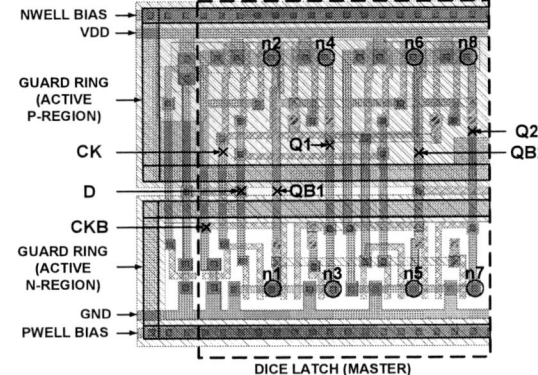

Figure 4.7. Close-up of "DICE Latch".

Table 4.1 lists the performance parameters from our circuit simulations. Each flip-flop was drawn in Cadence Virtuoso using 180nm CMOS technology, and post-layout parasitic extraction was performed on each layout design using a Cadence Assura RCX extraction script provided by the fabrication facility. We used Cadence Spectre to simulate the flip-flops using the extracted netlists. The simulation setup is shown in Fig. 4.10. For each simulation, the supply voltage was set at 1.8V, and two 40-MHz non-overlapping clocks drove the master and slave stages of the flip-flops. The speed and power performance of "LEAP-DICE FF" are comparable to "BASIC FF" and "DICE FF". "LEAP-DICE FF" (Fig. 4.8) incurs 40% additional area cost compared to "DICE FF" (Fig. 4.6) due to associated intra-cell routing.

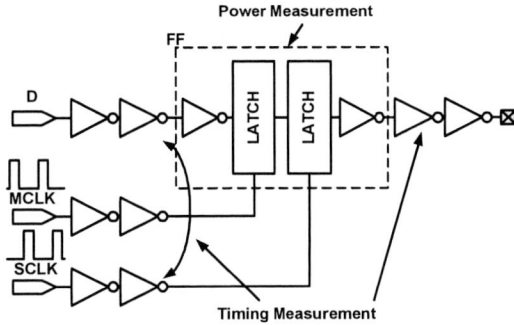

Figure 4.10. Simulation test bench for post-layout timing and power measurements. All inverters are minimum-sized.

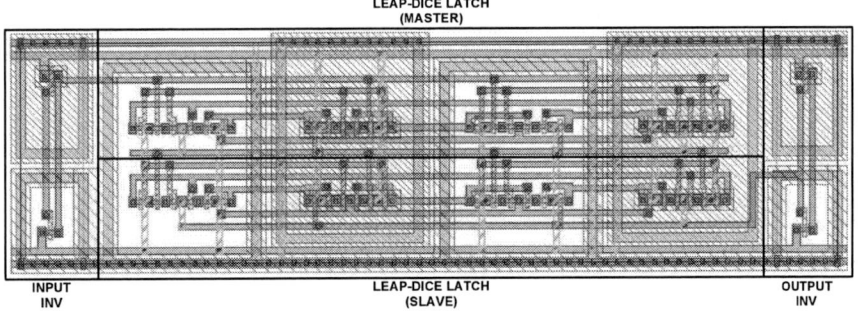

Figure 4.8. Layout of "LEAP-DICE FF".

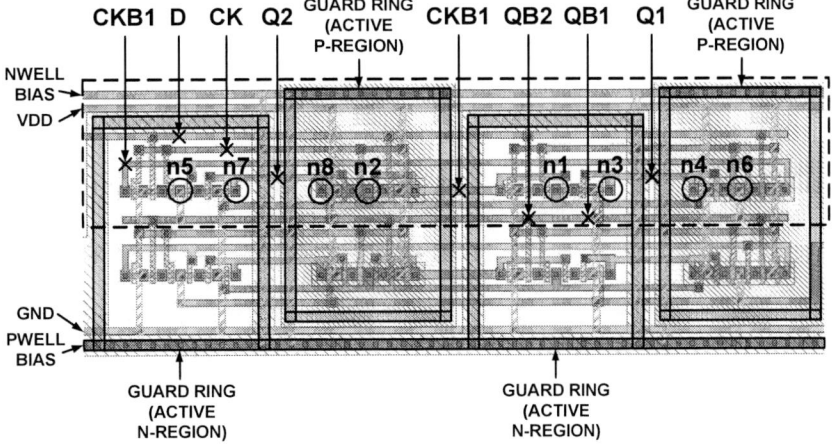

Figure 4.9. Close-up of "LEAP-DICE".

978-1-4244-5430-3/10 $26.00 © 2010 IEEE

Figure 4.11 shows our 5mm x 5mm test chip fabricated using an 180nm CMOS process from National Semiconductor Corporation. For each flip-flop design, we implemented a scan chain with 2,304 to 4,608 flip-flops. Each scan chain has its own I/O and core supplies so that each chain can be independently tested (Fig. 4.12). The non-overlapping clocks for the master and slave stages of the flip-flops, MCLK and SCLK respectively, are supplied from an external source.

TABLE **4**.1 NORMALIZED POST-LAYOUT SIMULATION PERFORMANCE FOR VARIOUS FLIP-FLOP DESIGNS AT **40** MHz AND **1**.8V SUPPLY.

Design	Transistor Count	Layout Area	Power	Average Clock to Output Delay
BASIC FF	24	1.00	1.00	1.00
DICE FF	52	1.67	1.50	1.06
LEAP-DICE FF	**52**	**2.33**	**1.54**	**1.07**

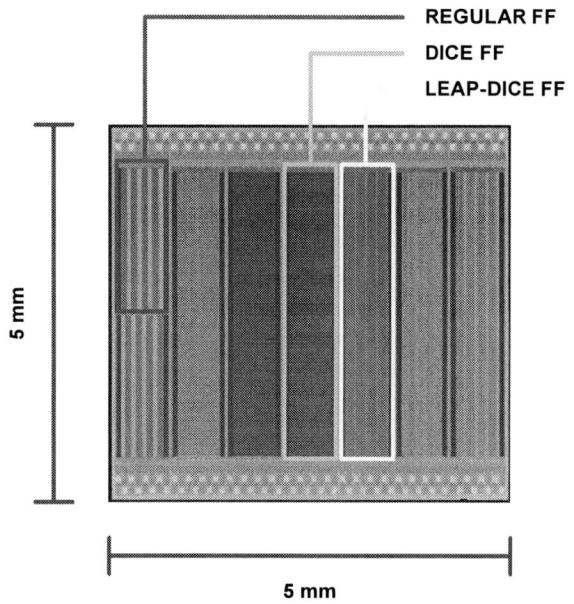

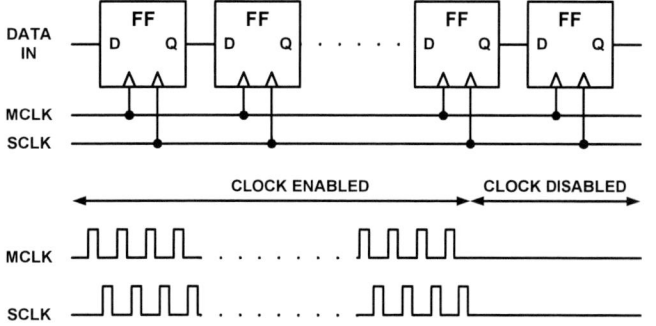

Figure 4.11. Die photograph of the 5x5mm test chip.

Figure 4.12. Test chip clocking scheme, with a scan chain of flip-flops using two separate external clocks, MCLK and SCLK, clocking the master and slave stages of the flip-flop, respectively.

V. RADIATION EXPERIMENT SETUP AND RESULTS

For radiation testing purposes, a test board containing only the test chip socket, decoupling capacitors and line matching resistors was placed in front of the beam. The socket board passes the test signals through two parallel 68-pin 30-feet SCSI cables to a CPLD (*Complex Programmable Logic Device*) board that provides the supply voltages for the test chip and buffers various test signals and clocks generated by an adjacent FPGA board. The FPGA board can generate various signal patterns and vary the test clock frequency, and can be commanded by a PC via the Ethernet. Fig. 5.1 shows the experimental test setup.

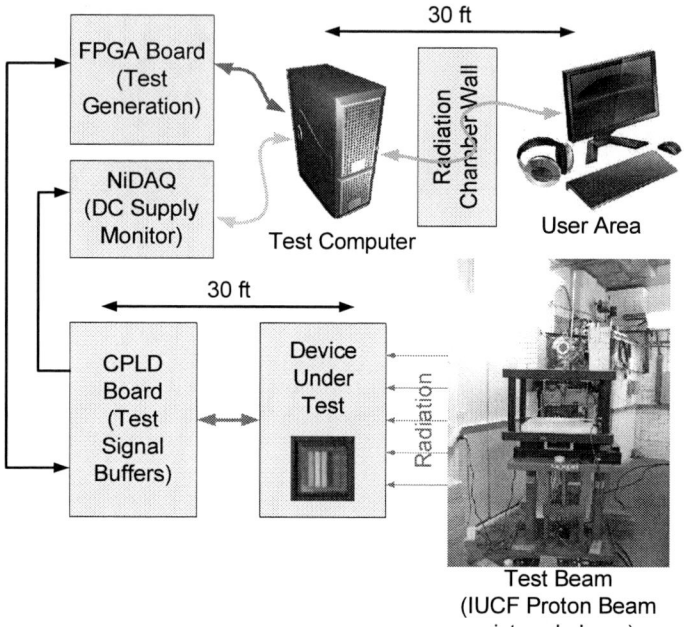

Figure 5.1. Experimental setup for single event testing.

Two types of tests were conducted for this study following single event testing guidelines from the JEDEC89A standard [JEDEC06]:

1. *Static Testing:* a pre-defined data pattern is first serially shifted into the flip-flops through two two-phase non-overlapping clocks. Two types of data patterns used in this test are "*all-1*" (all bits are 1) and "*all-0*" (all bits are 0). After the data is loaded into the flip-flops, the clocks are disabled such that both master and slave stages of the flip-flops retain their own data values. The test chip is then exposed to the beam at normal incidence. After irradiation for a set *fluence* (the number of particles passing through normalized by unit area), the beam is turned off and the flip-flops contents are serially shifted out by enabling the clocks again. The number of errors in the read data pattern is then counted to compute the *error cross-section* given by the following equation:

$$error\ cross\ section = \frac{number\ of\ errors\ counted}{number\ of\ flip-flops \times fluence}$$

TABLE 5.1. ERROR COUNTS FOR VARIOUS FLIP-FLOP DESIGNS FROM 200 MEV, 1V SUPPLY PROTON IRRADIATION AT INDIANA UNIVERSITY CYCLOTRON FACILITY (IUCF).

Test Type		BASIC FF (2304 FFs)		DICE FF (4608 FFs)		LEAP-DICE FF (2304 FFs)	
		Error Count	Dose Mrad[Si]	Error Count	Dose Mrad[Si]	Error Count	Dose Mrad[Si]
Static	All-0	852	0.2	42	2	4	2
	All-1	956	0.2	28	2	3	2
Dynamic (50 kHz)	All-0	8,915	2.0	28	2	4	2
	All-1	8,867	2.0	32	2	3	2

TABLE 5.2. CONFIDENCE INTERVALS FOR STATIC ERROR CROSS-SECTION OF VARIOUS FLIP-FLOP DESIGN AT 1V SUPPLY FROM 200 MEV PROTON IRRADIATION AT INDIANA UNIVERSITY CYCLOTRON FACILITY (IUCF).

Design	Pattern	Error Cross-Section	95% Confidence Interval		90% Confidence Interval	
			Min.	Max.	Min.	Max.
BASIC FF	All-0	1.10×10^{-13}	9.61×10^{-14}	1.26×10^{-13}	9.83×10^{-14}	1.24×10^{-13}
	All-1	1.24×10^{-13}	1.09×10^{-13}	1.41×10^{-13}	1.11×10^{-13}	1.38×10^{-13}
DICE FF	All-0	2.72×10^{-16}	1.96×10^{-16}	3.68×10^{-16}	2.07×10^{-16}	3.52×10^{-16}
	All-1	1.82×10^{-16}	1.21×10^{-16}	2.62×10^{-16}	1.29×10^{-16}	2.49×10^{-16}
LEAP-DICE FF	All-0	5.19×10^{-17}	1.41×10^{-17}	1.33×10^{-16}	1.77×10^{-17}	1.19×10^{-16}
	All-1	3.89×10^{-17}	8.02×10^{-18}	1.14×10^{-16}	1.06×10^{-17}	1.01×10^{-16}

TABLE 5.3. CONFIDENCE INTERVALS FOR DYNAMIC ERROR CROSS-SECTION OF VARIOUS FLIP-FLOP DESIGN AT 1V SUPPLY, 50 KHZ FROM 200 MEV PROTON IRRADIATION AT INDIANA UNIVERSITY CYCLOTRON FACILITY (IUCF).

Design	Pattern	Error Cross-Section	95% Confidence Interval		90% Confidence Interval	
			Min.	Max.	Min.	Max.
BASIC FF	All-0	1.16×10^{-13}	1.13×10^{-13}	1.18×10^{-13}	1.14×10^{-13}	1.18×10^{-13}
	All-1	1.15×10^{-13}	1.13×10^{-13}	1.17×10^{-13}	1.13×10^{-13}	1.17×10^{-13}
DICE FF	All-0	1.82×10^{-16}	1.21×10^{-16}	2.62×10^{-16}	1.29×10^{-16}	2.49×10^{-16}
	All-1	2.07×10^{-16}	1.42×10^{-16}	2.93×10^{-16}	1.51×10^{-16}	2.79×10^{-16}
LEAP-DICE FF	All-0	5.19×10^{-17}	1.41×10^{-17}	1.33×10^{-16}	1.77×10^{-17}	1.19×10^{-16}
	All-1	3.89×10^{-17}	8.02×10^{-18}	1.14×10^{-16}	1.06×10^{-17}	1.01×10^{-16}

2. *Dynamic Testing*: in contrast to static testing, the predefined data pattern is actively shifted into the flip-flops and shifted out while the test chips are irradiated. The test chips are also irradiated for target fluence but with a set clock frequency. Error cross-section is also computed for dynamic testing, and "all-1" and "all-0" patterns are used.

Two sets of tests were performed on separate chips from the same batch. A spread energy (pink) spectrum neutron test was first conducted at the Los Alamos National Laboratory (LANL)/ICE House facility on Sept 14-17, 2009. The neutron flux and energy spectrum offered at LANL are approximately 3×10^8 times the neutron flux at New York City ground level [JEDEC 06]. We used a 3-inch diameter neutron beam where the center of the beam strikes the silicon chip surface at 90° incident angle, and no latchup was observed. To maximize the upset rate, the supply voltage of the flip-flops was set to 0.8V, the lowest operating voltage for all flip-flops. Nonetheless, no upsets were observed for both DICE FF and LEAP-DICE FF designs in this experiment. The upset rate during static testing for the BASIC FF design was approximately one error in 2304 flip-flops per neutron fluence of $8.5 \times 10^8/cm^2$ at 0.8V. This translates to 7.1×10^{-3} FITs at New York City sea level.

We conducted a second set of tests at the Indiana University Cyclotron Facility on Oct 12-16, 2009 and Dec 14-18, 2009. Multiple test chips were irradiated at normal incidence using a 200 MeV proton beam up to a maximum dose of 2 Mrad[Si]. The core voltage values were set at 1V, 1.4V and 1.8V while the I/O voltage value was set at 1.8V. Again, no latchup was observed during the entire duration of tests. Due to substantial total dose effects at higher supply core voltages (including exponential increase in leakage current as well as transistor threshold shifts), we observed that soft error rates for all designs increased significantly after the total dose exceeded 300 Krad[Si] for core supplies of 1.4V and 1.8V. For tests conducted at a core voltage of 1V, the observed soft error rates remained stable throughout the experiment. Therefore, we report results for tests conducted at 1V in this paper. The raw error counts from actual experiments are reported in Table 5.1. The resulting error cross-sections and their confidence intervals (95% and 90% using chi-squared statistics) are shown in Table 5.2 and 5.3.

Based on results from both static and dynamic testing, we can conclude that the "LEAP-DICE FF" design encounters 2,000X fewer errors on average compared to the baseline "BASIC FF" design, and 5X fewer errors on average than the "DICE-FF" design.

978-1-4244-5430-3/10 $26.00 © 2010 IEEE

VI. CONCLUSION

The LEAP (**L**ayout Design through **E**rror **A**ware Transistor **P**ositioning) principle enables the design of the LEAP-DICE sequential element with significantly improved soft error resilience compared to our reference DICE sequential element. LEAP-DICE retains the circuit design of DICE, but uses a new layout design. Radiation experiment results using 180nm test chips demonstrate that our LEAP-DICE flip-flop design encounters 5X fewer errors on average compared to our reference DICE flip-flop and 2,000X fewer errors on average compared to a conventional D flip-flop. This improvement comes at a negligible power and performance cost compared to DICE. However, the flip-flop-level area cost increases by 40% compared to our DICE flip-flop. Future research directions include evaluation of LEAP in advanced technologies (beyond 180nm) and generalization of LEAP for arbitrary designs.

ACKNOWLEDGEMENT

This research was supported in part by the Defense Threat Reduction Agency (contract no. HDTRA1-09-P0011), the Air Force Research Laboratory (contract no. AFRL FA8718-05-C-0027) and the National Science Foundation. We also thank National Semiconductor Corporation for fabrication support of our test chip, and Benjamin Mossawir and Jumie Yuventi of Stanford University for their participation in the radiation experiments conducted at LANL and IUCF.

REFERENCES

[Amusan 07] Amusan, O.A., L.W. Massengill, M.P. Baze, B.L. Bhuva, A.F. Witulski, S. DasGupta, A.L. Sternberg, P.R. Fleming, C.C. Heath, and M.L. Alles, "Directional Effects on Single Event Charge Sharing in a 90 nm CMOS Latch", *IEEE Trans. Nucl. Sci.*, vol. 54, pp.2584-2589, Dec. 2007.

[Amusan 08] Amusan, O.A., L.W. Massengill, M.P. Baze, B.L. Bhuva, A.F. Witulski, J.D. Black, A. Balasubramanian, M.C. Casey, D.A. Black, J.R. Ahlbin, R.A. Reed and M.W. McCurdy, "Mitigation Techniques for Single-Event-Induced Charge Sharing in a 90-nm Bulk CMOS Process," *Proc. IEEE Intl. Rel. Phys. Symp.*, pp.468-472, 2008.

[Baumann 05] Baumann, R.C., "Radiation-Induced Soft Errors in Advanced Semiconductor Technologies", *IEEE Trans. Dev. Mat. Rel.*, vol. 5, no. 3, pp.305-316, Sept. 2005.

[Baze 08] Baze, M.P., B. Hughlock, J. Wert, J. Tostenrude, L. Massengill, O. Amusan, R. Lacoe, K. Lilja and M. Johnson, "Angular Dependence of Single Event Sensitivity in Hardened Flip-Flop Designs," *IEEE Trans. Nucl. Sci.*, vol. 55, no. 6, pp.3295-3301, Dec. 2008.

[Black 05] Black, J.D., A.L. Sternberg, M.L. Alles, A.F. Witulski, B.L. Bhuva, L.W. Massengill, J.M. Benedetto, M.P. Baze, J.L. Wert and M.G. Hubert, "HBD layout isolation techniques for multiple node charge collection mitigation," *IEEE Trans. Nucl. Sci.*, vol. 52, no. 6, pp.2536-2541, Dec. 2005.

[Calin 96] Calin, T., M. Nicolaidis and R. Velazco, "Upset Hardened Memory Design for Submicron CMOS Technology," *IEEE Trans. Nucl. Sci.*, vol. 43, no. 6, pp.2874-2878, Dec. 1996.

[Clein 99] Clein, D., "CMOS IC Layout: Concepts, Methodologies, and Tools", *Boston MA: Newnes Press*, pp.93-99, 2000.

[Dodd 03] Dodd, P.E. and L.W. Massengill, "Basic Mechanisms and Modeling of Single-Event Upset in Digital Microelectronics," *IEEE Trans. Nucl. Sci.*, vol. 50, no. 3, pp.583-602, June 2003.

[Gill 09] Gill, B., N. Seifert, and V. Zia, "Comparison of Alpha-particle and Neutron-induced Combinational and Sequential Logic Error Rates at the 32nm Technology Node," *Proc. IEEE Intl. Rel. Phys. Symp.*, pp.199-205, 2009.

[IP 08] Robust Chip Inc. patent pending US2009/0184733

[JEDEC 06] JEDEC Standard, "Measurement and Reporting of Alpha Particles and Terrestrial Cosmic Ray-Induced Soft Errors in Semiconductor Devices", *JEDEC89A*, JEDEC Solid State Technology Association, New York, October 2006.

[Karnik 04] Karnik, T. and P. Hazucha, "Characterization of Soft Errors Caused by Single Event Upsets in CMOS Processes", *IEEE Trans. Dep. Sec. Comp.*, vol. 1, no. 2, pp.128-143, April-June 2004.

[Koga 96] Koga R., "Single-Event Effect Ground Test Issues," *IEEE Trans. Nucl. Sci.*, vol. 43, no. 2, pp.661-670, April 1996.

[Lilja 09] Lilja K., "Single Event Cross-Section and Error-Rate Prediction for Digital Logic Using Accurate Simulation", *SEE Symposium*, La Jolla, 2009

[Mavis 02] Mavis D. G. and P. H. Eaton, "Soft Error Rate Mitigation Techniques for Modern Microcircuits", *Proc. IEEE Intl. Rel. Phys. Symp.*, pp.216-225, 2002.

[Meaney 05] Meaney P.J., S.B. Swaney, P.N. Sanda, and L. Spainhower, "IBM z990 Soft Error Detection and Recovery", *IEEE Trans. Dev. Mat. Rel.*, vol. 5, no. 3, pp. 419-427, Sept. 2005.

[Mitra 05] Mitra S., N. Seifert, M. Zhang, Q. Shi and K.S. Kim, "Robust System Design with Built-In Soft Error Resilience", *IEEE Computer*, vol. 38, no. 2, pp.43-52, Feb. 2005.

[Narasimham 08] Narasimham, B., "Quantifying the Effect of Guard Rings and Guard Drains in Mitigating Charge Collection and Charge Spread", *IEEE Trans. Nucl. Sci.*, vol. 55, no. 6, pp.3456-3460, Dec. 2008.

[Olson 05] Olson B.D., D.R. Ball, K.M. Warren, L.W. Massengill, N.F. Haddad, S.E. Doyle and D. McMorrow, "Simultaneous Single Event Charge Sharing and Parasitic Bipolar Conduction in a High-Scaled SRAM Design", *IEEE Trans. Nucl. Sci., vol. 52.*, pp. 2132-2136, Dec. 2005.

[Olson 07] Olson B.D., O.A. Amusan, S. Dasgupta, L.W. Massengill, A.F. Witulski, B.L. Bhuva, M.L. Alles, K.M. Warren, and D.R. Ball, "Analysis of Parasitic PNP Bipolar Transistor Mitigation Using Well Contacts in 130 nm and 90 nm CMOS Technology", *IEEE Trans. Nucl. Sci.*, vol. 54, no. 4, pp. 894-897, Aug. 2007.

[RCI 10] The ACCURO simulator from RCI [www.robustchip.com].

[Seifert 06] Seifert N., P. Slankard, M. Kirsch, B. Narasimham, V. Zia, C. Brookreson, A. Vo, S. Mitra, B. Gill and J. Maiz, "Radiation-Induced Soft Error Rates of Advanced CMOS Bulk Devices," *Proc. IEEE Intl. Rel. Phys. Symp.*, pp.215-225, 2006.

[Seifert 07] Seifert N., B. Gill, M. Zhang, and V. Ambrose, "On the Scalability of Redundancy based SER Mitigation Schemes", *Proc. Intl. Conf. on Int. Ckt. Des. and Tech.*, pp. 1-9, May 30-June 1 2007.

[Shuler 05] Shuler R. L., C. Kouba, and P.M. O'Neill, "SEU Performance of TAG Based Flip-Flops", *IEEE Trans. Nucl. Sci.*, vol. 52, no. 6, Dec. 2005.

Alpha-Particle-Induced Soft Errors and Multiple Cell Upsets in 65-nm 10T Subthreshold SRAM

Hiroshi Fuketa[†*], Masanori Hashimoto[†*], Yukio Mitsuyama[†*], and Takao Onoye[†*]

[†]Dept. Information Systems Engineering, Osaka University, JAPAN [*]JST, CREST

{fuketa.hiroshi,hasimoto,mituyama,onoye}@ist.osaka-u.ac.jp

Abstract—**This paper presents measurement results of alpha-particle-induced soft errors and multiple cell upsets (MCUs) in 65-nm 10T SRAM with a wide range of supply voltage from 1.0 V to 0.3 V. We reveal that the soft error rate (SER) at 0.3 V is eight times higher than SER at 1.0 V, and the ratio of MCUs to the total upsets increases as the supply voltage decreases. The SER and ratio of MCUs with body-biasing are also described. In addition, we investigate an impact of manufacturing variability on the soft error immunity of each memory cell. In our measurement, a distinct influence of manufacturing variability is not observed even in subthreshold region.**

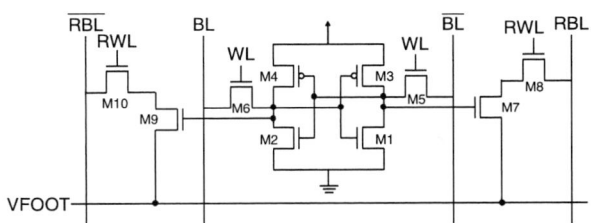

Fig. 1. Structure of the 10T memory cell. WL is only asserted at the write operation and RWL is asserted at the read operation.

I. INTRODUCTION

Subthreshold circuits, which operate at a lower supply voltage than threshold voltage, are expected to be used for ultra-low power applications, such as a sensor-node processor [1], [2]. On the other hand, soft error immunity of subthreshold circuits has become a concern because the ultra-low voltage operation reduces the energy required to cause upsets [3], [4]. Especially, the soft error rate (SER) in SRAM, which often characterize the SER of the entire circuit, must be carefully examined before adopting subthreshold circuits for practical applications.

According to [5], the neutron-induced SER in SRAM increases by 18% for every 10% reduction in the supply voltage. Reference [6] reports a trend that decrease in the supply voltage makes the alpha-particle-induced SER dominant in SRAM. However, these measurements were just performed between the nominal supply voltage and 0.8 V. As for soft errors in subthreshold region, single event transient (SET) was recently analyzed with ring oscillators in [3]. On the other hand, the SRAM SER in subthreshold region has not been measured as far as the authors know.

This paper is the first work to measure the alpha-particle-induced SER in subthreshold region using a newly designed 65-nm 10T SRAM. Measurement results show that 0.3 V operation increases SER eightfold compared to 1.0 V. In addition, information on frequency of multiple cell upsets (MCUs) is important for error prevention using error checking and correction (ECC) technique. In this paper, we also investigate MCUs in subthreshold region. We reveal that although the ratio of MCUs to the total upsets remarkably increases at lower voltage, conventional ECC technique is still effective in our 10T subthreshold SRAM. Additionally, the dependency of SER on body-bias is evaluated, and measurement results indicate that the SER and the ratio of MCUs are less sensitive to the body-bias voltage when the supply voltage is 0.4 V.

Furthermore, we examine the variation in soft error immunity of each memory cell. Reference [7] indicates that the charge required to cause an upset, that is, critical charge varies bit by bit due to manufacturing variability, especially threshold voltage variation, and the variation in critical charge becomes larger as the supply voltage is lowered. In this paper, we investigate the influence of manufacturing variability on the soft error immunity in subthreshold region. Measurement results show that the distribution of the number of SEUs in each memory cell is well explained by the spatial randomness of alpha-particle hits without taking into account critical charge variation due to manufacturing variability.

II. SRAM STRUCTURE

In order to measure the SRAM SER over a wide range of supply voltage, SRAM that can operate even in subthreshold region is required. On the other hand, traditional 6T SRAM, which is used in commercial off-the-shelf SRAMs, is scarcely functional in subthreshold region due to weak writability and read instability [8], [9]. Many types of memory cell structure have been proposed to overcome those problems in the low-voltage operation [9]–[13]. One of the potential solutions is to add a few transistors dedicated to a read port [10]. Based on this technique, 10T [9], [11] and 8T [12], [13] SRAMs achieve subthreshold operation.

In this paper, 10T SRAM was newly designed and fabricated. Fig. 1 depicts the structure of designed memory cell. This memory cell is based on the 8T memory cell proposed in [12]. We improved the 8T memory cell such that a differential read operation can be achieved by adding two NMOSs (M9 and M10). VFOOT is used to ensure the correct read operation in subthreshold region [12]. VFOOT of the accessed word is set to low and those of the unaccessed

978-1-4244-5430-3/10 $26.00 © 2010 IEEE   213

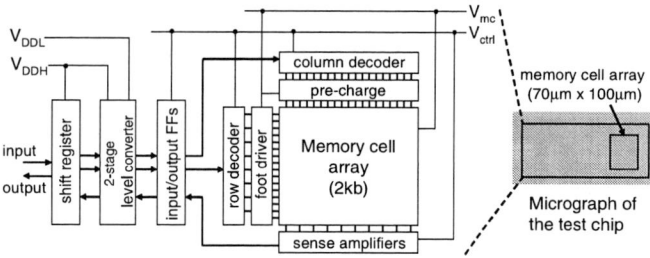

Fig. 2. Block diagram of the SRAM circuit and micrograph of the test chip.

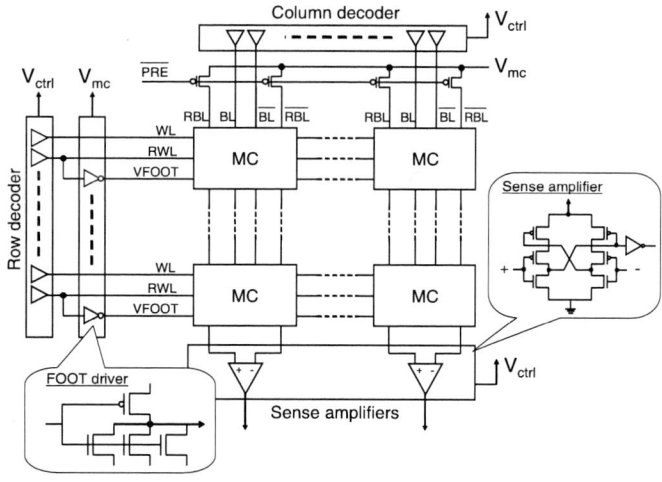

Fig. 3. Details of the memory cell (MC) array and its control logic.

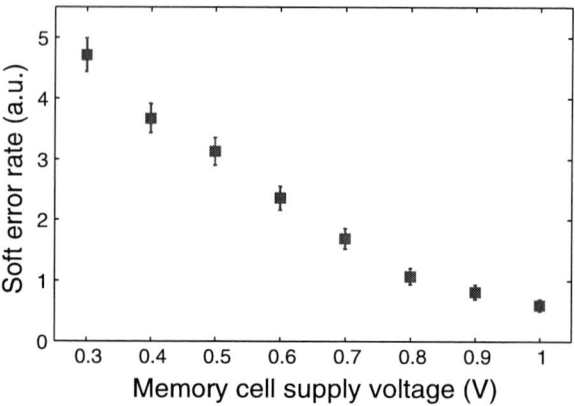

Fig. 4. SERs as a function of the supply voltage of the memory cell array (V_{mc}). Each error bar indicates $\pm 3\sigma$, where σ is defined as the square root of the number of the observed upsets.

III. EXPERIMENTAL RESULTS

This section describes measurement results. We used an Am-241 foil, whose flux is 9×10^9 cm^{-2}h^{-1}, as an alpha particle source. The main peak energy of the alpha particle is 5.49 MeV. The foil was placed immediately above the die according to [15].

A. Sort Error Rate and Multiple Cell Upset

Fig. 4 shows SERs as a function of the supply voltage of the memory cell array (V_{mc}). In this experiment, we first wrote zero to all bits in the SRAM. Then, all stored data was read every 10 seconds, and we checked whether each bit was flipped or not to evaluate the SER. The operation frequency in read operation was 5 kHz. Fig. 4 indicates that the SER increases as the supply voltage is reduced, and the SER at $V_{mc} = 0.3$ V is eight times higher than that at $V_{mc} = 1.0$ V.

Next, we focus on the multiplicity of soft errors. Especially, multiple upsets belonging to the same word are critical, because the conventional SEC-DED (Single Error Correction, Double Error Detection) ECC cannot correct two or more errors in a word. As explained in Fig. 5, the occurrence of multiple upsets in adjacent bits depends on the written data pattern [16]. In our memory cell layout, the intra-word upsets might occur in data pattern "01" and "10". Therefore, we periodically wrote the data pattern "1010···" to all words and checked the stored data until around 2000 soft errors were observed. Note that two errors originated from different particles might be misjudged as an MCU in this measurement. We, however, estimated the probability of this misjudgment, and the number of such "pseudo" MCUs was below two in this setup.

Fig. 6 lists the number of single bit upsets (SBUs) and MCUs and plots the ratio of MCUs to the total number of upsets as a function of the memory cell voltage (V_{mc}). All of the observed MCUs were two-bit ones. The ratio of MCUs dramatically increases when V_{mc} becomes less than 0.6 V. Interestingly, on the other hand, multiple upsets belonging to the same word were observed only once in $V_{mc} = 1.0$ V,

words remain high. The area of 6T (M1–M6) occupies 80% of the total area (10T) because the size of the cross-coupled inverters (M1–M4) was increased to mitigate threshold voltage variability and to retain the correct data in all bits even in subthreshold region.

Fig. 2 shows the block diagram of the SRAM circuit. A test chip was fabricated in a 65-nm CMOS process. 2kb (128 row × 16 column) cells with triple-well structure are implemented, and are verified to be functional from 1.0 V to 0.3 V. This time, any bit-interleaving technique is not implemented in our design. Fig. 3 illustrates the details of the memory cell array and its control logic. Different supply voltages are applied to the memory cell array and the control circuit which consists of the FFs for input/output data and the row/column decoder. The supply voltage of the control circuit V_{ctrl} is higher than the memory cell voltage V_{mc}. Since the word line (WL) and bit line (BL) drivers operate at V_{ctrl}, strong writability can be achieved, which is similar to so-called "boosted word line" technique [12], [14]. Throughout this paper, V_{ctrl} is set to $V_{mc} + 0.1$ V. Signals at V_{ctrl} are converted to the nominal supply voltage V_{DDH} via V_{DDL} by the 2-stage level shifter. For example, a typical configuration at $V_{ctrl} = 0.4$ V is $V_{DDL} = 0.6$ V and $V_{DDH} = 1.2$ V.

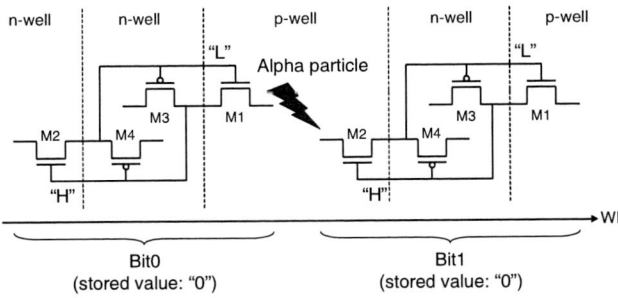

(a) intra-word upsets do not occur

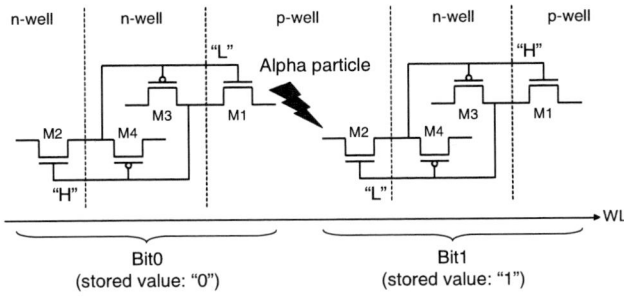

(b) intra-word upsets might occur

Fig. 5. Occurrence of multiple upsets in adjacent bits depends on the written data pattern [16]. Only cross-coupled inverters (M1–4 in Fig. 1) are illustrated in this figure. In our memory cell layout, intra-word upsets might occur in data pattern "01" and "10", and do not occur in "00" and "11".

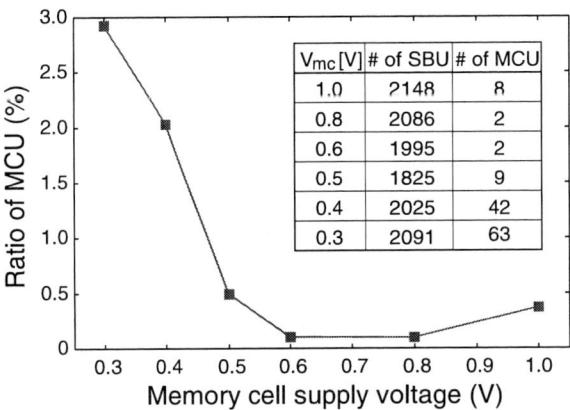

V_{mc} [V]	# of SBU	# of MCU
1.0	2148	8
0.8	2086	2
0.6	1995	2
0.5	1825	9
0.4	2025	42
0.3	2091	63

Fig. 6. The number of single bit upsets (SBUs) and multiple cell upsets (MCUs), and the ratio of MCUs to the total upsets (SBU + MCU) as a function of the supply voltage of the memory cell array (V_{mc}).

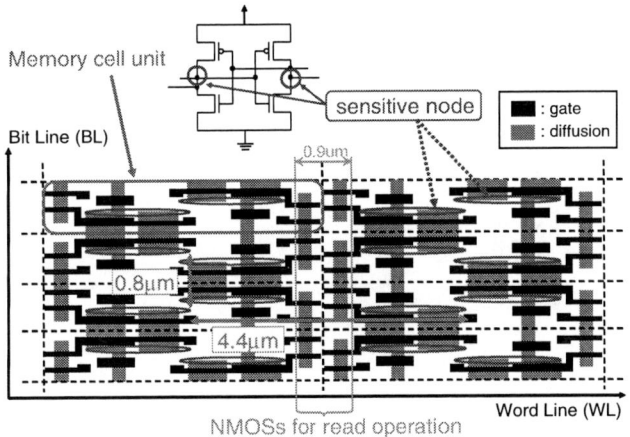

Fig. 7. Layout of the memory cell array. Blue ellipses indicate sensitive nodes for alpha particles. The NMOSs for the read operation (M7–10 in Fig. 1) are located between the adjacent columns.

and all other MCUs occurred in inter-word adjacent bits. This means the conventional ECC technique is still effective for robust operation in subthreshold region.

We here discuss the reason why the MCUs only arose in inter-word adjacent bits in subthreshold operation. Fig. 7 illustrates the layout of the memory cell array. The drains of P/NMOSs in the cross-coupled inverter pair are sensitive to alpha particles. The distance between the sensitive nodes of the adjacent rows is 1/5 shorter than that of the adjacent columns. In the designed 10T cells, the NMOSs for the read operation (M7–10 in Fig. 1) are located between the adjacent columns, which makes horizontally-adjacent cells distant. Thus, adjacent bits in different words along bit lines are more likely to upset.

Next, we measure the dependence of the SER and MCU on the body-bias voltage of the memory cells at $V_{mc} = 0.4$ V and 1.0 V. Fig. 8 shows the measurement results. When V_{mc} is 1.0 V, the SERs at 1.0 V-RBB (reverse body bias) and at 0.3 V-FBB (forward body bias) increase by more than 30% compared to the SER at ZBB (zero body bias), whereas when V_{mc} is 0.4 V, the SERs at 1.0 V-RBB and 0.3 V-FBB are higher by below 10% than the SER at ZBB. In addition, although FBB raises the ratio of MCUs to the total upsets at $V_{mc} = 1.0$ V, body-bias does not affect the ratio so much at $V_{mc} = 0.4$ V. The SER and the ratio of MCUs are less sensitive to the body-bias at 0.4 V than those at 1.0 V.

B. Variation in Soft Error Immunity

In this section, we investigate the influence of the critical charge variation due to manufacturing variability on the soft error immunity. We measured the number of soft errors in each memory cell for around 16 hours (the total number of errors was 300K) when the memory cell voltage (V_{mc}) is 0.3 V. Fig. 9 shows the distribution of the number of soft errors in each memory cell in a single chip. This figure indicates that the number of errors varies bit by bit.

In order to clarify whether this variation is caused by statistical variation due to the spatial randomness of particle hits or by the critical charge variation, we performed a Monte Carlo simulation with the following procedure:

1) Let P_i be the occurrence probability of errors in the i th memory cell ($0 \leq i \leq 2047$).
2) An integer number n and a real number p are randomly generated. n and p satisfy $0 \leq n \leq 2047$ and $0 \leq p < 1$, respectively.

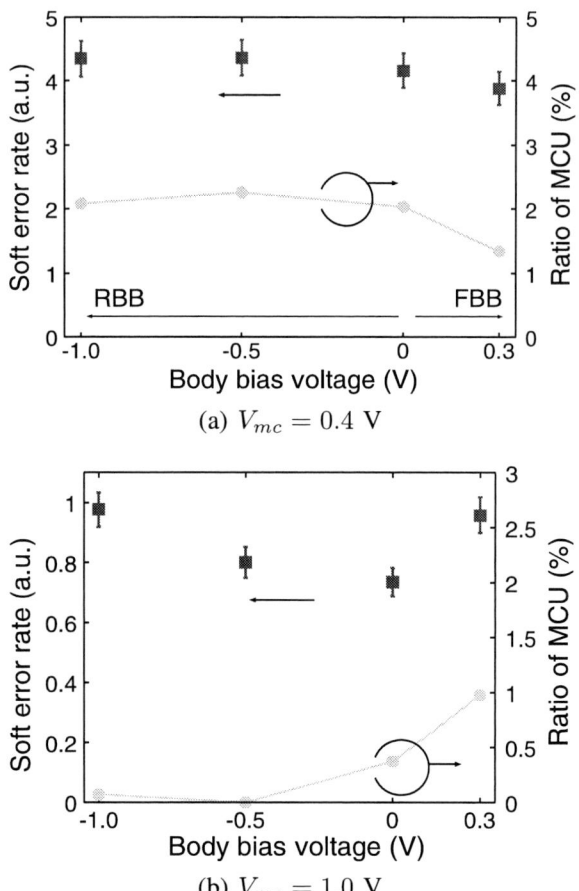

(a) $V_{mc} = 0.4$ V

(b) $V_{mc} = 1.0$ V

Fig. 8. Dependence of the SER and the ratio of MCUs on the body-bias voltage of the memory cells. "RBB" and "FBB" denote reverse body-bias and forward body-bias, respectively. Each error bar indicates $\pm 3\sigma$.

3) If p is less than P_n ($p < P_n$), we consider that an upset occurs in the n th memory cell.

4) Steps 2) and 3) are repeated until the total number of upsets reaches 300K. Consequently, the number of errors in each memory cell is obtained.

Fig. 10 shows the simulated distribution of the number of errors in each memory cell. Here, we assume that the occurrence probability of errors $P = \{P_0 P_1 \cdots P_{2047}\}$ is normally distributed with mean μ_P and standard deviation σ_P. In this simulation, μ_P is set to 0.5. Fig. 10-(a) depicts the simulated distribution in the case where σ_P is zero, that is, the occurrence probability of errors in each memory cell is the same. Fig. 10-(b) shows the distribution when σ_P/μ_P is 0.1. The distribution of Fig. 10-(a) is normally distributed, and its standard deviation is equivalent to the square root of the mean ($\sqrt{147.4} = 12.14$). This means that Fig. 10-(a) obeys the Poisson distribution.

If the variation in the number of errors is caused only by the spatial randomness of particle hits, the occurrence probability of errors in each memory cell is the same ($\sigma_P = 0$). In this case, the distribution of the number of errors follows the Poisson distribution as shown in Fig. 10-(a). On the other hand, if the critical charge fluctuation influences the variation

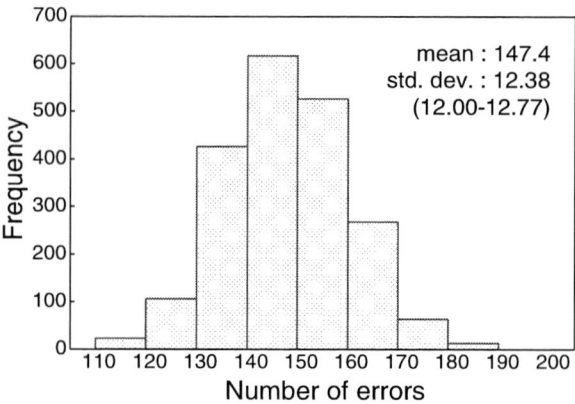

Fig. 9. Measured distribution of the number of soft errors in each memory cell in a single chip ($V_{mc} = 0.3$ V). The total number of errors is 300K. The values in parentheses represent the 95% confidence interval of the standard deviation.

in the number of errors in addition to the spatial randomness of particle hits, the distribution of the number of errors becomes different from the Poisson distribution like Fig. 10-(b).

The distribution of Fig. 9 is normally distributed, and its standard deviation is close to that of Fig. 10-(a). This implies that the both distributions are almost identical and follow the Poisson distribution. This means that the variation shown in Fig. 9 is explained by the spatial randomness of particle hits, and the influence of the critical charge variation was not distinguishably observed even at $V_{mc} = 0.3$ V.

IV. CONCLUSION

In this paper, we presented alpha-particle-induced SERs and MCUs in subthreshold region of a designed 10T SRAM in a 65-nm CMOS process. The measurement results showed that the SER increases as the supply voltage is lowered and SER at 0.3 V is eight times higher than SER at 1.0 V. We also pointed out that the ratios of MCUs to the total upsets in subthreshold region are much higher than those in super-threshold (nominal supply voltage) region, yet the conventional ECC is still valid because intra-word multiple upsets were not observed in subthreshold region. We additionally revealed that the SER and the ratio of MCU are less sensitive to body-bias at 0.4 V supply voltage. Furthermore, we investigated the influence of the critical charge variation due to manufacturing variability on the soft error immunity in each memory cell. Measurement results indicated that the influence of the critical charge variation was not recognizably observed even in subthreshold region.

ACKNOWLEDGMENTS

This research was performed by the authors for STARC as part of the Japanese Ministry of Economy, Trade and Industry sponsored "Next-Generation Circuit Architecture Technical Development" program. The VLSI chip in this study has been fabricated in the chip fabrication program of VLSI Design and Education Center (VDEC), the University of Tokyo in collaboration with STARC, e-Shuttle, Inc., and Fujitsu Ltd. This study was partly supported by NEDO.

978-1-4244-5430-3/10 $26.00 © 2010 IEEE

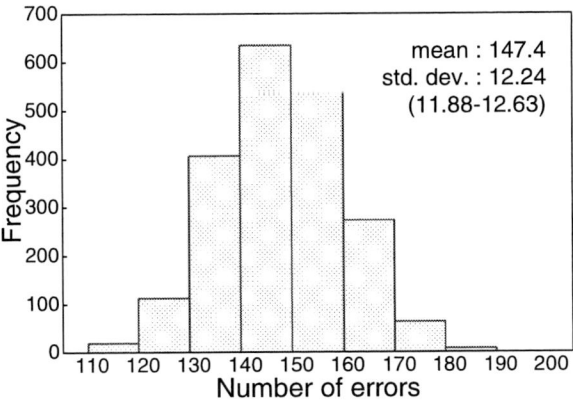

(a) $\sigma_P = 0$ (the occurrence probability of errors in each memory cell is the same)

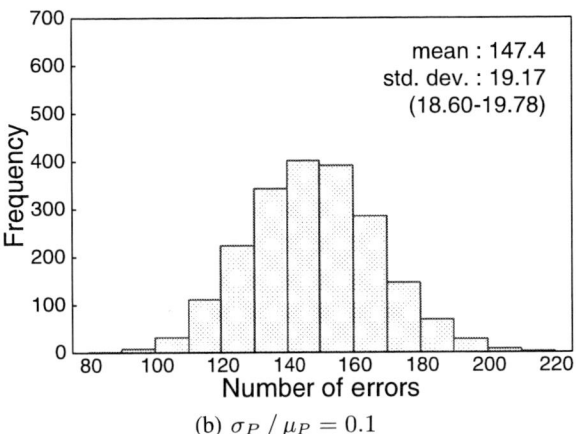

(b) $\sigma_P / \mu_P = 0.1$

Fig. 10. Simulated distribution of the number of errors in each memory cell. The total number of errors is 300K. The values in parentheses represent the 95% confidence interval of the standard deviation. The simulation assumes that the occurrence probability of errors is normally distributed with mean μ_P and standard deviation σ_P ($P \sim N(\mu_p, \sigma_P^2)$).

REFERENCES

[1] M. Seok, S. Hanson, Y. Lin, Z. Foo, D. Kim, Y. Lee, N. Liu, D. Sylvester, and D. Blaauw, "The Phoenix Processor: A 30pW Platform for Sensor Applications," in *Symp. VLSI Circuits Dig. Tech. Papers*, pp. 188–189, 2008.

[2] N. Ickes, D. Finchelstein, and A. Chandrakasan, "A 10-pJ/instruction, 4-MIPS Micropower DSP for Sensor Applications," in *Proc. ASSCC*, pp. 289–292, 2008.

[3] M.C. Casey, B.L. Bhuva, S.A. Nation, O.A. Amusan, T.D. Loveless, L.W. Massengill, D.McMorrow, J.S. Melinger, "Single-event effects on ultra-low power CMOS circuits," in *Proc. IRPS*, pp. 194–198, 2009.

[4] R. Garg and S.P. Khatri, "3D Simulation and Analysis of the Radiation Tolerance of Voltage Scaled Digital Circuit," in *Proc. ICCD*, pp. 498–504, 2009.

[5] P. Hazucha, T. Karnik, J. Maiz, S. Walstra, B. Bloechel, J. Tschanz, G. Dermer, S. Hareland, P. Armstrong, and S. Borkar, "Neutron Soft Error Rate Measurements in a 90-nm CMOS Process and Scaling Trends in SRAM from 0.25-mm to 90-nm Generation," in *IEDM Tech. Dig.*, pp. 21.5.1–21.5.4, 2003.

[6] Y. Tosaka, H. Ehara, M. Igeta, T. Uemura, H. Oka, N. Matsuoka, and K. Hatanaka, "Comprehensive Study of Soft Errors in Advanced CMOS Circuits with 90/130 nm Technology," in *IEDM Tech. Dig.*, pp. 38.3.1–38.3.4, 2004.

[7] A. Balasubramanian, P.R. Fleming, B.L. Bhuva, O.A. Amusan, and L.W. Massengill, "Effects of Random Dopant Fluctuations (RDF) on the Single Event Vulnerability of 90 and 65 nm CMOS Technologies," *IEEE Trans. Nucl. Sci.* vol. 54, no. 6, Dec. 2007.

[8] M. Yamaoka, N. Maeda, Y. Shinozaki, Y. Shimazaki, K. Nii, S. Shimada, K. Yanagisawa, and T. Kawahara, "Low-Power Embedded SRAM Modules with Expanded Margins for Writing," in *ISSCC Dig. Tech. Papers*, pp. 480–481, 2005.

[9] B. H. Calhoun and A. Chandrakasan, "A 256-kb 65-nm Sub-threshold SRAM Design for Ultra-Low-Voltage Operation," *IEEE J. Solid-State Circuits*, vol. 42, pp. 680–688, Mar. 2007.

[10] L. Chang, D.M. Fried, J. Hergenrother, J.W. Sleight, R.H. Dennard, R.K. Montoye, L. Sekaric, S.J. McNab, A.W. Topol, C.D. Adams, K.W. Guarini, and W. Haensch, "Stable SRAM Cell Design for the 32 nm Node and Beyond," in *Symp. VLSI Technology, Dig. Tech. Papers*, pp. 128–129, 2005.

[11] T.H. Kim, J. Liu, J. Keane, and C.H. Kim, "A High-Density Subthreshold SRAM with Data-Independent Bitline Leakage and Virtual Ground Replica Scheme," in *ISSCC Dig. Tech. Papers* pp. 330–331, 2007.

[12] N. Verma and A. Chandrakasan, "A 65nm 8T Sub-Vt SRAM Employing Sense-Amplifier Redundancy," in *ISSCC Dig. Tech. Papers*, pp. 328–329, 2007.

[13] T.H. Kim, J. Liu, and C.H. Kim, "A Voltage Scalable 0.26V, 64kb 8T SRAM with Vmin Lowering Techniques and Deep Sleep Mode," in *Proc. CICC* pp. 407–410, 2008.

[14] I.J. Chang, J.J. Kim, S.P. Park, and K. Roy, "A 32 kb 10T Sub-Threshold SRAM Array With Bit-Interleaving and Differential Read Scheme in 90 nm CMOS," *IEEE J. Solid-State Circuits*, vol. 44, no. 2, pp. 650–658, Feb. 2009.

[15] JEDEC standard JESD89, "Measurement and Reporting of Alpha Particles and Terrestrial Cosmic Ray-Induced Soft Errors in Semiconductor Devices."

[16] N. Mikami, A. Oyama, H. Kobayashi, H. Usui, and J. Kase, "A Novel Technique for Mitigating Neutron-Induced Multi-Cell Upset by means of Back Bias," in *Proc. IRPS*, pp. 187–191, 2008.

SEILA: Soft Error Immune Latch for Mitigating Multi-node-SEU and Local-clock-SET

Taiki Uemura, Yoshiharu Tosaka, Hideya Matsuyama and Ken Shono.
Fujitsu Microelectronics Ltd.,
1500 Mizono, Tado, Kuwana, Mie, 511-0192, Japan
phone: +81-594-24-2187, e-mail: uemura.taiki@jp.fujitsu.com

Chihiro J. Uchibori
Fujitsu Laboratories of America. Inc.,
Sunnyvale, CA, 94085-5401, USA

Keiji Takahisa, Mitsuhiro Fukuda and Kichiji Hatanaka.
Osaka University
10-1 Mihogaoka, Ibaraki, Osaka, 567-0047, Japan

Abstract—**We have developed a robust latch for achieving high reliability in LSI. The latch can attenuate multi-node single-event-upset (MNSEU) and single event transient on local-clock (SETLC). The robust latch has Dual-clock-buffers (DCB) and Double-height-cell (DHC) technologies. Results on neutron acceleration experiments show that DHC can dramatically attenuate MNSEU and DCB can protect almost SETLC of the latch. In addition, we investigate optimum design in well structure.**

Keywords: soft-error, charge sharing, SEU, SET, MNSEU, SETLC, local-clock.

I. INTRODUCTION

Radiation-induced soft-error rates (SERs) increase in modern LSI chips. Error correction code (ECC) is very effective countermeasure for soft-errors (SEs). However, it is difficult to apply ECC to logic circuits. SERs in logic circuits are significant problems in LSIs.

Redundancy-based latches having more than two storage nodes are used for decreasing SER in logic. [1][2][3][4] Storage data in these latches can not be corrupted by collecting charge on one node. They, however, can be corrupted by collecting charge on two or more nodes. There are four possible mechanisms of MNSEU as shown in Table 1. [5]

TABLE 1. POSSIBLE MECHANISM OF MNSEU. [5]

Symbol	Name
(A)	Successive hits by one ion
(B)	Multi hits by multi-ion
(C)	Drift/diffusion charge (charge sharing)
(D)	Parasitic bipolar action

Guard rings on latch cells can mitigate MNSEU. [6] [7] The cell height of unit cells in recent design is small such as nine to twelve grid height (one grid is half pitch [8]). Well-tap(tie)-less cell will be mainstream because of achieving large width of transistors with small height. In these design trends, it is difficult to use guard rings in commercial products.

Latches are controlled by clock signals distributed by clock trees. When single-event-transient (SET) occurs on clock buffers in clock trees, the buffer sends out the wrong signal to latches. This leads rewriting the data at unexpected timing. If input data is different with retention data in the latch, the retention data is rewritten the different data at the time. Gate size of last buffers on clock trees is smaller than upstream buffers in the trees. Small buffer is sensitive to SET. Generally, most latch cells have clock buffers at local (i.e. local-clock). We need to consider especially SETLC in latch. Contribution of SETLC increases as technologies advances. [9] Conventional robust latches can not mitigate SETLC.

In this paper, we have developed new technology to enhancing SER mitigation efficiency for attenuating MNSEU and SETLC in latches. For mitigating MNSEU, we develop design scheme of placement of critical area and cancelling area. We design the redundancy based latch with double height for achieving the design scheme. This technique is high affinity with existing standard cell libraries for commercial products. For mitigating SETLC in latches, we also design redundancy based clock buffer by splitting clock buffer of latch. If noise is occurred on one split clock buffer, retention data of the latch is not upset. Mitigation efficiency has been evaluated by using neutron acceleration test. In addition, we investigate optimum design in well structure through the experiment.

978-1-4244-5430-3/10 $26.00 © 2010 IEEE   218

II. MITIGATION TECHNIQUE

We have developed a soft-error immune latch (SEILA). The latch has inter-lock type storage structure as shown in Fig. 1. Data corruption on one node can not corrupt storage data on the storage structure. [1][4] The latch also includes two technologies, double-height-cell (DHC) and dual-clock-buffer (DCB) technologies for enhancing SER mitigation efficiencies. DHC technology can protect from MNSEU. DCB technology protects from corrupting storage data by SETLC on latches.

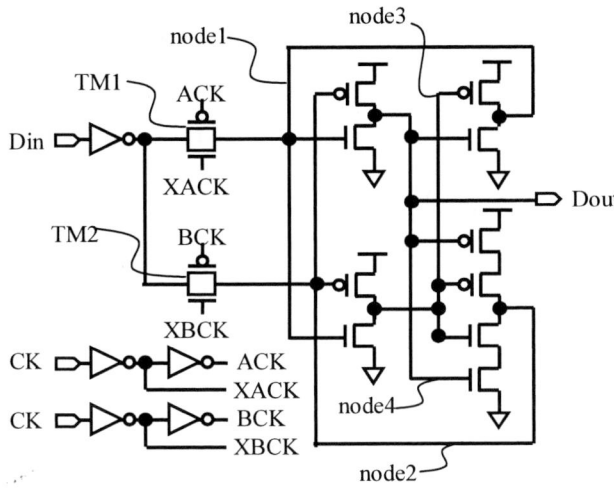

Figure 1. Schema of SEILA

A. Dual clock-buffer (DCB) technology

First, we evaluate SETLC of normal un-robust latches as shown in Fig. 2 through neutron acceleration test. This acceleration test has been carried out using spallation neutron beam in research center for nuclear physics (RCNP) at Osaka Univ. as shown in Fig. 3. We use three kind of the test chips, 65nm un-robust latches, 45nm un-robust latches, and 65nm robust latches [4]. Each chip includes more than 200kbit latches. Sixteen test chips are simultaneously irradiated.

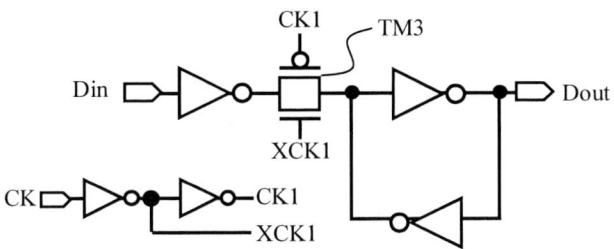

Figure 2. Schema of normal un-robust latch.

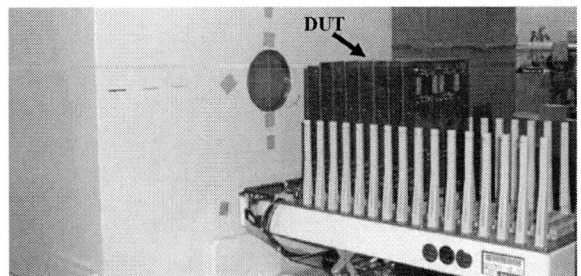

Figure 3. Setup of neutron acceleration experiment.

We have performed with following conditions;

TEST1: Clock signal (CK1 in Fig.2) is kept low after the data is written to the latch. The transfer gate (TM3 in Fig2) between the data retention node and the input terminal of the latch is off-state. The data in Din is kept the same as the retention data in the latch during neutron irradiation. Even when the gate is erroneously open, the data in the latch is not corrupted. SER only due to single-event-upset (SEU) on the retention data nodes is evaluated after irradiation in TEST1.

TEST2: With the transfer gate off condition, the data in Din is kept opposite to the retention data in the latch during neutron irradiation. When the gate is erroneously open, the data in the latch is corrupted. SER due to SEU on the retention data nodes and SETLC is evaluated after irradiation in TEST2.

SETLC can be evaluated from the difference between SERs in TEST1 and TEST2.

Figure 4 and 5 show SER of SEU on data node and SETLC in 65nm and 45nm-latches. Error bars in these figure show statistical error variations of 95% confidence level. SER on the vertical-axis on Fig. 4, 5 and 6 are the same scale. SER of SETLC is ten to fifteen percent of SER of the SEU in 65nm un-robust latches. Contribution of SETLC is not so large in 65nm technologies. That, however, is not negligible in 45nm technologies as shown in Fig. 5.

Generally, SER is different on DATA0 and DATA1 because data retention part of latch consists of un-balanced feed-back circuit and SER contribution is different between PMOS and NMOS. [10] In this experiment, SER difference between DATA0 and DATA1 in 45nm un-robust latch is larger than that in 65nm un-robust latch. The cause might be that beta ratio (size ratio on PMOS and NMOS) is large different between the 65nm and 45nm latches.

Figure 6 shows SERs of the SEU and SETLC in a 65nm robust latch [4]. Error bars in this figure show statistical error variations of 95% confidence level. This robust latch is inter-lock type latch, not SEILA. This latch is not included DCB and DHC technologies shown below. Schema of data retention part on the robust latch is the same with SEILA as shown in Fig. 1. The latch has only one local-clock buffer. The buffer connects to two transfer gates (TM1 and TM2 in Fig. 1).

978-1-4244-5430-3/10 $26.00 © 2010 IEEE   219

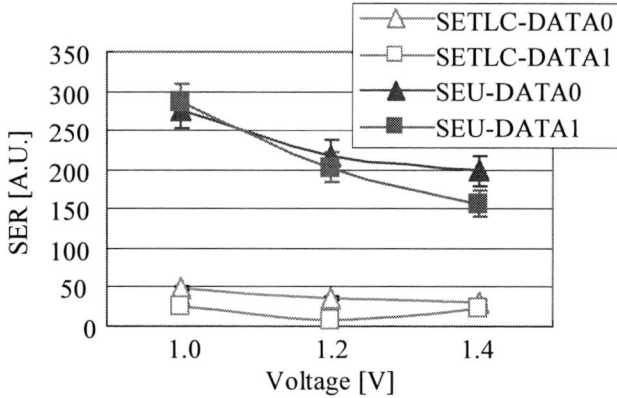

Figure 4. SER of SEU on data node and SETLC in retention data of all latches are 0 and 1 on un-robust latch as shown in Fig. 2 manufactured with 65nm technology. The error bars show statistical error variations of 95% confidence level.

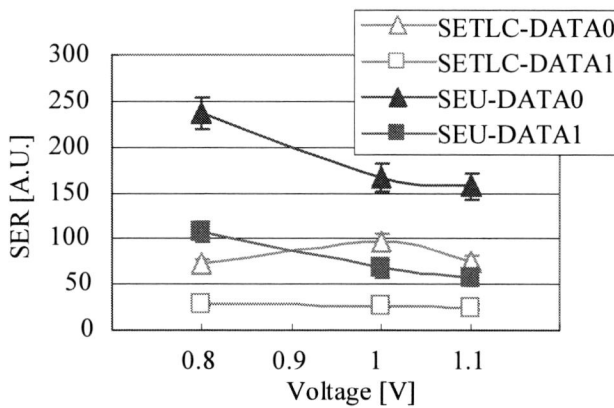

Figure 5. SER of SEU on data node and SETLC in retention data of all latches are 0 and 1 on un-robust latch as shown in Fig. 2 manufactured with 45nm technology. The error bars show statistical error variations of 95% confidence level.

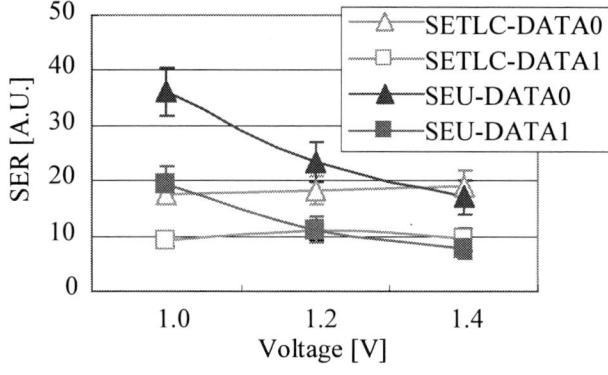

Figure 6. SER of SEU on data node and SETLC in retention data of all latches are 0 and 1 on inter-lock type robust latch [4] manufactured with 65nm technology. The error bars show statistical error variations of 95% confidence level.

SER of SEU decreases as operation voltage increases because electrical capacity increases as the voltage increases. The voltage change affects not only to electrical capacity but also to setup-time (necessary time for writing data) of latches. The setup-time shortens as the voltage increases. SER of SETLC increases as setup-time shorten. As a result, SER of SETLC is not so changed with the voltage changing.

SER of SETLC is not so changed from the SER of 65nm un-robust latches to 65nm robust latches although SER of SEU on retention data node decreases. The robust latch can mitigate SEU on retention data node, however, cannot mitigate SETLC. The contribution of SETLC is not negligible in robust latches on 65nm. So when we design high reliability LSI, we need to consider SETLC problem in not only 45nm but also 65nm technologies.

We develop DCB for mitigating SETLC. A latch with DCB has two clock-buffers and two transmission gates. Each clock-buffer controls each transmission gates independently. One transmission gate passes data to one storage node. Unless data are written simultaneously into both the two storage nodes, storage data of the latch is not changed. Therefore, if a noise is occurred in one clock-buffer in SEILA, storage data of the latch is not upset. DCB can protect SETLC of latches with small overhead (four transistors). When we change from 2X clock-buffer (consisting of two NMOS and two PMOS) to two 1X clock-buffers (consisting of a NMOS and a PMOS) in latch, the cell size is little changed.

If noise is occurred on clock buffers upstream from local-clock buffers (i.e. clock-trees), all latches connected with the clock buffer must be upset. Generally, buffer size in clock-trees is larger than that on local-clock. Additionally, a node in clock-trees is connected to many buffers and the node has larger electrical capacitance than that in local-clock. Therefore, large charge is necessary for changing voltage on clock-trees. As a result, SERs due to SET on clock-trees are very smaller than that on local-clock. We have not observed SET on clock-trees in the experiment.

B. Double height cell (DHC) technology

Collecting charge on one node can not corrupt storage data in inter-lock type latches. If charge is collected on critical two nodes, storage data of the latch can be corrupted. SER of MNSEU on the inter-lock type latch can be more than ten percent of SER of SEU on un-robust latch. [4]

When charge is collected in drain areas of PMOS and NMOS on the same node, voltage on the node is not so changed. [11] The area of PMOS connected to the same node is cancelling area for the area of NMOS.

We have estimated voltage variation using device and circuit simulation [4] in collecting charge on more than one drain area. When noises are occurred on two critical areas (NMOS on node1 and NMOS on node 2 in Fig. 1) in inter-lock type latch, retention data on the latch is upset as shown in

Fig. 7 (A). When noises are occurred on the two critical areas and canceling area (NMOS on node1, NMOS on node 2, and PMOS on node 3 in Fig. 1), the retention data is not upset as shown in Fig. 7 (B).

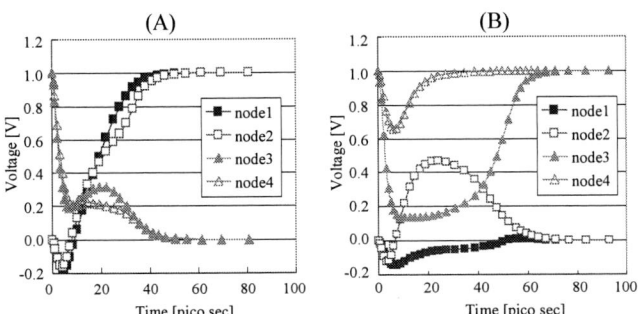

Figure 7. (A) Voltage variation when noises are occuerd on two critical drain areas (NMOS on node1 and NMOS on node 2 in Fig. 1). (B) Voltage variation when noises are occuerd on the two critical areas and one canncelling area (NMOS on node1, NMOS on node 2 and PMOS on node 3 in Fig. 1).

MNSEU due to (A) and (C) in Table 1 is attenuated by putting the cancelling area in position between the two critical areas as shown in Fig. 8. In this placement, one ion must not penetrate the two critical areas at the same time unless it penetrates the cancelling area. In addition, when charge is shared on the two critical areas, it must be shared also on the cancelling area.

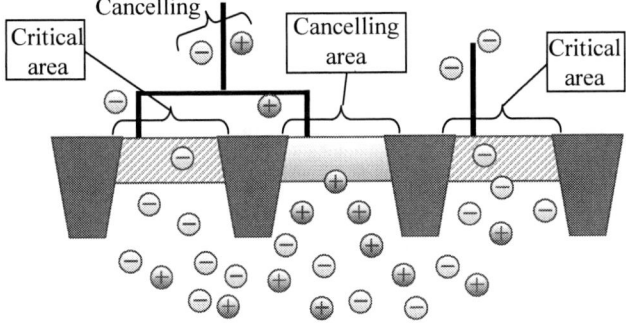

Figure 8. Cancelling area between two critical areas. Charge must not be collected in the two critical areas at the same time unless it is collected in the cancelling area.

For achieving the scheme of the area placement, we have designed the latch cell layout with double height. One critical area is put in upper side, another is put in lower side, and cancelling area is put between the two critical areas. The area size of DHC is not changed from single height cell, though area form is changed as shown in Fig. 9. DHC can mitigate MNSEU not only due to (A) and (C) in Table 1 but also parasitic bipolar effect as shown (D) in Table 1 because of well separation.

We also separately put two clock-buffers with the same way as shown in Fig. 10, for attenuating not only MNSEU but also SET in the two clock-buffers.

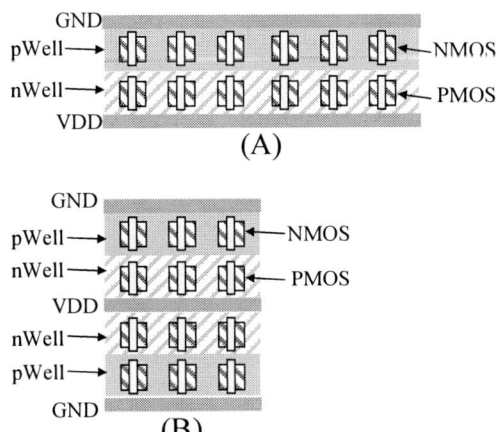

Figure 9. (A)single height cell and (B) Double height cell (DHC). Area sizes are not different on two cell height, though area shapse are different.

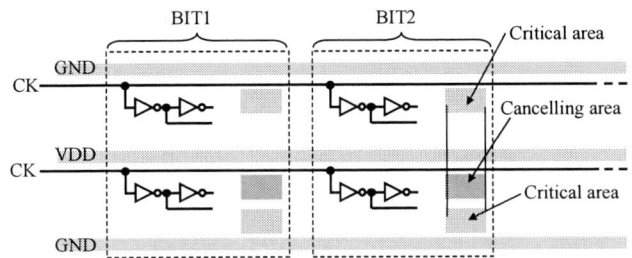

Figure 10. Double-Height-Cell (DHC) technology. Critical areas and canncelling areas are put as Fig. 8. Each height has a clock signal line. Latches have two clock buffers (DCB) and each clock signal line on the height is connected to each clock buffers on the height.

III. ACCELERATION TEST

We evaluate SER on SEILA and an un-robust latch as shown in Fig. 2 using spallation neutron beam in RCNP as shown in Fig. 3. We perform the experiment on 16 test chips including arrays of the un-robust latch and SEILA. The chip includes more than 200kbit and twin-well and triple-well area, designed with 65nm technology as shown in Fig. 11. PMOS is placed in the middle of the SEILA as shown in Fig. 9

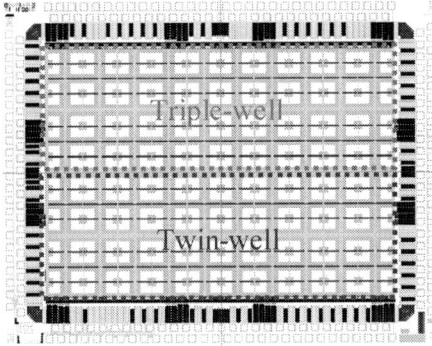

Figure 11. A layout of the test chip. The test chip includes twin-well and triple areas.

978-1-4244-5430-3/10 $26.00 © 2010 IEEE

Table 2 and Fig. 12 show SER in the un-robust latch as shown in Fig. 2 and SEILA. SERs in Fig.12 are average value of twin-well and triple-well. These SER is not included SETLC because these are experiment results under TEST1 condition.

SEILA can protect 99.3% of SEU. A robust latch on the same circuit without DHC can only protect 90% of SEU at 1V of operation voltage. [4] The DHC technology achieves very high mitigation efficiency against SEU with no area penalty. Well-structure dependence (i.e. twin and triple-well) of mitigation efficiency is a little on SEILA as shown in Fig. 13.

TABLE 2. SER OF LATCHES AT 1.0 V OPERATION VOLTAGE NORMALIZED WITH SER ON AN UN-ROBUST LATCH. THESE SERs ARE AVERAGE VALUE IN NEAR AND FAR POSITION LATCHES AS SHOWN IN FIG. 14.

	Normalized SER
Un-robust latch	1
Robust latch without DHC [4]	0.1020
SEILA (Twin-well)	0.0068
SEILA (Triple-well)	0.0073

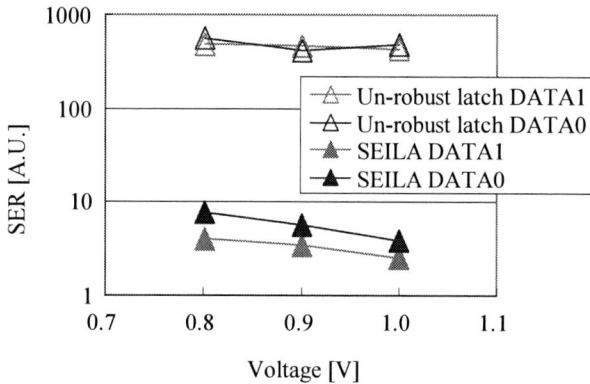

Figure 12. SER of the un-robust latch and SEILA. These SERs are average value in near and far position latches as shown Fig. 14.

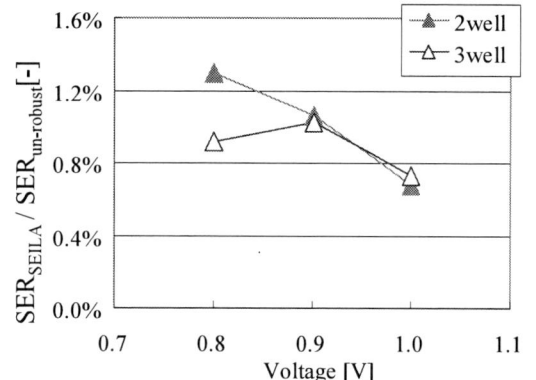

Figure 13. Mitigation efficiency. Ratio of SER on un-robust latch and SER on SEILA. These SERs are average value in near and far position latches as shown Fig. 14.

Well-tap(tie)-less cells are mainstream because of achieving large width of transistors with small height in recent design trend. Latches in the test chip are also well-tap-less

cells and need well-taps on out of latch cells. In the test chip, well-tap is put as shown in Fig. 14. There are latches which are near and far position from well-taps. The SERs in Table 2, Fig. 12 and 13 are average value at the near and far position. SER on far position latches are higher than on near position for both un-robust and SEILA as shown in Fig. 15. SERs in this figure are average value on DATA0 and DATA1. It is easer to change potential on well at far position from well-taps. Therefore, bipolar effects at far from well-tap are lager than that at near from well-tap. The position dependence may show that a major cause of MNSEU is parasitic bipolar effect as shown in Table 1 (D) and DHC can particularly mitigate the bipolar effect.

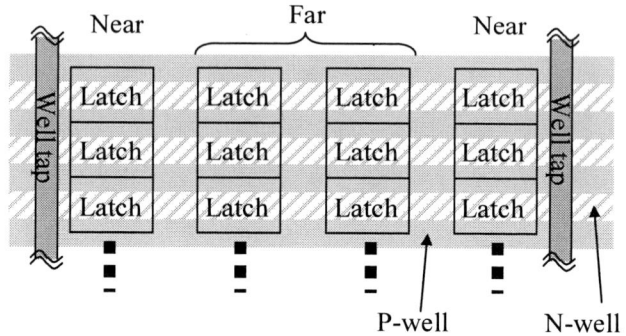

Figure 14. Layout of latches and well-taps in the test chips. There are four latches between well-taps. There are latches which are near and far position from well-taps.

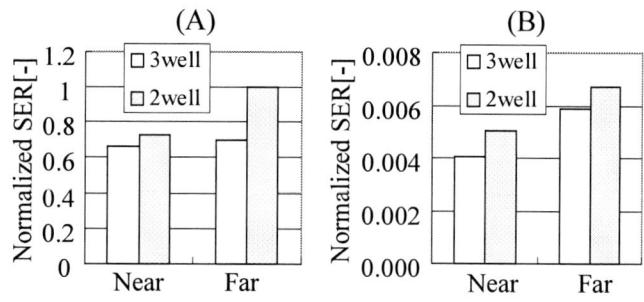

Figure 15. SER on (A) a un-robust latch and (B) SEILA on near and far position from well taps as shown in Fig. 14. These value are normalized with SER of un-robust latch on far position in twin-well. These SERs are average value on DATA1 and DATA0. SER on far position latches are higher than on near position.

In triple-well, electrons can diffuse from n-well to deep n-well and holes can not diffuse from p-well as shown in Fig. 16. An n-well connects other well-contacts in other n-well through deep n-well. It is difficult to change N-well potential and it is easy to change P-well potential in triple-well. Therefore, bipolar effects on PMOS are smaller than that on NMOS in triple-well

In twin-well, it is difficult to change P-well potential and it is easy to change N-well potential. Therefore, bipolar effects on NMOS are smaller than that on PMOS in twin-well Bipolar effects on PMOS in triple-well are larger than that on twin-well, and the effect on NMOS in triple-well are smaller than that on twin-well.

978-1-4244-5430-3/10 $26.00 © 2010 IEEE   222

PMOS is placed in the middle of the SEILA as shown in Fig. 9. All PMOS in the SEILA are included in one n-well. Therefore SER in triple-well is lower than SER in twin-well as shown in Fig. 15.

If NMOS is placed in the middle of SEILA, the efficiency on twin-well will be higher than triple-well. Optimum well layout design is different by well-structures. If SEILA is designed with triple height cell (THC), all PMOS and all NMOS in the SEILA are not included in one well. SEILA designed with THC will achieve higher mitigation efficiency than that with DHC.

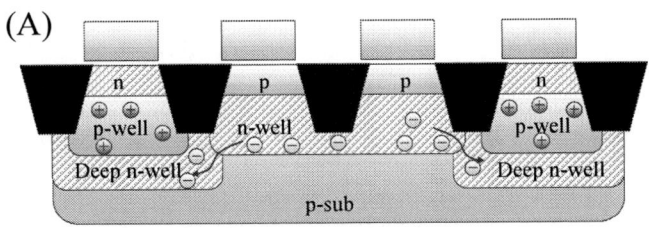

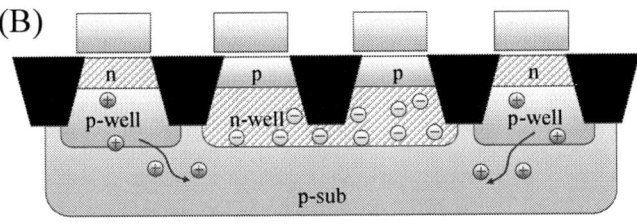

Figure 16. Electrons can diffuse from n-well to deep n-well and holes can not diffuse from p-well in (A) triple-well. An n-well connects other well-contacts in other n-well through deep n-well in triple-well. In (B) twin-well, electrons can not diffuse from n-well and holes can diffuse from p-well to p-substrate. An p-well connects other well-contacts in other p-well through p-substrate in twin-well.

We also evaluate SETLC on SEILA with the neutron acceleration test. Figure 17 shows error counts on TEST1 and TEST2 as we describe in session of "DCB technology" on SEILA applied DCB. Error bars in this figure show statistical error variations of 68% confidence level. There is no significant difference on TEST1 and TEST2 conditions. These results show that DCB can protect almost SETLC.

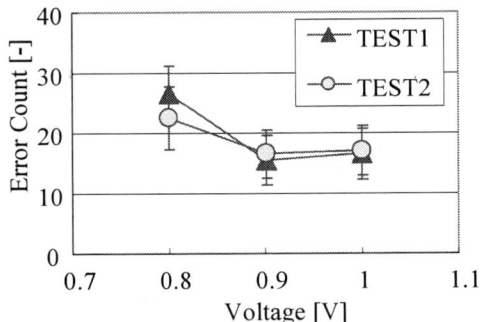

Figure 17. Error count in TEST1 and TEST2 for evaluating SEU on data node and SETLC in SEILA manufactured with 65nm technology. The error bars show statistical error variations of 68% confidence level.

IV. CONCLUSIONS

We have developed a new robust latch, SEILA for protecting many soft-error problems. SEILA includes DCB and DHC technologies. These technologies are high affinity with existing standard cell libraries for commercial products. SEILA can protect 99.3% SEU and almost SETLC. Optimum well layout design is different by well-structures. SEILA designed with triple height cell (THC) will achieve higher mitigation efficiency than that with DHC. DCB and DHC (THC) can be also used in other redundant base robust latches such as BISER [3] with little penalties.

ACKNOWLEDGE

The authors would like to thank several people. The latches were designed by Mr. A. Nishiwaki, Mr. M. Komaki, and Mr. N. Kakizawa of Fujitsu VLSI. The test chips are designed with advisements of Mr. M. Igeta of Fujitsu microelectronics. The acceleration test is carried out with Mr. R. Tanabe, Mr. T. Tsuruta and Mr. T. Kato of Fujitsu microelectronics. Fruitful discussions with Mr. H. Shimizu of Fujitsu microelectronics, Mr. H. Yamashita, Mr. T. Nakada and Mr. Y. Satsukawa of Fujitsu, are greatly appreciated. Finally, we would like to thank Dr. D. Kobayashi of Japan Aerospace Exploration Agency and Dr. S. Onoda of Japan Atomic Energy Agency for grateful discussion.

REFERENCES

[1] T. Calin, et. al., "Upset hardened memory design for submicron CMOS technology", *IEEE Trans. Nucl. Sci.,* vol. 43, no. 6, pp. 2874-2878, 1996.

[2] P. Hazucha, et. al., "Measurements and analysis of SER tolerant latch in a 90-nm dual-Vt CMOS process", *IEEE CICC*, pp. 617 - 620, 2003.

[3] M. Zhang, et. al., "Sequential Element Design With Built-In Soft Error Resilience," *IEEE Trans. on VLSI Systems,* Vol. 14, Issue 12, pp. 1368- 1378, 2006.

[4] T. Uemura, et. al., "Using low pass filters in mitigation techniques against single-event transients in 45nm technology LSIs" *IOLTS08*, pp. 117-122.

[5] E. Ibe, et. al., "Spreading Diversity in Multi-cell Neutron-Induced Upsets with Device Scaling," IEEE CICC 2006, pp. 437-444 (2006).

[6] B. D. Olson, el. al., "Analysis of parasitic PNP bipolar transistor mitigation using well contacts in 130 nm and 90 nm CMOS technology", Trans. Nuc. Sci. Vol. 54 No. 4. 2007.

[7] O. A Amusan, el. al., "Mitigation techniques for single-event-induced charge sharing in a 90-nm bulk CMOS process", Trans. Nuc. Sci. Vol. 54 No. 6. 2007.

[8] International Technology Roadmap for Semiconductors (ITRS) 2009 , (http://public.itrs.net/).

[9] N. Seifert et. al., "On the Scalability of Redundancy based SER Mitigation Schemes" *ICICDT07*, pp. 1-9.

[10] T. Heijmen, et. al., "A Comparative Study on the Soft-Error Rate of Flip-Flops from 90-nm Production Libraries", IEEE Trans. IRPS2006. pp. 204-211.

[11] Y. Kawakami, et. al. "Investigation of Soft Error Rate Including Multi-Bit Upsets in Advanced S U M . Using Neutron Irradiation Test and 3D Mixed-Mode Device Simulation", Tran. IEDM04, pp. 945-948, 2004.

Evolving MEMS Qualification Requirements

Andrew Olney

Worldwide Reliability and Product Analysis Department
Analog Devices, Inc.
Wilmington, Massachusetts USA
phone: 001-781-937-2362, andrew.olney@analog.com

Abstract—In the early 1990s, Microelectromechanical Systems (MEMS) products were qualified using a suite of application-specific stress tests that were previously developed and used for mechanical sensors. For example, automotive companies required MEMS accelerometers to pass the same stress tests as mechanical accelerometers, including shock, vibration, temperature cycling, acceleration, and operating life tests. Semiconductor companies augmented these traditional stress tests with other tests to assess the robustness of MEMS products, such as cyclic sensor deflection, ESD, and latch-up tests. The proliferation of MEMS products and applications, especially in consumer electronics, poses major challenges to MEMS suppliers qualifying products for a wide range of applications and operating environments. Unlike automotive companies, many consumer electronics companies do not have sensor-specific reliability requirements; instead, they increasingly impose the same stress test requirements specified for all other ICs. Products with MEMS elements directly exposed to the outside world (such as microphones) may have intrinsic failure mechanisms that preclude passing all industry-standard stress tests. In addition, these standard stress tests do not adequately assess the reliability of MEMS products in terms of sensor robustness to failure mechanisms such as stiction, fracture, and particles. Thus, MEMS suppliers and their customers sometimes need to develop and agree upon a set of stress tests and acceptance criteria that are tailored to a particular category of MEMS products. This paper will focus on the qualification requirements for four categories of MEMS products used in diverse applications and operating environments, and then will discuss likely future trends and challenges in qualifying MEMS products.

Keywords-MEMS, FMEA, IC, qualification, reliability

I. INTRODUCTION

When semiconductor companies first began qualifying MEMS products approximately twenty years ago, no qualification standards existed for doing so. At the time, the largest market for accelerometers was for automotive airbag deployment modules. Most automotive Original Equipment Manufacturers (OEMs) using accelerometers did not have MEMS-specific reliability requirements; they expected that the emerging MEMS accelerometers would pass the same stringent set of stress tests used to qualify mechanical accelerometers. These component-level stress tests typically included: mechanical shock testing; variable-frequency vibration testing; drop testing; temperature cycle testing; and constant acceleration testing. Specific test conditions were negotiated and agreed upon between MEMS suppliers and automotive OEMs, and were developed to ensure that MEMS accelerometers would remain fully functional through module assembly processes and automotive assembly processes, and, most importantly, would be highly reliable through at least 20 years of application in harsh under-the-hood operating conditions. In addition, semiconductor companies used Failure Modes and Effects Analysis (FMEA) to identify other appropriate MEMS-specific stress tests, such as conducting millions of cycles of sensor deflection testing to ensure that sensors would not mechanically fail due to fatigue or other mechanisms. Finally, a subset of standard semiconductor qualification tests was used to qualify MEMS accelerometers, such as: high temperature operating life test, low temperature operating life test, and ESD and latch-up robustness tests.

As a specific example of the above qualification approach, in the early 1990s, when Analog Devices, Inc. (ADI) was developing the ADXL50, ADI's first MEMS product, the reliability qualification plan was developed using the following inputs:

- The ADXL50 50 g Monolithic Accelerometer with Signal Conditioning data sheet [1];

- The design, fabrication process, and package FMEAs;

- ADI's general specification for qualifying ICs;

- JEDEC Standard JESD34, "Failure-Mechanism-Driven Reliability Qualification of Silicon Devices" [2];

- Details on OEM manufacturing and application environments for mechanical accelerometers;

- Automotive OEM reliability requirements for mechanical accelerometers.

Potential MEMS accelerometer failure mechanisms, including sensor stiction, sensor impediment due to particles, sensor warpage, and sensor fracture, were identified, and stress tests to accelerate these mechanisms were developed. The resulting qualification plan, which was successfully executed on the single-chip ADXL50 accelerometer, is provided in Table 1 [3]. The ADXL50 samples were fabricated on a BiCMOS process with an integrated polysilicon MEMS sensor consisting of a center member with fingers interdigitated between fixed plates. The samples were assembled in a 10-pin hermetic TO-100 package using a low-stress die attach material. Samples were randomly pulled from three different wafer fabrication and assembly lots for stress tests that were selected to accelerate the potential failure mechanisms listed above.

978-1-4244-5430-3/10 $26.00 © 2010 IEEE

TABLE I.   MEMS ACCELEROMETER QUALIFICATION PLAN FROM 1993

Stress Test	Qualification Vehicle: ADXL50 50 g Accelerometer in a TO-100 Package		
	Conditions	No. of Lots	Sample Size per Lot
Mechanical shock	0.5 ms pulsed shocks at 500 g to 1500 g conducted at -40 °C, +25 °C, and +105 °C in 3 package spatial axes	3	90
Variable frequency vibration	50 g pulses applied from 20 Hz to 2000 Hz (logarithmically swept) at +25 °C	3	45
Mechanical drop	4000 g minimum shocks from 1.0 m drops onto concrete in 3 package spatial axes	3	45
Thermal shock	1000 cycles -65 °C to +150 °C (liquid-to-liquid) per MIL-STD-883 Method 1011 Cond. C [4]	3	45
Temperature cycle plus constant acceleration	1000 cycles -65 °C to +150 °C (air-to-air) per MIL-STD-883 Method 1010 Cond. C followed by 30,000 g acceleration in the +z axis (perpendicular to the die surface) [4]	3	77
High temperature operating life	2000 hours at +125 °C T_A with V_S = 6.0 V per MIL-STD-883 Method 1005 [4]	3	116
Low temperature operating life	1000 hours at -55 °C T_A with V_S = 6.0 V	1	45
High temperature storage	1000 hours at +150 °C T_A per MIL-STD-883 Method 1008 [4]	1	45
Low temperature storage	1000 hours at -40 °C T_A	1	45
Mechanical life test	1000 hours at +25 °C and V_S = 6.0 V with ~3.6 million cycles of maximum sensor deflection	1	30
Human Body Model ESD test	Up to 2000 V per MIL-STD-883 Method 3015 [4]	1	12
Latch-up test	Per JEDEC Standard JESD17 [5]	1	6

II.   EVOLUTION OF MEMS QUALIFICATION REQUIREMENTS

As more semiconductor companies have entered the MEMS business and qualified products during the past 15 years, qualification requirements have evolved, but they are not standardized. This is not surprising given the very wide range of MEMS fabrication processes, packages, product categories, and end-customer applications. As the number and variety of customers for MEMS products proliferated, semiconductor companies tailored qualification plans to match the reliability expectations for a particular category of products. It became clear that no "one size fits all" qualification plan could be applied to all MEMS products. However, qualification testing is becoming increasingly standardized for particular families of MEMS products. Qualification plans are now better aligned with industry-standard semiconductor reliability stress tests since such tests have proven effective at predicting field reliability performance of both ICs and MEMS products. The demonstrated ability of MEMS products to pass most or all the same stress tests that ICs pass has accelerated the use of MEMS products in a wide range of electronics applications. The following sections discuss some of the reliability considerations for MEMS microphones, MEMS gyroscopes, and MEMS switches.

III.   QUALIFICATION OF MEMS MICROPHONES

MEMS microphones have been commercially available for nearly a decade and are offered by a number of semiconductor companies. As with traditional electret microphones, MEMS microphone applications include cell phones, headsets, digital cameras and camcorders, handheld music players, teleconferencing systems, and noise cancellation systems. Most MEMS microphones are offered in surface-mount Land Grid Array (LGA) packages consisting of a laminate base with a metal lid. A sound port (round hole) is made through the laminate base to directly expose the MEMS microphone inside

the package to sound waves. While this package design optimizes the acoustical performance of the microphone, it presents additional challenges in terms of board manufacturing as well as qualification and reliability performance. For example, most MEMS microphones require dry-pack, and product data sheets and application notes provide details on how to properly mount these packages without contaminating or otherwise degrading the MEMS element [6, 7].

Due to their design, MEMS microphones generally cannot pass all industry-standard IC stress tests. Any immersion in a liquid or prolonged exposure to 100% RH can result in the very thin diaphragm sticking to the thicker backplate due to capillary action, rendering the microphone non-functional unless this stiction recovers. Thus, the qualification suite for MEMS microphones does not include autoclave or other testing at 100% RH. MEMS microphones are generally designed to survive very severe mechanical abuse and extremely high sound levels, making them suitable for most consumer and industrial applications. For example, the ADMP421 omni-directional microphone has mechanical shock and Sound Pressure Level (SPL) Absolute Maximum Ratings of 20,000 g and 160 dB, respectively [6]. A key stress test for MEMS microphones is temperature-humidity-bias (THB), which is generally conducted for 1000 hours at +85°C and 85% RH with bias applied as per JESD22-A101C [8]. While an electrical potential is applied across the MEMS diaphragm and backplate, this stress test directly exposes these critical structures to high temperature and humidity in a non-condensing environment. THB testing should be augmented by saliva resistance testing to simulate human spittle entering the sound port during microphone use. Unfortunately, however, no industry standard exists for such testing. Potential THB and saliva resistance testing failure mechanisms include bond pad corrosion as well as diaphragm warping, deflection, or stiction to the backplate.

TABLE II. TYPICAL MEMS MICROPHONE QUALIFICATION PLAN

Stress Test	Qualification Vehicle: ADMP421 Microphone in a Land Grid Array Cavity Package with Sound Port		
	Conditions	No. of Lots	Sample Size per Lot
Mechanical shock - powered[a]	0.1 ms pulsed shocks at 10,000 g conducted at +25 °C in +x, -x, +y, -y, +z, -z axes	3	25
Endurance life test - powered[a]	96 hours with a continuous Sound Pressure Level of 130 dB	1	42
Temperature cycle	1000 cycles -40 °C to +125 °C (air-to-air) with 1 cycle per hour per JESD22-A104D [9]	3	77
Mechanical drop	10 drops from 1.2 m conducted at +25 °C in 3 package spatial axes	3	25
Mechanical shock + vibration, variable frequency + constant acceleration	Per MIL-STD-883 Method 5005 Group D Subgroup 4: Shock per Method 2002 Cond. B (1500 g, 0.5 ms, half-sine, 5 shocks in +x, -x, +y, -y, +z, -z axes); Vibration per Method 2007 Cond. B (50 g, 20-2000 Hz, 3 axes); Acceleration per Method 2001 (30,000 g, 1 min., +y axis) [4]	3	24
Early life failure rate	48 hours at +125 °C T_A with V_{DD} = 3.6 V per JESD22-A108C [10]	3	667
High temperature operating life[b]	1000 hours at +125 °C T_A with V_{DD} = 3.6 V per JESD22-A108C [10]	3	77
Low temperature operating life[b]	1000 hours at -40 °C T_A with V_{DD} = 3.6 V per JESD22-A108C [10]	3	77
High temperature storage	1000 hours at +150 °C T_A per JESD22-A103C [11]	3	77
Low temperature storage	1000 hours at -40 °C T_A per JESD22-A119 [12]	3	45
Temperature-humidity-bias[b]	1000 hours at +85 °C T_A and +85% RH with V_{DD} = 3.6 V per JESD22-A101C [8]	3	77
Human Body Model ESD test	±500 V, ±1000 V, ±1500 V, and ±2000 V per ANSI/ESD STM5.1-2007 [13]	1	12
Charged Device Model ESD test	±125 V, ±250V, and ±500 V per ANSI/ESD STM5.3.1-2009 [14]	1	15
Latch-up test	Per JESD78B [15]	1	6

a. These stress tests are conducted with the MEMS microphone samples mounted to a Printed Circuit Board.

b. The samples for these stress tests were solder heat preconditioned per IPC/JEDEC J-STD-020D [16].

Semiconductor companies also conduct other specific stress tests to assess the reliability of the MEMS diaphragm and backplate, as well as the overall package. These tests include: endurance life testing in which the microphone is continuously exposed to an extreme SPL; drop testing (generally from a height of at least 1 m); powered mechanical shock testing; and a sequence of mechanical stress tests, such as those specified in MIL-STD-883 Method 5005.14 Group D Subgroup 4: mechanical shock; vibration, variable frequency; and constant acceleration [4]. Finally, the qualification suite for MEMS microphones includes additional appropriate industry-standard IC stress tests. For example, Table II provides the qualification plan for ADI's open-market MEMS microphones. These microphones are composed of a sensor die and a CMOS ASIC die in a 3-, 5-, or 6-terminal LGA package with a sound port. The sensor is fabricated on a modified CMOS process and consists of a thin polysilicon diaphragm suspended by springs above a thicker silicon back plate.

IV. QUALIFICATION OF MEMS GYROSCOPES

As with their predecessor MEMS accelerometers, MEMS gyroscopes are used to sense motion, but in this case the motion involves the angular rate of an object. Unlike MEMS accelerometers, where the MEMS element is static (not in motion) when the accelerometer is at rest, MEMS gyroscopes typically incorporate a resonator circuit that continuously drives portions of the MEMS structures to resonance whenever power is applied. The resonating sensor structures establish a velocity element that then produces a Coriolis force in response to angular rate. The presence of continuously resonating MEMS structures results in continuous self-testing of much of the overall MEMS sensor, which is useful for identifying any unexpected sensor behavior during normal operation. Due to their design, gyroscopes are particularly sensitive to micron-scale particles as well as package stress. Susceptibility to particles can be minimized by wafer-level capping of the MEMS elements in a clean room environment. Susceptibility to package stress can be addressed by assembling gyroscopes in hermetic ceramic packages or in cavity plastic packages [17, 18]. These packages do not result in any solid material directly in contact with the sensor die surface, thus eliminating potential sources of die stress including point stress from mold compound fillers, and bulk stress from coefficient of thermal expansion differences between the die and mold compound.

An important measure of the component-level robustness of a MEMS sensor is its maximum shock rating. This rating provides an indication of the susceptibility of a MEMS sensor to damage during manufacturing operations or due to mishandling such as dropping the sensor component onto a work floor. (Once in most customer applications, MEMS sensors are buffered from extreme shock events, and thus they are rarely subjected to shocks that can cause damage.) The ratio of the sensor mass to the anchor area on gyroscopes is typically higher than that for accelerometers and microphones. Due to the mass to anchor ratio, as well as other geometric factors, MEMS microphones and MEMS accelerometers

TABLE III.   TYPICAL MEMS GYROSCOPE QUALIFICATION PLAN

Stress Test	Qualification Vehicle: ADXRS610 Gyroscope in a Ceramic BGA Package		
	Conditions	No. of Lots	Sample Size per Lot
Mechanical shock - powered[a]	0.5 ms pulsed shocks at 2,000 g conducted at +25 °C in 3 package spatial axes	3	25
Mechanical drop	10 drops from 1.2 m conducted at +25 °C in 3 package spatial axes	3	25
Mechanical shock + vibration, variable frequency + constant acceleration	Per MIL-STD-883 Method 5005 Group D Subgroup 4: Shock per Method 2002 Cond. B (1500 g, 0.5 ms, half-sine, 5 shocks in +x, -x, +y, -y, +z, -z axes); Vibration per Method 2007 Cond. B (50 g, 20-2000 Hz, 3 axes); Acceleration per Method 2001 (10,000 g , 1 min., +y axis) [4]	3	25
Temperature cycle	1000 cycles -55°C to +125 °C (air-to-air) per JESD22-A104D [11]	3	77
Early life failure rate	48 hours at +150 °C T_A with $AV_{CC} = V_{DD} = 6.0$ V per JESD22-A108C [10]	3	667
High temperature operating life	1000 hours at +150 °C T_J with $AV_{CC} = V_{DD} = 6.0$ V per JESD22-A108C [10]	3	77
Low temperature operating life	1000 hours at -40 °C T_A with $AV_{CC} = V_{DD} = 6.0$ V per JESD22-A108C [10]	3	77
High temperature storage	1000 hours at +150 °C T_A per JESD22-A103C [11]	3	77
Low temperature storage	1000 hours at -40 °C T_A per JESD22-A119 [12]	3	45
Human Body Model ESD test	±500 V, ±1000 V, ±1500 V, and ±2000 V per ANSI/ESD STM5.1-2007 [13]	1	12
Charged Device Model ESD test	±125 V, ±250 V, and ±500 V per ANSI/ESD STM5.3.1-2009 [14]	1	15
Latch-up test	Per JESD78B [15]	1	6

a. This stress test is conducted with the MEMS gyroscope samples mounted to a Printed Circuit Board.

generally pass mechanical shock stress conditions that are much more severe than those used to qualify MEMS gyroscopes. MEMS gyroscopes, such as the ADXRS610 ± 300 degree/second angular rate sensor, typically have shock survival ratings of 2,000 g in any axis, both when powered and unpowered [19]. In comparison, MEMS accelerometers, such as the ADXL335 ± 3 g, 3-axis accelerometer, typically have shock survival ratings of 10,000 g in any axis, both when powered and unpowered [20]. Other than the lower shock levels applied, qualification plans for MEMS gyroscopes are similar to those for current-generation MEMS accelerometers. Table III provides the qualification plan used to release ADI's MEMS gyroscopes in hermetic ceramic packages. The qualification samples were fabricated on a BiCMOS process with an integrated polysilicon MEMS sensor. The samples were assembled in a 32-ball hermetic Ball Grid Array (BGA) package using a low-stress die attach material. Qualification plans for MEMS gyroscopes in cavity plastic packages are similar, but include industry-standard stress tests for non-hermetic packages, i.e., autoclave along with temperature-humidity-bias (THB) and/or highly accelerated stress test (HAST).

V.   QUALIFICATION OF MEMS SWITCHES

MEMS switches and MEMS relays differ from the preceding MEMS products in that they are not sensing real-world phenomena, but rather are responding to an electrical stimulus. They offer advantages over conventional semiconductor switches and mechanical switches due to their combination of negligible off-state leakage current and capacitance, minimal on-state resistance, and extremely fast switching times. Although their designs vary significantly, a

MEMS switch generally consists of a metal beam with an anchored end comprising the source terminal, and an opposite end with a metal tip suspended above the drain terminal. Application of a positive voltage at the gate (relative to the source) causes the suspended beam to be electrostatically attracted to the gate, causing the beam tip to contact the drain. The thicknesses and compositions of the metal layers used for both the beam and the switch terminals are carefully selected to maximize electrical performance and durability while minimizing contact resistance, beam stiction, beam deformation, and material wear.

As with other MEMS products, mechanical robustness of MEMS switches is paramount to their overall quality, reliability, and robustness levels. Therefore, the set of qualification tests for MEMS switches should include drop testing and a sequence of mechanical stress tests, such as those specified in MIL-STD-883 Method 5005 Group D Subgroup 4. Depending on their intended application, MEMS switches must pass all electrical data sheet limits after millions or even billions of switching cycles at their rated maximum current levels. Qualification plans for MEMS switches therefore include endurance life tests to check for failure mechanisms including contact resistance degradation, material fatigue, and stiction. Table IV shows the qualification plan that ADI will use to qualify RF MEMS switch products that are currently in development. These products consist of a MEMS switch die with a silicon cap over it and a CMOS ASIC die, with both die packaged in a 24-terminal Leadframe Chip-Scale Package (LFCSP), also commonly referred to as a Quad Flat No-Lead (QFN) package. Since this is a plastic package, the qualification plan includes industry-standard stress tests for non-hermetic packages, i.e., autoclave, THB, and HAST.

978-1-4244-5430-3/10 $26.00 © 2010 IEEE

TABLE IV.   TYPICAL MEMS SWITCH QUALIFICATION PLAN

Stress Test	Qualification Vehicle: RF MEMS Switch in a 24-Terminal Leadframe Chip-Scale Package		
	Conditions	*No. of Lots*	*Sample Size per Lot*
Endurance life test - powered[a]	10^{10} actuation cycles conducted at +85 °C with V_{DD} = 3.6 V	1	25
Mechanical drop	10 drops from 1.2 m conducted at +25 °C in 3 package spatial axes	3	25
Mechanical shock + vibration, variable frequency + constant acceleration	Per MIL-STD-883 Method 5005 Group D Subgroup 4 [4]	3	25
Temperature cycle[b]	1000 cycles -55°C to +125 °C (air-to-air) per JESD22-A104D [9]	3	77
Early life failure rate	48 hours at +125 °C T_A with V_{DD} = 3.6 V per JESD22-A108C [10]	3	667
High temperature operating life[b]	1000 hours at +125 °C T_A with V_{DD} = 3.6 V per JESD22-A108C [10]	3	77
Low temperature operating life	1000 hours at -40 °C T_A with V_{DD} = 3.6 V per JESD22-A108C [10]	3	77
High temperature storage	1000 hours at +150 °C T_A per JESD22-A103C [11]	3	77
Low temperature storage	1000 hours at -40 °C T_A per JESD22-A119 [12]	3	45
Autoclave[b]	168 hours at +121 °C T_A, +100% RH, and 2 Atmospheres per JESD22-A102 [21]	3	77
Temperature-humidity-bias[b]	1000 hours at +85 °C T_A and +85% RH with V_{DD} = 3.6 V per JESD22-A101C [8]	3	77
Highly accelerated stress test[b]	96 hours at +130 °C T_A, +85% RH, and 2 Atmospheres with V_{DD} = 3.6 V per JESD22-A110C [22]	3	77
Human Body Model ESD test	±500 V, ±1000 V, ±1500 V, and ±2000 V per ANSI/ESD STM5.1-2007 [13]	1	12
Charged Device Model ESD test	±125 V, ±250 V, and ±500 V per ANSI/ESD STM5.3.1-2009 [14]	1	15
Latch-up test	Per JESD78B [15]	1	6

a. This stress test is conducted with the MEMS switch samples mounted to a Printed Circuit Board.

b. The samples for these stress tests were solder heat preconditioned per IPC/JEDEC J-STD-020D [16].

VI.   LOOKING TO THE FUTURE

As MEMS products penetrate into more and more applications, OEMs and Contract Manufacturers (CMs) using MEMS products will increasingly expect them to have quality, reliability, and robustness levels that are comparable to standard ICs. This will ensure that the incorporation of MEMS products into system designs will not degrade the overall reliability of end applications. OEMs/CMs will therefore require that MEMS products pass all the stress tests specified in their General Procurement Specifications (GPS's) for semiconductor components. ADI's review of twenty-five GPS's from a broad range of automotive, consumer, communications, and industrial companies showed significant overlap in the reliability qualification requirements for ICs. (Company-specific GPS's for semiconductor components are generally covered by non-disclosure or other confidentiality agreements, so they are not listed in the References section.) Table V lists the most common component-level stress tests, associated conditions, and sample sizes based on this review of GPS's. While many MEMS products are capable of passing most or all these stress tests, this stress test suite does not include MEMS-specific stresses such as powered mechanical shock and mechanical drop that are required to assess the robustness and reliability of MEMS elements. Thus, OEMs/CMs that rely solely on GPS requirements for procuring MEMS products may not achieve the quality, reliability, and robustness levels they expect. Consequently, OEMs/CMs need

to work directly with MEMS suppliers to identify and agree upon additional stress tests to demonstrate that MEMS products will not degrade through PCB-level and system-level assembly and test operations, and will remain reliable in *specific* end-customer applications. These manufacturability and reliability assessments require that MEMS suppliers develop technical expertise in board manufacturing processes and OEM applications, and that OEMs/CMs share detailed information about their specific manufacturing processes and applications with MEMS suppliers. This mutual working relationship will become even more critical as OEMs/CMs increasingly expect MEMS suppliers to conduct qualification work at the PCB-level rather than at the component-level.

Challenges remain in developing and releasing reliable, robust, and cost-effective MEMS products that can pass all the qualification tests previously outlined, including those specified in GPS's for semiconductor components. As this paper has shown, qualification tests have evolved as MEMS products and applications have proliferated, reliability-related failure mechanisms were discovered, and appropriate stress tests were developed. This process continues, though consumer customers usually have less developed return systems than automotive customers that systematically return rejects to MEMS suppliers for Failure Analysis (FA). MEMS suppliers need consumer customers to modify their systems to facilitate the return of rejects in a manner that preserves unknown failure mechanisms, especially those in new

TABLE V.   TYPICAL OEM & CM REQUIREMENTS FOR QUALIFYING INTEGRATED CIRCUITS

Stress Test	Component-Level Stress Tests for ICs (without Embedded EEPROM)		
	Conditions	No. of Lots	Sample Size per Lot
Highly accelerated stress test (HAST)[a] ----------- or -----------	96 hours at +130 °C T_A, +85% RH, and 2 Atmospheres with $V_{DD} \geq V_{MAX}$ or 264 hours at +110 °C T_A, +85% RH, and 2 Atmospheres with $V_{DD} \geq V_{MAX}$ per JESD22-A110C [22] ----------- or -----------	3	25 to 77
Temperature-humidity-bias[a]	1000 hours at +85 °C T_A and +85% RH with $V_{DD} \geq V_{MAX}$ per JESD22-A101C [10]	3	25 to 77
Unbiased HAST for laminate-based packages[a]	96 hours at +130 °C T_A, +85% RH, and 2 Atmospheres or 264 hours at +110 °C T_A, +85% RH, and 2 Atmospheres per JESD22-A118	3	25 to 77
Autoclave for leadframe-based packages[a]	168 hours at +121 °C T_A, +100% RH, and 2 Atmospheres per JESD22-A102 [21]	3	25 to 77
Temperature cycle[a]	1000 cycles -55°C to +125 °C (air-to-air) per JESD22-A104D [9]	3	25 to 77
Early life failure rate	48 hours at +125 °C per JESD22-A108C [10]	3	400 to 800
High temperature operating life	1000 hours at $\geq$ +125 °C T_J with $V_{DD} \geq V_{MAX}$ per JESD22-A108C [10]	3	25 to 77
Low temperature operating life	1000 hours at -40 °C T_A with $V_{DD} \geq V_{MAX}$ per JESD22-A108C [10]	3	25 to 77
High temperature storage	1000 hours at +150 °C T_A per JESD22-A103C [11]	1	25 to 77
Low temperature storage	1000 hours at -40 °C T_A per JESD22-A119 [12]	1	25 to 77
Human Body Model ESD test	Minimum 1000 V or 2000 V per JESD22-A114F [23]	1	12
Charged Device Model ESD test	Minimum 500 V per JESD22-C101E [24]	1	9
Latch-up test	Per JESD78B [15]	1	6

a. The samples for these stress tests were solder heat preconditioned per IPC/JEDEC J-STD-020D [16].

applications, so that FA is able to identify new failure mechanisms. Based on the FA results, qualification stress test suites and reliability monitor programs can be enhanced to minimize field failures. In conjunction with this largely reactive feedback loop, predictive MEMS research is required, particularly in the area of packaging, and more specifically the interactions between MEMS elements and ever cheaper and smaller form-factor packages. Packaging of MEMS has always presented technical and reliability challenges [25, 26]. This will continue to be the case, especially since consumer customers prefer to use standard IC packages with no special handling or manufacturing requirements.

As the critical dimensions of MEMS features enter the nanoscale regime and MEMS performance levels increase, MEMS sensors will be increasingly susceptible to stiction and particle-related failure mechanisms. To address these challenges, ongoing research is required in the areas of stiction mechanisms, anti-stiction coatings, and particle movement over time. As MEMS sensors with wafer-level caps increasingly migrate to industry-standard encapsulated plastic packages to support high-volume, cost-effective manufacturing, research will also be required into ultra-low-stress mold compounds and die attaches that ideally provide Moisture Sensitivity Level 1 performance. Finally, for MEMS elements that are particularly susceptible to ESD damage, further research is needed into protection methodologies that have minimal or no impact on the overall electrical performance of MEMS products. This is especially challenging in the case of MEMS elements that have no associated circuitry on the same die, thus precluding the use of standard ESD protection circuitry.

VII. CONCLUSIONS

Due to their functional requirements, widely varying manufacturing processes, unique failure mechanisms, and diverse end-customer applications, MEMS products cannot be qualified using a single set of standard stress tests. MEMS packaging, which is critical to the overall quality and reliability of MEMS products, is a key consideration when developing qualification plans. MEMS products in hermetic packages can generally pass most or all industry-standard stress tests for such packages, though their performance through mechanical shock, mechanical drop, and constant acceleration testing needs to be extensively characterized to determine the susceptibility of the MEMS elements to stiction and physical damage. MEMS products where the sensor is capped and then packaged in plastic packages likewise can generally pass industry-standard stress tests for such packages. However, MEMS products where the sensor is directly exposed to the outside world, such as with most microphones, generally cannot pass stress tests where the product is exposed to condensing moisture levels that can induce stiction.

While no single set of stress tests is appropriate for qualifying all types of MEMS products, standards are evolving for qualifying particular families of such products. The tables in this paper show typical qualification plans for various categories of MEMS products, and these plans should show a significant degree of commonality regardless of the specific manufacturer for each product category. Many MEMS products will pass most or all standard stress tests for semiconductors, including high and low temperature operating life test, high and low temperature storage, temperature

978-1-4244-5430-3/10 $26.00 © 2010 IEEE

cycling, and ESD and latch-up tests. Thus, these stress tests should serve as the foundation for all MEMS qualification plans, especially since these stress tests are required by MEMS users that increasingly expect MEMS products to pass all the same tests that standard ICs will pass. These baseline stress tests are then augmented by an appropriate set of mechanical tests such as drop, vibration, and powered mechanical shock testing to demonstrate the MEMS products will remain reliable through shipping and handling, board- and system-level manufacturing, and end-customer field application. Finally, MEMS products require endurance life testing where the MEMS elements are subjected to millions or billions of deflections/actuations to ensure they will function reliably beyond the intended life time of the products. Passing all these stress tests will remain challenging as MEMS sensors migrate to nanoscale feature sizes and are increasingly assembled in standard plastic packages. Thus, ongoing research is required in areas such as failure mechanisms, anti-stiction coatings, packaging, and ESD protection for MEMS products.

ACKNOWLEDGMENTS

The author gratefully acknowledges support from the following colleagues at Analog Devices: Mike Aldrich, Maurice Brodeur, Brad Gifford, James Griffin, David Grosjean, James Molyneaux, Carolyn Pipitone, Dave Watson, and Jay Yakura.

REFERENCES

[1] "ADXL50 Monolithic Accelerometer with Signal Conditioning" data sheet, Analog Devices, Inc., 1996. http://www.analog.com.

[2] JEDEC Standard JESD34, "Failure-Mechanism-Driven Reliability Qualification of Silicon Devices," The JEDEC Solid State Technology Association, March 1993 (rescinded November 2004). http://www.jedec.org.

[3] S. Bart J. Chang, T. Core, L. Foster, A. Olney, S. Sherman, and W. Tsang., "Design rules for a reliable surface micromachined IC sensor," 1995 IEEE International Reliability Physics Symposium, pp. 311-317.

[4] MIL-STD-883, "Department of Defense, Test Method Standard, Microcircuits." http://www.dscc.dla.mil.

[5] JEDEC Standard JESD17, "Latch-up in CMOS Integrated Circuits, The JEDEC Solid State Technology Association, August 1988 (rescinded February 1999). http://www.jedec.org.

[6] "ADMP421 Omnidirectional Microphone with Bottom Port and Digital Ouput" data sheet, Analog Devices, Inc., 2009. http://www.analog.com.

[7] "SPM0404HD5H-PB 'Mini' SiSonic™ Microphone Specification – Halogen Free" data sheet, Knowles Accoustics, 2009. http://www.knowles.com.

[8] JEDEC Standard JESD22-A101C, "Steady-State Temperature Humidity Bias Life Test," The JEDEC Solid State Technology Association, March 2009. http://www.jedec.org.

[9] JEDEC Standard JESD22-A104D, "Temperature Cycling," The JEDEC Solid State Technology Association, March 2009. http://www.jedec.org.

[10] JEDEC Standard JESD22-A108C, "Temperature, Bias, and Operating Life" The JEDEC Solid State Technology Association, June 2005. http://www.jedec.org.

[11] JEDEC Standard JESD22-A103C, "High Temperature Storage Life," The JEDEC Solid State Technology Association, November 2004. http://www.jedec.org.

[12] JEDEC Standard JESD22-A119, "Low Temperature Storage Life," The JEDEC Solid State Technology Association, November 2004. http://www.jedec.org.

[13] ESD Association Standard Test Method ANSI/ESD STM5.1-2007, "Electrostatic Discharge Sensitivity Testing – Human Body Model (HBM) Component Level," Electrostatic Discharge Association, 2007. http://www.esda.org.

[14] ESD Association Standard Test Method ANSI/ESD STM5.3.1-2009, "Electrostatic Discharge Sensitivity Testing – Charged Device Model (CDM) Component Level," Electrostatic Discharge Association, 2009. http://www.esda.org.

[15] JEDEC Standard JESD78B, "IC Latch-up Test," The JEDEC Solid State Technology Association, December 2008. http://www.jedec.org.

[16] Joint IPC/JEDEC Standard J-STD-020D, "Moisture/Reflow Sensitivity Classification for Nonhermetic Solid State Surface Mount Devices," The JEDEC Solid State Technology Association, March 2008. http://www.jedec.org.

[17] M. Zimmerman, L. Felton, E. Lacsamana, and R. Navarro, "Next generation low stress plastic cavity package for sensor applications," 2005 IEEE Electronics Packaging Technology Conference, pp. 231-237.

[18] J. S. Lee et al., "A cost effective MEMS cavity packaging technology for mass production," IEEE Transactions on Advanced Packaging, Vol. 32, No. 2, pp. 453-460, May 2009.

[19] "ADXRS610 ±300 °/sec Yaw Rate Gyro" data sheet, Analog Devices, Inc., 2007. http://www.analog.com.

[20] "ADXL335 Small, Low Power 3-Axis ±3 g Accelerometer" date sheet, Analog Devices, Inc., 2010. http://www.analog.com.

[21] JEDEC Standard JESD22-A110C, "Accelerated Moisture Resistance – Unbiased Autoclave," The JEDEC Solid State Technology Association, June 2008. http://www.jedec.org.

[22] JEDEC Standard JESD22-A110C, "Highly Accelerated Temperature and Humidity Stress Test (HAST)," The JEDEC Solid State Technology Association, January 2009. http://www.jedec.org.

[23] JEDEC Standard JESD22-A114F, "Electrostatic Discharge (ESD) Sensitivity Testing Human Body Model (HBM)," The JEDEC Solid State Technology Association, December 2008. http://www.jedec.org.

[24] JEDEC Standard JESD22-C101E, "Field-Induced Charged Device Model Test Method for Electrostatic Discharge Withstand Thresholds of Microelectronic Components," The JEDEC Solid State Technology Association, December 2009. http://www.jedec.org.

[25] T.-R. Hsu, "Reliability in MEMS packaging," 1996 IEEE International Reliability Physics Symposium, pp. 398-402.

[26] T. M. Tanner, "MEMS reliability: Where are we now?" Microelectronics Reliability, Vol. 49, pp. 937-940, 2009.

978-1-4244-5430-3/10 $26.00 © 2010 IEEE

EFM study of injected charges in the silicon nitride of an electrostatic actuated MEMS

Nowodzinski Antoine, Bloch Didier, Koszewski Adam, Toussaint Thibaut
Microsystems Characterisation and Reliability Laboratory
CEA-Leti
Grenoble France
33 –438781142, antoine.nowodzinski@cea.fr

Abstract— **In this paper we introduce an original Electrostatic Force Microscopy (EFM) -based methodology which allows a rapid and easy calculation of the charge density value of electric charges injected in a silicon nitride dielectric layer for conditions which prevail during MEMS switch actuation. The calculation of the number of trapped charges is reproducible and in agreement with the results already published in the literature. A physical model is suggested which describes the charge spot behavior inside the silicon nitride. This model can explain the pull-in voltage shift due to dielectric charging in a RF-switch MEMS.**

Keywords-EFM; dielectric; charging;MEMS; reliability

I. INTRODUCTION

Dielectric charging is a key failure mechanism for electrostatically actuated MEMS switches. The main consequences are a shift of the pull-in voltage and possible stiction. Although many efforts have been made in this field [1,2,3,10], no clear explanation concerning the physics and no valuable predictive model has been proposed so far.

In this paper we introduce an original Electrostatic Force Microscopy (EFM)-based methodology. To our knowledge, it is the first time that voltage measurements are made using EFM calibration and a procedure checking the possibility to neglect the tip-surface capacitance is applied in order to perform charge density evaluation. This methodology enables assessment of a silicon nitride dielectric layer for conditions prevailing during MEMS switch actuation. This approach allows the assessment of the quantity of electric charges present on the dielectric layer and the interpretation of the geometrical behavior of the spot of charges obtained by EFM imaging.

The calculation of the number of trapped charges is reproducible and in agreement with the results already published in literature. A physical model is suggested which describes the charge spot behavior inside the silicon nitride and can explain pull-in voltage shift due to dielectric charging in an RF-switch MEMS. This work contributes to a better understanding of dielectric charging issues in MEMS devices using silicon nitride dielectric layers

II. SAMPLE DESCRIPTION

The tested sample is a MEMS capacitive switch fabricated at the CEA-Leti [8]. Fig. 1 describes the device used. The stack is composed of a platinum bottom electrode and a 260 nm-thick Si_3N_4 dielectric layer.

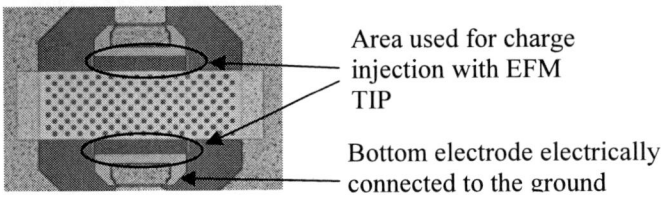

Figure 1.   Sample description

The charge injection procedure is performed by applying an electric field on specific areas located on the edges of the aluminum membrane of the switch (Fig. 1). The bottom electrode and EFM tip are electrically connected to a voltage source used for charge injection.

We know that pull-in and pull-out voltage shifts can be widely influenced by non-uniform charging or a non-uniform air-gap between the dielectric and the mobile membrane [16]. Injecting charge in close vicinity of the MEMS device mobile membrane results in a much more precise dielectric charging characterization than the usual functional test with MEMS [3,10].

III. EFM MEASUREMENTS PROTOCOL, CHARGE INJECTION

For our experiments, a highly doped diamond-coated conductive tip is used. The radius of the tip curvature is approximately 200 nm: such a large radius of curvature appears to allow more stable measurements and to be less sensitive to topographic artifacts due to large amounts of injected charges [4].

The electric charges are injected in the dielectric layer through the diamond coated EFM tip. This procedure, as well as conventional EFM measurements, is well known and described in the literature [6,7,8,9]. EFM measurements are based on phase shift measurements of a sustained oscillation of a conductive tip during scan at a few tens of nanometers above the surface sample. Usually, EFM is used to detect charges and not to quantify charges. This is because the link between phase shift measurements and charge voltage is proportional to $\delta C^2/\delta^2 z$, C being the tip-surface capacitance and z the tip-surface distance. The quantity $\delta C^2/\delta^2 z$ is difficult to calculate, and it is only possible to approximate it [14].   Moreover, most

978-1-4244-5430-3/10 $26.00 © 2010 IEEE

of the time, the EFM tip capacitance is high enough to influence the voltage of the injected charges. Fig. 2 shows the electrical equivalent circuit and explains why the voltage of the injected charges can be influenced by the tip capacitance because $Vq=Q/(C_{tip}+C_{SiN})$.

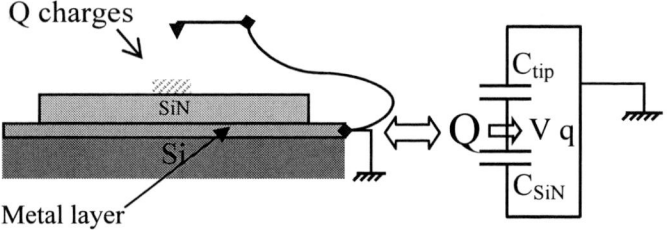

Figure 2. Electrical equivalent circuit of deposited charges during EFM measurements

Kelvin probe force microscopy (KPFM) has also been used to quantify the charge voltage [11, 12]. The KPFM technique uses a built-in voltage generator in order to equalize the voltage between the AFM tip and the surface potential. But most of the time with KPFM it is difficult to measure voltages as high as 10V, whereas the pull-in or the pull-out voltage shift due to dielectric charging in MEMS can reach 15V[13] or more [10].

In this paper, we use a methodology based on EFM measurements that permits charge spot voltage measurements up to 15V. This methodology uses an EFM calibration in order to convert the EFM phase shift into the potential of a charged spot. Then, the voltage variation as a function of the tip-surface height is measured on a spot of charge in order to find the height for which the tip capacitance can be neglected.

A. EFM calibration

When electrostatic forces influence the EFM tip, the apparent stiffness of the EFM tip lever and then its resonance frequency are modified. Girard [17] gives an approximation of the expression of $\Delta\varphi$, the phase shift of the tip oscillation due to an applied electrostatic force F:

$$\Delta\varphi=Q/k_0 \, grad(F) \qquad (1)$$

where k_0 is the stiffness and Q the quality coefficient of the lever. Girard also gives the relationship between grad(F) and the voltage V between the tip and the sample surface:

$$Grad(F)= \delta C^2/\delta^2 z \, V^2 \qquad (2)$$

It is then possible to deduce:

$$\Delta\varphi= Q/k_0 \, \delta C^2/\delta^2 z \, V^2 \qquad (3)$$

The EFM calibration consists of measuring $K=Q/k_0 \, \delta C^2/\delta^2 z$ for each tip-surface height to be used during the EFM measurement. To do this, we measure the induced phase variation when we apply a voltage ramp between the EFM tip and a conductive layer. The parabolic coefficient of the curve $\Delta\varphi=f(V)$ is an experimental measurement of $Q/k_0 \, \delta C^2/\delta^2 z$.

Fig. 3 gives an example of EFM calibration and shows that the relationship between $\Delta\varphi$ and V is parabolic as predicted by (3).

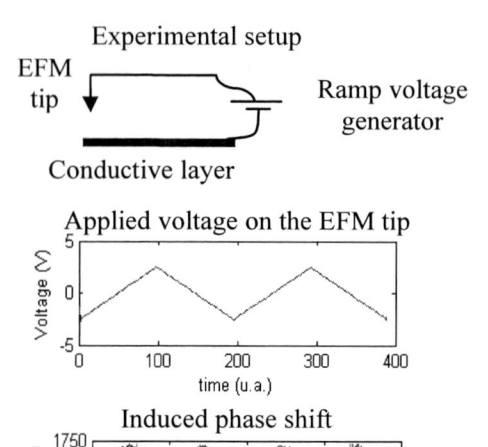

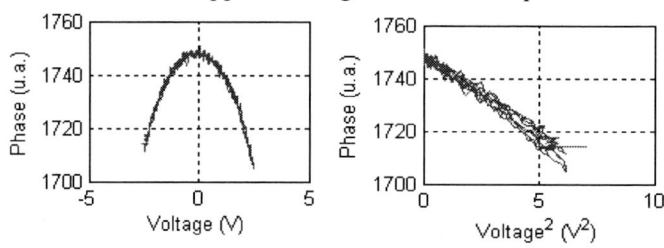

Experimental relationship between the induced phase shift and applied voltage on the EFM tip

Figure 3. Calibration example at one tip-surface height

It is useful to perform the experimental measurement of K for several tip-surface heights. For heights ranging between 50nm and 300 nm, Fig. 4 shows that K for the conductive diamond coated tip can be fitted by $K=K_0 z^{-\gamma}$.

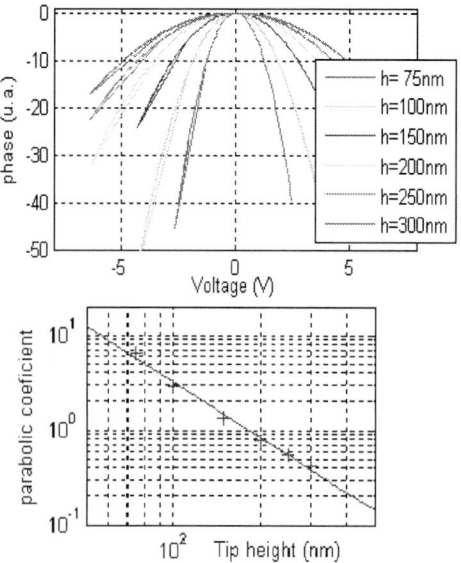

Figure 4. EFM calibration results for several tip-surface heigth

Equation (3) shows that the relationship between $\Delta\varphi$ and V is fully dependent on $\delta C^2/\delta^2 z$. It is then obvious that the calibration can be very affected by a tip shape modification due to tip wear, by the charge injection procedure, or by tip contamination. This explains why measurements were much more stable with the 200-300 nm diamond coated tip radius than with the 10-20 nm conventional EFM tip radius. Nevertheless, we always calibrated the diamond coated tip before and after EFM measurement sessions, in order to check for any anomalies.

B. Tip-surface height variation to neglect tip capacitance

This part of our procedure is used to find the height for which the tip capacitance can be neglected in order to be able to calculate the charge density deposited on the surface from the charge spot voltage measurement. The dielectric permittivity and thickness are known, and at the appropriate distance, the spot density of charge can be easily calculated: $\sigma = \varepsilon E$, with σ being the charge density, ε the permittivity of the dielectric, and E the electric field on the dielectric.

Considering Fig. 2, the voltage V_q of q deposited charges is influenced by C_{tip} and C_{SiN}, because:

$$V_q = q/(C_{tip} + C_{SiN}) \qquad (4)$$

Hence, from (4) when the EFM tip is raised, the measured voltage should increase until C_{tip} can be neglected in comparison with C_{SiN}. Fig. 5 shows an example of the voltage evolution of a spot of injected charge versus tip-surface height. On this figure, it is possible to see that the charge voltage is increasing when the tip-surface height increases up to 100 nm. It means that below 100 nm, the tip capacitance cannot be neglected.

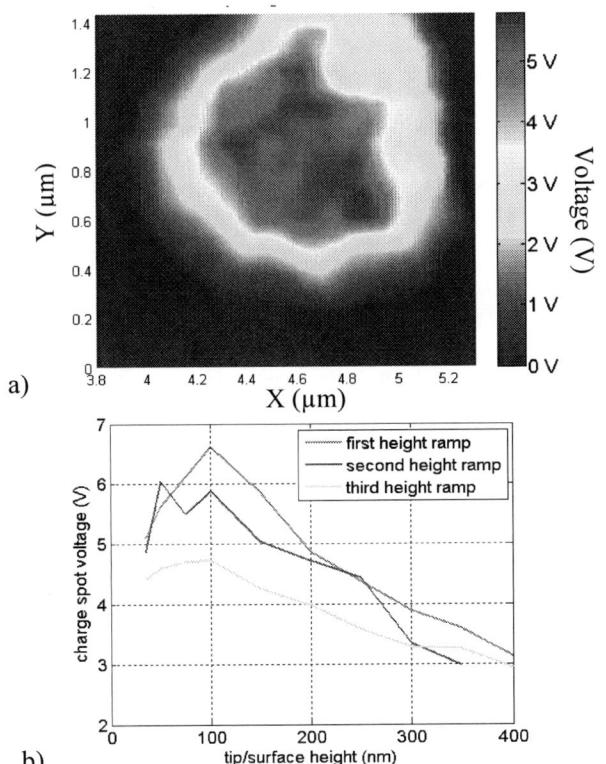

a)

b)

Figure 5. a) Charge spot image at a tip-surface height of 50 nm b) Voltage measurement according tip-surface height of a spot charge.

In Fig. 5, for a distance larger than 100 nm, a voltage decrease is observed. This decrease is due to:

- A noticeable voltage decay due to charge relaxation. The voltage decay is roughly 1V between each measurement at 100 nm, and we can assume that this decay is not negligible between the measurements performed at 100 nm and 400 nm for the three height ramps.

- An increase of the diameter of the tip/surface interaction that induces voltage measurement average on a surface larger than the charge spot detail (Fig. 6).

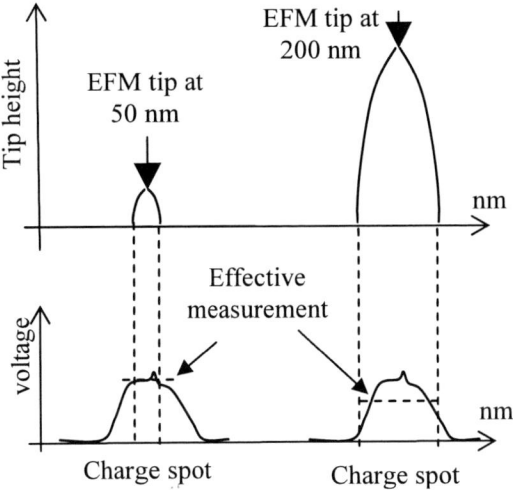

Figure 6. Averaging influence due to tip height

C. Calibration application on EFM measurement

For each experiment, a few hundred EFM pictures are post treated according to the phase-voltage conversion table obtained during the calibration step. Data such as the density of trapped charges, the behavior of the maximum spot charge density, and the charge spot area are extracted as a function of time. The charge spot area is obtained by measuring the spot area at its half maximum height.

IV. EXPERIMENTAL RESULTS

Measurements are performed for batches of charges injected during 300 seconds under an applied electric field ranging between 190kV/cm and 4.2MV/cm for positive polarization or ranging between -190kV/cm and -4.2MV/cm. Fig. 7 shows an example of charge spot evolution.

The AFM used is placed inside a quasi hermetic chamber under nitrogen flux in order to dry the atmosphere. The humidity is controlled with a data logger. For all the measurements the relative humidity was between 2% and 4%. Before the measurements, the sample is dried in an oven at 120°C and introduced hot inside an airlock next to the AFM chamber. Hence, the sample cools down slowly under dry atmosphere.

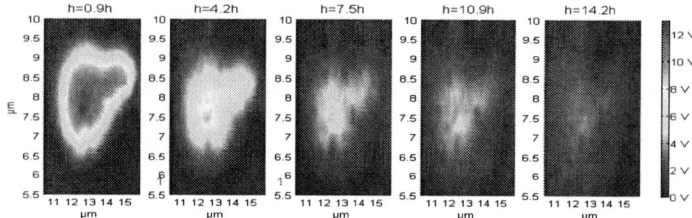

Figure 7. Pictures in false colors of the voltage evolution of an electric charge spot injected in the dielectric layer with an electric field of -2.7 MV/cm

Fig. 8a and 8b shows evolution examples of the maximum voltage of the charge spot with time, for negative and positive tip polarizations ranging between 1.2MV/cm and 4.2MV/cm during injection. During the first hours of each experiment, the charge spot voltage decreases with time and fits an exponential law whose time constant τ can be estimated. This behavior can be used to determine the voltage of the charge spot immediately after injection. The value σ of the injected charge density is obtained through the relation $\sigma = \varepsilon E$. It is here assumed that electric charges at t=0 are spread on or located close to the surface of the dielectric [10]. Each charge injection experiment leads to a different exponential law, depending on the value of the electric field.

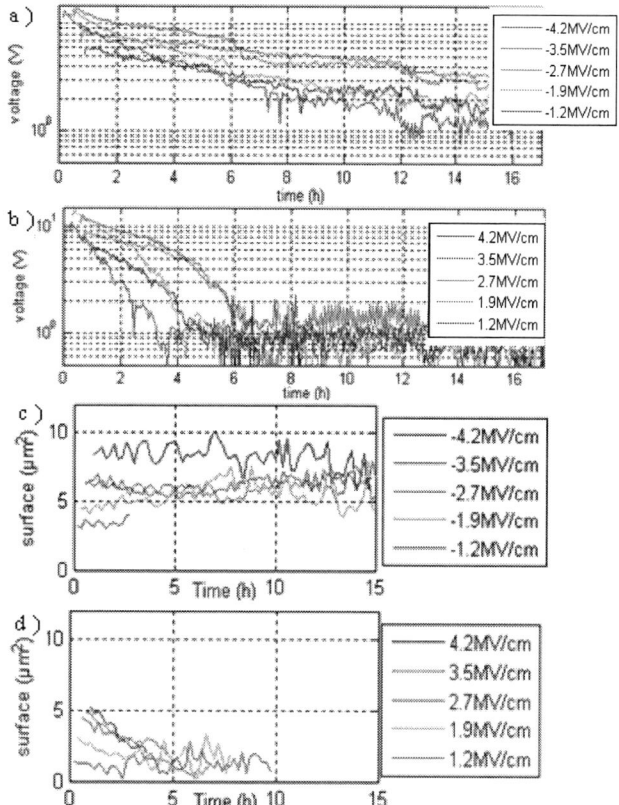

Figure 8. Maximum of each spot of charge versus time for b) positive and a) negative injection. Surface at half height versus time for each spot of charge for d) positive and c) negative injection.

Fig. 9a represents the initial charge densities injected as a function of the electric field. Fig. 9b represents the electric field dependence of the time constant. These measurements have been performed on the same device over a period of 36 days.

The consistency of the measurements confirms the acceptable reproducibility of the charge injection/relaxation mechanisms and gives us confidence in the reliability of the EFM measurement.

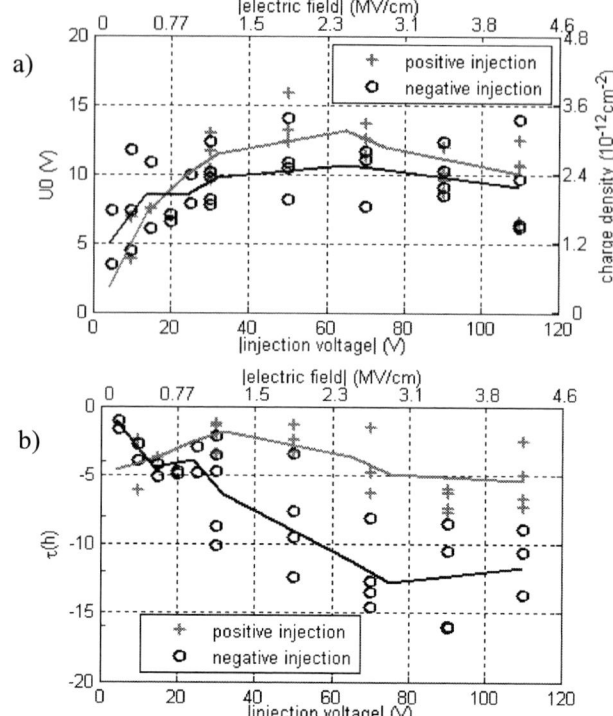

Figure 9. a) Initial charge spot voltage of the injected spots of electric charges as a function of the electric field; b) Voltage decrease of the charge spots maximum versus the injection voltage, considering: $U(t)=U_0\exp(-t/\tau)$. The lines are drawn to guide the eye.

Results introduced in Fig. 9 may be compared to literature results obtained with KPFM measurements on a 300nm thickness silicon nitride layer [11]. Taking into account the dielectric thickness and the electric charge injection time, the results are comparable for electric charge injection voltages lower than 10 Volts.

Fig. 9a also shows that there exists a maximum charge spot voltage value of $\approx$ 13 Volts. This maximum value is reached for an injection voltage of roughly 70 V (2.7MV/cm). This result may suggest that the maximum number of injected charges is limited by the number of available traps in the dielectric material, or by material intrinsic conduction mechanisms near the breakdown electric field.

As the tip-dielectric distance selected is equal to 200nm, the tip capacitance can be fully neglected. The calculation of the density of charges gives a value of 2.4×10^{12} cm^{-2} for positive injection. Similar values are measured in the literature with the same dielectric and the same wafer but with a different measurement method [10].

Fig. 8a and 9b exhibit the smaller relaxation times for positive polarization (vs. negative) during charge injection. It is here assumed in both cases (negative or positive polarization during charge injection) that electrons are the main charge carriers. Two cases are possible: if electrons are injected through the bottom electrode (positive tip polarization), electrons are trapped inside the dielectric, close to the metal-dielectric interface. The proximity of the bottom electrode allows a rapid charge relaxation. If electrons are injected through the EFM-tip (negative tip polarization), electrons have to cross through the dielectric thickness to reach the bottom electrode, thus increasing the diffusion time.

The conduction mechanism of our sample is determined by Poole-Frenkel (PF) conduction or space charge limited conduction (SCLC) [10]. Fig. 9a shows a small difference between positive and negative injection, especially for electric field superior to 1 MV/cm. Moreover, fig. 9b depicts for negative charges injection a clear increase of τ when the injection electric field increases. This could be due to trap saturation increase when the injection voltage increases. This trap saturation prevents PF conduction that is much less resistive than space charge limited conduction and could explain the τ increase and the difference between trapped charge density for positive and negative injection.

Fig. 10 reports the initial spreading and the surface behavior of the spot of charges. It can be observed that the higher the injection voltage, the larger the spot area. This phenomenon is due to the local saturation of the number of electron traps, which characterizes the changeover from a PF-type to a SCLC-type conduction mechanism, and which leads to a dramatic resistivity increase directly below the injection zone. Hence, charged species spread on the edges of the high-resistivity zone in order to reach trap-free areas.

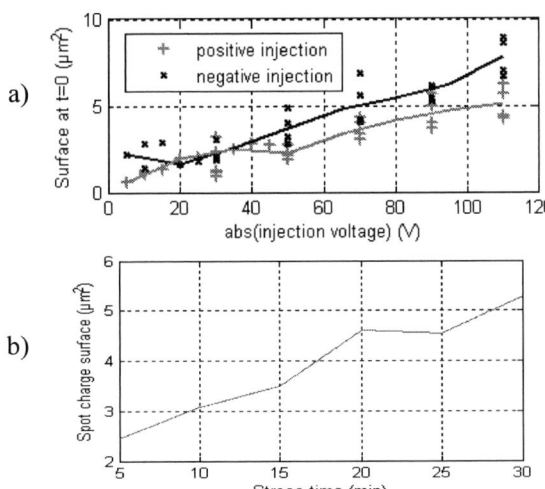

Figure 10. Initial spreading of injected charges versus a) injection voltage and b) injection time. The lines are drawn to guide the eye.

Fig. 11 illustrates by diagrams our assumption about charge injection and relaxation mechanism:

Step 1: Charge injection start up. Injection occurs mainly by PF conduction because of the low level of trapped charges:

Step 2: Charge injection saturation start up. Due to trap saturation by charges, PF conduction is troubled just under the tip and starts to be of the SCLC type. Because of the low SCLC conduction under the tip, the charges start to spread through the surface by PF conduction.

Step 3: For negative polarization on the EFM-tip, as for step 2 the charge mainly spreads through the surface. For positive polarization, the charge spreading reaches a few micrometers and starts to relax through the bottom because charges are located far from the electric field of the EFM tip.

Step 4: Charge relaxation occurs. For positive injection, electrons stay close to the bottom electrode, making relaxation easier than for electrons injected by the EFM-tip. This explains the difference between time relaxations for positive injection and negative injection.

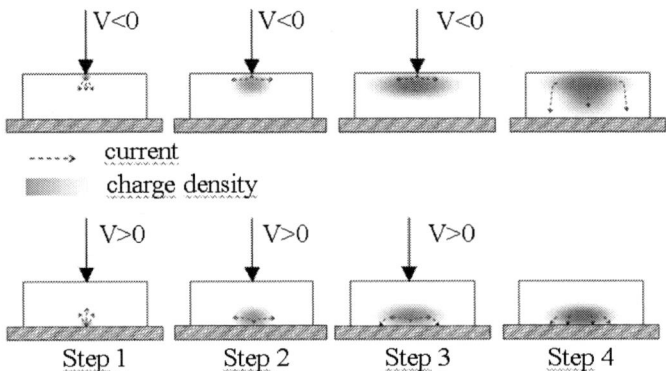

Figure 11. Charge injection and relaxation mechanism

V. COMPARISON WITH FUNCTIONAL TEST

The functional tests are obtained by capacitance versus applied voltage as described in [3]. This functional test has been realized on switches belonging to the same wafer as those used for the EFM measurements. A PAV200 (SUSS Microtech) has been used in order to perform measurements under nitrogen. Fig. 12 depicts pull-in shift versus holding time (membrane at the down state) for two stress voltages and for three MEMS that have representative behavior of a batch of several tens. This pull-in shift is clear evidence of the reliability issue for this batch of electrostatically actuated MEMS switches that can lead to membrane stiction. The comparison with the previous EFM measurements should help to understand the failure mechanism.

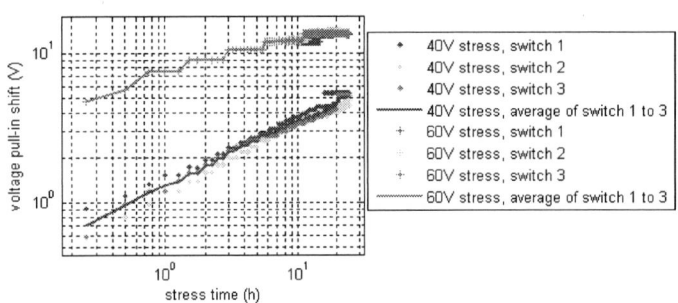

Figure 12. Pull in shift due to stress.

Fig. 12 shows that the voltage shift seems to reach a maximum of 14V when the switch is stressed at 60V. For a 40V stress voltage the pull-in shift stays below 5V, but it could be because it takes much more time to reach 14V voltage pull-in with a 40V stress voltage.

First, we can compare pull-in voltage shift due to stress (Fig. 12) and the EFM-calibrated measurements represented Fig. 9a. The EFM-calibrated measurements show that the maximum charge spot voltage is roughly 13V and the functional test shows that the maximum pull-in voltage shift is around 14V, suggesting that charge injection produces the pull-in shift.

Secondly, we can compare the 10 hour - 60V stress required to reach the maximum pull-in shift and 5 minutes required to reach 13V with EFM-tip charge injection. We know that the membrane induces non-homogenous injection [12] due to holes in the membrane or to an irregular surface on the membrane. The resultant pull-in shift is roughly the average voltage of all charge injection spots on the dielectric surface. We assume that the time required to reach the maximum pull-in voltage shift is the time required to homogenize dielectric charging. Fig. 13 sums up our assumption regarding the mechanism for charge injection homogenization.

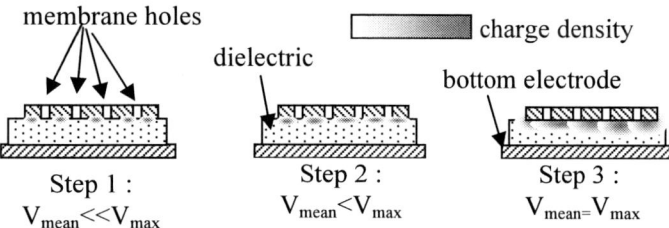

Figure 13. Charge homogenization mechanism, Vmax is maximum voltage of each spot of charge, and Vmean is the pull-in voltage shift.

VI. CONCLUSION

An EFM-based methodology has been implemented in order to study charge diffusion mechanisms in silicon nitride dielectric layers used for electrostatically actuated MEMS switches. In conditions prevailing for switch actuation, it has been shown that the trapped charge density is limited to 2.5×10^{12} cm^{-2}, similar to the values given in the literature. This charge density is found to decrease more rapidly for EFM tip positive polarization during injection, which lets us assume that electrons are the main charge carriers. Poole Frenkel and space charge limited conduction mechanisms explain the increase of the charge spot geometrical characteristics with the applied electric field used for charge injection.

The coherence and the reproducibility of the results described here, give us good confidence in using calibrated EFM for voltage measurements. Moreover, a comparison with functional tests show that the EFM-calibrated measurements are very useful in interpreting pull-in shifts of electrostatically actuated MEMS switches.

ACKNOWLEDGMENT

This work has been realized in framework of the PIDEA project FULL CONTROL with the support of the French Minister of Finance.

VII. REFERENCE

[1] G. Rebeiz, RF MEMS: theory, design and technology, Willey-Interscience, New Jersey, 2003.

[2] P. Czarnecki; X. Rottenberg; P. Soussan; P. Nolmans; P. Ekkels; P. Muller, "New Insights into Charging in Capacitive RF MEMS Switches", Proc. 46th IRPS, Phoenix, 2008, pp. 496-505.

[3] F.Souchon, PL. Charvet, C. Maeder-Pachurka, M. Audoin Dielectric Charging Sensitivity on MEMs Switches", Proc. Transducers & Eurosensors '07, Lyon, 2007, pp. 363-366.

[4] M. J. Yan and G. H. Bernstein "Apparent height in tapping mode of electrostatic force microscopy", Ultramicroscopy, 2006, vol. 106 pp. 582–586.

[5] L. Portes; P. Girard; R. Arinero; M. Ramonda, "Force gradient detection under vacuum on the basis of a double pass method", Review of Scientific Instruments, 2006, vol. 77, p. 96101.

[6] J.E. Stern; B.D. Terris; H.J. Mamin; D. Rugar, "Deposition and imaging of localized charge on insulating surfaces using a force microscope." Appl. Phys. Lett. vol. 53, pp. 2717, 1988.

[7] R. Dianoux, " Injection et detection de charges dans des nanostructures semiconductrices par Microscopie à Force Atomique ", " Injection and detection of charges inside semiconductor nanostructure by Atomique Force microscopie" thesis, 2004, université Joseph Fourier - Grenoble 1

[8] J. Lambert, "Etude de la dynamique de charges par microscopie à force électrostatique : exemple des isolants à grande constante diélectrique", "Study of the charges dynamics by electrostatic force microscopie" thesis 23-05-2003 at université Paris 7 – Denis Diderot.

[9] R.A. Oliver, "Advances in AFM for the electrical characterization of semiconductors", Rep. Prog. Phys. vol. 71, p. 76501, 2008.

[10] A. Koszewski; F. Souchon; T. Ouisse; "Conduction and trapping in RF MEMS capacitive Switches with a SiN Layer"; Proc. of ESSDERC 2009, Athens, pp. 339-342.

[11] A. Belarni, M. Lamhamdi, P. Pons, L. Boudou, J. Guastavino, Y. Segui, "Kelvin probe microscopy for reliability investigation of RF-MEMS capacitve switches", Microelectronics Reliability, 2008, vol. 46, pp. 1232-1236.

[12] R.W. Herfst, P.G. Steeneken, J. Schmitz, A.J.G. Mank, and M. van Gils "Kelvin probe study of laterally inhomogeneous dielectric charging and charge diffusion in RF MEMS capacitive switches" , 2008 IEEE International Reliability Physics Symposium IRPS., pp. 492-495.

[13] Rodolf W. Herfst, Peter G. Steeneken, H. G. A. (Bert) Huizing, and Jurriaan Schmitz , "Center-Shift Method for the Characterization of Dielectric Charging in RF MEMS Capacitive Switches", IEEE Trans. Semi. Manuf., May 2008, Vol. 21, No. 2, pp. 148-153.

[14] R. Dianoux, F. Martins, F. Marchi, C. Alandi, F. Comin, and J. Chevrier, "Detection of electrostatic forces with an atomic force microscope: Analytical and experimental dynamic force curves in the nonlinear regime", Phys. Rev. B vol. 68, pp. 45403, 2003

[15] M.Lamhamdi1, L.Boudou, P.Pons, J.Guastavino, A. Belarni, M. Dilhan, "Si3N4 thin films properties for RF-MEMS reliability investigation" Transducers '07 & Eurosensors XXI. 2007 14th International Conference on Solid State Sensors, Actuators and Microsystems. 2007: pp. 579-82.

[16] Xavier Rottenberg, Ingrid De Wolf, Bart K. J. C. Nauwelaers, Walter De Raedt, and Harrie A. C. Tilmans, "Analytical Model of the DC Actuation of Electrostatic MEMS Devices With Distributed Dielectric Charging and Nonplanar Electrodes", Journal of MEMS, 2007, vol. 16, no. 5, pp. 1243-1253.

[17] P. Girard, M. Ramonda, D. Saluel, "Electrical contrast observations and voltage measurements by Kelvin probe force gradient microscopy" J. Vac. Sci. Technol. B , 2002,vol. 20.4, pp. 1348-1355.

A Novel Low Cost Failure Analysis Technique for Dielectric Charging Phenomenon in Electrostatically Actuated MEMS Devices

U. Zaghloul[1,2], F. Coccetti[1,2,4], G.J. Papaioannou[1,2,3], P. Pons[1,2] and R. Plana[1,2]

[1]CNRS ; LAAS ; 7 avenue du colonel Roche, F-31077 Toulouse, France
[2]Université de Toulouse ; UPS, INSA, INP, ISAE ; LAAS ; F-31077 Toulouse, France
Phone : (33) – (5) –(61336817), E-mail: usama.zaghloul@laas.fr

[3]University of Athens, Solid State Physics, Panepistimiopolis, Zografos, Athens, Greece
[4]Novamems, 10 av. De l' Europe, F-31520, Toulouse, France

Abstract—**This work presents a novel failure analysis technique for the dielectric charging phenomenon in electrostatically driven MEMS devices. The new reliability assessment methodology makes use of Kelvin Probe Force Microscopy (KPFM) and it targets in this specific work thin PECVD silicon nitride films for electrostatic capacitive RF MEMS switches. The proposed technique took advantage of the AFM tip to simulate charge injection through asperities then measure the induced surface potential. The impacts of bias amplitude, bias polarity, and bias duration employed during charge injection have been explored. Various parameters have also been investigated: dielectric film thickness, substrate nature, and SiN material deposition conditions. FTIR and XPS material characterization techniques have been used to determine the chemical bonds and compositions, respectively, of the SiN films being investigated. The required samples for this technique consist only of thin dielectric films deposited over planar substrates and no photolithography steps are required. Therefore, the proposed methodology provides a low cost and quite fast solution compared to the currently available methods.**

Keywords-failure analysis; Kelvin Probe Force Microscopy; dielectric charging; silicon nitride; electrostatic MEMS.

I. INTRODUCTION

Electrostatically actuated Micro-Electro-Mechanical-Systems (MEMS) technology has already emerged as an enabling technology for a new generation of high-performance RF components such as RF MEMS switches, tunable capacitors and inductors. In addition, RF MEMS components can be fully integrated with monolithic microwave integrated circuits and therefore can potentially lead to systems with small size, lighter weight, low power consumption and mass production [1]. Among the mentioned RF MEMS devices, it is believed that the MEMS switch is the key device due to its unique RF performance comparing to the current existing devices. However, the commercialization of the electrostatic MEMS capacitive switch is still hindered by reliability issues, especially the dielectric charging phenomenon [2].

The dielectric charging effect arises from dipole orientation and the charge injection, displacement, and trapping that occurs under the strong electric field during the down state of MEMS switches [3]. The charge injection was reported to follow the Poole-Frenkel effect, as found from the dependence of the shift of pull-out voltage on the actuation bias [4]. In SiN material it has been shown that "hole injection" introduces metastable traps, which give rise to asymmetrical current-voltage characteristic, observed in Metal-Insulator-Metal (MIM) capacitors with electrode material symmetry [5]. On the other hand, the relaxation time of charged SiN MIM capacitors was reported to increase with the applied electric field intensity for both low frequency (LF) and high frequency (HF) deposited PECVD dielectric films [6]. Additionally, the stored charge was found to increase almost linearly with increasing film thickness for LF SiN material while there was no clear effect of the dielectric film thickness on the stored charge in the case of HF material [6]. Recently, charging-related lifetime tests of capacitive RF-MEMS switches showed that substrate charging can highly affect the reliability and it can even dominate the well known interposer dielectric charging [7].

A variety of characterization methodologies have been employed in order to assess the dielectric charging phenomenon. For example, current measurements in Metal-Insulator-Metal (MIM) capacitors have been introduced in [6, 8, 9, 10] while capacitance and/or voltage measurements in MEMS capacitive switches are reported in [3, 4, 11, 12]. These methods, although extremely useful, are quite expensive and time consuming since they require the fabrication of the complete MIM or MEMS devices including many levels of photolithography. These methods also lead to results that depend strongly on the nature of the device under test. Thus, in MIM capacitors the discharge takes place under a short circuit condition and the injected charges are collected by the injecting electrodes. In contrast with MEMS, the injected charges are collected only by the bottom electrode because the top electrode, the bridge, is in up-state.

978-1-4244-5430-3/10 $26.00 © 2010 IEEE

Recently, Kelvin Probe Force Microscopy (KPFM) has been employed and shown to be a promising method because it effectively simulates the charging and discharging conditions of MEMS capacitive switches providing qualitative information on dielectric free surface charge distribution and discharging process [13-17]. The evolution of the deposited charges by the AFM tip showed that the charges do not diffuse across the surface and the decay of the charge was attributed to diffusion towards the bottom electrode and trapping in the bulk of the dielectric [13, 14]. Finally, through the comparison between the surface potential and SEM mapping, experiments performed on MEMS switches after pull-down stress reveal the effects of the holes in the top electrodes, which translate to no injected charge found at that position [14]. The surface potential distribution was found to be more confined in thinner dielectric films when the same bias is applied during charge injection, independent of the substrate nature [16]. Furthermore, the decay time constant is found to be larger in thinner SiN dielectric films than in the thicker ones when the same bias is applied during charge injection, independently of the substrate nature [17].

The novelty of this work is to introduce a new failure analysis technique for the dielectric charging problem. Based on the results obtained from the proposed methodology, a selection criterion for a dielectric film which minimizes the dielectric charging will be available. The new methodology took advantage of the KPFM tip to simulate charge injection through asperities which exists in MEMS switches and then measure the induced surface potential. The required samples for this technique consist only of thin dielectric films deposited over planar substrates and no photolithography steps are required. Therefore, this technique provides a low cost and quite fast solution comparing to the currently available methods. In comparison to our previous work [16, 17], KPFM has been employed in this study to inject charges over SiN surfaces under different electric field intensities and for various charge injection durations. Moreover, a large variety of investigated test structures will be used. Also, the effect of relative humidity during charge injection as well as during measurements has been considered. Finally, material characterization has been performed on a variety of samples and the relation between the SiN chemical composition and the obtained results from the KPFM technique will be discussed.

This paper is organized as follows. First the layer structures of the investigated samples as well as the KPFM experiment conditions and parameters will be presented. This is followed by a detailed section for the results and discussion based on a comparative study between different samples and considering the material characterization. Finally, a comprehensive conclusion of the study will be addressed.

II. EXPERIMENTAL DETAILS

The effect of various parameters has been investigated in order to reach a selection criterion to minimize the dielectric charging phenomenon. The investigated samples are listed in Table 1. First, SiN films with different thicknesses have been deposited in order to study the effect of the thickness. The impact of the underneath physical layer over which the dielectric is deposited has also been investigated through depositing SiN films on evaporated gold (Evap-Au), electro-deposited gold (ECD-Au), titanium (Ti) layers, and over bare silicon substrates (Si). The main reason behind investigating the ECD-Au and Ti layers is that these layers are normally used in real RF MEMS switches: the ECD-Au is required for RF MEMS switches operating at lower frequency while the Ti layer is used as an adhesion layer between the Au and the SiN dielectric layers. Finally, the effect of the SiN deposition conditions on the charging/discharging processes has been investigated.

The deposition of the SiNx layers has been carried out using a (PECVD) reactor (STS, Multiplex Pro-CVD). For this reactor, a pair of 8" size upper and bottom electrodes made of stainless steel was used within the deposition chamber that is kept at high vacuum during the deposition process. The main feature of this system is its duality of RF operation modes, namely LF at 380 KHz and HF at 13.56 MHz. Also, the operating power can be adjusted to range from 0 to 1000 W in the LF mode and from 0 to 600 W in the HF mode. Other features of the system include the pressure ranging from 0 to 2000 mTorr; SiH_4 flow rates ranging from 0 to 500 sccm; NH_3 flow rates ranging from 0 to 360 sccm and N_2 flow rates varying from 0 to 5000 sccm. The detailed PECVD deposition conditions which have been employed in this study are listed in Table 2.

The KPFM experiments have been performed using a Digital Instrument Dimension-3100 with Nanoscope IV controller, and the SCM-PIT conductive tips have been used. Charges are injected in single points over the SiNx surface by applying voltage pulses of defined amplitude, Up and duration, Tp to the AFM tip in tapping mode (Fig. 1a). The Nanoscript software available from Veeco has been employed to control the motion of the tip during the charge injection step. Then the analysis of the results was performed by measuring the surface potential peak, Us at t=0, and the full width at half maximum (FWHM) of single charge points directly after the charge injection step (Fig. 1b). Moreover, the decay of Us versus time has been measured and analyzed. As the measured surface potential varies from one AFM tip to another due to the difference in the intrinsic characteristics of each tip, the same tip has been used in all measurements in order to be able to compare the results obtained from different samples precisely. The KPFM measurement parameters are optimized using the scheme presented in [18]. The optimum value of the lift scan height parameter, which controls the separation between the tip and sample surface during the surface potential scanning, in our experiments is found to be 20 nm.

TABLE 1. DESCRIPTION OF THE SAMPLES UNDER DISCUSSION. THE SAMPLE NAMES ARE DESIGNATED AS FOLLOWS: THE DEPOSITION FREQUENCY (HF OR LF) FOLLOWED BY THE SAMPLE INDICATION (SUBSTRATE LAYER), E.G.: HF-EVAP-AU MEANS HF SIN DEPOSITED OVER EVAPORATED GOLD.

Sample indication	HF-Evap-Au	HF-ECD-Au	HF-Ti	HF-Si	LF-Evap-Au	LF-ECD-Au	LF-Ti	LF-Si
SiNx (nm)	100→400	200, 300	200	200, 300	100→400	200, 300	200	200, 300
Ti (nm)	-	-	50	-	-	-	50	-
Electrodeposited Au(μm)	-	2.5	-	-	-	2.5	-	-
Evaporated Au (nm)	200	200	200	-	200	200	200	-
Ti (nm)	50	50	50	-	50	50	50	-
Low resistivity Silicon (μm)	500	500	500	500	500	500	500	500

TABLE 2. PECVD DEPOSITION CONDITIONS FOR LF AND HF SIN SAMPLES.

Parameter		Low Frequency SiN (LF)	High Frequency SiN (HF)
Gas flow rate (sccm)	SiH_4	21	18
	NH_3	15	40
	N_2	1960	1200
Radio frequency (RF) power (W)		185	20
Chamber pressure (mTorr)		650	1000
Radio frequency mode		380 KHz	13.56 MHz
Substrate temperature (°C)		200	200

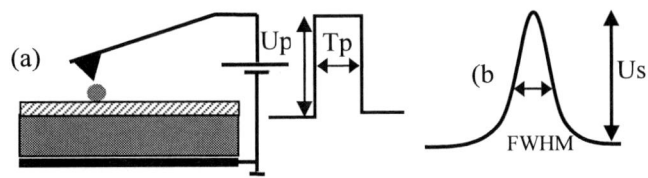

Figure 1. The procedure of the KPFM based failure analysis technique and result analysis (a) charge injection in single point, (b) resulting surface potential distribution.

The removal of the water-related layer from both surfaces on the sample and the AFM tip is reported to be very important to improve the reliability of the KPFM measurements [19]. Moreover, at low humidity ≈10% RH trapped surface charges could be detected in the KPFM measurements while such charging effects become less prominent at high humidity ≈80% RH [18]. The adsorbed water layer on the sample surface is considered to shield the surface potential [20]. Also, the possible water film and the excess surface charge which exists over the SiN dielectric surface at high relative humidity levels are found to contribute to the fast charge decay as well as the lateral charge redistribution [21]. It is believed that the external contribution of relative humidity can only be quantified by comparison with KPFM measurements in perfect vacuum and/or with shielded SiN samples, which were not available during our KPFM experiments. For these reasons, the investigated SiN samples in this study have been dehydrated at 150°C for 45 minutes under vacuum just before performing the KPFM experiments. In addition, surface potential measurements have been performed under dry air flow (relative humidity ≈ 6%). Finally, it should be pointed out that the KPFM experiments

have been performed several times for different samples and the error percentage in the measurement results presented in this article is kept to less than 6%.

III. RESULTS AND DISCUSSION

The parameters of the pulse applied to the AFM tip during the charge injection step have been investigated for different SiN samples. This includes the pulse duration, Tp, the pulse amplitude, Up, and bias polarity. Charges have been injected for different durations ranging from 0.1 msec to 100sec while the pulse amplitude Up has been changed from 10V to 70V using both positive and negative bias. Fig. 2 shows an example of the KPFM surface potential images obtained for a HF SiN sample deposited over evaporated gold layer (HF-Evap-Au) with 200nm thickness, under different pulse amplitudes and for different pulse durations. It is clear from the figure that the induced surface potential increases with increasing either the applied pulse amplitude or the pulse duration.

A. Effect of dielectric thickness

Fig. 3 depicts the measured surface potential (Us at t=0) versus the applied pulse amplitude, Up, for the HF-Evap-Au samples with different film thicknesses. It is evident that for each film thickness, the higher the Up value is, the higher the measured Us and hence the more trapped charge there will be. It is also evident that Us is smaller for thicker dielectric films for a given Up. For each film thickness, the increase of measured Us with the applied Up can be attributed to the fact that the injected charge density increases with the applied electric field intensity [22] which becomes larger for higher Up amplitudes. For a given pulse duration Tp, there are two main factors which determine the amount of trapped charge for SiN films with different thicknesses, which are the electric field intensity and the material stoichiometry. With thinner dielectric films, the electric field intensity is higher when the same Up is applied. Therefore, injected charge density and consequently the measured Us is expected to be larger for thinner dielectric films, for a given Up and Tp. In addition, the measured Us is found to be asymmetric when positive or negative Up is applied as highlighted in the figure. This can be attributed to different penetration depths for positive and negative charge carriers in a dielectric film [23]. In our investigated SiN samples, the total amount of negative charge that can be stored is higher than that of positive charge for a given Up and Tp.

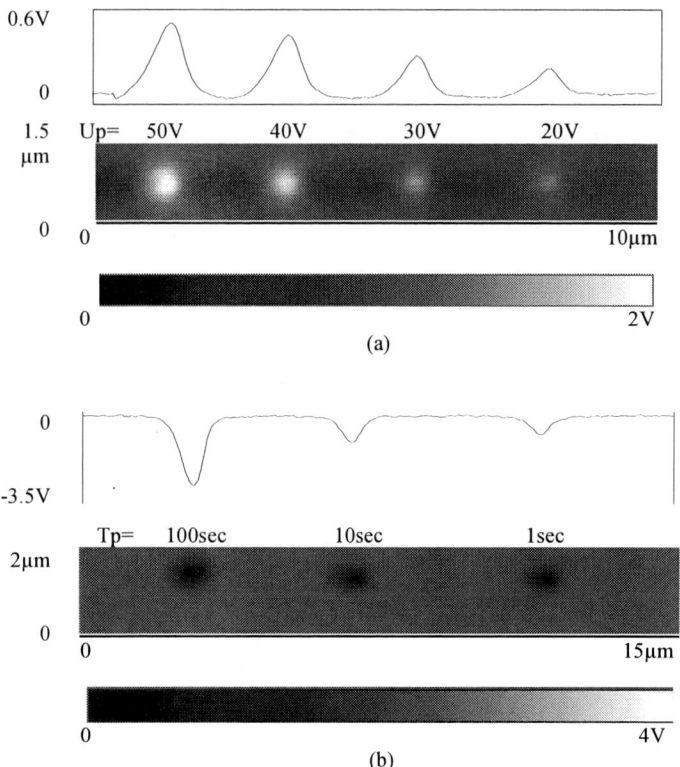

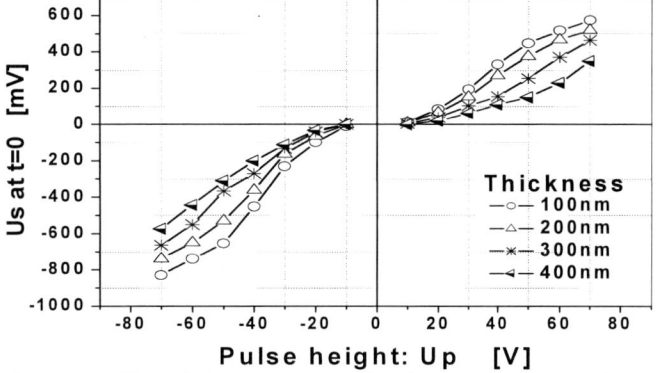

Figure 2. KPFM images for HF-Evap-Au SiN sample with 200nm thickness. Charges have been injected using (a) different pulse amplitudes Up with Tp=1sec and (b) different pulse durations Tp with Up=-40V.

In order to further investigate the impact of the dielectric film thicknesses, charges have been injected for different time durations, Tp, for the same HF-Evap-Au samples with different film thicknesses and the results are shown in Fig. 4. For a given Up, the measured Us is found to increase with increasing the charge injection duration, Tp, for all SiN film thicknesses. This finding is supported by the study of charging in dielectricless capacitive MEMS switches [24] where the drift in the pull in voltage is found to be directly proportional to the duty cycle value of the periodic square positive signal stress that have been used. In addition, initial modeling of dielectric charging shows that the amount of injected charge increases with the charging time [25].

Figure 3. Effect of Up on Us at t=0 for the HF-Evap-Au samples with different film thicknesses, Tp=1 msec.

Furthermore, from Fig. 4 thinner dielectrics are found once again to have larger Us and hence higher charge trapping than thicker ones, for the different investigated pulse durations. Finally, it should be pointed out that the same remarks concerning the effect of the dielectric film thickness under different charge injection conditions (Fig. 3, Fig. 4) have been observed for low frequency SiN samples deposited over evaporated Au (LF-Evap-Au) with different thicknesses.

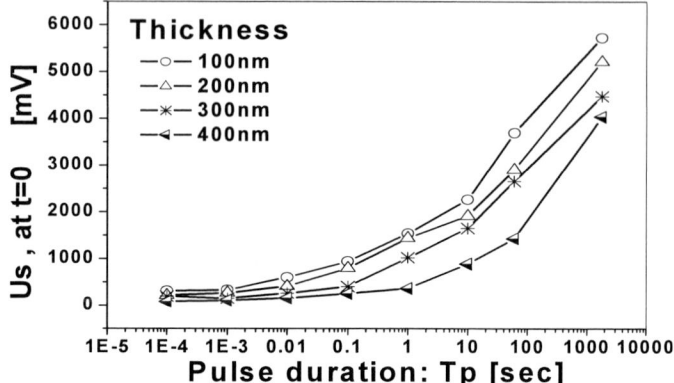

Figure 4. The surface potential Us at t=0 versus the pulse duration, Tp for HF-Evap-Au samples with different thicknesses with Up=40V.

The chemical composition of the SiN material for HF-Evap-Au samples with different thicknesses have been addressed in order to further understand the effect of the SiN film thickness on the charging/discharging processes. Two material characterization techniques have been applied; FTIR (Fourier transform Infra-Red spectroscopy) and XPS (X-ray photoelectron spectroscopy). First, FTIR spectroscopy has been performed in order to provide information about the chemical bonds of the dielectric film and their variations. Then, XPS experiments have been performed to provide information about the different chemical bonds and valence state in the SiN surface and then in depth through sputtering. Here it should be pointed out that the XPS technique is not sensitive to hydrogen or helium.

The FTIR spectra for HF-Evap-Au samples with different thicknesses are presented in Fig. 5. Such spectra reveal that there is a strong Si-H bond at around $2159cm^{-1}$ and N-H bond at around $3355 \ cm^{-1}$, independent of the SiN film thickness. The XPS results are summarized in Table 3 and highlight that the surface of the HF-Evap-Au samples with different film thicknesses are all silicon rich (N/Si is in the range of 0.56 to 0.73) with consistent concentrations for different elements. The 300nm SiN sample is an exception as its SiN surface has a higher carbon contamination (23.1%) compared to the other samples. Furthermore, the difference in the N/Si ratio between the surface and the bulk of the 100nm SiN film is relatively small, which is also expected to be the case for the bulk of the other samples with different thicknesses. Hence, both FTIR and XPS results lead to the conclusion that the HF-Evap-Au samples with different thicknesses have the same material stoichiometry. Actually, this was expected as these samples have been deposited using the same deposition conditions but for different durations. Furthermore, in spite of the higher carbon contamination on the surface of the 300nm SiN sample compared to the other samples, the KPFM Us measurement shows smooth transition from one thickness to another including the 300nm sample. This confirms the previous explanation that the main dominant parameters in determining the measured surface potential based on the KPFM technique for SiN samples with different thickness are the applied electric field intensity and the charge injection duration.

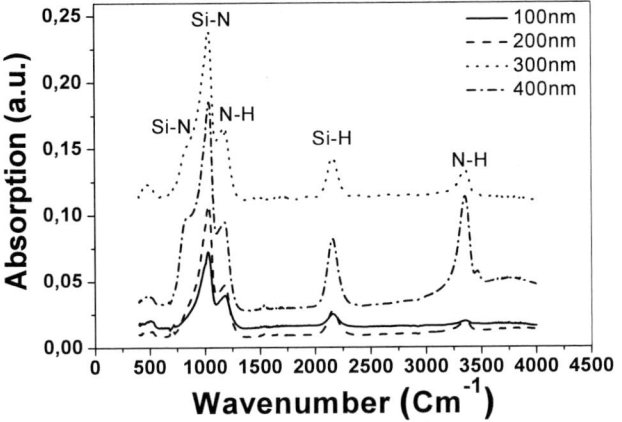

Figure 5 FT-IR spectra of HF-Evap-Au samples with different thickness at an angle of 70°.

TABLE 3: SURFACE AND BULK QUANTIFICATION OF XPS (% ATOMIC) FOR HF-EVAP-AU AND HF-SI SAMPLES WITH DIFFERENT THICKNESSES (X-RAY SOURCE AL OR TWIN MG MONOCHROMATOR, EP=20 EV, CORRECTION SCOFIELD).

Sample	Film thickness	Type	C	O	N	Si
			at %			
HF-Evap-Au	100nm	surface	8.6	23.6	27.6	40.2
HF-Evap-Au	200nm	surface	6.6	24.0	29.2	40.1
HF-Evap-Au	300nm	surface	23.1	25.9	18.3	32.6
HF-Evap-Au	400nm	surface	7.8	30.2	23.6	38.4
HF-Evap-Au	100nm	bulk	-	-	42	58
HF-Si	400nm	surface	13.8	28.6.	21.4	36.0
HF-Si	100nm	surface	9.0	29.2	23.9	37.8
HF-Si	100nm	bulk	-	-	41.5	58.5

The decay of the induced surface potential with time has been investigated for charges which had been injected under different Up. This set of experiments has been performed for two different pulse durations, Tp=1sec and Tp=1000sec. Fig. 6a presents the surface potential decay resulting from charge injection for Tp=1sec under different Up, for the sample HF-Evap-Au with 200nm thickness. As shown, the potential relaxation follows the stretched exponential law $\exp[-(t/\tau)^\beta]$ where τ is the process time constant and β ($0 \le \beta \le 1$) is the stretch factor. This remark is found to apply as well when a longer pulse duration (Tp=1000sec) is used. As already discussed in previous publications [3, 6] the stretch factor constitutes an index of charge collection complexity and the lower the value of β the more complex is the charge collection.

The decay time constants versus the applied Up for both investigated pulse durations Tp are plotted in Fig. 6b. It is obvious from the figure that the decay time constant increases with increasing Up, for both Tp=1sec and Tp=1000sec. This can be mainly attributed to the Poole-Frenkel effect [26]. According to that, when the applied electric field intensity increases, charges from shallower states will be easily released, leaving charges trapped in deeper ones which are characterized by larger relaxation times. At the same time, β is found to decrease with increasing electric field intensity. For the charging process at higher Up, the contribution of the Poole–Frenkel effect is expected to lead to a more complex process, hence to deviate from simple exponential law, which means the decrease of β. In the case of discharge, charge decay will be performed in the presence of internal electric fields generated by charge gradients. Then, the carrier removal will be expected to also follow a more complex process since the shallow states have been previously emptied by the charging under high electric fields. Here it is worth mentioning that a close trend for both the decay time constants, τ, and the stretch factors, β, versus the electric field intensity, has been obtained using the C/DCT (Charging/ Discharging Current Transients) methodology applied for MIM capacitors [6]. Note that the decay time constant for Tp=1000sec is found to be larger than in the case of Tp=1sec as presented in Fig. 6b. This may be partially supported by the initial modeling of dielectric charging which shows that the amount of injected charge increases with the charging time [25].

Based on this analysis for both the amplitude of the surface potential, Us, and the decay time constant, τ, thicker dielectric material which features smaller Us (hence smaller trapped charge) and shorter time constants (hence faster charge decay), is expected to be more robust with respect to the dielectric charging phenomenon when compared to thinner films, when charges are injected using the same conditions (same Up and Tp). Consequently, MEMS switches employing thicker SiN dielectric films are expected to have less dielectric charging and therefore longer lifetime. Certainly, one cannot use a very high dielectric thickness for electrostatic MEMS devices especially for capacitive RF MEMS switches which operate mainly based on the high capacitance ratio between the "on" and "off" states (C_{on}/C_{off}). As the dielectric thickness increases, C_{on}/C_{off} decreases resulting in degradation in the RF performance of the switch. Therefore, for capacitive MEMS switches the selection of the dielectric thickness should be a compromise between the dielectric charging phenomenon and the RF performance. Another way to look at this is that based on the previous results concerning the impact of both Up and Tp, it is recommended to actuate the capacitive MEMS switch using smaller bias amplitude and/or for shorter duration, in order to minimize the dielectric charging.

978-1-4244-5430-3/10 $26.00 © 2010 IEEE

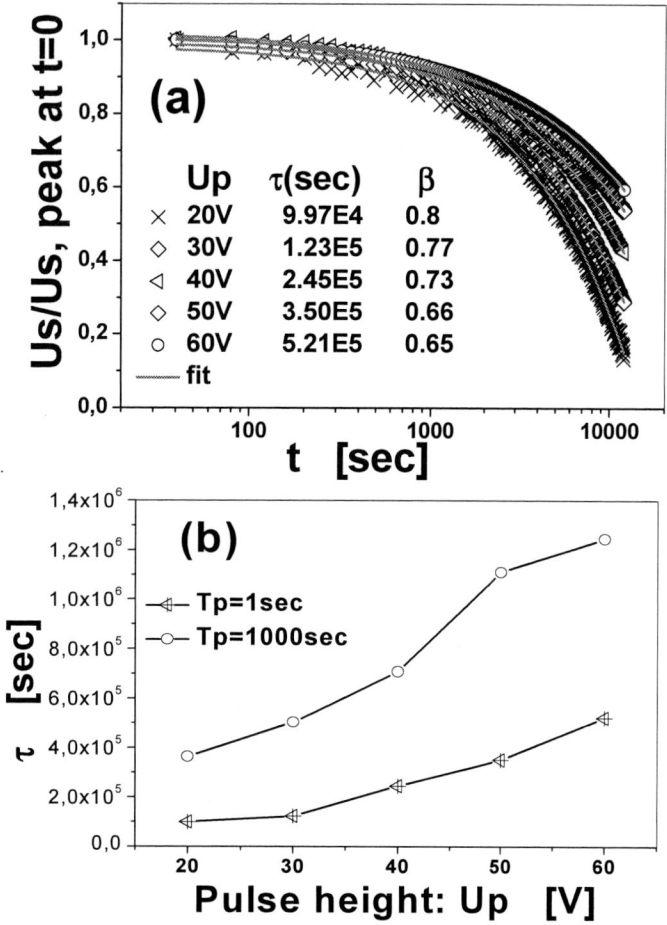

Figure 6. The decay of surface potential under different Up and Tp for the HF-Evap-Au samples with 200nm film thickness. (a) Surface potential decay under different Up, Tp=1sec. (b) Calculated decay time constants versus the applied pulse amplitude Up, for two different pulse durations (Tp= 1sec and Tp= 1000sec).

B. Effect of substrate and material deposition conditions

The influence of the material deposition conditions and the substrate is presented in Fig. 7 and Fig. 8. Here it should be highlighted that the SiN samples with the same thickness in this study have been deposited at the same time and using the same deposition conditions. For a given Tp, the samples HF-Evap-Au, HF-ECD-Au and HF-Ti are found to have almost the same Us (Fig. 7a) and the same FWHM (Fig. 7b) under different Up. Additionally, these samples are found to have very close values for the decay time constants τ (Fig. 8) when the same Up and Tp are applied. This indicates that for the HF material the conductivity of the underneath metal layer has a minimal effect on the charging and discharging processes. It is also evident from the results that the HF-Si samples feature smaller Us, larger FWHM values and finally longer decay time constants, when compared to the HF samples deposited over metal (HF-Evap-Au, HF-ECD-Au, HF-Ti). The same observations can be made for LF SiNx samples excluding the fact that the FWHM values for LF material deposited over metal features large dispersion at higher Up which is probably

due to the LF material defects. The density of defects is expected to be larger in the case of LF material in comparison to the HF material due to the high RF power used during the LF film deposition (see Table 2) which results in a very high deposition rate for the LF film when compared to the deposition rate for the HF material.

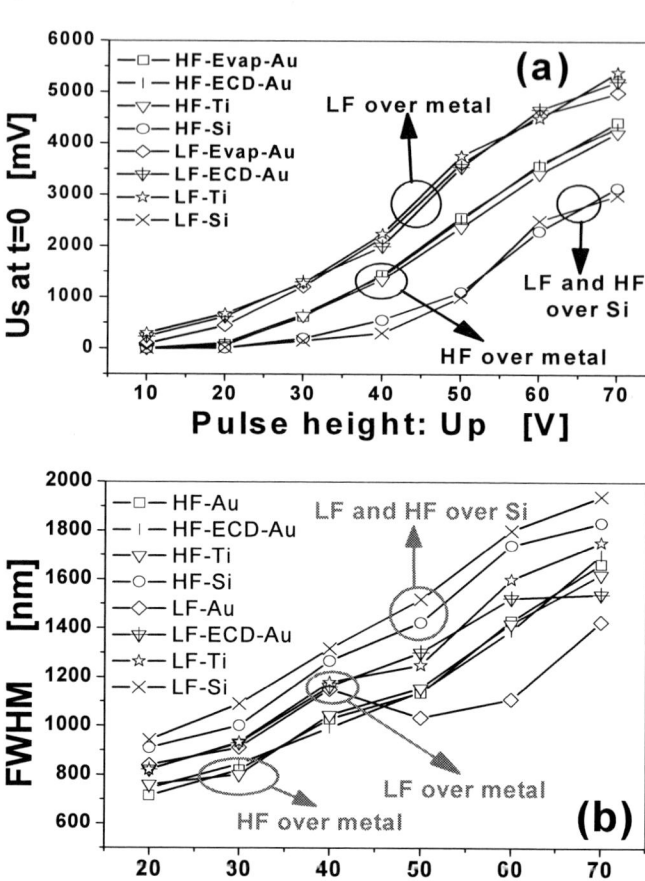

Figure 7. Effect of Up on (a) Us at t=0, (b) FWHM, for the 200nm SiN samples which feature different deposition conditions over various substrates, Tp=1sec.

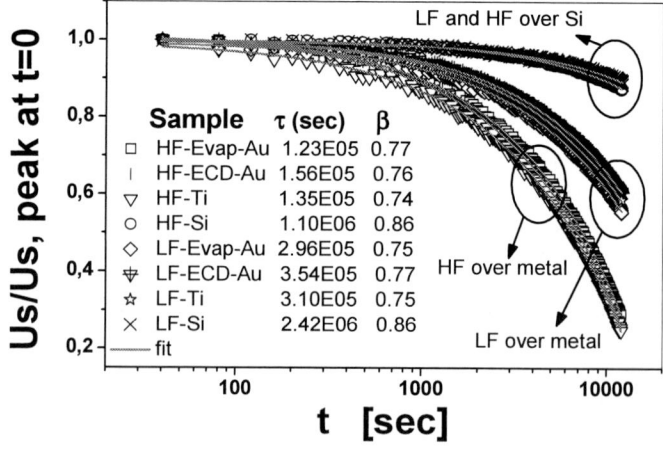

Figure 8. Surface potential decay for the 200nm SiN samples which feature different deposition conditions over various substrates, Up=30V and Tp=1sec.

In order to further understand the difference between the behavior of SiN samples deposited over metal layers and those deposited over bare silicon, FTIR and XPS techniques have been also performed for the SiN samples with different thicknesses deposited over bare silicon substrates, HF-Si. The FTIR spectra for the HF-Si samples are presented in Fig. 9. From the figure, it is obvious that the SiN samples deposited over silicon has a strong Si-H bond very close to $2159 cm^{-1}$ and N-H bond at around $3362 cm^{-1}$, which are very close to what have been found for HF-Evap-Au samples (see Fig. 5). In other words, the SiN samples deposited over metal and over silicon substrates have close FTIR absorptions, independent of the SiN film thickness. XPS measurements for HF-Si samples (see Table 3) highlight that the chemical compositions of SiN films deposited over metal layer or over silicon substrate are almost the same with the ratio of nitrogen to silicon around 0.72 for the SiN bulk. Hence, from both FTIR and XPS measurements, the stoichiometry of both HF-Evap-Au and HF-Si are almost the same. Therefore, the difference in the measured Us, FWHM and decay time constants τ for both samples is attributed mainly to the impact of the space charge region which exists in the case of SiN samples deposited over bare silicon substrate. As reported in our previous work [17], the space charge region which exists in the HF-Si and LF-Si samples contributes to the shape of the surface potential distribution during the charging step and also affects the discharging process.

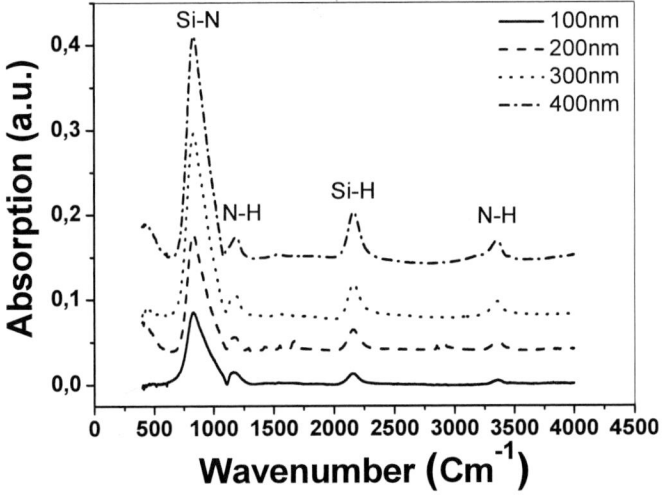

Figure 9. FT-IR transmission spectra for HF-Si samples with different thicknesses.

Comparing the HF and LF results (Fig. 7, 8) reveals that the HF SiN have smaller measured Us (hence smaller trapped charges), smaller FWHM values (hence less charge spreading over the dielectric surface) and finally shorter decay time constants (hence faster charge decay), when compared to the LF material. This can be attributed to the difference in stoichiometry and resistivity and finally to the difference in defect concentration and distribution in both materials. As presented in [9], LF SiN material has smaller resistivity and is more Si rich when compared to the HF one. Based on these

results, the proposed KPFM failure analysis technique reveals that the HF material is more reliable with respect to the dielectric charging compared to the LF material. We could further conclude that MEMS switches employing HF SiN as a dielectric layer would be expected to have longer lifetimes than the switches employing the LF material.

These findings further recommend the usage of Us and the decay time constant τ extracted from the KPFM measurements in a combined approach as a method for the figure-of-merit for dielectrics being used in electrostatic capacitive MEMS switch. This is because the KPFM method allows the monitoring of the dependence of charging on the polarity of the applied bias (Fig. 3). Moreover it allows the monitoring of the dependence of the charging on the nature and roughness of the substrate on which the dielectric film is deposited (Fig. 7a, b), the dielectric film thickness (Fig. 4) as well as the dielectric material composition (Fig. 7a, b). Finally, it allows the monitoring and prediction of the dependence of decay/relaxation time constant on the already mentioned parameters (Fig. 6 and 8).

The impact of the present work on the study of real MEMS devices becomes obvious taking into account that the KPFM method simulates the local charging caused by asperities which are present on the surface of a MEMS suspended electrode. The proposed technique cannot simulate the non flat MEMS electrodes and the additional field emission induced charging due to the small gap in the pull-down state. Yet, it allows the monitoring of the charge decay process over different positions on the dielectric film surface [14]. Also, the KPFM based methodology provides the possibility of monitoring the distribution of the charging process and its dependence on factors such as the composition of the dielectric film and nature of substrate as well as the polarity of applied bias, duration of charging, etc. These factors are key issue parameters for the reliability and the lifetime prediction of RF-MEMS capacitive switches.

An important issue that is worth mentioning here is the possible impact on the dielectric charging behavior due to the processing steps that follow the SiN deposition, during the fabrication of real MEMS devices. For example, in MEMS switches the air gap is normally made by etching away a sacrificial layer material which is deposited on top of the dielectric film. As the etchant has to stop on the SiN film, it may affect the charging behavior of the dielectric material. The SiN samples which have been investigated in this work consist only of thin dielectric films deposited over planar substrates and did not go through any photolithography steps or further chemical treatments. Consequently, the possible impact of further chemical treatments on the SiN charging behavior has not been addressed and it requires further investigation. An initial estimation related to this effect could be extracted from our previous work [17] where KPFM was used to study the impact of the substrate over which SiN material will be deposited on the dielectric charging. In that study, two groups of samples were explored. The first group of SiN films was implemented in a real MEMS switch and hence was exposed to further processing steps while the second group of samples was not exposed to any chemical treatment after the dielectric deposition. The impact of the substrate on the charging/discharging behavior of both groups of samples was found to be the same. The only existing difference was related to the exact values of both decay time constants and stretch factors for both groups. It was found that the first group of samples features larger decay time constants and stretch factors compared to the second one.

IV. CONCLUSIONS

A new failure analysis methodology has been proposed for the dielectric charging phenomenon in electrostatically driven MEMS devices. The technique is based on KPFM surface potential measurements and the investigated dielectric material is PECVD silicon nitride films in view of application in electrostatic capacitive RF MEMS switches. Different parameters have been explored including the impact of the dielectric film thickness, the substrate nature and finally the material deposition conditions. The impact of the pulse duration, pulse amplitude and pulse polarity which have been employed during the charge injection step, have been also studied. Finally, the chemical composition of SiN films for different samples has been investigated through FTIR and XPS. The results from these techniques have been then correlated to the KPFM surface potential measurements.

The electric field intensity along with the pulse duration applied during the charge injection step are found to be the main dominant parameters in determining both the amplitude as well as the distribution of the induced surface potential. Hence, for a given bias amplitude and charge injection duration, thinner dielectric films are found to have larger induced surface potential and hence higher charge trapping. Thus, thicker SiN films would be preferred for MEMS switches for less dielectric charging, considering the C_{on}/C_{off} ratio degradation. Additionally, the relaxation time is found to be higher when charges are injected using larger bias amplitudes due to the Poole-Frenkel effect.

When SiN films are deposited over metal layers, the conductivity of the employed metal layer is found to have almost no effect on the charging/discharging of the dielectric. Hence, the induced surface potential amplitude, distribution and decay with time, are found to be almost the same for SiN films deposited over evaporated Au, electrodeposited Au and titanium, when the same charge injection conditions are applied. On the other hand, the space charge region which exists in SiN films deposited over bare silicon substrate, is found to affect the surface potential amplitude, distribution and decay. For example, charges are found to decay faster in SiN films deposited over metal layers comparing to the dielectric films deposited over silicon substrates. Finally, high frequency PECVD SiN films are found to be more resistant to the dielectric charging comparing to the low frequency ones. For the same charge injection conditions, the HF SiN features smaller surface potential amplitude, more confined potential distribution and faster charge decay, when compared with the LF material.

ACKNOWLEDGMENT

This work has been partially supported by the following projects: the French ANR project FAME (PNANO-059), POLYNOE project funded by European Defense Agency (B-0035-IAP1-ERG), and the SYMIAE project funded by the Fondation STAE.

REFERENCES

[1] Yao, J. J., "RF MEMS from a Device Perspective", Journal of Micromechanics and Microengineering, Vol. 10, pp.R9-R38, 2000.

[2] J. Wibbeler, G. Pfeifer, and M. Hietschold, "Parasitic charging of dielectric surfaces in capacitive microelectromechanical systems (MEMS)," Sens. Actuators A, Phys., vol. 71, no. 1/2, pp. 74–80, Nov. 1998.

[3] G. Papaioannou, M. Exarchos, V. Theonas, G. Wang, and J. Papapolymerou, "Temperature Study of the Dielectric Polarization Effects of Capacitive RF MEMS Switches", IEEE Trans. on Microwave Theory and Techniques, vol. 53, pp. 3467-3473, 2005.

[4] S. Melle, D. De Conto, D. Dubuc, K. Grenier, O. Vendier, J.L. Muraro, J.L. Cazaux and R. Plana, "Reliability Modeling of Capacitive RF MEMS", IEEE Tans. on Microwave Theory and Techniques Vol. 53, pp. 3482, 2005.

[5] K.J.B.M. Nieuwesteeg, J. Boogaard, G. Oversluizen and M.J. Powell, " Current-stress induced asymmetry in hydrogenated amorphous silicon n-i-n devices", J Appl. Phys. Vol 71, pp. 1290–1297, 1992.

[6] U. Zaghloul, G. Papaioannou, F. Coccetti, P.Pons and R. Plana, "Dielectric Charging in Silicon Nitride Films for MEMS Capacitive Switches: Effect of Film Thickness and Deposition Conditions", Journal of Microelectronics Reliability 49, 1309-1314, 2009.

[7] X. Yuan, , J. C. M. Hwang, D. Forehand, and C. L. Goldsmith, "Modeling and Characterization of Dielectric-Charging Effects in RF MEMS Capacitive Switches", International Microwave Symposium 2004.

[8] R. Daigler, E. Papandreou, M. Koutsoureli, G. Papaioannou, J. Papapolymerou, "Effect of deposition conditions on charging processes in SiNx: Application to RF-MEMS capacitive switches", Microelectronic Engineering 86 , pp. 404–407, 2009.

[9] M. Lamhamdi, J. Guastavino, L. Boudou, Y. Segui, P. Pons, L. Bouscayrol, R. Plana: Charging-Effects in RF capacitive switches influence of insulating layers composition. Microelectronics Reliability 46(9-11): pp. 1700-1704 , 2006.

[10] E. Papandreou, M. Lamhamdi, C.M. Skoulikidou, P. Pons, G. Papaioannou and R. Plana, "Structure dependent charging process in RF MEMS capacitive switches", Journal of Microelectronics Reliability 47, pp 1812–1817, 2007.

[11] R.W. Herfst, H.G.A. Huizing, P.G. Steeneken and J. Schmitz, "Characterization of dielectric charging in RF MEMS capacitive switches", IEEE International Conference on Microelectronic Test Structures (ICMTS-06), pp 133-136, 2006.

[12] J. Ruan, G.J. Papaioannou, N. Nolhier, N. Mauran, M. Bafleur, F. Coccetti and R. Plana, "ESD failure signature in capacitive RF MEMS switches", Microelectronics Reliability 48, pp. 1237–1240, 2008.

[13] M. Lamhamdi, L. Boudou, P. Pons, J. Guastavino, A. Belarni, M. Dilhan, Y. Segui and R. Plana, "Si3N4 thin films properties for RF-MEMS reliability investigations", The 14th International Conference on Solid-State Sensors, Actuators and Microsystems, Lyon, France, June 10-14, pp 579-582, 2007.

[14] R.W. Herfst, P.G. Steeneken, J. Schmitz, A.J.G. Mank and M. van Gils, "Kelvin probe study of laterally inhomogeneous dielectric charging and charge diffusion in RF MEMS capacitive switches", 46th Annual International Reliability Physics Symposium, Phoenix, pp. 492-495, 2008.

[15] Belarni, M. Lamhamdi, P. Pons, L. Boudou, J. Goustavino, Y. Segui, G. Papaioannou and R. Plana," Kelvin probe microscopy for reliability investigation of RF-MEMS capacitive switches", Journal of Microelectronics Reliability 48, pp. 1232-1236, 2008.

[16] U. Zaghloul, A. Abelarni, F. Coccetti, G. Papaioannou, R. Plana and P. Pons, "Charging processes in silicon nitride films for RF-MEMS capacitive switches: The effect of deposition method and film thickness", MRS Fall Meeting, Boston, December 1-5, 2008.

[17] U. Zaghloul, A. Abelarni, F. Coccetti, G. Papaioannou, L. Bouscayrol, P. Pons and R. Plana, "A Comprehensive Study for Dielectric Charging Process in Silicon Nitride Films for RF MEMS Switches Using Kelvin Probe Microscopy", the 15th International Conference on Solid-State Sensors, Actuators, and Microsystems, Denver, pp. 1120-1125, 2009.

[18] H. O. Jacobs, H. F. Knapp, and A. Stemmer, "Practical aspects of Kelvin probe force microscopy," Rev. Sci. Instrum., vol. 70, pp. 1756-1760, 1999.

[19] S. Ono, M. Takeuchi, and T. Takahashi, "Kelvin probe force microscopy on InAs thin films grown on GaAs giant step structures formed on (110) GaAs vicinal substrates", Appl. Phys. Lett. 78, 1086, ,2001.

[20] H. Sugimura, Y. Ishida, K. Hayashi O. Takai, and N. Nakagiri, "Potential shielding by the surface water layer in Kelvin probe force microscopy", Appl. Phys. Lett. 80, 1459, 2002.

[21] U. Zaghloul, G.J. Papaioannou, F. Coccetti, P. Pons and R. Plana, "Effect of Humidity on Dielectric Charging Process in Electrostatic Capacitive RF MEMS Switches Based on Kelvin Probe Force Microscopy Surface Potential Measurements", MRS Fall meeting, DOI: 10.1557/PROC-1222-DD02-11, 2009.

[22] R. Ramprasad, "Phenomenological theory to model leakage currents in metal–insulator–metal capacitor systems", Physica status solidi. (B), Vol. 239, No. 1, pp. 59-70, 2003.

[23] G.M. Sessler (Ed.), Electrets, (Springer topics in applied physics Vol.33), Springer Verlag, Berlin, 2nd ed., 1987.

[24] D. Mardivirin, D. Bouyge, A. Crunteanu, A. Pothier, P. Blondy, " Study of residual charging in dielectric less capacitive MEMS switches", IEEE MTT-S Int. Microwave Symp. Dig., 2008.

[25] G. Papaioannou, F. Coccetti and R. Plana, "On the Modeling of Dielectric Charging in RF-MEMS Capacitive Switches", 10th topical meeting on silicon monolithic integrated circuits in RF Systems (SIRF), New Orleans, pp.108-111, 2010.

[26] Simmons JG, "Poole–Frenkel effect and Schottky effect in metal–insulator– metal systems", Phys Rev;155:657–60, 1967.

Accelerated Testing of RF-MEMS Contact Degradation through Radiation Sources

A. Tazzoli[*], M. Barbato, V. Giliberto, G. Monaco, S. Gerardin, P. Nicolosi, A. Paccagnella, and G. Meneghesso

Department of Information Engineering, University of Padova
Via Gradenigo 6/B, 35131, Padova, Italy
[*]Tel: +39 049 8277664; e-mail: augusto.tazzoli@dei.unipd.it

Abstract — **This work aims to propose a novel method to accelerate the lifetime of ohmic RF-MEMS switches by means of radiation exposure. Experimental results of proton and γ-ray irradiation were compared to cycling stresses, obtaining similar degradation in electrical performances. Electrical measurements, RF simulations, and AFM analysis of surface roughness were carried out to verify the proposed method.**

Keywords: RF-MEMS, accelerated testing, radiation

I. INTRODUCTION

Micro-Electro-Mechanical-Systems (MEMS) for radio frequency (RF) applications are acquiring an increasing number of admirers day after day thanks to their proven better performance than actual state-of-the-art solid state devices, and are becoming true contenders for future low-power, high-frequency, wired reconfigurable networks and wireless communication systems [1]-[2]. In many RF applications, a single MEMS device can replace and outperform an entire solid-state circuit; in other applications, a judicious combination of MEMS with active devices will result in smart communicating circuits and systems. Examples of RF-MEMS are 3D-suspended inductors, resonators, varactors, and switches. RF-MEMS devices in general exhibit almost zero power consumption, extremely good RF linearity, and high quality factor (Q) for tuning. Linearity metrics of input intercept second and third order harmonics (IIP2 and IIP3) both can achieve better than 70 dBm. RF-MEMS switches in particular achieve very low insertion loss (IL) of lower than 0.1 dB (up to 100 GHz) while maintaining high isolation (>20 dB). From a technology standpoint, they employ a low-cost fabrication process compared to other RF and microwave solid-state integrated circuits (e.g., RF Silicon on Insulator (SOI), Gallium Arsenide (GaAs), etc.). Several families of micromachined devices have been flooding the market for quite some time, the big examples of which being ink-jet heads, accelerometers and gyroscopes for both navigation systems and entertainment, pressure sensors, microphones, and micro-mirrors for video projection [3].

Despite the relative technological simplicity of RF-MEMS (Westinghouse and IBM demonstrated the first silicon based MEMS switch in the late 1960's), their reliability has not yet completely assessed because of the lack of standardized processes and near absence of accelerated lifetime tests, compared to traditional silicon solid state technology. Furthermore, since one of the most promising application of RF-MEMS devices is to be found in space applications (e.g., redundancy switches to substitute traditional bulky and heavy coaxial relays), typical reliability problems of solid state devices are compounded by space related problems, including the influence of radiation sources.

The study of different radiation effects on micromachined devices started to appear in the literature some years ago, in particular concerning both accelerometers and RF switches [4]-[15]. Shea made a very rich and complete review on the sensitivity of MEMS devices in 2009 [16], with the following main results: (i) silicon does not show heavy mechanical degradation for radiation doses lower than 100 Mrad; (ii) MEMS devices can fail at doses between 20 krad and 10 Mrad; (iii) in general radiation sensitivity depends strongly on the sensing or actuation principle, device design, and materials. Probably the most common result (iv) is that the failure or degradation is mainly due to the radiation-induced trapped charge in dielectrics.

In this work we show experimental results of proton and gamma ray irradiation effects on electrostatically actuated dielectric-less RF-MEMS switches. Confirming preliminary results shown in our previous work [17], the first result is the degradation of the insertion losses of tested devices, and an almost negligible dielectric charging. The second result concerns the comparison of the degradation induced by cycling and radiation stresses, with the aim of understanding if radiation stresses could be used as an accelerating mechanism in the study of MEMS reliability.

II. DEVICE DESCRIPTION AND MEASUREMENT SETUP

This work was based on the characterization and reliability analysis of dielectric-less electrostatically actuated ohmic RF-MEMS clamped-clamped switches, manufactured by FBK-IRST (Trento, Italy). The technology utilized for the fabrication of these devices consists of an eight-mask surface micromachining process [18]. The process allows the electrodeposition of two gold layers of different thicknesses for movable bridges with different mechanical properties and microstrip/coplanar waveguide lines. The suspended membrane is realized using a 3-µm thick photoresist as sacrificial layer. The bridge release is done with a modified plasma ashing process in order to avoid sticking problems. Polysilicon actuator lines, designed in an interdigitated configuration, are 400 nm below the gold contact points to avoid short circuit issues with the RF line, and the SiO_2 layer was removed over

978-1-4244-5430-3/10 $26.00 © 2010 IEEE

them in order to reduce charge trapping issues. Figure 1 shows a picture of tested switches (a), and a sketch of the technological process (b). Tested configurations were both series (c) and shunt (d) devices. Since behaviors of the series and shunt switches were almost identical, here we will show the results of the characterization of shunt switches only.

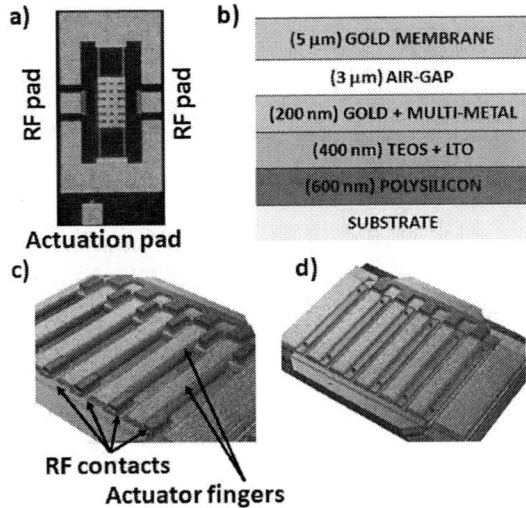

Figure 1. Picture of tested shunt switches (a); schematic process description (b); optical profilometer images after bridge removal of series (c), and shunt (d) topologies.

The effect of γ-rays with two different doses and dose rates (in SiO_2) was studied using two Cobalt-60 γ-cells. Doses were 60 krad @ 2.9 krad/h, and 1.2 Mrad @ 54 krad/h. We have used the proton irradiation facilities at INFN laboratory in Legnaro, Italy, in order to carry out 2 MeV proton stress at increasing doses of 1, 10, and 30 Mrad (on SiO_2) on on-wafer devices (radiation impinged RF-MEMS switches perpendicularly from the top of the structure). Note that the dose is the amount of energy deposited per unit of mass in the target material (measured in rad, 1 rad = 100 erg/g). The dose reported here was calculated using the energy of the particles at the surface, so it is strictly valid only at the surface. As a particle loses energy into a material, it slows down, changing its linear energy transfer (the rate at which it transfers energy to the target material) and so the dose [19], but in this case the structure is very thin compared to the range of the radiation, making the surface approximation reasonable. More details on Monte Carlo (TRIM) simulation of displacement damage for 2 MeV protons on tested devices was shown in [17].

Although doses in the Mrad range are probably greater than the doses typically collected by devices during their entire life [20], even in space applications, such high doses were applied with the aim of using them for reliability investigation purposes.

The setup adopted to characterize tested switches is based on a vector network analyzer (Agilent HP8753), for the measurement of the RF-performances, and a source meter (Keithley 2612) to bias the device and measure the current drained by the actuator structure. The measurement starts from 0 V up to $+V_{MAX}$, decreases down to $-V_{MAX}$, and finally comes

back to 0 V, see Figure 2a. This procedure leads to the traditional hysteresis-like diagram reported in Figure 2b, from which it is possible to extract the actuation voltage ($|V_{ACT}|$) the release voltage ($|V_{REL}|$), and the S_{11} and S_{21} parameters. During such measurements, the applied RF signal was a continuous wave (CW) sinusoidal signal at 6 GHz, 0 dBm. Furthermore, considering the graph's symmetry or translation, it is possible to study the presence of charge trapping or redistribution phenomena. More details on the setup are reported in [21].

Cycling stresses were carried out using a solid state pulser HP 8114A, whereas the evaluation of the resistance variation induced by cycling and radiation stresses was done using a Keithley 2612 source meter. To conclude, failure analysis was carried out using an atomic force microscope (Park System XE-70), and an optical profilometer (Polytec MSA-500, for both static topography and laser doppler vibrometer dynamic measurements).

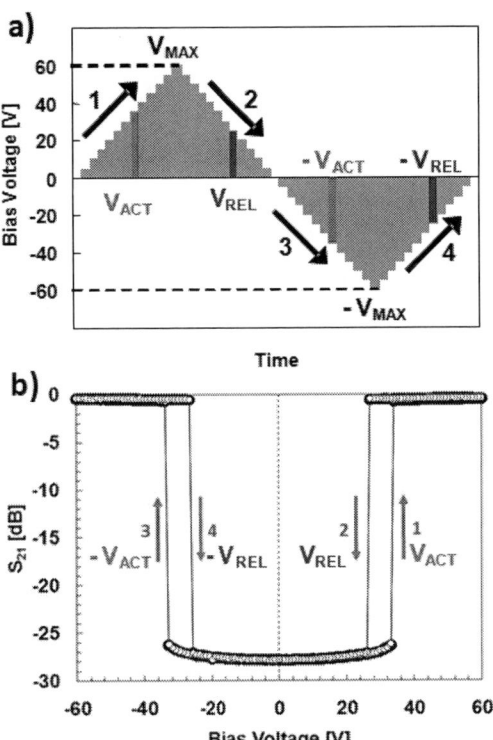

Figure 2. Bias signal waveform adopted for the characterization of the electrical ($\pm V_{ACT}$, $\pm V_{REL}$) and rf performances (S-parameters) of a series switch.

III. EXPERIMENTAL RESULTS

An excerpt of the evolution of $S_{21}(V_{BIAS})$ of shunt switches stressed with γ-rays at doses of 60 krad @ 2.9 krad/h, and 1.2 Mrad @ 54 krad/h is reported in Figures 3 and 4, respectively. The behavior shown just after the radiation stress was almost identical to the fresh one, but it starts to gradually change days after the radiation stress (devices were left at ambient temperature inside wafer holders).

Measurements have been repeated after several days from the stress, obtaining a heavy degradation of the S-parameters,

978-1-4244-5430-3/10 $26.00 © 2010 IEEE

indicative of a degradation of the metal-to-metal contact (increase of the switch resistance).

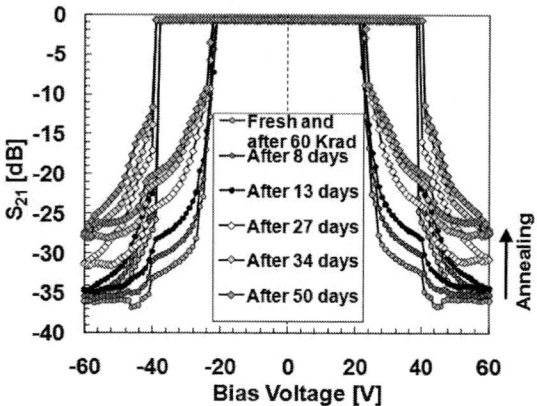

Figure 3. Degradation of $S_{21}(V_{BIAS})$ of a shunt switch after γ-ray stress (60 krad @ 2.9 krad/h) - f_{rf} = 6 GHz, P_{rf} = 0 dBm.

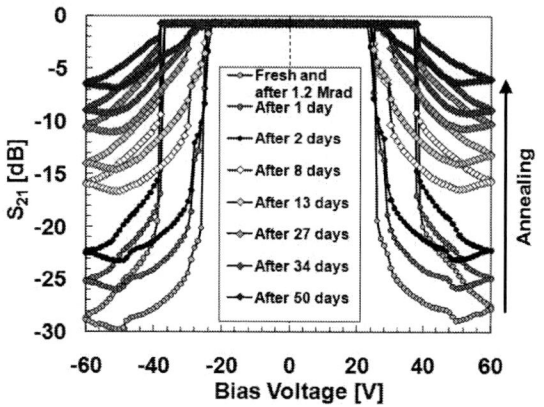

Figure 4. Degradation of $S_{21}(V_{BIAS})$ of a shunt switch after γ-ray stress (1.2 Mrad @ 54 krad/h) - f_{rf} = 6 GHz, P_{rf} = 0 dBm.

Note that the devices show a symmetric behavior for both positive and negative bias voltages, with the actuation voltage almost fixed at +39 V, and the release voltage constant at about +24 V, hence no charge trapping or charge redistribution issues are expected on tested devices after γ-ray irradiation stresses, which is contrary to the results of other experiments presented in the literature on RF-MEMS switches [8]-[15].

In order to better analyze the evolution of the RF parameters during the aging time, Figure 5 shows the degradation of the S_{21} parameter @ +60 V as a function of time for the two dose and dose rates.

From the comparison of the curves, we can extrapolate the following results: (i) the higher dose (1.2 Mrad) induced a heavy degradation of the S_{21} parameter, symptomatic of a strong degradation of the metal-to-metal contact of such ohmic switches; (ii) S_{21} degradation at higher dose seems to follow a square root evolution up to 10 days, and then continuing to degrade with a linear trend, in contrast to the almost linear trend shown by 60 krad exposed sample from just after the irradiation; (iii) the linear trends of both the curves have the same slope, a good indication that the failure mechanism,

leading to the increase of the contact resistance, is possibly the same.

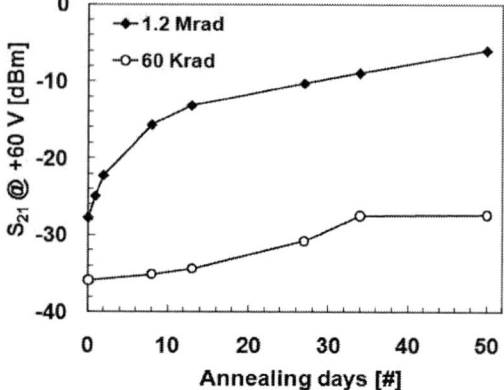

Figure 5. Degradation of S_{21} measured @ +60 V as a function of time for the two γ-ray stresses (values taken from Figures 3 and 4).

More comments on the proposed degradation mechanism, related to hydrogen bubble formation, will be found in the next section.

In order to verify that the dielectric layers were not impaired by γ-ray irradiation, we have submitted stressed devices to a continuous actuation stress, about 60,000 s long. Such a test is very useful to determine the robustness of switches to continuous actuation, e.g., when used as a redundancy switch, where the device must stay in the actuated state for a long time. As described in [22], the shift, or narrowing, of the S-parameters vs. bias voltage curves can furnish useful information on the propensity of tested devices to fall into stiction issues. Comparing the behavior shown by fresh and irradiated devices it is also possible to evaluate any eventual degradation of the oxide over (not in this case) or around the actuator structure. Figure 6 shows the results of such a stress, comparing the narrowing of the positive actuation voltage (normalized to the initial value) of a fresh and a 1.2 Mrad γ–ray exposed sample continuously actuated @ V_{BIAS} = +50 V. Both devices showed the same behavior, another piece of evidence that charge trapping or oxide degradation was not induced by gamma irradiation stresses. An identical behavior was shown by 60 krad γ-ray irradiated devices.

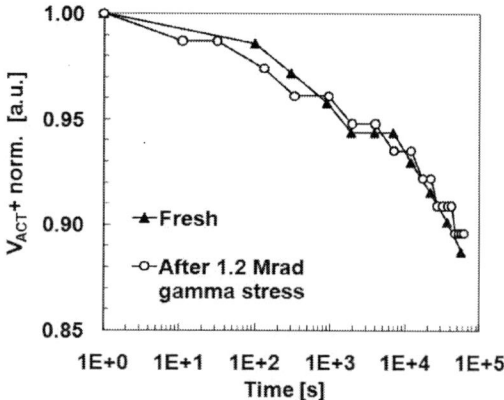

Figure 6. Comparison of the normalized positive actuation voltage variation of a fresh and 1.2 Mrad gamma ray stressed switch.

The degradation of the RF performances of γ-ray stressed switches, without evidence of charge trapping issues, is very similar to the behavior shown by similar samples exposed with 2 MeV protons, as shown in Figure 7.

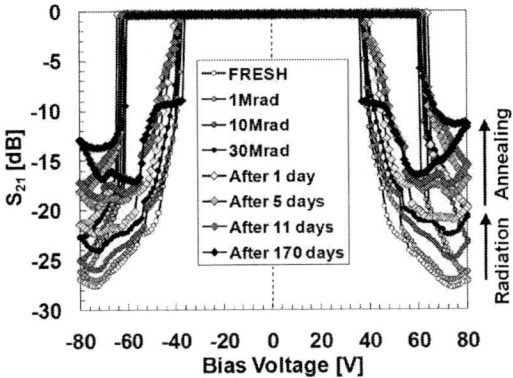

Figure 7. Degradation of S_{21} (V_{BIAS}) of a shunt switch during the proton irradiation stress and the successive days of annealing at room temperature.

S-parameters vs. bias voltage was measured after 1, 10, and 30 Mrad proton irradiation, and after successive days of aging carried out at room temperature.

Like after gamma ray stresses, protons induced a strong degradation of the RF performance, but almost no variation of actuation and release voltages, as well as of the symmetry of the curves. It is believed that the RF degradation is not due to an increase of the coplanar waveguide (CPW) or substrate losses, since repeating the measurements at different frequencies from 100 kHz to 6 GHz (also on simple CPW irradiated test structures) do not show any frequency-related dependence (more details can be found in [17]). Furthermore, in the off-state (0 V) RF parameters do not show any change. The interesting point is that both gamma ray and proton stresses induce a degradation of the RF parameters similar to the one induced by cycling stress, as shown in Figure 8 after 10^3, 10^4, 10^5, and 10^6 cycles at +60 V.

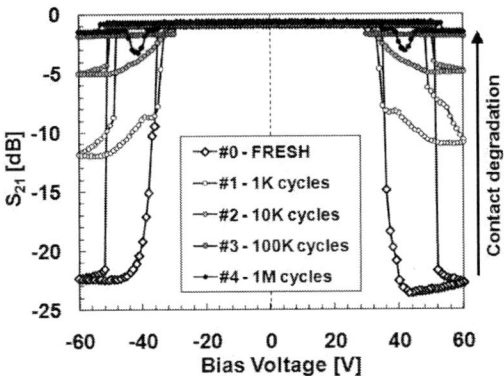

Figure 8. Degradation of S_{21}(V_{BIAS}) of shunt switches during cycling stress after 1k, 10k, 100k, and 1Meg cycles at +60 V.

The degradation of the switch RF performance (in on-state) induced by the cycling stress can be explained by the increase of the resistance of the metal-to-metal contact between the suspended gold membrane and the bottom gold layer. This hypothesis was verified measuring the contact resistance of cycled samples at selected intervals with a Keithley 2612 source meter, applying a DC signal of 100 mV. Measured behavior is shown in Figure 9.

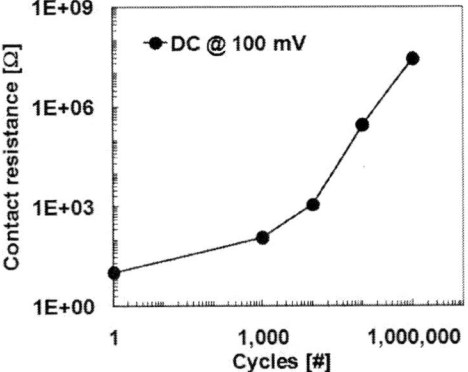

Figure 9. Degradation of the shunt switch contact resistance (measured in DC regime @ 100 mV) due to cycling.

There is little work in the literature on the degradation of the ohmic contact of MEMS devices, but the degradation (increase) of the contact resistance can be explained by considering the collision between the suspended membrane and the bottom contact, leading to material removal and an increase in the surface roughness of the contacting metals. A good description of the nanoscale behavior of MEMS contacts can be found in [23]. In order to understand the influence of the contact resistance on the RF performance of a shunt switch, we have carried out simplified simulations with a freely available RF simulator (Portview). Despite the simplicity of the model, in which the contact points of the RF-MEMS are modeled as a capacitor and a resistor connected between the center of the transmission line and ground, (see the inset in Figure 10) simulated data are in good agreement with experimental results. Figure 10 shows the decrease of the insertion loss (as expected for a shunt switch) with the increase of the shunt resistance value, and simulated points are in good agreement with both S_{21} values shown in Figure 8 and resistance values reported in Figure 9, shown with triangles in Figure 10 (the difference between simulated and measured results is due to a difference in the estimated shunt parasitic capacitance value).

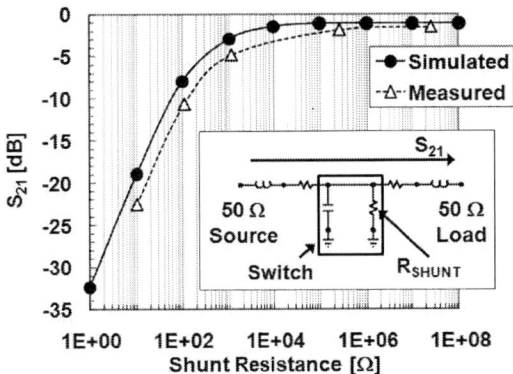

Figure 10. Simulation of the S_{21} parameter of a shunt switch as a function of the contact resistance (R_{SHUNT}).

The experimental evidence points to considering the physical degradation of the metal-to-metal contact due to both radiations and cycling stress. In order to understand the cause of the contact degradation induced by radiation exposure, we acquired 10 μm x 10 μm wide AFM topographies (Park System XE-70) of an untreated sample (see Figure 11) and a proton irradiated sample (see Figure 12).

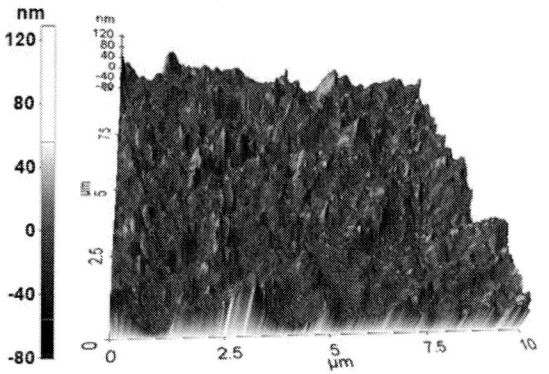

Figure 11. 10 x 10 μm AFM topography of an untreated gold layer. Measured roughness is Rq = 28.4 nm.

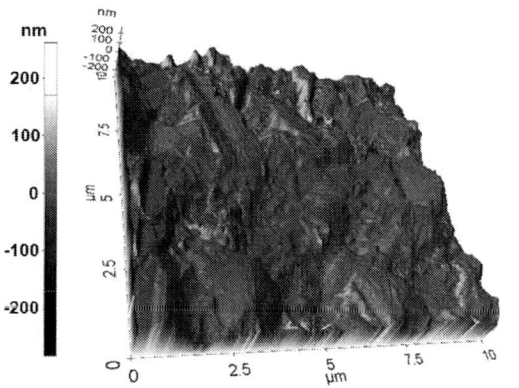

Figure 12. 10 x 10 μm AFM topography of an irradiated gold layer. Measured roughness is Rq = 87.4 nm (note the different z-axis scale with respect of Figure 11).

The AFM image was taken in non-contact mode, which seemed to offer better response than contact mode because of the softness of the material.

There are many different roughness parameters in use; here we adopted the traditional root mean square value to define the profile roughness. The untreated part showed a roughness Rq of about 28.4 nm, whereas the irradiated sample about 87.4 nm, another piece of evidence suggesting radiation stress induced mechanical degradation of the gold layer. Similar results were obtained analyzing the γ-ray irradiated and cycled samples.

IV. DISCUSSION

The comparison of the electrical and RF performance of irradiated (both protons and γ-rays) samples and cycled ones leads us to conclude that the degradation of the electrical performances is very similar, i.e., it is due to the degradation of the metal-to-metal contact between the suspended membrane and the bottom gold layer. If confirmed, this result could be very important, because it could be possible to age MEMS switches and other types of contact-based microstructures by means of radiation sources. Even if at first blush irradiation may not seem convenient because the lower availability of radiation facilities compared to traditional aging techniques of solid state devices (using thermal chambers and/or biasing devices at high electric fields), radiation could be used to stress in parallel (especially with γ-rays inside gamma cells) a huge number of devices directly at wafer level, avoiding expensive packaging solutions necessary to bias and stress in parallel a large number of devices. Protons cannot process in parallel, but they could instead be used to accelerate the degradation of the metal contacts of switches using high dose levels and dose rates.

The result that radiation stresses induce a degradation of the gold layer (an increase of the surface roughness leading to an increase in the contact resistance) is in good agreement with the theory that during irradiation (both protons and gamma rays) hydrogen gas is formed in the sample.

The hydrogen ion (H^+) is in fact the lightest and most common mobile ion impurity in semiconductor devices, and, unlike other heavier ions such as sodium and potassium which have been reduced to extremely low densities in state-of-the art commercial manufacturing facilities, hydrogen remains present in microelectronics processing [24]. It is present in cleaning procedures, film depositions, etches, high-temperature anneals, etc. Inevitably, a relatively large density of hydrogen is present in even the driest and cleanest of all Si-based semiconductor manufacturing processes [25]. Clearly, MEMS facilities are not immune from such problems, rather they can be more prone because of the lower quality required by micromachined structures compared to state-of-the-art MOS devices.

The problem is that not all of the hydrogen introduced during the semiconductor (or MEMS) processing sequence is tightly bound, and Si–H and/or other bonds can be broken when a device is exposed to ionizing radiation [26]. Furthermore, the presence of other contaminant elements, such as oxygen, carbon, etc., in the gold layer can make the material more prone to degradation induced by radiation.

The formation of hydrogen gas can easily cause the formation of vesicles, strong enough to break the lattice of the oxide and then of the metal, leading to an increase of the surface roughness of the stressed gold layer, and hence of the contact resistance [27]. There are few works in the literature studying the effects of hydrogen on metal surfaces, but bubble formation (blisters) was studied in detail in aluminum layers in [28]. Despite the different materials, we think the mechanism described can be adapted also to devices tested in this work, constituted by gold and Al/Ti/Al alloy layers.

To conclude, the process of lattice breakdown evolves in time, explaining the worsening device behavior with time, and it is probably limited by the diffusion properties of hydrogen through all the layers (Si, SiO₂, Al/Ti/Al, and Au, see Fig. 1b) of tested RF-MEMS switches.

CONCLUSIONS

In this work we have characterized by means of electrical, RF, and topography measurements the effects of γ-ray and proton irradiation on the reliability of ohmic RF-MEMS switches. The main result was the heavy degradation of the RF performance, but contrary to what is typically seen in the literature, negligible charge trapping or redistribution issues. Furthermore, the degradation of the metal-to-metal contact induced by radiation and cycling stresses was investigated, obtaining similar results, mainly due to the increase of the roughness of stressed surfaces, as verified by AFM measurements. This result, applicable to other families of micromachined devices, could be used to accelerate lifetime testing of contacts, by enabling the simultaneous stressing of a huge number of devices directly at wafer level with γ-rays, or reducing the stressing time by using a high dose of protons. Both of these solutions would obviate the need for expensive packaging solutions.

REFERENCES

[1] G. Rebeiz, J. B. Muldavin, "RF MEMS switches and switch circuits", IEEE Microwave Magazine, Dec. 2001, pp. 59-71.

[2] J. Ehmke, et al., "Rf MEMS devices: a brave new world for rf technology", Emerging Technologies Symposium: Broadband, Wireless Internet Access, April 2000, pp. 1-4.

[3] J. W. Gardner, V. K. Varadan, O. O. Awadelkarim, "Microsensors, MEMS, and smart devices", John Wiley & Sons, LTD, 2001.

[4] A. R. Knudsen, et al., "The effects of radiation on MEMS accelerometers", IEEE Trans. Nucl. Sci., vol. 43, no. 6, 1996, p. 3122.

[5] C. I. Lee, A. H. Johnston, W. C. Tang, and C. E. Bames, "Total dose effects on micromechanical systems (MEMS): accelerometers", IEEE Trans. Nucl. Sci., vol. 43, no. 6, 1996, p. 3127.

[6] L. P. Schanwald, et al., "Radiation effects on surface micromachined comb drives and microengines", IEEE Trans. Nucl. Sci., vol. 45, no. 6, 1998, p. 2789.

[7] L. D. Edmonds, G. M. Swift, and C. I. Lee, "Radiation Response of a MEMS Accelerometer: An Electrostatic Force", IEEE Trans. Nucl. Sci., vol. 45, Dec. 1998, pp. 2779-2788.

[8] R. N. Schwartz et al., "Gamma-ray radiation effects on RF MEMS switches", Proceedings of the 2000 IEEE Microelectronics Reliability and Qualification Workshop, October 2000, paper IV.6.

[9] S. S. McClure, et al., "Radiation effects in micro-electro-mechanical systems (mems): rf relays", IEEE Transaction on Nuclear Science, vol. 49, 2002, pp. 3197.

[10] M. Exarchos, E. Papandreou, P. Pons, M. Lamhamdi, G. J. Papaioannou, R. Plana, "Charging of radiation induced defects in rf mems dielectric films", Microelectronics Reliability, 2006, doi:10.1016/j.microrel.2006.07.045

[11] J. R. Caffey, P. E. Kluditis, "The effects of ionizing radiation on microelectromechanical systems (MEMS) actuators: Electrostatic, Electrothermal, and Bimorph", IEEE 17th Micro Electro Mechanical Systems, September 2004, pp. 133–136.

[12] J. Ruan, et al., "Alpha particle radiation effects in RF MEMS capacitive switches", Microelectronics Reliability, Vol. 48, August 2008, pp. 1241–1244.

[13] A. Crunteanu, et al., "Gamma radiation effects on RF MEMS capacitive switches", Microelectronics Reliability, Vol. 46, September 2006, 1741–1746.

[14] V. G. Theonas, M. Exarchos, G. J. Papaioannou and G. Konstantinidis, "RF MEMS dielectric sensitivity to electromagnetic radiation", Sensors and actuators A: physical, Vol. 132, November 2006, pp. 25–33.

[15] G. J. Papaioannou, V. Theonas, M. Exarchos, G. Konstantinidis, "RF MEMS Sensitivity to Radiations", 34th European Microwave Conference, Vol. 1, October 2004, pp. 65–68.

[16] H. R. Shea, "Radiation sensitivity of microelectromechanical system devices", J. Micro/Nanolith. MEMS MOEMS 8(3), 031303 (Jul–Sep 2009), pp. 1-11.

[17] A. Tazzoli, G. Cellere, E. Autizi, V. Peretti, A. Paccagnella, G. Meneghesso, "Radiation sensitivity of ohmic rf-mems switches for spatial applications", Proceedings of IEEE MEMS 2009, pp. 634-637.

[18] F. Giacomozzi, et al., "Electromechanical aspects in the optimization of the transmission characteristics of series ohmic rf-switches", Proceedings of MEMSWAVE 2004, pp. C25-C28.

[19] A. Holmes-Siedle, L. Adams, "Handbook of radiation effects", 2nd edition, Oxford University Press, 2002.

[20] J. Mazur, "The Radiation environment outside and inside a spacecraft", IEEE Nuclear and Space Radiation Effects Conference, Section II, 2002.

[21] A. Tazzoli, V. Peretti, G. Meneghesso, "Electrostatic discharge and cycling effects on ohmic and capacitive rf-mems switches", IEEE Transaction on Device and Materials Reliability, vol. 7, no. 3, Sept. 2007, pp. 429-437.

[22] A. Tazzoli, et al., "Evolution of Electrical Parameters of Dielectric-less Ohmic RF-MEMS Switches during Continuous Actuation Stress", IEEE ESSDERC 2009, September 2009, pp 343-346.

[23] B. D. Jensen, L. W. L. Chow, H. Kuangwei, K. Saitou, J. L. Volakis , K. Kurabayashi, "Effect of nanoscale heating on electrical transport in RF MEMS switch contacts", Journal of Microelectromechanical Systems, Vol. 14, Issue 5, Oct. 2005, pp. 935-946.

[24] E. H. Nicollian, J. R. Brews, "MOS (Metal oxide semiconductor) physics and technology", New York: Wiley; 1982.

[25] R. Gale, et al., "Hydrogen migration under avalanche injection of electrons in Si MOS capacitors", J. Appl Phys 1983;54:6938–42.

[26] D. M. Fleetwood, "Effects of hydrogen transport and reactions on microelectronics radiation response and reliability", Microelectronics Reliability, 42 (2002), pp. 523–541.

[27] L. Kogut, K. Komvopoulos, "Electrical contact resistance theory for conductive rough surfaces", Journal of Applied Physics, 94 (5), 2003, pp. 3153-3162.

[28] P. Rozenak, "Hemispherical bubbles growth on electrochemically charged aluminum with hydrogen", International Journal of Hydrogen Energy, 32 (2007), pp. 2816-2823.

X-ray Computed Tomography for Non-Destructive Failure Analysis in Microelectronics

Mario Pacheco, Deepak Goyal

Intel Corporation, M/S CH5-263, 5000 W. Chandler Blvd., Chandler, AZ 85226 USA
mario.pacheco@intel.com (480)552-5772

Abstract

In this paper a review of the development of x-ray computed tomography (CT) for non-destructive failure analysis in microelectronics is presented. The general operation principle, key design considerations and technical challenges faced by x-ray CT technology are discussed. A comparison between 2D and 3D x-ray imaging capabilities is presented, and critical failure analysis case studies that are hard or not possible to isolate by alternative methods are also discussed, as well as its unique progressive testing capability.

Introduction

The assembly package development roadmap has been increasing interconnect complexity, component density, the number of stacked dice, and material composition, while at the same time reducing critical dimensions. As a result of these trends, the isolation and root cause analysis of defects has become increasingly challenging for traditional analytical tools and techniques. In addition, reduced time-to-information and non-destructive approaches have become critical factors, introducing more challenges into the analytical tools development roadmap. One of these analytical capabilities is 2D x-ray imaging, which is also one of the more extensively used tools in the semiconductor industry to isolate and analyze defects in a non-destructive fashion. Some of these capability challenges are the detection of metal migration and dendrite growth, solder join non-wetting and cracking at both first and second level interconnect, wirebonds shorting and breakage, etc. One of the x-ray techniques identified as potentially feasible to fulfill these defect detection gaps was x-ray CT [1]. However, the standard assembly imaging techniques of x-ray CT had important limitations in terms of defect detection capability and time-to-data, and thus it needed to be extended into the next generation [2]. Key technical drivers for x-ray CT are the need to non-destructively and quickly detect micron and sub-micron sized defects.

In this paper, we review key technical challenges faced by x-ray CT technology, leading technical developments that have overcome some of these challenges, as well as application development for critical defect detection case studies.

X-ray CT Operation Principle

There are several possible configurations to achieve the challenges of the key drivers; however, all of them relay of a basic operation principle of using an x-ray source to radiate the object at different tilt angles, with a rotating stage providing such angular displacement in equally spaced angles, and a detector collecting the 2D x-ray images at each angle. This is schematically shown in Fig. 1. All 2D images are mathematically superimposed and processed to obtain a three dimensional map of the sample, as it is shown schematically in Fig. 1b. Information is typically displayed as a cross-section, slice view and 3D model.

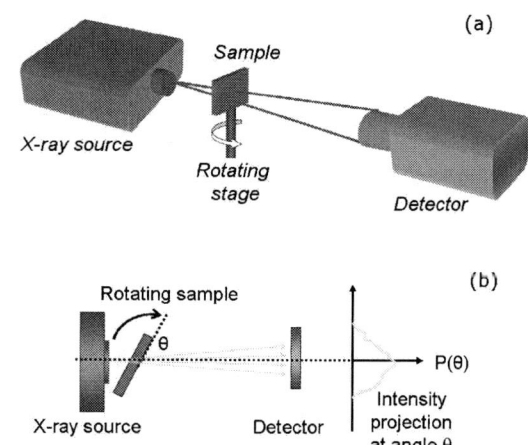

Figure 1: (a) Basic schematic of the x-ray CT setup consisting of x-ray source, rotating stage and detector. (b) The projections at each angle of the intensity of collected 2D x-ray images are superimposed and mathematically processed to generate the three dimensional image.

Key Design Factors in X-Ray CT

In this section we'll review the critical design considerations for X-Ray CT systems that are targeted for failure analysis in microelectronics applications

Resolution
X-ray image resolution depends on several factors, however geometric magnification, which is given by the ratio between sample-to-source and sample-to-detector distances, is one of the most critical ones, given by the following expression:

978-1-4244-5430-3/10 $26.00 © 2010 IEEE   252

$$M = \frac{D_1 + D_2}{D_1},$$

where D_1 and D_2 are the sample-to-source and sample-to-detector distances, respectively (Fig. 2). The total distance between x-ray source and detector is fixed by tool setup dimensions, thus the sample-to-source distance strongly limits system magnification. Smaller sample-to-source distances provide higher magnification and thus better spatial resolution.

One of the key technical challenges is the fact that in order to obtain the highest possible resolution, the sample has to be positioned very close to the x-ray source and since the sample has to fully rotate for 2D images collection the sample hits the x-ray source, as illustrated in Fig. 2. In order to avoid sample to source collision, the sample has to be positioned far enough from the source to freely rotate without hitting it, thus limiting the achievable resolution. One simple solution is to trim the sample under test so that it does not snag. However such sample preparation consumes time, it is invasive and non-practical in most cases. Resolution also depends on the spot size of the x-ray source; in general, the smaller the spot size the higher the resolution.

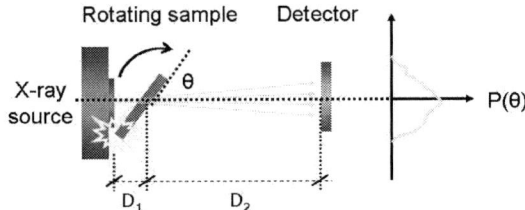

Figure 2: Conventional x-ray systems face a technical challenge: the sample has to freely rotate without hitting the x-ray source, thus it has to be positioned far enough to avoid hitting the source; however, increasing D_1 deteriorates magnification and thus limits the achievable resolution.

This technical issue is one of the key fundamental limitations of conventional x-ray CT technology in microelectronics applications. Limited angle scanning routines could be used as a way around to avoid sample to source crashing; however, the missing 2D images at high tilt angles have considerable amount of information of the defects, especially if the best way to visualize a specific defect is by using virtual cross sections, rather than virtual plane views. Design solutions to overcome these issues may require innovative x-ray optics configurations.

Maximum Energy
The maximum energy of the x-ray source is important in microelectronics applications because the copper content in the samples may highly absorb the x-ray radiation and thus directly impact the required exposure time to get a good quality two dimensional x-ray image and therefore the total tomography scanning time. This is especially important when there are solder joints in the field of view or even worse if they are located in the region of interest. In general, x-ray CT

systems with higher maximum energy will provide better x-ray penetration that will allow scanning highly absorbing samples. In addition, the higher the energy used to scan a specific sample, the faster the exposure time and thus total data collection time. For most microelectronic applications an x-ray source with maximum energy higher than 120-150kV is recommended.

It is important to keep in mind that based on current x-ray source technology the higher the maximum energy the bigger the spot size, so in order to take advantage of maximum energy in microelectronics applications at micron level resolution, the x-ray CT system needs to have a means to compensate for such resolution deterioration.

Data Acquisition Throughput Time
Data acquisition time depends on several factors, e.g. x-ray source energy as was mentioned earlier, sample density, sample size, detector-to-sample and source-to-sample distances, detector system sensitivity and the required resolution to detect the feature of interest. All these factors define the required minimum exposure time for each two dimensional image, thus contributing to total data acquisition time. The number of two dimensional images that needs to be used to obtain the required image quality in the tomography results for a given case study adds to the total data acquisition time. In many cases long data acquisition times could be show stoppers, especially when an alternative destructive or non-destructive method is available. Developing approaches to achieve data collection and 3D image reconstruction times less than 10 minutes is very important for x-ray CT technology to succeed in integrated circuit (IC) packaging applications.

Data Reconstruction
All x-ray CT suppliers have focused efforts on decreasing the data reconstruction time being achieving improvement from several hours to a few minutes in only a few years. The development of more efficient reconstruction and image processing algorithms has been crucial to improving the time to data. On the other hand, the availability of faster computing technologies have also been an important factor in reducing reconstruction processing time.

Image Quality
Regarding image quality, the greater the quality and the number of the two dimensional images used for reconstruction the higher the quality of the reconstructed tomography data. Depending on the specific case, the analyst can adjust the number of collected 2D images according to the required resolution and tomography image quality. Several filtering techniques to enhance the image quality have been developed, and the failure analyst could always find the best one to use for specific applications without introducing image processing artifacts.

Requirements for X-ray CT Systems in Microelectronics
Considering the diverse spectrum of package sizes as well as all key design factors described above, a suitable x-ray CT system for failure analysis in microelectronics needs to meet the following highlights: (a) must have a way to compensate for magnification / resolution deterioration due to large sample

sizes, (b) a maximum energy of 150KV or higher to deal with highly absorbing cooper features, and (c) fast data acquisition / reconstruction throughput time down to 10 minutes of scanning time to compete with alternative failure analysis methods.

The X-ray CT Microscope

The microelectronic case studies that are presented in this paper were analyzed using the x-ray ct microscope, model MicroXCT-200, manufactured by Xradia [3,4].

One of the key components of this system is an x-ray detector with proprietary x-ray optics and optical design that provides significant reduction of the dependence that magnification has with respect to the source-to-sample distance. The design achieves resolutions in the range of 1.5 to 2 microns without the need of placing the sample close to the source, thus allowing full sample rotation required for 3D x-ray imaging without sample-source collision for samples up to 4 inches in diameter (Fig. 3).

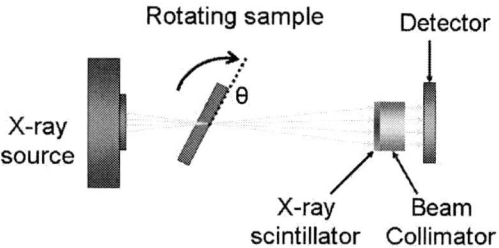

Figure 3. Xradia Micro XCT tool configuration.

The system also uses an x-ray tube that has a maximum voltage and power output of 150kV and 10W, respectively. This high voltage allows for effective penetration through highly absorbing samples with features like solder balls. The proprietary detection scheme also helps reduce the impact on resolution of increased spot size produced by a higher x-ray source energy.

2D X-ray Vs X-ray CT

2D x-ray imaging is a matured and widely used analytical method for failure analysis in microelectronics. One of the important reasons, besides demonstrated capability for defect detection, is its real time nature; in most cases, the user interface makes it easy to obtain good quality images without the need of intensive training, not to mention the good amount of automated routines that can be used for the detection of specific defects. On the other hand, some image quality and resolution improvements may be expected in the near future as a consequence of detector technology development.

The main technical challenges for 2D x-ray technology in microelectronics arise with the development of smaller and more complex microelectronic packages. Analysts often find that the combination of small defect dimensions with a lot more interfering features in the field of view makes it increasingly difficult and sometimes impossible to image certain defects in non-destructive fashion. Such samples often require preparation that may produce artifacts. This fundamental limitation of 2D x-ray imaging technology is a common roadblock that has been driving the development efforts of x-ray CT for failure analysis in microelectronic devices.

Microelectronics Failure Analysis Applications

In this section we will review different applications in which x-ray CT has demonstrated capability to detect defects that are hard or impossible to detect by alternative methods [5].

First Level Interconnect Solder Bump Non-Wet Detection

Non-wet detection in solder balls/bumps at first level interconnects (FLI) using 2D x-ray is based on the rounded shape of solder ball indicating non-wetting issues. When the shape is such that it is difficult to differentiate between passing and failing bumps due to 2D x-ray imaging limitations, this type of non-wet is known as "invisible". Figure 4 shows the physical cross section pictures of both passing and non-wet failing solder bumps. This figure also shows that the best 2D X-Ray image obtained using high-magnification at an oblique view angle cannot detect these types of non-wet solder bumps.

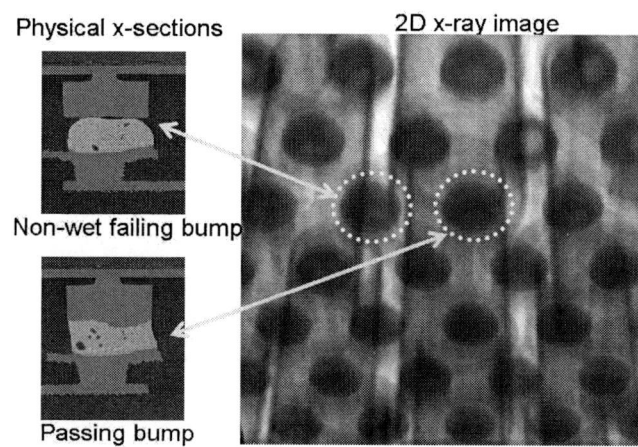

Figure 4: On the left, physical cross section view of both a non-wet failing solder bump and a passing one. On the right, the best 2D x-ray image of those two bumps indicating that they cannot be detected.

It is possible to isolate the electrical structure that has non-wet solder bumps by electrical testing for either dead open or high-resistance electric failure modes. The technical gap is to further non-destructively isolate failing bump locations in the failing electric structure.

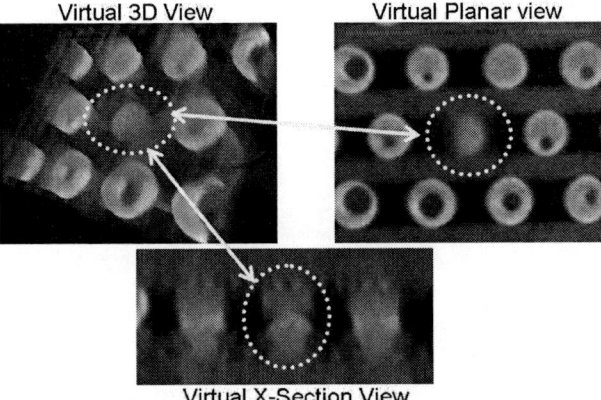

Virtual 3D View   Virtual Planar view

Virtual X-Section View

Figure 5: Virtual 3D model, planar and cross section views of a solder bump non-wet that cannot be detected using conventional 2D x-rays.

The objective of this application development was to establish a fast solution to detect solder bump non-wets along the failing structures using x-ray CT. The results are shown in Fig. 5 where the non-wet is clearly visible using either virtual cross section or planar grinding tomographic views. The typical scanning time for solder bump non-wet detection in first level interconnects is 15-20 minutes per data set of 3.5mm^2.

Wirebond Breaks in Multi-Stacked Packages

Wirebond breaks in multi-stacked packages are difficult to detect with conventional 2D x-ray equipment due to both the overlapping of wirebonds in the field of view as well as resolution limitations. Figure 6a shows the best 2D x-ray lateral image of a package, with several broken wirebonds. These breaks have different gaps and the ones that are less than 5 microns are very difficult to detect. This sizing required the development of a 3D x-ray imaging solution to image cracks in the range of 3-5 microns.

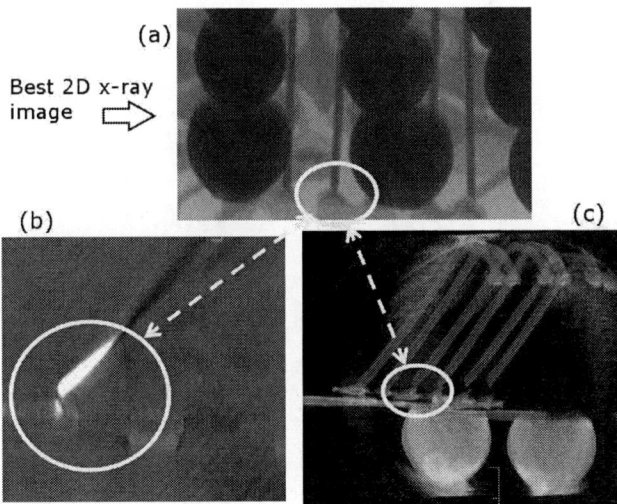

Figure 6. 6a shows the best 2D x-ray image is unable to detect the wire break. Figures 6b and 6c show the results obtained with x-ray CT. The wire breaks are clearly seen using either virtual cross section (Figure 6b) or 3D model view (Figure 6c).

Figures 6b and 6c show the results obtained with x-ray CT. The breaks are clearly seen using either virtual cross section (Fig. 6b) or a rendered 3D image (Fig. 6c). Typical scanning times are in the range of 12 to 20 minutes with a field of view of 3.5mm^2.

Wirebond Shorting in Multi-Stacked Packages

Another case study involves electrical shorting caused by wirebonds touching each other in either single die or multi-stacked packages. Figures 7a and 7b show the 2D x-ray images, lateral and top views, of the suspected shorted wirebonds from a two-dice stacked package. In this figure it is clear that the wirebonds overlap in the x-ray image field of view, but they do not provide conclusive data of actual physical contact. Root cause analysis would typically require a destructive technique such as chemical decapsulation in order to get physical access to the failing wirebonds. The main problem when using a physical de-processing technique is the risk of releasing pressure that could potentially hold the wirebonds together; thus root cause key data could be lost. This was the case of the unit shown in Fig. 7. It belonged to a group of units that failed with electric shorts and recovered during failure analysis. Chemical decapsulation and laser deprocessing were used without success in keeping the failure mode intact.

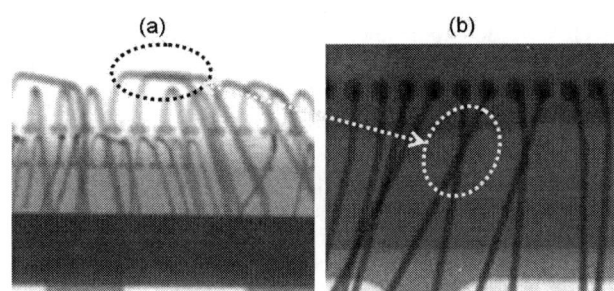

Figure 7. 2D x-ray images showing overlapping wirebonds in a dual-stacked package: (a) lateral view, and (b) top view.

A unit was scanned using x-ray CT. Figure 8a shows the rendered 3D image of this unit; the yellow circle highlights the clear physical contact between two wirebonds of the failing structure. Once the tomography data is reconstructed, the virtual rendered 3D image (model) can be rotated 360° to any possible viewing angle with a spatial image resolution better than 2 microns. This provides conclusive information on whether there is physical contact or not. Figure 8b shows the virtual cross section view of the potential area of physical contact. The two wirebonds appear to be touching each other within the 2 micron resolution of the virtual cross-section view. Once the exact location of the physical contact was found, the unit was physically cross-sectioned for correlation purposes only (Fig. 8c). As can be seen, correlation between the virtual Fig. 8b and physical Fig. 8c images is very clear.

978-1-4244-5430-3/10 $26.00 © 2010 IEEE

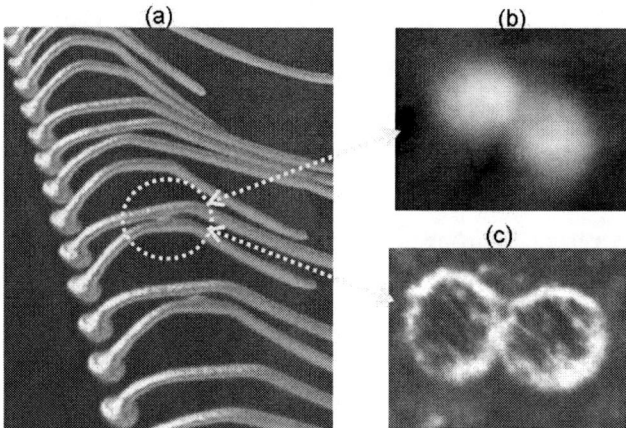

Figure 8. x-ray CT results: (a) Virtual 3D model clearly showing physical contact between failing wirebonds, (b) virtual cross section of x-ray image at location of physical contact, confirming information of virtual 3D model, and (c) optical image of physical cross-section showing that it correlates very well with virtual x-ray image.

Solder Bump Open Detection

In another interesting case study, a unit failed with a solder bump open. A Time Domain Reflectometry (TDR) tool could not isolate the defect location. 2D x-ray was also used to search for potential defective solder bumps without success. The unit was scanned using x-ray CT to search for any defective solder bump along the failing structure that could have caused the open failure.

A routine was created to perform hi-resolution scanning in the whole chain of solder bumps, which required 10 scans. Each scan took about 20 minutes, with a field of view of about 2mm square. As can bee seen in Figs. 9a-b, the defect was localized along the failing structure as low-volume solder that could not be seen with 2D x-ray due to the presence of a capacitor that blocked radiation thus masking-out the defect (Fig. 9a). The virtual plane view is shown in Fig. 9b, in which the defect can be clearly seen. For comparison purposes, Fig. 9a shows the top view 2D x-ray image indicating the area (solid yellow rectangle) where the low volume solder bump was found. Note that it is not detectable at all with 2D x-ray.

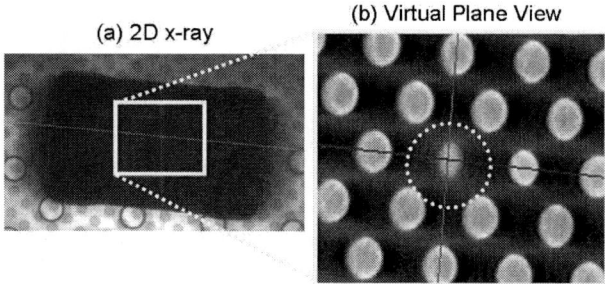

Figure 9. (a) 2D x-ray image of area of interest, notice defect cannot be seen, (b) Virtual planar grinding view, yellow circle highlights defective solder bump.

Figure 10a shows the virtual cross section of a defective low-volume and a passing solder bumps. The unit was physically

deprocessed for correlation purposes; Fig. 10b shows the the physical cross-section. Notice the good correlation between virtual and physical images.

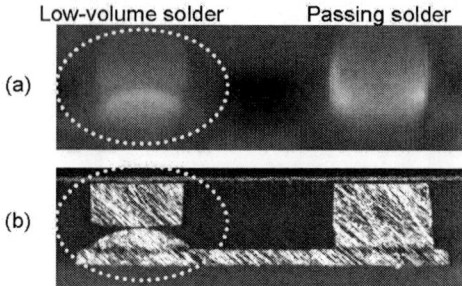

Figure 10. Comparison between (a) virtual and (b) physical cross section views of both failing low-volume and passing solder bumps.

Solder Bump Cracking Detection

Cracks in FLI solder bumps could appear in units under thermal stress for reliability evaluation. Electrical test cannot isolate partial cracks but just full solder joint cracks and the isolation resolution depends on the number of solder bumps between test points. Destructive techniques like Dye and Pull produce artifacts and cannot be used in packages with underfill. Physical cross-sectioning is a tedious and time consuming process that requires more than 3hrs for about ten C4 bumps.

The objective of this application development is to establish a non-destructive method to detect cracks in solder bumps. Figures 11a and 11b show the virtual cross section and planar grinding views of solder bumps that were obtained for a sample size of ~6cm^2, without any sample preparation required.

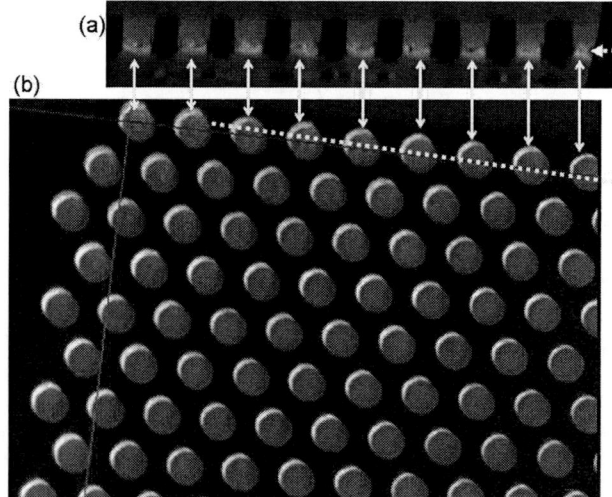

Figure 11. (a) Virtual cross section and (b) planar grinding of corner solder bumps. Yellow dotted line on plane view indicates line of cross section.

Figure 12 shows a comparison between virtual planar and cross sectional views for both a failing solder bump from the corner of the die and a passing solder bump from the center of the die. As can be seen in Fig. 12, x-ray CT can provide a clear finger print of cracking issues, either using a virtual plane or a cross-sectional view. In this slice the surface of a good solder joint looks smooth and uniform; whereas the slice surface of a failing bump looks uneven. A similar situation occurs with the virtual cross-sectional views.

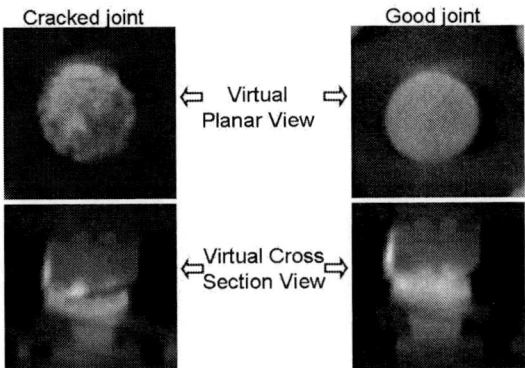

Figure 12. X-ray image comparison between a failing solder bump (left images) and a passing solder bump (right images): virtual planar views (top) and virtual cross sectional views (bottom).

Progressive Testing

One of the compelling capabilities of x-ray CT is the possibility of performing defect inspection during reliability testing to provide feedback to predictive reliability models, as well as at different steps during process development. In order to demonstrate progressive testing capability, the solder bump area of a unit without underfill was imaged before and after thermal stressing. The absence of underfill helped in cracking the solder bumps during thermal stress. A routine was created to scan the units with high resolution, with a data acquisition time of about 90 minutes. Figure 13a shows the virtual planar view image of the same solder bump area obtained prior to thermal stress to verify that solder bumps were not cracked and Fig. 13b shows a similar view after thermal stress displaying that solder joints with a finger print of bump cracking compared to the good joint and cracked joint show in Fig. 12.

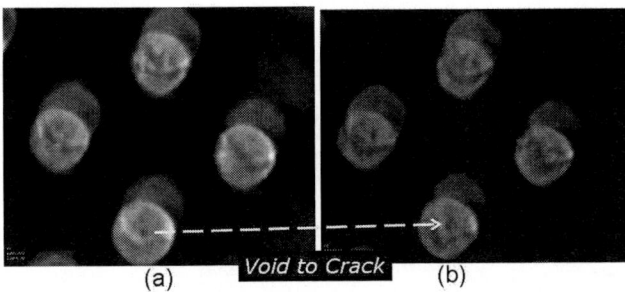

Figure 13. Comparison between virtual planar view of solder bumps (a) before and (a) after the applied thermal stress.

X-ray CT has several advantages over physical deprocessing. For instance, using physical cross sectioning the analyst usually gets information from about 10-20 bumps, with

limited information from a single cross-section, with a throughput time of 3-4hrs. X-ray CT can obtain full information of both virtual plane and cross sectional views of 60 to 100 solder bumps (between 6-10 rows) in about 30 to 90 minutes, depending on how x-ray-absorbing the sample is. Table 1 summarizes a comparison between physical and virtual methods for bump cracking applications.

Comparison Matrix		
Parameter	Physical	XRay CT
Analyzed bumps	10-20	60-100
Number of rows	1	6-10
Xsections/bump	1	Multiple
Time-to-information	3-4h	1-2h
Plane views	Cannot	Multiple
Progressive test?	No	Yes

Table 1. Comparison matrix between physical and x-ray CT virtual methods for bump cracking detection.

Ball Grid Array (BGA) Micro Voiding

Figure 14(a) shows a metallic socket in which a non-destructive characterization of voiding in solder balls is required. 2D x-ray imaging can provide some information about voiding; however it is limited to a single top view with overlapping details as shown in Fig. 14(b). This application requires obtaining a 2D x-ray image of the whole sample to select the solder balls to be inspected with x-ray CT. Once the area of interest has been determined, higher resolution scans can be performed in those areas to fully characterize voids as indicated in Figs. 14(c-d). Figure 14(c) show a virtual cross section view of a solder ball, indicating planes where voids are formed. Virtual planar views can be obtained at any plane level to detect its voiding, as can be seen in Figs. 14(d-f). Typical data collection time is around 15 minutes per region of interest of about 4mm^2.

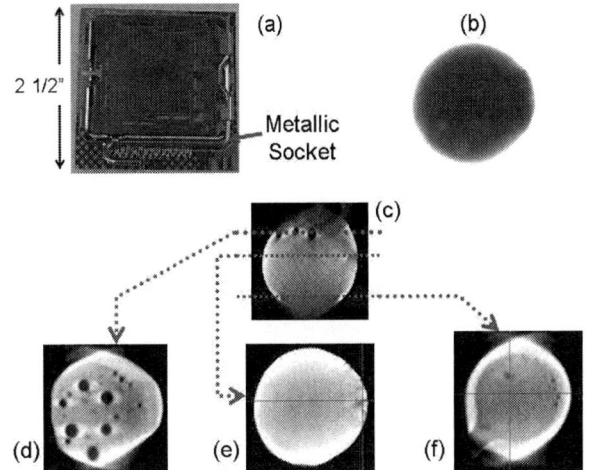

Figure 14. (a) Board sample with metallic socket. (b) Best 2D x-ray image of voids. (c) Virtual cross section of solder ball and (d)-(f) virtual plane views of voids.

BGA Solder Joint Cracking Detection

Another compelling capability of x-ray CT is the detection of solder ball cracking in the second level interconnects BGA package. In this case, the typical sample dimensions are in the range of 1-2 inches wide. Figure 15 shows a planar view indicating solder ball cracking issues (a), and also a virtual cross section confirming the existence of cracks and their plane location. Scanning times for this application using a 4X objective lens is around 30 to 60 minutes, with a field of view of ~4mm^2.

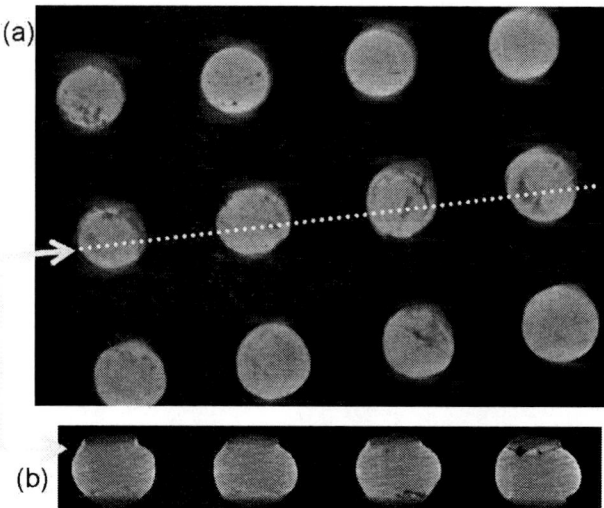

Figure 15. (a) Virtual plane view and (b) virtual cross section view of solder balls corresponding to yellow-doted line after thermal cycle stress testing, indicating typical finger print of solder joint cracking.

Conclusions

This paper presented a review of x-ray CT in failure analysis of microelectronics, with emphasis on the general operation principle, key design considerations, challenges for both 2D and 3D x-ray imaging, and compelling failure analysis case studies. The small dimensions of the defects with increased interfering features in the field of view makes it increasingly difficult and sometimes impossible to image certain defects in a non-destructive fashion, which is a fundamental limitation of 2D x-ray imaging technology and the roadblock that has been driving the development of x-ray CT technology in microelectronics. A critical x-ray CT design factor is the use of proper schemes to compensate for magnification deterioration due to large sample sizes so the tool's highest resolution can be achieved regardless of sample dimensions. A maximum energy of x-ray source of 150KV or higher is recommended to cover the diverse spectrum of microelectronics applications. On the applications side, progressive testing was also discussed as it appears to be a unique property of x-ray CT that can be important in microelectronics, especially for reliability assessment and process development. Further development in the x-ray CT technology roadmap should include improving spatial resolution to sub-micron levels to intercept the failure analysis needs in next generation package technologies, reducing time-to-data to levels of 10 minutes, increasing usability, and customizing data inspection.

References

1. D. Goyal, "X-Ray Tomography for Electronic Packages", ISTFA, 2000.
2. D. Goyal, Z. Fu, J. Thomas, "3D X-ray Computed Tomography for Electronic Packages", ISTFA, 2003.
3. Scott, F. Duewer, S. Kamath, A. Lyon, D. Trapp, S. Wang, W. Yun, "A Novel X-ray Microtomography System with High Resolution and Throughput for Non-Destructive 3D Imaging of Advanced Packages", ISTFA, 2004, pp. 94-98.
4. S. Wang, F. Duewer, D. Scott, W. Yun, "X-Ray Microscopy for NDE Micro- and Nano-structures", Proc. of SPIE, Vol 5766, pp40-48, 2005
5. M. Pacheco, D. Goyal, "New Developments in High-Resolution X-ray Computed Tomography for Non-Destructive Defect Detection in Next Generation Package Technologies", ISTFA 2008, pp 30-35

Impact on Device Performance and Monitoring of a Low Dose of Tungsten Contamination by Dark Current Spectroscopy

F. Domengie[1,2*], J. L. Regolini[1], D. Bauza[2], and P. Morin[1]

[1]STMicroelectronics, Silicon Technology Development,
850 rue Jean Monnet, 38926 Crolles Cedex, France
[2]IMEP-LAHC, Micro Nano Electronic Devices Group,
Minatec, 3 rue Parvis Louis Néel, BP 257, 38016 Grenoble Cedex 1, France
*phone: +33 (0) 4 38922182, e-mail address: florian.domengie@st.com

Abstract—**The dark current in CMOS Image Sensors induced by deliberate contamination with tungsten ion implantation is studied with the Dark Current Spectroscopy (DCS) technique. We obtain quantized dark current peaks associated to the donor level of W in silicon. Accounting for rigorously Schockley-Read-Hall formalism and Poisson distribution of metal atoms, the technique allows to check the generation rate and the fingerprint of this deep level. We use this information for the study of a very low level of accidental contamination that produced an increase in the white pixel number of the sensors. This contamination is then identified as coming from W clusters having an average number of 30 atoms and impacting a limited number of pixels. A hypothesis on the origin of this contamination is proposed and confirmed by snapshots of the pixels in the dark environment.**

Keywords-dark current; contamination; tungsten; image sensors; deep level; clusters

I. INTRODUCTION

Because of significant advantages such as lower power consumption, CMOS technology compatibility, high speed imaging, and signal processing functions integration possibility, CMOS image sensors (CIS's) have been adopted in many large volume applications, such as digital cameras, automotive applications, mobile phones, optical mouse, and webcams [1]. However dark current levels in CIS's are still more important than those in Charge-Coupled Device-based sensors (CCD's). With CIS's pixel size reduction down towards 1 μm and sensor performance increase, greater efforts in reduction of process-induced defects are necessary [2]. Dark current is a parasitic current due to carriers not generated by photons and is an important parameter that impacts the technology yield. Dark current in CIS's is mainly originated from the process-induced defects in which metallic contamination in silicon plays an important role [3]. Indeed, metals introduce deep levels in photodiodes silicon bandgap that generates non acceptable dark current levels. Therefore a serious challenge is to monitor and identify lower and lower doses of metallic contamination.

Dark Current Spectroscopy (DCS) is a very promising technique that has been developed to detect this contamination. From the first time observation of the quantized dark current in 1987 [4], DCS was used to detect metallic contamination in CCD sensors [5] or to characterize radiation induced defects [6]. In a previous study we showed that DCS could detect gold and tungsten introduced voluntarily by implantation in CIS's wafers with sensitivities as low as a few 1E8 /cm^3 and with this technique we observed, for the first time, the quantized dark current peaks related to the W donor deep level in silicon [7].

A real case of unknown contamination during the fabrication process that increased the dark current level and could not be identified with usual inline detection techniques is presented. In order to identify this unknown contamination a protocol has been elaborated which consists in (i) deliberate contamination of CIS's by implantation and (ii) determination of the contaminant deep level parameters by DCS related to this voluntary contamination. These results are applied to study the unintentional low-dose contamination issue and determine its characteristics. Finally, the sensitivity of DCS technique is compared to that of other techniques.

II. DARK CURRENT SPECTROSCOPY CONSIDERATIONS

Different metal contaminants form deep levels in silicon bandgap [8]. In depleted regions, they will emit electrons to the conduction band and holes to the valence band. The main parameters determining their electrical activity and the generation rate are their activation energy E_t, their capture cross-section for electrons σ_n and their capture cross-section for holes σ_p.

The Shockley-Read-Hall [9] model for generation rate of a deep level is used to calculate the dark current produced by a number of contaminant atoms. The electric field enhancement factor $\lambda(E)$ is also considered. Indeed, the generation rate is known to be enhanced by the combination of two major phenomena: energy barrier lowering due to Poole-Frenkel effect and phonon-assisted tunnelling mechanism [10]. Finally,

978-1-4244-5430-3/10 $26.00 © 2010 IEEE

dark current produced by a defined number of atoms N in one pixel can be expressed as:

$$I_N = qN\lambda(E) \frac{\sqrt{\sigma_n \sigma_p}\sqrt{v_{th,n} v_{th,p}}\, n_i}{2\cosh\left(\dfrac{E_t - E_i}{kT} + \dfrac{1}{2}\ln\left(\dfrac{\sigma_n v_{th,n}}{\sigma_p v_{th,p}}\right)\right)}, \quad (1)$$

with n_i the intrinsic carrier concentration, E_i the intrinsic level, $v_{th,n}$ the thermal velocity for electrons, $v_{th,p}$ the thermal velocity for holes, k the Boltzmann constant, T the temperature and q the electronic charge. Some pixels do not contain any impurity atom and have a dark current due to the intrinsic sources common to all pixels. Statistically, a fraction of them contains one atom and will exhibit an additional dark current due to the thermal generation of this center in the pixel volume. Other pixels have two metal atoms and therefore twice the dark current with regard to one impurity centre is generated etc. This will result in observation of quantized dark current peaks.

The distribution of contaminant atoms over the pixel matrix is considered as random and is expected to follow a Poisson distribution [5]. Other distributions have been evaluated to give a better fit in an interesting discussion proposed by Baer in [11], but for our study the use of a Poisson distribution has been found satisfactory:

$$P_N(X) = (X^N / N!)\exp(-X), \quad (2)$$

where X is the average number of atoms per pixel. Volume concentration of the contaminant is easily estimated from X value knowing that the pixel volume is about 1.2 μm^3 for the image sensors evaluated in this study.

Quantized dark current peaks result from a given deep level, from a number of contaminant atoms per pixel associated to this deep level, and from a peaks shape as discussed below. The peaks position resulting from this deep level is calculated from (1) as N times the generation rate of one contaminant atom. First peak that represents uncontaminated pixels is centered on the intrinsic dark current value I_0 and this value has to be added to the dark current calculated to obtain the N-th peak position. Number of pixels in each peak is linked to the statistical distribution of the metal atoms over the pixels matrix, which we consider as a Poisson distribution from (2). Finally peaks are considered as gaussian curves centered on the dark current calculated and the amplitude of which is related to the mean concentration of the contaminant in the pixels volume, proportional to the average number of atoms per pixel X. For each peak there is a dispersion in the dark current values and this results in a peaks width that is set to fit the experimental histogram. It was observed to increase with temperature and with N. Therefore, for some dark current values, contributions from several peaks add up and final dark current histogram does not always show clear quantized dark current peaks.

III. EXPERIMENTAL DETAILS

The pixel technology is based on a 1.75μm pixel pitch and a 4T-type pinned photodiode with 1.75 equivalent shared transistors per pixel [12]. W contamination is ion implanted with three different doses in CIS's wafers through a sacrificial oxide before the gate oxide is processed. A first dose (dose 1) is implanted, then 10 times this dose, and finally 100 times the first dose. Indeed, among all potential and dangerous contaminants, W has often been reported [13], [14]. After the process is finished, histograms of number of pixels versus the dark current are plotted for temperatures between 30°C and 65°C. Integration time for pixels dark current measurement is determined carefully. If it is too long, pixels dark current saturates while if it is too short, it decreases the resolution obtained on dark current measurement. With optimized conditions, quantized dark current peaks are obtained that correspond to the number of pixels containing 0, 1, 2, 3…contaminant atoms. Peaks spacing evolution with temperature permits to estimate the generation rate parameters of the contaminant which will be used to identify the accidental contamination studied hereafter.

Dark current measurements are normalized and we define a white pixel as a pixel that presents a dark current greater than 13 dark current code at 45°C. A white pixel appears bright in a picture taken in the dark.

Complementary Deep-Level Transient Spectroscopy (DLTS) measurements are carried out on N+/P diodes that have been specially included into the CIS's wafers. A -3.0 V reverse voltage and a 100 μs filling pulse with +0.1 V forward bias are used for capacitance transients.

IV. RESULTS

A. Voluntarily Contaminated CIS's

The histogram measured on W implanted wafers for the lowest dose shows individual peaks distributed along x-axis (dark current code). They appear clearly at 45°C, 55°C and 65°C in the histogram presented in Fig. 1. Our goal is to correlate this histogram that is the result of quantized dark current peaks with the W deep level. For that, we used the activation energy and the capture cross-sections for holes and electrons of the donor level of W in silicon found in the

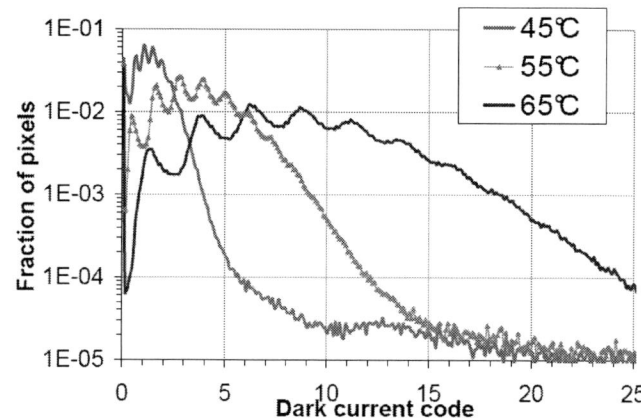

Figure 1. Dark current histogram of W ion implanted CIS's at 45°C, 55°C and 65°C (dose 1).

literature: $E_t = 0.41$ eV above the top edge of the valence band, $\sigma_p = 5.0E\text{-}16$ cm² and $\sigma_n = 4.8E\text{-}15$ cm² [8], [15], [16]. An electric field enhancement factor $\lambda(E) = 1.7$ is introduced for the fit. A value for this factor between 1.5 and 2 is realistic considering the electric field in our photodiodes that is estimated as a few 1E4 V/cm [10], [17].

Table I indicates the dark current generated by one W atom in the photodiodes from 40°C to 65°C. These values are calculated from (1). The fingerprint of W contamination is obtained with this dark current produced by one W atom on donor state within a temperature range. Dark current peaks are then simulated successfully. An example is shown in Fig. 2 for the 65°C temperature.

The average concentration of W over the pixels can be estimated by adjusting X in (2) to fit the experimental data. The use of an average number of 3.1 atoms per pixel allows the simulated peaks to fit the experimental histogram. This corresponds to a concentration of W atoms of 2.7E12 /cm³. For higher doses, quantization of dark current peaks is lost as the average number of atoms per pixel is more important and peaks merge.

pixel number increase. However it is difficult to identify the process step and the impurity responsible for this white pixel number excursion. Fig. 3 shows the results of the white pixel number increase compared to the reference value for each of the three doses of implanted W and for the sensors affected by this accidental and unknown contamination. Although the white pixel number increase is not so large compared to the case of W ion implantation, the unknown contamination brought is sufficient to show a clear fingerprint in the dark current histogram. Fig. 4 shows DCS histogram from this unknown contamination case, compared with the histogram corresponding to the W low dose implanted sensors measured at 60°C. It is difficult to point out any correlation between these two histograms at first glance. The dark current main peak created by the unknown contamination is centred on 50 dark current code whereas W implanted case shows a main distribution centred on 6 dark current code which includes quantized dark current peaks. Histogram from reference not contaminated sensors is plotted on the same figure.

Nevertheless, as we are going to demonstrate, the dark current histogram can be associated to W contamination taking into account two considerations.

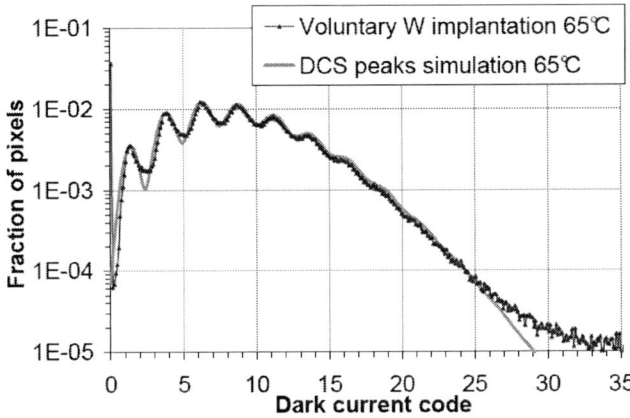

Figure 2. Dark current histogram of W ion implanted CIS's at 65°C, and simulation of the dark current peaks.

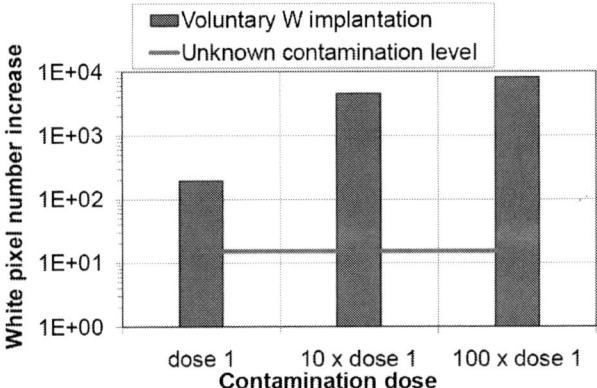

Figure 3. White pixel number increase compared to that of the reference not contaminated CIS's, for W ion implanted three doses and accidentally contaminated CIS's, at 45°C.

TABLE I. DARK CURRENT GENERATED BY ONE W ATOM BETWEEN 40°C AND 65°C. IT IS EQUIVALENT TO THE QUANTIZED DARK CURRENT PEAKS SPACING THAT APPEARS IN THE HISTOGRAMS.

Dark current generated by one W atom on donor level (dark current code)					
40°C	*45°C*	*50°C*	*55°C*	*60°C*	*65°C*
0.30	0.47	0.72	1.09	1.64	2.43

B. Accidentally Contaminated CIS's

This methodology developed in last subsection has been applied to the study of a casual contamination issue. For some of the CIS's wafers accidentally contaminated, the number of white pixels has been increased by 15 compared to the reference wafers. An unintentional metallic contamination from a process step is a good candidate to explain this white

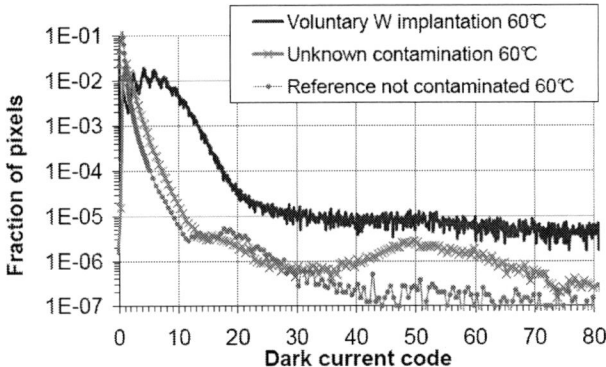

Figure 4. Dark current histogram of W ion implanted, accidentally contaminated and reference not contaminated CIS's at 60°C.

First, an average concentration of 30 W atoms per pixel results in a simulated dark current histogram that follows very closely the x-axis position of the histogram obtained (Fig. 5). Indeed, using deep level parameters of W donor level used for the fit in Fig. 2 and an average number of atoms per pixel equal to about 30, the dark current histogram position along the x-axis can be fitted for temperatures between 30°C and 65°C. This histogram is the combination of the quantized dark current peaks.

Secondly, dark current histogram simulated has to be normalized with a factor of 0.0006 to fit the experimental results and especially the peaks height. Random distribution of contaminant atoms follows a Poisson statistics but it only applies to about 0.06 % of the total number of pixels. This is quite clear evaluating in Fig. 4 the integration of the peak issued from the unknown contamination and that of the peaks related to W ion implantation which impacts all pixels. The result is shown in Fig. 5 with the simulation of the dark current histogram considering an average concentration of 30 W atoms on donor state in about 0.06% of the pixels. Simulated histograms are superimposed on each of the experimental histograms. The quantized dark current peaks are too close to each other to be visible individually so that they form the broad peak we observe at 50 dark current code at 60°C.

Then it is clear that W individual atoms are responsible for this dark current peak. However they are here in a much larger number in affected pixels (between ≈ 16 and 45 with a maximum at 30) than in the case of the intentional contamination. Contamination by W clusters can explain this fact, i.e. that a small fraction of the pixels is contaminated by a large number of W atoms. Once the clusters reached the pixel surface they diffused into individual W atoms during the subsequent thermal treatments. Assuming an average value of 30 W atoms in 0.06 % of the pixels is equivalent to a global concentration in the sensor of less than 1.6E10 /cm³. In Fig. 6 the full width at half maximum of the individual peaks composing that in Fig. 5 at 60°C is reduced to show how this large peak can be decomposed in its individual components. It results from a finite number of W atoms in different pixels, from 16 to 45 atoms for the visible peaks. Each atom produces

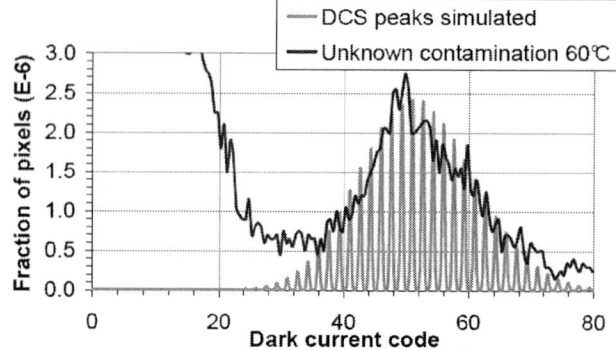

Figure 6.   Dark current histogram of acidentally contaminated CIS's at 60°C and simulation of the dark current peaks obtained from W contamination, that shows individual peaks associated to a number of W atoms between 16 and 45.

a generation current according to what was previously observed from the individual W atoms fingerprint in DCS as showed on Fig. 2 and Table I. The W clusters emission by a cathode gun is one of the realistic hypotheses for such results. Luksha and Tsybin [18] showed that clusters of a few tens of atoms are likely to be emitted by cathode guns. They confirmed this phenomenon in the case of Mo and Pt elements. We can reasonably state that the same occurs with W.

To present complementary evidence of this contamination origin, snapshots of pixels in the dark environment were taken, from image sensors affected by this fortuitous contamination and from image sensors deliberately contaminated by the W ion implantation over the whole pixels matrix at the lowest dose. Signals of pixels were integrated for 200 ms in the dark at 45°C temperature. Gain was adapted to show better contrast between all the pixels of the same picture. Fig. 7 is a clear illustration of the two ways the pixels have been contaminated. In the case of uniform contamination brought by the W implantation that induces a random Poisson-type distribution of atoms, all the pixels are affected and present different levels of gray from darker to whiter ones. This level of gray is directly associated with the number of W contaminant atoms contained into the pixels. In the other case that of random clusters contamination affecting a limited number of pixels, it appears that a large number of pixels remained unaffected i.e. black on the picture. Only a few of them (about 1000 to 2000 ppm) appear white. These white pixels contain a large number of W atoms each time due to one cluster, with an average number of 30 as demonstrated before.

V.   COMPARISON WITH DLTS AND SENSITIVITY

DLTS measurements on N+/P test structures embedded on the CIS's wafers are limited by a sensitivity in the order of 1E12 /cm³ as doping concentration of P region is about 5E16 /cm³. Measurement on W ion implanted CIS's wafers gave a peak around 220 K (Fig. 8). From the resulting Arrhenius plot, an activation energy of 0.41 eV above the top edge of the valence band is obtained, consistent with the W donor level already identified by DCS. W concentration deduced from

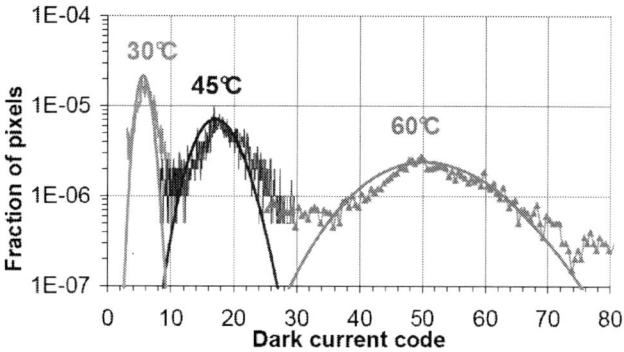

Figure 5.   Dark current histogram of accidentally contaminated CIS's at 30°C, 45°C and 60°C, and the simulated histogram for each temperature.

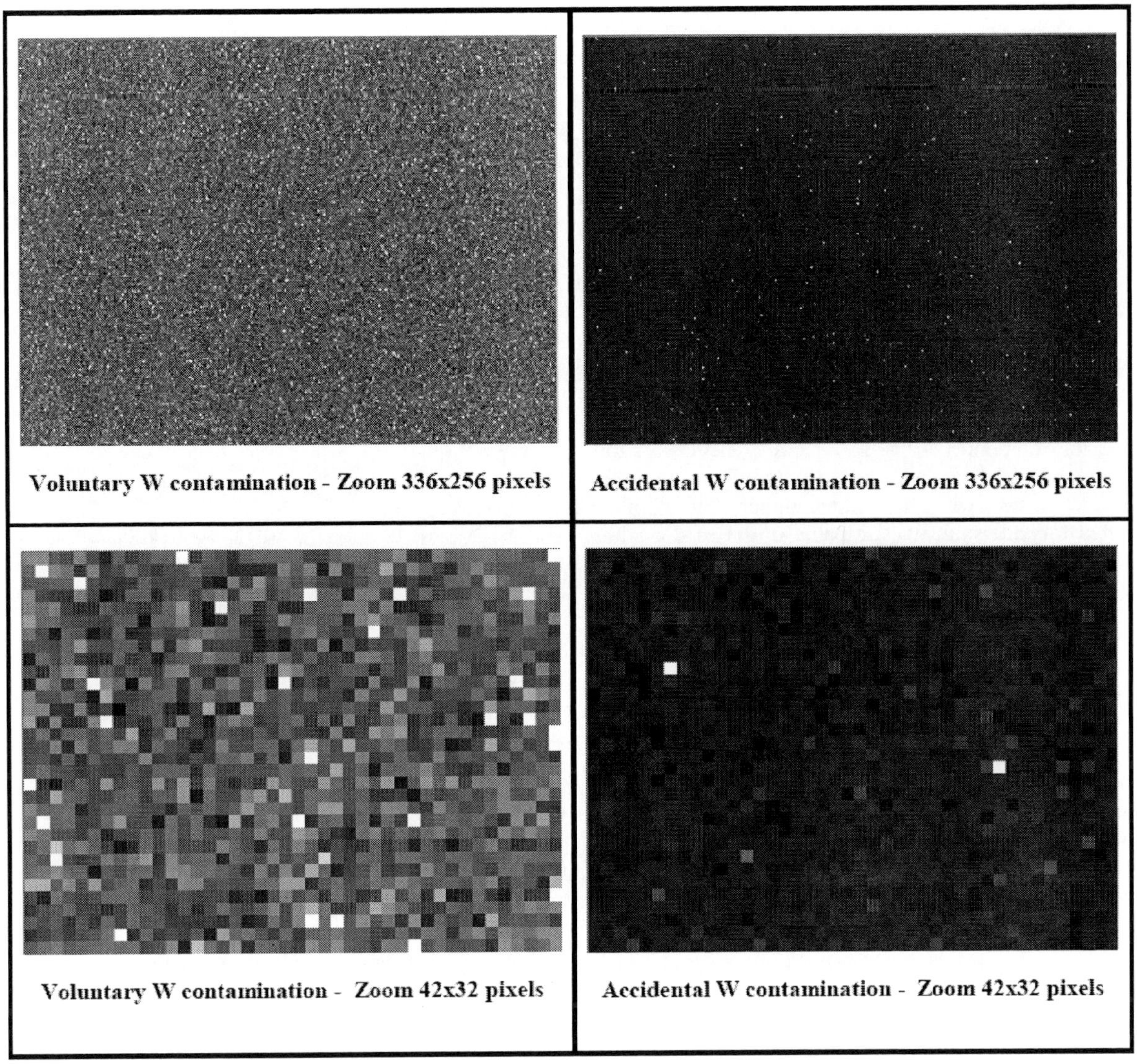

Figure 7. Snapshot of pixels from W ion implanted and accidentally contaminated wafers taken in the dark at 45°C and for 200 ms integration time. Two levels of zoom are shown.

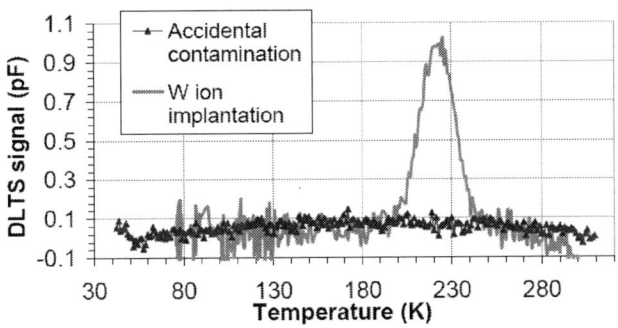

Figure 8. DLTS temperature scan on N+/P diodes from the W ion implanted and accidentally contaminated wafers.

DLTS peak height is 8E12 /cm^3, in the order of magnitude of the concentration found from DCS. But same measurements done on the accidentally contaminated CIS's wafers did not result in a DLTS peak and gave a flat baseline in the temperature scan. To compare with DCS, the sensitivity of which is in the order of a few 1E8 /cm^3, traditional techniques allowing detection of contaminant in volume like DLTS, Secondary Ion Mass Spectrometry (SIMS) or Neutron Activation Analysis (NAA), have sensitivities between 1E12 /cm^3 and 1E15 /cm^3.

Table II compares sensitivities and concentrations obtained both from DCS and DLTS measurements.

TABLE II. COMPARISON OF SENSITIVITIES AND OF W CONCENTRATIONS FOUND IN LOW DOSE-ION IMPLANTED AND ACCIDENTALLY CONTAMINATED SENSORS, IN CM^{-3}.

Technique	Sensitivity	W concentration - ion implanted	W concentration - accidental
DCS	1.0E8	2.7E12	1.6E10
DLTS	1.0E12	8.0E12	Not detected

VI. CONCLUSION

CIS's contaminated voluntarily by W have been studied by DCS with a Shockley-Read-Hall model combined with considerations on Poisson statistics and on electric field. We have determined the fingerprint of W donor level in silicon as the generation rate visible from the DCS peaks spacing and considering the W donor deep level parameters. Using this protocol, the source of an unknown contamination could be identified as W clusters with an average number of 30 atoms. It has permitted to understand the probable origin of the white pixels increase. Such a case, i.e. contamination by a large number of metal atoms in a small fraction of the pixels has not been reported previously. DCS is further confirmed as a powerful method for studying very low level metallic contaminations in CIS's. This technique combines advantages to identify with an outstanding sensitivity a contaminant through the accurate determination of its deep level parameters, and to provide spatial and statistical information about the contamination that is brought during the process. The impurity responsible for the white pixels increase can directly be investigated as DCS can track the pixels where an important dark current is generated.

ACKNOWLEDGMENT

Authors wish to thank N. Virollet and C. Augier from STMicroelectronics for their help in dark current measurements.

REFERENCES

[1] A. El Gamal and H. Eltoukhy, "CMOS image sensors", *IEEE Circuit. Devic.*, vol. 21, pp. 6-20, May-June 2005.

[2] J. C. Ahn, C.-R. Moon, B. Kim, Y. Kim, M. Lim, W. Lee, H. Park, K. Lee, K. Moon, J. Yoo, Y. J. Lee, B. J. Park, S. Jung, J. Lee, T.-H. Lee, Y. K. Lee, J. Jung, J.-H. Kim, T.-C. Kim, H. Cho, D. Lee, and Y. Lee,

"Advanced image sensor technology for pixel scaling down toward 1.0μm", in *IEDM Tech. Dig.*, Dec. 2008, pp. 1-4.

[3] L. Jastrzebski, R. Soydan, H. Elabd, W. Henry, and E. Savoye, "The effect of heavy metal contamination on defects in CCD image sensors", *J. Electrochem. Soc.*, vol. 137, pp. 242-249, Jan. 1990.

[4] R. D. McGrath, J. Dory, G. Lupino, G. Ricker, and J. Vallerga, "Counting of deep-level traps using a charge-coupled device", *IEEE Trans. Electron Devices*, vol. 34, pp. 2555-2557, Dec. 1987.

[5] W. C. McColgin, J. P. Lavine, and C. V. Stancampiano, "Probing metal defects in CCD image sensors", in *Proc. Mater. Res. Soc. Symp.*, vol. 378, 1995, pp. 713-724.

[6] C. Tivarus and W. C. McColgin, "Dark current spectroscopy of irradiated CCD image sensors", *IEEE Trans. Nucl. Sci.*, vol. 55, pp. 1719-1724, June 2008.

[7] F. Domengie, J. L. Regolini, and D. Bauza "Study of metal contamination in CMOS image sensors by dark current and deep level transient spectroscopies", *J. Electron. Mater.*, in press.

[8] K. Graff, *Metal Impurities in Silicon-Device Fabrication*. Berlin: Springer, 2000.

[9] D. K. Schroder, "The concept of generation and recombination lifetimes in semiconductors", *IEEE Trans. Electron Devices*, vol. 29, pp. 1336-1338, Aug. 1982.

[10] J. Bogaerts, B. Dierickx, and R. Mertens, "Enhanced dark current generation in proton-irradiated CMOS active pixel sensors", *IEEE Trans. Nucl. Sci.*, vol. 49, pp.1513-1521, June 2002.

[11] R. L. Baer, "A model for dark current characterization and simulation", in *Proceedings of the SPIE*, vol. 6068, 2006, pp. 37-48.

[12] M. Cohen, F. Roy, D. Herault, Y. Cazaux, A. Gandolfi, J. P. Reynard, C. Cowache, E. Bruno, T. Girault, J. Vaillant, F. Barbier, Y. Sanchez, N. Hotellier, O. LeBorgne, C. Augier, A. Inard, T. Jagueneau, C. Zinck, J. Michailos, and E. Mazaleyrat, "Fully optimized Cu based process with dedicated cavity etch for 1.75μm and 1.45μm pixel pitch CMOS image sensors", in *IEDM Tech. Dig.*, Dec. 2006, pp. 127–130.

[13] R. B. Liebert, G. C. Angel, and M. Kase, "Tungsten contamination in BF2 implants", in *Proceedings of the 11th International Conference on Ion Implantation Technology*, Austin, TX, USA, June 16-21, 1996, pp. 135-138.

[14] Y. Borde, A. Danel, A. Roche, A. Grouillet, and M. Veillerot, "Estimation of detrimental impact of new metal candidates in advanced microelectronics", *Solid State Phenomena*, vol. 134, pp. 247-250, 2008.

[15] S. Diez, S. Rein and S. W. Glunz, "Analysing defects in silicon by temperature –and injection- dependent lifetime spectroscopy (T-IDLS)", in *Proceedings of the 20th European Photovoltaic Solar Energy Conference*, Barcelona, Spain, June 6-10, 2005, pp. 1216-1219.

[16] T. Ando, S. Isomae, and C. Munakata, "Deep-level transient spectroscopy on p-type silicon crystals containing tungsten impurities", *J. Appl. Phys.*, vol. 70, pp. 5401-5403, Nov. 1991.

[17] A. F. Tasch, Jr. and C. T. Sah, "Recombination-generation and optical properties of gold acceptor in silicon", *Phys. Rev. B*, vol. 1, pp. 800-809, 1970.

[18] O. I. Luksha and O. Yu. Tsybin, "Emission of atomic particles from the surface of the thermionic cathode of an electron gun", in *Sov. Phys.-Tech. Phys.*, vol. 37, pp. 1041-1043, 1992.

Electromigration of NiSi Poly Gated Electrical Fuse and Its Resistance Behaviors Induced by High Temperature

Han-Byul Kang[1], Jongwoo Park[1]*, Gun-Rae Kim[1], Hyun-Woo Park[2], Woon-Hak Lee[2] and Joo-Byoung Yoon[2]

Technology Reliability[1], System LSI division, Samsung Electronics
Defect Analysis[2], Memory division, Samsung Electronics
San #24 Nongseo-Dong Giheung-Gu, Yongin-City, Gyeonggi-Do, Korea 446-711
jongwoo.s.park@samsung.com (email); 82-31-209-1344 (phone); 82-31-209-4312 (fax)

Abstract

Insight is given on improved behaviors of the programmed NiSi polygated electrical fuse (eFuse) during the high temperature storage (HTS) test. By using a noble transmission electron microscopy (TEM) that includes scanning transmission electron microscopy (STEM), energy dispersive x-ray spectrometry (EDS), electron energy loss spectrometry (EELS) and nano-beam electron diffraction (NBED), microstructural behavior and phase transition of NiSi in the fuse link are painstakingly investigated before and after HTS test. It is found that improved post-resistance of eFuse is attributed to the low temperature growth of Ni_3Si_2 induced by HTS test at 250°C, which is microscopically proven by both ex-situ and in-situ TEM. In fact, Ni agglomeration, in which Ni resides around void formed in the fuse link, plays an important role of this cystallization. As results, the root causes of improved post-resistance of eFuse are qualitatively substantiated with respect to dynamic phase transformation and microstructural change in the fuse link. (Keywords: eFuse, TEM, post-resistance, electromigration).

I. INTRODUCTION

In modern CMOS technology, NiSi polygated electrically programmable fuse (eFuse) is popular and widely used for redundancy implementation and electronic chip identification. In the case for redundancy implementation, the row or column that consists of bad cells is replaced by the redundant array segments using eFuse [1-3]. For the application of electronic chip identification, information contains wafer numbers and X-Y coordinates of the chip by using a string of eFuse. As known, fuse programming has been attributed to electromigration (EM), agglomeration and other mechanisms. Programmed eFuses are electrically high resistance, while the unprogrammed remain in a low resistance state. There are two kinds of method for fuse programming. One is the EM driven programming to increase electrical resistance of fuse, which is governed by silicide migration in polycrystalline silicon. The other one is to use thermal rupture, which bursts out the fuse by localized self-heating induced by very high current flow [3, 4]. On the comparative bases, the rupture based fuse programming is capable of producing high resistance, but unpredictable microstructure can appear then often results in poor yields and unstable resistance behaviors during reliability evaluations [4-5]. Thus, a steadfast methodology is necessary

for fuse programming to ensure for eFuse reliability. In fact, resistance behaviors of the programmed eFuse stressed under high temperature storage (HTS) test inherently rely on the kinetics of EM, Ni silicide migration from the cathode to anode. Even though eFuse is completely programmed within an optimized program window proven by both electrical characterization and yield validation with a statistical certainty, its post-resistance often varies under the given stress condition, which is a major reliability concern. However, the formation structure and phase transformation of Ni silicide have not been fully investigated. Moreover, intimate relationships between microstructures and electrical characteristics of programmed eFuse are a little studied.

In this paper, insight is given on the elucidation of improved resistance (V_{min} reduction) by using a noble TEM analysis that includes a variety of microscopy, such as scanning transmission electron microscopy (STEM), energy dispersive X-ray spectrometry (EDS), electron energy loss spectrometry (EELS) and nano-beam electron diffraction (NBED). Furthermore, in-situ TEM that enables dynamic structural analysis and direct observation with the resolution of the atomic scale of imaging and diffraction is employed then compared with ex-situ electron microscopy in order to investigate microstructural change and phase transformation of blown eFuse in the link under HTS test.

II. EXPERIMENTAL

NiSi fuse is blown by a programming transistor with a gate bias (V_{gate}) and constant voltage on the anode ($V_{fsource}$). During programming, a constant current tends to flow at the anode. V_{gate} and $V_{fsource}$ vary in a range of 1.7~1.9V to produce 3×3 program window in an increment of 0.1V with different programming pulses of 5μs and 10μs. To monitor physical degradation of the programmed eFuse, resistance is measured as a function of HTS test conditioned at 250°C for 168h in wafer level. After blowing within the programming window, a microscopy is conducted to confirm EM driven fuse programming. For the cross-sectional TEM sample preparation that requires uniform thickness, minimal contamination and large observable area, a focused ion beam is carefully conducted on the eFuse demonstrating increases in post-resistance during HTS test. In sequence, the samples are thoroughly examined by TEM equipped with STEM-EDS and EELS. Since the image contrast is sensitive to the mass thickness of the specimen, STEM imaging that enables to observe the elemental distribution is employed. Moreover, NBED technique is chosen for crystallographic confirmation of the phase transformation on the fuse link after HTS test.

In order for temperature-induced phase transformation and microstructural evolution, in-situ TEM is used for direct observation of dynamic microstructural change with the resolution of an atomic scale in heated stage. The in-situ TEM is performed at 300keV in a JEOL 3011 with a LaB6 filament. EM-31050 JEOL heating holder is used to resistively heat the sample and a Pt/Pt-13% Rh thermocouple is connected to the furnace. The heating rate is ~10°C/minute and the specimen is heated up to 250°C. In addition to normal photographic recording, the images from Gatan 622 TV rate camera system are viewed with the magnification of approximately 0.1~1 million and then recorded on a video tape with a time resolution of 30 frames/second.

III. RESULTS AND DISCUSSION

A. Vmin behaviors of before and after the HTS stress

Figure 1 shows Vmin behaviors of programmed eFuses before and after the HTS test conditioned at 250°C for 168h. As shown, there are little changes in Vmin for the eFuses below 0.6V even after HTS test. However, Vmin between 0.6 and 0.94V deviated from the major population decreases after HTS stress (marked as an arrow in Fig. 1). This indicates that the resistance of blown eFuse increases after HTS test. In fact, changes in eFuse resistance are hardly observed due to relatively higher thermal stability of NiSi. Note that when eFuse is programmed by EM mode, Ni migration from the cathode to anode, its resistance rapidly augments up to 10 kΩ. It is, however, reported that the resistance of eFuse is subjected to be varied in the presence of the rupture and void or the remnant of Ni in the fuse link based on the program modes [3, 4]. Again, it is obvious that that Vmin decreases with increasing after 168h of HTS test at 250°C.

B. Ex-situ TEM analysis after the HTS test

To investigate improved Vmin behaviors during HTS test, the programmed eFuse that reveals Vmin reduction is carefully characterized by ex-situ TEM. In Fig. 2(a), void exists in the tail of fuse link near the anode side. The fuse link between the metal contacts (CA) shows a lighter grey in contrast. This is an indicative of Ni migration from the cathode to anode through the fuse link. As shown, the darker tone appears in the middle of fuse link between the metal contacts. According to the contrast mechanism of TEM imaging of mass-thickness contrast, the fuse link is likely comprised of the element having higher atomic numbers. It is, in turn, hardly difficult to expect that Ni exists in the link even after blowing. Indeed, the darker spots highlighted by a dotted square in the middle of fuse link imply Ni. The enlargement of Ni spots is shown in Fig. 2(b). As shown, both lattice and Moire fringe are obvious. Furthermore, the selective area diffraction pattern (SADP) reveals multiple diffractions as shown in the inlaid of Fig. 2(b). Both high resolution image and SADP support that the darker

area in the fuse link represents the crystalline. However, the causes of void formation and crystalline structure are not known yet. In sequence, how much Ni is migrated, whether void is grown and why crystalline exits in the fuse link toward the cathode instead of the anode side are interesting subjects to be investigated.

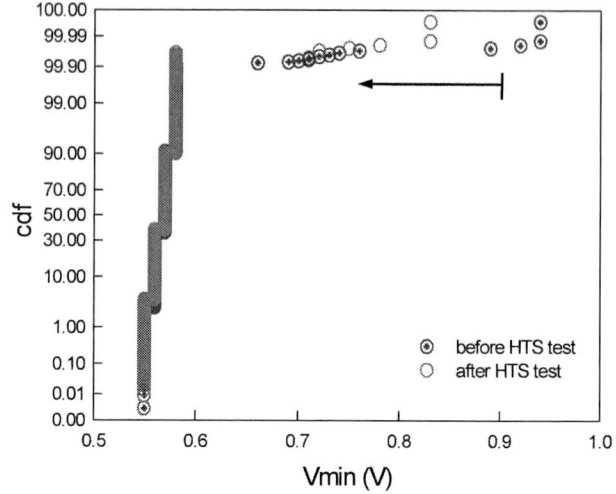

Figure 1. Vmin shift behaviors of the programmed eFuse before and after the HTS test conditioned at 250C for 168h. The arrow indicates the reduction of Vmin after the HTS test.

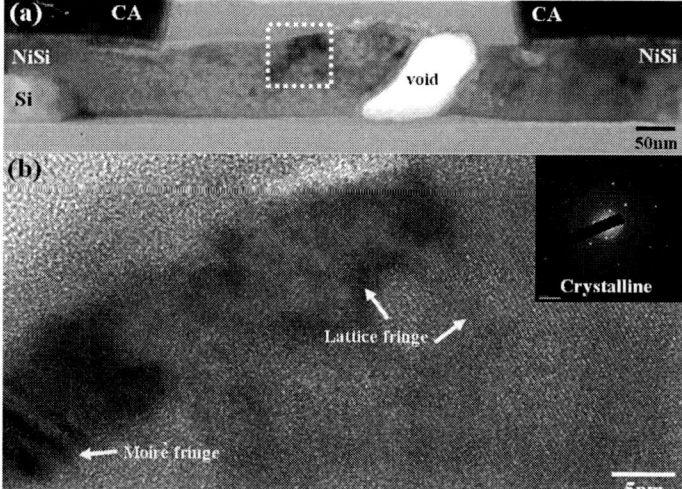

Figure 2. Results of ex-situ TEM analyses on the improved Vmin of eFuse after the HTS test at 250°C for 168h: (a) the cross-sectional TEM micrograp and (b) the HR-image and SADP obtained from the area highlighed with a square in Fig. 2(a). Based on the contrast, The darker area in the link reveals crystalline. Note that CA represents metal contact.

The STEM-HAADF and EELS analyses are taken for analyzing chemical composition and distribution in the fuse link. The results of STEM-HAADF and EELS elemental mapping conducted on Fig. 2(a) are shown in Fig. 3(a-c). It is shown that the fuse link consists of Ni and Si based on HAADF image [see Fig. 3(a)]. In addition, Fig. 3(b) and (c) of EELS analyses supports that Si and Ni are widely distributed in the fuse link and more NiSi exists underneath the anode. As such, one might argue that insufficient programming current results in partially programmed eFuse in the appearance of residual Ni peak from the fuse link. Such that Ni silicide is not fully migrated from the cathode to anode. However, voiding in the fuse link resulting from temperature rising due to excessive programming current flow implies an indicative of over programming. Hence, it is reasonable that the residual Ni in the fuse link is a natural phenomenon as long as the programming window produces a solid yield demonstration with a statistical certainty and robustness in process throughout reliability assessments.

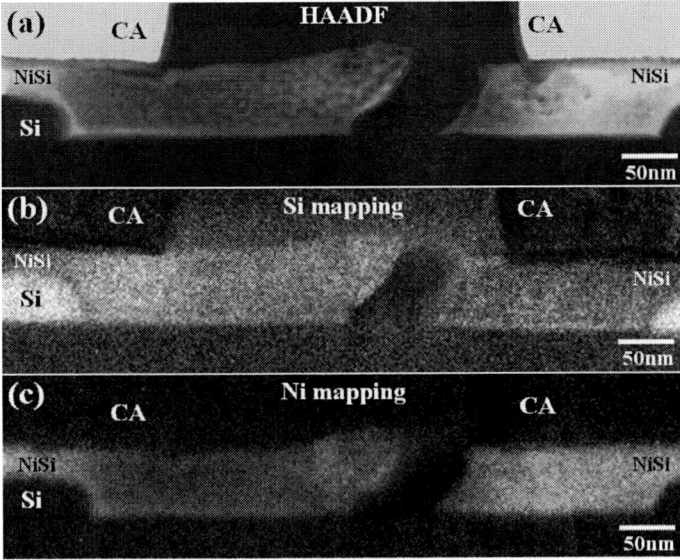

Figure 3. Results of STEM images and EELS analyses taken from Fig. 2(a): (a) HAADF image, (b) EELS Si mapping and (c) EELS Ni mapping. Si and Ni are widely distributed in the programmed fuse link.

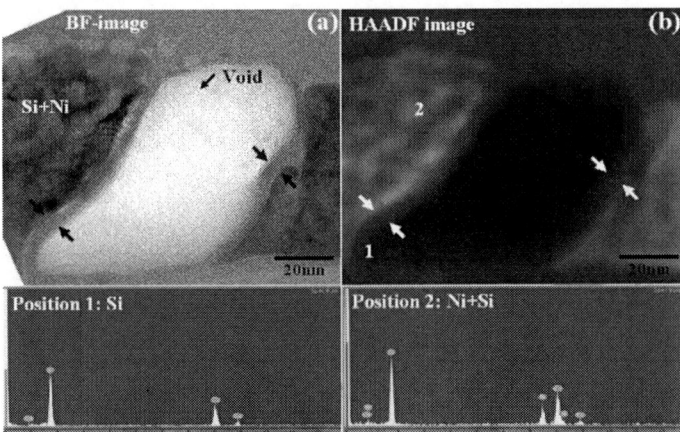

Figure 4. Results of STEM-EDS analyses on void formation in the fuse link: (a) BF-image and (b) HAADF image. Position 1 (Si peak) and Position 2 (Ni and Si peak).

Figure 4 shows STEM-EDS results on void in the programmed fuse link. As shown in Fig. 4(a) of the bright filed (BF) image, void is surrounded by thin line of Si marked by the arrows. Comparatively, it is much brighter than the fuse link in contrast. The physical feature of void is more evident with HAADF image as shown in Fig. 4(b). Seemingly, Ni is somehow agglomerated around the void in the fuse link. Note

that only Si peak is arisen from the Position 1, while the Position 2 shows Si and Ni peak. As such, void formed in the fuse link is primarily surrounded by Si.

C. In-situ TEM analysis before and after the HTS test

In-situ study particularly for temperature-induced phase transformation and structural evolution is of important for understanding structural stability and transformation coupled with electrical characterization. Indeed, in-situ TEM that can provide the resolution of an atomic scale is an ideal approach for direct observation of dynamical microstructural change in a heated environment. Hence, the usage of in-situ TEM is inevitable to unravel ex-TEM results for the reduction of Vmin during HTS test. Figure 5 is the sequential images from in-situ TEM analysis at 250°C. After blowing, void forms in the programmed fuse link near the anode. Suggested that void has not been nucleated by HTS test. With increasing time at elevated temperatures, the dark spots nucleate and then become vivid near a void in the fuse link. In turn, the dark spots aggravate then later cluster with increasing heat duration. In contrast, a void seems not to be grown in size. As shown, in-situ TEM analysis enables phase transition in the fuse link to be microscopically explicated. This evidences that the crystalline shown in Fig. 2(a) directly results from HTS test at 250°C.

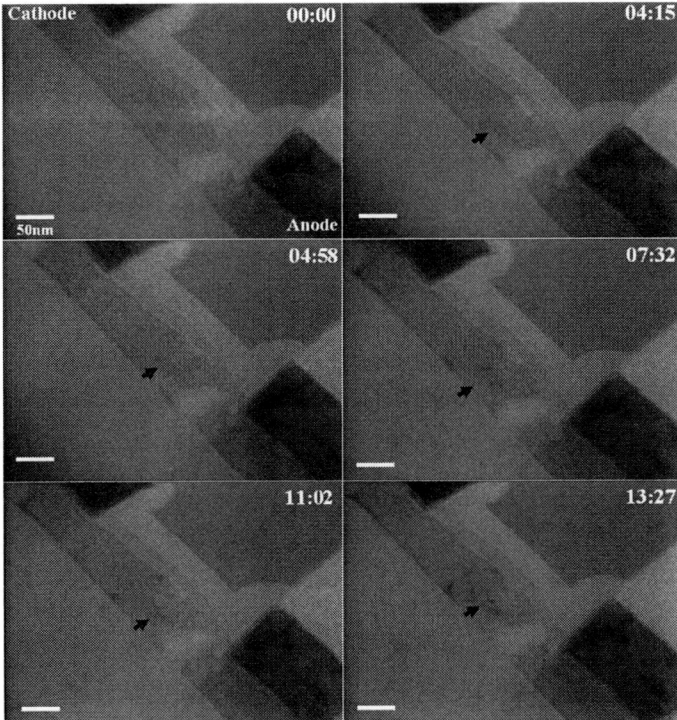

Figure 5. The sequential in-situ TEM images of the blown fuse link at 250°C as a function of time.

Figure 6(a-b) shows microstructural change of the blown fuse after 19 minutes at 250°C from in-situ TEM. The discrepancy between two images is obvious and articulated by the arrows. After heating, clustering of dark spots on the left side of void is evident. This is, in fact, the crystallization proven by the diffraction pattern in Fig. 6(a-b). Even though weak Ni peak is caught by the SADP before heat treatment, the fuse link between metal contacts is the amorphous state [see Fig.6 (a)]. In Fig. 6(b), the diffraction pattern of dark spots induced by heat treatment is in a good agreement with the ex-situ results shown in Fig. 2. The fact is that in-situ TEM, which intends to mimic effects of HTS test on the blown fuse, replicates the crystallization formed in the fuse link. As such, HTS test at 250°C triggers the crystallization in the fuse link as Ni is not fully eliminated due to void formation, which can play as a physical barrier and discontinuity hindering continuous Ni migration to the anode. Hence, it is legitimate to conclude that the crystallization of blown eFuse caught by the ex-situ TEM is attributed to HTS test at 250°C. Again, there are little changes in the geometrical feature of void. It means that the effect of void on eFuse reliability wound be negligible.

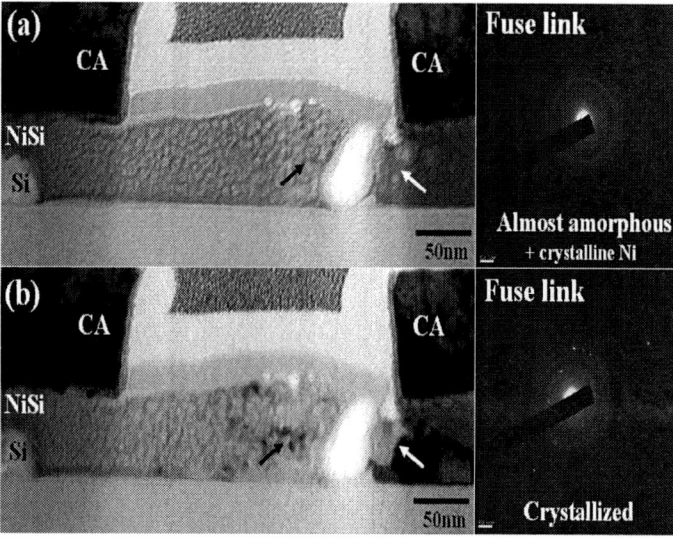

Figure 6. Microstructural changes of blown fuse before and after heat treatment in in-situ TEM: (a) before and (b) after 19 minutes of heat treatment.

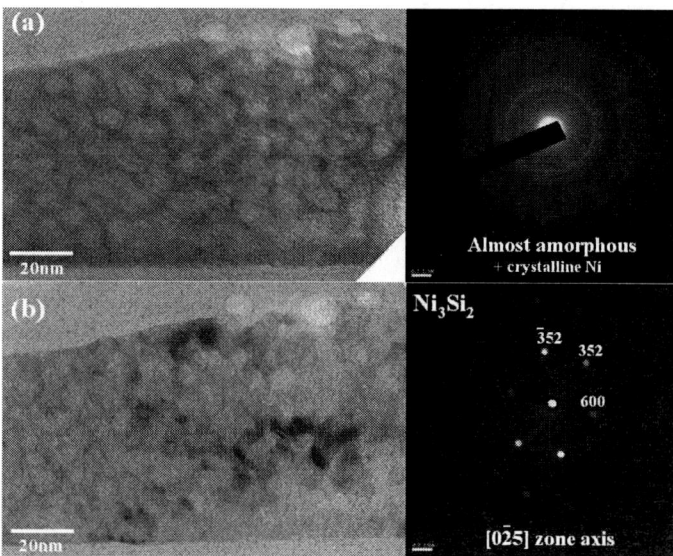

Figure 7. SADP and NBED analyses of the blown fuse link : (a) before and (b) after heat treatement from in-situ TEM.

To further identify the crystallization induced by heat treatment, the NBED technique is adopted. Dark spots are easily noticeable after heat involved in comparison to before heat exposure as shown in Fig. 7(a-b). The NBED reveals sharp diffraction patterns after heat exposure [see Fig. 7(b)]. Whereas, no diffraction pattern appears on the pristine blown fuse before heat exposure as shown in Fig.7 (a). This indicates that the fuse link is in the amorphous state. As shown in Fig. 7(b), it is manifested that the crystallization appeared on the blown eFuse during HTS test at 250°C is Ni_3Si_2. Suggested that HTS test at 250°C results in the low temperature growth of Ni_3Si_2.

In addition, the results of HR image and fast Fourier transformation (FFT) after heat treatment are shown in Fig. 8(a-b). Neither lattice nor Moire fringe found from the cathode side of the fuse link consisting of ~92 at. % of Si and ~8 at. % of Ni. Indeed, diffused halo ring from the FFT analysis represents the amorphous state [see Fig. 8(a)]. In contrast, Fig. 8(b) shows clear lattice fringes from the darker spots near void, which is composed of ~66 at. % of Si and ~34 at. % of Ni. Proven that Ni_3Si_2 forms near void in the fuse link, in which the fragment of Ni resides near the anode side.

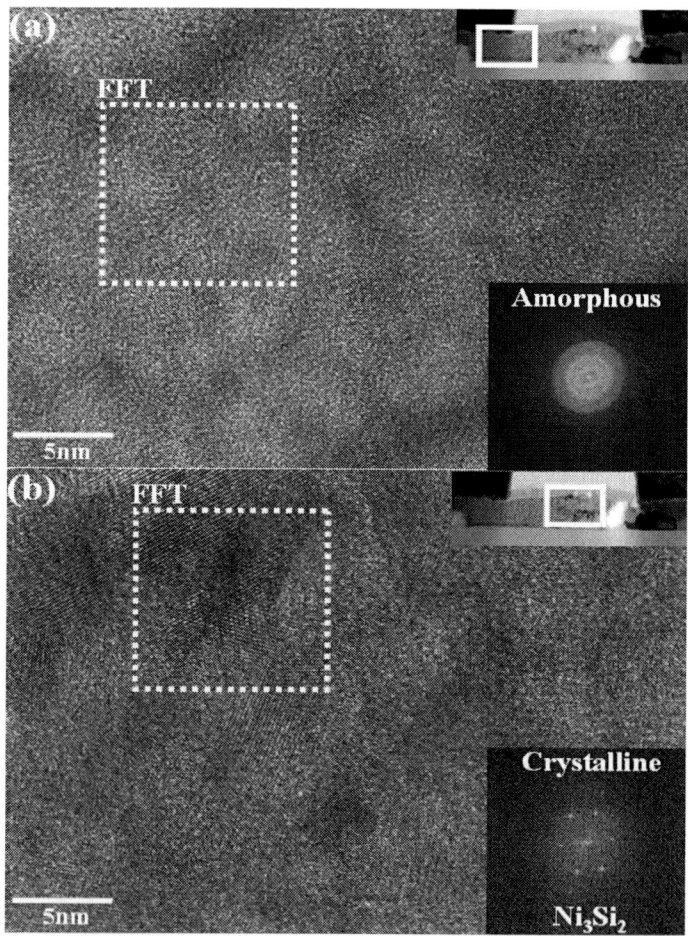

Figure 8. HR- image and FFT results of the blown eFuse from in-situ TEM: microstructural changes (a) at the cathode and (b) at darker spots near void,in the fuse link.

[6-7]. It is found that Ni_3Si_2 can be formed at lower temperature involved than any other Ni slicide. However, it has the highest resistance. In regard to 250°C of HTS test condition, Ni_3Si_2 formation appeared on the fuse link of blown eFuse is quite feasible from practical perspectives.

TABLE I.   FORMATION TEMPERATURE AND RESISTANCE OF NI SILICIDE

	Ni_2Si	Ni_3Si_2	$NiSi$	$NiSi_2$
Formation temp. (°C)	225~325	< 200	350~400	< 750
Resistivity (µ Ω cm)	25	~70	~15	~50

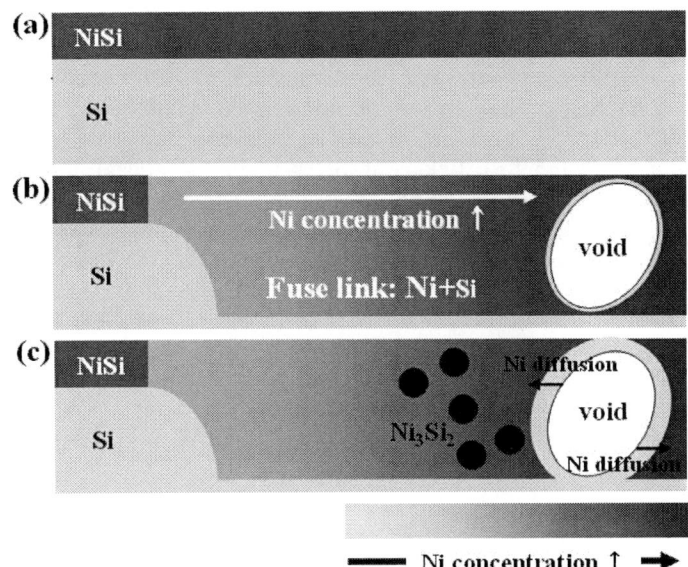

Figure 10.   *The schematic diagram of physical mechanism of EM driven Ni silicide migaration responsible for improved Vmin of the blown eFuse during HTS test: (a)before blowing, (b)after blowing and (c) after HTS stress.*

Figure 10 intends to illustrate the physical mechanism and microscopic model of EM based eFuse for the reduction of Vmin during HTS test. Recall that EM initiates from the cathode to the anode then Ni silicide is subjected to pile up underneath the anode. However, when void forms in the fuse link during programming, the entire amount of Ni silicide is not immaculately migrated from the cathode to anode because voiding plays a role as physical discontinuity [see Fig. 10(a)→(b)]. Such that the remnant of Ni exists in the fuse link, unless EM based blown fuse perfectly removes Ni through the fuse link without voiding or rupture. Hence, the cathode is more prone to be in the amorphous state. As such, the probability of the crystallization of Ni_3Si_2 near void in the cathode side in the fuse link becomes higher during HTS test in the presence of void. Proven that after heat exposure, Ni diffusion from the adjacent area of void to the fuse link triggers Ni agglomeration, contributing to improved Vmin as illustrated in Fig. 10 (b→c). In turn, similar structural change in void is reported in [8]. As results, it is reasonable to conclude that the decrease of Vmin of blown eFuse during HTS test is because of the low temperature growth of Ni_3Si_2 formation facilitated by Ni agglomeration in the fuse link. Although a successful corroboration between the ex-situ and

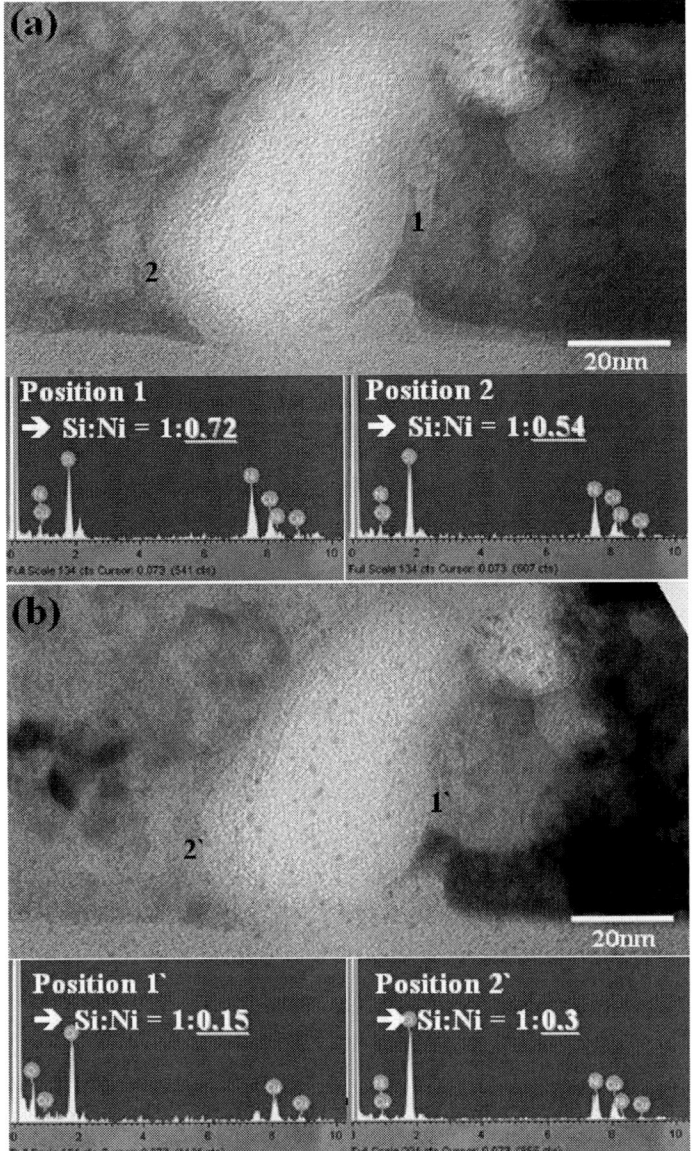

Figure 9.   *Spatial structure of void and its STEM-EDS results (a) before and (b) after heat treatment in in-situ TEM.*

Figure 9 exhibits in-situ TEM micrographs focused on the void formed in the fuse link before and after heating. The perimeter of void particularly for the edges marked by 1 and 2 seem to be alleviated after heat treatment. In fact, the ratio of Si to Ni peak detected from STEM-EDS on the Positions of 1 and 2 is dramatically diminished. This in-situ TEM result is well agreed with ex-situ as shown in Fig. 4. It is found that void is rarely grown in size during HTS test. This implies that one of reliability concerns inherently linked to the growth kinetics of void during HTS test would be negligible. As demonstrated, in-situ TEM enables to physically understand and explain Vmin behaviors of blown eFuse in a manner of intrinsically structural consistent and cohesively microscopic relevant.

TABLE I contains the type of Ni silicide that can be formed under the given temperatures and its corresponding resistances

978-1-4244-5430-3/10 $26.00 © 2010 IEEE   269

in-situ TEM qualitatively explicates the reduction of Vmin and negligible effect of void on blown fuse reliability during HTS test, the propensity of initial Vmin behaviors on void or rupture formation in the fuse link is the area ripe for further study.

IV. CONCLUSION

Phenomenological behaviors of the programmed eFuse during HTS test, improved post-resistance leading Vmin reduction, are meticulously characterized by both ex-situ and in-situ TEM analyses. It is found that the post-resistance augment of blown fuses during HTS test is attributed to the low temperature growth of Ni_3Si_2 facilitated by Ni diffusion in the fuse link at elevated temperatures. In fact, in-situ TEM successfully replicates the phase transition of NiSi crystalline and microstructural changes occurred in the blown fuse during HTS test. It is also shown that Ni agglomeration near void formed in the fuse link contributes to Ni_3Si_2 crystallization during HTS test. This implies that such microstructural change is cohesively adhered to EM based eFuse programming. Finally, the causes of improved resistance during HTS test and intimate relationships between microstructures and electrical characteristics of programmed eFuse are elucidated.

REFERENCES

[1] C. Kothandaraman, S. K. Iyer, and S. S. Iyer, "Electrically programmable fuse (eFuse) using electromigration in silicides", IEEE Electron Device Lett., Vol. 23, no. 9, 2002, pp. 523–525.

[2] T. Sasaki, N. Otsuka, K. Hisano, and S. Fujii, "Melt-segregate-quench programming of electrical fuse", IEEE 43rd Annual International Reliability Physics Symposum Procedings, pp 347-351, 2005.

[3] Chunyan E. Tian, Dan Moy, Chuck Le, et al, "Reliability Investigation of NiPtSi Electrical Fuse With Different Programming Mechanisms", IEEE TRANS. DEVICE MATERIALS RELIABILITY. Vol. 8, no. 3, 2008, pp. 536-541.

[4] B. Ang, S. Tumakha, et al, "NiSi Polysilicon Fuse Reliability in 65-nm Logic CMOS Technology", IEEE TRANS. DEVICE MATERIALS RELIABILITY. Vol. 7, no. 2, 2007, pp. 298-303.

[5] Deok-kee Kim, Anthony Domenicucci, and Subramanian S. Iyer, "An investigation of electrical current induced phase transformations in the NiPtSi/polysilicon system", J. Appl. Phys. Vol. 103 , 2008, 073708.

[6] Karen Maex and Marc Van Rossum, Properties of Metal Silicides, London, INSPEC press, 1995.

[7] P.Gas, F.M. dheurie, et al., "Formation of intermediated phases, Ni_3Si_2, Pt_6Si_5: Nucleation, identification, and resistivity", J. Appl. Phys. Vol. 50 ,1986, pp. 3458-3466.

[8] T. S. Doorn, "A Detailed Qualitative Model for the Programming Physics of Silicided Polysilicon Fuses", IEEE Electron Device, Vol. 54, no. 12, 2007, pp. 3285–3291.

DETERMINATION OF THE LOCAL ELECTRIC FIELD STRENGTH BY ENERGY DISPERSIVE PHOTON EMISSION MICROSCOPY

T Geinzer [a], R Heiderhoff [a], JCH Phang [b], LJ Balk [a, b]

[a] Department of Electronics, Faculty of Electrical, Information and Media Engineering, University of Wuppertal, Germany
[b] Centre for Integrated Circuit Failure Analysis and Reliability (CICFAR), National University of Singapore, Singapore
phone: +49 202 439 1928; fax: +49 202 439 1804; e-mail: geinzer@uni-wuppertal.de

ABSTRACT

Energy-dispersive Photon Emission Microscopy (PEM) allows the local electron temperature distribution to be characterized with high accuracy and sensitivity. The suitability and potential of this new technique for failure analysis and reliability investigation of semiconductor devices are demonstrated by the spatially resolved analysis of non-uniform breakdowns. With state-of-the-art devices in nanometer dimensions the charge carrier transport is analyzed in order to determine the electric field strength distribution.

INTRODUCTION

As a non-destructive technique, Photon Emission Microscopy (PEM) is a favored method for failure analysis and reliability investigation of semiconductor devices [1]. The Scanning Near-Field Photon Emission Microscopy (SNPEM) technique is able to resolve the nanometer dimensions of state-of-the-art processors and integrated circuits [2]. However, for failure analysis and reliability investigation, the localization of the defects is not the only decisive factor for a detailed understanding of the activities inside the device. Another important factor is the locally dominating electric field strength.

The charge carrier temperature which is dependent on the electric field strength has been determined for devices with active junctions in the micrometer range from the slope of the semi-logarithmic energy distribution of the visible electromagnetic radiation [3]. Spectral PEM analysis based on Wavelength-Dispersive PEM (WDPEM) allows the charge carrier temperature T_c to be determined. However, due to the poor signal-to-noise ratio (SNR) of the emitted photons, spatially resolved distributions of T_c have not been achieved [4].

The new approach of Energy-Dispersive PEM (EDPEM) circumvents the disadvantage of WDPEM. EDPEM is able to detect the T_c of a hot spot and has the potential to determine the accurate electric field strength distribution of state-of-the-art devices with nanometer dimensions.

LUMINESCENCE MECHANISMS WITHIN SEMICONDUCTOR DEVICES

The charge carrier distribution, particularly of electrons, is analyzed in order to interpret photon emission from semiconductor devices. The density of electrons versus energy n(W) inside a semiconductor is defined as the integral over the product of density of states S(W) and the energy distribution function f(W):

$$n = \int_{W_c}^{\infty} S(W) \cdot f(W)\, dW .$$ (1)

W_c is the conduction band energy and S(W) is calculated from the electronic band structure. The Fermi-Dirac-distribution is valid for the energy distribution function f(W), but the high energy tail can also be described by the Maxwell-Boltzmann distribution:

$$f_{MB}(W) = e^{\frac{-W}{k_B \cdot T_e}}$$ (2)

where k_B is the Boltzmann constant and T_e is the electron temperature. Electrons with high temperature are commonly known as "hot" electrons.

Electrons can reduce their temperature which implies loss of energy by emission of light. In silicon, electroluminescence is generated by means of recombinations involving both carrier types as well as relaxations involving only one type of carrier. These processes can further be separated into direct transitions in which a photon conserves energy and momentum or indirect transitions in which the photon only provides the energy according to [5]. These various luminescence mechanisms are summarized in Figure 1.

The emission spectra of the different transitions differ significantly from each other in the amount and the distribution of photons. Akil et al. [6] developed a multi-mechanism model for hot electrons in silicon devices and divided the near-infrared (NIR) and visible radiation into three zones which in each case is dominated by a different transition.

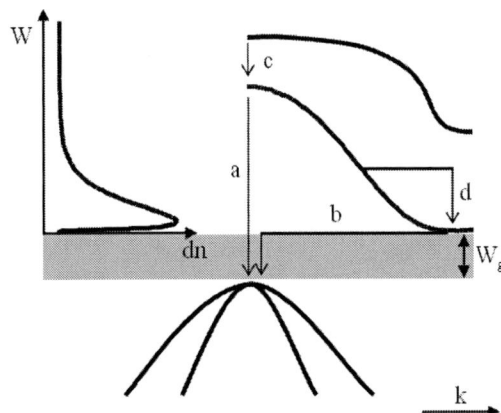

Figure 1: Relaxations and recombinations in indirect semiconductors according to [5]: (a) direct recombination, (b) indirect recombination, (c) direct relaxation, and (d) indirect relaxation.

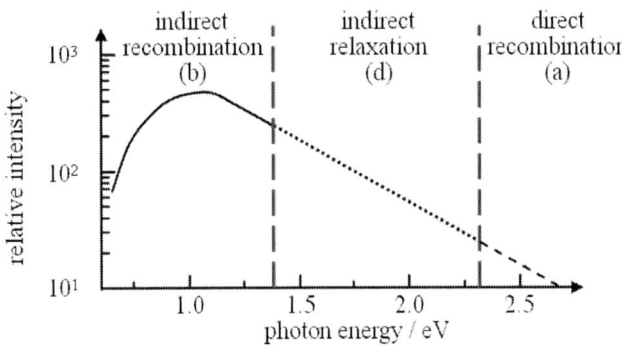

Figure 2: Photon emission of dominant relaxations and recombinations according to Figure 1 and [6]

Figure 2 indicates as leading emission the indirect recombination for photon energies below 1.4 eV, the indirect relaxation between 1.4 eV and 2.3 eV, and the direct recombination above 2.3 eV. The indirect relaxation in the intermediate range is also often called bremsstrahlung. The slope β of the semi-logarithmic energy distribution of this bremsstrahlung from hot electrons is inversely proportional to the electron temperature T_e [3]:

$$T_e \propto \frac{1}{\beta} \,. \tag{3}$$

For silicon devices, the emission efficiency over the visible range is very low. Cynoweth and McKay [7] reported 1 photon over the visible spectrum for every 10^8 electrons crossing the pn junction even in avalanche breakdown. In avalanche breakdown, multiplication of hot electrons takes place by impact ionization.

ENERGY DISPERSIVE PHOTON EMISSION MICROSCOPY

The new approach of EDPEM is introduced to determine the electron temperature sensitively and accurately. The main difference between WDPEM and EDPEM is illustrated in Figure 3 by means of the semi-logarithmic photon energy distribution. The notation of WDPEM and EDPEM is according to the notation of wavelength-dispersive and energy-dispersive X-Ray spectroscopy.

For WDPEM in Figure 3 (a), the slope of the straight line is calculated from the intensities I of at least two discrete points. In practice, a wavelength tunable monochromator or optical band-pass filters are used for the accurate measurement of two or more points. After correcting for the dispersion characteristics, the values can be transformed into the energy domain. For an accurate energy value, a small fragment of the total available photons is used. For a better SNR, optical filters with a wide band-pass are used to average the energy value.

For EDPEM, the number of photons above the threshold energy N is integrated by use of a optical high-pass filter for energy-dispersive PEM (b). The threshold energy can be set exactly and a significant amount of the total available photons is detected which improves SNR. The evidence that both WDPEM and EDPEM identify the same slope can be shown mathematically. The slope β is calculated for WDPEM according to:

$$\beta = \frac{\ln(I_1) - \ln(I_2)}{h\nu_1 - h\nu_2} \,. \tag{4}$$

For EDPEM, the quantities N can be described as:

$$N_1 = \int_{h\nu_1}^{\infty} \alpha \, e^{-\beta \cdot h\nu} \, dh\nu = \frac{\alpha}{\beta} e^{-\beta \cdot h\nu_1} \,, \tag{5}$$

which results into the ratio of N_1 and N_2 as follows:

$$N_1 / N_2 = e^{\beta(h\nu_2 - h\nu_1)} \,. \tag{6}$$

Consequently, the slope β is as follows:

$$\beta = \frac{\ln(N_1) - \ln(N_2)}{h\nu_2 - h\nu_1} \,. \tag{7}$$

In the next section, the EDPEM technique is used to resolve the electron temperature distribution of a hot spot with better resolution and sensitivity.

COMPARISON OF WAVELENGTH DISPERSIVE AND ENERGY DISPERSIVE PEM

Both WDPEM and EDPEM measurements are performed on a hot spot of the base emitter junction of a silicon photo-transistor which is driven into avalanche breakdown. Figure 4 (a) shows the optical image of the top view of the silicon photo-transistor (DUT) and the schematic cross section. The DUT is inserted in the prototype SNPEM system shown in Figure 4 (b). The SNPEM allows an improvement of the spatial resolution based on the near-field collection mode of the Scanning Near-field Optical Microscope (SNOM). The improvement of the spatial resolution causes a further reduction of the amount of photons available. The emission of the DUT is guided by a sharpened uncoated glass fiber. The dominating bremsstrahlung emission is discriminated by optical filters. A commercial photomultiplier tube (PMT) is used for photon detection and the PMT's spectral response is nearly uniform within the visible range.

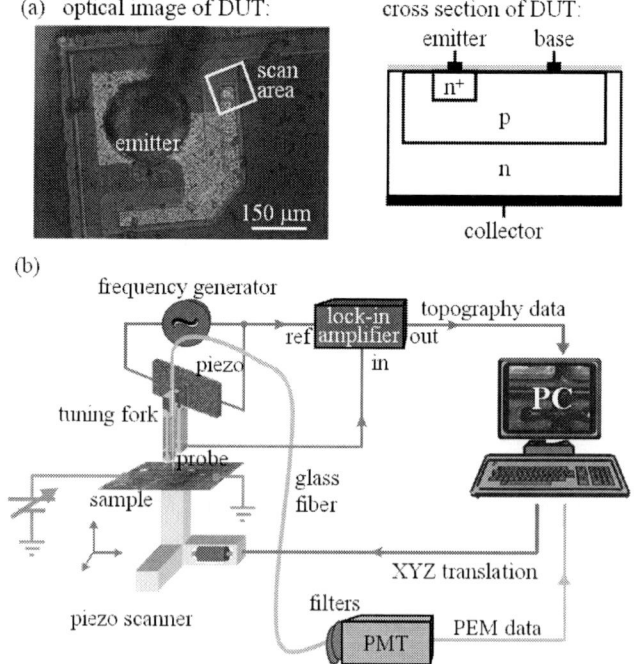

Figure 4: Description of the DUT (a) and measurement setup of SNPEM (b)

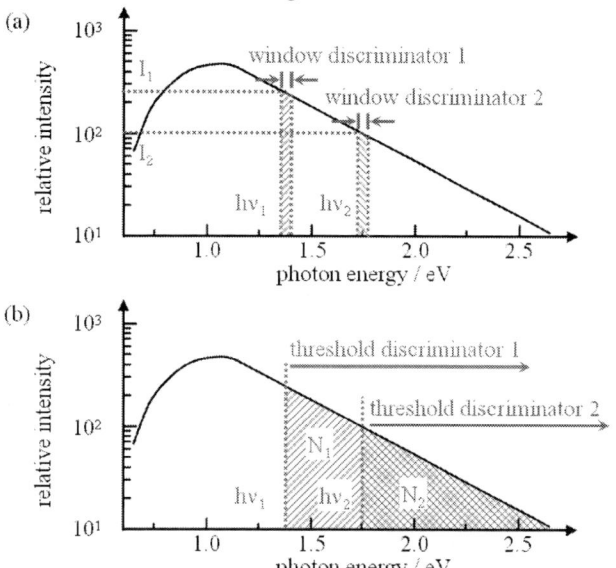

Figure 3: Comparison of the electron temperature determination using wavelength-dispersive (a) or energy-dispersive (b) PEM

In both cases, the measurement setup is identical for a direct comparison between WDPEM and EDPEM. Two optical band-pass filters are chosen to discriminate the panchromatic photon emission. On the one hand, the optical band-pass filter 1 allows the photon emission in the optical bandwidth between 1.5 eV and 2.1 eV to be measured, and, on the other hand, filter 2 transmits photons between 2.1 eV and 3.1 eV. For WDPEM results, the two experimental intensities I_1 and I_2, discriminated by the optical band-pass filters, are computed as in equation (4). For EDPEM, the intensities I_1 and I_2 are added to N_1 and the quantity N_2 corresponds to I_2. These quantities are computed according to the slope in (7).

With hot electrons, the calculated slope β is indirect proportional to the electron temperature T_e as mentioned above. A comparison of the qualitative results for resolving a hot spot for visible PEM, for WDPEM and EDPEM is illustrated in Figure 5. The photon emission in the visible range (a) shows a dominant peak with cliffy flanks. In contrast to (a), the results of the electron temperature distribution in (b) and (c) do not show this sharp peak. For physical interpretation, it is mandatory to analyze the electron temperature distribution detected by both techniques (b) and (c). A significant increase in the SNR is evident with energy-dispersive PEM (c) besides the possibility of higher spatial resolution.

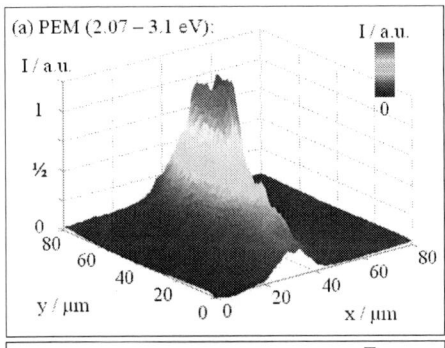

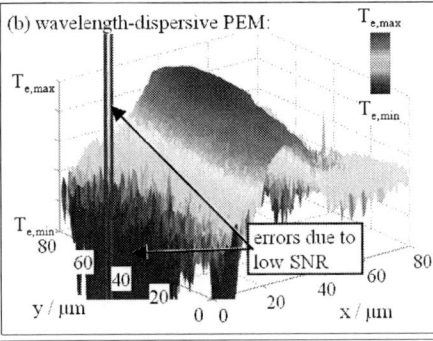

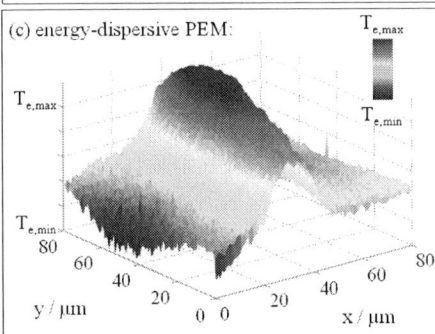

Figure 5: Comparison of PEM (a) and the qualitative electron temperature distribution by wavelength-dispersive (b) and energy-dispersive (c) PEM

TABLE 1
COMPARISON OF RELATIVE FAILURES FOR TYPICAL T_E = 2000 K

relative failure for T_e = 2000 K	energy-dispersive	wavelength-dispersive	
	filter 1: 1.5 - 2.1 eV *filter 2: 2.1 - 3.1 eV*	*filter 1: 1.5 - 2.1 eV* *filter 2: 2.1 - 3.1 eV*	*filter 1: 1.5 - 1.6 eV* *filter 2: 2.1 - 2.2 eV*
N(hv) = 10^6	2.36 %	97.6 %	23.7 %
N(hv) = 10^3	6.57 %	97.9 %	24.5 %

For the total amount of one million photons per second (N(hv) = 10^6) Gauß' error analysis provides a measuring inaccuracy of 2.36 % for the EDPEM compared to 97.6 % for WDPEM for quantitative T_e analysis. The amount of 10^6 photons is a value typical of an emission area of 1 µm. If the emission area is further confined to ensure better spatial resolution, the total amount of photons also decreases. The total amount of one thousand photons per second (N(hv) = 10^3) leads to a failure of 6.57 % for (c) and 97.9 % for (b). The significant difference in the measurement inaccuracy is due to the wide bandwidth of the optical filters which are effectively high-pass filters. For (b), the wide bandwidth falsifies the energy value whereas for (c) only the abruptness of the filter edge is decisive. The accuracy of (b) can only be improved by reducing the filter bandwidths, e.g. to 0.1 eV. The comparisons are given in Table 1.

DETERMINATION OF THE LOCAL ELECTRIC FIELD STRENGTH BY ENERGY DISPERSIVE PEM

The influence of bias on electrons has to be analyzed for the indirect determination of the local electric field strength E inside semiconductor devices. When electrons are accelerated by an electric field, the Maxwell-Boltzmann distribution function in (2) differs from the equilibrium through an increase in the electron temperature [8] as follows:

$$T_e \propto W_{el} = q \int \vec{E} d\vec{s} \qquad (8)$$

where W_{el} is the energy benefit electrons obtained by an electric field along the way $\vec{s}$, and q is the elementary charge.

An essential material parameter is the mean free path of charge carriers for state-of-the-art semiconductor devices. The dimension of the mean free path d_{mfl} compared to the extension of the electric field strength E decides whether charge carriers have to be described by ballistic or by classic transport. Within semiconductor devices, charge carriers are affected by the force of an electric field typically ruling in a space charge region with the width w_{SCR}. Therefore, under the condition of $d_{mfl} > w_{SCR}$ the transport is ballistic whereas under $d_{mfl} < w_{SCR}$ it is classic.

Under the ballistic condition ($d_{mfl} > w_{SCR}$) in the case of a symmetric and abrupt pn junction the integral of (8) simplifies to:

$$T_e \propto W_{el} = q \frac{E_{max} \cdot w_{SCR}(E)}{2} \qquad (9)$$

where E_{max} is the maximum electric field strength at the doping boundary and the space charge region width w_{SCR} is dependent on the electric field strength E. The space charge region width of an abrupt pn junction is proportional to the square root of the voltage difference between junction voltage U_J and applied voltage U. Furthermore, this voltage difference is proportional to the square of the maximum electric field strength. Finally, the relation between w_{SCR} and E_{max} is directly proportional:

$$w_{SCR}(E) \propto \sqrt{U_J - U} \propto \sqrt{E_{max}^2} \propto E_{max}. \qquad (10)$$

Inserting (10) into (9) results in:

$$T_e \propto E_{max}^2 \approx E^2 \qquad (11)$$

for the ballistic transport.

Under the classic transport condition ($d_{mfl} < w_{SCR}$) charge carriers scatter and they obtain an energy benefit W_{el} by an electric field also following (8). Under $d_{mfl} < w_{SCR}$ the electric field can be assumed as constant simplifying the integral to

$$T_e \propto W_{el} \overset{E\,const.}{=} q \cdot E \cdot v_{drift}(E) \cdot \tau_W \qquad (12)$$

where v_{drift} and τ_W are the drift velocity and the relaxation time. In connection with the electric field dependence of the drift velocity the band structure of the semiconductor determines the type of scattering. According to [9], generally three scattering mechanisms are considered in silicon: elastic scattering by acoustic phonons and inelastic scattering by optical phonons or by impact ionization.

At low electric field strengths, the scattering is mainly by acoustic phonons and the drift velocity is proportional to the electric field strength. Therefore, (12) results again in:

$$T_e \propto E^2 . \qquad (13)$$

At high electric field strengths, the scattering by optical phonons or by impact ionization gets more important. The drift velocity saturates according to [10] for $E > 10^6$ V/m and is independent of the electric field. When the drift velocity becomes field independent, charge carriers are named "hot" carriers. For avalanche breakdown in a space charge region the electric field strengths are of the order of those for "hot" carriers. Inserting the saturated drift velocity into (12) leads to the new proportionality of

$$T_e \propto E \qquad (14)$$

for classic transport at high electric field.

Balkan [11] provides representative values for the decision factor "mean free path length" which allows ballistic and classic transport to be differentiated. In silicon, the mean free path length decreases with carrier energy from around 40 nm (for electrons at room temperature) to 2 nm for energies around 1.5 eV (hot electrons). Thus, hot electrons guarantee the continued description of classic transport down to very small feature sizes of nanometer dimensions.

Further, the mentioned charge carrier energy is an important value for determining the energy balance within charge carrier transport. For the three general scattering mechanisms in silicon devices, the initial charge carrier must have enough energy so that it can produce the interaction products of acoustic and optical (> 0.05 eV) phonons or of an electron hole pair by impact ionization (> 1.1 eV). The ability of an electron to generate scattering interaction products can be seen in the electronic band structure. In Figure 6, this issue is illustrated on the basis of the indirect conduction band structure of silicon in [100] up to the band minimum. Based on the data of the conduction band structure the group velocity v and the effective mass m_e^* of the electron can be calculated, as shown in Figure 6. The different zones are shown in the colors yellow, green, and red referring to ballistic, classic low and high electric field transport, respectively.

The zones can be interpreted as follows: An electron of the valence band normally remains in the minimum of the band structure. An electric field leads to an energy increase and initially the electron follows the parabolic dispersion relation of a free electron (red). In the next zone (yellow), the electron has enough energy to start scattering by acoustic phonons for classic transport at a low electric field. The last zone (blue) is the classic transport at high electric fields where even scattering by optical phonons or by impact ionization gets more important and the effective electron mass can become infinite. This means that the momentum must be given to the crystal by scattering [12] as well.

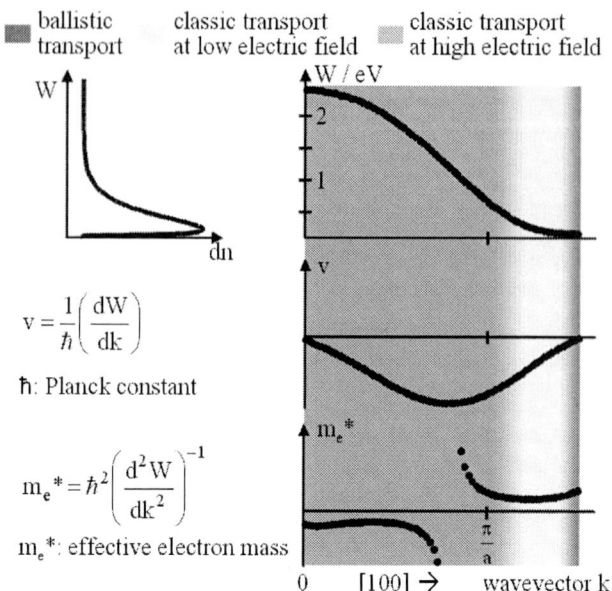

ballistic transport — **classic transport at low electric field** — **classic transport at high electric field**

$$v = \frac{1}{\hbar}\left(\frac{dW}{dk}\right)$$

$\hbar$: Planck constant

$$m_e^* = \hbar^2 \left(\frac{d^2W}{dk^2}\right)^{-1}$$

m_e^*: effective electron mass

Figure 6: Description of ballistic and classic transport in the electronic band structure

On the basis of the relation between the electron temperature and the electric field, EDPEM is used to characterize the base emitter junction of the DUT driven into avalanche breakdown. The results are shown for two different biasing conditions. Figure 7 demonstrates the expected increase of the electric field strength with higher device biasing. Furthermore, the main hot spot is expanded and a second small hot spot appears in the top left edge of Figure 7 (b).

Because of the excellent accuracy of energy-dispersive PEM (see again Table 1), it is possible to observe the expected relationship between the determined electric field and the biasing voltage for an abrupt pn junction even of the whole image for the first time. The electric field strength varies with the square root of the voltage difference between junction voltage U_J and applied voltage U according to (10). Figure 8 shows that this relation is fulfilled for moderate biasing above -8.5 V for the two lines indicated in Figure 7 (b). The square of the Pearson's correlation factor R^2 is above $R^2_{crit} = 0.92$ indicating the relation between electric field and biasing voltage of an abrupt pn junction. Only at the sides R^2 drops due to the original low emission for both lines (a) and (b).

Below -8.5 V, the correlation of this relation between electric field and voltage is not always fulfilled for both lines. This phenomenon is illustrated in Figure 9. Looking first at Figure 9 (a) it can be seen that the electric field follows the relation for an abrupt pn junction indicated by a good correlation above $R^2_{crit} = 0.92$.

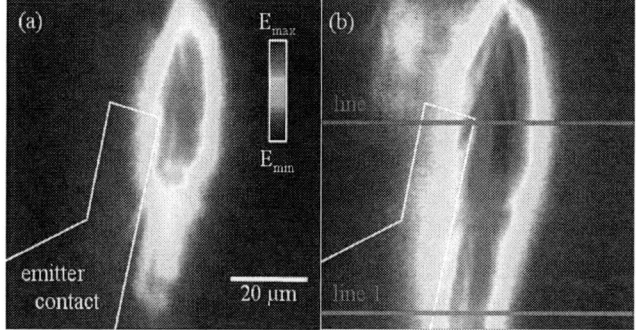

Figure 7: Determination of the electric field strength distribution for -8.4 V and -2 mA (a) as well as -12 V and -8 mA (b)

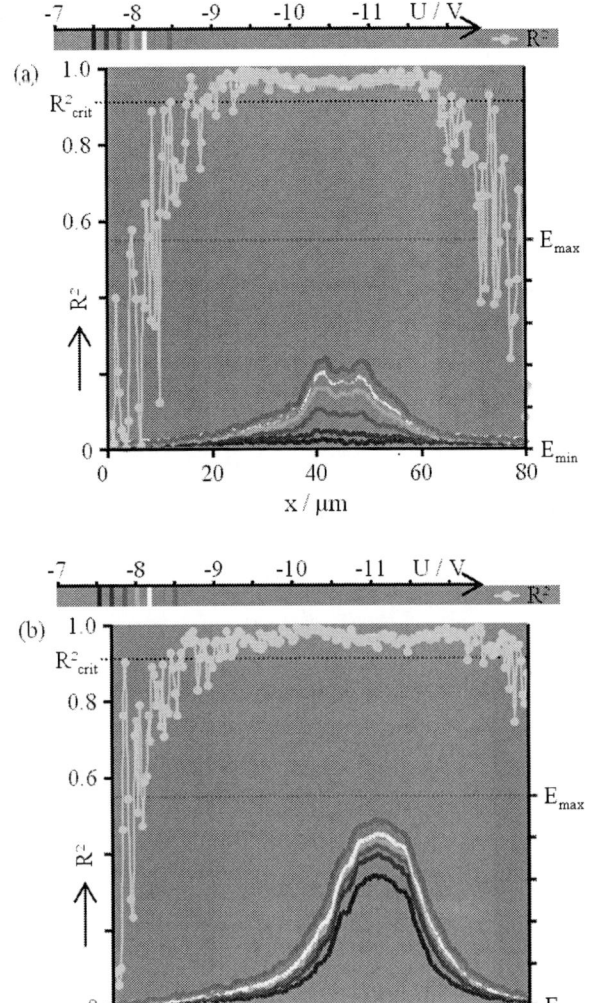

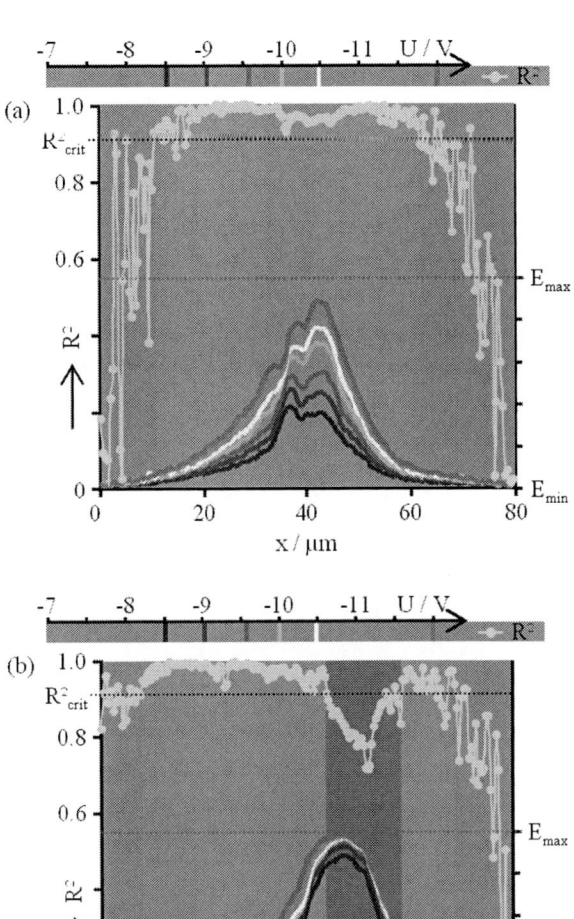

Figure 8: Correlation between electric field strength and voltage for an abrupt pn junction at line 1 (a) and line 2 (b) indicated in Figure 7 (b) for U > -8.5 V

Figure 9: Correlation between electric field strength and voltage for an abrupt pn junction at line 1 (a) and line 2 (b) indicated in Figure 7 (b) for U < -8.5 V

Only at the sides R^2 drops due to the originally low emission. In Figure 9 (b) the situation is different, as there is a critical area where R^2 is below 0.92 even within original high emission. The electric field strength saturates due to the enormous increase of mobile charge carriers by avalanche multiplication. The generated electron hole pairs partially compensate the space charge built up of localized, ionized donors and acceptors and lower the electric field strength. With further increase of biasing, the DUT reaches a crucial multiplication factor most likely within this critical area. This crucial multiplication leads to a malfunction of the whole device by thermal overload of the critical area. It is especially remarkable that the critical area is not symmetric around the maximum electric field strength. For panchromatic PEM, one would assume that the critical area is symmetric around the emission peak.

CONCLUSION

In this work, the charge carrier transport of a silicon photo-transistor is analyzed in order to determine the electric field strength distribution.

EDPEM allows the electric field strength distribution to be determined with a higher accuracy and sensitivity compared to WDPEM. The hot spot of the passivated base-emitter junction has been analyzed. The determined electric field strengths follow the behavior of an abrupt pn junction for different biasing. At high biasing, the electric field strength saturates indicating a crucial factor of avalanche multiplication. This critical area will lead to a malfunction of the whole device by thermal overload in case of higher biasing.

These non-uniform local breakdowns can be spatially resolved for the first time which demonstrates the potential of EDPEM for failure analysis and reliability investigation. Even state-of-the-art devices can be analyzed because hot electrons guarantee the continued description of classic transport down to very small feature sizes of nanometer dimensions. For these high-integrated devices, the breakthrough voltages are much lower. As a consequence, the luminescence shifts from the visible spectrum more to the NIR and the total amount of photons decreases. Due to the optimized SNR, energy-dispersive SNPEM is able to identify the local electric field strength with highest spatial resolution and accuracy.

REFERENCES

[1] JCH Phang, DSH Chan, SL Tan, WB Len, KH Yim, LS Koh, CM Chua, LJ Balk, "A Review of Near Infrared Photon Emission Microscopy and Spectroscopy", Proceedings of 12th IPFA, 2005, pp. 275-281.

[2] D Isakov, AAB Tio, T Geinzer, JCH Phang, Y Zhang, LJ Balk, "Near-field detection of photon emission from silicon with 30 nm spatial resolution", Microelectronics Reliability, 2008, Vol. 48, pp. 1285-1288.

[3] A Toriumi, M Yoshimi, M Iwase, Y Akiyama, K Taniguchi, "A study of photon emission from n-channel MOSFET's", IEEE Trans. Electron Devices, 1987, Vol. 34, No. 7, pp. 1501-1507.

[4] A Glowacki, P Laskowski, C Boit, P Ivo, E Bahat-Treidel, R Pazirandeh, R Lossy, J Würfl, G Tränkle, "Characterization of stress degradation effects and thermal properties of AlGaN/GaN HEMTs with photon emission spectral signatures", Microelectronics Reliability, 2009, Vol. 49, pp. 1211-1215.

[5] J Bude, N Sano, A Yoshii, "Hot-carrier luminescence in Si", Physical Review B, 1992, Vol. 45, No. 11, pp. 5848-5856.

[6] N Akil, SE Kerns, DV Kerns, Jr., A Hoffmann, JP Charles, "A multimechanism model for photon generation by silicon junctions in avalanche breakdown", IEEE Trans. Electron Devices, 1999, Vol. 46, No. 5, pp. 1022-1028.

[7] AG Chynoweth, KG McKay, "Photon Emission from Avalanche Breakdown in Silicon", Phys. Rev., 1956, Vol. 102, pp. 369-376.

[8] K Hess, Advanced Theory of Semiconductor Devices, IEEE Press: New York, 2000, chapter 5.

[9] W Mönch, "On the Physics of Avalanche Breakdown in Semiconductors", Physica Status Solidi (b), 1969, Vol. 36, No. 1, pp. 9-48.

[10] C Jacoboni, C Canali, G Ottaviani, A Alberigi Quaranta, "A review of some charge transport properties of silicon", Solid-State Electronics, 1977, Vol. 20, pp. 77-89.

[11] N. Balkan, Hot electrons in semiconductors: physics and devices, Clarendon Press: Oxford, 1998, chapter 2.

[12] C Kittel, Introduction to solid state physics, 8th ed., Wiley: Hoboken, NJ, 2005, chapter 8.

Explosion Phenomenon of High Resistance Via During TEM Sample Preparation Using FIB

Pan Liu, Irene Tee, Soo Sien Seah, Chi Wen Soo, Ye Chen, Zhi Qiang Mo
Failure Analysis department, GLOBALFOUNDRIES Singapore Pte Ltd,
60 Woodlands Industrial Park D, Street 2, Singapore 738406
Phone: +65-6360-3430, Fax: +65-6360-4592, E-mail: liupan@globalfoundries.com

Abstract— **High-resistance via explosion is a new damage phenomenon which is induced during TEM sample preparation using FIB. Two methods that will prevent this phenomenon are introduced. In practice, these methods have been effective in avoiding high-resistance via explosion.**

Keywords- Via explosion, TEM sample preparation, FIB, EM, High resistance

I. INTRODUCTION

The resistance measurement methodology is important to monitor semiconductor manufacturing and production reliability test, especially in evaluating multi-interconnect layers [1]. For wafer accept test (WAT), the common test structures for the back-end of line monitor include via chain, Kevin via and stack Kevin via structure. For reliability test, the common test structures include electro-migration (EM) and stress-migration (SM) structures. The major failure modes of these test structures are high resistance, open and short to other nearby lines. The high-resistance failure mode is more common than other failure modes. Therefore, knowing the root cause of high resistance via is important for the improvement of production yield and reliability.

Focused ion beam (FIB) is used to mill and view high resistance precision via structure as well as prepare transmission electron microscopy (TEM) samples [2]. With the shrinkage of the critical dimension (CD) of integrated circuits (IC), the CD of via becomes smaller and smaller. FIB is used to mill and view the microstructure in failure analysis. However, the artifacts and damage induced by FIB have been reported in many articles including curtain effect, re-deposition and ion beam induced amorphous layer [3,4]. The ion beam induced amorphous layer can be observed with the use of high voltage ion beam for TEM sample preparation. Thus, a low kV ion beam is used to remove the amorphous layer as part of the process of TEM sample preparation.

In this article, a new FIB damage phenomenon, called via explosion, is reported when a high resistance via structure is milled by FIB. The explosion causes damage to the area of interest which makes the determination of the root cause impossible. In the next section, we describe the explosion phenomenon during TEM sample preparation using FIB. Finally, in the last section, we give an explanation of this phenomenon and several methods to avoid this damage.

II. EXPERIMENT AND PHENOMENON

In general, the high resistance via means that the resistance is about ten times of the normal via in the resistance testing, but not totally open. To prepare this kind of TEM samples, the standard process is as follows. We coat a protection layer which can be metal, carbon or silicon oxide material, and this layer thickness is about 0.5μm using ion/electron beam deposition over the area of interest of about 1μm by 10μm. An ion current of 100~300pA is normally used, followed by trench milling, U-cut and finial fine milling. The final thickness of the sample is below 100nm. Finally, we observe the sample in the TEM.

We observed the explosion phenomenon in different electrical test (ET) and EM test structures. The images of high-resistance via explosion are shown in figure 1. In figure 1A, a six Al layer stack via structure explosion is because the tungsten via and Al metal melted during sample preparation. In figure 1B, the explosion occurred at Dual-Damascene structure copper via1 and metal2. In figure 1C, an EM test structure explosion occurred at copper via4 and copper metal5 layer. The EM test structure is a M4-M5-M4 stack. From the image, via4 is destroyed during the explosion.

From these images, the explosion damage induced by TEM sample preparation using FIB is very serious; especially in figure 1B where the 45-degree shear stress splits the oxide layer and the copper diffusion is sent to the gap. Once explosion occurs at the high resistance via, the area of interest is totally damaged. In fact, this explosion phenomenon became more serious after low k and ultra-low k materials were introduced into the process at the back-end of line. It has been observed that the explosion phenomenon occurs only with via structures that show high resistance during electrical testing.

III. RESULT AND DISCUSSION

From the above via explosion phenomenon, the damages are very serious. The area of interest is totally damaged, and the high-resistance via is destroyed. What is the mechanism and how can the damage be avoided?

We studied the TEM sample preparation process carefully. During FIB milling, an anomaly was found at the high-resistance via where the explosion occurred. Figure 2A and B show the FIB and TEM images respectively at the same Kevin via2 structure. The high-resistance via2 with evidence that the explosion has already occurred was observed in the FIB image. Hence, the explosion occurred during FIB sample preparation or before, and not during TEM imaging.

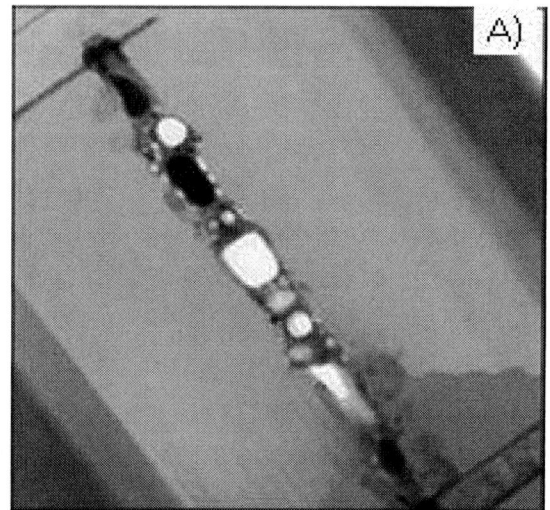

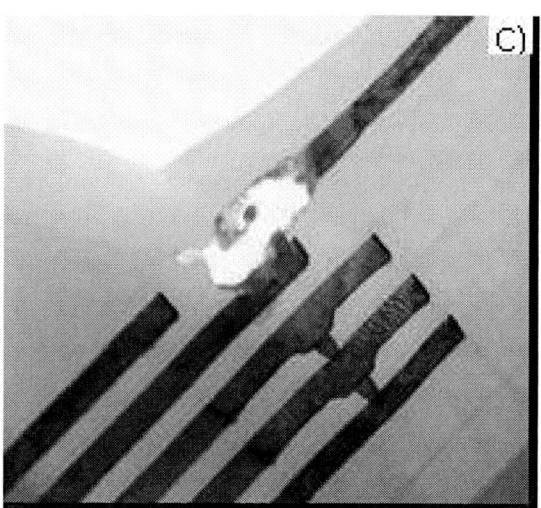

Figure 1. TEM images of high resistance via explosion phenomenon. (A) TEM image of six Al layer stack via structure; (B) TEM image of via1 and metal2; (C) TEM image of EM test structure

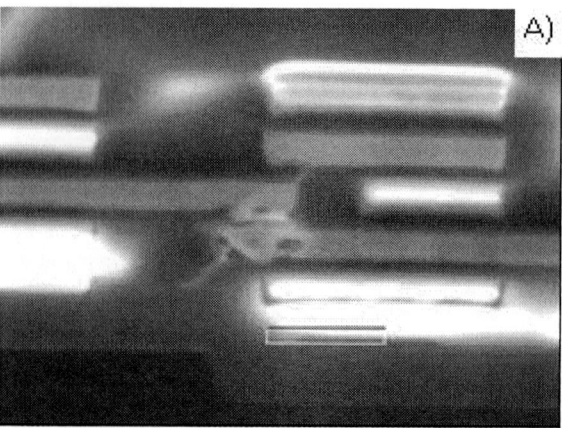

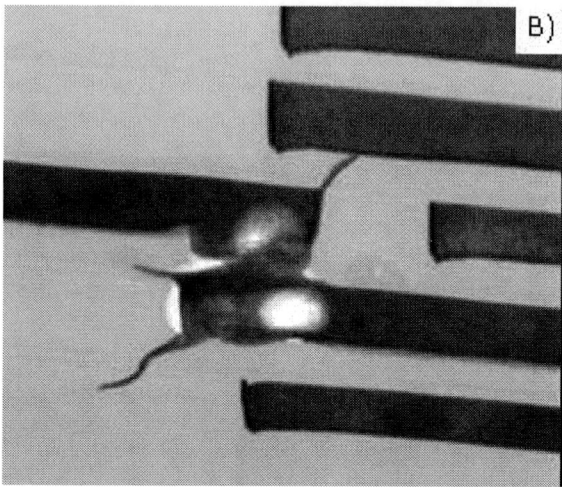

Figure 2. (A) FIB and (B) TEM images of Kevin via2 explosion of via1 and metal2 explosion of copper via;

In fact, not all high-resistance via samples suffer explosion, but via samples with explosions were all of high resistance. Therefore, we suspect that the FIB ion current induces the via explosion. At the same time, we found a similar explosion profile in the time dependent dielectric breakdown (TDDB) hard breakdown case. Figure 3 shows a TDDB hard breakdown example. The mechanism of high resistance via explosion is similar to the dielectric breakdown to a certain extent.

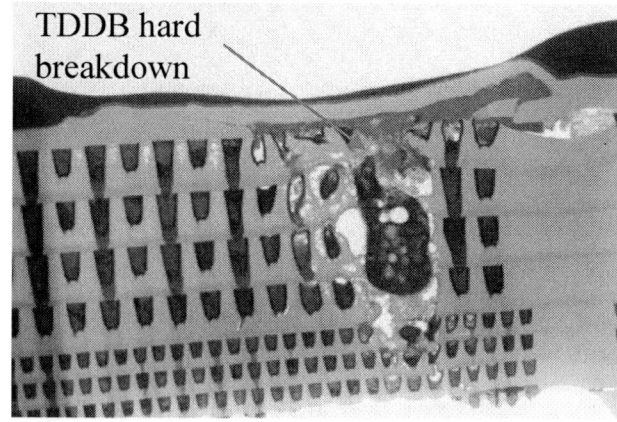

Figure 3. TEM image of stack via TDDB hard breakdown

The TDDB hard breakdown mechanism [5,6,7] is that the local electric field stresses the dielectric film and at the weak points the stress accumulated is sufficiently high to cause avalanche multiplication phenomena. For high resistance via, a similar breakdown phenomenon would occur under FIB ion/electron beam irradiation in the process of TEM sample preparation. Therefore, we can hypothesize that this breakdown and explosion phenomena in the high resistance via is based upon the dielectric breakdown mechanism. Firstly, during the process of scanning the sample surface with the electron beam or ion beam for imaging, or depositing a layer to protect the area of interest, the high resistance via will suffer a electron field stress or high current density at weak points which is only the liner (Ti/TiN or Ta/TaN very thin layer less than 50A) in some cases. Actually, this effect is like charging since charging is worse at the high resistance via. Secondly, this charging will lead to discharge when the buildup reaches a critical value. The buildup will depend on electron-ion current. The discharge will lead to damage as in electrostatic discharge (ESD). At the first stage of discharge, the liner seems to function as a conductive path in TDDB soft breakdown. Under continuing stress, this very thin liner will break and the capacitor-like structure will be formed in the via because the void separates two piece of metals of copper via. Lastly, if the high resistance via cannot discharge the accumulated charge from E/I beam continued scanning, the TDDB hard breakdown or ESD process would happen at the high resistance via.

In fact, the void in the high resistance via can be regarded as an air gap between copper layers and it has extreme low breakdown voltage compared to the dielectric breakdown voltage in TDDB due to the fact that k value of air is 1 while ultra-low k material is 2.45 .

According to the above analysis, reducing the ion current through the high resistance via will reduce the probability of high-resistance via explosion. Furthermore, charging can be reduced if we ground the bond pad which links the high resistance via since the ion current will be conducted to the ground because the ground structure resistance is much lower than that of the high resistance via. That is, the ion current bypasses the high resistance via to ground. It follows that no current or very low current passes through the high resistance via to avoid the high-resistance via explosion phenomenon.

To avoid the high-resistance via explosion, two methods are given to conduct the ion/electron current to the ground in order that the charge is released. The first method is to dig a deep trench near the bond pad which connects to the substrate by using FIB before the protection layer deposition, followed by filling the deep trench and linking the bond pad to the deep trench with Pt deposition using FIB.

As the sketch in figure 4 shows, the deep trench is linked to the silicon substrate and the deposition layer is connected to the deep trench. As a consequence, the charge induced by the FIB will be conducted to the substrate, without passing the high-resistance via.

A second method is to coat a conducting metal layer using Precision Etching Coating System (PECS). The laser mark aligning to the high resistance via is needed to identify the high resistance via during FIB milling. After that, a copper belt is used to conduct the sample surface to the FIB holder. In the sketch in figure 5, the sample is adhered to the carbon conductor adhesive layer, and the whole sample surface is coated

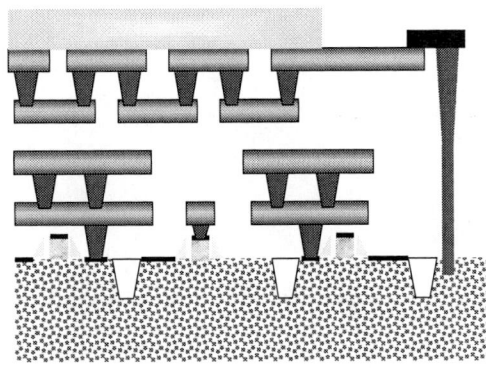

Figure 4. Schematic illustration of using deep trench to conduct charge and ion current to the ground

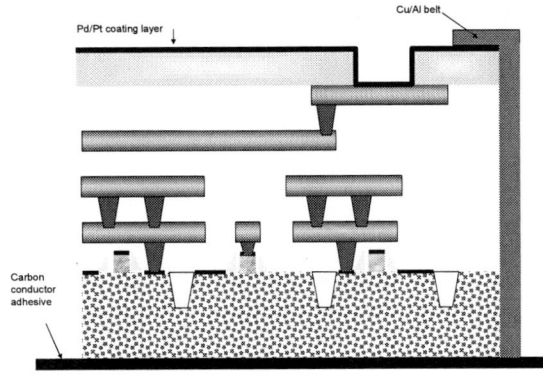

Figure 5. Schematic illustration of using coating metal layer to conduct charge and ion current to the ground

with Pd 200nm and connected to the carbon conductor adhesive by a copper belt.

Using the two methods, the charge induced by FIB is conducted to the ground. The high resistance via becomes safe, and thus will avoid explosion. Three TEM images are shown in figure 6A, B C, whose structures are the same as those of figure 1A, B 1C respectively. No via explosion was found after the bond pad was grounded.

Recently, we found some specially designed ET structures in which the top metal line flip up close to the testing via. Even though we performed grounding method to link all the bond pads, the via explosion still took place. In order to improve the success rate, we modified the milling structure after analyzing the mechanism of the via explosion. The flipped metal in the layout is kept as shown in figure 7A, the accumulated charge can be released through the grounded bond pads. Otherwise, if the flipping metal line is cut, the grounding method has no role to play. So, in practice, we make trench bigger than trench B, and make sure that trench B does not touch the flipping metal line. Due to keeping the flipping metal line, we should take care in the subsequent milling process as the metal line is close to the trench edge, which is shown in figure 7B.

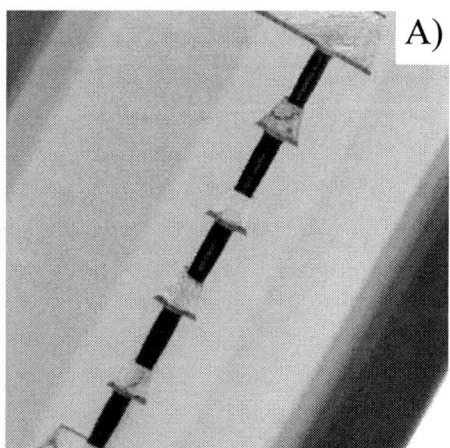

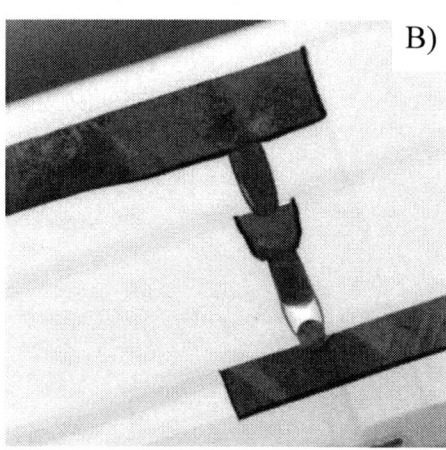

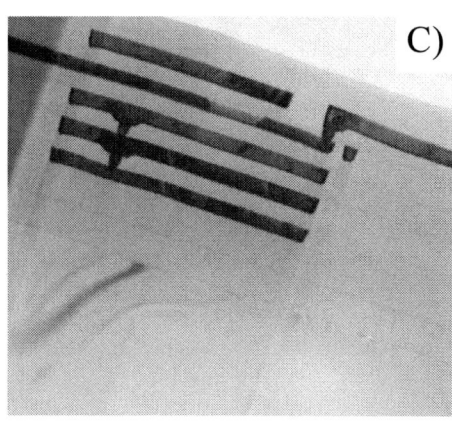

Figure 6. TEM images of high resistance via after grounding pads to the substrate. (A) TEM image of six Al layer stack via structure which is the same as figure 1A. (B) TEM image of via1 and metal2 of copper via which is the same as figure 1b; C) TEM image of EM test structure which is the same structure as figure 1C

We found that for 65nm technology node and beyond, the incidence of via or contact explosions is high. We combined deep trench link with coating to avoid the explosion phenomenon. In fact, any process which introduces charges, such as ESD, electron/ion beam, should be avoided to improve the TEM sample preparation quality further.

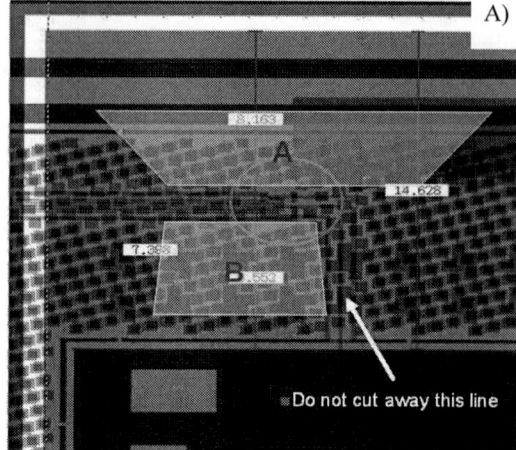

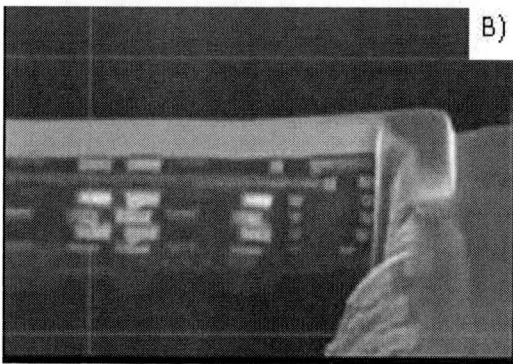

Figure 7. Flip up metal line layout TEM cutting method. (A) Trench cut layout ;(B) FIB image after trench cut

IV. CONCLUSIONS

High-resistance via explosion is a common issue at the back-end of line. The via explosion changes the profile of original via. To avoid high resistance via explosion, grounding methods and trench cutting policy are introduced. These methods are useful in keeping the original profile of the via.

REFERENCES

[1]. S.Wolf, *Silicon Processing for the VLSI Era, Vol. 4: Deep-Submicron Process Technology*, Lattice Press (Sunset beach CA, 2002) pp 573-602

[2].] Pan Liu, K. Li, Eddie ER and Siping Zhao "Plane-view Transmission Electron Microscopy for advanced integrated circuit," *2006 IEEE international conference on semiconductor electronics*, Kuala Lumpur, Nov, 2006, pp. 630-633.

[3]. GAO Qiang, ZHANG Mark, NIOU Chorng , LI Ming , CHIEN Kary "Sidewall damage induced by FIB milling during TEM sample preparation," *42nd Reliability Physics Symposium Proceedings*, April. 2004 Phoenix AZ pp 613-614.

[4]. Rubanov S, Munroe P.R "Investigation of the structure of damage layers in TEM samples prepared using a focused ion beam," *Journal of Materials*

Science Letters, Volume 20, Number 13, 1 July 2001, pp. 1181-1183(3)

[5]. Michael G. Pecht, Riko Radajcic, Gopal Rao, *Guidebook for managing silicon Chip reliability*, CRC (New York,1999), pp.25-29.

[6]. F. Chen, M. Shinosky, B. Li, J. Gambino, S. Mongeon, P. Pokrinchak, J. Aitken, D. Badami , M. Angyal, R. Achanta, G. Bonilla, G. Yang, P. Liu, K. Li, J. Sudijono, Y.Tan, T. J. Tang, C. Child "Critical Ultra Low-k TDDB Reliability Issues For Advanced CMOS Technologies", *IEEE CFP09RPS-CDR 47th Annual International Reliability Physics Symposium, Montreal,* 25-29 April 2004, pp464-475

[7]. S. Yokogawa, D. Oshida, H. Tsuchiya, T. Taiji, T. Morita, Y. Tsuchiya, and T. Takewaki , "Prediction of early failuer due to non-visual defct on time dependent dielectric breakdown of low-k dielectrics: experimental verification of a yield-reliability model", *IEEE CFP08RPS-CDR 46th Annual International Reliability Physics Symposium,* Phoenix, 2008, pp 144 - 149

GATE OXIDE EFFECT ON WAFER LEVEL RELIABILITY OF NEXT GENERATION DRAM TRANSISTORS

Yu Gyun Shin, Kab-Jin Nam, Heedon Hwang, Jeong Hee Han, Sangjin Hyun, Siyoung Choi, and Joo-Tae Moon
Process Development Team, Semiconductor R&D Center, Samsung Electronics Co., Ltd.
San #16, Banwol-dong, Hwasung-City, Gyeonggi-Do, Korea 445-701, phone: +82-31-208-0435, email: yugyun.shin@samsung.com

ABSTRACT

Wafer level reliability (WLR) issues of DRAM cell and peripheral transistors are discussed. Since the 70 nm technology node, recessed transistors have been accepted for assuring data retention time of DRAM cell transistors. Various recessed transistor structures suggest that the most important issue in reliability, in addition to optimizing data retention time, is the elimination of local regions of concentrated electric fields. In this paper, by modulating the cell gate oxidation process, local field concentration is effectively reduced. Particularly, the introduction of a radical oxidation process can create cell transistors that are more immune to Fowler-Nordheim (F-N) stress, which can degrade interface quality during cell transistor operation. On the other had, for DRAM peripheral transistors, for DRAM peripheral transistors, which currently use dual poly-Si gates and SiON dielectrics, high-k/metal gate (HK/MG) structure are expected to be adopted at the 20 nm technology node for improved equivalent oxide thickness (EOT) scaling. The high thermal budget of a conventional DRAM manufacturing process can significantly impact HK/MG WLR issues. However, we have evaluated reliability characteristics for HK/MG WLR on DRAM cell and peripheral devices, and concluded that WLR issues will not be critical for operation. [*Keywords:* DRAM, transistor, gate oxide, reliability]

INTRODUCTION

Wafer level reliability (WLR) for DRAM must be classified for two different types of transistors: cell transistors and peripheral transistors. In order to increase data retention time, DRAM cell transistors have evolved from two-dimensional planar transistors to three-dimensional recessed-channel array transistors (RCAT) and spherical SRCAT since the 70 nm technology node. [1] On the other hand, DRAM peripheral transistors have planar structures in order to obtain high on-currents. Scaling is performed by reducing the equivalent oxide thickness (EOT). The trends are summarized in Table 1. However, it is important to consider thermal processes which, unlike the case for other logic transistors, may result in reliability issues of the DRAM storage capacitors. Thus it is important to evaluate the effect of heat for both three-dimensional cell transistors and planar peripheral transistors. In Part I, the effect of oxidation process (wet and radical oxidation) on reliability will be investigated. In Part II, peripheral and cell transistors incorporating high-k/metal gate (HK/MG) stacks for EOT scaling will be examined.

EXPERIMENTAL DETAILS

SRCAT devices were used for the cell transistors. Peripheral transistors with a conventional planar structure were also fabricated with similar integration processes. In Part I, wet oxidation or radical oxidation was used to form the gate oxide. The design rule of the DRAM cell transistors was 60 nm and a dual poly-Si process was used for the gate electrodes. All WLR measurements were performed at 25°C-125°C. [2]

TABLE 1. DRAM CELL AND PERIPHERAL TRANSISTOR TRENDS AT DIFFERENT TECHNOLOGY NODES.

	~90nm	~70nm	~60nm	~40nm
Cell Tr.				RCAT SRCAT RFIN
Peri. Tr.	EOT ~35Å	~25Å	~20Å	<20Å

For the peripheral and cell devices discussed in Part II, a HK/MG gate stack was implemented. An interface layer (IL) was formed, followed by the deposition of HfSiON for the high-k layer. A plasma nitridation (PN) and post-nitridation anneal (PNA) were subsequently performed. [3] The gate consisted of a TiN layer, followed by poly-Si. The HK/MG gate stack formation was also performed for some of the DRAM cell transistors as well for reliability evaluation.

PART I: OXIDATION EFFECT ON WLR FOR DRAM TRANSISTORS

The results of the WLR measurements for peripheral transistors are shown in Figure 1. To fairly compare cell and peripheral transistors, a thick gate oxide was used having the same EOT~5 nm in both cases. To apply hot carrier injection (HCI) stress, the $I_{sub,max}$ condition was used. For Fowler-Nordheim (F-N) stress, only a gate voltage was applied while the other terminals of the transistor were grounded. For the off-state stress, only a drain voltage was applied while the other terminals of the transistor were grounded. From the results in Figure 1, HCI stress has the largest effect on thick transistors, which may be due to the high lateral electric field. In addition, the trends between radical and wet oxidation were similar, with the type of stress also having a similar tendency. The effect of F-N stress was lower than the off-state stress which was lower than the HCI stress.

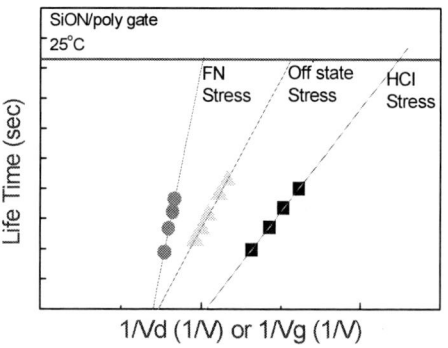

FIGURE 1. WLR LIFETIME OF DRAM PERIPHERAL TRANSISTORS.

For peripheral transistors, Vth and sub-threshold swing increase with a lowering of the drain current for both radical and wet oxidation processes were observed after F-N stress. Both show that the Vth shift is most likely to be caused by the same mechanism of electron traps particularly in the bulk. [12] However, the results for recessed cell transistors are different than those of the peripheral transistors. The shift in Id-Vg characteristics for radical and wet oxidation after F-N stress for the cell transistor is illustrated in Figure 2. For cell transistors with radical oxides after F-N stress, the Vth shift and degradation of sub-threshold slope has a similar trend to that of the peripheral transistors. However, the characteristics of the wet oxidation cell transistors are different than those of the peripheral transistors. Unlike the peripheral transistors, cell transistors with wet oxides show a reduction in Vth with increasing F-N stress time. From this Vth shift, it may be concluded that the wet oxide in cell transistors exhibit a hole trapping phenomenon in the bulk oxide after F-N stress. [4, 10]

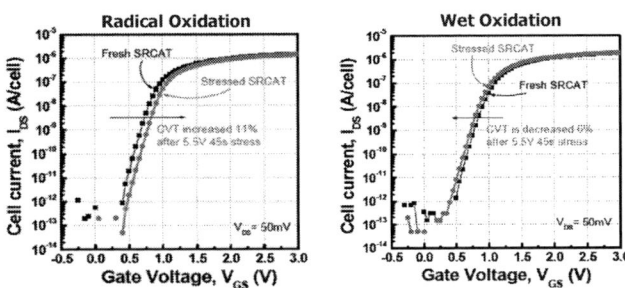

FIGURE 2. ID-VG DEGRADATION WITH F-N STRESS TIME FOR PERIPHERAL TRANSISTORS WITH RADICAL OXIDATION AND WET OXIDATION. [2]

The Id-Vg curves when both stress and measurement temperatures were 125°C were also performed. For radical and wet oxidation, the degradation of F-N stress was different. Although the stress conditions were exactly the same as those in Fig. 2, the Vth shifts in both the radical and wet oxidations had the same trend, but the degradation was more severe. Particularly for wet oxides, the change in Vth was so severe that the WLR lifetime could not be properly calculated. The worsening of Id-Vg characteristics for wet oxides will also lead to the worsening of refresh times.

Figure 3 summarizes the cell and peripheral WLR characterization with radical oxidation. When incorporated into the cell transistors, the HCI stress characteristics are better than those for the peripheral transistors. It believed to be due to the reduction in lateral field for the longer cell transistor channel lengths. Yet, there is no significant difference in F-N stress results for cell and peripheral transistors. When considering the operation of a DRAM cell, the operation voltage of the source and drain are much smaller than the gate voltage, so F-N stress in DRAM operation is the most important stress method of the three considered. Additionally, "burn-in" can occur when a high voltage is applied to the gate where degradation from high electrical fields can happen. Thus, F-N stress of the gate oxide is more critical than hot carrier effects when evaluating WLR.

Figure 4 shows the Id-Vd curves before and after F-N stress for both radical and wet oxidation. Since the gate voltage is less than the threshold voltage, the Id-Vd curves can show both the interface leakage current at the drain and the depletion region leakage along with the GIDL leakage. For drain voltages less than a certain level, the interface leakage at the channel with the depletion region leakage can be determined, while drain voltages over that certain level show the GIDL component. For the radical oxidation process, the Id-Vd

curves show no significant difference before and after F-N stress. However, for the wet oxidation process, there was a large increase in drain current after F-N stress. This may be explained by the wet oxidation process having a large increase in interface leakage.

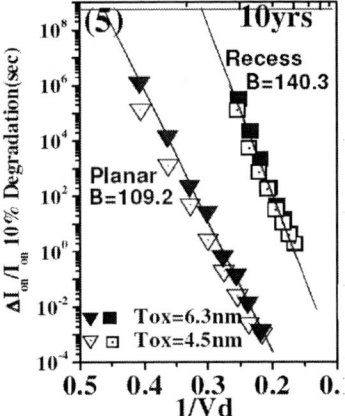

FIGURE 3. WLR LIFETIME OF DRAM CELL (RECESS) AND PERIPHERAL (PLANAR) TRANSISTORS. [4]

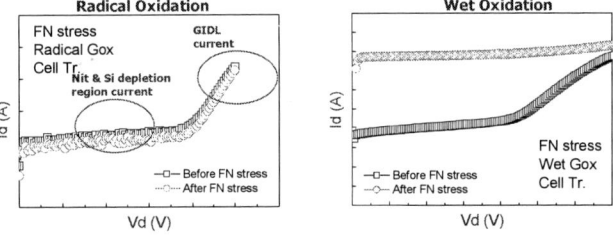

FIGURE 4. COMPARISON OF ID-VD LEAKAGE BEFORE AND AFTER F-N STRESS. FOR RADICAL OXIDATION AND WET OXIDATION.

The ability to recover from F-N stress was also tested. After stress, Id-Vg curves were measured at different time intervals. After a certain time frame, the obtained Id-Vg curves were no different than that of the immediately stressed device. Therefore, no recovery occurred and the F-N caused permanent damage to the interface. In addition, there was a difference between oxidation processes for cell transistors where there was no difference in peripheral transistors. To determine the mechanism explaining the disparity, the anode hole injection (AHI) was adopted as shown in the energy band diagrams in Figure 5. When the transistor is in inversion, an electron from the channel tunnels through the gate oxide and impacts the gate electrode. [5] The energy from this mechanism causes the formation of back-scattered holes. These holes then are trapped at the substrate/oxide interface, resulting in degradation. For a DRAM recessed cell transistor, the band diagram is shown in Figure 5(b). The electrical field near the gate is strongest (the electrical field can be determine by the slope of the bandgap) while the field near the channel/dielectric interface is lessened. Therefore, the holes are less likely to be trapped with this field compared to a planar peripheral transistor.

Furthermore, TOF-SIMS material analysis results shown in Figure 6 present evidence that there is more hydrogen present in the wet oxide dielectric. Especially for the substrate interface, the total number of Si-H or Si-O bonds contribute to number of possible hole traps. If there is a large H component within the dielectric, then the possibility of hole trapping increases. Particularly for recessed transistors, the lower electrical field at the substrate interface may not be large enough for trapped holes to escape. [6]

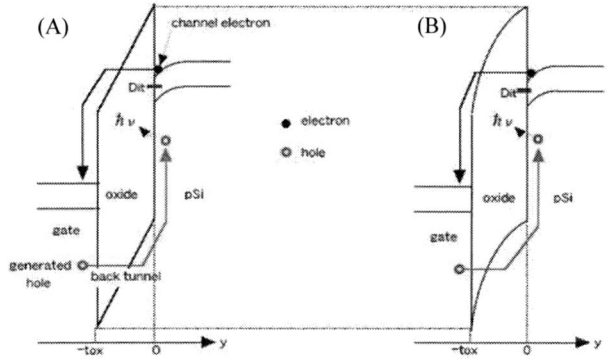

FIGURE 5. SUGGESTED DEGRADATION MECHANISM FOR (A) PERIPHERAL (PLANAR) AND (B) CELL (RECESSED) TRANSISTORS. [2]

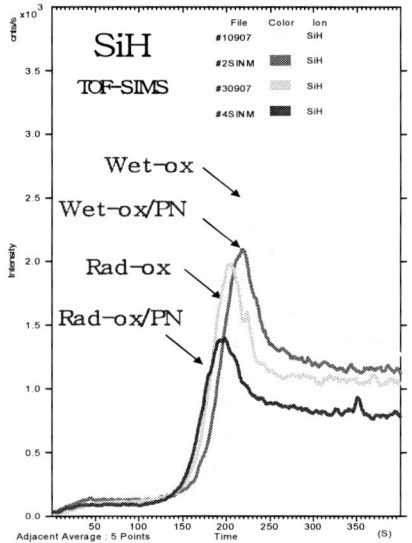

FIGURE 6. COMPARISON OF SI-H CONTENT WITH DIFFERENT OXIDATION METHODS AND PLASMA NITRIDATION. [2]

PART II. WLR OF DRAM CELL AND PERIPHERAL TRANSISTORS WITH HK/MG

The TDDB characteristics for peripheral thin oxide transistors were also performed. For PMOS transistors, devices with SiON/poly-Si gate stack structures show a large quantity of soft fail errors. HK/MG stacks, although having similar EOT, had no soft fail errors. This may be due to the physical thickness of the oxide is much thicker for the HK/MG and because the metal gate has less electrical resistance than the poly-Si gate. [7]

Table 2 summarizes the life time safe voltage values for both NMOS and PMOS thin transistors. For 10 year lifetime, the SiON/poly stack is similar or slightly worse than the HK/MG stack. Also, since typical operation voltages will be less than 1 V, there is still some margin for reliability. It is assumed that the reason for the slightly improved HK/MG is because of the larger physical thickness. [11]

TABLE 2. SUMMARY OF TDDB . [11]

		SiON		HfSiON	
		thin ox	thick ox	thin ox	thick ox
Electrical properties	EOT	1.7nm	3.6nm	1.4nm	4.1nm
	CET	2.9nm	4.6nm	3.1nm	5.4nm
	Vth	0.26V	0.55V	0.28V	0.80V
10yr life time voltage	TDDB	2.1V	3.4V	2.2V	4.2V
	HCI (Vg=Vd)	1.3V	1.8V	1.5V	1.8V
	HCI (Isub,max)	1.1V	1.4V	1.3V	1.4V
	PBTI	>3.3V	>3.7V	2.2V	>4.0V

To determine if the HK/MG process integration is the cause of TDDB Weibull slope degradation, s-parameters were evaluated. [8] This was done by applying inversion stress until breakdown occurs and then measuring the source and drain current in accumulation mode to estimate the location of the breakdown. Previous reports show that HK breakdown occurs near the center of the devices, whereas breakdown of SiON gate oxides occur at the overlapped source/drain areas, as shown in Figure 7. [13] This phenomenon was explained as process-related due to the spacer and thin physical thickness of the SiON film. However, our films have completely opposite results. For the samples with SiON/poly, the s-parameter shows a random distribution whereas for the HK/MG stack structure, the breakdown occurs at the overlapped area near the source. From this, it can be noted that the HK/MG integration process is not optimized. Particularly, since the overlap area is the location of the breakdown, gate etch-related processes must be improved.

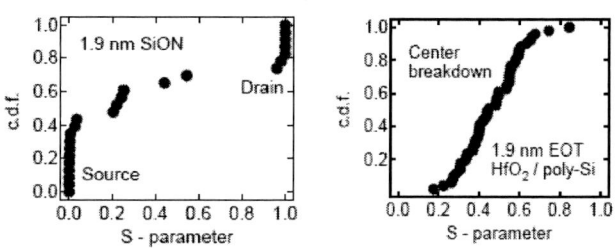

FIGURE 7. S-PARAMETER COMPARISON FOR SiON/POLY-SI AND HK/MG STACKS. [13]

The PBTI life time of HK/MG stacks for NMOS thin transistors was also investigated. For SiON/poly-Si stacks, oxide breakdown occurred before any significant Vth degradation so PBTI could not be evaluated. Despite incorporating HK/MG is the stack, the safe voltage observed was acceptably high for 10-year lifetime. This result, along with TDDB results, is very reasonable and it can be concluded that PBTI is not a critical reliability risk. [11]

It is also important to consider fast traps, particularly in high-k gate oxides. The effect of F-N stress of pulsed Id-Vg measurements for PMOS thin transistors was determined and it was concluded that for initial Id-Vg curves, there was negligible hysteresis. Therefore, in either gate stack of HK/MG or SiON/poly-Si, there was no evidence of the existence of fast traps. However, after F-N stress was applied, hysteresis was only observed for the HK/MG stack due to less immunity of stress. [9]

978-1-4244-5430-3/10 $26.00 © 2010 IEEE

Figure 8 shows the comparison of negatively biased temperature instability (NBTI) lifetime for the two different gate stacks. SiON/poly-Si gate stack has a slightly higher safe voltage for 10-year lifetime than that of the HK/MG. The NBTI characteristics are consistent with TDDB results. It is believed that the difference is process induced and the characteristics of the HK/MG stack can further be improved with optimization.

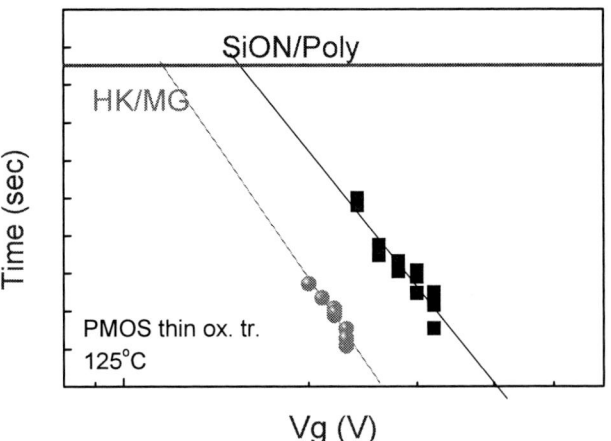

FIGURE 8. NBTI LIFETIME OF PMOS THIN OXIDE TRANSISTORS.

GIDL currents of the cell transistors significantly impact refresh characteristics. The comparison of GIDL currents for HK/MG and SiON/poly-Si gate stacks are shown in Figure 9. It was observed that GIDL levels were slightly larger for HK/MG. This may be due to the increase in electric field in the overlapped area for metal gate. Thus, it is important to optimize the gate etch process of the cell transistor.

Figure 10 shows the refresh characteristics of the HK/MG stack for cell transistors. Although the dynamic refresh characteristics are at acceptable levels, it is slightly worse that conventional SiON/poly-Si gate stacks. However, static refresh characteristics of HK/MG show twice the improvement of SiON/poly-Si stacks. Thus, it believed that the incorporation of HK/MG in DRAM cell transistors will not be an issue.

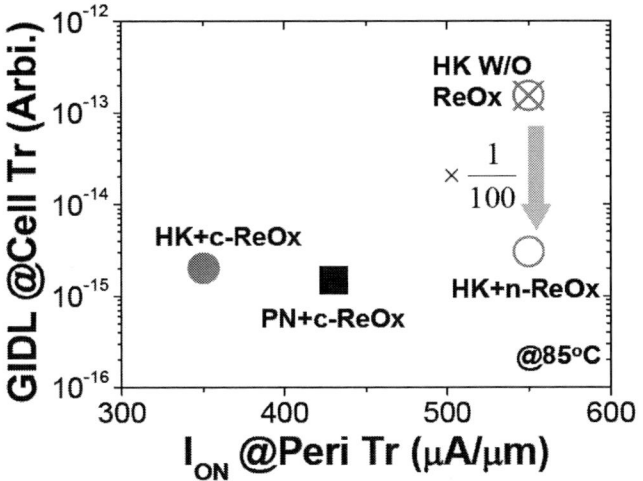

FIGURE 10. GIDL CURRENT VARIATION BEFORE AND AFTER F-N STRESS FOR DRAM CELL TRANSISTORS. [3]

CONCLUSIONS

WLR results with different trends were observed for DRAM cell and peripheral transistors. For peripheral (planar) transistors, the oxidation method to form the gate oxide, whether it be radical or wet oxidation, did not have a significant difference when F-N stress was applied. However, in cell transistors with wet oxidation, the Vth shifted negatively. This is believed to be caused by the three-dimensional structure of the cell transistor where the low electric field at the Si/oxide interface did not allow for the escaping of trapped holes. Also, HK/MG stacks incorporated in DRAM peripheral and cell transistors were evaluated. For peripheral transistors, HK/MG showed acceptable NBTI, PBTI, and TDDB reliability levels for 10-year life time. Additionally, HK/MG stacks in cell transistors resulted in acceptable refresh times. It is believed that the HK/MG stack can be further improved through process integration and optimization.

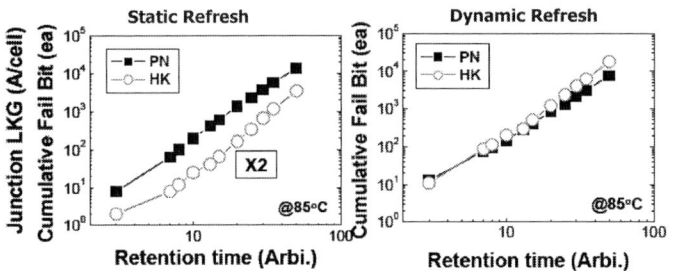

FIGURE 10. COMPARISON OF REFRESH TIME FOR HK/MG AND SiON/POLY-Si GATE STACKS FOR DRAM CELL TRANSISTORS. [3]

REFERENCES

[1] J. Y. Kim, et al., "S-RCAT Technology for 70nm DRAM feature size and beyond," in *Proceedings of the VLSI Symp.*, 2005, pp. 34-35.

[2] S. Y. Lee, et al., "New Charge Trapping Phenomena in Recessed-Channel-Array-Transistor (RCAT) after Fowler-Nordheim Stress," in *Proceedings of IDIDCT*, 2008, pp. 9-12.

[3] S. Hyun, et al., "Improvement of Performance and Data Retention Characteristics of Sub-50nm DRAM by HfSiON Gate Dielectric," in *Proceedings of VLSI Symp.*, 2007, pp. 184-185.

[4] J. Y. Seo, et al., "Investigation of Hot Carrier Degradation in Grooved Channel Structure nMOSFETs: SRCAT," in *Proceedings of IRPS*, 2006, pp. 723-724.

[5] K. F. Schuegraf, et al., "Hole Injection SiO2 Breakdown Model for Very Low Voltage Lifetime Extrapolation," in *IEEE Transactions on Electron Devices*, Vol. 41, No. 5, 1994, pp. 761-767.

[6] T. Hori, *Gate Dielectrics and MOS ULSIs*, Springer, 1997, pp. 161.

[7] B. Kaczer, et al., "Implications of progressive wear-out for lifetime extrapolation of ultra-thin SiON films," in *IEDM Technical Digest*, 2004, pp. 713-716.

[8] R. Degraeve, et al., "Relation between Breakdown Mode and Breakdown Location in Short Channel NMOSFETs and its Impact on Reliability Specifications," in *Proceedings of IRPS*, 2001, pp. 360-366.

[9] M. Houssa, et al., "Electrical Properties of High-k Gate Dielectrics: Challenges, Current Issues, and Possible Solutions," in *Materials and Engineering*, No. 51, 2006, pp. 37-85.

978-1-4244-5430-3/10 $26.00 © 2010 IEEE

[10] H. J. Cho, et al., "Reliability of Recess-channel Gate Cell Transistor under Gate Induced Drain Leakage Stress and Positive Bias Fowler-Nordheim Gate Stress," in *Proceedings of ESSDERC*, 2006, pp. 415-418.

[11] K. J. Nam, et al., "Investigation of hot carrier effects in n-MOSFETs thick oxide with HfSiON and SiON gate dielectrics," in Proceedings of IRPS, 2007, pp. 622-623.

[12] A. Shanware, et al., "Evaluation of positive bias stress stability in Hf-SiON gate dielectrics," in Proceedings of IRPS, 2003, pp. 208-213.

[13] R. Degraeve, et al., "Measurement and statistical analysis of single trap current-voltage characteristics in ultra thin SiON," in Proceedings of IRPS, 2005, pp. 360-365.

Reliability Characterization of 32nm High-K and Metal-Gate Logic Transistor Technology

Sangwoo Pae, Ashwin Ashok[2], Jingyoo Choi[3], Tahir Ghani[1], Jun He, Seok-hee Lee[1], Karen Lemay[2], Mark Liu[1], Ryan Lu, Paul Packan[1], Chris Parker[1], Richard Purser[2], Anthony St. Amour[1], and Bruce Woolery

Intel Corporation, LTD Q&R, [1]Portland Technology Development, [2]D1D Manufacturing, [3]D1C Q&R
Hillsboro, Oregon, 97124, USA, e-mail: sangwoo.pae@intel.com

Abstract—**High-K (HK) and Metal-Gate (MG) transistor reliability is very challenging both from the standpoint of introduction of new materials and requirement of higher field of operation for higher performance. In this paper, key and unique HK+MG intrinsic transistor reliability mechanisms observed on 32nm logic technology generation is presented. We'll present intrinsic reliability similar to or better than 45nm generation.**

Keywords-component; high-k dielectrics, metal-gate transistor, TDDB and BTI, process charging, Burn-in

I. INTRODUCTION

Logic CPU and SOC products with transistors made of HK+MG have been introduced at the 45nm technology node. 2nd generation of HK+MG transistors on 32nm node is also already in production containing more than billion transistors; in continuous support of Moore's law [1-2]. The Hf-based HK+MG transistors enabled very aggressive EOT scaling (0.9-1.0nm), supporting traditional 0.7x Tox scaling while demonstrating significant reductions (25-1000X) in gate leakage currents. This comes with more than 30% higher operating E-fields than 65nm node (which used Poly/SiO$_2$ gate stack) and as a result, reliability became challenging – both from the standpoint of introduction of new materials and requirement of higher field of operation for higher performance.

Key HK+MG reliability mechanisms and characterization work is detailed in this paper with focus on 32nm HK+MG transistors. Reliability results of 32nm transistors are also compared to our previously published 45nm results demonstrating intrinsic reliability similar to better than 45nm generation. Such superb intrinsic HK+MG transistor reliability was achieved by fully optimized process and also from the two generations of HK+MG technology learning.

II. EXPERIMENT DETAILS AND PROCESS

The transistors studied for reliability characterizations are from high performance digital logic transistors used in 32nm CPU products. 2nd generation of HK+MG transistors provide record drive currents at the tightest gate pitch (112.5nm) reported for any 32nm or 28nm logic technology [2]. It uses scaled HK dielectric of 9Å EOT with replacement metal gate flow and higher channel strain for improved transistor drives.

The gate stack has been fully optimized for performance, reliability, and yield. A 32nm process and transistor characteristics are detailed in [1-2]. Figure 1 shows cross section of the NMOS and PMOS devices. The introduction of raised NMOS source/drain (S/D) enables reduced device resistance. The proximity of the PMOS SiGe region to the channel continues to be reduced for enhanced channel strain.

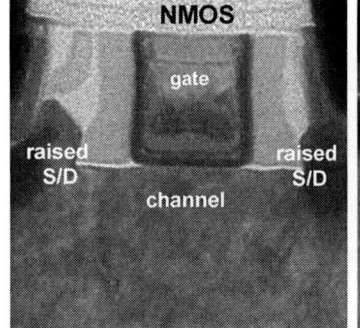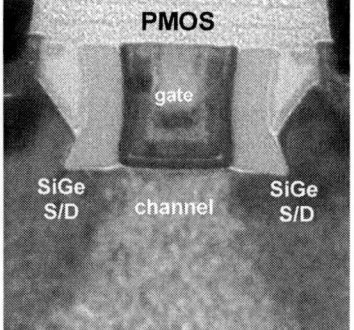

Figure 1. Cross section of NMOS and PMOS devices showing raised S/D regions for reduced parasitic resistance [2].

All typical transistor stresses for dielectric breakdown and Bias-temperature instability are performed in inversion with S/D and wells grounded and at hot temperature (T=90-125°C), otherwise noted. All device level reliability results are directly from the high performance transistors having a physical gate length of 30nm [2].

III. BREAKDOWN (TDDB) BEHAVIORS

Aggressive EOT scaling and higher E-field of operation increases reliability requirement for safe voltage operation on products made of HK+MG transistors. N-FET is the reliability limiter for the dielectric breakdown (TDDB). It exhibits larger gate tunneling currents than P-FETs in operation when Hf-based HK dielectric with band edge work functions are used as metal gate electrodes. In N-FET, physical band alignment of Si-channel with respect to the transitional interfacial layer (IL) and HK makes it easy for the direct tunneling (DT) of electrons from the substrate.

Figure 2 shows 32nm N-FET inversion TDDB vs. E-field showing superior oxide breakdown (supporting 10-15% higher E-field) performance relative to 45nm [1]. Transition of Field acceleration factor (EAF) to higher value at lower field has been attributed to the change in carrier tunneling mechanism in N-FETs. Exponential fit of E-field model was used to extrapolate. If power-law fit is used, as proposed by others, a small voltage gain is expected with the presence of low voltage data used in our study. SILC is not observed in fully optimized 32nm process as in 45nm generation [3]. Figure 3 summarizes the inversion TDDB lifetime distributions for logic N-FET. The modulation of TDDB distribution comes from the intended process splits. Improvement in Weibull shape factor (or beta) can be achieved with optimized skew, demonstrating superb intrinsic property of the gate dielectric film for the given sub-1nm EOT thickness – showing beta up to ~2. Hard breakdown (HBD) criteria was used as the TDDB failure criteria similar to our previous study on 45nm node [3].

With more Tox scaling, some reduction in P-FET inversion lifetime (TDDB) took place but is still more reliable than N-FETs at both stress and at end-of-life as shown in Figure 4. The low voltage lifetime data in Figure 4 exceeded 2.5 months of stress with unfailing units (as indicated by the arrows pointing upward). TDDB data was collected from large arrayed logic miller capacitors. The tightness of the distributions (as shown in Figures 3 and 4) suggests no extrinsic issues on gate dielectric in terms of defect learning.

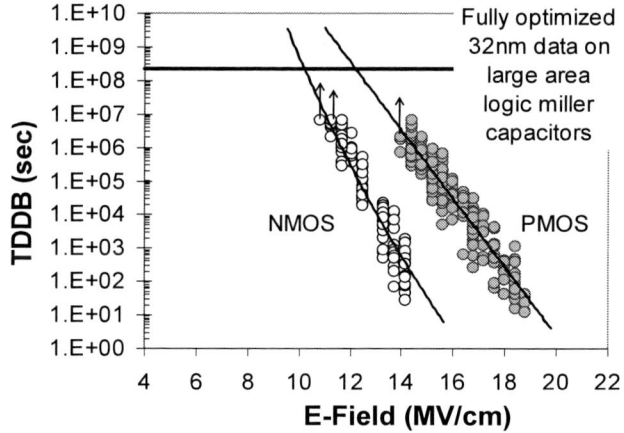

Figure 4. Most recent data of N-FET vs. P-FET TDDB vs. E-field. N-FET TDDB is reliability limiter. Low voltage stress, done through packaged units, is >2.5months. Arrows at low E-field indicates unfailed units.

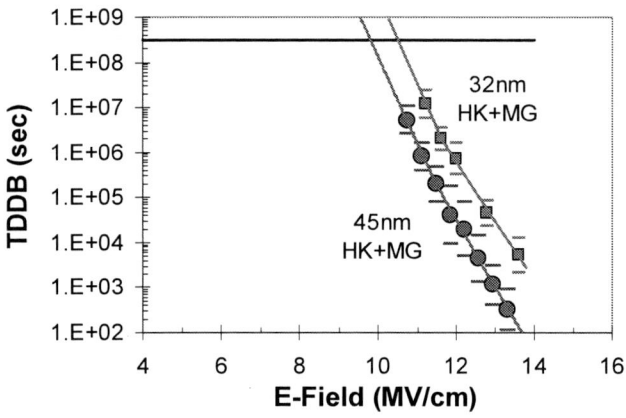

Figure 2. N-FET TDDB vs. E-field. E-field is calculated using V_G over inversion T_{ox}. 45nm and 32nm transistors have EOT of 1.0nm and 0.9nm respectively. 32nm TDDB supporting 10-15% higher intrinsic E-field at same lifetime.

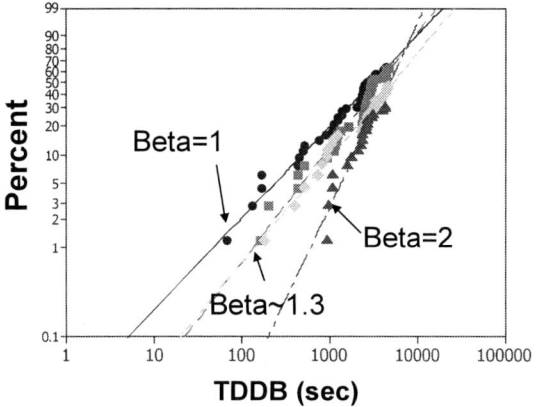

Figure 3. N-FET TDDB lifetime distributions on Weibull plot. Beta is modulated with process skews. All dielectric thickness (both HK & IL) is kept same.

The Weibull beta on P-FET is generally lower than N-FET and did not modulate significantly with the process changes shown for N-FETs in Figure 3. This implies the differences in the breakdown mechanism associated with process skews between N-FET vs. P-FET. As the process splits did not modulate the IL and it also showed polarity dependence, we attribute the P-FET breakdown to IL-limited degradation, consistent with literature reports [4]. To better understand the breakdown behavior, we've conducted the inversion TDDB stress on discrete transistors and measured gate leakage (I_G), delta-V_T shift and %G_m (linear peak transconductance, G_m) degradation at the stress readout intervals. The stress was performed in SMS (stress, measure, and then stress) fashion with fixed delay time. Readout voltages were kept low and minimized in time so that no additional degradation occurred during the readout phases. Figure 5 shows substantial change in V_T and well-correlated degradation of peak %G_m (measured in the linear region with V_{ds}=50mV) consistent with degradation close to the interface in P-FET during TDDB stress. This is very similar to what had been discussed in [5] for P-FET NBTI degradation. This behavior is very different on N-FET TDDB stress (shown in Figure 6). In P-FET, I_G vs. stress time change is relatively small suggesting no SILC generation. D_{it} (interface state trap) measurements (not shown) suggest large defect generation near the dielectric interface region. On N-FETs, V_T increase and change was not correlated to %G_m degradation during TDDB stress (as shown in Figure 6). This suggests that defect generation is farther away from the IL. Deterioration of

978-1-4244-5430-3/10 $26.00 © 2010 IEEE

HK bulk is suggested as the degradation mechanism of N-FET breakdown (BD) for the gate dielectric stack under study. The BD electrical behavior and findings are very similar to N-FET PBTI degradation mechanisms [5]. This analogy suggests a strong polarity dependence of the degradation mechanisms. Also note that P-FET can sustain much larger V_T-shift than the N-FET prior to BD, due to trap location being closer to the interface, which would impact V_T increase more strongly.

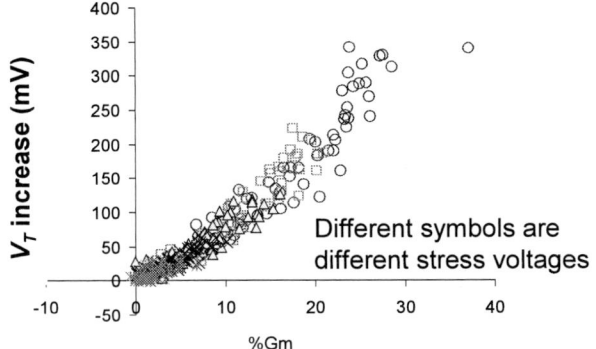

Figure 5. P-FET: V_T increase vs. peak G_m degradations characterized prior to hard breakdown for large V_T shifts.

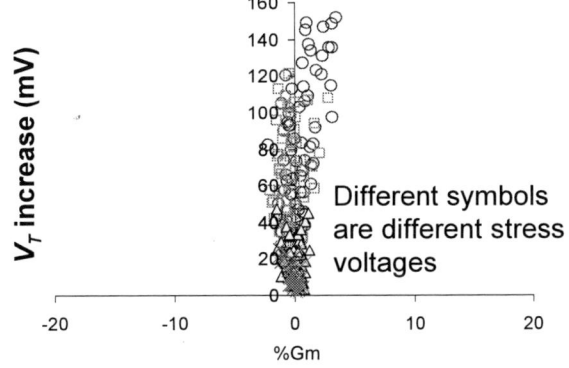

Figure 6. N-FET: V_T increse vs. Peak G_m degradation graph analogues to Figure 7. %Gm, implying degradation happening close to the interface, is not showing much movement until V_T shift is large (>100mV).

Conducting an experiment with varying dielectric thickness of interest and understand its dependency on reliability is one way to assess the location of traps that are being generated during stress. We've conducted studies with varying HK thicknesses, while maintaining the IL thickness the same. P-FET showed strong dependence on HK thickness while N-FET showed relatively weaker dependence with fixed stress voltage. P-FET TDDB behavior vs. HK thickness trend is consistent with P-FET discussions from previous section that traps are generated near the IL region. As HK thickness increases, E-field is reduced as stress voltage was kept constant, thereby reducing the trap generation rate. On N-FETs, much weaker N-FET TDDB dependence on HK was observed. Thicker HK, if not fully optimized, can result in larger volume of trap creation, resulting in relatively weaker thickness dependence.

We've also conducted transistor edge stress to study intrinsic reliability performance. The stress evaluation of gate edge is important since electrical over-/under-shoot (OS/US) can happen in I/Os. Any margin available will help mitigate the reliability concerns related to these OS/US cases, if any. Stress is performed with high bias on the drain (one side of S/D), while putting other terminals to ground. The resulting TDDB lifetime is substantially higher than those of typical inversion stress. Figure 7 shows sufficient margin is available. This improvement in lifetime can be explained by the flat-band offset in accumulation reducing the gate E-field coupled with large Si to IL/HK barrier height causing less tunneling carriers.

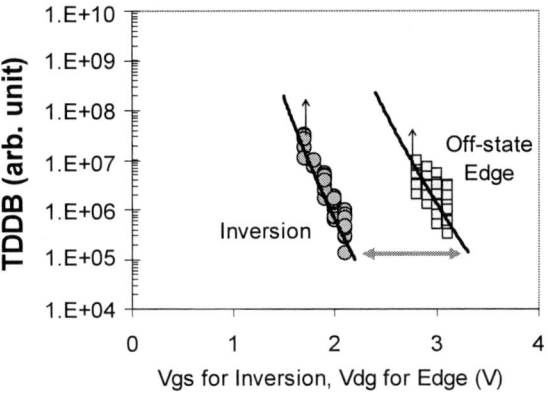

Figure 7. N-FET inversion vs. Edge stress TDDB results.

IV. TRANSISTOR DEGRADATION

Bias-temperature instability (BTI) degradation is observed on both N-FET and P-FET transistors made of HK+MG. It has been discussed that N-FET PBTI degradation at high field is due to bulk-HK trap creation (analogues to TDDB discussion in previous section) while P-FET NBTI mechanism was consistent with NBTI mechanism seen in SiO$_2$ at the interface [5]. In Figure 8, BTI degradation vs. duty cycle dependence is shown on both N-FET and P-FET. The magnitude of stress gate voltage was kept same while duty cycle in the gate pulse was varied. Note that there's a large difference in the characterized BTI behavior between N vs. P-FETs [6]. From the measured electrical stress data, we can deduce similar conclusion about the location of traps. Very fast P-FET recovery is similar to what has been shown in SiO$_2$ or SiON PMOS NBTI. In other reports, it is discussed that significant NBTI recovery could happen within few micro-seconds. This implies that the trap creation and recovery are happening near the interface. NMOS BTI recovery, on the other hand, was much slower. This is consistent with the picture that traps are generated relatively deeper into the bulk HK (and also deeper in energy band).

Figure 9 shows that N-FET PBTI can be further reduced in 32nm with respect to 45nm with further scaling of HK dielectric and with process improvements. Figure 10 shows

978-1-4244-5430-3/10 $26.00 © 2010 IEEE

PMOS NBTI can be also co-optimized to achieve similar degradation levels with scaled HK vs. 45nm generations.

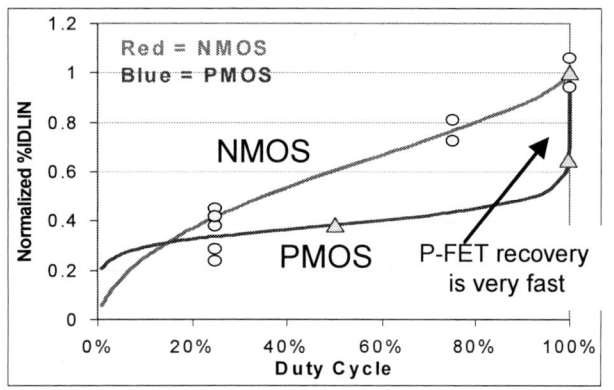

Figure 8. BTI degradation vs. Duty cycle dependence. P-FET shows very fast recovery as duty cycle is reduced from DC (100% duty cycle) stress to AC stress.

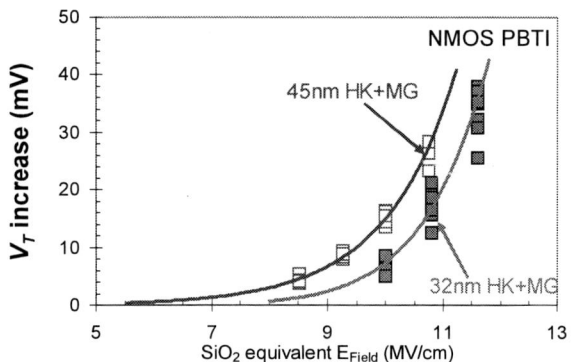

Figure 9. N-FET BTI (V_T increase) vs. E-field.

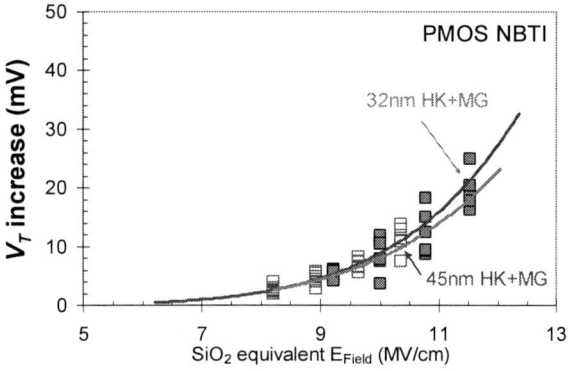

Figure 10. P-FET BTI (V_T increase) vs. E-field.

Ring Oscillators (RO) that were embedded onto products were stressed in either AC or DC during Burn-in (BI). DC stress in RO only exercises BTI stress since RO is locked (not running) while both BTI and hot carrier injection (HCI) effects can happen during AC stress (as it is free-running). Similar levels of degradation between AC and DC were obtained which suggests HCI degradation was not an issue in AC stress during BI. BI stress is highly accelerated and exceeded EOL stress for transistor degradations. The unique band structure of our HK+MG material significantly lowers levels of gate injection of the generated hot carriers resulting in overall reduction in HCI degradation. With the low level of pre-existing defects and traps of fully optimized HK+MG process, HCI degradation was not seen until a very high bias stress was applied thus it is relatively low risk. Figure 11 shows device level HCI stress done on N-FETs compared to previous results, demonstrating lifetime similar to better than 45nm.

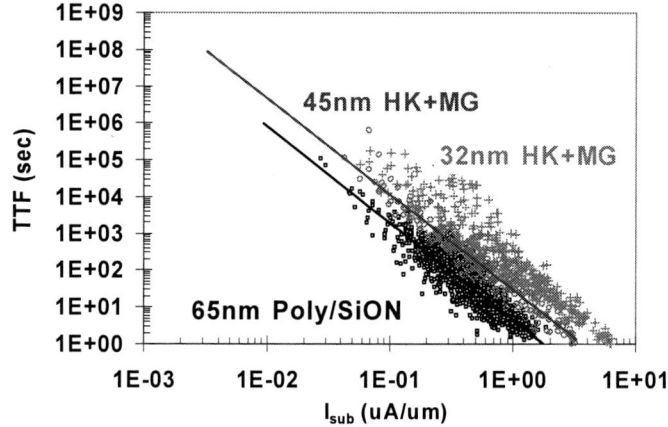

Figure 11. N-FET hot carrier lifetime (TTF) vs. substrate current showing technology comparisons. 32nm and 45nm shows comparable HCI results and is substantially better than 65nm SiO2.

V. PROCESS CHARGING

Process charging risk has been declining with the scaling of SiO_2 dielectrics preventing the charge build-up across gate dielectric (due to large direct tunneling leakage) [7]. The standard approach to protect against such plasma-induced damage is to provide a discharge path in the form of diodes or transistors. Design rules (DR) are defined to ensure sufficient protection to prevent any transistor damage during processing. With the large reduction in gate leakage that HK dielectrics provide, the charging rules must therefore be tightened to more historical levels. Significantly reduced gate leakages on P-FETs pose more process charging risk than N-FET. Figure 12 shows P-FET showing no charging issues on DR structures having cumulative antenna for all the metal layers. Still large margin is available (A ratio > DR) while showing charging signal (time0 V_T shift and tails) on structures with very high antenna ratios (B ratio >> DR) far beyond the DR. Transistor TDDB, and BTI stresses evaluated on DR charging structures show no latent damage issues and is consistent with intrinsic reliability model predictions.

978-1-4244-5430-3/10 $26.00 © 2010 IEEE   290

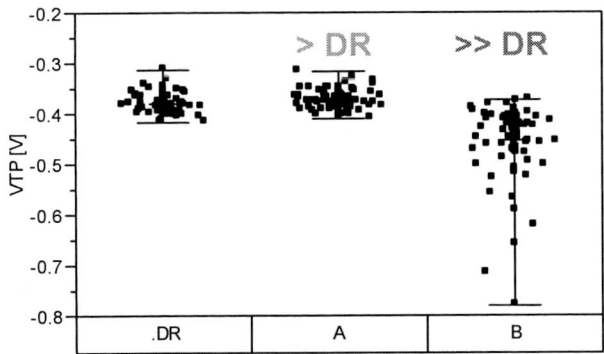

Figure 12. P-FET V_T measurements vs. various antenna sizes showing Design Rule (DR) vs. structures far exceeding DRs.

VI. MANUFACTURING TRANSISTOR RELIABILITY

Quick reliability testing for both ramped-VBD (breakdown voltage and correlation to TDDB) and BTI have been implemented on both N- & P-FETs similar to what had been done on 45nm node [8]. Correlation of our baseline vs. quick stress at several different stress conditions show excellent correlations (example is shown on Figure 13 for N-FET BTI). The quick stress capability provides high volume manufacturing (HVM) Fab reliability monitor and gauge for its health.

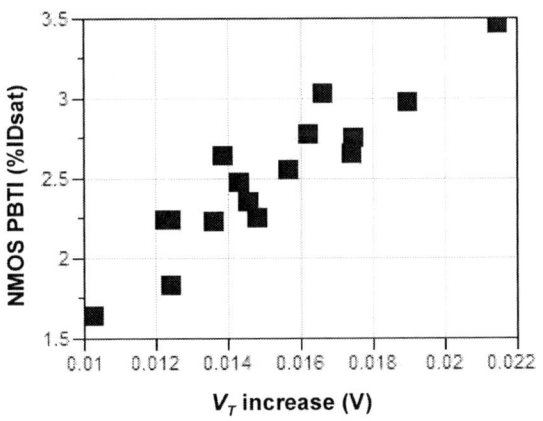

Figure 13. Quick stress (x-axis) to long stress (y-axis) comparisons for N-FET BTI example. Correlation can be made either in %ID or delta-V_T shifts.

Product high voltage BI that exercises the gate dielectric wearout is conducted and compared to transistor TDDB based Gate Oxide reliability model. Large margin in product data exists and this can be largely attributed to differences between the discrete transistors vs. circuit level to breakdown tolerance (Figure 14). Figure 15 shows that with reduction in gate leakage (I_G) on HK with respect to transistor drive currents (I_D), the functionality of transistors are not immediately lost even after BD. This, however, depends on the device size and breakdown hardness [9]. %ID degradation behavior continues to exhibit power-law dependence with time slope ~0.17-0.2

for N-FETs. The product BI data shows intrinsically superb reliability on product level that exceeds operating lifetime. It also demonstrates that there's circuit level reliability margin that can be leveraged with accurate circuit level reliability modeling.

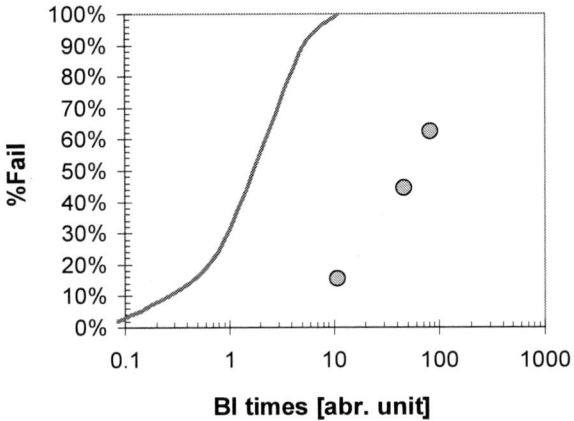

Figure 14. Product Burn-In data that excercises Gate Oxide wearout vs. Transistor based TDDB comparisons.

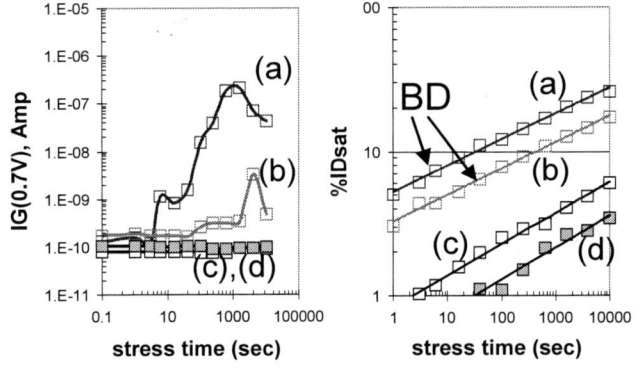

Figure 15. Transistor characteristics are maintained after BD due to high I_{Dsat} to I_G leakage ratios. Power-law time slope remains unchanged post-BD. (a)-(d) shows stresss done at different conditions.

VII. DISCUSSION

Technology trend of higher operating E-field continues to make reliability challenging as we scale technology. N-FET and P-FET breakdown based on its electrical characteristics and process dependencies were discussed for 32nm technology generations. In conclusion, P-FET TDDB mechanism is IL-limited while N-FET TDDB is modulated by the bulk HK. These findings are similar to those reported for BTI mechanisms and optimization of both layers is needed on CMOS process. Validation of key FE reliability on products made of HK+MG transistors showed no issues, supporting well beyond specified operating lifetime.

ACKNOWLEDGMENT

Authors would like to acknowledge Sanjay Natarajan (PTD 32nm program manager) for valuable inputs and support. Svetlana Torodov (Fab32 Q&R) for reliability discussions. Jason Xu (LTD Q&R) for technical reviews and management support.

REFERENCES

[1] S. Natarajan et al., "A 32nm Logic Technology Featuring 2nd-Generation High-k + Metal-Gate Transistors, Enhanced Channel Strain and 0.171um^2 SRAM Cell Size in a 291Mb Array," in *IEDM Tech. Dig.*, 2008.

[2] P. Packan et al., "High Performance 32nm Logic Technology Featuring 2nd Generation High-K + Metal Gate Transistors," in *IEDM Tech. Dig.*, 2009, pp. 659-662.

[3] S. Pae et al., "Characterization of SILC and Its End-of-Life Reliability Assessment on 45nm High-K and Metal-Gate Technology," in *International Reliability Physics Symp. Proc.*, 2009, pp. 499-504.

[4] K. Torii et al., "Physical model of BTI, TDDB and SILC in HfO$_2$-based high-k gate dielectrics," in *IEDM Tech. Dig.*, 2004, pp. 129-132.

[5] S. Pae et al., "BTI reliability of 45 nm high-K + metal-gate process technology," in *International Reliability Physics Symp. Proc.*, 2008, pp. 352-357.

[6] S. Ramey et al., "Frequency and recovery effects in high-k BTI degradation," in *International Reliability Physics Symp. Proc.*, 2009, pp. 1023-1027.

[7] M. Alavi et al., "Effect of MOS device scaling on process induced gate charging," in *Plasma Process-Induced Damage Symp. Proc.*, 1997, pp. 7-10.

[8] R. Kasim et al., "Reliability for manufacturing on 45nm logic technology with high-k + metal gate transistors and Pb-free packaging," in *International Reliability Physics Symp. Proc.*, 2009, pp. 350-354.

[9] B. Kaczer et al., "Impact of gate-oxide breakdown of varying hardness on narrow and wide nFET's," in *International Reliability Physics Symp. Proc.*, 2004, pp. 79-83.

RELIABILITY STUDIES ON A 45NM LOW POWER SYSTEM-ON-CHIP (SOC) DUAL GATE OXIDE HIGH-K / METAL GATE (DG HK+MG) TECHNOLOGY

C. Prasad[2], P. Bai[1], S. Gannavaram[1], W. Hafez[1], J. Hicks[2], C.-H. Jan[1], J. Lin[1], M. Jones[2], K. Komeyli[1], R. Kotlyar[3], K. Mistry[1], I. Post[1], C. Tsai[1]

[1]Logic Technology Development, [2]Logic Technology Development Quality & Reliability, [3]Design and Technology Solutions
Intel Corporation, 5200 N.E. Elam Young Pkwy, Hillsboro, OR 97124, USA
Primary Author Contact: chetan.prasad@intel.com, 503-613-7265 (phone)/503-613-1068 (fax)

ABSTRACT

In this paper, we present extensive reliability characterization results for a novel dual gate 45nm HK+MG technology. BTI, HCI and TDDB degradation modes on the Logic and I/O transistors are studied and excellent reliability is demonstrated. Emphasis is placed on the importance of process optimizations to support robust I/O transistors while maintaining the high performance and reliability of Logic transistors. Monitoring of reliability for HVM and collateral reliability are also addressed.

INTRODUCTION AND BACKGROUND

System-on-chip (SoC) products have experienced strong growth in the mobile, handheld and consumer device market due to a rapid increase in both functionality and computational power enabled by scaling; however, the Si technology features and performance requirements for these highly integrated low power products differ significantly from those aimed at high performance CPU applications. In 2007, Mistry *et al.* [1] reported a revolutionary 45nm gate stack technology with HK+MG processing where high CMOS performance was achieved coupled with large reductions in gate leakage and power. Robust gate and transistor reliability [2-5] as well as solid integrated-vehicle reliability and stable high-volume manufacturing trends [6] were demonstrated on this gate stack. In 2008, this 45nm technology was optimized for SOC applications with the integration of a dual gate oxide (DG) HK+MG flow to enable low standby power, high performance and support for high voltage legacy as well as contemporary I/O interfaces [7]. A novel hybrid SiO_2 and HK+MG architecture was incorporated for the I/Os and this work describes results for these transistors in detail.

DEVICE FABRICATION AND MEASUREMENTS

Transistors in this DG HK+MG technology can be classified as either *Logic* or *I/O* transistors. The *Logic* transistors are versions of those reported in [1] optimized to support high performance at a low standby power – they enable core logic as well as I/O interfaces. The *I/O* transistors use a novel hybrid SiO_2 and HK+MG architecture and enable 3.3V legacy I/O interfaces as well as 5V safe operation [7]. The *Logic* transistors were fabricated with Hafnium-based High-k dielectric and a thin SiO_2 interfacial layer. The *I/O* transistors have a hybrid SiO_2 + HK dielectric stack with a high quality thermal SiO_2 layer that is thicker than the interfacial layer of the *Logic* transistors [7]. All the transistor types use different band edge metal gate electrodes to enable full CMOS operation [1, 8]. The basic process steps for DG HK+MG flow deviate from the single gate flow only for the *I/O* transistor dielectric and S/D extensions and are otherwise identical.. *Logic* transistors have gate lengths of 40nm and widths of 0.45um with an EOT of 1.0nm, while the I/O devices have longer gate lengths of 200nm and widths up to 0.9um; all the values specified above are drawn lithographic dimensions. TEM images of both types of transistors are shown in Figure 1, and the difference in SiO_2 layer thickness as well as lithographic dimensions can be

clearly seen. All transistor degradation parameters were evaluated under quasi-DC conditions in stress-measure-stress (SMS) mode with a fixed stress-to-measure delay, unless stated otherwise. All TDDB lifetimes are referenced to "hard" breakdown events unless specified otherwise. For comparison purposes, *Logic* and *I/O* transistors fabricated on a mature 65nm SiO_2+Poly process technology [9, 10] are presented as references.

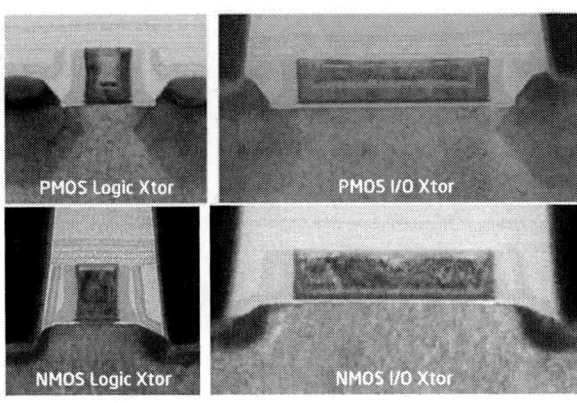

FIGURE 1. TRANSMISSION ELECTRON MICROGRAPH IMAGES OF NMOS AND PMOS LOGIC AND I/O TRANSISTORS [7]

TRANSISTOR DEGRADATION RESULTS

BTI degradation results for the 45nm single gate flow has been extensively treated in prior works [3-5] where it was reported that the BTI degradation mechanism differed for NMOS vs. PMOS on HK+MG flows due to the band structure asymmetry and trap distributions of the dual dielectric system. This work discusses the BTI degradation of the *Logic* and *I/O* transistors of the DG HK+MG flow and the impact of the hybrid SiO_2 + HK dielectric stack and it is conclusively demonstrated that NMOS PBTI is dominated by bulk trapping, while PMOS NBTI is explained by the conventional SiO_2-like interface state degradation model. A key observation is the similarity of BTI behavior between *Logic* and *I/O* transistors in the DG HK+MG flow, indicating that the presence of a significantly thicker SiO_2 layer in the hybrid stack of the *I/O* transistors does not modify the basic underlying physics of the NBTI and PBTI degradation mechanisms..

Figure 2 shows correlation plots of saturation drive degradation as a function of peak g_m degradation, where the change in g_m is an indicator of interface damage and mobility degradation. It is observed that NMOS transistors exhibit no g_m degradation whereas PMOS transistors show g_m degradation that is well-correlated to the drive degradation. This points to mobility reduction though interface state generation for PMOS devices. The lack of g_m degradation on NMOS indicates that the damage or traps caused by PBTI are not close to the interface, but are located remotely in the bulk region, and that drive degradation is dominated by a threshold voltage shift mechanism.

978-1-4244-5430-3/10 $26.00 © 2010 IEEE

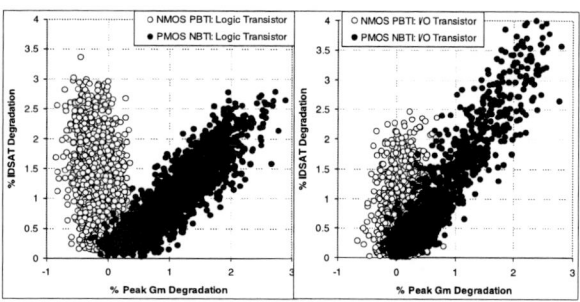

FIGURE 2. DRIVE DEGRADATION AS A FUNCTION OF PEAK GM DEGRADATION INDICATES THAT NMOS PBTI IS PRIMARILY BULK TRAP DRIVEN WHEREAS PMOS NBTI IS INTERFACE STATE DRIVEN. I/O AND LOGIC TRANSISTORS SHOW IDENTICAL BEHAVIOR.

These conclusions are further supported by the correlation plots for linear drive degradation as a function of threshold voltage shift shown in Figure 3. For a given threshold voltage shift caused by the PBTI or NBTI degradation, the PMOS devices always shows higher linear drive loss, consistent with the presence of an additional interface damage component that affects channel carrier transport. Qualitatively similar signals are seen for the Logic and I/O transistors in Figure 2 as well as Figure 3.

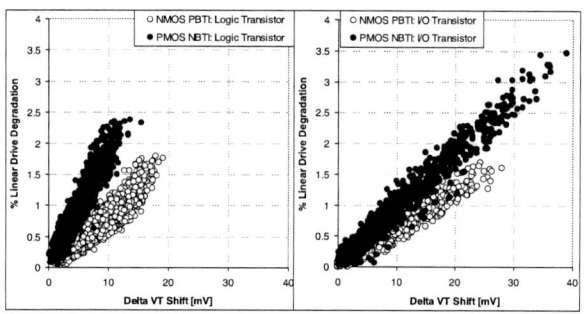

FIGURE 3. ADDITIONAL LINEAR DRIVE DEGRADATION AT EQUIVALENT THRESHOLD VOLTAGE SHIFT ON PMOS TRANSISTORS INDICATES MOBILITY DEGRADATION DRIVEN BY INTERFACE STATES.

Charge pumping (CP) measurements collected over a range of 1 kHz – 1 MHz are shown in Figure 4, and they provide further evidence to the location of the trap states. The CP data reveal that there is negligible D_{IT} increase for NMOS PBTI, whereas PMOS NBTI is accompanied by a moderate D_{IT} increase that scales consistently with VT shift. Since CP, even at 1 kHz, does not probe much beyond the Si-SiO$_2$ interface, this confirms the interface vs. bulk trap origin for NBTI vs. PBTI respectively. Furthermore, it is interesting to note that CP data from Logic and I/O transistors collapse onto common lines when D_{IT} change is normalized – indicating that the impact of the interface state generation on the channel is very similar for the Logic and I/O transistors, which is not unexpected.

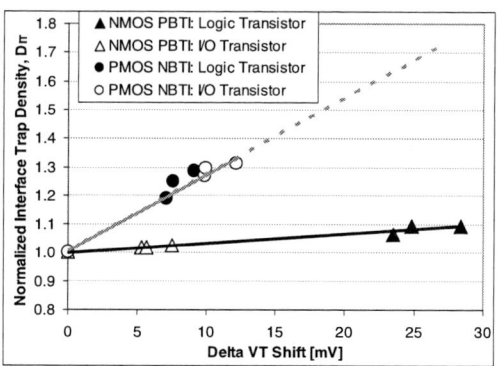

FIGURE 4. CHARGE PUMPING CONFIRMS INTERFACE V.S. BULK TRAP MODELS FOR BTI DEGRADATION ON NMOS AND PMOS

Pae et al. [3] report the magnitude of NMOS PBTI degradation for a single gate HK+MG flow to be quite similar in magnitude to the PMOS NBTI degradation at a use end of life condition. For the DG HK+MG flow, similar N/P degradation ratios are also seen on the Logic transistors as expected. The I/O transistors exhibit a different N/P degradation ratio skewed towards higher PMOS degradation (as shown in Table 1).

HK+MG Flow	Logic N vs. P	I/O N vs. P
Single Gate	0.40 / 0.60	n/a
Dual Gate	0.40 / 0.60	0.20 / 0.80

TABLE 1. RATIO OF N VS. P DEGRADATION (AT USE) OF LOGIC AND I/O TRANSISTORS ON SINGLE AND DUAL GATE HK+MG FLOWS.

To illustrate the origin of the difference in the relative degradation ratios, a 1-D simulation of the NMOS band structure is shown in Figure 5. At use biases, electrons see a barrier consisting of the HK and SiO$_2$ layers for both the Logic and I/O transistors, but at higher applied fields ("Va"), the transport through the HK layer transitions to F/N tunneling As fields are further increased ("Vc"), eventually the HK layer is "transparent" to energetic electrons and the tunneling barrier is determined either by the SiO$_2$ interfacial layer in Logic transistors or by the hybrid stack thick SiO$_2$ layer in I/O transistors. As transport transitions through these regimes, the charge injection characteristics modify the PBTI sensitivity to applied field in Logic transistors. Due to the robustness of the thick SiO$_2$ layer in the hybrid stack of the I/O transistors, such changes in PBTI behavior are not resolvable until even higher fields, where the thick SiO$_2$ of the hybrid stack itself enters the F/N tunneling regime ("Vd"), and a corresponding inflection in the gate stack leakage is observed (shown in Figures 10 and 13). Correspondingly, there is a non-negligible PBTI effect at very high fields on the I/O transistors but this contribution diminishes rapidly, becoming negligible at operating fields, as shown in Figure 6. As a consequence; PMOS NBTI is the dominant degradation mode for these I/O transistors. However, it is important to note that the scaling within the hybrid stack, whether it is achieved through HK layer scaling vs. SiO$_2$ layer scaling, will significantly modify the N/P degradation balance due to conflicting effects of bulk volumetric traps and tunneling fluence driven injection, and these tradeoffs are important to comprehend during process development. Figure 6 (left plot) demonstrates that I/O transistors on the DG HK+MG flow have PMOS NBTI reliability that is similar to or better than that of the reference mature 65nm SiO$_2$+Poly technology.

978-1-4244-5430-3/10 $26.00 © 2010 IEEE

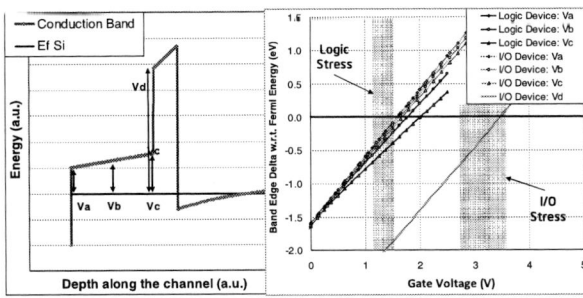

FIGURE 5. 1-D NMOS BAND SIMULATION OF *LOGIC* AND *I/O* BAND STRUCTURE DEMONSTRATING DT TO F/N TRANSITIONS THROUGH SCALING OF INDIVIDUAL CONDUCTION BAND SET-POINTS. SHADED REGIONS DENOTE THE RANGE OF *LOGIC* AND *I/O* STRESS BIASES.

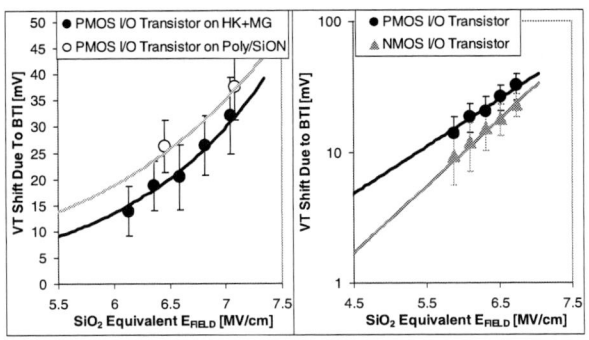

FIGURE 6. *I/O* TRANSISTORS ON HK+MG FLOW SHOW SIMILAR OR BETTER RELIABILITY TO REFERENCE 65NM SiO$_2$+POLY FLOW. NMOS PBTI DEGRADATION IS A RELATIVELY MUCH SMALLER CONTRIBUTION AND TENDS TO BE NEGLIGIBLE AT USE-LIKE BIASES

As a part of the AC testing suite, fast pulse measurements were performed on *I/O* transistors to assess fast trapping and data are shown in Figure 7. The lack of hysteresis observed in the traces with both 1.0us and 10us rise and fall times demonstrates the high quality of the hybrid stack with no evidence of fast trapping or shallow trap states.

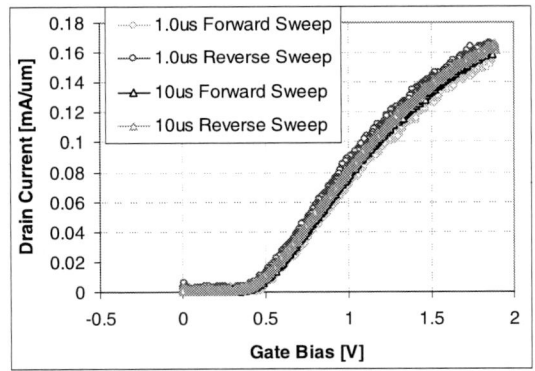

FIGURE 7. PULSE DATA FROM NMOS *I/O* TRANSISTORS AT V$_{DS}$ = 0.1V WITH T$_{RISE}$ = T$_{FALL}$ = 1.0US AND 10US CONFIRMS THE LACK OF FAST TRAP ISSUES

Figure 8 demonstrates that NMOS HCI degradation performance on DG HK+MG *I/O* transistors is identical to that of the reference SiO$_2$-Poly reference devices – which is as expected for an interface centric mechanism. The presence of the HK layer does not significantly modulate the charge injection at the drain, consistent with our understanding of hybrid stack band bending at high V$_{DS}$.

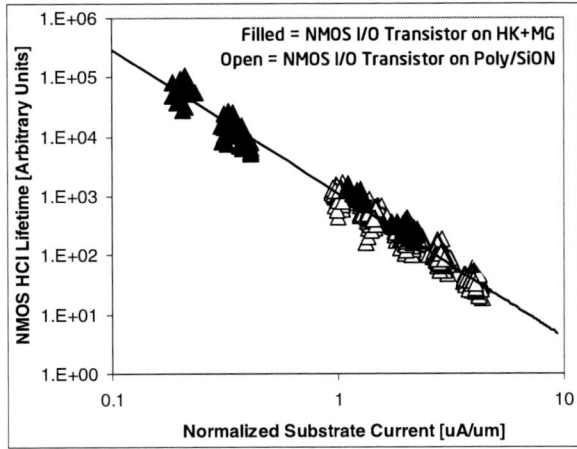

FIGURE 8. HCI DEGRADATION ON NMOS *I/O* TRANSISTORS IS IDENTICAL FOR HK+MG AND SiO$_2$+POLY PROCESS FLOWS

Figure 9 emphasizes the interfacial nature of HCI damage, where changes in interface state density (D$_{IT}$) caused by HCI degradation on NMOS and PMOS *Logic* as well as *I/O* transistors are shown. Corresponding D$_{IT}$ changes caused by PBTI degradation are shown for comparison purposes. All types of transistors evaluated (NMOS and PMOS, *Logic* and *I/O*) show large relative D$_{IT}$ increases as compared to the reference PBTI cases – as expected for hot carrier induced damage. Unlike the behavior seen for BTI data however, the normalized interface trap density does not collapse onto a single fit line. This is not unexpected as the threshold voltage shift is not a complete indicator of HCI damage – and evaluating this trend as a function of linear drive loss or mobility loss may show more universal behavior.

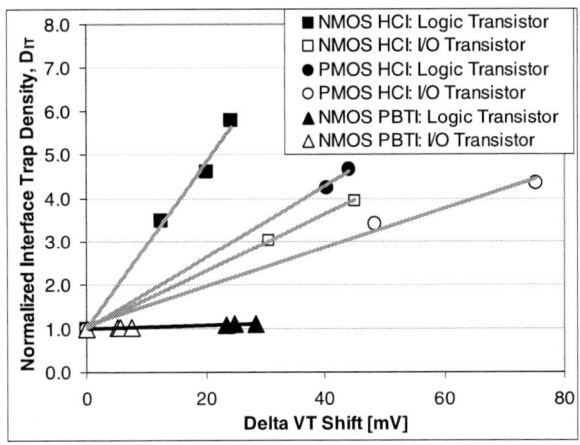

FIGURE 9. CHARGE PUMPING DATA SHOWS A CLEAR INCREASE IN NORMALIZED D$_{IT}$ CAUSED BY HCI DAMAGE, EMPHASIZING THE INTERFACE-CENTRIC NATURE OF THIS DAMAGE.

DIELECTRIC DEGRADATION RESULTS

In this section and subsequently through this work, NMOS inversion and accumulation will be denoted as N_{INV} and N_{ACC} and the corresponding PMOS counterparts as P_{INV} and P_{ACC} respectively. The HK+MG band structure of a single gate flow has been discussed in detail in several publications [2-4] and its intrinsic asymmetry results in better TDDB lifetime for N_{ACC}/P_{INV} than N_{INV}/P_{ACC}. The same band physics are confirmed to apply on the DG HK+MG flow and its effects are observed in both *Logic* and *I/O* transistors as shown in Figure 10, where relative leakage and dielectric strengths are reported. The *Logic* transistor Stress Induced Leakage Current (or SILC) behavior is very similar to the optimized stack data reported by Prasad *et al.* [2] and Pae *et al.* [11]. In these works, the SILC behavior was linked to the bulk trap density in the HK layer – and optimizations of the HK layer quality were shown to strongly mitigate the SILC behavior [2, 11]. Others in literature [12, 13] have reported this effect and postulated the cause as an oxygen vacancy defect mode in the bulk HK layer.

In the *I/O* transistors of the DG HK+MG flow, the high-quality optimized HK-layer with low trap densities as well as the thick SiO_2 layer of the hybrid stack strongly mitigates the effects of SILC degradation. As a consequence, SILC degradation effects and polarity dependent SILC effects that were reported for non-optimized HK+MG stacks [2, 11] are not observed in the optimized DG HK+MG flow, as shown by the *I/O* transistor SILC data in Figure 11. As discussed previously [2], high SILC degradation is unsuitable for product applications, as it can have an unacceptable impact on long-term standby power stability.

The nature of breakdown events on the *I/O* transistors is observed to be very abrupt or "hard" – this is similar to *Logic* transistors on the HK+MG flow as well as *I/O* transistors on the 65nm SiO_2+Poly flow but unlike *Logic* devices on the 65nm SiO_2+Poly flow, which exhibit less abrupt, "softer" BDs. This effect can be attributed to the physics of breakdown in a MG vs. Poly architecture and the overall physical thickness of the respective gate stacks.

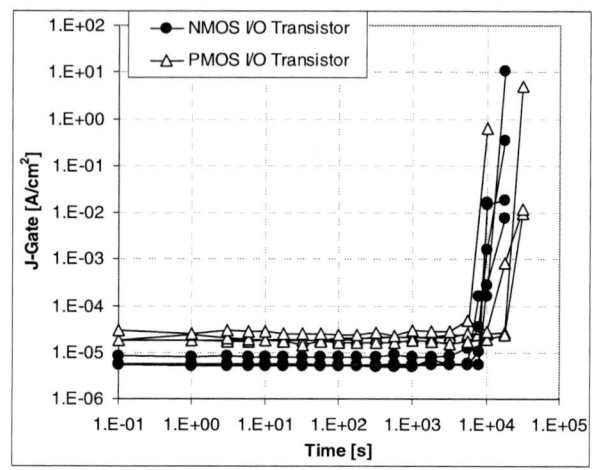

FIGURE 11. DATA DEMONSTRATES THAT *I/O* TRANSISTORS EXHIBIT NO SILC DEGRADATION AND HARD BD, AS EXPECTED FOR HIGH-QUALITY HYBRID SiO_2 AND HK+MG GATE STACKS

A challenge of the DG architecture, for which many process steps are shared, is ensuring robust reliability requirements are satisfied for both the *Logic* and *I/O* transistors and Figure 12 shows a comparison of TDDB performance for a very early unoptimized flow (designated *Initial*) with the final optimized DG HK+MG flow (designated *Final*). A large improvement in overall TDDB lifetime was achieved on both NMOS and PMOS *I/O* transistors through proper process optimizations that addressed three primary aspects of the hybrid stack architecture – individual layer quality and stoichiometry, uniformity across wafer and geometric attributes of the transistors. It is important to note that the TDDB lifetime gain on the *I/O* transistors was achieved with no negative impact to *Logic* transistor performance and reliability of *I/O* transistor performance. The magnitude of improvement achieved underlines the critical importance of simultaneous optimization of the Logic and I/O devices.

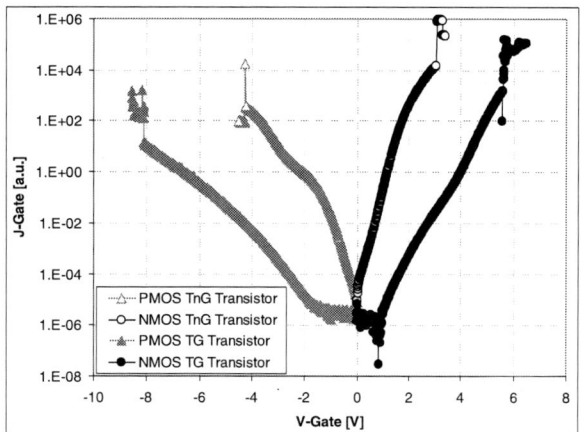

FIGURE 10. RELATIVE MAGNITUDES OF GATE LEAKAGE DENSITY AND DIELECTRIC BREAKDOWN STRENGTH ARE SHOWN FOR NMOS AND PMOS *LOGIC* VS. *I/O* TRANSISTORS INDICATING THE INFLUENCES OF CORRESPONDING BAND STRUCTURE ASYMMETRY.

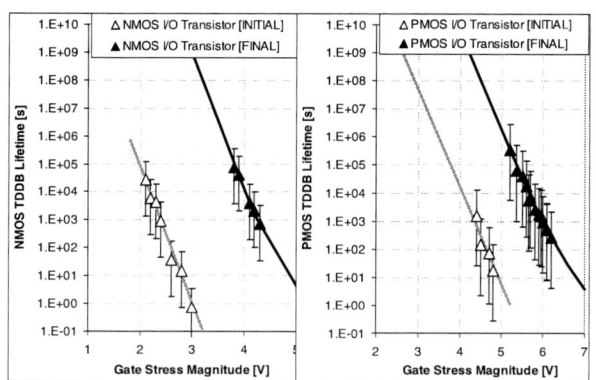

FIGURE 12. *INITIAL* AND *FINAL* TDDB LIFETIMES SHOWN FOR *I/O* TRANSISTORS EMPHASIZES THE IMPACT OF CRITICAL PROCESS FLOW OPTIMIZATIONS PERFORMED ON THE HYBRID DG HK+MG FLOW TO ENABLE ROBUST RELIABILITY BEHAVIOR.

978-1-4244-5430-3/10 $26.00 © 2010 IEEE

Voltage scaling transitions associated with changes in charge injection corresponding to a tunneling regime transitions from direct tunneling (DT) to Fowler-Nordheim (F/N) tunneling have been reported previously with extensive long term stress data in the single gate HK+MG flow [2] as well as in SiO2-Poly flows [14, 15]. Similar results are observed for the *I/O* transistor gate stack, and Figure 13 shows the gate stack leakage data presented on an F/N-tunneling plot, identifying the transition points. These inflections correlate the transitions in the NMOS 1-D band structure simulations of *Logic* and *I/O* shown in Figure 5. Corresponding composite empirical fits with exp(-kE) and exp(n/E) behavior are used to fit the *I/O* transistor data from Figures 12 and 15.

Figure 14 compares the TDDB reliability of *I/O* transistors on the DG flow to those on a mature 65nm SiO2-Poly reference. *I/O* transistor reliability is matched on the NMOS and supports 20% higher operating bias on the PMOS relative. Corresponding *Logic* transistor reliability on the DG flow is identical to that reported for the single gate HK+MG flow [1, 2, 3, 11].

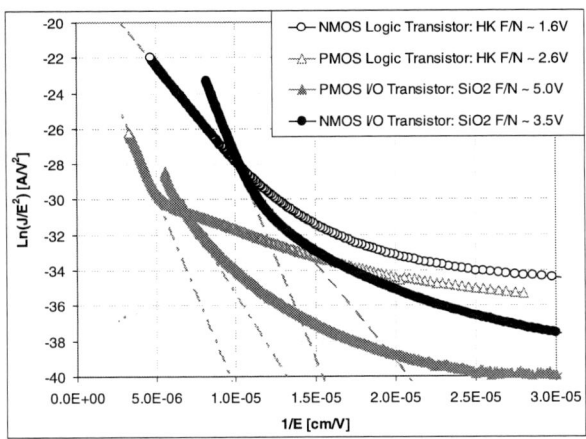

FIGURE 13. BOTH *LOGIC* AND *I/O* TRANSISTOR GATE STACKS EXHIBIT TRANSITIONS TO F/N REGIME AT HIGH BIAS CONDITIONS, CONSISTENT WITH 1-D BAND STRUCTURE SIMULATIONS.

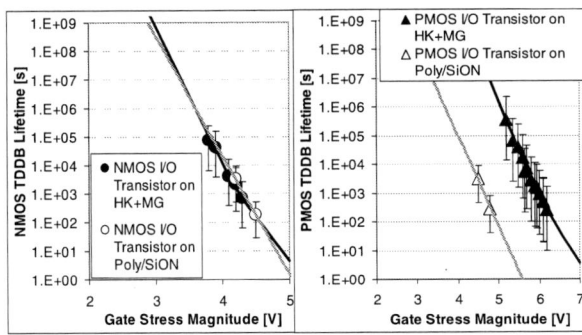

FIGURE 14. NMOS AND PMOS LIFETIME AS A FUNCTION OF VOLTAGE FOR *I/O* TRANSISTORS CONFIRMS RELIABILITY PERFORMANCE EQUIVALENT OR BETTER THAN THAT OF THE REFERENCE 65NM POLY/SiON TECHNOLOGY. OPERATING VOLTAGES UP TO 2.5V AND 3.3V CAN BE SUPPORTED ON THE NMOS AND PMOS I/O TRANSISTORS RESPECTIVELY [7]

PROCESS-CHARGING INDUCED DAMAGE EFFECTS

As gate leakage is significantly reduced with the HK+MG process flow, it is important to consider potential impact of process-charging induced damage or PID. Such damage is accrued during plasma deposition and/or etch process steps where metallization connected to the transistors acts as an antenna collecting and funneling charge through the gates, potentially damaging them if the charge transferred is excessive. To prevent PID effects caused by excessive charge transfer, layout design rules (DR) are defined such that they limit the ratio of antenna sizes with respect to the "victim" gate area and mandate adequate protection against PID.

Conventional SiO2 technologies usually exhibit PID signals through the presence of tails in the gate leakage distribution. In the DG HK+MG process flow, NMOS transistors exhibit similar tails in the gate leakage distribution with no tails in the threshold voltage distributions. However, PMOS transistors show a distinct signature with a tail in the threshold voltage distribution but no tail in the gate leakage distribution. This unique effect is attributed to the relative robustness of the PMOS gate stack, which results in significant charge trapping, and associated threshold voltage shifts, during the plasma processing steps prior to actual dielectric breakdown.

Understanding this effect and its manifestations is crucial to the definition of design rules that provide adequate protection for transistors from PID in high-volume manufacturing. Figure 16 illustrate these PID characteristics on NMOS and PMOS for 3 different *Logic* gate stack test structures: a reference structure with no antenna (and hence no possibility of PID), a structure designed to meet the defined DR set, and a structure that violates the DR limits by a factor of 25 ~ 30. The data demonstrates that the defined DRs are adequate to prevent transistors from PID. Similar results and margins to their respective DRs were observed on the *I/O* transistors.

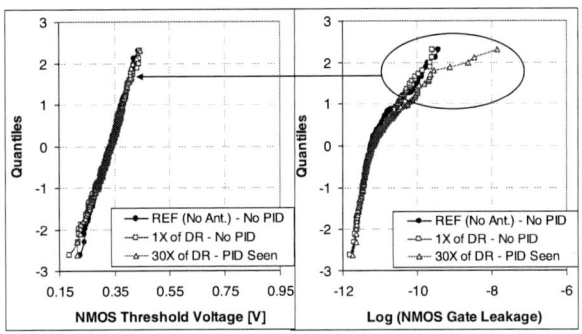

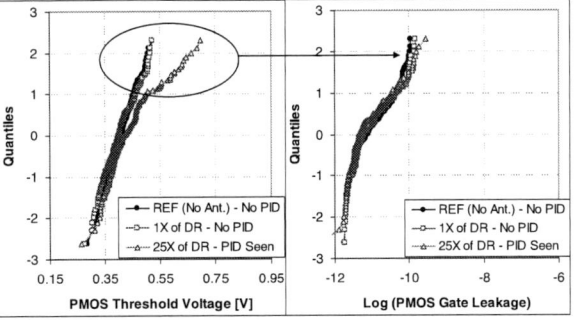

FIGURE 15. NMOS [TOP] AND PMOS [BOTTOM] GATE LEAKAGE AND THRESHOLD VOLTAGE DISTRIBUTIONS ON *LOGIC* TRANSISTORS SHOW THE IMPACT OF PROCESS-CHARGING INDUCED DAMAGE AND MITIGATION THROUGH DR DEFINITION

PRODUCTION BASELINES AND TRENDS

Figure 16 shows the factory production trend of *Logic* and *I/O* transistor reliability metrics similar those detailed in [6] over a period of 6 months. These metrics of TDDB performance are measured on every wafer at the End-Of-Line test. The stability of this metric underlines the excellent stability of the DG HK+MG dielectric reliability in high-volume manufacturing (HVM) conditions These granular trending data [6] allow Fab personnel to react rapidly to variations as well as identify opportunities for continuous improvement.

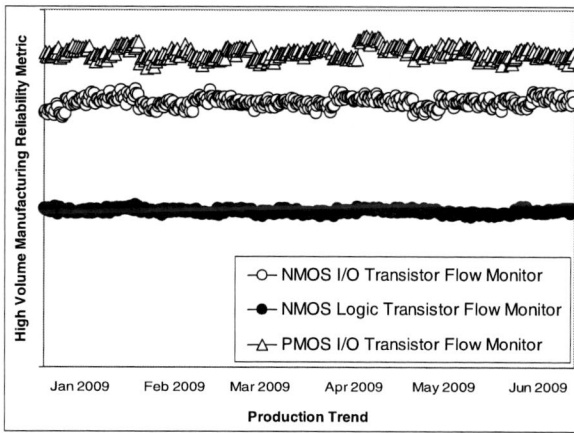

FIGURE 17. FACTORY PRODUCTION TREND OVER A PERIOD OF 6 MONTHS IS SHOWN FOR THE QUICK-STRESS (QS) RELIABILITY METRICS FOR *LOGIC* AND *I/O* TRANSISTORS. EACH SYMBOL CORRESPONDS TO A COMPOSITE METRIC FOR 1 WAFER.

CONCLUSION

Transistor and dielectric reliability were studied extensively on a low power 45nm DG HK+MG process technology. The similarity of degradation physics between logic and I/O transistors has been demonstrated and the critical importance of simultaneous optimization was emphasized. Excellent reliability referenced to SiO_2-Poly was achieved and manufacturing data presented confirms very good process stability.

REFERENCES

[1] Mistry K. *et al*, IEDM Tech Dig, pp.247-50 (2007)
[2] Prasad C. *et al*, 46th IRPS Sym Proc, pp.667-8 (2008)
[3] Pae S. *et al*, 46th IRPS Sym Proc, pp.352-7 (2008)
[4] Hicks J. *et al*, Intel Tech Journal, vol.12 iss.2 (2008)
[5] Ramey S. *et al*, 47th IRPS Sym Proc, pp.1023-7 (2009)
[6] Kasim R. *et al*, 47th IRPS Sym Proc, pp.350-4 (2009)
[7] Jan C.-H. *et al*, IEDM Tech Dig, pp.1-4 (2008)
[8] Auth C. *et al*, VLSI Tech Sym, pp.128-9 (2008)
[9] Bai P. *et al*, IEDM Tech Dig, pp.657-60 (2004)
[10] Jan C.-H. *et al*, IEDM Tech Dig pp.60-3 (2005)
[11] Pae S. *et al*, 47th IRPS Sym Proc pp.499-504 (2009)
[12] Torii K. *et al*, IEDM Tech Dig, pp.129-32 (2004)
[13] Zhang J. *et al*, IRW Final Report, pp.92-95 (2002)
[14] Hu C., Lu Q., 37th IRPS Sym Proc, pp.47-51 (1999)
[15] McPherson J. *et al*, Trans Elec Dev, vol.50 pp.1771-8 (2003)

Re-consideration of Influence of Silicon Wafer Surface Orientation on Gate Oxide Reliability from TDDB Statistics Point of View

Yuichiro Mitani

Advanced LSI Technology Laboratory, Corporate R&D Center,
Toshiba Corporation.
8 Shinsugita-cho, Isogo-ku, Yokohama 235-8522, Japan
Phone: (+81) –(45)- 776-5943, mitani@amc.toshiba.co.jp

Akira Toriumi

Department of Materials Science,
The University of Tokyo
7-3-1, Hongo, Bunkyo-ku, Tokyo 113-8656, Japan

Abstract— **Recently, in order to achieve higher performance and higher density in both CMOS/Logic and flash memories, some three-dimensional structures have been paid much attention. In these structures, the Si surfaces with the surface orientation except (100) are used for the channel of transistors/cells. The reliability depending Si surface orientation has been previously reported and worse reliability of the gate oxides has been suggested. From the viewpoint of the reliability assurance of the devices, the statistical distribution of the degradation is also important. In this paper, we focus on the influence of Si wafer surface orientation on the gate oxide reliability, in particular, focusing the statistical distribution of TDDB. As a result, not only average t_{BD} (50%-t_{BD}) but also Weibull slope for the gate oxide grown on (111) or (110) Si surface are less than those for conventional (100) Si surface. From time-dependent SILC characteristics, it is suggested that larger generated defect size invokes the small Weibull slope compared to SiO_2 on (100). It is expected that the SiO_2 on (111) or (110) Si surface involves the fragile SiO_2 structures, which cause both higher defect generation rate and larger generated defect size.**

Keywords-component; Gate Oxide, Reliability, TDDB, Wafer Surface Orientation, Weibull distribution

I. INTRODUCTION

Issues concerning the reliability of gate dielectrics constitute one of the most serious challenges in the scaling of ULSI. Understanding of gate oxide degradation mechanism is essentially important to realize highly-reliable devices as well as the device performance. Recently, in order to achieve superior performance and higher density in both CMOS/Logic and flash memories, some three-dimensional structures have been paid much attention [1-7], including FinFET for CMOS [3], the nanowire transistors [4,5] and polycrystalline silicon channel for high-density flash memories [6,7]. In these structures, the Si surfaces with the surface orientation except (100) are used for the channel of transistors/cells. The

reliability depending Si surface orientation has been reported previously [3, 8-10] and worse reliability of the gate oxides grown on the Si surfaces except (100) orientation has been suggested. However, the influence of Si surface orientation on the statistical distribution of reliability has yet to be fully discussed yet. This is one of the most important issues in terms of reliability assurance of the devices. In general, the lifetimes of gate oxide reliability at ppm-order cumulative failure are predicted from the extrapolation of the experimental results, which are measured under accelerated conditions (i.e. at higher temperature and stress voltage than actual operating condition). Therefore, wide distribution of t_{BD} (i.e. small Weibull slope β) is a serious concern from the viewpoint of device reliability including the lifetime prediction and the yield. That is, not only the average t_{BD} but also its statistical distribution should be considered in terms of the improvement of the dielectric breakdown characteristics of gate oxide.

In this paper, by focusing on the statistical distribution of TDDB for the gate oxide grown on (100), (110) and (111) Si surfaces, we discuss the degradation mechanism of gate dielectrics depending on the Si-wafer orientation.

II. DEVICE FABRICATION

The devices used in this study were n-channel MOSFETs fabricated on (100), (110) and (111) of Si surface orientation. The gate oxides were grown in dry O_2 ambient. Each wafer was pre-cleaned and oxidized under the same condition. The gate oxide thickness was 4 – 9 nm.

III. EXPERIMENTAL RESULTS AND DISCUSSION

Fig. 1 shows the Weibull distribution of t_{BD} for the gate oxides grown on (100), (110) and (111) Si surfaces having gate oxide thickness of ~6nm. Each device was stressed under the constant-current condition. Note that t_{BD} values for (111) and (110) are about one-order smaller than that for (100) surface in this case. Fig. 2 shows the average (at cumulative failure probability of 50%) t_{BD} values as a function of oxide thickness (t_{OX}). The average t_{BD} for (111) and (110) are less marked than that for (100), irrespective of t_{OX}. In particular, 50%-t_{BD} value for the oxide grown on Si(111) changes dramatically depending t_{OX}. These results are almost the same with the previous reports [8-10].

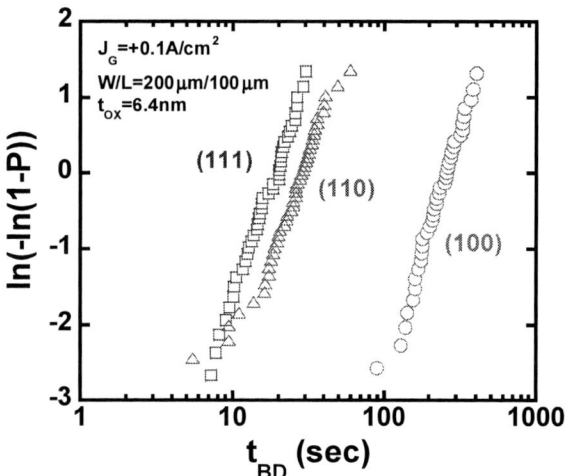

Fig.1 TDDB distribution for the gate oxides grown on the (100), (110) and (111) Si surfaces having the gate oxide thickness of ~6nm. The devices are stressed under constant-current condition of J_G=+0.1A/cm². t_{BD} values for the gate oxide on (110) and (111) are lower than that for conventional (100) surface.

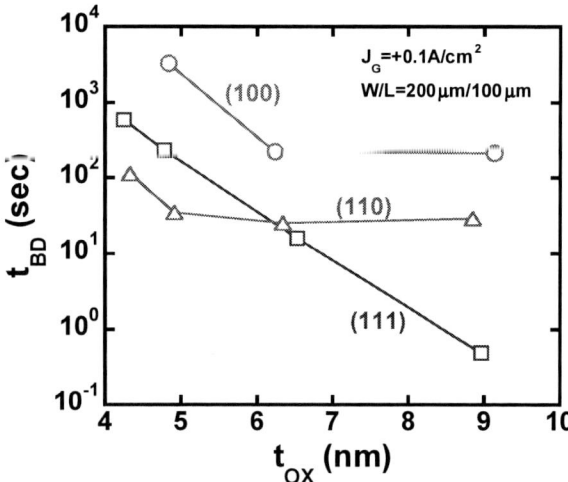

Fig.2 Average t_{BD} values (t_{BD} at cumulative failure probability of 50%) as a function of oxide thickness (t_{OX}). It should be noted that the average t_{BD} values for (110) and (111) are less marked than that for (100) Si surface, irrespective of t_{OX}.

Furthermore, we focus on the Weibull slope β as shown in Fig. 3. Weibull slope β strongly depends on t_{OX} and the data reported by several research groups including our experimental data are plotted on the same line in the case of the gate oxide grown on (100) Si surface. However, it should be noted that Weibull slope β for the gate oxide on (111) and (110) are

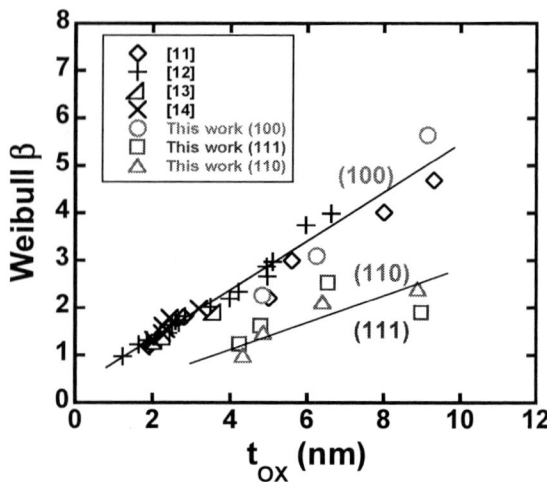

Fig.3 Weibull slope β for the gate oxides grown on the (100), (110) and (111) Si surfaces. The experimental data reported previously from the other research groups [11-14] are also plotted. Our data for the gate oxide on (100) Si surface are plotted on the same line to the previous reported data. However, in the cases of (110) and (111), the values of Weibull β are smaller compared to those for (100).

clearly smaller than that on (100). In addition, t_{OX} dependence of Weibull β is gradual compared to that on (100).

As shown in Fig. 2, t_{BD} values depend on t_{OX}. Therefore, in order to check weather the variability of t_{OX} is responsible for the small Weibull β for the oxide grown on Si(110) and Si(111), the t_{OX} variability is compared among the three oxide films, as shown in Fig. 4. The t_{OX} variability of 6nm-thick gate oxides is eliminated within 0.2nm in all cases, indicating that the t_{OX} variability does not affect on the TDDB distribution.

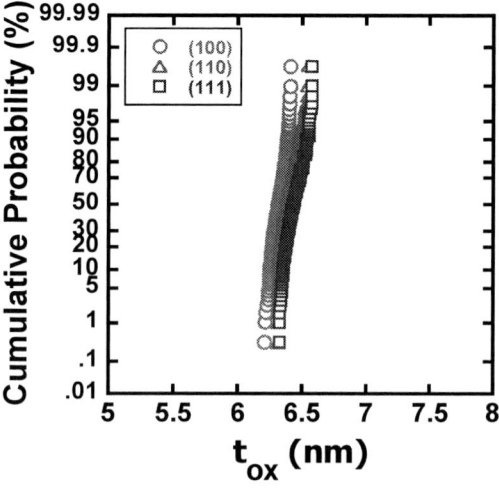

Fig.4 t_{OX} variability for the gate oxide on (100), (110) and (111). t_{OX} was estimated from C-V measurement. In all cases, the variability is eliminated within 0.2nm. This result implies that the t_{OX} variability does not affect on the TDDB distribution shown in Figs. 1 and 3.

978-1-4244-5430-3/10 $26.00 © 2010 IEEE

Figs. 5(a) and (b) show the J-V characteristics and J_{SUB}/J_G, which are estimated using the carrier separation method. Here, J_{SUB}/J_G is regarded as the hole generation efficiency at the anode interface [15]. The J-V characteristics and J_{SUB}/J_G correspond among the gate oxide on (100), (110) and (111) Si surfaces in each t_{OX}. That is, under the same oxide thickness, the same amount of holes per unit stress time is injected into the gate oxide among (100), (110) and (111) if the oxide thickness is identical among the three gate oxides.

Fig. 6 shows the Weibull distributions of hole fluence (Q_P) contrasting the gate oxide grown on (111) with that on (100). J_G, J_{SUB} and J_{SUB}/J_G correspond among the three gate oxides as shown in Fig. 5. Nevertheless, Q_P for (111) is less marked than that for (100) as shown in Fig. 6. From these results, the gate oxide grown on the Si surfaces except (100) orientation is expected to be more fragile compared to that on (100).

The less reliability of the gate oxide grown on (110) and (111) compared to that on (100) has been reported previously. And besides, it is said that the worse qualities of Si/SiO₂ interface and bulk SiO₂ cause the less reliability. Fig. 7 shows the generated interface-state densities under constant-current F-N stressing for the gate oxide grown on respective Si surface.

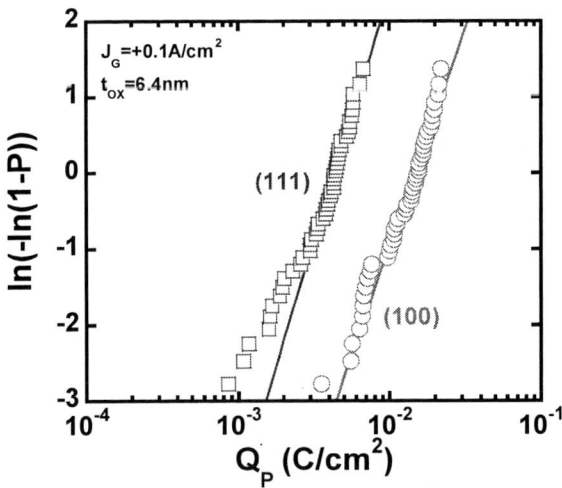

Fig.6 Weibull distribution of Q_P for the gate oxide on (100) and (111). Q_P values were estimated from the integration of JSUB during constant-current stressing. Although the same oxide thickness and the quantum yield at the anode interface, the difference of Q_P between the gate oxide on (100) and that on (111) is clear.

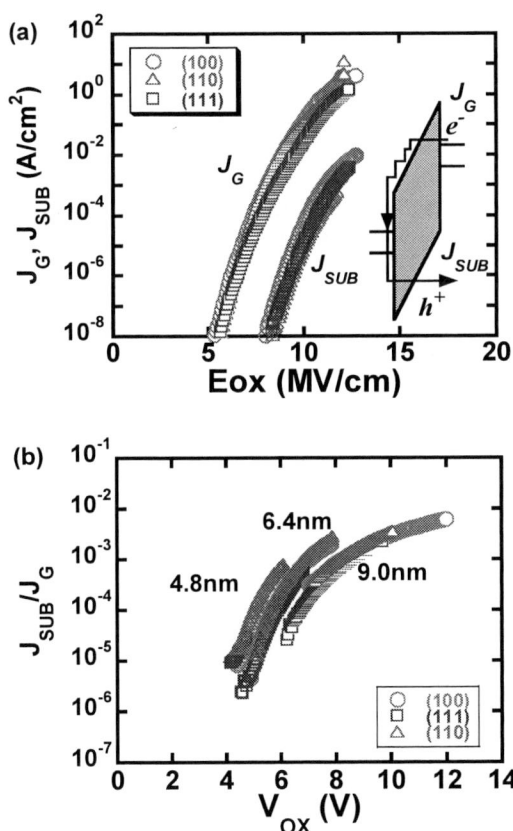

Fig.5 J-V characteristics (a) and J_{SUB}/J_G (b) for the gate oxide grown on (100), (110) and (111) Si surface. J_{SUB} were measured using the carrier separation method and J_{SUB}/J_G means the hole generation efficiency at anode interface [20]. J_G, J_{SUB} and J_{SUB}/J_G almost coincide among the three devices.

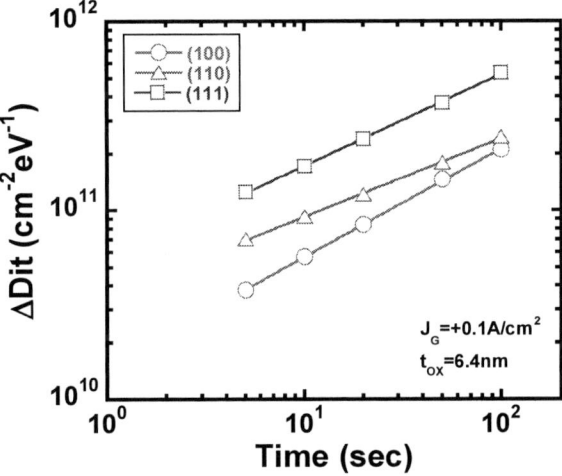

Fig.7 Generated interface-state density ΔDit for the gate oxide on (100), (110) and (111). Dit values were measured by using the conventional charge-pumping method [16]. ΔDit values for (110) and (111) are larger than that for conventional (100) Si surface

The values of ΔDit for the gate oxide on (110) and (111) are larger than that for (100).

It has been reported that higher interface-state generation deteriorates TDDB characteristics [17, 18]. According to the previous reports, higher interface-state generation during the stress forcing accelerates the defect generation in the bulk and, consequently, leads the shorter t_{BD}. As shown in Fig. 7, the interface-state generation for the gate oxide grown on the Si surfaces except (100) orientation is more marked. Therefore, it can be expected that the variability of ΔDit causes the smaller Weibull slope β in the case of (110) and (111).

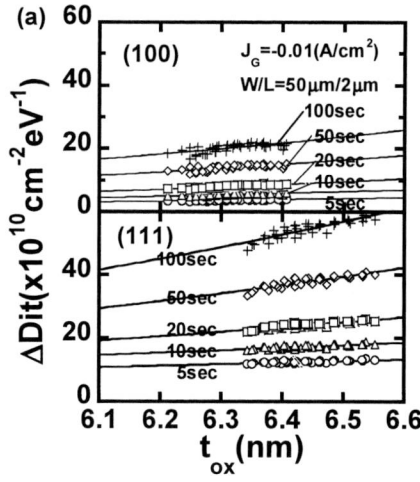

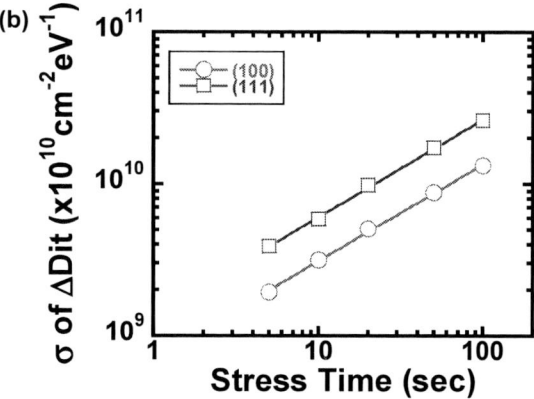

Fig.8 ΔDit as a function of t_{ox} (a) and standard deviation σ (b) of the ΔDit for the gate oxide on (100) and (111). The values of standard deviation σ are estimated based on the fitting curves of ΔDit-t_{ox} correlation. Note that σ of ΔDit increases monotonously with increasing stress time and that for the oxide on (111) is larger than that on (100).

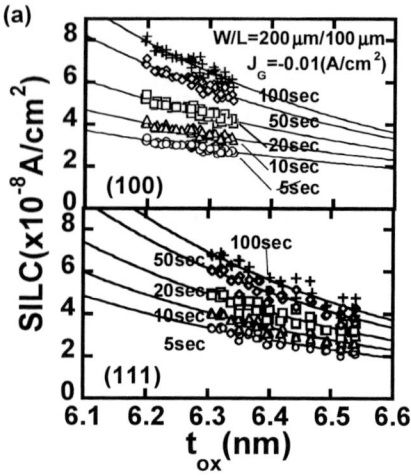

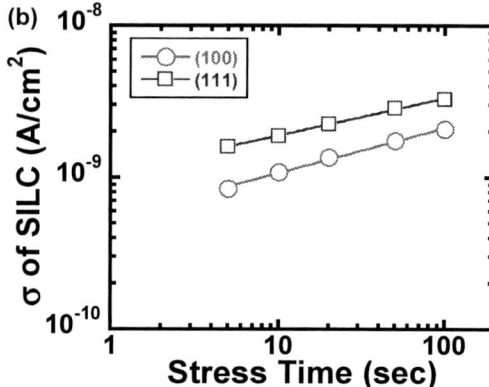

Fig.9 SILC as a function of t_{ox} (a) and standard deviation σ (b) of the SILC for the gate oxide on (100) and (111). The values of standard deviation σ are estimated based on the fitting curves of SILC-t_{ox} correlation. Note that σ of SILC also increases monotonously with increasing stress time and that for the oxide on (111) is larger than that on (100).

Fig. 8 shows the variability of ΔDit for the gate oxides on (100) and (111) Si surfaces. The values of standard deviation σ are estimated based on the fitting curves of ΔDit-t_{ox} correlation as shown in Fig. 8(a). It can be seen in Fig. 8(b), the value of σ for ΔDit increases monotonously with stress time. Furthermore, it should be noted that these σ for (111) are larger than those for (100). In the same way, the variability of SILC, which relates deeply to the dielectric breakdown, is evaluated as shown in Fig. 9(a). The value of σ for SILC also increases monotonously with stress time as shown in Fig. 9(b).

If the variability of ΔDit is responsible to that of SILC, the correlation between σ of ΔDit and σ of SILC of the gate oxide on Si(111) maybe coincides with that on Si(100). Fig. 10 shows the correlation between σ of ΔDit and σ of SILC for (100) and (111). The correlation of σ for (111) does not coincide with that for (100) and the data for (111) is higher than those for (100). That is, the variability of SILC for the gate oxide on (111) Si surface is deteriorated compared to that on (100) even though the same ΔDit variability is observed.

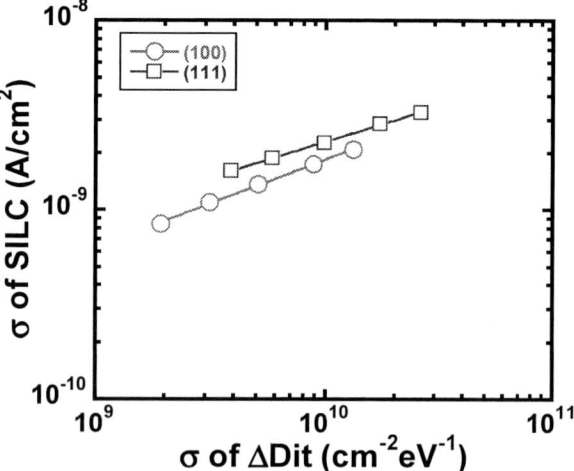

Fig.10 Correlation between σ of ΔDit and σ of SILC. The correlation of σ for (111) does not coincide with that for (100) even under the same σ of ΔDit. This result implies that the variability of SILC is not dominated by that of the interface-state generation.

This result implies that the variability of the interface-state generation does not cause directly the variability of the defect generation in bulk. Therefore, the difference of Weibull β between the gate oxide on (100) and that on (111) Si surfaces cannot be explained only by the variability of ΔDit.

It has been reported that Weibull distribution of dielectric breakdown in SiO₂ can be mathematically represented by the cell-based percolation model [19-22]. According to this model, as schematically shown in Fig. 11, a cubic cell with a lattice constant a_0 is defined in the oxide bulk and the gate oxide is assumed to be divided in this cell. a_0 is ascribed to physical defect diameter. During stressing, defects are generated at random in the gate oxide and the cell changes to conductive state. Dielectric breakdown occurs when all the cells in one straight vertical column are connected from electrode to substrate. Here, Weibit (W_{BD}) is represented by

$$W_{BD}(\lambda) = \ln\left\{\ln\frac{1}{1-F(t)}\right\} = \ln\left(\frac{A_{OX}}{a_0^2}\right) + \left(\frac{t_{OX}}{a_0}\right)\ln(\lambda(t)) \quad (1),$$

where A_{OX} is the oxide area and $\lambda(t)$ is the defect generation probability. $\lambda(t)$ is described by stress-induced leakage current (SILC) and this is represented by

$$\lambda(t) = \xi \cdot t^\gamma \quad (2),$$

where γ corresponds to the slope of the stress time dependence of SILC. From these equations, Weibull β is equal to

$$\beta = \gamma \cdot \frac{t_{OX}}{a_0}. \quad (3)$$

From the equation (3), if the oxide thickness is fixed, γ and a_0 modulate Weibull β. Fig. 12 shows the SILC for the gate oxide grown on (100) and (111) Si surface. SILC for the gate oxide on (111) is larger than that on (100). However, the slopes of the time dependence of SILC (i.e. γ in the equation (2)) correspond between (100) and (111).

As mentioned above, Weibull β of t_{BD} can be represented by using the defect size (a_0) according to the cell-based percolation model. Therefore, from these results, it can be concluded that the larger variability of SILC and the smaller Weibull β for the gate oxides grown on the Si surfaces except (100) orientation are consequent upon the larger defect size (a_0).

Previous reports show that the SiO₂ film density on (110) is higher [23], and diluted HF etching rate for (111) is larger compared to that for (100) [9, 23]. Momose et al. suggested that the SiO₂ film on (110) substrate includes a lot of sub-oxides, because of the faster oxidation rate [23]. Fig. 12 shows SiO₂ film density on (100), (110) and (111) measured by grazing incidence X-ray reflective (GI-XR) technique. The transition region near at Si/SiO₂ interface is wider and film density is higher in the case of (110). This result implies that the gate oxide on (110) involves the strained SiO₂ structure.

We have reported that strained SiO₂ structures can be relaxed by fluorine incorporation and consequently, Weibull slope β increases [11].

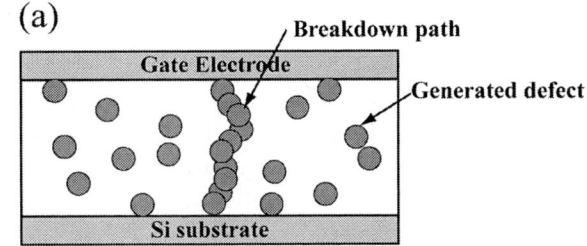

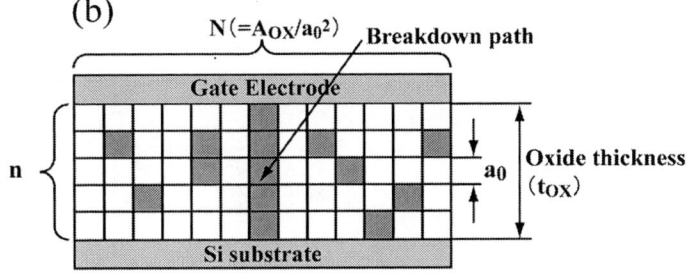

Fig.11 Schematic diagrams of dielectric breakdown phenomenon based on percolation model [16]. (a) percolation model of dielectric breakdown in SiO₂ (b) analytic "cell-based" percolation model

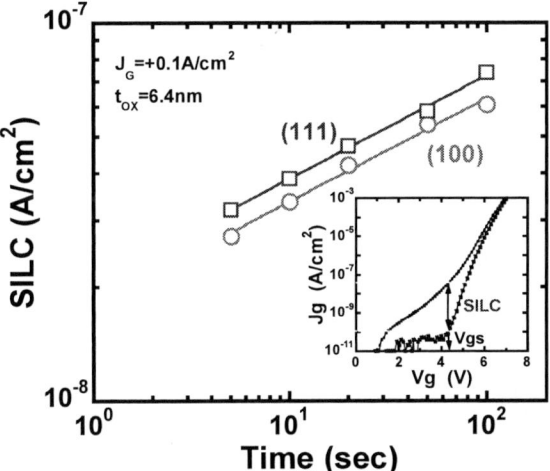

Fig.12 SILC for the gate oxide on (100) and (111) under the constant-current stress condition. SILC for the gate oxide on (111) is larger than that on (100). However, the slopes of the time dependence of SILC coincide between (100) and (111). The SILC is evaluated as the increase of the gate current at the sensing gate voltage (Vgs: Vg at which the gate current reaches 10^{-10} A/cm²) after the constant-current stressing.

Therefore, it is inferred that inherent strained SiO₂ structures cause the larger defect size.

On the other hand, in the case of the SiO₂ film on (111), the transition region is narrower, though film density coincides with that on (100). Tatsumura et al. calculated the SiO₂ structure on Si(111) and indicated that the SiO₂ structure on

Si(111) has less strain than that on Si(100), but involves a larger number of dangling bonds [24]. In the case of High-k gate dielectrics, it has reported that oxygen vacancies (V_O) attract each other and these are condensed when electrons are injected into High-k, which causes larger defect size and smaller Weibull β value [25]. In the same way, it is expected that a large number of Si dangling bonds is activated by F-N stressing and apparent defect size becomes larger in the case of (111). Thus, we infer that the fragile SiO_2 structures involved in the oxide on the Si surfaces except (100) orientation invokes larger defect size and smaller Weibull slope β.

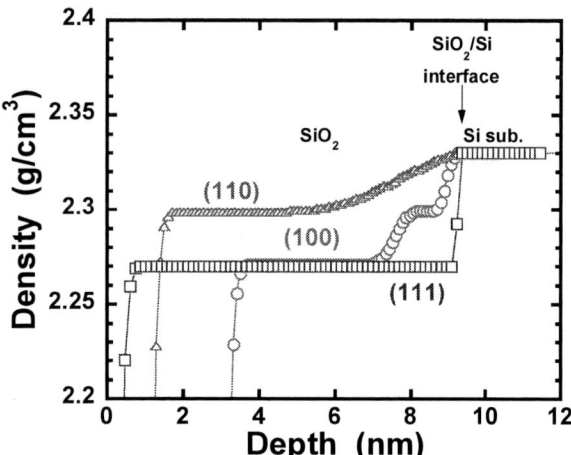

Fig.13 Film density measured by grazing incidence X-ray reflective (GI-XR) technique. The SiO_2 on (110) has a wider transition layer near the Si/SiO_2 interface and higher film density compared to that on (100). On the other hand, the film density grown on (111) Si surface almost coincides with that on (100), but transition layer is narrower.

IV. CONCLUSIONS

In this paper, the influence of Si wafer surface orientation on the gate oxide reliability was investigated. In particular, we focused on the statistical distribution of TDDB. As a results, not only the 50%-t_{BD} values but also the Weibull β of the gate oxide grown on (100) and (111) Si surface are smaller than that on (100). From the experimental results of the time-dependent SILC and the correlation between the variability of ΔDit and SILC, it was suggested that the defect size (a_0) generated in these SiO_2 during F-N stressing is larger than that for the gate oxide on a conventional Si(100). It is inferred that the fragile SiO_2 structures included in the oxide on (110) and (111) deteriorates not only the average values but also the statistical distributions of the deterioration of the gate oxides.

ACKNOWLEDGMENT

The authors would like to thank I. Koizumi and Prof. K. Kajita for their support of the measurements. The authors would like to also thank Drs. K. Kato, Y. Nakasaki, I. Hirano, S. Fukatsu and A. Masada for their thoughtful discussions and comments. In addition, I thank Dr. J. Koga for encouragement and support throughout this work.

REFERENCES

[1] J. M. Hergenrother, D. Monroe, F. P. Klemens, A. Kornblit, G. R. Weber, W. M. Mansfield, M. R. Baker, F. H. Baumann, K. J. Bolan, J. E. Bower, N. A. Ciampa, R. A. Cirelli, J. I. Colonell, D. J. Eaglesham, J. Frackoviak, H. J. Gossmann, M. L. Green, S. J. Hillenius, C. A. King, R. N. Kleiman, W. Y.-C. Lai, J. T.-C. Lee, R. C. Liu, H. L. Maynard, M. D. Morris, S.-H. Oh, C.-S. Pai, C. S. Rafferty, J. M. Rosamilia, T. W. Sorsch, H.-H. Vuong, "The vertical replacement-gate (VRG) MOSFET: A 50-nm vertical MOSFET with lithography-independent gate length," in IEEE International Electron Devices Meeting Technical Digest, 1999, pp. 75-78.

[2] D. Hisamoto, W.-C. Lee, J. Kedzierski, H. Takeuchi, K. Asano, C. Kuo, E. Anderson, T.-J. King, J. Bokor, C. Hu, "FinFET-a self-aligned double-gate MOSFET scalable to 20 nm," IEEE Trans. Electron Devices, ED-47, 2000, p.p.2320-2325.

[3] S.H. Tang, L. Chang, N. Lindert, Y.-K. Choi, W.-C. Lee, X. Huang, V. Subramanian, J. Bokor, T.-J. King, C. Hu,"FinFET-a quasi-planar double-gate MOSFET," in IEEE International Solid-State Circuits Conference (ISSCC). Digest of Technical Papers, 2001, p.p. 118-119.

[4] S. D. Suk, S.-Y. Lee, S.-M. Kim, E.-J. Yoon, M.-S. Kim, M. Li, C. W. Oh, K. H. Yeo, S. H. Kim, D.-S. Shin, K.-H. Lee, H. S. Park, J. N. Han, C. J. Park, J.-B. Park, D.-W. Kim, D. Park and B.-I. Ryu, "High Performance 5nm radius Twin Silicon Nanowire MOSFET(TSNWFET): Fabrication on Bulk Si Wafer, Characteristics, and Reliability, " in IEEE International Electron Devices Meeting Technical Digest, 2005, pp. 717-720.

[5] M. Li, K. H. Yeo, S. D. Suk, Y. Y. Yeoh, D.-W. Kim, T. Y. Chung, K. S. Oh and W.-S. Lee, "Sub-10 nm Gate-All-Around CMOS Nanowire Transistors on Bulk Si Substrate," in Symposium on VLSI Technology Digest of Technical Papers, 2009, p.p. 94-95.

[6] H. Tanaka, M. Kido, K. Yahashi, M. Oomura, R. Katsumata, M. Kito, Y. Fukuzumi, M. Sato, Y. Nagata, Y. Matsuoka, Y. Iwata, H. Aochi and A. Nitayama, "Bit Cost Scalable Technology with Punch and Plug Process for Ultra High Density Flash Memory," in Symposium on VLSI Technology Digest of Technical Papers, 2007, p.p. 14-15.

[7] J. Kim, A. J. Hong, S. M. Kim, E. B. Song, J. H. Park, J. Han, S. Choi, D. Jang, J.-T. Moon, and K. L. Wang, "Novel Vertical-Stacked-Array-Transistor (VSAT) for ultra-high-density and cost-effective NAND Flash memory devices and SSD (Solid State Drive)," in Symposium on VLSI Technology Digest of Technical Papers, 2009, p.p. 186-187.

[8] K. Nakamura, K. Ohmi, K. Yamamoto, K. Makihara, T. Ohmi, "Silicon Wafer Orientation Dependence of MOS Device Reliability," in Ext. Abst. Solid State Devices and Materials (SSDM), 1993, p.p.585-587.

[9] T. Ohmi, K. Matsumoto, K. Nakamura, K. Makihara, J. Takano, and K. Yamamotoa, "Influence of silicon wafer surface orientation on very thin oxide quality," J. Appl. Phys. 77, 1995, p.p. 1159-1164

[10] H. S. Momose, T. Ohguro, S. Nakamura, Y. Toyoshima, H. Ishiuchi and H. Iwai, "Study of wafer orientation dependence on performance and reliability of CMOS with direct-tunneling ;gate oxide," in Symposium on VLSI Technology Digest of Technical Papers, 2001, p.p 77-78.

[11] Y. Mitani, H. Satake, and A. Toriumi, "Impact of Deuterium and Fluorine Incorporation on Weibull Distribution Dielectric Breakdown in Gate Dielectrics," ECS Transactions, Volume 19, 2009, p.p 227-242.

[12] E. Y. Wu, J. Sune and W. Lai, "On the Weibull shape factor of intrinsic breakdown of dielectric films and its accurate experimental determination. Part II: experimental results and the effects of stress conditions," IEEE Trans. Electron Devices, 49, 2002, p.p. 2141-2150.

[13] P. E. Nicollian, A. T. Krishnan, C. A. Chancellor, and R. B. Khamankar, "The Traps that cause Breakdown in Deeply Scaled SiON Dielectrics," in IEEE International Electron Devices Meeting Technical Digest, 2006, pp. 743-746.

978-1-4244-5430-3/10 $26.00 © 2010 IEEE

[14] G. Ribes, S. Bruyère, D. Roy, M. Denais, J-M. Roux, C. Parthasarathy, V. Huard, G. Ghibaudo, "New extensive MVHR breakdown models," in IEEE International Reliability Physics Symposium proceedings, 2006, p.p. 621-622.

[15] I. C. Chen, S. Holland, K. K. Young, C. Chang, and C. Hu, "Substrate hole current and oxide breakdown," Appl. Phys. Lett., 49, 1986, p.p. 669-671.

[16] G. Groeseneken, H. E. Maes, N. Beltran, and R. F. De Keersmaecker, "A reliable approach to charge-pumping measurements in MOS transistors," IEEE Trans. Electron Devices, ED-31, 1984, p.p. 42-53.

[17] Y-. T Huang, A. Pinto, C-. T. Lin, C-. H. Hsu, M. Ramin, M. Seacrist, M. Ries, K. Matthews, B. Nguyen, M. Freeman, B. Wilks, C. Stager, C. Johnson, "PMOSFET Reliability Study for Direct Silicon Bond (DSB) Hybrid Orientation Technology (HOT)," IEEE Electron Dev. Lett., 28, 2007, p.p. 815-817.

[18] Y. Mitani, T. Yamaguchi, H. Satake, A. Toriumi, "Reconsideration of hydrogen-related degradation mechanism in gate oxide," in IEEE International Reliability Physics Symposium Proceedings, 2007, p.p. 226-231.

[19] R. Degraeve, G. Groeseneken, R. Bellens, M. Depas, H. E. Maes, "A consistent model for the thickness dependence of intrinsic breakdown in ultra-thin oxides," in IEEE International Electron Devices Meeting Technical Digest, 1995, pp. 863-866.

[20] J. Sune, "New physics-based analytic approach to the thin-oxide breakdown statistics," IEEE Electron Device Lett., 22, 2001, p.p. 296-298.

[21] M. A. Alam, J. Bude, B. Weir, P. Silverman, A. Ghetti, D. Monroe, K. P. Cheung, and S. Moccio, "An anode hole injection percolation model for oxide breakdown-the "doom's day" scenario revisited," in IEEE International Electron Devices Meeting Technical Digest, 1999, p.p. 715-718.

[22] A. T. Krishnan and P. E. Nicollian, "Analytic extension of the cell-based oxide breakdown model to full percolation and its implications," in IEEE International Reliability Physics Symposium proceedings, 2007, p.p. 232-239.

[23] H. S. Momose and S. Yoshitomi, "Effects of Si Channel Orientation on MOSFET Characteristics," in Proc. 26th International Conference on Microelectronics (MIEL), 2008, p.p. 137-144.

[24] K. Tatsumura, T. Watanabe, D. Yamasaki, T. Shimura, M. Umeno, I. Ohdomari, "Large-scale atomistic modeling of thermally grown SiO/sub 2/ on Si(111) substrate," Jpn. J. Appl. Phys. 43, 2004, p.p. 492-497.

[25] I. Hirano, Y. Nakasaki, S. Fukatsu, A. Masada, M. Goto, K. Nagatomo, S. Inumiya, K. Sekine and Y. Mitani, "Impact of Metal Gate Electrode on Weibull Distribution of TDDB in HfSiON gate dielectrics," in IEEE International Reliability Physics Symposium proceedings, 2009, p.p. 355-361.

Photovoltaic Module Reliability Studies at the Florida Solar Energy Center

Neelkanth G. Dhere, Shirish A. Pethe, Ashwani Kaul

Florida Solar Energy Center, University of Central Florida
1679 Clearlake Road
Cocoa, FL 32922
Phone: +1-321-638-1442, e-mail: dhere@fsec.ucf.edu

Abstract— **The accelerated tests currently carried out on PV modules reduce the infant mortality as well as improve the production techniques during the manufacture of PV modules. However, they do not completely duplicate the real world operating conditions of PV modules. Hence it is essential to deploy PV modules in the field for extended period of time in order to estimate the degradation, if any, as well as to elucidate the degradation mechanisms. Moreover, PV modules should be tested by specially designed tests in harsh climates. In this paper some of the results obtained on a-Si:H modules from various US companies is discussed.**

Keywords- Photovoltaic, a-Si:H, PVUSA Regression, High Voltage Bias.

I. INTRODUCTION

Large scale usage of PV technology will help to reduce fossil fuel consumption for electricity generation that in turn will offer means to minimize greenhouse gas emissions and to minimize the risks of climate change. Development of cost efficient photovoltaic modules having a useful lifetime of 30 year is the goal of the U.S. Department of Energy. So far, the issues with long term performance have not been resolved completely. However, the reliability and durability of PV modules has improved consistently over the recent years. Accelerated aging tests have been useful to anticipate failure modes and mechanisms and for establishing quality standards. However, the acceleration testing alone is not able to predict the various possible degradation modes and mechanisms in the PV modules. Studies have shown that most of the degradation mechanism and reliability issues in PV cells have been determined by the tests carried out on field-deployed modules. Moreover, exposure testing of PV modules can be used for determination of PV performance degradation rates. Despite the progress made in this area, additional research is required in order to establish stringent quality standards to fulfill the stated goal. Florida Solar Energy Center (FSEC) has been closely working with the PV industry and national laboratories for studying the performance and reliability of PV modules. Thin-film PV modules of leading US manufacturers are being tested in the hot and humid climate at FSEC [1]. Performance analysis of PV modules is being carried out systematically using the test methodology developed at FSEC. Such detailed testing of field deployed PV modules helps in studying and elucidating the correlation between PV performance and the various meteorological parameters such as solar irradiance, relative humidity, ambient temperature, etc. This paper discusses the general methodology employed in performance analysis of thin film a-Si:H PV modules and presents some of the results obtained during the three and half year test period.

Photovoltaic modules deployed in a field may operate at a high voltage relative to ground determined by position of the module in the overall array circuit. In a grid-connected photovoltaic system the cells may be as much as ±600 volts with respect to ground in USA and as high as ±1000 volts with respect to ground in Europe and elsewhere, i.e., with respect to the frame of the module. This voltage gradient between cells to frame gives rise to leakage currents across them. High leakage currents can lead to electromigration and degradation, and thus become important issue for durability and safety. Thus long-term effects of exposure to high voltage in the field must be studied to achieve desired service lifetime for PV modules [2-4]. One of the functions of the module encapsulation is to confine the generated electrical energy to the module circuitry. The energy that dissipates from the module circuitry through the encapsulation to the ground is termed leakage current. Leakage current may be composed of charge carriers that move under influence of voltage and concentration gradients through the insulation, reacting with it and the cell and frame metals to produce corrosion products. Leakage current levels are also determined in large part by the electrical conductivity of the insulation. This conductivity varies greatly with changing environmental conditions of temperature and relative humidity [5-6]. Therefore, high voltage bias testing of various PV modules in the hot and humid climate of Florida was undertaken at FSEC.

II. METHODOLOGY

Degradation in an electronic device is detected by deterioration in its output characteristics. In the case of PV modules, degradation results in a decrease in power generated by the PV module. It is therefore essential that continuous monitoring and analysis of PV parameters be carried out. Taking this fact into consideration, the outdoor monitoring and testing of PV modules in hot and humid climate of Florida was initiated with a project from National Renewable Energy Laboratory (NREL).

978-1-4244-5430-3/10 $26.00 © 2010 IEEE

A. Outdoor Monitoring of PV Modules

The PV modules are divided in two sets; modules in each set are connected in series so as to build maximum open circuit voltage of less than ±600 V with respect to ground and are connected across a fixed load resistor and a current measuring shunt. The value of fixed load resistor is estimated from the rated voltage and current values. It has been verified by earlier studies that fixed resistive loading is an effective and economic technique for loading PV modules during outdoor exposure testing, provided, the resistance value is near to the ratio of voltage to current at maximum power point under standard test conditions [7]. To understand the effect of biasing on PV performance, the modules in two arrays are connected with positive and negative bias with respect to ground. PV modules are visually inspected and photographed before installation to compare any signs of degradation with the initial condition of module. Continuous recording of PV parameters as well as the meteorological parameters is carried out and analyzed on a daily as well as monthly basis. Such continuous data when acquired over an extended period of time facilitates in estimating the PV performance degradation rates. The continuous monitoring of meteorological parameters aids in elucidating the correlation of ambient parameters such as solar irradiance, relative humidity and ambient temperature on PV performance.

The data is screened for any discrepancies and Power (V*I) versus Irradiance graphs are daily plotted. After the data for an entire month is available, it is plotted over time as a function of normalized power and for irradiance values above 600 W/m². Photovoltaic system performance monitoring by regressions under Performance Test Conditions (PTC) using a protocol developed by Photovoltaics for Utility Systems Applications (PVUSA) [8] is the commonly used method because PV module and array temperature measurements are not needed and the magnitude of power obtained by PTC is much closer to the actual operational power values. PVUSA regression analysis is carried out using a simplified version suited to our requirements of Labview code [9] originally provided by NREL. As per the PVUSA regression analysis, power is considered to be a function of irradiance, temperature and wind speed. However, for the analysis presented here wind speed is assumed to be constant as an approximation. Hence power is considered a function of irradiance and temperature and is represented mathematically as,

$$P = C_1.E + C_2.E^2 + C_3.T.E \qquad (1)$$

Here, P, E, and T represent power, irradiance, and temperature respectively, while C_1, C_2, C_3, are constants for a given month. A LABVIEW program was developed to estimate the three coefficients using the least square method. These coefficients are then used to calculate the PTC power (E = 1000 W/m² and T = 20 °C) for that month.

B. Current-Voltage (I-V) Measurements

Current-voltage characteristics of the module arrays are also taken to substantiate the results obtained with the outdoor data monitoring of the module arrays. This is carried out by using a curve tracer. Standard PV performance parameters such as open circuit voltage, short circuit current, fill factor, peak power, voltage and current at peak power are obtained from these measurements. The data is normalized at 1000 W/m² and 25°C using temperature coefficients provided by the module manufacturer. Variation in the PV parameters is a good sign of degradation in PV modules. Analysis of the current-voltage measurements carried over an extended period of time serves as a technique by itself in estimating the PV performance degradation rates. The analysis using current-voltage measurements complements the analysis based on data obtained from outdoor monitoring of PV modules. Moreover, the periodic current-voltage measurements facilitate in deciding the correct value of fixed load resistor used during the outdoor monitoring of PV modules. Degradation, if any, can be detected by a drop in the generated power. However, with periodic current-voltage measurements it might be possible to predict which of the PV parameters is degrading and consequently can facilitate in determining the actual cause of degradation.

C. High Voltage Bias Testing

Commercial grid-connected PV modules are subjected to high voltage biases of up to ±600 V in USA and up to ±1000 V in Europe and elsewhere. Thus, PV modules are not only expected to withstand the harsh climatic conditions but also the high voltage biasing that they are subjected to. When a PV module is biased to say +600 V it means that the PV cells are at 600 V with respect to ground. In case of framed PV modules the frames as well as the mounting rails on which the modules are mounted are connected to ground. This leads to a voltage gradient from the cell to the frame which is at ground potential. The voltage gradient gives rise to leakage current and leakage current is directly proportional to the voltage gradient and hence directly proportional to the biasing voltage. High leakage currents can lead to electromigration and degradation, and thus become important issues for durability and safety. Long-term effects of exposure to high voltage in the field thus become important aspect for study to achieve the desired service lifetime for photovoltaic module. For this reason individual modules are biased at +600 V and -600 V and the leakage current along with the relative humidity and ambient temperature is continuously recorded. For an accurate prediction of the degradation modes and mechanisms in PV modules It is essential to collect and analyze the data over extended period of time. Moreover, a statistical approach is essential to analyze the large data collected over this extended period of time.

III. RESULT AND DISCUSSIONS

The a-Si:H PV modules were deployed at the FSEC Outdoor facility in November 2003 and removed in November 2007. However, PV performance data is available only for a period of about 30 months. Data during the initial period of testing was not available due to several issues regarding the data acquisition system and because significant amount of time was lost during the hurricane season of 2004. This was one of worst ever hurricane seasons in Florida and the data could not be recorded during this time frame. Hurricane protectors were built and tested after these unprecedented five hurricanes (Bonnie, Charley, Frances, Ivan and Jeanne) hit Florida in

978-1-4244-5430-3/10 $26.00 © 2010 IEEE

2004. The usage of the hurricane protectors during the subsequent hurricane seasons resulted in minimal loss of data as well as damage to the PV modules.

Figure 1 shows the raw data collected for the positive biased array during a sample two year period from July 2005 to June 2007. It includes the data for all values of irradiance after the total data collected was verified and all the erroneous data points were removed from the data-set. Similar graph is obtained for the negative biased array. The average annual energy yield calculated from the data collected through continuous monitoring of PV performance is approximately 1385 kWhr/kWp/Yr that is comparable to value of 1200 kWhr/kWp/Yr available in literature for similar climatic conditions [10]. Based on this raw data the coefficients for calculation of PTC power are estimated and the consequent PTC power for that month is calculated. For example, in the month of July 2006 the calculated PTC power was 649.24 W and 643.83 W for the positive and negative biased module arrays, respectively. Similarly, PTC powers were calculated for all the months. A PTC power trend was plotted from the calculated PTC powers. The trend line will be a straight horizontal line if there is no degradation. It may be noted that since the data for the initial period could not be used for statistical analysis, the initial Staebler-Wronski degradation could not be recorded.

As is well-known, a-Si:H PV modules typically degrade slightly during the winter when the temperatures are low and recuperate during the summer when the temperatures are high. The result is a significant decrease and increase in the power respectively during the winter and summer. Therefore, in order to make a reasonable comparison of the performance, it is essential to consider multiples of 12-month periods as shown in Figures 2 and 3. Moreover, it is critical that the resistive load during this test period be kept constant for the regression analysis to be accurate and further for accurate prediction of the degradation rates.

Figures 2 and 3 show that over the 24 month test period, both the positive and negative arrays for either type of modules show degradation. The deviation from the trend-line is mainly due to the seasonal variations and cannot be considered as an error. Using the trend lines, the total degradation in one year for a two year period can be calculated as follows;

Equation of line: $y = m.x + c$; where m is the slope of line and c is the intercept over the Y axis indicating the power value at the beginning of the test period.

$$\% \text{ Degradation/year} = \frac{(m \times 12)}{C} \times 100 \qquad (2)$$

The annual degradation for the positive and negative array of Module type A calculated from the data over the 24 month period is 0.6%/year and 0.5 %/year, respectively. The annual degradation for the positive and negative array of Module type B calculated from the data over the 24 month period is 0.74%/year and 0.259 %/year, respectively. The uncertainty in slope of the trend-line will directly translate to uncertainty in the degradation rate. From the calculated standard error in the slope the uncertainty in degradation rate is approximately 2.9 %/year.

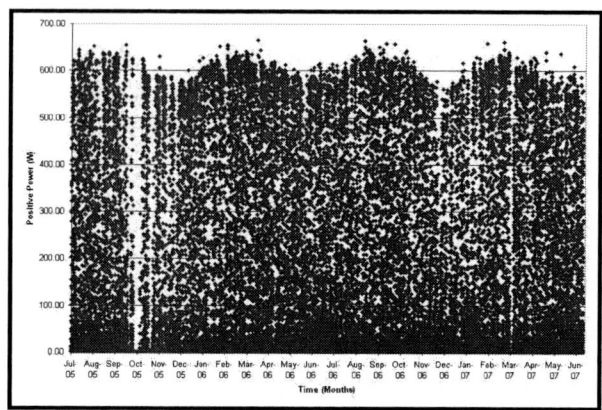

Figure 1. Raw data for the positive biased a-Si:H module array from July 2005 to June 2007

A complementary study of current-voltage characteristics was carried out for comparison with the outdoor data monitoring of the module arrays. Current-voltage characteristics of the module arrays taken on a regular basis provide important PV parameters such as short circuit current, open circuit voltage, fill factor, peak power, etc. Peak power and short circuit current values are normalized to 1000 W/m², 25 °C. This is done by calculating the power values at 1000 W/m² and then multiplying them with the known values of module temperature coefficient. Figures 4 and 5 show the trend for normalized peak power obtained from periodic current-voltage measurements carried out during the period from May 2005 to April 2007 for the positive and negative arrays, respectively.

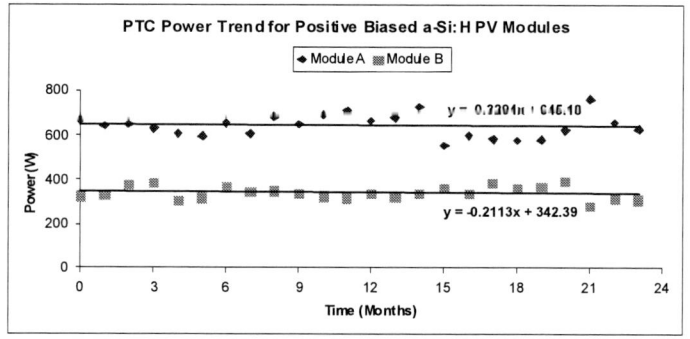

Figure 2. PTC Power Trend for Positive biased a-Si:H PV Modules

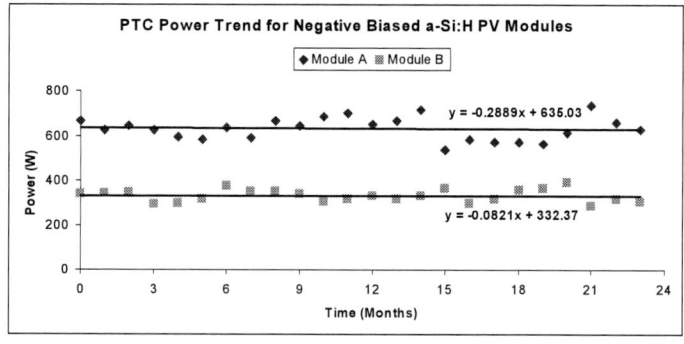

Figure 3. PTC Power Trend for Negative biased a-Si:H PV Modules

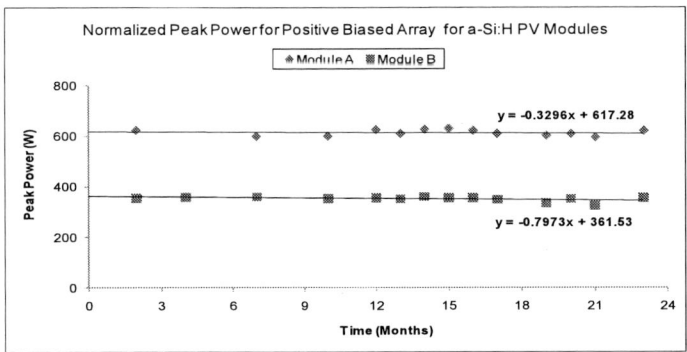

Figure 4. Normalized Peak Power Trend for Positive biased a-Si:H PV Modules

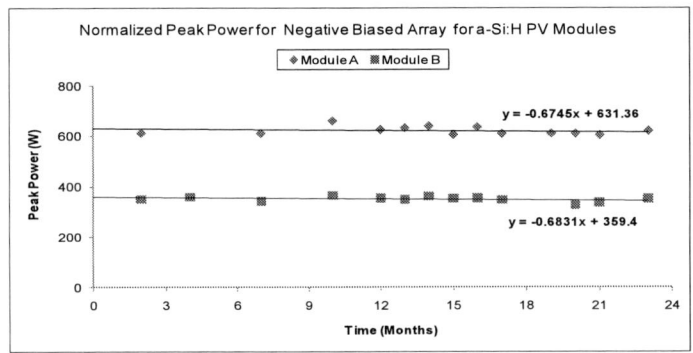

Figure 5. Normalized Peak Power Trend for Negative biased a-Si:H PV Modules

In case of Module type A, the annual degradation calculated from the current-voltage measurements carried out over the period from May 2005 to April 2007 is 0.6%/year for positive array and 1.3%/year for negative array and for Module type B it is 2.65% for positive array and 2.3% for negative array.

The annual degradation calculated from the comparison of PTC power over a two year period and from the comparison of normalized peak power calculated from periodic current-voltage measurements over similar time frame is comparable. However, a complete picture for the performance of PV modules cannot be obtained as the data during the initial period could not be recorded. Moreover, the current-voltage measurements for several months during the two year period are not available. Hence it is necessary to continue the outdoor testing of PV modules along with the periodic current-voltage measurements for longer period of time.

Figure 6 shows the typical variation of leakage current with respect to relative humidity for a +600 V and -600 V biased PV module on a cloudy day. It can be seen from the graph that the leakage current is directly proportional to the relative humidity. The leakage current data when collected over an extended period of time will enable to statistically estimate the activation energy of the leakage current and also aid in predicting an accurate model for the leakage current pathways [4].

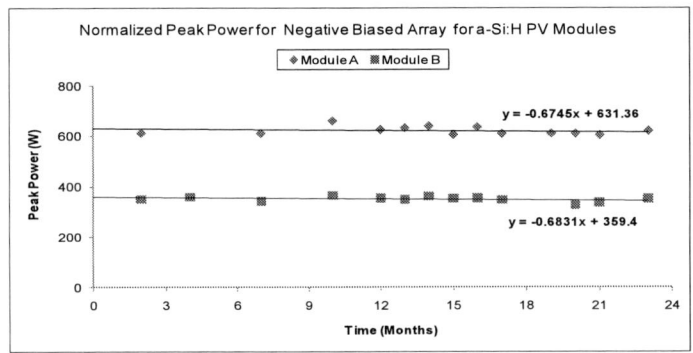

Figure 6. Typical variation of leakage current with respect to relative humidity on a cloudy day

Eight earlier generation a-Si:H modules were installed on the FSEC high voltage test bed in December 2001. The photovoltaic circuits of the modules were biased individually at ±600 V, ±300 V, and ±150 V. No external bias was applied to two modules. One of the first observations made during this test was that the leakage currents were proportional to the applied biases i.e. the value of leakage currents for module biased with +600V was approximately twice the value of leakage currents for module biased with +300V and was approximately four times the value of leakage currents for module biased with +150V. It is essential to model the leakage current pathways in order to determine the possible degradation mechanisms. Therefore, Arrhenius graphs of leakage current versus the inverse of the absolute module temperature were plotted in two distinct relative humidity ranges viz. 35-37% and 95-97%. The semi-logarithmic plots for +600 V and -600 V are shown in figure 7. It can be seen that that the logarithm of leakage current is inversely proportional to the reciprocal of the module temperature. Thus the photovoltaic module leakage conductance is thermally stimulated with a characteristic activation energy that depends on relative humidity.

TABLE I. ACTIVATION ENERGY DIFFERENT VOLTAGES FOR DISTINCT RH RANGES

Module Bias	Activation Energy	
	RH 35-37%	RH 95-97%
+600 V	0.445	0.683
-600 V	0.456	0.639

At high relative humidity, the activation energies are higher than those at low relative humidity for all the modules biased with different voltages. The activation energy can be calculated as per following equation;

$$i(RH,V,T) = i_o(RH,V)\exp\left[\frac{-E_A(RH)}{kT}\right] \quad (3)$$

Earlier generation a-Si:H modules were tested under similar condition. These modules were tested for 30 months and significant power loss was observed in all modules biased at different voltages. Moreover, the power loss was directly proportional to the

978-1-4244-5430-3/10 $26.00 © 2010 IEEE

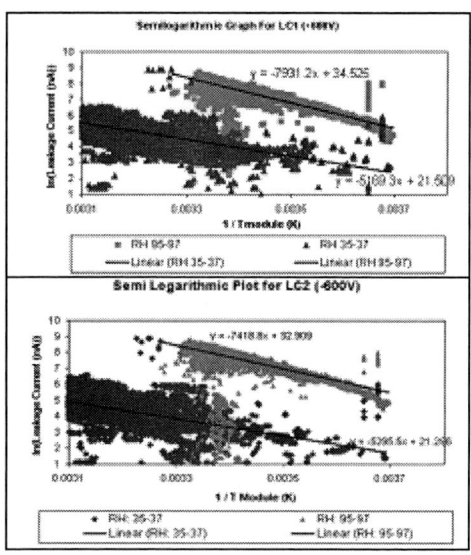

Figure 7.   Semi logarithmic Arrhenius plots for leakage current

biased voltage. Table II shows the percentage power drop for modules biased at various biasing voltages at two stages of the test. It can be seen that the power loss is not linear with time and as the modules are biased longer the rate at which the power loss occurs increases [11].

TABLE II.   PERCENTAGE REDUCTION IN POWER FOR HIGH VOLTAGE BIASED EARLIER GENERATION AMORPHOUS SILICON MODULES

Bias Voltage	After 12 months	After 18 months
-150 V	11.4%	21%
-300 V	40.7%	85.88%
-600 V	N/A	98%

It was observed that the module with high bias voltage – 600 V had been damaged extensively with a complete disintegration of the module within 24 months. Pairs of superstrate and bottom samples were extracted by coring from locations of damaged regions of the module biased at –600 V. The cored samples shown in figure 8 showed very little or no adhesion to the glass. This shows that the corrosion is associated with the delamination of SnO$_2$:F layer from the glass for the negatively biased modules. Visual inspection of the sample extracted from the superstrate side showed plain clear glass without any vestiges of the TCO and the rest of the cell.

Figure 8.   Cored samples from the bottom side (left) and from the superstrate side (right) of the second a-Si:H Module biased at –600 V

X-Ray Energy Dispersive Spectroscopy (XEDS) analysis of this sample showed the presence of silicon, oxygen and sodium, which are normal elements in soda-lime glass thus confirming the absence of a TCO coating and remainder of the cell. Visual inspection of the sample extracted from the bottom side showed a complete cell remaining on the EVA. XEDS analysis of this sample showed the presence of tin, oxygen and silicon on the cell. Thus the SnO$_2$:F TCO with the cell had been transferred on to the EVA which then resulted in its curling and cracking and complete destruction.

IV.   CONCLUSIONS

The PVUSA regression model is an effective tool for studying the performance loss in PV module arrays. Performance loss predictions by this method are found to be very accurate and consistent. Regression analysis is complemented by current voltage characteristics in order to study the performance of PV arrays. Using these tools the performance analysis of a-Si:H PV arrays has been carried out and it has shown minimal degradation over the given period of two years. The a-Si:H PV modules connected in series to form an array across a fixed load resistor were tested continuously for over 30 months. Average annual energy yield of approximately 1385 kWhr/kWp/Yr was calculated. The annual degradation of less than 1% was estimated by the PTC power method. The annual degradation rates estimated from the PTC power trend are complemented by the degradation rate estimated from the normalized peak power trend obtained from periodic current-voltage measurements. The annual degradation rates estimated from the normalized peak power trend are in the range of 1-2%. The discrepancy between the annual degradation rates estimated by the two methods can be explained by the fact that the number of data point used in the calculation in case of the normalized peak power is fewer than that used in case of PTC power. Hence, it can be concluded that it is essential to carry out long term outdoor testing of PV modules to predict the lifetime of PV modules. The long term outdoor testing of PV modules aids in determining the approximate degradation rate for PV modules, thus, providing means to approximately forecast the amount of degradation in PV modules under actual operating conditions. Moreover, it is helpful in detecting long term degradation, if any, and understanding the correlation of PV performance and the meteorological parameters. As a rule of thumb, each three-year study will increasingly assure approximately nine-year segments of useful life time. Hence to guarantee a US DOE goal of 30 years useful lifetime it is necessary to continue the outdoor testing of a-Si:H PV modules. The activation energy for the leakage current was calculated and it was observed that the activation energy was higher for higher relative humidity. However, it is crucial to collect extensive data to verify the activation energy in various relative humidity ranges. Delamination became clearly visible in a-Si:H PV modules fabricated using earlier generation SnO$_2$:F TCO layers biased at –600 V, –300 V and –150 V after 8, 15 and 27 months respectively. There was complete loss of adhesional strength in the damaged area. Initially the a-Si:H cell in the degraded region delaminated and transferred itself entirely from the superstrate glass to the bottom glass/EVA surface, then curled and cracked and finally was destroyed. For the long term stability of PV modules of any technology it is essential to optimize the packaging scheme. Therefore, it is essential to test the packaging of PV modules in terms of their long term durability. High voltage bias testing has proven to be a good testing method for determining packaging durability. The high voltage bias testing of field deployed PV modules is a more realistic accelerated test as compared to the damp heat test with or without high- voltage bias because it exposes the modules to real solar irradiance, relative humidity and temperature cycles. Therefore, it is recommended that the high voltage bias testing in the field should be considered as a reliable metric in determination of the long term performance of PV modules.

ACKNOWLEDGMENT

This work was supported by the NREL and the Thin-Film Partnership program under the contract # ZDJ 33360002. The authors would like to thank Keith Emery from NREL for providing with the original Labview program for carrying out the PVUSA regression analysis. The authors are thankful to Bolko von Roedern, Carl Osterwald, Harin Ullal and Ken Zweibel from NREL and colleagues from various US thin-film PV Companies that provided the a-Si:H PV modules and for their useful suggestions and discussions.

REFERENCES

[1]   A. Kaul, B. Kumar, S. Khatri, and N. G. Dhere, "Outdoor Monitoring High and High Voltage bias Testing of Thin Film PV Modules", Poster Presentation, Florida AVS, 2006.

[2]   N. G. Dhere, S. M. Bet and H. P. Patil, "High voltage bias testing of thin film PV modules", 3rd World Conference on Photovoltaic Energy Conversion, 2003.

[3]   J.A.del Cueto and T. J. McMahon, "Analysis of leakage currents in photovoltaic modules under high voltage bias in the field", Progress in Photovoltaics Research and Applications 2002, Vol. 10, pp.15-28.

[4]   N. G. Dhere, V. V. Hadagali and S. M. Bet, "Leakage currents pathways, magnitudes and their correlation to humidity and temperature in high voltage biased thin film PV modules", Proc. 19th European Photovoltaic Solar Energy Conference, Paris, 2004.

[5]   G. R. Mon and R. Ross, "Electrochemical degradation of a-Si:H photovoltaic modules", Proc. 18th IEEE PV Specialists Conference, Las Vegas, 1985, pp. 1142-1149.

[6]   G. Mon, L. Wen, R. G. Ross, Jr. and D. Ardent, "Effects of temperature and moisture on module leakage currents", Proc. 18th IEEE PV Specialists Conference, Las Vegas, 1985, pp. 1179-1185.

[7]   C.R. Osterwald, J. Adelstein, J.A. del Cueto, W. Sekulic, D. Trudell, P. McNutt, R. Hansen, S. Rummel, A. Anderberg, and T. Moriarty, "Resistive Loading of Photovoltaic Modules and Arrays for Long-Term Exposure Testing", Prog. Photovolt: Res. And Appl. 14(6), 2006, pp. 567-575.

[8]   [6]   R. N. Dows and E. J. Gough, "PVUSA procurement, acceptance, and rating practices for photovoltaic power plants", PG&E Co.Report #95-30910000.1, 1995.

[9]   Neelkanth G. Dhere, Ashwani Kaul, Bhaskar Kumar, Santosh D. Khatri, Shirish A. Pethe, "Performance Assessment of Thin-Film Photovoltaic Modules in Hot and Humid Climate of Florida", Conference Proceedings ASES 2007.

[10]   C. Jardine and K. Lane, "PV-COMPARE: Relative Performance of Photovoltaic Technologies in Northern and Southern Europe," Proceedings of the PV in Europe Conference, 2002, pp. 1057-1060.

[11]   N.G. Dhere, V. V. Hadagali and K. Jansen, "Performance degradation analysis of high voltage biased thin film PV modules in hot and humid conditions" Proc. 31st IEEE PV Specialists Conference, Orlando, 2005, pp. 507-510.

Intrinsic Reliability of Amorphous Silicon Thin Film Solar Cells

M. A. Alam, S. Dongaonkar, Karthik Y., S. Mahapatra, D. Wang[*], M. Frei[*]

Purdue University, School of Electrical and Computer Engineering
465 Northwestern Avenue, West Lafayette, Indiana 47907, USA
* Applied Materials, Santa Clara, CA
Phone: 765-494-5988 E-mail: alam@purdue.edu

1. Introduction and Motivation

Although solar cells made of single crystalline silicon (c-Si) have long dominated the market for their efficiency and reliability, there has been a growing interest in exploring other (somewhat) less efficient materials (e.g., amorphous Si or α-Si:H, CIGs, CdTe) that offer improved price/kWh, through significantly lower manufacturing costs. If the reliability challenges associated with these materials could be understood and addressed, the economic viability of these PV options would improve dramatically.

The reliability issues of α-Si PV could either be extrinsic (e.g., humidity, glass breakage, wiring fault)[1] or intrinsic (e.g., light induced degradation). In this article, we will discuss the physics and technological origin of three intrinsic reliability concerns for α-Si solar cells: (i) shunt conduction, (ii) shadow degradation and hot-spot generation, and (iii) light induced degradation. Over the years, these issues have been discussed in the literature by many groups, although the context of the discussion was often isolated, and the approach frequently empirical. Here we take a comprehensive approach to show that these reliability issues are usually not independent, but can and should be understood in a broader framework. Moreover, many of the reliability concerns of α-Si:H solar cells are shared by other low-cost, thin-film PV materials (e.g. CIGS, polymer) deposited by solution processing or chemical vapor deposition. Therefore, a good understanding of the reliability issues for a-Si solar cells will also address broader questions regarding the reliability of other thin film technologies.

2. Background Information

2. 1 Basics of PV Operation

The textbook description of the operation of a solar cell is as follows: The Sun radiates at 6000°C and the part of the '6000°C' blackbody spectrum that is not absorbed by various molecules in the atmosphere and reaches earth surface can generate electron-hole pairs within the p-n junction of a solar cell (see Fig. 1a). If the charges can be separated before they self-annihilate through recombination, the collected photo-induced carriers can deliver power to an external load. Assuming superposition is satisfied [2],

$$I_T = I_L - I_0\left(e^{(qV - IR_S)/nk_BT} - 1\right) \quad (1);$$

where I_L is rate of generation of electron-hole pairs (after accounting for various optical losses due to reflection and free carrier absorption), R_s is the series resistance, and n is the ideality factor. ($n = 1$ for classical diode, while $n = 2$ for diodes dominated by recombination in the space charge region).

For a variable load R_L, the maximum power delivered to the load is given by (see Fig. 1b for definitions)

$$P_{max} = I_{max} \times V_{max} = V_{oc} \times I_L \times FF; \quad (2)$$

where the open circuit voltage V_{oc} (as shown in Fig. 1b) is given by

$$V_{oc} = \left(\frac{kT}{nq}\right) ln\left(1 + \frac{I_L}{I_0}\right), \quad (3)$$

and the fill factor FF reflects the power lost to series resistance, R_s and other effects. (see Fig. 1b, bottom). Typically one would string 100s of such diodes in series as shown in Fig. 2 to increase the output voltage.

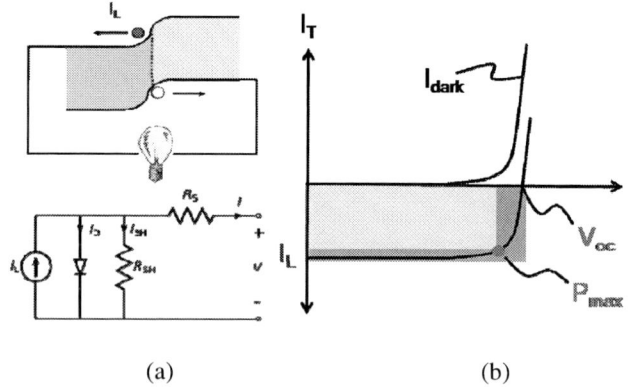

(a)   (b)

Fig. 1: (a) Classical diode as solar cell, and equivalent circuit of actual solar cell. (b) Schematic diagram showing light and dark IV characteristics. Superposition is assumed. The red-dot signifies maximum power point.

2.2 Basics of PV Reliability

Regardless of the material used for the solar cells, i.e., crystalline or amorphous, the reliability issues can be analyzed using Eq. (1). For a given value of I_L, Eq. (3) suggests that V_{oc} (and therefore the output power) decreases with increase in the ideality factor n, 'dark current' pre-factor I_0. A time dependent shift in these parameters as a consequence of operating conditions constitutes the reliability issues of a technology. Obviously, the reliability issues that influence these parameters depend on the material type and the configuration of the solar cell. As a result, the reliability concerns appear very different for various technologies; e.g., while light induced degradation is an important reliability concern for a-Si:H solar cells, the issue is less important for c-Si solar cells. The goal of this paper is to understand how various reliability issues of a-Si:H affect Eq. (2) and how they relate to the common reliability issues for other solar cell technologies.

2.3 Thin Film Technology

The fabrication and installation costs of large area macroelectronic devices like solar cells can be minimized only if they are processed with low energy input in high-efficiency equipment using inexpensive and durable substrate to reduce the use of raw materials to the minimum. While crystalline silicon requires 100s of μm thick active layer[3], thin film solar cells use only 100s of *nm* of intrinsic material sandwiched between 10-20nm thick, heavily doped p and n layers. Fluorinated Tin Oxide (SnO:F) or Indium Tin Oxide (ITO) is used as transparent front electrode, while Al acts as back electrode. The cells are connected in series using laser scribes on TCO and a-Si:H layers (see Fig. 2). This long string of series-connected cells produces a module output of ~100V at ~1A. The quality and thickness of the ITO and Al determine the optical properties and series resistance of the system.

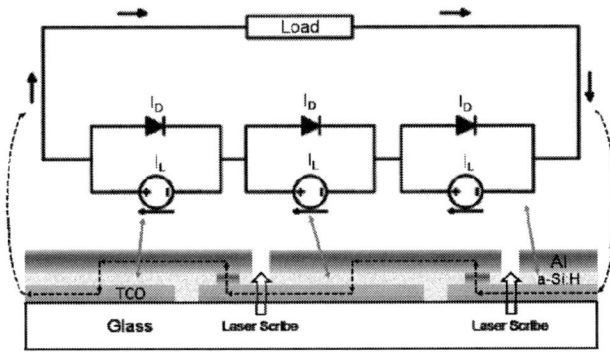

Fig. 2: Schematic diagram showing the series connected solar cells supplying a load in an equivalent circuit picture (top) and the actual device layout schematic (bottom), with the arrows indicating the current paths.

The economical use of materials in thin film technologies,

like a-Si:H or CIGS, is not due to any fundamental innovation, but rather related to the generic wavelength dependent absorption in semiconductors. Typically, a-Si:H has larger bandgap (~1.72 eV) compared to c-Si (~1.1 eV); given that absorption in silicon scales with density of states and is higher at shorter wavelengths, only a thin film of crystalline or amorphous silicon is necessary to completely absorb (E > 1.72 eV) part of the solar spectrum (discussed in more detail in Fig. 5). While the manufacturing cost is greatly reduced by low-temperature vapor phase deposition of thin-films on inexpensive glass substrate, there are some unique reliability concerns of thin film solar cells like light induced degradation, shunt leakage, etc.

3. Intrinsic Reliability Issues of a-Si:H Solar Cells

3.1 Shunt Leakage

Since the features of the dark current (I-V characteristics without sunlight shining on) dictate V_{oc} (Eq. (3)), measurement of this characteristics offers insight into ultimate device efficiency. While dark I-V of thin film PV at sufficiently large forward biases appears purely classical (with $1 < n < 2$), one of the most interesting and universal feature is an anomalous leakage component at low forward and reverse bias conditions. Although this feature has long been accounted for in equivalent circuit models with a parallel ohmic shunt resistance, R_{SH} (see Fig. 1(a), bottom), and has often been eliminated/reduced by so called 'shunt busting', the physical origin of the phenomena and the universality of leakage across various thin-film technologies have remained unclear [4].

We find that this anomalous current is defined by four characteristics; namely, current-voltage symmetry ($I_{anom}(V) = I_{anom}(-V)$); voltage nonlinearity ($I_{anom} \sim V^{\delta+1}$); temperature insensitivity ($I_{anom}(T_1) \sim I_{anom}(T_2)$), and large statistical fluctuation in the leakage current magnitude from sample to sample. These observations cannot be explained by intrinsic device properties (e.g., defect distributions), but rather lead us to attribute this anomalous leakage to generalized space-charge limited current through localized metal-semiconductor-metal structures, and described by of $I_{anom} \sim aV^{\delta+1}/L^{\delta+2}$, where $\delta \sim 1-2$ (to be published). These localized structures should not be confused with pin-holes related to inadequate deposition of Si, but are likely to arise from surface non uniformity of TCO coated glass substrates. Recent mapping of the localized light-spots by lock-in thermometry appears to support the conclusion[5]. Other experiments involving metal-a-Si-metal resistive memories are consistent with the hypothesis[6]. Indeed, this leakage must be reduced to a level so that it does not degrade V_{oc} and becomes a reliability concern for thin film technologies. Fortunately, however, shunt conductance may be reduced by improved surface planarization[7], blocking layers[8], shunt busting[4], etc. The choice of a specific approach depends on application.

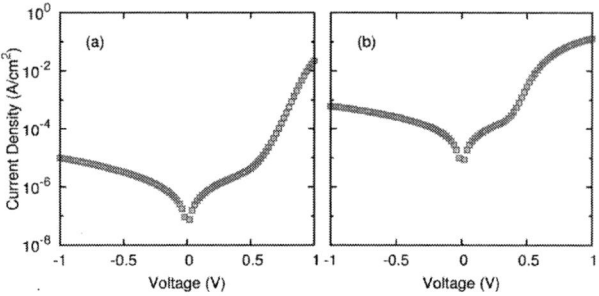

Fig. 3: Dark IV showing anomalous leakage currents in (a) a-Si:H p-i-n solar cells; (b) Organic BHJ cells from [7].

3.2 Shadow Degradation

Shadow degradation is related to the requirement that N solar cells are series connected in a module to achieve large $V_{oc}^{mod} = NV_{oc}^{cell}$. If the various structural elements of the assembly (e.g., antennas, solar paddles, booms, etc. in satellite applications) or other natural elements (e.g. clouds, leaves, trees, dirt, etc. for terrestrial applications) cast a shadow on one or several elements of the solar module, the current in the affected cells is reduced. Current continuity dictates the voltage be redistributed in such a way among the N-cells so that $V_L = (N - N_s)V_1 - N_sV_s$, where V_L is the voltage across the load, V_1 is the voltage generated across each of the $N - N_s$ illuminated cells of the array and V_s is the voltage developed across the shadowed part (N_s cells). The negative sign of V_s reflects the fact that the affected cells are reversed biased in the Zener tunneling or Avalanche breakdown mode (see Fig 4(a)). Indeed, shadowing not only eliminates the cells from power generation, but dramatically degrades the power output from the rest of the system by reducing the output voltage.

One may wonder if shadowing, like radiation induced soft errors in CMOS circuits, is a transient effect, and if the system might be restored to its pristine performance once the shadow is lifted (e.g., with change in orientation of the satellite with respect to the Sun, or cleaning of the solar panels). Indeed, most papers in the literature treat shadowing as a power management problem[9], with no discussion of long term consequences. The large reverse bias endured by the shadowed α-Si:H cells however will depassivate SiH and/or weak Si-Si bonds, create mid-gap defects, reduce recombination lifetime τ, and increase I_0. The corresponding shift in V_{oc} reflects the integrated duration of the shadowing (and the corresponding stress) endured by the cells. The change in the maximum power point would reduce power output of the system. Details of the system level effects will be discussed elsewhere.

Initial experiments with "shadow stress" demonstrate that degradation may be described by a power law, i.e., $(\Delta I(t) \sim B_s t^s)$ with $s \sim 1/4$ (see Fig. 4(b)), and the voltage acceleration is exponential (to be published). These two equations are sufficient to predict the effect of shadow degradation in large scale solar cell installations.

The problem of shadow degradation might be addressed both at device as well as circuit levels. At device level, the reduction of Zener voltage reduces reverse stress on the affected cell and reduces defect generation for a given duration of shadow. If cell redesign is difficult or expensive, circuit/system solutions are appropriate.

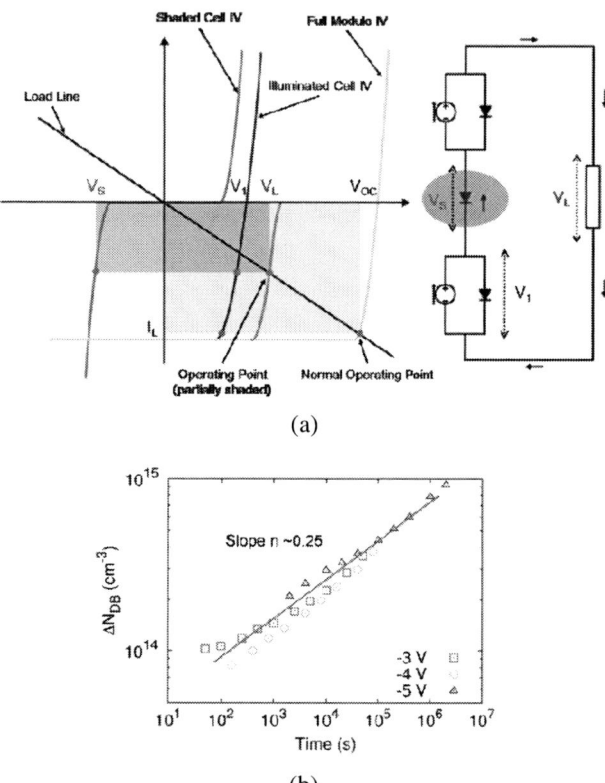

Fig. 4: (a) Schematics showing the operating points before and after shading; identifying voltages across individual cells (V_1), reverse voltage developed across shaded device (V_s), and the corresponding load voltage (V_L). (b) Power law time dependent degradation of solar cells under different reverse bias stresses that has been scaled by constant factors to highlight the power-exponent of degradation.

Recall that for SRAM memory, use of redundant arrays in post-Si phase dramatically improves yield by allowing re-mapping of defective cells (due to process related V_T fluctuation or fluctuation in for V_{DD} leading to READ or WRITE failures)[10]. Similar approaches may be appropriate for PV modules as well[11]. For example, a significant reduction in V_L (see Eq. (3)) would indicate onset of shadowing; a sweep across the elements of the module would determine the affected row; and a standby redundant array would then be switched in to bypass the affected cell. Obviously, once the shadow is lifted, the same redundant cell can be released and can be reconfigured such that they can be used by other shadowed cells. Simulations show that small degree of redundancy and management overhead result in large improvements [11].

In some cases, rapid defect generation in cells affected by intense shadows may lead to catastrophic failure of solar cells by a process called the "hot-spot" formation[12]. Hot spots may delaminate the mirrors and destroy the cells. Like 'hard breakdown' in thick gate dielectrics, hot-spot formation and propagation is likely to involve a positive feedback and interplay between temperature and current[13, 14]. Like the early literature on gate dielectric breakdown, there is considerable debate regarding whether the hot-spot is related to pre-existing defects like etch pits as discussed in Section 3.1 (thermal images seems to indicate such possibility) or intrinsic defect generation terminated by a runaway process. Further work is needed to identify the mechanism of hot-spot generation.

3.3 Light Induced Degradation

Light induced degradation (LID) involves time dependent reduction in output current of a solar cell, under solar illumination. Since the early definitive studies of LID in 1977s by Staebler and Wronski[15], this notorious reliability issue has been studied in depth; and several features are known thereof; (i) the degradation follows a power law of the form $\Delta I = G^{\frac{n}{2}} t^n$ with $n \sim 1/3$ as the power exponent [16] (see Fig. 6a), (ii) there is evidence of some recovery once light is removed, (iii) the degradation increases with light intensity, (iv) many studies suggest dissociation of Si-H complexes during light soaking, (v) degradation is reduced if H is replaced by isotope Deuterium, and finally, (vi) and most intriguingly, thin-films are susceptible to it, while the effect, although present, is considerably suppressed in crystalline silicon solar cells[17].

The last three observations appear to be contradictory; as the backplane of most c-Si solar cells involve large Si-SiO$_2$ interface passivated by SiH bonds, otherwise minority carrier recombination will increase I_0 and decrease V_{oc} and hence the efficiency. If light induced dissociation of SiH bonds creates defects in thin -Si:H solar cells, why isn't LID degradation a similar concern for c-Si solar cells? After all, in c-Si cells, light bounces many times between the electrodes for full absorption[3].

To resolve this puzzle, recall that absorption length of high energy photons are relatively small (~10s of nm) and it is low-energy, near gap, radiation that reaches the bottom

Si/SiO$_2$ interface of c-Silicon solar cells (See Fig. 5). These *low energy photons* ($hf \sim 1\,eV$) that reach the back-interface cannot dissociate SiH bonds. In contrast, the distance over which the *high energy photons* are absorbed in a-Si solar cells are replete with SiH bonds, and therefore it is hardly surprising that Si-H bonds dissociation is a concern. In short, film thickness of solar cells, bandgap of the PV material and the light-induced degradation are related by fundamental consideration and must be understood within this comprehensive framework. It is important to note that LID is not absent in c-Si, but involves dissociation processes (e.g. of B-H complexes) that can be initiated by lower energy photons.

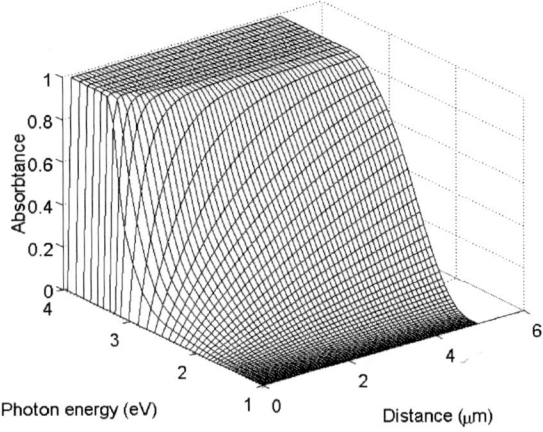

Fig. 5: Surface plot showing absorbance of crystalline silicon vs. thickness (μm) and photon energy (eV).

There have been several models proposed to explain the time dependent kinetics of LID issue. Some involve time dependent recombination of electron-hole pairs breaking weak Si-Si bonds[16], while others have explicitly attributed the defect generation to light induced dissociation of SiH bonds[18, 19]. LID continues to be an active topic of research, therefore more work is needed before a robust theory is developed. If the time-exponent of LID is robust, as several group have reported, a reaction-diffusion based model might be appropriate. Below we offer a simple derivation and numerical solution of the concept. Detailed prediction of the model needs to be explored by experiments.

Let us assume that the small wavelength photons of flux G have sufficient energy to dissociate the weak Si-H bonds (initial concentration N_0) at the rate of K_F. The broken dangling bonds N_{DB} remain fixed in space to act as recombination centers, while the freshly released atomic H diffuse laterally through the film. These two populations of mobile H and static dangling bonds interact throughout the volume of the thin film, with opportunities for repassivation (rate constant K_R). In sum, therefore, we can write the rate of defect creation as -

$$\frac{dN_{DB}}{dt} = k_F N_0 G - k_R N_{DB} N_H. \tag{4}$$

In addition to repassivation, H can react with each other and be lost from the kinetics of the problem by forming molecular H_2. Therefore the evolution of H is given by

$$\frac{dN_H}{dt} = \frac{dN_{DB}}{dt} - k_H N_H^2. \tag{5}$$

The coupled equations can be solved numerically (Fig. 6), although the following analytical solution provides additional insight. After an initial transient, the rate of change of the system can be presumed to be slow, compared to the fluxes sustaining them. Therefore, the forward dissociation and reverse repassivation of the Si-H bonds are evenly balanced as

$$k_F N_0 G \sim k_R N_{DB} N_H; \tag{6}$$

and $dN_{DB}/dt \sim N_{DB}/t$ so that $N_{DB}/t \sim k_H N_H^2$. Taken together, we find

$$N_{DB(t)} = (3k_H)^{\frac{1}{3}} \left(\frac{k_F N_{0G}}{k_R} \right)^{\frac{2}{3}} t^{\frac{1}{3}}. \tag{7}$$

Reassuringly, this theoretical result compares well with the experimental results from literature, as shown in Fig. 6. The scaling of degradation rate (observation 2 above) as a function of flux G (for concentrator PV applications, for

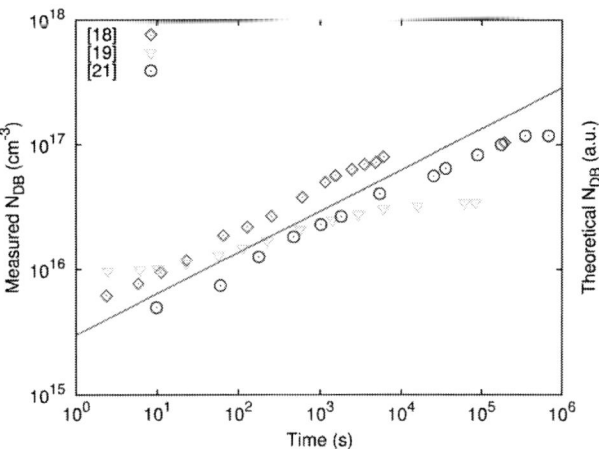

Fig. 6: The numerical solution of reaction diffusion equations anticipates the $t^{1/3}$ time dependence trends (right axis in a.u.) of measured DB density (left axis in cm^{-3}) from [18, 19, 21]. The theoretical predictions need to be confirmed with additional experiments.

example) is confirmed and once the Sun is down, relaxation is expected. The analogy of LID and NBTI is obvious[20],

and has been noted by many groups. This should motivate a NBTI like thorough study of solar cell degradation, with the full understanding that in addition to SiH bond-dissociation, other bond dissociation processes (e.g. weak Si-Si bonds, B-O complexes) may also contribute to LID.

4. Conclusion

In this paper, we have discussed three intrinsic reliability issues of thin-film -Si:H solar cells; space charge limited shunt conduction through localized metal-semiconductor-metal structures; shadow degradation in series connected cells in a module, and light induced degradation. Despite their distinct external manifestation, these intrinsic reliability issues appear to share common physical phenomena. For example, the light induced and the shadow degradation may be related because they are described by very similar time-exponents (see Fig. 4c and 6a). While the physics of G are different (e.g. photon induced dissociation for LID and (possibly) electron-hole recombination induced dissociation for shadow degradation), it is likely that they both break SiH bonds and are subsequently follow similar diffusive kinetics. Finally, analogies to CMOS reliability; e.g., shunt conduction related to non uniform conduction through oxides, shadow degradation to bulk defect generation and TDDB in gate dielectric, and light induced degradation to NBTI in PMOS transistors; may help illuminate many aspects of the degradation processes.

Acknowledgements

Fig. 5 is taken from unpublished work by Mohammad Ryyan Khan, a graduate student in Prof. Alam's group. We gratefully acknowledge Applied Materials for samples, the Birck Nanotechnology Center for characterization facilities, and the Network of Computational Nanotechnology for computational resources. The work is supported by grants from Applied Materials and Columbia EFRC.

References

[1] C. R. Osterwald and T. J. McMahon, "History of Accelerated and Qualification Testing of Terrestrial Photovoltaic Modules: A Literature Review," *PROGRESS IN PHOTOVOLTAICS*, vol. 17, pp. 11-33, Jan 2009.

[2] F. A. Lindholm, J. G. Fossum, and E. L. Burgess, "Application of the Superposition Principle to Solar-Cell Analysis," *IEEE Transactions on Electron Devices*, vol. 26, pp. 165-171, 1979.

[3] M. A. Green, J. H. Zhao, A. H. Wang, and S. R. Wenham, "Very high efficiency silicon solar cells -

Science and technology," *IEEE Transactions on Electron Devices,* vol. 46, pp. 1940-1947, Oct 1999.

[4] T. R. Johnson, G. Ganguly, G. S. Wood, and D. E. Carlson, "Investigation of the causes and variation of leakage currents in amorphous silicon p-i-n diodes," Warrendale, PA, USA, 2003, pp. 381-6.

[5] O. Kunz, J. Wong, J. Janssens, J. Bauer, O. Breitenstein, and A. G. Aberle, "Shunting Problems Due to Sub-Micron Pinholes in Evaporated Solid-Phase Crystallised Poly-Si Thin-Film Solar Cells on Glass," *Progress In Photovoltaics,* vol. 17, pp. 35-46, Jan 2009.

[6] J. W. Seo, S. J. Baik, S. J. Kang, Y. H. Hong, J.-H. Yang, L. Fang, and K. S. Lim, "Evidence of Al induced conducting filament formation in Al/amorphous silicon/Al resistive switching memory device," *Applied Physics Letters,* vol. 96, pp. 053504-3.

[7] M. D. Irwin, J. Liu, B. J. Leever, J. D. Servaites, M. C. Hersam, M. F. Durstock, and T. J. Marks, "Consequences of Anode Interfacial Layer Deletion. HCl-Treated ITO in P3HT:PCBM-Based Bulk-Heterojunction Organic Photovoltaic Devices," *Langmuir,* 2009.

[8] J. D. Hwang and C. H. Chou, "On the origin of leakage current reduction in TiO[sub 2] passivated porosus silicon Schottky-barrier diode," *Applied Physics Letters,* vol. 96, pp. 063503-3.

[9] J. Feldman, S. Singer, and A. Braunstein, "Solar-Cell Interconnections and the Shadow Problem," *Solar Energy,* vol. 26, pp. 419-428, 1981.

[10] A. Agarwal, B. C. Paul, H. Mahmoodi, A. Datta, and K. Roy, "A process-tolerant cache architecture for improved yield in nanoscale technologies," *IEEE Transactions on Very Large Scale Integration (Vlsi) Systems,* vol. 13, pp. 27-38, Jan 2005.

[11] D. Nguyen and B. Lehman, "A reconfigurable solar photovoltaic array under shadow conditions," *Apec 2008: Twenty-Third Annual Ieee Applied Power Electronics Conference and Exposition, Vols 1-4,* pp. 980-986, 2008.

[12] J. Bauer, J. M. Wagner, A. Lotnyk, H. Blumtritt, B. Lim, J. Schmidt, and O. Breitenstein, "Hot spots in multicrystalline silicon solar cells: avalanche breakdown due to etch pits," *Physica Status Solidi-Rapid Research Letters,* vol. 3, pp. 40-42, Mar 2009.

[13] M. A. Alam, B. E. Weir, and P. J. Silverman, "A study of soft and hard breakdown - Part I: Analysis of statistical percolation conductance," *IEEE Transactions on Electron Devices,* vol. 49, pp. 232-238, Feb 2002.

[14] M. A. Alam, B. E. Weir, and P. J. Silverman, "A study of soft and hard breakdown - Part II: Principles of area, thickness, and voltage scaling," *IEEE Transactions on Electron Devices,* vol. 49, pp. 239-246, Feb 2002.

[15] D. L. Staebler and C. R. Wronski, "Reversible Conductivity Changes in Discharge-Produced Amorphous Si," *Applied Physics Letters,* vol. 31, pp. 292-294, 1977.

[16] M. Stutzmann, W. B. Jackson, and C. C. Tsai, "Light-Induced Metastable Defects in Hydrogenated Amorphous-Silicon - a Systematic Study," *Physical Review B,* vol. 32, pp. 23-47, 1985.

[17] J. Schmidt, "Light-Induced Degradation in Crystalline Silicon Solar Cells," Brandenburg, Germany, 2004, pp. 187-196.

[18] H. R. Park, J. Z. Liu, and S. Wagner, "Saturation of the Light-Induced Defect Density in Hydrogenated Amorphous-Silicon," *Applied Physics Letters,* vol. 55, pp. 2658-2660, Dec 18 1989.

[19] C. Godet, "Metastable hydrogen atom trapping in hydrogenated amorphous silicon films: A microscopic model for metastable defect creation," *Philosophical Magazine B-Physics of Condensed Matter Statistical Mechanics Electronic Optical and Magnetic Properties,* vol. 77, pp. 765-777, Mar 1998.

[20] M. A. Alam, "A critical examination of the mechanics of dynamic NBTI for PMOSFETs," *2003 IEEE International Electron Devices Meeting, Technical Digest,* pp. 345-348, 2003.

[21] T. Shimizu, "Staebler-Wronski effect in hydrogenated amorphous silicon and related alloy films," *Japanese Journal of Applied Physics Part 1-Regular Papers Brief Communications & Review Papers,* vol. 43, pp. 3257-3268, Jun 2004.

Correlations of Capacitance-Voltage Hysteresis with Thin-film CdTe Solar Cell Performance During Accelerated Lifetime Testing

David S. Albin and Joseph A. del Cueto

National Renewable Energy Laboratory (NREL)
1617 Cole Boulevard, M.S. 3219
Golden, CO 80401
Phone: +1-303-384-6550, e-mail: david.albin@nrel.gov

Abstract—**In this paper we present the correlation of CdTe solar cell performance with capacitance-voltage hysteresis, defined presently as the difference in capacitance measured at zero-volt bias when collecting such data with different pre-measurement bias conditions. These correlations were obtained on CdTe cells stressed under conditions of 1-sun illumination, open-circuit bias, and an acceleration temperature of approximately 100 °C.**

Keywords- Photovoltaic, Cadmium Telluride (CdTe), Capacitance Voltage Hysteresis, Transient Ion Drift (TID), Transient Capacitance, Deep Level Transient Spectroscopy (DLTS).

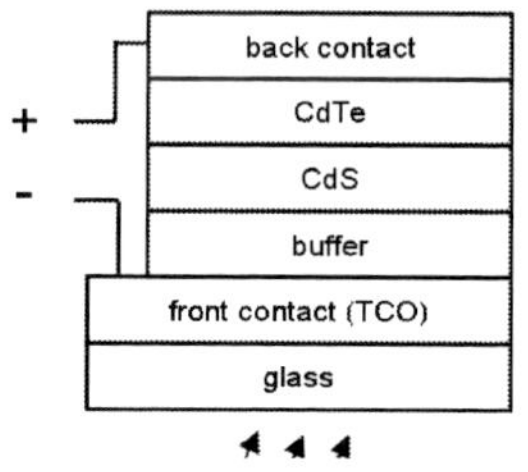

Fig. 1. Basic CdS/CdTe Solar Cell Design.

I. INTRODUCTION

Polycrystalline CdS/CdTe thin film solar cells have demonstrated small-area, laboratory efficiencies of 16.5% [1]. The highest reported efficiency for CdTe modules in [2] is 10.9% though this value is nearly a decade old. Higher performance levels for industrial products based on CdTe are likely but not openly disseminated for strategic reasons. In addition to considerable research addressing efficiency, recent work has focused on the intrinsic durability of these thin film semiconductor devices. The effects of back contact and doping strategies have been heavily researched for example [3, 4, 5]. This is driven by the inherent difficulty in fabricating ohmic contacts on wide band gap, p-type CdTe. More recently, the detrimental effects of localized shunts [6], the general effects of polycrystalline thin film micro-nonuniformities [7], and cell fabrication details revealed through factorial, design-of-experiment research [8] have been openly presented.

The basic structure of a thin film CdS/CdTe solar cell is that of a glass superstrate design in which light passes through a conducting/insulating (buffer) oxide film layer stack deposited on glass. Most laboratory and industrial cells use tin-oxide (SnO_2) for these layers, though there is considerable interest in more advanced stannate materials (Cd_2SnO_4 and $ZnSnO_x$). Once transmitted through the glass/TCO/buffer superstrate, light is then absorbed in the n-CdS/p-CdTe heterojunction structure which provides the field necessary for separating photo-generated carriers. A back contact structure completes the cell. A schematic of this basic design is shown in Figure 1.

In order to ascertain the long term reliability of modules based upon CdS/CdTe, cells of this basic design were exposed to 1-sun illumination under open-circuit, V_{oc}, bias and acceleration temperatures of 60 – 120 °C for times exceeding 1000 hours [9]. Under field-use conditions, series-connected cells nominally see voltages somewhat less than V_{oc}, thus, open-circuit conditions represent an additional form of acceleration. In this study, two dominant degradation mechanisms were identified in the temperature range studied. As shown in Fig. 2(a), between 60-80 °C, an activation energy of 2.94 eV was measured and was attributed to S-outdiffusion from the CdS layer into the CdTe based upon a reported value of 2.8 eV for bulk diffusion of S in CdTe [10]. This assertion was also based upon the observation of Kirkendall-like voiding of the CdS layer in cells that underwent stress. In the temperature range 100-120 °C, an activation energy of 0.63 eV was determined which agreed well with the reported value of 0.67 eV for Cu diffusion in CdTe [11].

Since cell efficiency, $\eta\%$, is determined by the equation:

$$\eta\% = {V_{oc} \times J_{sc} \times FF} / {\phi_{inc}} \tag{1}$$

(where Φ_{inc}, the incident power density is typically normalized to a solar value of 100 mW/cm^2), a correlation analysis $\Delta\eta\%$ versus changes in V_{oc}, short-circuit current density, J_{sc}, and fill factor, FF, during stress testing as a function of stress temperature was performed. This analysis is shown in Figure 2(b).

978-1-4244-5430-3/10 $26.00 © 2010 IEEE

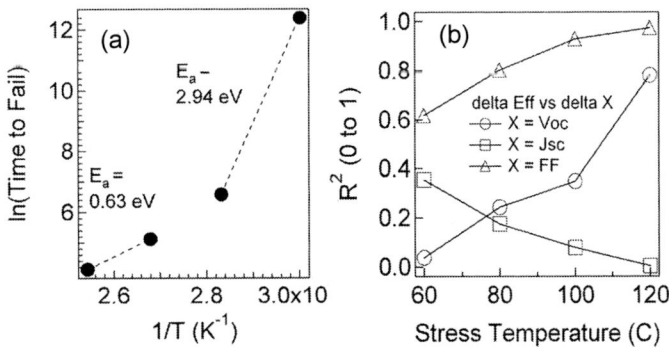

Fig. 2. Degradation activation energies (a) and correlation analysis (b) of how efficiency (Eff) changes versus changes in V_{oc}, J_{sc}, and FF.

The moderate correlation of $\eta\%$ with J_{sc} seen at lower stress temperatures is due to reduced optical attenuation associated with S-outdiffusion from the CdS. The most important variable affecting $\eta\%$ at all temperatures, approaching a near ideal correlation at 120 °C, was FF. FF represents the efficiency by which photons absorbed in a solar cell are collected by a combination of field and diffusion-limited collection mechanisms. Within the space-charge, photo-generated electrons and holes are swept towards the n and p-sides of the junction respectively by the built-in field. In the quasi-neutral region, carrier diffusion length determines whether they are collected. Recombination of carriers before collection is the greatest impediment to improving the overall performance, primarily V_{oc}, for these cells.

The fundamental current-density, J, and voltage, V, behavior of a solar cell is represented by:

$$J = J_{SCR} + J_{QNR} + \left(\frac{V - JR_s}{R_{sh}} \right) - J_{ph} \quad (2)$$

where J_{SCR}, J_{QNR}, and J_{ph} represent recombination currents in the space-charge and quasi-neutral regions of the cell, as well as the photo-generated current, while R_s and R_{sh} represent series and parallel (shunt) resistance losses. J_{SCR} and J_{QNR} are further represented by the following:

$$J_{QNR} = J_{01} \left(e^{q(V-JR_s)/kT} - 1 \right) \quad (3)$$

$$J_{SCR} = J_{02} \left(e^{q(V-JR_s)/2kT} - 1 \right) \quad (4)$$

where J_{01} and J_{02} are further dependent upon minority carrier transport properties.

In the third quadrant, maximum power output is achieved by maximizing the term, J_{ph}, and minimizing the first three terms in (2), often referred to collectively as the "forward" current. Each of the forward current terms contributes to decreased cell performance. It should be noted that the recombination currents, J_{QNR}, and J_{SCR} are themselves dependent upon resistive effects as shown in (3) and (4).

The loss parameters J_{QNR}, J_{SCR}, R_s, and R_{sh} can be determined graphically or by direct modeling with programs like Pspice. Using the latter approach, the percent contribution each parameter contributes to the forward current (and thus loss) in the power quadrant for a laboratory made 14.4% CdS/CdTe solar cell is shown in Fig. 3.

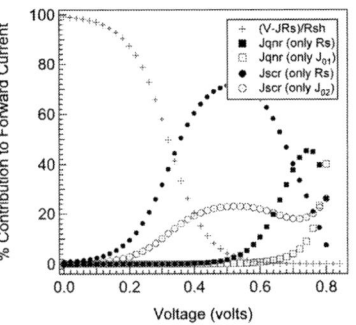

Fig. 3. Percent contribution to forward current losses modeled for a 14.4% CdS/CdTe cell.

The results shown in Fig. 3 use a model-fitted value of 3 ohms*cm^2 for R_s which is a reasonable upper value observed during stress testing of these cells [9]. As seen in this figure, recombination occurs mostly in the space-charge except near V_{oc} where recombination in the quasi-neutral region, i.e., between the depletion width and back contact begins to dominate. Note that resistive contributions for both effectively go to zero at V_{oc} where J = 0 in equations (3) and (4).

The effect of addressing recombination within the quasi-neutral region is shown in Fig. 4.

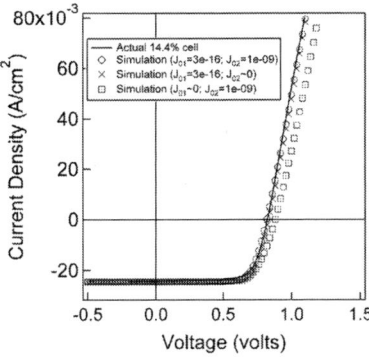

Fig. 4. The effect of reducing quasi-neutral recombination on the V_{oc} of CdTe solar cells (all J terms given in A/cm^2).

Shown in this figure is the Pspice model simulation used to extract the loss mechanisms shown in Fig. 3 along with actual cell data in a conventional J-V diagram. In this case J_{01} and J_{02} equal 3e^{-16} and 1e^{-09} A/cm^2 respectively. Having obtained a good fit, the model is then perturbed by simulating conditions where recombination is removed in either the space-charge (J_{02} ~ 0) or quasi-neutral (J_{01} ~ 0) regions. As can be deduced from Fig. 4, improving CdTe material quality within the depletion width (space-charge) does not improve cell performance. However, an additional 60 mV (0.82 to 0.88) can result if recombination in the quasi-neutral region is reduced.

The results of this fundamental calculation make understanding the depletion width position important when determining why FF changes during stress testing of cells. With this goal in mind, a new technique for collecting capacitance-voltage (C-V) data quickly during accelerated stress testing was developed [12]. This technique collects capacitance data first in a reverse direction (rev) voltage scan followed by a subsequent, forward direction (fwd) scan. This approach yields two distinct C-V curves as shown in Fig. 5 for representative cells with and without Cu added intentionally as a dopant during back contact fabrication.

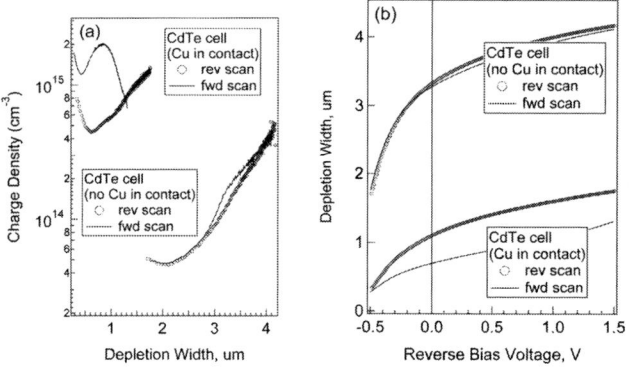

Fig. 5. Mott-Schottky plot (a) and corresponding depletion width vs. bias diagram (b) for CdS/CdTe cells with and without intentional Cu.

Fig. 5(b) clearly shows how Cu affects the space-charge depletion width, W_d. The strong decrease in W_d with Cu reflects an obvious increase in ionized acceptors, N_a^- in the CdTe possibly as Cu_{Cd} or as a paired, defect complex [13]. The decrease in W_d (since it can be so easily explained by doping) is not surprising. What is less obvious is the large degree of hysteresis associated with the introduction of Cu. When Cu is not intentionally added (Cu is well-known to be a naturally occurring trace impurity in CdTe for instance), we see little indication of hysteresis within our experimental error. This has been confirmed in every cell made in which Cu was intentionally absent. The presence of Cu introduces significant hysteresis in which W_d measured during the second (fwd) scan is always lower than the value of W_d determined during the first (rev) scan. Hysteresis in C-V measurements on polycrystalline thin film cells is also regularly reported by others [14, 15] in which the authors attribute this to the presence of deep states. The basis for this possibility is explained in more detail in [12].

Recently however, capacitance transients were used by Enzenroth, et al. [16] and Lyubomirsky, et al. [17] to determine the diffusion parameters of mobile Cu_i^+ ions in CdTe and CuInSe$_2$ cells and materials based upon a transient ion drift (TID) method first developed by Heiser and Mesli [18]. In particular, Enzenroth used the TID approach to quantify an increase in mobile Cu_i^+ as a function of increased Cu added during cell fabrication. The presence of mobile charge, in particular, Cu_i^+, is thus plausible as an explanation for the effects shown in Fig. 5.

II. EXPERIMENTAL DESIGN

Two sets of CdS/CdTe cells were fabricated in which the only difference between sets were the TCO/buffer structure used. In one set, a conventional bi-layer SnO$_2$ structure consisting of undoped (insulating) and doped (conducting) SnO$_2$/Corning 7059 borosilicate glass superstrates were used to grow CdS/CdTe devices. SnO$_2$ layers were grown by chemical vapor deposition (CVD) of tetramethyltin with bromotrifluoromethane (CBrF$_3$) added when F-doping was required. Cells using these superstrate structures will be referred to as cSnO$_2$/iSnO$_2$ cells. In the other set, cadmium (CTO) and zinc (ZTO) stannate materials were used as the conducting and buffer layer oxides. Stannate superstrate structures were fabricated by sputtering CTO and ZTO onto unheated borosilicate glass substrates with subsequent 650 $^{\circ}$C anneals in He used to obtain the best optical and electrical properties. CdS and CdTe layers were deposited by chemical bath deposition (CBD) and close-spaced sublimation (CSS) respectively. Both sets used Cu-doped graphite prior to Ag paste metallization. Further details regarding the fabrication of cells can be found in references [1] and [19].

Performance data using standard J-V scans were made on cells after fabrication and during subsequent stress testing with a current-calibrated Oriel solar simulator. C-V measurements were performed in the dark at room temperature using an Agilent 4294A Precision Impedance Analyzer operated manually at 100 kHz with a 50 mV oscillation voltage. Capacitance data was collected by scanning voltage in two directions. Immediately upon applying a voltage of +0.5 V forward bias, capacitance was measured as voltage was quickly swept (~3 s) in a reverse (rev) direction to -1.5 V where bias was maintained for exactly 5 m. During the subsequent forward (fwd) sweep back to +0.5 V, capacitance data was again collected.

For stress testing, cells were placed glass-side up, under an Atlas CPS+ solar light source (~AM 1.5; 1-sun) in machined Al blocks designed to keep the cells at V_{oc} bias. Cell temperature was set at 100 $^{\circ}$C. At times equal to 1, 4.4, 10, 28, 73, and 115 hrs cells were removed and allowed to relax in the dark for 12-24 hrs. After measurements of J-V and C-V, cells were again placed under stress. Some problems with temperature control were encountered during this test. It is very likely that actual stress temperatures exceeded 100 $^{\circ}$C though the design of the Al blocks insured that all cells were at identical temperature.

III. RESULTS AND DISCUSSION

The performance of cells, both initially as well as during stress testing are discussed in [12]. The uniformity of initial performance in cells grown on the SnO$_2$-based superstrates was very good. The highest performance achieved with this substrate was V_{oc} = 0.832, J_{sc} = 23.2, FF = 71.8, and η% = 13.8. In contrast, considerable variation in performance was observed when using the CTO/ZTO superstrates. Some cells exhibited very high R_s due to cracking of the CTO/ZTO layers.

However, the best performance ($V_{oc} = 0.827$, $J_{sc} = 24.6$, FF = 71.1, $\eta\% = 14.5$) was obtained using the CTO/ZTO substrate.

CTO/ZTO cell durability was also inferior to the $cSnO_2/iSnO_2$ cells. This has been a somewhat consistent observation when testing CTO/ZTO cells fabricated at NREL. The durability of CTO/ZTO cells is determined however by both TCO and cell processing conditions and some discussion of this is presented in [12].

Of interest to this paper was the correlation observed between C-V hysteresis and cell performance during stress. Hysteresis is defined as the difference in W_d at $V = 0$ between reverse and forward direction scans, i.e., $W_{d,rev} - W_{d,fwd}$. Fig. 6 summarizes the variation of hysteresis with stress time as well as the correlation between V_{oc} and FF with hysteresis.

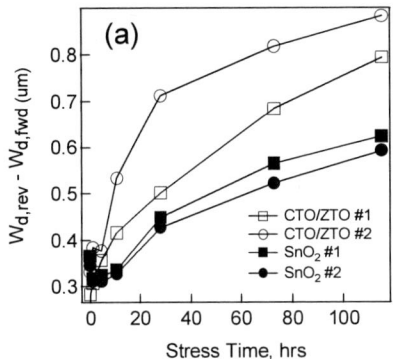

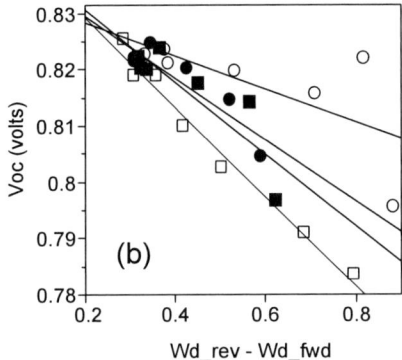

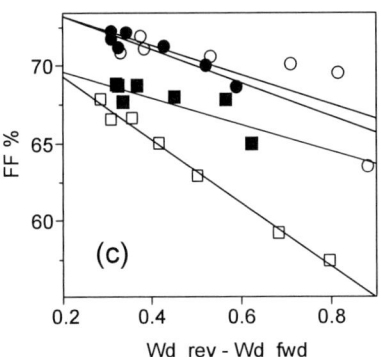

Fig. 6. Correlations during CdS/CdTe cell stress testing: (a) hysteresis vs. stress time, (b) V_{oc} vs. hysteresis, and (c) FF vs. hysteresis.

During stress, hysteresis was observed to increase as cell performance decreased. This increase with stress is shown in Fig. 6(a). The correlation of hysteresis with both V_{oc} and FF (shown in Fig. 6(b) and 6(c) respectively) was also apparent. The correlation coefficient, R^2 of V_{oc} with hysteresis for CTO/ZTO cells #1 and #2, and SnO_2 cells #1 and #2 were 0.98, 0.46, 0.75, and 0.82. The same values for FF were 0.99, 0.58, 0.63, and 0.87. Similar correlations with either $W_{d,fwd}$ or $W_{d,rev}$ were not nearly as good and did not show the monotonic behavior shown in Fig. 6(b) and 6(c).

The origin of capacitance hysteresis during stress is not presently clear. Recent TID research suggesting an ionic basis is further supported by the early work by Snow, et al, who used hysteresis to measure ionic transport in insulting films [20]. In the context of this experiment, an ionic explanation would infer an important result. Since cells fabricated using the different TCO structures used identical Cu-doped back contacts, then the additional hysteresis shown in Fig. 6(a) for CTO/ZTO cells must be associated with the introduction of additional ions. In this case, the likely source would be from either the CTO or ZTO layers. C-V measurements are capable of detecting changes in charge below 10^{15} cm^{-3} and thus, this technique might be a useful way to measure ionic changes in the space-charge of cells in a non-destructive, and easily implementable way during stress testing. For example, the durability, in particular, chemical reactivity of new TCOs and buffers could be evaluated in such a fashion prior to costly scale-up of such structures in module manufacturing. Similarly, the effectiveness of various diffusion barriers to mitigate Na$^+$ diffusion from soda-lime glasses might also be evaluated in such a fashion.

Regardless, an understanding of how best to interpret pre-bias dependent determinations of either $W_{d,fwd}$ or $W_{d,rev}$ are important since again, this is a key metric in understanding FF. The correlation of $\eta\%$ with FF is extremely high for both initial performance as well as performance during stress testing.

IV. SUMMARY

An easy-to-implement, quick, and non-destructive technique was demonstrated for obtaining capacitance hysteresis measurements in thin-film polycrystalline solar cells. Initial results on CdS/CdTe solar cells stressed at approximately 100 °C show a strong correlation between capacitance hysteresis (defined in this study as the difference in depletion width at V=0) with cell V_{oc} and FF was shown. The origin of this hysteresis, whether electronic or ionic, is unclear though past work suggests an ionic nature. The technique shows potential for being an important diagnostic tool for understanding why FF in these cells change with stress.

ACKNOWLEDGMENT

The authors thank Drs. Jian Li, Tim Gessert, and Xiaonan Li of NREL for discussions concerning C-V measurements, CTO/ZTO, and SnO_2 respectively. We acknowledge the support of the U.S. Department of Energy under Contract No. DOE-AC36-08G028308 with NREL.

REFERENCES

[1] X. Wu, "High efficiency polycrystalline CdTe thin-film solar cells," Solar Energy 77, 2004, pp. 803-814.

[2] M. A. Green, K. Emery, Y. Hishikawa, and W. Warta, "Solar Cell Efficiency Tables (Version 34)," Prog. Photovolt: Res. Appl., 17, 2009, pp. 320-326.

[3] K. Dobson, I. Visoly-Fisher, G. Hodes, and D. Cahen, "Stabilizing CdTe/CdS Solar Cells with Cu-Containing Contacts to p-CdTe," Advanced Materials, 13 (9), 2001, pp. 1495-1499.

[4] C. Corwine, A. Pudov, M. Gloeckler, S. Demtsu, and J. Sites, "Copper inclusion and migration from the back contact in CdTe solar cells, " Sol. Energy Mater. Sol. Cells, 82, 2004, pp. 481-489.

[5] D. Albin, D. Levi, S. Asher, A. Balcioglu, "Precontact surface chemistry effects on CdS/CdTe solar cell performance and stability," Proc. 28th IEEE Photovoltaics Specialists Conference, New York, 2000, pp. 583-586.

[6] T. McMahon, T.J. Berniard, and D.S. Albin, "Nonlinear shunt paths in thin-film CdTe solar cells," J. Appl. Phys., 97, 2005, pg. 054503.

[7] V.G. Karpov, A.D. Compaan, and D. Shvydka, "Effects of nonuniformity in thin-film photovoltaics," Appl. Phys. Lett., 80 (22), 2005, pp. 4256-4258.

[8] D.S. Albin, S.H. Demtsu, and T.J. McMahon, "Film thickness and chemical processing effects on the stability of cadmium telluride solar cells," Thin Solid Films, 515, 2006, pp. 2659-2668.

[9] D.S. Albin, "Accelerated Stress Testing and Diagnostic Analysis of Degradation in CdTe Solar Cells," in Reliability of Photovoltiac Cells, Modules, Components, and Systems, edited by Neelkanth G. Dhere, Proceedings of SPIE Vol. 7048 (SPIE, Bellingham, WA, 2008) 70480N.

[10] B. McCandless, M. Engelmann, and R. Birkmire, "Interdiffusion of CdS/CdTe thin films: Modeling x-ray diffraction line profiles," J. Appl Phys. 89(2), 2001, pp. 988-994.

[11] H. Woodbury and M. Aven, "Some Diffusion and Solubility Measurements of Cu in CdTe," J. Appl. Phys. 39(12), 1968, pp. 5485-5488.

[12] D.S. Albin, R.G. Dhere, S.C. Glynn, J.A. del Cueto, and W.K. Metzger, "Degradation and Capacitance-Voltage Hysteresis in CdTe Devices," in Reliability of Photovoltaic Cells, Modules, Components, and Systems II, edited by Neelkanth G. Dhere, Proceedings of SPIE Vol. 7412 (SPIE, Bellingham, WA, 2009) 74120I.

[13] S. Wei and S.B. Zhang, "Chemical trends of defect formation and doping limit in II-VI semiconductors: The case of CdTe," Phy. Rev. B 66, 2002, pp. 15521.

[14] M. Wimbor, A. Romeo, and M. Igalson, "Electrical characterization of CdTe/CdS photovoltaic devices," Opto-Electronics Review, 8(4), 2000, pp. 375-377.

[15] S. Hegedus and W.N. Shafarman, "Thin-Film Solar Cells: Device Measurements and Analysis," Prog. Photovolt: Res. Appl., 12, 2004, pp. 155-176.

[16] R. A. Enzenroth, K. L. Barth, and W.S. Sampath, "Transient Ion Drift Measurements of Polycrystalline CdTe PV Devices," 4th IEEE World Conference Photovoltaic Energy Conversion, Hawaii, 2006, pp. 449-452.

[17] I. Lyubomirsky, M.K. Rabinal, and D. Cahen, "Room-temperature detection of mobile impurities in compound semiconductors by transient ion drift," J. Appl. Phys. 81(10), 1997, pp. 6648-6691.

[18] T. Heiser and A. Mesli, "Determination of the Copper Diffusion Coefficient in Silicon from Transiet Ion-Drift," Appl. Phys. A 57, 1993, pp. 325-328.

[19] D. Rose et al., "Fabrication Procedures and Process Sensitivities for CdS/CdTe Solar Cells," Prog. Photovolt: Res. Appl. 7, 1999, 331-340.

[20] E.H. Snow, A.S. Grove, B.E. Deal, and C.T. Sah, "Ion Transport Phenomena in Insulating Films," J. Appl Phys. 36, 1965, pp. 1664-1673.

Solar Cell Interface Stability Probed by Charge Extraction

R. L. Graham, C. E. France, S. A. Carter and G. B. Alers

Physics Department
University of California
Santa Cruz, CA 95064, USA
phone: (831)459-3657, RLGraham@ucsc.edu, GAlers@ucsc.edu

Abstract— **Electrical properties of CdTe Schottky solar cells were investigated during exposure to air. Trap states were probed and characterized using charge extraction. Mechanisms for degradation are discussed, and it is argued that degradation in this case is due to instability at the interface between CdTe and the metal electrode.**

Keywords- photovoltaic, thin film, CdTe, solar cell, reliability, encapsulation

I. INTRODUCTION

The reliability of solar cells made with a cadmium telluride (CdTe) absorber layer and a Schottky junction was investigated to determine the effects of degradation in air due to an oxygen-rich environment. Electrical properties of solar cell devices with and without encapsulation were studied and compared to the behaviors of an ideal solar cell.

A good solar cell requires efficiency in the absorption of all incoming light, the excitation of valence electrons across the energy band gap, and the transport of all excited charges between the conduction band and the electrodes. CdTe has been known as a good absorber material since the 1950's because it has a direct band gap of 1.5 electron volts (eV), and therefore utilizes much of the spectrum of light coming from the sun. The excitation of an electron across the energy gap to the conduction band leaves behind a vacancy (hole) in the valence band; this electron-hole pair are collectively known as an exciton. Before the electron and hole recombine and return to the ground state, the exciton pair must be separated and the electron and hole must be transported to opposite electrodes. The extracted electrons produce the output current desired from the solar cell.

Efficient charge transport can be impeded by defects states, or traps, between the valence and conduction band. A trap slows down transport to the electrode, such that the electron and hole are more likely to recombine than to conduct to the electrodes. Traps increase the recombination rate of excitons, and therefore decrease the external quantum efficiency (EQE) and the overall output current. EQE is determined by the number of incoming photons of light versus the number of outgoing electrons. Trap states near or at the interfaces at the front and back contact can be current-limiting by impeding the transport of electrons and holes.

Thin film CdTe solar cells are a good candidate for making an efficient and low cost solar panel because CdTe has high optical absorption and only needs to be 300 nanometers (nm) thick to absorb all the incoming light above the 1.5 eV band gap. Many questions remain, however, about the proper functioning of the cell and how the interfacial interactions affect the current-voltage characteristics. In this paper we use a simple charge extraction method [1] to probe electronic trap states at the Al/CdTe interface in a Schottky junction solar cell. The Schottky junction CdTe solar cell is formed with a CdTe absorber layer deposited onto a transparent front-electrode, which is annealed in an oxygen atmosphere, and then an Al back contact is thermally evaporated onto the CdTe. Prior to depositing the metal back-electrode, the CdTe is stable and will not degrade due to the presence of oxygen. After deposition of the back-electrode, however, the electrical properties of the CdTe solar cell begin to degrade significantly upon exposure to air.

II. FABRICATION OF THE CdTe SOLAR CELL

Solar cells were produced using a patterned indium-tin-oxide (ITO) transparent electrode on a glass substrate. The CdTe absorber layer was formed on the ITO with a spin-cast CdTe nanoparticle solution, followed by a coating of cadmium chloride ($CdCl_2$). The $CdCl_2$ treatment is well-documented [2] and facilitates grain growth in the CdTe film. The nanoparticle film is then sintered at 400 °C to form a uniform thin film ~200 nm thick [3]. Efficiencies as high as 5% have been achieved with this very simple solution-based fabrication technique. Aluminum electrodes were deposited by evaporation to form the back-contacts. Normally, air sensitive devices would be encapsulated to prevent the ingress of water and oxygen. In this case a worst case condition was studied, with one solar cell device having no encapsulation. A second device, fabricated in an identical way and at the same time as the first device, went through a final fabrication step where the device was encapsulated with epoxy and covered with glass, leaving only the edge of the electrodes exposed. The final device structures are shown in Fig. 1. The final evaporation of Al, encapsulation of the second device, and the initial electrical testing of both devices were performed in an oxygen-free nitrogen-purged glove box.

978-1-4244-5430-3/10 $26.00 © 2010 IEEE

Device 1

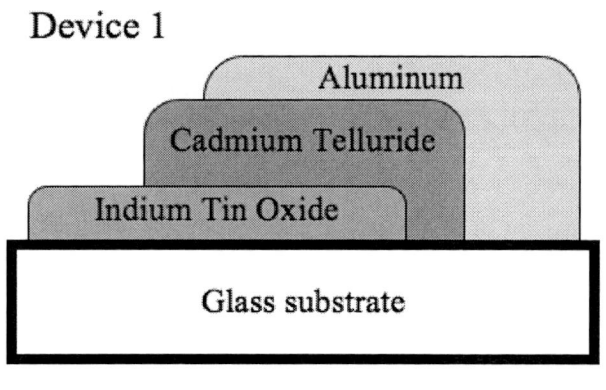

Device 2

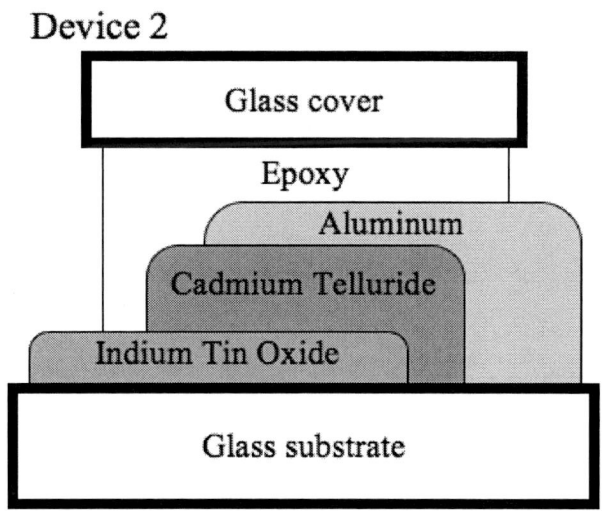

Figure 1. Structure information for each device. Device 1 is unencapsulated, and device 2 is encapsulated with epoxy and glass.

III. ELECTRICAL PERFORMANCE DURING DEGRADATIONS

Initial observation occurred before the completed devices were exposed to air. Then the solar cells were exposed to air, and electrical properties were measured at equal time intervals as the cells were in the continuous presence of oxygen. Fig. 2 shows current and voltage characteristics under light bias during the first several hours of degradation. The short-circuit current (at zero voltage) of Device 1 had completely degraded within 80 minutes of exposure to oxygen-rich air, while the short-circuit current of Device 2 showed only slight degradation during observation for over 600 minutes.

The active region is defined as the region where current density is negative and voltage is positive. The voltage in the active region remained steady throughout the degradation, which suggests that the band gap remained continuous and that no significant change was occurring within the CdTe absorber layer. The current in the active region decreased during degradation, however, which suggests an increase in the series resistance of the devices.

There are two types of resistance in a solar cell, a series resistance in series with the device and a shunt resistance in parallel with the device. An ideal solar cell has zero series resistance and has infinite shunt resistance. The series

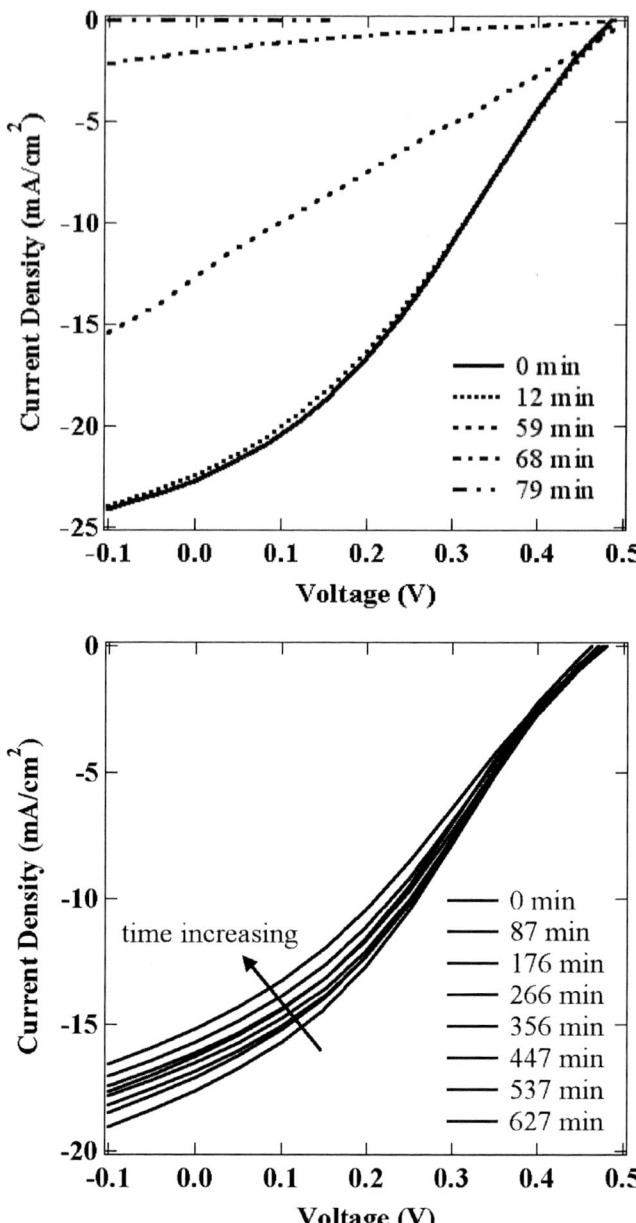

Figure 2. Current density versus applied voltage for Device 1 (top) and Device 2 (bottom). Note different time scales. The devices are exposed to air starting at *time* = 0.

resistance of the cell was calculated from the slope of the current versus voltage at zero current for each measurement under light bias. Fig. 3 shows that the series resistance in Device 1 increased several orders of magnitude, while the series resistance for Device 2 increased by less than 20%. The shunt resistance was calculated from the slope of the output current of the cell in response to a reverse voltage bias with no light illumination, and it did not change significantly for either device throughout the observation.

978-1-4244-5430-3/10 $26.00 © 2010 IEEE

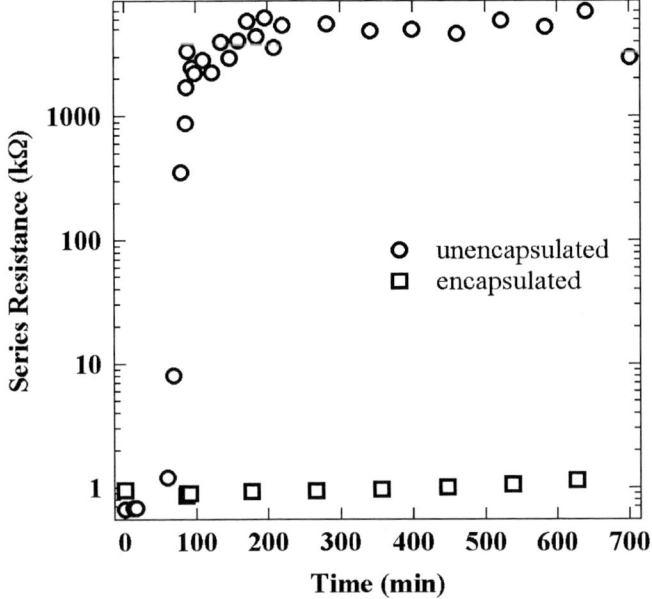

Figure 3. Series resistance versus time for each device. The devices are exposed to air starting at *time* = 0.

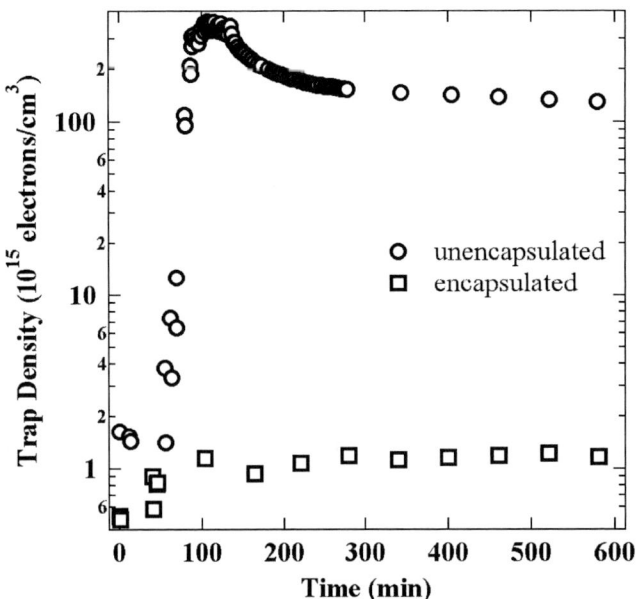

Figure 5. Density of trapped charges versus time for each device. The devices are exposed to air starting at *time* = 0.

IV. CHARGE EXTRACTION

When a solar cell is biased with voltage and without illumination, a dark current will flow that populates the trap states within the CdTe layer and at the interfaces between CdTe and the electrodes. The trapped charges generate an internal electric field that opposes the field due to the externally applied voltage. When the voltage bias is switched off, then the charges are no longer trapped and can flow freely out of the system.

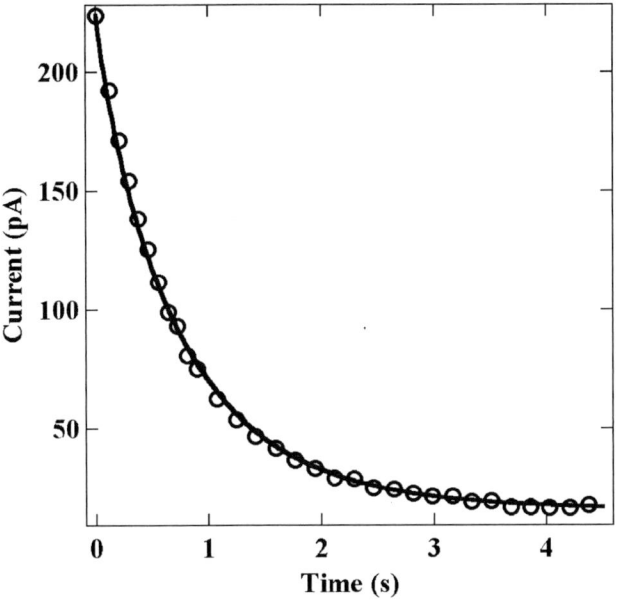

Figure 4. Example of current transient in response (markers) to switching off the applied voltage, compared to the stretched exponential in (1) with $\beta = 0.9$, $\tau = 0.72$, $I_0 = 2.063 \times 10^{-10}$A, and $I_{OFFSET} = 1.58 \times 10^{-11}$A. The applied voltage is switched off, and the cell is put into short circuit conditions, at *time* = 0.

For these experiments, the cell was biased in the dark with a voltage near the open circuit voltage, and then it was quickly switched to short circuit conditions. The switching time is defined as *time* = 0 for that charge extraction experiment, and the corresponding current output is shown in Fig. 4. Starting at *time* = 0, a transient current is observed that decays over a time period of seconds due to the emission of trapped charges. The time scale for the RC decay in this cell is less than 1 ms, which is much faster than the decay observed due to the de-population of traps.

The only mechanism that can produce current on these time scales is deep level traps. The time dependence of the relaxation current has a stretched exponential form,

$$I(t) = I_0 \exp[-(t/\tau)^\beta] + I_{OFFSET}. \qquad (1)$$

Fig. 4 compares the current transient from a charge extraction experiment to a stretched exponential with $\beta = 0.9$, $\tau = 0.72$, $I_0 = 2.063 \times 10^{-10}$A, and $I_{OFFSET} = 1.58 \times 10^{-11}$A. This is similar to the transient response observed with amorphous silicon [4]. Integration over time of the transient current yields a value of the total charge that was trapped in the defects of the cell, and from that value the density of traps (trapped charges per volume) can be found.

Fig. 5 shows the density of trap states over time, beginning with the cell's introduction to air. During the degradation process for Device 1, the density of trap states, and thus the total number of defects in the cell, increased several orders of magnitude larger than its initial value. During the degradation process for Device 2, the density of trap states doubled from the initial measurement. The trap density is normalized by the full film thickness of 200 nm.

V. DISCUSSION

The increase in trap states correlates with the increase in series resistance for the cell. We believe that this increase in trap states is due to the formation of a poor-quality oxide layer at the CdTe/Al interface. Singh, *et al.* [5], modeled the degradation of a CdS-CdTe solar cell by assuming that oxygen diffuses through the aluminum and reacts with excess tellurium at the CdTe-aluminum interface, which produces an insulating $CdTeO_3$ layer between the absorber layer and the electrode, and they found correlation between this model and their experimental results. As the thickness of an oxide layer increases, the number of trap states also increases. An insulating oxide barrier between the CdTe and the aluminum impedes transport of the charges and increases the series resistance of the cell, which is consistent with the results observed in this study. The diffused oxygen would continue to react until all the excess tellurium has reacted. Based on the data for Device 1 in Figures 3 and 5, it is likely that the excess tellurium was used up between 0 minutes and about 100 minutes. After 100 minutes, the series resistance leveled off to a fairly constant value, suggesting that the thickness of the oxide layer was no longer increasing.

The trap density in Device 1 increased several orders of magnitude between 0 minutes and 100 minutes, and then it began to decrease and leveled off at about 1/3 of the maximum value for the rest of the measurements. The series resistance in Device 1 also increased several orders of magnitude between 0 minutes and 100 minutes, and then it remained steady for the rest of the measurements. These trap density and resistance behaviors are consistent with the oxide formation during the first 100 minutes and annealing of the oxide layer during the remaining observation. As the oxide layer anneals, grains grow and defects (trap states) are removed. The oxide layer is so much more resistive than the aluminum, leaving the series resistance unchanged by the annealing.

VI. CONCLUSION

Comparison of Devices 1 and 2 shows the benefit of the encapsulation. Trap density and series resistance in Device 1 increased several orders of magnitude from their initial values, while the trap density in Device 2 doubled and the series resistance in Device 2 increased by 20%. The encapsulation of Device 2 was effective, but not perfect. Electrical measurements required access to the electrodes, which meant the edges of those electrodes were exposed to air. The oxygen could still diffuse through the aluminum in the encapsulated device, though the diffusion distance was increased by many orders of magnitude (from about 50 nm to a few mm).

An unencapsulated CdTe Schottky solar cell was shown to be unstable during exposure to air. The density of trap states was seen to increase significantly during the degradation process, indicating that the total number of defects in the cell increased due to the presence of oxygen in the air. The series resistance also increased significantly, suggesting the formation of an oxide layer at the CdTe interface due to oxygen diffusion through the aluminum. The overall degradation of the unencapsulated device is likely due to instability at the interface between the CdTe and the aluminum electrode.

ACKNOWLEDGMENT

R.L.G. thanks UCSC and GAANN for financial support. R.L.G. and G.B.A. thank Steve Weinzierl and Orpheous Nelson of Keithley Instruments for their technical support.

REFERENCES

[1] B. C. O'Regan and J. R. Durrant, "Measuring charge transport from transient photovoltage rise times. A new tool to investigate electron transport in nanoparticle films," J. Phys. Chem. B 110, 8544, 2006.

[2] B. E. McCandless, L. V. Moulton, and R. W. Birkmire, "Recrystallization and sulfur diffusion in CdCl₂-treated CdTe/CdS films," Prog. Photovolt. Res. Appl. 5, 249–260, 1997.

[3] I. E. Anderson, A. J. Breeze, J. D. Olson, L. Yang, Y. Sahoo, and S. A. Carter, "All-inorganic spin-cast nanoparticle solar cells with nonselective electrodes," Appl. Phys. Lett. 94 (6), 063101, 2009.

[4] R. I. Hornsey, K. Aflatooni, and A. Nathan, "Reverse current transient behavior in amorphous silicon Schottky diodes at low biases," Appl. Phys. Lett. 70 (24), 3260, 1997.

[5] V. P. Singh, O. M. Erickson, and J. H. Chao, "Analysis of contact degradation at the CdTe-electrode interface in thin film CdTe-CdS solar cells," J. Appl. Phys. 78 (7), 4538, 1995.

Reliability Aspects of Organic Light Emitting Diodes

Thomas Riedl
Institute of Electronic Devices
University of Wuppertal
Wuppertal, Germany
+49–202- 439-1965, t.riedl@uni-wuppertal.de

Thomas Winkler, Hans Schmidt, Jens Meyer*, Daniel Schneidenbach, Hans-Hermann Johannes, Wolfgang Kowalsky
Institut für Hochfrequenztechnik
Technische Universität Braunschweig
Braunschweig, Germany

Thomas Weimann, Peter Hinze
Physikalisch Technische Bundesanstalt
Braunschweig, Germany

Abstract—Various functional elements in organic optoelectronics devices are sensitive to oxygen an moisture. Thus, without an encapsulation the lifetime of organic light emitting diodes or organic solar cells does not meet the requirements of a serious application. We will discuss the particular challenges to form reliable, dense, pin-hole free thin-film barriers on top of organic devices. Specifically, atomic layer deposition (ALD) will be shown to be a highly attractive technique, that allows to operate at temperatures below 100 °C. Particularly, the use of nanolaminates - multilayer structures of two alternating oxide materials - have evolved as a promising candidate to seal organic devices with a powerful moisture barrier. We will discuss the preparation technology of these barrier layers, the characterization of their gas permeation rate as well as their performance in real OLED devices. With a proper ALD barrier, the lifetime of an OLED operated at 1 000 cd/m^2 can be increased to be well in excess of 20 000 h. For top-emitting or transparent OLEDs, the optical properties of the thin-film encapsulation layer can be used to concomitantly tune the light extraction efficiency of the devices.

Keywords-organic light emitting diodes, encapsulation, atomic layer deposition, nanolaminate, top-emitter

I. INTRODUCTION

The field of organic optoelectronics has matured significantly in recent years. First products, like displays based on organic light emitting diodes (OLEDs) have already entered the marked. Owing to their outstanding quantum efficiency, OLEDs enable highly efficient displays, which e.g. allow for extended battery lifetimes in mobile applications. For a wide range of specialized applications, transparent OLED displays are considered as a unique opportunity for novel human-machine interfaces. A further branch of application will use OLEDs for ambient lighting. With a share of 19% of global energy consumption by general lighting, OLEDs as highly efficient luminaire seed

the prospect for significant savings in the near future. As a result, OLEDs are expected to enter a multi-billion dollar market. While there is a continuous effort to improve efficiency and to lower manufacturing cost, at the same time, reliability of the devices and systems will become a topic of paramount importance.

II. THIN-FILM BARRIER LAYERS

A. General Requirements

Gas diffusion barriers are important for various applications ranging food-packing to flat panel displays or solar cells grown on plastic substrates. Owing to the sensitivity of many organic semiconductors to oxygen and moisture, OLEDs state a particularly demanding application [1]. Moreover, low function electrode materials (e.g. Al) or electron injection layers based on Li, Cs, etc. which are often used in OLED structures add to the sensitivity of the devices against ambient gases [2, 3]. Without encapsulation a rapid degradation and a severely limited lifetime of the devices has been reported [4]. A very commonly quoted figure for upper limits of the water vapor transmission rate (WVTR) or a particular barrier layer in order to reach a minimum OLED lifetime of 10 000 h is 10^{-6} g/(m^2 day). This value is based on an estimate of the amount of water required to corrode the reactive cathode material in a typical OLED [5]. Aside from water transmission, for the oxygen transmission rate (OTR) typical values to reach similar OLED lifetimes have been reported in the region of 10^{-5} cm^3/(m^2 day) - 10^{-3} cm^3/(m^2 day) [6].

B. State of the Art Techniques

As of yet, organic devices are often encapsulated with a glass lid which is applied under inert atmosphere.

---

* Present address: Department of Electrical Engineering,
Princeton University,
Princeton, NJ, USA

Additionally some getter material is included to minimize the residual moisture in the cavity between organic device. This approach suffers from substantial costs and its incompatibility with flexible or transparent devices. Therefore, thin-film encapsulation strategies using metal oxide or nitride layers such as Al_2O_3, SiO_2, TiO_2, or SiN are vigorously pursued.

Early work on thin-film permeation barriers was based on single layer systems prepared by plasma enhanced chemical vapor deposition (PECVD) or sputtering. Poor barrier properties with a WVTR of about 0.3 g/(m^2 day) have to be attributed to structural imperfections in the films or limited chemical stability [7]. As an improvement, multilayer barriers have been proposed by various groups [8, 9]. Specifically, Yan et al. have reported on graded multi-layers consisting of SiO_xC_y and SiO_xN_y, prepared by PECVD [9]. Thereby, substantially lowered WVTR values on the order of 10^{-5} g/(m^2 day) have been achieved which are close to meeting the demands of OLEDs discussed above. A severe drawback of these approaches is the use of plasma based deposition techniques. Consequently, the application of these barriers on top of organic devices may be problematic due to potential plasma damages to the brittle organics during the deposition process. Therefore, these barriers have thus far predominantly been used to form barrier layers on sufficiently stable polymer substrates without further (opto-)electronic functionality.

C. Thin-Film Barriers by Atomic Layer Deposition

An alternative technique which promises highly uniform thin film coatings without the a-priori need for plasma enhancement is atomic layer deposition (ALD). ALD relies on the sequential exposition of a surface that needs to be coated to a metal-organic precursor and a reactant (H_2O, O_3, NH_3 etc.) [10]. ALD is based on a self-limitation of the reaction on the sample surface rather than in the gas phase. A precise control of the layer thickness and an outstanding conformity are notable advantages of this technique. ALD has thus far predominantly been used in the preparation of ultra-dense high-k dielectric layers for microelectronic applications. Owing to the reactivity of the precursors, ALD allows for the deposition of very dense films even at low temperatures (< 100 °C). Consequently, ALD has been envisaged as promising technique to prepare encapsulation layers on top of organic electronic devices. Early reports on ALD barriers based on single Al_2O_3 layers have evidenced low permeation rates [11, 12]. A very low WVTR of 6.5×10^{-5} g/(m^2 day) at 60 °C ambient temperature has been reported for Al_2O_3 films which were grown at 120 °C [12]. However, deposition temperatures beyond 100 °C might be critical because glass transition temperatures for many functional OLED materials.

Moreover, recent results have shown that Al_2O_3 films are easily corroded by exposure to H_2O vapors [13]. Therefore, neat Al_2O_3 barriers may provide excellent sealing properties, which are subsequently degraded by the formation of pinhole defects as a result of corrosion of the Al_2O_3 by H_2O. Consequently, for increased reliability, multilayer ALD barriers are mandatory in which efficient sealing properties

are accompanied by elevated chemical robustness [13, 14]. In this paper we will predominantly focus on results obtained from either neat Al_2O_3 or Al_2O_3/ZrO_2 nanolaminate barriers which have been prepared by 80°C [14].

D. Barrier Properties

For Al_2O_3 layers prepared by the well established trimethylaluminum(TMA)/water ALD process [15], an optimum processing temperature of around 250°C has been found before. In this case the relative dielectric constant is about $\varepsilon \approx 9$. At lowered processing temperatures, a dielectric constant of $\varepsilon \approx 7.9$ hints to a reduced film density. Figure 1 shows the structural properties of the Al_2O_3 or Al_2O_3/ZrO_2 nanolaminate layers depicted by transmission electron microscopy. In Fig. 1a) defects can be identified as tiny voids leading to more sponge-like morphology. Dillon et al. have shown by FTIR spectroscopy that there is an accumulation of Al-OH species due to incomplete precursor reaction in films prepared with a TMA/H_2O ALD process at growth temperatures below 450 °K (177 °C) [16].

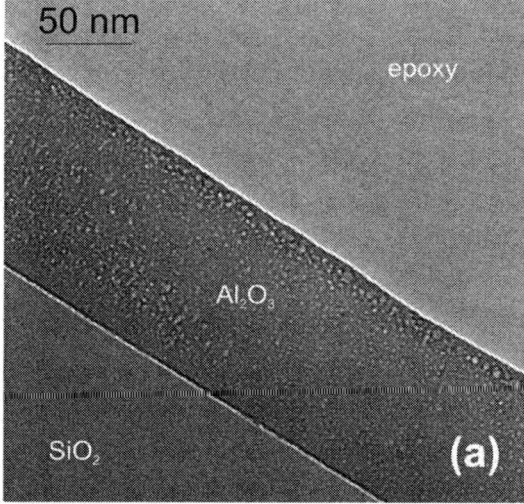

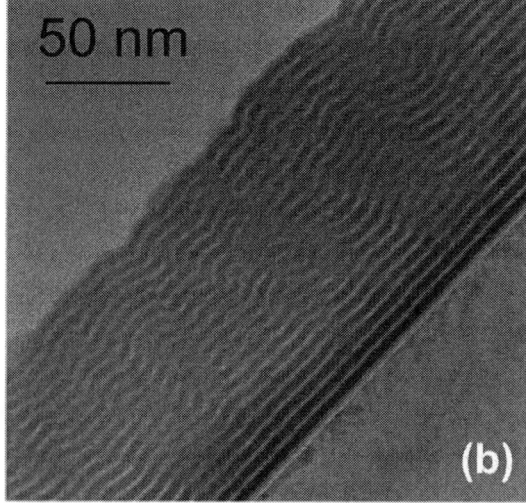

Figure 1 Transmission electron microscopy (TEM) images of a neat Al_2O_3 layer (a) and a Al_2O_3/ZrO_2 nanolaminate (b) prepared by ALD at 80°C.

It thus appears reasonable that small amounts of Al-OH species are piled up in the Al_2O_3 film at our growth temperatures of 80 °C. As a result, we would expect the Al_2O_3 films with a reduced packing density will lead to a higher density of permeation channels for gaseous species. As opposed to that, no voids are found in the nanolaminate (Fig. 1b) prepared at 80°C as well. The reason for the absence of voids in this case is still under investigation. A possible explanation may be linked to the Zr-precursor Tetrakis(dimethylamido)zirconium(IV) (TDMA(Zr)) used in this case. This substance has a lower vapor pressure than TMA. Therefore residues of TDMA(Zr) may remain in the reactor during the entire process and may function as a scavenger for H2O. Thus a pile-up of H2O in the layers is suppressed. Strictly speaking, with our assumption of the TDMA(Zr) precursor staying in the reactor, we leave the ALD "window" and approach the regime of chemical vapor deposition. This is partially reflected in Fig. 1b, where we see that the nanolaminate structure is perfect with sharp and well defined interfaces between the sub-layers only during the initial stages of the deposition process. With increasing number of sub-layers the nanolaminate evolves in a wavy structure. The origin of this change in the growth is not clarified yet. We believe that the onset of parasitic CVD is the reason for this effect. If the purging time between precursor doses is increased, the ideal NL structure can be preserved (not shown here). Longer purging times, however, directly lead to overall longer processing times. Thus, in our studies we deliberately did not increase purging times.

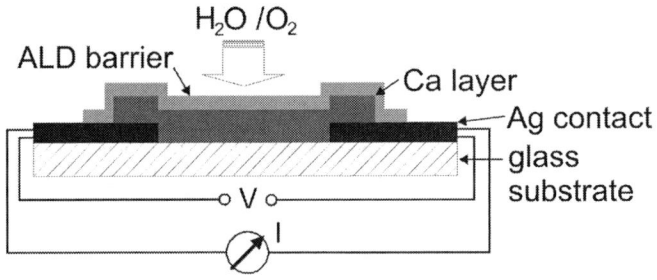

Figure 2. Setup of a sensor to electrically measure the permeation rates of a barrier layer by the corrosion of a Ca sensor.

For the measurement of the permeation rates a very sensitive technique is required. For example, the most widely used, commercially available test equipment by MOCON™ industries, is limited to rates in the range of 5×10^{-3} g/(m^2 day) for water vapor. A more sensitive permeation measurement method based on radioactive tritium-containing water (HTO) as tracer material can be used. Here, the theoretical detection limit has been shown to be as low as 2.4×10^{-7} g/(m^2 day) [17]. Alternatively, a method introduced by Paetzold et al [18]. The method allows for the measurement of WVTR the order of 10^{-6} g/(m^2 day) by *in-situ* monitoring of the conductivity of corroding Ca films. A possible sensor setup for this technique is shown in Fig. 1. A thin Ca pad is deposited between two Ag electrodes. The entire sensor is coated with the barrier material to be

measured. Specifically, the sensor detects the resistance changes due to the chemical reaction of moisture or oxygen with Ca. For accelerated aging experiments the sensors are typically stored under controlled condition in a climate cabinet at elevated temperature/humidity.

In an initial set of experiments, Ca sensors have been coated by 130 nm of Al_2O_3 or 130 nm of an Al_2O_3/ZrO_2 nano-laminate by ALD at 80 °C reactor temperature. For the nanolaminate 20 sub-cycles in the ALD process have been used for Al_2O_3 and ZrO_2. The thickness of the sub-layers is on the order of 2.6 nm and 3.6 nm for Al_2O_3 and ZrO_2, respectively. Upon storage of the sensors at 70°C /70%rh a linear decay of the Ca conductivity vs. time was detected. From the Al_2O_3 data, a permeation rate of 9.9×10^{-5} g/(m^2 day) for water was estimated. A very low WVTR was derived in the case of the Al_2O_3/ZrO_2 nanolaminate of 4.7×10^{-5} g/(m^2 day). By variation of the temperature in the climate cabinet we were able to determine the activation energy E_A for the permeation of water/oxygen through our barriers. For the E_A of the WVTR through the nanolaminate we obtained a value of 92 kJ/mol. Using the nanolaminate E_A data and our WVTR determined above, we estimate an ultra-low WVTR for our nanolaminate barrier layers at room temperature on the order of 5×10^{-7} g/(m^2 day). This low value is even below the resolution of our experiment.

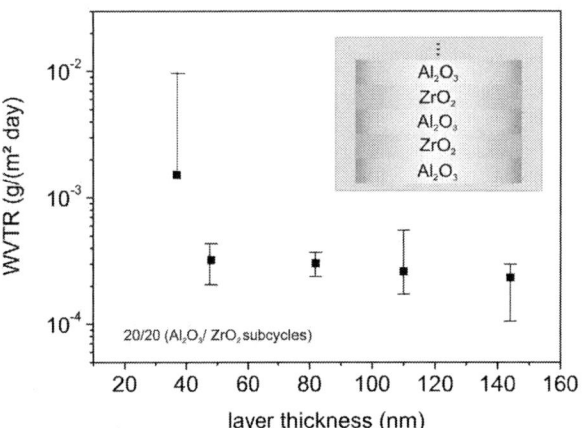

Figure 3. WVTR for NL barrier layers with varied thickness (here measured at 80 °C/80% rh). The inset sketches the NL layer sequence.

An important question related to the minimum thickness of the barrier in order to achieve a reasonable WVTR. Increased manufacturing throughput is directly related to lowered fabrication costs. Thus, thinner barrier layers not only lead to smaller tact times but also reduce costs in the sense that less precursor chemicals are required. Therefore, the above experiment has been repeated with a varied thickness of the nanolaminate (parameters in the climate cabinet 80°C/80%rh). As can be seen in Fig. 3, the WVTR does not substantially depend on the layer thickness as long as the thickness does not fall below a critical value of about 40 nm. Below 40 nm a dramatic increase in WVTR is encountered. Previous reports on Al_2O_3 or SiO_2 barriers

978-1-4244-5430-3/10 $26.00 © 2010 IEEE   329

have demonstrated a similar threshold-like characteristic for the WVTR vs. barrier thickness [19, 20]. On the other hand, the leveling off of the WVTR at ~3×10^{-4} g/(m^2 day) towards thicker barriers is substantially consistent with gas permeability through pinhole defects in barrier films [20].

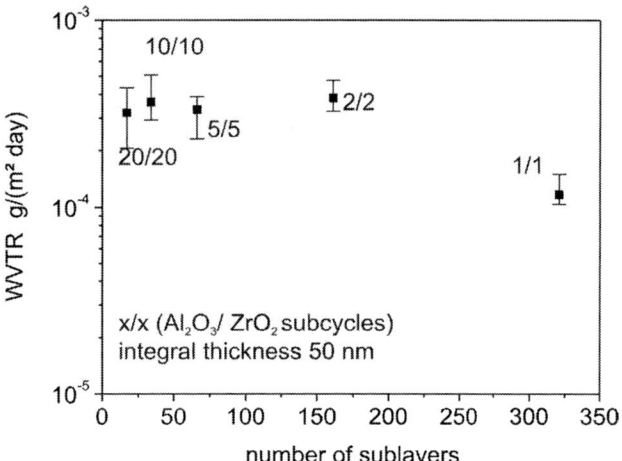

Figure 4 WVTR for varied thickness of sub layers in the nanolaminates. The overall thickness of 50 nm was kept constant in each case.

In a further study, the composition of the nanolaminate has been varied. Specifically, the thickness of the individual sub-layers has been varied via the number ALD cycles used for the sub-layers. The overall barrier thickness was kept constant at 50 nm in this case. The WVTR data is plotted in Fig. 4. Notably, the WVTR does not vary substantially from the 20/20 nanolaminate to the 1/1 nanolaminate, in which there is an alteration of one monolayer Al$_2$O$_3$ and ZrO$_2$ until 50 nm are reached. There appears to be a slight tendency, that in the well intermixed 1/1 sample a lower WVTR rate is found. One could speculate that the formation of mixed phase of Zr-Al-O, a so called aluminate phase, may be promoted in this case. Zirconium-Aluminates have previously been associated with increased film densities as opposed to that of the individual materials [21].

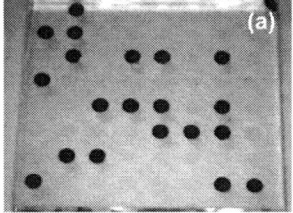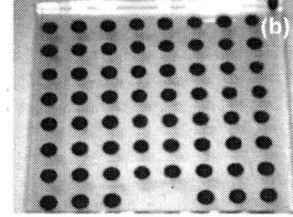

Figure 5 Photographs of Ca pads on a glass substrate encapsulated with 100 nm thick layers of neat Al$_2$O$_3$ (a) and Al$_2$O$_3$/ZrO$_2$ nanolaminate (b) after 160 h stored under 70°C / 70 % rh. .

Aside from low permeation rates on small areas like that of our permeation sensors, the encapsulation layers also must bring about the ability to form homogeneous films without statistical local defects or paths of elevated water/oxygen permeation on significantly larger areas. To assess this property of an encapsulation, one can use a number of Ca pads on a carrier as sensor which may be encapsulated. The probability of statistical defects can be derived simply by counting the degraded vs. the non-degraded Ca pads. In a specific experiment, an array of 64 individual Ca pads (diameter of 1 mm, thickness 100 nm) has been encapsulated with either Al$_2$O$_3$ or alternatively with an Al$_2$O$_3$/ZrO$_2$ nanolaminate. The samples have been photographed after 160 h storage in a climate cabinet (70°C/70%rh) as shown in Figure 5. Obviously, there is some statistical variation in the permeation rate of the thin-film encapsulation which leads to the observed failure statistics of the Ca pads in the case of the Al$_2$O$_3$ encapsulation. Very remarkably, if the Ca sensors are covered by the Al$_2$O$_3$/ZrO$_2$ nanolaminate, the failure rate of the Ca pads stored for 160 h under the same conditions as above is only about 3 % Fig. 5b. For both experiments identical substrates have been used. Therefore, pinhole defects due to dust particles or other sources of roughness should be the same for both sensor substrates and can thus not explain the difference between neat Al$_2$O$_3$ and ZrO$_2$/Al$_2$O$_3$ encapsulating layers. We, therefore, conclude that the structural imperfection of the low-temperature Al$_2$O$_3$ layers as evidenced in Fig. 1a and the previously encountered corrosion of neat Al$_2$O$_3$ is the reason for the high failure rate. On the contrary, vastly defect free barrier layers can be obtained with the multilayer approach. Here, structural integrity and chemical stability pave the way towards reliable thin film encapsulation layers.

III. ENCAPSULATION OF OLEDs

A. Conventional OLED Structures

Thus far, the barriers have only been used to protect metallic Ca pads. The application of ALD barriers on top of OLEDs will be discussed in this subsection. In a previous report on OLEDs encapsulated by ALD, Chang *et al.* have found that TMA attacks the double bonds in the conjugated OLED polymer poly[1-methoxy-4-(2'-ethyl-hexyloxy)-2,5-phenylenevinylene] (MEH-PPV). The resulting chemical defects cause efficient quenching of the desired electro-luminescence of the device. For the present study OLEDs with a layer sequence shown in Fig. 5 have been used. Here, the small organic molecules have been deposited by evaporation from Knudsen cells at high-vacuum (10^{-8} mbar). The OLED structure is based on a state of the art *p-i-n* OLED with an emitter system (IHF-TE-15) [22]. The anode and cathode side of the OLED comprise chemically doped organic charge injection/transport layers [23, 24].
The encapsulation process was carried out at 80°C in analogy to our discussion above. Specifically, we have used neat Al$_2$O$_3$ and ZrO$_2$/Al$_2$O$_3$ encapsulating layers. For reference on sample has been encapsulated by a conventional glass lid with getter.

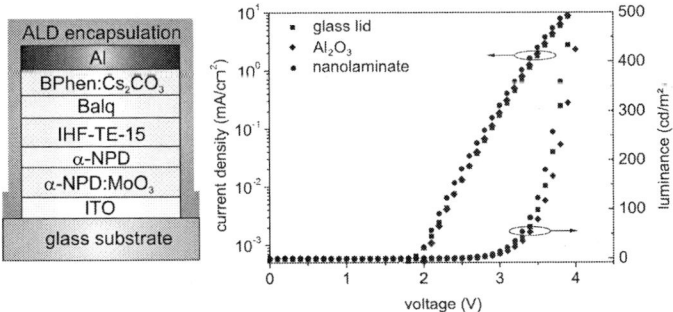

Figure 6. (left) Layer sequence of an OLED structure with ALD encapsulation. (right) *L-I-V* characteristics of encapsulated OLEDs measured directly after fabrication.

It is important to note, that the ALD process does not deteriorate the OLED luminance-current-voltage characteristics (Fig. 6). There is no significant difference in the *L-I-V* characteristics between the ALD passivated samples compared to the reference device. It may be speculated, that the 100 nm thick Al top-electrode facilitates nucleation of the ALD film and prevents the reactive precursors from penetrating the organic layer system. Moreover, obviously the processing temperature of 80°C does not affect the doped layers and does not lead to substantial diffusion of the dopant molecules.

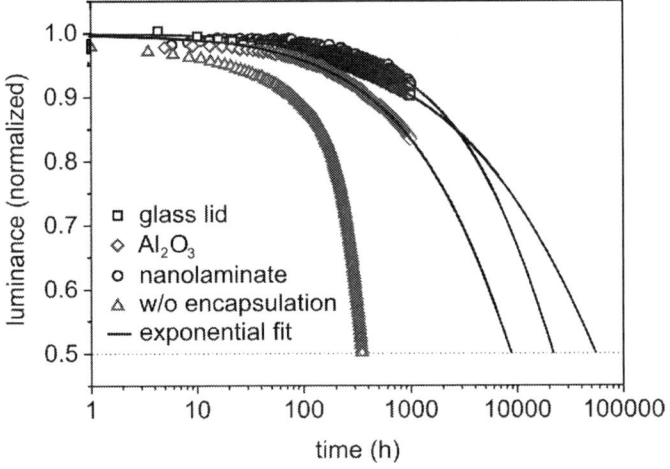

Figure 7. Normalized luminance vs. time characteristics of neat and encapsulated OLEDs. Starting luminance 1000 cd/m².

A typical lifetime test of and OLED is conducted at a fixed current density. With an initial brightness of 1 000 cd/m² the decay of the luminance level over time is recorded. T_{50} is the time after that the initial luminance level has decayed to 50% of its initial value. In the present case the encapsulated samples along with an un-encapsulated reference have been studied (Fig. 7). It is important to note, that for the neat OLED a rapid decay of the luminance is found with $T_{50} = 350$ h. The active area at this stage already shows significant black areas, which indicate severe degradation of the device. For the encapsulated OLEDs, the lifetime test has been stopped after 1 000 hours of operation and a model based on

a stretched exponential decay can be fitted to derive an estimate for T_{50} [25]. For the glass lid encapsulated device $T_{50} = 55\,000$ h can be extracted. The ALD passivated devices reach 8 700 h and 22 000 h for the Al_2O_3 and ZrO_2/Al_2O_3 encapsulating layers, respectively [26]. In view of the uncertainty associated with the extrapolation of a 1 000 h measurement to a timescale of 10 000 h, it can be stated that the NL encapsulation allows for a substantial lifetime close to that of the reference OLEDs. It is essential to note, that the active OLED area in case of the device encapsulated with the nanolaminate is essentially free from black spots (not shown here). This is in favorable agreement with the vast absence of statistical defects in this encapsulation system.

B. Top-Emitting and Transparent OLEDs

As opposed to conventional OLEDs where the light is emitted via the transparent glass substrate, in top-emitting devices the light is emitted via the top electrode. Top-emitting OLEDs are favorable for active matrix OLED displays, where the opaque Si based backplane electronics positioned under each pixel substantially compromises the amount of light coupled out for bottom emitting structures [29]. Recently, transparent thin film transistors based on metal-oxide semiconductors have evolved as a powerful replacement for a-Si backplane technology, which would allow for efficient driver electronics that does not interfere with the light emission of bottom emitting OLEDs [30, 31]. At the same time, entirely see-through active matrix OLED displays become feasible [32-34].

In view of these prospects, the encapsulation of top-emitting or entirely transparent OLEDs has to be considered. As opposed to the previous case, where only the barrier functionality of the encapsulation layer was required, in this case also the optical properties of the thin-film encapsulation is of paramount importance. Ideally, the dielectric properties of the barrier layer can be tuned to improve the light extraction from the top emitting OLED. Figure 8 shows the setup of a top-emitting OLED with an encapsulation layer on top. Again, the layer sequence is based on a state of the art device (c.f. Fig. 6) with the peculiarity of a 15 nm thin Aluminum top cathode that allows for some out-coupling of light. The bottom electrode consists of 100 nm Ag which is reflective. *I-V* characteristics of the fresh and encapsulated OLEDs are shown in Fig. 8 as well. No substantial change due to the ALD process is encountered. A closer look at Fig. 8 reveals some increased current density in the encapsulated devices. A similar behavior is observed for OLEDs treated at elevated temperatures (like those used in an ALD process), leading to slightly increased current densities. The quantum efficiency, on the other hand, does not change. It may be speculated that annealing of the chemically doped $BPhen:Cs_2CO_3$ layer may lead to a more efficient carrier injection. Nevertheless, the microscopic reason is not clarified, yet.

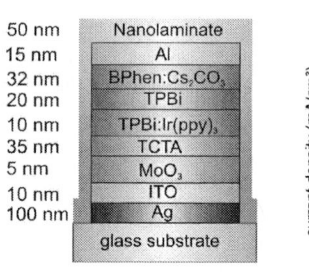
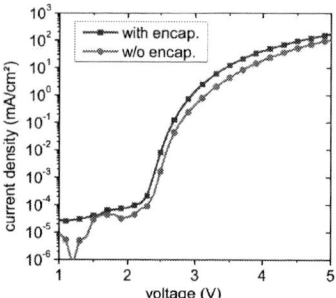

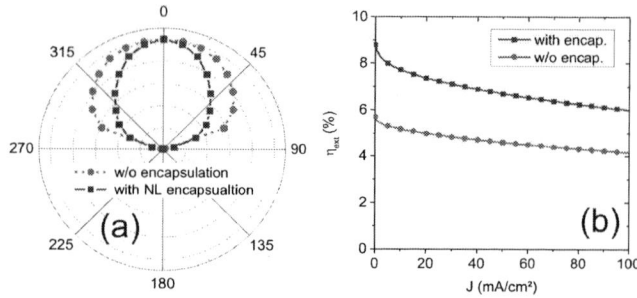

Figure 8. (left) Layer sequence of a top-emitting OLED structure with a 50 nm thick nanolaminate barrier layer on top. (right) I-V characteristics of the encapsulated OLED vs. neat devices directly after preparation.

Figure 9. Angular resolved light emission from the OLEDs (a) and external quantum efficiency η_{ext} vs. current density for the neat and encapsulated top-emitter (b).

Note, the layer sequence of the un-encapsulated devices has been optimized for maximum light extraction. A maximum current efficiency of 27.5 cd/A (at 100 cd/m^2) has been determined. Upon careful design of the optical properties of the encapsulation layer (thickness and composition) the amount of light extraction can be further increased. Specifically, in the present case a ZrO_2/Al_2O_3 nanolaminate (20/20 sub-cycles) structure with a thickness of 50 nm has been used. Under 0° observation the current efficiency almost doubles and reaches a value of 45 cd/A. It is, of course, well known, that the capping layer may substantially alter the angular dependence of the top-emitter. This is also observed in the case of our devices (Fig. 9a). The radiation pattern narrows towards the forward (0°) direction. One could speculate that this changed emission pattern is the sole reason for the increased efficiency found under 0° observation. While this would still be a substantial improvement for some display applications, for lighting the amount of photons extracted over the entire half-space has to be considered. To this end the OLEDs have been studied in an integrating sphere which allows us to determine the external quantum efficiency of the both OLED structures. While for the optimized OLED without encapsulation layer an external efficiency η_{ext} on the order of 4-6% is derived, the OLED with tuned encapsulation layer shows an approximated 40% increased external efficiency (Fig. 9b). In both devices η_{ext} is limited by the semi-transparent 15 nm thick Aluminum top-electrode. Substantially increased external efficiencies are expected for metal-oxide based top electrodes (e.g. ITO or Al:ZnO).

IV. CHALLENGES

Today, ALD is still suspected to be a deposition technique that is slow and not suitable for continuous manufacturing environments. In microelectronics, ALD has established its position as fabrication technique of high-k dielectrics. However, the required film thickness in this case is on the order of a few nm. Moreover microelectronic manufacturing is wafer based.

In organic (opto-)electronics, large area substrates are envisaged. Ultimately, roll-to-roll processing of organic devices and systems is desired. Aside from dedicated equipment, like in-line ALD tools, novel precursor chemicals are required, which particularly allow for high-speed processing of barrier layers even at temperatures below 100°C. One of the major strengths of ALD is the parallel processing of large batches. Therefore, the coated area/time can be large. Recently, Kodak has published an atmospheric ALD process which does not require vacuum equipment and is compatible with being integrated in a roll-to-roll process [27].

The most demanding application for gas permeation barriers, will be future flexible devices, e.g. rolled up OLED displays or foldable organic photovoltaic cells. Typically, the inorganic barrier layers lack flexibility and concepts have to be developed to harvest the favorable sealing properties of the oxide barriers while at the same time allow for repeated bending of these barriers. An interesting approach aims at a combination of ALD and molecular layer deposition, an ALD analogue for the preparation of organic thin films [28]. Thereby, organic layers may be interposed between oxide layers, mechanically decoupling the layers thereby increasing the critical strain associated with film cracking.

[1] F. Papadimitrakopoulos, X.-M. Zhang, D. L. Thomsen, III, and K. A. Higginson, "Inhibition of dark spots growth in organic electroluminescent devices", Chem Mater, vol. 8, pp.1363-1365, July 1996.

[2] M. K. Fung, Z. Q. Gao, C. S. Lee, and S. T. Lee, "Inhibition of dark spots growth in organic electroluminescent devices", Chem. Phys. Lett., vol. 333, pp. 432-436, Jan. 2001.

[3] M. Schaer, F. Nüesch, D. Berner, W. Leo, and L. Zuppiroli, "Water Vapor and Oxygen Degradation Mechanisms in Organic Light Emitting Diodes", Adv. Funct. Mater, vol. 11, pp. 116-121, Apr. 2001.

[4] P. E. Burrows, V. Bulovic, S. R. Forrest, L. S. Sapochack, D. M. McCarty, and M. E. Thompson, "Reliability and degradation of organic light emitting devices", Appl. Phys. Lett. vol. 65. pp. 2922-2924, Dec 1994.

[5] P. E. Burrows, G. L. Graff, M. E. Gross, P. M. Martin, M. Hall, E. Mast, C. Bonham, W. Bennet, L. Michalski, M. S. Weaver, J. J. Brown, D. Fogarty, and L. S. Sapochack, "Gas permeation and lifetime tests on polymer-based barrier coatings", Proc. SPIE, vol. 4105, pp. 75-83, 2001.

[6] G. Nisato, M. Kuilder, P. Bouten, L. Moro, O. Philips, and N. Rutherford, "Thin Film Encapsulation for OLEDs: Evaluation of

---

We gratefully acknowledge financial support by the German Federal Ministry for Education and Research BMBF (FKZ 13N9152, 13N10316) and the Deutsche Forschungsgemeinschaft (DFG).

978-1-4244-5430-3/10 $26.00 © 2010 IEEE

Multi-layer Barriers using the Ca Test", SID Symp. Dig. Tech. Papers vol. 34, pp. 550-553, May 2003.

[7] A. S. da Silva Sobrinho, M. Latrèche, G. Czeremuszkin, J. E. Klemberg-Sapieha, and M. R. Wertheimer, J. Vac. Sci. Technol. A vol. 16, pp. 3190-3198 Nov. 1998.

[8] G. L. Graff, R. E. Williford, and P. E. Burrows, "Mechanisms of vapor permeation through multilayer barrier films: Lag time versus equilibrium permeation", J. Appl. Phys. vol. 96, 1840-1849, Aug. 2004.

[9] M. Yan, T. W. Kim, A. G. Erlat, M. Pellow, D. F. Foust, J. Liu, M. Schaepkens, C. M. Heller, P. A. McConnelee, T. P. Feist, and A. R. Duggal, "A Transparent, High Barrier, and High Heat Substrate for Organic Electronics", Proc. IEEE, vol. 93, pp. 1468-1477, Aug. 2005.

[10] S. Suntola, "Surface chemistry of materials deposition at atomic layer level", Appl. Surf. Science, vol. 100-101, pp. 391-398 July 1996.

[11] A. P. Ghosh, L. J. Gerenser, C. M. Jarman and J. E. Fornalik, "Thin-film encapsulation of organic light-emitting devices", Appl. Phys. Lett., vol. 86, 223503, May 2005.

[12] P. F. Carcia, R. S. McLean, M. H. Reilly, M. D. Groner and S. M. George, "Ca test of Al_2O_3 gas diffusion barriers grown by atomic layer deposition on polymers", Appl. Phys. Lett., vol. 89, 031915, July 2006.

[13] A. A. Dameron, S. D. Davidson, B. B. Burton, P. F. Carcia, R. S. McLean, and S. M. George, "Gas Diffusion Barriers on Polymers Using Multilayers Fabricated by Al_2O_3 and Rapid SiO_2 Atomic Layer Deposition ", J. Phys. Chem. C, vol. 112, pp. 4573-4580, Mar. 2008.

[14] J. Meyer, P. Görrn, F. Bertram, S. Hamwi, T. Winkler, H.-H. Johannes, T. Riedl, and W. Kowalsky, "Al_2O_3/ZrO_2 nanolaminates as ultra-high gas diffusion barriers - a strategy for reliable encapsulation of organic electronics", Adv. Mater., vol. 18, pp. 1845-1849, May 2009.

[15] R. Puurunen, "Surface chemistry of atomic layer deposition: A case study for the trimethylaluminum/water process", J. Appl. Phys., vol. 97, 121301, Jun. 2005.

[16] A. C. Dillon, A. W. Ott, J. D. Way, S. M. George, "Surface chemistry of Al_2O_3 deposition using $Al(CH_3)_3$ and H_2O in a binary reaction sequence", Surface Science, vol. 322, pp. 230-242, Jan. 1995.

[17] R. Dunkel, R. Bujas, A. Klein, and V. Horndt, "Method of Measuring Ultralow Water Vapor Permeation for OLED Displays", Proc. IEEE, vol. 93, pp. 1478-1482, Aug. 2005.

[18] R. Paetzold, A. Winnacker, D. Henseler, V. Cesari, K. Heuser, "Permeation rate measurements by electrical analysis of calcium corrosion ", Rev. Sci. Instrum., vol. 74, pp. 5147-5150, Dec. 2003.

[19] M. D. Groner, S. M. George, R. S. McLean, and P. F. Carcia, "Gas diffusion barriers on polymers using Al_2O_3 atomic layer deposition", Appl. Phys. Lett. vol. 88, 051907, Jan. 2006.

[20] A. S. da Silva Sobrinho, G. Czeremuszkin, M. Latrèche, and M. R. Wertheimer, "Defect-permeation correlation for ultrathin transparent barrier coatings on polymers", J. Vac. Sci. Technol. A, vol. 18, pp. 149-157, Jan./Feb. 2000.

[21] W. F. A. Besling, E. Young, T. Conard, C. Zhao, R. Carter, W. Vandervorst, M. Caymax, S. De Gendt, M. Heyns, J. Maes, M. Tuominen, S. Haukka, "Characterisation of ALCVD Al_2O_3–ZrO_2 nanolaminates, link between electrical and structural properties ", J. Non-Cryst. Solids, vol. 303, 123-133, May 2002.

[22] D. Schneidenbach, S. Ammermann, M. Debeaux, A. Freund, M. Zöllner, C. Daniliuc, P. G. Jones, W. Kowalsky, and H.-H. Johannes, "Efficient and Long-Time Stable Red Iridium(III) Complexes for Organic Light-Emitting Diodes Based on Quinoxaline Ligands", Inorg. Chem., vol. 49, pp. 397-406, Jan. 2010.

[23] K. Walzer, B. Maennig, M. Pfeiffer and K. Leo, "Highly Efficient Organic Devices Based on Electrically Doped Transport Layers", Chem. Rev., vol. 107, pp. 1233-1271, Mar. 2007.

[24] J. Meyer, S. Hamwi, S. Schmale, T. Winkler, H.-H. Johannes, T. Riedl, and W. Kowalsky, "A strategy towards p-type doping of organic materials with HOMO levels beyond 6 eV using tungsten oxide", J. Mater. Chem., vol. 19, pp. 702-705, Jan. 2009.

[25] C. Féry, B. Racine, D. Vaufrey, H. Doyeux, and S. Ciná, "Physical mechanism responsible for the stretched exponential decay behavior of aging organic light-emitting diodes", Appl. Phys. Lett., vol. 87, 213502, Nov.2005.

[26] J. Meyer, D. Schneidenbach, T. Winkler, S. Hamwi, T. Weimann, P. Hinze, S. Ammermann, H.-H. Johannes, T. Riedl, and W. Kowalsky, "Reliable Thin Film Encapsulation for Organic Light Emitting Diodes grown by Low-Temperature Atomic Layer Deposition", Appl. Phys. Lett. , vol. 94, 233305, Jun. 2009.

[27] D. H. Levy, D. Freeman, S. F. Nelson, P. J. Cowdery-Corvan, and L. M. Irving, "Stable ZnO thin film transistors by fast open air atomic layer deposition", Appl. Phys. Lett., vol. 92, 192101, May 2008.

[28] D. C. Miller, R. R. Foster, Y. Zhang, S.-H. Jen, J. A. Bertrand, Z. Lu, D. Seghete, J. L. O'Patchen, R. Yang, Y.-C. Lee, S. M. George, and M. L. Dunn, "The mechanical robustness of atomic-layer- and molecular-layer-deposited coatings on polymer substrates", J. Appl. Phys., vol. 105, 093527, May 2009.

[29] T. Dobbertin, O. Werner, J. Meyer, A. Kammoun, D. Schneider, T. Riedl, E. Becker, H.-H. Johannes, and W. Kowalsky, "Inverted hybrid organic light-emitting device with polyethylene dioxythiophene-polystyrene sulfonate as an anode buffer layer", Appl. Phys. Lett., vol. 83, pp. 5071-5073 , Dec. 2003.

[30] P. Görrn, P. Hölzer, T. Riedl, W. Kowalsky, J. Wang, T. Weimann, P. Hinze, and S. Kipp, "Stability of transparent zinc tin oxide transistors under bias stress", Appl. Phys. Lett., vol. 90, 063502 , Feb. 2007.

[31] T. Riedl, P. Görrn, P. Hölzer, and W. Kowalsky, "Ultra-high long-term stability of oxide-TTFTs under current stress", phys. stat. sol. (RRL), vol. 1, pp. 175-177, Jul. 2007.

[32] P. Görrn, M. Sander, J. Meyer, M. Kröger, E. Becker, H.-H. Johannes, W. Kowalsky, and T. Riedl, "Towards See-Through Displays: Fully Transparent Thin-Film Transistors Driving Transparent Organic Light-Emitting Diodes", Adv. Mater. vol. 18, 738-741, Mar. 2006.

[33] J. Meyer, T. Winkler, S. Hamwi, S. Schmale, H.-H. Johannes, T. Riedl, and W. Kowalsky, "Transparent Inverted OLEDs with Tungsten Oxide Buffer Layer ", Adv. Mater. vol. 20, pp. 3839-3843, Sept. 2008.

[34] T. Riedl, P. Görrn, and W. Kowalsky, "Transparent Electronics for See-Through AMOLED Displays", IEEE/OSA J. Displ. Technol., vol. 5, pp. 501-508, Dec. 2009.

Light, Bias, and Temperature Effects on Organic TFTs

N. Wrachien, A. Cester, N. Bellaio, A. Pinato, M. Meneghini, A. Tazzoli, G. Meneghesso

Department of Information Engineering
University of Padova
Via Gradenigo 6B, Padova – Italy
phone: +39-0498277625, fax: +39-0498277699; e-mail: wrachien@dei.unipd.it

K. Myny, S. Smout, J. Genoe

imec
Kapeldreef 75, 3001 Leuven, Belgium

Abstract— **In this work we present the instabilities observed in organic Thin-Film Transistors when subjected to stress test in different bias, temperature and illumination conditions. C-V measurements, indicate the presence of two distinct trapping phenomena. Appreciable charge trapping can be achieved using relatively high biases for long times (1000s). Illumination strongly enhances charge trapping only under positive gate biases, while it has no effect on the charge trapping/detrapping under negative gate bias. Charge detrapping is thermally activated coherently with trapping/detrapping at the SiO_2/pentacene interface from hydrogenoid species. A first order model explaining the observed relaxation kinetics is also presented.**

Keywords-component; Organic thin film transistors, TFT, organic electronics, charge trapping, traps, reliability.

I. INTRODUCTION

Organic thin-film transistors (OTFT) are attracting much attention due to their recent performance improvements [1]-[4]. In particular, many publications reported very good results in terms of mobility, subthreshold slope, and I_{ON}/I_{OFF} ratio, making OTFTs a very compelling candidate as possible replacement for their inorganic counterparts in some specific applications. As a representative example, since their first appearances in the literature, OTFTs gained several orders of magnitude in terms of carrier mobility [5]-[8]. OTFTs favorably compare with inorganic thin-film transistor due to their much lower cost and ease of manufacture. In fact, many organic semiconductors can be deposited at very low temperatures, eliminating the need for very high and unpractical thermal budgets [9]-[10]. In addition organic semiconductor deposition techniques are well suited for large area devices, such as LCD or AMOLED devices. Another advantage of organic semiconductors over their inorganic counterpart, is the compatibility with many plastic substrates, which are flexible, much more robust, inexpensive and much lighter than glass. This latter advantage, would allow a further reduction of the finished device cost and it would open the door to a broad field of applications, which can be hardly achieved using inorganic semiconductors, such as flexible displays,

smart textiles, or various kind of disposable electronic devices like smart labels or cheap lab-on-a-chip.

Unfortunately, many issues are still affecting OTFTs, such as the sensitivity to external environment factors (temperature, humidity, light) [11]-[12], the poor stability and the shorter lifetime if compared to silicon TFTs.

Many publications reported the most recent advancements on new structures, materials or deposition techniques for organic transistors [5]-[10]. However, still few publications have addressed the stability of the electrical characteristics of OTFTs subjected to bias [13]-[16]. The bias stability is an issue in several applications, but it is crucial especially for AMOLED displays [17]-[22]. From this application viewpoint, the light sensitivity is also a concern, because it might enhance the bias effects. Moreover, even the temperature might enhance the effects of bias. At this regard, very few authors in the literature explored the effects of temperature and bias on the stability of organic thin film transistors.

The aim of this work is to investigate the effects of bias combined with illumination and temperature on the OTFT electrical characteristics. We also provide a first order model, which allows to determine the energetic trap density distribution.

II. EXPERIMENTAL AND DEVICES

Throughout this work, we analyzed p-type bottom-contact bottom-gate pentacene OTFTs (see Fig. 1a). The gate dielectric

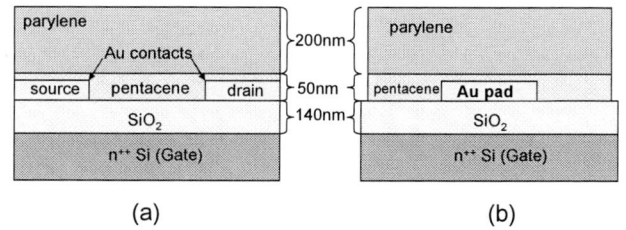

Figure 1. Cross section of the devices used throughout this work: (a) the organic thin-film-transistor; (b) the organic capacitor.

This work was partially supported by Progetto di Ateneo 2009 – Università di Padova, Italy (Project Number CPDA083941).

consists of a 140-nm thermal SiO_2 layer grown over an n^{++} silicon substrate, which also acts as gate contact. A 50-nm pentacene layer was deposited by evaporation. Source and drain contacts are made of 30-nm gold layer. The gold layer was deposited by sputtering and patterned using a standard lift-off technique. A 200-nm parylene layer is employed as encapsulation layer and it has been deposited using chemical vapor deposition at a temperature below 40°C. The devices

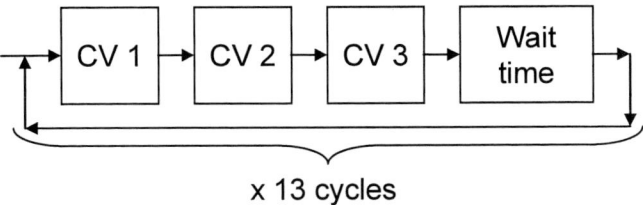

x 13 cycles

Figure 2. Experimental procedure followed for the C-V measurements. Each cycle consists of 3 double sweep C-Vs, followed by a wait time. Illumination has been turned on only for cycles 3 and 4.

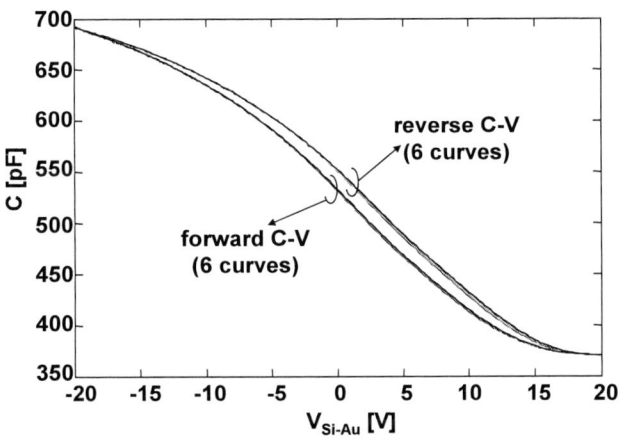

Figure 3. First 6 double sweep C-V performed in dark conditions (cycle 1 and cycle 2). The 6 curves are almost overlapping.

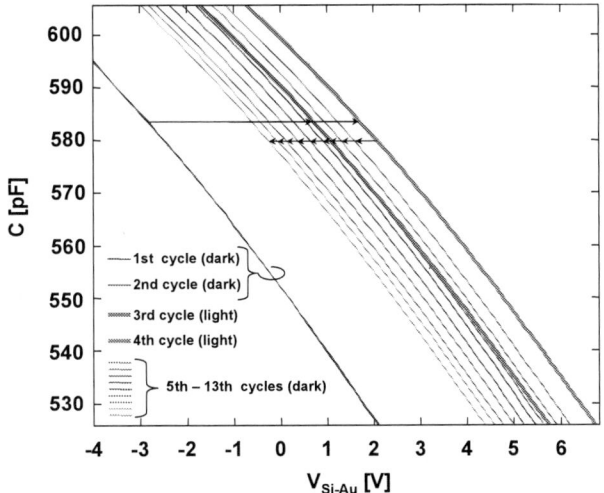

Figure 4. Zoom of the 3rd reverse C-V sweep of each cycle. When C-Vs are performed in dark, negligible changes occur (1st and 2nd cycle). Strong variation occurs, if the C-Vs are performed under illumination (3rd and 4th cycle). If the C-Vs are performed again in dark, (cycles 5-13), the C-Vs progressively move toward negative voltages.

feature a hole mobility between 0.5 and 0.7 $cm^2V^{-1}s^{-1}$ in ohmic region. C-V measurements have been performed at f=1kHz on a 2.8-mm^2 area capacitor, featuring the same vertical structure of the OTFT (see Fig. 1b); the pentacene layer around the gold pad was removed to achieve the desired capacitor area.

In the followings, we will refer as "bias" to the silicon (i.e. gate) to pentacene voltage.

III. RESULTS

A. C-V and I-time measurements on organic capacitors

Capacitors are useful tools to investigate device stability. In fact, fast interface states can be detected from the stretch of the C-V measurement. The presence of slow traps (i.e. with response times comparable or larger than the C-V measurement time) can be detected from the hysteresis of double sweep C-V measurement or even performing subsequent C-V measurements. The presence of long-lived trapped charge can

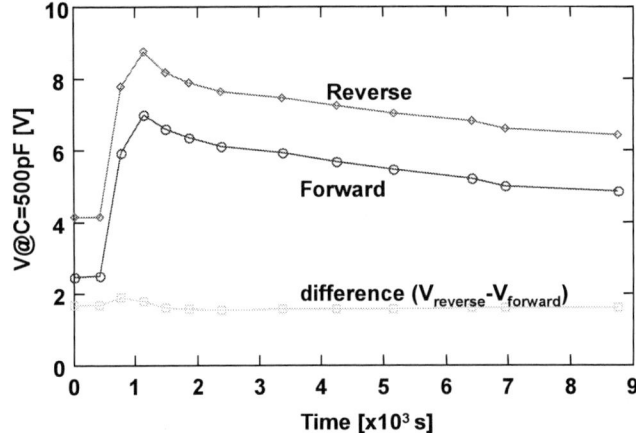

Figure 5. Voltages extrapolated from the third C-V of each cycle, at C=500pF. The C-V hysteresis (squares) is almost constant, regardless the shift on the C-V.

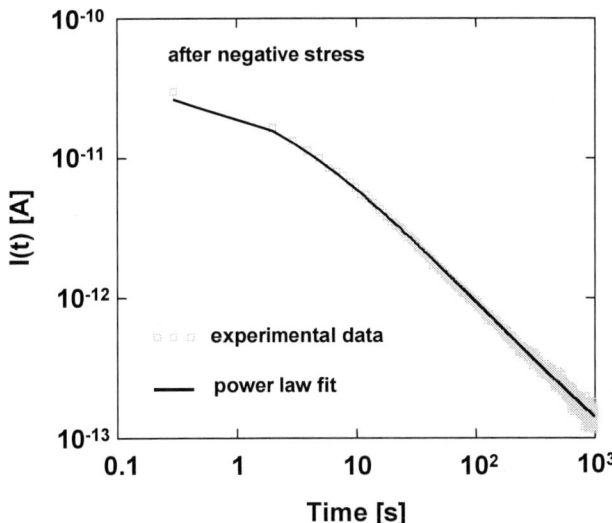

Figure 6. Evolution of the current measured after a 1000-s -30-V stress on a organic capacitor. Symbols are experimental data, while the solid line is the power-law fit (see Section IV).

be detected also from the rigid shift of the C-V after a constant stress. In particular, we performed the experimental procedure described in Fig. 2. Each measurement cycle consists of 3 consecutive double-sweep C-V measurements. The first two cycles were performed in dark conditions and they are plotted in Fig 3. The 3rd and 4th cycle were performed under white light illumination, and all the other cycles were performed in dark. White light from a halogen light bulb was chosen in order to achieve a wide and continuous energy spectrum. While this choice does not allow us to establish any photon energy dependence, it is still useful to determine if charge trapping can be enhanced by light. The light power supply was chosen so that the optical power intensity was 0.1mW/cm^2. A zoom of the transition region of the C-Vs of the cycles 1 through 13 is plotted in Fig 4. For clarity, we show for each cycle only the third reverse C-V sweep. In Fig 5, we plotted, for each cycle, the voltage required to achieve a capacitance of 500pF, on the third forward ($V_{forward}$) and reverse ($V_{reverse}$) C-V sweep.

Remarkably, the C-Vs performed in dark (see Fig 3) show a hysteresis as large as 1.7V, indicating the presence of traps with response time comparable to the time taken by the C-V measurement. Furthermore, despite the hysteresis, all the double sweep C-Vs of Fig 3 overlap each other.

On the contrary, when C-Vs are taken under illumination, a large rightward shift occurs on both the forward and the reverse C-V after the 3rd and 4th cycles (see Figs. 4 and 5), indicating that the double sweep C-V measurement induced a net negative long-lived trapped charge. During cycles 5-13 (performed in dark) the C-V begins to shift leftward indicating that the negative trapped charge slowly decreases.

Noticeably, the hysteresis in Fig. 5 is almost constant, regardless the shift on the C-V curve (i.e. the negative trapped charge), suggesting that the traps responsible for the C-V hysteresis are not the same traps responsible for the rigid rightward shift on the C-Vs.

We have also subjected the capacitor to a 1000-s -30-V constant negative bias stress. After that, we sampled the capacitor current for 1000 seconds (I-time measurement). The current evolution of the I-time measurement is shown in Fig. 6. The same procedure has been repeated using a 1000-s +30-V positive bias, with a similar evolution. As it will be discussed in Section IV, the capacitor current is excellently fitted with a power law, which decreases with time.

B. Bias, light and temperature effects on OTFT

While the experimental procedure of Fig. 2 provides some clues about charge trapping and traps density, it cannot be usefully exploited to perform more accurate investigations. In particular, the C-V measurements take a considerable amount of time, hence its effect on the relax kinetics cannot be neglected, at least in the first 1000 seconds. Furthermore, the bias, which induces the trapped charge, is not constant during the C-V. Hence, we performed the experimental procedure shown in Fig. 7, for a much more accurate investigation on the role of light, temperature, and bias. In particular, we subjected the O-TFT to a 1000-s gate stress, with source and drain grounded (stress phase). The gate voltage during the stress phase was either -30V (negative stress) or +30V (positive

stress) and the stress phase was performed in dark conditions. Just after the stress, we subjected the device to a relax phase, by applying no or moderate gate bias with source and drain grounded. The gate biases employed during the relax phases were either 0V, +5 or +10V, for the relax phases performed after the negative stress, and either 0, -5V or -10V for the relax phases performed after the positive stress. The relax phase was performed either in dark or under illumination, using LEDs with different wavelengths. The LED biases were calibrated so that they emitted the same optical power. These relax phases were periodically interrupted to perform the I_D-V_{GS} measurements. For these I_D-V_{GS} measurements, the V_{GS} sweep range was chosen so that the devices operated in the linear

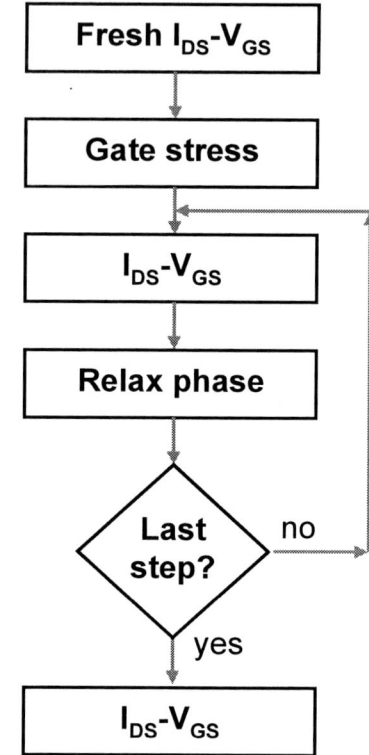

Figure 7. Experimental procedure used for the charge relaxation experiments described in section III-B.

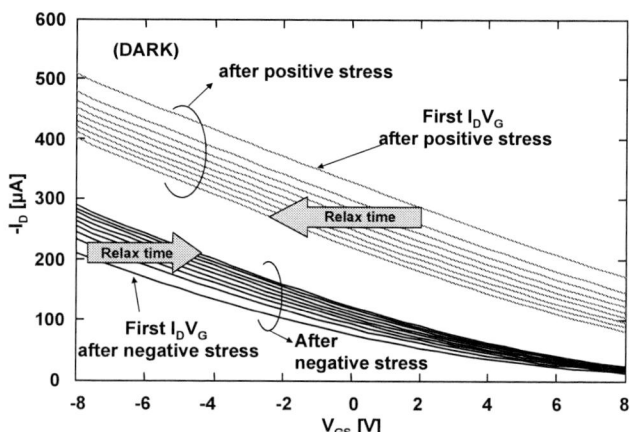

Figure 8. I_D-V_{GS} curves measured during the charge relaxation experiments performed after the 1000-s gate stress.

region and the corresponding I_D was high enough to allow accurate and fast I_D measurements, without inducing a strong impact on the relax kinetics. The total time taken from the I_D-V_{GS} is less than 1 second.

Fig. 8 shows the I_D-V_{GS} curves of a device taken during the relax phase after positive and negative stress, in dark. The I_D-V_{GS} (in linear region) are almost parallel suggesting no modifications in the transconductance (and hence the carrier mobility), as already observed in [16],[23]. In the following, we define the gate voltage at which the drain current is 200μA as V^*. Fig. 9 shows the time evolution of the V^* variation with respect to its initial value ($V^*(t)-V^*(0)$), extrapolated from the I_D-V_{GS} measured during the various relax phases. The lines are the fits, which will be discussed in Section IV.

Noticeably, if the relax phases is performed under 0-V bias, the kinetics are independent on the illumination wavelength. Negative gate bias accelerates the relax kinetics but, again,

there is no dependence on the illumination wavelength. On the contrary, light has a strong impact when a positive gate bias is applied during the relax phase after the negative stress.

C. Effects of temperature

We also analyzed the effects of temperature, performing the same relaxation experiments described in Fig. 7 at different temperatures, from 30°C to 60°C. The maximum temperature was chosen so that any organic semiconductor degradation phenomena are avoided.

Remarkably, at least within our range, the temperature has a small impact on the transconductance. In fact, the transconductance variation between 30°C and 60°C is less than 4%, hence, as a first approximation, its impact on the relax kinetics (if any) can be neglected. In particular, we found

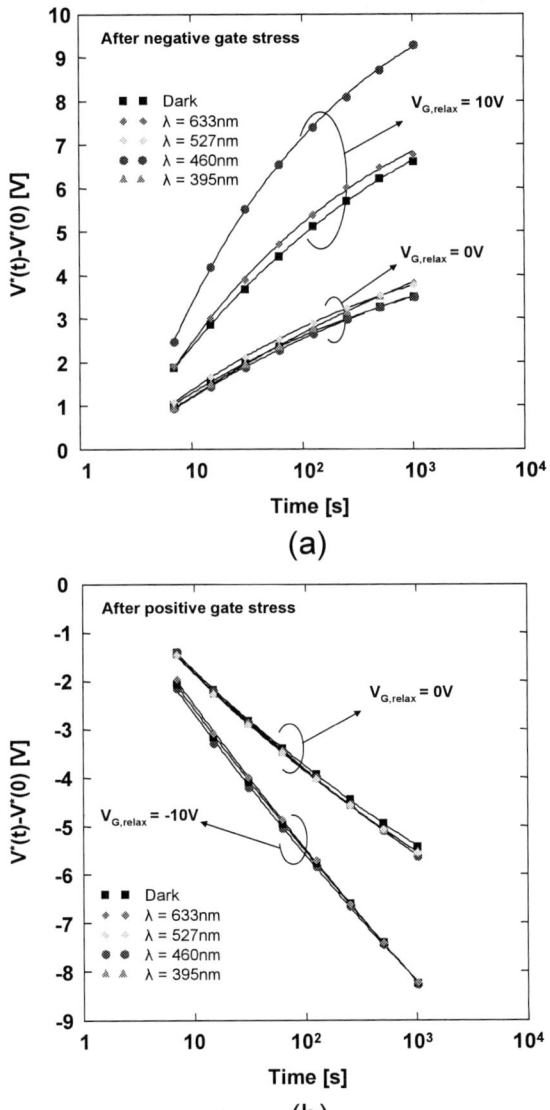

Figure 9. Variation of the V^* extrapolated from the I_D-V_{GS} measured during the relax phases, under different illumination conditions: after the negative gate stress (a) and after the positive gate stress (b).

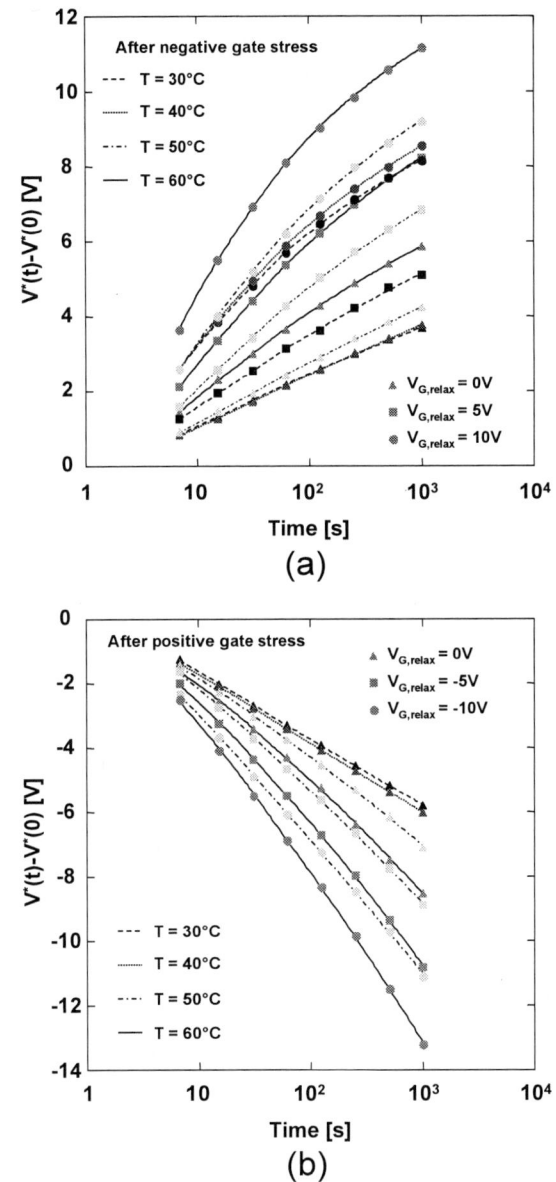

Figure 10. Variation of the V^* extrapolated from the I_D-V_{GS} measured during the relax phases, at different temperatures: after the negative gate stress (a) and after the positive gate stress (b).

6.31μS and 6.56μS at 30°C and 60°C, respectively. The results are summarized in Fig. 10. As expected, we found that the temperature accelerates the charge relaxation kinetics, especially under bias.

To avoid temperature effects on the stress and measurements, we also followed a different experimental setup. We have stressed the devices at a constant -30V gate bias, at 27°C. Then, just after the stress, we measured the I_D-V_{GS}, characteristics. Later, we stored the device on a climatic chamber for 40 minutes, at a constant temperature ranging from 30°C to 60°C. Due to the very small sample size and thermal mass, we expect that the thermal transient can be neglected over the total 40-minutes annealing time. During this annealing time, we left the device floating. After the 40-minutes, we forced the device to cool-down to 27°C and we measured again the I_D-V_{GS}. In Fig. 11, we plot the ratio between the currents after and before annealing. The I_D-V_{GS} curves measured before and after annealing are almost parallel, indicating that no mobility variation occurred and the threshold voltage variation was the responsible for the decreased current. Fig. 11 confirms that the charge relaxation kinetics is thermally activated, with an average activation energy of 213meV.

IV. DISCUSSIONS

The results of Figs. 3-11 indicate that moderately high biases induce charge trapping, whose sign depends on the bias polarity. In particular, with positive gate-to-pentacene bias, a net negative trapped charge contribution can be observed, while, if a negative gate-to-pentacene bias is employed, positive charge is trapped.

In the literature some authors reported charge trapping in organic TFT, both polymer and small-molecule based [13]-[16]. In principle charge can be trapped at: the bulk of the oxide, at the pentacene/oxide interface, in the bulk of the pentacene layer. Due to the relatively small electric field, we argue that the charge trapped in the bulk of the SiO_2 is negligible. In fact, during our stress and relax kinetics, the oxide electric field is in the range of 0.7-2.1MV/cm, which is a very low value to achieve appreciable charge injection [24]-

[25]. Traps (or even border traps) at the dielectric/pentacene interface might account for the hysteresis measured during the C-V measurements. Finally, some authors reported about trap in the organic semiconductor [17],[26]-[27] or bias-induced charged defects [27], which, in turn, move in the organic semiconductor, inducing long-lived threshold voltage variations.

In the following we will discuss about the combined effects of light and bias (Section IV-A), the temperature effects (Section IV-B) and we will provide a first order model for the relaxation kinetics (Section IV-C).

A. Light and bias effects

The C-V measurements (see Figs. 3-5) clearly highlight the presence of traps featuring different response times. On one hand, we observed a hysteresis (see $V_{forward}$-$V_{reverse}$ of Fig 5), which is almost constant regardless the measurement condition (dark or under illumination). Taking into account the C-V measurement and sampling times, we argue that the response time of the traps responsible for the C-V hysteresis is within 10-100 seconds. In fact, traps faster than the sampling-time would only stretch-out the C-V curve, whereas traps much slower than the total C-V measurement-time would not be affected by the C-V measurement, hence, they would not give any contribution on the total trapped charge, which is reflected on the C-V position.

On the other hand, if the double-sweep C-V is performed under illumination, both the forward and the reverse curves are shifted rightward (see $V_{forward}$ and $V_{reverse}$ in Fig 5), indicating that negative charge has been trapped under the combined action of light and bias. Remarkably, the presence of this trapped charge does not affect the hysteresis width, which is constant. Once the light is removed, a very slow relax kinetics can be measured on the $V_{forward}$ and $V_{reverse}$ (see Fig 5, after 1140 s). The timescale of this kinetics is much slower than the C-V measurement time and from Fig. 5 we argue that these trap, which are responsible for the almost rigid shift of both $V_{forward}$ and $V_{reverse}$, have a response time at least two orders of magnitude slower than the traps responsible for the C-V hysteresis. This explains why the hysteresis is almost unaffected by the particular value of $V_{forward}$ or, equivalently, $V_{reverse}$: by the time the full double-sweep C-V is performed (120 s), very small net charge is trapped in dark, and the position of the forward-sweep C-V relative to the reverse-sweep C-V depends almost only on the traps with response time within 10-100s.

From the C-V measurements, we also argue that light mostly enhances negative charge trapping. Negative charge trapping (or equivalently, hole neutralization) occurs when the gate-to-pentacene bias is positive (see position of forward-sweep C-V with respect the reverse-sweep C-V). This also confirmed by Fig 9a. In fact, from Fig. 9a we noticed that trapped hole neutralization is strongly enhanced under blue light if a positive bias is applied.

From the data of Fig. 5, we estimate that an interface trap density of $2.6 \cdot 10^{11} cm^{-2}$ can account for the 1.7-V hysteresis, while the 4.6V shift on both the forward and reverse C-V is induced by a negative trapped charge density of $7.0 \cdot 10^{11} cm^{-2}$ (if

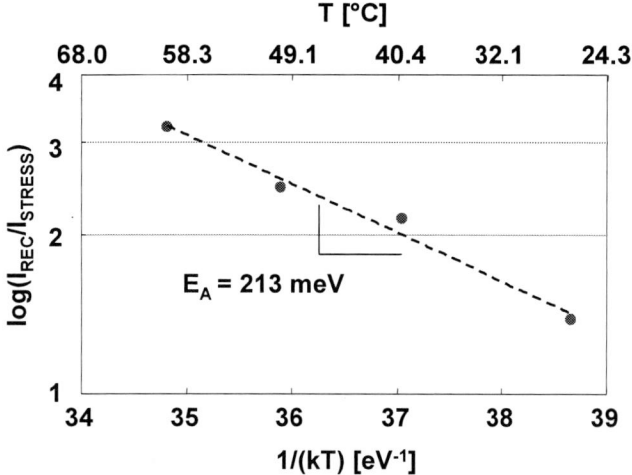

Figure 11. Arrhenius plot of the drain current recovery induced by a 40-min annealing performed after the negative stress.

all the charges are supposed concentrated at the SiO_2/pentacene interface). In about two hours, only half of these traps are effectively discharged.

Coming now to the stress, we argue that, due to the electric field direction, most of the trapped charges are located near the pentacene/SiO_2 interface. When a high enough positive bias is applied, electrons are trapped. Photons with enough energy increase the effects of positive bias, because of the excess of photogenerated electrons. In fact, it is well known [28] that photons with energy larger than the energy gap can photogenerate electron-hole pairs, which may either be trapped or they may neutralize the trapped holes. Conversely, if a high enough negative bias is applied, an excess of positive charge is trapped. At least with our optical intensity ($\sim 0.1 mW/cm^2$), light has a minimal impact because holes are majority carrier and the excess of photogenerated holes is negligible. Hence, there is also negligible light wavelength dependence under negative bias.

In the same way, light accelerates positive charge neutralization if a positive bias is applied. In fact photogenerated electrons drift toward the interface under the action of the positive bias, neutralizing the trapped holes [16]. Again, at least under low level of illumination intensities, light does not accelerate electron neutralization under negative bias, because the excess of photogenerated holes is negligible with respect to the equilibrium hole density value.

The effect of light under positive bias explains why the C-V measurements performed under white light feature a strong

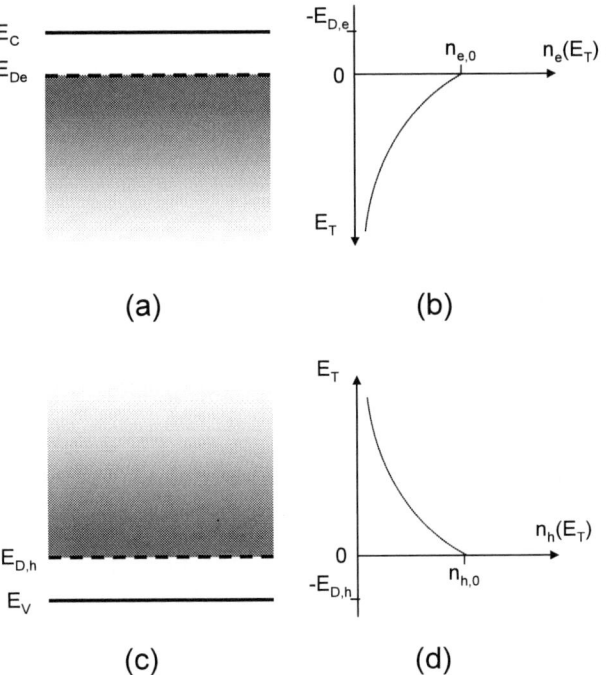

Figure 12. Visual representation of the electron trap density below the conduction band (LUMO band). Darker zones correspond to higher density (a). Sketch of the exponential energy trap density assumed for electrons in Section IV-C (b). Visual representation of the hole trap density above the valence band (HOMO band). Darker zones correspond to higher density (c). Sketch of the exponential energy trap density assumed for holes in Section IV-C (d)

rightward shift. In fact, when the measurements are performed in dark, the positive and negative gate biases induce almost the same absolute value of V^* variation (around 4V in 30s, see Fig. 9). However, when a positive relax bias is applied under light (e.g. λ=460nm in Fig 9), a 6-V V^* variation occurs, whereas the V^* variation with negative relax bias is -4V, independent of the illumination condition.

B. Effects of temperature

Figs. 10-11 indicate that temperature has a strong impact on the charge detrapping kinetics, especially if we consider the very small temperature variation (only 30°C). The temperature might accelerate the relaxation kinetics in several ways. First, it is well known that the time to escape from a trap exponentially depends on the temperature [25],[29]. Secondly, the mobility of charged mobile defects might be increased by the temperature. The bias might also accelerate the relaxation kinetics adding a drift contribution to the thermal-activated diffusion of mobile charged defects, which move away from the interface toward the source/drain contacts.

The 213-meV activation energy extrapolated in Fig. 11 perfectly correlates with the 0.2-eV activation energy associated with the electron transfer from/to hydrogenoid species [30]. These species may derive either from H atoms used for dangling bond passivation or from H_2O contaminants within the parylene encapsulation layer, which eventually migrate toward the SiO_2/pentacene layer during aging.

C. Relax kinetics

According with literature [15], all the relax kinetics of Figs. 9-10, fit with a power law. In particular, we found that our experimental data excellently fit with the following power law:

$$\Delta V^* (t) = V^* (t) - V^* (0) = C \left[1 - \left(\frac{t}{t_0} \right)^\gamma \right]. \qquad (1)$$

In the following we give a first order model that allows to quantitatively estimate the mean trap energy level.

In principle, the trap energetic distribution within the pentacene layer might be position dependent. Furthermore, V^* also depend on the actual trapped charge position. For sake of simplicity, we will assume that all the trapped charges are located near the interface between SiO_2 and pentacene. Of course, this is only a first order approximation, since traps might also be located in the bulk of the pentacene, as already stated before and as it has been also reported in [31]. Still, the calculated interface trap density might be viewed as an equivalent trap density, which accounts both for the interface and for the pentacene bulk trap density.

According with several papers in literature, we expect that traps in the in the polycrystalline pentacene are energetically distributed between the HOMO and LUMO energy level (see Figs. 12a and 12b). In particular, we assume that there is an exponential distribution of deep and slow traps starting at depth $E_{D,e}$ below the LUMO and it exponentially fades with a energy constant U. Following these assumptions, we write the trap density per unit area and per unit energy as (assumed concentrated at the SiO₂/pentacene interface):

$$d_e\left(E_T\right) = d_{e,0} \cdot e^{-E_T/U}, \qquad (2)$$

Where E_T is the trap depth with respect to $E_{D,e}$ (see Fig. 12b). Of course, (2) is only a first order approximation and holds only for a limited E_T range. A similar distribution was hypothesized in [32].

Shallow traps are supposed energetically concentrated at $E_{D,e}$ and we suppose that such shallow traps quickly exchange electrons/hole in time much shorter than 2 seconds, i.e., the time elapsed by the end of the stress pulse and the beginning of the I-time sampling. For this reason, we neglect the contribution of the shallow traps in the hysteresis and in the slow C-V curve displacement during the relax period. The same hypotheses and considerations holds true for the HOMO level (see Figs. 12c and 12d).

In a first approximation, we also assume that, after the positive stress, all the traps energetically located below the Fermi level are filled with an electron, i.e. the filled trap density per unit area and per unit energy is $n_e(E_T) = d_e(E_T)$:

We suppose the emission time τ of a trap located at energy E_T+E_D below the LUMO (or above the HOMO) exponentially depends on the trap depth, i.e.,

$$\tau\left(E_T\right) = \tau_0 \cdot e^{E_T/(KT)}, \qquad (3)$$

Where τ_0 depends also on the temperature, bias and $E_{D,e}$.

Now, at the time t, the number of detrapped electrons is:

$$N_e\left(t\right) = \int_0^{E_T(t)} n_e\left(E_T\right) dE_T, \qquad (4)$$

Where $E_T(t)$ is found solving (3) for E_T.

Using (2) and (4) we find:

$$N_e(t) = \frac{d_{e,0}}{U}\left[1 - \left(\frac{t}{\tau_0}\right)^{-\frac{KT}{U}}\right]. \qquad (5)$$

Assuming the effect of $N_e(t)$ as if it were concentrated at the SiO₂/pentacene interface, the corresponding $\Delta V^*(t)$ describing the relax kinetics after the positive gate stress is:

$$\Delta V^*(t) = -\frac{q}{C_{ox}} \cdot \frac{d_{e,0}}{U}\left[1 - \left(\frac{t}{\tau_0}\right)^{-\frac{KT}{U}}\right], \qquad (6)$$

which has the same form of (1).

The same procedure holds true for the case of trapped holes (N_h) after a negative stress. Of course, in this case the resulting sign is opposite to (6).

Our results indicate that U is within the range of 150-400meV.

Incidentally, the relax current measured on a capacitor also fits with a power law:

$$i(t) = i_0\left(\frac{t}{t_i} + 1\right)^{-D} + i_k, \qquad (7)$$

In (7), the first term is the transient component, responsible for the current kinetics at least for the first 1000 seconds. On the contrary the constant i_k accounts for both the capacitor and the instrumentation/cables/setup leakage, which do not play any role in the charge trapping kinetics. The same constant might also account for a much slower kinetics (such as those induced by the motion of hydrogenoid species), which induces negligible effects at least in the first 1000s, but it could play a key-role if much longer timescales are considered. The constant i_k will considered equal to zero hereafter.

After simple calculation (see appendix) we derive the trap energetic distribution:

$$n\left(E_T\right) = \frac{i_0 \cdot \tau_0}{q \cdot KT} \cdot \frac{e^{E_T/(KT)}}{\left(\frac{\tau_0}{t_i} e^{E_T/(KT)} + 1\right)}. \qquad (8)$$

In our fits, we found that $\tau_0 \cong t_i = 2s \pm 0.3s$, hence, if $E_T/(KT)$ is high enough, the right term denominator of (8) can be approximated as $\left((t_0/t_i)e^{E_T/(KT)}\right)^D$. Using this approximation, (8) has the form of (2). In other words, if we assume that (3) holds, and if the measured current fits with (7), then the $n(E_T)$ has, with good approximation, the exponential form hypothesized in (2).

APPENDIX

It is straightforward to derive the following result from (4):

$$i(t) = \frac{dQ}{dt} = q\frac{dN\left(E_T(t)\right)}{dt} = q \cdot n_e\left(E_T\right)\frac{dE_T(t)}{dt}, \qquad (9)$$

From (9), one can obtain:

$$n\left(E_T\right) = \frac{i(t)}{q} \cdot \frac{1}{\dfrac{dE_T(t)}{dt}} = \frac{i(t)}{q} \cdot \frac{t}{KT}, \qquad (10)$$

Now, using (8) and substituting t with (3), we find $n(E_T)$:

$$n\left(E_T\right) = \frac{i_0 \cdot \tau_0}{q \cdot KT} \cdot \frac{e^{E_T/(KT)}}{\left(\frac{\tau_0}{t_i} e^{E_T/(KT)} + 1\right)}. \qquad (11)$$

V. CONCLUSIONS

In this work, we showed the combined effects of bias, light and temperature on organic thin-film transistors. Relatively high biases induce a noticeable charge trapping, which manifests itself as a rigid shift of the transfer characteristics. Such shift can be larger than 10V if a ±30V bias is applied for 1000 seconds. However, no bias-induced effects can be appreciated on the device transconductance.

978-1-4244-5430-3/10 $26.00 © 2010 IEEE

Light enhances both the electron trapping and hole neutralization kinetics, if a positive bias is applied after the negative gate-to-source stress, whereas the effects of light are negligible under zero or negative bias after the positive gate-to-source stress. Still, we do not exclude that with a much higher optical intensity light could also enhance hole trapping or electron neutralization.

Temperature increases the charge detrapping rate, and the calculated activation energy suggests that the motion of hydrogenoid species may have a strong contribution on the relaxation kinetics.

Finally, we presented a first order model, which provides excellent fits of the charge relaxation kinetics, in all the operating conditions.

REFERENCES

[1] D.J., Gundlach, C.-C., Kuo, S.F. Nelson, and T.N. Jackson, "Organic thin film transistors with field effect mobility >2 cm 2/V-s," 57th Annual Device Research Conference Digest, 1999, pp.164-165, 1999.

[2] S. K. Park, C.-C. Kuo, J.E. Anthony, and T.N. Jackson, "High mobility solution-processed OTFTs," IEEE International Electron Devices Meeting, 2005. IEDM Technical Digest, pp. 4, 5-7 Dec. 2005.

[3] C.-C. Kuo; M.M. Payne, J.E. Anthony, and J.E. Jackson, "TES anthradithiophene solution-processed OTFTs with 1 cm²/V-s mobility," IEEE International Electron Devices Meeting, 2004. IEDM Technical Digest. , pp. 373-376, 13-15 Dec. 2004.

[4] D.J. Gundlach,H. Klauk, C.D. Sheraw, C.-C. Kuo; J.-R. Huang; and T.N. Jackson, "High-mobility, low voltage organic thin film transistors,"

[5] Z. Bao, A. J. Lovinger, and A. Dodabalapur ,"Organic field-effect transistors with high mobility based on copper phthalocyanine," Appl. Phys. Lett. 69, 3066 (1996).

[6] R. Parashkov, E. Becker, G. Ginev, T. Riedl, H.-H. Johannes, and W. Kowalsky, "All-organic thin-film transistors made of poly(3-butylthiophene) semiconducting and various polymeric insulating layers," J. Appl. Phys. 95, 1594 (2004).

[7] Fang-Chung Chen, Chih-Wei Chu, Jun He, and Yang Yang, "Organic thin-film transistors with nanocomposite dielectric gate insulator," Appl. Phys. Lett. 85, 3295 (2004).

[8] H. Klauk, M. Halik, U. Zschieschang, G.Schmid, and W. Radlik, "High-mobility polymer gate dielectric pentacene thin film transistors," J. Appl. Phys. 92, 5259 (2002).

[9] T.N. Jackson, "Organic thin film transistors-electronics anywhere," International Semiconductor Device Research Symposium, 2001, pp.340-343, 2001.

[10] K. Kudo, "Recent progress on organic thin film transistors and flexble display applications," Nanotechnology Materials and Devices Conference, 2006. NMDC 2006. IEEE , vol.1, pp.290-291, 22-25 Oct. 2006.11 IRPS 09

[11] C. Pannemann, T. Diekmann, and U. Hilleringmann, "On the degradation of organic field-effect transistors," Microelectronics, 2004. ICM 2004.

[12] Y. Qiu, Yuanchuan Hu, Guifang Dong, Liduo Wang, Junfeng Xie, and Yaning Ma, "H$_2$O effect on the stability of organic thin-film field-effect transistors," Appl. Phys. Letters, (83), 8,2003, pp. 1644-1646.

[13] Hsiao-Wen Zan, and and Shin-Chin Kao "The Effects of Drain-Bias on the Threshold Voltage Instability in Organic TFTs ," IEEE Elec. Dev. Lett, vol 29, no. 2, Feb. 2008, pp .155-157.

[14] C. Erlen, F. Brunetti, P. Lugli, M. Fiebig, S. Schiefer, and B. Nickel, "Trapping Effects in Organic Thin Film Transistors," in Proc. of Sixth IEEE Conference of Nanotechnology, June 2006, pp. 82-85.

[15] A. Salleo, F. Endicott, and R. A. Street, "Reversible and irreversible trapping at room temperature in pol(thiophene) thin-film transistors,", Appl. Phys. Lett. Vol. 86, 263505, 2005.

[16] M. Debucquoy, S. Verlaak, S. Steudel, K. Myny, J. Genoe, and P. Heremans, "Correlation between bias stress instability and phototransistor operation of pentacene thin-film transistors," Appl. Phys. Lett. 91, 103508 (2007).

[17] Jae-Hoon Lee, Sang-Geun Park, Sang-Myeon Han, Min-Koo Han, and Kee-Chan Park ,"New PMOS LTPS–TFT pixel for AMOLED to suppress the hysteresis effect on OLED current by employing a reset voltage driving," Solid-State Electronics 52 (2008) 462–466

[18] Z. Tang, M.S. Park, S.H. Jin, C.R. Wiem, "Drain bias dependent bias temperature stress instability in a-Si:H TFT,", Solid-State Electronics, vol. 53 (2009), pp. 225–233.

[19] Jae-Hoon Lee, Woo-Jin Nam, Kwang-Sub Shin, and Min-Koo Han, "Hysteresis phenomenon of hydrogenated amorphous silicon thin film transistors for an active matrix organic light emitting diode," Journal of Non-Crystalline Solids vol. 352 (2006) pp. 1719–1722.

[20] Taek Ahn, Hye Jung Suk, Jongchan Won, Mi Hye Yi, "Extended lifetime of pentacene thin-film transistor with polyvinyl alcohol (PVA)/layered silicate nanocomposite passivation layer," Microelectronic Engineering, vol. 86. (2009), pp 41–46

[21] Juan Li, Chunya Wu, Jianpin Liu, Shuyun Zhao, Zhiguo Meng, Shaozhen Xiong, and Lizhu Zhang, "A new instability phenomenon in microcrystalline silicon thin film transistors", Journal of Non-Crystalline Solids, vol. 352 (2006) 1715–1718.

[22] Bong-Hyun You, Jae-Hoon Lee, and Min-Koo Han, "Polarity Balanced Driving Scheme to Suppress the Degradation of Vth in a-Si:H TFT Due to the Positive Gate Bias Stress for AMOLED," IEEE journ. Display Technology, vol. 3, no. 1, mar. 2007, pp. 40-44.

[23] Tae Ho Kim, Chung Kun Song, Jin Seong Park, and Min Chul Suh, "Constant Bias Stress Effects on Threshold Voltage of Pentacene Thin-Film Transistors Employing Polyvinylphenol Gate Dielectric", IEEE Elec. Dev. Lett. vol. 28, nO. 10, Oct. 2007, pp. 874-876.

[24] M. Lenzlinger, E .H. Snow, "Fowler-Nordheim tunnelling into thermally grown SiO$_2$", J. Appl. Phys., Vol. 40, pp. 278-283, 1969.

[25] P. Hesto, in Instability on Silicon Devices, Ed. G. Barbottin and A. Vapaille, North Holand, Vol. 1, 1986.

[26] John E. Northrup and Michael L. Chabinyc, "Gap states in organic semiconductors: Hydrogen- and oxygen-induced states in pentacene," Phys. Rev. B 68, 041202(R) (2003).

[27] D.V. Lang, X. Chi, T. Siegrist, A.M. Sergent, and A. P. Ramirez, "Bias-Dependent Generation and Quenching of Defects in Pentacene," Phys. Rev. Lett., vol. 93, no. 7, 076601, 2004.

[28] Tse Nga Ng, M.L. Chabinyc, R.A. Street, and A. Salleo, "Bias Stress Effects in Organic Thin Film Transistors," *45th annual. IEEE International Reliability physics symposium, 2007. proceedings.*, pp.243-247, 15-19 April 2007.

[29] W. B. Jackson, J. M. Marshall, and M. D. Moyer, "Role of hydrogen in the formation of metastable defects in hydrogenated amorphous silicon", Phys. Rev. B 39, 1164–1179 (1989).

[30] K. Vanheusden, S. P. Karma, R. D. Pugh, W. L. Warren, D.M Fleetwood, and R. A. B. Devine, "Thermally activated electron capture by mobile proton in SiO$_2$ thin films", Appl- Phy- Lett. 72, pp. 28, 1998.

[31] V. Nádaždy, R. Durný, J. Puigdollers, C. Voz, S. Cheylan, and M. Weis, "Defect states in pentacene thin films prepared by thermal evaporation and Langmuir–Blodgett technique," Journal of Non-Crystalline Solids, Vol. 354, May 2008, pp 2888-2891.

[32] A. Rose, "Space-Charge-Limited Currents in Solids", Physical Review, Vol. 97, no. 6, Mar. 1955, pp 1538-1544.

RELIABILITY OF (100) AND (110) ORIENTED SINGLE-GRAIN SI TFTS WITHOUT SEED SUBSTRATE

Tao Chen, Ryoichi Ishihara, and C.I.M Beenakker

Delft Institute of Microsystems and Nanoelectronics Technology (DIMES)

Delft University of Technology

Delft, the Netherlands

Phone: (0031) –152786474, e-mail address: echosteve@hotmail.com, t.chen@tudelft.nl

Abstract— **We report high performance (100) and (110) oriented single-grain TFTs below 600°C by orientation controlled μ-Czochralski process. Due to surface and in-plane orientation control, the uniformity approaches to the SOI counterpart. Electron mobilities are 732cm²/Vs for (100) and 630cm²/Vs for (110). Devices show stable performance under gate and drain stress respectively. After applying electrical stress on gate and drain for 1000s respectively, the electron mobility has not deteriorated for (100) SG-TFT and (110) SG-TFT.**

Keyword; Location and orientation control, single-grain Si TFT

I. INTRODUCTION

The μ-Czochralski process[1] has realized single-grain (SG) Si TFTs with the electron mobility of 600cm²/Vs which are comparable to the SOI counterpart. However, variation in the mobility was as large as 20% due to the random crystallographic orientation, resulting in effective mass[2] variation between the grains. It is therefore important to control both the surface and in-plane orientation of the location-controlled grains. One solution is to control the crystallographic orientation by growing a seed before the μ-Czochralski process. We have reported that Metal-Induced-Lateral-Crystallization (MILC) can grow both (100) and (110) surface orientation controlled poly-Si on SiO₂ by using anisotropic tensile stress[3]. We have used the MILC poly-Si as a seed for the μ-Czochralski process and realized both (100)[4] and (110)[5] surface orientation- and location- con-trolled Si grains. In this paper, we report initial electrical characteristics of both surface and in-plane orientation controlled (100) and (110) SG-TFTs without any seeding substrate and also the reliability on electrical stress. Electron and hole mobilities are optimized by the surface and in-plane orientation along the current flow of (001)/ [1$\bar{1}$0] and (011)/ [100], respectively. With those configurations, we have realized electron mobilities of 732cm²/Vs for (100) SG-TFT and 630cm²/Vs for (110) SG-TFT, respectively, which surpass the SOI counterpart. Nickel concentration is below SIMS detection limit. We have studied reliability of (100) and (110) SG-TFTs by positive and negative gate bias or drain bias. After applying the electrical stress on gate or drain for 1000s respectively, the

electron mobility has not been deteriorated for (100) and (110) SG-TFT.

II. EXPERIMENT

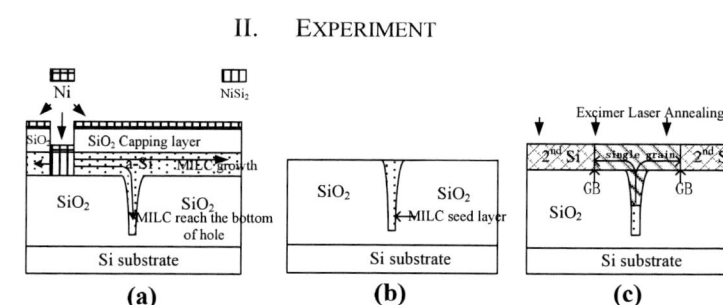

Fig.1 (a)-(c) Cross section structure of process

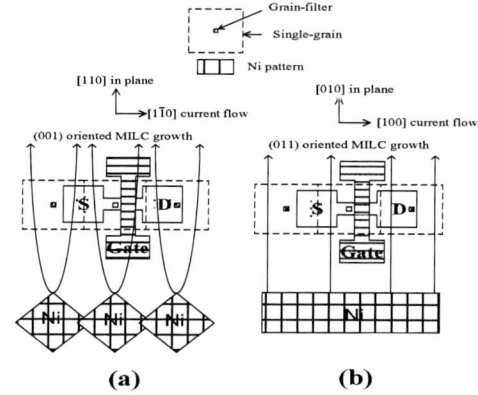

Fig.2 Planar view of Ni pattern design for (100) and (110) single-grain TFT

The (100) and (110) seeds were formed by the MILC as follows: a 15nm Ni was deposited on 250nm a-Si in a rectangular shaped SiO₂ window and further annealed at 600°C for 4h. The (100) or (110) orientated poly-Si seeding layer was selected, using the orientation dependence on the position with respect to the Ni pattern.[3] Figure 1 shows cross-sectional structure for the orientation and location-control of the grains. A grid of 1.2μm deep cavities (grain-filter) with a diameter of 100nm was formed in 1200nm thick SiO₂ on a c-Si wafer; the bottom of the hole is separated from the substrate with 845 nm thick SiO₂. A 250nm thick a-Si was deposited by LPCVD at

978-1-4244-5430-3/10 $26.00 © 2010 IEEE   342

545°C. The thick Ni was formed at 20μm away from the matrixes of the grain-filter, letting the MILC poly-Si grows into the grain filter and reaches to the bottom of the filter. The (100) or (110) orientated poly-Si seeding layer was selected by positioning the grain filter near the corner or the side of the Ni pattern, respectively[3]. For the (100) orientation, the corner was repeated on a line with a Ni pattern of zigzag shape[3] to increase the area of the (100) oriented grains (Fig. 2(a)). Then the 250nm MILC poly-Si layer was etched away by a dry etching while keeping the MILC poly-Si inside the grain-filter as a seed (Fig.2 (b). After that, a 2nd a-Si layer with a thickness of 250nm was deposited by LPCVD and the sample was crystallized by excimer-laser at a substrate temperature of 450°C. The laser irradiation energy is 1500mJ/cm². During laser irradiation, (100) or (110) oriented grains epitaxially grew from the seed in the grain filter (Fig.1(c)). After pattering the Si into island, a 80nm thick SiO_2 was deposited as a gate oxide with PECVD by TEOS at 350°C. After that, it was annealed at 400°C in H_2 to passivate interface defects. The source and drain were doped with 10^{16} ions/cm² by phosphorous implantation at 30keV for n-channel and 10^{16} ions/cm² by boron at 20keV for p-channel. Then the dopants are activated by excimer laser at 300mJ/cm². The channel width and length was 2μm for both p- and n- channel SG-TFTs. The SG-TFTs with (100) surface orientation has the current flow direction of [1$\bar{1}$0] while the TFTs with (110) orientation has the flow direction of [010].

III. RESULTS AND DISCUSSION

A. Initial Status of device

Figure 3 plots the pole-figure of EBSD of the location-controlled grains grown from the grain filters positioned at the corner (a) and the side (b) of the Ni area. It can be seen that surface orientation of (100) and (110) was obtained for the grains at the corner (a) and the side (b) respectively. The in-plane orientation was also controlled for the both cases. The (100) oriented grains have the in-plane orientation of [110] along the direction perpendicular to the zigzag line of the Ni, while the (110) oriented grains have [010] along the direction perpendicular to the side. Table.1 shows the uniformity of the TFTs have been improved due to (100) and (110) orientation control, when compared with the randomly oriented SG-TFTs. The mobility variation is around 7% for the (100) n-channel and 3.7% (110) p-channel TFTs, which are approaching those of the SOI counterpart. For n-channel, the (100) TFTs show a high mobility of 732cm²/Vs, while the (110) TFTs show a mobility of 630cm²/Vs before electrical stress. We have also studied reliability on those devices. The maximum mobility is achieved around the threshold voltage. Fig.4 and 5 show the transfer characteristic of (110) and (110) n-channel device under 6V gate bias, respectivity, with a drain voltage of 0.05V for 100s and 1000s. The electrical field across the gate is 0.88MV/cm.

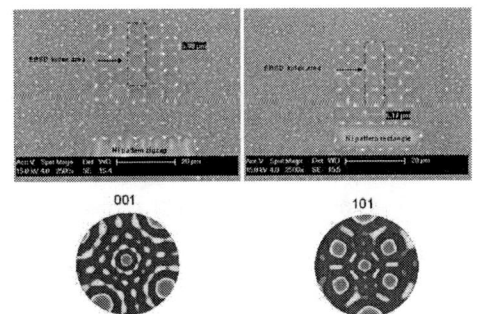

Fig.3 SEM and pole-figure of (100) and (110) oriented single grains after laser crystallization

	Mobility(cm²/Vs⁻¹)	Variation (%)
<100>	732	6,9
<110>	630	3,6
SOI<100>	727	2,4
random orientation	597	16,9

Table.1 Uniformity of (100) and (110) SG TFTs together with SOI and random orientation

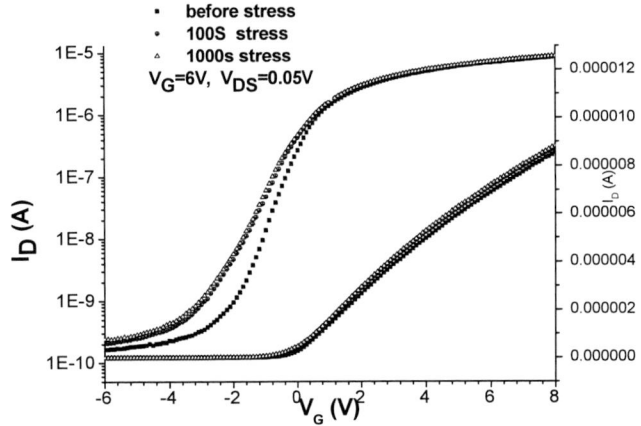

Fig.4 Transfer characteristic of (110) n-channel device under 6V gate bias with drain voltage of 0.05V

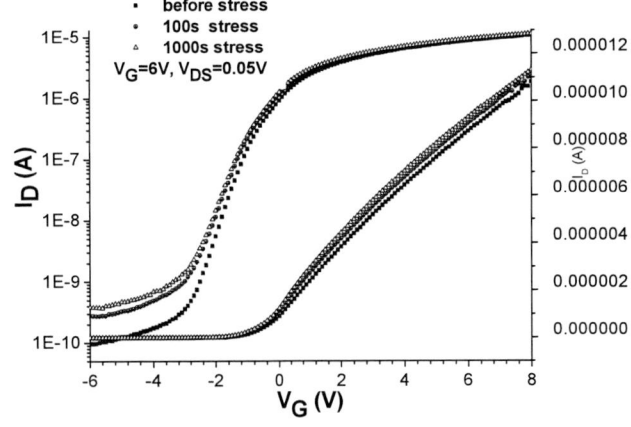

Fig.5 Transfer characteristic of (100) n-channel device under 6V gate bias with drain voltage of 0.05V

B. Positive gate bias effect

After the gate is stressed with a voltage of 6V for 100s and 1000s, for n channel, no positive shift has been observed for both (110) and (100) SG-TFT. The mobility of (110) and (100) SG-TFTs haven't been deteriorated. The substhreshold swing (S) and the off-current for (110) and (100) SG-TFTs have been slightly increased. Those are different than poly-Si.[7] The S of (100) SG-TFT is more stable than (100) under positive gate bias which is same as single crystalline Si[8]. The leakage current has increased more for (100) SG-TFT than (110) SG-TFT. Moreover, the V_{th} has a negative shift. Thus we can conclude that hot holes injection in the gate oxide, accompanied by acceptor states generation at upper half of the bandgap at the intereface[9].

C. Positive drain bias effect after gate bias

After the (110) and (100) n-channel device is stressed by gate, we continue to put 9V bias on drain with gate voltage of 2V on the same device. Fig.6 and 7 show the transfer characteristic of (110) and (100) n-channel device under 9V drain bias with V_G of 2V for 10s, 100s and 1000s. With the drain bias stress, subthreshold swing is improved from 0.726V/dec to 0.630V/dec and from 0.871V/dec to 0.630V/dec for (100) and (110) SG-TFTs, respectively. Also leakage current has decreased and a slightly positive shift of V_{th} has been observed. Based on the experiment results, we concluded that the positive shift of V_{th} is due to hot electrons generated by the high electrical field[10] around drain recombining with the trapped hot holes generated by the previous gate bias stress. And the improvement of S and leakage current is due to self heating[11] which passivate the interface defects during electrical stress.

D. Positive drain bias effect under different voltage

To further investigate the effect of hot electrons effect caused by high electrical field around the drain, we applied 12V, 15V, 20V, 25V and 30V on drain with gate voltage of 2V for 1000s on new device which has not been stressed under positive gate bias. From Fig.8, we see neither V shift nor S improvement below 25V drain bias stress. A negative V_{th} shift together with deteriorate S has been observed when the drain voltage is higher than 25V. Leakage current and on current has also increased under 30V drain bias. These are due to generation of electron-hole pairs by impact ionization under high drain voltage[12].

E. Negative gate bias effect

Furthermore, we have performed the stress experiment by negative gate bias (off-state). Fig.9 and 10 show (110) and (100) n-channel device under gate voltage of -10V with drain voltage of 0.05V for 100s and 1000s. From Fig.9 and 10 we see until 100s only positive shift of transfer characteristic for (110) and (100) n- channel due to electrons tunneling into oxide from source and drain. And S has been deteriorated from 0.722V/dec to 0.792V/dec and from 0.617V/dec to 0.738V/dec for (110) and (100) n- channel after 1000s stress. Leakage current has deteriorated from 40nA to 70nA only for (110) SG-TFT after 1000s stress. V_{th} has a positive shift of 0.3V and 0.6V for (110) and (100) respectively. With increasement of stress time under high gate negative bias, more electrons can gain enough energy

tunneling from drain to channel and generated trapped states at mid-gap [13]. And (110) SG-TFT has higher leakage current increase than (100) after negative gate bias. Fig. 11 and 12 summarize the mobility and S change under different stress conditions. Based on these result, we can conclude that hot holes caused deterioration of device and negative shift of V_{th} by positive gate stress while hot electrons caused deterioration of device and positive shift of V_{th} by drain bias and negative gate bias.

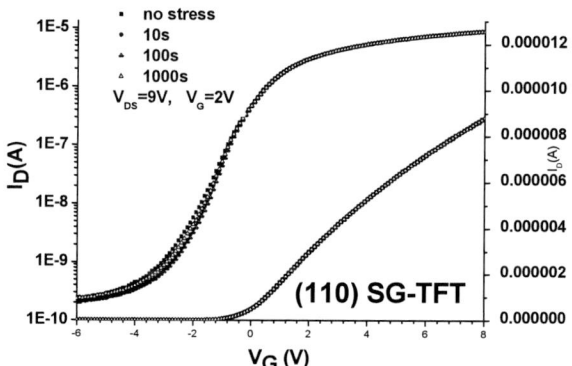

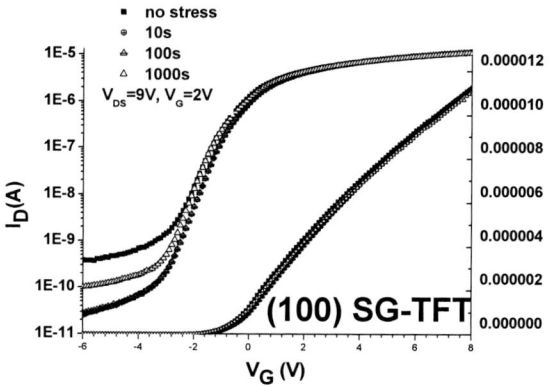

Fig.6 transfer characteristic of (110) and (100) n-channel device under 9V drain bias for 10s, 100s and 1000s

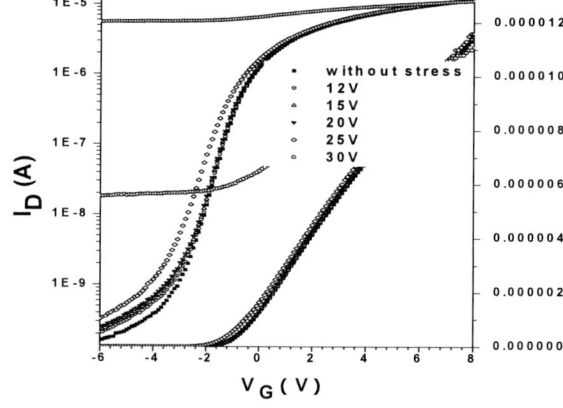

Fig.8 Transfer characteristic of n-channel device under different drain bias for 1000s

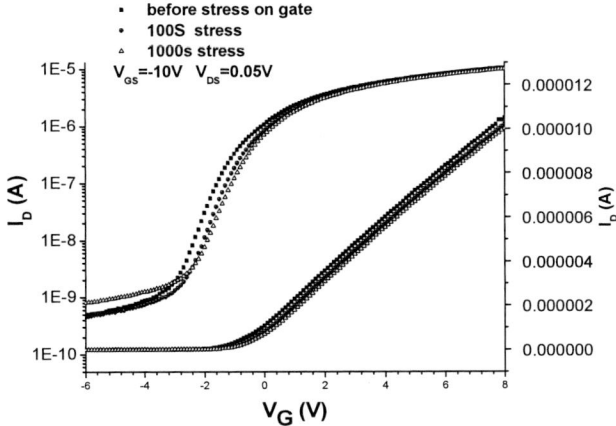

Fig.9 Transfer characteristic of (110) n-channel device under negative gate bias 10V and drain voltage of 0.05V

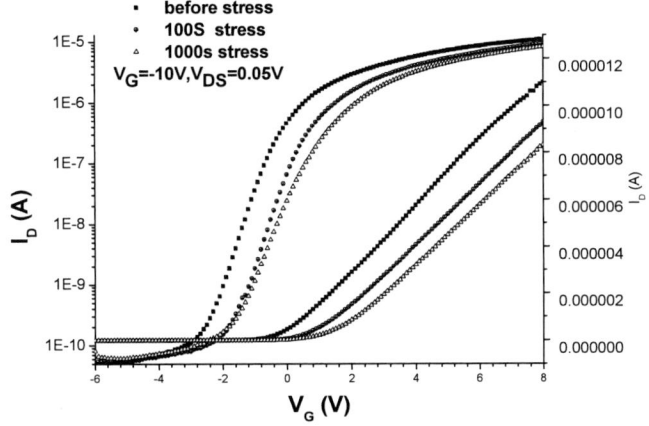

Fig.10 Transfer characteristic of (100) n-channel device under negative gate bias 10V and drain voltage of 0.05V

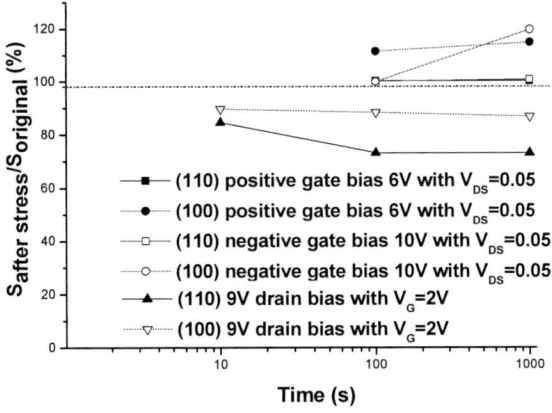

Fig.11 Mobility of (100) and (110) n-channel device after different stress

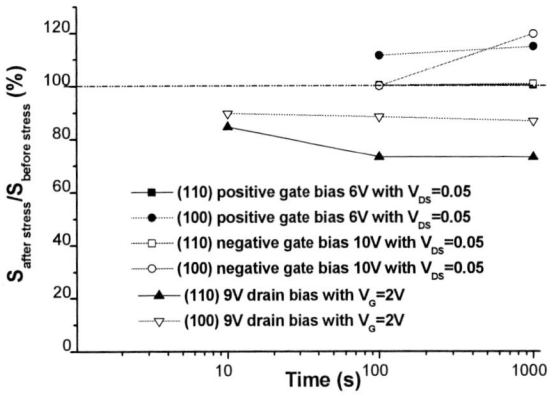

Fig.12 Subthreshold change of (100) and (110) n-channel device after different stress

IV. SUMMARY

We have studied n-channel (100) and (110) SG-TFTs under positive gate stress, drain stress and off-state stress. Under positive gate stress, hot holes dominate the deterioration of device with a negative shift of the V_{th}. Under positive drain bias, hot electrons dominate the deterioration of the device with a positive shift of the V_{th}. Under negative gate bias, electrons tunneling through the drain to channel dominate the deterioration of the device with a positive shift of the V_{th}. SG-TFT shows more stable performance than poly-Si TFT under different stress conditions.

ACKNOWLEDGMENT

This research work is supported by Dutch STW project.

REFERENCES

[1] V.Rana, Ryoichi Ishihara, Yasushi Hiroshima, Daisuke Abe, Satoshi Inoue, Tatsuya Shimoda, Wim Metselaar,and Kees Beenakker, "Dependence of Single-Crystalline Si TFT Characteristics on the Channel Position Inside a Location-Controlled Grain" IEEE Trans. Electron Devices, vol. 52, no. 12, pp. 2622-2628, December 2005

[2] Tai.Sato, Yoshiyuki Takeishi, Hisashi Hara, "Mobility Anistropy of Electrons in Inversion Layers on Oxidized Silicon Surfaces" Physical Review, vol. 4, no. 6, pp. 1950-1960, Septmber, 1971

[3] Shimada Hiroyuki, Ryoichi. Ishihara, Tao Chen, "半導体装置の製造方法、電気光学素子、および電子機器" Japanese Patent, 2008-129544, May 16th, 2008

[4] Chen Tao, R. Ishihara, Wim Metselaar, Kees Beenakker , Meng Yue Wu, "Location and Crystallographic-Orientation Control of Si Grains through Combined MILC and μ-Czochralski process", Japan Journal of Applied Physics, vol 47, no. 3, pp. 1880-1883, March, 2008

[5] Tao Chen, Ryoichi Ishihara, Wim Metselaar, Kees Beenakker, "<110> Orientation and Location Controlled Si Grains Through Combined MILC and μ-Czochralski Process" Proceeding of International Display Workshop, pp. 2011-2012, December 2007

[6] Tao Chen, Ryoichi Ishihara, Baiano Alessandro Kees Beenakker, "Highly Uniform Single-Grain Si TFTs Inside (110) Orientated Large Si Grains" Proceeding of International Display Workshop, pp.1599-1600, December 2008

978-1-4244-5430-3/10 $26.00 © 2010 IEEE

[7] Giannis P. Kontogiannopoulos, Filippos V. Farmakis a, Dimitrios N. Kouvatsos, George J. Papaioannou, Apostolos T. Voutsas, "Hot-carrier stress induced degradation of SLS ELA polysilicon TFTs – Effects of gate width variation and device orientation" Solid State Electronics, 52 pp. 388-393, 2008

[8] Masaaki Kinugawa,Toshiroh Usami and JUN ICHI MATSUNAGA "Effects of Silicon Surface Orientation on Submicron CMOS Devices" International Electron Device Meeting, pp581-584, 1985

[9] Doyle B. The hot carrier effect. In: Chang CY, Sze SM, editors. ULSI devices. John Wiley & Sons, Inc; 2000. p. 281-3 chapter 6

[10] Yukiharu URAOKA, Tomoaki HATAYAMA, Takashi FUYUKI, Tetsuya KAWAMURA and Yuji TSUCHIHASHI, "Hot Carrier Effects in Low-Temperature Polysilicon Thin-Film Transistors" Japan Journal of Applied Physics, Vol. 40 pp.2833-2836, 2001

[11] Satoshi INOUE, Mutsumi KIMURA and Tatsuya SHIMODA "Analysis and Classification of Degradation Phenomena in Polycrystalline-Silicon Thin Film Transistors Fabricated by a Low-Temperature Process Using Emission Light Microscopy", Japan Journal of Applied Physics, Vol. 42 pp.1168-1172 (2003)

[12] Marina Valdinoci, Luigi Colalongo, Giorgio Baccarani, Guglielmo Fortunato, A. Pecora, and I. Policicchio, "Floating Body Effects in Polysilicon Thin-Film Transistors" IEEE Transactions on Electron Devices, Vol. 44, No. 12, 1997

[13] D N Yaung, Y K Fang, K C Huang, C Y Chen, Y J Wang, C C Hung, S G Wuu and M S Liang, "Mechanism of device instability for unhydrogenated polysilicon TFTs under off-state stress" Semicond. Sci. Technol, 15 pp.888–891(2000)

Non-uniform Threshold Voltage and Non-saturating Drain Current in Amorphous-Si TFT after Saturation-Mode Bias Temperature Stress

C. R. Wie and Z. Tang

Department of Electrical Engineering
State University of New York at Buffalo
Buffalo, NY, 14260, USA
phone: 1–716-645-1023, e-mail address: wie@buffalo.edu

Abstract—We investigated the degradation effects on a-Si:H thin film transistors under a saturation-mode bias temperature stress. The simulated I-V characteristics using a non-uniform channel profile of V_t, with a maximum at source and minimum at drain, agreed very well with the measured data. The C-V characteristics were simulated using ATLAS based on this non-uniform V_t profile, which agreed well with data. The non-saturating reverse configuration I_d-V_{ds} characteristics was explained using the channel length modulation effect. Moreover, the device V_t extracted from linear I_d-V_{gs} characteristics can be modeled as a weighted average of the non-uniform V_t channel-profile. For short-channel devices, the experimental C-V data and ATLAS simulation show that the channel profiles of V_t, deep (NGA) and tail (NTA) state densities are affected significantly by the self heating effect during stress, where NGA (NTA) shows a flat maximum (minimum) level from source to mid channel.

Keywords- a-Si:H TFT; Bias-temperature stress; Non-saturating drain current; Short-channel; Non-uniform Threshold voltage shift; Self-heating effect

I. INTRODUCTION

The research interest of a-Si:H TFTs has recently moved from active-matrix liquid crystal display (AMLCD) to active-matrix organic light emitting diode display (AMOLED) significantly. In AMLCD, the TFT is used as a switch, controlled by a pulsed gate bias with duty cycles ≤0.1%, making the circuit fairly stable and nearly free from a large threshold voltage (V_t) shift. In contrast, the current driver TFT in an AMOLED pixel operates under a DC bias with a comparable magnitude in both the gate and drain voltages, and the transistor V_t limits the OLED current directly. In the gate pixel circuit of AMLCD, the measured V_t will be the same as the actual V_t because the channel after stress is nearly uniform. However, in the AMOLED driver circuit where the driver TFT operates in a saturation mode, the drain bias reduces the gate overdrive stress non-uniformly along the channel (*i.e.*, overdrive stress is maximum at Source and minimum at Drain and varies continuously) and therefore, it results in a non-uniformly distributed V_t profile [1]. The measured V_t will be an averaged value over the entire channel. It is necessary to learn the behavior and correctly model the V_t in a current driver circuit of AMOLED and in general, in the a-Si:H TFTs used in an analog circuit. This will help the circuit designer to predict

the TFT lifetime in AMOLED more accurately. In addition, when the channel length decreases to below about 10 μm with a gate width of 400μm, a non-saturating I_d-V_{ds} characteristic, analogous to the "kink effect", became apparent [2]. The self-heating effect also became important in the virgin, short channel devices. They both contribute to the non-saturating drain current. The cause for non-saturating drain current characteristic (often referred to as kink effect erroneously) in a-Si:H TFT is not well understood yet. We recently argued that this non-saturating drain current can be explained reasonably well by the channel length modulation (CLM) and the non-uniform threshold voltage profile $V_t(y)$ [2].

II. EXPERIMENTS

The a-Si:H TFTs used in the experiments had an inverted staggered structure with symmetrical source and drain terminals. The gate insulator silicon nitride (SiN_x) had a thickness of 430nm and the gate metal consisted of Mo/Al/Mo. All metal layers were sputter deposited, and all dielectric layers were PECVD deposited [3]. Both the stress experiments and the electrical measurements were performed on a temperature-controlled hotchuck, which was mounted on a probestation, all within an electromagnetic interference (EMI)-shielded dark box. The stress experiments and the measurements were conducted at an elevated temperature, such as 55°C or 100°C.

The I-V transfer characteristics in linear mode (V_{ds}=0.1V) and in saturation mode (V_{ds}=20V), as well as the output characteristics, were measured before and after stress using the HP4145B semiconductor parameter analyzer. I-V characteristics were measured in the forward (in which source and drain terminals were the same between the stress and the measurement) and reverse (in which source and drain terminals are interchanged) configurations. The gate-to-drain capacitance (C_{gd}) and gate-to-source capacitance (C_{gs}) were measured, before and after the stress, using the Agilent E4980A LCR meter with the high signal terminal connected to the gate, the low signal terminal connected to the drain or source and the unused terminal connected to a physical ground. For the virgin samples, the I-V data were measured after the C-V data. For the stressed samples, the I-V data were measured intermittently throughout stress and the C-V data was measured once after all stress experiments were completed.

III. RESULTS

Figure 1 shows the C-V characteristics at 10kHz measured before (C_{gs0} and C_{gd0}) and after (C_{gs} and C_{gd}) a saturation-mode BTS at $V_{GS}=30V=V_{DS}$ at $55^{\circ}C$ for 900s [2].

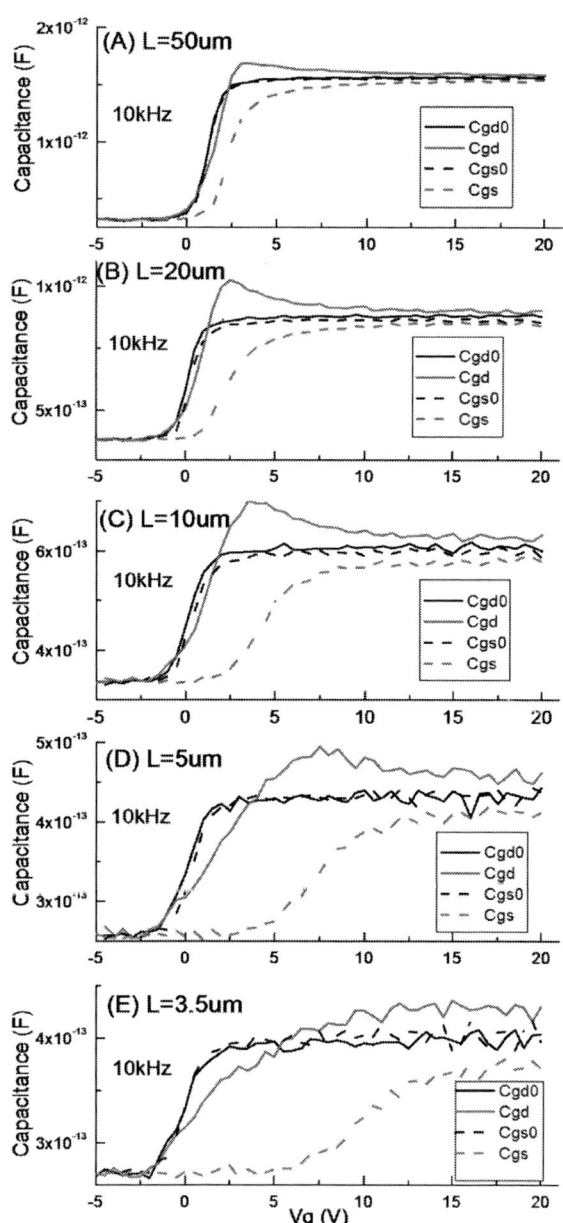

Fig. 1 C_{gs}-V_g (dash) and C_{gd}-V_g (solid) data at 10 kHz before (C_{gs0} and C_{gd0}) and after (C_{gs} and C_{gd}) the 900s BTS at $V_{gs}=30V=V_{ds}$ and $55^{\circ}C$ [2]. $W=400\mu m$.

The C_{gs0} and C_{gd0} data taken before stress closely overlap each other as expected from the sample symmetry and the uniformity of channel. After the stress, however, the C_{gs}-V_g data is right-shifted more than the C_{gd}-V_g data, clearly indicating that the V_t is higher at the Source side than at the Drain side of the channel [1,2]. The C-V data also shows that the maximum V_t-shift, found at the source side, increases with the decreasing channel length. The I_d-V_{ds} data at $V_{gs}=12.5V$, measured after the BTS, are shown in Figure 2 as symbols. The calculated curves (line) will be discussed later in this

section. We have shown that both the C-V and I-V data can be explained by considering the following [1,2,4]: (i) the saturation-mode BTS induces a continuously varying threshold voltage along the channel, $V_t(y)$ where $y=0$

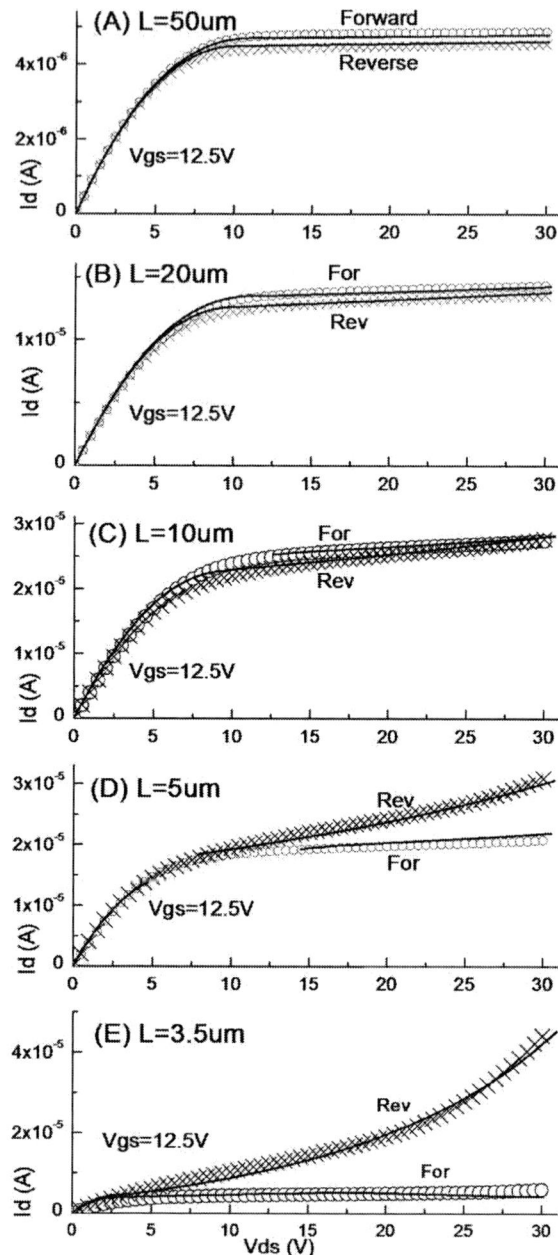

Fig.2 Experimental (symbol) and calculated (solid) forward and reverse output characteristics after a saturation-mode BTS at $V_{gs}=30V=V_{ds}$ for 900s. Part of the linear-saturation transition region calculations are not shown [2]. $W=400\mu m$.

is source and $y=L$ is drain with L=channel length; (ii) for long-channel ($L\geq 20\mu m$), both C-V and I-V data can be well explained by only the non-uniform $V_t(y)$ profile which is shown in Fig.3. Particularly, the relative level of drain current (*forward* is higher than *reverse*) in Fig.2(A) and (B) is explained by this $V_t(y)$ profile.; (iii) for short-channel ($L \leq 10\mu m$), the increasingly non-saturating reverse characteristics can be explained by the $V_t(y)$ profile and channel-length modulation (CLM); and (iv) again in the short channel samples, the self-heating effect of the BTS significantly altered the $V_t(y)$ profile, in which the final $V_t(y)$ profile is a combined product of the square-root function overdrive stress profile of Fig.4(A), which has maximum at the source end, and the channel-profile of SHE temperature which has maximum near the middle of channel [4]. In addition to the above four main results, we also measured the V_t-shift values as a function of the stress V_{DS} voltage.

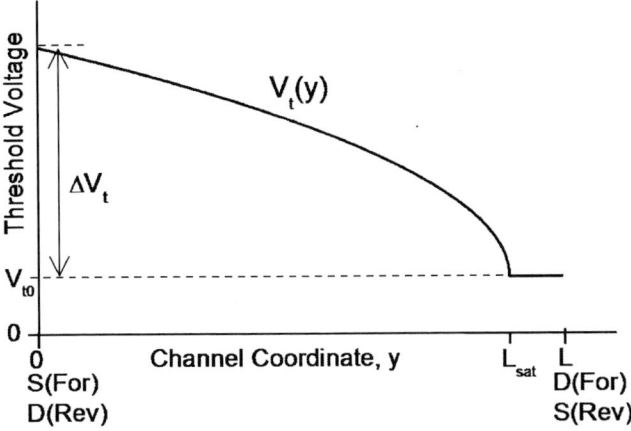

Fig.3 A sketch of the non-uniform threshold voltage profile along the channel [2].

1. Long-channel (L>10µm) a-Si:H TFT after saturation-mode BTS

We follow the defect pool model of Powell [5] and derive the channel V_t-profile after the saturation-mode stress. First, we assume that the final V_t-profile will follow the initial channel profile of electron density induced by the stress voltage. Fig.4 shows a gate overdrive stress (V_{ov}) profile corresponding to the channel potential (V_{str}) due to the V_{DS} stress voltage with an initial, uniform threshold voltage [1]. This overdrive stress will produce a charge density profile as given by $Q_{str}(y) = C_G V_{ov}(y)$ $= C_G(V_{GS} - V_{t0} - V_{str}(y))$ and according to the defect pool model, the V_t-shift, $\Delta V_t(y)$, is proportional to $Q_{str}(y)$. Consequently, the non-uniform $V_t(y)-V_{t0}$ follows $V_{ov}(y)$ [1]. The resulting non-uniform threshold voltage profile is shown in Fig.4(B). The negative overdrive stress beyond the pinch-off point is neglected. The position-dependent $V_t(y)$, can be found from the uniform stress BTS data (at V_{DS}=0V), i.e., ΔV_t as a function of $V_{GS}-V_{t0}$ given by $\Delta V_t = B(V_{GS}-V_{t0})+C$ where B and C are found from the data as Fig.5(a) shows[3]. Here, this ΔV_t is constant throughout the channel because V_{DS}=0V. When V_{DS} $\neq$ 0V, by replacing the gate overdrive stress $V_{GS}-V_{t0}$ of the uniform stress (V_{DS}=0V) by the local overdrive stress $V_{ov}(y)=V_{GS}-V_{str}(y)-V_{t0}$ of the non-uniform BTS ($V_{DS}\neq$0V), the V_t-shift can be found as follows: The threshold voltage extracted from a triode-mode (V_{ds}=0.1V) I_d-V_{gs} data is approximately a simple channel average of $V_t(y)$ [3]:

$$< \Delta V_t > \cong \frac{1}{L} \int_0^L \Delta V_t(y) dy \qquad (1)$$

This may be written as

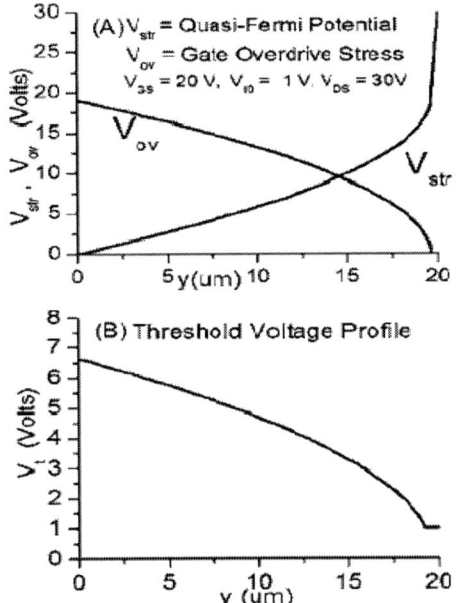

Fig.4 (A) An example profile of quasi-Fermi potential and gate overdrive stress under a saturation-mode bias stress. (B) An example of the resulting nonuniform profile of V_t[1].

$$< \Delta V_t > = \frac{W\mu}{L I_D} \int_0^{V_{DS}} \Delta V_t(V) Q_{str}(V) dV \qquad (2)$$

Here, V is the same as V_{str} mentioned earlier. For a triode mode stress, this yields

$$< \Delta V_t > = \frac{2}{3} B \frac{(V_{GS}-V_{t0})^3 - (V_{DS}-V_{t0})^3}{(V_{GS}-V_{t0})^2 - (V_{DS}-V_{t0})^2} + C \qquad (3)$$

And for a saturation-mode stress,

$$< \Delta V_t > = \frac{2}{3} B (V_{GS} - V_{t0}) + C \qquad (4)$$

For C=0, the above results reduce to those of ref.[6]. Figure 5(b) shows a reasonable agreement between data and calculation based on Eqs.(3) and (4). It also shows that it is correct to interpret that the role of channel potential (due to the stress V_{DS} voltage) is to reduce the gate overdrive stress. The data shows an initially decreasing ΔV_t with the increasing stress V_{DS} and a saturated ΔV_t at a higher V_{DS}. The validity of the $V_t(y)$ profile in Fig.3 was further confirmed by comparing the calculated I-V data against the experimental data. The drain current was calculated in the gradual channel approximation (GCA) and the position dependent field-effect mobility $\mu_{FE}(y)$ [1].

$$I_d = \frac{W C_G}{L'} \int_0^{V_d'} dV (V_{gs} - V_t(y) - V(y)) \mu_{FE}(y) \qquad (5)$$

Here, L'=L and V_d'=V_{ds} for the triode-mode measurement and L'=L_{sat} and V_d'=V_{dsat} for the saturation-mode measurement, where L_{sat} is the distance between source and pinch-off point and V_{dsat} is the quasi-Fermi potential at the pinch-off point. The calculated (solid lines) linear-mode and saturation-mode I_d-V_{gs} and the output characteristic curves are compared with the experimental data (symbols) in Fig.6, for the forward (source and drain terminals are the same between stress and measurement) and reverse (source and drain are interchanged) characteristics.

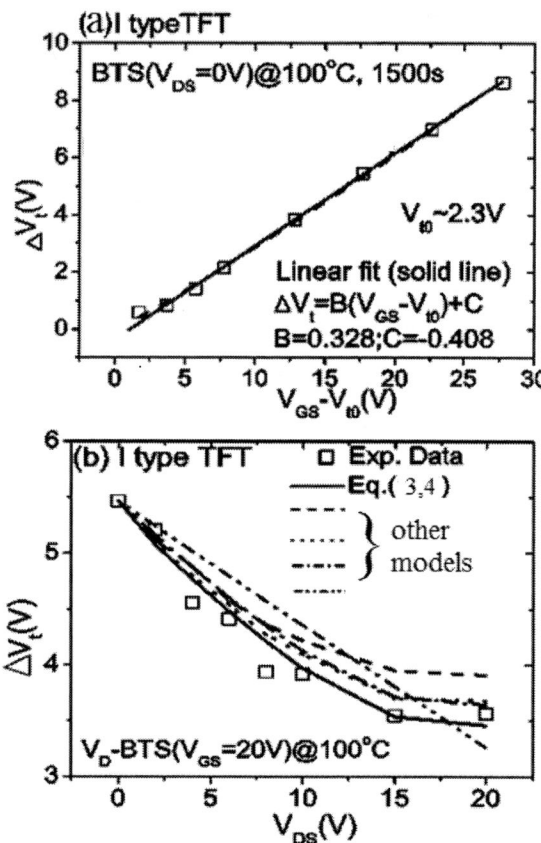

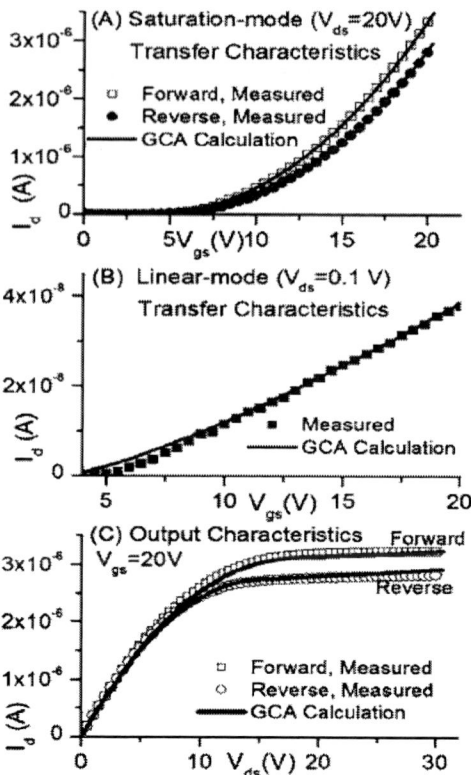

Fig.6 The GCA-calculated (A) saturation, (B) linear transfer characteristics, (C) output characteristics were compared with the measured data (symbol) in forward and reverse configuration [1]. W/L=400µm/20µm.

Fig.5 Both (a) the uniform stress and (b) the non-uniform stress results [3]. Other models include ref. [6] and are discussed in detail in [3]. W/L=400µm/20µm.

The good agreement between the calculated I-V characteristics and the experimental data indicates that the non-uniform V_t-profile of Fig.3 is indeed correct. In addition, the difference between the forward and reverse saturation currents (see Fig. 6(c)) is due to the V_t-profile alone, which has maximum at the source-end for *forward* but has maximum at the drain-end for *reverse* [1].

The C-V characteristics were simulated by ATLAS. Since the V_t shift after BTS is caused by the deep level defect states generated by the BTS, the peak density of states parameter (N_{GA}) in the ATLAS TFT density of states model [7] was used as the key variable to simulate the non-uniform V_t profile.

Figure 7 shows C-V and density of states (DOS) profiles before and after a uniform (V_{DS}=0V) BTS at V_{GS}=30V and 55°C. The voltage-shifted C-V curve after BTS of Fig.7(a) can be fitted with the increased deep state density (NGA) (dashed line in Fig.7(b)). The N_{GA} channel-profile of Fig.8(a) was made such that it follows the $V_t(y)$ profile of Fig.3 [4]. Both simulated curves of C_{gs} and C_{gd} (solid lines) agreed well with the experimental data as Fig.8(b) shows.

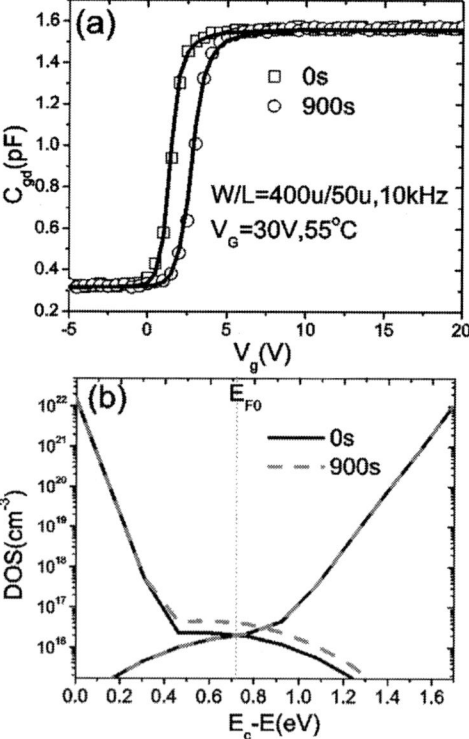

Fig.7(a) The measured C_{gd} data (symbols) and the ATLAS simulation results (lines); (b) DOS energy profile before (solid) and after (dashed) the stress. Bias temperature stress with V_G=30V was applied for 900-seconds at 55°C [4]. W/L=400µm/50µm.

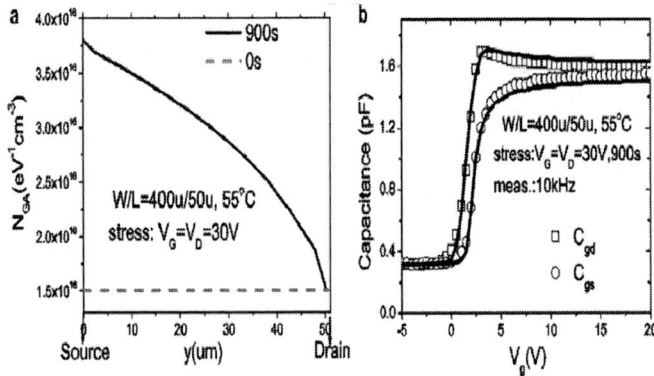

Fig.8 (a) NGA profile used in atlas simulation (b) the measured c_{gd} and c_{gs} curves (symbol) were compared with the simulated curves (solid lines) [4]. W/L=400μm/50μm.

2. Short-channel (L≤10μm) Non-saturating I_d-V_{ds} Characteristics

For short-channel, three things needed to be considered: source and drain contact resistance R_{ds}, channel length modulation (CLM), and self-heating effect (SHE) [2]. After the BTS, however, the SHE could be ignored because of the increased V_t which reduces the current level, which in turn decreases the self heating effect to an insignificant level. The R_{ds} and CLM were incorporated into the GCA calculation, Eq.(5), with the following replacement of parameters [2]: $L' = L$ and $V'_d = V_{ds} - I_d R_{ds}$ for triode mode; $L' = L - \Delta L$ and $V'_d = V_{dsat} - I_d R_{ds}$ for saturation mode where $V_{dsat} = V_{gs} - V_t(L')$; and R_{ds} is the sum of source and drain contact resistances plus the bulk resistance between the external source (or drain) contact and the end of intrinsic channel, given as $R_{ds} = R_s + R_d = 2R_s$. For a non-uniform sample with $V_t(y)$ as shown in Fig.3, the forward I_d-V_{ds} has a large V_{dsat} because of the small $V_t(y=L)$ at the drain. The reverse I_d-V_{ds} has a small V_{dsat} because of the large $V_t(y=0)$ at drain of the reverse configuration (see Fig.3). Therefore, a large forward V_{dsat} means a small CLM effect ($\Delta L = \lambda'(V_{ds}-V_{dsat})$), and the small reverse V_{dsat} means a large CLM effect. Figure 2 (D) and (E) show a large (super-linear) increase of the reverse current over the forward current. Eq.(5) was numerically integrated assuming that the channel potential $V(y)$ is also a square-root function [1,2]. Here, ΔV_t and λ' were varied to fit the experimental data.

In addition to the difference in the *CLM* effect, the different behavior of the reverse-forward current also comes from the opposite behavior of $V_t(L')$ as the drain bias increases. The reverse $V_t(L')$ decreases while the forward $V_t(L')$ increases as the pinch-off point ($y=L'$) moves further into the channel from drain. The changing $V_t(L')$ alone will cause the forward current to decrease and the reverse current to increase as the drain bias is increased, as Fig.8 shows (the dashed curves without CLM). The CLM effect adds a rising current to both forward and reverse currents, but by a different amount as was pointed out earlier. The *CLM* effect adds a more rapidly rising current in the *reverse* direction, as Fig.8 shows (solid lines). This combined *CLM* and changing $V_t(L')$) effect seems to fit the experimental data reasonably well in both the *forward* and

reverse directions. The difference between the forward and reverse characteristics in the rising currents due to CLM is much more striking in the *L=3.5μm* sample as Fig.8(B) shows (curves-1 versus curves-3).

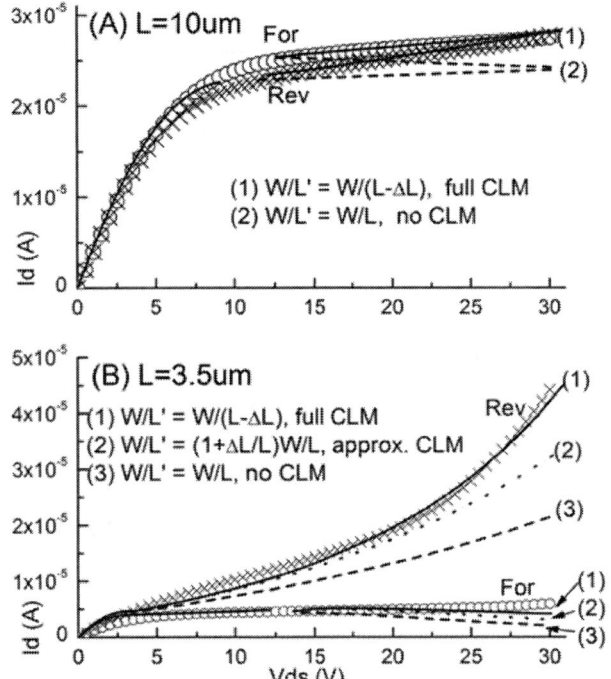

Fig.9 The effect of changing $V_t(l')$ in the drain current. (A) L=10um, (B)L=3.5um. the linear-saturation transition region where the fit quality was unreliable is omitted. W=400μm.

In addition, the shorter the channel length and the larger the ΔV_t, the $V_t(L')$ changes more rapidly with the increasing V_{ds}. This is evident by comparing the curves-2 of Fig.9(A) with the curves-3 of Fig.9(B) which are due to the changing $V_t(L')$. It should also be mentioned here that in calculating the forward current for L=3.5μm of Fig.9(B), the pinch-off happens at a large drain bias of $V_{dsat}=V_{gs}-V_t(drain) \approx 12V$. For $V_{ds} < V_{dsat}$, the device is in triode mode and the effective threshold voltage for the continuous channel of triode mode is a weighted average of $V_t(y)$ over the channel length [3]. Therefore, for this region ($V_{ds}<V_{dsat}$), a constant effective $V_{t,eff}$ was assumed, which value was then determined by fitting the forward current data over $0<V_{ds}<V_{dsat}$. The $V_{t,eff}$ thus obtained was about 9.5 V, which is midway between the simple average (7.7V) of $V_t(y)$ and the maximum V_t (about 11.5V) for L=3.5μm [2].

3. Short-channel (L <10 μm) and Effect of Self-Heating on $V_t(y)$ Profile

It was found that the self-heating, if present during the saturation-mode BTS, produces a significant effect on the $V_t(y)$ profile. This is because the temperature rise due to SHE is not uniform across the channel. We first analyzed the C-V data using ATLAS simulation based on the assumed NGA and NTA profiles. The assumed profiles will be justified later.

978-1-4244-5430-3/10 $26.00 © 2010 IEEE   351

Figure 10 shows the experimental (symbol) and simulated (solid) C-V data where the ATLAS C-V simulation was based on the NGA and NTA profiles shown in Fig.11.

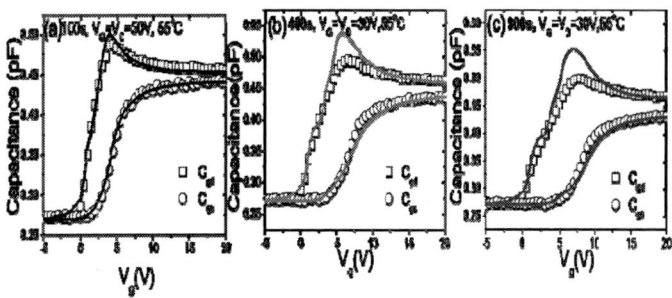

Fig.10 Calculated and experimental C_{gd}-V_g ($\square$) and C_{gs}-V_g (O) data. Calculated curves are by ATLAS device simulator based on the channel-profile of deep Gaussian acceptors (NGA) and shallow tail acceptors (NTA) as shown in Figure 11 [4]. $W/L=400\mu m/5\mu m$.

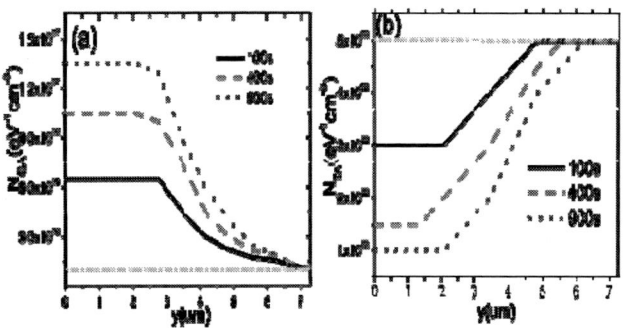

Fig.11 The channel profiles of deep (NGA) and shallow (NTA) defectsused in the ATLAS simulation of C-V curves of Fig10 [4].

The flat portion in the NGA profile of Fig.11(a) extended from Source into the mid channel. This is believed to be due to the self-heating effect by the BTS. The SHE is known to increase the device temperature non-uniformly along the channel with a maximum temperature in the middle of the channel [8]. Channel profiles of the gate overdrive stress voltage and of the temperature (due to SHE) are shown in Fig.12. When the two effects (i.e., saturation-mode BTS and SHE) are combined, the NGA profile of Fig.11(a) results, which is relatively flat from source to the mid channel. Note that in the regions where NGA increases with the increasing stress time (between source and mid channel), there is a decrease in the NTA (Fig.11(b)). This is probably because the deep states (NGA, the dangling bonds) are created by breaking the tail states (NTA, the weak bonds) [5]. However, the decrease in NTA was orders of magnitude higher than the increase in NGA.

Considering the $V_t(y)$ profile due to the saturation-mode bias stress (which has a maximum stress at source, decreasing toward drain, following a square-root profile, see Fig.4(A)) and the SHE temperature profile during stress (with maximum near mid channel), which were shown in Figure 12, the resulting NGA profile of Fig.11(a) is conceivable.

Finally, the dependence of C-V curve on the tail state density (NTA) profile is shown in Fig.13. The flat portion of capacitance at the higher V_g range is affected sensitively. A uniform NTA, with the NGA profile given by Fig.11(a), gives rise to the C-V curve shown by the dashed line; while the stepped-up NTA profile, shown in the inset of Fig.13, with the stepped-down NGA profile of Fig.11(a), produced the C-V curve shown as solid line Fig.13. It is this high V_g portion C-V data that allowed determination of NTA profile.

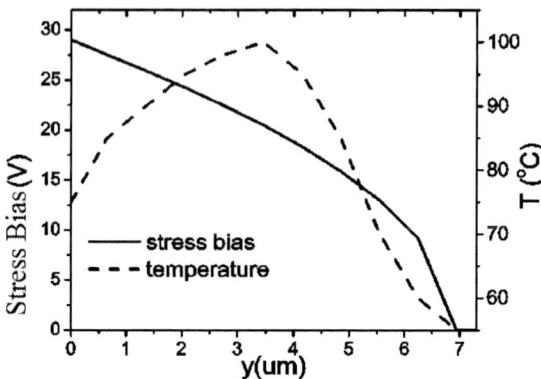

Fig.12 Qualitative channel profiles of bias stress and SHE-induced temperature in a L_d=5-μm (drawn width) device which shows about $7\mu m$ L_{eff} [4].

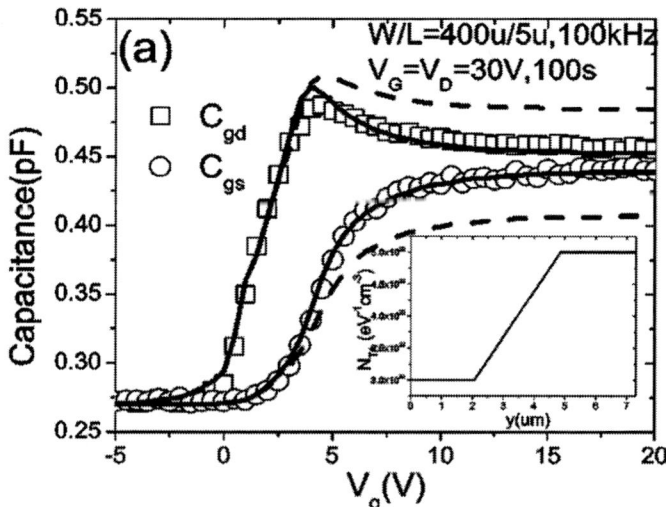

Fig.13 The dashed lines were simulated with a uniform constant NTA from source to drain; and the solid line was simulated with the NTA channel-profle shown in the inset [4].

IV. SUMMARY

Amorphous Si TFTs in analog circuits, operating in saturation mode, experience a bias stress that is not uniform across the channel. As opposed to the interpretations in the literature about the effect of the saturation-mode BTS, such as the maximum channel V_t controlling the device operation [9], a two-region model similar to the hot-carrier degraded MOS transistor [10] or the total charge and averaged-V_t approach [6], we have developed a position-dependent V_t channel-profile

(Fig.3) and showed its validity by analyzing the I-V and C-V data. We also developed a channel-length modulation model to explain the non-saturating output characteristics in short-channel devices [2]; showed that the self-heating effect of the BTS significantly modifies the V_t and DOS channel-profiles in short-channel devices (Fig.11) [4]; and developed a transistor parameter extraction method for such short-channel a-Si:H TFTs [11]. The contact resistance R_{sd}, and CLM and SHE parameters were incorporated into the parameter extraction method for short-channel a-Si:H TFTs [11]

ACKNOWLEDGMENT

The authors would like to thank Dr. Mun-Soo Park of Samsung Electronics for providing the samples.

REFERENCES

[1] C.R. Wie, Z. Tang, M.S. Park, "Non-uniform Threshold Voltage Profile in a-Si:H TFT Stressed Under Both Gate and Drain Biases", *J. Appl. Phys.*, 104, 114509, (2008)

[2] C.R. Wie, "Non-saturating Drain Current Characteristic in Short-Channel Amorphous Silicon Thin Film Transistors", *IEEE Trans. Electron Dev, to appear*, Apr. 2010

[3] Z. Tang, M.S. Park, S.H. Jin and C.R. Wie, "Drain bias dependent bias temperature stress instability in a-Si:H TFT" *Solid-State Electron.*, *53*(2) pp.225-233, (2009)

[4] Z. Tang, and C.R. Wie, "Capacitance-voltage characteristics and device simulation of bias temperature stressed a-Si:H TFTs" *Solid-State Electron.*, 54(3) pp.259–267, (2010)

[5] M.J.Powell. C. van Berkel, and J.R.Hughes, "Time and temperature dependence of instability mechanisms in amorphous silicon thin-film transistors", Appl.Phys.Lett., 54(14) 1323-1325 (1989)

[6] K.S. Karim, A. Nathan, M. Hack, W.I. Milne, "Drain-Bias Dependence of Threshold Voltage Stability of Amorphous Silicon TFTs", IEEE Electron Dev. Lett. 25(4) 2004, pp.188-190

[7] *ATLAS User's Manual* 2008

[8] S. Inoue, H. Ohshima, and T. Shimoda, "Analysis of Degradation Phenomenon Caused by Self-Heating in Low-Temperature-Processed Polycrystalline Silicon Thin Film Transistors", *Jpn.J. Appl. Phys.* 41(Pt1, No.11A), 6313-6319 (2002)

[9] Y.Kaneko, A. Sasano, and T. Tsukada, "Characterization of instability in amorphous silicon thin-film transistors", J.Appl.Phys.69,.7301-7305 (1991)

[10] R.Shringarpure, S.Venugopal, L.T.Clark, D.R.Allee, and E.Bawolek, "Localization of Gate Bias Induced Threshold Voltage Degradation in a-Si:H TFTs", IEEE Electron Dev.Lett. 29(1) 93-95 (2008).

[11] Z.Tang, M.S.Park, S.Jin, and C.R.Wie, "Parameter Extraction of Short-Channel a-Si:H TFT Including Self-Heating Effect and Kink Effect", IEEE Trans. Electron Dev., *to appear*, May 2010.

New Insight into the TDDB and Post Breakdown Reliability of Novel High-κ Gate Dielectric Stacks

K.L. Pey[1,*], N. Raghavan[1], X. Li[1,2], W.H. Liu[1], K. Shubhakar[1], X. Wu[1] and M. Bosman[2]

[1]Division of Microelectronics, School of Electrical & Electronic Engineering,
Nanyang Technological University (NTU), Singapore – 639798.
[2]Institute of Microelectronics (IME), Singapore Science Park II, Singapore – 117685.
*Ph: (+65) 6790-6371, Fax: (+65) 6792-0415, E-mail: – eklpey@ntu.edu.sg

Abstract — **In order to achieve aggressive scaling of the equivalent oxide thickness (EOT) and simultaneously reduce leakage currents in logic devices, silicon-based oxides (SiON / SiO₂) have been replaced by physically thicker high-κ transition metal oxide thin films by many manufacturers starting from the 45nm technology node. CMOS process compatibility, integration and reliability are the key issues to address while introducing high-κ at the front end. In this study, we analyze in-depth the reliability aspect of high-κ dielectrics focusing on both the time-dependent-dielectric breakdown (TDDB) and the post breakdown evolution stage. Electrical characterization, physical failure analysis, statistical reliability modeling as well as atomistic simulations have all been used to achieve a comprehensive understanding of the physics of failure in HK and the associated microstructural defects and failure mechanisms. The role played by different gate materials ranging from poly-Si → FUSI → metal gate and different HK materials (HfO₂, HfSiON, HfZrO₄) is also investigated. Based on the results obtained, we emphasize the need and propose a few approaches of design for reliability (DFR) in high-κ gate stacks.**

Keywords – Breakdown recovery, Grain boundary, High-κ dielectric, Interfacial layer, Metal gate, Post breakdown, Random telegraph noise (RTN), Time dependent dielectric breakdown (TDDB).

I. INTRODUCTION

For the past three decades, silicon dioxide (SiO₂) and recently oxynitride (SiON) have been the standard choice for gate oxide material in silicon-based complementary metal-oxide-semiconductor (CMOS) technology. Reliability of these oxynitride based MOS transistors was very good considering the continued process optimization involved in growing these films, eliminating extrinsic failure modes and ensuring a low defect density [1]. With continued downscaling however, the thickness of SiON had to be reduced to as low as 12Å, which is equivalent to 3-4 monolayers of atoms. These ultra-thin gate oxides showed very high intrinsic stress induced leakage currents (SILC) resulting in increased power dissipation during circuit operation [1]. To circumvent this high leakage problem and enable equivalent oxide thickness (EOT) downscaling at the same time, high-κ (HK) dielectrics turned out to be a more feasible solution. The physically thicker HK layer with a lower EOT helped to address the leakage issue. However, the introduction of an unoptimized HK thin film deposition process has given way yet again to extrinsic failure issues due to poor

(defective) interfaces with the gate and Si substrate and introduced microstructural defects such as grain boundaries [2, 3] and dislocations that are detrimental from the reliability point of view. There is a need for a holistic study on HK reliability and its dependence on material and microstructure as most studies in the past have been purely based on electrical characterization [4]. Unless physical failure analysis tools are employed to unearth the physical morphology and chemical composition of the defect in the failed device, conclusions from purely electrical studies remain highly speculative.

It is with this motivation that we present here a detailed investigation into the reliability of HK thin films combining and correlating the various electrical, statistical, physical analysis and atomistic simulation results to fully understand the physics of failure. Use of the various failure analysis tools help in identifying the root cause of failure which can later be used as an input to design for reliability (DFR) initiatives to fine tune and optimize the processing conditions so that process-induced defects could be eliminated. It can be confirmed from our studies that HK reliability is highly influenced by the surrounding materials (gate, substrate and presence of interfacial SiOₓ layer) as well.

The layout of the paper is as follows. Section II presents the electrical characterization results for various gate stacks. The failure analysis investigations on devices subjected to TDDB are presented in Section III focusing primarily on the transmission electron microscopy (TEM) – electron energy loss spectroscopy (EELS) and scanning tunneling microscopy (STM) tools. Following these qualitative studies, quantitative lifetime assessment using statistical models and their implications are dealt with in Section IV. Based on the analysis in the above sections, the inferences of these experimental results on HK gate stack technology is presented in Section V, emphasizing the need of design for reliability (DFR) in the front-end gate-first / gate-last HK process. Finally, we conclude the study highlighting the scope for further studies to be carried out relating to HK breakdown and reliability.

II. ELECTRICAL CHARACTERIZATION

Various devices with different material gate stacks have been tested for this study. Table I lists out the details of the devices tested and their dielectric thickness values. In order to understand the pattern of degradation and failure in HK stacks, different electrical test algorithms have been developed and

---

This work is supported by the A*STAR Institute of Microelectronics (IME) and Institute of Materials Research and Engineering (IMRE), Singapore.

978-1-4244-5430-3/10 $26.00 © 2010 IEEE

applied. We shall describe the various test algorithms used and the information that can be extracted from them. The three main algorithms are – (a) Two-step sequential HK→IL breakdown [5], (b) K-cycle multiple stage constant voltage stress (K-CVS) [6] and (c) Successive constant voltage stress (SCVS) test [7]. While the first one is used for TDDB stage, the second one is confined to the post-BD conduction study. The SCVS test spans both the TDDB and post-BD regime.

TABLE I. DETAILS OF THE VARIOUS DEVICES TESTED AND THEIR DIELECTRIC THICKNESS VALUES

Sample #	Gate Stack	Dielectric Thickness
A	Poly-Si – SiON	$t_{ox} = 16\text{Å}$
B	Poly-Si – HfO_2 – SiO_x - Si	$t_{HK} = 44\text{Å}$ $t_{IL} = 6\text{-}8\text{Å}$
C	NiSi – SiON – Si	$t_{ox} = 14\text{Å}$
D	NiSi – HfSiON – SiO_x – Si	$t_{HK} = 25\text{Å}$ $t_{IL} = 12\text{Å}$
E	TiN/TaN – $HfZrO_4$ - SiO_x	$t_{HK} = 25\text{Å}$ $t_{IL} = 12\text{Å}$
F	TiN - $HfLaO_x$	$t_{HK} = 12\text{Å}$
G	CeO_2 (Blanket) – n-Si	$t_{HK} = 30\text{Å}$
H	Sc_2O_3 / La_2O_3 / SiO_x / n-Si (Blanket – STM study)	$t_{Sc2O3} = 30$ Å $t_{La2O3} = 40$ Å $t_{SiOx} = 10$ Å

A. Two-Step Sequential HK-IL Breakdown

Although a pure HK film is deposited on the Si substrate, the annealing conditions involved during the CMOS process flow give rise to growth of a thin defective sub-oxide interfacial layer (IL) of SiO_x ($x < 2$) sandwiched between the HK and Si substrate. The presence of the IL is detrimental to EOT scaling but helps to improve adhesion of the HK which does not have a good interface with the Si substrate due to lattice and thermal mismatch issues [3]. Given the presence of the IL layer, it is necessary to evaluate the role of the HK and IL in the overall reliability of this dual layer dielectric gate stack. In general, the stress voltage applied for TDDB assessment in HK stacks is quite high [8] and for such stress conditions, once one of the dielectrics fail, the electric field across the surviving dielectric layer could exceed its breakdown field causing it to fail abruptly almost instantly. Therefore, single stage CVS is not a good tool to assess the "intrinsic" reliability of the bi-layer gate stack.

We have developed a two-step stress CVS methodology as shown in Fig.1 that enables successive controlled BD of the HK and IL by carefully tuning the compliance levels (I_{g1} and I_{g2}) and stress levels (V_{g1} and V_{g2}), where $V_{g2} < V_{g1}$. Poole-Frenkel conduction (specific to HK material) characteristics are used as a yardstick to investigate whether the HK or IL broke down first. The analysis revealed that for substrate injection stress conditions, HK is almost always the first layer to break down. Further details on this methodology and associated results are presented in [5]. In Section IV, the application of this two-step breakdown trend will be made use

of for quantitative statistical analysis of the lifetime of HK and IL layers. Fig.2 illustrates the two-step TDDB initiated in some devices of Samples B and D and Fig.3 is the corresponding I_g-V_g plot in these samples for fresh device, after 1-layer HK BD, after 2-layer HK+IL BD and progressive BD (~100μA compliance). Note that the FUSI sample shows a unique trend of leakage current recovery (about 2-3 orders of magnitude) after the 2-layer TDDB breakdown.

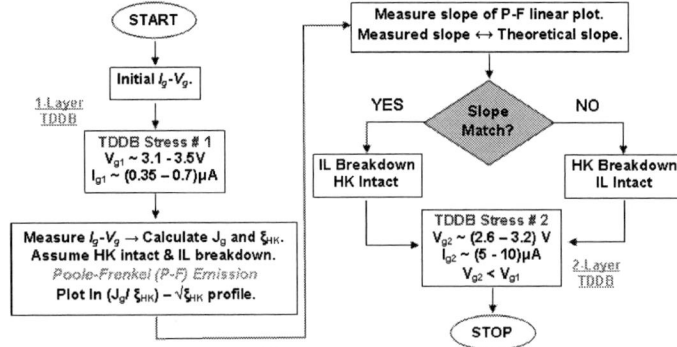

Figure 1. Electrical test algorithm to initiate a two-step sequential TDDB HK→IL breakdown and use of Poole-Frenkel emission slope as a criteria to determine the layer which breaks down first [5].

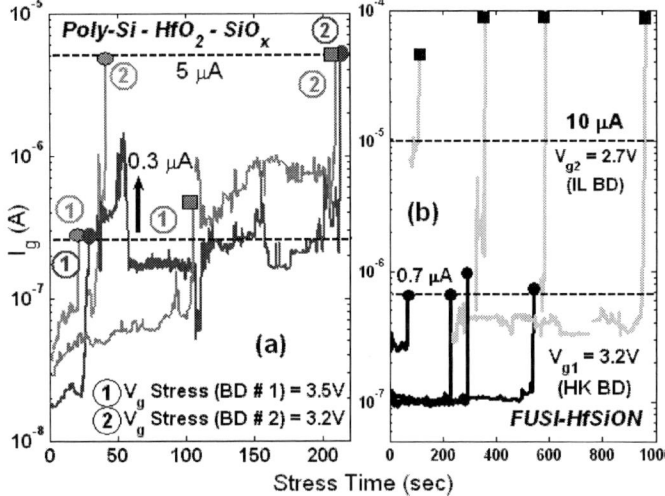

Figure 2. Observation of two-step sequential TDDB HK→IL breakdown in Samples B and D using the proposed test algorithm in Fig.1.

B. K-cycle Multiple Stage Constant Voltage Stress (K-CVS)

This algorithm involves application of a staircase-like voltage stress profile starting from very low operating gate voltage to incrementally higher gate stress conditions in the post-BD regime of nanodevices with the compliance set to a very high value. The purpose of this approach is to investigate the presence of random telegraph noise (RTN) fluctuations (also known as digital breakdown – Di-BD) at low gate voltages and the eventual transition to the analog breakdown (An-BD) [6] phase wherein there is a monotonic increase in the post-BD leakage current due to significant structural degradation of the dielectric caused by silicon epitaxial defects such as dielectric breakdown induced epitaxy (DBIE) [9] or percolation path dilation. It is believed that there is a critical voltage (V_{CRIT}) corresponding to this digital to analog transition, which is a strong function of the oxide thickness, as evidenced by past investigations on SiON dielectrics [10]. The Di-BD RTN phase provides for an additional reliability margin for the HK gate stack as device performance characteristics in this regime are moderate and within acceptable limits. Fig.4 illustrates the K-CVS leakage current trends in Sample B and Fig.5 shows explicitly the RTN trends in various HK gate stacks confirming that it is a universal phenomenon in the post-BD phase of HK materials.

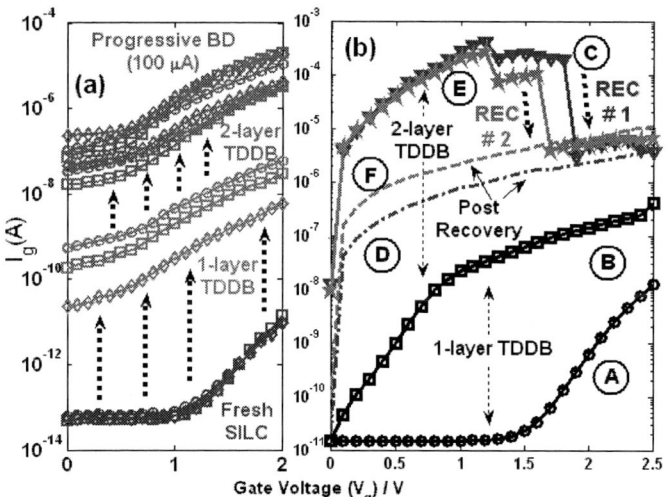

Figure 3. I_g-V_g plot for different stages of breakdown in (a) poly-Si and (b) FUSI gated high-κ samples. Note in (b), the unique trend of repeated leakage current recovery (sweeps C and E) and subsequent low leakage high resistance state trends (sweeps D and F) in the FUSI sample after a bi-layer TDDB breakdown.

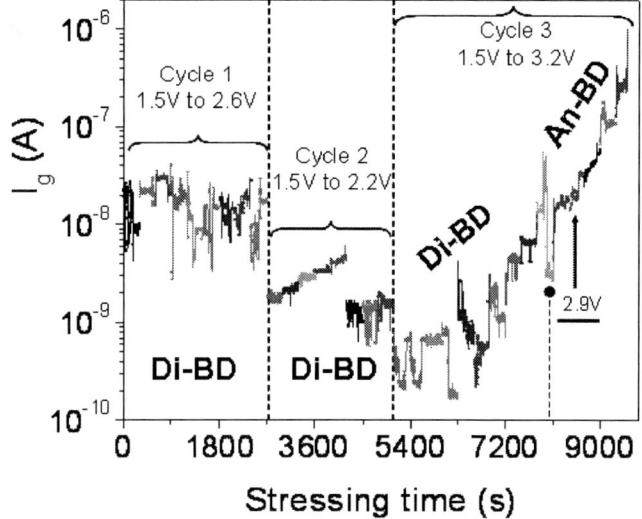

Figure 4. K-cycle multiple stage CVS in Sample B device for accumulation mode stress condition. Notice the random gate current fluctuations in cycles 1 and 2 (digital-BD), but a monotonic increase in current during cycle 3 for V_g>2.8V corresponding to analog-BD evolution. The critical voltage governing the Di-BD→An-BD transition is 2.9V.

It may be noted in Fig.5 that the RTN fluctuations for single layer oxynitride based dielectrics have clearer and fewer discrete current steps for the Poisson transition. However, the bilayer stacks in general show a more noisy pattern which may be attributed to the physically thicker stack (more traps in the

percolation path), different trap charge capture and emission properties of the HK and IL layers respectively and the presence of a high concentration of pre-existing process induced traps across the dielectric. As illustrated by Fig.6 [11], the RTN regime can be quite long and therefore, plays a significant role in extending the lifetime of HK gate stacks. Note that the clear observation of RTN signature is an indication of the "softness" of the BD event. In the analog regime, where microstructural defects are predominant, the RTN fluctuations appear as a background noise and are not clearly detectable as can be seen in the last cycle of Fig.4.

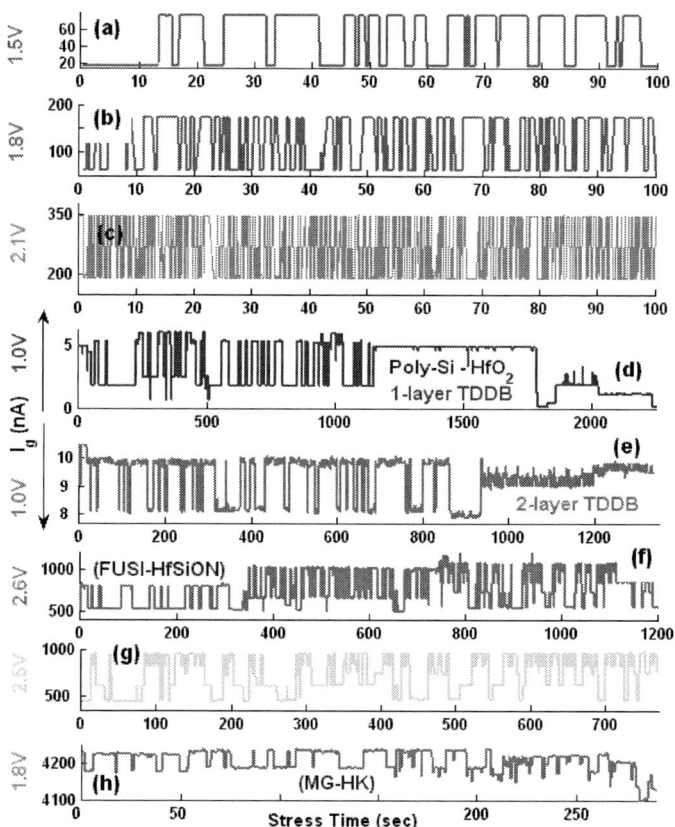

Figure 5. Post breakdown inversion stress RTN Di-BD fluctuations observed in various gate stacks at different test conditions – (a) – (c) → Sample A, (d), (e) → Sample B, (f), (g) → Sample D and (h) → Sample F. The corresponding gate stress values are shown in the ordinate axis. The common unit for the y-axis is "nA". Plots (d) and (e) correspond to RTN after 1-layer and 2-layer TDDB BD in the poly-HfO₂ based stack respectively. Plots (f) and (g) show the RTN noise after 1-layer TDDB and after post-HBD unipolar recovery in the FUSI-HfSiON sample.

C. Successive Constant Voltage Stress (SCVS)

A high stress voltage is applied to accelerate the TDDB and post-BD failure; however, the compliance levels (I_{gl}) are gradually relaxed from very low values to very high values as shown in Fig.7 [7]. When the test halts upon reaching compliance at each setting, device performance is evaluated. The purpose of this test is to investigate in detail the variation in device performance with breakdown hardness and observe any anomalies, such as the dielectric breakdown recovery in NiSi-based gate stacks (Samples C & D) (Fig.7).

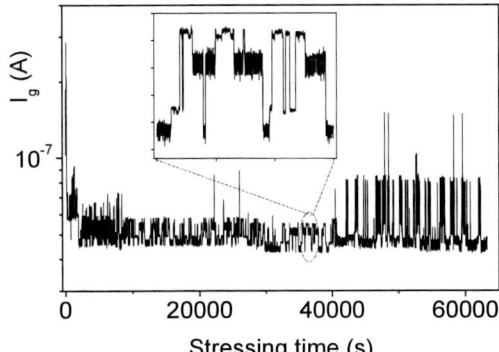

Figure 6. Prolonged observation of RTN noise in Sample D for more than 15 hours at a stress condition of $V_g = 1.3V < V_{CRIT}$ [11].

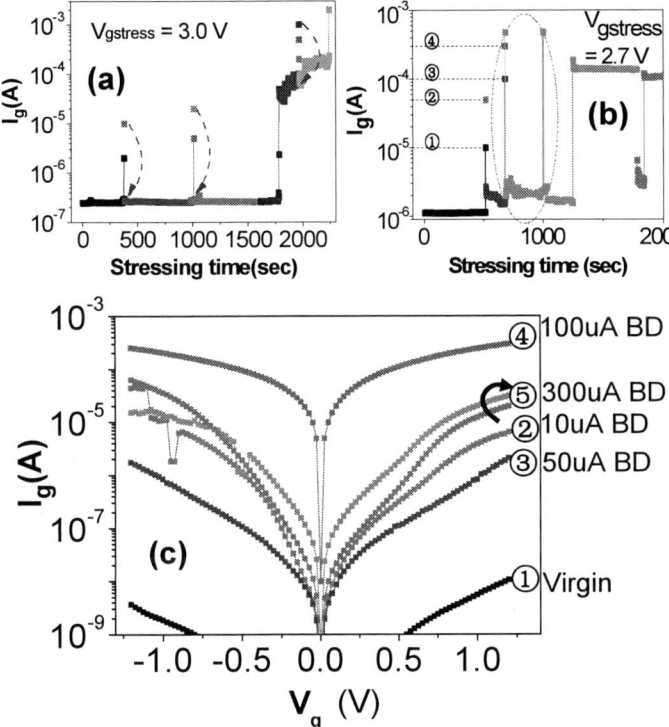

Figure 7. **(a)** Sample D subjected to SCVS at V_g=3.0V, I_{gl} is relaxed from 2µA, 10µA, 20µA and then to 2mA, with I-V measurement (such as I_g–V_g, I_d–V_d and I_s–V_s) taken upon reaching each I_{gl}. After I-V measurement, very low I_g, similar to pre-BD leakage, is observed when stress is resumed after I_{gl}=10µA, 20µA and 1mA, meaning that recovery in I_g happens during post-I_{gl} (i.e., I_g–V_g, I_d–V_d and I_s–V_s) measurement. **(b)** Similarly for Sample C pMOSFET under SCVS of 2.7V, switching behavior is observed i.e., I_g decreases to its original pre-BD value after I_{gl}=50µA and 300µA. **(c)** An I_g-V_g sweep confirming I_g switching to low value after I_{gl}=50µA, 300µA in (b) [7].

III. PHYSICAL FAILURE ANALYSIS

As discussed in the beginning, to understand the root cause of failure and design approaches to prolong device reliability, it is imperative to perform in-depth failure analysis study on the post-BD device and uncover the physics of failure and driving forces behind the failure mechanism. We make use of the STEM-EELS / energy dispersive X-ray spectroscopy (EDS) and STM high resolution analysis tools in this study on various gate stacks and identify the material dependent modes of failure.

A. Transmission Electron Microscopy

Fig.8 presents the HRTEM images of the breakdown site for (a) poly-Si gate stack, (b) NiSi FUSI stack and (c) & (d) TaN metal gate stack where the location of imaging is based on the breakdown location determined electrically [12]. The insets in the figure are the high angle annular dark field (HAADF) images. A summary of the failure defects and their driving forces is provided in Table II. Each of these stacks show a different defect signature. Failure in poly-Si gated materials involves a protrusion of the silicon epitaxy from the substrate into the oxide assisted by the electron wind force through the small cylindrical percolation path (~20-50 nm in diameter) [13] and thermomigration [14]. This epitaxial defect is called DBIE. Fig.8(b) shows migration of highly diffusive Ni from the S/D contacts towards the substrate assisted by temperature and concentration gradients. This dielectric breakdown induced metal migration (DBIM) [7] can lead to channel short for the case of a hard breakdown, which can be detrimental to device performance. As for the MG-HK devices, metal (Ta) filamentation is clearly seen wherein a punchthrough of the gate material occurs through the dielectric into the substrate [15]. This is caused by ionic and hole migration at high joule-heating temperature and electric field after the dielectric breakdown. Similar filamentation and gate metal (Ni) spiking has also been observed in some of the FUSI stacks.

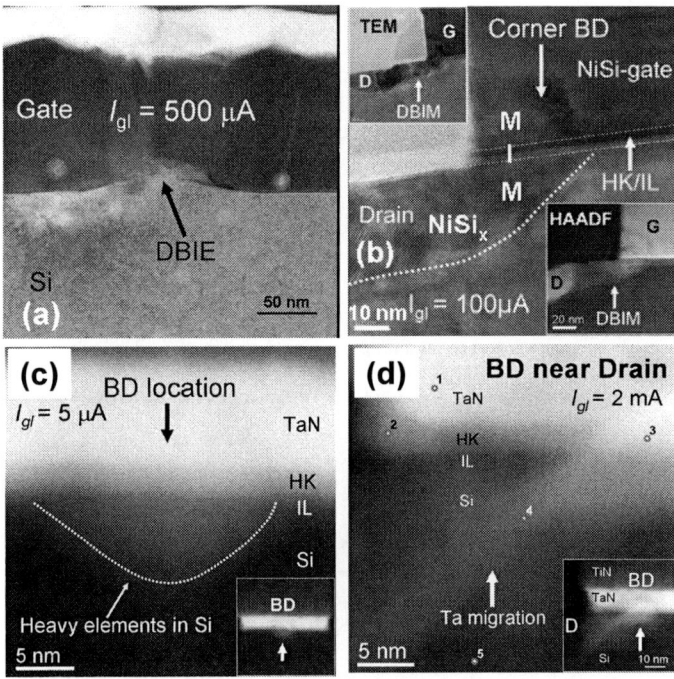

Figure 8. HRTEM image of dielectric BD in various gate material stacks – (a) Sample A, (b) Sample D with NiSi$_x$ phase formation and (c) & (d) Sample E. Clearly, the physics of failure is determined by the gate material.

B. Chemistry of Breakdown Filament – EELS / EDS

Having located and imaged the failure site, the EELS / EDS technique is used to find out the elemental composition and confirm the material which diffuses / migrates after the dielectric BD event. As an illustration, Fig.9 is the EELS result for one of the FUSI samples which showed Ni spiking from the gate into the substrate [16] and Fig.10 is the EDS spectra [15] corresponding to the filamentation in the TaN-based MG-HK stack of Fig.8(d).

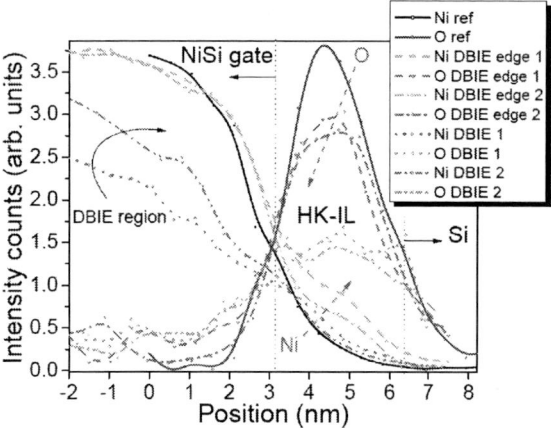

Figure 9. EELS image of Ni spiking from gate to substrate in Sample D. The Ni composition profile shifts to the right with respect to the reference fresh device profile indicating Ni punchthrough. Furthermore, the percolated region is O-deficient (comprising oxygen vacancies) and the NiSi gate is serving as a reservoir for oxygen ions storage due to its gettering capability [16].

TABLE II.   FAILURE DEFECTS OBSERVED IN VARIOUS MATERIAL GATE STACKS AND THEIR DRIVING FORCES

Gate Stack	Failure Defect in Oxide	Driving Forces
Samples A & B	DBIE Si nanowire filament Percolation Path Dilation Grain Boundary assisted failure in Sample B.	Current Density. Thermomigration. Grain Boundary – faster diffusion and enhanced conduction path.
Samples C & D	DBIE. DBIM (Metal migration in Ni/Co/Ti silicides). Ni spiking.	Temperature Gradient. Concentration Gradient. Ionic Conduction Electron Wind Force.
Samples E & F	Metal filamentation. Metal migration into substrate.	Ionic Conduction. Hole Migration.

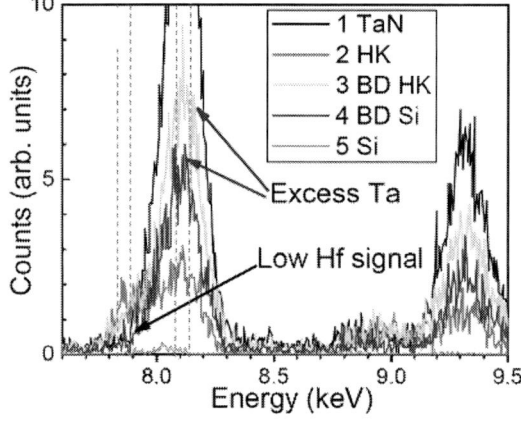

Figure 10. EDS spectra of filament for TaN MG-HK device (Sample E). The breakdown region in the dielectric and the substrate beneath the breakdown location show a drastic increase in the Ta count indicating Ta to be the main constituent of the filament [15].

C. Scanning Tunneling Microscopy (STM)

A very useful tool for localized BD study is the STM. This technique may be used to probe the conduction of HK blanket samples (without gate deposited) at very localized sites of interest. It may also be used to probe the process induced and stress induced traps in the HK as well as the IL depending on the stress polarity [17] and infer the energy and location of these traps in the gate stack [18]. Current induced tunneling spectroscopy (CITS) in conjunction with the topography image is used to identify the defective regions in the dielectric showing enhanced trap assisted tunneling (TAT). Fig.11 shows the (a) topography and corresponding (b) CITS map of a 3nm CeO_2 HK dielectric (without any IL layer) in a 30nm × 30nm area at an STM bias condition of +3.5V (Sample G). The contour of bright contrast (marked by the dotted line) observed in both these maps is indicative of high conduction pathways that are likely to be grain boundaries (GB) in the polycrystalline HK film. These GBs are defective inter-grain interfaces with high density of dangling bonds (process induced traps) and serve as low resistance paths for leakage conduction and percolation [3]. Grain boundaries therefore play a major role in determining the reliability of novel HK gate stack materials. From Fig.11, the diameter of these grains range from 15nm to 25nm.

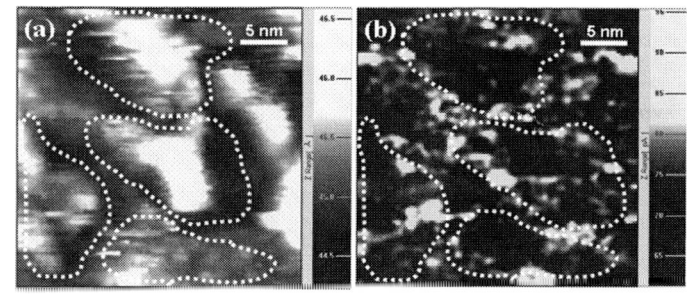

Figure 11. Bright contrast patches observed in (a) topography and (b) CITS images indicative of the grain boundary contours in a CeO_2 HK dielectric (Sample G), uniformly stressed at an STM bias of +3.5V.

The RTN noise has also been detected using the STM technique as illustrated by Fig.12 in a Sc_2O_3(3nm)/La_2O_3(4nm)/SiO_x(1nm)/n-Si sample at +4V substrate injection [19]. This figure shows the digital variations in the tip height with time in a constant voltage constant current imaging (CCI) mode which is analogous to the standard digital current noise fluctuations at CVS.

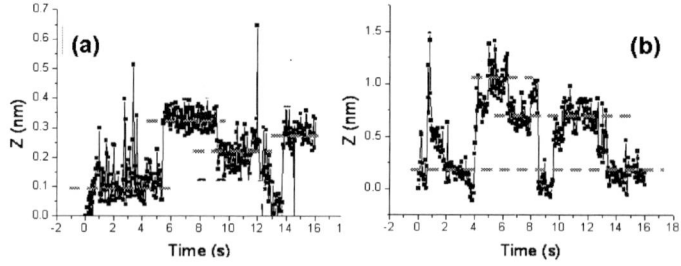

Figure 12. Observation of digital fluctuations in the tip height of the STM probe at two different locations of the Sc_2O_3/La_2O_3/SiO_x sample (Sample H) indicative of RTN leakage in the CCI mode [19].

It is clear from the results above that failure analysis can be a very useful technique to understand the material kinetics and physics of failure for HK gate stack dielectric breakdown. We now briefly look into the statistical lifetime assessment methodology for HK gate stacks.

IV. STATISTICAL LIFETIME ASSESSMENT

Statistical models for accelerated life test (ALT) and reliability extrapolation are well established for oxide reliability and have been widely applied to SiON-based devices [20] – [23]. Although these models were directly extended to be applied for HK gate stack lifetime studies [8, 24], it is hard if not inappropriate to use the same model for an intrinsic reliability assessment of the bi-layer HK-IL stack. The presence of a dual layer complicates the analysis in this case and application of a single stage CVS results in the second surviving layer experiencing a field exceeding its critical breakdown E-field after 1-layer TDDB, causing it to abruptly breakdown.

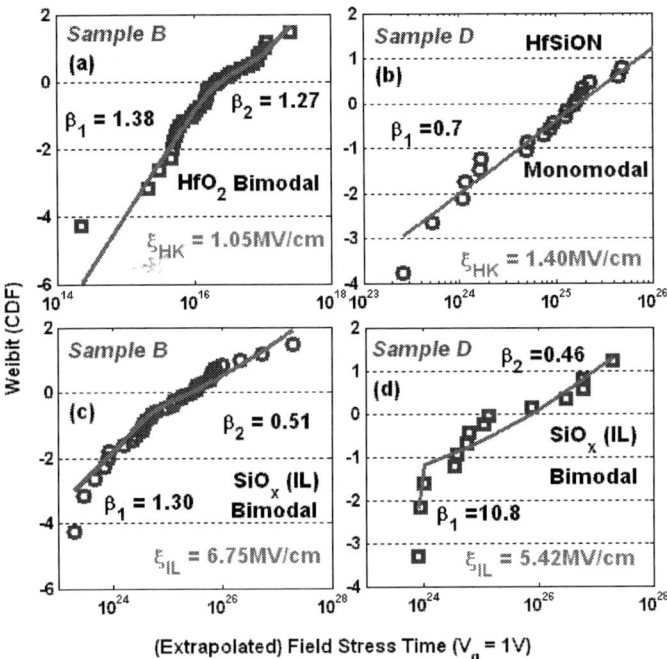

Figure 13. Weibull probability plot of extrapolated operating condition lifetime for the HK and IL layers using the CDM-IPL model for (a) & (c) Sample B and (b) & (d) Sample D.

To assess the intrinsic reliability, it is therefore necessary to develop an algorithm [5] as detailed in Section II.A and Fig.1 where $V_{g2} < V_{g1}$ and $I_{g1} \ll I_{g2}$ (compliance setting). Using the proposed two-step TDDB methodology, since the voltage loading is "time-variant" for the surviving second layer dielectric, the *cumulative damage model* (CDM) [25] is used to model separately the reliability of the HK and IL layers. We use the Poole-Frenkel emission mechanism as a yardstick to detect the layer which breaks down first. In general, HK is almost always the first layer to break down [5] given its lower breakdown field ($\xi_{BD-HfO2}$ = 4-7MV/cm and $\xi_{BD-SiO2}$ = 15MV/cm) [26]. The mathematical details of the statistical model are discussed in detail in [27] and shall not be presented

here. It is sufficient to take note that the CDM model enables us to determine the Weibull slope of the HK and IL layers separately {β_{HK}, β_{IL}} and also infer their mean lifetime {η_{HK}, η_{IL}} and power law exponent in the inverse power law (IPL) ALT model using the maximum likelihood optimization routine [25].

Fig.13 is the extrapolated Weibull plot at V_g = 1V for both the HK and IL layers in Sample B (polycrystalline HfO$_2$) and Sample D (amorphous HfSiON – 60% HfO$_2$ + 40% SiO$_2$) using the CDM-IPL model mentioned above. Since the HK and IL thickness are different for the two samples, as a common parameter of comparison, we use the electric field across the HK and IL layers. It is interesting to note from the lifetime plot (refer to plots (a) and (b)) that the polycrystalline HfO$_2$ has a lifetime about 9 orders lower than the amorphous HfSiON although the ξ-field across the HfO$_2$ is much lower. Moreover, HfO$_2$ shows bimodal failure distribution while HfSiON is clearly monomodal. This is a clear indicator that the HK microstructure plays a significant role in determining the lifetime of the gate stack. The bimodal distribution for the HfO$_2$ stack is indicative of the presence of grain boundary assisted failure (~63% of samples show early GB failure) and bulk failures (percolation through the grain). Fig.14 shows a TEM image as an evidence of GB-assisted HfO$_2$ breakdown [28] where the DBIE defect prefentially nucleates adjacent to the GB, which is an easier diffusion pathway. Since the STM results in the previous section reveal the average grain size to range between 15-25nm, it is possible that for advanced CMOS technology nodes (32nm and 22nm), some of the logic devices may not have any GB in them. However, for 45nm and older technologies, a large proportion of the devices are likely to have percolating GB severely limiting the lifetime of HfO$_2$-based stacks.

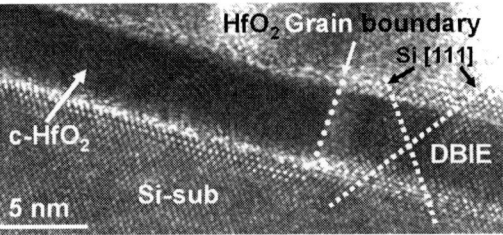

Figure 14. TEM cross-section of a failed device clearly shows the grain boundary site failure in the HK layer along with other microstructural DBIE transformations [28].

Comparing the IL and HK layers for samples B and D (refer to Figs.13(a) & (c)) and Figs.13(b) & (d))), amorphous HK and IL are equally robust with similar lifetime; however poly-HfO$_2$ has a very low lifetime (~10 orders of magnitude) compared to its IL layer. Therefore, the IL layer serves as a "savior" to the whole gate stack in extending the device and circuit lifetime. Removal of the IL layer may prove to be very detrimental to the device reliability although it may help improve the drive current, I_{ON}/I_{OFF} ratio, subthreshold slope etc... [29] The IL layers in samples B and D compare very well (refer to Figs.13(c) and (d)). At a weibit value of 0 corresponding to 63.2% failure, the IL layer in sample B has a slightly lower η as it experiences a higher ξ-field. The IL

978-1-4244-5430-3/10 $26.00 © 2010 IEEE   359

layers show a bimodal behavior as well which could correspond to extrinsic and intrinsic failure modes since these very thin sub-oxides can be highly defective (early failures).

In general, therefore, amorphous HK materials such as HfSiON have superior TDDB robustness. But, from a device scaling point of view, HfO$_2$ is preferred since it has a higher κ value (~20-30) compared to HfSiON (~14-17), thereby enabling more aggressive scaling of the EOT. The analysis above only pertains to a small-area device level study. It is necessary to extend these results to the circuit level by using the area scaling law [30]. Area scaling is only applicable to the HK layer as it is the first layer to BD; subsequent IL failure is localized at the HK BD spot. As explained in [27], circuit level data projection clearly suggests that circuit failures can only occur by mutliple HK BD events for both amorphous and polycrystalline HK stacks. Sequential HK→IL breakdown seldom occurs as the IL failures are localized and very robust as observed in Fig.13. Having performed the electrical, physical and statistical analysis on a variety of gate stacks, the following section focuses on the inferences and implications for future HK technology from this holistic study.

V. DESIGN FOR RELIABILITY IN HIGH-K GATE STACKS

We analyze the practical implications of the experimental results above on future HK gate stack technology and propose approaches that would help prolong the gate stack lifetime. There are four key issues to consider – (A) Recovery of dielectric breakdown, (B) Role of grain boundary, (C) Feasibility of zero interfacial layer (ZIL) solution, (D) Role of gate electrode on the post breakdown reliability margin.

A. Recovery of Dielectric Breakdown

In FUSI gated stacks, as shown in Fig.3(b), recovery in the leakage current at the HBD stage is consistently observed. Both unipolar (Fig.15) [31] and bipolar modes [7] of recovery have been found. This recovery is analogous to the resistive switching in RRAM [32]. While unipolar recovery is governed by joule heating assisted phase transition at a critical temperature (T_{CRIT}), bipolar recovery has an additional driving force of "drift" where the O^{2-} ions which are stored in the gate electrode during inversion stress, can be flushed back to neutralize the oxygen vacancy traps (V_0^{2+}) that constitute the percolation path thereby shutting it "off". The post recovery device performance characteristics such as I_g-V_g and I_d-V_d show proper transistor action (Fig.16).

We may make use of this recovery phenomenon to "repair" a failed device which has suffered a HBD and in a practical circuit, all that is needed to achieve this is a simple trigger of an I_g-V_g sweep, as in Fig.15. However there are two main restrictions to its implementation – (a) this recovery is successfully observed only in NiSi FUSI gate stacks. MG (TaN) stack can only recover for compliance levels, $I_{gl} < 2\mu A$ [7] and poly-Si gated devices do not show any recovery trend at all. Therefore, for the TaN/TiN based MG-HK stacks that are currently used, if the BD events are "soft", it is possible to initiate a partial recovery and repair the circuit. Whether the MG stacks show a soft BD or abrupt HBD with irreversible

metallic filamentation paths at normal operating conditions is still not clear. (b) Only when the BD hardness is high (low percolation resistance), the recovery voltage is moderately low such that $V_{REC} < V_{BD}$. Here, V_{REC} is the gate voltage needed to cause a 1mA gate leakage which is the typical current level at which recovery occurs in FUSI stacks (Fig.15). However, as explained at the end of Section IV, real circuit may suffer multiple "soft" HK BD events without any IL failure. In such cases, the percolation resistance is very high and $V_{REC} >> V_{BD}$ implying that the whole gate stack may break down even before the 1mA current level is reached during the I_g-V_g sweep. If the IL layer is eliminated from the gate stack, then the HK BD events are sufficient to cause a low resistance percolation path and achieving recovery is then plausible.

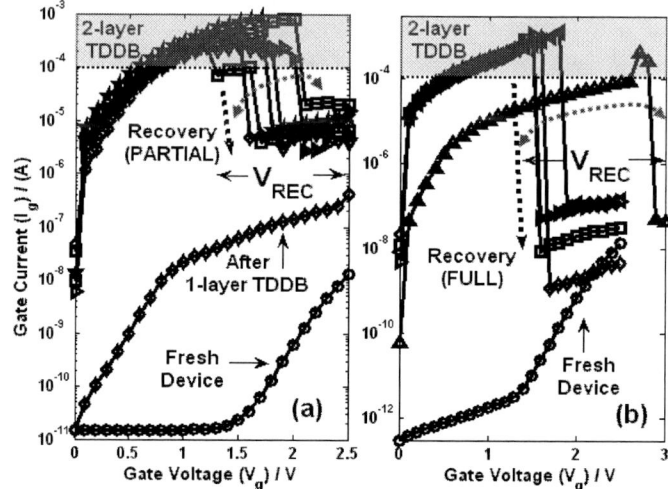

Figure 15. Repeated observations of partial and full recovery of gate leakage current during I_g-V_g sweep after a 100μA compliance controlled HBD in Sample D. Partial recovery involves leakage drop by 2-3 orders of magnitude while full recovery corresponds to 5-7 orders leakage reduction such that the recovered current is almost as good as the fresh device leakage value [31].

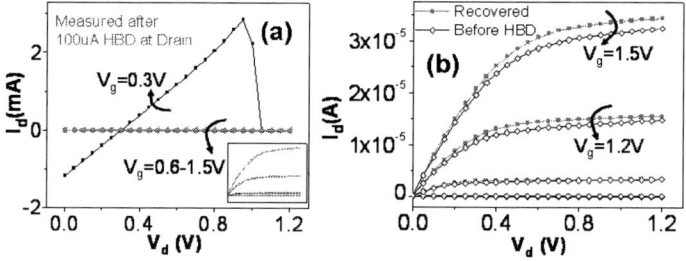

Figure 16. **(a)** I_d-V_d measurement after Sample D was stressed in inversion mode to 100μA. HBD near the drain was found. "Switch off" of the BD path happens at V_g=0.3V and V_d=1.0V during an initial sweep of I_d-V_d. For example, the I-V curves show a HBD ohmic characteristic from $0 \leq V_d < 1.0$V while at V_g=0.3V, but the transistor becomes functional at V_d=1.0V corresponding to a reverse bias of V_{gd} = -0.7V (bipolar recovery). The insert shows following sweep (i.e., after the switch off) of I_d-V_d with V_g=0.6-1.5V. **(b)** Comparison of I_d-V_d characteristics before BD and after "switching off" of the BD path. The transistor "recovered" without much degradation [7].

B. Role of Grain Boundary and High-κ Microstructure

Based on the STM, TEM and statistical results in Sections III and IV, grain boundary (linked chain of process induced defects) plays a major role in determining the reliability of HK gate stacks. The microstructure of the HK material during deposition and subsequent annealing needs to be precisely controlled so as to avoid or minimize the grain boundary density so that early HK failures can be avoided. Interaction of oxygen vacancies with monoclinic HfO_2 was recently reported by McKenna and Shluger [33]. Using first principle calculations, they showed that there is a strong driving force for vacancy segregation to the GB where diffusion barriers are reduced relative to the bulk. The activation barrier for V_0^{2+} diffusion along a twin boundary was found to be ~0.57eV, while it is ~0.71eV in the bulk. Moreover, diffusion of V_0^{2+} towards the GB has a barrier of 0.42eV, while the reverse direction for vacancy movement away from the GB shows a very high barrier of 1.14eV [33]. This clearly points to the fact the GB is an efficient sink for oxygen vacancies. These results are further substantiated by Eq.(1) [33], where f_{GB} and f_{bulk} represent the fraction of GB and bulk sites occupied by vacancies, E_S is the vacancy segregation energy (with respect to bulk), k_B is the Boltzmann constant and T is the Kelvin temperature. Since $E_S < 0$ for GB width, $d_{GB} < 7$Å, there is a higher concentration of vacancies segregated at the GB interface rather than the bulk. The maximum cluster density of V_0^{2+} at the GB is only limited by coulombic repulsion forces.

$$\frac{f_{GB}}{f_{bulk}} = \exp\left[\frac{-E_S}{k_B T}\right] \quad (1)$$

To avoid the segregation and clustering of oxygen vacancies along GB, the process flow (HK deposition + subsequent annealing conditions) needs to be optimized so as to achieve an amorphous microstructure that can be expected to be more robust to TDDB failures.

C. Feasibility of Zero Interfacial Layer (ZIL) Solution

To achieve the best EOT value, "zero" IL solution has been recently proposed by Huang *et al.* [29]. They reported improved subthreshold slope, I_{on}/I_{off} and lower interface state density with the implementation of the ZIL scheme. From a device performance perspective, ZIL solution is favorable, although carrier mobility may be adversely affected. However, from a reliability point of view, since the IL layer is the "savior" helping to prolong the TDDB lifetime, removal of this layer can paralyze the device reliability. ZIL solution may be feasible only if the final microstructure of the HK layer is amorphous and the metal gate does not show filamentation. This is however very difficult to achieve at present. If filamentation can be avoided, then the post BD RTN noise in the HK layer may provide for an additional reliability margin. The presence of the filament causes a catastrophic uncontrolled BD leading to a "dead" transistor.

It is desirable to have an ultra-thin IL layer (but not ZIL) so that the digital BD fluctuations in the sub-oxide can provide sufficient reliability margin. Post-BD statistical analysis on 16Å SiON has shown that the Di-BD stage after TDDB can provide a lifetime enhancement as much as 212 years [11]. Moreover, as shown in Fig.17, as V_{CRIT} saturates for $t_{ox} < 16$Å

and $V_{op} \ll V_{CRIT}$, presence of an ultra-thin IL provides a very good post-BD reliability.

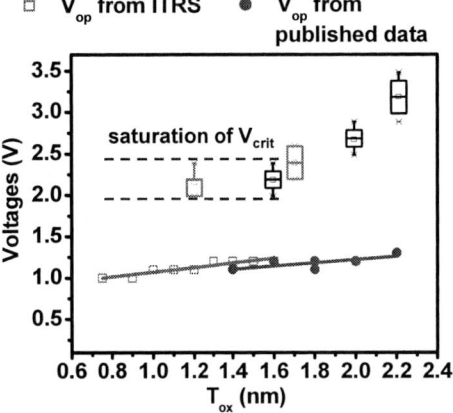

Figure 17. Variation of critical voltage for digital to analog BD transition, V_{CRIT}, for different SiON thickness. Note the saturation of V_{CRIT} as oxide thickness is scaled down below 16Å. The difference of the operating voltage, V_{OP} and V_{CRIT} is the voltage margin for prolonged post-BD reliability.

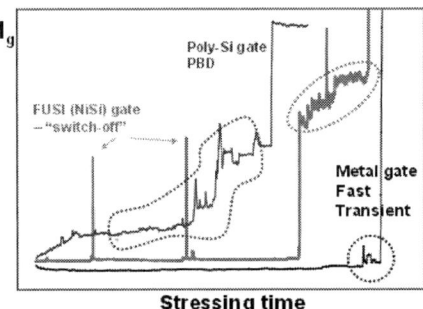

Figure 18. Post breakdown time evolution of leakage current for poly-Si, FUSI and MG based HK stacks. The duration of progressive BD (before HBD) for MG << FUSI << poly-Si gated devices [7].

D. Role of Gate Electrode on Post Breakdown Reliability

It is well documented in the literature on conventional oxynitride based devices that the initial stages of the post-TDDB lifetime when RTN noise is significant without much degradation in the device performance characteristics, should also be accounted for in the quantitative reliability assessment [34]. The same holds true for HK gate stacks as well. The post-BD reliability of the HK stack however largely depends on the material used as the gate electrode. Fig.18 shows an experimental trend of post-BD leakage evolution in Samples B, D and E with poly, FUSI and MG electrodes respectively [7]. The general trend of progressive BD duration (gradual increase in leakage current) follows the order – MG << FUSI << poly-Si as indicated by the dotted contours in the figure. Although the RTN duration during the ALT stress condition in MG-HK stacks seems to be much lower than the other gate stacks, at operating voltage conditions, this RTN phase could be sufficient enough to provide an enhancement in lifetime for at least a few years, thereby enabling us to satisfy the standard reliability specification of 10 years. Fig.19 shows six devices belonging to Sample E, that further confirm that MG-HK devices also show a notable progresive BD phase before the ultrafast transient BD occurs [15].

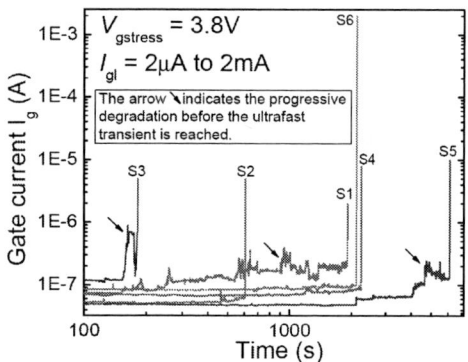

Figure 19. Time evolution of the gate leakage current I_g for six n-MOSFET devices (Sample E) (denoted as S1 to S6) under $V_{gstress}$ = 3.8V CVS with a compliance current limit I_{gl} varying from 2μA to 2mA. Progressive degradation is initially observed prior to the ultrafast transient BD, as indicated by the arrows [15].

A possible explanation of the trends observed in Figs.18 and 19 is given by the schematic in Fig.20 [7]. The mobile oxygen (O^{2-}) ions and V_0^{2+} vacancies are created during the trap formation in the dielectric. In a poly-Si stack, the O^{2-} ions react with the Si gate electrode ($Si + O \rightarrow SiO_x$), while in the FUSI and TaN gate, they do not react with the gate electrode, as oxidation of these gate materials is thermodynamically unfavorable. The gate electrode functions as "reservoir" for storage of O^{2-} ions in this case. During enhanced joule heating and/or reverse polarity bias sweep, these O^{2-} ions are driven back to the percolation path to passivate the dangling bonds and deactivate the traps thereby shutting-off the percolation path. This explains the observed recovery in FUSI in Fig.18.

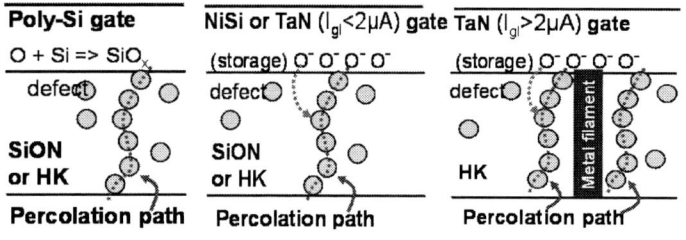

Figure 20. Schematic showing the kinetics of mobile O^{2-} ions, filamentation and effect of different gate materials on the post-BD leakage evolution. [7].

TABLE III.   DEFECT FORMATION ENERGY AND BANDGAP COLLAPSE FOR DIFFERENT GATE MATERIAL PENETRATION INTO HAFNIUM OXIDE.

Material	Defect Formation Energy (E_f)	Bandgap
HfO_x, $1V_o$	6.22 – 6.40 eV	---
HfO_x, Ta	9.92 eV	0 eV
HfO_x, 1Ni	6.92 eV	1.587 eV
HfO_x, 1Ti	7.53 eV	0.296 eV

As for the poly-Si gate device, the leakage current gradually increases due to percolation path dilation and effective oxide thinning due to Si DBIE. The final stage of HBD in poly-Si stack can be attributed to gate-drain or gate-source short due to lateral DBIE propagation or formation of Si nanoclusters. In the MG-HK stack, for $I_g > 2μA$, Ta filaments begin to nucleate in the dielectric [15] and grow rapidly shorting the gate and substrate thus leading to abrupt HBD and a very short span of

post-BD phase. Some devices belonging to the FUSI stack also show Ni spiking [16] and punchthrough the dielectric leading to HBD after the progressive-BD phase. The dielectric BD recovery is effective only for a limited number of cycles as complete recovery of the BD path or filament is not always possible. The accumulated damage during the multiple BD-REC events causes eventual failure of the FUSI stack as well.

The above electrical test results are further supported by *Ab-Initio* simulation results in Table III performed using supercell approach [35], where the defect formation energy (E_f) and bandgap are calculated for four different cases. Oxygen vacancies are the first to be generated prior to any metal filamentation in the HK as it has the lowest defect formation energy. Ta-based metal gate materials show catastrophic BD with limited progressive BD as the bandgap falls to zero (fully conducting) upon Ta migration into HfO_2. Ni-based gate stack has a relatively longer progressive BD phase and shows good recovery trends than Ta, as the FUSI-based HK bandgap shrinks to a non-zero value of 1.587eV.

VI. CONCLUSION

A comprehensive study on different material HK gate stacks and their impact on TDDB and post-BD reliability has been accomplished. The need for a diversified study on HK breakdown from an electrical, physical, statistical and theoretical perspective and the correlation of these results has been emphasized. The synergy of all these results led us to propose novel approaches to design for reliability in future HK technology such as amorphous microstructure, recovery of dielectric breakdown, use of ultra-thin interfacial layer etc...

Further studies are needed to gain more in-depth understanding on the fundamentals and failure kinetics in HK materials. This includes studying effect of other substrate materials apart from Si (e.g. SiGe, GaAs, InGaAs etc...) on the lifetime and failure mode, determination of critical voltage for the HK dielectric film (without IL), statistical study on ZIL devices, trap energy and location analysis using 1/f noise and RTN data [36] and physical analysis of FUSI/MG device after recovery. It is also essential to develop better circuit reliability models for HK stacks which can be used to accurately predict circuit lifetime based on device level failure data.

ACKNOWLEDGMENT

We would like to thank all current and past graduate students of the Gate Oxide Reliability Research Group, NTU for their diligent involvement and contributions to this work. We would like to acknowledge the sample support provided by our international collaborators - Interuniversity Microelectronics Centre (IMEC) - Belgium, Tokyo Institute of Technology (TIT) - Japan, Global Foundries (GF) - Singapore and SEMATECH - US. We would also like to thank the A*STAR Institute of Microelectronics (IME) and Institute of Materials Research and Engineering (IMRE) for provision of unconditional support for failure analysis facilities. The financial support and research grant provided by the Ministry of Education (MOE), Singapore is also very much appreciated.

978-1-4244-5430-3/10 $26.00 © 2010 IEEE

REFERENCES

[1] E.Y. Wu, J.H. Stathis and L.K. Han, "Ultra-thin oxide reliability for ULSI applications", *Semiconductor Science and Technology*, Vol. 15, pp.425-435, (2000).

[2] M.N. Jones, Y.W. Kwon and D.P. Norton, "Leakage current behavior of HfO_2 thin films", *ECS Proceedings*, Vol. 2003-22, pp.131-142, (2003).

[3] G.D. Wilk, R.M. Wallace, J.M. Anthony, "High-κ gate dielectrics: Current status and materials properties considerations", *Applied Physics Review, Journal of Applied Physics*, Vol. 89, No. 10, pp.5243-5275, (2001).

[4] Y.H. Kim and J.C. Lee, "Reliability characteristics of high-κ gate dielectrics", *Microelectronics Reliability*, Vol. 44, pp.183-193, (2004).

[5] N. Raghavan, K.L. Pey and X. Li, "Detection of high-κ and interfacial layer breakdown using the tunneling mechanism in a dual layer dielectric stack", *Applied Physics Letters*, Vol. 95, 222903, (2009).

[6] V.L. Lo, K.L. Pey, C.H. Tung and D.S. Ang, "A critical voltage triggering irreversible gate dielectric degradation", *IEEE International Reliability Physics Symposium (IRPS)*, pp.576-577, (2007).

[7] W.H. Liu, K.L. Pey, X. Li and M. Bosman, "Observation of switching behaviors in post-breakdown conduction in NiSi-gated stacks", *IEEE International Electron Device Meeting (IEDM)*, pp.135-138, (2009).

[8] R. Degraeve, T. Kauerauf, M. Cho, M. Zahid, L-Å. Ragnarsson, D.P. Brunco, B. Kaczer, Ph. Roussel, S. De Gendt and G. Groeseneken, "Degradation and breakdown of 0.9 nm EOT SiO_2 / ALD HfO_2 / metal gate stacks under positive constant voltage stress", *IEEE International Electron Device Meeting (IEDM)*, (2005).

[9] T.A.L. Selvarajoo, R. Ranjan, K.L. Pey, L.J. Tang, C.H. Tung and W. Lin, "Dielectric-breakdown-induced-epitaxy: A universal breakdown defect in ultrathin gate dielectrics", *IEEE Transactions on Device and Materials Reliability*, Vol. 5, No. 2, pp.190-197, (2005).

[10] V.L. Lo, "Theoretical study of breakdown in ultrathin gate dielectrics in nanoscale MOSFETs", *PhD Thesis*, Nanyang Technological University (NTU), (2008).

[11] N. Raghavan, X. Wu, X. Li, W.H. Liu, V.L. Lo and K.L. Pey, "Post breakdown reliability enhancement of ULSI circuits with novel gate dielectric stacks", *IEEE International Symposium on Integrated Circuits*, pp.505-513, (2009).

[12] F. Crupi, T. Kauerauf, R. Degraeve, L. Pantisano and G. Groeseneken, " A novel methodology for sensing the breakdown location and its application to the reliability study of ultrathin Hf-silicate gate dielectrics", *IEEE Transactions on Electron Devices*, Vol. 52, No. 8, pp.1759-1765, (2005).

[13] X. Li, C.H. Tung and K.L. Pey, "The physical origin of random telegraph noise after dielectric breakdown", *Applied Physics Letters*, 94, 132904, (2009).

[14] C.H. Tung, K.L. Pey, L.J. Tang, M.K. Radhakrishnan, W.H. Lin, F. Palumbo and S. Lombardo, "Percolation path and dielectric-breakdown-induced-epitaxy evolution during ultrathin gate dielectric breakdown transient", *Applied Physics Letters*, Vol. 83, No. 11, pp.2223-2225, (2003).

[15] X. Li, K.L. Pey, M. Bosman, W.H. Liu and T. Kauerauf, "Direct visualization and in-depth physical study of metal filament formation in percolated high-κ dielectrics", *Applied Physics Letters*, Vol. 96, 022903, (2010).

[16] N. Raghavan, X. Li, W.H. Liu, X. Wu, M. Bosman and K.L. Pey, "Metal filamentation switching in FUSI-gated $HfSiON$-SiO_x MIS device – A new approach using breakdown location dependent switching for high density embedded RRAM in LOGIC", *IEEE VLSI Technology Symposium*, Submitted, (2010).

[17] D.S. Ang, Y.C. Ong, S.J. O'Shea, K.L. Pey, C.H. Tung, T. Kawanago, K. Kakushima and H. Iwai, "Polarity dependent breakdown of the high-κ/SiO_x gate stack: A phenomenological explanation by scanning tunneling microscopy", *Applied Physics Letters*, Vol. 92, 192904, (2008).

[18] Y.C. Ong, "Analysis of high-dielectric constant gate stack reliability for nanoscale CMOS devices application via scanning tunneling microscopy", *PhD Thesis*, Nanyang Technological University (NTU), (2008).

[19] K.L. Pey, W.H. Liu, V.L. Lo, Y.C. Ong, X. Li and C.H. Tung, "Role of interfacial oxide on digital breakdown in high-κ gate stacks", Unpublished, (2010).

[20] J. Sune, E.Y. Wu, D. Jimenez and W.L. Lai, "Statistics of soft and hard breakdown in thin SiO_2 gate oxides", *Microelectronics Reliability*, Vol. 43, pp.1185-1192, (2003).

[21] M.A. Alam, R.K. Smith, B.E. Weir and P.J. Silverman, "Uncorrelated breakdown of integrated circuits", *Nature*, Vol. 420, pp.378, (2002).

[22] T. Pompl and M. Rohner, "Voltage acceleration of time-dependent breakdown of ultra-thin gate dielectrics", *Microelectronics Reliability*, Vol. 45, pp.1835-1841, (2005).

[23] A. Kerber, T. Pompl, M. Rohner, K. Mosig and M. Kerber, "Impact of failure criteria on the reliability prediction of CMOS devices with ultrathin gate oxides based on voltage ramp stress", *IEEE Electron Device Letters*, Vol. 27, No. 7, pp.609-611, (2006).

[24] N.A. Chowdhury, G. Bersuker, C. Young, R. Choi, S. Krishnan and D. Misra, "Breakdown characteristics of nFETs in inversion with metal / HfO_2 gate stacks", *Microelectronic Engineering*, Vol. 85, pp.27-35, (2008).

[25] W. Nelson, "Accelerated Testing: Statistical Models, Test Plans and Data Analyses", *John Wiley & Sons*, (1990).

[26] J.W. McPherson, J. Kim, A. Shanware, H. Mogul and J. Rodriguez, "Trends in the ultimate breakdown strength of high dielectric-constant materials", *IEEE Transactions on Electron Devices*, Vol. 50, No. 8, pp.1771-1778, (2003).

[27] N. Raghavan, K.L. Pey, W.H. Liu and X. Li, "New statistical model to decode the reliability and Weibull slope of high-κ and interfacial layer in a dual layer dielectric stack", *IEEE International Reliability Physics Symposium (IRPS)*, Accepted, (2010).

[28] R. Ranjan, K.L. Pey, C.H. Tung, L.J. Tang, G. Groeseneken, L.K. Bera and S. De Gendt, "A comprehensive model for breakdown mechanism in HfO_2 high-κ gate stacks", *IEEE International Electron Device Meeting (IEDM)*, pp.725-728, (2004).

[29] J. Huang, D. Heh, P. Sivasubramani, P.D. Kirsch, G. Bersuker, D.C. Gilmer, M.A. Quevedo-Lopez, M.M. Hussain, P. Majhi, P. Lysaght, H. Park, N. Goel, C. Young, C.S. Park, C. Park, M. Cruz, V. Diaz, P.Y. Hung, J. Price, H.H. Tseng and R. Jammy, "Gate first high-κ/metal gate stacks with zero SiO_x interface achieving EOT=0.59nm for 16nm application" *IEEE VLSI Technology Symposium*, Kyoto, pp.34, (2009).

[30] Y.H. Kim, K. Onishi, C.S. Kang, H.J. Cho, R. Nieh, S. Gopalan, R. Choi, J. Han, S. Krishnan and J.C. Lee, "Area dependence of TDDB characteristics for HfO_2 gate dielectrics", *IEEE Electron Device Letters*, Vol. 23, No. 10, pp.594-596, (2002).

[31] N. Raghavan, K.L. Pey, W.H. Liu, X. Wu and X. Li, "Unipolar recovery of dielectric breakdown in fully silicided high-κ gate stack devices and its reliability implications", *Applied Physics Letters*, Submitted, (2010).

[32] I.H. Inoue, S. Yasuda, H. Akinaga and H. Takagi, "Nonpolar resistance switching of metal/binary-transition-metal oxides/metal sandwiches: Homogeneous/inhomogeneous transition of current distribution", *Physical Review B*, Vol. 77, 035105, (2008).

[33] K. McKenna and A. Shluger, "The interaction of oxygen vacancies with grain boundaries in monoclinic HfO_2", *Applied Physics Letters*, Vol. 95, 222111, (2009).

[34] G. Bersuker, N. Chowdhury, C. Young, D. Heh, D. Misra and R. Choi, "Progressive breakdown characteristics of high-κ/metal gate stacks", *IEEE International Reliability Physics Symposium (IRPS)*, pp.49-54, (2007).

[35] D.B. Laks, C.G. Van de Walle, G.F. Neumark, P.E. Blochl and S.T. Pantelides, "Native defects and self compensation in ZnSe", Vol. 45, No. 19, pp.10965-10978, (1992).

[36] F. Martinez, C. Leyris, G. Neau, M. Valenza, A. Hoffmann, J.C. Vildeuil, E. Vincent, F. Boeuf, T. Skotnicki, M. Bidaud, D. Barge and B. Tavel, "Oxide traps characterization of 45nm MOS transistors by gate current RTS noise measurements", *Microelectronic Engineering*, Vol. 80, pp.54-57, (2005).

High-K gate stack breakdown statistics modeled by correlated interfacial layer and high-k breakdown path

G. Ribes[1], P. Mora[1], F. Monsieur[1], M. Rafik[1], F. Guarin[2], G. Yang[3], D. Roy[1], W.L. Chang[2], J. Stathis[2]

[1]STMicroelectronics, Hopewell Junction, NY 12533 USA & Crolles, 850 rue Jean Monnet, BP16 38926, France;
[2]IBM Hopewell Junction, NY 12533 & Thomas J. Watson Research Center, Yorktown Heights, NY 10598;
[3]Global Foundries Singapore;
Phone: (1) 845-894-1414; E-mail: gribes@us.ibm.com

Abstract — **We show that a model in which the breakdown of the interfacial layer induces a correlated breakdown in the high-K, at the same location, provides a good model of the high-K/IL gate stack statistics. We discuss of the implication of this model on the lifetime projection.**

Keywords: High-K metal gate dielectrics; breakdown; reliability; CMOS

I. INTRODUCTION

High-K dielectric stacks with Hf-based dielectric are being used as a replacement for conventional SiON dielectrics for sub-45nm technology nodes. A number of studies suggest that the ultra-thin interfacial layer IL determines the high-K/IL stack breakdown and that the high-K layer has little impact on the breakdown statistics [1]. This is in agreement with the small Weibull slopes commonly observed for high-K dielectrics. In a recent paper, these observations have been explained by a percolation model with uncorrelated percolation paths in the high-K and in the interfacial layer. The percolation paths are generated by different defect generation rates in the high-K layer and interfacial layer [2]. In this paper, we will show that a model in which the breakdown of the interfacial layer induces a correlated breakdown in the high-K at the same location, provides a good model of the high-K/IL gate stack statistics. We will show that this mechanism is more likely to control the breakdown statistics in the experimental range (below $10^4 um^2$).

II. EXPERIMENTAL RESULTS

In previous work [3] it has been shown that data from 0.006 um^2 to 0.6 um^2 can be explained by uncorrelated breakdowns [2] for $1um^2$ to $1000um^2$ the accuracy of the model is questionable. In this study PMOS high-K devices have been stressed in inversion. Breakdown lifetimes have been collected for larger area ($0.033um^2$ to $300um^2$) with gate current≤1E-4A and scaling perfectly with area. The I-V curves are plotted Figure.1. The breakdown is characterized by a so called "hard breakdown" and detections are shown Figure 2. Because the breakdown is not progressive the detection is obvious. Hence no progressive extension can be suspected. According to Fig.3 it can be observed that area scaling works for area from 0.033 to $33um^2$. Furthermore the lifetime distribution does not follow a pure Weibull distribution. Hence this data confirm the

previous work showing that vertical area scaling is accurate for small area ($<33um^2$).

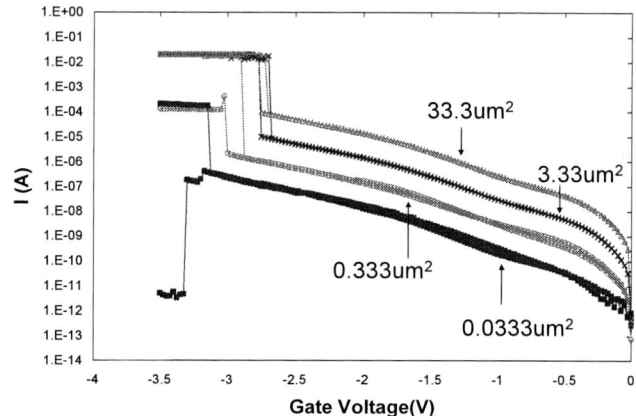

Figure 1. *PMOS Gate current for different area from 0.033um2 to 333um2*

Two mechanisms can explain this kind of lifetime distribution in a bi-layer stack with different defect generation rate $R(s^{-1})$ and/or critical defect density N_{BD}:

- Two uncorrelated connecting breakdown paths generated randomly on the high-K and interfacial layer area [2].
- A breakdown path initiated in one layer induces locally a dramatic increase of defect generation rate. This results in a correlated breakdown path in the second layer. The mathematics of this mechanism is very similar to progressive breakdown.

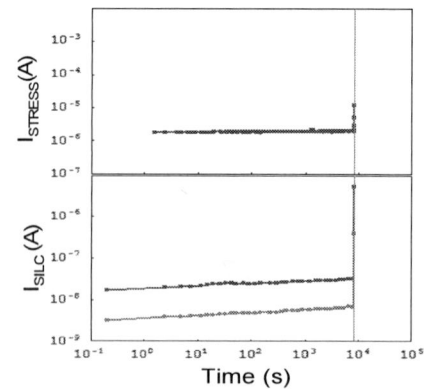

Figure 2. Typical TBD detection on PMOS high-K devices characterized by hard breakdowns

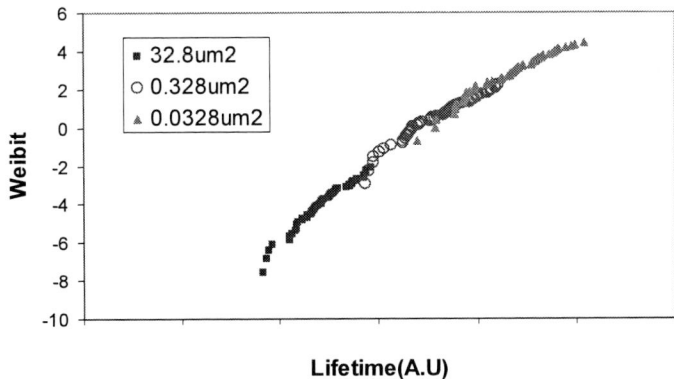

Figure 3. *PMOS Device A: IL=XÅ and HK=X+3Å experimental high-K breakdown lifetime obtained by vertical area scaling between 32.8um² and 0.0328um²*

In this paper we demonstrate that a breakdown initiated in the interfacial layer followed by a correlated breakdown in the high-K can fully explain the statistics and dominate it in the experimental range.

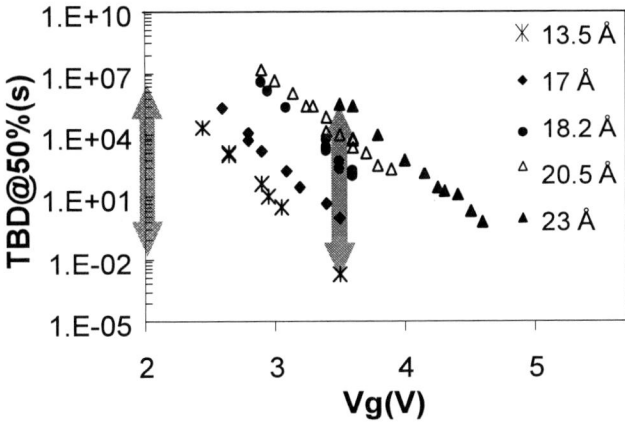

Figure 4. *SiON gate oxide thickness lifetime dependence obtained on SiON thickness from 13.5 Å to 23 Å.*

First we evaluate which is the driving mechanism with respect to various experimental stress conditions. At the first order we suppose that shorting electrically a 10A IL will increase the gate current of the high-K stack with dependence similar to SiON gate current thickness dependence.

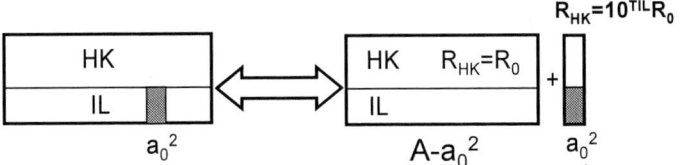

Figure 5. *Schematic of correlated breakdown path between high-K and IL*

In Fig.4, the SiON gate dielectrics time to breakdown is shown for various SiON thicknesses. As can be seen, the TDDB increases by about 1 decade per Å. This result means that if a breakdown path is initiated in a 10 Å interfacial layer IL, the defect generation rate in the high-K, $R_{HK}(s^{-1})$, increases by 10^{TIL} in the IL percolation path region (Fig.5). According to this assumption the T_{BD} ratio between uncorrelated breakdown in the HK itself and correlated breakdown in the high-k layer can be given as follow:

$$\frac{T_{BD\,uncorrelated}}{T_{BD\,correlated}} = \left(\frac{a_0^2}{A - a_0^2} \right)^{\frac{1}{\beta}} \cdot 10^{ToxIL} \qquad (2)$$

Where a_0 is the defect size, A is the device area, ToxIL is the thickness of the IL in Angstroms and β the Weibull slope.

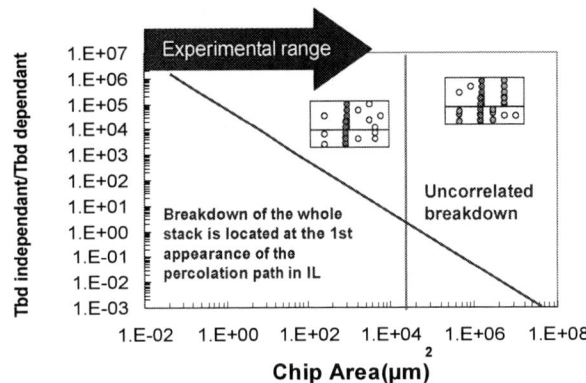

Figure 6. *T_{BD} ratio between uncorrelated breakdown and correlated breakdown with respect to device area.*

When the ratio is greater than 1, the correlated breakdown is more favorable. Fig.6 shows the TBD ratio between uncorrelated breakdown and correlated breakdown with respect to device area. The calculation has been done using a Weibull slope equal to $\beta=1$ which is roughly the value obtained experimentally and a_0=20 Å [5, 6]. With these assumptions, it is shown that below 10^4 um² the correlated breakdown is more favorable. That means for the device areas used in typical experiments the correlated breakdown is predicted to drive the breakdown. It has to be noted that in this simple calculation in the case of uncorrelated breakdown, the need to align two uncorrelated breakdowns has not been taken into account. This implies that above 10^4um² uncorrelated breakdown could drive the statistics, but considering the size of a chip and the size of the percolation path, the probability to align 2 uncorrelated breakdowns is very small, and thus the threshold area is far above 10^4um².

III. DISCUSSION

It has been reported by many authors that breakdown is initiated in the interfacial layer [1,7,8]. In order to model the correlated breakdown time, we consider that the time to breakdown of the stack is given by a convolution of the time to break the interfacial layer IL($T_{BD\,IL}$) and the time to extend the breakdown path in the high-K layer at the same location,

($T_{BDHKcorrelated}$). In this case the time to breakdown of the IL follows Poisson statistics. The correlated breakdown of the high-K layer is localized above the IL breakdown path; its localization does not correspond to the HK weakest point. So it does not follow a Poisson area scaling. Hence the time to breakdown of the bi-layer can be expressed as a convolution ($\oplus$) of the time to break the IL (T_{BDIL}) and the time to extend the BD path through the high-K:

$$T_{BDHK+IL} = T_{BDIL} \oplus T_{BDHKcorrelated}$$

$$T_{BDHK+IL} = T_{0IL}.Vg^{-n}.A^{\frac{1}{\beta IL}}.(-\ln(1-F))^{\frac{1}{\beta IL}} \oplus T_{correlated_{HK}}.Vg^{-nhk}.(-\ln(1-F))^{\frac{1}{\beta HK}}$$

(3)

In the case of high-K initiated breakdown, the mathematics remains the same, only the fitting parameters will change. Fig.7 and 8 illustrate typical data and modeling. Using a Monte Carlo simulation, IL layer and high-K layer lifetime distributions are generated and convolved numerically.

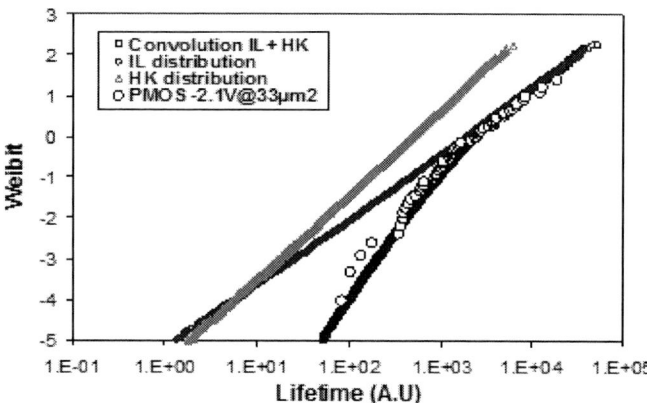

Figure 7. *Interfacial layer and high-K correlated breakdown comparison with experimental data (Device A)*

To test the theory in detail, we compared it against a large set of experimental conditions (different voltages, different gate areas and two IL and HK thicknesses). According to Fig.7 and Fig.8, we can see that very good agreement is obtained with large experimental data for gate voltage ranging between -2.7V and -2.1V and a large set of areas from 0.033 to 33um². The parameters T_{0IL}, n, β, $T_{correlatedHK}$, n_{hk} for this device (A) are listed in the table I.

TABLE I.   MODEL PARAMETERS FOR DEVICE A AND B

Parameters	Value for Device A	Value for Device B
T_{0IL}	7.93x10⁻⁶ (A.U)	2.88x10⁻² (A.U)
β_{IL}	0.7	0.77
n	50	50
$T_{correlated\,HK}$	1.03x10¹(A.U)	900 (A.U)
β_{HK}	0.9	0.9
nhk	42	42

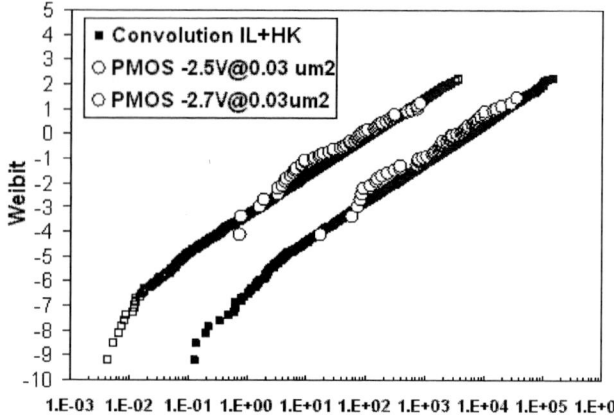

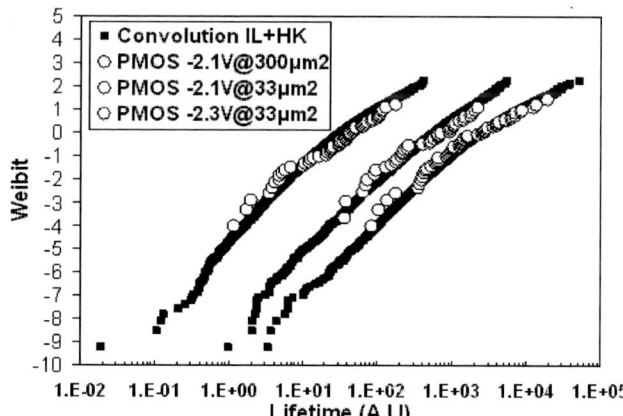

Figure 8. *High-K gate stack convoluted breakdown model and experimental data device A ranging between -2.7V and -2.1V and a large set of areas from 0.03 to 300um²*

Fig.9 shows experimental and theoretical data for a second device (B). A good agreement with the model is obtained for the parameters listed in the table above (Device B). Furthermore according to MVHR (Multi Vibrational Hydrogen Release) model [4] for high-K gate dielectrics as for SiON gate oxide the degradation rate $R_{SiON\,or\,HfSiON}$ depends on the gate current Ig:

$$R_{SiONorHfSiON} \propto Ig^{3.3}$$

(4)

The parameters for device B have been calibrated according to the MVHR predictions: $T_{0ILB}=T_{0ILA}*(J_A/J_B)^{3.3}=7.93x10^{-6}$ *2858=2.88x10⁻² using the experimental gate leakage currents:
Device A gate leakage density at 2.6V $J_A(A/um2)=1.8x10^{-6}$
Device B gate leakage density at 2.6V $J_B(A/um2)=1.62x10^{-7}$
Device B has a thicker IL by 2 Å so the breakdown path in the interfacial layer takes more time, but when the breakdown appears the degradation rate increase by 2 orders of magnitude with respect to degradation rate in device A. However, the high-K thickness is thicker for device B by 4 Å so the resulting degradation rate - taking into account the complete stack thickness (IL+HK) - should be roughly lower by 2 orders of magnitude. The $T_{0correlatedHK}$ for device B and device A follow this ratio.

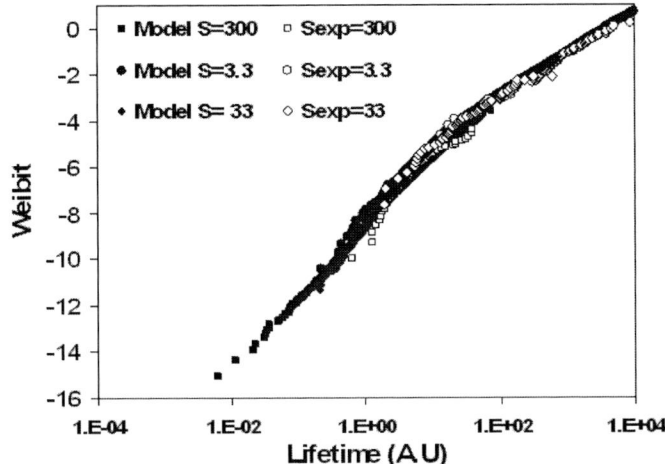

Figure 9. *PMOS Device B: IL=X+2Å and HK=X+7Å experimental high-K breakdown lifetime obtained by vertical area scaling between 3.3um² and 300um² and model comparison*

It has to be noted that there is a threshold area for which the IL time to breakdown becomes smaller than $T_{BDHKcorrelated}$. For these typical cases which are difficult to obtain experimentally the area scaling is not preserved. Nevertheless according to previous work [9], multiple BDs in the IL may occur. Hence different correlated BD paths in the high-k can be induced.

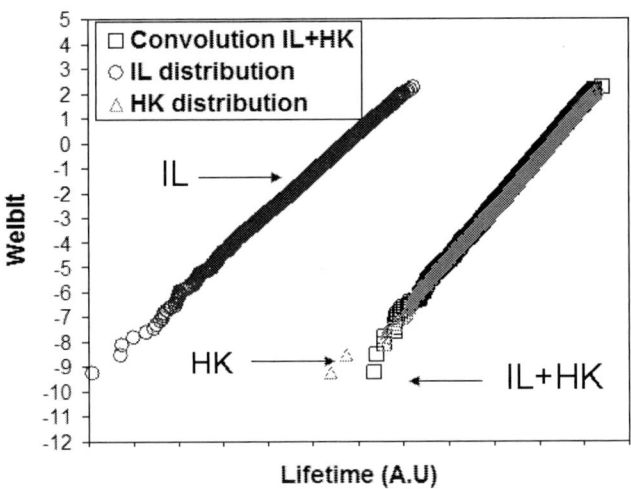

Figure 10. *IL + HK convolution at typical chip area (0.1cm²) and at typical Vdd*

The occurrence of multiple breakdown events can lead to a shorter time to breakdown of the complete stack, especially for high failure rate. In order to determine the impact of these events on the statistics, we have extrapolated our model to typical chip area (Fig.10). It is shown that at typical chip area of 0.1cm² the time to obtain the correlated BD is large with

respect to the IL breakdown. In this case multiple BD events in the IL can occur before the HK BD event. Based on the analytical formula provided by Sune et al. [9], it is possible to express the Kth BD distribution with respect to the Weibit and the Weibull slope:

$$\Delta_A(K, W_{spec}) = (K!)^{\frac{1}{K\beta}} \exp(-\frac{W_{spec}}{\beta}\frac{(K-1)}{K}) \qquad (5)$$

Where $\Delta_A(K, W_{spec})$ is the ratio between the 1st time to BD and the Kth, for a given Weibit (Wspec=ln(-ln(1-F))) and β (the Weibull slope).

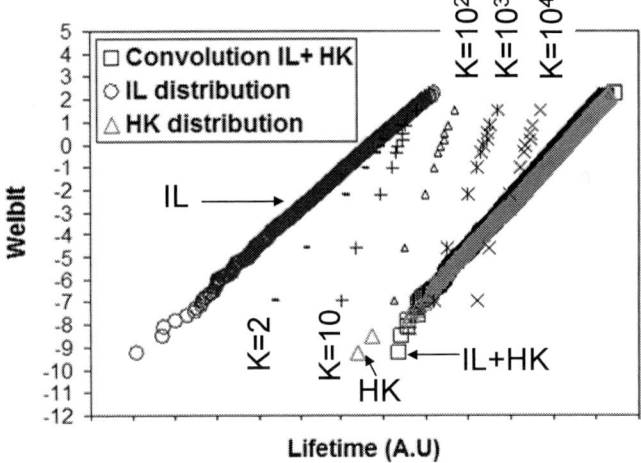

Figure 11. *Determination of the number of breakdowns in the IL before that the correlated HK breakdown induced by the 1st IL breakdown append*

TABLE II.   NUMBER OF BREAKDOWN IN THE IL FOR A GIVEN FAILURE PERCENTILE

Failure percentile (%)	Weibit	Number of BD
0.1	-6.91	10^3
10	-2.25	10^4
1	-4.60	3.10^3
5	-2.97	7.10^3
0.01	-9.21	10^2

Based on this expression we can determine the distribution of the Kth BD in the IL for any K (Fig.11). The maximum number of IL BD events, for a given failure rate (FR) or W_{spec}, corresponds to the highest $\Delta_A(K, W_{spec})$ for which the multiple BD time is less than the convoluted IL+HK time. For example, for FR=0.01% (Weibit=-9.2), a maximum of 10^2 BDs will occur in the IL, corresponding an IL time to multiple breakdown equivalent to the time to breakdown IL+HK. Nevertheless it is possible that one of them creates a correlated BD in the high-k with a shorter time than the one induced by the 1st IL BD. In order to evaluate this possibility, based on formula (5), we determine at FR=0.01% the time to BD of each Kth IL BD with K=1, 2...10^2 and associate randomly a time

978-1-4244-5430-3/10 $26.00 © 2010 IEEE

from the HK distribution (red triangle). The minimum time of the 10^2 IL+HK BDs is plotted (Fig.12). We can generalize this calculation for different failure percentile as illustrated in Fig.12 and table II.

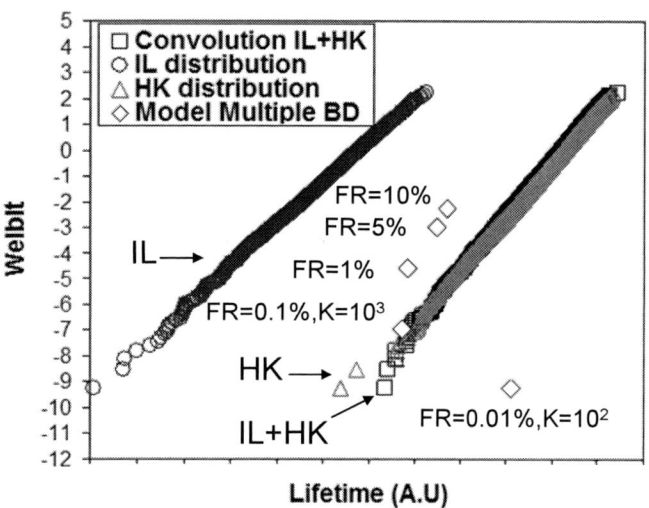

Figure 12. *Comparison between the time to BD distribution induced by a correlated BD related to the 1st IL BD (black) and the minimum time to BD correlated to K earlier IL breakdowns at a given failure percentile (blue).*

According to the Fig.12 the multiple breakdown mechanism does not drive the breakdown of the stack below a failure percentile of FR=0.1%. In a typical application the failure percentile of interest are lower than FR=0.1%. That means that our model neglecting the impact of multiple breakdowns is still relevant for projection at low percentile (FR<0.1%). It has to be noted that uncorrelated breakdown could drive the BD for large area ($A>>10^4 um^2$) but this mechanism is not possible at experimental conditions. This statement shows that different possible mechanisms can drive the statistics of the bi-layer stack breakdown. It is shown that the mechanism which drives at experimental level is not automatically the one which drives at product level. Hence in order to assess the high-k stack lifetime it is mandatory to keep a conservative methodology.

IV. CONCLUSION

In conclusion, we have seen that breakdown initiated in the interfacial layer followed by a correlated localized breakdown path in the high-K layer can explain the statistic of bi-layer time to breakdown. It is shown that this mechanism is more favorable in the experimental device area range (A<10^4um2). The model presented in this paper predicts the complete breakdown statistics with very good accuracy for two different gate stacks. Furthermore the model parameters are physically explained by MVHR model. It is shown also that this model can be used to assess the lifetime of a product at low percentile (FR<0.1%). Nevertheless it is highlighted that uncorrelated breakdown could drive the breakdown statistics for large product area. This possibility make the product lifetime assessment difficult because the mechanism which drives the statistics at experimental level could be different that the one at

product level. Hence a conservative methodology should be used.

REFERENCES

[1] G. Bersuker et al., IEDM, pp. 791-794, 2008.
[2] T. Nigam et al IRPS 2009
[3] A. Kerber et al IRPS 2009
[4] G. Ribes et al ECS 2009
[5] J. Stathis J. Appl. Phys, Vol.86, No.10, 1999
[6] G. Ribes et al IRPS 2006
[7] G. Ribes et al IRPS 2005
[8] B. Linder et al IRPS 2009
[9] J. Sune et al TED. Vol.51, No10, 2004

Acknowledgement: This work was performed at the IBM Microelectronics Div. Semiconductor Research & Development Center, Hopewell Junction, NY 12533

We are grateful for helpful discussions with Jordi Sune, Tanya Nigam, Ernest Wu, Eduard Cartier, Andreas Kerber.

Impact of charge trapping on the voltage acceleration of TDDB in metal gate/high-k n-channel MOSFETs

A. Kerber
GLOBALFOUNDRIES Inc.
1101 Kitchawan Road, Yorktown Heights, NY, 10598, USA
phone: +1 (914) 945 1607, **email:** Andreas.Kerber@globalfoundries.com

A. Vayshenker
Semiconductor Research and Development Center (SRDC)
IBM Systems and Technology Division, Hopewell Junction, NY, 12533, USA

D. Lipp
GLOBALFOUNDRIES Inc.
Wilschdorfer Landstr. 101, 01109 Dresden, Germany

T. Nigam
GLOBALFOUNDRIES Inc.
1050 East Arques Street, Sunnyvale, CA, 94085, USA

E. Cartier
T.J. Watson Research Center, IBM
1101 Kitchawan Road, Yorktown Heights, NY, 10598, USA

Abstract — The root cause for the increase in the TDDB voltage acceleration with decreasing stress voltage in metal gate/high-k n-channel FETs is investigated. Using DC and AC stress methodologies, the effect could be linked to charge trapping in the high-k gate dielectric. Furthermore, a correction for charge trapping is proposed, which results in a single power law voltage dependence for all stress conditions.

Keywords- high-k dielectrics, metal gate, SILC, TDDB, oxygen vacancies

I. INTRODUCTION

The voltage acceleration is a crucial parameter for time-dependent dielectric breakdown (TDDB) lifetime predictions. In a recent TDDB study for metal gate / high-k (MG/HK) CMOS devices [1], it was demonstrated that the voltage acceleration changes from a 1/E-dependence in the high-field regime to a E-dependence in the low-field regime and the transition point was attributed to the change in the tunneling behavior from FN to direct tunneling.

In this contribution, we present a detailed study on the impact of charge trapping on TDDB in MG/HK nFET devices. The study was motivated by recent reports which showed that threshold voltage shifts (ΔV_T) and stress induced leakage currents (SILC) during positive bias temperature stress at

typical TDDB stress voltages can be very large in MG/HK stacks [2, 3] and thus may impact the defect generation in dual layer MG/HK stacks, modulating the TDDB values. By comparing the device degradation during a bipolar AC with the degradation during a DC stress, strong differences in the charge trapping behavior are observed and it is shown that these differences are likely responsible for the deviation of the voltage acceleration from a simple power law behavior.

II. EXPERIMENTAL

N-channel MG/HK MOSFET devices with a Hafnium based gate dielectric were fabricated utilizing a conventional CMOS process flow on SOI substrates. TDDB tests were performed at $125^{O}C$ using a fast wafer-level data acquisition setup [4] for DC and AC testing. In addition, module-level stress was performed. Stress times ranged from ~ 1 ms to $\geq 10^5$ s, and test structure areas ranged from 0.008 μm^2 up to 166 μm^2.

The voltage acceleration of TDDB was examined using a constant voltage stress (CVS, DC stress) procedure with intermittent low-voltage current sensing as shown in the upper panel of Fig. 1. In addition to DC stress, a bipolar AC stress methodology was also applied [5] as illustrated in the lower panel of Fig. 1. The SILC at a lower sense voltage was monitored periodically within 1 ms as an aid for BD detection. Additionally, on individual FETs, intermittent I_d

978-1-4244-5430-3/10 $26.00 © 2010 IEEE

measurements were used to monitor the threshold voltage change during stress. For I_d measurements, the stress was interrupted for t_{delay}=1ms and the time interval between I_d measurements was increased with increasing stress time to minimize the perturbation to the stress.

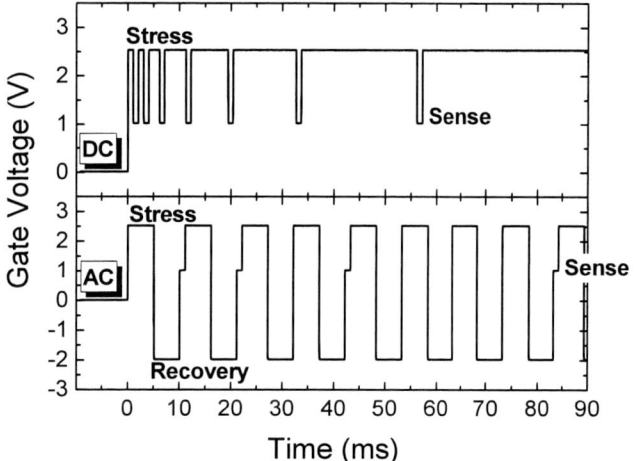

Figure 1. Voltage time traces for the DC Constant Voltage Stress (upper panel) and the bipolar AC stress where the bias is altered between stress and recovery (lower panel). Intermittend low voltage I_d sensing is performed to track the evolution of the threshold voltage shift.

With the bipolar AC stress method, the gate voltage was altered between the stress bias and a fixed recovery bias of -2 V at a frequency of 100 Hz and with a duty cycle of 50 %. The low voltage sense measurement was performed after the recovery cycle with t_{delay} = 1 ms for TDDB testing, as indicated in Fig. 1. For threshold voltage shift measurements sensing was performed before and after recovery cycle. As for DC measurements, the interval between sense measurements was increasing with stress time.

The threshold voltage shift, ΔV_T, during stress was used as a quantitative measure for the charge trapping in the gate stack during TDDB measurements. Details on the ΔV_T extraction procedure can be found in [6]. In short, the voltage shift is determined by projecting the drain current value at the sense voltage onto the pre-stress I_d-V_g characteristics, using an interpolation procedure.

For comparison, the V_T shifts during TDDB stressing were also measured for conventional poly Si / SiON gate stacks.

III. RESULTS AND DISCUSSION

Typical current time traces during DC and bipolar AC stress are shown in Fig. 2. As can be seen for bipolar AC stress SILC formation is strongly suppressed compared to the DC stress condition. These results are consistent with earlier SILC studies [2, 3]. The SILC was attributed to trap-assisted tunneling (TAT) near oxygen vacancies [3]. During positive bias stress the oxygen vacancies may trap negative charge opening TAT pathways, leading to the current increase during stress. When a negative bias is applied to the gate stack, the vacancies are being discharged, simultaneously resulting in a threshold voltage [7] and SILC [3] recovery. A large SILC increase prior to breakdown can compromise breakdown

detection. As can be seen from the data in Fig. 2, bipolar AC stressing provides a mean to decouple SILC from dielectric breakdown. AC stress furthermore can be used to study the breakdown transient in MG/HK n-channel MOSFETs in absence of strong charge trapping and SILC, providing a means to investigate the existence of a progressive wear out component. In this study the focus is only on the TDDB voltage acceleration and thus a hard breakdown criteria of a few µA at sense condition was used to determine the acceleration behavior for DC and bipolar AC stress.

The measured gate voltage, V_g, dependence of the TDDB lifetime (t63%) under DC stress conditions is summarized in Fig. 3 (solid symbols). As can be seen, the t63% values cannot be fitted by a single power-law voltage dependence [5]. For V_g > 2.2 V, a power law exponent of n = -54.8 is extracted, while for V_g < 2.2 V the value increases to n = -62. This change in slope is consistent with data reported in [1].

For comparison, values for t63% using AC stress are also shown (open symbols). These data can be fitted to a single power law. However, a significantly lower voltage exponent of n = -40.3 is obtained. In addition, the failure times for AC stress are reduced at all stress voltages when compared to the DC stress, consistent with previous reports [5, 9].

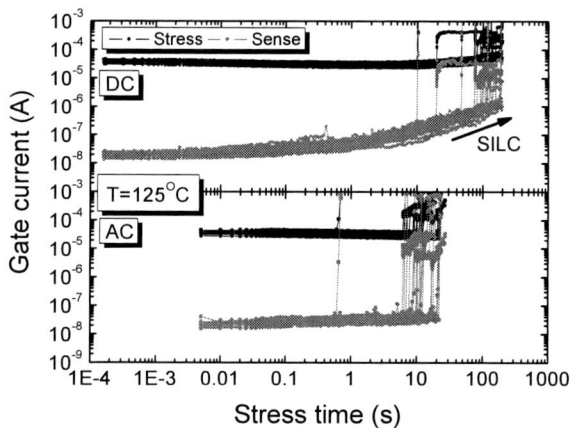

Figure 2. Current time traces during DC (upper panel) and bipolar AC (lower panel) stress at V_g = 2.2 V. Note that SILC formation is suppressed for bipolar AC compared to DC stress.

The impact of TDDB stressing on ΔV_T is summarized in Fig. 4, where the variations of ΔV_T with stress time for DC stress (upper panel) and for AC stress (lower panel) are compared. As can be seen from Fig. 4, the charge trapping under DC and AC conditions is very different. Under DC stress, the values for ΔV_T, rapidly increases beyond 100 mV and the time evolution follows a power law with a shallow power-law time exponent, n = 0.12, as compared to n = 0.17 at lower stress voltages [10]. During AC stressing the voltage shift (measured after the recovery cycle) is significantly lower than for DC stress, demonstrating that the recovery cycle effectively removes trapped charges from the high-k layer as discussed in Ref. [2]. The time evolution of ΔV_T during AC BTI testing is dependent on the measurement details like stress bias,

978-1-4244-5430-3/10 $26.00 © 2010 IEEE

recovery bias and duty cycle. In contrast to the DC stress, the time exponent for AC stress changes from n = 0.19 at short times to n = 0.45 near dielectric breakdown (for 1.9 V stress). To some extent the increase in time exponent can be attributed to trans-conductance degradation (not shown) which is typically not reported for n-channel MG/HK devices stressed in inversion.

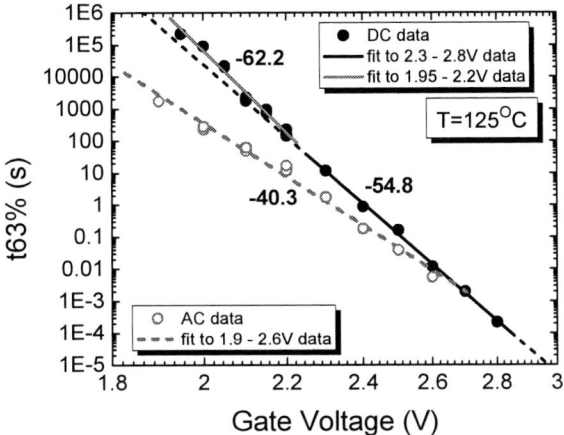

Figure 3. nFET DC TDDB voltage acceleration fitted to a power law dependence. Note the increase in acceleration factor with decreasing stress voltage. For AC stress (100 Hz, 50 % duty cycle and $V_{Recovery}$ = -2 V) a significantly lower acceleration factor (n = - 40.3) is obtained and no increase with decreasing stress voltage is observed.

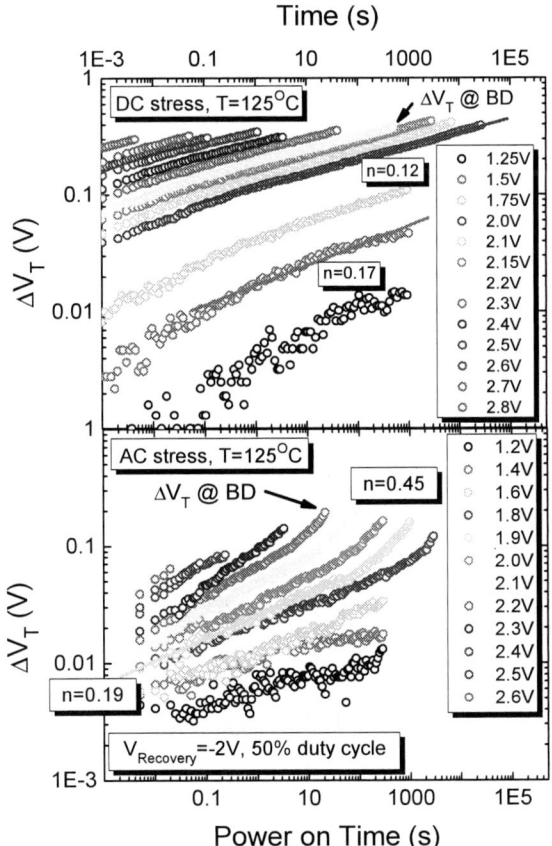

Figure 4. Threshold voltage shift during DC (upper panel) and AC (lower panel) TDDB stress. The AC shift was determined after the recovery cycle.

From the ΔV_T-versus-time traces shown in Fig. 4, the voltage shifts prior to dielectric breakdown (ΔV_T @ BD) for all measurement conditions are summarized in Fig. 5. Up to 10 samples were used per stress conditions. For DC stress, ΔV_T @ BD progressively increases with decreasing stress voltage, reaching shifts of up to 500 mV at the lowest voltage tested. In sharp contrast, ΔV_T for AC stressing strongly decreases with decreasing stress voltage below ~ 2.2 V. Above 2.2 V, shifts are quite comparable to the DC values and the "true" charge state is measured when sensing is performed prior to the recovery cycle (solid symbols). If ΔV_T is measured after the recovery cycle, effective discharging is observed because of the discrete nature of the AC stress.

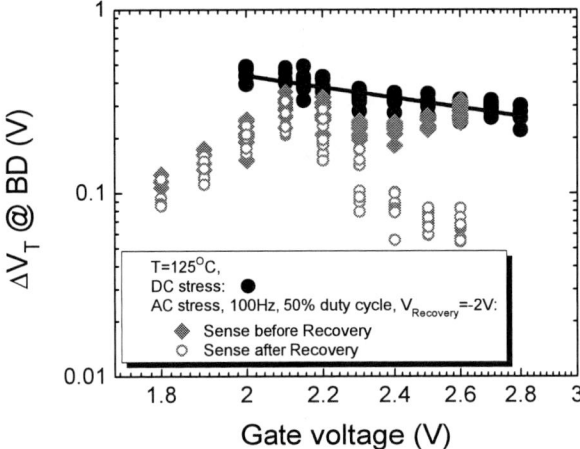

Figure 5. Threshold voltage shift prior to dielectric breakdown for DC and AC stress. Note the increase in voltage shift with decreasing stress voltage for DC stress. For AC stress voltage shifts are shown for sensing prior and after recovery cycle.

For conventional poly-Si/SiON gate stacks similarly large voltage shifts prior to dielectric breakdown are measured as shown in Fig. 6. Furthermore, no significant reduction in TDDB is observed when the AC stress procedure is applied (data not shown), suggesting stable trapped charge.

Since the charging behavior during DC and AC stress is significantly different for MG/HK nFETs, it can be expected that a correction to the stress condition may result in a universal acceleration behavior. In Fig 7 the t63% failure times form Fig. 1 are plotted against the corrected stress voltage (V_{Stress} - ΔV_T @ BD) using the threshold voltage shifts from Fig. 5. As can be seen, with this correction the failure times for DC and AC stress coincide and the change in voltage acceleration at low stress voltages is no longer present, resulting in a single TDDB voltage acceleration factor of n = -40 for all measurement conditions. A power law model following,

$$t63_{Use} = t63_{Stress} \times \left(\frac{V_{Use}}{V_{Stress} - \Delta V_T} \right)^n$$

with n = -40, can be applied to n-channel MG/HK device TDDB voltage acceleration.

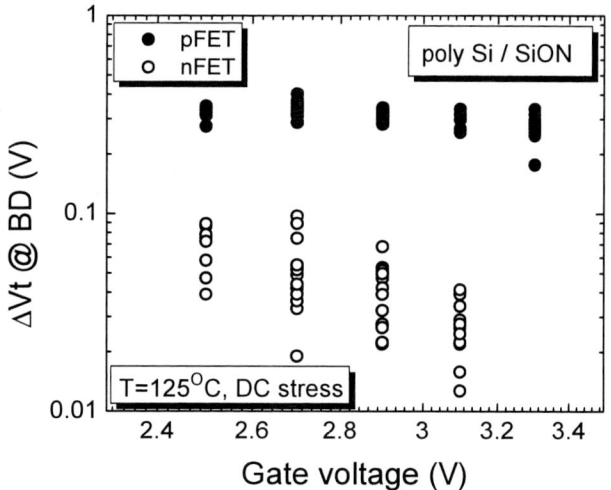

Figure 6. Threshold voltage shift prior to dielectric breakdown for conventional poly-Si/SiON using a DC stress procedure. Note that similarly large voltage shift can be observed for conventional gate stacks as for MG/HK at typical TDDB stress conditions.

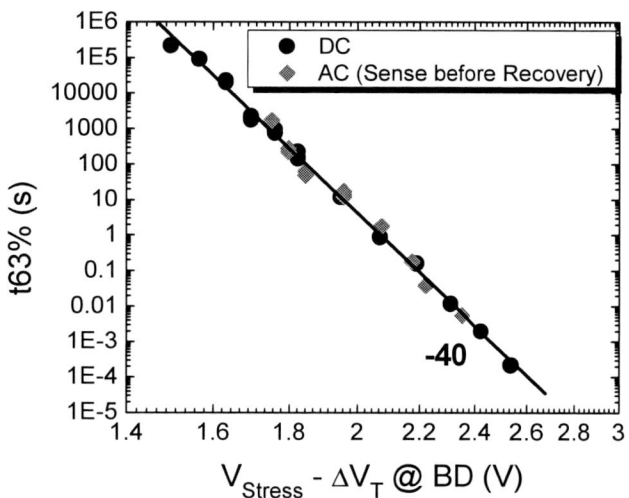

Figure 7. nFET TDDB voltage acceleration corrected for charge trapping. When plotting the failure time versus $V_{Stress} - \Delta V_T$ @ BD using the data shown in Fig. 3 a good agreement between DC and AC stress is obtained.

Since charge trapping at typical TDDB stress conditions has a significant impact on the voltage acceleration the lifetime projection for DC and bipolar AC stress are different. For an accurate assessment of TDDB lifetimes, the voltage acceleration and the failure distributions have to be determined simultaneously. Failure distributions have been studied in [5, 9]. According to Ref. [5], the failure distribution remains unchanged for bipolar AC stress based on sub-μm^2 devices, whereas others [9] reported an increase in the Weibull slope. As SILC formation is suppressed for bipolar AC stress larger areas can be tested in this mode and the failure distribution can be extended to verify the shape of the distribution towards product areas and low failure percentiles which may compensate for the reduction in the acceleration factor.

IV. CONCLUSION

Based on extensive TDDB testing of MG/HK n-channel FETs under DC and AC stress condition, the apparent increase in voltage acceleration with decreasing stress voltage was directly linked to charge trapping. It was shown that a universal acceleration behavior is obtained by accounting for charge trapping during TDDB testing. These results confirm the applicability of the power law for MG/HK n-channel devices.

ACKNOWLEDGMENT

This work was performed by the Research Alliance Teams at various IBM Research and Development facilities.

REFERENCES

[1] C. Prasad, M. Agostinelli, C. Auth, M. Brazier, R. Chau, G. Dewey, T. Ghani, M. Hattendorf, J. Hicks, J. Jopling, J. Kavalieros, R. Kotlyar, M. Kuhn, K. Kuhn, J. Maiz, B. McIntyre, M. Metz, K. Mistry, S. Pae, W. Rachmady, S. Ramey, A. Roskowski, J. Sandford, C. Thomas, C. Wiegand, J. Wiedemer, "Dielectric Breakdown in a 45 nm High-K/Metal Gate Process Technology", International Reliability Physics Symposium, pp. 667-668, 2008.

[2] F. Crupi, R. Degraeve, A. Kerber, D. H. Kwak, and G. Groeseneken, "Correlation between stress-induced leakage current (SILC) and the HfO_2 bulk trap density in a SiO_2/HfO_2 stack," International Reliability Physics Symposium, pp. 181–187, 2003.

[3] E. Cartier, A. Kerber, "Stress Induced Leakage Current and Defect Generation in nFETs with HfO_2/TiN Gate Stacks during Positive Bias Temperature Stress", International Reliability Physics Symposium. pp. 486-492, 2009.

[4] A. Kerber, M. Kerber, "Fast Wafer Level Data Acquisition for Reliability Characterization of sub-100 nm CMOS Technologies", International Integrated Reliability Workshop Final Report, pp. 41-45, 2004.

[5] A. Kerber, E. Cartier, B.P. Linder, S.A. Krishnan and T. Nigam, "TDDB failure distribution of metal gate / high-k CMOS devices on SOI substrates", International Reliability Physics Symposium, pp. 505-509, 2009.

[6] A. Kerber, K. Maitra, A. Majumdar, M. Hargrove, R. J. Carter, and E. Cartier, "Characterization of fast relaxation during BTI stress in conventional and advanced CMOS devices with HfO_2/TiN Gate Stacks," *IEEE Trans. Electron Devices*, Vol. 55, No. 11, pp. 3175–3183, 2008.

[7] A. Kerber, E. Cartier, L. Pantisano, R. Degraeve, T. Kauerauf, Y. Kim, A. Hou, G. Groeseneken, H. E. Maes, and U. Schwalke, "Origin of the threshold voltage instability in SiO_2/HfO_2 dual layer gate dielectrics," *IEEE Electron Device Lett.*, Vol. 24, No. 2, pp. 87–89, 2003.

[8] E. Wu, A. Vayshenker, E. Nowak, J. Sune, R.-P. Vollertsen, W. Lai, D. Harmon, "Experimental evidence of TBD power-law for voltage dependence of oxide breakdown in ultrathin gate oxides", *IEEE Trans. Elec. Dev.*, Vol. 49, No. 12, pp. 2244–2253, Dec. 2002.

[9] M. Kerber, R. Duschl, H. Reisinger, S. Jakschik, U. Schröder, T. Hecht, S. Kudelka, "Influence of Charge Trapping on AC Reliability of high-k dielectrics ", International Reliability Physics Symposium, pp. 585-586, 2004.

[10] A. Kerber, S.A. Krishnan and E.A. Cartier, "Voltage Ramp Stress for Bias Temperature Instability Testing of Metal Gate/High-k Stacks", *IEEE Electron Device Letters*, Vol. 30, No. 12, pp. 1347-1349, 2009.

Mechanism of high-k dielectric-induced breakdown of the interfacial SiO₂ layer

G. Bersuker, D. Heh, C. D. Young, L. Morassi[1], A. Padovani[1], L. Larcher[1], K. S. Yew[2],

Y. C. Ong[2], D. S. Ang[2], K. L. Pey[2], W. Taylor

SEMATECH, 2706 Montopolis Dr., Austin, TX 78741, USA; gennadi.bersuker@sematech.org

[1]DISMI Università di Modena e Reggio Emilia and IU.NET, 42100 Reggio Emilia, Italy

[2]Nanyang Technological University, Singapore 639798

Abstract— **A mechanism of degradation and breakdown in high-k/metal gate transistors was investigated. Based on the electrical test, physical analysis, and modeling results, we propose that the breakdown path formation/evolution in the interfacial SiO₂ layer is associated with the growth of an oxygen-deficient filament facilitated by the grain boundaries of the overlaying high-k film. The model allows reproducing SILC temperature dependency and its exponential increase from the fresh through soft and progressive breakdown phases.**

Keywords - breakdown, high-k dielectrics, interfacial layer

I.   INTRODUCTION

Aggressive gate stack scaling makes its breakdown (BD) stability a central reliability concern. Understanding the breakdown mechanism in high-k/metal gate stacks is complicated by their multilayer structure, which includes, in addition to the high-k dielectric (HK), an interfacial SiO₂ layer (IL) about 1nm thick. Stress-induced degradation of the dielectric stack may originate from defect generation in both the HK and SiO₂ layers, which is expected to be controlled by different mechanisms. Hence, the observed time dependency of device characteristics represents a convolution of the contributions associated with each active degradation mechanism. Since relative contributions of these mechanisms to the overall degradation may vary with the stress conditions (voltage/temperature), a failure to separate these contributions and identify the origin of the leading degradation component— HK or IL—would lead to an erroneous projection of reliability.

Previous studies demonstrated a strong correlation between the degradation of the IL in high-k stacks under stress (as measured by the frequency-dependent charge pumping method, f-CP) [1,2] and an increase of the stress-induced leakage current (SILC). On the other hand, SILC evolution during stress correlates to the gate stack transition through the soft and progressive BD phases [3,4], suggesting that the IL degradation might be primarily responsible for the overall degradation of the high-k gate stack (at least, when T<125°C). However, mechanisms of the BD path formation in the IL as well as the specific influence of the HK were not addressed. In this study, by combining electrical test, physical analysis, and simulation results, we propose that the BD path formation/evolution in the

IL is associated with the growth of an oxygen-deficient filament facilitated by the grain boundaries (GBs) of the overlaying high-k film.

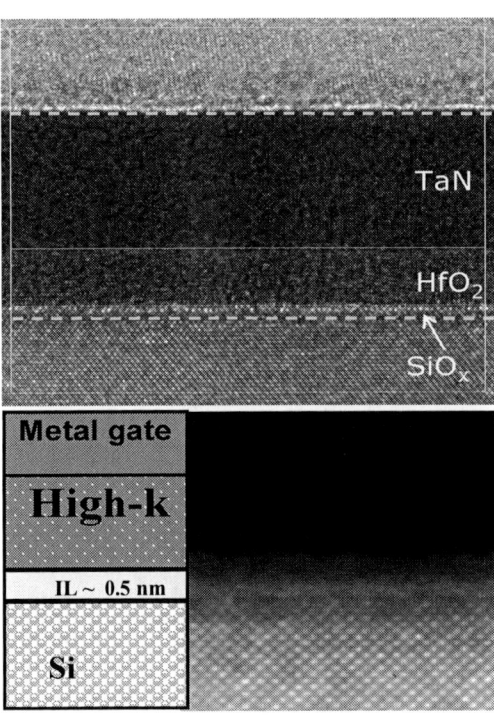

Fig. 1. TEM images of the IL-engineered (a) 0.3nm IL/4nm HfO₂ and (b) 0.5nm/2nm HfO₂ MOSFET gate stacks.

II.   VERIFICATION OF IL-RELATED SILC ORIGIN

As was previously reported, the evolution of SILC during stress was successfully simulated in a "conventional" high-k stack with a relatively thick IL, 1.1nm SiO₂/3nm HfO₂, using the dependency of the electron trap density in the IL on stress time as obtained by f-CP measurements [1]. To verify that SILC is an appropriate characteristic for monitoring IL degradation even in very thin ILs, HfO₂ transistors with a

978-1-4244-5430-3/10 $26.00 © 2010 IEEE   373

highly scaled IL, Fig.1, were fabricated using an IL-engineered process [5]. The gate stacks included 2nm HfO_2 and 0.5nm IL, and 4nm HfO_2 and 0.3nm IL (IL physical thicknesses were estimated from transmission electron microscopy [TEM] and electrical data), with equivalent oxide thickness (EOT) values of 0.6nm and 0.86nm, respectively. The slopes of the stress time dependencies of the transistor characteristics and interface trap density (estimated by the high frequency CP, which probes defects near the substrate) are similar, as shown in Fig.2,

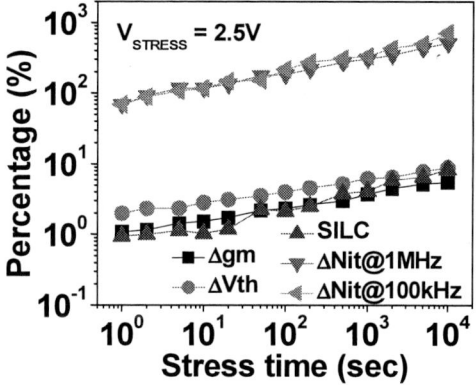

Fig. 2. Relative change of the characteristics of the 0.3nm IL/4nm HfO_2 gate stack transistor during stress

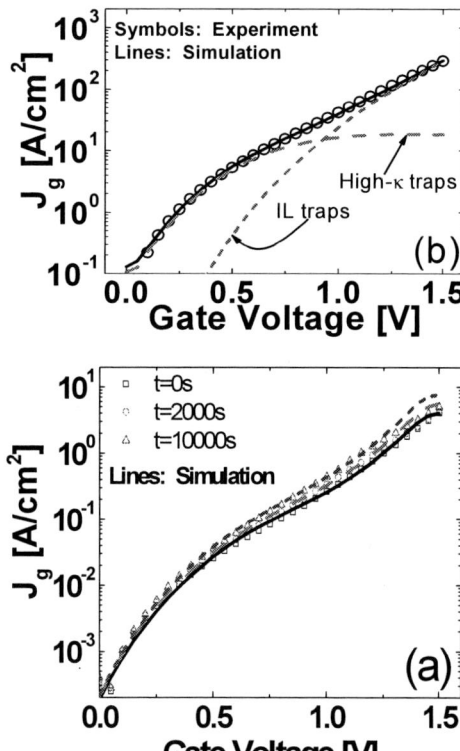

Fig. 3 (a) Measured (symbols) and simulated (lines) leakage current in (a) 0.3nm IL/4nm HfO_2 and (b) 0.5nm IL/2nm HfO_2 gate stack transistors at different stress times. Broken lines in (b) show the current components due to either bulk high-k or IL defects. In (b), SILC is negligible due to high gate leakage. Parameters used in simulations are shown in Table I (The values of the IL trap energy $E_{T,OX}$ were adjusted by the IL band offset decrease listed in Table II).

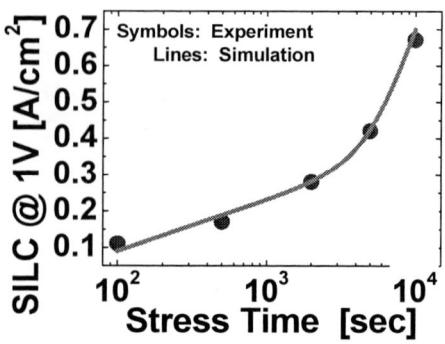

Fig. 4 Evolution of max SILC ($@V_g=1V$) in Fig. 3(a) during stress.

Table I. Defect parameters used in SILC simulations
$E_{T,OX}$ = 2.4 -2.8 eV– IL trap energy (wrt the bulk SiO_2 conduction band edge)
$E_{rel, ox}$ = 0.36 eV – IL trap relaxation energy
$\sigma_{0,IL}$ = 2E-14 cm^2 - IL trap x-section
$\sigma_{0,HK}$ = 1E-14 cm^2 – HK trap x-section
$N_{t,HK}$ = 2E19 cm^{-3} – HK trap density
$E_{T,HK}$ = 1.3-1.6 eV – HK trap energy

SiO_x IL physical thickness	3A (4nmHfO_2)	5A (2nmHfO_2)
IL band offset	1.3 eV	2 eV
IL band gap	4.3 eV	6.3 eV
IL k-value	9	7.5
Total stack EOT	0.86 nm	0.6 nm

Table II. IL characteristics in scaled stacks used in SILC simulations in Fig.3

Table III. Extracted energy characteristics (in eV) of IL defects in 1.1nm SiO_2/3nm HfO_2 stack

parameters \ degradation stage	FRESH	SBD	PBD
E_{rel}, relaxation energy	0.36	1.6	0.9
E_0, total energy differ.	-0.12	0.11	0.1
E_d, defect energy	2.6	3.1	3.1
E_B, activation energy	0.15	0.25	0.095

suggesting that the gate stack instability, specifically SILC, originates from IL degradation. This is further supported by the fact that the low and high frequency CP data sets are identical, indicating that no defects generation was detected in the bulk of the dielectric stack.

To verify the IL origin of the stress-generated traps, SILC gate leakage-gate voltage (I_g-V_g) data were simulated using the electron transport model, which considers a multi-phonon trap-assisted tunneling mechanism, including random defect generation and barrier deformation induced by the charged traps [6]. We took into account that in ultra-thin SiO_2 films, the band gap and band offset diminish while the dielectric constant increases as the SiO_2 becomes thinner, Table II [7]. In these simulations, the measured IL trap generation data, Fig. 2, were used as an input. The gate current is calculated by accounting for direct tunneling and TAT contributions due to both single- and multi-trap conductive paths [6]. The current driven by a single trap is calculated by

$$I_{trap_j} = \frac{q}{\tau_{c,j} + \tau_{e,j}} \cdot \quad (1)$$

where q is the electron charge and $\tau_{c,j}$ and $\tau_{e,j}$ are the time constants of the electron capture and emission of electrons by and from the j-th trap, respectively. $\tau_{c,j}$ and $\tau_{e,j}$ are calculated by summing up the phonon time constant contributions, $\tau_{c,j,n}$ and $\tau_{e,j,n}$, corresponding to the discrete phonon energies $E_{j,n}$ in the trap:

$$E_{j,n} = E_{C,j} + n \cdot \hbar\omega_0$$

$$\tau_{c,j} = \left(\sum_n N(E_{j-1,n}) \cdot f(E_{j-1,n}) \cdot P_T(E_{C,j} - E_{j-1,n}, F_{j-1,j}, D_{j-1,j}) \cdot Ca_{j,n} \right)^{-1}$$

$$\tau_{e,j} = \left(\sum_n N(E_{j+1,n}) \cdot P_T(E_{C,j} - E_{j,n}, F_{j,j+1}, D_{j,j+1}) \cdot Ep_{j,n} \right)^{-1} \quad (2)$$

$E_{C,j}$ is the j-th trap electronic energy level vs. the conduction band edge; $\hbar\omega_0$ is the effective phonon energy; N is the density of states; f is the Maxwell–Boltzmann occupation probability; P_T is the tunnel probability, which depends on the distance D and the electric field F between subsequent traps, respectively; and $Ca_{j,n}$ and $Ep_{j,n}$ are the capture and the emission rates of the electron traps, respectively. The $Ca_{j,n}$ and $Ep_{j,n}$ factors account for the lattice relaxation process, which depends on the electron-phonon coupling [8, 9]:

$$Ca_{j,n} = \frac{(4\pi)^2 \sigma_t^{1.5}}{E_{g,0}\pi^{1.5}} \frac{q^2 F^2 \hbar}{2m_e} \left(\frac{f_B+1}{f_B}\right)^{\frac{m}{2}} \exp[-2(2f_B+1)] \cdot I_n\left(2S\sqrt{f_B(f_B+1)}\right)$$

$$Ep_{j,n} = Ca_{j,n} \exp\left(-\frac{n \cdot \hbar\omega_0}{kT}\right) \quad (3)$$

$$f_B = 1/\exp(\hbar\omega_0/kT) + 1$$

Here I_n is the modified Bessel function of the order n, while f_B is the Bose function providing the phonon occupation number; σ_t is the trap capture cross-section; and $E_{G,0}$ is the energy bandgap of the dielectric. The Huang-Rhys factor S accounts for the structural relaxation associated with the electron capture or emission, $E_{relax} = S\, \hbar\omega_0$ (S is the number of phonons required for the system to reach its ground state). The capture probability peaks when the relaxation energy E_{relax} is equal to the difference between the initial (before electron trapping) and final (after trapping) total energies of the system, $E_0 = P\, \hbar\omega_0$, where P is the number of the phonons that should be emitted (or absorbed) to accommodate this energy difference. Therefore, an interplay between E_{relax} and E_0 determines the electron trapping efficiency and, hence, the leakage current. The physical meaning of these parameters is discussed below (see Fig. 8).

In both scaled gate stack sets, the I_g-V_g dependencies, Fig.3a,b, and SILC evolution, Figs. 3(a) and 4, were

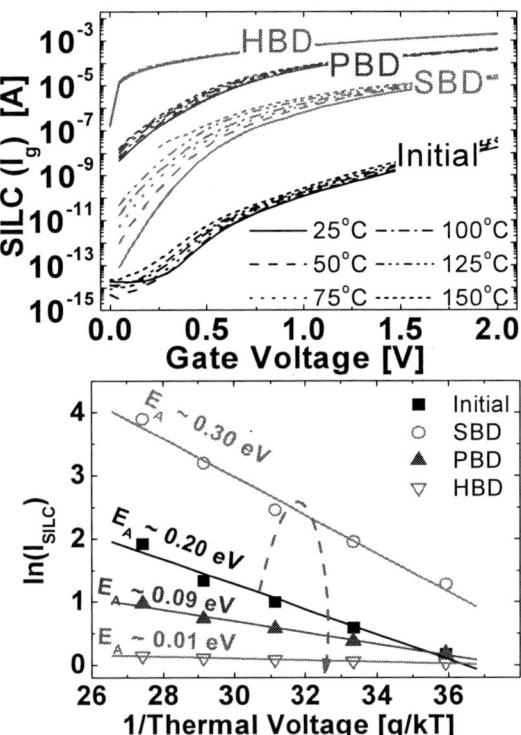

Fig. 5. (a) Temperature dependency and (b) activation energy of SILC measured before stress, after SBD, during PBD phase, and after hard BD in 1.1nmSiO₂/3nm HfO₂/TiN transistors

successfully reproduced using a *single* set of trap parameter values (see Table I) and accounting for the *stress-induced trap generation exclusively in the IL*. The values reported in Table II precisely match those obtained by simulating a conventional 1.1 nm IL high-k stack [1].

Results confirmed the association of SILC with the IL degradation.

III. SiO₂ IL BREAKDOWN MODEL

To identify specific defects in the SiO₂ IL that could be responsible for the observed SILC variation and, presumably, IL degradation, temperature-dependent current-voltage (I-V) characteristics were collected on fresh devices and during stress after the soft BD (SBD), progressive BD (PBD) and hard BD (HBD) phases, Fig. 5a. The corresponding activation energy values were extracted and are plotted in Fig. 5b. The activation energies exhibit a peculiar characteristic of, first, sharply increasing after SBD and then gradually decreasing with further dielectric degradation. This feature needs a plausible explanation within the breakdown model.

Further discussion is based on the recent TEM/electron energy-loss spectroscopy (EELS) results [10] revealing that SBD in a thin SiO₂ gate dielectric is associated with the formation of a highly oxygen-deficient filament. On the other hand, IL/high-k stacks are known to have a high density of as-processed O-vacancies in the IL region adjacent to the high-k film [11]. Thus, one can expect that such an oxygen-deficient region would initiate the formation of a BD path in the IL near the high-k film. Based on these facts, we propose a model that assumes that a highly *oxygen-deficient filament formed during a SBD event* (as in SiO$_2$ [10]) *consumes only a portion of the IL oxide adjacent to the high-k*, Fig. 6. Within this BD region, the band gap is effectively collapsed due to a high density of localized states associated with unpaired Si 3p electrons (since oxygen is lacking, similar to the well-studied oxygen-deficient transitional SiO$_2$ layer near the SiO$_2$/Si interface). These unoccupied states help transfer the electrons through the dielectric.

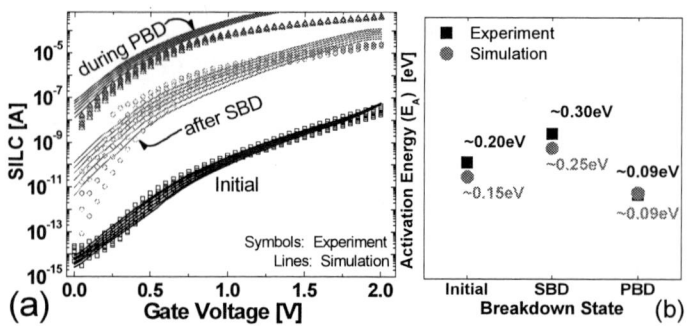

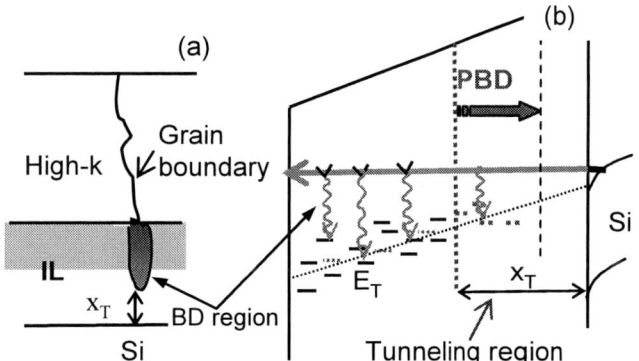

Fig. 7. (a) Measured (symbols) and simulated (lines) SILC (data in Fig. 5) and (b) SILC activation energy at different phases of BD in 1nmSiO$_2$/3nm HfO$_2$/TiN transistors. The measured PBD SILC at higher V_g is affected by series resistance causing mismatch with the simulations.

Fig. 6. Schematic illustration of the proposed IL BD model. (a) Formation of the O-deficient filament in IL in the result of SBD. Shaded area indicates high density of pre-existing O vacancies in IL. (b) Band energy diagram of the SBD region. PBD is associated with propagation of the BD filament toward Si substrate accompanied by generation of additional O-vacancy states (dotted lines). The waving line indicates trap relaxation due to an electron capture. x_T is the minimum electron tunneling distance through the unbroken IL portion, E_T is the energy of the relaxed trap. (See on a GB role below)

Based on this model, we simulated leakage current using the above approach, Eqs. (1-3). Simulations quantitatively describe both SILC temperature dependency and exponential current increase at different stages of BD, Fig.7a, as well as the corresponding SILC activation energies, Fig.7b. The energy characteristics of the trapping centers extracted by fitting the electron transport equations to the temperature-dependent leakage current data in Fig. 5a are presented in Table III (Note that the E_b values there are calculated using Eqs. (2-3)).

To gain better understanding of the major factors defining the evolution of the TAT current during stress, we discuss an electron capture process using a simplified form of Eq. (2-3). The characteristic capture cross-section of the trapping process, which determines the magnitude of the TAT current in Eq. (1), in most cases, can be approximated by

$$\sigma_c = \sigma_0 \times \exp\left(-\left(\frac{x_T}{\lambda}\right)\right) \times \exp\left(-\left(\frac{E_B}{kT}\right)\right) \qquad (4)$$

Here, σ_0 is the effective electronic capture cross-section (which incorporates the constant factors of Eq. (3)). The first exponential term describes the electron tunneling from the channel to the trap (P_T in Eq. (2)), x_T is the trap distance from

the substrate corresponding to the electron tunneling distance, and λ is the characteristic electron tunneling length. The second exponential term, which originates from the I_n and f_B dependencies in Eq. (3), describes the lattice re-arrangement required to accommodate the trapped charge. E_B is the associated energy barrier, which is defined by the following expression (as can also be deduced using the schematic in Fig.8):

$$E_B = (E_{relax} - E_0)^2 / 4 E_{relax} \qquad (5)$$

Here, E_0 is the change of the total energy of the system caused by the electron trapping at the defect, as in Eq. (3): (*dielectric/semiconductor with the electron in the channel*) - (*dielectric/semiconductor with the electron in the trap*), see Fig. 8. This energy barrier strongly depends on the trap relaxation energy E_{relax}, which is determined by the trapped charge-induced displacements of the lattice atoms around the defect.

According to the simulations, the stress-induced increase of the TAT current and its temperature dependency before SBD can be assigned to the generation of relatively shallow (~2.6 eV) IL traps (with a density on the order of 10^{19} cm^{-3}), which are characterized by a low relaxation energy upon electron capture (~0.36 eV), see Table III. Electrons can be transferred rather quickly (~10^{-6}sec @V_g=1.5V) via these defects, thus allowing for a high observed leakage current.

After SBD, the nature of the defects supporting the TAT current has changed as follows from their energy characteristics, Table III. An exponential current increase after SBD, Fig.6a, is due to a much shorter electron tunneling distance through the unbroken portion of the dielectric to the states in the BD region (smaller x_T, ~ 0.35 nm, in Eq. (4), see also Fig. 6). This is accompanied by a larger λ (due to a smaller energy band offset, which is characteristic of thinner SiO$_2$ films, see Table II) than the pre-SBD TAT current controlled by the electron transfer through the isolated, randomly distributed traps. An increase in the average relaxation energy of these traps and their deeper energy states results in higher current activation energy E_B, Table III, as calculated by Eq. (5). Overall, the characteristic time of the electron transfer through these traps, the density of which increased to $\sim 10^{20}$ cm^{-3}, was reduced to $\sim 10^{-11}$ sec at V_g=1.5V.

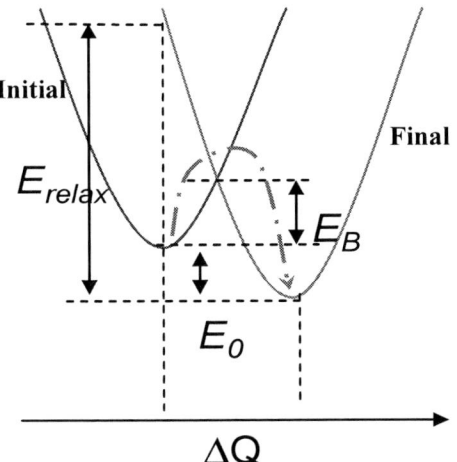

Fig.8. Schematic of the total energy diagram corresponding to the electron transition from the channel (Initial state) to the trap in the dielectric (Final state). E_{relax} is the energy associated with the charge trapping-induced displacements (ΔQ) of the dielectric atoms, positions of which are described by the generalized coordinate Q. The diagram shows that the charge trapping, in general, is associated with transitioning of the system over the energy barrier (E_B), even in the case of the resonant charge trapping (E_0=0).

The PBD phase is associated with further propagation of the BD filament towards the substrate (further x_T decrease, ≤0.2 nm) and its higher overall oxygen deficiency. The latter is expected to lower the characteristic vibration frequency ω due to a less dense structure of the oxygen-depleted BD region, which, in turn, leads to a lower relaxation energy, E_{relax} = S $\hbar\omega$, and, subsequently, lower leakage current activation energy E_B, Eq. (5). However, as follows from the unchanged value of the defect energy E_d, the post-SBD evolution of the leakage current through the PBD phase is supported by the generation of similar defects, indicating that PBD is mostly a quantitative rather than qualitative evolution of SBD.

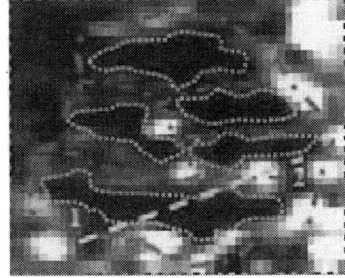

Fig. 9 Constant Current Tunneling Spectroscopy map of the 5nm HfO$_2$/1nm IL stack (annealed at 1000°C) at V_{sub}=-5V. Dotted lines (in green) outline GBs.

Identifying the physical structure of the SiO$_2$ defects based on their energy characteristics will be discussed elsewhere.

Leakage current activation energies at various phases of the stress-induced gate stack degradation calculated (Eq. (5)) using the trap parameters (Table III) are in excellent agreement with the experimental values, Fig. 7b.

IV. EFFECT OF HIGH-K GB ON BD PATH FORMATION IN IL

Conduction through a gate stack of 5nm HfO$_2$ and 1.1 nm IL was studied at the nanoscale level using scanning tunneling microscopy (STM) in ultrahigh vacuum. [12]. The bias voltage V_s was applied to the substrate. Constant imaging tunneling spectroscopy (CITS) was performed to examine local conduction through the gate stack. The feedback circuit was temporarily disabled after the preset tunneling current was achieved to allow V_s sweep from +5V to −5V at every pixel of the scanned area. The tunneling spectra thus form the CITS current map as shown in Fig. 9. Granular dark regions (corresponding to the low conductive area) are identified as the HK grains and the brighter borders (correspond to the high conductive area) as grain boundaries (GBs) [12]. The tunneling spectra of the grain and GBs are extracted before and after ramp voltage stress (RVS) to compare I-V as shown in Fig. 10(a). In the negative V_s regime, the tunneling current, which is sensitive to the emitting electrode barrier height, is observed to

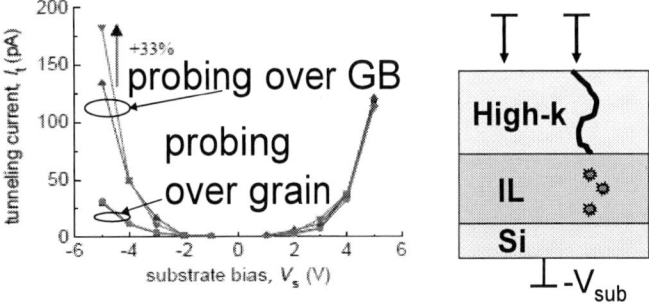

Fig. 10 (a) Initial and post-stress tunneling currents measured over the grain and GB. (b) Schematic of the STM probing over the grain and GB, and stress-induced defects generated in IL

be much higher at the GB, indicating a higher conduction through the IL underneath the GB [12]. After five cycles of RVS, the tunneling current at the GB increases significantly more (~33%) than at the grain (~4%). This shows that the IL beneath the GB is highly susceptible to degradation under electrical stress, Fig. 10(b), due to lower GB resistance (vs. the bulk of the grain) and, subsequently, enhanced electric field across the IL.

V. CONCLUSION

Using scaled-IL HK gate stack transistors, it was re-confirmed that SILC in MIS high-k devices under stress in inversion is controlled by IL degradation. A proposed physical model describes the evolution of IL degradation through SBD and PBD as a propagation of a highly oxygen-deficient filament from the IL/high-k region toward the substrate. The model allows reproducing SILC temperature dependency and its exponential increase from fresh through PBD phases. The stress-induced defects in the IL are found to be preferentially generated in the regions underlying the high-k grain boundaries, lower resistance of which results in a higher electric field across the adjacent IL regions.

REFERENCES

[1] G. Bersuker et al., IEDM, 2008
[2] S. Sahhaf et al., IRPS,2008
[3] G. Bersuker et al., IRPS,2007
[4] N. Chowdhury et al., Microel.Eng. 8,27,2008
[5] J. Huang.,VLSI 2009
[6] L.Larcher, TED, 50, 2003
[7] F.Giustino et al., J.Phys:17 S2065,2005
[8] C.H. Henry et al., Phys. Rev. B, 15, 989,1977
[9] W. B. Fowler et al., Phys. Rev. B, 41, 8313, 1990
[10] X. Li et al., APL 93, 262902, 2008; ibid. IEDM, 2008
[11] G.Bersuker., JAP,100,094108,2006
[12] Y. C. Ong, et al., APL 92, 2008; ibid. APL 91, 2007

Characterization of millisecond-anneal-induced defects in SiON and SiON/Si interface by gate current fluctuation measurement

Tsunehisa Sakoda[1], Keita Nishigaya[1], Tomohiro Kubo[1], Mitsuaki Hori[1], Hiroshi Minakata[1], Yuko Kobayashi[1], Hiroko Mori[1], Katsuji Ono[1], Katsuto Tanahashi[1], Naoyoshi Tamura[1], Toshifumi Mori[1], Yoshiharu Tosaka[1], Hideya Matsuyama[1], Chioko Kaneta[2], Koichi Hashimoto[1], Masataka Kase[1] and Yasuo Nara[1]

[1]Fujitsu Microelectronics Ltd.
1500, Mizono, Tado-cho, Kuwana-shi, Mie-ken, 511-0192, Japan
+81-594-24-2645, t.sakoda@jp.fujitsu.com

[2]Fujitsu Laboratories Ltd.
10-1, Morinosato-Wakamiya, Atsugi, Kanagawa 243-0197, Japan

Abstract—In this paper, we have investigated bulk trap and interface trap density (D_{it}) caused by millisecond annealing (MSA) using gate current fluctuation (GCF) and charge pumping measurements. We show that the high energy flash lamp annealing (FLA) creates the GCF with a long duration time and it is critical issue to get a stable SRAM operation. FLA creates interface traps localized at the gate edge of MOSFET.

Keywords-component; gate current fluctuation; charge pumping; Dit; millisecond-anneal; flash lamp annealing; laser annealing; SRAM reliability; V_{min}

I. INTRODUCTION

Millisecond annealing such as laser-spike annealing (LSA) and FLA are promising technology on the current high performance MOSFET and many papers have been reported [1-4]. These were intended to improve the transistor performance. However, it has been reported that D_{it} was generated by the thermo-mechanical stress during the millisecond annealing fast thermal gradient [2-4], so recovery annealing or sub 10 millisecond annealing after millisecond annealing is necessary [2]. Millisecond annealing also creates the bulk trap which can induce trap-assisted tunneling with GCF. The GCF induces V_{min} shift of SRAM [5] and large current fluctuation increases SRAM bit failure rate [6]. Therefore it is important to investigate bulk trap for reliability and yield. However there have been few reports which investigate the D_{it} and bulk trap generated by millisecond annealing systematically.

In this paper, we investigated the impact of various condition of millisecond annealing on defect generation in gate dielectric using gate current fluctuation measurement.

II. EXPERIMENTAL

Using 45nm node transistor with 1.1 nm thickness of SiON gate, D_{it}, negative bias-temperature instability (NBTI) lifetime and GCF were measured [7]. Fig. 1 shows the process flow for the device sample preparation. The samples have a poly-Si/SiON layered structure and millisecond annealing was performed for extension and source/drain annealing. Gate current and capacitance of millisecond annealing sample is almost same as without millisecond annealing sample.

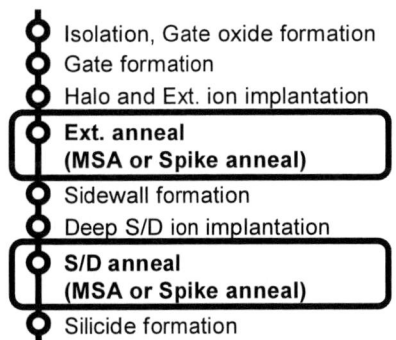

Fig. 1 Process flow for the device sample preparation. MSA was performed for extension and S/D annealing.

FLA with variable powers and assist temperatures was used (Fig. 2).

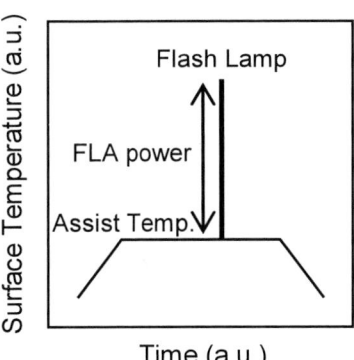

Fig. 2 Surface temperature simulated from assist temperature and flash power.

The non-patterned Si wafers with 1.1 nm SiON film and annealing was prepared for measuring the density profile of gate dielectric by high luminance x-ray reflectivity (XRR) analysis based on the synchrotron radiation at Spring8 [8].

The charge pumping measurement was performed at 2.5 MHz using the MOSFET with gate width and length of 20 and 0.09 μm respectively. NBTI was measured at 125 °C using pMOSFET by fast spot measurement. Measurement delay is under 10 millisecond. NBTI criterion is $\Delta I_d/I_{d0} = 10\%$, and we used power model for extrapolation of NBTI life time. We measured four samples per stress condition and four stress conditions were applied. GCF was measured by the timing chart sequence shown in Fig. 3(a) with nMOSFET array connected in parallel. We measured 52 chips per sample in parallel to count GCF. The measurement condition was the stress voltage (V_{stress}) of 1.8 V, the measuring voltage ($V_{measure}$) of 1.0 V at 150 °C. Fig. 3 (b) is actually measured example of GCF.

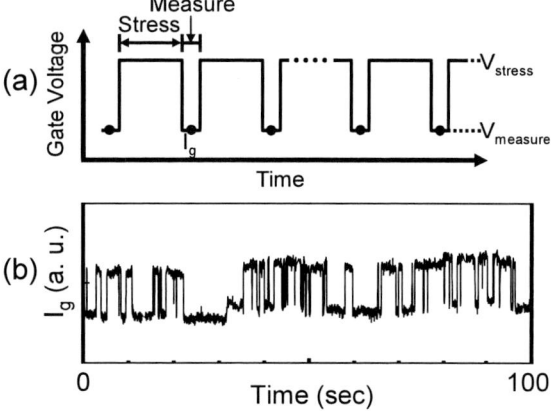

Fig. 3 Timing chart sequence for GCF measurement and a example of GCF.

Fig. 4 shows the Fourier transform of gate current, so the Lorentz noise dominates the gate noise below 0.1 Hz. We determined that the sampling interval is set to 10 s for characterizing the single defect and slow trap. The GCF count is defined as the number of gate current jump which variation is over 1 nA.

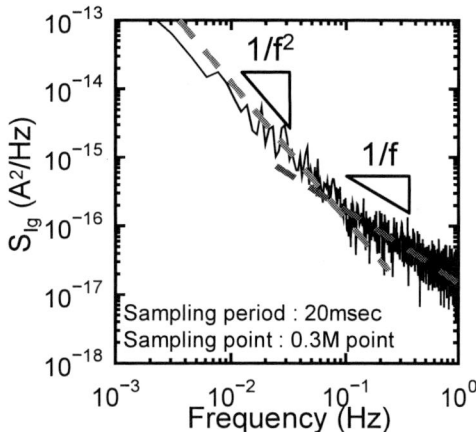

Fig. 4 The Fourier transform of gate current. The Lorentz noise dominates the gate noise below 0.1 Hz.

III. RESULTS AND DISCUSSION

The charge pumping current (I_{cp}) of nMOSFETs which performed FLA was measured (Fig. 5). I_{cp} increases with increasing the annealing temperature of FLA. I_{cp} of high FLA power sample was larger than the others. These phenomena are induced by thermo-mechanical stress and consistent with the reports [3]. We also measured NBTI life time to confirm the impact of interface trap (Fig. 6). NBTI life time of sample of high FLA power was shorter for annealing temperature than low FLA power.

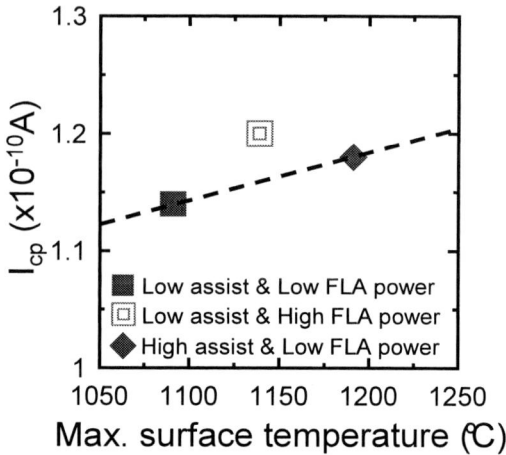

Fig. 5 I_{cp} of nMOSFETs with performed FLA. I_{cp} of high FLA power sample was larger than the others.

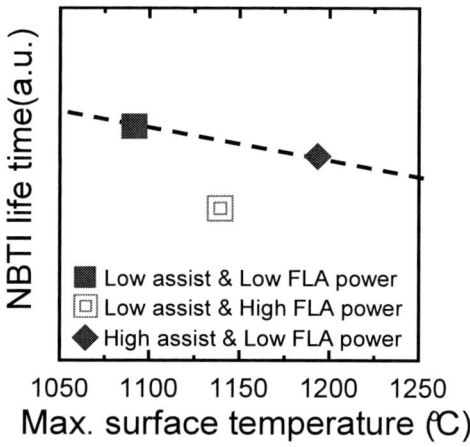

Fig. 6 NBTI lifetime of pMOSFETs. NBTI life time of high FLA power was shorter for annealing temperature than low FLA power.

Fig. 7 shows the relation of GCF count and I_g which was measured for the samples of various thickness of gate dielectric and adopted various condition of FLA, in particular, various annealing temperature (Fig. 7). GCF count increases with I_g because trap assisted tunneling probability depends on the thickness of gate dielectric. High FLA power samples causes large GCF counts even in same I_g, and it is seemed to be independent of annealing temperature. Fig. 5 and 8 show that I_{cp} increases with maximum surface temperature in the case of same FLA power and GCF count was constant if FLA power was low. CGF measurement is sensitive to bulk trap and I_{cp} is sensitive to interface trap. These results suggest that FLA creates traps in bulk and interface because I_{cp} and GCF count increased in the case of high FLA power.

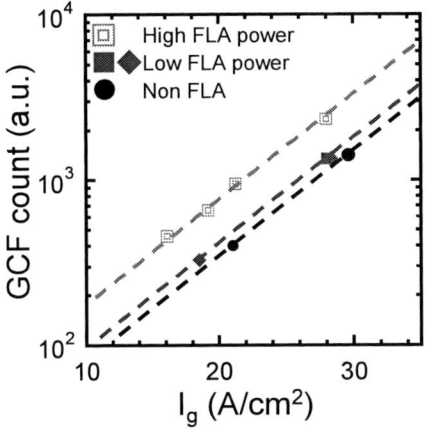

Fig. 7 Dependence of I_g on GCF count. GCF count of high power FLA was larger than low power or non FLA.

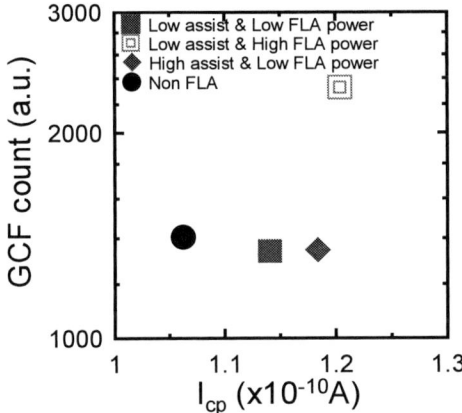

Fig. 8 Relationship between GCF count and Dit. Dit increases with temperature and GCF increase with FLA power.

Large I_g jump causes SRAM bit failure when the cell is selected (Figs. 9 and 10). Moreover, I_g jump with long duration time can increase SRAM bit failure because the probability of I_g jumps at the accessed cell increases.

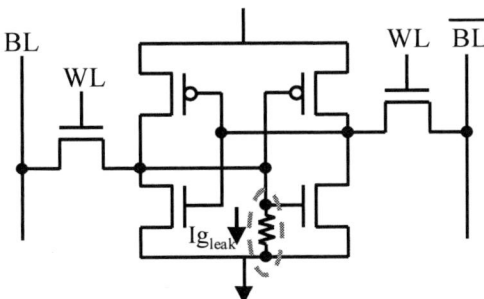

Fig. 9 Schematic of 6T SRAM cell with gate leakage. Gate leakage in nMOS driver transistor is the most sensitive factor to SRAM bit failure when the cell is selected.

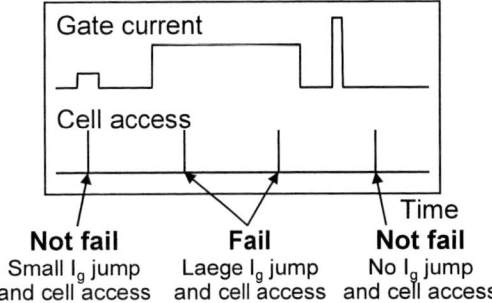

Fig. 10 Relationship between GCF and SRAM cell access timing. If large I_g jump and cell access occurred at the same time, SRAM bit failure will occur.

GCF counts were characterized by duration time and amplitude of I_g jump (Figs. 11 and 12). Large and long I_g jump was observed with high FLA power samples. Therefore FLA power must be optimized, because large I_g jump and long duration time cause SRAM bit failure.

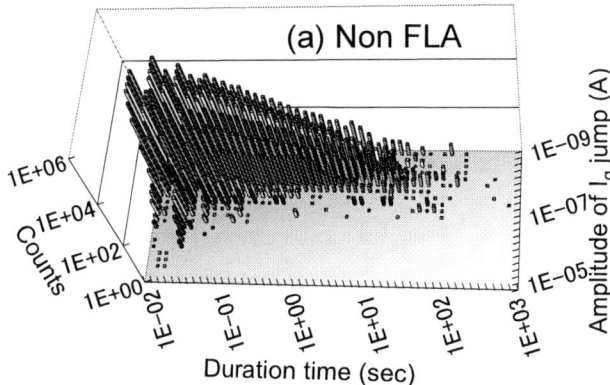

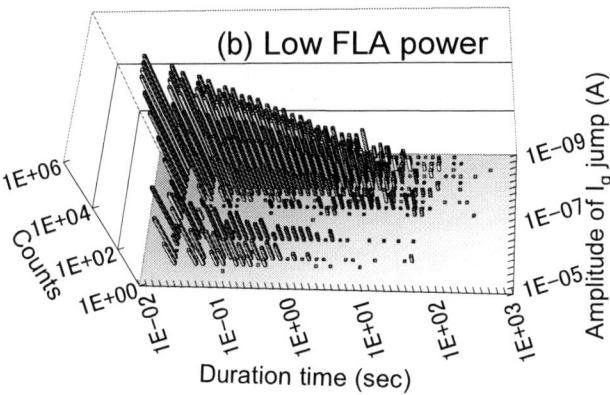

Fig. 11 Relationship between duration time, I_g jump size and GCF count of (a) non FLA sample and (b) low FLA power sample.

978-1-4244-5430-3/10 $26.00 © 2010 IEEE   381

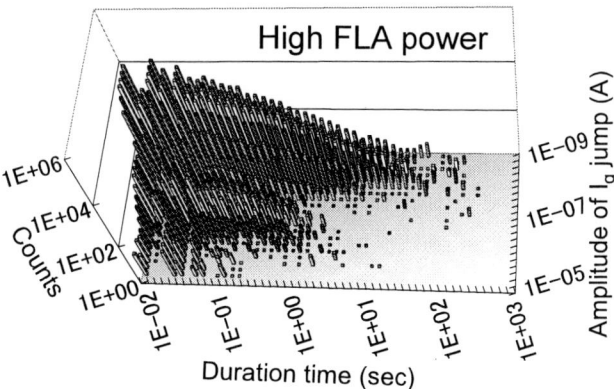

Fig. 12 Relationship between duration time, I_g jump size and GCF count of high FLA power. Large I_g jump was observed with high FLA power sample. Large I_g jump and long duration time can cause SRAM bit failure.

Depth profile of density was measured by x-ray reflectivity analysis using non-patterned Si wafer with SiON film which was subjected to FLA (Fig. 13). Density of FLA samples was lower than non FLA sample at Si/SiON interface and bulk SiON about 0.5 nm from interface. These results indicate that FLA generates defects following by structure change at SiON/Si interface and bulk SiON. We also suggest that the trap level does not depend on FLA power, because the time constant and defect position were not changed.

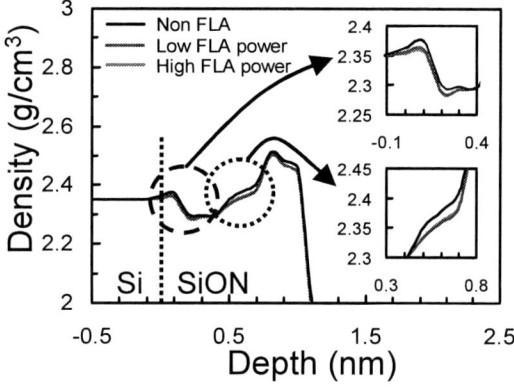

Fig. 13 Depth profile of density measured by XRR. Density of FLA samples was lower than non FLA sample at Si/SiON interface and bulk SiON.

We measured I_{cp} and GCF count using LSA and FLA sample (Figs. 14 and 15). The LSA samples were better NBTI and less GCF count. Laser scan type millisecond annealing including LSA might cause smaller thermo-mechanical stress than FLA [3].

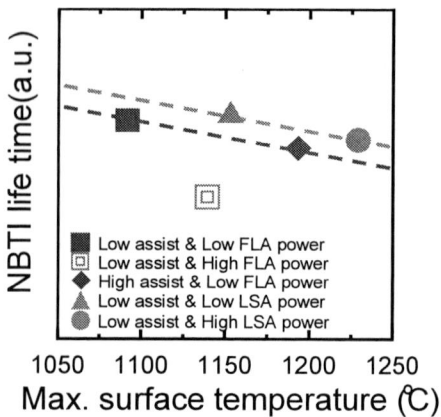

Fig. 14 NBTI life time of pMOSFETs performed FLA and LSA. LSA samples have longer lifetime than the others.

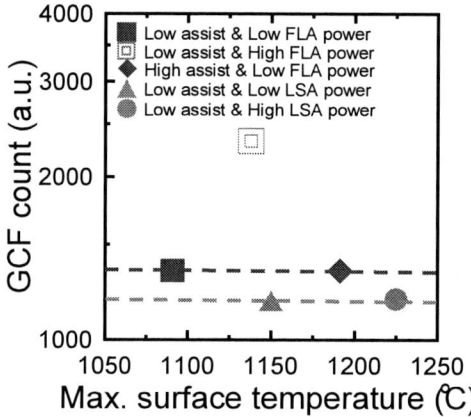

Fig. 15 GCF count of nMOSFETs performed FLA and LSA. LSA sample was less GCF count than FLA sample.

Also we suggest the generation of SiON gate stack defects such as H releasing from the shorter wave length characteristics of Xe flash lamp of FLA as shown in Fig. 16 [9].

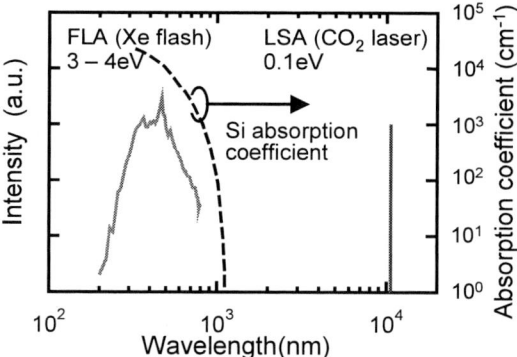

Fig. 16 The spectrum of Xe flash lamp (FLA) and CO_2 laser (LSA). FLA has short-wavelength component that have high photon energy. High energy photon disassociate Si-H bonds at SiON/Si interface.

978-1-4244-5430-3/10 $26.00 © 2010 IEEE   382

Gate length dependency of I_{cp} and GCF count was measured (Figs. 17 and 18). Both I_{cp} and GCF count of FLA sample were larger than non FLA sample at short gate length. These results indicate that defects were generated near the gate edge by FLA as shown in Fig. 19(a) and (b).

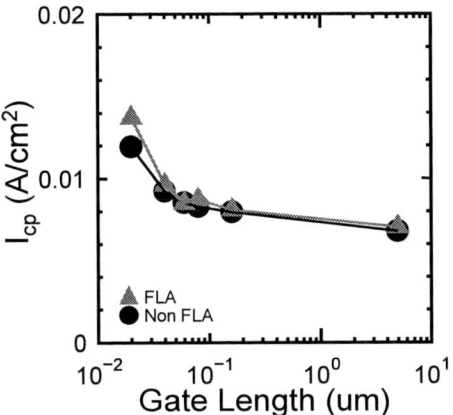

Fig. 17 Dependence of L_g on I_{cp}. I_{cp} increase with short L_g.

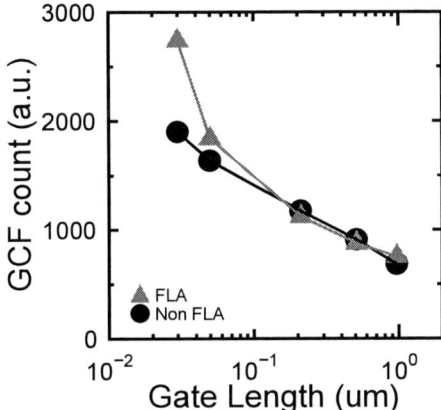

Fig. 18 Dependence of L_g on GCF count. GCF count also increase with short L_g.

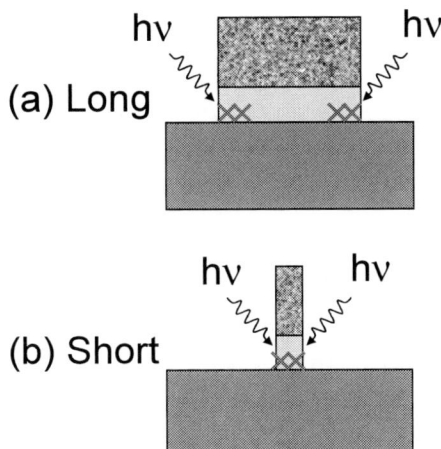

Fig. 19 Short MOSFET (b) is more affected than Long MOSFET (a) by UV light that enters obliquely because almost all UV light is absorbed by poly-Si electrode.

IV. CONCLUSION

We have investigated that trap generation in gate dielectric and interface caused by millisecond annealing using gate current fluctuation and charge pumping measurement. We show that the high energy FLA creates the GCF with a long duration time and it is critical issue to get a stable SRAM operation. The investigation of GCF count clarifies successfully to distinguish the defect generation of FLA and LSA and indicates the defect generation at the gate edge.

ACKNOWLEDGMENT

The authors would thanks to Mr. S. Doi of Fujitsu laboratories Ltd. for the helpful discussion and XRR measurements.

REFERENCES

[1] T. Yamamoto, T. Kubo, T. Sukegawa, E. Takii, Y. Shimamune, N. Tamura, T. Sakoda, M. Nakamura, H. Ohta, T. Miyashita, H. Kurata, S. Satoh, M. Kase and T. Sugii, "Junction Profile Engineering with a Novel Multiple Laser Spike Annealing Scheme for 45-nm Node High Performance and Low Leakage CMOS Technology", IEDM Tech. Dig., pp.143-144 (2007).

[2] Takashi Onizawa, Shinich Kato, Takayuki Aoyama, Yasuo Nara and Yuzuru Ohji, "A Proposal of New Concept Milli-second Annealing: Flexibly-Shaped-Pulse Flash Lamp Annealing (FSP-FLA) for Fabrication of Ultra Shallow Junction with Improvement of Metal gate High-k CMOS Performance", Symp. VLSI Tech., pp.110-111 (2008)

[3] S. Severi, E. Augendre, S. Thirupapuliyur, K. Ahmed, S. Felch, V. Parihar , F. Nouri, T. Hoffman, T. Noda, B. O'Sullivan, J. Ramos, E.San Andrés, L. Pantisano, A. De Keersgieter, R. Schreutelkamp, D. Jennings, S. Mahapatra, V.Moroz, K. De Meyer, P. Absil, M. Jurczak, S. Biesemans, "Optimization of Sub-Melt Laser Anneal: Performance and Reliability", IEDM Tech. Dig., pp.859-861 (2006)

[4] Pankaj Kalra, Prashant Majhi, Dawei Heh, Gennadi Bersuker, Chadwin Young, Nikhil Vora, Rusty Harris, Paul Kirsch, Rino Choi, Man Chang, Joonmyoung Lee, Hyunsang Hwang, Hsing-Huang Tseng, Rajarao Jammy, and Tsu-Jae King Liu, "Impact of Flash Annealing on Performance and Reliability of High-κ/Metal-Gate MOSFETs for sub-45nm CMOS", IEDM Tech. Dig., pp.353-356 (2007)

[5] M. Agostinelli, J. Hicks, J. Xu, B. Woolery, K. Mistry, K. Zhang, S. Jacobs, J. Jopling, W. Yang,B. Lee, T. Raz, M. Mehalel, P. Kolar, Y. Wang, J. Sandford, D. Pivin, C. Peterson, M. DiBattista,S. Pae, M. Jones, S. Johnson and G. Subramanian, "Erratic Fluctuations of SRAM Cache Vmin at the 90nm Process Technology Node", IEDM Tech. Dig., pp.671-674 (2005)

[6] Tsunehisa Sakoda, Naoyoshi Tamura, Shiqin Xiao, Hiroshi Minakata, Yusuke Morisaki, Keita Nishigaya, Takashi Saiki, Toshiyuki Uetake, Toshio Iwasaki, Hideo Ehara, Hideya Matsuyama, Hiroshi Shimizu, Koichi Hashimoto, Masayoshi Kimoto, Masataka Kase and Kazuto Ikeda, "1st quantitative failure-rate calculation for the actual large-scale SRAM using ultra-thin gate-dielectric with measured probability of the gate-current fluctuation and simulated circuit failure-rate", Symp. VLSI Tech., pp.27-28 (2007)

[7] T. Miyashita, K. Ikeda, Y. S. Kim, T. Yamamoto, Y. Sambonsugi, H. Ochimizu, T. Sakoda, M. Okuno, H. Minakata, H. Ohta, Y. Hayami, K. Ookoshi, Y. Shimamune, M. Fukuda, A. Hatada, K. Okabe, T. Kubo, M. Tajima, T. Yamamoto, E. Motoh, T. Owada, M. Nakamura, H. Kudo, T. Sawada, J. Nagayama, A. Satoh, T. Mori, A. Hasegawa, H. Kurata, K. Sukegawa, A. Tsukune, S. Yamaguchi, K. Ikeda, M. Kase, T. Futatsugi, S. Satoh, and T. Sugii, "High-Performance and Low-

Power Bulk Logic Platform Utilizing FET Specific Multiple-Stressors with Highly Enhanced Strain and Full-Porous Low-k Interconnects for 45-nm CMOS Technology", IEDM Tech. Dig., pp.251-254 (2007)

[8] SPring-8 Research Frontiers 2001B/2002A p92

[9] C. Kaneta, T. Tamasaki, T. Uda, "Reaction of atomic hydrogen with the Si(100)/SiO$_2$ interface defects", Conf. Phys. Semicond., pp.419-420 (2000)

Gate Dielectric Reliability in the Sub Threshold Regime

Paul E. Nicollian, Cathy A. Chancellor, and Anand T. Krishnan

Advanced CMOS Technology-Design Integration
Texas Instruments Incorporated
Dallas, Texas USA
Phone: (972) - 995-2820, email: nicollian@ti.com

Abstract—**The dielectric reliability of devices operating in the high bias, low gate current sub threshold state that can occur in analog and mixed signal designs is investigated using substrate hot carrier injection to accelerate breakdown. The complexities of this stress mode are elucidated and a reliability projection methodology is presented.**

Keywords - Reliability, Breakdown, TDDB, Sub-threshold, Dielectric, Oxide, SiON, Hydrogen

I. INTRODUCTION

The high cost of technology development has led to the proliferation of silicon-on-chip applications with multiple transistor elements to attain additional capability through design rather than scaling. In cascoded design schemes where transistors are used as voltage dividers, this can result in over-design voltages being applied to one or more device terminals. An example is shown in Fig. 1, where 3.3 V is applied to the drain of a 1.8 V (EOT = 2.7 nm) NFET and the device operates in the low drain current sub threshold (depletion/weak inversion) regime. With an applied gate voltage of 2.6 V, the source voltage will float to approximately $V_G - V_T = 2.2$ V in the sub threshold state. The drain voltage feeding into the next transistor stage will then be 2.2 V.

If the bias conditions shown in Fig. 1 resulted in the device operating in the on-state (strong inversion), breakdown would be driven by $V_{GS} = 0.4$ V, so the failure rate would be negligible. However, the surface charge is a quadratic rather than exponential function of band bending in depletion and weak inversion [1] since there are not a sufficient number of electrons to screen the bulk field. Accordingly, breakdown of the transistor that is operating in the sub threshold state will be driven by $V_{GB} = 2.6$ V, which predicts high failure rates in some applications. This would require incurring the cost of extra mask sets to define 3.3 V transistor components if the sub threshold operating conditions shown in Fig. 1 are truly unreliable. However, I_G is low in weak inversion since it is supply limited, which raises the possibility that reliability could be favorable since breakdown is both energy and fluence driven [2]. The problem is how to accelerate breakdown in this low I_G, low E_{OX}, and high V_{GB} deep depletion regime.

In this work, we use substrate hot electron stress (SHE) to develop breakdown models for sub threshold operation.

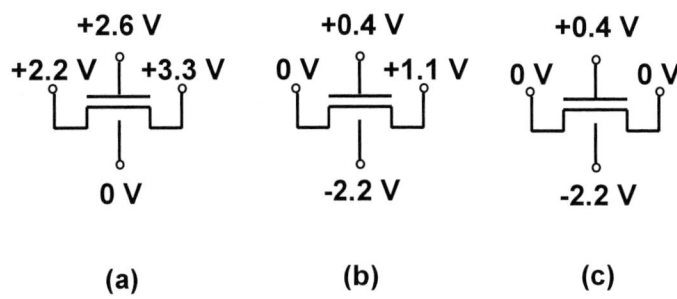

Figure 1. (a) Example of the biasing of a cascoded device operating in depletion/weak inversion. (b) Simplified equivalent biasing scheme. (c) Bias configuration for 3-terminal reliability projection.

II. RESULTS

An idealized cross section of the SHE device structure is shown in Fig. 2. It is a 4-terminal MOSFET or 3-terminal gated diode (with $V_D = V_S$) with an external PN junction that is forward biased to supply electrons. The oxide field, silicon field, and electron fluence can be separately controlled respectively by V_G, V_B, and I_{INJ} [3]. SHE has been used to study charge trapping [4], to separate the effects of energy and field on breakdown [5], to investigate vibrational excitation (VE) of silicon-hydrogen bonds as a mechanism for trap generation [6], and to profile trap distributions in high-k films [7]. As V_{GD} is small, we are primarily interested in modeling the effects of high V_{GB} in this work. We use the 3-terminal configuration as shown in Fig. 1(c) and refer to the combined source-drain terminal as the drain with a bias V_D.

The gate current versus stress time for uninterrupted stress (UIS) is shown in Fig. 3. SHE increases I_G by more than seven orders of magnitude and can accelerate breakdown even in this low E_{OX} sub threshold state. Due to the low E_{OX} and the low free surface charge density during stress, the first SBD event cannot be accurately detected for UIS because (i) I_{SBD} is difficult to resolve from the background current, (ii) I_G is initially a decreasing function of time due to charge trapping since de-trapping rates are reduced at low E_{OX} [8], and (iii) the application of a substrate bias further "softens" breakdown due to low surface electron density, which reduces the energy (since it is proportional to the square of the electron density) that can be dissipated during the transient formation of the SBD

978-1-4244-5430-3/10 $26.00 © 2010 IEEE

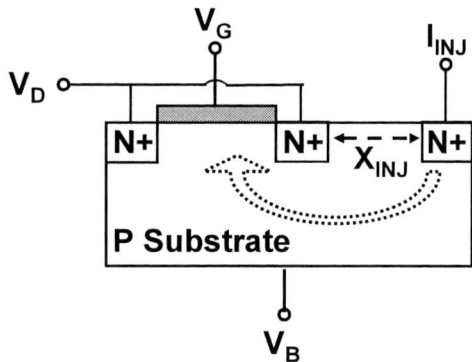

Figure 2. Idealized cross section of an SHE test structure in 3-terminal configuration, where $V_D = V_S$. X_{INJ} is the separation between the injector and NFET drain.

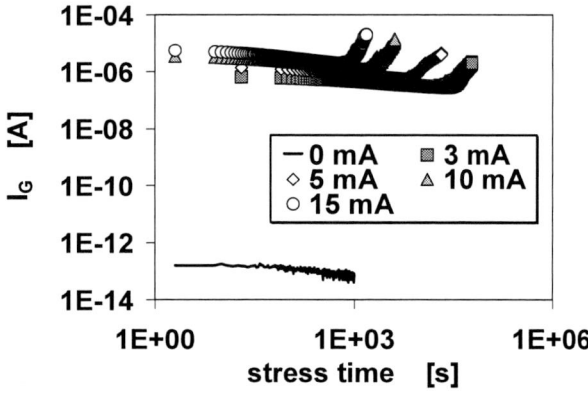

Figure 3. Gate current versus stress time for UIS with injector current as a parameter. $V_G = +3.0$ V, $V_D = +3.0$ V, $V_B = -2.2$ V.

path [9]. The first SHE SBD event can be detected in a conventional SILC measurement where only the gate is biased, so the stress must be periodically interrupted to perform a sense operation. For UIS, only the onset of wear out is measured.

When trap generation during electrical stress results from desorption of silicon-hydrogen (Si-H) bonds through multi-electron vibrational excitation (VE), the general form of the voltage and current dependence of the trap generation efficiency $\zeta(V,I)$ is given by [10], [11]

$$\zeta(V,I) = cI_O^{M-1}V^N, \qquad (1)$$

where c is a constant, I_O is the total current impinging on the interfaces and M is the number of electrons required to desorb one Si-H bond. The charge to breakdown is proportional to $1/\zeta(V,I)$ so that for VE, the *voltage scaling* of Q_{BD} will follow the TDDB Power Law Model [11], [12]. From (1), when only one electron is involved in the hydrogen release process, $\zeta(V,I)$ becomes independent of current and breakdown will be controlled only by fluence and energy.

Q_{BD} versus I_{INJ} for UIS is shown in Fig. 4 for the off-state, weak inversion/depletion, and on-state conditions. In all cases, $Q_{BD} \sim I_{INJ}^{-1}$. From (1), the mechanism that follows this current dependence is two electron VE of Si-H bonds. Fig. 5 shows that the injector has little effect on the time-0 activation energies for transport, which are nearly zero as expected for band-to-band-tunneling (BTBT) in the substrate. During stress, $I_{INJ} \gg I_B(BTBT)$ so that $I_O \alpha I_{INJ}$ whereas under operating conditions, $I_{INJ} = 0$ so that $I_O \alpha I_B(BTBT)$. Accordingly, I_O will be proportional to the sum of $I_{INJ} + I_B(BTBT)$. For 2-electron VE, the *current scaling* relationship from stress to operating conditions will be

$$Q_{BD} \alpha [I_{INJ} + I_B(BTBT)]^{-1}. \qquad (2)$$

. The proportionality relationship in (2) applies whether trap generation occurs solely at the polysilicon-SiON interface, solely at the Si-SiON interface, or simultaneously at both interfaces. It can be seen that the $I_B(BTBT)$ term prevents the charge to breakdown from becoming infinite when the injector is disabled. Otherwise, it would not be possible to use (2) to extrapolate Q_{BD} to operating conditions

The Weibull slopes (β's) for the first SBD event are shown in Fig. 6 and the corresponding bulk and interface trap generation power law exponents "m", where $N(Q) = bQ^m$ are shown in Fig. 7. β drops sharply when $V_{GD} > 0$. Since β depends on the stress condition, a reliability assessment cannot be obtained for this experiment. There is a corresponding drop in the bulk trap generation power law exponent from approximately 1/2 to 1/6, which is numerically consistent with the observed change in β as predicted by the cell based percolation model [13]. It has been shown that the application of a substrate bias can modify the trap generation reaction, which results in a change in the trap generation power law exponent "m" and corresponding Weibull slope [14]. m = 1/2 means that a positively charged species such as H^+ is a product of the bulk trap reaction [15] whereas m = 1/6 corresponds to a dimerized species such as H_2 [16]. The similarity of the V_G dependence of the Weibull slope (Fig. 6) and bulk trap generation (Fig. 7) shows that bulk traps control SBD in this stress regime. No clear systematic trend is observed for the reaction that generates interface traps.

In LV-SILC measurements where the stress is periodically interrupted and the current is sensed at $V_G \sim V_{FB}$ [17], I_G and I_D sense traps at the polysilicon-SiON interface. I_G and I_B sense traps at the Si-SiON interface [18]. The I_D increase is due to tunneling from polysilicon-SiON interface traps into the pwell conduction band. The electrons subsequently diffuse out the drain contact as shown in Fig. 8(a)-(b). The increase in I_B is due to tunneling from either the polysilicon conduction band or from traps at the polysilicon-SiON interface into Si-SiON interface traps, followed by recombination with holes in the pwell as shown in Fig. 8(c)-(d). A tunneling/recombination mechanism between metal gate Fermi Level and p++ substrates was first proposed in the 1960's to explain the high conductance in samples that were known to have high as-processed interface trap densities [19]. $I_D(t)/I_D(0)$ and $I_B(t)/I_B(0)$ are shown in Fig. 9 and Fig. 10 respectively. For $V_{GD} > 0$, more traps are generated at the poly-SiON interface compared to

978-1-4244-5430-3/10 $26.00 © 2010 IEEE

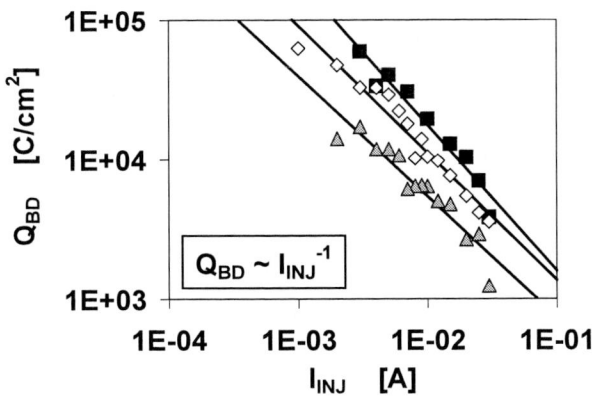

Figure 4. Q_{BD} versus injector current for UIS with $V_B = -2.0$ V. Squares: $V_G = +3.0$ V, $V_D = +3.0$ V (off-state). Diamonds: $V_G = +3.4$ V, $V_D = +3.0$ V (depletion/weak inversion). Triangles: $V_G = +3.4$ V, $V_D = +1.0$ V (on-state).

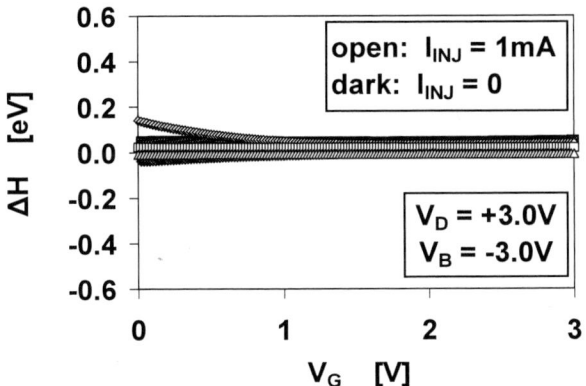

Figure 5. Time-0 activation energies for I_G (diamonds), I_D (squares), and I_B (triangles). $\Delta H \sim 0$ for all terminal currents.

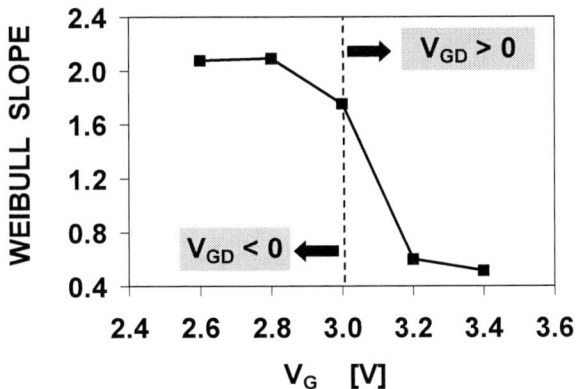

Figure 6. Weibull slope versus gate stress voltage. $V_D = +3.0$ V and $V_B = -2.2$ V during stress.

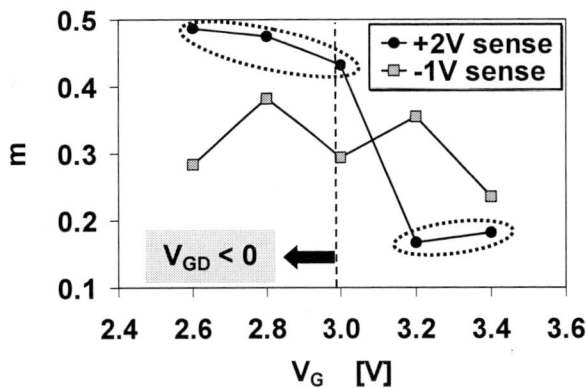

Figure 7. Trap generation power law exponents versus gate stress voltage with $V_D = +3.0$ V and $V_B = -2.2$ V during stress. Only V_G is biased during the sense operation. $V_G = +2.0$ V senses bulk traps and $V_G = -1.0$ V senses interface traps.

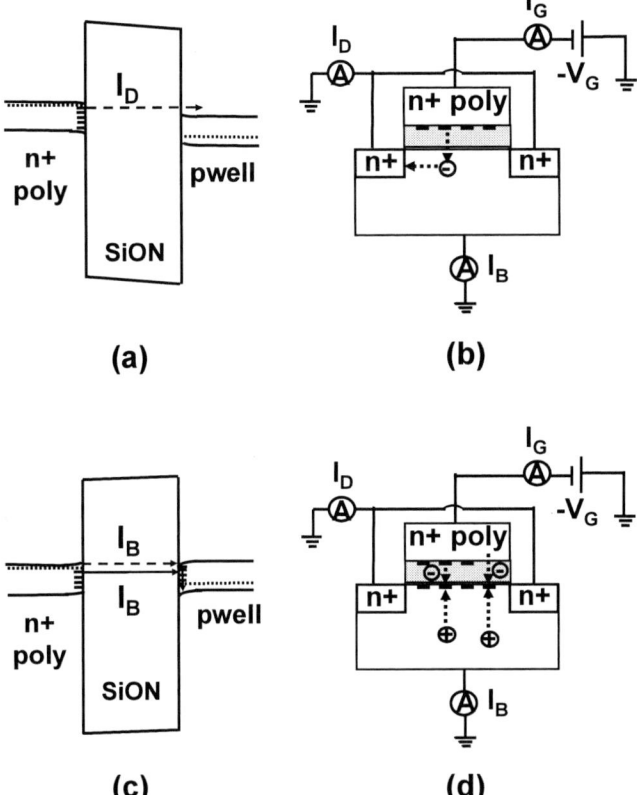

Figure 8. Band diagrams and device cross sections illustrating the current flow for LV-SILC for an NMOS device sensed at $V_G \sim V_{FB}$. (a), (b) Tunneling from n+ polysilicon interface traps into the pwell conduction band. The electrons subsequently diffuse out the drain contact and give rise to LV-SILC peaks in the drain and gate currents. (c), (d) Tunneling from either polysilicon interface traps or polysilicon conduction band into pwell interface traps, followed by recombination with holes. This process results in LV-SILC peaks in the substrate and gate currents. After Nicollian, ref. [18].

978-1-4244-5430-3/10 $26.00 © 2010 IEEE   387

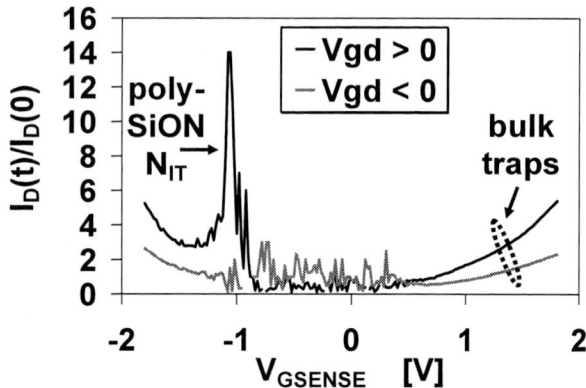

Figure 9. $I_D(t)/I_D(0)$ versus sense voltage after 1,000 C/cm² fluence. Only V_G is biased during the sense operation. V_D = +3.0 V and V_B = -2.2 V during stress. For V_{GD} > 0, V_G = +3.4 V during stress whereas for V_{GD} < 0, V_G = +2.6 V during stress.

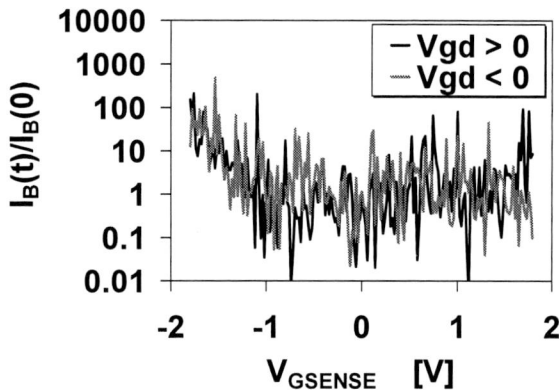

Figure 10. $I_B(t)/I_B(0)$ versus sense voltage after 1,000 C/cm² fluence. Only V_G is biased during the sense operation. V_D = +3.0 V and V_B = -2.2 V during stress. For V_{GD} > 0, V_G = +3.4 V during stress whereas for V_{GD} < 0, V_G = +2.6 V during stress.

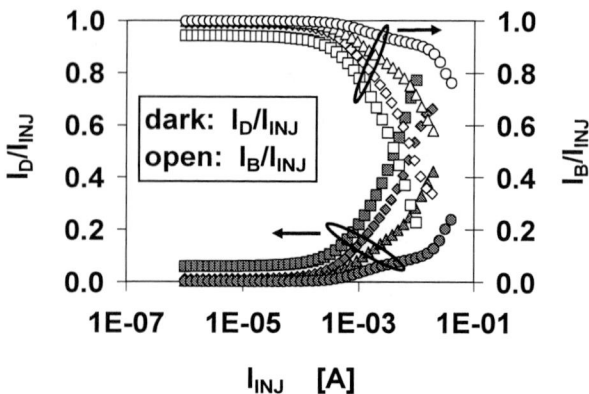

Figure 11. I_D/I_{INJ} (y1-axis) and I_B/I_{INJ} (y2-axis) versus I_{INJ} with X_{INJ} as a parameter. V_G = +3.4 V, V_D = +3.0 V, and V_B = -2.2 V. $X_1 < X_2 < X_3 < X_4$. Squares: X_1. Diamonds: X_2. Triangles: X_3. Circles: X_4.

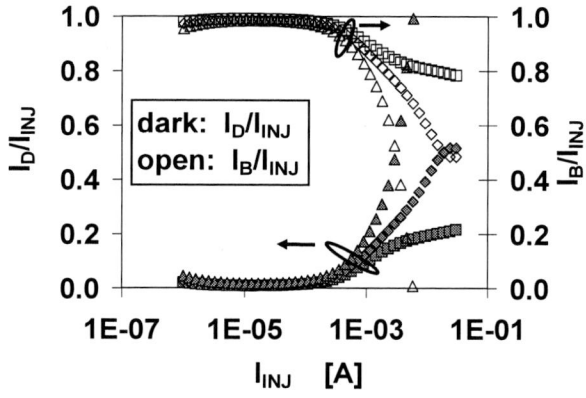

Figure 12. I_D/I_{INJ} (y1-axis) and I_B/I_{INJ} (y2-axis) versus I_{INJ} at fixed X_{INJ} with V_B as a parameter. V_G = +3.4 V and V_D = +3.0 V. Squares: V_B = -1.0 V. Diamonds: V_B = -2.0 V. Triangles: V_B = -3.0 V.

V_{GD} < 0. The species liberated increases the density of bulk traps. No significant effect of V_{GD} on Si-SiON interface traps is observed, but this could be a resolution issue due to the low magnitude of I_B.

The cause for this puzzling behavior arises from the parasitic lateral NPN bipolar transistor formed between the injector (emitter), pwell (base), and drain (collector). Fig. 11 shows the ratio of I_D/I_{INJ} and I_B/I_{INJ} versus I_{INJ} with X_{INJ} as a parameter. At low I_{INJ}, I_B is primarily due to electrons from the forward biased injector diffusing out the substrate contact so that $I_B \sim I_{INJ}$. At high I_{INJ}, I_B drops sharply. This is a high level injection effect where the large concentration of electrons in the pwell results in an increase in the electric field, which enhances impact ionization in the drain space charge region. The created holes diffuse out the substrate contact, which oppose the electron flow and reduce I_B. The generated electrons increase I_D. Fig. 12 shows that at fixed X_{INJ}, increasing V_B also

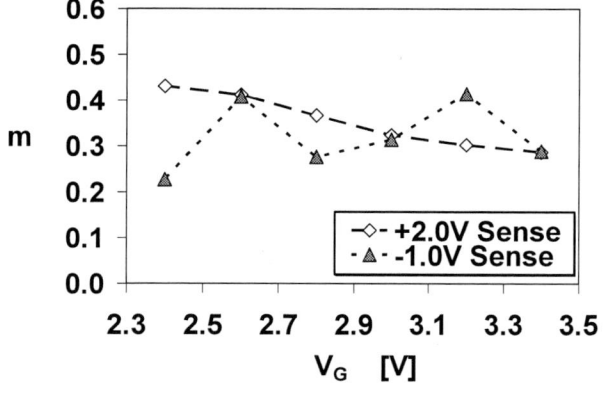

Figure 13. Trap generation power law exponents for constant V_{GD} = +0.4 V stress with V_B = -2.2 V, where parasitic bipolar effects have been minimized. V_G = +2.0 V senses bulk traps and V_G = -1.0 V senses interface traps.

978-1-4244-5430-3/10 $26.00 © 2010 IEEE

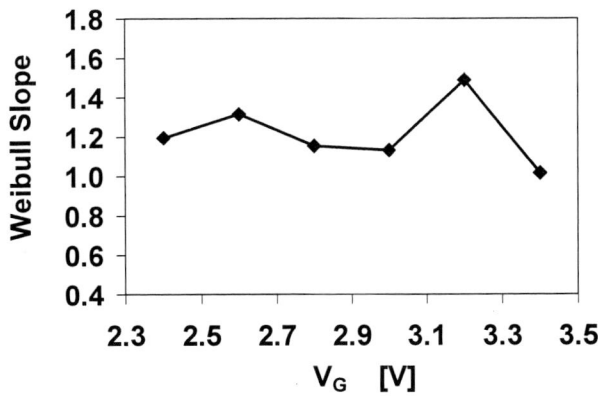

Figure 14. Weibull slopes for constant $V_{GD} = +0.4$ V SHE stress with $V_B = -2.2$ V where parasitic bipolar effects have been minimized.

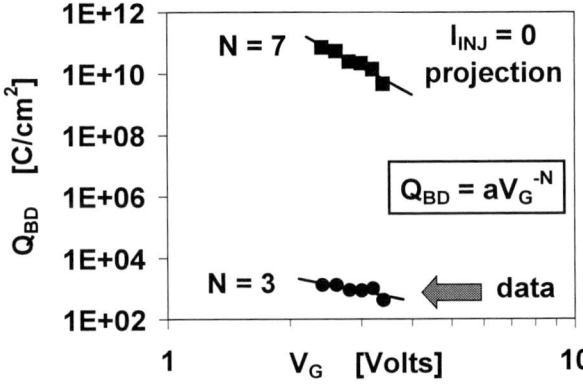

Figure 15. Q_{BD} versus V_G for constant $V_{GD} = +0.4$ V stress at $V_B = -2.2$ V where parasitic bipolar effects have been minimized.

increases high injection effects. The behavior seen in Fig. 11 and Fig. 12 is known as the Reverse Base Current Effect [20]. Some of the carriers that are generated through this process will impinge on the interfaces and affect trap generation. The value of I_{INJ} where this becomes significant decreases with decreasing X_{INJ} (the base width of the NPN is X_{INJ} in the idealized structure; it is larger when trench isolation separates the injector from the drain). Accordingly, the stress must be carefully optimized to obtain a robust reliability model. More generally, these parasitic bipolar effects need to be considered for any experiment where SHE is applied.

Trap generation power law exponents and Weibull slopes are shown in Fig. 13 and Fig. 14 respectively for an optimized constant $V_{GD} = +0.4$ V stress. β is independent of stress voltage and there are no abrupt changes in the trap generation reactions. Q_{BD} for the stress data, and current scaled to operating conditions using (2) are shown in Fig. 15. SHE accelerates Q_{BD} by about 7 orders of magnitude, corresponding to 14 orders of magnitude for t_{BD}. The V_G dependence of $I_B(BTBT)$ at fixed V_{GD} results in a higher TDDB Power Law Model "N" for *projected* Q_{BD}. Nonetheless, the N = 7 TDDB Power Law Model exponent is lower than conventional on-state stress, indicating that V_B has more influence on the energy of the carriers that generate bulk traps than V_G. Projected to operating voltage, temperature, gate area, and cumulative fail fraction, gate dielectric reliability requirements can be met for sub threshold operation for an optimized SHE experiment that minimizes parasitic bipolar effects.

III. CONCLUSIONS

The dielectric reliability of devices operating in the low current sub threshold state was investigated using substrate hot carrier injection to accelerate breakdown. The trap generation mechanism was found to be due to two-electron vibrational excitation of silicon bonds for all conditions analyzed where an injector was utilized including the off-state, on-state, and sub threshold state. Consequently, extrapolating charge to breakdown to operating condition requires scaling of both current and voltage. A quantitative model was developed to enable reliability projections.

Applying SHE for accelerated TDDB stress introduces complexities in the data analysis. Under the high injection conditions that can arise during SHE stress, the parasitic lateral NPN bipolar transistor formed between the injector (emitter), pwell (base), and drain (collector) can give rise to the Reverse Base Current Effect that results in data that are not useful for reliability projections because the carriers created through the resulting impact ionization modify the reactions that lead to the generation of the trap states that cause breakdown. This in turn causes the Weibull slope to become stress voltage dependent. In general, parasitic bipolar effects should be carefully considered in *any* experiment where SHE is employed.

When properly optimized, SHE stress yields meaningful reliability assessments which showed that it is possible to meet reliability requirements for sub threshold operation; thus saving the cost of adding high voltage components.

ACKNOWLEDGMENT

The authors would like to thank Srikanth Krishnan, Steven Zuhoski, and James Ondrusek for their support of this work. We also appreciate useful discussions with Vijay Reddy and Gianluca Boselli.

REFERENCES

[1] S. M. Sze, *Physics of Semiconductor Devices*, 2nd edition, New York: Wiley, 1981, ch. 7.

[2] D. J. DiMaria, "Explanation for the Polarity Dependence of Breakdown in Ultra Thin Silicon Dioxide Films", *Appl. Phys. Lett.*, vol **68**, pp. 3004 – 3006, 1996.

[3] D. J. DiMaria, "Defect Generation under Substrate Hot Electron Injection into Ultra Thin Silicon Dioxide Layers", *J. Appl. Phys.*, vol. **86**, pp. 2100 – 2109, 1999.

[4] W. D. Zhang, J. F. Zhang, M. Lalor, D. Burton, G. V. Groeseneken, and R. Degraeve, "Two Types of Neutral Traps Generated in the Gate Silicon Dioxide", *IEEE Trans. Electron Devices*, vol. **49**, pp. 1868 - 1875, 2002.

[5] E. M. Vogel, J. S. Suehle, M. D. Edelstein, B. Wang, Y. Chen, and J. B. Bernstein, "Reliability of Ultra Thin Silicon Dioxide under Combined Substrate Hot-Electron and Constant Voltage

Tunneling Stress", *IEEE Trans. Electron Devices*, vol. **47**, pp. 1183 – 1191, 2000.

[6] G. Ribes, S. Bruyere, M. Denais, D. Roy, and G. Ghibaudo, "MVHR (Multi-Vibrational Hydrogen Release): Consistency with Bias Temperature Instability and Dielectrics Breakdown", in *Proc. IRPS*, pp. 377 – 380, 2005.

[7] R. O'Connor, L. Pantisano, R. Degraeve, T. Kauerauf, B. Kaczer, Ph. J. Roussel, and G. Groeseneken, "SILC Defect Generation Spectroscopy in HfSiON using Constant Voltage Stress and Substrate Hot Electron Injection", in *Proc. IRPS*, pp. 324 – 329, 2008.

[8] D. J. DiMaria, E. Cartier, and D. Arnold, "Impact Ionization, Trap Creation, Degradation, and Breakdown in Silicon Dioxide Films on Silicon", *J. Appl. Phys.*, vol. **73**, pp. 3367 – 3384, 1993.

[9] S. Lombardo, F. Crupi, and J. H. Stathis, "Softening of Breakdown in Ultra Thin Gate Oxide nMOSFETs at Low Inversion Layer Density", in *Proc. IRPS*, pp. 163 – 167, 2001.

[10] B. C. Stipe, M. A. Rezaei, W. Ho, S. Gao, M. Persson, and B. I. Lundqvist, "Single-Molecule Dissociation by Tunneling Electrons", *Phys. Rev. Lett.*, vol. **78**, pp. 4410 – 4413, 1997.

[11] J. Suñé and E. Y. Wu, "Mechanisms of Hydrogen Release in the Breakdown of SiO2-Based Gate Oxides", in *IEDM Tech. Dig.*, pp. 339 – 402, 2005.

[12] E. Y. Wu, et al., "Voltage-Dependent Voltage-Acceleration of Oxide Breakdown for Ultra-Thin Oxides", in *IEDM Tech. Dig.*, pp. 541 – 544, 2000.

[13] J. Suñé, "New Physics-Based Analytical Approach to the Thin-Oxide Breakdown Statistics", *IEEE Electron Device Lett.*, vol. **22**, pp. 296 – 298, 2001.

[14] P. E. Nicollian, A. T. Krishnan, C. Bowen, S. Chakravarthi, C. A. Chancellor, and R. B. Khamankar, "The Roles of Hydrogen and Holes in Trap Generation and Breakdown in Ultra-thin SiON Dielectrics", in *IEDM Tech. Dig.*, pp. 403 – 406, 2005.

[15] S. Ogawa and N. Shiono, "Generalized Diffusion-Reaction Model for the Low-Field Charge-Buildup Instability at the Si-SiO2 Interface", *Phys. Rev. B*, vol. **51**, pp. 4218 – 4230, 1995.

[16] S. Chakravarthi, A. T. Krishnan, V. Reddy, C. F. Machala, and S. Krishnan, "A Comprehensive Framework for Predictive Modeling of Negative Bias Temperature Instability", in *Proc. IRPS*, pp. 273 – 282, 2004.

[17] P. E. Nicollian, M. Rodder, D. T. Grider, P. Chen, R. M. Wallace, and S. V. Hattangady, "Low Voltage Stress-Induced-Leakage-Current in Ultra Thin Gate Oxides", in *Proc. IRPS*, pp. 400 – 404, 1999.

[18] P. E. Nicollian, A. T. Krishnan, and V. K. Reddy, "Two-Trap Model for Low Voltage Stress-Induced Leakage Current in Ultra Thin SiON Dielectrics", *J. Appl. Phys.*, vol. **104**, pp. 053718-1 – 053718-9, 2008.

[19] W. E. Dahlke and S. M. Sze, "Tunneling in Metal-Oxide-Silicon Structures", *Solid-State Electron.*, vol. **10**, pp. 865 – 873, 1967.

[20] P. F. Lu and T. C. Chen, "Collector-Base Junction Avalanche Effects in Advanced Double-Poly Self-Aligned Bipolar Transistors", *IEEE Trans. Electron Devices*, vol. **36**, pp. 1182 – 1188, 1989.

Soft Error Assessments for Servers

K Paul Muller, Pia N. Sanda

Systems and Technology Group
International Business Machines Corp.
Poughkeepsie, NY. USA
Phone: 001-845-435-8359

Abstract — In order to assess the soft error rate (SER) of a server, it is important to not only quantify the soft error contribution of the individual semiconductor components, but also to account for derating and for SER mitigation like hardening and shielding. Derating describes the fact that not every soft error has an impact. A large number of soft errors vanish based on electrical, logical or timing considerations. They have no impact. Additionally, a server can, to a large degree, be protected from the impact of soft errors by implementing error detection and correction means. In these cases the impact of the soft error is limited to the extra compute time needed for the correction. Summing up the SER contributions from transistors and circuits results in the so-called raw soft error rate, a rate which describes just the bottom layer of the system stack. Powerful protection mechanisms at higher layers can reduce that rate by several orders of magnitude. Awareness of this vertical interaction across the different layers in the system stack leads to servers optimized for robustness.

Keywords: soft error rates, SER protection, derating, system assessments, server, cross-layer optimization

I. INTRODUCTION

Energetic particles have the ability to propagate through the active area of a semiconductor chip and leave trails of electron/hole pairs. These particles have either cosmic origin or are emitted from the materials used in chip fabrication or packaging. If the electron/hole pairs are generated in a p/n-junction, a current and voltage pulse is the result. Above a certain critical limit, the pulse has the ability to inadvertently switch a transistor temporarily from non-conducting to conducting or to change the potential of a node. In a case where this transistor or the node is part of a storage element, e.g. an SRAM cell or a latch, this can lead to an upset, i.e. a change in the stored value. This is called a soft error induced single event upset (SEU). If the transistor is part of a logic cone, it can lead to a temporary change of the calculated value (soft error transient, SET). SEUs, and to lesser degree SETs, are well studied in the literature and understood. However, for an accurate assessment of a server, soft error mitigation and all mechanisms which can contribute to derating have to be taken into account. Mitigation describes techniques which reduce particle bombardment of sensitive circuitry or which make the circuitry less susceptible to soft errors. Derating describes the fact that the impact of a soft error is reduced either through the implementation of error detection and correction, or through vanishing. Error detection and correction is designed into a server, not only for handling soft errors but also hard errors. A broad variety of approaches are known in the industry. In a

server, the errors are detected by hardware and firmware checkers. At system level, which would include the software layer, one could also include software checkers. The following discussion focuses on hard- and firmware and excludes the software. The way an error is treated after detection differs and ranges from instantaneous correction, to instruction retry, to checkstop. In contrast to a hard error, a soft error does not damage the hardware, but the rate of soft errors, particularly if they are not mitigated or derated, can be much higher than the rate of hard errors.

The derating factor is defined as the ratio of all soft error events to the events with an impact. The inverse is called the architectural vulnerability factor. A large number of SEUs and SETs have no impact, neither on the server configuration, nor the outcome of a calculation. This factor is used to describe both, the degree of protection through error detection and correction, as well as the probability that a non-detected soft error vanishes. Great effort is put on the accurate quantification of that probability.

A particle induced soft error is always caused at the lowest layer of the system stack, namely at transistor level. It can be addressed at that layer, or in some cases more efficiently at one of the higher layers. The optimization of this cross-layer interaction results in the most robust servers.

II. CATEGORIES

A. Soft Error Categories

In any server, hard and soft errors contribute to the overall failure rate and have to be addressed appropriately. Besides particle induced soft errors, there are other errors which are soft in nature, e.g. errors caused by noise, cross talk, temporary threshold modulations, or voltage droop. In this paper, we are concerning ourselves only with particle induced soft errors in semiconductor chips.

Besides the two categories, vanished errors and corrected errors, both with little or no impact, there are three categories with impact: (a) Silent data corruption (SDC) (b) system failure, and (c) system degradation. In the case of silent data corruption the data are affected. They either become incorrect, or correct data get stored at an incorrect location or they get fetched from an incorrect location. SER induced system failure is characterized by the fact that the entire system becomes temporarily non-functional. And we talk about SER induced system degradation if individual components of a system fail. They are then either re-initialized, or failover mechanisms make use of backup solutions. Since energetic particles can impact semiconductor circuitry, much effort in today's high end servers is put into avoiding SDC and system failure, and to provide protection as well as failover and backup solutions.

At IBM, a hierarchical approach to recovery has a long tradition. Nearly 30 years ago, the desire was to take action at the lowest possible level [1]. As we will show in the following sections, today we have tools at hand which enable efficient cross-layer optimization at an early stage without compromising reliability.

B. Chip Categories and Protection

The semiconductor chips in a complex server can be categorized in the following way: (a) computational, (b) transactional and (c) storage chips. Computational chips are microprocessors and chips dedicated to cryptography. Transactional chips are chips which enable transmission of data. E.g. I/O hub chips, memory controllers, and network adapters. Storage chips can be SRAM, DRAM, flash memory chips, etc. For a mainframe server, the number of chips can easily exceed 2,000. A supercomputer may have close to 20,000 processor chips alone, excluding IO and memory chips. On the other hand, a chip can combine more than one category. In fact, all categories can be found on one chip in the case of a system-on-a-chip.

The error detection and recovery means vary between the three different chip categories. For computational chips predictive parity [2], self test [3], and residue checking [4,5] can be employed. For the transactional chips, a variety of protocols have been developed to address relatively high bit error rates in digital data transmission in general. They include error checking and can be used to detect soft errors as well. Typically, additional data beyond the payload data for the application are transmitted in order to allow for the checking. Checksum in the TCP/IP protocol [6] is used for both, error detection in the header of a packet as well as in the payload data. Cyclic redundancy checking (CRC) is used as part of the PCI Express (TM) protocol for checking on the header information (16 or 32 bit LCRC) [7]. Additionally, the payload can be protected by invariant end-to-end CRC. Other examples are RAID [8], and T10-Data Integrity Fields (T10-DIF) [9]. These are very powerful detection mechanisms. The probability of transmitting errors undetected can easily be suppressed by four orders of magnitude or more [10].

For storage chips, SER protection can easily be achieved through parity and error correction code (ECC). Most implementations of ECC code today allow for double bit detection, but only single bit correction. However, an energetic particle can cause multi bit upsets, either by hitting several transistors in close proximity sequentially, or by generating several daughter particles in a spallation event [11], which then hit a multitude of transistors. With higher integration density in future technologies, the SER sensitive nodes will move physically closer together and an increase in dual, triple, and quadruple bit upsets will be observed. Physical separation of the nodes by interleaving and/or more complex ECC codes will mitigate this effect.

III. IMPACT TREE

As mentioned above not every soft error has an impact. Once it occurred, e.g. in form of a latch flip or an SRAM cell flip, it is in many cases detected by parity, checksum and the like. If the error is detected and corrected it is recovered. In this case there is no impact on the server or the application other than the extra time necessary for the recovery. If the first recovery attempt is not successful, there may be additional recovery attempts. If it is detected, but in the end not recoverable through any of the available means, then there is an impact on the server. In a compartmentalized server, the component in which the error occurred can be checkstopped and re-initialized. If the error is detected but no recovery means are available, neither for the component nor for the system, then, in most cases, a system checkstop and re-initialization would be advisable.

If the error is not detected, it still may be the case that there is no impact on the server or the application data [12]. In many cases the error propagates for some period of time, but then vanishes. It can vanish through electrical derating, timing derating, utilization derating, or logic derating. Electrical derating occurs if the soft error induced pulse is attenuated and fades away. Timing derating occurs if the error arrives at the next latch outside of the time window for latching. Utilization derating occurs if a portion of a chip is not utilized. And logic derating occurs if the combination of logic gates disallows the propagation of the error. Derating is discussed in more detail in section IV. The error can also vanish, if it occurred in a branch prediction, which is not used later. Another example of an error vanishing is the case where the latch or the memory cell, in which the error resides, is re-written before the erroneous bit is read. Exact quantification of the percentage of errors vanishing is important in the SER assessment of servers.

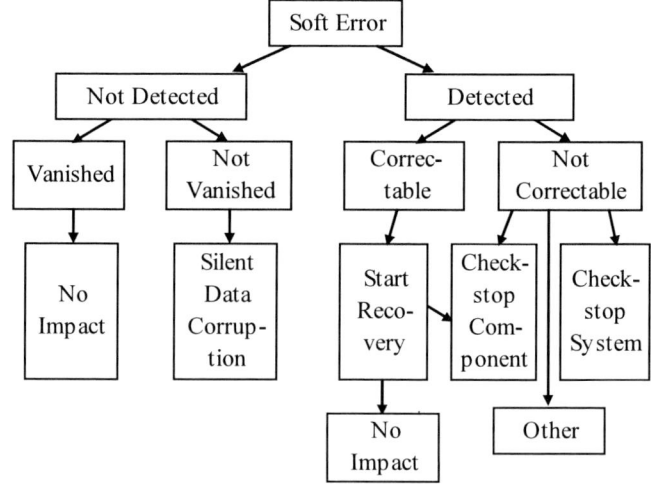

Fig. 1: Impact Tree of Soft Errors

Fig. 1 shows the impact tree of soft errors. A similar tree is published in [13] showing how a faulty bit impacts a microprocessor chip. In Fig. 1 the branches are comprehensive and the impact of the detected error is tracked beyond the microprocessor chip, namely to the point of component recovery or component or system checkstop. Checkstop is the typical step after an uncorrectable error has been detected, but other approaches are possible also, e. g. keeping the system running through a manual override. Note: Once an error has been detected, it cannot vanish any more. And vice versa, a vanished error cannot be detected any more.

IV. TIME LINE

It is desirable to carry out an SER assessment of a chip as early as possible. This means during the design phase and prior to fabricating the chip. That way, any possible shortcomings of the design become visible early enough so that they can be addressed appropriately. A final validation of the SER assessment of a chip can be done once hardware is available, e.g. in form of a thorium foil or radio active underfill experiment [14], or a proton or neutron beam experiment. By definition, during the design phase the design is not finalized, and any SER assessment during that phase is preliminary. If significant design modifications have to be captured, an iterative SER assessment may be required. Fig. 2 depicts the process. Depending upon the degree of completion of the design, different tools can be used for the SER assessment.

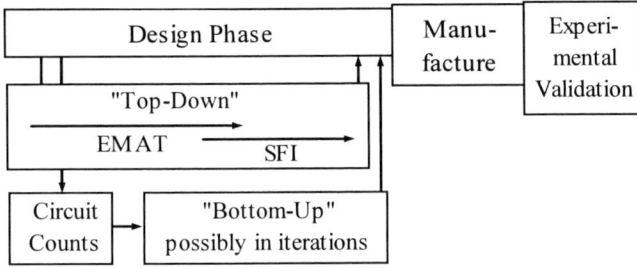

Fig. 2: Time line of soft error assessments, SFI = Statistical Fault Injection, EMAT = Early Microarchitectural Assessment Tool

A. "Bottom-Up" Approach

As soon as circuit count predictions for SRAM cells, DRAM cells, and flip-flips (latches) are available, one can start calculating the SER contribution from these circuits. This may be called "Bottom-Up" assessment. Since summing up the SER contributions from these circuits typically leads to a fairly high and in many cases unacceptable soft error rate, the bottom-up assessment is usually followed by bottom-up mitigation. A variety of known SER mitigation techniques can be implemented. As an example, blocking layers can shield sensitive circuitry from alpha particles [15], latches can be hardened through the implementation of dual interlock [16], and in the case of SOI, through the implementation of series transistors [17]. These techniques come with power, area, or processing overhead. In addition to mitigation, protection can be obtained through the implementation of detection and correction means. Most prominent at this layer are parity and ECC. However, the knowledge of the presence of any cross-layer interaction in form of higher layer protection allows for an optimized design point. A cost effective robust solution is possible [18]. In cases in which a large number of circuits are affected by the mitigation and/or protection, the "Bottom-Up" assessment may be iterative.

B. Electrical Derating and Timing Derating

In order for an SET to be latched at the next stage, the transient must be large enough that it propagates. If the SET pulse is small, it can be attenuated when propagating through the gates, and finally die out. This is called electrical derating. Note: The opposite can happen also, the pulse may become larger and wider through amplification. Assuming that the SET is not electrically derated, it still may vanish through timing derating. In order for the SET to have an impact, it needs to be present at the next latch at the exact time when the clock edge triggers the latching. If the SET arrives too early or too late, the transient will not be captured in the latch and will thus have no impact [19]. SPICE simulations can be used to make timing predictions in general and can also be used in this case. A broad variety of logic gate combinations need to be analyzed in order to obtain a statistically meaningful derating factor. Electrical and timing derating are limited to the transistor and circuit layers. The following sections describe interactions with higher layers.

C. Utilization Derating

Additional mitigation comes from the fact that not all parts of a chip are utilized all the time, which leads to a utilization derating factor. A part of a chip which is not in use for computing, transmitting or storing application data for a specific period of time, cannot contribute to SDC for that period of time. It can still contribute to system failure or system degradation. In order to quantify the utilization derating factor, one can make use of performance simulation data. Performance simulation is typically carried out during the design phase as well. It makes performance predictions for specific anticipated workloads using statistics on the residency of relevant application data. These statistics are one-to-one applicable to the quantification of the utilization derating. An early microarchitectural assessment tool like PHASER [20] helps to automate this process. Since utilization derating takes application conditions into account, it originates in a higher layer and is part of the so-called "Top-Down" assessment.

D. Logic Derating and Statistical Fault Injection

As mentioned above, logic derating occurs if the propagation of the soft error is disallowed based on the logic combination of the gates. E.g. if the error reaches an AND-gate of which the other input is a "0", the output remains a correct "0" and the error cannot propagate. Statistical fault injection (SFI) is a tool which allows for verification and quantification of protection mechanisms, as well as quantification of logic derating including transparency derating [21]. Transparency derating accounts for the fact that an SEU within a latch can be corrected by incoming data as long as the transmission gate is open (transparent).

In order to carry out SFI simulations, a chip or a subsection thereof together with an architectural verification program is logically represented in a simulator, preferably a hardware accelerated simulator. Randomly, a latch is selected and the data stored in it inverted (flipped), hereby emulating an SER

978-1-4244-5430-3/10 $26.00 © 2010 IEEE

event. The soft error is then tracked for a period of time, e.g. one million clock cycles. At that time, it typically falls into one of the following four categories: (a) Vanished (b) detected and corrected (c) detected but checkstopped, (d) SDC. With a large number of random latch flips one can quantify not only the degree of detection and protection, but also the logic derating factor. This assessment is also a "Top-Down" activity since it originates in the architectural layer.

V. CONCLUSION

Depending upon the presence or absence of SER mitigation techniques as well as detection and correction mechanisms, the soft error rates of servers can vary by many orders of magnitude. In order to achieve SER resilient servers, a combination of SER robust circuitry and error detection combined with recovery is implemented. Some very effective cross-layer protection mechanisms are available for trans-actional chips, which can efficiently be utilized for SER protection. Moreover, the different types of derating can help lower the soft error rate. With the knowledge of the cross-layer interaction and the help of simulation tools, the derating factors can be determined with the required accuracy.

Future technology generations are expected to be similarly soft error sensitive as current 65nm and 45nm generations regarding SEUs. The complexity of chips measured in transistors per unit area will continue to increase. This will allow us to produce substantially more complex servers. This complexity increase will primarily drive an increase in soft errors in the future, and will require an increase in SER protection accordingly. Due to closer physical proximity of sensitive nodes, an increase in multi bit upsets is expected, which needs to be addressed at the same time.

ACKNOWLEDGMENT

Thanks for fruitful discussions and critical review go to Stefanie Chiras, Luiz Alves, William Clarke and Larry Wissel.

REFERENCES

[1] M. Y. Hsiao, W. C. Carter, J. W. Thomas, W. R. Stringfellow, "Reliability, Availability, and Serviceability of IBM Computer Systems: A Quarter Century of Progress", IBM J. Res. & Dev. Vol. 25, No 5, pp. 453-468, September 1981

[2] S. Vassiliadis, E. Schwarz, M. Putrino, "Parity Predict for 34-bit Adders with Selection", Proceedings of the 1988 IEEE Southern Tier Technical Conference, Oct. 1988

[3] J.-L. Rainard, Y.-J. Vernay, "A 16-bit self testing multiplier", IEEE Journal of Solid-State Circuits, Vol. SC-16, No. 3, June 1981

[4] T. R. N. Rao, "Biresidue Error-Correcting Codes for Computer Arithmetic", IEEE Transactions on Computers, Vol. C-19, No. 5, May 1970

[5] S. Wei, K. Shimizu, "Error detection of Arithmetic Circuits Using a Residue Checker with Signed-Digit Number System", Proceedings of the 2001 IEEE International Syposium on Defect and Fault Tolerance in VLSI Systems, October 2001

[6] C. Kozierok, "The TCP/IP Guide: A Comprehensive, Illustrated Internet Protocols Reference", No Starch Press, Oct. 2005

[7] R. Buduk, D. Anderson, T. Shanley, "PCI Express System Architecture", Addison-Wesley Professional, 2003, p. 221

[8] D. A. Patterson, G. Gibson, R. H. Katz, "A Case for Redundant Inexpensive Disks (RAID)" Proceedings of the 1988 ACM SIGMOD international conference on Management of data, Chicago, Illinois, pp. 109-116, 1988

[9] D. Nagle, M. E. Factor, S. Iren, D. Naor, E. Riedel, O. Rodeh, J. Satran, "The ANSI T10 object-based storage standard and current implementations", IBM J. Res. & Dev., Vol. 52, No. 4/5, p. 401, 2008

[10] T. C. Maxino, P. J. Koopman, " The Effectiveness of Checksums for Embedded Control Networks", IEEE Transactions on Dependable and Secure Computing, Vol. 6, No. 1, Jan.-Mar. 2009

[11] K. P. Rodbell, D. F. Heidel, H. H. K. Tang, M. S. Gordon, and C. E. Murray, "Low-Energy Proton-Induced Single-Event-Upsets in 65 nm Node, Silicon-on-Insulator, Latches and Memory Cells", IEEE Transactions on Nuclear Science, Vol. 54, No. 6, pp. 2574-2479, December 2007

[12] P. N. Sanda, J. W. Kellington, P. Kudva, R. Kalla, R. B. McBeth, J. Ackaret, R. Lockwood, J. Schumann, C. R. Jones, "Soft-error resilience of the IBM POWER6 processor", IBM J. Res. & Dev., Vol. 52, No. 3, pp. 275-283, May 2008

[13] S. Mukherjee, J. Emer, S. K. Reinhardt, "The Soft Error Problem: An Architectural Perspective", Proceedings of the 11th Int'l Symposium on High-Performance Computer Architecture (HPCA-11 2005)

[14] J. D. Ackaret, R. B. Bhend, D. F. Heidel, P. N. Sanda, Naoko Pia, S. B. Swaney, J. Jones, T. H. Zabel, "System and method for accelerated detection of transient particle induced soft error rates in integrated circuits", US patent 7084660, August 2006

[15] C. Cabral, M. Gordon, K. Rodbell, "Method for reduction of soft error rates in integrated circuits", US patent 7601627, October 2009

[16] T. Calin, M. Nicolaidis, R. Velazco, "Upset hardened Memory Design for Submicron CMOS Technology", IEEE Transactions on Nuclear Science, Vol. 43, No. 6, December 1996

[17] J. Warnock, L. Sigal, D. Wendel, K. P. Muller, J. Friedrich, V. Zyuban E. Cannon, AJ. KleinOsowski, " POWER7 (TM) Local Clocking and Clocked Storage Elements", to be published at ISSCC February 2010, San Francisco

[18] S. Mita, K. Brelsford, "Cross-Layer Resilience Challenges: Metrics and optimization", Design and Test in Europe, DATE2010, March 2010 Dresden, Germany

[19] AJ. Kleinosowski, E. H. Cannon, J. A. Pellish, P. Oldiges, L. Wissel, „Design Implications of Single Event Transients in a Commercial 45 nm SOI Device Technology", IEEE Transactions on Nuclear Science, Vol. 55, No. 6, pp. 3461-3466, December 2008

[20] J. A. Rivers, P. Bose, P. Kudva, J-D. Wellman, P. N. Sanda, E. H. Cannon, and L. C. Alves, "PHASER: Phased methodology for modeling the system-level effects of soft errors", IBM J. Res. & Dev. Vol. 52, No 3, pp. 293-306, 2008

[21] P. Kudva, J. W. Kellington, P. N. Sanda, R. McBeth, J. Schumann, R. Kalla, "Fault Injection Verification of IBM POWER6 Soft Error Resilience", Proceedings of the Workshop on Architectural Support for Gigascale Integration, San Diego, CA, June 2007

Contribution of Low-Energy (< 10 MeV) Neutrons to Upset Rate in a 65 nm SRAM

Brian D. Sierawski, Kevin M. Warren, Robert A. Reed, Robert A. Weller,
Marcus M. Mendenhall, Ronald D. Schrimpf
Institute for Space and Defense Electronics
Vanderbilt University
Nashville, Tennessee 37212
1-615-343-9833, brian.sierawski@vanderbilt.edu

Robert C. Baumann, Vivian Zhu
Texas Instruments
Dallas, TX 75243

Abstract—**Predictions of single and multiple cell upsets in a 65 nm bulk CMOS SRAM are presented for the low-energy (< 10 MeV) portion of the NYC neutron spectrum. Scattering is identified as a significant nuclear mechanism for this regime and the consequence for multiple bit upset is discussed. The contribution is compared to the full spectrum.**

I. Introduction

In terrestrial environments, the single event upset (SEU) failure rate is collectively attributed to alpha emissions, thermal neutron capture products, and the fast neutron component of the cosmic ray spectra. SEU from fast neutrons occur as the result of a number of different mechanisms including elastic and inelastic scattering, absorption, and spallation. Reaction cross sections, as well as the species of secondary fragments, vary with the incident neutron energy. Quantifying the failure rate for a given radiation environment requires an inclusive treatment of each mechanism.

The calculation of a failure rate given an atmospheric or white neutron source requires an assumption, or direct measurement, of the low-energy threshold for a neutron induced upset. For previous generations of devices, this threshold has been characterized to be 10 MeV as is stated in the JEDEC Standards document 89A [1].

The atmospheric neutron spectrum introduces a reliability concern as a large portion of the incident neutrons are below 10 MeV. The relative abundance of these neutrons warrants an assessment of their effects. A number of publications have cited the importance of the low-energy threshold in determining an accurate estimate of the part failure-in-time (FIT) rate.

In [2], neutron cross sections for energies below 10 MeV were shown to be significant to determining the FIT rate. The authors show that the threshold can make a large difference in the predicted SER and measured the SEU cross section for SRAMs as small as the 0.13 μm technology node with mono-energetic neutron beams. In addition, a series of publications [3][4][5] which have examined the thresholds for various reaction channels, the effects of various materials, and have used Monte Carlo simulations to determine the impact to the ground-level SEU FIT rate over various technology nodes.

In scaled SRAM devices the probability of more than one cell upsetting from a single event, defined as a multiple cell upset (MCU), becomes important to quantify. The increased SEU susceptibility and decreased separation of bits increases the likelihood that several bits will be affected by the energy deposition of a single ionizing particle. More than one upset within the same data word results in a multiple bit upset (MBU). Although error correcting codes (ECC) and bit-interleaving can be effective measures to reduce the MBU rate, they can increase memory complexity and access time, and may not be suitable for small memories such as register files.

The determination of an energy threshold is difficult because it requires a low-energy mono-energetic neutron beam. Our methodology for predicting an SEU response, especially for environments where testing is impractical, is to model the charge collection mechanisms based on well-characterized radiation sources and use radiation transport codes to simulate the initial energy deposition for the prediction environment. In this work we use a single event upset model that has been calibrated to heavy ion SEU cross section data, which has been shown to reproduce proton data, to predict the neutron response. We will then examine the mechanisms leading to neutron-induced upsets below 10 MeV and evaluate their impact on MCU through Monte Carlo simulation.

II. Experimental

The 4 Mbit SRAM used in this experiment was a test array fabricated in a commercial 65 nm bulk CMOS process. The SRAM consisted of 6-transistor, high density cells. The overlayers were approximately 5 μm thick and the device was bonded as a chip-on-board to allow for front side testing. For heavy ion and proton testing, the test board was biased at 1.2 V and operated by a low cost Field Programmable Gate Array (FPGA) tester designed by NASA Goddard Space Flight Center.

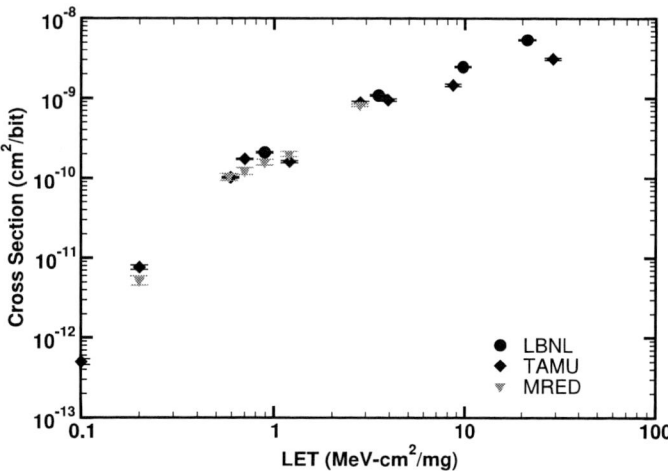

Fig. 1. Heavy ion SEU cross sections measured at LBNL and TAMU compared to MRED simulated cross sections [6].

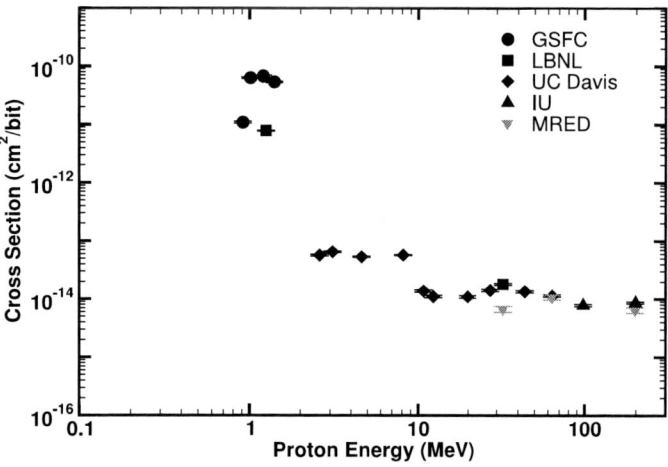

Fig. 2. Simulated and experimental proton SEU cross sections. Model provides good agreement at high energy and captures the steep increase in cross section at low energy [6].

Heavy ion data (Fig. 1) were collected using cyclotrons at Lawrence Berkeley National Laboratories (LBNL) and Texas A&M. Particular attention was given to characterizing the SEU cross section for low linear energy transfer (LET) particles, as it has been shown that devices with low-LET thresholds can be upset from the direct ionization of a single proton [7][8][6]. Measurements show significant contributions ($> 10\%$) to the event cross section from multiple cell upsets for particles with LET greater than 3 MeV-cm²/mg.

Proton single event upset data, shown in Fig. 2, were collected over a wide energy range at four facilities. The cyclotron at Indiana University was used to obtain SEU cross sections at 198 and 98 MeV. A 63.3 MeV proton beam at UC Davis was used (with degraders) to obtain SEU cross section data as low as 19.8 MeV and a 14.6 MeV primary beam was degraded to obtain data down to 2.6 MeV. The low-energy (< 2 MeV) data were acquired at the NASA GSFC Van de Graaff facility. The LBNL cyclotron was used to confirm the

TABLE I
TERRESTRIAL NEUTRON FIT RATE

Voltage	Counts/Mbit	FIT/Mbit (E_{th} = 10 MeV)
0.9	107	514
1.0	113	374
1.2	315	310
1.5	78	249

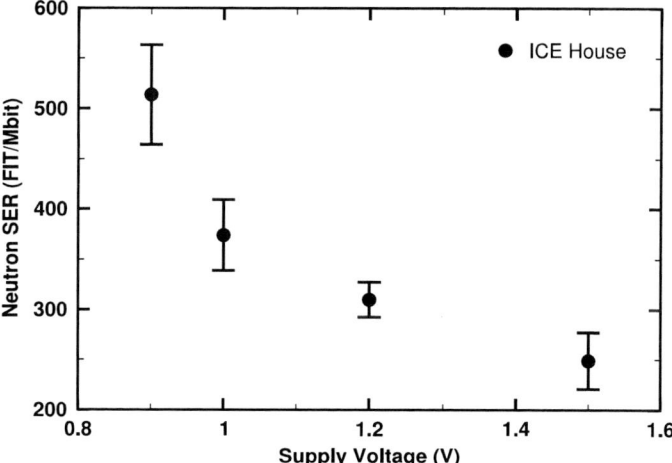

Fig. 3. Predicted New York City FIT rate as a function of supply voltage as determined by neutron accelerated testing at the LANL ICE House.

measured proton cross section at 32.5 MeV and a 6 MeV H_2^+ beam was broken up and degraded through air to obtain a proton spectrum with a peak at 1.2 MeV. The proton cross-section curve demonstrates the expected characteristics when direct ionization from protons is one of the upset mechanisms. The three orders of magnitude increase in cross section going from high energy (> 10 MeV) to low energy (< 2 MeV) indicates a transition from upsets due to nuclear events to direct ionization. At higher energies, the upset cross section is consistent with SEU from spallation reactions. At low energies, the proton cross section is on the order of the dimensions of the physical features of the cell, indicating that these devices are susceptible to SEU from proton direct ionization.

Neutron accelerated testing was done at the ICE House at Los Alamos National Laboratory. The white neutron spectrum approximates the atmospheric spectrum down to 1 MeV [1]. Terrestrial neutron-induced failures were calculated as a function of operating voltage (Fig. 3) and are summarized in Table I. FIT rates were computed with the JESD89A recommended energy threshold E_{th} = 10 MeV. Accelerated test and field failure rates are well correlated.

III. MODELING AND SIMULATION

The simulation tool used for this investigation is MRED (Monte Carlo Radiative Energy Deposition, developed at Vanderbilt University) [9]. MRED is a GEANT4 application [10] and has been successfully used for predicting SER in thermal and fast neutron environments [11][12]. A three-dimensional

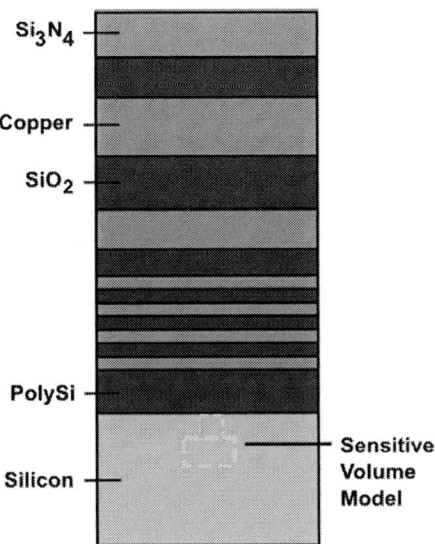

Fig. 4. Schematic drawing (not to scale) of three-dimensional material structure used in radiation transport simulations. Material thicknesses were obtained for the process technology.

TABLE II
MULTIPLICITY OF EVENTS

Distance (μm)	Counts (LANL)	Exp. Prob. (LANL)	MRED Prob. (> 1 MeV)	MRED Prob. (< 10 MeV)
0.48	631	92	89	98
0.96	46	6.7	9	2
1.44	11	1.6	1.2	-
1.92	1	0.1	0.3	-
2.4	1	0.1	0.1	-

IV. CONTRIBUTION TO SOFT ERROR RATE

The model was used to predict the single and multiple cell error rate in both the full New York City neutron spectrum and the portion of the spectrum below 10 MeV. All interaction mechanisms, including scattering and absorption, were included in the physics models for all simulations. The histograms of flux-scaled events are reverse integrated to obtain the soft error rate as a function of critical charge values. The FIT rate represents the integral of events depositing a given charge from infinity to Q. Fig. 5 shows the predicted FIT rate for the SRAM in the NYC neutron environment. The error rate for this part is indicated by the vertical dashed line marked 'Q_{crit}'. The curves at this point indicate the rate of events generating charge exceeding Q_{crit}. The predicted single bit upset rate is 10X larger than the rate of MCU involving two bits indicated as '2-MCU'. Similarly, the predicted rates for errors of multiplicity up to '7-MCU' are given as well. Events upsetting eight or more bits are aggregated into the rate for '8-MCU+'. For the simulated fluence, no events exceeded Q_{crit} in more than five cells.

Fig. 6 shows the predicted failure rate for the SRAM when only considering the flux of neutrons below 10 MeV. The simulated rate for the low-energy portion of the NYC spectrum contributes approximately 13% of the rate for this device. This result is similar to reported values in [5]. Multiple cell upsets in this regime contribute to less than 1% of the SEU rate. Simulated events using the NYC spectrum > 1 MeV (Table II) agree well with observed MCU from accelerated testing. In these environments, MCU contributes approximately 9% of the total events. This probability falls to 2% for the low-energy portion of the spectrum. Further, no more than two cells were simultaneously upset from a single neutron event in this set of simulations.

V. DISCUSSION

The charge collection process that leads to a single event upset can originate from the energy deposition of spallation secondaries, absorption particle emissions, or scattering recoils. The importance of elastic scattering as a mechanism for low-energy neutron-induced upsets is discussed in [5][13]. The reason for this is seen in Fig. 8, which summarizes the contribution of the mechanisms to the neutron total cross section. The cross section for elastic events is large compared to the other mechanisms at neutron energies below 10 MeV. Many of the upsets seen in simulation were the result of elastic scatters.

multiplanar material stack (Fig. 4) has been used to approximate the back-end-of-line. A weighted sensitive volume model of a cell has been constructed and calibrated to heavy ion and high-energy (> 32.5 MeV) proton single event upset cross sections in Fig. 2. The model has been shown to predict the cross sections at lower energies including the contributions of direct ionization from protons in [6]. The critical charge for this cell is 1.3 fC.

For MCU (multiple cell upsets not necessarily distributed within the same word) analysis, an array of device models has been placed in accordance with the cell pitch within a multiplanar stack structure representing the silicon substrate and including SiO_2 and copper in the back-end-of-line materials. The energy deposition at each device is computed for each incident particle event. Histograms are maintained for each multiplicity of upset and labeled SBU, 2-MCU, 3-MCU, up to 8-MCU+. On an event-by-event basis the application reads the energy deposited in each sensitive volume model representing an SRAM bit. The list of energies is sorted from highest to lowest, e.g., E_1, E_2, E_3, E_4. For SBU, the weight of the event is added to the E_1 energy bin and also subtracted from the E_2 energy bin. The motivation for this scheme is evident in the reverse-integrated results. There exists a range of critical deposited energies such that no bits have upset for a given event and progressively smaller values of critical energy will classify the same event as a single-bit upset, two-bit upset, three-bit upset, etc. For $E > E_1$, no bits are upset. For $E_2 < E \leq E_1$ we know only one bit should be considered as upset. If $E < E_2$ at least two bits should be considered upset. A similar analysis is performed for the 2-MCU and larger histograms.

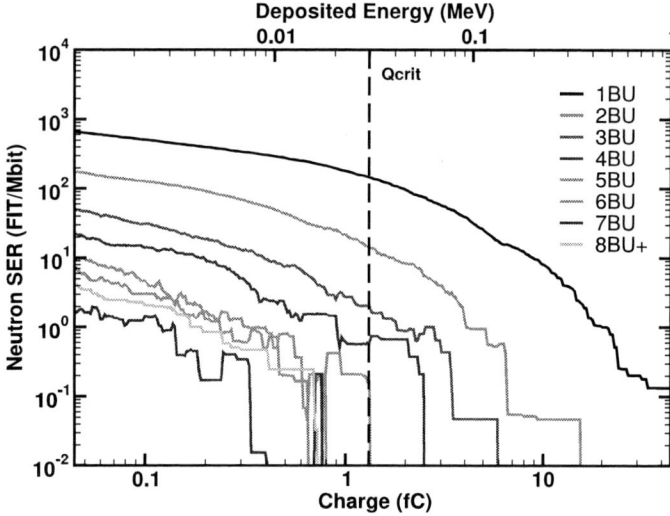

Fig. 5. Simulated failure rate in NYC terrestrial neutron environment for upset multiplicity.

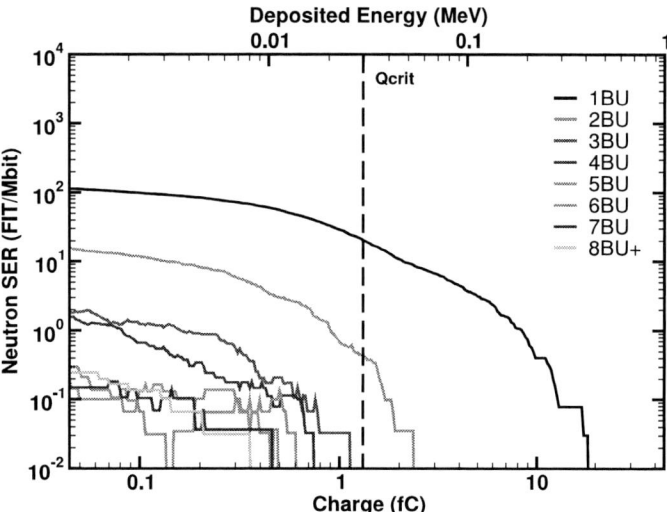

Fig. 6. Simulated failure rate for neutrons less than 10 MeV in NYC terrestrial environment for upset multiplicity.

We infer from the data that a large portion of the single bit upsets from low-energy neutrons in the ground-level neutron spectrum are the result of ionization from recoils. The kinetic energy imparted to the recoiling particle can be computed according to Eq. 1 given the initial neutron kinetic energy E_n and mass m and the scattering angle θ of the recoil with mass M.

$$E_{recoil} = \frac{4mM}{(m+M)^2} E_n cos^2\theta \qquad (1)$$

For 10 MeV neutrons, the maximum energy of a silicon recoil is 1.3 MeV and for oxygen 2.2 MeV. The range of these recoils is 1 and 2.4 μm with LETs of 6 and 6.5 MeV-cm^2/mg, respectively. In this case charge generation is localized to the vicinity of the interaction. These short range particles are therefore unable to affect a high multiplicity of cells at this

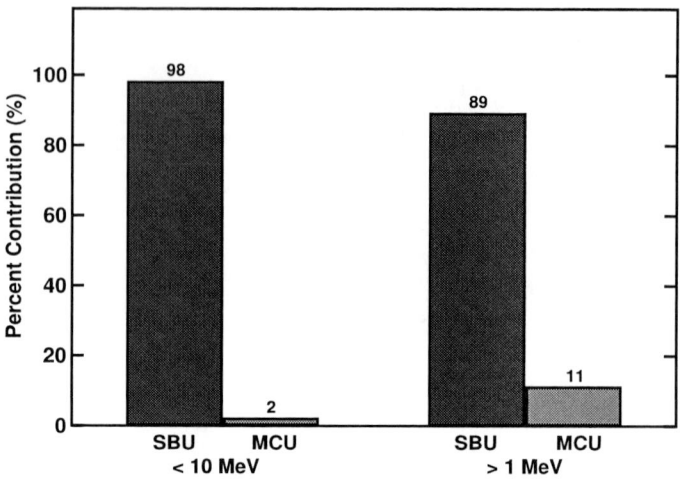

Fig. 7. Contribution of "< 10 MeV" and "> 1 MeV" NYC neutrons to single and multiple bit upset rate.

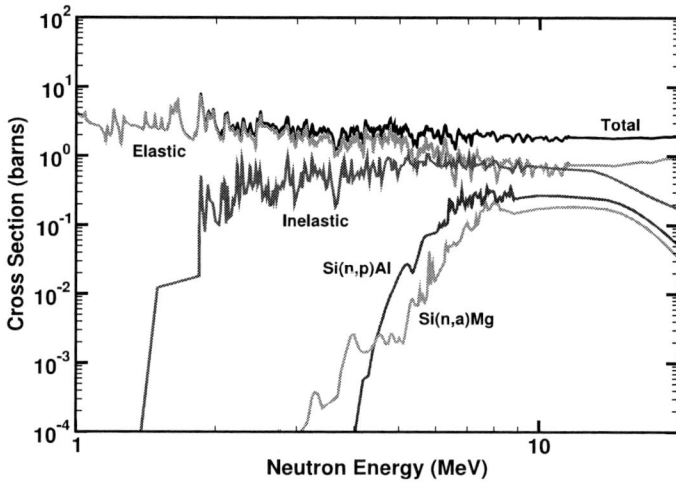

Fig. 8. Contribution of reaction channels to neutron total cross section for a silicon target as reported by the Evaluated Nuclear Data File (ENDF) [14].

technology node. This consequence is seen in Fig. 6 as the flux of neutrons below 10 MeV contributes to the single bit error rate, but very little to MCU. Error correcting codes and bit interleaving are therefore effective at reducing the impact of neutrons in this energy region.

VI. CONCLUSIONS

Neutrons below 10 MeV constitute a large portion of the terrestrial spectrum and their contribution to the total error rate, both through single and multiple cell upsets, should be assessed. It has been shown that elastic scattering is responsible for a large contribution of single event upsets from fast neutrons below this energy threshold. Assuming thermal neutron capture does not contribute to errors, multiple cell upsets, and hence multiple bit upsets are less likely for the low-energy portion of the spectrum.

ACKNOWLEDGMENT

This work was supported in part by the Defense Threat Reduction Agency grants HDTRA1-08-1-003 and HDTRA1-08-1-0034. The authors would also like to thank the NASA Electronic Parts and Packaging Program (NEPP), especially Ken Label and Mike Xapsos, for their support in the testing and characterization of the part.

REFERENCES

[1] *JESD89A: Measurement and reporting of alpha particle and terrestrial cosmic ray-induced soft errors in semiconductor devices.* JEDEC Solid State Technology Association, 2006.

[2] Y. Yahagi, E. Ibe, Y. Takahashi, Y. Saito, A. Eto, M. Sato, H. Kameyama, M. Hidaka, K. Terunuma, T. Nunomiya, and T. Nakamura, "Threshold energy of neutron-induced single event upset as a critical factor," in *Reliability Physics Symposium Proceedings, 2004. 42nd Annual. 2004 IEEE International*, April 2004, pp. 669–670.

[3] G. Gasiot, V. Ferlet-Cavrois, J. Baggio, P. Roche, P. Flatresse, A. Guyot, P. Morel, O. Bersillon, and J. du Port de Pontcharra, "SEU sensitivity of bulk and SOI technologies to 14-MeV neutrons," *IEEE Trans. Nucl. Sci.*, vol. 49, no. 6, pp. 3032–3037, Dec 2002.

[4] D. Lambert, J. Baggio, V. Ferlet-Cavrois, O. Flament, F. Saigne, B. Sagnes, N. Buard, and T. Carriere, "Neutron-induced SEU in bulk srams in terrestrial environment: Simulations and experiments," *IEEE Trans. Nucl. Sci.*, vol. 51, no. 6, pp. 3435–3441, Dec. 2004.

[5] J. Baggio, D. Lambert, V. Ferlet-Cavrois, P. Paillet, C. Marcandella, and O. Duhamel, "Single event upsets induced by 1-10 mev neutrons in static-RAMs using mono-energetic neutron sources," *IEEE Trans. Nucl. Sci.*, vol. 54, no. 6, pp. 2149–2155, Dec. 2007.

[6] B. D. Sierawski, J. A. Pellish, R. A. Reed, R. D. Schrimpf, K. M. Warren, R. A. Weller, M. H. Mendenhall, J. D. Black, A. D. Tipton, M. A. Xapsos, R. C. Baumann, X. Deng, M. J. Campola, M. R. Friendlich, H. S. Kim, A. M. Phan, and C. M. Seidleck, "Impact of low-energy proton induced upsets on test methods and rate predictions," *IEEE Trans. Nucl. Sci.*, vol. 56, no. 6, pp. 3085–3092, Dec. 2009.

[7] K. Rodbell, D. Heidel, H. Tang, M. Gordon, P. Oldiges, and C. Murray, "Low-energy proton-induced single-event-upsets in 65 nm node, silicon-on-insulator, latches and memory cells," *IEEE Trans. Nucl. Sci.*, vol. 54, no. 6, pp. 2474–2479, Dec. 2007.

[8] D. Heidel, P. Marshall, K. LaBel, J. Schwank, K. Rodbell, M. Hakey, M. Berg, P. Dodd, M. Friendlich, A. Phan, C. Seidleck, M. Shaneyfelt, and M. Xapsos, "Low energy proton single-event-upset test results on 65 nm SOI SRAM," *IEEE Trans. Nucl. Sci.*, vol. 55, no. 6, pp. 3394–3400, Dec. 2008.

[9] R. A. Weller, M. H. Mendenhall, R. A. Reed, R. D. Schrimpf, K. M. Warren, B. D. Sierawski, and L. W. Massengill, "Monte carlo simulation of single event effects," *IEEE Trans. Nucl. Sci.*, in press 2010.

[10] S. Agostinelli *et al.*, "GEANT4-a simulation toolkit," *Nuclear Instruments and Methods in Physics Research A*, vol. 506, pp. 250–303, Jul. 2003.

[11] K. M. Warren, B. D. Sierawski, R. A. Weller, R. A. Reed, M. H. Mendenhall, J. A. Pellish, R. D. Schrimpf, L. W. Massengill, M. E. Porter, and J. D. Wilkinson, "Predicting thermal neutron-induced soft errors in static memories using TCAD and physics-based monte carlo simulation tools," *Electron Device Letters, IEEE*, vol. 28, no. 2, pp. 180–182, Feb. 2007.

[12] K. Warren, J. Wilkinson, R. Weller, B. Sierawski, R. Reed, M. Porter, M. Mendenhall, R. Schrimpf, and L. Massengill, "Predicting neutron induced soft error rates: Evaluation of accelerated ground based test methods," in *Reliability Physics Symposium, 2008. IRPS 2008. IEEE International*, 27 2008-May 1 2008, pp. 473–477.

[13] Y. Arita, M. Takai, I. Ogawa, and T. Kishimoto, "Influence of elastic scattering on the neutron-induced single-event upsets in a static random access memory," *Japanese Journal of Applied Physics*, vol. 43, no. 9A/B, pp. L1193–L1195, 2004. [Online]. Available: http://jjap.ipap.jp/link?JJAP/43/L1193/

[14] M. Chadwick, P. Obložinský, M. Herman *et al.*, "ENDF/B-VII.0: Next generation evaluated nuclear data library for nuclear science and technology," *Nuclear Data Sheets*, vol. 107, no. 12, pp. 2931–3118, December 2006. [Online]. Available: http://www.sciencedirect.com/science/journal/00903752

Scaling Trends of Neutron Effects in MLC NAND Flash Memories

S. Gerardin, M. Bagatin, A. Paccagnella

Department of Information Engineering - University of Padova, Padova, Italy,
via Gradenigo 6B, phone: +39-049-8277786; fax: +39-049-8277699; e-mail: simone.gerardin@dei.unipd.it

G. Cellere

Applied Materials Baccini, Treviso, Italy

A. Visconti, S. Beltrami

Numonyx R&D - Technology Development, Agrate Brianza, Italy

C. Andreani

Department of Physics - University of Roma – Tor Vergata, Roma, Italy

G. Gorini

Department of Physics - University of Milano – Bicocca, Milano, Italy

C.D. Frost

Rutherford Appleton Laboratory - ISIS, Didcot, UK

Abstract— We investigate atmospheric neutron effects on floating gate cells in MLC NAND Flash memories. Loss of information is shown to occur especially at the highest program levels, but to an extent that does not challenge current error correction capabilities. We discuss the physical mechanisms and analyze scaling trends, which show a rapid increase in sensitivity for decreasing feature size. A large spread in the cross section is visible from vendor to vendor for comparable feature size.

Keywords- Floating-gate Cells; Soft Errors; Atmospheric Neutrons; Error Correction Codes;

I. INTRODUCTION

It was 1962 when Wallmark and Marcus predicted that the smallest achievable feature size of CMOS technology was 10 μm due to various issues, one of which was sensitivity to atmospheric neutrons [1]. In more recent times, Butt and Alam cast dark shadows on the scaling of Floating Gate (FG) memories for the exact same reason [2].

Soft errors at sea level, originating from scattered particles in the atmosphere or alpha-emitting contaminants in chip materials, are a known source of disturbances in SRAMs and, to a lesser extent, in DRAMs [3]. Relatively less is known about the sensitivity of FG memories. An extensive literature covers the effects of heavy ions on FG cells [4]-[5], but little data obtained with particles matching the terrestrial neutron environment are available [6].

In general, technology scaling is bringing about devices where smaller and smaller amounts of charge are used to store information, which could make them increasingly sensitive to ionizing radiation. Actually, the scenario is more complicated than that, because, even though less charge is required to upset a more scaled SRAM cell, the cell area reduction causes the charge collection efficiency to diminish with shrinking cell sizes. As a result, experimental data show a constant bit error rate as a function of the feature size [3], or even a decrease starting from the 90-nm node [7].

This trend must be coupled with the tendency to include an increasing amount of memory in integrated circuits, usually resulting in a growing chip (system) sensitivity to atmospheric neutrons and alpha particles [3].

The reduction in storage charge is particularly evident in state-of-the-art NAND Flash with Multi-Level Cell (MLC) architectures, where only few hundred electrons separate two contiguous levels. In these memories, the loss (or trapping) of just a few carriers due to ionizing radiation can easily upset a bit.

NAND flash memories require the use of Error Correction Codes (ECC) because they are affected by a number of mechanisms which can lead to non-zero bit error rate: program disturb, quantum-level noise, erratic tunneling, stress induced leakage current, read disturb, and detrapping-induced retention [8]. The purpose of this work is to understand whether atmospheric neutrons are to be added to that list, and whether mandatory ECC should be made more powerful because of neutrons. The experimental data set presented in this contribution extends for the first time the analysis of neutron effects down to the 48-nm node. As we will see, the results are very reassuring: the atmospheric neutron threat is well under control.

The paper is organized as follows: Section II details the features of the devices used in this work. Section III presents the experimental results obtained both with neutrons and heavy ions. The simulation results on the interactions of neutrons with the back-end materials are illustrated in Section IV. Section V discusses the main factors determining the neutron sensitivity of floating gate memories as a function of the feature size: characteristics of the charged secondaries produced by neutrons, electric field, back-end materials, and peripheral circuitry.

978-1-4244-5430-3/10 $26.00 © 2010 IEEE

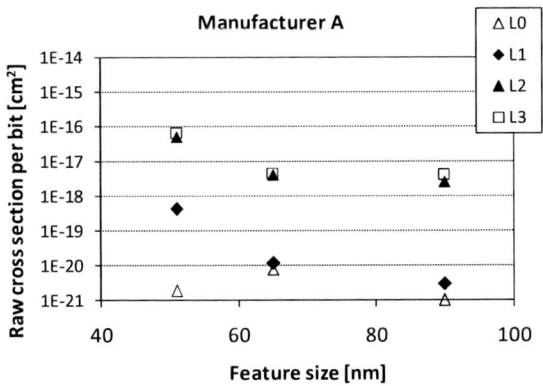

Figure 1. Cross section for raw bit errors induced by wide-spectrum neutrons in MLC NAND Flash memories from manufacturer A.

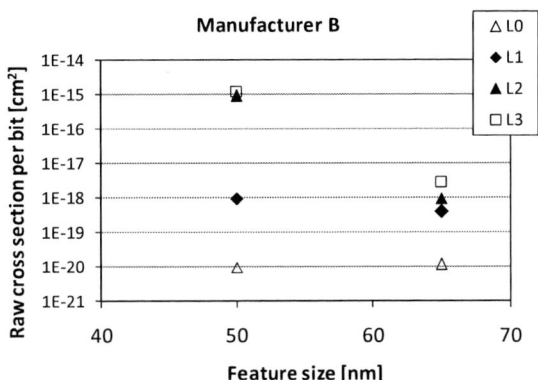

Figure 2. Cross section for raw bit errors induced by wide-spectrum neutrons in MLC NAND Flash memories from manufacturer B.

II. EXPERIMENTAL SET-UP AND DEVICES TESTED

Throughout this work we used commercial MLC NAND Flash memories manufactured by three different vendors. The feature size of the tested devices ranges from 90 nm to 48 nm (Table I). In Numonyx samples, measurement of the cell threshold voltage (V_{th}) was possible thanks to the availability of reserved test modes. On the other two sets of samples, only the number of raw digital errors was measured (without the application of any error correction code). In the following, the cell levels are numbered from the lowest to the highest V_{th} as L0, L1, L2, L3.

Irradiation was performed at the VESUVIO line of the ISIS facility at the Rutherford Appleton Laboratory, in Didcot, UK, using a neutron beam, which reasonably reproduces the terrestrial environment, with several orders of magnitude of acceleration [9].

Devices were measured before and after irradiation, and left unbiased at room temperature during the exposure to neutrons. For each irradiation thousands of errors were gathered, assuring the statistical significance of the results. Unirradiated reference devices were kept alongside the irradiated ones and measured at the same time.

The Los Alamos Neutron Science Center (LANSCE)-equivalent neutron fluence on the samples was $1.68 \cdot 10^{10}$ n/cm^2, corresponding to 148000 years at New York City. The LANSCE-equivalent flux can be used to compare the experimental data obtained at the ISIS facility with those at LANSCE [9], the facility that comes the closest to matching the JEDEC JESD89A [10] reference spectra. This correlation

TABLE I. DEVICES USED IN THIS WORK

Vendor	Architecture	Feature size [nm]	Capacity [Gbit]
A	NAND MLC	90	8
A	NAND MLC	65	4
A	NAND MLC	51	32
B	NAND MLC	65	8
B	NAND MLC	50	32
Numonyx	NAND MLC	60	8
Numonyx	NAND MLC	48	16

is necessary, because ISIS-VESUVIO features a different spectrum than LANSCE.

In addition, a few samples were irradiated with heavy ions at the SIRAD (SIlicon RADiation) facility of the Istituto Nazionale di Fisica Nucleare (INFN) Legnaro National Laboratories, Italy [11]. Four different ions were used: O, Si, Ni, Ag with a Linear Energy Transfer (LET), ranging from 3 to 28 MeV·cm^2/mg and energy between 100 and 200 MeV. All irradiations with heavy ions were performed at normal incidence and at room temperature.

III. EXPERIMENTAL RESULTS

Fig. 1 shows the raw bit cross section (i.e. with no error correction code), σ, of errors attributable to neutrons, as a function of the feature size for MLC NAND Flash memories manufactured by Vendor A. The bit cross section is defined as the number of errors divided by the fluence and by the total number of bits. The most affected bits are those belonging to the two highest levels, L2 and L3. For these bits, σ increases more than one order of magnitude in two generations. The cross section for the smallest feature size is less than 10^{-16} cm^2, a factor of 100 smaller than that for a typical SRAM cell [12]. Fig. 2 shows a similar figure for the MLC NAND Flash memories manufactured by Vendor B. The increase from one generation to the other is stronger, more than three orders of magnitude. The absolute value of σ is higher, as well, at 10^{-15} cm^2, for the worst-case pattern and smallest feature size. In all cases, the upsets are from one level to the next one with lower V_{th}.

σ can be used to calculate the raw bit error rate on the field for a certain location, using data on the neutron flux. Assuming a value of 13 n · cm^{-2} · hour^{-1} at NYC [10], the raw bit error rate due to atmospheric neutrons over a period of ten years in the worst case (Vendor B, L3) is $1.13 \cdot 10^{-9}$. This is not enormous, but comparable to other mechanisms of bit errors [8]. The uncorrectable bit error rate is the probability that more errors will be generated in an ECC codeword than can be corrected. Assuming an ECC scheme capable of correcting 8 bits and an ECC codeword of 539 bytes, as prescribed by the manufacturer datasheet, the uncorrectable bit error rate will be, according to the binomial distribution, much less than $1 \cdot 10^{-18}$. At avionics altitudes, the neutron flux can be 300 times higher, resulting in a worst-case raw bit error rate of $3.4 \cdot 10^{-7}$, which, translated to uncorrectable bit errors, is still

Manufacturer A

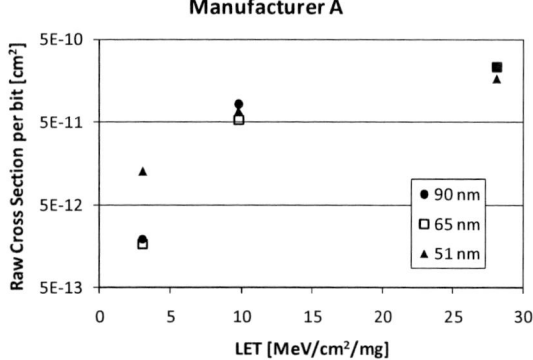

Figure 3. Heavy-ion raw cross section for L3 in MLC NAND Flash memories from manufacturer A.

Manufacturer B

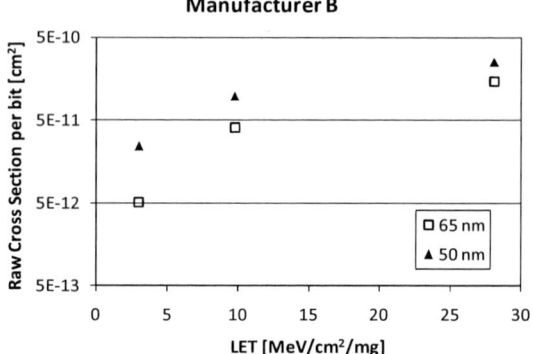

Figure 4. Heavy-ion raw cross section for L3 in MLC NAND Flash memories from manufacturer B.

well below $1 \cdot 10^{-18}$. The other tested memories exhibit much lower raw bit error rates.

Memories from Vendors A and B were also exposed to monoenergetic heavy-ion beams at the SIRAD facility, which emulate (with some limitations as we will see) the secondary particles produced by neutrons. The experimental results are shown in Figs. 3-4, which illustrate the bit cross section as a function of the ion LET, for Vendor A and Vendor B, respectively. We recall that the Linear Energy Transfer is a measure of the energy (hence the number of electron-hole pairs) deposited by an ionizing particle per unit of length. As seen in the figures, for a given vendor, at low LET (3 MeV·cm²/mg), the raw bit cross section grows with decreasing feature size, regardless of the manufacturer. At higher LET, the dependence is much less pronounced, or the larger feature size devices can even display a lower σ with respect to the smaller one. It is also interesting to observe that, at high LET, the absolute value of the cross section is very similar between the two vendors, whereas larger differences appear at low LET. It is interesting to note that the cross section (which is a measure of the sensitive area of the cell) at high LET can be larger than the size of the floating gate.

V_{th} distributions were measured, in addition to the number of errors, before and after neutron irradiation in another set of samples from Numonyx. Fig. 5 shows the cross section for raw bit errors (lower V_{th} program and erase levels show less or no errors). Increasingly longer tails appear in the distributions (Fig. 6), for increasing cell V_{th}. Fig. 7 compares the V_{th} distributions before and after irradiation for two different technology nodes. The two devices were irradiated with the same fluence ($1.68 \cdot 10^{10}$ n/cm²), so it is apparent how both the number of cells in the tail and the maximum V_{th} shift increase with decreasing feature size.

IV. GEANT4 SIMULATIONS

We developed an application based on the Geant4 toolkit [13] to simulate the interactions of neutrons with the chip materials. Geant4 is a collection of C++ classes, which contains a variety of utilities and physics processes to model the interaction of radiation with matter. It includes both electromagnetic and hadronic interactions.

The geometry and materials of the Numonyx samples were modeled in detail in three dimensions. A 2D drawing of the examined structure (not to scale) is shown in Fig. 8. The ISIS

neutron energy spectrum (Fig. 9) was also carefully reproduced in the simulations.

The output of the simulations provides the number, species, energy, and trajectory of the particles that cross the floating gates depositing energy along their paths. An event is defined as the deposition of energy inside a FG.

To reduce the variance on the rarest events the low- and high- energy parts of the ISIS spectrum were simulated separately. This allowed us to reduce the total amount of events to be simulated, while producing enough high-energy neutrons events to obtain accurate statistics for rare events. Afterwards the results were combined, applying the proper weight (given by the ratio between high and low energy neutrons in the ISIS spectrum) to the results coming from the different simulations.

Geant4 treats energy deposition by ionization as the sum of a continuous deposition and discrete events, whose production depends on the range cut set by the user. When the ionization process produces electrons whose range is above the cut, these electrons are tracked, otherwise the energy deposition associated with them becomes part of the continuous component. The range cut for electrons was set to 1 nm, so that energy deposition in the FG is possible not only for ions directly crossing the FG, but also for particles passing close enough to generate electrons, which in turn can reach the FG.

Fig. 10 shows the number of the secondary particles crossing the FG as a function of the effective LET for the 48-nm and the 60-nm Numonyx parts. The values were normalized to match the experimental conditions of Figs. 5-7, in terms of exposed cells and received fluence, thus allowing a direct comparison. The effective LET was calculated as the deposited energy divided by the density of Silicon and by the FG thickness. Because of the larger geometrical size, the total number of events (the integral of the curves in Fig. 10) is larger in the 60-nm parts than in the 48-nm ones. The shapes of the two curves are different because of the different thicknesses of the materials the particles cross before reaching the floating gate. As seen in Fig. 10, the distributions are linear in a log-lin scale and the maximum LET is around 10 MeV · cm²/mg for both technologies. The smallest LET bin, below 1 MeV · cm²/mg, contains a lot more particles than the other bins, due to the large amount of generated protons and alpha particles. These lighter particles can travel a longer distance than those with higher LET, which have lower initial energy and lose the energy faster along the way.

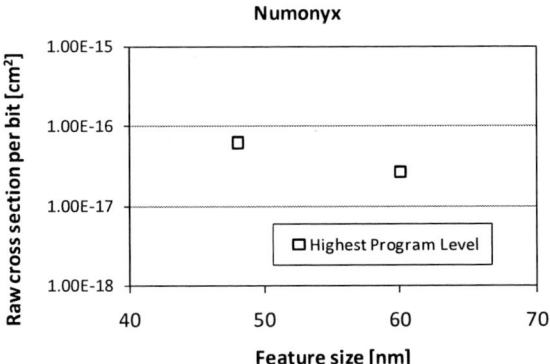

Figure 5. Cross section for L3 raw bit errors induced by wide-spectrum neutrons in MLC NAND Flash memories from Numonyx.

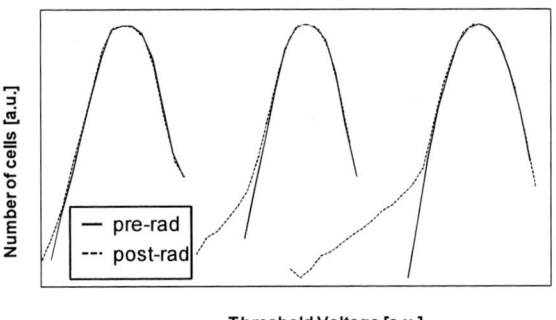

Figure 6. Threshold voltage distribution for a 60-nm MLC NAND Flash memory irradiated with wide-spectrum neutrons.

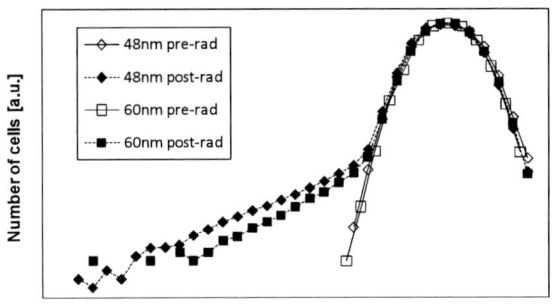

Figure 7. Comparison between the threshold voltage distributions before and after wide-spectrum neutron irradiation for two MLC NAND Flash memories with different feature sizes.

Fig. 11 shows the energy distribution of the particles with LET above 1 MeV·cm²/mg produced by the neutrons. As seen, the energies are fairly limited, most of the particles are below 10 MeV. This is an important aspect, because the structure of the e-h track generated by the ions is strongly dependent on the energy, in addition to LET [4]. Furthermore, low energies mean that the neutron triggered events are not likely to affect many bits, and that tracks such as those observed in [14] are extremely unlikely with neutrons. We must remark, though, that we only simulated the conditions found in our experiments - in particular, normal incidence. It may be possible that with different angles, multiple effects occur, because the directions of the secondaries are a function of the imping-ing neutron direction.

Another interesting correlated aspect is shown in Fig. 12, which reports the locations in the chip where ionizing secon-daries originate during exposure to neutrons. Most of the events are due to particles generated in the proximity of the worldines, whereas, as expected, the other layers give a de-creasing contribution as their distance from the floating gate increases. The events come for the most part from the oxide isolating the wordlines, rather than from the wordlines them-selves.

Our simulation results show that the largest part of the events is due to reactions between neutrons and Silicon or Oxygen atoms. Events with Tungsten and other heavier mate-rials in the back-end are much more rare, and contribute to the total number of events for less than 1%.

V. DISCUSSION

As we have discussed in the previous section, neutrons in-teract with the chip materials through elastic or inelastic nuc-lear reactions that generate charged secondary particles (heavy ions). These byproducts can cross (or pass nearby) FG's, in-ducing charge loss [4] and, to a lesser extent, charge trapping in the oxides surrounding the FG [15]. This translates into a shift in the threshold voltage of the hit cells, which, when large enough to take the cell beyond the reference voltage, causes a raw bit error. Obviously this is not enough to cause an error visible to the user, because mandatory ECC schemes have to be implemented by the application.

A. Secondaries and Threshold Voltage Shift

The tails in Figs. 6 and 7 are approximately linear in a log-lin scale, meaning that the distribution of threshold voltage shifts

induced by neutron byproducts is exponential. This distribu-tion arises from the fact that the generated secondaries have a wide spread in energy, range, and LET. The number of events observed in our simulations reasonably matches the number of cells in the tail in Fig. 7 (the difference is below 30%). It is also consistent with the number of observed errors. To make this comparison, one needs to set a threshold LET (possibly dependent on energy), above which the charge loss and trap-ping are high enough to cause an error. Some clues can be obtained from the heavy-ion data of Figs. 3-4, even though an exact value cannot be extracted. For sure, particles with an LET of 3 MeV·cm²/mg can cause errors in all the technologies we tested. It is also clear enough from the heavy-ion data that, if all the curves follow the usual Weibull fit, the threshold value decreases with technology scaling. This means that an increasingly higher percentage of the secondaries produced by the impinging neutrons will give rise to raw bit errors.

The dependence of the cross section on the LET in Figs. 3-4 is very similar to that of SRAMs and can be interpreted in the same way. In SRAMs, it is attributed to the fact that a cell is characterized by regions with different collection efficien-cy. As a result, particles with low LET are able to produce an upset only if they hit a region with high collection efficiency, whereas high-LET ions cause upset just by crossing a sensi-tive region, no matter its collection efficiency. In floating gate cells, the same interpretation implies that charge loss/trapping is maximum for strikes in the center of the floating gate (the cross section for low-LET particles is smaller than the FG area). Strikes outside of the FG can also lead to errors, pro-

978-1-4244-5430-3/10 $26.00 © 2010 IEEE

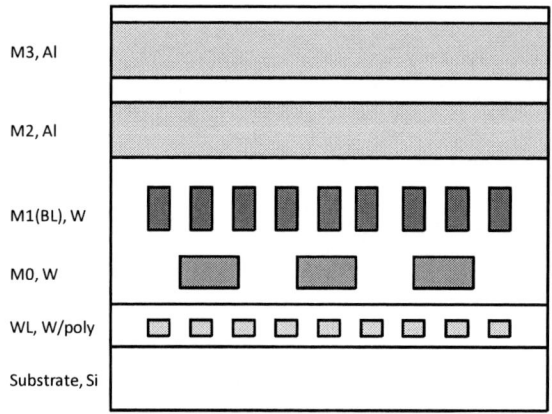

Figure 8. Drawing (not to scale) of the structure simulated for the assessment of neutron byproducts. WL = Wordline, BL=Bitline, Mx = Metal x. M0 is present only in 48-nm parts.

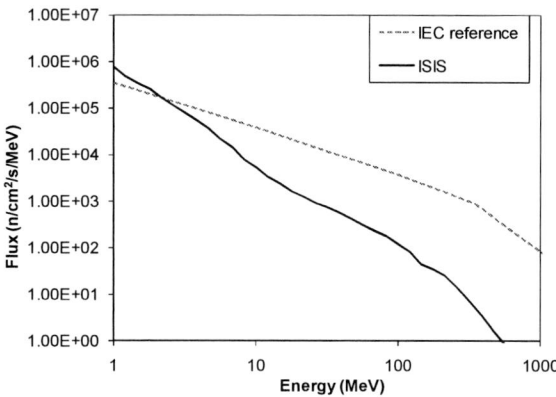

Figure 9. Neutron ISIS spectrum as compared to the IEC reference [10] multiplied by a factor of 10^8.

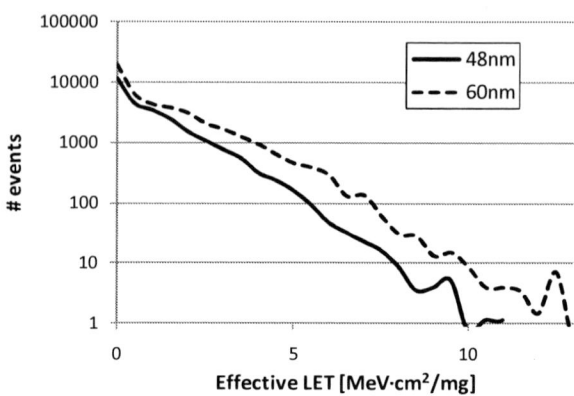

Figure 10. Simulated effective LET distribution of ISIS neutron byproducts crossing FG in 48-nm and 60-nm Numonyx parts.

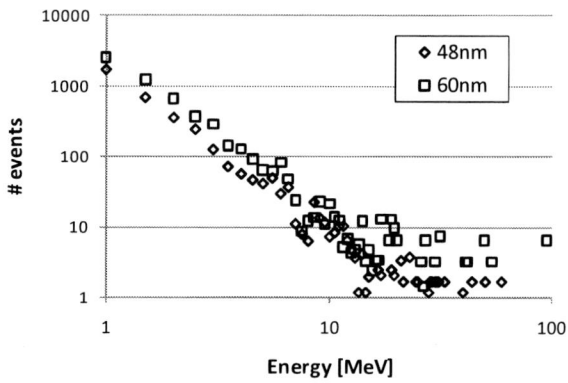

Figure 11. Simulated energy distributions of ISIS neutron byproducts with effective LET > 1 MeV·cm²/mg crossing floating gates in 48-nm and 60-nm Numonyx parts.

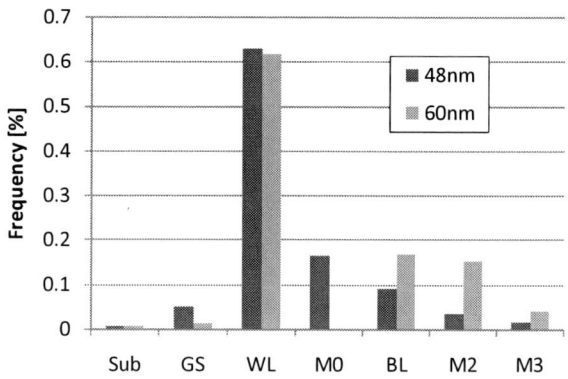

Figure 12. Simulated origin location distributions of the neutron byproducts with effective LET > 1 MeV·cm²/mg crossing floating gates in 48-nm and 60-nm Numonyx parts. Sub = Substrate, GS = Gate stack, WL = Wordline, BL = Bitline, Mx = Metal x. Each layer contains also the isolation up to the next layer going from the substrate to the passivation. Note that M0 is not present in 60-nm parts.

vided the impinging particle has a sufficiently high LET (the cross section for high-LET particles is larger than the FG area). On top of this we must add that LET is an average concept, and that large fluctuations in the energy deposition for nominally identical particles are possible, especially when we are dealing with very small volumes, where the number of interactions between the particle and the target material is small [16]. As a result, it is possible that at low LET, only "strongest" events (i.e. those with the largest energy deposition) are able to trigger an error, thus reducing the cross section. Scaling of course increases this variability.

Concerning Figs. 3-4, two different behaviors at high LET can be observed for the two manufacturers. The cross section for Vendor A decreases for decreasing feature size (Fig. 3). In this case, the geometrical factor (reduced area) wins over the reduced noise margin, and the smaller cell is less sensitive. On the contrary, NAND devices from Vendor B (Fig. 4) exhibit a larger cross section for the more scaled devices.

Extensive studies on the impact of heavy ions have shown that the amount of induced charge loss is more or less linear with the impinging particle linear energy transfer, at least when the energies of the different particles are not too different. In turn, the LET depends on the mass and energy of the particle: the higher the mass, the higher the LET.

However, no simple relation can be found between the experimental V_{th} distribution of cells in the tail and the simulated distribution of the effective LET of the secondaries. For instance, a linear relationship between LET and threshold voltage shift (which is commonly observed in heavy-ion results) does not produce a satisfactory match (not shown). The situation is indeed more complex because in addition to the LET, also the energy of the particle crossing the FG has a huge impact on the amount of charge loss, as demonstrated for instance in [4].

As illustrated in Fig. 11, for the most part, the energies of the secondaries are limited (less than 10 MeV) and far below those used for heavy ion-testing. This fact limits the applicability of heavy-ion testing to understanding the effects of neutron secondary by-products. Following the arguments of [4], a lower energy ion produces a track with a smaller diameter, because of the lower energy of the delta electrons emitted normal to the ion trajectory, and this translates into a higher threshold voltage shift, because of the higher conductivity of the transient conductive path.

Concerning the raw production of neutrons (i.e. independently of the induced threshold voltage shifts), Fig. 10 shows limited differences, apart from the geometrical scaling (of course the number of events is proportional to the exposed area if we keep constant the particle fluence and the number of irradiated cells). The larger V_{th} tail in the 48-nm parts than in the 60-nm parts (Fig. 7) is due to the fact that, for a given ion LET, the shift is more severe for the smaller feature size.

B. Back-end Materials

Although neutron nuclear interactions with Si and O are by far the most common, byproducts originating from reactions with high-Z materials employed in the CMOS flow (such as Tungsten) can be very important, because they generate high-LET particles. However, our simulations show that these reactions account for only a very small number of events. They might have been significant a few generations ago, when the threshold LET for an error was far higher, so that these events were the only ones able to produce an error.

Given the relatively low threshold LET necessary to produce an error, changes in materials (for instance the replacement of Tungsten) in future technology nodes are not expected to play a major role on the sensitivity to neutrons (as long as Silicon and Oxygen will be the two most common materials).

C. Vendors

In principle, all the analyzed memories from the three vendors share the same architecture, so it is at first surprising to see such large differences in the raw bit error rate. Differences in the materials surrounding the FG or in the back-end can alter the spectrum of particles impinging on the FG. Yet, this does not appear as the main driving factor: the memories show different sensitivities even when irradiated with heavy ions. These observations support the possibility that the spectrum of particles generated by neutrons is the same in all the devices, but the FG sensitivity changes, even for nominally identical feature sizes, or the read window margin changes.

D. Electric field

The dependence of charge loss on the electric field (E) has been investigated in detail for heavy ions. A larger E increases the discharging current triggered by the heavy ion [17]. This translates into a higher charge loss, and explains very well the level dependence of our data. In an alternative view, the electric field modulates the transient carrier flux across the oxide barriers, impacting the amount of charge loss [2].

In addition to the data on the raw single bit sensitivity, this dependence is evident also on the V_{th} distribution of Fig. 6, where a longer tail appears on the highest program level. The lowest program level and the erased level are practically immune from errors. The fact that the neutron sensitivity increases with the electric field is somehow helpful. The reason is that several other factors that affect reliability tend to have the same dependence on E. In this way, a custom ECC designed to exploit the level dependence can efficiently act also for neutron–induced errors.

No scaling is in sight for the electric field, at least as long as the floating-gate architecture will be maintained. Therefore no reduction (nor increase) of the neutron sensitivity can be expected from this front.

E. Peripheral Circuitry

In addition to the sensitivity of the floating gate array, one should also consider the impact of the peripheral circuitry. In fact, the read-out protocol of NAND memories includes the transfer of data from the floating gate matrix to a temporary storage, called page buffer. This page buffer is an array of latches, which in principle is susceptible to soft errors [15]. The microcontroller that manages the operation of the Flash memory contains state information, which can be upset by radiation [5]. Furthermore, Single Event Transients (SET), can originate in combinational or analog blocks. These errors have been observed with heavy ions, but, so far, not with neutrons. However, the heavy-ion data are reassuring: high-LET ions are needed to produce these events [5]. This is due to the fact that the NAND periphery is realized using thick oxides for cost reasons (fewer masks), and large cells, which therefore do not mirror the scaling trends of SRAM cells and logic circuits. In addition, no change is expected at least in the near future, since there are no plans to change these features even when faster access protocols will be implemented.

VI. CONCLUSIONS

We showed that neutrons can induce raw data loss in FG memories. The neutron raw bit error rate increases with decreasing feature size. This reflects the simple fact that the charge used to store a bit decreases with each new generation, whereas the ion track remains constant. In one of the tested devices, the raw bit error for the most sensitive level increased by three orders of magnitude with just a generation shrink. The exact scaling trend is difficult to predict, as shown by the large differences from vendor to vendor, due to the large mix of parameters that play a role in determining it. In spite of this, the numbers are reassuring. Only in some conditions does the neutron threat appear to be of some significance, but nowhere large enough to defy current ECC schemes.

VII. ACKNOWLEDGEMENT

The authors are greatly indebted to Pietro Guzzi, Numonyx, Agrate Brianza, who provided the technological and layout information used in the GEANT4 simulations. We would also like to thank the people from the Numonyx failure analysis lab for decapping the samples for the heavy-ion irradiation. Finally, we kindly acknowledge Serena Mattiazzo, Devis Pantano, and Mario Tessaro, INFN, Padova, for their invaluable help with the SIRAD irradiation facility.

REFERENCES

[1] J.T. Wallmark, S.M. Marcus,, "Minimum Size and Maximum Packing Density of Nonredundant Semiconductor Devices,", *Proceedings of the IRE, 50*, pp. 286-298, 1962,

[2] N.Z. Butt and M. Alam, "Modeling single event upsets in Floating Gate memory cells", *Proceedings of IEEE International Reliability Physics Symposium (IRPS) 2008*, pp. 547-555, April 2008

[3] R. C. Baumann, "Radiation-induced soft errors in advanced semiconductor technologies", *IEEE Trans. On Dev. Mat. Rel., 5*, pp.305-316, 2005

[4] G. Cellere et al., "Effect of Ion Energy on Charge Loss From Floating Gate Memories", *IEEE Transactions on Nuclear Science*, 55 (4), pp. 2042–2047, Aug. 2008

[5] F. Irom and D. Nguyen,, "Single Event Effect Characterization of High Density Commercial NAND and NOR Nonvolatile Flash Memories", *IEEE Transactions on Nuclear Science, 54 (6)*, pp. 2547-2553, Dec. 2007

[6] G. Cellere et al., "Neutron-induced soft errors in advanced Flash memories", *2008 IEEE International Electron Devices Meeting (IEDM) Tech. Digest*, pp. 357-360, Dec. 2008

[7] N. Siefert, B. Gill, K. Foley, P. Relangi, "Multi-cell upset probabilities of 45nm high-k + metal gate SRAM devices in terrestrial and space environments" *Proceedings of IEEE International Reliability Physics Symposium (IRPS), 2008*, pp. 181-186, April 2008

[8] N. Mielke, et al., "Bit Error Rate in NAND Flash Memories", *Proceedings of IEEE International Reliability Physics Symposium (IRPS), 2008*, pp. 9-19, April 2008

[9] S. Platt, Z. Torok, C.D. Frost, S. Ansell, "Charge-Collection and Single-Event Upset Measurements at the Isis Neutron Source", *IEEE Transactions on Nuclear Science, 55 (4)*, pp. 2126-2132, Aug. 2008

[10] JEDEC Standard JESD89A "Measurement and Reporting of Alpha Particle and Terrestrial Cosmic Ray Induced Soft Errors in Semiconductor Devices", Oct. 2006.

[11] J. Wyss, D. Bisello, D. Pantano, "SIRAD: An irradiation facility at the LNL tandem accelerator for radiation damage studies on semiconductor detecttos and electronics devices and systems," *Nucl. Instruments Methods Phys. Res. A*, vol. 462, pp. 426-434, 2001

[12] A. Hands, C. S. Dyer, F. Lei, "SEU Rates in Atmospheric Environments: Variations Due to Cross-Section Fits and Environment Models", *IEEE Transaction on Nuclear Science 56 (4)*, pp. 2026-2034, Aug. 2009

[13] S. Agostinelli et al., "Geant4: a simulation toolkit," *Nuclear Instruments and Methods in Physics Research Section A, 506*, 250-303, 2003

[14] G. Cellere, A. Paccagnella, A. Visconti, M. Bonanomi, R. Harboe-Sørensen, A. Virtanen, "Angular dependence of heavy ion effects in Floating Gate memory arrays", *IEEE Transaction on Nuclear Science 54 (6)*, pp. 2371-2378, Dec. 2007

[15] M. Bagatin, et al., "Key Contributions to the Cross Section of NAND Flash Memories Irradiated with Heavy Ions", IEEE Transactions on Nuclear Science, 55 (6), pp. 3302-3308, Dec. 2008

[16] R.A. Weller, A.L. Sternberg, L.W. Massengill, R.D. Schrimpf, D.M. Fleetwood, "Evaluating Average and Atypical Response in Radiation Effects Simulations", *IEEE Transactions on Nuclear Science, 50 (6)*, pp. 2265-2271, Dec. 2003

[17] G. Cellere, A. Paccagnella, L. Larcher, A. Visconti, M. Bonanomi, "Subpicosecond conduction through thin SiO_2 layers triggered by heavy ions", Journal of Applied Physics, vol. 99, 2006.

SEE test and modeling results on 45nm SRAMs with different well strategies

Gilles Gasiot, Member IEEE, Slawosz Uznanski and Philippe Roche
Technology R&D/Central CAD and Design Solutions
STMicroelectronics
, 850, Rue J. Monnet 38926 Crolles Cedex, France,
+33 476 925 660; fax: +33 476 925 678; gilles.gasiot@st.com

Abstract—**This paper presents heavy ion experimental results on SRAMs processed with 45nm bulk technology. Experiments were analyzed for Multiple Cells Upset (MCU) occurrence. The tested device was especially designed for MCU studies. In order to limit their spread it embeds different well strategies: usage of triple well layer and several densities of well ties.**

Keywords-heavy ions, well engineering, Multiple Cell Upsets, triple well

I. INTRODUCTION

Multiple Cell Upsets (MCU) are topological multiple upsets which can be corrected by classical Error Correction Codes. Multiple Bit Upsets (MBU) are logical multiple upsets uncorrectable by simple Error Correction Code (ECC) scheme since the bits in error belong to the same bit-word. Most studies related to MCU have identified the bipolar amplification as source of their occurrence [1]. The bipolar amplification is facilitated by high well resistance (figure 1). The usage of triple well layer increases the P-Well resistance and MCU occurrence. In this paper the tested device embeds different well strategies (usage of triple well layer and several densities of well ties) to modulate the well resistance and to quantify its effect on MCU spread, ion cross sections and error rate in orbit.

II. EXPERIMENTAL DETAILS

A. Devices under test

SRAMs were manufactured in 45nm commercial CMOS technology. The testchip has a total of 24Mb of single port SRAMs. Two different bitcell designs were embedded. First one has an area of $0.252\mu m^2$ (High Density: HD) while the other is $0.299\mu m^2$ (Standard Density). SRAM arrays were processed with or without triple well layer. Additionally for the bitcells processed with triple well four well tie frequencies were implemented. Table 1 synthesizes main feature of the device under test. It is noteworthy that the reference well tie frequency corresponds to the usual frequency i.e. without design modification.

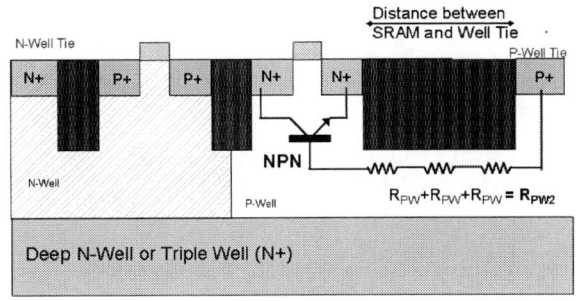

Figure 1. Schematic cross section of cmos inverter with triple well showing parasitic bipolar.

TABLE 1. CONTENT OF TESTCHIP UNDER TEST

Capacity	Bitcell area	Triple well usage	Well tie frequency
2Mb	$0.299\mu m^2$	No	Reference
2Mb	$0.299\mu m^2$	Yes	Ref.
512k	$0.299\mu m^2$	Yes	Ref. x2
512k	$0.299\mu m^2$	Yes	Ref. x4
512k	$0.299\mu m^2$	Yes	Ref. x8
6Mb	$0.252\mu m^2$	Yes	Ref.

B. Irradiation facilities

Experimental testings were performed at the RADEF [2] (RADiation Effect Facility) cyclotrons. The RADEF facility is located in the Accelerator Laboratory at the University of Jyväskylä, Finland (JYFL). Heavy ions used at the RADEF facility have stopping ranges in silicon much larger than the whole stack of back-end metallization and passivation layers (~10μm). The design-of experiment included different test patterns and supply voltages. The test procedure is compliant with ESA test specification ESCC 25100 [3].

III. EXPERIMENTAL TEST RESULTS

No Single Event Latchup (SEL) was recorded during the experiments whose worst test condition SEL-wise was with an ion LET of 60MeV/mg/cm², at 125°C and at nominal voltage plus 10%.

This work was supported in part by the European Space Agency (ESA) and the European Space Research and Technology Centre (ESTEC) under contract no. 18799/04/NL/AG, COO-4.

978-1-4244-5430-3/10 $26.00 © 2010 IEEE

C. 45nm compared to 65nm

Figure 2 shows a comparison of experimental heavy cross-sections for Single Port SRAMs processed in 65nm and 45nm. For LET below 10Mev/cm².mg both SRAMs have the same cross-sections. From 15Mev/cm².mg LET, the 45nm SRAM has a lower cross-section than the 65nm. In 45nm, the increase in MCU rate is counterbalanced by the feature size decrease. Moreover, it was verified that in 45nm the cross section variation between standard density and high density single port SRAMs is limited to 2% on average over the LET range.

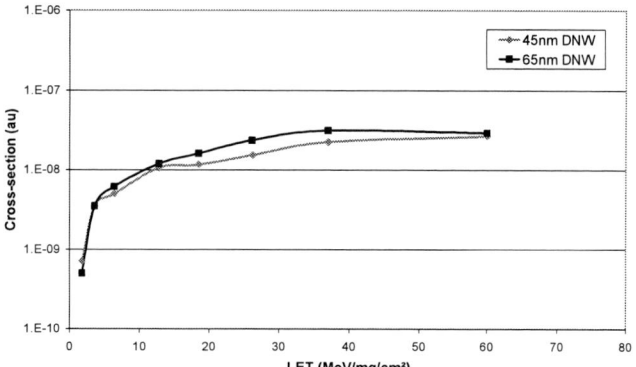

Figure 2. Cross section as a function of LET for 45nm and 65nm technologies processed with triple well for a standard density bitcell.65nm data from [4].

D. Effect of well tie frequency on SEU sensitivity

Figure 3 synthesizes experimental cross-sections measured for 0.299μm² bitcells presented in table 1. It is noteworthy that for a given well-tie frequency (reference frequency) the usage of triple well layer worsen the cross-section by a factor x3.6 on average. This was already observed in 65nm and explained by MCU rate increase due to parasitic bipolar amplification [1]. Moreover, for SRAM arrays processed with triple well, the higher the well tie frequency the lower the cross-section. The cross-section is reduced by a factor x1.5 on average between the reference and the maximum frequency. It turns out that the cross-section of the SRAM array without triple well is decreased by a factor x2.5 compared to with triple well and maximum well tie frequency. Figure 4 and 5 show MCU distribution for SRAM arrays with triple well for respectively reference and maximum well-tie densities. The higher the well ties density, the lower the MCU order. This confirms that parasitic bipolar amplification is the key parameter driving MCU response and that higher well tie frequency is able to reduce well resistance and therefore to mitigate bipolar amplification.

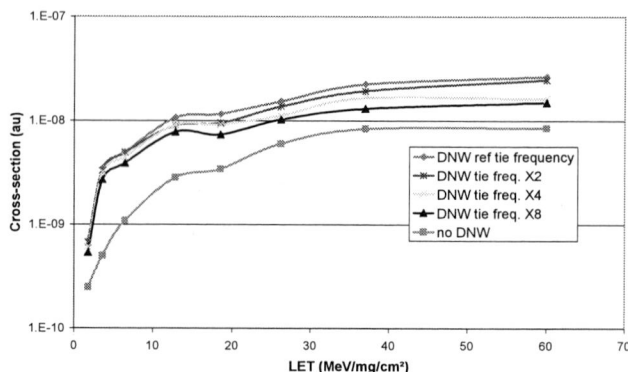

Figure 3. Cross section as a function of LET. 4 bitcells were processed with triple well and different density of well ties. 1 bitcell was processed without triple well with the standard well tie density.

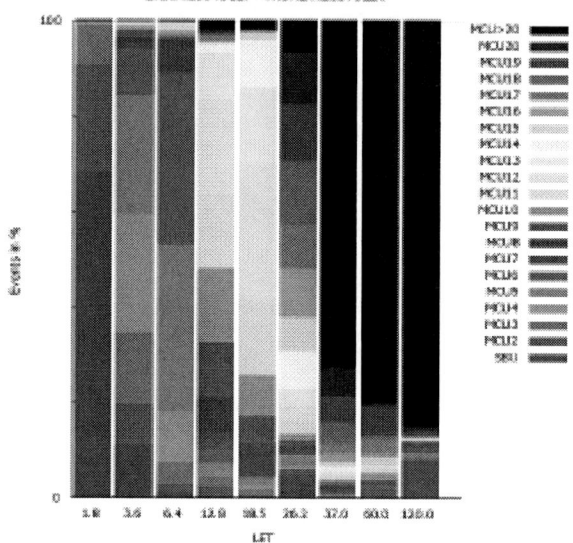

Figure 4. MCU distribution for SRAM array with reference well tie frequency.

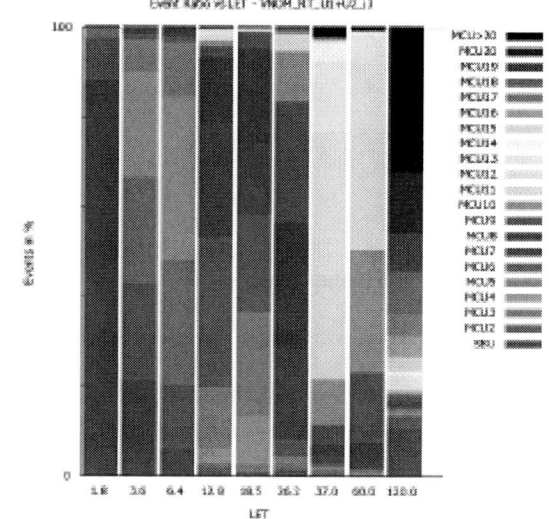

Figure 5. MCU distribution for SRAM array with well tie frequency multiplied by x8 compared to reference.

IV. Discussion and Modeling

Previous part has experimentally demonstrated that increasing the well ties density is efficient to reduce the cross section by up to a factor x3. The main drawback of increasing the well tie density is the surface increase. As a matter of a fact, surface initially used for bitcells is progressively filled-up by well ties as the tie density increases. Table 2 synthesizes the area overhead for the different well tie strategies experimentally tested.

Beyond the heavy ion cross sections, upset rate in orbit are of the utmost importance for system designers to assess the reliability of their system in its real mission environment. Upset rates in orbit were therefore derived from experimental cross sections with web-based tool CREME96 [6]. Upset rate were computed for a GEO orbit considering an Al shielding of 100 mils and solar quiet conditions (maximum heavy ions flux). The obtained rates are synthesized in table 2. As expected the higher the density of well ties the lower the upset rate. This rate can be decreased by up to a factor x2. However, the lowest upset rate is obtained for the SRAM processed without triple well (rate decreased by a factor x5). The upset rate improvement is not constant with well ties density increase. The first doubling of well tie density allows reducing the upset rate by 17% while increasing the area by 3%. The second doubling of well ties density further reduces the upset rate by 19% while increasing the area by 7%. The third doubling leads to the smallest upset rate decrease (10%) while having the highest area penalty (13%). The highest upset rate improvement with minimum area penalty is measured for a well tie density twice larger than reference. As expected the minimum upset rate is measured with the highest well tie density while it induces a large area penalty.

TABLE 2. SILICON AREA OVERHEAD AND RELATIVE UPSET RATE IN ORBIT FOR THE DIFFERENT WELL STRAP STRATEGIES (NORMALIZED TO 100 FOR DNW AND REFERENCE WELL TIE DENSITY).

Well strategy	Area overhead (%)	Relative upset rate in GEO orbit
DNW Ref.	-	100
DNW Ref. x2	3	83
DNW Ref. x4	10	64
DNW Ref. x8	23	54
no DNW	-	18

In previous part it was shown that a higher density of well-ties has led to lower MCU occurrence when a triple well layer was used. This part explores the feasibility of further reducing the MCU occurrence when a triple well layer is not used. This is also done by increasing the well tie density and verified by means of 3D TCAD simulations. All 3-D SRAM structures were built using a methodology described in [5] and the tool suite of Sentaurus Synopsys package. In this abstract, results are shown for two 3D structures accounting for the maximum and minimum frequency of well ties available in the silicon tested in this work (figure 6). Figure 7 compares the drain collected current after an ion of 5MeV/cm².mg striking a PMOS source. It shows that even without triple well, increasing the density of well ties is efficient to decrease the bipolar amplification in wells. The authors believe that this lower bipolar amplification will reduce MCU occurrence, ion cross section and upset rate in orbit. However, TCAD

simulations are deterministic and can not be used to assess the anticipated reduction in ion cross section and upset rate.

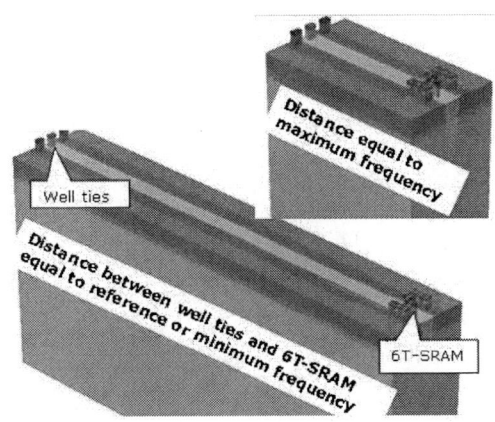

Figure 6. 3D TCAD structures whose 6-transistor bitcell is located far from the well ties to account for bipolar amplification [1]. Two distances are considered: reference and maximum=ref. x 8.

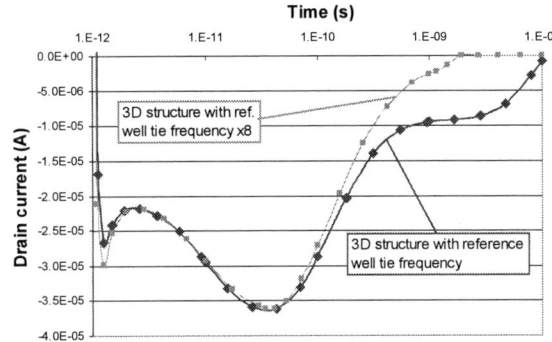

Figure 7: PMOS Drain collected currents after ion strike in 3D structures without triple well. 3D structures with reference well tie frequency has more bipolar amplification compared to structures with tie frequency multiplied by x8.

V. Conclusions

Heavy ion testings were carried out on 45nm SRAMs processed with and without triple well layer. The test vehicle embeds design variation of well tie frequency on parts processed with triple well. It was shown that increasing the well tie frequency by a factor x8 allows decreasing the MCU occurrence and therefore upset cross-section. Moreover, the upset rate in orbit can be decreased by up to a factor x2 for the highest well tie density. With triple well, the most efficient well tie density is to double the reference density (maximum upset rate reduction compared to area penalty). 3D TCAD modeling has also shown that without triple well a well tie frequency increase allows to decrease bipolar amplification and therefore MCU occurrence, ion cross section and upset rate in orbit.

Acknowledgement

The authors would like to thank Brice Lhomme and Bertrand Borot for their design efforts. We greatly appreciate the support by the RADEF teams.

REFERENCES

[1] G.Gasiot et al., "Multiple Cell Upsets as the Key Contribution to the Total SER of 65nm CMOS SRAMs and its Dependence on Well Engineering," 44th Annual International NSREC 2007, Honolulu, July 2007.

[2] A. Virtanen et al., "High Penetration Heavy Ions at the RADEF Test Site," presented at the RADECS, 2003.

[3] Single Event Effects Test Method and Guidelines, European Space Agency, ESA/SCC Basic Specification No.25100, 1995

[4] D. Giot et al., "Heavy Ion Testing and 3D Simulations of Multiple Cell Upset in 65nm Standard SRAMs," IEEE TNS 2007.

[5] Ph. Roche et al., IEEE Trans. on Nuc. Sci., Vol. 45, Issue: 6, pp. 2534–2543, Dec 1998.

[6] CREME96 web-based code: https://creme96.nrl.navy.com. CREME stands for Cosmic Ray Effects on Micro Electronics.

FIDELITY OF ENERGY SPECTRA AT NEUTRON FACILITIES FOR SINGLE-EVENT EFFECTS TESTING

S. P. Platt [1], A. V. Prokofiev [2], and Cai Xiao Xiao [1]

[1] University of Central Lancashire, [2] The Svedberg Laboratory

School of Computing, Engineering and Physical Sciences, University of Central Lancashire, Preston, Lancs. England.
Tel: +44 1772 893341 Fax: +44 1772 892915 Email: spplatt@uclan.ac.uk, xxcai1@uclan.ac.uk

The Svedberg Laboratory, Uppsala University, Uppsala, Sweden.
Tel: +46-18-471 3850 Fax: +46-18-471 3833 Email: alexander.prokofiev@tsl.uu.se

Abstract—We quantify the fidelity of neutron beam spectra for single-event effects accelerated testing. Beam spectra are folded with representative SEE cross-section curves and errors in predicted SEE rates are determined. The use of quasi-monoenergetic neutrons to complement spallation neutron sources is described. Errors can be controlled to insignificant levels even when the maximum neutron energy is below 200 MeV.

INTRODUCTION

Accelerated testing for single-event effects induced by cosmic-ray neutrons is a well established technique [1], [2]. Currently, the benchmark facility for tests of this kind is the 800 MeV spallation source at the LANSCE ICE House [3]–[5]. The LANSCE beam provides a simulation of the neutron energy spectrum arising due to cosmic ray interactions in the atmosphere, through the spallation of a tungsten target under the influence of an 800 MeV proton beam, leading to a neutron flux around eight orders of magnitude greater than that normally experienced at mid-latitude sea-level locations.

A number of similar test facilities are available [6]–[12]; their number has proliferated in recent years due to increased demand for testing of this kind. One such facility is at The Svedberg Laboratory (TSL) where an established quasi-monoenergetic neutron (QMN) capability [7] has recently been supplemented by a 180 MeV spallation source, ANITA [8], [9]. The facility at TRIUMF has been developed over recent years to provide several complementary spallation sources [10], [11]. New facilities are either under active development or under consideration for future developments [13]–[15].

Inevitably, each facility simulates the neutron field induced by cosmic rays with limited fidelity. In order to predict system reliability in service from the results of accelerated testing some means of assessing this fidelity is desirable.

NATURAL AND SYNTHETIC NEUTRON FIELDS

One of the key characteristics of SEE-inducing neutron fields is the distribution of neutrons by energy, defined by the differential neutron flux, $\phi(E)$. Gordon et al [16], [17] measured the neutron field at several locations in the USA

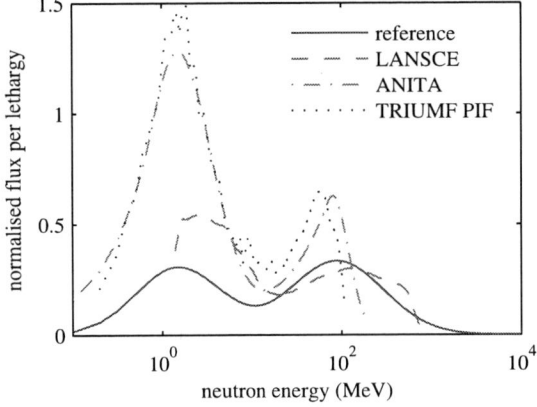

FIG. 1: NEUTRON SPECTRA, NORMALISED TO INTEGRAL FLUX ABOVE 10 MEV. SEE THE APPENDIX FOR A DETAILED EXPLANATION OF THIS FIGURE.

and parameterised a model for a reference spectrum at New York City. Following JEDEC [1], we also adopt this as our reference. This reference field, normalised to the total flux above 10 MeV, is shown in Fig. 1. The equilethargic form used in Fig. 1 ensures that equal areas under the curve enclose equal integral flux despite the logarithmic x-axis, as described in the appendix. Fig. 1 also shows spectra for the synthetic fields available at the LANSCE ICE House [18], TSL ANITA, and TRIUMF PIF (BL2C, 116 MeV protons [11]). Although these "white" fields have broadly similar spectra, differences among them are clear. For example, the component in the reference field which is above the LANSCE cut-off energy is small, whereas the corresponding component above the ANITA and TRIUMF PIF cut-off energies is larger. Each synthetic field exaggerates the evaporation peak around 1 MeV; the exaggeration is greater for ANITA and TRIUMF PIF than for LANSCE. ANITA and TRIUMF PIF fields are very similar.

However, Fig. 1 does not tell us the significance of these differences either between any two synthetic fields or between a synthetic field and the natural one. Their implications for the accuracy of reliability predictions cannot be determined without some additional considerations.

978-1-4244-5430-3/10 $26.00 © 2010 IEEE

SEE CROSS-SECTIONS AND SEE RATES

The SEE response of a device to neutrons can in principle be described by a cross-section function, $\sigma(E)$. An expected SEE rate, λ, could be calculated by folding the cross-section function with the differential flux in the reference field:

$$\lambda = \int_0^\infty \sigma(E)\phi_r(E)dE \tag{1}$$

In practice $\sigma(E)$ cannot be measured because of the absence of suitable monoenergetic fields, and so the purpose of accelerated testing in white neutron beams is to use measurable quantities to estimate an integral cross-section in a synthetic field and apply that to the corresponding integral flux in the reference field to predict the SEE rate in service.

The integral cross-section of an event type of interest to neutrons above a threshold energy E_0 is

$$\sigma_{E_0} = \frac{\int_0^\infty \sigma(E)\phi(E)dE}{\int_{E_0}^\infty \phi(E)dE} \tag{2}$$

The best estimate of this quantity for a synthetic field ϕ_s is

$$\sigma_{s,E_0} \approx \frac{N}{\Phi_{s,E_0}} \tag{3}$$

assuming N events to have been observed during a period T in which the integral fluence received is

$$\Phi_{s,E_0} = \int_0^T \int_{E_0}^\infty \phi_s(E,t)dEdt \tag{4}$$

The fluence Φ_{s,E_0} is normally provided by facility beam monitoring and is typically known to an uncertainty of about 10%. The threshold energy is commonly taken to be 10 MeV, under the assumption that the contribution of lower energy neutrons to the event rate is small [1].

The standard procedure [1] for determining the SEE rate in service from accelerated test data amounts to an equation of the integral cross-section in the synthetic field to that in the reference field:

$$\sigma_{s,E_0} \approx \frac{N}{\Phi_{s,E_0}} \approx \sigma_{r,E_0} \tag{5}$$

and to apply that to the integral flux in the reference field to estimate the event rate

$$\lambda_r = \sigma_{r,E_0}\int_{E_0}^\infty \phi_r(E)dE \approx \frac{N}{\Phi_{s,E_0}}\int_{E_0}^\infty \phi_r(E)dE \tag{6}$$

One source of error in the estimation of λ_r is the limited fidelity of the synthetic field used for the accelerated test; for example because of a limited upper neutron energy. Such limits cannot be entirely avoided and may be significant. The fractional error in the prediction of λ_r due to discrepancies between ϕ_r and ϕ_s is:

$$\varepsilon_s = \frac{\sigma_s - \sigma_r}{\sigma_r} = \frac{\int_0^\infty \sigma(E)\phi_s'(E)dE}{\int_0^\infty \sigma(E)\phi_r'(E)dE} - 1 \tag{7}$$

where ϕ' represents the differential flux normalised to that above the threshold energy, E_0. If the reference and synthetic fields are known (cf. Fig. 1) we can use (7) with suitable

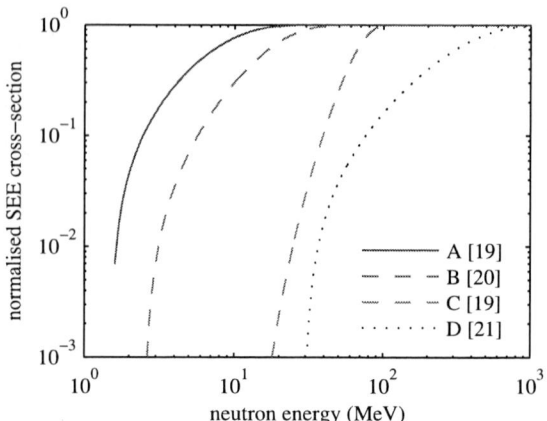

FIG. 2: SELECTED SEE CROSS-SECTION FUNCTIONS

TABLE I: SELECTED SEE CROSS-SECTION WEIBULL FIT PARAMETERS

	E_t/MeV	W/MeV	S
A [19]	1.5	6.3	1.2
B [20]	2.5	15.34	1.44
C [19]	12	56	3.1
D [21]	30	300	1.2

cross-section functions $\sigma(E)$ to quantify the fidelity of a given neutron spectrum to the reference spectrum in terms of the error component, ε_s.

Cross-section curves

We have selected some cross-section functions from the literature [19]–[21] These were determined by the original authors by measurement using a mixture of monoenergetic neutron fields at lower energies and proxy protons at high energies. Weibull fits are used to represent them:

$$\sigma(E) = 1 - \exp\left[-\left(\frac{E - E_t}{W}\right)^S\right] \tag{8}$$

The fitting parameters (Table I) were either taken from tabulated values (case B [20]) or else determined from published curves (cases A, C & D [19], [21]. As our analysis is not influenced by the absolute value of the cross-section the curves are shown normalised to their saturation values as shown in Fig. 2.

Case A [19] represents the single-event upset (SEU) cross-section for an SRAM with an especially low threshold energy, $E_t = 1.5$ MeV. Case B [20] describes the SEU cross-section of a more typical SRAM. Case C [19] represents the cross-section function for SEU in a DRAM with a higher threshold energy. Case D [21] describes the energy dependence of single-event latch-up (SEL) in the SRAM of case B, and is characterised by a high onset threshold energy and slow rise (described by parameters W and S) to a plateau value. Between them, cases A and D represent curves close to the extremes expected for most phenomena in most devices of interest.

978-1-4244-5430-3/10 $26.00 © 2010 IEEE

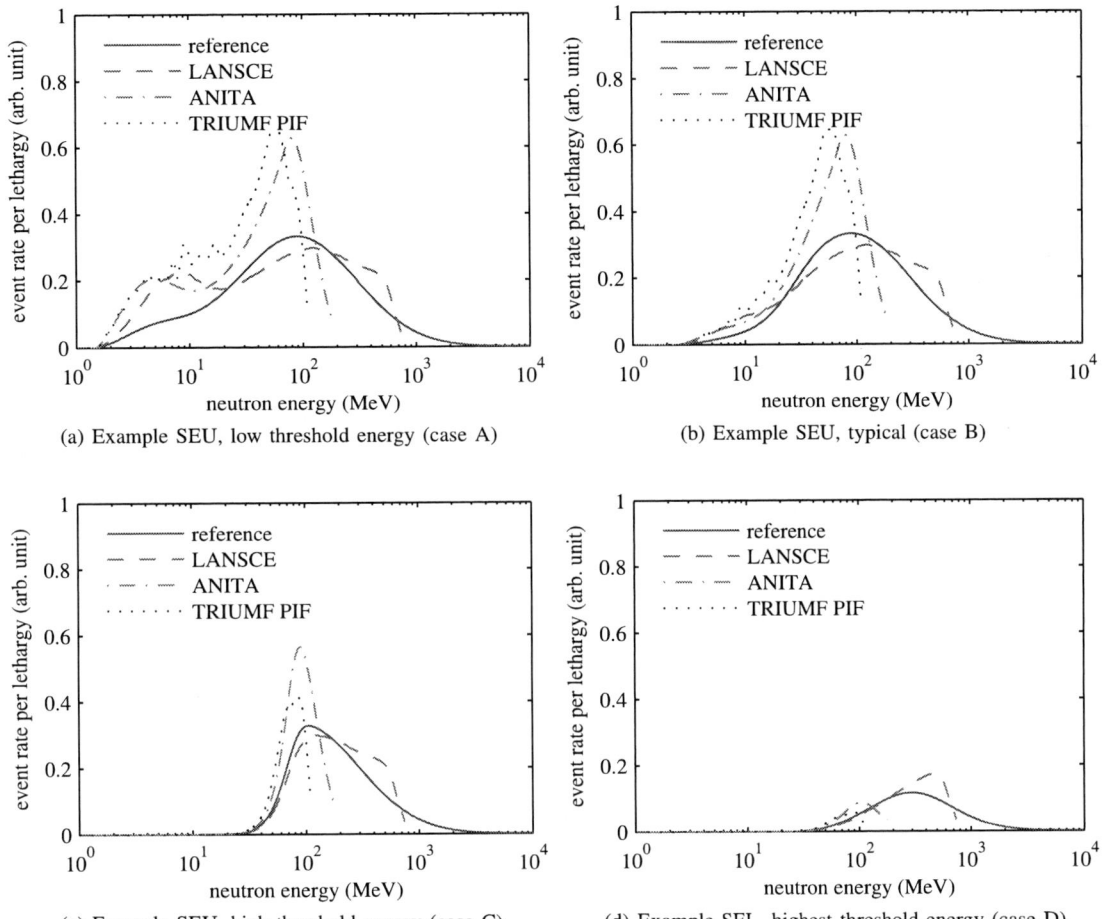

(a) Example SEU, low threshold energy (case A)

(b) Example SEU, typical (case B)

(c) Example SEU, high threshold energy (case C)

(d) Example SEL, highest threshold energy (case D)

FIG. 3: CROSS-SECTION FUNCTIONS FOLDED WITH NEUTRON SPECTRA. THE CONTRIBUTION OF A PORTION OF EACH NEUTRON FIELD TO THE SEE RATE IS PROPORTIONAL TO THE CORRESPONDING AREA UNDER EACH CURVE.

Influence of neutron spectra on SEE rates

Fig. 3 shows the result of folding each of the selected cross-section fits with each of the neutron spectra considered here. That is, it plots the function $E\sigma(E)\phi(E)$ versus energy, E, exploiting the equilethargic representation to illustrate to what extent each part of each spectrum is effective in inducing SEE in each case. In addition, the fractional error in predicted SEE rate due to discrepancies in spectral shapes has been determined for ANITA (ε_A), TRIUMF PIF (ε_P) and LANSCE (ε_L) using (7) and tabulated in Table II for each test case.

For case A, all synthetic fields suffer from an excess of events from neutrons in the evaporation peak. Apart from any other source of error, this leads to a predicted overestimate of SEU rate in this device of between 10% and 20%. This is likely to be similar in magnitude to other sources of error (for example those arising from uncertainty in received fluence or from event counting statistics). For case B, all predictions have negligible errors due to the shape of the neutron spectrum. This is because the cross-section function saturates at a sufficiently low level that it is effectively constant over the region in which the form of the synthetic fields differs from that of

TABLE II: FRACTIONAL ERRORS (%) IN SEE RATE ESTIMATES. FOR EACH OF THE REPRESENTATIVE CROSS-SECTION FITS, THE CONTRIBUTION TO UNCERTAINTY IN THE MEASURED SEE RATE DUE TO LIMITED BEAM FIDELITY (7) IS SHOWN FOR LANSCE (ε_L), TRIUMF PIF (ε_P), ANITA (ε_A), AND ANITA WITH QMN ADJUSTMENT ($\varepsilon_{A'}$)

	ε_L	ε_P	ε_A	$\varepsilon_{A'}$
A	12.4	18.1	15.6	4.7
B	0.7	−4.0	−0.3	0.0
C	−1.5	−55.0	−26.8	4.5
D	1.2	−84.4	−69.1	26.5

the reference fields. Case B is typical of several cross-section functions of interest reported in the literature [19], [20], [22] For case C, with a higher energy threshold and higher saturation energy, LANSCE provides a good estimate of the SEE rate in service, whereas ANITA or TRIUMF PIF testing would lead to an underestimate which might limit the accuracy of the prediction, and which might benefit from an adjustment procedure, as described below. For case D, the lower maximum energies at ANITA and TRIUMF PIF lead to a significant underestimate of the SEL rate in this device.

978-1-4244-5430-3/10 $26.00 © 2010 IEEE

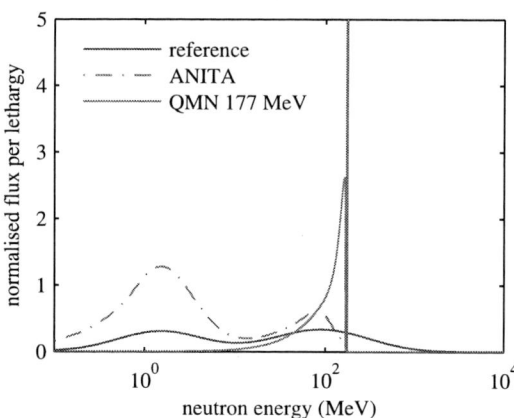

FIG. 4: NORMALISED NEUTRON SPECTRA, INCLUDING THE
TSL QMN 177 MEV FIELD. THE QMN 177 MEV SPECTRUM IS
NORMALISED TO THE FLUX IN THE NARROW PEAK, WHICH CONTAINS
APPROXIMATELY 40% OF THE QMN FLUX AND EXTENDS TO OVER 35
UNITS ON THE Y-AXIS ON THIS SCALE.

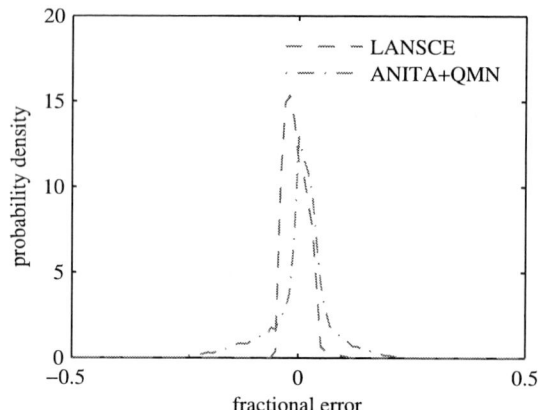

FIG. 6: CONTRIBUTION TO ERROR IN PREDICTED SEE
EVENT RATES DUE TO SPECTRAL DISCREPANCIES IN
SYNTHETIC NEUTRON FIELDS. ASSUMES WEIBULL FITS WITH
PARAMETERS IN THE RANGE OF FIG. 2 AND TABLE I.

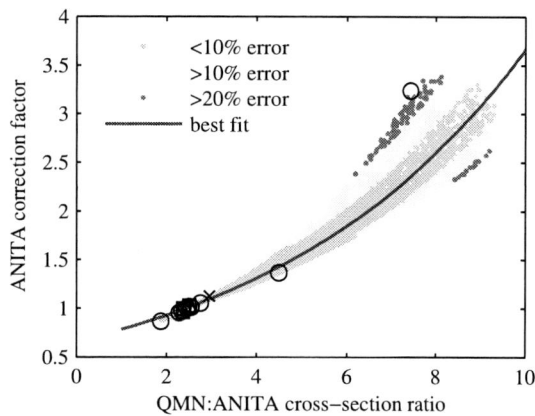

FIG. 5: FACTORS REQUIRED TO ADJUST ANITA DATA.
ASSUMES WEIBULL FITS WITH PARAMETERS IN THE RANGE OF FIG. 2
AND TABLE I. GREEN POINTS REPRESENT SAMPLES FROM THE
PARAMETER SPACE FOR WHICH ADJUSTMENT ACCORDING TO THE BEST
FIT IS ACCURATE TO WITHIN 10%; YELLOW AND RED POINTS REPRESENT
CORRESPONDING ERRORS LESS THAN AND GREATER THAN 20%,
RESPECTIVELY. BLACK SYMBOLS DESCRIBE APPLICATION OF THE
TECHNIQUE TO REPRESENTATIVE CROSS-SECTION FUNCTIONS.

Adjustment using quasi-monoenergetic neutrons: For devices of interest with low threshold energies for neutron-
induced SEE the upper energy limits at TRIUMF PIF and
ANITA are unlikely to lead to significant errors in cross-
section measurement. For other devices or phenomena the
influence of these limits may be significant. In such cases
measurements can be adjusted using a quasi-monoenergetic
neutron (QMN) beam.

Fig. 4 shows the spectrum of a 177 MeV QMN beam
available at TSL. The QMN field contains substantial flux
in the region underestimated by the ANITA field. The QMN
and ANITA fields are complementary, and QMN neutron

irradiations can be used to control uncertainties in results
from ANITA irradiations through an adjustment procedure,
as follows. We can determine an adjustment factor, k, such
that $\sigma_r = k\sigma_A$. Fig. 5 illustrates the required adjustment as a
function of the (measurable) ratio of integral cross-sections
in QMN (σ_Q) and ANITA (σ_A) beams. We used a set of
10 000 cross-section curves, expressed as Weibull functions
with parameters (E_t, W, and S) chosen randomly according to
a uniform distribution so as to lead to curves bounded by the
extreme cases (A and D) shown in Fig. 2. This distribution
does not necessarily represent the distribution of SEE cross-
section functions in devices of interest, but it does populate
the parameter space within which most such curves could be
expected to fall. Fig. 5 also shows a best-fit exponential to the
distribution:

$$k = Ae^{B\sigma_Q/\sigma_A} \qquad (9)$$

The fit parameters are $A = 0.651$, $B = 0.175$.

The adjustment of (9) has been applied to the results
obtained from the example SEE fits of Table I. The resulting
fractional errors $\varepsilon_{A'}$ (7) are included in the final column of
Table II. We see that the adjustment improves the SEE rate
estimates significantly. In the three SEU cases (A..C) the
resulting errors are insignificant (less than 5%). In the extreme
case (SEL, case D) the error due to is also much reduced. The
positions in Fig. 5 corresponding to these four cases are shown
as open circles.

Fig. 6 shows the distribution of the contribution to errors in
predicted SEE rates due to spectral discrepancies in the LAN-
SCE beam and the ANITA beam with the QMN adjustment
applied. For both cases the results assume the distribution of
Weibull fits used to generate the data set of Fig. 5.

Application to the Imaging SEE Monitor: Although Weibull
fits have been widely assumed to represent neutron SEE
curves, they provide, at best, a crude approximation to the

978-1-4244-5430-3/10 $26.00 © 2010 IEEE

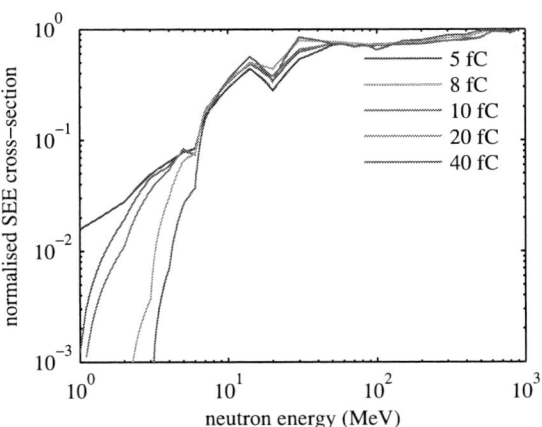

FIG. 7: SIMULATED ISEEM CROSS-SECTION FOR EVENTS EXCEEDING THE SPECIFIED CHARGE LEVELS

TABLE III: FRACTIONAL DISCREPANCIES (%) IN EVENT RATE ESTIMATES FOR CROSS-SECTIONS SIMULATED USING ISEEM (7)

	ε_L	ε_P	ε_A	$\varepsilon_{A'}$
5	3.0	−7.3	0.2	1.1
8	7.3	1.0	3.2	4.9
10	7.2	0.9	3.3	4.3
20	6.0	−0.4	0.9	1.4
40	5.8	−3.2	−2.5	−4.0

general form of cross-section functions. For example, there is evidence both from simulations and from measurements that at least some phenomena in some devices exhibit local maxima and minima in their cross-section curves [22], [23]. We have previously used a CCD-based Imaging SEE Monitor (ISEEM) to investigate effects in a variety of neutron fields, including making benchmarking comparison between neutron beams [8], [12]. Fig. 7 shows simulated [24] cross-sections for events in ISEEM from which charge exceeding thresholds in the range from 5 fC to 40 fC is collected. The form of these curves is similar to those reported for measurements and simulations of devices of interest, for example showing the same structure attributed to the onset and peak cross-sections of important interactions including $Si(n,\alpha)Mg$ and $Si(n,p)Al$ reactions [25]. Applying cross-sections of these forms to the procedure of (7) results in the data of table III. The tabulated data include the results of the adjustment procedure (9), also illustrated by open squares in Fig. 5. In all cases the discrepancies between simulated and reference neutron fields are below 10% for cross-sections of this complex form.

CONCLUSION

When testing electronic devices for neutron-induced SEE due to cosmic radiation it is advantageous to mimic the natural radiation field as closely as possible. However, all white neutron beams used to simulate atmospheric cosmic radiation for accelerated SEE testing have some limit to their fidelity. This is particularly so for those fields at which the limiting energy is 200 MeV or less. It is useful to have some means of quantifying the error in SEE rate inferred from measurements made at such facilities. In particular it us useful to know under what conditions this component of error is likely to be insignificant compared to other errors in such measurements (in practice, less than about 10% would certainly be insignificant and somewhat larger errors are likely to be tolerable).

We have described a simple technique for assessing this fidelity by folding with selected cross-section curves, and demonstrated this by using selected Weibull fits as well as by using the cross-section of the Imaging SEE Monitor as a proxy for devices of interest.

In many cases, especially those of modern devices with low energy thresholds for the onset of neutron-induced SEE (e.g. below 10 MeV), a low upper energy to a neutron field (e.g. below 200 MeV) is not likely to be a significant limitation. In such cases measurements of integral cross-sections at facilities such as ANITA (180 MeV) and TRIUMF PIF (BL2C, 116 MeV) are unlikely to be affected by discrepancies between the beam spectrum and that of the natural atmospheric radiation field.

In other cases, for example those where the SEE cross-section is still a strong function of neutron energy above about 10 MeV, it is possible to apply an adjustment factor derived from measurements in a complementary quasi-monoenergetic neutron field. This complementary technique is readily possible at TSL where a quasi-monoenergetic source is available from the same 180 MeV proton beam as that used for ANITA. In the example cases shown here, this adjustment technique is shown to be capable of reducing the effect of spectral discrepancies to insignificant levels for many cases of interest.

In still other cases, especially those in which the threshold neutron energy for an effect of interest is known or suspected to be high (for example above about 100 MeV) a low upper energy limit in a synthetic neutron field might limit the accuracy of a cross-section measurement. If an accurate cross-section measurement is required (rather than a screening test) it becomes advantageous to select the most faithful field. Of those facilities considered here the preferred one for this kind of measurement would be the LANSCE ICE House, leading to an acceptable contribution to measurement uncertainty in almost all likely cases (Fig. 6).

The technique described here would also enable the fidelity of neutron spectra at future SEE test facilities (e.g. [14], [15]) to be quantified in a comparable manner.

APPENDIX
EXPLANATION OF THE LETHARGY PLOT

In the neutron SEE field, when plotting neutron energy spectra (e.g. Fig. 1, Fig. 4) or functions derived from these (e.g. Fig. 3), it is normally necessary to use a logarithmic scale for energy to cover the wide range of neutron energies of interest (perhaps as much as 12 orders of magnitude, from meV to GeV). Most commonly log-log scales have been used to accommodate wide ranges of both dependent and independent variables. The disadvantage of this approach is that the nonlinear scales make it more difficult to interpret graphical data. More recently a different representation has become more common, most notably in the latest issue of the relevant JEDEC standard [1]. This representation, which is well established in the field of neutron spectrometry [16], [26], [27], is alternatively described as a "lethargy plot" or "equilethargic representation".

Neutron lethargy is a dimensionless measure of the relative energy of a neutron, or of a portion of a neutron ensemble:

$$L = \ln \frac{E}{E_0} \tag{10}$$

The value of the reference energy, E_0 is arbitrary. The differential neutron flux per lethargy is

$$\frac{d\Phi(E)}{dL} = \frac{d\Phi(E)}{dE}\frac{dE}{dL} = E\frac{d\Phi(E)}{dE} = E\phi(E) \tag{11}$$

and it is this quantity which is plotted directly in Fig. 1 and Fig. 4 and after weighting by SEE cross-section functions in Fig. 3. (Here $\Phi(E)$ is the integral flux, for example in $n\,cm^{-2}\,s^{-1}$, at energy E and $\phi(E)$ is the corresponding differential quantity, for example in $n\,cm^{-2}\,s^{-1}\,MeV^{-1}$.)

The usefulness of the lethargy plot is that the distortion resulting from multiplying differential flux by energy (and presenting the result on a linear y-axis) compensates for that involved in presenting energy on a logarithmic x-axis. As a result, the area under a differential spectrum between two energies E_1 and E_2 as shown on a lethargy plot is proportional to the integral flux between the two energies:

$$\int_{\log E_1}^{\log E_2} E\phi(E)\,d\log E = \int_{L_1}^{L_2}\frac{d\Phi(E)}{dL}\,dL = \Phi(E_2) - \Phi(E_1) \tag{12}$$

Alternatively, we can consider that we plot differential flux per unit lethargy on a linear y-axis, and consider the logarithmic representation of energy on the x-axis as being a linear representation of lethargy; integrating flux per lethargy over lethargy gives us the integral flux between the two lethargy points corresponding to the limiting energies.

REFERENCES

[1] *Measurement and Reporting of Alpha Particle and Terrestrial Cosmic Ray-Induced Soft Errors in Semiconductor Devices*, JEDEC Solid State Technology Association Std. JESD89A, Oct. 2006. [Online]. Available: http://www.jedec.org/download/search/jesd89a.pdf

[2] *Test Method for Beam Accelerated Soft Error Rate*, JEDEC Solid State Technology Association Std. JESD89-3A, Nov. 2007. [Online]. Available: http://www.jedec.org/download/search/jesd89-3A.pdf

[3] P. W. Lisowski, C. D. Bowman, G. J. Russell, and S. A. Wender, "The Los Alamos National Laboratory spallation neutron sources," *Nucl. Sci. Eng.*, vol. 106, p. 208, 1990.

[4] S. A. Wender *et al.*, "A fission ionization detector for neutron flux measurements at a spallation source," *Nucl. Inst. Meth. Phys. Res., A*, vol. 336, no. 1–2, pp. 226–231, Nov. 1993.

[5] B. E. Takala, "The ICE House: Neutron testing leads to more-reliable electronics," *Los Alamos Science*, no. 30, pp. 96–103, 2006.

[6] T. Nakamura, M. Baba, E. Ibe, Y. Yahagi, and H. Kameyama, *Terrestrial Neutron-Induced Soft Errors in Advanced Memory Devices*. World Scientific, 2008.

[7] A. V. Prokofiev *et al.*, "The TSL neutron beam facility," *Radiat. Prot. Dos.*, vol. 126, no. 1-4, pp. 18–22, 2007.

[8] A. V. Prokofiev, J. Blomgren, S. P. Platt, R. Nolte, S. Röttger, and A. N. Smirnov, "ANITA – a new neutron facility for accelerated SEE testing at The Svedberg Laboratory," in *Proc. Int. Rel. Phys. Symp*, 2009, pp. 929–935.

[9] A. V. Prokofiev *et al.*, "Characterization of the ANITA neutron source for accelerated SEE testing at The Svedberg Laboratory," in *Proc. IEEE Radiation Effects Data Workshop*, 2009, pp. 166–173.

[10] E. Blackmore, P. Dodd, and M. Shaneyfelt, "Improved capabilities for proton and neutron irradiations at TRIUMF," in *2003 IEEE Radiation Effects Data Workshop Record*, 2003, pp. 149–155.

[11] E. W. Blackmore, "Development of a large area neutron beam for system testing at TRIUMF," in *2009 IEEE Radiation Effects Data Workshop Record*, 2009.

[12] S. P. Platt, Z. Török, C. D. Frost, and S. Ansell, "Charge-collection and single-event upset measurements at the ISIS neutron source," *IEEE Trans. Nucl. Sci.*, vol. 55, no. 4, pp. 2126–2133, Aug. 2008.

[13] S. Ansell and C. D. Frost, "A design of an irradiation beamline for Target Station 2, ISIS," in *Proc. 9th European Conf. Radiat. Effects Compon. Syst. (RADECS 2007)*, Sep. 2007, paper F-1.

[14] C. D. Frost, S. Ansell, and G. Gorini, "A new dedicated neutron facility for accelerated SEE testing at the ISIS facility," in *Proc. IEEE International Reliability Physics Symposium*, 2009, pp. 952–955.

[15] L. Dominik, E. Normand, M. J. Dion, and P. Ferguson, "Proposal for a new integrated circuit and electronics neutron experiment source at oak ridge national laboratory," in *Proc. IEEE International Reliability Physics Symposium*, 2009, pp. 940–947.

[16] M. S. Gordon *et al.*, "Measurement of the flux and energy spectrum of cosmic-ray induced neutrons on the ground," *IEEE Trans. Nucl. Sci.*, vol. 51, pp. 3427–3434, Dec. 2004.

[17] M. S. Gordon *et al.*, "Correction to "Measurement of the flux and energy spectrum of cosmic-ray induced neutrons on the ground"," *IEEE Trans. Nucl. Sci.*, vol. 52, pp. 2703–2703, Dec. 2005.

[18] B. E. Takala, Mar. 2006, personal communication.

[19] Y. Yahagi *et al.*, "Threshold energy of neutron-induced single event upset as a critical factor," in *Proc. Int. Rel. Phys. Symp.* IEEE, 2004, pp. 669–670.

[20] A. Hands, C. S. Dyer, and F. Lei, "SEU rates in atmospheric environments: Variations due to cross-section fits and environment models," *IEEE Trans. Nucl. Sci.*, vol. 56, no. 4, pp. 2026–2034, Aug. 2009.

[21] C. S. Dyer, S. N. Clucas, C. Sanderson, A. D. Frydland, and R. T. Green, "An experimental study of single-event effects induced in commercial SRAMs by neutrons and protons from thermal energies to 500 MeV," *IEEE Trans. Nucl. Sci.*, vol. 51, no. 5, pp. 2817–2824, Oct. 2004.

[22] J. Baggio, D. Lambert, V. Ferlet-Cavrois, P. Paillet, C. Marcandella, and O. Duhamel, "Single event upsets induced by 1-10 MeV neutrons in static-RAMs using mono-energetic neutron sources," *IEEE Trans. Nucl. Sci.*, vol. 54, pp. 2149–2155, Dec. 2007.

[23] D. Lambert *et al.*, "Analysis of quasi-monoenergetic neutron and proton SEU cross sections for terrestrial applications," *IEEE Trans. Nucl. Sci.*, vol. 53, no. 4, pp. 1890–1896, Aug. 2006.

[24] X. X. Cai, S. P. Platt, and W. Chen, "Modelling neutron interactions in the Imaging SEE Monitor," *IEEE Trans. Nucl. Sci.*, vol. 56, no. 4, pp. 2035–2041, Aug. 2009.

[25] "ENDF/B-VII Incident-Neutron Data," Los Alamos National Laboratory. [Online]. Available: http://t2.lanl.gov/data/neutron7.html

[26] P. Goldhagen *et al.*, "Measurement of the energy spectrum of cosmic-ray induced neutrons aboard an ER-2 high-altitude airplane," *Nucl. Inst. Phys. Res., A*, vol. 476, p. 4251, 2002.

[27] *Reference neutron radiations – Part 1: Characteristics and methods of production*, International Organization for Standardization Std. ISO 8529-1:2001(E), 2001.

978-1-4244-5430-3/10 $26.00 © 2010 IEEE

MOBILE AND STABLE HYDROGEN SPECIES IN THE INTERFACE LAYER
BETWEEN POLY SILICON AND GATE OXYNITRIDE

Ziyuan Liu, Shuu Ito, Shoichi Hiroshima, [2]Shin Koyama, [2]Mariko Makabe,
[3]Markus Wilde and [3]Katsuyuki Fukutani

Test and Analysis Engineering Division, [2]Process Technology Division, NEC Electronics Corporation,
1753 Shimonumabe, Nakahara-ku, Kanagawa 211-8668, Japan. Phone: +81-44-435-1438; Fax: +81-44-435-1873; Email: z.liu@necel.com
[3]Institute of Industrial Science, University of Tokyo and CREST-JST

ABSTRACT

The diffusion behavior of hydrogen contained in the surface layer of oxynitrides serving as models for poly-Si/oxynitride interfaces in MOS transistors was studied with H depth profiling by nuclear reaction analysis. The poly-Si/oxynitride interface is found to contain mobile and stable H species. The mobile H species tends to desorb in vacuum at room temperature. A TDDB improvement caused by resting in air for more than 24 h prior to the post nitridation annealing is attributed to the reduction of mobile H species. Eliminating the mobile H from the gate interface is thus suggested to improve the reliability of oxynitride dielectrics.

INTRODUCTION

It is well recognized that hydrogen (H) plays an important and intricate role in the breakdown and instability of gate dielectrics. On one hand, H species present in the oxide contribute to the superior charge-to-breakdown (Qbd) quality of 'wet' oxides compared to 'dry' oxides; on the other hand H is involved in oxide instabilities such as negative-bias-temperature instability (NBTI) and hot-carrier instability (HCI) [1-6]. These effects are related to the diffusivity of H and to its ability to passivate/ depassivate dangling bonds [7, 8]. Accordingly, the resistance against H diffusion of the respective layers and interfaces that constitute the MOS stack becomes essential. We have repeatedly observed that MOS device degradation is connected to significant H accumulation in the oxide/Si interface region [5, 8] and realized that the H diffusion should trace back to a source of mobile H. We already demonstrated that a specific ultrathin oxynitride functions as a potential H-storage layer [9] and showed that building an oxynitride layer as a H diffusion barrier in front of the oxide/Si interface improves the reliability [10]. In this study we focus on the source of the diffusing H species that approach the oxide/Si interface.

A qualitative model explaining gate dielectrics degradation and breakdown by H release has been proposed a long time ago and has widely been debated [11-13] but only little experimental evidence exists to directly confirm it. This is possibly related to the experimental difficulty to observe mobile H at buried interfaces. We therefore applied H depth profiling by nuclear reaction analysis (NRA) to probe the H diffusion and found in combination with thermal desorption spectroscopy (TDS) that two kinds of H, referred to as mobile and stable H species, are coexisting in the interface region between the poly Si gate and the oxynitride dielectric of MOS transistors [14, 15]. As it will be shown, this situation is similar to, although not the same, as that of H stored in air-exposed oxynitride layers on Si_3N_4/SiO_2 stacks [9]. NRA confirms that the mobile H, if stimulated by energetic carriers, migrates across the gate films and relocates to the SiO_2/Si interface. The mobile H is found to desorb easily from the oxynitride surface after fabrication by plasma nitridation. We hence propose that the TDDB improvement due to air exposure prior to post-nitridation annealing is related to a reduction of the mobile H. Thus eliminating the mobile H from the electrode/dielectric interface is expected to improve the reliability of the oxynitride dielectrics.

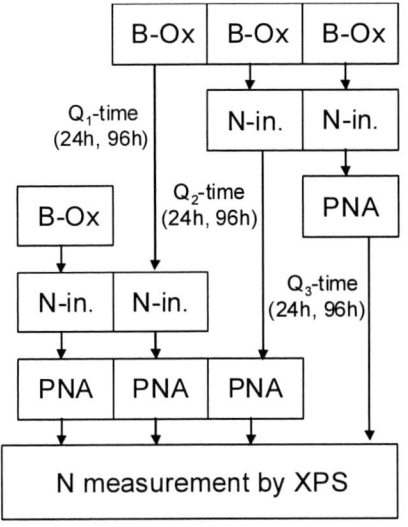

FIGURE 1. EXPERIMENTAL SCHEME TO DETERMINE THE Q-TIME INFLUENCE ON THE NITROGEN CONTENT IN THE GATE OXYNITRIDES.

EXPERIMENTAL

Devices under test were nMOSFETs with 1.6 nm oxynitride gate dielectrics. As shown in Fig. 1, the gate oxynitrides were fabricated in a 3-step procedure: base oxidation (B-Ox), plasma-nitridation (N-in), and post-nitridation annealing (PNA). Each time interval (Q-time) between the process steps was arranged as <2 h (ref.), 24 h, and 96 h, respectively. After the final PNA the nitrogen content of the oxynitride films with split Q-time intervals was measured by X-ray photoelectron spectroscopy (XPS). The influence of the Q-time on the Time-Dependent Dielectric Breakdown (TDDB), gate leakage current (Ig), the threshold voltage (Vt), and the channel current (Ion) were evaluated at room temperature by electrical measurements.

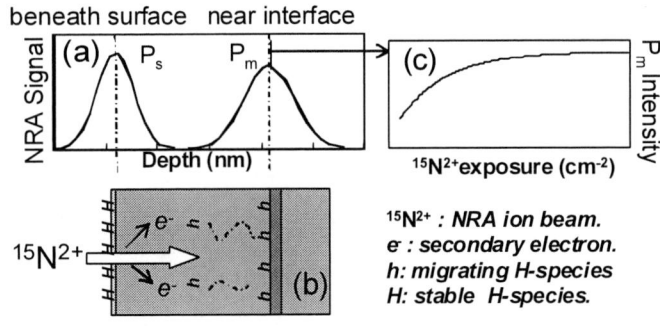

FIGURE 2. DETECTION PRINCIPLE OF MIGRATING H SPECIES CONTAINED IN THE SURFACE LAYER. NRA DISTINGUISHES BETWEEN MOBILE (P_M) AND STABLE (P_S) H SPECIES.

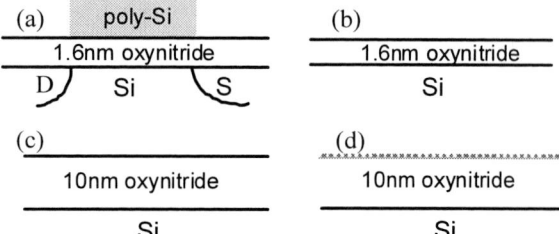

FIGURE 3. SCHEMATICS OF THE n-MOSFET (A), THE 1.6 NM OXYNITRIDE FILMS FOR XPS MEASUREMENT (B), THE 10 NM OXYNITRIDE FILMS FOR NRA H-MEASUREMENTS (C), AND 10 NM OXYNITRIDE COVERED BY AN ULTRA-THIN POLY SI LAYER FOR TDS MEASUREMENTS (D).

With the same process, 10 nm thick oxynitride films were fabricated separately for H depth profiling analysis by $^1H(^{15}N,\alpha\gamma)^{12}C$ NRA (Fig. 2 (a)). Irradiation with ^{15}N ions at ~6.4 MeV induces H-redistribution in silicon oxide and in Si nitride films [8]. The effect is caused by a secondary electron-induced H-release (Fig. 2 (b)) and results in H-accumulation in the oxide/Si interface region (Fig. 2 (c)). Rather than showing the original H distribution in the oxynitride films, NRA therefore distinguishes between mobile (P_m) and stable (P_s) H species [9]. By observing the P_m intensity after saturation of the H relocation process the proportion of mobile H can be estimated. TDS and Time-of-Flight Secondary Ion Mass Spectrometry (TOF-SIMS) were applied to study H desorption from the oxynitride and the oxynitride film composition, respectively. Figure 3 illustrates the cross-section of (a) the tested nMOSFET device and the films evaluated in this study, (b) the gate oxynitride chosen for XPS measurements, (c) the thicker oxynitride prepared for NRA H-measurements, (d) the thicker oxynitride film covered with an ultra thin poly Si layer prepared for TDS measurements.

RESULTS AND DISCUSSION

A. Oxynitride surface layer on nitride/oxide stacks

1. Hydrogen storage layer

A specific thin (<1 nm) oxynitride layer of Si_2N_2O-like composition, formed just below the surface of $Si_3N_4/SiO_2/Si$ stacks during N_2-annealing near 1000°C, was found to have highly efficient H-storage properties [9]. Figure 4 shows NRA H-depth profiles of

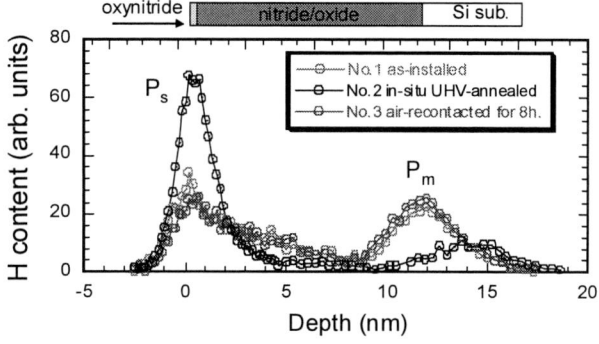

FIGURE 4. NRA H DEPTH PROFILES FOR N2-ANNEALED NITRIDE/OXIDE STACK. THE SPECIMENS DIFFER ONLY IN THE CONDITION IN VACUUM PRIOR TO THE NRA MEASUREMENT.

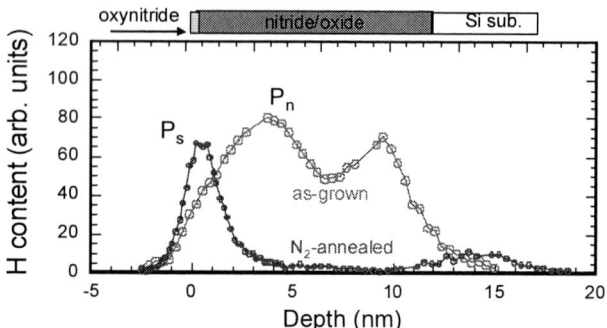

FIGURE 5. NRA H-PROFILES FOR AS-GROWN AND N2-ANNEALED NITRIDE/OXIDE STACK.

such N_2-annealed nitride/oxide stacks. The specimen No. 1-3 originated from the same wafer but received different in-situ treatments in the vacuum chamber before the NRA measurements. Condition No. 1 refers to the sample as it was installed from air. In condition No. 2 the specimen was first heated in UHV at 750°C under a pressure of 10^{-5} mbar for 5 min, without admitting air to the sample before the NRA measurement. In contrast, in condition No.3 the 750°C vacuum-annealed specimen was exposed to air again for 8 hours prior to NRA. The H depth profiles exhibit two distinctly peaked signals. The peak (P_s) centered at a depth <1.0 nm is attributed to H near the Si_3N_4 surface, while the peak (P_m) close to the SiO_2/Si interface corresponds to relocated H, which was initially present in the Si_3N_4 but migrated to the SiO_2/Si interface under the NRA ion irradiation [9]. Figure 4 indicates that after the N_2-anneal H is predominantly present near the surface of the Si_3N_4 films. In case of the specimens that had air contact prior to NRA (No. 1 and 3), this region mainly contains mobile H species that relocate to the SiO_2/Si interface under NRA ion irradiation as indicated by the pronounced P_m peaks. In contrast, the P_s peak in the H-depth profile of the vacuum-annealed sample without air contact (No.2) is narrow, nearly symmetric, and centers slightly beneath the surface. The P_m peak is much smaller. This demonstrates the formation of a ~0.5 nm thin near-surface layer that contains H species at a concentration of ~10^{20}cm^{-3}, which are stable against ion beam-induced relocation. Before air admission, the penetration of this stable H from the storage layer into the bottom oxide is strongly suppressed. These NRA results combined with layer structure and composition analysis demonstrate that a thin, thermally stable, yet air-sensitive Si_2N_2O layer is formed just beneath the native surface oxide layer on N_2-annealed Si_3N_4 [9]. This Si_2N_2O layer can store H species, which are resistant to energetic electron damage because apparently their diffusion into the bottom oxide is strongly suppressed.

The formation of the surface H-storage layer during the N_2-anneal becomes evident by comparing the NRA H depth profiles of as-grown and N_2-annealed nitride/oxide stacks shown in Figure 5. The as-grown nitride film exhibits a broad P_n feature at the center of nitride layer, which indicates that the H distribution in the nitride film is nearly homogeneous. The H signal observed near the oxide/Si interface corresponds to mobile and hence relocated H. After N_2-annealing, the nitride film as well as the bottom oxide are almost completely swept clear of H. Only a sharp peak P_s remains near the surface, which represents a thin layer holding H at a high concentration. We refer to this near-surface layer as 'H-storage layer' [9] and believe that it is highly desirable for MOS stacks because of its ability to retain stable, i.e., relocation-resistant H species.

978-1-4244-5430-3/10 $26.00 © 2010 IEEE   418

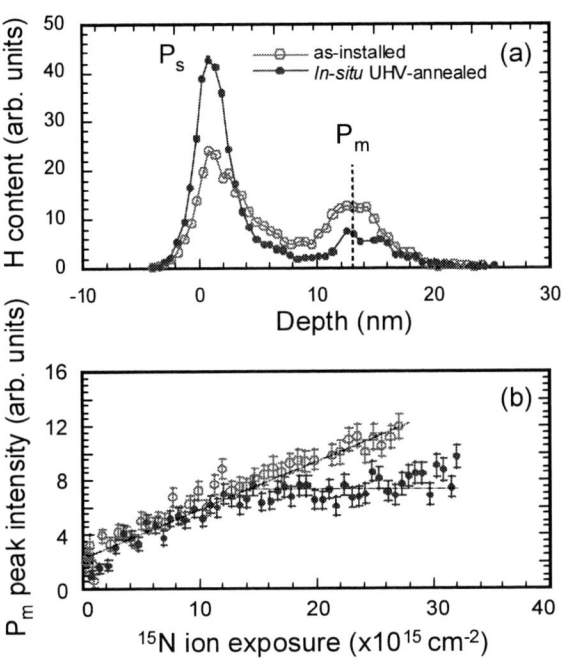

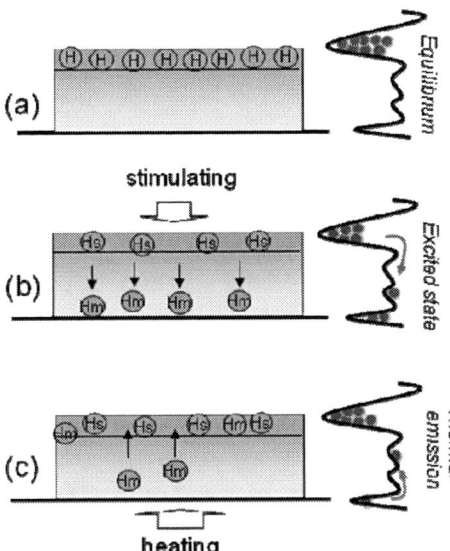

FIGURE 6. THE NRA H-PROFILE (A) AND THE MOBILE H (P$_M$) INTENSITY VARIATION WITH NRA ION EXPOSURE (B). THE DATA ARE OBSERVED FROM THE SAME SAMPLE, MEASURED AS INSTALLED (OPEN CIRCLES) AND AFTER IN SITU VACUUM ANNEAL (FULL CIRCLES), RESPECTIVELY.

2. Mobile and stable hydrogen species

Figure 6 shows the H depth profile and H diffusion kinetics obtained from the N$_2$-annealed nitride/oxide stack with different *in situ* treatments in the NRA vacuum system. Again, the (P$_s$) and (P$_m$) peaks seen in Fig. 6 (a) represent the stable and mobile H species, respectively. Figure 6 (b) provides clear evidence that the mobile H observed near the SiO$_2$/Si interface initially resided in the Si$_2$N$_2$O-like surface layer but migrated to the SiO$_2$/Si interface due to the NRA ion irradiation. Note that the proportion of mobile H (P$_m$)

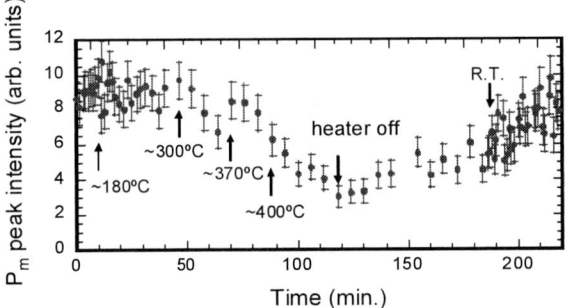

FIGURE 7. SATURATED MOBILE H INTENSITY VARIATION OF THE IN SITU ANNEALED NITRIDE/ OXIDE STACK AT DIFFERENT TEMPERATURES.

FIGURE 8. ILLUSTRATIONS OF THE H TRANSPORT BACK AND FORTH BETWEEN THE H-STORAGE LAYER AND THE OXIDE/SI INTERFACE. THE H MIGRATION BETWEEN BINDING SITES IN THE NEAR-SURFACE H-STORAGE LAYER AND THE OXIDE/SI INTERFACE DEPENDS ON THE TEMPERATURE AND THE STIMULATION BY ENERGETIC CHARGE IRRADIATION. RIGHT SIDE SHOWS CORRESPONDING ENERGY LEVELS OF H IN THE STORAGE LAYER, THE DIELECTRIC INTERIOR, AND THE SI-INTERFACE, RESPECTIVELY.

significantly decreased while the stable H (P$_s$) increased due to *in situ* vacuum heating at 770°C. The specific oxynitride layer apparently contains two kinds of H species (mobile and stable) of different mobility, which can be converted into each other by thermal stimulation or air contact. The important message for the processing optimization is that the annealing is capable of stabilizing the mobile H, implying that an air-degraded H storage layer can be repaired.

Figure 7 shows the variation of the mobile H peak (P$_m$) intensity while the sample is heated under continuous NRA observation. The measurement starts (t=0) from the P$_m$ peak saturation level at room temperature and monitors the H signal at different temperatures. The mobile H intensity localized in the oxide/Si interface decreases between ~370 °C and ~400 °C. After a significant reduction of the H intensity above 400°C the heater was stopped. Note that the reduced P$_m$ H intensity recovers back to its saturation level after the heater is turned off and the sample cools to room temperature. Since the P$_m$ intensity reflects the balance of the H transport back and forth between the H storage layer and the oxide/Si interface, this result means that H is preferably bound in the H storage layer, but migrates into the oxide/Si interface when the sample is irradiated by energetic charges. The heating process transfers H back to the H-storage layer. Based on these results, we propose a model for the reversible H migration as illustrated in Fig. 8. Beside schematic stages of the H-relocation as observed by NRA, (a), (b), and (c), the energy levels of the migrating H species at specific locations within the oxynitride/Si stack are illustrated on the right side. H initially resides in the most stable sites in the oxynitride H-storage layer near the surface. Energetic electrons (e.g., from the NRA ion irradiation) excite H into a mobile state that allows diffusion within the stack interior to trap sites near the Si interface. Heating thermally activates the H escape from the interfacial traps into the mobile diffusion state and hence

978-1-4244-5430-3/10 $26.00 © 2010 IEEE   419

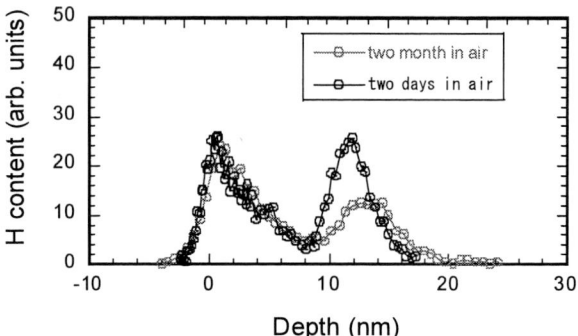

FIGURE 9. NRA H-PROFILES OBSERVED FROM THE SAME NITRIDE/OXIDE SPECIMEN STORED IN AIR FOR TWO DAYS AND TWO MONTHS, RESPECTIVELY.

promotes its return back to the near-surface storage layer. This characteristic is important because oxynitride dielectrics usually endure both electrical and temperature stress.

The thermal reduction of the interfacial trap population as proposed in our model occurs around 370-400°C (Fig. 7), which is well consistent with the thermal stability of H species incorporated near SiO_2/Si interfaces [14]. Recently, T. Grasser et al. presented a triple-well model to explain the H release from Si-H bonds associated to bias temperature instability, where the primary H bonding sites are assumed to be randomly distributed in the amorphous dielectric network [16]. Our NRA results, however, strongly suggest that the mobile H originates mainly from the ultra thin H-storage layer rather than from a homogeneous H distribution in the bulk dielectric.

NRA data show further that the amount of mobile H species is reduced during storage in air atmosphere (Fig. 9). In contrast, no change of the stable H peak height was observed even after 2 months of storage in air. The partial reduction of the mobile H species is possibly due to desorption.

B. H species in the poly-Si/oxynitride interface

The poly-Si/oxynitride interface resembles the above described H-storing surface oxynitride layer on the N_2-annealed nitride/oxide stack in the fact that it apparently contains stable and mobile H species as well. Using the concepts established in section A, we first present evidence for the coexistence of stable and mobile H species at the poly-Si/ oxynitride interface. Implications of the mobile H species for the electrical properties of the respective MOS transistors and possible chemical differences of the H species to the nitride/oxide system are then discussed in sections C and D.

1. Stable hydrogen species

Figure 10 shows TDS spectra of H_2 and SiO species obtained from the as-fabricated oxynitride films (a), (c) formed by plasma-nitridation, and from the same material that underwent deposition of an ultra thin poly-Si film (b), (d). The deposited Si mimics the poly-Si interface in intense reactive contact with the oxynitride as evidenced by TOF-SIMS, where the surface Si is indistinguishable from the underlying oxynitride. The TDS data reveal that H predominantly desorbs below 600 °C (β_1, β_2) for the as-fabricated oxynitride, but an additional γ peak at a temperature above 1100°C appears after poly-Si nucleation. This indicates the ability of the poly-Si interface to retain thermally highly stable H species. Note that this stable H species desorbs combined with SiO species above

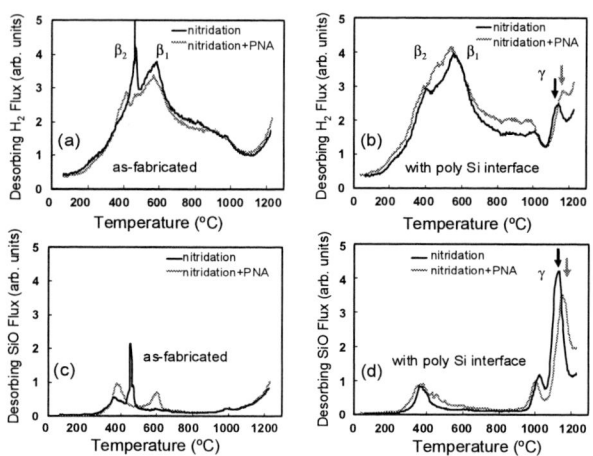

FIGURE 10. TDS OF H2 AND SIO SPECIES OBSERVED FROM 10 NM OXYNITRIDE FILMS IN AS-FABRICATED CONDITION (A), (C), AND AFTER NUCLEATION OF A ULTRA-THIN POLY SI LAYER (B), (D).

1100°C. H bonds in ordinary Si-H, N-H, and O-H units cannot explain such high desorption temperatures. We suggest instead that the stability of these H species is attributable to the high thermal stability of the interfacial layer matrix, rather than to the H bond strength itself.

2. Mobile H species and their volatility on the plasma-nitrided oxynitride surface

Figure 11 compares H depth profiles of the 10 nm plasma-nitrided oxynitride with different residence times in vacuum prior to NRA. These measurements were performed 72 h after plasma nitridation and without PNA, i.e., they reflect the oxynitride condition within the later Q_2-interval (Fig. 1). H profiles after poly-Si interface nucleation are qualitatively very similar. We present the plasma-nitrided bare oxynitride data here in order to discuss the H-behavior during the Q_2-time, which strongly influences the MOS electrical properties (see below). The depth profiles exhibit two peaked signals labeled P_s and P_m, representing stable and mobile H species, respectively. The emergence of the P_m peak intensity during ion exposure (shown in the insert of Fig. 11) indicates clearly that the relocating mobile H initially resides in the oxynitride surface layer. The P_m peak decrease from Fig. 11 (a) to (b) reveals that this

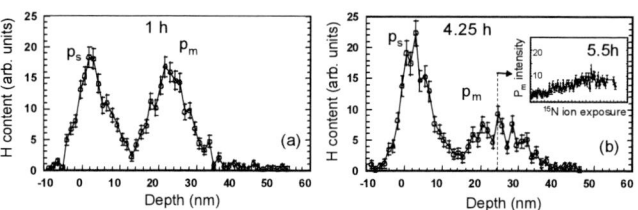

FIGURE 11. NRA H DEPTH PROFILES OF 10 NM OXYNITRIDE FILM DURING THE Q2-INTERVAL. THE VACUUM RESIDENCE TIMES PRIOR TO THE NRA MEASUREMENT WERE 1 H (A) AND 4.25 H (B). THE INSERT SHOWS THE P_M PEAK INTENSITY VARIATION SEEN IN (B) AS FUNCTION OF NRA ION EXPOSURE. THE HORIZONTAL AXIS INDICATES THE DEPTH ALONG A GLANCING INCIDENCE ANGLE OF 65 DEGREES VERSUS THE SURFACE NORMAL.

978-1-4244-5430-3/10 $26.00 © 2010 IEEE   420

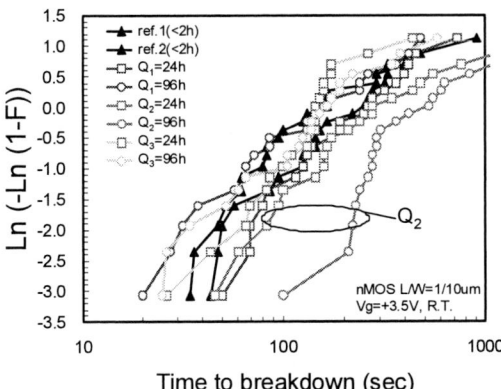

FIGURE 12. WEIBULL PLOTS OF TDDB OBTAINED FROM NMOS. STRESS CONDITION: VG=+3.5 V AT ROOM TEMPERATURE.

mobile H shows a strong tendency to desorb from the oxynitride surface at room temperature. The mobile H in the plasma-nitrided oxynitride dielectric thus behaves notably different than the relocating H species contained in the desirable Si_2N_2O oxynitride layer on Si_3N_4, which apparently were convertible into stable H by vacuum annealing (Figs. 4 and 6).

C. Influence of mobile hydrogen on the electrical properties

1. TDDB improvement during the Q_2-interval

Figure 12 shows Weibull plots of the TDDB data obtained from nMOS devices fabricated with different Q-times (Fig. 1). Obviously, only the Q_2-time impacts significantly on the capacitor life time under constant voltage stress. The TDDB improves markedly as the Q_2-time increases, especially from 24 h to 96 h. No such effects are observed after the respective Q_1 and Q_3-intervals. The Q-time dependence of the N content in the oxynitride films is shown in Figure 13. The N content decreases from 4% to 2.8% during the initial 24 h of Q_2-time, but becomes stable afterwards. It thus appears that the later TDDB improvement is caused by a different factor than the N content reduction. Recall that the NRA H depth profiles in Fig. 11 revealed the tendency of the mobile H species to desorb at room temperature from the plasma-nitrided surface. This strongly suggests that the TDDB improvement after increased Q_2-times is related to a

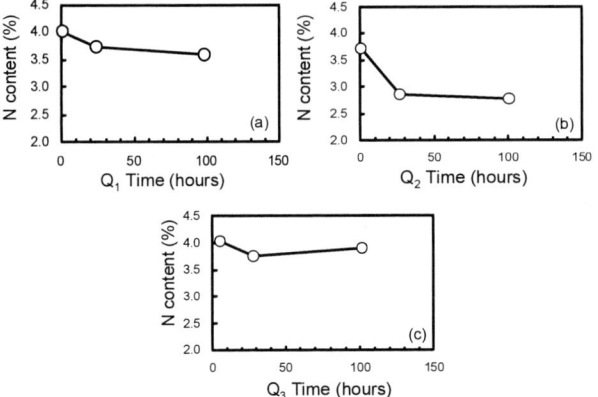

FIGURE 13. VARIATION OF THE N CONTENT REMAINING IN THE OXYNITRIDE FILMS AFTER DIFFERENT Q-TIMES AS MEASURED BY XPS.

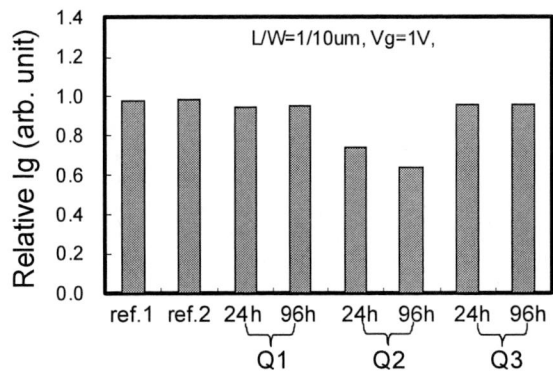

FIGURE 14. THE GATE LEAKAGE CURRENT (IG) COMPARED FOR NMOS WITH DIFFERENT Q-TIMES.

reduction of the mobile H species from the oxynitride surface during the Q_2-interval.

2. Gate leakage reduction and Vt shift with Q-time

Figure 14 shows the gate leakage current compared for different Q-times. Only the Q_2-time influences the gate leakage, and the leak current reduction still continues between 24h and 96h. This observation combined with the above XPS and NRA results implies that N as well as mobile H residing in the oxynitride appears to be involved in the formation of the gate leak path.

Figure 15 compares the Vt and Ion of the nMOSFETs for the different Q-times. Obviously, Vt shift and Ion degradation were only affected during the initial 24 h of Q_2-time; no further change is observed afterwards. This Q_2-time-dependence differs from the TDDB improvement as well as from the Ig reduction and appears to correlate with the N content reduction seen in Fig. 13, regardless of

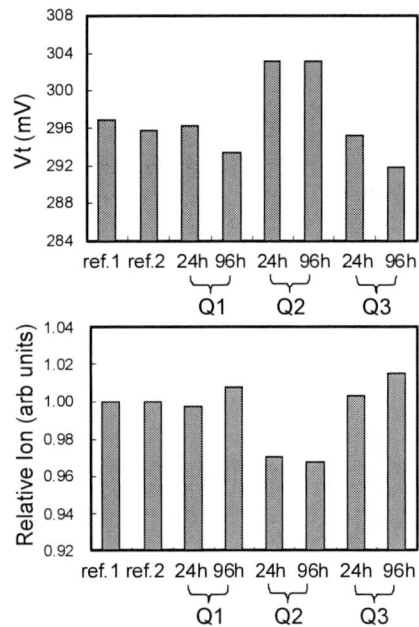

FIGURE 15. COMPARISON OF VT AND ION FOR NMOS WITH DIFFERENT Q-TIMES. THE INITIAL 24H OF Q_2-TIME CAUSE A Vt SHIFT AND A CONCOMITANT ION DECREASE.

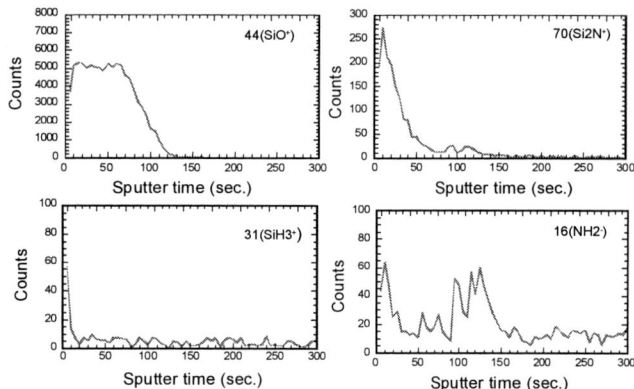

FIGURE 16. COMPOSITION DEPTH PROFILE OF THE PLASMA OXYNITRIDED FILM WITH PNA COVERED BY A POLY-SI INTERFACE LAYER, OBSERVED BY TOF-SIMS COMBINED WITH AR-SPUTTERING.

the ongoing H species desorption. The Vt and Ion shift can be explained by charge generation near the oxide/Si interface. Chen et al. have demonstrated that N-incorporation by plasma nitridation induces a negative Vt shift in nMOS due to positive fixed-charge buildup [17]. Our observations agree with this interpretation, as the notable Vt increase during the Q_2-interval is probably caused by the reduction of the N-induced positive charge as the N content decreases. The reduced Ion (Fig. 15) is a direct consequence of the increased Vt value, which reflects a positively shifted I-V curve [18].

D. **Nature of the mobile H species**

In order understand the possible chemical nature of the mobile H species contained in the poly-Si/oxynitride interface, composition profiles of the PNA treated plasma oxynitride with a poly-Si interface layer were analyzed by TOF-SIMS combined with Ar sputtering. Three typical depth profile shapes were observed (Fig. 16). The m/e=44 (SiO^+) species profile, a square-like curve, reflects the matrix structure of the base oxide film/Si substrate. The m/e=70 (Si_2N^+) and m/e=31 (SiH_3^+) profiles have mountain-like curves, revealing that N and H are localized near the surface. Only the m/e=16 (NH_2^-) species profile exhibits a weak but clear interface peak, which is accompanied by a small near-surface signal. Since NRA ion irradiation causes a secondary electron-induced H migration into the Si interface region, Ar sputtering can be expected to induce a similar penetration of mobile species into the Si substrate interface. Thus the NH_2 signal in the lower oxide/Si interface region probably reflects N-H species that diffused from the top interface during the TOF-SIMS analysis. We therefore suggest in turn that the mobile H species contained in the poly-Si/oxynitride interface may, at least partially, be related to N-containing species, implying that also the interfacial nitrogen chemistry should be controlled.

Realizing that the oxynitride/Si interface contains mobile and stable H species is important. These H species can be released from the layer under device operating stress, then diffuse toward the oxynitride/Si substrate interface as illustrated in Fig. 17. The diffusing H species may react with the SiO_2 network, leaving leakage paths across the dielectric along their track, which eventually result in breakdown. This model is consistent with Buchanan et al., who demonstrated that H release and transport induced by hot-electrons in Al gate MOS stacks results in H buildup at the substrate interface and generates electrically active defects [19]. Their article supports our interpretation that TDDB correlates with mobile H species.

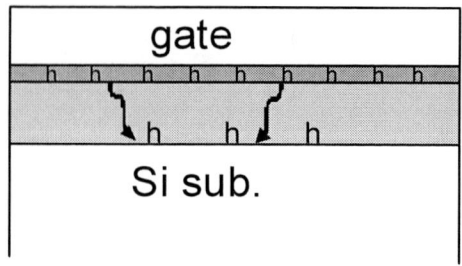

FIGURE 17. SCHEMATIC OF HYDROGEN DIFFUSION FROM THE H-STORAGE LAYER EXISTED IN THE POLY SI/ OXYNITRIDE INTERFACE, TOWARD THE BOTTOM SI INTERFACE. THE DIFFUSION TRACK RESULTS IN THE FORMATION OF LEAKAGE PATH IN THE OXYNITRIDE DIELECTRIC.

According to our experimental observation of the H diffusion and with reference to the percolation model explaining gate dielectric breakdown, we thus suggest that building a poly-Si/dielectric interface without mobile H species is the essential key to achieve high reliability MOS stacks with oxynitride gate dielectrics.

CONCLUSIONS

Hydrogen residing at the interface between poly-Si and oxynitride was investigated for plasma-nitrided gate oxide. We found that this interface contains mobile and stable H species, and the mobile H species tends to desorb from the oxynitride surface. The TDDB improvement after increased air storage following nitridation is due to the reduction of mobile H species. Optimizing the poly-Si/dielectric interface by eliminating the mobile H species is suggested to improve the reliability of devices applying oxynitride dielectrics.

ACKNOWLEDGMENTS

The author, Z. Liu, wishes to thank Y. Taniguchi, T. Ishiyama, T. Ide of NEC Electronics Corporation for their warm support. Many thanks are to N. Hirashita and K. Maejima of ESCO, Ltd. for helpful TDS measurements. Thanks are given to K. Arima, S. Asada of NEC Fabserve Ltd. for wafer preparation. The authors are grateful for assistance in the MALT tandem accelerator operation by H. Matsuzaki and C. Nakano at the University of Tokyo.

This work was partially supported by New Energy and Industrial Technology Development Organization of Japan (NEDO).

REFERENCES

[1] A. Toriumi, H. Satake, 'The boundary between hard and soft breakdown in ultra thin silicon dioxide films', Materials Research Society Symposium Proceedings, Boston, Vol. **592**, 1999, pp. 323-9.

[2] G. J. Gerardi, E. H. Poindexter, P. J. Caplan, M. Harmatz, W. R. Buchwald, and N. M. Johnson, 'Generation of P_b Centers by Negative Corona Stress Across the Si/SiO₂ Interface' J. Electrochem. Soc. **136**, 1989, pp. 2609.

[3] D. K. Schrodera and Jeff A. Babcock, 'Negative bias temperature instability: Road to cross in deep submicron silicon semiconductor manufacturing', J. Appl. Phys. **94**, 2003, pp. 1.

[4] C. E. Blat, E. H. Nicollian, and E. H. Poindexter, 'Mechanism of negative-bias-temperature instability', J. Appl. Phys. **69**, 1991, pp. 1712.

[5] Z. Liu, S. Fujieda, K. Terashima, M. Wilde and K. Fukutani, 'Hydrogen Redistribution Induced by Negative-Bias-Temperature Stress in Metal-Oxide-Silicon Diodes', Appl. Phys. Lett. **81**, 2002, pp. 2397.

[6] Y. Nissan-Cohen and T. Gorczyca,, 'The Effect of Hydrogen on Trap Generation, Positive Charge Trapping, and Time-Dependent Dielectric Breakdown of Gate Oxides', IEEE Electron Dev. Letters, **9**, 1988, pp. 287.

[7] L. Dori, J. H. Stathis and J. A. Tornello 'Moderate-temperature anneal of 7-nm thermal SiO_2 in O_2- and H_2O-free atmosphere: Effects on Si-SiO_2 interface-trap distribution' J. Appl. Phys. **70**, 1991, pp. 1510.

[8] Z. Liu, S. Fujieda, F. Hayashi, M. Shimizu, M. Nakata, H. Ishigaki, M. Wilde, and K. Fukutani, 'Influence of hydrogen permeability of liner nitride film on Program/Erase endurance of split-gate type flash EEPROMs', IEEE proc. of IRPS, **2007**, pp. 190.

[9] Z. Liu, S. Ito, M. Wilde, K. Fukutani, I. Hirozawa, and T. Koganezawa, 'A hydrogen storage layer on the surface of silicon nitride films', Appl. Phys. Lett. **92**, 2008, pp. 192115.

[10] Z. Liu, S. Ito, T. Ide, M. Nakata, H. Ishigaki, M. Makabe, M. Wilde, K. Fukutani, H. Mitoh and Y. Kamigaki, 'Indications for an ideal interface structure of oxynitride tunnel dielectrics', IEEE Proc. of IRPS, **2009**, pp. 902.

[11] P. E. Nicollian, A. T. Kfishnan, C. A. Chancellor, R. B. Khamankar, S. Chakravarthi, C. Bowen, and V. K. Reddy, 'The current understanding of the trap generation mechanisms that lead to the power law model for gate dielectric breakdown,' IEEE Proc. of IRPS, **2007**, pp. 197.

[12] D. J. DiMaria, 'Impact ionization trap creation, degradation, and breakdown in silicon dioxide films on silicon', J. Appl. Phys. **73**, 1993, pp. 3367.

[13] D. J. DiMaria and J. W. Stasiak, 'Trap Creation in Silicon Dioxide Produced by Hot Electrons', J. Appl. Phys. **65**, 1989, pp. 2342.

[14] M. Wilde, M. Matsumoto, K. Fukutani, Z. Liu, K. Ando, Y. Kawashima and S. Fujieda, 'Influence of H_2-annealing on the hydrogen distribution near SiO_2/Si.(100) interfaces revealed by in situ nuclear reaction analysis', J. Appl. Phys., **92**, 2001, pp. 4320.

[15] N. Hirashita, M. Kinoshita, I. Aikawa, and T. Ajioka 'Effects of surface hydrogen on the air oxidation at room temperature of HF-treated Si (100) surfaces', Appl. Phys. Lett., **56**, 1990, pp.451.

[16] T. Grasser, B. Kaczer and W. Goes, 'An Energy-level Perspective of Bias Temperature Instability', IEEE Proc. of IRPS, **2008**, pp. 28.

[17] Chien-Hao Chen, Yean-Kuen Fang, Shyh-Fann Ting, Wen-Tse Hsieh, Chih-Wei Yang, Tzu-Hsuan Hsu, Mo-Chiun Yu, Tze-Liang Lee, Shih-Chang Chen, Chen-Hua Yu, and Mong-Song Liang 'Downscaling Limit of Equivalent Oxide Thickness in Formation of Ultrathin Gate Dielectric by Thermal-Enhanced Remote Plasma Nitridation', IEEE Transactions on Electron Dev., **49**, 2002, pp. 840.

[18] S. M. Sze, Physics of semiconductor devices, 2nd ed., New York, Wiley, 1981, pp. 475.

[19] D. A. Buchanan, A. D. Marwick, and D. J. DiMaria, L. Dori, 'Hot-electron-induced hydrogen redistribution and defect generation in metal-oxide-semiconductor capacitors', J. Appl. Phys. **76**, 1994, pp. 3595.

Novel TDDB Mechanism for p-FET accelerated by Hydrogen from HfSiON Film

Izumi Hirano, Koichi Kato, Yasushi Nakasaki, Shigeto Fukatsu and Yuichiro Mitani
Advanced LSI Technology Laboratory, Corporate R&D Center,
Toshiba Corporation
8, Shinsugita-cho, Isogo-ku, Yokohama 235-8522, Japan
phone: (+81) -(45)-776-5950, izumi3.hirano@toshiba.co.jp

Masakazu Goto, Seiji Inumiya, Katsuyuki Sekine and Motoyuki Sato*
Semiconductor Company, Toshiba Corporation
8, Shinsugita-cho, Isogo-ku, Yokohama 235-8522, Japan
*present affiliation selete

Abstract—**Time Dependent Dielectric Breakdown (TDDB) in p-FETs with HfSiON/SiO₂ gate stacks under negative bias stress has been studied. It is shown that the shape parameter of Weibull distribution of T_{bd}, β, is very small value independent of gate electrode materials. This small β seems to arise from the interface layer (I.L.) breakdown. Further experimental result reveals the existence of additional interface layer degradation mechanisms due to hydrogen in HfSiON. The reduction of hydrogen amount in high-k dielectrics leads to the longer-term reliability in metal-gate /high-k gate stacks.**

Keywords-component; High-k dielectrics, HfSiON, Reliability, TDDB, p-FET, Hydrogen

I. INTRODUCTION

Time dependent dielectric breakdown (TDDB) is one of the major concerns before reaching the practical use of metal-gate/high-k gate stacks and these mechanisms have been discussed [1-6]. From the reliability assurance viewpoints, the gate stacks with much larger Weibull slope, β, for time to breakdown (T_{bd}) is essential because the smaller β leads to shorter TDDB lifetime. We have reported that the size of defects created during stress, which strongly influences the β, is determined by carrier balance flowing into the gate stack in n-FETs [1].

However, the breakdown behavior for n-FETs and p-FETs in MG/HK are quite different and especially small β was observed in p-FET, as shown in Fig.1. It is important to elucidate the origin of the small β in MG/HK devices in p-FET.

In this paper, we report on Weibull distribution of T_{bd} for metal-gate/high-k devices with various gate electrode materials and with different interface layer thickness. We have found that the breakdown of the p-FETs is attributed to the breakdown of interfacial SiO₂ layer. Furthermore, another hydrogen atoms induced degradation seems to exist, which differ from hydrogen release model [7-9].

II. EXPERIMENTAL AND THEORY

The devices used in this study were n-/p-FETs with poly-Si gate/HfSiON and metal-gate/HfSiON stacks. The gate dielectrics HfSiON (Hf/(Hf+Si)=50%, [N]=20 at. %) were fabricated by MOCVD method. The areas of gate electrode were $10^{-7} \sim 10^{-8}$ cm². The seven kinds of gate electrode materials were prepared, as show in Table 1 (A) ~ (F). The gate dielectrics HfSiON below these electrodes were fabricated by the same process. In addition, the four samples of poly-Si/HfSiON/SiO₂ stacks with different I.L. thickness and ordinary SiON were prepared, as shown in Table 1 (G) ~ (K).

TABLE I.   SAMPLES USED IN THIS WORK. THE THICKNESS OF DIELECTRICS WAS THE TARGET THICKNESS AT FILM DEPOSITION. (*TA-RICH TAC **TA-RICH TAC)

sample	gate electrode	thickness of dielectrics	
		HfSiON	I.L.
(A)	poly-Si		
(B)	TiN		
(C)	*TaCx-1 3nm	2.3 ~ 2.5nm	SiO₂ 0.8nm
(D)	*TaCx-1 5nm		
(E)	*TaCx-1 10nm		
(F)	**TaCx-2 10nm		
(G)			SiO₂ 1.0nm
(H)	poly-Si	4.0nm	SiO₂ 2.0nm
(I)			4.0nm
(J)			SiON 3.0nm
(K)		—	SiON 3.0nm

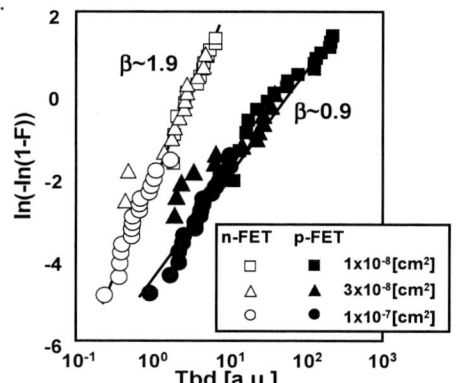

Fig.1 Area scaled Weibull plots of T_{bd} for n-FET and p-FET with TaC/HfSiON (sample F) under inversion stress.

978-1-4244-5430-3/10 $26.00 © 2010 IEEE

The TDDB measurements were performed under constant voltage stress at 125°C. The electron and hole currents injected during TDDB tests were independently measured by using carrier separation technique. Interface states (D_{it}) were measured by charge-pumping method.

Hydrogen atom reactions with hydrogen terminated interface defects were examined with first principles calculations.

III. RESULTS AND DISCUSSION

A. Weibull distribution of HfSiON/SiO₂ stacks

As we reported previously [10, 11], the breakdown of the n-FETs is triggered by the defect creation in HfSiON film, and the size of the defects is dependent on carrier balance flowing into the stack, leading to a strong correlation between Weibull β and gate electrode material (Fig.2(a)). In p-FETs, however, Weibull β are independent of gate electrode materials and almost constant around the value of 0.9 (Fig.2 (b)) although the balance of injected carrier, $J_{hole}/J_{electron}$, is largely varied with the gate electrode materials, as shown in Fig.3.

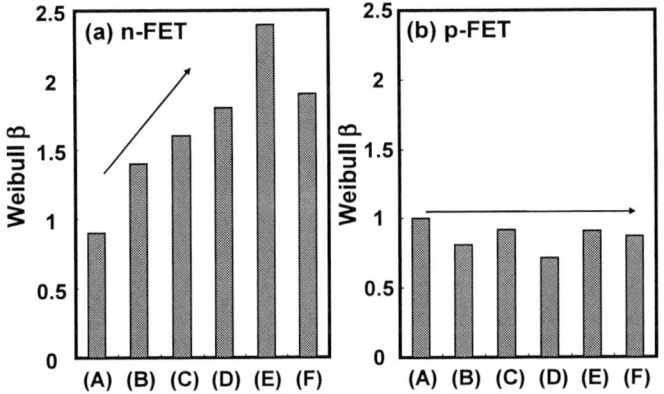

Fig.2 Gate electrode dependences of Weibull β in (a) n-FETs [1] and (b) p-FETs under inversion bias stress.

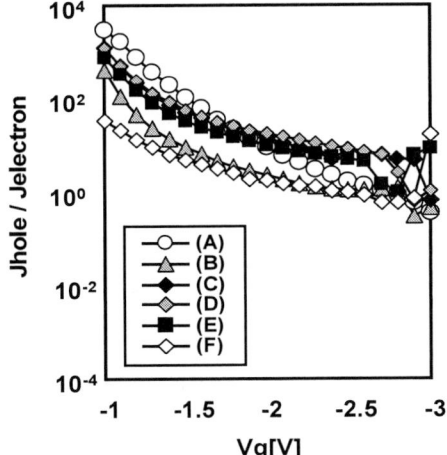

Fig.3 $J_{hole}/J_{electron}$ for each sample in p-FET. J_{hole} was measured by J_{sd} and $J_{electron}$ was measured by J_{sub}.

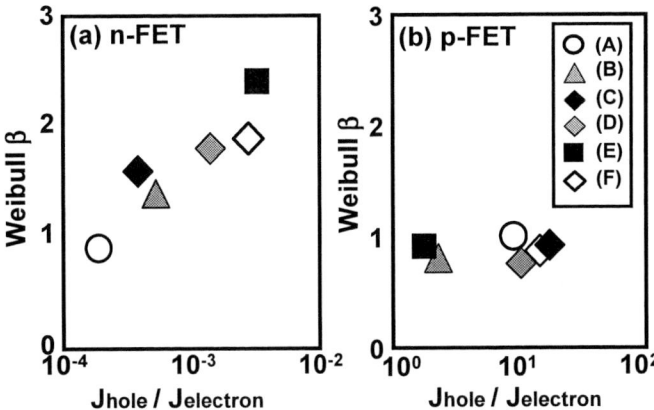

Fig. 4 Relation between Weibull β of T_{bd} and $J_{hole}/J_{electron}$ in (a) n-FET [10] and (b) p-FET.

Fig.4 shows the correlation between $J_{hole}/J_{electron}$ and Weibull β in n-FETs and p-FETs. Weibull β in p-FETs are independent of $J_{hole}/J_{electron}$ although those in n-FETs are strongly depends on $J_{hole}/J_{electron}$. These results reveal that the defect creation in HfSiON film does not dominate over the breakdown in p-FETs.

To clarify the breakdown mechanisms in detail, we focused on the dependence of the T_{bd} distribution of interfacial SiO₂ layer thickness. Fig.5 shows the Weibull β for n-FETs and p-FETs with HfSiON/SiO₂ stacks as a function of the thickness of interfacial SiO₂ layer. The Weibull β for ordinary SiO₂ stacks with various SiO₂ thicknesses, which were previously reported in [12], are also plotted. It is found that the Weibull β for p-FETs with HfSiON/SiO₂ stacks are in good agreement with the trend for ordinary SiO₂ stacks, although the Weibull β for n-FETs with HfSiON/SiO₂ stacks show a tendency different from ordinary SiO₂ stacks. Therefore, it may suggest that the breakdown for the p-FETs is arises from the breakdown of interfacial SiO₂ layer.

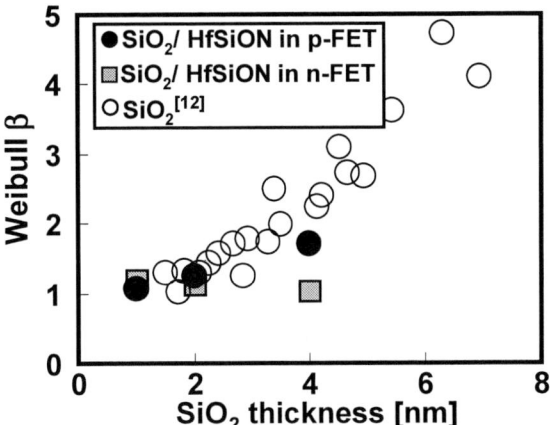

Fig.5 SiO₂ thickness dependence of Weibull β in p-FET with HfSiON/SiO₂ stacks. The Weibull β for conventional SiO₂ stacks with various thicknesses, which were previously reported [12], are also plotted.

B. Degradation mechanisms in p-FET with HfSiON/SiO₂

In breakdown of thin SiO₂, hydrogen release (HR) model have been widely accepted [7-9]. According to the model, T_{bd} can be described by following equation. i.e. power law.

$$T_{bd} = A \cdot Vg^{\gamma} \qquad (1)$$

Where Vg is applied voltage and γ is acceleration factor. Recently, extended HR model, Multi-vibrational hydrogen release (MVHR) model, have been reported [13-16].

From this model, T_{bd} can be given by following equation.

$$T_{bd} \propto Vg^{-\left[n(fin)\frac{Eth}{n\sigma}+n(lg)\frac{Eth}{n\sigma}\right]} \qquad (2)$$

where

fin : inelastic part of the current

Eth : desorption energy

σ : stretching mode energy of the bond

n : number of quanta

n(fin) and *n(Ig)* are power law exponents of *fin* and *Ig* versus *Vg*.

In this work, we used the reported value [3] for *n(fin)*, Eth, σ and *n* while *n(Ig)* was obtained from our experimental data.

Fig.6 shows the relation between acceleration factor γ obtained by experimental data and that calculated by MVHR model for the sample (A) ~ (H) and (K). γ of HfSiON/SiO₂ stacks (sample(A)~(H)) obtained from experimental results are smaller than that calculated by MVHR model, although γ of SiON (sample (K)) obtained from experimental data is in good agreement with that of MVHR model. These results indicate that additional degradation of SiO₂ (or SiON) with small voltage dependence occurs due to HfSiON stacked on SiO₂.

Next, we investigated the influence of HfSiON on the interface SiO₂ layer degradation. Diffused nitrogen from HfSiON into SiO₂ interface layer may enhance the interface states (D$_{it}$) generation of high-k/SiO₂ stacks [17]. In order to eliminate the influence of nitrogen diffusion from HfSiON, we use the HfSiON/SiON and SiON. Here, both SiON are fabricated by same process and N concentration is around 20 at %. Initial interface states of both samples are almost equal, as shown in Fig.7(a). After electrical stress, the interface degradation of HfSiON/SiON stack (sample (J)) is higher than that of SiON (sample (K)) as shown in Fig 7(b). Furthermore, the total fluence of the holes to dielectric breakdown, Q$_p$, in HfSiON/SiON stack (sample (J)) is smaller than that in SiON (sample (K)) (Fig.8). From these results, the deposited HfSiON seems to enhance the degradation of SiO₂ (or SiON) interface layer.

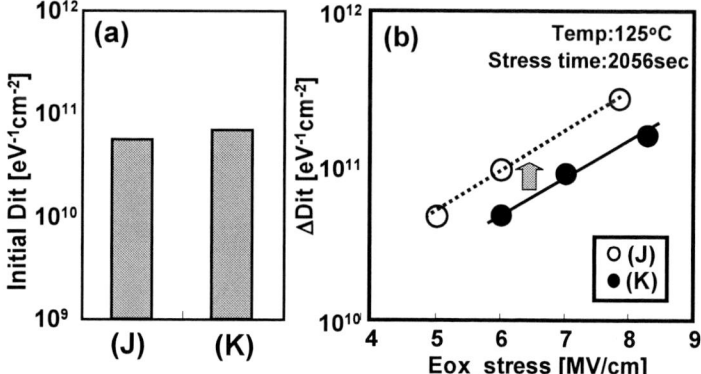

Fig.7 (a) Dit for SiON (sample (K)) and HfSiON/SiON stacks (sample (J)) before stress. (b) ΔDit for SiON and HfSiON/SiON stack after stress. ΔDit for HfSiON/SiON stack is higher than that for SiON.

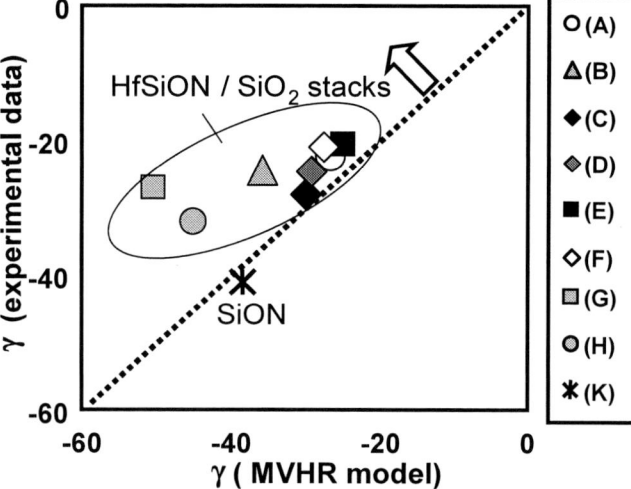

Fig.6 Relation between acceleration factor γ obtained by experiments and these calculated by MVHR model [3]. γ of SiO₂/HfSiON stacks obtained from experiments are smaller than these obtained from MVHR model.

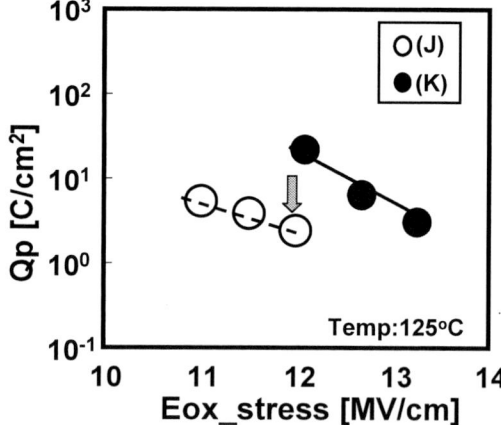

Fig.8 Eox_stress dependence of Qp for SiON (sample (K)) and HfSiON/SiON stack (sample (J)).Qp for HfSiON/SiON stacks is smaller than that for SiON.

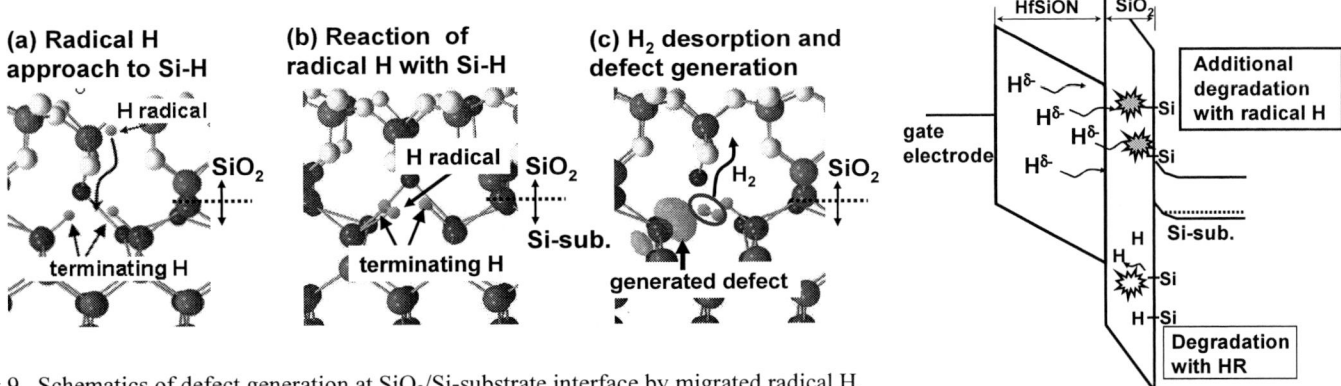

(a) Radical H approach to Si-H

H radical
terminating H
SiO$_2$

(b) Reaction of radical H with Si-H

H radical
terminating H
SiO$_2$
Si-sub.

(c) H$_2$ desorption and defect generation

H$_2$
generated defect
SiO$_2$

Fig.9   Schematics of defect generation at SiO$_2$/Si-substrate interface by migrated radical H.
(a)   Radical H migrates to near Si-H bond.
(b)   Radical H reacts with Si-H.
(c)   Formation and desorption of H$_2$. Defect (oxygen vacancy) generate at SiO$_2$/Si-substrate
       interface. These reaction takes place with little energy barrier.

Fig.10   Schematic diagram of additional degradation in SiO$_2$ I.L. by migrated radical H from HfSiON for p-FET.

From first principle calculations, as schematically shown in Fig.9, when radical hydrogen migrates near Si-H bond, it reacts with Si-H to form H$_2$. As a result, defects (oxygen vacancy) generate near SiO$_2$/Si-substrate interface. Furthermore, it was reported that interstitial hydrogen in HK is slightly negatively charged [18]. Under inversion stress in p-FETs, negatively-charged hydrogen in HfSiON migrates from HfSiON to lower SiO$_2$/Si substrate interface by electrical field. Migrated H breaks the Si-H bond and, as a result, degradation of SiO$_2$ interface layer will be enhanced. Migration of hydrogen atoms can be key factor of additional degradation (Fig.10).

Finally, in order to demonstrate the effects of hydrogen in HfSiON on degradation of MG/HK in p-FET, we investigate the degradation of HfSiON/SiO$_2$ stack with different hydrogen content, namely, high-H HfSiON and low-H HfSiON. Hydrogen content was changed by using different source gases of HfSiON. Fig.11 shows the SIMS depth profile of hydrogen in both samples. Hydrogen in low-H HfSiON is approximately an order of magnitude lower than that in high-H HfSiON.

Here, we used the p-FETs with 1nm SiO$_2$ I.L. The thicknesses of high-H HfSiON are 4nm and 6nm, and the thickness of low-H HfSiON is 4nm. Fig.12 shows the comparison of interface degradation (ΔDit) between high-H HfSiON and low-H HfSiON under negative bias stress (6MV/cm). ΔDit of low-H HfSiON is smaller than that of high-H HfSiON. Fig.13 shows the T_{bd} of low-H HfSiON and high-H HfSiON under negative bias stress (9MV/cm). T_{bd} of low-H HfSiON was longer than that of high-H HfSiON though the β for high-H HfSiON and low-H HfSiON was almost the same value as shown in inset of fig.13. These results indicate that the degradation of SiO$_2$ interface layer was substantially suppressed and the lifetime of p-FET has been dramatically improved by reduction of hydrogen content in HfSiON. Furthermore, fig.14 shows the acceleration factor γ obtained from experimental results of high-H HfSiON and low-H HfSiON. In these samples, γ calculated by MVHR model is approximately -50 as shown in hatched region in fig.14. A closer fit between γ (experimental) to γ (MVHR model) is obtained by lowering hydrogen amount in HfSiON, as shown in fig.14.

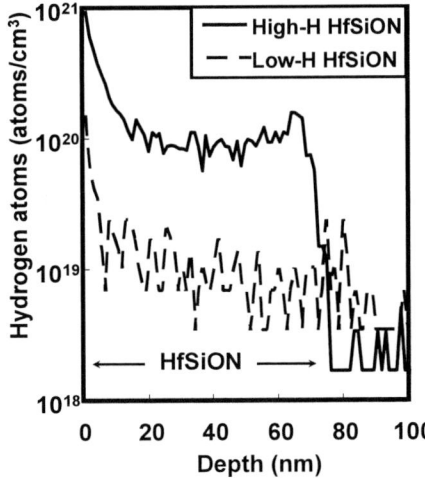

Fig.11   SIMS depth profile of hydrogen in high-H HfSiON and low-H HfSiON. H in low-H HfSiON is around one order of magnitude lower than that in high-H HfSiON.

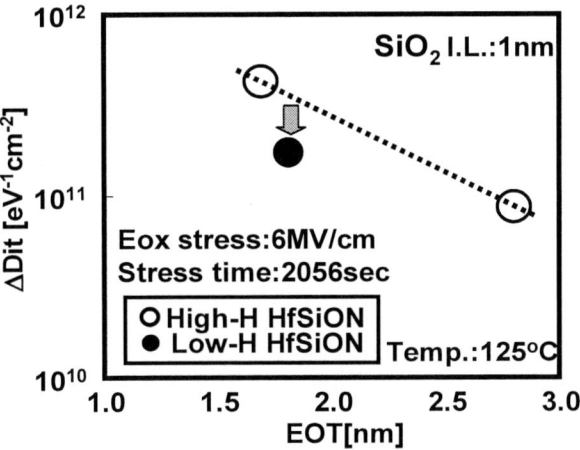

Fig.12 ΔDit for high-H HfSiON sample and low-H HfSiON sample. ΔDit of low-H HfSiON was lower than that of high-H HfSiON.

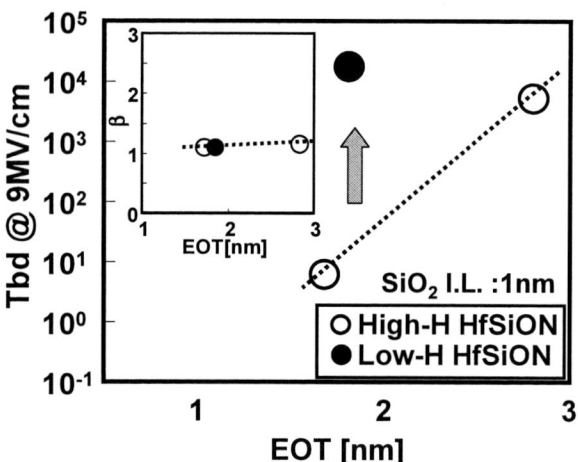

Fig.13 Tbd of high-H HfSiON and low-H HfSiON. Tbd of low-H HfSiON was longer than that of high-H HfSiON. Inset shows the EOT dependence of β. β for high-H HfSiON and low-H HfSiON was almost same value.

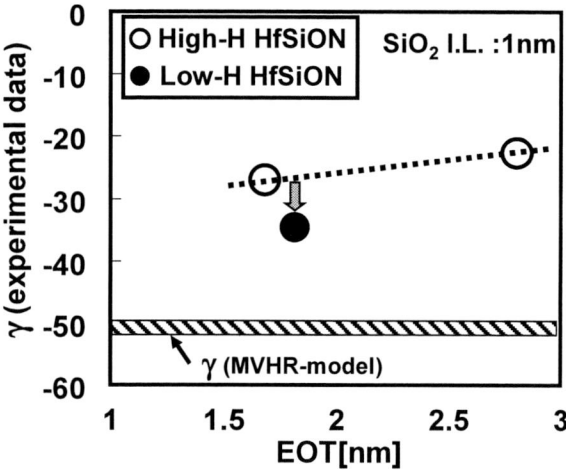

Fig.14 Acceleration factor γ from experimental of high-H HfSiON and low-H HfSiON. Hatched region shows the calculated γ by MVHR model. γ of low-H HfSiON was nearer that from MVHR-model than that of high-H HfSiON.

IV. CONCLUSIONS

We investigated the Weibull distribution of T_{bd} for MG/HfSiON in p-FETs. It was found that Weibull β in p-FETs is hardly influenced by gate electrode materials and injected carrier balance, suggesting another important mechanism rules over. The detailed experimental analyses revealed that the SiO_2 interface layer controls the T_{bd} distribution of $HfSiON/SiO_2$ stacks in p-FETs. The additional interface layer degradation mechanisms related to hydrogen in HfSiON exists. The controlled hydrogen in high-k dielectrics leads to better long-term reliability in metal-gate/ high-k gate stacks.

ACKNOWLEDGMENT

The authors would like to thank Drs. S.Kawanaka, A. Azuma, K.Nakajima, K.Yoshikawa, K.Nagatomo, A.Masada, M. Miyata and A. Nishiyama for useful discussions and continuous supports.

REFERENCES

[1] K. Torii, H. Kitajima, T. Arikado, K. Shiraishi, S. Miyazaki, K.Yamabe, M. Boero, T. Chikyow, K Yamada" Physical model of BTI,TDDB and SILC in HfO2- based high-k gate dielectrics" *International Electron Devices Meeting* 2005 p.129

[2] R. Degraeve, T. Kauerauf, M. Cho, M.Zahid, L-A. Ragnarsson, D. P. Brunco, B. Kaczer, Ph. Roussel, S De Gendt, and G.Groeseneken, "Degradatiom and breakdown of 0.9nm EOT SiO2/ALD HfO2/metal gate stacks inder positive Constant Voltage Stress" *International Electron Devices Meeting* 2005 p.408

[3] T. Yamaguchi, I.Hirano, R.Iijima, K.Sekine, M. Takayanagi, K. Eguchi, Y. Mitani and N. Fukushima, "Thermochemical understanding of dielectric breakdown in HfSiON with current acceleration" *International Reliability Physics Symposium* 2005 p.67

[4] K. Okada, H. Ota, T.Nabatame, and A.Toriumi, "Dielectric breakdown in High-k gate dielectrics: mechanism and lifetime assessment", p.36 International Reliability Physics Symposium 2007

[5] I. Hirano, T.Yamaguchi, Y.Nakasaki, K. Sekine, and Y. Mitani "Influence of traps and carriers on reliability in HfSiON/SiO2 stacks" *Transaction of Device and Materials Reliability* 2009 p.163

[6] T.Nigam, A.Kerber and P. Peumans, "Accurate model for time-dependent dielectric breakdown of high-k metal gate stacks" *International Reliability Physics Symposium* 2009 p.523-30

[7] J. Sune and E.Y.Wu "Hydrogen-Release Mechanisms in the Breakdown of Thin SiO2 Films" *Phys. Rev. Let.* vol.92, no.8 (2004) 087601.

[8] V. Huard, M.Denais and F. Monsieur "Hydrogen release and defect generation rate in ultra-thin oxides" *International Integrated Reliability Workshop Final Report* 2004 p.4

[9] K.Ohgata, M. Ogasawara, K.Shiga, S.Tsujikawa E.Murakami, H.Kato, H.Umeda and K.Kubota "Universality of Power-Law Voltage dependence for TDDB Lifetime in Thin Gate Oxide PMOSFETs" *International Reliability Physics Symposium* 2005 p.372

[10] I. Hirano, Y. Nakasaki, S. Fukatsu, A. Masada, M. Goto, K. Nagatomo, S. Inumiya, K. Sekine and Y. Mitani "Impact of Metal Gate Electrode on Weibull Distribution of TDDB in HfSiON gate dielectrics" *International Reliability Physics Symposium* 2009 p.355

[11] Y. Nakasaki, I. Hirano K. Kato and Y. Mitani, "Atomic-scale theory on current-assisted thermochemical degradation mode and its filed acceleration via charge trapping of O vacancy in HfSiO4" *Microelectron Eng*, vol.86, p1901-1904 (2009)

[12] E. Y. Wu , J. Sune, W. Lai, A. Vayshenker, E. Nowak and D. Harmon,"Critical reliability challenges in scaling SiO2-based dielectric to its limit" *Microelectronics Reliability* 43 (2003) p.1175-1184

[13] G. Ribes, M. Rafik, D.Barge, S. Kalpat, M. Denais, V.Huard and D.Roy, "New Insight on Percolation Theory and The Origin of Oxide Breakdown Thickness and Process Deposition Dependence" International Reliability Physics Symposium 2007 p.578

[14] G. Ribes, S. Bruyere, D. Roy, M. Denais, J-M. Roux, C. Partharathy, V.Huard and G. Ghibaudo "New Extensive MVHR Breakdown Models" *International Reliability Physics Symposium* 2006 p.621

[15] G. Ribes, S. Bruyere, M. Denais, F. Monsieur, V.Huard, D.Roy, and G.Ghibaudo "Multi-vibrational hydrogen release: Physical origin of Tbd, Qbd power-law voltage dependence of oxide breakdown in ultra-thib gate oxides"*Microelectronics Reliability* vol.45, no.12, p1842 (2005)

[16] M. Rafik, G.Ribes, D.Roy, and G. Ghibaudo, "New understanding on the breakdown of high-k dielectric stacks using multi-vibrational hydrogen release model" *International Reliability Physics Symposium* 2007 p.44-48

978-1-4244-5430-3/10 $26.00 © 2010 IEEE

[17] X. Garros, M. Casse, M. Rafk, C. Fenouillet-Beranger, G. Reimbold, F. Martin, C. Wiemer and F. Boulanger, "Process dependence of BTI reliability in advanced HK MG stacks" *Microelectronics Reliability* (2009) p.982-988

[18] J. F. Kang, H. Y. Yu, C. Ren, M.-F.Li, D. S. H.Chan, H. Hu, and H. F. Lim and W.D. Wang. D. Gui, and D, -L. Kwong, "Thermal stability of nitorgen incorporated in HfNxOy gate dielectrics prepared by reactive sputtering" *Appl. Phys.Lett.* vol.84 no.9 (2004) p.1588

Buckling, Wrinkling and Debonding in Thin Film Systems

S. Goyal, K. Srinivasan, G. Subbarayan *and T. Siegmund
School of Mechanical Engineering, Purdue University
West Lafayette, IN 47907
Email: ganeshs@purdue.edu

February 26, 2010

Abstract—Thin films bonded to substrates commonly occur in semiconductor dielectric stacks. In these systems, many times, the mismatch in the coefficient of thermal expansion between the films and the substrate result in significant compressive stresses during processing. These compressive stresses may lead to instabilities such as buckling or wrinkling, possibly resulting in debonding of the films. In general, at the present time, the mechanics of buckling and wrinkling leading to debonding in thin film systems are not well understood. A further significant challenge to modeling these systems is that the conditions under which the instabilities occur depend on possible plastic deformation in metal films (in turn dictated by processing temperatures) as well as the presence of geometric features such as vias in the film stack. In this paper, we review analytical derivations of conditions under which buckling and wrinkling induced debond occur in thin film systems. We describe the film interface and (where appropriate) the compliant substrate using a bi-linear Cohesive-Zone fracture model. We utilize the developed theory to estimate interfacial fracture toughness of weakly bonded thin film systems through a newly proposed non-contact, thermally driven, patterned buckling delamination test. The proposed test does not need a weakened region of the interface to initiate crack, nor does it require mechanical loading; it relies on inducing a compressive stress due to heating of the film on a thick silicon substrate. The test is demonstrated on a model system consisting of Al/Su-8/Si stack. Finally, we review wrinkling of thin film systems, and analyze the impact of pattern features on the propensity of thin films to wrinkle sooner than predicted for blanket films.

Keywords: buckling; wrinkling; debond; energy minimization; thin film

---

*Corresponding author

1 Introduction

Thin films are commonly used in several applications including semiconductor devices, magnetic storage media, and surface coatings [1]. In the currently popular Cu/Low-k technology, the dielectric stacks consist of films deposited layer by layer with intervening etch-stop layers. The reliability of these devices containing thin films is critically dependent on the integrity of the film's interface with the substrate. Often, the choice of film material, geometrical features, and processing conditions may induce undesirable instabilities in the film resulting in buckling or wrinkling, which in turn may lead to debond initiation followed (upon further loading) by debond progression. Examples of instability induced debond initiation and progression are shown in Fig. 1.

Typically, semiconductor thin film systems are subjected to compressive stresses induced by thermal expansion mismatch between the film and the silicon substrate under processing conditions. Possessing a substantially lower thermal expansion coefficient, thicker geometry, and greater elastic modulus value (as a result significantly stiffer than the film), the silicon substrate is nearly rigid, but dictates the deformation of the film. Under the induced compressive stress, the films buckle, wrinkle and/or debond. These phenomena may also be complicated in semiconductor systems by the fact that the film may deform plastically prior to buckling or wrinkling due to the high processing temperatures. In general, there is relatively little understanding of the mechanics that govern thin film debond initiation, progression, and wrinkling in the literature. There is even less of an understanding of these phenomena as they apply to thin films in semiconductor systems. The goal of this paper is to provide a broad review of the fundamental mechanics governing these phenomena and discuss its application to semiconductor systems. Below, we briefly survey the mechanics literature that is relevant for an understanding of the physics governing buckling, debonding, propagation and wrinkling for elastic-plastically behaving films.

Relatively few studies have addressed the process of buckling or wrinkling-induced debond initiation in thin film systems, although a number of studies pertaining to debond propagation exist [2–4]. Commonly, to study propagation, the existence of an interface flaw of a specified size and location is assumed. Debond initiation is assumed to occur in this flawed region of the interface, and the critical buckling stress for debond initiation is then the Euler buckling load corresponding to a free-standing film of width equal to the initial width of the flaw. However, none of the above studies account for the stiffness of the interface and therefore, the above studies do not completely describe the mechanics of debond initiation. This gap in existing studies was addressed in our prior work [5] by modeling the interface through a cohesive zone fracture model and then deriving the analytical descriptions of the process of debond initiation. In our prior work, a non dimensional parameter termed as the "Foundation Compliance Ratio" was identified to be of importance in describing the mechanics of debond initiation. We also developed the analytical formalism that enables one to estimate the cohesive zone model parameters for a given interface through observations of buckling and debond initiation temperatures.

A debond initiated by either buckling or wrinkling instability will propagate on further loading. While such a propagation has been studied in the literature for elastic films [2–4], typically, the processing temperatures are sufficiently high to cause semiconductor metal films to undergo plastic deformation. In our prior work [6], we derived the conditions for the propagation of debond in a film undergoing elastic-plastic deformation. We further proposed an experimental procedure to estimate the fracture toughness of the interface between the film and the substrate by studying the propagation of the debond. The elasto-plastic model of debond propagation was applied to estimate the toughness of an interface between Aluminum and SU8 films by using the experimental results obtained from the non-contact thermally-driven buckling delamination test.

Wrinkling is also an instability, which is commonly observed in residually compressed stiff films deposited on relatively compliant substrates. The formation of wrinkles in stiff films on compliant substrates has been observed in several experiments [7–11]. The phenomenon of wrinkling is widely studied and several models have been proposed to explain the mechanics of wrinkling [12–20]. Debond induced by wrinkling were observed in [21] and in [22]. In general, wrinkling phenomenon has a detrimental effect on processing semiconductor thin films (see Figure 2). Thus, in this paper, based on our prior work [23], we demonstrate the impact of patterned substrates on the propensity of thin films to wrinkle.

2 Mechanics of Buckling-Induced Debonding

In this section, we briefly review the mechanics governing buckling or wrinkling-induced debond initiation in thin films. The discussion here follows the detailed theory presented in [5]. Consider a film of width $2d$, which is fixed at both ends and bonded to a rigid substrate, similar to the condition illustrated in

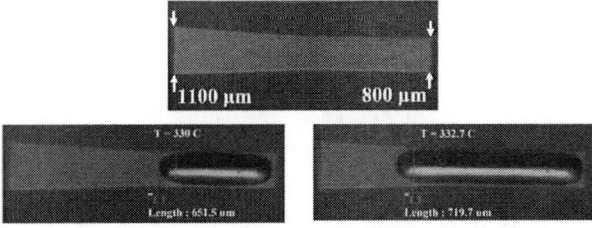

(a) Progression of instability induced debond in a funnel-shaped line [6].

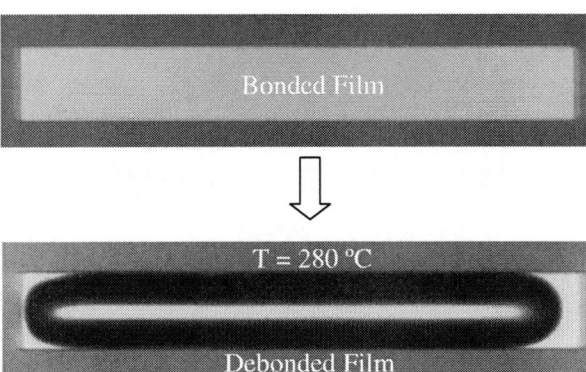

(b) A buckled aluminum metal line of thickness $t = 2.5$ μm, width=800 μm and length=6 mm .

Figure 1: Examples of compressive stresses leading to instability-induced debonding in thin film systems [5].

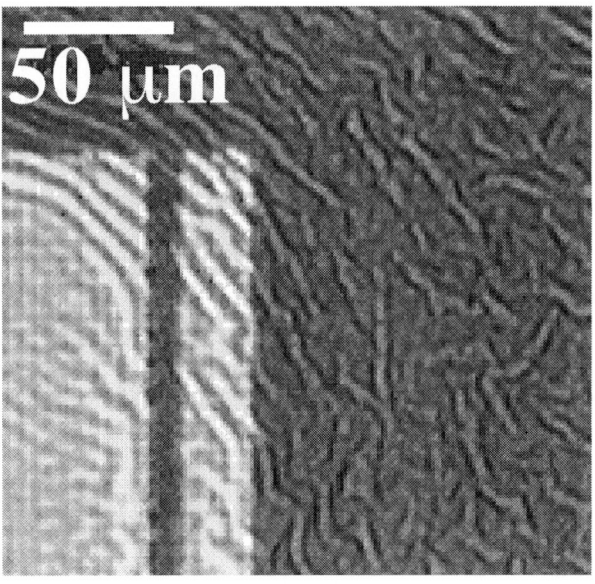

Figure 2: An example of wrinkling in semiconductor thin film systems [23].

Fig. 1(b). Upon heating from a stress free reference state, the film is compressively loaded due to differential expansion relative to the substrate. This compressive loading causes the film to buckle or wrinkle and then possibly debond from the substrate. Such a film is schematically illustrated in Fig. 3. The film is assumed to be loaded equibiaxially by a thermal compressive strain $\hat{\varepsilon} = \Delta\alpha\Delta T$ imposed in x and y directions, where $\Delta\alpha$ is the CTE difference between the film and the substrate and ΔT is the applied temperature excursion. However, the film is assumed to preferentially buckle in the x direction.

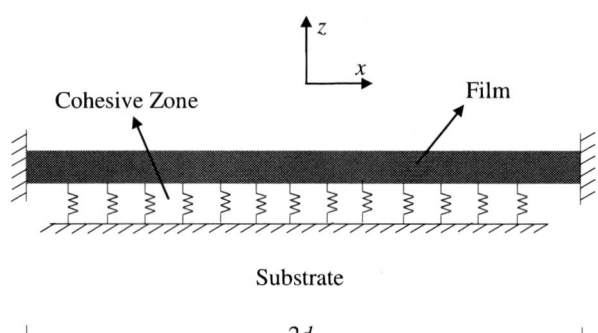

Figure 3: Bonded film on a non-linear foundation.

In the theoretical derivation, both the interface between the film and the rigid substrate (in the case of buckling) or the stiffness of the compliant film (underlying a stiffer film causing wrinkling) and the interface between the compliant film and a rigid substrate is modeled as a cohesive foundation with a bi-linear traction-separation law (TSL) shown in Fig. 4. The traction-separation relation gives the out-of-plane tractions, T_n, as a function of separation, Δ_n, between the film and the substrate. A bi-linear TSL is introduced and described by the following parameters: cohesive (interfacial) strength, σ_{max}; cohesive (interfacial) energy Γ; and cohesive lengths δ_c and δ_{max}. The cohesive length δ_c represents the separation at which the cohesive strength, σ_{max} is reached and damage initiates. The cohesive length δ_{max} represents the critical separation beyond which the foundation is fully damaged and loses its ability to support an applied traction.

The subsequent analyses of film instability and the instability induced debonding is based on the minimization of total energy. The total energy, U^t, is:

$$U^t = U^f + U^i \tag{1}$$

where, U^f is the stored energy in the film calculated using von Karman plate theory [5] and U^i is the energy stored in the interface.

We use a Rayleigh-Ritz approximation for the out-of-plane displacements, $w(x,y)$, and in-plane displacements, $u(x,y)$, of the film and determine the configuration that minimizes the total energy given

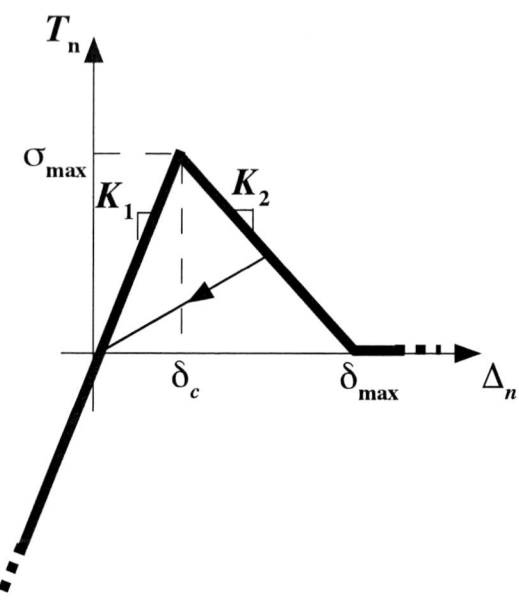

Figure 4: Traction-separation relation for the cohesive foundation.

by Eq. (1). The trial function for the out-of-plane displacement is assumed to be of the form:

$$w(x,y) = \frac{w_0}{2}\left[1 + \cos\left(\frac{\pi x}{d}\right)\right] \text{ for } -d \leq x \leq +d \tag{2}$$

Similarly, the in-plane displacements $u(x,y)$ obtained from the in-plane equilibriun equation are of the form:

$$u(x,y) = \frac{w_0^2 \pi}{32d}\sin\left[\frac{2\pi x}{d}\right] \text{ for } -d \leq x \leq +d \tag{3}$$

For a film configuration with non-zero out-of-plane displacements, three distinct regions may develop in the interface depending on the extent of interfacial separation as shown in Fig. 5. These regions are classified on the basis of the energy provided to interface as being recoverable, partially recoverable or unrecoverable. These regions are termed the intact region (Region 1, $w(x,y) < \delta_c$), the damaged region (Region 2, $\delta_c < w(x,y) < \delta_{max}$), and the debonded region (Region 3, $w(x,y) > \delta_{max}$) respectively. With the assumed trial function for the out-of-plane displacement, Eq. (2), the size of these regions may be evaluated (see [5]).

We define several non-dimensional quantities prior to describing the governing principle leading to the conditions that cause debonding to initiate: $\hat{\varepsilon} = \Delta\alpha\Delta T$, $\kappa = K_1 t/E$ and $r = \delta_{max}/\delta_c$, $\xi = d/t$, $\bar{d}_i = d_i/d$, $\bar{\delta}_c = \delta_c/t$, $\bar{\delta}_{max} = \delta_{max}/t$, $\bar{w}_0 = w_0/t$. Using these, it is possible to derive the expression for

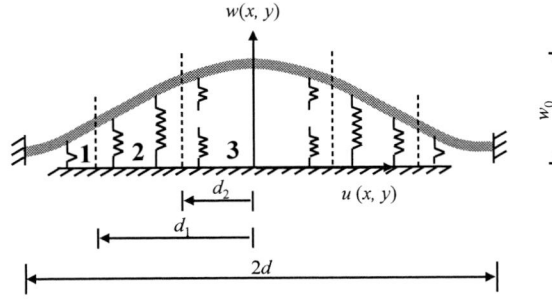

Figure 5: Illustration of different regions resulting from cohesive zone interface after buckling.

the energy quantities as [5]:

$$\bar{U}^f = 1 - \frac{1}{(1+\nu)\,\hat{\varepsilon}^2}\left[\frac{\pi^2 \bar{w}_0^2\,(1+\nu)\,\varepsilon_0}{16\xi^2} - \frac{\pi^4 \bar{w}_0^4}{512\xi^4} - \frac{\pi^4 \bar{w}_0^2}{192\xi^4}\right]$$
(4a)

$$\bar{U}_1^i = \frac{(1-\nu)\,\kappa \bar{w}_0^2}{4\hat{\varepsilon}^2}\left[\frac{3\,(1-\bar{d}_1)}{4} - \frac{1}{\pi}\sin\left(\pi\bar{d}_1\right) - \frac{1}{8\pi}\sin\left(2\pi\bar{d}_1\right)\right]$$
(4b)

$$\begin{aligned}
\bar{U}_2^i &= \frac{(1-\nu)\,\kappa}{2\hat{\varepsilon}^2}\left[\frac{3\,(\bar{d}_1-\bar{d}_2)\,\bar{w}_0^2}{8\,(1-r)}\right.\\
&\quad + \frac{\bar{w}_0^2}{2\pi\,(1-r)}\left[\sin\left(\pi\bar{d}_1\right) - \sin\left(\pi\bar{d}_2\right)\right]\\
&\quad + \frac{\bar{w}_0^2}{16\pi\,(1-r)}\left[\sin\left(2\pi\bar{d}_1\right) - \sin\left(2\pi\bar{d}_2\right)\right]\\
&\quad - \frac{\kappa\bar{\delta}_{\max}\,(\bar{d}_1-\bar{d}_2)\,\bar{w}_0}{(1-r)}\\
&\quad + \frac{\bar{\delta}_{\max}\bar{\delta}_c\,(\bar{d}_1-\bar{d}_2)}{(1-r)}\\
&\quad \left. - \frac{\bar{\delta}_{\max}\bar{w}_0}{\pi\,(1-r)}\left[\sin\left(\pi\bar{d}_1\right) - \sin\left(\pi\bar{d}_2\right)\right]\right]
\end{aligned}$$
(4c)

$$\bar{U}_3^i = \frac{(1-\nu)\,\kappa\bar{d}_2\bar{\delta}_{\max}\bar{\delta}_c}{2\hat{\varepsilon}^2}$$
(4d)

The film buckling temperature is obtained by minimizing the total energy of the film and the interface

with respect to $\bar{w}_0$. The solution obtained thus is:

$$\bar{w}_0 = 0; \qquad \Delta T < \Delta T_b$$
(5a)

$$\Delta T_b = \frac{1}{2\Delta\alpha}\left[\frac{\pi^2}{6\,(1+\nu)\,\xi^2} + \frac{6\,(1-\nu)\,\xi^2\kappa}{\pi^2}\right]$$
(5b)

where, ΔT_b is the temperature excursion at which buckling occurs. Clearly, the buckling temperature depends on the interfacial stiffness, aspect ratio of the film, and the expansion mismatch between the film and the substrate. Further, one can define a non-dimensional "interfacial compliance ratio" that governs the nature of post buckling response [5]:

$$C_f = \pi\left[\frac{1}{36\kappa(1-\nu^2)}\right]^{\frac{1}{4}}\left[\frac{1}{\xi}\right]$$
(6)

This ratio can in turn be used instead to define the buckling temperature:

$$\Delta T_b = \frac{1}{2\Delta\alpha}\sqrt{\frac{\kappa\,(1-\nu)}{(1+\nu)}}\left[C_f^2 + C_f^{-2}\right]$$
(7)

Eq. (7) indicates that the buckling temperature is minimized when C_f is unity. For $\Delta T \geq \Delta T_b$, the solution for $\bar{w}_0$ obtained by minimizing the energy is analytically solvable as long as $w_0 \leq \delta_c$:

$$\bar{w}_0 = 2\sqrt{\frac{1}{3}\left[1 + C_f^{-4}\right]\left[\frac{\Delta T}{\Delta T_b} - 1\right]} \qquad w_0 \leq \delta_c$$
(8)

Now, using the condition $w_0 = \delta_c$ at the initiation of damage, an expression for the temperature excursion, ΔT_{db}, at damage initiation following buckling is obtained:

$$\Delta T_{db} = \Delta T_b\left[1 + \frac{3\delta_c^2}{4t^2\left(1 + C_f^{-4}\right)}\right]$$
(9)

Depending on the value of the interface compliance ratio C_f, the post-buckling response of the film varies (see Figure 6). When $C_f < 1$, there is a sudden, unstable increase in the film deflection at ΔT_{db}. During this event the film debonds over much of the pattern width. On the other hand, when $C_f \geq 1$, the out of plane deflection gradually increases after damage initiation. The knowledge of the buckling temperature and damage initiation temperature can be utilized to design a test where the post-damage response is maximized (for maximum observability). Such a test can then be used to determine the cohesive-zone parameters defining the interface. Thus, from observations of ΔT_b, the interfacial stiffness K_1 can be inferred. From Eq. (5b), the non-dimensional stiffness parameter κ is obtained as [5]:

$$\kappa = \frac{\pi^2 t^2}{3d^2\,(1-\nu)}\left[|\Delta\alpha\Delta T_b| - \frac{\pi^2 t^2}{12(1+\nu)d^2}\right]$$
(10)

978-1-4244-5430-3/10 $26.00 © 2010 IEEE

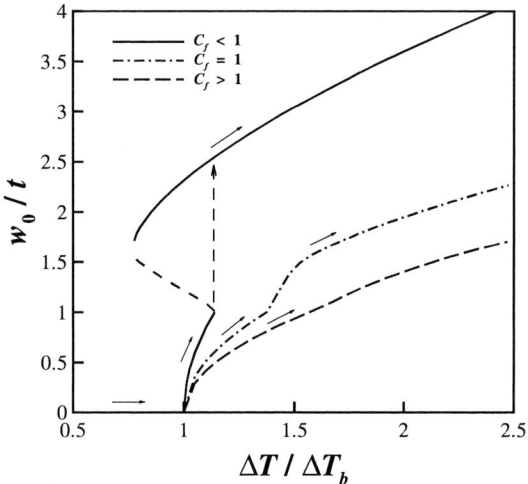

Figure 6: Buckling and debonding response for systems characterized by $\kappa = 8.5714 \times 10^{-6}$, $\delta_c/t = 1$, $r = 1.5$, $E = 70$ GPa, $\nu = 0.3$. Three pattern widths are considered such that $C_f = 0.6897$, 1.0 and 1.333. For each case ΔT is normalized with the respective ΔT_b.

In the above equation, when the term within the square brackets is zero, the interfacial stiffness is zero, which corresponds to the case of a free standing film. Similarly, interfacial strength can be inferred once ΔT_{db} is known. Following Eq. (9), and employing a value of κ as obtained from Eq. (10) one obtains the normalized cohesive strength, $\bar{\sigma}_{max} = \sigma_{max}/E$, as [5]:

$$\bar{\sigma}_{\max} = \frac{\pi^2 t^2}{3d^2(1-\nu)}\left[\Delta\alpha\Delta T_b - \frac{\pi^2 t^2}{12(1+\nu)d^2}\right.$$
$$\left. \times \sqrt{\frac{4}{3}\left[1+C_f^{-4}\right]}\left[\frac{\Delta T_{db}}{\Delta T_b}-1\right]\right] \quad (12)$$

3 Debond Propagation Model

We now describe the mechanics of debond propagation in an elastic-plastically deforming film. This model differs from the model developed by [24] as it takes into account plasticity in the film. It is assumed that the film is under a plastic state after debond initiation has occured. An example of debond propagation was shown in Figure 1(a) wherein the debond propagated after buckling caused debond initiation over the portion of the film with constant width.

To model debond propagation, consider a partially-debonded film column of width $2D$ containing a debonded region $2d$ wide as shown in Figure 7. The film is assumed to be made of a linear elastic strain-hardening material with Young's Modulus E and tangent modulus E_T. The film is assumed to be loaded in compression by a stress σ. The analysis begins by considering a debond of width $2d < 2D$. The stress in the film when the debond initiates is assumed to be σ_d. The stress σ_d is potentially greater than the yield strength σ_Y of the strip. Therefore, the film is in a plastic state at the point when the debond initiates and remains so as the debond further propagates due to an increase in stress magnitude beyond the debond stress σ_d. The fracture toughness ahead of the debonded region is Γ.

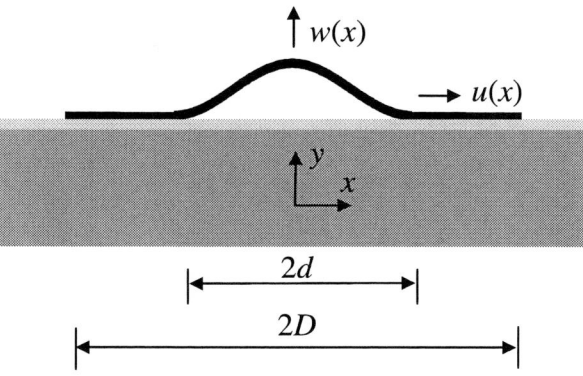

Figure 7: One-dimensional Model of Debond Propagation.

As with debond initiation, the propagation of the debonded region is modeled through the minimization of the total free energy in the partially debonded film and the interface system. The total free energy consists of portions corresponding to the bonded film, debonded film, and the energy overcome in causing the debond:

$$\psi = \psi_b + \psi_d + 2\Gamma d$$
$$= Et\int_d^D (\varepsilon_b^e)^2 dx + Et\int_0^d \left[\varepsilon_b^e - \left(u_x + \frac{w_x^2}{2}\right)\right]^2 dx$$
$$+ EI\int_0^d w_{xx}^2 dx + 2\Gamma d \quad (13)$$

where ε_b^e is the elastic strain in the bonded film and is given by:

$$\varepsilon_b^e = \frac{\sigma}{E} = \frac{1}{E}\left[\sigma_Y + \left(\varepsilon_b^t - \frac{\sigma_Y}{E}\right)E_T\right] \quad (14)$$

Assuming the same form of out-of-plane deflections, $w(x)$ and in-plane deflections $u(x)$ as in Section 2 and substituting in Eq. (13), the expression for total free energy is obtained. Minimizing the free energy with respect to w_0, one obtains the following the peak out-of-plane displacement of the debonded region as [6]:

$$\bar{w}_0 = \sqrt{\frac{4}{3}\left[\frac{\varepsilon_b^e}{\varepsilon_{cr}} - 1\right]} \tag{15a}$$

$$\varepsilon_{cr} = -\frac{\pi^2 I}{d^2 t} \tag{15b}$$

where, $I = t^3/12$ is the cross-sectional inertia. Further, minimizing the free energy with respect to d the fracture toughness of the debonded region is obtained as:

$$\Gamma = \frac{Et(\varepsilon_b^e)^2}{2}\left[1 + 3\frac{\varepsilon_{cr}}{\varepsilon_b^e}\right]\left[1 - \frac{\varepsilon_{cr}}{\varepsilon_b^e}\right] \tag{16}$$

When the toughness is greater than the right hand side, crack will not propagate, and the right hand side is to be interpreted as the energy release rate.

4 Elastic-Plastic Estimates of Energy Release Rate

An aluminum film patterned on to a SU8 subtrate was shown in Fig 1(a). The fim-substrate system was heated until debond initiated at a temperature of 330° in the region of uniform width after which it propagated into the region of varying width. The crack length of the propagating region was measured and fit to the elastic-plastic model discussed in the above section. The crack length measurements in the varying width region for the propagating debond from the experiments conducted on the funnel shaped patterns were used. The fit of the crack length measurements to those predicted by Eq. (16) is shown in Figure 8. For purposes of comparison, a similar fit was also obtained assuming an elastic behavior for the film. This fit is shown in Figure 9. The fracture toughness estimated from the elastic-plastic fit was 0.30 J/m². On the other hand, fracture toughness obtained from the elastic fit was 9 J/m². From a comparison of the fits shown in Figures 8 and 9, clearly the elastic-plastic model provides a much better fit to the experimental data than the one based on elastic behavior.

5 Wrinkling in Thin Films

We define wrinkled films as those containing a repeating deformation pattern with a wavelength equal to λ (Fig. 10). We note that the cohesive foundation model now represents the combination of the compliant substrate and the interface which bonds the film to the substrate. The initial stiffness of the cohesive foundation K_1 is therefore dominated by the substrate response, while the cohesive strength corresponds to the failure of the interface bond (or the failure of the substrate). We describe the out-of-plane displacements of the wrinkled film using a trial func-

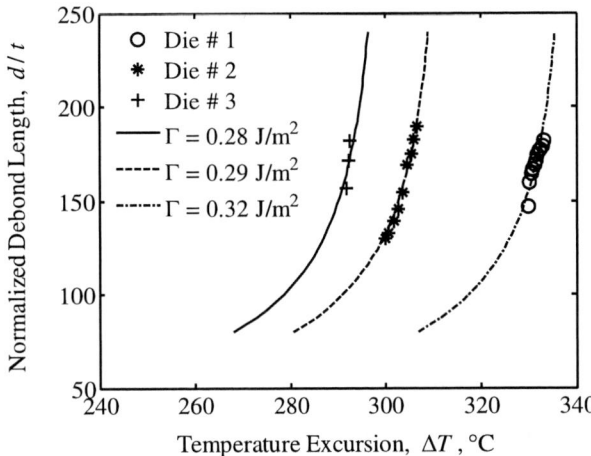

Figure 8: Estimate of interfacial fracture toughness using the elasto-plastic model.

tion of the following form, which ensures a deflection of $+w_0$ at the peaks and a deflection of $-w_0$ at the valleys:

$$w(x,y) = w_0 \cos\left[\frac{2\pi x}{\lambda}\right] \tag{17}$$

To identify the appropriate trial function corresponding to in-plane displacements, we again make use of the in-plane displacement equilibrium condition to obtain [5]:

$$u(x,y) = \frac{w_0^2}{4\lambda}\cos\left[\frac{4\pi x}{\lambda}\right] \tag{18}$$

The amplitude and wavelength of a wrinkled film are such as to minimize the energy at the wrinkling temperature. In a wrinkled film, since both the out-of-plane displacements and in-plane displacements are periodic with wavelengths equal to λ and 2λ, we focus our attention on a span of a single wavelength in the wrinkled film to determine the energy per unit area in the wrinkled film.

The energy stored in the film may be described as before using von Karman plate theory:

$$\bar{U}^f = 1 - \frac{1}{(1+\nu)\,\hat{\varepsilon}^2}\left[\frac{\pi^2\bar{w}_0^2(1+\nu)\,\hat{\varepsilon}}{16\xi^2} - \frac{\pi^4\bar{w}_0^4}{512\xi^4} - \frac{\pi^4\bar{w}_0^2}{768\xi^4}\right] \tag{19}$$

where, the wavelength is normalized as $\xi = \lambda/4t$. Also, the energy stored elastically in the foundation over a span of length λ upon wrinkling of the film pattern is:

$$U^i = U_1^i = \frac{1}{2}\int_{-\lambda/2}^{\lambda/2} K_1 w^2 dx \tag{20}$$

978-1-4244-5430-3/10 $26.00 © 2010 IEEE

Minimizing the total energy of the wrinkled film and the foundation with respect to w_0 and λ, one obtains the λ as:

$$\lambda = \pi t \left[\frac{4}{3\kappa(1-\nu^2)} \right]^{\frac{1}{4}} \tag{21}$$

and the wrinkling temperature, ΔT_w is given by:

$$\Delta T_w = \frac{1}{\Delta \alpha} \sqrt{\frac{\kappa(1-\nu)}{3(1+\nu)}} \tag{22}$$

One can also define a "foundation compliance ratio" for wrinkled films similar to the buckled films:

$$C_f = \pi \left[\frac{1}{192\kappa(1-\nu^2)} \right]^{\frac{1}{4}} \left[\frac{4t}{\lambda} \right] \tag{23}$$

However, for wrinkled films, the ratio C_f is always unity, unlike in the case of buckled films where the ratio is determined by the pattern width to film thickness. The peak deflection of the wrinkled film is given by:

$$\bar{w}_0 = \sqrt{\frac{2}{3} \left[\frac{\Delta T}{\Delta T_w} - 1 \right]} \tag{24}$$

The above equations for ΔT_w, λ and w_0 are identical to those obtained by Huang et. al. [15] for wrinkling of films on thin substrates.

Finally, for a compliant film of Young's modulus E_c and thickness t_c, the out-of-plane stiffness K of the elastic foundation is given by the relation [15]:

$$K = \left[\frac{(1-\nu_c)}{(1-2\nu_c)(1+\nu_c)} \right] \frac{E_c}{t_c} \tag{25}$$

where ν_c is the Poisson's ratio of the compliant film. Using the above foundational stiffness, the wrinkling temperature derived in Eq. (22) can be expressed as:

$$\Delta T_w = \frac{1}{\Delta \alpha} \left[\frac{(1-\nu)(1-\nu_c)}{3(1-2\nu_c)} \left(\frac{E_c}{E} \right) \left(\frac{t}{t_c} \right) \right]^{1/2} \tag{26}$$

6 Wrinkling on Patterned Substrates

In this section we study wrinkling on Patterned Substrates. Such a study is of direct relevance to the microelectronic industry in the design and fabrication of stiff film-compliant film systems on patterned substrates. The system under consideration is shown in Figure 11(a). A stiff film of Young's Modulus E, thickness t and coefficient of thermal expansion (CTE) α is bonded to a compliant film of Young's modulus E_c, which in turn is bonded to a rigid *patterned* substrate. The compliant film has a thickness $t_c = t_1$ within vias and a thickness $t_c = t_2$ above pillars $(t_1 > t_2)$. A square unit cell of side L, as shown in Figure 11(a) and 11(b), is considered. Patterned features are captured by varying the thickness of the underlying compliant film in a single square region of

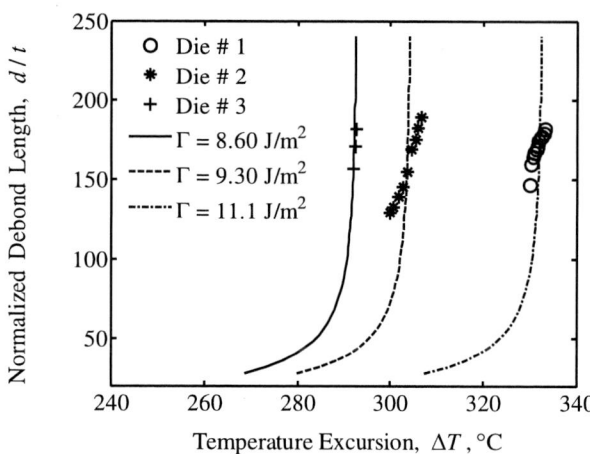

Figure 9: Estimate of interfacial fracture toughness using the elastic model.

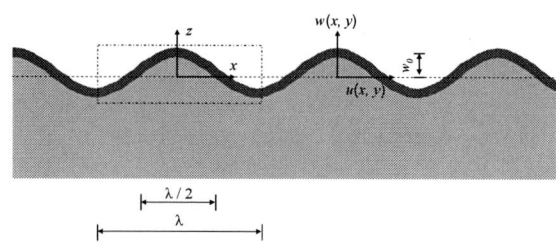

Figure 10: System under consideration for film wrinkling.

size L_p, as shown in Figure 11(b).

For the system under consideration, the foundation stiffness is a spatially varying quantity on account of the varying thickness of the compliant film. Let K_1 and K_2 be the stiffness of the foundation in regions of thickness t_1 and t_2 respectively. K_1 and K_2 are obtained from Eq. (25) by substituting $t_c = t_1$ and $t_c = t_2$ respectively. Then the spatially varying foundation stiffness is written as:

$$K(x_1, x_2) = K_1 + [(K_2 - K_1) \times \Pi(x_1, x_2)] \quad (27)$$

where $\Pi(x_1, x_2) = 0$ within vias ($t_c = t_1$) and $\Pi(x_1, x_2) = 1$ above pillars ($t_c = t_2$). The second term on the R.H.S of Eq. (27) captures the spatial variation in the foundation stiffness and is absent in existing approaches.

The spectral method used by Huang et al. [15, 25] was extended to study the wrinkling of stiff-compliant films on patterned substrates. The equation for out-of-plane equilibrium of the wrinkled film was transformed into Fourier Space and a numerical viscosity parameter was used to stabilize the computation. The equation for out-of-plane equilibrium after the introduction of viscosity parameter is given by:

$$F(w^{n+1}) = \frac{ik_\beta F(N^n_{\alpha\beta}, w^n_{,\alpha}) - [(K_2 - K_1)F(\Pi \times w^n)]}{(K_1 + \frac{t^3 \bar{E}}{12} k^4 + \zeta)} + \frac{\zeta F(w^n)}{(K_1 + \frac{t^3 \bar{E}}{12} k^4 + \zeta)} \quad (28)$$

where w^n and w^{n+1} are the out-of-plane displacements at the nth and $(n+1)^{\text{th}}$ iterations respectively. Note that the viscosity ζ does not have a physical basis and is introduced solely for improving numerical convergence. The above equation was updated at each iteration using the in-plane and out-of-plane equilibrium equations from the von Karman plate theory to obtain solution at every step as in [15, 25].

The reader is referred to [23] for details of the numerical solution procedure behind the simulations presented here. The solutions were obtained iteratively at each grid point. The compressive load is applied as a uniform initial strain at each grid point, $\hat{\epsilon}_{11} = \hat{\epsilon}_{22} = \Delta\alpha\Delta T$, where $\Delta\alpha$ is the mismatch in the coefficient of thermal expansion (CTE) between the stiff film and the substrate and ΔT is the applied temperature difference relative to a stress-free temperature.

The parameters used in the simulations are as follows: A modulus ratio of $E/E_c = 10^5$ and film thickness ratios of $t_c/t = t_1/t = 50$ within vias and $t_c/t = t_2/t = 10$ above pillars was considered. A grid with $N = 2^7 = 128$ points was used to discretize a square unit cell of side $L = 1500t$. A constant shear stiffness for the compliant film of $S = 0.75K_1$ was employed in all simulation cases. Also, in all cases, an initial imperfection is applied to the out-of-plane deflection w at the start of the simulation. This initial imperfection is a sum of all the possible eigen-modes and has a magnitude smaller than $0.001t$. All simulations were run until the out-of-plane deflection w

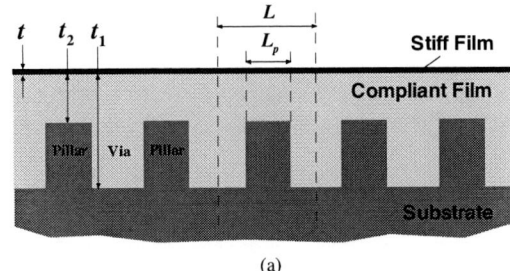

(a)

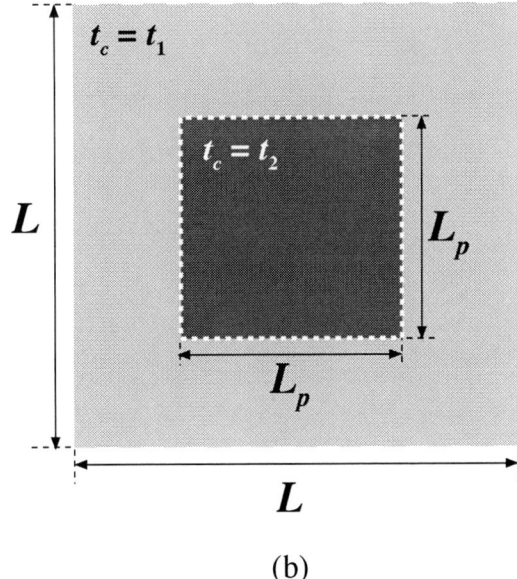

(b)

Figure 11: (a) Geometry of film stack, (b) unit cell geometry - uniform

attains convergence.

Results from simulations are summarized in Figure 12. Simulations for the reference case of blanket films are presented in Column 1 ($t_c = t_1 = 50t$) and Column 3 ($t_c = t_2 = 10t$) of the figure. Using the analytical model of Eq. (26), the critical strains for film wrinkling for these cases are $\Delta\alpha\Delta T_w^{(1)} = 2.20 \times 10^{-4}$ (for $t_c = t_1$) and $\Delta\alpha\Delta T_w^{(2)} = 4.92 \times 10^{-4}$ (for $t_c = t_2$). Note that this gives a ratio of the critical temperatures as $\Delta T_w^{(2)}/\Delta T_w^{(1)} = 2.24$. This would imply that for the applied loading in the range $\Delta T/\Delta T_w^{(1)} = 1.02 - 2.96$, considered in Figure 12, all simulation cases with $t_c = t_1$ should show wrinkling (as $\Delta T > \Delta T_w^{(1)}$ for $\Delta T/\Delta T_w^{(1)} \geq 1.02$), and for $t_c = t_2$, only the cases $\Delta T/\Delta T_w^{(1)} \geq 2.25$ should show wrinkling (as $\Delta T > \Delta T_w^{(2)}$ for $\Delta T/\Delta T_w^{(1)} \geq 2.25$). This is indeed the case as can be seen from the results in Column 1 and Column 3 of Figure 12. Also from the results in Column 1 it can be seen that close to the critical load, the wrinkling has a checkerboard mode. This changes to herringbone and labyrinth wrinkling modes for higher applied loads. Consider

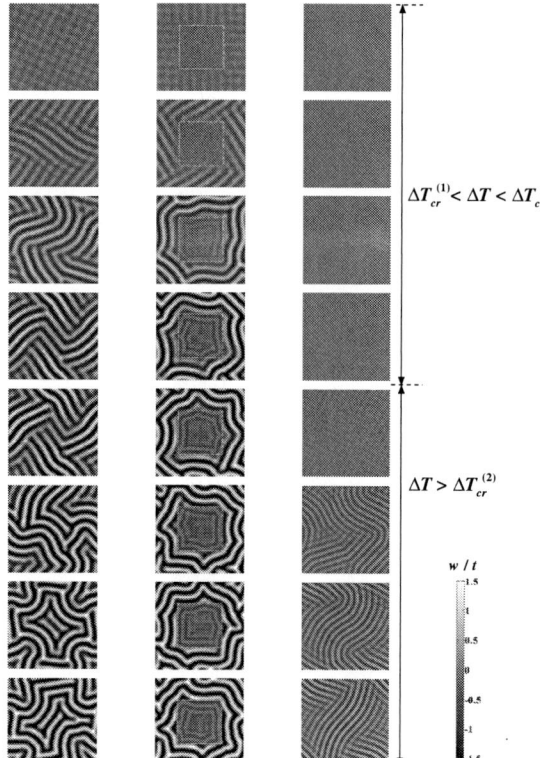

Figure 12: Wrinkling on a patterned substrate with uniformly arranged pillars. Here, ΔT_{cr} refers to the wrinkling temperature ΔT_w corresponding to the film thickness.

the results in Column 2 of Figure 12, which are for a patterned substrate with uniformly spaced pillars ($L_p/L = 1/2$). Qualitatively, the introduction of patterns on the substrate significantly changes the nature of wrinkling compared to the blanket film solutions in Column 1 and Column 3. For cases $\Delta T/\Delta T_w^{(1)} = 1.02 - 2.05$, regions above pillars are not expected to wrinkle based on the analytical solution of Eq. (26). However, as can be seen from the results in Column 2, wrinkling is observed even in regions above pillars. For cases, $\Delta T/\Delta T_w^{(1)} \geq 2.25$, regions above pillars should undergo wrinkling based on the the analytical solution of Eq. (26). However, for these cases the wrinkling amplitudes above pillars are significantly lower than the corresponding blanket film solutions in Column 3.

7 Summary

In the present paper, we analytically described the conditions for buckling, wrinkling and buckling-induced debond initiation as well as progression. The mechanics of such phenomena are relatively less studied, especially when the films undergo plastic deformation. We used a bilinear traction-separation law and analytically described the temperature excursions corresponding to onset of instabilities and of damage. Once the two critical temperature excursions - for buckling and debonding onset - are known from experiment, the cohesive stiffness and cohesive strength are obtained. Together with methods described in [6] for the determination of the interfacial fracture toughness, a complete description of the film/interface/substrate system is then available. We demonstrated the derivations by applying them to develop a non-contact, buckling-induced debond test that enabled us to measure fracture toughness of Al/SU-8 film on Si substrate.

Wrinkling of blanket films on substrates is known to be size-independent as it depends only on the ratios of moduli and film thickness. We demonstrated that in the presence of patterned features this is no longer the case. The size of the pattern introduces a length scale into the problem and this length scale interacts with other dimensions like the wrinkling wavelength in determining the nature of the solution for wrinkling on patterned wafers.

The fundamental mechanics that govern the buckling, wrinkling and debonding phenomenon reviewed in the present paper is of increasing relevance to the semiconductor industry as alternative dielectric materials are explored in the quest for low-k dielectric systems.

Acknowledgments

The financial support from the Semiconductor Research Corporation (SRC) through Grant No. 1292.011 is gratefully acknowledged.

References

[1] L.B. Freund and S. Suresh. *Thin Film Materials: Stress, Defect Formation and Surface Evolution.* Cambridge University Press, New York, 2003.

[2] J.W. Hutchinson, M.D. Thouless, and Liniger E.G. Growth and Configurational Stability of Circular, Buckling-Driven Film Delaminations. *Acta Metallurgia et Materialia40(2)*, 40(2):295–308, 1992.

[3] J.W. Hutchinson and Z. Suo. Mixed mode cracking in layered materials. *Advances in applied mechanics*, 29:63–191, 1992.

[4] M.W. Moon, K.R. Lee, KH Oh, and J.W. Hutchinson. Buckle delamination on patterned substrates. *Acta Materialia*, 52(10):3151–3159, 2004.

[5] S. Goyal, K. Srinivasan, G. Subbarayan, and T. Siegmund. On Instability-Induced Debond Initiation in Thin Film Systems. *Engineering Fracture Mechanics*, 2010. To appear, doi:10.1016/j.engfracmech.2010.02.001.

[6] S. Goyal, K. Srinivasan, G. Subbarayan, and T. Siegmund. A non-contact thermally-driven buckling delamination test to measure interfacial fracture toughness of thin film systems. *Thin Solid Films*, 518:2058–2064, 2010.

[7] N. Bowden, S. Brittain, A.G. Evans, J.W. Hutchinson, and G.M. Whitesides. Spontaneous formation of ordered structures in thin films of metals supported on an elastomeric polymer. *Nature*, 393:146–149, 1998.

[8] W.T.S. Huck, N. Bowden, P. Onck, T. Pardoen, J.W. Hutchinson, and G.M. Whitesides. Ordering of spontaneously formed buckles on planar surfaces. *Langmuir*, 16(7):3497–3501, 2000.

[9] F. Iacopi, S.H. Brongersma, and K. Maex. Compressive stress relaxation through buckling of a low-κ polymer-thin cap layer system. *Applied Physics Letters*, 82(9):1380–1382, 2003.

[10] P.J. Yoo, K.Y. Suh, H. Kang, and H.H. Lee. Polymer elasticity-driven wrinkling and coarsening in high temperature buckling of metal-capped polymer thin films. *Physical Review Letters*, 93(3):034301, 2005.

[11] J.R. Serrano, Q. Xu, and D.G. Cahill. Stress-induced wrinkling of sputtered sio2 films on polymethylmethacrylate. *Journal of Vacuum Science and Technology A*, 24(2):324–327, 2006.

[12] H.G. Allen. *Analysis and Design of Structural Sandwich Panels.* Permagon Press, Oxford, 1969.

[13] E. Cerda and L. Mahadevan. Geometry and physics of wrinkling. *Physical Review Letters*, 90(7):074302, 2003.

[14] X. Chen and J.W. Hutchinson. Herringbone buckling patterns of compressed thin films on compliant substrates. *Journal of Applied Mechanics*, 71:597–603, 2004.

[15] Z. Huang, W. Hong, and Z. Suo. Nonlinear analysis of wrinkles in a film bonded to a compliant substrate. *Journal of the Mechanics and Physics of Solids*, 53:2101–2118, 2005.

[16] D. Lee, N. Triantafyllidis, J.R. Barber, and M.D. Thouless. Surface instability of an elastic half space with material properties varying with depth. *Journal of the Mechanics and Physics of Solids*, 56(3):858–868, 2008.

[17] J. Song, H. Jiang, Z.J. Liu, D.Y. Khang, Y. Huang, J.A. Rogers, C. Lu, and C.G. Koh. Buckling of a stiff thin film on a compliant substrate in large deformation. *International Journal of Solids and Structures*, 2008.

[18] B. Audoly and A. Boudaoud. Buckling of a stiff film bound to a compliant substrate, Part I: Formulation, linear stability of cylindrical patterns, secondary bifurcations. *Journal of the Mechanics and Physics of Solids*, 56(7):2401–2421, 2008.

[19] B. Audoly and A. Boudaoud. Buckling of a stiff film bound to a compliant substrate, Part II: A global scenario for the formation of herringbone pattern. *Journal of the Mechanics and Physics of Solids*, 56(7):2422–2443, 2008.

[20] B. Audoly and A. Boudaoud. Buckling of a stiff film bound to a compliant substrate, Part III: Herringbone solutions at large buckling parameter. *Journal of the Mechanics and Physics of Solids*, 56(7):2444–2458, 2008.

[21] H. Mei, R. Huang, J.Y. Chung, C.M. Stafford, and H.H. Yu. Buckling modes of elastic thin films on elastic substrates. *Applied Physics Letters*, 90:151902, 2007.

[22] D. Vella, J. Bico, A. Boudaoud, B. Roman, and M.P. Reis. The macroscopic delamination of thin films from elastic substrates. *Proceedings of the National Academy of Science*, 106(27):10901–10906, 2009.

[23] K. Srinivasan, S. Goyal, T. Siegmund, G. Subbarayan, and Q. Lin. Thermally-Induced Wrinkling in Thin-Film Stacks on Patterned Substrates. *IBM Journal of Research and Development*, 53(3):12:1–12:10, 2009.

[24] J.W. Hutchinson and Z. Suo. Mixed mode cracking in layered materials. *Advances in Applied Mechanics*, 29:63–191, 1992.

[25] Z. Huang, W. Hong, and Z. Suo. Evolution of wrinkles in hard films on soft substrates. *Physical Review E*, 70:030601, 2004.

Reliability of Microelectronics Packaging in the Era of EnergyWise and Borderless Networks

Li Li and Jie Xue
Technology & Quality, CVCM
Cisco Systems Inc.
170 West Tasman Drive
San Jose, California 95134, USA
Phone: +1 408 527-0801, Email: lili2@cisco.com

Abstract—**In an ever-increasingly connected world, the boundaries or obstacles to accessing information are being torn down by business necessity, personal preferences and technical innovations. The Borderless Network is creating the ability for customers to work anywhere, with any device using services and applications like video, collaboration, with a secure, reliable and seamless communication experience. The customer will also experience reduced energy consumption from network attached devices thanks to innovations made in the microelectronic devices, systems that enable the next generation of smart and EnergyWise networks. In this paper, challenges and opportunities in reliability of microelectronics packaging for the next generation of EnergyWise, Borderless networks will be reviewed. The impact of materials, manufacturing processes and field environmental conditions on the packaging reliability will be discussed with one case study.**

Keywords-EnergyWise; low power design; reliability; microelectronics packaging; temperature and humidity effect

I. INTRODUCTION

We're living in an increasingly connected world. The next generation of smart, borderless networking is coming and will create the ability for customers to work anywhere, with any device using services and applications like video, collaboration, with a secure, reliable and seamless communication experience [1]. But in addition to ubiquitous broadband connecting devices wherever they are, we also need to pay attention to the energy required to power the whole network and all the devices attached to the network.

The Environmental Protection Agency (EPA) calculated that the energy consumed by the nation's servers and data centers was estimated at 61 billion kilowatt-hours (kWh) in 2006, which accounted for 1.5 % of total U.S. electricity consumption at a cost of about $4.5 billion. This consumption is two times larger than what was consumed in 2000 and is projected to double again by 2011 unless efficiencies are implemented, according to the EPA [2]. The EPA report to congress on server and data center energy efficiency also estimated the power and cooling infrastructure accounted for 50% of data center total energy consumption.

To enable the next generation of smart and EnergyWise networks, the microelectronic components designed into the computing, networking and storage equipment need to be low-power and yet with high reliability. They also need to be able to withstand different field environmental conditions and sometimes very harsh field conditions. From system design perspective, the energy efficient designs will drive less redundancy in hardware or trade-off of design efficient with redundancy in reliability requirements.

In this paper we will review the challenges and opportunities in reliability of microelectronics packaging for the next generation of EnergyWise, Borderless networks. We will also discuss the impact of materials, manufacturing processes and field environmental conditions on the packaging reliability and the responses from the reliability community for meeting those challenges.

II. INDUSTRY TRENDS IN "GREEN IT"

A. "Green" Data Center

To limit climate changes, many believe that greenhouse gas emissions must be reduced by 25 gigatonnes (Gt) to 30 Gt of CO_2 equivalents (CO_2e) annually by 2030. Much attention has been placed upon individual devices and data center power consumption within the Information Technology (IT) vendor community in an effort to reduce carbon footprints and energy cost. Measures that have been put in place at the data center, system and operating system (OS) levels include:

- Air-side and water-side economizer
- Better data center design and layout
- More efficient power and cooling
- Consolidation and virtualization
- Throttle to match required performance
- New architectures to allow idle CPUs and memory
- Power aware resource management and scheduling

B. Leveraging Network Infrastructure

Clearly there has to be better ways to leverage IT to not only reduce its own power consumption but also the power consumption of non-IT devices. For example, in a typical

978-1-4244-5430-3/10 $26.00 © 2010 IEEE

commercial building, lighting plus heating and cooling represents some 66% of total electrical energy consumption while IT represents between 25 and 30%. Within IT, desktop computers, printers, etc., consume 50%, data centers draw 30% while networks represent 10% of electrical energy consumption. There is a broader networked-based approach to address the remaining power consumed which can deliver far greater gains in power efficiency and cost reduction [3].

The networked-based solution to power management is leveraging the connectivity of all devices within the network. In short the network touches every device. Today these devices are IT-based, including computers, storage, printers, access points, cameras, phones, special network appliances such as firewalls, mobile devices, and increasingly, TVs and other non-IT electronics. The network has a unique position to monitor, distribute commands and most importantly control the power consumption of the devices it connects. This concept is straightforward for devices that obtain their power from network switches via Power over Ethernet (PoE) such as wireless LAN access points (AP), IP phones, Ethernet/IP-based video surveillance cameras, etc., as the network is their source of power. But the networked approach to power management concept can be extended to non-PoE IT devices such as computers, digital signage, printers, storage, fax machines, etc. The concept can be extended further to non-IT systems such as building controls, lighting, elevators, 24/7 monitoring systems, HVAC-sensors, fire/smoke sensors, etc. This is the idea of an EnergyWise network [3].

In addition to power consumption information gathering and reporting, EnergyWise network provides the means to control energy consumption by time of day. With a simple energy policy of shutting down devices for an average of four or five hours a day, a company will gain significant power usage savings. In essence, EnergyWise can utilize the network to act as a distributed programmable thermostat changing the power consumption of devices and business control systems based on time of day. The network keeps time and thus can be the timer for connected devices.

C. Reducing Power At Chip Level:

Innovations have also been made at the chip level to reduce the power of the devices in order to save energy while still meeting the performance requirements. These include:

- Low-power chip designs, e.g., multi-core, multi-threaded microprocessor designs

- Decreased leakage current to reduce stand-by power

- Throttle to match required performance using idle/sleep states

- New architectures to allow idle CPUs and memory in the systems

While the EnergyWise network is promising, we should not ignore the potential reliability implications. A better understanding of these reliability concerns is very important in making the right design trade-offs and in reconciling such conflicting development goals as high performance and high reliability vs. energy efficient.

III. RELIABILITY IMPLICATIONS

A. Expanded Operating Environmental Envelop

To provide greater flexibility in facility operations, particularly with the goal of reducing energy consumption in data centers, the American Society of Heating, Refrigerating and Air-Conditioning Engineers (ASHRAE) recommended to expand the data center environmental envelop in its 2008 Environmental Guidelines for Datacom Equipment [4].

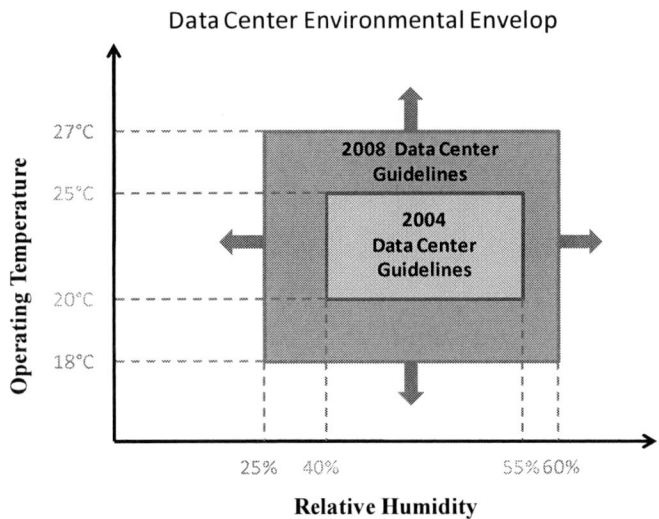

Figure 1. Expanded data center environmental envelop.

The expanded operating environmental envelop for data centers is shown in Figure 1. It was expanded for both the temperature and humidity conditions. For other applications such as those at the edge of the network, similar trends have also been seen, e.g.

- Network equipment that is used to be in a controlled environment is now subjected to uncontrolled environment and sometimes outdoor conditions with no active convection cooling.

- Increased power and temperature cycling due to device, equipment/port power on/off as part of the EnergyWise policy.

- Increased power and temperature cycling due to throttling performance with devices in idle/sleep states.

B. Impact on Chip and Package Reliability

There are several reliability implications from the expanded environmental envelop. It will force servers, switches, storage and other IT equipment to operate at a higher temperature and a high humidity environment. This in turn will cause the IT equipment to draw more power and the cooling fans run faster. The high temperature environment will increase the risks of component failures due to insufficient cooling and high operating temperatures. The higher humidity field conditions will increase the risks of component failures including electro-chemical migration, growth of conductive anodic filaments (CAF) in Printed Circuit Boards (PCB) and interfacial delamination in component packages. A summary of the impact on component reliability is shown in Table 1.

Table 1. Effect of Field Environmental Conditions on Component Reliability

Environmental Conditions (Stress)	Failure Mechanism Examples
Temperature	<ul><li>Thermal degradation</li><li>Thermal break down</li><li>Material aging</li><li>Thermal/stress migration</li><li>Electromigration</li></ul>
Temperature Cycling and thermal shock (Time Dependent Thermo-mechanical Stress)	<ul><li>Dielectric film cracking</li><li>Fractured bonds and interconnects</li><li>Solder joint fatigue/cracking</li><li>Chip cracking</li><li>Interfacial delamination</li><li>Via/Plated Through Hole (PTH) cracking</li></ul>
Humidity (Moisture Concentration)	<ul><li>Moisture induced swelling, delamination</li><li>Corrosion</li><li>Electro-chemical migration</li><li>Conductive anodic filaments (CAF)</li><li>Electrostatic discharge</li></ul>

A case study is included in this paper to illustrate the potential increase of component failure risks due to operating environmental condition changes and the opportunities that the reliability community is facing to better understand the failure mechanism acceleration under different conditions and to make design improvement or trade-off to mitigate the increase of potential risks.

IV. CASE STUDY: ENVIRONMENTAL EFFECTS ON PLASTIC ENCAPSULATED SILICON DEVICES

A. Moisture Induced Failure Mechanism

When a silicon device is assembled into a plastic package and subjected to highly accelerated temperature and humidity conditions such as autoclave test at 121°C, 100%RH; Highly Accelerated Temperature and Humidity Stress Test (HAST) at 130°C, 85%RH; and Temperature Humidity Biased (THB) test at 85°C, 85%RH; delamination at the die-to-encapsulating material and the die-to-die attach pad interfaces is the major failure mode. This type of failure is usually caused by hygroswelling of the plastic encapsulating material, which often is the precursor to the "popcorn" cracking during solder reflow process. On the other hand, moisture induced delamination in the dielectric film stacks of the silicon devices has been reported in the literature as well [5]. The moisture

absorbed can weaken the chemical bonds within the dielectric film. It can also reduce the interfacial strength and induce hygroscopic stresses causing moisture induced damage in the dielectric films. The delamination typically starts at the edge of the die and propagates towards the center of the die.

The effects of temperature and moisture on the adhesive and cohesive strength of dielectric films and the interfaces that are composed of those materials commonly used in silicon devices have been studied by researchers in the field [6]. For example, the barrier-SiO$_2$ interfaces found in typical Al or Cu interconnect structures have been found to be susceptible to subcritical debonding similar to stress-corrosion cracking in bulk glasses. It was also determined that, at least over the temperature range investigated, it is the moisture content and not the temperature that has a large effect on reducing the effective work of adhesion [6].

It has been reported that doped silicon oxides absorb and react with water causing reliability problems in microelectronic devices. The absorption and reaction of water with doped and undoped oxides as well as the effect of annealing have been studied extensively [7].

As the semiconductor industry migrates from oxide based dielectrics to Cu interconnect and low-k dielectrics, both the bulk and interfacial properties of the mechanically fragile low-k dielectrics must be carefully characterized in terms of the effects of the environment on their overall cohesive and adhesive strengths. As the dielectric constant is reduced, the cohesive strength of the dielectric is reduced as well. Given that both will depend heavily on the density of Si–O bonds in the dielectric, a linear relationship has been shown [6].

B. Device Reliability Evaluation

For plastic encapsulated silicon devices, the temperature and moisture effects on the dielectric films are investigated using accelerated environmental stress testing. By simultaneously applying temperature, humidity, and voltage bias, the accelerated stress tests can be used to understand moisture related failure mechanisms in the silicon devices and their plastic packages. The tests are conducted to further evaluate the sensitivities to moisture and temperature induced damages in the dielectric films of the plastic packaged silicon devices.

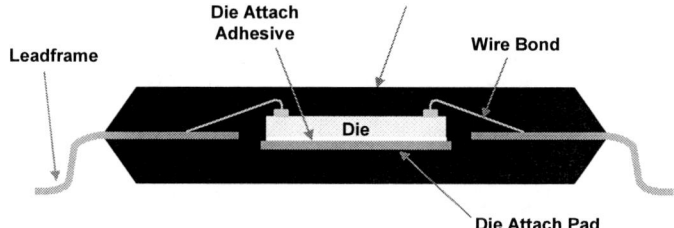

Figure 2. A schematic of LQFP package.

Several devices and packages were used as test vehicles for this study. These included both PBGA and LQFP packages. For each package and each test condition, 45 samples were selected and tested. A schematic of the LQFP package used is shown in Figure 2.

The test vehicles were subjected to such highly accelerated temperature and humidity conditions as the autoclave at 121°C, 100%RH; HAST at 130°C, 85%RH, and THB at 85°C, 85%RH. The accelerated environmental stress conditions and test durations are summarized in Table 2. No bias was used since the focus of this investigation is mainly on the temperature and humidity effects.

Table 2. Accelerated Environmental Stress Conditions

Environmental Stress	Test Condition	Test Duration (hour)
Autoclave	121°C, 100%RH	96
HAST	130°C, 85%RH	384
THB	85°C, 85%RH	1000

Both the autoclave and the HAST are stringent stress tests. The failures occurred early during the autoclave and HAST tests. At the 48 hour readout point, electrical open failure was found for one of the test vehicles. Delamination in the dielectric film stacks of the encapsulated silicon device was identified as the failure mode. The delamination started at the edge of the die and propagated towards the center of the die. A schematic showing the dielectric film delaminated above the poly-gate is given in Figure 3.

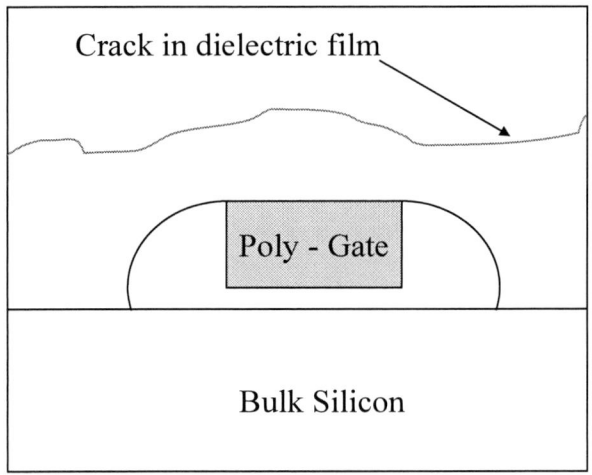

Figure 3. Dielectric film delamination during HAST, autoclave and temperature and humidity (85°C/85%RH) tests.

Through failure and statistical analysis, the delamination of dielectric films was concluded as moisture induced failure in the silicon device.

Figure 4 contains the Weibull analysis plots for the HAST and THB testing results. Note that the slopes are almost identical. The final Weibull analysis results for the HAST and THB are summarized in Table 3. The results revealed both the temperature and humidity effects on the failure mechanism.

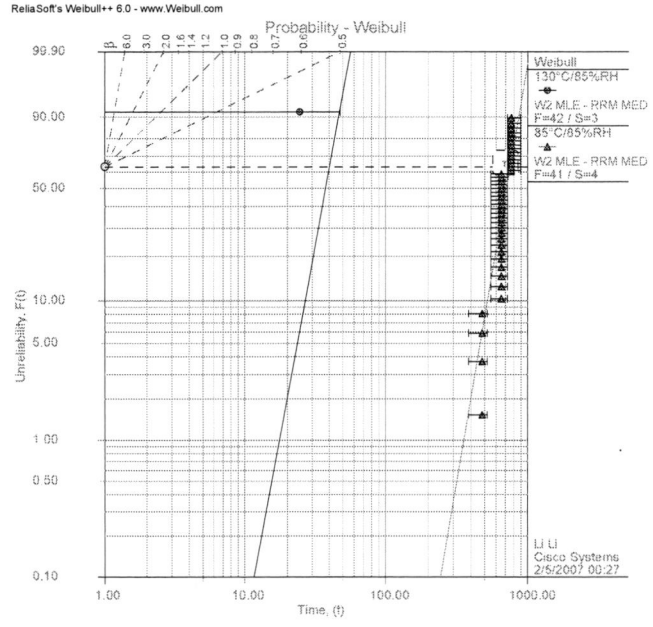

Figure 4. Weibull distributions of the HAST and THB testing results.

Table 3. Summary of Weibull analysis for HAST and THB testing for one of the test vehicles.

Environmental Stress	Sample Size	Characteristic Life (hour)	Slope
HAST 130°C, 85%RH	45	40	5.59
THB 85°C, 85%RH	45	754	6.13

C. Field Reliability Modeling

The acceleration of moisture induced failure in the dielectric films was modeled using an empirical reliability model. For accelerated environmental (temperature and humidity) effects, the unbiased Peck model is expressed as [8]:

$$t_f = \tau_0 \cdot RH^{-n} \exp\left(\frac{E_a}{RT}\right) \qquad (1)$$

where RH is the relative humidity, R is the Boltzmann constant, T is the absolute temperature (°K) and E_a is the activation energy. The results of autoclave test at 121°C, 100%RH; HAST at 130°C, 85%RH, and THB test at 85°C, 85%RH, were used to fit the Peck model.

By fitting Equation 1 with the accelerated stress testing results, the two parameters in Equation 1 were determined [9]. Once the two parameters are identified, the acceleration factors and hence reliability of the silicon devices under typical product use or storage environment conditions can be assessed using the validated Peck model.

D. Impact of Higher Temperature and Humidity

The effects of temperature and humidity on the component reliability are shown in Figure 5 and Figure 6.

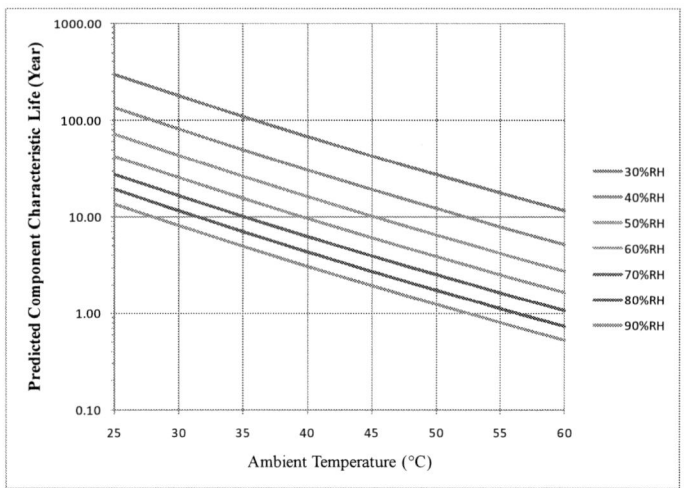

Figure 5. Effect of ambient temperature on component reliability for a given ambient relative humidity level (%RH).

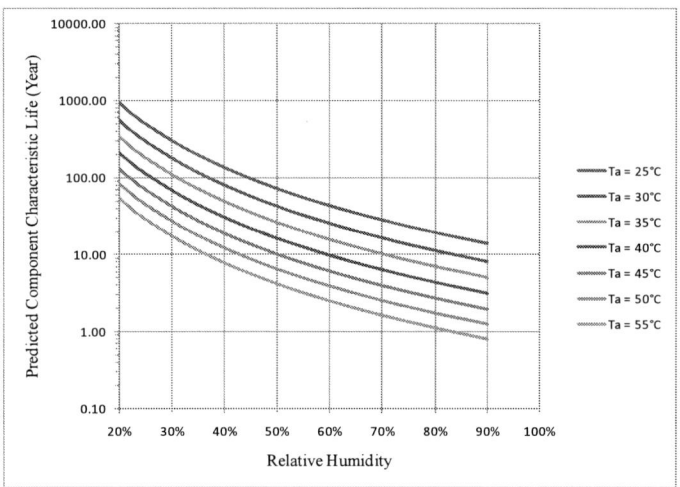

Figure 6. Effect of relative humidity on component reliability for a given ambient temperature (T_a).

It is clear as shown in both Figure 5 and Figure 6, the predicted component reliability deceases exponentially as temperature and humidity level increases. For example, an increase in temperature from 40°C to 45°C together with an increase in humidity from 40%RH to 50%RH can cause about 66% reduction in the predicted component field life.

When we deal with the failure of moisture-induced dielectric film delamination, the localized moisture concentration at the die level needs to be quantified. When the component is in non-operating state such as in the power-off or in the storage state for a relative long period of time, the die surface humidity will be the same as the ambient humidity.

However, when the device is operating and dissipating power, it has been shown in [8] that the die surface humidity will only be a fraction of the ambient humidity. Self-heating by the device power dissipation will drive die temperature rise above the ambient temperature. The ratio of die surface humidity to the ambient is determined by the saturated water vapor pressure at ambient and the die surface. The die self-heating effect is shown in Figure 7. The die surface humidity level can be adjusted accordingly when we assess the operating component reliability using Equation 1.

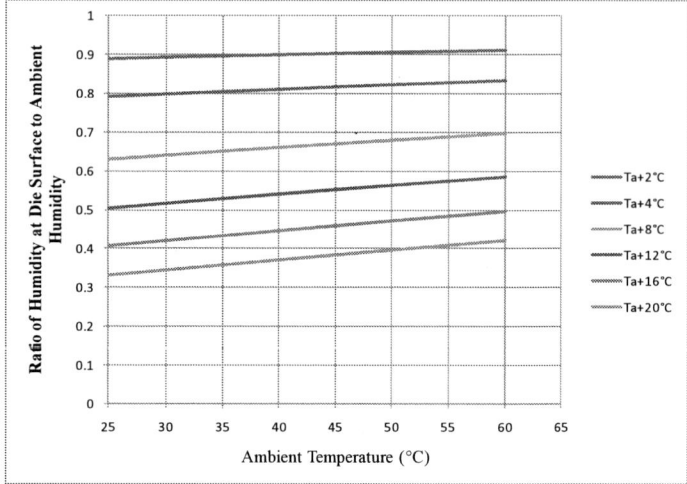

Figure 7. Ratio of humidity at die surface to ambient humidity for a given die temperature rise above ambient temperature ($T_a + dT$).

E. Opportunities for the Reliability Community

The expanded operating environmental envelop can be a challenge to the delivery of high performance, high reliability and energy efficient systems that our customers demand. On the other hand, it also creates opportunities for the reliability community to focus or re-focus on:

- Better understanding field conditions and failure mechanism acceleration

- Exploring component qualification based on physics-of-failure in addition to the traditional stress-test-driven qualification methodology to truly understand the component application envelop

- Making design improvement and right design trade-off

- Better field reliability monitoring and data analysis

- Learning from low-power computing applications

The following is a good example which demonstrated making design improvement in the die "seal ring" for the devices discussed in this case study, the moisture penetration can be sealed out of the die active area and hence the component reliability to the moisture induced delamination was greatly improved.

Usually the die "seal ring" consists of an array of metal vias and contacts along the die periphery. The structure and density of the metal vias and contact are very important to employing the "seal ring" structure as a moisture barrier. When the density of the vias and contacts is low, the "seal ring" will not be effective.

On the other hand, the most effective "seal ring" design merges all the vias into a continuously filled trench. The two types of "seal ring" designs are shown schematically in Figure 8.

Finite element modeling showed [9] that the full trench design served well as a moisture barrier by keeping the moisture concentration in the area bounded by the "seal ring" well below the critical level, after 1000 hours exposure to the 85°C/85%RH conditions. Temperature and humidity stress testing also confirmed the improved component reliability with the full trench "seal ring" design. To prevent moisture penetrating into the active area of the packaged silicon device, a full trench, redundant "seal ring" design is needed.

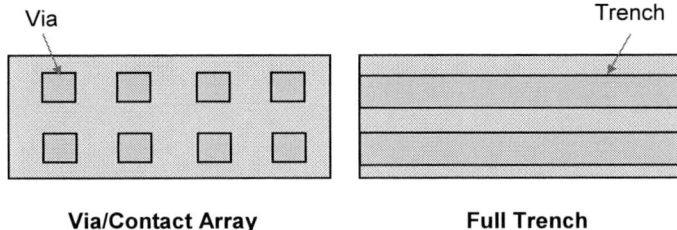

Figure 8. A comparison of die "seal ring" designs (top view).

V. SUMMARY

Innovations have been made at various levels to make the network smart and energy efficient. While the EnergyWise network is promising, we should be aware of the potential reliability implications.

A case study is used to illustrate the potential increase of component failure risks due to the operating environmental condition changes. It also showed that the reliability community can respond to the challenges while focusing on better understanding the failure mechanism acceleration under different field use conditions and on making energy efficient and reliability aware design improvement to mitigate the potential increase of risks.

REFERENCES

[1] http://www.ciscoknowledgenetwork.com/borderless/index.php

[2] U.S. Environmental Protection Agency, "Report to Congress on Server and Data Center Energy Efficiency", August, 2007.

[3] http://www.cisco.com/en/US/solutions/ns726/intro_content_energywise.html.

[4] American Society of Heating, Refrigerating and Air-Conditioning Engineers, "2008 ASHRAE Environmental Guidelines for Datacom Equipment," 2008.

[5] M. C. Bost, et al, "Method of forming a guard wall to reduce delamination effects," US Patent 5,270,256.

[6] M. W. Lane, et al, "Environmental Effects on Cracking and Delamination of Dielectric Films," IEEE Trans-Device and Materials Reliability, Vol. 4, No. 2 (2004), pp. 142-147.

[7] A. G. Thorsness and A. J. Muscat, "Moisture Absorption and Reaction in BPSG Thin Films," Journal of The Electrochemical Society, Vol. 140, No. 12 (2003), pp. 219-228.

[8] C. G. Shirley and C. Hong, "Optimal Acceleration of Cyclic THB Test for Plastic-Packaged Devices," in Proc. IRPS Conf., 1988, pp. 90–95

[9] L. Li, et al, "Environmental Effects on Dielectric Films in Plastic Encapsulated Silicon Devices," Proc. 57th ECTC, pp.755-760, 2007.

Comparison of two calibration methods for a package stress measurement testchip

Christian Djelassi, Helmut Köck and Michael Glavanovics

Kompetenzzentrum für Automobil- und Industrieelektronik (KAI)
Europastrasse 8, 9524 Villach, Austria
+434242-34890-04, christian.djelassi@k-ai.at

Abstract— **Analog and digital ICs are built up with bipolar or MOS (Metal Oxide Semiconductor) transistors. Such miniaturized transistor structures can fail when the thermo-mechanical stress becomes critical. The main aim of this work is to analyze the thermo-mechanical stress in integrated devices. Different approaches have been proposed to determine the thermo-mechanical stress effect in an IC device, based on stress-dependent properties of resistors, bipolar transistors or MOS transistors [1], [2]. In this paper two calibration methods for stress sensitive on-chip CMOS transistors are presented. One benefit of the described structures is that they can be produced in a standard semiconductor process. Another benefit is that electrical parameters of compact MOS transistors, in particular their charge carrier mobility, are highly stress dependent [1]. The reliability of the presented calibration methods is verified by measurements and simulations.**

Keywords: silicon stress measurement, package stress, CMOS stress sensor, beam bending calibration

I. INTRODUCTION

The increasing complexity of smart power systems in the automotive sector leads to a huge number of ICs. To guarantee the safety and reliability of the application and devices, respectively different investigations are required. One important part of the reliability issue is the packaging process, due to the fact that the packaging assembly process of the semiconductor device can stress the chip considerably. The many temperature cycles over the lifetime of the device further stresses the chip, especially in automotive applications. This thermo-mechanical stress may influence the functionality of the device and lead to an application failure.

To analyze this stress we designed a Package Interaction Testchip chip (PIT) with on-chip stress sensitive sensors. Those sensors are PMOS and NMOS transistors connected as current mirrors with special arrangement, which enables the possibility to analyze different stresses. Both, the PMOS sensor which is sensitive to shear stress and the NMOS sensor sensitive to normal stress are measurement devices based on the piezo-resistive effect. Those stress sensitive chips will be mounted in different package types, and subjected to temperature cycling to investigate the induced stresses. However, before we can analyze the measured stresses, the sensors have to be calibrated. In this paper we show two different calibration systems, an indenter and a beam bending

system. The presented calibration systems are verified by measurements and FEM simulations.

A. General Stress Definition

The average stress on an infinitesimal area expressed by the quotient of force per area, assuming that Σ approaches a finite value for Lim $\Delta A \to 0$ [3]:

$$\Sigma = \lim_{\Delta A \to 0} \frac{\Delta F}{\Delta A} = \frac{dF}{dA} \tag{1}$$

The stress consists of a normal component, called normal stress (σ) and a tangential component, called shear stress (τ). On a 3D body like in Figure 1, the stress is a second rank tensor which may be represented in Cartesian coordinates [3],[4]:

$$\sigma_{ij} = \begin{pmatrix} \sigma_{xx} & \tau_{xy} & \tau_{xz} \\ \tau_{yx} & \sigma_{yy} & \tau_{yz} \\ \tau_{zx} & \tau_{zy} & \sigma_{zz} \end{pmatrix} \tag{2}$$

Application of moment equilibrium conditions at a point leads to [3],[4]:

$$\tau_{xy} = \tau_{yx} \qquad \tau_{yz} = \tau_{zy} \qquad \tau_{xz} = \tau_{zx} \tag{3}$$

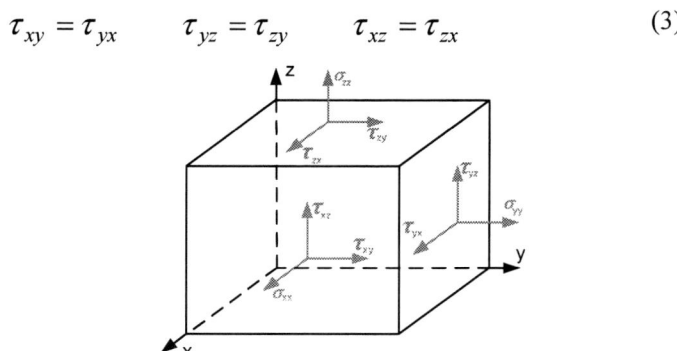

Figure 1 - Stress Definition

Most standard materials show a linear behavior relating stress to strain over a limited range and are fully reversible. For a force acting on a body and leading to a change in length, Hooke's law states [3],[4]:

$$\sigma = \frac{\Delta l}{l} \cdot E = \varepsilon \cdot E \tag{4}$$

978-1-4244-5430-3/10 $26.00 © 2010 IEEE

where E is Young's modulus. For anisotropic material like silicon, the crystal structure (Figure 2) causes directional dependencies of E, i.e. different E components for different crystallographic directions.

B. Piezoresistance Effect

The piezoresistive effect describes a change of the electrical resistance due to mechanical stress, which is highly distinctive for semiconductors, such as silicon [5], [6], [7]. The piezo-resistive effect of silicon can be attributed to the change of the inter-atomic distances due to thermo-mechanical stress. This deformation affects the band gap, i.e. the energy difference between the valence and the conduction band. The mobility of charge carriers depends on the band gap, causing the piezo-resistive effect, but also on temperature, an effect that has to be compensated if one wants to measure stress. Moreover, the change of inter-atomic distance depends on the specific orientation of the atomic bonds within the silicon crystal [4]. The quantitative relation between mechanical stress and mobility for silicon was explored by Kanda [8].

To describe the resistance of silicon, the following equation may be used [4]:

$$R = \rho_0 \cdot \frac{L}{A} = \frac{1}{qn\mu}\frac{L}{WH} \tag{5}$$

where n is the density of free charge carriers, μ their mobility and q is the charge per carrier, L, W and H are the length, width and height of the resistor, respectively. To calculate the change in resistance due to stress, the differential form of (5) is used and simplified [9] to:

$$\frac{\Delta R}{R} = \frac{R(\sigma \neq 0) - R(\sigma = 0)}{R(\sigma = 0)} = \frac{\Delta L}{L} - \frac{\Delta W}{W} - \frac{\Delta H}{H} - \frac{\Delta \mu}{\mu}$$

$$\frac{\Delta R}{R} \cong -\frac{\Delta \mu}{\mu} \tag{6}$$

It may be assumed that for silicon the change of geometrical dimensions L, W and H are negligible in comparison to the change of mobility μ [9]. Finally, the change of specific resistance (ρ) can be calculated as [4], [7]:

$$\frac{\Delta \rho}{\rho} = \pi_{ij} \cdot \sigma_{ij}$$

$$\frac{1}{\rho}\begin{bmatrix} \Delta\rho_{11} \\ \Delta\rho_{22} \\ \Delta\rho_{33} \\ \Delta\rho_{23} \\ \Delta\rho_{13} \\ \Delta\rho_{12} \end{bmatrix} = \begin{bmatrix} \pi_{11} & \pi_{12} & \pi_{12} & 0 & 0 & 0 \\ \pi_{12} & \pi_{11} & \pi_{12} & 0 & 0 & 0 \\ \pi_{12} & \pi_{12} & \pi_{11} & 0 & 0 & 0 \\ 0 & 0 & 0 & \pi_{44} & 0 & 0 \\ 0 & 0 & 0 & 0 & \pi_{44} & 0 \\ 0 & 0 & 0 & 0 & 0 & \pi_{44} \end{bmatrix} \cdot \begin{bmatrix} \sigma_{xx} \\ \sigma_{yy} \\ \sigma_{zz} \\ \tau_{yz} \\ \tau_{xz} \\ \tau_{xy} \end{bmatrix} \tag{7}$$

π_{ij} is the resistor orientation dependency 6x6 matrix, referring to the silicon crystal structure (Figure 2). The piezo-coefficient π_{11} represents the longitudinal, π_{12} the transversal stress and π_{44} the shear stress component. Equation (7) is only valid if the applied stress is described in the silicon crystal coordinate system (see Figure 2). Otherwise, a coordinate transformation is required.

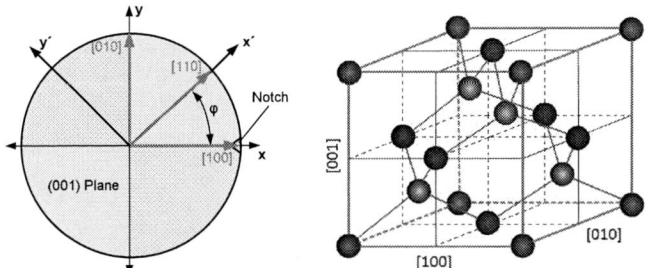

Figure 2 - Used wafer orientation (l) and structure of a silicon crystal (r)

First investigations to define the piezo-coefficients (π) of silicon were done in the 1960s for n- and p-type devices. Table 1 shows the first order coefficients for n- and p-type of silicon material.

Table 1 - First order piezo-resistance coefficients from [8]

	$\pi_{11}\left[\frac{10^{-11}}{Pa}\right]$	$\pi_{12}\left[\frac{10^{-11}}{Pa}\right]$	$\pi_{44}\left[\frac{10^{-11}}{Pa}\right]$
n-type Si	-102.2	53.4	-13.6
p-type Si	6.6	-1.1	138.1

It has to be mentioned that the values from Table 1 are only valid for slightly doped silicon material (~10^{16} cm^{-3}). Table 1 shows that n-type Si is more sensitive to normal stress (π_{11} and π_{12}), whereas p-type Si is most sensitive to shear stress (π_{44}), in particular when rotated 45° with respect to the silicon [100] crystal axis [1], [9], [10], [11]. The graphical representation of the piezo-coefficients (Figure 3) shows that the n-type Si has its sensitivity maximum along the [100] silicon lattice axis, whereas the p-type shows its maximum in the [110] direction, which is in our case rotated by $\pm$ 45° relative to the main wafer orientation. Hence, a coordinate transformation (φ=45°) of the π_{ij} piezo-coefficients has to be done:

$$\frac{1}{\rho}\begin{bmatrix} \Delta\rho_{11} \\ \Delta\rho_{22} \\ \Delta\rho_{33} \\ \Delta\rho_{23} \\ \Delta\rho_{13} \\ \Delta\rho_{12} \end{bmatrix} = \begin{bmatrix} \pi'_{11} & \pi'_{12} & \pi'_{13} & 0 & 0 & 0 \\ \pi'_{12} & \pi'_{11} & \pi'_{13} & 0 & 0 & 0 \\ \pi'_{13} & \pi'_{13} & \pi'_{33} & 0 & 0 & 0 \\ 0 & 0 & 0 & \pi'_{44} & 0 & 0 \\ 0 & 0 & 0 & 0 & \pi'_{44} & 0 \\ 0 & 0 & 0 & 0 & 0 & \pi'_{66} \end{bmatrix} \cdot \begin{bmatrix} \sigma'_{xx} \\ \sigma'_{yy} \\ \sigma'_{zz} \\ \tau'_{yz} \\ \tau'_{xz} \\ \tau'_{xy} \end{bmatrix} \tag{8}$$

$$\pi'_{11} = \frac{1}{2}(\pi_{11} + \pi_{12} + \pi_{44}), \pi'_{12} = \frac{1}{2}(\pi_{11} + \pi_{12} - \pi_{44})$$

$$\pi'_{13} = \pi_{12}, \pi'_{33} = \pi_{11}, \pi'_{44} = \pi_{44}, \pi'_{66} = \pi_{11} - \pi_{12}$$

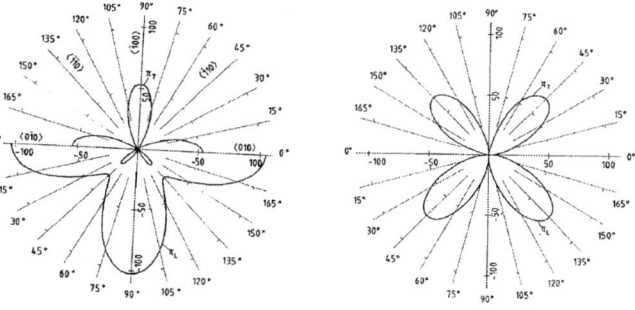

Figure 3 – Piezo-resistive coefficients for n-Si(l) and p-Si(r) from [4]

The whole set of transformation equations is derived in [10], [12]. According to Figure 3, a longitudinal stress sensitive n-type sensor should be placed along the silicon main direction [100] and a shear stress sensor along the [110] direction.

C. Piezo-MOS Effect

As mentioned before the stress cells are MOS devices. In principle, a MOS transistor may be regarded as a voltage controlled resistor. The drain-to-source current can be controlled by the gate voltage. In general, MOS transistors have three possible operation regions, the weak ($V_{GS} < V_{th}$), the moderate and the strong inversion mode. These modes depend on the forced voltages at the transistor interfaces. The moderate inversion mode is called linear region ($V_{GS} > V_{th}$ and $V_{DS} < (V_{GS}-V_{th})$), and the strong inversion mode also referred as saturation region ($V_{GS} > V_{th}$ and $V_{DS} > (V_{GS}-V_{th})$)[13].

$$\text{Linear region: } I_D = \beta(V_{GS} - V_{th} - \frac{V_{DS}}{2})V_{DS} \qquad (9)$$

$$\text{Saturation region: } I_D = \frac{\beta}{2}(V_{GS} - V_{th})^2 \qquad (10)$$

$$\text{Gain factor: } \beta = \mu_i C_{ox} \frac{W}{L} \qquad (11)$$

where V_{th} is the threshold voltage and μ_i is the mobility.

In the strong inversion region, the drain current is - in first approximation - controlled only by the gate voltage. For a defined gate and drain voltage, the current depends on the gain factor, see (11). Former investigations have found that the influence of width and length changes due to thermo-mechanical stress is negligible in contrast to the mobility change [9], [10], [11]. Furthermore, it is shown in [9], [10], [11] that the mobility change due to mechanical stress has its maximum effect in the saturation region (10):

$$\frac{\Delta I_d}{I_d} = \frac{I_d(\sigma \neq 0) - I_d(\sigma = 0)}{I_d(\sigma = 0)} = \frac{\Delta\mu}{\mu} + \frac{\Delta W}{W} - \frac{\Delta L}{L} - \frac{\Delta V_{th}}{V_{gs}}$$

$$\frac{\Delta I_d}{I_d} \cong \frac{\Delta\mu}{\mu} \qquad (12)$$

Equation (12) states that stress affected current changes are, in good approximation, only an effect of variation in charge mobility. Also the temperature dependency reduces in (12) because only the current differences are significant. The mobility change of the MOS transistor can here be interpreted as a change in channel resistivity [11]. Therefore, the channel resistivity is strongly affected by thermo-mechanical stress [9], [10], [11] and the piezo-coefficients from Table 1 (which are the same as for lightly doped silicon resistors [1], [11]) may consequently be applied to calculate the current I_d shift.

II. STRESS SENSOR AND LAYOUT

The stress sensor cells are MOSFET (metal oxide semiconductor field effect transistor) devices. Table 1 and Figure 3 indicate that for in-plane normal stress measurements an n-type transistor oriented along the crystal main axis [100] should be used, whereas for shear stress measurements, a p-

type transistor 45° rotated along the [110] axis would be preferable.

If the device current applies along the [100] direction the mobility change due to stress according to (12) and (7), is [7], [10]:

$$\frac{\Delta\mu}{\mu_0}\Big|_{[100]} = \frac{\Delta\rho_{11}}{\rho} + \frac{\Delta\rho_{12}}{\rho} = \pi_{11}\sigma_{xx} + \pi_{12}\sigma_{yy} + \pi_{12}\sigma_{zz} + \pi_{44}\tau_{xy} \qquad (13)$$

Equation (13) includes normal and shear stress components, as expected from (7). These components are:

➤ $\pi_{11}\sigma_{xx}$, representing the longitudinal in-plane normal stress component

➤ $\pi_{12}\sigma_{yy}$, representing the transversal in-plane normal stress component

➤ $\pi_{12}\sigma_{zz}$, representing the off-plane normal stress component

➤ $\pi_{44}\tau_{xy}$, representing the shear stress component

By rotating the device 90°, to the [010] direction, now the mobility change becomes:

$$\frac{\Delta\mu}{\mu_0}\Big|_{[010]} = \frac{\Delta\rho_{22}}{\rho} + \frac{\Delta\rho_{12}}{\rho} = \pi_{12}\sigma_{xx} + \pi_{11}\sigma_{yy} + \pi_{12}\sigma_{zz} + \pi_{44}\tau_{xy} \qquad (14)$$

A possible circuit to compare the stress sensitivity of the [100] and [010] direction is a current mirror, where the transistors are arranged along the silicon lattice axis but 90° rotated to each other (Figure 4). The differential combination of the two orientations, besides compensating for temperature dependency in first order, yields:

$$\frac{\Delta I_d}{I_d} = \frac{\Delta\mu}{\mu_0}\Big|_{[100]} - \frac{\Delta\mu}{\mu_0}\Big|_{[010]} = -(\pi_{11} - \pi_{12}) \cdot (\sigma_{xx} - \sigma_{yy}) \qquad (15)$$

These directions correlate directly to the wafer coordinate system (Figure 2); therefore, no coordinate transformation needs to be performed.

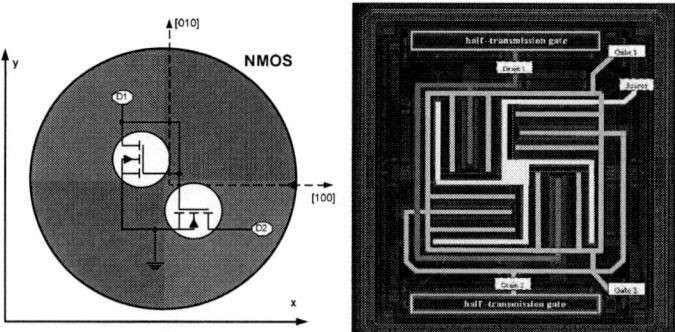

Figure 4 - NMOS current mirror wafer orientation (l) and final NMOS stress cell (r)

Hence, equation (15) shows the final correlation between the current ratio of the NMOS stress cell (Figure 4) and the difference of in-plane normal stresses.

Another important stress factor is the shear stress, where the PMOS transistors are most sensitive (Table 1). The PMOS device is most stress sensitive when placed 45° rotated to the notch, i.e. along the [110] silicon crystal axis [1], [9], [10], [11]. Again a current mirror (Figure 5) is the best practical solution. If the current flows along the [110] direction the mobility change due to thermo-mechanical stress will be:

$$\left.\frac{\Delta\mu}{\mu_0}\right|_{[110]} = \frac{\Delta\rho_{11}}{\rho} + \frac{\Delta\rho_{12}}{\rho} = \pi_{11}\sigma'_{xx} + \pi_{12}\sigma'_{yy} + \pi_{12}\sigma'_{zz} + \pi_{44}\tau'_{xy} \quad (16)$$

The thermo-mechanical stress influence on the mobility along the [$\bar{1}$10] direction is:

$$\left.\frac{\Delta\mu}{\mu_0}\right|_{[\bar{1}10]} = \frac{\Delta\rho_{22}}{\rho} + \frac{\Delta\rho_{12}}{\rho} = \pi_{12}\sigma'_{xx} + \pi_{11}\sigma'_{yy} + \pi_{12}\sigma'_{zz} + \pi_{44}\tau'_{xy} \quad (17)$$

The stress-related current change of a p-type MOS device mirror is now the difference between (16) and (17), which is:

$$\frac{\Delta I_d}{I_d} = \left.\frac{\Delta\mu}{\mu_0}\right|_{[110]} - \left.\frac{\Delta\mu}{\mu_0}\right|_{[\bar{1}10]} = -(\pi_{11}-\pi_{12})\cdot(\sigma'_{xx}-\sigma'_{yy}) \quad (18)$$

However, the equation above cannot be used to adequately describe the PMOS devices because of two reasons:

➤ The p-type piezo-coefficients π_{11} and π_{12} (Figure 3) have to be rotated by 45° so that the piezo-coefficients share the global coordinate system (Figure 2)

➤ The stress components σ'_{xx} and σ'_{yy} also have to be rotated by 45° to share the global coordinate system (Figure 2)

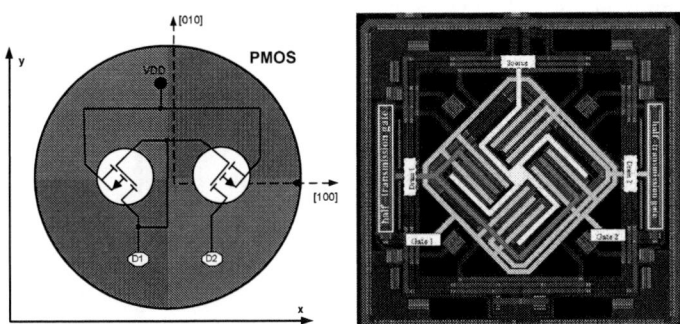

Figure 5 - PMOS current mirror wafer orientation (l) and final PMOS stress cell (r)

The 45° rotation of the piezo-coefficients π_{ij} is displayed in (8). The 45° rotation of π_{11} - π_{12} yields the π'_{66} component, which represents the shear coefficient π_{44} of the shear stress τ_{xy}. A rotation of 45° of the stresses $(\sigma'_{xx} - \sigma'_{yy})$ yields:

$$\sigma'_{xx} = \frac{1}{2}(\sigma_{xx}+\sigma_{yy}) + \frac{1}{2}(\sigma_{xx}-\sigma_{yy})\cos 2(45°) + \tau_{xy}\sin 2(45°)$$

$$\sigma'_{yy} = \frac{1}{2}(\sigma_{xx}+\sigma_{yy}) - \frac{1}{2}(\sigma_{xx}-\sigma_{yy})\cos 2(45°) - \tau_{xy}\sin 2(45°) \quad (19)$$

$$\sigma'_{xx} - \sigma'_{yy} = 2\cdot\tau_{xy}$$

The final equation relating the measured current ratio of the PMOS stress cell and the shear stress is this:

$$\frac{\Delta I_d}{I_d} = -\pi_{44}\cdot 2\cdot\tau_{xy} \quad (20)$$

A. Measurements

For practical stress measurements, we have designed two test chips with each 32 PMOS and 32 NMOS stress cells placed on different positions on a chip (Figure 6). To select individual stress cells, an analog multiplexer is placed on each stress chip. This multiplexer consists of a digital 5-bit address decoder (Figure 1) and two analog "transmission gates (Figure 4 and Figure 5) for each stress cell.

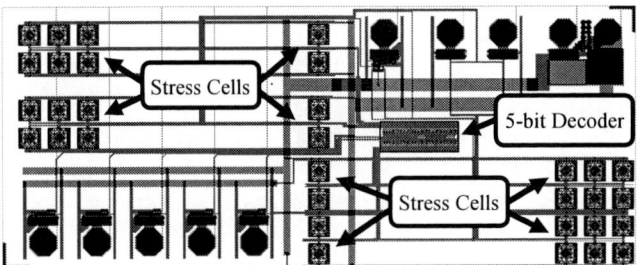

Figure 6 - Layout of one PMOS stress cell chip

III. INDENTER CALIBRATION

To verify the functionality of the stress cells and to calibrate them, a mechanical test setup was required. The first approach includes a force gauge and a blade-shaped indenter to generate a local stress on the chip surface next to the stress cells and measure their electrical response.

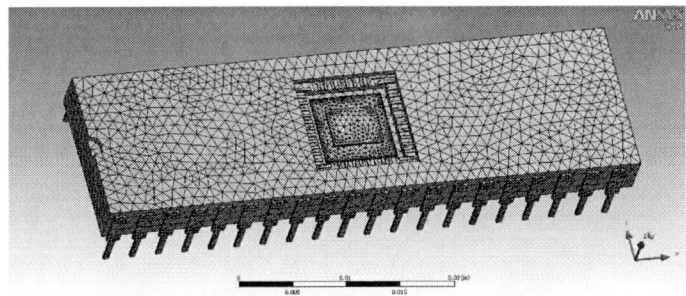

Figure 7 – FEM Model of PIT including the ceramic package

To elaborate the mechanical stress state within the chip, generated by the indenter, an FEM (Finite Element Method) simulation with ANSYS [14] was performed. For this purpose a model with the according material properties has been designed (Figure 7) and the simulation results have been compared to the measurements (Figure 8). The indenter calibration was performed for NMOS (σ_{xx} - σ_{yy}) and PMOS (τ_{xy}) stress cells to analyze the different stress states.´

In the second indenter calibration method we have used a ball shaped indenter, directly acting between the stress cells. Again, an FEM simulation was performed and compared with the measurement (Figure 9). Both indenter calibration methods show a good agreement between measurements and simulations. However, due to the fact that the indenter position and entrance angle cannot be defined exactly, an advanced method, the beam bending calibration, was developed.

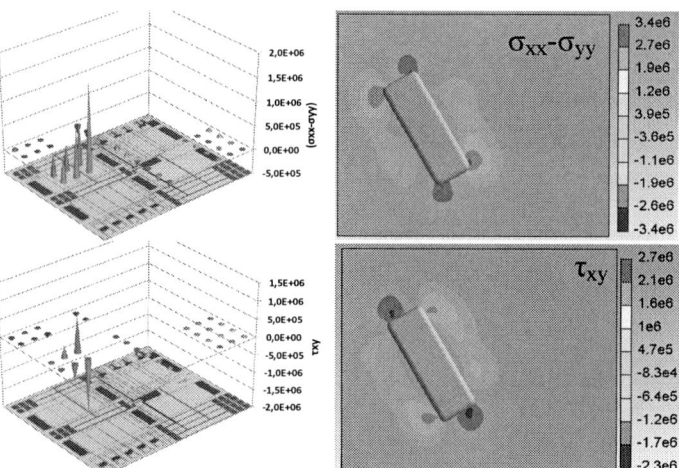

Figure 8 - Stress measurements (l) and FEM simulations results (r) of manually stressed chip with a blade-shaped indenter

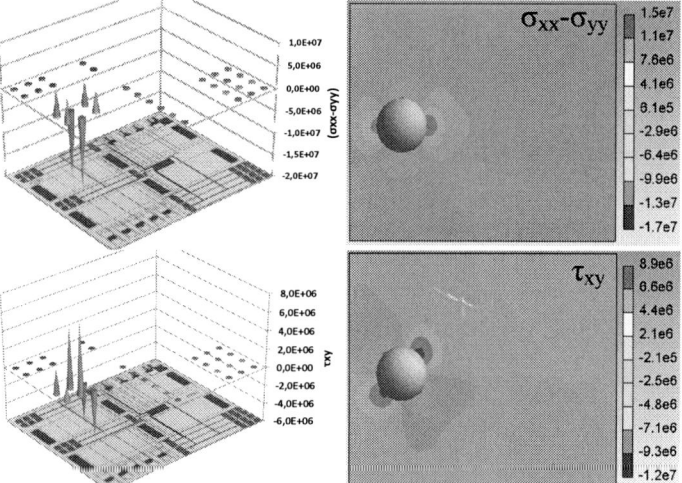

Figure 9 - Stress measurements (l) and FEM simulations results (r) of manually stressed chip with a ball shape indenter

IV. BEAM BENDING CALIBRATION

The cantilever or beam bending calculation is a well-known procedure. Traditionally, beam bending calculations are used in construction engineering. Here they will be applied to calculate a special uniaxial stress state in a silicon wafer. First, a simplified model (Figure 10) of the beam bending is needed [3]. Based on this model, the two dimensional stress in a strip of silicon can be calculated (Figure 11).

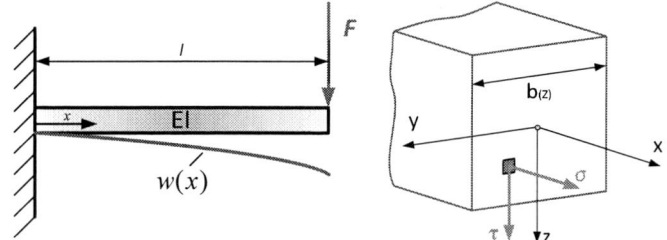

Figure 10 - Schematic of a 2D beam bending (l) and stress orientation of a beam (r)

First, the deflection curve is calculated to determine the displacement in each point. To calculate the deflection curve of a beam the Euler-Bernoulli beam theory is used. Solving the four differential equations of the simplified model (Figure 1) yields [3]:

$$w'' = -\frac{F}{EI} \cdot (-x + l) \tag{21}$$

where E is Young's modulus and I is the geometrical moment of inertia. Integration of w'' yields the beam deflection in z direction [3]:

$$w(x) = \frac{Fl^3}{6EI} \cdot \left(-\frac{x^3}{l^3} + 3\frac{x^2}{l^2} \right) \tag{22}$$

With $w(x)$ the deflection in each point under load is now known. However, for the present application it is more necessary to determine the stress distribution. As outlined before, there are two types of stress, normal stress (σ) and the shear stress (τ) (Figure 10 and Figure 11). Those stresses can be calculated as [3]:

$$\sigma = \frac{M}{I} \cdot z = \frac{F}{I} \cdot z(-x + l)$$

$$\tau = \frac{VQ}{Ib} = \frac{3Q}{2bh} \left(1 - \frac{4z^2}{h^2} \right) \tag{23}$$

where σ is the normal stress, F is the force, I is the geometrical moment of inertia, z is the distance from the neutral line to the edge of the beam (Figure 11), τ is the shear stress, Q being the static moment, V is the total shear force at the point of interest and Ib the geometrical moment of inertia (Figure 11).

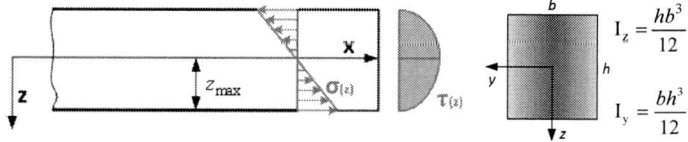

Figure 11- Stress distribution of a silicon beam cross section (l) and its geometrical moment (r)

For a practical approach again a mechanical test setup is needed [15]. In Figure 12 the final construction and the concept drawing of the bending beam system with the main parts are displayed.

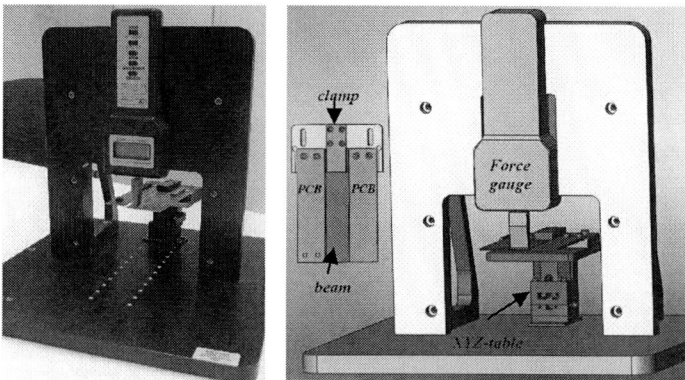

Figure 12 – Bending beam system (l) and its CAD drawing (r)

978-1-4244-5430-3/10 $26.00 © 2010 IEEE   450

In our case the bending beam is a silicon strip cut from an 8 inch wafer (Figure 13). To achieve a connection to one of the stress-sensing chips, two PCBs with gold pads and connectors (Figure 13) were designed and mounted left and right next to the silicon strip. The electrical connections between the chip and the PCBs were done with gold bond wires (Figure 13).

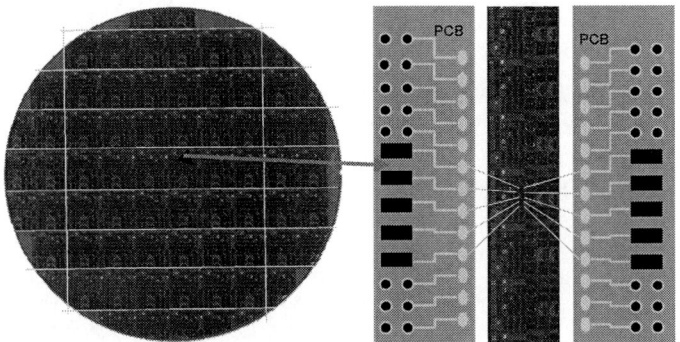

Figure 13 - 8" wafer with sawing subscription (l) and schematic of electrical connections for beam bending (r)

For the first experiment we forced 0.08 N at the free edge of the beam and measured the deflection (Figure 14). For verification we also calculated it based on (22) and simulated it with ANSYS. Both, the calculated and simulated deflection are in good agreement with the measurement (Figure 14).

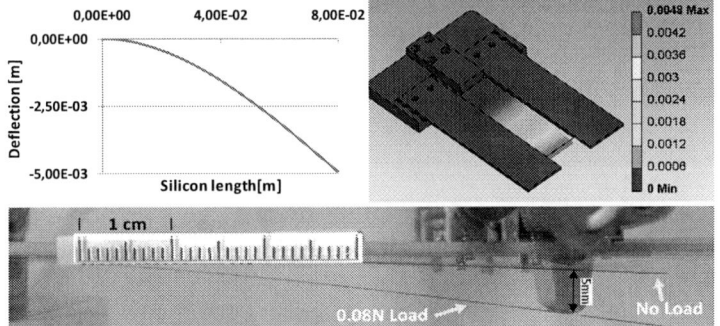

Figure 14 – Beam bending above calculation (l), simulation above (r) and measurement bottom

Next we measured the electrical response at different loadings and created for each loading again FEM simulations (Figure 15). Unfortunately, the shear stress becomes zero at the chip surface (Figure 11) [3] in a symmetrical beam bending, which was confirmed also by the PMOS cells that show no change of the electrical response under uniaxial bending. Hence, a symmetrical beam bending under uniaxial stress cannot be used to calibrate the shear stress sensor directly. Therefore a small modification of the beam bending system was introduced and FEM simulated, which was to fix the silicon beam also on the third corner and force a load on the free fourth corner (Figure 15).

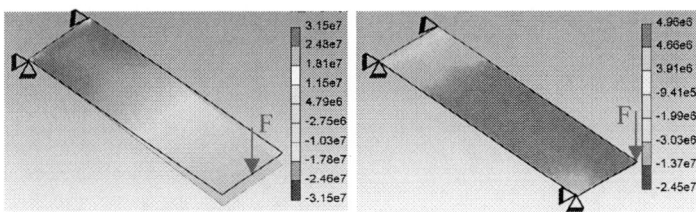

Figure 15 – FEM simulation of σ_{xx}-σ_{yy} for a symmetrically loaded beam bending (l) and τ_{xy} for the three point fixed (warped) beam bending (r)

A. Discussion

The electrical responses of the sensor cells are nearly linear with respect to stress caused by mechanical distortion (Figure 16). It is of interest to find the quantitative relation between stress due to the applied loading force and sensor signal ($\Delta I/I$). Corresponding mechanical FEM simulations were performed (Figure 15) and compared to the previously described measurements (Figure 16).

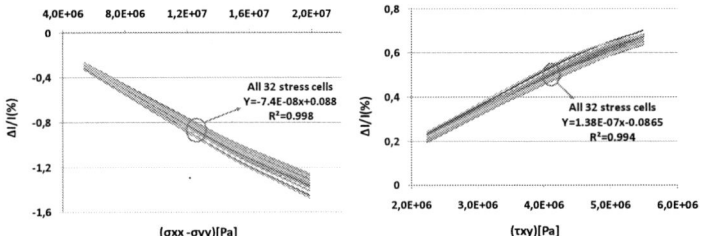

Figure 16 – Measured relation between $\Delta I/I$, σ_{xx}-σ_{yy} for the symmetrically loaded beam bending (l) and τ_{xy} for the three point fixed (warped) beam bending (r)

Figure 16, respectively (24) and (25) represent the thus established relationships between the sensor signals ($\Delta I/I$) and the stresses.

$$\frac{\Delta I}{I}_{NMOS} = 7.4 \cdot 10^{-10} \, Pa \tag{24}$$

$$\frac{\Delta I}{I}_{PMOS} = 13.8 \cdot 10^{-10} \, Pa \tag{25}$$

The observed stress coefficients for NMOS and PMOS are about two times smaller than the literature values [8] of $-(\pi_{11}^{n}-\pi_{12}^{n}) = 15.5 \cdot 10^{-10}$ Pa and $2 \cdot \pi_{44}^{p} = 27.6 \cdot 10^{-10}$ Pa for low doped Si-resistors (computed from Table 1).

The deviation of the measured values may be attributed to two different effects, which are difficult to distinguish:

➤ The error introduced by the transmission gates (Figure 4 and Figure 5) in series to the actual sensors, because their stress sensitivity adds to the asymmetry of the current mirror.

➤ Previous investigations have also shown that the piezo-resistive coefficients depend on the sensor doping concentrations. A comparison of different measured piezo-resistive coefficients for heavily and low doped n- and p-type Silicon is shown in Table 2.

Table 2 – Comparison of measured and literature values of piezoresistive coefficients for different doping concentrations

Type	Lightly doped Si [8]	Heavily doped Si [16]	Heavily doped Si [17]	Presented
	($\times 10^{-11}$ Pa^{-1})			
π_{11}^{n}	-102.2	-45.4	-26.7	
π_{12}^{n}	53.4	24.2	14.2	
$-(\pi_{11}^{n}-\pi_{12}^{n})$	155.6	69.6	40.9	74.5
π_{44}^{p}	138.1	62.1	72.2	68.9

Comparing Table 2 to the above presented calibration and simulation results it can be concluded that our stress cells are suitable for measuring thermo-mechanical stress. However, the observed stress coefficients may be influenced by the sub-optimal multiplexer architecture.

CONCLUSION

In this paper we have introduced two calibration methods for a thermo-mechanical stress measurement test chip. The test-chips consist of piezo-resistive MOS sensors which are sensitive for thermo-mechanical normal and shear stress. Both, the indenter and beam bending calibration methods are verified by measurements and simulations. The indenter calibration shows a good correlation between FEM simulation and measurement but an accurate calibration is not possible due to the manual positioning handicap.

An improved calibration method has been introduced with the beam bending system, which can be used for an accurate calibration of the measurement signal versus mechanical stress. However, an outcome of the beam bending calibration is that our analog multiplexer, more precisely the transmission gates are also stress sensitive and may limit the accuracy of the measurement results. Based on the presented data from this paper a new test chip replacing the multiplexer with an innovative sensor arrangement has been designed. The new sensor chip yields promising results which will be published soon.

ACKNOWLEDGMENT

We thank our industry partner Infineon Technologies Austria for testchip circuit design and layout as well as for providing silicon samples. The contributions of Thomas Detzel, Michael Nelhiebel, Stefan Wöhlert, Bernhard Meldt and Javad Zarbakhsh for project management and coordination are gratefully acknowledged. Theoretical considerations and FEM simulations were supported by the expert know-how of Vladimir Kosel and Josef Schweda. This work was jointly funded by the Federal Ministry of Economics and Labour of the Republic of Austria (contract 98.362/0112-C1/10/2005 and the Carinthian Economic Promotion Fund (KWF) (contract 18911 | 13628 | 19100).

REFERENCES

[1] Richard C. Jaeger, Jeffrey C. Suhling, Ramanathan Ramani, Arthur T. Bradley, and Jianping Xu, "CMOS Stress Sensors on (100) Silicon," *IEEE JOURNAL OF SOLID-STATE CIRCUITS,* vol. 35, January 2000.

[2] Jeffrey C. Suhling and R. C. Jaeger, "Silicon Piezoresistive Stress Sensors and Their Application in Electronic Packaging," *IEEE SENSORS JOURNAL,* vol. 1, June 2001.

[3] Gross, Hauger, Schröder, and Wall, *Technische Mechanik 2 - Elastostatik* vol. 9. Berlin Heidelberg New York: Springer ISBN 978-3-540-70762-2, 2007.

[4] Ulrich Mescheder, *Mikrosystemtechnik - Konzepte und Anwendungen* vol. 2. Stuttgart Leibzig Wiesbaden: B. G. Teubner, ISBN 3-519-16256-3, Year 2004.

[5] P. W. Bridgeman, "The effect of homogenous mechanical stress on the electrical resistance of crystals," *Physical Review,* vol. 42, pp. 858-863, 1932.

[6] P. W. Bridgeman, "The effect of pressure on the electrical resistance of certain semiconductors," *Proc. Amer. Acad. Arts Sci.,* vol. 79, pp. 125-179, 1951.

[7] Charles S. Smith, "Piezoresistance Effect in Germanium and Silicon," *Physical Review,* vol. 1, April 1954.

[8] Yozo Kanda, "A Graphical Representation of the Piezoresistance Coefficients in Silicon," *IEEE TRANSACTIONS ON ELECTRON DEVICES,* vol. 29, January 1982.

[9] Yonggang Chen, "CMOS STRESS SENSOR CIRCUITS," Graduate Faculty of Auburn University Alabama: Dissertation, 2006.

[10] Josef Schweda, "Analysis of Shift Mechanisms in BiCMOS Hall Devices with Scope of a Precise SPICE Simulation Model," Fakultaet fuer Elektrotechnik und Informationstechnik TU Vienna: Dissertation, 2007.

[11] Richard C. Jaeger, Ramanathan Ramani, Jeffrey C. Suhling, and R. W. Johnson, "Effects of Stress-Induced Mismatches on CMOS Analog Circuits," in *VLSI Technology, Systems and Applications,* 1995.

[12] Gökhan Kizilirmak, "Frei applizierbare MOSFET-Sensorfolie zur Dehnungsmessung," Fakultät für Elektrotechnik und Informationstechnik RTWH Aachen: Dissertation 2007.

[13] Kurt Hoffmann, *System Integration - From Transistor Design to Large Scale Integrated Circuits* Munich: John Wiley & Sons, ISBN 0-470-85407-3, Year 2004.

[14] ANSYS Inc. : Web page, http://www.ansys.com/.

[15] Christian Djelassi, "Development of an Automated Measurement System for a Smart Power Technology Testchip to Characterize Mechanical Package Stress," Carinthia University of Applied Sciences Villach: Diploma Thesis, 2009.

[16] Kuo Tian, Zheyao Wang, Min Zhang, and L. Liu, "Design, Fabrication, and Calibration of Piezoresistive Stress Sensor on SOI Wafers for Electronic Packaging Applications," *IEEE TRANSACTIONS ON COMPONENTS AND PACKAGING TECHNOLOGIES,* vol. 32, June 2009.

[17] David W. Peterson, James N. Sweet, Steven N. Burchett, and A. Hsia, "Stresses From Flip-Chip Assembly and Underfill; Measurements with the ATC4.1 Assembly Test Chip and Analysis by Finite Element Methodt," in *Electronic Components and Technology Conference,* Proc. 1997 47th ECTC.

Reliability Evaluation Methodology for Chip-Package Interaction

Die Corner Edge Failure Mechanism

Melida Chin and Amit Marathe
GLOBALFOUNDRIES
1050 E Arques, Sunnyvale, California 94085, USA
phone: +1-408- 462-4127, email: melida.chin@globalfoundries.com

Abstract— **A novel reliability qualification methodology to evaluate chip-package interaction (CPI) is presented. Experimental and finite element analysis results are combined in a new protocol that provides solutions to address the need of CPI test vehicles for pre-qualification purposes, more accurate CPI risk projection, and more flexibility for productization.**
[Keywords: CPI, crack driving force, crackstop, reliability]

I. INTRODUCTION

Chip-package Interaction (CPI) is of significant importance for flip-chip packages that comprise low-k interlayer dielectrics (ILD), large dies, organic packages, and, lately, lead free bumps [1,2,3]. Massive delamination failures can be seen at the die corner edge if a crackstop is not included in the back-end-of-line (BEOL) of the die. If lead-free bumps are used as the controlled-collapse chip connect (C4), near-bump delamination failures arise near the bump and break the BEOL interconnect. The former CPI failure mechanism, die corner edge, is well understood and recommendations to minimize the risks have been widely published [4,5]. The latter CPI failure mechanism, near bump, is a relatively recent issue related to the new regulation of replacing lead with lead-free materials, along with adding ultra-low-k (ULK) as ILD, [6].

The focus of this paper is on CPI die corner edge, in which, even though it is well understood, reliability qualification can be challenging. Unlike failure mechanisms such as stress migration, time-dependent dielectric breakdown, and electromigration, CPI is more product-package driven and requires matured fab processes and a particular package design to start any qualification test. To address the challenge of short times to complete the qualifications, [7] proposed to use temperature shock test instead of temperature cycling test to reduce stress times by a factor of at least 28. However, the lack of proper test vehicle for pre-qualification purposes has not been addressed yet. The pre-qualification of the CPI is generally difficult because there is no chip from the node to be evaluated. Finite element analysis (FEA) and hybrid look-ahead test vehicles, which are in many cases from a previous node with some modifications to resemble the upcoming node, are usually used to evaluate CPI during the early stages of the technology node development cycle. FEA results are many times used to determine critical knobs, their trends, and,

therefore, the splits in a set of experimental tests. Unfortunately, with the increasing options for packaging, chip applications, and shorter timelines in the technology roadmaps, the number of splits can be too large to cover in the required time frame, or the parameter to be evaluated could be available much later during the qualification process. In this paper, we present a new protocol to combine FEA results with experimental tests, in which the FEA results are used more effectively to design the experimental tests. Besides enabling the experimental tests to cover a larger range of critical knobs in a shorter time, this novel methodology will provide solutions to address the need of CPI test vehicles for pre-qualification purposes, more accurate CPI risk projections, and more flexibility for productizing a variety of product configurations and technology nodes.

II. ANALYSIS

Critical parameters in the CPI die corner edge are die size and aspect ratio, ILD, assembly, underfill, package, lidded/lidless, and crackstop. Among these parameters, the crackstop is by far the most important because for the most challenging requirements - large die, ULK ILD, and organic and lidded package - a proper crackstop structure can arrest any crack growing from the edge of the die [5,8,9]. The fracture toughness of the crackstop, which is measured experimentally, is usually compared to the crack driving force in the flip-chip package, which is computed via FEA. The methodology herein is based on our capability to compute accurately crack driving force behavior, measure accurately crackstop fracture toughness, validate our CPI FEA model, and modulate our CPI accelerated reliability tests, and on our thorough understanding of these accelerated reliability tests.

A. Crack Driving Force (G)

The crack driving force (G) is the strength that a crack flaw, usually originated during mechanical saw dicing, has to grow along the BEOL. This G depends on the crack flaw size, and it is compared to the fracture toughness (Gc) of the BEOL thin film interfaces, cohesive strength of the BEOL thin-films, and Gc of the crackstop (Fig. 1). When ULK is one of the ILDs in the BEOL, G is usually much higher than Gc of the interfaces and cohesive strength of the BEOL thin films (case shown in

978-1-4244-5430-3/10 $26.00 © 2010 IEEE

Fig. 1). Gc of the crackstop, on the other hand, is set to be higher than G so the crack growth can be arrested. Crack driving force values are usually estimated by FEA methods. A schematic of the location of the crack flaw, crackstop, and the strategy we have used to compute the value of G via FEA are shown in Fig. 2. As shown in Fig. 1, G increases sharply with the length of the crack up to a threshold value; after that, G increases very slowly. G also depends on the change of temperature (ΔT). This is because CPI is driven mainly by large thermal strains, which are a result of a coefficient of thermal expansion (CTE) mismatch between the die and the package. The ΔT is an important parameter to determine G. During the assembly process, when the die is attached to the package, both parts are heated to temperatures above 100°C then cooled to room temperature. It is during the cool-down process that large thermal strains build up between the die and the package. The larger the cool-down ΔT, the larger G is expected to be. The family of G vs. crack length curves for a set of ΔT is shown in Fig. 3.

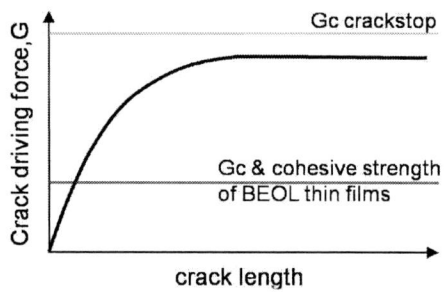

Figure 1. G versus crack length curve compared to fracture toughness of crackstop and BEOL lowK thin films

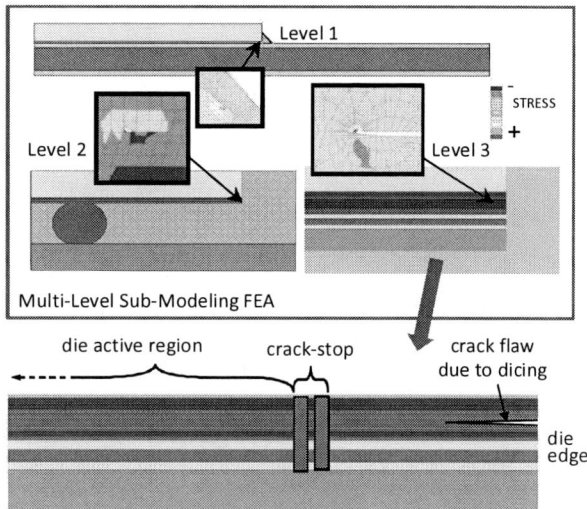

Figure 2. Schematic of the location of the expected crack flaw, crackstop, and strategy to compute G from FEA model

B. Crackstop Fracture Toughness (Gc)

The crackstop is a stack-up from backside contact all the way to top metal and around the perimeter of the chip. No active circuitry is connected to the crackstop. It has two main purposes: serve as a moisture oxidation barrier and as a crack stopper. The full stack-up characteristic allows the crackstop to serve as a moisture oxidation barrier, whereas a high via contact density, a large number of via bars, and vias arranged in a staggering manner can contribute to a high Gc in the crackstop, which in turn can arrest the growth of a crack. Many efforts and methods have been reported to accurately measure experimentally the value of Gc: four-point bend test (FPBT), double-cantilever beam (DCB), modified edge lift-off test (mELT), and others [5,9]. Fig. 4 shows a SEM picture of one DCB test. It is clearly shown that the crack is completely arrested by the crackstop.

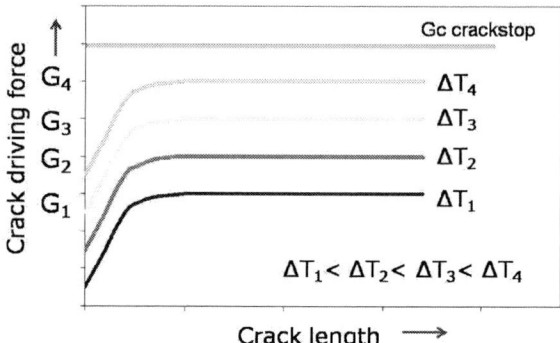

Figure 3. Family of G versus crack length curves
The larger ΔT, the larger is G

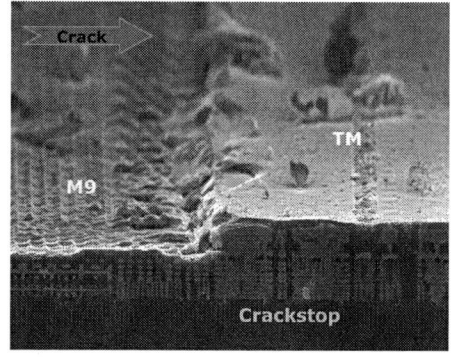

Figure 4. Crack arrested by crackstop in DCB test

C. Finite Element Analysis (FEA) Model

The FEA model used to evaluate the methodology presented uses the multi-level sub-modeling technique in which the package, the die corner edge, and the thin films in the BEOL (from 10^{-2}m to 10^{-9}m) are all evaluated in the same model. A schematic of the model is shown in Fig. 2. The multi-level sub-modeling approach interconnects independent models (level 1, level 2 and level 3 in Fig. 2) by exporting boundary conditions from one level to another (e.g., from level 1 to level 2). A two-dimensional (2D) plain-strain model was chosen because the CPI die corner edge failure mechanism occurs mainly at the corner edge of the die, where the assumption of the third dimension being much larger than the other two fits well. The model also includes some variations of each level (level 1 could be lidded or lidless). A similar FEA model is described in more detail in [4]. In that paper, comprehensive results of G values and the effect of different parameters are widely discussed.

Validation of the FEA Model - The reliability qualification methodology proposed requires that the FEA model be validated with experimental data. To validate the model, a simple test can be carried out. For a given flip-chip package type, two tests are run (see Fig. 5). In the first test, flip-chip package sample 1 is stressed at ΔT_1. In the second test, flip-chip package sample 2 is stressed at ΔT_2. In the latter test, ΔT_2 is increased until failure is detected. Sample 2 will be carefully monitored during the test to determine whether the sample has failed or not. Once failure is detected, the value of ΔT_2 is recorded. Then the FEA model is used to evaluate the values of G_1 and G_2. If G_1 and G_2 are the lower and upper bound of Gc (whose value has been determined from experimental measurements), then the model is validated. Some iterations are needed, so G_1 and G_2 are very close to the value of Gc. The flip-chip package must be equipped with proper crack sensor structures that can facilitate monitoring of crack growth in the BEOL and far-back-end-of-line (FBEOL).

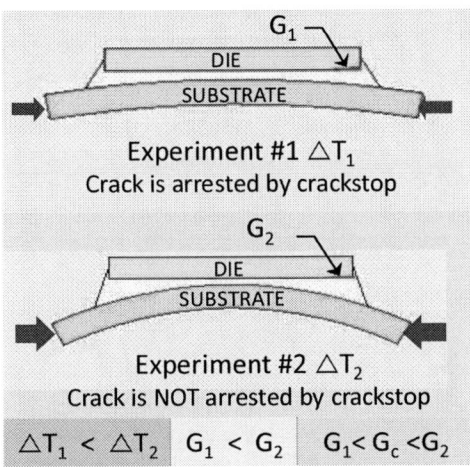

Figure 5. Schematic of FEA CPI validation method

D. CPI Accelerated Reliability Tests

Current recommended reliability qualification tests for flip-chip CPI are temperature cycling (TC), high-temperature storage life (HTSL), temperature humidity bias (THB), and highly accelerated temperature and humidity stress (HAST). The TC test accelerates damage caused by thermal-mechanical stress as a result of CTE mismatch and dimensional differences. This test is directly related to die corner edge failure mechanism. Although the specifications of the TC test (ΔT, number of cycles, and sample size) are based on solder joint fatigue data, the large number of cycles required in the test (500 to 2,300) assures that the crack flaw has grown to its maximum length possible (from the edge of the die to the crackstop), or at least to the threshold value where G is stable. The increase of the crack length with respect to an increase of number of cycles is given by the Paris Law:

$$\frac{da}{dN} = C(\Delta \sigma Y \sqrt{\pi a})^m \qquad (1)$$

where a is the crack length, N is the number of load cycles, C and m are material constants, Y is a geometry constant, and σ

is the uniform tensile stress perpendicular to the plane of the crack, and which is proportional to the ΔT in the test. Efforts to determine the exponent m are still in progress, but by driving the crack flaw to its maximum length possible or threshold value, it can be assumed that G is being driven to its maximum possible value. If the sample has not failed, this means that the G maximum is smaller than Gc.

III. METHODOLOGY

CPI risk assessment is based on the gap between Gc and G. Fig. 6 shows one of our CPI risk assessment Gc-G for a particular flip-chip package (as predicted, CPI was not an issue during the evaluation tests of the actual flip-chip package.) The smaller the gap, the riskier the CPI design. For two different scenarios of Gc and G (e.g., node 45 nm and node 32 nm), the same gap Gc-G represents equivalent CPI risk. If Gc values are known, and G can be computed via FEA for any given ΔT and package-die combination, it is possible to drive G such that Gc-G gaps are equivalent. Equivalent CPI risk of two different flip-chip packages, 1 and 2, can be written as:

$$Gc_1 - G_1' \cong Gc_2 - G_2 \qquad (2)$$

where Gc_1 and Gc_2 are the fracture toughness of the crackstop in flip-chip 1 and 2, respectively; G_2 is the crack driving force in flip-chip 2 computed via FEA under the regulatory TC ΔT; and G_1' is the crack driving force in flip-chip 1 under a $\Delta T'$, such that Eq.(2) holds true.

Assume flip-chip 1 is available for testing, and flip-chip 2 is not, and that they are each from different technology nodes. Now, if Eq. (2) is held true, flip-chip 1 could be used to evaluate flip-chip 2. Since flip-chip 2 is not available for testing, Gc_2 and G_2 are fixed based on experimental measurements of Gc and FEA computations of G. Gc_1 is known from experimental measurements, so Eq. (2) can be solved for G_1'. This G_1' is the value that flip-chip 1 needs to see to have the same CPI risk as flip-chip 2. This value of G_1' is used in the FEA model to find the value of $\Delta T_1'$ that will drive the crack driving force to the G_1' value during the experimental tests. Fig. 7 shows the flow of the methodology, and Fig. 8 shows the behavior of crack driving force of flip-chip 1 throughout the tests when the same sample used to evaluate flip-chip 1 is used to evaluate flip-chip 2. Notice that if the package in flip-chip 1 is organic, then $\Delta T_1'$ might be limited to the stress test ranges defined by JEDEC [10], temperature cycling conditions B, C or G.

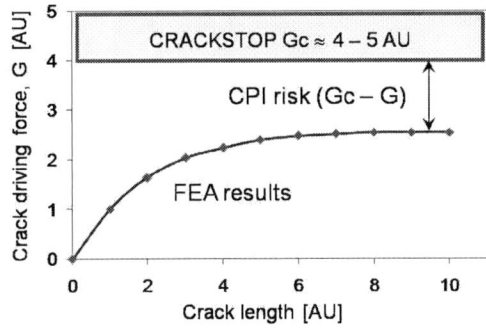

Figure 6. CPI risk assessment case of a particular flip-chip package - as predicted, CPI was not an issue during qual tests

IV. CASE STUDY AND DISCUSSON

A test case involving two flip-chips, 1 and 2, was studied. The goal was to pre-qualify flip-chip 2 by testing a possibly higher risk existing test vehicle, flip-chip 1. Flip-chip 1 and flip-chip 2 were from different technology nodes and had different properties. The most relevant properties and the relative comparison between flip-chip 1 and flip-chip 2 are shown in Table I. From the relative comparison, a decrease or increase of Gc-G gap could be defined (last column in Table I.) Half of the property comparison shows a decrease effect and the other half an increase. FEA simulations were then run to evaluate the values of G for flip-chip 1 and flip-chip 2. It was found that G_1 is slightly larger than G_2, that each one is smaller than their respective Gc values, and that the Gc_1-G_1 gap was larger than the Gc_2-G_2 gap. Since flip-chip 1 was at lower risk, it was apparently not a feasible test vehicle for testing flip-chip 2. To bridge this gap, the parameter ΔT was then introduced in the analysis through Eq. (2). The measured values of Gc_1 and Gc_2, and that of G_2 from FEA, were plugged in Eq. (2) to find G_1'. FEA simulations were then run to find $\Delta T'$ that would induce a G value of G_1' in the flip-chip 1. This $\Delta T'$ was then later used as the stress load in a TC test in flip-chip 1 samples that have completed the regulatory TC tests (see Fig. 8, arrows from G_0 to G_1'). These samples passed the new test. Later on, it was found that actual flip-chip 2 samples passed their corresponding die corner edge TC tests as well.

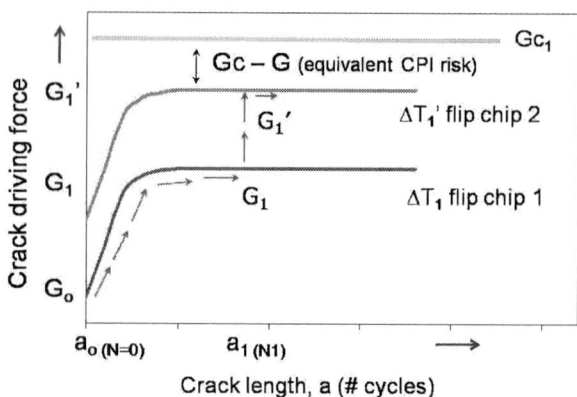

Figure 8.   Schematic behavior of G_1 during TC test (from G_0 to G_1')

TABLE I.   CASE STUDY: FLIP-CHIP 1 AND FLIP-CHIP 2

Property	Flip-chip 1 compared to Flip-chip 2	
	Property	*Effect on Gc-G gap*
Die size	Larger	Decrease
Package substrate thickness	Thicker	Decrease
Type of ILD	Mechanically stronger	Increase
Crackstop Gc	Stronger	Increase

V. CONCLUSIONS

By introducing the Gc-G gap, ΔT, and Eq. (2) in the combined FEA and experimental tests analyses, it is possible to evaluate a larger range of CPI critical parameters per given CPI test vehicle. Accurate measurements of fracture toughness of crackstops and accurate FEA models that have been carefully validated are crucial for the implementation of the methodology. As a result of this protocol, FEA is used more effectively to design experimental qualification tests, CPI risk projections are more accurate, lack of CPI test vehicles for pre-qualification is solved, and productization is more flexible to account for the large variety of product configuration and technology nodes.

ACKNOWLEDGMENT

Sincere thanks go to Dmytro Chumakov for providing the SEM picture in Fig. 4 and for the interesting discussions regarding measurements of crackstop's fracture toughness. We also thank Frank Kuechenmeister for coordinating the TC tests on flip-chip 1 and flip-chip 2, Richard Blish for the encouragement to pursue this work and for reviewing it, and Mahidhar Rayasam for participating in the discussions.

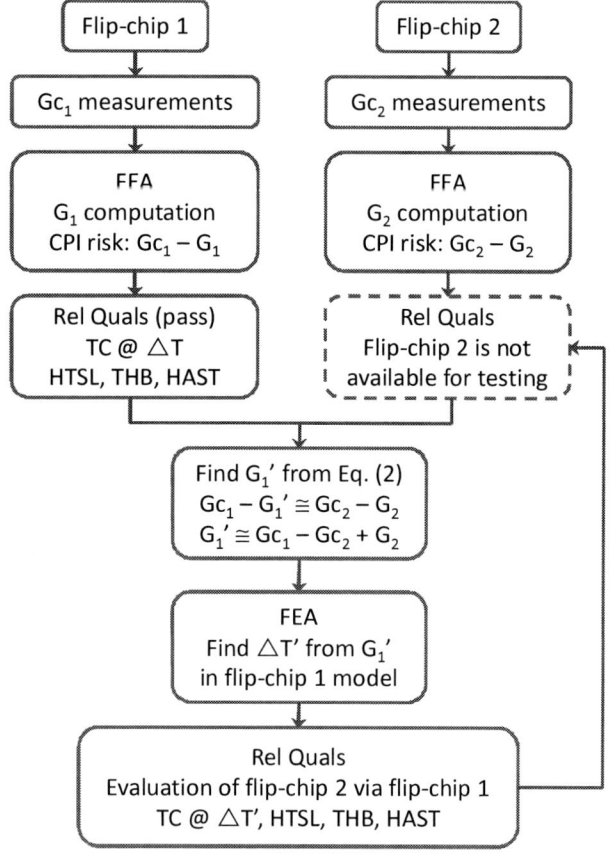

Figure 7.   Flow chart of reliability evaluation methodology

REFERENCES

[1] J. Auersperg, D. Vogel, M.U. Lehr, M. Grillberger and B. Michael, "Crack and damage evaluation in low-k BEoL stacks under chip package interaction aspects," in *Thermal, Mechanical and Multi-Physics Simulation and Experiments in Microelectronics and Microsystems*, 2009.

[2] C.J. Uchibori, X. Zhang, P.S. Ho and T. Nakamura, "Chip package interaction and mechanical reliability impact on Cu/ulta low-k

interconnects in Flip Chip package," in *Solid-State and Integrated-Circuit Technology International Conference*, 2008.

[3] Z. Zhang, C.J. Zhai and R.N. Master, "3D fracture mechanics analysis of underfill delamination for flip chip packages," in *IEEE ITherm Conference*, 2008.

[4] X. H. Liu, T.M. Shaw, M.W. Lane, E.G. Liniger, B.W. Herbst and D.L. Questad, "Chip-package-interaction modeling of ultra low-k/copper back end of line," in *IEEE International Interconnect Technology Conference*, 2007.

[5] T. M. Shaw, E. Liniger, G. Bonilla, J.P. Doyle, B. Herbst, X.H. Liu and M.W. Lane, "Experimental determination of the fracture toughness of crack stop structures," in *IEEE International Interconnect Technology Conference*, 2007.

[6] M. Chin and A. Marathe, "Near-bump chip package interaction failure mechanism in flip-chip packages," *IMAPS Device Packaging Conference*, 2010, in press.

[7] W. Kanert and R. Pufall, "Investigation of chip-package interaction – looking for more acceleration in product qualification tests," *IEEE International Reliability Physics Symposium*, 2009.

[8] C.J. Zhai, U. Ozkan, A. Dubey, Sidharth, R.C. Blish II and R.N. Master, "Investigation of Cu/low-k film delamination in flip chip packages," in *IEEE Electronic Components and Technology Conference*, 2006.

[9] D. Chumakov, F. Lindert, M.U. Lehr, M. Grillberger and E. Zschech, "Fracture toughness assessment of patterned Cu-interconnect stacks by dual-cantilever-beam (DCB) technique," *IEEE Transactions on Semiconductor Manufacturing*, 2009, vol. 22, no. 4.

[10] "Stress-Test-Driven Qualification Integrated Circuits." *JEDEC Solid State Technology Association*, Standard JESD47F, January 2007.

A Failure Levels Study of Non-Snapback ESD Devices for Automotive Applications

Yiqun Cao [1, 2], Ulrich Glaser [1], Stephan Frei [2] and Matthias Stecher [1]

[1] Infineon Technologies, Am Campeon 10, 85579, Neubiberg, Germany
phone: +49-89-234-63936; e-mail: yiqun.cao@infineon.com
[2] Technische Universität Dortmund

Abstract—**Snapback ESD devices suffer from increasing danger when the protected ICs experience ESD events in powered up states. To ensure more reliable ESD protections, non-snapback ESD structures are gaining more importance in the field of automotive ESD design. Two types of on-chip non-snapback ESD devices, pn-diodes and active FET structures are investigated in this work regarding their failure levels. Characteristics of the ESD devices as well as electrical SOA of an nLDMOS are evaluated and discussed in detail with TCAD electro-thermal simulation, SPICE circuit simulation and mainly TLP measurements. Comparison of the efficiency of different ESD protections considering ESD window is also given, delivering the basic idea of choosing the right ESD devices in automotive applications.**

Keywords: non-snapback, on-chip ESD, failure levels, pn-diode, bigFET, DMOS, SOA, TLP, SPICE, TCAD

I. INTRODUCTION

System-level ESD has gained more and more importance in the field of automotive applications. ESD test on system-level introduces a new concern because it is testing for events that can occur in the system in either powered up or powered off states [1]. When the device is powered up, the power supply is available to cause damage with its practically unlimited energy. Thus, a snapback on-chip ESD protection element could face serious danger if its holding voltage is below the supply voltage. In fact, non-snapback devices become nowadays particularly attractive in power technologies to give reliable and also area efficient ESD solutions in the automotive industry.

Failure levels of semiconductor devices are introduced in Wunsch-Bell criterion [2], providing a chance to estimate failure power or energy concerning different pulse durations. However, the approximated thermal model is no longer valid for modern power technologies. Robustness of high voltage (HV) or power MOS transistors is often limited not only thermally but also electrically [3]. As a matter of fact, electrical safe operating area (SOA) usually determines the operating boundary of DMOS transistors under very short, high power pulses like ESD when the transistor is used as ESD protection. It is thus desirable to investigate and systematically compare the failure levels of various non-snapback ESD structures. The deep understanding of the ESD capabilities of such ESD devices considering ESD window should be one of the prerequisites to improve the concepts of ESD engineering in HV IC design.

Two types of on-chip ESD protection approaches including conventional pn-diodes and active clamping circuits are mainly used to realize non-snapback ESD devices [4]. In section II pn-diodes as ESD protection are fully characterized with TLP and very fast (vf) TLP measurements. Particular effects in the junction area are demonstrated with assistance of device simulation. In section III two active clamping structures are introduced, giving the examples of implementing active ESD devices. Circuit-level design procedure is shown with SPICE simulation as well. I-V characteristics especially electrical SOA of the used bigFET are discussed in detail utilizing TCAD simulation. Various non-snapback devices concerning their failure levels are evaluated and compared with experimental results in section IV. ESD protection efficiencies of both types of ESD devices are also discussed, followed by conclusions.

II. PN-DIODE

Pn-diodes are often designed as vertical diodes which provide very efficient use of silicon area since very robust vertical junctions are available in automotive technologies such as BCD (Bipolar-CMOS-DMOS). Reverse-biased pn ESD diodes utilizing avalanche breakdown focus mainly on the engineering of doping profiles along the pn junction deep inside the silicon.

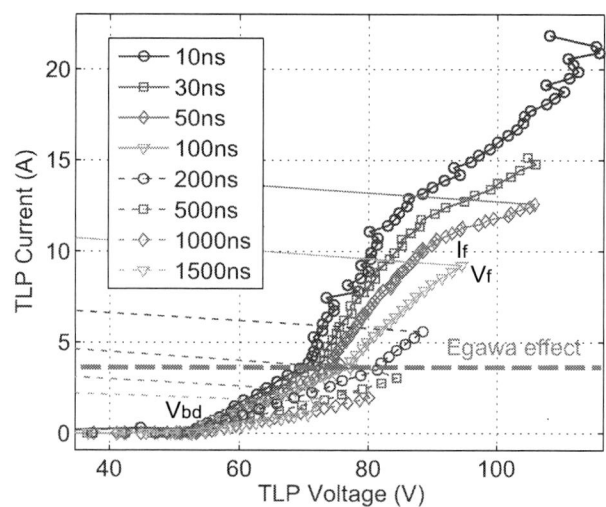

Figure 1: TLP I-V characteristic of a 45V voltage class pn-diode named s9e45k2 tested with various pulse widths from 10ns to 1500ns and same rise-time of 1ns. Parameter s9 indicates the specific technology, where e45 the voltage class and k2 the size of the device.

978-1-4244-5430-3/10 $26.00 © 2010 IEEE

Figure 1 shows the TLP measurement results for a 45V class pn-diode. The TLP voltage and current are generated by averaging the whole pulse duration in order to take self-heating of junction especially for long pulses into account. The rise-time is not relevant as it didn't change the results in a practical manner. Note that V_f, I_f and V_{bd} indicate failure voltage, failure current and breakdown voltage, respectively. TLP test with 5ns pulse width was also performed but not explicit plotted since the device did not fail within the test hardware limitation of 30A maximum current. For pulse widths smaller than 200ns, the current density in the pn junction reaches the level where Egawa effect [5] takes place and changes the slope of I-V characteristic of the device. The slope changing points according to the tests with different pulse durations generate the nearly horizontal dashed line (Figure 1), which confirms the current density level is the main condition to enable Egawa effect around the pn junction area. To give a detailed observation of this phenomenon, 1-D TCAD simulation is sufficient to show qualitatively the behavior of the pn-diode. Figure 2 shows the simulation results where the second local maximum of electric field appears with the higher current density.

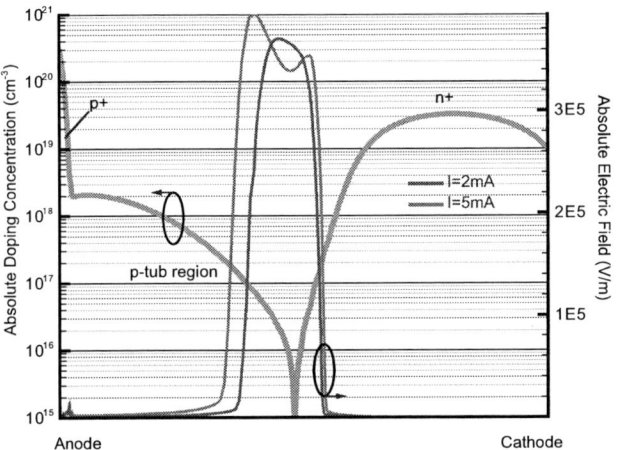

Figure 2: 1-D TCAD simulation for the pn-diode in the reverse-biased condition shows the differences in the electric field with or without Egawa effect. Doping profile is shown as well. With 2mA current biasing, only one local electric field maximum exists along the diode from anode to cathode, where double peaks are observed at 5mA bias current. The higher hole and electron current densities lead to the space charge modification, resulting in double peaks of electric field.

Egawa effect surely improves the ESD performance of the diode since it can lower the total on-resistance regarding ESD (R_{on}) of the device, resulting in improvement of on-chip area efficiency. In this work, several pn-diodes implemented for two voltage classes were characterized using the same TLP test method discussed above. If one simply defines R_{on} as $(V_f - V_{bd})/I_f$, the dependence of R_{on} area product on the pulse duration can be derived from the measurement results as depicted in Figure 3. As expected, the diodes with the same breakdown voltage have nearly the same R_{on} referring to unit area. However, the pn-diodes in higher voltage classes have to be designed larger to sustain the same R_{on}. The weak doping concentration in p-tub region [4] increases the resistance in the current path significantly while adjusting a higher breakdown voltage. This results in less area efficiency of 45V devices with a factor of approximately 1.8 comparing to 25V diodes.

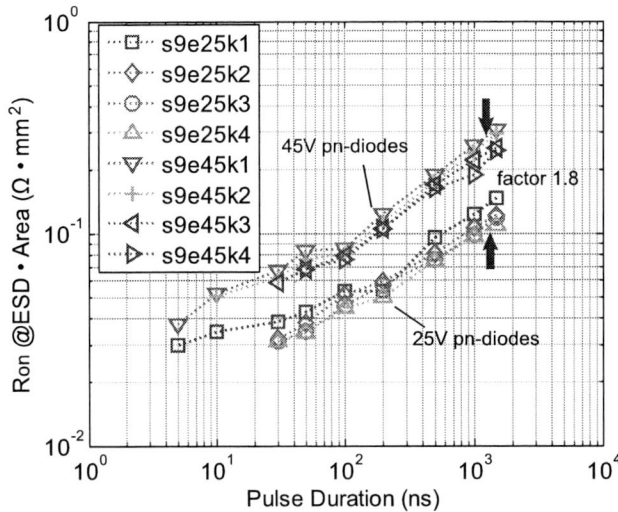

Figure 3: R_{on} area product versus pulse duration for two groups of pn-diodes in 25V and 45V voltage classes. k1 to k4 denotes chip area of the devices with the relation k4≈2k2≈4k1.

As a matter of fact, in high voltage applications ESD design windows often have upper limits of several tens of volts as constraints. Hence very high V_f is out of concern for real ESD design. Effective R_{on} in terms of ESD window rather than thermal V_f is a more practical measure which probably will reduce the area efficiency of pn-diodes. Nevertheless, failure power or energy versus pulse duration is interesting for the investigation of junction robustness. Figure 4 shows the characteristics of energy and power to failure of various ESD diodes. Overlapped lines imply the fact that failure levels of the pn-diodes are relatively independent on the voltage classes within the given technology and proper designs. Linear power and energy to failure per area in double logarithm plot basically confirm the Wunsch-Bell theory. However, the slope is approximately -0.8 in the reality in double logarithm plot rather than -0.5 according to Wunsch thermal model.

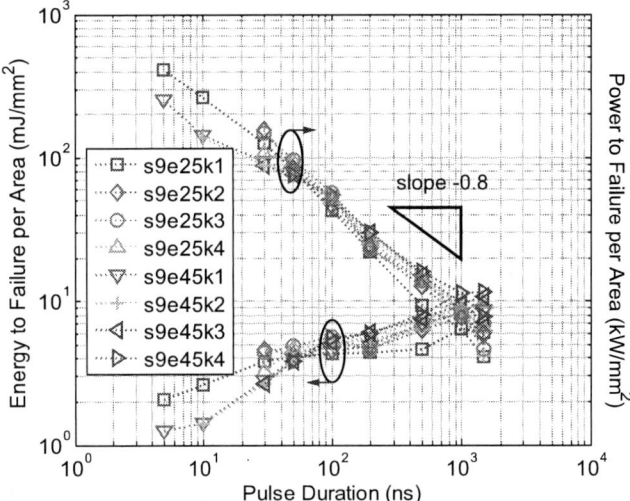

Figure 4: Energy and power to failure per area versus pulse duration for two groups of pn-diodes in 25V and 45V voltage classes.

III. ACTIVE CLAMPING STRUCTURE

Another common approach for non-snapback on-chip ESD protection utilizes an active controlled bigFET [4]. RC triggered HV-MOS ESD circuits have been applied in the advanced CMOS world for years [7]. To avoid the risks of wrong dynamic triggering due to EOS disturbances, static triggering concept is preferred in automotive HV ESD design field.

A. Design and evaluation of active clamping structures

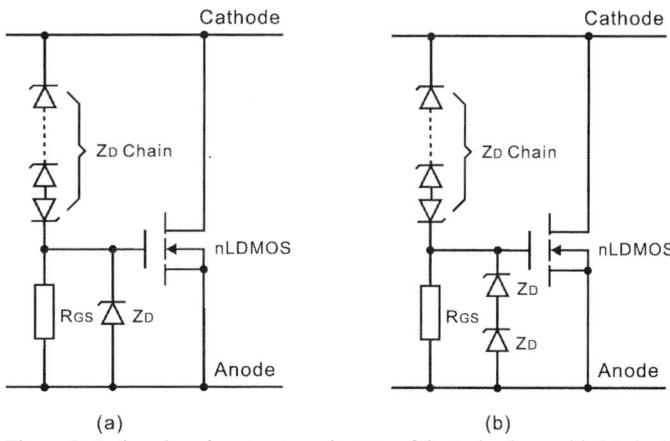

Figure 5: Active clamping structures in state of the technology with (a) single or (b) two in serial connected Zener diodes as gate protection. An active ESD structure can be considered as an ESD device with two terminals cathode and anode as in a pn-diode.

Among many possibilities allowing transient gate-biasing of the bigFET during ESD stresses, two examples employing Zener diodes chain or Zener clamping as breakdown voltage definition circuit are demonstrated in Figure 5. Note that the protection circuits are designed and fabricated in the BCD technology with the bigFET implemented using an n-type lateral DMOS (nLDMOS). Type A has one Zener diode as gate protection while type B engages two of them. With combination of forward- and reverse-biased Zener diodes in the Zener chain, various static V_{bd} of the active clamping structures can be flexibly implemented. This is true as long as the designed V_{bd} is considerably smaller than the avalanche breakdown voltage of nLDMOS with zero gate-source voltage (V_{gs}). In these two test circuits, V_{bd} was designed for ESD protection in 25V voltage class. The trigger circuit part occupies certainly much less area comparing to the nLDMOS but is robust to withstand the clamping voltage without being firstly damaged. As designed, V_{gs} of nLDMOS can be temporarily biased up to one or two Zener voltages in type A or type B, respectively.

Different as in product design, ESD design on circuit-level usually suffers from the lack of proper simulation models which would accurately describe electro-thermal behavior of circuit elements under high power and short time conditions. Therefore the expectation of the outputs of SPICE simulation should not be too high. In this design, the I-V characteristic of the nLDMOS was extracted with DC measurements. Thus electrical SOA regarding short pulses in high drain-source voltage range is not included in the SPICE models. Figure 6 shows the simulation results of the active structures.

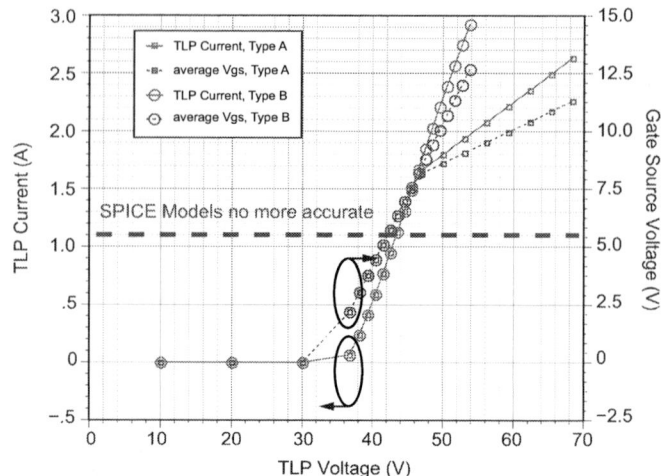

Figure 6: Simulated TLP characteristics of type A and type B circuitries in SPICE. TLP current and voltage as well as gate-source voltage of the nLDMOS are calculated via average windows. Above certain current level (dashed line), simulation only delivers qualitatively the I-V characteristics of the structures.

Breakdown voltage of the active clamps and gate-source voltage of the nLDMOS can be well simulated while the TLP current can only be qualitatively shown. Limiting V_{gs} in type A with one Zener diode directly affects the total current conducted through the bigFET.

As mentioned earlier, using an nLDMOS as the main circuit element, the physical limitation of ESD capability of the active triggered circuits is given by its SOA. I-V points at the onset of snapback or triggering of the inherent bipolar transistor for different V_{gs} give the boundary in which the bigFET is able to sustain ESD currents. In the given BCD technology, non-uniform snapback of DMOS often results in filament formation and local burn-out [6], allowing the trigger voltage and current to be treated as failure voltage and current immediately. Further, the powered on ESD protection aspect also supports this consideration of seeing snapback directly as failure.

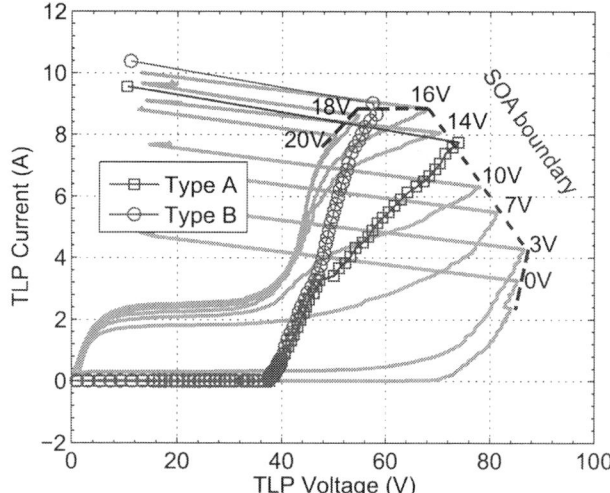

Figure 7: SOA characteristics of nLDMOS at different gate voltages and I-V curves of two active triggered circuits using the same nLDMOS are recorded with standard TLP measurements (10ns rise-time, 100ns pulse duration).

The pulsed I-V characteristics of an nLDMOS and designed active circuits with V_{bd} of 35V at room temperature tested with standard TLP are shown in Figure 7. Note that the SOA boundary indeed restricts the snapback and failure points of these active devices. Type B device shows an advantage against type A, confirming the simulation results. Besides, with the given SOA, V_{gs} of type A and type B at failure points can be readily estimated using the plot without extra measurements (approx. 13V and 17V).

B. Failure levels study on actively controlled and single nLDMOS

In order to study the failure levels of active structures, TLP tests with various pulse durations were applied to the active ESD devices. Figure 8 gives the performance of a type B device employing a smaller nLDMOS as in Figure 7. Vf-TLP measurements in TDRs setup were used for 5ns pulse width tests. Apart from the pn-diodes, as pulse widths were scaled

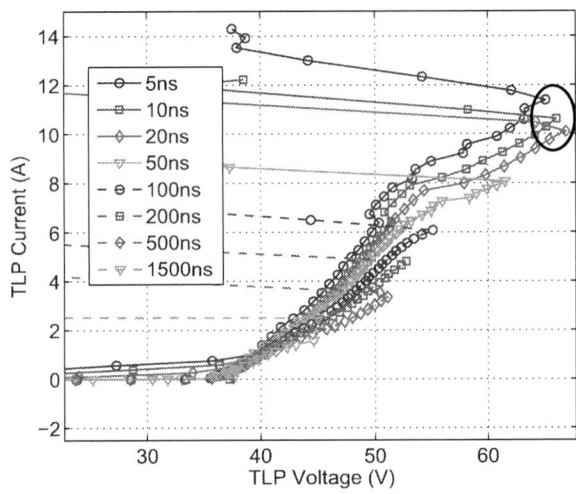

Figure 8: TLP I-V characteristic of a 25V voltage class active clamping structure named e25 type B tested with various pulse widths from 5ns to 1500ns. The crowding of failure points tested with below 20ns long pulses is indicated comparing to the case of the pn-diode in Figure 1.

down below 20ns the snapback points start crowding together with less deviation of V_f and I_f. In other words, with less energy content of the TLP pulses, failure power levels of the active structure do not increase significantly. Note that some other active devices including type A and type B on different breakdown voltages are also characterized with TLP and vf-TLP, showing the same effect.

As already discussed, ESD capability of proper designed active clamping structures relies on the bigFET, in this case the nLDMOS transistor. To understand the phenomenon of less increasing of failure power at short pulse durations, electrical SOA of the nLDMOS is investigated applying stresses with various pulse durations instead of the standard TLP with 100ns pulse width. Two groups of measurements were performed for different V_{gs} as depicted in Figure 9. Note that for tests with 2.5ns and 5ns pulse durations, the failure points forming pure electrical SOA remain nearly unchanged. This can be used to explain the fact that failure power of active ESD structures is

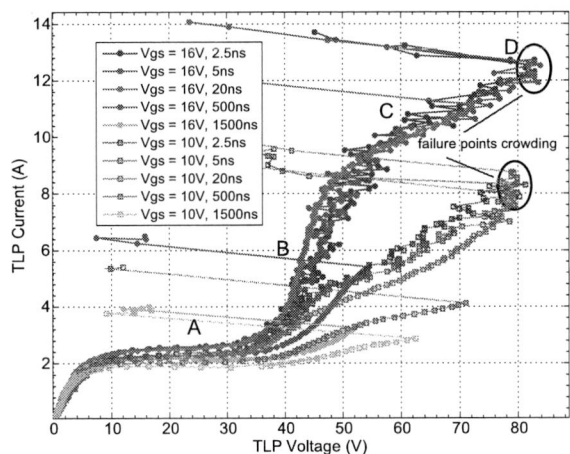

Figure 9: Electrical SOA of nLDMOS for Vgs=10V and 16V with different pulse durations. Pulses shorter than 20ns no longer enhance the SOA significantly. A to D indicates four different regions of I-V characteristics where the transistor gate is biased with high voltage.

saturated as pulse duration getting shorter. It is worth to mention that the SOA measurements were repeated also with different rise-times (100ps and 1ns), delivering the same phenomenon. No significant difference regarding rise-times was observed. This implies the triggering of snapback has no significant dependence on dV/dt or dI/dt.

In the active clamping structures, the gate voltage of nLDMOS should be biased on a relative high level to facilitate a higher failure current. This encourages the further analysis of the used nLDMOS under high voltage gate-biasing with 2-D TCAD simulation involved. As marked in Figure 9, A to D represents four regions which are of concern for ESD designers:

- Region A: nLDMOS works in saturation region as a usual MOSFET, where the channel is pinched off with total electron current flowing through the

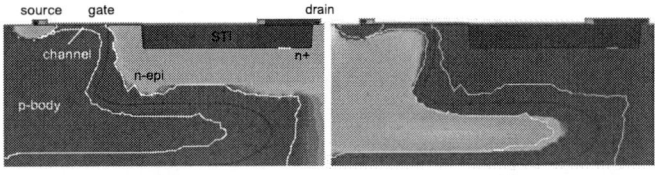

Figure 10: Simulation of (a) electron current density and (b) hole current density in region A. Source and body contacts were shorted to ground while the gate was biased on 10V. The same simulation setup was also applied to Figure 11 to Figure 13.

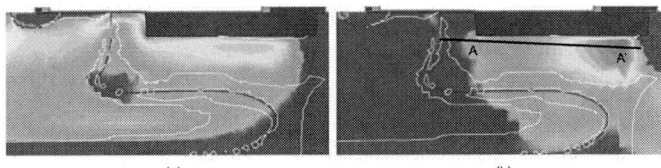

Figure 11: Simulation of (a) hole current density and (b) impact ionization amplitude in region B. Significant hole current starts flowing to the body contact in comparison with region A (Figure 10 (b)). The delocalization of the impact ionization maximums as well as electric field peaks along the cut line AA' represents Kirk effect.

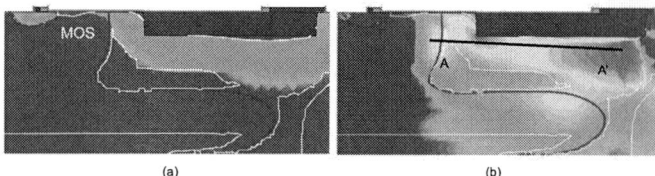

Figure 12: Simulation of (a) electron current density and (b) impact ionization amplitude in region C. Electron current is distributed similar as in region A. Along the cut line AA', three local maximums of impact ionization can be observed.

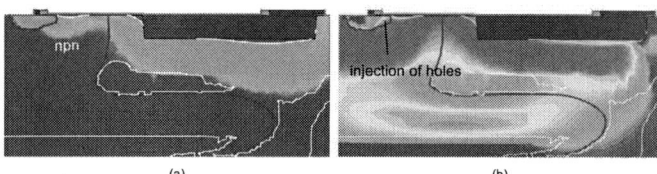

Figure 13: Simulation of (a) electron current density and (b) hole current density in region D. Comparing to the operation as MOSFET in Figure 10 (a), the npn bipolar transistor is now dominating the nLDMOS behavior leading to snapback.

narrow channel to the drain contact. No additional hole current through n-epi layer and p-body region to the body contact can be observed (Figure 10).

- Region B: In this region, avalanche multiplication occurs accompanied with strong local electric field [8] [9]. The electrons generated due to impact ionization flow to drain contact while the holes flow to the body, providing a significant addition on total current (Figure 11). The delocalization of the electric field peaks under high current density to the more highly doped drain region is called Kirk effect [10] after its first investigator in bipolar transistors, having a similar effect as so called Egawa effect in pn-diodes described earlier.

- Region C: As the generation of the holes keeps increasing, additional local maximums of impact ionization can be observed at p-body to n-epi junction area (Figure 12), exhibiting the main difference of region C and B. Note that no significant electron current flows from source through p-body to drain as the parasitic bipolar transistor is not triggered yet.

- Region D: This region is defined at the moment when snapback just happens. As hole current through p-body is sufficient large to build up an ohmic voltage drop on base-emitter junction, which is larger than the build-in voltage. The bipolar transistor is then triggered accompanied with significant electrons injected into p-body and holes into source (Figure 13).

In addition to the simulation results comparing the electric parameters in four regions inside the transistor, transient simulation using the commercial SDEVICE simulator by SYNOPSYS was also performed for the used nLDMOS in electro-thermal mode.

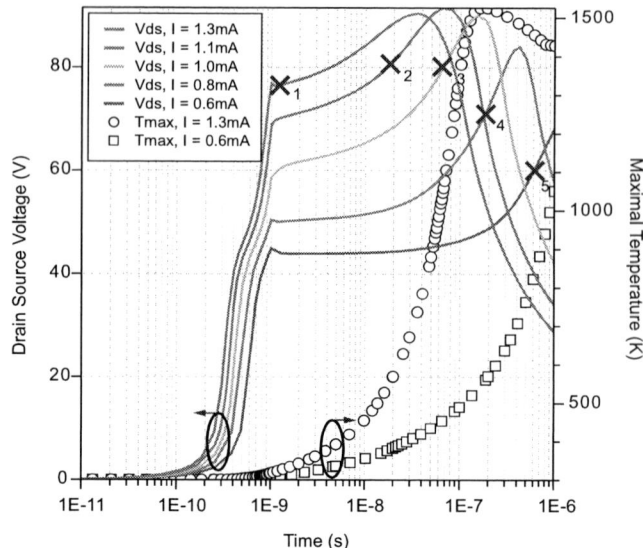

Figure 14: Transient simulation of drain-source voltage corresponding to various drain-source current pulses with rise-time of 1ns. In case current reaches a critical level, snapback takes place at very beginning of the pulse (indicated by X_1), where the self-heating of the nLDMOS is not significant.

Figure 14 depicts the waveforms of voltage and maximal lattice temperature of the LDMOS with V_{gs} of 10V under various drain-source current biasing. X_1 to X_5 denotes the time instants when the triggering of the npn-transistor takes place. These specific time points were extracted by examining the electron current density in the device cross section (Figure 13 (a)). A time instant was then recorded, when a large amount of electrons are not only found in the MOS channel but also in the p-body region. This means the device was driven into the critical region D. In the measurements, triggering of the parasitic bipolar transistor is accompanied immediately with voltage snapback due to current filamentation (Figure 9). In the simulation however, the voltage snapback of an ideal homogenous device occurs after a certain time period (tens of nanoseconds shown in Figure 14) beyond the bipolar triggering. The simulated maximal voltage peaks observed in the transient waveforms are therefore not usable to identify the snapback or device failure in practice. With X_1 to X_5 as the

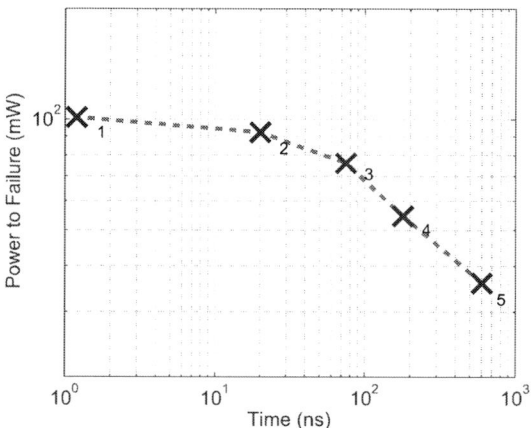

Figure 15: Simulated power to failure (triggering of the bipolar transistor) of the nLDMOS at Vgs=10V. X1 to X5 indicates the same failure points shown in Figure 14.

failure points of the studied nLDMOS, power failure levels can be plotted in Figure 15. Note that the same effect comes out as described in the investigation of SOA (Figure 9). Again, power to failure of nLDMOS as well as the active ESD protection structures using nLDMOS as the bigFET cannot increase significantly by shortening the pulse width. Furthermore, a dynamic triggering of the inherent bipolar transistor due to dV/dt or dI/dt is not observed in the device simulation as well.

IV. COMPARISON OF FAILURE LEVELS AND DISCUSSION

As the major failure mechanisms of two types of non-snapback ESD devices are already detailed discussed in the former sections, systematic comparison of the protection devices is performed in order to achieve a better overview.

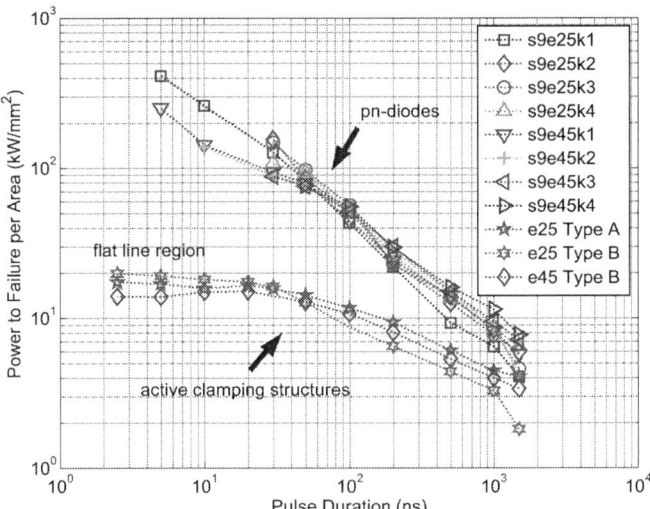

Figure 16: Failure power levels per unit area of various non-snapback devices with different dimensions and V_{bd}. Pn-diodes show overall advantage in terms of the current capability per unit area.

Figure 16 compares the failure power levels per unit area of the investigated non-snapback ESD devices calculated with $I_f \cdot V_f$ /area. Very good scalability is found in the pn-diodes while certain dependences on trigger types are seen in the active ones due to gate-biasing issues. For short pulse durations, the curves of failure power of active circuits become flat confirming the phenomenon simulated for nLDMOS in Figure 15.

Robust junction of pn-diodes enables higher power to failure levels due to the fact that the heat generated mainly in junction area can be dissipated into all directions inside the device, where the hotspots within a DMOS based active clamping structure are located in the surface area. Besides, the inherent bipolar snapback and accompanying current filamentation are the major constraint of the active devices in the short pulse range. In contrast to this, ruggedness of pn-diodes is only limited thermally rather than electrically.

However, as emphasized earlier, absolute maximum value of current conducting capability or power/energy to failure

might not be the right measure of evaluating the area efficiency of ESD devices. ESD window or the maximum allowed voltage (V_{max}) defined by the protected IC circuitries should be considered in each ESD design. The ESD window could be enlarged by e.g. employing a secondary protection concept using serial on-chip resistance with extra area cost. But this is often not possible in the real applications. Therefore, effective R_{on} provides a better comparison of the area efficiency

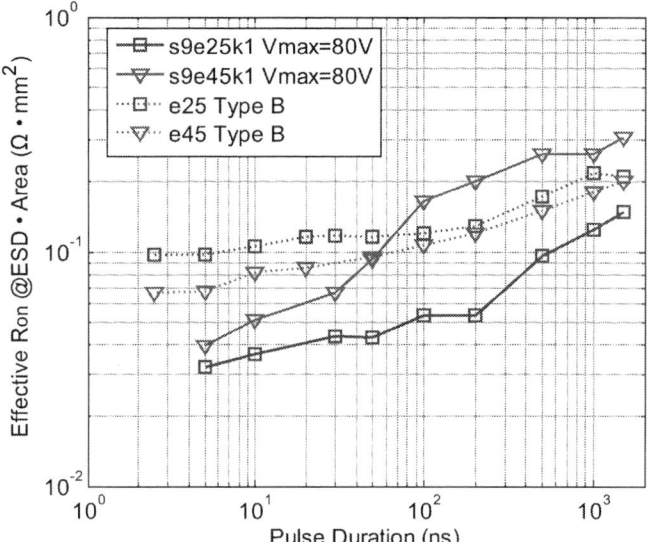

Figure 17: Effective R_{on} area product with V_{max}=80V. 80V is already an optimistic value in HV ESD design windows of 25V, 45V or 55V voltage classes. For short pulse durations, pn-diodes still have more area efficiency.

regarding ESD. Figure 17 illustrates effective R_{on} over pulse duration of several devices with V_{max} assumed with 80V. Note that R_{on} area product of active structures does not change with higher breakdown voltage compared to the pn-diodes (Figure 3). In 45V voltage class, e45 type B is already leading in terms of area efficiency with lower effective R_{on} for pulse duration larger than 50ns. The benefit of a proper active clamping design could be even higher when an ESD device with higher V_{bd} is required.

In addition to the effective R_{on} as discussed above, several other advantages of active clamping structures or drawbacks of pn-diodes are summarized as following:

- Zener triggered active clamping structures give the possibility of defining V_{bd} in a more flexible way where pn-diode demands process design which is much more time consuming to generate a new device with another breakdown voltage.

- For higher breakdown voltages needed in certain applications, pn-diodes are only realizable on extra cost of area efficiency.

- Process fluctuation and temperature dependence of pn-diode is usually much higher than the existing Zener diode and nLDMOS in the active structures.

978-1-4244-5430-3/10 $26.00 © 2010 IEEE   463

Among the device comparison and characterization, behavior of these two types of ESD devices under realistic ESD pulses needs to be studied further. In order to address the impact of different failure mechanisms for non-snapback devices in a practical manner, a novel human metal model (HMM) tester based on coaxial line technique [11] is used as the ESD test method which can deliver the same shape of current pulse defined in system-level ESD and provide a excellent resolution of transient voltage and current measurements [12].

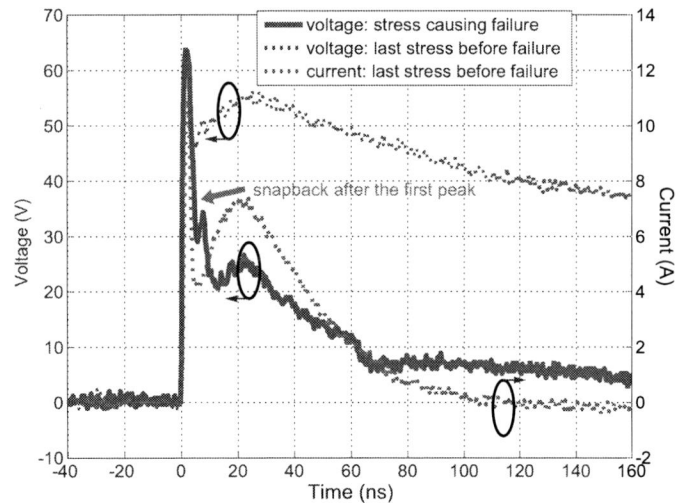

Figure 18: Active device e25 type B tested with HMM. Snapback observed right after the initial peak causing damage of the device. This device alone can sustain a 3.5kV system-level ESD discharge (150pF/330Ω).

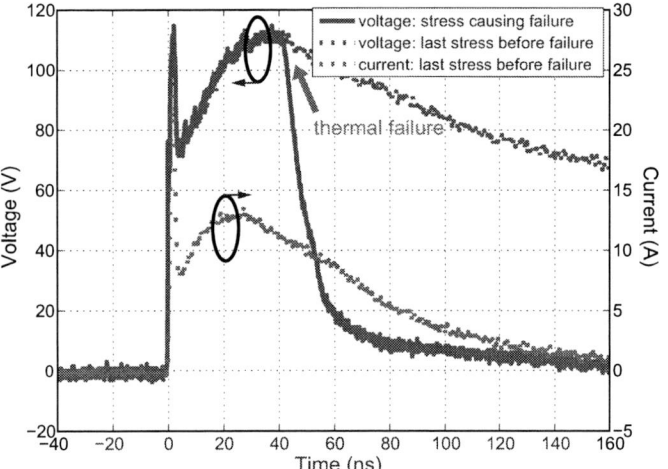

Figure 19: Pn-diode s9e25k2 tested with HMM. Thermal failure observed in the broad peak region. This device alone can sustain a 6kV system-level ESD discharge (150pF/330Ω).

Figure 18 and Figure 19 show the waveforms confirming that the decisive damage due to snapback actually comes from the initial peak of ESD pulses for active structures, while the pn-diode burned out by certain energy content of the broad peak. For the active device, the calculated first peak power per unit area is located on the flat line region shown in Figure 16. On the other hand, the pn-diode is not damaged by the first peak because the peak power is below the power to failure level at the corresponding pulse duration.

Again, applying system-level ESD pulse on the single ESD devices delivers the characterizations but does not reflect the real ESD protection capabilities. For example, a circuit protected with the tested pn-diode would probably not survive a 6kV GUN discharge since the clamping voltage of the diode exceeds 110V, which is already beyond the ESD window. Nevertheless, from the effective R_{on} point of view, the first peak of the HMM pulse in the range of several nanoseconds and the broad peak in the range of tens of nanoseconds both imply the fact that the pn-diode could be in favor in common system-level ESD applications compared to the active ESD protection in 25V voltage class as exhibited in Figure 17.

V. CONCLUSIONS

Conventional avalanche pn-diodes with designed breakdown voltage for HV ESD utilize robust vertical junction deep inside the silicon and deliver inherently very good ESD ruggedness with high area efficiency. Egawa effect occurs above certain current density, providing an extra advantage of using pn-diodes. Characteristics of active ESD structures rely mainly on the used bigFET, in this case, the nLDMOS. The saturation effect of failure power levels while shortening the pulse length is studied carefully both from test and simulation results.

In this work, a novel aspect of evaluating non-snapback ESD devices in a wide range of pulse duration is presented based on Wunsch-Bell theory. However, Wunsch theory has its limitation in practical use because of the simplified thermal models. Additionally, ESD hardness of active ESD protections is not able to be estimated using Wunsch theory. In short-pulse range, active devices actually suffer from electrical failure rather than a thermal one. The slope 0.5 is thus no longer applicable according to Figure 16. For the evaluation of ESD efficiency considering ESD window, Figure 17 provides the possibility of choosing ESD devices properly against ESD pulses depending on their pulse lengths. HMM tests with system-level ESD like pulse waveforms were performed on the pn-diode and the active device as an example. It was shown the pn-diode in 25V voltage class is more efficient according to the effective R_{on} in the first peak and the broad peak ranges.

Generally, the much lower area efficiency of nLDMOS in short-pulse region implies that pn-diodes furnish a better approach for protecting ESD pulses with a short-pulse part such as IEC 61000-4-2, HMM, CDE and etc. Active ESD devices would be of less interest in these cases since the dangerous bipolar triggering can be launched by a critical power level, resulting in overall sensitivity of active clamps on peak power. More detailed, on a certain voltage level across drain-source of nLDMOS, the current density in the body region will reach a sufficient value to trigger the parasitic bipolar transistor, driving the MOS transistor into destructive snapback mode. On the other hand however, within the given ESD window, active clamping structures become eventually more area efficient compared to pn-diodes. Active ESD solution is hence a very attractive alternative for ESD design against pulses such as HBM. Furthermore, if one considers

978-1-4244-5430-3/10 $26.00 © 2010 IEEE

flexibility, cost of development, voltage class and other factors, design of active ESD clamping will be more beneficial and even necessary. With elaborate analysis of the nLDMOS as well as its SOA in this work, the improvement of active ESD clamps could be achieved with different methods such as better gate-biasing techniques (see the comparison of active clamps type A and type B). Being aware of the ESD limitation of a DMOS transistor due to its SOA, designers are encouraged to develop new DMOS transistors by improving their SOA [13] especially the trigger current and finally boost the total ESD capability of the active clamps.

Above all, characterizations and failure level studies for two types of non-snapback devices are carried out in this work. Experimental and simulation results show various failure mechanisms and I-V characteristics of the protection devices, which are of great importance in high voltage ESD design. Systematic comparison also shows the drawbacks and advantages of the various ESD devices, helping to make proper decisions on ESD concepts according to applications.

ACKNOWLEDGEMENT

The authors would like to thank Esmark Kai and Howard Tang for fruitful discussions and mentoring the paper. This work was supported by the MEDEA+ project 2T205 SPOT-2.

REFERENCES

[1] SP5.6-2008, "Electrostatic discharge sensitivity testing - human metal model (HMM)," in *ESDA Tech. Rep.*, to be published in 2009.

[2] D.C. Wunsch and R.R. Bell, "Determination of threshold failure levels of semiconductor diodes and transistors due to pulse voltages," in *IEEE Trans. Nuc. Sci.*, 1968, NS-15 244-259.

[3] P. Moens et al., "On the electrical SOA of Integrated Vertical DMOS Transistors", in *IEEE Electron Device Letters*, Vol. 26, N° 4, April 2005, pp. 270-272.

[4] M. Mergens et al., "ESD Protection Considerations in Advanced High-Voltage Technologies for Automotive," in *EOS/ESD Symposium*, 2006, pp. 54-63.

[5] H. Egawa, "Avalanche characteristics and failure mechanism of high voltage diodes," in *IEEE Transactions on Electron Devices,* Volume 13, Issue 11, Nov 1966, pp.754 - 758.

[6] M. Mergens et al., "Analysis of lateral DMOS power devices under ESD stress conditions", in *IEEE Transactions on Electron Devices*, Volume 47, Issue 11, Nov. 2000, pp. 2128 – 2137.

[7] M. Stockinger et al., "Boosted and Distributed Rail Clamp Networks for ESD Protection in Advanced CMOS Technologies," in *EOS/ESD Symposium*, 2003.

[8] M. Denison, "Single Stress Save Operating Area of DMOS Transistors integrated in Smart Power Technologies," Dissertation, published by Shaker Verlag, 2004, pp. 40-47.

[9] A. Amerasekera and C. Duvvury, "ESD in Silicon Integrated Circuits", *2nd Edition*, published by John Wiley & Sons, 2002, pp. 356–357.

[10] C.T. Kirk, Jr., "A theory of transistor cutoff frequency (ft) falloff at high current densities", in *IRE Transactions on Electron Devices*, Volume ED-9, Mar. 1962, pp. 164 – 174

[11] E. Grund et al., "Delivering IEC 61000-4-2 current pulses through transmission lines at 100 and 330 ohm system impedances," in *EOS/ESD Symposium*, Sept. 2008, pp. 132–141.

[12] Y. Cao et al., "A TLP-based Human Metal Model ESD-Generator for Device Qualification according to IEC 61000-4-2," in *APEMC*, Apr. 2010, to be published.

[13] A. Podgaynaya et al., "Improvement of the Electrcal Safe Operating Area of a DMOS Transistor during ESD Events," in *IRPS*, Apr. 2009,pp. 437-442.

Understanding Transient Latchup Hazards and the Impact of Guard Rings

Farzan Farbiz and Elyse Rosenbaum
Department of Electrical and Computer Engineering,
University of Illinois at Urbana-Champaign
Urbana, Illinois, USA
ffarbiz2@illinois.edu

Abstract—**An experimental study of transient latchup is conducted. Measurements are performed on test structures fabricated in 90-nm and 130-nm CMOS technologies. The worst case testing conditions differ for static and transient latchup. Device simulation is used to understand the measurement results. P-well and N-well guard rings are evaluated under transient test conditions.**

Keywords—Latchup; Guard rings

I. INTRODUCTION

Latchup refers to the undesired triggering of a parasitic PNPN, located between the power supply and the ground rail of an integrated circuit. Figure 1 illustrates how a parasitic PNPN is formed in a bulk-Si CMOS inverter. Under normal circuit operating conditions, this PNPN should be in its high impedance state. If it is triggered into the low impedance mode, the resulting localized current conduction may result in circuit malfunction or even permanent damage to the part. Latchup may be triggered by a voltage perturbation at one of the terminals of the parasitic PNPN (referred to as *internal* latchup), or it may be triggered by a perturbation applied elsewhere in the circuit, i.e., outside the PNPN (referred to as *external* latchup). After the perturbation, or *trigger source*, is removed, the PNPN will remain in the low impedance state if its holding voltage is lower than the supply voltage [1]. Therefore, latchup is a serious reliability concern as long <u>any</u> of circuit blocks within an IC have a supply voltage that is greater than the holding voltage of a parasitic PNPN, roughly 1.2V [2].

The JEDEC latchup test standard, JESD78B, describes several (static) tests that may trigger latchup [3]. Among them, the *I-tests* are relevant to external latchup, which is the focus of this work. In the negative [positive] I-test, negative [positive] current is injected at a signal pin; the injected current is the trigger source. The chip is powered up during testing, and the current drawn from each of its power supplies is recorded both before and after the application of the trigger source. If the current drawn from any of the power supplies has increased appreciably after the trigger source is removed, latchup has occurred. The smallest valued injection current that causes latchup is called the latchup trigger current and is denoted I_{trig}.

The current source used for standardized I-tests [3] has a slow rise-time (5μs-5ms) and a long pulse-width (10μs-1s), which is why this is classified as static testing. Under

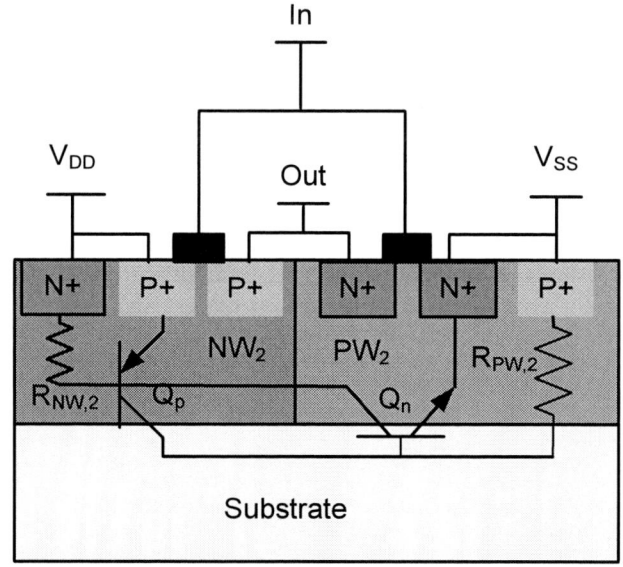

Figure 1. Device cross section of a CMOS inverter. The parasitic bipolar transistors form a PNPN that could be triggered on and cause latchup.

real-world use conditions, latchup may be caused by powered-on ESD events such as cable discharge events (CDE) [4] — such events are characterized by their short duration (hundreds of nanoseconds) and fast rise-time (a few nanoseconds). A variety of non-standardized procedures have been proposed for TLU testing. One such test is described in a recommended practice document published by the ESD Association [5]; in this test, the stimulus—a negative voltage pulse—is applied directly to the supply terminal of the PNPN device, placing it in the category of internal latchup testing. More relevant to this work is the transient I-test [6][7], in which the trigger source is a current pulse with short pulse duration and fast rise-time. The details of the experimental setup for the transient I-test are given in Section II.

In the following sections of the paper, I_{trig} values obtained from positive and negative transient I-tests are compared. The efficacy of guard rings under transient testing conditions is evaluated. Henceforth, I_{trig} measured during a negative I-test will be referred to as *negative I_{trig}*, and I_{trig} measured during a positive I-test will be referred to as *positive I_{trig}*. Furthermore, only the magnitude of I_{trig} is reported, i.e., both negative I_{trig} and positive I_{trig} are reported as positive numbers.

978-1-4244-5430-3/10 $26.00 © 2010 IEEE

II. Experimental Setup and the Test Structures

Figure 2 illustrates the experimental set-up used to study TLU in this work. A high power pulse generator, with output impedance of 50Ω, is connected to the signal pad of the test structure, labeled I/O, via a rise-time filter (RTF) and a 50Ω resistive matching network. RTF sets the rise-time of the pulse, t_r, to the desired value, which is 7 ns, unless otherwise noted. The pulse-width, T_{PW}, is variable. The use of the matching network results in cleaner pulse waveforms by eliminating reflections [8]. The current injected into the signal pin, I_{inj}, is calculated by measuring the voltage drop across R_S of the matching network (refer to Fig. 2). I_{DD} denotes the current provided by the power supply connected to the PNPN, or *victim*. Latchup is said to have been triggered if I_{DD} exceeds 2mA after the trigger source has been removed. Before the trigger source is applied, $I_{DD} < 0.1$ mA.

In a well designed IC, the latchup trigger current is high, 100s of mA. Such high levels of current injection can be provided by forward-biased PN junctions, and the test structures for the positive and negative I-tests reflect this. (Note that in a typical CMOS IC, many PN junctions will be found at the signal pads.) Figure 3 contains cross-sectional representations of the test structures used in this work. The test structures were fabricated in 90 and 130-nm CMOS technologies. In Fig. 3, victim *orientation* refers to the relative placement of PW_2 and NW_2, with respect to the substrate current injector.

In advanced CMOS technologies, different power supply voltages are used with I/O and core transistors. The supply voltage is 2.5V for the I/O and 1V for the core in the 90-nm technology used in this work; these numbers are 3.3V and 1.2V for the 130-nm technology. Parasitic PNPN are found in both I/O and core logic circuits, meaning that the victims of Fig. 3 could have V_{DD} of 1V, 1.2V, 2.5V or 3.3V. I_{trig} is only a weakly decreasing function of V_{DD} for $V_{DD} \geq 1.5$V [9]; note

that latchup cannot be sustained after the removal of the trigger source if $V_{DD} \leq 1.2$V. Therefore, V_{DD} was set to 1.5V in all the experiments. V_{DDIO} was also set to 1.5V, arbitrarily, because I_{trig} is insensitive to V_{DDIO}.

It is important to note that during a negative [positive] I-test, the PN junction injects minority [majority] carriers into

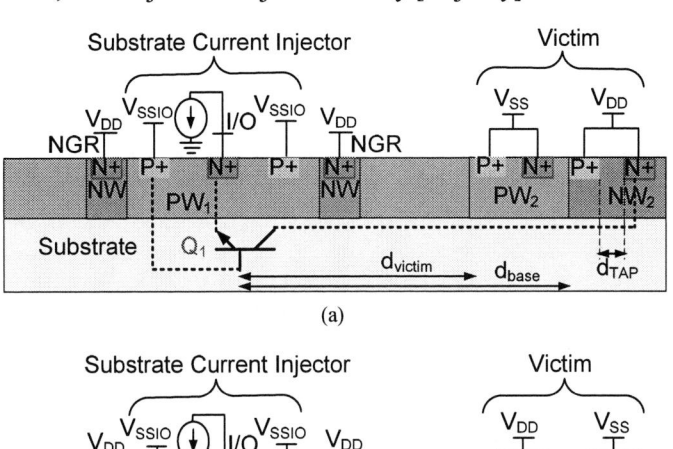

(a)

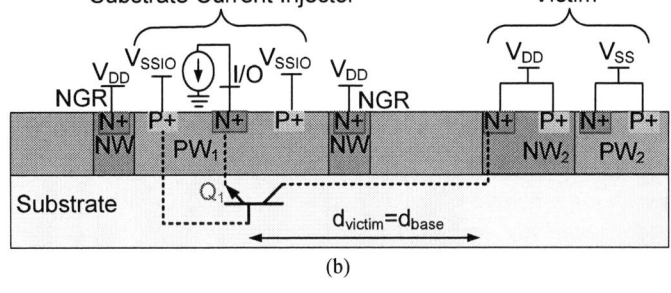

(b)

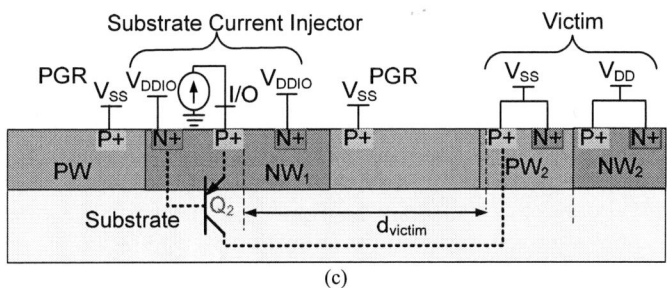

(c)

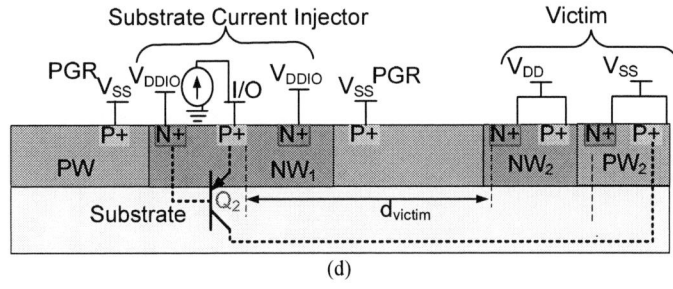

(d)

Figure 3. Test structure cross-sections. Each consists of a substrate current injector and a PNPN (victim). The injectors have 4 fingers (only one is shown). Key layout spacings are highlighted in the figures; d_{TAP}=40 µm and d_{victim}=5 µm, unless otherwise noted. d_{victim} is measured from the victim to the right-most finger of the injector. Structures (a) and (b) are used for negative I-tests; (c) and (d) are used for positive I-tests. The victims in (a) and (c) are $0°$ oriented, which means that the P-well of the victim (PW_2) is closer to the injector than is its N-well (NW_2). (b) and (d) show $180°$ oriented victims, in which the relative positions of PW_2 and NW_2 are swapped. N-well guard rings (NGR) and P-well guard rings (PGR) are 1 µm wide.

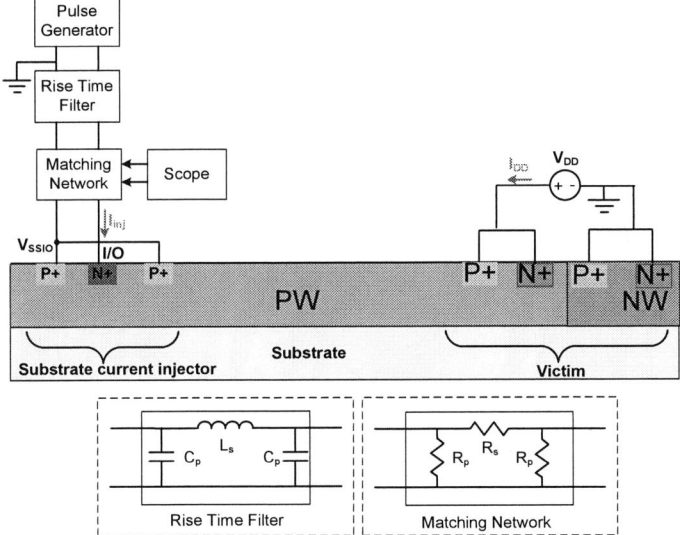

Figure 2. Experimental set-up. The rise-time filter adjusts the rise-time of the pulse injected into the device under test. I_{inj} is calculated from the measured voltage drop across the matching network. I_{DD} is the current through the dc power supply.

the substrate. It is normal practice to surround a minority carrier injector with an N-well guard ring (NGR), to prevent the electrons from reaching the victims. Similarly, majority carrier injectors are surrounded by P-type guard rings (PGR)— comprised of a P+ diffusion inside a P-well. The guard ring efficacy may be assessed by measuring I_{trig} both with and without the guard ring activated. An inactive GR is left floating; in this work, an active NGR is connected to V_{DD}, and an active PGR is connected to V_{SS}.

III. RESULTS AND DISCUSSIONS

A. Pulse-width dependence

The pulse-width dependence of I_{trig} is investigated by changing T_{PW} of the trigger source. The measurement results are shown in Figs. 4 and 5. The results shown in Fig. 4 confirm previous observations [6][7]: negative I_{trig} is a decreasing function of T_{PW}. These data also show that for a fixed substrate current injector to victim spacing (d_{victim}), negative I_{trig} is virtually identical for 90 and 130-nm CMOS technologies. The data of Fig. 5 reveal that positive I_{trig} is also a function of pulse-width; a comparison of Figs. 4 and 5 indicates that I_{trig} becomes independent of T_{PW} on a significantly shorter timescale under positive test conditions than under negative test conditions. The different time dependencies observed under positive and negative test conditions suggest that these time dependencies are not intrinsic to the victim. This hypothesis is confirmed by applying a pulsed overvoltage to the V_{DD} terminal of the victim so as to trigger internal latchup. The measurement results of Fig. 6 show that the latchup trigger voltage is independent of T_{PW} on a time scale ranging from less than 100ns up to 100µs.

The pulse-width dependence of negative I_{trig} is attributed to the non-quasi-static behavior of the parasitic NPN transistor Q_1, formed by the N^+ region of the injector, the P-substrate, and NW_2 (refer to Fig. 3). This is confirmed by an experiment. The potential at the emitter of Q_1 is pulled below zero, resulting in emitter current. The steady-state value of I_E is just a little bit less than the value of I_{trig} obtained from the static I-test. The

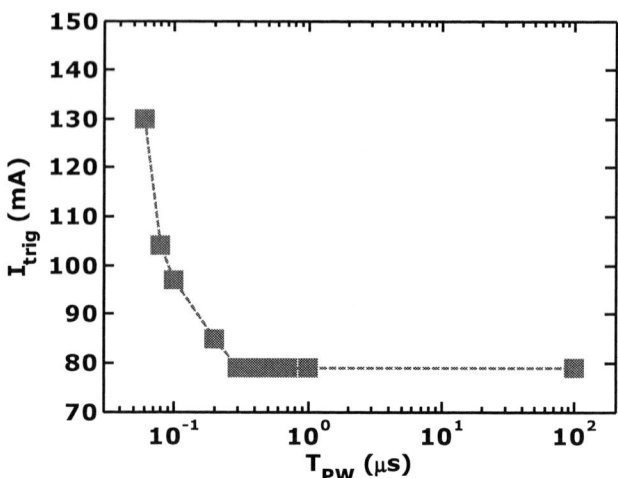

Figure 5. I_{trig} from positive I-test. Guard rings are inactive. 0° oriented victim. 130-nm CMOS technology.

measured rise time for I_E is 7ns. The collector current $I_{C,Q1}(t)$ is monitored. As shown in Fig. 7, $I_{C,Q1}$ approaches steady-state far more slowly than does I_E. Latchup is triggered when $I_{C,Q1}$ is large enough to forward bias the base-emitter junction of the victim, i.e., when

$$I_{C,Q1} = I_{NW}^{crit} . \qquad (1)$$

In (1), I_{NW}^{crit} is the minimum amount of current that has to be collected by NW_2 to trigger latchup. Based on Figure 7, if the current injected at the IO pad is just slightly higher than the static I_{trig}, it will take about 10µs for $I_{C,Q1}$ to reach I_{NW}^{crit}. Therefore, I_{trig} should be a decreasing function of T_{PW} for $T_{PW} \leq 10µs$, consistent with the data of Fig. 4.

$I_{C,Q1}$ increases slowly as a result of the large transit time for minority carriers in the substrate. The transit time is affected by recombination in the base region of Q_1. One may show this

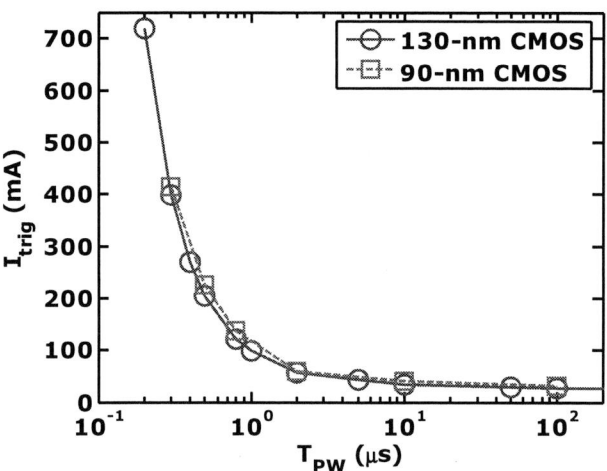

Figure 4. Negative I-test. 0° oriented victims. I_{trig} is a decreasing function of T_{PW}. 90-nm and 130-nm CMOS technologies. NGR are inactive.

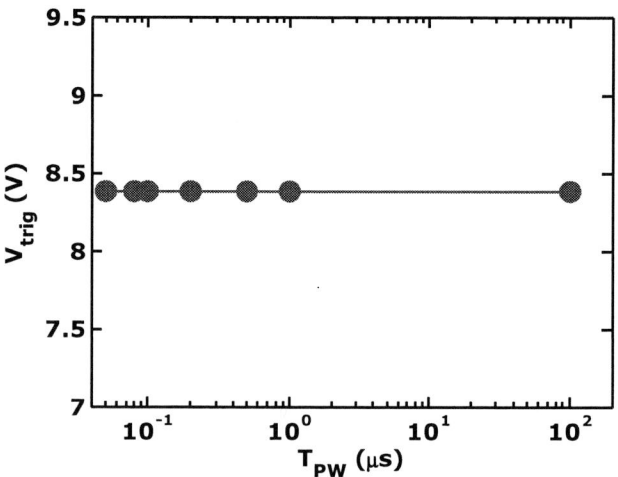

Figure 6. V_{trig} for internal latchup vs. T_{PW}. The victim is shown in Fig. 2. The terminals of the substrate injector are left floating. A trigger source with pulse-width of T_{PW} and rise-time of 7ns is placed in series with the dc supply, V_{DD}. 130-nm CMOS technology.

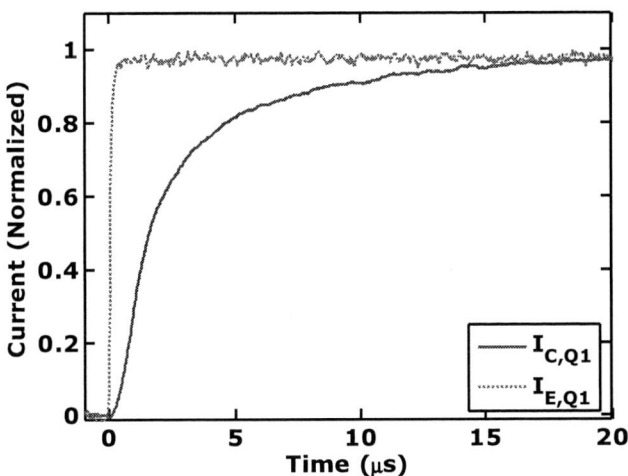

Figure 7. Collector and emitter current of Q_1 during a negative I-test for a $0°$ oriented victim. 130-nm CMOS technology.

mathematically by solving the diffusion equation in the base region of Q_1 to obtain an analytical expression for $I_{C,Q1}(t)$. A closed form solution cannot be obtained if one attempts to model the non-uniform, three-dimensional geometry of Q_1. Here, we formulate and solve the diffusion equation for a simplified NPN transistor that has uniform geometry in two dimensions. This transistor is shown in Fig. 8. The base length in the x-direction is d_{base}. Under low-level injection conditions, the diffusion equation in the base is written as

$$\frac{\partial n_p}{\partial t} = D_n \frac{\partial^2 n_p}{\partial x^2} - \frac{n_p}{\tau_n}. \tag{2}$$

In (2), D_n is the diffusion constant for electrons in the substrate and J_n and τ_n are the current density and the carrier lifetime, respectively, of electrons in the substrate (base of Q_1). Partial differential equation (2) is solved by the Laplace transform method. Taking the Laplace transform of (2) and solving for $N_p(s,x)$, one gets

$$N_p(s,x) = C_1 e^{-\sqrt{\frac{s+\frac{1}{\tau_n}}{D_n}}x} + C_2 e^{\sqrt{\frac{s+\frac{1}{\tau_n}}{D_n}}x} \tag{3}$$

and

$$J_n(s,x) = qD_n \frac{\partial N_p(s,x)}{\partial x}$$
$$= qD_n \sqrt{\frac{s+\frac{1}{\tau_n}}{D_n}} \left(-C_1 e^{-\sqrt{\frac{s+\frac{1}{\tau_n}}{D_n}}x} + C_2 e^{\sqrt{\frac{s+\frac{1}{\tau_n}}{D_n}}x} \right). \tag{4}$$

One may set C_2 to zero, as long as the solution will not be used to find the current in devices with $d_{base} << \sqrt{D_n\tau_n}$, the diffusion length for electrons in the substrate. C_1 is found by considering the boundary condition at the base edge x=0,

$$J_n(t,0) = J_E \cdot u(t), \tag{5}$$

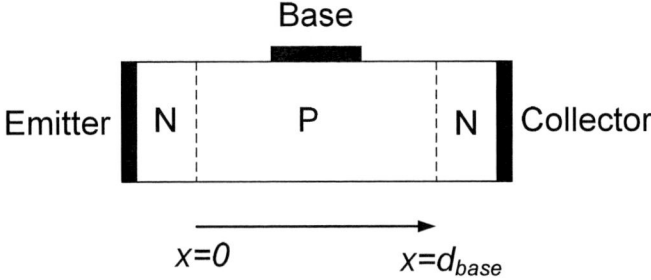

Figure 8. A 1-d NPN is used to simplify the derivation of the diffusion equation.

where u(t) is the unit step function and J_E is the emitter current density. A unit step function is used in (5) because the source driving the emitter has a very fast rise-time, justifying the step function approximation which greatly simplifies the algebra. Eq. (5) can be rewritten in the Laplace domain,

$$J_n(s,0) = \frac{J_E}{s}. \tag{6}$$

(6) is substituted in (4), yielding

$$C_1 = -\frac{1}{qD_n\sqrt{\frac{s+\frac{1}{\tau_n}}{D_n}}x} \cdot \frac{J_E}{s}. \tag{7}$$

The final expression for $J_n(s)$ thus becomes

$$J_n(s,x) = \frac{J_E}{s} e^{-\sqrt{\frac{s+\frac{1}{\tau_n}}{D_n}}x}. \tag{8}$$

The collector current density, $J_C(t)$, is calculated by setting $x=d_{base}$ in (8) and then taking the inverse Laplace transform;

$$J_C(t) = J_E \frac{D_n d_{base}}{2\sqrt{\pi}D_n^{1.5}} \int_0^t \frac{e^{-\frac{t}{\tau_n}} \cdot e^{-\frac{d_{base}^2}{4D_n t}}}{t^{1.5}} dt. \tag{9}$$

Finally, one may write an expression for the common base current gain, α.

$$\alpha(t) = \frac{J_C}{J_E} = \frac{I_C}{I_E} = \frac{D_n d_{base}}{2\sqrt{\pi}D_n^{1.5}} \int_0^t \frac{e^{-\frac{t}{\tau_n}} \cdot e^{-\frac{d_{base}^2}{4D_n t}}}{t^{1.5}} dt. \tag{10}$$

From (1), latchup is triggered when the current collected by NW_2 of the victim is equal to I_{NW}^{crit} . This N-well region is analogous to the collector region of the transistor in Fig. 8. I_{trig} is related to I_{NW}^{crit} by the NW collection efficiency, α_{NW}. Reasonably assuming that α_{NW} has the same functional form as does α derived above, one obtains the following expression for the pulsewidth-dependent I_{trig}.

978-1-4244-5430-3/10 $26.00 © 2010 IEEE

$$I_{trig} = \frac{I_{NW}^{crit}}{C \int_0^{T_{PW}} \dfrac{e^{-\frac{t}{\tau_n}} \cdot e^{-\frac{d_{base}^2}{4D_n t}}}{t^{1.5}} dt} . \qquad (11)$$

In (11), the constant C is a function of the emitter and collector areas, and other material and geometric constants. Eq. (11) is plotted in Fig. 9; the model predicts that I_{trig} should be a decreasing function of T_{PW} for $T_{PW} \leq 10\mu s$, consistent with the measurement data. Generally, in a technology with a moderate or high resistivity substrate, τ_n will be large and negative I_{trig} will be a strong function of T_{PW}.

The pulse-width dependence of positive I_{trig} (see Fig. 5) may be understood by considering current conduction through Q_2 (refer to Fig. 3(c)). Figure 10 shows that $I_{C,Q2}$ saturates after 200ns, which is consistent with the behavior of I_{trig} in Fig. 5. $I_{C,Q2}$ reaches steady-state much faster than does $I_{C,Q1}$ because Q2 has a shorter base width than Q_1 Therefore, the pulse-width dependence of negative I_{trig} is more pronounced than that of positive I_{trig}.

B. Effect of trigger source rise-time

During transient test conditions, displacement current will augment the carrier injection into the substrate. Recall that the PN junction current injectors have an associated capacitance. Displacement current is a majority carrier current. Referring to Fig. 3(c) or (d), let I_{disp} denote the amount of displacement current injected into the I/O when the pad voltage is raised from V_{DDIO} to V_{inj} with a rise-time of t_{edge}; for the test structure of Fig. 3(a) or (b), let I_{disp} denote the amount of displacement current injected into the I/O when the pad voltage is lowered from 0 to $-V_{inj}$ with a fall-time of t_{edge}. Because I_{disp} is a decreasing function of t_{edge}, fewer carriers are injected into the device as t_{edge} increases. Therefore, one might expect I_{trig} to be an increasing function of t_{edge}. However, both positive and negative I_{trig} are insensitive to dV/dt, as indicated by the data in Figs. 11 and 12. These results are explained below.

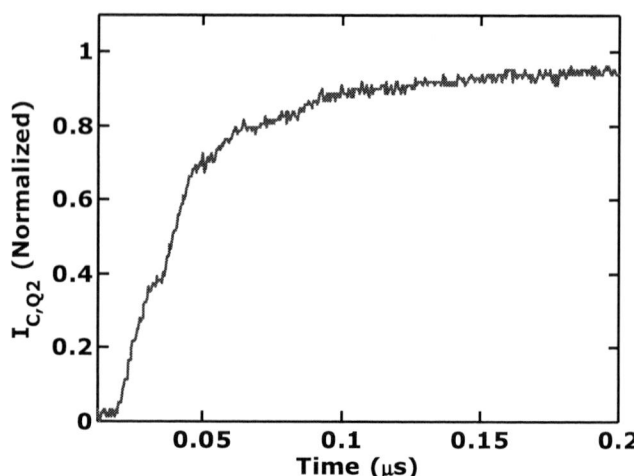

Figure 10. Collector current of Q_2 during a positive I-test for a 0° oriented victim. 130-nm CMOS technology.

During a negative I-test, latchup is triggered by electrons, whereas the displacement current in the substrate is a hole current. In fact, during a negative I-test, substrate hole current has the wrong polarity to trigger latchup [10]. Thus, negative I_{trig} is insensitive to t_{edge}, as was shown in Fig. 11.

Figure 13 shows parasitic PNP Q_2, which controls substrate current injection during a positive I-test. The substrate current is equal to the collector current of Q_2, $I_{C,Q2}$. The displacement current component of $I_{C,Q2}$ is denoted as $I_{C,Q2,disp}$. R_{NW} is small ($< 3\Omega$) because the substrate current injector was, in fact, an ESD protection diode, which has a small on-resistance. It follows that $R_{NW}(C_{BC} + C_{BE}) \ll t_{edge}$, and therefore

$$\frac{I_{C,Q2,disp}}{I_{disp}} \simeq \frac{C_{BC}}{C_{BC} + C_{BE}} e^{-\frac{t_{edge}}{R_{NW}(C_{BC}+C_{BE})}} \ll 1 . \qquad (12)$$

The inequality in (12) indicates that only a small fraction of the displacement current at the I/O pin is injected into the

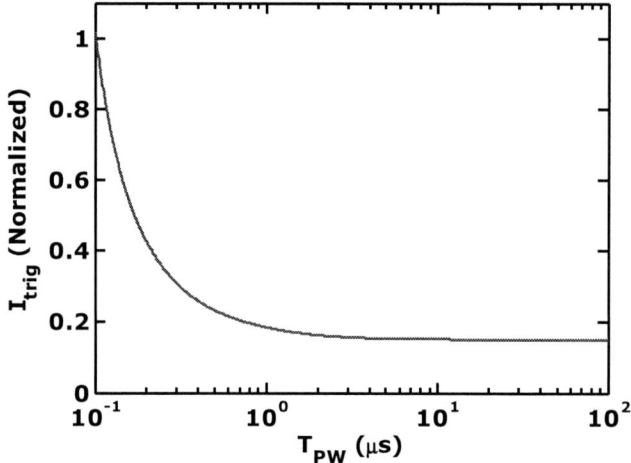

Figure 9. I_{trig} vs. T_{PW} as predicted by eq. (11). $D_n = 30$ cm^{2}s^{-1} and $\tau_n = 3\mu s$, reasonable values for 90 and 130-nm technologies [10]. I_{trig} becomes constant for $T_{PW} \geq 10\mu s$, similar to the data of Fig. 4.

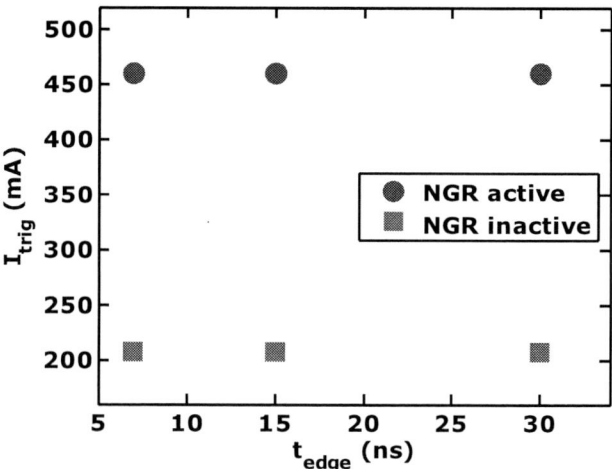

Figure 11. Negative I-test. I_{trig} is insensitive to t_{edge}. $T_{PW}=500$ns. 0° oriented victims. 130-nm CMOS.

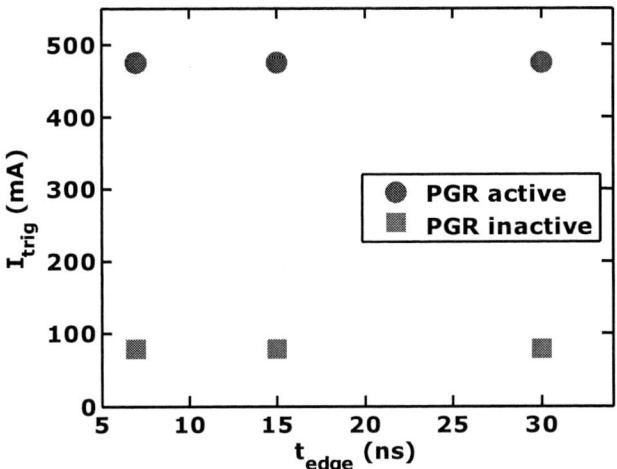

Figure 12. Positive I-test. I_{trig} is insensitive to t_{edge}. T_{PW}=1μs. $0°$ oriented victims. 130-nm CMOS.

substrate. This is why positive I_{trig} is insensitive to t_{edge}.

C. Orientation of the victim

In previous works, it was claimed that $0°$ victim orientation (Fig. 3(a)) provides the lowest value of negative I_{trig} [6][7]. Figures 14 and 15 indicate that this is generally not a correct assertion. It's true only under static test conditions in the absence of guard rings. In most cases, the $180°$ oriented victim (Fig. 3(b)) has the lower trigger current; this can be attributed to the smaller base width of Q_1, d_{base}.

The device simulation results of Fig. 16 may be used to further understand the effects of orientation. The value of I_{trig} depends both on the fraction of current injected at the I/O pad that gets collected by NW_2 of the victim (i.e., α_{NW}) and on the direction of current flow within NW_2. The current must be directed such that it lowers the N-well potential in the vicinity of the P^+ diffusion, thus forward biasing the PN junction, a necessary step to trigger latchup. When NGR are used (Fig. 16(a)), current flows vertically through NW_2, regardless of orientation, and the $180°$ oriented victim always has lower I_{trig} due to the smaller d_{base} and consequently larger α_{NW}. Figures

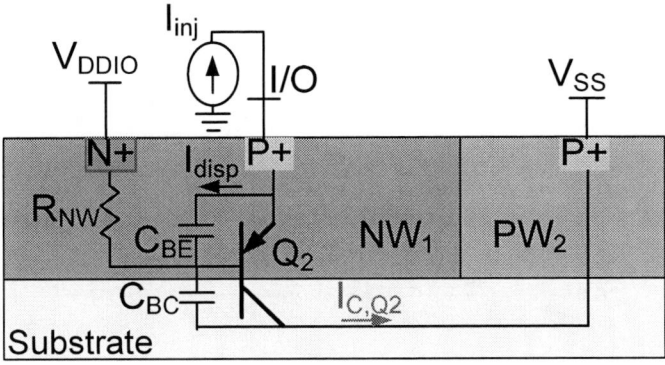

Figure 13. Parasitic PNP Q_2 is the substrate current injector during a positive I-test (see Figs. 3(c)(d)). R_{NW} is the N-well resistance. C_{BE} and C_{BC} are the base-emitter and the base-collector junction capacitances, respectively. Only the P+ diffusion of the victim is shown.

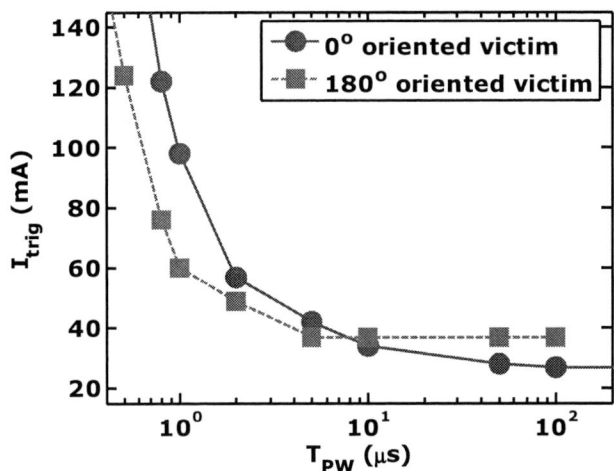

Figure 14. Negative I-test with inactive NGR. Only for large T_{PW} does the $0°$ oriented victim have smaller I_{trig} than the $180°$ oriented victim. 130-nm CMOS technology.

16(b) and 16(c) illustrate current flow in structures without GR. For the $180°$ oriented victim (Fig. 16(b)), the portion of the current that flows laterally between the injector and NW_2 does not assist in lowering the N-well potential in the vicinity of the P^+ diffusion (see Fig. 16(d)). For the $0°$ oriented victim (Fig. 16(c)), all of the current in NW_2 helps to lower the potential near the P^+ diffusion (see Fig. 16(e)). Thus, for the case of no NGR and long T_{PW}, I_{trig} of the $0°$ oriented victim is lower than that of the $180°$ oriented victim. However, as the stress pulse-width is made small, eq. (11) indicates that the number of carriers collected by NW_2 of the $0°$ oriented victim (long d_{base}) becomes a decreasing fraction of the number collected by NW_2 of the $180°$ oriented victim (shorter d_{base}). Therefore, for short stress durations, the $180°$ oriented victim has the lowest I_{trig}, as shown in Fig. 14.

Figure 17 examines the effect of orientation on positive I_{trig}. For the positive I-test, the $180°$ victim orientation (Fig. 3(d)) provides the lower I_{trig}, regardless of T_{PW}; only the GR active

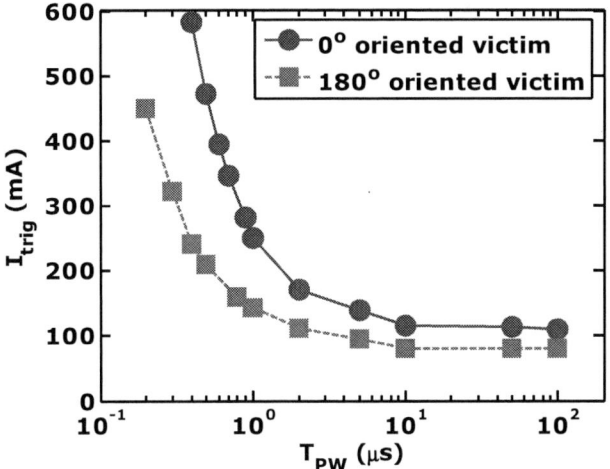

Figure 15. Same experiment as for Fig. 14, except that the NGR are active. The $180°$ oriented victim has the smallest I_{trig}, regardless of T_{PW}.

978-1-4244-5430-3/10 $26.00 © 2010 IEEE   471

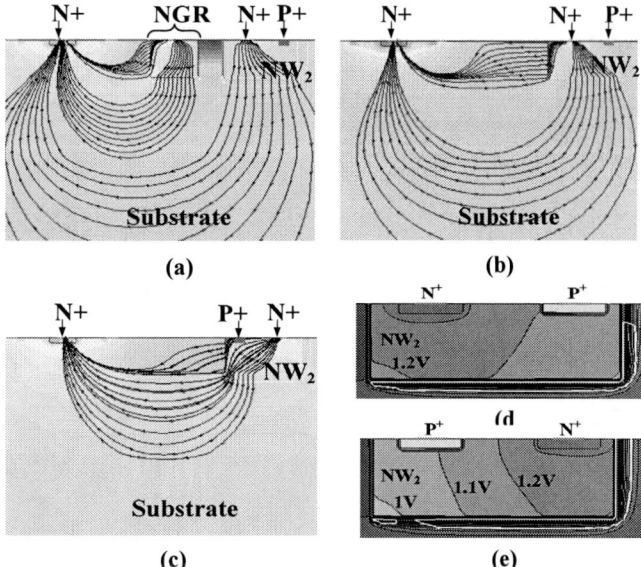

Figure 16. Simulated current flow during a negative I-test for (a) 180° oriented victim with active NGR, (b) 180° oriented victim without NGR, and (c) 0° oriented victim without NGR. Figs. (d) and (e) show potential contours for cases (b) and (c), respectively. I_{inj}= 3 mA/μm in all of the simulations.

case is examined. In [9], it is shown that in a positive I-test

$$I_{trig} = \frac{1}{\alpha_{Q2}} \cdot \frac{V_{BE,on}}{R_{PW,2}}, \qquad (13)$$

where α_{Q2} is the common-base current gain of Q_2, $R_{PW,2}$ is the P-well resistance shown in Fig. 1, and $V_{BE,on}$ represents the on-state base-emitter voltage drop of Q_2. (13) seems to suggest that positive I_{trig} will be independent of both spacing (d_{victim}) and orientation; the first of these predictions is consistent with measurement data [10]. The orientation effect seen here (*cf.* Fig. 17) is an artifact of the test structure design. Since the test structures contain guard rings, (13) is modified to account for their presence,

$$I_{trig} = \frac{1}{\alpha_{Q2}} \cdot \frac{1}{f_{PGR}} \cdot \frac{V_{BE,on}}{R_{PW,2}}, \qquad (14)$$

where f_{PGR} is the fraction of injected carriers that is collected by the PGR. Note that I_{trig} is a decreasing function of $R_{PW,2}$. In the test structures with a 0° oriented victim, the PGR are placed in the same well as is the victim (PW2). In these test structures, the PGR not only increase I_{trig} by collecting some of the excess holes from the substrate, they also increase I_{trig} by decreasing $R_{PW,2}$.

D. Guard ring efficiency under TLU testing

N-well [P-well] guard rings increase negative [positive] I_{trig}, thus improving latchup resilience; however, Fig. 18 shows that the relative benefit of NGR decreases for short stress durations. In the earlier dataplots, e.g. Fig. 4, it was shown that I_{trig} is higher for short T_{PW}; we have previously shown that NGR efficiency drops under high-level injection conditions [11]. Taken together, these two observations explain why NGR raise I_{trig} by a smaller percent as T_{PW} decreases.

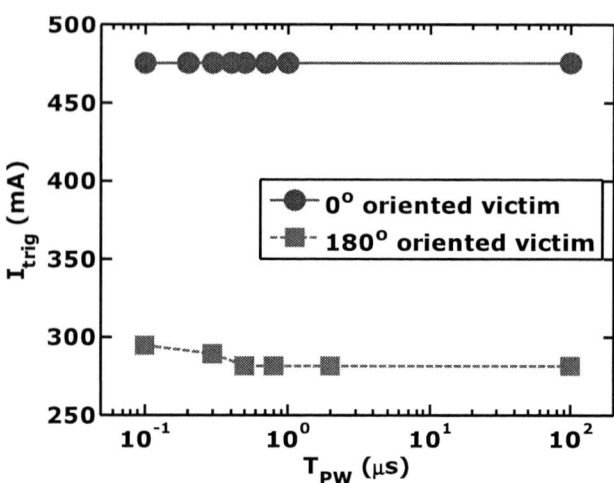

Figure 17. Positive I-test with active PGR. The 180° oriented victim has the lower I_{trig}. 130-nm CMOS technology.

E. Triple well technology

Triple well can be used to reduce the latchup hazard due to substrate hole injection, which occurs during a positive I-test. In the 130-nm technology, placing the victim inside a deep N-well was found to increase I_{trig} by almost a factor of 2. However, a comparison of the data in Figs. 4 and 19 shows that placing the victim inside a deep N-well enhances its susceptibility to negative TLU; that is, for a given d_{victim}, I_{trig} is lower when the victim P-well lies within a deep N-well. A similar observation was made about static latchup [12]. Triple well raises the latchup threat because the deep N-well provides an additional collection area for electrons.

F. Negative I-test vs. Positive I-test

Previously, it had been reported that the negative I-test provides the lowest I_{trig}, that is, it's the worst case test condition [12]. However, the data of Figs. 20 and 21 show that this is too

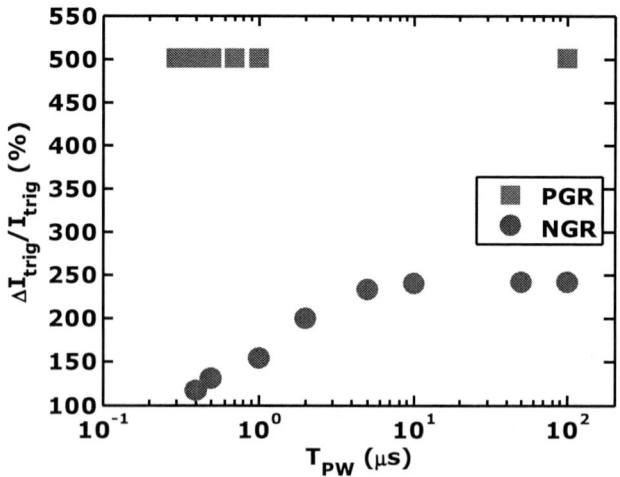

Figure 18. $\Delta I_{trig}/I_{trig}$(GR inactive) where $\Delta I_{trig} \equiv I_{trig}$(GR active) - I_{trig}(GR inactive). 0° oriented victims. Circular data markers are for a negative I-test; square ones are for a positive I-test. 130-nm CMOS technology.

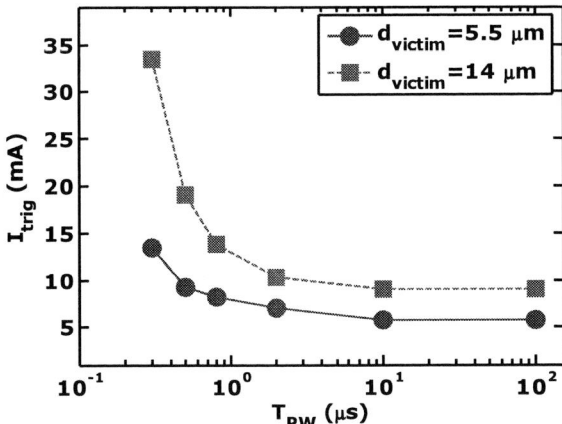

Figure 19. Negative I-test on $0°$ oriented victims built using triple well. NGR are inactive. 130-nm CMOS technology.

general a claim. These figures compare the values of I_{trig} obtained from positive and negative I-tests, both when the GR are inactive (Fig. 20) and when they are active (Fig. 21). If T_{PW} is large, as is the case for static latchup testing, the negative I-test does yield the smallest I_{trig}. However, for stress durations less than about 500ns, the positive I-test yields a lower value of I_{trig}.

IV. CONCLUSIONS

Negative current injection is worst case during static latchup testing [7], which is generally used for product qualification. However, real world stresses, such as cable discharges, are transient, in which case positive current injection is worst case.

A $0°$ oriented victim is most susceptible to latchup only during static testing and if guard rings are not used. Otherwise, a $180°$ oriented victim will have a lower latchup trigger current. For the case of a negative I-test, this is because of the proximity of victim N-well to the substrate current injector (larger α_{NW}); for the case of a positive I-test, it is due to the

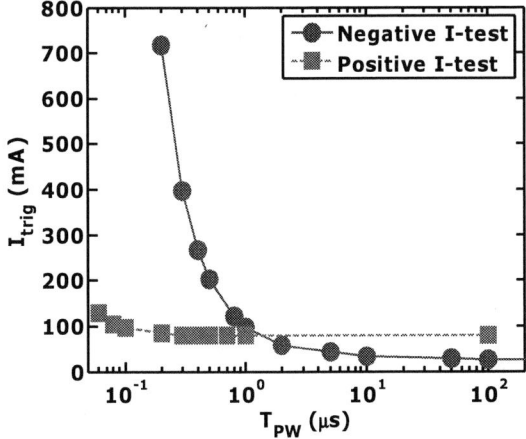

Figure 20. I_{trig} from positive and negative I-tests. Guard rings are inactive. $0°$ oriented victims. 130-nm CMOS technology.

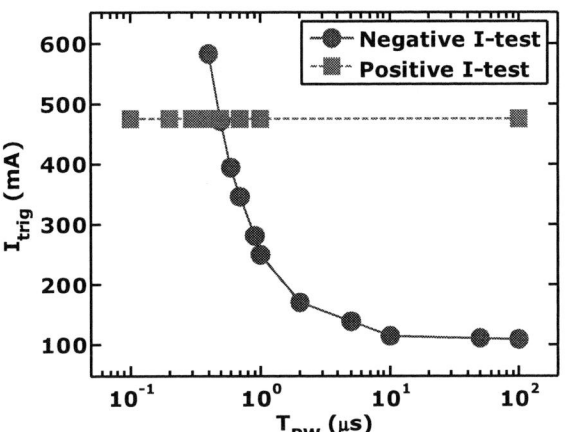

Figure 21. Same experiment as for Fig. 20, except the guard rings are active.

direction of current flow within the victim P-well.

N-type guard rings become less efficient in preventing TLU as the pulse-width of the injected current decreases. Thus, they must be evaluated under short transient pulses relevant to TLU.

ACKNOWLEDGMENT

The authors gratefully acknowledge UMC for fabricating the test structures through the UMC University Program. This work was supported by Semiconductor Research Corporation.

REFERENCES

[1] R. Troutman, *Latchup in CMOS Technology: The Problem and Its Cure*, Kluwer Academic Publishers, NY, 1986.

[2] W. Moms, "Latchup in CMOS," *Proceedings of International Reliability Physics Symposium*, pp. 76-84, 2003.

[3] "IC Latch-Up Test, JESD78B Standard", http://www.jedec.org, pp. 1-19, 2008.

[4] Telecommunications Industry Association (TIA), Category 6 Cabling: Static Discharge Between LAN Cabling and Data Terminal Equipment, Category 6 Consortium, Dec. 2002.

[5] "Transient Latch-up testing- Component Level Supply Transient Stimulation", ANSI/ESD SP5.4-2004, ESD Association, 2004.

[6] K. Domanski et al., "Development strategy for TLU-robust products," *Proceedings of ESD/EOS Symposium*, pp. 299-307, 2004.

[7] D. Kontos et al., "External latchup characterization under static and transient conditions in advanced bulk CMOS technologies," *Proceedings of International Reliability Physics Symposium*, pp. 358-363, 2007.

[8] S. Joshi and E. Rosenbaum, "Transmission line pulsed waveform shaping with microwave filters," Proc. EOS/ESD Symp., pp. 364-371, 2003.

[9] F. Farbiz and E. Rosenbaum, "Analytical modeling of external latchup," *Proceedings of ESD/EOS Symposium*, pp. 338-346, 2007.

[10] F. Farbiz and E. Rosenbaum, "Modeling of majority and minority carrier triggered external latchup," *Proceedings of International Reliability Physics Symposium*, pp. 270-277, 2008.

[11] F. Farbiz and E. Rosenbaum, "Guard ring interactions and their effect on CMOS latchup resilience," *IEEE Electron Device Meeting*, pp. 345-348, 2008.

[12] D. Kontos et al., "Investigation of external latchup robustness of dual and triple well design in 65nm bulk CMOS technology," *Proceedings of International Reliability Physics Symposium*, pp. 145-150, 2006.

978-1-4244-5130-3/10 $26.00 © 2010 IEEE

A Novel TCAD-Based Methodology to Minimize the Impact of Parasitic Structures on ESD Performance

Nicholas Olson, Gianluca Boselli*, Akram Salman* and Elyse Rosenbaum

Department of Electrical & Computer Engineering
University of Illinois at Urbana-Champaign
Urbana, IL USA
naolson@illinois.edu

*Texas Instruments Inc.
Dallas, Texas USA

Abstract—During an ESD event, breakdown of a parasitic bipolar transistor can lead to chip failure. For a variety of ESD protection networks, it is demonstrated that TCAD simulations correctly predict the stress level at which failure occurs due to bipolar breakdown. A procedure to characterize the interaction between any two N-type diffusions and the ESD cells to which they are connected is presented.

Keywords- Electrostatic Discharge; ESD; Parasitic Bipolar; TCAD

I. INTRODUCTION

It is well established that the breakdown of a parasitic NPN during an ESD event can lead to chip failures [1]. Parasitic NPNs may be formed between any pair of two N-type diffusions. One such scenario is illustrated in Figure 1; in this figure, the ground net is tied to the substrate, which forms the base of the parasitic NPN. During an ESD event from pin POS to pin NEG, the substrate is at a positive potential with respect to pin NEG, thus forward biasing the base-emitter junction of any parasitic NPNs connected between the two pins. This forward bias condition lowers the breakdown voltage, V_{t1}, of the parasitic bipolar transistor [2]. Parasitic NPNs are often formed between circuit blocks and it may not be possible to engineer all the parasitic devices to safely operate in breakdown; therefore, the ESD engineer's objective is to avoid parasitic bipolar breakdown. Achieving this objective is non-trivial because the breakdown voltage varies with the layout.

Multiple strategies for preventing parasitic NPN breakdown [2, 3, 4] have been disclosed in the literature, with the most common being to space the N-regions farther apart. The challenge is to optimize the spacing so as to ensure a robust design without unnecessarily sacrificing silicon area. The breakdown voltage depends on the location of the substrate taps, the doping profiles of the N-regions and the base, the ESD clamp characteristics and the orientation of the N-regions with respect to each other; thus, there is not a single value for the optimum spacing. Process complexity increases the number of variables. For example, in a modern BiCMOS process, there may be over 10 types of N-regions (acting as collectors/emitters) and more than 3 types of P-regions (acting as bases), leading to hundreds of different types of parasitic devices. It is necessary to develop a methodology to efficiently and accurately determine the interaction between any two N-type regions and the ESD cells connected to them, without the

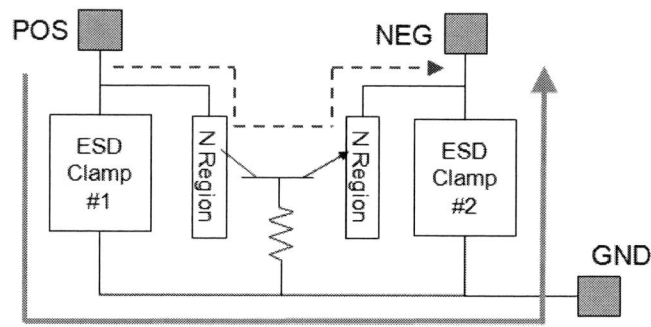

Figure 1. Example ESD network. Solid arrow depicts intended path for ESD current and dashed arrow represents a parasitic path, for the case of positive stress on pad POS with respect to pad NEG.

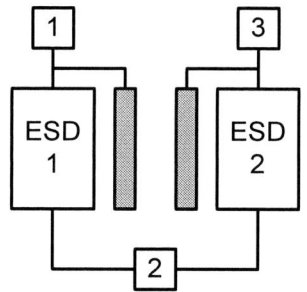

Figure 2. Test structure. Shaded regions are N-regions.

need to run hundreds of test structures. This paper presents such a methodology; it requires only a limited number of 2D TCAD simulations to be performed.

II. MEASUREMENTS

A. Test Structures

Test structures consisting of two ESD clamps and a built-in NPN were fabricated in a state-of-the-art BiCMOS process. Figure 2 depicts a conceptual view of the test structure layout. Each ESD clamp consists of a snapback device and a diode for reverse conduction. Pad 2 is connected to the P+ guardrings that surround the ESD clamps. A variety of N-type and P-type diffusions are available in this process technology and are listed in Table I. This paper presents results only for a subset of these diffusions. Multiple N-type diffusions may be combined to form a single N-region (for example, see Figure 3B). The

978-1-4244-5430-3/10 $26.00 © 2010 IEEE

test structures vary in terms of the N-regions used to form the emitter and collector of the parasitic NPN and also the spacing between the two N-regions. Figures 3A and 3B show two of the different NPNs that were investigated. High voltage N-Well (HV-NW) is a N-well diffusion that is deeper than the normal N-well and can reach the N-buried layer, NBL. N-SINKER is similar to the HV-NW, except it has a higher doping concentration.

B. Testing Method

Pulsed I-V characteristics were obtained using a TLP system with a 100 ns pulse-width and 10 ns rise-time. First, pulses are applied between pads 1 and 2, and then between pads 2 and 3 to generate the I-V curves (taken to failure) of the stand-alone ESD clamps. Next, the pulsed I-V characteristic is measured between pad 1 and pad 3 on a fresh structure. This exercises the whole ESD network with the parasitic NPN in parallel. Finally, the first two I-V curves are summed; that is, current is plotted as a function of $V_{12} + V_{23}$. The resulting curve is what is expected if there were no parasitic device and is hereafter referred to as the "summed I-V."

The summed I-V is compared with the I-V curve measured between pad 1 and pad 3. If the pad 1 to pad 3 measurement produced failure at a lower current, it is concluded that the parasitic bipolar was driven into breakdown and the test is classified as a fail. This is a conservative definition of failure since in a normal design process, an I_{TARGET} would be specified as the maximum expected current through the ESD network. Therefore, the pass threshold would be lowered from the ESD protection device failure level.

C. Results

For a given pair of N-regions, the pass/fail results are tabulated as a function of the spacing and the type of ESD clamp. Examples are given in Tables IIA and IIB, which contain the measurement results for the specific test structures shown in Figures 3A and 3B. Tables IIA and IIB are only a small portion of the full results grid, which contains data for test structures that contain all allowed combinations of ESD clamps and diffusions from Table I. In Table II, ESD1 through ESD4 denote four different bipolar-based snapback ESD clamps that have increasingly larger trigger voltages; the higher trigger voltages are achieved by cascoding the protection devices. The pulsed I-V characteristics of these ESD clamps are shown in Figure 4. The data in Table II indicate that, as expected, the likelihood of failure increases when the ESD clamp trigger voltage is increased and/or the N-region spacing is decreased. The data in Table II also show that the addition of higher doped diffusions is detrimental. This is unsurprising since higher doped junctions are known to breakdown at lower voltages.

Table II shows that the minimum spacing needed to avoid parasitic NPN triggering is variable, depending on the type of clamp and N-diffusion. To avoid applying overly conservative spacing rules to all configurations, a comprehensive matrix of ESD cells and parasitics needs to be created and the corresponding failure levels determined. Given the size of the matrix, it is NOT feasible to do this at the silicon level. Thus, in the next section, a TCAD-assisted procedure for characterizing critical elements in the matrix of ESD cells and parasitics is presented.

TABLE I. LIST OF AVAILABLE DIFFUSIONS THAT MAY FORM PARASITIC NPN DEVICES. THE TWO N-DIFFUSIONS FORMING THE NPN MAY BE OF THE SAME OR DIFFERENT TYPES .

N Diffusions	P Diffusions
N+	P+
N-Well	P-Well
High Voltage Well	P-sub
N-SINKER	Moat
NBL	PBL

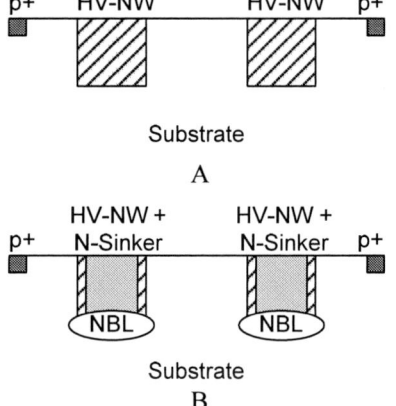

Figure 3. Cross section of parasitic structures. A: HV-NW. B: HV-NW+NBL+N-SINKER.

TABLE II. PASS/FAIL GRID. HORIZONTAL LINES (GREEN)=PASS. VERTICAL LINES (RED)=FAIL. NO LINES (GRAY)=NO DATA AVAILABLE. TABLES A AND B EACH CONTAIN DATA FOR A FIXED PAIR OF N-REGIONS WITH VARIABLE SPACING (I.E. BASE WIDTH OF THE NPN). IN BOTH EXPERIMENTS, THE MAXIMUM SPACING IS IDENTICAL AND ALL OTHER SPACINGS ARE NORMALIZED WITH RESPECT TO THIS VALUE. CIRCLED CASE WILL BE REFERENCED LATER.

Primary Protection Device	Device in Diode Operation	HV-NW to HV-NW				
ESD1	ESD2					
ESD2	ESD1					
ESD3	ESD4					
ESD4	ESD3					
Normalized Spacing		0.22	0.27	0.44	0.72	1

A

Primary Protection Device	Device in Diode Operation	HV-NW+NBL+N-Sinker to HV-NW+NBL+N-Sinker				
ESD1	ESD2					
ESD2	ESD1					
ESD3	ESD4					
ESD4	ESD3					
Normalized Spacing		0.23	0.36	0.44	0.55	0.72	1

B

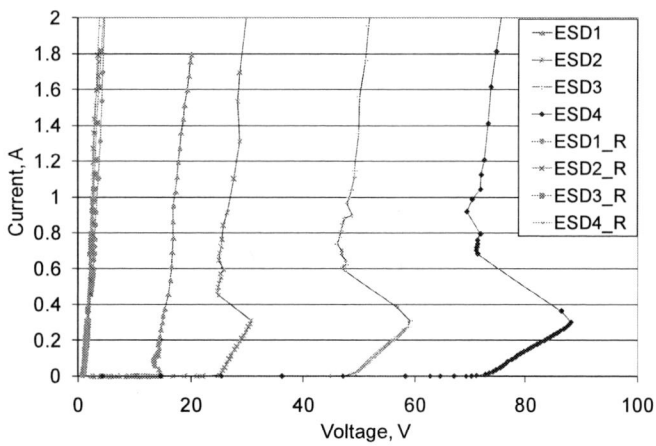

Figure 4. TLP I-Vs of clamps ESD1 through ESD4 used in Table II. The added R in the legend indicates a clamp was tested in reverse operation mode.

III. TCAD-ASSISTED BREAKDOWN PREDICTION

TCAD simulations have been shown to be useful in analyzing ESD networks with parasitic bipolar transistors [5]. This suggests it may be possible to develop structure-specific spacing rules without needing to fabricate an unreasonable amount of test structures. In the rest of this section, a method for using TCAD simulations to predict NPN breakdown and to develop spacing rules is presented.

A. Methodology

First, 2D quasi-static TCAD simulations are used to obtain the stand-alone I-V characteristics of the NPN devices of interest. The substrate bias is varied and the collector voltage is ramped up. As indicated in Figure 5, the potential at the base of the parasitic bipolar transistor will be less than the applied substrate bias voltage, and this potential difference is a function of the substrate tap placement. Figure 6 shows an example set of simulated I-V curves. Using the data of Figure 6, a plot of V_{t1} vs. substrate bias can be generated. The next step is to combine this data with the summed I-V. An example summed I-V is shown in Figure 7. Figure 7 also shows the I-V data for the second ESD clamp alone, this is equivalent to a plot of V_{sub}. Figure 8 shows the NPN V_{t1}, obtained from TCAD simulation, and the voltage drop across the ESD network, obtained from the summed I-V, both plotted as a function of the substrate bias. The voltage drop across the ESD network is, of course, V_{CE} of any parasitic NPN that is connected in parallel. As long as V_{CE} is less than the NPN V_{t1}, the NPN will not enter breakdown. The breakdown point is reached when the curves cross each other. The current at this point is denoted as the failure current, I_{fail}. I_{fail} is obtained by locating the point $V_{CE} = V_{t1}$ in Figure 8 and then finding the current at this V_{CE} in the dataplot of Figure 7. Figure 9 summarizes the complete flow of this method for predicting the failure point of any ESD network which contains a parasitic BJT.

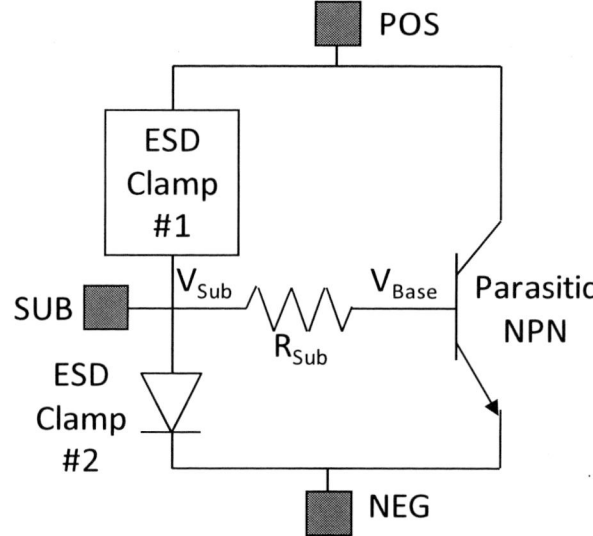

Figure 5. ESD network showing bias condition of parasitic NPN. V_{sub} (potential at the substrate taps, i.e., P+ guard rings) and V_{base} are at different potentials. In the TCAD simulations, V_{sub} is set externally; thus, it is important to include substrate taps into the TCAD simulions since R_{sub} will affect the results.

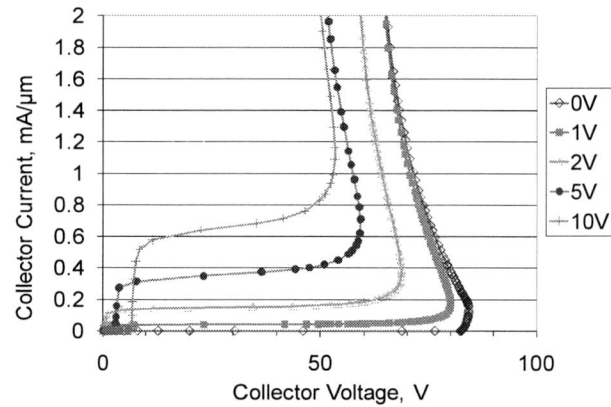

Figure 6. TCAD simulated I-V characteristics of a parasitic NPN. Substrate bias values in legend.

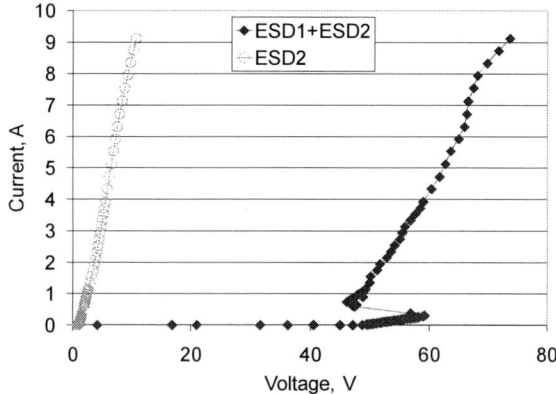

Figure 7. Pulsed I-V data of ESD clamps summed together and the second ESD clamp by itself. The voltage across ESD2 is equal to the substrate bias when the full test structure is measured from pad 1 to pad 3.

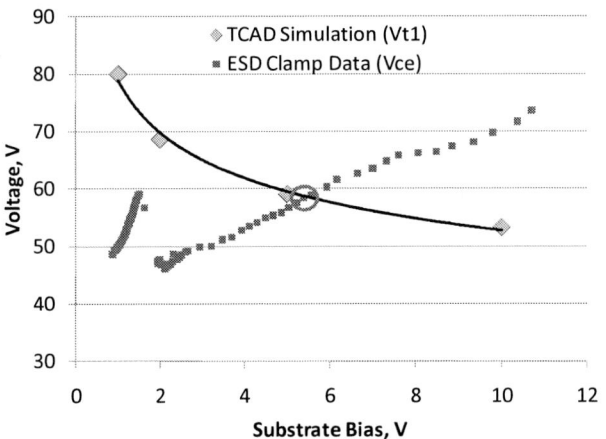

Figure 8. Simulated NPN V_{t1} from Figure 6 is plotted along with the expected collector voltage from Figure 7, all with respect to the substrate bias. Estimated failure point circled.

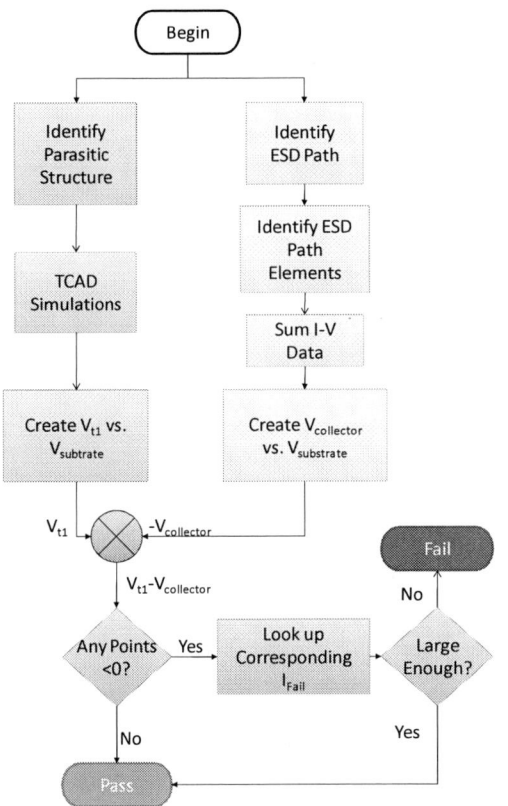

Figure 9. Methodology to calculate failure from TCAD simulation along with the pulsed I-V data for the ESD network.

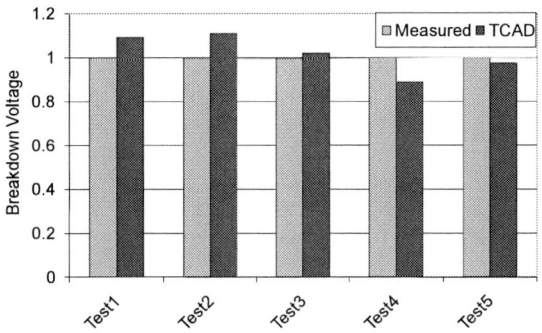

Figure 10. Failure voltage measured using TLP with that obtained from TCAD simulations using the described methodology. Values normalized to failure level of TLP measurements. Test 1 is the same structure as the circled case in Table II.

B. Results and Discussion

The above described procedure was applied to several different test structures and the results are shown in Figure 10. The test structures included two different ESD clamps, two different types of N-regions, and several different spacings. The estimated failure points derived from the TCAD simulations agree well with the failure currents and voltages measured using TLP. This confirms that TCAD simulation data may be combined with ESD clamp data to estimate failure levels when any arbitrary NPN is located between the anode and cathode of the ESD network.

Without loss of generality, the methodology can be applied using ESD clamp I-V curves obtained directly from simulation, making this approach applicable for pre-silicon evaluation. A few stand-alone NPN test structures would be needed to calibrate the process deck. When doing any of the TCAD simulations proposed here, it is important to use an accurate cross-section in a well calibrated process.

Accurate simulation of the breakdown voltage is not only dependent on how well the process has been simulated but also on the grid used for TCAD electrical simulations. It was found that truncating the cross-section to a too shallow depth below the silicon surface can alter the results. In practice, one can run multiple TCAD simulations, increasing the vertical extent of the simulation grid each time, until the parasitic NPN behavior no longer changes; at this point, a sufficiently large device cross-section is being simulated. In this study, it was found that extending the simulation grid to a depth over 2x that of the deepest diffusion was required.

To construct Figure 8, only a few V_{t1} values were simulated, but intermediate values can be interpolated with a good degree of accuracy using the solid curve shown. The solid curve has the form $y=Ax^B$. However, note that the solid curve is not extended down to zero substrate bias. At zero bias, the NPN breakdown mechanism changes to reverse PN junction breakdown. It is important to do several TCAD simulations for low substrate bias values, since extrapolated data from simulations obtained at only higher substrate voltages can introduce too much error.

IV. ANALYSIS

TCAD simulations may be used to provide additional insight into the relation between N-region type and breakdown voltage. Figures 11 and 12 depict TCAD results for the parasitic structures defined in Figure 3A and 3B, respectively; they show impact ionization rates near the collector of each parasitic NPN. Figure 11 shows that for the NPN of Figure 3A, the most impact ionization occurs inside the HV-NW. The addition of the N-SINKER and NBL layers (Figure 12) shifts the location of maximum impact ionization to the PN junction. Figure 12 shows increased impact ionization along the entire junction although the highest concentration is at the NBL to substrate interface. This is seen more clearly on a structure with just HV-NW and NBL, Figure 13. Information about impact ionization and current density, obtained from TCAD simulations, provides guidance as to how to inhibit breakdown of the parasitic NPN. For example, if breakdown occurs near the surface, adding an implant or some type of shallow isolation, if available, may help to increase the breakdown voltage of the parasitic NPN. If the breakdown occurs deeper in the substrate, as seen in the cases presented here, the solution may be to change the P-region profile below the surface. For this process, a PBL diffusion was available that was at the desired depth.

Indeed, additional TCAD simulations confirm that breakdown of the parasitic NPN may be inhibited by adding a P+ buried layer (PBL) in between the two N-regions, but not extending all the way to the N-regions. The PBL is helpful because it is located in the substrate at the depth where the parasitic bipolar has its highest current density. The PBL degrades the bipolar's current gain and increases the breakdown voltage. Figure 14 shows the simulated current density in devices with and without PBL at similar collector voltage levels (but different collector current levels). Figure 15 compares the simulated I-V curves for these two devices, and Table III provides TLP measurement results which confirm the simulation results. For comparison, the second row in Table III is the same structure as the circled case in Table II. In all cases, the addition of the PBL increases the I_{fail} value.

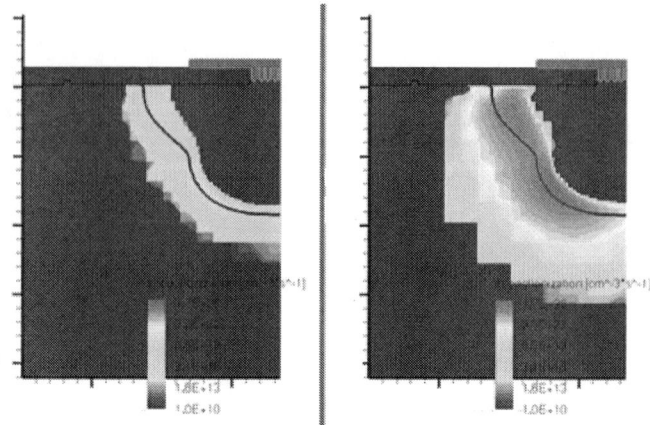

Figure 12. Simulated impact ionization rate near the collector of the structure in Figure 3B. Plot on the left is at a point where $I_C<I_{t1}$ and the plot on the right is at a point where $I_C>I_{t1}$. The highest level of impact ionization is at the NBL/substrate interface near the bottom left of the N-region.

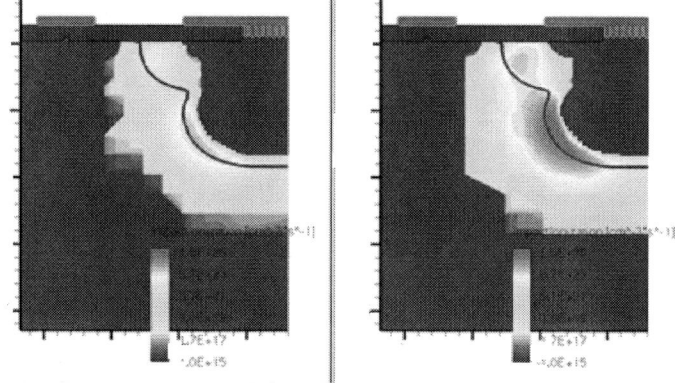

Figure 13. Simulated impact ionization rate near the collector of a structure with HV-NW and NBL. Plot on the left is at a point where $I_C<I_{t1}$ and the plot on the right is at a point where $I_C>I_{t1}$. The highest level of impact ionization is at the NBL/substrate interface near the bottom left of the N-region.

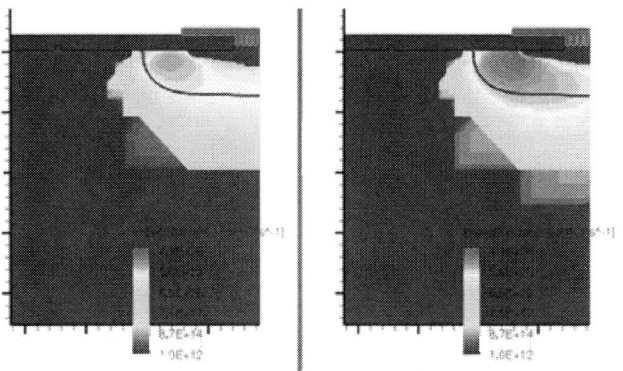

Figure 11. Simulated impact ionization rate near the collector of the structure in Figure 3A. Plot on the left is for a bias point where $I_C<I_{t1}$ and the plot on the right is is for $I_C>I_{t1}$. The highest level of impact ionization is inside the HV-NW near the actual contact where there is a shallow region of higher doping.

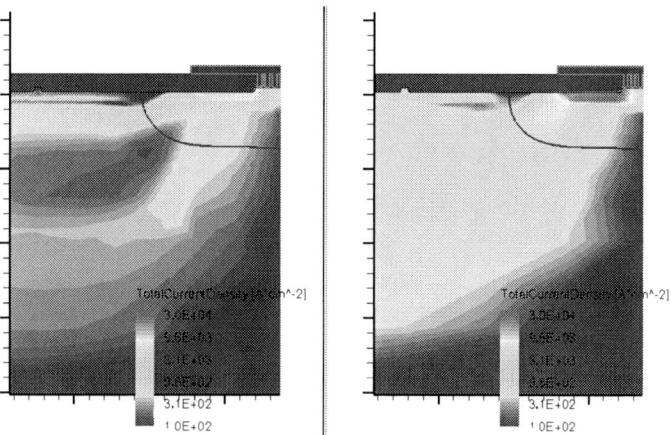

Figure 14. Total current density of a HV-NW based parasitic NPN. Cross section on the left is with PBL and cross section on the right is without PBL. The PBL interrupts the normal current path and changes the bipolar's behavior.

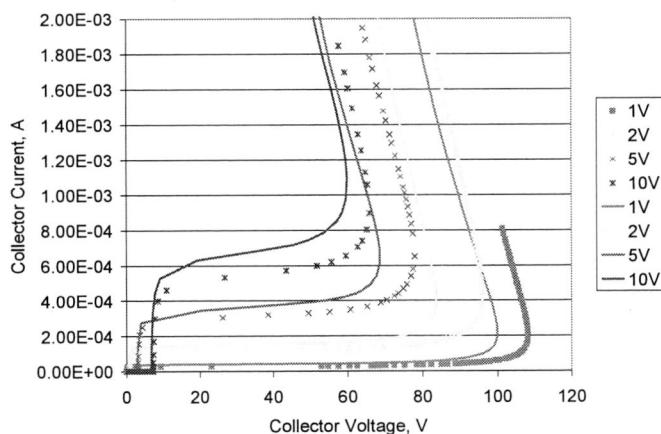

Figure 15. TCAD simulation showing benefit of adding PBL in between collector and emitter of the parasitic NPN. Solid curves are without PBL. Dotted curves are with PBL. Collector and emitter are HV-NW diffusions.

TABLE III. TLP DATA FOR TWO DIFFERENT ESD NETWORKS AND TWO DIFFERENT VALUES OF THE PARASITIC NPN BASE WIDTH (I.E., N-REGION SPACING). Δ IFAIL IS THE DIFFERENCE BETWEEN THE NO PBL CASE AND THE PBL CASE. THE ADDITION OF A PBL BETWEEN THE N-REGIONS INCREASES IMPEDES BIPOLAR BREAKDOWN AND IMPROVES THE FAILURE CURRENT.

ESD Network	Spacing (Normalized)	No PBL I_{fail} (A)	PBL I_{fail} (A)	ΔI_{fail} (A)
A	0.63 (No PBL) 0.66 (PBL)	0.177	0.957	+ 0.78
A	1	0.198	2.608	+ 2.41
B	0.63 (No PBL) 0.66 (PBL)	1.959	3.243	+ 1.28
B	1	3.074	5.701	+ 2.63

V. CONCLUSIONS

Parasitic NPN transistors can cause ESD failures. A time-efficient 2D TCAD-based methodology for accurately determining the interaction between any two N-type diffusions and the ESD cells connected to them has been demonstrated. The methodology is shown to drastically reduce the number of test structures needed to create N-diffusions spacing rules, even prior to silicon fabrication.

TCAD simulations have also been utilized to analyze the behavior of the parasitic bipolar. By identifying the breakdown location and the region of highest base current density, design changes can be made to inhibit the NPN breakdown. A successful demonstration of this process was shown; the TCAD-informed addition of a PBL layer raised the failure current of a particular test structure.

REFERENCES

[1] D. Trémouilles et al., "Latch-Up Ring Design Guidelines to Improve Electrostatic Discharge (ESD) Protection Scheme Efficiency," IEEE Journal of Solid-State Circuits, vol. 39, no. 10, pp. 1778-1782, 2004.

[2] A. Amerasekera, C. Duvvury, "ESD in Silicon Integrated Circuits", John Wiley and Sons, 1995.

[3] U. Glaser et al., "Base pushout driven snapback in parasitic bipolar devices between different power domains," IEEE International Reliability Physics Symposium, pp. 387-392, 2004.

[4] L. Cerati et al., "Novel Technique to Reduce Latch-up Risk due to ESD protection devices in smart power technologies," EOS/ESD Symposium, pp. 32-38, 2006.

[5] K. Esmark (private communication), 2007.

On the Differences Between 3D Filamentation and Failure of N & P Type Drain Extended MOS Devices Under ESD Condition

Mayank Shrivastava[1], S. Bychikhin[2], D. Pogany[2], Jens Schneider[3], M. Shojaei Baghini[1], Harald Gossner[3], Erich Gornik[2], V. Ramgopal Rao[1]

[1]Center for Excellence in Nanoelectronics, Department of Electrical Engineering, Indian Institute of Technology-Bombay Mumbai-400076, India, mailto:mayank@ee.iitb.ac.in; mailto:rrao@ee.iitb.ac.in
[2]Institute for Solid State Electronics, Vienna University of Technology, mailto:dionyz.pogany@tuwien.ac.at
[3]Infineon Technologies AG, P.O. Box 80 09 49, D-81609 Munich, Germany, mailto:Harald.Gossner@infineon.com

Abstract- **We present differences in the ESD failure mechanisms, intrinsic behavior and various phases of filamentation of STI type DeNMOS and DePMOS devices using detailed 3D TCAD simulations, TLP and TIM experiments. The impact of localized base-push-out, power dissipation because of space charge build-up, regenerative bipolar triggering and various events during the current filamentation are compared. Measurements show that the absence of base push out in DePMOS device leads to ~2.5X higher I_{T2} as compared to DeNMOS.**

Index Terms- **DEMOS, ESD Failure, space charge build-up, Filamentation.**

I. INTRODUCTION

Low Voltage (LV) NMOS device is widely used as a ESD clamp because of its high failure current value (I_{T2}), whereas the Drain extended NMOS (DeNMOS) devices have been found to be extremely vulnerable towards the ESD event [1]-[4]. Recently we presented a complete picture of STI type DeNMOS device failure and current filamentation using Transient Interferometric Mapping (TIM) experiments and 3D TCAD simulations [5][6]. In the past, LV PMOS devices have shown lower I_{T2} as compared to LV NMOS. Intuitively, Drain extended PMOS (DePMOS) devices should perform even worse under the ESD conditions, whereas our measurement results show a ~2.5X higher I_{T2} for DePMOS device, as compared to DeNMOS. This gives an indication towards a different filamentation and failure mechanism involved in DePMOS devices. We observed that the drain extension region doping is slightly higher for DePMOS device (P-Well doping) as compared to DeNMOS (N-Well doping), which makes the DePMOS device less prone to base push out or charge modulation. This work compares the 3D filamentation and device failure mechanisms involved in STI type DeNMOS and DePMOS devices, which has not been reported before.

II. DEVICE AND EXPERIMENTAL SETUP

Thin gate oxide, double finger DeNMOS and DePMOS device with STI under the gate-drain overlap, as shown in Fig. 1a & 1b, are processed in a state-of-the-art 65 nm node CMOS technology. The electrical scheme for device stressing is given in Fig. 1c. TIM method monitors the temperature and free-carrier concentration induced changes in the silicon

refractive index with a 1.5μm areal resolution and 3ns time resolution [7]. During the scanning, 30 stress pulses are applied at every scan point. A well calibrated process and device simulation (for low currents) deck is used for 3D TCAD simulation, as described in our previous papers [5][6][8]. It is worth mentioning that simulations here underestimate (by 15%) the I_{T2} values, which is due to slight differences in the simulated P-Well profile with the actual SIMs profile. Nevertheless, TCAD setup can still capture the physical phenomena behind current filamentation and device failure accurately.

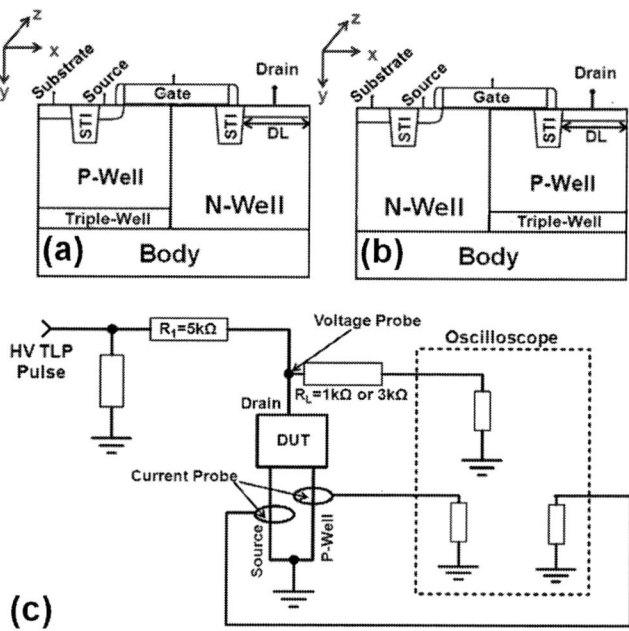

Fig. 1: Schematic of (a) DeNMOS and (b) DePMOS device. Figure show half structure only, whereas it was originally fabricated on silicon as in folded geometry/double finger (W= 2 x 5 μm) using state of the art 65nm CMOS process. (c) Electrical scheme for device stressing and TIM experiments.

III. ON THE DIFFERENCE OF DEVICE INTRINSIC BEHAVIOR

The TLP characteristics (Fig. 2) show that the DeNMOS device undergoes failure at very low currents (~1.2mA/μm), whereas a DePMOS device survives up to ~2.5X higher currents (~2.9mA/μm). The higher I_{T2} is attributed to a

978-1-4244-5430-3/10 $26.00 © 2010 IEEE

slightly higher doping in the drain extension region of DePMOS as compared to DeNMOS (Inset, Fig. 2). Higher doping in the drain extension region survives the base push out driven filamentation [5][6], which eventually leads to higher failure threshold. It is worth mentioning that besides a lower hole mobility, both the devices have almost an equal on-resistance at lower currents, whereas DePMOS starts outperforming at higher currents.

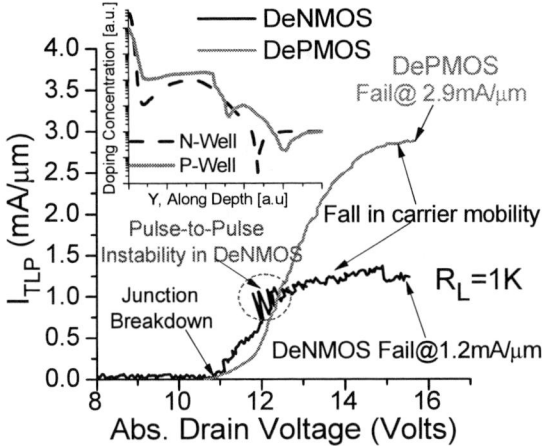

Fig. 2: Measured TLP characteristics of DeNMOS and DePMOS device at lower load line (i.e. R_L=1kΩ). Figure shows (i) junction breakdown at 10.7V, (ii) pulse-to-pulse instability and base pushout driven early failure in DeNMOS (iii) ~2.5X higher I_{T2} of DePMOS device as compared to DeNMOS (iv) Inset shows Drain extension region doping profile of DeNMOS and DePMOS (i.e. N-Well and P-Well respectively)

Differences in the device behavior before failure can be explained as follows-

(a) Junction Breakdown: Initially the current flow was dominated by junction breakdown, which occurs at 10.7V in both the devices (Fig. 2).

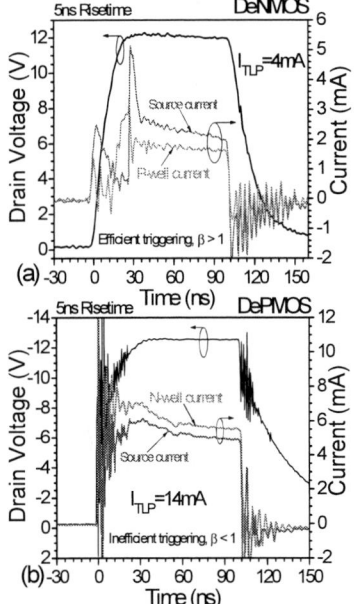

Fig. 3: Measured transient voltage/current plots during 100ns HBM event. (a) DeNMOS device triggers efficiently at lower TLP currents (4mA) whereas (b) DePMOS device show inefficient bipolar triggering even at higher TLP currents (14mA).

(b) Bipolar Triggering: Fig 3 shows that parasitic bipolar turns on efficiently (β > 1) at lower currents (4mA) in DeNMOS device, whereas DePMOS shows inefficient bipolar triggering (β < 1) even at higher TLP currents (14mA).

(c) Pulse-to-Pulse Instability: DeNMOS device has shown a pulse-to-pulse instability, whereas DePMOS device does not show a similar instability. Pulse-to-pulse instability was previously linked with bipolar turn-on and base push-out phenomena [5][6]. This difference is attributed to inefficient bipolar (PNP) turn on and absence of base push out.

(d) Base push out: Fig. 4 shows that DeNMOS device suffers from high electric fields underneath the drain diffusion [5], which is resulted from base push out and leading to early thermal failure. On the other hand, DePMOS device does not suffer from base push out and can withstand higher currents.

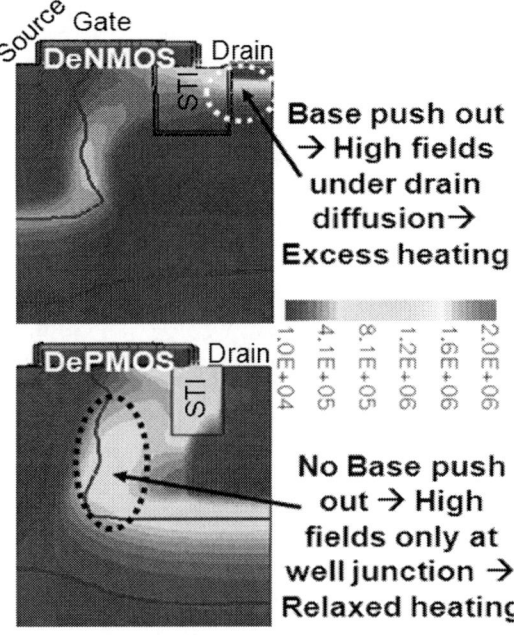

Fig. 4: Electric Field Plot (V/cm). DeNMOS device suffers from high electric field under drain diffusion because of base push out, which leads to early thermal failure. Whereas DePMOS device does not suffer from base push out and eventually survives up to higher currents.

IV. ON THE DIFFERENCE OF 2D v/s 3D BEHAVIOR

Fig. 5 shows a distinct 2D and 3D behavior of DeNMOS device, which is attributed to localized charge modulation [6]. Onset of filamentation and thermal run away at very low temperature (600 K) signifies the leading role of base push out on the device failure (Inset Fig. 5) mechanism. Furthermore, Fig. 6 shows identical 2D and 3D behavior of DePMOS device, which is attributed to the absence of base push out. This eventually leads to a significantly higher value of critical temperature (1050 K) for the onset of filamentation and thermal run away (Inset Fig. 6) and provides higher value of failure currents.

978-1-4244-5430-3/10 $26.00 © 2010 IEEE   481

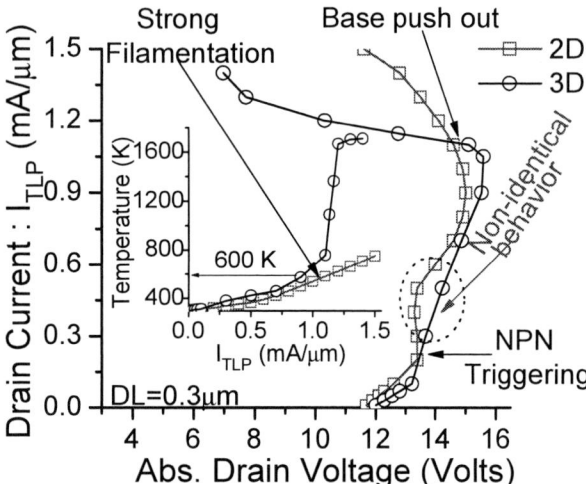

Fig. 5: TLP and self heating characteristics of DeNMOS device extracted from 2D and 3D simulations. Figure shows non-identical 2D v/s 3D behavior, whereas inset shows the onset of filamentation and thermal run away at very low temperature (600 K)

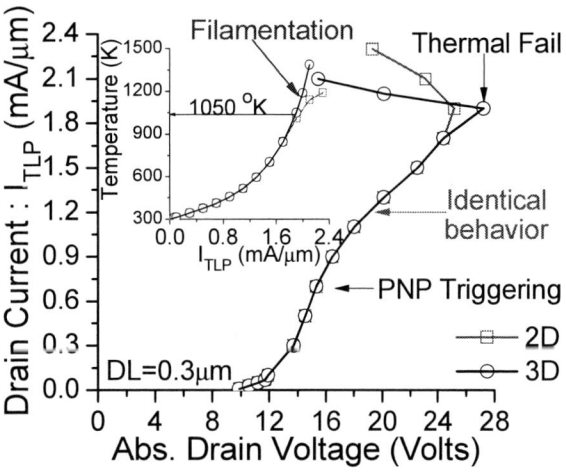

Fig. 6: TLP and self heating characteristics of DePMOS device extracted from 2D and 3D simulations. Figure shows identical 2D v/s 3D behavior, whereas inset shows the onset of filamentation and thermal run away at very high temperature (1050 K).

V. ON THE DIFFERENCES OF FILAMENTATION AND THERMAL FAILURE

Fig. 7a shows uniform current flow across the width of DePMOS device, even at higher TLP stress. Furthermore, uniform current flow at source side shows the absence of regenerative PNP action, which was found to be a cause of faster filamentation after base push-out. Fig. 7b shows that a strong filamentation exists in DeNMOS device, even at very low TLP stress values. Current shrink at drain side is because of elevated temperature, whereas shrink at source side is attributed to regenerative NPN action [5]. Figure 8 shows that DePMOS device has a hot spot at Nwell-to-Pwell junction, whereas it resulted underneath the drain diffusion in DeNMOS device. This is attributed to absence of base push-out in DePMOS device, which keeps the high electric field

region near to the Well junction. (1) Lower current densities and (2) higher silicon area available for thermal diffusion at Nwell-to-Pwell junction is leading to relaxed self heating in DePMOS device as compared to DeNMOS, which eventually provides ~2.5X higher I_{T2}.

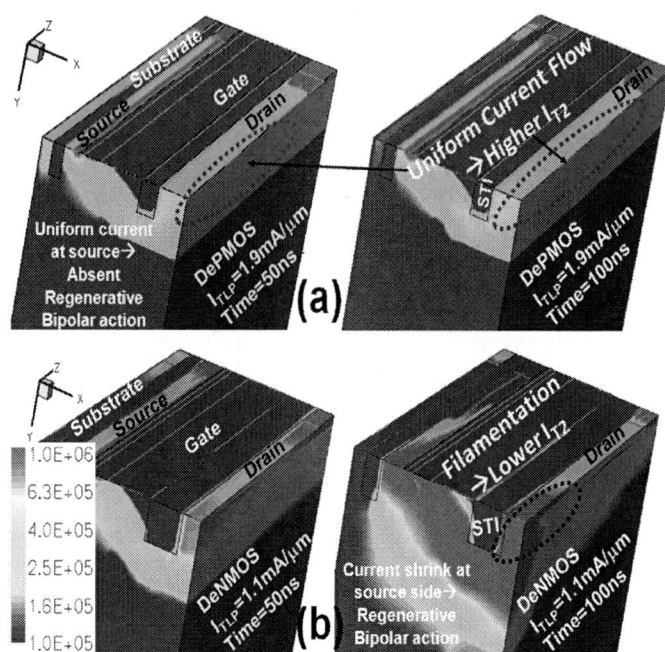

Fig. 7: Current Density Plot (A/cm²) for (a) DePMOS and (b) DeNMOS devices at different time.

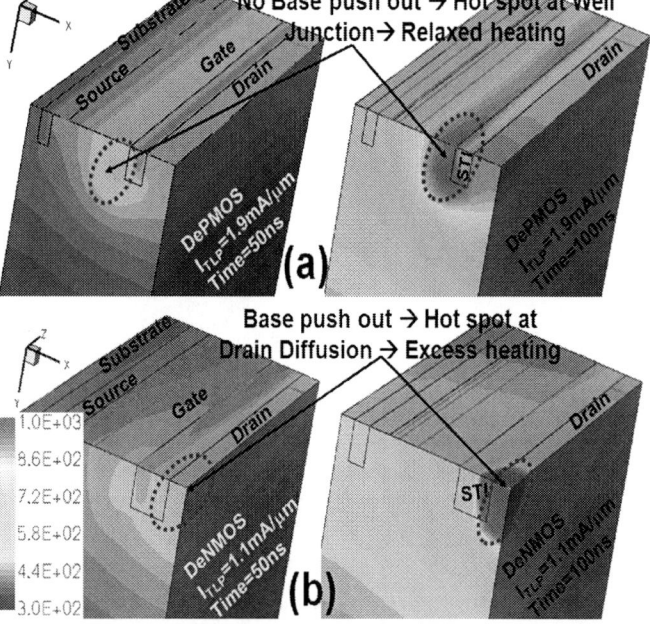

Fig. 8: Temperature Plot (K) for (a) DePMOS and (b) DeNMOS devices at different time. DePMOS device has hot spot at Nwell-to-Pwell junction, whereas it is underneath the drain diffusion in DeNMOS device.

VI. ON THE DIFFERENCES OF TIM BEHAVIOR

Fig 9 shows ~2X higher phase shift during the TLP stress (14mA), extracted from TIM measurements, in DeNMOS as compared to DePMOS device. This further validates the presence of excess heating in DeNMOS device, which leads to lower I_{T2}. Furthermore, DeNMOS device also shows significant heating (positive phase shift) during initial times (0-40ns), which is attributed to space charge build-up and current discharge after the onset of base push out [5][6]. Whereas, absence of base push out in DePMOS device leads to uniform phase shift with respect to time, i.e., a linear rise in temperature as a function of time.

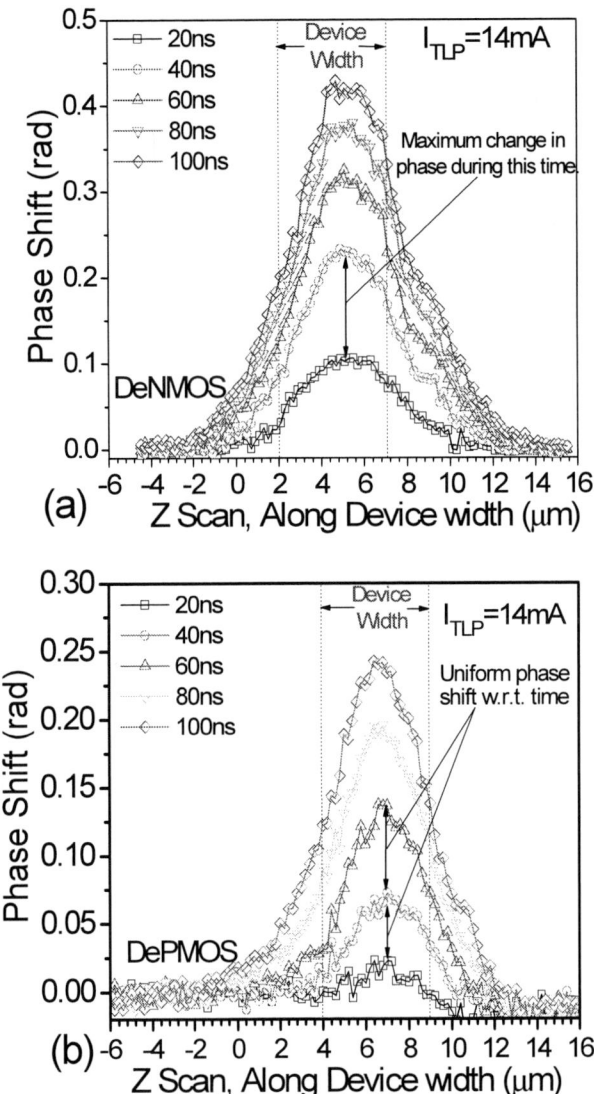

Fig. 9: Transient development of the TIM signal for (a) DeNMOS and (b) DePMOS devices (Z scan @ X': where X' is a point on X axis at which maximum phase shift occurs during X scan). DeNMOS device results ~2X higher phase shift as compared to DePMOS, which validates the presence of excess heating in DeNMOS device. Unlike to DePMOS, DeNMOS device also shows significant heating (positive phase shift) during initial times (0-40ns), which is attributed to space charge build-up and current

VII. DEVICE BEHAVIOR AT HIGHER LOAD

Fig. 10 shows that both the devices fail at almost an equal TLP current when they are stressed at higher load (R_L=3KΩ). Relatively DeNMOS device shows a significant improvement (~3X) in I_{T2} as compared to DePMOS (30%). This behavior is attributed to the difference in failure mechanism of these devices. At higher load DeNMOS survives localized charge modulation, which increases the critical temperature for the onset of current filamentation. This eventually improves the failure threshold significantly. On the other hand, this critical temperature for DePMOS device is much higher, which therefore provides only a narrow window for improvement (30%) when the filamentation is survived using a higher load.

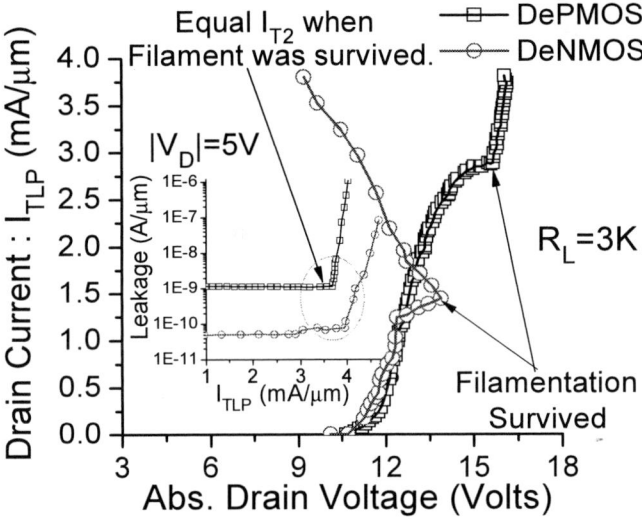

Fig. 10: Measured TLP characteristics of DeNMOS and DePMOS device at higher load line (i.e. R_L=3kΩ). Larger load survives the filamentation and leads to higher (almost equal) I_{T2} for both the devices.

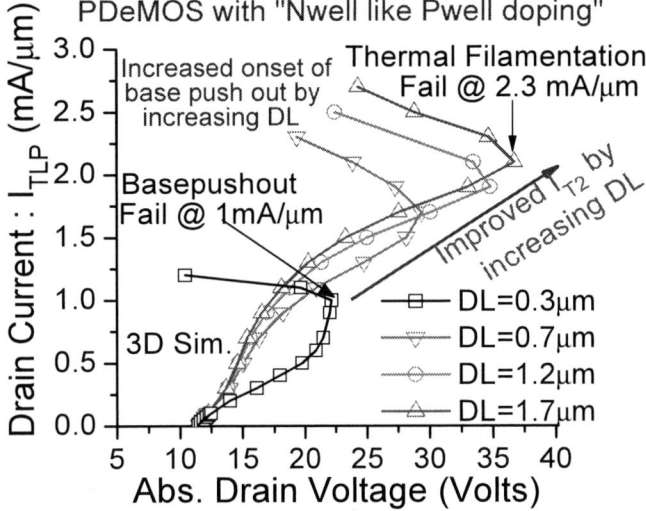

Fig. 11: TLP characteristics of DePMOS device *with N-Well like P-Well doping profile*, extracted from 3D simulations. DePMOS device (DL=0.3μm) with lower drain extension doping fails at 1mA/μm, which was further improved up-to 2mA/μm by increasing DL.

978-1-4244-5430-3/10 $26.00 © 2010 IEEE   483

VIII. DePMOS WITH N-WELL LIKE P-WELL DOPING

By changing the drain extension region doping with *N-Well like doping profile*, DePMOS device (DL=0.3μm) has been seen to fail at 1mA/μm (Fig. 11). This further validates that lower drain extension region doping causes the base push out driven early filamentation and failure. Further, the same device shows a ~2X improvement in I_{T2} when the drain diffusion length (DL) is increased, which is attributed to the shift of the onset of base push out towards higher currents [5]. This also gives an indication that failure threshold and ESD window of standard STI DeNMOS can also be improved by changing the N-Well profile in order to mitigate the onset of base push-out.

IX. CONCLUSION

We demonstrate that the base push out in DeNMOS causes a very low critical temperature for the onset of current filamentation, which eventually results in a lower I_{T2}. However, absence of base push out in DePMOS device leads to ~2.5X higher I_{T2} as compared to DeNMOS. DePMOS devices do not show pulse-to-pulse instability and have identical 2D v/s 3D behavior, which is attributed to the absence of localized charge modulation and regenerative bipolar triggering. One can expect a width scaling due to the uniformity of current. TIM experiments have further validated the point that DePMOS devices have a relaxed self heating. Unlike the DeNMOS device, DePMOS devices have no space charge build-up and current discharge during initial time. In spite of a slightly worse voltage clamping capability per width, DePMOS is a better device option for power clamps, as compared to DeNMOS. It can also be concluded from this work that along with increasing the drain diffusion length, changing the N-Well profile in order to mitigate the onset of base push-out, is also a key for improving the ESD performance of the standard STI DeNMOS.

REFERENCES

[1] P. L. Hower, J. Lin, S. Haynie, S. Paiva, R. Shaw and N. Hepfinger, "Proceedings of International Symposium on Power Semiconductor Devices and ICs, ISPSD '99, pp. 55-58.

[2] P. L. Hower, J. Lin and Steve Merchant, "Snapback and Safe operating area of LDMOS Transistors", Proceedings of International Electron Device Meeting, pp. 193-196, 99.

[3] P. Moens, S. Bychikhin, K. Reynders, D. Pogony and M. Zubeidat, " Effects of hot spot hopping and drain ballasting in Integrated Vertical DMOS devices under TLP stress", Proceedings of International Reliability Physics Symposium, pp. 393-398, 2004.

[4] Gianluca Boselli, Vesselin Vassilev and Charvaka Duvvury, " Drain Extended NMOS High Current behavior and ESD protection strategy for HV application in sub 100nm CMOS technologies", Proceedings of International Reliability Physics Symposium, pp. 342-347, 2007.

[5] Mayank Shrivastava, Jens Schneider, Maryam Shojaei Baghini, Harald Gossner, V. Ramgopal Rao, "A New Physical Insight and 3D Device Modeling of STI Type DENMOS Device Failure under ESD Conditions", IEEE International Reliability Physics Symposium (IRPS), 2009, pp 669-675.

[6] Mayank Shrivastava, S. Bychikhin, D. Pogany, Jens Schneider, M. Shojaei Baghini, Harald Gossner, Erich Gornik, V. Ramgopal Rao, "Filament Study of STI type Drain extended NMOS Device using Transient Interferometric Mapping", IEEE International Electron Device Meeting (IEDM), 2009.

[7] M. Litzenberger, C. Fürböck, S. Bychikhin, D. Pogany and E. Gornik, "Scanning heterodyne interferometer setup for the time resolved thermal and free carrier mapping in semiconductor devices", IEEE Trans Instrumentation Measurement Vol. 54, 2005, pp. 2438–2445.

[8] Mayank Shrivastava, Jens Schneider, Maryam Shojaei Baghini, Harald Gossner, V. Ramgopal Rao, "Highly resistive body STI: n-DEMOS: An optimized DEMOS device to achieve moving current filaments for robust ESD protection", IEEE International Reliability Physics Symposium (IRPS), 2009,2009, pp 754-759.

Predictive Simulation of CDM Events to Study Effects of Package, Substrate Resistivity and Placement of ESD Protection Circuits on Reliability of Integrated Circuits

Vrashank Shukla, Nathan Jack and Elyse Rosenbaum
Department of Electrical and Computer Engineering
University of Illinois at Urbana Champaign
1308 W Main St., Urbana, IL, 61801 USA
vshukla2@illinois.edu

Abstract— **Power domain crossing circuits, also known as internal I/O's, are susceptible to gate oxide damage during charged device model (CDM) events. Circuit-level simulations of internal I/O circuits along with elements representing the package, electro-static discharge (ESD) circuits and the substrate, elucidate the roles of the package, power clamp placement, back-to-back diode placement and the decoupling capacitors in determining the amount of stress at the internal I/O circuits. This paper presents an internal I/O model that can be used for CDM simulations. The effects of power and ground bus resistance, substrate resistivity, decoupling capacitance, local ESD clamp at the gate of the receiver and the presence of local back-to-back diodes are investigated. The paper further contains design recommendations for preventing CDM failures in the internal I/O circuits.**

Keywords- CDM; Simulation; Internal I/O

I. INTRODUCTION

During a field induced charge device model (FICDM) test, a packaged chip, called a device under test (DUT), is placed upside-down on a charge plate as shown in Figure 1. The charge plate potential is raised to a predetermined pre-charge voltage and then a pogo pin attached to a ground plate is lowered until it makes contact with a pin of the DUT. Charge flows through the pogo pin to the DUT in order to support the potential difference between the grounded component and the field charge plate. The charge flows through the lowest impedance paths that lie between the "zapped pin" and the charge storage sites throughout the component. Simulation of CDM events in integrated circuits requires modeling the plural charge storage locations and discharge paths. To simulate the discharge current waveform, the netlist must include elements representing the package, ESD circuits and substrate [1][2]. If full-chip modeling is undertaken, one can also simulate the CDM waveforms at power domain crossing circuits, known as internal I/Os. These circuits may be susceptible to gate oxide damage during CDM events [3][4]. Previous work on full-chip CDM simulation to study the power domain crossing circuits did not include the substrate model [5]. In this paper, we show that the substrate resistivity plays an important role in determining the stress at the power domain crossing circuits.

Ref. [4] introduced the concept of a structured cross-domain signal interface in which anti-parallel diodes APD (i.e. back-to-back diodes) and power clamps are placed near the driver and the receiver of the internal I/O. In practice, this requires the internal I/O to be placed close to the pad ring because this is the preferred location for placing a power clamp due to its large size. This impacts the chip floorplanning and limits the blocks between which signals may be passed. This work uses circuit simulation to explore whether a cross-domain signal interface may safely be located far from the pad ring and the primary ESD protection, if the primary protection is augmented by small APD and decoupling capacitors placed near the internal I/O. In [4], APDs are also used between power rails, a strategy not preferred by designers due to a power sequencing problem. In this work, protection devices are therefore not placed directly between different power busses. Simulation results compare favorably with measurements obtained on a CDM testchip; this provides confidence in the results of the predictive simulations.

II. MODELING METHODOLOGY

A. Capacitive Model of Charge Storage

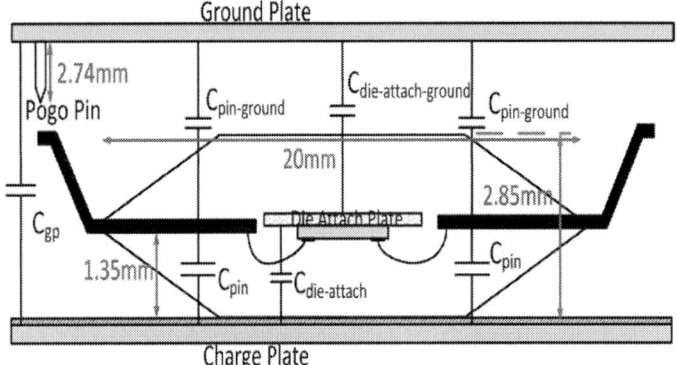

Figure 1. QFP Package that was modeled in this work. Important capacitances formed by the parts of the components with the charge plate and ground plate of the FICDM tester are shown.

978-1-4244-5430-3/10 $26.00 © 2010 IEEE

The amount of charge transferred during a CDM event depends on the capacitors formed by the metallic parts of the DUT with the plates of the CDM tester. The capacitance between the plates of the tester also contributes to the amount of charge storage. A lumped 3-capacitor model [6] is sufficient to calculate the total amount of charge; subsequently, the peak current may be found by roughly estimating the series impedance of the discharge paths [7]. Although this method for estimating the peak current is sufficient to predict the reliability of I/O circuits, distributed modeling of charge storage is needed to estimate the stress on the internal circuitry, particularly the power domain crossing circuits. Therefore, a distributed capacitance model of the part is needed.

In this work, the various capacitances between the DUT and the tester plates were calculated using CSURF, a 3-D capacitance modeling tool [8]. The actual dimensions of the package and the die were drawn using the graphical user interface of CSURF. The package dielectric constant was entered and capacitances were extracted. The capacitance between the on-chip power/ground interconnect metal and the charge plate was estimated by extracting the area of the interconnect metal using a layout extraction tool. At each metal level, the area of the metal exposed to the charge plate was extracted. A customized layout extraction rule set was written to perform the extraction.

This paper presents the model of a cavity-up quad flat package (QFP). The QFP is placed in a deadbug position on an FICDM charge plate as shown in Figure 1. In a QFP with a large number of pins or long pin leads, package pins are the primary charge storage locations. Packages that have a small die attach plate in comparison to the package size can have long pin leads. This can lead to a large pin capacitance. For very thin packages, the bondpads and the on-die interconnect will be closer to the charge plate and the capacitance formed by them with the charge plate may become significant. However, for a small die in a thicker package, the capacitances formed by the interconnect metal and the bondpads can be negligibly small. For packages with a metal heat sink, the capacitance between the heat sink and the charge plate will be significant, and the connection of the heat sink to the rest of the on-die elements will decide the path of discharge for the charges stored in the heat sink. For example, the heat sink could be downbonded to a ground pin of a particular power domain. In that case, zapping a pin belonging to another power domain will lead to a higher current flowing across this domain compared to zapping the pin downbonded to the heat sink.

B. Schematic Model Including On-chip, Package and Substrate Elements

A schematic representation of the CDM simulation model for a QFP package is shown in Figure 2.

Parasitic resistance, inductance and mutual capacitance of the package pins can be obtained from the package data sheets. The parasitic resistance and the inductance of the bondwires can be calculated once the die is packaged and bondwire lengths are known. Alternatively, these values may be estimated with reasonable accuracy before packaging, if the die size and the package size are known.

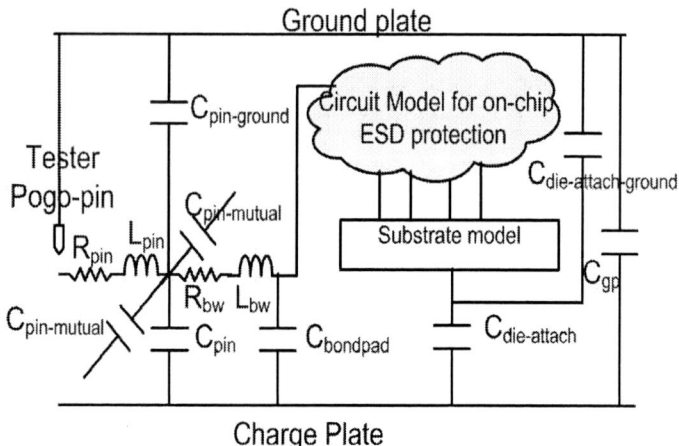

Figure 2. High-level representation of CDM simulation model showing interconnection of substrate, package and on-chip models

The charge stored on $C_{die-attach}$ flows through the substrate into the on-chip ground bus through the substrate contacts. To understand the flow of charge through the substrate and its impact on the on-chip circuitry, it is important to model the substrate. The chip substrate was divided into a 3-D grid of boxes, each represented by 6 resistors, as suggested in [1]. A script was written to create the substrate resistor network, given the substrate size and the grid size. The grid size and the chip size determine the number of nodes in the substrate model. The backside of the substrate is connected to $C_{die-attach}$, and the top side of the substrate is connected to the substrate contacts of VSS busses as shown in Figure 3.

Figure 3 corresponds to the particular case in which the die attach plate is connected to the die substrate using conductive glue (silver filled epoxy). If, instead, the die is attached using an insulating glue, the $C_{die-attach}$ should be distributed across the bottom-most grid boxes of the substrate in the x-y plane.

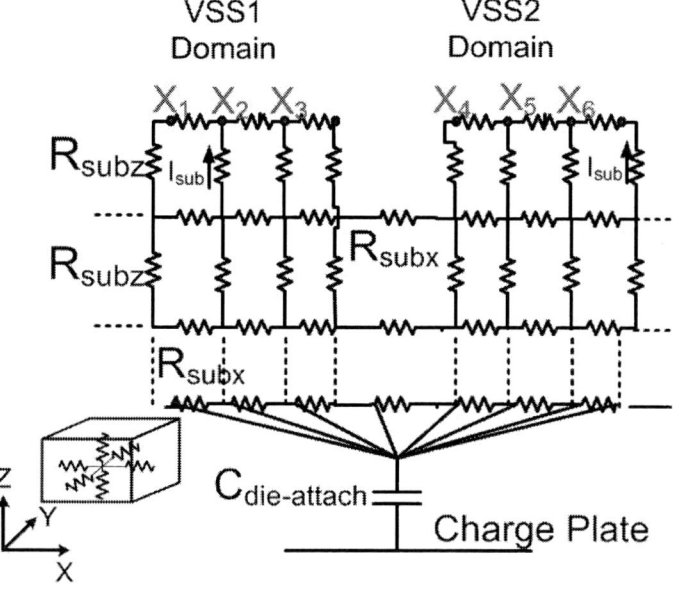

Figure 3. 2-D slice (in X-Z plane) of 3-D substrate resistor grid. X1, X2 … Xn are the substrate contacts. The resistors in the top row model the Vss busses. A sample grid-box with 6 resistors is shown on the lower left corner.

Capacitive coupling is stronger along the periphery of the substrate than in the center due to fringing fields, and thus the capacitors connected to the outermost grid boxes would be larger than those in the interior. The top-side of the substrate is connected to the substrate contacts. In this work, substrate contacts were automatically extracted from the layout and a reduction algorithm was used to merge the substrate contacts that were in the same grid box of the substrate model. The locations of the substrate contacts were extracted using a customized layout extraction rule set. A script was written to identify the top-most grid box of the substrate model that contained each of the substrate contacts. Multiple substrate contacts contained within the same grid box were connected together.

The circuit-level model is connected to the package and the substrate models. The substrate contacts provide connection to the ground busses. Bondpads provide connection to the package model. The circuit-level model contains ESD protection devices at each pad, parasitic resistance of power and the ground busses, and decoupling capacitors between the power and the ground busses. In this work, ESD protection includes dual diodes at the signal pads and active clamps as the power clamps [9]. The circuit-level model should also contain any core circuitry of interest, e.g. internal I/O. The capacitance between the interconnect metal and the charge plate, if significant, should be included in the netlist.

Figure 4 illustrates an internal I/O on the test chip described in Section III. One can observe the connections between the circuit-level model, the package model and the substrate model.

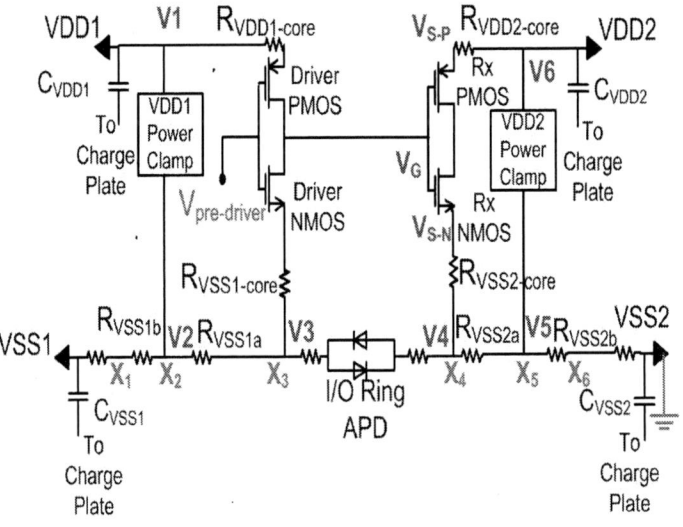

Figure 4.   Internal I/O circuit. Driver in VDD1 domain; receiver in VDD2 domain. The substrate contacts (X1, X2, etc.) are shown. Refer to Fig. 3 for substrate connections. C_{VDD1}, etc., are pin capacitors

III.   DESCRIPTION OF TEST CASES AND SIMULATION SETUP

Simulation studies were conducted on two 90-nm test cases: Testcase A and Testcase B. Testcase A is a small chip that is used only for predictive (i.e., "what if") simulations. Cadence's Spectre was used as the simulation engine. In Testcase A, a 500μm x 500μm die is modeled as being inside a 34-pin QFP package. The package height is 2.85mm and its

other dimensions are 20mm x 16mm. Its die attach plate is 6mm x 6mm. A capacitance model was extracted as described in section II, and the values are tabulated in Table I. For the 34-pin package, the total pin capacitance amounts to 3.74pF, which is much larger than the next largest capacitor, $C_{die-attach}$. The bondpad and the interconnect capacitance were found to be very small and were neglected in subsequent analyses. Testcase A uses a high resistivity substrate (ρ=15Ω-cm), unless otherwise noted.

TABLE I.   VALUES OF CAPACITORS EXTRACTED FOR THE QFP PACKAGE USING 3-D CAPACITANCE EXTRACTION TOOL. REFER TO FIG. 2 FOR COMPONENT NAMES

Component Name	Value	Component Name	Value
C_{pin}	0.11pF	$C_{die-attach}$	1.33pF
$C_{pin-ground}$	60fF	$C_{die-attach-ground}$	160fF
C_{gp}	20pF	$C_{bondpad}$	0.08fF

The pad assignments for Testcase A are illustrated in Figure 5. The power and ground busses for each domain run along the periphery of the chip. The VSS1 bus is continuous while the VSS2 bus runs only in the VDD2 domain section of the I/O ring. The parasitic resistance grid of the core power bus and that of the core ground bus were created using a script and are shown in Figure 5. The substrate contacts were created with a regular grid and are shown as red dots in Figure 5. Note that the VDD1 domain is larger compared to the VDD2 domain. The contents of each pad cell are shown in Figure 6. Decoupling capacitance of 10pF is placed in each of the pad cell. Bus resistance of each power/ground bus is 0.25Ω per pad cell.

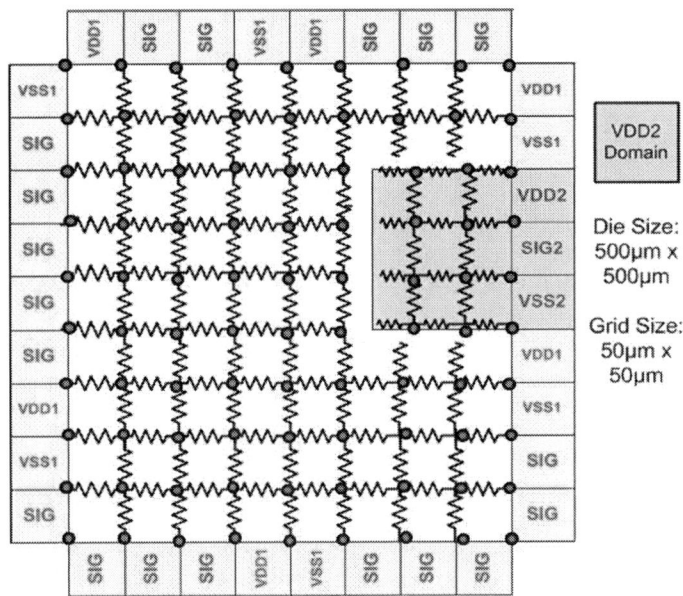

Figure 5.   Pad arrangement for Testcase A. VDD1/VSS1 domain is larger with higher pin count and larger number of substrate contacts. Ground bus model for the core is also shown.

978-1-4244-5430-3/10 $26.00 © 2010 IEEE   487

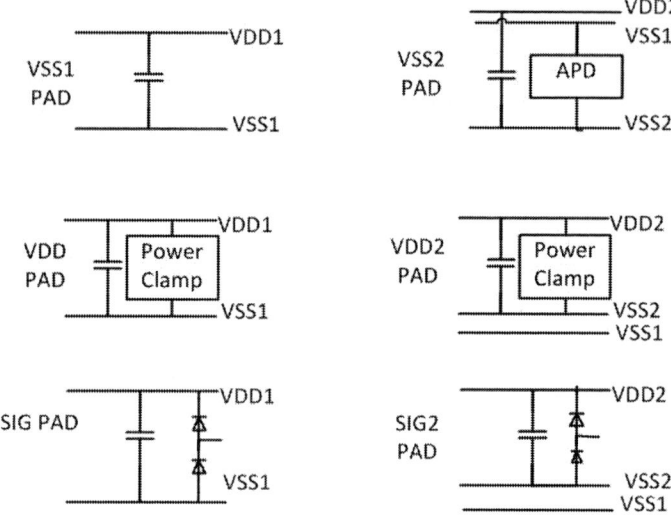

Power/Ground Bus Resistance per cell = R_{bus}

Figure 6. Block diagram of each pad cell used in the pad-ring described in Figure 5

Figure 4 illustrates one of the internal I/Os for Testcase A. The driver is in the VDD1 domain and the receiver is in the VDD2 domain. Note that the respective grounds, VSS1 and VSS2, are connected by back-to-back diodes in the pad ring. The transistors in the internal I/O circuits have thin gate oxides which can conduct significant current under CDM conditions; therefore, gate tunneling current models valid in the high voltage regime were used in the simulations.

In the CDM simulations of Testcase A, the charge plate is brought to 300V and then a VSS2 pin is grounded by the pogo pin. The paper focuses on the stress induced at an internal I/O whose receiver is in the VDD2 domain, such as that illustrated in Fig. 4. During any CDM event in which the rail clamps turn on, the digital logic will be activated; in the simulations, it is assumed that the input to the driver is pulled low.

Testcase B is a CDM test chip that was fabricated and tested; this chip was packaged in the 100-pin QFP package described in section II. The total pin capacitance for this package is large (11pF). The die size is 2mm x 4mm. The test chip is built on a high resistivity substrate (ρ=15Ω-cm). The test chip has 64 pads and multiple internal I/Os similar to the one illustrated in Figure 4. The 2 domains VDD1 and VDD2 are equally sized. In Section V of this paper, simulation results for Testcase B are compared with data obtained using capacitively coupled TLP (cc-TLP) [10] and very fast TLP (VF-TLP) testers.

IV. SIMULATION RESULTS AND DISCUSSIONS

In this section, simulation results for Testcase A are presented. Recall that the simulations are for a 300V CDM event with respect to a VSS2 pin. The currents through the APD and that at the grounded pin are plotted in Figure 7. A large fraction of the total current flows through the APD from VSS1 to VSS2. This is because most of the charge is stored in the pin capacitors and most pins are in the VDD1 domain.

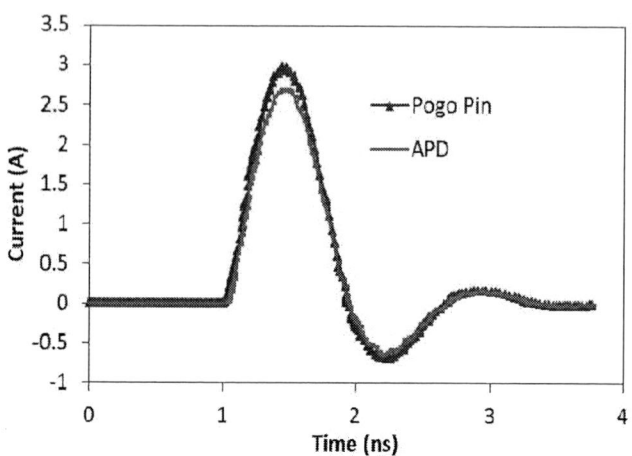

Figure 7. Simulated current waveforms at the grounded pin and at the APD. Current flows through the APD from VSS1 to VSS2.

The primary path for ESD current from the VDD1 to the VDD2 domain is through the VSS busses and the APD.

A. Effect of Substrate Resistivity

Simulations were performed for two cases: ρ_{sub} = 15Ω-cm (default case) and ρ_{sub} = 0.02Ω-cm In Figure 8, the peak V_{gs} of the receiver NMOS and that of PMOS are plotted as a function of the per cell power/ground bus resistance in the pad ring. Despite the internal I/O being off-path for the ESD current, large potential differences appear across the gate oxides of the receiver transistors, especially when high-resistivity substrate is used. During this CDM event, the NMOS is biased in inversion and the PMOS is biased in accumulation. The stress can be measured in terms of the voltage drop across the oxide, which is V_{GS} in inversion and V_{GB} in accumulation. However, since the source and body nodes of these transistors are tied together, there is no need to distinguish between V_{GS} and V_{GB}.

Using the node voltages introduced in Figure 4, the voltage stress across the receiver NMOS can be written as:

$$V_{gs\text{-}nmos} = V_G - V_{S\text{-}N} \approx V1 - V4 \qquad (1)$$

$$V1 - V4 = (V1 - V2) + (V2 - V3) + (V3 - V4) \qquad (2)$$

The right-most quantity in eq. (1) is approximately equal to $V_{gs\text{-}nmos}$ since the ESD current primarily flows through the pad ring and there is little voltage drop in the core busses, e.g. negligible potential drop through $R_{VSS2\text{-}core}$ or $R_{VDD1\text{-}core}$. In (2), the quantity (V1-V2) is the voltage drop across the power clamp. V2-V3 is the voltage drop in the VSS1 bus between the power clamp and the APD. V3-V4 is the voltage drop across the APD. Similarly the stress across the receiver PMOS can be written as:

$$V_{gs\text{-}pmos} = V_G - V_{S\text{-}P} \approx V1 - V6 \qquad (3)$$

$$V1 - V6 = V_{gs\text{-}nmos} + (V4 - V5) - (V6 - V5) \qquad (4)$$

The quantity (V4-V5) is the voltage drop in the VSS2 bus between the point at which the receiver source is connected to

978-1-4244-5430-3/10 $26.00 © 2010 IEEE   488

the VSS2 bus in the I/O ring and the VDD2 power clamp. V6-V5 is the voltage drop between VDD2 and VSS2; this quantity is positive because current flows from VDD2 to VSS2 when a VSS2 pin is zapped.

Equation (4) explains why the receiver PMOS, not just the NMOS, is stressed when VSS2 pin is zapped. $V_{gs-pmos}$ may be higher or lower than $V_{gs-nmos}$ depending upon the relative values of (V4-V5) and (V6-V5). If the quantity (V6-V5) is larger, the voltage stress on the receiver PMOS oxide will be smaller than the stress on the NMOS oxide. However, in this testcase, the VDD2 domain is small and the current flowing from VDD2 to VSS2 is small, leading to a very small voltage drop between VDD2 and VSS2. This leads to a significant voltage stress across the PMOS.

The oxide breakdown voltage of 90-nm thin oxide transistors was measured using a very fast transmission line pulsing (VFTLP) system with a pulse width of 10ns. The oxide breakdown voltage for an NMOS in inversion was about 4.8V, and the PMOS oxide breakdown voltage in accumulation region was about 5.2V. The simulation results of Fig. 8 indicate that oxide breakdown at the internal I/O is far more likely to occur when high resistivity substrate is used rather than low resistivity substrate.

To simulate the case of a chip built on low resistivity substrate (0.02Ω-cm), the substrate model was replaced with a new resistance grid, appropriate for the lower resistivity material. When low resistivity substrate is used, the charge from $C_{die-attach}$ flows through the substrate directly to the grounded VSS2 bus. Although the charge stored on the pins in the VDD1 domain will enter the VSS1 bus, much is diverted to the low resistivity substrate that is in parallel, thus bypassing the APD and the bus resistances; the reduced voltage drop across the VSS1 bus resistance and the APD yields a greatly reduced receiver V_{gs}. Therefore, CDM failures at internal I/O's are not expected to occur in ICs built on low-resistivity substrates.

B. Effect of Decoupling Capacitance

Increasing the amount of decoupling capacitance placed between the power and the ground busses reduces the stress on the internal I/O receiver gates, as shown in Figure 9.

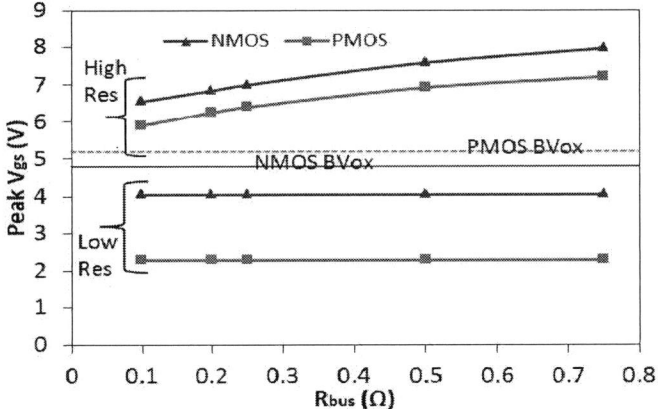

Figure 8. V_{gs} of the receiver transistors for low and high resistivity substrates.

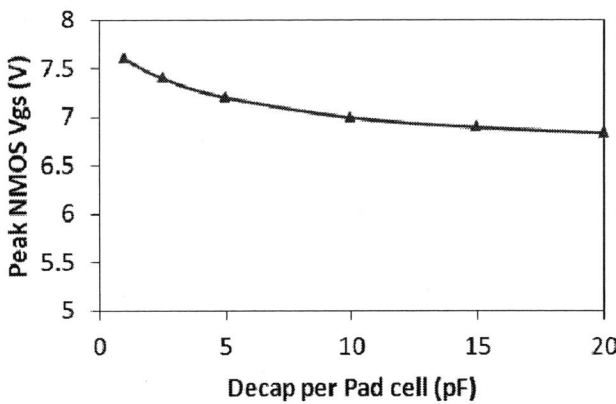

Figure 9. Effect of decoupling capacitance on the receiver NMOS V_{gs}.

Increasing the decoupling capacitance reduces the voltage drop between VDD1 and VSS1, which is the quantity (V1-V2) in (1). Displacement current through the decoupling capacitors ($I=C*dv/dt$) is not negligible in the CDM timescale. The beneficial properties of decoupling capacitors were noted in [4]; this work highlights the quantitative importance of including them in CDM simulation models. Even though decoupling capacitance reduces the stress across the receiver gates, it may not be sufficient to protect the receiver gates from oxide breakdown and hence, additional protection strategies are considered in the following sections.

C. Effect of Cross-Domain Power Clamp

A cross-domain power clamp may be used to reduce the stress at the internal I/Os, as shown in Figure 10. However, the cross-domain clamp is beneficial only if the VDD12 clamp shown in Figure 10 is placed near the VSS2 pad such that R_{VDD12} and R_{VSS12} are minimized. A plot of PMOS and NMOS V_{gs} as a function of the sum of R_{VDD12} and R_{VSS12} is shown in Figure 11. Though the cross-domain clamps reduce the voltage stress across the receiver gate oxides, depending upon the parasitic resistance of the power and the ground bus and the placement of the clamp, the voltage stress may not be low enough to protect the gate oxide. Designers concerned with noise isolation between the domains may not wish to use cross-domain power clamps.

D. Local APD

As shown in Figure 12, a small APD is placed between VSS1 and VSS2 in close proximity to the internal IO circuit in order to augment the APD in the I/O ring. R_{con} is the parasitic resistance of the ground bus connecting the local APD to the source of the driver NMOS or to the source of the receiver NMOS.

The local APD reduces the voltage drop between the VSS busses local to the driver and the receiver. The local APD carries very little current compared to the APDs in the I/O ring. As shown in Figure 13, the receiver V_{gs} is reduced significantly when using a local APD. R_{con} was set to 0.5 ohms in the simulations.

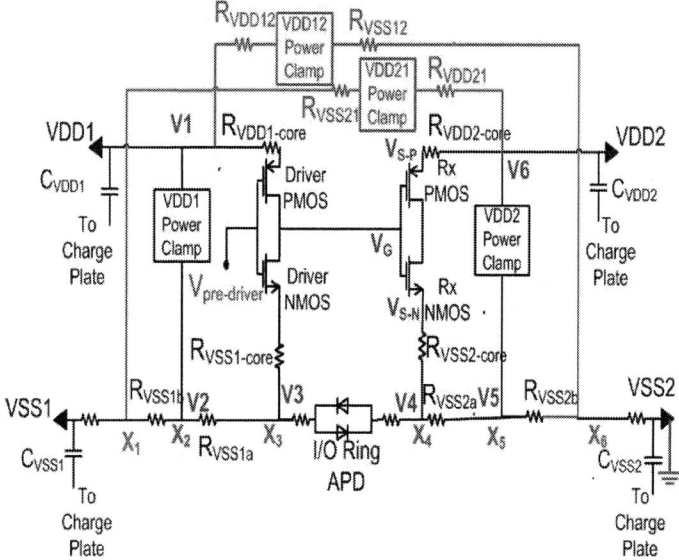

Figure 10. Internal I/O circuit with cross-domain power clamps between VDD1 and VSS2 and between VDD2 and VSS1

The data in Figure 13 show that the voltage stress can be further reduced by increasing the amount of decoupling capacitance in the vicinity of the driver and the receiver in addition to adding the local APD. Decoupling capacitors placed near the driver reduce the receiver NMOS V_{gs}. C_{local} placed near the driver reduces the local potential difference (V1'-V3'), while the local APD reduces the potential difference (V3'-V_{S-N}), thus the two reduce the potential difference, V1'-V_{S-N}. Decoupling capacitors placed near the receiver reduce the receiver PMOS V_{gs}. Specifically, C_{local} in the receiver domain (i.e. VDD2 domain) reduces the potential difference ($V_{S-P} - V_{S-N}$), which reduces the potential difference V1-V_{S-P}.

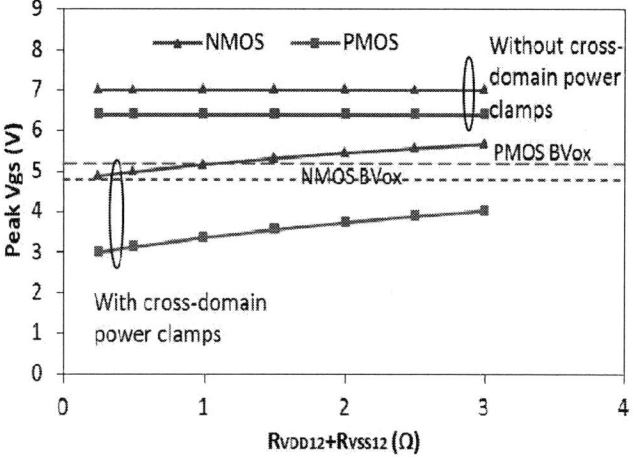

Figure 11. The cross-domain power clamps (VDD1-VSS2 and VDD2-VSS1) reduces V_{gs} of the receiver transistors. Resistance from the clamp to the respective pads is important

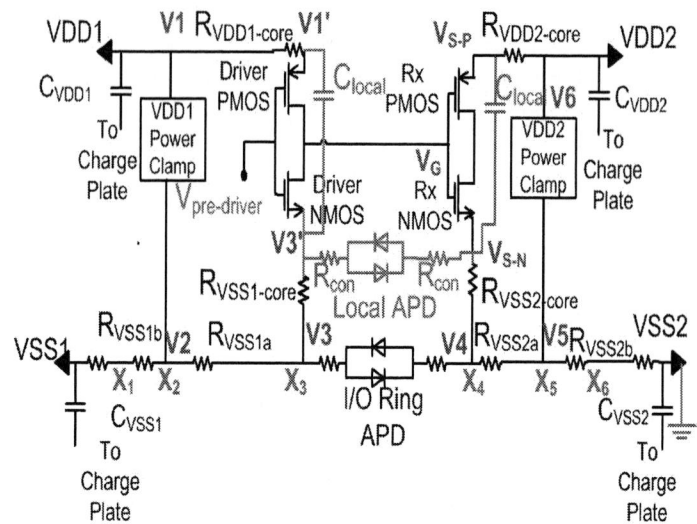

Figure 12. Internal I/O circuit with a local APD added close to the driver and the receiver, in addition to the APD in the I/O ring

Increasing the size of the APD devices in the pad ring does not provide the same benefit as does adding a local APD, as evidenced by the data shown in Figure 14. In these simulations, the parasitic resistance from the core VSS1 and VSS2 busses to the APD in the I/O ring ($R_{vss1-core}$ and $R_{vss2-core}$ in Figure 11) is 6 Ω each, and the parasitic resistance to the local APD (R_{con}) is 0.5 Ω.

Figure 15 is a plot of the peak V_{gs} of the receiver NMOS as a function of both the amount of local decoupling capacitance and the local APD perimeter. V_{gs} decreases as either the amount of decoupling capacitance is increased and/or the perimeter of the local APD is increased. To clamp the V_{gs} to a particular value at a given CDM stress level, one adjusts the diode perimeter or the value of C_{local}. For example, a local

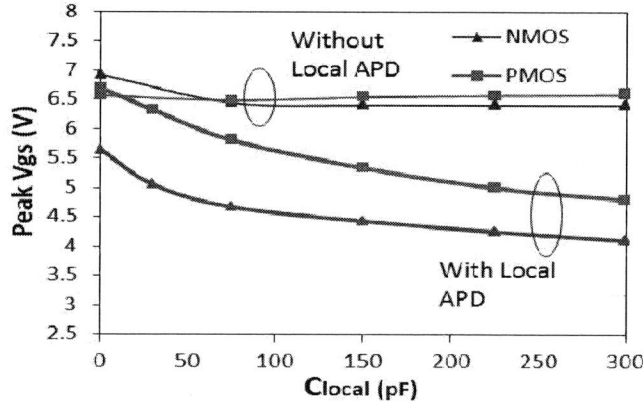

Figure 13. A local APD along with local decoupling capacitance greatly reduces Vgs of the receiver transistors. Local APD perimeter=42μm. APD perimeter in the I/O ring is 210 μm.

978-1-4244-5430-3/10 $26.00 © 2010 IEEE

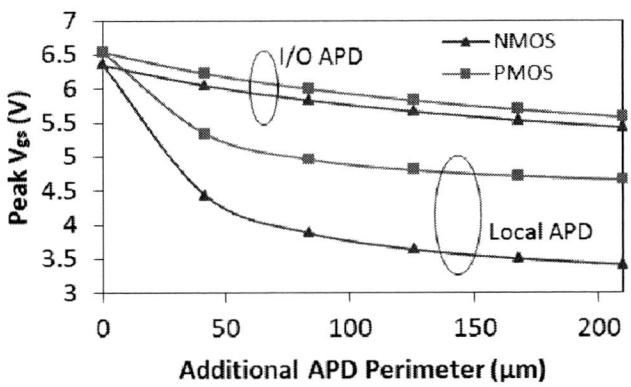

Figure 14. Effect of adding a local APD, close to the internal I/O, is compared with the effect of increasing the perimeter of the APD in the I/O ring. C_{local}=150pF in both cases, APD perimeter in the I/O ring=210μm. Local APD not present initially.

APD with a perimeter of 84μm needs 40pF of C_{local} to clamp the NMOS Vgs to about 4.5V. In order to get a further reduction of 0.3V, either the local APD perimeter can be increased by about 150%, to 210μm, or the value of C_{local} can be increased by about 50%, to 60pF. It might appear that adjusting C_{local} is the favorable approach, however, the absolute silicon area required to increase the amount of decoupling capacitance is much higher than that required to increase the local APD perimeter. This analysis suggests that, in light of the silicon area constraint, improved voltage clamping is best achieved by increasing the local APD perimeter rather than increasing the amount of local decoupling capacitance. However, in designs with a large amount of core decoupling capacitance, a small local APD may be sufficient to clamp the V_{gs} to a safe level.

The efficacy of the local APD also depends on the parasitic resistance between the APD terminals and the source terminals of the driver and receiver NMOS transistors, R_{con}. If R_{con} is comparable to the parasitic resistance from the internal I/O to the APD in the I/O ring ($R_{vss1-core}$ and $R_{vss2-core}$ in Figure 12), the local APD provides little benefit, as shown in Figure 16. The

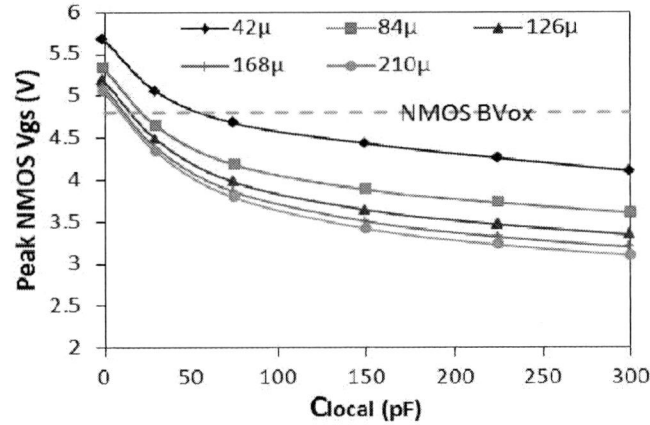

Figure 15. Peak V_{gs} across receiver NMOS as function of local decoupling capacitance for various perimeters of local APD

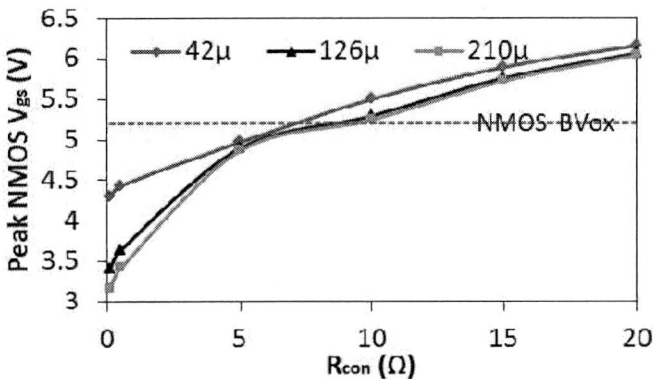

Figure 16. Effect of parasitic resistance connecting the driver and receiver transistors to the local APD terminals. Perimeter of APD in the I/O ring is 210μm. $R_{vss1-core}$ and $R_{vss2-core}$ are each 6 Ω, this is the parasitic resistance to the APD in the I/O ring. C_{local}=150pF

local APD is intended to be used when the internal I/O is located far the pad ring, in which case R_{con} will be smaller than the resistance out to the ring, e.g., $R_{vss1-core}$. In flip-chip designs with the power and ground bumps over the core region, the APDs in the I/O ring may behave as local APDs if $R_{vss1-core}$ is very small.

If placing an APD in the core region between, say, the analog and the digital grounds is a noise concern, the APD could be replaced by a circuit which is off during normal circuit operation and conducts current between the domains only during an ESD event [11].

E. Local Clamps

A local clamp may be placed at the receiver input, but this will increase the path delay unless the designer considers the capacitive loading due to the ESD clamp when sizing the driver. Simulation results show, however, that the CDM benefits of local clamps are significant. A receiver with a diode based local clamp is shown in Figure 17. The peak stress voltage across the receiver gate oxides is plotted in Figure 18 as a function of the series resistance R_{series}.

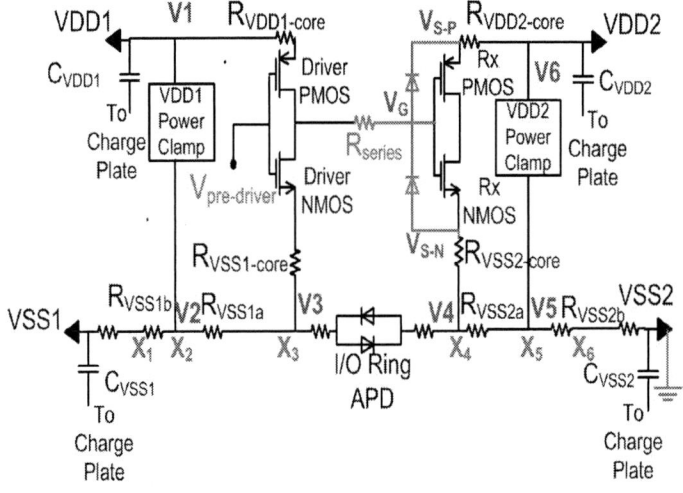

Figure 17. Internal I/O protected by a local clamp consisting of diodes across the gate-source terminals of the receiver transistors and a series resistor

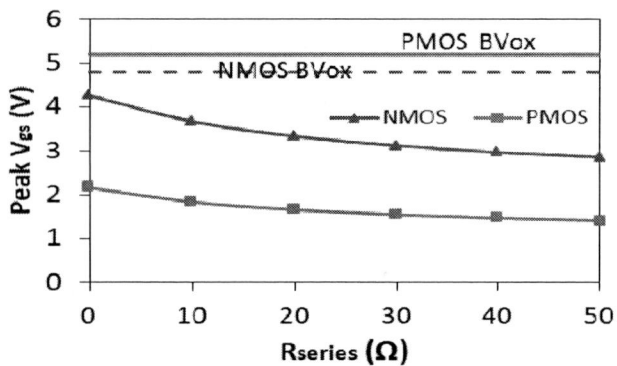

Figure 18. Voltage stress at the receiver transistors when there is a diode based clamp at the receiver input. Perimeter of each diode in the dual diode clamp is 42μm

The stress voltage is clamped to a lower value when R_{series} is increased. However, increasing R_{series} provides diminishing returns. The simulation results indicate that an R_{series} of 20 Ω provides sufficient margin to protect the receiver gate oxides against a 300V CDM event.

Table II compares the peak V_{gs} of the receiver transistors for all of the investigated protection strategies. From Table II, one can see that the local APD with local decoupling capacitance can clamp the voltage across the gate oxides below the gate oxide breakdown voltage (BVox). A local clamp at the receiver input will clamp the voltage to an even lower value, but there is a performance penalty due to an additional series resistor and the capacitance of the diodes. The driver transistors may need to be resized to meet the initial performance goals.

TABLE II. COMPARISON OF PEAK VOLTAGE STRESS ACROSS THE RECEIVER TRANSISTORS FOR DIFFERENT PROTECTION STRATEGIES. SIMULATION RESULTS.

ESD Protection Methods	Receiver Voltage Stress	
	RX NMOS Vgs (V)	RX PMOS Vgs (V)
Diode based Clamp at the gate(Diode perimeter=42μm, R_{series}=20Ω)	3.3	1.7
Local APD of 120μm perimeter with C_{local}=150pF,	3.64	4.8
Cross domain power clamp (VDD1-VSS2) with $R_{VDD12}+R_{VSS12}$=0.5Ω	5	3.11

V. SIMULATION RESULTS OF TESTCASE B

In this section, simulation results for Testcase B are compared with the actual measurement data to establish the validity of the simulation model and the methodology. A simulation model was created for the 64-pad CDM testchip embedded in 100-pin QFP package.

Waveforms at internal nodes of the CDM testchip were obtained using real-time probing at the die level and, also, peak voltages at internal nodes were recorded by voltage monitor circuits [10]. The simulation model was created for the die on the VFTLP tester. A VF-TLP stress of 80V was applied between the VDD1 and VSS2 pads of the design. Table III compares the measured and simulated values of peak NMOS V_{gs} at two internal receivers, one with and one without a local clamp at the input. This local clamp was a grounded-gate. The simulation results match well with the measurement data.

TABLE III. COMPARISON OF MEASURED AND SIMULATED VOLTAGE STRESS ACROSS THE RECEIVER NMOS GATE OXIDE UNDER VF-TLP STRESS

	Voltage Monitor Reading (V)	Simulated voltage (V)
No clamp at the gate	3.4	3
Local clamp at the gate	1.89	1.86

Table IV quantifies the effect of a local APD (without local decoupling capacitors) using both measurement and simulation. Here, the potential difference is measured between the sources of the driver and receiver NMOS devices. A cc-TLP stress was applied to the VSS2 pad near the receiver in the core. The peak discharge current was 1 A. The potential difference was measured between VSS1 and VSS2 probe pads placed near the driver and the receiver, respectively. The agreement between measurement and simulation results is reasonably good. The data in Table III and Table IV provide confidence in the results of the what-if studies presented in Section IV.

TABLE IV. COMPARISON OF CC-TLP MEASUREMENT RESULTS WITH THE SIMULATION RESULTS

	CCTLP Measurement (V)	Simulation Results (V)
Control Case	5.58	6.15
Internal I/O with local APD	5.05	5.45

The minor differences between the measurement and the simulation results result from known inaccuracies in the model for the core power/ground bus mesh and in the models of the ESD protection devices. However, both the measurement and the simulation results indicate that the stress at the receiver gate is lower in the case of the local APD placed near the internal I/O circuit.

VI. CONCLUSIONS

Internal I/O failures are less likely to occur with a low resistivity substrate. At an internal I/O, if the driver output is high, not just the receiver NMOS, but also the receiver PMOS can experience significant voltage stress when the receiver ground is zapped, especially if the charge storage capacitance of the receiver power domain is small. A local APD along with some local decoupling capacitance reduces the gate voltage stress at internal I/Os significantly with no performance penalty. However, the interconnect resistance between the local APD and the source terminals of the driver and receiver NMOS transistors has to be small in order for the local APD to effectively reduce the voltage stress at the receiver. Cross-domain clamps provide another approach to reduce the stress at the internal I/O; they are effective if placed close to the receiver ground pad.

ACKNOWLEDGMENTS

The authors gratefully acknowledge UMC for fabricating the test structures through the UMC University Program. This work is partially supported by the National Science Foundation (grant NSF ECCS 0725406). Vrashank Shukla is supported by a grant from the Semiconductor Research Corporation. Elyse Rosenbaum acknowledges a Micron Professorship Award.

REFERENCES

[1] M.S.B Sworariraj, Smedes T, Salm C., Mouthaan, A.J. and Kuper, F.G, "Role of package parasitics and substrate resistance on the CDM failure levels," *Microelectronics Reliability,* vol. 43, pp. 1569-1575, 2003.

[2] M. Etherton, "Charged Device Model (CDM) ESD in ICs", PhD dissertation, ETH Zurich, 2006.

[3] K. Watanabe, T. Hiraoka, K. Sato, T. Sei, K. Numata., "New protection techniques and testchip design for achieving high CDM robustness," *Proceedings of the EOS/ESD Symposium,* pp. 332-338, 2008.

[4] E. R. Worley, "Distributed gate ESD network architecture for inter-power domain signals," *Proceedings of the EOS/ESD Symposium,* pp. 238-247, 2004.

[5] J. Lee, Y. Huh, J. Chen, P. Bendix and S. M. Kang, "Chip level simulation for CDM failures in multi-power ICs," *Proceedings of the EOS/ESD Symposium,* pp. 456-464, 2000.

[6] R. Renninger, M-C. Jon, D. L. Lin, T. Diep and T. L. Welsher, "A field induced charged device model simulator," *Proceedings of the EOS/ESD Symposium.* pp. 59-71, 1989.

[7] B. C. Chou, T. J. Maloney and T. W. Chen, "Wafer-level charged device model testing," *Proceedings of the EOS/ESD Symposium,* pp. 115-124, 2008.

[8] T.T. Lu, Z.Y. Wang, and W. J. Wu, "Hierarchical Block Boundary-Element Method (HBBEM): a fast field solver for 3-D capacitance extraction," *IEEE Trans. On Microwave Theory Tech.,* vol. 52, no. 1, pp. 10 - 19, 2004.

[9] C. Torres, J. Miller, M. Stockinger, M. Akers, M. Khazhinsky and J. Weldon, "Modular, portable and easily simulated ESD protection networks for advanced CMOS technologies," *Proceedings of the EOS/ESD Symposium.,* pp. 82-95, 2001.

[10] N. Jack, "Measurement and test methods for charged device model electrostatic discharge", M. S Thesis, Univ. of Illinois, Urbana, 2009.

[11] Private correspondence with Dr. C. Duvvury, Texas Instruments Inc.

Thin-Film Photovoltaics: What are the Reliability Issues and Where Do They Occur?

James R. Sites
Colorado State University
Department of Physics
Fort Collins, CO 80523 USA

Abstract-Several aspects of the typical thin-film poly-crystalline solar-cell structure can be susceptible to reliability issues. Potential reliability issues for CIGS and CdTe cells are discussed and their probable physical causes are presented. These possible problems include the metallic back contacts, the grain boundaries in polycrystalline materials, the primary cell junctions, the transparent front contacts, and the edges delineating the cells. A general principle is that reliability issues are always localized to specific areas and never spread uniformly over a cell.

INTRODUCTION

The three thin-film photovoltaic materials that have currently achieved industrial commercialization are amorphous silicon (a-Si), Cu(In,Ga)Se$_2$ (CIGS), and cadmium telluride (CdTe). This focus here will be on the latter two, though some of the reliability issues apply to a-Si as well. In general, the completed thin-film cell structure consists of a metallic back contact, two semiconductor layers of different materials that form the p-n junction, and a transparent front contact. Most often, these layers are deposited on glass, but in some cases a flexible metal or plastic is used. Some thin-film cells are fabricated starting with the metallic contact, as with CIGS (substrate configuration) and others starting with the transparent contact, as with CdTe (superstrate configuration). These two configurations are shown in Fig. 1. In either case, a second piece of glass with an appropriate sealant is commonly used to protect the completed cell from moisture.

BACK CONTACT

CIGS and CdTe are both p-type materials, which means that an ohmic back contact can be difficult to achieve, especially for the higher-band-gap CdTe. For CIGS cells (left in Fig. 1), the back contact is generally made with molybdenum, and such a contact is now generally stable and has low resistance.

The CdTe situation, however, is more prone to difficulty. The usual back-contact approach involves the use of copper to form a layer of copper telluride or heavily-doped CdTe. This layer, however, is metastable, and there is the possibility for diffusion of copper away from the contact area, deterioration of the contact quality, and hence, a reliability issue for the solar cell. The rate of such deterioration depends on temperature, but also on the concentration of copper, the chemistry with neighboring atoms, the porosity of the CdTe, and the electric-field profile within the CdTe.

Experimental CdTe studies of back-contact deterioration are often done at elevated temperatures to accelerate any changes, and these studies are often referred to as stress tests. Fig. 2 shows the effect on the current-voltage (J-V) curves of one CdTe cell fabricated at NREL after it was held at 100°C under illumination and open-circuit bias for varying periods of time [1]. Note that this is an extreme temperature for solar-cell operation and is likely to accelerate any J-V changes by a factor of 100-1000 [2]. In many cases cell performance under such stress is reasonably stable for a month or more, and with a plausible model, one can credibly predict little or no change in performance over decades of field operation.

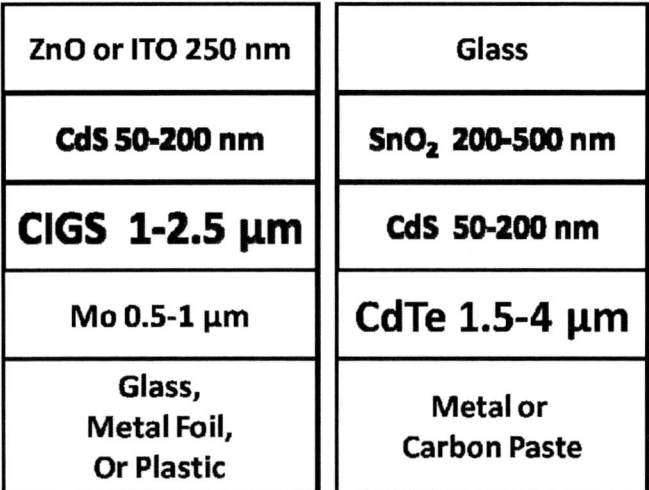

Fig. 1. Typical substrate and superstrate configurations for CIGS and CdTe solar cells. In both cases, light would enter the cell from the top.

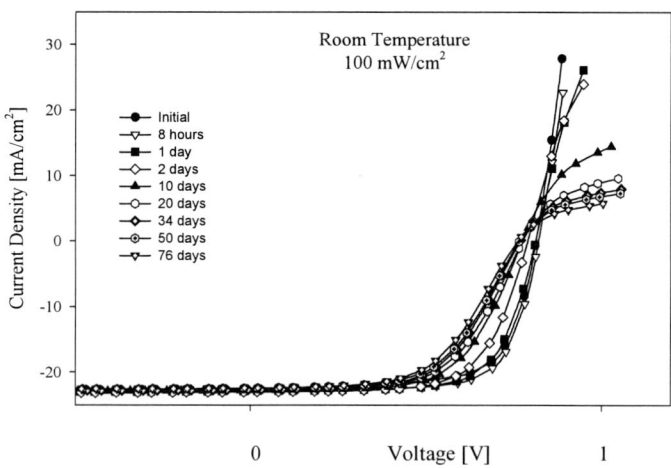

Fig. 2. Progressive changes over 76 days for a CdTe solar cell held at open-circuit and 100°C [From Ref. [1].

The primary performance parameter that changes in Fig. 2 is the fill-factor, but it is clear that the curves are also becoming significantly current-limited in the first quadrant. This latter effect is often referred to as "rollover" and is characteristic of a second diode barrier with polarization opposite the primary diode barrier [3]. In fact, J-V curves calculated with a judicious choice of the back-contact barrier give good fits to the data in Fig. 2. The highly plausible physical model is that copper diffusion away from the back contact allows the contact barrier to increase towards its non-copper value and cell performance is compromised. Close inspection of Fig. 2 also shows that the voltage is also somewhat reduced. If the barrier height becomes still larger, the voltage decrease becomes more significant, and eventually the current will also degrade.

When there are stress-driven performance changes, temperature and bias conditions, as well as the porosity of the material, will affect their rate, as shown in Fig. 3. Greater change at higher temperature is intuitive, but the greater change seen at higher biases (OC for open-circuit) may be less so. The plausible physical explanation here is that the diode-junction field for CdTe extends to near the back contact and opposes the diffusion of positive Cu ions away from the contact. In forward bias, however, the field is both weaker and farther from the back contact, and hence less effective.

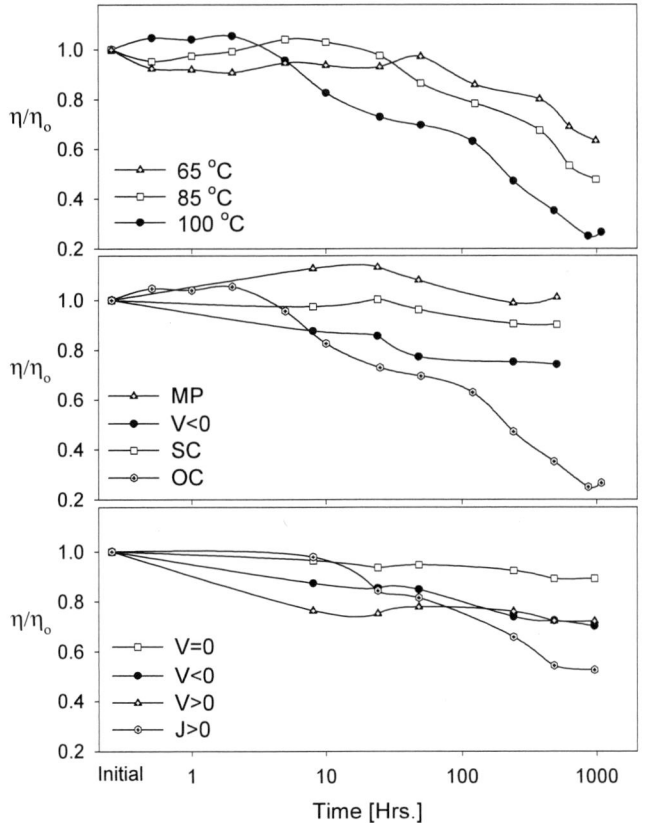

Fig. 3. CdTe efficiency vs. stress time. Top: three temperatures at open circuit (light). Middle (light) and bottom (dark): four biases at 100°C.

The changes observed in Figs. 2 and 3 do not occur uniformly across a solar cell. The spatial variations can be seen through LBIC (light-beam-induced current) measurements, which are maps of the photocurrent response from a focused light spot that is systematically stepped over a designated cell area [4]. Fig. 4 is an example of LBIC data taken before and after a CdTe cell was subjected to elevated-temperature stress. The field of view is half a millimeter. Before the stress, a single non-uniformity in photocurrent, about 10 μm in diameter, is visible. Afterwards, that non-uniformity remains present, but a number of other small areas also exhibit reduced photocurrent. The likely physical explanation is that copper is diffusing away from the back contact preferentially along grain boundaries in the CdTe material.

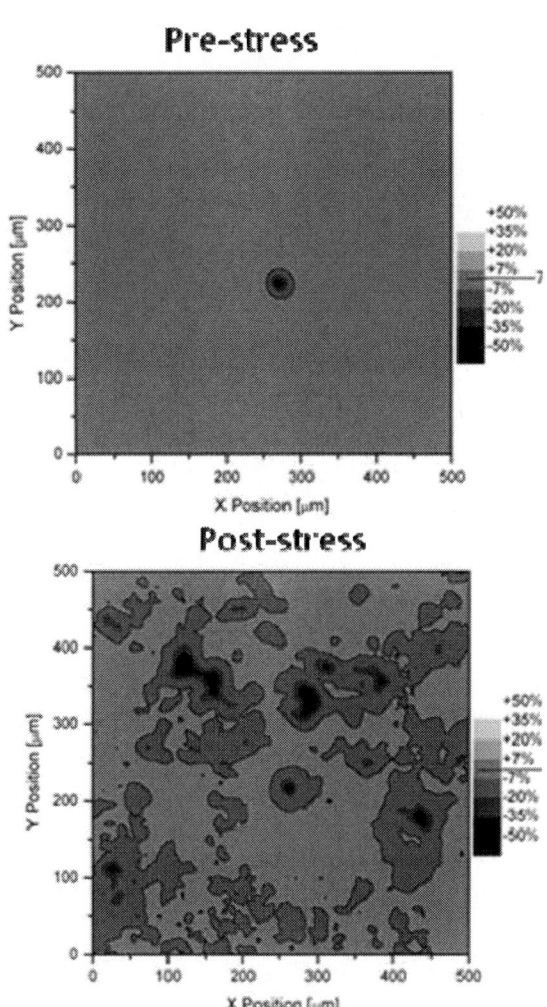

Fig. 4. CdTe photocurrent maps before and after elevated-temperature stress.

GRAIN BOUNDARIES

A schematic of the granular structure of polycrystalline thin-film solar cells is shown in Fig. 5. For both CIGS and CdTe, the typical dimension of the grains is the order of 1 μm, and the grains tend to be columnar, or longer in the growth direction than perpendicular to it. Grain boundaries, as mentioned above, are potential paths for atomic diffusion. Copper diffusion is an issue for the CdTe back contact, but in the extreme case, there may be shunting paths along the grain boundaries.

In addition to providing diffusion paths for copper and other atoms in thin-film cells, however, the grain boundaries inherently present in both CIGS and CdTe solar-cell absorbers (see Fig. 5) are a potential source of recombination for minority carriers (electrons in p-type material). This recombination can limit the collection of photogenerated electrons, but more importantly, it can enhance the forward diode current and reduce the cell's voltage.

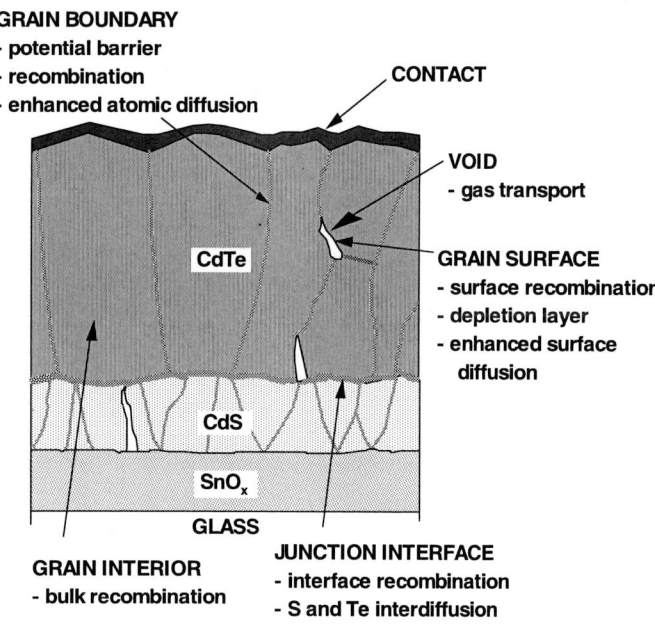

GRAIN BOUNDARY
- potential barrier
- recombination
- enhanced atomic diffusion

CONTACT

VOID
- gas transport

GRAIN SURFACE
- surface recombination
- depletion layer
- enhanced surface diffusion

GRAIN INTERIOR
- bulk recombination

JUNCTION INTERFACE
- interface recombination
- S and Te interdiffusion

Fig. 5. Schematic drawing of CdTe solar cell showing grain structure, roughness, and other physical non-uniformities. CIGS is similar.

Single-crystal solar cells can have a minority-carrier lifetime the order of a microsecond. In contrast, the highest-efficiency CIGS cells have a minority-carrier lifetime that is stable and approximately 50 ns in magnitude [5], but that also is large enough to not significantly reduce voltage. The best CdTe cells, however, have a lifetime of 2 ns, generally less, [6], and their lifetime is not always stable over time. In particular, the diffusion of copper out of the back-contact region can lower the CdTe lifetime, and hence affect the cell performance in a third way, in addition to back-contact changes and potential shorting. Such changes may continue to do so over extended times, and again the details will depend on the temperature and electric-field history of the cell as well as the CdTe porosity and the amount of copper available.

PRIMARY CELL JUNCTION

The diode junction for both CIGS and CdTe solar cells is generally a heterojunction formed with n-CdS and the p-type absorber material, though other materials similar to CdS have also been used successfully. CdS in particular tends to form a junction with a relatively modest density of interfacial states. Hence, it is often referred to as a buffer layer. The CdS itself does not generally contribute to the photocurrent, because its density of gap states is relatively high. Hence, the short-

wavelength photons that are absorbed in the CdS contribute very little to the photocurrent, and the thickness of the CdS layer is generally kept small to minimize photon absorption. Fig. 6 shows the dramatic difference in short-wavelength CdTe quantum efficiency (QE). The QE of CIGS shows a very similar pattern when CdS thickness is varied.

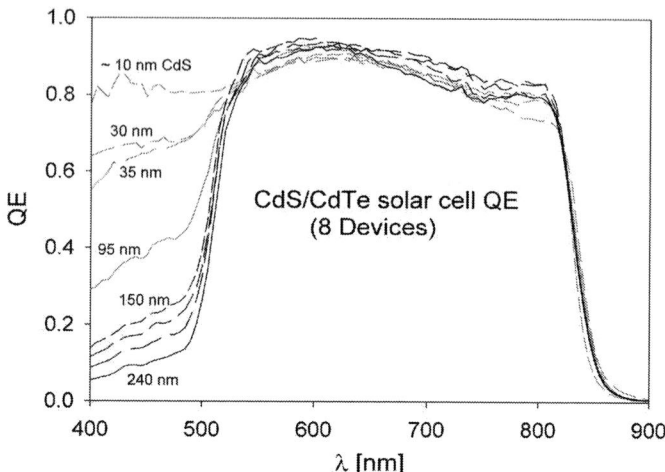

Fig. 6. Variation in QE response with CdS thickness. From Ref. [7].

Thin CdS, however, raises another potential reliability issue. A thin layer is susceptible to pinholes, and a pinhole in the CdS will put the absorber in direct contact with the transparent contact. Locally, this contact results in a diode region with a reduced turn-on voltage, which will reduce open-circuit voltage overall cell performance. Fig. 7 from Ref. [7] shows the steady increase in current at smaller CdS thicknesses, as well as the fairly dramatic decrease in voltage near 100 nm.

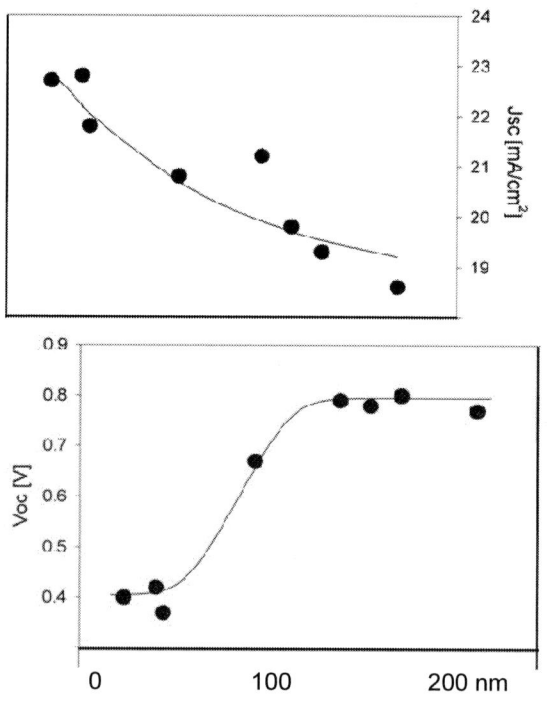

Fig. 7. CdTe cell current and voltage vs. CdS thickness.

The voltage reduction with thin CdS is accentuated by the roughness of the layers, as suggested by Fig. 5. Because of this roughness, there are small areas of very thin CdS, which are essentially shunts between the transparent contact and the absorber. In this case also, the problem areas are not distributed uniformly over the solar cell. Furthermore, there is a strong possibility that the larger current flow through these regions will lead to further deterioration of the partially shunted area, which becomes an additional reliability issue. Fig. 8 again uses LBIC to illustrate the uniformity difference between CdTe cells with thicker and thinner CdS. A similar effect is seen when thin CdS is used with CIGS, but the smaller band gap of CIGS means that there is proportionally less current gain by thinning CdS.

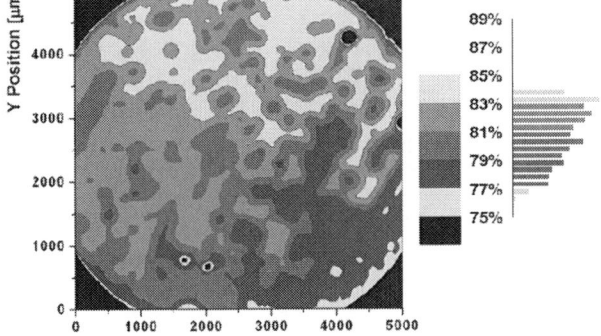

Fig. 8. LBIC comparison of CdTe cells with thin and intermediate-thickness CdS. From Ref. [7].

The actual CdS/absorber interface is generally less of an issue for cell performance and reliability. Although the interface may have a significant number of interface states, thin-film junctions are almost always "one-sided", which means that the central part of the depletion region where the recombination probability is large, will not be near the materials interface. Hence, forward recombination current is likely to be primarily determined by the bulk absorber material.

TRANSPARENT FRONT CONTACTS

Front contacts to thin-film solar cells must be sufficiently transparent that most photons will reach the solar-cell absor-

ber. Generally, the front contacts for thin-film cells consist of various transparent conducting oxides (TCO's) that are doped to improve their conductivity, but not doped to such a degree that their transparency is seriously compromised. Typically as shown in Fig. 1, ZnO or ITO is used with CIGS, and SnO_2 with CdTe. In many cases a double TCO layer is used, where that adjacent to the CdS has a sufficiently high resistivity to mitigate the pin-hole effect of thin CdS.

The primary front-contact reliability issue, however, is the possibility of severe degradation of a TCO when it exposed to moisture. Some of the most common TCO's are particularly vulnerable to moisture, and hence part of TCO research involves a search for materials that are less vulnerable. For the time being, however, great care is required to seal a thin-film module so that moisture cannot easily penetrate to the individual solar cells. In this case, the greater concern is with CIGS cells, where the TCO is the topmost layer that is most exposed following cell fabrication.

With either CIGS or CdTe, or a-Si, a second piece of glass is typically placed over the active part of the module, with an encapsulating material in between. When a flexible substrate is used, the potential for moisture ingress is larger, and the concern for TCO deterioration greater. Similarly, moisture-related deterioration is more likely when modules are deployed in hot, damp climates. Major non-uniformities in a TCO layer, as well as changes in these non-uniformities, can often be observed visually. As with other types of degradation, deterioration of the TCO is likely to begin locally, then spread with time over larger areas.

CELL DELINEATION

Thin-film photovoltaic cells that are incorporated into modules are vulnerable to additional local damage, generally in the form of shunts, when the edges of individual cells are defined. These edges are often formed by laser scribing as the cells that comprise the module cells are delineated. As with other reliability issues, small performance problems immediately after fabrication can become much more severe over a period of time. Shunting at cell edges is also a situation where the information gained from measuring the performance of an entire module is limited, so again it is helpful to be able to have mapping techniques available that enable specific changes to be followed.

TRACKING CHANGES

With each of the reliability issues described, it is necessary to have measurements that can track the changes and whenever possible pinpoint where in the cell or module the changes occur. The topmost layer of a CIGS cell can be tracked optically, and other changes can be tracked with LBIC measurements as illustrated in Fig. 5. The LBIC technique relates directly to actual cell operation, especially since it can be repeated with different bias voltages, but it tends to be relatively slow. Alternatively, a snapshot of electroluminescence (EL), photoluminescence (PL), or infrared temperature from a biased cell or module, can now be done straightforwardly with a CCD camera and can be tracked over time with relatively low overhead. Techniques for relating such data to specific physical changes are not currently well developed, but this type of approach will likely gain widespread attention in the next few years. A very good review that compares different mapping techniques that may become increasingly useful for tracking local degradation is given in Ref. [8].

SUMMARY AND CONCLUSIONS

Several parts of the typical thin-film solar-cell structure are vulnerable to reliability issues:

(1) Ohmic back-contact to a p-type absorber can be difficult to achieve. The common solution of using copper to improve the CdTe contact produces metastabilities where copper diffuses to other layers of the cell and results in degradation of cell performance.

(2) Grain boundaries in either CIGS or CdTe can cause a variety of reliability issues including diffusion paths for foreign atoms, possibly leading to shunting, reduction in photocurrent collection, and excessive forward recombination current. In each case the problem can become more severe over time.

(3) Because of roughness and interdiffusion, the primary junction of either a CIGS or CdTe cell can be by-passed when the CdS layer is made quite thin, a typical stategy to increase current. Here also weak spots can increase in severity over time.

(4) The transparent front contact, particularly with CIGS cells, is vulnerable to moisture ingress, and it too can deteriorate over time.

(5) Scribe lines for cell delineation when fabricating a CIGS or CdTe thin-film module may also produce weak spots that cause increasing performance problems over time.

All of the potential reliability issues discussed here (back-contact, grain boundaries, primary cell junction, front contact, and cell delineation) are the result of local non-uniformities in cell structure. It is critical to track cell performance changes over time at the cell and module level, but it is also important to design tests to identify the nature, location, and magnitude of the changes. These tests, if at all possible, should include techniques to measure spatial variations that occur, often randomly, across the area of the cell.

ACKNOWLEDGMENTS

The author is grateful to his colleague Dr. Alan Fahrenbruch and several former students who have contributed to the reliability studies over the past 15 years. These students include Gunther Stollwerck, Jason Hiltner, Samuel Demtsu, Alan Davies, and Lei Chen. Financial support for much of the work presented here was provided by the U.S. National Renewable Energy Laboratory.

REFERENCES

[1] S.E. Asher, F.S. Hasoon, T.A. Gessert, M.R. Young, P. Sheldon, J.F. Hiltner, and J.R. Sites, "Determination of Cu in CdTe/CdS Devices Before and After Accelerated Stress Testing," Proc. 28th IEEE PV Specialists Conf., p. 479 (2000).

[2] J.F. Hiltner and J.R. Sites, "Stability of CdTe Solar Cells at Elevated Temperatures: Voltage, Temperature, and Cu Dependence," AIP Conf. Series **462**, 170 (1998).

[3] G. Stollwerck and J.R. Sites, "Analysis of CdTe Back–Contact Barriers," Proc. European Photovoltaic Solar Energy Conf. **13**, 2020 (1995).

[4] J.F. Hiltner and J.R. Sites, "Local Photocurrent and Resistivity Measurements with Micron Resolution," Proc. 28th IEEE PV Specialists Conf, p. 543 (2000).

[5] I. L. Repins, W.K. Metzger, C.l. Perkins, J.V. Li, and M.A. Contreras, "Measured Minority-Carrier Lifetime and CIGS Device Performance," Proc. 34th IEEE PV Specialists Conf., p. 978 (2009).

[6] W.K. Metzger, D. Albin, D. Levi, P. Sheldon, X, Xi, B.M. Keyes, and R.K. Ahrenkiel, "Time-Resolved Photoluminescence Studies of CdTe Solar Cells," J. Appl. Phys. **94**, 3549-3555 (2003).

[7] A.R. Davies, A.E. Enzenroth, W.S. Sampath, and J.R. Sites, "All-CSS Processing of CdS/CdTe Thin-Film Solar cells with Thin CdS Layers," Proc. Mat. Res. Soc. **1012**, 157-162 (2007).

[8] S.W. Johnston, N.J. Call, B. Phan, and R.K. Ahrenkiel," Applications of Imaging Techniques for Solar Cell Characterization," Proc. 34th IEEE PV Specialists Conf., p. 276 (2009).

THERMOREFLECTANCE IMAGING OF DEFECTS IN THIN-FILM SOLAR CELLS

D. Kendig, G. B. Alers, and A. Shakouri
Baskin School of Engineering
University of California, Santa Cruz
Santa Cruz, CA 95064 USA
+1-(510)-565-5257, dkendig@ucsc.edu

Abstract—We have identified and characterized various defects in thin-film *a*-Si and CIGS solar cells with sub-micron spatial resolution using thermoreflectance imaging. A megapixel silicon-based CCD was used to obtain noncontact thermal images simultaneously with visible electroluminescence (EL) images. EL can be indicative of pre-breakdown sites due to trap assisted tunneling and stress induced leakage currents. Physical defects appear at reverse bias voltages of 8V in a-Si samples. Linear and nonlinear shunt defects are investigated as well as electroluminescent breakdown regions at reverse biases as low as 4.5V. Pre-breakdown sites with electroluminescence are investigated.

Keywords-thermoreflectance; lock-in; shunts; thin-film; solar cell; defects; thermal imaging

I. INTRODUCTION

Thin-film photovoltaic production is increasing rapidly, and this calls for new, versatile characterization techniques to obtain high resolution images of the microscopic features of these devices. Many different techniques have previously been developed in order to characterize solar cells and their defects. One of those techniques is lock-in thermography (LIT). LIT is a well-established failure analysis tool and is typically done with infrared (IR) cameras. [1] LIT techniques provide a wide rang of applications for qualitative and quantitative analysis of solar cell parameters. IR-LIT can produce thermal images with 10 μK temporal resolution and 5-10 μm spatial resolution. Unfortunately, this resolution cannot always be obtained. Thin-film solar cells with glass superstrates/substrates can block direct imaging of the active region since typical glass used in fabrication is not transparent to IR light of 3-10 μm. [2] This limits the resolution of LIT to the thickness of the glass or ~3 mm. Thermoreflectance (TR) imaging is a LIT technique that uses the change in a materials reflectivity due to a change in temperature to obtain a high spatial resolution thermal image. [3,4,5] This technique is typically done with visible light and can produce 200-400 nm spatial resolution and 10 mK temperature resolution after averaging with the use of typical 12-bit CCDs. TR imaging can be used to see through materials such as glass (with visible light) or silicon (with near-IR light >1200nm) in order to obtain high resolution thermal images of areas of interest. Using a silicon-based CCD, visible electroluminescence (EL) and thermal images can be can taken simultaneously. This versatile, noncontact characterization tool can help locate a broad range of thermal and EL defects with sub-micrometer spatial resolution. This paper will focus on using the high spatial resolution of thermoreflectance imaging, along with its ability to simultaneously obtain visible electroluminescence, to locate and characterize various defects in photovoltaic cells and mini-modules.

A. Failures in Solar Cells

Solar cells are made of thin *p-n* junctions through a series of complicated and delicate processing steps. To obtain high efficiency cells with high yield, uniform material deposition must be made under highly controlled conditions, with some steps that require high temperatures. A large variety of defects can arise during any of the processing steps such as Schottky-type shunts, ohmic shunts, weak-diodes, avalanche and Zener breakdown regions, linear and nonlinear edge shunts, cracks and dislocations, and areas of high resistance. [6] Further damage from the environment can also occur, such as oxidation, delamination, or light-induced degradation (e.g. Staebler-Wronski effect). These defects can provide localized parasitic pathways for photogenerated current, thus limiting the power provided to the load. The local high current density at these defects produces heating and sometimes light which can be detected by LIT or infrared camera. Shunt defects are low resistance paths in parallel with the photovoltaic cell that bypass current from the load. Two general categories of shunt defects are linear and nonlinear(weak diode), shown in Fig 1. These categories can then be further divided into specific failure mechanisms (e.g. Zener, avalanche, weak-diode etc). Linear shunts are visible in the forward and reverse bias thermal image, and nonlinear shunts typically appear in forward bias. Recombination centers can also emit light when in forward or reverse bias, such as an avalanche breakdown region or impurity state. In avalanche breakdown, impact ionization due to electrons accelerated by high electric fields in the reverse-biased depletion region can emit visible white light. Under partial shading conditions some series connected cells will output reduced current and can become reverse biased as the higher current of the module is forced through the cells. Under these reverse biased conditions a great deal of heat can be dissipated in the shaded cells. [7] Typical single crystal silicon solar cells can undergo reversible breakdown at reverse biases of ~15V. However, localized breakdown sites can occur at voltages as low as -7V in multi-crystalline silicon. [8] Thin film photovoltaics have lower breakdown voltages due the higher electric field in the thin film and can show irreversible damage at localized hot spots under partial shading conditions. Damage and shunting from these localized hot spots will degrade the long term efficiency and reliability of the photovoltaic module.

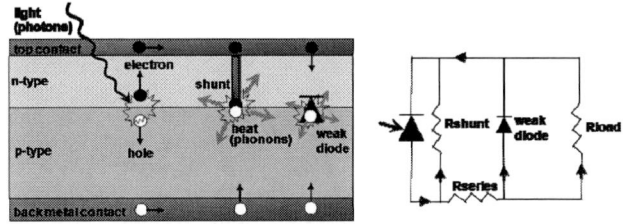

Figure 1. Diagram of electron and hole flow through linear shunts and nonlinear weak diode defects and the corresponding simplified circuit.

II. EXPERIMENTAL METHOD

Images were obtained with our custom-built silicon-based CCD thermoreflectance system. A 1-Megapixel Dalsa 1M60 CCD was used with custom LabVIEW software. Multiple algorithms can be used to obtain a thermal signal. [3,4,5] This setup used a differencing algorithm with precisely controlled phase delay. The current system is able to produce 20 ns time resolution and has been developed for transient thermal imaging.

978-1-4244-5430-3/10 $26.00 © 2010 IEEE

For a typical measurement, the camera images are collected at frequencies between 1 and 30 Hz. When the camera shutter is open, LED pulses are used to obtain normalized measurements of the initial reflected intensity and the intensity after heating. The change in reflectivity is then related to a change in temperature through a thermoreflectance coefficient obtained through calibration for each material and wavelength of light. Voltage pulses 2.5 ms wide, from 0 to 20V, at 10% duty cycle were used for excitation in forward and reverse bias. The pulse width and duty cycle were adjusted to limit the maximum temperature of the device to prevent damage from the measurement. Electroluminescent images were taken under the same bias conditions without the measurement LED on.

III. RESULTS

Images of thin-film a-Si, CIGS, and poly-Si solar cells were taken with forward and reverse bias pulses at voltages up to 20V. The images revealed linear ohmic shunts as well as nonlinear weak diode defects in forward bias. In reverse bias, visible electroluminescence defects were found at voltages as low as 4.5V. Thermal images also show cold regions due to decreased current flow and high resistances. Heating around the edge of the back aluminum contact in a-Si is due to high electric fields and leads to delamination of the cell edges at 20V in forward bias and -8V in reverse bias. Spot defects in the center of the device appear at 5V forward bias.

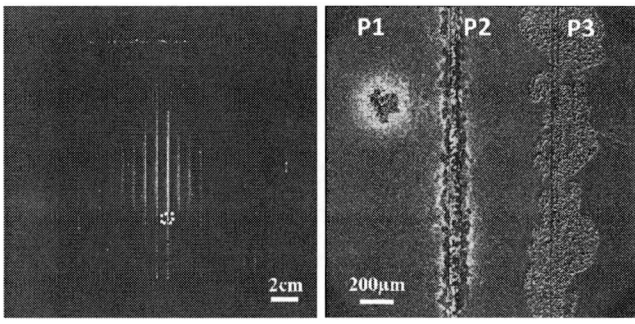

Figure 2. This is an image of a CIGS mini-module (left) and a high magnification image of a defect along the back P1 metal scribe line (right). This also shows nonuniform heating along the P2 scribe line

A. Scribe Line Defects

The low magnification image on the left in Fig. 2 reveals the general locations of strong defects at the scribe lines separating individual cells in a thin film CIGS module that had undergone high temperature stressing in dry heat. The images used 300s of averaging. These images were taken at high voltages to increase the thermal signal, while the duty cycle was used to limit overall substrate temperature. Optimization of the illumination source as well as longer averaging times will increase the thermal sensitivity and quality of the images. High magnification images obtained through the glass superstrate of the CIGS cells show that heating occurs at the P2 laser scribe line that delaminates the CIGS absorber layer from the neighboring cells. At this scribe line current flows through the transparent conducting oxide (TCO) to connect the bottom Mo contact of one cell with the top TCO contact of the next cell. P1 and P3 scribe lines are identifiable in the bright field image and delineate the Mo back contact layer and the TCO top layer respectively. Localized heating of the CIGS scribe line is the result of high resistivity TCO that occurred from oxidation during high temperature stressing. [9] Nonuniform heating of the scribe line that connects neighboring cells is visible in this image and correlates with

residual CIGS material that is left in the scribe line. Images with this resolution would not be possible with conventional infrared imaging.

B. Linear Defects

In low magnification images we can see ohmic shunts appear in both forward and reverse bias where the thermal signal at identical voltages in forward and reverse bias is the identical. Linear defects in a thin-film a-Si solar cell are shown in Fig. 3. These shunts can be caused by defects introduced in the manufacturing process such as an incompletely opened emitter at the edge of the cell or defects induced by stressing. [6] Aluminum impurities can create p^+-regions that form an ohmic contact with the base and a tunnel contact to the emitter. Impurity defects such as Al particles can also be locations of avalanche breakdown at reverse biases between -5V and -10V.

C. Nonlinear Defects

Nonlinear shunts shown in the thermal images typically appear in forward bias and not in reverse bias. These shunts show diode-like I-V characteristics and are sometimes known as "weak-diode" defects. [6] At these locations, the open circuit voltages of the cell are lower due to the lower effective shunt resistance of the cell. Nonlinear shunts can be caused by areas of higher concentration of recombination-active defects. Schottky-type shunts can be caused by direct contact between the p-type base material and metal. Nonlinear shunts are often found along the edges of the solar cell and along scribe lines. They also appear if the emitter metallization is deposited on a region where there is no or little emitter layer. In reverse-bias, weak diodes can appear as pre-breakdown defects in thermal and EL images. Avalanche breakdown occurs in some defects as low as -4.5V where these defects emit white light that is easily detectable by a silicon CCD.

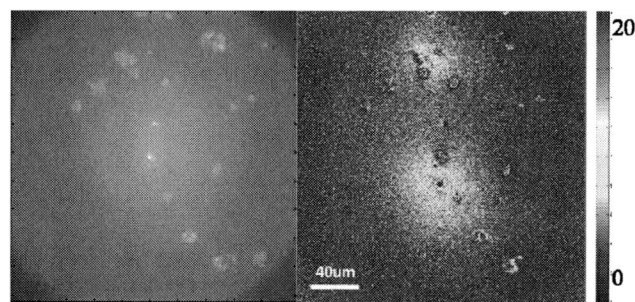

Figure 3. This thermal image at 7V reverse bias shows the details of the linear shunts in the a-Si sample. Not all of these physical defects conduct current. The scale bar is a rough calibration 0 to 20 K.

D. Resistive Defects

Cooler areas in the thermal images correspond to areas of decreased current flow. These areas have abnormally large resistances that can be caused by poor contacts with the base/emitter metallizations, microstructural defects, or other resistive defects such as oxidation. If the encapsulation materials fails to protect the active photovoltaic region or its metal contacts, oxidation of the interface or TCO layer from moisture can cause high series resistances. Oxidation and delamination typically appear along scribe lines in solar cell mini-modules. These high series resistances create nonuniform current paths that can lead to further destruction of the module. High series resistance can also be seen in the forward bias electroluminescence image. Fig. 2 shows how the oxidation of the Zn:Al:O transparent conductive oxide causes increased resistance.

978-1-4244-5430-3/10 $26.00 © 2010 IEEE   500

This increase in resistance causes large voltage drop across the module which is shown by the strong heating on the side of the module shown in Fig. 4. These resistances greatly reduce the fill factor and efficiency of the modules.

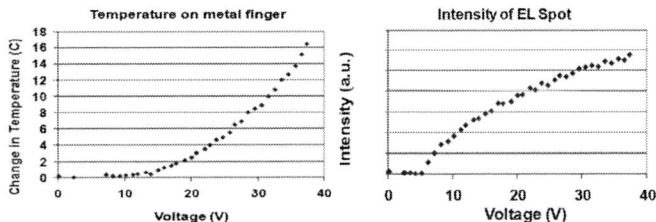

Figure 7: Typical Joule heating response from a thermal defect (Left) and the intensity light emitted by breakdown defect (Right)

E. Electroluminescent Defects

Many defects can be found through electroluminescence in reverse bias. These EL defects are typically sites of avalanche breakdown and emit white light that can be detected by a silicon CCD. Fig. 6 shows an avalanche defect with visible EL. We can see the dramatic difference between the voltage response of a thermal and electroluminescent defect in Fig. 7. These graphs show the thermal response of a defect along a metal finger and the light intensity of an EL defect. We have found EL spots in reverse bias voltages as low as 4.5V. By using a silicon CCD, we can easily detect these breakdown defects and distinguish them from Zener-type (tunneling) breakdown regions due to the emission of visible light.

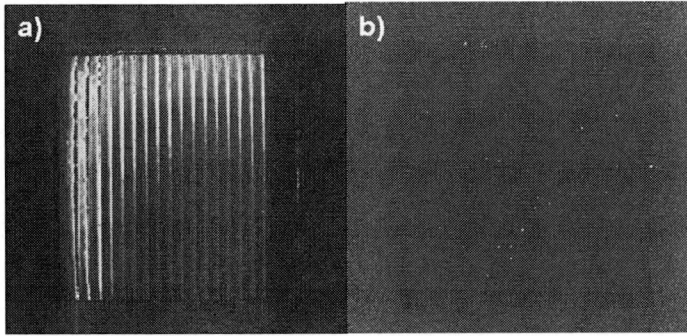

Figure 4. a) forward and b) reverse bias electroluminescence images of a CIGS solar cell mini-module with high resistance TCO.

Avalanche breakdown occurs at high reverse bias voltages typically around 50V for crystalline silicon with 10^{19} cm^{-3} doping concentrations. Avalanche breakdown causes high currents to flow through localized spots in the solar cell. In reverse bias thermally generated charge carriers are accelerated at very high voltages. When the kinetic energy is large enough to cause impact ionization (Fig. 5), avalanche carrier multiplication occurs causing a sharp increase in reverse bias current. This breakdown voltage is controlled by doping levels, and impurities can create local breakdown points where this can happen. Also, this avalanche current decreases with increasing temperature (negative temperature coefficient). Zener breakdown is due to the tunneling of electrons through the reverse biased p-n junction. High doping concentrations with a sharp transition region lead to greater tunneling. The probability of an electron tunneling is independent of temperature since tunneling occurs when an electron from the valance band of the p-side tunnels through to the conduction band of the n-side, the electron does not have a large change in energy compared to an electron from avalanche breakdown. This difference in the breakdown mechanism allows us to distinguish between avalanche and Zener breakdown regions by the detection of visible light.

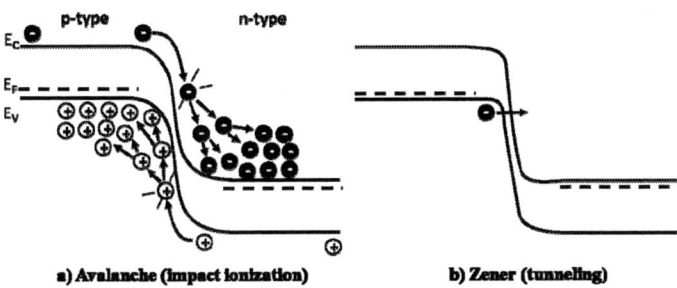

Figure 5: This figure shows the band diagram of two typical breakdown mechanisms in semiconductors

The high currents created by breakdown can lead to electromigration and irreversible damage to the device. Electromigration can move metal ions through this breakdown region and create a permanent linear shunt.

F. Quantitative Analysis

From these thermal images and the biasing parameters, other information such as current density, diode ideality factor, current multiplication factor, relative temperature coefficient, resistance, and local I-V characteristics measured thermally (LIVT) can be calculated. Thermal data has previously been used to help distinguish between Zener-type and avalanche-type breakdown defects. [7]

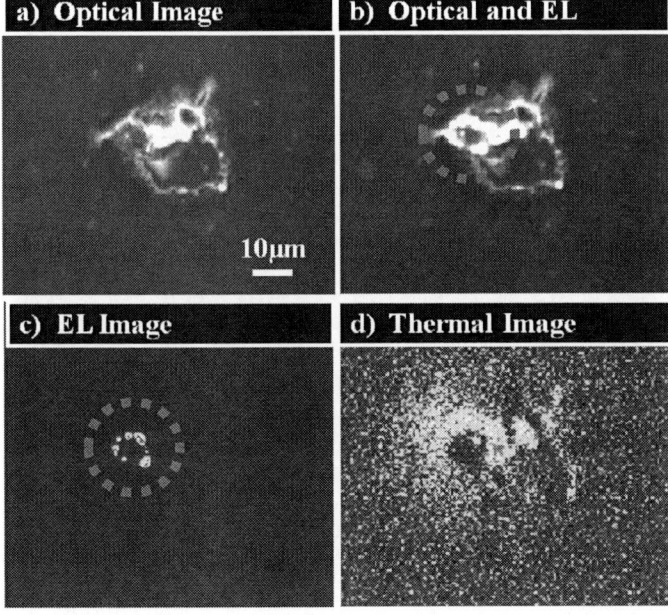

Figure 6. a) optical image of a defect in poly-Si, b) optical image with EL combined to show the location of the EL on the defect, c) EL signal in reverse bias, d) Thermal image of defect at 30V reverse bias

978-1-4244-5430-3/10 $26.00 © 2010 IEEE

IV. CONCLUSION

Thermoreflectance with a silicon CCD is a useful tool that can obtain visible electroluminescence and thermal images simultaneously while providing the optical image that shows the locations of defects with respect to the physical features of the solar cell. We have shown sub-micron thermal images of defects in solar cells that correlate with electroluminescence defects in reverse bias. Detailed thermal images of module scribe lines are obtained through 3mm of glass encapsulation with visible light. Thermal and electroluminescence images can be used to distinguish between avalanche and Zener breakdown effects. Electroluminescence in poly-Si was found at reverse bias voltages as low as 4.5V.

V. ACKNOWLEDGEMENTS

The authors would like to thank M. D. Kempe and K. N. Terwilliger of NREL and D. Tarrant of Shell Solar for the CIGS samples. This work was partially funded by the UC Discovery grant.

VI. REFERENCES

[1] J. Bauer, O. Breitenstein, and J.-M. Wagner, "Lock-in Thermography: A Versatile Tool for Failure Analysis of Solar Cells," *ASM International*, no. 3, pp. 6-12, 2009.

[2] D. Shvydka, J. P. Rakotoniaina, and O. Breitenstein, "Lock-in thermography and nonuniformity modeling of thin-film CdTe solar cells," *Applied Physics Letters*, vol. 84, no. 5, pp. 729-731, Feb. 2004.

[3] D. Lüerßen, J. A. Hudgings, P. M. Mayer, and R. J. Ram, "Nanoscale Thermoreflectance With 10mK Temperature Resolution Using Stochastic Resonance," in *Proceedings of the 21st IEEE SEMI-THERM Symposium*, San Jose, 2004.

[4] G. Tessier, S. Hole, and D. Fournier, "Quantitative thermal imaging by synchronous thermoreflectance," *Applied Physics Letters*, vol. 78, no. 16, pp. 2267-2269, Apr. 2001.

[5] J. Christofferson and A. Shakouri, "Thermoreflectance based thermal microscope," *Review of Scientific Instruments*, vol. 76, 2005.

[6] O. Breitenstein, J. P. Rakotoniaina, M. H. Al Rifai, and M. Werner, "Shunt Types in Crystalline Silicon Solar Cells," *Progress in Photovoltaics: Research and Applications*, vol. 12, pp. 529-538, Jul. 2004.

[7] O. Breitenstein, J. Bauer, J. Wagner, and A. Lotnyk, "Imaging Physical Parameters of Pre-Breakdown Sites by Lock-in Thermography Techniques," *Progress in Photovoltaics: Research and Applications*, vol. 16, pp. 679-685, Jul. 2008.

[8] M. Kasemann, W. Kwapil, M.C. Schubert, H. Habenicht, B. Walter, M. The, S. Kontermann, S. Rein, O. Breitenstein, J. Bauer, A. Lotnyk, B. Michl, H. Nagel, A. Schütt, J. Carstensen, H. Föll, T. Trupke, Y. Augarten, H. Kampwerth, R.A. Bardos, S. Pingel, J. Berghold, W. Warta, and S.W. Glunz, "Spatially resolved silicon solar cell characterization using infrared imaging methods," in *Proceedings of the 33rd IEEE Photovoltaic specialists conference*. 2008. San Diego, CA, USA.

[9] Kempe, M. D.; Terwilliger, K. M.; Tarrant, D.; , "Stress induced degradation modes in CIGS mini-modules," *Photovoltaic Specialists Conference, 2008. PVSC '08. 33rd IEEE*, vol., no., pp.1-6, 11-16 May 2008

SEAM and EBIC Studies of Morphological and Electrical Defects in Polycrystalline Silicon Solar Cells

L. Meng[1], D. Nagalingam[1], C.S. Bhatia[1,2], A.G. Street[3,4] and J.C.H. Phang[1,4]

[1]Centre for Integrated Circuit Failure Analysis and Reliability (CICFAR), National University of Singapore
[2]Institute of Materials Research and Engineering (IMRE), Singapore
[3]Qualcomm CDMA Technologies, San Diego, U.S.A.
[4]Inscope Labs Pte. Ltd., Singapore
Phone: +65-6516-2244, E-mail: lei.meng@nus.edu.sg

Abstract—**Morphological and electrical defects in solar cells are distinguished by combining Scanning Electron Acoustic Microscopy (SEAM) and Electron Beam Induced Current (EBIC) techniques. These techniques provide complementary information on grain boundaries and defects that affect solar cell performance in different ways.**

Keywords- morphological defects, electrical defects, solar cell, Scanning Electron Acoustic Microscopy, SEAM, Electron Beam Induced Current, EBIC

I. INTRODUCTION

Polycrystalline silicon materials are promising candidates for the manufacture of low cost, high efficiency and large area solar cells. The poor performance of these cells is attributed to the existence of morphological and electrical defects [1]. Morphological defects such as lamellar twins, stacking faults and double positioning twin boundaries affect the electrical performance of solar cells made from polycrystalline photovoltaic materials. The electrical properties of morphological defects may vary at different locations within the same defect [2]. Electrical defects without morphological features also affect the electrical performance of these cells.

SEAM is a morphological imaging technique that is used for subsurface imaging, depth profiling and material characterization of metals, doped silicon [3] and solar grade polycrystalline silicon [4]. However, the technique is not sensitive to non-morphological electrical defects. Alternatively, EBIC is an electrical defect characterization technique that detects recombination sites, doping level inhomogeneities [5] and electrical irregularities in solar cells [6]. However, EBIC requires a pn junction and the detection range is limited to a few diffusion lengths from the junction. Both techniques can be implemented in the Scanning Electron Microscope (SEM).

In this paper, a method is described to distinguish morphological and electrical defects in a polycrystalline silicon solar cell by combining these two techniques. This method can be used to detect subsurface anomalies in a solar cell and differentiate the properties of the defects which can lead to further improvements in process design and device performance.

II. EXPERIMENTAL SETUP

A Hitachi S2700 SEM is modified for both SEAM and EBIC imaging. Fig. 1 shows the block diagram of the SEAM and EBIC setup. An intensity-modulated electron beam is achieved with the beam blanking system that comprises a function generator, amplifier and blanking plates [7]. When the modulated electron beam irradiates the sample surface, localized heating induces acoustic waves that propagate through the sample and are detected by a lead zirconate titanate (PZT) piezoelectric transducer attached to the backside of the sample. The PZT transducer converts the acoustic waves into electrical signals, which are amplified and fed into a lock-in amplifier (LIA). The reference of the LIA is the same signal that is used to drive the blanking plates of the SEM. The LIA outputs comprise both SEAM amplitude (A) and phase (θ) signals. An active imaging system, SEMICAPS 2200A, is used to generate SEAM amplitude (A) and phase (θ) images by synchronizing the scan generator with the output of the LIA. EBIC imaging is performed with the beam blanking system turned off. The EBIC current is amplified with a low noise Current Amplifier. The same active imaging system is used to generate EBIC images from the output of the Current Amplifier.

978-1-4244-5430-3/10 $26.00 © 2010 IEEE

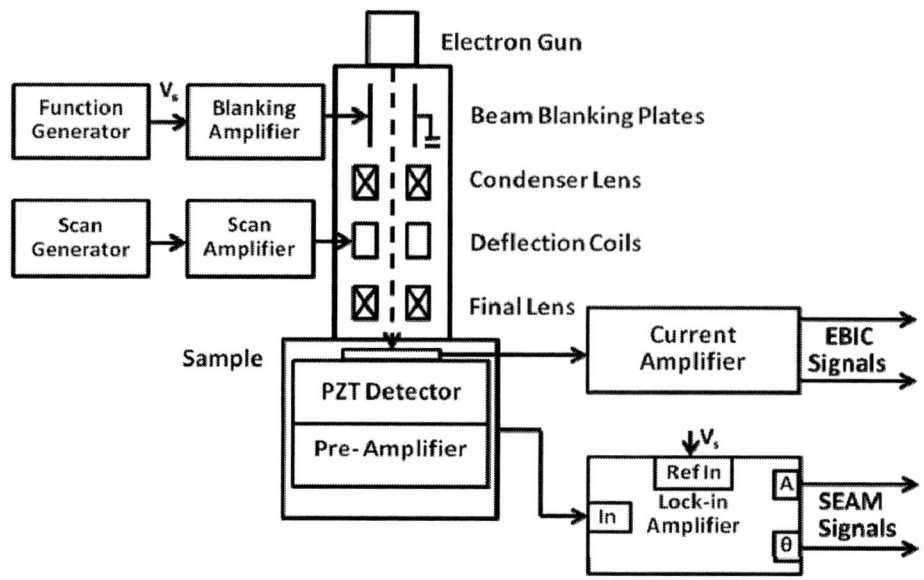

Figure 1. SEAM and EBIC experimental setup

Fig. 2 shows an isometric view of a polycrystalline silicon solar cell sample that is formed by growing a 0.6 μm thick n$^+$ silicon epitaxial layer on a 200 μm thick p$^-$ silicon substrate. The solar cell sample consists of several individual solar cells separated by 2 μm deep isolation trenches. As the modulated electron beam irradiates the sample surface, the beam energy is dissipated through inelastic scattering within the interaction volume and thermal waves are produced by the heat generated during this energy dissipation. The SEAM signal detection range is determined by both the beam energy dissipation and thermal-wave propagation. These thermal waves attenuate rapidly and in turn produce elastic acoustic waves [3], which propagates through the sample to the PZT detector for SEAM imaging.

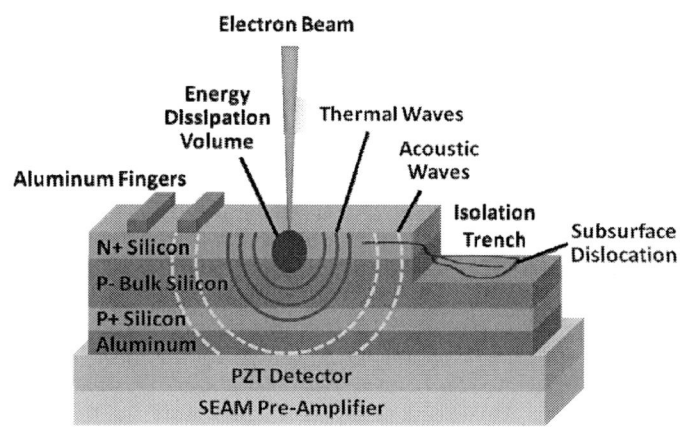

Figure 2. SEAM signal detection of the solar cell

The incident electron beam also generates electron-hole pairs, which are separated by the electric field by the pn junction to produce an EBIC signal. A morphological defect that extends from a solar cell into the isolation trench is illustrated in both Fig. 2 and 3. The part of the defect that is within several diffusion lengths of the pn junction is observed by EBIC. However, the defect that extends into the trench where there is no pn junction only appears in the SEAM image.

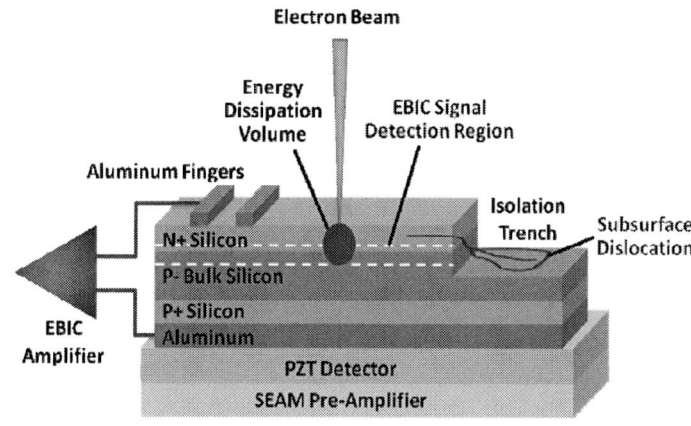

Figure 3. EBIC signal detection of the solar cell

III. RESULTS AND DISCUSSION

Fig. 4a shows the secondary electron (SE) image of the solar cell device depicted in Fig. 2 and 3 using 30 KeV electron beam energy. Fig. 4b shows the corresponding EBIC image using 20 KeV electron beam energy and Fig. 4c shows the SEAM amplitude image using 30 KeV electron beam energy and an electron beam modulation frequency of 261 KHz. Both the EBIC and SEAM images show strong contrast of similar defects that are not observed in the SE image. These defects are located within the EBIC signal detection range in the subsurface of the solar cell illustrated in Fig. 3.

978-1-4244-5430-3/10 $26.00 © 2010 IEEE

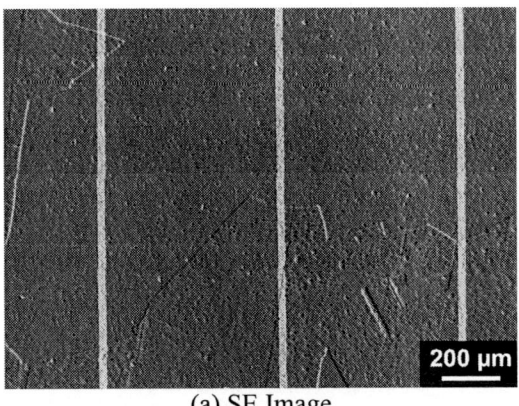

(a) SE Image

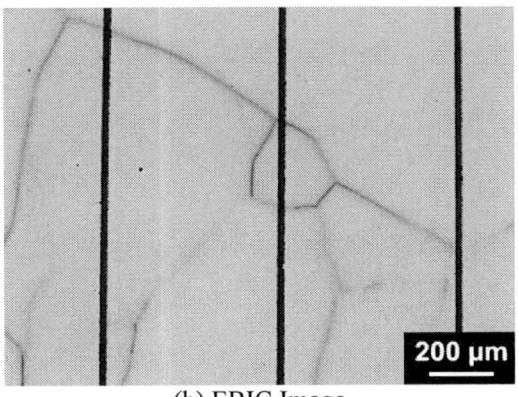

(b) EBIC Image

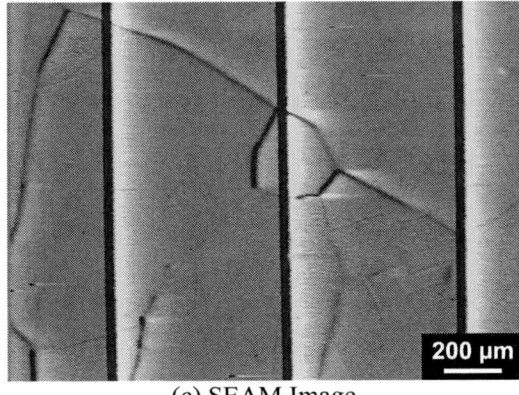

(c) SEAM Image

Figure 4. (a) SE image at 30 KeV; (b) EBIC image at 20 KeV; (c) SEAM amplitude image at 30 KeV and modulation frequency of 261 KHz.

Fig. 5a shows the SE image of the same device depicted in Fig. 4 using 20 KeV electron beam energy. This image provides surface topographic information. The bottom right half of the SE image corresponds to the isolation trench. Fig. 5b shows the corresponding EBIC image using 20 KeV electron beam energy. Fig. 5c and 5d show the corresponding SEAM amplitude and phase images respectively using 30 KeV electron beam energy and an electron beam modulation frequency of 363 KHz. Both the EBIC and SEAM images show similar defects except in the trench area where the defects

are evident in the SEAM image but not the EBIC image. This is expected as the pn junction is truncated at the trench for isolation purposes. The image shows that the defect extends into the substrate of the solar cell, which may have existed prior to the formation of the pn junction.

(a) SE Image   (b) EBIC Image

(c) SEAM Amplitude Image   (d) SEAM Phase Image

Figure 5. (a) SE image and (b) EBIC image at 20 KeV; (c) SEAM amplitude image and (d) SEAM phase image at 30 KeV and modulation frequency of 363 KHz.

Fig. 6a and 6b show the EBIC and SEAM images respectively at another location of the same solar cell. Fig. 6c shows the line profiles across defects at X-X' and Y-Y' for both images. The contrast of the signal is defined as the ratio of the signal intensity of the subsurface defects to the standard deviation of the surface background signals. After normalizing the intensity of surface background signals to zero for both line profiles, the subsurface defects show negative contrast. At X-X', the contrast of the defect from EBIC and SEAM line profiles is -11.0 and -7.0 respectively, showing that the defect is both morphological and electrical in nature. The magnitude of the EBIC and SEAM signal contrast cannot be compared directly due to the different imaging parameters used. However, at Y-Y', the contrast of the defect from the EBIC line profile shows a contrast of -9.5 while the defect contrast from the SEAM signal line profile is not observable. This defect is predominantly electrical in nature without morphological features.

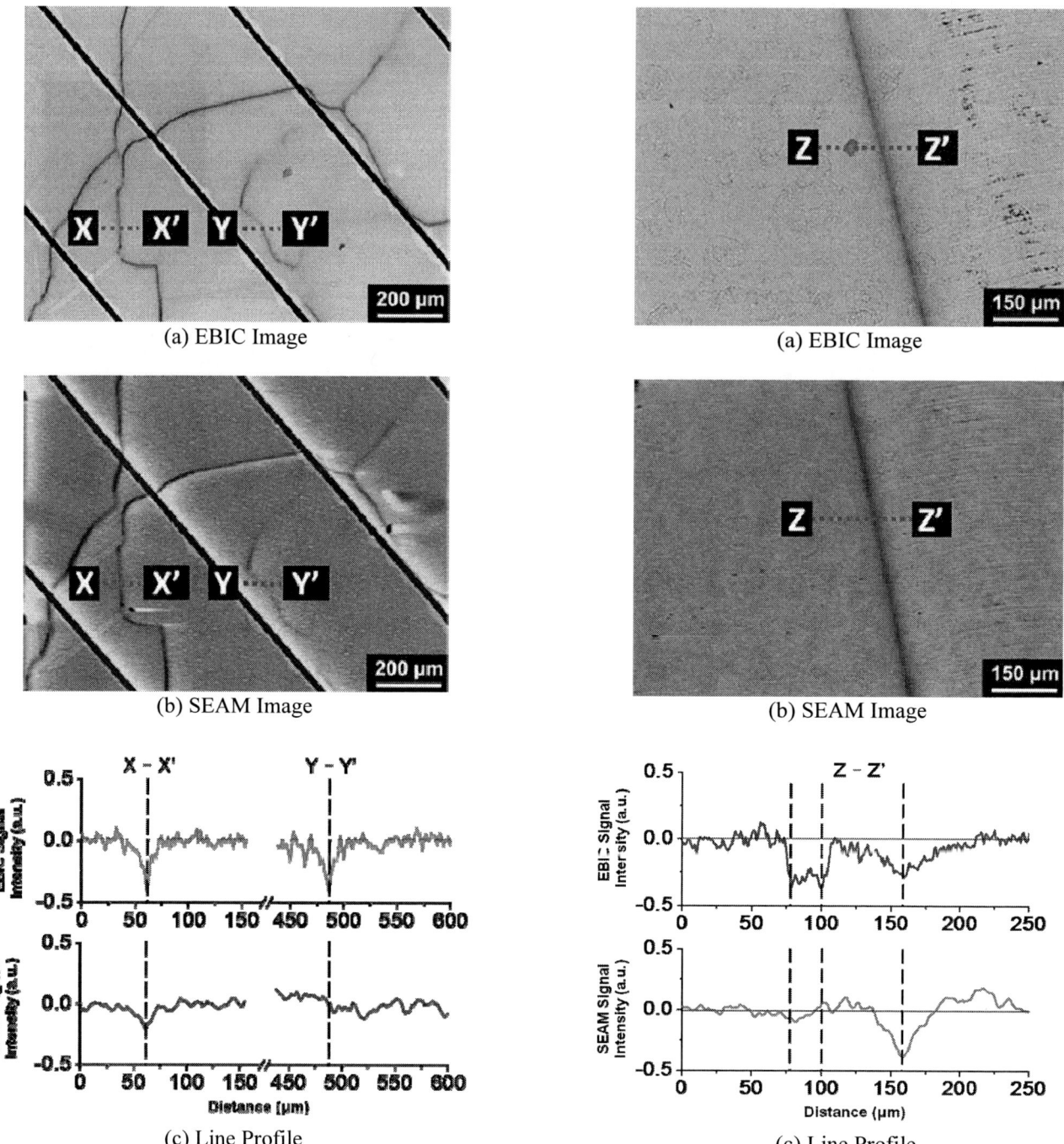

Figure 6. (a) EBIC image at 20 KeV; (b) SEAM amplitude image at 30 KeV and modulation frequency of 391 KHz; (c) Line Profile at X-X' and Y-Y' of both EBIC and SEAM images.

Figure 7. (a) EBIC image at 20 KeV; (b) SEAM amplitude image at 30 KeV and modulation frequency of 292 KHz; (c) Line Profile at Z-Z' of both EBIC and SEAM images.

978-1-4244-5430-3/10 $26.00 © 2010 IEEE

An experiment was performed to validate the ability of the SEAM and EBIC techniques to distinguish morphological and electrical defects. A reverse biased overstress was applied to a commercial solar cell to create defects at the pn junction [8]. As a result of applying a 1 KV pulse, the current-voltage (I-V) characteristics of the solar cell changed from a diode to a resistive short. Fig. 7a shows the EBIC image of the stressed solar cell indicating both a point defect that is induced by the electrical overstress and a line defect that was present in the unstressed solar cell. Both the point and line defects are not observed in the SE image. Fig. 7b shows the SEAM amplitude image of the same location where only the line defect is clearly observable and matches that imaged by EBIC, indicating that the defect is morphological in nature. Fig. 7c shows the normalized signal intensity line profiles across the defects at Z-Z' for both EBIC and SEAM images. The EBIC line profile shows the point defect with a width of 25 μm and a contrast of approximately -6.2 while the corresponding SEAM line profile does not show any appreciable contrast, which confirms that the point defect is evident in the EBIC image but absent in the SEAM image. This defect is an electrical defect without morphology that is not detectable by SEAM.

In order to verify that the SEAM and EBIC techniques can indeed distinguish the morphological and electrical defects, a Focused Ion Beam (FIB) cross-section was performed on the defects along Z-Z' shown in Fig. 7a and 7b after a 100 nm thick platinum layer was deposited on the sample surface. Fig. 8 shows a cross-sectional SE image at the location of the line defect where a dark oblique line contrast was captured on the cross-section of the line defect. There is no physical defect observed at the location of the point defect. It is clear that the point defect which is detected only by EBIC is a non-visual electrical defect, whereas the line defect captured by both SEAM and EBIC is both a morphological and electrical defect.

IV. CONCLUSION

SEAM and EBIC techniques are successfully applied to distinguish morphological and electrical defects in polycrystalline silicon solar cells. These techniques provide complementary information about defects that may affect solar cell performance in different ways. With this method it is possible to further investigate the properties of the defects and their effects on solar cell quality and performance.

ACKNOWLEDGMENTS

Samples for this project were provided by IBM Thomas J. Watson Research Center, Yorktown Heights, NY, U.S.A.,

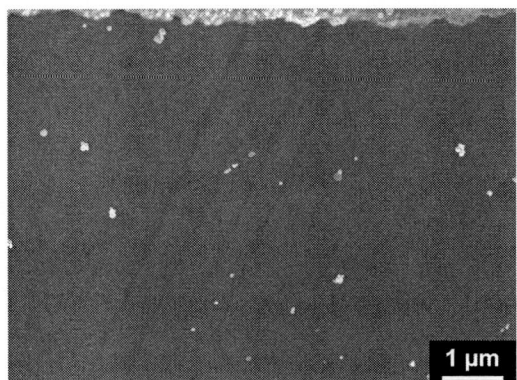

Figure 8. Cross-sectional SE image of the line defect shown in Figure 7a and 7b

under the NUS & IBM Joint Study Agreement # W0853529. Funding and support from Michael Campbell of Qualcomm CDMA Technologies, San Diego, U.S.A., under the NUS & Qualcomm Research Agreement # NAT-12150 is acknowledged. Funding for this research under Singapore National Research Foundation grant # R - 263 000 544 272 is acknowledged.

REFERENCES

[1] Y. Yan, K. M. Jones, C. S. Jiang, X. Z. Wu, R. Noufi, M. M. Al-Jassim, "Understanding the Defects in Polycrystalline Photovoltaic Materials", Physica B: Physics of Condensed Matter, vol. 401-402, pp. 25-32, 2007.

[2] Y. Ohshita, Y. Nishikawa, M. Tachibana, V. K. Tuong, T. Sasaki, N. Kojima, S. Tanaka, M. Yamaguchi, "Effects of Defects and Impurities on Minority Carrier Lifetime in Cast-grown Polycrystalline Silicon", Journal of Crystal Growth, vol. 275, pp. e491-e494, 2005.

[3] L. J. Balk, "Scanning Electron Acoustic Microscopy", Advances in Electronics and Electron Physics, vol. 71, pp. 1-73, 1988.

[4] L. J. Balk, N. Kultscher, "Nonlinear Scanning Electron Acoustic Microscopy", J. de Physique, vol. C2, Supplement No. 2, pp. 869-872, 1984.

[5] H. J. Leamy, "Charge Collection Scanning Electron Microscopy", Journal of Applied Physics, vol. 53, pp. R51-R80, 1982.

[6] O. Breitenstein, J. Bauer, M. Kittler, T. Arguirov, W. Seifert, "EBIC and Luminescence Studies of Defects in Solar Cells", Scanning, vol. 30, pp. 331-338, 2008.

[7] L. Meng, A. G. Street, J. C. H. Phang, "Subsurface Imaging of Multi-level Integrated Circuits using Scanning Electron Acoustic Microscopy", Int. Symp. Testing & Failure Analysis (ISTFA 2009), 15-19 Nov. 09, San Jose, California, U.S.A. pp. 27-32, 2009.

[8] R. L. Pease, J. R. Barnum, "Silicon Solar Cell Damage From Electrical Overstress", IEEE Transactions on Nuclear Science, vol. ns-29, n 6, pp. 1526-1532, 1982.

Photovoltaic (PV) cells characterization using advanced optical tools

Franco Stellari, Steven E. Steen, Kathryn C. Fisher, and Xiaoyan Shao
IBM T.J. Watson Research Center
1101 Kitchawan Rd., Yorktown Heights, NY 10598
phone: +1–914-945-3223, e-mail: stellari@us.ibm.com

Abstract—**In this paper we investigate the use of advanced imaging characterization tools for photovoltaic (PV) cell characterization. These tools were originally developed for CMOS VLSI circuits and offer very high sensitivity and spatial resolution. Optical imaging of PV cells using different tools and techniques will be compared and discussed.**

Keywords: Photovoltaic, Electro-Luminescence, Optical Beam Induced Resistance Change (OBIRCH), Laser Beam Induced Current (LBIC).

I. INTRODUCTION

Photovoltaic (PV) cells are essentially different types of diodes with very large areas that are optimized for absorbing light and converting it into electricity with the highest possible efficiency. Many different types of materials may be used for the cell fabrication, from more traditional ones such as Silicon to rare metals such as Cadmium Telluride, Copper Indium Selenide (CIS), GaAs, and organic materials/polymers [1]. PV cells can also be divided in bulk or thin film, depending on the thickness of the absorbing film.

Although PV cells are mostly known for absorbing light, thanks to the reciprocity principle, they can also be operated or stimulated to efficiently emit light. For example, it is well known that by forward biasing a diode, one can obtain Electro-Luminescence (EL) as in the case of Light Emitting Diodes (LEDs). Since metal contacts used for PV cells are designed to minimize obstruction to the incoming solar light, they also offer minimum blockage to optical characterization techniques. Based on this observation, it is natural that imaging characterization techniques may be used to evaluate the performance of PV cells, measure their reliability and degradation, localize problematic and defective areas, characterize materials and fabrication steps, *etc*. Therefore, optical characterization methods, such as Photo-Luminescence (PL) [2], Electro-Luminescence (EL) [3], Laser Beam Induced Current (LBIC) [4], Thermal Imaging [5], *etc*, are the natural way of characterizing PV cells of any type: mono- and multi-crystalline, amorphous silicon, CIGS, and CIS [6]. Most of these tools are designed for large area collection, such as for entire PV panel imaging, have relatively inexpensive detectors and optics to reduce cost, are very simple to operate and can offer many different levels of information depending on the sophistication of the analysis software and user experience.

Photo-Luminescence (PL) techniques [2] use a short laser pulse to create carriers in the cell substrate while a camera (usually a Charge Coupled Device, CCD, or an InGaAs camera) are used to measure the resulting photoemission, thus allowing measurement of cell uniformity [7], carrier life times [8], detection of local defects [9], *etc*. Most commercial techniques use a continuous (shorter wavelength) source and filter out that wavelength in the path to the detector. Electro-Luminescence (EL) [3] is another imaging technique but it requires electrical contact with the device (and therefore can only be used after metal contacts have been deposited). By applying bias (FB) or reverse bias (RB) to the PV cell one can obtain recombination or hot-carrier emission, respectively. This methodology can be used for detecting defects and shunts, as well as measuring the cell uniformity. Laser Beam Induced Current (LBIC) [4] techniques are also used for measuring cell uniformity and performance by scanning a laser beam, possibly at different wavelengths, across the PV cell and measuring the current generated by the cell and collected by the contacts.

Emission and laser-based techniques have also been developed and have established themselves as a standard way of testing and characterizing CMOS VLSI chips. Emission Microscopy (EMMI) can measure recombination light to detect latchup phenomena [10,11], or hot-carrier luminescence from switching and leaking gates to study many characteristics of the chip, such as voltage drops [12], logic state mapping [13], power supply noise [14], process variations [15], *etc*. Additionally, many laser stimulation techniques have been widely adopted to cause detectable alterations in the circuit under test and localize defects, such as resistive vias, openings in metal lines, weak devices, and so forth. All these advanced tools are designed to reach very high spatial resolution (down to few hundred nanometers), increase sensitivity of weak signals from the chip and have high stability for long acquisition times.

Key challenges with PV applications for these types of tools are the limited field of view available at the lowest possible magnification (typically from 0.8X-1.5X for macro lenses) and the limited space available inside the dark boxes used to block ambient and background light during the circuit testing. Although these are probably insurmountable problems for testing entire panels, we believe that the tools capabilities can be effectively applied to smaller PV cells during new material development, fabrication process refinement, *etc*. In fact, high-end microprocessor chips have also outgrown the maximum field of view of these tools for several years now [15]. Therefore, different methodologies have been developed already in CMOS VLSI chip testing that could be applicable to PV cell characterization. The automation of the tools allows the acquisition of sequences of images, following for example a raster scan pattern, as well as automated stitching of the images using cross correlation algorithms. As a result, high resolution images of large areas, exceeding the field of view of the tool, can be acquired from the device under test [15].

In this paper we will investigate and discuss advanced optical characterization tools for VLSI circuits and their usefulness in characterizing PV devices. Although most of these advanced tools are not engineered to allow a large field of view, as required for PV applications, their high sensitivity, high spatial resolution, and the large array of methodologies developed around them, may have a significant impact on PV devices characterization. These additional capabilities have the potential of furthering the understanding of material preparation, processing steps, defect localization, and failure mechanisms, especially during the development phase of new materials, device designs, or when process (fine) tuning is necessary for performance and yield improvement. Although, the higher cost of these tools has so far limited their penetration into PV applications, we would like to asses their full potential.

II. EXPERIMENT DESCRIPTION

All of the data that will be presented and discussed in the next sections has been acquired from small mono-crystalline PV cell samples that are commonly fabricated to investigate new manufacturing processes such as junction implants, metal finger deposition, structure optimizations, *etc.* The performance and quality of these cells is not representative of optimal fabrication processes but the cells serve as useful test samples for the evaluation of optical characterization tools and the development of new methodologies, which is the scope of this paper. The sample sizes range from 1 cm x 1 cm or 4 cm x 4 cm mono-crystalline PV cells with phosphorous implanted front junction, boron implanted back surface field, SiN anti-reflection coating, Al rear contact and miscellaneous front contacts followed by Cu plating.

The optical characterization of the samples relies on the use of Electro-Luminescence (EL) and different forms of laser stimulation techniques. EL was primarily chosen because our samples are processed through contact metallization, thus making EL measurements very easy to achieve with the emission tools available in our lab. Both forward bias (FB) and reverse bias (RB) EL imaging were used with particular attention paid to studying the metal contact deposition processes.

Laser stimulation is a very common and widely used family of techniques that allows very quick and effective localization of fails and defects for CMOS circuit debugging and failure analysis applications. For PV applications, we have used the Optical Beam Induced Resistance Change (OBIRCH) [16] technique. During the experiment, the device under test is biased at a constant voltage (current) while current (voltage) at the external terminals is measured with a low noise amplifier.

When the laser spot is scanned cross a device, an image is created with the intensity of each pixel corresponding to the value of the measured electric quantity (voltage or current). When the laser hits a region sensitive to carrier generation or temperature alterations, such as a defect, an increase or decrease of the local resistivity is generated and a corresponding bright or dark spot is recorded in the OBIRCH image. Contrary to the commonly used LBIC technique used for PV, laser stimulation in our advanced tools relies on the scanning capability of a Laser Scanning Microscope (LSM) instead of a translation stage to achieve a raster scanned image of the sample. This allows very fast image acquisitions with very high spatial resolution at the price of a smaller field of view. Additionally, only two specific laser wavelengths are commonly used and available for circuit testing applications: 1064 nm and 1340 nm. The former is short enough that electron-hole pairs can be created in the active area of integrated circuits but also long enough that it can penetrate through the relatively thick substrate of a heavily doped wafer that is commonly used for VLSI CMOS circuit fabrication. In our experimentation we have found that, although 1064 nm is out of the absorption spectrum of interest for PV applications, it can create a detectable amount of response from the PV cells that we have investigated. The 1340 nm laser is used to create local temperature hot-spots inside the sample so that temperature sensitive responses can be excited and detected through their effect on external variables such as the cell short circuit current or open circuit voltage. For example, this methodology could be used to localize resistive current paths and shunts.

Two different systems from Hamamatsu Photonics [17] were used during the experiments. The first was a Phemos 200 retrofitted with a Roper back-illuminated CCD camera. This was used for most of the EL measurements due to several advantages: larger dark box, extended translation stage, lowest magnification macro lens available (0.8X), and a top-down microscope that allows very simple positioning of the sample under the collection optics. During the experiments we found that the absence of a shutter in front of the CCD camera on the Phemos 200 was a key limitation of the current configuration of our tool. In fact, the readout of the sensor data for short acquisition of intense light sources may cause artifacts in the image. This is usually not a problem when imaging CMOS devices because of the much fainter nature of the emission signals. A simple solution that can be adopted for imaging PV cells is to reduce the bias of the cell and increase the acquisition time. The second tool used for our experiments was

Figure 1.   Small mono-crystalline solar cell mounted on a clip fixture.

a state of the art TriPhemos system with a Peltier-cooled InGaAs static camera and LSM-based Optical Beam Induced Resistance Change (OBIRCH) [16] laser stimulation. The advanced capabilities of this tool were mostly used to identify defective regions. Since the TriPhemos system is based on a bottom-up microscope, which is commonly used for CMOS circuit measurements requiring tester head docking, mounting small samples is a little bit more laborious. For this reason, PV cells were mounted on a Cu plate with a clip that allows the cell to be held in place and biased at the same time (Fig. 1).

III. ELECTRO-LUMINESCENCE TECHNIQUE

A. Forward Bias Electro-Luminescence

In order to compare the two tools, Forward Biased (FB) and Reverse Bias (RB) Electro-Luminescence (EL) from the PV cells were acquired using both the back-illuminated CCD and the Peltier-cooled InGaAs cameras. FB EL is based on recombination light emitted by the carriers crossing the large area diode junction of the cell like in an LED. For the best cells, the emission would be very uniform across the cell area, and its intensity would be increasing with the increased quality of the cell. However, if a shunt that bypasses the junction, or a pinching of the voltage, or a reduction of the current due to series resistance is resent, a reduction of the corresponding emission would be observed. This could be a global effect, a gradient, or localized to the proximity of the defect. Therefore, localizing regions of depressed FB EL is the key to isolate problematic areas.

Fig. 2 shows a false color image of the EL signal collected from a mono-crystalline Si PV cell using both cameras. Red colors correspond to higher EL signals while yellow, green and blue correspond to progressively weaker emission. Fig. 2(a) was acquired using the CCD and a 0.8X macro lens with the cell in FB at 500 mV. The small cell fits in a single snapshot and the shadowing due to the Cu clip is visible at the left hand side of the bottom metal bus bar. The mono-crystalline cell exhibits a good uniformity of EL signal across the entire area with a progressive decrease away from the contact and interrupted only by the two bus bars and thin Cu fingers. Some artifacts (red) are visible in the image due to defects in the

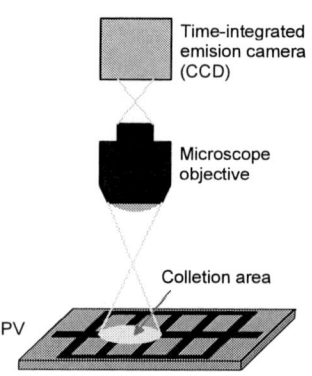

Figure 3.   Schematic diagram of the measurement setup. The collection area is raster scanned in order to cover the entire sample under test.

CCD detector. A few darker regions are located at the edges of the cell, suggesting some problems with the cell dicing process. Fig. 2(b) shows a FB EL image of the same cell at the same electrical conditions acquired using an InGaAs camera. In this case, a macro lens was not available on the TriPhemos system, so the stage had to be stepped with a raster pattern (Fig. 3) in order to reconstruct a composite image of the entire cell as shown in Fig. 4. If one neglects the shading of the Cu clip and some artifacts at the stitching edges of the individual 2.5X acquisitions, good spatial uniformity is also observed with this camera. There is a slight decrease in emitted light from the bottom-left corner (where the clip makes contact) to the top-right corner as previously observed with the CCD camera. This is probably due to a resistive IR drop across the fingers of the cell. The InGaAs camera image shows also a bright region around the cell that is not observed by the CCD. This difference between the two detector images, corresponding to the poorly cleaved edge, is possibly due to the longer wavelength of the emission caused by the recombination centers at the cell edge.

By comparing the two FB images one can conclude that the back-illuminated CCD system appears better suited for measuring the cell uniformity. In fact, the sensitivity to the recombination emission (at least for Si cells) is well matched while the lower detector noise, better detector spatial uniformity, larger field of view of collection optics are clear advantages. Additionally, 4 x 4 images of 30 sec each had to be acquired to measure the entire PV cell with the InGaAs camera, with a significant increase in the total measurement time

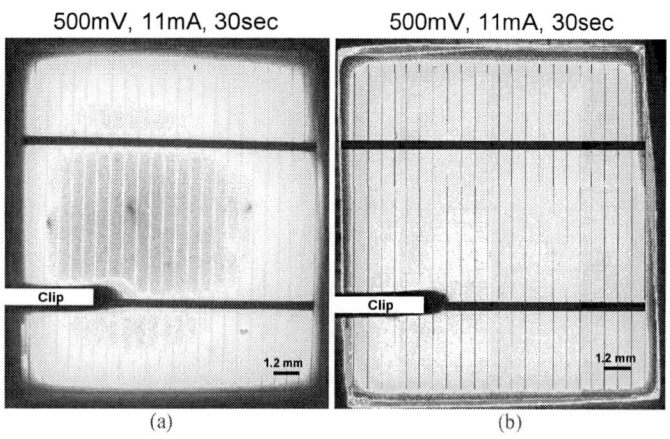

500mV, 11mA, 30sec   500mV, 11mA, 30sec

(a)   (b)

Figure 2.   Forward Bias (FB) Electro Luminescence (EL) images of a mono-crystalline Si cell acquired with two different detectors: (a) back-illuminated silicon CCD and (b) Peltier-cooled InGaAs camera.

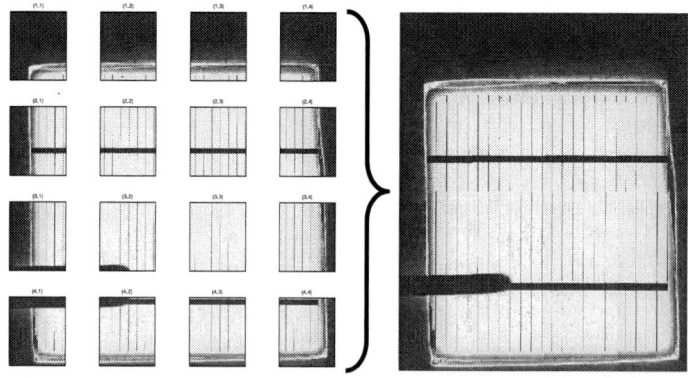

Figure 4.   Multiple acquisitions (e.g. 4 x 4) are combined by means of a cross-correlation algorithm to create a single EL image of the PV cell.

compared to the CCD system. On the contrary, the InGaAs camera may be better suited to localize defective regions such as the cleaved edges.

B. Reverse Bias Electro-Luminescence

EL images may also be taken with the same cameras but with the cell in reverse bias (RB). While the diode junction is in RB, very little current should be flowing in the cell if the voltage is kept below the breakdown voltage. However, if a preferred current path is available or the electric field across the depleted region is locally intensified, localized emission spots may be observed as shown in Fig. 5. The carriers in these regions are strongly accelerated by the high electric field, thus becoming "hot-carriers" and emitting luminescence with a broadband spectrum, usually located in the near-infrared region. This type of emission is a strong function of the energy acquired by the carriers by means of the electric field, contrary to the FB recombination emission case where the recombination emission wavelength is dominated by the bandgap energy. Therefore, in RB, one would expect to observe a linear dependence from the current (number of carriers crossing the RB junction) and exponential with the voltage (defining the electric field in the depleted region of the junction), similarly to what is observed during CMOS circuit testing [18,19]. The strong amplification caused by the field makes it easier to localize the spots but, at the same time, more difficult to quantify the flowing current.

Since the RB EL images show usually only a few emission spots, it may be sometimes difficult to precisely align them with the cell structures. Even when reflected light images are acquired by the setup, slight misalignment is possible. We found that a much better way to achieve precise alignment is to take a FB and RB EL image in succession without moving the stage and to overlay the latter (in false color) with the former (in gray scale). Unless the cell under test has very serious problems that do not permit the acquisition of the FB image, the FB EL has usually enough features (such as metal fingers) to easily localize the RB EL regions of interest on the sample.

Fig. 5 shows the RB EL images of the small PV cell acquired at -20 V reverse bias voltage and about 6 mA of

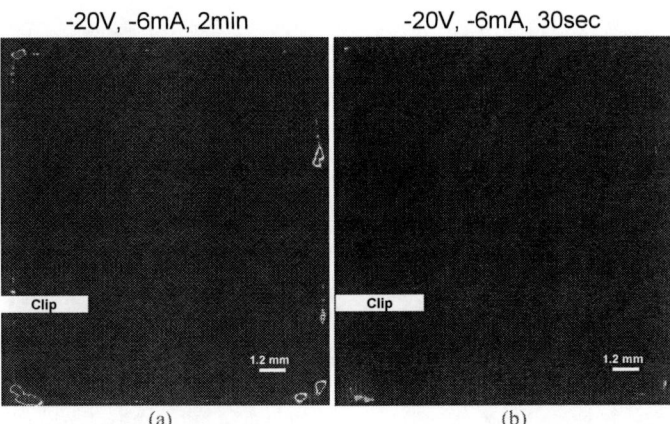

Figure 5. Reverse Bias (RB) Electro Luminescence (EL) images of the same PV cell (as appear in Fig. 2) acquired with two different detectors: (a) back-illuminated silicon CCD and (b) Peltier-cooled InGaAs camera.

reverse current using both types of detector. As previously observed in the FB EL data, the cell does not show significant problems in the active area and all the RB EL hot spots are located at the edge of the die, probably due to our inaccurate manual dicing of the sample. The image acquired with the CCD (Fig. 5(a)) exhibits significant differences compared to the one acquired with InGaAs camera (Fig. 5(b)). In particular, the InGaAs camera picks up fewer regions of interests: mainly the three corners and the right hand side of the two metal bus bars (obviously any possible emission spot under the clip would be masked by its presence). The spectral response of the InGaAs camera, shifted towards longer wavelengths compared to the CCD, makes it better suited to pick up hot-carrier emission. These locations are also in very good agreement with thermal laser stimulation data that will be discussed in the next section. It should also be noted that the InGaAs camera image (Fig. 5(b)) exhibits some artifacts (vertical and horizontal lines) organized as a grid. These are due to the stitching process required to image the entire cell at 2.5X magnification and are not related with defects in the PV cell.

As shown in the previous example, very useful information can be collected by using the RB EL method. In particular, we believe that the localization of the problematic areas can be achieved with much higher spatial accuracy compared to FB EL. At the same time, one must remember that the phenomena causing a dark region in a FB EL image and a bright spot in the RB EL may or may not be driven by the same underlying problem. Therefore, both types of images should be acquired. Additionally, it should be noted that for certain types of cells the RB EL may not be possible due to premature breakdown of the cell (e.g. with CIGS based cells). Additionally, maintaining the cell in a RB condition for a long time, especially when intense emission spots are present, may cause significant local degradation of the cell performance due to both thermal and impact ionization effects, to the limit of destroying the sample.

IV. OBIRCH TECHNIQUE

A. OBIRCH at 1064 nm

The TriPhemos tool also offers two separate laser wavelengths that can be used for the OBIRCH technique, which is described elsewhere [16]. A 1064 nm laser is commonly used for CMOS applications because it can penetrate silicon substrates but at the same time generate enough electron-hole pairs to alter the circuit behavior. Although the wavelength is longer than other LBIC tools used for PV, it is still able to create enough carriers that can be picked up at the cell contacts using a current amplifier. A Laser Scanning Microscope (LSM) was used to raster scan the laser to study the portion of the cell in the field of view.

Fig. 6(a) shows a 2.5X back-reflected LSM image of the bottom right portion of the cell shown in Fig. 2. Here, gray tones indicate different reflectivity in the materials of the cell: a lot of dust particles are visible on the cell surface that was left exposed in the lab for several weeks without any protection. Fig. 6(b), as a comparison, shows the corresponding OBIRCH image acquired by biasing the cell at a constant voltage (i.e. 0 V in this case). Here, the different gray levels of each pixel of the image represent different currents induced by the laser

Reflected light pattern image   OBIRCH 1064nm @ 0V

(a)   (b)

Figure 6. (a) Reflected light image of the bottom right corner of the cell shown in Fig. 2. (b) OBIRCH image of the same portion of the cell acquired using 1064 nm laser.

Reflected light pattern image   OBIRCH 1340nm @ -8V

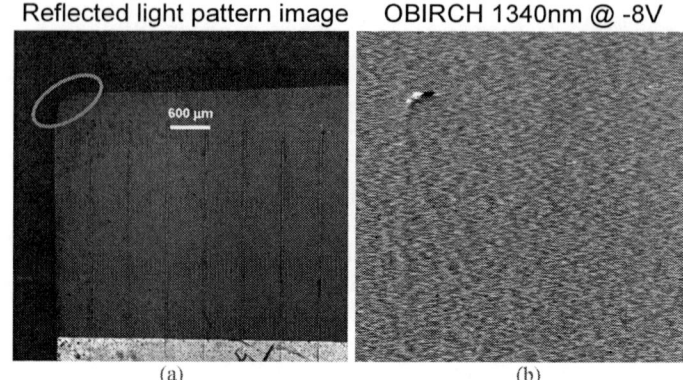

(a)   (b)

Figure 8. (a) Reflected light image of the top left corner of the cell. (b) OBIRCH image of the same portion f the cell acquired using 1340 nm laser.

beam for each corresponding scan position. Since there is no interaction of the laser with the dust particles, the image appears much cleaner.

The good uniformity across this portion of the cell confirms the data collected using the EL tools. However, fine granular patterns of lighter and darker grays may be also noticed between the fingers. Such pattern does not correspond to the dust particles but was later correlated to processing induced damage of the anti-reflective coating. This damage results in local reflectivity variations and can only be seen through local optical stimulation of the cell while reading out electrical response, highlighting the superior resolution of the OBIRCH system.

An opening in one of the metal Cu fingers is also clearly visible (red circle in Fig. 6(b)) thanks to the much cleaner image free of dust particles. With some attention, the metal opening can also be noticed as an interruption of the dark line corresponding to the finger shadowing in the previous EL images shown in Fig. 2. It is however interesting to note that the opening is not big enough to cause a significant depression of the FB EL signal in the region. By using micro probes, we were able to quantify the resistivity of the open to be ~40 Ω. Visual inspection showed that the evaporated portion of the metallization was still present and apparently thin enough to be partially transparent to both EL in FB light from the cell and OBIRCH stimulation.

During our experiments we also studied the effect of the cell bias on the OBIRCH image. We have found that increasing the FB of the cell does not lead to differences in the images, while RB at progressively larger voltages introduces artifacts in the OBIRCH image (Fig. 7).

B. OBIRCH at 1340 nm

The second wavelength available on our tool for OBIRCH measurements is a 1340 nm laser that is commonly used to induce local thermal variations to the device under test. In particular, the resistivity of shunts and current paths may be altered using the laser and the changes in induced current, at a fixed voltage bias, may be plotted as a function of the laser position. In order to see an effect of the thermal laser either a current or voltage bias needs to be applied to our test cell since the photon energy is below the band gap of the device (which would be true for most PV materials for this laser).

Fig. 8 shows an example of thermal laser stimulation acquired at -8 V RB compared to the reflected light image acquired with the same LSM. The top left corner of the cell shows a response to the thermal laser, in very good agreement with the previous RB EL images acquired, especially with the InGaAs camera.

Fig. 9 shows a detailed comparison between the thermal laser stimulation localization and the RB EL images acquired with the InGaAs camera. After localizing several bright emission spots near the right end of the bus bar using 2.5X magnification, additional 20X images were acquired with both

0V   -3V   -5V   -8V

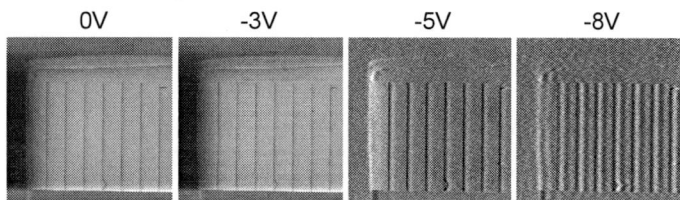

Figure 7. OBIRCH 1064 nm images acquired at different values of reverse voltage bias of the cell diode.

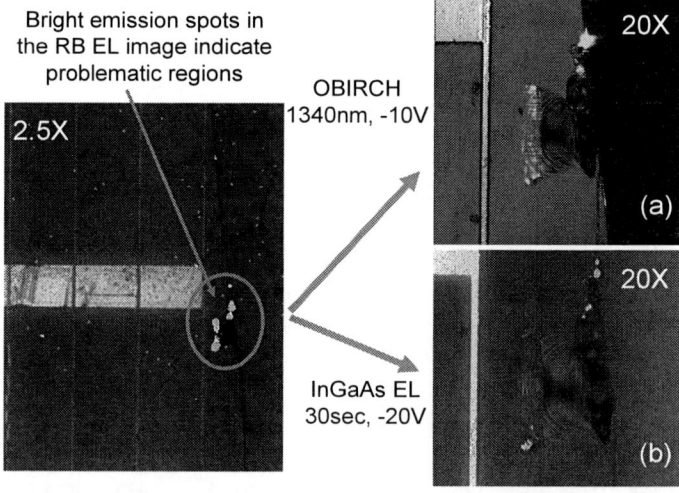

Figure 9. 20X magification images of the region at the right hand side of the bus bar using both (a) RB EL and (b) thermal OBIRCH at 1340 nm.

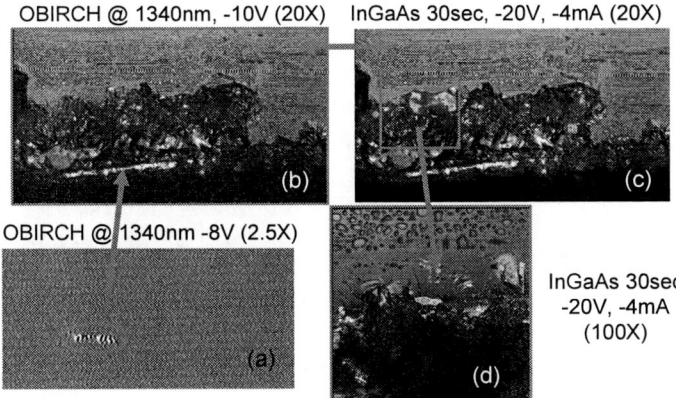

OBIRCH @ 1340nm, -10V (20X)   InGaAs 30sec, -20V, -4mA (20X)

OBIRCH @ 1340nm -8V (2.5X)

InGaAs 30sec
-20V, -4mA
(100X)

Figure 10. Starting from a low magnifcation OBIRCH image (a), high magnification (20X and 100X) images (b) (c) (d) of the defective area at the bottom left of the cell.

techniques. The correlation between the locations identified is very good and coincident with a chipped edge of the die.

Since both EMMI and OBIRCH characterization methodologies are based on advanced microscope systems with high magnification and spatial resolution capabilities, it is possible to obtain very detailed localization of defective areas. Fig. 10 shows an example of the combined used of both techniques. An initial low magnification (2.5X) thermal stimulation image (a) of the cell was used to highlight a problematic area at the bottom left corner of the cell. Subsequently, the magnification was increased to 20X and (b) laser and (c) EL data was acquired and compared. In this case, both the EL and the laser stimulation signals are overlaid to reflected light images of the sample to aid in the alignment with the cell structure. Additionally, (d) higher magnification EL images (100X) can be acquired to further narrow down the exact position of the shuts inside the larger defective area caused in this case by the hand cleaving of the cell.

V. APPLICATION EXAMPLES

A. Characterization of cell isolation

During the development of new PV processes, it is common to manufacture several cells of different characteristics, size, and layout on the same wafer. If the cells are small and

numerous, it is convenient to be able to measure their IV characteristics without cleaving them into individual parts. To measure the in-wafer IV characteristics of multiple cells the test cell can either be masked during the measurement or electrically isolated from the other cells by removing part or all of the emitter between the cells. It is common practice in solar cell manufacturing to use a laser or mechanical scribe to isolate the cell from edge shunts [20]. In the following example we show that the FB EL can be used to determine the effectiveness of such cell isolation techniques.

Two 200 mm wafers were sent through similar cell fabrication sequences to make multiple cells on the one wafer. The cells in Wafer A were isolated by removing the emitter in a 5 μm wide trench around the perimeter of each cell after the emitter formation and prior to nitride passivation (*trench isolation*). The cells in Wafer B were isolated by etching down through the nitride and the emitter at the end of the processing sequence everywhere except inside the cell perimeter (*mesa isolation*).

Fig. 11 shows the IV curves of a cell from Wafer A measured with and without aperture and a cell from Wafer B measured without aperture. The short circuit current density of the cell with trench isolation (from Wafer A) is impossibly high without aperture but is around the same level as the cell with mesa isolation without aperture when an aperture is applied exposing only the active cell. This indicates that the isolation trench is leaky: the current generated in the surrounding cells is being conducted into the cell under test. Fig. 12 shows the FB EL image of (a) the entire trench isolated cell (1 cm x 1 cm) and (b) the top left corner of the mesa isolated cell (2 cm x 2 cm). Additionally, Fig. 13 shows the intensity of the EL signal in a cross section corresponding to the black line in Fig. 12. The cell in Fig. 12(b) has actually been cleaved out of the original die but it still shows that there is no EL signal generated outside the isolation boundary (top left corner). The cell in Fig. 12(a), on the other hand, clearly shows that the cell has not been properly isolated by the trench since EL signal is being generated outside the isolation boundary.

FB EL may also be used to estimate the diffusion length of carriers similarly to what has been developed to test latchup propagation inside CMOS circuits [10,11]. In this case, a PV cell is forward biased and the EL signal is acquired from both

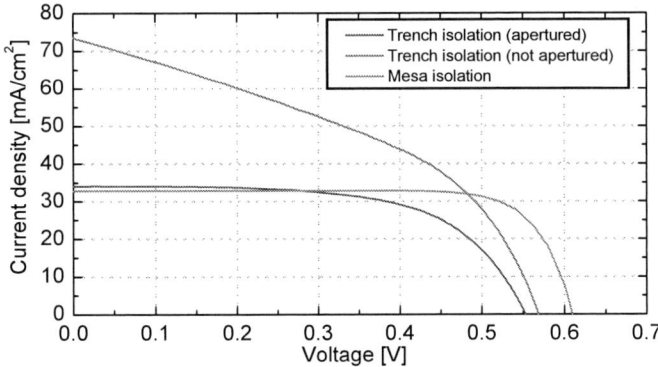

Figure 11. 1 sun IV curves of a trench isolated cell measured (dark blue) with and (light blue) without apeture and a mesa isoalted cell measured (red) without apeture.

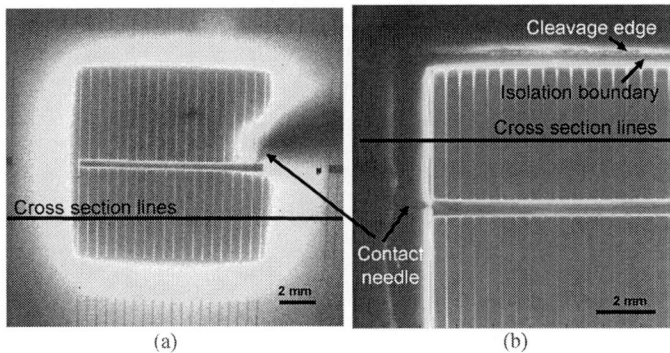

Figure 12. FB EL signal from two mono-crystalline silicon PV cells imaged in wafer: (a) trench isolated 1 cm x 1 cm cell and (b) mesa isolated 2 cm x 2 cm cell (only top left corner is visble) The black cross section lines correspond to the data shown in Fig. 13.

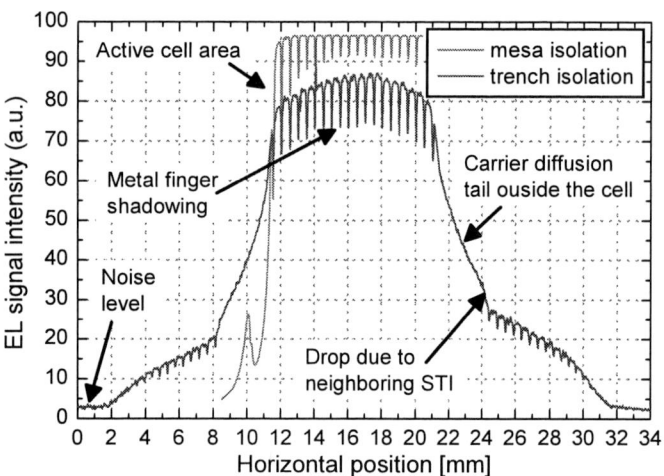

Figure 13. Cross section of the FB-EL data shown in Fig. 12 for both a trench isolated 1 cm x 1 cm cell (blue) and a mesa isolated 2 cm x 2 cm cell (red).

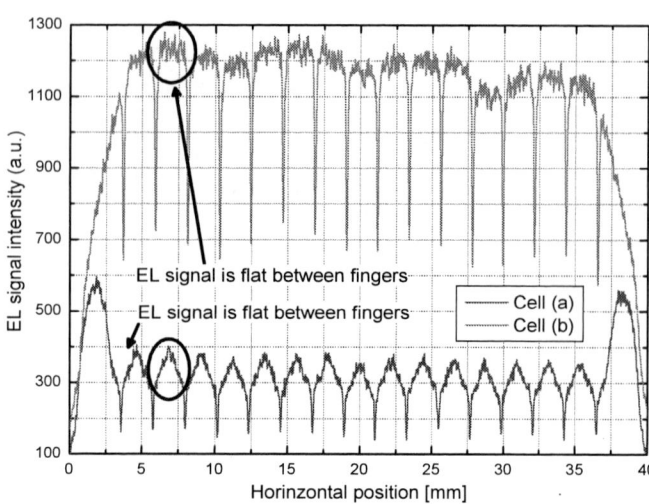

Figure 15. Cross section of the FB-EL from the two cells shown in Fig. 14.

the cell and the surrounding area as shown in Fig. 12. By plotting a cross section of the measured EL signal (Fig. 13), one can estimate the diffusion constant of the carriers as well as quantify the number of carriers that cross the trench enclosing the biased cell. Referring to the cross-section of Fig. 13, an exponential decay of the EL is observed in the region on the left and right hand side of the cell metal fingers. A steep reduction of the signal is also observed where the trench is located (see marking in Fig. 13). However, significant EL signal is still observed from the region corresponding to the neighboring cells: modulation of the emission by the metal fingers can even be easily observed. This clearly shows that there is a conduction path across the isolation trench since only the central cell here is biased.

B. High resolution EL patterns

FB EL images were also used to assess the impact of novel interface materials on device performance. Fig. 14 shows the FB EL image acquired from two single-crystal PV cells (4 cm x 4 cm) with the same underlying structure at the same voltage bias condition of 0.5 V. The only difference between the two cells is the silicide at the interface between Si and the Cu fingers of the top contact. Cell (a) has a much higher EL signal intensity than Cell (b) indicating that there is a

systematic problem with the interface material used for Cell (b). To confirm this, a quantitative cross section of the EL signal from each cell is shown in Fig. 15. At the edges of the die, where there are no contacts, both cells have a similar EL signal intensity. However, the emission from Cell (b) is quenched by the first finger and remains depressed between all the contacts.

A 10X magnification was used to magnify the detail between the metal fingers of both cells, as shown in the cross sections of Fig. 16. The dips in emission at both sides of the cross section correspond to the metal fingers and can be used as a reference. EL from Cell (a) shows a typical contact behavior: the EL signal is highest at the contact and decreases to minimum between the contacts. Cell (b) shows the opposite trend: the signal at the contacts is suppressed and is highest between the contacts. This effect is probably due to the metal used for the interface silicide in Cell (b) having diffused further from the contact than the metal used in Cell (a), thus causing junction leakage and/or increased emitter recombination. Whatever the cause, FB EL imaging clearly shows that the process used on Cell (a) is superior to that used on Cell (b).

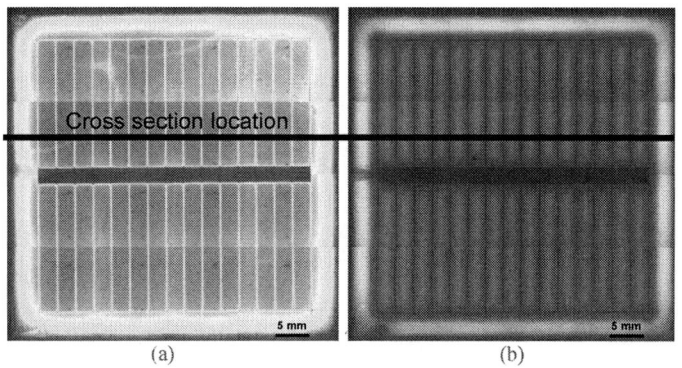

Figure 14. FB-EL images of two mono-crystalline Si cells (4 cm x 4 cm) with identical design but different interface materials under the front metal contacts: sample (a) shows a good response while sample (b) shows much lower emission around the contacts

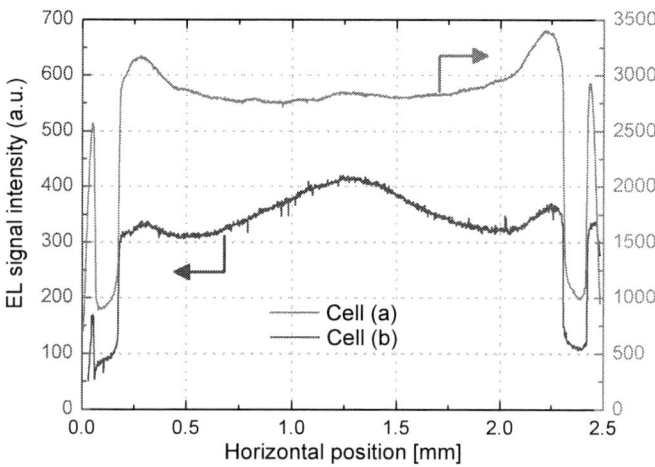

Figure 16. Cross section of the FB-EL signal between two adjacent metal contacts acquired at 10X from the cells shown in Fig. 14.

978-1-4244-5430-3/10 $26.00 © 2010 IEEE   514

RB EL @ -10V,-10mA

Figure 17. The RB-EL signal (false colors) acquired at -10 V (10 mA) is overlaid to the previously acquired FB-EL image of the cell (Fig. 14(b)).

A low magnification (0.8X) RB EL image of Cell (b) shows also two bright regions. Fig. 17 plots the RB EL image of the cell in false colors overlaid to the gray scale FB EL acquisition. It is immediate to notice that one of the bright RB EL spots corresponds to a reduction in FB EL signal in the lower-left corner of the cell. Higher magnification RB EL images (10X and 100X) overlaid to the reflected light image of the cell may be used to further pin point the location of the defective areas as shown in Fig. 18: (a) and (b) are 10X and 100X images of a defective spot near the bus bar; (c) is a 10X image of a defective location at the edge of the cell.

VI. CONCLUSIONS

In this paper we have discussed the use of advanced optical characterization techniques with elevated sensitivity and spatial resolution developed for CMOS circuit debugging and failure analysis for photovoltaic (PV) applications. These tools allow the detection of many degradation mechanisms that can occur in PV cells, thus providing very useful information for their characterization and optimization. In this paper we have presented and discussed the use of Electro-Luminescence (EL), in both Forward Bias (FB) and Reverse Bias (RB) conditions, as well as Optical Beam Induced Resistance Change (OBIRCH), using both 1064 nm and 1340 nm lasers. The combined use of these techniques may allow additional insight into the defect characteristics present in many types of PV

cells. Finally, additional examples and applications of the techniques have been presented and discussed.

ACKNOWLEDGMENT

P. Song, A. Weger, and R. John from IBM Research for their help in setting up the experiment, preparing the samples, and discussing the results; S. Kim, E. Chan, B. Roche, and T. Nakamura from Hamamatsu Photonics for tool installation and support.

REFERENCES

[1] "Solar Cell Efficiency Tables (Version 33), *Prog. Photovolt: Res. Appl.* 2009; 17:85–94.

[2] T. Trupke *et al.*, "Photoluminescence imaging of silicon wafers", *Appl. Phys. Lett.*, 2006.

[3] T. Fuyuki *et al.*, "Photographic surveying of minority carrier diffusion length in polycrystalline silicon solar cells by electroluminescence", *Appl. Phys. Lett.*, 2005.

[4] A. Kaminski *et al.*, "Light beam induced current and infrared thermography studies of multicrystalline silicon solar cells", *J. Phys.: Condens. Matter*, 2004, pp. 9-18.

[5] O. Breitenstein *et al.*,"The imaging of shunts in solar cells by infrared lock-in thermography", *European Photovoltaic Solar Energy Conference*, 2002, pp. 1499-1502.

[6] T. Kirchartza and U. Rau, "Electroluminescence analysis of high efficiency Cu(In,Ga)Se2 solar cells", *J. Appl. Phys.*, 102, 2007.

[7] P. Würfel *et al.*, "Diffusion lengths of silicon solar cells from luminescence images", *J. Applied Phys.*, 2007.

[8] E. Pink *et al.*, "Fast series resistance imaging using photoluminescence", *European Photovoltaic Solar Energy Conference*, 2007.

[9] T. Trupke *et al.*, "Luminescence imaging for fast shunt localization in silicon solar cells and silicon wafers", *Int. Workshop on Science and Technology of Crystalline Silicon Solar Cells*, 2006.

[10] F. Stellari *et al.*, "Study of critical factors determining latchup sensitivity of ICs using Emission Microscopy", *Int. Symp. for Testing and Failure Analysis (ISTFA)*, 2003, pp. 19-24.

[11] F. Stellari *et al.*, "Latchup analysis using Emission Microscopy", *European Symposium on Reliability of Electron Devices, Failure Physics and Analysis (ESREF)*, 2003, pp. 1603-1608.

[12] F. Stellari *et al.*, "Testing and diagnostics of CMOS circuits using Light Emission from Off-State Leakage Current", *IEEE Trans. on Electron Dev.*, vol. 51, no. 9, 2004, pp. 1455-1462.

[13] P. Song *et al.*, "A novel scan chain diagnostics technique based on light emission from leakage current", *Int. Test Conf. (ITC)*, 2004, pp. 140-147.

[14] F. Stellari *et al.*, "On-chip power supply noise measurement using Time Resolved Emission (TRE) waveforms of Light Emission from Off-State Leakage Current (LEOSLC)", *Int. Test Conf. (ITC)*, 2009.

[15] F. Stellari *et al.*, "Mapping systematic and random process variations using Light Emission from Off-State Leakage", *Int. Symp. on Physics and Reliability (IRPS)*, 2009, pp. 640-649.

[16] K. Nikawa and S. Inoue, "New Capabilities of OBIRCH method for fault localization and defect detection," *Asian Test Symposium (ATS)*, 1997, pp. 214-219.

[17] Hamamatsu Photonics, USA: http://sales.hamamatsu.com/en/products/ system-division/semiconductor-industry/failure-analysis.php

[18] F. Stellari *et al.*, "Tools for non-invasive optical characterization of CMOS circuits", *Int. Electron Dev. Meeting (IEDM)*, 1999, pp. 487-490.

[19] A. Tosi *et al.*, "Characterization of backside hot-carrier luminescence in scaled CMOS technologies", *Int. Symp. on Physics and Reliability (IRPS)*, 2006, pp 595-601.

[20] D. Kray *et al.*, "Study on the edge isolation of industrial silicon solar cells with waterjet-guided laser", *Solar Energy Materials and Solar Cells*, vol. 91, no. 17, 2007, pp. 1638-1644.

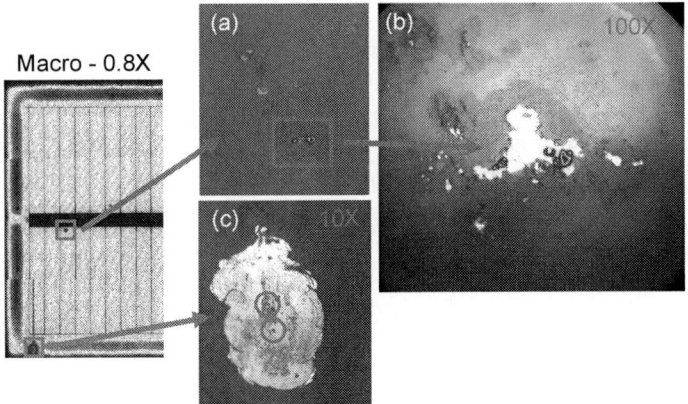

Figure 18. High magnification images of two hot spots identified by the RB-EL image acquired in Fig. 17.

Reliability Assessment of State-of-the-art GaN HEMT by Means of Cellular Monte Carlo Simulation

Fabio Alessio Marino, Diego Guerra, Stephen M. Goodnick, and Marco Saraniti,
Arizona State University, Tempe, 85281, Az,
Fabio.Marino@asu.edu

Abstract— Here we discuss how Monte Carlo Simulations can be exploited to investigate reliability issues in GaN high electron mobility transistor (HEMT) devices. In particular, we report simulation results for high-frequency, high-power state-of-the-art GaN HEMTs focusing our analysis on the effects of that threading edge dislocations on the DC and RF performance of state of the art technologies. A complete characterization of an InGaN back – barrier device has been performed, and the influence of dislocation density on device performance analyzed.. Furthermore, a device structure based on the N-face configuration is analyzed.

Keywords- Dislocations, GaN, high-electron mobility transistor (HEMT), high-frequency, Monte Carlo, numerical simulation.

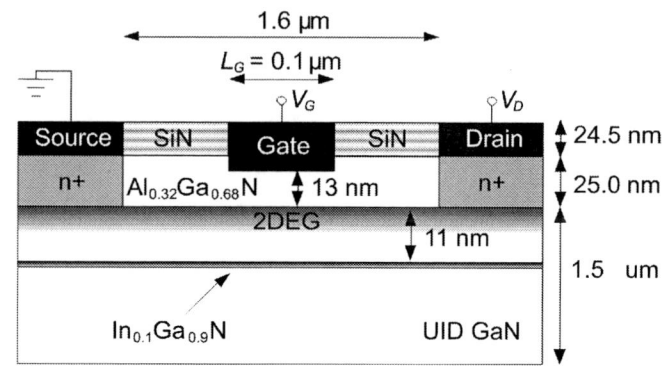

Figure 1 Schematic cross-section of the simulated InGaN back-barrier HEMT.

I. INTRODUCTION

Recently, AlGaN/GaN HEMTs with a current gain cut-off frequency f_T of about 180 GHz and a maximum oscillation frequency f_{max} of 230 GHz have been reported in literature [1,2]. This has been made possible by novel techniques, including the use of thin, high-Al-content barrier layers that compensate the decrease of the sheet channel charge concentration with decreasing of the barrier thickness [1]. An InGaN back-barrier has also been used in order to raise the conduction band in the buffer with respect to the GaN channel, consequently increasing the confinement of the carriers at the heterointerface [2].

Another emerging technology that seems to have several advantages is based on the N-face polarity growtj technique, which have the opposite polarization polarity of Ga-face material. To date, Ga-face devices have largely been investigated, mainly because of the more challenging fabrication of N- face high quality film. However, new high quality N-face nitride processes were recently developed [3,4], making the exploitation of the N-face reverse polarization appealing for future device engineering.

Material quality is still a limiting factor in the high frequency performance, and is therefore a very important issue for modern GaN device technology. Indeed, because of the lack of a suitable lattice-matched substrate, epitaxial layers of GaN can contain a high density of dislocations. So far, the substrate

of choice has been sapphire (Al_2O_3) and silicon carbide (SiC), and both have shown such problems.

Simulation-based analysis is important both for exploring new semiconductor devices and for understanding the static and dynamic behavior of existing devices. Most commercial simulators, such as DESSIS and MEDICI, are well designed for cubic materials like Si and GaAs. However, they are not tailored toward the unique physical models required for hexagonal GaN. In order to explore the reasons for various physical phenomena and to implement more realistic models to describe GaN based device behavior, Monte Carlo simulations [7] are extremely important.

Monte Carlo is now a well established method for studying semiconductor devices and is particularly well suited to highlighting physical mechanism and exploring material properties. Not surprisingly, the more completely the material properties are built into the simulation, up to and including the use of a full zone representation of the band structure, the more accurate is the method. Indeed, it is now becoming increasingly clear that phenomena such as transport properties in GaN HEMTs cannot be understood satisfactorily without using full band Monte Carlo.

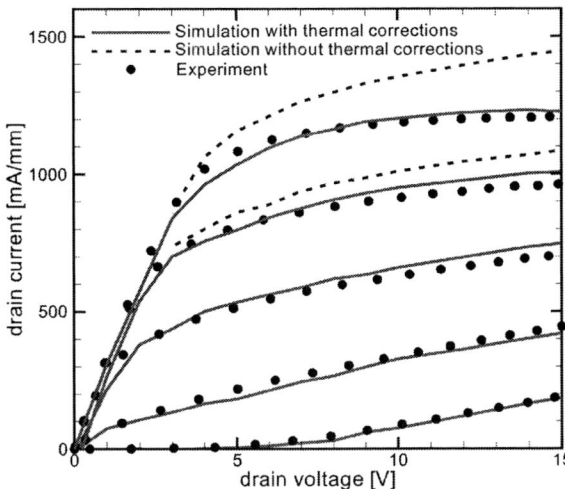

Figure 2 Comparison of the simulated (lines) and measured (dots) I_D - V_D characteristics of the HEMT structure shown in Fig. 1. Dashed lines represents the CMC results obtained with the thermal corrections (see text). The gate voltage V_G is ranged between 1 and -3V with a bias step of $V_G = 1V$.

In this paper, we first demonstrate the capability of Monte Carlo simulations in modeling state-of-the-art GaN HEMT characteristics, showing the agreement between simulation and experiment for two different device structures. In particular, we first focused our attention on the device layout including an InGaN back-barrier proposed by Palacios *et al.* [5]. We also show the results obtained by the analysis of on N-face device, comparing its performance with the one of a standard Ga-face HEMT. Finally, we use the approach of Weimann [6] to model within the Cellular Monte Carlo (CMC) method [8], the effect of dislocation scattering on the high-frequency performance ostate-of-the-art GaN HEMT power devices.

II. FULL BAND PARTICLE-BASED SIMULATION

The Monte Carlo particle-based method [7] provides a space-time statistical solution for the semi-classical charge transport equation. For more than three decades now, intensive research has been done to improve the performance of this method, leading to a simulation tool based on the so-called Cellular Monte Carlo approach [8].

As the electronics industry reduces the size of Si transistors to their fundamental limits, innovative device structures and new materials must be explored to continue the fast progress of information technology and telecommunications. A simulation tool able to model new materials with few adjustable parameters and device structures is therefore required.

The main characteristics of the CMC method are the discretization of the first Brillouin zone (BZ1) of the crystal lattice into an inhomogeneous grid, and the storage of a transition table containing the transition rates for a particle in a given cell in the momentum space. For memory efficiency issues, the symmetry of the crystal is fully exploited so that the

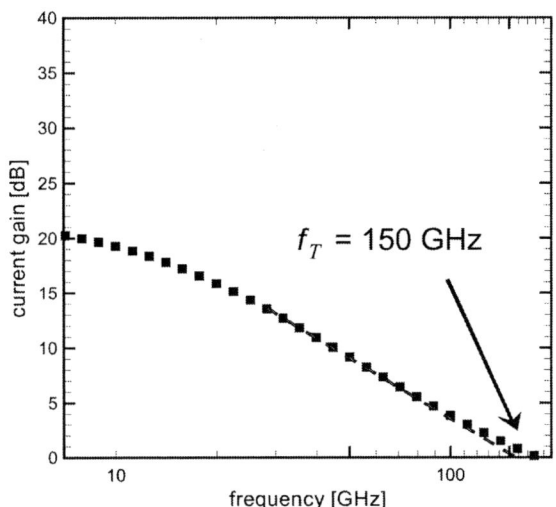

Figure 3 Current gain as a function of the frequency obtained by CMC simulation.

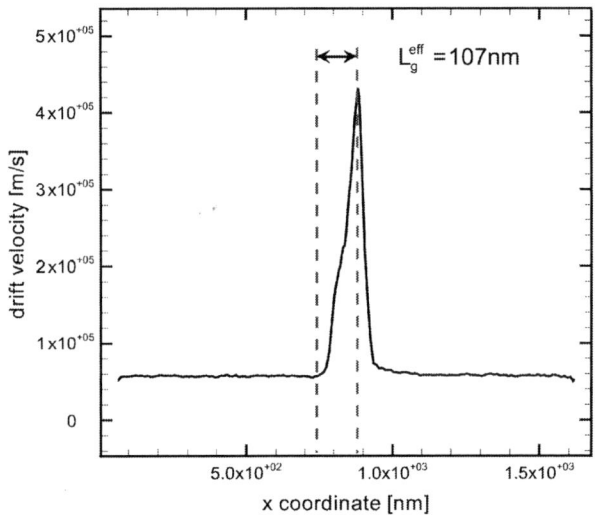

Figure 4 Electron drift velocity along the channel obtained by CMC simulation.

table is computed only for initial states within the irreducible wedge of the BZ1.

Material characteristics such as the electronic band-structure, the phonon spectra and the scattering rates are computed during an initialization phase. The initial carrier distribution inside the computational domain representing the device is also computed in this step, in general by imposing the neutrality condition on mobile carriers, fixed ions and other fixed charges. Then, the real-time carrier distribution is used to compute Poisson's equation and extract the electric field distribution. The field is kept constant for a short ballistic free-flight step and is used to update the carrier positions in the full phase space according to the electric field and scattering mechanisms. The duration of the time-step in which the field is considered constant is crucial in order to prevent numerical artifacts and fully resolve the plasma oscillations of the

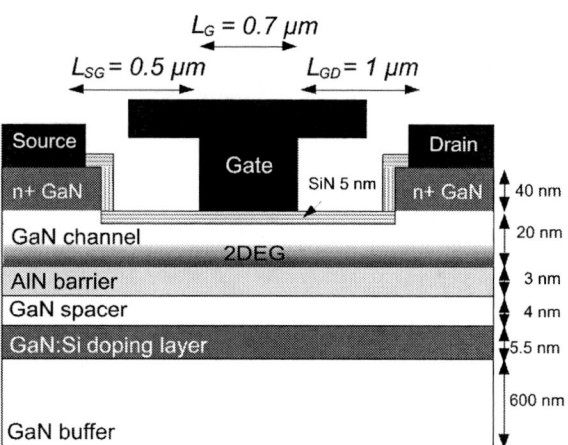

$L_G = 0.7 \ \mu m$

$L_{SG} = 0.5 \ \mu m$ $L_{GD} = 1 \ \mu m$

Figure 6 Schematic cross-section of the simulated N-face HEMT.

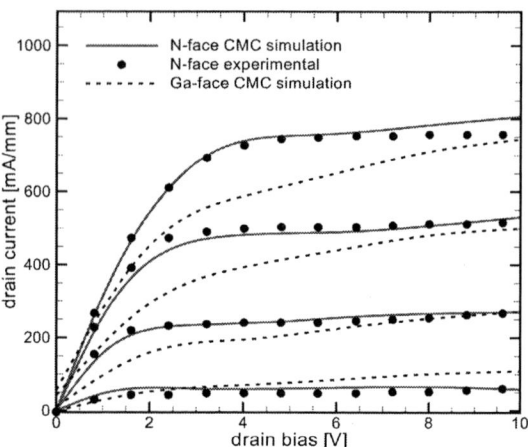

Figure 7 Comparison of simulated (lines) and measured (dots) I_D - V_D characteristics of the N-face HEMT. The gate voltage V_G is ranged between 2 and -1V with a bias step of $V_G = 1V$.

electron gas [7]. At the end of the free flight, a stochastic Monte Carlo procedure is used to select the scattering mechanisms based on their probability, as it was tabulated during the initialization step. The carrier position in momentum space is then updated after each scattering event.

The cycle described above is then repeated for the population of simulated carriers until a preset simulation time is reached. Finally, quantities of interest such as the average energy, the drift velocity, or the mobility are averaged over the ensemble of carriers at the end of each iteration.

The Cellular Monte Carlo (CMC) algorithm [8] was developed to reduce the high computational demand of a conventional full band EMC simulation, associated with the calculation of the final state in the full Brillouin zone at the end of the free flight. To do so, the CMC algorithm pre-calculates all the possible final momentum states for every initial state, and for all possible scattering mechanisms, and stores them in a rather large look-up table. The price for this reduction of the scattering process calculation is the size of the look-up table, which often requires storage larger than three gigabytes of RAM.

In the context of reliability, the CMC tool can be exploited in order to obtain information relevant to the performance of the most advanced GaN based devices. Indeed, threading dislocations effects on charge transport in new materials like InGaN and N-face GaN, and crystallographic damages due to hot electrons can be investigated with this powerful tool.

In this work, we focus on the effects of threading edge dislocations on the DC and RF performance of state-of-the-art GaN HEMT devices. Indeed, these dislocations and other point defects present in AlGaN/GaN HEMTs degrade the device performance and decrease the device long-term reliability [9]. The improvement of material quality of AlGaN/GaN HEMT heterostructure on SiC substrate plays, therefore, an important role for the further improvement of reliability performance in AlGaN/GaN HEMT technology [10].

III. GaN HEMT with InGaN back barrier

The layout of the InGaN back-barrier device described in this work is illustrated in Figure 1, and is composed of an AlGaN/GaN heterostructure with an InGaN back-barrier grown on a semi-insulating SiC substrate by metalorganic chemical vapor deposition [2].

A. Simulation setup

Figure 2 shows a comparison of the simulated (dashed lines) and the measured (dots) I_D-V_D characteristics of the HEMT. As it can be seen, the simulation results are in very good agreement with the experimental data in the low drain bias region. However, the simulated current is increasingly higher than the experimental one with increasing drain and gate bias. This discrepancy is due to self-heating effects under high power operation.

To account for these effects, thermodynamic [11] simulations were carried out with the commercial simulation program Sentaurus [12]. The temperature distribution within the device has been computed at $V_{GS} = 0$ V and 1 V for different drain biases. The results obtained by this analysis were used to set the number of phonons in the scattering tables used for CMC simulations. As it can be seen in Figure 2, the thermally corrected simulation results (solid lines) agree well with the experimental data.

B. RF analysis

The current gain as a function of the frequency of the simulated device is shown in Figure 3. This plot, representing the simulation results, is obtained by the Fourier decomposition (FD) method [13], a straightforward technique, where the perturbation is a small step voltage applied to one electrode of the device in steady state. A cut-off frequency f_T of 150 GHz is found from this analysis, which closely matches the 153 GHz measured experimentally [5].

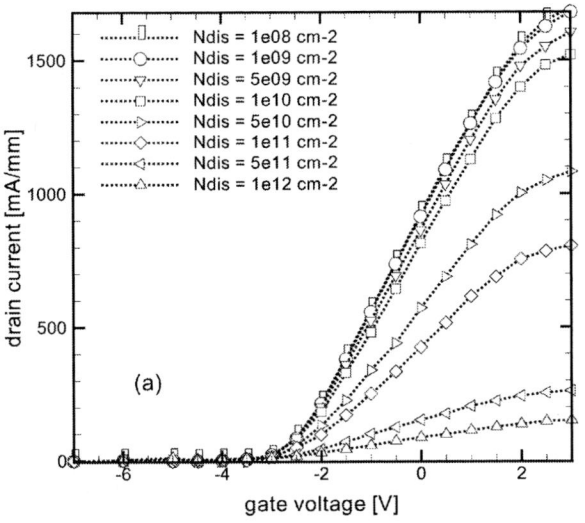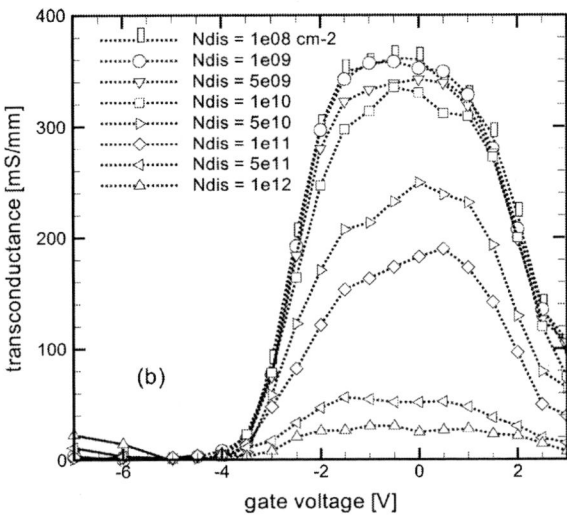

Figure 8 $I_D - V_G$ characteristics (a) and transconductance versu gate voltage curves (b) as obtained by simulations where the threading dislocation concentration ranges from $N_{dis} = 10^8 \, \text{cm}^{-2}$ up to $N_{dis} = 10^{12} \, \text{cm}^{-2}$.

To understand the relation of this frequency response to the electron dynamics in the channel, we have analyzed the cut-off frequency in terms of the average velocity under the gate [14].

$$ f_T = \left[2\pi \int_{L_{eff}} \frac{1}{v_{ave}(x)} dx \right]^{-1} , $$

where L_{eff} is the effective gate length and $v_{ave}(x)$ is the average velocity calculated along the channel. Good agreement with the experimental and simulated f_T is found from this expression using the simulated velocity, shown in Figure 4, and an effective gate length very close to the metallurgical gate length.

IV. N-FACE TECHNOLOGY

N-face HEMTs show several potential advantages over Ga-face devices. In particular, the wider-bandgap AlGaN layer, which is located below the GaN channel layer in N-face devices, provides a natural back-barrier to the channel electrons when the transistor is biased near pinch-off, improving carrier confinement particularly in deep submicrometer devices [15]. At the same time, the smaller bandgap of the top N-face GaN layer allows for easier access to the 2DEG channel than AlGaN, so non-alloyed ohmic contacts with low access resistance can be fabricated [16].

Recently, 150 nm gate length N-face metal-insulator semiconductor (MIS) HEMTs have been reported [17] with a current gain cut-off frequency f_T of about 47 GHz and a maximum oscillation frequency f_{max} of 81 GHz.

A. N-face device structure

The characterization of the N-face device structure proposed by Nidhi *et al.* [17] has been performed in order to evaluate the transport properties of the N-face technology. The layout of the simulated N-face device is illustrated in Figure 6,

and is composed of an GaN/AlN/GaN heterostructure grown by metal-organic chemical vapor deposition on a semiinsulating SiC substrate. A silicon-doped layer was used to supply charge to the channel and prevent modulation of slow responding trap states near the valence-band edge at the bottom AlN/GaN interface.

The material characteristics, e.g. the band-structure and phonon dispersion, are calculated using the geometrical basis formed by the primitive vectors of the reciprocal Bravais lattice. The rotation of the crystal needed to simulate N-face GaN is therefore performed by simply rotating these three primitive vectors.

Figure 7 shows a comparison of the simulated (solid lines) and the measured (dots) I_D-V_D characteristics of the N-face HEMT. As it can be seen, the simulation results are in very good agreement with the experimental data.

B. N-face vs Ga-face

In the same Fig. 7, is reported also the comparison with the I_D-V_D characteristics (dashed line) of a Ga-face device obtained from the N-face one by switching the GaN channel layer and the AlN barrier layer, and by removing the Si-doped layer. Furthermore, in order to obtain a fair comparison between the two different structures, we imposed the condition that the integral of the charge along the perpendicular direction to the channel is the same in both structures for a bias of $V_G = 0 \, \text{V}$ and $V_D = 0 \, \text{V}$.

As it can be seen, the N-face technology can achieve significantly better performance than the Ga-face technology. The drain current provided by the N-face device is, indeed, higher with respect to the one in the Ga-face configuration, especially at low drain voltage values where the increase is up to 20%. This is due to the higher confinement of the carriers in the N-face channel respect with the Ga-face configuration.

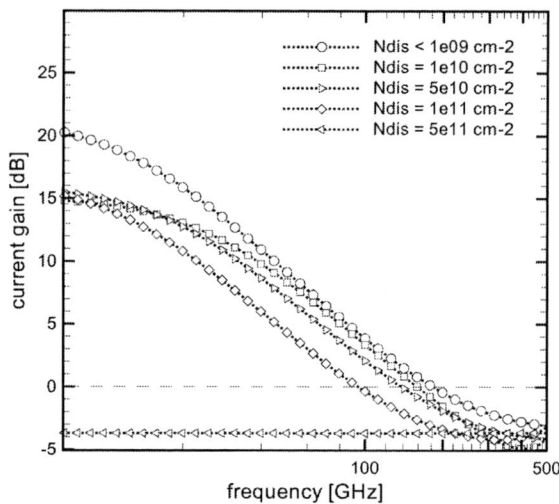

Figure 9 Current gain as a function of the frequency for different densities of threading edge dislocations.

V. EFFECT OF DISLOCATIONS ON DEVICE PERFORMANCE

In order to obtain a quantitative insight about the effects of dislocations on the HEMTs performance, the approach of Weimann *et al.* [6] to calculate the threading dislocation scattering rate, was implemented in our code and used to perform DC and frequency analysis at different dislocation concentrations in the InGaN back-barrier device.

A. DC analysis

Various degradation modes related to the mobility reduction caused by the dislocation scattering have been identified, including the lowering of the drain current and the transconductance peak value. Both these effects can be seen in Fig. 8, where the $I_D - V_G$ characteristics and the transconductance curves are reported as obtained by simulations at values of threading edge dislocations density in the range $N_{dis} = [10^8 - 10^{12}]$ cm^{-2}.

As it can be seen, when N_{dis} ranges between 10^8 and 10^{10} cm^{-2}, the threading defects do not significantly affect the device behavior, which shows a high drain current (> 1500 mA/mm). Also, the transconductance peak value is larger than 350 mS/mm and is almost unchanged by the defect scattering. In particular, if $N_{dis} < 10^9$ cm^{-2} the scattering from dislocations is negligible respect to the other scattering mechanisms.

However, when the dislocation concentration exceeds the "threshold" value of 10^{10} cm^{-2}, the transport properties start to degrade significantly, and both the drain current and the transconductance decrease by a factor of 2 or more. When N_{dis} is increased above 10^{11} cm^{-2} the device is completely compromised and ceases to work as a transistor.

B. Frequency analysis

The main results of the small signal rf analysis are illustrated in Fig. 9, where the current gain versus the frequency is plotted for different values of dislocation density.

As in Fig. 3, the operating point ($V_{GS} = 0$ V, $V_{DS} = 6$ V) is chosen.

The simulated cut-off frequency exhibits fairly large changes with the dislocation density, ranging from 150 GHz, corresponding to a concentration of $N_{dis} = 10^8 - 10^9$ cm^{-2}, down to 90 GHz, for $N_{dis} = 10^{11}$ cm^{-2}. Also in this case, the device performance is compromised above the critical concentration of $N_{dis} = 10^{11}$ cm^{-2}. Beyond this value, in fact, the current gain is almost zero for all frequencies and a cut-off frequency is not identifiable.

V. CONCLUSIONS

In this paper, we reported simulation results for high-frequency, high-power state-of-the-art GaN HEMTs, using a full band Cellular Monte Carlo (CMC) simulator, which includes the full details of the band structure and the phonon spectra, in order to show how this powerful tool can be used in reliability assessment of the new promising technology available nowadays. A complete characterization of an InGaN back-barrier device has been performed using experimental data to calibrate the few adjustable parameters of the simulator. Also, a state-of-the-art N-face structure has been analyzed, in order to evaluate the transport properties of the emerging N-face technology. Threading dislocation effects on the device transport properties were investigated as well. Our simulations indicate that the GaN HEMT frequency performance is directly related to the material quality. We also showed that, if the number of dislocations exceed the critical concentration of N_{dis} = 10^{11} cm^{-2}, the device transport properties are completely compromised.

ACKNOWLEDGEMENTS

This work was partially supported by the Air Force Research Laboratory, contract number FA8650-08-C-1395 (Controller: C. Bozada), and by the Arizona Institute for Nano-Electronics of Arizona State University. The author would like to thanks Prof. Tomas Palacios who generously provided us the experimental data for the InGaN device. F. A. Marino acknowledges Prof. Gaudenzio Meneghesso of the University of Padua, Italy, for his guidance.

REFERENCES

[1] M. Higashiwaki and T. Matsui, Japanese Journal of Applied Physics 44(16), L 475L 478 (2005).

[2] G. Simin, X. Hu, A. Tarakji, J. Zhang, A. Koudymov, S. Saygi, J. Yang, A. Khan, M. S. Shur, and R. Gaska, Japanese Journal of Applied Physics 40, L1142 – L1144 (2001).

[3] S. Rajan, A. Chini, M.H.Wong, J. Speck, and U. Mishra, Journal of Applied Physics 102, 044501 (2007).

[4] S. Keller, C. Suh, Z. Chen, R. Chu, N. Fichtenbaum, M.Furukawa, S. DenBaars, J. Speck, and U. Mishra, Journal of Applied Physics 103, 033708 (2008).

[5] T. Palacios, A. Chakraborty, S. Heikman, S. Keller, S. P. DenBaars, and U. K. Mishra, IEEE Electron Device Letters 27(1), L475 – L478 (2006).

[6] N. Weimann, L. Eastman, D. Doppalapudi, H. Ng, and T. Moustakas, Journal of Applied Physics 83, 3656– 3659 (1998).

[7] C. Jacoboni and P. Lugli, The Monte Carlo Method for Semiconductor Device Simulation, Wien, NewYork: Springer–Verlag, 1989.

[8] M. Saraniti and S. Goodnick, IEEE Transactions on Electron Devices 47(10), 1909–1915 (2000).

978-1-4244-5430-3/10 $26.00 © 2010 IEEE

[9] D.S. Green, S.R. Gibb, B. Hosse, R. Vetury, D.E. Grider and J.A. Smart, Journal of Crystal Growth, 272, 285-292 (2004).

[10] K. K. Chu, P. C. Chao, M. T. Pizzella, R. Actis, D. E. Meharry, K. B. Nichols, R. P. Vaudo, X. Xu, J. S. Flynn, J. Dion, and G. R. Brandes, Electronic Device letters, 25, 596-598 (2004).

[11] G. Wachutka, An extended thermodynamic model for the simultaneous simulation of the thermal and electrical behaviour of semiconductor devices (1989), p. 409414.

[12] Synopsys, Sentaurus Device User's Manual (2008).

[13] F. Schwierz and J. J. Liou, Modern Microwave Transistors, Theory, Design and Performance ,Wiley InterScience, 2003.

[14] R. Akis, J. Ayubi-Moak, N. Faralli, D. K. Ferry, S. M. Goodnick, and M. Saraniti, IEEE Electron Device Letters 29(4), 306–308 (2008).

[15] J. Chung, E. Piner, and T. Palacios, IEEE Electron Device Letters 30(2), 113 – 116 (2009).

[16] M. H. Wong, Y. Pei, T. Palacios, L. Shen, A. Chakraborty, L. S. McCarthy, S. Keller, S. P. DenBaars, J. S. Speck, and U. K. Mishra, Applied Physics Letters 91, 232–233 (2007).

[17] Nidhi, S. Dasgupta, Y. Pei, B. Swenson, D. Brown, S. Keller, J. Speck, and U. Mishra, IEEE Electron Device Letters 30(6), 599 – 601 (2009).

A study of the failure of GaN-based LEDs submitted to reverse-bias stress and ESD events

Matteo Meneghini, Augusto Tazzoli, Enrico Ranzato, Nicola Trivellin, Gaudenzio Meneghesso, and Enrico Zanoni
DEI- University of Padova, V. Gradenigo 6/B, 35135 Padova, Italy,
Phone: +390498277664, Fax: +390498277699, matteo.meneghini@dei.unipd.it

Maura Pavesi, Manfredo Manfredi,
University of Parma, Parma, Italy

Rainer Butendeich, Ulrich Zehnder, Berthold Hahn
OSRAM-OptoSemiconductors
Regensburg, Germany

Abstract—This paper describes an extensive analysis of the degradation of InGaN-based LEDs submitted to reverse-bias stress and Electrostatic Discharge events. Results described within the paper indicate that: *(i)* reverse-bias current flows through localized leakage paths, related to the presence of structural defects; *(ii)* the position of these paths can be identified by means of emission microscopy; *(iii)* reverse-bias stress can induce a degradation of the electrical characteristics of the devices (decrease in breakdown voltage); *(iv)* degradation is due to the injection of highly accelerated carriers through the active region of the LEDs, with the subsequent generation/propagation of point defects; *(v)* the localized leakage paths responsible for reverse-current conduction can constitute weak regions with respect to reverse-bias ESD events.

[*Keywords:* GaN, LED, degradation, reverse-bias]

I. INTRODUCTION

Over the last years, important efforts have been done in order to improve the performance and the reliability of InGaN-based Light-Emitting Diodes (LEDs). As a result, LEDs with efficiency in excess of 150 lm/W are currently available [1]: these devices have an high expected lifetime, and are therefore considered as excellent candidates for the realization of the next-generation light sources. However, recent studies [2-4] have highlighted that blue and green LEDs grown on sapphire can be particularly susceptible to reverse bias stress and Electro-Static Discharge (ESD) events: the active region of these devices can be highly defective - due to the inherently high dislocation densities, and to the difficulty of growing uniform quantum-wells with high indium content - and this can result in a limited robustness to high electric field and/or reverse-current levels. Despite the importance of this topic, only little information is available on the origin of LED failure under reverse-bias operation and ESD events [2-4].

Therefore, the aim of this paper is to describe a detailed investigation of the factors that limit the robustness of InGaN-based LEDs towards reverse-bias operation and ESD events. The results described in the following indicate that: *(i)* reverse current conduction occurs by tunnelling through localized structural defects; *(ii)* reverse-bias tunnelling determines

radiative recombination (reverse-bias luminescence, RBL), due to the injection of carriers in the quantum-well (QW) region; *(iii)* reverse-bias stress can induce the degradation of the electrical characteristics of the LEDs, due to the generation/propagation of point defects in proximity of pre-existing defective regions; *(iv)* reverse-bias degradation is induced by hot carriers, and the degradation rate is proportional to the stress current level; *(v)* ESD failures are determined by the shortening of the junction in proximity of one of the structural defects responsible for leakage current conduction.

II. EXPERIMENTAL DETAILS

The analysis was carried out on green LEDs emitting at 2.33 eV (532 nm), based on multi-quantum well. Device structure consists in a 5 μm GaN:Si buffer layer grown on sapphire, an active region containing green InGaN/GaN MQWs, a p-AlGaN electron barrier layer and a p-GaN contact layer (see [5] for details). After an initial electro-optical characterization, a number of devices with uniform characteristics (and an area of 290x290 μm^2) was submitted to reverse-bias stress tests and to reverse-bias ESD events. Reverse-bias stress tests were carried out by applying a constant current or a constant voltage to the devices. During stress time, devices were repeatedly characterized by means of electrical (current-voltage) and optical (emission microscopy, optical power) measurements, in order to achieve information on the degradation process. ESD tests were carried out by a Transmission Line Pulser - Time Domain Reflectometer (TLP-TDR): this system can simulate ESD events by generating 100 ns pulses with increasing voltage amplitude. The analysis described within this paper were carried out on a statistically relevant number of samples (>50 samples). Analyzed devices showed reproducible behavior: in the following, we report results obtained on representative samples.

III. RESULTS AND DISCUSSION

A. Electro-optical characterization

Before carrying out the stress experiments and the ESD robustness tests, the electro-optical characteristics of the devices were fully characterized by means of Current-Voltage-

Temperature (I-V-T), Emission Microscopy (EMMI), and Electroluminescence (EL) measurements. In Figure 1 we report the results of the (reverse-bias) I-V-T measurements carried out on one of the analyzed samples. Measurement data indicate that reverse current has a power-law dependence on voltage (I~V^n). This kind of dependence is typical for a soft-breakdown behavior [6]. Furthermore, the log-log slope of the (reverse-bias) I-V curves is nearly independent of temperature: this result suggests that that reverse-current conduction is mainly due to tunneling through structural defects (dislocations/V-defects) [6, 7]. In Figure 2 we report the results of emission microscopy measurements carried out before stress on one of the analyzed samples. EMMI measurements indicate that the LEDs can emit light when submitted to reverse-bias. The intensity of Reverse-Bias Luminescence (RBL) was found to be proportional to the reverse current density (Figure 2), and localized on a number of emissive spots randomly distributed on device area (see the inset of Figure 2). These spots correspond to the presence of preferential paths responsible for leakage current conduction, correlated to the presence of structural defects [6, 7].

In order to better understand the origin of reverse-bias luminescence, we have carried out Electroluminescence (EL) measurements under reverse-bias conditions (see Figure 3). EL analysis indicate that RBL has a broad emission peak, centered in the blue spectral region (around 2.58 eV, 480 nm). The emission wavelength (close to the emission of the InGaN-QWs) suggest that RBL is due to the recombination of the carriers injected by tunneling in the QW region. The blue-shift between forward and reverse-bias luminescence can be explained by the compensation of the Quantum Confined Stark Effect, due to the high reverse voltages and/or to fluctuations in the well width in proximity of the defects responsible for leakage conduction [8, 9].

The results described in this section indicate that reverse-current conduction does not take place on the whole device area, but only on localized leakage paths. These paths - related to the presence of structural defects - can represent weak regions with respect to reverse-bias stress or ESD events. In fact, under reverse-bias conditions, they can be crossed by a significant reverse current density, that can generate both gradual and catastrophic degradation. This is described in detail in the following.

B. Reverse-bias stress tests

After the electro-optical characterization described above, devices were submitted to stress under reverse-bias conditions: a set of LEDs was submitted to stress test under constant (negative) current, while another set of samples was submitted to stress test under constant (negative) voltage. In Figure 4 we report the reverse-bias emission vs voltage characteristics measured on one of the analyzed devices before and after stress at -1 mA dc (960 minutes).

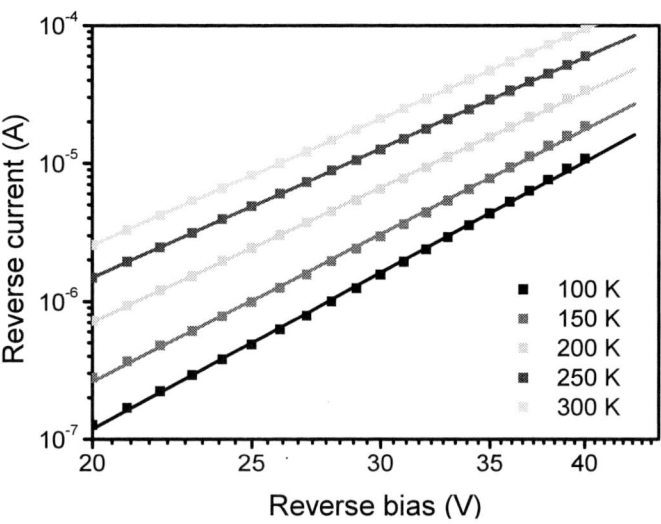

Figure 1: Current-Voltage characteristics measured at different temperature levels on one of the analyzed samples (log-log scale, reverse-bias region)

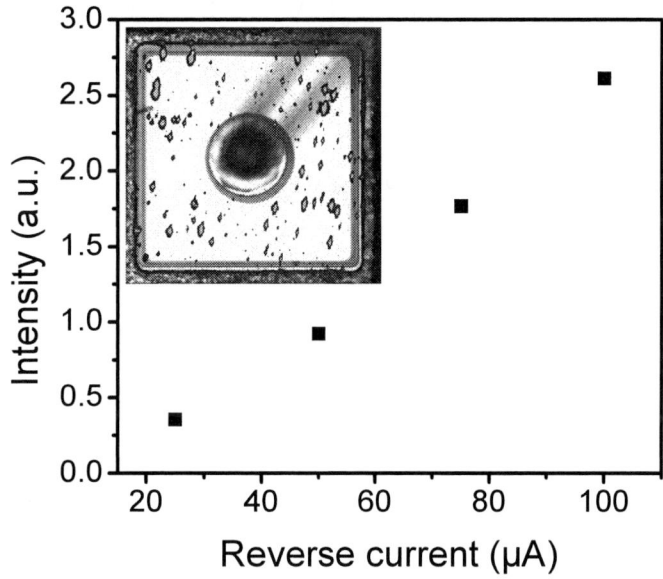

Figure 2: Dependence of the intensity of reverse-bias luminescence on the reverse-current level. Inset: false color emission microscopy image, showing the distribution of reverse-bias luminescence on device area

As can be noticed, stress induced a significant increase in the reverse-bias luminescence signal. Furthermore, an increase in the reverse-current of the LEDs was also detected after stress (see the I-V curves in Figure 5). Since both reverse-bias luminescence and reverse current are related to the presence of structural defects, these results suggest that stress determined the generation/propagation of defects in the active layer of the devices.

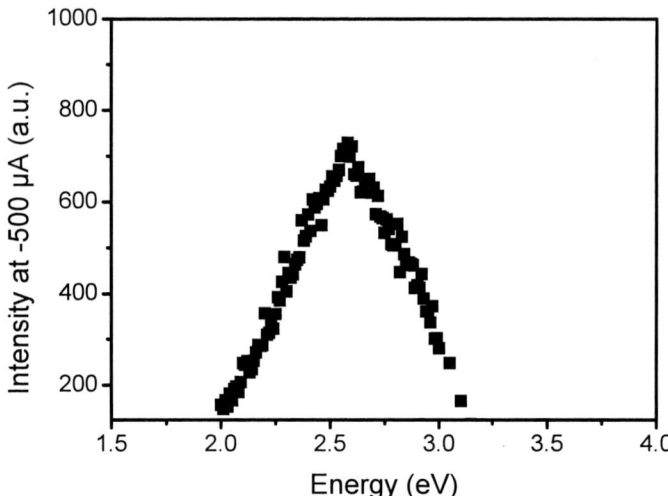

Figure 3: electroluminescence measurement carried out under reverse-bias on one of the analyzed samples. Measurement was taken at -500 μA, 300 K

On the other hand, stress did not significantly modify the optical characteristics of the LEDs in the forward-bias region: in Figure 6 we report the (forward bias) Light Output vs Injected Current (L-I) characteristics of one sample, as measured before and after stress at -1 mA dc (960 minutes). No significant optical power degradation can be measured as a consequence of stress. During stress time, reverse current and RBL were found to increase with similar kinetics (Figure 7 and 8), indicating a correlation. Both parameters increased with a t^k dependence on time (see Figure 8), with k in the range 0.2-0.3 depending on the analyzed sample.

The results described above indicate that reverse-bias stress can induce a significant degradation of the electrical and optical characteristics of GaN-based LEDs. Stress mainly influences the reverse-bias characteristics of the devices, and in particular induces an increase in reverse-leakage current, corresponding to a decrease in the breakdown voltage. In order to better understand the origin of the degradation process, and how the degradation kinetics depend on the different driving forces (electric field and reverse-current), we carried out ageing tests under different stress conditions. In particular, a set of LEDs was submitted to stress at constant (negative) voltage, while another set of devices was stressed under constant (negative) current. Several stress voltage and current levels were used to study the dependence of the degradation kinetics on the electric field and on the reverse-current level. In Figure 9 we report the current vs voltage characteristics measured (before and after stress) on a set of devices aged at different constant voltage levels. As can be noticed, moderate stress voltage levels (V<30 V) did not determine any strong degradation (i.e. any significant increase in reverse-leakage current) of the devices. On the other hand, an increase in the stress voltage beyond 30 V was found to determine a significant increase in the degradation rate. Figure 10 reports the current-voltage characteristics measured (before and after stress) on a set of devices aged at different constant current levels. As can be noticed, an increase in stress current level was found to determine an increase in the degradation rate for the analyzed devices. Figure 11 reports the dependence of the degradation rate on the stress voltage and current level:

degradation rate is defined as the relative reverse-current increase measured (at -25 V) after stress on samples aged under different conditions. As can be noticed, degradation rate has a superlinear dependence on the stress voltage level, and a linear dependence on the stress current level. This last result suggest that current plays a significant role in determining the degradation process: degradation is correlated to the injection of carriers in the active layer of the LEDs. Under reverse bias conditions, current flows only through localized paths, related to the presence of structural defects. The position of these paths can be identified by means of emission microscopy measurements, as shown in Figure 2. During stress, injected carriers can be significantly accelerated by the high electric field, thus achieving enough energy to interact with the lattice and induce the generation/propagation of point defects. This process can result in an increase in reverse-bias current and luminescence.

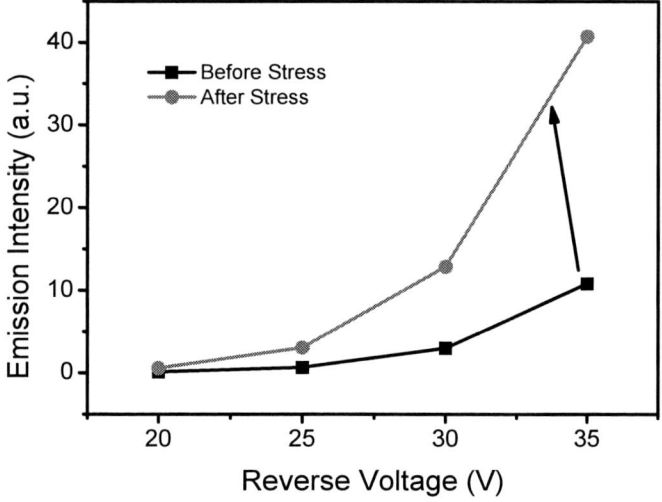

Figure 4: reverse-bias luminescence vs voltage characteristics measured on one of the analyzed samples before and after stress stress at -1 mA dc (960 minutes)

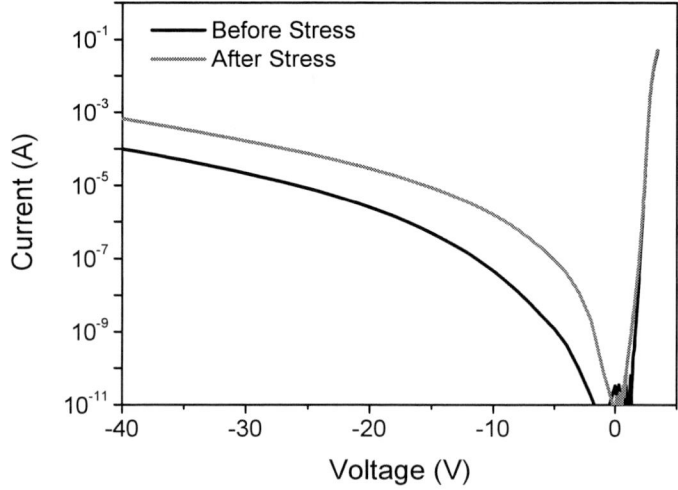

Figure 5: current-voltage characteristics measured on one of the analyzed samples before and after stress at -1 mA dc (960 minutes)

978-1-4244-5430-3/10 $26.00 © 2010 IEEE

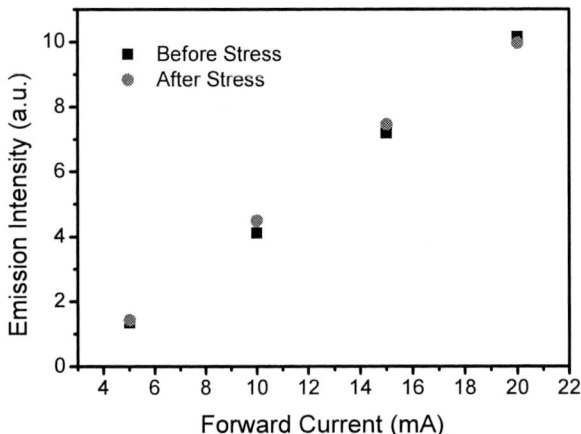

Figure 6: forward-bias optical power vs injected current characteristics measured on one of the analyzed samples, before and after stress at -1 mA dc

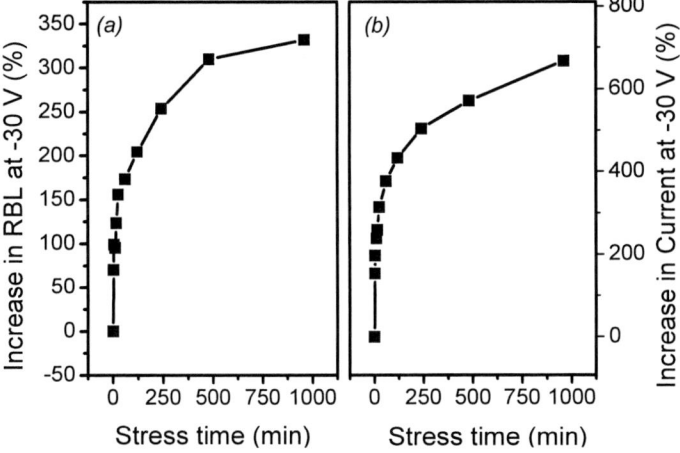

Figure 7: (a) increase in reverse-bias luminescence measured (at -30 V) during stress at -1 mA dc, on one of the analyzed samples. (b) increase in reverse-current measured (at -30 V) during stress at -1 mA dc, on one of the analyzed samples

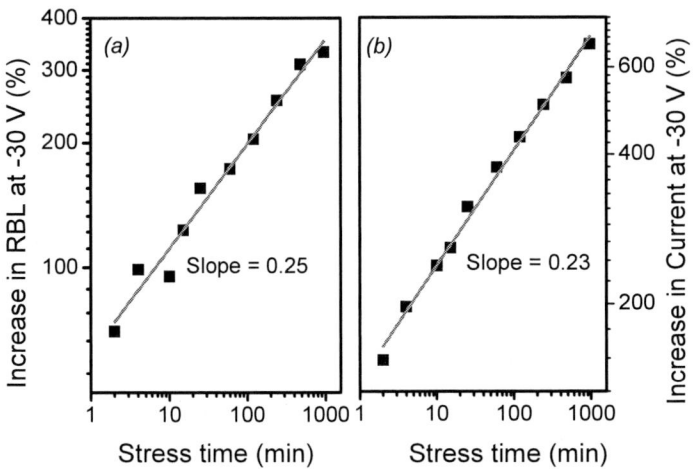

Figure 8: (a) increase in reverse-bias luminescence measured (at -30 V) during stress at -1 mA dc, on one of the analyzed samples (log-log scale). (b) increase in reverse-current measured (at -30 V) during stress at -1 mA dc, on one of the analyzed samples (log-log scale)

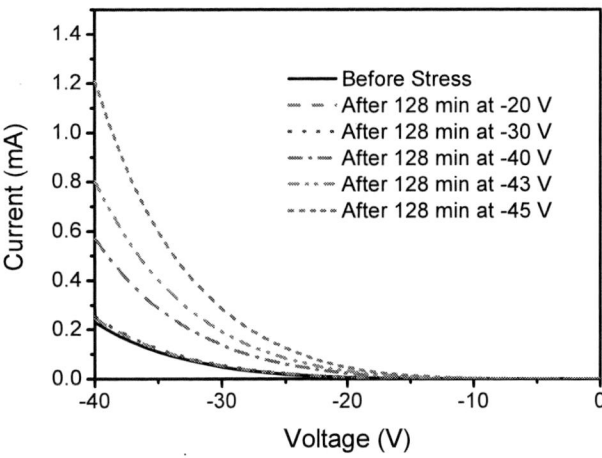

Figure 9: current vs voltage characteristics measured before and after stress at different voltage levels

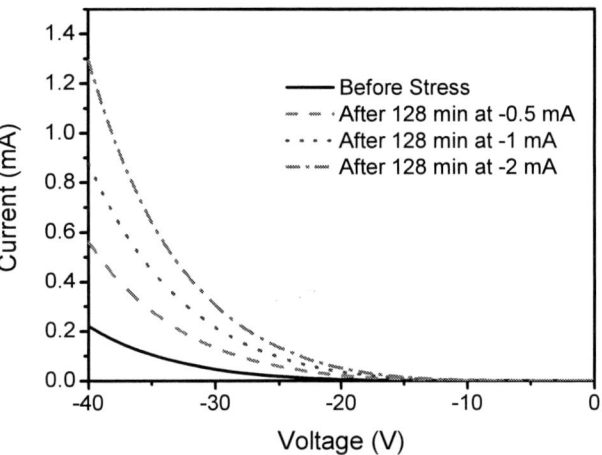

Figure 10: current vs voltage characteristics measured before and after stress at different current levels

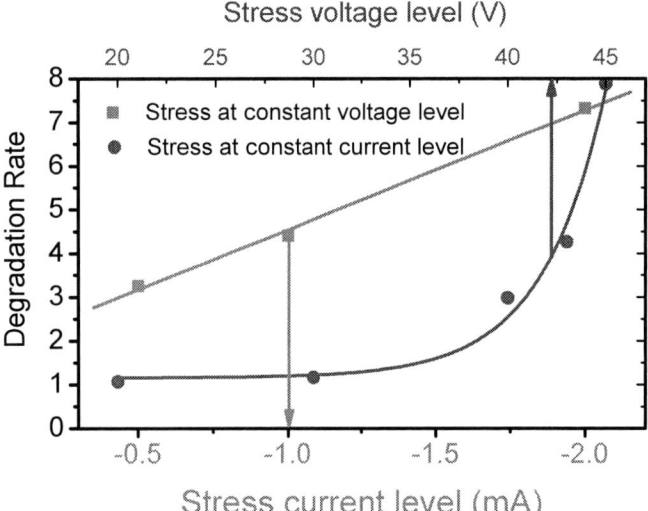

Figure 11: dependence of the degradation rate on the stress current level (red, x-axis on bottom) and on the stress voltage level (blue, x-axis on top). Solid curves are guides to the eye

C. ESD tests

As described above, the leakage paths responsible for reverse-current conduction can represent weak regions with respect to reverse-bias stress. Therefore, the presence of these paths can – in principle – influence the robustness of LEDs towards ESD events. In order to evaluate how the devices behave when submitted to ESD events, we have carried out an ESD testing campaign on a large number of LEDs. The ESD robustness tests were carried out by the TLP method, by applying pulses (100 ns duration) with increasing voltage amplitude to the LEDs. For each voltage pulse we have measured *(i)* the corresponding TLP current (Figure 12 (a)) and *(ii)* the leakage current after the pulse (Figure 12 (b)). After ESD failure, LEDs behave as short circuits: ESD damage interests a localized region, as demonstrated by Scanning Electron Microscopy (SEM) investigation (see a sample image in Figure 13). In most of the cases, damaged region is located in correspondence of one of the leaky paths identified (before stress) by emission microscopy. This result suggests that the leakage paths responsible for reverse current conduction constitute weak points with respect to ESD events, since they allow an extremely high current to flow through a small-size path. The presence of structural defects (V-defects, dislocations, …) can therefore strongly limit the ESD stability of the LEDs. Results indicate that even TLP pulses with (reverse) current smaller that the failure threshold can modify the electrical characteristics of the LEDs: in particular, reverse current levels greater than 10 mA can induce a decrease in the leakage current of the LEDs (Figure 12 (b)), corresponding to a certain improvement of the electrical characteristics of the samples. This effect can be possibly due to the annihilation of one or more leakage paths, due to the extremely high current densities injected through small-size defects.

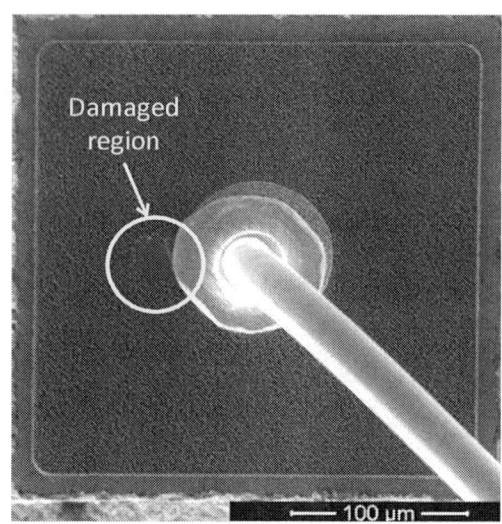

Figure 13: SEM image of one LED after ESD failure. The position of the damaged region is highlighted in the figure

IV. CONCLUSIONS

We have described an analysis of the origin of the failure of LEDs submitted to reverse-bias stress and ESD robustness tests. We have demonstrated the following relevant results: *(i)* under reverse-bias, tunneling is the dominant conduction mechanism, and current flows through localized defect-related paths; *(ii)* reverse-bias stress can induce an increase in the reverse current and RBL of the LEDs, due to the generation/propagation of defects in proximity of pre-existing leakage paths; *(iii)* reverse-bias degradation is induced by highly accelerated carriers, injected by tunneling in the active region of the LEDs; *(iv)* the presence of structural defects can strongly limit the ESD robustness of the LEDs. The measurements carried out within this work indicate that a significant improvement of the ESD and reverse-bias robustness of GaN-based LEDs can be obtained only by an accurate control of the defectiveness of the active layer.

ACKNOWLEDGMENT

This work was partially supported by the CARIPARO Foundation, and by the ALADIN project (Italian Ministry of Industry)

REFERENCES

[1] R. D. Dupuis and M. R. Krames, "History, Development, and Applications of High-Brightness Visible Light-Emitting Diodes", Journal of Lightwave Technology, 26, 9, 1154, 2008X. Cao et al., Microel. Reliab., vol. 43, pp. 1987-1991, 2003

[2] H. Kuan, "High Electrostatic Discharge Protection Using Multiple Si:N/GaN and Si:N/Si:GaN Layers in GaN-Based Light Emitting Diodes", Jpn. J. Appl. Phys. 47, 1544, 2008H. Kuan, Jpn. J. Appl. Phys. 47, 1544, 2008

[3] X.A. Cao, P.M. Sandvik, S.F. LeBoeuf, and S.D. Arthur, "Defect generation in InGaN/GaN light-emitting diodes under forward and reverse electrical stresses", Microelectronics Reliability 43, 1987-1991, 2003

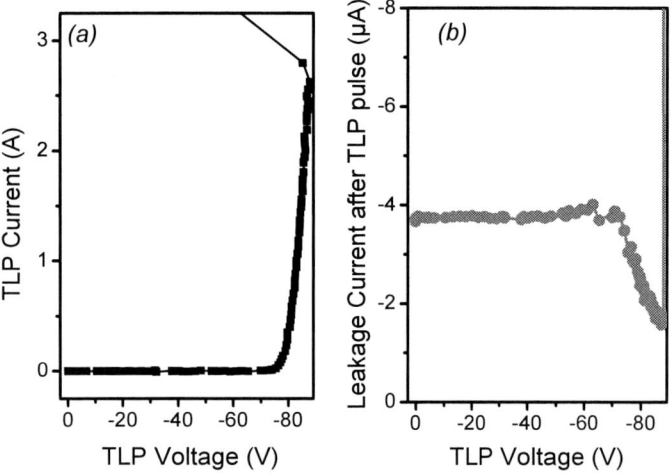

Figure 12: *(a)* I-V curves of one of the analyzed LEDs measured by the TLP setup. The maximum TLP current reached by the LEDs before failure was used as a parameter to define the ESD robustness of the LEDs (for this device it is equal to -2.6 A). *(b)* leakage current measured after each TLP pulse: TLP current levels greater than 10 mA can induce a decrease in the leakage current of the devices, possibly due to the annihilation of defect-related leakage paths

[4] S.-K. Jeon, J.-G. Lee, E.-H. Park, J. Jang, J.-G. Lim, S.-K. Kim, and J.-S. Park, "The effect of the internal capacitance of InGaN-light emitting diode on the electrostatic discharge properties", Appl. Phys. Lett. 94, 131106, 2009

[5] M. Peter, A. Laubsch, P. Stauss, A. Walter, J. Baur, and B. Hahn, "Green ThinGaN power-LED demonstrates 100 lm", Phys. Stat. Sol. (c) Vol. 5, No. 6, pp. 2050–2052, 2008

[6] Z.-Q. Fang, D.C. Reynolds, and D.C. Look, "Changes in Electrical Characteristics Associated with Degradation of InGaN Blue Light-Emitting Diodes", Journal of Electronic Materials, 29, 4, 448, 2000

[7] X.A. Cao, S.F. LeBoeuf, K.H. Kim, P.M. Sandvik, E.B. Stokes, A. Ebong, D. Walker, J. Kretchmer, J.Y. Lin, H.X. Jiang, "Investigation of radiative tunneling in GaN/InGaN single quantum well light-emitting diodes", Solid-State Electronics 46, 2291, 2002Y.

[8] Y. D. Jho, J. S. Yahng, E. Oh, and D. S. Kim, "Field-dependent carrier decay dynamics in strained $In_xGa_{1-x}N/GaN$ quantum wells", Phys. Rev. B, 66, 035334, 2002

[9] A. Hangleiter, F. Hitzel, C. Netzel, D. Fuhrmann, U. Rossow, G. Ade, and P. Hinze, "Suppression of Nonradiative Recombination by V-Shaped Pits in GaInN=GaN QuantumWells Produces a Large Increase in the Light Emission Efficiency", Phys. Rev. Lett. 95, 12, 127402, 2005

Temperature Assessment of AlGaN/GaN HEMTs: A Comparative study by Raman, Electrical and IR Thermography

N. Killat and M. Kuball
University of Bristol, H.H. Wills Physics Laboratory
Bristol BS8 1TL, United Kingdom
Phone: +44 (0)117 928 8750; fax: +44 (0)117 925 5624; e-mail: Nicole.Killat@bristol.ac.uk

T.-M. Chou, U. Chowdhury, and J. Jimenez
TriQuint Semiconductor Inc.
Richardson TX 75080, USA

Abstract—**The accuracy of different thermography techniques for the determination of AlGaN/GaN HEMT channel temperature was investigated. Micro-Raman thermography, a novel electrical testing method, and IR thermography were applied to measure the temperature in the active region of AlGaN/GaN HEMTs with different device geometries. Due to its accepted accuracy, micro-Raman thermography was performed on different devices in order to validate thermal simulation results. When compared to the validated thermal model, pulsed I-V measurements underestimated channel temperature to some degree, while IR thermography determined unrealistically low device temperatures.**

Keywords— **HEMT, device reliability, thermal management.**

I. Introduction

AlGaN/GaN high electron mobility transistors (HEMTs) are an important technology of high interest for various defense and commercial sector applications. During operation, HEMTs experience highly localized power dissipation within a submicron region adjacent to the gate contact, resulting in a highly localized temperature rise [1], [2]. The channel temperature (which is essentially a function of the device geometry, package, thermal conductivities of constituent materials, base-plate temperature, and dissipated power) causes thermally activated device degradation. Therefore, accurate knowledge of the channel temperature in elevated temperature life-tests is necessary for accurate extrapolation of the operational lifetime for these devices. Accurate channel temperature measurement also facilitates improvements in device and package design for better thermal management.

Raman thermography [3]-[5] is a non-invasive optical technique, based on Raman scattering spectroscopy, enabling temperature measurements with sub-micron spatial and nanosecond time resolution. However, it is currently rarely available in device research laboratories. Electrical temperature measurements [6] and IR thermography [4], [7] are well established and readily available in most semiconductor industry laboratories. Nevertheless, these methods have spatial resolution constraints, limiting their suitability for measurement of devices with micron/submicron source-drain spacing. In this work, we quantify the relative accuracy of the different device temperature assessment methods including thermal simulations and also comment on other aspects of the measurement techniques.

II. Experimental Setup

Measurements were performed on small dies of AlGaN/GaN/SiC HEMTs, containing FETs of the following gate peripheries: 4×100, 6×100, 12×100, 2×280 and 14×280 µm, bonded to a Cu-Mo carrier plate using Au-Sn brazing. The devices were operated at a constant drain-source voltage (V_{ds}) of 35 V and the gate-source voltage (V_{gs}) was varied to measure the power density dependence. Micro-Raman thermography was applied to determine the maximum device temperature occurring close to the drain side edge of the gate contact. Raman thermography uses scattered laser light to probe the energy of phonons in the device materials, which are dependent on temperature, enabling a 0.5-0.7 micron lateral spatial resolution to be achieved. Further details on micro-Raman thermography are described in [3], [5]. In addition, a novel electrical method, based on synchronized pulsed I-V measurements, was applied. For this method, the device is first calibrated by measuring its pulsed I_{DS}-V_{DS} characteristics at zero quiescent dissipated power and different base-plate temperatures. Then, pulsed I_{DS}-V_{DS} curves are obtained at different quiescent dissipated power and with the base-plate at room temperature. By comparing the two sets of curves, channel temperature at a given dissipated power can be estimated. More details on the electrical method can be found in [8]. This method shows several merits compared to other electrical temperature assessment methods. Particularly, this technique is found to be superior to Schottky barrier-based techniques that are standard in GaAs technology. These Schottky barrier-based techniques are often found unsuitable for measurement of GaN devices due to non-ideality of the junction and change of Schottky barrier properties during measurement. The pulsed I-V measurement setup probes electrical characteristics of the entire device, hence it measures an average temperature over the entire gate width, similar to the results reported in [3]. Standard IR thermography was also performed on the devices and is further described in [4]. The spatial resolution of the IR system used was 5-10 µm.

III. Results and Discussion

Maximum device temperature rise as function of dissipated power density for different device layouts assessed using Raman

---

This work was supported in part by the U.S. Defense Advanced Research Project Agency and Army research Laboratory, ARL contract no. W911QX-05-C-0087 under program manager Mark Rosker.

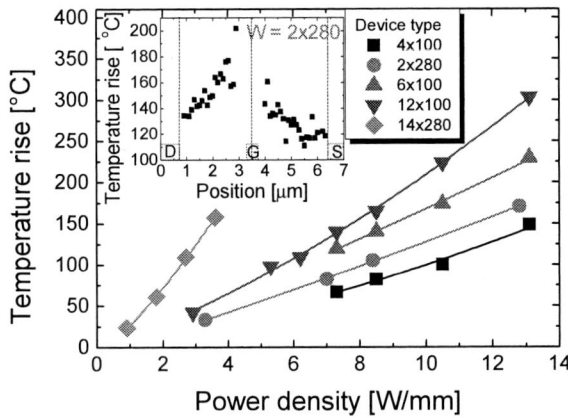

Figure 1. Peak temperature rise as function of power density measured by Raman thermography for different device geometries (base-plate temperature of 25 °C) at the drain edge of the gate contact. The inset shows the GaN temperature distribution over the source-drain gap of a 2×280 device with 6 μm source-drain seperation at a power density of 13.5 W/mm.

thermography is shown in Fig. 1. Larger devices show a significant increase in thermal resistance (R_{TH} = temperature rise/power density) due to thermal cross-talk between the gate fingers, in agreement with earlier results and theoretical expectation [9]. The inset in Fig. 1 shows the lateral temperature profile of the active region of one of the fingers of a 2×280 device. The profile indicates that the area of the highest channel temperature is located at the drain edge of the gate contact, where the highest power dissipation occurs. This is the point of measurement with respect to the gate edge of the central finger, where channel temperature was determined in the multi-finger devices shown in Fig. 1.

To assess the measurement repeatability of Raman thermography in an industrial setting, identical devices were measured over a period of two months. Fig. 2 shows the temperature rise of a 2×280 μm AlGaN/GaN HEMT as an example. A standard deviation of better than ±5 °C was achieved, confirming good repeatability throughout the multiple measurements performed at different times, on different dies and chips. These small variations in temperatures measured are attributed to small inaccuracies in the alignment/position of the

0.5 μm size Raman thermography measurement spot with respect to the gate edge (alignment accuracy of ~0.2 μm for point measurements) and small variations in thermal mounting of the die on the base-plate. The Raman thermography measurement accuracy was better than ±5 °C, after taking into account the stress contributions to the phonon shifts [10].

Results from thermal simulation were compared to Raman thermography in order to validate its accuracy. These results are displayed in Fig. 3. Both the device and package designs were included in the model. Temperature was averaged in the thermal simulation over a $0.75 \times 0.75 \times 2$ μm^3 area close to the drain-gate edge, to compare to the Raman measurement area. There is good agreement between Raman thermography and thermal simulation for different device geometries. A comparison of the thermal model with all the temperature assessment methods used is further displayed in Fig. 4. The thermal simulation performed for the devices investigated and with realistic material parameters shows an accuracy better than 2% with regard to Raman thermography. This is within the repeatability and measurement accuracy of the Raman thermography method (Fig. 2). Since the agreement with simulation data is very good with a wide range of device geometries, the simulation models and parameters used can be considered accurate and trustworthy for scaling to other device geometries. The agreement also establishes Raman thermography as a suitable tool for measurement of temperature at localized points of such devices.

Data from IR thermography measurements is plotted in Fig. 4 for comparison. This method significantly underestimates the device temperature, as IR thermography does not provide sufficiently high spatial resolution. The significant underestimation of temperature in high-performance devices using IR thermal imaging has also been reported by Sarua et al. [4] and Claflin et al. [11]. Typically IR spatial resolution is 5-10 microns – which is at best diffraction limited by the wavelength of the IR radiation used for the measurement. Since common electronic devices have a source-drain spacing in the order of 3 to 6 microns, this lack of resolution often causes IR to be inadequate for determination of channel temperature and hence failure prediction. Considering life-testing applications, inaccurate temperature determination using IR thermography gives unrealistic activation energies [12], affecting the accuracy of lifetime predictions. Moreover, comparison between different devices is challenging and unreliable due to the lack of spatial resolution. However, the IR temperature measurement technique is an extremely useful method for fast and comparative screening of devices, but the peak temperature determined by this technique needs to be considered with caution.

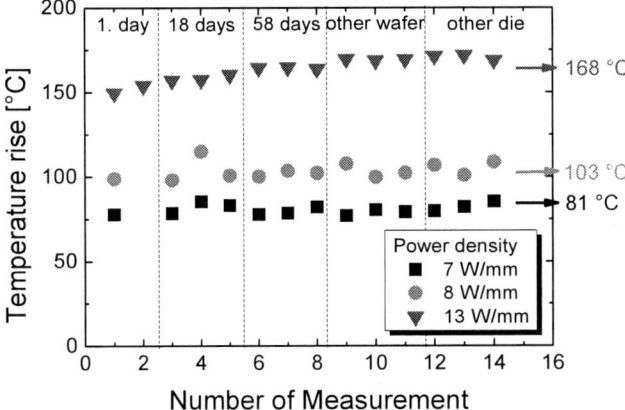

Figure 2. Repeatability of temperature measurements using Raman thermography, on a 2×280 μm AlGaN/GaN device over a time period of two months and on different dies and wafers.

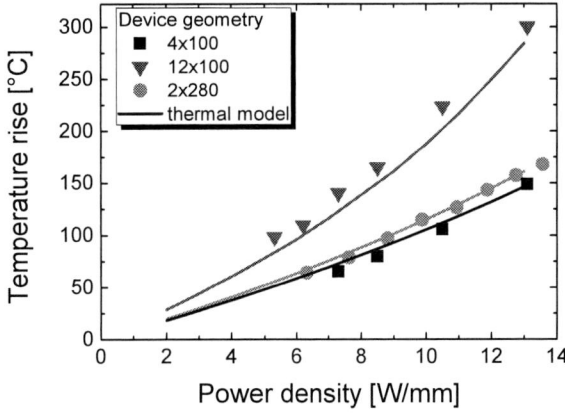

Figure 3. Comparison of thermal simulation with micro-Raman thermography results for different device geometries.

978-1-4244-5430-3/10 $26.00 © 2010 IEEE

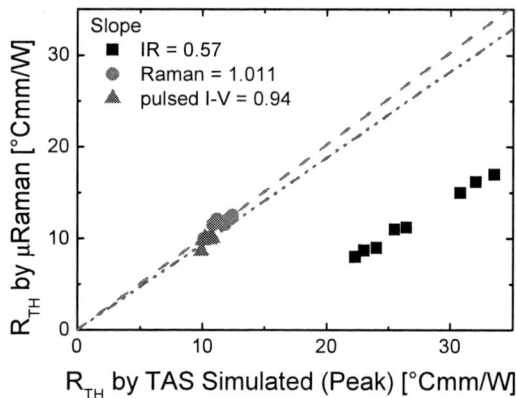

Figure 4. Comparison of Raman, electrical and IR thermography techniques with thermal simulation considering the thermal resistance R$_{TH}$. Thermal resistance was determined for power densities in the range of 7-13 W/mm

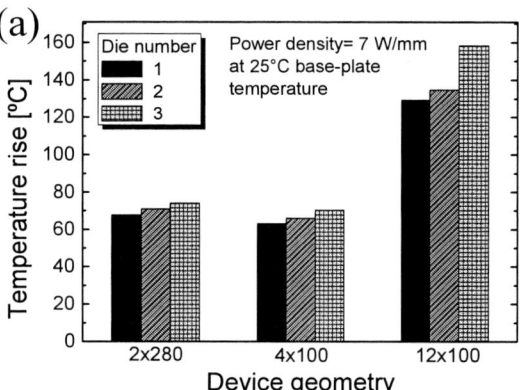

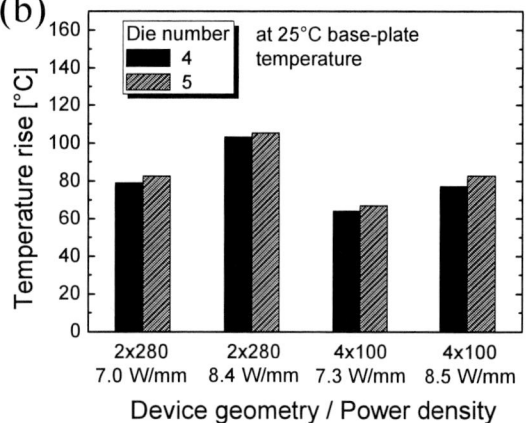

Figure 5. Die to die comparison for different devices with temperatures determined by (a) synchronized pulsed I-V measurements and (b) micro-Raman thermography.

Fig. 4 also displays results from the synchronized pulsed I-V measurement setup used. Using this method, device thermal resistance is underestimated by about 7%, with respect to the Raman results. As the thermal resistance from thermal simulation used (as horizontal axis in this graph) considers device temperature averaged over 0.75×0.75×2 μm^3 close to the drain-gate edge, the underestimation

presumably is a result of the fact that the electrical method only measures an average temperature over the entire gate width. However, it needs to be emphasized that the less than 10% difference to the Raman thermography data observed here is already a significant improvement in comparison to other electrical methods, which underestimated device temperature by up to 50% compared to Raman thermography [3]. This indicates that the pulsed electrical method averages temperature over the saturation velocity region, not over the whole source-drain gap as a typical DC electrical method would do. However, it needs to be noted that even the pulsed I-V method does not measure the peak channel temperature, i.e., some spatial averaging also takes place here. Since a micro-Raman setup is often not available in device research laboratories, pulsed I-V measurement provides a quick and more accessible way of evaluating thermal resistivity of these devices. Since this technique also demonstrates a high reproducibility, with temperatures varying in the range of 3%, the pulsed I-V setup is very suitable for comparing devices with nominally the same structure.

The synchronized pulsed I-V setup was used to compare different dies from one wafer. Fig. 5(a) shows the temperature rise for devices with three different gate peripheries, operated at a power density of 7 W/mm and measured on three different dies. For all the device geometries presented, die number 3 showed the highest temperature. This behavior was also observed for other power densities and as well confirmed by Raman thermography measurements performed on different device geometries and power densities, as shown in Fig. 5(b). As the difference in temperature rise and therefore in thermal resistance occurs independent from the device measured, it can be assumed that this difference is due to variations in the die attach, i.e., the brazing onto the carrier plate. The observed temperature variation between the dies of up to 20% represents a significant difference in thermal properties of the die/package assembly and would play an essential role in the reliability of the ultimately packaged HEMT.

IV. CONCLUSIONS

Three different thermography techniques were compared. Good agreement of temperature determined by Raman thermography with a simulation model for a wide variety of device geometries established that the simulation model used for these devices is accurate. Multiple temperature measurements demonstrated very good repeatability of the Raman thermography technique. Raman thermography temperature results were also compared to temperatures determined by electrical and IR thermography measurements. These other methods showed some underestimation of device peak temperature due to lateral spatial averaging. A combination of thermography methods, such as Raman and electrical thermography, used in conjunction with thermal simulations would be ideally suited for gaining accurate information on the thermal properties of devices.

REFERENCES

[1] R. Gaska, A. Osinsky, J. W. Yang, and M. S. Shur, "Self-heating in high power AlGaN-GaN HFETs," *IEEE Electron Device Lett.*, vol. 19, no. 3, pp. 89–91, March 1998.

[2] S. Nuttinck, B. K. Wagner, B. Banerjee, S. Venkataraman, E. Gebara, J. Laskar, and H. M. Harris, "Thermal analysis of AlGaN-GaN power HFETs," *IEEE Trans. Microw. Theory Tech.*, vol. 51, no. 12, pp. 2445–2452, December 2003.

[3] R. J. T. Simms, J. W. Pomeroy, M. J. Uren, Member IEEE, T. Martin, and M. Kuball, "Channel temperature determination in high-power AlGaN/GaN HFETs using electrical methods and Raman Spectroscopy," *IEEE Trans. Electron Devices*, vol. 55, no. 2, pp. 478-482, February 2008.

[4] A. Sarua, H. Ji, M. Kuball, M. J. Uren, T. Martin, K. P. Hilton, and R. S. Balmer, "Integrated micro-Raman/Infrared thermography probe for monitoring of self-heating in AlGaN/GaN transistor structures," *IEEE Trans. Electron Devices*, vol. 53, no. 10, pp. 2438-2447, October 2006.

[5] M. Kuball, G. J. Riedel, J. W. Pomeroy, A. Sarua, M. J. Uren, T. Martin, K. P. Hilton, J. O. Maclean, and D. J. Wallis, "Time-Resolved

978-1-4244-5430-3/10 $26.00 © 2010 IEEE

Temperature Measurement of AlGaN/GaN Electronic Devices using Micro-Raman Spectroscopy," *IEEE Electron Device Lett.*, vol. 28, no. 2, pp. 86-89, February 2007.

[6] J. Kuzmik, P. Javorka, A. Alam, M. Marso, M. Heuken, and P. Kordos, "Determination of channel temperature in AlGaN/GaN HEMTs grown on sapphire and silicon substrates using DC characterization method," *IEEE Trans. Electron Devices*, vol. 49, pp. 1496-1498, August 2002.

[7] J. D. McDonald and G. C. Albright, "Microthermal imaging in the infrared," Electron. Cooling, vol. 3, no. 1, pp. 26-29, January 1997.

[8] J. Joh, J. A. Del Alamo, U. Chowdhury, T.-M. Chou, H.-Q. Tserng, and J. L. Jimenez, "Method for estimation of the channel temperature of GaN high electron mobility transistors," *ROCS Workshop,* [Reliability of Compound Semiconductors Digest], pp. 87-101, October 2007.

[9] M. Kuball, S. Rajasingam, A. Sarua, M. J. Uren, T. Martin, B. T. Hughes, K. P. Hilton, and R. S. Balmer, "Measurement of temperature distribution in multi-finger AlGaN/GaN HFETs using micro-Raman spectroscopy," *Appl. Phys. Lett.*, vol. 82, no. 1, pp. 124–126, January 2003.

[10] T. Batten, J. W. Pomeroy, M. J. Uren, T. Martin, and M. Kuball, "Simultaneous measurement of temperature and thermal stress in AlGaN/GaN HEMTs using Raman scattering spectroscopy," *J. Appl. Phys.*, vol. 106, no. 9, 094509, November 2009

[11] B. Clafin, E. R. Heller, B. Winningham, J. E. Hoelscher, M. Bellot, K. Chabak, A. Crespo, J. Gillespie, V. Miller, M. Trejo, G. H. Jessen, and G. D. Via, "Accurate channel temperature measurement in GaN-based HEMT devices and its impact on accelerated lifetime predictive models," *CS MANTECH Conference,* May 2009

[12] E. R. Heller, "Simulation of life testing procedures for estimating long-term degradation and lifetime of AlGaN/GaN HEMTs," *IEEE Trans. Electron Devices*, vol. 55, no. 10, pp. 2554-2560, October 2008.

Analysis of Interface-Trap Effects in Inversion-Type InGaAs/ZrO$_2$ MOSFETs

L. Morassi, G. Verzellesi, A. Padovani, L. Larcher
DISMI, University of Modena and Reggio Emilia, Reggio Emilia, Italy
phone: +39 –0522-52-2605, giovanni.verzellesi@unimore.it

P. Pavan
DII, University of Modena and Reggio Emilia, Modena, Italy

D. Veksler, Injo Ok, G. Bersuker
SEMATECH, Austin, TX, USA

Abstract— Interface-trap effects are analyzed in inversion-type, self-aligned In$_{0.53}$Ga$_{0.47}$As and In$_{0.53}$Ga$_{0.47}$As/ In$_{0.2}$Ga$_{0.8}$As MOSFETs with ALD ZrO$_2$ gate dielectric. Interface-trap densities in the order of 10^{13} cm^{-2} eV^{-1} are required to explain the measured subthreshold slopes. For these D$_{it}$ values, donor-like interface traps are compatible with threshold-voltage values in the 0-0.15 V range as those observed in these devices. Moreover, the presence of donor-like interface traps can explain the negative V$_T$ shift induced by the inclusion of the In$_{0.2}$Ga$_{0.8}$As cap layer, as the result of the influence of interface traps located at the In$_{0.2}$Ga$_{0.8}$As/ZrO$_2$ interface on the inversion channel forming at the In$_{0.53}$Ga$_{0.47}$As/In$_{0.2}$Ga$_{0.8}$As interface.

Keywords: III-V MOSFETs; high-k dielectric; InGaAs; interface traps; numerical simulation.

I. INTRODUCTION

InGaAs is intensively being studied as one of the possible high-mobility materials for replacing Si in the channel of n-MOSFETs for the CMOS technology beyond the 22-nm node. The inversion-type, enhancement-mode InGaAs MOSFET is, in particular, one of the most attractive options as for the specific device structure to be adopted [1-3]. Traps at the interface between InGaAs and the high-k gate dielectric play a major role in the performance of these devices, influencing key parameters, such as threshold voltage (V$_T$), channel mobility, and sub-threshold slope (SS), and therefore indirectly impacting both I$_{ON}$ and I$_{OFF}$ currents as well as device speed. Interface-trap generation is also expected to be one of the most severe reliability-limiting factors. For these reasons, developing an in-depth understanding of interface-trap effects is important in the current stage of InGaAs MOSFET development, characterized by continuous technological improvements, as well as, prospectively, in view of the rapidly approaching phase, when reliability of these devices will start to be systematically tested and analyzed.

In this paper, we present results from a detailed analysis of the impact of interface-trap distributions on the electrical characteristics of inversion-type, self-aligned InGaAs

MOSFETs with ALD ZrO$_2$ gate dielectric. Two-dimensional device simulations [Dessis-8.0, Synopsys Inc.] are adopted to gain insight on the measured IV characteristics and to analyze the impact of different interface-trap distributions.

II. SAMPLES AND EXPERIMENTAL RESULTS

Devices under consideration have two different structures, one with the ZrO$_2$ gate dielectric directly deposited onto the In$_{0.53}$Ga$_{0.47}$As channel, the other with a 2-nm In$_{0.2}$Ga$_{0.8}$As cap layer interposed. The latter is intended for mobility improvement by outdistancing the inversion channel forming at the In$_{0.53}$Ga$_{0.47}$As/In$_{0.2}$Ga$_{0.8}$As interface from the dielectric surface. Figure 1 shows schematic cross sections for the two device types. Layer thicknesses are reported in Fig. 1. P-type doping (Zn) concentration is 5x10^{17} cm^{-3}. Gate length is 10 µm for both structures. More details on device fabrication can be found in [4].

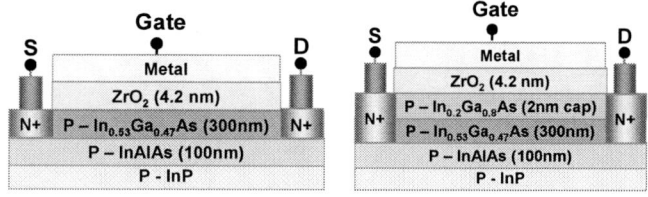

Fig. 1. Schematic cross sections of devices under study. Left figure: In$_{0.53}$Ga$_{0.47}$As/ZrO$_2$ MOSFET (uncapped device); Right figure: In$_{0.53}$Ga$_{0.47}$As/ In$_{0.2}$Ga$_{0.8}$As/ZrO$_2$ MOSFET (capped device).

Figures 2 and 3 show experimental drain-current (I$_D$) vs gate-source-voltage (V$_{GS}$) characteristics at V$_{DS}$=0.05 V for representative MOSFETs with and without the In$_{0.2}$Ga$_{0.8}$As cap in linear and semilog scales, respectively. Simulation data shown in Figs. 2 and 3 will be discussed in section III. As can be noted from Fig. 2, the device with the cap exhibit a larger I$_{ON}$ value. The same applies to the transconductance (not shown). This occurs as a result of the larger channel mobility, as confirmed by CV-split measurements (not shown), yielding peak mobilities of ≈ 1960 cm^2/Vs and ≈ 1500 cm^2/Vs for the capped and uncapped MOSFETs, respectively. Moreover, capped devices show a smaller V_T than devices without the cap (≈ 0.03 V against ≈ 0.13 V), see Fig. 2. SS is instead similar in the two device types down to the 10^{-6} A/mm drain-current regime, see Fig. 3. For smaller currents, drain reverse leakage current dominates in the MOSFET without cap, while, in the capped device, SS increases further down to the 10^{-8} A/mm regime, till I_D eventually reaches the reverse leakage current floor. Different values of drain reverse leakage current for the two device types are explained by the different electron lifetime characterizing wafers onto which devices with/without cap have been fabricated.

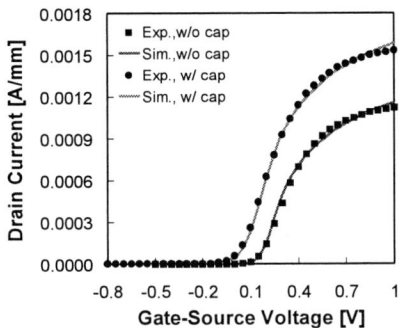

Fig. 2. Experimental (symbols) and simulated (lines) I_D vs V_{GS} characteristics at $V_{DS}= 0.05$ V for MOSFETs without and with the In$_{0.2}$Ga$_{0.8}$As cap.

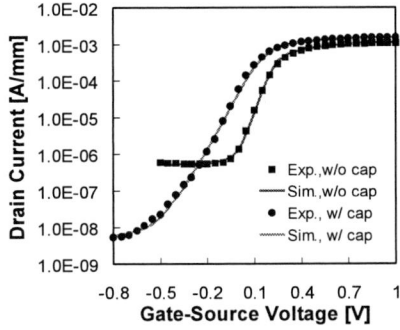

Fig. 3. Experimental (symbols) and simulated (lines) I_D vs V_{GS} sub-threshold characteristics at $V_{DS}= 0.05$ V for MOSFETs without and with the In$_{0.2}$Ga$_{0.8}$As cap.

III. SIMULATIONS RESULTS

Two-dimensional device simulations [Dessis-8.0, Synopsys Inc.] have been adopted to gain insight on the measured IV characteristics and to analyze the impact of different interface-trap distributions.

Nominal values have been adopted for the substrate p-type doping and for all geometrical parameters. The InGaAs doping-dependent low-field electron mobility has been set in accordance with literature data [5]. Source/drain doping distributions have been modeled by Gaussian profiles. The latter have been adjusted to reproduce source/drain sheet

resistances obtained from TLM measurements. Channel-mobility-model parameters have been tuned to fit measured C-V split data for the two device types (with and without cap). Source and drain contact resistances have been derived from TLM measurements. The ZrO$_2$ relative dielectric constant has been set to 21.

In agreement with [6], interface traps at the high-k-dielectric/InGaAs interface have been defined to be donor-like and their density (D_{it}) has been assumed to peak at the valence-band edge (E_V) and to decrease about linearly throughout the InGaAs bandgap. Maximum and minimum D_{it} values (at E_V and E_C, respectively) allowing experimental V_T and SS to be reproduced accurately are 7×10^{13} cm^{-2} eV^{-1} and 7×10^{12} cm^{-2} eV^{-1} for the In$_{0.53}$Ga$_{0.47}$As/ZrO$_2$, while they are 8×10^{13} cm^{-2} eV^{-1} and 1×10^{13} cm^{-2} eV^{-1} for the In$_{0.2}$Ga$_{0.8}$As/ZrO$_2$ device.

Simulated I_D vs V_{GS} characteristics at $V_{DS}=0.05$ V are compared with measurements in Figs. 2 and 3. As can be noted, a fairly accurate description of both above- and sub-threshold $I_D(V_{GS})$ behavior have been achieved for both device types. It is worth noticing that interface-trap densities in the order of 10^{13} cm^{-2} eV^{-1} are required to explain the measured subthreshold slopes. For these relatively-high D_{it} values, only donor-like interface traps are compatible with V_T values in the 0-0.15 V range as those observed in these devices, whereas acceptor-like traps would result in much larger V_T values. This will appear clearer in section III.B, where the effects of acceptor-like traps are described.

In addition, simulations suggest that the negative V_T shift observed in capped MOSFETs with respect to uncapped ones can be of fundamental nature (and not induced by process parameter variations). Simulations actually reproduce the observed V_T shift (see Figs. 2 and 3), provided that interface traps at the InGaAs/ZrO$_2$ interface are donor-like traps. If interface traps had instead been assumed to be acceptor-like, the V_T shift in capped devices would have been opposite to what experimentally observed. Explanation for this will be given in sections III.A and III.B.

In order to analyze the effect of interface traps more systematically, a simplified interface-trap distributions characterized by uniform D_{it}'s has been adopted. This simplification does not limit generality of conclusions drawn, as far as the different effects of donor- and acceptor-like traps are concerned. Results are described in Sections III.A and III.B for donor-like and acceptor-like traps, respectively.

A. Effects of Donor-Like Interface Traps

Figure 4 shows the simulated I_D vs V_{GS} subthreshold characteristics at $V_{DS}= 0.05$ V for a MOSFET without the In$_{0.2}$Ga$_{0.8}$As cap for different donor-like, uniform interface-trap distributions. As can be noted, in the uncapped MOSFET donor-like interface traps impact the subthreshold characteristics by increasing SS. They instead leave V_T virtually unchanged.

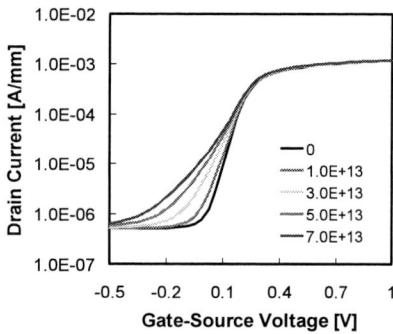

Fig. 4. Simulated I_D vs V_{GS} sub-threshold characteristics at V_{DS}= 0.05 V for a MOSFETs without the $In_{0.2}Ga_{0.8}As$ cap for different donor-like, uniform D_{it}.

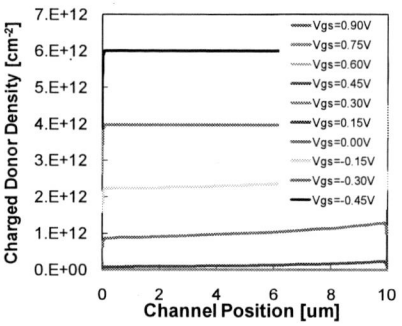

Fig. 5. Charged donor density along the ZrO_2-$In_{0.53}Ga_{0.47}As$ interface in the MOSFET without the $In_{0.2}Ga_{0.8}As$ cap at V_{DS}=0.05 V for different V_{GS} values.

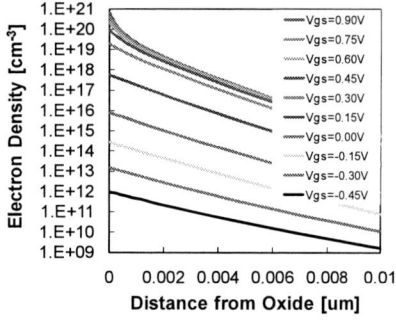

Fig. 6. Electron density along a vertical cut in the middle of the gate in the MOSFET without the $In_{0.2}Ga_{0.8}As$ cap at V_{DS}=0.05 V for different V_{GS} values.

This can easily be understood with the aid of Figs. 5 and 6, showing the charged donor density along the ZrO_2-$In_{0.53}Ga_{0.47}As$ interface (Fig. 5) and the electron density along a vertical cut in the middle of the gate (Fig. 6) in the uncapped MOSFET for different V_{GS} values. As a matter of fact, donor traps are mostly positively charged (i.e., unoccupied) in accumulation and weak inversion (see Fig. 5), as the electron density in the channel is small (see Fig. 6). Under these conditions, the larger D_{it} the higher I_D, thus explaining the SS increase shown in Fig. 4. As strong-inversion conditions are reached, the electron density is large (see Fig. 6) and traps are almost all neutral (i.e., completely occupied) (see Fig. 5), so that they can not impact I_D any more. For this reason V_T does not change with increasing D_{it}, see Fig. 4.

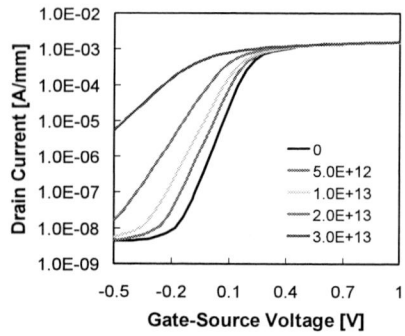

Fig. 7. Simulated I_D vs V_{GS} sub-threshold characteristics at V_{DS}=0.05 V for a MOSFETs with the $In_{0.2}Ga_{0.8}As$ cap for different donor-like, uniform D_{it}.

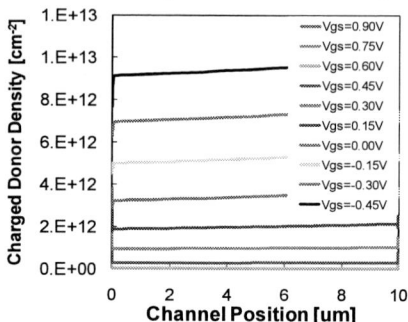

Fig. 8. Charged donor density along the ZrO_2-$In_{0.2}Ga_{0.8}As$ interface in the MOSFET with the cap at V_{DS}=0.05 V for different V_{GS} values.

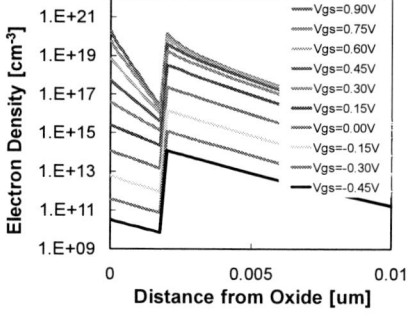

Fig. 9. Electron density along a vertical cut in the middle of the gate in the MOSFET with the $In_{0.2}Ga_{0.8}As$ cap at V_{DS}=0.05 V for different V_{GS} values.

Figure 7 shows the simulated I_D vs V_{GS} sub-threshold characteristics at V_{DS}= 0.05 V for a MOSFET with the $In_{0.2}Ga_{0.8}As$ cap for different donor-like, uniform interface-trap distributions. As can be noticed, in the capped device increasing the donor D_{it} shifts V_T to smaller values (besides increasing SS like for uncapped ones). This can be explained with the aid of Figs. 8 and 9, showing the charged donor density (Fig. 8) along the ZrO_2-$In_{0.20}Ga_{0.80}As$ interface and the electron density along a vertical cut in the middle of the gate (Fig. 9) in the capped MOSFET for different V_{GS} values. As can be noted, traps that are located at the $In_{0.2}Ga_{0.8}As$/ZrO_2 interface continue, in this case, to be modulated by the V_{GS} change (see Fig. 8) even after the channel at the $In_{0.53}Ga_{0.47}As$/$In_{0.2}Ga_{0.8}As$ interface has attained strong inversion (see Fig. 9). In this respect, donor-like traps at the dielectric-cap interface influence the inversion layer at the

978-1-4244-5430-3/10 $26.00 © 2010 IEEE

$In_{0.53}Ga_{0.47}As/In_{0.2}Ga_{0.8}As$ similarly to fixed, positive charges in the dielectric of an uncapped MOSFET, i.e. they shift V_T negatively. Donor-like traps at the $ZrO_2/InGaAs$ interface can therefore explain the negative V_T shift observed experimentally in capped MOSFETs with respect to uncapped ones (see Figs. 2 and 3).

B. Effects of Acceptor-Like Interface Traps

The effect of acceptor-like interface traps is illustrated by Figs. 10 and 11, showing simulated I_D vs V_{GS} subthreshold characteristics at V_{DS}=0.05 V for different acceptor-like, uniform interface trap distributions in a MOSFET without and with the $In_{0.2}Ga_{0.8}As$ cap, respectively.

As shown in Fig. 10, in devices without the $In_{0.2}Ga_{0.8}As$ cap, increasing the acceptor-like D_{it} increases SS and, at the same time, shifts V_T positively. Acceptor-like traps are in fact neutral (unoccupied) under accumulation, this making I_D independent on their density. As the MOSFET is pushed towards inversion, traps charge up negatively, so that the larger D_{it} the smaller I_D. D_{it} values in the order of 10^{13} cm^{-2} eV^{-1} are required to reproduce measured SS values. As shown in Fig. 10, these relatively-large D_{it} values would result in V_T values >0.5 V, that are much larger than the measured ones (see Figs. 2 and 3).

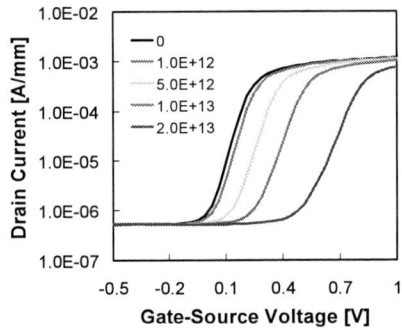

Fig. 10. Simulated I_D vs V_{GS} sub-threshold characteristics at V_{DS}= 0.05 V for a MOSFETs without the $In_{0.2}Ga_{0.8}As$ cap for different acceptor-like, uniform D_{it}.

In devices with the cap, similar considerations apply. As shown in Fig. 11, for the same D_{it}, the positive V_T shift is however larger than in the uncapped device, because traps located at the $In_{0.2}Ga_{0.8}As/ZrO_2$ interface continue to be modulated by the V_{GS} change, even after the channel at the $In_{0.53}Ga_{0.47}As/In_{0.2}Ga_{0.8}As$ interface has reached strong inversion. If (dominant) interface traps were acceptor-like traps in the devices under consideration, inclusion of the $In_{0.2}Ga_{0.8}As$ cap would shift V_T positively, which contrasts with what observed experimentally. On the contrary, as previously shown, donor-like interface traps can explain experiments. In these respects, our results are in agreement with conclusions drawn for $In_{0.65}Ga_{0.35}As/Al_2O_3$ MOSFETs in [6], providing further evidence of the donor-like nature of interface traps at the high-k/InGaAs interface and their role in allowing strong-inversion operation even in the presence of relatively large D_{it}.

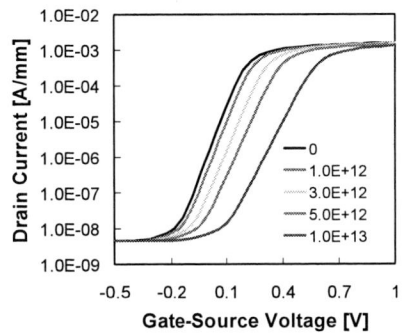

Fig. 11. Simulated I_D vs V_{GS} sub-threshold characteristics at V_{DS}= 0.05 V for a MOSFETs with the $In_{0.2}Ga_{0.8}As$ cap for different acceptor-like, uniform D_{it}.

IV. CONCLUSIONS

We have analyzed the impact of interface-trap distributions on the electrical characteristics of inversion-type, self-aligned InGaAs MOSFETs with ALD ZrO_2 gate dielectric by means of two-dimensional device simulations. Devices under consideration have two different structures, one with the ZrO_2 dielectric directly deposited onto the $In_{0.53}Ga_{0.47}As$ channel, the other with a 2-nm $In_{0.2}Ga_{0.8}As$ cap layer interposed. Interface-trap densities in the order of 10^{13} cm^{-2} eV^{-1} are required to explain the measured subthreshold slopes. For these D_{it} values, donor-like interface traps are compatible with V_T values in the 0-0.15 V range as those measured in these devices, whereas acceptor-like traps would result in much larger V_T values (> 0.5 V). Moreover, the presence of donor-like interface traps can explain the negative V_T shift induced by the inclusion of the $In_{0.2}Ga_{0.8}As$ cap layer, as the result of the influence of interface traps located at the $In_{0.2}Ga_{0.8}As/ZrO_2$ interface on the inversion channel at the underlying $In_{0.2}Ga_{0.8}As/In_{0.53}Ga_{0.47}As$ interface.

REFERENCES

[1] J. Q. Lin, S.J. Lee, H.J. Oh, G.Q. Lo, D.L. Kwong, D.Z. Chi, "Inversion-mode self-aligned $In_{0.53}Ga_{0.47}As$ n-channel metal-oxide-semiconductor field-effect transistor with HfAlO gate dielectric and TaN metal gate", IEEE Electr. Dev. Lett., vol. 29, no. 9, pp. 977-980, Sept. 2008.

[2] Y.Q. Wu, W.K. Wang, O. Koybasi, D.N. Zakharov, E.A. Stach, S. Nakahara, J.C.M. Hwang, P.D. Ye, "0.8-V supply voltage deep-submicrometer inversion-mode $In_{0.75}Ga_{0.25}As$ MOSFET", IEEE Electr. Dev. Lett., vol. 30, no. 7, pp. 700-702, July 2009.

[3] H.-C. Chin, X. Gong, X. Liu, Y.-C. Yeo, "Lattice-mismatched $In_{0.4}Ga_{0.6}As$ source/drain stressors with in situ doping for strained $In_{0.53}Ga_{0.47}As$ channel n-MOSFETs", IEEE Electr. Dev. Lett., vol. 30, no. 8, pp. 805-807, Aug. 2009.

[4] S. Koveshnikov, N. Goel, P. Majhi, H. Wen, M.B. Santos, S. Oktyabrsky, V. Tokranov, R. Kambhampati, R. Moore, F. Zhu, J. Lee, W. Tsai, "$In_{0.53}Ga_{0.47}As$ based metal oxide semiconductor capacitors with atomic layer deposition ZrO_2 gate oxide demonstrating low gate leakage current and equivalent oxide thickness less than 1 nm", Appl. Phys. Lett., **92**, 222904 (2008).

[5] EMIS Datareviews Series No. 8, "Properties of lattice-matched and strained Indium Gallium Arsenide", P. Bhattacharya Ed., 1993.

[6] D. Varghese, Y. Xuan, Y.Q. Wu, T. Shen, P.D. Ye, M.A. Alam, "Multi-probe interface characterization of $In_{0.65}Ga_{0.35}As/Al2O3$ MOSFET", IEEE IEDM Tech. Dig., 2008.

Degradation of III-V inversion-type enhancement-mode MOSFETs

N. Wrachien, A. Cester, E. Zanoni, G. Meneghesso
Department of Information Engineering
University of Padova
Via Gradenigo 6B, Padova – Italy
phone: +39-0498277625, fax: +39-0498277699; e-mail: wrachien@dei.unipd.it

Y.Q. Wu and P.D. Ye
School of Electrical and Computer Engineering
Purdue University
West Lafayette, IN 47906, U.S.A.

Abstract—**We performed gate ramp voltage stress on III-V InGaAs based MOSFETs. Stress induces trapped charge and it also leads to interface trap generation, which has detrimental effects on the subthreshold slope and on the transconductance. At high electric fields, before the hard breakdown, a very low-frequency high-current random telegraph noise appears at the gate, which seems to be not correlated with the soft breakdowns commonly observed in other devices.**

Keywords-stress; III-V MOSFET; reliability.

I. INTRODUCTION

In the past, III-V based MOSFETs have been evaluated as possible replacement for silicon N-MOSFETs in high speed VLSI-ULSI devices, which will face several scaling limits in the next decade [1-3]. One of the main limits of high speed VLSI-ULSI devices is the dynamic power dissipation, which depends on the square of the power supply voltage. From this viewpoint, III-V MOSFETs are very promising due to their much higher electron mobility with respect their silicon counterparts, which, in turn, translates in a much lower ON-state MOSFET resistance. Therefore, III-V MOSFETs allow to use much lower power supply voltages, while operating at the same switching speeds of their silicon counterpart, reducing the dynamic power consumption.

However, several issues affected III-V MOSFETs in the past, mainly due to the lack of a good gate dielectric. The native oxide forms a very poor semiconductor/dielectric interface, with an interface trap density so high that induces the pinning of the Fermi level, leading to very poor electrical characteristics [4-11]. Therefore, extensive research (see for instance [12-25]) was carried out to find materials and deposition techniques, which allowed the Fermi level unpinning.

Recently, several advancements have been achieved both in flatband and inversion mode III-V MOSFETs, reaching electron mobility exceeding 5000 $cm^2V^{-1}s^{-1}$ [17,26,27].

Despite these improvements, III-V MOSFETs have still many open issues, such as very exacerbated short channel effects [28], which negatively impact on the static power consumption. Moreover, the reliability of III-V MOSFETs is still an unexplored field. In this work, we show the results of gate ramp stress performed on InGaAs inversion-type enhancement-mode MOSFETs.

This is the first reliability investigation on InGaAs MOSFET devices. Even if some preliminary reliability data on GaAs devices, have been recently published in [29], this study indeed represent the first deep and systematic investigation of the reliability of III-V MOSFETs devices.

This work is organized as follows: in Section II we describe the devices and the experimental procedure; in Section III-A we show and discuss the gate current stress kinetics; in Section III-B we discuss the impact on the electrical characteristics and the degradation kinetics on the extrapolated parameters; finally,

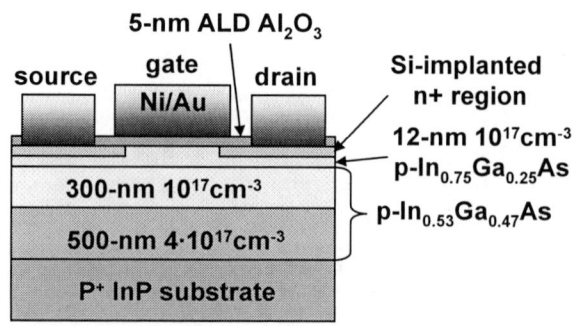

Figure 1. Cross section of the devices used throughout this work.

This work was partially supported by Progetto di Ateneo 2009 – Università di Padova, Italy (Project Number CPDA083941).
The work at Purdue University is supported by United States National Science Foundation and the SRC FCRP MSD Center.

in Section IV we draw our conclusions.

II. EXPERIMENTAL AND DEVICES

Throughout this work, we analyzed inversion-type enhancement-mode InGaAs with Al_2O_3 gate dielectric. Fig. 1 shows a schematic cross section of the analyzed devices. A 500-nm p-type $4\times10^{17}/cm^3$ buffer layer, a 300-nm p-type $1\times10^{17}/cm^3$ $In_{0.53}Ga_{0.47}As$ layer, and a 12-nm strained p-type $1\times10^{17}/cm^3$ $In_{0.75}Ga_{0.25}As$ channel were sequentially grown by molecular beam epitaxy over a p^+-InP substrate. The devices feature a 5-nm Al_2O_3 gate oxide. For further details, the interested reader may refer to [28]. The gate lengths (L) considered in this work were 150nm, 200nm, and 250nm. All the MOSFETs feature the same channel width W=5µm.

We performed gate ramp stress on devices with different channel lengths. The stress procedure is schematically depicted in Fig. 2a. The devices were initially characterized and they were subjected to a stress-characterization-relax-characterization loop, which lasted until device breakdown was detected. Each stress step consists of a 100-s constant voltage gate stress (CVS), with the other terminals grounded. The 200-s relax phase was introduced to neutralize any positive unstably trapped charge, which may affect the first characterization performed after each stress step, especially at the earlier stress steps (i.e. when the device is subjected to the lower stress voltages). During the relax phase, the gate was biased to -0.5V, while the other terminals were grounded.

Incidentally, in our samples, the drain to substrate diode features a quite large leakage, regardless the channel length. This leakage is drain voltage dependent and it is in the nA range even at V_{DB} as low as 0.2V. This current is quite high and it would not allow to appreciate the small variations, which occur at the very first stress steps, therefore, we extrapolated all our parameters from the source current, rather than drain current. Performance evaluation is outside the scope of this work, while our primary focus is to evaluate the stress impact and the degradation kinetics.

III. RESULTS AND DISCUSSIONS

A. Gate current stress kinetics

In Fig. 3 we plot the gate current evolution measured during the stress, on devices with different L. Each stress step was followed by a relax phase (as in Fig. 2a). Noticeably, the kinetics are almost identical, regardless the L value. Remarkably, the gate current evolution of Fig. 3 can be divided in three distinct zones. In Zone 1 (V_{STRESS}<2.6V-2.8V), the gate current decreases during each step (Fig 4a). In Zone 2, (i.e. for 2.6V-2.8V<V_{STRESS}<3.6V-3.8V) the current increases during each step (Fig 4b). Zone 3 is characterized by soft and hard breakdowns (Fig. 4c).

The decreasing gate current evolution of each stress step in Zone 1 (see Fig. 4a) suggests the presence of bulk traps in the

Al_2O_3 oxide, as also found in [30]. Since the trap density is finite, the gate current diminishes with time as the traps release their electrons, which then move toward the gate. Noticeably, those traps show a very slow response time. In fact, from Fig 4a, we argue that the response time of these traps is in the order of tens of seconds.

When the stress voltage is higher than 2.6V-2.8V (Zone 2), the gate current evolution starts to be increasing during each stress step (see Fig. 4b). This suggests that the increase of the gate current due to the formation of neutral traps in the dielectric dominates over the current relaxation observed in Fig. 4a.

Zone 3 (see Fig. 4c) features a very low frequency Random Telegraph Noise (RTN)-like evolution at rather high current values. It is well-known [31-33] that stressed oxides may feature gate leakage currents, which may be characterized by RTN. Still, in the literature, the RTN is observed at current

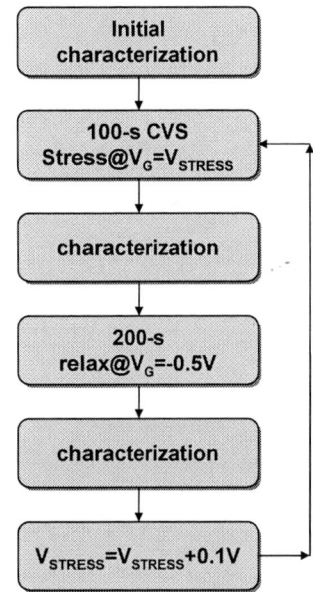

Figure 2. Experimental procedure adopted for the ramp stress with relax phase.

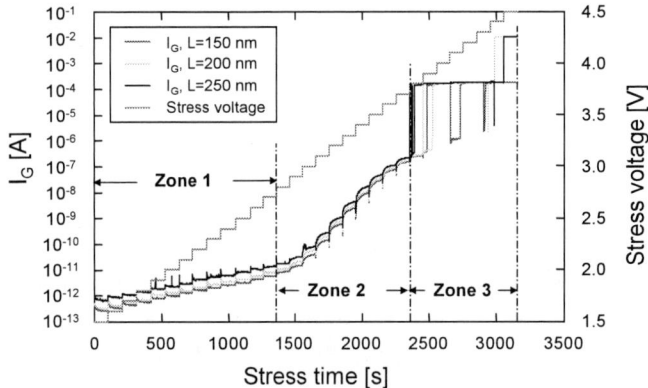

Figure 3. Evolution of the gate current measured during the stress. The stress kinetics has been divided in 3 zones featuring different behaviors (see zooms in Fig. 4).

values within the 1pA-100nA range, which are several orders of magnitude below the 100-µA range shown in Fig. 4c. Furthermore, it has been observed that the stronger the electric field, the higher is the RTN frequency [33], due to the onset of multilevel fluctuations induced by several separated leakage paths, while Fig. 4c (and the third zone of Fig. 3) has timescales in the 10-50s range and very high electric fields (exceeding 7MV/cm).

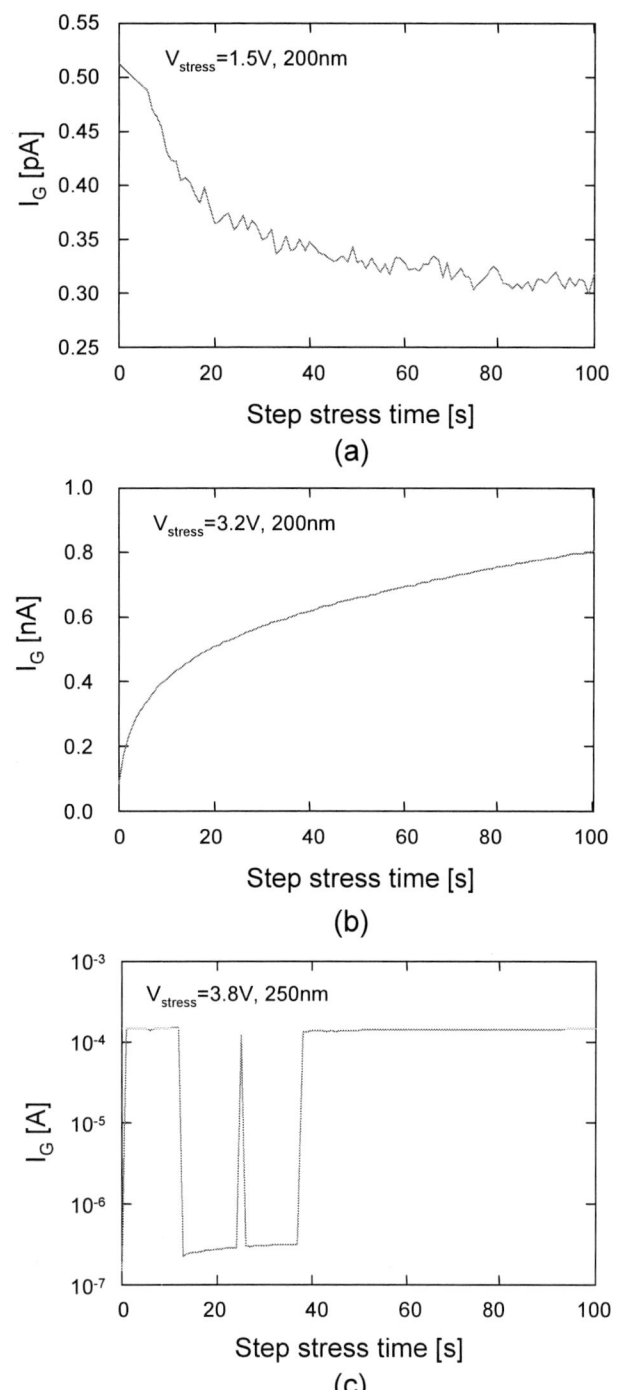

Figure 4. Zoom of 3 steps belonging to Zone 1 (a), Zone 2 (b) and Zone 3 (c) of Fig. 3.

Tentatively, we may attribute these low-frequency high-current fluctuations to the formation/rupture of a filament due to the high electric field, during the stress. The phenomenon is similar to what commonly happens in resistive random access memories, which works on the ability of some oxides (also Al_2O_3, see [34]) to reversibly switch their conductivity. The filament is generated by the very high gate electric fields and it induces a very high localized current flow, resulting in a local increase of the temperature. The high temperature may induce either a thermal runaway process, which further increases the gate current (hard breakdown), or it may lead to the filament rupture due to thermal redox and/or anodization [35]. Once the filament has been broken off, the gate current temporarily returns to a much smaller value, until the filament is opened again. The continuous random formation/rupture of the filament might be the responsible of the observed RTN-like I_G.

B. Effects on the I-V characteristics and degradation kinetics

In Fig. 5 we plot the I_S-V_{GS} and I_G-V_{GS} taken after various stress steps on a 250-nm gate-length device. Remarkably, the gate current (see Fig. 5b) features a noticeable permanent increase only for $V_{STRESS}>3.5V$ indicating that an Al_2O_3 electric field larger than ~6.5MV/cm (value calculated taking into account the flatband voltage and the semiconductor band bending) is required to appreciably degrade the gate dielectric, at least within 100s. Noticeably, this threshold V_{STRESS} value matches the gate-to-bulk voltage required to achieve Fowler-Nordheim (FN) Injection of electrons, which could enhance the dielectric degradation. Breakdown occurs at Al_2O_3 electric fields larger than ~7MV/cm, which is a value that is in good agreement with data reported in other works [36-37] for Al_2O_3.

In Figs 6-13 we plot the evolution of some of the most important MOSFET parameters during the electrical stress. The degradation kinetics is shown as a function of the applied gate voltages. All curves refer to the characteristics measured after the relax phase. In particular, we show the subthreshold slope (SS), the ON current (I_{ON}, measured with $V_{DS}=V_{GS}=1V$), the leakage current at $V_{GS}=-0.5V$ and $V_{DS}=1V$, the transconductance peak, and the threshold voltage (V_{th}). V_{th} has

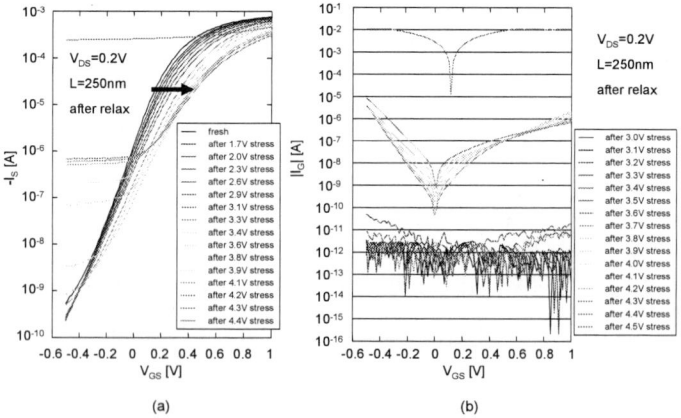

Figure 5. (a) Source-current- and gate-current- (b) gate-to-source-voltage-characteristics measured on a device with L=250nm, after each stress step, after relax.

been defined as the V_{GS} required to drive a drain current chosen so that $L \cdot I_D$ equals to 0.5 A·cm, to take into account the different W/L values.

Noticeably, the subthreshold slope increases with the stress voltage, (Fig. 6) indicating that new semiconductor/dielectric

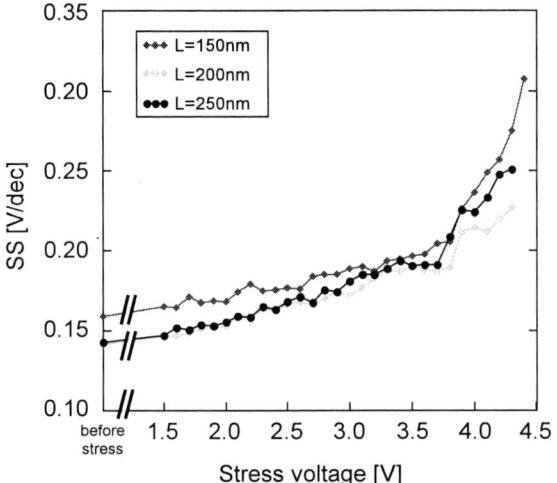

Figure 6.   Subthreshold slope extrapolated from the characterizations performed after the relax phases.

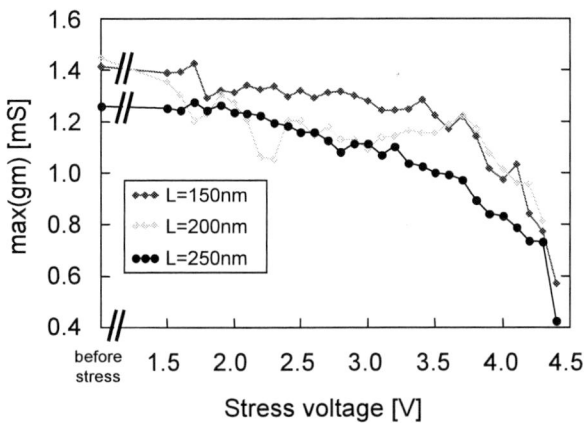

Figure 7.   Transconductance peak extrapolated from the characterizations performed after the relax phases. V_{DS}=0.2V.

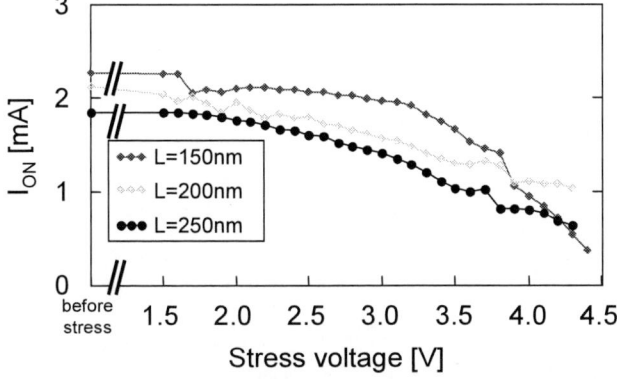

Figure 8.   ON-state source current, measured at V_{DS}=V_{GS}=1V after each relax phase.

interface traps are generated. The transconductance peak (Fig 7) and, consequently, I_{ON} (Fig 8), are both decreasing, with increasing stress voltage. Incidentally, there is an almost linear correlation between the inverse of the transconductance peak, normalized to its fresh value, and the subthreshold slope, normalized to its fresh value (Fig. 9). This suggests that the transconductance variation is due to the mobility degradation, which, in turn, is affected by the interface trap generation as widely confirmed in the literature (see for instance [38]).

Remarkably there is also an increase of the leakage current (Fig. 10) especially at the latter stress steps. The observation of the drain- (Fig. 11a), source- (Fig. 5a), bulk- (Fig 11b) and gate- (Fig. 5b) currents as a function of the gate voltage leads to the conclusion that the observed leakage current (measured at V_{GS}<0V) is a true drain-to-source current (i.e. it is a current flowing through the channel).

We ascribe the increased drain-to source leakage to two

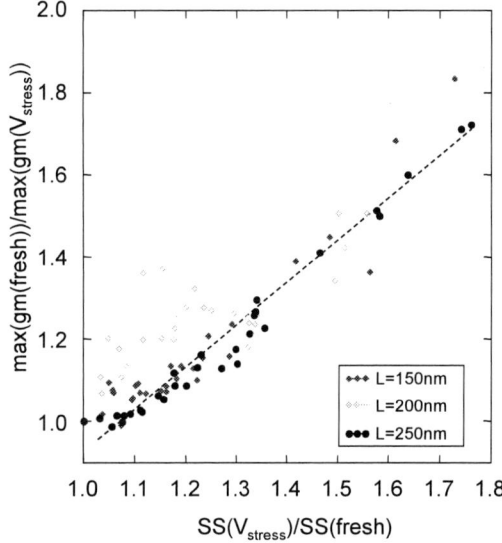

Figure 9.   Correlation between the subthreshold slope (normalized to its pre-stress value) and the inverse of the transconductance peak (normalized to its pre-stress value).

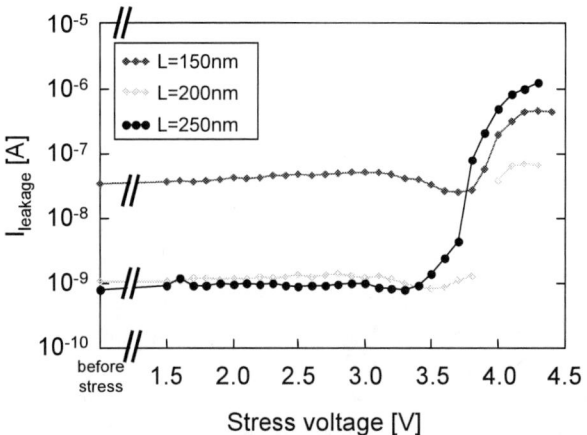

Figure 10. Source leakage current, measured at V_{DS}=1V, V_{GS}=-0.5V after each relax phase.

phenomena. First, positive charges might be trapped near the edges of the gate, where the electric fields are enhanced during stress by the sharp shape of the gate edges. The positive trapped charge in the Al_2O_3 near the gate edges might then induce a parasitic low-V_{TH} lateral MOSFET, increasing the drain-to-source leakage current. Noticeably, our devices do not feature any field or shallow trench isolation.

The second possible explanation for the onset of the drain-to-source current could be the Fermi level pinning at the channel surface, induced by the stress-induced additional interface traps. In fact, the observed leakage current is almost independent of V_{GS} (see Fig. 5a and Fig 11a). This conclusion is consistent to [39], where the minimum OFF-state leakage current on inversion-type III-V MOSFET was ascribed to the Fermi level pinning at the channel surface. The combined effects of Fermi level pinning and the parasitic lateral MOSFET might be the root cause for the strong increase of the drain-to-source leakage current measured for $V_{GS}<0$.

At this point, one might also wonder if the observed increased leakage current is indeed a drain-to-source current or, if it derives from the increased drain-to-substrate and source-to-substrate leakage currents, due to a possible stress-induced drain and source junctions' degradation. However, analyzing the I-V characteristics measured after each relax phase, we found that:

F1) the leakage current increase at the source is equal to the leakage current increase measured at the drain, and this relation holds for each V_{DS} at which the I-V characteristics have been taken, i.e. up to 1V (please note that for sake of simplicity, in Figs. 5 and 11 we show only the I-V taken at V_{DS}=0.2V). Of course, the value of the source/drain leakage increase depends on the V_{DS}.

F2) The difference (i.e. the algebraic sum) between the source and drain currents, measured after each stress step, is almost equal to the drain-to-bulk leakage current measured before the device was subjected to any stress.

If the origin of the increased leakage at the source and drain were the stress-induced source and drain junctions' degradation, we would expect a very large leakage dependence on the applied bias (i.e. V_{DB} and V_{SB}). However, as we stated in F1, we found that the increase of the leakage current at the source (which has V_{SB}=0 ±100μV) is equal to the increase of the leakage current at the drain (which has V_{DB} from 0.2V up to 1V). Furthermore, F2 suggests that the drain-to-bulk leakage current is unaffected by the stress. In other words F2 suggests that is the junction is not degraded by the stress.

One might also hypothesize that the drain and source junctions were not equally degraded by the stress. However, in that case, F1, would not hold for every V_{DS}. Furthermore, our devices are geometrically symmetrical and, during the stress, the source and drain were at the same potential (0V), thus we expect an uniform degradation (if any) at the source and drain junctions.

The comparison of Fig. 5b and Fig. 11b gives us also some information about the position of the soft breakdown. As expected, the breakdown is not preferably located near the drain or the source (because of the uniform stress setup, i.e. with $V_B=V_S=V_D$), and it manifests itself as a gate to bulk current, as confirmed by the diode-like I-V characteristics shown in Fig. 12b for $V_{STRESS} \geq 3.8V$.

Finally, V_{th} increases with increasing stress voltage as shown in Fig. 12. The increase of V_{th} arises from many contributions:

1) the subthreshold slope increase;

2) the reduction of the transconductance;

3) the negative charge trapping.

To assess how much charge has been trapped in the

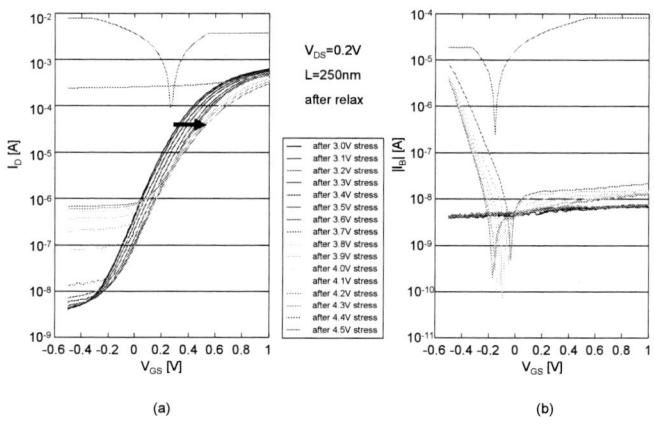

(a)   (b)

Figure 11. (a) Drain-current- and bulk-current- (b) gate-to-source-voltage characteristics measured on a device with L=250nm, after each stress step, after relax.

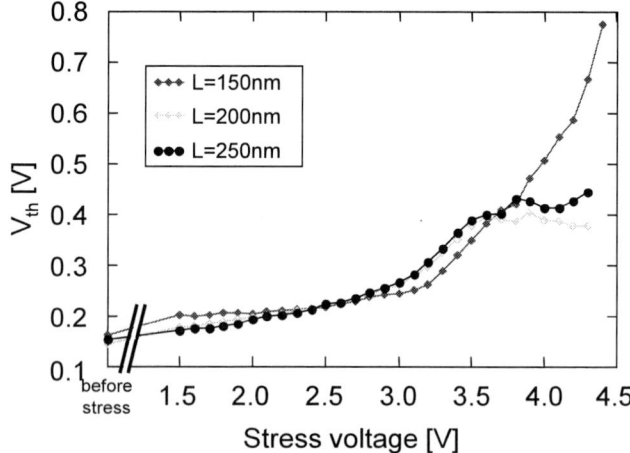

Figure 12. Threshold voltage extrapolated from the characterizations performed after the relax phases. V_{DS}=0.2V.

dielectric, we evaluated the midgap voltage variation using the well known subthreshold-midgap method [40-41]. The midgap voltage of fresh devices has been calculated analytically and it has been estimated to 45mV. In Fig. 13, we show the midgap voltage variation and the corresponding estimated dielectric trapped charge, as if it were concentrated in the center of the dielectric. Noticeably, the midgap voltage variation is smaller than V_{th}. The trapped charge at breakdown may vary between different samples, but it is within $3{\sim}9{\cdot}10^{12}$ electrons/cm^2.

IV. CONCLUSIONS

In this work we showed the results of gate ramp stresses performed on III-V inversion mode MOSFETs with different gate lengths. This is the first works that present systematic and deep investigation of electrical stress on InGaAs-based MOSFETs.

The interface trap generation reduces the ON-state current due to mobility and subthreshold slope degradation. Negative charge trapping also induces a threshold voltage increase, further reducing the ON-state current. Interface trap generation also may play a role in increasing the OFF-state leakage current, due to Fermi level pinning.

The increased OFF-state leakage current might be also induced by the onset of a parasitic lateral MOSFET due to positive charge trapping in the gate dielectric near the gate edges, where the electric field is locally enhanced.

A peculiar device soft breakdown is observed at Al_2O_3 electric fields around 7MV/cm. This soft breakdown is characterized by a very low-frequency RTN-like high gate current and it may be correlated to the continuous generation and rupture of a filament in the oxide as it happens in resistive memories.

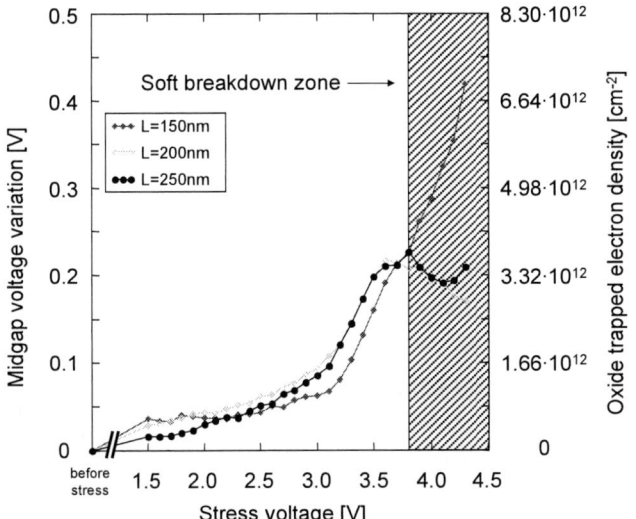

Figure 13. Evolution of the midgap voltage variation (left scale) and corresponding oxide trapped electron density variation induced by stress.

REFERENCES

[1] R. Chau, S. Datta, and A. Majumdar, "Opportunities and challenges of III–V nanoelectronics for future high speed, low power logic applications," in Proc. IEEE CSIC Tech. Dig., 2005, pp. 17–20.

[2] 'Process integration, devices, and structures' in 'International technology roadmap for semiconductors', 2007, ITRS. p. 3-4, www.itrs.net.

[3] Semiconductor Research Corporation website: www.src.org.

[4] H. Becke, R. Hall, and J. White, "Gallium arsenide MOS transistors," Solid-State Electronics., vol. 8, pp. 813–823, 1965.

[5] T. Ito and Y. Sakai, "The GaAs inversion-type MIS transistors," Solid-State Electronics., vol. 17, no. 7, pp. 751–759, 1974.

[6] B. Bayraktaroglu, E. Kohn, and H. L. Hartnagel, "First anodic-oxide GaAs M.O.S.F.E.T.S based on easy technological processes," Electronics Lett., vol. 12, no. 2, pp. 53–54, 1976.

[7] A. Colquhoun, E. Kohn, and H. L. Hartnagel, "Improved enhancement/ depletion GaAs MOSFET using anodic oxide as the gate insulator," IEEE Trans. Elec. Dev., vol. 25, pp. 375–376, Mar. 1978.

[8] T. Mimura, K. Odani, N. Yokoyama, and M. Fukuta, "New structure of enhancement-mode GaAs microwave M.O.S.F.E.T.," Electron. Lett., vol. 14, no. 16, pp. 500–502, 1978.

[9] T. Mimura, K. Odani, N. Yokoyama, Y. Nakayama, and M. Fukuta, "GaAs microwave MOSFETs," IEEE Trans. Elec. Dev., vol.25, pp. 573–579, June 1978.

[10] K. Kamimura and Y. Sakai, "The properties of GaAs–Al O and InP–Al O interfaces and the fabrication of MIS field-effect transistors," Thin Solid Films, vol. 56, pp. 215–223, 1979.

[11] G. G. Fountain, R. A. Rudder, S. V. Hattangady, R. J. Markunas, and J. A. Hutchby, "Demonstration of an n-channel inversion mode GaAs MISFET," in IEDM Tech. Dig., Dec. 1989, pp. 887–889.

[12] Y. C. Wang et al., "Advances in GaAs MOSFETs using Ga2O3 (Gd2O3) as gate oxide," in Proc. Mater. Res. Soc. Symp., 1999, vol. 573, pp. 219–225.

[13] M. Passlack, J. K. Abrokwah, R. Droopad, Z.Yu, C. Overgaard, S. I.Yi, M. Hale, J. Sexton, and A. C. Kummel, "Self-aligned GaAs p-channel enhancement mode MOS heterostructure field-effect transistor," IEEE Electron Device Lett., vol. 23, no. 9, pp. 508–510, Sep. 2002.

[14] M. Hale, S.I. Yi, J.Z. Sexton, A.C. Kummel, and M. Passlack, "Scanning Tunneling Microscopy and Spectroscopy of Gallium Oxide Deposition and Oxidation on GaAs(001)-c(2x8)/(2x4)", J. Chemical Physics, vol. 119, no. 13, pp.6719 -6728, 2003.

[15] Z. Yu, C. M. Overgaard, R. Droopad, M. Passlack, and J. K. Abrokwah, "Growth and physical properties of GaO thin films on GaAs(001) substrate by molecular-beam epitaxy," Appl. Phys. Lett., vol. 82, no. 18, pp. 2978–2980, May 2003.

[16] M. Passlack, R. Droopad, K. Rajagopalan, J. Abrokwah, R. Gregory, D. Nguyen, "High Mobility NMOSFET Structure with High-K Dielectric," IEEE Electron. Dev. Lett., vol. 26, no. 10, pp. 713-715, 2005.

[17] R. Droopad, K. Rajagopalan, J. Abrokwah, M. Canonico, and M. Passlack, "Ino 75Gao 25As Channel Layers with Record Mobility Exceeding 12,000 cm^2/Vs for Use in High-K Dielectric NMOSFETs," Solid State Electronics, vol. 50, no.7-8, pp. 1175-1177, 2006.

[18] Y. Xuan, H. C. Lin, P. D. Ye, and G. D.Wilk, "Capacitance-voltage studies on enhancement-mode InGaAs metal-oxide-semiconductor field-effecttransistor using atomic-layer-deposited Al2O3 gate dielectric," Appl. Phys. Lett., vol. 88, no. 26, pp. 263 518–263 520, Jun. 2006.

[19] K. Rajagopalan, J. Abrokwah, R. Droopad, and M. Passlack,"Enhancement-mode GaAs n-channel MOSFET," IEEE Electron Devices Lett., vol. 27, no. 12, pp. 959–962, Dec. 2006.

[20] I. Ok et al., "Self-Aligned n- and p-channel GaAs MOSFETs on undoped and p-type substrates using HfO2 and silicon interface passivation layer," in IEDM Tech. Dig., Dec. 2006, pp. 829–832.

[21] K. Rajagopalan et al., "1-μm enhancement mode GaAs n-channel MOSFETs with transconductance exceeding 250 mS/mm," IEEE Electron Devices Lett., vol. 28, no. 2, pp. 100–102, Feb. 2007.

[22] Y. Sun et al., "Enhancement-mode buried-channel $In_{0.70}Ga_{0.30}As/In_{0.52}Al_{0.48}As$ MOSFETs with high-k gate dielectrics," IEEE Electron Devices Lett., vol. 28, no. 6, pp. 473–475, Jun. 2007.

[23] H. C. Lin et al., "Enhancement-mode GaAs metal-oxide-semiconductor high-electron-mobility transistors with atomic layer deposited Al_2O_3 as gate dielectric," Appl. Phys. Lett., vol. 91, no. 21, pp. 212 101–212 103, Nov. 2007.

[24] Y. Xuan, Y. Q. Wu, H. C. Lin, T. Shen, and P. D. Ye, "Submicrometer inversion-type enhancement-mode InGaAs MOSFET with atomic-layerdeposited Al_2O_3 as gate dielectric," IEEE Electron Devices Lett., vol. 28, no. 11, pp. 935–938, Nov. 2007.

[25] Y. Xuan, Y. Q. Wu, T. Shen, T. Yang, and P. D. Ye, "High performance submicron inversion-type enhancement-mode InGaAs MOSFETs with ALD Al_2O_3, HfO_2, HfAlO as gate dielectrics," in IEDM Tech. Dig., Dec. 2007, pp. 637–640.

[26] M. Passlack, R. Droopad, K. Rajagopalan, J. Abrokwah, P. Zurcher; P. Fejes, "High Mobility III-V Mosfet Technology," . *IEEE Compound Semiconductor Integrated Circuit Symposium, 2006. CSIC 2006*, pp.39-42, Nov. 2006

[27] R.J.W. Hill, D.A.J. Moran, Li Xu, Zhou Haiping, D. Macintyre, S. Thoms, A. Asenov, P. Zurcher, K. Rajagopalan, J. Abrokwah, R. Droopad, M. Passlack, I.G. Thayne, "Enhancement-Mode GaAs MOSFETs With an $In_{0.3}Ga_{0.7}As$ Channel, a Mobility of Over 5000 $cm^2/V \cdot s$, and Transconductance of Over 475 $\mu S/\mu m$," *IEEE Electron Device Letters*, vol.28, no.12, pp.1080-1082, Dec. 2007

[28] Y.Q. Wu, W.K. Wang, O. Koybasi, D.N. Zakharov, E.A. Stach, S. Nakahara, J. Hwang, P.D.Ye, "0.8-V Supply Voltage Deep-Submicrometer Inversion-Mode $In_{0.75}Ga_{0.25}As$ MOSFET," *IEEE Electron Device Letters*, vol.30, no.7, pp.700-702, July 2009.

[29] Z. Tang, P.D. Ye, D. Lee, C.R. Wie, "Electrical measurements of voltage stressed Al_2O_3/GaAs MOSFET" Microelectronics Reliability, vol 47, pp. 2082-2087, 2007.

[30] D. Varghese, Y. Xuan, Y. Q. Wu, T. Shen, P. D. Ye, and M. A. Alam "Multi-probe Interface Characterization of $In_{0.65}Ga_{0.35}As/Al_2O_3$ MOSFET", in IEDM Tech. Dig., Dec. 2008, pp 379-382.

[31] A. Cester, L. Bandiera, G. Ghidini, I. Bloom, And A. Paccagnella, "Soft Breakdown Current Noise in Ultra-thin Gate Oxides," Solid-State Electronics, vol. 46, p. 1019-1025, 2002.

[32] T. Tomita, H. Utsunomiya, T. Sakura, Y. Kamakura and K. Taniguchi , A new soft breakdown model for thin thermal SiO_2 films under constant current stress. *IEEE Trans. Elec. Dev., vol* **46** no. 1, pp. 159–164, 1999.

[33] M. Depas, T. Nigam and M.M. Heyns , Soft breakdown of ultra-thin gate oxide layers. IEEE T Electron Dev 43 9 (1996), pp. 1499–1504.

[34] Chih-Yang Lin, Dai-Ying Lee, Sheng-Yi Wang, Chun-Chieh Lin, Tseung-Yuen Tseng, "Effect of thermal treatment on resistive switching characteristics in $Pt/Ti/Al_2O_3/Pt$ devices," Surface & Coatings Technology 203 (2008) 628–631.

[35] Kyung Min Kim, Byung Joon Choi, and Cheol Seong Hwang, "Localized switching mechanism in resistive switching of atomic-layer-deposited TiO_2 thin films," Appl. Phys. Lett. 90, 242906 2007.

[36] H. C. Lin, P. D. Ye, and G. D. Wilk," Leakage current and breakdown electric-field studies on ultrathin atomic-layer-deposited Al_2O_3 on GaAs," Appl. Phys. Lett. 87, 182904, 2005.

[37] M.D. Groner, J.W. Elam, F.H. Fabreguette, S.M. George, "Electrical characterization of thin Al_2O_3 films grownby atomic layer deposition on silicon and various metal substrates," Thin Solid Films 413 (2002) 186–197.

[38] S. C. Sun and J. D. Plummer, "Electron Mobility in Inversion and Accumulation Layers on Thermally Oxidised Silicon Surfaces," IEEE Trans. Electron Dev. ED-27, 1497 (1980).

[39] M. Passlack, "OFF-State Current Limits of Narrow Bandgap MOSFETs," IEEE Trans. Elec. Dev., vol. 53, 2006, pp. 2773-2778.

[40] P. S. Winokur, J. R. Schwank, P. J. McWhorter, P. V. Dressendorfer, and D. C. Turpin, "Correlating the Radiation Response of MOS Capacitors and Transistors," IEEE Trans. Nuc. Sci. w, 1453 (1984).

[41] P. J. McWhorter and P. S. Winokur, "Simple Technique for Separating the Effects of Interface Traps and Trapped-Oxide Charge in Metal-Oxide-Semiconductor Transistors," Appl. Phys. Lett. 48, 133 (1986).

978-1-4244-5430-3/10 $26.00 © 2010 IEEE

E- and $\sqrt{E}$-model too conservative to describe low field time dependent dielectric breakdown

K. Croes and Zs. Tőkei

Imec, Kapeldreef 75, B-3001 Leuven, Belgium

phone: (32)16/281621; fax: (32)16/281576; e-mail: Kristof.croes@imec.be

Abstract - **Extremely low field time dependent dielectric breakdown measurements were performed on single damascene structures with 90 and 50nm ½pitch integrated in a porous low-k material (k=2.5). We found with statistical significance that the E-model and the $\sqrt{E}$-model were too conservative to extrapolate our high field data to these low fields. Also, while soft breakdown does not occur at higher fields and wider spacings, we demonstrate that the time difference between soft and hard breakdown becomes significant in the 50nm ½pitch structures at normal operating fields. Besides this, we detected a change in distributional shape at these low fields and we argue that an extrinsic failure mode could be driving these failures or that the role of spacing variations across the wafer becomes more significant at lower fields.**

Keywords - BEOL, copper, low-k, TDDB, low field

I. INTRODUCTION

Time dependent dielectric breakdown (TDDB) of the intermetal dielectric (IMD) in Back-End-Off-Line (BEOL) interconnects [1] became a serious concern with the replacement of SiO_2 as IMD by low-k materials. These materials are generally weaker and tend to break faster. Typical lifetime criteria for such dielectrics are 0.01% failures after 10 years at normal operating fields. Reliability experiments are conducted at much higher fields and methods have to be developed to extrapolate the results obtained at high fields to normal operating fields.

Over the past years, a significant amount of studies have been done about the field dependence of the TDDB time to failure (TTF). The E-model, where ln(TFF) is assumed to be linearly dependent on the electric field E, has been widely proposed [2-4]. Theoretical justifications were based on the assumption that weak bonds in the dielectric can be broken by applying an external electric field [5] or that Cu+ diffusion and drift through the dielectric forms leakage passages and results in dielectric breakdown [3]. More recently, the $\sqrt{E}$-model (ln(TTF)~ $\sqrt{E}$) has been proposed [6-7]. This model was motivated by assuming dielectric degradation due to Cu-ion drift through the dielectric and on the observation that the two main conduction mechanisms (Poole Frenkel and Schottky Emission) have a $\sqrt{E}$-dependence of the leakage current. Lloyd et al. [8] proposed the impact damage model where ln(TTF)~A$\sqrt{E}$+B/E. The key assumption in this model is that damage induced to the IMD by electrons that are accelerated by the electric field is directly linked to failure. At lower fields, this model is dominated by the 1/E-term. Finally, G. Haase [9]

proposed a numeric model based on charges being trapped in the IMD. The model is less conservative than the $\sqrt{E}$-model, but more conservative than the 1/E-model.

Besides the above models that have been proposed for BEOL TDDB, also the pure 1/E-model (ln(TTF)~1/E) [10-11] and the power law (TTF~E^n) [12-13] are often considered as they have been widely studied for the application of gate oxide TDDB. Finally, some more recent studies argue that different models could exist at high and low fields[14-15].

The main drawback of the above mentioned studies is that the model validation was mainly done using indirect measurements (i.e. measurements at different temperatures, areas or spacings, studies of current versus voltage or current versus time curves, …). A direct measurement at normal operating temperatures has never been done.

This paper summarizes TDDB tests carried out at very low fields on structures with 90 and 50nm ½pitch at 100°C, which is close to typical normal operating temperature. The purpose of the tests was to investigate the TDDB-behavior at these low fields.

II. EXPERIMENTAL

Single damascene copper structures were integrated on 300mm wafers in a SiOCH CVD low-k IMD (k-value ~2.5, porosity ~25%). A TaN/Ta layer was used as barrier between copper and IMD. To prevent moisture absorption, the wafers were passivated with a 35nm SiCN, 300nm SiO_2 and 500nm Si_3N_4-stack.

Figure 1 shows the two types of test structures that have been used for this study: A) a 1cm long meander-fork structure with 90nm ½pitch and B) a 1cm long fork-fork structure with 50nm ½pitch. Both structures were printed using 193nm immersion lithography and patterned by using a metal hard-mask approach. The 90nm ½pitch structures were integrated

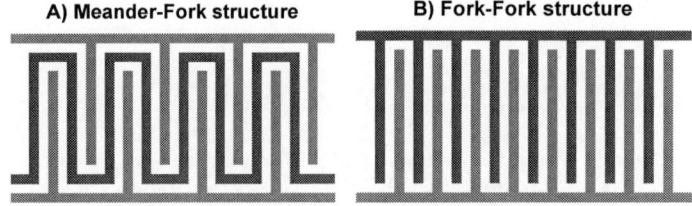

Figure 1. Different test structures used in this study.

with conventional patterning techniques [16], while the 50nm ½pitch structures were integrated using the double patterning approach [17]. The actual spacings of the integrated structures have been obtained using cross-sectional TEM. The spacings for the 90nm ½pitch structures were obtained to be 105nm, while those for the 50nm ½ pitch structures were 45nm.

To find the test voltages for the TDDB experiments, V-ramps have been performed at wafer level at 100°C. Figure 2 shows a typical IV sweep for each of the structures. The arrows indicate at what fields the TDDB tests were performed. These tests were performed at 100°C at wafer level or package level, depending on the failure times at a given field. For each structure, one condition was chosen on which both wafer level and package level tests were performed in order to validate that tests at wafer- and package level gave the same result. Figure 3 shows an example for the 90nm ½pitch structures where failure times obtained at wafer- and package level are compared. No differences are observed.

III. RESULTS

Figure 4 shows current versus time curves. The 90nm ½pitch structures show an abrupt current increase at failure at all fields (Figure 4A). Also the 50nm ½pitch structures show such an abrupt increase at high fields, but at lower fields a noisy behavior is observed before breakdown (Figure 4B). This behavior has been linked to soft breakdown (SBD) in earlier papers about gate oxide reliability and recently also in papers about BEOL TDDB [9,18-19]. When SBD occurs before hard breakdown (HBD), two different failure criteria can be used. Failure criterion 1 (FC1), we define failure as any significant current increase, which is higher than the system background, representing a leaky capacitor. Failure criterion 2 (FC2) is where the sample is completely shorted, meaning that the device is not functional anymore.

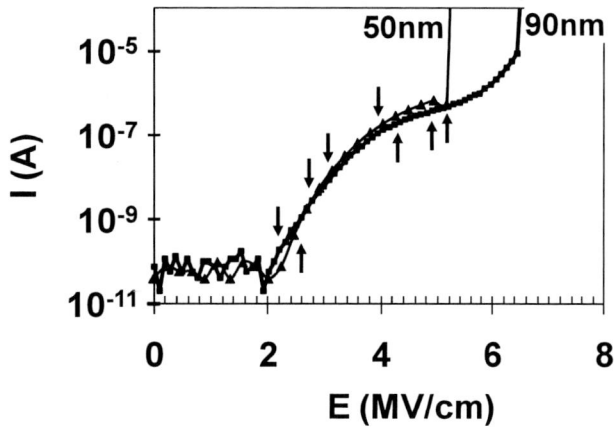

Figure 2. Typical IV curve for the two used structures. Arrows indicate the fields at which the TDDB-tests have been performed.

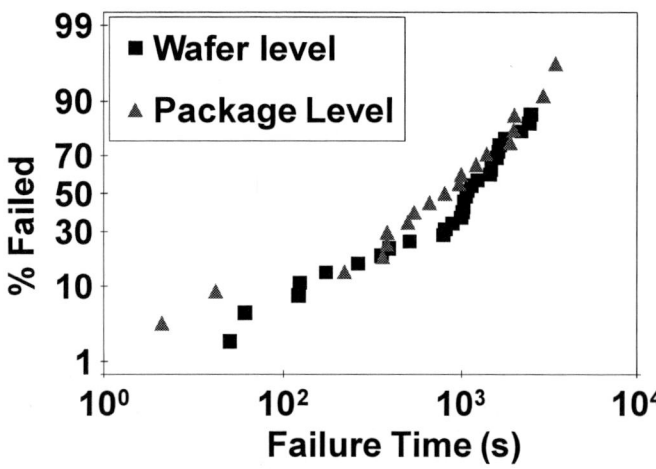

Figure 3. Lognormal probability plot of failure times obtained on 90nm ½pitch structures stressed at 4.3MV/cm at wafer- and package level.

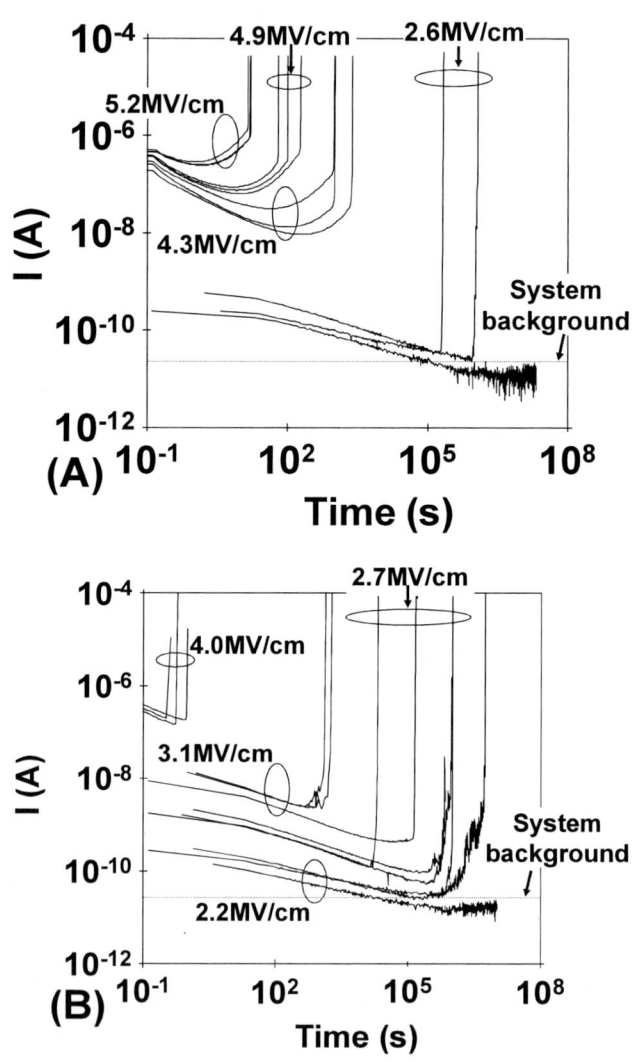

Figure 4. Current versus time curves for the (A) 90nm and (B) 50nm ½pitch structures for different fields. Noisy behavior before final breakdown is observed for small pitches at low fields.

978-1-4244-5430-3/10 $26.00 © 2010 IEEE

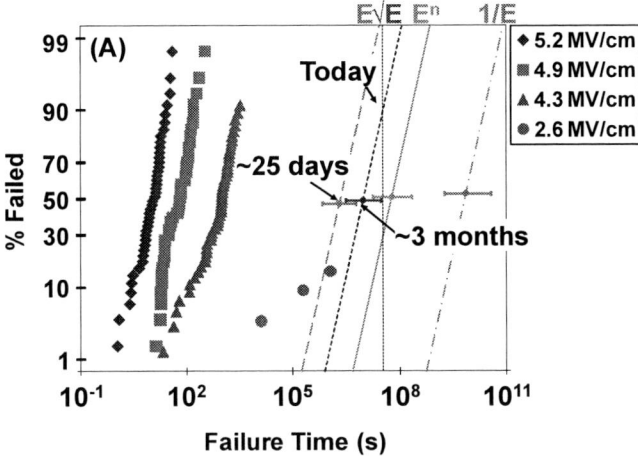

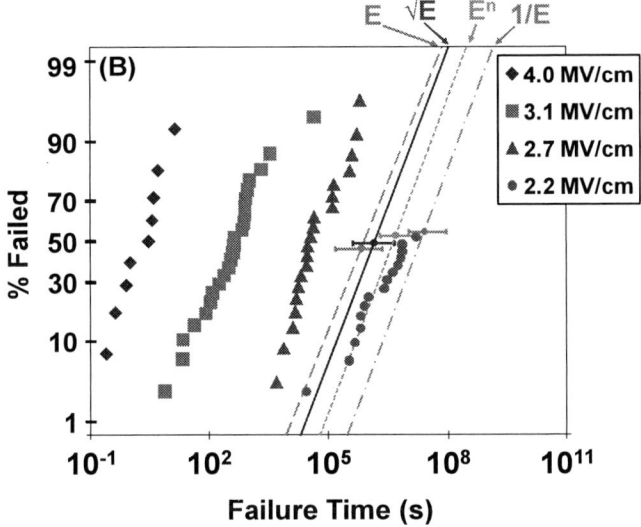

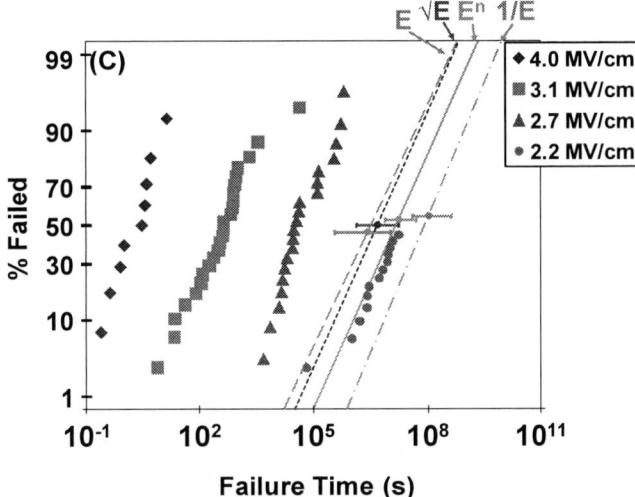

Figure 5. Lognormal probability plot of failure times: (A) 90nm ½pitch, (B) 50nm ½pitch using FC1 and (C) 50nm ½pitch using FC 2. Extrapolation to lower fields using the E-model, the √E-model, the power law and the 1/E-model has been done based on the three higher fields using maximum likelihood estimation.

Figure 5 shows the lognormal probability plots of the failure times for the 90nm (Figure 5A) and 50nm ½pitch structures at the different stress fields. For the 50nm ½pitch structures, both failure criteria have been used (Figure 5B shows FC1 and Figure 5C shows FC2). Using maximum likelihood fitting, the three highest conditions have been fitted using the E-model, the √E-model, the power law (indicated with E^n) and the 1/E-model. From this fit, an extrapolation is done to the lowest stress field for all percentiles and the 95% prediction interval of the extrapolated $t_{50\%}$ is indicated on the graph.

IV. DISCUSSION

A. E- and √E-model too conservative

Based on the high field data from the 90nm ½pitch-structures (Figure 5A), the E-model would have estimated 50% failures after ~25 days of testing at 3MV/cm, while the most optimistic prediction of the $t_{50\%}$ would be ~3 months (upper level of the 95% prediction interval). For the √E-model, the estimated $t_{50\%}$ at 3MV/cm is about 3 months. Today, after about 1 year of testing, we reached the upper limit of the 95% prediction interval of this model as well. At this moment, we only observed 15% failed samples, indicating that the upper limit of the 95% prediction interval obtained by using the √E-model gives a too conservative estimate of the real $t_{50\%}$.

The choice for calculating *prediction* intervals rather than *confidence* intervals on the extrapolated $t_{50\%}$ is important. An x% *confidence* interval of an estimate of a parameter A returns two values between which the real parameter A will fall. The probability that the real parameter A will fall outside of this interval is x%. An x% *prediction* interval returns two values between which a **future observation** will fall. If a future observation falls outside an estimated confidence interval, statistical evidence that the used estimate is wrong is not obtained, while this evidence is obtained when a future observation falls outside a prediction interval [20]. Since for our 90nm ½pitch-data, the observed $t_{50\%}$ at 2.6MV/cm is above the upper limit of the 95% prediction interval as estimated by both the E-model and the √E-model, we can conclude with statistical significance that both models are too conservative to explain our low field data.

With the 50nm ½pitch data using FC1 as failure criterion (Figure 5B), the observed $t_{50\%}$ at 2.2MV/cm is more than a factor 2 higher than the upper level of the 95% prediction interval of the $t_{50\%}$ as predicted by the √E-model. When using FC2 (Figure 5C), similar conclusions can be drawn: both the E-model and the √E-model are too conservative to explain our low field data.

As there is no common agreement on the choice of the distribution function of failures times for damascene BEOL TDDB, we wanted to exclude the dependency of our conclusions on this choice. Figure 6 shows the exact same data as shown on Figure 5B, but now on a Weibull scale. Also the fits of the different models were done using the Weibull distribution. The same conclusions as before can be made: both the E-model and the √E-model are too conservative to

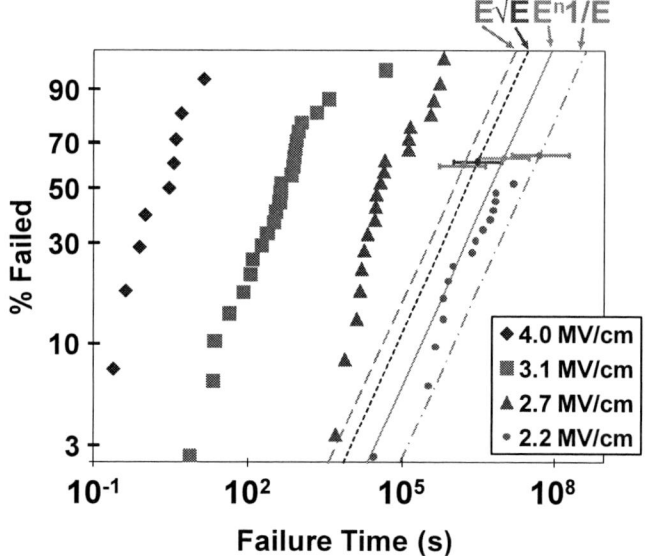

Figure 6. Weibull probability plot of the failure times obtained on the 50nm ½pitch structures using FC1. Extrapolation to lower fields using the E-model, the √E-model, the power law and the 1/E-model has been done based on the three higher fields.

explain our low field data. As such, the choice of the distribution function does not influence our conclusions.

In summary, we measured directly that the most optimistic prediction (i.e. the upper level of the prediction interval) of both the E- and the √E-model are too conservative to predict our low field data. This means that only lifetime models that predict longer lifetimes than the E- and √E-model can be valid alternatives to extrapolate data obtained at high fields. To our knowledge, the only models that predict such a behavior are proposed by Lloyd et al. [8] and Haase [9]. We argue that the theoretical arguments described in these papers are in good agreement with our experimental data.

Figure 7 shows the $t_{50\%}$ versus electric field for the 50nm ½pitch samples using FC1. Predictions of the different models were done using maximum likelihood estimation and are based on the three highest stress conditions (as in Figure 5B). The impact of a wrong model assumption is demonstrated. While the E-model and the √E-model predict median lifetimes in the order of 10-100 years at an electric field of 1.5MV/cm, less conservative models predict median lifetimes of several orders of magnitude higher.

B. Difference in failure times when considering SBD or HBD

To quantify the effect of the choice of the failure criterion on the lifetimes at normal operating conditions, the 50nm ½pitch data at all fields have been fitted using the E-model, the √E-model, the power law and 1/E-model using both failure criteria FC1 and FC2. Using these fits, extrapolation was done to lower fields. The ratio of the lifetimes between SBD and HDB increases significantly with decreasing field (Figure 8). For the most conservative lifetime model (the E-model), this ratio is about 3 to 5, depending on the normal operating field, while for a less conservative model, this ratio can go as high as 15 for the power law and even 950 for the 1/E-model.

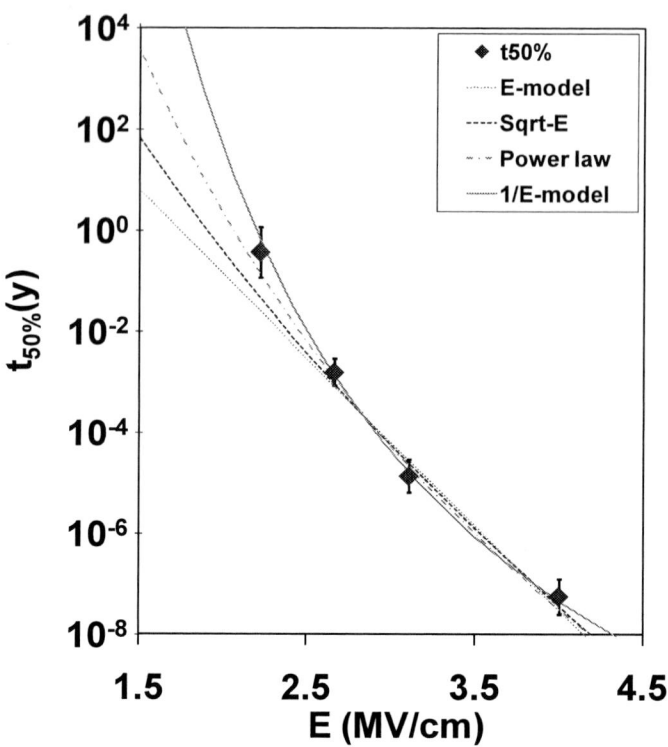

Figure 7. $t_{50\%}$ versus E-field for the 50nm ½ pitch samples using FC1. Predictions of the different models were done using maximum likelihood estimation and are based on the three highest stress conditions (as in Figure 5B).

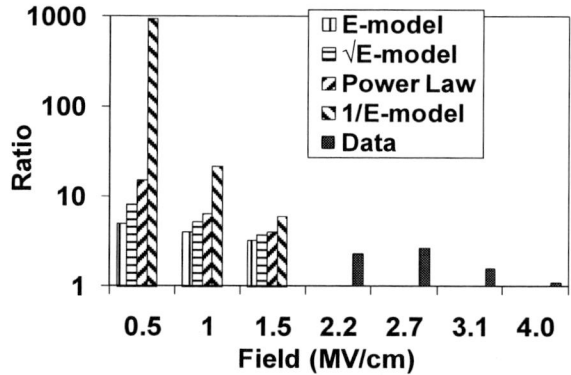

Figure 8. Estimated ratio between the $t_{50\%}$ obtained using FC1 and FC2. Data have been obtained by fitting individual stress conditions. Predictions to lower fields are based on maximum likelihood fits of the different models and failure criteria using all 4 stress conditions.

C. Change in distributional shape at low fields

Figure 5 suggests that the spread of the distribution of failure times becomes higher at lower fields. Additional to the simultaneous maximum likelihood fit of the three highest conditions of the failure times obtained from the 50nm ½pitch structures using FC1, Figure 9 shows a separate fit of the lowest stress condition. A significantly higher sigma compared to the fits at higher fields is observed.

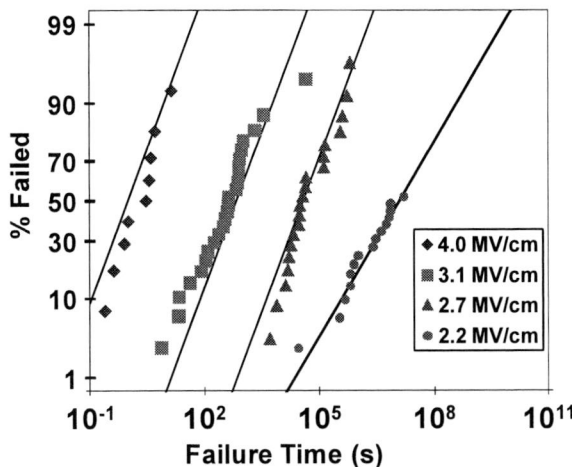

Figure 9. Lognormal probability plot of the failure times obtained using FC1 on the 50nm ½pitch structures. The three highest conditions were fitted using a simultaneous maximum likelihood fit while the lowest condition was fitted separately.

It is hard to judge whether the failed samples of the 2.6MV/cm-condition from the 90nm ½pitch samples are early failures or are due to an increased width of the distribution (Figure 5A). To confirm this, we used slightly higher fields to stress a companion wafer with the same structures. The result is shown in Figure 10. Still, both interpretations of a separate failure mode and an increased sigma are valid alternatives to interpret the distributional shape at the lowest field.

While Kim et al. [21] reported an early failure mode at extremely low fields, Vilmay et al. [22] argued that, based on simulations using the √E-model, spacing variations across the wafer would result in lower sigma's at normal operating fields. We performed similar simulations using all 4 lifetimes models. To estimate the spacing variation for our test material, we performed CD-measurements of the spacings after etch across the wafer. After the corrections with the TEM's as mentioned in section II, we found the spacings were normally distributed

with a mean of 45nm and a sigma of 3.6nm. The simulation approach was as follows:

1. Simulate a set of spacings from a normal distribution with mean=45 and σ=3.6.

2. For a certain voltage, generate the field for each spacing.

3. Based on the $t_{50\%}$ and acceleration factors obtained from the fits done in Figure 5B, calculate a failure time for each of the simulated fields.

4. Calculate the sigma of the logarithm of these failure times.

Note that the simulated sigma is only the result of the spacing variations, while any other sources of variation are excluded. Figure 11 shows the obtained sigma at different fields for the different lifetime models. The same decreasing trend is observed for the √E-model as obtained in [22]. Less conservative models, however, predict an increasing sigma with decreasing field. This is in agreement with our data, which is another argument that the E- and √E-model are too conservative to explain the lifetimes at lower fields.

The predicted increase of sigma when assuming the 1/E-model is significant and its impact on low percentiles at lower fields is investigated in Figure 12. The exact same lines as in Figure 7 are shown but now using the 0.01%-percentile ($t_{0.01\%}$) instead of the median ($t_{50\%}$). The extrapolation of the $t_{50\%}$ from Figure 7 to the $t_{0.01\%}$ was done using the simulated sigma values at the different fields (Figure 11). The differences between the different models become smaller at the 0.01%-percentile, but are still significant.

V. CONCLUSIONS

Extremely low field time dependent dielectric breakdown data have been obtained on damascene structures with 90 and 50nm ½pitch integrated in a porous low-k material (k=2.5). High field data have been used to predict the

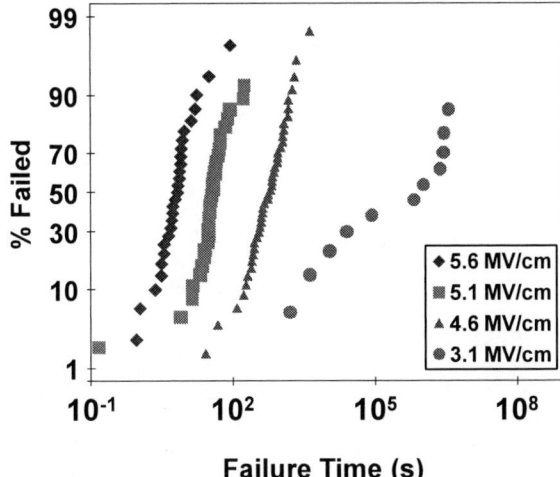

Figure 10. Lognormal plot of failure times obtained at 90nm ½pitch structures from a companion wafer as the one used in this study.

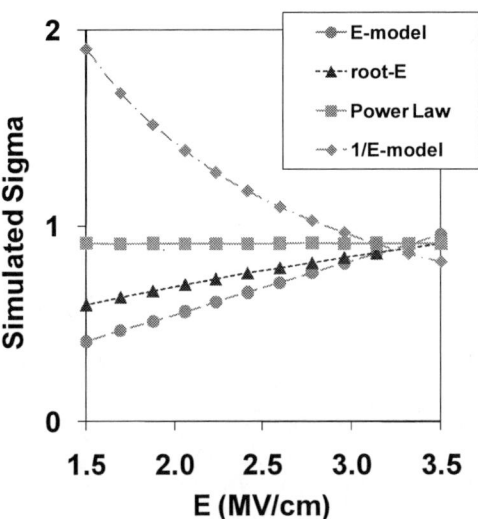

Figure 11. Simulated contribution of spacing variations across the wafer to the sigma of the lognormal distribution.

978-1-4244-5430-3/10 $26.00 © 2010 IEEE

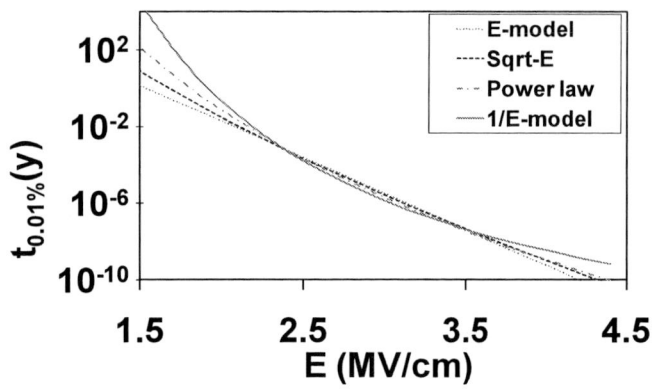

Figure 12. $t_{0.01\%}$ versus E-field for the 50nm ½ pitch samples using FC1. Prediction from the $t_{50\%}$-values shown in Figure 7 to the 0.01%-percentiles were done using the simulated sigma-values from Figure 11.

median time to failure at low fields. We found that even the most optimistic prediction of both the E- and the √E-model are too conservative to describe our low field data, meaning that only lifetime models that predict longer lifetimes than the E- and √E-model can be valid alternatives to extrapolate data obtained at high fields to normal operating fields. Besides this, we found that the difference in failure times between soft- and hard breakdown becomes significant at low spacings and low fields. Finally, we observed a change in distributional shape at lower fields and simulations showed that, assuming the less conservative models, part of this widening could be explained by spacing variations across the wafer.

ACKNOWLEDGMENT

The authors would like to thank Ellipsiz TestLab Pte Ltd for providing their stress and measurement facilities during the course of this study. Also, the participants of imec's barrier reliability meeting and the members of the interconnect and remo group are acknowledged for fruitful and stimulating discussions. In particular, Dr. Marianna Pantouvaki and Dr. Laureen Carbonel are acknowledged for supplying the test material, Dr. Mirko Scholtz and Dr. Dimitri Linten for their support in helping to resolve the packaging issues, Chris Wright for the operational work and Larry Zhao, Intel assignee at imec, and Dr. Michele Stucchi for the nice discussions.

REFERENCES

[1] R. Tsu, J. W. McPherson and W. R. McKee, "Leakage and breakdown reliability issues associated with low-k dielectrics in a dual-damascene Cu process", IEEE Int. Reliability Physics Symposium (IRPS), p. 348, 2000

[2] G. S. Haase, E. T. Ogawa and J. W. McPherson, "Breakdown Characteristics of interconnect dieletrics", IEEE Int. Reliability Physics Symposium (IRPS), p. 466, 2005

[3] W. Wu, X. Duan and J. S. Yuan, "A physical model of time-dependent dielectric breakdown in copper metallization", IEEE Int. Reliability Physics Symposium (IRPS), p. 282, 2003

[4] J. Kim, E. T. Ogawa and J. W. McPherson, "Time dependent dielectric breakdown characteristics of low-k dielectric (SiOC) over a wide range of test areas and electric fields", IEEE Int. Reliability Physics Symposium (IRPS), p. 399, 2007

[5] J. W. McPherson and H. C. Mogul, "Underlying physics of the thermochemical E model in describing low-field timedependent

dielectric breakdown in SiO2 thin films," Journal of Applied Physics, Vol. 84, p. 1513, 1998

[6] F. Chen, O. Bravo, K. Chanda, P. McLaughlin, T. Sullivan, J. Gill1, J. Lloyd, R. Kontra and J. Aitken, "A comprehensive study of low-k SiCOH TDDB phenomena and its reliability lifetime model development", IEEE Int. Reliability Physics Symposium (IRPS), p. 46, 2006

[7] N. Suzumura, S. Yamamoto, D. Kodama, K. Makabe, J. Komori, E. Murakami, S. Maegawa and K. Kubota, "A new TDDB degradation model based on Cu ion drift in Cu interconnect dielectrics", IEEE Int. Reliability Physics Symposium (IRPS), p. 484, 2006

[8] J. R. Lloyd, E. Liniger and T. M. Shaw, "Simple model for time-dependent dielectric breakdown in inter- and intralevel low-k dielectrics", Journal of Applied Physics, Vol. 98, p. 084109, 2005

[9] G. S. Haase, "A model for electric degradation of interconnect low-k dielectrics in microelectronic integrated circuits", Journal of Applied Physics, Vol. 105, p. 044908, 2009

[10] I. C. Chen, S. Holland and C. Hu, "A quantitative model for time-dependent breakdown in SiO2", IEEE Int. Reliability Physics Symposium (IRPS), p. 24, 1985

[11] J. W. McPherson, R. B. Khamankar and A. Shanware, "Complementary model for intrinsic time-dependent dielectric breakdown in SiO2 dielectrics", Journal of Applied Physics, Vol. 88, p. 5351. 2000

[12] J. Suñé and E. Wu, "A New Quantitative Hydrogen-Based Model for Ultra-Thin Oxide Breakdown", Digest of Technical Papers on Symposium on VLSI Technology, p.97, 2001

[13] E. Wu, J. Suñé, W. Lai, A. Vayshenker and D. Harmon, "A Comprehensive Investigation of Gate Oxide Breakdown of P+Poly/PFETs Under Inversion Mode", Technical Digest of the International Electron Devices Meeting (IEDM), p. 339, 2005

[14] F. Chen, J. Gambino, M. Shinosky, B. Li, O. Bravo, M. Angyal, D. Badami, and J. Aitken, "Correlation between I-V Slope and TDDB Voltage Acceleration for Cu/Low-k Interconnects", International Interconnect Technology Conference (IITC), p. 182, 2009

[15] N. Suzumura, S. Yamamoto, D. Kodama, H. Miyazaki, M. Ogasawara, J. Komori and E. Murakami, "Electric-field and temperature dependencies of TDDB Degradation in Cu/low-k damascene structures", IEEE Int. Reliability Physics Symposium (IRPS), p. 138, 2008

[16] M. Pantouvaki, H. Struyf, D. Hendrickx, N. Heylen, O. Richard and G. P. Beyer, "The impact of ash on TDDB of metal-hard-mask-etched Cu/low-k interconnects", ", Proceedings of the Advanced Metallization Conference (AMC), p. 733, 2008

[17] J. Van Olmen, A. Al-Bayati, G. P. Beyer, P. Boelen, L. Carbonell, C. Zhao, I. Ciofi, M. Claes, A. Cockburn, G. Druais, D. Hendrickx, N. Heylen, E. Kesters, S. Lytle, A. Noori, M. Op de Beeck, H. Struyf, Zs. Tőkei and J. Versluijs, "Integration of 50nm half pitch single damascene copper trenches in BDII by means of double patterning 193nm immersion lithography on metal hardmask", Proceedings of the Advanced Metallization Conference (AMC), p. 355, 2008

[18] K. Croes, G. Cannatá, L. Zhao and Zs. Tőkei, "Study of copper drift during TDDB of intermetal dielectrics by using fully passivated MOS capacitors as test vehicle", Microelectronics Reliability, Vol. 48, p. 1384, 2008

[19] F. Chen, B. Li, J. Gambino, S. Mongeon, P. Pokrinchak, J. Aitken, D. Badami, M. Angyal, R. Achanta, G. Bonilla, G. Yang, G., P. Liu, K. Li, J. Sudijono, Y. Tan, T. J. Tang and C. Child, "Critical Ultra Low-k TDDB Reliability Issues For Advanced CMOS Technologies", IEEE Int. Reliability Physics Symposium (IRPS), p. 464, 2009

[20] J. Neter, W. Wasserman and M. H. Kutner, "Applied Linear Statistical Models", IRWIN, 3rd Edition, 1990

[21] J. Kim, E. T. Ogawa and J. W. McPherson, "Time Dependent Dielectric Breakdown Characteristics of Low-k Dielectric (SiOC) Over a Wide Range of Test reas and Electric Fields", IEEE Int. Reliability Physics Symposium (IRPS), p. 399, 2007

[22] M. Vilmay, D. Roy, C. Monget, F. Volpi, J.-M.Chaix, "Copper line topology impact on the SiOCH low-k reliability in sub 45nm technology node", IEEE Int. Reliability Physics Symposium (IRPS), p. 606, 2009

978-1-4244-5430-3/10 $26.00 © 2010 IEEE

STUDY OF LEAKAGE MECHANISM AND TRAP DENSITY IN POROUS LOW-K MATERIALS

Gianni Giai Gischia*, Kristof Croes, Guido Groeseneken* and Zsolt Tőkei
IMEC
Kapeldreef 75, B-3001 Leuven, Belgium
phone: (0032)16-288-214, e-mail: giai.gischia.gianni@imec.be

* also at ESAT Department
K. U. Leuven
Kasteelpark Arenberg 10, B-3001 Leuven, Belgium

Valery Afanas'ev
Department of Physics and Astronomy
K. U. Leuven
Celestijnenlaan 200d, B-3001 Leuven, Belgium

Larry Zhao
Intel assignee at IMEC
Kapeldreef 75, B-3001 Leuven, Belgium

Abstract—The field and temperature dependence of the leakage current of low-k material is studied by using planar capacitors. First it is shown that our planar capacitors are suitable test vehicles to analyze the intrinsic properties of low-k materials. Then an evaluation of the trap density of the investigated low-k material is performed. Eventually different models such as Poole-Frenkel emission, Schottky emission and trap-assisted Fowler-Nordheim tunneling are analyzed in a temperature range from 80 K to 473 K and in a field range from 4.2 MV/cm to 6.6 MV/cm. It is found that Poole-Frenkel emission and Schottky emission do not fit the experimental data, whereas trap-assisted Fowler-Nordheim tunneling exhibits an adequate match between the extracted parameters and the theoretical predictions.

Keywords: conduction mechanism, leakage current, low-k, trap density

I. INTRODUCTION

The ongoing scaling of integrated circuits has required the use of porous low-k materials in order to replace SiO_2. When SiO_2 was used as inter-metal dielectric Back End of Line (BEOL) reliability was never a big concern. By contrast, the introduction of porous low-k materials, together with the decrease of critical dimensions, made interconnect dielectric reliability a major issue. Consequently a study of the mechanisms that cause low-k degradation is necessary.

In order to gain a detailed knowledge of the degradation mechanisms of porous low-k materials one of the first steps is to establish the leakage current mechanisms. In literature this topic has been addressed many times [1], [2], [3] but at the

moment there is a lack of uniform explanations of the phenomena. Some of the proposed conduction mechanisms take place through the traps which are present in the low-k material; therefore, an evaluation of the trap density and type is required, although, to our knowledge this parameter has been rarely investigated [4], [5].

A comprehensive assessment of the leakage current of low-k materials can be accomplished only if the intrinsic properties of the involved materials are analyzed. Damascene structures, which are the typical BEOL test vehicles, do not allow focusing on the fundamental properties of the low-k materials because the dielectric is affected by various process related issues, such as dielectric modification during patterning, field enhancement due to Line Edge Roughness (LER) and unwanted Cu residues after Chemical Mechanical Polishing (CMP). In order to cope with this problem low-k planar capacitors have been developed. They are test structures which allow focusing on the fundamental features of porous low-k materials since plasma etching on the dielectric is not required during the patterning whereas LER and CMP Cu residues do not affect these structures.

In this paper, it is first shown that the features of low-k planar capacitors scale with the area, indicating that they are proper devices to study the intrinsic properties of low-k materials. Then the trap density is assessed in order to achieve a more exhaustive knowledge of the dielectric. Finally the field and temperature dependence of the leakage current in low-k material is analyzed by comparing different models, such as Poole-Frenkel emission, Schottky emission and trap-assisted Fowler-Nordheim tunneling.

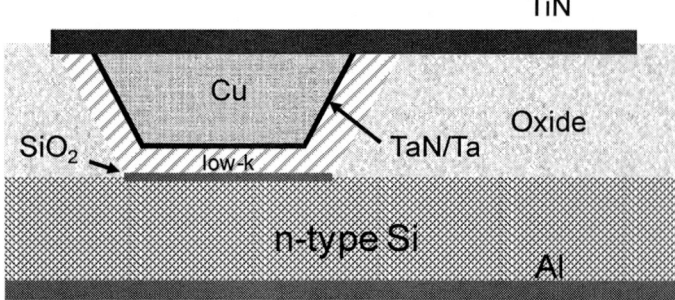

Figure 1. Cross section of low-k planar capacitor used in the experiments.

II. EXPERIMENTAL DETAILS

Low-k planar capacitors were used during the experiments. A schematic cross section is shown in Fig. 1. On top of a n-type silicon substrate first a 1nm SiO_2 layer is thermally grown in order to minimize interface effects between the substrate and the low-k material, followed by a CVD-deposition of a 60 nm low-k dielectric. The low-k deposition takes place after the patterning of the device, therefore the low-k is kept pristine and no plasma damage occurs on it. The dielectric material is a silica-based low-k with dielectric constant ε_r=2.5, porosity 25%, pore radius 0.9 nm and the following chemical composition: 40% H, 13% C, 29% O, 18% Si. Then a 6 nm layer of 6nm PVD TaN/Ta is deposited as metal barrier followed by the deposition of a Cu gate. The copper layer thickness after seed deposition, electroplating and CMP is approximately 300 nm. Finally, the devices are sealed with 30 nm of TiN which acts as passivation layer. On the back side a layer of 500 nm PVD Al is deposited in order to create an ohmic contact. More details about the test structure fabrication can be found in [6] and [7].

The characterization of the properties of low-k planar capacitors was performed through Time Dependent Dielectric Breakdown (TDDB) experiments. The conduction mechanism was studied through voltage sweeps with a ramp rate of 1 V/s. The trap density was estimated by using voltage sweeps and constant voltage measurements. A HP4142B DC power source was used for the TDDB experiments and a HP4156C semiconductor parameter analyzer was employed for the voltage sweeps and the constant voltage measurements. The experiments were performed in a temperature range from 80 K up to 473 K. From 80 K up to 275 K the measurements were performed in a low pressure chamber ($\sim 10^{-6}/10^{-7}$ mbar) cooled with liquid nitrogen. From 275 K up to 473 K the experiments were conducted in a chamber at atmospheric pressure purged with compressed air. In every experiment the voltage was applied on the TiN electrode, with the Al backside contact grounded.

III. RESULTS AND DISCUSSION

A. Poisson Area Scaling

When the dielectric breakdown is caused by randomly distributed defects, the time-to-breakdown is expected to scale

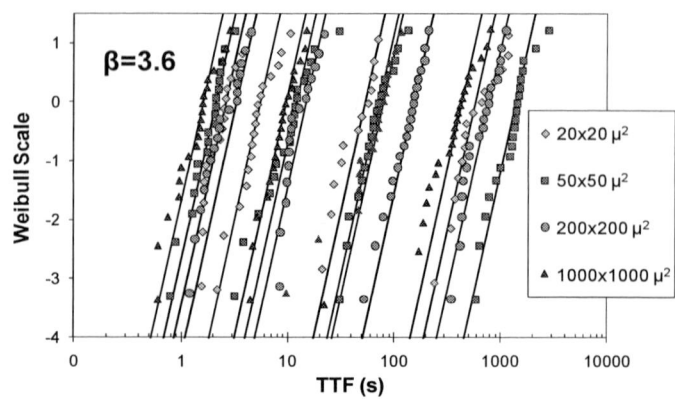

Figure 2. Experimental Weibull distributions of TDDB experiments. The solid lines are the extrapolated fit

with the area [8]. That means that Poisson area scaling, as given by (1) must be fulfilled:

$$\eta(A_1) = \eta(A_2)\left(\frac{A_2}{A_1}\right)^{\frac{1}{\beta}} \qquad (1)$$

where A_1 and A_2 are the areas of capacitors of different size and η and β are the location and the shape parameter of the Weibull distribution, respectively.

To check the validity of (1) TDDB experiments at 100°C were conducted on capacitors with the following areas: 20x20 μm^2 (applied voltages: 41 V, 39 V, 34 V, 32 V), 50x50 μm^2 (applied voltages: 40 V, 37 V, 34 V, 30 V), 200x200 μm^2 (applied voltages: 39 V, 36 V, 32 V, 30 V), 1000x1000 μm^2 (applied voltages: 40 V, 36 V, 32 V, 29 V). Fig. 2 shows the overall experimental Weibull distributions. η and β can be extracted by including all data on Fig. 2 in a single maximum likelihood fitting procedure. The extracted values at a reference area of 50x50 μm^2 and a reference voltage of 32 V are: η=439±14 s and β=3.6±0.1 within 95% confidence bounds. Another way to calculate β is performed by using the least square fit method. As shown in Fig. 3, the location parameter η is calculated singularly (with a reference voltage of 32 V) for each size and then it is plotted versus the area of the investigated capacitors.

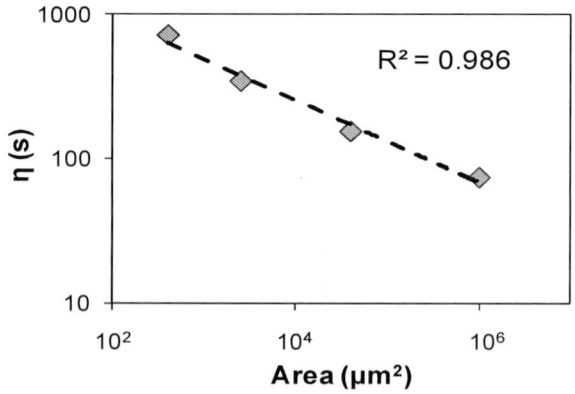

Figure 3. Least square fit of η as a function of the area of the capacitors. The slope of the fit is equal to $-1/\beta$.

978-1-4244-5430-3/10 $26.00 © 2010 IEEE

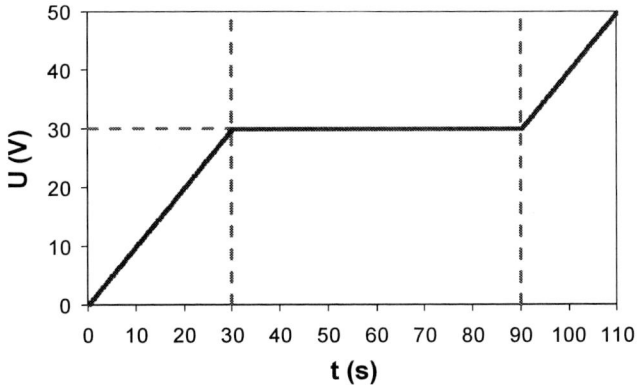

Figure 4. Illustration of the trap density evaluation method. Two voltage sweeps are alternated with a part at constant voltage.

The result of the fit is β=3.5±0.2. The calculated β values from maximum likelihood and least square fit are identical within the confidence bounds. Thus it is concluded that Poisson area scaling is fulfilled for the investigated low-k planar capacitors, indicating that they are suitable test vehicles to study the intrinsic properties of low-k materials.

B. Trap Density Estimation

An evaluation of the trap density has been obtained by performing two voltage sweeps alternated with a constant voltage stress part as depicted in Fig. 4. The first voltage sweep goes from 0 V to V_0. Then the voltage is kept constant at V_0 for a given time. Eventually, without dropping the voltage to 0 V, a second voltage sweep is executed starting from V_0. Both voltage sweeps are performed with a ramp rate of 1 V/s.

Fig. 5 describes the same experimental procedure of Fig. 4 by using a different representation: in this case, the current is plotted as a function of the applied voltage. From 0 V to 30 V the current increases due to the voltage increase. When the voltage is kept constant at 30 V, the traps are gradually filled and thus the current decreases. Eventually, when the second voltage ramp starts the current increases again. The additional voltage required to attain again the current value reached before the drop (ΔV), is a measure for the filled trap concentration, which then can be extracted as [5]:

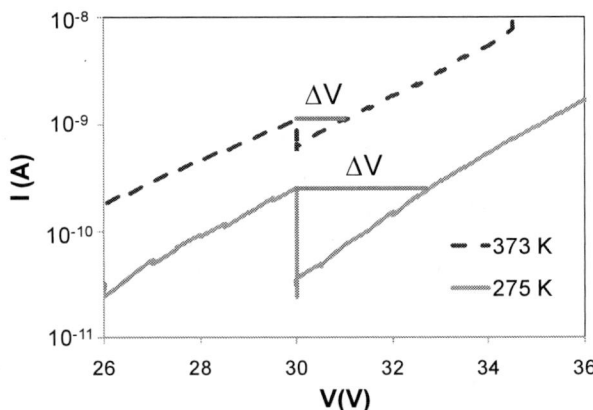

Figure 5. Trap density evaluation. During the part at constant voltage the traps are filled and thus the current decreases.

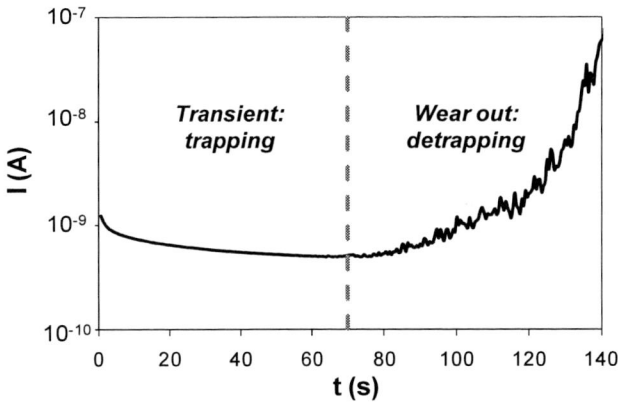

Figure 6. I-t curve measured at V=30 V and T=373 K. The vertical dashed line separates the current transient from the wear-out phase.

$$N_{trap}(1/cm^2) = \frac{2 \cdot \Delta V \cdot \varepsilon_0 \cdot \varepsilon_r}{q \cdot d} \qquad (2)$$

where d is the low-k thickness, q is the elementary charge, ε_0 is the vacuum permittivity and ε_r is the dielectric constant of the low-k material. (2) is based on the capacitance formula of a planar structure: $C=Q_{trap}/(\Delta V \cdot 2)$ where Q_{trap} is the overall charge related to the filled traps and the factor 2 is due to the assumption that the traps are in the middle of the low-k layer, hence the considered low-k thickness is halved.

To obtain a precise estimation of the trap density of the low-k material the proper time length of the part at constant voltage must be selected. In order to do that, prior to the trap density evaluation, I-t curves are measured by using the same temperature and the same voltage value of the constant voltage part (Fig. 6). The current trend in Fig. 6 can be divided into two parts: current transient and wear-out phase. During the current transient the traps are gradually filled and thus the current decreases, when all the traps are filled the current reaches the minimum. During the wear-out phase the detrapping effect has a relevant role and thus the current increases again. If the chosen time length of the part at constant voltage corresponds to the initial part of the current transient, not all the traps are filled and hence the trap density is underestimated.

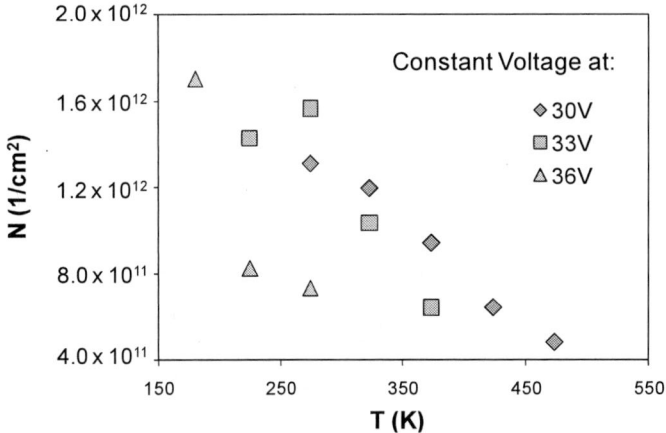

Figure 7. Trapped electron density as a function of the temperature.

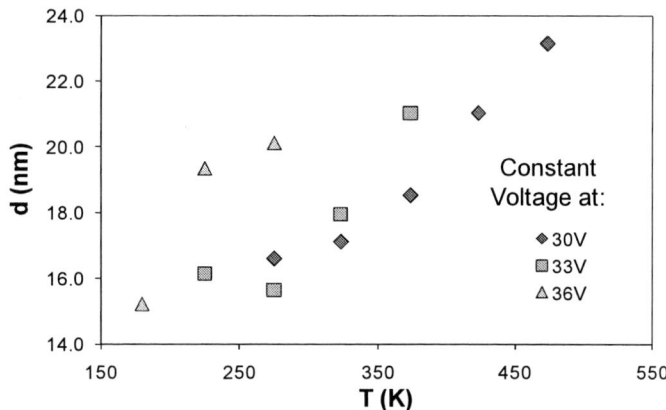

Figure 8. Mean distance between occupied traps. The result is obtained by assuming a uniform trap distribution.

If the chosen time length corresponds to the beginning of the wear-out phase, the detrapping effect affects the current and even in this case the trap density is underestimated. Therefore, the appropriate time length corresponds to the minimum of the current value.

The results of the trap density evaluation are shown in Fig. 7 for an investigated temperature range from 180 K up to 473 K and for three different voltage values of the constant voltage part (30 V, 33 V and 36 V). It is noted that when the temperature is increased or when the part at constant voltage is performed with a higher voltage value, the extracted trap density decreases: in both cases this trend can be explained by an increase of the detrapping effect.

The extracted trap density varies from 4.8×10^{11} cm^{-2} (obtained at 473 K with a constant voltage stress of 30 V) to 1.7×10^{12} cm^{-2} (obtained at 180 K with a constant voltage stress of 36 V). These results are in agreement with typical values found in literature [4]. Moreover, since the overall volume of the low-k material is known, the mean distance between traps (d_T) can be calculated if a uniform trap distribution is assumed. The result (Fig. 8) is in the range of 15 nm to 23 nm.

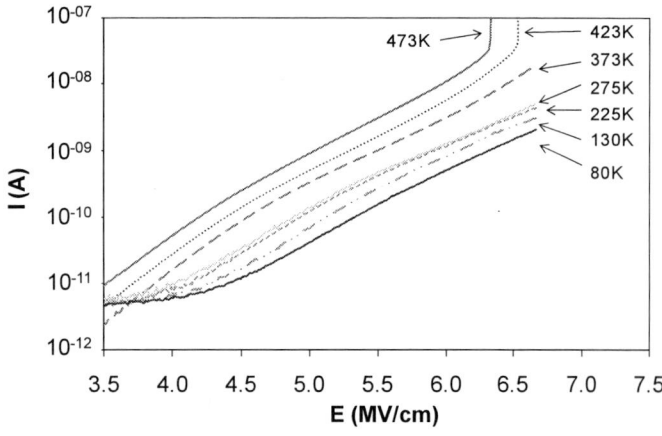

Figure 9. Current-field characteristics measured at different temperatures with planar capacitors of 100×100 μ^2 by using 1V/s voltage sweep rate.

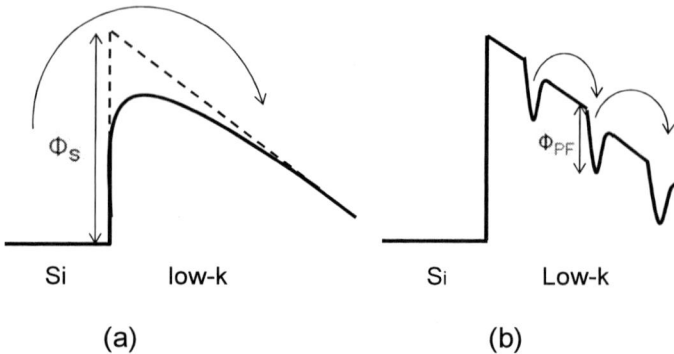

Figure 10. Band diagram illustrating Schottky emission (a) and Poole-Frenkel emission (b) mechanisms.

C. Schottky and Poole-Frenkel emission

The conduction mechanism is studied through voltage sweeps as shown in Fig. 9 by using capacitors with an area of 100×100 μm^2. The investigated field range is between 4.2 MV/cm and 6.6 MV/cm.

In literature both Schottky (3) and Poole-Frenkel emission (4) have been suggested as possible conduction mechanisms for dielectrics [9], [10].

$$J = A^* \cdot T^2 \cdot \exp\left(\frac{-q\phi_S}{kT}\right) \cdot \exp\left(\frac{\beta_S}{2kT} E^{1/2}\right) \quad (3)$$

$$J = \sigma_0 \cdot E \cdot \exp\left(\frac{-q\phi_{PF}}{kT}\right) \cdot \exp\left(\frac{\beta_{PF}}{kT} E^{1/2}\right) \quad (4)$$

β_S and β_{PF} can be expressed as:

$$\beta_{PF} = 2\beta_s = \sqrt{\frac{q^3}{\pi \varepsilon_0 \varepsilon_r}} \quad (5)$$

where A^* is the effective Richardson constant, T is the absolute temperature, q is the elementary charge, k is the Boltzmann constant, E the electric field, σ_0 is the low-field conductivity, Φ_S is the Schottky barrier height, Φ_{PF} is the Poole-Frenkel trap depth, ε_0 is the vacuum permittivity and ε_r is the dielectric constant of the low-k material.

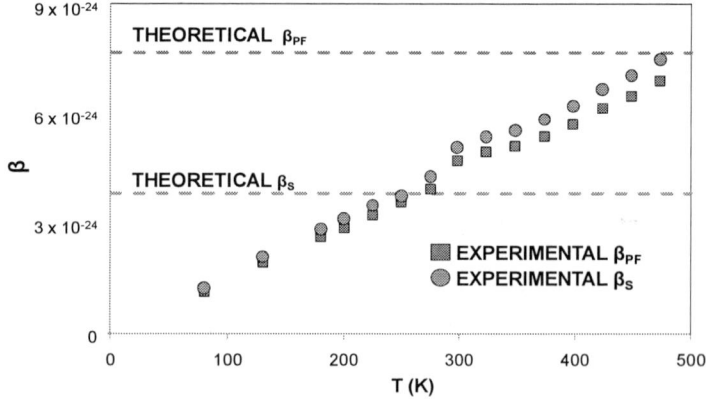

Figure 11. Comparison between theoretical and experimental β_{PF} and β_S.

978-1-4244-5430-3/10 $26.00 © 2010 IEEE

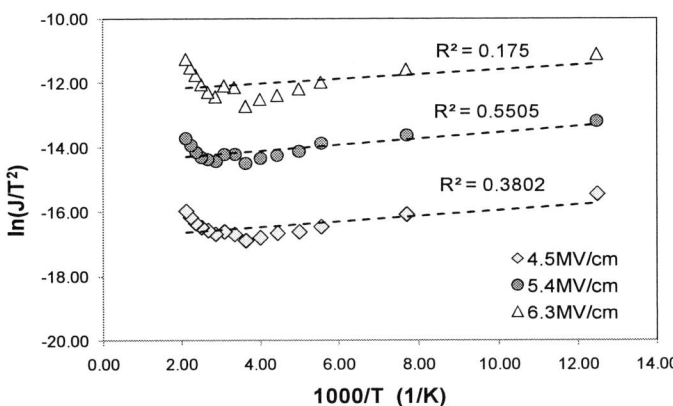

Figure 12. Schottky temperature dependence: the dashed line is the extrapolated linear fit.

Both Schottky and Frenkel models are associated with a barrier lowering in presence of an external electric field; in both cases the electrons emission requires thermal excitation. The Schottky emission takes place at the dielectric interface (Fig. 10 (a)) whereas the Poole-Frenkel emission takes place through traps in the bulk of the dielectric (Fig. 10 (b)).

It is noted from (5) that the theoretical values of β_S and β_{PF} can be easily calculated; with ε_r=2.5 the results are β_{PF}=7.68 x 10^{-24} J·m$^{1/2}$/V$^{1/2}$ and β_S=3.84 x 10^{-24} J·m$^{1/2}$/V$^{1/2}$. Furthermore β_{PF} and β_S can also be extracted from the experimental data by performing a linear fit of $\ln(J/T^2)$ as a function of $E^{1/2}$ for Schottky and a linear fit of $\ln(J/E)$ as a function of $E^{1/2}$ for Poole-Frenkel. β_{PF} and β_S are eventually extracted from the slope of the linear fit. A comparison between the theoretical and the experimental value of β_S and β_{PF} can identify the dominant conduction mechanism. The results are shown in Fig. 11: it is clear that neither of the models delivers an adequate match to the experiment in the entire temperature range.

The temperature dependence of the experimental results can be pointed out if the data in Fig. 9 are plotted for a fixed electric field value as a function of the temperature.

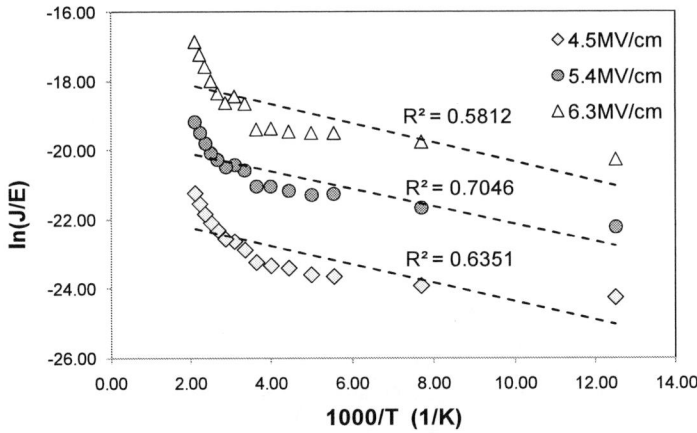

Figure 13. Poole-Frenkel temperature dependence: the dashed line is the extrapolated linear fit.

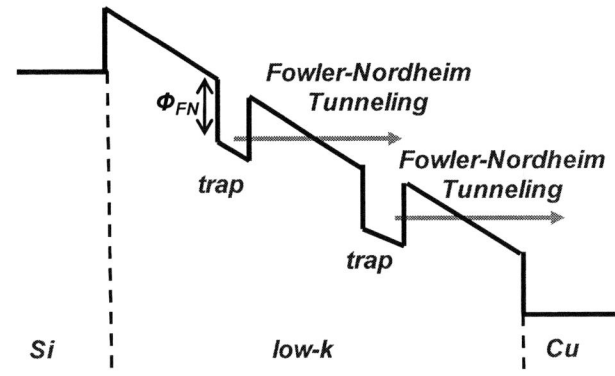

Figure 14. Band diagram showing trap assisted tunneling based on Fowler-Nordheim tunnelling.

For Schottky emission if $\ln(J/T^2)$ is plotted as a function of $1/T$ a linear dependence is expected. For Poole-Frenkel emission if $\ln(J/E)$ is plotted as a function of $1/T$ a linear dependence is expected. Fig. 12 and Fig. 13 show that there is no correlation between the theoretical prediction and the experimental data. As a result both models can be excluded as dominant leakage mechanism.

D. Trap-assisted Fowler-Nordheim tunneling

Another possible conduction mechanism is trap-assisted Fowler-Nordheim tunneling. This model is based on a series of Fowler-Nordheim tunneling transitions [9, 10] which occur through the trap levels in the low-k material (Fig. 14). The leakage current is assumed to be proportional to the Fowler-Nordheim current (6), as given by (7):

$$J_{FN} = \frac{q^3}{8\pi h \phi_{FN}} \cdot \frac{m}{m^*} \cdot E^2 \cdot \exp\left(-\frac{8\pi\sqrt{2m^*} \cdot \phi_{FN}^{3/2}}{3hq \cdot E}\right) \quad (6)$$

$$J_{TAT} \propto J_{FN} \quad (7)$$

where q is the elementary charge, E the electric field, Φ_{FN} is the Fowler-Nordheim trap depth, h is Planck's constant, m is the electron mass and m* is the effective electron mass in the low-k material, which is assumed to be the close to the effective electron mass in SiO$_2$ (m*=0.5m).

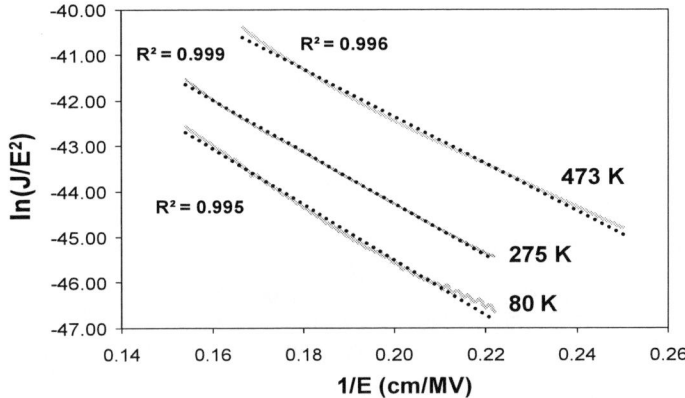

Figure 15. Fowler-Nordheim plot of the leakage current: the dotted line is the extrapolated linear fit.

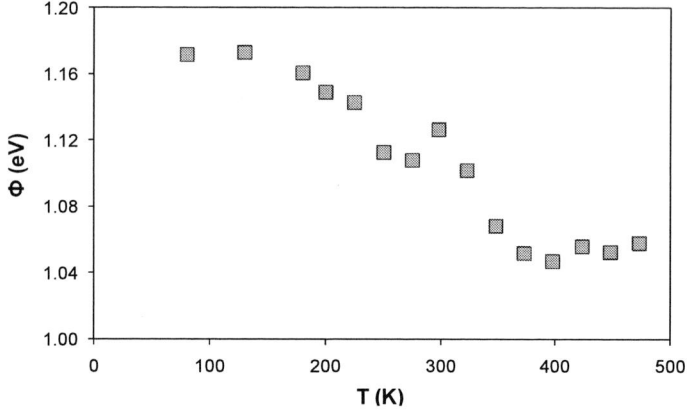

Figure 16. Fowler-Nordheim trap depth as a function of the temperature.

For trap-assisted Fowler-Nordheim tunneling if $\ln(J/E^2)$ is plotted as a function of $1/E$ (the so-called Fowler-Nordheim plot) a linear dependency is expected. Fig. 15 shows that the experimental data match the theoretical field dependence in the entire investigated temperature range.

By using the slope of the lines in Fig. 15 the Fowler-Nordheim trap depth Φ_{FN} can be calculated, according to (8).

$$\phi_{FN} = \left(-\frac{3hq}{8\pi\sqrt{2m^*}} \cdot slope \right)^{2/3} \qquad (8)$$

The results, shown in Fig. 16, indicate that Φ_{FN} varies between 1.05 eV and 1.17 eV. Taking into account the large height of the low-k interface barriers, which appears to be in excess of 4 eV [11], the obtained barrier height of about 1 eV suggests that the observed current is controlled by trap levels inside the insulating layer.

The extracted trap depth can therefore be employed to calculate the barrier transparency coefficient P_t, which for Fowler-Nordheim tunneling is described by (9). Assuming a trap-to-band tunneling mechanism P_t depends only on the trap depth and the electric field, whereas it is independent from the distance between the traps.

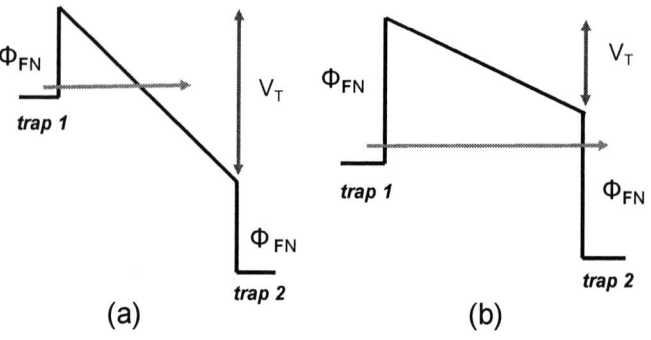

(a)   (b)

Figure 18. (a) Fowler-Nordheim (trap-to-band) tunneling: $V_T > \Phi_{FN}$. (b) Direct (trap-to-trap) tunnelling: $V_T < \Phi_{FN}$.

$$P_t = \exp\left(-\frac{8\pi\sqrt{2m^*} \cdot \phi_{FN}^{3/2}}{3hq \cdot E} \right) \qquad (9)$$

In the investigated field range, which is from 4.2 MV/cm to 6.6 MV/cm, P_t varies from 4.26×10^{-7} to 3.68×10^{-4}, as shown in Fig. 17. The time constant of Fowler-Nordheim detrapping can be calculated with the formula $(P_t \cdot \nu)^{-1}$ where ν is the attempt frequency which is assumed to be in the order of the Si-O bond vibration frequency, which is 10^{13} s^{-1} [12]. As a result, the time constant is in the range of 10^{-7}-10^{-10} s, indicating that a tunneling phenomenon is realistic.

The last parameter investigated is the barrier shape: Fowler-Nordheim tunneling occurs when the barrier is triangular (Fig.18 (a)), whereas if the barrier is trapezoidal a direct tunneling takes place (Fig.18 (b)). The necessary condition for a triangular barrier is $V_T > \Phi_{FN}$ where V_T is the voltage between two adjacent traps; if $V_T < \Phi_{FN}$ the shape of the barrier is trapezoidal. V_T can be written as $V_T = V_0 \cdot d_T / d$, where V_0 is the total voltage applied on the capacitor, d is the low-k thickness and d_T is the mean distance between traps, therefore the condition for a triangular barrier it is given by (10):

$$V_T = V_O \frac{d_\tau}{d} > \phi_{FN} \qquad (10)$$

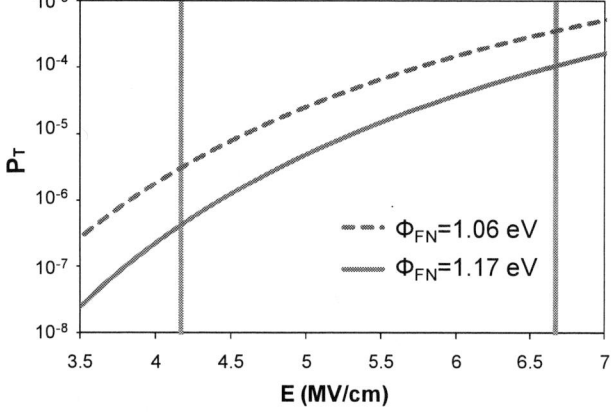

Figure 17. Fowler-Nordheim transmission coefficient of a triangular barrier with height Φ_{FN}. The two vertical lines correspond to the investigated field range (from 4.2 MV/cm up to 6.6 MV/cm). The two curves correspond to the minum and the maxium value of the trap depth found in our samples.

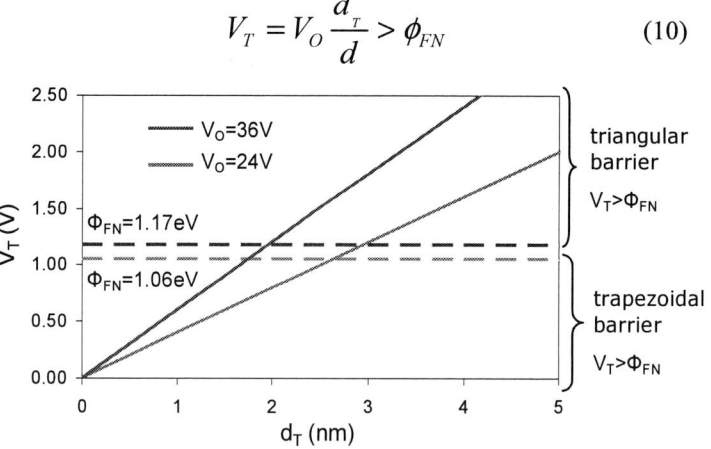

Figure 19. Barrier shape evaluation. The two obliques lines represent the results of (10) for $V_0=24$ V and $V_0=36$ V, which correspond to the maximum and the minum of the investigated field range, respectively. The two horizontal lines correspond to minimum and maximum value of the extracted trap depth.

978-1-4244-5430-3/10 $26.00 © 2010 IEEE

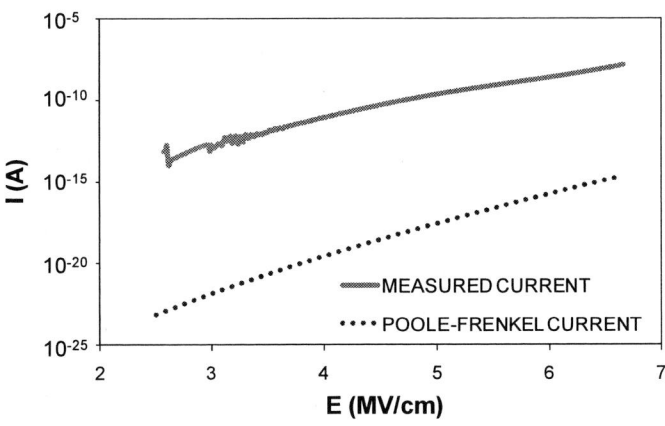

Figure 20. Comparison between measured current and current due to Poole-Frenkel emission at T=298K.

Fig. 19 shows that, in the investigated voltage range, if d_T is longer than 3nm the barrier is triangular. Since the calculated distance between traps is above 15nm (Fig. 8) it is concluded that the extracted barrier shape is consistent with the Fowler-Nordheim tunneling.

Moreover with the extracted parameters it is possible to estimate the magnitude of the leakage current due to the Poole-Frenkel emission. Both trap-assisted Fowler-Nordheim tunneling and Poole-Frenkel emission involve the traps inside the low-k material. Therefore if the calculated trap depth Φ_{FN} is used in (4) the Poole-Frenkel emission leakage current can be assessed. The low-field conductivity of the low-k material is assumed to be the same of SiO_2, which is $\sigma_0 = 5 \times 10^{-18}$ $m^{-1} \cdot \Omega^{-1}$ at 298 K [13]. The result is shown in Fig. 20: the leakage current due to the Poole-Frenkel emission is 9 orders of magnitude lower than the measured current, indicating once again that Poole-Frenkel is not the main conduction mechanism.

IV. CONCLUSIONS

It is proven that Poisson area scaling is fulfilled for the used low-k planar capacitors indicating that they are suitable test vehicles to study the intrinsic properties of the low-k materials.

The obtained trap density varies from 4.8×10^{11} cm^{-2} to 1.7×10^{12} cm^{-2}: the results are in agreement with typical values found in literature [4].

The leakage mechanism of the low-k material is studied in the field range of 4.2 MV/cm to 6.6 MV/cm and in the temperature range of 80 K to 473 K. In the experimental range Poole-Frenkel and Schottky emission are found not to be the dominant conduction mechanisms. In fact the experimental temperature and field dependences do not match the theoretical models.

In contrast, when trap-assisted Fowler-Nordheim tunneling is analyzed, every parameter obtained from the experimental data is consistent with the theoretical model. The appropriate field dependence is confirmed experimentally, the calculated trap depth leads to a realistic tunneling probability and the barrier shape obtained is triangular, as required by the Fowler-Nordheim model. It can thus be concluded that in the investigated field and temperature range the leakage current of low-k planar capacitors is dominated by trap-assisted Fowler-Nordheim tunneling.

REFERENCES

[1] J.W. Tringe et al., "Temperature-dependent current transport in low-k inorganic dielectrics", Journal of the Electrochemical Society, vol. 151 , pp. F128-F131, no5, 2004.

[2] Ou Y. et al., "Conduction Mechanisms of Ta/Porous SiCOH Films under Electrical Bias", Journal of the Electrochemical Society, vol. 155, no12, 2008.

[3] V. Verriere et al., "Dielectric conduction mechanisms of ULK/CU interconnects: Low field conduction mechanism and determination of defect density", IRPS, 2008, pp. 679–680.

[4] J. M. Atkin et al., "Photocurrent spectroscopy of low-k dielectric materials: Barrier heights and trap densities", Journal of Applied Phyics, vol. 103, issue 9, 2008.

[5] M. Vilmay et al., "Characterization of low-k SiOCH dielectric for 45 nm technology and link between the dominant leakage path and the breakdown localization", Microelectronic Engineering, vol. 85, issue 10, October 2008,pp. 2075-2078.

[6] L. Zhao et al., "A Novel Test Structure to Study Intrinsic TDDB of Copper/Low-k, IRPS, 2009.

[7] L. Zhao et al., "A New Perspective of Barrier Material Evaluation and Process Optimization" IITC, 2009, pp. 206-208.

[8] E. Y. Wu et al., "On the Weibull shape factor of intrinsic breakdown of dielectric films and its accurate experimental determination. Part I: theory, methodology, experimental techniques", IEEE Transactions on Electron Devices, vol. 49, issue 12, 2002, pp. 2131-2140.

[9] S. M. Sze, Physics of Semiconductor Devices, Oxford; New York: Wiley, 2006, ch. 4.

[10] K. C. Kao, Electrical transport in solids, 1981, Pergamon Press (Oxford, New York).

[11] S. Shamuilia et al., "Internal photoemission of electrons at interfaces of metals with low-k insulators", Appl. Phys. Lett. 89, 202909 (2006).

[12] A. M. Urbanowicz et al., "Damage Reduction and Sealing of Low-k Films by Combined He and NH_3 Plasma Treatment", Electrochemical and Solide State Letters, vol. 10, issue 10, 2007, pp. G76-G79.

[13] Srivastava, J. K.& Prasad, M. Electrical conductivity of silicon dioxide thermally grown on silicon, J.Electrochem. Soc. 132, 955–963 (1985).

New Voltage Ramp Dielectric Breakdown Methodology Based on Square Root *E* Model for Cu/Low-*k* Interconnect Reliability

Mingte Lin, James W. Liang, and K. C. Su
United Microelectronics Corporation
3, Li-Hsin Rd. II, Science-Based Industrial Park, Hsinchu, 30077, Taiwan
phone: 886-3-5679797, e-mail: mingte_lin@umc.com

Abstract— A new voltage ramp dielectric breakdown (VRDB) methodology based on square root (SQRT) *E* model for dielectrics reliability evaluation was developed. We conducted VRDB and time dependent dielectric breakdown (TDDB) experiments on dielectrics in Cu/low-*k* interconnect for reliability assessment. The experimental results show very good correlation between VRDB and TDDB data underlying SQRT *E* model. The electric field acceleration parameter of SQRT *E* model can be estimated with our methodology by applying dual ramping rate during VRDB test. Furthermore, we derived a relation to transform the shape factor of statistic distribution of VRDB breakdown voltage (*Vbd*) to shape factors expected for TDDB failure time distributions. As a result, our novel methodology allows to define a reliability criteria of *VBD* that is equivalent to meet a TDDB specification lifetime.

Keywords-low-k; dielectric; Cu; breakdown; reliability; VRDB

I. INTRODUCTION

As scaling down interconnect dimension in microelectronic devices, copper and low-*k* dielectrics were introduced as interconnect material to meet the *RC* reduction requirement for IC performance. Low-*k* dielectrics became a reliability concern due to their lower dielectric strength [1]. Time dependent dielectric breakdown (TDDB) is a long-term constant voltage stress which provides time to failure (*TTF*) as the dielectric reliability index, while voltage ramp dielectric breakdown (VRDB) is a short-term ramping voltage test monitoring the breakdown voltage or field (*Vbd* or *Ebd*) as the reliability index. The relation between *TTF* and *Vbd* was established for *E* model [2] and Power law [3]. Since the SQRT *E* model is considered to be the appropriate breakdown model for low-*k* dielectrics [4], it was straightforward to apply this particular model for VRDB test for low-*k* dielectrics reliability estimation. Previous research had studied the SQRT *E* VRDB model based on charge to breakdown concept [5]. However, a simple and complete VRDB reliability methodology, which can also consider the statistical aspect of *Vbd*, is still worth studying.

In this study, we will establish a new VRDB methodology based on SQRT *E* model based on damage law concept [2]. VRDB and TDDB experiments on low-*k* dielectrics are conducted for reliability evaluation to correlate *TTF* and *Vbd* results with the SQRT *E* VRDB model. Applying dual ramping rate VRDB tests, the electric field acceleration factor of SQRT

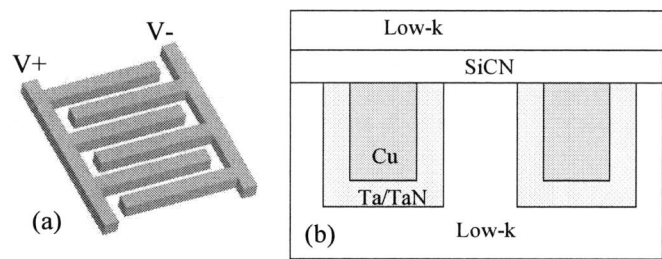

Fig. 1. (a) Comb test structure used for TDDB and VRDB tests. (b) Cross section of the test structure showing the respective interfaces.

E model will be extracted with our model. In order to allow a quantitative dielectric reliability estimation, a relation of Weibull distribution shape factor of *Vbd* from VRDB with that of TDDB *TTF* will be derived. Finally, we will develop a reliability specification of *Vbd* by SQRT *E* model to meet an equivalent TDDB operation lifetime criterion for low-*k* films with different *k* values.

II. EXPERIMENTAL DETAILS

VRDB and TDDB tests were performed on 300 mm wafers using 45 nm Cu/SiCOH low-*k* dual damascene process. Cu lines were electroplated into low-*k* dielectric trenches with TaN/Ta liners and coated with SiCN cap layers on top (Fig. 1). The leakage currents of test structures were monitored under voltage stress to detect the breakdown event. The test temperature is 110 °C. The test patterns are comb structures of Metal-2 lines embedded in SiCOH film (Fig. 1) with a total length of about 2.7 cm and a nominal space between metal lines of 70 nm. The sample size of each test condition is 20. Weibull statistic is used throughout this paper for its simple mathematical form and Poisson area scaling property [6].

III. RESULTS AND DISCUSSION

TDDB experiments were conducted with stress voltages from 22 to 25V, corresponding to 2.71 to 3.57 MV/cm electrical field. Fig. 2 (a) shows the Weibull distribution of TDDB *TTF* with shape factor of 0.979. *TTF* of low-*k* dielectric became shorter as larger stress fields are applied. VRDB tests were performed with two different ramping rates: 0.3 and 3

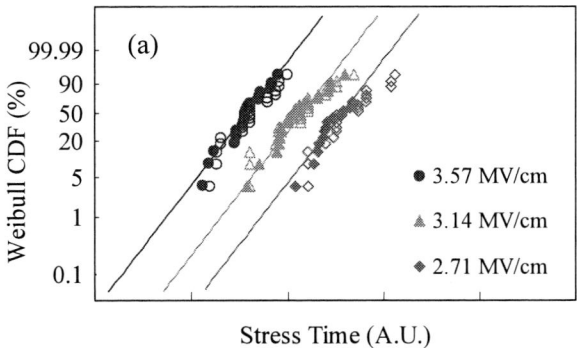

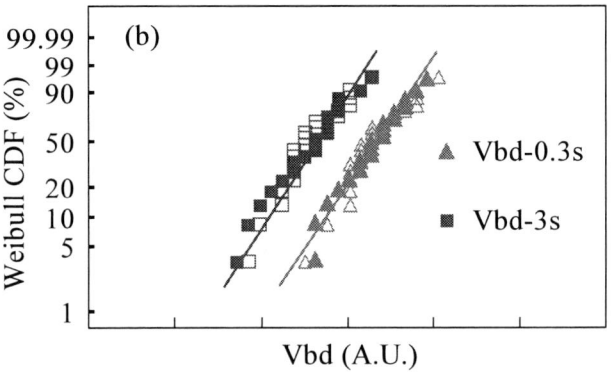

Fig. 2. (a) TDDB *TTF* distributions obtained for various electric fields (closed symbols). Open symbols are *TTF* converted from VRDB results. (b) VRDB *Vbd* distributions with ramping time steps 0.3 and 3 seconds (close symbols), respectively. Open symbols are *Vbd* converted from TDDB results. Solid lines are fitting lines for reference.

seconds at the same voltage step of 0.65 volt. Fig. 2 (b) shows the resulting Weibull distributions of VRDB *Vbd* having the similar shape factor of 6.62. *Vbd* became lower as longer time step (3 s) was applied.

Poole-Frenkel (PF) conduction has been reported to dominate the failure mechanism of low-*k* dielectric breakdown under bias [7, 8]. PF emission is described by

$$I = A \cdot E \cdot \exp\left(-\frac{q(\phi_B - \sqrt{qE / \pi k \varepsilon_o})}{k_B T} \right) \qquad (1)$$

where *I* is the leakage current, *A* is a constant, *E* is the applied electric field, *q* is the electronic charge, φ_B is the potential barrier, *k* is the relative dielectric constant, ε_o is the dielectric constant of free space, *T* is the absolute temperature, and k_B is the Boltzmann constant. Fig. 3 shows a typical I-V result from VRDB test in PF plot. The slope of fitting line in the PF plot was used to estimate the relative dielectric constant (k). This value is consistent with the k value of raw SiCOH film (*k-raw* =0.93*k*). The effective dielectric constant *k* is somewhat higher due to the SICN cap layer with higher dielectric constant and process effects. After verification of PF conduction, SQRT *E* model can be applied to properly describe low-*k* TDDB in our study. With the dominant exponential term of (1), the square root *E* model of TDDB is expressed as [7]

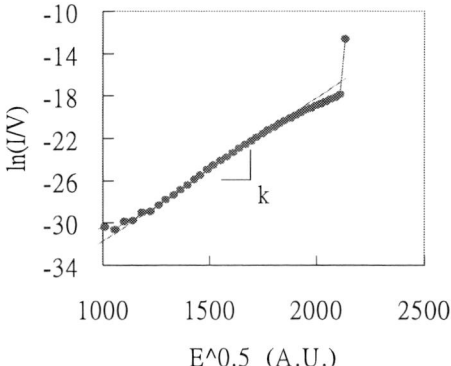

Fig. 3. IV characteristics of low-k dielectrics structures on PF plot. Relative dielectric constant (*k*) is estimated by the slope of IV data.

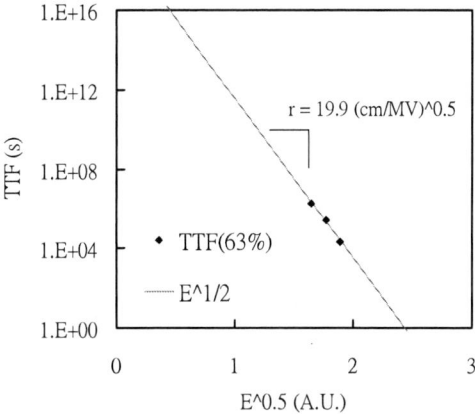

Fig. 4. TDDB time to failure (63%) of different stress voltages on SQRT *E* plot. Filed acceleration factor r is estimated with *TTF* slope.

$$TTF = B \cdot Exp(-r\sqrt{E}) \qquad (2)$$

where *TTF* is the time to failure, *B* is a constant, and *r* is the field acceleration factor. TDDB 63% *TTF* of each stress field are shown on SQRT *E* plot in Fig. 4 resulting in a field acceleration of *r* = 19.9 (cm/MV)$^{0.5}$. With the conduction and TDDB characteristics described above, SQRT *E* model is shown to be appropriate for our low-k breakdown model. Breakdown sites were identified and analyzed with OBIRCH and TEM. A typical breakdown is located at the top corner of metal line where leakage current crowded and Cu metal is forming a bridge (Fig. 5).

Following the damage law concept [2], we develop a correlation between SQRT *E* model TDDB lifetime and VRDB breakdown voltage. If a dielectric structure is stressed with an electrical field *E* in a time interval *dt*, its damage or defect (*dD*) created relatively to a critical breakdown damage (*Ddb*), can be defined according to [2] as:

$$dD \equiv \frac{dt}{B \cdot Exp(-r\sqrt{E})} \qquad (3)$$

978-1-4244-5430-3/10 $26.00 © 2010 IEEE

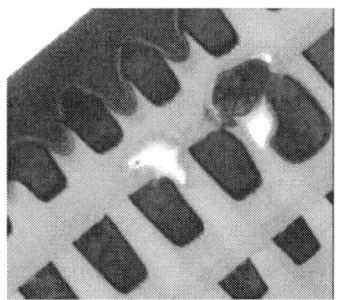

Fig. 5. Cross section of low-k dielectrics breakdown test sample. Typical breakdown located at the top corner of metal-2 line.

When the time interval is equal to *TTF*, the relative damage reaches the critical breakdown condition and the value of the breakdown damage (*Dbd*) becomes unity.

$$D_{bd} = \int_0^{TTF} \frac{dt}{B \cdot Exp(-r\sqrt{E})} = \frac{TTF}{B \cdot Exp(-r\sqrt{E})} = 1 \quad (4)$$

We now apply the damage law to the VRDB test. When the dielectric structure is under VRDB test with linear ramping rate $R = E/t$ (Fig. 6(a)), the damage function (3) for the breakdown condition can be expressed as

$$D_{bd} = \int_0^{Ebd/R} \frac{dt}{B \cdot Exp(-r\sqrt{E})} = 1 \quad (5)$$

where *Edb/R* is equal to the time to breakdown in a VRDB test with increasing electrical field *E(t)*. With a constant ramping rate, the time interval *dt* can be substituted by *dE/R* and hence

$$D_{bd} = \int_0^{Ebd} \frac{Exp(r\sqrt{E})dE}{B \cdot R} = 1 \quad (5a)$$

It is then convenient to define a time parameter *t0*, which represents the equivalent time to breakdown under (hypothetical) constant electrical field at *Ebd* according to Eq. (4) as:

$$t_o = B \cdot Exp(-r\sqrt{E_{bd}}) \quad (6)$$

The solution of (5a) combined with (6) results in

$$t_o = \frac{2}{Rr^2}(r\sqrt{E_{bd}} - 1 + Exp(-r\sqrt{E_{bd}})) \quad (7)$$

For typical values of *r* and *Ebd* relevant in our experiments, the first term (~50) on the right side of Eq. (7) dominates and *to* can be simplified as

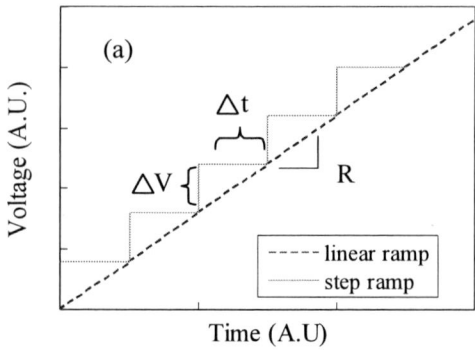

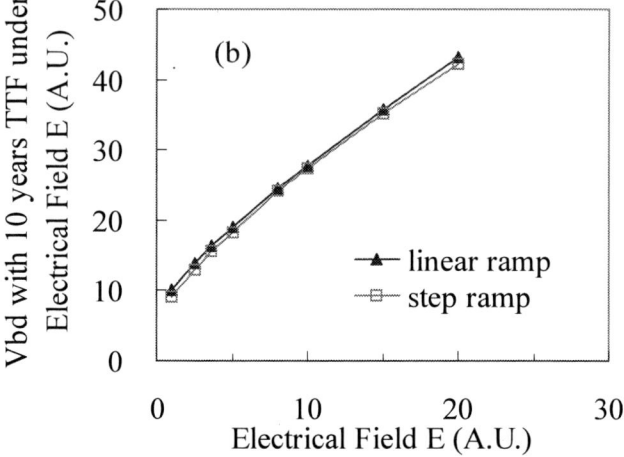

Fig. 6. (a) Voltage v.s. time plots of linear and step ramp VRDB test. (b) Vbd values calculated with (9) (linear ramp VRDB) and (11) (step ramp VRDB) for 10 years TDDB TTF with different electrical fields.

$$t_o \approx \frac{2}{Rr}\sqrt{E_{bd}} \quad (8)$$

With Eq. (4), (6), and (8), we can link the *Ebd* obtained in a VRDB test with *TTF* of TDDB under constant stress field *E* as

$$\ln(TTF) = \ln(\frac{2\sqrt{E_{bd}}}{Rr}) + r(\sqrt{E_{bd}} - \sqrt{E}) \quad (9)$$

If the dielectric structures are under VRDB test with step ramp (Fig. 6(a)), the VRDB damage law leads to

$$\sum_{n=1}^{Ebd/\Delta E} \frac{\Delta t}{B \cdot Exp(-r\sqrt{n\Delta E})} = 1 \quad (10)$$

where Δt is time step and ΔE is the corresponding electric field step.

Appling Eq. (2) to Eq. (10), we can link the breakdown field *Ebd* of stepped VRDB test to *TTF* of TDDB under stress field *E* as follows

978-1-4244-5430-3/10 $26.00 © 2010 IEEE

$$TTF = \sum_{n=1}^{Ebd/\Delta E} \Delta t \cdot Exp(r(\sqrt{n\Delta E} - \sqrt{E})) \quad (11)$$

With the TDDB tests performed on low-k dielectric structures, we can assess TTF under field E and field acceleration r. Ebd (or Vbd) of these structures under step VRDB test can be transformed from TTF with Eq. (11). In Fig.2b, the transformed Vbd (open symbol) are compared with the directly measured Vbd (close symbol). We can also reversely transform Vbd from VRDB test to TDDB TTF using Eq. (11). The transformed TTF (open symbol) and the measured TDDB TTF (closed symbol) are compared in Fig. 2(a). The transformed results show very good correlation between Vbd and TTF through the damage law transformation with SQRT E model.

The Ebd (Vbd) calculated with Eq. (9) for linear ramp VRDB test is consistent with the Ebd calculated with Eq. (11) for step ramp VRDB test. The Vbd results calculated with Eq. (9) and Eq. (11) are compared in Fig. 6(b), where the ramping rates are the same for both types of VRDB tests. The Vbd values are calculated for 10 years TTF with various electric fields E in Eq. (9) and (11). Their differences are rather small in the range of the voltage step unit.

In Fig.7 we compare the reliability estimations made by different breakdown models. As can be seen, a SQRT E model predicts longer TDDB lifetime (Fig. 7(a)) and lower Vdb value (Fig. 7(b)) required for 10 years TDDB lifetime than an E model [9]. Power law, which has been applied e.g. for ultra thin oxide breakdown [4], is the most optimistic.

To apply an Ebd specification with respect to a low percentile failure target, the statistic aspects of the Ebd distribution need to be considered. The relations of statistic distribution characteristics of TDDB and VRDB can also be estimated with the above transformation equations (9) and (11). In Weibull statistics the parameter W is defined by

$$W \equiv \ln(-\ln(1-F)) \quad (12)$$

where F is the cumulative probability of failure. TTF of SQRT E TDDB model can be expressed with Weibull distribution as

$$TTF = B \cdot Exp(-r\sqrt{E}) \cdot ((-\ln(1-F)))^{1/m}$$

$$\ln(TTF) = -r\sqrt{E} + W/m + \ln(B) \quad (13)$$

where m is the Weibull shape factor.

For TDDB test under constant field E, the Weibull shape factor m can be calculated as

$$m = dW / d(\ln(TTF)) \quad (14)$$

We consider the Weibull distribution of Vdb under VRDB test. Under practical condition (for our experiment example: Ebd = 4.5MV/cm, r =19.9 (cm/MV)$^{0.5}$, R=3.1E^{-2} MV/cm/s), the last term at the right side of Eq. (9) dominates and the

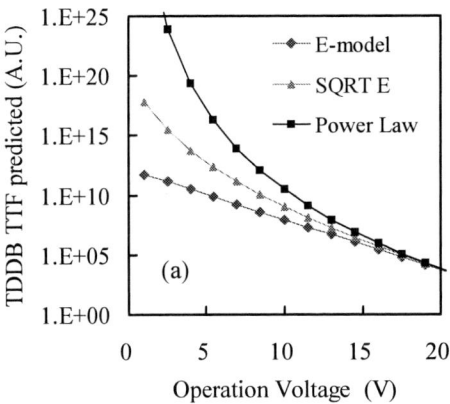

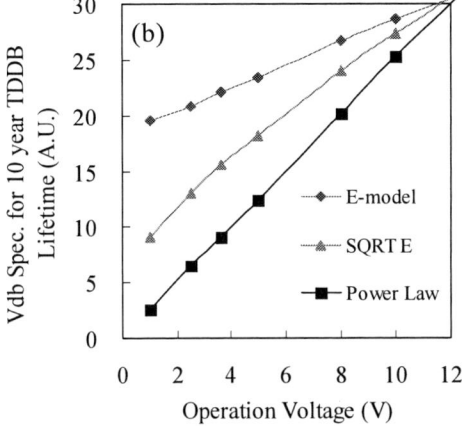

Fig. 7. (a) TDDB TTF prediction under operation voltage for different models. (b) Vdb criteria corresponding to different models and various TDDB lifetime requirements.

variations of Ebd and TTF under constant E due to statistics have following relation:

$$d(\ln(TTF)) = d(r\sqrt{E_{bd}}) \quad (15)$$

With Eq. (14) and (15), we can estimate the Weibull shape factor m' of Ebd (or Vbd equivalently) from shape factor m of TDDB TTF distribution by the following derivation:

$$m = \frac{dW}{d(\ln(TTF))} = \frac{dW}{d(r\sqrt{E_{bd}})} = \frac{1}{r} \cdot \frac{dW}{d(\sqrt{E_{bd}})}$$

$$m' \equiv \frac{dW}{d(\sqrt{E_{bd}})} = r \cdot m \quad (16)$$

Note that the Weibull shape factor m' of Ebd is defined with respect to $\sqrt{E_{bd}}$-scale. From our experimental results, the Weibull shape factor m of TTF and the field acceleration factor r from TDDB are determined to be m=0.979 and r=19.9 (cm/MV)$^{0.5}$, respectively. The resulting value $m \cdot r$ = 19.48 is then consistent with the calculated Weibull shape factor

978-1-4244-5430-3/10 $26.00 © 2010 IEEE

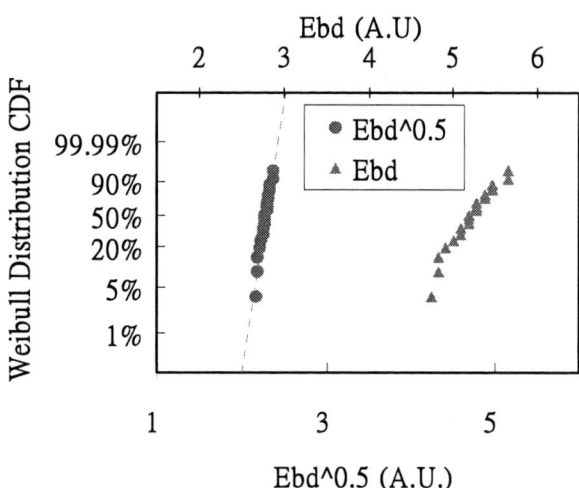

Fig. 8. Weibull distributions obtained for *Ebd* and Sqrt(*Ebd*) from VRBD test plotted with respect to Ebd and Ebd^0.5 as x-axis, respectively. The calculated Weibull shape factor (*m'*) for Sqrt(*Ebd*) distribution is 19.39.

m'=19.39 directly obtained from the measured $\sqrt{E_{bd}}$ - distribution in Fig. 8 - which is fitted very well underlying Weibull statistics.

In the next section, we describe how the field acceleration factor *r* in SQRT *E* model can be derived from a dual ramping rate VRDB test. For that purpose, we perform two VRDB tests with different ramping rate R_1 and R_2 to get two sets of Vbd distributions. According to Eq. (8) and the definition of the ramping rate R (with the same field step $\triangle E_i$ but different time steps $\triangle t_i$) there are 2 different times to breakdown t_{0i}:

$$t_{oi} = \frac{2\sqrt{E_{bdi}}}{R_i r} \quad , R_i \equiv \frac{\Delta E}{\Delta t_i} \quad , i = 1, 2. \quad (17)$$

And with (6) the two corresponding breakdown fields Ebd_i :

$$t_{o1} = B \cdot Exp(-r\sqrt{E_{bd1}})$$

$$t_{o2} = B \cdot Exp(-r\sqrt{E_{bd2}})$$

$$\frac{t_{o1}}{t_{o2}} = Exp(r(\sqrt{E_{bd2}} - \sqrt{E_{bd1}})) \quad (18)$$

With (17) and (18), the field acceleration factor r becomes:

$$\frac{\Delta t_1 \sqrt{E_{bd1}}}{\Delta t_2 \sqrt{E_{bd2}}} = Exp(r(\sqrt{E_{bd2}} - \sqrt{E_{bd1}}))$$

$$\ln(\frac{\Delta t_1}{\Delta t_2}) = r(\sqrt{E_{bd2}} - \sqrt{E_{bd1}}) + \ln(\frac{\sqrt{E_{bd2}}}{\sqrt{E_{bd1}}})$$

$$\ln(\frac{\Delta t_1}{\Delta t_2}) \approx r(\sqrt{E_{bd2}} - \sqrt{E_{bd1}})$$

$$r \approx \ln(\Delta t_1/\Delta t_2)/(\sqrt{E_{bd2}} - \sqrt{E_{bd1}}) \quad (19)$$

Since the Vbd results of step ramp VRDB is similar to linear ramp VRDB under our ramping rate condition (Fig. 6(b)), the field acceleration factor with dual step ramping rate VRDB can be approximated with Eq. (19). This can be shown for step ramp VRDB tests with different time steps $\triangle t_i$ following Eq. (11):

$$t_{o1} = \Delta t_1 \cdot \sum_{n=1}^{E_{bd1}/\Delta E} Exp(r(\sqrt{n\Delta E} - \sqrt{E_{bd1}}))$$

$$t_{o2} = \Delta t_2 \cdot \sum_{n=1}^{E_{bd2}/\Delta E} Exp(r(\sqrt{n\Delta E} - \sqrt{E_{bd2}}))$$

$$\frac{t_{o1}}{t_{o2}} = \frac{\Delta t_1}{\Delta t_2} \cdot \frac{\sum\limits_{n=1}^{E_{bd1}/\Delta E} Exp(r(\sqrt{n\Delta E} - \sqrt{E_{bd1}}))}{\sum\limits_{n=1}^{E_{bd2}/\Delta E} Exp(r(\sqrt{n\Delta E} - \sqrt{E_{bd2}}))} \quad (20)$$

When we calculate Eq. (20) with $\triangle t_1/\triangle t_2$ ~10 as in our experiments, the summation ratio term at the most right side is approximately equal to unity and Eq. (20) can be simplified as

$$\frac{t_{o1}}{t_{o2}} \approx \frac{\Delta t_1}{\Delta t_2} \quad (21)$$

Comparing (21) and (18), we can use the same equation (19) for step and linear ramp VRDB test.

The field acceleration factor *r* calculated according to (19) from dual ramp VRDB results is 19.1 (cm/MV)^0.5 (Fig. 2(b)). This value is consistent with the field acceleration *r*=19.9 (cm/MV)^0.5 determined in TDDB test (Fig. 4).

To demonstrate the SQRT *E* VRDB methodology for low-k dielectrics reliability assessment, we performed VRDB and TDDB tests on structures embedded in SiCOH films with different *k* values (*k1*<*k2*). Fig. 9 shows the *Vbd* results of these two films. The transformed *Vbd* from TDDB TTF through Eq. (11) were plotted to confirm again the good correlation with measured *Vdb*. The SiCOH film with *k1*<*k2* shows smaller *Vbd* strength. This is a consequence of higher Si-CH4/Si-O ratio, which implies smaller density and mechanical strength. This causes a weaker dielectric reliability as shown in our VRDB experiment.

The reliability specification for low-*k* dielectrics is usually defined as 10 years TDDB lifetime under e.g. 3.63 Volt operation condition. We can then calculate a *Vdb* specification for 10 years TDDB lifetime at 3.63 Volt operation voltage with

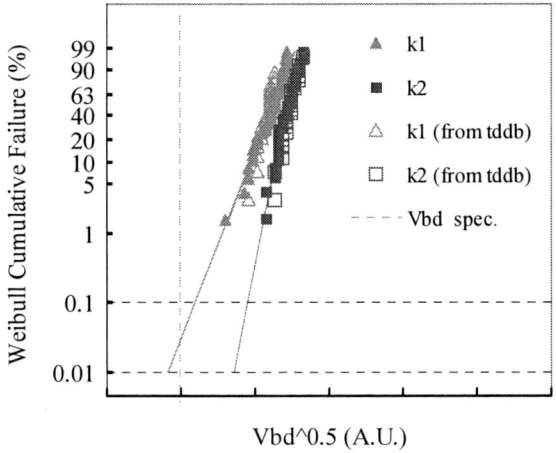

Fig. 9. Vbd distributions of dielectrics with different k values (k1<k2). The vertical dashed line is the Vdb specification that corresponds to 10 years TDDB lifetime under 3.63 Volt operation condition.

Eq. (9) or (11). This value is plotted as the vertical dashed line in Fig. 9. From the extrapolation lines of *Vdb* distributions, both *k1* and *k2* SiCOH films can pass the *Vdb* specification at 0.1% percentile. However, only *k2* SiCOH film can pass *Vdb* specification at 0.01% percentile.

IV. CONCLUSION

A new VRDB methodology based on SQRT *E* model for dielectrics reliability evaluation was established. TDDB and VRDB experiments were performed on low-k SiCOH dielectrics to verify this methodology. *Vdb* estimated with the SQRT *E* VRDB model from TDDB correlated well with *Vdb* measured by VRDB test. The relation between the Weibull shape factors of *Vdb* distribution and TDDB distribution was derived and confirmed by the experiments. The electric field acceleration factor *r* of SQRT *E* model can be estimated by applying dual ramping rate VRDB tests with our model. The estimated *r*-value was demonstrated to be consistent with the one extracted from TDDB results. A reliability specification for *Vdb* was derived to meet the required operation TDDB lifetime

criterion. We applied such specification of *Vdb* to different low-k dielectric films and demonstrated the effectiveness of such VRDB methodology to differentiate their reliability performances.

Since VRDB is a quick assessment test with test time usually less than 100 seconds, it is especially suitable for dielectric reliability evaluation for assessment purpose with various dielectric process split conditions or monitoring purpose with massive sample size. With the VRDB methodology derived above, we can replace time consuming TDDB test by quick VRDB tests to evaluate the reliability of dielectric the SQRT *E* breakdown model applied well.

REFERENCES

[1] E. T. Ogawa, K. Jinyoung, G.S Haase, H.C. Mogul, J.W. McPherson, "Leakage, breakdown, and TDDB characteristics of porous low-k silica-based interconnect dielectrics", IRPS 2003, pp. 166 – 172.

[2] A. Berman, "Time-Zero Dielectric Reliability Test by a Ramp Method", IRPS 1981, pp. 204 - 209.

[3] S.C. Fan, J.C. Lin, A.S. Oates, "Accurate Characterization on Intrinsic Gate Oxide Reliability using Voltage Ramp Tests", IRPS 2006, pp. 625 - 626.

[4] F. Chen, M. Shinosky, "Addressing Cu/Low-k Dielectric TDDB-Reliability Challenges for Advanced CMOS Technolo-gies", IEEE Trans. on Electron Devices, Vol. 56, No. 1, 2009, pp. 2-12.

[5] O. Aubel, M. Kiene, W. Yao, "New approach of 90nm low-k interconnect evaluation using a voltage ramp dielectric breakdown (VRDB) test", IRPS, 2005, pp. 483–489.

[6] F. Chen, P. McLaughlin, J. Gambino, E. Wu, J. Demarest, D. Meatyard, M. Shinosky, "The Effect of Metal Area and Line Spacing on TDDB Characteristics of 45nm Low-k SiCOH Di-electrics", IRPS, 2007, pp. 382-389.

[7] J. R. Lloyd, E. Liniger, and T. M. Shaw, "Simple model for time-dependent dielectric breakdown in inter and intra level low-k dielectrics J. Appl. Phys. 2005, 084109, pp. 98-102.

[8] Kok-Yong Yiang, Yao, H.W., Marathe, A., and Aubel, O., "New perspectives of dielectric breakdown in low-k intercon-nects", IRPS, 2009, pp. 476 – 480.

[9] G. S. Haase, E.T. Ogawa, J.W. McPherson, "Breakdown characteristics of interconnect dielectrics", IRPS, 2005. , pp. 466 – 473.

A Simple Electrical Method for Etch Bias and Process Reliability Determination

Kok-Yong Yiang, Melida Chin, and Amit Marathe
GLOBALFOUNDRIES Inc., Technology Reliability Development
1050 E Arques Ave, MS 143, Sunnyvale CA 94085, USA
Tel: +1-408-462-4135; E-mail: kok-yong.yiang@globalfoundriescom

Oliver Aubel
GLOBALFOUNDRIES Dresden Module One LLC & Co. KG, Quality and Reliability Engineering
Wilschdorfer Landstrasse 101, 01109 Dresden, Germany

Abstract— A fast and simple electrical method is developed to characterize the etch bias and post-patterned ILD breakdown strength of back-end-of-line (BEOL) interconnects, as well as the middle-of-line (MOL) contact/poly module. The method provides a timely and valuable monitoring mechanism for assessing lithography, etch, thin-film quality and process reliability windows.

Keywords- Etch bias, line-edge roughness (LER), VRDB.

I. INTRODUCTION

The aggressive scaling of CMOS technology has led to greatly reduced spacings between (a) interconnect wires and (b) the distance between contact and poly/gate. Limitations in photolithography, etch and chemical-mechanical polishing (CMP) at deep sub-micron technologies are making the control of line-edge roughness (LER), critical dimensions (CD), and overlay extremely challenging. These factors contribute to the overall etch bias (i.e., the difference in lateral dimensions between the etched image and mask image).

Poor etch bias control leads to large variability in metal, via, and intralevel dielectric (ILD) dimensions and impacts the statistical distributions of the major reliability mechanisms (i.e., electromigration (EM), stress migration (SM), and time-dependent dielectric breakdown (TDDB)). It is therefore crucial to have a fast monitoring methodology that can give timely metrics on etch bias to the process floor. Unfortunately, the characterization of etch bias conventionally relies either on inline SEM, which is restricted in field-of-view (FOV) coverage, or finite-element analysis, which has to be based on some pre-conceived probabilistic assumptions. [1] There is also no means of assessing the post-patterned quality of the thin-film stack, which is greatly dependent on the etch, clean and CMP chemistries. [2],[3]

In this paper, we demonstrate a fast and simple electrical method to measure the actual etch bias and breakdown strength (E_{bd}) of the intralevel dielectric (ILD) stack, without the aforementioned limitations. This method can also give very important insights into the statistical variabilities that are seen in EM, SM, and TDDB.

II. EXPERIMENTAL

We highlight two different types of test structure designs that can be used in the characterization of back-end-of-line (BEOL) and middle-of-line (MOL) etch bias.

For BEOL, it consists of typical via comb ($M_x/V_x/M_{x+1}$) test structures with various degree of via misalignment, as illustrated in Figure 1.

For the contact-to-poly (CA/PC) MOL module, the spacing between CA and PC is modulated by different PC widths, as shown in Figure 2. All CA and PC arrays are drawn at critical poly pitch (CPP) to prevent any unintentional overlay deviations.

Both the BEOL and MOL structures are fabricated on 45 nm wafer technology. Based on voltage-ramp dielectric breakdown (VRDB) experiments, the statistical median breakdown voltages (V_{bd}) of these structures were derived and used in the etch bias analysis.

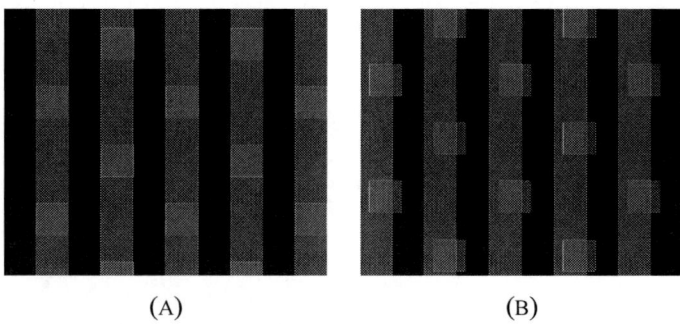

(A)   (B)

Figure 1. Via combs ($M_x/V_x/M_{x+1}$) with metals drawn at minimum pitch. On different structures, the vias are drawn with different degrees of misalignment: (a) misalignment = 0 nm, and (b) misalignment > 0 nm.

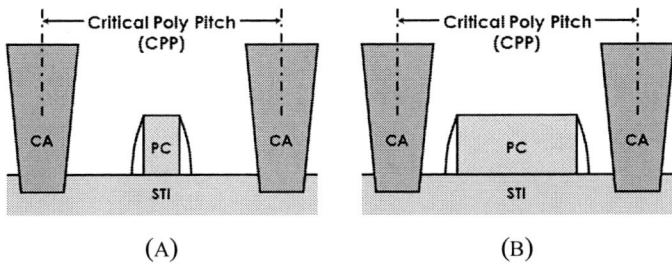

(A)   (B)

Figure 2. Contact-to-poly (CA/PC) test structures with the gate/poly (a) at minimum design width (i.e., wimpy = 0 nm), and (b) at wider widths (i.e., wimpy > 0 nm). All contact and poly arrays are drawn at critical poly pitch (CPP) to prevent unintentional overlay deviations.

III. RESULTS AND DISCUSSION

A. BEOL Via Comb Module

Figure 3 shows the V_{bd} distributions of the via combs with different degrees of via misalignment, plotted using lognormal statistics[a]. As expected, the greater the misalignment, the lower the V_{bd}, due to the reduced spacing between the via and nearest neighboring metal.

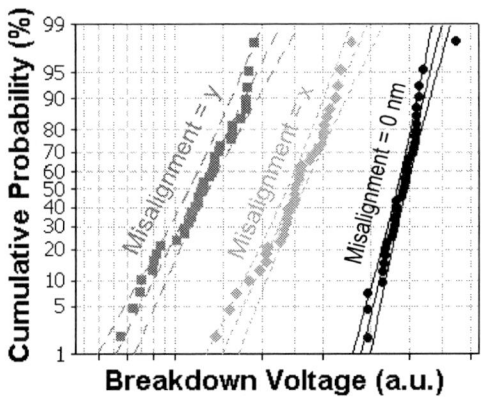

Figure 3. Breakdown voltage distributions of via comb structures having different degrees of via misalignments ($y > x > 0$ nm).

To derive the etch bias, we plot the median V_{bd} as a function of "as-drawn" via-to-metal spacing, as shown in Figure 4. From the linear relationship that ensues, we can extrapolate to an as-drawn spacing on the x-axis, which would yield an electrical "short" (i.e., $V_{bd} \to 0$ V). Reading off the x-intercept in this manner, we get the total etch bias on the via and neighboring metal, which is independent of the degree of via misalignment.

The simple derivation of the etch bias, based on these test structures and method, can be explained mathematically. Consider the schematic diagram in Figure 5, which illustrates the via comb structure with misaligned via. Δx_{via} and Δx_{metal} represent the individual etch bias contributions from the via and metal, respectively. x_{drawn} represents the as-drawn metal-to-metal spacing, and $\Delta x_{mis.via}$ (not labeled) is the amount of as-

---

[a] The choice of lognormal statistics over weibull is discussed elsewhere. [4],[5]

---

drawn via misalignment. Based on these, we can easily derive the effective via-to-metal spacing, x_{eff}:

$$x_{eff} = x_{drawn} - (\Delta x_{via} + \Delta x_{metal} + \Delta x_{mis.via}) \quad (1)$$

The nominal breakdown field, E_{bd}, of the ILD material between via and metal is:

$$E_{bd} = \frac{V_{bd}}{x_{eff}} = \frac{V_{bd}}{x_{drawn} - (\Delta x_{via} + \Delta x_{metal} + \Delta x_{mis.via})} \quad (2)$$

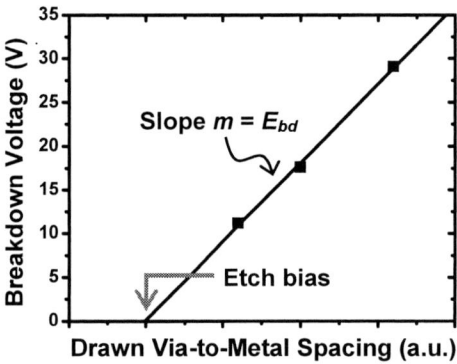

Figure 4. Median V_{bd} as a function of as-drawn via-to-metal spacing. The x-intercept value indicates the etch bias, where $V_{bd} \to 0$ V. The slope of the fitted line represents the nominal E_{bd}.

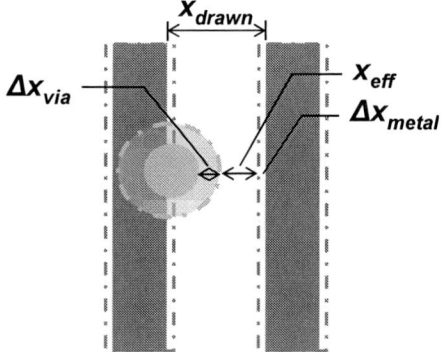

Figure 5. Schematic diagram of via comb structure with misaligned via. Δx_{via} and Δx_{metal} represent the individual etch bias contributions from the via and metal, respectively. x_{drawn} is the as-drawn metal-to-metal spacing. $\Delta x_{mis.via}$ (not labeled) is the amount of via misalignment as-drawn, and x_{eff} is the effective via-to-metal spacing after all etch biases are considered.

For any given temperature, the nominal E_{bd} is constant (since it is a material property); however, the complex geometries of a patterned interconnect (or MOL) structure can cause locally enhanced electric fields at the metal (or poly/gate) corners. This is validated by finite-element analysis (using ANSYS 12) as shown in Figure 6, and by various authors. [6],[7] Therefore, the nominal E_{bd} for a patterned BEOL/MOL ILD stack cannot simply be assumed from thin-film studies, and will be treated as an unknown in Eq. 2. V_{bd} on the other hand, is experimentally obtained, and is therefore a known entity. x_{drawn} and $\Delta x_{mis.via}$ are also known entities, since they are *as-drawn* dimensions.

To solve for the two unknowns, E_{bd} and total etch bias (Δx_{via} + Δx_{metal}), we need a minimum of two simultaneous equations (test structures); these can be in the form of (a) a standard structure with non-misaligned vias (i.e., $\Delta x_{mis.via}$ = 0 nm), and (b) another structure with misaligned vias (i.e., $\Delta x_{mis.via}$ > 0 nm).

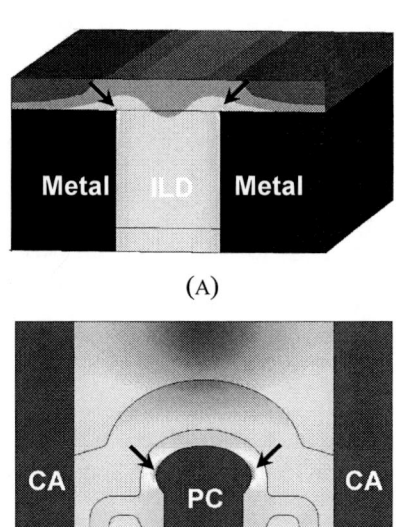

(A)

(B)

Figure 6. Finite-element analysis showing locally-enhanced electric fields (indicated by arrows) at (a) metal corners, and (b) PC corners.

Eq. 2 proves consistent across all three median V_{bd} values that were obtained experimentally for the different degrees of via misalignment. A nominal E_{bd} of about 9 MV/cm for the post-patterned ILD stack is obtained, which is typical for low-k SiCOH. Alternatively, the same E_{bd} value can be derived quickly from Figure 4 by simply calculating the slope of the fitted line.

B. MOL CA/PC Module

The same method can be used to determine the etch bias between CA and PC in the MOL module. Figure 7 shows the V_{bd} distributions and median V_{bd} values for a considerable range of *as-drawn* CA-to-PC spacings, measured at temperatures ranging from 25 to 200°C. All three fitted lines intersect the x-axis at almost the same value; since this value corresponds to the total etch bias afflicting both the CA and PC, it is therefore expected to be constant regardless of temperature.

When the total etch bias is taken into account (resulting in reduced CA-to-PC spacing), finite-element analysis reveals a locally enhanced electric field of about 11 MV/cm on the PC corners upon dielectric breakdown; this is close to the breakdown strength of the spacer material. In addition, the slopes of the fitted lines, representing the nominal E_{bd}, degrade progressively with increasing temperature; when plotted as a function of temperature, as shown in Figure 8, the excellent fit of E_{bd} degradation with temperature [b] gives a second-order validation of the high accuracy of this electrical method.

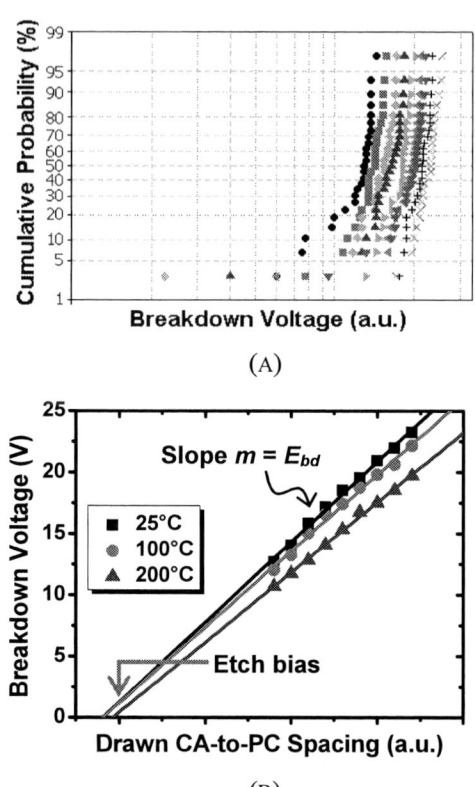

(A)

(B)

Figure 7. (A) Breakdown voltage (V_{bd}) distributions at 25°C and (B) median V_{bd} of CA/PC ILD with different CA-to-PC spacings *(as-drawn)*, at temperatures ranging from 25 to 200°C. x-intercept of (B) is the etch bias, where $V_{bd} \rightarrow 0$ V. The slopes of the fitted lines represent the nominal breakdown strength, E_{bd}. Decreasing slopes at high temperatures imply lower E_{bd}.

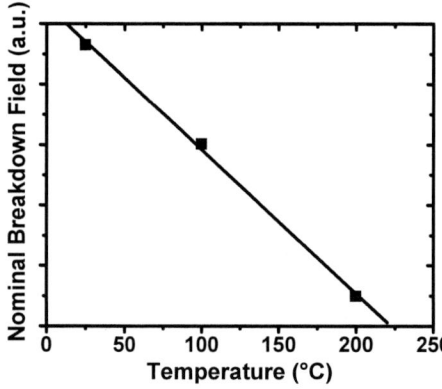

Figure 8. Nominal E_{bd} of the post-patterned CA/PC ILD, derived from the slopes of the fitted lines in Figure 7. Excellent fit demonstrates the accuracy of this electrical method.

---

[b] The nature of E_{bd} degradation with increasing temperature is explained in [8], and exemplified in [9],[10].

IV. CONCLUSION

A fast and simple electrical method was developed to derive the etch bias and breakdown strength (E_{bd}) of the post-patterned ILD. The etch bias can provide timely and valuable feedback in terms of lithography and etch process windows, whereas the E_{bd} is a good indicator of post-patterned thin-film quality.

The derivations of etch bias and E_{bd} are mutually independent and can be applied to both the BEOL and MOL modules on any ILD (i.e., dense or ultra-low-k films in the BEOL stack, and different spacer thicknesses and materials in the MOL module). In addition, this method can give very important insights into the statistical variabilities in EM, SM, and TDDB.

REFERENCES

[1] Michele Stucchi and Zsolt Tökei, "Impact of LER and Misaligned Vias on the Electric Field in Nanometer-scale Wires," in *Proceedings of IITC*, p. 174-176 (2008).

[2] Julien Michelon and Romano J.O.M. Hoofman, "Moisture Influence on Porous Low-k Reliability," *IEEE Trans. Dev. and Mat. Rel., Vol. 6 No. 2,* p. 169-174 (2006).

[3] P.S. Ho, J. Leu, and W.W. Lee, in "Low Dielectric Constant Materials for IC Applications," *New York: Springer-Verlag,* p. 1-21 (2002).

[4] J.R. Lloyd, E. Liniger, and T.M. Shaw, "Simple model for time dependent dielectric breakdown in inter- and intralevel low-k dielectrics," *J. Appl. Phys. Vol. 98 No. 8,* p. 084109.1-084109.6 (2005).

[5] Zs. Tökei *et al.,* "Impact of the barrier/dielectric interface quality on reliability of Cu porous-low-k interconnects," in *Proceedings of IRPS,* p. 326-332 (2004).

[6] K.Y. Yiang, T.S. Mok, W.J. Yoo, and Ahila Krishnamoorthy, "Reliability Improvement using Buried Capping Layer in Advanced Interconnects," in *Proceedings of IRPS,* p. 333-337 (2004).

[7] Andrew T. Kim *et al.,* "Line Edge Roughness of Metal Lines and Time-Dependent Dielectric Breakdown Characteristics of Low-k Interconnect Dielectrics," in *Proceedings of IITC,* p. 155-157 (2007).

[8] S.O. Kasap, in "Principles of Electrical Engineering Materials and Devices, Revised Edition," *New York: McGraw-Hill,* p. 550 (2000).

[9] Krishna Seshan, in "Handbook of Thin-film Deposition Processes and Techniques, Second Edition," *New York: Noyes,* p. 572 (2002).

[10] H. Zhou, F.G. Shi, B. Zhao, and J. Yota, "Temperature accelerated dielectric breakdown of PECVD low-k carbon doped silicon dioxide dielectric thin films," *App. Phys. A, Vol. 81,* p. 767-771 (2005).

COMPREHENSIVE INVESTIGATIONS OF COWP METAL-CAP IMPACTS ON LOW-K TDDB FOR 32NM TECHNOLOGY APPLICATION

F. Chen, M. Shinosky, B. Li, C. Christiansen, T. Lee, J. Aitken, D. Badami
IBM Microelectronics, Essex Junction, VT 05452, USA
Phone: 802-769-7917; Fax: 802-769-4139; E-mail: chenfe@us.ibm.com

E. Huang , G. Bonilla
IBM Thomas J. Watson Research Center, Yorktown Heights, NY 10598, USA

T.-M. Ko, T. Kane, Y. Wang, M. Zaitz, L. Nicholson, M. Angyal, C. Truong, X. Chen
IBM Microelectronics, Hopewell Junction, NY 12533, USA

G. Yang, S. B. Law, T. J. Tang
GLOBALFOUNDRIES Singapore, Hopewell Junction, NY 12533, USA

S. Petitdidier, G.Ribes
STMicroelectronics, Hopewell Junction, NY 12533, USA

M. Oh, C. Child
GLOBALFOUNDRIES, Inc, Hopewell Junction, NY 12533, USA

H. Sawada
Toshiba America Inc., Hopewell Junction, NY 12533, USA

A. Kolics, O. Rigoutat, N. Gilbert
Lam Research Corporation, Fremont, CA, USA

Abstract — as the current-carrying capability of a copper line is reduced due to interconnect dimension shrinkage, self-aligned CoWP metal-cap has been reported to be helpful to improve degraded electromigration (EM) reliability. However, adoption of the metal cap in general further exacerbates the already problematic low-*k* dielectric TDDB reliability at 32nm and beyond. This paper provides a comprehensive study of CoWP metal-cap impacts on low-*k* TDDB for 32nm technology application. It was found that CoWP could induce a severe degradation of low-*k* TDDB if its process is not optimized, and its impacts on dense low-k and porous ultra low-*k* (ULK) dielectrics were different. An optimized CoWP process with the least defect density could lead to an acceptable TDDB performance as compared to the control for both dense low-*k* and porous ULK dielectrics, while showing substantial improvements in EM and stress migration (SM).

Keywords - time-dependent dielectric breakdown, TDDB, low-k reliability, CoWP metal-cap, low-k defect density, EM, SM, leakage

I. INTRODUCTION

With the continuing aggressive scaling of interconnect dimensions and introduction of new lower *k* materials, the back-end-of-line (BEOL) interconnect reliability margins of TDDB, EM and SM mechanisms are greatly reduced. EM is of increasing concern at 32nm because wire cross-section scales by 70% from the 45nm to 32nm node but circuit voltage and liner thickness do not scale at the same rate, therefore, even greater current density is imposed for the interconnect wires at the 32nm node. In addition to the geometry shrinkage, process induced challenges such as Cu microstructure degradation could further exacerbate the EM problem at 32nm. Therefore, there is a great need for EM performance enhancement in order to let EM meet its target at 32nm and beyond. It has been reported that the EM degradation behavior of Cu interconnects could be significantly improved by increasing the bonding strength of the top interface [1-2]. Several approaches have been discussed including metal-cap and local alloying [3-5]. In all cases, the mass transport along the Cu/capping layer interface was reduced. Among metal-caps, one promising material is CoWP because of its ability to be selectively deposited on Cu. About 100x improvement in EM lifetime in both up stream and down stream modes of Cu interconnects within low-*k* dielectrics were achieved by using self-aligned CoWP metal-cap [6].

While providing significant EM reliability improvement by adopting CoWP metal-cap, CoWP metal-cap in general could further exacerbate the already problematic low-*k* dielectric TDDB reliability [7]. Low-*k* TDDB is commonly considered a big challenge at 45nm technology and beyond. Therefore, in order to minimize the detrimental impact from CoWP on TDDB, the CoWP cap must be selectively deposited on the copper but not on the low-*k* surface. The defects induced by CoWP process such as metallic particles and the lateral growth of CoWP toward low-*k* surface must be well controlled during CoWP integration. In this study, we carefully investigated the effects of electroless deposited (ELD) CoWP metal-cap on TDDB performance of dense low-k and porous ULK SiCOH films during the development of 32nm BEOL technology. More than 20 different CoWP process splits were analyzed during the investigation period. It was found that CoWP could induce a severe degradation of low-*k* TDDB if its process was not optimized, and its impact on dense low-k and porous ultra low-k (ULK) dielectrics was different. The degradation of TDDB was attributed to metallic particles arising from the electroless CoWP deposition process. Therefore, elimination of metallic contaminants after CoWP deposition became crucial for the successful implementation of CoWP metal-cap. Finally, we demonstrated that with careful CoWP process optimization, comparable leakage and TDDB performance with substantial improvements in EM and SM could be achieved for both dense low-k and porous ULK dielectrics at the 32nm technology node.

II. EXPERIMENTAL

A *k*=2.7 dense CVD SiCOH and a *k*=2.4 CVD porous ULK SiCOH dielectric were fabricated using a 32 nm CMOS process on 300mm wafers. As shown in Figure 1, Cu lines were uniformly capped with 4-5nm ELD CoWP layer [6]. The entire CoWP process included preclean, ELD, post cleaning, and drying. Both line only (serpentine-comb and comb-comb) and via-related structures with various sizes were used in this study. The actual mean spacing between lines and vias ranged from 20nm to 50nm for the different test structures and processed wafers, as measured by top-down SEM after wafer processing. The method of multiple time-steps to apply low sensing voltages to monitor stress-induced leakage current (SILC) was used during the constant voltage TDDB stress. Hard breakdown was defined as a sudden leakage current increase during stress, in this case, corresponding to a SILC current above 100nA. 0.1V steps with 1V/s ramp rate and 200ms delay at each step were used for voltage ramp dielectric breakdown (VRDB) test. All VRDB tests were performed at 125 °C and at wafer level. TDDB tests were performed at 125 °C as well but at both wafer and module levels.

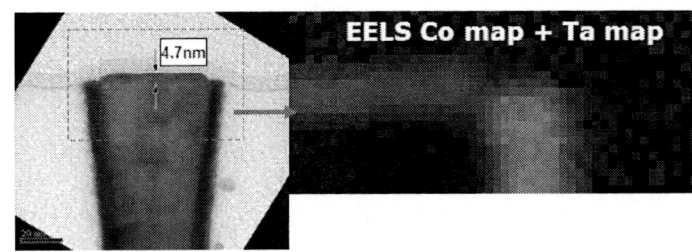

Figure 1: A TEM image of a 50 nm wide Cu line in ULK having a thin CoWP layer at the top with uniform coverage

III. RESULTS AND DISCUSSIONS

A. Early stage of CoWP Development

During the early CoWP development stage, we found that adding CoWP onto Cu surfaces imposed a great impact on dielectric TDDB characteristics. As shown in Figures 2 and 3, CoWP could degrade all TDDB characteristics including t_{BD}, voltage acceleration, and Weibull slopes for both dense low-k and porous ULK materials with comb-serpentine test structures of 1m long serpentine. ULK could be degraded more than dense low-k. However, with a smaller size test structure (100um long structure), CoWP had different influences for different dielectric materials. It had little impact on dense low-k dielectric TDDB while it still degraded ULK TDDB significantly at such small test structure. As shown in Figure 3, the Weibull slopes of CoWP and the control samples with dense low-k were comparable using 100um long test structures, while the Weibull slope of CoWP with porous ULK was still much worse than the porous ULK control using 100um long structures. This clear size dependent TDDB difference of dense low-k and porous ULK revealed that CoWP impact on TDDB could have a strong dependence on low-*k* materials.

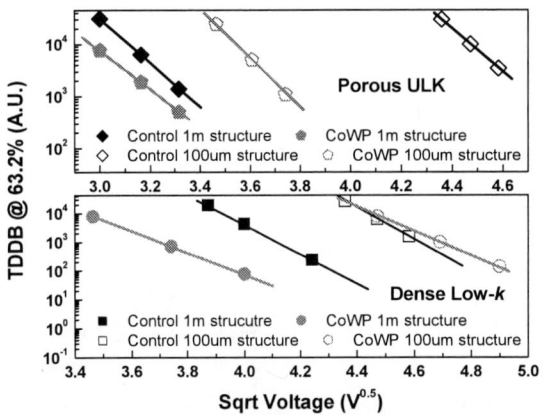

Figure 2: TDDB performance for CoWP and control at dense and porous low-*k* levels

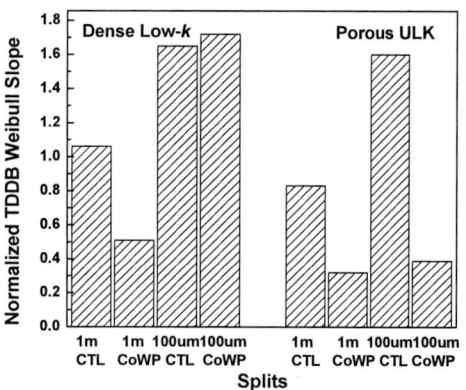

Figure 3: Weibull slope comparison for different sizes of structures

The degraded TDDB performance of CoWP wafers was found to be related to metallic particles arising from the CoWP plating process. Some examples of these defects observed immediately after Co deposition without optimized post CoWP deposition cleans are shown in Figures 4 and 5. Both EDX and Auger analysis confirmed that metallic particles were mainly Co particles, which caused increase leakage and even electrical shorts between interconnects.

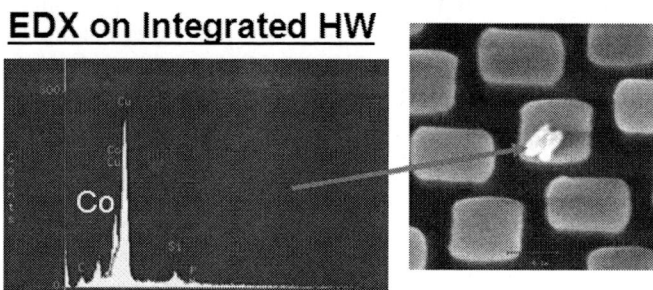

Figure 4: SEM and EDX of a metallic defect on an integrated hardware post CoWP deposition

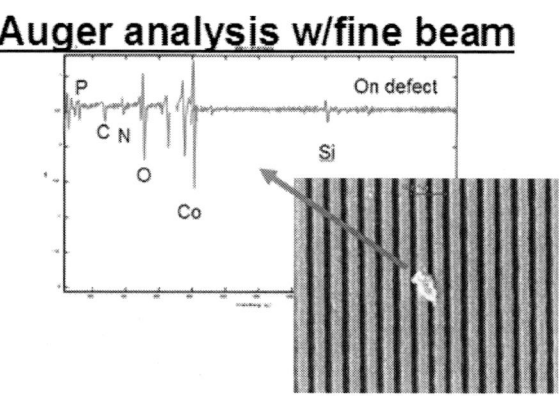

Figure 5: SEM and Auger analysis of a metallic particle on a comp-serpentine structure post CoWP deposition

Based on the square-root of E TDDB breakdown model [8], a physical picture to explain the observed TDDB degradation of dense low-k and porous ULK is illustrated in Figure 6 and equations 1 and 2. Because dense low-k has a relatively dense surface which has a better resistance to surface metallic contamination, its surface could possibly only be contaminated by limited Co particles after CoWP deposition, depending on its dielectric surface size. Consequently low-k TDDB exhibited an obvious area dependence. Dense low-k had poor TDDB on a larger size structure, but had a much improved TDDB on a smaller size structure as shown in Figures 1 and 2 due to its limited defect density. However, ULK, due to its porous surface, has a much higher susceptibility to Co contamination during CoWP deposition. On the other hand, the degradation of ULK intrinsic electrical properties, caused by surface damage formation during the integration process, could be severe. Such degradation could further exacerbate its TDDB problem as cobalt-based metal-cap was reported to be extremely sensitive to galvanic corrosion [9] if moisture entrapped by damaged ULK was present. Therefore, even with a very small testing structure size, ULK could still suffer a severe TDDB problem due to its extremely high susceptibility to entrapping Co particles at its surface and even gross sub-surface entrapment. Extremely high defect density could wash out the area dependence of TDDB.

The contribution of Co particles to TDDB can be described by equations 1 and 2:

$$C_{critical} - C_0(Co) = 1.12 \times C_{time} \left(D_0 \exp\left[-\frac{E_D}{k_B T} \right] \right)^{0.5} (t_{BD})^{0.5} \quad (1)$$

$$C_{time} \propto A^* T^2 \exp\left[\frac{\beta_s \sqrt{E} - \varphi_s}{K_B T} \right] \quad (2)$$

here $C_{critical}$ is the critical percolative defect density to trigger an ultimate low-k breakdown. It consists of a T0 Co defect density $C_0(Co)$ and is a function of a time dependent defect density C_{time} generated within a specific location between two interconnects during TDDB stress. C_{time} is proportional to electron fluence ($C_{Cu} \propto I/I_0$, where I_0 is an effective unit length) as shown in equation 2 based on the square-root of E breakdown model. As Co density is limited with a strong area dependence for dense low-k, while it is high enough to distribute Co everywhere for ULK, this $C_0(Co)$ difference could explain our observed difference of TDDB area dependence between dense low-k and porous ULK dielectrics. As shown in Figure 7, TXRF surface Co contamination analysis results of dense and ULK blanket monitoring wafers passed through CMP and CoWP supported our hypothesis as it clearly revealed the Co contamination differences between dense low-k and porous ULK as compared to the control after CoWP deposition.

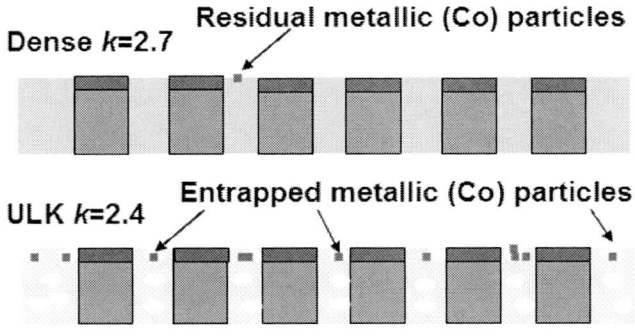

Figure 6: Proposed physical model for CoWP impact on TDDB

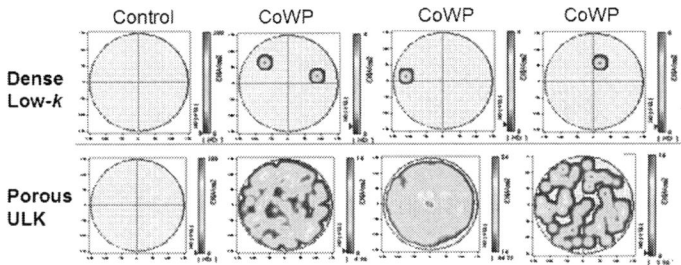

Figure 7: TXRF surface Co contamination results for blanket dense and porous low-*k* dielectrics with CoWP

Historically, metallic defects and/or residues leading to poor shorts yield and degraded TDDB behavior have been studied [7, 10, 11], and because of that, the implementation of CoWP metal cap layer was reported to be extremely challenging. Therefore, optimizing CoWP process has become a major task for successful CoWP implementation at 32nm with TDDB assurance in addition to the improved EM benefit. A way to maintain Co on the Cu surface for EM while removing residual particles at dielectric surface for TDDB was rigorously needed. As discussed above, some special focus needed to be placed on ULK dielectrics as their porous surface might entrap Co particles easily.

B. CoWP Process Optimization and low-k Defect Denstiy Reduction

The removal of metallic defects and contamination from the dielectric surface can be achieved by the optimization of CoWP processes. A careful selection of various chemistries and their sequence, with the understanding of integration dependent parameters, is critical for CoWP defectivity reduction and TDDB improvement.

Many aggressive CoWP wet process schemes have been developed including various process aspects of CoWP modules. In order to identify the best CoWP process quickly and accurately, we developed a comprehensive reliability testing methodology during the 32nm CoWP development period. It consists of two major tests – 1) quick VRDB test on the largest available structure (10m long structure) with large

sample size per wafer as a screen for quantitative defect density study, 2) constant voltage TDDB stress with various sizes of test structures (from 100um to 1m) on wafers passing the 1st test for final selection based on their lifetime and failure rate projections. Multiple wafers per split were also used to address wafer-to-wafer variation concern. With this methodology, not only the extrinsic and intrinsic breakdown characteristics but also the extrinsic defect density dependence on critical areas and baseline wafer-to-wafer variability could all be carefully evaluated for various CoWP wafers. The TDDB results used for the final CoWP process selection were mainly from 1m long structures. It should be noted that the effectiveness of product burn-in on BEOL TDDB defects is still not known yet. All BEOL TDDB breakdowns should be considered as legitimate wearout fails if they couldn't be detected and screened out at $1.5 \times V_{DD}$-125 °C product stress condition. Therefore, passing TDDB with a relatively larger structure is a conservative and more accurate (closer to real product use condition) method for a robust technology qualification.

Figure 8 is an example of VRDB results from different CoWP splits with ULK. It was found that different CoWP processes could modulate defect tails significantly, and the split 3 process was found to be the worst. Figure 9 illustrates the calculated defect density with 90% confidence levels from the wafers in Figure 8. The defect density was calculated based on the following equation [12]

$$C \leq 1 - \sum_{i=Y}^{N} \frac{N!}{i!(N-i)!}[\exp(-D_0 A)]^i [1-\exp(-D_0 A)]^{N-i} \quad (3)$$

where C is the confidence bound level, D_0 is the defect density, A is the area of test structure, N is the total sample size, and Y is the number of good samples. By further fine tuning the process of split 1, a good CoWP process producing limited defect density was finally identified as shown in Figure 9 for both dense low-*k* and porous ULK dielectrics, respectively. Further as shown in Figure 10, by aggressively changing some process parameters within the split 1 process, the defect density was barely changed at the dense low-*k* level, and was even slightly improved at the ULK level, suggesting that this process has a good manufacturing process window.

C. Reliability Performance of Optimized CoWP wafers

After the initial VRDB screening, constant voltage TDDB stresses were carried out on the promising splits for final process selection. As shown in Figures 11 and 12, wafers with the optimized CoWP process exhibit comparable TDDB Weibull slopes to the control wafers without CoWP for both dense and porous dielectrics. Furthermore, the TDDB kinetics of CoWP and the control wafers were studied. As shown in Figures 13 and 14, the optimized CoWP wafers overall maintained the TDDB kinetics with no degradation.

978-1-4244-5430-3/10 $26.00 © 2010 IEEE

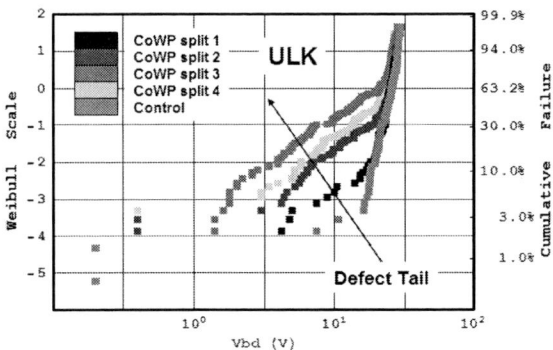

Figure 8: Vbd distributions of various CoWP splits

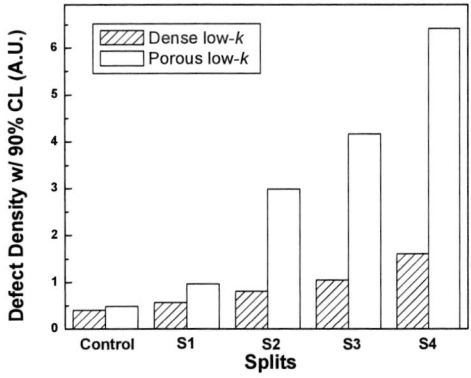

Figure 9: Calculated defect density results from wafers in Figure 8 by VRDB test

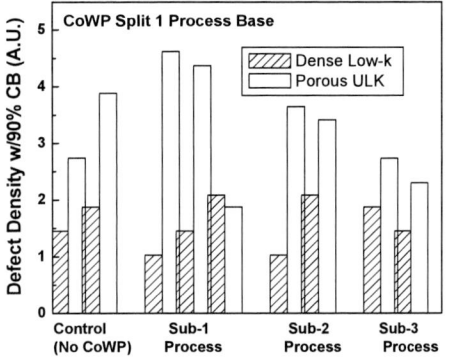

Figure 10: Calculated defect density results of various sub split 1 CoWP processes by VRDB test

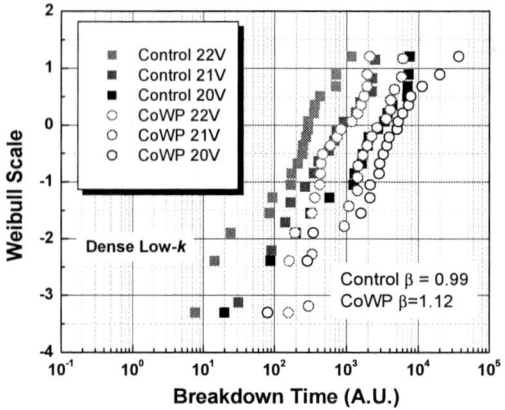

Figure 11: TDDB comparison between control and optimized CoWP wafer at dense low-*k* level

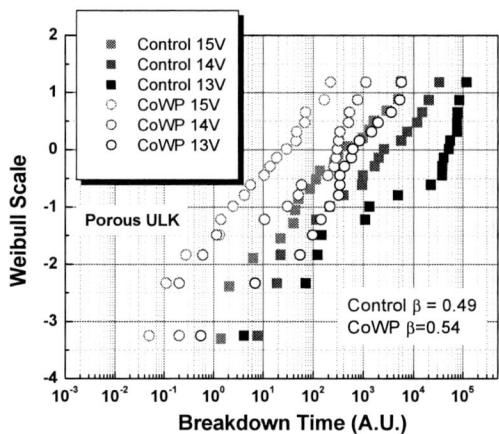

Figure 12: TDDB comparison between control and optimized CoWP wafers at porous ultra low-*k* level

Figure 15 illustrates the TDDB reliability index of various CoWP splits with subtle chemistry and sequence changes of split 1 CoWP process. The TDDB reliability index was defined as the product of TDDB Weibull slope and voltage acceleration factor, which were two important parameters for TDDB lifetime projection and failure rate determination. As shown in Figure 15, no significant variations from various CoWP splits were found, suggesting that a relatively wide CoWP process window could be established after elimination of some problematic CoWP steps. CoWP wafers overall have comparable TDDB reliability index to the control wafers.

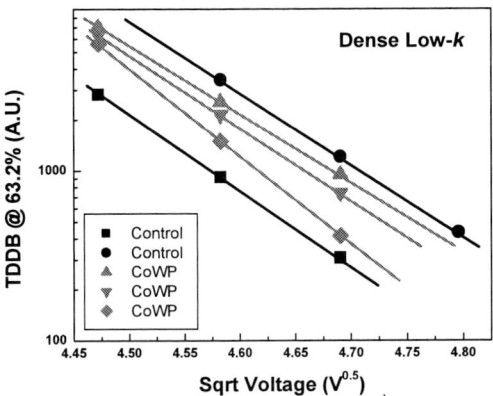

Figure 13: TDDB kinetics and t_{BD} of CoWP and the control wafers with dense low-k dielectric

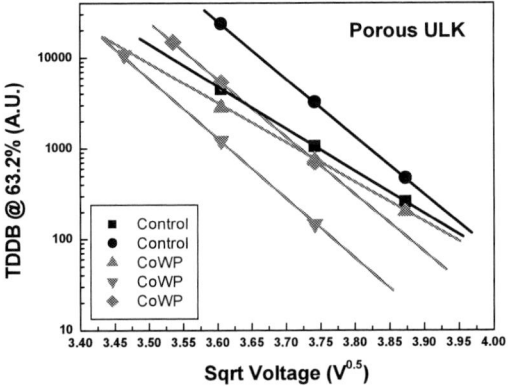

Figure 14: TDDB kinetics and t_{BD} of CoWP and the control wafers with porous ULK dielectric

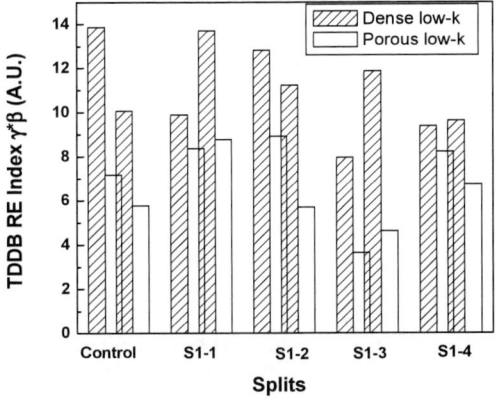

Figure 15: TDDB reliability index for CoWP and the control wafers with dense and porous low-k dielectrics

From Figure 15, it is obvious that porous ULK had overall worse TDDB performance index than dense low-k. The poor TDDB performance of ULK was attributed to its intrinsically weaker material properties, which made it more prone to have process induced defects. However, those defects were found not to be related to optimized CoWP process. As shown in Figures 16, 17 and 18, TDDB early fails of ULK with and without CoWP were found to fall into two general categories based on extensive post TDDB physical failure analysis – 1) a particle containing oxygen sitting at the top of Cu lines and depression of Cu lines possibly induced by CMP, and 2) A local O and Cu rich region between Cu lines at breakdown site. Such defects could cause enhanced local field with accelerated local Cu ionization (due to presence of oxygen nearby), which result in poorer TDDB. Both kinds of observed defects were obviously not related to CoWP. Therefore, for ULK material, other process induced defects may modulate ULK TDDB more than CoWP processing. It was concluded that a robust baseline process together with an optimized CoWP process was imperative to assure good TDDB performance at 32nm technology and beyond.

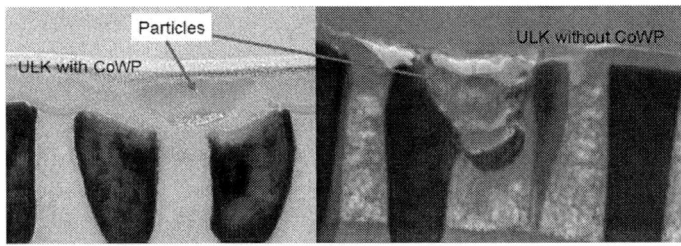

Figure 16: TEM results showing an early ULK TDDB fail signature for both control and CoWP wafers – a foreign particle containing oxygen sitting at the top of Cu lines

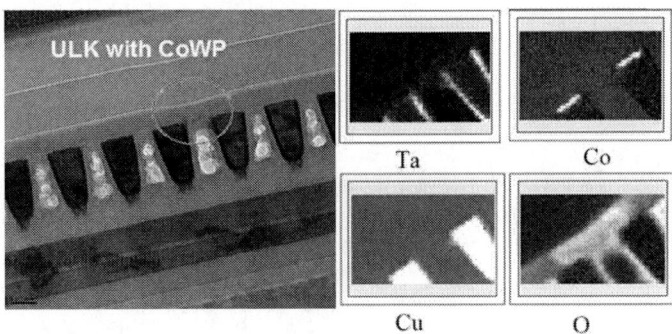

Figure 17: TEM and EELS results showing an early ULK TDDB fail signature for a CoWP wafer – a local O and Cu rich region between Cu lines. Gaps between lines are TEM artifacts.

One extra benefit of using CoWP was shown in Figures 19 and 20. Implementation of CoWP could reduce the wafer-to-wafer VRDB variation as compared to the control wafers for ULK film based on a large scale VRDB study. Such reduction of variability was attributed to the extensive cleans used before and after CoWP deposition, which could minimize the potential post CMP Cu corrosion when ULK became more hydrophilic due to its surface damage during Cu/low-k

978-1-4244-5430-3/10 $26.00 © 2010 IEEE

integration. Therefore adding CoWP could help to maintain V_{BD} (voltage-to-breakdown) for those severely damaged ULK wafers. On the other hand, adding CoWP could reduce the intrinsic V_{BD} of less damaged and damage free ULK wafers. So the net effect of adding CoWP is that CoWP potentially could reduce overall ULK wafer-to-wafer VRDB variability.

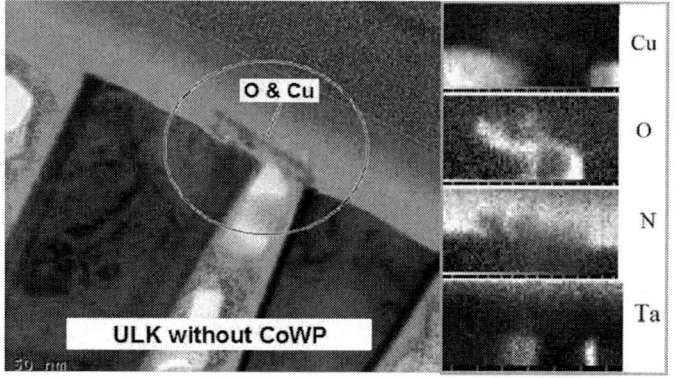

Figure 18: TEM and EELS results showing an early ULK TDDB fail signature for a control wafer – a local O and Cu rich region between Cu lines

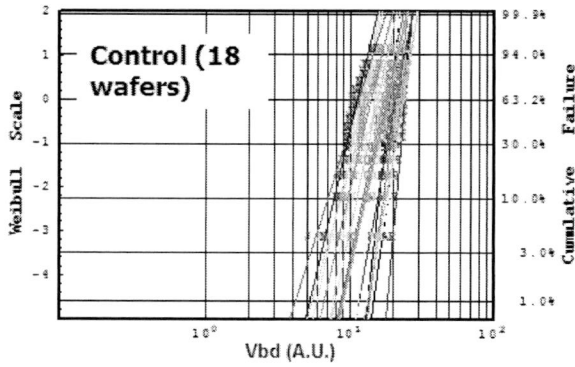

Figure 19: The variability study of 18 control wafers with ULK by VRDB showing a large wafer-to-wafer variation

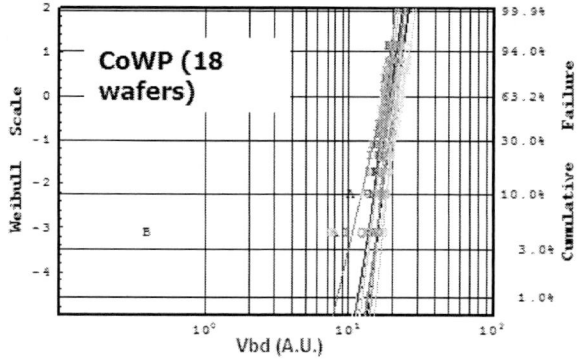

Figure 20: The variability study of 18 CoWP wafers with ULK by VRDB showing a reduced wafer-to-wafer variation as compared to the control wafers

Finally, an optimized CoWP process with substantial improvements of both stress migration (SM) and EM, with comparable TDDB performance was demonstrated at 32nm technology node. As shown in Figures 21 and 22, this CoWP process offered the best solution for both EM and SM improvements as compared to pure Cu due to CoWP's superior ability to effectively slow down the mass transport at Cu/capping layer interface as compared to pure Cu.

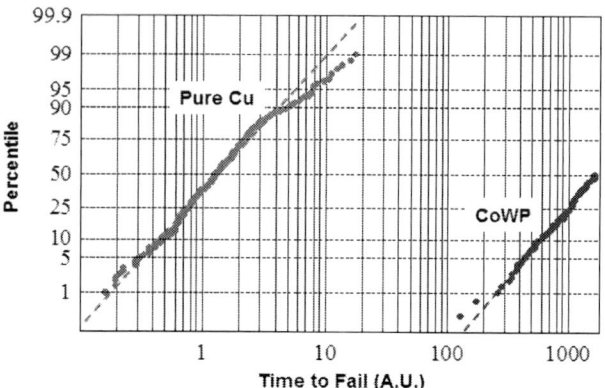

Figure 21: EM V1→ M2 upstream mode performance comparisons for CoWP and pure Cu wafers

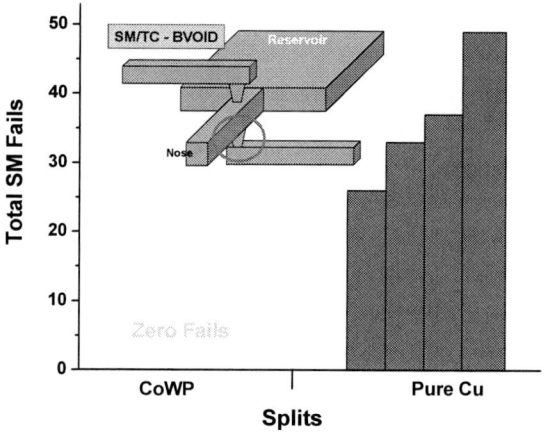

Figure 22: SM performance comparisons for CoWP and pure Cu wafers

IV. CONCLUSIONS

In conclusion, impacts from ELD CoWP on dense low-k and porous ULK TDDB have been thoroughly investigated, and a physical model was proposed. Post ELD CoWP cleaning is critical to assure robust extrinsic and intrinsic low-k TDDB. A comprehensive TDDB reliability evaluation method including defect density study by VRDB testing of 10m long structures with large sample size, and dielectric breakdown lifetime determination by constant voltage TDDB testing on 1m long structures has been proven to be very useful for CoWP process development. Finally, a reliable CoWP process for Cu with both dense low-k and porous ULK dielectrics was demonstrated. Comparable leakage and TDDB performance with substantial improvements in EM and SM were achieved with an optimized CoWP process.

ACKNOWLEDGMENT

The authors wish to acknowledge all the people working within IBM alliance programs for 32nm SOI and Bulk CMOS technology development. This work has been supported by the independent Bulk CMOS and SOI technology development projects at the IBM Microelectronics Division, Semiconductor Research & Development Center, Hopewell Junction, NY 12533.

REFERENCES

[1] M. W. Lane, E.G. Liniger, and J. R. Lloyd, J. Appl. Phys., vol. 93, pp.1417-1421, 2003

[2] E. Zschech, M.A. Meyer, and E. Langer, Proc. Mater. Res. Soc. Symp, 2004, vol. 812, pp. 361-372

[3] Y. S. Diamand, et al., Proc. Electrochem. Soc., 2000, vol. 99-34, pp.102

[4] C.K. Hu, et al., Appl. Phys. Lett., vol. 81, no. 10, pp. 1782-1784, 2002

[5] S. Yokogawa, Y. Kakuhara, H. Tsuchiya, and K, Kikuta, 45th Annu. IEEE Rel. Phys. Symp., 2007, pp. 117-121

[6] E. Huang et al., Proc. of Adv. Metallization Conference, 2009, pp. 33-34

[7] J. Gambino et al., Phys. & Failure Analysis of Integrated Circuits, 2007, IPFA, 14th International Symp. on the, pp 59-64.

[8] F. Chen, et al., Proc. 44th Annu. IEEE Rel. Phys. Symp, 2006, pp. 46-53

[9] C.I. Lang, et al., Semiconductor International, July 1, pp. 1-8 (2009)

[10] K-J Moon, et al., Proc. of Adv. Metallization Conference, 2009, pp. 105-106

[11] O. Aubel, et al., 46th Annu. IEEE Rel. Phys. Symp, 2008, pp. 675-676

[12] JEDEC Standard JESD35-A, pp 39, 2001

EM and SM Induced Degradation Dynamics in Copper Interconnects Studied Using Electron Microscopy and X-Ray Microscopy

Ehrenfried Zschech, René Hübner
Fraunhofer Institute for Nondestructive Testing
Dresden, Germany
phone: (49) –(351) 88815 543, e-mail address: ehrenfried.zschech@izfp-d.fraunhofer.de

Oliver Aubel
GLOBALFOUNDRIES Inc.
Dresden, Germany

Paul S. Ho
The University of Texas at Austin
Austin/TX, USA

Abstract—In addition to statistically relevant standard reliability tests and lifetime analysis, the study of solid-state physical degradation mechanisms for a limited number of representative samples is needed to understand weaknesses in the interconnect technology and to exclude reliability-related failure in Cu interconnects. We present dynamic studies of damage mechanisms in on-chip Cu interconnects caused by electromigration (EM) and stress migration (SM). Scanning electron microscopy (SEM) and synchrotron-based transmission X-ray microscopy (TXM) are applied to visualize the void evolution, electron backscatter diffraction (EBSD) in the SEM and conical dark-field (CDF) analysis in the transmission electron microscope (TEM) are applied to characterize the Cu microstructure.

In case of EM, our experiments show that voids are formed at interfaces and grain boundaries, often far away from vias in via/line interconnect structures. Due to the gradient of the electrical potential, Cu atoms migrate along weak pathways for material transport. Depending on the interface strength, voids that virtually move along interfaces or grain boundaries over large distances into the opposite direction of the current flow, i. e. toward the cathode end of the line, have been visualized. In case of SM, our experiments do not show void movement over large distances during the stress-induced voiding (SIV). In via/line interconnect structures we rather observe that voids are formed directly beneath the via, i. e. in wide Cu line at the edge of the via bottom. It is concluded that voids are originally formed at the site where eventually the catastrophic failure occurs. During SM tests, Cu atoms migrate from regions of low tensile (or high compressive) stress to regions of high tensile stress, and simultaneously, vacancies migrate along the stress gradient (within a

limited range of some μm) in the opposite direction, to the location where vias connect wide Cu lines. For both EM and SM, the driving forces for atomic transport depend strongly on the particular geometry of the tested structure, but also on interface bonding and metal microstructure.

Keywords: electromigration, stress migration, stress-induced voiding, degradation mechanism, microscopy

I. INTRODUCTION

Electromigration (EM) and stress migration (SM) are reliability concerns for damascene Cu interconnects in advanced integrated circuits [1-6]. Continuous shrinking of the dimensions of on-chip interconnects, the introduction of advanced backend-of-line (BEoL) manufacturing process steps, and various combinations of thin film materials result in microstructure changes of the metal interconnects (including the stress state), new types of interfaces, and so far unknown degradation phenomena during accelerated reliability tests [7]. Therefore, the effect of interconnect design and Cu microstructure on EM and SM resistance is an important issue that has attracted increasing attention. However, the correlation between Cu interconnect properties (particularly impurity chemistry of the electroplating bath, interconnect geometry, and Cu microstructure) and the degradation dynamics is not well understood.

"Post mortem" studies of failed EM and SM structures, mainly based on SEM images of Focused Ion Beam (FIB) cross-sections through the region of interest including the void that causes the catastrophic failure have been published [1,8]. In addition, several models have been proposed to describe the EM and SM/SIV mechanisms [1,9,10]. However, the experimental verification of such models and the direct study

978-1-4244-5430-3/10 $26.00 © 2010 IEEE

of the dynamics of degradation mechanisms in Cu interconnects require either in-situ imaging of voids or non-destructive observation of the void evolution at several stages of the degradation process. This approach was successfully applied for EM and SM studies [11-14]. In particular, these experiments have directly shown the real degradation mechanisms for the first time. In this paper, EM and SM degradation dynamics in damascene Cu interconnects are presented, based on SEM and TXM imaging during the degradation process. These studies are complemented by "post mortem" FIB cross-sections and high-resolution Cu microstructure studies applying SEM and TEM.

II. EXPERIMENT

A. Samples

For the experiments performed in this study multilevel damascene Cu/low-k interconnect stacks, which include the test structures, were extracted. For both EM and SM experiments, cross-section samples with the regions of interest were prepared using the FIB technique in such a way that the respective interconnect structures remain fully embedded in the insulating dielectric material (either SiO_2 or organosilicate glass, OSG).

For the EM studies, the samples with via/line test structures (a long line connected by two vias at each end to the upper metal level) were stressed in particularly designed chambers during the accelerated in-situ SEM and TXM experiments. The test structures were connected to bond pads. The temperature range was between 150 ^{0}C and 350 ^{0}C, and the current densities were up to 30 MA/cm^2. These conditions were chosen to perform the in-situ experiments in a reasonable period of time.

The SM test structures, a Cu via (diameter about 100 nm) connected to wide (> 1 μm) Cu plates (above and below the via), were stressed in vacuum ($p_{residual} \sim 10^{-2}$ mbar) at a temperature of 175 °C, a typical temperature for SM baking. The initial pre-stressing of the sample was performed for 50 hours. Subsequently, a series of TXM images was recorded after every five hours of baking at the same temperature.

For the in-situ SEM studies, the FIB cross-section was milled in such a way that about 50 to 100 nm dielectrics remains in front of the Cu interconnect structure, without disturbing any interfaces. The insulating dielectric thin film reduces the SEM image resolution for the buried interconnect structure to some nm. For the "post-mortem" EBSD analysis, the Cu via/line interconnect structure was cross-sectioned through the middle of the line using FIB milling.

Sample preparation for the TXM experiments was also performed by applying the FIB technique. A lamella containing the region of interest was thinned to a thickness of about 1-2 μm in such a way that the Cu via remains fully embedded in dielectric material and the FIB cuts are several 100 nm away from the region of the Cu line where the voiding is expected. For the "post mortem" TEM analysis, this sample was further thinned to a final thickness of about 60 nm in such

a way that the structure of interest is located within the lamella.

B. SEM and TXM experiments

In-situ SEM and TXM experiments allow the visualization of material transport in fully embedded damascene copper test structures. To obtain information about the principle degradation mechanisms of the individual samples, in-situ and ex-situ (after some stages of stressing) investigations using SEM and TXM were performed. In these experiments, cross-sections of fully embedded test structures were imaged to visualize the degradation dynamics after several periods of time.

In the in-situ SEM experiments, firstly described in [11], cross-sections with embedded dual-damascene Cu via/line interconnects were imaged continuously, while EM stressing is performed at elevated temperatures and current densities. For these experiments, the samples were mounted on modified 24-pin test chip packages and wire bonded to ensure electrical connection. This test chip was mounted on a custom-made heating stage of a modified Zeiss Gemini 1550 SEM tool. However, SEM experiments do not seem to be appropriate to study the SIV dynamics during baking since the FIB cut, which has to be located less than 100 nm in front of the Cu via, would intersect the Cu wide line (which is necessary to pronounce SM behavior) in a region where the voiding is expected.

Synchrotron-based TXM is a powerful imaging technique for both EM and SM induced void detection in Cu interconnects [12]. In the case of TXM, using photon energies between 0.5 and 2.0 keV, the sample thickness has to be 1-2 μm. Consequently, at least several 100 nm of wide line material is remaining in each direction of the Cu via structure, which is sufficient for SM experiments. The TXM studies were performed at the full-field soft X-ray microscopes installed at a bending magnet at the Advanced Light Source (ALS), Berkeley, CA (photon energy of 1.8 keV) [12] and at the an undulator beam line at the BESSY II electron storage ring, Berlin, Germany (photon energy of 0.5-0.8 keV) [14]. For Fresnel zone plates providing 20 nm lateral resolution the depth of field is about 1 μm for soft X-rays. Thus, all object details of an on-chip interconnect stack (with metal line dimensions << 1 μm) are simultaneously in focus, and as a consequence the images can be treated as magnified projections. By tilting the object perpendicular to the optical axis of the microscope, X-ray tomography can be performed. Using this 3-D imaging technique, it is possible to locate features as small as 20 nm in a Cu interconnect stack.

C. EBSD and CDF studies

EBSD is a proven analytical technique to characterize the microstructure of Cu interconnects, also at FIB-prepared cross-sections. Particularly inverse pole figure maps (IPF) are used to correlate void evolution and microstructure. This technique is applicable up to grain sizes of about 30 nm. However, in sub-100nm interconnect structures, new phenomena like agglomerates of small grains (dominantly at the lower half of the line) occur, which require analytical techniques with an improved spatial resolution.

Due to a resolution in the < 5 nm range, Conical Dark-Field (CDF) analysis in the TEM provides grain size distributions and complete grain orientation maps for grains < 30 nm. Consequently, it allows to characterize the Cu microstructure of regions with small grains (grain sizes << 100 nm) [15]. In a first step, centered dark-field images of a particular view of the specimen are recorded in a circular way for several tilt and rotation angles of the parallel electron beam. In a second step, spot diffraction patterns for every point of the view of interest are reconstructed using all recorded dark-field images. Based on the crystallographic structure of the polycrystalline material, the diffraction patterns are indexed. Inverse pole figure maps, and hence, the crystallographic orientations of individual Cu grains, particularly of the "neighbour grains" of a void, can be determined. This microstructure information can be used to discuss degradation phenomena in copper interconnect structures.

III. RESULTS

A. Electromigration

For dual-damascene Cu lines that were embedded in metal liners (typically Ta or Ta-based compounds or Ta/TaN layer stacks) at the bottom and at the sidewalls, and with surfaces covered by a dielectric material (typically SiN_x or $a-SiC_xH_y$ etch stop layers) after the CMP process, interfaces dominate usually the EM-induced mass transport, and therefore, degradation of Cu interconnects is a function of the bonding strength of the weakest interface [7]. Most of the previous EM studies have shown that the void growth is dependent on material transport along the Cu/dielectric interface but not dependent on material transport along the Cu/metal interface (the metal liner provides good adhesion, i. e., high bonding strength). That means, the top Cu/dielectrics interface has been found to be the weakest interface in most cases [2,11,16]. The reason that Cu/dielectrics interfaces are rapid pathways for mass transport is the weak bonding and the high degree of atomic disorder, and consequently, the resulting poor adhesion between Cu and the adjacent material from which the interface is composed.

Lane et al. established relationships between atomic transport along interfaces, adhesion and EM-induced drift velocity for BEoL structures [17]. TEM investigations of Cu/SiN_x and Cu/CoWP interfaces revealed distinct differences that provide the understanding of the relation between interface structure and bonding on the one hand side and the activation energy for atomic transport processes along these interfaces on the other hand, i. e., the root cause for the different EM behavior [18]. The Cu/SiN_x interface is usually characterized by an "intermixing layer" of about 2 nm thickness with a high degree of disorder. This "intermixing layer" can be explained with the plasma enhanced chemical vapor deposition process (PECVD) of SiN_x on top of the polished Cu. At the Cu/CoWP interface, however, lattice planes are visible in both metal films, sometimes without any interruption at the interface. This kind of interface with regions of coherence (i. e. with epitaxy-like growth) is much less disordered, and there does not exist any "intermixing layer".

The experimentally observed preferential voiding at the Cu/capping layer interface indicates the "weakness" of the top Cu/SiN_x interface, which is the fastest pathway for atomic transport. Under EM, some triple junctions at the interface will act as flux divergent sites for void nucleation. This depends on the overall balance of the mass fluxes along the two neighboring interfaces and the intersecting grain boundary. At the weak Cu/SiN_x interface where interfacial diffusion dominates mass transport, the flux divergence arises primarily from the difference of the diffusion fluxes along the interfaces of the two adjacent grains, and consequently, grain boundary diffusion is not important. In this case, most of the voids will firstly appear at the top interface and move toward the cathode, causing eventually via failure. This is consistent with experiments where most voids were formed initially at the Cu/SiN_x interface by a heterogeneous void nucleation - often far away from the cathode end. Void nucleation and growth in the line segment can further facilitate the relaxation of a tensile stress developed during EM near the cathode [7]. The individual processes for a weak Cu/cap layer interface can be described as in [11]: Cu/SiN_x interface diffusion and diffusion along the inner surface of a void, grain boundary diffusion and void agglomeration, as well as void growth and Cu redeposition.

The EM-induced degradation process for a sample with strengthened top interface is significantly different to that for samples with Cu/SiN_x top interface [18,19]. The CoWP-coated Cu lines – as an example for interconnects with strengthened top interface - show void formation at both the Cu/capping layer and at the Cu/liner interfaces, and subsequent movement of the voids. Since interfacial mass transport is significantly reduced due to the strong interface bonding, grain boundary diffusion becomes important in contributing to flux divergence and void nucleation. As a result, the Cu microstructure and the grain boundary orientation play a more important role than for the Cu/SiN_x structure. In some cases, the mass transport along this interface was reduced to an extent that the Cu/liner interface becomes the dominant diffusion path.

Figure 1 shows the SEM image of a cross-section via/line test structure with strengthened top interface, that failed during an in-situ SEM degradation experiment (at position D, see [19]) together with an inverse pole figure (IPF) map representing the crystallographic orientation of the individual grains relative to the wafer surface. The IPF map shows clearly that voids occur at triple junctions with neighbored Cu grains that have different crystallographic orientation. These triple junctions which are created by Cu grain boundaries and interfaces act as heterogeneous nucleation sites for void formation [18], i. e., the Cu microstructure is influencing already the first phase of the degradation process described above. Moreover, so-called slit voids that were observed at three positions during the in-situ SEM experiment coincide with vertical grain boundaries that span across the entire line.

978-1-4244-5430-3/10 $26.00 © 2010 IEEE

At position A for instance, such a slit void was formed suddenly during the in-situ SEM experiment, and it remained there until it merged with another void. Subsequently, the merged slit void moved to position B, where it again remained stationary. That means, grain boundary/interface intersection points can act as "void stoppers" (see also [20]), where voids are pinned for a relatively long period of time. On the cathode side of the grain boundaries reside either large grains (A and D) or a twinned region (B). In region C, where a faceted void was observed to remain stationary for some time, two grains are visible: a Cu<111>-oriented grain above the void position and a Cu<511>-oriented grain at the cathode side of the void. Such grain orientations seem to play a special role for the void pinning at triple junctions and void movement along an interface. Bower et al. discussed the mass transport as a function of the ratio between chemical potential differences (surface energy of inner surfaces of voids) and electric current flow. That means in particular, that increased EM driving forces can de-pin voids from triple junctions (acting as "void stoppers") [21].

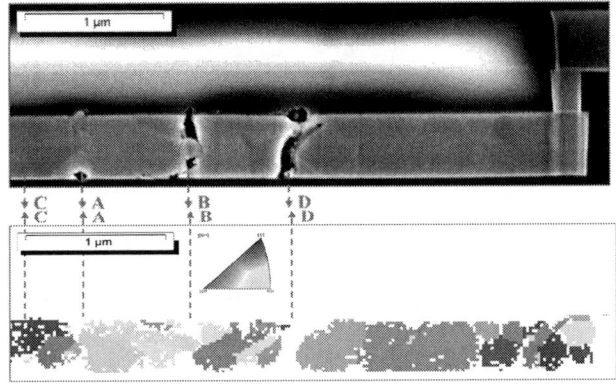

Figure 1: SEM/EBSD analysis of a Cu/CoWP interconnect sample after EM experiment [19].

B. Stress-induced voiding

A series of TXM images of the SM sample as described above was recorded to make the void growth process visible. The time series of TXM images (Fig. 2) shows the void evolution in Cu interconnect structures [6]. In Fig. 2 (b), after 90 hours of thermal treatment, an early stage of void formation is seen in the wide Cu line beneath the via, starting at the edge of the via bottom. The void continues to grow due to further thermal treatment, as clearly seen after 95 hours of thermal treatment in Fig. 2 (c). For this particular sample, the void is clearly after 100 hours of baking at a temperature of 175°C, as seen in Fig. 2 (d). No void movement was observed. To measure exactly the location of the void in the interconnect stack, we performed nano-tomography by tilting the sample in the TXM and recorded X-ray images at different viewing angles. To reveal the 3-D location of the stress-induced void, we reconstructed the tomographic data set, indicating clearly that the void is located beneath the via [6].

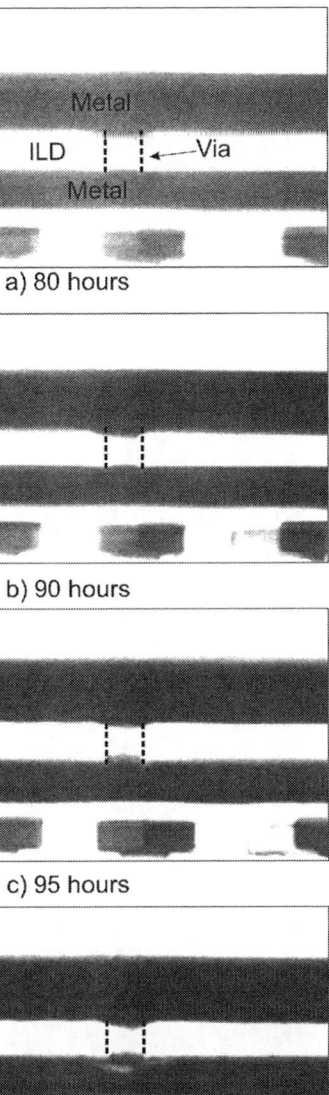

a) 80 hours

b) 90 hours

c) 95 hours

d) 100 hours

Figure 2: Time series of TXM micrographs showing the stress-induced voiding during the SM experiment. The sample was pre-stressed by heating at 175°C for 80 (a), 90 (b), 95 (c), and 100 (d) hours, respectively. Note that the void is clearly visible after baking at 175 °C for 100 h. The dashed lines represent the position of the via [6].

The location of the void formation and growth process depends strongly on the interconnect geometry as well as particular process variations and resulting configurations like via gouging [22]. In our study, the voids in Cu nucleated and grew where the via is connected to the bottom wide line.

Subsequently, i.e. after SM baking and the TXM experiments, the FIB-based serial sections technique was used to create a number of parallel cuts through the wide Cu line and the Cu via, and particularly the region where the void was formed. In coincidence with X-ray tomography data, the extension of the void is significantly larger than the via cross-section. Fig. 3 shows a series of SEM images recorded during FIB milling [6]. Stress-induced voids are typically observed at the edge of the via bottom, which supports the observations made by Gregoire et al. [23].

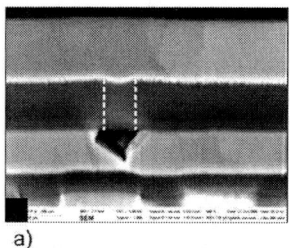

a)

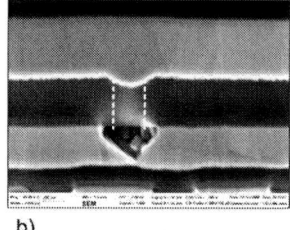

b)

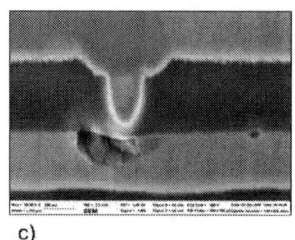

c)

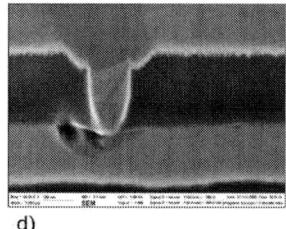

d)

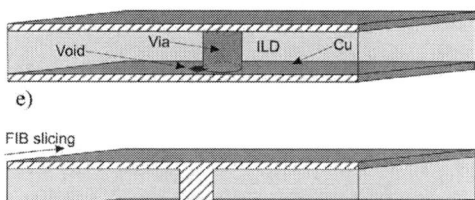

e)

Figure 3: (a-d) Series of SEM images recorded during FIB milling (serial section technique) of the SM test structure after the void evolution experiment in the TXM for 100 h at 175 °C. The SEM image series gives an impression about the 3-D shape of the void located close to the via. (e) Schematic representation of the void location [6].

Since the local microstructure plays an important role for the SM degradation process, the Cu grain orientations in the region of interest, i. e., in the surrounding of the stress-induced void, were analyzed using the CDF technique. A FIB target preparation was performed to fabricate a TEM sample with a thickness of about 60 nm in such a way that the void formed during SIV-based degradation and the surrounding Cu grains were partially included in the lamella. Fig. 4 shows an SEM image of the TEM lamella before final thinning from the backside, a TEM brightfield image and two inverse pole figure maps (including the orientation triangle) of the structure of interest [6]. The void formation depends on geometry and interface strength (barrier and etch stop layer processes and materials). In the particular case, the Cu microstructure information is provided for the region next to the void formed at the edge of the via bottom. This edge is the location of the maximum stress gradient, where eventually the SM damage occurs. The crystallographic orientations of all Cu grains, and particularly the "neighbour grains" of the void, could be determined, except for the small grains next to the void. Energy-filtered TEM, which is based on electron energy loss spectroscopy (EELS), shows an increased oxygen concentration in addition to the Cu signal for this particular interconnect region [6].

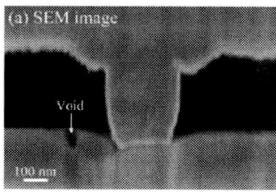

Figure 4a: SEM image of the TEM lamella before final thinning from the backside [6].

 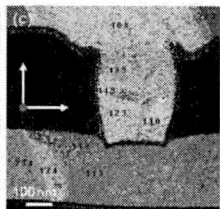 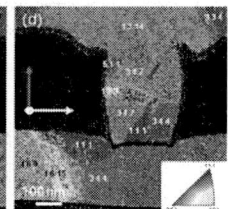

Figure 4b-d: TEM brightfield image (b) and two inverse pole figure maps (including orientation triangle) (c,d)) of the structure of interest, i. e., next to the void formed below the via that connects the wide line where the SIV-induced damage occurred [6].

The experimental results support the diffusion-flux model proposed by Ogawa et al. [1], which assumes an "active diffusion volume", i.e. a region where high diffusion rates induced by stress gradients and vacancy supersaturation at critical interconnect geometries (design) lead to void formation. The model implies that the vacancy source for void formation is highly localized although its nature has not been clarified. Interestingly, the grain structures in Figs. 4c and 4d show that the void was formed between a <111> grain and a high-index grain. The elastic anisotropy of Cu (E_{111}~$3E_{200}$) will give rise to a highly localized stress gradient at such grain junctions to drive vacancy agglomeration and void formation. This suggests a SM mechanism which is directly correlating void formation to local grain structures at critical line/via geometries. This can account for the SIV observed beneath the small vias that connect wide Cu lines as reported by Ogawa et al. [1].

IV. CONCLUSIONS

Two conditions are necessary to form and to grow a void: a vacancy source and a (directed) driving force. The dynamics of the Cu interconnect degradation is based on directed atomic transport caused by the specific driving forces in both cases, for EM and SM. However, the ranges of the driving forces, i. e. the gradients of the electrical potential and the stress gradient, are very different: for EM along the whole interconnect structure (several 10 μm to several 100 μm) and for SM within the relatively short range of the stress gradient (about 3 μm). Consequently, the SM mechanism is different from the EM mechanism where voids are usually formed aside the site of catastrophic failure, and where voids are (virtually) moving along relatively long distances of interconnect structures. In the case of SM, voids are formed at the site where the catastrophic failure eventually occurs, i. e. preferably where a small via connects to a wide line of an interconnect structure.

In summary, our dynamical EM degradation study showed that voids are often formed in via/line Cu interconnect structures far away from the point where eventually the catastrophic failure occurs. Void evolution includes virtual void movement along the interconnect lines, caused by the gradient of the electrical potential. Our SM-based degradation study showed that voids in via/line Cu interconnect structures are formed directly in the Cu wide line at the edge of the via bottom. This leads to the conclusion that voids are originally formed at the site of the maximum stress gradient where the catastrophic failure eventually occurs, i. e. preferably where a small via connects to a wide line of an interconnect structure. For wide lines, the reservoir of available vacancies, the so-called "active diffusion volume" (from which vacancies can diffuse and can contribute to the void formation), is larger, and therefore, larger voids grow in a certain period of time. The driving force is the stress gradient below the via bottom [24]. Results from grain structure analysis indicate that voids are formed between <111> and high-index Cu grains, where a high local stress gradient exists to drive vacancy agglomeration and void formation.

In both cases, EM and SM, the dynamics of the degradation processes is determined by the strength of interfaces and by the copper microstructure. The next step that has to be done is to consider interactions between simultaneously occurring EM and SM effects.

ACKNOWLEDGMENT

The authors thank Andreas Meyer and Eckhard Langer, both with Globalfoundries Inc., Dresden, Germany, and Peter Guttmann and Gerd Schneider, both with Helmholtz Zentrum Berlin, Germany, for performing the SEM and TXM experiments, as well as Yvonne Ritz, Fraunhofer Institute for Nondestructive Testing, Dresden, Germany, for careful sample preparation. Valuable discussions with Valery Sukharev, Mentor Graphics Inc., San Jose/CA, USA, are acknowledged.

REFERENCES

[1] E. T. Ogawa, J. W. McPherson, J. A. Rosal, K. J. Dickerson, T. C. Chiu, L. Y. Tsung, M. K. Jain, T. D. Bonifield, J. C. Ondrusek, W. R. McKee, "Stress-Induced Voiding under Vias Connected to Wide Cu Metal Leads", Proc. IEEE IPRS 2002, 312 (2002).

[2] E. T. Ogawa, K. D. Lee, V. A. Blaschke, P. S. Ho, "Electromigration reliability issues in dual-damascene Cu interconnections", Proc. IEEE IRPS 2002, 403 (2002).

[3] P. S. Ho, K. D. Lee, E. T. Ogawa, S. Yoon, X. Lu, "Impact of Low-k Dielectrics on Electromigration Reliability for Cu Interconnects", Proc. Characterization and Metrology for ULSI Technology, AIP Proc. 683, 533 (2003).

[4] R. C. J. Wang, C. C. Lee, L. D. Chen, K. Wu, K. S. Chang-Liao, "A Study of Cu/Low-k Stress-Induced Voiding at Via Bottom and its Microstructure Effect", Microelectronics Reliability 46, 1673 (2006).

[5] E. Zschech, P. S. Ho, D. Schmeisser, M. A. Meyer, A. V. Vairagar, G. Schneider, M. Hauschildt, M. Kraatz, V. Sukharev, "Geometry and Microstructure Effect on EM-Induced Copper Interconnect Degradation", IEEE TDMR 9, 20 (2009).

[6] E. Zschech, R. Huebner, D. Chumakov, O. Aubel, D. Friedrich, P. Guttmann, S. Heim, G. Schneider, "Stress-Induced Phenomena in Nanosized Copper Interconnect Structures Studied by X-Ray and Electron Microscopy", J. Appl. Phys. 106, 093711 (2009).

[7] V. Sukharev, E. Zschech, W. D. Nix, "A Model for Electromigration-Induced Degradation Mechanisms in Dual-Inlaid Copper Interconnects: Effect of Microstructure", J. Appl. Phys. 102, 053505 (2007).

[8] T. Oshima, K. Hinode, H. Yamaguchi, H. Aoki, K. Torii, T. Saito, K. Ishikawa, J. Noguchi, M. Fukui, T. Nakamura, S. Uno, K. Tsugane, J. Murata, K. Kikushima, H. Sekisaka, "Suppression of Stress-Induced Voiding in Copper Interconnects", Proc. IEDM 2002, 136 (2002).

[9] T. C. Huang, C. H. Yao, W. K. Wan, H. H. Lin, C. C. Hsia, M. S. Liang, "Numerical Modeling and Characterization of the Stress Migration Behavior upon Various 90 Nanometer Cu/Low-k Interconnects", Proc. IITC 2003, 207 (2003).

[10] Z. Suo, "Reliability of Interconnect Structures", Comprehensive Structural Integrity 8, 265 (2003).

[11] M.A. Meyer, M. Herrmann, E. Langer, E. Zschech, "In-situ SEM Observation of Electromigration Phenomena in Fully Embedded Copper Interconnect Structures", Microelectronic Engineering 64, 375 (2002).

[12] G. Schneider, G. Denbeaux, E.H. Anderson, B. Bates, A. Pearson, M.A. Meyer, E. Zschech, E.A. Stach, "Dynamical X-Ray Microscopy Investigation of Electromigration in Passivated Inlaid Cu Interconnect Structures", Appl. Phys. Lett. 81, 2535 (2002).

[13] V. Sukharev, E. Zschech, "A Model for Electromigration-Induced Degradation Mechanisms in Dual-Inlaid Copper interconnects: Effect of Interface Bonding Strength", J. Appl. Phys. 96, 6337 (2004), Microelectronic Engineering 82, 629 (2005).

[14] G. Schneider, P. Guttmann, S. Heim, S. Rehbein, D. Eichert, B. Niemann, "X-ray Microscopy at BESSY: From Nano-Tomography to Fs-Imaging", Proc. of the 9th International Conference on Synchrotron Radiation Instrumentation (Eds. J. Y. Choi, S. Rah), AIP Conf. Proc. 879, 1291 (2007).

[15] E. Zschech, R. Huebner, P. Potapov, I. Zienert, M. A. Meyer, D. Chumakov, H. Geisler, M. Hecker, H. J. Engelmann, E. Langer, „Process Control and Physical Failure Analysis for Sub-100nm Cu/Low-k Structures", Proc. IITC 2008, 67 (2008).

[16] A. V. Vairagar, S. G. Mhaisalkar, A. Krishnamoorthy, K. N. Tu, A. M. Gusak, M. A. Meyer, E. Zschech, "In-situ Observation of Electromigration-Induced Void Migration in Dual-Damascene Cu Interconnect Structures", Appl. Phys. Lett. 85, 2502 (2004).

[17] M. W. Lane, E. G. Liniger, J. R. Lloyd, "Relationship between Interfacial Adhesion and Electromigration in Cu Metallization", J. Appl. Phys. 93, 1417-1421 (2003).

[18] E. Zschech, H. J. Engelmann, M. A. Meyer, V. Kahlert, A. V. Vairagar, S. G. Mhaisalkar, A. Krishnamoorthy, M. Yan, K. N. Tu, V. Sukharev, "Effect of Interface Strength on Electromigration-Induced Inlaid Copper Interconnect Degradadtion: Experiment and Simulation", Z. f. Metallkde. 96, 996 (2005).

[19] M. A. Meyer, E. Zschech, "New Microstructure-Related EM Degradation and Failure Mechanisms in Cu Interconnects with CoWP Coating", AIP Conf. Proc. of 9th Int. Workshop on Stress Induced Phenomena in Metallization 945, 107-114 (2007).

978-1-4244-5430-3/10 $26.00 © 2010 IEEE

[20] H. P. Bonzel, in J. M. Blakely (Ed.), Surface Physics of Materials II, Academic, New York (1975).

[21] A. F. Bower, S. Shankar, "A Finite Element Model of Electromigration-Induced Void Nucleation, Growth and Evolution in Interconnects", Modelling Simul. Mater. Sci. Eng. 15, 923 (2007).

[22] A. H. Fischer, O. Aubel, J. Gill, T. C. Lee, B. Li, C. Christiansen, F. Chen, M. Angyal, T. Bolom, E. Kaltalioglu, "Reliability Challenges in Copper Metallizations Arising with the PVD Resputter Liner Engineering for 65nm and Beyond", Proc. IRPS 2007, 511 (2007).

[23] M. Gregoire, M. Juhel, P. Vannier, P. Normandon, "Post Electrochemical Cu Deposition Anneal Impact on Stress-Voiding in Individual Vias", Microelectronic Engineering 85, 2146 (2008).

[24] C. J. Zhai, H. W. Yao, P. R. Besser, A. Marathe, R. C. Blish, D. Erb, C. Hau-Riege, S. Taylor, K. O. Taylor, "Stress Modeling of Cu/Low-k BEoL – Application to Stress Migration", Proc. IEEE IRPS 2004, 714 (2004).

Effects of Cap Layer and Grain Structure on Electromigration Reliability of Cu/Low-k Interconnects for 45 nm Technology Node

L. Zhang, J. P. Zhou, J. Im, P. S. Ho

The University of Texas at Austin, 10100 Burnet Road Bldg. 160, Austin, TX 78758, USA
phone: (512)-471-8966; fax: (512)-471-8969; e-mail: ljzhang@mail.utexas.edu

O. Aubel, C. Hennesthal

GLOBALFOUNDRIES Dresden Module One LLC & Co. KG, 01109, Dresden, Germany

E. Zschech

Fraunhofer Institute for Non-Destructive Testing IZFP, Dresden, Germany

Abstract—The effects of cap layer and grain structure on electromigration (EM) reliability of Cu/low-k interconnects were investigated for the 45 nm technology node. Compared to the SiCN cap only, the CoWP capped samples showed a 40x lifetime improvement with a small lifetime variation (σ=0.34) at the M1 level. By tuning the process parameter, Cu lines of two different grain sizes were fabricated at the M2 level for both with and without the CoWP cap. The EM results showed that, for both caps, the Cu lines with the large grain structure had a longer EM lifetime compared with the small grain structure, and the EM enhancement of the metal cap was reduced for the small grain structure. Failure analysis revealed two failure modes for the SiCN cap, with void formation either at the via corner or in the trench away from the via; on the contrary, voids mostly formed several microns away from the via for the large grain CoWP cap. The difference in voiding locations for the two caps was attributed to the different interfacial mass transport rate. Implications of scaling effect on EM reliability were also discussed.

Keywords-Electromigration; Cu interconnect; CoWP cap; Grain structure; failure mode; Scaling

I. INTRODUCTION

Electromigration (EM) is a major reliability concern for Cu damascene interconnects and is becoming more important as technology scaling continues. Under EM, the mass transport induced by the electrical current can lead to void formation at the cathode end of the line to fail the interconnect structure. Prior to the 65 nm technology node, the Cu interconnects commonly had a bamboo-like grain structure where the mass transport under EM was dominated by the Cu/SiNx interface diffusion [1-4]. In this case, the EM lifetime was shown to degrade by about half for every new generation, even with the same current density [3]. However, a recent study [5] showed that at the 65 nm node, small grains agglomerated and mixed with large bamboo grains in Cu damascene lines narrower than 90 nm. This resulted in further degradation of the EM lifetime due to the increased contribution of the grain boundary

diffusion to mass transport [5]. As scaling continues, the aspect ratio and surface to volume ratio of Cu trench lines will increase further inducing more small grain presence. Therefore, the grain structure effect on EM lifetime has to be understood together with the cap layer effect in order to ensure the interconnect reliability for future technology node. Earlier studies have demonstrated that modification of the Cu/SiNx interface using a metal cap layer [6-8] or solute addition such as Al [9] can reduce the interfacial mass transport. However, the grain structure effect on EM reliability for the metal capped structures has not been reported.

In this paper, we first studied the EM behavior of M1 Cu interconnects capped with and without the CoWP layer. Then we investigated the grain structure effect on EM reliability by examining the EM lifetimes and failure modes of the M2 Cu lines with two different grain sizes for both with and without the CoWP cap. Finally, the impact of scaling effect on the EM lifetime due to the cap layer and the grain boundary was discussed.

II. EXPERIMENTAL DETAILS

The test structures were fabricated by GLOBALFOUNDRIES using the 45 nm technology process. All test samples are fully built wafers with five Cu layers and different interlayer dielectric (ILD) materials (dense and porous low-k dielectrics). Fig. 1 shows the two types of single-linked EM test structures used for this study. The type I structure consists of two-level interconnects with the M1 line as the current stressing line. The type II structure consists of three-level interconnects with the M2 line as the test line. The M1 test line is 72 nm wide and 110 nm thick; the M2 test line is 72 nm wide and 144 nm thick. Both test lines are 200 um in length. Two different cap layer configurations were applied to the type I structure at the M1 level: low-k SiCN with and without a CoWP cap. As for the M2 test line, two different grain sizes, large and small grains, were obtained by changing the current-profile during electroplating. In this way, four groups of samples were fabricated for the type II structure: Cu

978-1-4244-5430-3/10 $26.00 © 2010 IEEE

interconnects with large and small grains with and without a CoWP cap.

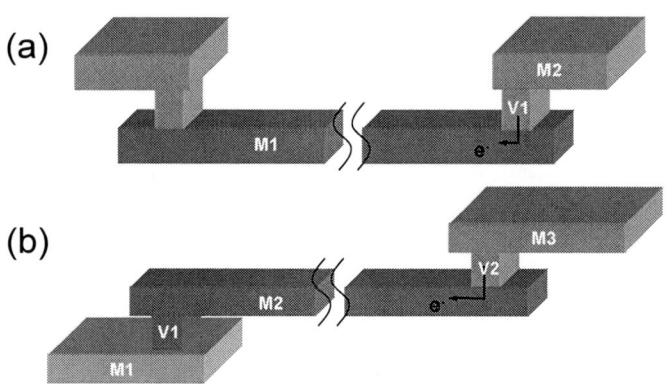

Figure 1. Schematic diagrams of EM test structures: a) two-level interconnects with M1 test line (Type I) and b) three-level interconnects with M2 test line (Type II).

During the EM test, the electron current passed downward along the M2-V1-M1 and M3-V2-M2 directions for type I and type II structures, respectively. The test current density varied from 1.03 MA/cm^2 to 1.34 MA/cm^2 while the temperature was kept constant at 330 °C. The resistance was monitored during the EM test and failure criteria were set at 10% resistance increase. Focused ion beam (FIB) and transmission electron microscopy (TEM) were used for failure analysis and microstructure characterization.

III. RESULTS

A. SiCN Cap vs. CoWP Cap

To investigate the effect of the capping materials, EM tests were performed on the two-level interconnect structures with and without the CoWP metal cap. Fig 2 shows the cross-sectional TEM image of Cu interconnect with the CoWP cap and the EDX line profile scanned across the Cu/cap interface from point A to point B. The EDX profile revealed that the CoWP cap was approximately 6 nm thick.

The cumulative distribution function (CDF) plot obtained at 330 °C and 1.34 MA/cm^2 is shown in Fig. 3. The test structure with the CoWP metal cap demonstrates a ~40x lifetime improvement compared with that without the metal cap. In addition, the samples with the CoWP cap show a tight EM distribution with a small standard deviation (σ=0.34) compared with results reported previously [6, 7, 10]. Longitudinal TEM was employed to examine the microstructure of Cu interconnects with and without the CoWP cap. As shown in Fig. 4, there is no significant difference in their grain structures. Thus the EM lifetime improvement can be attributed to the reduced atomic transport along the Cu/cap interface, indicating that interface modification using the metal cap is an effective way to improve EM reliability.

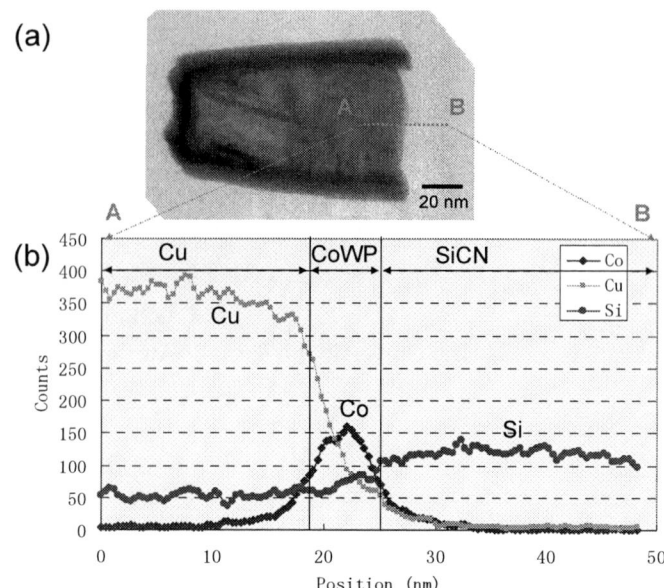

Figure 2. (a) Cross-sectional TEM image of M1 Cu trench with the CoWP cap; (b) EDX line profile scanned from point A to point B as indicated in (a).

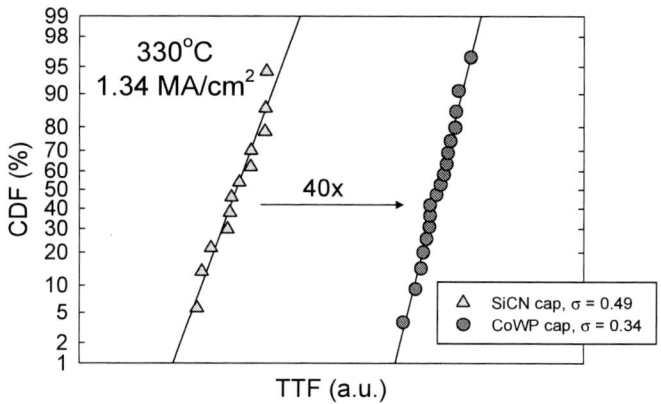

Figure 3. Cumulative distribution function (CDF) plot of the type I structure with and without the CoWP cap. EM test was preformed at T = 330 °C and j = 1.34 MA/cm^2.

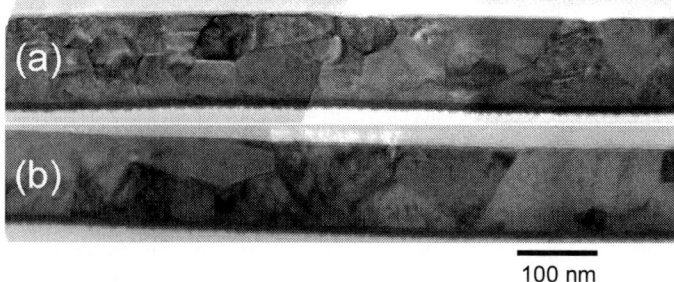

Figure 4. Longitudinal TEM images of M1 Cu lines with (a) and without (b) the CoWP cap.

B. Large Grain vs. Small Grain

The effect of grain structure on EM reliability was examined with large and small grains at the M2 level for the type II structure for both with and without the CoWP cap. The two types of grain structures were compared by using longitudinal scanning TEM (STEM) images as shown in Fig. 5, and the results from grain analysis are presented in Table I. Table I shows that the large grain structure has a significantly larger average grain size compared with the small grain structure, especially at the trench top. EM results obtained at 330 °C and 1.03 MA/cm^2 are shown in Fig. 6. The large grain structure is found to have a longer EM lifetime compared with the small grain structure for both caps. If we assume that the interfacial diffusion for the same cap is the same, independent of the grain structure, the degradation in EM reliability for the small grain structure can be attributed to the increase in mass transport along the grain boundary.

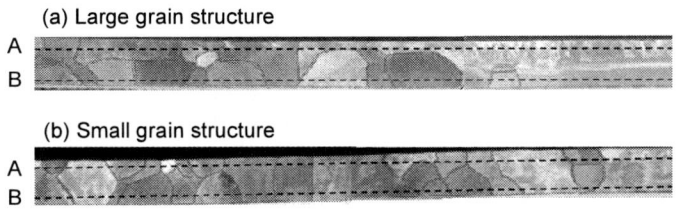

Figure 5. Longitudinal STEM images of M2 Cu lines with the CoWP cap: a) large grain, b) small grain.

TABLE I.   AVERAGE GRAIN SIZE FOR LARGE AND SMALL GRAIN STRUCTURES ALONG LINE A AND LINE B IN FIG. 5. THE GRAIN SIZE WAS CALCULATED BY AVERAGING ALONG THE DASHED LINE.

(* based on 5~10 um long Cu lines, 40~60 grains)

	Average grain size along line A (nm)*	Average grain size along line B (nm)*
Large grain structure	215	181
Small grain structure	123	126

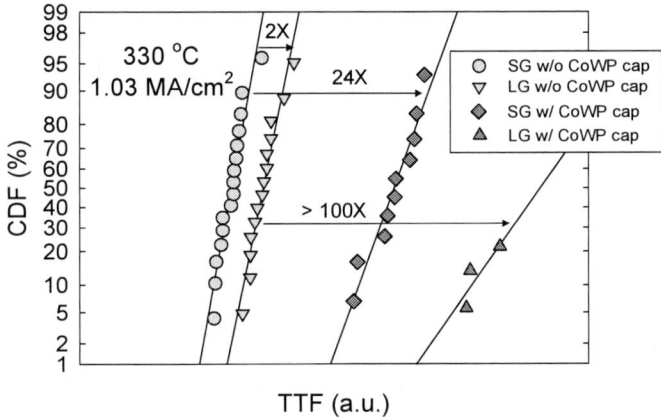

Figure 6.   Cumulative distribution function (CDF) plot for the type II structure with large and small grains for both with and without the CoWP cap. EM test was performed at 330 °C and 1.03 MA/cm^2. "SG" and "LG" refer to "small grain" and "large grain", respectively.

Comparing the EM lifetimes with and without the CoWP cap, the large grain structure reveals over 100x EM lifetime improvement, while the small grain structure exhibits a reduced improvement of 24x. This reduction in EM performance enhancement for the small grain structure can be attributed to the additional grain boundary diffusion contribution to the mass transport. This indicates that when the interfacial diffusion is reduced to a point that it contributes little to the mass transport, the grain boundary diffusion will play a more dominant role in controlling EM lifetime; a small variation in the grain size might induce a large difference in EM performance. In addition, the EM lifetime improvement by the CoWP cap for large grain M2 line is more significant than that of the M1 line, regardless of the differences in line thickness and test current density, indicating a better interface quality in suppressing interfacial diffusion for the higher metal level for the 45 nm technology process.

C. Failure Analysis

Failure analysis was performed using FIB and TEM for both types of structures. Two failure modes were observed as shown in Fig. 7 where mode I represented void formation at the cathode via corner and mode II represented void formation in the trench line far away from the via. For both modes, the Ta based liner remained intact and provided the electrical redundancy after void formation. Fig. 8 shows the progressive resistance increases for both cases with a small initial resistance increase corresponding to mode I and a large initial resistance jump for mode II. Overall mode II failures exhibited a slightly longer EM lifetime compared with mode I failures. Both types of failures were observed for SiCN capped samples for both large and small grains. However, only mode II failures were observed for large grain CoWP capped samples. The difference in the voiding location for these two types of caps can be understood by considering their different interfacial diffusivity. For the SiCN cap, the atomic diffusivity along the Cu/SiCN cap interface is large. The Cu atoms or voids can move readily along the interface until the voids eventually accumulate at the cathode via corner. However, for the CoWP cap, the interfacial diffusion is significantly suppressed by the metal cap. Instead of moving along the interface, micro-voids can be readily trapped at the local interface where either a large grain resides or an interface defect is present. The void will stay there and grow larger until it finally causes the interconnect failure.

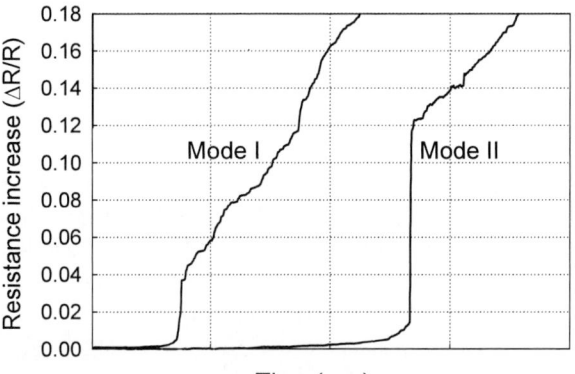

Figure 7.   Typical resistance traces of mode I and mode II failures: mode I with a small initial resistance increase and mode II with a large initial resistance jump.

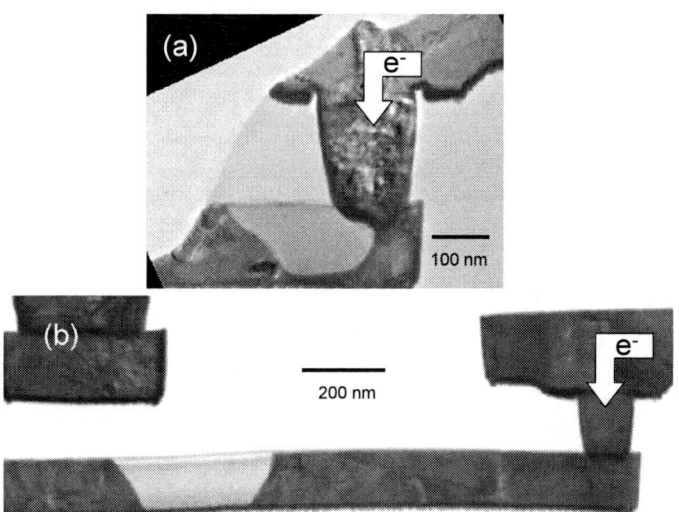

Figure 8. Cross-sectional TEM images of EM failed samples showing different voiding locations: (a) at the via corner, (b) in the trench away from the via, representing mode I and mode II failures in Fig. 7, respectively.

IV. DISCUSSION

To better understand the effects of cap layer and grain structure on EM reliability as scaling continues, we investigated the problem by considering the mass transport and EM lifetime in Cu interconnects. During EM, the mass transport driven by the EM driving force $F_e = Z_{eff}^* epj$ can be written as:

$$v_d = (\frac{D_{eff}}{kT})F_e = (\frac{D_{eff}}{kT})Z_{eff}^* e\rho j \qquad (1)$$

where v_d is the drift velocity; D_{eff} is the effective diffusivity; $Z_{eff}^* e$ is the effective charge; ρ is the metal resistivity; j is the applied current density; k is the Boltzmann constant and T is the absolute temperature. Two parameters in Eq. (1) determine the drift velocity of metal ions and in turn the EM lifetime. The first is the current density j which continues to increase as specified by the ITRS [11]. The second is the effective diffusivity D_{eff} of moving ions through various diffusion paths. In Cu damascene interconnects, there are two dominant diffusion paths: Cu/cap interface and Cu grain boundary. Accordingly, the effective diffusivity can be expressed as:

$$Z_{eff}^* D_{eff} = \frac{Z_N^* D_N \delta_N}{h} + \frac{Z_{GB}^* D_{GB} \delta_{GB}}{d} \qquad (2)$$

where δ_i is the effective width and D_i is the diffusivity of the metal ions. The subscripts N and GB refer to the Cu/cap interface and the Cu grain boundary, respectively. The factor h is the line thickness and d is the average grain size.

For simplicity, we consider the single damascene Cu interconnects where the void forms at the cathode end of the line to fail the interconnect. The EM lifetime τ can be expressed as:

$$\tau = \frac{\Delta L_{cr}}{v_d} = \frac{\Delta L_{cr} kT}{e\rho j(Z_N^* D_N \delta_N / h + Z_{GB}^* D_{GB} \delta_{GB} / d)}$$

$$= \frac{\Delta L_{cr} h kT}{Z_N^* D_N \delta_N e\rho j(1 + fgh/d)} \qquad (3)$$

where ΔL_{cr} is the critical void length to cause the failure, which is approximately the via size or line width w; f is a geometrical factor defined by the average orientation of the grain boundaries relative to the current flow. The g factor is defined as the ratio of the mass transport through the grain boundary vs. the cap interface and can be expressed as:

$$g = Z_{GB}^* D_{GB} \delta_{GB} / Z_N^* D_N \delta_N \qquad (4)$$

Equation (3) is written in a format to facilitate the discussion of cap layer and grain structure effects on EM lifetime. According to Eq. (3) and the model proposed by Hu et al. [3], the ratio of the median lifetime for each technology node relative to that of the 0.13 um technology is plotted in Fig. 9 as a function of the critical void volume $\Delta L_{cr}h$ (or cross-sectional area wh). Both grain boundary and interfacial diffusion contributions to mass transport are included here. The data points with a circular symbol are experimental data based on this study and the results provided by GLOBALFOUNDRIES.

For standard SiCN capped Cu interconnects with bamboo or near-bamboo microstructures, the f term is approaching to zero and can be neglected, thus the EM mass transport is predominantly through Cu/SiCN interface diffusion. For each new technology, if we assume the current density j remains the same, the EM lifetime degrades by half due to the scaling of the geometrical factor $\Delta L_{cr}h$ (wh), as shown by the reference line "1" in Fig. 9. However, after the CoWP metal cap is applied to the Cu interconnects, the atomic diffusion along the interface is significantly suppressed; the D_N term is reduced by a factor of 40~150 depending on the metal cap process. The scaling curve will shift up significantly as shown in Fig. 9, where we assumed a 40x lifetime improvement.

As scaling continues, more small grains emerge in the Cu lines after the 65 nm node, and a higher proportion of grain boundaries are aligned with the current flow. This will result in a larger f and a smaller d value, thus the fgh/d term will increase. Therefore, grain boundary diffusion will become important, which will accelerate the mass transport under EM and degrade the EM performance, as illustrated by the green dashed line labeled "2" in Fig. 9. Such EM lifetime degradation due to the presence of small grains can be more significant for the CoWP cap, as observed in our experimental results. This is due to the fact that with the suppression of interfacial diffusion by the CoWP cap, grain boundary diffusion will become more dominant. Therefore, a small change in the grain structure can result in a large change in the

978-1-4244-5430-3/10 $26.00 © 2010 IEEE

overall Cu diffusivity and in turn the EM lifetime. Nevertheless, the benefit of the CoWP cap over the SiCN cap remains significant even for the 32 nm technology node according to our estimate. However, to ensure EM reliability for the 22 nm technology node and beyond, the grain structure must be optimized together with the cap layer process.

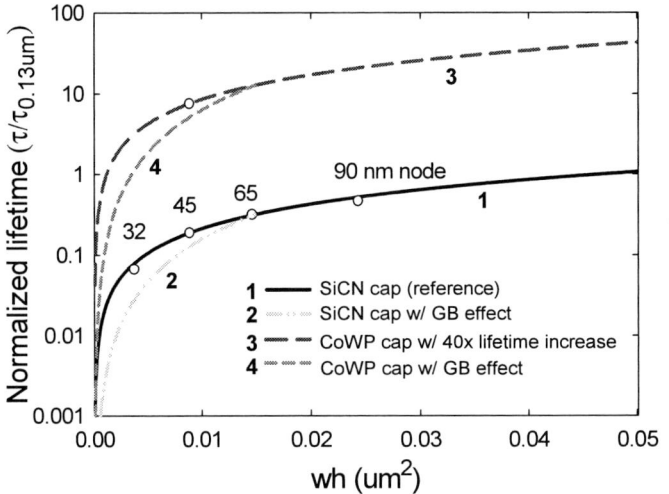

Figure 9.   Normalized EM median lifetime vs. *wh* (cross-sectional area) for various Cu interconnect generations considering different dominant Cu diffusion paths. The data points of open circle are experimental data based on this study and the results provided by GLOBALFOUNDRIES.

V.   CONCLUSIONS

The EM mass transport mainly occurs through interfacial diffusion and grain boundary diffusion for damascene Cu interconnects. For the 45 nm technology node, by implementing the CoWP metal cap, the EM performance at the M1 level is enhanced by a factor of ~40 with a small lifetime variation (σ=0.34). The EM results of different grain sizes with and without the CoWP cap at the M2 level show that grain boundary diffusion could become predominant in controlling EM reliability when more small grains are present, especially when the Cu/cap interface diffusion is suppressed by the metal capping. Failure analysis shows that voids mostly form in the trench away from the cathode via for the CoWP cap; in contrast, voids form in the trench both at the via corner and away from the via for the SiCN cap. Analysis of scaling effect on EM lifetime suggests that EM reliability is ensured by the application of the CoWP cap for the 32 nm technology. However, further investigation of interconnect processing to optimize the grain structure is critical to developing the 22 nm technology node and beyond.

ACKNOWLEDGMENT

The authors would like to acknowledge the support from GLOBALFOUNDRIES at Dresden Germany, and the Center for Advanced Interconnect Science and Technology of the Semiconductor Research Cooperation (SRC/CAIST).

REFERENCES

[1]   C. K. Hu, R. Rosenberg, and K. Y. Lee, "Electromigration path in Cu thin-film lines," *Applied Physics Letters,* vol. 74, p. 2945, 1999.

[2]   C. S. Hau-Riege and C. V. Thompson, "Electromigration in Cu interconnects with very different grain structures," *Applied Physics Letters,* vol. 78, pp. 3451-3453, 2001.

[3]   C. K. Hu, L. Gignac, and R. Rosenberg, "Electromigration of Cu/low dielectric constant interconnects," *Microelectronics Reliability,* vol. 46, pp. 213-231, 2006.

[4]   A. Von Glasow, A. H. Fischer, D. Bunel, G. Friese, A. Hausmann, O. Heitzsch, M. Hommel, J. Kriz, S. Penka, and P. Raffin, "The influence of the SiN cap process on the electromigration and stressvoiding performance of dual damascene Cu interconnects," in *IEEE International Reliability Physics Symposium Proceedings,* 2003 pp. 146-150.

[5]   C. K. Hu, L. Gignac, B. Baker, E. Liniger, R. Yu, and P. Flaitz, "Impact of Cu microstructure on electromigration reliability," *IEEE International Interconnect Technology Conference Proceedings,* pp. 93-95, 2007.

[6]   C. K. Hu, L. Gignac, R. Rosenberg, E. Liniger, J. Rubino, C. Sambucetti, A. Domenicucci, X. Chen, and A. K. Stamper, "Reduced electromigration of Cu wires by surface coating," *Applied Physics Letters,* vol. 81, p. 1782, 2002.

[7]   J. Gambino, J. Wynne, J. Gill, S. Mongeon, D. Meatyard, B. Lee, H. Bamnolker, L. Hall, N. Li, and M. Hernandez, "Self-aligned metal capping layers for copper interconnects using electroless plating," *Microelectronic Engineering,* vol. 83, pp. 2059-2067, 2006.

[8]   M. A. Meyer and E. Zschech, "New microstructure-related EM degradation and failure mechanisms in Cu interconnects with CoWP coating," in *Proc. Int. Workshop Stress Induced Phenom. Metallization,* 2007, p. 107.

[9]   M. A. Meyer, *Ph. D. Thesis, Brandenburg University of Technology, Cottbus, Germany,* 2007.

[10]   G. Dixit, D. Padhi, S. Gandikota, J. Yahalom, S. Parikh, N. Yoshida, K. Shankaranarayanan, J. Chen, N. Malty, and J. Yu, "Enhancing the electromigration resistance of copper interconnects," *Proceedings of the IEEE 2003 International Interconnect Technology Conference,* pp. 162-164, 2003.

[11]   http://www.itrs.net/reports.html.

Study of Stress Migration and Electromigration Interaction in Copper/Low-κ Interconnects

A. Heryanto[1], K.L. Pey[1,#], Y.K. Lim[2], W. Liu[2], J. Wei[3], N. Raghavan[1], J.B. Tan[2] and D.K. Sohn[2]

[1]School of Electrical and Electronic Engineering, Nanyang Technological University, Singapore - 639798
[2]GLOBALFOUNDRIES Singapore Pte Ltd, 60 Woodlands Industrial Park D Street 2, Singapore - 738406
[3]Singapore Institute of Manufacturing Technology, 71 Nanyang Drive, Singapore - 638075
[#]Phone : +65-6790-6371, Fax : +65-6793-3318, E-mail : eklpey@ntu.edu.sg

Abstract— **Stress migration (SM) and electromigration (EM) are key reliability concerns for advanced metallization in nanoscale CMOS technologies. In this paper, the interaction between these two mechanisms is studied in dual-damascene Cu/low-k interconnects. It is found that these mechanisms are not independent; EM failure time could be strongly affected by the presence of residual stress induced by SM, causing significant EM lifetime degradation. The reliability implication of the residual stress in copper interconnects on the EM is further investigated at the voiding site using transmission electron microscopy (TEM) failure analysis. A failure mechanism model is proposed to explain the lifetime degradation due to vacancy accumulation near the voiding via. The vacancy accumulation leads to higher tensile stress and shortens the time to reach the critical stress (σ_{crit}) for void nucleation.**

Keywords- Stress Migration; Stress Induced Voiding; Electromigration; Copper; Vacancy accumulation; Stress gradient; Finite Element Analysis; Critical stress.

I. INTRODUCTION

Stress migration (SM) or stress induced voiding (SIV) in Cu dual damascene structure was first reported in 2002 by Ogawa *et al.*, where SM was found under vias connected to wide bottom Cu metal lines [1]. Since then, SM has been studied by many groups to gain further understanding of the voiding mechanism [2-4]. However, for a more practical study relevant to the real operating conditions, it becomes imperative not only to understand the voiding mechanism, but also to analyze how the presence of residual stress would influence the electromigration (EM) behavior in Cu interconnects.

In general, stress migration is caused by the interaction between the thermomechanical stress in the interconnect systems and the diffusion of vacancies. The presence of these two factors gives stress migration an interesting dependence on temperature. Unlike EM, in which reliability degrades monotonically with increase of temperature, the temperature dependence curve for stress migration is bell-shaped. The peak of stress migration is at an intermediate temperature range, between 150 °C and 200 °C [1]. This can be explained by the fact that the tensile stress in the interconnect structure decreases almost linearly with the temperature. On the other hand, the vacancy diffusion is temperature dependent. It will increase exponentially with temperature.

In accelerated EM tests, the effect of SM may not be prominent because of the high temperature testing conditions that are typically close to the stress free temperature (SFT). As a result, the intrinsic hydrostatic stress in the metal lines is relaxed and SM is unlikely to occur. However in reality, SM may occur at the chip operating temperature, typically around 100-125°C, where the thermal stresses are considerably higher. This fact demonstrates the inadequacy of current reliability testing methodology, where the effect of SM and EM are studied separately. Hence, further work is needed to assess EM failure in the presence of residual stress in Cu interconnects in order to obtain a more realistic estimation of the interconnect reliability and lifetime.

This paper presents the study of the interaction of SM and EM in dual-damascene Cu interconnect structures with carbon-doped oxide, SiCOH, as the low-κ dielectric material. The focus of this paper is to study this interaction in the lower metal (M1) of the Cu interconnect system. The reliability implication of the SM effect on the EM is further investigated at the voiding site using TEM failure analysis. Three dimensional (3D) finite element analysis (FEA) simulation was carried out to study the SM behavior of Cu/Low-κ interconnects. A failure mechanism model for stress evolution and void formation is proposed to provide insight into the interaction between these two failure mechanisms.

II. EXPERIMENTAL DETAILS

Fig. 1 shows the schematic diagram of the via-chain test structure consisting of 28 plates with 56 via connections used to investigate the effect of SM on EM. The advantage of the via-chain test structure is that a relatively small population of the SM failures can be detected with small sample size.

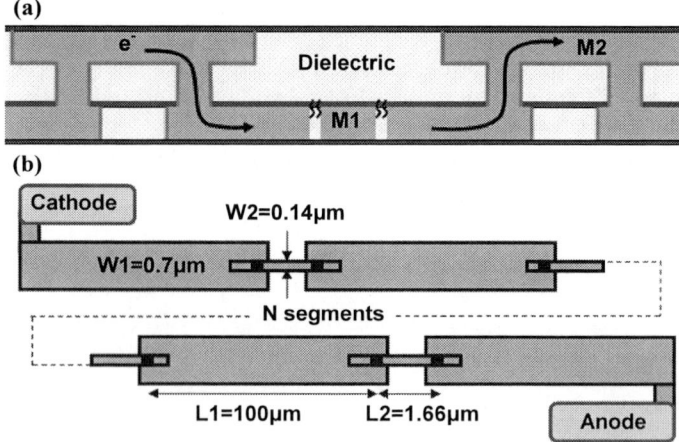

Figure 1. Schematic diagrams of the **(a)** side view and **(b)** top view of via-chain test structures used for the SM-EM interaction study.

The test samples were fabricated using a standard 45-nm CMOS process on 300 mm wafers, with dual damascene process scheme. To see the interaction in lower metal (M1), the interconnect system is designed such that M1 is susceptible to SM/EM failure while the chances of SM/EM failure at upper metal (M2) is suppressed tremendously. It has been reported by Ogawa et al. [1] that wide metal is more prone to SM failure compared to narrow metal lines since larger metal segments are able to supply more vacancies for voiding. Therefore the test structure was designed to have a wide M1 (width = 0.7 μm) and a narrow M2 (width = 0.14 μm). The via diameter was kept at 0.14 μm.

The concept of Blech length effect was used to localize the interconnect failure due to EM [5]. The test structure of M1 was designed sufficiently long, such that $(jL) > (jL)_c$ and hence EM failure is very likely to occur. While M2 was designed with very short length such that $(jL) < (jL)_c$, leading to immortality with respect to EM. The length of M1 and M2 is 100 μm and 1.66 μm, respectively, with a 1.15 μm metal extension in M2 to further improve the EM reliability (due to reservoir effect) [6]. Using an EM current density stress of 0.8 MA/cm² and 4 MA/cm² for M1 and M2, respectively, we can obtain $(jL)_{M1}$ = 8000 A/cm and $(jL)_{M2}$ = 664 A/cm. Thus, large value of $(jL)_{M1}$ and low value of $(jL)_{M2}$ imply that EM failures are probable to occur only in M1.

The SM/EM testing was performed at package level. Firstly, the samples were subjected to a thermal bake at 200°C for 1000 hours. A very small sensing current of less than 0.05 MA/cm² was used to perform in-situ resistance measurement. It was observed that there was no resistance increase for all the structures after the 1000-hours SM stressing. Immediately after the SM stress, the temperature was increased to the EM testing temperature of 350°C with a stressing current of 0.8 MA/cm² in M1. The failure criterion used was the first step of resistance jump.

III. RESULTS AND DISCUSSION

A. Failure Distribution

Fig. 2 shows the lognormal probability plot of the failure data for the EM tests with pre-SM stressed samples (i.e. Sample-1). The "EM-stress only" test was used as a control sample (i.e. Sample-2). The statistical data fitting for the failure distribution was obtained using the Expectation and Maximization (E&M) algorithm and Bayes' posterior probability theory [7]. This algorithm determines the distribution parameters by maximizing the log-likelihood function of the mixture distribution. This statistical method was developed to identify the number of underlying failure mechanisms embedded in a given set of test data and classify the units belonging to each of these different mechanisms.

From the statistical study, the failures of Sample-1 show bimodal distribution, indicative of the presence of two different failure mechanisms [8]. It can be further deduced that 90% of the test structures failed due to Failure Mechanism A (FM-A) and 10% of the failures correspond to FM-B. The median TTF of FM-A or $t_{50(1A)}$ = 61.13 hours with σ_{1A} = 0.23, while for FM-B, $t_{50(1B)}$ = 130.28 hours and σ_{1B} = 0.15. The earliest failures of Sample-1 tend to form a cluster which speculate to be due to

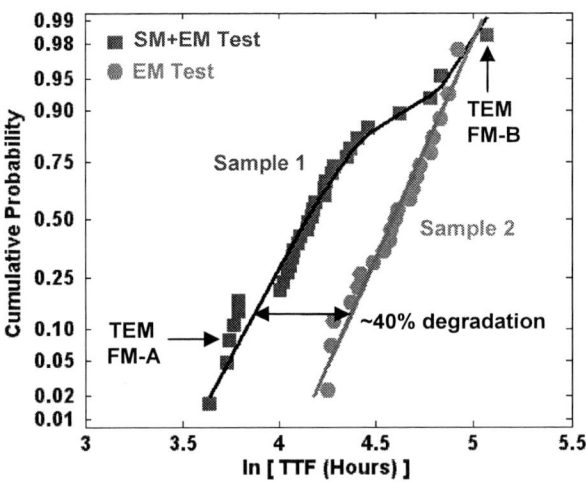

Figure 2. Lognormal probability plot of samples failed during SM+EM test compared to "EM-test only" failures. The data is fitted with E&M algorithm probability theory. It shows that the TTF of samples with SM+EM test has ~40% degradation.

extrinsic defect. The control sample shows a monomodal distribution trend with $t_{50(2)}$ = 98.62 hours and σ_2 = 0.196. The difference in the statistical parameters suggests that SM can potentially accelerate EM failure during circuit operation. When the SM effect in the metal line is dominant, the EM lifetime could be degraded by as much as ~40%. On the other hand, there is a small population where the EM lifetime is not greatly affected by SM, as observed in 10% of the samples.

B. Resistance Degradation Trends

The resistance degradation trends for Sample-1 and Sample-2 are shown in fig. 3 and fig. 4, respectively. It is observed that, in most cases (81.25%) Sample-1 has a catastrophic failure resistance trend where open-circuit failures occur abruptly. On the other hand, the resistance trend for the majority (85.71%) of the Sample-2 devices is a staircase-like trace which indicates that multiple, small random voids in the via-chain structure are occurring before a final failure. This suggests that the failure of Sample-1 is void-nucleation-limited which resulting from a void that nucleates directly under via. On the other hand, the failure of Sample-2 is void-growth-limited which results from a void that nucleates just downstream of the via but does not immediately uncover the via [9].

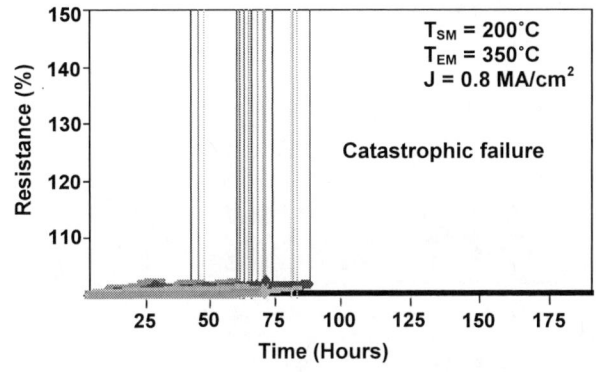

Figure 3. Resistance-time degradation trends for SM+EM test. It shows that most of the samples (i.e. Sample-1) under SM+EM test show abrupt resistance jump which indicates that a fatal void is formed below via.

978-1-4244-5430-3/10 $26.00 © 2010 IEEE

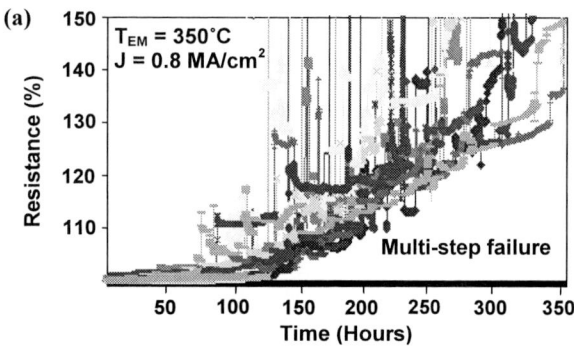

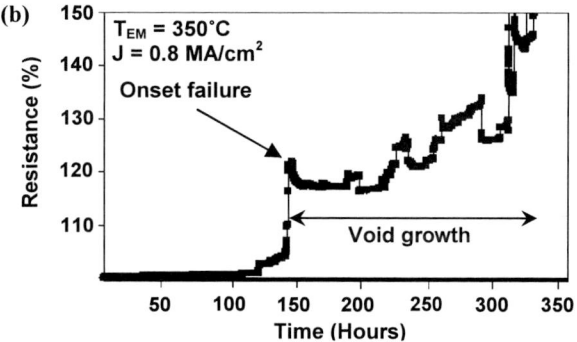

Figure 4. (a) Resistance-time degradation trends for the "EM-test only" samples (i.e., Sample-2). It shows that most of the samples have a step-like resistance characteristic which indicates occurrence of multiple failures. (b) Focusing on the observed trend in one of the degraded samples in (a).

C. Physical Failure Analysis

Physical analysis of the failed test structures from SM+EM samples, i.e. FM-A and FM-B group, were performed to examine the nature of the void formation process. In order to locate the failure site in the long connective chain of the vias in the multi-link structures after the SM/EM test, passive voltage contrast (PVC) technique was employed to isolate and locate the failures. PVC is an effective method to identify defective via in via-chain structures [10]. After that, TEM analysis was performed on the failed via-chain to study the voiding trends and the associated failure mechanisms. Fig. 5 shows that the devices failed in FM-A population are due to the formation of

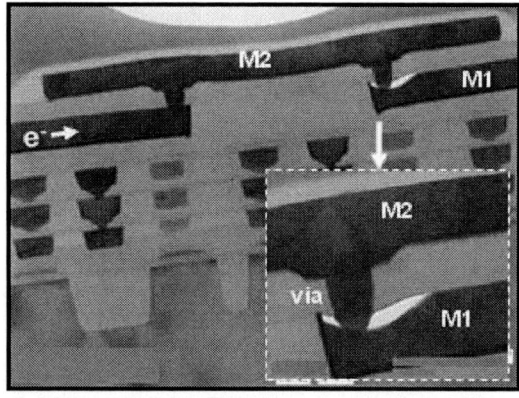

Figure 5. TEM analysis of a degraded sample after SM+EM test from the FM-A population (t_{50}=61.13-hours) where void formed below the via of the cathode end causing open circuit failure. The TTF is reduced because a small amount of Cu depletion around the via/metal interface is sufficient to cause a large resistance change.

voids exactly below the via at the M1 cathode side, leading to the abrupt failures shown in fig. 3. On the other hand, in FM-B, the failures occurred at the later stage where void forms in the trench of the metal line instead (see fig. 6).

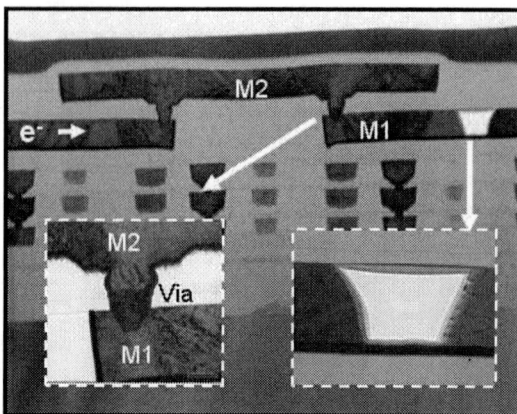

Figure 6. TEM analysis of a degraded sample after SM+EM test from the FM-B population (t_{50}=130.28-hours) where void forms in the trench of the Cu metal line. The TTF is enhanced because a large volume of Cu must be depleted to reach the resistance failure criteria.

D. Finite Element Analysis Simulation

FEA simulation is a good tool to study the SM behavior due to process-induced thermal stress profiles generated in the Cu/low-k interconnect structure and assess the most probable sites for vacancy accumulation. The simulation model of the test structure shown in fig. 1 was developed using ANSYS®. All of these models were implemented assuming isotropic material properties with linear elastic behaviors. Because of line symmetry, only a quarter of the model was simulated. The material properties used are given in Table 1 and the stress free temperature was set at 300°C. In order to evaluate the validity of the simulation model more quantitatively, the computed results have been compared against the data obtained from volume-averaged thermal stress measurements in Cu using the XRD technique as described in [11].

Fig. 7(a) shows the simulated global hydrostatic stress profile in the M1 line of the Cu/Low-κ interconnect system. It reveals that the insertion of a via in the Cu metal lead creates non-uniform stress in the system, where the tensile stress near the via is always lower compared to the Cu bulk. As reported in [12], the thermally generated stress gradient in the Cu interconnect drives the saturated vacancies from high stress regions towards low stress regions. This might be one of

TABLE I. MATERIAL PROPERTIES USED IN OUR SIMULATION MODEL[13,14]

Materials	CTE (10^{-6}/K)	Modulus (GPa)	Poisson's Ratio
Cu	17.7	104.2	0.352
Ta	6.5	185.7	0.342
Si_3N_4	3.2	220.8	0.27
SICOH	12	16.2	0.3

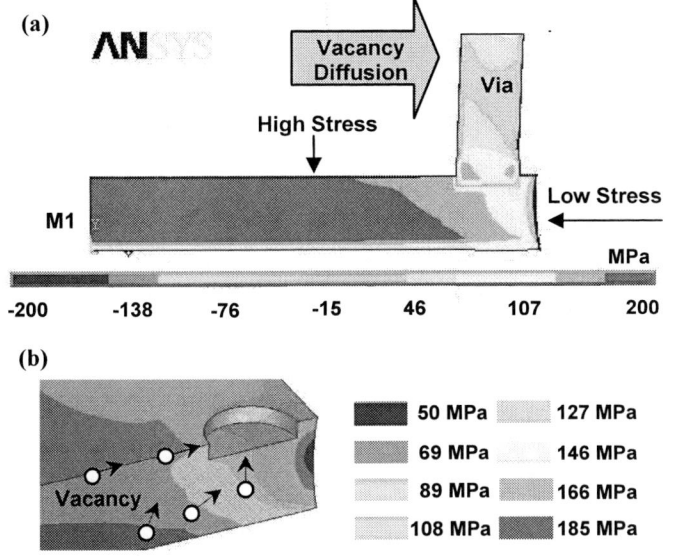

Figure 7. Hydrostatic stress simulation using FEA. (a) Global interconnect stress profile in the M1 test structure. The lower tensile stress near the via/metal edge enhances vacancy diffusion towards the via. (b) Local interconnect stress contours in the vicinity of the via. The non-uniform tensile stress near the via location leads to a prominent stress gradient that drives vacancies towards the via.

possible explanations as to why the SM void tends to form near the via location as reported in [1-4]. Fig. 7(b) further emphasizes that the tensile stress near the via location is non-uniform. It leads to a considerable stress gradient causing the vacancies to be driven towards the via.

E. Stress Migration and Electromigration Interactions

From the electrical results and physical failure defect signatures, we believe that during the SM test, the thermally generated stress gradient of copper interconnect drives the saturated vacancies to diffuse towards "weak" via(s) as shown in fig. 8(a). Assuming that the amount of vacancies accumulated near the vias after SM test are the same, the metal line become more tensile at both ends by almost the same increment (σ_{SM1} and σ_{SM2}) as illustrated in fig. 8(b). If failure occurs when a critical stress (σ_{crit}) is reached [15], this SM effect shortens the time-to failure (t_f) for the cathode end to reach a certain threshold for vacancy superstaturation, which in turn, leads to early void nucleation at this interface during the EM test. This void which functions as a sink of ensuing vacancies supplied by EM, continues to grow as shown in fig. 8(c). On the other hand, at the anode side, the EM mass flux will cancel out the vacancy accumulation. Therefore, the stress will be neutralized and during subsequent EM stressing, leading to progressive compressive stress due to Cu atom accumulation.

It is proposed that SM can influence EM when there is significant amount of vacancy accumulation due to SM in the cathode area which accelerates EM lifetime degradation (FM-A). Furthermore, the impact of SM on EM is observed to be very severe because it increases the probability of EM voiding failures under the category of catastrophic failure as shown in fig. 3 and fig. 5. On the other hand, the experimental result shows that there is a small population of Sample-1 whose the

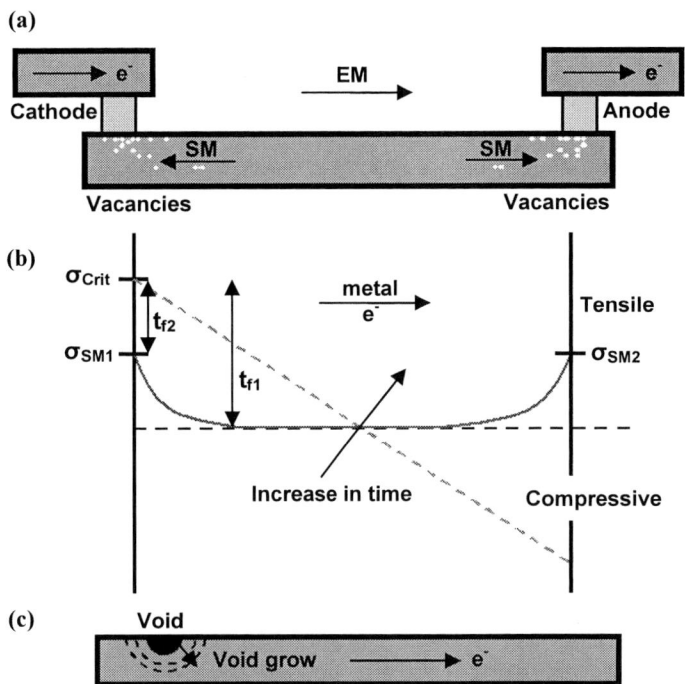

Figure 8. SM and EM interaction model. (a) Schematic diagram of the test structure after SM test. (b) Stress as a function of location along the M1 interconnect at different times. Due to the SM stress, the time to reach critical stress is shorter, $t_{f2} < t_{f1}$ (c) Void evolution process near the weaker via in the metal line during the subsequent EM test.

EM lifetime is not affected by the presence of any residual stress i.e. pre-SM stress (FM-B). The TEM analysis shows that the failure only occurs in the trench of the metal line and there is no void nucleation observed in the surrounding of the via near the cathode side (fig. 6). This failure is just one type of typical EM downstream failures as reported in [16, 17]. The mechanism behind this observed non-degradation might be due to less vacancy accumulation at the cathode side (i.e. a result of asymmetrical stress distribution during SM) and hence there is no stress enhancement after SM on the cathode side. It is worth to take note that, regardless of the concentration of vacancies induced by SM in the anode end, it will be completely replenished by Cu atoms during the EM stress, leading to eventual stress neutralization.

IV. CONCLUSION

The interaction of SM and EM is important as these two phenomena are expected to occur during chip operation, where the thermal stresses are considerably higher. Unfortunately, the typical accelerated EM testing conditions are at a relatively high temperature, typically close to the stress free temperature of the metal lines. As a result, the effect of intrinsic thermal stress in the metal lines is underestimated.

The influence of the residual stress in copper interconnects, simulated by pre-SM test in our experiment, on EM behavior in Cu interconnects is studied from a statistical, electrical and physical perspective. It was found that the EM failure time of the lower metal line could be strongly affected by the presence of the residual stresses at the cathode end. We believe that vacancy accumulation to a sufficiently high value in the cathode region due to SM will greatly affect the threshold for EM failures to occur in the metal line.

The effect of SM on EM will become even more prominent when the technology node is scaled down to the sub-45nm range. Smaller vias need a lower vacancy density to cause the metal line to reach the required critical tensile stress for void nucleation. In addition, the introduction of more porous low-κ materials lowers the critical nucleation stress [18]. This will reduce the time-to-failure for the cathode end to reach a vacancy threshold for supersaturation which, in turn, leads to a much earlier void nucleation at the interface.

ACKNOWLEDGMENT

This research is supported by the Singapore Economic Development Board (EDB) and GLOBALFOUNDRIES Singapore Joint Industry Postgraduate Program. The authors would like to thank Mr. C.K. Cheng from Institute of Microelectronics (IME), Singapore, for the SM/EM testing support, the Chartered FA team for performing the failure analysis and the School of Mechanical and Aerospace Engineering, NTU, for access to ANSYS simulation tools.

REFERENCES

[1] E. T. Ogawa, J. W. McPherson, J. A. Rosal, K. J. Dickerson, T. C. Chiu, L. Y. Tsung, M. K. Jain, T. D. Bonifield, J. C. Ondrusek, and W. R. McKee, "Stress-induced voiding under vias connected to wide Cu metal leads," IRPS 2002, pp. 312-21.

[2] K. Yoshida, T. Fujimaki, K. Miyamoto, T. Honma, H. Kaneko, H. Nakazawa, and M. Morita, "Stress-induced voiding phenomena for an actual CMOS LSI interconnects," IEDM 2002, pp.753-6.

[3] T. Oshima, K. Hinode, H. Yamaguchi, H. Aoki, K. Torii, T. Saito, K. Ishikawa, J. Noguchi, M. Fukui, T. Nakamura, S. Uno, K. Tsugane, J. Murata, K. Kikushima, H. Sekisaka, E. Murakami, K. Okuyama, and T. Iwasaki, "Suppression of stress-induced voiding in copper interconnects," IEDM 2002, pp.757-60.

[4] Y. K. Lim, Y. H. Lim, C. S. Seet, B. C. Zhang, K. L. Chok, K. H. See, T. J. Lee, L. C. Hsia, and K. L. Pey, "Stress-induced voiding in multi-level copper/low-k interconnects," IRPS 2004, pp. 240-5.

[5] I. A. Blech and C. Herring, "Stress generation by electromigration," Applied Physics Letters, vol. 29, pp. 131-3, 1976.

[6] W. Shao, A. V. Vairagar, C.-H. Tung, Z.-L. Xie, A. Krishnamoorthy, and S. G. Mhaisalkar, "Electromigration in copper damascene interconnects: reservoir effects and failure analysis," Surface and Coatings Technology, vol. 198, pp. 257-61, 2005.

[7] C.M. Tan and N. Raghavan, "Unveiling the electromigration physics of ULSI interconnects through statistics," Semiconductor Science and Technology, vol. 22, pp. 941-6, 2007.

[8] A. H. Fischer, A. Abel, M. Lepper, A. E. Zitzelsberger, and A. von Glasow, "Modeling bimodal electromigration failure distributions," Microelectronics Reliability, vol. 41, pp. 445-53, 2001.

[9] C. S. Hau-Riege, "An introduction to Cu electromigration," Microelectronics Reliability, vol. 44, pp. 195-205, 2004.

[10] G. B. Ang, Y. N. Hua, S. K. Loh, Yogaspari, and S. Redkar, "Application of passive voltage contrast and focused ion beam on failure analysis of metal via defect in wafer fabrication," IPFA 2001, pp. 107-11.

[11] S.H. Rhee, D. Yong, and P. S. Ho, "Thermal stress characteristics of Cu/oxide and Cu/low-k submicron interconnect structures," Journal of Applied Physics, vol. 93, pp. 3926-33, 2003.

[12] K.-N. Tu, J. W. Mayer, L. C. Feldman, "Electronic Thin Film Science," Macmillan Publishing Company, 1992.

[13] P. Jong-Min, P. Hyun, and J. Young-Chang, "Effect of low-k dielectric on stress and stress-induced damage in Cu interconnects," Microelectronic Engineering, vol. 71, pp. 348-57, 2004.

[14] A. Grill, V. Patel, K. P. Rodbell, E. Huang, M. R. Baklanov, K. P. Mogilnikov, M. Toney, and H. C. Kim, "Porosity in plasma enhanced chemical vapor deposited SiCOH dielectrics: a comparative study," Journal of Applied Physics, vol. 94, pp. 3427-35, 2003.

[15] M. A. Korhonen, P. Borgesen, K. N. Tu, and C.-Y. Li, "Stress evolution due to electromigration in confined metal lines," Journal of Applied Physics, vol. 73, pp. 3790-9, 1993.

[16] M. H. Lin, Y. L. Lin, K. P. Chang, K. C. Su, and T. Wang, "Copper interconnect electromigration behavior in various structures and precise bimodal fitting," Japanese Journal of Applied Physics, vol. 45, pp. 700-9, 2006.

[17] S.C. Lee and A. S. Oates, "Identification and analysis of dominant electromigration failure modes in copper/low-k dual damascene interconnects," IRPS 2006, pp.107-14.

[18] C. S. Hau-Riege, S. P. Hau-Riege, and A. P. Marathe, "The effect of interlevel dielectric on the critical tensile stress to void nucleation for the reliability of Cu interconnects," Journal of Applied Physics, vol. 96, pp. 5792-6, 2004.

Electromigration and stress-induced-voiding in dual damascene Cu/low-k interconnects: a complex balance between vacancy and stress gradients

K. Croes, C.J. Wilson, M. Lofrano, B. Vereecke, G.P. Beyer and Zs. Tőkei

Imec, Kapeldreef 75, B-3001 Leuven, Belgium
phone: (32)16/281621; fax: (32)16/281576; e-mail: Kristof.croes@imec.be

Abstract - **The influence of residual vacancy concentrations and stress gradients on electromigration both in the metal layer below and above copper via's with a diameter of 90nm integrated in low-k materials has been investigated. Variations in stress gradients and vacancy concentrations were created by applying different post-plating anneal conditions. The impact of these variations was quantified based on high temperature storage tests both at the optimum stress-induced-voiding temperature and around the copper stress free temperature. By linking the results from these high temperature storage tests to electromigration data, we observe residual vacancy concentrations contribute more to upstream electromigration, while downstream electromigration is more vulnerable to residual stress gradients.**

Keywords - BEOL, copper, stress-induced-voiding, electromigration, vacancy, stress, FEM

I.   INTRODUCTION

Electromigration and stress-induced-voiding (SIV) are the two main metal failure mechanisms in back-end-of-line copper interconnects [1-3]. Electromigration is the result of the movement of copper ions due to an electron wind caused by an electrical current [1], while SIV is caused by hydrostatic stress gradients in the copper [2].

In general, metal failure in back-end-of-line interconnects is the result of three forces acting on copper ions: electron wind, stress gradients and vacancy gradients. The electron wind is induced by an external electric field, where two forces act on the copper ions: a direct coulomb force which pulls the ions in the direction of the applied field and a force resulting from a momentum exchange with scattering electrons. As this latter dominates, the electron wind induces a force which results in a net material flow and ions move in a direction opposite to the field. This force is generally expressed as $F_e = Z^* e \rho j_e$, where e is the electronic charge, Z^* is the effective ion valence (taking both Coulomb force and electron wind into account), ρ is the resistivity and j_e is the current density. Stress gradients are the result of a mechanical stress due to a thermal mismatch between the metal and the surrounding materials; grain growth; vacancy annihilation; and re-crystallization [4]. Besides mechanical stress, stress gradients can also be the result of electron wind. Atoms which are driven out from the cathode end of the conductor cause a tensile stress and the drifted atoms accumulate at the anode end causing an increase in the atomic density and a higher compressive stress [1,5]. Stress gradients

act as an external driving force for atom movement which can be quantified by $F_{MECH} = \Omega \nabla \sigma$, where Ω is the atomic volume and $\nabla \sigma$ the hydrostatic stress gradient. As explained above, the term $\nabla \sigma$ needs to be subdivided in two factors: a factor relating to the residual stress gradients in the material due to mechanical stress without the presence of an electron wind, further referred to as $\nabla \sigma_R$ and the stress gradient induced by the electron wind, further referred to as $\nabla \sigma_{EM}$. Also a gradient in concentration of copper ions will result in a driving force for ion movement. This force can be quantified by $F_{DIFF} = -k_B T \nabla C / C$, where k_B is the Boltzmann constant, T the absolute temperature and C and ∇C the concentration and concentration gradient of copper ions, respectively. The Nernst-Einstein equation links the forces acting on copper ions to atomic velocity: $v_d = DF/k_B T$, where D is the diffusion coefficient and all other terms are defined above. The Einstein equation relates the drift velocity to mass migration. This equation is given by $J_{mass} = C v_d$. Filling in the above mentioned forces results in the generally accepted equation for mass transport [6]:

$$J_{mass} = \underbrace{-D \cdot \nabla C}_{1} + \underbrace{\frac{DC}{k_B T} \cdot Z^* q \rho j_e}_{2} - \underbrace{\frac{DC}{k_B T} \cdot \Omega \cdot \nabla \sigma}_{3}, \quad (1)$$

where term 1, 2 and 3 describe the influence of concentration gradients, electrical current and stress gradients, respectively.

Electromigration is mainly driven by term 2 and SIV by term 3, where $\nabla \sigma_{EM}$ is zero. The higher dependence of electron wind induced stress gradients $\nabla \sigma_{EM}$ on electromigration void growth compared to vacancy concentration gradients ∇C and residual stress gradients $\nabla \sigma_R$ is widely discussed. However, it needs to be fundamentally understood how both changes in $\nabla \sigma_R$ and ∇C will impact electromigration. The current paper will contribute to this understanding.

Since typical flows to integrate copper interconnects involve dual damascene processing, the failure modes of the metal lines below and above via's are quite different [7-8]. The metallic barrier at the via-bottom acts as a flux divergence point, making void formation in the via-region more likely. The processing and lay-out of via's make stress distributions, vacancy concentrations, diffusion paths, current crowding effects, etc. complex phenomena that have to be understood for both the degradation below and in/above the via. For example,

978-1-4244-5430-3/10 $26.00 © 2010 IEEE

downstream electromigration and SIV below the via are affected by the damage caused to the copper during etching the via onto the metal layer, while upstream electromigration and SIV in/above the via are more influenced by the barrier-coverage and the filling of the via [9]. To study all related phenomena, different test structures and test methods to evaluate electromigration and SIV have to be applied.

As higher temperatures increase ion diffusivity and higher current densities result in a stronger electron wind, both temperature and current density are acceleration factors for electromigration. The situation is more complex for SIV: a higher temperature gives a higher ion diffusivity, but a lower hydrostatic stress and stress gradient. To accelerate SIV, an optimum temperature has to be determined at which SIV happens the fastest [2,10]. Figure 1 summarizes the key findings of our earlier work [11]. It is shown that, even close to the stress free temperature of the copper, copper degradation still occurs. It is argued that at these temperatures, concentration gradients and not stress gradients dominate the degradation. The main argument was that degradation at the optimum SIV temperature had an activation energy close to interface diffusion (0.9eV), while the activation energy of the mechanism dominating the degradation close to the copper stress free temperature was close to grain boundary diffusion (1.2eV). Therefore, degradation during storage tests at different temperatures could be used as a quantification for either residual stress gradients (and thus SIV) or concentration gradients.

Many studies have been published either on electromigration or on degradation during high temperature storage, but studies that use storage related phenomena to study electromigration are rare. Earlier papers link stress migration to electromigration in Al interconnects [12,13] and a more recent study links storage tests at typical electromigration temperatures to electromigration [14].

In this paper, we summarize a detailed study where the failure characteristics of storage tests at the optimum SIV-temperature and close to the copper stress free temperature were used to interpret the electromigration failure mechanisms. We produced varying residual stress and concentration gradients in the copper by using different post-plating anneal conditions. Storage tests were used to validate existing literature data, to confirm the arguments made in [11] and to quantify the residual stress and concentration gradients in the produced test material. Finally, these results were linked to electromigration experiments and the different contributions of residual stress gradients and vacancy concentrations to upstream and downstream electromigration were evaluated.

II. EXPERIMENTAL

A. Test material

Dual damascene copper structures were integrated on 300mm wafers. Intermetal (IMD) and interlayer dielectric (ILD) was a SiOCH CVD low-k dielectric. The M1-IMD was with k-value ~3.0, porosity ~9% and E-modulus ~15GPa, while via-ILD and M2-IMD was with k-value~2.5, porosity~25% and E-modules ~9GPa. A 5/25nm SiCN/SiCO etch stop layer (E-modulus ~100 GPa) was used between the M1- and via/M2-layer. A 3/3nm TaN/Ta barrier was used between copper and dielectric. The wafers were passivated with a 35nm SiCN, 300nm SiO_2 and 500nm Si_3N_4-stack.

Studying the influence of concentration and stress gradients on electromigration requires test material where both parameters are varied. However, stresses and vacancies are difficult to measure, especially in integrated structures. Measuring stress in patterned lines requires high intensity and focused X-ray from advanced light facilities or the use of special test structures [15-18], while vacancies in blanket films have been measured by positron-annihilation lifetime spectroscopy [19-22].

In this study, different post-plating anneal conditions were applied to vary the residual stress and vacancy concentration. Post-plating anneals for longer times and higher temperatures will lead to higher stresses in small dimensions and to bigger grains [17-18,23]. Vacancy concentrations are also lowered by higher temperature post-plating anneals [19] and the increased grain size indirectly lowers the vacancy concentration [20]. We considered 3 different post-plating anneal conditions: 175°C for 20s (or short-low temperature), 175°C for 20min (or long-low temperature) and 250°C for 20min (or long-high temperature). TABLE I summarizes these conditions.

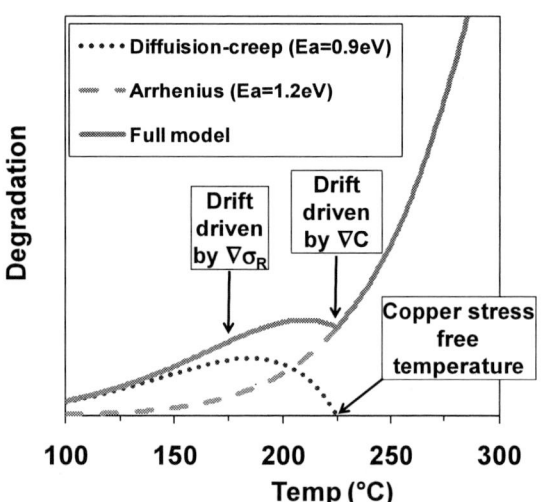

Figure 1   Model of degradation during high temperature storage as argued in [11]. Two different mechanisms were proposed: at lower temperatures, diffusion-creep and thus stresses and stress gradients were hypothesized to drive the degradation, while close to the copper stress free temperature, vacancy agglomeration was suggested to be the dominant degradation mechanism. For our test material, the copper stress free temperature was found to be around 225°C.

TABLE I   POST-PLATING ANNEAL CONDITIONS USED IN THIS STUDY.

Condition	short-low temperature	long-low temperature	long-high temperature
Time	20 s	20 min	20 min
T(°C)	175	175	250

B. Test structures and test approach

Electromigration was done using a single via structure connected to 120nm wide and 500µm long lines both at M1- and M2-level. Upstream and downstream electromigration was studied by applying the current in two different directions (Figure 2A). All experiments were done at package level at 300°C with a current density of 1.5MA/cm^2 through the via. A first jump in the resistance versus time curve was used as failure criterion as in [24]. To study degradation below a via, a 10x10µm M1-plane was designed with two 90nm via's positioned at two opposite edges of the plane. Each via was connected to a narrow M2-line (Figure 2B). This structure was inverted to study the voiding in a via (Figure 2C).

As discussed in [11] and mentioned in the introduction (Figure 1), degradation after high temperature storage occurs in two distinct temperature regions. In a first region, at a lower temperature, the stress gradients in the via-region dominate the degradation. For our test material, the temperature at which this degradation is maximum is around 175°C. In the second region, close to the stress free temperature (~225°C for our material), the residual vacancies dominate the degradation. To study both stress gradients and concentration gradients, wafer pieces with ~20-25 DIE's in total were stored at either 175 or 225°C. An absolute shift in resistance per via has been used as drift criterion. To improve measurement resolution, all measurements were done at a controlled chuck temperature of 40°C. Repetitive measurements were conducted to prove a measurement resolution <0.02 Ohm/via.

III. HIGH TEMPERATURE STORAGE TESTS

A. M1-plane structure

Figure 3 shows the drift/via of the M1-plane structures after 36 days storage at 175°C and 225°C. The 175°C condition is below the stress free temperature, thus degradation is expected to be dominated by hydrostatic stress gradients [2,11]. Literature suggests high temperature post-plating anneals [17] or longer anneals [18] leave a higher stress in the copper. A lower degradation of the short-low temperature post-plating anneal samples and the high degradation of the long-high temperature post-plating anneal is observed and thus confirms the literature data and the hypothesis that degradation during storage at 175°C is driven by residual stresses and stress gradients.

After 225°C storage, the samples with the short-low temperature post-plating anneal show a higher drift. We showed that in this storage region it is expected that the drift is mainly driven by residual vacancy gradients [11]. As for our material 225°C is close to the stress free temperature (Figure 1), it is unlikely that stress gradients alone are the driving force for this degradation. As tiny voids are more stable than the vacancies themselves, vacancies tend to agglomerate when going to a sufficiently high temperature [25]. The fact that the short-low temperature post-plating anneals show higher drift is due to the large amount of residual vacancies left in the copper

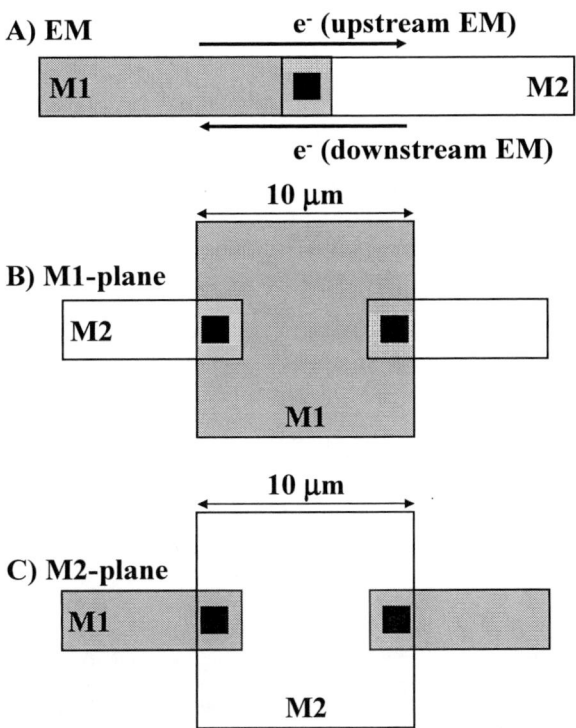

Figure 2   Test structures used to study A) upstream and downstream electromigration, B) SIV below a via and C) SIV in/above a via.

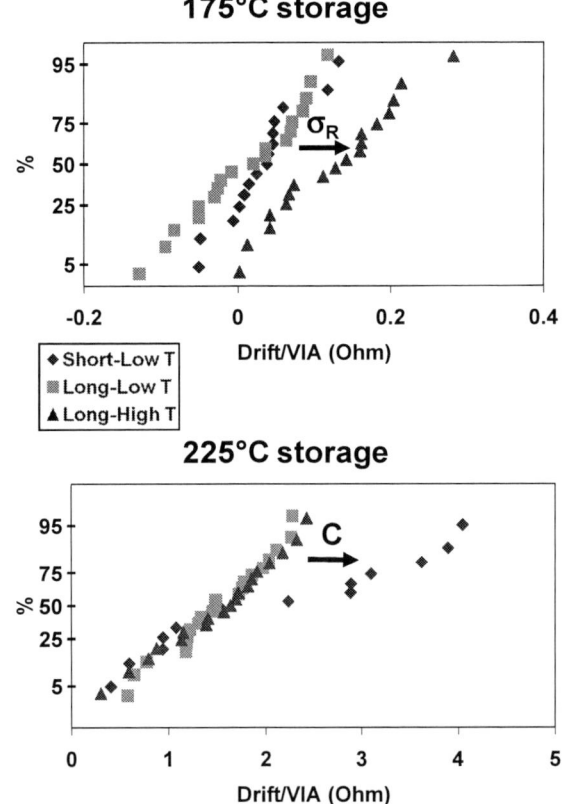

Figure 3   Drift/via of the M1- plane structures after high temperature storages of 36 days at 175°C and 225°C.

after this short low temperature post-plating anneal and confirms the vacancy driven degradation in this temperature region and is in agreement existing literature data [19,20].

B. M2-plane structure

Figure 4 shows the drift/via of the M2-plane structure after 36 days storage at 175°C and 225°C. As in the M1-plane structures, the long-high temperature post plating anneal shows the higher drift after 175°C storage and the short-low temperature post-plating anneal shows the higher drift after 225°C storage. This again is in agreement with existing literature data that argue higher stresses in the long-high and more vacancies in the short-low temperature post-plating anneals and validates our hypothesis of stress-driven degradation during 175°C storage and vacancy-driven degradation during 225°C storage.

The behavior of the M2-plane structures for the long-high temperature post-plating anneal condition is more surprising, as one would expect that the high anneal drives out vacancies to the surface before CMP, leading to smaller vacancy gradients (as observed for the M1-plane structure for this long-high temperature post-plating anneal). We postulate that the degradation of this M2-plane structure is affected by micro-voids that are left in the via after this aggressive post-plating anneal. Instead of moving to the surface, we believe the vacancies agglomerate and form micro-voids when going to a sufficiently high temperature [25]. To further investigate this,

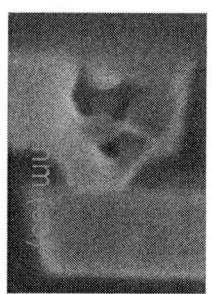

Figure 5 FIB/SEM picture of a via after a post-plating anneal of 350°C for 20min. A severe void is observed above the via.

we explored the effect of a more severe post-plating anneal. Figure 5 shows big voids are formed due to CMP in the via region after a 350°C-20min post-plating anneal. As a result of the severe micro-voiding, the copper in the via is easily pulled out during CMP. We believe that the degradation observed for the long-high temperature post-plating anneal after 225°C high temperature storage in the region above the via is related to this phenomenon.

C. Discussion of high temperature storage tests

A negative shift is observed for some samples. In [11], we showed that both structures with narrow lines and wide planes without the presence of a via showed a negative resistance drift. We suggest that another mechanism which is not related to via-degradation plays a role. For our current test material, this negative shift is the same for all our splits. As such, only relative shifts between the different splits should be considered.

As SIV is driven by stress gradients, we need to confirm whether the observed higher *stresses* after long-high temperature post-plating anneals in the literature also lead to higher *stress gradients*, which create the drift during high temperature storage at 175°C. For this, a 3D finite element model (FEM) of the M2-plane structure was created in ANSYS Multi-Physics. All materials were considered to be isotropic linear elastic solids. The material properties of each material were determined using internal data. Figure 6 shows the hydrostatic stress above the via when varying the copper stress free temperature as input parameter for the modeling. A higher stress free temperature leads to a higher stress at room temperature. Figure 7 shows the normalized stress above the

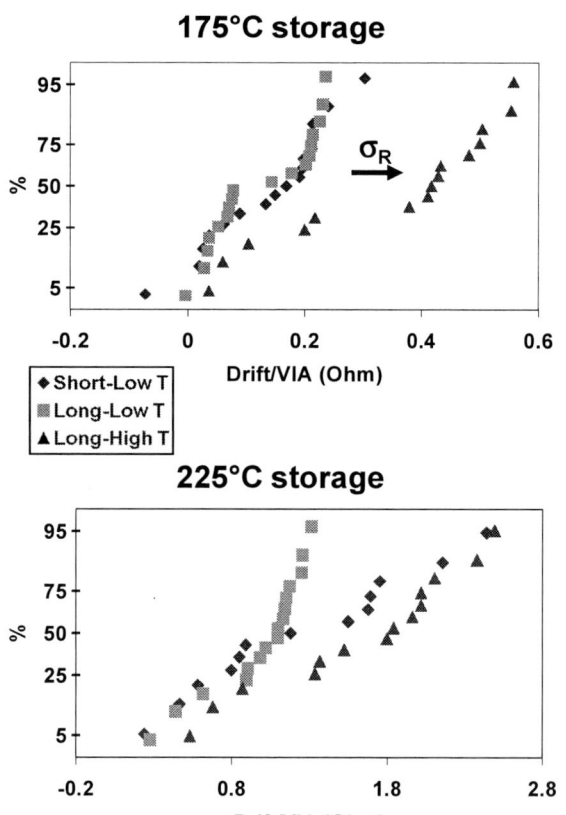

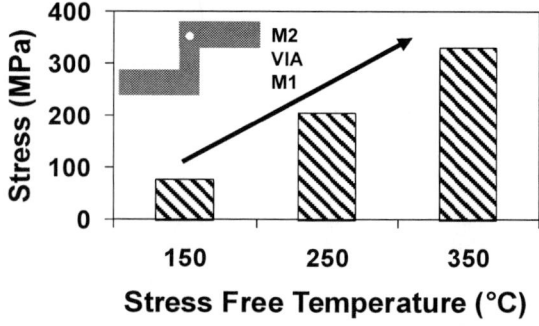

Figure 6 Hydrostatic stress with varying copper stress free temperatures as input parameter for the FEM. Stress above the via is shown for the particular case of the M2-plane structures (point where stress-value is taken is indicated with a white dot).

Figure 4 Drift/via of the M2-plane structures after high temperature storages of 36 days at 175°C and 225°C.

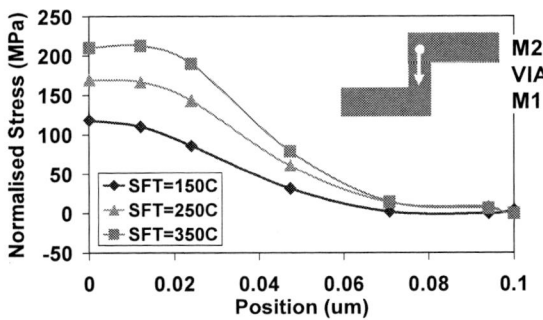

Figure 7 Normalized stress in a via for different copper stress free temperatures (SFT) as input parameter for the FEM. Normalization was done with respect to the start the white arrow.

via for these different stress free temperatures. Higher stress free temperatures, and thus higher stresses, lead to higher stress gradients. These higher gradients are therefore the main driving force for the shift of the samples with the long-high temperature post-plating anneal.

D. Summary of high temperature storage tests

Literature suggests longer and higher temperature post-plating anneals leave less vacancies and induce higher residual stresses in the material [17-20,23]. In addition, our earlier work hypothesized degradation during high temperature storage at 175°C is driven by residual stress gradients $\nabla\sigma_R$, while degradation during high temperature storage at 225°C is driven by vacancy gradients ∇C. In the current work, the existing literature and our earlier work are used together to propose the vacancy concentrations and stress gradients in our material, which we summarized in TABLE II.

IV. ELECTROMIGRATION TESTS

Figure 8 shows the failure times after upstream and downstream electromigration. The long-high temperature post-plating anneal samples show the weakest downstream electromigration and the short-low temperature post-plating anneal samples show the weakest upstream electromigration. Early failures were detected for the long-high temperature post-plating anneal samples when performing upstream electromigration.

The high temperature storage experiments showed residual stress gradients $\nabla\sigma_R$ were the most significant for the long-high temperature post-plating anneal and the short-low temperature post-plating anneal was affected by high vacancy concentrations and gradients ∇C. These trends can be used to

TABLE II   POST-PLATING ANNEAL CONDITIONS USED IN THIS STUDY AND INDICATION OF POSTULATED RESIDUAL VACANCY AND STRESS GRADIENTS.

	short-low temperature	long-low temperature	long-high temperature
Vacancies	↑	→	↓
Micro-voids in via	→	→	↑
Stress	↓	→	↑

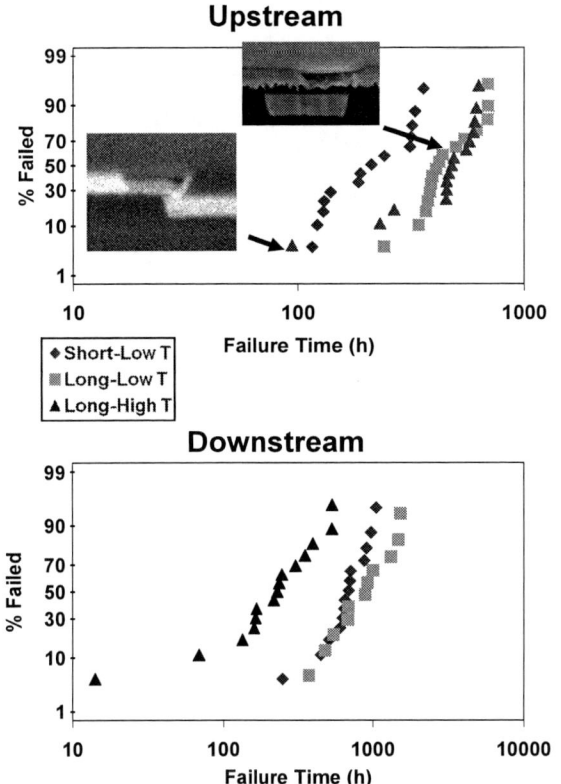

Figure 8   Failure times for upstream and downstream electromigration

explain the electromigration behavior. Hence, the larger residual stress gradients $\nabla\sigma_R$ reduce the downstream failure times of the samples with higher post-plating anneal conditions, and the higher residual vacancy concentrations ∇C lower the failure times of the devices with short post-plating anneals. Therefore, the HTS experiments suggest differences in vacancy concentrations ∇C contribute more to upstream electromigration, while downstream electromigration is more vulnerable to differences in residual stress gradients $\nabla\sigma_R$.

V. DISCUSSION

Electromigration is mainly driven by the strong electron wind j_e (term 2 of equation 1) and by the electron wind induced stress gradients $\nabla\sigma_{EM}$. Although these effects are higher compared to vacancy concentration gradients ∇C and residual stress gradients $\nabla\sigma_R$, the contribution of the latter two needs to be fundamentally understood.

We believe the higher influence of residual stress gradients $\nabla\sigma_R$ on downstream electromigration is explained by the fact that stress gradients below the via are mainly along the line, i.e. parallel to the top-interface that dominates electromigration [1]. This is schematically depicted in Figure 9. As the low temperature post-plating anneal is expected to give the lowest residual stress gradients $\nabla\sigma_R$, the downstream electromigration lifetime is superior. The high temperature anneal has the highest stress, thus the weakest downstream electromigration lifetime. The short-low temperature post-plating anneal is still expected to have a lower stress and good downstream

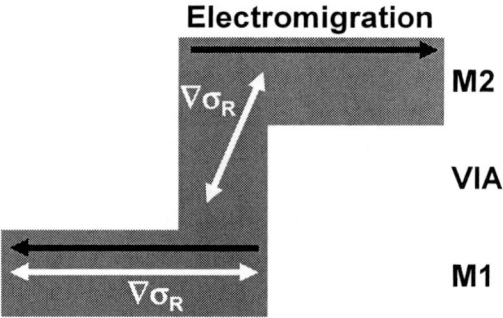

Figure 9 Direction of preferred electromigration diffusion paths (black arrows) and stress gradients built up in the line due to electromigration (white arrows). For downstream electromigration, the preferred diffusion path is parallel to the main stress gradients, while this does not hold for upstream electromigration.

electromigration, but is likely still affected by the large residual vacancy concentration. A schematic representation of the postulated contributions of the variations in residual stress gradients $\nabla\sigma_R$ and vacancy gradients ∇C for downstream electromigration is depicted in Figure 10-bottom.

For upstream electromigration, the relevant residual stress gradients are both present in the via and in the line above the via, i.e. the average stress vector is not parallel to the top-interface of the metal line above the via (Figure 9). We hypothesize the resulting residual stress gradients $\nabla\sigma_R$ have a less significant impact compared to downstream electromigration, making the contribution of vacancy gradients ∇C more dominant for upstream electromigration. In this case the long-high temperature post-plating anneal shows the longest upstream electromigration lifetime as the vacancy concentration is minimized. The short-low temperature post-plating anneal has the highest vacancy concentration remaining after deposition, thus the weakest upstream electromigration lifetime. The long-low temperature post-plating anneal is able to reduce the vacancy gradients, with an upstream electromigration lifetime approaching the long-high temperature post-plating anneal samples. The early failures of long-high temperature post-plating condition that occurred during upstream electromigration are explained by the micro-voids that are present in the via caused by this high post-plating anneal condition (same effect as in section III.B, Figure 5). This is confirmed by our failure analysis, where voiding above the via is found for the early failed devices, while the bulk failures show voids in the metal line away from the via (Figure 8-top). A schematic representation of the postulated contributions of the variations in residual stress gradients $\nabla\sigma_R$ and vacancy gradients ∇C for upstream electromigration is depicted in Figure 10-top.

This complex interaction between residual vacancy and stress gradients, the degradation after high temperature storage at different temperatures and electromigration is schematically summarized in Figure 11.

VI. CONCLUSIONS

We have performed a detailed study of the influence of the residual vacancy and stress gradients on electromigration.

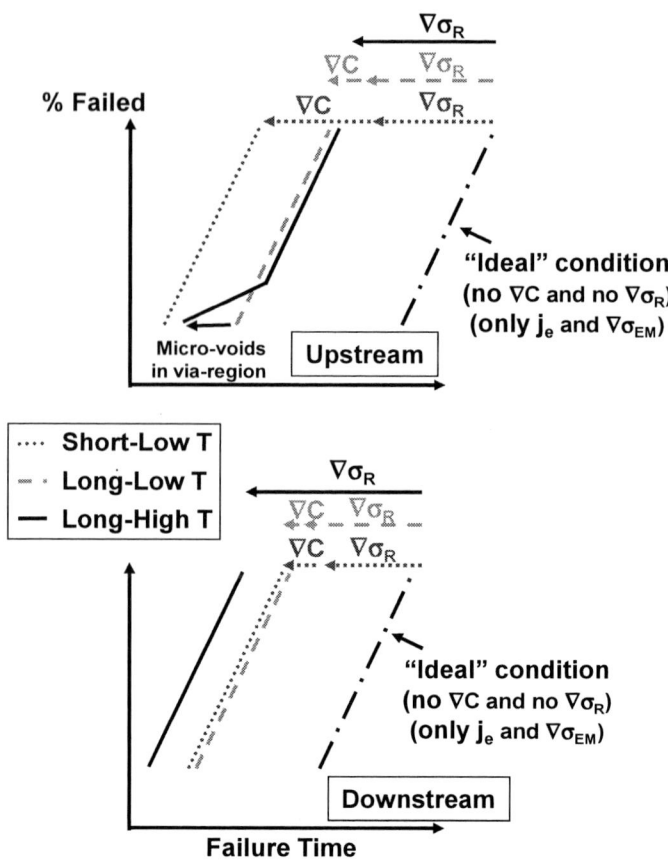

Figure 10 Qualitative sketch of the influence of residual vacancy and stress gradient variations on electromigration for up and downstream electromigration.

Residual vacancy concentration and stress gradients were varied by applying different post-plating anneal conditions. Although agreed that the main driving force for electromigration is the electron wind and the by the electron wind induced stress gradients, we observed that the contribution of residual vacancy and stress gradients could not be neglected as a secondary order effect. The high temperature storage experiments showed stress gradients were the most significant for the long-high temperature post-plating anneals, and the samples receiving a short-low temperature post-plating anneal were affected by high vacancy concentrations and gradients. We used these observations to explain the electromigration behavior. This showed the long-high temperature post-plating anneal degraded downstream electromigration, while the short-low temperature post-plating anneal was more detrimental for upstream electromigration. Based on our high temperature storage experiments, we suggest that the relative contribution of a higher a residual stress gradient is more severe for downstream electromigration, while the relative contribution of concentration gradients are more detrimental for upstream electromigration. We propose that the residual stress gradients accelerate electromigration below the via as they are parallel to the electron flow while this is not the case for upstream electromigration, and upstream electromigration is therefore more sensitive to vacancy gradients.

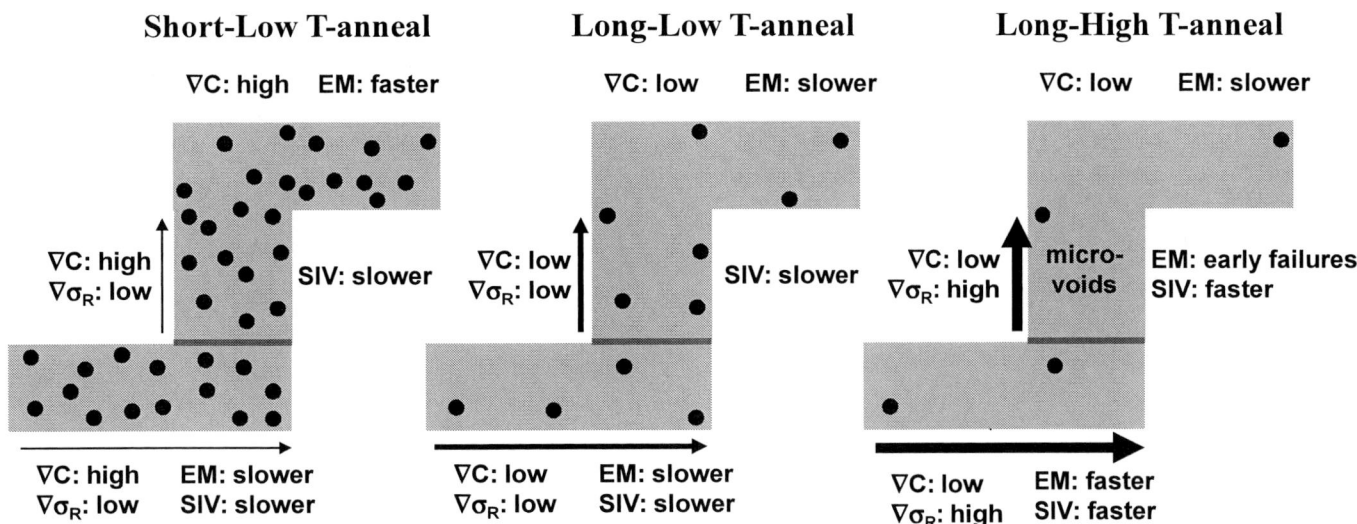

Figure 11   Schematic drawing of residual vacancy concentration and stress gradient (indicated by arrows) for different post-plating anneal conditions and their impact on electromigration and SIV.

ACKNOWLEDGMENT

The authors would like thank the participants of imec's electromigration meeting and the members of the interconnect and remo group for fruitful and stimulating discussions. In particular, Rudy Caluwaerts is acknowledged for the failure analysis work, Veerle Simons and Myriam Van De Peer for the operational work and Yuichi Miyamori, Sony assignee at imec, for the nice discussions.

REFERENCES

[1] C.-K. Hu, L. Gignac and R. Rosenberg, "Electromigration of Cu/low dielectric constant interconnects", Microelectronics Reliability, Vol. 46, p. 213, 2006

[2] E. T. Ogawa, J. W. McPherson, J. A. Rosal, K. J. Dickerson, T.-C. Chiu, L. Y. Tsung, M. K. Jain, T. D. Bonifield, J. C. Ondrusek and W. R. McKee, "Stress-Induced Voiding Under Vias Connected To Wide Cu Metal Leads", IEEE Int. Reliability Physics Symposium. (IRPS), p. 312, 2002

[3] X. He, "EM and SIV of Cu-lowk semiconductor interconnects - 65nmm interconnect technology and beyond", Journal of Nanoelectronics and Optoelectronics, Vol. 2, p. 115, 2007

[4] M. F. Doerner and W. D. Nix, "Stresses and deformation processes in thin films on substrates," Critical Reviews In Solid State and Materials Sciences, Vol. 14, 1988

[5] I. A. Blech and C. Herring, "Stress generation be electromigration", Applied Physics Letters, Vol. 29, p. 131, 1976

[6] J. R. Black, "Mass transport of aluminum by momentum exchange with conducting electrons", IEEE Int. Reliability Physics Symposium. (IRPS), p. 148, 1967

[7] C. L. Gan, C. V. Thompson, K. L. Pey, W. K. Choi, H. L. Tay, B. Yu and M. K. Radhakrishnan, "Effect of current direction on the lifetime of different levels of Cu dual-damascene metallization", Applied Physics Letters, Vol. 79, p. 4592, 2001

[8] A. Roy and C. M. Tan, "Probing into the asymmetric nature of electromigration performance of submicron interconnect via structure", Thin Solid Films, Vol. 515, p. 3867, 2007

[9] A. H. Fischer, A. von Glasow, S. Penka and F. Ungar, "Process optimization - the key to obtain highly reliable Cu interconnects", Proceedings of the International Interconnect Technology Conference (IITC), p. 253, 2003

[10] J. W. McPherson and C. F. Dunn, "A model for stress-induced metal notching and voiding in very large-scale-integrated Al-Si(1%) metallization," Journal of Vacuum Science and Techology, Vol. B5(5), p. 1321, 1987

[11] K. Croes, C. J. Wilson, M. Lofrano, Y. Travaly, D. De Roest, Zs. Tőkei and G. P. Beyer, "Time and Temperature Dependence of Early Stage Stress-Induced-Voiding in Cu/Low-k interconnects", IEEE Int. Reliability Physics Symposium. (IRPS), p. 457, 2009

[12] S. A. Lytle and A. S. Oates, "The effect of stress-induced voiding on electromigration", Journal of Applied Physics, Vol. 71, p. 174, 1992

[13] A. S. Oates, "Electromigration in Stress-Voided Al Alloy Conductors", IEEE Int. Reliability Physics Symposium. (IRPS), p. 297, 1993

[14] Y.-J. Park, K.-D. Lee and W. R. Hunter, "Observation and restoration of negative electromigration activation energy behaviour due to thermo-mechanical effects", IEEE Int. Reliability Physics Symposium. (IRPS), p. 18, 2005

[15] P. R. Besser, S. Brennan and J. C. Bravman, "An x-ray method for direct determination of the strain state and strain relaxation in micron-scale passivated metallization lines during thermal cycling," Journal of Materials Research, Vol. 9, p. 13, 1994.

[16] S.-H. Rhee, Y. Du and P. S. Ho, "Thermal stress characteristics of Cu/oxide and Cu/low - k submicron interconnect structures," Journal of Applied Physics, Vol. 93, p. 3926, 2003.

[17] C. J. Wilson, K. Croes, Zs. Tőkei, B. Vereecke, G. P. Beyer, A. G. O'Neill1 and A. B. Horsfall1, "Application of a nano-mechanical sensor to monitor stress in copper damascene interconnects", Applied Physics Express, Vol. 2, p. 096503, 2009

[18] C. J. Wilson, K. Croes, C. Zhao, T. H. Metzger, L. Zhao, G. P. Beyer, A. B. Horsfall, A. G. O'Neill and Zs. Tőkei, "Synchrotron measurement of the effect of line-width scaling on stress in advanced Cu/Low-k interconnects" Journal of Applied Physics, Vol. 106, p. 053524, 2009

[19] A. Uedono, T. Suzuki, T. Nakamura, T. Ohdaira and R. Suzuki, "Defects in Electroplated Cu and Their Impact on Stress Migration Reliability Studied using Monoenergetic Positron Beams", Japanese Journal of Applied Physics, Vol. 46, p. 1938, 2007

[20] K. B. Yin, Y. D. Xia, W. Q. Zhang, Q. J. Wang, X. N. Zhao, A. D. Li, Z. G. Liu, X. P. Hao, L. Wei, C. Y. Chan, K. L. Cheung, M. W. Bayes and K. W. Yee, "Room-temperature microstructural evolution of electroplated Cu studied by focused ion beam and positron annihilation

978-1-4244-5430-3/10 $26.00 © 2010 IEEE

lifetime spectroscopy", Journal of Applied Physics, Vol. 103, p. 066103, 2008

[21] S. Ogawa, T. Ohdaira, N. Hosoi, N. Tarumi, R. Suzuki and S. Saito, "Cu/Barrier Metal Stack Film Characterization for Reliability Estimation", Proceedings of the International Interconnect Technology Conference (IITC), p. 52, 2007

[22] F. Shoji, T. Ohdaira, I. Suzuki, M. Shimada, R. Suzuki and S. Ogawa, "Behaviors of vacancies in electro-plated Cu films by positron-annihilation lifetimes spectroscopy correlated with SIV phenomena", Proceedings of the Advanced Metallization Conference (AMC), p. 313, 2004 (published in 2005)

[23] P. R. Besser, A. S. Mack, D. B. Fraser and J. C. Bravman, "Finite-Element Modeling and X-Ray Measurement of Strain in Passivated Al Lines during Thermal Cycling,", Journal of the Electrochemical Society, Vol. 140, p. 1769, 1993

[24] J. Michelon, C. Bruynseraede, D. Tio Castro, Ph. Roussel, R. Hoofman and K. Maex, "Electromigration study of sub-100nm Cu-lines", Proceedings of the Advanced Metallization Conference (AMC), p. 253, 2004 (published in 2005)

[25] W. Li, C. M. Tan and N. Raghavan, "Dynamic simulation of void nucleation during electromigration in narrow integrated circuit interconnects", Journal of Applied Physics, Vol. 105, p. 014305, 2009

AUTHOR INDEX

Abe, Kenichi 683
Abeln, Glenn 1115
Abramowitz, Peter 125, 1115
Abu-Rahma, M. H. 1008
Ackermann, Markus 1006
Acosta, Antonio G. 689
Afanas'ev, Valery 549
Ahlbin, J. R. 198, 763, 1031
Ahn, Kun-Ok 975
Aichinger, Thomas 1063, 1073
Aitken, J. 566
Alam, Muhammad Ashraful 65, 312, 1069, 1091
Albin, David S. 318
Alers, G. B. 323, 499
Allee, David R. 644
Allmon, R. 188
Ambrose, V. 188
Amoroso, Salvatore M. 966
Anderson, Gary 125, 1115
Andreani, C. 400
Ang, D. S. 373, 1058
Angyal, M. 566
Aoulaiche, Marc 1095
Arigane, T. 1001
Arreghini, Antonio 731
Asheghi, Mehdi 99
Ashok, Ashwin 287
Atkinson, N. M. 198
Aubel, Oliver 117, 562, 574, 581, 926
Auluck, Kshitij 981
Baburske, Roman 162
Baccarani, Giorgio 881
Badami, D. 566
Bae, Kidan 104
Baek, Chang-Ki 94
Baek, Rock-Hyun 94
Bagatin, M. 400
Baghini, Maryam Shojaei 480, 841
Bai, P. 293
Balasubramanian, A. 188
Balk, L. J. 271
Banerjee, Kaustav 822
Barbato, M. 246
Barsky, M. E. 807
Bashir, Muhammad M. 895
Basker, V. S. 1099
Baumann, Robert C. 395
Bauza, D. 259
Beenakker, C. I. M. 342
Bellaio, N. 334
Beltrami, Silvia 400, 604
Bender, Hugo 712
Benvenuti, A. 615
Bergman, J. 813
Bersuker, G. 73, 373, 532

Besset, C. 55
Beyer, G. P. 591
Bhatia, C. S. 503
Bhuva, Bharat L. 198, 763, 1026, 1031
Biedenbender, M. 807
Bisht, Gaurav 981
Bisschop, Jaap 830
Bittel, B. C. 947
Blaauw, D. 676
Bloch, Didier 231
Block, T. R. 807
Bluet, J.-M. 139
Boit, C. 890
Boku, K. 750
Bonilla, G. 566
Borghs, Gustaaf 146
Boselli, Gianluca 474
Bosman, M. 354
Bounasser, Mounaim 203
Bourgeois, F. 129
Brar, B. 813
Bravaix, A. 55
Breuer, T. 890
Brewer, Forrest 853, 932
Briggs, Benjamin 80
Bronner, W. 129
Bru-Chevallier, C. 139
Bu, Huiming 1099
Burnett, David 125, 1115
Butendeich, Rainer 522
Butera, Geni 712
Buttari, D. 807
Bychikhin, S. 480
Cacho, Florian 655
Cagli, C. 620
Cai, Xiao Xiao 411
Calderoni, Alessandro 738, 970
Camozzi, Elisa 966
Campbell, J. P. 804
Cannon, Ethan H. 1019
Cao, Yiqun 458
Cao, Yu 650
Carter, S. A. 323
Cartier, E. 50, 369, 787, 1044
Cäsar, M. 129
Catthoor, Francky 1014
Cavelaars, Jan 724
Cellere, G. 400
Ceric, Hajdin 911
Cester, A. 334, 536
Cha, Nam Hyun 1004
Chancellor, Cathy A. 385, 670
Chang, C. S. 627
Chang, C. Y. 818
Chang, W. L. 364, 787

AUTHOR INDEX

Chang, Y. F. 627
Chao, Y. P. 960
Chappaz, C. 887
Chatterjee, Amitabh 853, 932
Chatterjee, I. 1031
Chatty, Kiran 846
Chen, An 84
Chen, Chih-Hsien 918
Chen, F. 566
Chen, K. C. 627, 960
Chen, K. F. 627
Chen, K. H. 818
Chen, Kuan-Fu 634
Chen, Kuang-Chao 634, 639, 951
Chen, M. J. 665
Chen, M. S. 627
Chen, Ming-Shiang 634
Chen, Tao..342
Chen, Wen-Yi 857
Chen, X. 566
Chen, Y. J. 627
Chen, Y. N. 89
Chen, Ye 277
Chen, Yin-Jen 634
Chen, Zuhui 977
Cheng, Cheng-Hsien 634
Cheng, Kai 146
Cheng, Kangguo 1099
Cheung, K. P. 804
Chevallier, Remy 655
Chikhaoui, W. 139
Child, C. 566
Chimeno, Alejandro 655
Chin, Melida 453, 562
Chiu, J. P. 960
Cho, Eun Suk 611
Cho, M. 1078, 1082
Cho, Moonju 1095
Cho, Myoung Kwan 975
Cho, Seokwon 975
Choi, Gil-Bok 94
Choi, Gil-Heyun 903
Choi, Hyun Ki 599
Choi, Hyun-Sik 94
Choi, Jeong-Hyuk 599, 611
Choi, Jingyoo 287
Choi, R. 1008
Choi, Siyoung 282, 903
Choi, Zungsun 903
Chong, Lit Ho 634
Chopra, Sanjeev 655
Chou, H. L. 170
Chou, T.-M. 528
Chou, Y. C. 807
Chou, Y. L. 960

Chouard, Florian R. 826
Chowdhury, U. 528
Christiansen, C. 566
Chu, B. H. 818
Chu, F. 750
Chua, E. C. 938
Chung, Chilhee 599, 611
Chung, Steve S. 1053
Cirba, C. 670
Coccetti, F. 237
Compagnoni, Christian Monzio 604, 966, 970
Croes, Kristof 543, 549, 591, 712
Curutchet, A. 139
Dadgour, Hamed F. 822
Dammann, M. 129
Das, Jo 146
DasGupta, S. 813
De Jaeger, J.-C. 139
Decoutere, Stefaan 146
Defrance, N. 139
Degraeve, Robin 26, 1078, 1095
Degroote, Stefan 146
Del Alamo, Jesús A. 134
Del Cueto, Joseph A. 318
Demirtas, Sefa 134
Demuynck, Steven 712
Denison, Marie...881
Deora, S. 1105
Depner, M. 750
Derkits, G. E. 879
Derluyn, Joff 146
Detcheverry, C. 111
Detzel, Thomas 911
Dhere, Neelkanth G......................................306
Di Sarro, James 846
Djelassi, Christian 446
Domengie, F. 259
Dongaonkar, S. 312
Donnet, David 724
Doris, Bruce 1099
Douglas, E. A. 818
Douvry, Y. 139
Doyen, L. 922
Drijbooms, Chris 712
Du, A. Y. 906
Du, G. A. 1058
Du, Pei-Ying 951
Dua, C. 139
Duvvury, Charvaka.......................................853
Eaton, P. H. 198
Eliason, J. 750
Eneman, G. 1082
Eng, D. C. 807
Enichlmair, H. 43
Fang, Z. 964

AUTHOR INDEX

Fantini, Paolo 738, 970
Faqir, M. 152
Farbiz, Farzan 466
Farina, Fabrizio 970
Ferro, M. 738
Fisher, Kathryn C. 508
Fleetwood, D. M. 813
Fleming, D. 879
Forbes, Keith 125
Forsythe, Eric 644
France, C. E. 323
Francis, Rick 117
Franco, Jacopo 26, 1082, 1095
Franey, J. P. 879
Frei, M. 312
Frei, Stephan 458
Frost, C. D. 400
Fugazza, Davide 743
Fujii, Shosuke 956
Fujiki, Jun 956
Fujisawa, Takafumi 683
Fujitsuka, Ryota 956
Fujiwara, Tetsuo 988
Fukatsu, Shigeto 424
Fuketa, Hiroshi 213
Fukuda, Mitsuhiro 218
Fukuda, Toshikazu 694
Fukutani, Katsuyuki 417
Fulde, Michael 826
Funayama, Kota 995, 1001
Futase, Takuya 988, 995
Gadlage, M. J. 198, 763
Galbiati, Nadia 966
Gannavaram, S. 293
Gao, B. 964
Gasiot, Gilles 407
Gauthier, Robert J. 846
Ge, L. 1008
Geinzer, T. 271
Genoe, J. 334
Gerardin, S. 246, 400
Germain, Marianne 146
Gertas, J. 750
Ghani, Tahir 287
Ghetti, A. 615
Ghidini, Gabriella 966
Ghidotti, Michele 604
Ghilardi, Tecla 966
Gilbert, N. 566
Giliberto, V. 246
Gill, B. 188
Gischia, Gianni Giai 549
Glaser, Ulrich 458
Glavanovics, Michael 446
Gnade, Bruce 644

Gnani, Elena 881
Gnudi, Antonio 881
Goes, W. 16
Goguenheim, D. 55
Goodnick, Stephen M. 516
Goodson, Kenneth 99
Gorini, G. 400
Gornik, Erich 480
Gossner, Harald 480, 841, 853
Goto, Masakazu 424
Goyal, Deepak 252
Goyal, S. 430
Graham, R. L. 323
Grasser, Tibor 7, 16, 26, 43, 1063, 1073, 1082
Green, Keith 689
Groat, J. 750
Groeseneken, Guido 26, 549, 712, 1078, 1082, 1095
Grossi, Alessandro 966
Grzegorczyk, Andrzej 724
Guarin, F. 364
Guérin, C. 55
Guerra, Diego 516
Gustin, Wolfgang 7, 175
Ha, Sungmok 104
Hafez, W. 293
Haggag, Amr 125, 1115
Hahn, Berthold 522
Han, B. M. 1008
Han, Jeong Hee 282
Han, Tzung-Ting 627, 634
Hashikawa, Naoto 988, 995
Hashimoto, Koichi 379
Hashimoto, Masanori 213
Hashimoto, T. 1001
Hatanaka, Kichiji 218
He, Jun 287
Heh, D. 373
Heiderhoff, R. 271
Heinze, Birk 162
Hendriks, Teun 1014
Hennesthal, C. 581
Heryanto, A. 586
Hicks, J. 293
Higman, Jack 1115
Hinze, Peter 327
Hirano, Izumi 424
Hiroshima, Shoichi 417
Hisamoto, D. 1001
Hitoshi IIda 874
Ho, Paul S. 574, 581
Hoel, V. 139
Hoffmann, Thomas Y. 926, 1095
Hofmann, Ralf 981
Holman, W. T. 198
Hong, Chonga 975

AUTHOR INDEX

Hong, Shih-Ping 951
Hori, Mitsuaki 379
Horita, K. 1001
Hsia, L. C. 906
Hsiao, Yi-Hsuan 951
Hsieh, E. R. 1053
Hsieh, Jung-Yu 639, 951
Hsieh, Kuang-Yeu 627, 951
Hsieh, Sunnys 170
Hsu, Chia-Lin 918
Hsu, Fang-Hao 951
Hsu, Tzu-Hsuan 634, 951
Hu, Y. Z. 1058
Huang, Climbing 918
Huang, E. 566
Huang, I-Jen 627, 634
Huang, Jyun-Siang 634
Huang, Rui 911
Huang, Yu-Hui 170
Huard, Vincent 33, 55, 655
Hübner, René 574
Hughes, Greg 799
Hurkx, G. A. M. 99
Hurley, P. K. 775
Hutter, H. 1063
Hwang, Heedon 282
Hyun, Sangjin 282
Ielmini, Daniele 620, 738, 743
Im, J. 581
Inaba, Yutaka 988
Inan, Umran S. 203
Inoue, M. 1001
Inumiya, Seiji 424
Ioannou, D. P. 1044
Ioannou, Dimitris E. 846
Ishigaki, T. 1049
Ishihara, Ryoichi 342
Islam, Ahmad Ehteshamul 65, 1069
Ito, Shuu 417
Iwai, Hidenao 874
Iwamatsu, T. 1001
Jack, Nathan 485, 835
Jagannathan, Hemanth 1099
Jain, Palkesh 698
Jammy, R. 73
Jan, C.-H. 293
Jeong, YeonJoo 975
Jeong, Yoon-Ha 94
Jha, N. K. 665
Jimenez, J. 528
Johannes, Hans-Hermann 327
Jones, M. 293
Jung, S. O. 1008
Jungemann, C. 175
Jurczak, Malgorzata 731

Kaapor, Neeraj 655
Kaczer, Ben 16, 26, 1082, 1095
Kakuhara, Y. 717
Kamino, Takeshi 988
Kane, T. 566
Kaneoka, T. 1001
Kaneta, Chioko 379
Kang, Han-Byul 265
Kang, J. F. 964
Kang, Ju Seong 1004
Kang, M. G. 1008
Karthik, Y. 312
Kase, Masataka 379
Kato, Koichi 424
Kauerauf, Thomas 712, 1078, 1095
Kaul, Ashwani 306
Kaureauf, Thomas 799
Keller, R. 175
Kendig, D. 499
Ker, Ming-Dou 857
Kerber, A. 50, 369
Kerst, U. 890
Khakifirooz, Ali 1099
Kiefer, R. 129
Killat, N. 528
Kim, Byoung Taek 611
Kim, D. 750
Kim, Dae Mann 94
Kim, Dong-Won 94
Kim, Gun-Rae 104, 265
Kim, Hyun 1004
Kim, Hyun Jung 611
Kim, Kinam 94
Kim, Min 104
Kim, SangBum 99
Kim, Seong Soo 599
Kim, Woo Sup 1004
Kim, Yong Seok 599
Kim, Yongshik 104
Kimura, S. 1049
King, S. 947
Ko, T.-M. 566
Kobayashi, Yuko 379
Köck, Helmut 446
Koh, Yohwan 975
Kolics, A. 566
Komeyli, K. 293
Konstanzer, H. 129
Kopf, R. 879
Koszewski, Adam 231
Kotlyar, R. 293
Kowalsky, Wolfgang 327
Koyama, Shin 417
Krause, J. 188
Krick, J. 670

AUTHOR INDEX

Krishnan, Anand T. 385, 670, 1122
Krishnan, Srikanth 650, 689, 698, 1122
Ku, Shaw-Hung 634
Kuball, M. 152, 528
Kubo, Tomohiro 379
Kufluoglu, Haldun 670
Kulkarni, Pranita 1099
Kulshrestra, Vishal 655
Kumashiro, Shigetaka 694
Kuper, Fred 111, 724
Kuroda, Rihito 683
Labat, N. 139
LaBel, Kenneth A. 768
Lacaita, Andrea L. 604, 620, 743, 966, 970
Lai, R. 807
Lai, Sheng-Chih 951
Lamontagne, P. 922
Langer, E. 890
Larcher, Luca 373, 532, 731
Laurin, L. 615
Lavizzari, Simone 743
Law, S. B. 566
Lee, Byoungil 99
Lee, Chi Kyoung 599
Lee, Dong Jun 599
Lee, Dong-Kyu 975
Lee, Eun-Kyu 80
Lee, Ho Seok 975
Lee, Hsiao-Heng Kelin 203
Lee, J. H. 665
Lee, Jeong-Soo 94
Lee, Jian-Hsing 182
Lee, Jong Myeong 903
Lee, L. S. 807
Lee, S. C. 705, 932
Lee, Sang-Hyun 94
Lee, Seok-hee 287
Lee, T. 566
Lee, Woon-Hak 265
Lee, Y. C. 627
Lee, Y.-H. 170, 665
Lemay, Karen 287
Lenahan, P. M. 43, 947, 1086, 1122
Leu, L. C. 818
Leung, D. L. 807
Leys, Maarten 146
Li, B. 566
Li, Junjun 846
Li, Li 440
Li, X. 354, 778, 964
Li, Yuan 724
Lian, Nan-Tzu 951
Liang, James W. 556
Liang, Z. 111
Liao, Jeng-Hwa 639

Liaw, M. H. 627
Lilja, Klas 203
Lim, K. Y. 73
Lim, Phyllis Shi Ya 1055
Lim, Se Young 1004
Lim, Sunme 104
Lim, Y. K. 586, 906, 938
Lin, Chung-Hsun 807, 1099
Lin, Hung Sung 861, 865, 870
Lin, J. 293
Lin, Jeh-Chieh 918
Lin, M. H. 705, 1053
Lin, Mingte 556
Lin, Shang-Wei 634
Lin, Shihuan 977
Lin, Wen-Chin 918
Lin, Y. H. 1053
Linscott, Ivan R. 203
Lipp, D. 369
Liu, Bin 1055
Liu, C. C. 170
Liu, H. 906
Liu, J. F. 938
Liu, Mark 287
Liu, Ning 1115
Liu, P. W. 1053
Liu, Pan 277
Liu, Tsu-Jae King 1117
Liu, W. 586, 906, 938
Liu, W. H. 354, 778
Liu, W. J. 964
Liu, Ziyuan 417
Lloyd, J. R. 943
Lo, Chester 627, 818
Lo, Patrick G. Q. 964
Lofrano, Melina 591, 712
Lu, Chih-Yuan 627, 634, 639, 951, 960
Lu, Kuan-Ting 918
Lu, Lei 1040
Lu, Ryan 287
Lu, Wen-Pin 627, 634
Lu, Xiaowei 1040
Lue, Hang-Ting 627, 634, 639, 951
Lutz, Josef 162
Lwin, Z. Z. 89
Ma, G. H. 1053
Ma, H. C. 960
Ma, Zhe 1014
Maconi, Alessandro 966
Mahapatra, Souvik 89, 312, 981, 1105
Mahatme, N. N. 1031
Maheta, V. D. 1105
Mairena, Andrew 1117
Maitra, Kingsuk 1099
Makabe, K. 120

AUTHOR INDEX

Makabe, Mariko 417
Makiyama, H. 1001
Malbert, N. 139
Mandich, M. L. 879
Manfredi, Manfredo 522
Marathe, Amit 117, 453, 562
Marca, Vincenzo Della 731
Marcon, Denis 146
Marino, Fabio Alessio 516
Marmiroli, Andrea 970
Marshall, A. 670
Mascellino, Evelyne 966
Massengill, L. W. 198, 763, 1031
Masuduzzaman, M. 1069
Matsuyama, Hideya 218, 379
Mauri, Aurelio 966
Medjdoub, Farid 146
Mei, X. B. 807
Mendenhall, Marcus H. 1026
Mendenhall, Marcus M. 395
Meneghesso, Gaudenzio 1, 246, 334, 522, 536
Meneghini, Matteo 1, 334, 522
Meng, L. 503
Mertens, Robert 146
Meyer, Jens 327
Miccoli, Carmine 604
Mikulla, M. 129
Milor, Linda 895
Minakata, Hiroshi 379
Miranda, E. 775
Mishra, Anand 655
Mishra, Rahul 846
Mishra, U. K. 152
Mistry, K. 293
Mitani, Yuichiro 299, 424
Mitra, Souvick 846
Mitra, Subhasish 203
Mitsuyama, Yukio 213
Mittl, S. 1044
Mizutani, M. 1001
Mo, Zhi Qiang 277
Moise, Ted 689, 750
Monaco, G. 246
Monsieur, F. 364
Moon, Joo-Tae 282, 903
Moosa, Mohamed 125, 1115
Mora, P. 364
Morassi, L. 373, 532
Mori, Hiroko 379
Mori, Toshifumi 379
Morin, P. 259
Morita, Y. 1049
Morton, David 644
Mottadelli, Riccardo 604
Mukherjee, S. 188

Mukhopadhyay, Gautam 981
Muller, K. Paul 391
Müller, S. 129
Murakami, E. 120
Myny, K. 334
Nagabhirava, Bhaskar 80
Nagalingam, D. 503
Nakamura, Hideyuki 694
Nakamura, Tomonori 874
Nakasaki, Yasushi 424
Nakkala, P. 887
Nam, Kab-Jin 282
Nara, Yasuo 379
Narasimham, B. 198
Nardi, F. 620
Nassif, N. 188
Nelhiebel, Michael 1063, 1073
Ney, David 655, 922
Ngan, P. 111
Nicholson, L. 566
Nicollian, Paul E. 385
Nicolosi, P. 246
Nigam, T. 369
Nikolic, Borivoje 1117
Nishigaya, Keita 379
Nowodzinski, Antoine 231
Oates, A. S. 705, 932
Obradovic, Borna 689
O'Connor, E. 775
O'Connor, Robert 799
Ogasawara, M. 120
Oh, M. 566
Ohmi, Tadahiro 683
Ok, Injo 532
Oki, A. K. 807
Oldiges, Philip J. 1099
Olney, Andrew 224
Olson, Nicholas 474
Ong, Tong-Chern 182
Ong, Y. C. 373
Ono, Katsuji 379
Onoye, Takao 213
Orita, Kenji 1
Ottogalli, F. 615
Oualli, M. 139
Ouchi, T. 120
Paccagnella, A. 246, 400
Pacheco, Mario 252
Packan, Paul 287
Padovani, Andrea 373, 532, 731
Pae, Sangwoo 287
Pan, JiFong 627
Pantelides, S. T. 813
Papaioannou, G. J. 237
Parihar, Sanjay 1115

AUTHOR INDEX

Park, Chan-Hoon 94
Park, H. 73, 1008
Park, Hong Sik 1004
Park, Hyung-Jin 1004
Park, Hyun-Woo 265
Park, Jongwoo...104, 265
Park, Junkyun 104
Park, Milim 975
Park, Noh Seok 1004
Park, Sukkwang 975
Park, Young-Joon 698
Parker, Chris..287
Parthasarathy, Chittoor 655
Paruchuri, Vamsi 1099
Pavan, Paolo 532, 731
Pavesi, Maura 522
Pearton, S. J. 818
Pei, Y. 152
Pellish, Jonathan A. 768
Pendharkar, Sameer 853, 881
Perng, Dung-Ching 918
Persin, Flore 655
Pethe, Shirish A. 306
Petitdidier, S. 566
Petitprez, E. 922
Pey, K. L. 89, 354, 373, 586, 778, 964
Phang, J. C. H. 271, 503
Piazza, M. 139
Pickholtz, J. 188
Pinato, A. 334
Pion, Emmanuel 655
Pirovano, A. 615
Plana, R. 237
Planes, Nicolas 655
Platt, S. P. 411
Pobegen, Gregor 1073
Pogany, D. 480
Poli, Stefano 881
Pons, P. 237
Post, I. 293
Prasad, C. 293
Prokofiev, A. V. 411
Puchner, S. 1063
Purser, Richard 287
Quay, R. 129
Quevedo-Lopez, Manuel 644
Rafik, M. 364
Raghavan, N. 354, 586, 778
Ragheb, T. 670
Ragnarsson, Lars-Åke 26, 799, 1095
Randriamihaja, Y. Mamy 55
Rangoni, Armando 966
Ranjan, R. 665
Ranzato, Enrico 522
Rao, V. Ramgopal 480, 841

Recchia, C. 188
Redaelli, An. 615
Reddy, Vijay 650, 670
Reed, Robert A. 395, 813, 1026
Reents, W. D. 879
Reggiani, Susanna 881
Regolini, J. L. 259
Reifenberg, John 99
Reisinger, Hans 7, 16, 26, 175
Relangi, Prasanthi 203
Remack, K. 750
Ren, F. 818
Renard, S. 55
Ribes, G. 364, 566
Riedl, Thomas 327
Riedlberger, E. 175
Riepe, K. 129
Rigoutat, O. 566
Robert, Vincent 655
Robl, Werner 911
Roche, Philippe 407
Rödle, T. 129
Rodriguez, John 689, 750
Rodriguez-Latorre, J. 750
Romain, Michael 1036
Rosenbaum, Elyse 466, 474, 485, 835
Roussel, J. 26
Roussel, Philippe 712, 1095
Roy, D. 55, 364
Ruelke, H. 890
Ruiz-Amador, Natalia 655
Rumyantsev, S. 73
Rupp, Roland 156
Ryan, J. T. 43, 1122
Ryan, S. 879
Sagong, Hyun Chul 94
Sahhaf, S. 1078
Sakoda, Tsunehisa 379
Salman, Akram 474
San, Tamer. 689
Sanda, Pia N. 391
Sandhya, C. 981
Saraniti, Marco 516
Sasse, Guido T. 830
Sato, Motoyuki 424
Sawada, H. 566
Schanovsky, F. 16
Schlünder, Christian 7
Schmidt, Hans 327
Schmitt-Landsiedel, Doris 826
Schneidenbach, Daniel 327
Schneider, Jens 480, 841
Schrimpf, Ronald D. 395, 763, 813, 1026
Scozzari, Claudia 966
Seah, Soo Sien 277

AUTHOR INDEX

Seetharaman, Sridhar 881
Seifert, N. 188
Seirawski, Brian D. 1026
Sekine, Katsuyuki 424, 956
Seo, Jae Yong 1004
Shakouri, A. 499
Shao, Xiaoyan 508
Sheldon, Douglas J. 759
Shen, X. 813
Sheshadri, Vijay B. 1026
Shi, Q. 188
Shiga, K. 1001
Shih, J. R. 170, 182, 665
Shimamoto, Y. 1001
Shin, Changhwan 1117
Shin, Yu Gyun 282
Shinosky, M. 566
Shono, Ken 218
Shrivastava, Mayank 480, 841
Shubhakar, K. 354
Shukla, Vrashank 485
Shuler, R. 1031
Shur, M. 73
Siegmund, T. 430
Sierawski, Brian D. 395
Simms, R. J. T. 152
Simoen, E. 26
Singh, P. 676, 981
Singh, P. K. 89
Sites, James R. 494
Sivatheja, M. 981
Smout, S. 334
So, Byung Se 1004
Sohn, D. K. 586, 906
Sölkner, Gerald 156
Song, Du Heon 599, 611
Song, Jai Hyuk 599, 611
Song, S. C. 1008
Song, Seung Hyun 94
Soo, Chi Wen 277
Spessot, Alessio 970
Spinelli, Alessandro S. 604, 966, 970
Spitzer, A. 175
Srinivasan, K. 430
Srividya, V. 1078
St Amour, Anthony 287
Stathis, J. 364
Stathis, James H. 50, 787, 1099
Stecher, Matthias 175, 458
Steen, Steven E. 508
Stellari, Franco 508
Street, A. G. 503
Su, K. C. 556
Subbarayan, G. 430
Suehle, J. 804

Sugawa, Shigetoshi 683
Sugii, N. 1049
Suh, Kang-Deog 599, 611
Summerfelt, Scott 689, 750
Sun, Kai 1040
Suñe, Jordi 792
Sury, C. 139
Suzuki, Hiroyoshi 683
Sylvester, D. 676
Takahisa, Keiji 218
Takeuchi, Kan 694
Tam, Nelson 1036
Tamura, Naoyoshi 379
Tan, J. B. 586, 906
Tanahashi, Katsuto 379
Tanaka, Katsuhiko 694
Tang, T. J. 566
Tang, Z. 347
Tanimoto, Hisanori 988, 995
Tapajna, M. 152
Taylor, W. 73, 373
Tazzoli, Augusto 246, 334, 522
Tee, Irene 277
Teo, Z. Q. 1058
Terada, Hirotoshi 874
Teramoto, Akinobu 683
Tessariol, Paolo 966
Tobimatsu, Hiroshi 995
Toh, Seng Oon 1117
Tokei, Zsolt 543, 549, 591, 712
Torii, K. 1001
Toriumi, Akira 299
Tortorelli, I. 615
Tosaka, Yoshiharu 218, 379
Tous, Santi 792
Toussaint, Thibaut 231
Treu, Michael 156
Trivellin, Nicola 1, 522
Truong, C. 566
Tsai, C. 293
Tsai, C. H. 1053
Tsai, C. T. 1053
Tsai, R. S. 807
Tsai, Teng-Chun 918
Tsai, Y. S. 665
Tseng, Joshua 1095
Tsuchiya, H. 717
Tsuchiya, R. 1001, 1049
Tsukamoto, Yasumasa 1117
Tsukasa, Matsuda 903
Uchibori, Chihiro J. 218
Udayakumar, K. R. 750
Uemura, Taiki 218, 694
Uznanski, Slawosz 407
Van Den Bosch, Geert 731

AUTHOR INDEX

Van Der Wel, P. J. 129
Van Dijk, K. 111
Van Houdt, Jan 731
Van Hove, Marleen 146
Vandelli, Luca 731
Vandevelde, Bart 712
Varghese, Dhanoop 1091
Vayshenker, A. 369
Veksler, D. 73, 532
Velamala, Jyothi B. 650
Ventrice, D. 738
Venugopal, Sameer M. 644
Vereecke, B. 591
Vermunt, Frank 1014
Verzellesi, G. 532
Vialle, Nicolas 655
Vincent, E. 55
Visalli, Domenica 146
Visconti, Angelo 400, 604
Volf, P. A. J. 111
Wagner, P.-J. 16
Waltereit, P. 129
Wang, D. 312
Wang, J. 1008
Wang, L. 750
Wang, Miaomiao 1099
Wang, Mingxiang 1040
Wang, Szu-Yu 627, 951
Wang, Tahui 960
Wang, Wayne 665
Wang, Y. 566, 1044
Wang, Z. R. 964
Warren, Kevin M. 395, 1026
Watabe, Shunichi 683
Wei, J. 586
Weimann, Thomas 327
Weir, Bonnie 1091
Weller, Robert A. 395, 1026
Wen, Shi-Jie 1026, 1036
Wen, Yong-Ru 857
Werner, Christoph 826
Wie, C. R. 347
Wilde, Markus 417
Wilson, Christopher J. 591, 712
Winkler, Thomas 327
Wise, Rick 881
Witulski, A. F. 198
Wojtowicz, M. 807
Wong, H.-S. Philip 99
Wong, Richard 1026, 1036
Woo, S. H. 1008
Woolery, Bruce 287
Wouters, Y. 922
Wrachien, N. 334, 536
Wu, Ernest Y. 792

Wu, J. Y. 918
Wu, Kenneth 170, 182, 665
Wu, L. 964
Wu, Ming-Tsung 951
Wu, Mong Sheng 861, 870
Wu, X. 354
Wu, Y. Q. 536
Xu, C. 879
Xue, Jie 440
Yamamoto, Hirohiko 988, 995
Yamauchi, Toyohiko 874
Yanagi, I. 1001
Yang, G. 364, 566
Yang, J. W. 1008
Yang, Ling-Wu 639, 951
Yang, Tahone 627, 639, 951
Yang, Yang 846
Yasuda, Naoki 956
Yau, A. 111
Ye, P. D. 536
Yeap, G. 1008
Yeh, Chun-Chen 1099
Yeh, Teng-Hao 634
Yeo, Kyoung Hwan 94
Yeo, Yee-Chia 1055
Yeoh, Yun Young 94
Yew, K. S. 373
Yiang, Kok-Yong 117, 562
Yokogawa, S. 717
Yoo, Jae-Yoon 104
Yoon, Joo-Byoung 265
Yoshimoto, H. 1001, 1049
Yoshioka, N. 120
Young, C. D. 73, 373
Yu, Bin 80
Yu, H. Y. 964
Yu, Tianhua 80
Yuri, Masaaki 1
Zaghloul, U. 237
Zaitz, M. 566
Zanoni, Enrico 1, 522, 536
Zehnder, Ulrich 522
Zeng, X. 938
Zhang, B. C. 906
Zhang, F. 804, 906
Zhang, L. 581
Zhang, W. 938
Zhao, K. 50
Zhao, Larry 549
Zhao, Y. H. 906
Zheng, Rui 650
Zhou, C. 750
Zhou, J. P. 581
Zhou, Xing 977
Zhu, Guojun 977

AUTHOR INDEX

Zhu, Vivian 395
Zous, Nian-Kai 634
Zschech, Ehrenfried 574, 581

CURRAN ASSOCIATES INC.
proceedings
.com

9781424454303

2010 IEEE International Reliability Physics Symposium (IRPS 2010)

Garden Grove (Anaheim), California, USA
2-6 May 2010

IEEE Catalog Number: CFP10RPS-POD
ISBN: 978-1-42445-430-3

2010 IEEE International Reliability Physics Symposium

(IRPS 2010)

Anaheim, California, USA
2-6 May 2010

Pages 599-1125

IEEE Catalog Number: CFP10RPS-PRT
ISBN: 978-1-4244-5430-3

Copyright © 2010 by the Institute of Electrical and Electronic Engineers, Inc
All Rights Reserved

Copyright and Reprint Permissions: Abstracting is permitted with credit to the source. Libraries are permitted to photocopy beyond the limit of U.S. copyright law for private use of patrons those articles in this volume that carry a code at the bottom of the first page, provided the per-copy fee indicated in the code is paid through Copyright Clearance Center, 222 Rosewood Drive, Danvers, MA 01923.

For other copying, reprint or republication permission, write to IEEE Copyrights Manager, IEEE Service Center, 445 Hoes Lane, Piscataway, NJ 08854. All rights reserved.

***This publication is a representation of what appears in the IEEE Digital Libraries. Some format issues inherent in the e-media version may also appear in this print version.**

IEEE Catalog Number:	CFP10RPS-PRT
ISBN 13:	978-1-4244-5430-0
Library of Congress No.:	1541-7026
ISSN:	82-640313

Additional Copies of This Publication Are Available From:

Curran Associates, Inc
57 Morehouse Lane
Red Hook, NY 12571 USA
Phone: (845) 758-0400
Fax: (845) 758-2633
E-mail: curran@proceedings.com
Web: www.proceedings.com

TABLE OF CONTENTS

ESREF BEST PAPER

ESREF A Review on the Reliability of GaN-Based Laser Diodes.. 1
Nicola Trivellin, Matteo Meneghini, Enrico Zanoni, Kenji Orita, Masaaki Yuri, Gaudenzio Meneghesso

SESSION 2A: TRANSISTORS: BTI, HOT CARRIER

2A.1 The Statistical Analysis of Individual Defects Constituting NBTI and Its Implications for Modeling DC- and AC-Stress 7
Hans Reisinger, Tibor Grasser, Wolfgang Gustin, Christian Schlünder

2A.2 The Time Dependent Defect Spectroscopy (TDDS) for the Characterization of the Bias Temperature Instability 16
T. Grasser, H. Reisinger, P.-J. Wagner, F. Schanovsky, W. Goes, B. Kaczer

2A.3 Origin of NBTI Variability in Deeply Scaled pFETs 26
B. Kaczer, T. Grasser, J. Roussel, J. Franco, R. Degraeve, L.-A. Ragnarsson, E. Simoen, G. Groeseneken, H. Reisinger

2A.4 Two Independent Components Modeling for Negative Bias Temperature Instability 33
Vincent Huard

2A.5 Recovery-Free Electron Spin Resonance Observations of NBTI Degradation 43
J. T. Ryan, P. M. Lenahan, T. Grasser, H. Enichlmair

SESSION 2B: TRANSISTORS BTI, HOT CARRIERS

2B.1 PBTI Relaxation Dynamics After AC Vs. DC Stress in High-k/Metal Gate Stacks 50
K. Zhao, J. H. Stathis, A. Kerber, E. Cartier

2B.2 Off State Incorporation into the 3 Energy Mode Device Lifetime Modeling for Advanced 40nm CMOS Node 55
A. Bravaix, C. Guérin, D. Goguenheim, V. Huard, D. Roy, C. Besset, S. Renard, Y. Mamy Randriamihaja, E. Vincent

2B.3 Mobility Enhancement Due to Charge Trapping & Defect Generation: Physics of Self-Compensated BTI 65
Ahmad Ehteshamul Islam, Muhammad Ashraful Alam

2B.4 Understanding Noise Measurements in MOSFETs: The Role of Traps Structural Relaxation 73
D. Veksler, G. Bersuker, S. Rumyantsev, M. Shur, H. Park, C. Young, K. Y. Lim, W. Taylor, R. Jammy

SESSION 2C: NANOELECTRONICS

2C.1 Reliability Study of Bilayer Graphene - Material for Future Transistor and Interconnect 80
Tianhua Yu, Eun-Kyu Lee, Benjamin Briggs, Bhaskar Nagabhirava, Bin Yu

2C.2 Failure Analysis of Resistive Switching Devices 84
An Chen

2C.3 Charging and Discharging Characteristics of Metal Nanocrystals in Degraded Dielectric Stacks.................... 89
Z. Z. Lwin, K. L. Pey, Y. N. Chen, P. K. Singh, S. Mahapatra

2C.4 Characterization of Gate-All-Around Si-NWFET, Including R_{sd}, Cylindrical Coordinate Based 1/f Noise and Hot Carrier Effects 94
Rock-Hyun Baek, Hyun-Sik Choi, Hyun Chul Sagong, Sang-Hyun Lee, Gil-Bok Choi, Seung Hyun Song, Chan-Hoon Park, Jeong-Soo Lee, Yoon-Ha Jeong, Chang-Ki Baek, Dae Mann Kim, Yun Young Yeoh, Kyoung Hwan Yeo, Dong-Won Kim, Kinam Kim

2C.5 Thermal Disturbance and Its Impact on Reliability of Phase-Change Memory Studied by the Micro-Thermal Stage 99
SangBum Kim, Byoungil Lee, Mehdi Asheghi, G. A. M. Hurkx, John Reifenberg, Kenneth Goodson, H.-S. Philip Wong

SESSION 2D: RELIABILITY IN MANUFACTURING

2D.1 Mature Processability and Manufacturability by Characterizing V_T and V_{MIN} Behaviors Induced by NBTI and AHTOL Test 104
Jongwoo Park, Sungmok Ha, Sunme Lim, Jae-Yoon Yoo, Junkyun Park, Kidan Bae, Gunrae Kim, Min Kim, Yongshik Kim

2D.2 Validating Foundry Technologies for Extended Mission Profiles 111
K. van Dijk, P. A. J. Volf, C. Detcheverry, A. Yau, P. Ngan, Z. Liang, F. G. Kuper

2D.3 Non-Destructive Current-Ramp Dielectric Breakdown (IRDB) for Fast BEOL Reliability Monitoring 117
Kok-Yong Yiang, Rick Francis, Amit Marathe, Oliver Aubel

2D.4 Modeling of Cu IMD-TDDB Caused by Extrinsic Defects 120
T. Ouchi, K. Makabe, M. Ogasawara, E. Murakami, N. Yoshioka

2D.5 Product Failures: Power-Law or Exponential Voltage Dependence? 125
Amr Haggag, Keith Forbes, Gary Anderson, Dave Burnett, Peter Abramowitz, Mohamed Moosa

SESSION 2E: COMPOUND SEMICONDUCTORS

2E.1 Reliability Status of GaN Transistors and MMICs in Europe 129
M. Dammann, M. Cäsar, H. Konstanzer, P. Waltereit, R. Quay, W. Bronner, R. Kiefer, S. Müller, M. Mikulla, P. J. van der Wel, T. Rödle, F. Bourgeois, K. Riepe

2E.2 Effect of Trapping on the Critical Voltage for Degradation in GaN High Electron Mobility Transistors 134
Sefa Demirtas, Jesús A. del Alamo

2E.3 Reliability Assessment in Different HTO Test Conditions of AlGaN/GaN HEMTs 139
N. Malbert, N. Labat, A. Curutchet, C. Sury, V. Hoel, J.-C. de Jaeger, N. Defrance, Y. Douvry, C. Dua, M. Oualli, M. Piazza, C. Bru-Chevallier, J.-M. Bluet, W. Chikhaoui

2E.4 High Temperature On- and Off-State Stress of GaN-On-Si HEMTs with In-Situ Si_3N_4 Cap Layer 146
Denis Marcon, Farid Medjdoub, Domenica Visalli, Marleen Van Hove, Joff Derluyn, Jo Das, Stefan Degroote, Maarten Leys, Kai Cheng, Stefaan Decoutere, Robert Mertens, Marianne Germain, Gustaaf Borghs

2E.5 Identification of Electronic Traps in AlGaN/GaN HEMTs Using UV Light-Assisted Trapping Analysis 152
M. Tapajna, R. J. T. Simms, M. Faqir, M. Kuball, Y. Pei, U. K. Mishra

SESSION 2F: HIGH VOLTAGE DEVICES

2F.1 Reliability of SiC Power Devices and Its Influence on Their Commercialization – Review, Status, and Remaining Issues 156
Michael Treu, Roland Rupp, Gerald Sölkner

2F.2 Effects of Negative Differential Resistance in High Power Devices and Some Relations to DMOS Structures 162
Roman Baburske, Josef Lutz, Birk Heinze

2F.3 Investigation of Monotonous Increase in Saturation-Region Drain Current During Hot Carrier Stress in N-Type Lateral Diffused MOSFET with STI 170
Yu-Hui Huang, J. R. Shih, Y. H. Lee, Sunnys Hsieh, C. C. Liu, Kenneth Wu, H. L. Chou

2F.4 Modeling the Lifetime of a Lateral DMOS Transistor in Repetitive Clamping Mode 175
E. Riedlberger, R. Keller, H. Reisinger, W. Gustin, A. Spitzer, M. Stecher, C. Jungemann

2F.5 Low-Side Driver's Failure Mechanism in a Class-D Amplifier Under Short Circuit Test and a Robust Driver Device 182
Jian-Hsing Lee, J. R. Shih, Tong-Chern Ong, Kenneth Wu

SESSION 3A: SOFT ERRORS

3A.1 On the Radiation-Induced Soft Error Performance of Hardened Sequential Elements in Advanced Bulk CMOS Technologies 188
N. Seifert, V. Ambrose, B. Gill, Q. Shi, R. Allmon, C. Recchia, S. Mukherjee, N. Nassif, J. Krause, J. Pickholtz, A. Balasubramanian

3A.2 Effect of Multiple-Transistor Charge Collection on SET Pulse Widths...198
    J. R. Ahlbin, M. J. Gadlage, N. M. Atkinson, B. L. Bhuva, A. F. Witulski, W. T. Holman, L. W. Massengill, P. H. Eaton, B. Narasimham

3A.3 LEAP: Layout Design through Error-Aware Transistor Positioning for Soft-Error Resilient Sequential Cell Design...203
    Hsiao-Heng Kelin Lee, Klas Lilja, Mounaim Bounasser, Prasanthi Relangi, Ivan R. Linscott, Umran S. Inan, Subhasish Mitra

3A.4 Alpha-Particle-Induced Soft Errors and Multiple Cell Upsets in 65-nm 10T Subthreshold SRAM...213
    Hiroshi Fuketa, Masanori Hashimoto, Yukio Mitsuyama, Takao Onoye

3A.5 SEILA: Soft Error Immune Latch for Mitigating Multi-Node-SEU and Local-Clock-SET...218
    Taiki Uemura, Yoshiharu Tosaka, Hideya Matsuyama, Ken Shono, Chihiro J. Uchibori, Keiji Takahisa, Mitsuhiro Fukuda, Kichiji Hatanaka

SESSION 3B: MICRO-ELECTRONIC SYSTEMS AND MEMS

3B.1 Evolving MEMS Qualification Requirements...224
    Andrew Olney

3B.2 EFM Study of Injected Charges in the Silicon Nitride of an Electrostatic Actuated MEMS...231
    Antoine Nowodzinski, Didier Bloch, Adam Koszewski, Thibaut Toussaint

3B.3 A Novel Low Cost Failure Analysis Technique for Dielectric Charging Phenomenon in Electrostatically Actuated MEMS Devices...237
    U. Zaghloul, F. Coccetti, G. J. Papaioannou, P. Pons, R. Plana

3B.4 Accelerated Testing of RF-MEMS Contact Degradation through Radiation Sources...246
    A. Tazzoli, M. Barbato, V. Giliberto, G. Monaco, S. Gerardin, P. Nicolosi, A. Paccagnella, G. Meneghesso

SESSION 3C: FAILURE ANALYSIS

3C.1 X-Ray Computed Tomography for Non-Destructive Failure Analysis in Microelectronics...252
    Mario Pacheco, Deepak Goyal

3C.2 Impact on Device Performance and Monitoring of a Low Dose of Tungsten Contamination by Dark Current Spectroscopy...259
    F. Domengie, J. L. Regolini, D. Bauza, P. Morin

3C.3 Electromigration of NiSi Poly Gated Electrical Fuse and Its Resistance Behaviors Induced by High Temperature...265
    Han-Byul Kang, Jongwoo Park, Gun-Rae Kim, Hyun-Woo Park, Woon-Hak Lee, Joo-Byoung Yoon

3C.4 Determination of the Local Electric Field Strength by Energy Dispersive Photon Emission Microscopy...271
    T. Geinzer, R. Heiderhoff, J. C. H. Phang, L. J. Balk

3C.5 Explosion Phenomenon of High Resistance Via During TEM Sample Preparation Using FIB...277
    Pan Liu, Irene Tee, Soo Sien Seah, Chi Wen Soo, Ye Chen, Zhi Qiang Mo

SESSION 3D: PROCESS AND INTEGRATION RELIABILITY

3D.1 Gate Oxide Effect on Wafer Level Reliability of Next Generation DRAM Transistors...282
    Yu Gyun Shin, Kab-Jin Nam, Heedon Hwang, Jeong Hee Han, Sangjin Hyun, Siyoung Choi, Joo-Tae Moon

3D.2 Reliability Characterization of 32nm High-K and Metal-Gate Logic Transistor Technology...287
    Sangwoo Pae, Ashwin Ashok, Jingyoo Choi, Tahir Ghani, Jun He, Seok-hee Lee, Karen Lemay, Mark Liu, Ryan Lu, Paul Packan, Chris Parker, Richard Purser, Anthony St. Amour, Bruce Woolery

3D.3 Reliability Studies on a 45NM Low Power System-on-Chip (SoC) Dual Gate Oxide High-K / Metal Gate (DG HK+MG) Technology...293
    C. Prasad, P. Bai, S. Gannavaram, W. Hafez, J. Hicks, C.-H. Jan, J. Lin, M. Jones, K. Komeyli, R. Kotlyar, K. Mistry, I. Post, C. Tsai

3D.4 Re-Consideration of Influence of Silicon Wafer Surface Orientation on Gate Oxide Reliability from TDDB Statistics Point of View...299
    Yuichiro Mitani, Akira Toriumi

SESSION 3E: PHOTOVOLTAIC DEVICES

3E.1 Photovoltaic Module Reliability Studies at the Florida Solar Energy Center.................306
Neelkanth G. Dhere, Shirish A. Pethe, Ashwani Kaul

3E.2 Intrinsic Reliability of Amorphous Silicon Thin Film Solar Cells 312
M. A. Alam, S. Dongaonkar, Y. Karthik, S. Mahapatra, D. Wang, M. Frei

3E.3 Correlations of Capacitance-Voltage Hysteresis with Thin-Film CdTe Solar Cell Performance During Accelerated Lifetime Testing 318
David S. Albin, Joseph A. del Cueto

3E.4 Solar Cell Interface Stability Probed by Charge Extraction 323
R. L. Graham, C. E. France, S. A. Carter, G. B. Alers

SESSION 3F: THIN FILM DEVICES

3F.1 Reliability Aspects of Organic Light Emitting Diodes 327
Thomas Riedl, Thomas Winkler, Hans Schmidt, Jens Meyer, Daniel Schneidenbach, Hans-Hermann Johannes, Wolfgang Kowalsky, Thomas Weimann, Peter Hinze

3F.2 Light, Bias, and Temperature Effects on Organic TFTs....................................334
N. Wrachien, A. Cester, N. Bellaio, A. Pinato, M. Meneghini, A. Tazzoli, G. Meneghesso, K. Myny, S. Smout, J. Genoe

3F.3 Reliability of (100) and (110) Oriented Single-Grain Si TFTs without Seed Substrate 342
Tao Chen, Ryoichi Ishihara, C. I. M. Beenakker

3F.4 Non-Uniform Threshold Voltage and Non-Saturating Drain Current in Amorphous-Si TFT After Saturation-Mode Bias Temperature Stress 347
C. R. Wie, Z. Tang

SESSION 4A: DEVICE DIELECTRIC BREAKDOWN

4A.1 New Insight into the TDDB and Post Breakdown Reliability of Novel High-k Gate Dielectric Stacks...................354
K. L. Pey, N. Raghavan, X. Li, W. H. Liu, K. Shubhakar, X. Wu, M. Bosman

4A.2 High-k Gate Stack Breakdown Statistics Modeled by Correlated Interfacial Layer and High-k Breakdown Path 364
G. Ribes, P. Mora, F. Monsieur, M. Rafik, F. Guarin, G. Yang, D. Roy, W. L. Chang, J. Stathis

4A.3 Impact of Charge Trapping on the Voltage Acceleration of TDDB in Metal Gate/High-k n-Channel MOSFETs 369
A. Kerber, A. Vayshenker, D. Lipp, T. Nigam, E. Cartier

4A.4 Mechanism of High-k Dielectric-Induced Breakdown of the Interfacial SiO_2 Layer...................373
G. Bersuker, D. Heh, C. D. Young, L. Morassi, A. Padovani, L. Larcher, K. S. Yew, Y. C. Ong, D. S. Ang, K. L. Pey, W. Taylor

4A.5 Characterization of Millisecond-Anneal-Induced Defects in SiON and SiON/Si Interface by Gate Current Fluctuation Measurement 379
Tsunehisa Sakoda, Keita Nishigaya, Tomohiro Kubo, Mitsuaki Hori, Hiroshi Minakata, Yuko Kobayashi, Hiroko Mori, Katsuji Ono, Katsuto Tanahashi, Naoyoshi Tamura, Toshifumi Mori, Yoshiharu Tosaka, Hideya Matsuyama, Chioko Kaneta, Koichi Hashimoto, Masataka Kase, Yasuo Nara

4A.6 Gate Dielectric Reliability in the Sub Threshold Regime 385
Paul E. Nicollian, Cathy A. Chancellor, Anand T. Krishnan

SESSION 4B: SOFT ERRORS

4B.1 Soft Error Assessments for Servers 391
K. Paul Muller, Pia N. Sanda

4B.2 Contribution of Low-Energy (< 10 MeV) Neutrons to Upset Rate in a 65 nm SRAM 395
Brian D. Sierawski, Kevin M. Warren, Robert A. Reed, Robert A. Weller, Marcus M. Mendenhall, Ronald D. Schrimpf, Robert C. Baumann, Vivian Zhu

4B.3 Scaling Trends of Neutron Effects in MLC NAND Flash Memories 400
S. Gerardin, M. Bagatin, A. Paccagnella, G. Cellere, A. Visconti, S. Beltrami, C. Andreani, G. Gorini, C. D. Frost

4B.4 SEE Test and Modeling Results on 45nm SRAMs with Different Well Strategies 407
Gilles Gasiot, Slawosz Uznanski, Philippe Roche

4B.5 Fidelity of Energy Spectra at Neutron Facilities for Single-Event Effects Testing.............411
S. P. Platt, A. V. Prokofiev, Xiao Xiao Cai

SESSION 4C: PROCESS AND INTEGRATION RELIABILITY

4C.1 Mobile and Stable Hydrogen Species in the Interface Layer Between Poly Silicon and Gate Oxynitride 417
Ziyuan Liu, Shuu Ito, Shoichi Hiroshima, Shin Koyama, Mariko Makabe, Markus Wilde, Katsuyuki Fukutani

4C.2 Novel TDDB Mechanism for p-FET Accelerated by Hydrogen from HfSiON Film 424
Izumi Hirano, Koichi Kato, Yasushi Nakasaki, Shigeto Fukatsu, Yuichiro Mitani, Masakazu Goto, Seiji Inumiya, Katsuyuki Sekine, Motoyuki Sato

SESSION 4C: ASSEMBLY AND PACKAGING

4C.3 Buckling, Wrinkling and Debonding in Thin Film Systems..............430
S. Goyal, K. Srinivasan, G. Subbarayan, T. Siegmund

4C.4 Reliability of Microelectronics Packaging in the Era of EnergyWise and Borderless Networks 440
Li Li, Jie Xue

4C.5 Comparison of Two Calibration Methods for a Package Stress Measurement Testchip..............446
Christian Djelassi, Helmut Köck, Michael Glavanovics

4C.6 Reliability Evaluation Methodology for Chip-Package Interaction - Die Corner Edge Failure Mechanism 453
Melida Chin, Amit Marathe

SESSION 4D: ESD AND LATCHUP

4D.1 A Failure Levels Study of Non-Snapback ESD Devices for Automotive Applications..............458
Yiqun Cao, Ulrich Glaser, Stephan Frei, Matthias Stecher

4D.2 Understanding Transient Latchup Hazards and the Impact of Guard Rings 466
Farzan Farbiz, Elyse Rosenbaum

4D.3 A Novel TCAD-Based Methodology to Minimize the Impact of Parasitic Structures on ESD Performance..............474
Nicholas Olson, Gianluca Boselli, Akram Salman, Elyse Rosenbaum

4D.4 On the Differences Between 3D Filamentation and Failure of N & P Type Drain Extended MOS Devices Under ESD Condition 480
Mayank Shrivastava, S. Bychikhin, D. Pogany, Jens Schneider, M. Shojaei Baghini, Harald Gossner, Erich Gornik, V. Ramgopal Rao

4D.5 Predictive Simulation of CDM Events to Study Effects of Package, Substrate Resistivity and Placement of ESD Protection Circuits on Reliability of Integrated Circuits..............485
Vrashank Shukla, Nathan Jack, Elyse Rosenbaum

SESSION 4E: PHOTOVOLTAIC DEVICES

4E.1 Thin-Film Photovoltaics: What are the Reliability Issues and Where Do They Occur?..............494
James R. Sites

4E.3 Thermoreflectance Imaging of Defects in Thin-Film Solar Cells..............499
D. Kendig, G. B. Alers, A. Shakouri

4E.4 SEAM and EBIC Studies of Morphological and Electrical Defects in Polycrystalline Silicon Solar Cells..............503
L. Meng, D. Nagalingam, C. S. Bhatia, A. G. Street, J. C. H. Phang

4E.5 Photovoltaic (PV) Cells Characterization Using Advanced Optical Tools 508
Franco Stellari, Steven E. Steen, Kathryn C. Fisher, Xiaoyan Shao

SESSION 4F: COMPOUND SEMICONDUCTORS

4F.1 Reliability Assessment of State-of-the-Art GaN HEMT by Means of Cellular Monte Carlo Simulation 516
Fabio Alessio Marino, Diego Guerra, Stephen M. Goodnick, Marco Saraniti

4F.2 A Study of the Failure of GaN-Based LEDs Submitted to Reverse-Bias Stress and ESD Events 522

Matteo Meneghini, Augusto Tazzoli, Enrico Ranzato, Nicola Trivellin, Gaudenzio Meneghesso, Enrico Zanoni, Maura Pavesi, Manfredo Manfredi, Rainer Butendeich, Ulrich Zehnder, Berthold Hahn

4F.3 Temperature Assessment of AlGaN/GaN HEMTs: A Comparative Study by Raman, Electrical and IR Thermography 528

N. Killat, M. Kuball, T.-M. Chou, U. Chowdhury, J. Jimenez

4F.4 Analysis of Interface-Trap Effects in Inversion-Type InGaAs/ZrO$_2$ MOSFETs 532

L. Morassi, G. Verzellesi, A. Padovani, L. Larcher, P. Pavan, D. Veksler, Injo Ok, G. Bersuker

4F.5 Degradation of III-V Inversion-Type Enhancement-Mode MOSFETs 536

N. Wrachien, A. Cester, E. Zanoni, G. Meneghesso, Y. Q. Wu, P. D. Ye

SESSION 5A: INTERCONNECT AND BEOL DIELECTRICS

5A.1 E- and Square Root E-Model Too Conservative to Describe Low Field Time Dependent Dielectric Breakdown 543

K. Croes, Zs. Tökei

5A.2 Study of Leakage Mechanism and Trap Density in Porous Low-k Materials 549

Gianni Giai Gischia, Kristof Croes, Guido Groeseneken, Zsolt Tökei, Valery Afanas'ev, Larry Zhao

5A.3 New Voltage Ramp Dielectric Breakdown Methodology Based on Square Root E Model for Cu/Low-k Interconnect Reliability 556

Mingte Lin, James W. Liang, K. C. Su

5A.4 A Simple Electrical Method for Etch Bias and Process Reliability Determination 562

Kok-Yong Yiang, Melida Chin, Amit Marathe, Oliver Aubel

5A.5 Comprehensive Investigations of CoWP Metal-Cap Impacts on Low-k TDDB for 32nm Technology Application 566

F. Chen, M. Shinosky, B. Li, C. Christiansen, T. Lee, J. Aitken, D. Badami, E. Huang, G. Bonilla, T.-M. Ko, T. Kane, Y. Wang, M. Zaitz, L. Nicholson, M. Angyal, C. Truong, X. Chen, G. Yang, S. B. Law, T. J. Tang, S. Petitdidier, G. Ribes, M. Oh, C. Child, H. Sawada, A. Kolics, O. Rigoutat, N. Gilbert

SESSION 5B: INTERCONNECT AND BEOL DIELECTRICS

5B.1 EM and SM Induced Degradation Dynamics in Copper Interconnects Studied Using Electron Microscopy and X-Ray Microscopy 574

Ehrenfried Zschech, René Hübner, Oliver Aubel, Paul S. Ho

5B.2 Effects of Cap Layer and Grain Structure on Electromigration Reliability of Cu/Low-k Interconnects for 45 nm Technology Node 581

L. Zhang, J. P. Zhou, J. Im, P. S. Ho, O. Aubel, C. Hennesthal, E. Zschech

5B.3 Study of Stress Migration and Electromigration Interaction in Copper/Low-k Interconnects 586

A. Heryanto, K. L. Pey, Y. K. Lim, W. Liu, J. Wei, N. Raghavan, J. B. Tan, D. K. Sohn

5B.4 Electromigration and Stress-Induced-Voiding in Dual Damascene Cu/Low-k Interconnects: A Complex Balance Between Vacancy and Stress Gradients 591

K. Croes, C. J. Wilson, M. Lofrano, B. Vereecke, G. P. Beyer, Zs. Tokei

SESSION 5C: MEMORY

5C.1 The New Scaling Limitation of the Floating Gate Cell in NAND Flash Memory 599

Yong Seok Kim, Dong Jun Lee, Chi Kyoung Lee, Hyun Ki Choi, Seong Soo Kim, Jai Hyuk Song, Du Heon Song, Jeong-Hyuk Choi, Kang-Deog Suh, Chilhee Chung

5C.2 Investigation of the Threshold Voltage Instability After Distributed Cycling in Nanoscale NAND Flash Memory Arrays 604

Christian Monzio Compagnoni, Carmine Miccoli, Riccardo Mottadelli, Silvia Beltrami, Michele Ghidotti, Andrea L. Lacaita, Alessandro S. Spinelli, Angelo Visconti

5C.3 Optimal Cell Design for Enhancing Reliability Characteristics for Sub 30 nm NAND Flash Memory 611

Eun Suk Cho, Hyun Jung Kim, Byoung Taek Kim, Jai Hyuk Song, Du Heon Song, Jeong-Hyuk Choi, Kang-Deog Suh, Chilhee Chung

5C.4 Impact of the Current Density Increase on Reliability in Scaled BJT-Selected PCM for High-Density Applications 615

An. Redaelli, A. Pirovano, I. Tortorelli, F. Ottogalli, A. Ghetti, L. Laurin, A. Benvenuti

SESSION 5D: MEMORY

5D.1 Trade-Off Between Data Retention and Reset in NiO RRAMs 620
D. Ielmini, F. Nardi, C. Cagli, A. L. Lacaita

5D.2 Chip-Level Reliability Study of Barrier Engineered (BE) Floating Gate (FG) Flash Memory Devices 627
Hang-Ting Lue, JiFong Pan, C. S. Chang, Szu-Yu Wang, Y. F. Chang, Y. C. Lee, M. H. Liaw, Y. J. Chen, K. F. Chen, Chester Lo, I. J. Huang, T. T. Han, M. S. Chen, W. P. Lu, T. Yang, K. C. Chen, Kuang-Yeu Hsieh, Chih-Yuan Lu

5D.3 Source/Drain Dopant Concentration Induced Reliability Issues in Charge Trapping NAND Flash Cells 634
Yin-Jen Chen, Lit Ho Chong, Shang-Wei Lin, Teng-Hao Yeh, Kuan-Fu Chen, Jyun-Siang Huang, Cheng-Hsien Cheng, Shaw-Hung Ku, Nian-Kai Zous, I-Jen Huang, Tzung-Ting Han, Tzu-Hsuan Hsu, Hang-Ting Lue, Ming-Shiang Chen, Wen-Pin Lu, Kuang-Chao Chen, Chih-Yuan Lu

5D.4 Performance and Reliability Optimizations of BE-SONOS NAND Flash Using SiON Bandgap-Tuning Tunneling Barrier 639
Jeng-Hwa Liao, Jung-Yu Hsieh, Hang-Ting Lue, Ling-Wu Yang, Tahone Yang, Kuang-Chao Chen, Chih Yuan Lu

SESSION 5E: CIRCUIT RELIABILITY

5E.1 Flexible Electronics: What Can It Do? What Should It Do? 644
Sameer M. Venugopal, David R. Allee, Manuel Quevedo-Lopez, Bruce Gnade, Eric Forsythe, David Morton

5E.2 On the Bias Dependence of Time Exponent in NBTI and CHC Effects 650
Jyothi B. Velamala, Vijay Reddy, Rui Zheng, Srikanth Krishnan, Yu Cao

5E.3 Managing SRAM Reliability from Bitcell to Library Level 655
Vincent Huard, Remy Chevallier, Chittoor Parthasarathy, Anand Mishra, Natalia Ruiz-Amador, Flore Persin, Vincent Robert, Alejandro Chimeno, Emmanuel Pion, Nicolas Planes, David Ney, Florian Cacho, Neeraj Kaapor, Vishal Kulshrestra, Sanjeev Chopra, Nicolas Vialle

5E.4 Prediction of NBTI Degradation for Circuit Under AC Operation 665
Y. S. Tsai, N. K. Jha, Y.-H. Lee, R. Ranjan, Wayne Wang, J. R. Shih, M. J. Chen, J. H. Lee, K. Wu

5E.5 An Extensive and Improved Circuit Simulation Methodology for NBTI Recovery 670
Haldun Kufluoglu, V. Reddy, A. Marshall, J. Krick, T. Ragheb, C. Cirba, A. Krishnan, C. Chancellor

SESSION 5F: CIRCUIT RELIABILITY

5F.1 Adaptive Sensing and Design for Reliability 676
P. Singh, D. Sylvester, D. Blaauw

5F.2 Statistical Evaluation of Dynamic Junction Leakage Current Fluctuation Using a Simple Arrayed Capacitors Circuit 683
Kenichi Abe, Takafumi Fujisawa, Hiroyoshi Suzuki, Shunichi Watabe, Rihito Kuroda, Shigetoshi Sugawa, Akinobu Teramoto, Tadahiro Ohmi

5F.3 Scaling Reliability and Modeling of Ferroelectric Capacitors 689
Antonio G. Acosta, John Rodriguez, Borna Obradovic, Scott Summerfelt, Tamer San, Keith Green, Ted Moise, Srikanth Krishnan

5F.4 Measurement of Neutron-Induced Single Event Transient Pulse Width Narrower Than 100ps 694
Hideyuki Nakamura, Katsuhiko Tanaka, Taiki Uemura, Kan Takeuchi, Toshikazu Fukuda, Shigetaka Kumashiro

SESSION 6A: INTERCONNECT AND BEOL DIELECTRICS

6A.1 New Electromigration Validation: Via Node Vector Method 698
Young-Joon Park, Palkesh Jain, Srikanth Krishnan

6A.2 Electromigration Mechanisms in Cu Nano-Wires 705
M. H. Lin, S. C. Lee, A. S. Oates

6A.3 Degradation and Failure Analysis of Copper and Tungsten Contacts Under High Fluence Stress 712
Thomas Kauerauf, Geni Butera, Kristof Croes, Steven Demuynck, Christopher J. Wilson, Philippe Roussel, Chris Drijbooms, Hugo Bender, Melina Lofrano, Bart Vandevelde, Zsolt Tokei, Guido Groeseneken

6A.4 Effective Thermal Characteristics to Suppress Joule Heating Impacts on Electromigration in Cu/Low-k Interconnects 717
S. Yokogawa, H. Tsuchiya, Y. Kakuhara

6A.5 Assessing the Degradation Mechanisms and Current Limitation Design Rules of SiCr-Based Thin-Film Resistors in Integrated Circuits..724
Yuan Li, David Donnet, Andrzej Grzegorczyk, Jan Cavelaars, Fred Kuper

SESSION 6C: MEMORY

6C.1 Role of Holes and Electrons During Erase of TANOS Memories: Evidences for Dipole Formation and Its Impact on Reliability 731
Luca Vandelli, Andrea Padovani, Luca Larcher, Antonio Arreghini, Geert Van den bosch, Malgorzata Jurczak, Jan Van Houdt, Vincenzo Della Marca, Paolo Pavan

6C.2 Reset Current Distributions in Phase Change Memories..738
A. Calderoni, M. Ferro, D. Ventrice, D. Ielmini, P. Fantini

6C.3 Random Telegraph Signal Noise in Phase Change Memory Devices 743
Davide Fugazza, Daniele Ielmini, Simone Lavizzari, Andrea L. Lacaita

6C.4 Reliability of Ferroelectric Random Access Memory Embedded within 130nm CMOS 750
J. Rodriguez, K. Remack, J. Gertas, L. Wang, C. Zhou, K. Boku, J. Rodriguez-Latorre, K. R. Udayakumar, S. Summerfelt, T. Moise, D. Kim, J. Groat, J. Eliason, M. Depner, F. Chu

SESSION 6E: EXTREME ENVIRONMENTS

6E.1 Electronic Failures in Spacecraft Environments..759
Douglas J. Sheldon

6E.2 Single Event Transient Pulse Width Measurements in a 65-nm Bulk CMOS Technology at Elevated Temperatures 763
M. J. Gadlage, J. R. Ahlbin, B. L. Bhuva, L. W. Massengill, R. D. Schrimpf

6E.3 Practicality of Evaluating Soft Errors in Commercial Sub-90 nm CMOS for Space Applications..................768
Jonathan A. Pellish, Kenneth A. LaBel

POSTER PRESENTATIONS

BD - DEVICE DIELECTRIC BREAKDOWN POSTERS

BD.1 Analysis of the Breakdown Spots Spatial Distribution in Large Area MOS Structures 775
E. Miranda, E. O'Connor, P. K. Hurley

BD.2 New Statistical Model to Decode the Reliability and Weibull Slope of High-k and Interfacial Layer in a Dual Layer Dielectric Stack..778
N. Raghavan, K. L. Pey, W. H. Liu, X. Li

BD.3 Role of Interface Layer in Stress-Induced Leakage Current in High-k/Metal-Gate Dielectric Stacks..787
W. L. Chang, J. H. Stathis, E. Cartier

BD.4 A Compact Analytic Model for the Breakdown Distribution of Gate Stack Dielectrics 792
Santi Tous, Ernest Y. Wu, Jordi Suñe

BD.5 Time Dependent Dielectric Breakdown and Stress Induced Leakage Current Characteristics of 8Å EOT HfO$_2$ N-MOSFETs..799
Robert O'Connor, Greg Hughes, Thomas Kaureauf, Lars-Åke Ragnarsson

BD.6 Frequency-Dependent Charge-Pumping: The Depth Question Revisited 804
F. Zhang, K. P. Cheung, J. P. Campbell, J. Suehle

CD - COMPOUND SEMICONDUCTORS POSTERS

CD.1 Progressive Schottky Junction Reaction Induced Degradation in Pt-Sunken Gate InP HEMT MMICs for High Reliability Applications 807
Y. C. Chou, D. L. Leung, M. Biedenbender, D. Buttari, D. C. Eng, R. S. Tsai, C. H. Lin, L. S. Lee, X. B. Mei, M. Wojtowicz, M. E. Barsky, R. Lai, A. K. Oki, T. R. Block

CD.2 Electrical Stress Induced Degradation in InAs – AlSb HEMTs..813
S. DasGupta, R. A. Reed, R. D. Schrimpf, D. M. Fleetwood, X. Shen, S. T. Pantelides, J. Bergman, B. Brar

CD.3 InAlAs/InGaAs MHEMT Degradation During DC and Thermal Stressing 818
E. A. Douglas, K. H. Chen, C. Y. Chang, L. C. Leu, C. F. Lo, B. H. Chu, F. Ren, S. J. Pearton

CR - CIRCUIT RELIABILITY

CR.1 A Built-In Aging Detection and Compensation Technique for Improving Reliability of Nanoscale CMOS Designs 822
Hamed F. Dadgour, Kaustav Banerjee

CR.2 A Test Concept for Circuit Level Aging Demonstrated by a Differential Amplifier 826
Florian R. Chouard, Christoph Werner, Doris Schmitt-Landsiedel, Michael Fulde

CR.3 The Hot Carrier Degradation Rate Under AC Stress 830
Guido T. Sasse, Jaap Bisschop

EL - ESD AND LATCHUP POSTERS

EL.1 ESD Protection for High-Speed Receiver Circuits 835
Nathan Jack, Elyse Rosenbaum

EL.2 On the Failure Mechanism and Current Instabilities in RESURF Type DeNMOS Device Under ESD Conditions 841
Mayank Shrivastava, Jens Schneider, Maryam Shojaei Baghini, Harald Gossner, V. Ramgopal Rao

EL.3 Characterization of High-k/Metal Gate Stack Breakdown in the Time Scale of ESD Events 846
Yang Yang, James Di Sarro, Robert J. Gauthier, Kiran Chatty, Junjun Li, Rahul Mishra, Souvick Mitra, Dimitris E. Ioannou

EL.5 Robust High Current ESD Performance of Nano-Meter Scale DeNMOS by Source Ballasting 853
Amitabh Chatterjee, Forrest Brewer, Harald Gossner, Sameer Pendharkar, Charvaka Duvvury

EL.6 A Bending N-Well Ballast Layout to Improve ESD Robustness in Fully-Silicided CMOS Technology 857
Yong-Ru Wen, Ming-Dou Ker, Wen-Yi Chen

FA - FAILURE ANALYSIS POSTERS

FA.1 Isolating Marginally Defective Gate Using Photoperturbation Induced via a C-AFM Laser Beam 861
Hung Sung Lin, Mong Sheng Wu

FA.2 A Novel Sample Preparation Technique for Visualizing Invisible Defects Embedded in Poly Gate 865
Hung Sung Lin

FA.3 A Case Study of High Temperature Pass Analysis Using Thermal Laser Stimulation Technique 870
Hung Sung Lin, Mong Sheng Wu

FA.4 High Spatial and Temporal Resolution Thermal Imaging for LSI Circuits with Phase Microscopy 874
Tomonori Nakamura, Hidenao Iwai, Toyohiko Yamauchi, Hirotoshi Terada, Hitoshi Iida

FA.5 Reliability of Electronic Equipment Exposed to Chlorine Dioxide Used for Biological Decontamination 879
G. E. Derkits, M. L. Mandich, W. D. Reents, J. P. Franey, C. Xu, D. Fleming, R. Kopf, S. Ryan

HV - HIGH VOLTAGE DEVICES POSTERS

HV.1 Analysis of HCS in STI-Based LDMOS Transistors 881
Susanna Reggiani, Stefano Poli, Elena Gnani, Antonio Gnudi, Giorgio Baccarani, Marie Denison, Sameer Pendharkar, Rick Wise, Sridhar Seetharaman

IC - INTERCONNECT AND BEOL DIELECTRICS POSTERS

IC.1 Lifetime Extrapolation for Electromigration Tests at Wafer Level with a Dedicated Device 887
C. Chappaz, P. Nakkala

IC.2 Ultra-Low-k Dielectric Degradation Before Breakdown 890
T. Breuer, U. Kerst, C. Boit, E. Langer, H. Ruelke

IC.3 Analysis of the Impact of Linewidth Variation on Low-K Dielectric Breakdown 895
Muhammad M. Bashir, Linda Milor

IC.4 Effect of Pre-Existing Void in Sub-30nm Cu Interconnect Reliability 903
Zungsun Choi, Matsuda Tsukasa, Jong Myeong Lee, Gil-Heyun Choi, Siyoung Choi, Joo-Tae Moon

IC.5 Study of Upstream Electromigration Bimodality and Its Improvement in Cu Low-k Interconnects 906
W. Liu, Y. K. Lim, F. Zhang, H. Liu, Y. H. Zhao, A. Y. Du, B. C. Zhang, J. B. Tan, D. K. Sohn, L. C. Hsia

IC.6 Modeling of Stress Evolution of Electroplated Cu Films During Self-Annealing 911
Rui Huang, Werner Robl, Thomas Detzel, Hajdin Ceric

IC.7 The TDDB Failure Mode and Its Engineering Study for 45nm and Beyond in Porous Low k Dielectrics Direct Polish Scheme 918
Chia-Lin Hsu, Kuan-Ting Lu, Wen-Chin Lin, Jeh-Chieh Lin, Chih-Hsien Chen, Teng-Chun Tsai, Climbing Huang, J. Y. Wu, Dung-Ching Perng

IC.8 Resistance Trace Modeling and Electromigration Immortality Criterion Based on Void Growth Saturation 922
P. Lamontagne, D. Ney, L. Doyen, E. Petitprez, Y. Wouters

IC.9 Practical Considerations of Process Corner Evaluation for Deep-Sub Micron Technology Nodes Using the Example of Its Impact on Electromigration 926
Oliver Aubel, Thomas Hoffmann

IC.10 Investigating the Electro-Thermal Origin of Breakdown in Low-K/Cu Dielectrics Under Short Duration over Stressed Pulsed Regime 932
Amitabh Chatterjee, Forrest Brewer, S. C. Lee, A. S. Oates

IC.11 Study of Electric Field-Based Lifetime Projection Method in IMD TDDB 938
W. Zhang, X. Zeng, W. Liu, Y. K. Lim, J. F. Liu, E. C. Chua

IC.12 On the Physical Interpretation of the Impact Damage Model in TDDB of Low-k Dielectrics 943
J. R. Lloyd

IC.13 Reliability and Performance Limiting Defects in Low-k Dielectrics for Use as Interlayer Dielectrics 947
B. C. Bittel, P. M. Lenahan, S. King

MY - MEMORY POSTERS

MY.1 A High-Endurance (>100K) BE-SONOS NAND Flash with a Robust Nitrided Tunnel Oxide/Si Interface 951
Szu-Yu Wang, Hang-Ting Lue, Tzu-Hsuan Hsu, Pei-Ying Du, Sheng-Chih Lai, Yi-Hsuan Hsiao, Shih-Ping Hong, Ming-Tsung Wu, Fang-Hao Hsu, Nan-Tzu Lian, Jung-Yu Hsieh, Ling-Wu Yang, Tahone Yang, Kuang-Chao Chen, Kuang-Yeu Hsieh, Chih-Yuan Lu

MY.2 Transition of Erase Mechanism for MONOS Memory Depending on SiN Composition and Its Impact on Cycling Degradation 956
Shosuke Fujii, Jun Fujiki, Naoki Yasuda, Ryota Fujitsuka, Katsuyuki Sekine

MY.3 Use of Random Telegraph Signal as Internal Probe to Study Program/Erase Charge Lateral Spread in a SONOS Flash Memory 960
Y. L. Chou, J. P. Chiu, H. C. Ma, Tahui Wang, Y. P. Chao, K. C. Chen, C. Y. Lu

MY.4 Bias Temperature Instability of Binary Oxide Based ReRAM 964
Z. Fang, H. Y. Yu, W. J. Liu, K. L. Pey, X. Li, L. Wu, Z. R. Wang, Patrick G. Q. Lo, B. Gao, J. F. Kang

MY.5 Reliability Constraints for TANOS Memories Due to Alumina Trapping and Leakage. 966
Salvatore M. Amoroso, Aurelio Mauri, Nadia Galbiati, Claudia Scozzari, Evelyne Mascellino, Elisa Camozzi, Armando Rangoni, Tecla Ghilardi, Alessandro Grossi, Paolo Tessariol, Christian Monzio Compagnoni, Alessandro Maconi, Andrea L. Lacaita, Alessandro S. Spinelli, Gabriella Ghidini

MY.6 Variability Effects on the V_T Distribution of Nanoscale NAND Flash Memories 970
Alessio Spessot, Alessandro Calderoni, Paolo Fantini, Alessandro S. Spinelli, Christian Monzio Compagnoni, Fabrizio Farina, Andrea L. Lacaita, Andrea Marmiroli

MY.7 NAND Flash Reliability Degradation Induced by HCI in Boosted Channel Potential 975
Milim Park, Sukkwang Park, Seokwon Cho, Dong-Kyu Lee, YeonJoo Jeong, Chonga Hong, Ho Seok Lee, Myoung Kwan Cho, Kun-Ok Ahn, Yohwan Koh

NA - NANOELECTRONICS POSTERS

NA.1 Interface-Trap Modeling for Silicon-Nanowire MOSFETs 977
Zuhui Chen, Xing Zhou, Guojun Zhu, Shihuan Lin

NA.2 Applicability of Dual Layer Metal Nanocrystal Flash Memory for NAND 2 or 3-Bit/Cell Operation: Understanding the Anomalous Breakdown and Optimization of P/E Conditions 981
Pawan Singh, C. Sandhya, Kshitij Auluck, Gaurav Bisht, M. Sivatheja, Ralf Hofmann, Gautam Mukhopadhyay, Souvik Mahapatra

PI - PROCESS AND INTEGRATION RELIABILITY POSTERS

PI.1 Pattern-Independent, Fine-Morphology Ni-Pt Silicide Formation by Partial Conversion with Low Metal-Consumption Ratio 988
Takuya Futase, Takeshi Kamino, Naoto Hashikawa, Yutaka Inaba, Tetsuo Fujiwara, Hirohiko Yamamoto, Hisanori Tanimoto

PI.2 Disconnection of NiSi Shared Contact and Its Correction Using NH$_3$ Soak Treatment in Ti/TiN Barrier Metallization 995
Takuya Futase, Kota Funayama, Naoto Hashikawa, Hiroshi Tobimatsu, Hirohiko Yamamoto, Hisanori Tanimoto

PI.3 Analysis of Statistical Variation in NBTI Degradation of HfO$_2$/SiO$_2$ FETs 1001
H. Yoshimoto, D. Hisamoto, Y. Shimamoto, R. Tsuchiya, I. Yanagi, T. Arigane, K. Torii, K. Funayama, T. Hashimoto, H. Makiyama, K. Horita, T. Iwamatsu, K. Shiga, M. Mizutani, M. Inoue, T. Kaneoka

RM - RELIABILITY IN MANUFACTURING POSTERS

RM.1 Method of Deciding Burn-In Stress Voltage in Conceptual Design Phase 1004
Jae Yong Seo, Noh Seok Park, Hyung-Jin Park, Hong Sik Park, Woo Sup Kim, Se Young Lim, Hyun Kim, Nam Hyun Cha, Ju Seong Kang, Byung Se So

RM.2 Device-Level Reliability Simulation for High Temperature Applications of a Modular CMOS Foundry Process 1006
Markus Ackermann

RM.3 Accurate Projection of V$_{ccmin}$ by Modeling "Dual Slope" in FinFET Based SRAM, and Impact of Long Term Reliability on End of Life V$_{ccmin}$ 1008
H. Park, S. C. Song, S. H. Woo, M. H. Abu-Rahma, L. Ge, M. G. Kang, B. M. Han, J. Wang, R. Choi, J. W. Yang, S. O. Jung, G. Yeap

SE - SOFT ERRORS POSTERS

SE.1 System-Level Analysis of Soft Error Rates and Mitigation Trade-Off Explorations 1014
Zhe Ma, Francky Catthoor, Frank Vermunt, Teun Hendriks

SE.2 Soft Errors from Neutron and Proton-Induced Multiple-Node Events 1019
Ethan H. Cannon

SE.3 Effects of Multi-Node Charge Collection in Flip-Flop Designs at Advanced Technology Nodes ... 1026
Vijay B. Sheshadri, Bharat L. Bhuva, Robert A. Reed, Robert A. Weller, Marcus H. Mendenhall, Ron D. Schrimpf, Kevin M. Warren, Brian D. Seirawski, Shi-Jie Wen, Ricky Wong

SE.4 Analysis of Soft Error Rates in Combinational and Sequential Logic and Implications of Hardening for Advanced Technologies 1031
N. N. Mahatme, I. Chatterjee, B. L. Bhuva, J. Ahlbin, L. W. Massengill, R. Shuler

SE.5 Thermal Neutron Soft Error Rate for SRAMs in the 90nm-45nm Technology Range 1036
ShiJie Wen, Richard Wong, Michael Romain, Nelson Tam

TF - THIN FILM DEVICES POSTERS

TF.1 Evaluation of Self-Heating and Hot Carrier Degradation of Poly-Si Thin-Film Transistors Using Charge Pumping Technique 1040
Xiaowei Lu, Mingxiang Wang, Kai Sun, Lei Lu

XT - TRANSISTORS: BTI, HOT CARRIER POSTERS

XT.1 PBTI Response to Interfacial Layer Thickness Variation in Hf-Based HKMG nFETs 1044
D. P. Ioannou, E. Cartier, Y. Wang, S. Mittl

XT.2 HCI and NBTI Including the Effect of Back-Biasing in Thin-BOX FD-SOI CMOSFETs 1049
T. Ishigaki, R. Tsuchiya, Y. Morita, H. Yoshimoto, N. Sugii, S. Kimura

XT.3 The Understanding of Strain-Induced Device Degradation in Advanced MOSFETs with Process-Induced Strain Technology of 65nm Node and Beyond 1053
M. H. Lin, E. R. Hsieh, Steve S. Chung, C. H. Tsai, P. W. Liu, Y. H. Lin, C. T. Tsai, G. H. Ma

XT.4 Effect of Strain on Negative Bias Temperature Instability of Germanium p-Channel Field-Effect Transistor with High-k Gate Dielectric 1055
Bin Liu, Phyllis Shi Ya Lim, Yee-Chia Yeo

XT.5 A Robust Ultrafast Switching Methodology for Device Parameter Characterization of Bias-Temperature Instability 1058
  Y. Z. Hu, D. S. Ang, Z. Q. Teo, G. A. Du

XT.6 Impact of Hydrogen on Recoverable and Permanent Damage Following Negative Bias Temperature Stress 1063
  T. Aichinger, S. Puchner, M. Nelhiebel, T. Grasser, H. Hutter

XT.7 A Multi-Probe Correlated Bulk Defect Characterization Scheme for Ultra-Thin High-k Dielectric 1069
  M. Masuduzzaman, A. E. Islam, M. A. Alam

XT.8 Dependence of the Negative Bias Temperature Instability on the Gate Oxide Thickness 1073
  Gregor Pobegen, Thomas Aichinger, Michael Nelhiebel, Tibor Grasser

XT.9 Interpretation of PBTI/TDDB Predicted Lifetime Based on Trap Characterization by TSCIS in V_{th}-Adjusted Transistors 1078
  S. Sahhaf, R. Degraeve, V. Srividya, M. Cho, T. Kauerauf, G. Groeseneken

XT.10 Improvements of NBTI Reliability in SiGe p-FETs 1082
  J. Franco, B. Kaczer, M. Cho, G. Eneman, G. Groeseneken, T. Grasser

XT.11 A Model for NBTI in Nitrided Oxide MOSFETs without Hydrogen or Diffusion 1086
  P. M. Lenahan

XT.12 A Generalized, I_B-Independent, Physical HCI Lifetime Projection Methodology Based on Universality of Hot-Carrier Degradation 1091
  Dhanoop Varghese, Muhammad Ashraful Alam, Bonnie Weir

XT.13 Positive and Negative Bias Temperature Instability on Sub-Nanometer EOT High-k MOSFETs 1095
  Moonju Cho, Marc Aoulaiche, Robin Degraeve, Ben Kaczer, Jacopo Franco, Thomas Kauerauf, Philippe Roussel, Lars Å. Ragnarsson, Joshua Tseng, Thomas Y. Hoffmann, Guido Groeseneken

XT.14 Hot-Carrier Degradation in Undoped-Body ETSOI FETs and SOI FINFETs 1099
  Miaomiao Wang, Pranita Kulkarni, Kangguo Cheng, Ali Khakifirooz, V. S. Basker, Hemanth Jagannathan, Chun-Chen Yeh, Vamsi Paruchuri, Bruce Doris, Huiming Bu, Chung-Hsun Lin, James H. Stathis, Kingsuk Maitra, Philip J. Oldiges

XT.15 NBTI Lifetime Prediction in SiON p-MOSFETs by H/H2 Reaction-Diffusion(RD) and Dispersive Hole Trapping Model 1105
  S. Deora, V. D. Maheta, S. Mahapatra

XT.16 Product NBTI Distribution and Voltage Dependence - Impact of Relaxation and Droops 1115
  Amr Haggag, Ning Liu, Peter Abramowitz, Mohamed Moosa, Gary Anderson, David Burnett, Sanjay Parihar, Glenn Abeln, Jack Higman

XT.17 Analysis of the Relationship Between Random Telegraph Signal and Negative Bias Temperature Instability 1117
  Yasumasa Tsukamoto, Seng Oon Toh, Changhwan Shin, Andrew Mairena, Tsu-Jae King Liu, Borivoje Nikolic

XT.18 Energy Resolved Spin Dependent Trap Assisted Tunneling Investigation of SILC Related Defects 1122
  J. T. Ryan, P. M. Lenahan, A. T. Krishnan, S. Krishnan

Author Index

New Scaling Limitation of the Floating Gate Cell in NAND Flash Memory

Yong Seok Kim, Dong Jun Lee, Chi Kyoung Lee, Hyun Ki Choi, Seong Soo Kim, Jai Hyuk Song,
Du Heon Song, Jeong-Hyuk Choi, Kang-Deog Suh* and Chilhee Chung.

NAND Flash Process Architecture Team, Semiconductor Business Division, Samsung Electronics Co.,
San #24, Nongseo-Dong, Giheung-Gu, Yongin-City, Gyunggi-Do 446-711, Korea
*School of Information and Communication Engineering, Sungkyunkwan University, Suwon, 440-746, South Korea.
Tel) 82-31-209-3546, Email) yongseok7.kim@samsung.com

Abstract— **As the scaling in NAND Flash Memory is progressed, the various interferences among the adjacent cells are more and more increased and the new phenomenon which is ignored until now has to be considered. In this paper, we will introduce the new program interference phenomenon which is generated between the program word line and the adjacent word lines along the bit-line. This new program interference is that the Vth's of the adjacent word lines along the bit-line are decreased while a word line is programming. Because this phenomenon is severely aggravated as the gate space is decreased, we have to consider this program interference for the future technology nodes.**

Keywords- Scaling Limitation, interference, Floating Gate Type NAND Flash

I. INTRODUCTION

In many papers, the scaling limitations of the floating gate NAND flash cell are listed and most of all are related with the increment of the interference among the neighboring cells and the decrement of coupling ratio [1]. Especially, the cell to cell interference, which has been attributed to the parasitic capacitance coupling effect, is a biggest barrier of MLC operations [2]. This cell-to-cell coupling between the floating gates has been studied analytically by involving the size of the cell transistor and the operation bias conditions [3] and recently, the new concept of the direct field effect interference, which is generated between the adjacent channels of cell transistors, is introduced [4]. Despite of these scaling problems, the scaling of floating gate type NAND flash has been continued to sub-30nm through Error Code Correction and program algorithm progress [5], [6]. In this paper, we will introduce the new program interference, which has never been mentioned in the published paper so far. This program interference is the Vth reduction of the adjacent high state cells while the adjacent cells are programming. In the way that this program interference is due to the charge loss by the leakage current, it is different from the well-known interference by the cell-to-cell coupling effect, which has to do with the potential induced by the charge in neighbor floating gate. But because this new phenomenon is generated between the program word line and the adjacent word lines along the bit-lines, we call this phenomenon the program interference. This program interference is very much increased as the space between the

adjacent cells along the bit-line is decreased and is more serious in the fresh cells than in the cycled cells.

II. EXPERIMENT RESULTS

As shown in Fig.1, when WL(n) is programmed to the high state, the high program voltage is applied to the control gate of WL(n) and the pass voltage is to other control gates for inhibiting the programming of the unselected strings and for turning on the cells of the selected string [7]. Also, conventionally, NAND flash device is programmed from lower word line which is closer to the CSL(Common Sense Line) to higher word line which is closer to the BL(Bit Line). Although we had observed this program interference phenomenon in the main chip under the low pass voltage for the first time, we had conducted experiments with the test modules located beside the main cell, which is 35nm technology node. The test module is easy to control and trace each cell. Also the test modules with different design rule can be tested under the same condition. The test modules of n-type floating gate cell have 4 bit-lines and 64 word-lines as shown in Fig. 1. The control signals of the test modules are generated from external signal generators and measured by HP4083.

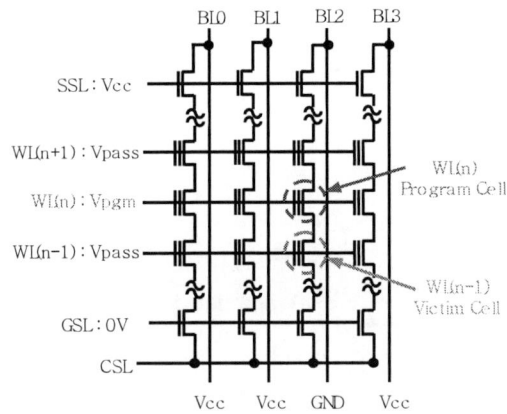

Figure 1. The schematic diagram for the test module with 4 bit-line and the basic program condtion for the local boosting scheme.

978-1-4244-5430-3/10 $26.00 © 2010 IEEE

At first, we erased all word lines in the test module and programmed only the WL(n-1) of BL2, which the target Vth of the erased cell is about -3V and the Vth of the programmed cell is about 4.7V. Next, we applied Vpgm to the WL(n) and Vpass to all other word lines and then we monitored the Vth with the stress time. The bias condition of this program stress is same as the condition for the normal program except for the amplitude of the voltage and the time. The Fig. 2 is a measured result for a single cell. At the first stage of the program stress, the program stress cell, WL(n), is being programmed and the Vth of the victim cell, WL(n-1), is being increased due to the cell-to-cell coupling between the program stress cell and the victim cell. Next, the Vth of the program stress cell is saturated, which is due to the leakage of the inter-poly dielectric layer, and the Vth of victim cell is deceased. Although this program interference don't distinguish between the WL(n+1) and WL(n-1), we call the WL(n-1) the victim cell because the program is done from the low word line to the high word line. In this paper, we will focus on the victim cell's behavior. The Fig. 3 is the measured results for 40 cells, where the Vth reductions have a wide distribution. In fact, at the main cell, this program interference phenomenon are observed as the under tail of the Vth distribution when all word line are programmed to high state in the multi-level cell operation. Because we are unfamiliar with this new phenomenon, we have confirmed whether it is dependent upon the back pattern, which means the Vth of other word lines except the program stress cell and the victim cell, and the sequence of program. But this phenomenon is only related to the adjacent cells along the bit-line. We have confirmed that this phenomenon doesn't distinguish between the upper word line, WL(n+1), and the under word line, WL(n-1). To understand this phenomenon, we have investigated the dependence of the program stress voltage, the pass voltage, and the initial Vth of the victim cell. As shown in Fig. 4, we have measured the Vth of victim cell with the various program stress voltage. To magnify the difference of voltage between the neighboring gates, we have applied the low pass voltage, which is 4.5V. The decrement of the victim cell's Vth is getting larger with the program stress voltage increase. We have measured the same cells for the program stress voltage 24V, 26V, and 28V after initializing the cell and the each Vth in Fig. 4 is the average values for 20 cells.

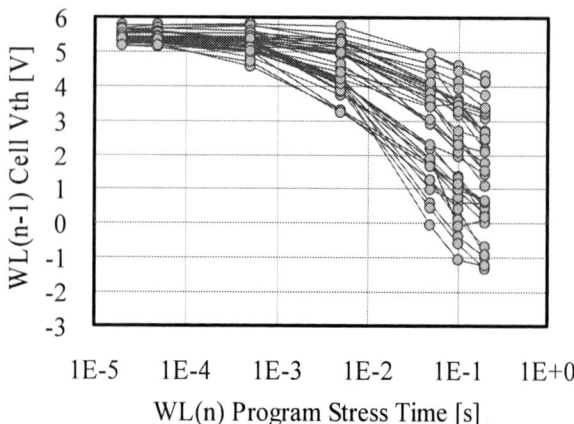

Figure 3. The Vth reduction of the victim cells, WL(n-1), with the program stress times for 40 cells. The program stress voltage is 26V and the pass voltage is 4.5V.

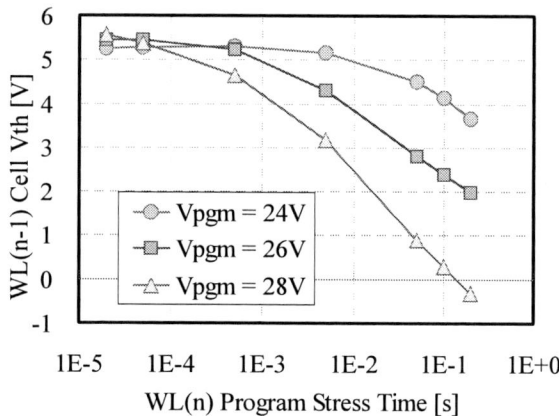

Figure 4. The Vth reduction of the victim cell, WL(n-1), with the program voltages, which are 24, 26, and 28V. The pass voltage is 4.5V.

The Fig. 5 is the measured results of the victim cell's Vth with the various Vpass. We have measured at the same cells for the pass voltage 4.5V, 7V, and 10V after initializing the cell repeatedly, and the results of each WL(n-1) Vth in Fig. 5 are the average of 20 cells. Contrary to the program voltage experiment, the decrement of the victim cell's Vth is getting larger as the pass voltage decreases. This is the reason why the low pass voltage is used when the experiment for this program interference is performed. The potential difference between the control gate and the floating gate can be changed by altering the initial cell Vth of the victim cell. The charge loss of the victim cell decreases as the initial cell Vth of the victim cell decreases. The results in Fig. 6 are measured at the same cell after the Vth of the cells is adjusted again. Although the initial Vth's of the victim cell are different, it appears as if the victim cell's Vth converges as program stress time increases, as shown in Fig. 6. Some cells exactly converge into a single values and other only gathered at a range. This convergence is well observed at high program stress voltage and in cells with small spaces between the gates.

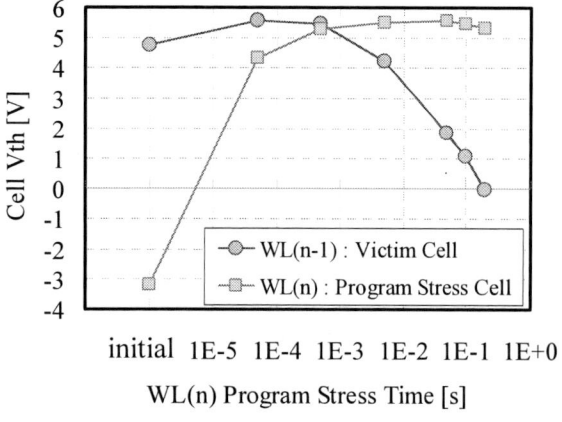

Figure 2. The Vth's of WL(n-1) and WL(n) with the program stress time for cells. The program stress voltage is 28V and the pass voltage is 4.5V.

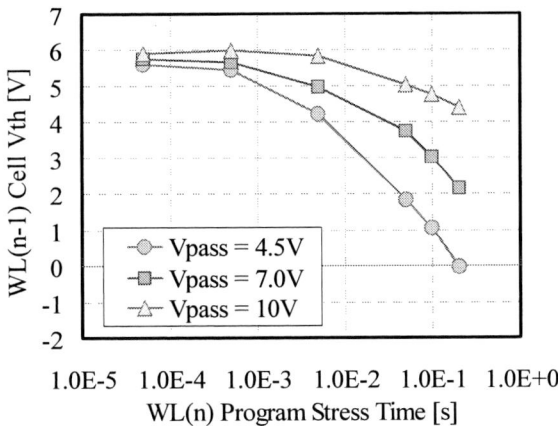

Figure 5. The Vth's reduction of WL(n-1) with the Vpass. The program voltage is 28V and Vpass is 4.5, 7, 10V, respectively.

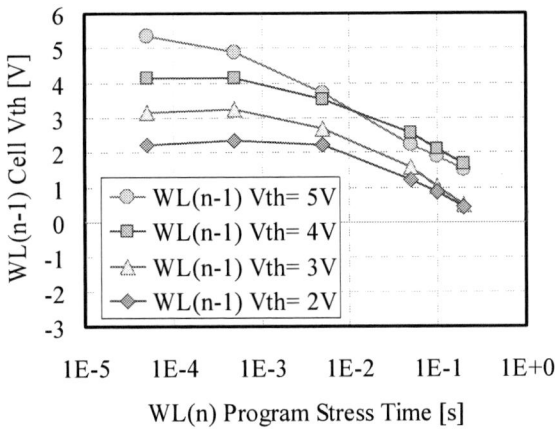

Figure 6. The Vth's reduction of WL(n-1) with the initial Vth. The program stress voltage is 26V and the pass voltage is 4.5V. This result is the measured result for a single cell.

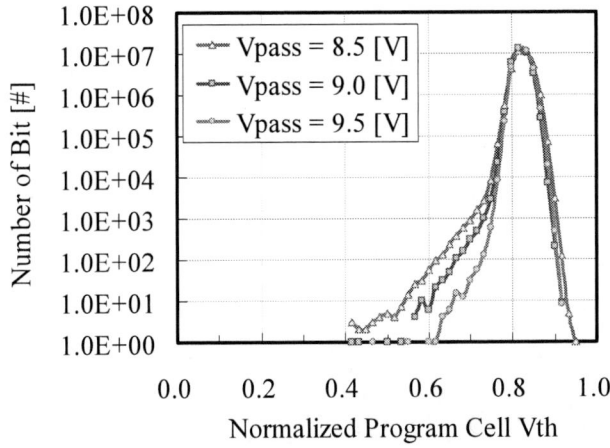

Figure 7. The distribution of program cell Vth after program of the main chip, where Vpass is 8.5, 9.0, 9.5V, respectively.

Until now, we have shown the results under the extreme program condition of the TEST modules. The program time of the main chip is shorter and Vpass in the main chip is higher than the program stress condition given for the TEST Module. So we cannot observe this phenomenon in the main cell more than Sub-40nm. Fig. 7 has shown the normalized Vth distribution after all word lines are programmed to high state(ex. 4.7V) from lower word line to higher word line at the main chip in order. The gate design rule of the main chip is sub-30nm. The under tail of Fig.7 is generated at the WL(n-1) when the WL(n) is programmed. The final word line without the upper word line no has the under tail of the Vth distribution. As the pass voltage is increasing, the tail of Vth distribution is decreasing. In the main cell, the dependence upon the program voltage and the initial Vth of WL(n-1) are the same tendency as the results in the test modules. Because the reduction of Vth in the test module have a very large distribution, as shown in Fig. 3, the program interference phenomenon in the main cell is appeared as a tail.

III. DISCUSSION

It is important to know that the variation rate of the victim cell's Vth with the program stress voltage in Fig. 4 and with the pass voltage in Fig. 5 are different. The variation rate of the Vth's reduction for the program stress voltage is from 0.9 to 1.5 and the variation rate of the Vth's reduction for the pass volatge is from 0.65 to 0.8. From these results, we first assumed that the decrement of the victim cell's Vth is caused by the leakage current due to the potential difference between the control gate of the program word line and the floating gate of the victim cell. The coupling ratio of the sub-30nm node used for this experiment is around 0.68 coming from 3D - simulation. The result of Fig. 6 also supports that this program interference is caused by the potential difference between the control gate and the floating gate. Because the Vth of the victim cell in Fig.2 is decreasing after the one of the program stress cell is saturated, the potential difference between the floating gates is not the cause of the charge loss. Even if it is so that the program interference is due to the potential difference between the control gate and the floating gate, is it FN tunneling current? The Fig. 8 is a simulation result of the electric field at t=0 for 30nm node. The simulation was performed in 3D-structure with practical dimension reflecting doping on Si channel, poly-Si floating gates, and control gates.

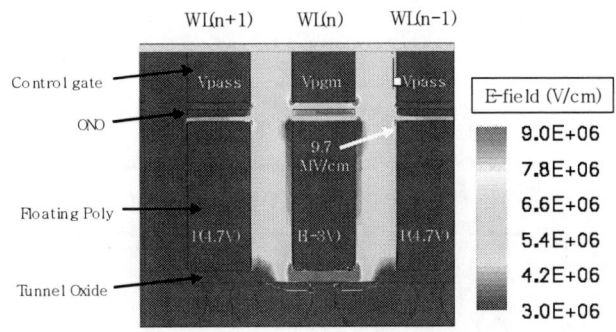

Figure 8. The simulation result of the electric field distribution under the program operation at t=0, where the Vpgm is 24V and the Vpass is 8.5V.

The target Vth's of the floating gates are adjusted from Id-Vg curve by controlling charges of the floating gates. The maximum electric field is formed between the top edge of the floating gate and the bottom edge of the control gate. We only have confirmed that the edge field is large enough to generate FN tunneling current through the simulation. Although more simulations have to be performed to prove the type of the leakage current, the measured results show that the Vth's reduction of the victim cell is similar one when the floating gate cell is erased by FN tunneling current. In the main cell, this program interference is decreased with the P/E cycle as shown in Fig. 9. Because the program interference mentioned in this paper is caused by the leakage current between the floating gate and the control gate, the charge trapped in the tunnel oxide have nothing to do with it directly. The result in Fig. 9 is only due that the program voltage to adjust the same Vth is decreased as the P/E cycle is progressed [8]. After 100 P/E Cycle and 1000 P/E Cycle, the program speed is faster 0.3V and 0.5V, respectively, than the initial program speed. Fig. 10 shows the 20th measured results for the program interference at the same cell repeatedly.

If the leakage current is related with the trap assisted tunneling, the measured results have to show any tendency. So we guess that this leakage current is supposed to be the FN like tunneling current.

IV. SCALING EFFECT

We have measured these Vth's reductions in the NAND test modules with different gate design rules, which are from sub-30nm to sub-50nm. Fig. 11 shows that the space between the adjacent WL gates is very critical to this phenomenon. The y-axis in Fig. 11 is the amount of the Vth reduction after 0.2s of program stress. The each value of Fig. 11 is the mean of 20 points. Although the Vth's reductions are measured at different program voltage according to the gate design rule, all of the measured results are simply generalized with the electric field between the control gates. The normalized results with the electric field are given in Fig. 12, where the y-axis is the decrement of the victim cell's Vth after program stress for 0.2s.

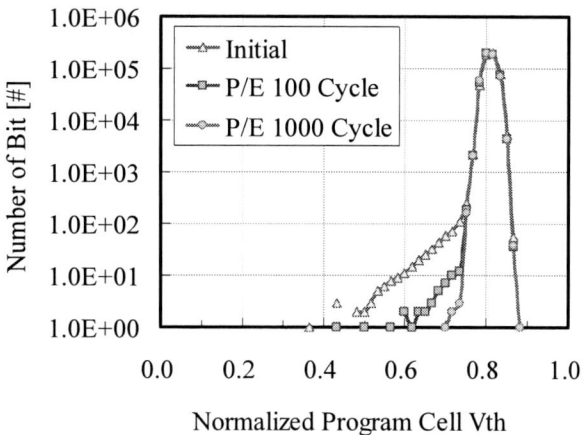

Figure 9. The distribution of program cell Vth after P/E cycle at the main chip , where the program speed is faster 0.3V and 0.5V, respectively, than the initial program speed after 100 P/E Cycles and 1000 P/E Cycles.

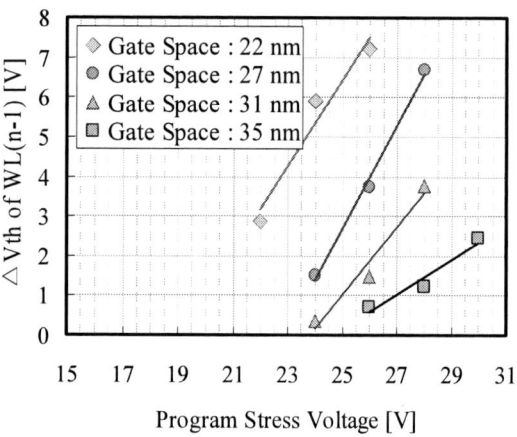

Figure 11. The Vth's reduction of WL(n-1) for samples with different gate space after 0.2s of program stress, the pass voltage is 4.5V.

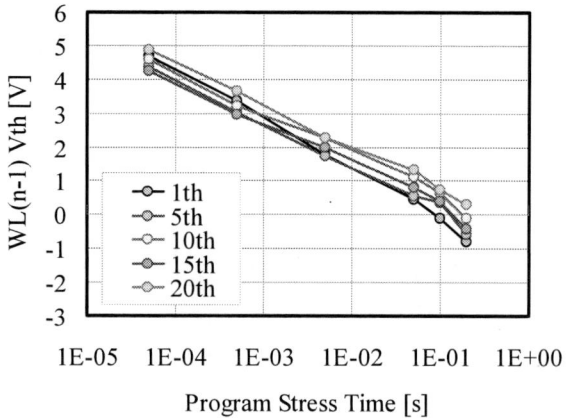

Figure 10. The 20th measured results for the program interference at the same cell in the test module repeatedly.

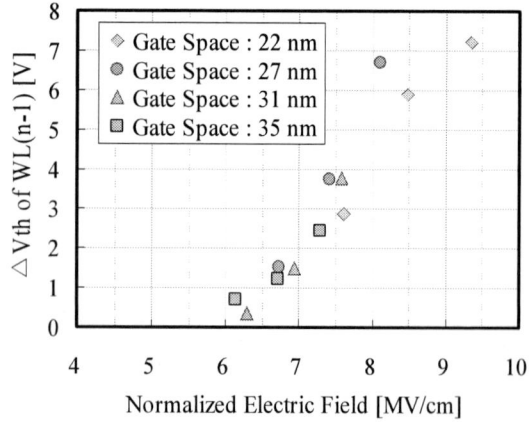

Figure 12. The Vth's reduction of WL(n-1) for samples with different gate space. The x-axis is the electric field between the control gate of WL(n) and the WL(n-1).

978-1-4244-5430-3/10 $26.00 © 2010 IEEE

The reason for using the electric field between the control gates instead of the electric field between the WL(n)'s control gate (program stress cell) and the WL(n-1)'s floating gate (victim cell) is because it is difficult to use the representative value. As shown in Fig. 8, the electric field of the edge is susceptible to the shape and profile. If the coupling ratio of the floating gate cell is similar, the electric field between the control gates is a simple and proper parameter for representing this phenomenon in the industries. The Vth reduction of WL(n-1) is measured above 6.0 MV/cm without the gate design rules. This means that the Vth reduction of WL(n-1) is general phenomenon related to the electrical field between the floating gate and the adjacent control gate along the bit-line.

V.   CONCLUSION

We have introduced a new scaling limitation of the floating gate cell in NAND Flash memory, which is the program interference. The charge loss of WL(n-1), when the WL(n) is programming, is attribute to the leakage current between WL(n)'s control gate and WL(n-1)'s floating gate. By extension, we have generalized this phenomenon with the electric field between the control gates. Because this program interference is rapidly increased as the space between the gates is decreased, the program speed and program operation condition should be determined with considering how this charge loss will be handled in the following nodes.

REFERENCES

[1]   Chih-Yuan Lu, Tao-Cheng Lu, and Rich Liu, "Non-Volatile Memory Technology-Today and Tomorrow", in Proceedings of 13th IPFA, 2006, pp. 18-23.

[2]   T.Cho, Y. Lee, E. Kim, J.Lee, S. Choi, S. Lee, D. Kim, W. Han, Y. Lim, J. Lee, J. Choi, and K. Suh, "A dual-mode NAND flash memory: 1-Gb multilevel and high-performance 512-Mb single-level modes," Solid-State Circuits, IEEE, 2004, pp. 895-900.

[3]   K. Park, M. Kang, D. Kim, S. Hwang, B. Choi, Y. Lee, C. Kim, and K. Kim, "A zeroing cell-to-cell interference page architecture with temporary LSB storing and parallel MSB program scheme for MLC NAND flash memories," Solid-State Circuits, IEEE, 2008, pp. 919-928.

[4]   Mincheol Park, Keonsoo Kim, Jong-Ho Park, and Jeong-Hyuck Choi, "Direct field effect of neighboring cell transistor on cell-to-cell interference of NAND flash cell arrays." IEEE Elrctron Deivce Letters, 2009, pp. 174-177.

[5]   A. Nakamura, H. Moriya, T. Terano, H. Kosaka, A. Hashiguchi, K. Nomoto, I. Fujiwara, and T. Oda, "Narrow distribution of threshold voltage in 3-Mbit MONOS memory-cell array with F-N channel write and direct/F-N tunneling erase operation as a single transistor structure," IEEE Trans. Eectron Devices, 2004, pp. 895-900.

[6]   Won-Seong Lee, "Future Memory Technologies", in Solid-State and Integrated-Circuit Technology, ICSICT 9th international Conference, 2008, pp. 1-4.

[7]   Kang-Deog Suh, Byung-Hoon Suh, Young-Ho Lim, et. al., "A 3.3V 32 Mb NAND flash memory with incremental step pulse programming scheme", Solid State Circuits, IEEE, 1995, pp. 1149-1156.

[8]   T. Kamigaichi, F.Arai, H. Nitsuta, M. Endo, K. Nishihara, H. Takekida, et. al., "Floating Gate Super Multi Level NAND Flash Memory Technology for 30nm and Beyond", IEDM, IEEE, 2008, pp. 1-4.

Investigation of the Threshold Voltage Instability after Distributed Cycling in Nanoscale NAND Flash Memory Arrays

Christian Monzio Compagnoni*, Carmine Miccoli*, Riccardo Mottadelli*, Silvia Beltrami[†],
Michele Ghidotti*, Andrea L. Lacaita*[‡], Alessandro S. Spinelli*[‡], and Angelo Visconti[†]

* Dipartimento di Elettronica e Informazione, Politecnico di Milano–IU.NET,
piazza L. da Vinci 32, 20133 Milano, Italy, e-mail: monzio@elet.polimi.it
[†] Numonyx, R&D - Technology Development, via C. Olivetti 2, 20041 Agrate Brianza (MI), Italy
[‡] IFN-CNR, Milano, Italy

Abstract—**This paper presents a detailed experimental investigation of the cycling–induced threshold voltage instability of deca-nanometer NAND Flash arrays, focusing on its dependence on cycling time and temperature. When the array is brought to a programmed state after cycling, instability mainly shows up as a negative shift of its threshold voltage cumulative distribution, increasing with time and resulting from partial recovery of cell damage created in the previous cycling period. The threshold voltage loss displays a strong dependence not only on the tunnel oxide electric field during retention, but also on the cycling conditions. In particular, performing cycling over a longer time interval or at higher temperatures delays the threshold voltage transients on the logarithmic time axis. The delay factor is studied as a function of the cycling duration and temperature on 60 and 41 nm technologies, extracting the parameter values required for a universal damage–recovery metric for NAND.**

Keywords: Flash memories, program/erase cycling, semiconductor device reliability, semiconductor device modeling.

I. INTRODUCTION

Electrical stress due to repeated program/erase (P/E) cycles is a main source of threshold voltage (V_T) instabilities in deca-nanometer NAND Flash memories, determining the damage of the tunnel oxide and of its interface with cell substrate [1]–[4]. Instabilities are, first of all, the result of the damage itself, giving the possibility, for example, of enhanced random telegraph noise (RTN) fluctuations with respect to the fresh device as a consequence of the larger trap density in the tunnel oxide [5]–[7]. In addition to that, damage recovery also represents a source of statistical displacements for V_T, resulting from charge detrapping from the tunnel oxide and interface state annealing [3], [8]–[15]. Besides a statistical dispersion, these displacements have a non-zero average value and can be easily investigated by monitoring the array V_T cumulative distribution since the end of cycling.

Despite damage recovery mainly represents a critical reliability issue for NAND data retention *after* cycling, its quantitative impact on V_T should be assessed carefully considering the possibility for it to take place also *during* cycling. In fact, while P/E cycles create cell damage, this can be

recovered during the time elapsing in between the cycles. As a consequence, the total amount of cell damage contributing to V_T instabilities since the end of cycling depends not only on the number of P/E cycles that were previously performed but also on the duration of the cycling period. Fast–cycling tests where P/E cycles are performed in quick succession do not correctly reproduce the damage conditions occurring in real device operation, with cycling distributed over a longer timescale. In particular, fast–cycling experiments represent a worst case for V_T instabilities which, however, can hardly be considered realistic.

A clear reduction of V_T instabilities due to charge detrapping during data retention has been shown for NOR Flash memories when distributed– instead of fast–cycling is adopted [14], [15]. Results have shown not only that the increase of the cycling time (t_{cyc}) allows more damage recovery to take place *during* cycling, then reducing V_T instabilities *after* cycling, but also that the same damage recovery can be obtained increasing the cycling temperature (T_{cyc}). This gives the possibility to design fast–cycling experiments reproducing the damage conditions at the end of time–distributed cycling tests. All these results have never been assessed for deca-nanometer NAND technologies, displaying different damage and recovery physics as a consequence of the different P/E mechanisms and of the different cell design [3], [4], [11]–[13].

In this work we present a detailed experimental investigation of the V_T instability due to damage recovery in NAND Flash arrays, focusing on its dependence on cycling time and temperature. Monitoring the array cumulative distribution from the programmed state, we show that damage recovery gives rise to V_T–loss transients having a strong dependence not only on the tunnel oxide electric field during data retention but also on the cycling conditions. In particular, the V_T–loss transient is delayed along the logarithmic time axis when t_{cyc} or T_{cyc} are increased. The delay factor is studied over a wide range of t_{cyc} and T_{cyc} on 60 and 41 nm NAND technologies, extracting the parameter values required for a universal damage–recovery metric (UDM) for NAND.

978-1-4244-5430-3/10 $26.00 © 2010 IEEE

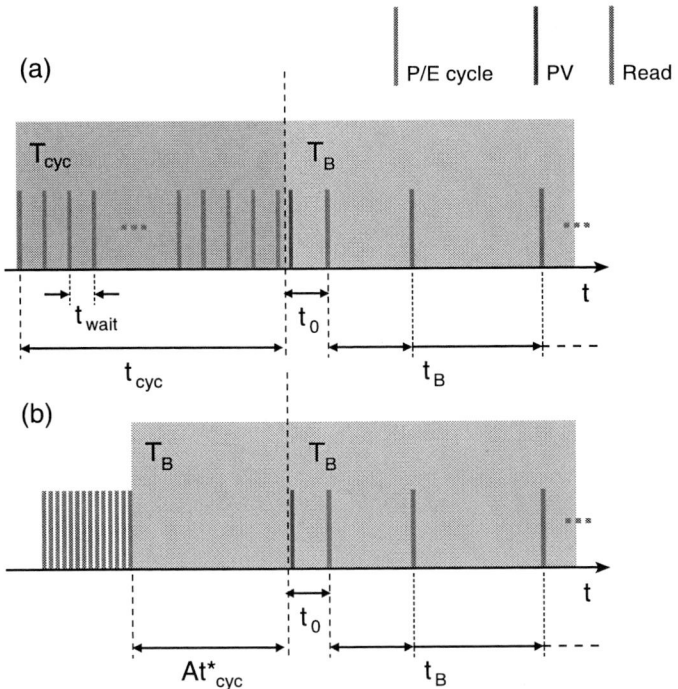

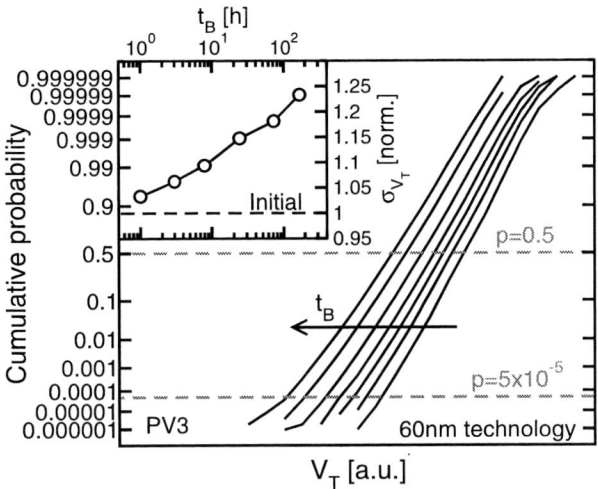

Fig. 1. Schematics for the experimental procedure used in this work to investigate cycling–induced V_T instabilities (a) and equivalent model for distributed cycling (b).

Fig. 2. V_T cumulative distribution at the first read operation after PV and after increasing t_B. Cells were programmed to the highest V_T level by PV after cycling with $t_{cyc} = 0.6$ h and $T_{cyc} =$RT. Inset shows the distribution standard deviation, normalized to the initial value, as a function of t_B.

II. EXPERIMENTAL AND THEORETICAL BACKGROUND

Fig. 1a schematically shows the experimental procedure most commonly adopted to test V_T instabilities after cycling on multi-level NAND devices: 1) a certain number N of P/E cycles is performed in a time $t_{cyc} = N \cdot t_{wait}$ (t_{wait} is a constant delay between cycles); 2) a program-and-verify (PV) algorithm brings the cells to a certain programmed V_T level; 3) V_T is monitored at logarithmically–spaced times t_B since the first read operation, performed after a delay t_0 since the end of cycling. Note that the V_T monitoring phase corresponds to a data retention experiment at temperature T_B, which may be room temperature (RT) or, more generally, a selected bake temperature. In the latter case, bakes are periodically interrupted and the device cooled to room temperature for V_T reading.

Despite the previous test may appear quite simple, many experimental details have a non-negligible impact on the results and should be specified when investigating deca-nanometer NAND devices. First of all, the data pattern created on the NAND block during cycling should be defined: in order to maximize damage creation for fixed number of P/E cycles, we used a uniform cycling pattern, moving all the cells in the block from the erased to the highest programmed V_T level. Second, how the increase of t_{cyc} is obtained in the experimental procedure should be clarified, *i.e.* inserting time delays after block erase or after block program or after both. Results presented in this work were obtained increasing t_{cyc} by adding delays after cells programming. Finally, the data

pattern created by the PV operation and used to monitor the V_T instabilities is a last important piece of information: we used uniform patterns where all the cells in the NAND block were raised to one of the three programmed V_T level. The possibility for a background pattern sensitivity of data retention from the programmed state was in fact recently shown for deca-nanometer NAND Flash memories in the fresh state [16], and cannot be excluded for the V_T–loss transients after cycling.

The amount of cell damage present at the end of cycling in the experimental test of Fig. 1a is the result of damage creation by P/E cycles and damage recovery during the time elapsing in between the cycles. Assuming that damage creation by P/E cycles depends neither on t_{wait} nor on T_{cyc} and that damage recovery during cycling can be reproduced by a bake period of duration proportional to t_{cyc} at temperature T_{cyc} after damage has been created [14], [15], the testing procedure of Fig. 1a is equivalent to that of Fig. 1b. In this latter experimental test the same cell damage existing in Fig. 1a prior to the PV operation is obtained by a fast cycling at RT and a subsequent damage recovery period of duration $A \cdot t_{cyc}^*$, where A is a constant to be determined from experiments. In order to deal with damage recovery at a single temperature, the time t_{cyc}^* was introduced, corresponding to the time at T_B that is required to have the same damage recovery taking place in a time t_{cyc} at T_{cyc}:

$$t_{cyc}^* = t_{cyc} \cdot e^{E_A(1/kT_B - 1/kT_{cyc})} \qquad (1)$$

where an Arrhenius law of activation energy E_A was used for the time conversion. Assuming now that V_T has a logarithmic decrease due to damage recovery since the end of the damage creation period, the following formula holds for the V_T variation (ΔV_T) resulting in a time t_B since the first read operation in the experimental test of Fig. 1b and, in turn, of Fig. 1a [15]:

$$|\Delta V_T| = \alpha \ln\left(1 + \frac{t_B}{t_0 + At_{cyc}^*}\right) = \alpha \ln\left(1 + \frac{t_B}{t_B^*}\right) \qquad (2)$$

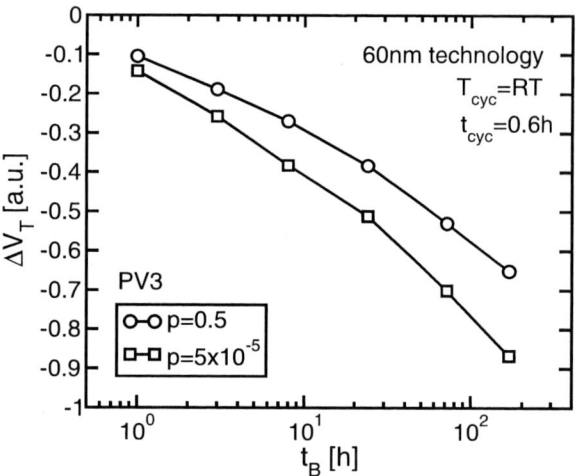

Fig. 3. V_T–loss transients extracted from the cumulative distributions of Fig. 2 at $p = 5 \times 10^{-5}$ and $p = 0.5$.

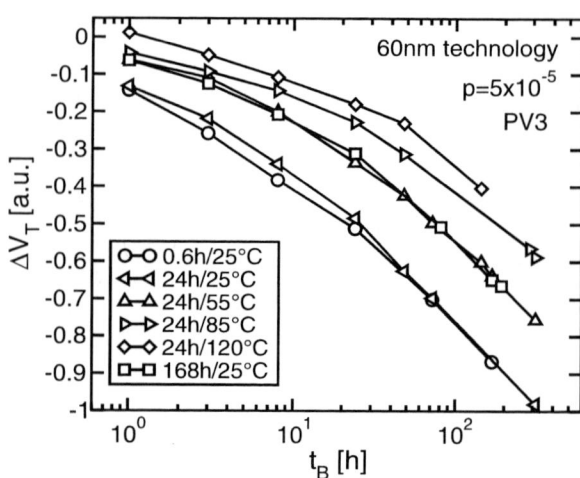

Fig. 5. ΔV_T transients from PV3 ($p = 5 \times 10^{-5}$) after cycling tests with different t_{cyc} and T_{cyc}.

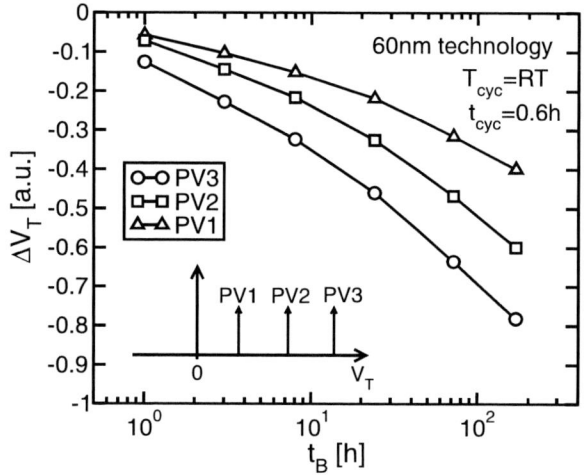

Fig. 4. V_T–loss transients at $p = 0.5$ for different PV levels after cycling with $t_{cyc} = 0.6$ h and T_{cyc} =RT. Inset shows V_T placement for the different PV.

where α gives the magnitude of the logarithmic decrease of V_T due to partial damage recovery and $t_B^* = t_0 + A t_{cyc}^*$. From the t_B^* definition, lower V_T–loss transients should result from longer t_{cyc} and higher T_{cyc}.

III. DISTRIBUTED–CYCLING RESULTS ON 60 NM TEST-CHIPS

We applied the experimental procedure of Fig. 1a on 60 nm multi-level NAND test-chips. For each cycling test, $N = 10$k P/E cycles were performed on a block of cells, then monitoring data retention at T_B =RT after the block was uniformly programmed by the PV algorithm to a selected V_T level. The total delay between the first read operation and the end of cycling was $t_0 \simeq 0.8$ h, as required by an unoptimized experimental set-up. Fig. 2 shows the V_T cumulative distribution at the first read operation and for increasing t_B up to 1 week, after cycling with $t_{cyc} = 0.6$ h, T_{cyc} =RT and PV to the highest V_T

level. In order to minimize RTN effects on the V_T distribution during data retention experiments [17], the PV operation was performed by an incremental step pulse programming (ISPP) algoritm with loose step amplitude [18]–[20], resulting in the nearly gaussian distributions of Fig. 2. As a result of damage recovery as retention time proceeds, the distribution displays a negative shift and a slight increase of its standard deviation (see inset), similarly to what previously observed on NOR devices in [14], [15]. We monitored the distribution V_T–loss considering two reference probability levels, namely, $p = 0.5$ and $p = 5 \times 10^{-5}$, corresponding to the average and the edge shift of the distribution. Due to the slight increase of the spread with t_B, a faster V_T–loss is observed for $p = 5 \times 10^{-5}$ than for $p = 0.5$, as shown in Fig. 3.

Fig. 4 shows that the ΔV_T transients display a strong dependence on the cell V_T level during data retention, with larger V_T–loss observed when moving from the lower (hereafter, PV1) to the intermediate (PV2) to the higher (PV3) V_T level. This result highlights a strong dependence of damage recovery on the electric field in the cell tunnel oxide, which may be attributed both to enhanced detrapping efficiency for higher electric fields [9] and to a backpattern dependence in presence of string effects. As a consequence of the strong field dependence, the amount of P/E damage directly recovered during cycling should depend not only on t_{cyc} and T_{cyc}, as discussed in the previous section, but also on the data pattern present in the array during cycling delays.

In order to explore the dependence of the data retention V_T–loss on cycling time and temperature, Fig. 5 shows results obtained for t_{cyc} and T_{cyc} ranging, respectively, from 0.6 h to 168 h and from RT to 120°C. A monotonous reduction of ΔV_T clearly appears when, for fixed $t_{cyc} = 24$ h, the cycling temperature is increased, confirming that a larger damage recovery takes place directly during cycling when this is performed at higher temperatures. For fixed T_{cyc} =RT, a reduction of the V_T–loss is, instead, evident only when t_{cyc} is

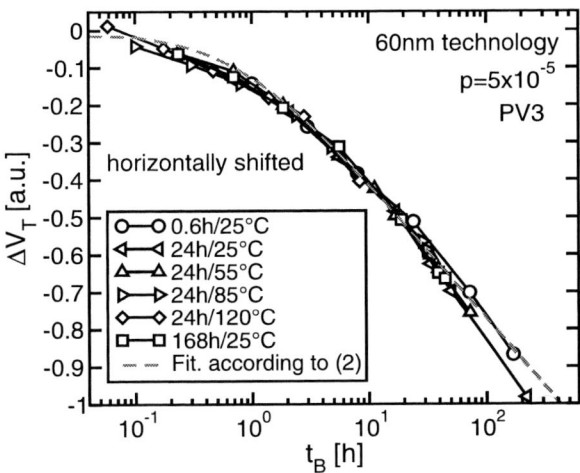

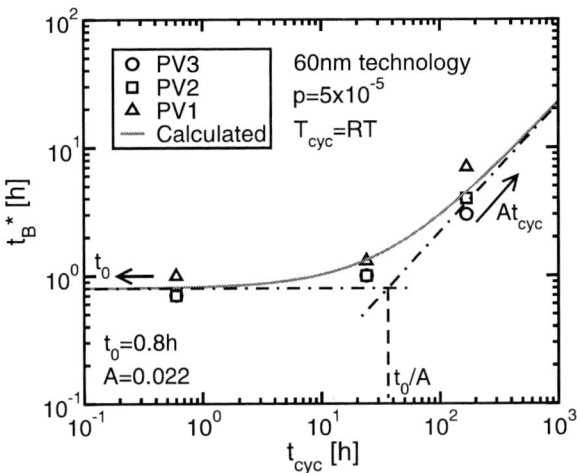

Fig. 6. Same as in Fig. 5, but with the curves horizontally shifted to overlap with the $t_{cyc} = 0.6$ h, $T_{cyc} =$ RT transient. Dashed line is a fitting of experimental data according to (2).

Fig. 8. t_B^* dependence on t_{cyc} for the 60 nm test-chip in the case $T_{cyc} =$ RT.

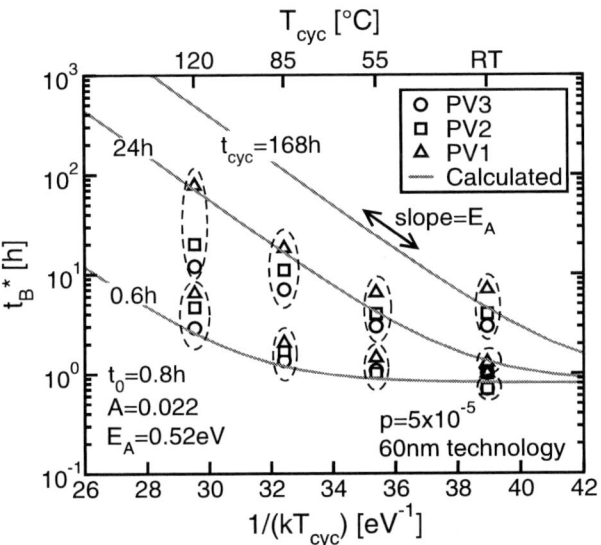

Fig. 7. Arrhenius plot for cycling for the 60 nm test-chip.

increased to 168 h, with quite similar ΔV_T transients observed for $t_{cyc} = 0.6$ h and 24 h. The quantitative assessment of the improvements given by longer t_{cyc} and higher T_{cyc} on the V_T–loss can be obtained after noting that the transients for the different cycling conditions are only horizontally shifted on the logarithmic t_B axis, as shown in Fig. 6 and as predicted by (2). This result, valid also when considering $p = 0.5$ or different PV levels, reveals that the delay between the curves represents the main parameter needed to quantify the effect of cycling time and temperature on V_T instabilities during data retention.

Figs. 5-6 confirm that (2) gives a formally correct expression to describe the effect of cycling time and temperature on the following V_T–loss transients. This expression includes two main free parameter that have to be extracted from

experimental data, namely, α and t_B^*. The former gives the shape of the logarithmic ΔV_T transient and, due to the overlap of the curves after a horizontal shift shown in Fig. 6, depends neither on t_{cyc} nor on T_{cyc}. Figs. 3-4 reveal instead that α depends on the probability level used to monitor the V_T–loss (larger α for $p = 5 \times 10^{-5}$ than for $p = 0.5$) and on the PV level during data retention (the higher the PV the larger α).

For a selected PV and p, once α is extracted by fitting the experimental data as shown in Fig. 6, t_B^* can be obtained for the different cycling conditions. In fact, the same fitting operation can be used to extract t_B^* for the fastest cycling curve (0.6 h at RT), then extracting t_B^* for all the other cycling conditions from the delay of their ΔV_T transient with respect to this curve. Results are shown in Fig. 7 as a function of $1/kT_{cyc}$, referring to $p = 5 \times 10^{-5}$. We defined this graph as the Arrhenius plot for cycling, showing a characteristic time for the data retention ΔV_T transients as a function of the reciprocal of the *cycling* temperature and not of the *retention* temperature, which is always equal to RT. Experimental data can reasonably be reproduced by the theoretical definition of t_B^* given in Section II (lines in Fig. 7), allowing the extraction of $E_A = 0.52$ eV, $t_0 = 0.8$ h and $A = 0.022$ independently of the PV level and p. Note that the extracted value of t_0 well matches the experimental delay between the end of cycling and the first read operation on our 60 nm NAND test-chip.

Experimental data and extracted theoretical trends in Fig. 7 show that for fixed t_{cyc}, t_B^* grows with T_{cyc} in the large T_{cyc} regime, where the slope of the t_B^* curve is given by E_A, while reaches a constant value equal to t_0 for low T_{cyc}. The transition from the high to the low T_{cyc} regime depends on the t_{cyc} value, with longer cycling times allowing reaching the T_{cyc} sensitive regime at lower temperatures. The dependence of t_B^* on t_{cyc} is shown in Fig. 8 for the case of $T_{cyc} =$ RT, revealing a linear increase of t_B^* with t_{cyc} for cycling times longer than t_0/A and a saturation to t_0 below this value. Note that the saturation of t_B^* to t_0 in Figs. 7-8, obtained when t_0 is larger than $A \cdot t_{cyc}^*$, prevents the direct observation of the

978-1-4244-5430-3/10 $26.00 © 2010 IEEE

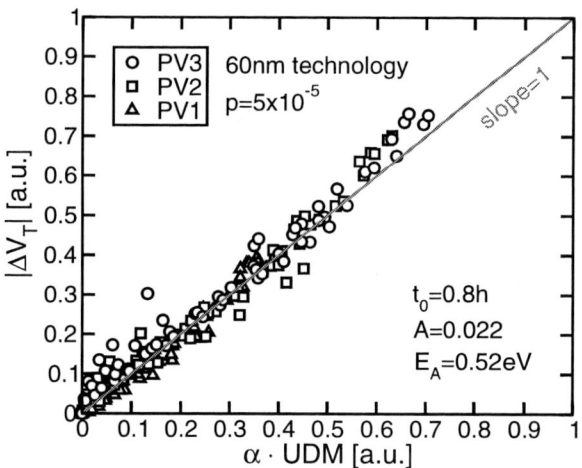

Fig. 9. ΔV_T transients as a function of $\alpha \cdot$ UDM for the 60 nm test-chip in the case $p = 5 \times 10^{-5}$. Experimental data referring to all the explore t_{cyc}, T_{cyc} and PV levels are reported.

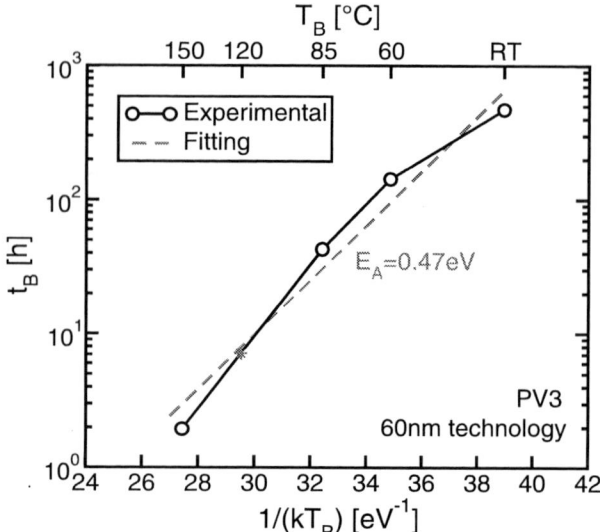

Fig. 10. Arrhenius plot for data retention on the 60 nm test-chip.

Technology node	E_A
90 nm	1.1 eV
70 nm	1.1 eV
60 nm	0.47–0.52 eV
48 nm	1.0–1.2 eV
41 nm	1.0–1.2 eV
32 nm	1.0–1.2 eV

TABLE I
ACTIVATION ENERGY EXTRACTED FOR DIFFERENT NAND TECHNOLOGIES WITH FEATURE SIZE RANGING FROM 90 NM TO 32 NM.

dependence of the V_T–loss transient on t_{cyc} and T_{cyc} for very low cycling times and temperatures.

Finally, note that previous results not only confirm that V_T instabilities are reduced when cycling time and temperature are increased, but allow a quantitative evaluation of the distributed–cycling effects in the case of NAND. From these results, the logarithmic term in (2) assumes the role of UDM for NAND:

$$ UDM = \ln\left(1 + \frac{t_B}{t_0 + At_{cyc}^*}\right) \qquad (3) $$

giving the time dynamics of V_T instabilities due to damage recovery irrespective of the data retention PV and the monitored p level (whose dependence is included in the coefficient α). The validity of the extracted parameters for the UDM definition is further confirmed in Fig. 9, showing a good linear relation with slope 1 for all the ΔV_T transients at $p = 5 \times 10^{-5}$ as a function of $\alpha \cdot$UDM. A similarly good linear relation was also obtained referring to all the V_T–loss transients at $p = 0.5$.

IV. TECHNOLOGY DEPENDENCE

The value for E_A obtained from the distributed–cycling results on our 60 nm test-chip is in good agreement with what obtained from the Arrhenius plot for data retention, shown in Fig. 10. In this case, the same cycling experiment ($t_{cyc} = 0.6$ h, T_{cyc} =RT) was used, than monitoring cell V_T at different T_B and extracting t_B as the time to reach a selected V_T–loss. A linear fitting of the semi-log plot of Fig. 10 results into $E_A \simeq 0.47$ eV, confirming that this activation energy describes the damage recovery transients for the investigated test-chip. The lower E_A with respect to the 1.1 eV usually reported for detrapping [14], [15], [21] is attributed to an excessively thin tunnel oxide used for this test-chip and to a non-optimized oxide/silicon interface, resulting into a significant contribution of direct tunneling on charge detrapping. Due to the larger activation energy of thermal

with respect to tunneling detrapping, the former mechanism is expected to dominate the charge loss at high T_B, with the latter getting more and more important as T_B is reduced [22]. As a result, the activation energy for the damage recovery process should depend on the temperature regime which is investigated by the bake experiments, decreasing when moving from high to low T_B. This allows the observation of activation energies that are lower than the 1.1 eV value that is typical of pure thermal detrapping, as in the case of Fig. 10. As a further confirmation of this physical picture, note that experimental data in Fig. 10 display a sub-linear behavior in the semi-log plot, revealing that a weak decrease of the locally-defined E_A (local slope of the $\ln(t_B)$ vs. $1/kT_B$ curve at each point) is yet detectable when T_B decreases in the temperature range explored in this figure.

The $E_A \simeq 0.5$ eV represents, for our technologies, a unique feature of the 60 nm test-chip, which was not observed in all other NAND technologies we investigated, ranging from the 90 nm to the 32 nm feature size. As shown in Table I, all other technologies displayed E_A in the 1–1.2 eV interval, representing a more usual range for the activation energy of damage recovery. As an example, Fig. 11 shows the Arrhenius plot for data retention on our 41 nm and 32 nm technologies, allowing the extraction of an activation energy in the 1–1.2 eV range. This result was also confirmed on the 41 nm technology

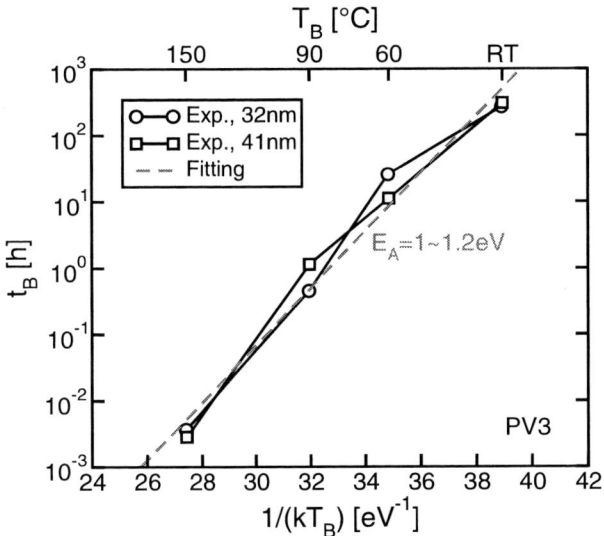

Fig. 11. Arrhenius plot for data retention on different NAND technologies.

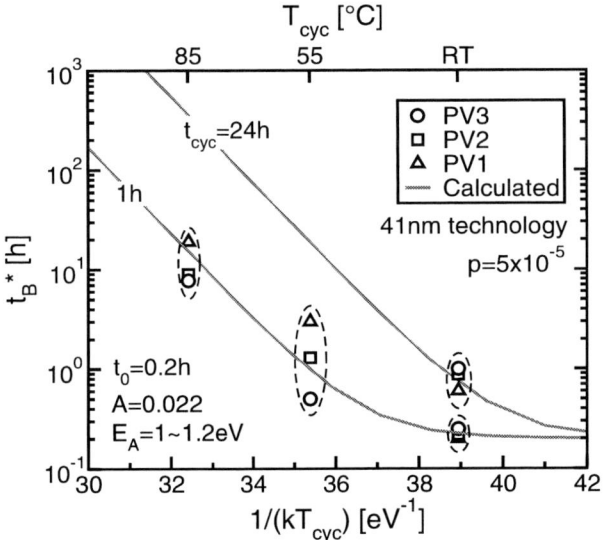

Fig. 12. Same as in Fig. 7, but for the 41 nm technology.

by distributed–cycling experiments similar to those presented in Section III, as shown by the Arrhenius plot for cycling of Fig. 12. Note that the same value $A = 0.022$ extracted for the 60 nm test-chip is obtained also for this technology, while a different $t_0 = 0.2$ h was used, mainly due to a different experimental set-up.

V. CONCLUSIONS

We presented for the first time a comprehensive investigation of distributed–cycling effects on data retention V_T instabilities of NAND technologies, quantitative assessing how the V_T–loss transients are modified as a function of cycling duration, cycling temperature, electric field during retention and probability for V_T extraction. Similarly to what previously reported on NOR devices, results on NAND technologies

display a reduced data retention V_T–loss when cycling time or temperature are increased. Moreover, the 1.0–1.2 eV range was shown to be the most typical range for the temperature activation of the damage recovery process, both in distributed cycling and post-cycling retention. These results should be considered when designing distributed–cycling and post-cycling retention qualification stress tests for NAND Flash.

VI. ACKNOWLEDGMENTS

Authors would like to thank P. Cappelletti, E. Camerlenghi and R. Bez from Numonyx for discussions and support. This work has been partially supported by ENIAC under the MODERN Project 120003 and by MIUR under the FIRB Project No. RBIP06YSJJ.

REFERENCES

[1] S. Yamada, Y. Hiura, T. Yamane, K. Amemiya, Y. Ohshima, and K. Yoshikawa, "Degradation mechanism of Flash EEPROM programming after program/erase cycles," in *IEDM Tech. Dig.*, pp. 23–26, 1993.
[2] Y.-B. Park and D. K. Schroder, "Degradation of thin tunnel gate oxide under constant Fowler-Nordheim current stress for a Flash EEPROM," *IEEE Trans. Electron Devices*, vol. 45, pp. 1361–1368, June 1998.
[3] M. Park, K. Suh, K. Kim, S.-H. Hur, K. Kim, and W.-S. Lee, "The effect of trapped charge distributions on data retention characteristics of NAND Flash memory cells," *IEEE Electron Device Lett.*, vol. 28, pp. 750–752, Aug. 2007.
[4] A. Fayrushin, K. Seol, J. Na, S. Hur, J. Choi, and K. Kim, "The new program/erase cycling degradation mechanism of NAND Flash memory devices," in *IEDM Tech. Dig.*, pp. 823–826, 2009.
[5] C. Monzio Compagnoni, A. S. Spinelli, S. Beltrami, M. Bonanomi, and A. Visconti, "Cycling effect on the random telegraph noise instabilities of NOR and NAND Flash arrays," *IEEE Electron Dev. Lett.*, vol. 29, pp. 941–943, August 2008.
[6] K. Fukuda, Y. Shimizu, K. Amemiya, M. Kamoshida, and C. Hu, "Random telegraph noise in Flash memories - model and technology scaling," in *IEDM Tech. Dig.*, pp. 169–172, 2007.
[7] H. Kurata, K. Otsuga, A. Kotabe, S. Kajiyama, T. Osabe, Y. Sasago, S. Narumi, K. Tokami, S. Kamohara, and O. Tsuchiya, "Random telegraph signal in Flash memory: its impact on scaling of multilevel Flash memory beyond the 90-nm node," *IEEE J. Solid-State Circuits*, vol. 42, pp. 1362–1369, 2007.
[8] M. Kato, N. Miyamoto, H. Kume, A. Satoh, T. Adachi, M. Ushiyama, and K. Kimura, "Read-disturb degradation mechanism due to electron trapping in the tunnel oxide for low-voltage flash memories," in *IEDM Tech. Dig.*, pp. 45–48, 1994.
[9] R. Yamada, Y. Mori, Y. Okuyama, J. Yugami, T. Nishimoto, and H. Kume, "Analysis of detrap current due to oxide traps to improve flash memory retention," in *Proc. IRPS*, pp. 200–204, 2000.
[10] R. Yamada, T. Sekiguchi, Y. Okuyama, J. Yugami, and H. Kume, "A novel analysis method of threshold voltage shift due to detrap in a multilevel Flash memory," in *2001 Symp. VLSI Tech. Dig.*, pp. 115–116, 2001.
[11] J.-D. Lee, J.-H. Choi, D. Park, and K. Kim, "Degradation of tunnel oxide by FN current stress and its effects on data retention characteristics of 90 nm NAND Flash memory cells," in *Proc. IRPS*, pp. 497–501, 2003.
[12] J.-D. Lee, J.-H. Choi, D. Park, and K. Kim, "Data retention characteristics of sub-100 nm NAND Flash memory cells," *IEEE Electron Device Lett.*, vol. 24, pp. 748–750, 2003.
[13] J.-D. Lee, J.-H. Choi, D. Park, and K. Kim, "Effects of interface trap generation and annihilation on the data retention characteristics of Flash memory cells," *IEEE Trans. Device and Materials Reliab.*, vol. 4, pp. 110–117, March 2004.
[14] N. Mielke, H. Belgal, I. Kalastirsky, P. Kalavade, A. Kurtz, Q. Meng, N. Righos, and J. Wu, "Flash EEPROM threshold instabilities due to charge trapping during program/erase cycling," *IEEE Trans. Device and Materials Reliab.*, vol. 4, pp. 335–344, Sept. 2004.
[15] N. Mielke, H. Belgal, A. Fazio, Q. Meng, and N. Righos, "Recovery effects in the distributed cycling of Flash memories," in *Proc. IRPS*, pp. 29–35, 2006.

[16] C. Monzio Compagnoni, A. Ghetti, M. Ghidotti, A. S. Spinelli, and A. Visconti, "Data retention and program/erase sensitivity to the array background pattern in deca-nanometer NAND Flash memories," *IEEE Trans. Electron Devices*, vol. 57, pp. 321–327, Jan. 2010.

[17] C. Monzio Compagnoni, R. Gusmeroli, A. S. Spinelli, A. L. Lacaita, M. Bonanomi, and A. Visconti, "Statistical model for random telegraph noise in Flash memories," *IEEE Trans. Electron Devices*, vol. 55, pp. 388–395, Jan. 2008.

[18] G. J. Hemink, T. Tanaka, T. Endoh, S. Aritome, and R. Shirota, "Fast and accurate programming method for multi-level NAND EEPROMs," in *1995 Symp. VLSI Tech. Dig.*, pp. 129–130, 1995.

[19] C. Monzio Compagnoni, R. Gusmeroli, A. S. Spinelli, and A. Visconti, "Analytical model for the electron-injection statistics during programming of nanoscale NAND Flash memories," *IEEE Trans. Electron Devices*, vol. 55, pp. 3192–3199, Nov. 2008.

[20] C. Monzio Compagnoni, M. Ghidotti, A. L. Lacaita, A. S. Spinelli, and A. Visconti, "Random telegraph noise effect on the programmed threshold-voltage distribution of Flash memories," *IEEE Electron Dev. Lett.*, vol. 30, pp. 984–986, Sept. 2009.

[21] "JEDEC Standard JEP122E: Failure mechanisms and models for semiconductor devices," tech. rep., JEDEC Solid State Technology Association, March 2009.

[22] C. Monzio Compagnoni, A. S. Spinelli, and A. L. Lacaita, "Experimental study of data retention in nitride memories by temperature and field acceleration," *IEEE Electron Dev. Lett.*, vol. 28, pp. 628–630, July 2007.

Optimal Cell Design for Enhancing Reliability Characteristics for sub 30 nm NAND Flash Memory

Eun Suk Cho, Hyun Jung Kim, Byoung Taek Kim, Jai Hyuk Song,
Du Heon Song, Jeong-Hyuk Choi, Kang-Deog Suh* and Chilhee Chung.

NAND Flash Process Architecture Team, Semiconductor Business Division, Samsung Electronics Co.,
San #24, Nongseo-Dong, Giheung-Gu, Yongin-City, Gyunggi-Do 446-711, Korea
*School of Information and Communication Engineering, Sungkyunkwan University, Suwon, 440-746, South Korea.
Tel) 82-31-209-9195, Email) happy.cho@samsung.com

Abstract— **One of the critical scaling barriers in sub 30 nm NAND Flash technology node is an abrupt threshold voltage drop of cell transistors by short channel effect. It increases program voltage which leads, in turn, to fatal reliability issues. A simple way to relieve the short channel effect is increasing the channel boron concentration. However it degrades endurance characteristics by deteriorating boosting efficiency on inhibit operation. In this paper, we present an optimal cell design for the improved reliability characteristics in the level of mass production for the future NAND Flash with floating gate cells.**

Keywords- Reliability, SCE, Coupling Rario, Floating Gate, NAND Flash

I. INTRODUCTION

It has been predicted that a charge-trapping-type cell is a promising candidate for developing NAND Flash memory of sub 40 nm technologies. Because the conventional NAND with floating gate (FG) cells has scaling issues such as structural limit, cell-to-cell interference and low coupling ratio (C/R). [1]-[4] However, it has been plausible to extend a development of FG NAND until sub 30 nm with smart controller having various compensation algorithms such as randomized program method and ECC [5]. On the other hand, as the NAND Flash is scaled down under 40 nm, the neutral threshold voltage (V_{TH}) of unit cell transistor, which can be called fresh cell (not cycled) V_{TH}, is decreased dramatically so that the program voltage (V_{PGM}) at V_{TH} = 4.5 V and T_{PGM} = 100 μs continues to show an upward tendency mainly driven by short channel effect (SCE) [Fig.1]. Then, V_{TH} distribution of programmed cell transistors on program operation get broader because of cell interference induced by increased incremental step pulse program (ISPP) loop. As a result, reliability such as endurance and hot temperature data retention (HTDR) characteristics are degraded. [Fig.2] In general, as the gate length of cell transistor decreases, it is necessary to increase the boron doping concentration of substrate to prevent punch-through between source and drain. However, high boron doping concentration of NAND cells can not only deteriorate boosted channel potential by large junction leakage current on unselected Bit Line (BL), but also aggravate program disturb severely at the worst test pattern like check board patterns, especially for sub 32 nm node. [Fig.3][Fig.4] In this paper, we present an optimal cell design which prevents undesirable increment of V_{PGM} by SCE without increasing channel boron concentration.

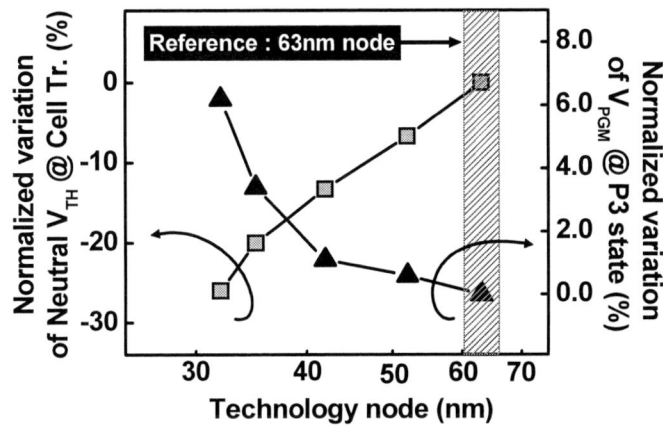

Figure 1. Normalized neutral V_{TH} and V_{PGM} of NAND cell transistor for each technology node. As the scaling down of NAND Flash device proceeds, the neutral V_{TH} of cell transistor without program/erase operation tends to decrease and higher V_{PGM} is required.

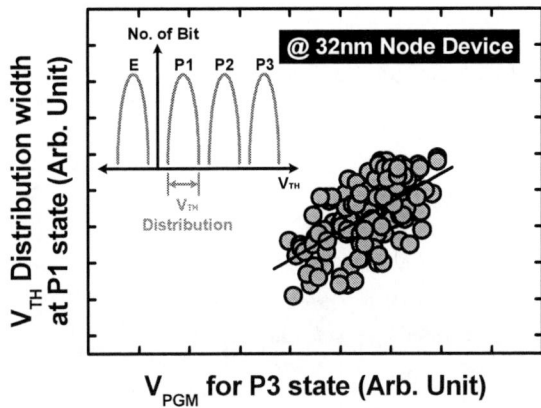

Figure 2. Verified V_{TH} distribution of programed cells with P1 state as a function of V_{PGM} for P3 state. Cells with high V_{PGM} shows a broader V_{TH} distribution.

978-1-4244-5430-3/10 $26.00 © 2010 IEEE   611

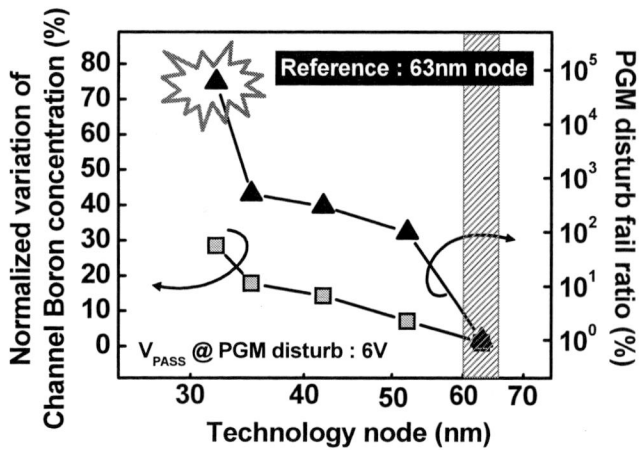

Figure 3. Channel boron doping concentration and program disturbance characteristics for each technology node. It was predicted to degrade program disturbance characteristics drastically at sub 32nm technology node.

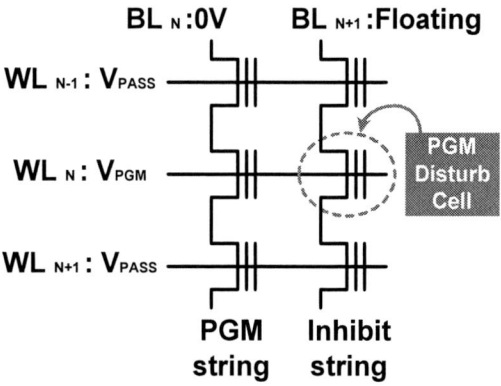

Figure 4. Schematic for program disturbance condition.

II. CELL STRUCTURES

Fig. 5 shows a simplified process flow and key parameters of the NAND Flash cell of 32 nm size. Half pitch of active and gate are 42 nm and 32 nm, respectively. Co-Salicide was adopted for control gate to reduce sheet resistance of word line (WL). The height of the floating gate and the equivalent oxide thickness (EOT) of the inter-poly dielectric (IPD) was processed with various conditions for an optimal cell design.

Implantation for Transistor	Half Pitch of Active	42nm
Tunnel oxidation		
F-poly deposition	Half Pitch of Gate	32nm
Active pattern		
STI gap fill & Recess	Height of Floating Gate	Split (Ref:90nm)
ONO & C-poly deposition		
Gate pattern & Co Silicidation	Height of Control Gate	120nm
MEOL / BEOL	EOT of IPD (ONO)	Split (Ref:13nm)

(a)   (b)

Figure 5. (a) Simplified process flow and (b) key parameters of the NAND Flash cell.

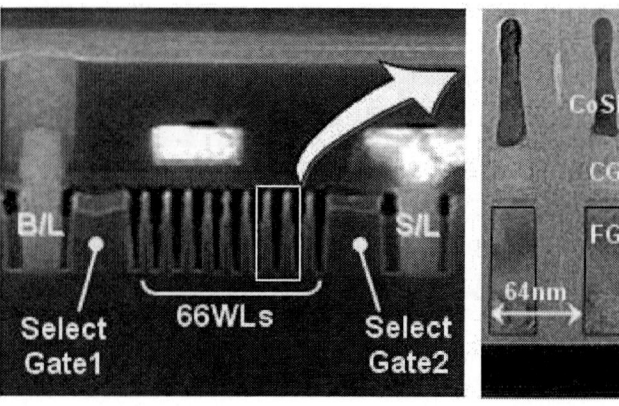

Figure 6. Cross sectional SEM and HRTEM Image of NAND cell string on bit line direction. It consists of 66 word-lines(WLs) including two dummy WLs between select gates. The half pitch of gate is 32nm.

Fig. 6 shows a cross sectional SEM and HRTEM images of NAND cell string on bit line direction. NAND string consists of 66 WLs including two dummy WLs between select gates.

III. RESULTS AND DISCUSSION

The major reliability characteristics of NAND flash, such endurance and HTDR characteristics after program (PGM) / erase (ERS) cycling operation can be improved largely by the following two methods: One is to narrow the initial distribution of the programmed cell V_{TH}, and another is to strengthen the immunity of dielectrics such as tunnel oxide and IPD during PGM/ERS stress.

Tightening the initial V_{TH} distribution of cells is very critical to achieve better reliability margin along with enhancing the quality of dielectrics. The factors that can drive narrow V_{TH} distribution of the programmed cells are higher cell string current, smaller cell-to-cell interference and lower PGM disturb. In terms of cell string current, a smaller cell string current is considered as higher V_{TH} state during read-sensing operation. As a result, the V_{TH} distribution of erased cells is shifted to higher and endurance fail bits after cycling are increased. [Fig.7]

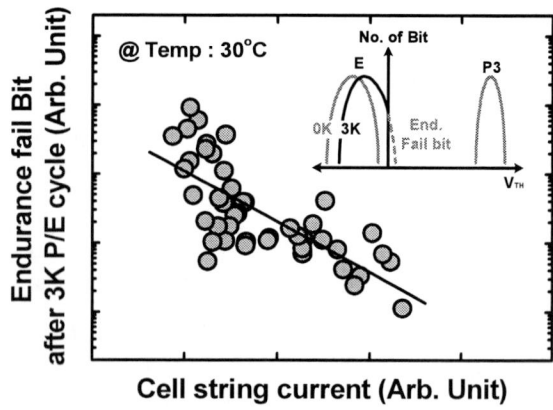

Figure 7. Endurance fail bit after 3K program/erase cycles as a function of cell string current. The smaller cell string current induces larger endurance fail bits.

978-1-4244-5430-3/10 $26.00 © 2010 IEEE   612

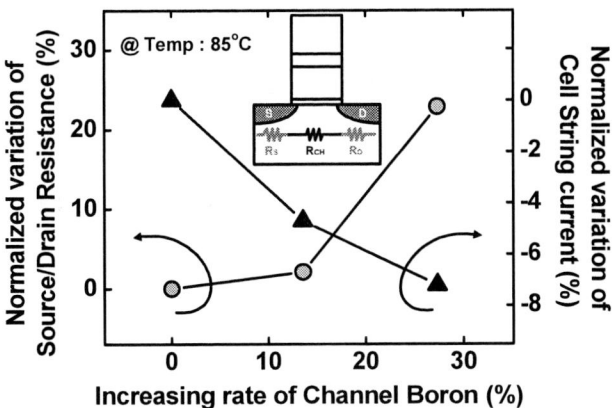

Figure 8. Variation rate of source/drain resistance and cell string current as a function of channel boron concentration. Reduction of boron doping concentration results in an enhanced string current.

To achieve higher cell string current, the channel boron doping concentration of the cell transistor can be decreased. It reduces the resistance at source/drain and the cell string current increase as a result. In addition, the reduction of the channel boron doping concentration can improve the PGM disturb by the reduction of leakage current at source/drain junction.[Fig.8] Therefore, the reduction of channel boron concentration is very effective way to improve endurance failures with the increased cell string current.

However, the reduction of the channel boron doping concentration results in a lower neutral V_{TH} of cells and a higher V_{PGM}. Therefore, more electrons are necessary for the same program state and HTDR characteristic is degraded by increased electron trap at tunnel oxide and IPD layer. [Fig.9] Thus, in order to keep the HTDR characteristics as same, it is essential to raise the C/R (Coupling Ratio) of cell transistor to lower V_{PGM}. On the other hand, reduction of the channel boron doping concentration results in a lower neutral V_{TH} of cells and a higher V_{PGM}. Therefore, more electrons are necessary for the same program state and HTDR characteristic is degraded by increased electron trap at tunnel oxide and IPD layer. [Fig.9] So, it is essential to raise the C/R of cell transistor for lowering the increased V_{PGM}.

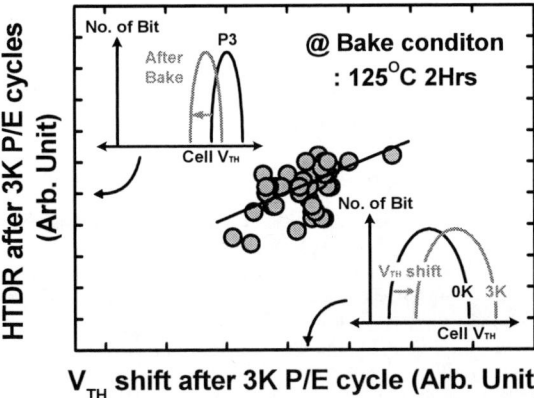

Figure 9. HTDR characteristics as a function of V_{TH} shift after 3K program/erase cycles. Oxide traps worsen the HTS characteristics.

As shown in equation (1), a direct method to get a larger C/R of cell transistor is increasing C_{IPD}, that is, raising the height of the FG and scaling the EOT of the IPD layer.

$$C / R = \frac{C_{IPD}}{C_{IPD} + C_{TUNNEL} + C_{S / D}} = \frac{C_{IPD}}{C_{TOTAL}} \quad (1)$$

However, the increase of the height of FG raises cell-to-cell interference accordingly, and to scale down the EOT of the IPD layer may results in fatal reliability issues (as discussed earlier on wide initial V_{TH} of programmed cell) by electron tunneling through the IPD layer (ONO) during PGM/ERS operation, respectively. To address the concerns all together, it is necessary to increase the cell C/R by proper combination of raising FG height and scaling of the EOT of the IPD, which leads to minimize the electric field through the IPD during the PGM/ERS operations.

As shown in [Fig 10.], [Fig 11.], and [Fig 12.], we extracted various combinations of the FG height and the ONO EOT from 3-D TCAD simulation for enlarging the cell C/R. We simulated the cell-to-cell interference and applied electric field to IPD layer especially for three cases (group A, B, C).

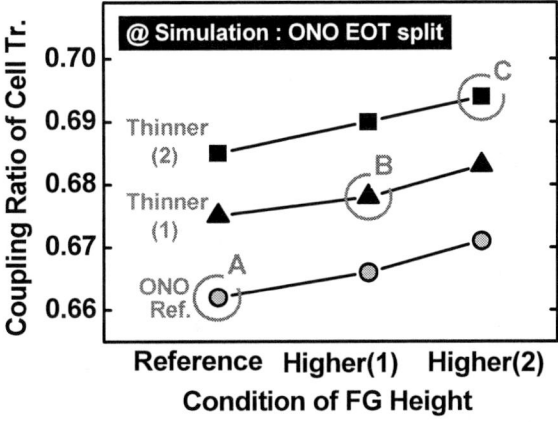

Figure 10. Simulated coupling ratio of the cell transistors according to the height of floating gate and the EOT of the ONO layer

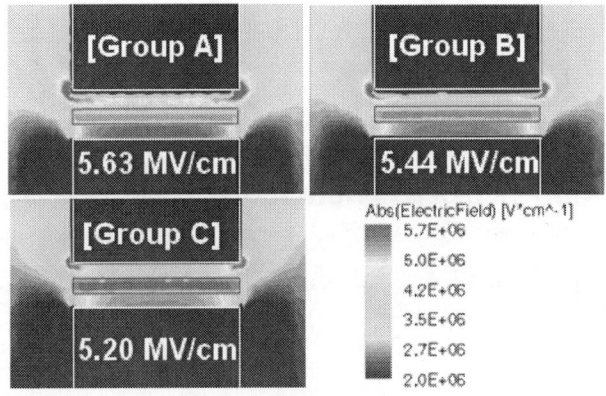

Figure 11. Electric field comparison through ONO layer for program operation based on 3-dimensional TCAD simulation.

978-1-4244-5430-3/10 $26.00 © 2010 IEEE   613

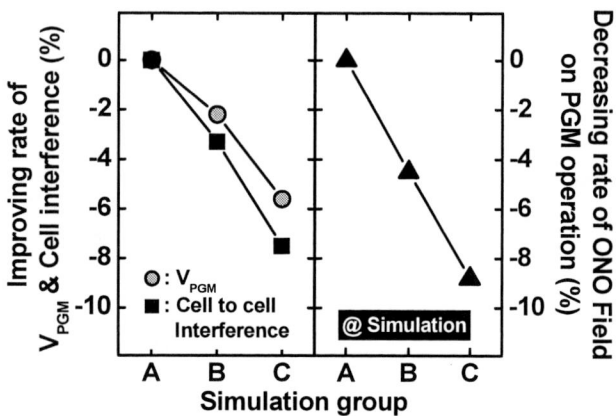

Figure 12. Simulation results regarding V_{PGM}, cell-to-cell interference and electric field though ONO layer on program.

As the cell C/R is up, the V_{PGM}, cell to cell interference and the electric field through IPD on program operation tends to decrease. However, reliability of the "group C" is degraded comparing with that of reference group. Aggressive scale down of the IPD layer induced a wider initial distribution of cell V_{TH} (not shown). Therefore "group B" is considered as a promising cell structure for optimal reliability characteristics.

By applying "group B" condition to the 32Gb NAND flash device with 11% lowered channel boron doping concentration, the initial distribution of cell V_{TH} and cell-to-cell interference have improved by 5.3% and 6.7%, respectively. [Fig.13] It means that the enlarged cell C/R of "group B" compensates sufficiently the decrement of neutral V_{TH} of cell by SCE. Finally, the endurance and HTS characteristics after 3K PGM /ERS cycles have enhanced 73% and 9.1% simultaneously. [Fig.14] [Fig.15]

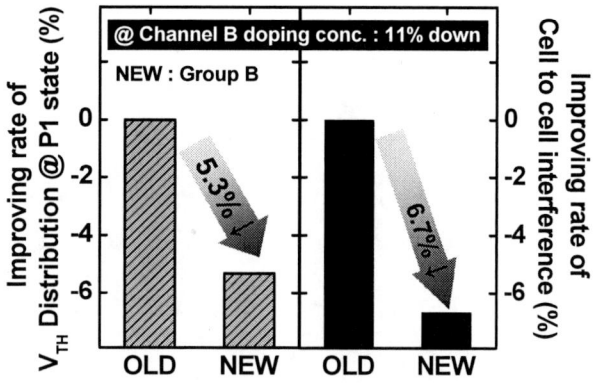

Figure 13. Improving rate of initial distribution of cell V_{TH} and cell-to-cell interference in condition of NEW process.

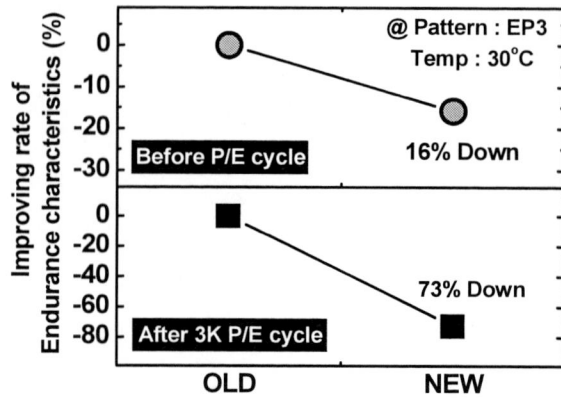

Figure 14. Improving rate of enduarnce characteristics before/after 3K program/erase cycles for the NEW process condition. Enduarance fail bit at E/P3 test pattern decreased by 73%.

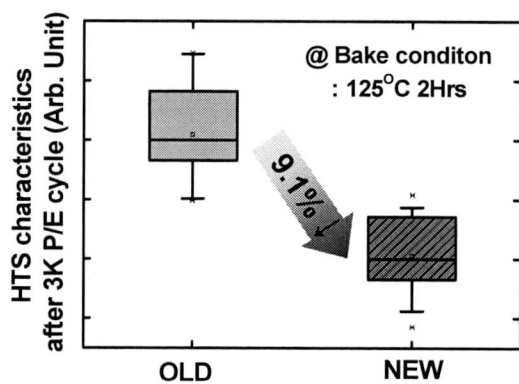

Figure 15. Comparison of HTS characteristics of the old and NEW process. The HTS characteristics imporved ~9%.

IV. CONCLUSION

We proposed an optimal cell design of NAND Flash with FG cells for sub 30m devices by raising the C/R with combination of FG height and the IPD EOT. It prevented the dropping of program speed due to the lower channel boron doping concentration for narrow V_{TH} distribution on PGM operation. Therefore, the reliability characteristics such as endurance and HTS for mass production could be successfully improved for 32 Gb NAND Flash with 32 nm technology node.

REFERENCES

[1] Kinam Kim, and Jungdal Choi, "Future outlook of NAND Flash Technology for 40nm node and beyond", NVSMW, 2006, pp. 9-11

[2] Jae-Duk Lee, Sung-Hoi Hur, and Jung-Dal Choi, "Effects of floating-gate interference on NAND flash memory cell operation", EDL, Volume 23, Issue 5, May 2002, pp. 264-266.

[3] Chandra Mouli, Kirk Prall, and Ceredig Roberts, "Trend in memory technology-Reliability Perspectives, Challenges and Opportunities", IPFA, 2007, pp. 130-134.

[4] Mincheol Park, Keonsoo Kim, Jong-Ho Park, and Jeong-Hyuck Choi, "Direct field effect of neighboring cell transistor on cell-to-cell interference of NAND Flash cell arrays", EDL, Volume 30, Issue 2, Feb 2009, pp. 174-177.

[5] Yohwan Koh, "NAND flash scaling beyond 20nm", IMW, 2009, pp. 1-3.

Impact of the Current Density Increase on Reliability in Scaled BJT-selected PCM for High-Density Applications

A. Redaelli, A. Pirovano, I. Tortorelli, F. Ottogalli, A. Ghetti, L. Laurin and A. Benvenuti

Numonyx, R&D Technology Development
Via C. Olivetti 2,
Agrate Brianza, 20041, Italy
andrea.redaelli@numonyx.com

Abstract— Some of the largest semiconductor companies involved in the non volatile memory business have demonstrated that Phase-Change Memory (PCM) technology has today reached the product maturity at 90 and 65 nm nodes and 45 nm platform is under development. In this approaches the architectural choice for large density arrays decoding relies on silicon diodes or BJT-selected PCMs (BJT-PCM), thus defining it as the mainstream PCM non-volatile memory technology. However to continue the PCM technology scaling roadmap the current density required to program the storage element will increase linearly with the lithography reduction, becoming of the order of tens of *MA/cm²* in the BJT selector and of hundreds of *MA/cm²* in the storage element at ultra-scaled lithographic nodes. In the ITRS 2008 the maximum current density to program a PCM cell has been recognized as the main physical mechanism that can impact the reliability of scaled PCM devices and it can be a serious show-stopper for this technology beyond the 32 nm technology node. Aim of this paper is to investigate the impact of the increasing current density on the functionality and reliability of scaled BJT-PCM architecture down to the 16 nm node

Keywords: Phase Change Memory, Reliability, BJT selector, PCM endurance, heater degradation.

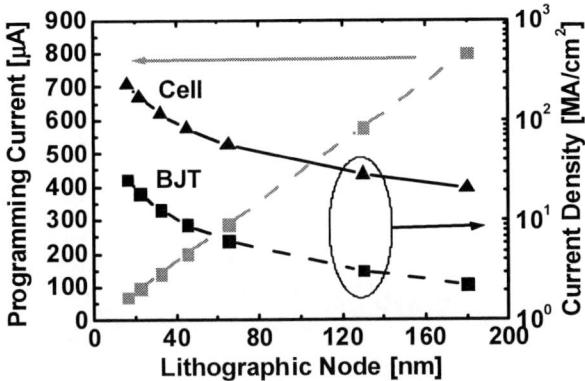

Fig.1: EXPECTED SCALING TREND FOR PCM PROGRAMMING CURRENT (LEFT AXIS) AND CURRENT DENSITIES IN THE STORAGE ELEMENT AND BJT (RIGHT AXIS).

I. INTRODUCTION

Despite Phase-Change Memory (PCM) technology has nowadays reached the product maturity at 90 and 65 nm technology [1, 2] and BJT-selected PCMs (BJT-PCM) are expected to become a mainstream non-volatile memory technology at 45 nm[3], to continue the PCM technology scaling roadmap the current density required to program the storage element will increase linearly with the lithography reduction. This is reported clearly in Fig. 1 where the current required by a PCM [1] to be programmed is reported on the left axis as a function of lithographic node. As already reported in other publications [4,5], the programming current scales linearly if an isotropic scaling is performed. On the right axis of Fig. 1 the current density sustained by both the storage element and the selecting element during the programming operation is shown. As a consequence of the linear decrease of the programming current, the current density increases with the lithographic node for both the phase change and the selecting elements. The difference in the current density in the two devices comes from the different area of the conductive path. Assuming a BJT selected PCM cell based on the µTrench [1] or "Wall" [3] architecture, the BJT is typically defined as half-pitch squared active areas, while the active region of the storage element is realized by the intersection of a GST line fabricated at the lithographic half-pitch with a thin film heater that can be even 10 times smaller than the lithographic capability. From Fig. 1 it can be seen that the current density becomes of the order of 20 *MA/cm²* in the BJT selector and in the order of 200 *MA/cm²* in the storage element at 16 nm node. As a consequence, the electrical stress due to the interaction between carriers and the lattice of active materials as well as the thermal stress due to the high temperatures reached with such high current densities can of course cause damages during the life of both the devices. ITRS recognizes that such high programming current density could be the main physical mechanism impacting the reliability of scaled PCM devices and it can be a serious show-stopper for this technology beyond the 32 nm technology node [6]. Aim of this paper is to investigate the impact of the increasing current density on the functionality and reliability of scaled BJT-PCM architecture down to the 16 nm node.

978-1-4244-5430-3/10 $26.00 © 2010 IEEE

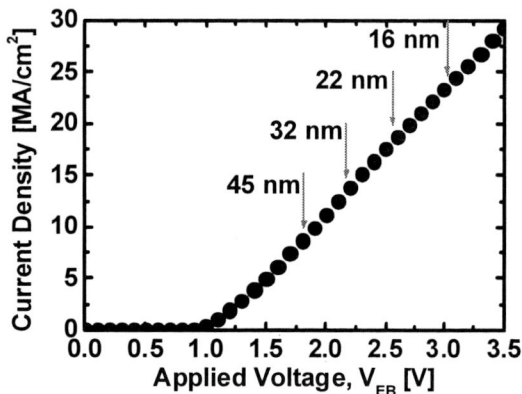

Fig. 2: EXPERIMENTAL DIRECT CURRENT DENSITY-VOLTAGE (J-V) CHARACTERISTIC OF A 45NM BJT SELECTOR. CURRENT DENSITIES REQUIRED TO PROGRAM DOWNSCALED CELLS ARE ALSO REPORTED.

II. BJT SELECTOR SCALING

The bipolar transistors used for phase change memory selection purpose is engineered mainly to provide low leakage and high forward current, while the gain of the transistor is not a major concern for the technology. This means that only a part of the current is collected by the common collector, while the remaining part of current goes into the base region. In this sense the emitter-base junction plays a major role, thus it is the object of this characterization. Fig. 2 reports a typical current density-voltage (J-V) curve measured on a 45 nm node BJT selector [3]. In the same picture the current density values required to program the PCM cell at the 32, 22, and 16 nm nodes are reported in accordance with the PCM scaling rules [4, 5], showing that the BJT selector is able to provide the required current density by incrementing the emitter-base voltage drop. It is worth noting that in the range of interest the current density is limited mainly by the series resistance along the current path. This resistance is constituted by the intrinsic emitter and base resistances (quasi neutral silicon regions) and

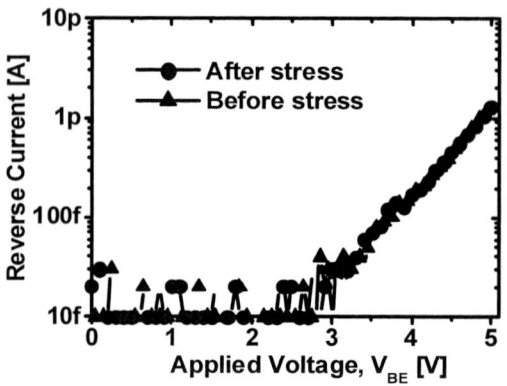

Fig.4: REVERSE CURR.-VOLTAGE (I-V) CHARACTERISTIC OF A 45NM BJT BEFORE AND AFTER A DC STRESS OF 100 S AT $25MA/CM^2$.

from extrinsic contributions (mainly silicon/silicide contact resistances formed before tungsten contacts definition). To assess the reliability impact, 45nm BJTs have been stressed for 100 s at 25 MA/cm^2, a current density value that exceeds the projected value at 16 nm and for a time that corresponds to about 10^9 programming cycles. Fig. 3 reports the current-voltage electrical characteristics in forward biased condition before and after the stress. It is worth noting that basically the electrical characteristic does not change and the current density for programming the storage element can be easily reached without any additional voltage drop. Practically no damage occurred to the junction functionality and to the resistance values along the current path. However to effectively prove that the junction is not damaged at all, a characterization of the reverse bias condition is needed. Fig. 4 reports the emitter current under reverse biased condition as a function of the base-emitter voltage before and after the DC stress. Also in this case no degradation of the BJT performances is detected, thus confirming that the emitter-base junction is not degraded by the very high current density required by the PCM programming

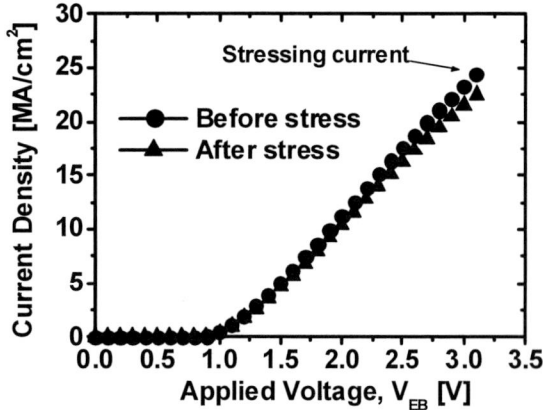

Fig.3: CURRENT DENSITY-VOLTAGE (J-V) CHARACTERISTIC OF A 45NM BJT SELECTOR BEFORE AND AFTER A DC STRESS OF 100 s AT $25MA/CM^2$.

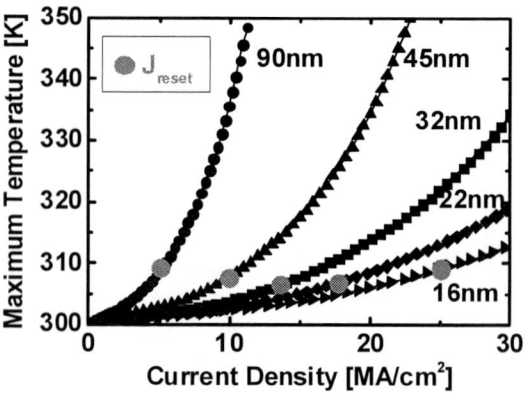

Fig. 5: SIMULATED MAXIMUM TEMPERATURE IN THE BJT SELECTOR DURING A DC STRESS PERFORMED WITH THE CURRENT DENSITY REPORTED IN THE X-AXIS.

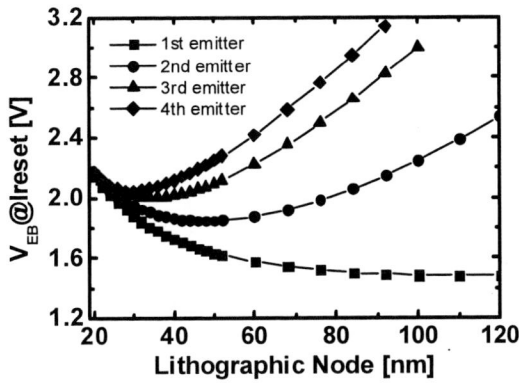

Fig.6: VOLTAGE DROP ON THE BIPOLAR REQUIRED TO PROGRAM THE STORAGE ELEMENT AS A FUNCTION OF THE LITHO NODE.

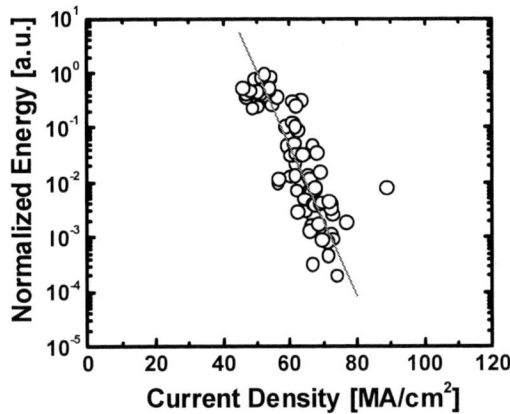

Fig.8: NORMALIZED FAILURE ENERGY AS A FUNCTION OF THE CURRENT DENSITY.

operation down to the 16nm technology node.

It is interesting to note that these accelerated stress experiments correctly reproduce the current density reached in scaled BJT, but the temperature reached inside the device can be altered by the thermal boundaries that cannot be correctly scaled in the experiment. To investigate the impact of the temperature on the scaling projection, the maximum temperature in the BJT as a function of the current density has been numerically simulated for a BJT with the same geometry of the devices electrically characterized and for scaled devices. As reported in Fig. 5, the same current density value makes the scaled devices always colder than the larger ones due to the closer proximity of the thermal boundaries. Experimental data collected on the 45 nm BJT thus represent a worst case condition for scaled devices due to the higher temperature reached during this accelerated stress experiment. Another aspect that is not taken into account in the experiment of Fig. 2 is the possible variation of the BJT base and emitter resistances due to the device scaling. Fig. 6 shows the voltage drop required to drive the PCM cell programming current as a function of the technology node and considering a plurality of

emitters for each base contact. The detailed architecture can be found in [3]. While the voltage drop of the emitter close to the base contact increases with the scaling due to the growth of the contact resistance, the far emitter shows an opposite trend, taking advantage from the reduction of the absolute distance from the base contact in scaled technologies. It follows that in scaled technologies the voltage drop marginally increases, but there is much more room for multi-emitters-per-base-contact architectures.

III. PCM CELL SCALING

To assess the effect of current density increase with scaling on cell reliability, several experimental data were collected on PCM cell integrated on 180, 90 and 45 nm CMOS platforms. For these tests, several writing and erasing operations have been repeatedly carried out (cycling) through an Agilent 81110 pulse generator with a constant 50 ns box reset pulse at a current $I_{reset\psi}$ and 150 ns box set pulse at a current I_{set}, never changed during the experiment. The cell failure is defined when the programmed set resistance increases by a factor 3 with respect to the minimum set level or when the reset

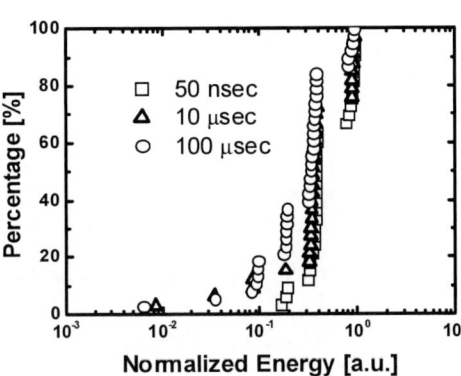

Fig.7: NORMALIZED FAILURE ENERGY DISTRIBUTION OBTAINED FOR DIFFERENT CYCLING PULSE WIDTHS.

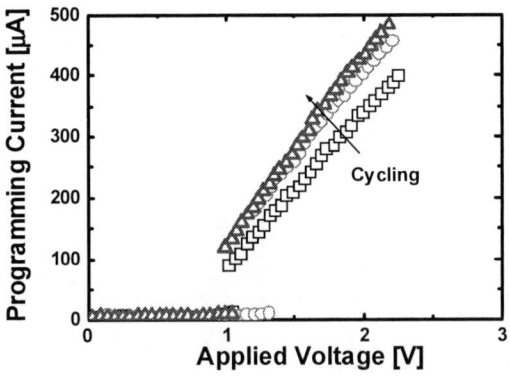

Fig.9: CURRENT-VOLTAGE (I-V) CHARACTERISTICS EVOLUTION DURING CYCLING.

978-1-4244-5430-3/10 $26.00 © 2010 IEEE

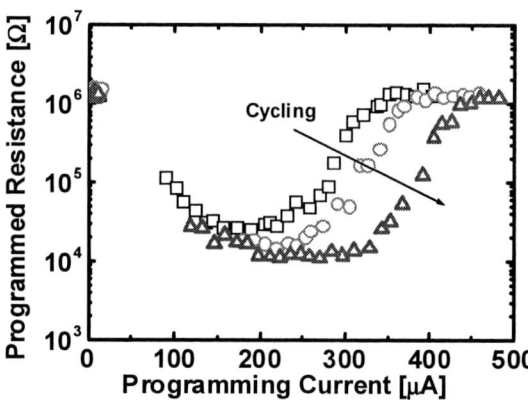

Fig.10: PROGRAMMING CHARACTERISTICS (R-I) EVOLUTION DURING CYCLING, SHOWING THE I_{RESET} SHIFT TOWARDS HIGHER CURRENTS.

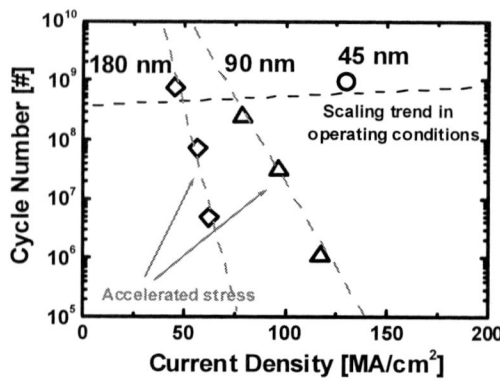

Fig. 12: CYCLES NUMBER AS A FUNCTION OF THE CURRENT DENSITY FOR 3 DIFFERENT LITHO NODES.

resistance decreases by a factor 3. To investigate the nature of the device degradation, accelerated stress experiments were carried out with increasing programming currents, I_{reset}, measuring several cells of the same wafer (about 40). The goal of this characterization was to asses the phenomenology of the failure mechanism considering as major players the current density used and the duration of the programming pulse, thus defining the energy delivered to the device. The energy delivered to each cell was calculated as the cumulative sum of the energy provided during each cycle computed as the product of programming voltage, current, and pulse duration for both the reset and set pulses (all the delivered pulses are rectangular). The cumulative sum of the energy is performed till the failure of the device occurs. It is worth noting that eventual changes in the electrical characteristics of the cells (reset and set voltages and currents) are properly considered since their values are updated at each readout (three times per decade). Fig. 7 reports the failure energy, E, normalized to its maximum value, as a function of the reset pulse width. It is important to note that the energy at the failure, E, does not

depend on time. This means that the physical mechanism behind is cumulative with time and the effect of coming back to the equilibrium with the programming current/voltage pulse does not take any role.

Fig. 8 shows the normalized failure energy as a function of the reset pulse current density. It is worth noting that the failure energy is exponentially activated by the current density, and thus the current density seems to play a major role for the failure during cycling. In order to better understand the physical mechanism we monitor the electrical characteristics of the PCM during cycling. Results are reported in Fig. 9 and 10 where the current-voltage (I-V) and programming curve (R-I) are reported for 3 different cycles. In Fig. 9 it is clear that the heater resistance decreases with time, suggesting that a modification of the heater material is occurring. Consistently, Fig. 10 shows a decrease in the set resistance (at fixed set current and pulse length) as well as an increase in the programming current. Note that the set resistance change cannot be fully explained by the decrease in the heater resistance. Large part of the set resistance decrease may be due to a change in the GST material composition [7] and thus in the crystallization speed of Sb-rich resulting alloy [8]. The cell failure, on the other hand, can be ascribed to a decreasing of the on-state differential resistance (*i.e.*, the heater resistance), that actually implies a reduced capability to heat up the GST during the reset pulse thus leading to an increase of the saturation reset current, I_{sat}. If the chip circuitry is not able to follow the programming current increase of the PCM device, the bit fails remaining in the set state. The behavior of Fig. 9 and Fig. 10 has been observed in all the tested PCM devices realized with 180nm, 90nm and 45 nm lithographic nodes. To understand if the decrease in the heater resistance is directly driven by the current density or induced by the temperature reached during programming, the same experiment has been repeated on devices without the GST, with the heater in direct contact with the top electrode, where a lower temperature is expected due to the closer proximity of the top metallic cold thermal boundary. To compare the results with the previous experiments, the device failure was defined by monitoring the decrease in the heater resistance as in Fig. 9. Fig. 11 shows that devices without GST has lower heater degradation than standard PCM

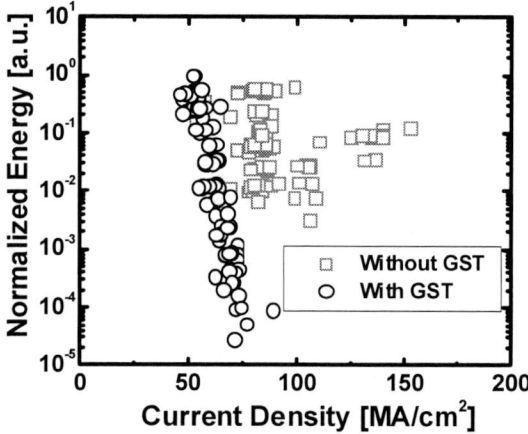

Fig. 11: NORMALIZED FAILURE ENERGY AS A FUNCTION OF CURRENT DENSITY FOR DEVICE WITH AND WITHOUT GST.

devices, thus suggesting that the degradation mechanism is mainly driven by the higher temperatures reached in accelerated stress experiments and not directly by current density. This is definitively pointed out in Fig. 12 where the device endurance is reported as a function of programming current density for three different technology nodes. It is worth nothing that at a fixed lithographic node the overall cycle life is strongly reduced with the current density increase. On the other hand, no impact is observed on scaled devices where the higher current density values are needed to reach the same GST melting temperature.

IV. Conclusions

An extended experimental analysis of the impact of the increase of the current density with scaling on both selector and PCM cell reliability was presented. It was shown that no effect on BJT and storage element reliability is expected in scaled BJT-PCM architectures down to the 16 nm technology node. Temperatures higher than in standard operating conditions were demonstrated to be responsible for the PCM cell endurance degradation observed in accelerated tests carried out at high current densities.

Acknowledgment

The authors acknowledge Prof. Daniele Ielmini and colleagues Fabio Pellizzer and Eleonora Brini for fruitful discussions.

References

[1] F. Pellizzer et al., *Symp. on VLSI Tech.*, pp. 150–151, 2006.

[2] Oh J. H. et al., IEDM Tech. Dig., 49-52, 2006

[3] G. Servalli, *IEDM Tech. Dig.*, pp. 113-116, 2009.

[4] U. Russo et al., *Transactions on Electron Devices*, vol. 55, n 2, pp. 506-514, 2008

[5] A. Pirovano, *IEDM Tech. Dig.*, pp. 699–702, 2003.

[6] ITRS 2008 update, http://www.itrs.net/Links/2008ITRS/Home2008.htm.

[7] C. Kim et al., Applied Physics Letters, (94) 193504-1/193504-3

[8] M. Boniardi et al., Accepted to IMW 2010

TRADE-OFF BETWEEN DATA RETENTION AND RESET IN NiO RRAMs

D. Ielmini, F. Nardi, C. Cagli and A. L. Lacaita[§],

Dipartimento di Elettronica ed Informazione and IU.NET, Politecnico di Milano, piazza L. da Vinci 32, 20133 Milano, Italy.
Email: ielmini@elet.polimi.it [§] also with CNR-IFN, Politecnico di Milano, Milano, Italy.

ABSTRACT

NiO-based resistive-switching memory (RRAM) is a promising new technology for high-density non-volatile storage. The main obstacles to practical application in nonvolatile memories are the variability of program/erase voltages, the large and hardly scalable programming current and the cell reliability. We have investigated data retention in RRAM samples with NiO as active switching material. Temperature-accelerated bake experiments show that data retention limited by oxidation of the conductive filament (CF) obey an Arrhenius law, while the retention time decreases for decreasing size of the CF. The results are interpreted by a physical model for CF dissolution in an oxidizing environment, which can be applied to both data retention extrapolation at long times and low temperatures, and the reset operation in the ns/ms regime. The model is verified with experimental data, and the tradeoff between data retention and reset current is finally discussed.

[*Keywords*: Resistive-switching memory (RRAM), non-volatile memory, reliability estimation, reliability modeling]

INTRODUCTION

Resistive-switching memory (RRAM) is an emerging storage concept with significant potential applications in the area of non-volatile and high density memories [1, 2]. RRAM is a two-terminal resistive memory where the change of resistivity in an active switching material, usually a transition metal oxide, is exploited [1, 3]. Two schemes of resistance switching have been shown, depending on the switching material and the polarity of program/erase: in unipolar devices, both program and erase can be performed at the same bias polarity, *e.g.* positive voltage, while bipolar switching requires a change of bias polarity between program and erase. RRAM generally displays a large resistance window compatible with multilevel operation and attracts interest thanks to good potential scalability and integration capability. On the other hand, concerns have been raised regarding a relatively large programming current and reliability issues, including cycling endurance and data retention.

Fig. 1 shows typical current-voltage ($I - V$) characteristics for a RRAM cell with NiO as insulating switching material, which is considered in this work. The $I - V$ curves were collected for a memory cell with Pt top and bottom electrode and large area (about $1600~\mu m^2$) [4]. The curves were measured after the so-called forming operation, that is a dielectric breakdown of the original NiO layer resulting in the formation of a localized conductive filament (CF) characterized by low resistance [1, 3]. The figure shows $I - V$ curves for both the low-

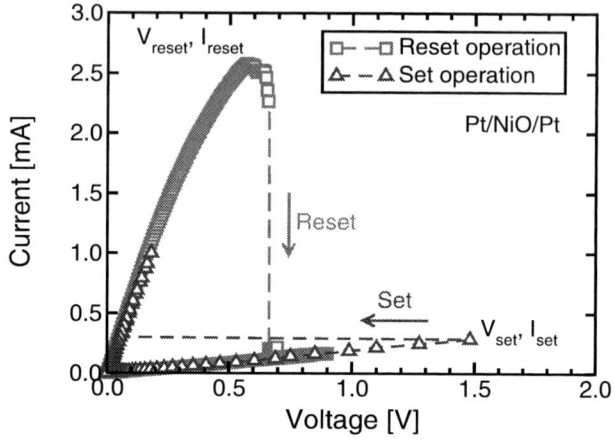

Figure 1: MEASURED $I - V$ CURVES SHOWING THE SET AND RESET TRANSITIONS FOR THE RESET AND SET STATES, RESPECTIVELY. A UNIPOLAR RRAM CELL WITH NiO AS ACTIVE SWITCHING MATERIAL WAS USED [4].

resistance (set) state and the high-resistance (reset) state. In the set state the CF is fully connecting the top and bottom electrodes, thus shunting the high-resistance NiO layer, while in the reset state the CF is partially or fully discontinued, resulting in a high resistance. The figure also shows the two operations for the transitions between the set and the reset states. The application of a sufficiently high voltage and current to the set state results in a sudden drop of the current, *i.e.* increase of resistance, marking the transition to the reset state. This is referred to as reset operation, and is physically explained as a dissolution of the CF by solid-state diffusion and oxidation [3, 5]. The application of a large voltage (usually larger than that required for the reset operation) to the reset state results in a sudden drop (snap back) of voltage, marking the transition to a lower resistance. This is referred to as set operation, and is explained as a localized chemical reduction activated by the temperature and/or the current density, capable of restoring the metallic-like CF responsible for the low resistance. The local high temperature and current density allowing for the set transition are possible through a threshold switching, that is an electronic transition to a high-conductivity state in the insulating material [6]. Extensive dynamic analysis of set and reset processes are in agreement with these physical models in NiO RRAM cells [7].

As indicated in Fig. 1, the reset current is generally large, usually in the $0.1 - 1$ mA range. This is due to the rather inef-

978-1-4244-5430-3/10 $26.00 © 2010 IEEE

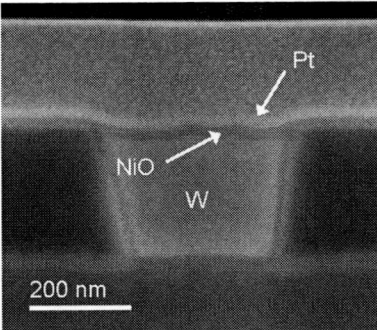

Figure 2: TEM PICTURE OF A NiO RRAM CELL STUDIED IN THIS WORK.

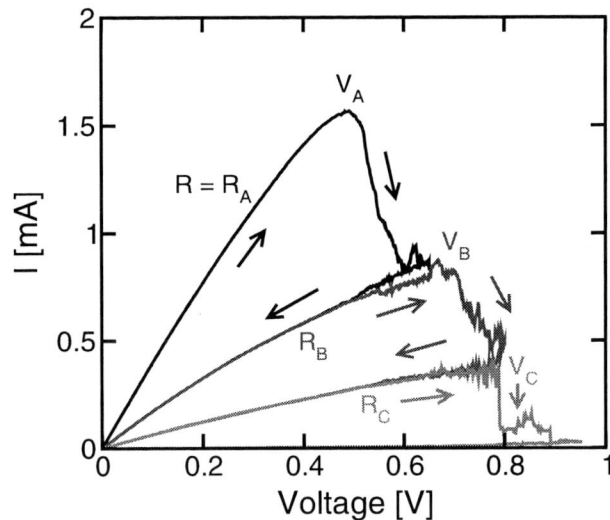

Figure 3: MEASURED $I - V$ CURVES SHOWING THE SET/RESET TRANSITIONS (a) AND THE INCREMENTAL RESET ALGORITHM ENABLING FINE TUNING OF R (b).

ficient Joule heating in a high-conductivity CF. In fact, the low thermal resistance enhances thermal conduction and makes it difficult to reach high temperatures needed to activate diffusion and oxidation in the CF. At the same time, the low electrical resistance results in a high current flowing through the CF. The large reset current limits high-density memory application, since it requires relatively large and/or optimized select devices. To limit the reset current, it has been proposed that the CF size should be limited, allowing for a larger electrical resistance hence a smaller dissipated current during reset [3]. Experimental verification of this idea have been provided, with the use of select transistor capable of limiting the supply current during the set transition and, consequently, the size of the resulting CF [5, 8, 9]. However, the possible reliability implications of small CF have not been addressed to date.

This work addresses the trade-off between reset current reduction by CF size control and reliability. Temperature-accelerated retention experiments for variable resistance states show that data retention obeys an Arrhenius law and that retention times degrades for increasing resistance and decreasing CF size. This is explained by a size-dependent diffusion model. Physical models for CF dissolution controlling data retention and reset transitions are developed, allowing for a quantitative assessment of the tradeoff between reset and retention time in NiO RRAMs.

EXPERIMENTS

Fig. 2 shows a cross section SEM picture of the RRAM cell considered in this work. The cell consists of a MIM structure with a W bottom electrode with diameter from 0.18 to 1 μm, an active NiO film deposited by atomic layer deposition (ALD) with thickness $t_{NiO} = 20$ nm, and a top Pt electrode [10, 11]. The area of the device can be estimated by the bottom contact size, namely in the range from 0.032 to 1 μm^2. However, the electrical characteristics were not found to depend significantly on the sample area[11]. This is because the switching behavior in NiO is confined at the CF, thus the CF area controls both the $I - V$ curves in the low resistance state and the reset voltage and current. Also, since the CF is usually not completely switched off at reset, the reset state also suffer from a certain localization. Only the initial resistance in the pre-forming state was found to

depend predictably on the sample area, *i.e.* the pre-forming resistance scale with the inverse of the sample area [11].

To allow for a broad investigation of the programming and reliability for different programmed states, we changed the resistance of the cell by a partial reset algorithm as shown by the $I - V$ curves in Fig. 3. Applying a first reset sweep (black curve in the figure) to the set state with initial resistance of about 0.2 kΩ, the onset of reset occurred at a reset voltage $V_A = 0.5$ V. At this voltage, the resistance was found to increase gradually, instead of the sharp reset transition usually found in large area samples (see *e.g.* Fig. 1). The first sweep in the figure was stopped at a voltage V_B of about 0.65 V, resulting in an increase of resistance to about 0.5 kΩ. A second sweep (blue curve in the figure) was applied to this cell state and interrupted at a slightly larger voltage $V_C = 0.8$ V, yielding a further increase of resistance to a value of about 1 kΩ. more generally, the partial reset algorithm allowed to tune the resistance from about 50 (full set state) to 10^8 Ω (full reset state) by changing the stopping voltage in the reset sweep. Similar programming procedures were previously shown for bipolar HfO_2 based RRAMs [12].

NUMERICAL MODEL

Multiple states in Fig. 3 can be interpreted as due to a different size and/or a different composition of the CF, as shown in Fig. 4: The full set state (a) corresponds to a metallic CF with relatively large diameter [7]. Partial reset can result in a continuous CF with smaller diameter (b) or in a dispersed CF (c) consisting of residual conductive elements (*e.g.* excess Ni and O vacancies, shown as blue dots in Fig. 4). Due to the discontinuous nature of the CF in Fig. 4c, the resistivity has increased being dictated by Poole-Frenkel transport mechanism at localized states instead of carrier drift which is typical for metals or highly doped semiconductors. Finally, the full reset state (d)

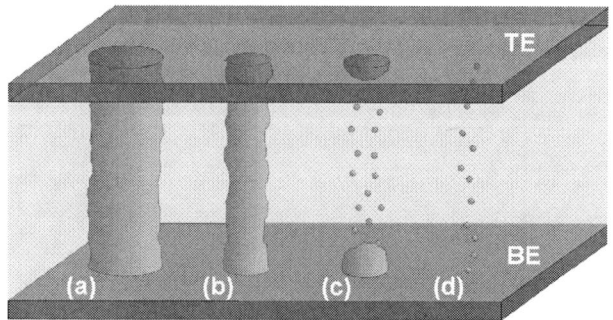

Figure 4: SCHEMATIC FOR RRAM MULTIPLE STATES: FULL-SET STATE (a), INTERMEDIATE STATE WITH SMALL METALLIC CF (b), INTERMEDIATE STATE WITH SEMICONDUCTIVE CF (c) AND FULL RESET STATE (d).

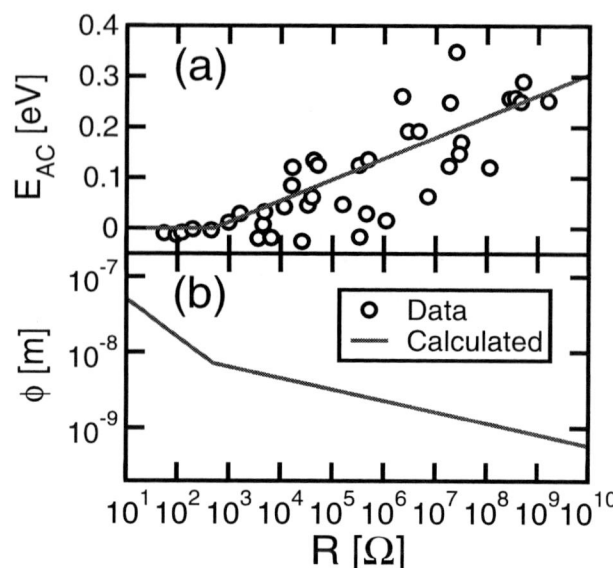

Figure 5: MEASURED E_{AC} WITH ITS PIECEWISE LINEAR FIT (a) AND CALCULATED CF DIAMETER ϕ (b) FROM EQ. (1), AS A FUNCTION OF R.

consists of quasi-stoichiometric NiO with negligible concentration of localized states. Thus the resistance in this state is expected to approach the pre-forming value of the sample.

To support this physical picture for different programmed states in our samples, we measured the temperature dependence of resistance. From $R - T$ data, we collected the activation energy E_{AC} for conduction as the slope of resistance in the Arrhenius plot. Fig. 5a shows the measured E_{AC}: the activation energy is small and negative for low resistance, indicating a metallic conduction, while E_{AC} becomes positive at about $R_{metal} = 0.5$ kΩ and increases for increasing R. A negative or zero activation energy is consistent with the picture of metallic-like continuous CF in Fig. 4a and b. On the other hand, a positive E_{AC} indicates a semiconductor-like transport and can be understood by the picture in Fig. 4c and d. Based on this picture, in fact, E_{AC} might indicate the difference between the Fermi level, marking the highest occupied energy level for localized states, and the conduction or valence mobility edge, where carriers must at least be thermally excited to contribute to conduction [13]. Data for E_{AC} in the figure display a significant spread, which can be explained by the combined dependence of R on size and material composition. For instance, a similar value of high resistance can be obtained by a relatively narrow CF with metallic nature (low resistivity) and low E_{AC}, or by a relatively large CF characterized by semiconductor-like conduction (high resistivity) and large positive E_{AC}. Note also that an E_{AC} value can only be attributed to metallic-states R within a small temperature range, since these states do not obey a strict Arrhenius dependence on temperature.

The extracted E_{AC} allows to estimate the effective CF area A by the formula:

$$R = \rho_0 exp\left(\frac{E_{AC}}{kT}\right)\frac{t_{NiO}}{A}, \qquad (1)$$

where k is the Boltzmann constant and $\rho_0 = 100$ $\mu\Omega$cm [5] is the pre-exponential constant for resistivity. The NiO layer thickness t_{NiO} was used to identify the CF length in Eq. (1) according to the schematic in Fig. 4. Fig. 5b shows the extracted CF diameter $\phi = (4A/\pi)^{1/2}$ obtained from Eq. (1) using the piecewise-linear fitting of E_{AC} in Fig. 5a, where

$E_{AC} = 0$ was assumed for metallic states ($R < R_{metal}$) while $E_{AC} = E_{AC0}\log(R/R_{metal})$ with $E_{AC0} = 0.0178$ eV was assumed for semiconductor-like states ($R > R_{metal}$). The resulting CF diameter ϕ changes only by a factor 20 in the investigated R window, from 20 nm at $R = 100$ Ω to about 1 nm at 3×10^8 Ω. Note again that ϕ should be viewed as an effective value for the CF size, taking into account the possible irregular shape of CFs, possibly including multiple branches and non-uniform cross section. Also, conduction may be largely non-uniform at relatively large resistivity as a result of percolation effects: In this case, the extracted effective area could yield an estimate for the conduction path in the percolation process.

NUMERICAL MODEL

Although decreasing the CF size results in a smaller I_{reset} [5, 8, 9], it also raises a concern in terms of CF stability. In fact, T-activated diffusion and oxidation of small CF can result in data loss at relatively low T/short times. This is in agreement with previous results for bipolar CuO-based RRAMs, showing that the set state obtained with relatively high current limit and displaying relatively low resistance was less sensitive to elevated-temperature bake [14]. A similar report on $Cu : MoO_x/GdO_x$ capacitors showed that the set states with high resistance were affected by a shorter retention time [15]. To assess data-loss effects for variable size of the CF, we performed high-T annealing experiments on RRAM cells. To handle the large spread of retention behavior in our samples, we performed the annealing at a fixed time and temperature on a large set of samples, usually including 50 cells programmed at approximately the same initial resistance. The resistance of the cells was measured at room temperature before and after bake, and after each bake the cells were set again and subjected

978-1-4244-5430-3/10 $26.00 © 2010 IEEE

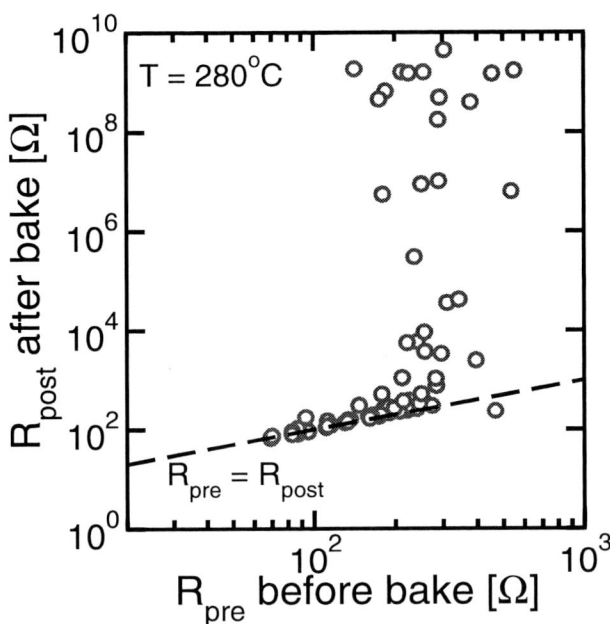

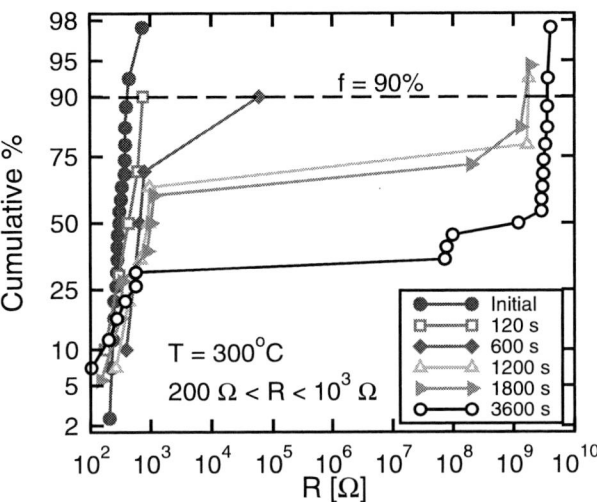

Figure 6: CORRELATION PLOT OF R MEASURED BEFORE (R_{pre}) AND AFTER BAKE (R_{post}).

Figure 7: CUMULATIVE DISTRIBUTIONS OF R FOR INCREASING ANNEALING TIMES AT 300C, STARTING FROM A SET STATE RESISTANCE BETWEEN 0.2 AND 1 KΩ.

to a new bake experiment for a different time and/or temperature. Sufficiently long annealing times were used in our bake experiments, to ensure that a steady-state, constant temperature throughout the annealing chamber and the samples was reached during the experiment.

Fig. 6 shows the correlation plot of the post-bake resistance R_{pre} as a function of pre-bake resistance R_{post}, for an annealing temperature $T = 280°C$. The dashed line indicating $R_{pre} = R_{post}$ and corresponding to negligible effects of the annealing on resistance is also shown for reference. From the figure, one may note that R_{post} is generally larger than R_{pre}, as a result of oxidation and dissolution of the CF. Most importantly, cells with relatively large R are weaker against thermal reset, as indicated by the larger tendency to move to a high R_{post}. This demonstrates the size-dependent data loss effects in our samples, thus posing a reliability issue for scaled RRAMs with small CF.

Fig. 7 shows cumulative distributions of measured R before annealing (initial) and after bake experiments for increasing annealing times at 300°C. Cells were programmed at an initial R between 0.2 and 1 kΩ. Annealing results in a shift to higher resistance, usually close to the reset state. This indicates a sharp transition between the set and reset state, also consistent with the generally small voltage range where the reset transition takes place on the $I - V$ curve. The amount of cells belonging to the moving tail increases for increasing time. This behavior is similar to the previously observed data-loss process in phase-change memories, and can be explained by the relatively large retention time distribution and by the sharp transition from set to reset state as the CF is thermally dissolved [16].

From the cumulative distributions in Fig. 7, the average evolution of resistance with time can be obtained, similarly to the

established procedure in Flash memories [17]. The cumulative distributions were analyzed at a constant percentile f (e.g. see the horizontal line corresponding to $f = 90\%$ in Fig. 7). In correspondence of the intersections between the cumulative distributions and the horizontal line at constant f, resistances values were extracted and plotted as a function of the corresponding annealing time. Fig. 8 shows the resulting R as a function of time for increasing percentile $f = 25, 50, 75$ and 90%, obtained from the cumulative distributions in Fig. 7. The results indicate relatively abrupt transitions from the set to the reset state, where the transition time increases for decreasing f. This is clearly the consequence of the ordering procedure inherent to the cumulative distribution plot. Note in particular that no resistance transition could be detected for $f = 25\%$ within the experimental time range considered.

From Fig. 8, one can define a retention time τ_R as the time for a $10\times$ increase of R (10 kΩ in the figure). Fig. 9a shows the Arrhenius plot of retention times τ_R at variable f from 25% to 90%. The measured retention time τ_R obeys the Arrhenius law with an activation energy for retention $E_{AR} = 1.21$ eV for $f = 50\%$, in good agreement with previous results (1.4 eV) of voltage-driven reset studies [7]. Extrapolation indicates a 10-years-retention T of 110°C for $f = 50\%$. However, τ_R decreases at low f, confirming that statistical studies are mandatory for RRAM reliability assessments. From a careful inspection of the figure, one may notice that more data points are available at large f as compared to small f: This is due to the already mentioned shift of τ_R to longer times for decreasing f, hence away from the statistical tail of moving cells. Longer experimental time would obviously result in more point for τ_R at smaller percentiles.

Fig. 9b shows the Arrhenius plots for three different ranges of R at $f = 90\%$. Data show that τ_R decreases for increasing R, hence decreasing CF size, at a given temperature. Although τ_R is reported for $f = 90\%$, similar results may be expected

978-1-4244-5430-3/10 $26.00 © 2010 IEEE

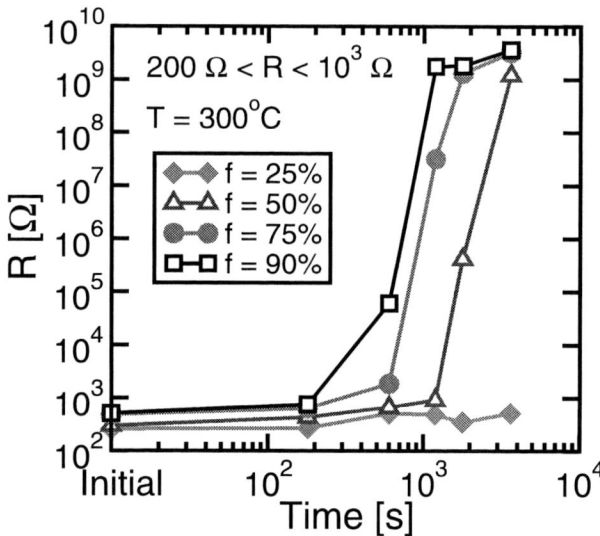

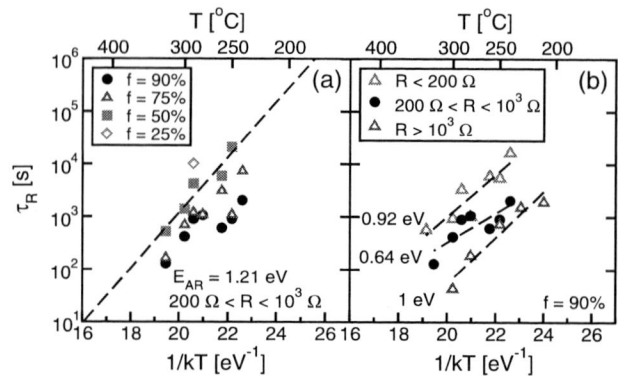

Figure 9: ARRHENIUS PLOT OF τ_R FOR INCREASING PERCENTILE f AND EXTRAPOLATION AT 10 YEARS (a) AND ARRHENIUS PLOT OF DATA RETENTION TIME τ_R FOR INCREASING R AT $f = 90\%$ (b).

Figure 8: TIME EVOLUTION OF RESISTANCE AT CONSTANT PERCENTILE f, FROM FIG. 7.

for smaller f, e.g. 50%, due to the similar behavior at different f in the Arrhenius plot of Fig. 9a. The results in Fig. 9b demonstrate that smaller CFs are affected by a shorter τ_R. The extracted activation energy E_{AR} in the figure is between 0.64 and 1 eV, with no clear dependence on initial resistance. Note however that the extracted E_{AR} may be affected by a measurement spread due to the variability of data loss in memory cells. More data along an extensive temperature range are needed for a better and more reliable evaluation of the activation energy as a function of the cell resistance.

RETENTION-RESET TRADEOFF

To explain the R-dependence of reliability, we modeled CF oxidation by reaction-diffusion in cylindrical coordinates. Diffusion of Ni atoms/ions from the CF was assumed the limiting step in the reaction-diffusion process of oxidation. This is in agreement with results on thermal oxidation of Ni films [18]. The following diffusion equation in cylindrical coordinates was solved numerically:

$$\frac{\partial n}{\partial t} = \frac{D}{r} \frac{\partial}{\partial r} r \frac{\partial n}{\partial r}, \qquad (2)$$

where r is the radial coordinate, n is the concentration of conductive elements (e.g. Ni atoms/ions) and D is their diffusivity. The CF was assumed with cylindrical shape as in Fig. 4 with infinite length to reduce the numerical problem to 2D. The retention time was defined as the time for the Ni concentration in the middle of the CF to reduce by a factor 10 or 100. Fig. 10 shows the calculated τ_R, as a function of the initial CF diameter ϕ in the simulation. The retention time increases with ϕ^2, as evident from the slope 2 in the loglog plot. Simulation results in the figure were obtained for the same value of D, hence at a fixed temperature.

To describe both the temperature dependence (Arrhenius model) and the geometrical dependence on ϕ^2, we introduced

the compact formula:

$$\tau_R = \tau_{R0} \left(\frac{\phi}{\phi_0} \right)^2 exp \left(\frac{E_{AR}}{kT} \right), \qquad (3)$$

where τ_{R0} and ϕ_0 are constants. Fig. 11a shows the experimental τ_R at 250°C, obtained from Arrhenius regressions in Fig. 9b, and results by Eq. (3), using $E_{AR} = 1.2$ eV. The calculated CF diameter ϕ from Fig. 5b was used in Eq. (3). These results suggest that the CF size ϕ should be maximized and R minimized for best data retention.

Since both reset and data-loss are due to CF-oxidation process, improving retention negatively impacts reset and vice versa. Fig. 11 shows measured V_{reset} (b) and I_{reset} (c) as a function of R. Fig. 11b also show calculation results by:

$$V_{reset} = \left(\frac{R}{R_{th}} (T_{reset} - T_0) \right)^{1/2}, \qquad (4)$$

where R_{th} is the effective thermal resistance, T_0 the room temperature and T_{reset} the critical T for reset obtained from Eq. (3) as:

$$T_{reset} = \frac{E_{AR}}{k \log \left(\frac{\tau_{reset}}{\tau_{R0}} \left(\frac{\phi_0}{\phi} \right)^2 \right)}, \qquad (5)$$

where τ_{reset} indicates a specific reset time [7]. V_{reset} first decreases with R, due to the decreasing ϕ, then increases for increasing R above about 1 kΩ. The V_{reset} rise is due to the decrease in the ratio R_{th}/R for decreasing CF size, as a result of the parallel thermal conduction out of the CF through the surrounding NiO [5]. On the other hand, the electrical resistance scales with the inverse CF area, with no significant contribution from the NiO phase. Note that the increase of reset voltage for decreasing CF size was already shown by reset simulations [5]. Note in Fig. 11b that fast reset, e.g. at 10 ns, requires a larger voltage, because a larger temperature is needed to accelerate diffusion and oxidation in a shorter time, according to the Arrhenius law in Eq. (3). The reset current was calculated by:

$$I_{reset} = V_{reset}/R, \qquad (6)$$

978-1-4244-5430-3/10 $26.00 © 2010 IEEE

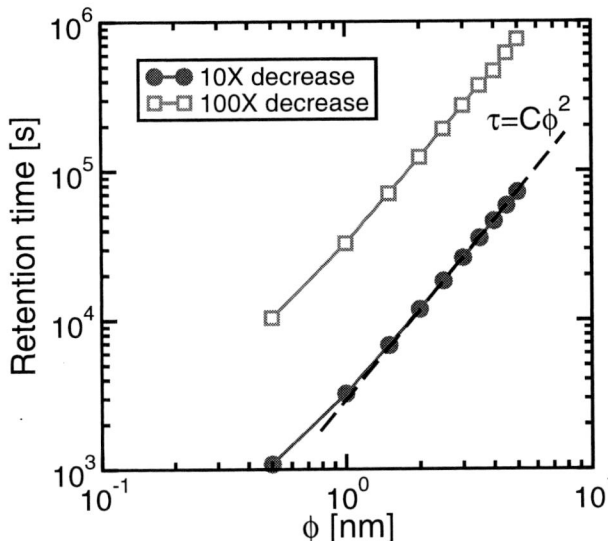

Figure 10: CALCULATED τ_R AS A FUNCTION OF ϕ, ACCORDING TO REACTION-DIFFUSION SIMULATIONS IN RADIAL COORDINATES.

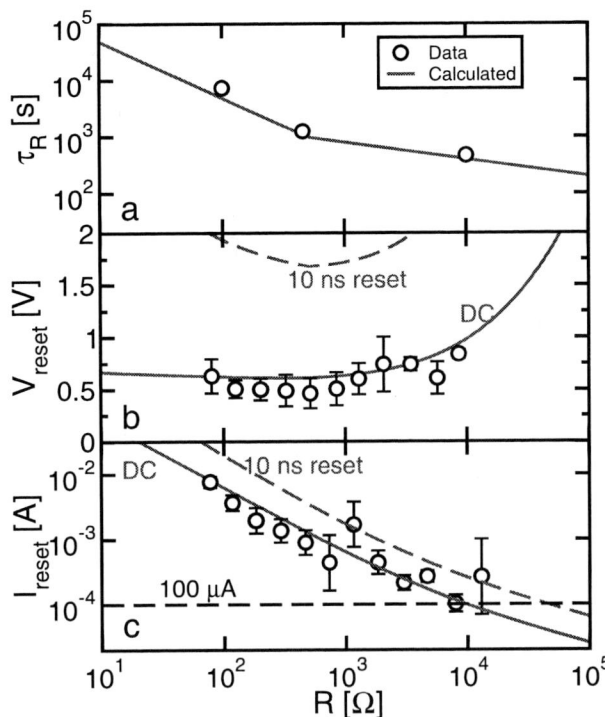

Figure 11: MEASURED AND CALCULATED τ_R (a), V_{reset} (b) AND I_{reset} (c), AS A FUNCTION OF R. BOTH DC AND PULSED (10 NS) RESET CONDITIONS ARE SHOWN.

thus neglecting the non-linear behavior in the $I - V$ curves due to the thermally-induced resistance increase in metallic CFs [3].

DC reset with $I_{reset} < 100$ μA, *i.e.* a possible reference for low-current reset [8], requires $R > 10^4$ Ω, yielding a relatively poor data retention in Fig. 11a. Pulse-mode (10 ns) reset condition at 100 μA yields an even larger $R = 5 \times 10^4$ Ω. A better tradeoff between τ_R and I_{reset} may require further studies aimed at an optimization of active/electrode materials, cell structure and set/reset algorithms.

CONCLUSIONS

This work provides and extensive characterization of data retention in RRAM cells as a function of annealing time, annealing temperature and initial resistance. From a physical interpretation of RRAM states and a statistical data-retention characterization, we link retention time to CF size. Numerical simulations based on the reaction-diffusion model are used to provide a quantitative explanation of experimental findings. From these results, the tradeoff between reset current and data retention in RRAM was addressed.

ACKNOWLEDGMENTS

The authors acknowledge S. Spiga and E. Cianci (MDM) for sample preparation, A. Vigani for experimental assistance and EU for partial support (project Emma FP6-033751).

REFERENCES

[1] I. Baek, M. Lee, S. Seo, M. Lee, D. Seo, D.-S. Suh, J. Park, S. Park, H. Kim, I. Yoo, U.-I. Chung, and J. Moon, "Highly scalable nonvolatile resistive memory using simple binary oxide driven by asymmetric unipolar voltage pulses," in *IEDM Tech. Dig.*, 2004, pp. 587–590.

[2] R. Waser and M. Aono, "Nanoionics-based resistive switching memories," *Nature Mater.*, vol. 6, pp. 833–840, 2007.

[3] U. Russo, D. Ielmini, C. Cagli, and A. L. Lacaita, "Filament conduction and reset mechanism in NiO-based resistive-switching memory (RRAM) devices," *IEEE Trans. Electron Devices*, vol. 56, pp. 186–192, 2009.

[4] U. Russo, D. Kamalanathan, D. Ielmini, A. L. Lacaita, and M. N. Kozicki, "Study of multilevel programming in programmable metallization cell (PMC) memory," *IEEE Trans. Electron Devices*, vol. 56, pp. 1040–1047, 2009.

[5] U. Russo, D. Ielmini, C. Cagli, and A. L. Lacaita, "Self-accelerated thermal dissolution model for reset programming in unipolar resistive-switching memory (RRAM) devices," *IEEE Trans. Electron Devices*, vol. 56, pp. 193–200, 2009.

[6] D. Ielmini, C. Cagli, and F. Nardi, "Resistance transition in metal oxides induced by electronic threshold switching," *Appl. Phys. Lett.*, vol. 94, p. 063511, 2009.

[7] C. Cagli, F. Nardi, and D. Ielmini, "Modeling of set/reset operations in NiO-based resistive-switching memory RRAM devices," *IEEE Trans. Electron Devices*, vol. 56, pp. 1712–1720, 2009.

[8] Y. Sato, K. Tsunoda, K. Kinoshita, H. Noshiro, M. Aoki, and Y. Sugiyama, "Sub-100 − μA reset current of nickel oxide resistive memory through control of filamentary conductance by current limit of MOSFET," *IEEE Trans. Electron Devices*, vol. 55, pp. 1185–1191, 2008.

[9] K. Kinoshita, K. Tsunoda, Y. Sato, H. Noshiro, S. Yagaki, M. Aoki, and Y. Sugiyama, "Reduction in the reset current in a resistive random access memory consisting of NiO$_x$ brought

about by reducing a parasitic capacitance," *Appl. Phys. Lett.*, vol. 93, p. 033506, 2008.

[10] S. Spiga, A. Lamperti, C. Wiemer, M. Perego, E. Cianci, G. Tallarida, H. L. Lu, M. Alia, F. G. Volpe, and M. Fanciulli, "Resistance switching in amorphous and crystalline binary oxides grown by electron beam evaporation and atomic layer deposition," *Microelectron. Eng.*, vol. 85, pp. 2414–2419, 2008.

[11] A. Demolliens, C. Muller, D. Deleruyelle, S. Spiga, E. Cianci, M. Fanciulli, F. Nardi, C. Cagli, and D. Ielmini, "Reliability of NiO-based resistive switching memory (ReRAM) elements with pillar W bottom electrode," in *International Memory Workshop*, 2009, pp. 25–27.

[12] H. Y. Lee, P. S. Chen, T. Y. Wu, Y. S. Chen, C. C. Wang, P. J. Tzeng, C. H. Lin, F. Chen, C. H. Lien, and M.-J. Tsai, "Low power and high speed bipolar switching with a thin reactive ti buffer layer in robust HfO_2 based RRAM," in *IEDM Tech. Dig.*, 2008, pp. 297–300.

[13] D. Ielmini, "Threshold switching mechanism by high-field energy gain in the hopping transport of chalcogenide glasses," *Phys. Rev. B*, vol. 78, p. 035308, 2008.

[14] T.-N. Fang, S. Kaza, S. Haddad, A. Chen, Y.-C. Wu, Z. Lan, S. Avanzino, D. Liao, C. Gopalan, S. Choi, S. Mahdavi, M. Buynoski, Y. Lin, C. Marrian, C. Bill, M. VanBuskirk, and M. Taguchi, "Erase mechanism for copper oxide resistive switching memory cells with nickel electrode," in *IEDM Tech. Dig.*, 2006, pp. 789–792.

[15] J. Park, J. Yoon, M. Jo, D.-J. Seong, J. Lee, W. Lee, J. Shin, E.-M. Bourin, and H. Hwang, "Effect of filament resistance on retention characteristics of ReRAM," in *as discussed at 2009 IEEE SISC, Arlington, VA*, 2009.

[16] D. Mantegazza, D. Ielmini, E. Varesi, A. Pirovano, and A. L. Lacaita, "Statistical analysis and modeling of programming and retention in PCM arrays," in *IEDM Tech. Dig.*, 2007, paper 46.6.

[17] D. Ielmini, A. S. Spinelli, A. L. Lacaita, and A. Modelli, "Equivalent cell approach for extraction of the SILC distribution in Flash EEPROM cells," *IEEE Electron Device Lett.*, vol. 23, pp. 40–42, Jan. 2002.

[18] S. A. Makhlouf, "Electrical properties of NiO films obtained by high-temperature oxidation of nickel," *Thin Solid Films*, vol. 516, pp. 3112–3116, 2008.

Chip-Level Reliability Study of Barrier Engineered (BE) Floating Gate (FG) Flash Memory Devices

Hang-Ting Lue, JiFong Pan[+], C.S. Chang[+], Szu-Yu Wang[*], Y.F. Chang[+], Y. C. Lee[+], M. H. Liaw[+], Y. J. Chen[*], K. F. Chen[*], Chester Lo[*], I. J. Huang[*], T. T. Han[*], M.S. Chen[*], W. P. Lu[*], T. Yang[*], K. C. Chen[*], Kuang-Yeu Hsieh, and Chih-Yuan Lu

Macronix International Co., Ltd., Emerging Central Lab, [+]Product Engineering Center, [*]Technology Development Center
16 Li-Hsin Road, Hsinchu Science Park, Hsinchu, Taiwan.
E-mail: htlue@mxic.com.tw

Abstract—Floating gate (FG) devices using barrier-engineered (BE) tunneling dielectric have been studied both theoretically and experimentally. Through WKB modeling the tunneling efficiency of various multi-layer tunneling barriers can be well predicted. Experimental results for FG devices with oxide-nitride-oxide (ONO) U-shaped barrier are examined to validate our model. Furthermore, a large-density array (1Mb) was studied to provide chip-level reliability understandings. Finally, these results are compared with barrier engineered charge-trapping (CT) devices. Our results suggest that BE FG device is not promising in terms of serious reliability degradation and tail bits. Moreover, the speed enhancement is not better than using the conventional gate-coupling ratio (GCR) improvement or tunnel oxide scaling. On the other hand, CT devices do not have GCR and it need BE tunneling barrier to solve the erase and retention dilemma. We also prove that BE-SONOS device is immune to tail bits due to the nature of discrete trapped charge storage.

Keywords-component; Floating gate; barrier engineer (BE), Reliability, Tunneling, Modeling, charge-trapping memory.

I. INTRODUCTION

Barrier engineered (BE) tunneling barrier was widely considered as an attracting way to increase the programming/erasing (P/E) speed of Flash memory devices [1-6]. A schematic diagram of tunneling current density (J) versus electric field (E) of a tunneling barrier is illustrated in Fig. 1. In general, a "steeper" J-E curve is preferred for a Flash memory device so that a higher tunneling current density is achieved at high field during programming, while leakage current at low field is suppressed in order to reduce the program/read disturb and retention leakage.

The first original concept was proposed by Likharev [1], where a crested barrier was proposed and simulated (Fig. 2). The crested barrier can be approximated by a multi-layer "low-ϕ_B/high-ϕ_B/low-ϕ_B" stack, where ϕ_B is the barrier height of electron tunneling from Si conduction band. Simulation results indicated that the tunneling current can be steeper than a single tunnel oxide. However, Likharev's model did not consider the dielectric constant changes when barrier height is altered. In fact, barrier height is often correlated to dielectric constant (K) for most materials. Materials with a smaller ϕ_B often have a higher K. The higher-K layer in the multi-layer stacks has a lower E field than the other layers,

leading to a degraded tunneling efficiency, and the overall tunneling curve is not as ideal as expected.

Later, a "VARIOT" [2] concept was proposed (Fig. 3), in which multi-layer low-K and high-K materials can be stacked together to form a "U-shape" or double barrier. A few experimental results were demonstrated [3]. However, reliability seems to be the most critical issue since high-K materials often contains a lot of traps, leading to endurance and retention problems.

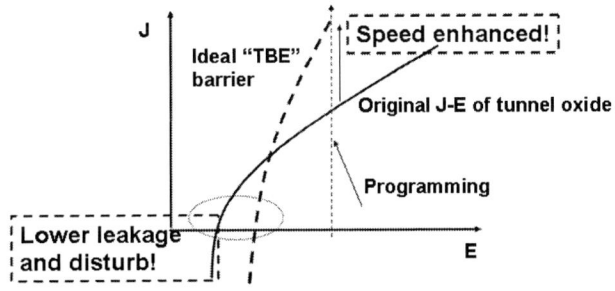

Fig. 1 A schematic diagram to illustrate the J-E curves of a tunneling barrier (Tox). The solid line is the J-E curve of a conventional single tunnel oxide. The dash line is an "ideal" tunneling barrier which has a steeper J-E curve. At the programming voltage, the tunneling current is higher so that programming speed is faster. On the other hand, at lower voltage, J is smaller so that both program and read disturb are suppressed. The lower leakage current is also important to avoid retention loss.

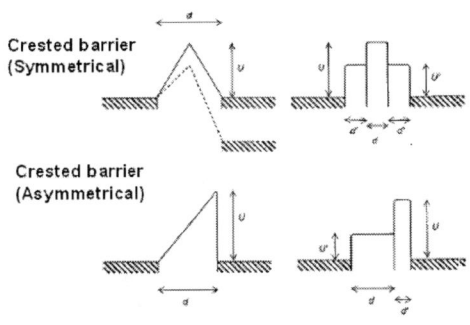

Fig. 2 Crested barrier proposed by Likharev [1]. Both symmetrical and asymmetrical barriers were proposed to modify the tunneling barrier. A multi-layer "low-ϕ_B/high-ϕ_B/low-ϕ_B"barrier was also proposed to simulate the crested barrier shape.

Contacting author: Hang-Ting Lue, htlue@mxic.com.tw

978-1-4244-5430-3/10 $26.00 © 2010 IEEE

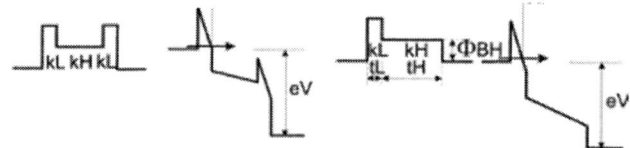

Fig. 3 "VARIOT" proposed in [2]. Multi-layer barrier with low-K or high-K materials can be designed to form a U-shape barrier.

The difficulty of tunnel barrier (Tox) engineering for floating gate memory can be briefly summarized in the followings:

(1) The desired small tunneling barrier (ϕ_B) in a BE Tox is unfortunately often accompanied with a higher K. However, a higher K is not preferred in the tunneling barrier because it decreases the electric field. The overall J-E curve is not as steep as predicted by the crested barrier.

(2) The direct contact of a low ϕ_B material (or a high-K material) on silicon is impractical, because it naturally introduces a thin interfacial oxide. In terms of this processing issue, the U-shaped barrier by a low-K/high-K/low-K seems more practical because SiO_2 can serve as the first tunnel oxide. However, it should be mentioned that the U-shaped barrier doesn't have a steeper tunneling barrier than the conventional tunnel oxide. There is also thickness limitations (O1 or O2 <3nm) due to the band offset effect, which will be discussed later.

(3) Usually one cannot find a perfect high-K layer that has sufficiently low trap density. The overall tunneling current (J-V curve) are affected by some inelastic trap assisted tunneling, leading to non-ideal J-E curves. Furthermore, the traps in tunneling barrier easily lead to reliability problems such as endurance degradation and retention loss.

(4) FG Flash memory is prone to the erratic behaviors due to randomly generated tunnel oxide traps during P/E cycling. This effect can not be monitored by only a few single cells but must be studied through a large-density memory chip. For BE FG devices, this effect would be more critical since the tunnel dielectrics often have more defects. However, no chip-level reliability was ever reported before.

In this wok, we provide the first chip-level study of BE FG memory device (90nm node) to understand the detail reliability physics and statistics. Finally, BE charge-trapping (CT) device (75nm node) will be compared.

II. U-SHAPED ONO TUNNELING BARRIER

An oxide-nitride-oxide (ONO) barrier is studied in this work. Since nitride has a much smaller bandgap than oxide, the ONO barrier thus resembles a U-shaped tunneling barrier (Fig. 4). Note that the nitride must be sufficiently thin (<3nm) to minimize the charge-trapping [4, 7], while the total BE Tox stack must be sufficiently thick to avoid serious charge loss and stress induced leakage current (SILC) [8, 9]. On the other hand, the first tunnel oxide (O1 for +FN and O2 for –FN) must

be sufficiently thin to provide enhanced tunneling (to be explained). Considering the above factors, a typical ONO thickness is designed as O1/N/O2=7/2/2 nm, respectively. Since there are no new high-K materials and metal gate, the devices are easily integrated in a conventional Flash memory process. A 90 nm node 4Mb test chip is evaluated for the BE FG devices. Although the use of a 90nm node technology is not advanced, the conventional STI processes do not cut through the tunnel oxide (not SA-STI) and hence it can avoid tunnel oxide damage and provide a better reliability study.

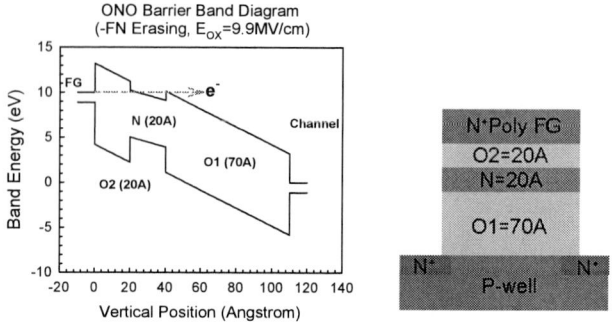

Fig. 4 The band diagram and a schematic diagram of the ONO barrier. At high field during –FN, the electrons see the U-shaped barrier because nitride has a lower bandgap than oxide. The U-shaped barrier reduces the effective tunnel barrier thickness and enhances the tunneling efficiency. Note that the band offset is more significant in the valence band.

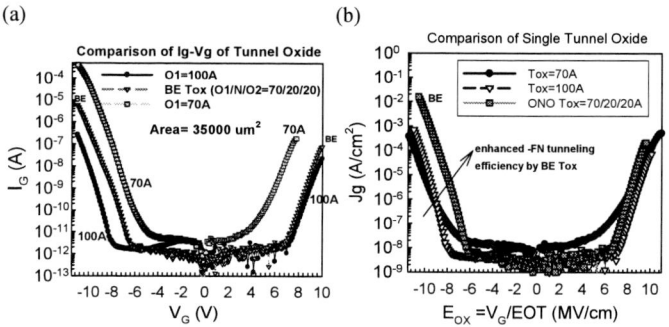

Fig. 5 (a) Ig-Vg characteristics of tunnel oxide capacitor for various Tox. Thin (7 nm) Tox shows the largest tunneling current density. (b) When normalized to Jg-E, where E is the electric field of tunnel oxide (O1 for BE Tox). Both thick and thin Tox has similar +./-FN efficiency. BE Tox has enhanced efficiency at –FN. However, the +FN has no apparent enhancement.

The experimental results of a tunnel oxide (Tox) capacitor for a thick (10 nm), thin (7 nm), and BE Tox (O1/N/O2=7/2/2 nm) barriers are shown in Fig. 5(a). In Ig-Vg plot, the thin Tox shows the largest tunneling current. But this is simply due to the EOT difference. For a fair comparison, it should be normalized to a J-E curve, where E is the electric field in the tunnel oxide. The results are shown in Fig. 5(b). It shows that both thick and thin Tox have similar J-E curves. This is reasonable because the FN tunneling equation is only dependent on E field and independent of thickness:

$$J = \alpha \cdot E_{ox}^2 \ \exp\left(-E_C / E_{ox}\right) \qquad (1)$$

On the other hand, BE Tox indeed shows higher tunneling current than a single Tox at –FN. However, at +FN, the difference is not significant.

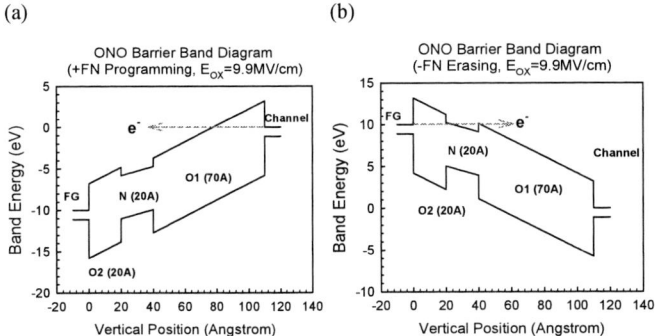

(a) (b)

Fig. 6 Calculated band diagram for the ONO barrier under (a) +FN and (b) –FN operation. At +FN, the electrons from Si channel only sees the O1 triangular barrier. At –FN, electrons from FG see a U-shaped barrier. Due to the band offset effect, the effective tunneling thickness is reduced.

The asymmetrical tunneling behavior can be explained in Fig. 6. At high field during +FN (Fig. 6(a)), electrons tunnels from channel, and it only sees the triangular barrier contributed by the thick O1. Therefore, it is independent of N and O2 and the J-E curve is the same with a single Tox.

On the other hand, at –FN operation (Fig. 6(b)), the electron tunnel from the FG and it will see the U-shaped barrier since O2 is sufficiently thin. The band offset effect screens the barrier of N and O2 and hence the tunneling efficiency is enhanced.

In the previous work [10], we have developed an analytical equation to model the tunneling current density of multi-layered barrier engineered tunneling dielectric using the WKB approximation, as illustrated in Eq. (2):

$$J = \frac{q^3}{8\pi h} \frac{m_s}{m_{O1}} E_{O1}{}^2$$

$$\times \left(\phi_{B,O1}{}^{1/2} - \left(\phi_{B,O1} - qE_{O1}T_{O1} \right)^{1/2} + \frac{\varepsilon_{N1}}{\varepsilon_{O1}} \sqrt{\frac{m_{N1}}{m_{O1}}} \left(\phi_{B,N1} - qE_{O1}T_{O1} \right)^{1/2} \right.$$
$$\left. + \frac{\varepsilon_{O2}}{\varepsilon_{O1}} \sqrt{\frac{m_{O2}}{m_{O1}}} \left(\phi_{B,O2} - qE_{O1}T_{O1} - qE_{N1}T_{N1} \right)^{1/2} \right)^{-2}$$

$$\times Exp \left(-\frac{8\pi\sqrt{2m_{O1}}}{3hqE_{O1}} \left(\phi_{B,O1}{}^{3/2} - \left(\phi_{B,O1} - qE_{O1}T_{O1} \right)^{3/2} + \frac{\varepsilon_{N1}}{\varepsilon_{O1}} \sqrt{\frac{m_{N1}}{m_{O1}}} \left(\phi_{B,N1} - qE_{O1}T_{O1} \right)^{3/2} \right.\right.$$
$$\left.\left. + \frac{\varepsilon_{O2}}{\varepsilon_{O1}} \sqrt{\frac{m_{O2}}{m_{O1}}} \left(\phi_{B,O2} - qE_{O1}T_{O1} - qE_{N1}T_{N1} \right)^{3/2} \right) \right)$$

(2)

The required parameters are the barrier height of O1, N1, O2 and effective mass. We use 3.2 eV, and 2.2 eV for oxide and nitride, respectively, which are reasonable parameters for the well-known materials. In Eq. (2), the current density is described in terms of first tunnel oxide E field (E_{O1}).

Figure 7(a) shows that WKB simulation basically can reproduce the experimental result. The slightly deviation from theoretical model may be due to some non-ideal trap-assisted tunneling [10] effects.

Figure 7(b) shows that the experimental –FN tunneling current is enhanced when O2 is thinner. On the other hand, when O2 is thicker than 3 nm, the difference between the BE Tox and a single Tox is smaller.

Figure 8(a) shows the WKB simulation can well explain the experimental result, where thinner O2 indeed shows enhanced tunneling efficiency. When O2 is greater than 3 nm, the J-E curves merge with the single tunnel oxide, consistent with the explanation in Fig. 6. The above results simply conclude that the first tunnel oxide must be sufficiently thin (<3nm) to offer an effective BE Tox. On the other hand, it is difficult to make both thin O1 and O2 due to reliability constraint (to be discussed later). Thus asymmetrical +/- FN tunneling would limit the capability of BE FG devices.

Figure 8(b) shows the simulation result of using other high-K materials for the U-shape barrier. Our simulation results indicate that other high-K material does not offer significantly improved efficiency than ONO barrier. The reason is that higher-K material will decrease the electric field and thus not favor the tunneling. Another reason is that SiN already offers very small electron barrier height.

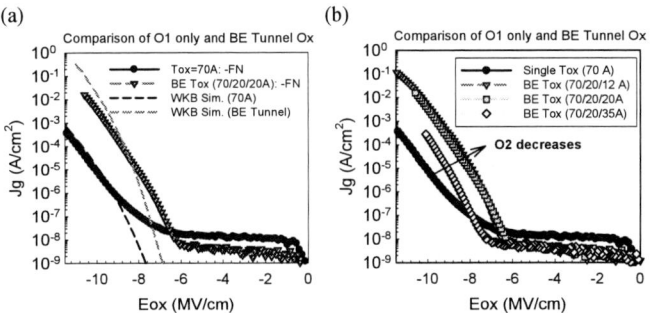

(a) (b)

Fig. 7 (a) J-E curve simulation of the BE Tox. The WKB simulation excellent reproduces the experimental result. (b) Experimental J-E curve of various BE Tox. A thin O2 (<3nm) is necessary for enhanced –FN tunneling.

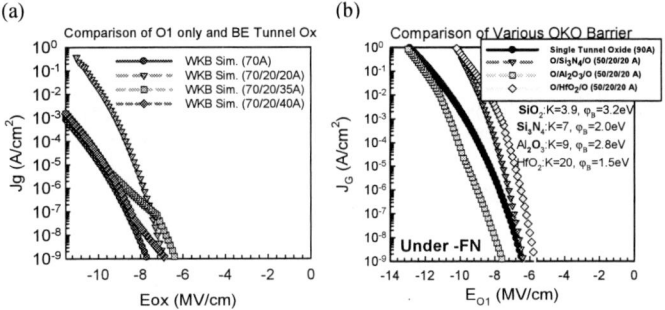

(a) (b)

Fig. 8 (a) J-E curve simulation of the BE Tox with various ONO thickness. The simulation is consistent with the experimental result in Fig. 7. (b) J-E curve simulation of BE Tox with various OKO barrier, where K can be SiN, HfO2, Al2O3.

III. BE FG DEVICE CHARACTERISTICS AND CHIP-LEVEL RELIABILITY

The 90nm FG devices are measured using the above mentioned Tox's. The +/- FN programming/erasing comparison are briefly summarized in Fig. 9. For a fair

978-1-4244-5430-3/10 $26.00 © 2010 IEEE

comparison, the thick Tox (10 nm) and BE Tox (O1/N/O2=7/2/2 nm) have very close EOT so that the tunnel oxide E field are the same. The results show that –FN erasing of BE FG indeed is faster than a single Tox. However, the +FN programming do not have difference. The results are very consistent with Fig. 5.

On the other hand, Figs. 9(c) and (d) indicate that thinner Tox (7 nm) shows a much faster speed enhancement. This is because the tunnel oxide E field is enhanced when Tox thickness is reduced (assuming the same gate voltage and GCR).

The above results simply raise a critical question that reducing Tox thickness seems to be more efficient than using a BE Tox.

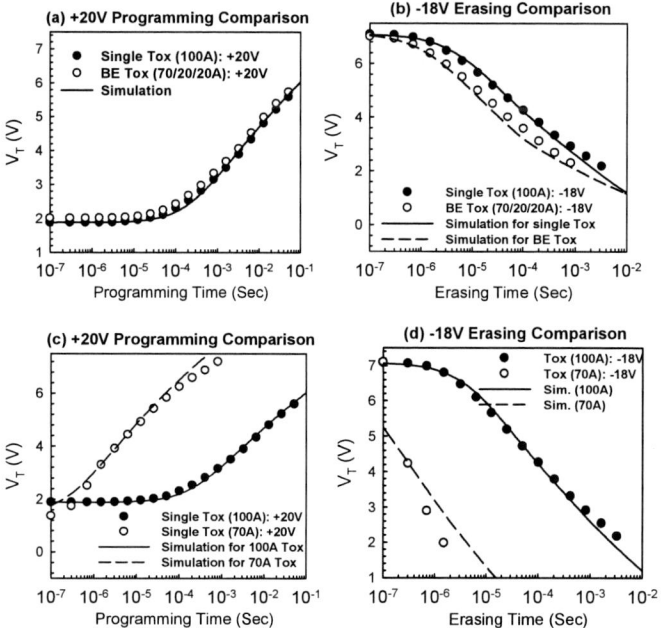

Fig. 9   (a) +FN programming comparison of single Tox (10 nm) and BE Tox (O1/N/O2=7/2/2 nm). (b) –FN erasing comparison of single Tox (10 nm) and BE Tox (O1/N/O2=7/2/2 nm). (c) +FN programming comparison of thick Tox (10nm) and thin Tox (7 nm). (d) -FN erasing comparison of thick Tox (10 nm) and thin Tox (7 nm). The dotted symbols are experimental results. The lines are simulation using the J-E curve model in Figs. 5 and 7. The effective GCR is around 0.65 in our device.

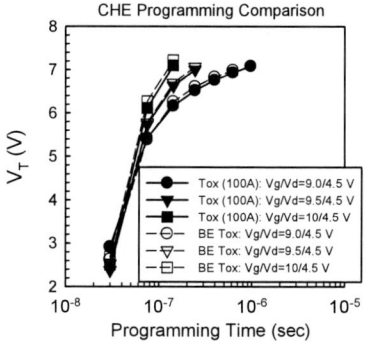

Fig. 10   CHE programming comparison of single Tox (10 nm) and BE Tox (O1/N/O2=7/2/2 nm).

Figure 9 also shows the simulation results for each device. The J-E curve model in Figs. 5 and 7 can excellently reproduce the experimental Vt-time for FG devices. Therefore, the tunneling current analysis (J-E or J-V) on a capacitor can well predict the BE FG characteristics.

We also compare the channel hot electron (CHE) injection methods for a single Tox and a BE Tox, as illustrated in Fig. 10. BE Tox does not improve the CHE efficiency. This is because CHE is operated under lower vertical field (<5MV/cm) and there is no such band offset effect. Besides, CHE injection efficiency is more related to substrate hot carrier generation.

(a)   (b)

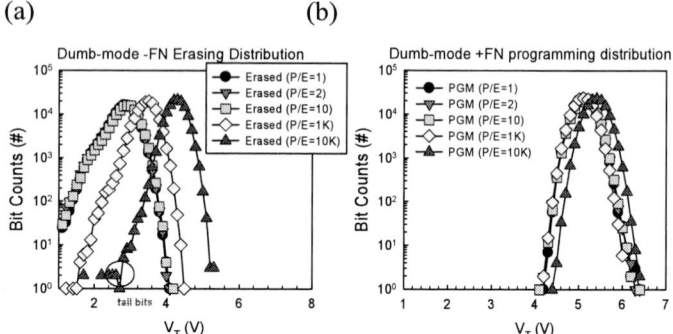

Fig. 11   Chip-level P/E distribution of a BE Tox during 10K cycling stress. (a) Dumb-erasing (-16V 5msec), and (b) dumb-programming (+18V 10msec) are carried out.

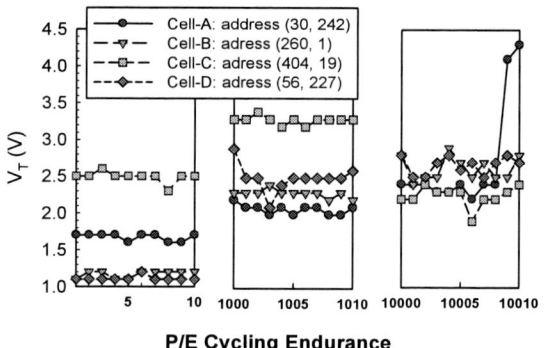

P/E Cycling Endurance

Fig. 12   Bit tracking during 10K P/E cycling for a few tail bits illustrated in Fig. 11(a). P/E=1 to 10, 1K to 1K+10, 10K to 10K+10 are recorded. The erratic behavior is clearly observed.

The test chip P/E distribution of the BE FG device with BE Tox (O1/N/O2=7/2/2 nm) are shown in Fig. 11. Even at low P/E cycling, the erase distribution is wide and certain tail distribution is observed. After 10K P/E cycling stress, erase degradation is significant. On the other hand, the programmed state has little degradation after 10K cycling stress, and the distribution is normal.

This again manifests the asymmetry of +/-FN operation of BE Tox. –FN erasing sees the U-shape barrier and is therefore more sensitive to the defect/trap at BE Tox. The erase degradation comes from the accumulated trapped electrons in the BE Tox, leading to decreased electron tunneling. On the other hand, +FN programming only sees a triangular barrier of the thick O1 and is therefore less sensitive.

978-1-4244-5430-3/10 $26.00 © 2010 IEEE

We also observe that a few tail bits happened after 10K cycling. We have particularly trace the tail bits during PE cycling, as shown in Fig. 12. In general, all cells have erase degradation and Vt goes higher after cycling. However, most cells often show erratic behavior (instable Vt during cycling), such as Cell-C.

The erratic behavior is a common behavior for all FG memory devices due to the randomly generated tunnel oxide traps. A single defect or leakage path in the tunnel oxide will drain the entire charge, leading to erratic behaviors. For BE Tox, this effect becomes more severe.

The chip-level retention is compared in Fig. 13. Thin Tox (7nm) shows a significant tail bits after a 10K P/E cycling stress and room-temperature retention test. This just manifests the well-know SILC [9] issue that limits the tunnel oxide scaling. On the other hand, BE Tox shows suppressed tail bits, possibly due to the thicker stack height that suppresses TAT.

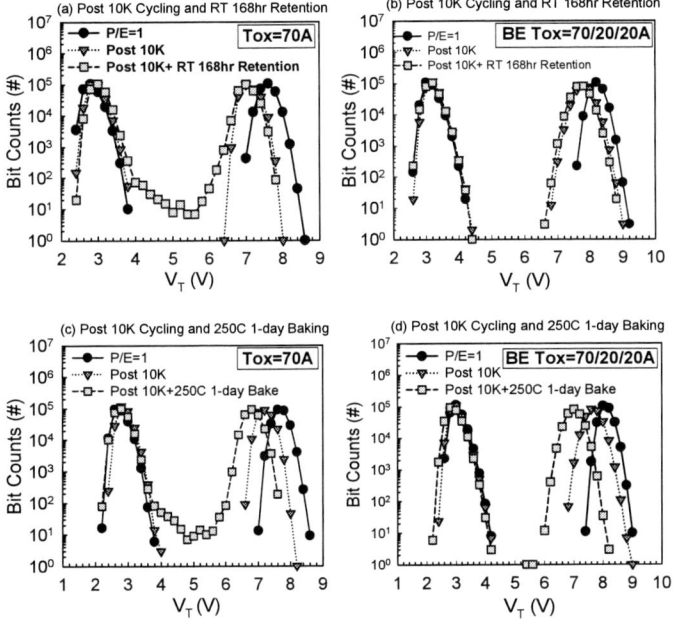

Fig. 13 **(a)** Room-temp. (RT) retention of 70A tunnel oxide after 10K cycling. **(b)** Room-temp (RT) retention of BE Tox. **(c)** 250C retention of Tox=70A. **(d)** 250C retention of BE Tox. The tail bits of thin tunnel oxide come from the high-Vt charge loss.

IV. COMPARISON WITH BE-SONOS CHARGE-TRAPPING DEVICES

Unlike FG devices, charge-trapping (CT) device do not have gate coupling ratio (GCR) design. In Fig, 14, the planar SONOS device has equal electric fields at top and bottom oxides. This will lead to a large gate injection and small memory window during erase. The conventional SONOS [12] must use an ultra-thin (<2nm) tunnel oxide to offer efficient direct tunneling and provide an asymmetrical tunneling efficiency for bottom and top oxides. However, such thin tunnel oxide causes unacceptable retention loss [13].

The difficulty of SONOS can be illustrated in Fig. 15. Erase to lower than initial Vt (or V_{FB}) is not possible with O1>2.5nm. However, with O1<2.5nm, the retention is too poor for a practical application. The physical reason can be explained by the hole direct tunneling calculation in Fig. 16. Since oxide has a large hole barrier height, the hole FN tunneling with a thick O1 (>3nm) is negligibly small under a reasonable erase field (<13MV/cm). In fact, SONOS with a thick tunnel oxide is erased by electron de-trapping, which is still very slow. To offer a suitable hole direct tunneling current, O1 must be thinner than 2nm. However, direct tunneling also causes large leakage at low field, leading to poor retention.

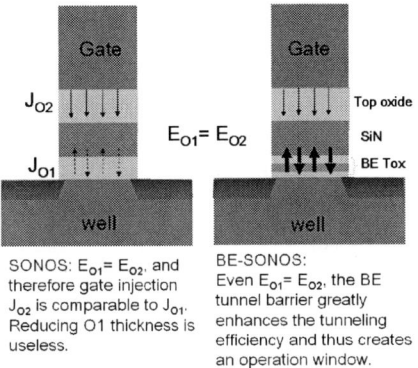

Fig. 14 Schematic diagram to illustrate the issue of CT SONOS device. Because there is no GCR design, the top oxide electric field is equal to that of the tunnel oxide. This leads to large gate injection and small memory window.

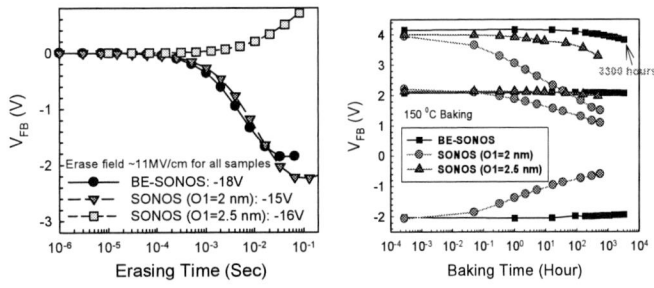

Fig. 15 (a) Erasing comparison of SONOS with O1=2, 2.5 nm and BE-SONOS. When O1>2.5nm, erase of SONOS becomes impossible. (b) Retention comparison. SONOS has poor data retention for O1<=2.5nm.

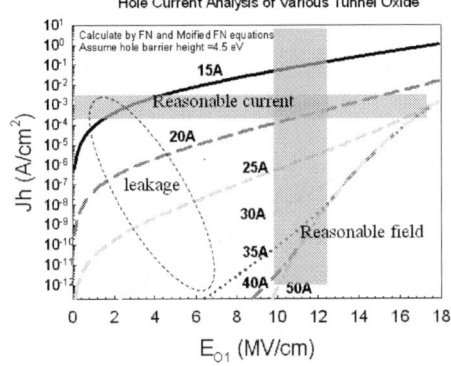

Fig. 16 Calculated hole tunneling current of a single tunnel oxide for various thickness. Due to the large hole barrier height (>4.5eV). hole FN tunneling is impossible. Only when ultra-thin oxide is used, hole direct tunneling is provided. However, direct tunneling has a large leakage current at low field, leading to retention and disturb issues.

978-1-4244-5430-3/10 $26.00 © 2010 IEEE

The solution of SONOS is to introduce a BE Tox to offer efficient hole tunneling injection and suppressed low-field leakage for good retention. The structure is the so-called BE-SONOS [4]. Figure 17(a) shows the band diagram of ONO barrier under –FN erase. The band offset effect greatly reduces the hole barrier of N1 and O2 thus improves the hole tunneling efficiency. Figure 17(b) shows that the BE ONO barrier creates more than 6 orders of magnitude larger hole tunneling current than a thick tunnel oxide. The efficiency gain by BE barrier is much more significant that the electron tunneling in Figs. 7-8.

It should be mentioned that hole injection was not allowed for the conventional thick (>3nm) tunnel oxide because of the large hole barrier height (>4.5eV). It is only allowed by an ultra-thin oxide (to offer direct tunneling) or the BE (such as ONO) barrier. However, only BE barrier offers good retention and fast erase simultaneously.

For SONOS (or MONOS) with a thick tunnel oxide, the erase mechanism mainly comes from the electron de-trapping. Compared with the MONOS, the hole current density of BE ONO barrier is more than 2 orders of magnitude larger than the electron-de-trapping current (Fig. 17(b)).

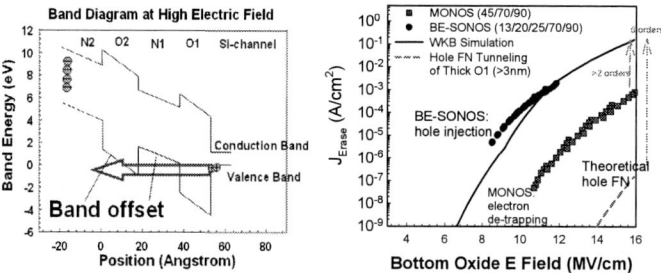

Fig. 17 (a) Band diagram of BE ONO barrier under –FN erasing. A band offset is introduced. (b) The calculated hole tunneling current of BE ONO barrier. Experimental data of hole current from BE-SONOS and electron de-trapping current from MONOS are compared.

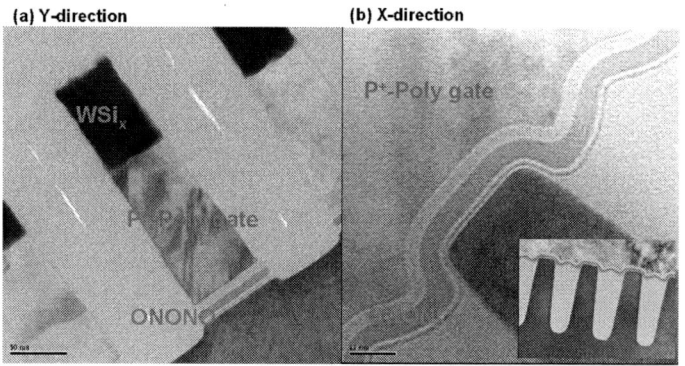

Fig. 18 A 75 nm BE-SONOS NAND array. Both X (STI) and Y (WL) directions have half pitch =75nm.

We have fabricated BE-SONOS charge-trapping NAND Flash test chip in a 75nm (half-pitch) technology to study the chip-level reliability and the statistical distribution.

Figure 18 shows the TEM cross-sectional view of the device. Both X (STI) and Y (WL) directions have half pitch

equal to 75nm. The device is fabricated in a near-planar STI structure.

Figure 19 shows the dumb-mode (without any P/E verify) programming/erasing Vt distribution of the BE-SONOS NAND array with two different O1 process (pure oxide O1 and nitrided O1). Although there is a certain erase degradation and Vt roll-up, no single tail bit is observed and all bits follow a normal Gaussian distribution.

The erase degradation mainly comes from Dit generation of Si/O1 and can be improved by using nitrided O1 [14], as shown in Fig. 19(b).

Figure 20 illustrates the post-cycled retention behavior of BE-SONOS NAND. Again, there is no single tail bit observed. In Fig. 20(b), with suitable engineered device, BE-SONOS possess excellent data retention even with an ultra-thin O1 (1.3nm). This indicates that charge-trapping device can tolerate much thinner and non-perfect tunneling barrier than FG devices.

The above results suggest that charge-trapping device have very different reliability from the FG devices. The major difference is that charges are stored in discrete traps, thus the device is naturally immune to the SILC problem in FG device, where a single trap in tunnel oxide would causes entire charge leakage in FG. Therefore, CT devices do not have tail bit behavior, as demonstrated in previous NROM[TM] technology [15].

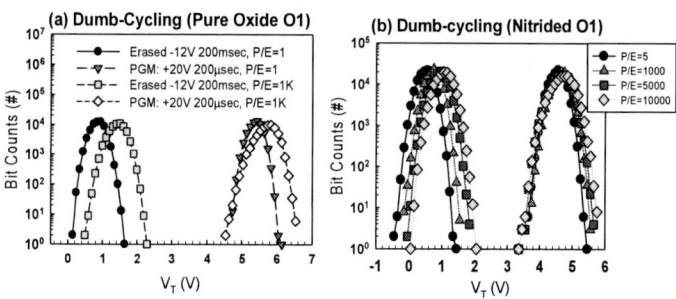

Fig. 19 (a) Dumb PE cycling (without any P/E verify) of BE-SONOS with pure oxide O1. (b) Dumb cycling of BE-SONOS with nitrided O1. Erase degradation is improved by nitrided O1. For both devices, dumb-cycling do not create a single tail bit. Edge WL's (WL0 and WL31) are excluded to avoid processing differences.

(a)   (b)

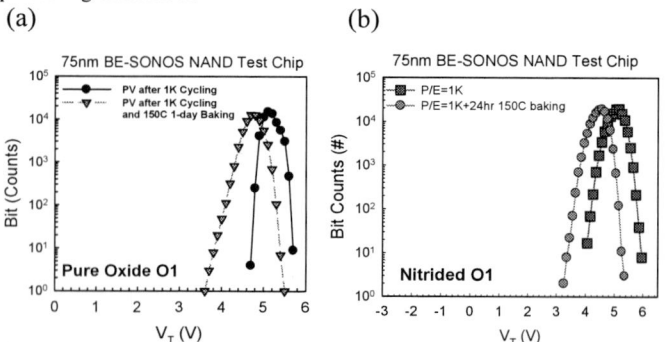

Fig. 20 Retention of programmed state after 1K P/E cycling and 150C high-temperature baking test for one block (128kb). (a) Pure oxide O1 (by ISSG process), (b) Nitrided O1. Although charge loss is observed after P/E cycling, no single tail bit is observed and the distribution is normal.

V. Conclusions

Through our detail analysis we suggest that the barrier engineering (BE) is not very promising for FG devices because:

(1) Increasing GCR or Tox scaling are more effective ways to improve the speed.

(2) U-shaped BE Tox must be designed with first tunnel oxide (O1 for +FN and O2 for –FN) thinner than 3 nm for a sufficient band offset effect. Besides, it shows asymmetrical +/- FN tunneling behavior.

(3) FG is prone to tail bit behavior and this effect is more critical if a non-perfect dielectric is used in the BE Tox. The thin O1 or O2 together with reliability constraint of SILC limits the feasibility of BE FG devices.

It should be mentioned that when FG eventually scales to a planar structure with very thin FG stack height where GCR is very low, the use of BE Tox may still create more operational window. However, the reliability still remains the most critical concern.

On the other hand, charge-trapping devices do not have GCR and it strongly requires an efficient hole injection method to erase the deeply-trapped electrons (de-trapping is slow). Thus the barrier engineering of charge-trapping devices is more related to hole tunneling rather than electron tunneling. This is a major difference from the BE FG devices.

Since both efficient hole injection and good retention are impossible by a single tunnel oxide, the advantage and importance of BE barrier is much more evident.

Chip-level demonstration shows that there is no single tail bit during programming/erasing and retention tests for BE-SONOS devices. The most important advantage of CT device is that it is discrete trapped-charge storage. Thus CT is a more promising showplace for the barrier engineering (BE).

Acknowledgment

The authors would like to thank many colleagues in technology development center (TD) and product engineering (PE) for their supports in BE-SONOS research.

References

[1] K. K. Likharev, "Layered tunnel barriers for nonvolatile memory devices", Applied Physics Letters (APL), vol. 73, no. 15, pp. 2137-2139, 1998.

[2] B. Govoreanu, P. Blomme, M. Rosmeulen, J. Van Houdt, and K. De Meyer, "VARIOT: A Novel Multilayer Tunnel Barrier Concept for Low-Voltage Nonvolatile Memory Devices", IEEE Electron Device Letters, pp. 99-101, 2003.

[3] P. Blomme, J. V. Houdt, and K. D. Meyer, "Write/Erase Cycling Endurance of Memory Cells With SiO2/HfO2 Tunnel Dielectric", IEEE Trans. Electron Devices, VOL. 4, NO. 3, pp. 345-352, 2004

[4] H. T. Lue, S.Y. Wang, E.K. Lai, Y.H. Shih, S.C. Lai, L.W. Yang, K.C. Chen, J. Ku, K.Y. Hsieh, R. Liu, and C.Y. Lu, "BE-SONOS: A Bandgap Engineered SONOS with Excellent Performance and Reliability", Tech. Digest 2005 International Electron Devices Meeting, pp. 547-550, 2005.

[5] S. C. Lai, H. T. Lue, M. J. Yang, J. Y. Hsieh, S. Y. Wang, T. B. Wu, G. L. Luo, C. H. Chien, E. K. Lai, K. Y. Hsieh, R. Liu and C. Y. Lu, "MA BE-SONOS: A Bandgap Engineered SONOS using Metal Gate and Al2O3 Blocking Layer to Overcome Erase Saturation", IEEE Non-Volatile Semiconductor Memory Workshop (NVSMW), pp-88-89, 2007.

[6] S. C. Lai, H.T. Lue, C.W. Liao, Y.F. Huang, M.J. Yang, Y.H. Lue, T.B. Wu, J.Y. Hsieh, S.Y. Wang, S.P. Hong, F.H. Hsu, G.L. Luo, C.H. Chien, K.Y. Hsieh, R. Liu and C.Y. Lu "An Oxide-Buffered BE-MANOS Charge-Trapping Device and the Role of Al2O3", NVSMW-ICMTD, pp. 101-102, 2008.

[7] H.T. Lue, P.Y. Du, S.Y. Wang, E.K. Lai, K.Y. Hsieh, R. Liu, and C.Y. Lu, "A Novel Gate-sensing and Channel-sensing Transient Analysis Method for Real-time Monitoring of Charge Vertical Location in SONOS-Type Devices and its Applications in Reliability Studies", International Reliability Physics Symposium (IRPS), session 3A-4, pp. 177-183, 2007.

[8] K. Naruke, S. Taguchi, and M. Wada, "Stress induced leakage current limiting to scale down EEPROM tunnel oxide thickness", Tech. Digest 1988 International Electron Devices Meeting, pp. 424-427, 1988.

[9] F. Arai, T. Maruyama, an R. Shirota, "Extended data retention process technology for highly reliable Falsh EEPROMs of 106 to 107 W/E Cycles", International Reliability Physics Symposium (IRPS), pp. 378-382, 1998.

[10] H. T. Lue, S. C. Lai, T. H. Hsu, P. Y. Du, S. Y. Wang, K. Y. Hsieh, R. Liu, and C. Yuan Lu, "Understanding Barrier Engineered Charge-Trapping NAND Flash Devices With and Without High-K Dielectric", International Reliability Physics Symposium (IRPS), pp. 874-882, 2009.

[11] Daniele Ielmini, Alessandro S. Spinelli, Member, IEEE, Andrea L. Lacaita, Senior Member, IEEE, and Alberto Modell, "A Statistical Model for SILC in Flash Memories", IEEE TRAN. ON ELECTRON DEVICES, VOL. 49, NO. 11, pp1955-1961, 2002.

[12] M. White, "On the go with SONOS", IEEE Circuits and Designs, pp.22-31, 2000.

[13] S. Y. Wang, H. T. Lue, E. K. Lai, L. W. Yang, T. Yang, K. C. Chen, J. Gong, K. Y. Hsieh, R. Liu, and C. Y. Lu, "Reliability and Processing Effects of Bandgap Engineered SONOS (BE-SONOS) Flash Memory", International Reliability Physics Symposium (IRPS), session 3A-3, pp. 171-176, 2007.

[14] S. Y. Wang, H. T. Lue, T. H. Hsu, P. Y. Du, S. C. Lai, Y. H. Hsiao, S.P. Hong, M. T. Wu, F. H. Hsu, N. T. Lien, C. P. Lu, J. Y. Hsieh, L. W. Yang, T. Yang, K. C. Chen, K. Y. Hsieh, and C. Y. Lu, "A High-Endurance (>100K) BE-SONOS NAND Flash with a Robust Nitrided Tunnel Oxide/Si Interface", International Reliability Physics Symposium (IRPS), 2010, in publication (this conference).

[15] M. Janai, "Data retention, endurance and acceleration factors for NROM devices", Proceeding of the International Reliability Physics Symposium (IRPS), pp. 502-505, 2003.

Source/Drain Dopant Concentration induced Reliability issues in Charge Trapping NAND Flash Cells

Yin-Jen Chen, Lit Ho Chong, Shang-Wei Lin, Teng-Hao Yeh, Kuan-Fu Chen, Jyun-Siang Huang, Cheng-Hsien Cheng, Shaw-Hung Ku, Nian-Kai Zous, I-Jen Huang, Tzung-Ting Han, Tzu-Hsuan Hsu, Hang-Ting Lue, Ming-Shiang Chen, Wen-Pin Lu, Kuang-Chao Chen, and Chih-Yuan Lu

Macronix International Company Ltd., No. 16, Li-Hsin Road, Science Park, Hsin-Chu, Taiwan, R. O. C
Phone: +886-3-5786688-78062 E-mail: nkzou@mxic.com.tw

Abstract

Source/Drain (S/D) dopant concentration related reliability issues including erase speed degradation, sub-threshold swing (SS) increase, and program/erase (P/E) cycling induced low threshold voltage (V_T) state drift and on-state current (I_{ON}) reduction are carefully examined in charge trapping (CT) NAND flash memories. Residual charges above S/D junctions has been identified as a dominant factor and cell performances are greatly improved with increasing S/D dosages. Moreover, a new program disturbance behavior, which possibly originates from junction leakage or breakdown induced hot carriers injection, is observed. Simulation results confirm that a high lateral junction field occurs at a program-disturbed cell once its S/D is fully depleted. Although optimizing S/D dosage can ease this situation, it is still a possible obstacle for further device scaling.

Keywords-component: dopant concentration, erase speed degradation, sub-threshold swing (SS) increase, program/erase (P/E) cycling, on-state current (I_{ON}) reduction, program disturbance

Introduction

Trapping storage cells have been studied as an candidate to replace floating gate (FG) flash memories due to the immunity to interference, simple fabrication process flow [1-3], and easy transformation to 3D structures [4,5]. Recently, reliability issues related to S/D doping concentration have been mentioned for both FG [6] and CT [7] devices. Due to the locally stored nature of a non-conductive storage node, detailed review of the impact of S/D doping on the performance of a CT-NAND cell is necessary. For example, channel hole injection is utilized as an erase method for a CT device.

Unfortunately, as shown in Fig. 1, the hole density near junctions is always lower than the electron density, which means the stored electrons in this area are hard to neutralize. Once more electrons are accumulated after successive P/E cycles, possible degradations arise and its impact will be different according to the S/D dopant concentration. Another concern is the performance of self-boosting (SB). Fig. 2 demonstrates the NAND cell operating conditions for programming. Bit A is the programmed bit and bit B is the inhibited bit. The channel potential (V_{CH}) of the selected bit-line (BL) always keeps ground and a large vertical voltage drop across the gate stack results in F-N electron injection. In the unselected BL, the channel potential is initially precharged by V_D=Vcc and is then coupled by gate voltage (V_{PGM} and Vpass). When $V_{CH,ini}$ increases to the condition of $V_{CH,ini}$ > Vcc+$V_{T,SSL}$, SSL becomes cut off and the potential is quickly boosted to achieve "inhibit" operation. According to the literature [8], the global SB becomes local SB if the adjacent channels are cut off. As the device dimensions shrink, local SB could occur because of the reduction of junction dosage and junction depth [9]. Although the V_{CH} can be boosted more efficiently under local SB, a large potential difference exists between channel and junction regions (Fig. 7 of ref [9]). Such a high lateral field will be a potential risk from reliability point of view. In this paper, we explore the reliability risk from the impacts of S/D dosage on erase V_T distributions; SS increment, I_{ON} reduction, and endurance performance of a NAND string are described. Second, a high V_{CH} induced additional disturbances are demonstrated and explained by simulations.

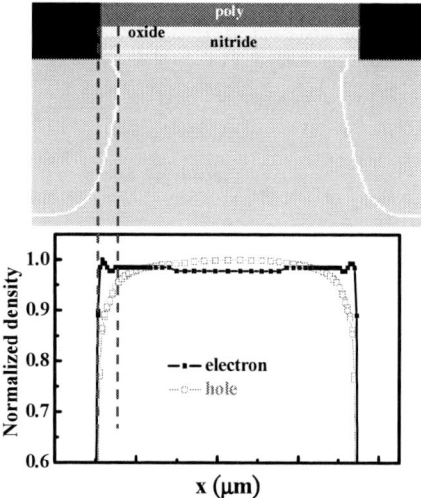

Fig.1 Simulated channel electron and channel hole density during program and erase. Electron density is larger than hole density near junctions

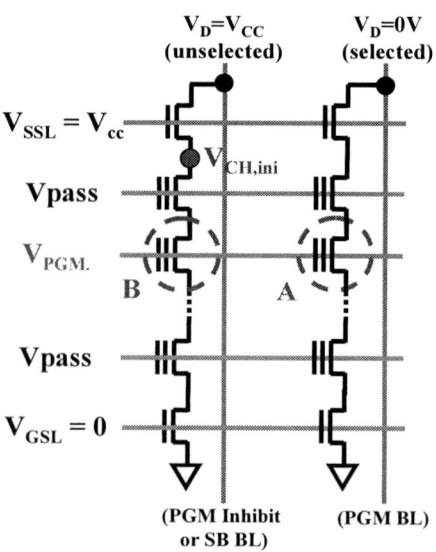

Fig.2 Bias conditions for NAND cell programming. A is the programmed bit and B is the self-boosting bit (or inhibited bit). When $V_{CH,ini}$ is raised up to $V_{CH,ini}$ > Vcc+$V_{T,SSL}$, SSL of the unslected BL is cut-off and the channel potential is boosted more efficiently.

978-1-4244-5430-3/10 $26.00 © 2010 IEEE

Experimental results and Discussions

BE-SONOS cells with different S/D dosage are fabricated by 70nm technology. The effective oxide thickness (EOT) is about 14nm. 32 cells connected serially compose a NAND string. Operation conditions are the same for all wafers. The V_T of a cell is defined as the required gate voltage to reach a certain read current.

(a) Impacts on Cell Performances

Fig. 3 shows the V_T distribution between strings with different dosages and no significant difference in fresh state is observed. After

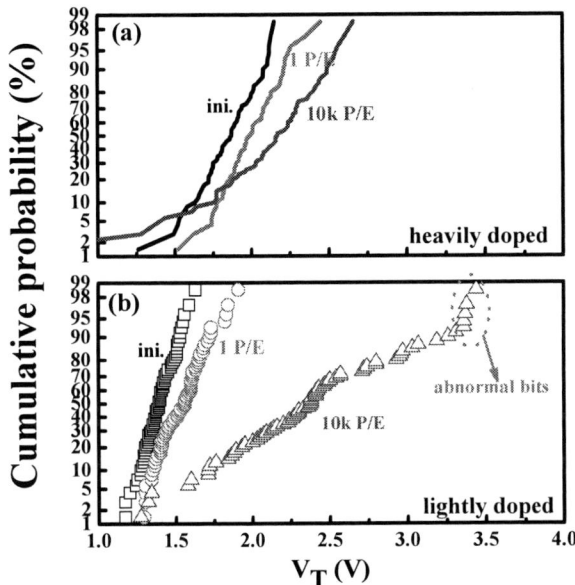

Fig.3 S/D dopant effect on VT distribution after stress. Abnormal bits are easily observed in lightly S/D doped cells.

100k P/E cycling stress, a wider distribution is observed for the lighter dosage case. Since heavily doped sample only shows minor change, interface states generation is not responsible for this huge degradation. The cycling evolution of Id-Vg curves is shown in Fig. 4 and serious degradation on SS and more than 30% reduction in I_{ON} are observed in a lightly doped sample. The cycling endurance of a single cell and a 32-cells string were measured respectively and shown in Fig. 5. Severe erase V_T rolled up is observed in the P/E cycled NAND string. After 100k cycling, the net window of a single cell is still ~3V since the high V_T state drifts up at the same time. However, the net window ~0V when we merely consider 32-cells string. To further understand the degradation, SS and I_{ON} are plotted against P/E numbers in Fig. 6. I_{ON} reduction occurs in both programmed and erased states but SS degradation only happens in erased state and is recovered after programming. The suspected root cause is schematically depicted in Fig. 7. Initially, no charges exist above junctions and thus no degradation on SS and I_{ON}. Since the electron and hole densities are not uniformly distributed beneath the storage node as illustrated in Fig. 1, negative charges are easily accumulated above S/D junctions, causing negative field to deplete the overlap dopant and thus I_{ON} reduces no matter in programmed or erased states. These localized charges will also induce SS degradation at erased state due to enhanced drain induced barrier lowering [10].

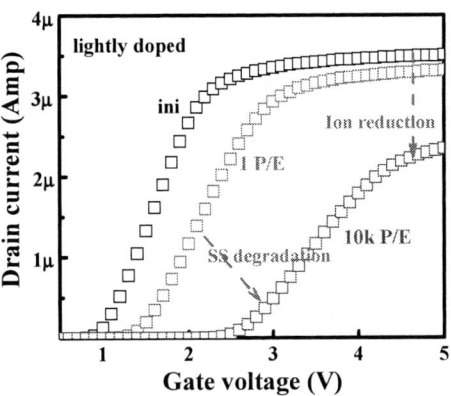

Fig.4 Comparison the Id-Vg behaviors of the lightly doped cell among initial, 1P/E and after 10k P/E stress. Serious I_{ON} and SS degradation are observed after stress.

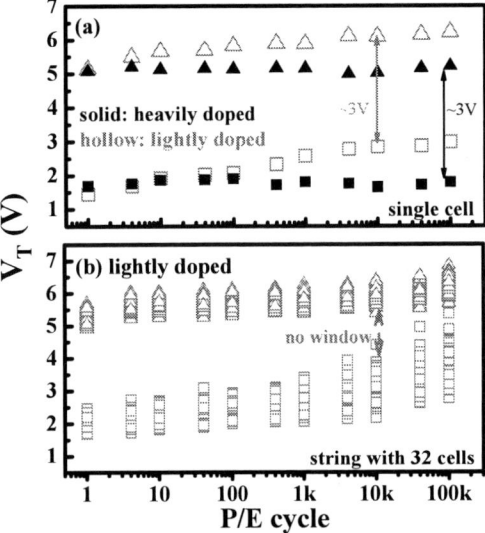

Fig.5 (a) S/D dopant effect on endurance behavior on single cell. (b) Wider VT variation on endurance if all 32 cells within a string are stressed.

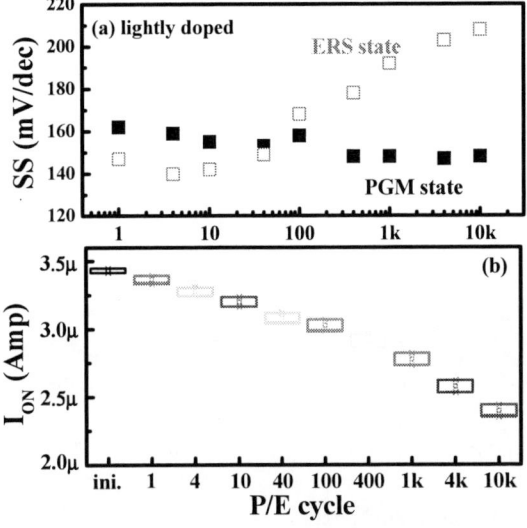

Fig.6 (a) Comparison the SS degradation between ERS and PGM state during P/E stress. (b) I_{ON} degradation with P/E cycle number.

978-1-4244-5430-3/10 $26.00 © 2010 IEEE

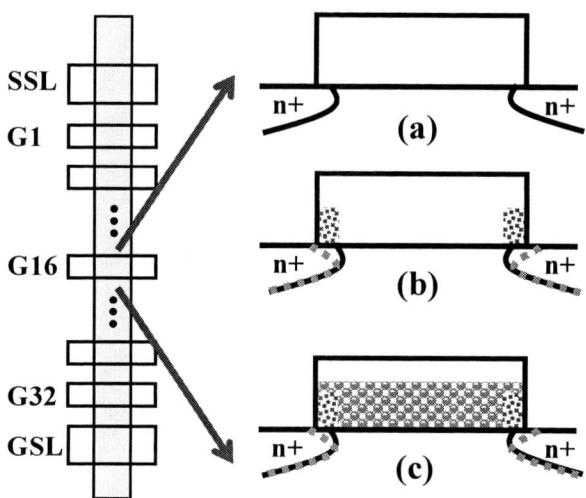

Fig.7 The cell is on (a) initial state, (b) erase state after P/E stress, and (c) program state after P/E stress. S/D might be depleted due to the residual charges (blue) above junctions. Then uniform programmed charges can mask the residual charges (red) induced SS degradation.

But this effect is insignificant once the V_{CH} is also increased. That is why SS is recovered after program operation. In a lightly doped S/D sample, this phenomenon becomes pronounced since its n+ junction is much easier to be depleted. The role of residual charges in erase speed is also examined in Fig. 8(a) and (b). Unlike a heavily doped case, the erase speed becomes 100X slower after P/E stresses for a cell with a lightly doped S/D. This also implies that the residual negative charges are located at the overlay region, which results in erase difficulty. To further confirm our suspicions, a gate-induced drain leakage (GIDL) method is employed to directly measure the residual charges during P/E stress. In heavily doped junctions, larger GIDL is observed, but the influence of the charges is relative smaller to lightly doped cases. Thus, in Fig. 8(c), the increase in GIDL with P/E cycles is easily observed in the lightly doped case.

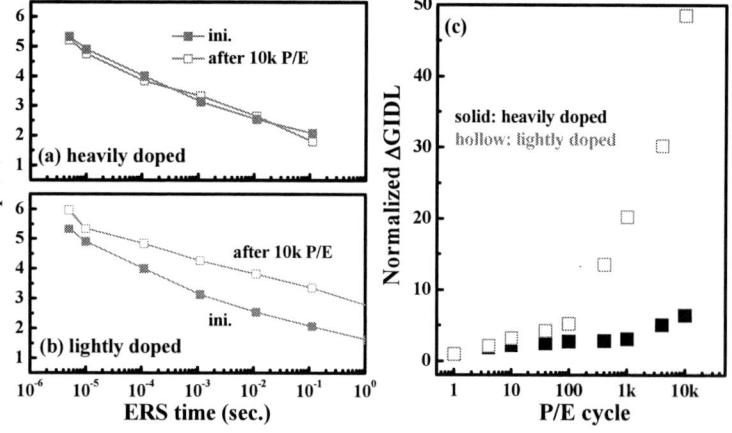

Fig.8 Comparison dopant effect on erase speed in (a) and (b). Normalized GIDL shift between heavily and lightly doped cells (c).

(b) New Program Disturbance

The behaviors of program disturbances between the heavily and lightly doped cells are compared in Fig 9 with a fixed total

programming time of 100µs. The heavily doped device follows a typical disturbance behavior, which gets worse with increasing programming time/shot [11]. Reducing SB ability due to junction leakage is the reason for this behavior. The lightly doped sample, however, shows a reverse trend. In Fig. 9(c), the increase of SS is observed in disturbed cells. The SS further degrades once these cells are erased. This implies that the disturb V_T mainly comes from charges located above S/D junction. To further investigate the program disturbance behavior in lightly doped case, the dependences of Vpass and V_{PGM} on disturbance are measured under a fixed

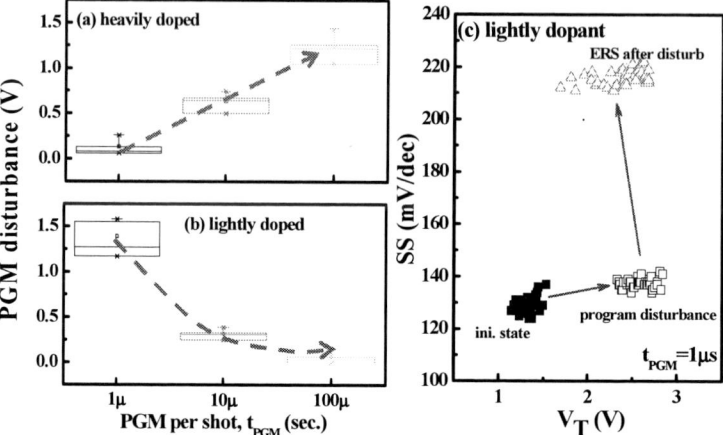

Fig.9 Dopant effect on program disturbance in heavily doped (a) and lightly doped (b) cells. After disturbance, SS is degraded in the lightly case (c).

V_{PGM}=20V or Vpass=10V, respectively (Fig. 10(a)). Result shows that the disturbed ΔV_T has a weak dependence on Vpass but gets larger with increasing V_{PGM}. This means that the boosting V_{CH} is not affected by Vpass but only determined by V_{PGM}, which is consistent to the behavior of a local SB case. In other words, the V_{CH} below the

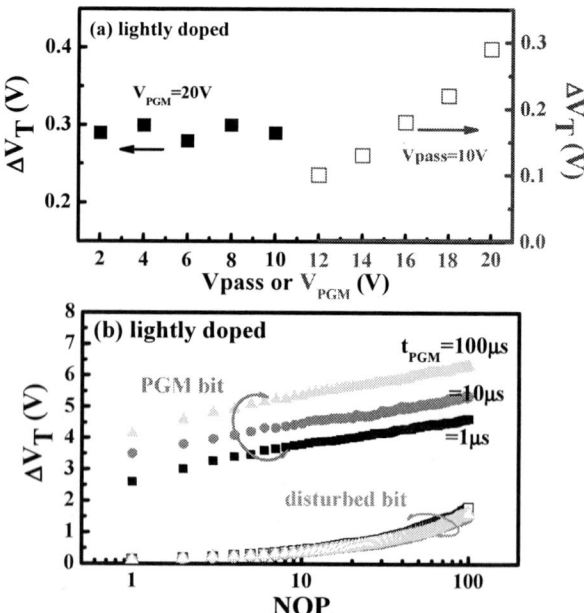

Fig. 10 (a) Vpass and VPGM dependence on disturbance. (b) NOP effect on disturbance with difference t_{PGM}.

978-1-4244-5430-3/10 $26.00 © 2010 IEEE

disturbed bit is only modulated by V_{PGM} because Vpass effect is cut by the depleted S/D junction. In Fig. 10(b), the disturb ΔV_T is plotted against the number of program shots (NOP). For the programmed bit, as expected, ΔV_T raises with increasing program time/shot. However, ΔV_T of disturbed bit is only related to NOP and is independent to program time/shot. To further explain this unique disturbance, a transient simulation is performed for a NAND string with a lightly doped S/D, as shown in Fig. 11. Although V_{CH} is boosted highly by a fully depleted S/D for a disturbed cell, it release and get to saturate rapidly (<1μs), implying that the most efficient disturbed time locates within 1μs. This is the reason that the ΔV_T of disturbed bit is rarely dependent on program time/shot in our case. The simulated lateral fields of a disturb cell is shown in Fig. 12. Though this high V_{CH} can lower the vertical field, it greatly increases the probability of junction breakdown. When junction leakage or breakdown occurs, the generated excess electrons will be accelerated by the lateral field and injected toward the nitride layers above S/D junctions and thus induces undesirable V_T shifts. This is very similar to the programming

mechanism in [12]. To fine tune the doping concentration, an optimized region with lower conventional disturbance (induced by vertical field) and smaller new disturbance (caused by lateral field) is obtained, as shown in Fig. 13.

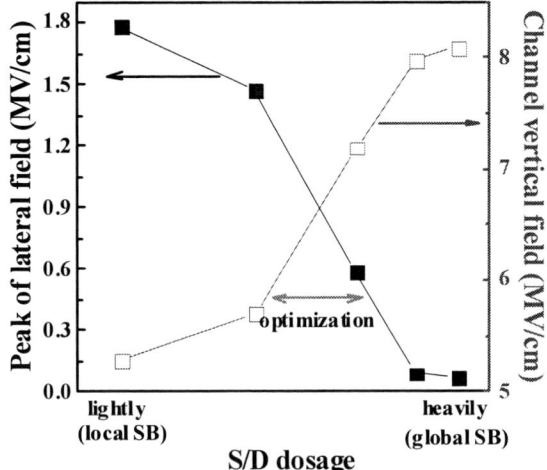

Fig.13 Self-boosting simulation for the the lateral field (new disturbance) and vertical field (conventional disturbance) under different S/D dopant.

Conclusions

In this paper, two different mechanisms resulting in reliability concerns for NAND flash that are both influenced by S/D doping concentration are investigated. The first part talking about the accumulated charges above the lightly-doped S/D junctions after P/E cycling will cause a serious SS increment, I_{ON} reduction, and window closure. Erase speed degradation and the increase of GIDL current also give another evidence to confirm that the root cause truly caused by residual charges inducing a depleted S/D junction. The second part illustrates that this depleted S/D can increase the channel potential and reduce the vertical voltage drop between gate and channel, but a large lateral field near the junctions occurs in the meantime. Once the junction has leakage or breakdown, excess carriers will be injected into the S/D junctions and cause a new kind of program disturbance. Increasing the S/D doping can ease such disturbance.

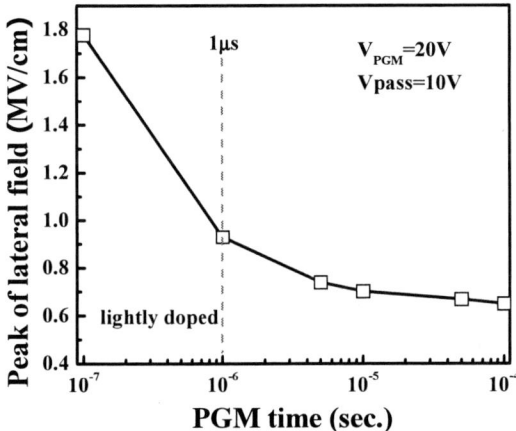

Fig.11 The transient behavior in peak of lateral field during programming. Field gets to saturate when PGM time>1μs.

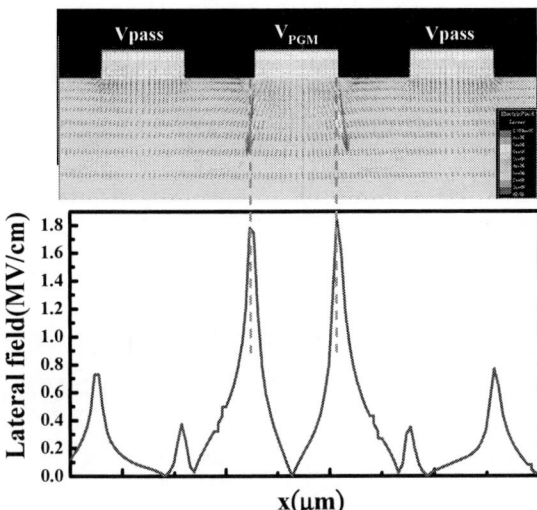

Fig.12 Self-boosting simulation for the channel field in a string with lightly doped cells. In program disturbed bit, large field is observed near the junctions

References

[1] T. Wang et. al., "Reliability Models of Data Retention and Read-Disturb in 2-bit Nitride Storage Flash Memory Cells (Invited Paper)," *IEDM Tech. Dig.*, pp. 169-172, 2003.

[2] H. T. Lue et. al., "Scaling evaluation of BE-SONOS NAND Flash beyond 20 nm", *Symp. VLSI Tech.,* pp. 116-117, 2008.

[3] C. H. Lee et. al., "Cell Endurance Prediction from a Large-area SONOS Capacitor," *Proc. Int. Reliability Phys. Symp.*, pp. 347-348, 2009.

[4] C. Friederich et. al., "Multi-level p+ tri-gate SONOS NAND string arrays," *IEDM Tech. Dig.*, pp. 1-4, 2006

[5] S. Jung et. al., "Three Dimensionally Stacked NAND Flash Memory Technology Using Stacking Single Crystal Si Layers on ILD and TANOS Structure for Beyond 30nm Node," *IEDM Tech. Dig.*, pp. 1-4, 2006.

[6] Y. Koh, "NAND Flash Scaling beyond 20nm," *IEEE International Memory Workshop*, pp. 3-5, 2009.

[7] M. F. Beug et. al., "Anomalous Erase Behavior in Charge Trapping Memory Cells," *IEEE International Memory Workshop*, pp. 121-123, 2008.

[8] T. Cho et. al., "A New Self-Boosting Phenomenon by Soure/Drain Depletion Cut-off in NAND Flash Memory," *IEEE International Memory Workshop*, pp. 39-41, 2007.

[9] D. Oh et. al., "A Dual-Mode NAND Flash Memory: 1-Gb Multilevel and High-Performance 512-Mb Single-Level Modes," *IEEE JOURNAL OF SOLID-STATE CIRCUITS,* pp. 1700-1706, 2007.

[10] L. Larcher et. al., "Impact of Programming Charge Distribution on Threshold Voltage and Subthreshold Slope of NROM Memory Cells," IEEE Trans. Elec. Dev., Vol. 49, pp.1839-1946, 2002.

[11] S. Satoh et. al., "A Novel Gate-Qffset NAND Cell (GOC-NAND) Technology Suitable for High-Density and Low-Voltage-Operation Flash Memories" IEDM Tech. Dig., pp. 271-274, 1999.

[12] J. Y. Wu et. al., "A NAND-type Flash Memory Using Impact Ionization Generated Substrate Hot Electron Programming ($\gg$ 20MB/sec) and Hot Hole Erasing" IEDM Tech. Dig., pp. 87-90, 2007.

Performance and Reliability Optimizations of BE-SONOS NAND Flash Using SiON Bandgap-Tuning Tunneling Barrier

Jeng-Hwa Liao, Jung-Yu Hsieh, *Hang-Ting Lue, Ling-Wu Yang, Tahone Yang, Kuang-Chao Chen, and Chih Yuan Lu

Macronix International Co. Ltd, Technology Development Center,
*Macronix International Co. Ltd, Emerging Central Lab.,
16 Li-Hsin Road, Hsinchu Science Park, Hsinchu, Taiwan.
Corresponding author: Jeng-Hwa Liao; Tel: +886-3-5786688 ext. 78168; Fax: +886-3-5789087; E-mail: jhliao@mxic.com.tw

Abstract—Bandgap-tunable SiON (oxynitride) tunnel barrier is developed to optimize the performance and reliability of BE-SONOS NAND Flash devices. The HTO O2 layer of the ONO tunnel barrier is replaced by SiON thin films with various refractive index (n) and thickness. We found that with n $\leq$ 1.72, SiON can provide excellent data retention comparable to conventional BE-SONOS. On the other hand, the erase speed can be greatly improved by using SiON O2 when n > 1.50. We suggest that hole barrier lowering by SiON is the root cause of such erase speed improvement. Our results suggest an optimal 35Å SiON O2 layer with n = 1.63 for the best program/erase speed and data retention performance. Finally, a 75 nm BE-SONOS NAND device is fabricated to validate the performances of this concept.

I. INTRODUCTION

Recently, floating gate (FG) NAND Flash memory has scaled down rapidly, and the nearly yearly doubling of density has stimulated new applications with ever lower cost. However, continuing this phenomenal scaling faces many fundamental obstacles, such as FG interference, geometrical limitation that threatens the gate coupling ratio (GCR), and the few-electron statistical limit. It is forecasted that charge-trapping Flash (CTF) devices [1-3] will continue the Flash memory scaling and further propel into 3D-stackable memory devices.

Several potential candidates of CTF devices were proposed [1-3]. BE-SONOS is one of the most promising candidates [2, 3]. In this device, a modulated ONO "U-shaped" tunneling barrier is used to replace the traditional single tunnel oxide. Efficient hole tunneling current can be provided at high electric field due to the band offset, while excellent retention can be provided at low field due to a thicker ONO barrier that prevents the direct tunneling leakage.

The most crucial element for BE-SONOS is the bottom O1/N1/O2 tunneling barrier. The previous BE-SONOS devices use conventional oxide for O1 and O2, where O1 is thermally grown ultra-thin oxide, while O2 uses a CVD (typically high-temperature LPCVD: HTO) process. While this structure provides excellent reliability, the erase speed is limited by the balance of hole injection and gate electron injection rates. In this work, we further improve the barrier property by engineering the O2 layer – by replacing the O2 with an oxynitride (SiON).

SiON can be fabricated with a wide range of dielectric constant (k is between oxide and nitride) as well as barrier height. We expect by reducing the hole barrier height we can improve the erase speed. Meanwhile, this new O2 layer must have sufficient barrier height to prevent charge loss. SiON is an ideal adjustable material for engineering the tunneling barrier.

II. SAMPLE DESCRIPTIONS

We chose to engineer the O2 layer of the O1/N1/O2 barrier because: (1) replacing O1 by a CVD process instead of thermal oxidation may introduce a higher interface trap density (D_{it}) and thus is not desirable, and (2) N1 (pure nitride) already has minimal hole barrier height, thus changing it does not improve the tunneling speed. Various BE-SONOS devices were fabricated, as shown in Table 1. The O2 layer was replaced by SiON thin films with various thickness and refractive index (n) between 1.50 and 1.72. Optical refractive index is a very useful indicator for the properties of SiON. For the reference, n=1.46 for pure oxide, and 2.0 for pure nitride.

	O1 / N1 / O2 (or SiON) / N2 / O3 (BE-SONOS film stack)
STD	13 / 20 / 25 / 60 / 60
SiON 1	13 / 20 / 25(n=1.63) / 60 / 60
SiON 2	13 / 20 / 25(n=1.72) / 60 / 60
SiON 3	13 / 20 / 35(n=1.50) / 60 / 60
SiON 4	13 / 20 / 35(n=1.63) / 60 / 60
SiON 5	13 / 20 / 35(n=1.72) / 60 / 60

Table 1 BE-SONOS NAND devices studied in this work. Various CVD SiON is used for the O2 layer of tunneling barrier. "n" is the index of refraction of SiON.

The SiON layer was deposited in a single wafer RTP CVD module, with substantially smaller thermal budget compared to the standard HTO O2. After O2 deposition the trapping layer (N2) and the top oxide O3, which is thermally converted from nitride, were fabricated. A heavily doped P$^+$-poly gate is used for all the samples in this work.

III. RESULTS AND DISCUSSIONS

A. Dielectric constant (k) and index of refraction (n):

Figure 1 illustrates the relationship between the dielectric constant (k) and the refractive index (n) of the SiON films. All the dielectric constant of SiON films were normalized to that of Si_3N_4 film. The k value is in between 3.9 to 7, and increases with increasing refractive index of the SiON film from 1.46 to 2.0, as predicted. A typical condition for SiON is k=5.0 and n=1.63.

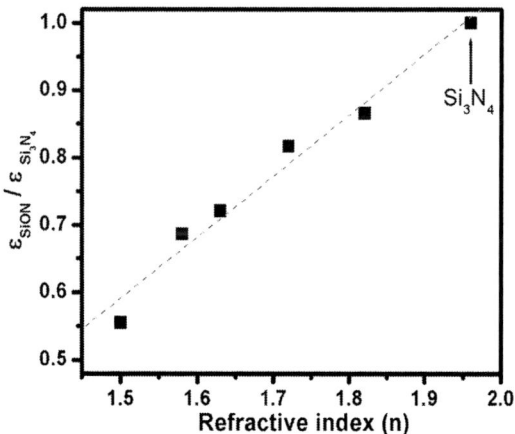

Fig. 1 Refractive index and dielectric constant of SiON. "n" is in between 1.5 to 2.0. Dielectric constant (K) is in between 3.9 to 7. All the dielectric constant of SiON films were normalized to that of Si_3N_4 film. The dash line is a straight line fitted to the data.

B. Capturing efficiency of SiON with various n:

The charge trapping efficiency of different SiON trapping layers in SONOS (O/N/O = 70/70/90 Å) devices were evaluated using incremental step pulse programming (ISPP) analysis [4] and shown in Figure 2. ISPP method is to apply a constant voltage step (ΔV_{PGM}) after successive +FN programming. It is often plotted as V_t (or V_{FB}) versus V_{PGM}. Our previous analysis indicates that the ISPP slope is strongly correlated to the capture efficiency [4]. Theoretical model predicts that a nearly fully-capturing property would give the ISPP slope very close to 1. On the other hand, a lower capture efficiency results in a lower ISPP slope. The SiON layers with $n \leq 1.63$ exhibit negligible charge-trapping behavior because the corresponding ISPP slopes are close to zero.

Therefore, SiON with n <= 1.63 is nearly "trap-free" and can be used in the barrier engineered tunneling dielectric.

C. Programming/Erasing comparison of BE-SONOS with various SiON O2:

The program and erase characteristics of BE-SONOS with various SiON O2 were shown in Figure 3. For a fair comparison, we normalize the initial electric field (V_G/EOT) to adjust for the EOT difference between the samples. The programming speed increases slightly with increasing the refractive index of SiON layer. However, the erase speed can be largely improved for SiON layer with n > 1.50, which could

be due to the increased hole tunneling current density during erasing. The reason will be explained next.

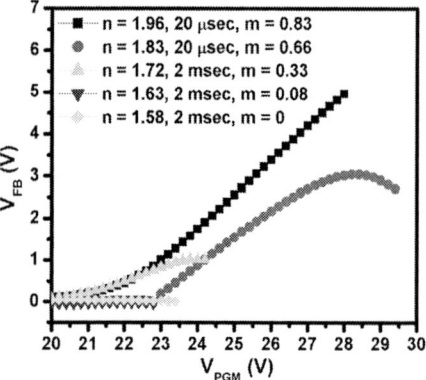

Fig. 2 ISPP programming of SONOS (O/N/O=70/70/90 Angstrom) with various SiON trapping layer. where "m" represents the maximum ISPP slope of each SONOS. When n ≤1.63, SiON becomes trap-free, leading to small ISPP slope (~0).

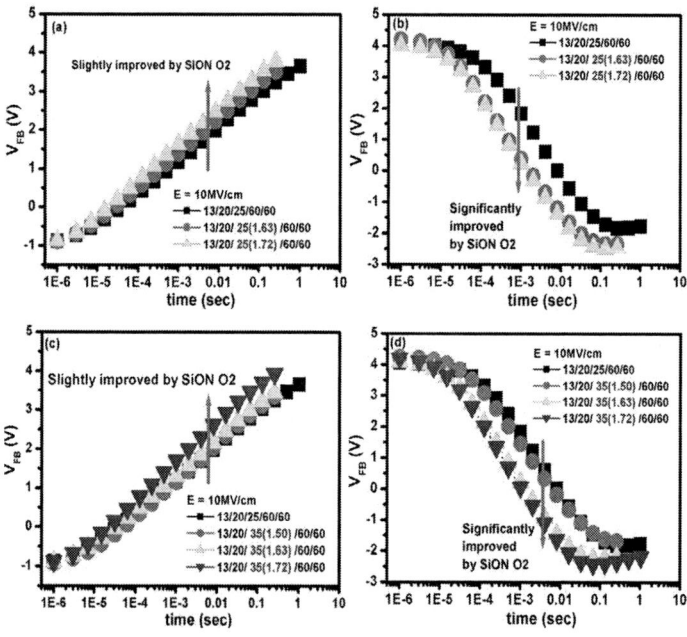

Fig. 3 The program and erase characteristics of BE-SONOS for (a) (b) 25A and (c) (d) 35A SiON O2 layers with different refractive index. Standard BE-SONOS with pure oxide O2 is also compared. SiON greatly improves the erase speed, but only slightly improves the programming speed.

D. Detailed erasing comparison:

Figure 4 shows the detailed erasing speed comparison and the corresponding hole current density using transient analysis. Transient analysis can transform the erase transient (V_{FB}-time) curve into a J-E_{TUN} curve, where J is the erase transient current density and E_{TUN} is the instantaneous tunnel oxide electric field [5]. Figure 4(a) and 4(b) indicate that increasing O2 from 25 to 35 Å does not change the erase speed. Figure 4(c) and 4(d) show that the hole current density is enhanced by SiON O2 with n > 1.50. The difference between n=1.63 and 1.72 is minor.

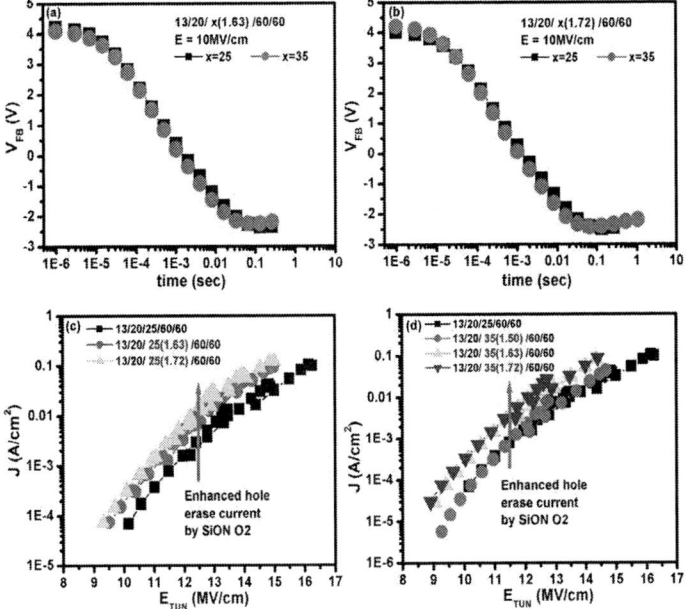

Fig. 4 The erase speed of BE-SONOS with various SiON thickness at (a) n=1.63 and (b) n=1.72. Transient analysis of the erase hole current density is also compared in (c) and (d).

E. WKB Model Simulation and Explanation:

Figure 5(a) shows the calculated band diagram of various ONO barrier under the same E field (10 MV/cm in O1). The shadow area (inside O1, N1 and O2) corresponds to the WKB tunneling barrier. N1 barrier is almost entirely screened at higher field due to the band offset. However, a small portion of O2 barrier still exists. Increasing O2 thickness does not change the WKB barrier since the triangular barrier of O2 only depends on E field. On the other hand, if the O2 barrier height is reduced (by SiON), the O2 barrier becomes less significant and the WKB tunneling probability is significantly enhanced. Note that the barrier reduction is expected to saturate when barrier is further lowered (when O2 barrier completely disappears due to the band offset). This explains why when n becomes >1.63 there is no further improvement.

Figure 5(b) shows the detailed WKB tunneling model fitting for the −FN erasing using the previous model [6]. Our model predicts that the SiON O2 hole barrier is approximately 1eV lower than that of the pure oxide HTO O2.

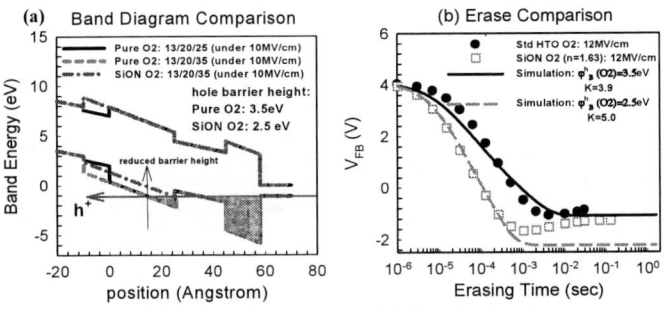

Fig. 5 (a) The band diagram to explain the enhanced erase speed. The shadow area corresponds to the WKB tunneling barrier. SiON with lower hole barrier height reduces the O2 barrier. (b) The erase

simulation of BE-SONOS. About 1eV hole barrier height lowering is suggested to be the root cause of erase speed improvement.

F. ISPP programming:

The ISPP programming of BE-SONOS with various SiON O2 is shown in Fig. 6. ISPP slopes are all close to 0.9. This indicates that tunnel barrier engineering does not change the ISPP performances [6].

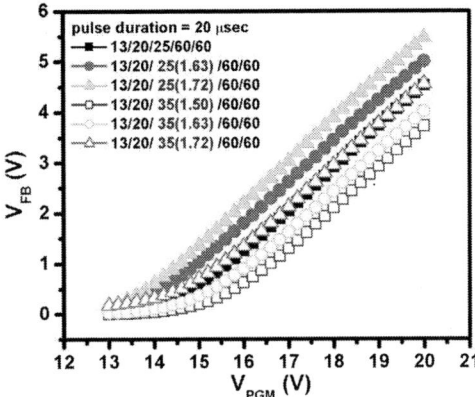

Fig. 6 ISPP programming for all BESONOS devices with the same pulse duration. ISPP slopes are close to 0.9, and independent of SiON O2.

G. P/E Cycling endurance:

Fig. 7 shows the cycling endurance of BE-SONOS devices. All the samples are subjected to the same stress voltage during cycling. There is no obvious P/E cycling endurance degradation by SiON O2.

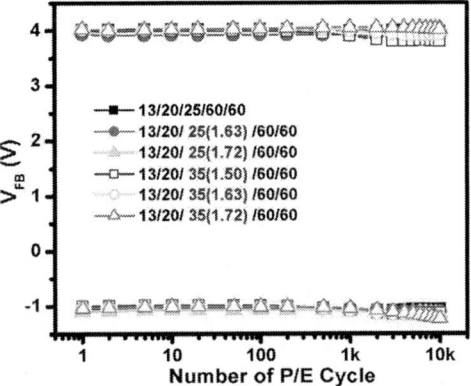

Fig. 7 The dumb-mode P/E cycling endurance of all BE-SONOS under the same stress voltage.

H. Retention comparison:

Figures 8(a) and 8(b) compare the retention of 25 and 35 Å O2, respectively. The retention performance was measured under 150°C baking. For the 25 Å O2, SiON shows worse retention than pure oxide. This is expected, since the lower bandgap SiON also offers less protection against leakage. The retention performance can be further improved by increasing the SiON thickness to 35A.

978-1-4244-5430-3/10 $26.00 © 2010 IEEE   641

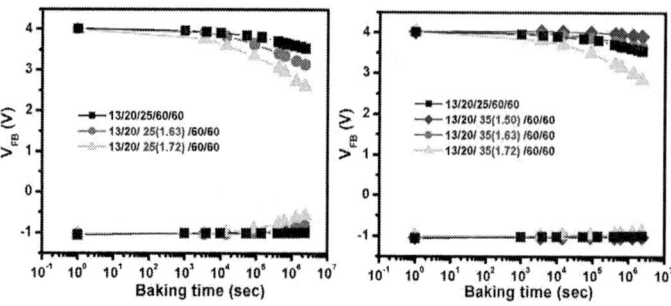

Fig. 8 The high-temperature (150C) retention of BE-SONOS devices for (a) 25A and (b) 35A SiON O2 layers with different refractive index.

IV. 75 NM NAND DEVICE DEMONSTRATION

We have successfully applied the SiON O2 (35 Å and n=1.63) in a 75 nm (half-pitch) BE-SONOS NAND device, as shown in Fig. 9. The typical programming and erasing characteristics of the BE-SONOS NAND device (measured in the center WL of the 32-WL NAND string) is shown in Fig. 10. More than 6V memory window can be obtained within typical P/E time (+20V 200usec for the programming, and -15V 10msec for the erasing). Figure 11 compares the erasing speed with BE-SONOS using HTO O2. By using SiON O2, the erasing speed is indeed improved and erase saturation is lower, thus creating more operational window.

Note that there is no new high-K material or metal gate and our device only uses matured materials including oxide, nitride, SiON, and P⁺-poly gate. All the processes are carried out by the conventional mass-production tools in IC manufacturing so that the device is reliable and repeatable. This is a very important advantage of BE-SONOS device.

(a)   (b)

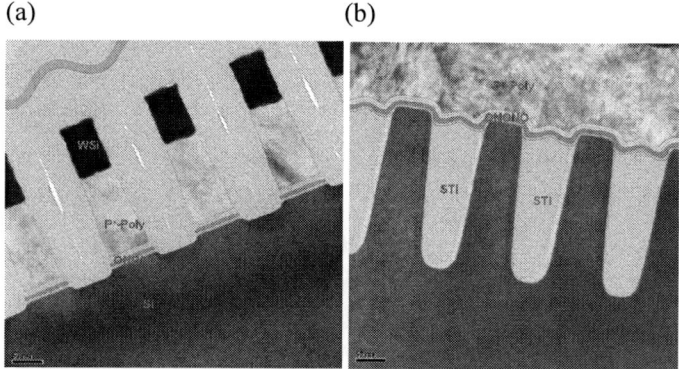

Fig. 9 TEM picture of the 75nm BE-SONOS NAND device with SiON O2. The pitch is 150nm in both channel length and width direction.

The dumb-mode (without any P/E verify) P/E cycling endurance is shown in Fig. 12. It shows reasonable endurance up to 10K cycling. The slightly roll-up of Vt is caused by the interface state (Dit) generation [7], as illustrated by the degraded subthreshold slope of IV curves in Fig. 12 (b). Dit effect is caused by the tunnel oxide degradation, and we expect the endurance can be further improved by combining the nitrided O1 to strengthen the interface of Si/O1 [7].

Figure 13 show the retention distribution of the devices. After 150C baking, it shows good retention and well preserve the memory window.

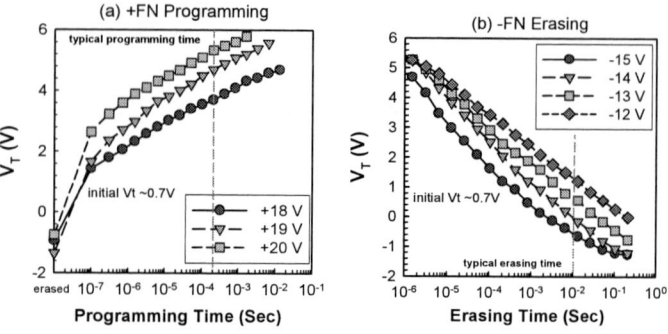

Fig. 10 (a) +FN programming and (b) –FN erasing characteristics of the 75nm BE-SONOS NAND device with SiON O2. The device is the center WL measured in a 32-WL NAND string.

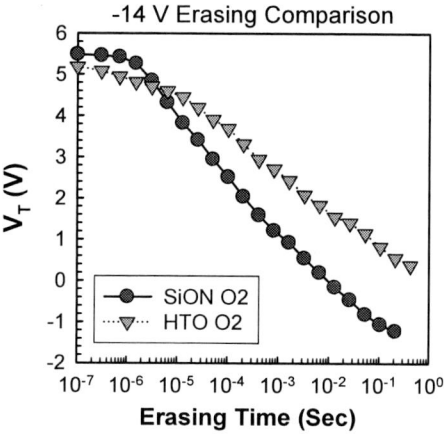

Fig. 11 –FN erasing speed comparison of 75nm BE-SONOS with various O2 thin film. SiON O2 shows faster erase speed and lower erase saturation than te standard HTO O2 (pure oxide). P⁺-poly gates are used for both devices.

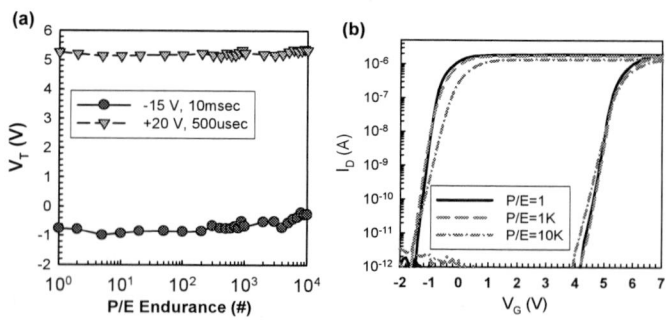

Fig. 12 (a) Dumb-mode P/E cycling endurance (with constant P/E conditions) of the 75nm BE-SONOS NAND with SiON O2, and (b) The corresponding IV curves measured in a 32-WL NAND string (with Vpass=7V) during P/E cycling.

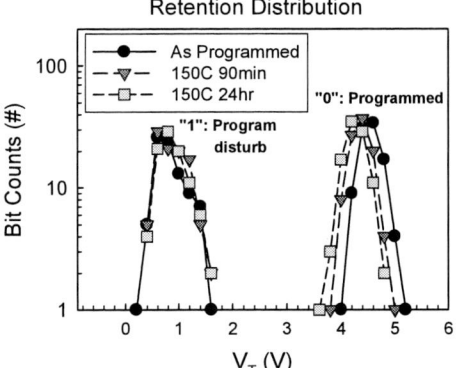

Fig. 13 Retention distribution of many 75nm BE-SONOS NAND devices with SiON O2. A SLC checker board pattern is designed in a NAND mini array with hundreds of memory cells. Conventional self-boosting method is employed and a successful SLC operation is shown. After 150C baking, it shows good retention.

V. CONCLUSIONS

A bandgap-tunable SiON tunneling barrier is explored and successfully applied in BE-SONOS devices. Erase speed and retention trade-off can be obtained by suitably adjusting the SiON properties. Our results show that a 35 Å SiON with n=1.63 optimizes both erase speed and retention properties.

REFERENCES

[1] Y. Shin, J. Choi, C. Kang, C. Lee, K.T. Park, J.S. Lee, J. Sel, V. Kim, B. Choi, J. Sim, D. Kim, H.J. Cho and K. Kim, "A Novel NAND-type MONOS Memory using 63nm Process Technology for Multi-Gigabit Flash EEPROMs", Tech. Digest 2005 International Electron Devices Meeting, pp. 327-330, 2005.

[2] H.T. Lue, S.Y. Wang, E.K. Lai, Y.H. Shih, S.C. Lai, L.W. Yang, K.C. Chen, J. Ku, K.Y. Hsieh, R. Liu, and C.Y. Lu, "BE-SONOS: A Bandgap Engineered SONOS with Excellent Performance and Reliability", Tech. Digest 2005 International Electron Devices Meeting, pp. 547-550, 2005.

[3] S. Y. Wang, H. T. Lue, E. K. Lai, L. W. Yang, Tahone Yang, K. C. Chen, J. Gong, K. Y. Hsieh, R. Liu, and C. Y. Lu, "Reliability and Processing Effects of Bandgap Engineered SONOS (BE-SONOS) Flash Memory", IEEE 45th Annual International Reliability Physics Symposium (IRPS), pp.171-176, 2007.

[4] H. T. Lue, et al, "Study of incremental step pulse programming (ISPP) and STI edge effect of BE-SONOS NAND flash," International Reliability Physics Symposium (IRPS), 2008, pp. 693-694.

[5] H.T. Lue, Y. H. Shih, K. Y. Hsieh, R. Liu, and C.Y. Lu, "A transient analysis method to characterize the trap vertical location in nitride-trapping devices", IEEE Electron Device Letters, vol. 25, pp.816-818, 2004.

[6] H. T. Lue, S. C. Lai, T. H. Hsu, P. Y. Du, S. Y. Wang, K. Y. Hsieh, R. Liu, and C. Y. Lu, "Understanding Barrier Engineered Charge-Trapping NAND Flash Devices With and Without High-K Dielectric", International Reliability Physics Symposium (IRPS), 2009, pp. 874-882.

[7] S. Y. Wang, H. T. Lue, T. H. Hsu, P. Y. Du, S. C. Lai, Y. H. Hsiao, S. P. Hong, M. T. Wu, F. H. Hsu, N. Z. Lien, C. P. Lu, J. Y. Hsieh, L. W. Yang, T. Yang, K. C. Chen, K. Y. Hsieh, R. Liu, and C. Y. Lu "A High-Endurance (>100K) BE-SONOS NAND FLASH With a Robust Interface of Si / Nitrided Tunnel Oxide", International Reliability Physics Symposium (IRPS), 2010, in publication (this conferecne).

978-1-4244-5430-3/10 $26.00 © 2010 IEEE

Flexible Electronics:
What can it do? What should it do?

Sameer M. Venugopal and David R. Allee
Flexible Display Center
Arizona State University
7700 S River Parkway, Tempe, AZ, USA, 85284
Phone: (001) – (480) - 727- 8986, sameermv@asu.edu

Manuel Quevedo-Lopez and Bruce Gnade
Department of Material Science and Engineering
University of Texas at Dallas
Richardson, TX, USA

Eric Forsythe and David Morton
Army Research Laboratory, M/S AMSRD-ARL-SE-EO,
2800 Powder Mill Rd, Adelphi, MD, USA, 20783-1145

Abstract—The development of low temperature, thin film transistor processes that has enabled flexible displays also presents opportunities for flexible electronics. A variety of flexible digital and analog electronics have been demonstrated, although typically of modest performance. We review the state-of-the-art in flexible electronics followed by a discussion of the constraints, remaining challenges and realistic potential applications of thin film transistors and flexible integrated systems.

Keywords- flexible electronics, thin film transistors

I. INTRODUCTION

Flexible electronics began with flexible solar cells which were silicon based but thinned down to improve their efficiency. Basically, flexible electronics deals with circuits developed using thin film transistors (TFT). The first TFTs were reported in 1968 by Brody and colleagues [1]. During the 1980s, the active matrix TFT backplanes were fabricated for display applications. This led to a huge success for the LCD industry. The amorphous silicon based backplanes were inexpensive to make and uniform across a large area.

Throughout the last four decades, there has been continuous improvement in developing TFTs on flexible substrates such as paper, polyimide, Mylar, stainless steel etc. Flexible solar cells have been extensively researched, and many of the manufacturing plants today develop amorphous silicon solar cells using roll-to-roll processes [2].

During the 1980s, the liquid-crystal display (LCD) industry began using amorphous silicon (a-Si:H) active matrix TFT backplanes for LCD displays. After an R&D effort of more than a decade, today we see flat panel displays being used in televisions, computer monitors, cellular phones, mp3 players etc.

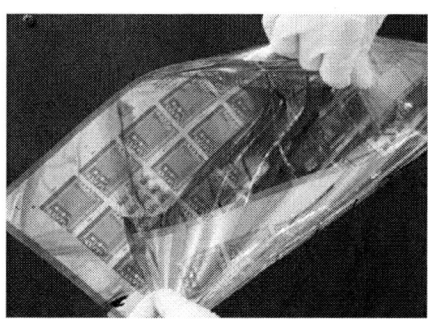

Fig. 1. Amorphous silicon backplane on Gen II PEN Substrate

Thin film transistor based circuits on flexible polyimide substrates were first demonstrated by Constant et al. in 1994 [3]. Since then several companies and research groups have demonstrated circuits and displays on flexible substrates using a-Si:H, organic materials, mixed oxide TFT and also hybrid organic/inorganic CMOS technologies [4-7]. The research effort is still ongoing to develop new materials and processes to manufacture flexible displays. Several electro-optic materials have been identified such as E-Ink Corp's electrophoretic ink, Kent Display's cholesteric material, organic light emitting diodes (OLED) etc. Fig. 1 shows an array of a-Si:H TFT backplanes for 1.1 inch diagonal flexible displays on heat stabilized polyethylene naphthalate (PEN) developed at Flexible Display Center at Arizona State University [8]. The size of the panel is 370 mm x 470 mm (Gen II). The panel is bonded to a glass carrier before it is processed in the Gen II pilot line. After the completion of the process, it is de-bonded from the carrier.

978-1-4244-5430-3/10 $26.00 © 2010 IEEE

II. CURRENT STATUS OF THIN FILM ELECTRONICS

In this section, we will discuss the current trends in flexible electronics and the different TFT technologies available today to develop circuits and displays on large area substrates.

A. Amorphous Silicon Technology

Hydrogenated amorphous silicon (a-Si:H) TFTs are the workhorse of today's active matrix LCD displays. Today, the LCD manufacturers are manufacturing panels on Gen 10 mother-glass which is 2880 mm x 3080 mm. These LCD panels are processed at higher temperatures which are not compatible with flexible substrates such as PEN, PET, etc. However, during the last decade, several research groups have shown progress in the development of flexible displays using a-Si:H TFT backplanes [4, 9, 10]. Fig. 2 shows a 4.1 inch active matrix OLED display on PEN using a-Si:H TFTs developed at the Flexible Display Center at ASU [11].

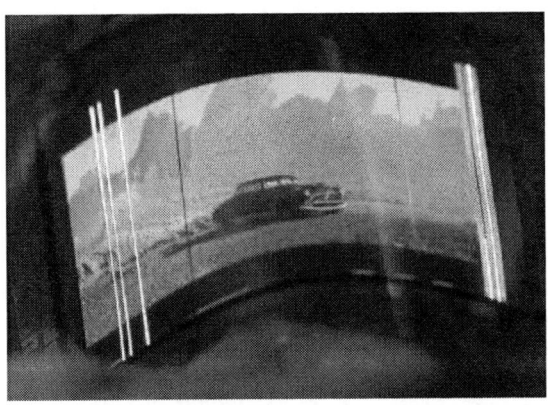

Fig. 2. A 4.1" QVGA OLED display on PEN

Materials and manufacturing processes for OLED displays are continuously evolving. The main driving force for OLED display is its emissive characteristics, good color saturation and clarity. It is also sunlight readable unlike many LCD technologies. The main limitation in using a-Si:H TFTs for OLED displays is the threshold voltage (Vt) shift of the TFTs. Due to electrical stress, the Vt of the TFTs increase over time which reduces the drive current of the TFTs. This will degrade the brightness of the OLEDs. Also, the lifetime of the OLED materials is limited due to moisture intake through the PEN substrate. Barrier coating is necessary to prevent this degradation.

Currently, there are a few ways of manufacturing flexible displays on plastic substrates – Surface Free Technology by Laser Annealing (SUFTLA), Electronics on Plastic by Laser Release (EPLaR), Bond-Debond method [12,13,4]. All these are low temperature processes which give rise to higher threshold voltage variations in a-Si:H TFTs. In order to reduce Vt shifts, the process temperature has to be increased and thus requires a substrate which can sustain higher temperatures. Princeton University researchers along with their industrial partners have shown high temperature plastics [14] which can be processed at 250 - 280 °C. Processing free standing PEN substrates is still a challenge as the size of the substrate increases.

Another important area is the integration of drivers for flexible displays on the substrate. Although, there are multiple research groups and industrial partners working on flexible TFT backplanes, very few have shown functional integrated electronics on flexible substrates to drive these displays [15, 16]. Sarnoff worked on high temperature TFT process on glass substrates for a number of years and developed integrated row and column drivers for LCD displays. Several Vt compensation techniques were used in this design.

Row drivers are relatively easier to integrate using amorphous silicon technology. The Vt shift does not affect the row drivers as much as it does to the column drivers since the row drivers are "on" only for a small period of time in a frame. If the display is running at 60 Hz, then the frame time is 0.166 seconds and for a QVGA display, the row time is only 69 µs. For column drivers, it is more important that the TFTs function with good stability since they are "on" throughout the frame time. Some of the work done on integrated drivers using low temperature amorphous silicon process on flexible substrates have been published by our group (FDC) in recent years [16, 17]. Fig 3 shows the circuit of a single column driver which can drive an electrophoretic display with 3 output levels. This circuit has been shown to be functional on stainless steel as well as PEN substrates. A 64x64 display with integrated column drivers was demonstrated as described in [17]. Due to its low mobility and high Vt shift, a-Si:H TFTs are not a good candidate for developing integrated source drivers for video rate displays. However, for bistable displays such as electrophoretic and cholesteric displays, a-Si:H TFT based integrated drivers can be used in applications which require only occasional image updates such as advertising , map applications, point of sale labels etc.

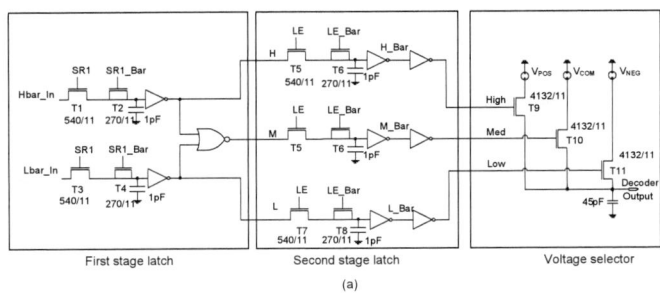

Fig. 3. Schematic of a single column of integrated source drivers for electrophoretic displays on PEN substrates.

B. Polysilicon Technology

Poly-Si TFTs are processed at higher temperature using laser re-crystallization of a-Si:H material and can have mobilities greater than 100 cm^2 V^{-1} s^{-1}. The threshold voltages of these TFTs are very stable and can be made in both varieties – n-type and p-type. Hence, Poly-Si TFTs can be used to develop display backplanes as well as CMOS digital circuits [18]. However, the process and substrate costs are

978-1-4244-5430-3/10 $26.00 © 2010 IEEE

comparatively higher and hence restrict the use of these TFTs in higher end applications such as high resolution displays in smart phones and high-end radio frequency tags.

C. Organic Thin Film Transistors

Organic TFTs can be manufactured using a number of organic semiconductors such as Pentacene, TIPS Pentacene, etc. These semiconductors can be processed at low temperatures using solution processes or vacuum evaporated processes such as spin coating, and ink-jet printing. Roll-to-roll processing may bring down the cost of production. The Pentacene based TFTs are p-type with carrier mobilities ranging from 0.1 to 5 cm^2 V^{-1} s^{-1} [19,20] which will allow them to be used in low speed applications such as active matrix electrophoretic displays. The OTFT is sensitive to air and hence its performance degrades over time when exposed to the environment. Barrier coating is required to protect it from exposure. Some of the research done at Stanford University shows that it is possible to develop mixed signal analog to digital converters using organic TFTs [21].

D. Single Crystal Silicon on Flexible Substrates

It is possible to develop single crystal silicon circuits on flexible substrates with mobilities greater than 500 cm^{2}V^{-1}s^{-1} and response frequencies greater than 500 MHz [22, 23]. In this technique, a semiconducting micro/nanomaterial known as microstructured silicon (μs-Si) is printed using dry transfer or solution based techniques onto plastic substrates to produce high performance TFTs [24].

E. Mixed Oxide Thin Film Transistors

Mixed oxide thin film transistors such as IZO, IGZO have been shown to provide better mobility, higher current densities and better stability compared to a-Si:H TFTs [25-27]. Another feature of mixed oxide TFTs is that they are transparent. Hence, there is much interest to develop transparent electronics on large area flexible substrates.

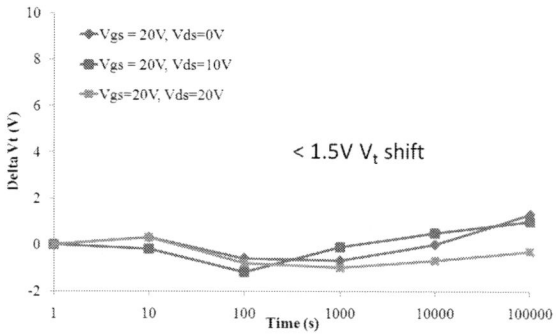

Fig. 4. Vt shift of low temperature IZO TFTs under DC bias stress

Simple digital inverters were built using low temperature (180 °C) IZO TFTs and stressed electrically for more than 350 hours. Fig 4 shows the variation of Vt over time under positive DC stress for these TFTs and Fig 5 shows the stability of an inverter with an AC stress.

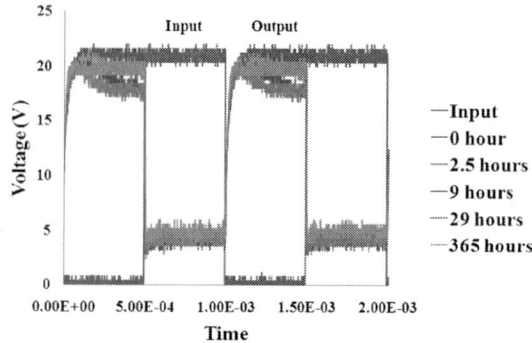

Fig. 5. Digital inverter using IZO TFTs stressed for 365 hours

F. Hybrid (CMOS) Technology

Complementary Metal Oxide Semiconductor technology has several advantages over nMOS only (a-Si:H, IZO, IGZO) or pMOS only (Pentacene) technologies. By including n-type TFTs and p-type TFTs on the same substrate, it is possible to implement CMOS circuits which reduce power consumption, leakage currents and improve the gain of the digital logic circuits. Some of the work done in this area is presented in [20, 21]. Research done at Flexible Display Center in collaboration with University of Texas at Dallas has shown that these CMOS logic circuits are more stable compared to a-Si:H TFT circuits. This is because the Vt of the a-Si:H shifts in positive direction with electrical stress while that of organic TFTs shift negative (Fig 6). In this technology, we have integrated a-Si:H TFTs and Pentacene TFTs on PEN substrate and successfully demonstrated a CMOS column driver for electrophoretic displays [28].

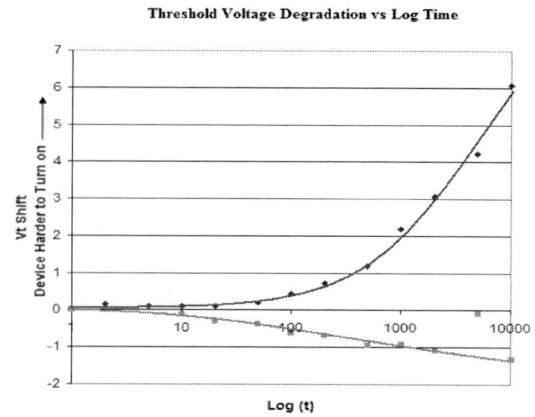

Fig. 6. Vt shift – a-Si:H nMOS and Pentacene pMOS

III. APPLICATIONS

Flexible electronic circuits can be used in several applications which require large area, rugged environments and have to be conformal to the device or structure. As discussed above, currently, the active matrix displays using flexible TFT backplanes is the most attractive application for industries since it does not require high speed devices comparable to single crystal silicon. For the future, however, it is important to have a vision with new applications which can only be done using flexible TFT based circuits. Some of the applications are discussed below.

A. Large Area Detectors

In the medical field, x-ray imaging is one of the most important tool to determine a patient's health. Digital x-ray imaging is gaining acceptance and has a long term economical advantage by having patients' records in digital format. By developing these x-ray imaging sensors on flexible substrates, it is possible to have portable equipments for field use such as in battlefield and for medical emergencies such as earthquakes, floods, etc. Potentially, a stretcher can have an integrated x-ray detector which sends the images wirelessly to the hospital servers for immediate use. This technology can be further expanded to detect neutrons, and gamma rays for security purposes in ports of entry and other public transport stations.

B. Prognostics and Diagnostics

Today, there are several autonomous machines which help in reconnaissance for the military and civilian applications. For prognostics and diagnostics, these machines have to be decommissioned temporarily and moved to a testing facility. This exercise is prohibitively expensive. Flexible electronics in the future can potentially be conformal to the surface of these vehicles/machines and the tests can be done onsite reducing the cost significantly. For example, flexible sensors can be put on the wings of an airplane to monitor structural health, air flow, etc.

C. Smart textiles

Recently, there is increase interest in smart textiles for health monitoring, entertainment and display applications. These "smart textiles" can be embedded with multiple sensors and display devices for monitoring heart rate, stress, monitoring toxic gases in the environment etc. Fig. 7 shows a concept of a smart textile with weave able "smart threads". Each "smart thread" is basically a shift register with a small display pixel and possibly a sensor, which can be used to transfer data from one end to the other. Data from all the "smart threads" can be read at the edge of the textile for further data processing.

D. Flexible Antennas

Transceivers are an important part of any wireless device. Future applications require some form of communication which requires high performance TFTs as well as integrated antennas. Researchers at ASU have shown flexible antennas fabricated on PEN substrate in the same fabrication line as the flexible displays [30].

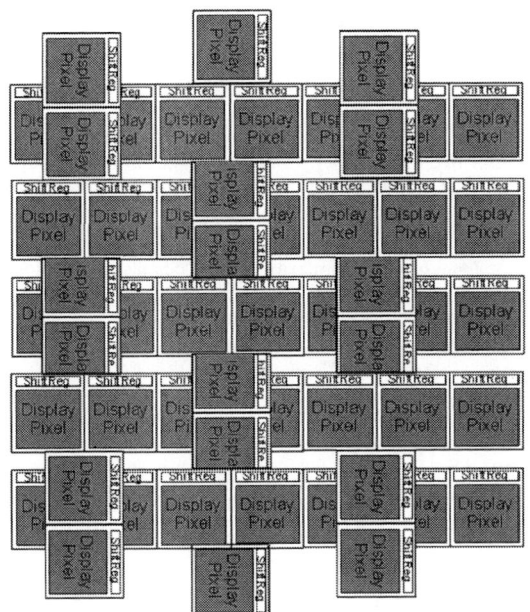

Fig. 7. Smart textile concept

E. Blast sensors/dosimeters

In recent years, more and more soldiers are experiencing brain trauma due to exposure to a large number of blasts in the battlefield. In order to understand the effect of blasts on human brain, researchers are developing blast sensors/dosimeters which can measure the blast wave as well as the direction from which it came from [31].

F. Organic Photoreceptors

From past two decades, significant amount of research has been done on organic photoreceptors and organic imaging systems for xerography [32, 33]. However, much effort is needed to develop organic photoreceptors on flexible substrates and integrate on to roll drums.

IV. CONCLUSION

This paper gives a brief overview of how the field of flexible electronics has evolved over the years and what the future holds for large area, flexible, rugged, low power electronics. Flexible displays have been the main focus so far for the industry. Several new manufacturing techniques are being developed from vacuum processing to ink-jet printing to roll-to-roll process. This paper introduces to the different TFT technologies being researched such as amorphous silicon, polysilicon, single crystal silicon, mixed oxide and organic TFTs. Some of the applications which can be developed on flexible substrates have been introduced.

ACKNOWLEDGMENT

We would like to acknowledge U.S. Army Research Labs for their continued support and funding for the development flexible electronics and displays at Flexible Display Center at Arizona State University.

REFERENCES

[1] Brody, T. Peter , "The thin film transistor - a late flowering bloom," *IEEE Transactions on Electron Devices*, vol. 31, no.11, pp. 1614–1628, Nov 1984.

[2] William S. Wong and Alberto Salleo, Flexible electronics: materials and applications, Springer Science and Business Media, 2009

[3] A. Constant, S. G. Burns, H. R. Shanks, C. Gruber, A. Landin, D. Schmidt, C. Thielen, F. Olympie, T. Schumacher, and J. Cobbs, "Development of thin film transistor based circuits on flexible polyimide substrates," *Electrochemical Society Meeting*, 9-14 October 1994, Miami, Florida. Published Proceedings. Invited Paper

[4] S.M. O'Rourke, D.E. Loy, C. Moyer, E.J. Bawolek, S.K. Ageno, B.P. O'Brien, M. Marrs, D. Bottesch, J. Dailey, R. Naujokaitis, J. P. Kaminski, D. R. Allee, S. M. Venugopal, J. Haq, N. Colaneri, G. B. Raupp, D.C. Morton and E.W. Forsythe, "Direct fabrication of a-Si:H thin film transistor arrays on plastic and metal foils for flexible displays," Proc. 26th Army Science Conference, Dec. 2008

[5] M. G. Kane et al, "Analog and digital circuits using organic thin-film transistors on polyester substrates," *IEEE Electron Device Letters*, vol. 21, no. 11, Nov 2000

[6] Randy Hoffman, Tim Emery, Bao Yeh, Tim Koch, Warren Jackson, "Zinc Indium Oxide thin-film transistors for active-matrix display backplane," *Proc. of Society for Information Display*, 2009

[7] S. Gowrisanker, M.A, Quevedo-Lopez, H. N. Alshareef, B. Gnade, S. Venugopal, R. Krishna, K. Kaftanoglu, D. Allee, "Low temperature integration of hybrid CMOS devices on plastic substrates," *Proc. Flexible Electronics and Displays Conference*, Phoenix, Arizona, Feb 2-5, 2009

[8] Flexible Display Center, Arizona State University [online] http://flexdisplay.asu.edu

[9] John K. Borchardt, "Developments in organic displays," *Materials Today*, Volume 7, Issue 9, September 2004, Pages 42-46, ISSN 1369-7021, DOI: 10.1016/S1369-7021(04)00401-8

[10] S.E. Burns et. al, "Flexible active matrix displays," *Proc. Society for Information Display*, vol. 36, no. 1, pp. 19-21, 2005

[11] D. E. Loy, Y. K. Lee, C. Bell, M. Richards, E. J. Bawolek, S. K. Ageno, C. Moyer, M. Marrs, S. M. Venugopal, J. P. Kaminski, N. Colaneri, S. M. O'Rourke, J. Silvernail, K. Rajan, R. Ma, M. Hack, and J. J. Brown, "Active matrix PHOLED displays on temporary bonded polyethylene naphthalate substrates with 180°C a-Si:H TFTs," *Proc. Society for Information Displays*, San Antonio, TX, May 31-June 5, 200

[12] Sumio Utsunomiya, Satoshi Inoue, and Tatsuya Shimoda, "Low-temperature poly-Si TFT transferred onto plastic substrates by using surface free technology by laser ablation/annealing (SUFTLA)," *J. Soc. Inf. Display*, vol. 10, no.1, March 2002

[13] Ian French, David George, Thierry Kretz, Francois Templier, and Herbert Lifka, "Flexible displays and electronics made in AM-LCD facilities by the EPLaR process," *SID Symposium Digest*, vol. 38, no. 1, pp. 1680-1683, May 2007

[14] H. Bahman, K.H. Cherenack, A. Z. Kattamis, K. Long, S. Wagner, J.C Sturm, "Highly stable amorphous-silicon thin-film transistors on clear plastic," *Applied Physics Letters*, vol. 93, no.3, 2008

[15] H. Lebrun, F. Maurice, J. Magarino, and N. Szydlo, " AMLCD with integrated drivers made with amorphous silicon TFTs", *Society for Information Display, Journal of*, vol. 3, no. 4, pp. 177-180, Dec 1995

[16] S.M. Venugopal and D. R. Allee, "Integrated a-Si:H source drivers for 4" QVGA electrophoretic display on flexible stainless steel substrate," *IEEE Journal of Display Technology*, Vol. 3, No. 1, March 2007, pp. 57-63

[17] Sameer M. Venugopal, Rahul Shringarpure, David R. Allee and Shawn M. O'Rourke, "Integrated a-Si:H Source Drivers for Electrophoretic Displays on Flexible Plastic Substrates", *Proc. 7th Annual Flexible Electronics & Displays Conference*, Phoenix, Arizona, pp. 1-5 Jan 2008

[18] Po-Chin Kuo, Abbas Jamshidi-Roudbari, and Miltiadis Hatalis, "Electrical characteristics and mechanical limitation of polycrystalline silicon thin film transistor on steel foil under strain", *J. Appl. Phys.* vol. 106, no. 11, pp. 114502, Dec 2009

[19] Dirk Zielke, Arved C. Hubler, Ulrich Hahn, Nicole Brandt, Matthias Bartzsch, Uta Fugmann, Thomas Fischer, Janos Veres, and Simon Ogier, "Polymer-based organic field-effect transistor using offset printed source/drain structures", *Appl. Phys. Lett.* vol. 87, pp.123508, 2005, DOI:10.1063/1.2056579

[20] Manuel Quevedo, S. Gowrisanker, H.N. Alshareef, B. E. Gnade, D Allee, S Venugopal, R Krishna, and K Kaftanoglu, "Novel materials and integration schemes for CMOS-based circuits for flexible electronics," *Meet. Abstr. - Electrochem. Soc.* 902, 2408 (2009),

[21] W. Xiong, U. Zschieschang, H. Klauk, and B. Murmann, "A 3V, 6b Successive Approximation ADC using Complementary Organic Thin-Film Transistors on Glass," to appear, *ISSCC Dig. Techn. Papers*, Feb. 2010

[22] Jong-Hyun Ahn et. al., "High-speed mechanically flexible single-crystal silicon thin-film transistors on plastic substrates", *IEEE Electron Device Letters*, vol. 27, no. 6, pp. 460-462, June 2006

[23] Dae-Hyeong Kim et. al., "Ultrathin Silicon Circuits With Strain-Isolation Layers and Mesh Layouts for High-Performance Electronics on Fabric, Vinyl, Leather, and Paper," *Advanced Materials*, vol. 21, no. 36, pp. 3703-3707, Sept. 2009

[24] E. Menard, K. J. Lee, D.-Y. Khang, R. G. Nuzzo, and J. A. Rogers, "A printable form of silicon for high performance thin film transistors on plastic substrates", *Appl. Phys. Lett.*, vol.84, no. 26, pp. 5398-5400, June 2004, DOI:10.1063/1.1767591

[25] P. G"orrn, M. Sander, J. Meyer, M. Kr"oger, E. Becker, H.-H. Johannes, and W. K. T. Riedl, "Towards see-through displays: Fully transparent thin-film transistors driving transparent organic light-emitting diodes," *Adv. Mater.*, vol. 18, pp. 738–41, 2006.

[26] U. Ozgur, Ya. I. Alivov, C. Liu, A. Teke, M. A. Reshchikov, S. Dogan, V. Avrutin, S.-J. Cho, and H. Morkoc, "A comprehensive review of ZnO materials and devices," *J. Appl. Phys.* Vol. 98, pp. 041301, 2005.

[27] Sameer M. Venugopal, Edward Bawolek, Michael Marrs, Korhan Kaftanoglu, Aritra Dey, James R. Wilson, Anil Indluru, Terry L. Alford, David R. Allee and Shawn O'Rourke, "Effect of bias stress on Indium Zinc Oxide thin film transistors manufactured at low temperature", *Proc. 9th annual Flexible Electronics and Display Conference*, Phoenix, AZ, Feb 2010

[28] D. R. Allee, S. Venugopal, R. Krishna, K. Kaftanoglu, M. Quevedo-Lopez, S. Gowrisanker, A. Avendano-Bolivar, and B. Gnade, "Flexible CMOS and electrophoretic displays," invited paper, *Society for Information Displays*, International Symposium, Digest of Technical Papers, San Antonio, TX, May 31-June 5, 2009

[29] [online] http://www.warwickaudiotech.com/index.php

[30] B. Gnade, M. Quevedo, D. Allee, S. Venugopal, C. Balanis, T. Jackson, H. McHugh, K. Baugh, E. Forsythe, and D. Morton, "Flexible integrated sensor systems," invited paper, *Special Operations Forces Industry Conference*, Tampa, FL, June 2-4, 200

[31] Daniel J, Garner S, Nga N T, Knights J, Arias A. C, "Sensing blast events with flexible sensor tapes", *Proc. 9th annual Flexible Electronics and Displays Conference*, Phoenix, AZ, Feb 2010.

[32] Borsenberger P.M, Weiss D.S, Organic Photoreceptors for imaging systems, Marcel Dekker, Inc., New York, 1993

[33] Borsenberger P.M, Weiss D.S, Organic Photoreceptors for Xerography, Marcel Dekker, Inc., New York, 1998

Sameer Venugopal (PhD, 2007, Arizona State University) is a Display Design Engineer at Flexible Display Center at Arizona State University. His research interests include circuit design using novel thin film devices for flexible electronics and backplane design for flexible displays. He has authored/co-authored over 20 peer reviewed scientific publications.

David R. Allee (PhD, 1990, Stanford University) is an Associate Professor of Electrical Engineering at Arizona State University. He is currently Director of Research for Backplane Electronics at the Flexible Display Center at Arizona State University, and is investigating a variety of flexible electronics circuit applications. He has co-authored over 70 archival scientific publications.

Manuel Quevedo-Lopez is a Research Professor in the Department of Materials Science and Engineering at the University of Texas at Dallas. His research interests include novel materials and devices for flexible electronics. He has authored/coauthored over 75 peer review papers and holds 5 US issued patents with 7 more pending.

Bruce Gnade (PhD, 1982, Georgia Institute of Technology) is currently Vice-president for Research at the University of Texas at Dallas and the Distinguished Chair in Microelectronics in the Electrical Engineering Department also at the University of Texas at Dallas. He has 69 U.S. patents, 54 foreign patents, and approximately 150 peer reviewed publications. Dr. Gnade research interests include novel materials and devices for flexible electronics.

Eric Forsythe (Ph.D. 1996, Stevens Institute of Technology) is a staff physicist at the Army Research Laboratory. He is the Team Leader for Display Technologies and is an Associate Program Manager for the Army's Flexible Display Center. His research activities include, organic based light emitting device for flexible displays, organic based thin film transistors and photovoltaics. He has authored/coauthored more than 45 papers.

David Morton is a staff physicist at the Army Research Laboratory and the Program Manager for the Army's Flexible Display Center.

978-1-4244-5430-3/10 $26.00 © 2010 IEEE

On the Bias Dependence of Time Exponent in NBTI and CHC Effects

[1]Jyothi B. Velamala, [2]Vijay Reddy, [1]Rui Zheng, [2]Srikanth Krishnan, [1]Yu Cao

[1]Department of Electrical Engineering,
Arizona State University, Tempe, AZ 85287, USA
[2]Texas Instruments, Dallas, TX 75243, USA
E-mails: [1]{jvelamal, rui.zheng.1, ycao}@asu.edu, [2]{vreddy, s-krishnan1}@ti.com

ABSTRACT

NBTI and CHC are two leading reliability concerns. Their degradation rate, which is represented by the time exponent (*n*), varies with multiple factors, such as the measurement method and bias voltages (i.e., different *n* for sub-threshold or linear current). Such a variation significantly affects the long-term prediction of circuit lifetime. By investigating the underlying mechanisms and silicon data, we conclude that the bias dependence is due to *intrinsic device non-linearity*. With a unified aging model of threshold voltage (V_{th}) shift, different time exponents in different operation regions are consistently explained. The proposed solution captures the change of *n* under various supply voltages (V_{dd}), as validated with silicon data from transistors and RO measurement. It helps improve the accuracy in reliability prediction, reducing unnecessary design margins. Based on the result, the device and circuit lifetime is expected to be enhanced operating at lower V_{dd} due to the reduction in the time exponent.

Keywords: Temporal degradation, NBTI, CHC, time exponent, reaction-diffusion.

I. INTRODUCTION

The temporal degradation of the device plays a key factor in limiting the circuit lifetime. Negative bias temperature instability (NBTI) and Channel hot carrier (CHC) effects are two primary aging mechanisms in scaled CMOS design [1-3]. NBTI occurs due to the generation of interface traps at the Si-SiO_2 interface when a PMOS is stressed with negative bias while CHC causes the generation of interface charges at the gate-oxide interface near the drain end when the NMOS gate switches. The degradation due to CHC is permanent while the degradation due to NBTI can be recovered. In reality, such recovery property of NBTI makes it more difficult to predict the lifetime due to the uncertainty in dynamic operation.

The degradation due to both NBTI and CHC can be modeled based on the general reaction-diffusion (RD) and trapping/detrapping mechanisms [3,10-13] which are dependent on the process and design parameters like V_{th}, V_{dd}, duty cycle etc. Both effects manifest themselves as the temporal degradation of device parameters, especially V_{th} and mobility [3] leading to poor drive current, lower noise margin and shorter device and circuit lifetime. In order to investigate

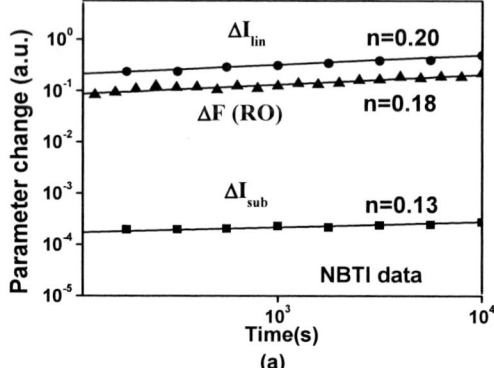

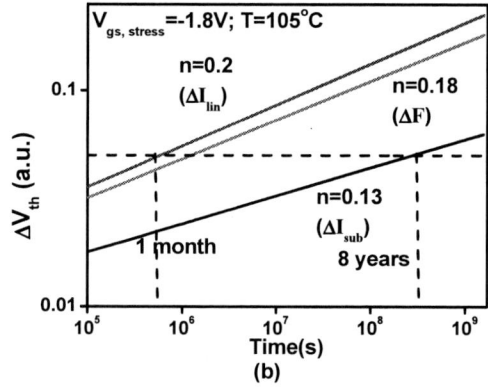

Figure 1. (a) Different *n* from different extraction methods in NBTI; (b) A wide variation of lifetime prediction from various extraction methods, assuming a 50mV V_{th} shift at the end of the lifetime.

the degradation due to NBTI and CHC effects, several models for the threshold voltage shift have been previously proposed [3,10,13]. Although the underlying reliability mechanisms are still under the debate [2], their degradation rate, such as V_{th} shift (ΔV_{th}), usually follows a power law of the stress time (t): $\Delta V_{th} \propto t^n$, where the time exponent is a characteristic of NBTI (*n*~0.16) or CHC (*n*~0.5) [1-3, 8-10]. The diffusing species determines the time exponent in NBTI: it is widely accepted that H_2 is the diffusing species resulting in *n*~0.16 against the H atom diffusion resulting in *n*~0.25 [3]. The exact value of *n*

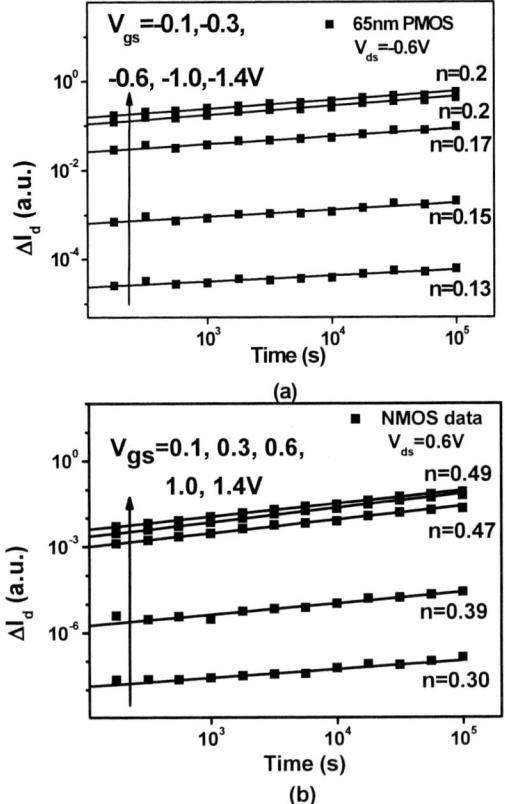

Figure 2. The change of time exponents in different bias regions: (a) NBTI ($V_{gs,stress}=-1.8V$) (b) CHC ($V_{gs,stress}=1.7V$)

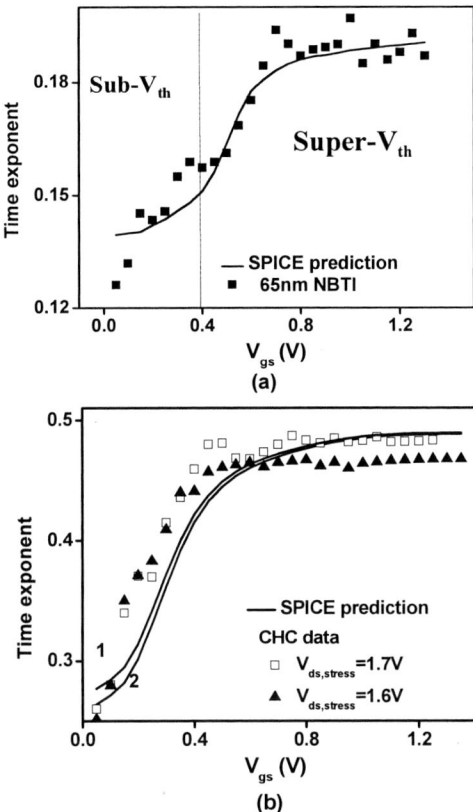

Figure 3. SPICE simulation with V_{th} aging model [3] predicts the increase of n from sub-V_{th} to super-V_{th} region, approaching the theoretical expectations: (a) NBTI and (b) CHC.

is essential to extrapolate the amount of degradation in the long term.

In practice, there exist various extraction methods to obtain the value of n, such as the OTF measurement that monitors the change in linear current (I_{lin}) [4], the constant current definition of V_{th} from sub-threshold current (I_{sub}) [5], and the frequency (F) measurement from a ring oscillator (RO) [6]. These methods assume a linear dependence between the change of measured parameters and ΔV_{th} and thus, their extracted n should be the same. However, Fig. 1a illustrates that even from the same fabrication technology and the same stress condition [7], the values of n from different extraction methods do not match each other. Specifically, I_{sub} has a lower n than that of I_{lin}, while n of RO frequency is between them.

Although the deviation between the extracted n and the theoretical expectation (e.g., 0.16 in NBTI) can be explained by different reliability mechanisms, such as trapping/ detrapping [2], its bias dependence cannot be attributed to this reason. Worse more, even a small difference in n leads to a significant change in the long-term prediction of circuit lifetime (Fig. 1b), complicating the protection strategy and the amount of guardbanding.

To clarify this concern and improve the accuracy of circuit reliability prediction, this paper investigates the reason for the bias dependence of n in NBTI and CHC. It is concluded that

such bias dependence is due to the inherent non-linear behavior of a transistor.

II. TRANSISTOR LEVEL DEPENDENCE ON BIAS VOLTAGES

Experimental data were first collected from scaled transistors [7], which are stressed with high voltages (NBTI for the PMOS devices and CHC for the NMOS devices) and high temperature (i.e., 105°C). A set of IV curves are measured using stress-measure-stress method (SMS). Figure 2 presents the degradation of drain current (I_d) under various bias voltages (V_{gs} and V_{ds}), for both NBTI and CHC. It is evident that in both cases, the time exponent goes up with higher V_{gs}. For instance, n increases from 0.13 to 0.2 when V_{gs} changes from -0.1V to -1.4V for a PMOS device; CHC in a NMOS device exhibits a similar shift, from $n=0.30$ to $n=0.49$. Meanwhile, n is relatively independent on V_{ds}, in both NBTI and CHC.

The change of n is correlated to the intrinsic property of a transistor: in the strong-inversion (super-V_{th}) region, I_d is a linear function of V_{th}, because of either the property of the linear region or strong velocity saturation in the saturation region [3, 13]; in this case, I_d degrades at the same rate of V_{th}.

978-1-4244-5430-3/10 $26.00 © 2010 IEEE

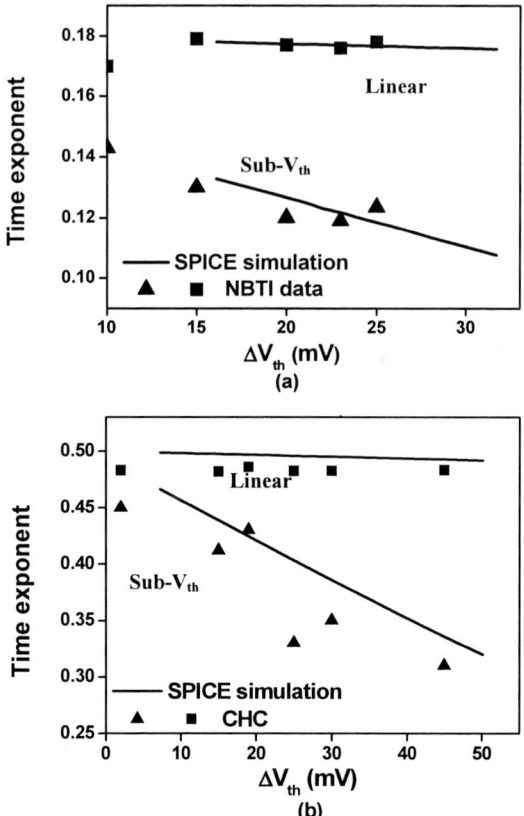

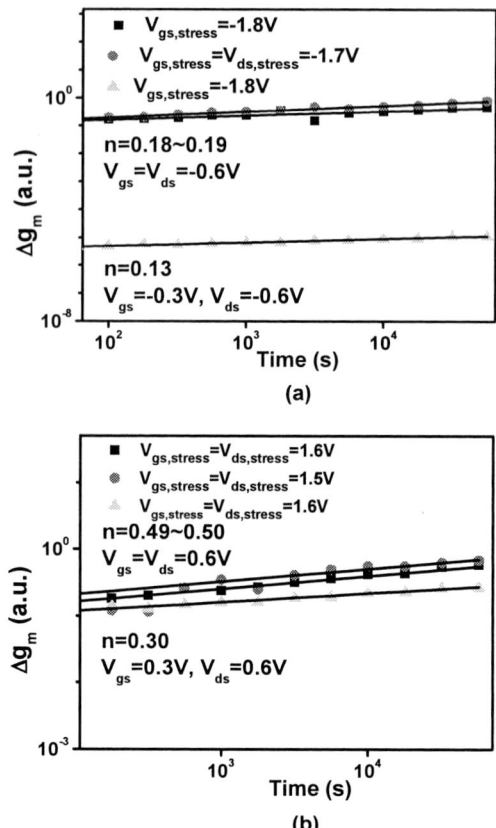

Figure 4. The change of n further depends on the amount of V_{th} shift: (a) NBTI and (b) CHC.

Figure 5. The validation of SPICE model with silicon data of g_m degradation: (a) NBTI and (b) CHC.

Therefore the time exponent of ΔI_d is the same as the time exponent of ΔV_{th} in linear and saturation regions. However, in the sub-V_{th} region, I_d is an exponential function of V_{th} and thus:

$$I_d \propto \exp(-V_{th}/s) \propto 1 + \left(\frac{\Delta V_{th}}{s}\right) + \left(\frac{\Delta V_{th}}{s}\right)^2 + \ldots \quad (1)$$

where s is the sub-V_{th} swing. The non-linear terms distort the relationship between ΔI_d and ΔV_{th}, reducing n to a lower value.

To verify the reasoning above, SPICE simulations are conducted (Fig. 3) inducing an increase in the absolute threshold voltage in both PMOS and NMOS. By including the aging model of $\Delta V_{th} \propto t^n$ [3, 13] to a nominal device model, where n is fixed at the value from the linear region, the changing behavior of n is successfully reproduced, as shown in Fig. 3. In the super-V_{th} region, n is close to the theoretical expectation (0.16 for NBTI and 0.5 for CHC) confirming the linear relationship between ΔI_d and ΔV_{th}. As V_{gs} decreases, the non-linear terms become more significant resulting in gradual reduction of n with a relative independence on V_{ds}.

Furthermore, Eq. (1) implies that the non-linear terms become stronger when ΔV_{th} is larger. Therefore, the reduction of n in the sub-V_{th} region is more pronounced with a larger ΔV_{th}. Figure 4 confirms this behavior, using data from

different stress time and gate voltages, such as that in Fig. 3b. It indicates that when a device is stressed for a longer period, the time exponent in the sub-V_{th} region decreases with the stress time. On the other hand, n remains as a constant in the super-V_{th} region, because of the linear dependence of I_d on V_{th}.

In a brief summary, the characterization of n should be conducted in the super-V_{th} region. Other methods, such as the I_{sub}-based V_{th} extraction, may suffer from the intrinsic non-linearity of the transistor and underestimate the degradation rate. With an appropriate ΔV_{th} model, the degradation rate of IV in all regions can be correctly predicted (Figs. 2-4). Also an increase in V_{th} with time, lowers the time exponent of the drain current in sub-V_{th} region. It also supports the prediction of other device aging properties. Figure 5 presents an example of the shift in trans-conductance (g_m) in both NBTI and CHC effects. With ΔV_{th} from the linear region ($n \sim 0.20$ for NBTI and $n \sim 0.50$ for CHC), the proposed method well handles different n values of Δg_m under different bias voltages during the measurement. Also, the operation of the device close to the sub-threshold region decreases the time exponent of ΔI_d, thereby significantly reducing the degradation rate.

978-1-4244-5430-3/10 $26.00 © 2010 IEEE

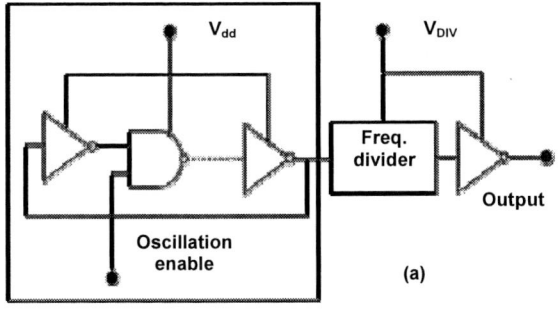

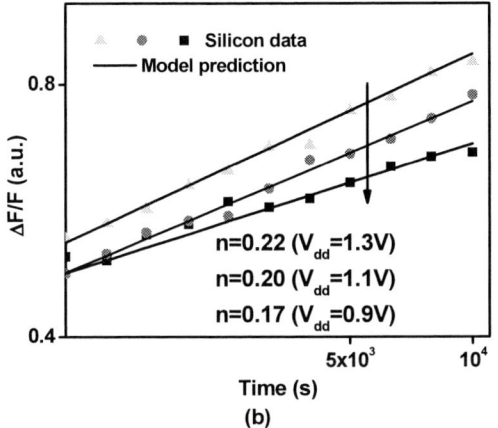

Figure 6. n of RO frequency degradation changes with V_{dd}: (a) the test structure of a RO [7], (b) n value decreases at lower V_{dd}.

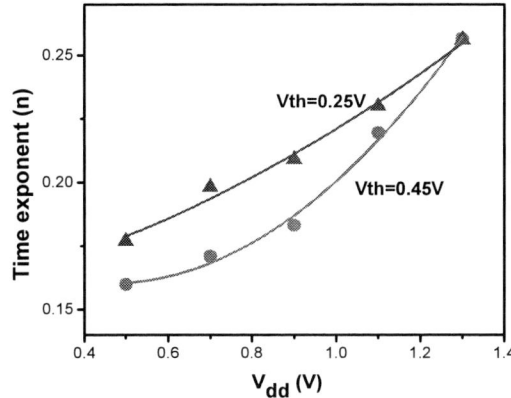

Figure 7. Simulation results of n change with V_{dd} in an 11 stage RO under two different PMOS V_{th} values using 65nm PTM [14].

III. IMPLICATION ON CIRCUIT RELIABILITY

The appropriate extraction of V_{th} degradation provides a solid basis for circuit reliability analysis. In order to reduce the power consumption, today's circuit design usually involves dynamic scaling of the supply voltage (DVS). Dynamic switching in digital circuits involving operating parameters like supply voltage (V_{dd}) and switching activity (α) can influence the degradation rate due to the change of n. The decrease in the supply voltage involved in DVS further reduces n enhancing the circuit lifetime. This section further investigates the characteristic of n at the circuit level.

Figure 6a presents the test structure used to investigate the behavior of n in a ring oscillator. Frequency change induced due to NBTI is monitored in the test as the data is collected at regular intervals for 10000 seconds at 105°C. The performance degradation of RO under different V_{dd} can be managed through V_{DIV} pin.

In a digital circuit, V_{gs} and V_{ds} usually switch from 0 to V_{dd} due to the rail-to-rail switching. Consequently, the degradation rate at the circuit level is a compound of the linear and non-linear response. Such a behavior results in the n of RO between that of I_{lin} and I_{sub} (Fig. 1a). Furthermore, the exact n of RO depends on the relative time in which a transistor stays in each region: when V_{dd} is higher, the transistor spends more time in the saturation region and thus, n

is higher; at lower V_{dd}, n decreases since the transistor stays relatively longer in the linear and sub-V_{th} region (Fig. 3).

Figure 6b validates this behavior by comparing the RO data [7] with SPICE simulation with a unified ΔV_{th} model [3]. As shown in Fig. 6b, our method well captures the change of n under various operating V_{dd}'s. In particular, it indicates that the aging rate of a digital circuit decreases under lower V_{dd}: $n=0.17$ at $V_{dd}=0.9V$ as compared to $n=0.22$ at $V_{dd}=1.3V$. Figure 7 further compares the change of n with supply voltage in an 11 stage RO with PMOS under two different threshold voltages. In the circuit with fast PMOS transistors (i.e., lower V_{th}), n is larger than that with slow PMOS transistors (i.e., higher V_{th}), This is because a lower V_{th} leads to a faster switching and thus, transistor-switching stays longer in the saturation region, allowing it to reach a higher value of n compared to that with slow PMOS transistor. Furthermore, rate of n change is lower with faster PMOS transistors, while the rate of n change is higher if slow PMOS transistors are used. This behavior is due to the larger voltage headroom (V_{dd}-V_{th}) of a fast PMOS transistor which reduces the sensitivity to V_{th}. Therefore, the usage of a slow PMOS transistor reduces the time exponent. It improves the circuit lifetime besides reducing the leakage.

Moreover, the difference in n with V_{dd} could lead to more than 30X difference in the prediction of circuit lifetime, as shown in Fig. 8. Therefore, correct modeling of the bias dependence of n is essential to accurate reliability prediction, avoiding pessimistic guardbanding in low-power design practice.

IV. CONCLUSIONS

In conclusion, the time exponent of drain current degradation in NBTI and CHC effects varies with bias voltages. Based on experimental data and SPICE simulations, this behavior is consistently explained by intrinsic device non-linearity, particularly the exponential dependence of I_d on V_{th} in the sub-V_{th} region. Therefore, it is more appropriate to characterize the aging model in the super-V_{th} region, in order

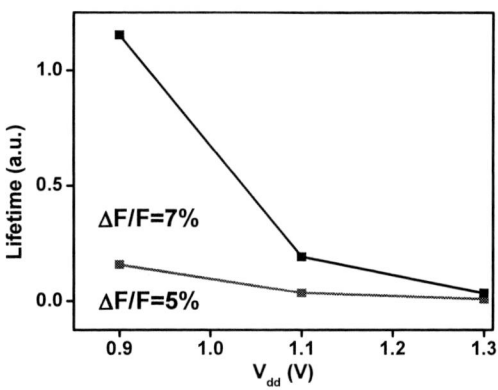

Figure 8. Change of n with V_{dd} leading to dramatic difference in prediction of lifetime

to correctly predict the long-term aging. The changing nature of n implies that in dynamic switching, circuit degradation rate is influenced by operating parameters, such as V_{dd} and signal switching rate. The reduction in the supply voltage helps improve the circuit lifetime by lowering the value of n. Our RO data and simulation method confirm this phenomenon, facilitating designers to accurately predict circuit reliability in low-power applications.

ACKNOWLEDGMENT

The authors would like to acknowledge the support by SRC-1609.

REFERENCES

[1] S. V. Kumar, C. H. Kim and S. S. Sapatnekar, "Adaptive techniques for overcoming performance degradation due to aging in digital circuits", *Asia and South Pacific Design Automation Conference*, pp. 284-289, 2009.

[2] M. A. Alam, "A critical examination of mechanics of dynamic NBTI for PMOSFETs", *International Electron Devices Meeting*, pp. 14.4.1-14.4.4, 2003.

[3] W. Wang, V. Reddy, A. T. Krishnan, R. Vattikonda, S. Krishnan and Y. Cao, "Compact modeling and simulation of circuit relaibility for 65nm CMOS technology", *IEEE Transactions on Device and Materials Reliability*, vol. 7,no. 4, pp. 509-517, 2007.

[4] M. Denais, et al., "On-the fly charecterization of NBTI in ultra-thin gate oxide PMOSFET's", *International Electron Devices Meeting*, pp. 109-112, 2004.

[5] X. Zhou, K. Y. Lim and D. Lim, "A simple and unambiguous definition of threshold voltage and its implications in deep-submicron MOS device modeling", *IEEE Transactions on Electron Devices*, pp. 807-809, 1999.

[6] V. Reddy, "Impact of negative bias temperature instability in digital circuit reliability", *International Reliability and Physics Symposium*, pp. 248-253, 2002.

[7] R. Zheng, et al., "Circuit aging prediction for low-power operation", *Customs Integrated Circuits Conference*, pp. 427-430, 2009.

[8] D. K. Schroder and J. A. Babcock, "Negative bias temperature instability: Road to cross in deep submicron silicon semiconductor manufacturing," *Journal of Applied Physics*, vol. 94, no. 1, pp. 1-18, 2003.

[9] K. Kunhyuk, M. A. Alam, K. Roy, "Estimation of NBTI degradation using IDDQ measurement", *International Reliability and Physics Symposium*, pp. 10-16, 2007.

[10] M. A. Alam, S. Mahapatra, "A comprehensive model of PMOS NBTI degradation," *Microelectronics Reliability*, vol. 45, pp. 71--81, 2005.

[11] K. L. Chen, S. A. Saller, I. A. Groves, and D. B. Scott, "Reliability effects on MOS transistors due to hot carrier injection", *IEEE Transactions on Electron Devices*, vol. ED-32, no. 2, pp. 386-393, 1985.

[12] C. Hu, S. C. Tam, F. C. Hsu, E K. Ko, T. Y. Chan, and K. W. Terrill, "Hot-electron induced MOSFET degradation model-model, monitor, and improvement," *IEEE Transactions on Electron Devices*, vol. ED-32, no. 2, pp.375-385, 1985.

[13] S. Bharadwaj, W. Wang, R. Vattikonda, Y. Cao and S. Vrudhula, "Predictive modeling of the NBTI effect for reliable design," *Customs Integrated Circuits Conference*, pp. 189-192, 2006.

[14] Y. Cao, W. Zhao, " Predictive technology model for nano-CMOS design exploration", *International Conference on Nano-Networks*, 2006.

Managing SRAM reliability from bitcell to library level

Vincent Huard[1]*, Remy Chevallier[2]#, Chittoor Parthasarathy[3], Anand Mishra[3], Natalia Ruiz-Amador[1], Flore Persin[1],
Vincent Robert[1], Alejandro Chimeno[2], Emmanuel Pion[1], Nicolas Planes[1], David Ney[1], Florian Cacho[1], Neeraj
Kapoor[3], Vishal Kulshrestha[3], Sanjeev Chopra[3], Nicolas Vialle[4]

[1] Technology R&D, Electrical Characterization and Reliability team, STMicroelectronics, Crolles, France
[2] Technology R&D, Central CAD and Design Solutions, STMicroelectronics, Crolles, France
[3] Technology R&D, Central CAD and Design Solutions, STMicroelectronics, Noida, India
* phone: + 33(0)4-38-92-29-07, e-mail: vincent.huard@st.com
# phone: + 33(0)4-76-92-63-25, e-mail: remy.chevallier@st.com
[4] Apache Design Solutions, Paris, France
e-mail: Nicolas@apache-da.com

Abstract— **Static Random Access Memories (SRAMs) are present nowadays in all CMOS products in large quantities. Besides, they are often very challenging both on process side (due to small dimensions) and on design side (due to performance request). As a consequence, managing their reliability is of prime importance, though it is quite complex due to their overall complexity. This paper demonstrates a full reliability-based design flow for SRAM libraries including both Front-End degradation modes (NBTI, PBTI and HCI) as well as Back-End degradation modes (Electromigration). Large experimental datasets on various technologies and SRAM bitcells have been used all along the paper to show clear Silicon-CAD correlation evidences, demonstrating the efficiency and accuracy of the developed flow.**
Keywords- SRAM, NBTI, V_{MIN}, HCI, PBTI, Design, Library

I. INTRODUCTION

In previous works [1-2], Static Random Access Memory (SRAM) intrinsic reliability has been addressed from a transistor level viewpoint with a specific focus on Bias Temperature Instability (BTI). From that approach, it is possible to extrapolate some behaviors for a single bitcell and potentially up to an array of bitcells. Nevertheless, SRAM libraries really introduced in a product present a dedicated control logic wrapped around the bitcells array making the prediction of their behavior far more complex. In this study, the intrinsic reliability of SRAM bitcell will be studied first enlarging previous works to all intrinsic degradation modes both in Front-End and Back-End parts of the process. In a second time, the intrinsic reliability aspects of the whole SRAM library will be discussed based on both long term silicon experiments and ageing simulations.

II. SRAM BITCELL RELIABILITY

In current designs, SRAM bitcells are optimized based on performance and density requirements. High performance and high reliability cells, necessary in general purpose processors, require a larger area while denser cells with medium performance are preferred in low power applications.

In advanced technology nodes, both variability and reliability are important barriers to SRAM bitcell scaling [3-4]. The resulting reduction of transistor geometries has caused stronger inter-die variations in process parameters (such as V_T, L, W...) and intra-die variations in the number and spatial distribution of dopants in the channel region. Among them, the random placement of dopants is more of a concern since it cannot be eliminated by a tight control of the manufacturing process. In addition, NBTI degradation introduces systematic V_{TP} drift in correlation with increased intra-die variability. As a consequence, it is necessary to optimize not only a SRAM bitcell for time zero yield but also for reliability degradation.

A. Reliability modeling

The reliability models will be described here below as well as the global approach (including required methodologies) to extract model parameters. The parameter ΔD chosen to represent the degradation forms the bridge between the physical degradation and the evolution of MOS parameter. The models are made linear with respect to time for being suitable for integration during circuit simulation. The boundaries of the integrals represent the window during which the stress assessment is made (during circuit simulation).

Negative Bias Temperature Instability (NBTI) modeling

Applying a careful look to all available NBTI data in literature, we have noted the following characteristics of NBTI degradation, which are common to all groups:

- Conventional methodology loses some amount of degradation. In this context, either OTF or Fast methodologies are required to capture nearly all the degradation
- Recovery is present post NBTI stress
- There is some permanent damage – referred to as 'lock-in' damage [3] which cannot be recovered. This is the value of the degradation after long times of recovery.

It is widely admitted that the threshold voltage V_{th} is the best monitor of NBTI degradation. As a consequence, V_{th} is taken as the degradation parameter ΔD_NBTI.

We have recently developed a new approach to model NBTI degradation [5] based on the coexistence of two components [6-8]:

- A permanent part ΔD_P which is related to Si-H bond breaking.
- A recoverable part ΔD_R which is most likely due to trapping/detrapping of holes in the oxide.

Based on a large dataset of experimental evidences, we conclude that interface traps are created during NBTI degradation which constitutes at the end the so-called 'permanent damage'. Though interface traps recovery has been recently demonstrated, it occurs for time period longer than the

978-1-4244-5430-3/10 $26.00 © 2010 IEEE

product use in the field [9]. Consequently, the interface traps creation is only stress time dependent.

Permanent part D$_P$	$\Delta D_P \approx K_P V_g{}^a t_s{}^b$

On the contrary, the recoverable part is stress duty dependent (i.e. dependent on both the stress and the recovery times). It has been already reported that this component is weakly temperature activated and strongly dependent on the nitrogen dose [9 and references therein]. It was also recently shown that the recovery dynamics is strongly voltage dependent [10], which cannot be explained by Reaction-Diffusion model predicting gate bias independent recovery [11]. All these facts lead to the conclusion that the recoverable part is related to hole trapping/detrapping phenomenon. Following the pioneer work of Tewksbury [12], it is possible to show that the recoverable part can be described by:

Recoverable part D$_R$	$\Delta D_R \approx K_R V_g{}^c \ln\left(1 + \dfrac{t_s.\tau_e}{t_r.\tau_c}\right)$

where t_s stands for 'stress' time and t_r for 'recovery' time. In this case, 'recovery' means the time period when the gate voltage is lowered below the gate voltage of the previous activity phase.

It is worth noticing that experimental voltage acceleration exponent c presents typical values higher than one (c>1) as expected from direct tunneling theory. Recently, experimental values were convincingly supported by the theoretical description of Multi-Phonon Field-Assisted Tunneling (MP-FAT) as the leading mechanism of hole trapping-detrapping [10,14].

As a consequence, the global degradation (and recovery) is modeled by assuming a cumulative and independent contribution of both components in such a way that $\Delta D = \Delta D_P + \Delta D_R$. Fig. 1 shows the good agreement of this approach to model both stress and recovery phases (black lines).

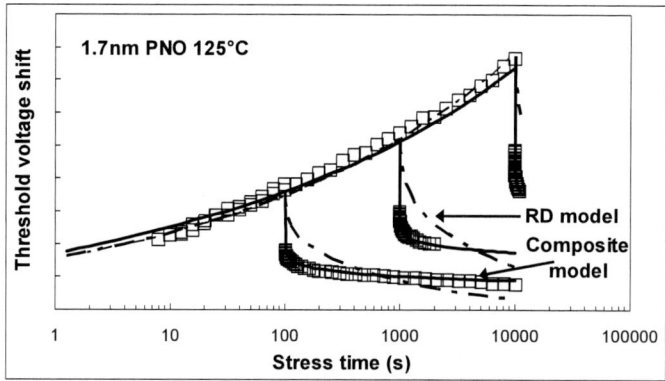

Fig. 1: Composite NBTI (full line) and RD (dashed line) models benchmarked vs experiments for both stress and recovery phases.

Finally, the results of our quasi-static modeling approach are benchmarked with AC stress conditions which would be more representative of digital circuitry. First, we do confirm the lack of frequency dependence from 1 Hz up to 2 Ghz based both on AC-stressed dummy inverter (i.e. within ring oscillator surrounding) and real ring oscillators. Fig. 2 shows both the absence of frequency dependence for experimental results but also the good agreement of the model. This result is true not only for one intra-cycle duty factor but it is also the case for

duty factor ranging from 100% (pure DC case) down to 0.1% (cf. figure 3).

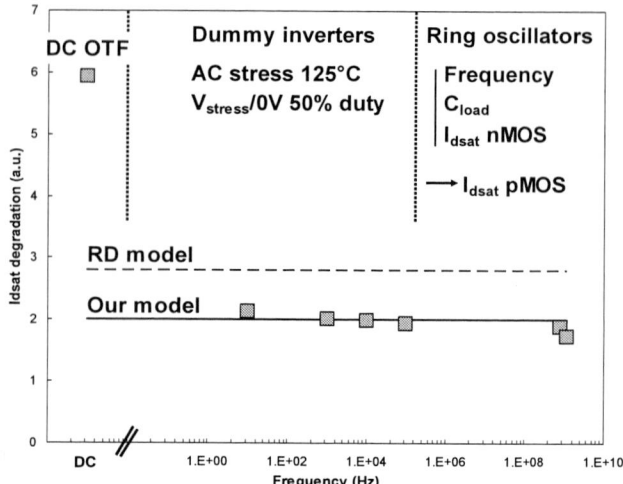

Fig. 2: Frequency impact on Idsat shift during AC stress compared with model predictions (lines).

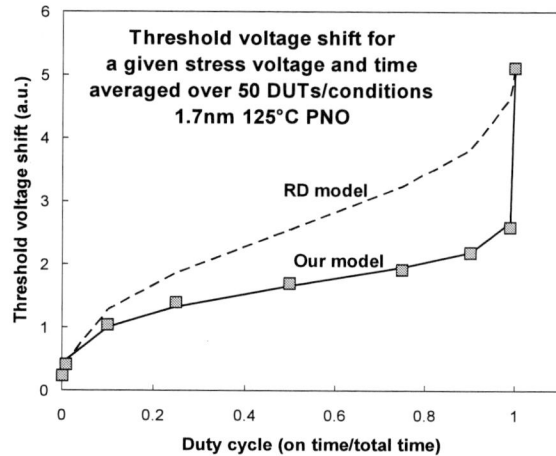

Fig. 3: Composite NBTI (full line) and RD (dashed line) models benchmarked vs experiments for both stress and recovery phases.

Hot Carrier Injection (HCI) modeling

Hot-Carrier Injection degradation presents a renewed interest in the more recent nodes where high level of device reliability is difficult to achieve at high temperature as a function of supply voltage V_{DD}. This point is mainly explained by a continuous increase in lateral electric field since 120nm node. Both digital and analog applications require HCI modeling of the whole V_{gs}/V_{ds} design space described by devices either during transitions in between two logic levels or in analog mode. Recent experimental HCI analysis allowed separating the contributions of three independent modes. The first mode is related to carriers bringing individually enough energy to break the Si-H bond. The second mode is related to moderate carrier energy range where Electron-Electron Scattering plays a role to promote a single carrier to high energy and allow Si-H bond breaking [15]. Altogether, these two modes can be considered as Channel Hot Carrier (CHC) modes since they are related to carrier energy. Finally, a third

978-1-4244-5430-3/10 $26.00 © 2010 IEEE

mode in low energy range was recently attributed for the first time without ambiguity to Multiple Vibrational Excitation (MVE) [16-17]. In this configuration, the degradation is lead mostly by the number of carriers "hitting" the bond and in a less important way by the acquired energy. This mode should be considered as Channel Cold Carrier (CCC) mode since the degradation is no longer related to carrier energy. This description in three modes allows modeling the HCI degradation in the whole V_{gs}/V_{ds} design space (cf. fig. 4).

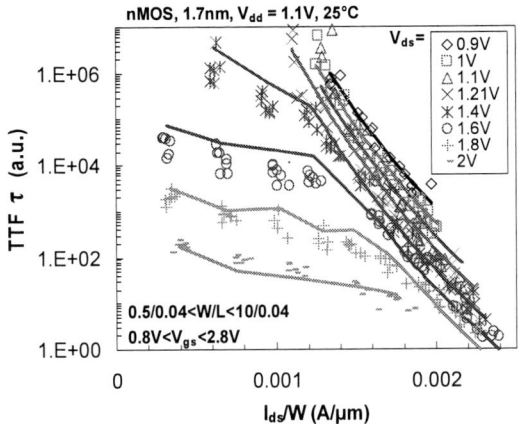

Fig. 4: HCI modelling (lines) at RT compared to experimental dataset covering the whole Vgs/Vds design space

Understanding accurately the underlying physics of each mode as extensively described in [17] allows also describing the HCI degradation behavior over a wide range of temperature at the cost of no additional fit parameters (cf. fig. 5).

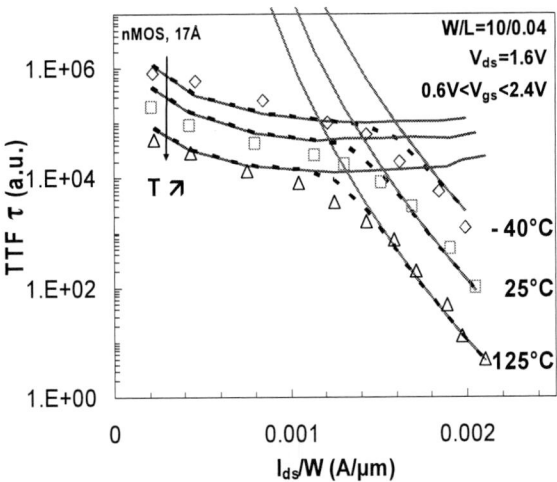

Fig. 5: HCI modelling (blue line: high-energy modes; green line: low-energy modes; dotted line: full model) at different temperatures compared to experimental dataset.

The amount of HCI degradation (Age) impacting the device can thus be normalized whatever the stress conditions as (see also [17] for more details):

Age	$Age = \dfrac{t}{\tau} = t \cdot \left[C_1 \cdot \left(\dfrac{I_{ds}}{W}\right)^{a_1} \cdot \left(\dfrac{I_{bs}}{I_{ds}}\right)^m + C_2 \cdot \left(\dfrac{I_{ds}}{W}\right)^{a_2} \cdot \left(\dfrac{I_{bs}}{I_{ds}}\right)^m + C_3 \cdot V_{ds}^{1/2} \cdot \left(\dfrac{I_{ds}}{W}\right)^{a_3} \cdot \exp\left(\dfrac{-E_{emi}}{k_b T}\right) \right]$

Degradation mapping to compact models

ELDO simulator has been enhanced to communicate with a proprietary Application Programming Interface named UDRM [18] (fig. 6), in which the previously described stress models are encoded in a flexible way. In addition, parameter update equations are also provided, as mapped out of fresh and degraded I-V curves.

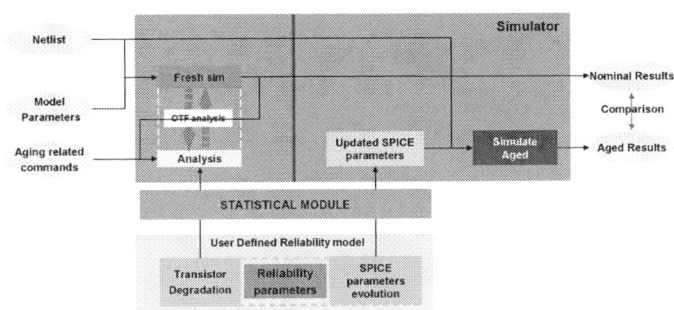

Fig. 6: Schematics of the reliability simulation flow. Standard BSIM and PSP equations are equally supported.

Recent works have proposed a statistical reliability modeling based on experimental datasets composed of both logic and SRAM transistors (especially small size transistors) [19-20]. It was demonstrated that parameters drifts distributions do not follow normal distributions, as it was mathematically expected since [21]. Skellam distributions allow reproducing adequately the high V_{TP} shifts tails (in the case of NBTI degradation) (cf. figure 7).

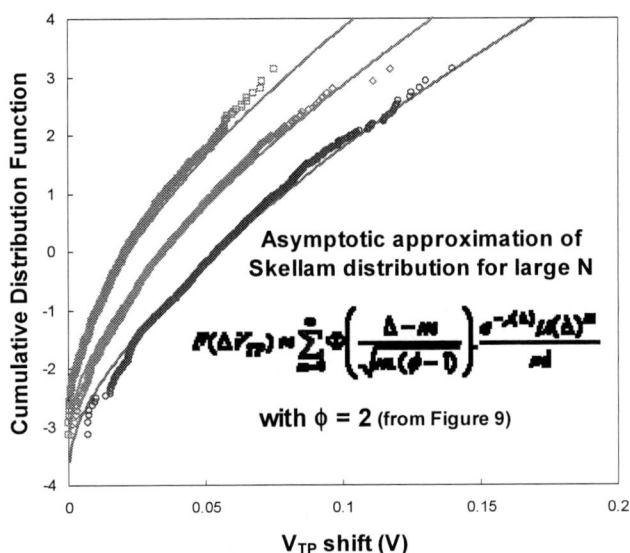

Fig. 7: Comparison of experimental V_{TP} shift cumulative distributions (symbols) for SRAM-sized transistors measured at three different readout times with the asymptotic approximation of Skellam distribution for large N (lines).

This non-normality was recently reproduced through TCAD simulations assuming randomly placed defects within the transistor channel [22]. Considering the case of NBTI degradation, the non-normality of V_{TP} shifts distribution might have a different impact at product level depending on the situation. The worst-case situation is SRAM arrays where

978-1-4244-5430-3/10 $26.00 © 2010 IEEE

millions of transistors are involved, thus enhancing the potential discrepancy between normal approximation and high V_{TP} shifts tail. As a conclusion, normal distribution approximation for V_{TP} shifts is a good approximate for a low number of transistors but might have a significant impact, though not severe in situations such as SRAM arrays where millions of transistors are involved.

Since statistics is of prime importance for SRAM analysis, we have developed a new Statistical Module (cf. figure 6), which can exchange with UDRM in order to achieve timely, efficient statistical reliability simulations.

B. Front-End Reliability

In this section, the tolerance of SRAM bitcells to both NBTI and PBTI degradation will be discussed. A conventional six transistors SRAM bitcell schematic is shown in Figure 8. It consists of a cross-coupled inverter pair (two "pull-down" (PD) nMOS and two "pull-up" (PU) pMOS) and two access (PG) nMOS transistors that couple the inverter pair to the bitlines.

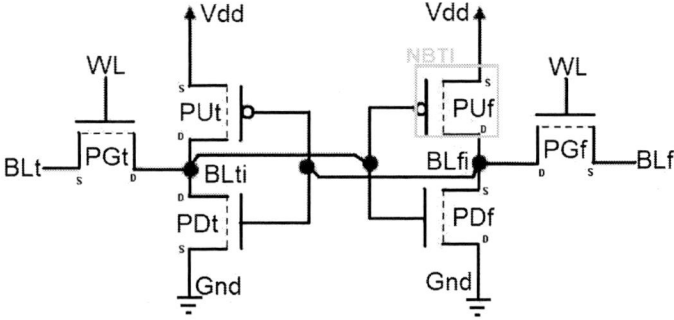

Fig. 8: 6 transistor Static Random Access Memory (SRAM) bitcell schematic. Shaded transistor suffers NBTI degradation due to long-term storing of '0' at BLTi node and '1' at BLFi node.

The transistors in a bitcell are designed in a way to reach targeted stability and yield, which requires a fine tuning of relative strengths of all transistors. A functional bitcell needs to fulfill several tasks, as storing data and being writable fast enough to pass product specifications. Some metrics have been developed to quantify each of these requirements:

- Static Noise Margin (SNM) shows how stable the bitcell is during a reading operation. A large SNM corresponds to a stable bitcell.

- Write Margin (WM) is the voltage that causes the bitcell to flip when the Bitline (BL) voltage is swept from "1" to "0". The external circuitry should apply a voltage lower than WM to successfully write the cell. Thus, a larger WM is desirable.

- Read current (I_{cell}) is the current discharged through the Bit-line (BL) when the cell is accessed. This parameter determines how fast the bitcell can be read, i.e. how performant is the bitcell. Therefore, a larger current is desirable.

Using Monte-Carlo simulations, it is possible to accurately reproduce by simulations the electrical behavior observed on silicon for single isolated bitcells (cf. figure 9), including the whole distributions.

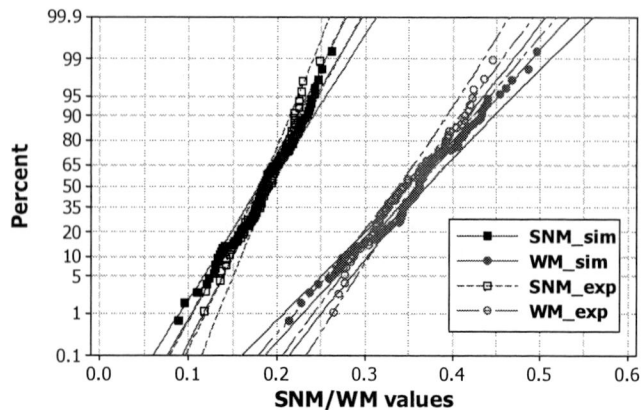

Fig. 9: Cumulative distributions of SNM (squares) and WM (circles) values for both experimental measurements at 25°C (open symbols) and electrical simulations (filled symbols) for 45nm HD bitcell.

It is important to notice that the experimental mean value variations as a function of power supply as well as the standard deviations (cf. figure 10) are well reproduced by simulations in the practical range of applications.

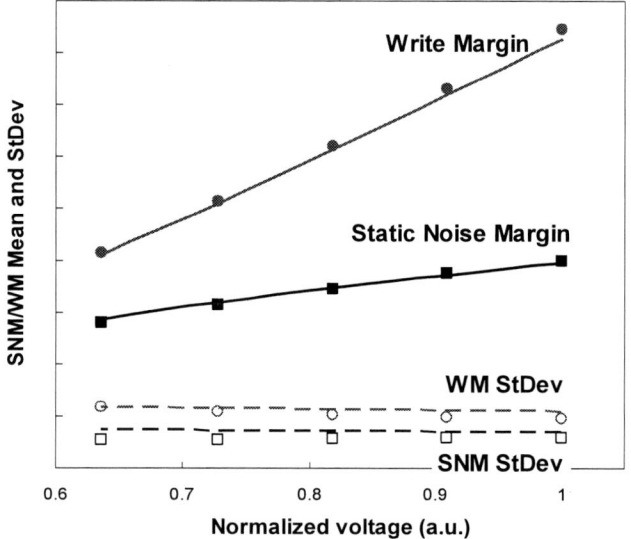

Fig. 10: Experimental mean values and standard deviations of SNM (squares) and WM (circles) dependence with respect to supply voltage are well reproduced by electrical simulations for 45nm HD bitcell.

From electrical analysis of a bitcell, it is expected that SNM value decreases when V_{TP} pMOS is increased. It has been observed both experimentally and through simulations using our in-house Design-in Reliability tools [23] that SNM shift linearly with respect to V_{TP} shift (Figure 11), inline with previous findings [24].

978-1-4244-5430-3/10 $26.00 © 2010 IEEE   658

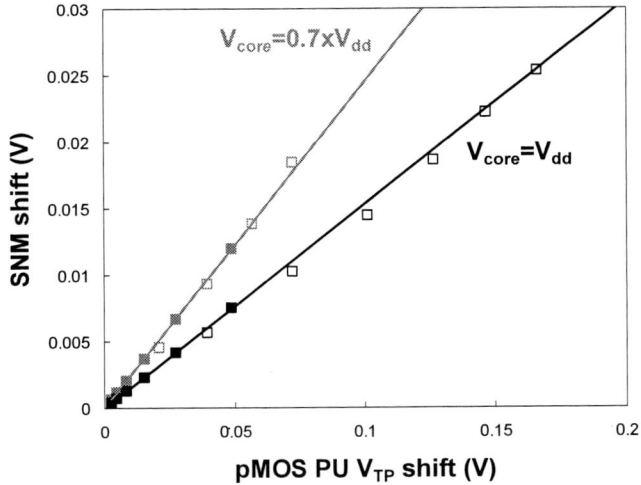

Fig. 11: SNM increases linearly with pMOS PU V_{TP} shift, as experimentally measured on 65nm HD SRAM bitcell (open symbols) and simulated using our Design-In Reliability tools (filled symbols and lines). SNM sensitivity to similar V_{TP} shift is greater when V_{core} decreases.

The sensitivity of SRAM bitcell during READ operation to NBTI degradation can thus be written as:

$$\Delta SNM = -\frac{\partial SNM}{\partial V_{TP}}(V_{core}).\Delta V_{TP} = -dSNM.\Delta V_{TP}$$

where the correlation parameter dSNM (positive value) is dependent on the core voltage V_{core} supplying the array. This dependence needs to be modeled to further analyze the NBTI degradation of real product SRAM arrays.

In a NBTI weakened bitcell, the degraded PU pMOS will tend to less resist to a bitcell flip when in concurrence with PD nMOS. As a consequence, the bitcell WM is increasing with NBTI degradation. In a similar way to READ operation, it is possible to define the sensitivity of SRAM bitcell during WRITE operation as:

$$\Delta WM = \frac{\partial WM}{\partial V_{TP}}(V_{core}).\Delta V_{TP} = dWM.\Delta V_{TP}$$

where the correlation parameter dWM (positive value) is dependent on the core voltage V_{core} supplying the array. This dependence needs to be modeled to further analyze the NBTI degradation of real product SRAM arrays.

Due to the difficult implementation, electrical reliability simulations so far focused solely on the impact of parameters mean drifts. Nevertheless, linking single bitcell analysis to electrical properties of large arrays with up to billions of SRAM bitcells request more than solely mean electrical simulations.

Using our newly introduced reliability statistical module, it is possible to address the reliability-related variability introduced by aging processes. Figure 12 shows that NBTI-induced variability translates directly into increased SNM and WM standard deviations. It is worth noticing that, though mean WM is improved (i.e. increased) by NBTI (as discussed above), its standard deviation is more degraded (i.e. larger) in this case than SNM ones.

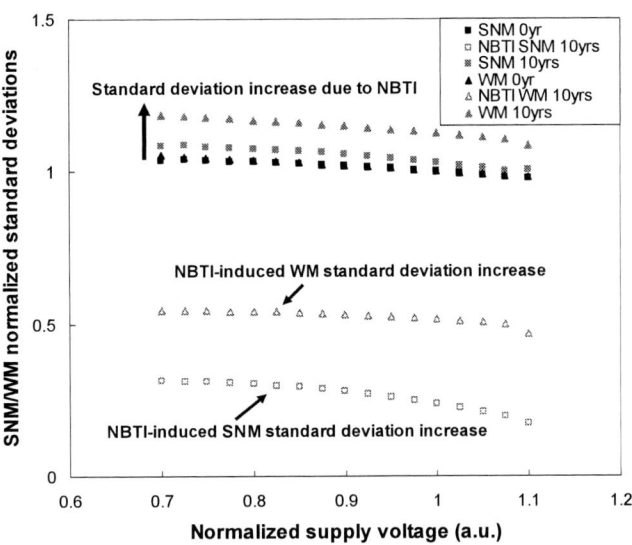

Fig. 12: Electrical statistical simulation results for 45nm HD bitcell showing SNM and WM standard deviations increase due to NBTI. In spite of the fact that mean WM is improved (i.e. increased) by NBTI, its standard deviation is more degraded than SNM ones.

We will here follow the experimental approach we had proposed previously [20] to determine the number of faulty cells present in a memory cut. This approach has two main advantages: (1) to open the way to a direct link between bitcell/device and SRAM arrays levels and (2) to make quantitative predictions at the product end-of-life whatever the memory cut considered. This set of advantages can be further used to check if six-sigma confidence requirements are fulfilled and/or support process development to have realistic reliability guardband. The core element of this approach is the evaluation of the number of faulty cells in a memory cut at a given supply voltage. This number corresponds to a bitcell failure probability (F_{bit}).

Following electrical simulations, it is possible to show that NBTI does not degrade the read current I_{cell}, since the read current is only determined by the relative strengths of PG and PD nMOS. As a consequence, the modification of the number of faulty cells can be related to either a read or a write failure. The method to determine the read failure probability will be thoroughly described here either from experimental result or electrical statistical simulations. The same approach will be followed to obtain the write failure probability.

As mentioned previously, the bitcell read failure probability F_{bit_read} is directly related to SNM values. SNM is subject to random fluctuations in a similar way than device parameters such as threshold voltage V_T. Figure 13 shows SNM distributions for 45nm HD SRAM bitcell for various supply voltages. When SNM value is equal to zero, a read failure occurs when the bitcell is addressed. This criterion allows determining the probability for which the read failure occurs at a given supply voltage.

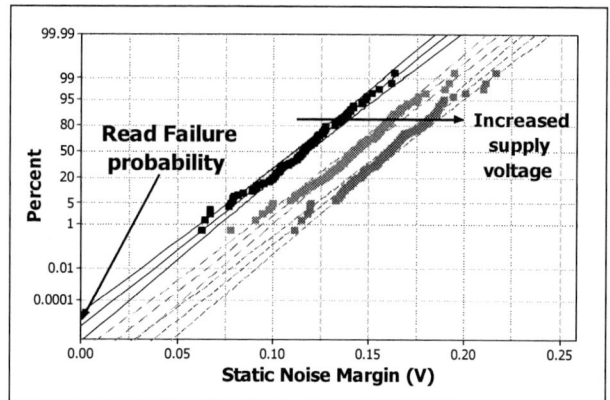

Fig. 13: Static Noise Margin distributions for 45nm HD bitcell for three different supply voltages. Read failure probability is extracted when SNM value equals defined criterion, in this case zero value.

In a similar way, it is possible to define a criterion for the Write Margin WM which translates the operational time window allowed for writing. If the WM is below this criterion, the bitcell will take longer than the operational time window to be written correctly, i.e. a write failure occurs. This criterion is more design/operating test dependent than the SNM criterion and must be simulated.

Figure 14 shows the Read (F_{bit_read}) and Write (F_{bit_write}) failure probabilities as simulated for 45nm HD isolated bitcells using statistical monte-carlo approach. Since a bitcell has either a read or a write failure, the two failure probabilities are independent and can be straightforwardly added. The resulting F_{bit} is the expected total bitcell failure probability as a function of the supply voltage (also seen as the number of faulty cells).

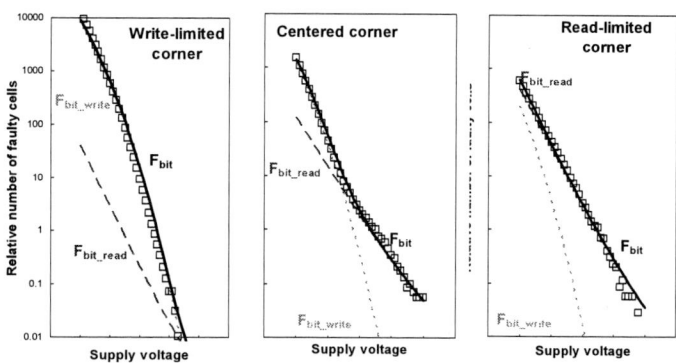

Fig. 14: Relative number of faulty cells measured experimentally in real SRAM arrays (open symbols) is in agreement with the electrical statistical simulations of 45nm HD bitcell (lines).

The predictive simulations of the number of faulty cells in SRAM arrays show good agreement with non-degraded bitcell, i.e. at yield level. These simulations, as shown above, only require the knowledge of SNM and WM distributions. In a previous section of this paper, the dependence of SNM and WM to NBTI degradation was thoroughly examined. By combining these two approaches, simulations can be run to predict the evolution of the number of faulty cells during the operating life of SRAM arrays. Figure 15 shows an example of the output of these simulations. We first simulated the impact of mean V_{TP} shift degradation, i.e. considering that NBTI does not add additional variations to the final V_{TP} distributions (dashed lines). Composite model [4,8] was used to simulate

accurately the mean V_{TP} shift value as a function of transistor/bitcell activity. In a second step, the increased dispersion induced by NBTI degradation is introduced (full lines). In the read-limited part (i.e. high supply voltage part), the mean V_{TP} shift increases the number of faulty cells, with respect to degraded SNM. Additional introduction of NBTI-induced dispersion further increase the number of faulty cells. In the write-limited part, the mean V_{TP} shift introduces a decrease of the number of faulty cells (towards dashed lines), which is linked to the improvement of WM. Nevertheless, this decrease is further compensated by the introduction of the NBTI-induced dispersion.

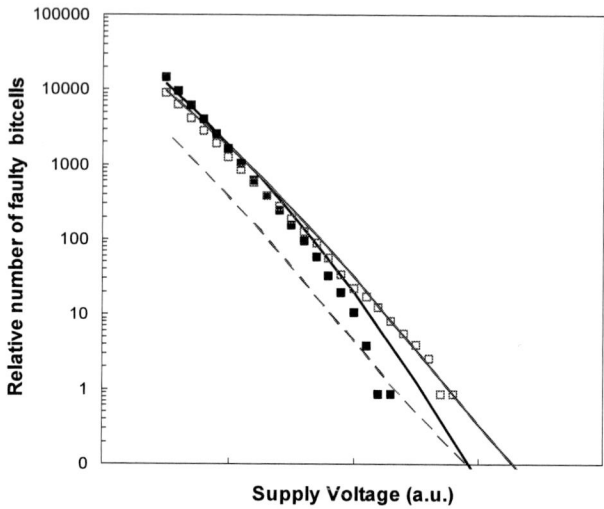

Fig. 15: Simulated relative number of faulty bitcells for fresh 45nm HD bitcell (black line), for NBTI-degraded bitcell without dispersion modification (dotted line) and NBTI-degraded bitcell with increased dispersion (blue line). Degradation has been simulated to reproduce experimental setup (i.e. voltage supply, temperature amd stress time). Experimental results obtained on large arrays are shown for comparison (symbols). Process corner is write-limited in this case.

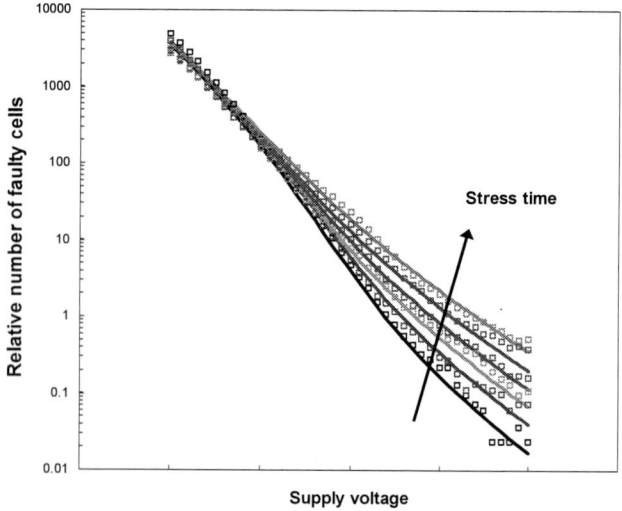

Fig. 16: Simulations of the relative number of faulty cells (lines) compare accurately with experimental measurements made on large SRAM (symbols) for various readout times up to hundreds of hours.

978-1-4244-5430-3/10 $26.00 © 2010 IEEE

This electrical simulation approach can be shown to be efficient to reproduce several stress times (cf. figure 16), providing confidence that our statistical reliability simulation setup is robust enough to reproduce SRAM arrays behaviours. By simulating the bitcell failure probability, this approach allows also being independent of the SRAM array size and thus can accurately reproduce experimental results even for large, product-sized SRAM arrays.

This approach was also demonstrated to be robust while simulating process corner impact vs aging. Figure 17 shows median VMIN values for large arrays (several Mbytes) as simulated using our procedure described above and compared to silicon results.

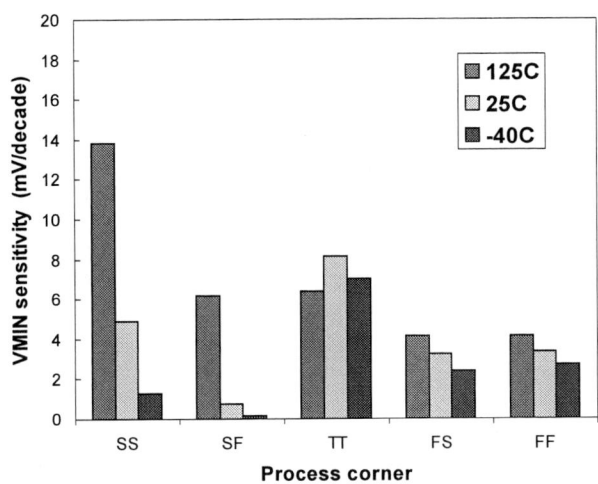

Fig. 18: Simulations of the dependence of V_{MIN} drift due to NBTI as a function of the number of SRAM bitcells in the array. As expected, the greatest the number of SRAM bitcells, the greatest is the V_{MIN} drift. Nevertheless, not all the process corner neither the temperature behaviours are similar.

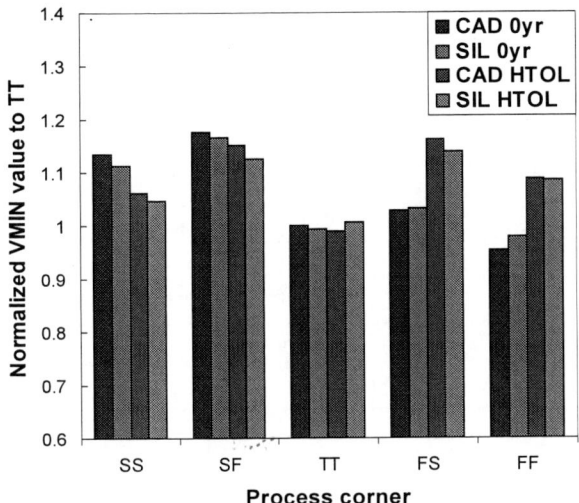

Fig. 17: Simulations (CAD) and experimental results (SIL) of normalized V_{MIN} values for large arrays of 45nm HD bitcell both for non-stressed SRAM test vehicles (0yr) and after HTOL stress (HTOL) as a function of process corners. Some corners are showing degradation while some others are showing improvement. In all cases, our simulation setup can reproduce quantitatively the observed evolutions.

Using similar procedure, it is possible to evidence the V_{MIN} drift dependence as a function of SRAM array size. As expected, the greatest the SRAM array (in terms of number of bitcells), the greatest the VMIN drift induced by electrical aging. Nevertheless, it is worth noticing that this cut size dependence is strongly dependent on the process corner but also on the characterization temperature (cf. figure 18).

This procedure can also be used to evaluate potential reliability risks in advance to silicon yield for advanced nodes (32nm here) and for mitigating the impact of new degradation modes (PBTI in HK) (cf. figure 19).

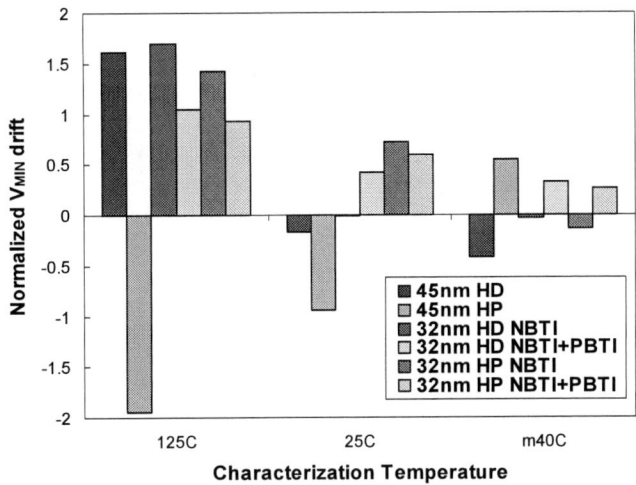

Fig. 19: Comparison of V_{MIN} drift for a given SRAM array size for various technologies, SRAM bitcells and also showing the combined effect of NBTI and PBTI in HK.

II. SRAM LIBRARY RELIABILITY

A. SRAM IP overview

The impact of intrinsic reliability can have significant impact on Static Random Access Memories (SRAM) libraries as they are more susceptible to functional failure. Besides, SRAM libraries are performance bottlenecks in high-performance VLSI circuits, and occupy a majority of on-chip silicon area while requiring good tolerance throughout the life of usage. A SRAM library is typically divided in 4 main blocks as shown in figure 20. The operation in the clock cycle is computed in the control block. When a read or a write cycle is performed, the address is chosen by the decoder block before selecting the right word inside the memory array. In parallel, the input/output block is either collecting the data from the memory array for a read cycle, or collecting from outside of the memory the data which will be saved in the memory array for a write cycle.

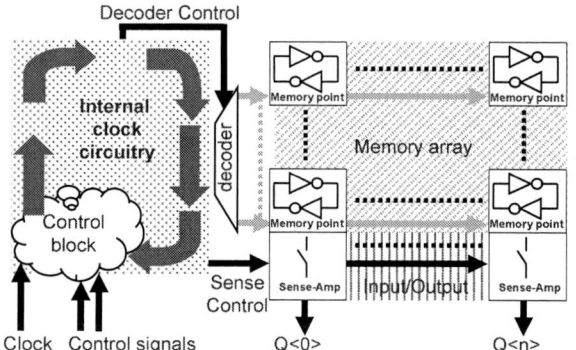

Fig. 20: SRAM library schematics including wrapped control logic.

In order to reach the highest level of performance specific design techniques are implemented in SRAM libraries: dynamic logic for control and decoder, sense-amplifiers for reading the data in the bit-cell. However, this strategy yields the generation of enabling signals like internal clock circuitry which mimics the longest timing path.

B. Front-End Reliability

The addition of control logic around the SRAM bitcells array can be thought to add an impact of the yield loss at V_{min}. Actually, though dedicated access to both control logic and SRAM arrays, we have checked that the yield loss is solely controlled by the SRAM array (cf. figure 21).

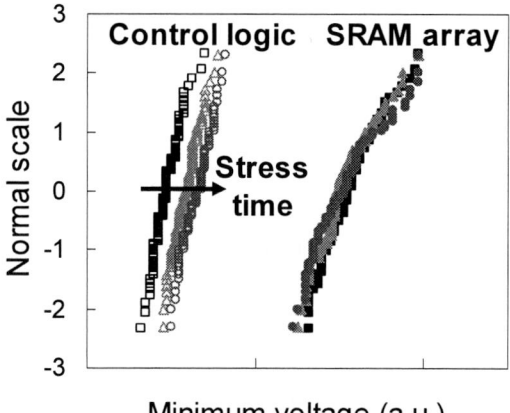

Fig. 21: V_{MIN} library is controlled by V_{MIN} array solely. NBTI impacts differently the array (small change when bitcells are centered), while control logic is monotonously increased.

Nevertheless, to support the automatic integration of SRAM IPs into a semi-custom design flow, not only the functionality at V_{MIN} should be maintained but the speed performance as well. An accurate forecast of the performance loss is needed to assure that adequate IPs timing characterizations are provided upwards into the design flow.

Timing performances are not only related to bitcell properties (like I_{cell}) but also to the whole SRAM IP including the control logic. We first wanted to be able to evaluate

Identify applicable sponsor/s here. *(sponsors)*

accurately which parts of the critical path are responsible for most of the degradation within the IP. For that purpose, reliability electrical simulations were performed on several SRAM libraries using our in-house reliability modeling solution described above. In parallel, dedicated test vehicles are built in order to cross-check simulation results with silicon results. These test vehicles embarked specific structures which allow at-speed tests. Two kind of analysis were performed. First, an automated timing monitor embarked on silicon allows performing accurate timing characterization. Though useful, this timing monitor does not provide information on which parts of the critical path are most impacted by transistor aging parameters drift. Another possibility is to proceed to ebeam analysis on a statistically relevant set of test vehicles prior to and after the electrical stress. This ebeam analysis has been done on dedicated pads connected to inner nets at the interfaces of main blocks in order to perform correlations at block level. Figure 22 shows that excellent silicon to CAD correlation has been obtained in the prediction of SRAM critical path aging.

OPERATION	TIMING DEFINITIONS	SILICON	AGING SIMULATIONS
		Delta Aged-Fresh (ps)	
WRITE 0 AT	CK to PAD1	35	40
BOTTOM ROW	PAD1 to PAD2	30	43
READ 0 AT	CK to PAD1	30	34
BOTTOM ROW	PAD1 to PAD2	20	21
	PAD2 to Q	0	1
WRITE 1 AT	CK to PAD1	30	34
TOP ROW	PAD1 to PAD2	40	42
READ 1 AT	CK to PAD1	30	33
TOP ROW	PAD1 to PAD2	25	21
	PAD2 to Q	5	3

Fig. 22: Timing drifts monitored by ebeam technique on 45nm HP SRAM IP are well reproduced by aging simulations

It is worth pointing out that the overall contribution of the bitcell speed is negligible and the whole performance degradation is driven by the control logic timing path. One of the main conclusions we obtained from reliability simulations at that point is that both NBTI and HCI are contributors of the timing path degradation. This result can be efficiently demonstrated by silicon measurements. Two sets of electrical stresses on identical SRAM IPs were done. The first set was dedicated electrical stresses using external clock at slower frequency (~ 100 kHz). In this configuration, we expect the HCI contribution to be negligible due to its strong frequency dependence, while NBTI degradation should dominate due to its frequency independence (cf. figure 2). On another hand, a second set of electrical stresses were performed at-speed to trigger not only NBTI but also HCI degradation. Figure 23 shows that though low-frequency results are well explained by NBTI-only reliability simulations, it is not the case for at-speed tests where HCI degradation needs to be taken into account to explain the whole degradation.

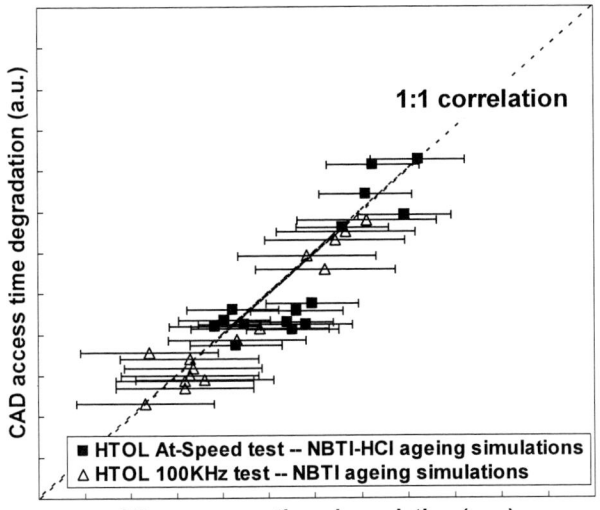

Fig. 23: At-Speed HTOL access time degradation can be explained by combining NBTI and HCI ageing simulations while NBTI only simulations are needed for 100kHz HTOL tests. It is worth noticing the good 1:1 correlation between silicon results and ageing simulations on complex critical paths.

C. Back End Reliability

The continuous downscaling of CMOS technologies imposes more and more constraints on the Back-End of the Line (BEOL) part. As illustrated with figure 24, the current consumption in integrated circuits (ICs) is continuously increasing, making the ICs more and more likely to suffer electromigration (EM) failure. One way to prevent EM degradation is by defining current limits.

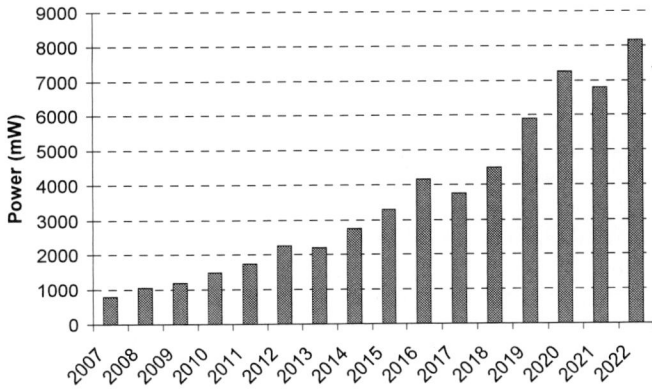

Fig. 24: ITRS 2007 System Drivers: SOC customer stationary power consumption trends

Nevertheless, the thread is growing up and cannot be longer ignored by designers, who need to reach the highest performance while keeping a good level of reliability. However, tackling this degradation mode by designers requires to compute current flow in complex IPs which is a real challenge considering the number of interconnects and the overall complexity.

1) Back End Reliability Strategy

To avoid early failure of chips, rigorous reliability rules are implemented in design foundations for each failure mechanism, these rules depending of the process capability and chip mission profile. Especially for EM, designers have to check 3 types of current rules on all drawn lines.

• the dc current (i_{dc}) generated directly from black equation [25]. Respecting this rule ensures a minimum operating time-period without EM failure for a given temperature junction.

• the root mean square current (i_{rms}) is given to limit Joule heating in metal lines which can accelerate EM phenomenon. This rule is established from thermal model [26-27].

• the peak current rule (i_{peak}) is necessary to avoid local metal melting due to instantaneous excessive dissipated power.

The downscaling of interconnect cross-sections induces a strong limitation of the dc current level permitted to designers. A simple way to get around this limitation is to take advantage of the Blech effect for short lines, in which a stress-induced backflow balances the EM flux [28]. Today, the use of current rules depending on L becomes fundamental to preserve the global benefits of the node change.

2) Back-End Reliability Verification Strategy

The challenge of complex analog block verification is to reach the highest level of accuracy in an acceptable simulation runtime. Indeed, on one side, the flow based on very accurate extracted netlist and dynamic vectors [29] is providing a good level of accuracy but is not able to model very complex physical phenomenon just as Blech effect. As a consequence, over design is mandatory to pass this flow. On the other side, another verification strategy [30] is possible by modeling the EM physical effects but cannot support full dynamic vectors analysis. In this case, pessimism on currents computed are introduced which induces over design. Our goal is to associate an accurate modeling of physical rules by including Blech effect and dynamic vector simulation to mimic the behavior of the custom block in the final chip.

3) Back-End Reliability Verification applied on SRAM

First of all, spice simulations [31-32] are performed to pick-up transistor activities. Then, in a second time, EM verification checks are computed using Totem provided by Apache [33], and taking into account the Blech effect and relax I_{dc} current rules in short lines, as shown by figure 25. A schematic description of this flow is given for a SRAM (45 nm node).

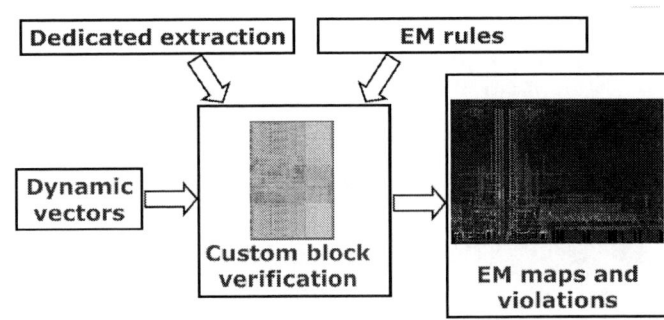

Fig. 25: Bottom-up verification flow for Electromigration.

Using this flow, it is possible to localize the potential weaknesses in the overall layout and thus identifies specifically the design part to work on.

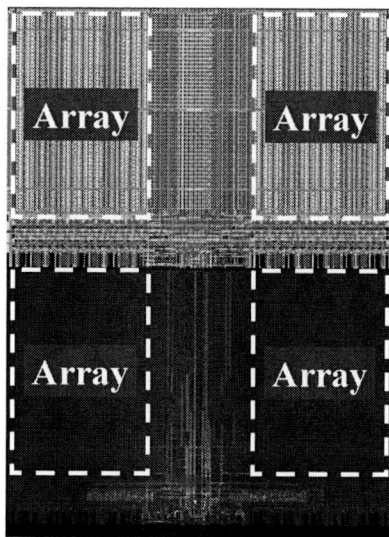

Fig. 26: Top figure represents SRAM library layout under analysis by our EM checker flow. Bottom figure represents the EM limits violations results. SRAM arrays are showing no EM risks which are localized in control logic.

A first analysis (cf. figure 27) revealed a weakness on a power wire used to supply 64 simultaneous switching buffers by duplicating power connections and considering the Blech effect, the design was optimized to assure a reliable SRAM.

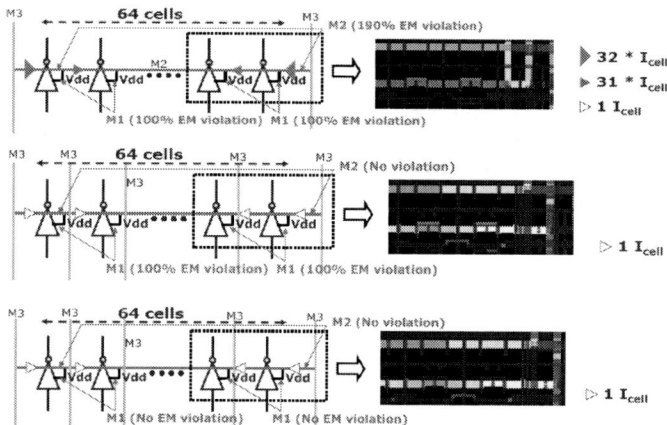

Fig. 27: Top: Initial EM verification of C45 SRAM: 2 EM violations. Mid: EM verification with design fix: M2 violation is solved. Bottom: EM verification with design fix and Blech filter enabled: M1 violation solved

EM checking is usually done on achieved IP (eg in table1), for which only limited layout modifications are possible. To anticipate potential EM weakness, during early stages of SRAM design, a hierarchical analysis flow has been developed.

TABLE I. RUN TIME FOR SRAM BLOCK VERIFICATION

Block	Size (nb devices)	Spice simulator	Run time	techno
Associative-SRAM (128 words, 22 bits)	25 600	Hsim	1h30	C65
Associative SRAM (2048 words, 19 bits)	311 200	Hsim	4h30	C65
SRAM (256 words, 32 bits)	60 000	Hsim	26min	C45

4) Improve EM Rules When Silicon is Not Available

For advanced technologies, the required current to supply particularly aggressive design can be higher than I_{dc} limit in some nets which thus require particular attention. When EM limits are violated for specific layout configurations not addressed by silicon testing, Finite-Element simulations are performed. A multiphysics model [34] has been developed based on the understanding the physics of atomic transport during EM in such particular configurations. It allows assessing if classical current rules provided by DRM are still relevant in these specific cases, further filtering out false violations.

II. CONCLUSION

This paper studies the role of the different blocks composing a SRAM library as found in a product. At front-end side, the good accuracy of aging simulations versus silicon results allows on one side predicting the yield (or V_{MIN}) of the SRAM bitcells array and on another hand optimizing the control logic design without the cost of additional silicon verifications. At back-end side, a full automated flow was developed for the designers. This was made possible by the close collaboration between reliability experts, SRAM designers, and TCAD experts in partnership with the vendor. Overall, we have evidenced in this study all the elements needed in order to insure adequate reliability-performance trade-off for customers in the SRAM libraries commonly found in all products.

REFERENCES

[1] Bansal, A. *et al.,* IEEE IRPS (2009)

[2] Huard, V. *et al.,* IEEE IRPS (2009)

[3] Kapre, R. *et al.,* IEEE IRPS (2007)

[4] Mukhopadhay, S. *et al.,* IEEE CICC (2006)

[5] Rangan, S. *et al.,* IEEEE IEDM Proc. (2003) 41

[6] Huard, V. *et al.,* IEEE IEDM Proc. (2007)

[7] Reisinger, H. *et al.,* IEEE IRPS (2006) 448-453

[8] Schroder, D. *et al.,* J. Appl. Phys. 94 (2003)

[9] Huard, V. *et al.,* IEEE IRPS (2006) 733-734

[10] Huard, V. *et al.,* IEEE IRPS (2010)

[11] Parthasarathy, C.R. *et al.* Micro. Rel., 46 (2006) 1464-1471

[12] Grasser, T. *et al.,* IEEE IRPS Proc. (2007) 268-280

[13] Tewksbury, T., MIT PhD (1992).

[14] Grasser, T. *et al.,* IEEE IRPS Proc. (2009)

[15] Rauch, S.E. TDMR, vol. 1, (2001)

[16] Guerin, C. *et al.,* IEEE IRPS Proc. (2008)

[17] Bravaix, A. *et al.,* IEEE IRPS Proc. (2009)

[18] Huard V. *et al.,* IEEE TDMR vol. 7, (2007)

[19] Rauch, S., IEEE TDMR (2007)

[20] Huard, V. *et al.,* IEEE IRPS Proc. (2008)

[21] Skellam, J. G., J. Roy. Statistical Soc. (1946)

[22] Bukhori, M. F. *et al.,* IEEE TED (2010)

[23] Huard, V. *et al.,* IEEE IRPS (2008)

[24] Krishnan, A. T. *et al.,* IEEE IEDM (2007)

[25] Black, J. R. *Proc. IEEE,* vol. 57, no. 9, pp. 1587–1594, Sep. 1969.

[26] Bilotti, A.A., 1974, Vol. 21, pp. 217-226

[27] D. Ney, D. Girault, V. Federspiel, X., IEEE IRW Final report, 2003.

[28] Blech I. A., *J. Appl.Phys.,* vol. 47, no. 4, pp. 1203–1208, Apr. 1976

[29] Sangameswaran, S. Yamauchi, S. IEEE DCAS, pp. 211-214

[30] G. Jerke, G. and Lienig, J. , IEEE TCADICS (2004)

[31] Users Manual for Eldo spice simulator www.mentor.com

[32] User Manual for Hsim spice simulator, www.synopsys.com

[33] User manuals for Totem/Redhawk, www.apache-da.com

[34] F. Cacho, *et al.,* Eurosime (2008)

PREDICTION OF NBTI DEGRADATION FOR CIRCUIT UNDER AC OPERATION

Y.S. Tsai*, N. K. Jha, Y.-H. Lee, R. Ranjan, Wayne Wang, J.R. Shih, M. J. Chen, J.H. Lee and K. Wu

Technology Q&R Division, TSMC, 9, Creation Rd. 1, Hsinchu Science Park, Hsinchu, Taiwan 300-77

*email: ystsaia@tsmc.com

ABSTRACT

A model predicting the negative bias temperature instability (NBTI) reliability of high performance nitrided oxides is developed from discrete p-type metal-oxide-semiconductor field effect transistor (PMOSFET) data and verified with ring oscillator degradation in various frequencies for up to 1GHz. Based on the experimental data and the simulation results, hole traps generation is considered to be major factor for AC NBTI degradation. An AC/DC NBTI improvement factor of around 10 has been observed at low frequency of 0.01Hz while it is significantly larger (~10000) at 1GHz frequency range. It is established that the measurement techniques are very crucial for accurate NBTI reliability estimation.

INTRODUCTION

P-type metal-oxide-semiconductor field effect transistor (PMOSFET) negative bias temperature instability (NBTI) degradation is a key mechanism limiting integrated circuit (IC) lifetime for the advance technology having the thin nitrided-gate or high-k/metal-gate oxide. Since the NBTI degradation is strongly modulated by the charge trapping/detrapping, and/or the fast trap relaxation and re-filling, an accurate AC NBTI model is required for the circuit degradation projection. Extensive studies have been done on the NBTI relaxation and its implication on the frequency dependent AC/DC NBTI improvement factor [1]-[5], however, most of the papers emphasized on the relaxation component [6]-[9] and only briefly mentioned about the re-filling of the traps. In addition, the AC NBTI papers were primarily based on comparatively slow measurements, in which data may get affected by the data acquisition method [10], [11]. This work first address the effect of the measurement techniques on the experimental results to explain the AC/DC NBTI improvement factor discrepancy observed from various publications [4], [6], [12]-[15]. Then we establish a NBTI device physical degradation model, which is capable to explain our experimental data with various frequencies (0.01Hz to 1GHz) and duty cycles (0.01-99.9%). Once a model is established and validated, the paper discusses how it can be used to assess reliability impact from the changes in the device processing, use conditions, and the circuit design.

EXPERIMENTAL

The devices and ring oscillators used were fabricated with the industry VLSI processes. The equivalent oxide thickness (EOT) is around 19Å. Poly critical dimension (CD) for the device is around 30-100 nm. The back-end process has 9-layer Cu for interconnects. PMOSFET NBTI stress was conducted at various voltages (V_{gstr}), frequencies, and duty cycles to properly characterize the AC/DC improvement factors. In this study, we focus on the drive current change (%ΔI_{dsat}) to minimize the measurement delay that induces degradation recovery. The devices used in this study have been characterized by ultra-fast Transient I-V (TIV) measurement setup with measurement delay being as fast as less than 1μs [16], [17]. The devices are stressed under AC conditions (high and low cycles) with two types of measurement methodology, a) the stressed devices are measured immediately after high stress cycle within 1μsec delay defined as "Mode-A" in Fig. 1, and b) "Mode-B" is defined as the devices measured after recovery time of half of the total AC period as shown in Fig. 1.

To compare the measurement condition impact on frequency dependence of NBTI degradation, the I_{dsat} degradation behavior for different frequencies during Mode-A and Mode-B measurement conditions are shown in Figs. 2 and 3, respectively. It has been observed that the I_{dsat} degradation for Mode-A is frequency dependent, while Mode-B measurement methodology shows frequency independent behavior. Thus we can experimentally resolve the discrepancy in NBTI frequency dependent or frequency independent degradation behavior as reported in the literature [5], [12]-[15].

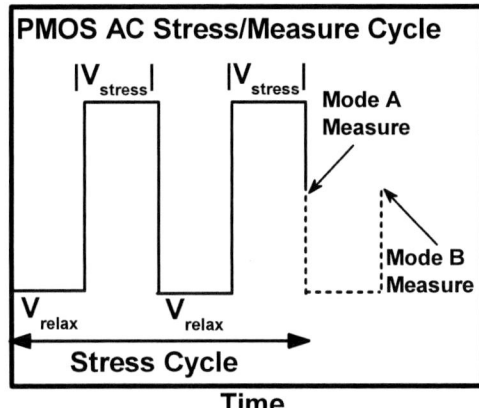

Fig. 1 Measurement technique – with limited delay after the stress (Mode-A) and with a relatively longer delay (Mode-B). The limited delay is less than 1μs using the fast measurement setup, while the long delay can be in the range between 10 μs to msec depends on the AC stress frequency.

NBTI RELAXATION AND RE-FILL

Depending on the NBTI degradation measurement methodology, different aspects of overall degradation would be characterized. It is therefore essential to carry out rigorous evaluation of possible components in the overall degradation. There are many hole trapping parameters that can influence the overall degradation. It is also necessary to understand various type of trapping defects and their overall generation, trapping/detrapping behavior. Three different types of positive traps have been considered to cause the hole trapping/neutralization of traps under NBTI stress conditions [18]. Namely "As Grown Hole Traps" (AHT), "Cyclic Positive Charges" (CPC) or shallow traps, and "Anti-Neutralization Positive Charges" (ANPC) or deep traps as described by Zhang *et al.* [18]. The exact mechanism of ANPC type of trap generation is already well described by Grasser *et al.* [19]. The scope of the present work is to see the NBTI behavior of above-mentioned traps under AC stress and subsequently its impact on overall circuit degradation. In this context it is important to revisit Figs. 2 and 3. As we can see that lower frequency results in higher I_{dsat} degradation or higher hole trapping after stress cycle (Mode A characterization condition) but almost all of these charges detrap after recovery half cycle or under "Mode B" characterization implying that CPC or shallow trap is dominant hole trapping mechanism under 0.1Hz and above AC stress conditions. This means that we are measuring "CPC+ANPC" charge magnitude under "Mode A" measurement conditions (Fig. 2), while degradation due to "ANPC only" is observed under "Mode B" measurement condition (Fig. 3). This also implies that CPC trap generation is strongly dependent on frequency while ANPC trap generation is frequency independent and it only depends on the cumulated stress time.

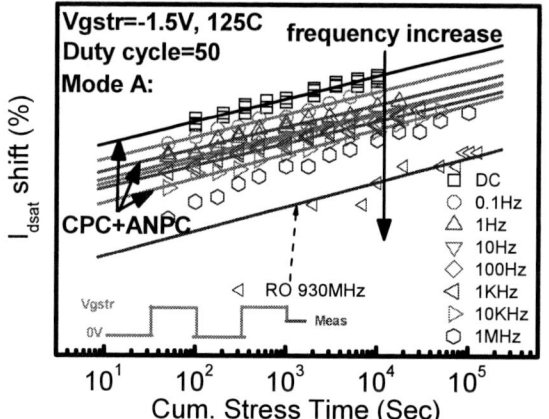

Fig. 2 NBTI degradation as a function of the AC frequency - Data collection within 1-µs after AC stress with a gate stress voltage (Vgstr) of −1.5V at 125°C of temperature. The frequency dependent I_{dsat} degradation is dominated by the generation of cyclic positive charge (CPC), while anti-neutralization positive charges (ANPC) generation is frequency independent.

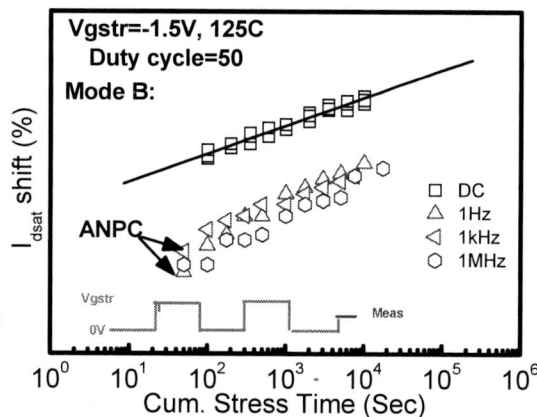

Fig. 3 NBTI degradation as a function of the AC frequency - Data collection with some recovery (half of pulse period). I_{dsat} degradation is only dominated by the ANPC trapped charges or slow-states.

To further distinguish the I_{dsat} degradation behavior under AC and DC stress conditions, the I_{dsat} degradation and relaxation with time are plotted in Fig. 4 with Figs. 5 and 6 show the relaxation rate during recovery and trap refill rate during restress, respectively. The relaxation and refill rates are the same irrespective of total stress time indicating that trapping and detrapping mechanisms to be similar in nature. So the charges defined as CPC are the dominant hole trapping mechanism for the devices with the DC and AC stress conditions under study. On the other hand, ANPC kind of traps seems to be not neutralizing or neutralizing very slowly (which we could not observe within limited measurement time frame).

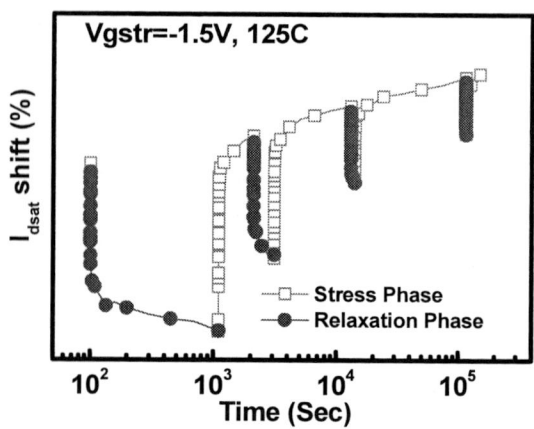

Fig. 4 A typical AC NBTI stress of the PMOSFET. Data showed a fast degradation relaxation after the bias (relaxation phase) and a fast re-fill of the traps (and/or new defect generation) when a stress bias is re-applied.

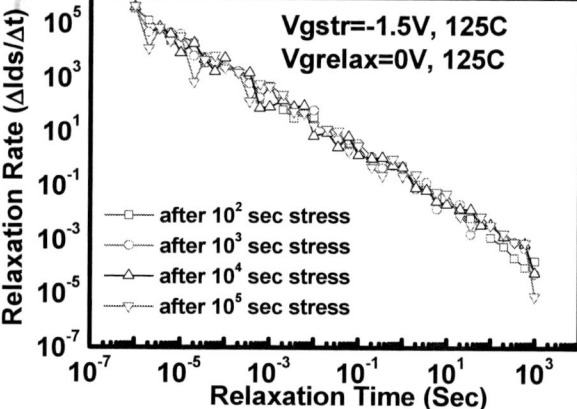

Fig. 5 Relaxation rate as a function of the relaxation time using data from Fig. 4. Data showed that relaxation rate is independent of the number of trapped charges and defects generation. The relaxation slope is around -1.

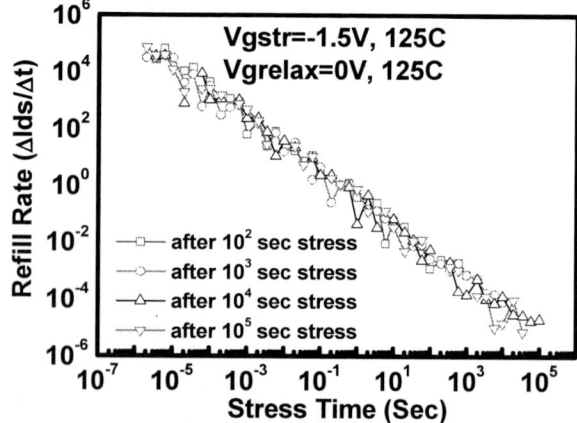

Fig. 6 The re-fill rate of the NBTI AC stress based on data from Fig. 4. The re-fill rate (with slope ~ 1) is almost the same as the relaxation rate (Fig. 5) – implies the fast-refilling of the shallow-traps.

To further illustrate the role of trapping and detrapping in overall NBTI degradation mechanism, I_{dsat} degradation as a function of duty cycle for different frequencies have been plotted for Mode-B and Mode-A in Figs. 7 and 8, respectively. The cumulative stress time is constant for all the measurements in Figs. 7 and 8. For Mode-B degradation, the duty cycle dependence is independent of frequency. Mode-A degradation has a similar duty cycle dependence compared to Mode-B; however, Mode-A degradation shows a frequency dependence, which is not observed with Mode-B measurement condition. This again implies that trap generation is frequency dependent but all captured holes detrap fast enough, so they can't be measured during Mode-B conditions. Implying that CPC or shallow traps are the dominant hole trapping and detrapping mechanism for different duty cycles also. The equal trapping and detrapping time constants suggest that the holes can elastically tunnel to and from the generated CPC traps. We can model resulting threshold voltage (V_t) and I_{dsat} degradation shift as a function of stress time (t_{stress}) and recovery time ($t_{recovery}$) [20]:

$$\Delta I_{dsat} = \frac{qD_{ot}x_o}{C_{ox}(V_g - V_{t0})}(E_F - E_{F0})\ln\left(1 + \frac{\tau_{oe}t_{stress}}{\tau_{oc}t_{recover}}\right)\ ......(1)$$

In Eq. (1), x_o is the characteristic tunneling depth, D_{ot} is trap density in oxide, and τ_{oc} and τ_{oe} are capture and emission time constant of traps. To calculate the duty cycle dependence of I_{dsat}, D_{ot} oxide trap density for unstressed PMOSFET is extracted by $1/f$ noise spectral density [21] as shown in Fig. 9. The model Eq. (1) shows good correlation with the experimental data for Mode B type of degradation as shown in Fig. 7.

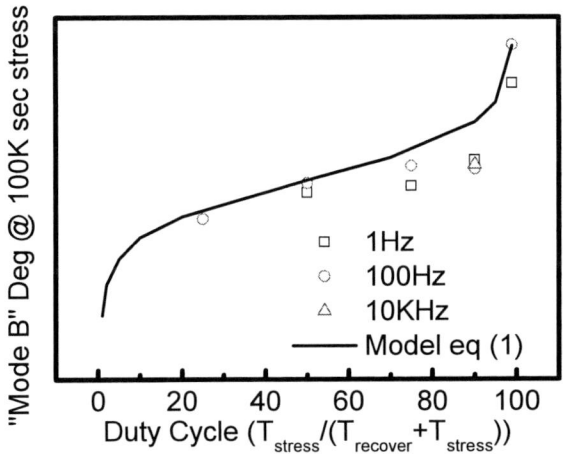

Fig. 7 Plot of the degradation of Mode B as a function of the duty cycle and frequency, with model equation (1) fits experimental data well.

The model Eq. (1) is unable to explain the frequency dependent I_{dsat} degradation data of Mode A as observed in Fig. 8. To explain I_{dsat} degradation during Mode A, we need to consider the different trapping and detrapping times for traps located at variable distance from the interface. The trap occupation probability P_T for a trap located at distance x from the interface due to different stress and recovery time can be simulated as [20]:

$$P_T = g(x, t_{stress})(1 - g(x, t_{recover}))............................(2)$$

The simulated trap occupation probabilities using Eq. (2) are shown in Fig. 10, where $g(x,t_s)$ and $g(x,t_r)$ are trapping and detrapping probability at depth x from the Si/SiO$_2$ interface.

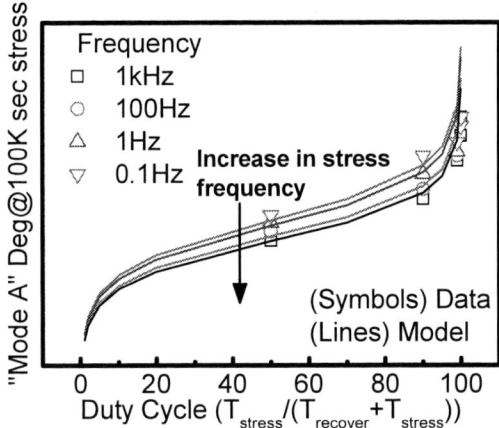

Fig. 8 Plot of the degradation for Mode A as a function of duty cycle and frequency, the trapping/detrapping model in (1) has been modified to include variation of D_{ot} as a function of frequency as Eq. (4). The model would be described in the following section.

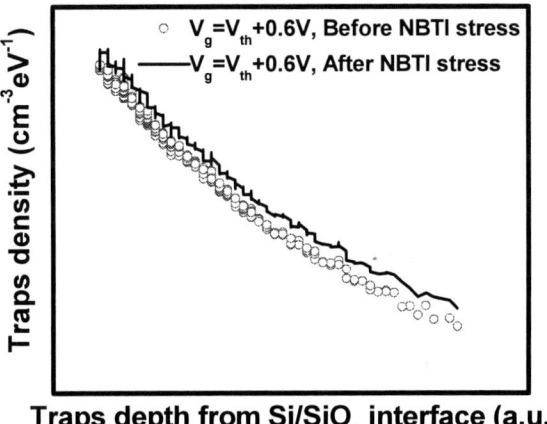

Fig. 9 Oxide trap density at different depths for PMOSFET before and after NBTI stress has been calculated through $1/f$ noise measurements. The trap density increases with stress (shown as line).

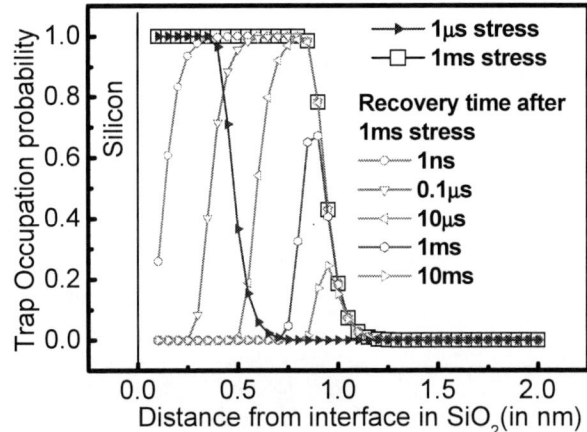

Fig. 10 Trap occupation probability (P_T) in oxide at different stress and recovery voltages. The probability of trap occupancy is larger at larger distance from interface for low frequency stress.

Hence, the total I_{dsat} degradation considering trapping/detrapping from existing traps at variable distance x from Si/SiO$_2$ interface can be modeled as:

$$\Delta I_{dsat} = \frac{q x_0 D_{ot}}{C_{ox}(V_g - V_{t0})} \int_0^{t_{ox}} \left(1 - \frac{x}{t_{ox}}\right) P_T \, dx \dots\dots\dots(3)$$

Simulated data using Eq. (3) assuming constant D_{ot}, does not show any frequency dependence for equal cumulative stress times as shown in Fig. 11 (open symbols). The filling of traps shows a saturation behavior with time suggesting that at long enough stress times all the traps at characteristic depth would be filled. So we can only get frequency independent I_{dsat} degradation by considering simple elastic hole trapping/detrapping model. One of the factors that is being overlooked in the model till now is the generation of new traps with stress time. The possibility of generated traps with stress has been verified with $1/f$ noise measurements before and after NBTI stress. The extracted bulk traps from $1/f$ noise data does show increase in oxide traps with stress (shown as line in Fig. 9). The D_{ot} is extracted after long enough stress time so that extracted D_{ot} has signature of CPC in addition to ANPC kind of generated traps. With inclusion of D_{ot} as a variable with function of time, the frequency dependent I_{dsat} degradation for Mode-A can be evaluated by:

$$\Delta I_{dsat} = \frac{q x_0}{C_{ox}(V_g - V_{t0})} \int_0^{t_{ox}} D_{ot}(t_{stress}, T)\left(1 - \frac{x}{t_{ox}}\right) P_T \, dx..(4)$$

In Eq. (4), T is defined as total time period of the AC stress pulse. The resulting I_{dsat} degradation are shown in Fig. 11 as solid symbols.

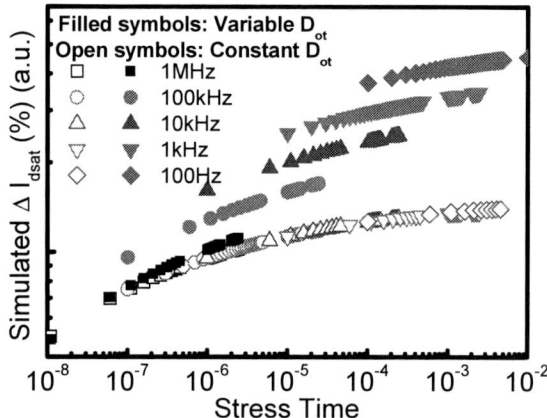

Fig. 11 I_{dsat} Degradation with time does not show any frequency dependencies if trap generation with time is not included. Simulation including trap generation with time results in frequency dependence of degradation as observed with Mode-A measurement methodology. The open symbols are similated ΔI_{dsat} with constant D_{ot} for all frequencies, which does not show any frequency dependence. Assuming increase in D_{ot} with time shows ΔI_{dsat} as a function of frequency with smaller frequency results in larger degradation as shown by solid symbols in Fig. 11.

AC/DC MODEL AND CIRCUIT APPLICATION

With the understanding developed in last section, a semi-empirical model to calculate the AC to DC ratio at different frequencies for Mode-A conditions are plotted in Fig. 12. The AC to DC ratio suggests

an improvement factor of around 10 for minimum frequency of 0.01 Hz and 50% duty cycle, essentially due to frequency dependence of CPC kind of trap generation. On the other hand, Mode-B has a much higher AC/DC factor at ~1000X level.

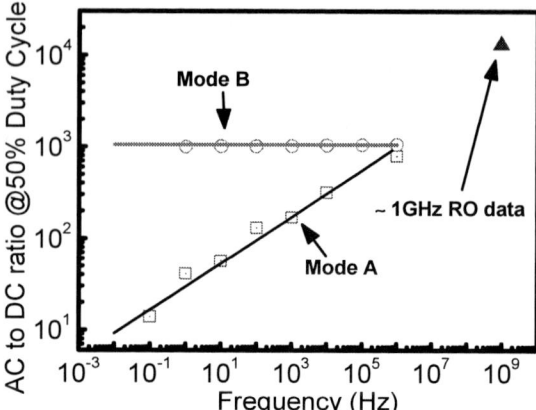

Fig. 12 The AC to DC NBTI improvement ratio for AC signal with 50% duty-cycle. The model lines extrapolated from semi-empirical models are also shown. The AC/DC improvement ratio is about 10 for frequency of 0.01Hz for Mode-A, while it is nearly 1000 for Mode-B measurements upto 1MHz. The RO circuit measurements under NBTI stress shows AC/DC improvement ratio ~10000 at 1GHz freqency.

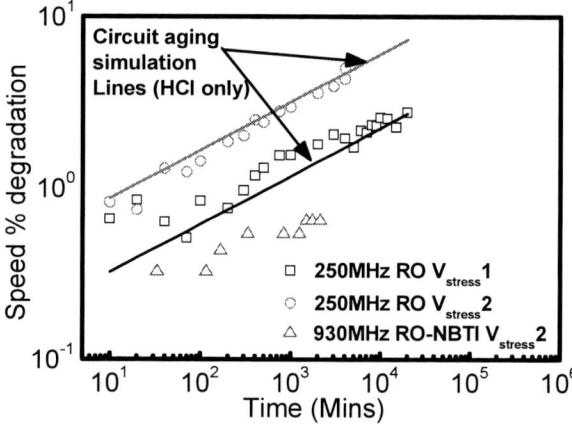

Fig. 13 The speed degradation for nominal channel length ring oscillator (RO) with 250MHz of frequency and large channel length ring-oscillator (RO-NBTI) having frequency of 930MHz is shown. $V_{stress}1$ and $V_{stress}2$ are applied stress voltage, where $V_{stress}2 > V_{stress}1$. The nominal channel length RO data matches very well with "HCI only" circuit aging simulation as NBTI contribution is observed to be less significant during high frequency circuit operation. The high frequency NBTI degradation is much lower as shown by RO-NBTI circuit degradation data stressed with $V_{stress}2$ as shown by hallow triangle ($\triangle$) in the figure. The HCI+NBTI (not shown here) circuit aging simulation with NBTI degradation having frequency independent AC to DC ratio extracted from DC data overestimate RO degradation.

To further explore the role of hole trapping and detrapping in determining actual circuit degradation at high frequency, ring oscillator (RO) circuits are stressed with high voltage. Two type of RO circuit is designed. One RO is designed with larger channel lengths NMOSFET and PMOSFET, such that possibility of hot carrier injection (HCI) degradation is excluded, we can call it RO-NBTI. Another RO is designed with on-rule channel length devices, which would be simply called as RO. The RO and RO-NBTI oscillator is stressed at $V_{stress}2$ with temperature of 150°C to further amplify the NBTI degradation conditions. Fig. 13 shows the performance speed shift of both RO and RO-NBTI circuit. The on-rule RO shows much higher speed

978-1-4244-5430-3/10 $26.00 © 2010 IEEE

degradation compared to RO-NBTI largely due to higher HCI degradation. The observed NBTI degradation for RO-NBTI is much lower compared to RO speed degradation data stressed at same voltage (V_{stress}2). The AC to DC NBTI lifetime ratio extracted from RO-NBTI is projected to be more than 10000 (for frequency near 1GHz) as shown in Fig. 12. The related I_{dsat} degradation percent for PMOSFET under RO-NBTI stress (930MHz) is also plotted in Fig. 2. The RO is also stressed under lower voltage (V_{stress}1) to study the impact of stress voltage on speed degradation. As expected higher stress voltage results in higher speed degradation for RO as shown in Fig. 13. The RO degradation is simulated with conventional aging simulation tool. RO degradation with data matches well with circuit aging simulation results, if only HCI factor is considered in the overall degradation. With both NBTI+HCI aging factors are considered in circuit aging simulations, the simulated degradation overestimate experimental data by a large margin (not shown here). This is mainly due to the current AC NBTI aging model, which is extracted with DC degradation data with very small AC to DC ratio. Hence, there is a need to implement more accurate AC/DC factor considering frequency dependent hole traps generation, in the circuit aging model simulation tool to calculate the effective I_{dsat} drift and the subsequent speed degradation under the circuit usage conditions. It would ensure a sufficient build-in design margin for product reliability.

CONCLUSION

This paper utilizes ultra fast NBTI degradation measurements to establish NBTI lifetime ratio under different frequencies AC stress compared to DC stress. A NBTI AC degradation model is established which takes into consideration, the generation of CPC or fast traps during NBTI stress. The relaxation and re-fill of the CPC or fast states can explain the frequency and duty-cycle dependent NBTI degradation. The current slow measurements essentially can measure only slow traps and interface states, which are frequency independent. A minimum AC/DC NBTI improvement factor ~10 has been observed even at frequency as low as 0.01Hz. The physics and measurement-based AC degradation model can be used for the more accurate projection of the circuit NBTI degradation.

ACKNOWLEDGMENT

Authors would like to thank TSMC Research and Development (R&D) team for providing the wafer and to Dr. N. S. Tsai for his guidance and support. We would like to give special thanks to Mr. Naresh K. Emani for his valuable input regarding MATLAB simulations. We are also thankful to Mr. T.Y. Yew for circuit measurements.

REFERENCES

[1] T. Yang, C. Shen, M. F. Li, C. H. Ang, C. X. Zhu, Y. -C. Yeo, G. Samudra, S. C. Rustagi, M. B. Yu, and D.-L. Kwong, "Fast DNBTI Component in p-MOSFET with SiON Dielectric," *IEEE Electron Dev. Lett.*, vol. 26, pp. 826-828, 2005.

[2] V. Huard, C. Parthasarathy, N. Rallet, C. Guerin, M. Mammase, D. Barge, and C. Ouvrard, "New Characterization and Modeling Approach for NBTI Degradation from Transistor to Product Level," in *Proc. IEDM*, 2007, pp. 797-800.

[3] T. Nigam, "Pulse-Stress Dependence of NBTI degradation and Its Impact on Circuits," *IEEE Trans. Dev. Mat. Rel.*, vol. 8, pp. 72-78, 2008.

[4] Y. Mitani, "Influence of nitrogen in ultra-thin SiON on negative bias temperature instability under AC stress," in *Proc. IEDM*, 2004, pp. 117-120.

[5] R. Fernandez, B. Kaczer, A. Nackaerts, S. Demuynck, R. Rodriguez, M. Nafria, and G. Groseneken, "AC NBTI studied

in the 1 Hz–2 GHz range on dedicated on-chip CMOS circuits," in *Proc. IEDM*, 2006, pp. 337-340.

[6] D. Ielmini, M. Manigrasso, F. Gattel, and G. Valentini, "A unified model for permanent and recoverable NBTI based on hole trapping and structure relaxation," in *Int. Rel. Phys. Symp.*, 2009, pp. 26-32.

[7] T. Grasser, W. Gos, V. Sverdlov, and B. Kaczer, "The Universality of NBTI Relaxation and its Implications for Modeling and Characterization," in *Int. Rel. Phys. Symp.*, 2007, pp. 268-280.

[8] C.R. Parthasarathy, M. Denais, V. Huard, G. Ribes, E. Vincent, and A. Bravaix, "New Insight into Recovery Characteristics Post NBTI Stress," in *Int. Rel. Phys. Symp.*, 2006, pp. 471-477.

[9] S. Mahapatra, M. A. Alam, P. B. Kumar, T. R. Dalei, and D. Saha, "Mecahnism of Negative Bias Temperature Stability in CMOS devices: Degradation, Recovery and Impact of Nitrogen," in *IEDM Tech. Dig. 2004*, pp. 105-108.

[10] H. Reisinger, "Fast Measurement Technique for Determination of Degradation due to NBTI," in *Int. Rel. Phys. Symp.*, 2008 (tutorial).

[11] H. Reisinger, U. Brunner, W. Heinrigs, W. Gustin, and C. Schlü nder, "A Comparison of Fast Methods for Measuring NBTI Degradation," *IEEE Trans. Dev. Mat. Rel.*, vol. 7, pp. 531-539, 2007.

[12] A. T. Krishnan, C. Chancellor, S. Chakravarthi, P. E. Nicollian, V. Reddy, A. Varghese, R. B. Khamankar, and S. Krishnan, "Material Dependence of Hydrogen Diffusion: Implications for NBTI Degradation," in *IEDM Tech. Dig.*, 2005, pp. 691-694.

[13] S. S. Tan, T. P. Chen, C. H. Ang,, and L. Chan, "A New Waveform Dependent Lifetime Model for Dynamic NBTI in PMOS Transistor," in *Int. Rel. Phys. Symp.*, 2004, pp. 35-39.

[14] W. Abedeer, and W. Ellis, "Behavior of NBTI under AC Dynamic Circuit Conditions," in *Int. Rel. Phys. Symp.*, 2003, pp. 17-22.

[15] M. A. Alam, "A Critical Examination of the Mechanics of Dynamic NBTI for PMOSFETs," in *IEDM Tech. Dig.*, 2003, pp. 345-348.

[16] H. Reisinger, U. Brunner, W. Heinrigs, W. Gustin, and C. Schlü nder, "Analysis of NBTI Degradation and Recovery Behavior Based on Ultra-Fast Vt-Measurements," in *Int. Rel. Phys. Symp.*, pp. 448-453, 2006.

[17] Agilent, *B1500 WGFMU module user manual*

[18] J. F. Zhang, Ce. Z. Zhao, A. H. Chen, G. Groseneken, and R. Degraeve, "Hole Traps in Silicon Dioxides- Part 1: Properties," *IEEE Tran.on Electron Dev.*, pp. 1267-1273, 2004.

[19] T. Grasser, B. Kaczer, W. Goes, Th. Aichinger, Ph. Hehenberger, and M. Nelhiebel, "A Two-Stage Model for Negative Bias Temperature Instability," in *Int. Rel. Phys. Symp.*, pp. 33-43, 2009.

[20] T. Tewksbury, "*Relaxation Effects in MOS Devices due to Tunnel Exchange with Near-Interface Oxide Traps,* " Ph.D thesis, MIT, 1992.

[21] H. D. Xiong, D. Heh, S. Yang, X. Zhu, M. Gurfinkel, G. Besuker, D. E. Ioannou, C. A. Richter, K. P. Cheung, and J. S. Suehle, "Stress-Induced Defect Generation in HfO_2/SiO_2 Stacks observed by using Charge-Pumping and Low Frequency Noise Measurements," in *Int. Rel. Phys. Symp.*, pp. 319-323, 2008.

An Extensive and Improved Circuit Simulation Methodology For NBTI Recovery

Haldun Kufluoglu, V. Reddy, A. Marshall, J. Krick, T. Ragheb, C. Cirba, A. Krishnan, C. Chancellor

Texas Instruments, TDI

13121 TI Blvd, Dallas, TX, 75243

Email: h-kufluoglu@ti.com

Abstract—**A feasible computational framework that enables improved predictability of NBTI degradation within commercially available tools is discussed. The NBTI model is used for both delay correction in transistor characterization data and real-time circuit operation where recovery is present. The complementary nature of implementation is readily incorporated into existing model extraction and verification tools. The method provides significantly enhanced accuracy in simulations when compared to circuit data, yet retains practicality and flexibility.**

I. INTRODUCTION

Accurate characterization of Negative Bias Temperature Instability (NBTI) is essential for integrated circuit lifetime assessment. In recent years, although considerable emphasis has been given to minimizing NBTI recovery artifacts in transistor measurements and speculating on physical aging mechanisms, conceptual gaps in circuit-level modeling are rarely addressed [1]-[7]. Such modeling ambiguities are particularly important in cost-conscious CMOS development where the number of masks is intentionally reduced and the resulting transistors are often operated at aggressive -yet less familiar- biasing schemes. Possible sources of modeling inaccuracies arise from: 1) recovery contamination in PMOS transistor measurements, 2) extraction of SPICE-based transistor models from limited degradation data [8],[9], 3) lack of real-time computation methods that account for recovery and history effects in common circuit simulation tools, and 4) improper handling of recovery in product-level test design and data assessment. Without these proper guidelines, it is a challenge to build reliable circuits while, at the same time, avoid over-design to compensate for degradation. In order to rectify these issues, we report an extensive and practical methodology to correctly estimate the NBTI degradation while not deviating from the industry-standard circuit design flow.

II. NBTI CHARACTERIZATION AND MODELING

Two major NBTI characterization techniques, On-The-Fly (OTF) and Stress-Measure-Stress (SMS), are illustrated in Fig 1. The OTF minimizes the recovery, however the degradation is measured at a single condition [7]. Often, the stress/measure condition is outside the real operating regime; furthermore, the transistor could be used in vastly different applications in digital and analog circuits whose performance criteria correlate with specific transistor metrics other than V_{TH}, I_{DLIN} or I_{DSAT} [10]. For a scalable model that can calculate the

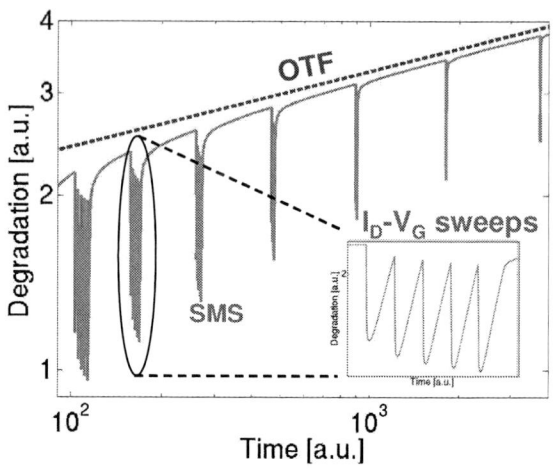

Fig. 1. SMS TECHNIQUE MEASURES FULL $I_D - V_{GS}$ (INSET) BUT TIME SLOPE AND MAGNITUDE ARE AFFECTED BY DELAY COMPARED TO OTF METHOD. EACH $I_D - V_{GS}$ SWEEP IS DONE AT A DIFFERENT V_{DS}.

degradation at broader range of operating conditions, family of $I_D - V_{D/G}$ data need to be measured intermittently during stress. As Fig. 2 shows, when OTF degradation data is used for SPICE-level aging model, it is difficult to predict the transistor response over the whole operating regime. This is because the degradation depends not only on the stress, but also on the measurement, i.e., terminal voltages. The Stress-Measure-Stress (SMS) method provides a family of $I_D - V_{D/G}$ data but only in the expense of relaxation between stress and measure cycles (Fig. 1). The measured time-slope (in the log-log plot), and magnitude of the degradation -important parameters used for reliability projections- suffer from the delay and therefore the SMS requires delay correction on the data along with the knowledge of the exact experimental design in terms of the stress durations, the delays and the voltages applied.

Several theories have been proposed to explain the NBTI stress and recovery trends in transistors. Such a model is summarized in Fig. 3. The details will be published elsewhere since the Reaction-Diffusion (R-D) model is not the main point in this paper. The R-D framework can be used to simulate the recovery and activity behavior. This improved model utilizes spatial variation in hydrogen diffusion constants (D_{H_2}) in the dielectric which is illustratively divided into separate regions

978-1-4244-5430-3/10 $26.00 © 2010 IEEE

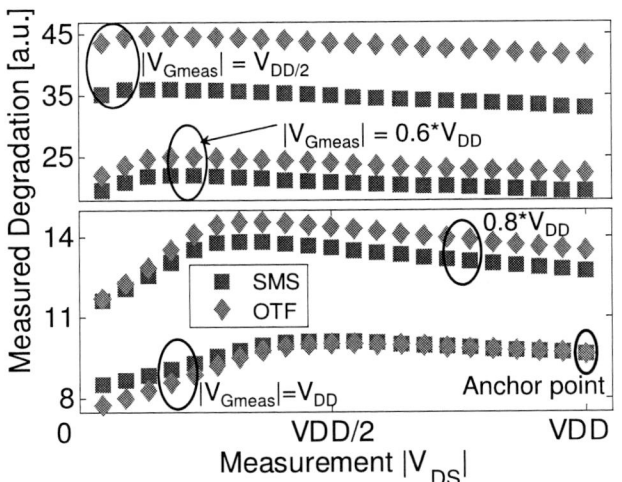

Fig. 2. SINGLE POINT OTF CANNOT CAPTURE THE FULL TRANSISTOR CHARACTERISTICS TO MEASUREMENT BIAS COMPARED TO SMS. BOTH ARE AFTER THE SAME STRESS CONDITION AND ANCHORED AT THE ΔI_{DSAT} ($V_{G,MEAS} = V_{D,MEAS} = V_{DD}$).

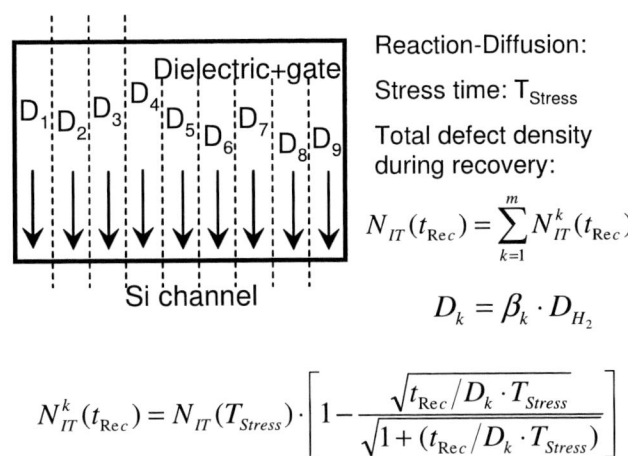

Reaction-Diffusion:

Stress time: T_{Stress}

Total defect density during recovery:

$$N_{IT}(t_{Rec}) = \sum_{k=1}^{m} N_{IT}^k(t_{Rec})$$

$$D_k = \beta_k \cdot D_{H_2}$$

$$N_{IT}^k(t_{Rec}) = N_{IT}(T_{Stress}) \cdot \left[1 - \frac{\sqrt{t_{Rec}/D_k \cdot T_{Stress}}}{\sqrt{1 + (t_{Rec}/D_k \cdot T_{Stress})}}\right]$$

Fig. 3. R-D MODEL WITH PARALLEL DIFFUSION PATHWAYS CAN EXPLAIN LOG(TIME) TREND DURING RECOVERY. APPROXIMATIONS ARE DERIVED FROM THE BASIC R-D THEORY [16]. D_K IS A MULTIPLE OF HYDROGEN DIFFUSION CONSTANT, D_{H_2}.

based on the diffusion rate. Each section brings an effective time constant as in Fig. 4 (inset). The parallel interaction of these passivation pathways give rise to log(time) type of recovery (solid line in Fig. 4). This approach is simulated to compare with experimental recovery trend in Fig. 5; it is from the full numerical solution and not from the analytical expressions of Fig. 3. The measurement pulses that were used to sense the amount of degradation during recovery are also considered in simulation. The pulses do not affect the slope behavior but change the magnitude only slightly when the pulse duration is short. In Fig. 6, the outcome of the improved R-D simulation agrees well with multiple recovery data across 6 orders of magnitude in time for PMOSFETs that were stressed and recovered at the same conditions. The model is also capable of capturing the slope changes in recovery through diffusivity distribution which could appear due to process variations. Finally, in Fig. 7, the AC stress duty cycle dependence reported in recent literature ([11]) is successfully estimated by this new R-D based model while the frequency independence is retained (not shown). On the down side, this additional complexity impacts the computation time; such a simulation scenario takes several minutes just for a single transistor.

III. COMPACT RECOVERY MODELING

Despite the capabilities of the aforementioned R-D based, and in general all the models based on detailed numerical calculations, they are seldom practical when it comes to circuit design tools because of the computational requirements as the number of transistors increase. Furthermore, the analytical approximations derived from such theories are inadequate since they are directed to very specific cases in circuit operation [12]. These limitations are overcome by a new model that focuses on the general properties of NBTI degradation and recovery rather than microscopic mechanisms. This method

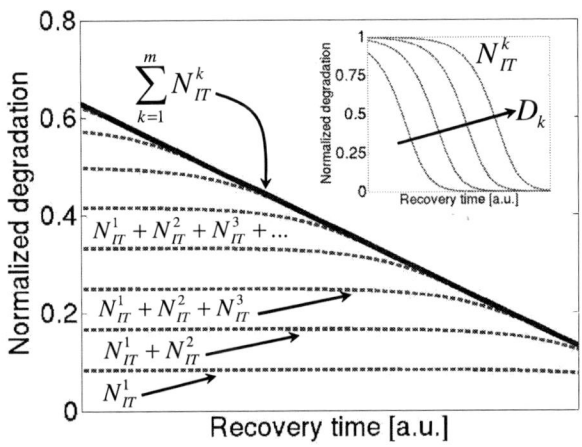

Fig. 4. RECOVERY IN EACH SEGMENT (DASHED LINES AND INSET) IN FIG. 3 ARE SUMMED AND THE DISTRIBUTED DIFFUSION YIELDS LOG(TIME) (SOLID LINE) WHOSE SLOPE IS DETERMINED BY THE DISTRIBUTION OF THE DIFFUSION CONSTANTS, D_K.

does not claim the presence or the lack of certain physical phenomena. It is assumed that observable trends resulting from the source of degradation are quantifiable in an encapsulated form. This makes the implementation simpler, more efficient in computation time, and is not restricted in the value of the time-slope or transistor activity [2], i.e., flexible for capturing data behavior.

A. Formulation

Efficient recovery calculation can be performed by the use of a convolution integral as in (1).

$$D(t) = h(t) \otimes g(t) = \int_0^t h(\tau) \cdot g(t - \tau) d\tau \tag{1}$$

where $\otimes$ is the convolution operator. Equivalently, $\mathcal{D}(\omega) = \mathcal{H}(\omega) \cdot \mathcal{G}(\omega)$ in frequency domain, or

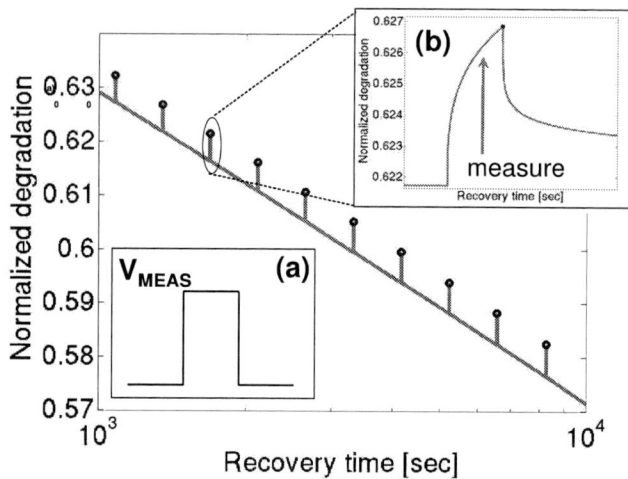

Fig. 5. R-D MODEL IS SIMULATED TO GET THE RECOVERY TREND IN COMPARISON TO THE DATA. SHORT MEASUREMENT PULSES (INSET (A)) THAT WERE PLACED LOGARITHMICALLY IN TIME ARE ALSO INCLUDED. CIRCLES ARE THE MEASURED DEGRADATION VALUES. INSET (B) SHOWS THE DETAILED TREND DURING MEASUREMENT; WITH EACH PULSE, DEGRADATION INCREASES SLIGHTLY WITHOUT CHANGING THE SLOPE. THE TREND IS FROM THE FULL ITERATIVE SOLUTION, AND NOT OF THE ANALYTICAL DERIVATION.

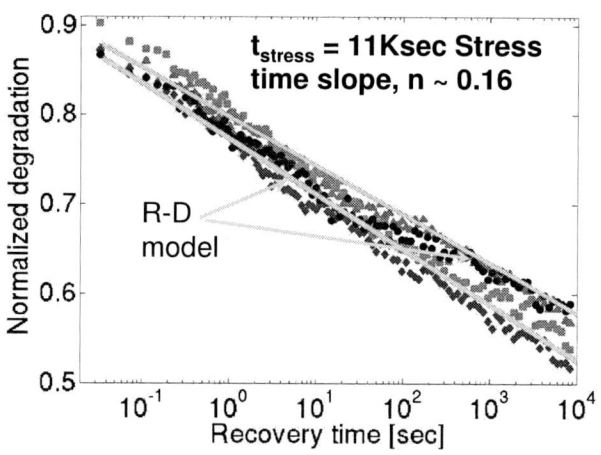

Fig. 6. NUMERICAL R-D RESULTS (LINES) COMPARE VERY WELL WITH RECOVERY DATA (SYMBOLS) OVER 6 DECADES IN TIME. SIMULATION REPEATS DATA MEASUREMENT TECHNIQUE (CIRCLES IN FIG. 5). SLOPE CHANGE IN MODEL DUE TO D_{H_2} DISTRIBUTION.

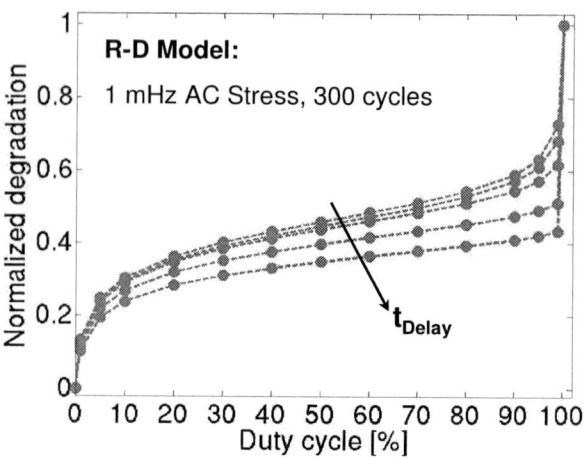

Fig. 7. IMPROVED R-D MODEL ALSO PREDICTS THE DUTY CYCLE DEPENDENCE REPORTED IN LITERATURE. THE t_{delay} IS THE DURATION AFTER AC STRESS BEFORE MEASUREMENTS AND THE DEGRADATION DECREASES WITH INCREASING t_{Delay}.

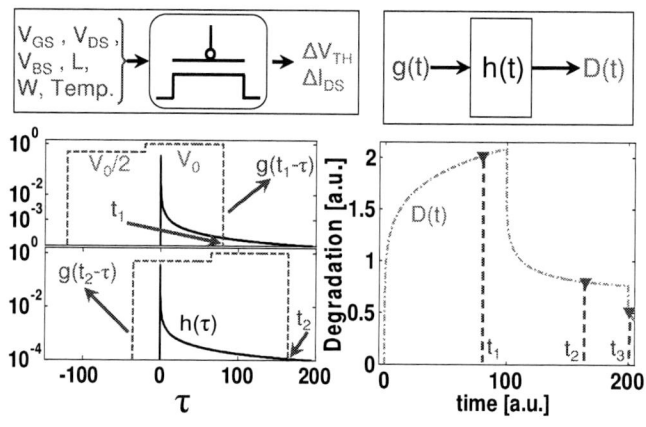

Fig. 8. TERMINAL VOLTAGES AND AMBIENT CONDITIONS ARE INPUTS TO SYSTEM WHOSE OUTPUT IS THE DEGRADATION. THE RESPONSE RECOVERS WHEN THE DEGRADATION RATE GETS SMALLER (TRANSITION TO $g(\tau) = V_0/2$). t_1 AND t_2 ARE TIME POINTS DURING STRESS AND RECOVERY, RESPECTIVELY. THE DEGRADATION RELAXES MUCH FASTER WHEN THE $g(\tau)$ DROPS TO 0 ($t = t_3$).

$\mathcal{D}(s) = \mathcal{H}(s) \cdot \mathcal{G}(s)$ using the Laplace transform, with $\mathcal{D}(\cdot)$, $\mathcal{H}(\cdot)$, and $\mathcal{G}(\cdot)$ as the tranform pairs of $D(t)$, $h(t)$, and $g(t)$ in either domain, respectively. Here, only time-domain is considered. Above, the $D(t)$ represents the degradation as a response to the input function, $g(t)$, which is a combination of terminal voltages, temperature, time, and device geometry ($g(time, V_{GS}, V_{DS}, V_{BS}, L_{channel}, W_{channel}, Temperature)$). Notice that (1) allows both degradation and recovery under the given conditions. The $g(t - \tau)$ term signifies the memory in the system, i.e., the overall response depends on the past history of stress and relaxation [13]. Fig. 8 shows the

convolution process with two different stress levels. When the $g(t)$ becomes smaller (V_0 to $V_0/2$), the degradation rate decreases and $D(t)$ begins to recover. The entire trend is governed by the amount of overlap between $h(\tau)$ and $g(t - \tau)$. Because of the history effect, the initial stress contributes to later stages, even into recovery, but its weight is reduced over time due to the monotonically decreasing $h(t)$.

The $h(t)$ is the response function of the degradation mechanism and it is assumed to be solely due to NBTI. It can be obtained by considering a DC stress for the $g(t)$, e.g., $= V_0 \cdot u(t)$, $u(t)$ being the unit step function. After

978-1-4244-5430-3/10 $26.00 © 2010 IEEE

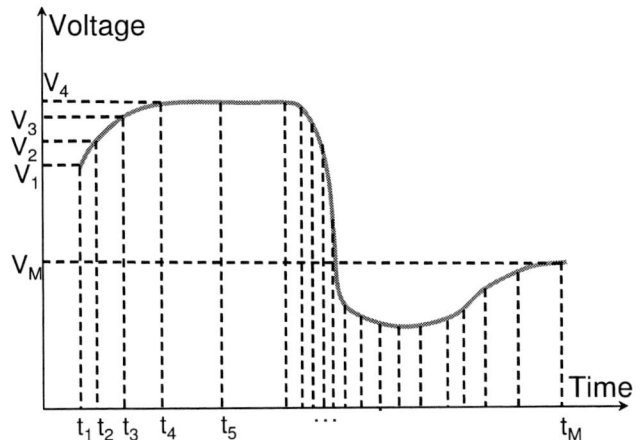

Fig. 9. TYPICAL SPICE OUTPUT OF A TRANSIENT HAS VARYING SAMPLING INTERVALS BASED ON THE RATE OF CHANGE IN VOLTAGES. THE WAVEFORM CAN BE APPROXIMATED BY A SUM OF PULSES WITH MAGNITUDE V_J AND DURATION Δt_j.

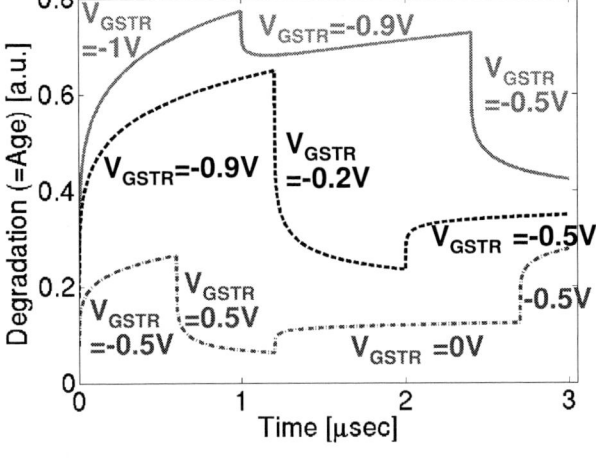

Fig. 10. THE CONVOLUTION-BASED (NOT R-D BASED) MODEL IS IMPLEMENTED TO FACILITATE REAL-TIME DEGRADATION AND RECOVERY CALCULATIONS IN CIRCUIT SIMULATORS. EXAMPLES OF PMOS BEHAVIOR ARE SHOWN. V_{GSTR} IS STRESS V_{GS}.

the experimental characterization of the DC degradation, the time-dependence can be modeled analytically and then the response function can be extracted by accounting for the $g(t)$ which has a constant magnitude for DC stress. This work does not impose any specific $h(t)$ -the choice is left to the end user- but rather focuses on the method of computation once appropriate functions are chosen. Similarly, the particular voltage, temperature, and L/W dependences in $g(t)$ are also user defined and will not be detailed in this work.

Although (1) provides a compact method for modeling, it is not directly applicable in CAD tools. Not only the simulation output in time domain is of discrete nature, but also the time steps are not uniform. For instance, SPICE uses adaptive sampling, i.e., smaller time increments in fast transients and bigger time steps in slowly varying conditions as illustrated in Fig. 9 . The convolution integral above needs to be modified accordingly. For clarity, the input $g(t)$ in the discussion below is taken as a linear function of a single terminal voltage but can be extended for multiple terminals, temperature, and geometry. The $g(t)$ can be approximated with pulses as

$$g(t) \approx V_1 \cdot [u(t) - u(t - t_1)] + V_2 \cdot [u(t - t_1) - u(t - t_2)] + \quad (2)$$
$$\cdots + V_M \cdot [u(t - t_{M-1}) - u(t - t_M)].$$

In order to evaluate (1) in a feasible way, one of the properties of convolution can be employed, that is

$$x(t) \otimes \frac{dy(t)}{dt} = \frac{dx(t)}{dt} \otimes y(t). \quad (3)$$

For the step function, $d[u(t - t_j)]/dt = \delta(t - t_j)$. Another function, $H(t) = \int_0^t h(q)dq$, can be introduced to be used in (3) and is interpreted as the DC degradation in response to $u(t)$ input. Since $\int x(\tau)\delta(t - \tau)d\tau = x(t)$, (1) becomes

$$D(t) \approx V_1 \cdot H(t) + (V_2 - V_1) \cdot H(t - t_1) + \cdots + \quad (4)$$
$$(V_M - V_{M-1}) \cdot H(t - t_{M-1}) - V_M \cdot H(t - t_M).$$

Under DC stress, the $V_{j+1} - V_j = 0$ and so $D(t) \propto H(t)$ due to the first term. In the case of recovery where $V_{j+1} < V_j$, the negative $V_{j+1} - V_j$ term reduces the magnitude of degradation.

The sum in (4) is calculated by generating a vector of time, and at each time step, the function $H(t - t_j)$ is evaluated with $H(t - t_j) = 0$ for $t < t_j$. Finally, each term is multiplied by its weight which is determined by the input function, $g(t)$, and added to the degradation obtained in the previous step. Depending on the choice of the $g(t)$, the weights can vary non-linearly with voltage and temperature. Nonetheless, with sufficiently good sampling rate, these quantities change slowly in a particular time step, therefore can be further approximated as linear terms.

Fig. 10 demonstrates the outcomes of the new model to several stress patterns. Although all patterns end with the same condition, the final degradation depends on the stress history; the memory effect in the results is clear. The model captures the trend when the recovery voltage is slightly less than the stress condition (top: -0.9V after -1V) where the degradation recovers initially but increases at a slower rate afterwards, the impact of reverse polarity (bottom: 0.5V after -0.5V) in which the amount of recovery increases, and different stress and relaxation times during the patterns. For simulation purposes, trends during all voltages including $|V_{GSTR}| < |V_{TH}|$ are assumed to measurable although on a real transistor, those results would be too noisy.

The convolution based methodology can also be utilized to correct for measurement inaccuracies due to delay. As shown in Fig. 11, the degraded characterization data is used to fit a SPICE model at each condition. Besides different stress voltages, temperatures, and transistor geometries, the data contains a set of measurement conditions (sweeps of V_{GS} and V_{DS}) in order to represent the impact of aging accurately. As long as the stress and measurement conditions are known, the input function $g(t)$ can be tailored and by comparing with

978-1-4244-5430-3/10 $26.00 © 2010 IEEE

the measured data, no-delay cases can be extrapolated through backward calculation.

B. Application to CAD

It is useful to briefly review how degradation is calculated in a typical CAD tool. In this work, although Relxpert ([14]) is used model the degradation behavior, the framework mentioned in the previous section is generally applicable to other CAD software as well as directly on SPICE output. As given in Fig. 11, the circuit netlist contains the time=0 transistor characteristics and the degradation model. The SPICE simulation generates the voltage waveforms for all transistors in the circuit. After this, in each time step (Δt), rate of aging $dAge/dt$ is calculated based on the voltages, temperature, and transistor geometry (Fig. 10). Next, the total age for each transistor is calculated as a sum of incremental aging in each time step, i.e.,

$$TotalAge = \sum_{j}^{M} \Delta t_j \cdot \frac{d}{dt} Age_j. \qquad (5)$$

The default mode of computation in Relxpert only allows degradation, the age never decreases even if the voltage is reduced during operation. One reason is that the default mode has no memory of stress history, the tool cannot judge if degradation should recover or not.

The add-on methodology presented in Section III-A can be implemented to enhance the capabilities of CAD. The effective degradation in (4) is substituted into (5) as in Fig. 11 (dotted lines), and thus reflected in final aging results of each transistor in the circuit. Other CAD tools might require different paths of calculation but the recovery methodology would still be similar to the flow shown here. [1]

IV. VERIFICATION

With all the pieces in hand, the completed methodology is presented in Fig. 11. Transistor SMS data are obtained at several different stress and measure conditions (V_{GS}, V_{DS}, V_{BS}, temperature), and geometries (channel length, width). Next, the $I_D - V_{D/G}$ data are used to optimize BSIM4 transistor models [15] across subthreshold, linear and saturation regions. The tool aims to minimize the RMS error between degradation data and the SPICE model. Fig. 12 shows RMS error distribution for all the data after the optimization; most fits yield small errors. The procedure is automated and therefore fast; fits 1590 $I_D - V_{D/G}$ sweeps in about 2 minutes. This efficient optimization tool also allows fine-tuning the fits to improve the model. The data is then used for Relxpert aging model extraction which is further refined for delay correction with the new NBTI model. This step also simulates the $I_D - V_{D/G}$ measurement sweeps (Fig. 1) besides stress and relaxation periods. The model is incorporated into Relxpert Ultrasim Reliability Interface (URI) equations to include the recovery

[1]The default version of Relxpert URI is not functional enough to handle the calculation in (4). In this work, the script is duplicated on a separate platform and the modifications given in Section III-A are added to demonstrate the proof-of-concept.

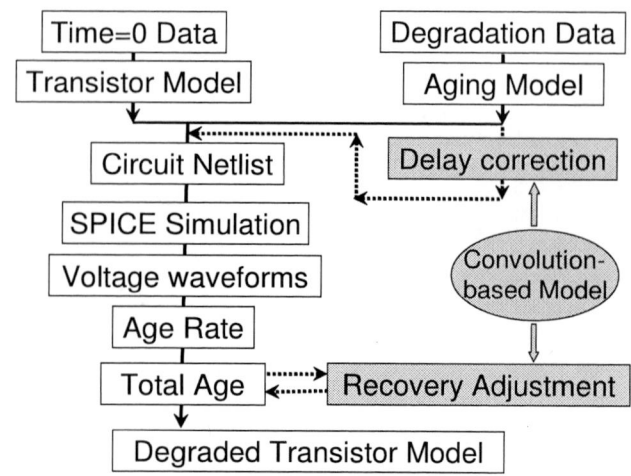

Fig. 11. THE COMPLETE AND PRACTICAL METHODOLOGY USES THE NEW MODEL TO CORRECT FOR DELAY IN PMOS DATA AND IS BUILT INTO RELXPERT URI EQUATIONS FOR REALISTIC ESTIMATION. SOLID LINES SHOW THE DEFAULT AGE CALCULATION WHEREAS DOTTED PATHS (AND SHADED BOXES) PROVIDE THE DELAY CORRECTION IN TRANSISTOR DATA AND URI CALCULATIONS.

in circuit operation as in Fig. 13. This important modification allows accurate real-time calculation of degradation and avoids the overestimation of the default Relxpert method and even the delay-corrected model. This flow is suitable for both digital and analog circuits and therefore, applicable to other simulation tools as well.

This comprehensive flow is applied to simulate the degradation of an on-chip ring oscillator (ROSC) which runs in the GHz frequency range. The degradation in ROSC frequency is measured OTF through an integrated divider on the same chip. The SPICE simulation contains the RC parasitics extraction on top of NBTI and Hot Carrier (HCI) Relxpert models. Fig. 14 compares the data with Relxpert results with and without the methodology. Overall, HCI effect is negligible at this V_{DD}. Without the delay correction to the degraded transistor $I_D - V_{D/G}$, Relxpert predictions are seen to be very pessimistic. When the corrections are performed to the PMOS NBTI characteristics (time slope, n, changes from ~0.26 to ~0.14) but still using the default Relxpert version, the degradation is overestimated due to lack of relaxation in the computation (Fig. 13, inset). The advantage of the URI is obvious with recovery activated; the degradation is bounded by the mid- and bottom-level (see points 2 and 3 in Fig. 13) results with the latter getting very close to the lifetime observed experimentally.

V. CONCLUSION

A robust and accurate simulation approach that addresses both PMOS-level measurement delay artifacts and real-time degradation/recovery calculation in circuits has been presented. This methodology is incorporated and verified in Relxpert tool, and compared with on-chip ring oscillator data. The backbone of this new flow is the unique NBTI model that

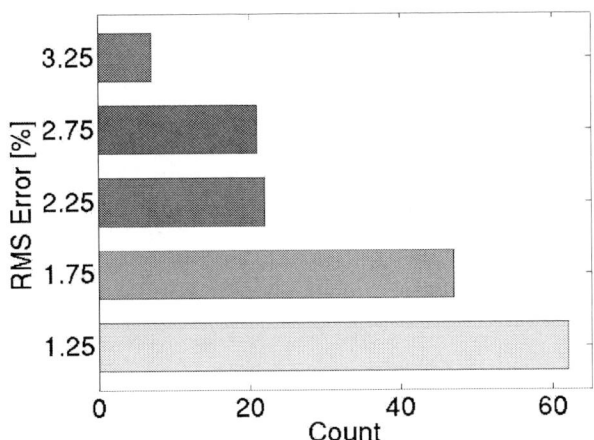

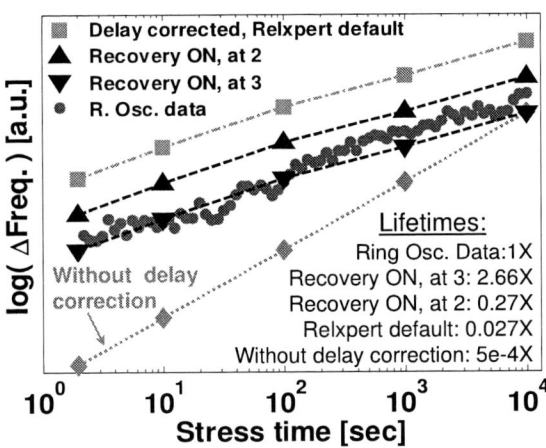

Fig. 12. BSIM4 SPICE MODEL IS FIT TO DEGRADED $I_D - V_{G/D}$ DATA. ENTIRE PROCESS IS AUTOMATED AND USED FOR AGED MODEL EXTRACTION. THE OPTIMIZATION HAS 159 DEGRADED DATASETS WITH 10 $I_D - V_{G/D}$ SETS EACH. TOTAL OPTIMIZATION TIME IS APPROXIMATELY 2 MINUTES.

Fig. 14. METHODOLOGY APPLIED TO ON-CHIP RING OSCILLATOR DATA. THE NEW MODEL BOUNDS THE DATA AND LIFETIME ESTIMATION IS MUCH BETTER. 2 AND 3 REFER TO RECOVERY LEVELS IN FIG. 13.

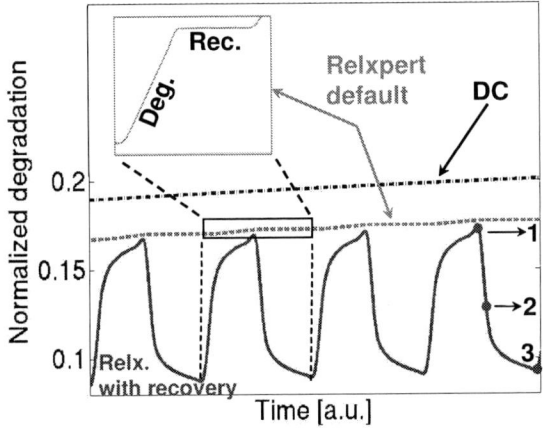

Fig. 13. ONLY WITH THE NEW MODEL, RELXPERT URI CAN CALCULATE RECOVERY IN REAL-TIME. OVERESTIMATION IS AVOIDED.

[3] J. Campbell et al., Proc. IEEE IRPS, pp.442-447, 2006.
[4] S. Zafar et al., VLSI Symp. Tech. Dig., pp.208-209, 2004.
[5] S. Mahapatra et al., Proc. IEEE IRPS, pp.1-9, 2007.
[6] B. Kaczer et al., Proc. IEEE IRPS, pp.381-387, 2005.
[7] M. Denais et al., Proc. IEEE IEDM, pp.109-112, 2004.
[8] W. Wang et al., IEEE Trans. DMR, pp.509-517, 2007.
[9] B. Paul et al., Proc. DATE, pp. 780-785, 2006.
[10] A. Marshall, ITC Tutorial, 2005.
[11] B. Kaczer et al., Proc. IEEE IRPS. pp.20-27, 2008.
[12] V. Huard et al., Proc. IEEE IRPS, pp.624-633, 2009.
[13] H. Kufluoglu et al., Proc. IEEE IRPS, pp. 690-691, 2007
[14] Cadence Relxpert Manual
[15] BSIM4 Manual
[16] M.A. Alam, Proc. IEEE IEDM, pp. 346-349, 2003.

is as capable as detailed calculations without the computational demands but simultaneously more veritable and scalable compared to analytical approximations. This balance enables true NBTI recovery integration into industry-standard circuit simulators, permitting improved use of the design space that is tightened by several variability sources in modern CMOS technology.

ACKNOWLEDGMENT

The authors would like to thank the colleagues at TDI SPICE Modeling Lab and J. Ondrusek, J. Carulli, and L. Salmon for support and collaboration.

REFERENCES

[1] T. Grasser et al., Proc. IEEE IRPS, pp.28-38, 2008.
[2] H. Kufluoglu et al., IEEE Trans. ED vol.54, pp.1101-1107, 2007

ADAPTIVE SENSING AND DESIGN FOR RELIABILITY

P. Singh, D. Sylvester, D. Blaauw

University of Michigan, Ann Arbor

email: {prsingh, sylvester, blaauw}@umich.edu

ABSTRACT

Chip lifetime degradation due to oxide break down is a major concern for today's designers. We review existing methods to solve the gate oxide reliability issues and also introduce an *in situ* degradation monitoring technique. This technique allows early detection of oxide degradation and makes a system aware of its reliability. When used in conjunction with reliability management schemes, it minimizes existing pessimistic reliability margins and allows an improvement in device performance.

Keywords: reliability, oxide degradation, sensors

INTRODUCTION

Technology scaling has resulted high levels of integration which has enabled sustained growth of the electronics industry. Advanced process nodes meet the performance requirements of today's electronic systems. However, it is becoming increasingly difficult to meet chip lifetime specifications while maintaining high yield. Modern technology scaling has abandoned the constant field paradigm, resulting in steady increases in gate oxide electric field in the past few generations. High electric fields increase the rate of degradation of the gate oxide, making gate oxide degradation a major bottleneck in sustaining the functionality of a chip throughout its lifetime [1]. This makes it harder to meet the reliability targets in advanced process nodes.

Designers have traditionally used static reliability management to meet reliability specifications. This method limits the supply voltage to a fixed maximum value for all fabricated chips such that a chip lifetime of the required level is attained. The lifetime of a chip, however, is a statistical variable due to the innate randomness in the degradation process. Moreover, process variation, fluctuations in environmental conditions such as voltage and temperature, and state dependence of the oxide degradation also add to the randomness in lifetime [2]. Hence the voltage limit is set so that the *weakest* chip meets the lifetime requirements under worst-case conditions. Hence, many chips will fail much later than the desired lifetime. The pessimism of static reliability management creates reliability slack on a chip-by-chip basis that can be traded off with performance to provide a significant performance benefit in cutting edge technology nodes [3].

To enable this tradeoff, a system needs to be aware of its reliability status. Such a system is then capable of dynamically adjusting the operating conditions of the chip (in particular, supply voltage and the maximum temperature limit) to achieve peak performance benefits while just meeting the lifetime specification. This approach has been referred to as Dynamic Reliability Management (DRM) [4, 5, 6]. These systems use on-chip sensors to get information that can be used to compute or predict the reliability state of the chip. DRM can be implemented in three ways: model based, degradation sensor based, and *in situ* monitoring based, as proposed in this paper.

In the degradation model-based approach, on-chip voltage and temperature sensors are employed to sense the operating conditions of the chip. The data from these sensors is fed to a degradation model which is used to compute the expected reliability state of the chip [4]. DRM systems based on this approach must account for inaccuracies in the sensors and the degradation models themselves by adding margins that make this approach more conservative. In addition, this approach does not address any innate sources of variation in lifetime such as those due to process variation, state-dependence and the inherent randomness of oxide breakdown, and hence has to be used with considerable margins.

The degradation sensor-based approach obviates the need for voltage and temperature sensors as well as degradation models. Instead, it relies on special sensors that directly monitor the degradation of replicated transistor oxides. Degradation sensors are distributed across the core in large numbers so that the sensors experience the same environmental conditions as the devices in the actual circuit. Degradation data from these sensors is then used to estimate the degradation of the actual devices in the chip [2]. Since such degradation sensors experience the same process conditions as the actual devices and also do not incur model inaccuracies, they can operate with tighter margins. However, degradation sensors do not account for lifetime variations due to innate randomness of degradation mechanisms, process mismatch between the replicated oxides that are monitored and the functional oxides of devices in the chip, and state dependence of stress. Thus, considerable margins remain.

The *in situ* monitoring-based DRM scheme employs a direct approach to monitor degradation. In this methodology the actual devices used in the circuit are measured directly to determine their degradation. This approach addresses all sources of lifetime variation and hence provides the most accurate observation of degradation, resulting in almost complete margin elimination. The key challenge is to achieve this with minimal invasiveness and overhead.

This paper reviews the static reliability management technique and the DRM schemes based on degradation models and degradation sensors. We also discuss a proposed *in situ* monitoring technique that allows the DRM system to be more aggressive with voltage scaling and hence minimizes reliability margins.

STATIC RELIABILITY MANAGEMENT

Breakdown of a MOSFET's gate oxide is a statistical process due to the innate randomness of the degradation mechanism. Moreover, factors including device process variation, fluctuations in operating conditions (voltage and temperature), and state dependence of stress further add to the randomness in device degradation. As a result, process engineers have to consider all these factors at design time before imposing a supply and temperature limit on the operating conditions of the devices. Fig. 1 shows the simulated lifetime distribution of an ensemble of chips under different process, voltage, and temperature conditions.

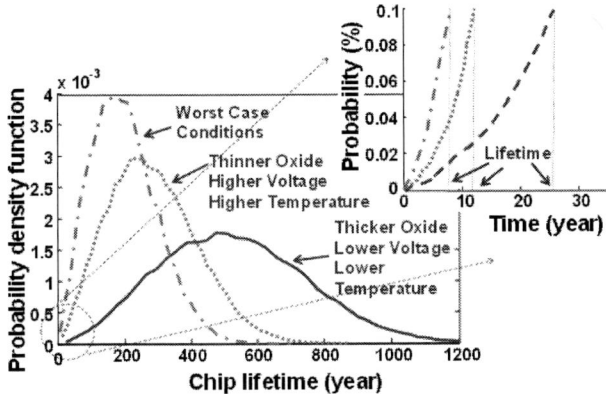

Fig. 1 Lifetime distribution of chips under different process, voltage and temperature (PVT) conditions. The desired lifetime of 10 years is not met under worst PVT conditions.

A percolation-based oxide degradation model was used in this simulation [7] . The figure shows that as the voltage and temperature are lowered and the oxide thickness reduced, the mean and spread in the lifetime of the chips decrease [3]. To ensure that a very low number of the chips fail (the desired yield) during the in-field operation of the chips, worst process, voltage, and temperature conditions are assumed to arrive at a lifetime distribution, which further reduces the mean lifetime of the chips. In reality these conditions will not be experienced by all chips at all times, making such assumptions very conservative.

If a lifetime of ten years and a yield of 99.9% is desired, then not more than 0.1% of the chips should fail at the end of ten years. As shown in Fig. 1 this does not hold true under the conservative worst process and operating condition assumptions. Hence, the voltage would have to be scaled down until the above mentioned criterion is met. This results in reduced performance of the chips and a lifetime much greater than the specification for many chips, or a *reliability slack,* owing to pessimistic assumptions. This slack can potentially be traded with performance enhancement by increasing the supply voltage. However, this requires that the system is self-aware of its reliability.

These systems can use DRM schemes to dynamically manage the reliability and the performance of the system. To employ DRM schemes the chip must be able to compute its reliability state. This can be accomplished by feeding temperature and voltage readings from sensors into a reliability model. However, accurate calibration of such a model with every technology node is non-trivial and resource intensive. Hence, in this paper we focus on DRM schemes that use reliability sensors as well as a proposed scheme that uses *in situ* reliability monitoring. We discuss each in the following two sections.

SENSOR-BASED RELIABILITY MONITORING

To employ DRM schemes, the chip must be able to compute its reliability state. The shortcomings of the model-based approach are partially overcome by degradation sensor-based reliability management. The idea is to use degradation sensors that consist of test devices exposed to the same voltage and temperature conditions as the core devices. Hence, these sensors are placed close to the core circuits to be monitored. However, the variation in degradation due to innate randomness cannot be captured by only a couple sensors; hundreds or even thousands of sensors are required to capture the statistics of degradation [2]. This imposes tight constraints on the area and the power consumption of these sensors. Moreover, it is important that these sensors track degradation very early after the onset of degradation so that the DRM scheme has enough time and flexibility to manage the chip reliability and react in time to changes in the degradation state.

Oxide degradation sensors have been proposed in [8, 9]. Fig. 2 shows the schematic of an oxide degradation sensor proposed in [8].

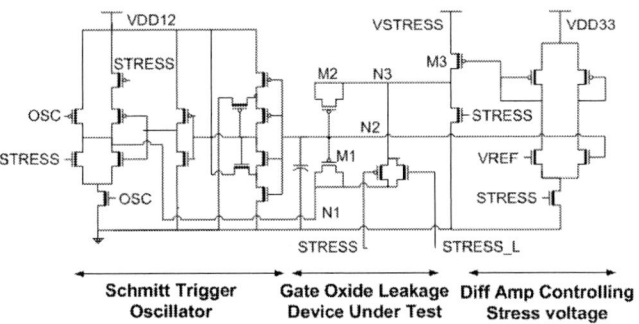

Fig. 2 Gate-oxide degradation Sensor schematic.

The sensor consists of test devices M1 and M2, which are exposed by the stress voltage Vstress, which is the regular supply voltage of the chip or can be a higher voltage for accelerated testing. When a measurement is taken, the senor outputs the frequency of a ring oscillator, the delay of which is determined by the gate leakage of M1 and M2. Hence the frequency output of this sensor is directly proportional to the gate leakage of M1 and M2. As the gate oxide of M1 and M2 degrades, their gate leakage and the frequency output of the sensor rise. The digital output of the sensor makes it easy to process its data. The sensor has the area of 21 NAND3 gates, making it amenable for use in large numbers.

Fig. 3 shows degradation data collected using the sensor in [8]. The degradation captured by the sensors is gradual, allowing the DRM scheme to manage the operating conditions and control the degradation rate of the chip. For a small sample of 16 sensors, the gate leakage degradation varies from 5-40% over a die. The sensor data is used to collect statistics of degradation, which are then used to compute an upper and lower bound on the degradation. In a chip consisting of millions of transistors, this variation is much larger and hence there is a considerable difference between the upper and lower bounds of degradation. To avoid unexpected failures, the DRM scheme assumes the upper bound on degradation for all devices and adds some additional margin to this to determine the acceptable maximum supply voltage setting.

978-1-4244-5430-3/10 $26.00 © 2010 IEEE

IN SITU RELIABILITY MONITORING

The sensor-based DRM scheme addresses the environmental conditions (voltage/temperature) and global process conditions among the sources of lifetime variation. To account for the other factors, such as local process variations and the innate randomness of the degradation, pessimism is introduced while computing the reliability state of the chip. This pessimism can be eliminated by the *in situ* monitoring of the degradation of the actual circuits as opposed to predicting their degradation using degradation sensors that lie alongside the circuit. This technique removes uncertainty due to local process, voltage, temperature conditions, the state dependence of degradation, as well as the innate randomness of the degradation process.

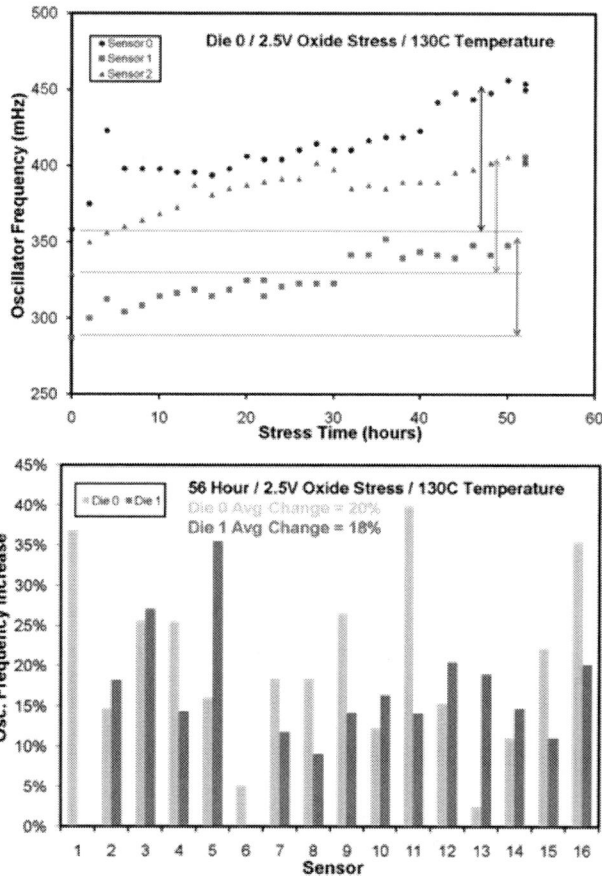

Fig. 3a (Top) Gradual gate-oxide degradation captured by the degradation sensor. Fig. 3b (Bottom) The gate current degradation varies from 5% to 40%.

In situ monitoring can be implemented in a number of ways. Delay has been proposed as a metric for degradation sensing; however, it is difficult to detect small changes in gate delay. In addition, the delay is sensitive to environmental conditions. Due to these factors a delay sensor's threshold to flag the onset of degradation must be set high to prevent any false alarms. This in turn would delay the degradation detection which may not leave enough time to take corrective action before the chip fails. Moreover, our silicon measurements show that a gate's delay might actually decrease with degradation, making the delay sensor-based *in situ* monitoring unreliable at times.

Gate leakage is the most direct measure of gate oxide degradation. However, it is difficult to measure the gate current in the presence of background currents such as subthreshold leakage, band-to-band tunneling, and gate-induced drain leakage. In addition, voltage and temperature strongly impact the total measured current, further complicating the measurement of changes in gate current.

In our proposed approach the key in detecting oxide degradation is sensing the change in the I_g-V_{gs} characteristics of a degraded device. In [10] the leakage of a degraded gate oxide is modeled as follows:

$$I_g = K \, (V_{gs})^P, \qquad (1)$$

where I_g is gate current and V_{gs} is gate-source voltage.

Fig. 4 illustrates how the values of K and P vary with degradation in time and consequently how the gate leakage increases. The values of K and P are based on measurements reported in [10].

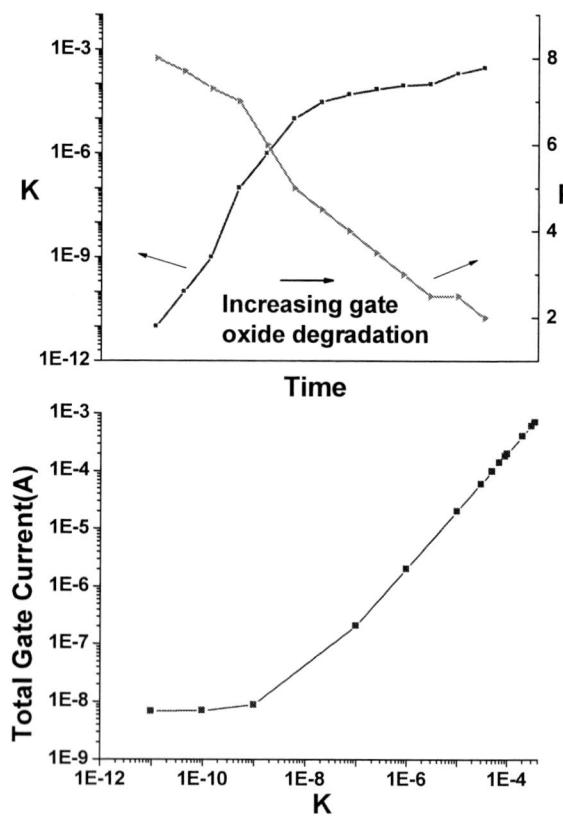

Fig. 4a (Top) Increasing gate-oxide degradation can be modeled as an increase in the value of K and decrease in the value of P. Fig. 4b (Bottom) The gate current increases by orders of magnitude due to gate-oxide degradation.

As the gate oxide degrades, defects, or trap sites, are formed in the oxide. The defects are at lower energy levels than the barrier height introduced by the insulator and hence the defects alter the exponential dependence of the gate current on voltage across the gate oxide.

978-1-4244-5430-3/10 $26.00 © 2010 IEEE

Due to this phenomenon the I_g-V_{gs} characteristics of the device become more linear as a device degrades, which is illustrated by progressively straighter lines in Fig. 5. It is this key behavior that we use to detect the degradation of a device. We monitor this behavior for a cluster of gates to reduce monitoring overhead.

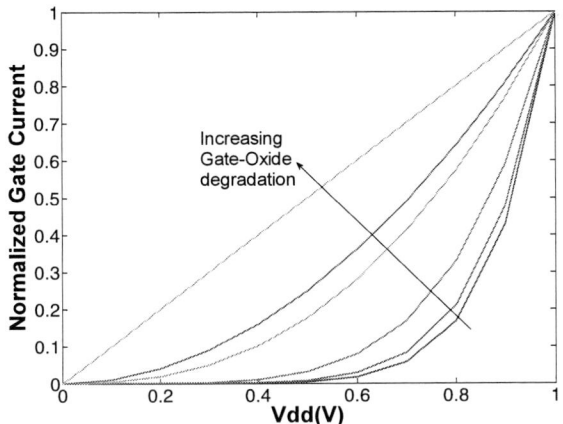

Fig. 5 The non linear nature of the Ig-Vgs characteristics of the gate-oxide becomes more linear with degradation.

As introduced in [11], the proposed *in situ* monitoring implementation leverages the prevalence of MTCMOS-based designs with PMOS header switches, a common technique to reduce standby power [12] with relatively low overhead.

Fig. 6 shows a circuit block partitioned into clusters, each connected to the power supply through a standard high-Vt MTCMOS switch and a weak PMOS device (WP) with a controllable gate voltage, Vbias.

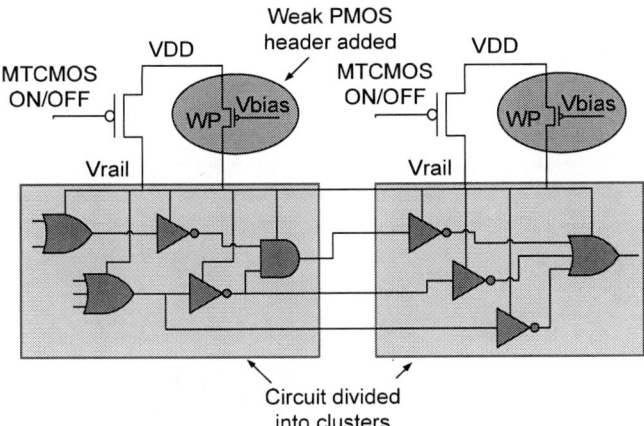

Fig. 6 The *in situ* monitoring technique is implemented by dividing a circuit into clusters using MTCMOS headers. Weak PMOS headers are added to monitor the conductance of the clusters.

The design is periodically taken offline and tested for oxide degradation by sweeping the gate voltage of WP from 0 to VDD using an on-chip DAC, while the virtual rail voltage is recorded using an on-chip ADC. The resulting Vbias vs. Vrail (V/V) curve is then analyzed to detect the onset of oxide breakdown (OBD). Initially, both oxide leakage and sub-threshold leakage are strongly non-linear with

supply voltage, resulting in a characteristic "hockey stick" curve. However, as the device degrades the V/V curve flattens (Fig. 7).

Based on this key behavior shift, we define a figure of merit called the degradation voltage angle (DVA) that measures the angle of a straight line fitted to the V/V curve over the 90 – 10% Vrail interval. As a gate degrades, the oxide displays more linear resistive behavior and a sharp drop in DVA is observed (Fig. 8).

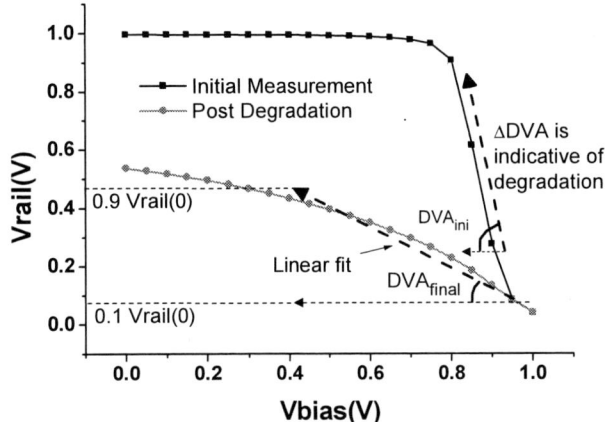

Fig. 7 The nature of Vrail vs. Vbias curve changes with degradation. DVA is defined to quantify this change in behavior.

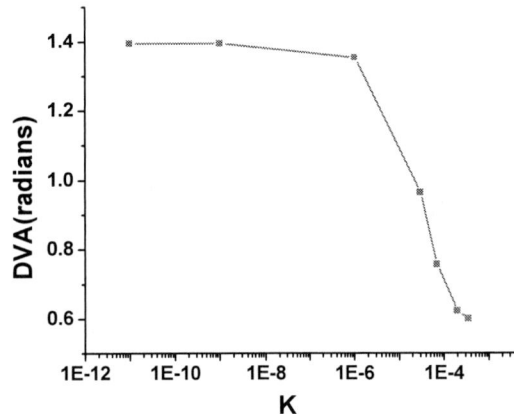

Fig. 8 DVA drops sharply with degradation which flags the onset of breakdown.

The technique was implemented in two test chips fabricated in 65nm CMOS. The nominal supply voltage in this 65nm technology is 1V. The first chip applies the technique to individual gates (INVERTER, NAND, NOR) and XOR parity trees (gate count ranging from 64-1024 gates) for a detailed study of the OBD effect. The stress voltage of 4V at a temperature of 125C is used for all the experiments on this test chip. The second chip implements a FIR filter to demonstrate applicability to larger circuit blocks. The technique can be applied to larger designs, in which case the overhead is further amortized. The stress voltage of 3V at a temperature of 165C is used for all the experiments on this test chip.

Fig. 9 shows the measured V/V, DVA, and gate delay curves for an inverter. We define 15% drop in DVA as the detection point of OBD onset. In this case the proposed technique detects the onset of gate-oxide degradation with as little as a 3% increase in the delay of an individual gate.

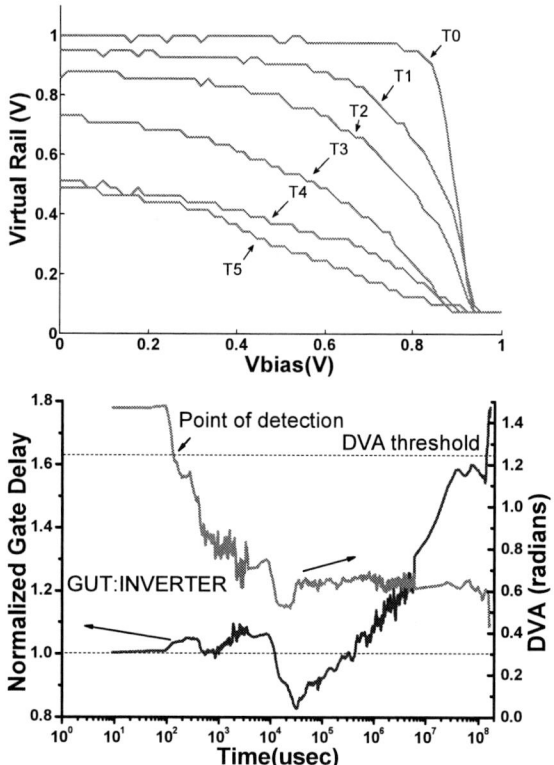

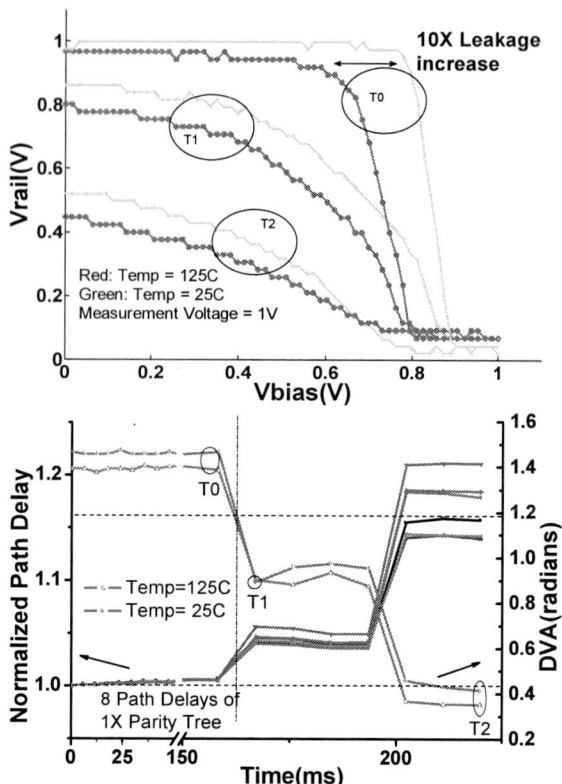

Fig. 9a (Top) Silicon measurement of Vrail vs Vbias curve for a stressed INVERTER at different points of degradation.
Fig. 9b (Bottom) Silicon measurement of DVA and Delay of a stressed INVERTER.

This illustrates that delay monitors like the one proposed in [13, 14, 15] will not be able to detect degradation until the delay is severely affected. Impact of stress on delay shows non-monotone behavior, at times resulting in faster gate delays. This is expected and is caused by the suppression of voltage swing under certain failure modes [10]. The corresponding degradation is captured by the DVA. Eventually the gate delay increases to 20X and then fails completely.

Since the subthreshold current is a strong function of temperature, the leakage of the non-failing gates can overwhelm the change in the gate leakage of the failing gate. To determine the sensitivity of the V/V curve to temperature, Fig. 10 shows the V/V, DVA, and path delay curves for an XOR tree consisting of 64 gates at 25C and 125C. There is a 10X change in subthreshold leakage for this 100C change in temperature, which changes the V/V curves significantly. However, the DVA metric changes by only 7% (less than the picked failure detection threshold of 15%), showing that DVA provides a robust measure of gate oxide degradation across temperature. The excellent robustness of DVA metric eliminates the need to calibrate or compensate for temperature, which would increase the complexity of the approach.

Fig. 10a (Top) Silicon measurement of Vrail vs Vbias curves of a 64 gate XOR parity tree at 25C and 125C at three different points in degradation showing immunity of DVA measure to environmental conditions.
Fig. 10b (Bottom) Measured DVA (at 25C and 125C) and Delay of a stressed 64 gate XOR parity tree.

Fig. 11 shows the effectiveness of this technique as the cluster size varies.

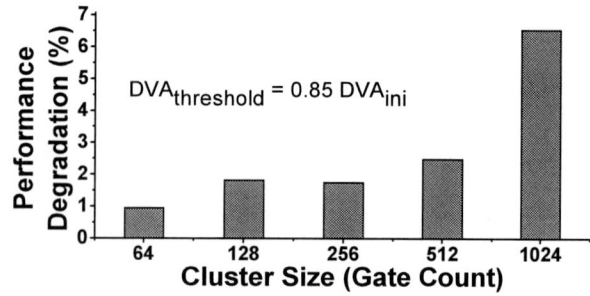

Fig. 11 Measurements show that the time to detection of onset of degradation increases with cluster sizes larger than 512 gates.

As the cluster size increases past 512 gates the failure detection is delayed since the leakage increase of the failing gate(s) is masked by the background leakage of the non-failing gates.

Fig. 12 shows a large variation in time to onset of OBD for 63 inverters, illustrating the statistical nature of oxide degradation.

978-1-4244-5430-3/10 $26.00 © 2010 IEEE   680

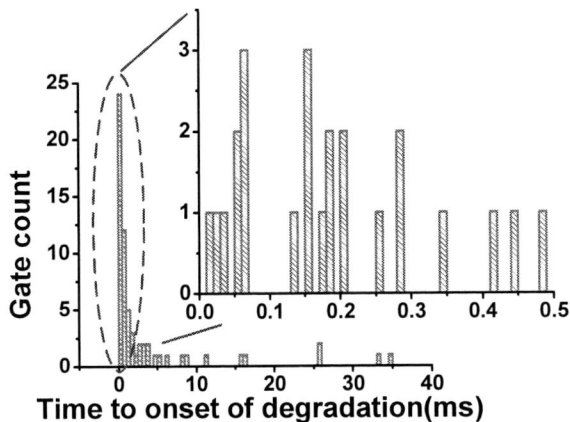

Fig. 12 Measurements show a large variation in the time to onset of gate-oxide degradation.

The second test chip applied the approach to a 16-bit, 8-tap FIR filter consisting of approximately 7K gates (Fig. 13).

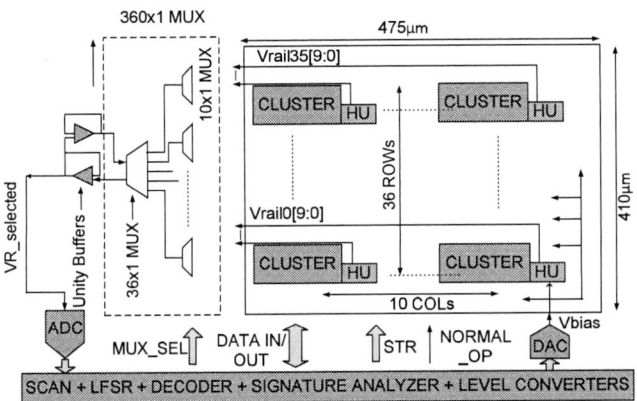

Fig. 13 *In situ* monitoring enabled FIR Filter architecture.

The FIR is divided into 360 clusters of ~20 gates placed into 36 rows and 10 columns using an automated design flow. To monitor each of the 360 virtual rails (VRs), a low leakage 360x1 two-stage mux is used. Since the VRs are driven by small leakage currents, it is extremely important to isolate the selected VR. To this end, a unity gain buffer mirrors the voltage seen on the selected VR onto the other non-selected VRs. The design area overhead of implementing this technique is 17% compared to a design without MTCMOS and 5% compared to a standard MTCMOS design. The overhead can be reduced by increasing the cluster size, which reduces the number of header devices and VRs.

Fig. 14 shows the clusters flagged for onset of degradation over time. Out of 360 clusters, 141 clusters were stressed and monitored. The first detected cluster failure corresponds to a performance degradation of the FIR by 0.5%.

All the silicon measurement results presented show that the *in situ* technique successfully monitors the degradation state of the actual devices in the chip and detects the onset of OBD very early in the chip's life. This provides the DRM controller with sufficient time to manage the chip's reliability, eliminating nearly all reliability

slack and allowing the maximum performance from the whole ensemble of the chips, irrespective of PVT conditions.

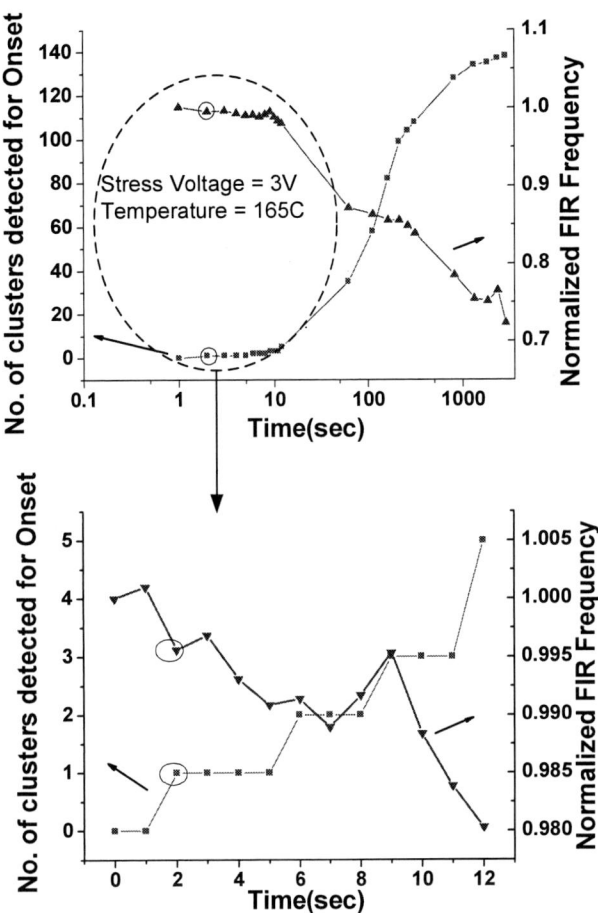

Fig. 14a (Top) The performance of the FIR degrades as clusters are detected with onset of gate-oxide degradation.
Fig. 14b (Bottom) The performance degradation is 0.5% when the first cluster is flagged for onset of degradation.

CONCLUSION

With technology scaling, the feature size of devices has reduced considerably while their supply voltage has not decreased proportionally. This yields performance gains at the cost of reduced device reliability. Gate oxide degradation is a major reliability threat to devices in advanced process nodes. To meet chip lifetime specifications, designers traditionally employ a "static reliability management" technique that does not address the sources of lifetime variations: 1) innate randomness in the degradation process, 2) process variation in the devices, 3) fluctuations in environmental conditions, and 4) circuit state. Hence, this approach tends to be very pessimistic and wastes considerable reliability slack that can be traded to recover performance losses.

Reliability-aware systems enable margin reduction by using different DRM schemes that are based on degradation models, degradation sensors, or *in situ* monitoring of the degradation of the actual core devices on the chip. Among these three methods, only the

978-1-4244-5430-3/10 $26.00 © 2010 IEEE

in situ monitoring approach addresses all the sources of lifetime variation and gives the most accurate degradation measurement.

We proposed a method for *in situ* monitoring of the gate oxide degradation of the *actual* devices of the core. The method enables early detection of the gate oxide degradation, providing sufficient time for a DRM controller to manage chip reliability while maximizing the performance gain from the whole ensemble of chips, irrespective of the sources of lifetime variations.

ACKNOWLEDGEMENTS

The authors would like to thank NSF and GSRC for funding support and ST Microelectronics for fabrication support.

REFERENCES

[1] J. H. Stathis, "Gate Oxide Reliability for Nano-Scale CMOS," in *International Conference on Microelectronics*, 2006, pp. 78-83.

[2] E. Karl, D. Blaauw, and D. Sylvester, "Analysis of System-Level Reliability Factors and Implications on Real-Time Monitoring Methods for Oxide Breakdown Device Failures," in *Proceedings of IEEE* International Symposium on Quality Electronic Design *(ISQED),*2008, pp. 391-395.

[3] Cheng Zhuo, David Blaauw, and Dennis Sylvester, "Post-Fabrication Measurement-Driven Oxide Breakdown Reliability Prediction and Management," in *Proceedings of IEEE/ACM International Conference on Computer-Aided Design,* San Jose, November 2009, pp.441-448.

[4] E. Karl, D. Blaauw, D. Sylvester, T. Mudge," Multi-Mechanism Reliability Modeling and Management in Dynamic Systems," in *IEEE Transactions On VLSI Systems*, April 2008, pp. 476-487.

[5] J. Srinivasan, S. V. Adve, P. Bose, and J. A. Rivers, "The case for lifetime reliability-aware microprocessors," in *Proceedings of 31st Annual International Symposium on Computer Architecture*, 2004, pp. 276–287.

[6] Z. Lu, W. Huang, M. R. Stan, K. Skadron, and J. Lach, "Interconnect lifetime prediction under dynamic stress for reliability-aware design," in *Proceedings of IEEE/ACM International Conference Computer-Aided Design*, 2004, pp. 327–334.

[7] R. Degraeve, et. al, "A consistent model for intrinsic breakdown in ultra-thin oxides," *International Electron Devices Meeting*, Dec. 1995, pp. 863-866.

[8] E. Karl, P. Singh, D. Blaauw, D. Sylvester, "Compact In-Situ Sensors for Monitoring Negative-Bias-Temperature-Instability Effect and Oxide Degradation," in *IEEE International Solid-State Circuits Conference (ISSCC) Dig. Tech. Papers*, 2008, pp.410-411.

[9] J. Keane, S. Venkatraman, P. Butzen, C. H. Kim, "An array-based test circuit for fully automated gate dielectric breakdown characterization," in *IEEE Custom Integrated Circuits Conference (CICC),*Sept. 2008, pp. 121-124.

[10] R. Rodriguez, J. H. Stathis, and B. P. Linder, "Modeling and experimental verification of the effect of gate oxide breakdown on CMOS inverters," in *Proceedings of IEEE International Reliability Physics Symposium*, Dallas, TX, 2003, pp. 11–16.

[11] P. Singh *et al.*," Early Detection of Oxide Breakdown Through In Situ Degradation Sensing", in *IEEE International Solid State Circuits Conf. (ISSCC) Dig. Tech. Papers*, Feb. 2010, pp. 190–191.

[12] G. Gerosa *et al.*, "A sub 1 W to 2 W low power IA processor for mobile internet devices and ultra-mobile PCs in 45 nm high-k metal gate CMOS," in *IEEE International Solid-State Circuits Conference (ISSCC) Dig. Tech. Papers*, Feb. 2008, pp. 256–257.

[13] A. Drake *et al.*, "A distributed critical-path timing monitor for a 65 nm high performance microprocessor," in *IEEE International Solid-State Circuits Conference (ISSCC) Dig. Tech. Papers*, Feb. 2007, pp. 398–399.

[14] J. A. Blome, S. Feng, S. Gupta, and S. Mahlke, "Online timing analysis for wearout detection," in *2nd Workshop on Architectural Reliability(WAR-2)*, Dec 2006.

[15] T.W. Chen, K. Kim, Y. Kim and S. Mitra, "Gate-Oxide Early Life Failure Prediction," in *IEEE VLSI Test Symposium*, 2008.

978-1-4244-5430-3/10 $26.00 © 2010 IEEE

Statistical Evaluation of Dynamic Junction Leakage Current Fluctuation Using a Simple Arrayed Capacitors Circuit

Kenichi Abe, Takafumi Fujisawa, Hiroyoshi Suzuki, Shunichi Watabe, Rihito Kuroda, Shigetoshi Sugawa
Graduate School of Engineering
Tohoku University
Sendai, Japan
Phone: +81-22-795-3977, e-mail: k-abe@fff.niche.tohoku.ac.jp

Akinobu Teramoto, [†]Tadahiro Ohmi
New Industry Creation Hatchery Center, [†]WPI Research Center
Tohoku University
Sendai, Japan

Abstract—We investigate statistical behaviors of steady-state p-n junction leakage currents at source/drain of MOSFET devices (I_{leak}s) and dynamic fluctuations of I_{leak}s using a newly developed test circuit. The test circuit can acquire the leakage currents from 28,672 n^+-p diodes in 7.7 s with 10 times averaging with the range from 0.1 fA to 23 fA. We demonstrate that two normal distributions exist in the steady-state (time averaging) I_{leak} distributions, which have different temperature dependency. A distribution of the activation energy which extracted from temperature dependence of I_{leak} is also revealed. Dynamic fluctuation of I_{leak} can be measured precisely with a simple configuration to execute pseudo parallel sampling among numerous samples for a long time. It can clarify a positive correlation between mean values of I_{leak} ($<I_{leak}>$) and amplitudes of quantum fluctuation of I_{leak} (ΔI_{leak}).

Keywords – p-n junction leakage current; random telegraph signal (RTS); MOSFET; dynamic random access memory (DRAM); retention time; test circuit;

I. INTRODUCTION

Reverse bias leakage current (I_j) of the p-n junction at drain of field emission transistor (FET) is a key characteristic to govern a retention time of dynamic random access memory (DRAM) [1], and it strongly relates to standby power consumption of a highly scaled microprocessor or a system-on-chip (SoC). The density of I_j has increased due to strengthening of electric field in the depletion layer at the drain side. It is caused by non-constant field device scaling with increase in substrate doping concentrations and introduction of new silicide materials and shallower structures, which are adopted to resolve short channel effects and to reduce source/drain and contact resistances. Nevertheless the importance of statistical evaluation of junction leakage currents has been growing, a measurement technique of junction leakage currents is still unsatisfactory on its measurement time. The measurement of a very small current as I_j needs a long time generally and the

time will become longer as the source/drain area shrinks. Therefore, a p-n junction test pattern with large area is often used to evaluate I_j even in a submicron technology. However, the I_j values measured by the large area pattern cannot reflect correctly geometric effects, such as areal, peripheral and corner leakage current components [2, 3], of an actual device size in a circuit (that is typically the smallest processible dimension of the technology). Moreover, the result is averaged out and the mean value of measured I_j has no information about extreme I_js, so that the leakage current, which is a very small but anomalous large for the actual device, cannot be detected. Some studies have pointed out that local and rare anomalous large leakage current is related to gate induced drain leakage (GIDL) and STI mechanical stress [4-6]. Furthermore, the anomalous leakage currents sometimes show dynamic fluctuation with random transition between bistable current states similar to random telegraph signal (RTS) [7-9], and this phenomenon makes it increasingly difficult to evaluate the I_j. These problems are not negligible for DRAM to determine an appropriate refresh rate which is dominated by the worst value of leakage current of the pass transistor.

In this work, we propose a new measurement scheme using an arrayed capacitor current-voltage conversion circuit, which is simple and needs only one or two interconnect layers, can be fabricated by any process technologies not only for DRAM, but also for other memories, logic and analog applications without special cares. It enables us to acquire femto ampere ordered leakage currents from more than 28,000 samples in 7.7 s with 10 times averaging. Moreover, pseudo parallel sampling among numerous p-n diodes with two simple shift registers and very long observation (up to 3,800 s) also enable us to grasp dynamic and quantum fluctuations of I_j with extremely varied time constants. We can analyze static and dynamic behaviors of I_j in FET precisely under same conditions of actual products (highly scaled junction area and dense arrangement comparable to memory applications) using this measurement technique.

978-1-4244-5430-3/10 $26.00 © 2010 IEEE

II. METHODOROGY

A. Test Circuit

Fig. 1 shows a schematic of a proposed test structure. The structure consists of two shift registers, current source transistors, a large array of unit cells including devise under test (DUT), and an output amplifier. This is based on same concepts of our previous works [10-18], which include current-voltage conversion by a capacitor and fast readout of a signal voltage corresponding to the leakage current using source follower amplifiers in each cell. A unit cell is composed of three transistors as shown in Fig. 2. The measured leakage current (I_{leak}) which can flow through the gate of M2, the source of M1, and an interconnect between M2 gate and M1 source. When the gate insulator is thick enough to the gate leakage components and the interconnect is short enough to ignore the leakage through the interlayer dielectric film, the dominant components of I_{leak} consist of the junction leakage of M1 source and the subthreshold leakage of M1. Additionally, V_{DS} can be reduced to 0 V and the subthreshold leakage is eliminated sufficiently. The measurement principle is represented by a simple equation as $I_{leak} = C_0 * dV_j/dt$, where C_0 is the capacitance connected to the source of M1 and V_j is the voltage at the source of M1. The capable range of I_{leak} is determined by the capacitance value of C_0, the sampling frequency and the voltage swing of V_j. The jitter of the sampling frequency is small enough to be negligible and what is worrisome is the dispersion of C_0. The capacitor of current-voltage conversion is expressed in the summation of the junction capacitance of M1 (C_j), the gate capacitance of M2 (C_g) and other parasitic capacitances (C_{para}). Careful design and optimization reduce C_{para} negligibly and C_g will be lowered due to Miller effect. So, it can be simplified that $C_0 \sim C_j + (1-A_V)C_g$ where A_V is the gain of M2 source follower amplifier. Figs. 3 show the procedure of the measurement and measured V_j curves as a function of time.

B. Sample Fabrication and Measurement Setup

To conform the principle of this measurement, we have designed the circuit including 28,672 unit cells which consist of n-MOSFETs (n$^+$-p diodes at source/drain, Ti silicidation, area = 3.89 μm^2, perimeter = 8.8 μm) and fabricated it by 0.35 μm logic CMOS process (Metal 1 layer, EOT = 7.4 nm, LOCOS isolation). The junction capacitance C_{sb1} is calculated and measured as 2.3 fF when V_j equals 2.0 V. It can be realized that the swing of V_j achieves 1.0 V and the sampling frequency

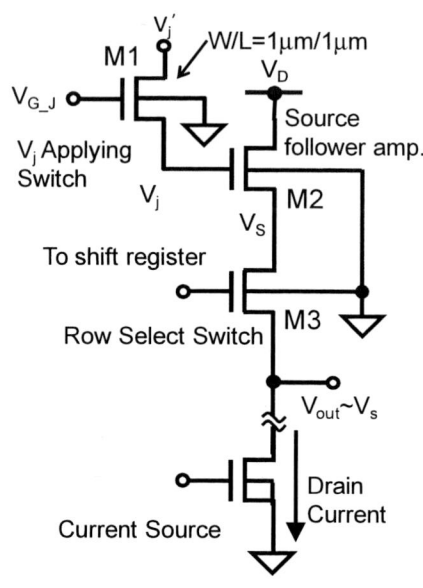

Figure 2. Schematic of the unit cell. M1 has a p-n junction to be measured at the source contact. M2 works as a voltage buffer for Vj. M3 operate as a switch to select the vertical address.

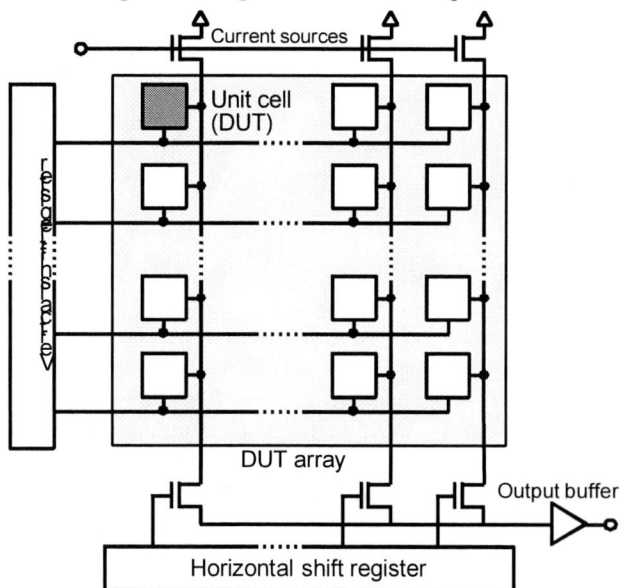

Figure 1. Schematic of the proposed test circuit. The circuit consists of a spatially-arrayed unit cells, two shift registers, column current sources and an output amplifier. The array includes 128 x 224 (=28,672) cells.

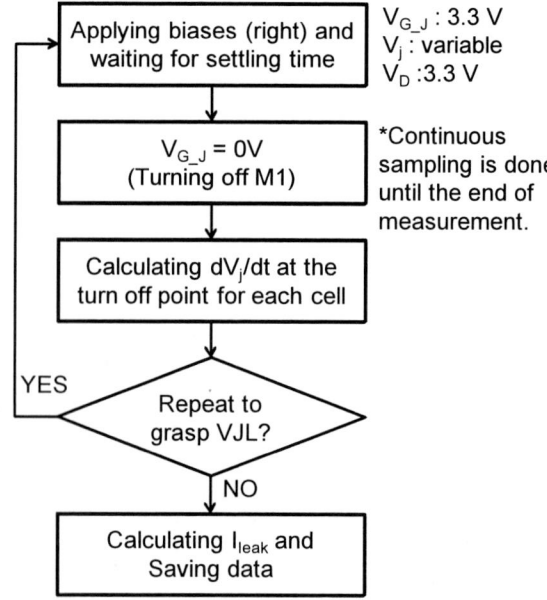

Figure 3. (a) Flow chart of the I_{leak} measurement. The test circuit is driven by pulses from FPGA and bias voltages controlled by Agilent B1500A.

978-1-4244-5430-3/10 $26.00 © 2010 IEEE

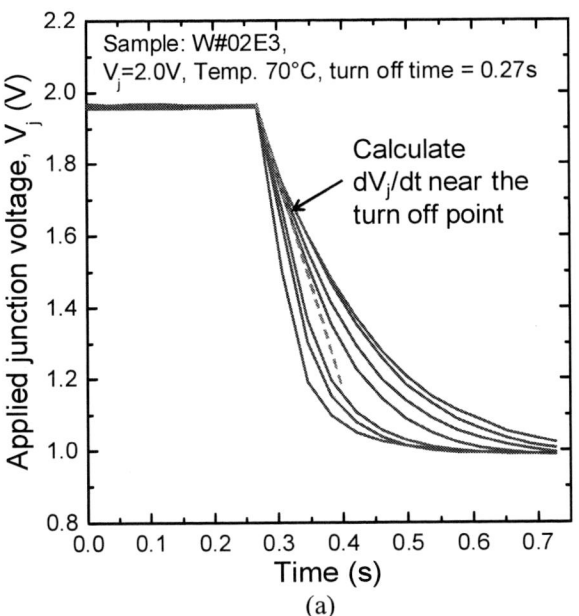
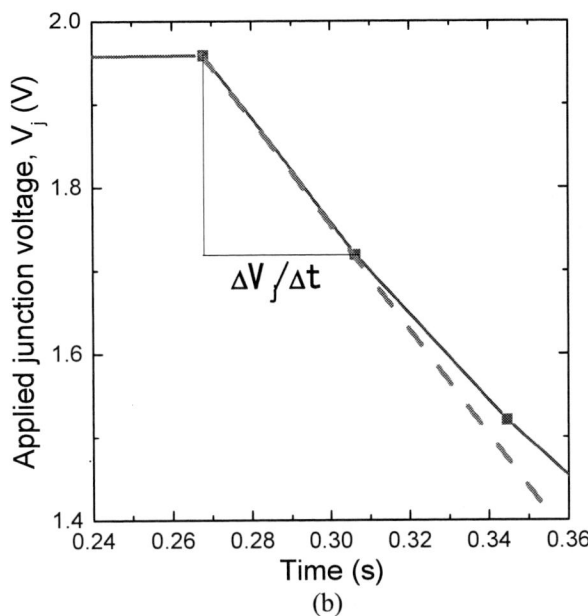

(a)   (b)

Figure 4. (a) V_j vs. time curves for typical cells including ones with maximum and minimum slope. (b) Enlarged figure of (a) at $\Delta V_j/\Delta t$ extracted point. Values of V_j decrease from the turn off point depending on I_{leak} of each sample. The nonlinearity of V_j comes from the dependency of the junction capacitance (C_{sb1}) on V_j, but it is negligible near the turn off point.

equals 1.31 Hz by employing a CMOS imager evaluation system [12]. Consequently, this enables us to measure the I_{leak} with the range from 1×10^{-16} to 2.3×10^{-14} A. We used a manual prober with a chuck stage temperature controller (MX-1105HUI, Vector Semiconductor Co., Ltd.) and B1500A semiconductor parameter analyzer (Agilent Technologies) for applying dc biases.

III. RESULTS AND DISCUSSION

Fig. 5 shows the temperature dependence of the I_{leak} distribution in the range of 10 °C to 75 °C. Measured values

more than detecting upper limit (23 fA) are counted as 23 fA. The distributions consist of two parts. The main distributions hold 95 – 99 % of total cells for each temperatures and the secondly distributions correspond to very minor parts (the exponential tail parts). However, the minor parts are a critical problem for product reliabilities and performances such as DRAM. As the temperature increases, the ratio of the main distribution decreases and the ratio of the secondly distribution drastically increases. Arrhenius plots for total cells belonging to the main distributions are shown by Fig. 6 which is depicted only by three lines with minimum, median and maximum currents at 50 °C to avoid less visibility. The distribution of

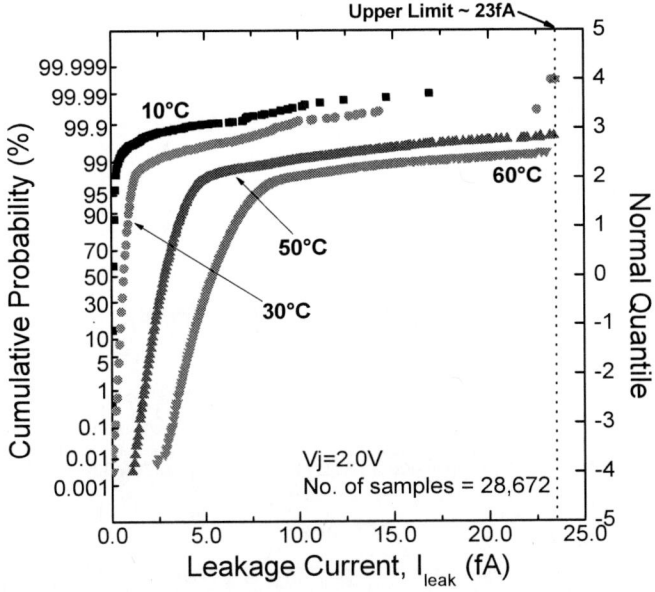

Figure 5. Temperature dependence of I_{leak} cumulative distributions. The ratio of the main distribution reduces and the second distribution becomes the dominant component as temperature increases.

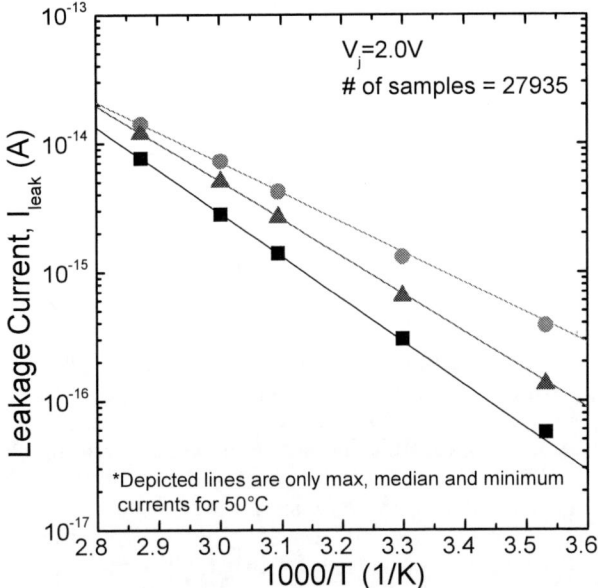

Figure 6. Arrhenius plots of I_{leak} (max, median, min). The measurement temperature range is from 10 °C to 75 °C.

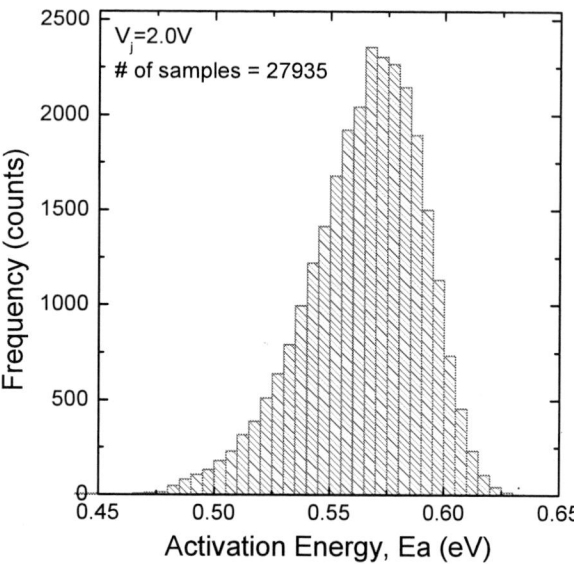

Figure 7. The distribution of activation energy Ea extracted by Fig. 6. The measurement temperature range is from 10°C to 75 °C. The activation energies are distributed around a half of bandgap energy of Si (0.56 eV).

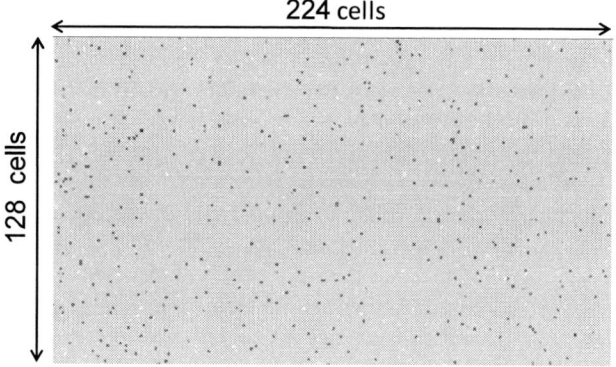

Figure 8. I_{leak} color map in the spatial array. Green dots indicate small I_{leak} cells and red dots indicate large I_{leak} cells. Large I_{leak} cell distributes randomly in the arrayed pattern.

the activation energies (E_a) extracted from Fig. 6 is shown by Fig. 7. The values of E_a are distributed around a half of bandgap energy of Si at room temperature (0.56 eV). As a result, the p-n junction leakages of most cells are dominated even by generation-recombination current for this sample in the temperature range of 10 °C to 75 °C. Fig. 8 shows the spatial distribution of I_j by colored map. Large I_j cells distribute randomly in the array and the two divided normal distributions do not comes from the systematic variation of the fabrication process.

Variable junction leakage (VJL) or variable retention time (VRT), which is discrete fluctuation of I_{leak} similar to random telegraph signal (RTS) has been reported [8, 19]. Figs. 9(a)-(c) indicate VJL waveforms measured by pseudo parallel long sampling [12]. In the literature, time constants of VJL which are characterized as transition intervals are spread to more than hundreds of seconds [8, 20]. However, it is clarified that VJLs with short time constants less than 1 s remaining large amplitude exists infrequently through our analysis. Fig. 10 shows a peculiar VJL waveform with three states. Possible origins for this phenomenon are that two individual defects

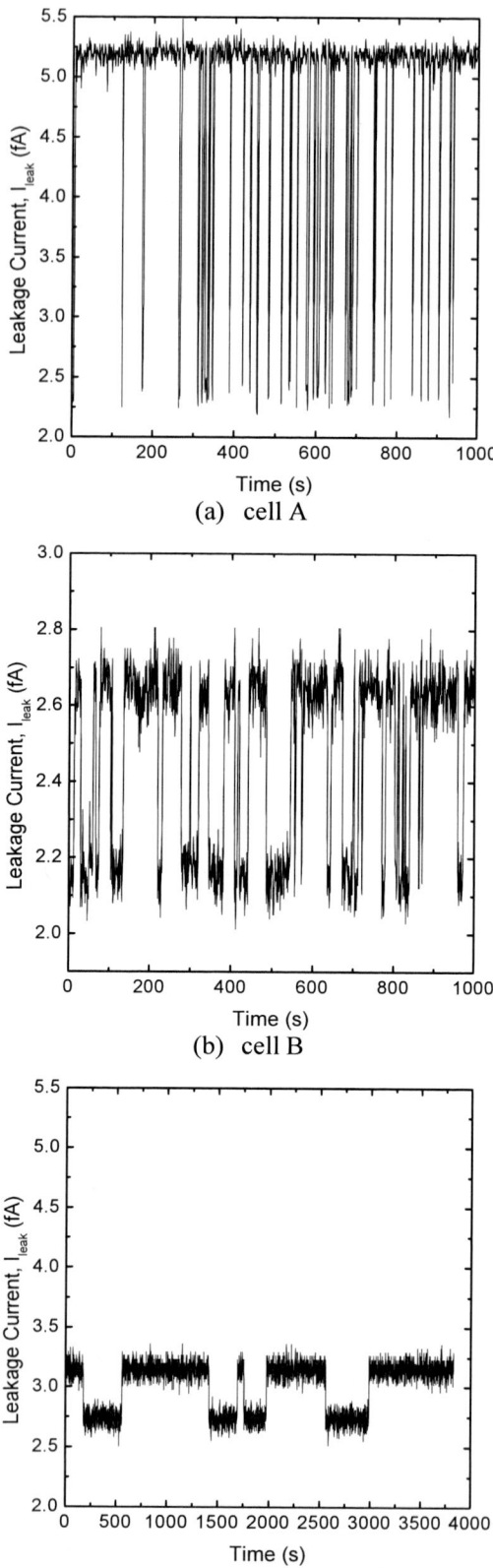

(a) cell A

(b) cell B

(c) cell C

Figure 9. Typical waveforms for detected variable junction leakages (VJLs) with various time constants (mean transition interval) and amplitudes. Time constants spread widely and multi level VJL has been observed. The measurement conditions: V_j = 2.0V, Temperature = 50°C, Sampling frequency = 1.31Hz.

interact with each other or one individual trap which is related to divacancy-oxygen defect [19, 21, 22] has three stable states, but further investigations need to be examined. Fig. 11 indicates the temperature dependence of VJL for a particular cell. The amplitude of VJL (ΔI_{leak}) which is defined as peak-to-peak value within the sampling span and the transition rate increases as the temperature increases. This thermal activated behavior agrees with a work regarding VRT in DRAM [20]. Fig. 12 shows a scatter plot between mean values of I_{leak} ($<I_{leak}>$) and ΔI_{leak}. There is a weak positive correlation between $<I_{leak}>$ and ΔI_{leak}. This shows that it is easier for the cell with many defects to cause VJL with large ΔI_j. It is suggested that the defects increasing $<I_{leak}>$ and the defects causing VJL behave in the same manner fundamentally. Fig. 13 shows the V_j dependence of σI_{leak} distributions. We define σI_{leak} as the standard deviation of I_{leak} within 5000 sampling points. The tail part corresponds to the cells regarding large VJL and its V_j dependency is not really great. On the other hand, Fig. 14 indicates the temperature dependence of σI_{leak} distributions. In contrast, the tail part increases drastically as the temperature increases. Therefore, the statistical behavior of VJL is strongly affected by the temperature rather than the reverse bias voltage.

IV. CONCLUSION

We demonstrated that the statistical evaluation of 28,672 n^+-p diodes by the newly proposed method, which allows us to measure small junction leakage currents for numerous samples in a short time with the simple circuit to be fabricated by any process technologies such as memory, logic and analog applications. It shows that two normal distributions exist in the mean I_{leak} distributions, which have different temperature dependency. The temperature dependency of I_j is statistically investigated and the distribution of the activation energy is successfully extracted. And also, VJL can be measured precisely with a simple coordination and we reveals that there are the typical and peculiar VJLs and the positive correlation between $<I_{leak}>$ and ΔI_{leak} exists. Because of its simplicity, this test structure is very useful for the evaluation of p-n junction leakage not only for the transfer gate of DRAM cell but also any other circuits.

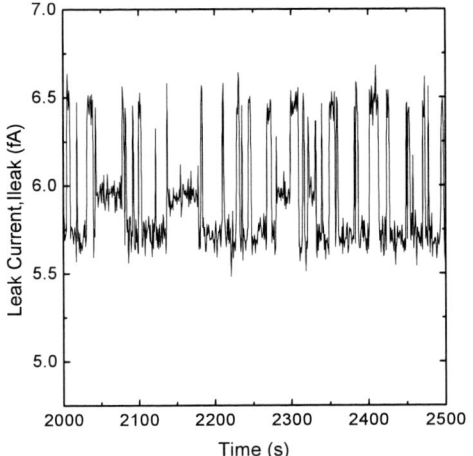

Figure 10. A peculiar VJL with three stable states. The measurement conditions: $V_j = 2.0V$, Temperature = 60°C, Sampling frequency = 1.31Hz.

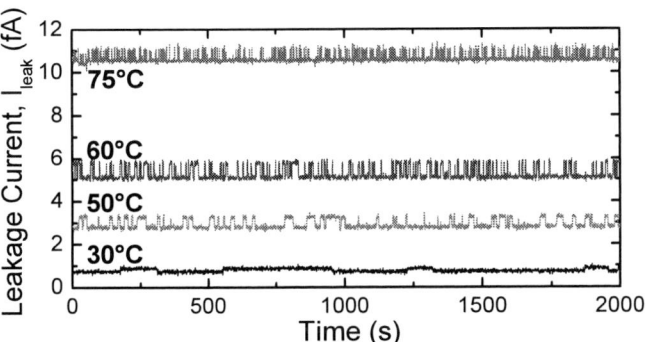

Figure 11. Temperature dependence of VJL. The measurement conditions: $V_j = 2.0V$, Sampling frequency = 1.31Hz.

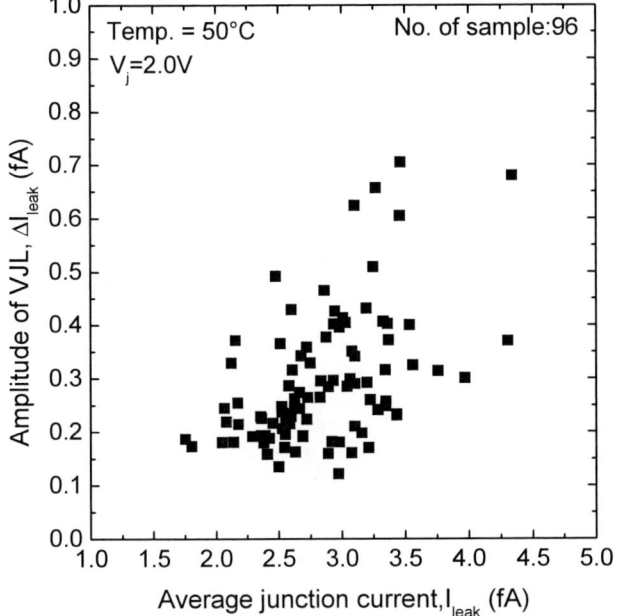

Figure 12. Correlation diagram between mean values of I_{leak} ($<I_{leak}>$) and amplitude of VJL (ΔI_{leak}). The positive correlation appears.

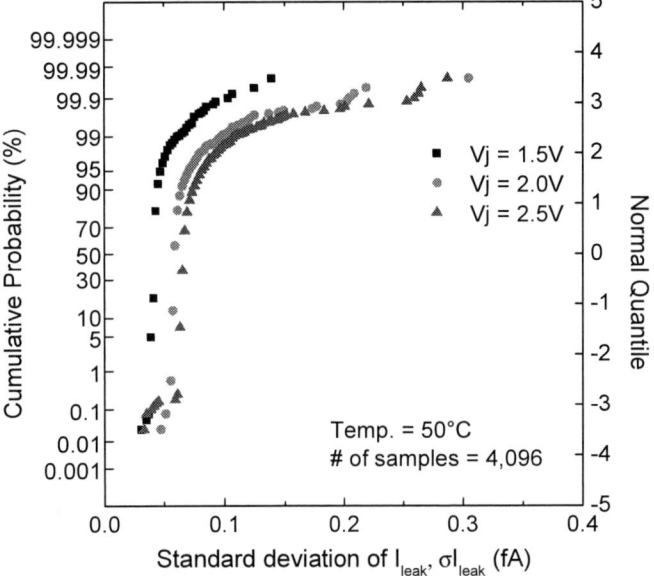

Figure 13. V_j dependence of σI_{leak} distributions. Total noise does not increase so much with increase in V_j.

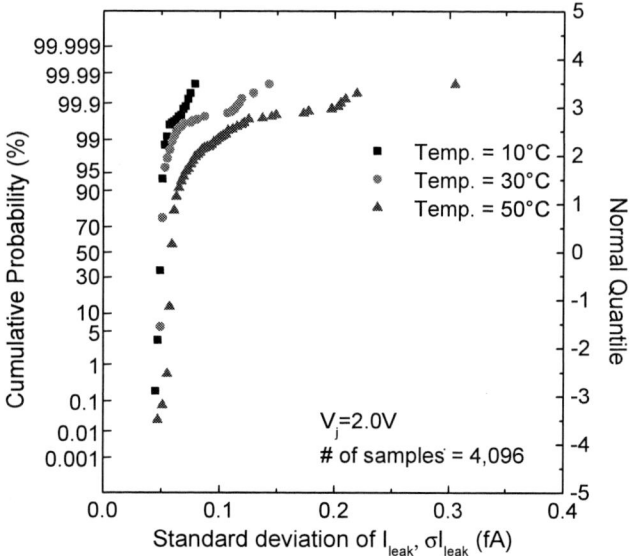

Figure 14. Temperature dependence of σI_{leak} distributions. The σI_{leak} increases drastically with increase in temperature.

ACKNOWLEDGMENT

The authors would like to thank Asahi Kasei Microdevices Co. Ltd. and Dr. M. Toita for the sample fabrication and technical supports.

REFERENCES

[1] T. Hamamoto, S. Sugiura, and S. Sawada, "On the retention time distribution of dynamic random access memory (DRAM)," IEEE Trans. Electron Devices, vol. 45, pp. 1300-1309, 1998.

[2] H.-D. Lee and J.-M. Hwang, "Accurate extraction of reverse leakage current components of shallow silicided p+-n junction for quarter- and sub-quarter-micron MOSFET's," IEEE Trans. Electron Devices, vol. 45, pp. 1848-1850, 1998.

[3] A. Poyai, I. Rittaporn, E. Simoen, C. Claeys, and R. Rooyackers, "Shallow trench isolation dimensions effects on leakage current and doping concentration of advanced p-n junction diodes," Materials Science and Engineering B, vol. 114-115, pp. 372-375, 2004.

[4] T. Y. Chan, J. Chen, P. K. Ko, and C. Hu, "The impact of gate-induced drain leakage current on MOSFET scaling," IEDM Tech. Dig., pp. 718-721, 1987.

[5] K. Saino, S. Horiba, S. Uchiyama, Y. Takaishi, M. Takenaka, T. Uchida, Y. Takada, K. Koyama, H. Miyake, and C. Hu, "Impact of gate-induced drain leakage current on the tail distribution of DRAM data retention time," IEDM Tech. Dig., pp. 837-840, 2000.

[6] H. Daewon, C. Changhyun, S. Dongwon, K. Gwan-Hyeob, C. Tae-Young, and K. Kinam, "Anomalous junction leakage current induced by

STI dislocations and its impact on dynamic random access memory devices," IEEE Trans. Electron Devices, vol. 46, pp. 940-946, 1999.

[7] D. S. Yaney, C. Y. Lu, R. A. Kohler, M. J. Kelly, and J. T. Nelson, "A meta-stable leakage phenomenon in DRAM charge storage - Variable hold time," IEDM Tech. Dig., pp. 336-339, 1987.

[8] Y. Mori, K. Ohyu, K. Okonogi, and R. Yamada, "The origin of variable retention time in DRAM," IEDM Tech. Dig., pp. 1034-1037, 2005.

[9] H. Kim, K. Kim, T.-K. Oh, S.-Y. Cha, S.-J. Hong, S.-W. Park, and H. Shin, "RTS-like Fluctuation in Gate Induced Drain Leakage Current of Saddle-Fin Type DRAM Cell Transisor," IEDM Tech. Dig., pp. 271-274, 2009.

[10] S. Watabe, S. Sugawa, A. Teramoto, and T. Ohmi, "New statistical evaluation method for the variation of metal-oxide-semiconductor field-effect transistors," Jpn. J. Appl. Phys., vol. 46, pp. 2054-2057, Apr 2007.

[11] Y. Kumagai, A. Teramoto, S. Sugawa, T. Suwa, and T. Ohmi, "Statistical evaluation for anomalous SILC of tunnel oxide using integrated array TEG," Proc. Int. Reliability Physics Symp., pp. 219-224, 2008.

[12] K. Abe, S. Sugawa, S. Watabe, N. Miyamoto, A. Teramoto, Y. Kamata, K. Shibusawa, M. Toita, and T. Ohmi, "Random Telegraph Signal Statistical Analysis using a Very Large-scale Array TEG with 1M MOSFETs," Symp. VLSI Technology Dig., pp. 210-211, 2007.

[13] S. Watabe, S. Sugawa, K. Abe, T. Fujisawa, N. Miyamoto, A. Teramoto, and T. Ohmi, "A Test Structure for Statistical Evaluation of Characteristics Variability in a Very Large Number of MOSFETs," ICMTS, pp. 114-118, 2009.

[14] T. Fujisawa, K. Abe, S. Watabe, N. Miyamoto, A. Teramoto, S. Sugawa, and T. Ohmi, "Accurate Time Constant of Random Telegraph Signal Extracted by a Sufficient Long Time Measurement in Very Large-Scale Array TEG," ICMTS, pp. 19-24, 2009.

[15] T. Fujisawa, K. Abe, S. Watabe, N. Miyamoto, A. Teramoto, S. Sugawa, and T. Ohmi, "Statistical Analysis of Time Constant Ratio of Random Telegraph Signal with Very large-Scale Array TEG," SSDM, pp. 28-29, 2009.

[16] K. Abe, A. Teramoto, S. Watabe, T. Fujisawa, S. Sugawa, Y. Kamata, K. Shibusawa, and T. Ohmi, "Impact of Channel Doping Concentration on Random Telegraph Signal Noise," SSDM, pp. 30-31, 2009.

[17] A. Teramoto, Y. Kumagai, K. Abe, T. Fujisawa, S. Watabe, T. Suwa, N. Miyamoto, S. Sugawa, and T. Ohmi, "Stress-induced leakage current and random telegraph signal," Journal of Vacuum Science & Technology B, vol. 27, pp. 435-438, Jan-Feb 2009.

[18] K. Abe, Y. Kumagai, S. Sugawa, S. Watabe, T. Fujisawa, A. Teramoto, and T. Ohmi, "Asymmetry of RTS characteristics along source-drain direction and statistical analysis of process-induced RTS," 2009 IEEE International Reliability Physics Symposium, pp. 996-1001, 2009.

[19] K. Ohyu, T. Umeda, K. Okonogi, S. Tsukada, M. Hidaka, S. Fujieda, and Y. Mochizuki, "Quantitative identification for the physical origin of variable retention time: A vacancy-oxygen complex defect model," IEDM Tech. Dig., pp. 389-392, 2006.

[20] P. J. Restle, J. W. Park, and B. F. Lloyd, "DRAM variable retention time," IEDM Tech. Dig., pp. 807-810, 1992.

[21] T. Umeda, K. Okonogi, K. Ohyu, S. Tsukada, K. Hamada, S. Fujieda, and Y. Mochizuki, "Single silicon vacancy-oxygen complex defect and variable retention time phenomenon in dynamic random access memories," Appl. Phys. Lett., vol. 88, p. 253504, 2006.

[22] T. Umeda, Y. Mochizuki, K. Okonogi, and K. Hamada, "Electrically detected magnetic resonance of ion-implantation damage centers in silicon large-scale integrated circuits," J. Appl. Phys., vol. 94, pp. 7105-7111, 2003.

Scaling Reliability and Modeling of Ferroelectric Capacitors

Antonio G. Acosta
Department of Electrical and Computer Engineering
University of Florida
Gainesville, FL 32601-6211, USA
1-352-392-7657, agacosta@ufl.edu

John Rodriguez, Borna Obradovic[*], Scott Summerfelt, Tamer San, Keith Green[*], Ted Moise, and Srikanth Krishnan
Analog Technology Development
[*]Analog Technology Development SPICE Modeling Lab
Texas Instruments Incorporated
Dallas, Texas, 75265, USA

Abstract—We report on reliability properties of MOCVD PZT ferroelectric capacitors as a function of film thickness. Data is presented for fatigue, thermal depolarization, and imprint. It is important to be able to model these parameters as they can significantly affect the switching polarization, which in turn affects the signal margin of an FRAM circuit. A ferroelectric SPICE model is presented that can be used to accurately simulate hysteresis and switching polarization behavior. This model agrees with experimental data and can be used to simulate FRAM circuit behavior through "end of life".

Keywords-FRAM; SPICE model; reliability; ferroelectric capacitor;

Embedded ferroelectric random access memory (FRAM) has become a viable non-volatile memory (NVM) for applications requiring low power consumption, fast write times, and high cycling endurance [1-3]. Over the past decade the technology has progressed to the point that high density ferroelectric memory is currently in production at the 130-nm node [4]. Further scaling of the technology can enable an increase in storage capacity leading to new applications for FRAM [5]. However, there are several challenges that must first be overcome. The success of FRAM as a commercial non-volatile memory has been largely due to its integration into complementary metal-oxide-semiconductor (CMOS) technology. Therefore, FRAM must be able to scale along with CMOS technology if it is to continue as a practical NVM. One challenge is to maintain the reliability properties of the ferroelectric capacitor as the PZT thickness is scaled. These include the polarization loss due to read/write cycling, or fatigue, and data retention [6]. Data retention can be significantly affected by thermal depolarization (TD). This effect causes the polarization in a ferroelectric capacitor to decrease as the ambient temperature increases toward the Curie point [6].

Also, as FRAM technology advances further, there is a need to design FRAM memory circuits that achieve the required performance and target reliability specifications. Thus, an additional challenge lies in the development of accurate models that can be used to simulate the hysteresis and switching properties of ferroelectric capacitors, including the effects of relaxation, imprint and thermal depolarization.

In this work we report on FRAM capacitor reliability properties as a function of thickness scaling. Experimental data is presented which shows the effects of film thickness, fatigue, thermal depolarization, and imprint on ferroelectric capacitor behavior. These results indicate a need for an accurate model that can be used to simulate FRAM signal margin. A ferroelectric SPICE model based on a modified Preisach model (for hysteresis effect) is also presented, including a procedure for modeling the FRAM circuit reliability. Simulation results using our model closely resemble those obtained experimentally.

I. EXPERIMENTAL DETAILS

A. Device Fabrication

The ferroelectric capacitors studied were processed with PZT thicknesses of 50-nm, 60-nm, 70-nm, and 90-nm. The 70-nm and 90-nm thick capacitors were fully integrated through a CMOS process with a total capacitor area of $1979\mu m^2$ and $3471\mu m^2$, respectively. The PZT films studied in this paper were deposited by MOCVD on Ir bottom electrodes with Ir/Ir-oxide top electrodes. Additional fabrication and device details are described in [1].

B. Experiment Set-up

1) Switched and non-switched polarization: A modified Sawyer-Tower circuit [7], with the load capacitor replaced by a 50-ohm resistor, was used to measure the switched and non-switched polarization (Pns), using P-U-N-D pulses as described in [6]. An HP 8110A Pulse Generator was used to apply voltage pulses of 1.65V and 2µs pulse width to the ferroelectric capacitor. The subsequent voltage on the load resistor was measured with a Tektronix TDS 540C Oscilloscope and the load current was calculated. These load

Texas Instruments, Inc.

978-1-4244-5430-3/10 $26.00 © 2010 IEEE

currents were then integrated over time to calculate the polarization charge density. For testing the switching polarization as a function of temperature, the devices were brought to the appropriate temperature using a thermal chuck. This chuck was capable of cooling the wafer down to -40°C and heating up to 125°C.

2) Cycling endurance: For fatigue measurements, the capacitor was cycled at +/- 1.8V for the 90-nm films and +/- 1.5V for the 50- and 60-nm films at 500kHz and 1200ns pulse widths. The total number of cycles was 1×10^{10}. Switched polarization (Psw) was measured incrementally throughout the cycling process using the aforementioned P-U-N-D method at 1.5V for all capacitors.

3) Data Retention: Data retention can be best described by three effects: polarization relaxation, imprint, and thermal depolarization, which have been previously discussed. Experiments were conducted that included these three effects. Polarization relaxation refers to a reversible decrease in polarization shortly after the capacitor has been polarized. Imprint is the preference of the ferroelectric capacitor to stabilize in one of the two remanent polarization states. In these experiments the polarization was measured using the same experimental set-up described above; however the pulse sequence differed from the P-U-N-D method. Fig. 1 shows a sample plot of the pulse and load voltage waveforms that were used to calculate polarization. Polarization charge density was calculated by integrating the magnitude of the current through the load resistor, both for on-pulse (P) and after-pulse (Pa) terms.

A detailed description of the test sequence used to measure all of these effects is shown in Fig. 2. Parameters are collected for both time-zero (initial Psw properties, pre-bake) and through the lifetime (post-bake) of the device. The same-state and opposite-state [6] polarization was measured at time-zero, and after bakes of 12 and 100 hours at 175°C, or the

equivalent of 10 years at 85°C and 105°C. In order to capture the imprint and depolarization effects using the very first read pulse, different capacitors were used for each read voltage at each polarity.

II. RESULTS AND DISCUSSION

A. Experimental Measurements

The switched polarization of the 70-nm and 90-nm films as a function of read voltage is shown in Fig. 3. The polarization is read approximately 1s after it is set to allow for relaxation. The switched polarization is invariant in this range of PZT film thickness, indicating residual stresses and interfacial layers between the ferroelectric and electrodes as well as film processing play a minor role. Since the coercive field remains constant, the voltage at which the polarization saturates scales accordingly with the thickness. The effects of thermal depolarization [6] can be seen in Fig. 4 where the switching properties of the 70nm films are shown as a function of temperature. The polarization decreases gradually as the ambient temperature increases towards the Curie point (~430°C), where the films undergo a phase transition from the ferroelectric tetragonal state to paraelectric cubic state. The switchable polarization decreases while the non-switching polarization increases. In a similar

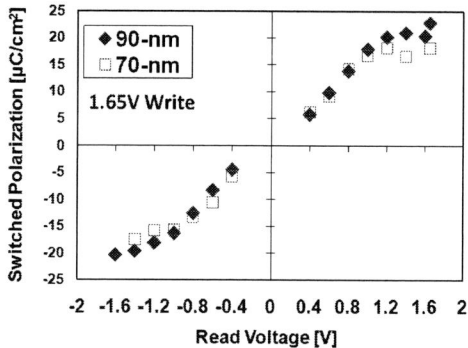

Fig. 3. Switched polarization comparison of ferroelectric capacitor with 90-nm and 70-nm thick PZT film at room temperature.

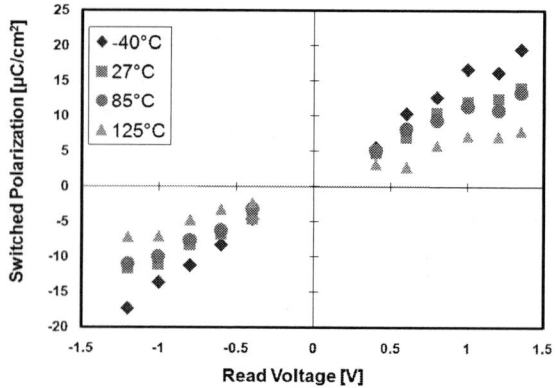

Fig. 4. Switched polarization, Psw, as a function of temperature for 70-nm PZT film capacitor.

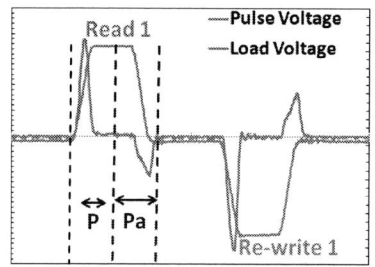

Fig. 1. Sample read/re-write data 1 waveforms with >1s delay between pulses. Bake step may be included prior to data read. Re-write pulse simulates FRAM circuit operation.

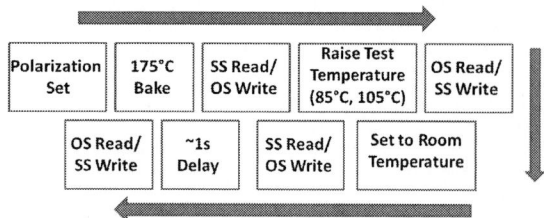

Fig. 2. Test sequence for imprint data measurement.

manner to Psw, the coercive field (Ec), which represents the electric field at which maximum polarization switching occurs, also decreases with temperature. This reduction of Ec lowers the voltage at which the Psw saturates; this can be seen in Fig. 2 where the Psw saturation voltage varies from ~1.2V at -40°C to less than 1V at 125°C.

One of the major challenges of FRAM reliability is polarization loss due to read/re-write cycling, known as fatigue [8]. Fig. 5 shows the normalized switched polarization as a function of cycles for different PZT film thicknesses. The trend is very similar for the 50-nm and 60-nm films while the 90-nm capacitor is slightly different. In all cases, the polarization loss is <10% after 1×10^{10} cycles for all three films, demonstrating that cycling reliability is not negatively impacted by scaling the film thickness to 50-nm. Cycling as a function of temperature is shown in Fig. 6 for the 90-nm film. The normalized cycling data is very similar at 25°C and 85°C and is just slightly worse at 125°C indicating that temperature does not strongly accelerate the read/write cycling fatigue for these films.

B. Modeling and Simulation

An important reliability parameter for FRAM is data retention. Ferroelectric capacitors are prone to polarization margin loss after exposure to high temperatures for long periods of time [9]. The bake effect causes an imprint shift in the hysteresis curve along the voltage axis which in turn reduces the Psw magnitude as a function of read voltage [6]. These bake effects can potentially be impacted by the PZT thickness [10]. Since the Psw is directly related to the circuit signal margin it is critically important to develop models that

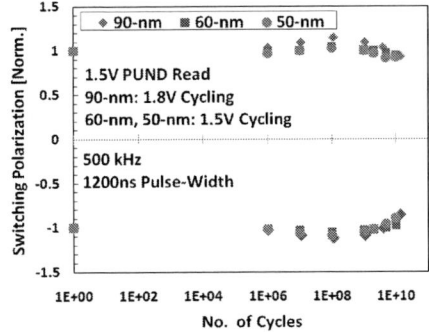

Fig. 5. Fatigue of ferroelectric capacitors at different PZT thickness up to 1×10^{10} cycles. Polarization loss is similar at different film thickness.

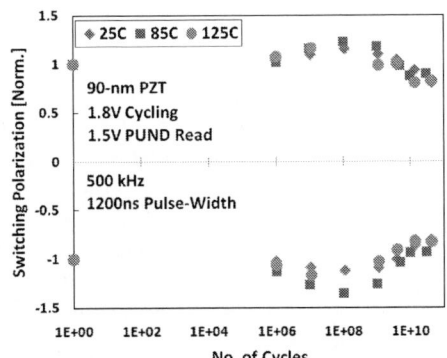

Fig. 6. Fatigue properties as a function of temperature. Temperature does not strongly accelerate fatigue for the different PZT films.

can accurately simulate the polarization as a function of read voltage through equivalent "life use" bake conditions.

To this end, we present a new SPICE model for FRAM circuit simulation that can be used to simulate ferroelectric polarization switching pre/post imprint stress. The model represents the ferroelectric capacitor as a collection of near-ideal Preisach domains [11], each with negative and positive switching thresholds (V_a and V_b). The behavior of the model is determined by calibrating the domain distribution parameters to the measured polarization data. The joint probability distribution function of the V_a and V_b thresholds is modeled using a 2-variable correlated Pearson IV function. The model parameters are the peak position, sigma, skew, and kurtosis (for V_a and V_b), as well as the correlation coefficient. In addition to the distribution parameters, non-ferroelectric capacitance and relaxation fraction are also extracted. A modified version of the relaxation model proposed by Kuhn et al. [12] is used. A bounded Levenberg-Marquardt algorithm is used to perform the fit [13].

The calibration is performed separately for the nominal (time-zero) and imprinted (positive and negative bake) capacitors. A comparison of the calibrated model to the measured data is illustrated in Figs. 7-9. Psw and Pns refer to the switching and

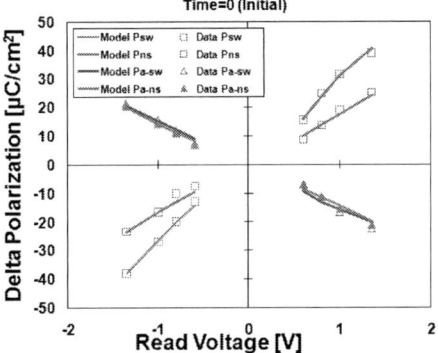

Fig. 7. Calibration of domain parameters to measured data for nominal ferroelectric capacitors.

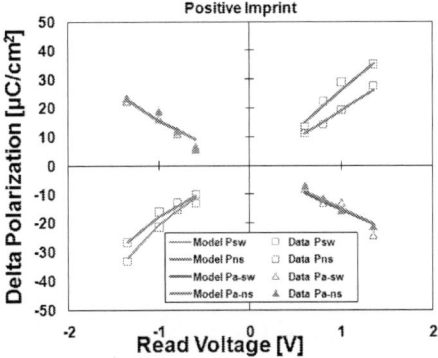

Fig. 8. Calibration of domain parameters to measured data for positive imprint.

non-switching on-pulse polarization, respectively. Pa-sw and Pa-ns correspond to the after-pulse terms for switched and non-switched polarization measurements, respectively. Both the data and the model indicate ~10% switched polarization reduction for the 10 year, 85°C equivalent bake. The non-switched polarization remains constant. Both of these trends are consistent with the measured results reported previously in an FRAM device [14]. The good match between the model, the capacitor bake data and the FRAM device gives confidence the calibrated SPICE model can be used to accurately simulate the imprint effects during FRAM circuit design. Using data from the calibrated model the corresponding Psw vs. read voltage curves can then be constructed using SPICE simulations. As shown in Fig. 10 the switching polarization was simulated for various operating temperatures. The data very closely matches the experimental results shown in Fig. 4, validating the accuracy of our model across wide voltage and temperature range.

The model has been used to simulate the ferroelectric switching properties using a capacitive voltage divider circuit similar to that shown in Fig. 11. This is done for two complementary states (BL and BLB) of polarization in a bit-cell, as in a 2T-2C cell design, where the data signal and its complement are input to a sense amplifier to detect the data state. The simulation results for a Data 1 BL (Data 0 BLB) before and after bake are shown in Fig. 12. As the plate line voltage increases to 1.5V, the ferroelectric capacitors

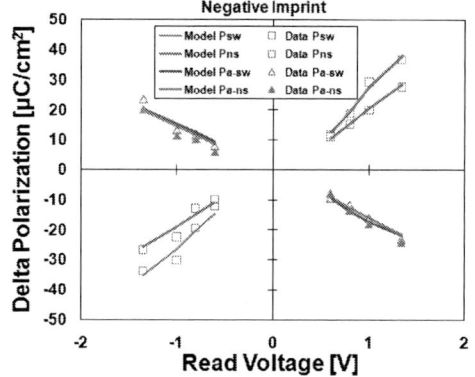

Fig. 9. Calibration of domain parameters to measured data for negative imprint.

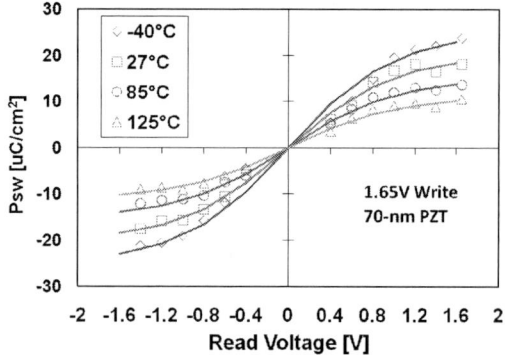

Fig. 10. SPICE simulation of switching polarization at different operating temperatures. The simulation data (solid lines) very closely matches experimental results (open symbols) from Figure 4.

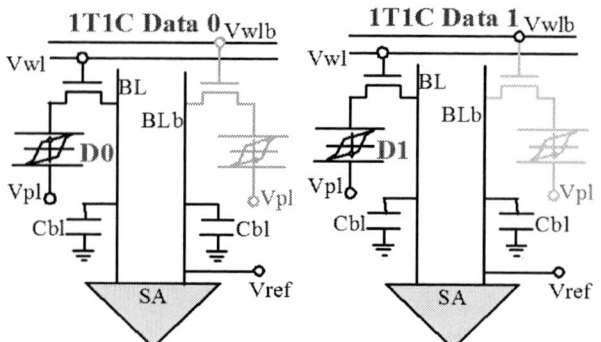

Fig. 11. FRAM circuit used to simulate ferroelectric capacitor behavior.

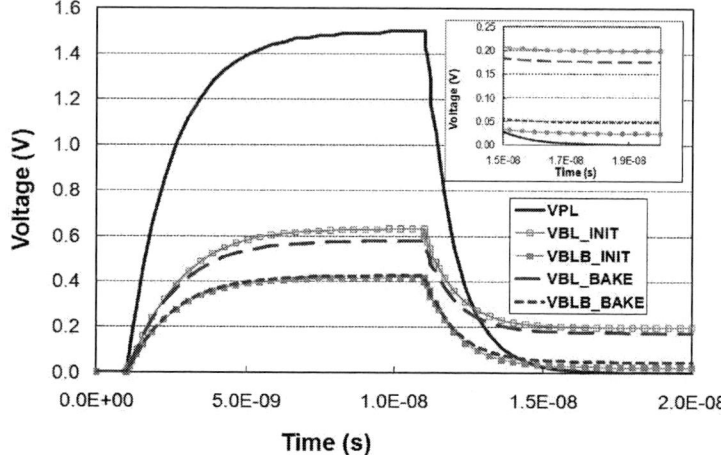

Fig. 12. Simulated voltage polarization waveforms of ferroelectric test circuit. The insi is a zoomed-in view of the waveforms showing after-pulse behavior. Data 0 wi relaxation clearly results in non-zero VBL.

charge up the BL and BLB load capacitances. The net charge remaining on the load capacitances after the plate line returns to ground creates the voltage signals which can then be detected by a sense amplifier. As shown in Fig. 12, the primary impact of the 10 year equivalent 85°C bake is to reduce the BL voltage by 10% (0.2V initial vs. 0.18V after bake). This demonstrates the model can be used to accurately simulate ferroelectric capacitor behavior in a circuit, including the effects of imprint, which is of fundamental importance for designing reliable FRAM circuits. The SPICE model is useful for assessing the signal margins

III. SUMMARY AND CONCLUSIONS

We have shown that the electrical and reliability properties of MOCVD PZT ferroelectric capacitors are not significantly altered by process technology and thickness scaling. We also presented a new SPICE model that can be used to accurately simulate hysteresis and switching properties as a function of voltage, including imprint effects resulting from high temperature bakes. The model shows very good agreement with the measured data and can be used to simulate circuit

978-1-4244-5430-3/10 $26.00 © 2010 IEEE

behavior at time-zero and through the expected life use conditions, enabling accurate prediction of FRAM performance as the technology is adopted in new applications.

REFERENCES

[1] T. Moise, *et al.*, "Demonstration of a 4Mb, high density ferroelectric memory embedded within a 130nm, 5LM Cu/FSG logic process," in *Proc. IEEE IEDM,* pp. 535-538, 2002.

[2] S. Aggarwal, *et al.*, "MOCVD Pb(Zr, Ti)O$_3$ thin films for ferroelectric memories," *Int. Symp. on Integrated Ferroelectrics Conference Presentation,* 2003.

[3] K.R. Udayakumar, *et a.l,* "High density embedded ferroelectric memories," in *Fall MRS Meeting Presentation,* 2002.

[4] www.ramtron.com

[5] K.Kim and S.Y. Lee, "Innovation in 1T1C FRAM technologies for ultra high reliable mega density FRAM and future high density FRAM," *Integrated Ferroelectrics,* vol. 81, pp. 77-88, 2006.

[6] J. Rodriguez, *et a.l.* "Reliability properties of low-voltage ferroelectric capacitors and memory arrays," *IEEE Trans. Dev. Mat. Rel.,* pp. 436-449, 2004.

[7] C.B. Sawyer and C.H. Tower, "Rochelle salts as a dielectric," *Phys. Rev.,*vol.35, p. 269, 1930.

[8] J. Rodriguez, *et al.*, "Empirical model for fatigue of PZT ferroelectric memories," *Proc. Rel. Phys. Symp.,* pp. 39-44, 2002.

[9] S. Traynor, "Polarization as a driving force for imprint," in *Proc. Int. Symp. Integrated Ferroelectrics,* 1998 .

[10] S.Y. Lee and K. Kim, "Future 1T1C FRAM technologies for highly reliable, high density FRAM," *Proc. IEEE IEDM,* pp. 547-550, 2002.

[11] A.T. Bartic, D.J. Wouters, H.E. Maes, J.T. Rickes and R.M. Waser, "Preisach model for the simulation of ferroelectric capacitors," *J. Appl. Phys.,* vol. 89, pp. 3420-3425, 2001.

[12] C. Kuhn , H. Honigschmid, O. Kowarik, "A new physical model for the relaxation in ferroelectrics," *Proc. ESSDERC,* pp.164-168, 2000.

[13] K. Levenberg, "A method for the solution of certain non-linear problems in least squares," *Q. Appl. Math.,* vol. 2, pp. 164-168, 1944.

[14] J. Rodriguez, et al., "Reliability demonstration of a ferroelectric random access memory embedded within a 130nm CMOS process," *Proc. IEEE NVMTS,* pp. 63-65, 2007.

Measurement of Neutron-induced Single Event Transient Pulse Width Narrower than 100ps

Hideyuki Nakamura, Katsuhiko Tanaka, Taiki Uemura, Kan Takeuchi, Toshikazu Fukuda and Shigetaka Kumashiro

MIRAI-SELETE

Sagamihara Office, NEC Sagamihara Plant, 1120 Shimokuzawa, Sagamihara 229-1198, Japan

e-mail: nakamura.hideyuki@selete.co.jp

Abstract— **A novel SET pulse measurement circuit is proposed which can detect pulses narrower than 100ps. Alternation of SET pulses during the propagation through the chain of target cells is minimized, which is attributed to small chain length (typically 20). This circuit configuration contributes to obtaining pulse distribution similar to that observed in actual circuit in use. Distribution of SET pulse width measured by our circuit through the white neutron beam testing agrees well with that estimated by computer simulation.**

Keywords-SET; single event transient; neutron; combinational logic; SEE; single event effects; soft errors

I. INTRODUCTION

Single event phenomena can be a major threat to reliability in today's ICs. For instance, soft error rate due to the single event upsets (SEUs) in memory cells already reaches unacceptable level for the specific use that requires high reliability. Most of the soft errors in memory cells caused by SEU can be cured by error correcting code (ECC). For upsets of flip-flops, redundancy techniques such as triple modular redundancy (TMR) can reduce the error rate.

Recently, another type of single event phenomena that occurs in combinational-logic circuits gets much attention, which is called single event transient (SET). In the logic circuits, an erroneous pulse signal raised by incidence of ionized particle propagates to a latch such as a flip-flop, and might be latched depending on the clock signal. It is predicted that soft error due to SET phenomenon becomes significant as device dimension is scaled down [1] mainly due to the increase in the clock frequency. Although the technique to filter SET pulses is proposed [2], it is necessary to know the SET pulse widths in order to optimize the filter.

One way to measure the SET pulse width is to simultaneously latch the output signals of each inverter in the inverter chain where an SET pulse propagates and detect a series of latches that do not have normal status [3-5]. Although many SET pulse measurements using such a circuit have been reported, only few of them adopted white neutron beam which is preferable to heavy ion beam since the spectrum of white beam resembles that observed at the terrestrial level. According to the previous measurement [6] using white neutron beam, SET pulse widths within the range between 300ps and 4ns were observed. However, use of available white neutron beam usually imposes a restriction that quite long chain composed of

several hundreds of target cells is necessary as a source of SET pulse in order to obtain sufficient number of SET pulses. Because of this restriction, observed pulse might be altered during the propagation through the chain. The purpose of this work is to measure the SET pulse which is not much affected by such alternation, and to obtain the distribution of SET pulse width that would be seen in actual circuits in use. For this purpose, expected pulse width distribution is estimated through computer simulation firstly, and a novel measurement circuit is designed based on the predicted SET pulse width. This measurement circuit includes target cell chains, each of which has reasonable length, and pulse detection circuits that can capture pulses narrower than 100ps.

II. CALCULATION OF PULSE WIDTH DISTRIBUTION

Fig. 1 illustrates the flow to calculate the SET pulse width distribution. Firstly, current response of single MOS transistor to the incidence of ionized particle is analyzed by three-dimensional device simulation for various incident particles (i.e. for typical positions of incidence and LETs), and the database of several hundreds of current responses is constructed. This current database is utilized to estimate the actual current response in the primitive circuit cells. From the estimated current response, SET pulse width is derived by circuit simulation. By applying the process of estimating current response and calculating the pulse width to various incident events of ionized particles, distribution of SET pulse width is obtained. Calculation of SET pulse utilizing the current database and the circuit simulator is managed by soft-error assessment tool TFIT [7].

The estimation process of current response is much faster than analysis by mixed-mode TCAD device simulation, and calculated current response and SET pulse agree well with

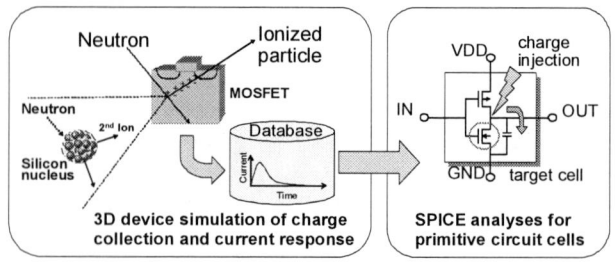

Figure 1. Simulation flow to calculate SET pulse width.

This work was supported by NEDO.

978-1-4244-5430-3/10 $26.00 © 2010 IEEE

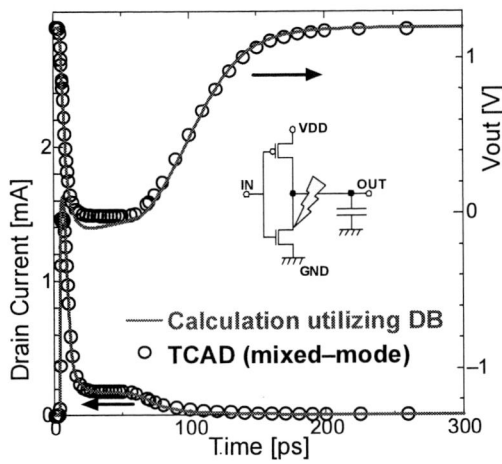

Figure 2. Comparison between the method utilizing the current response database and 3D mixed-mode circuit-device simulation for the calculation of current and voltage curves.

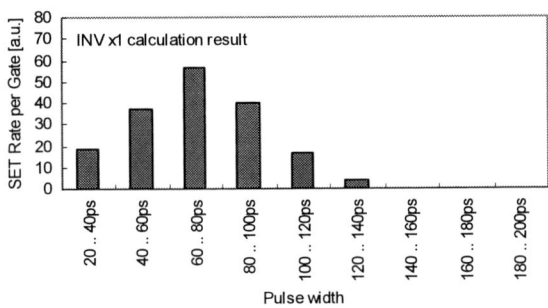

Figure 3. SET pulse-width distribution for inverter cell of 90nm technology node.

TCAD simulation as shown in Fig. 2 [8]. Fig. 3 shows the calculated pulse width distribution for inverter cell of 90nm technology node. It is clearly seen that SET pulse wider than 200ps hardly occurs and pulses narrower than 100ps occupies a large part of SET pulses. This implies that one should prepare a certain measurement circuit that can detect such short pulses in order to measure such SET pulse that occurs in circuits in practical use.

III. MEASUREMENT CIRCUIT DESIGN

Two types of measurement circuits are integrated into our test chip. One (circuit 1) is based on the circuit which is commonly used so far. Another one (circuit 2) aims at capturing SET pulse that does not travel too much in the chain, and hence keeps short width. Use of white neutron beam is assumed in our circuit design since the goal of this work is to measure the SET pulses similar to those in actual circuit environment. As a consequence, at least one million gates should be incorporated into the chip for each target cell in order to obtain sufficient number of SET pulses.

A. Circuit 1 (Conventional Circuit)

Circuit 1 is illustrated in Fig. 4, which is a measurement circuit similar to that used in [3]. This circuit consists of 511 chains each of which has 1000 inverters connected in series as sources of SET pulse, an OR tree to bundle outputs of 511 chains, and a chain of pair of inverter and flip-flop which captures SET pulse by self-trigger to measure pulse width.

Since it is predicted by SPICE calculation that the pulse narrower than 100ps cannot propagate through 1000 stage inverter chain as shown in Fig. 5, this measurement circuit detects only sufficiently wide pulses.

In order to avoid attenuation of SET pulse in the OR tree and to distinguish SET pulses raised in the OR tree, amplifiers to extend pulse width by about 300ps are inserted between inverter chains and the OR tree.

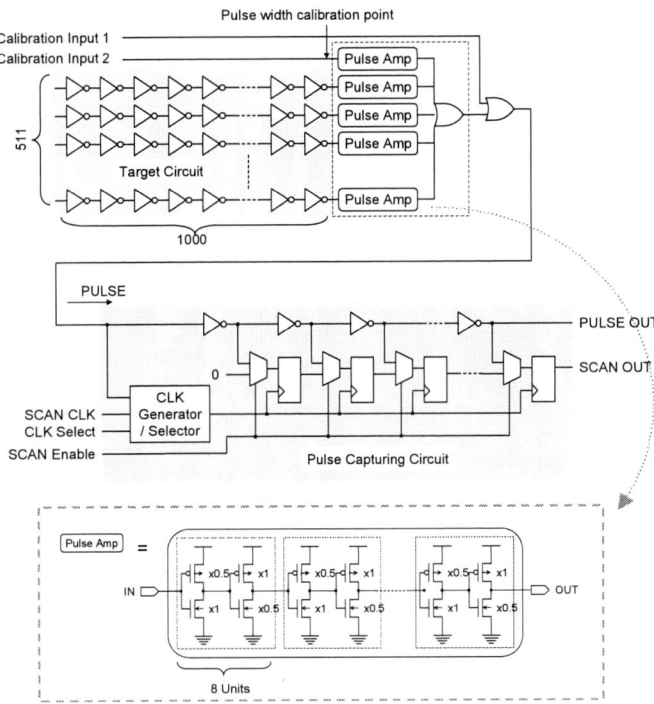

Figure 4. Overview of measurement circuit 1.

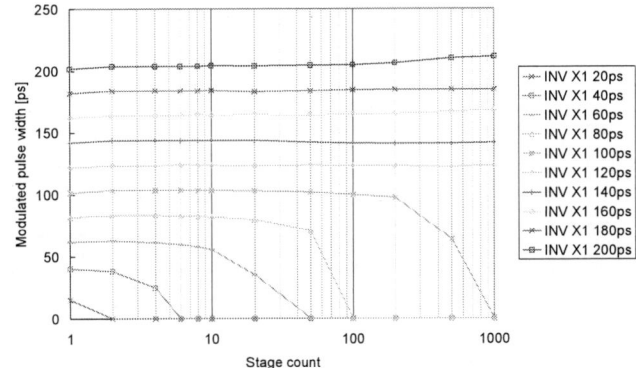

Figure 5. Attenuation characteristics of pulses in a inverter chain calculated by SPICE.

B. Circuit 2 (Proposed Circuit)

Circuit diagram of circuit 2 is shown in Fig. 6. This circuit is intended for detecting very narrow SET pulses coming from chain of typically 20 target inverters, whose stage count is comparable to that in the practical logic circuits. In order to avoid any side effects originating in the OR-tree, one pulse detection circuit is connected to one target inverter chain. Since one million gates (i.e. 50,000 inverter chains) are required to obtain statistically meaningful results, the pulse detection circuit must be compact as much as possible.

In order to make detectable pulse as narrow as possible, a Set-Reset latch (SR-latch) using a feedback loop of two NAND gates is adopted. Minimum detectable pulse width is 63ps in case that the circuit is fabricated by 90nm process. A flip-flop circuit used in a practical logic circuit is also a similar feedback loop. Therefore, roughly speaking, only a pulse wider than this minimum width can be caught by the flip-flop, and sensitivity of the detection circuit by SR-latch would be enough. Combining the plural SR-latches with different detection sensitivity, distribution of pulse width can be obtained. SR-latches with three different detection sensitivities are combined, taking into account the trade-off between pulse width resolution and the size of the pulse detection circuit.

SR-latches L1a and L1b, which have the same highest detection sensitivity, form redundant configuration in order to exclude SEU in SR-latch itself. In addition, four SR-latches are put in order of L1a-L2-L3-L1b in the layout, which contributes to reducing probability of multi-cell-upset of L1a and L1b. There are two input-buffering inverters for four SR-latches. L1a and L2b are connected to different ones, which prevents latches from sensing SET raised in the buffering inverter.

The sensitivities of L2 and L3 are adjusted by load capacitances connected to the retention nodes. When the circuit detects a pulse wider then L2 sensitivity, L1a, L1b and L2 are activated, and therefore signal D2 is activated. For a even wider pulse than L3 sensitivity, all four SR-latches are activated, and hence signal D3 is activated. D0 is prepared to find out nonfunctional chips due to the stuck failures of any SR-latches. D0 may be activated during neutron testing by SEU in SR-latches or SET in two input-buffering inverters. Table I shows the output signal patterns and the corresponding detected phenomena. The pulse detection sensitivities shown here have been calculated by SPICE.

TABLE I.   OUTPUT SIGNAL PATTERN AND CORRESPONDING EVENT

Signal	D1	D2	D3	D0	Event	SET pulse width
State	0	0	0	0	none	-
	1	0	0	1	SET	63.0 - 85.9ps
	1	1	0	1	SET	85.9 - 108.5ps
	1	1	1	1	SET	Over 108.5ps
	0	0	0	1	SEU	-

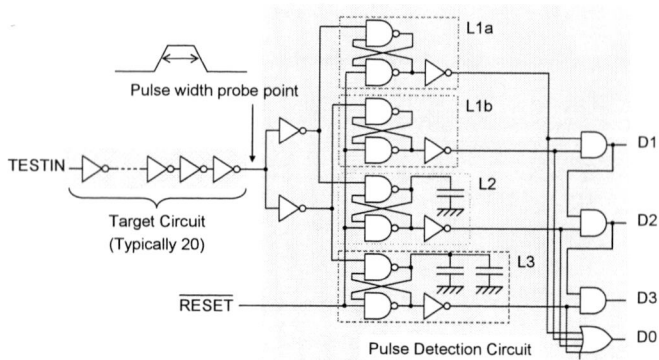

Figure 6.   Overview of measurement circuit 2.

Table II shows variation of the target circuit for measurement circuit 2. Typically 50,000 pairs of 20 stage target inverter chain and pulse detection circuit are integrated on one test chip. Each output signals D0-D3 coming from 50,000 detection circuits are gathered by OR logic, since the read-reset cycle during the neutron test can be made shorter enough than the expected time period between SETs.

TABLE II.   VARIATION OF THE TARGET CIRCUIT FOR MEASUREMENT CIRCUIT 2

Name	Number of chain stages	Function	Drive strength	Copies
INV x1	20	inverter	1	50,000
INV x1	40	inverter	1	10,000
INV x1	80	inverter	1	10,000

IV.   EXPERIMENTAL RESULTS

The test chip incorporating measurement circuit 1 and 2 was fabricated in 90nm process. Experiment by the white neutron beam was performed at the TRIUMF irradiation facility. The test method basically follows JEDEC standard JESD89A.

A. Circuit 1

Fig. 7 shows experimental results obtained by measurement circuit 1. Amount of pulse width expansion by the amplifier is subtracted from the pulse width. Therefore, observed pulses with negative width are considered as pulses stemming from the amplifier or the OR tree. SETs coming from target inverter chains are distributed up to 600ps. This result is consistent with the previous work [6], however, it is quite different from our calculation results. Considering that a pulse narrower than 100ps cannot propagate through a long inverter chain, any test result from the test chip taking this kind of target circuit configuration will be similar to this result.

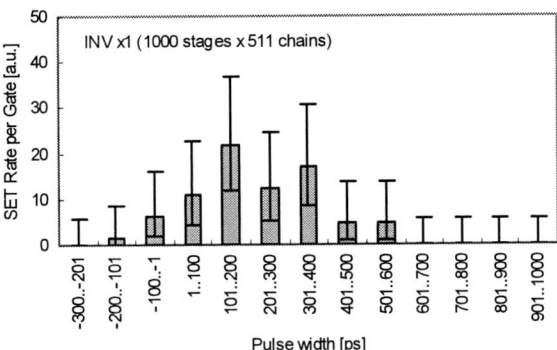

Figure 7. Experimental results obtained by measurement circuit 1. Error bars is relatively large since irradiation time was shorter than that initially scheduled.

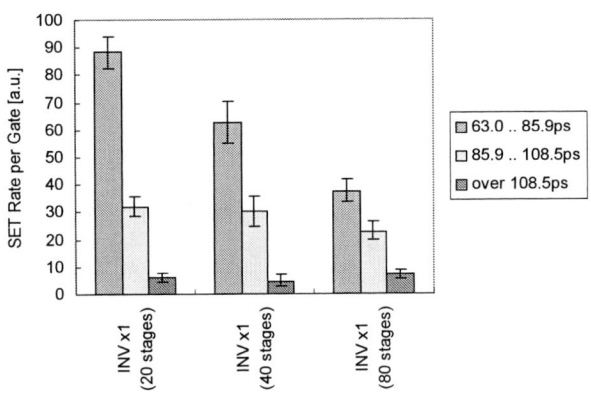

Figure 8. Experimental results obtained by measurement circuit 2.

B. Circuit 2

Fig. 8 shows experimental results obtained by measurement circuit 2. It is clearly seen that many pulses narrower than 100ps are detected, which are not observed in the result of circuit 1. Regarding an inverter cell, it is found that most of the SET pulses consist of the pulses narrower than 108ps. As the inverter chain gets longer, generation rate of pulse per gate becomes smaller, especially for short SET pulses. This is because attenuation of SET pulse occurs for the pulse narrower than 100ps even in the relatively short inverter chain whose number of stages is below 100, as shown in Fig. 5. The result of circuit 1 can be interpreted as an ultimate case. Therefore, it is important to determine the configuration of the target circuit carefully referring to the actual logic circuits in use.

Fig. 9 shows the distribution of SET pulse width calculated for the same target circuit as the measurement circuit 2 by our simulation flow. Shape of pulse width distribution is quite similar to that actually measured. SET generation rates of calculated and measured results agree within the factor of 1.5.

V. CONCLUSIONS

It is shown that SET test result depends greatly on the design of the target circuit. Unless the target circuit which has configuration similar to actual logic circuits, observed pulse width will be much different from actual one. In the realistic logic circuits, number of gates to latch would be at most 10 or so, and hence it is expected that most of the generated SET pulses will be narrower than 100ps. Therefore, our novel pulse detection circuit that can detect pulse width below 100ps would be highly useful to know the actual pulse width distribution in the logic circuits.

ACKNOWLEDGMENT

This work was supported by NEDO. The authors would like to thank people in iRoC technologies for their effort in experiment and for their support in designing measurement circuit 1.

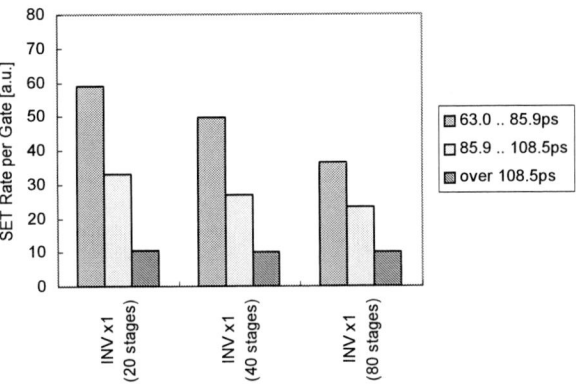

Figure 9. Calculation results for the same target circuit as the measurement circuit 2.

REFERENCES

[1] P. Shivakumar, M. Kistler, S.W. Keckler, D. Burger, and L. Alvisi, "Modeling the effect of technology trends on the soft error rate of combinational logic", International Conference on Dependable Systems and Networks, pp.389–398, 2002.

[2] M. Nicolaidis, "Time redundancy based soft-error tolerance to rescue nanometer technologies", VLSI Test Symposium, pp.86–94, 1999.

[3] M. Nicolaidis and R. Perez, "Measuring the width of transient pulses induced by ionizing radiation", 41st Annual International Reliability Physics Symposium, pp.56–59, 2003.

[4] B. Narasimham et al., "On-chip characterization of single-event transient pulsewidths", IEEE Transactions on Device and Materials Reliability, Vol.6, pp.542–549, 2006.

[5] T. Makino et al., "LET dependence of single event transient pulse-widths in SOI logic cell", IEEE Trans. Nucl. Sci., Vol.56, pp.202–207, 2009.

[6] B. Narasimham et al., "Neutron and alpha particle-induced transients in 90nm technology", 46th Annual International Reliability Physics Symposium, pp.478–481, 2008.

[7] http://www.iRoCtech.com

[8] M. Hane et al., "Synthetic soft error rate simulation considering neutroninduced single event transient from transistor to LSI-chip level", 2008 International Conference on Simulation of Semiconductor Processes and Devices, pp.365–368, 2008..

New Electromigration Validation: Via Node Vector Method

Young-Joon Park, Palkesh Jain*, and Srikanth Krishnan
Analog Technology Development/ASIC*
Texas Instruments Inc.
13121 TI Boulevard, MS364, Dallas, TX 75243, USA/
IN::Bangalore - 560 093:Bangalore-Bagmane TechPk, India*
Tel: 214-567-5379; e-mail: yjpark@ti.com

Abstract— Electromigration (EM) is a traditional reliability concern, aggravated recently due to intense shrinking of wire sizes and the increase in the number of interconnections on a system-on-chip (SoC). Thereby, it challenges the state-of-the-art in design, physics, process and CAD processes. To that regard, we propose a new EM check method named Via Node Vector method. The conventional EM check ignores the lead EM interaction in circuits and only checks the local current densities. The new method addresses the EM interactions and checks the EM reliability at the lead connection sites (called via node). It converts the electrical current density of each lead into an *effective* current density for the EM interaction consideration. For this, we introduce three new factors: length (F_L), width (F_W), and interaction (F_B). The effective current density *divergence* at a via node is derived as an addition of the effective current densities of all the interacting leads, which is a close proxy to represent the physical atomic flux divergence at the via node. This *divergence* at a via node can then be readily compared with the technology EM spec. The proposed method is successfully applied to 28nm node IPs and shows up to ~4X higher safe operating frequency than the conventional method allows. Additionally, it successfully identifies risky sites missed by conventional check. The Via Node Vector Method will provide higher performance with reliability in designing advanced digital and analog circuits.

Keywords—electromigration; electromigration interaction; circuit electromigration; electromigration check

I. INTRODUCTION

Electromigration (EM) is atomic flux driven by electrical current and its *divergence* at a site leads to interconnect failure, either in form of increasing lead resistance or shorting out the circuits. The EM divergence is net atomic flux at a site between IN and OUT electromigrated atoms. Even if the amount of atomic flux is very large along an interconnect, there can be no atomic flux *divergence*; i.e. EM damage, if the atomic flux is uniform along the interconnect (Figure 1).

IN atomic flux (EM) = OUT atomic flux (EM)

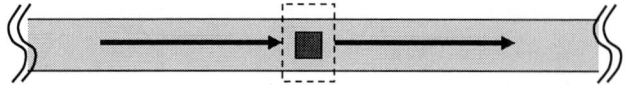

FIGURE 1. AS THE 'IN' AND 'OUT' ATOMIC FLUX ARE THE SAME, THE ATOMIC FLUX DIVERGENCE ALONG THE INTERCONNECT IS ZERO, RESULTING IN NO EM DAMAGE.

While we all know that atomic flux divergence is the cause for EM failures, the present industry practice to confirm the EM reliability is to check the local current density of interconnects [1]. However, the local current density is related to the amount of the atomic flux but does not correctly represent the divergence of the atomic flux. In limited cases, such as the case when an interconnect is terminated by via or contact, the atomic flux itself is the divergence at the lead ends as the atomic flux is completely blocked and cannot migrate to the upper or lower layer.

Accordingly, it is the intent of this work to propose a more physically justified way to manage the EM phenomenon. The best practice to check EM reliability is to limit the atomic flux divergence at a site where the flux meets, such as vias, contacts, lead merge, or division points, as these sites are most susceptible to divergence and eventually to EM damages. These sites are referred to as "via node" in this paper. The flux divergence at non-via nodes would also occur, but the divergence effect at those sites is possibly less than at via nodes.

The experimental observation of the current divergence effect (electrical interaction effect) at lead connection sites on the EM lifetime has been well reported [2-5]. It is also understood that, when the electrical divergence at a site is high, the EM lifetime is short. (Note that the electrical divergence referred to in this paper is the current divergence only among the connected leads, excluding the current to (or from) via. The atomic flux to (or from) via is blocked by the diffusion barrier layer, and there will be no atomic flux to (or from) via. Thus, we may consider only the divergence effect among the connected leads for the EM lifetime consideration. See Figure 2.)

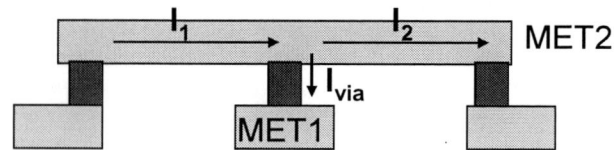

$$I_1 + I_2 + I_{via} = 0$$

$$I_1 + I_2 = \text{electrical divergence for EM damage}$$

FIGURE 2. THERE IS NO ELECTRICAL DIVERGENCE IN A CIRCUIT IF WE INCLUDE THE CURRENT TO OR FROM VIA. WHEN WE CONSIDER THE ELECTRICAL DIVERGENCE FOR EM DAMAGES, IT IS THE DIVERGENCE ONLY AMONG THE CONNECTED LEADS, EXCLUDING CURRENT FLOW TO OR FROM VIA.

If the atomic flux is purely determined by electrical currents and is constant during product lifetime, the electrical

978-1-4244-5430-3/10 $26.00 © 2010 IEEE

current divergence would be sufficient to predict EM reliability. However, the EM divergence at a via node is affected by additional factors as it is an atomic mass flow divergence. Among other things, a connected lead's dimensions and the topology of the connections can impact the EM divergence. If the interconnections to a via node are physically short, the atomic flux from the lead to a via node is not the same during the whole lifetime, even though the electrical current flowing through it is constant. The atomic flux may decrease with time and can be zero eventually due to the back stress effect while the current continues to flow [6, 7]. Thus the pure electrical current is not sufficient to represent the total atomic flux to a via node during the entire lifetime.

The dimensional impact of leads on atomic flux divergence is demonstrated in Figure 3. The same current j flows through all the leads. Thus the electrical current divergence at both via node A and B is zero as $j - j = 0$. This is also true for the atomic flux divergence at the via node during the initial state of the stressing as initial atomic flux is not limited by the back stress effect and can be proportional to the electrical current. However, as time goes on, the atomic flux coming from the lead $Q2$ into via node B becomes smaller than that into the lead Q from the via node B. This decrease is due to the short-length induced early-onset of back stress effect in lead $Q2$, which further results in the generation of mechanical stress at the via node B. On the other hand, equal lengths of the leads Q and $Q1$ enforce zero atomic flux divergence (and mechanical stress) over the lifetime at the via node A. Consequently, at steady state, the via node B has a finite tensile stress while the via node A does not, as shown in Figure 3, due to the different dimensions of the attached leads of $Q1$ and $Q2$.

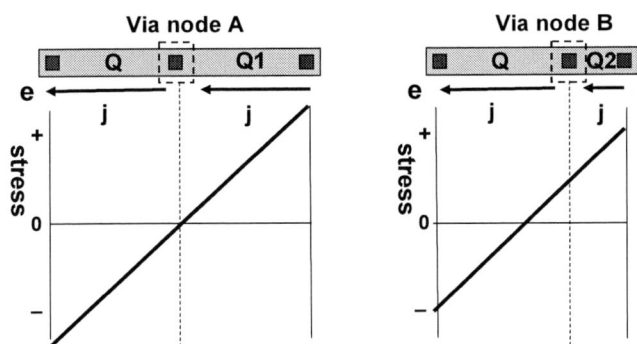

FIGURE 3. LEAD $Q2$ IS SHORTER THAN LEAD Q. THUS THE ATOMIC FLUX FROM $Q2$ INTO THE VIA NODE B CANNOT LAST AS LONG AS THE ATOMIC FLUX INTO Q FROM THE VIA NODE B. EVENTUALLY, THERE WILL BE STRESS GENERTION AT VIA NODE B DUE TO THE IMBALANCE IN THE ATOMIC FLUX EVEN WITH THE SAME ELECTRICAL CURRENTS IN BOTH LEADS.

The criticality of these non-electrical effects on the atomic flux divergence warrants its inclusion in the circuit EM reliability assessment. To address these effects, we propose a new EM check named Via Node Vector Method, which encompasses these non-electrical factors and estimates the atomic flux divergence at via nodes. It converts the electrical current density of each lead into an *effective* current density for the EM interactions by applying three new factors—length (F_L), width (F_W) and interaction (F_B)—to electrical current density. It is "effective" as it represents the atomic flux impact to via node during the overall lifetime period. All the effective

current densities from connected leads to a via node in question will be added to get an effective current density *divergence* at the via node. The *divergence* will be compared with the EM spec. We demonstrate this new method on the standard cells of a 28 nm technology node and show that it allows higher operation frequencies and detects more relevant violations than the conventional method.

II. EXPERIMENTS

The EM modules used in this research have two connected leads: main and attached. (In the example case of Figure 3, lead Q is the main lead and one of the leads $Q1$ or $Q2$ is the attached lead, respectively.) The two leads are connected serially through a middle via. The main lead is 20 um long and 50 nm wide while the attached leads vary in length (1 um or 20 um) and in width (50 nm or 200 nm).

We applied current of j, $2j$, or $2.5j$ as illustrated in each figure. All tests have been done at 300°C. When the currents in the attached leads are different from that in the main lead, the difference is supplied from the middle via. The electron flow direction in the attached lead is also a controlled factor which may be opposite to or be the same as the main lead. (Arrows represent the electron flow directions.) The EM failure criterion is 20% resistance increase. The resistance for both the main and the attached leads is measured separately during the testing. The cross sectional SEM and FIB images of the EM failure sites show the fatal void morphology.

III. RESULTS

A. Inaccuracy of conventional EM check method

Figure 4 shows the EM lifetimes of two connected leads, X and Y. The anode end of lead Y is connected to the cathode end of lead X which is stressed at $2j$, while lead Y is at j. The width and length of the two leads are the same, thus 2X more current flows through X than Y.

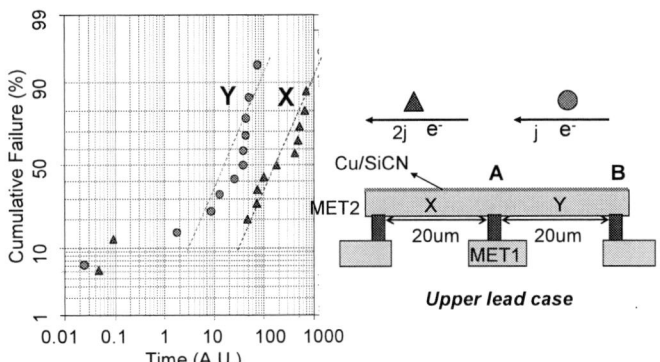

FIGURE 4. LEAD Y FLOWS AT HALF THE CURRENT DENSITY OF LEAD X, BUT FAILS FASTER THAN LEAD X; CONTRARY TO THE CONVENTIONAL EM EXPECTATION. THE NEW VIA NODE VECTOR METHOD EXPLAINS IT WELL WITH THE DIVERGENCE EFFECT.

The conventional method expects faster failure for lead X with $2j$, but experiments show that lead Y (with only j) fails ~10X earlier. The inadequacy of the conventional approach to explain this phenomenon is due to the lack of consideration of the lead interactions. The conventional check assumes the two leads are isolated in terms of EM reliability, but they are physically connected and their atomic flux indeed interacts.

978-1-4244-5430-3/10 $26.00 © 2010 IEEE

Thus, the actual divergence at the via node A and via node B is not purely determined by the local current density of the each lead X and Y, respectively. (Cathode via node is the most susceptible EM failure site due to voiding. The divergence at via node A determines the lead X EM reliability while that at B determines the lead Y.) The lead EM interactions also affect the divergences at the via nodes.

Lead X has a current flow of $2j$ but is supplied a current of j from lead Y. So, the electrical divergence at via node A is just j. (In this case, the dimensions of the two leads are the same so that the non-electrical effect is minor and the electrical consideration is good.) Thus the EM damaging rate (or atom depletion rate from the cathode via of lead X) is similar to that of the cathode via of an isolated single lead flowing j.

Lead Y has a current flow of j, but the current keeps flowing into X with a larger amount of $\sim 2j$. The atomic flux from lead Y is not accumulated at via node A, but will be transferred into lead X at a faster rate due to the higher current density of $2j$. It will eventually increase the atom depletion rate from via node B and make the EM lifetime of lead Y short. A conservative electrical way to consider this effect is to add the two current densities, which makes the current divergence from via node B $3j$, meaning that the EM lifetime of lead Y is similar to that of an isolated single lead having $3j$ current density. Thus lead Y fails much earlier than lead X. (The 3X larger current density now makes $\sim$10X shorter lifetime due to the short length effects where the apparent current exponent can be larger than one [8].)

The role of lead X in lead Y's EM lifetime is, in other words, to increase the atomic depletion time. Without lead X, the stress profile of lead Y will reach a steady state quickly and atomic depletion can happen only during initial current stressing. However, due to lead X, the atomic deletion from the via node B is extended and the steady stress profile occurs later with larger atomic depletion. If the length of lead Y is sufficiently long, lead X does not affect lead Y EM lifetime, and the EM failure at the via node B will happen before reaching the steady state profile.

The conventional EM check is good to estimate the EM reliability when a lead is isolated and terminated by via or contact at both ends. However, this experimental result clearly shows that it cannot correctly predict the EM reliability of circuit leads due to the lack of lead interaction consideration, thereby posing potential risk to reliability assessment of the system. The lead interaction consideration estimates the divergence at via node A as j while that at the via node B as $3j$, consistent with the experimentally observed EM lifetime behavior.

B. Via Node Vector EM check

Motivated by theoretical understanding and experimental observations, we formulate the Via Node Vector Method by combining all the EM interaction effects at a via node. The EM interaction is atomic flux interaction that is affected by the lead geometry, the layout topology, and the connecting interface property, in addition to the electrical flow. To comprehend these effects, we convert the electrical lead current density, j_{lead} ($=j_{avg}$), to an EM effective current density, j_{eff}. The j_{eff} indicates the atomic flux impact from a lead to a via node over whole service period. As an example, if the

atomic flux from a lead to a via node changes with time, the j_{eff} represents the 'time-averaged' atomic flux from the lead to the via node.

For this conversion of j_{lead} to j_{eff}, we introduce three new factors: the length factor (F_L), the width factor (F_W), and the interaction factor (F_B). Each factor is intended to address three different non-electrical impacts on the atomic flux of a lead. The j_{eff} is a product of these three factors multiplied by the j_{lead}. ($j_{eff} = F_L \times F_W \times F_B \times j_{lead}$). The polarity (vector) sum of the j_{eff}s of all the leads connected to a via node is the effective current density divergence of the via node, $j_{eff.div}$. We can then compare this with the allowed current density, $j_{EM.spec}$. (See Figure 5)

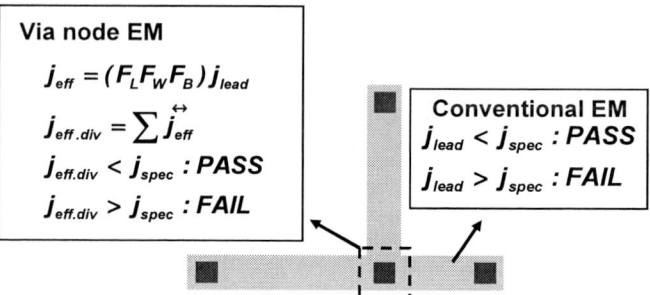

FIGURE 5. THE VIA NODE VECTOR METHOD COMPARES THE $j_{eff.div}$ OF A VIA NODE WITH THE $j_{EM.spec}$ WHILE THE CONVENTIONAL METHOD COMPARES THE j_{lead} TO THE $j_{EM.spec}$. THE VIA NODES ARE MOST SUSCEPTIBLE TO THE EM FAILURES, AND THE $j_{eff.div}$ SHOULD BE SMALLER THAN THE $j_{EM.spec}$.

For a given via, two sets of lead connections are possible: one for leads above the via and the other for leads below the via. Both of the two lead connections need to be checked separately to ensure the EM reliability of the via node.

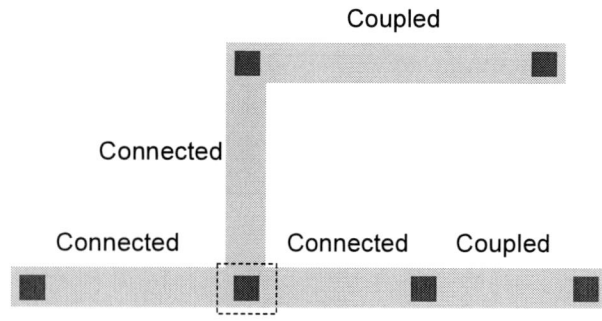

FIGURE 6. CONNECTED LEADS ARE DIRECTLY ATTACHED TO A VIA NODE WHILE COUPLED LEADS ARE INDIRECTLY LINKED TO A VIA NODE THROUGH CONNECTED LEADS. WHEN THE CONNECTED LEADS ARE SHORT, THE COUPLED LEADS CAN AFFECT THE EM RELIABILITY OF A VIA NODE IN QUESTION.

In a real circuit, there are both 'connected' leads—ones directly connect to a via node—and 'coupled' leads—ones that do not directly attach to the via node. (See Figure 6). Depending on the connected lead's length, the coupled leads may or may not affect the EM reliability of the via node in question. If the connected lead is sufficiently short, the atomic flux from the coupled leads affects the via node EM reliability by affecting the stress generation in the via node side of the connected lead. In such cases we need to add all the j_{eff} values including those from the coupled leads. On the other hand, when the connected leads are sufficiently long, we can ignore the effect from couple leads. In such cases, the EM failure of the via node happens when the stress profile near the via is localized and far earlier than reaching a steady state; thus they are not affected by the stress generation in the coupled leads.

In this paper, we explore the three non-electrical factors—length (F_L), width (F_W), and interaction (F_B)—only for the connected lead case as follows.

C. F_L: the length factor

Elaborating on the previous discussion, long attached leads serve as large atomic reservoirs, so the EM flux from the lead to a via node may last until failure. However, the EM flux from a short attached lead will not last long. The EM interaction of the short lead is limited to an early lifetime even though the electrical current flows during the whole lifetime. We comprehend this EM flux interaction difference on top of the electrical interaction with a length factor: $F_L = L_i / L_{max}$ where L_i=each lead length and L_{max}=the maximum length among leads connected to a via node *or* a threshold length. Besides the linear normalization used in this work, other ways to incorporate length factors are also possible, depending on the EM failure modes.

L_{max} is defined as the maximum length among leads connected to a via node or a threshold length. The threshold length is used to incorporate the back stress effect. A lead which is longer than the threshold length is able to supply the atomic flux to the via node during the whole lifetime without being limited by the back stress effect. In such a case, we assign the length factor as one. For example, if the threshold length is 50 um, all the connected leads >50 um are assigned a length factor of 1. This means that the L_{max} can be the threshold length and the maximum F_L is one, in case that the maximum length is larger than the threshold length.

If we use the threshold length as L_{max} for the case when all the connected leads are shorter than the threshold length, it is an aggressive EM check as the $j_{eff.div}$ of a via node becomes smaller than that when we use the maximum length among the connected leads as L_{max}. This smaller $j_{eff.div}$ implies that the via node has a better EM reliability due to the short length effects. (Note that the threshold length is a function of the current waveforms.)

Figure 7 shows the F_L and the polarity impact on the EM lifetime. The left-side leads (main leads) of all three cases have the same conditions: 20 um long, 50 nm wide, single via ended, and with 2.5j. The right-side lead is a connected lead attached to the main lead. The attached leads have the same current density of 2.5j, but the length or polarity is different. First of all, conventional EM check estimates the same EM reliability for the main leads, regardless of the attached lead conditions, as it does not comprehend any interaction. However, the experiment shows that the EM lifetime of the main lead is vastly different depending on the topology. The main lead EM lifetime is determined by the $j_{eff.div}$ at the middle via node serving as the cathode via to the main leads.

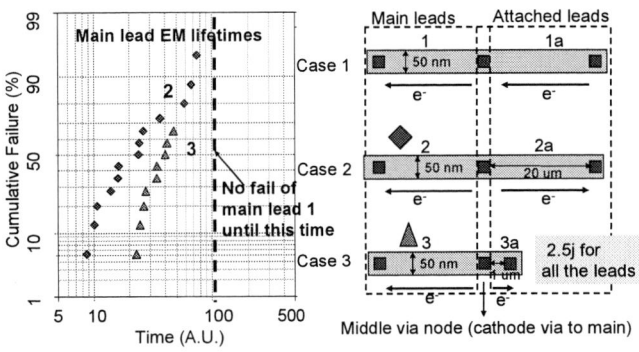

FIGURE 7. THE MET2 IS EM TESTED BY MET1 FEEDER: UPPER LEAD CASE. THE MAIN LEADS SHOW DIFFERENT LIFETIMES DEPENDING ON THE POLARITY AND F_L OF THE ATTACHED LEAD. THE EM LIFETIME OF MAIN LEAD *1* IS VERY LONG AS THE ATTACHED LEAD CAN SUPPLY THE ATOMIC FLUX TO THE MIDDLE VIA NODE. THE MAIN LEAD *2* SHOWS THE SHORTEST LIFETIME AS THE ATTACHED LEAD *2a* IS LONG AND KEEPS OUT-FLUXING FROM THE MIDDLE VIA NODE.

Case 1 has the longest EM lifetime (no fail is observed until testing is terminated) since the attached lead *1a* supplies the same effective current density with the main lead to the middle via node as the F_L for both the two connected leads (main and attached) is the same. Thus, the $j_{eff.div}$ of the middle via node, which is the cathode via of the main lead, is ~ 0.

Case 2 has the attached lead *2a* with out-going electron flow from the cathode via of the main lead *2* and the F_L for both leads is the same as 1 (=20/20).(The maximum length of 20 um is used as L_{max}.) So the $j_{eff.div}$ of the cathode via is ~2X of 2.5j (2.5j +2.5j = 5j). Thus the EM lifetime is short.

Case 3 has the attached lead *3a* having out-going electron flow from the middle via, but the length is very short (1 um) so the j_{eff} from the attached lead is very small due to small F_L (=1/20). So the $j_{eff.div}$ of the cathode via is ~[2.5j + (1/20) 2.5j], which is much smaller than the 5j for case *2*. Thus the EM lifetime of the main lead in case *3* is much longer than that in case *2*.

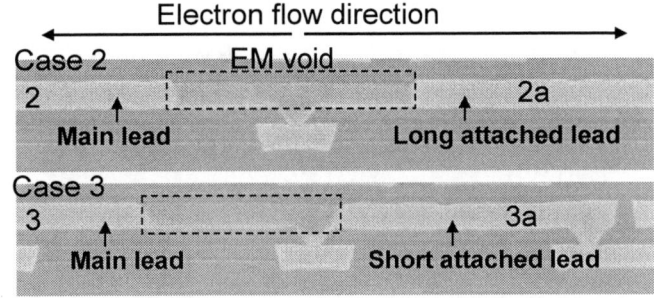

FIGURE 8. THE EM VOID DOES NOT GROW TOWARD THE SHORT ATTACHED LEAD *3a* EVEN WITH THE FAVORABLE ELECTRICAL CURRENT, WHILE THE LONG ATTACHED LEAD *2a* SHOWS VOID GROWTH.

Figure 8 shows cross-sectional images around the middle via node (cathode via to the main lead) of cases *2* and *3*. As expected, in case *3*, the EM induced void grows only toward the main lead while in case *2* it grows symmetrically toward both main and the attached leads. The difference is the attached lead length, not the electrical current; the attached lead *2a* is as long as the main lead while the *3a* lead is short. Thus, in case *3*, the void grows only in main lead *3* as the

978-1-4244-5430-3/10 $26.00 © 2010 IEEE

atomic flux toward the short lead *3a* happens only in the early stage of the stress conditions and the total amount of the flux is limited. We address this effect in calculation of j_{eff} of the lead by applying the F_L factor (1/20) to the current which makes the current impact that much smaller. As the critical void volume should be depleted by only the main lead 3 atomic flux, it takes a longer time to fully deplete over the middle via node.

Above experiments clearly establish that a) attached lead length has a first-order impact on the EM lifetime of the main lead and b) $j_{eff.div}$ of a via node determines the EM lifetime of the node. Furthermore, the via node with effective current divergence of $j_{eff.div}$ can be treated as equivalent to an isolated long lead having the $j_{eff.div}$ as electrical current density.

D. F_W: the width factor

Lead width is a primary variable in conventional EM checks since it directly governs the current density. Additionally, from the Via Node Vector Method perspective, if the widths of connected leads are different, the total EM flux coming from or going into a lead becomes proportional to the (current density) x (width). We thereby define the width factor, F_W, as W_i/W_{min} where W_i = each lead width and W_{min}= the minimum width among them.

Figure 9 shows the F_W effect. Case *4* is the same as case *2* except the 4X wide attached lead. Thus, the $j_{eff.div}$ would increase due to the 4X larger F_W of the attached lead *4a* having the out-going electron flow, which is the same as the main lead. Thus the lifetime of the main lead decreases further.

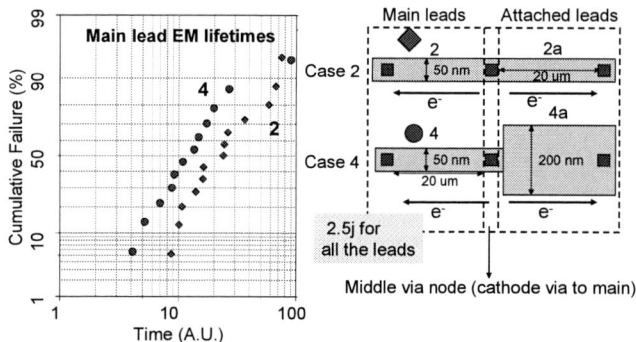

FIGURE 9. MET2 IS EM TESTED AS UPPER LEAD WITH MET1 FEEDER. THE WIDER ATTACHED LEAD *4a* MAKES THE MAIN LEAD *4* EM LIFETIME SHORT DUE TO LARGER F_W EFFECT MAKING THE $j_{eff.div}$ AT THE MIDDLE VIA NODE LARGER THAN THE CASE 2.

The $j_{eff.div}$ can create a fatal void among one of the leads connected to the via node. The worst case of EM failure happens when the void nucleates at a minimum width lead because the fatal void volume is the smallest. To address this worst case, the F_W factor converts the current density of each lead into the effective current density when the current would flow through the minimum width lead. For a given current density, the wider connected lead will have a larger atomic flux impact to the void.

E. F_B: the interaction factor

If the atomic diffusion is perfectly continuous through a via node without experiencing any mobility degradation, we define that as a perfect interaction (F_B=1). Otherwise, the leads interact partially ($0<F_B<1$). The interaction factor, F_B, deals with this impact. Even though the leads are perfectly connected for the electrical flow, the connection efficiency for the atomic flux may not be perfect.

The perfect flux interaction of F_B=1 is almost true for upper lead cases of Cu interconnects (connections above via as shown in Figure. 4) where we consider the via node EM reliability for the connected leads above the via nodes. In this case, the Cu interconnects are connected over the via node through the Cu/SiCN interface which is the fast diffusion path for EM. On the other hand, the $0<F_B<1$ is true for lower lead cases (connections below via as shown in Figure. 10) where the 'good' via bottom (Cu/Ta) interface slows down the atomic flux through the leads.

The EM lifetime difference between Figures 4 and 10 shows the F_B effect. The EM test conditions in Figure 10 are the same as those in Figure 4. The only difference in Figure 10 is that it is a lower lead case while Figure 4 is an upper lead case; thus the F_B is different. Figure 4 shows ~10X difference in lifetime between lead X and Y, while Figure 10 shows only ~1.5X difference between lead M and N. The smaller F_B of Figure 10, as the atomic flux is partially blocked by Cu/Ta interface, limits the atomic flux supply to the middle via node and increases its $j_{eff.div}$. Thus lead M cannot have a much longer EM lifetime than lead N, compared to the difference between leads X and Y in the upper lead cases in Figure 4. The F_B for this lower lead case is estimated as 0.6.

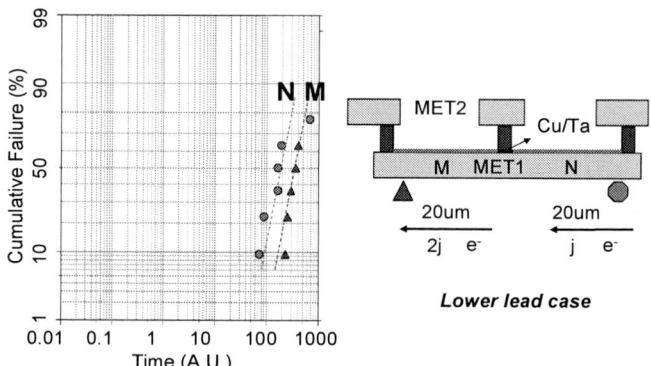

FIGURE 10. THE SMALL F_B (~0.6) OF THIS LOWER LEAD CASE MAKES THE EM LIFETIME DIFFERENCE BETWEEN LEAD M AND N SMALL COMPARED TO THE UPPER LEAD CASE (FIGURE 4). THIS DIFFERENCE CLEARLY SHOWS THAT WE NEED TO ESTIMATE ATOMIC FLUX DIVERGENCE, RATHER THAN PURE ELECTRICAL DIVERGENCE, AT A VIA NODE FOR EM VALIDATION.

This difference clearly shows that the EM interaction is atomic diffusion interaction and we have introduced the interaction factor, F_B, to address the connecting interface property. It is worthy to note that the F_B can also be different according to the lead width to the via width for the lower lead case. If the lead width is exactly the same as the via width, the via bottom Cu/Ta interface efficiently blocks the atomic flux interaction through it; but if the lead width is much wider, the lead can be partially connected through the fast Cu/SiCN interface, which will increase the F_B.

IV. DISCUSSION: VIA NODE VECTOR METHOD IMPLEMENTATION ON THE 28 NM TECHNOLOGY STANDARD CELLS

Having formulated the new Via Node Vector Method, we implement that in the in-house EM checking suite; details of which are kept out of scope for this work. We apply both the conventional and the Via Node Vector Method EM check to the standard cells of the 28nm technology node to see the impact. The comparison metric is the maximum EM safe frequency of the standard cell, when everything else is kept constant—operating load, input slews and voltages. Figure 11 shows the EM safe maximum frequency of each cell for a given design and load to drive. The cell current is proportional to CVf, where C=capacitance, V=voltage, and f=frequency, and the higher frequency means the higher allowance in EM currents.

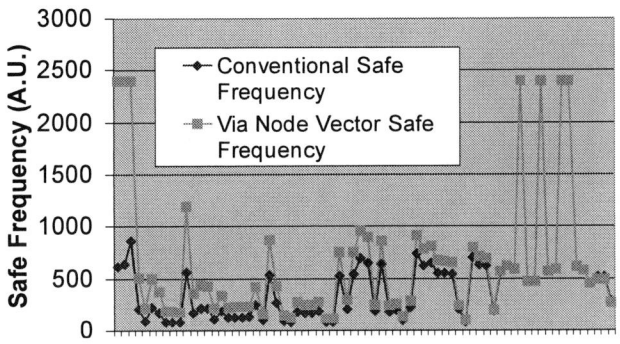

(A) EM SAFE FREQUENCIES OF ASIC CELLS

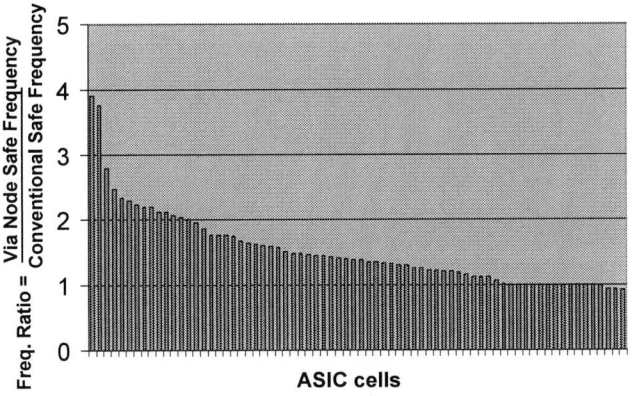

(B) SAFE FREQUENCY RATIOS BETWEEN VIA NODE VECTOR AND CONVENTIONAL EM CHECK METHOD

FIGURE 11. THE NEW VIA NODE VECTOR METHOD ALLOWS UP TO ~ 4X HIGHER FREQUENCIES THAN THE CONVENTIONAL METHOD.

As can be seen, for most of the cells, the Via Node Vector Method EM check allows higher safe operating frequencies—up to ~4X more than the conventional method. This means the leads are connected in such a way to reduce the $j_{eff.div}$. For example, if a via node has two connected leads and the electron flow in one lead is going into the via node while that in another lead is going out, the $j_{eff.div}$ can be smaller. In this case the local electrical current density may be allowed more

while meeting the EM rules at the via node. This is, in essence, the advantage of the new Via Node Vector Method.

Even if a $j_{eff.div}$ at a via node meets the EM spec, we cannot allow unlimited local current density in a lead. For example, if two leads are connected to a via node and 100X j of the EM spec is going in and out, then the $j_{eff.div}$ at the via node is zero, if the F_L, F_W, and F_B are the same for the two leads. Even in this case, EM failure can initiate at the body part of each line and we need to limit this possibility. This line body failure can be evaluated with the bow tie test structure which does not have via or contact at both ends of the lead. We may need to check the local current density with more relaxed rules than the conventional EM spec.

A detailed comparison of the conventional and the Via Node Vector Method EM check is shown in Figure 12, where the exact EM violation numbers for each cell are compared at a fixed frequency, load, and design condition. As expected, the Via Node Vector Method EM check produces less violations than the conventional check. It is also noted that the Via Node Vector EM violations are not always a subset of those from conventional checking. Additionally, the Via Node Vector Method EM check finds violations which were not detected by the conventional method. These sites would have a majority of the current going into or out of the via node, so the $j_{eff.div}$ violates the EM spec even though the local lead current density meets the spec. These sites are the actual threats in EM reliability and the Via Node Vector Method identifies them successfully.

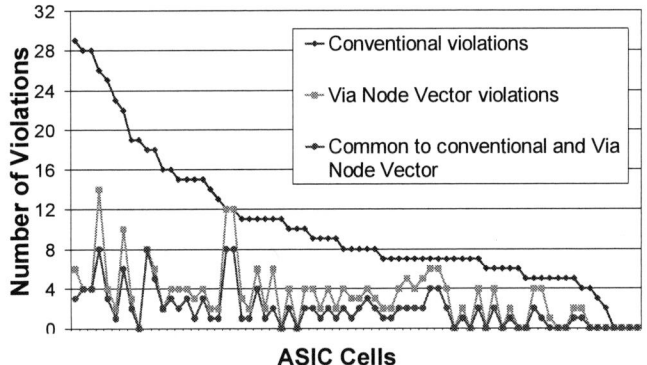

FIGURE 12. THE NEW VIA NODE VECTOR METHOD FINDS CRITICAL VIOLATIONS THAT WERE NOT DETECTED BY THE CONVENTIONAL METHOD, GUARANTEEING SUPERIOR RELIABILITY WITH HIGHER ENTITLEMENT.

V. CONCLUSIONS

Conventional EM checks, which do not encompass atomic flux divergence and lead interaction, merely check the local current densities of each lead; thereby incurring the loss of either entitlement or integrity. This problem is solved through formulation of the new Via Node Vector Method, which successfully addresses the interaction effects, efficiently controls the divergence effect, and allows maximum performance. It will provide us with more flexibility and higher reliability in designing advanced digital and analog circuits. Application of the new method on 28nm standard cells results in up to 4X higher EM performance, while simultaneously identifying the real problematic locations within the circuits.

978-1-4244-5430-3/10 $26.00 © 2010 IEEE

REFERENCES

1. EDA tools. Example EDA tool vendors: Synopsis, Cadence Design Systems, Mentor Graphics, Zuken Inc, and Magma Design Automation.
2. Stefan P. Hau-Riege and Carl V. Thompson, "Electromigration Saturation in a Simple Interconnect Tree," J. Appl. Physics, V. 88, No 5. pp. 2382-5 (2000).
3. Stefan P. Hau-Riege and Carl V. Thompson, "Experimental Characterization and Modeling of the Reliability of Interconnect Trees," J. Appl. Physics, V. 89, No 1. pp. 601-9 (2001).
4. C. L. Gan et al., "Experimental Characterization and Modeling of the Reliability of Three-terminal Dual-damascene Cu Interconnect Trees," J. Appl. Physics, V. 94, No 2. pp. 1222-7 (2003).
5. Frank L. Wei, Christine S. Hau-Riege, Amit P. Marathe, and Carl V. Thompson, "Effects of Active Atomic Sinks and Reservoirs on the Reliability of Cu/low-k Interconnects," J. Appl. Physics, V. 103, No 8 084513-1 to 11 (2008).
6. I. A. Blech, "Electromigration in Thin Aluminum Films on Titanium Nitride," J. Appl. Physics, V. 47, pp. 1203-8 (1976).
7. I. A. Blech and C. Herring, "Stress Generation by Electromigration," Appl. Phys. Lett. V. 29, pp. 131-3 (1976).
8. Young-Joon Park, Ki-Don Lee, William R. Hunter, "A Variable Current Exponent Model for Electromigration Lifetime Relaxation in Short Cu Interconnects," *Proc. of International Electron Devices Meeting (IEDM)* pp. 4.2.1-4.2.4. (2006).

Electromigration Mechanisms in Cu Nano-Wires

M. H. Lin, S. C. Lee, and A. S. Oates

Taiwan Semiconductor Manufacturing Company. Ltd.,
9, Creation Rd., Hsinchu, Taiwan 30077
tel:+886-3-566-6090, email:aoates@tsmc.com

ABSTRACT

In this article, we propose a new drift velocity technique to measure electromigration at temperatures of 125°C to directly assess electromigration transport at use conditions. We present measurements of the temperature of the drift velocity of Cu conductors with small and large polycrystalline grain size. A significant grain size dependence of drift velocity was found, indicating a large flux of atoms through grain boundaries when the fraction of polycrystalline segments is a significant fraction of the conductor length. A physical model is proposed to explain the drift velocity behavior in Cu nano-wires and we determine the diffusion parameters from grain boundary diffusion and interface diffusion.

INTRODUCTION

Electromigration transport mechanisms in IC thin film conductors have been extensively studied because of the potential for electromigration-induced void formation to limit reliability. For Al conductors grain boundary transport dominate electromigration when the conductor width is larger than the grain size. As feature sizes are reduced and grain structure approaches a bamboo configuration, electromigration transport is limited to bulk transport [1]. In contrast, Cu damascene conductors used in advanced IC technologies exhibit multiple modes of transport. Electromigration along the Cu/dielectric cap interface [2-4], the Cu/metal trench-liner interface [5], grain boundaries [6], dislocations, and the surfaces of voids at the Cu/dielectric interface [7,8] have all been suggested to occurs under various conditions of fabrication process, temperature and Cu geometry. In particular, recent studies of narrow Cu conductors have indicated a potential for grain boundaries to strongly influence electromigration as feature sizes are reduced due to limitations of grain growth in damascene trenches. Determining the dominant mode(s) of electromigration at circuit operating conditions is critical to accurate reliability estimation since the rate of mass transport determines failures times due to voiding [9]. Moreover, how such transport mechanisms are affected by reduction in the geometry of the Cu conductor, as well as changes in the material interfaces (e.g. replacement of dielectric cap layer with metallic) of the Cu damascene structure are important considerations for the viability of the continued scalability of Cu interconnects.

In this study we propose a new measurement method of the Cu electromigration drift velocity (v_d) that enables measurements of electromigration at temperatures typical of circuit operation. We are thus able, for the first time to our knowledge, to directly assess the transport mechanisms that operate at the conditions of circuit operation.

This paper is organized as follows: First, we demonstrate a new technique. Second, since electromigration mass transport phenomena are likely to increasingly involve grain boundary transport as conductor geometry shrink. We develop a predictive model of v_d as a function of Cu microstructure. As will become apparent, accurate characterization of v_d requires the knowledge of the degree of polycrystalline of the Cu conductor. We, therefore, verify a model to approximate the length of polycrystalline segments within a damascene conductor as a function of its width. In the result section we present experimental data for v_d, using both conventional as well as our newly proposed method, to verify our modeling approach, and demonstrate that electromigration is dominated by transport at the Cu/dielectric interface and grain boundaries for geometries of technological interest. Our modeling allows us to accurately determine the activation energies for diffusion along Cu grain boundaries, and along the Cu/capping interface.

EXPERIMENTAL DETAILS

Electromigration tests were carried out at the package level. Samples were fabricated using Cu dual-damascene process on 300 mm wafers. The film stacks in the test structures were fabricated using a dual-damascene process in which both vias and trenches were etched into the low-k dielectric followed by backfilling with a liner, a Cu seed, and electroplated Cu. The excess Cu and liner films were removed using a chemical mechanical polishing process. The TaN/Ta liner and Cu seed layers were deposited sequentially by using physical vapor deposition (PVD). All test structures were fabricated SiN based or metallic film capping layers. The inter-level dielectric was a CVD low k material. A two-level interconnects structure was tested,

978-1-4244-5430-3/10 $26.00 © 2010 IEEE

where via studs are located at both ends of metal line. Current densities in the range of 0.6 to 2.0×10^6 A/cm^2 (defined with respect to the cross section of the metal line) were used to limit the increase in temperature due to Joule heating to less than 3°C. To determine v_d we required accurate value of the temperature coefficient of resistance (TCR) of the trench barrier TaN/Ta. Our result shows the resistance of barrier layer versus temperature down to chip operation temperature are nearly temperature independent. This was determined to be between −259 to −414 ppm/°C i.e. negligible compared to that for Cu of 2500 ppm/°C. We, therefore, assume that the barrier resistivity is effectively temperature independent relative to the TCR for Cu, which is 2500 ppm/°C.

1. DRIFT VELOCITY TECHNIQUE AT LOW TEMPERATURES

The technique we use to determine v_d was originally developed by English and Kinsbron for Al-0.5%Cu films [10] and subsequently was adapted to determine v_d of circuit interconnects [11]. The basic principle is straightforward and involves measurement of the rate of resistance increase, dR/dt, of a conductor resulting from the motion of a void front along the conductor with time. The current density of the measurements was chosen to be a factor of an order of magnitude large than j$_c$ [12]. So that influence from j$_c$ on v_d is minimal. For a damascene structure consisting of a trench with a metallic barrier layer on 3 sides, and for a constant v_d, dR/dt is given by:

$$\frac{dR}{dt} = (\frac{\rho_b}{t_{bb}w} + \frac{\rho_b}{2t_{bs}h})v_d \qquad (1)$$

where ρ_b is the resistivity of barrier layers, and t_{bb} and t_{bs} is the bottom barrier thickness and sidewall thickness. Fig. 1 shows a typical example of resistance versus time evolution under electromigration stress of a test structure consisting of a single Cu conductor terminated by vias at each end. In this case a void is formed by electromigration in the metal trench connecting the vias, and a linear increase with time is clearly observed as the void grows demonstrating that v_d is constant for such void growth. However, a practical limitation to the application of this technique arises below about 200C, because very long times are required to obtain reliable values of dR/dt. To overcome this limitation, we propose to increase the sensitivity of the resistance measurement by using test structures consisting of serial chains of identical via units (multi-links) as shown schematically in Fig. 2. We have shown previously that electromigration in each link of such structures is identical to that which occurs in test structures consisting of single links. Assuming that electromigration-induced voiding can be made to occur in each link of an N-link structure, then magnitude of the dR/dt at any temperature is increased by a factor of N. Note that since we are only interested in determining dR/dt, this technique does not require identical

initial resistance change, dR, in each link. All that is required is that the dR corresponds to void growth along the conductor length. Thus, the temperature limitation to v_d determination can be almost entirely removed, and now the only limitation to the temperature range of v_d measurement results from the ability to accurately measure dR itself. In practice there are some potential sources of error in this approach to measure v_d: the occurrence of multiple voiding sites in links, will increase the absolute value of v_d although the temperature dependence will be unaffected provided the number of voids is temperature independent; voids can also occur inside vias, in addition to, or instead of in trenches. Since via-voids occur with relatively low probability with appropriate optimization of the via and trench processing [13], increasing the number of links in the structures can in principle reduce this error to a negligible value. As we discuss below, both these sources of error appear to have negligible impact on v_d, and so we proceed by assuming that each link has a single void in the trench region.

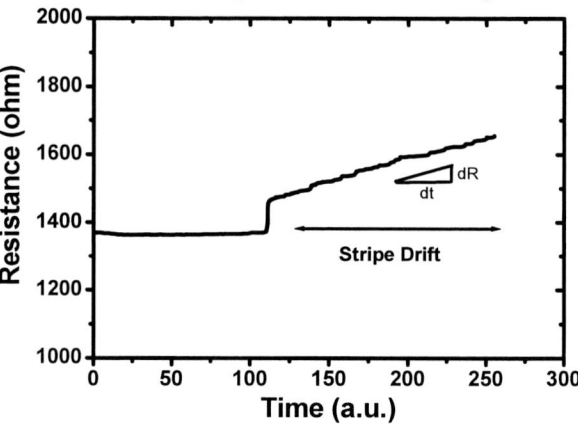

FIGURE 1. An example of resistance vs. time evolution, only progressive resistance increase range was used to calculate drift velocity.

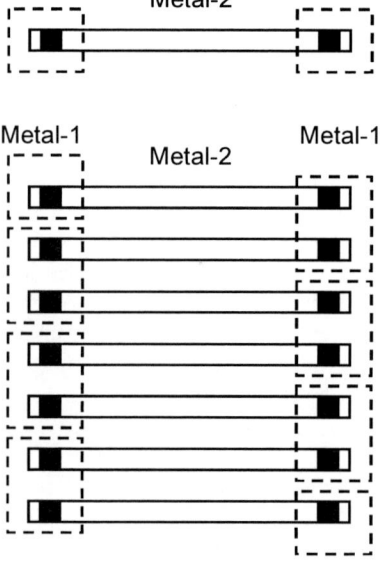

FIGURE 2. Schematic single and multi-link test structures.

To determine v_d, we first need to perform a "pre-stress" experiment with a sufficiently high temperature to ensure that voiding occurs in each link of the N-link structure. The v_d measurement begins after this pre-stress by changing the stress temperature to the desired value and then measuring dR/dt. Stress current density is chosen to minimize Joule heating. Cu dual damascene structures typically exhibit multiple modes of voiding due to electromigration. We use a test structure that is specifically designed to produce voids in trenches by using a 2 level test structure with narrow upper conductor width. We have previously shown that in the case of upstream electron flow, voids may also form inside vias, but the probability of via-voiding is ~2% in our samples and so here we assume that all voids occur in trenches [13].

We performed experiments with test structures consisting of 50 identical links, and compared the results with single-link structures, which have been used extensively to characterize v_d [14]. The magnitude of dR/dt of the 50-link structure is very close to 50 times that of the single-link (Table 1), implying that voiding in each of the links 50- and single-link structures is identical. Consequently, single-link and multi-link structures give identical v_d estimates, as shown in Fig. 3, confirming that the multi-link approach provides accurate estimates of v_d using the 50-link structure.

We extended drift measurements to temperatures typical of circuit operation. Fig. 4 shows an example of v_d measured in the range 300 to 125 °C. In principle, even lower temperatures may be accessed with a larger link number.

TABLE 1. dR/dt of multi-link divide by N is very close to single-link. Here N=50.

Number of links	N=1 (single-link)	N=50 (multi-links)
dR/dt	0.068	3.5

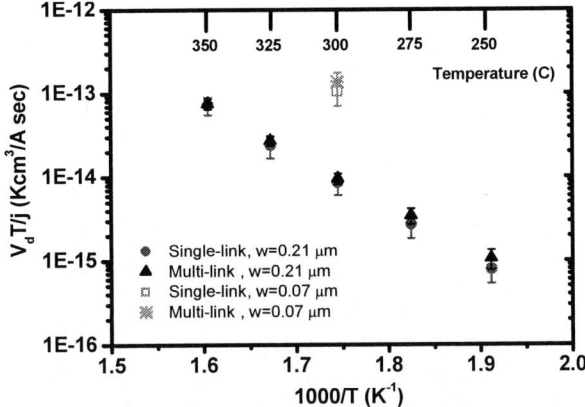

FIGURE 3. The comparison of v_d for single and multi-link at the same stress conditions over limited range of temperature.

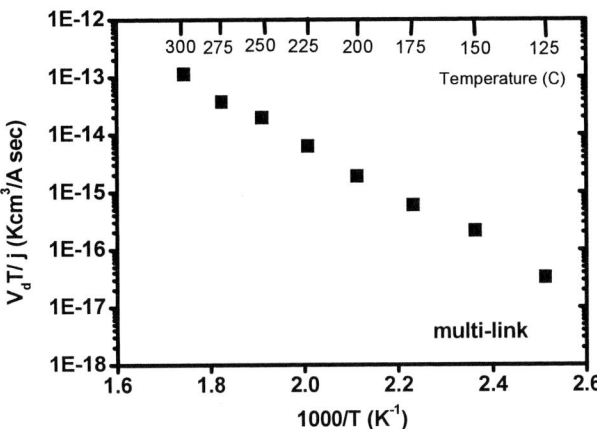

FIGURE 4. Using our new technique, v_d measured in the range 300 to 125°C.

2. Modeling of Electromigration Transport in Cu Damascene Conductors

2.1 Cu Drift Velocity (v_d):

In this section we develop a model of v_d for a Cu nano-wires incorporating the impact of grain structure. The Cu line is assumed to be capped by a dielectric layer, and the damascene trench to be lined on 3 sides by TaN-based barriers, since this is the most technologically relevant situation.

The flux of atoms due to electromigration is given by:

$$ J = \frac{ND_{eff}}{kT} eZ * \rho(j - j_c) = Nv_d \qquad (2) $$

where D_{eff} is the effective diffusivity of the atoms, $eZ*$ is the effective charge for electromigration, ρ is the resistivity of the conductor, N is the atom density, kT has the usual meaning, v_d is the average drift velocity of the atoms in the conductor, j is the current density, and j_c is the critical current density for electromigration. The critical current density is defined by $(j_cL)=\Omega\Delta\sigma/eZ*\rho$, where L is the stripe length, Ω is the atomic volume, and $\Delta\sigma$ is the maximum stress gradient that the conductor can sustain without plastic deformation. The effective diffusivity D_{eff} will have contributions from all possible diffusion paths in the conductor. The D_{eff} can be characterized as [15]:

$$ D_{eff} = n_B D_B + \sum_{j}^{n} D_{gbj}(\frac{\delta_{gbj}}{d}) + D_s\delta_s(\frac{1}{h}) $$
$$ + D_I[\delta_I(\frac{2}{W} + \frac{1}{h})] + D_p\rho_{disl}A^2 \qquad (3) $$

978-1-4244-5430-3/10 $26.00 © 2010 IEEE

where the subscripts B, gb, S, I, and p denote bulk, grain-boundary, top surface, barrier interface, and pipe parameters, respectively; δ denotes grain-boundary or interface width, w is the line width, h is the line thickness, d is Cu grain size, and ρ_{disl} is the dislocation density, A^2 is the cross section area of dislocation, and n_B, $\delta_I(2/W+1/h)$, δ_S/h, and δ_{gb}/d are the fractions of atoms diffusing along the bulk, interface, surface, and grain boundary, respectively. In addition to these mechanisms, it has been observed that electromigration can occur by motion of voids along the Cu/dielectric cap interface. This process involves diffusion of Cu atoms along the clean Cu surface of voids. The reported activation energy for this surface self-diffusion of Cu is 0.45 eV [8]. Because of the multiplicity of potential mass transport pathways in Cu, simplifications are required to make the description of v_d tractable. It is known that lattice diffusion is significantly less than grain boundary diffusion and so we neglect it here [2]. For Cu/dielectric interface diffusion, we assume that diffusion along single-crystal regions is much faster than regions where grain boundaries intersect the Cu surface. Recent studies have shown that Cu atoms moving along surfaces are significantly slowed at grain boundary triple points (i.e. where grain boundaries intercept the surface) [16]. Measurements of electromigration along the Cu/TaN interface have indicated a large activation energy (~1.4eV) [17], and so we neglect contributions to electromigration from this interface.

2.2 Modeling of Drift velocity for Polycrystalline Conductor

We begin by assuming an interconnect line is composed of single and polycrystalline segments in series, as shown in Fig. 5. Large arrows represent major EM transport paths. In single crystal segments, surface diffusion is the dominant transport path, while in polycrystalline segments, grain boundary diffusion dominates transport. We quantify the proportion of the total line length of polycrystalline segments as the parameter p. This parameter was used to explain the effects of microstructure on Al interconnect mass transport [18]. However, Cu interconnects are fabricated through a damascene process. Since it is typical that the thickness of the electroplated Cu film before chemical mechanical polishing (CMP) step exceeds the conductor width, the grain size is limited by the deposited Cu thickness rather than by the metal width [19]. A method to calculate p is presented to the following section. We determine the overall average drift velocity along unit single and ploy-crystalline segments with consideration of proportion of polycrystalline to be:

$$ J = cv_d = pcv_{GB} + (1-p)cv_S \qquad (4) $$

Here J is the atomic flux, c is the atomic density, and v_d is the average drift velocity we measured. v_{GB} is drift velocity along grain boundary, v_S is drift velocity along surface, and p is proportion of the total line length of polycrystalline segments.

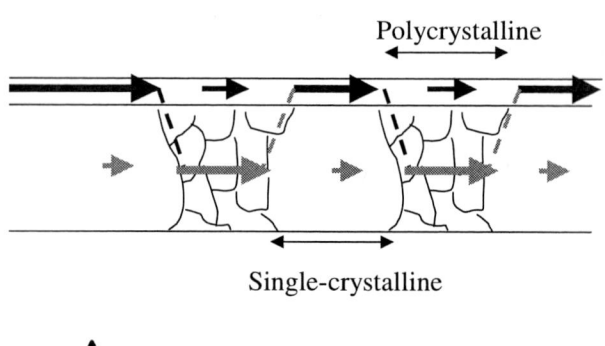

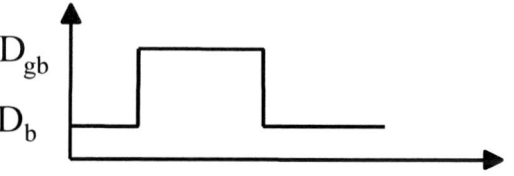

FIGURE 5. Schematic of the interconnect line is composed of single and polycrystalline segments in series. The arrows represent EM transport paths.

2.3 Calculation of Polycrystalline Segment Length Proportion

It is necessary to know the total polycrystalline segment length, p, to calculate v_d. Here we define L_p and L_s as the mean polycrystalline and single-crystal length for a given linewidth w. Using the assumption of log-normal grain size distribution [19], the total lengths of polycrystalline and single-crystal for a given metal line width w can be represented in closed form expressions (see Appendix).

$$ L_p = \frac{D_{50}}{W} e^{2\sigma^2} [1 + erf(\frac{\ln(W/D_{50}) - 2\sigma^2}{\sqrt{2}\sigma})] \qquad (5) $$

$$ L_s = D_{50} [\frac{1}{2} + \frac{1}{2} erf(\frac{\ln(W/D_{50})}{\sqrt{2}\sigma})] \qquad (6) $$

The proportion of the conductor length occupied by polycrystalline segments is then given by [20]:

$$ p = \frac{L_p}{L_p + L_s} \qquad (7) $$

TEM bright field images of grain structure as a function of conductor width were obtained to determine the accuracy of this model. Fig. 6 shows p measured using the mean lineal intercept method as a function of w. There is good quantitative agreement with equation 7, and in our subsequent study we will use this expression to determine grain structures for our Cu conductors.

978-1-4244-5430-3/10 $26.00 © 2010 IEEE

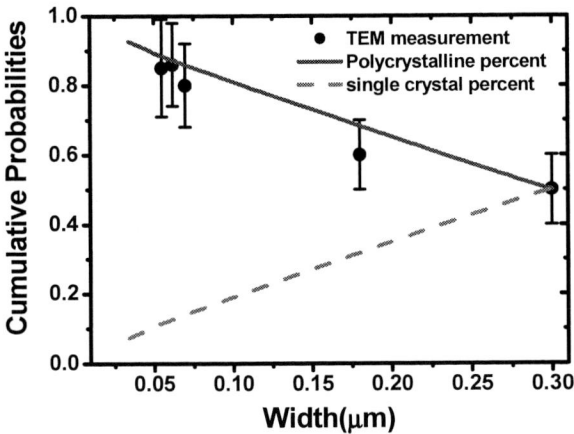

FIGURE 6. Predicted values of p as a function of width (w), compared with measurements by TEM.

RESULTS

Comparison of Experimental v_d with Model Predictions for a Dielectric Cap Layer

Fig. 7 plots v_d at 300 °C as a function of p for single-link structures with a dielectric cap layer. There is an increase of over an order of magnitude in v_d as p decreases from ~0.9 to 0.5. The increase in v_d is due to changes in the relative proportion of grain boundary and Cu/dielectric transport. Equation (4) accurately describes the experimental data using a single set of diffusion parameters, which are listed in table 2. The samples for this experiment were all obtained from a single wafer to minimize variations in the thickness of the trench liner, which can influence the absolute value of v_d through the sheet resistance, R_s. The E_a values of 0.84 and 0.95 eV are required for grain boundaries and the Cu/dielectric interface respectively. Grain boundary diffusion activation energies have been determined to be in the range 0.78 – 0.84 eV on bulks samples, and our estimate is consistent with this range within experimental uncertainty. Lloyd et al [21] also reported a similar value of 0.84 eV for polycrystalline damascene Cu conductors with similar trench and dielectric cap materials. Direct measurements of the activation energy for dielectric interface diffusion are not available, and our data represents an unambiguous determination of this value for the Cu/dielectric system.

We now consider the transport mechanism(s) that are involved in Cu damascene conductors. First, from the magnitude of the measured E_a, and the fact that the anticipated v_d due to void motion along the Cu surface at 125 °C is around three orders of magnitude higher than our directly measured value [8], we conclude that void motion along the Cu surface not the rate limiting factor in electromigration transport or voiding failure at circuit operating conditions. We infer that electromigration must occur either along the dielectric or metallic interfaces, or within the conductor itself.

Fig. 8 shows the measured temperature dependence of v_d as a function of p value. In this case only single link test structures were used and so the temperature range of measurements is limited to 175°C. An excellent quantitative fit to the whole data set is obtained using identical parameters as those used in fig. 7. The extrapolated temperature data of fig. 8 also show that the apparent E_a, defined as the average over the temperature range 300 to 100 °C, is quite sensitive to p, with changes of about 0.1 eV occurring for the relatively small range of p 0.6 to 0.9. The importance of grain boundaries to v_d is particularly evident for the narrowest lines, where v_d is relative large, and E_a can reach as low as 0.84 eV for conductors with large polycrystalline proportions. An experimental example of this sensitivity is shown in fig. 9, using the data shown in figure 4. Included in this figure is the model prediction using the parameters in table 2. The model prediction to this data is in good agreement with the experimental data, but now because the conductor is almost totally polycrystalline (p = 0.95) apparent Ea=0.84 eV is both observed and predicted.

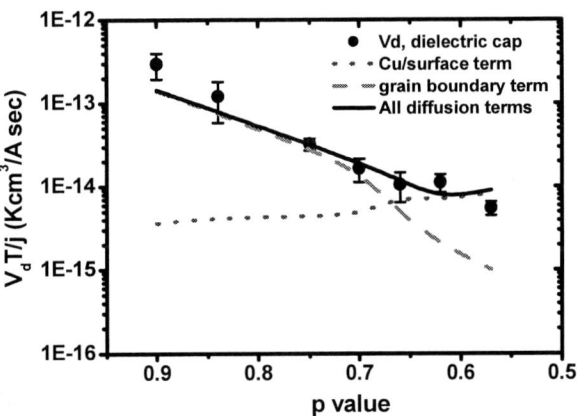

FIGURE 7. Drift velocity at 300C as a function of p (symbols), together with model of diffusion paths (lines).

TABLE 2. Diffusion path parameters used in model prediction

Diffusion path	Surface diffusion	Grain Boundary
Ea (eV)	0.95	0.84
$Z^*D_0\delta$ (cm^3/sec)	2.5×10^{-9}	4×10^{-10}

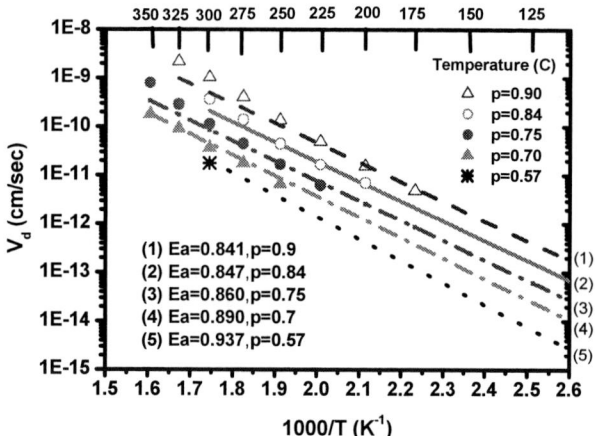

FIGURE 8. Measured temperature dependence of v_d as a function of p value.

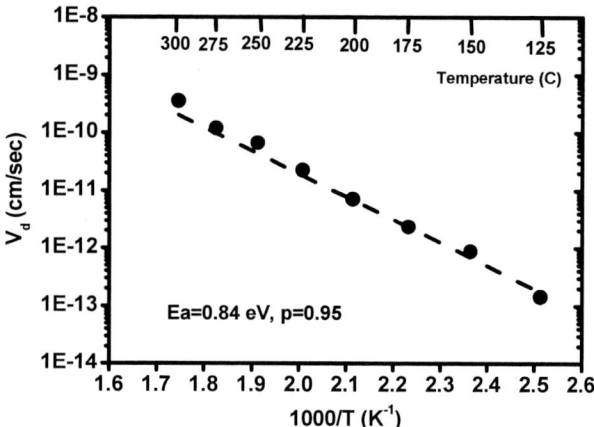

FIGURE 9. Experimental normalized drift velocity (symbols), together with model prediction (lines).

DISCUSSION

Our data indicate that electromigration transport in damascene Cu conductors can be accurately described by considering contributions from grain boundaries and capping interfaces for the most commonly used materials used in circuit fabrication. We find that the activation energy for electromigration is, in principle, dependent on p value because of the differing contributions of grain boundaries and interface mechanisms with temperature. One may argue that the contribution of transport path along Cu/barrier (TaN/Ta) or dislocation pipe should be estimated before being ignored. First, we examine the contribution of electromigration along the TaN/Ta interface, which we have thus far assumed to be negligible in our analysis. One reason to reconsider this pathway is the observation of a deviation between our model prediction and experimental data at the higher p (≥ 0.75) at temperatures above 300°C (fig. 8). Since we do not observe this deviation at lower p (=0.7), it is possible that

the mechanism involved is grain size dependent, which suggests that it is not lattice diffusion. This leaves the possibility that diffusion along the Cu/(TaN/Ta) metallic interface is contributing to high temperature electromigration. While we have no data for dislocation densities in our samples, recent measurements by others of dislocations in Cu conductors using synchrotron techniques have indicated much lower densities of 10^{11} cm^{-2} at widths in the range of 0.5 µm [7]. To explain the high temperature deviation by dislocation pipe diffusion would require dislocation densities larger than 10^{15} cm^{-2}. Unless dislocation densities are a very strong function of width, pipe diffusion appears to be an unlikely explanation. Furthermore, given that dislocation pipe diffusion has activation energy of 1.2 eV, it is improbable that it has a major contribution at circuit operating temperatures. We can, therefore, suggest that our measurements of v_d confirm the most fundamental assumptions of electromigration reliability engineering: that it is a reasonable approximation to use a single activation energy, determined at accelerated conditions, to extrapolate data to circuit operating temperatures.

CONCLUSIONS

We have developed a new method to measure v_d of with significantly increased sensitivity to dR. We model v_d as a function of Cu microstructure, and show that grain boundaries dominate when microstructure consists of a large fraction of polycrystalline segments. How to control microstructure of Cu nano-wire and block grain boundary diffusion will be the critical key factor for interconnects beyond 45 nm technology nodes.

ACKNOWLEDGEMENTS

The authors thank S.Y. Chiang, and Y.C. Sun for their support of this work. They also wish to acknowledge the support of the TSMC Failure Analysis Division in providing TEM images of Cu line morphologies.

REFERENCES

[1] C. K. Hu, M. B. Small, K. P. Rodbell, C. Stanis, and P. Blauner, Appl. Phys. Lett. **62**, p. 1023, 1993.
[2] C. K. Hu, R. Resenberg, and K. Y. Lee, Appl. Phys. Lett. **74**, p. 2945, 1999.
[3] C. K. Hu, R. Rosenberg, H. S. Rathore, D. B. Nguyen, and B. Agarwala, *in Proc. Int. Interconnect Technology Conf. (IITC)*, p. 267, 1999.
[4] C. S. Hau-Riege, Microelectronics Reliability **44**, p.195, 2005.
[5] J. Proost, T. Hirato, T. Furuhara, K. Maex, and J.-P. Celis, J. Appl. Phys. **87** p.2792, 2000.

[6] C. K. Hu, L. Gignac, B. Baker, E. Liniger, and R. Yu, in *Proc. Int. Interconnect Technology Conf. (IITC)*, p. 93, 2007.

[7] A. S. Budiman, C. S. Hau-Riege, P. R. Besser, A. Marathe, Y.-C. Joo, N. Tamura, J. R. Patel, and W. D. Nix, in *Proc. Int. Reliability Physics Symp.(IRPS)*, p.122, 2007.

[8] Z.-S. Choi, R. Monig, and C. V. Thompson, J. Appl. Phys., **102**, 083509, 2007.

[9] M. Hauschildt, et al, J. Appl. Phys., **101**, 043523, 2007.

[10] A. T. English and E. Kinsborn, J. Appl. Phys., **54**, 268, 1983.

[11] A. S. Oates, J. Appl. Phys., **70**, 5369, 1991.

[12] A. S. Oates and M. H. Lin, *Trans. Dev. Mat. Rel.*, vol. 9, p.244, 2009.

[13] A.S. Oates and S. C. Lee, Microelectronics Reliability, **46**, p.1581, 2006

[14] S. C. Lee and A. S. Oates, in *Proc. Int. Reliab. Phys. Symp.(IRPS)*, p.107, 2006.

[15] C. K. Hu, L. Gignac, and R. Rosenberg, Microelectronics Reliability 46, p.213, 2006.

[16] K. C. Chen, et al, Science, 321, 1066, 2008.

[17] C. K. Hu et al. Appl. Phys. Lett. **84**, p. 4986, 2004.

[18] K. Y. Fu, J. Appl. Phys., **69**, 2656, 1991.

[19] A.von Glasow, A. H. Fisher, and A. E. Zitzelsberger, in *Proc. Int. Reliab. Phys. Symp.(IRPS)*, pp126-131, 2003.

[20] M. L. Dreyer, K. Y. Fu, and C. J. Varker, J. Appl. Phys. **73** p.4894, 1993.

[21] J. R. Lloyd, M. W. Lane, E. G. Liniger, C. K. Hu, T. M. Shaw, and R. Rosenberg, *Trans. Dev. Mat. Rel.*, vol. **5**, p.113, 2005.

APPEDIX

Here we define L_p and L_s as the mean polycrystalline and single-crystal length for a given linewidth W. We assume a grain is represented by a square with size D_i. Schematic diagram for simplification of Lp calculation is shown below:

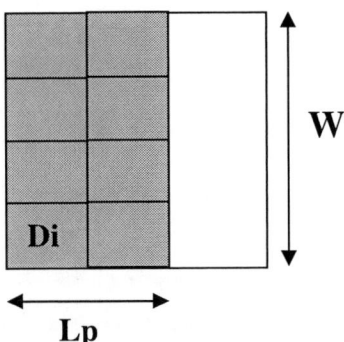

The area of a polycrystalline segment L_pW is proportional to the sum of the incremental grain areas $D_i^2N(D_i)$, where $N(D_i)$ is the number of grain having a characteristic length D_i. Lp can be then expressed as:

$$L_p = \frac{1}{W}\sum_i D_i^2 N(D_i) \tag{A1}$$

For a single-crystal segment with average length Ls, the Ls is proportional to $D_{50}N(D_i)$. The segment length Ls can be expressed as:

$$L_s = D_{50}\sum_i N(D_i) \tag{A2}$$

We assume Cu grain size distribution follows a lognormal distribution [9]. having a median grain size D_{50}. Defining $x=D/D_{50}$, the probability density function of $f(x)$ is given by:

$$f(x) = f(\frac{D}{D_{50}}) = \frac{1}{\sqrt{2\pi}\sigma(D/D_{50})}\exp\frac{-[\ln(D/D_{50})]^2}{2\sigma^2} \tag{A3}$$

Assuming all Cu grains are symmetric with a characteristic length D_i. Equations (A1) and (A2) can be rearranged as

$$L_p = \frac{1}{W}\int_0^W x^2 f(x)d(x) \tag{A4}$$

$$L_s = D_{50}\int_0^W f(x)d(x) \tag{A5}$$

(A4) and (A5) can be solved to give:

$$L_p = \frac{D_{50}}{W}e^{2\sigma^2}[1+erf(\frac{\ln(W/D_{50})-2\sigma^2}{\sqrt{2}\sigma}] \tag{A6}$$

$$L_s = D_{50}[\frac{1}{2}+\frac{1}{2}erf(\frac{\ln(W/D_{50})}{\sqrt{2}\sigma})] \tag{A7}$$

The proportion of the conductor length occupied by polycrystalline segments is then given by:

$$p = \frac{L_p}{L_p+L_s} \tag{A8}$$

Degradation and Failure Analysis of Copper and Tungsten Contacts under High Fluence Stress

Thomas Kauerauf, Geni Butera, Kristof Croes, Steven Demuynck, Christopher J. Wilson, Philippe Roussel, Chris Drijbooms, Hugo Bender, Melina Lofrano, Bart Vandevelde, Zsolt Tőkei, Guido Groeseneken*

imec
Kapeldreef 75, B-3001 Leuven, Belgium
phone: +32 1628 1148, e-mail: kauerauf@imec.be
*also at KU Leuven, ESAT Department, Leuven, Belgium

Abstract - **The reliability of Cu and W contacts under high fluence stress mimicking source/drain contacts in the on-state of a transistor is evaluated. We use Kelvin structures to study the contact degradation and to determine the lifetime as a function of voltage and temperature. Failure analysis reveals significant damage created in the proximity of the contacts. It is concluded that not electromigration alone, but also Joule heating of the contact and the contact interfaces triggers failure.**

Keyword: plug, Cu, W, barrier reliability

I. INTRODUCTION

The concept of Cu contacts is being considered for advanced technology nodes due to advantages related to lower contact resistance and reduced tool cost [1-3]. The use of Cu, however, requires a diffusion barrier in order to guarantee sufficient reliability and yield, and continuous optimization of this barrier is necessary to reduce the contact resistance and ensure further contact scaling [4]. So far, the reliability of different Cu/barrier combinations has mainly been studied by (i) gate oxide breakdown [5-8] or (ii) on electromigration (EM) structures [9].

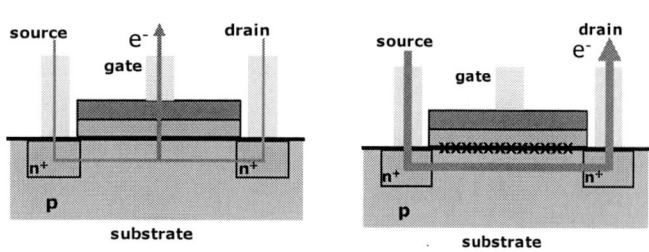

Fig. 1. During gate oxide tests the current flow through the contacts is limited (left), while in the transistor on-state the current through S/D contacts can be significantly higher (right). In this case, however, the gate dielectric will degrade and prevent controlled contact stress experiments.

In this work we are investigating the degradation of source/drain (S/D) contacts under high fluence, mimicking an on-state of a transistor. To our opinion, this is a crucial test

since both the increased drive current and reduced contact size will lead to significant current density increase in S/D contacts of future scaled devices, and the higher aspect ratio compared to gate contacts makes the barrier more vulnerable [10]. The most realistic device to study high fluence stress in S/D contacts is obviously a transistor (Fig. 1). The single S/D contact transistors presented in [8] would allow high current density, but the predominant degradation of the gate dielectric leading to V_t shifts and reduced drive current does not allow a controlled high fluence stress experiment.

Measuring through pn-junctions would avoid oxide degradation, but either the large voltage drop over the Si or multiple contacts leading to low current densities prevent their use. In our opinion, the ideal structures for high fluence experiments in contacts are therefore integrated four terminal Kelvin structures: the current is forced through a single contact (Fig. 2), the low resistivity connection allows for a controlled experiment and the contact resistance can be monitored during stress.

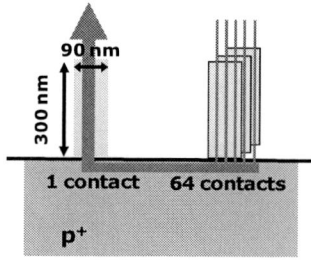

Fig. 2. Layout and schematics of a Kelvin structure typically used for resistance measurements.

II. EXPERIMENTAL

The Kelvin structures we tested were 90 nm diameter (bottom 40-50 nm, top 80-90 nm) high aspect ratio M1 to p+ active area contacts processed in 65 nm CMOS technology (Fig. 3). To benchmark the degradation and failure mechanism we studied both W and Cu contacts, with the latter having a 10nm PVD TaN diffusion barrier. The resistance of a single contact is about 22 Ω with Cu and 42 Ω with W.

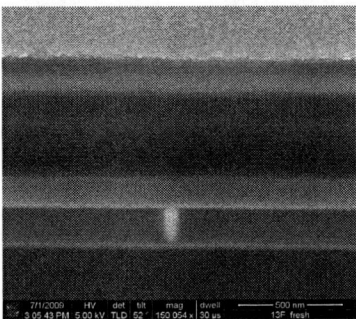

Fig. 3. SEM cross-sections of unstressed Kelvin structures. On the top a Cu contact with 10 nm PVD TaN diffusion barrier is shown. On the bottom a W contact and the typical proximity of unstressed contacts.

Initial I-V characteristics of the Kelvin structures show a linear current increase, implying that the structure can be seen as a simple resistor (Fig. 4). In agreement, the current through 150 nm contacts is larger compared to 90 nm contacts and at elevated temperature (200°C) a lower current is measured. For the 90 nm contacts, a sudden drop in current around 1.3 V can be observed.

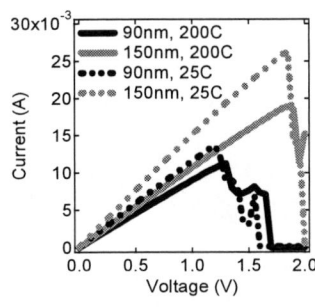

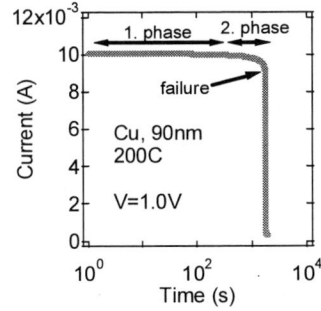

Fig. 4. I-V curves of 90 nm and 150 nm Kelvin structures at 25°C and 200°C. With the 90 nm contact the current drops around 1.3 V.

Fig. 5. I-t curve of a 90 nm Cu Kelvin structure. Time-to-failure is triggered if the current drops 10% below the initial value.

To stress at a realistic scenario the degradation of the contacts was studied by constant voltage stress (CVS) in which the current through the contact is monitored over time. The measurements were done on wafer level at a temperature of 25°C and 200°C. For each stress condition approximately 10 fresh 90 nm contact Kelvin structures were tested and to simplify the experiment we used a two-point probe technique without measuring the resistance simultaneously.

III. RESULTS AND DISCUSSION

When applying a constant voltage over a Kelvin structure and measuring the current over time, two phases of degradation are present: during the first the current slowly decreases, while during the second an accelerated current reduction is observed (Fig. 5). To avoid triggering failure during the first phase, a current reduction of 10% from the initial value is defined as the criterion.

The stress conditions are chosen such that failure of the Kelvin structures under stress occurs above the time resolution of the parameter analyzer (V_{stress}=1.3 V) but within reasonable testing time (V_{stress}=0.8 V). It is important to notice that at these stress voltages a current of ~10 mA is flowing through a single contact (Fig. 6). Even when considering the wider top part of the contact (90 nm), this represents a current density larger than 150 MA/cm^2.

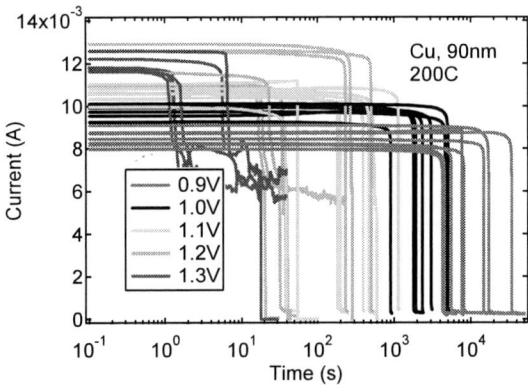

Fig. 6. A set of I-t traces measured at a temperature of 200°C on 90 nm Cu Kelvin structures for different stress voltages.

A. Electrical stress

To investigate the voltage dependence of the time-to-failure (t_f), CVS on Kelvin structures with Cu or W contacts was applied. From the I-t traces measured on the Kelvin structures with Cu contacts shown in Fig. 6, several observations can be made: (i) As expected, at lower voltage the current is reduced and t_f increases; (ii) On a log-time scale the length of the first phase is always a fixed fraction of the total failure time t_f. This suggests that both degradation phases are related to the same mechanism; (iii) Even though some devices stressed at different voltage carry similar current, the ones stressed at higher voltage systematically fail earlier; (iv) At the same stress voltage in most cases a larger t_f is observed for the contacts with higher current (Fig. 7), being an indication that the failure mechanism is related to resistance and Joule heating.

978-1-4244-5430-3/10 $26.00 © 2010 IEEE   713

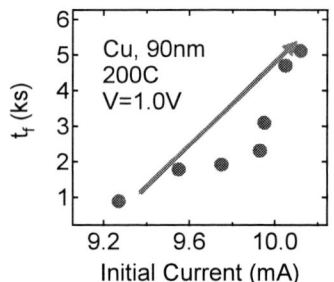
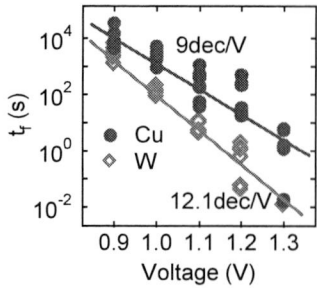

Fig. 7. At the same stress voltage the Kelvin structure with higher initial current exhibits a longer lifetime.

Fig. 8. Voltage acceleration of t_f for the Cu and W Kelvin structures. When stressed at the same voltage Cu contacts fail later.

B. Failure analysis

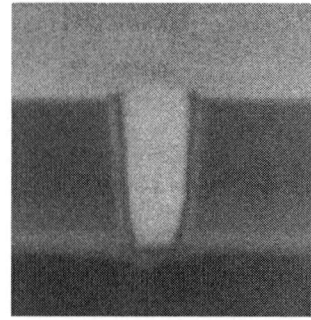
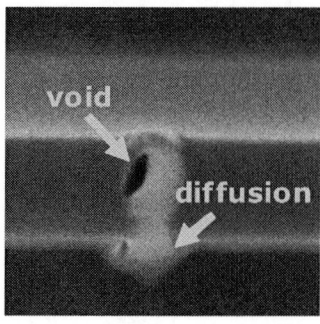

Fig. 10. SEM images of stressed W (left) and Cu (right) contacts. While with W contacts only minor damage at the bottom is visible, the Cu contact shows voids and diffusion into the Si.

After determining the t_f for both Cu and W contacts for at least 8 samples per stress voltage, an exponential dependence of t_f on the stress voltage is found (Fig. 8). Although at the same stress voltage the current through Cu contacts compared to W contacts is higher (10 mA vs. 8 mA), the t_f with Cu contacts is longer. This can be attributed to the lower resistivity of the Cu leading to reduced Joule heating (see next section). The voltage acceleration of W contacts, however, is higher (Fig. 8), suggesting a longer lifetime at low stress voltages.

In a further experiment, the temperature dependence of t_f with Cu contacts was studied, and at the same stress voltages the lifetime at 25°C appears to be lower than at 200°C. This is, however, entirely due to the lower resistivity of the contact at lower temperatures (Fig. 4). When plotting the t_f vs. the initial current for both temperatures, no additional degradation at higher ambient temperature is observed (Fig. 9). This implies that when testing contacts at typical electromigration (EM) experiment temperatures of 300°C with CVS, the obtained lifetime might actually be overestimated.

Even when the stress was immediately stopped after a 10% current reduction, from the FIB/SEM failure analysis on stressed samples, considerable damage is observed (Fig. 10). The W contacts themselves are almost unchanged and only minor extensions into the silicide below are detected. The Cu contacts, however, exhibit significant voiding and Cu silicide is formed. Since Cu silicide formation through TaN barrier requires temperatures ~550°C, significant heat generation on the bottom of the contact is expected.

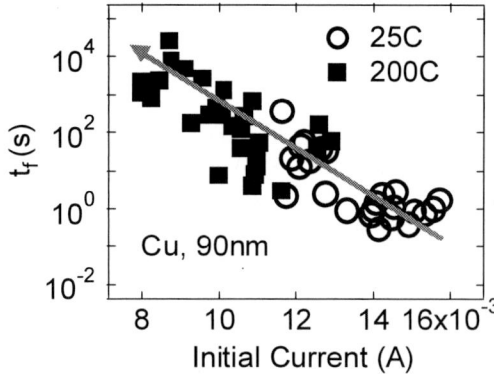

Fig. 9. The t_f vs. the initial stress current of 90 nm Cu contacts measured at 25°C and 200°C. The lifetime is not reduced when stressing at higher temperature.

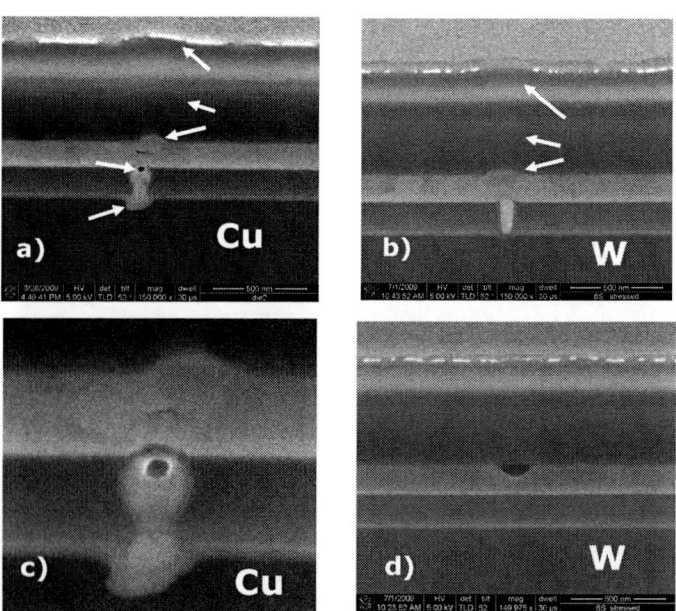

Fig. 11. The proximity of stressed contacts. Independent of the contact material, severe damage above the contact in the M1 is observed and the distortion extends to the layers above.

Common for both contact materials is severe damage in the M1 layer (Fig. 11). In the Cu grain-boundaries appear (Fig. 11a and 11c) and in several cases large voids are created (Fig. 11a and 11d). Moreover, above the contact a distortion is formed which extends to the overlying layers (Fig. 11a and 11b). Creating such kind of damage requires considerable forces and possible explanations are electromigration and/or Joule heating. Although the contact dimensions are much smaller than in typical EM tests [11, 12], the Blech threshold product of about 4500 A/cm is above the values reported for interconnects [11]. Since the damage also occurs with W where no voids in the contact are present, our assumption is that local Joule heating causes significant temperature increase.

C. Thermo-mechanical simulations

To identify the failure mechanism, various thermo-mechanical simulations using an axis symmetric finite element model (FEM) have been carried out. In a first attempt, a very simple model with a Cu wire surrounded by SiO_2 was simulated. For a current density of 157 MA/cm^2, corresponding to a current of 10 mA and 90 nm wire diameter, a temperature increase of 344 K due to Joule heating was calculated. Considering the bottom of a contact with 50 nm diameter (Fig. 3), the same current results in a temperature increase of 3650 K. This value seems to be very high, but is still significantly lower than one would expect from the $\Delta T=0.32*J^2$ relation obtained experimentally on Cu interconnects [13].

In a second attempt a more realistic model of cylindrical 90 nm Cu and W contacts with the layers and dimensions shown in Fig. 12 was built. Based on electrical measurement with 90 nm contacts, a total electrical resistance of 22 Ω with Cu and 42 Ω with W is obtained. This compares to a total resistance of 5 Ω and 12 Ω, respectively, calculated using the resistivity of the individual components. This reveals that the dominant resistance contribution, and power loss, has to be attributed to the interface between the metal and the NiSi. It was shown in [4, 14] that this interface resistance strongly depends on the cleaning process and to include the effect of the interface resistance a contact element was included in the FEM.

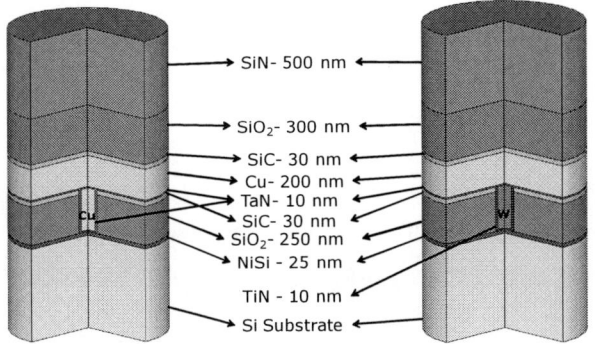

Fig. 12. The various layers of the axis symmetric finite element model (FEM) which was used for the thermo-mechanical simulations of the Cu and W contacts under electrical stress.

In contrast to the results of the simple Cu wire, the simulations of the more realistic 90 nm contact structures (Fig. 12) suggest that significant heat is only generated in the interface between the barrier and the NiSi (Fig. 13). For a current of 10 mA a maximum temperature increase of $\Delta T=92K$ for Cu and $\Delta T=211$ K for W contacts is determined. The area and volume of the M1 layer is significantly larger than the contact and the M1 might act as a gigantic cooling element reducing the overall temperature. A maximum temperature increase of 211 K, however, seems not high enough to explain the severe damage observed.

We extended the simulations by introducing voids at different positions in the contact similar to the observations in the stressed samples (Fig. 11). This results in very high local current densities, leading to a calculated temperature rise $\Delta T>1000K$.

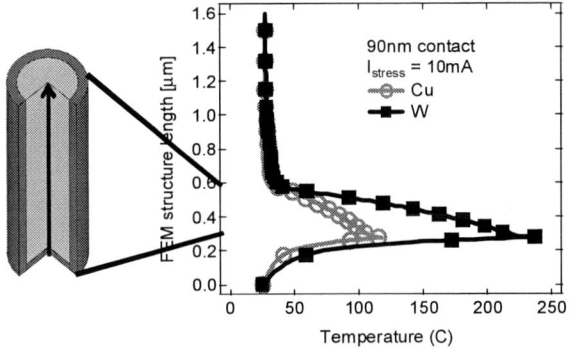

Fig. 13. The simulated temperature along 90 nm Cu and W contacts shows maximum heat generation at the silicide interface.

IV. CONCLUSIONS

High fluence stress in Cu and W contacts is studied on Kelvin structures and we propose this method for dedicated contact reliability evaluation. We demonstrate similar failure times for both filling materials and that time-to-failure exponentially depends on the voltage. After stress severe damage in the contact proximity is observed, originating from local heating at the contact interfaces. Therefore independent of the filling materials, for scaled contacts interface engineering is necessary to guarantee sufficient reliability.

REFERENCES

[1] S. Demuynck, A. Nackaerts, G. Van den bosch, T. Chiarella, J. Ramos, Zs. Tőkei, J. Vaes, N. Heylen, G.P. Beyer, M. Van Hove, T. Mandrekar, R. Schreutelkamp, "Impact of Cu contacts on front-end performance: a projection towards 22nm node", IITC, pp. 178-180, 2006.

[2] R. Islam, S. Venkatesan, M. Woo, R. Nagabushnam, D. Denning, K. Yu, O. Adetutu, J. Farkas, Tab Stephens and T. Sparks, "A 0.20 pm CMOS technology with copper-filled contact and local interconnect", VLSI Technology, pp. 22-23, 2000.

[3] M. Inohara, T.Fujimaki, K.Yoshida, K.Miyamoto, T.Katata, J.Wada, A.Sakata, K.Kinoshita, F.Matsuoka, "Copper Filling Contact Process to Realize Low Resistance and Low Cost Production fully compatible to SOC devices", IEDM, pp. 4.6.1 - 4.6.3, 2001.

[4] S. Demuynck, "Patterning & metallization options for advanced contact module integration", SSDM, pp. 795-796, 2009.

[5] G. Van den bosch, S. Demuynck, Zs. Tőkei, G. Beyer, M. Van Hove, G. Groeseneken, "Impact of copper contacts on CMOS front-end yield and reliability", IEDM, pp. 1-4, 2006.

[6] Soon-Cheon Seo, Chih-Chao Yang, Chao-Kun Hu,a Andreas Kerber,b Dave Horak, Karen Petrillo, Steve Holmes, Susan Fan, Veeraraghavan S Basker, Bala Haran, Christian Lavoie, Lahir Adam, James Demarest, Philip Flaitz,c Zhibin Ren,c Andreas Knorr,d Donald Canaperi, Matthew Smalley, Jason Cummings, Tuan Vo, James Kelly, Sanjay Mehta, Dae-Gyu Park,c Jean Wynne, Terry Spooner, Daniel Edelstein, Vamsi Paruchuri, Bruce Doris, Dale McHerron, "Copper contact metallization using Ru-based barrier liners for 45 nm and beyond", AMC, pp. 31-32, 2008.

[7] T. Kauerauf, S. Demuynck, Zs. Tőkei, K. Croes, G. Beyer, G. Groeseneken, "Reliability analysis of Cu contacts with various diffusion barriers", AMC, pp. 197-198, 2008.

[8] T. Kauerauf, S. Demuynck, G. Butera, J. Bogan, Zs. Tőkei, G. Groeseneken, "On the reliability of Cu contacts for the 32nm technology node and beyond", SSDM, pp. 799-800, 2009.

[9] K. Wang, C.J. Wilson, A. Cuthbertson, R. Herberholz, H.P. Coulson, A.G. O'Neill, A.B. Horsfall, "Influence of barriers on the reliability of dual damascene copper contacts", IRPS, pp. 677-678, 2008.

[10] ITRS, www.itrs.net

[11] E.T. Ogawa, K.-D. Lee, V.A. Blaschke, Paul.S. Ho, "Electromigration reliability issues in dual-damascene Cu interconnections", IEEE Transactions on reliability, vol. 51, no. 4, pp. 403-419, 2002.

[12] A.S. Oates, M.H. Lin, "Electromigration Failure Distributions of Cu/Low-k Dual-Damascene Vias: Impact of the Critical Current Density and a New Reliability Extrapolation Methodology", IEEE Transactions on device and materials reliability, vol. 9, no. 2, pp. 244-254, 2009.

[13] K.D. Lee, "Electromigration Critical Length Effect and Early Failures in Cu/oxide and Cu/low k Interconnects", PhD thesis, pp. 35, 2003.

[14] I. Vos, D. Hellin, S. Demuynck, O. Richard, T. Conard, J. Vertommen, W. Boullarta, "A Novel Concept for Contact Etch Residue Removal", ECS Transactions, vol. 11, no. 2, pp. 403-407, 2007.

EFFECTIVE THERMAL CHARACTERISTICS TO SUPPRESS JOULE HEATING IMPACTS ON ELECTROMIGRATION IN CU/LOW-K INTERCONNECTS

S. Yokogawa, H. Tsuchiya, and Y. Kakuhara

Advanced Device Development Division, NEC Electronics Corporation
1120 Shimokuzawa, Sagamihara, Kanagawa 229-1198, Japan
phone: +81-42-771-4274; fax: +81-42-771-0916; e-mail: shinji.yokogawa@necel.com

Abstract— We investigated the thermal characteristics of Cu/low-k interconnects to suppress the impact of Joule heating (JH) on electromigration. The thermal time constants in multilayered Cu/low-k interconnects were experimentally investigated for the first time on the basis of the transient thermal response. Furthermore, we analyzed the use of direct radiation through stacked contact/via to suppress the temperature increase from JH. The impact of JH on the critical product of the current density and the line length was also investigated.

Key words; Electromigration; Joule heating; Transient thermal response; Thermal conduction; Critical product

I. INTRODUCTION

The impact of the temperature increase due to Joule heating (JH) on the reliability of a recent Cu/low-k material is substantial and cannot be disregarded [1, 2]. Because low-k dielectrics, generally with a higher degree of porosity, are characterized by poorer thermal conductivity, JH is enhanced by the resistivity increase. With scaling, these trends induce lifetime degradation.

In contrast, the operating current density inevitably increases because of the scaling scenario. This increase in current density causes not only a temperature increase because of JH but also an accelerated degradation of electromigration (EM). Furthermore, the EM lifetime decreases as the cross-sectional area of the line decreases [3–5]. To overcome the problem of EM degradation, several advanced process technologies, labeled "EM boosters," have been proposed. However, the tradeoff for improving the EM is an increase in resistivity [6, 7], which contributes to the temperature increase due to JH.

Recently, we reported that an enhancement in the reliability of 32-nm node technology without an excessive increase in Cu line resistance occurs when Cu 0.5-wt% Al seed technology is used with annealing at high temperature over a short time period [8]. The annealing property of this seed technology results in an acceleration of the Cu grain growth, which leads to a suitable Al concentration in the Cu line. With a post-electroplating annealing condition of 350 °C for 30 s, Al accumulates at the Cu/SiCN interface. The EM lifetime improves because of the suppression of Cu diffusion, which in turn results from the accumulation of Al at the top surface of the Cu interconnects.

In addition, we compared Ti as a barrier metal (Cu/Ti) and CVD-Co capping (Co/Cu/Ta/TaN) [9]. Both Cu interconnects were close enough to show longer EM lifetimes and larger lifetime activation energies. The residual resistance of the Cu lines was measured cryogenically; the measurements showed that the residual resistance of Cu/Ti is higher than that of Co/Cu/Ta/TaN and Cu/Ta/TaN, which indicates that more impurities are present in Cu/Ti and that the grain size in Cu/Ti is smaller. The results of physical analysis indicate that a large amount of Ti doped into the Cu line segregates mostly at the grain boundaries on the Cu surface. Co capping suppresses Cu diffusion through the whole on the Cu surface. However, Ti doped into the Cu line suppresses Cu diffusion through the grain boundaries on the Cu surface. Because the grain boundary density increases on the Cu surface as the interconnects shrink, suppressing grain boundary diffusion on the Cu surface is a highly effective method to strengthen the EM reliability in narrow lines with small grains.

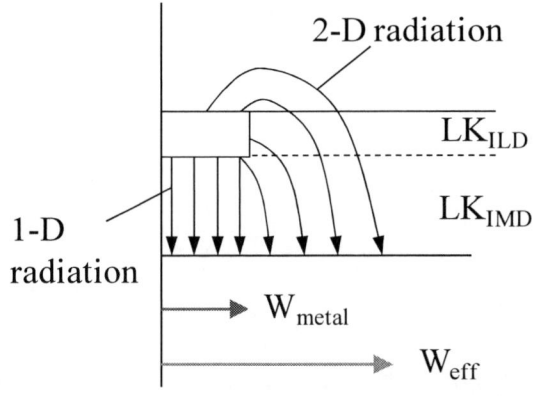

Figure 1. Schematic of radiation from metal to Si substrate.

To achieve a higher current density limit along with long-lifetime EM boosters, a method for controlling JH needs to be developed. In CMOS circuits, many interconnects carry pulsed (unidirectional) direct current (PDC) and pulsed alternating current (PAC). The former is carried by local VDD and GND lines, and the latter is carried by clock and signal lines. The pure alternating component of the current only contributes to temperature rise because of JH. Meanwhile, the average component of the current contributes to both EM and JH. In general, the "design-rule current" is classified into two parts [10]. The first part J_{avg} is interpreted to be the mean time average value of the current. The second part, J_{rms}, is set to limit the temperature rise due to JH. The J_{rms} part will not only prevent an acceleration of EM induced by the J_{avg} part but also effectively control the increase in line resistance that is caused by the JH-temperature increase. Both parts depend somewhat on the current waveform and the frequency [11–17]. To achieve a highly reliable interconnect system, both parts must be considered in the waveform and layout designs.

Last year, we reported the effect of JH on the EM lifetime in Cu/low-k interconnects [18]. A high-frequency pulse current reduces JH and therefore improves the EM lifetime. The temperature increase is primarily due to one-dimensional (1D) radiation (see Fig. 1). The passive lines in the layer do not act as radiation activators but rather as a compounding factor.

In the present article, we discuss the thermal characteristics required of interconnects to suppress the impact of JH on EM. The thermal time constants of JH in multilevel Cu/low-k interconnects are experimentally investigated for the first time by measuring the transient thermal response. The contributions of thermal conduction and thermal capacitance to the pulse effect are also discussed. Furthermore, we analyze the "upside-down-chimney effect," which occurs when direct radiation reaches the Si substrate through the stacked contact/via (S-C/V), to evaluate the suppression of the contribution to the temperature increase from the passive lines. The impact of JH on the critical product of current density and line length [19] is also investigated.

II. EXPERIMENTS

A. Sample preparation

Test structures were prepared with multilevel 40 nm-node technology (Fig. 2: lines/space = 66/66 nm). Dual-damascene test structures with five metal layers were fabricated to evaluate JH and EM reliability. Electrochemically plated Cu was used to fill the damascene structure after the physical vapor deposition (PVD) of the Ta/TaN and CuAl seed layers. The Cu overburden and the Ta-based films were removed by chemical-mechanical polishing. Finally, the top surface of the metal layer was passivated with a SiCN/SiC bilayer low-k barrier cap.

B. Joule heating measurements

Two types of test structures were used to evaluate the temperature increase due to JH. For thermal-response measurements, a straight Cu line surrounded by the passive lines was used. A fast waveform generator measurement unit (Agilent B1530A) was used to measure the resistance profile due to various PDC waveforms (Fig. 3), and the temperature profile in the line was transformed using the temperature coefficient of resistance (TCR), which was obtained using DC current with negligible JH. The thermal response observed from the Cu/low-k interconnect was analyzed with a simple model proposed by Li et al. [20].

The upside-down-chimney effect was evaluated by using meandering lines of S-C/V on a Si substrate. Figure 4 shows the schematics of the test structures used to measure line resistivity and JH. The JH level without S-C/V effects was estimated with these structures, as shown in Fig. 4(a). In the same layer, the meandering lines mutually heat each other. They contain electrically passive lines that alternate with active lines. The upside-down-chimney effect was investigated with the structure depicted in Fig. 4(b). The S-C/Vs connect the passive lines to the Si substrate with various densities. The area ratios of the S-C/V over the entire area of the structure were 0, 2.6, 7.8, and 23 ppm and 1.1%. The passive lines and S-C/Vs do not contribute to the electrical resistance of the active line. Consequently, a radiation effect through the S-C/V can be evaluated by comparing them. A temperature rise due to a measured DC was estimated using the TCR.

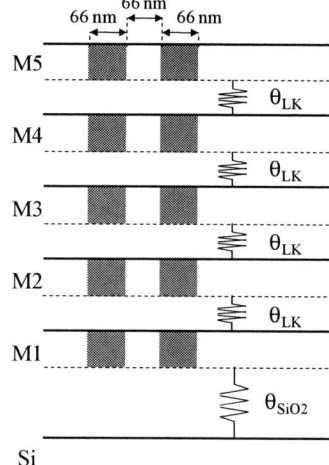

Figure 2. Cross-sectional schematic diagram of the thermal resistance for multilevel interconnects. The inter-layer dielectric (ILD) mainly contributes to the temperature increase because of JH [18].

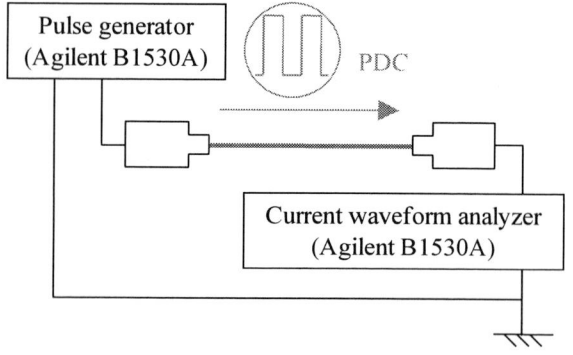

Figure 3. Overview of setup for measuring the dynamic characteristic of increase in temperature of line due to JH.

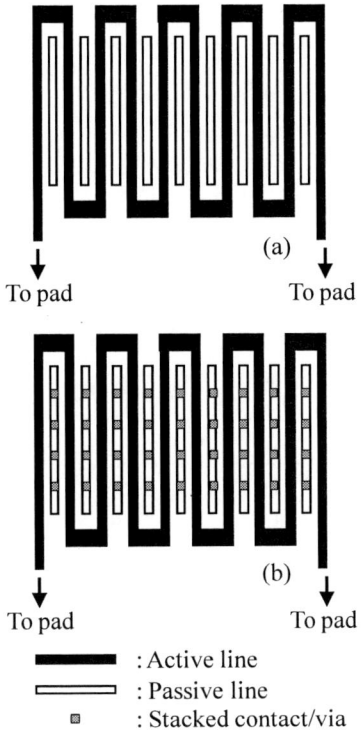

To pad   To pad

         (b)

To pad   To pad

▬▬▬ : Active line

▭▭ : Passive line

▣ : Stacked contact/via

Figure 4. Schematic of the test structure for evaluating upside-down chimney effect.

C. Electromigration test

EM tests were performed at 300 °C with constant current stress. Multilink test structures that have 20 segments terminated with vias [21] were used to evaluate the backflow effect [19], and the downstream EM lifetime was evaluated. Both the M1 line width and the via diameter were 66 nm. The M1 line lengths were 10, 30, 50, or 100 μm. Dense lines were prepared as test segments in the structures as shown in Fig. 5. The dense lines are comprised of active lines and passive lines. The active lines are surrounded by many passive lines with minimum lines/space.

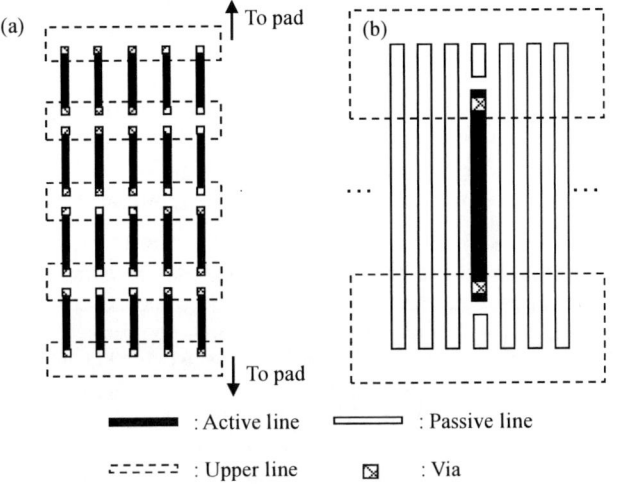

▬▬▬ : Active line   ▭▭ : Passive line

▭▭▭▭ : Upper line   ▣ : Via

Figure 5. Schematics of EM test segments. (a) Image of entire test structure. (b) Enlarged view of the segments.

The failure time is defined as the time when the first via fails, with the failure criterion being an abrupt resistance increase of 0.5%. EM lifetime distributions are analyzed by a lognormal plot of the cumulative failure probability per via using the cumulative hazard method [21, 22].

III. RESULTS AND DISCUSSIONS

A. Thermal time constants in Cu/low-k

We measured the thermal response of the Cu/low-k interconnects. The transitional-resistance response under PDC was measured at sampling rates from 10^3 to 10^6 s^{-1} at room temperature. The maximum and minimum pulse-generator output voltage was 5 V and 50 mV, respectively. The maximum current under the maximum voltage output was approximately 7.7 mA, which generates a line temperature increase due to JH. The frequency of the PDC was 10 kHz to 5 MHz, and the on-duty cycle was 50%. The waveform of the estimated line resistance in this measurement is due to JH. Figure 6 shows an example of this measurement.

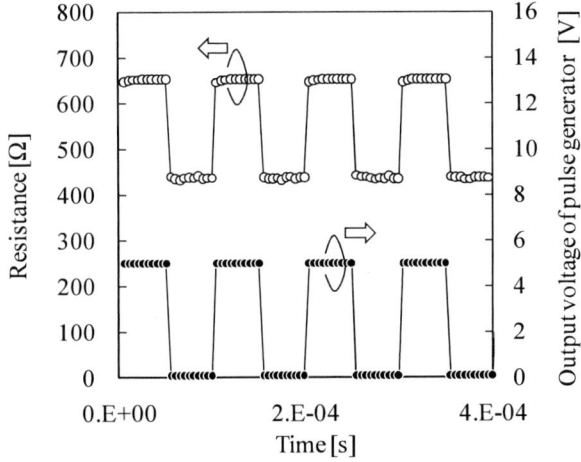

Figure 6. Figure 6. Example of pulse-generator output voltage pulse and line resistance estimated by the current waveform in M5. The maximum and minimum pulse-generator output voltage is 5 V and 50 mV, respectively.

Several researchers have reported a slight delay in the thermal response at high frequency PDC in Al lines [7–9]. Similarly, for above 1 MHz PDC, we confirm a remarkable delay in the thermal response of the resistance. Figure 7 depicts schematically the temperature increase of the line due to JH. In the dynamic characteristic region, which occurs immediately after current pulse-on, the temperature profile distorts. Eventually, the temperature saturates at the equilibrium temperature under DC. As the PDC frequency increases, the pulse-on time approaches the thermal time constant, and it eventually becomes less than the constant value. In the high-frequency region, the temperature profile will eventually converge to a constant value. Consequently, we expect the EM lifetime to improve as a result of reducing the increase in the JH temperature. The results of measuring the EM lifetime with high frequency PDC agree with the estimation [18].

978-1-4244-5430-3/10 $26.00 © 2010 IEEE   719

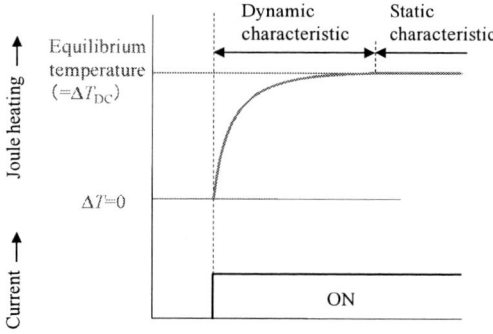

Figure 7. Schematic of the dynamic characteristics of the line temperature increase due to JH.

We analyzed the observed thermal response of the Cu/low-k interconnect with a simple model proposed by Li *et al.* [20]. The JH during the on and off period can be expressed as

$$\Delta T_{ON}(t) = \Delta T_{DC} - \Delta T_1 \cdot \exp\left(-\frac{t}{\tau_{th}}\right), \quad (1)$$

$$\Delta T_{OFF}(t) = \Delta T_2 \cdot \exp\left(-\frac{t}{\tau_{th}}\right), \quad (2)$$

respectively, where $\Delta T_{ON}(t)$ is the JH during the on period, ΔT_{DC} is the JH under DC, ΔT_1 is the JH at the beginning of the on period, $\Delta T_{OFF}(t)$ is the JH during the off period, ΔT_2 is the JH at the end of the on period, and τ_{th} is the thermal time constant of the measured structure. The results of the calculation indicate that after several pulse periods, the temperature profile during the on and off period will reach the steady state. The only parameter used to fit the model to the observed temperature profiles is τ_{th}.

Examples of the observed temperature changes and fits from the thermal analytical model are shown in Fig. 8. The fifth-layer (M5) line was measured with 10 kHz or 1 MHz PDC. The parameter τ_{th} is taken to be 2.6×10^{-7} s. For a frequency of 10 kHz [Fig.8(a)], the PDC period is much larger than τ_{th} and temperatures during the on and off periods reach their respective constant values. For a frequency of 1 MHz [Fig.8(b)], the PDC period approaches τ_{th} and the aforementioned remarkable delay in the thermal response is observed. If the PDC period is much less than τ_{th}, the temperature during the on and off periods will be a similar and approximately constant value determined by the product of ΔT_{DC} and on-duty ratio [20].

The parameter τ_{th} depends on the thermal coefficient of the measured structure. Because the 1D radiation mainly contributes to JH, a change in τ_{th} for each layer in the multilevel interconnects is an important quantity that is needed to evaluate the impact of JH on the EM lifetime. The temperature rise due to JH is linearly dependent on the power density. In particular, the temperature increase of the upper layer is larger than that of the lower layer [18]. Therefore, the static temperature increase per unit power density ($\Delta T_{DC}/D_{Power}$) is an effective quantitative scale to discuss the impact of JH on multilayer interconnects.

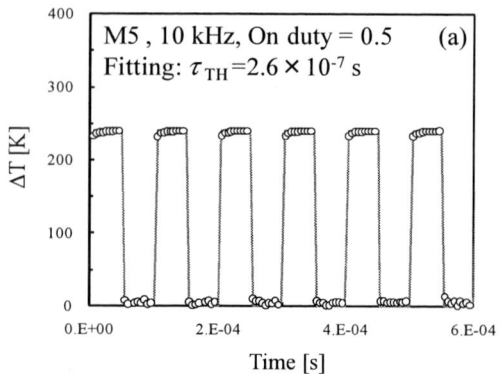

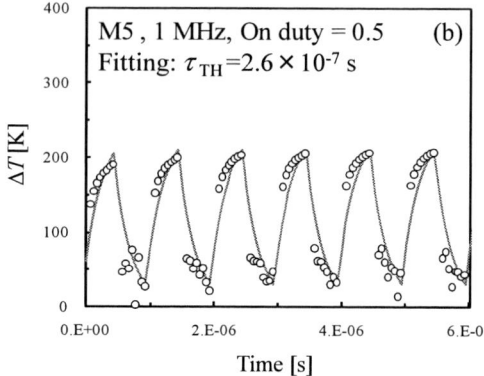

Figure 8. Figure 8. Temperature profiles obtained over several periods (a) of 10 kHz PDC and (b) of 1 MHz PDC. The solid curves were calculated using the thermal analytical model given in the text.

The thermal time constants of each layer were estimated and are shown in Fig. 9. The x axis represents $\Delta T_{DC}/D_{Power}$ and the y axis denotes the time constant. These two factors are linearly dependent. This result suggests that the thermal capacitance per unit area is the same for layers M1–M5. As shown by the equations in Fig. 9, it can be expressed based on an analogy with an RC circuit. The ratio $\Delta T_{DC}/D_{Power}$ represents a proportionality factor between heat flow and temperature increase, and it is equivalent to the thermal resistance, $R_{th}S_{TEG}$, where R_{th} is a thermal resistance per unit area and S_{TEG} is the area of the structure about the 1D radiation. These results suggest that the thermal time constant depends mainly on the thermal resistance of the structures. In other words, the thermal capacitance per unit area is the same for each layer. These results suggest that the reduction in the thermal resistivity of the line effectively suppresses the impact of JH.

In addition, the estimated time constant is approximately 1/10 that of the reported value for Al interconnects [20, 24, 25], but it agrees with their different thermal conductivities (Cu: 398 $W \cdot m^{-1} \cdot K^{-1}$; Al: 237 $W \cdot m^{-1} \cdot K^{-1}$ [26]). These results suggest that the time required to attain a peak temperature in Cu interconnects tends to be shorter than that required in Al interconnects. Similarly, some EM boosters may affect the time constant, so it is necessary to investigate the material dependence of JH to determine the future development of interconnects.

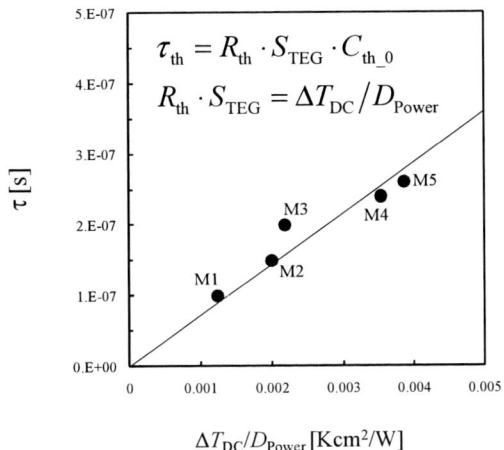

Figure 9. Estimated thermal time constants of the structures for each metal layer.

In addition, the thermal time constants of various layout lines can be estimated from $\Delta T_{DC}/D_{Power}$, because the thermal capacitance per unit area is the same for each layer. Temperature profiles for PDC or PAC due to JH can be estimated with the help of the time constants. This provides exact information to find a hot spot on circuits.

B. Direct radiation through stacked contact/via

To reduce the temperature increase due to JH, the thermal resistivity of the line must be reduced. Numerous studies have proposed that vias contribute to the radiation [1, 27], and because the thermal conductivity of Cu is much larger than that of any dielectric, vias connected to different levels transfer heat to a heat sink. After finding the hot spot, we connect the hot active line to the Si substrate by S-C/V, which reduces JH by the via effect. However, it also increases the circuit capacitance. For a method to reduce JH, it is desirable to have no impact on the circuit capacitance. Additionally, the passive lines in the same layer do not act as radiation activators but rather as compounding factors [18]. Consequently, direct radiation through S-C/V for suppressing the temperature increase due to the passive lines is an actual and reasonable method to reduce JH.

The upside-down-chimney effect is expected to reduce the thermal resistivity of the focused structure. In particular, when the passive lines act as a compounding factor, the upside-down-chimney effect may be used as a method to suppress JH. Figure 10 shows the test structures that are used to investigate the upside-down-chimney effect under DC. The area ratios of the S-C/Vs are 0, 2.6, 7.8, and 23 ppm, and 1.1%. The dependence of the temperature increase on power is shown in Fig. 10. A remarkable upside-down-chimney effect is observed only for the area ratio of 1.1%. This result suggests that a substantial number of S-C/Vs per unit area is required to suppress JH.

The impact of the thermal capacitance of the structure on the thermal time constant is not well understood. However, it is indispensable for an accurate estimation of JH and will be investigated in future work.

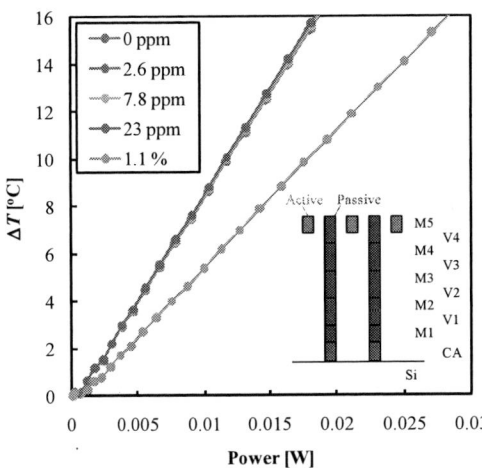

Figure 10. Figure 10. Dependence of upside-down-chimney effect on area ratio for passive line.

C. Joule heating impact on the Blech's critical product

For short lines, a comparatively large current density can be permitted in circuit design because of the backflow effect [19]. In the steady state, the following relation describes the balance of EM and stress-induced backflow:

$$(jL)_{th} = \frac{\Omega\Delta\sigma}{eZ^*\rho}, \qquad (3)$$

where j is the current density, L is the line length, Ω is the atomic volume, $\Delta\sigma$ is the stress difference within the line, eZ^* is the effective charge, and ρ is the metallic resistivity. This critical product represents the EM threshold product of the current density and line length. If the product is smaller than the result of Eq.(3), EM-induced void nucleation and growth does not occur. For Cu interconnects, no linewidth dependence [4], no temperature dependence [15], no current-waveform dependence [16], and no impurity-concentration dependence [28] are reported for the critical product. These results indicate that the critical product is robust under any condition. In particular, no temperature dependence means that the critical product is stable even with JH.

The critical products under various JH conditions were evaluated at 300 °C and with DC. The current densities were set at three levels. There is no JH for the first set point of the current density, and a low and high JH are estimated for the second and third current-density set points. Under these conditions, the observed lifetime for various line lengths provides the estimation of the critical product. The results were analyzed by the simple plot proposed by Wang and Filippi [29].

Figure 10 shows the comparison of the results for various levels of JH. The estimated critical products do not depend on the amount of JH, which agrees with the previous results of no temperature dependence [15]. The critical product is estimated to be 9330 ± 2080 A/cm. These results suggest that the impact of JH on the critical product is negligible even for design rules for short lines.

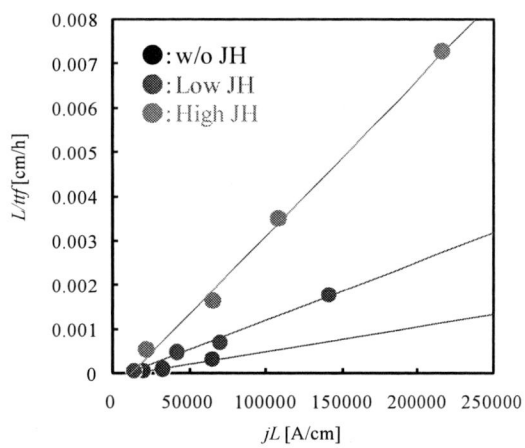

Figure 11. Effect of JH on backflow effects of down-stream EM lifetime.

IV. CONCLUSIONS

We experimentally investigated several thermal characteristics with the aim of understanding and suppressing the impact of JH on EM. We measured the thermal time constants for the first time in Cu/low-k interconnects. The thermal time constants measured for each metal layer depend mainly on the thermal resistance of the structures, which suggests that reducing the thermal resistivity of the interconnect structures is significantly effective in suppressing the impact of JH on EM. The upside-down-chimney effect of the S-C/Vs is one of the most effective ways to control JH, and the temperature increase due to passive lines can be reduced by the S-C/Vs. The impact of JH on the critical product is found to be negligible, which implies that the short-line EM design rule under the critical product limit does not depend significantly on JH. These results can be used to provide several suggestions for the layout of advanced interconnects.

ACKNOWLEDGMENT

The authors would like to thank to H. Takizawa for S-TEM observations and EM experiments. The authors would like to thank N. Nakamura and I. Akiba of NEC Electronics Corporation for research support.

REFERENCES

[1] S. Im, N. Srivastava, K. Banerjee, and K.E. Goodson; "Scaling Analysis of Multilevel Interconnect Temperatures for High-Performance ICs", *IEEE Transactions on Electron Devices*, Vol.52, pp.2710-2719 (2005).

[2] J.W. McPherson; "Reliability Challenges for 45nm and Beyond", *Proc. of 2006 DAC*, pp.176-181 (2006).

[3] C.-K. Hu, D. Canaperi, S. T. Chen, L.M. Gignac, B. Herbst, S. Kaldor, M. Krishnan, E. Liniger, D. L. Rath, D. Restaino, R. Rosenberg, J. Rubino, S.-C. Seo, A. Simon, S. Smith, W.-T. Tseng; "Effects of Overlayers on Electromigration Reliability Improvement for Cu/Low-k Interconnects", *Proc. of 2004 International Reliability Physics Symposium* (2004).

[4] S. Yokogawa and H. Tsuchiya; "Scaling Impacts on Electromigration in Narrow Single-Damascene Cu Interconnects", *Japanese Journal of Applied Physics*, Vol.44, No.4A, pp.1717-1721 (2005).

[5] C.-K. Hu, L. Gignac, B.Baker, E. Liniger, R. Yu, and P. Flaitz; "Impact of Cu Microstructure on Electromigration Reliability", *Proceedings of International Interconnect Technology Conference*, pp.93-95 (2007).

[6] S. Yokogawa, K. Kikuta, H. Tsuchiya, T. Takewaki, M. Suzuki, H. Toyoshima, Y. Kakuhara, N. Kawahara, T. Usami, K. Ohto, K. Fujii, Y. Tsuchiya, K. Arita, K. Motoyama, M. Tohara, T. Taiji, T. Kurokawa, and M. Sekine; "Trade-Off Characteristics Between Resistivity and Reliability for Scaled-Down Cu-Based Interconnects", *IEEE Transactions on Electron Devices*, Vol.55, pp.350-357 (2008).

[7] S. Yokogawa, Y. Kakuhara, H. Tsuchiya, and K. Kikuta; "Analysis of Al Doping Effects on Resistivity and Electromigration of Copper Interconnects", *IEEE Transactions on Device and Materials Reliability*, Vol.8, pp.216-221 (2008).

[8] M. Iguchi, S. Yokogawa, H. Aizawa, Y. Kakuhara, H. Tsuchiya, N. Okada, K. Imai, M. Tohara, K. Fujii, T. Watanabe; "Optimization of Metallization Processes for 32-nm-node Highly Reliable Ultralow-k (k=2.4)/Cu Multilevel Interconnects Incorporating a Bilayer Low-k Barrier Cap (k=3.9)", *2009 International Electron Device Meeting*, pp.36.1.1-36.1.4 (2009).

[9] Y. Kakuhara, S. Yokogawa, and K. Ueno; "A Comparison of Lifetime Improvements in Electromigration between Ti Barrier Metal and CVD Co Capping", to be published in Japanese Journal of Applied Physics.

[10] J. Tao, N.W. Cheung, and C. Hu; "Electromigration Failure Under Bidirectional Current Stress," *Proc. of Fourth International Workshop on Stress Induced Phenomena in Metallization*, AIP Conf. Proc. 418, H. Okabayashi, S. Shingubara, P.S. Ho Eds., pp.201-211 (1997).

[11] J. Tao, K.K. Young, N.W. Cheung, and C. Hu; "Comparison of Electromigration Reliability of Tungsten and Aluminum Vias Under DC and Time-Varying Current Stressing," *Proc. 30th International Reliability Physics Symposium*, pp.338-343 (1992).

[12] K. Hiraoka and K. Yasuda; "The Enhancement of Electromigration Lifetime under High Frequency Pulsed Conditions", *IEICE Transactions Fundamentals*, Vol.E77-A, pp.195-203 (1994).

[13] A.T. English, K.L. Tai, and P.A. Turner; "Electromigration in Conductor Stripes under Pulsed DC Powering", *Applied Physics Letters*, Vol.21, pp.397-398 (1972).

[14] B.K. Liew, N.W. Cheung, and C. Hu; "Electromigration Interconnect Lifetime under AC and Pulsed DC Stress," *Proc. 27th International Reliability Physics Symposium*, pp.215-219 (1989).

[15] S. Yokogawa; "Electromigration Reliability of Damascene Copper Interconnects", *Proceedings of 2003 Advanced Metallization Conference*, pp.259-269 (2004).

[16] S. Yokogawa, "Electromigration Behavior of Single-Damascene Cu Line under Pulsed-Current", *IEICE Transactions on Electronics (Japanese Edition)*, Vol.J.88-C, pp.253-260 (2005), in Japanese.

[17] L.M. Ting, J.S. May, W.R. Hunter, and J.W. McPherson; "AC Electromigration Characterization and Modeling of Multilayered Interconnects," *Proc. 31th International Reliability Physics Symposium*, pp.311-316 (1993).

[18] S. Yokogawa, Y. Kakuhara, and H. Tsuchiya; "Joule heating Effects on Electromigration in Cu/Low-k Interconnects", *Proc. of 2009 International Reliability Physics Symposium*, pp.837-843 (2009).

[19] I.A. Blech, "Electromigration in Thin Aluminum Films on Titanium Nitride", *Journal of Applied Physics*, Vol.47, pp.1203-1208 (1976).

[20] Z. Li, G. Wu, Y. Wang, Z. Li and Y. Sun, "Numerical Calculation of Electromigration Under Pulse Current with Joule Heating", *IEEE Transactions on Electron Devices*, Vol.46, pp.70-77 (1999).

[21] H. Tsuchiya and S. Yokogawa, "Electromigration Lifetimes and Void Growth at low Cumulative Failure Probability", *Microelectronics Reliability*, Vol.46, pp.1415–1420 (2006).

[22] W. Nelson, *Applied Life Data Analysis*, New York, John Wiley & Sons (1982).

[23] K. Nikawa, and S. Inoue, *DRIP (Defect Recognition and Image Processing in Semiconductors) VII*, Templin, 1997/9/7, Proc. DRIP VII, Inst. Phys. Conf. series No.160, Inst. Physics Publ., Bristol and Philadelphia, pp. 37-46 (1998).

978-1-4244-5430-3/10 $26.00 © 2010 IEEE

[24] W. Wu, J.S. Yuan, S.H. Kang, and A.S. Oates, "Electromigration Subjected to Joule heating under Pulsed DC Stress", Solid-State Electronics, Vol.45, pp.2051-2056 (2001).

[25] T. Furusawa, K. Hinode, and Y. Homma, "Pulsed current electromigration mechanism—Instantaneous temperature profile model," in *Extended Abstracts of the 22nd Conference on Solid State Devices and Materials*, pp. 255-258 (1990).

[26] S.P. Murarka, I.V. Verner, and R.J. Gutmann, *Copper –Fundamental Mechanisms for Microelectronic Applications*, New York, John Wiley & Sons (2000).

[27] K. Ramakrishna, M. Gall, P. Justison, and H. Kawasaki; "Prediction of Maximum Allowed RMS Currents for Electromigration Design Guidelines", *Proc. of Seventh International Workshop on Stress-Induced Phenomena in Metallization*, AIP Conf. Proc. 741, P.S. Ho, S.P. Baker, T. Nakamura, C.A. Volkert Eds., pp.156-164 (2004).

[28] S.Yokogawa and H. Tsuchiya; "Effects of Al doping on the electromigration performance of damascene Cu interconnects", *Journal of Applied Physics*, Vol.101, pp.013513-1-6 (2007).

[29] P.-C. Wang and R.G. Filippi; "Electromigration Threshold in Copper Interconnects", *Applied Physics Letters*, Vo.78, pp.3598-3600 (2001).

Assessing The Degradation Mechanisms and Current Limitation Design Rules of SiCr-based Thin-film Resistors in Integrated Circuits

Yuan Li[1], David Donnet[1], Andrzej Grzegorczyk[1], Jan Cavelaars[1] and Fred Kuper[1,2]

[1]NXP Semiconductors, Gerstweg 2, 6534AE Nijmegen, The Netherlands, , [2] University of Twente, The Netherlands
phone: +31-24-3535357; e-mail: Y.Li@nxp.com

Abstract— The degradation of SiCr-based thin-film resistors under current and temperature stress and the Joule heating in the resistors are experimentally investigated to set current limitation design rules. Degradation mechanisms, the failure modes, and the impact of the stress test on Temperature Coefficient of Resistance (TCR), are studied with the use of various test structures stressed under different conditions (temperature, current density and direction), followed by optical inspections, Infra-Red imaging and SEM/TEM cross-sections. Electromigration (EM) is found to be the dominating degradation mechanism, but the EM process differs from that in commonly used interconnects. Current accelerating factor and the equivalent activation energy are determined for data extrapolation. To avoid errors in Joule-heating determination from TCR, integrated temperature sensors are employed. Current limitation design rules are deduced based on the EM and Joule heating results.

Keywords-SiCr; integrated resistor; electromigration; Joule heating; design rule

I. INTRODUCTION

To optimize the design of RF ICs integrated resistors made of materials with high precision resistivity, low Temperature Coefficient of Resistance (TCR), low noise and good matching performances are needed [1]. SiCr-based materials meet these criteria and are therefore attractive alternatives to poly-silicon for integrated thin-film resistors in RF technologies. Despite many investigations carried out on SiCr-based materials in material sciences [2-9], information on reliability of SiCr-based integrated thin-film resistors in IC processes has rarely been reported. In the work of Brynsvold *et al* [10] intensive investigations are carried out on the "wear-out" of SiCr-based resistors under current and temperature stress in their process technology. However no predictions of lifetimes under use conditions are provided because of the lacking of a proper degradation model. Therefore no current limitation design rules can be deduced based on their experimental results.

In this work, a series of experiments are performed with the use of test structures of SiCr-based thin-film resistors in an IC process flow of NXP Semiconductors. Firstly investigations are carried out to clarify whether the degradation takes place in the SiCr film or in the SiCr–W plug interface, based on which further experiments can be designed. Samples are stressed under various current densities and temperatures, including

varying the current flow directions. Attention is paid to the degradation mechanisms and failure modes, aiming to define a model for data extrapolation and prediction of lifetime under use conditions. Joule heating of the SiCr film is characterized with built-in temperature sensors, to avoid errors in Joule-heating determination from TCR, which is small and quadratic with temperature for SiCr-based material. The final goal of the work is to characterize the reliability of the SiCr-based resistors and to define current limitation design rules by taking into account the degradation process and Joule heating effects.

II. THE EXPERIMENTS

In NXP's QUBIC4+ BiCMOS RF process SiCr thin film resistors are integrated between two AlCu interconnection metal levels, with the electrodes of the resistors being connected to the metal layer above through W via plugs. The SiCr film is deposited with a sputtering system, with a sheet resistance of 270 Ohms per square. Three sets of test structures are used in this work: In the first set of test structures a SiCr resistive film of 160 µm long and 8 µm wide is used. The number of via's used to connect the resistive film to the AlCu electrodes is varied between 1 and 8. This set of test structures is used to determine where the observed degradation takes place. The second set of test structures consists of a 200 µm long and 9.8 µm wide SiCr line. 33 via's in 3 columns are employed to connect the SiCr film to the AlCu electrodes in order to minimize the impact of any possible degradation related with the via's. This set of test structures is meant to study the degradation in the SiCr film. In the third set of test structures a relatively narrow (0.5 µm) AlCu interconnect line is used as the temperature sensor for Joule heating characterization. The temperature sensor is placed either above or beneath, and along the axis of, the SiCr film. The width of the SiCr films is varied from 2 µm to 8 µm, with the length varied accordingly to ensure a constant number of squares of the film. In all these test structures the SiCr resistors and the AlCu sensors are constructed such that they enable 4-probe testing. In Fig. 1 one of the test structures is shown in cross-section as an example. The test structures are packaged in ceramic housings to allow reliability testing under highly elevated temperatures.

The electrical and temperature stresses are carried out with the use of a commercial EM tester. TCR is characterized before

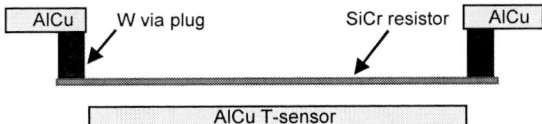

Figure 1. Cross-sectional schematic description of a SiCr test structure with an AlCu line beneath it as the temperature sensor.

and after the stress tests for some of the test runs to study the impact of current and temperature stress on TCR. Sample size is 15 or more per group of samples and per stress condition. Stress current and oven temperature are varied in a range within which the failure mode does not change with the stress conditions. A resistance change of 2% is taken as the failure criterion. The choice of such a failure criterion is based on the accuracy requirement of the product to the SiCr resistors. Failure modes are studied by optical inspection, Infrared (IR) imaging and SEM/TEM cross-sections. For Joule heating characterization the TCR of the AlCu temperature sensor is calibrated beforehand with the EM tester. The Joule heating is then measured as a function of current flow in the SiCr resistors.

III. THE RESULTS AND DISCUSSIONS

A. The general degradation phenomenon

The general degradation phenomenon is that the resistance of the Device Under Test (DUT) decreases during the current and temperature stress rather than increases as what is normally observed for interconnects. More precisely, the relative degradation curve shows an initial small increase (<2%) of the resistance shortly after the stress started and for a relatively short period, followed by a steady decrease of resistance, nearly perfectly linear with time, down to less than 90% of the original values (Fig. 2). And then a sudden deeper decrease occurs, followed by an abrupt increase of the resistance to infinite, suggesting that an open is created. As our failure criterion is set at 2% resistance change, for lifetime determination only the region of resistance decrease on the degradation curve is important. Therefore this work will focus on resistance decrease of SiCr resistors.

The decrease of resistance of SiCr under current and temperature stress is attributed to silicon migration, $CrSi_2$ grain formation and growth, and possibly $CrSi_2$ reorientation etc. in literature [10]. But no theory could explain our results reasonably well. In fact it is reported in another paper that the formation of $CrSi_2$ even causes resistance increases in SiCr-based materials [3]. On the other hand, because not much is known about the SiCr-W interface, it is speculated that any change of the property of this interface might have played a role in the observed resistance decrease process. Therefore it is important to determine where, in the SiCr film or at the SiCr-W plug interface, the degradation takes place. It is important to know where the degradation takes place also because it affects the definition of current density and DUT temperature in the experimental design and data interpretation.

B. The degraded element

To determine where the degradation takes place, test structures with the same SiCr layout (W x L = 8 μm x 160 μm) but with different number of via's, 1, 3 or 8 via's, respectively, are stressed. The via's are placed in one column, perpendicular to the length of the resistors at each end of the film which connect the SiCr film to the AlCu electrodes. Oven temperature is set to 240 °C in this experiment. Test structure 1 (with 8 via's) and 2 (with 3 via's) are stressed with a current of 3 mA, and test structure 3 (with only 1 via) is stressed with a current of 1 mA. Their relative degradation curves are shown in Fig. 3. Comparing test structure 1 with test structure 2, where the current densities in the SiCr film are the same while the current densities in the via's are different, it can be seen that their degradation curves overlap. This can be seen more clearly from their cumulative time-to-failure (TTF) distribution graph (Fig. 4). On the other hand, when the results of test structure 2 and 3 are compared, where the current density in each via is the same while the current density in the SiCr film differs by a factor of three, the degradation of test structure 3 with a lower current density in SiCr is much slower. In summary, the decrease of resistance depends on the current density in the SiCr film and is

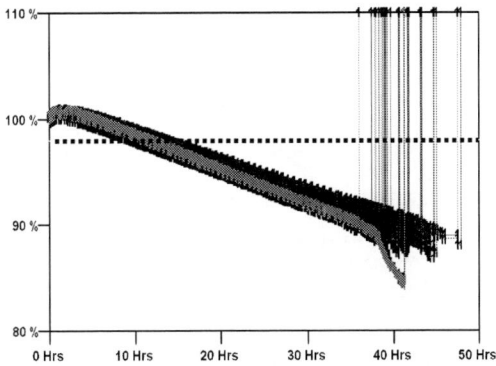

Figure 2. Typical relative degradation curve in time. T_{DUT}= 386.5 °C, I_{stress} = 0.61 mA/μm line width. Dashed line refers to the 2% resistance change failure criterion

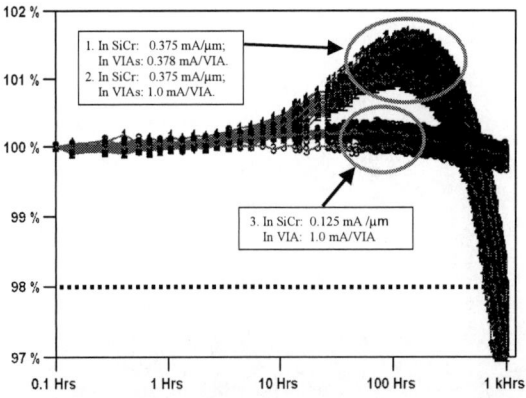

Figure 3. Relative degradation curves of 3 test structures for comparison: test structures 1 and 2 are stressed with the same curent density in the SiCr film but different current density in via's; and test structures 2 and 3 with the same curent density in the via's but different current density in the SiCr film. Refer to the insets for details. Dashed line refers to the 2% resistance change failure criterion. Oven temperature: 240 °C. X-axis in Log scale.

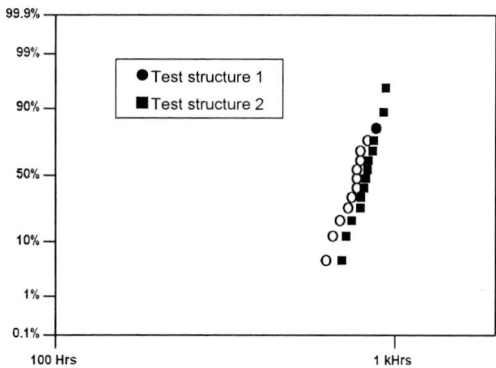

Figure 4. Cumulative failures in time in LogNormal distribution of test structure 1 and 2 with their relative degradation curves shown in Fig. 3.

independent of the current density in the via's.

The observed resistance decrease certainly cannot be explained by the formation of voids in the test structure. Resistance decrease could be caused by extrusion. However, no extrusion is observed at all. Supposing the resistance decrease is caused by the improvement of contact between W-plug and SiCr, it would be difficult to imagine that the improvement at the eight via's of test structure 1 could be just equal to that in the three via's of test structure 2. In addition, optical inspection reveals a discoloration in the SiCr film (for details see next section) at the anode side of the film. The length of the discolored area is found to be proportional to the decrease of resistance. The observed discoloration and the fact the resistance decrease depends on the current density in the SiCr

film, suggest that the degradation takes place in the SiCr film and not in the via or at the SiCr–W plug interface.

C. The degradation mechanism and failure modes

Further investigations are carried out under various current flows in the SiCr film and under various oven temperatures, in order to build a degradation model for lifetime extrapolation. In these experiments the SiCr test structure of 9.8 μm x 200 μm (W x L) is used. The stress conditions and the results of TTF fitting to LogNormal distribution are summarized in Table 1, where $TTF_{50\%}$ is the median point of TTF distribution, T_{DUT} is the temperature of the DUT, and σ is the standard deviation of TTF in natural logarithms scale. In general, the TTF decreases with the increase of current flow and DUT temperature. A remarkable characteristic of the TTF distributions is that the sigma's are very small (< 0.1), in contrast to those for interconnect lines which are normally 0.2 to 0.7 or even higher.

After stress test the samples are inspected with optical microscope as the first step to study the failure mode. It is observed that part of the SiCr resistors becomes darker at their anode side (see Fig. 5a). The length of the discolored area is found to be proportional to the percentage of resistance decrease due to stress. IR imaging while current flows in the SiCr film confirms that the darkly colored section of the resistor is cooler and therefore has a lower resistivity (Fig. 5b). Based on these results it can be concluded that the resistance decrease is related to the discoloration of the resistor. Whether any other change in the rest of the film plays a role as well is not clear yet.

To understand better what the color change means, current flow direction is changed for small groups of samples (three to six samples per group) at different stages of a stress experiment. Small sample sizes are acceptable for this semi-quantitative study because the differences among samples in their degradation behaviors are small, indicated by the small sigma as shown in Table 1. It is found that reversing the current flow direction can trigger a recovery process for the observed degradations at any stage of the degradation process (see Fig. 6, one sample per group is shown as examples). Some of the

TABLE I.   STRESS CONDITIONS AND THE RESULTS.

Fail/ samples	I_{stress} (mA)	T_{DUT} (°C)	$TTF_{50\%}$ (hrs)	σ	$I_{allowed}$ (mA/μm)	Life (yrs)
28/28	6	386.5	10.8	0.096	17.2	97K
15/15	5	333.9	159.3	0.099	17.2	97K
14/15	4.6	335.2	184.5	0.071	17.7	105K
15/15	4.6	355.0	76.9	0.047	18.2	113K

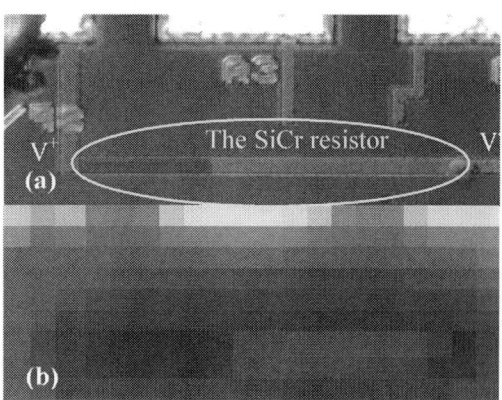

Figure 5. Color change on SiCr (a) is shown to correspond to a decrease in resistance by IR image (b) where darker means a lower temperature under current stress.

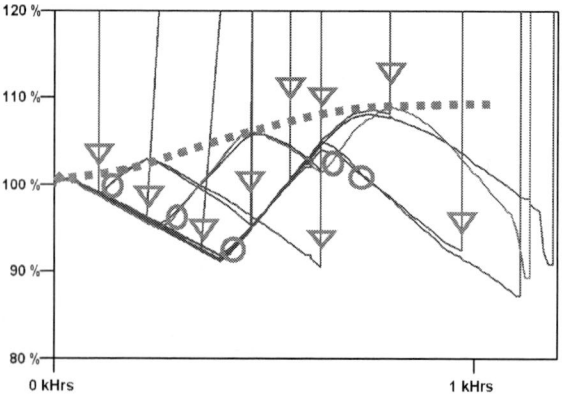

Figure 6. Relative degradation and the recovery with reversed current flow at different stage of the degradation process. Circles indicate a change in current flow direction. Triangles indicate the termination of current stress with those samples for visual inspections.

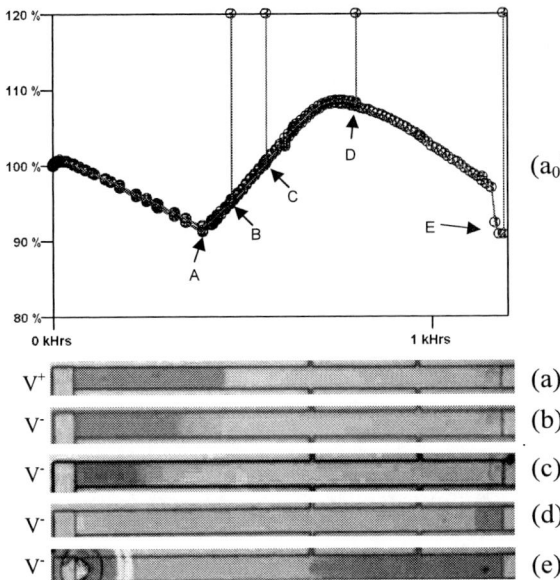

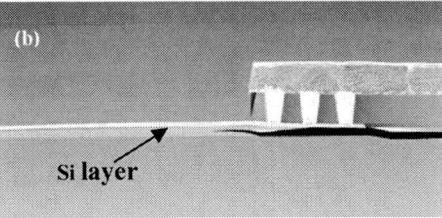

Figure 7. Impact of current flow reversing on the observed discoloration in SiCr film at different stages of the degradation process. Refer to text for details.

Figure 8. The "bubble" at the cathode end of the SiCr film after stress (a) and the TEM cross-section showing the presence of a Si layer under the SiCr film (b).

samples are terminated with current stress at different stages of the degradation process for visual inspections at the end of the experiment. It is assumed that with no stress current through the resistors no further degradation or any recovery would take place because the temperature of the SiCr film is much lower due to lack of Joule heating. Such an assumption has been proven to be valid by experiments. Some of the inspection results are shown in Fig. 7. In Fig. 7a0 the relative degradation curves are shown for a number of samples which belong to the same group originally. The optical micrograph of a sample terminated with current stress at stage A is shown in Fig. 7a. After stage A the current flow direction is reversed for all the other samples. Pictures of samples at stage B, C, D and E are given in Fig. 7b, c, d and e, respectively. Comparing the pictures at stage B, C, D and E with that at stage A, it is seen that with the reversed current flow the discolored area at the original anode side of the test structures becomes shorter and shorter. Thereafter color change takes place at the original cathode side, which is now the anode side after the change of the current direction. And the discolored area becomes larger and larger as the stress continues.

Based on the characteristics that the degradation process is dependent on current density, current flow directions and DUT temperature, and it is reversible, it can be concluded that the color change and the degradation of the resistors (resistance decrease) are caused by electromigration (EM).

What is migrating causing the color change and resistance decrease in SiCr-based films is an important point in this case. In conventional EM process in interconnects the matrix material of the metal interconnect migrates in the direction of electron flow. The failure mode is dominated by void formation in the upstream side of the metal lines, especially at locations where the EM path is blocked by W-plugs or barrier layers in which no EM can take place. In literature, resistance decrease in SiCr-based films is attributed to migration of Si, or

CrSi$_2$ grain formation and growth etc., in the SiCr material [10]. However, the results described so far cannot be explained by any of those speculations except for Si EM. But question still remains when EM of Si is considered as the degradation mechanism in SiCr because in that case the EM process would be similar as in the commonly used interconnects, based on the fact that Si is the dominating element in the SiCr film. If the dominating material migrates voids would be expected at the cathode end of the test structures. The fact is that voids have never been observed in the SiCr film at the cathode end of the film in all our experiments.

Instead of void formation at the cathode end of the SiCr films a "bubble" always appears there (Fig. 8a). TEM cross-section shows that the bubble reflects the presence of a Si layer under the SiCr film (Fig. 8b). The presence of such a Si layer changes the level of SiCr in the vertical direction, alters the morphology of the surface of the device and causes mechanical stress in the surrounding dielectrics. As such, it appears as a bubble under optical microscope. As the stress time is extended, delamination in this Si layer will take place, forming a gap between the SiCr film and the dielectrics beneath it. The formation of the gap between the SiCr film and the dielectrics underneath enhances the Joule heating effect so that the SiCr film becomes very hot locally. Because the TCR of the SiCr resistor is negative when the temperature is above 180 °C, such a temperature increase will result in an additional resistance decrease. In addition, the Si layer might short cut partly the SiCr film next to it at higher temperatures as well. These two effects can explain the deeper resistance decrease (low part of the curve in Fig. 2 and stage E in Fig. 7a0) before an open is formed. When the stress continues an open will form finally. The location of the void/open is not at the end of the SiCr film as is commonly observed for interconnects. Instead, the open is formed where the edge of the bubble crosses the SiCr film (Fig. 9). The shape of the open, which takes the form of the edge of

978-1-4244-5430-3/10 $26.00 © 2010 IEEE   727

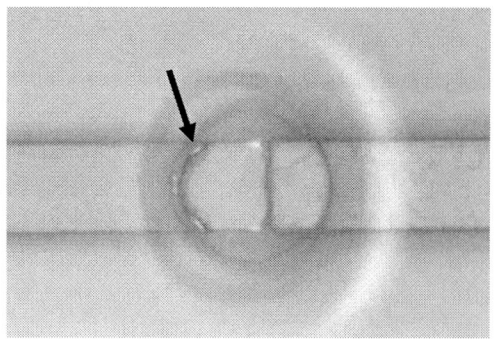

Figure 9. An open in the SiCr film, in the shape conforming the edge of the "bubble".

the bubble (Fig. 9), confirms that the open is the result of the bubble/delamination. Similar delamination has been reported in literature [10], where it is attributed to a wire bonding triggered damage.

The observations mentioned above suggest that EM in this case does not only mean migrations of material(s) within the SiCr film, but also indicate, even more importantly, that material(s) outside the SiCr film are involved and migrated along the SiCr film from the cathode side to the anode side of the film. The element which migrates is most likely oxygen. Auger analysis shows that the discolored part of the resistors indeed contains much more oxygen. A hypothesis on the mechanism of EM of oxygen in SiCr is shown schematically in Fig. 10. Taking oxygen away from the dielectric, basically SiO_2, around the cathode end of the SiCr film leaves only elementary Si there (Fig. 10a, area A), which causes volume shrink of the dielectric locally, resulting in cracking/delamination within area A. The oxygen pumped into the SiCr film migrates with the electron flow and accumulated at the anode side of the SiCr film (area B, Fig. 10a). As current flow direction is reversed the oxygen accumulated in the SiCr film in area B will migrate back to area A where it came from (Fig. 10b). If this process continues all the accumulated extra oxygen in the SiCr film will become exhausted, thereafter oxygen from the dielectrics around the original anode end of the film (area C, Fig. 10c) will be pumped into the SiCr film and migrated to and accumulated in the film at the original cathode side of the film (area D, Fig. 10c). As a result, a

"bubble" is formed at the original anode end.

Besides the EM degradation which dominates, there seems to be a slower degradation process resulting in gradual resistance increase as well. This could be seen by the increasing trend of the maximum points where a recovery process has completed and a new migration process in the reversed direction has started as shown in Fig. 6 (dashed line). The difference in slope between resistance increase (with a higher slope) and resistance decrease as is seen in Fig. 6 suggests also that a secondary degradation process is in place.

The lifetime of EM induced degradation can be described as a function of current density and operation temperature by the well-known Black's equation

$$TTF_{50\%} = AJ^{-n} \exp(E_a/kT), \qquad (1)$$

where A is a pre-factor, J is the current density, T is the temperature, n is the current accelerating factor, E_a is the thermal activation energy and k is the Boltzmann constant. Fitting the $TTF_{50\%}$ data obtained at different currents and T_{DUT} as summarized in table 1 with Black's equation, a current accelerating factor of 2.53 and an activation energy of 1.46 eV are obtained. For a required lifetime of 3000 operation hours at 150 °C and a 0.1% failure rate at the end-of-life, the allowed current per micrometer film width are calculated and given in Table 1. An $I_{allowed}$ of 17.2 mA/μm or more is obtained. This can be considered as the EM determined current limitation design rule.

From the fundamental point of view the story is far not to its end. It is reported in literature that introduction of more oxygen during the deposition of SiCr-based films increases the resistivity of the films [2]. This does not seem to be supported by our findings. It is possible that the reactions of oxygen with Si and/or Cr during the deposition and after the deposition and formation of $CrSi_2$ crystal grains are different. During deposition more SiO_2 can be formed with more oxygen sources, resulting in an increase of the resistivity of the deposited SiCr-O films [2]. It is also reported that $CrSi_2$ is a semiconductor material with a narrow band gap [3]. Oxidization of deposited SiCr films could occur only in a few nanometer layer thickness [2]. Extra oxygen brought into the film after the deposition and formation of $CrSi_2$ may play a role in coupling the $CrSi_2$ grains so that the resistivity becomes lower, rather to form SiO_2 to increase the resistivity. Furthermore, it is not well understood why the Si layer is formed under the SiCr film but not above it. Further investigations are needed to fully understand the mechanism.

D. SiCr TCR stability

In literature, the decrease of resistance of SiCr-based resistor during electrical and temperature stress is sometimes explained by a decrease of TCR (becoming more negative) due to the stress test so that the resistance value measured at elevated temperature decreases. To be sure the data obtained in this work, which are measured in situ during the stress test at high temperature, do not involve big errors due to that effect, TCR is characterized before and after a stress test run. A quadratic but relatively flat TCR as a function of temperature is obtained before the stress test, varying from roughly +125

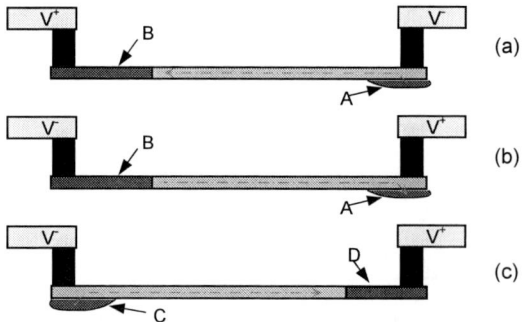

Figure 10. Schematic illustration of the EM process in SiCr involving dielectrics around it. (a) electron flows from right to left; (b) and (c) eletron flows from left to right. Refer to the text for details.

ppm/°C in average in the temperature range from 50 °C to 155 °C, to about −80 ppm/°C in average from 155 °C to 220 °C. After more than 600 hour stress, TCR changes to around 155 ppm/°C in the temperature range from 40 °C to 195 °C, and to about −85 ppm/°C from 195 °C to 225 °C. It is still quadratic, but the turning point from positive to negative shifted from about 155 °C before the stress test to about 195 °C after the stress test (see Fig. 11). This means that TCR of our SiCr film does change due to stress test but the change is not enough to explain the observed resistance decrease. Fig. 11 shows clearly that the resistance values are indeed lower after the stress test even if the measurements are done around room temperature.

It is reported in literature that introduction of more oxygen during the deposition of the SiCr film would decrease the TCR of the material to more negative [2]. While $CrSi_2$ crystallization due to annealing or stress would increase the TCR. Concerning the impact of oxygen on TCR it seems to be not in agreement with our findings of oxygen migration and accumulation in SiCr. A similar argument as for the impact of oxygen content on resistance decrease might apply. Further investigations are needed to make it clearer. TCR becomes more positive with the stress test is well in line with that is reported in literature [10].

E. Joule heating in SiCr resistor

The TCR of the SiCr-based material is small and quadratic. Large errors can occur when Joule heating is calculated from TCR measurement if the fitting is made with a software designed for metal interconnect of which the resistance increases linearly with temperature. Therefore it is necessary to calibrate the Joule heating with an alternative method. In this work AlCu lines of 0.5 μm in width are integrated in the test structure as the temperature sensors to measure the Joule heating in SiCr resistors. Choice of such a narrow AlCu line as the temperature sensor is based on the criteria that the AlCu line should not cause significant changes to the thermal resistance between the SiCr resistor and the Si substrate and it should not have a by-function as a heat sink etc. The error of temperature measurement with the use of AlCu sensor is dependent on the accuracy of the temperature of the oven in which the AlCu sensor is calibrated, which is specified at ±1.5%.

Joule heating for SiCr resistors with a film width of 2 μm, 4 μm and 8 μm are measured. For all film widths, at the same current density, a lower temperature is measured with the sensor under the SiCr film than above the SiCr film, although the sensor under the SiCr film is much closer to the SiCr film. This indicates that most of the thermal energy is passed to the Si substrate. Similar to normal metal lines, the measured line temperature increases with the line width for a constant current flow density. Because the sensor is not in contact with the SiCr film, the error of temperature measurement must be bigger for narrower resistors if their widths are in the same order as the distance between the SiCr films and the sensors. That is, the temperature for narrow lines could be underestimated. For this reason, the test structure with an 8 μm wide SiCr line is paid more attention in this study.

Joule heating is characterized at room temperature and at an oven temperature of 80 °C. No clear differences are found. This is understandable because the TCR of the SiCr material is small (~ 125 ppm/°C) so that Joule heating can be considered to be independent of the ambient temperature in a wide temperature range within the measurement error.

Corrections are made by taking into account the distances between the sensor and the SiCr film, and between the SiCr film and the Si substrate, assuming a constant thermal gradient between the SiCr film and Si substrate. The Joule heating of three samples measured at room temperature as a function of current density per line width are shown in Fig. 12. It can be seen in Fig. 12 that the SiCr resistor is already very hot even if the current flow is less than 1 mA per micrometer line width.

Suppose a 10 °C temperature increase is allowed, a current limitation of 0.12 mA per micrometer film width is determined. Considering the possible impact of the over heat of SiCr on the other elements around the SiCr resistors, such as the metal interconnects or transistors underneath, the 10 degree limitation is not too tight. This current limitation determined from Joule heating is more than two decades smaller as compared with the current limitation due to EM given in Table 1. This means that Joule heating is the limiting factor rather than EM for SiCr resistors. This current limitation is recommended as part of the design rule for SiCr resistor design.

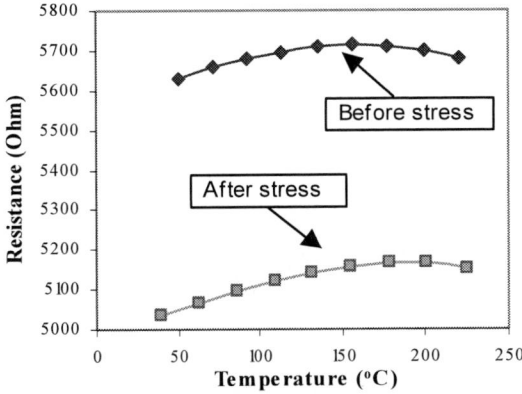

Figure 11. Resistance change with temperature of a sample before and after a stress test.

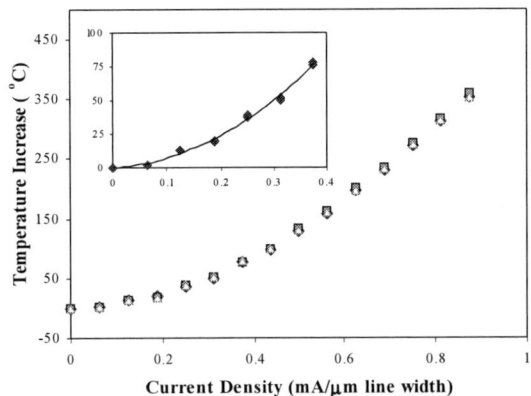

Figure 12. Joule heating in SiCr as a function of current density per micrometer line width, measured on 3 samples.

With the recommended current limitation the lifetime of SiCr due to EM is found to be longer than 90k years. The results are given in Table 1 (last column) as well.

IV. CONCLUSIONS

The degradation of SiCr-based thin-film resistors under current and temperature stress and the Joule heating effect are experimentally investigated. The main degradation is a current induced resistance decrease. The change of resistance is found to occur in the SiCr film itself and not at the SiCr–W plug interface. Two degradation processes are identified: EM, which dominates and leads to resistance decreases; and another process that results in a relatively small resistance increase. The EM induced degradation is found to be completely reversible. The observed phenomena are explained by a new EM mechanism, in which the initial material of the SiCr film does not migrate. Instead, oxygen from the material surrounding the SiCr film is pumped into and migrates along the SiCr film. Resistance decrease is resulted from oxygen accumulation in the SiCr resistor. Current accelerating factor and the equivalent activation energy are determined to enable data extrapolation from experimental conditions to use conditions. The change of TCR due to a current and temperature stress is less than 25%. Joule heating is found to be the limiting factor for current flow in the SiCr film. A current limitation design rule is proposed accordingly. At the proposed design current limitation the lifetime of SiCr due to electrical stress is determined to be more than 90k years.

ACKNOWLEDGMENTS

The authors are grateful to Eelco de Koning for his support in designing some of the test structures for the investigations. One of the authors, Yuan Li, would like to thank Som Nath for his encouragement to conduct the investigation and his critical review of the paper with valuable inputs for improvements. The support of the Dutch government for the knowledge workers project Resilience for Automotive is greatly appreciated.

REFERENCES

[1] R. Dekker, "Silicon process technology innovations for low-power RF applications", Proce. 28th European Solid-State Device Research Conference, pp. 71-80, 1998.

[2] Y. Narizuka, T. Kawahito, T. Kamei and S. Kobayashi, "Properties of high-resistivity Cr-Si-O thin-film resistor", IEEE Trans. Compon., Hybrids, Manuf. Technol., 11(4), pp. 433–438, Dec. 1988.

[3] F. Nava, T. Tien and K.N. Tu, "Temperature dependence of semiconducting and structural properties of Cr-Si thin film", J. Appl. Phys. 57 (6), pp. 2018-2025, Mar. 1985.

[4] R. Glang, R.A. Holmwood and S.R. Herd, "Resistivity and structure of Cr-SiO cermet films", J. Vac. Sci. Technol. 4 (4), pp. 163-170. Jul. 1967.

[5] E. G. Colgan, B.Y. Tsaur and J.W. Mayer, "Phase formatiion in Cr-Si thin-film interactions", Appl. Phys. Lett. 37(10), pp. 938-940. Nov. 1980.

[6] A.T. Burkov, H. Vinzelberg, J, Schumann, T. Nakama and K. Yagasaki, "Strongly nonlinear electronic transport in Cr-Si composite films", J. Appl. Phys. 95(12), pp. 7903-7907. Jun. 2004.

[7] A. P. Botha, R. Pretorius and S. Kritzinger, "Determination of the diffusion species and mechanism of diffusion during $CrSi_2$ formation, using ^{31}Si as a marker, Appl. Phys. Lett. 40(5), pp. 412-414, Mar. 1982.

[8] K. Affolter, X.A. Zhao and M-A. Nicolet, "Transition-metal silicides formed by ion mixing and by thermal annealing: which species moves?", J. Appl. Phys. 58(8), pp. 3087-3093, Oct. 1985.

[9] T. Hirano and M. Kaise, "Electrical resistivities of single-crystalline transition-metal disilicides", J. Appl. Phys. 68(2), pp. 627-633. Jul. 1990.

[10] R.R. Brynsvold and K. Manning, "Constant-current stressing of SiCr-based thin-film resistors: initial "wearout" investigation", IEEE Trans. Dev. & Materia. Reliability, 7(2), pp.259-269, Jun. 2007.

Role of Holes and Electrons During Erase of TANOS Memories: Evidences for Dipole Formation and its Impact on Reliability

Luca Vandelli, Andrea Padovani, Luca Larcher
DISMI (DIpartimento di Scienze e Metodi dell'Ingegneria)
Università di Modena e Reggio Emilia
42122 - Reggio Emilia, Italy
0039–0522-522638, andrea.padovani@unimore.it

Antonio Arreghini, Geert Van den bosch, Malgorzata Jurczak , Jan Van Houdt Senor member IEEE
imec
Leuven, Belgium

Vincenzo Della Marca, Paolo Pavan
DII (Dipartimento di ingegneria dell'informazione)
Università di Modena e Reggio Emilia
Modena, Italy

Abstract— The systematic investigation of the role played by electrons and holes during the erase operation of TANOS memories by means of charge separation experiments and physics-based simulations is reported for the first time. We determined a dominance of electrons back-tunneling in the first part of the transient, and dominance of holes in the second part. Good agreement is reached between experimental and simulated data. In addition we demonstrate for the first time the formation of a vertical charge dipole in TANOS devices, whose polarity depends on the P/E operation sequence. This dipole severely affects the program and erase performances and the retention of mild programmed and erased states, which is a concern especially for multilevel applications.

*Keywords—*TANOS memories, TANOS erase, nitride, charge-trapping devices, charge separation, charge dipole.

INTRODUCTION

The scaling of Floating Gate based nonvolatile memories is reaching its limit with the 22 nm technology nodes, requiring a replacement technology. The most promising and less disruptive evolution is represented by charge trapping memories: SONOS and especially the more recent TANOS stacks provide perfect adaptability to existing NOR/NAND FLASH arrays and good compatibility with the planar CMOS technology. Moreover they reduce the parasitic interference with neighboring cells and virtually eliminate the SILC effect [1].

However many aspects of these devices still need to be thoroughly studied in order to solve their remaining issues: Program (P) and Erase (E) operations are more complicated to

be optimized than in Floating Gate transistors as they require to understand not only injection phenomena through the tunneling barrier, but also the capture/emission of carriers from traps located in the bandgap of the Silicon Nitride (SiN) layer. In particular, the physical mechanisms of the erase operations have not been unambiguously understood yet, because two carrier species, electrons (e^-) and holes (h^+), may be involved [2][3]; the relative contribution of these carriers was only roughly evaluated before, and mainly on old SONOS and NROM devices [4][5]. The injection of two distinct carrier species may generate physically separated distributions in the nitride [4][6], i.e. a *charge dipole*, which can lead to severe reliability issues [7][8], not thoroughly investigated until now.

In this paper, we evaluate the role played by e^- and h^+ during the erase operation of TANOS memories by means of charge separation experiments [9][10] and simulations. Then, we demonstrate for the first time the formation of a vertical charge dipole in TANOS devices whose polarity and entity depends on the P/E operation sequence. This dipole severely affects the retention of mild programmed or erased states, and causes distortions in the P/E curves, therefore representing a serious reliability concern especially for multi-level memory applications.

EXPERIMENTAL SETUP AND DEVICES

Fig. 1 schematically represents the band diagram of a TANOS stack during an erase operation and the fluxes of charge involved: the negative gate bias bends the tunnel oxide electrical barrier increasing the probability of h^+ tunneling from the substrate (flux 1); these carriers are then captured by the neutral or negatively charged traps, i.e. the traps filled by the e^-

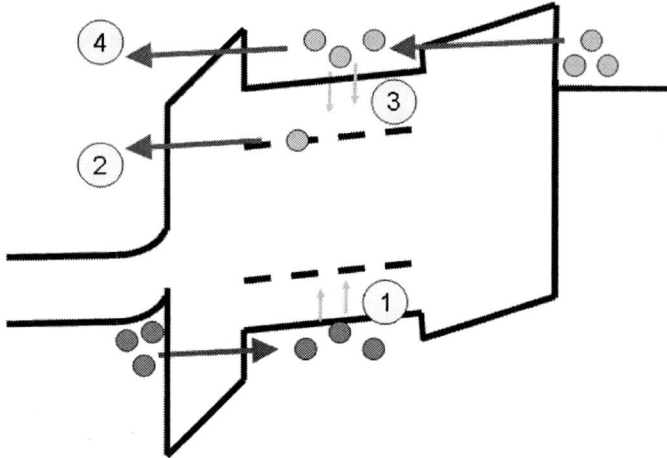

Figure 1. Band diagram of a TANOS structure under erase and main current flows involved.

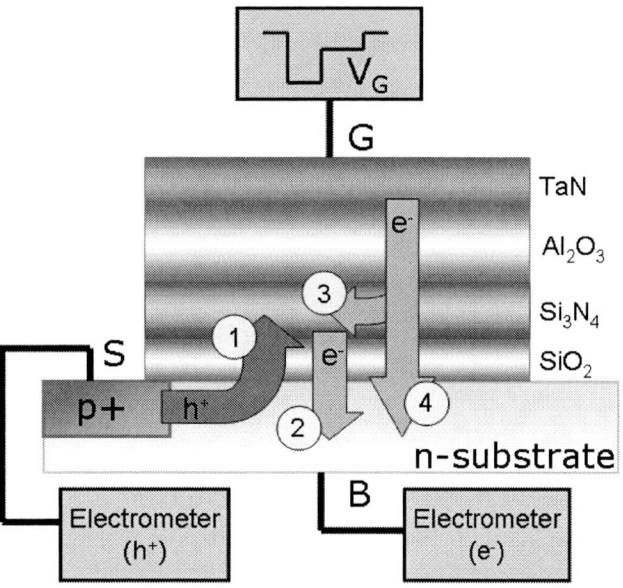

Figure 2. Setup for the carrier separation experiment during erase of pMOS TANOS devices and current flows involved.

injected in the trapping layer during a previous programming operation. At the same time, the traps filled by e^- can emit their charge via Direct Tunneling toward the substrate (flux 2).

Both contributions result in a reduction of the threshold voltage (V_T). In addition, when low threshold voltages are reached, usually below the neutral V_T, the electric field increases in the blocking oxide, allowing an undesired injection of e^- from the gate (flux 3): part of these carriers will be trapped, resulting in the erase saturation phenomenon [3], while the most energetic ones will directly reach the substrate (flux 4).

In order to separate the fluxes 1 and 2 (Fig. 1) responsible for the V_T reduction, hence effectively contributing in erasing the device, we developed a charge separation experimental setup. As illustrated in Fig. 2, we connected two Coulomb-meters at the source and bulk terminals of pMOS TANOS devices. In this way, we measured separately the charge

flowing at the source terminal (h^+, flux 1) and at the substrate (e^-, fluxes 2 and 4); the measurements can be considered reliable only when flux 4 is negligible, hence when the device V_T is not approaching the erase saturation conditions. As shown in Fig. 2, a pulse generator connected at the Gate applies three-level pulses (V_G): after issuing a high voltage to erase the device, the gate is biased for some seconds to a charge reading condition, i.e. a slightly negative bias, whose value is equivalent to the V_T variation induced by the erase pulse. In this way it is possible to compensate for the contribution of the displacement charge caused by the variation of the electric field in the gate capacitor during the erase transient [11]. Further, in order to minimize the spurious recombination in the substrate of electrons back tunneling from the charge trapping layer, we adopted a pMOS structure. In this way the thin hole inversion layer created applying a negative gate bias, can be easily crossed by the electrons back tunneling from the stack, and their contribution can be cleanly measured by the bulk coulomb-meter [11].

All our measurements were performed on TANOS pMOS 3-terminal capacitors; we used large area ($50 \times 50 \ \mu m^2$) devices in order to inject large packets of charge, above the noise level of the electrometers, i.e. few pC. All devices feature a gate stack formed by a 4 or 5 nm of SiO_2 tunnel oxide, a 6 nm of next-to-stoichiometric SiN, 12 nm of Al_2O_3 (contracted about 15% after high temperature PDA crystallization) and 10 nm of TaN metal gate.

MODELING

In order to gain a deeper understanding on the evolution of e^-/h^+ charge during TANOS erase, we simulated TANOS operations and charge separation experiments using an extended version of the model in [12].

To model the time evolution of the trapped electron/hole charge during TANOS program and erase operations, the nitride layer is discretized in energy and space into a matrix of bins. For simplicity, uniform spatial and energy distributions are considered for both electron ($E_{T,e}=1.8\text{-}2.8eV \pm 0.1eV$) and hole traps ($E_{T,h}=2.0\text{-}2.8eV\pm0.2eV$). At the beginning of each simulation time step the Poisson equation is solved to calculate the electric fields and the potential profile across the multilayer stack. The density of the free e^-/h^+ in the Conduction/Valence Band (CB/VB) and of the e^-/h^+ trapped into nitride defects are then calculated by solving self-consistently drift-diffusion (DD) and charge balance (CBE) equations [12]. Electron and hole trapping processes are described through Shockley-Read-Hall (SRH) theory [13], [14], assuming an instantaneous thermalization of e^-/h^+ when entering the nitride CB/VB [12]. Field-induced trap energy barrier lowering, Thermal Emission (TE) [15], Poole-Frenkel (PF) [16] and Trap-to-Band Tunneling (TBT) mechanisms are simultaneously considered for e^-/h^+ emission. For simplicity, no trapping is considered in tunnel oxide and alumina as we verified that this assumption does not lead to significant errors for the devices considered here.

Electron and hole current densities flowing through bottom oxide and alumina layer during program and erase are

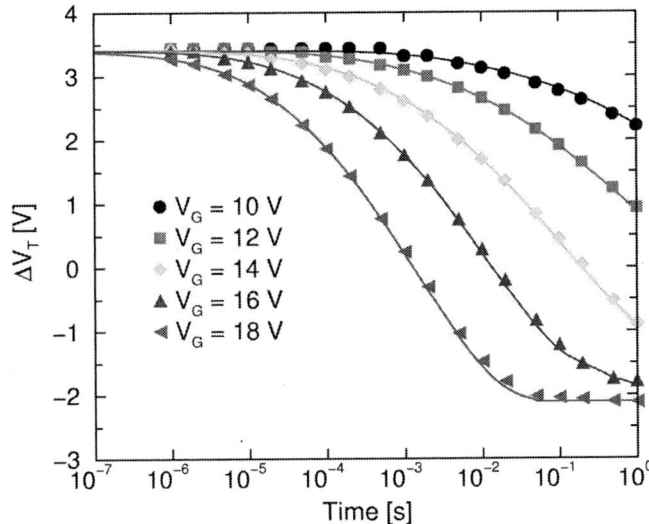

Figure 3. Erase V_T shifts measured (symbols) and simulated (lines) for different erase voltages.

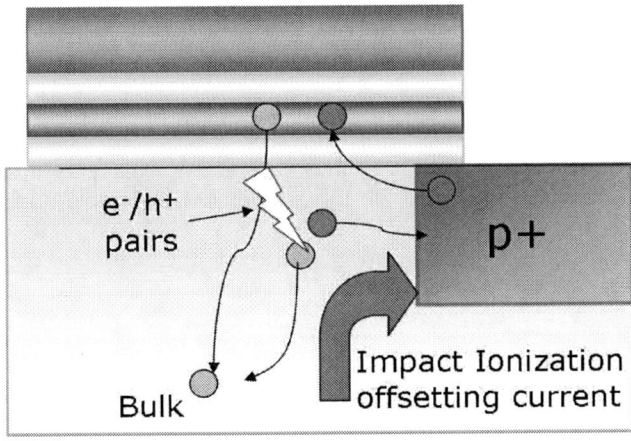

Figure 4. Current flows generated by impact ionization induced by hot electron injection in a p-type substrate.

calculated using the model reported in [17], which assumes a multi-phonon trap-assisted tunneling (TAT) conduction

mechanism and accounts for charge quantization effects at the Si/SiO_2 interface. During erase, the different current contributions depicted in Fig. 1 are calculated separately, thus allowing the simulation of charge separation experiments. Experimental P/E transients measured at different V_G were used to calibrate the model parameters, which agree with those reported in the literature [12]. Model simulations reproduce accurately the experimental erase transients, as shown in Fig. 3.

CHARGE SEPARATION EXPERIMENTS

Charge separation experiments under erase are affected by Impact Ionization (II) [18]. The e^- reaching the substrate may have a high enough energy to generate $e^- - h^+$ pairs that are separated by the electric field across the source – bulk junction: e^- drift toward the substrate terminal, whereas h^+ reach the source terminal; as illustrated in Fig. 4, a spurious net current contribution superimposes the clean measurement. To properly account for $e^- - h^+$ pairs generated by impact ionization, a

corrective factor, representing the average number of pairs generated by one e^-, was introduced. This factor is usually referred to as Quantum Yield (QY) [18]. In the Silicon it is a known function of the kinetic energy (E_e) of the injected e^- [19]. We obtained an approximate value of the QY simply estimating the average value of E_e, assuming a complete thermalization of the e^- in the nitride and neglecting their energy loss due to phonon scattering in the SiO_2. The QY values obtained vary with the erase voltage applied and during the transient, and range between 1.1 and 1.9, indicating that the effect is absolutely not negligible. Fig. 5 shows the measured currents of e^- and h^+ as a function of the threshold voltage shift from the fresh state (ΔV_T) corrected for the II contribution. Devices have been programmed to $\Delta V_T = 3.4$ V and then erased with different V_G ranging from -10 V to -18 V. h^+ current, due to h^+ tunneling through the SiO_2 (flux 1 in Fig. 1), decreases as the devices are erased, because of the reduction of the electric field in the tunnel oxide (F_{ox}). Instead, the e^- current follows a

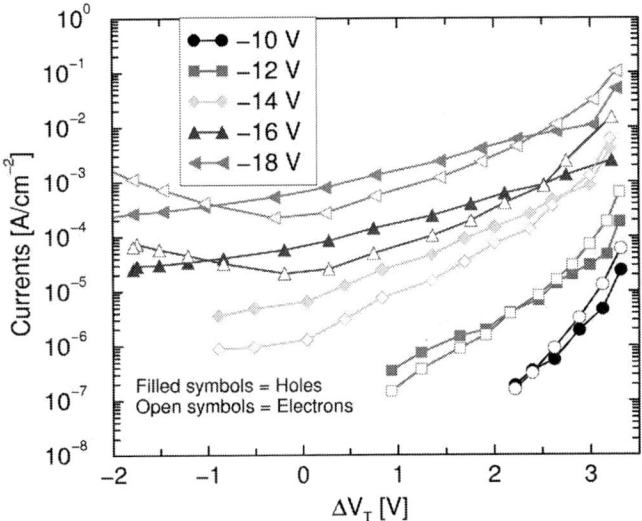

Figure 5. Measured electron (open symbols) and hole (filled symbols) current during erase.

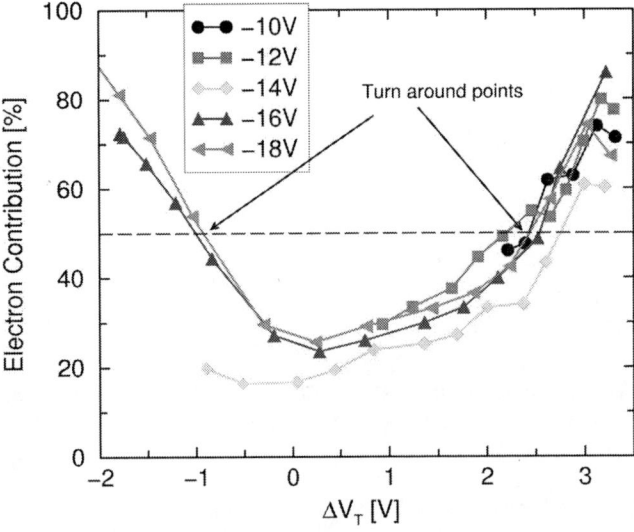

Figure 6. Measured percentage contribution of electrons to the total current during erase.

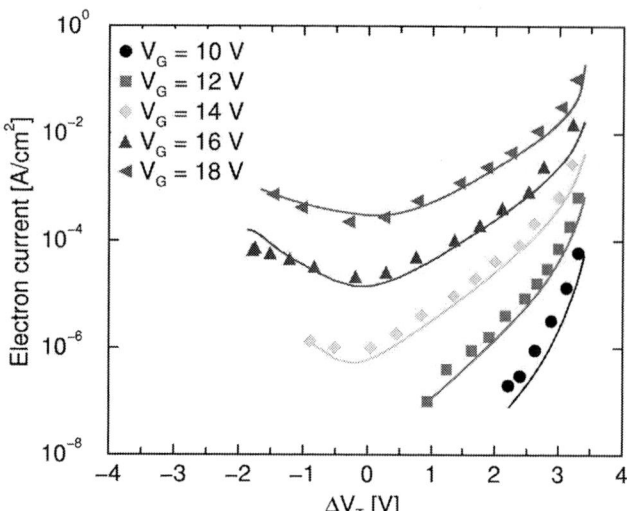

Figure 7. Electron current densities simulated (lines) and measured (symbols) through charge separation experiments as a function of the threshold voltage shift during erase.

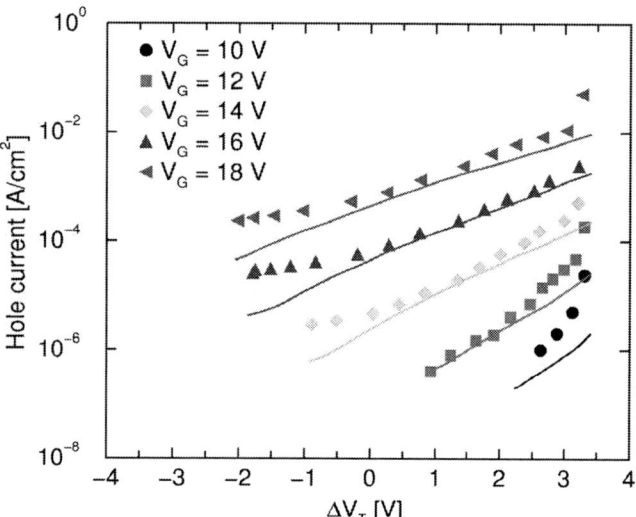

Figure 8. Hole current densities simulated (lines) and measured (symbols) through charge separation experiments as a function of the threshold voltage shift during erase.

more complex dynamics: in the first part of the transient detrapping dominates (flux 2) and thus the current decreases very fast due to the reduction of F_{ox} and to the concurrent removal of the e^- charge trapped; the high Al_2O_3 field at negative ΔV_T forces e^- injection from the gate (flux 4), and the e^- current starts increasing again.

Fig. 6 shows the percentage contribution of e^- to the total erase current. The trend is similar for every V_G: e^- prevail at strong programmed states and for negative ΔV_T, whereas h^+ injection prevails for mild programmed states. Two turnaround points, where the dominant carrier species changes, are present, interestingly located at similar ΔV_T for every V_G. Qualitative understanding of e^- and h^+ contributions is quantitatively confirmed for the first time by model simulations in Fig. 7 and 8, respectively, accurately reproducing the experimental data.

CHARGE DIPOLE INVESTIGATION

The results of charge separation experiments shown in the previous section clearly demonstrate that both e^- detrapping and h^+ injection mechanisms contribute to the erase of a previously programmed TANOS device, so that both carriers may be present in the nitride at the end of the erase operation, as recently observed in SONOS devices [2]. The concurrent presence of physically separated e^- and h^+ distributions in the nitride layer after subsequent P and E operations may lead to the formation of a dipole, which has been identified as one of origins of the retention degradation observed in nitride-based charge trapping devices [7], [8], [20].

In order to investigate and quantify this potential TANOS reliability issue, we performed high temperature (200°C) retention experiments on fresh and reset devices, i.e. devices whose V_T is brought back to its fresh value after a sequence of P+E or E+P operations. Results are shown in Fig. 9. While as expected the characteristics of fresh devices are flat, indicating that the initial charge in the nitride is stable, devices in the reset state show a significant V_T variation of up to 300mV, whose polarity depends on the reset sequence: positive for P+E (circles), negative for E+P (squares). Therefore, when the cell is brought back to its fresh V_T after a given P/E sequence, a significant amount of both e^- and h^+ charge is still trapped in the nitride. Moreover, the different polarity of the shifts observed for P+E and E+P sequences, see Fig. 9, is an evidence that the charge is not uniformly distributed across the nitride. Instead, it indicates the formation of a vertical charge dipole, i.e. a misalignment of e^- and h^+ trapped charge distributions.

Further confirmations come from the high temperature retention experiments performed on slightly programmed (erased) devices shown in Fig. 10. As expected, perfect retention is observed when only e^- (h^+) are present in the nitride, thus confirming that the losses shown in Fig. 9 are related to the concurrent presence of e^- and h^+ charges. Moreover, since the results in Fig. 10 indicate that charge leakage in presence of a small amount of carriers is negligible, the retention losses in Fig. 9 are determined by the

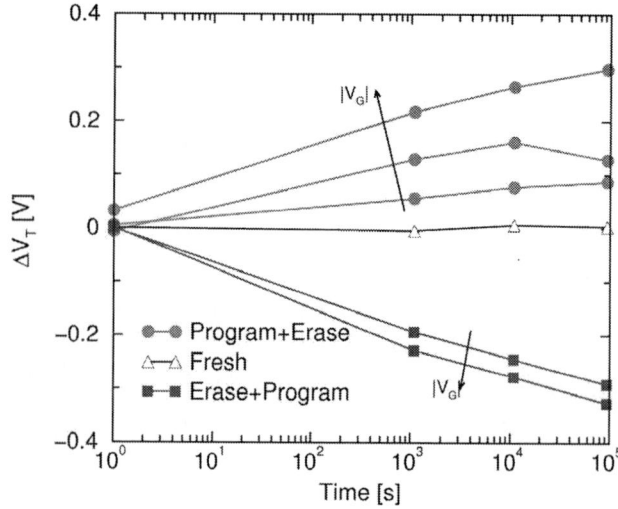

Figure 9. 200 °C retention of a fresh device (black) and of devices resetted with different combinations of P+E (red) and E+P (blue) pulses.

recombination of nitride e^- and h^+ charges driven by the internal dipole electric field.

This is confirmed also by retention experiments analogous to the ones in Fig. 9 performed at room temperature, when the charge leakage is negligible because the Thermal Emission is inactive. As shown in Fig. 11, V_T variations are still present, featuring the same dependence on the reset sequence, but as expected the time dependence is slower, because the charge redistribution is accelerated by the temperature.

In conclusion, the retention behavior observed in Fig. 9 is fully consistent with the formation of a charge dipole in the nitride, whose orientation depends on the last operation performed. This behavior can be understood by considering the simple model depicted in Fig. 12. In the case of the P+E sequence sketched in Fig. 12(b)-(d), when an erase operation is performed on a previously programmed device, the pre-

existing electron charge is not fully detrapped. Therefore, at the end of the E operation separated e^- and h^+ distributions are generated near the SiN/Al₂O₃ and SiN/SiO₂ interfaces, respectively, (Fig. 12(c)). During high temperature retention experiments, this charge tend to recombine under the action of the dipole electric field, thus reducing the amount of e^- and h^+ in the nitride (Fig. 12(d)). Since the charge close to the SiO₂ (i.e. h^+) has a higher impact on V_T, see equation in Fig. 12, the overall effect of the recombination is an increase of the threshold voltage, coherently to what observed in Fig. 9. The same concept applies to the case of the E+P sequence sketched in Fig. 12(e)-(g), where the opposite dipole orientation gives place to a V_T shift of opposite polarity.

In order to verify the simple model illustrated in Fig. 12 we reproduced the P+E and E+P sequences shown in Fig. 9 using our TANOS model. Fig. 13(a) and 13(b) shows the final distribution of the net trapped charge obtained in the two cases by using the same simulation parameters used to reproduce TANOS program (not shown) and erase transients (see Fig. 3).

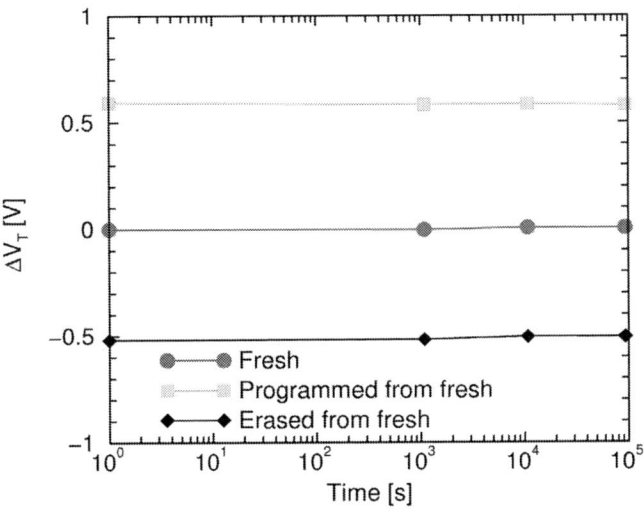

Figure 10. Retention of a fresh (red), slightly programmed (green) and slightly erased (black) devices.

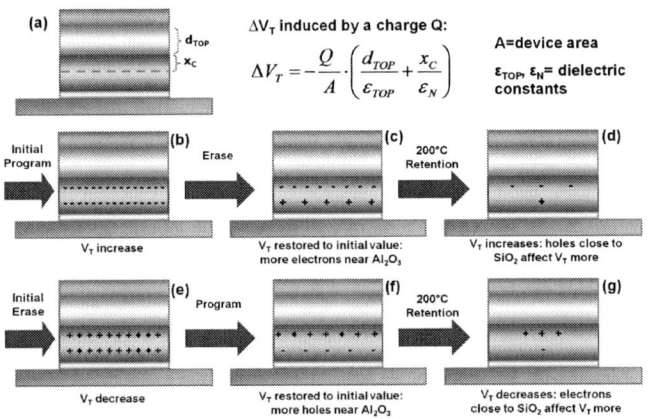

Figure 12. Schematic representation of the dipole orientation for the P+E and E+P cases and equation to compute ΔV_T induced by a charge Q.

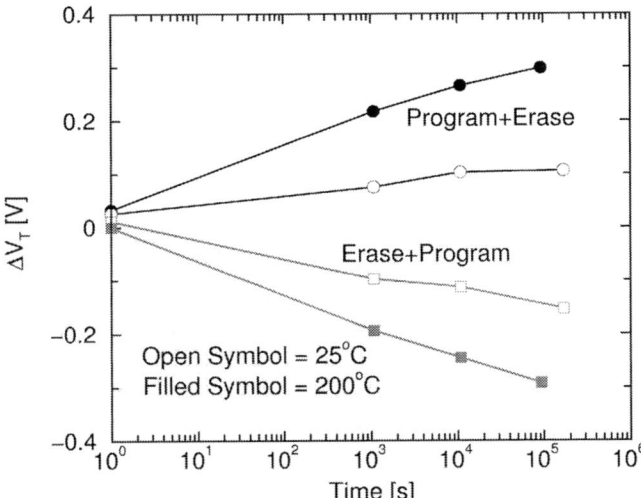

Figure 11. 200 °C (full symbols) and room temperature (open symbols) retention of devices resetted with P+E (black) and E+P (red) pulses.

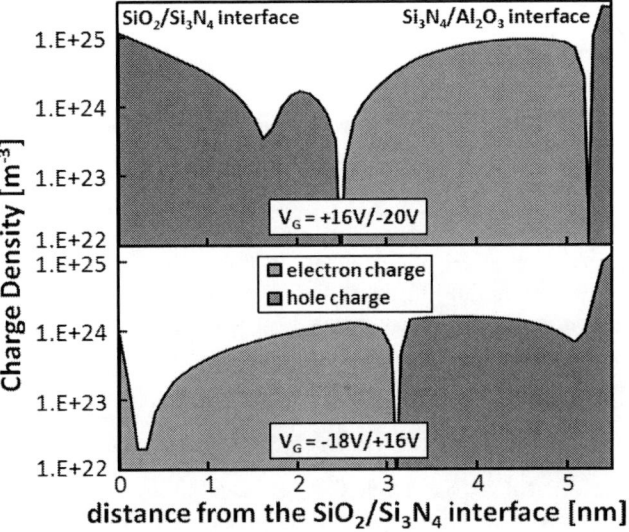

Figure 13. Net charge distribution simulated for P+E (a) and E+P (b) reset devices

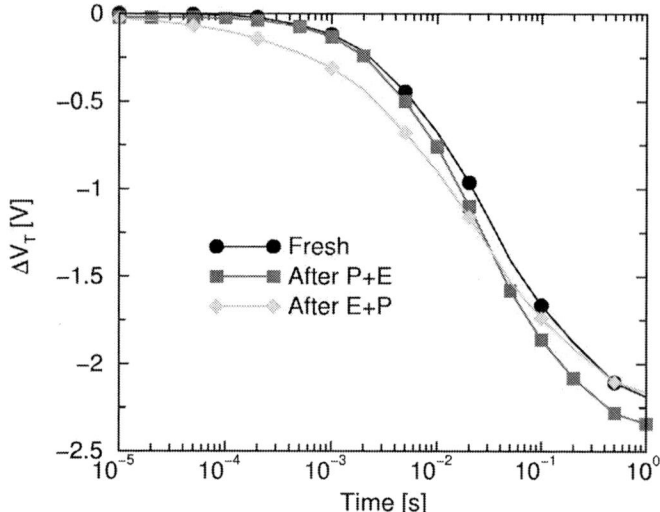

Figure 14. -16V erase-from-fresh V_T characteristics measured on a fresh device (black), after P+E (red) and E+P (green) reset.

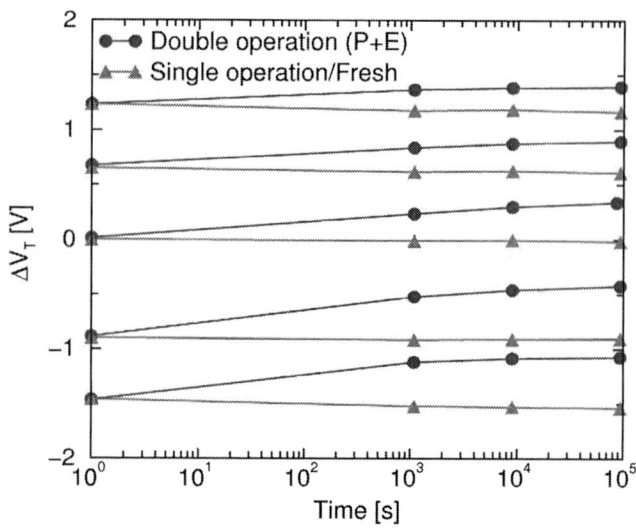

Figure 16. Comparison of retention of a device written with a single operation and a double operation (P+E).

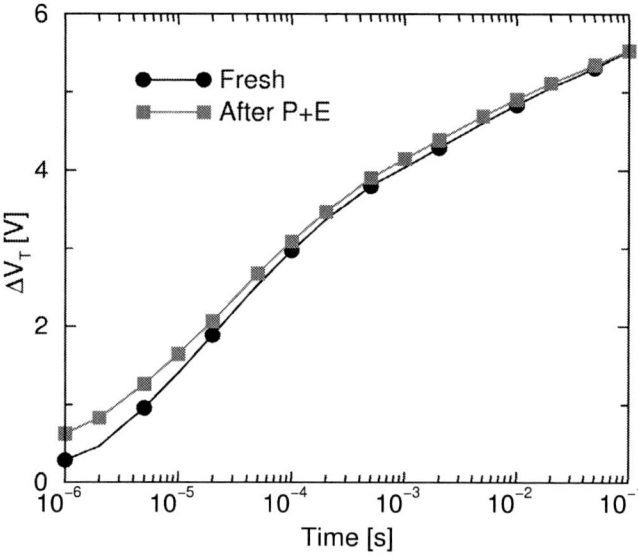

Figure 15. 16V program-from-fresh V_T characteristics measured on a fresh device (black) and after P+E (red)

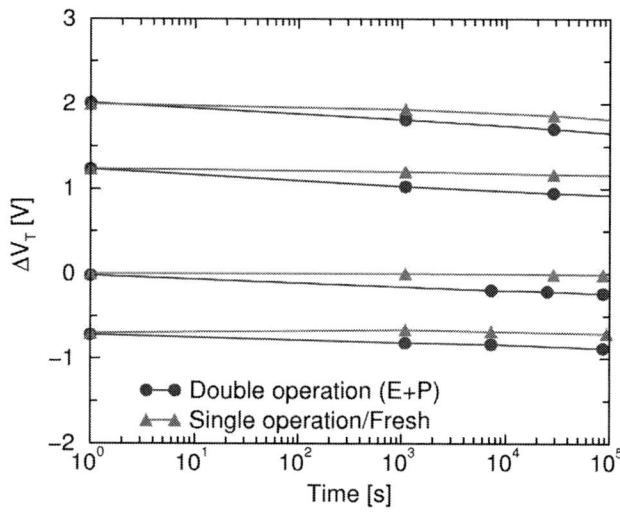

Figure 17. Comparison of retention of a device written with a single operation and a double operation (E+P).

These results prove that a charge dipole is created in the nitride, with an orientation that depends on the last operation performed. Moreover, the orientation of the dipole agrees with the simplified model illustrated in Fig. 12, thus confirming that the retention behaviour shown in Fig. 9 is determined by the recombination of nitride e^- and h^+ charges driven by the dipole field.

RELIABILITY CONSIDERATIONS

The vertical charge dipole that is created in the nitride layer after a reset operation is a serious reliability concern for TANOS memories. On one hand the presence of a dipole affects subsequent P and E operations, altering the normal operation of the device. To better understand this point we

compared in Fig. 14 and in Fig. 15 respectively the -16V erase–from–fresh and 16V program-from-fresh transients measured on P+E (squares) and E+P (diamonds) reset devices. The presence of a dipole severely distorts the characteristics with respect to the fresh ones (circles), also shifting the erase saturation level: this may lead to a spread of the V_T distribution in a TANOS array, with serious consequences on its reliability, especially in the framework of multi-level applications.

Moreover, the dipole leads to charge losses that affect the retention of the device. In order to understand the impact of this effect and the ΔV_T range in which the dipole influences the performances of the device, we performed the following experiment: as shown in Fig. 16 and 17, we compared the retention transients of devices brought to a certain ΔV_T through single-pulse operations (triangles), hence injecting only one kind of carriers, and double-pulse (P+E in Fig. 16, E+P in Fig. 17, circles) operations, hence injecting both h^+ and e^-. The retention characteristics depend on how the starting state is reached: this is due to the presence of a dipole, generated by double-pulse sequences. In agreement with the previous results, the dipole formed after P+E operations causes positive V_T variations. For programmed states this effect compensates the loss due to charge leakage, improving the retention; for erased states retention is instead severely worsened especially for negative V_T, causing a shift of up to 0.5V in the case of ΔV_T=-1V. Vice versa, the dipole formed after E+P operations causes negative V_T variations, mostly affecting the retention of mild programmed states. The range in which the dipole seems to be active extends from ΔV_T=-1V to 2V, and the impact on retention is absolutely not negligible for NAND multi-level applications, where negative V_T and mild programmed states are logical states of the cell. This demands for an optimization of program/erase voltages and pulse times to minimize this effect.

CONCLUSIONS

In this work we presented a detailed investigation of the erase operation of TANOS devices. First, we used charge separation experiments and numerical simulations to investigate the role played by electrons and holes. We found that electron detrapping dominates the first part of the transient, whereas hole injection prevails after the removal of the initial trapped electron charge. Then, we demonstrated for the first time the formation of physically separated electron and hole distributions, i.e. of a vertical charge dipole, whose polarity has been correlated to the last operation of P/E sequence applied to the cell. Finally, we evaluated the impact of the charge dipole on TANOS reliability. On one hand, the dipole influences successive program or erase operation, introducing a variability on the final V_T. On the other hand, it strongly degrades the retention of mild programmed and erased states thus representing a serious reliability concern for multilevel applications.

ACKNOWLEDGMENT

This work was supported in part by the E. U. under FP7 Research Contract 214431 "GOSSAMER" and by imec's Industrial Affiliation Program on Advanced Flash Memory

REFERENCES

[1] Y. Park, J. Choi, C. Kang, C. Lee, Y. Shin, B. Choi, J. Kim, S. Jeon, J. Sel, J. Park, K. Choi, T. Yoo, J. Sim, K. Kim, "Highly Manufacturable 32Gb Multi – Level NAND Flash Memory with 0.0098 μm² Cell Size using TANOS (Si - Oxide - Al2O3 - TaN) Cell Technology", IEDM Technical Digest, pp. 29-32, 2006.

[2] P.-Y. Du, H.-T. Lue, S.-Y. Wang, T.-Y. Huang, K.-Y. Hsieh, R. Liu, and C.-Y. Lu, "A Study of Gate-Sensing and Channel-Sensing (GSCS) Transient Analysis Method Part II: Study of the Intra-Nitride Behaviors and Reliability of SONOS-Type Devices", IEEE Transaction on Electron Devices, Vol. 55, N. 8, pp. 2229-2237, 2008.

[3] G. Wang, M. H. White, "Characterization of scaled MANOS nonvolatile semiconductor memory (NVSM) devices", Solid State Electronics, Vol. 52, p. 1491-1497, 2008.

[4] H.-T. Lue, P.-Y. Du, S.-Y. Wang, K.-Y. Hsieh, R. L. and C.-Y. Lu, "A Study of Gate-Sensing and Channel-Sensing (GSCS) Transient Analysis Method—Part I: Fundamental Theory and Applications to Study of the Trapped Charge Vertical Location and Capture Efficiency of SONOS-Type Devices", IEEE Transaction on Electron Devices, Vol. 55, N. 8, pp. 2218-2228, 2008.

[5] Larcher, L.; Pavan, P.; Eitan, B., "On the physical mechanism of the NROM memory erase", IEEE Transaction on Electron Devices,Vol. 51, N. 10, pp. 1593-1599, 2004.

[6] P.-Y. Due et al., "Study of gate-injection operated SONOS-type devices using the gate-sensing and channel-sensing (GSCS) method", Proc. of IRPS, pp. 288-293, 2009.

[7] A. Shappir, Y. Shacham-Diamand, E. Lusky, I. Bloom, B. Eitan, "Lateral charge transport in the nitride layer of the NROM non-volatile memory device", Microelectronic Engineering, Vol. 72, pp. 426-433, 2004.

[8] C. Kang, J. Choi, J. Sim, C. Lee, Y. Shin, J. Park, J. Sel, S. Jeon, Y. Park and K. Kim, "Effects of Lateral Charge Spreading on the Reliability of TANOS (TaN/AlO/SiN/Oxide/Si) NAND Flash Memory", Proc. of IRPS, p.167, 2007.

[9] A. Roy, M. H. White, "Electron and Hole Charge Separation With a Dual Channel Transistor-", IEEE Transaction on Electron Devices, Vol. 36, N. 11, p. 2604, 1989.

[10] Bachhofer, H.; Reisinger, H.; Bertagnolli, E., "Transient conduction in multidielectric silicon–oxide–nitride–oxide semiconductor structures", Journal of Applied Physics, Vol. 89, p. 2791, 2001.

[11] A. Arreghini, F. Driussi, E. Vianello, D. Esseni, M. J. van Duuren, D. S. Golubovic, N. Akil and R. van Schaijk, "Experimental Characterization of the Vertical Position of the Trapped Charge in Si Nitride-Based Nonvolatile Memory Cells", IEEE Transaction on Electron Devices,Vol. 55, N. 5, pp. 1211-1219, 2008.

[12] A. Padovani, L. Larcher, D. Heh and G. Bersuker, "Modeling TANOS Memory Program Transients to Investigate Charge-Trapping Dynamics", IEEE Electron Device Letters, Vol. 30, N. 8, , pp. 882-884, 2009.

[13] W. Shockley and W. T. Read, Physical Review, Vol. 87, p. 835, 1952.

[14] R. N. Hall, "Electron-Hole Recombination in Germanium", Phys. Rev. vol. 87, p. 387, 1952.

[15] P. J. McWhorter, S. L. Miller, T. A. Dellin, "Modeling the memory retention characteristics of silicon-nitride-oxide-silicon nonvolatile transistors in a varying thermal environment", Journal of Applied Physics, Vol. 68, p. 1902, 1990.

[16] J. Frenkel, "On Pre-Breakdown Phenomena in Insulators and Electronic Semi-Conductors", Physical Review, Vol. 54, pp. 647-648, 1938.

[17] A. Padovani, L. Larcher, S. Verma, P. Pavan, P. Majhi, P. Kapur, K. Parat, G. Bersuker, and K. Saraswat, "Statistical modeling of leakage currents through SiO2/high-k dielectric stacks for non-volatile memory applications," Proc. of IRPS, pp. 616-620, 2008.

[18] C. Chang, C. Hu, R. W. Brodersen, "Quantum yield of electron impact ionization in silicon", Journal of Applied Physics, Vol. 57, p. 302, 1985.

[19] R. C. Alig, S. Bloom, and C. W. Struck "Scattering by ionization and phonon emission in semiconductors", Phys. Rev. B vol. 22, no. 12, p. 5556, 1980.

[20] A. Padovani, L. Larcher and P. Pavan, "Hole Distributions in Erased NROM Devices: Profiling Method and Effects on Reliability", IEEE Transaction on Electron Devices, Vol. 55, N. 1, pp. 343-349, 2008.

RESET CURRENT DISTRIBUTIONS IN PHASE CHANGE MEMORIES

A. Calderoni [1], M. Ferro [2], D. Ventrice [1], D. Ielmini [2] and P. Fantini [1]

[1] Numonyx, R&D – Technology Development, via Olivetti 2, 20041 Agrate Brianza, Milano, Italy
e-mail: alessandro.calderoni@numonyx.com

[2] Politecnico di Milano-IU.NET, Dip. Elettronica e Informazione, Piazza L. da Vinci 32 – 20133 Milano – Italy

ABSTRACT

Abstract - **In this work a new analytical transport model for the readout region of amorphous GST is proposed. The model is employed to assess, through Monte Carlo (MC) simulations, the sources of variability responsible for the width and shape of the readout current distributions of reset bits. The correlation between transport mechanisms and the statistical spread is highlighted, also considering the reset pulses dependence. Furthermore, the temperature effect on the reset bits is addressed. The statistical characterization results are discussed within the framework of the proposed model.**

Keywords: Phase Change Memories (PCM), Non-Volatile Memory modeling, variability effects, chalcogenide, reliability modeling.

INTRODUCTION

Among the emerging memories alternative to the Flash memory concept, the GST(Ge2Sb2Te5)-based phase change Memory (PCM) represents today a promising candidate for next Non-Volatile Memory (NVM) technology generations [1]. To develop a mature PCM technology optimized programming algorithms for amorphization (reset process) or crystallization (set process) and methods to control variability effects are essential [2]. From this standpoint, the physical understanding and modeling of the transport properties in the readout region of PCM devices are strongly needed. Carrier transport and electrical switching in amorphous GST are highly debated in the literature: recent models have interpreted these phenomena in terms of drift-diffusion in band-like states [3] or small polaron conduction [4]. More recently, a thermally-activated hopping mechanism has been proposed to describe the subthreshold carrier transport and threshold switching in the PCM amorphous state [5]. Thermally activated hopping was shown to account for several electrical characterization results, such as current-voltage curves, threshold voltage and activation energy for conduction. Very recently, thermally-activated hopping was shown to coherently explain *1/f* noise, which was experimentally observed in amorphous GST, by means of a trap-level fluctuation picture [6, 7]. Thermally activated hopping at high field was shown to naturally account for threshold switching, as the result of carrier energy gain and the consequent negative differential resistance [8]. An extension of the thermally-activated hopping model to distributed energy barriers in the amorphous region was shown to explain the thickenss dependence of electrical parameters of the reset-state cells, allowing for scaling projections of resistance window and noise [6, 9].

Although the conduction model can account for the main experimental features of PCM resistance and switching, some aspects need to be further addressed. In particular, the logarithm of subthreshold current is sometimes reported to increase with the square root of the voltage (Poole-Frenkel model) [10, 11], instead of the usual linear dependence on voltage (Poole model) [5]. Therefore, a *unified* hopping conduction model able to describe both Poole (P) and Poole-

Frenkel (PF) characteristics is strongly needed. In this work we develop a new analytical model for I-V curves in reset-state PCM cells. The model is extended to a statistical framework, allowing to describe the distribution spread for varying temperature and programming conditions. The new model can be fruitfully applied to investigate the reliability figure related to the reset-induced read window budget.

TRANSPORT MODEL

The basic physical picture is that of electrons being thermally excited from a positively-charged localized state. The applied electric field is responsible for a barrier lowering that carrier must overcome. On the basis of this picture, two different *I-V* relationships are obtained according to either *i*) a single isolated Coulomb potential well or *ii*) two or more Coulombic traps interacting in the trap-to-trap hopping (see case A and B, respectively, in Fig. 1). Mechanisms *i*) and *ii*), namely the PF and P laws, lead to dependences of the potential barrier lowering ΔU on the square root or linear as a function of voltage, respectively.

The PF and P laws are commonly used to describe the low and high trap density regime respectively [7]. Figs. 2a) and 2b) report the *I-V*

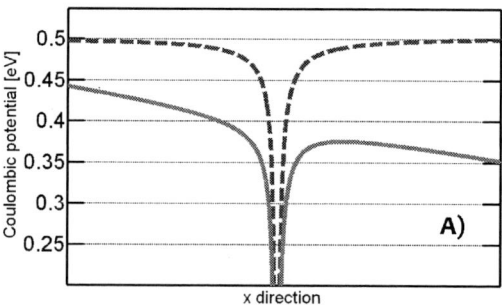

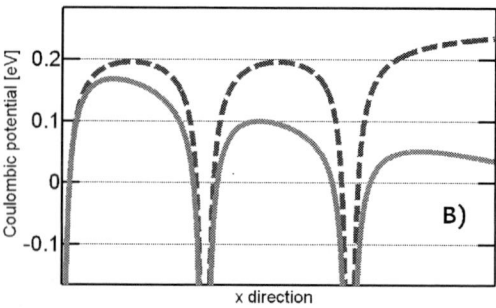

FIG. 1: A) CALCULATED POTENTIAL PROFILE CONSIDERING ONLY ONE DONOR TRAP WITHOUT (DASHED LINES) AND WITH (CONTINUOUS LINES) AN APPLIED FIELD. B) CALCULATED POTENTIAL WELLS FELT BY TWO TRAPS WITHOUT (DASHED LINES) AND WITH (CONTINUOUS LINES) AN APPLIED FIELD.

978-1-4244-5430-3/10 $26.00 © 2010 IEEE

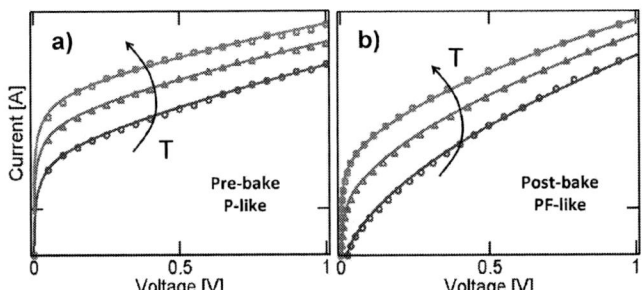

FIG. 2: EXPERIMENTAL (SYMBOLS) I-V CURVES FOR THREE TEMPERATURES MEASURED PRE-BAKE A) AND POST-BAKE B). CONTINUOUS LINES ARE THE SIMULATED CURVES OBTAINED BY USING P- A) AND PF- B) EQUATIONS, RESPECTIVELY.

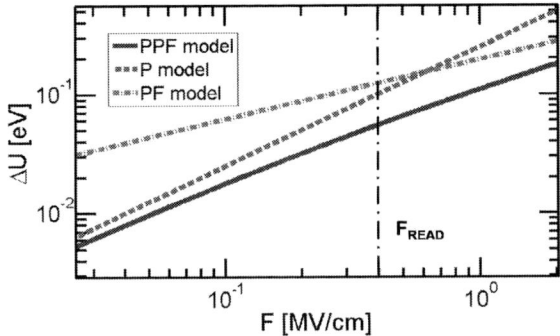

FIG. 3: CALCULATED BARRIER LOWERING ACCORDING TO POOLE (P), POOLE-FRENKEL (PF) AND HERE PROPOSED POOLE/POOLE-FRENKEL (PPF) MODEL. VERTICAL LINE IS THE TYPICAL READOUT FIELD.

curves for three different temperatures for reset state cells measured before and after 1 hour annealing at 400K, respectively. The annealing accelerates the onset of the drift phenomenon that is ascribed to a structural relaxation (SR) process, which reduces the density of localized point defects, hence of localized states in the amorphous volume [6, 11]. Experimental curves can be quantitatively explained by the P (Fig. 2a) and PF (Fig. 2b) transport laws, coherently with the occurrence of P and PF regime due to the reducing trap density. Therefore the reset-state bit population consists, in general, of the coexistence of P and PF-like cells, depending on both the trap density in amorphous GST and the applied electrical field [5, 9]. This highlights the necessity for a *unified* model picturing both I-V relationships (P and PF).

Based on this finding we propose a combined P/PF transport model obtained as a mixture of barrier lowering models through the Mathiessen's rule, giving the barrier lowering $\Delta U_{P/PF}$ by:

$$\Delta U^{-1}_{P/PF} = \Delta U^{-1}_P + \Delta U^{-1}_{PF} \qquad (1)$$

where ΔU_P and ΔU_{PF} are the barrier-lowering functions according to P and PF models, respectively. From Eq. (1), a close expression for $\Delta U_{P/PF}$ as a function of bias (V_A) is obtained as:

$$\Delta U_{P/PF} = q \frac{K_P \cdot K_{PF} \cdot V_A}{K_P \sqrt{V_A} + K_{PF}} \qquad (2)$$

where the costant K_P is given by $K_P = dz/2u_a$; dz is the average distance between adjacent traps, u_a is the amorphous GST thickness and the constant K_{PF} is given by $K_{PF} = (q/(\pi\varepsilon u_a))^{1/2}$, where ε represents the dielectric constant of amorphous GST. The constants K_P and K_{PF} are related to the P and PF barrier lowering, respectively. In fact, Eq. (2) yields $\Delta U_{P/PF} = qFdz/2$ for $V_A \to 0$, coherently with the P model, while we have $\Delta U_{P/PF} = (q^3 V_A/(\pi\varepsilon u_a))^{1/2}$ for $V_A \to \infty$ in agreement with the PF model describing an isolated Coulombic well.

Fig. 3 shows the calculated barrier lowering obtained from P, PF and the mixed P/PF expressions as a function of electric field. The vertical dashed line represents the electric field in a reset-state PCM cell at a typical readout bias (400 mV) and assuming $dz = 6$nm, $u_a = 20$nm and $\varepsilon_R = 15$. The typical read-mode electric-field lies in the middle of P and PF-like regions, thus the proposed model provides an effective description for the overall bias range.

Using the mixed P/PF model for the barrier lowering of Eq. (2) within the analytical equation for the current [5] leads to the following current equation:

$$I_{P/PF} = 2qAN_T \frac{dz}{\tau_0} e^{-\left(\frac{E_A}{kT}\right)} \sinh\left(\frac{q}{kT} \frac{K_P K_{PF} V_A}{K_P \sqrt{V_A} + K_{PF}}\right) \qquad (3)$$

where N_T is the total density of traps, A is the GST/Heater interface area, E_A is the zero-field barrier energy, T the sample temperature and τ_0 the attempt-to-escape time constant [5]. Now, we consider an arbitrary cell measured after a certain time from programming but without any accelerated bake. It is expected to show a blending behavior between P and PF. Fig. 4 shows calculated and measured I-V curves at various temperatures, demonstrating the excellent agreement between the proposed P/PF combined model and data in the overall bias range. This is a signature of the capability of the mixed P/PF model to correctly follow the barrier lowering both in the low- and high-field region.

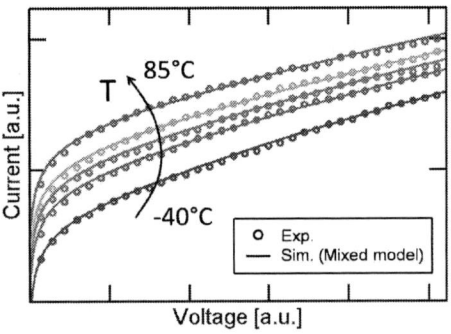

FIG. 4: EXPERIMENTAL (DOTS) AND SIMULATED (CONTINUOUS LINES) I-V CURVES FOR VARIOUS TEMPERATURES. SIMULATIONS ARE OBTAINED BY USING THE PROPOSED MIXED POOLE/POOLE-FRENKEL MODEL (EQ.2).

STATISTICAL MODEL AND EXPERIMENTAL CHARACTERIZATION

We next consider the statistical description of the readout currents of a PCM memory array. A Monte-Carlo (MC) approach has been employed to investigate the current distributions of amorphous state bits. MC simulations were run starting from the aforementioned single cell description in Eq. (3) and assigning Gaussian spreads to cell parameters to describe the intra-die dispersion. Both technology-induced variability of geometry (A, u_a) and physical spread of amorphous material properties (N_T, dz, E_A) were considered. Cell geometrical spreads were estimated from in-line process control measurements. A correlation between the trap concentration (N_T) and the average traps distance (dz) was assumed [12]. Reset current distributions from the BJT-selected PCM arrays in 90 nm technology node have been experimentally characterized [13]. Spreads, associated to the heater and the BJT selector were also introduced for a realistic comparison with the experimental results [14]. However, in contrast to what shown for the set state [14], heater and BJT selector were found to contribute negligibly to the width and shape of read current distributions, because the highly-resistive amorphous GST mainly controls the current distribution. We underline that, due to the exponential relationships in Eq. (3), a Gaussian spread of parameters in that equation results in a log-normal distribution of the current. Therefore, calculated and measured distributions were plotted on a logarithmic scale to highlight the log-normal shape.

STRUCTURAL RELAXATION

The metastable nature of amorphous phase causes a spontaneous process of structural relaxation toward the thermodynamic equilibrium state. This tendency is accomplished by an atomistic-level defect-annihilation process, leading to higher resistance and activation energy (E_A) [15, 16]. As a consequence, the electronic properties of the amorphous region change with time.
In particular, the reset-state resistance displays a steady increase with time, obeying to the power-law equation controlling the so-called drift phenomenon:

$$R(t) = R_0 \left(\frac{t}{t_0} \right)^{\nu} \qquad (4)$$

where R_0 and t_0 are constants, t is the time after the programming (reset) pulse, and ν is a characteristic exponent, controlling the resistance drift rate [15]. According to Eq. (3), the read resistance window increases with time, with beneficial consequences on memory reliability. For optimum monitoring and prediction of the impact of drift on resistance distributions with time, it is essential to model the evolution of the current statistics due to SR.

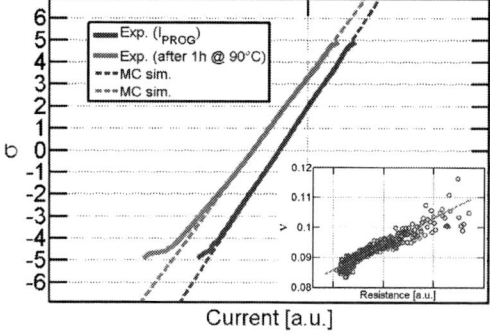

FIG. 5: EXPERIMENTAL CUMULATIVE DISTRIBUTIONS COLLECTED BOTH BEFORE AND AFTER BAKE (LINES). DASHED LINES ARE THE MC SIMULATIONS. INSET: CORRELATION BETWEEN THE DRIFT POWER LAW EXPONENT AND THE INITIAL RESISTANCE.

Fig. 5 shows the cumulative distributions of currents in the reset state measured before and after 1 hour annealing at 400K to accelerate the drift phenomenon. MC simulations were run increasing the E_a parameter, promoted by N_T lowering as a consequence of the SR, as the only difference between pre- and post-drift situation [7]. It can be

observed that simulations (dashed lines) well capture the distribution drift and broadening. Within our conduction model, cells moved from P to PF mechanism according to the results in Fig. 2. The distribution widening after annealing can be explained by the correlation between the drift characteristic exponent ν and the initial cell resistance reported in the inset of Fig. 5 [16, 17]. Namely, ν increases for increasing resistance, thus high-resistance cells gain resistance at a faster rate compared to cells with a relatively small initial resistance. This results in the broadening of the resistance distribution shown in the figure.

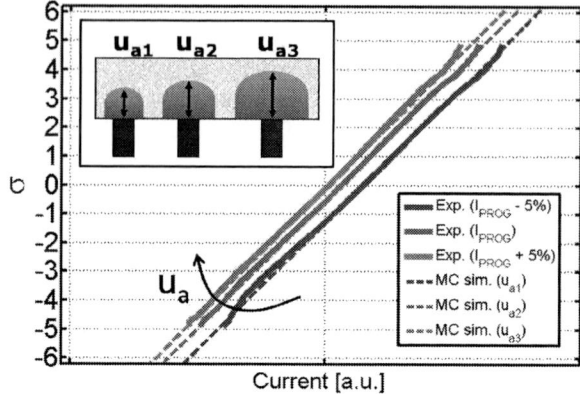

FIG. 6: EXPERIMENTAL AND SIMULATED CUMULATIVE DISTRIBUTIONS, IN LOG-LOG SCALE, FOR THREE DIFFERENT PROGRAMMING PULSES. MC SIMULATIONS WERE OBTAINED INCREASING THE AMORPHOUS DOME THICKNESS (u_A) WITH PULSE INCREASING AS SKETCHED IN THE INSET.

PROGRAMMING PULSE

The choice of a specific programming algorithm requires that both the average and the standard deviation of the reset current distribution are consistent with the read window budget. Therefore, the impact of programming pulses on the reset distribution must be well investigated and modeled. Programming pulse intensity allows to tightly modulate the amorphous GST thickness as sketched in the inset of Fig.6. This parameter is important since, together with the planar technological control of the cell, u_a defines the amorphous active region controlling the resistance, subthreshold slope and threshold voltage [5, 8]. Recently, it was shown that the amorphous thickness can be well controlled by appropriate choice of the cell architecture [18].
Fig. 6 shows the cumulative distributions of experimental currents in the RESET state obtained by different single pulse programming currents (continuous lines). MC simulations (dashed lines) are also shown. As a first step, we fitted the variability accounting for the currents distribution related with the typical I_{prog} pulse. Then, the distributions corresponding to a lower (-5%) and a higher (+5%) programming current were calculated by modulating only the amorphous thickness (u_a). A good agreement with the experimental data can be observed. This demonstrates the capability of our model to sense the programming pulse effects and to reliably predict the fail probability as a function of the programming current. In fact, cells need to be programmed above a minimum resistance (or, equivalently, below a maximum current) to avoid the risk merging the set and reset distributions. Improving the reset current allows to push the statistical distribution beyond this threshold, thus reducing the need for program/verify loops. Fig. 7 shows the fail probability as a function of the programming current, pointing out the controllability of resistance distribution by the reset current.

978-1-4244-5430-3/10 $26.00 © 2010 IEEE

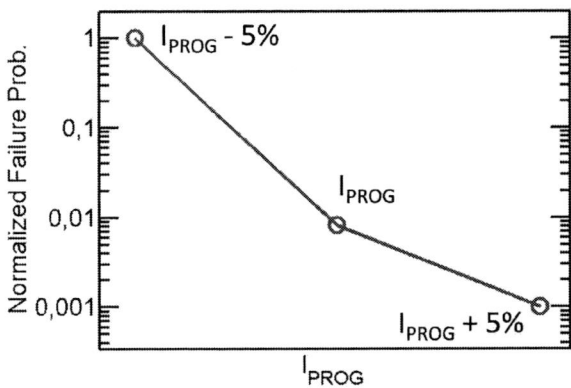

FIG. 7: FAILURE PROBABILITY EXTRACTED FROM FIG.5 AS A FUNCTION OF THE PROGRAMMING CURRENT WHEN A RESET VERIFY CURRENT IS FIXED.

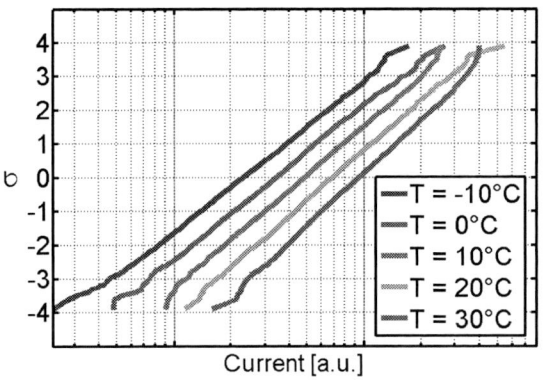

FIG. 9: SIMULATED CURRENT CUMULATIVE DISTRIBU- TION SHOWING THE SHIFT AND WIDTH MODULATION AS A FUNCTION OF TEMPERATURE.

MC simulations (dashed lines) were obtained by fitting, as first step, the variability providing the currents distribution related to the typical I_{prog} pulse. Further simulated distributions, corresponding to a lower (-5%) and a higher (+5%) programming current, were derived only modulating the amorphous thickness (u_a). A good fitting with the experimental data can be observed. This shows the capability of our model to sense the programming pulse effects, then to predict, in a controlled way, the fail probability with the energy spent in the programming operation. The fail probability as a function of the programming current is reported in Fig. 7 and obtained through the intercept of a horizontal cut-line with the distributions at a defined standard deviation of Fig. 6.

TEMPERATURE DEPENDENCE

In the framework of a thermally assisted hopping model describing transport in amorphous GST, the impact of temperature on the reset cell distribution needs to be well understood. In agreement with equation (2) the readout current of PCM cells in the reset state show an Arrhenius-like dependence on temperature as depicted by measured resistance for three different cells in Fig. 8. The cells were chosen to represent the average and borderline behavior of array cells respectively displaying typical, relatively high and relatively low

activation energy. The latter two cells are referred to as SLOW and FAST cells coherently with a steeper or smoother currents decreasing in the Arrhenius plot with respect to the typical (TYP) behavior as a consequence of higher or lower E_A values. Note that E_A increases for increasing resistance, as a result of percolation effects in the non-uniform disordered chalcogenide phase. This implies that a T increase results in a narrower distribution width. This phenomenon is captured through the MC simulated distribution for increasing temperature as shown in Fig. 9.

In addition to the resistance dependence of activation energy, we found that the pre-exponential factor for resistance exponentially depends on E_A. This dependence goes under the name of Meyer-Neldel rule (MNR) which is broadly verified for a number of amorphous materials. MNR was recently verified in amorphous GST for both resistance [19] and phase relaxation/crystallization [20]. While MNR is known as an empirical result since 1937 [21] many conflicting theories have been proposed as a theoretical explanation [22, 23] describing thermally activated process for which a measured property X follows an Arrhenius behavior:

$$X = X_0 \cdot \exp\left(-\frac{E_A}{kT}\right) \qquad (4)$$

When E_A varies, X_0 compensates the E_A variation according to the relation:

$$X_0 = X_{00} \cdot \exp\left(\frac{E_A}{kT_0}\right) \qquad (5)$$

being X_{00} and T_0 characteristic constants of the material. So, the MN term counterbalance the effect of an E_A variation in the X value definition. This contributes to the narrowing effect on the readout current distribution in Fig. 9. Fig. 10 reports the pre-factor of the conductibility ($X_0=\sigma_0$, in equation 4) as a function of E_A parameter for a number of analytical cells measured in 90 nm technology node. A linear fitting of the experimental data allows to verify the occurrence of the MN-compensation law concerning the transport mechanism in the amorphous GST and to extract the MN temperature T_0 [20].

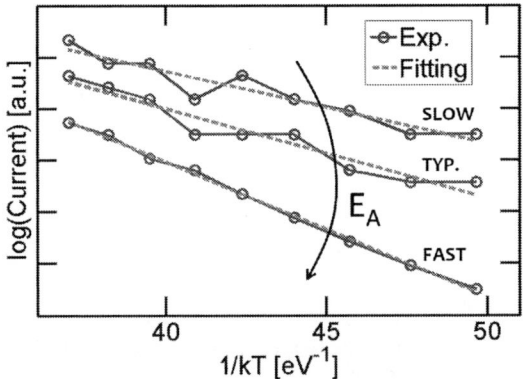

FIG. 8: ARRHENIUS PLOT OF THE RESET CURRENT (IN LOG SCALE) RELATED TO THREE CELLS WITH DIF- FERENT E_A. DIFFERENT CURRENT SPREAD AS A FUNCTION OF TEMPERATURE IS OBTAINED POSSIBLY RECALLING THE MEYER-NELDEL RULE [18].

978-1-4244-5430-3/10 $26.00 © 2010 IEEE

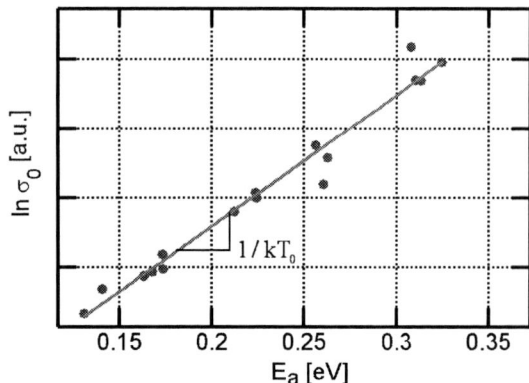

FIG. 10: SIMULATED CURRENT CUMULATIVE DISTRIBUTION SHOWING THE SHIFT AND WIDTH MODULATION AS A FUNCTION OF TEMPERATURE.

CONCLUSIONS

We have presented a unified transport model for the amorphous state allowing a description of the I-V curves for the reset state in both high and low trap-density regimes. The model was extended on a statistical basis to correctly account for the readout current distributions as a function of SR phenomena. The dependence of the programming-fail probability on the reset pulse amplitude was investigated. Also, the impact of temperature on the reset state distribution was addressed, evidencing a MNR in the transport properties of PCM cells. Our statistical model provides a useful tool to assess the reliability figure related to the reset-induced read window budget.

ACKNOWLEDGMENT

The authors would like to thank P. Cappelletti, E. Camerlenghi, R. Bez and A. Marmiroli from Numonyx and A.L. Lacaita from Politecnico di Milano for discussions and support.

REFERENCES

[1] G. Servalli , *"A 45nm Generation Phase Change Memory Technology"*, in IEDM Tech. Dig. pp. 113-116, 2009, Baltimora (ML).

[2] F. Bedeschi, R. Fackenthal, C. Resta, E.M. Donze, M. Jagasivamani, E.C. Buda, F. Pellizzer, D.W. Chow, A. Cabrini, G. Calvi, R. Faravelli, A. Fantini, G. Torelli, D. Mills, R. Gastaldi, G. Casagrande, *"A Bipolar-Selected Phase Change Memory Featuring Multi-Level Cell Storage"*, IEEE J. Solid State Circuits 44, Page: 217, ISSN: 0018-9200 (2009).

[3] Agostino Pirovano, Andrea L. Lacaita, Fabio Pellizzer, Sergey A. Kostylev, Augusto Benvenuti, Roberto Bez, *"Low-Field Amorphous State Resistance and Threshold Voltage Drift in Chalcogenide Materials"*, IEEE Transactions on Electron Devices, vol. 51, no. 5, 0018-9383 (2004).

[4] S. A. Baily, David Emin, Heng Li, *"Hall mobility of amorphous Ge₂Sb₂Te₅"*, Solid State Communications 139, 161-164 (2006).

[5] D. Ielmini and Y. Zhang, *"Analytical model for subthreshold conduction and threshold switching in chalcogenide-based memory devices"*, J. Appl. Phys. 102, 054517 (2007).

[6] P. Fantini, G. Betti Beneventi, A. Calderoni, L. Larcher, P. Pavan and F. Pellizzer, *"Characterization and Modeling of Low-Frequency Noise in PCM device"*, in IEDM Tech. Dig. pp. 219-222, 2008, San Francisco (CA).

[7] G. Betti Beneventi, A. Calderoni, P. Fantini, L. Larcher, P. Pavan, *"Analytical model for low-frequency noise in amorphous chalcogenide-based phase-change memory devices"*, Journal of Applied Physics, Volume 106, Issue 5, pp. 054506-054506-8 (2009).

[8] D. Ielmini, *"Threshold switching mechanism by high-field energy gain in the hopping transport of chalcogenide glasses"*, Phys. Rev. B 78, 35308 (2008).

[9] D. Fugazza, D. Ielmini, S. Lavizzari, and A. L. Lacaita *"Distributed-Poole-Frenkel modeling of anomalous resistance scaling and fluctuations in phase-change memory (PCM) devices"* in IEDM Tech. Dig. pp. 723-726, 2009, Baltimore (ML).

[10] Y.H. Shih, M.H. Lee, M. Breitwisch, R. Cheek, J.Y. Wu, B. Rajendran, Y. Zhu, E.K. Lai, C.F. Chen, H.Y. Cheng, A. Schrott, E. Joseph, R. Dasaka, S. Raoux, H.L. Lung, and C. Lam *"Understanding Amorphous States of Phase-Change Memory Using Frenkel-Poole Model"*, in IEDM Tech. Dig. pp. 753-756, 2009, Baltimore (ML).

[11] D. Ielmini, D. Sharma, S. Lavizzari and A. L. Lacaita, *"Reliability impact of chalcogenide-structure relaxation in phase change memory (PCM) cells – Part I: Experimental study"*, IEEE Trans. Electron Devices 56, 1078-1085 (2009).

[12] S. Lavizzari, D. Ielmini, D. Sharma and A. L. Lacaita, *"Reliability impact of chalcogenide-structure relaxation in phase change memory (PCM) cells – Part II: Physics-based modeling"*, IEEE Trans. Electron Devices 56, 1078-1085 (2009).

[13] F. Pellizzer, A. Benvenuti, B. Gleixner, Y. Kim, B. Johnson, M. Magistretti, T. Marangon, A. Pirovano, R. Bez, and G. Atwood, *"A 90nm phase change memory technology for stand-alone non-volatile memory applications"*, in VLSI 2006, pp. 122–123.

[14] D. Ventrice, A. Calderoni, A. Spessot, P. Fantini, A. Sanasi, S. Braga, A. Cabrini and G. Torelli, *"Statistical Modeling of Bit Distributions in Phase Change Memories"*, Proc. of 38th European Solid-State Device Research Conference, ESSDERC. Athens. 14-18 Sept. 2009. pp 157-160.

[15] D. Ielmini, S. Lavizzari, D. Sharma and A. L. Lacaita, *"Physical interpretation, modeling and impact on phase change memory (PCM) reliability of resistance drift due to chalcogenide structural relaxation"*, in IEDM Tech. Dig. pp. 939-942, 2007, Washington (DC).

[16] D. Ielmini, Andrea L. Lacaita, and D. Mantegazza, *"Recovery and Drift Dynamics of Resistance and Threshold Voltages in Phase-Change Memories"* IEEE Transactions on Electron Devices, vol. 54, no. 2, 308-315 (2007).

[17] M. Boniardi, D. Ielmini, S. Lavizzari, A. L. Lacaita, A. Redaelli and A. Pirovano, *"Statistical and scaling behavior of structural relaxation effects in phase change memory (PCM) devices"*, Proc. IRPS, 122-127 (2009).

[18] Byoung-Jae Bae, SangBum Kim, Yuan Zhang, Youngkuk Kim, In-Gyu Baek, Soonoh Park, In-Seok Yeo, Siyoung Choi, Joo-Tae Moon, H.-S. Philip Wong, and Kinam Kim, *"1D Thickness Scaling Study of Phase Change Material (Ge2Sb2Te5) using a Pseudo 3-Terminal Device"*, in IEDM Tech. Dig. pp. 93-96, 2009, Baltimora (ML).

[19] Semyon D. Savransky and Ilya V Karpov, *"Investigation of SET and RESET States Resistance in Ohmic Regime for Phase-Change Memory"*, Mater. Res. Soc. Symp. Proc. Vol. 1072 (2008).

[20] D. Ielmini and M. Boniardi, *"Common signature of many-body thermal excitation in structural relaxation and crystallization of chalcogenide glasses"*, Applied Physics Letters 94, 091906 (2009).

[21] W. Meyer and H. Neldel, Z. Tech. Phys. 12, 588 (1937).

[22] A. Yelon, B. Movaghar, H. M. Branz, *"Origin and consequences of the compensation (Meyer Neldel) law"*, Physical Review B, volume 46, number 19, (1992).

[23] Y. F. Chen and S. F. Huang, *"Connection between the Meyer-Neldel rule and stretched-exponential relaxation"*, Physical Review B. vol. 44, n. 24 (1991).

Random Telegraph Signal Noise in Phase Change Memory Devices

Davide Fugazza,* Daniele Ielmini,* Simone Lavizzari,* Andrea L. Lacaita* †

* Dipartimento di Elettronica e Informazione
Politecnico di Milano–IU.NET
piazza L. da Vinci 32, 20133 Milano, Italy.
phone +39 02 2399 4007, e-mail: fugazza@elet.polimi.it

† also with IFN-CNR, Politecnico di Milano, 20133 Milano, Italy.

Abstract— **Reliability in phase-change memory (PCM) devices is mainly related to the metastable nature of the amorphous phase, affected by crystallization and structural relaxation (SR) processes. More recently, low-frequency noise has attracted interest both as a valuable investigation tool of the microscopic properties of the chalcogenide material and as a possible reliability topic for future technology nodes. Moreover the recent appearance of random telegraph-signal noise (RTN) as a result of cell downscaling supports the need of a deeper insight into these phenomena. This work presents for the first time RTN in PCM, describing both frequency and time dependences of the noise. We analyze (i) the R dependence with the aid of a distributed Poole-Frenkel (DPF) conduction model and (ii) the voltage and temperature dependences, discussing possible physical origins of RTN in phase-change memories.**

Keywords – phase change memory (PCM); non-volatile memory; random telegraph noise; random resistance network (RRN); Meyer-Neldel law.

I. INTRODUCTION

Among the emerging nonvolatile memory concepts, phase change memory (PCM) holds a privileged position due to the degree of technology maturity, supported by the good scaling potential and a broad application range [1]–[3]. Its working principle relies on the ability of a chalcogenide material, typically $Ge_2Sb_2Te_5$ (GST), to attain a reversible crystalline-to-amorphous phase change as a result of the application of fast electrical pulses. The cell can be programmed in two states with different resistance: The high resistance (reset) state is associated to the amorphous phase, while the low-resistance state is obtained in correspondence to a polycrystalline phase of the chalcogenide material.

The low-frequency noise has been recently addressed by experimental and theoretical investigations [4]–[6]. In this frame, the study of further aspects, such as random telegraph noise (RTN), which was experimentally evidenced for the first time in PCM devices as a result of cell-size downscaling [7], can support the technology and its continuous development [8], [9]. In addition, RTN studies may strongly contribute to the understanding of the structural stability in the amorphous phase [10], [11] and the conduction mechanism in the amorphous chalcogenide phase [7], [12].

This work presents an experimental and numerical analysis of RTN on single cells. RTN is studied for variable initial resistance in both the time and frequency domain, demonstrating the Poisson nature of the two-level noise. RTN is then interpreted as due to a structural fluctuation at a bistable defect, which affects the local conductivity in a percolation chain. Simulations based on random resistance network (RRN), namely a standard approach to describe conduction [13], [14] and current fluctuations in amorphous solids [7], [15]–[17], are shown as a function of the thickness of the amorphous region, in agreement with experimental results. The temperature and voltage dependences of the noise are finally discussed with reference to the physical nature of the bistable defect responsible for RTN.

II. EXPERIMENTS

We carried out experiments on PCM devices fabricated by Numonyx with a 90 and 180 nm technology [18], [19]. Fig. 1 shows measured current-voltage $(I - V)$ characteristics for a PCM cell with a 90 nm technology for the reset state ($R \cong 10$ MΩ), the set state ($R \cong 10$ kΩ) and an intermediate state ($R \approx 600$ kΩ, cell A). The intermediate state was obtained by controlling the programming current pulse [20]: In fact, a proper modulation of the programming pulse amplitude allows to finely tune the resistance by different thicknesses u_a of the programmed amorphous volume [7].

Fig. 2 shows the power spectral density (PSD) of the current S_I divided by the square of the bias current I^2 (*normalized PSD in the following*) as a function of frequency. The set, reset and intermediate states A and B are considered. Prior to the electrical noise measurement, the cells were subjected to a sufficiently long annealing at 345 K to accelerate the SR of the amorphous phase [10], [11], in order to quench any resistance drift effect during noise measurement. The PSD was evaluated in the sub-threshold regime (read current below the threshold point in Fig. 1). For set and reset states, S_I/I^2 displays a characteristic $1/f$ behavior [5], which can be explained as a superposition of several individual Lorentzian ($1/f^2$) contributions with distributed characteristic frequencies [21], [22]. The normalized PSD is larger for reset than for set state,

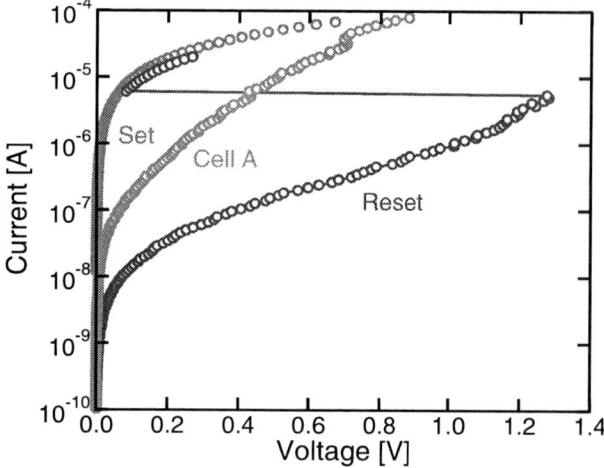

Fig. 1. Measured I-V curves for three different programmed states: low resistive (SET), an intermediate one (CELL A) and the high resistive state (full RESET).

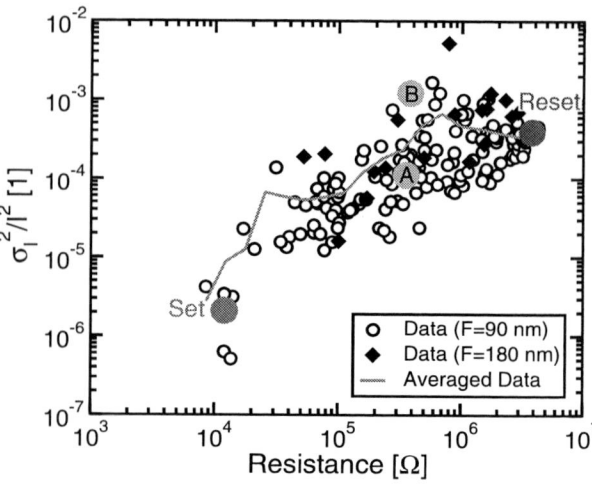

Fig. 3. Measured normalized variance $\sigma_I{}^2/I^2$ and its running average as a function of the resistance R read @100 nA. Integrals of spectra of Fig. 2 are highlighted.

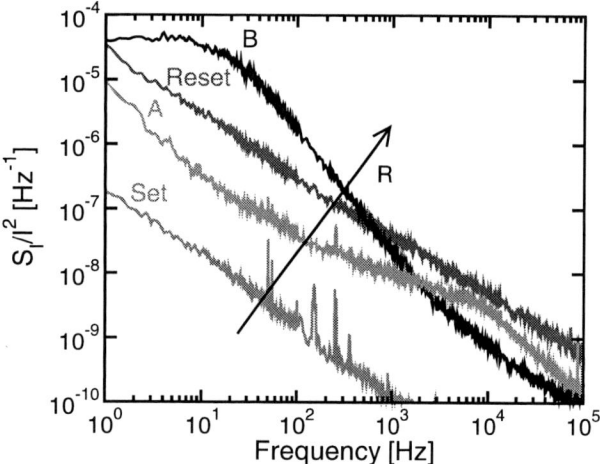

Fig. 2. Normalized PSD of two intermediate states (A, B) showing the appearance of clear Lorentzian components vs. SET and RESET $1/f$-like spectral behavior.

with no significant dependence on the applied bias. On the other hand, intermediate states A and B display clear "bumps" affecting in the $1/f$ shape: This can be attributed to a relatively small number of individual Lorentzian contributions, whose envelope does not result in a purely straight $1/f$ behavior [7]. In particular, the large bump for state B, clearly evidencing a plateau followed by a $1/f^2$ region at larger frequencies, can be considered as a signature for a dominant RTN contribution to noise. While the noise level for state A was intermediate between the set and reset state, state B displays a larger noise, despite both states A and B had the same programmed resistance.

Fig. 3 shows the normalized standard deviation $\sigma_I{}^2/I^2$, obtained integrating the normalized PSD in the frequency

range from 1 to 4×10^4 Hz. Two different technologies (180 and 90 nm) and several resistance states were characterized. The four states (set, reset, A and B) considered in Fig. 2 are marked for reference and the average normalized variance is shown as a solid curve. The normalized variance increases for increasing R. This power-law dependence is also consistent with what predicted by percolation theory and previously shown by numerical simulation on random resistance networks in disordered solids [15], [23]. Note the relatively large spread of data in correspondence of intermediate resistances, which agrees with the different noise level at the same R for states A and B in Fig. 2.

Fig. 4 shows the measured I as function of time for a cell displaying a dominant Lorentzian contribution (bump) in the PSD in the frequency domain. The current was measured at room-temperature. Two transition times can be defined, namely τ_{12} for the transition from low- to high-current level and τ_{21} for the transition from high- to low-current level, as indicated in the figure. We performed a statistical analysis on the collected waveform in the time domain, to extract the probability distributions of τ_{12} and τ_{21}. These are shown in an histogram representation in Fig. 5. Both distributions show a clear exponential shape, evidencing the Poisson nature of the conductance fluctuation.

III. RTN ANALYSIS AND MODELING

RTN characteristics can be understood in the framework of the conduction mechanism in amorphous chalcogenides [7], [12], [24]. Due to the strong structural disorder, amorphous chalcogenides lack a long-range lattice periodicity, thus cannot be described by conventional band structure models usually applied to crystalline semiconductors. The amorphous-semiconductor band structure is characterized by having a mobility gap with a large concentration of localized states [24], [25]. The Fermi level is usually pinned in the middle of

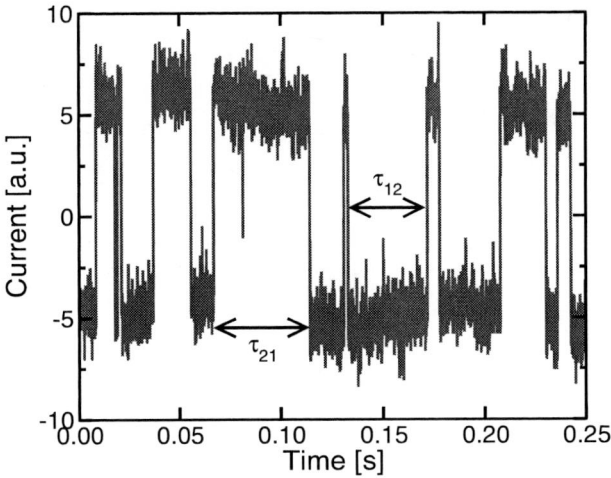

Fig. 4. Measured room temperature current vs. time for cell B in Fig. 2, exhibiting a clear RTN.

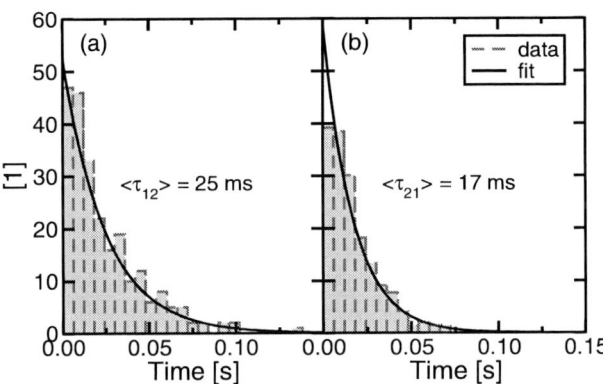

Fig. 5. Poisson distributions of the transition times τ_{12} and τ_{21} from RTN analysis of the waveform reported in Fig. 4.

the mobility gap, therefore carriers need to be thermally excited close to the band edges to effectively contribute to electrical transport [24]. As a result, both the voltage and the temperature dependence of the current can be explained by Poole-Frenkel (PF) mechanism [12]. More recently, the analysis of the thickness dependence of $I - V$ characteristics in PCM devices have evidenced a distributed Poole-Frenkel (DPF) conduction mechanism, where carrier conduction is localized at percolation paths due to the spread of activation energy for PF transport [7]. Since the emission activation energies are dictated by the local structural defect responsible for a localized state, fluctuations of bistable defects along the hopping percolation path can result in a significant two-level variation of current [7].

We used the DPF model to calculate the current density and time-fluctuation in the amorphous GST. The activation energy for thermally-assisted hopping among localized states within the mobility gap was assumed randomly distributed with a uniform distribution between 0 and 1 eV. The random-

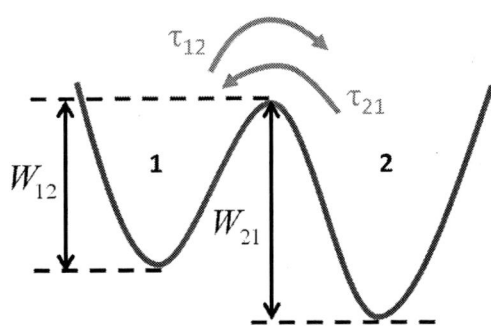

Fig. 6. Two-level free-energy diagram for defect fluctuation. The asymmetry in the two barriers results in different transition rates between the energy potential minima.

resistive network describing the hopping current was solved, assuming that each mesh element has a dimension of 1 nm³ and a resistance R_i, given by [7]:

$$R_i = R_0 \, e^{\frac{E_i}{kT}}, \qquad (1)$$

where E_i is the activation energy for hopping at the i-th mesh element (localized state). In order to model RTN, we extended the model assuming that each mesh element can randomly switch between two states (1 and 2), according to the two-level free-energy diagram in Fig. 6. For instance, the transition time τ_{12} is given by:

$$\tau_{12} = \tau_0 \, e^{\frac{W_{12}}{kT}}, \qquad (2)$$

where W_{12} is the energy barrier in the figure. The transition time τ_{21} is similarly linked to the energy barrier W_{21}. The two states in Fig. 6 correspond to two different local configuration of the structural defect, each associated to different values of the hopping energy barrier. As a result, the local resistance R_i fluctuates over time according to Eq. 1.

To illustrate the DPF model for RTN, we ran Monte Carlo simulations in two dimensions. Fig. 7a shows one of the randomly selected maps of activation energies E_i in a 2D sample (high/low E_i is indicated by dark/light gray) with $u_a = 15$ nm. Switching of each mesh element for increasing time was simulated according to Eq. 2, then the resistance map and the current distribution were calculated at each time step. Figs. 7b and c show the current distribution across the sample (dark/light gray for low/high current density) at two different times during a Monte Carlo simulation. A single percolation path carries most of the current in Fig. 7b, as a result of the non-uniform activation energy distribution [7]. The switching of a dominant trap in the percolation path results in the onset of a second percolation path in Fig. 7c, leading to a nearly 60% increase of the current.

Fig. 8 shows the calculated current (a) and σ_I^2/I^2 (b) as a function of sample thickness u_a, from 3D simulations at constant sample width and applied voltage. The current decreases for increasing thickness, however the current drop is much steeper than what expected from a simple ohmic behavior,

978-1-4244-5430-3/10 $26.00 © 2010 IEEE

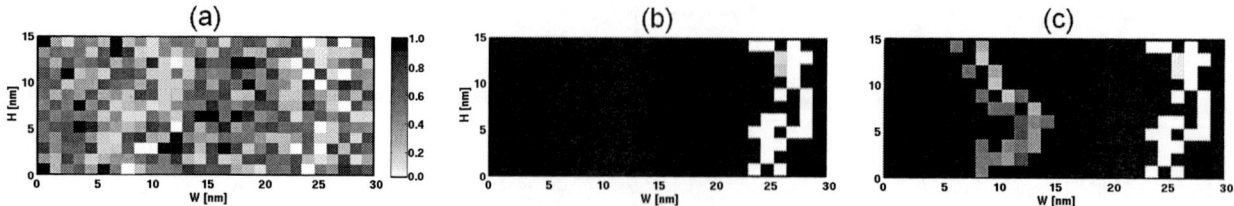

Fig. 7. Simulated 2D current distributions (b, c) at two different times for a 15 nm amorphous thickness, given the activation energy map reported in (a). Trap switching through time causes the onset/disappearance of a second percolation path (c) in parallel to the single path shown in (b).

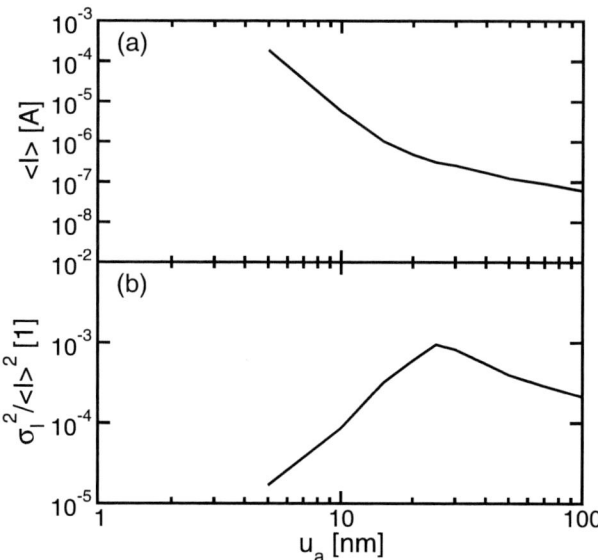

Fig. 8. Simulated average current (a) and normalized variance σ_I^2/I^2 (b) as a function of the amorphous thickness by the means of the 3D DPF Monte Carlo model.

i.e. $I \propto u_a^{-1}$. The thickness dependence of the current is mainly due to the increase of the conduction activation energy for increasing thickness [7], [26]. In fact, percolation paths become longer for increasing u_a, thus the probability for the percolation chain to include a trap with relatively high energy barrier increases with u_a. This is similar to percolation effects in discrete trap memories, where the threshold voltage in the programmed state increases for increasing channel length, due to the increased probability for finding a blocking chain of negatively-charged nodes along the channel [27].

The bell-shaped behavior of the calculated normalized variance in Fig. 8b can be interpreted in terms of variation of the aspect ratio of the amorphous volume. For small u_a, conduction occurs through multiple parallel percolation paths, thus resulting in an averaging of the individual fluctuation contributions [7], [26]. Consequently, the uncorrelated noise contributions are averaged out and thus σ_I^2/I^2 decreases for decreasing u_a. For thick samples, the current is instead confined into a single percolation path (see *e.g.* Fig. 7). For increasing u_a, multiple fluctuations within the same relatively-

long trap chain are again averaged out, resulting in a decreasing noise for increasing u_a. The intermediate case, where one/few short percolating paths carry most of the current, is the most affected by noise. Note that this is in qualitatively agreement with the average behavior of the measured σ_I^2/I^2 in Fig. 3, indicating a maximum around $R = 7 \times 10^5 \ \Omega$.

IV. PHYSICAL ORIGIN OF RTN

The fluctuating traps responsible for RTN appear to be located within the amorphous GST material, instead of sitting at the contacts and/or surrounding dielectrics. In fact, we have observed that the reset operation, consisting of a temperature increase over the melting point, followed by a fast quenching of the liquid phase into a disorder amorphous state, generally suppresses RTN traps and generates new ones. Moreover, RTN can spontaneously disappear during moderate/high T annealing, which is known to affect the metastable amorphous GST structure by means of SR effects [10], [11]. Thus fluctuating centers belong to the amorphous GST volume, as assumed in our model. In further support of this hypothesis and in order to gain more insight into the nature of the fluctuating defects responsible for RTN, we conducted an analysis of the T- and V-dependence of single Lorentzian contributions to the noise.

A. Temperature dependence

Fig. 9a shows measured S_I/I^2 for a cell state affected by a Lorentzian contribution. The PSDs were collected for increasing T, indicating a clear temperature activation of the RTN characteristic frequency, which shows an increase for increasing T. A further increase in temperature then resulted in the nearly classic flicker behavior. The spectra were analyzed and decomposed into $1/f$ and $1/f^2$ components to obtain the RTN characteristic time τ_P, related to transition times τ_{12} and τ_{21} by:

$$\tau_P = \frac{\tau_{12}\tau_{21}}{\tau_{12} + \tau_{21}}. \qquad (3)$$

Extracting the best least-square fit of S_I/I^2 and multiplying it by the frequency, we obtained the curves reported in Fig. 9b: Here, Lorentzian components clearly arise as bumps over the flat $1/f$ background function, which exhibits a weak dependence on T. The curves provide a better view of the increasing characteristic frequency of RTN for increasing temperature. The inset in Fig. 9a shows the Arrhenius plot of τ_P, which agrees with the Arrhenius function:

$$\tau_P = \tau_{P0} \ e^{\frac{E_{AP}}{kT}}, \qquad (4)$$

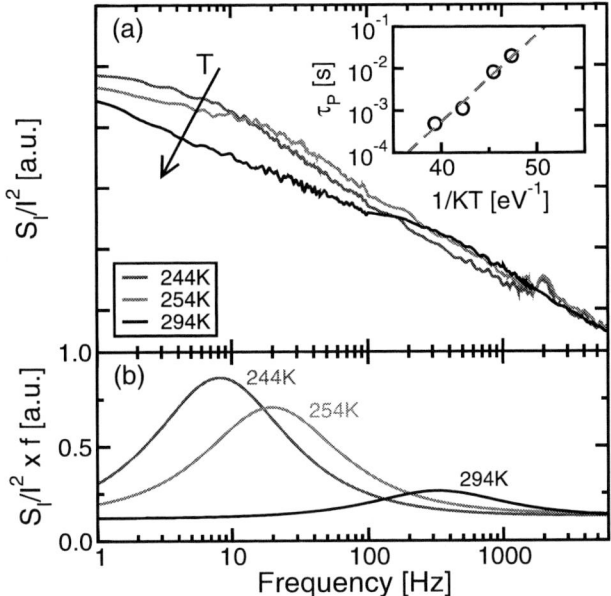

Fig. 9. (a) Smoothed single Lorentzian spectra for various T; Arrhenius plot of τ_P activation is shown in the inset. (b) least-square fits of the measured normalized PSD multiplied by frequency.

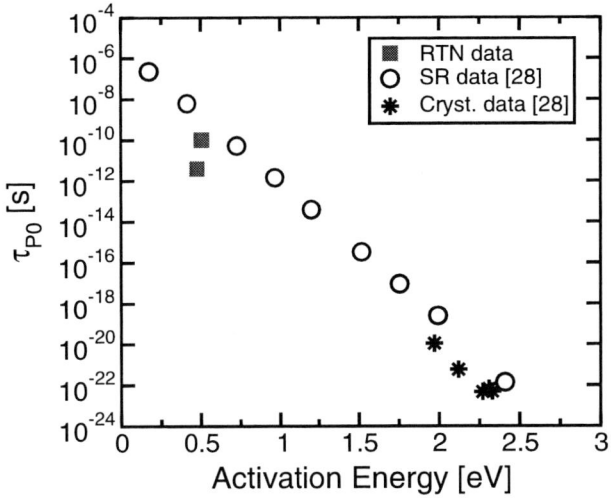

Fig. 10. Extracted τ_{P0} pre-exponential factor vs. the related activation energy, compared to SR and crystallization data in amorphous GST [28].

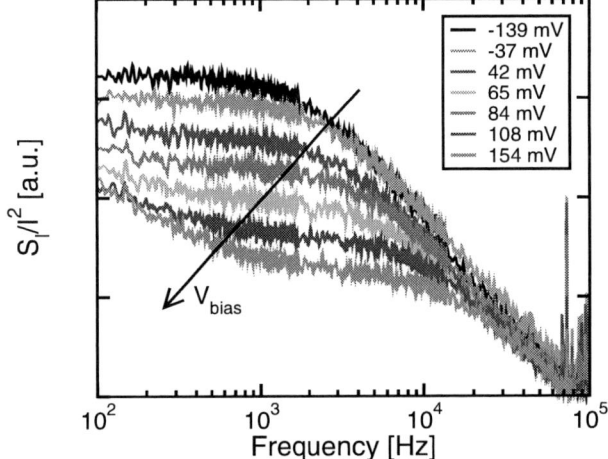

Fig. 11. Normalized single Lorentzian PSD for an intermediate state for different applied voltage values.

where $E_{AP} = 0.46$ eV is an activation energy and τ_{P0} is a pre-exponential constant. Fig. 10 shows τ_{P0} as a function of the extracted activation energy E_{AP}, for two RTN characteristics in our samples. RTN data are compared with similar plots for SR and crystallization in GST, indicating an exponential (Meyer-Neldel) law [28]. RTN parameters are in reasonable agreement with previous SR data: This confirms our previous assumption that RTN results from a bistable SR effect at structural defects in the amorphous GST. In this regard, the two-level energy profile in Fig. 6 differs from the similar one observed in SR processes [10], [11] by the similarity between the forward and reverse energy barriers (W_{12} and W_{21}). This results in a sequence of reversible transitions between the two equally-stable states, originating RTN, instead of an irreversible transition between a metastable and a stable state, which is typical of SR processes.

B. Voltage dependence

Fig. 11 shows a representative case of single RTN spectral dependence on the bias voltage V, also for applied fields of opposite polarity. A voltage increase from negative to positive values clearly results in a monotonic shift to higher frequencies of the characteristic pole frequency f_P and hence in a lower τ_P.

In order to get a deeper insight in the phenomenon, the transition times τ_{12} and τ_{21} were separately characterized by time-domain analysis of the collected waveforms. The obtained results are reported in Fig. 12, which shows the extracted τ_{12} and τ_{21} as a function of the applied voltage: For increasing applied electric field τ_{21} increases, while τ_{12} decreases. The nearly exponential dependence on voltage of

both the transition times results in an exponential increase of the characteristic frequency with V, in good agreement with the values extracted by the spectral analysis performed in the frequency domain.

For the case reported in Figs. 11 and 12, the voltage dependence of the two transition rates is explained by a field-induced barrier-lowering effect of the two-state energy profile, as sketched in Fig. 13. An applied voltage shifts the energy of state 2 with respect to state 1, resulting in an exponentially shorter τ_{12} and longer τ_{21}, as found in the experimental data reported in Fig. 12. This effect can be thus interpreted as polarizability of the bistable defect which modifies the energy barriers for relaxation.

In Fig. 11 the increasing applied voltage also results in a lowering of the Lorentzian plateau S_0, which changes approx-

978-1-4244-5430-3/10 $26.00 © 2010 IEEE

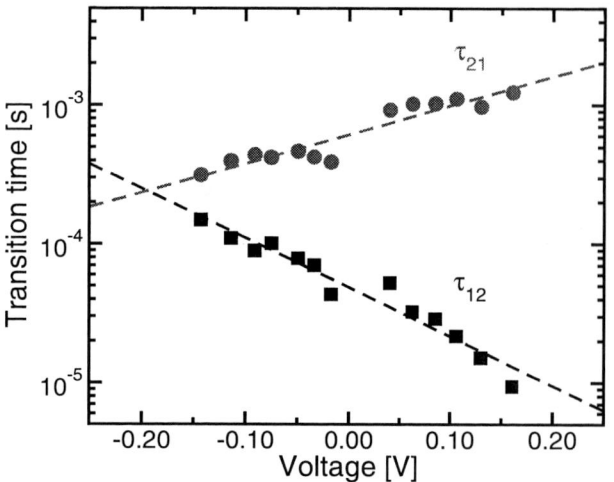

Fig. 12. Extracted transition times for cell in Fig. 11, both showing a nearly exponential behavior with the applied bias.

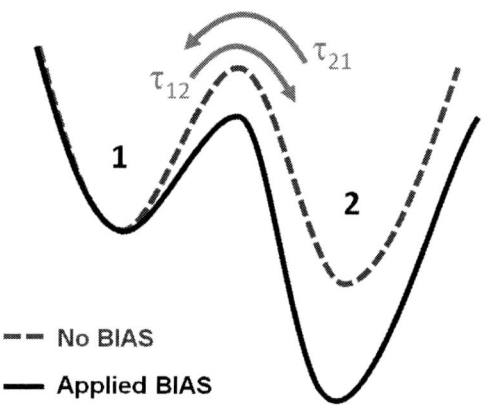

Fig. 13. Pictorial view of voltage modulation of the two-level energy barrier for the case reported in Figs. 11 and 12.

imately according to $S_0 \propto f_P^{-2}$. As a result, the Lorentzian contribution to the normalized noise decreases with voltage. This is opposite to the bias-independent normalized PSD usually observed for the $1/f$ spectra [4], [5]. To reproduce such an invariant $1/f$ spectral behavior, one may expect that all individual Lorentzian components would obey the relation $S_0 \propto f_P^{-1}$. To account for this disagreement, we speculate that Lorentzian noise contributions may obey to a different relationship $S_0 \propto f_P^{-\alpha}$, where α is statistically distributed around -1. Indeed, different voltage dependences with distributed α were observed for different RTN components which were experimentally studied.

In conclusion, the obtained results indicate that the GST polarization strongly affects the fluctuation dynamics of the single fluctuating center belonging or near to the main percolation chain.

V. CONCLUSIONS

We reported a detailed investigation of RTN noise in phase-change memory cells. The study showed the current fluctuation dependence on the programmed resistance both experimentally, by varying the programmed amorphous volume, and numerically, by the means of a RTN Monte Carlo model based on a distributed Poole-Frenkel approach for conduction. The discrete two-level resistance fluctuations were investigated in the time and frequency domain. The observed thermally and voltage activated kinetics was interpreted as due to fluctuations of the atomic structure between two metastable configurations, affecting the conductivity at percolation paths.

ACKNOWLEDGMENTS

The authors acknowledge Numonyx for providing experimental samples and financial support by Intel (Project 34523) and MIUR (FIRB-RBIP06YSJJ).

REFERENCES

[1] S. Lai, "Non-volatile memory technologies: The quest for ever lower cost," in *IEDM Tech. Dig.*, 2008, pp. 11–16.

[2] R. Bez, "Chalcogenide pcm: a memory technology for next decade," in *IEDM Tech. Dig.*, 2009, pp. 89–92.

[3] G. Servalli, "A 45nm generation phase change memory technology," in *IEDM Tech. Dig.*, 2009, pp. 113–116.

[4] P. Fantini, G. Betti Beneventi, A. Calderoni, L. Larcher, P. Pavan, and F. Pellizzer, "Characterization and modelling of low-frequency noise in PCM devices," in *IEDM Tech. Dig.*, 2008, pp. 219–222.

[5] G. Betti Beneventi, A. Calderoni, P. Fantini, L. Larcher, and P. Pavan, "Analytical model for low-frequency noise in amorphous chalcogenide phase-change memory devices," *J. Appl. Phys.*, vol. 106, p. 54506, 2009.

[6] M. Nardone, V. I. Kozub, I. V. Karpov, and V. G. Karpov, "Possible mechanisms for 1/f noise in chalcogenide glasses: A theoretical description," *Phys. Rev. B*, vol. 79, p. 165206, 2009.

[7] D. Fugazza, D. Ielmini, S. Lavizzari, and A. L. Lacaita, "Distributed-Poole-Frenkel modeling of anomalous resistance scaling and fluctuations in phase-change memory (PCM) devices," in *IEDM Tech. Dig.*, 2009, pp. 723–726.

[8] N. Tega, H. Miki, H. Yamaoka, H. Kume, T. Mine, T. Ishida, Y. Mori, R. Yamada, and K. Torii, "Impact of threshold voltage fluctuation due to random-telegraph noise on scaled-down sram," in *Proc. IRPS*, 2008, pp. 541–546.

[9] A. Ghetti, M. Bonanomi, C. Monzio Compagnoni, A. S. Spinelli, A. L. Lacaita, and A. Visconti, "Physical modeling of single-trap rts statistical distribution in flash memories," in *Proc. IRPS*, 2008, pp. 610–615.

[10] D. Ielmini, D. Sharma, S. Lavizzari, and A. L. Lacaita, "Reliability impact of chalcogenide-structure relaxation in phase-change memory (PCM) cells – Part I: Experimental Study," *IEEE Trans. Electron Devices*, vol. 56, pp. 1070–1077, 2009.

[11] S. Lavizzari, D. Ielmini, D. Sharma, and A. L. Lacaita, "Reliability impact of chalcogenide-structure relaxation in phase-change memory (PCM) cells - Part II: Physics-based modeling," *IEEE Trans. Electron Devices*, vol. 56, pp. 1078–1085, 2009.

[12] D. Ielmini and Y. Zhang, "Analytical model for subthreshold conduction and threshold switching in chalcogenide-based memory devices," *J. Appl. Phys.*, vol. 102, p. 054517, 2007.

[13] M. Pollak, "A percolation treatment of dc hopping conduction," *J. Non-Cryst. Solids*, vol. 11, pp. 1–24, 1972.

[14] V. Ambegaokar, B. I. Halperin, and J. S. Langer, "Hopping conductivity in disordered systems," *Phys. Rev. B*, vol. 4, no. 8, pp. 2612–2620, 1971.

[15] R. Rammal, C. Tannous, P. Breton, and A.-M. Tremblay, "Flicker ($1/f$) noise in percolation networks: a new hierarchy of exponents," *Phys. Rev. Lett.*, vol. 54, no. 15, pp. 1718–1721, 1985.

[16] L. M. Lust and J. Kakalios, "Dinamical percolation model of conductance fluctuations in hydrogenated amorphous silicon," *Phys. Rev. Lett.*, vol. 75, no. 11, pp. 2192–2195, 1995.

[17] K. M. Abkemeier and D. G. Grier, "Topological disorder and conductance fluctuations in thin films," *Phys. Rev. B*, vol. 54, no. 4, pp. 2723–2727, 1996.

[18] F. Pellizzer, A. Pirovano, F. Ottogalli, M. Magistretti, M. Scaravaggi, P. Zuliani, M. Tosi, A. Benvenuti, P. Besana, S. Cadeo, T. Marangon, R. Morandi, R. Piva, A. Spandre, R. Zonca, A. Modelli, E. Varesi, T. Lowrey, A. Lacaita, G. Casagrande, P. Cappelletti, and R. Bez, "Novel μtrench phase-change memory cell for embedded and stand-alone non-volatile memory applications," in *Symp. on VLSI Tech. Dig.*, 2004, pp. 18–19.

[19] F. Pellizzer, A. Benvenuti, B. Gleixner, Y. Kim, B. Johnson, M. Magistretti, T. Marangon, A. Pirovano, R. Bez, and G. Atwood, "A 90nm phase change memory technology for stand-alone non-volatile memory applications," in *Symp. VLSI Technology, Dig. Tech.*, 2006, pp. 122–123.

[20] D. Ielmini, A. Lacaita, A. Pirovano, F. Pellizzer, and R. Bez, "Analysis of phase distribution in phase-change nonvolatile memories," *IEEE Electron Device Lett.*, vol. 25, pp. 507–509, July 2004.

[21] P. Dutta and P. M. Horn, "Low-frequency fluctuations in solids: 1/f noise," *Rev. Mod. Phys.*, vol. 53, no. 3, pp. 497–516, 1981.

[22] M. B. Weissman, "1/f noise and other slow, nonexponential kinetics in condensed matter," *Rev. Mod. Phys.*, vol. 60, pp. 537–571, 1988.

[23] A.-M. Tremblay, S. Feng, and P. Breton, "Exponents for $1/f$ noise near a continuum percolation threshold," *Phys. Rev. B*, vol. 33, no. 3, pp. 2077–2080, 1986.

[24] D. Ielmini, "Threshold switching mechanism by high-field energy gain in the hopping transport of chalcogenide glasses," *Phys. Rev. B*, vol. 78, p. 035308, 2008.

[25] N. F. Mott and E. A. Davis, *Electronic processes in non-crystalline materials*. Clarendon Press, Oxford, 1979.

[26] A. Yakimov, N. Stepina, and A. Dvurechenskii, "Spontaneous fluctuations of variable-range hopping current in amorphous silicon microstructures," *Phys. Lett. A*, vol. 179, no. 2, pp. 131–134, 1993.

[27] D. Ielmini, C. Monzio Compagnoni, A. S. Spinelli, A. L. Lacaita, and C. Gerardi, "A new channel percolation model for V_T shift in discrete-trap memories," in *Proc. IRPS*, 2004, pp. 515–521.

[28] D. Ielmini and M. Boniardi, "Common signature of many-body thermal excitation in structural relaxation and crystallization of chalcogenide glasses," *Appl. Phys. Lett.*, vol. 94, p. 091906, 2009.

Reliability of Ferroelectric Random Access Memory Embedded within 130nm CMOS

J. Rodriguez, K. Remack, J. Gertas, L. Wang, C. Zhou, K. Boku, J. Rodriguez-Latorre,
K. R. Udayakumar, S. Summerfelt, T. Moise
Texas Instruments Inc.
Dallas, Texas USA
jrz@ti.com

D. Kim, J. Groat, J. Eliason, M. Depner, F. Chu
RAMTRON International Corporation
Colorado Springs, Colorado USA

Abstract—We present results of a comprehensive reliability evaluation of a 2T-2C, 4Mb, Ferroelectric Random Access Memory embedded within a standard 130nm, 5LM Cu CMOS platform. Wear-out free endurance to 5.4×10^{13} cycles and data retention equivalent of 10 years at 85°C is demonstrated. The results show that the technology can be used in a wide range of applications including embedded processing.

Keywords- embedded memory; ferroelectric memory reliability; cycling endurance; data retention; high-temperature operating life; sof-error rate.

I. INTRODUCTION

Ferroelectric memory (FRAM) offers unique capabilities which make it an attractive non-volatile memory choice for many applications. Distinguishing features include fast, RAM-like, write speeds, low voltage, low power write operation, high cycling endurance lifetime and architectural flexibility.

The low voltage operation, low write power, and fast write capability is of specific interest for ultra low power mobile/wireless electronics, battery powered measurement applications, and other new areas including intelligent, battery-less sensors. The ability to operate as a universal memory (memory block dynamically partitioned as cache, code, or data storage) provides flexibility for embedded microcontroller and system on chip devices. Non-volatile information storage coupled with high endurance enables FRAM to be a replacement for battery-backed SRAM, for example, in data logging systems.

The FRAM operates at 1.5V, and unlike floating gate devices, does not require the use of charge pumps. As with all non-volatile memories, the reliability challenges include write/read cycling endurance, data retention, and high temperature operating life. In this report we present results in these key areas demonstrating the reliability of the FRAM technology.

II. BACKGROUND

A. Process Description

The ferroelectric memory is embedded within a standard 130nm CMOS process with 5 layers of copper metallization, where the ferroelectric capacitor module is integrated between contact (CONT) and metal 1 (MET1). The process details have been reported previously [1], [2].

A single-mask capacitor stack etch defines the memory storage elements within the circuit. A second mask defines the bi-level VIA plug that connects MET1 to either the top electrode of the capacitor or to the contact. A representative scanning-electron micrograph of the integrated device is shown in Fig. 1.

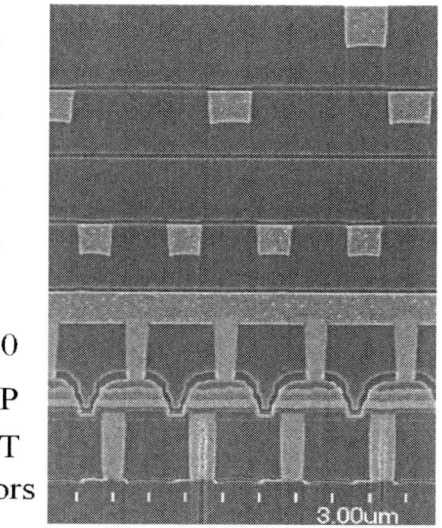

Figure 1. Cross-sectional scanning electron micrograph showing the FRAM module integrated between CONT and MET1 in a 5LM Cu/FSG process.

B. Device Description

The FRAM device was designed to operate in an 8Mb, 1T-1C or 4Mb, 2T-2C mode [3]. In this work we evaluated the memory reliability operating as a 4Mb, 2T-2C device. The bit behavior as a function of electrical stress conditions was analyzed by measuring the internal 8Mb bit signal distributions. All the results reported here were obtained from tests on assembled devices in a standard TSOP II package.

An optical die micrograph of this device is shown in Fig. 2. A centrally located spine separates two 4Mb, 1T-1C blocks. Each 4Mb block consists of eight 512Kb sections, and each section is made up of sixteen 32Kb segments. The device incorporates row, column, and bit level redundancy. The area in an embedded application, estimated by removing the I/O pad ring area from the stand-alone device is < 12 mm^2.

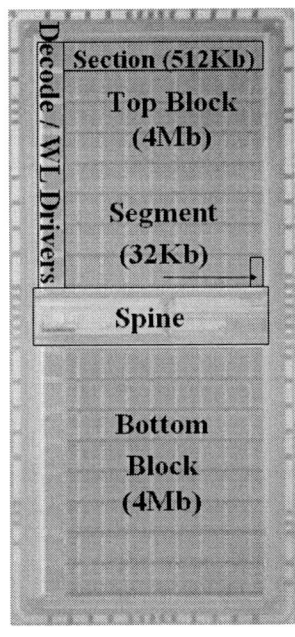

Figure 2. Optical micrograph of the FRAM device used in this evaluation.

Included in the device is a ~4Kb configuration memory which resides in 2T-2C FRAM rows located within each section. The configuration memory stores programmable features including core and word-line regulator voltages, speed trim, and redundancy information.

C. 1T-1C and 2T-2C FRAM Circuit Operation

A circuit depicting the 1T-1C bit-cell is shown in Fig. 3 together with a table showing the circuit bias conditions for data write and read operations. During data write, the plate line (PL) is either biased high (to write logic '0', stored at Pr+) or set to ground (to write logic '1', stored at Pr-). If reading logic '0', the capacitor polarization state will not change (per the table, the polarity is the same for both write and read '0'), and the only displacement charge density

detected is related to the linear or "non-switching" portion of the hysteresis loop, shown in Fig. 4. If reading logic '1', the polarization state will switch orientation (the read polarity on the capacitor is opposite to that of the write '1' polarity as shown in the table), and hence the detected signal will consist of the sum of a non-linear polarization switching charge and the linear charge. The logic '1' signal level therefore is larger than logic '0'. During data read, the BL is left floating so that it may rise to a voltage level dependent on the polarization state of the capacitor.

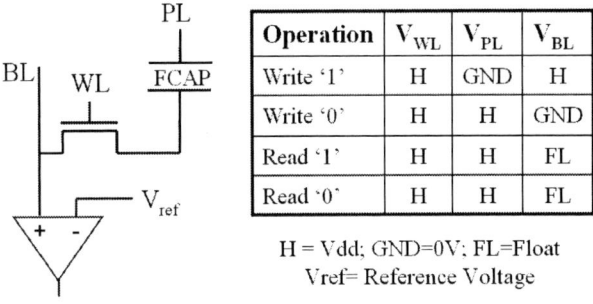

Operation	V_{WL}	V_{PL}	V_{BL}
Write '1'	H	GND	H
Write '0'	H	H	GND
Read '1'	H	H	FL
Read '0'	H	H	FL

H = Vdd; GND=0V; FL=Float
Vref= Reference Voltage

Figure 3. FRAM 1T-1C bit-cell circuit with bias conditions shown for writing and reading data.

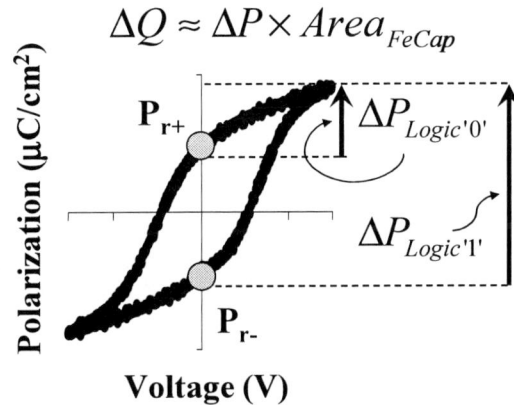

Figure 4. Hysteresis loop showing the definitions of the logic states.

A representative distribution of '0' and '1' bit signals is shown in Fig. 5, with a generic definition of signal margin window shown as the separation between the highest '0' bit and the lowest '1' bit on the left, and an example of a 1T-1C margin definition on the right. In a 1T-1C circuit, the reference voltage level is selected to be of a magnitude in between the highest '0' and lowest '1' signal levels. During data read, the V_{BL} and the reference voltage magnitudes are compared to each other using a sense amplifier. V_{BL} values below the reference are sensed as logic '0', whereas V_{BL} values greater than V_{ref} are sensed as logic '1'. A special test mode allows the V_{ref} to be varied over a wide range to determine the distribution of bit signals. This is done by measuring all bit addresses repeatedly for each V_{ref} step, and recording the number of bits "failing" at a given V_{ref} magnitude.

978-1-4244-5430-3/10 $26.00 © 2010 IEEE

In 2T-2C operation, the true data and its complement are stored in each bit-cell. The V_{ref} signal in Fig. 3 is replaced by the complement BL signal from an additional 1T-1C cell storing the complement logic state. This provides a larger signal margin at the expense of increased area. In both 2T-2C and 1T-1C, the data reads are "destructive". Consequently, every read operation is followed by a data restore. Additional FRAM circuit operation details are available in literature reports as found in [3], [4] and [5].

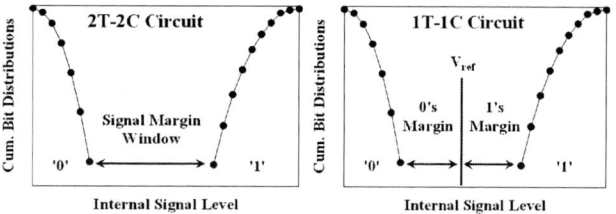

Figure 5. Definitions of FRAM signal margin in a 2T-2C and 1T-1C circuit. The margin increases with increasing separation between the highest '0' and lowest '1' bit signals. For a given separation distance, the 2T-2C margin is greater than the margins in a 1T-1C circuit.

III. RESULTS AND DISCUSSION

A. Data Retention

Reliable FRAM operation depends strongly on the logic data states having sufficient signal separation for accurate data sensing. There are two key mechanisms which need to be considered when evaluating FRAM retention reliability as follows.

1) Thermal Depolarization (TD): A reduction of the spontaneous polarization occurs as the ambient temperature of the ferroelectric film increases towards the Curie temperature [6], which for bulk ceramics with composition used in this process is ~430°C [7], [8]. Polarization decays gradually as a function of temperature, following an approximately second-order temperature dependence reported previously [8]. The depolarization effect is not strongly time dependent; it occurs when the sample temperature equilibriates with the stress temperature. When the sample is cooled to room temperature, a reduced signal margin is detected at the first data read, dependent on the maximum exposure temperature. The depolarization effect is not permanent; the polarization is restored after the first read/restore operation [8].

2) Imprint: Ferroelectric memories experience imprint [8], [9] when exposed to high temperatures for long periods of time after data is written. The imprint mechanism is historically the most important retention reliability mechanism for FRAMs. A high temperature bake of a bit-cell in one logic state (the "same-state") can *strengthen that state*, while it can *weaken its ability to store the complement logic state* ("opposite-state"). The imprint reduces the switchable polarization, which in turn lowers the signal margin in the circuit via a reduction of the '1' signal. The '0' signal is fairly stable in after-pulse sensing architectures

[8], [10]. After a period of time at high temperature, e.g., 1,000 hours at 125°C, the '1' signal magnitude can decrease, illustrated in Fig. 6, and may cause read fails for bits with the lowest signal levels. For example, a read failure can occur when the 2T-2C signal separation for a given bit has reduced to a magnitude below the sense amplifier detection limit. The data in Fig. 6 is from a 2T-2C data bake, where bits are physically storing both 0's and 1's, intended to represent a typical data memory use condition with mixed states.

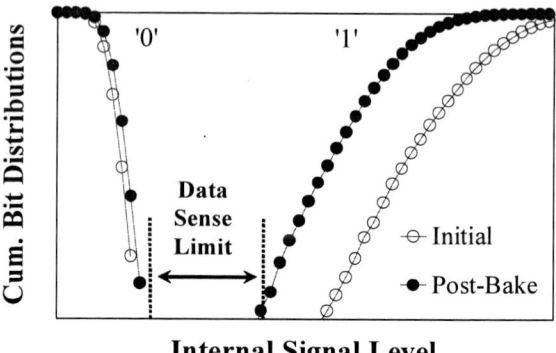

Figure 6. Data '1' signal level reduces uniformly with bake stress causing signal margin loss.

The kinetics of signal margin loss were quantified by monitoring the bit distributions (measured at room temperature) as a function of bake time over a temperature range of 85°C to 175°C. The bake was performed in a 2T-2C '0' data pattern, so that physically half of the bits were baked as Data '1' and the other half as Data '0', representing a mixture of data states. The lowest '1' bit signal measured at the different read-points is shown in Fig. 7. Signal margin reduction activation energy of 1.4eV is extracted from these experimental results. The equivalent margin loss expected after ten years at 85°C operation is achieved by accelerating the test using a 1,000 hour bake at 125°C.

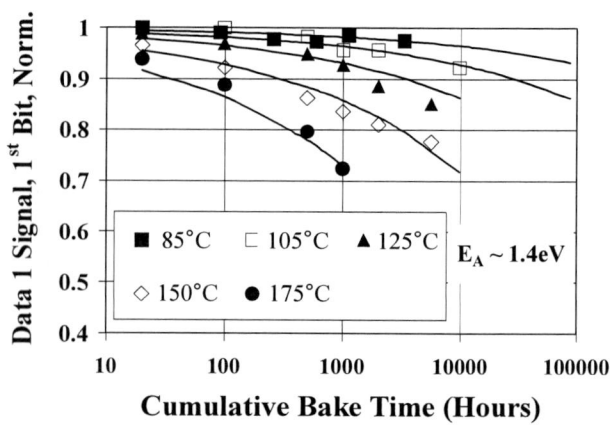

Figure 7. 1.4eV activation energy is extracted by monitoring lowest '1' bit-cell signal vs. time with bake temperature as a paramter.

These measured data shown in Fig. 7 are fit with an empirical power law model with a time exponent of ~0.3, shown in Fig. 8. A reduced acceleration factor at 175°C indicates there is a change in imprint kinetics at very high temperatures. This effect is similar to what has been reported previously on test capacitors baked at temperatures above 170°C [8].

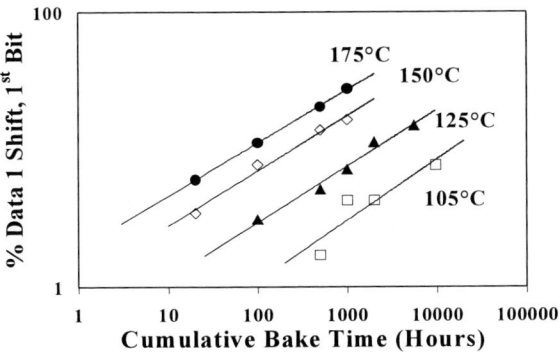

Figure 8.   The % degradation follows a power law trend vs. bake time.

A generalized data retention test flow for FRAM is shown in Fig. 9. The procedure begins by baking an initial data state, referred to as the "same-state," or SS, at a high temperature for a specified amount of time. For example, the read points typically follow a logarithmic time scale. After reading the SS, the complement data state, referred to as the "opposite-state," or OS, is written and stressed with a thermal depolarization (TD) bake. The OS depolarization bake is performed at the maximum operating temperature rating for the application (e.g., 85°C for an industrial temperature rating). Since the depolarization mechanism is not time dependent, a bake time of 30 minutes is specified to ensure enough time elapses for the samples to equilibrate at the desired temperature. Following the OS data read, the SS is re-written and the high temperature bake continues to the next read-point in the bake test.

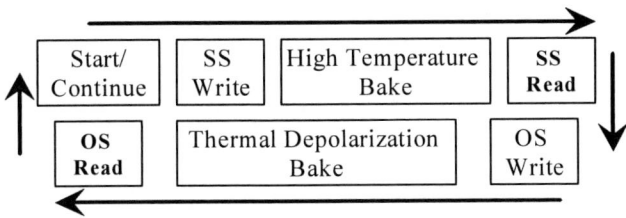

Figure 9.   FRAM Data retention test procedure is illustrated.

The procedure described above and detailed in Fig. 9 is optimized to identify potential imprint issues, especially in memories where the stored information may change during normal use, as can be expected in data storage applications. By design, the amount of time the bits spend at high temperature in the SS is maximized. Keeping the bits primarily in the SS during the bake tests can induce an imprint condition which will be observed as an OS read fail. This fail mechanism is slightly different from other non-volatile memory fail modes. If an imprint condition occurs,

the SS will actually be strengthened by the HT bake, but the probability of an OS read fail increases. Note that the data in Fig. 7 was collected as a special case of the procedure outlined in Fig. 9. The bit distributions were measured as the last step of the read-point tests, after both the SS and OS data reads were completed.

Extensive bake tests were used to quantify long-term data retention as a function of initial signal margin, an example of which is shown in Fig. 10. Successful results for both SS and OS data retention were obtained for units with initial margin beyond a critical minimum. Units with insufficient signal margin window at the beginning of the bake test either failed the SS read, a consequence of the thermal depolarization at the HT bake step, or after a period of time failed the OS read as imprint reduced the margin (see Fig. 7) below the sensing limits as illustrated in Fig. 6.

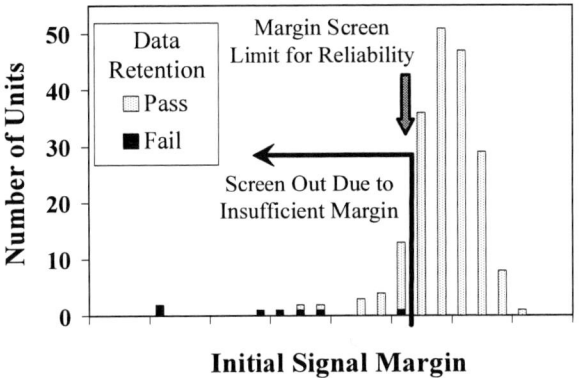

Figure 10. Margin screens for reliability were established via extensive bake tests at 125°C. Initial signal margin is critical for long term reliability.

These experimental results, repeated through extensive learning cycles, led to the definition of a margin screen limit for reliability, illustrated in Fig. 10, which includes 1) a built-in tolerance for the expected signal drift that occurs during typical use conditions in the field; 2) tolerance for the depolarization at the maximum operating temperature rating in the field; and 3) data sensing circuitry parameters.

In embedded memory applications, program, code and device trim information is often written during wafer-level test and stored through high-temperature assembly steps. With the advent of lead-free solder reflow processes a typical thermal profile requires exposure to temperatures in the range of 245°C - 260°C for a short amount of time, ≤ 2 minutes [11]. As a result of this high-temperature stress, the signal level of the capacitor is reduced through the thermal depolarization effect. Appropriate design safeguards need to be applied to ensure that the FRAM retains data through this step. It is also critical to include these high temperature steps as part of the pre-conditioning stresses during qualification tests [12]. The full polarization margin is restored naturally during the first read/restore operation following those assembly steps. FRAM data retention reliability through pre-conditioning tests was confirmed on >400 devices.

B. Write/Read Cycling Endurance

As described above, FRAM data access requires a restore operation after the data is read. As a consequence, the number of switching cycles during normal use can be relatively high, $>10^{10}$, or higher, depending on the specific application. Two distinct test patterns have been defined to evaluate the endurance properties. First, an intrinsic test was implemented which accesses 128 bits up to a total 5.4×10^{12} cycles in a thirty day stress period. This intrinsic test accesses five different rows simultaneously, each at a different duty cycle, in order to cover a cycling range of $5.4 \times 10^{9} - 5.4 \times 10^{12}$ during the thirty day stress period. The second pattern, a "full-chip" test, accesses all 8Mb a total of 10^{8} cycles within the thirty day stress period.

To separate the effects of cycling wear-out mechanisms from imprint induced signal degradation, the endurance tests are run at 25°C. This is reasonable since the polarization wear-out in thin film ferroelectric capacitors resulting from data cycling is strongly accelerated by voltage, but is not strongly dependent on temperature [13], [14]. Temperature, therefore, is not a useful parameter to accelerate cycling degradation. Furthermore, cycling tests run at high temperature can potentially induce imprint related signal loss. For example, the 8Mb full-chip cycling test accesses all 8,192 rows, with 16 accesses per each row of 1,024 bits. Since the bit access duty cycle is very low, $\sim 1/(8,192 \times 16)$, the bits are under a bias a total of 20 seconds during the thirty day stress period. Consequently, the results will essentially be dominated by the equivalent of a "thirty day bake stress" if performed at high ambient temperature, potentially masking any endurance wear-out effects.

Table I summarizes the endurance results measured on >1300 total assembled units with the standard thirty day cycling tests.

TABLE I.   CYCLING ENDURANCE TEST RESULTS

Cycling Endurance Results Pass/Fail 30 Days Test @25°C	
Description	No. of Chips Pass/Fail
Intrinsic, 128 bits each at 5.4×10^{9} - 5.4×10^{12}	969/0
Full-Chip, 8Mb to 10^{8}	360/0
Total No. of Devices Stressed, P/F	1329/0

Test arrays fabricated with this process were reported previously to show no polarization reduction, relative to the initial margin, after 10^{13} cycles [8]. For characterization purposes, we evaluated 37 packaged devices with the intrinsic test pattern for an extended stress period of three hundred days, achieving a maximum 5.4×10^{13} switching cycles, with all units passing the post-cycling functional tests. As shown in Fig. 11, where the cycled bits are contrasted with non-cycled bits from the same group of devices, we measure no signal margin degradation through that amount of data cycling stress.

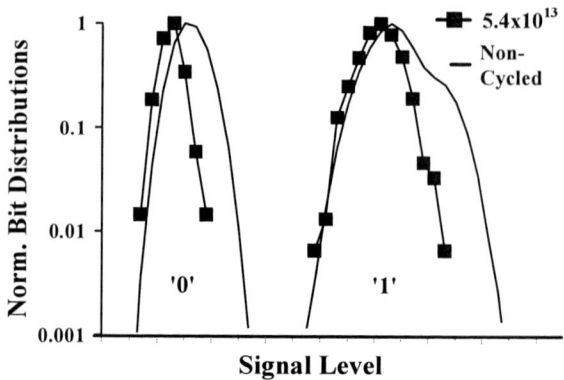

Figure 11. Signal distributions are shown comparing non-cycled and cycled bits after 5.4×10^{13} cycles at 25°C ambient. No degradation in the intrinnsic signal margin window is observed after 5.4×10^{13} cycles.

An additional intrinsic cycling test was run on a group of 150 packaged parts as a function of operating voltage, shown in Fig. 12, up to 5.4×10^{12} cycles. All 150 units passed the post-cycling functional tests. Each 0.1V operating voltage increase leads to ~ 10x reduction in the number of cycles required to reach an equivalent signal level, as indicated by the demarcation line (arbitrarily set) in Fig. 12. Based on the voltage acceleration, the intrinsic lifetime at 1.5V operation is estimated to be greater than 5×10^{14} cycles.

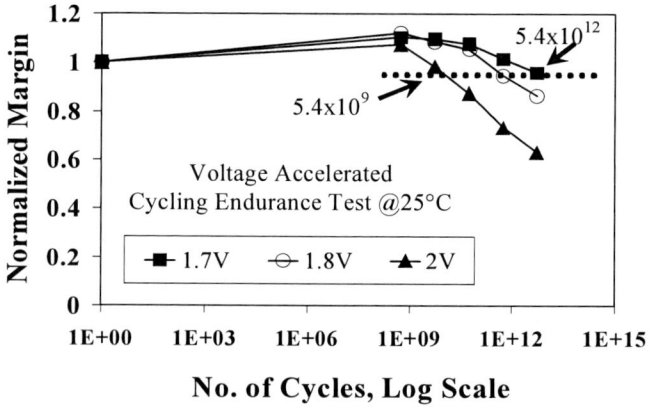

Figure 12. Cycling wear-out rate is strongly accelerated by voltage. Cycling life is estimated greater than 5×10^{14} cycles @1.5V.

The signal margin pre/post 10^{8} cycles measured on 160 devices tested with the full 8Mb chip cycling pattern is shown in Fig. 13, where, contrary to the degradation reported for other electrodes, an increased margin is observed after cycling for our optimized ferroelectric stack. The margin increase, measured in the signal level change for the lowest '1' bit, is in the range of $\sim 5\%$ through 10^{8} cycles at nominal operating voltage, and is attributed to the activation of pinned domain walls or the movement of initially passive domains [8]. A detailed review of domain nucleation effects and polarization switching kinetics can be found in texts [7] and literature references, for example see [15] - [18]. The

978-1-4244-5430-3/10 $26.00 © 2010 IEEE

lowest signal bits (<10,000 bits, Fig. 13 inset) benefit from this domain activation mechanism during cycling. Referring back to the voltage acceleration results shown in Fig. 12, which also show a margin increase before the onset of polarization fatigue, there appear to be two competing mechanisms during data cycling. First, initially passive domains get activated as a consequence of the bipolar voltage cycling. Second, an onset of polarization margin reduction occurs, especially at high voltages, as the data cycling induces defects which inhibit domain switching [17], [19] and [20].

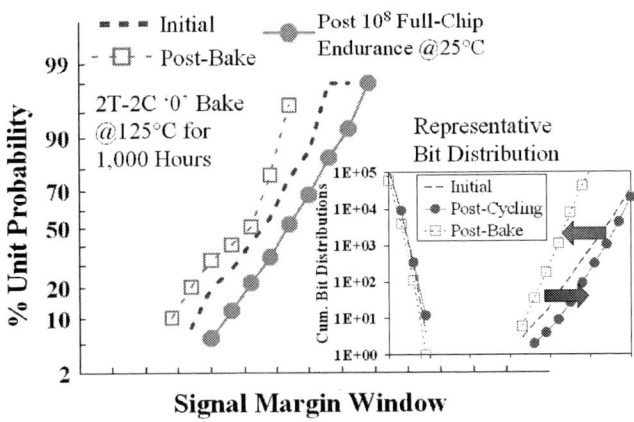

Figure 14. Post-cycling bake results in uniform margin reduction similar to the bake effects on non-cycled parts.

C. High Temperature Operating Life (HTOL)

HTOL patterns exercise the FRAM in a manner analogous to an SRAM, with a variety of bit access patterns including solids, scan and march patterns. In total, the bits experienced $\sim 10^7$ access cycles during the 1,000 hour life test, operated at 1.8V, 125°C stress conditions.

As stated in the previous discussions on cycling and imprint effects, the signal margin is impacted strongly by the bake stress. From a margin window reduction perspective, the 1,000 hours of 125°C HTOL stress represents the equivalent of 10 years operation at 85°C. The HTOL stress voltage is 0.3V above the 1.5V nominal operation. Since each 0.1V operating voltage increase leads to a 10x reduction in the equivalent number of cycles at normal operating voltage, (see Fig. 12 and the discussion on the resulting voltage acceleration), from an endurance point of view the 10^7 accesses during 1,000 hours of HTOL stress at 1.8V is equivalent to 10^{10} cycles at normal 1.5V operation at 25°C.

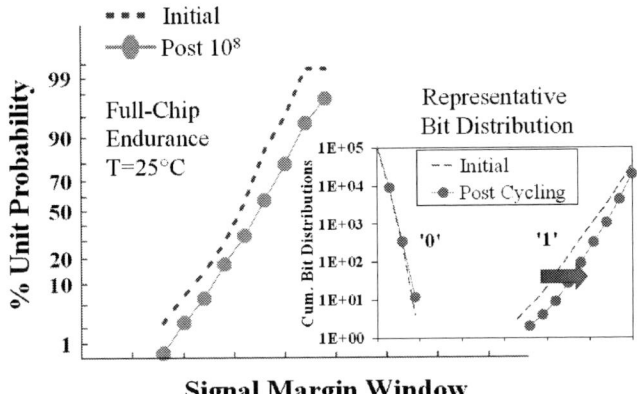

Figure 13. Full chip (8Mb) cycling endurance test@25°C to 10^8 cycles results in increased signal margin.

Data retention reliability was evaluated after both the intrinsic and full-chip cycling tests. The FRAM data retention is not impacted negatively by the cycling as shown in Table II, which summarizes the data retention results for both cycled and non-cycled units totaling 1,271 parts. Performing cycling tests prior to retention can have a beneficial effect on the FRAM since the signal margins tend to increase after cycling.

Burn-in oven monitor signals were used to validate the expected FRAM data at intermittent points during the stress. In addition, read-point tests were used to confirm device functional parameters through the entire stress period. Preliminary learning cycle tests identified "stuck" bits as an extrinsic mode, observed during both HTOL and Cycling Endurance tests. New screens for those bits at wafer probe were proven effective at eliminating that extrinsic mode during subsequent life tests. Table III summarizes the Early Life Failure Rate evaluations to 168 hours on over 3800 units. Subgroups of those units totaling over 1400 were then evaluated through the entire 1,000 hour operating life test as summarized in Table IV.

TABLE II.   DATA RETENTION RESULTS FOR THE 4MB FRAM.

Data Retention Pass/Fail 125°C, 1,000 Hours SS Bake, with 30 Minutes, 85°C OS Depolarization Bake at each Read-Point	
Description	No. of Chips Pass/Fail
Non-Cycled	794/0
Cycled	477/0
Total No. of Devices Stressed, P/F	1271/0

The bit distributions measured on 50 representative units baked after cycling show consistent margin reduction behavior as non-cycled units, shown on Fig. 14 + inset. The room temperature data retention also passed successfully after both intrinsic and full-chip cycling, reported in an earlier publication [21]. Based on these results, we conclude the FRAM endurance tests do not create anomalous defective behavior for retention in this range of switching cycles.

TABLE III.   EARLY FAILURE RATE TEST RESULTS

Early Life Fail Rate Pass/Fail Results 125°C, 1.8V Operation to 168 Hours	
Description	No. of Chips Pass/Fail
Non-Cycled	3437/0
Cycled	417/0
Total No. of Devices Stressed, P/F	3854/0

TABLE IV.   4MB HIGH TEMPERATURE OPERATING LIFE TESTS.

HTOL Pass/Fail 125°C, 1.8V Operation to 1,000 Hours	
Description	No. of Chips Pass/Fail
Non-Cycled	991/0
Cycled	417/0
Total No. of Devices Stressed, P/F	1408/0

Bit distribution signal levels (averaged from a 30 unit sample size) measured during the HTOL read-points are shown in Fig. 15 at various cumulative stress times. The '1' signal increases through the first ~500 hours of stress before showing a decrease at 750 hours. A detailed examination of the bit signals in Fig. 16 shows that bits on the low edge of the '1' distribution exhibit a signal margin increase similar to the increase observed after cycling tests. At longer stress times, the bake effect begins to dominate and the 1's distribution uniformly drifts downward, similar to that observed during retention bake tests, a consequence of the imprint mechanism. The repeated access has a positive effect on the 0's signal as it reduces slightly, lessening the net impact on signal margin.

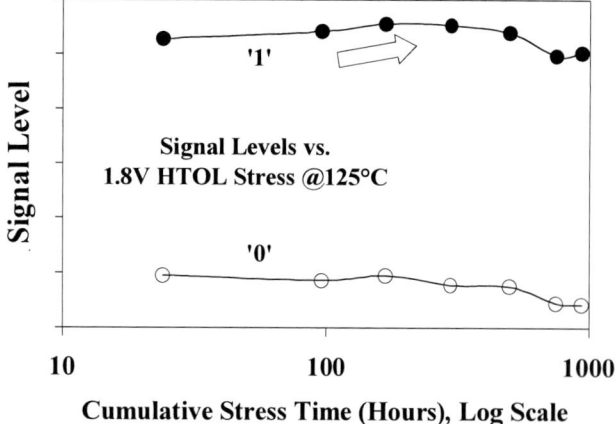

Figure 15. Margin initially increases at 125°C HTOL before the high temperature exposure reduces the Data 1 signal from imprint effects.

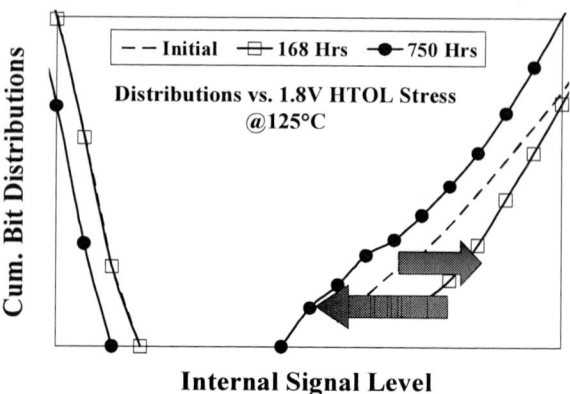

Figure 16. Data '1' signal distribution initially increases during 125°C HTOL stress. High Temperature effects dominate at long stress times, causing a reduction in the 1's signal via the imprint effect. The 0's bits tend to reduce with repeated access and cycling.

The high temperature operating life test in a sense combines the effects of endurance and retention bake into one test. An increased margin initially results from the bit accesses and switching. At long (~750 hours) stress time the high temperature causes imprint signal reduction in a manner similar to a data retention bake. By screening to appropriate margins levels during probe, reliability during normal operation can be assured as demonstrated with these accelerated life test results.

It is worth noting that in microcontroller devices, the FRAM may be partitioned as program, code and data memory. Any program or code partition will operate in "same-state" mode as it is not expected to change during normal use. During stress tests, it stores the instructions used by the microcontroller during high temperature operating life tests. The data partition, however, is expected to be updated during normal operation, so it is exercised as an SRAM at HTOL, and both SS/OS are tested during retention bake.

D. Soft-Error Rate (SER)

Ferroelectric memory devices exhibit a relatively high immunity to radiation effects since information is stored as a remnant polarization and not as an electronic charge. Switching the polarization requires application of an electric field to the capacitor, so the ferroelectric memory element is not disturbed by neutron strike events.

The 4Mb and another FRAM device of comparable density were evaluated at the Neutron Beam Facility at Los Alamos National Laboratories totaling over one-hundred thirty hours of beam time. All testing followed the JEDEC JESD-89A protocol [22]. Two types of experiments were conducted: 1) retention mode (static) which consisted of data write, power down, static retention time, power-up and data read while exposed to radiation, and 2) a dynamic mode where all the bits were accessed using the periphery circuits.

The retention mode test yielded a FIT rate <0.051/Mb with a 90% confidence interval. Dynamic access of a checkerboard pattern resulted in a FIT rate < 0.16/Mb at 90%

confidence, more than three orders of magnitude lower than standard 6T-SRAM devices.

A detailed analysis determined that errors result from sensitivities within the periphery logic. This is consistent with the observation of negligible FIT rate for static stress mode.

E. Prospects for 1T-1C Designs

As illustrated in Fig. 5, for a given margin window, the 1T-1C circuit margin is less than that of a 2T-2C circuit since the data sensing reference signal has a magnitude in between the Data 0 and Data 1 distribution end-points as shown in Fig. 17. Bake evaluations confirm that bits along the distribution edge (illustrated in Fig. 17) are the first to fail data retention for units with insufficient initial margin at a given stress condition. Reliable operation is ensured for units screened to margin levels that include 1) an operating tolerance window for the operating temperature depolarization and data sensing circuit operation and 2) sufficient signal margin to sustain the expected signal drift (see Fig. 7) during normal use in the field.

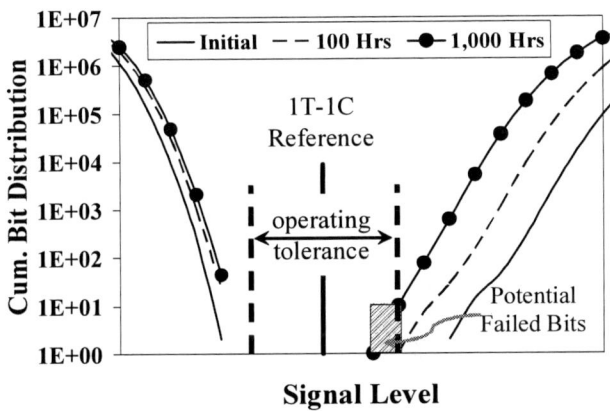

Figure 17. 1T-1C reliability is ensured by quantifying the circuit operating tolerance window and screening to a sufficient margin that accommodates the use requirements.

IV. CONCLUSIONS

Comprehensive reliability test results were presented for a 4Mb, 2T-2C FRAM device. Detailed bit distributions from all key non-volatile memory test results were shown including: data retention, cycling endurance, and high temperature operating life. The intrinsic cycling lifetime at 1.5V operation is greater than 5×10^{14} cycles. Data retention and operating life are not impacted negatively by pre-cycling the FRAM. The retention and operating life has been validated for a 10 year 85°C specification using the 1,000 hour, 125°C equivalent stress.

The development of robust processes and screen test conditions helped eliminate extrinsic issues observed during preliminary evaluations. Using signal margin as a screen parameter ensures reliable retention and operating life. In total, >6800 units have been tested with robust results in all key reliability areas. Using appropriate design guidelines, the FRAM reliably stores data through high temperature stresses required for lead-free solder processing. The margin screen methodology has been applied to 1T-1C evaluations with successful results. Reliable FRAM operation is ensured by designing in a signal margin window that accommodates the time/temperature profile and the depolarization at the use conditions.

SER tests of the FRAM confirm the FIT rate is dramatically lower than that of standard 6T-SRAM devices.

ACKNOWLEDGMENTS

We gratefully acknowledge the management support for this work, including Dr. Srikanth Krishnan and Dr. Venu Menon of Texas Instruments. We thank Bob Landers, Jerry Elkind, Larry Zhang, Hugh McAdams, Sudhir Madan and David Toops for SER experimental results and analysis. We thank Dr. Robert Baumann for additional insights and analysis of the SER test results. Numerous individuals have contributed to the FRAM project, especially process engineering staff at the Texas Instruments Kilby Center and DMOS5 fab, and we gratefully acknowledge their contributions.

REFERENCES

[1] T. Moise, et al., "Demonstration of a 4Mb, High Density Ferroelectric Memory Embedded within a 130nm, 5LM Cu/FSG Logic Process," IEEE Int. Electron Devices Meeting Digest, pp. 535 – 538, December, 2002.

[2] K.R. Udayakumar, et al., "Manufacturable High-Density 8Mb One Transistor-One Capacitor Ferroelectric Random Access Memory Embedded within a Low-Power 130nm Logic Process," Japan Journal of Applied Physics, Vol. 47, No. 4, pp. 2710-2713, 2008.

[3] J. Eliason, et al., "An 8Mb 1T1C Ferroelectric Memory with Zero Cancellation and Micro-Granularity Redundancy," IEEE Custom Integrated Circuits Conference Digest, pp. 427 – 430, 2005.

[4] Ali Sheikholeslami, and P. Glenn Gulak, "A Survey of Circuit Innovations in Ferroelectric Random Access Memories," Proceedings of the IEEE, Vol. 88, No. 5, May 2000.

[5] Hugh P. McAdams, et al., "A 64-Mb Embedded FRAM Utilizing a 130-nm 5LM Cu/FSG Logic Process," IEEE Journal of Solid-State Circuits, Vol. 39, No. 4, April 2004.

[6] Jack C. Burfoot and George W. Taylor, Polar Dielectrics and their Applications. Los Angeles: University of California, 1979.

[7] M. E. Lines and A. M. Glass, Principles and Applications of Ferroelectrics and Related Materials. Oxford: Clarendon, 1977.

[8] J. Rodriguez, et al., "Reliability Properties of Low-Volage Ferroelectric Capacitors and Memory Arrays," IEEE Transactions on Device and Materials Reliability, Vol. 4, No. 3, pp. 436 – 449, 2004.

[9] S. Traynor, "Polarization as a Driving Force in Accelerated Retention Measurements on Ferroelectric Thin Films," Proceedings of the IEEE Int. Symposium on the Applications of Ferroelectrics, pp. 15-18, August,1998.

[10] J. Rodriguez, et al., "Reliability Demonstration of a Ferroelectric Memory Embedded within a 130nm CMOS Process," Proceedings of the IEEE Non-Volatile Memtory Technology Symposium, pp. 64 – 66, November, 2007.

[11] "Moisture/Reflow Sensitivity Classification for Nonhermetic Solid State Surface Mount Devices," IPC/JEDEC Joint Industry Standard, J-STD-020D.1, March 2008.

[12] "Pre-conditioning of Nonhermetic Surface Mount Devices Prior to Reliability Testing," JEDEC Solid State Technology Association Standard JESD22A113E, March 2006.

[13] F. Chu, "A Mathematical Model for the Fatigue Behavior of Ferroelectric Thin Films for Memory Applications," Integrated Ferroelectrics, 58:1381-1393, 2003.

[14] J. Rodriguez, et al., "Empirical Model for Fatigue of Ferroelectric Memories," IEEE Int. Reliability Physics Symposium, pp. 39 – 45, 2002.

[15] H. Orihara, S. Hashimoto, and Y. Ishibashi, "A Theory of D-E Hysteresis Loop Based on the Avrami Model," J. Phys. Soc. Jpn., Vol. 63, pp. 1031-1035, 1994.

[16] V. Shur, E. Rumyantsev, S. Makarov, "Kinetics of phase transformations in real finite systems: Application to switching in ferroelectrics," J.Appl. Phys., Vol. 84, p. 445, 1998.

[17] N. Setter, et al., "Ferroelectric thin films: Review of materials, properties, and applications," J. Appl. Phys., Vol. 100, 051616, 2006.

[18] A.K. Tagantesev, at al., "Non-Kolmogorov-Avrami switching kinetics in ferroelectric thin films," Phys. Rev. B, Vol. 66, 214109; 2002.

[19] A.K. Tagantsev, et al., "Polarization fatigue in ferroelectric films: Basic experimental findings, phenomenological scenarios, and microscopic features," J. Appl. Phys., Vol. 90, No. 3, pp. 1387-1402, 2001.

[20] M. Dawber, K.M. Rabe, J.F. Scott, "Physics of thin-film ferroelectric oxides," Reviews of Modern Physics, Vol. 77, pp. 1083-1130, 2005.

[21] J. Rodriguez, et al., "Reliability Characterization of a Ferroelectric Random Access Memory Embedded within 130nm CMOS," Proceedings of the IEEE Int. Symposium on the Applications of Ferroelectrics, Vol. 1, pp. 1 - 2, February, 2008.

[22] "Measurement and Reporting of Alpha Particle and Terrestrial Cosmic Ray-Induced Soft Errors in Semiconductor Devices," JEDEC Solid State Technology Association Standard JESD89A, Oct. 2006.

Electronic Failures in Spacecraft Environments

Douglas J. Sheldon

Jet Propulsion Laboratory
California Institute of Technology
Pasadena, California, USA
818-393-5113 douglas.j.sheldon@jpl.nasa.gov

Abstract—**This paper will provide a review of data describing electronic part failures in spacecraft environments. The details of the spacecraft environment and their individual contributions to the overall failure population will be described. The paper will also present an analysis of new data regarding electronic part failures during spacecraft assembly and test. This data provides an indication of incoming part level quality and insight into handling practices.**

Keywords-component; radiation, reliabiity, electronic parts, failure, spacecraft

I. INTRODUCTION

Electronic parts are critical to all aspects of spacecraft implementation. Electronic parts used in spacecraft are subjected to significant amounts of additional screening and evaluation. These additional processes are designed to eliminate incoming quality problems that would cause infant mortality failures as well as evaluating the long-term failure mechanisms of the parts to ensure that there is sufficient operating margin to meet mission requirements before the parts begin to fail. Despite these significant efforts, failures of electronic parts still occur, sometimes with catastrophic results for the spacecraft. The first portion of this paper will discuss electronic parts failures in the operational phase of spacecraft missions. The second part of the paper will address parts failures seen in the assembly and test phases of spacecraft development and design.

In terms of how it relates to possible failures of electronic parts, the operational spacecraft environment can broadly be defined in two main areas, radiation and reliability. Radiation covers a wide array of ionizing phenomena. Ionizing radiation includes trapped protons and electrons, galactic cosmic rays, and solar particle events. These types of ionizing radiation have a very broad range of energies and densities and determining their effects requires very accurate and detailed knowledge of the mission trajectory. Reliability is a broad reference to long-term operation and possible degradation of an electronic part due to other environmental effects besides ionizing radiation. These can include the thermal stresses that the part is subjected to, variations in voltage and current that supply the part and occasionally more sophisticated effects due to Ionospheric plasmas for example.

Studying electronic part failures in operational spacecraft means understanding how spacecraft fail and then being able to relate the failure to the electronic part. Often this is a very complex undertaking and with no guarantee of 100% certainty.

Spacecraft failures can occur due to wide range of environmental conditions, incorrect human intervention, and excessive stresses beyond design tolerance, systematic design faults or random failures. The definition of failure in a spacecraft can range from a complete loss of ability to control or communicate to degradations of specific instrument or subsystem performance. Electronic part failures can contribute to this entire range of failures.

Spacecraft rarely carry sufficient instrumentation to identify the failure of a discrete part. Instead, a component or system is first identified as the source of failure. Then ground based testing on a variety of simulators and test benches is performed to mimic the failure as accurately as possible. Spacecraft engineering often go to heroic levels to find work arounds and alternative operating modes to continue to provide some level of functionality and performance.

II. HISTORICAL REVIEW

A 1996 NASA [1] review of over 100 failures and anomalies occurring from 1974 through 1994 was published. This review categorized failures by the details of the space environments. In this study 33% of the failures and anomalies were caused by ionizing radiation. Radiation was the second largest category. The largest contributor to spacecraft failures was spacecraft interactions and degradations with Ionospheric and magnetospheric plasmas (36%). Ionizing radiation failures manifested as single event upsets (SEU), bit errors, and degradation of materials and performance. Over 42% of the total radiation failures were SEU related.

SEU effects can produce a wide range of errors and anomalies. For example, SEUs were systematically recorded on TDRS-1 (Tracking and Data Relay Satellite) from 1984 to 1990. This allowed correlations to be drawn between those upsets and the space environment [2]. During the transfer orbit, the first anomalous responses were observed in the Attitude Control System (ACS). These anomalies were traced to state changes in the Random Access Memory (RAM) in the ACS caused by SEUs. The most serious ACS anomalies were considered mission threatening by operators because they could cause the satellite to tumble. Ground control was required to maintain the satellite's proper attitude. The first verified proton induced latch up was seen on the Precision Range and Range Rate Equipment (PRARE) instrument on the ESA Remote Sensing Satellite (ERS) following a transient high current [3]. The failure was found to have occurred close to the center of the South Atlantic Anomaly. Ground tests showed certain memories to be sensitive to proton-induced latch-ups.

978-1-4244-5430-3/10 $26.00 © 2010 IEEE

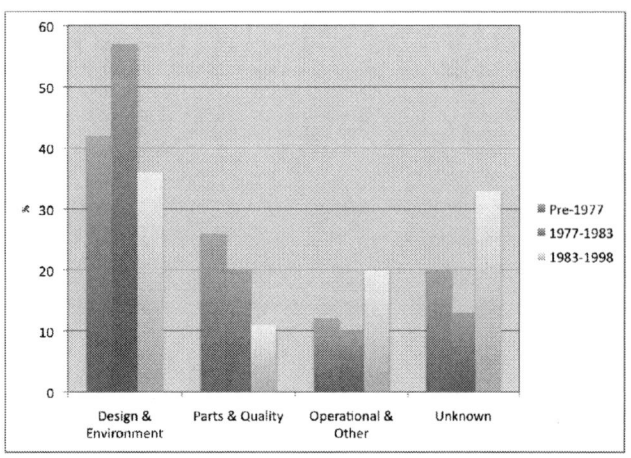

Figure 1. RAND study on spacecraft failures.

In 1998 the RAND Corporation [4] did a similar study of spacecraft failures and organized the failures into categories of design and environment, parts and quality, operational, and finally unknown. This is one of the first studies to identify parts specifically. The data from this report has been re-plotted in figure 1. Environmental and design related effects are the single largest category of failures. Over the 20 years studied in this report, environmental and design related effects cause between 40 to almost 60% of the failures. This category includes radiation effects.

The study also shows an important trend in decreasing failure percentages due to parts quality. The quality of electronic parts increased enormously in the 1980's and 1990's as part defect rates dropped by orders of magnitude. This is due to process improvements and the adoption of systematic quality engineering and statistical process control method. Another important trend that was noted was that the severity of the failures decreased over this time frame. In the pre-1977 time frames, almost 50% of the total number of failures was classified as significant.

In 2008 a study was done on 156 on-orbit failures from 1980 to 2005 [5]. In this study failures were categorized as electrical, mechanical, software and unknown. The largest percentage of failures was electrical failures. These represented 45% of the total. Mechanical failures were 32% and software failures only 6%. The overall space environment (radiation, solar storms, meteorites, etc.) caused 16% of the total failures. An important conclusion from this study was that 41% of all failures occur in the first year of the mission. These early life failures represent a significant opportunity for improving spacecraft performance.

A statistical analysis of on-orbit failures from 1990 to 2008 was completed in 2009 [6]. The author's Kaplan-Meier estimates show a reliability of 96% after two years on orbit. The spacecraft reliability drops to 94% after 6 years and then to 90% for extended times in excess of 12 years. This more rigorous result re-enforces the preceding point of a conceptual "bathtub curve" of spacecraft reliability.

Infant mortality failures dominate spacecraft reliability. Once this early life failure period is over, the probability of a spacecraft not only completing its initial mission but going on to perform an extended mission is quite high. One of the main historical approaches to address infant mortality is to "burn-in" the system to reduce failures. This is not possible for a spacecraft as it is the final end user platform. This means that improved systems engineering practices must be used during the design, test and assembly stages of a spacecraft. These improved practices include more robust design and materials choices and more accurate screening procedures. Electronic parts failures still represent a double-digit percentage of the overall spacecraft failures. Radiation related failures of electronic parts remain a major contributor to on orbit anomalies. The overall high levels of reliability of modern electronic parts provide the long term, predictable performance seen in the overall decreasing failure rate of modern spacecraft systems. The next section of this paper will discuss failures of electronic parts prior to launch. This will be used to provide a possible contributing explanation to the high levels of infant mortality in spacecraft.

III.   PRE-LAUNCH FAILURES

Failures of electronic parts in spacecraft environments can be divided into two major categories, pre-launch and post launch failures. These two categories allow for an evaluation of the entire electronic parts quality and reliability process.

Pre-launch failures can be further organized into escapes and implementation failures. Escapes are failures that were not caught in manufacturer's screening and production flows. These escapes provide a quantitative measure of a vendor's quality control and quality systems. Implementation failures refer to parts that initially passed all incoming quality metrics yet were subjected to stresses at some point during the overall development cycle that caused them to fail.

JPL flight parts are based on minimum QML-Q qualification standards. Depending upon mission requirements, additional screening levels such as QML-V may be required. Also custom qualification plans are sometimes developed to support the infusion of new technologies. Examples of these new technologies include certain kinds of non-volatile memories and radio frequency components.

JPL experience in pre-launch failures encompasses all electronic part types and technologies. From discrete transistors and passive components up to the most complex CMOS VLSI device, our experience has shown that no one part type seems to fail more than another. Failure modes for in the pre-launch phase are shown in figure 2. Figure 2 shows that "short" failures occur almost 3X as often as "open" failures. Short failures have a similar failure occurrence as "other" and "functional". Failure modes are dominated by short and functional failures, as they are 60% of the total population. Parts failing open only occur about 10% of the time. "Other" in figure 2 refers to such failures as visual defects. Functional failures mean the part has does not meet data sheet parameters yet has both functionality and no detectable physical or material defects.

From a root cause point of view, manufacturer's workmanship is the number one reason for pre-launch

978-1-4244-5430-3/10 $26.00 © 2010 IEEE

electronic parts failures at JPL. This is shown in Figure 3. This means additional work is required on both the

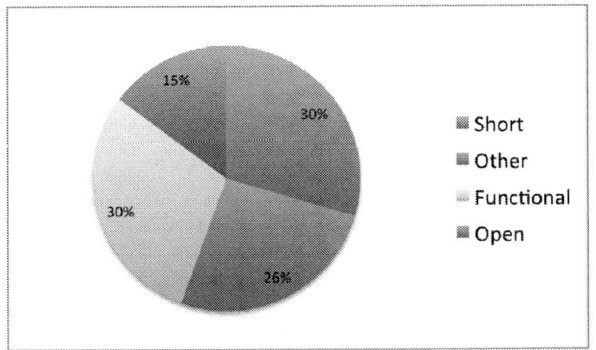

Figure 2. Failure Modes for JPL Pre-Launch Parts

manufacturing and spacecraft developer sides. There are more areas and opportunities for improvement in the product screening and design processes to help drive down workmanship failures. The actual implementation of space grade parts may need to be re-examined under certain circumstances. Part per billion quality metrics can only practically resolved with parts that come from extremely high volume commercial manufacturing lines. Continuous improvements in supplier QA and JPL IQA required. EOS/ESD (Electrical Overstress and Electrostatic discharge) failures reflect challenging spacecraft assembly and test environment.

Component reliability data often assume ideal handling and processing. For a commercial system, human hands often never touch components. Entire systems are assembled by automated processing equipment. Spacecraft applications remain custom in the sense that human technicians handle parts and components throughout the process. The effects of part handling and processing on manufacturer predictions of reliability are not well correlated.

All electronic parts are screened for minimum 2000V HBM ESD. Parts with exceptions are noted and discouraged. The

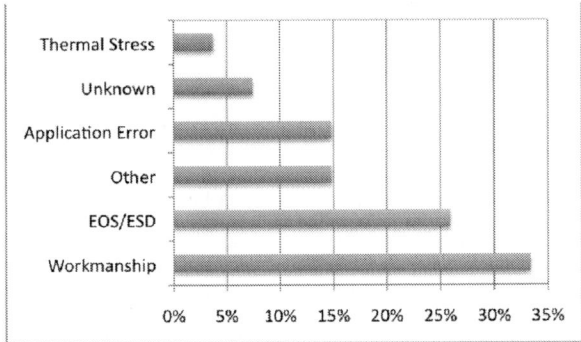

Figure 3. Root Cause of JPL Pre-Launch Failures

fact that there are continued ESD failures during spacecraft assembly and test indicate lack of correlation between integrated circuit ESD models and system ESD design. An additional level of insight may be obtained by examing the Cable Discharge Event or CDE for ESD events [7].

Ethernet/network users have developed the CDE to account for long and often untwisted pair cables. The model features a large amount of charge with low source impedance versus the standard high impedance/low capacitance models (MM, HBM, CDM). The CDE is a discharge that occurs when a cable is connected to a piece of electronic equipment. It occurs because there is a differential between the charge on a cable to be connected and the equipment that it is being plugged into. It is this differential that causes the discharge. The CDE discharge has a large spike followed by an oscillatory decay. Charge can accumulate due to triboelectricity and electromagnetic induction. These effects are combined with common materials that are used in spacecraft assembly that have very low dielectric leakage. This provides the ability to retain charge for long periods of time.

The CDE model provides a next step in the analysis of pre-launch electronic part failures. A continued evolution and sophistication in understanding and preventing ESD/EOS events in electronic parts will result in substantial reductions in spacecraft infant mortalities. This existence of latent defects in semiconductors continues to be a very active area of research [8]. Latent defects can provide a supporting contribution to overall part failure and weakness during normal spacecraft operation.

IV. CONCLUSIONS

Over 40 years of spacecraft failure analysis has shown that electronic parts play a significant contributing role. Radiation effects remain an important contributor to on orbit anomalies and interruptions. The continued evaluation of radiation degradation and possible mitigation schemes is critical to reducing these failures. Increasingly sophisticated electronic parts will require increasingly sophisticated radiation models and analysis.

The existence of a strong infant mortality condition in post-launch spacecraft failures reflects the need for improvement in design, screening and test procedures. Electronic parts used for spacecraft still have measurable instances of incoming quality issues. Any escapes from screening these parts will translate directly into increasing the overall infant mortality probabilities for the overall spacecraft system. The necessary additional human handling of custom spacecraft contributes to ESD and EOS style failures in electronic parts. Models that take overall system configuration and possible latent defect conditions into account must be used to both accurately describe and then improve the reliability of electronic parts used in spacecraft.

REFERENCES

[1] K.L. Bedingfield, R.D. Leach, and M.B. Alexander, "Spacecraft System Failures and Anomalies Attributed to the Natural Space Environment", NASA Reference Publication 1390, August 1996.

[2] Wilkinson, D., "TDRS-1 Single Event Upsets and the Effect of the Space Environment," *IEEE Transactions on Nuclear Science*, Vol. 38, No. 6, December 1991.

[3] Adams, L.,"A Verified Proton Induced Latch-Up in Space," *IEEE Transactions on Nuclear Science* NS-39, pp. 1804-1808.

[4] Liam P. Sarsfield, "The Cosmos on a Shoestring: Small Spacecraft for Earth and Space Science", RAND Corporation, 1998.

[5] M, Tafazoli, "A study of on-orbit spacecraft failures", Acta Astronautica 64 (2009) 195–205.

[6] Jean-Francois Castet and Joseph H. Saleh, "Satellite Reliability: Statistical Data Analysis and Modeling", Journal of Spacecraft and Rockets, Vol. 46, No. 5, September–October 2009.

[7] Rich Brooks, "A Simple Model For a "Cable Discharge Event", IEEE802.3 Cable Discharge Ad-hoc, March 2001.

[8] M. Diatta, E. Bouyssou, D. Trémouilles, P. Martinez, F. Roqueta, O. Ory, M. Bafleur, "Failure mechanisms of discrete protection device subjected to repetitive electrostatic discharges (ESD)", Microelectronics Reliability 49 (2009) 1103–1106

Single Event Transient Pulse Width Measurements in a 65-nm Bulk CMOS Technology at Elevated Temperatures

M. J. Gadlage, J. R. Ahlbin, B. L. Bhuva, L. W. Massengill, and R. D. Schrimpf

Vanderbilt University
Nashville, TN
matthew.j.gadlage@vanderbilt.edu

Abstract— **Soft errors are fast becoming a significant reliability issue for advanced technologies due to lower drive currents and higher operating frequencies. Single-event transients (SETs), a precursor for soft errors, show a strong dependence on operating temperature. In this work, heavy-ion induced SET pulse widths measured in a 65-nm bulk CMOS technology at temperatures ranging from 25° to 100° C with an autonomous SET capture circuit are presented. Experimental results for SETs induced in an inverter chain indicate an increase in average SET pulse width as a function of operating temperature. Unique SET test structures were also designed to differentiate between SETs induced in an nMOS transistor and those induced in a pMOS transistor. SET widths induced in a pMOS transistor increase more with temperature than SETs induced in an nMOS transistor.**

Keywords- soft error; single event; single event transient; SET; SER; pulse width; temperature; radiation environment

I. INTRODUCTION

A key factor affecting soft-error rates is the time duration of transient signals induced by energetic particles. For combinational logic circuits, the time duration of these transients determines the probability that they will arrive at a storage cell during a latching edge of the clock signal (and thus be recorded as an error). The parameters that determine the transient pulse width, such as charge generation and collection processes, are strong functions of operating temperature. Thus, for electronic circuits operating in extreme environments where both temperature and radiation are of concern, effects of temperature changes on the soft-error rate are of vital importance. Recent work has shown that an enhancement in the parasitic-bipolar action at elevated temperatures is the dominant mechanism that causes SET pulse widths to increase with temperature [1-3]. In bulk CMOS processes with a p-substrate and an n-well, simulation work has shown that the parasitic-bipolar effect is worse in pMOS transistors than in nMOS transistors, which results in larger pulse widths for SETs induced in pMOS transistors [4-5]. Because of difficulties associated with SET width measurements, separate measurement of SETs induced in either pMOS transistors or nMOS transistors has not been reported before. In this paper, SET pulse widths at elevated temperatures are measured for nMOS and pMOS transistors separately using a novel autonomous SET pulse width measurement circuit. The SET measurements in the nMOS and pMOS transistors are also compared to SETs measured in a long inverter chain.

II. SET MEASUREMENT STRUCTURES

Three test structures to characterize SET pulses were fabricated in an IBM 65-nm bulk CMOS technology. Each of the test structures consists of a target circuit in which the SETs are generated, followed by an on-chip measurement circuit. The measurement circuit is identical for all target circuits used in this study. The measurement portion of the test structure measures the SET pulse width in terms of inverter stage delays. The measurement circuit is based on the principle that within an inverter chain, an SET pulse will affect a number of inverters that is directly related to the pulse width. The measurement circuit was first described by Narasimham et al. [6] and has been implemented and tested successfully in multiple technologies [7-9]. In the implementation of the measurement circuit used here, 80 inverter stages are connected to latches to store the number of inverter stages affected by an SET. An SET pulse detection circuit is used to trigger the latches. With an individual inverter stage delay of 25 ps at room temperature, this circuit allows measurement of SET pulses ranging from 25 ps to 2 ns with a 12.5 ps measurement resolution.

The three target circuits used in this work consist of (1) a minimum sized and spaced linear 1000-inverter chain, (2) an "N-hit" circuit, and (3) a "P-hit" circuit. Figs. 1 and 2 show the basic blocks of the "N-hit" and "P-hit" target circuits. The "N-hit" ("P-hit") target circuit consists of four chains of 100 NAND (NOR) gate/inverter blocks "OR"-ed together to form a single output. In both circuits, individual ion strikes on the inverters are unable to propagate through the logic chain due to logic masking. (Logic masking is a term used to describe a situation in which a signal such as an SET is unable to propagate through a combinational logic block due to the state of the remaining logic. For example, in a two-input NAND gate, if one input is at a logic "0" the output will always be at a logic "1" regardless of the state of the other input to the gate.) In the "N-hit" circuit stage shown in Fig. 1, an SET generated by an ion hit in one of the inverters will not propagate through the NAND gate. Only an ion hit on an nMOS transistor in the NAND gate will propagate through the chain. Therefore, only an SET generated in the "OFF" nMOS transistor connected to the output of the NAND gate will be measured. All other SETs will be blocked and will not be measured. The "P-hit"

978-1-4244-5430-3/10 $26.00 © 2010 IEEE

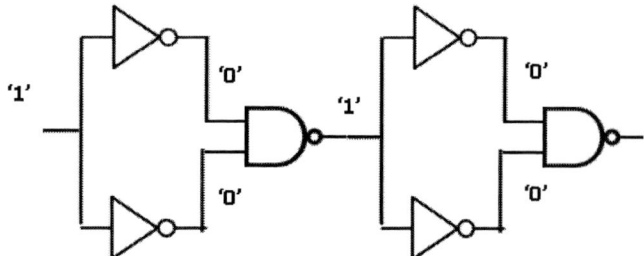

Fig. 1: Schematic of two of the blocks of "N-hit" target circuit. The target circuit used in this work consisted of four linear chains of 100 of these combinational logic blocks "OR"-ed together to form a single output.

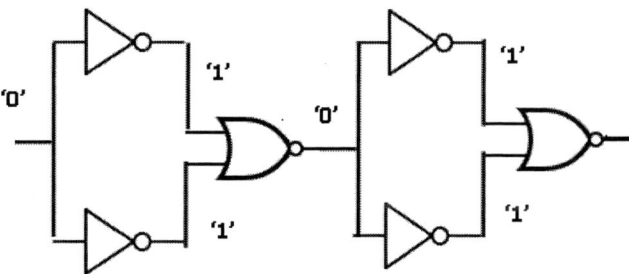

Fig. 2: Schematic of two of the blocks of "P-hit" target circuit. The target circuit used in this work consisted of four linear chains of 100 of these combinational logic blocks "OR"-ed together to form a single output.

target circuit works in a similar manner, with the NAND gates replaced by NOR gates.

One important item to take note of is the spacing of the two inverters in the "N-hit" and "P-hit" circuits. If the inverters are spaced close together in the layout, it may be possible for an ion strike to create a simultaneous SET on each inverter. If this were to happen, an SET may be able to propagate through either the NAND or the NOR gate, and as a result the circuit would no longer allow only hits on nMOS or pMOS devices to be measured. To ensure that a transient was not induced on both inverters by a single ion, the inverters were placed on top and bottom of the NAND/NOR gates with a separation of 3.5 μm as shown in Fig.3.

III. HEAVY ION EXPERIMENTAL RESULTS

The SET-measurement structures were tested at elevated temperature with heavy ions at the Lawrence Berkeley National Lab Cyclotron facility using xenon ions with an LET (linear energy transfer) of 58.8 MeV-cm^2/mg. The temperature of the device under test (DUT) was controlled through a resistive heater attached to the package of the DUT, and temperature measurements were taken using a sensor also attached to the package. Ion exposures were carried out at temperatures of 25°, 50°, and 100° C. The temperature reported in the following section is the package temperature as measured by the temperature sensor. Variations in inverter stage delays for this temperature range were recorded using a ring oscillator that was designed using the same inverter stages used in the measurement circuit. The ring oscillator frequency was measured at the temperatures used for the heavy ion experiment to determine the individual stage delay of the measurement circuit.

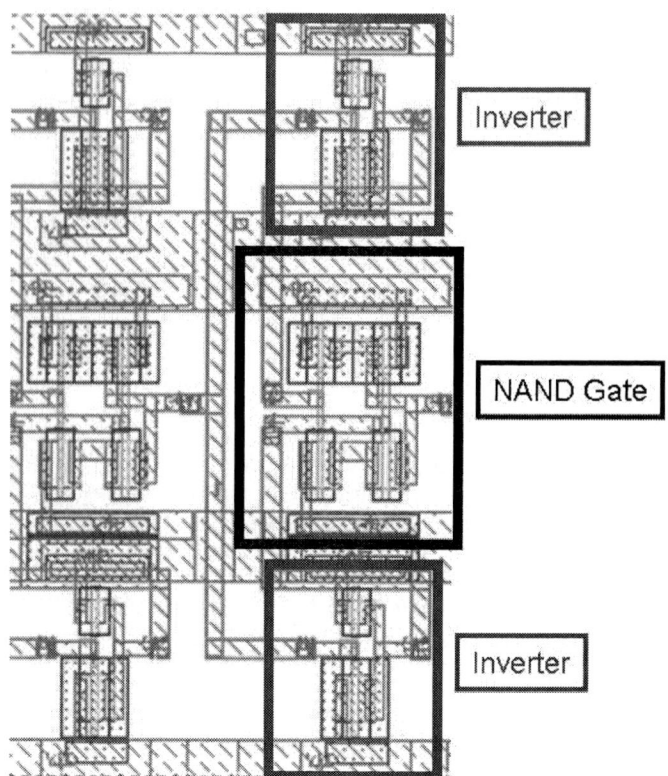

Fig. 3: Layout of two of the blocks of "N-hit" target circuit. The spacing between the two inverters needs to be large enough to ensure that an ion can not induce an SET on both at the same time.

The inverter stage delay increased linearly with temperature from approximately 25 ps at 25° C to 34 ps at 100° C.

For comparison purposes, the room temperature SET cross section for the inverter chain, "N-hit" and "P-hit" target circuits is shown in Fig. 4. The plotted SET cross section is simply the number of measured SETs divided by the total fluence of ions normalized to one logic block. For the inverter chain, the cross section is plotted per inverter, while for the "N-hit" and "P-hit" circuit the cross section is plotted per one NAND/NOR-inverter block combination. As can be seen in the plot, the threshold LET to create a measurable SET for the "N-hit" and "P-hit" circuits is much larger than that for the inverter chain circuit. Also of note is that the cross section for the "P-hit" circuit is larger than that of the "N-hit" circuit. One reason for this is that the size of the sensitive pMOS transistor in the "P-hit" circuit is much larger than the sensitive nMOS transistor in the "N-hit" circuit. The W/L ratio of the sensitive pMOS transistor is 1.3 μm/50 nm, while the W/L ratio of the sensitive nMOS transistor is 400 nm/50 nm. This means that the drain area of the sensitive pMOS transistor was over four times as large as the area of the sensitive nMOS transistor.

Histograms of the measured SET pulse width distributions at three different temperatures for exposures to xenon ions for the 1000-inverter chain target circuit are shown in Fig. 5. As can be seen in the histograms, the SET pulse width distribution clearly shifts towards longer SET widths as the temperature increases. More than an 80% increase in the average pulse width is observed as the temperature increases from 25° C to 100° C (as shown in Fig. 6). The longest measured SET pulse width increased from 200 ps to 304 ps. The total ion fluence was 10^8 ions/cm^2 for each exposure. The number of SET's measured for the total fluence of ions was 139 at 25° C, 163 at 50° C, and 235 at 100° C. The increase in the number of transients measured suggests that the sensitive volume (i.e., the area around each transistor that can collect enough charge to generate an SET) increases with temperature. SET measurements using inverter chains in 90-nm and 130-nm bulk technologies showed no increase in the number of SET events with temperature [2-3].

SETs measured for the "P-hit" circuit is about an order of magnitude larger than for the "N-hit" circuit (this is also shown in the cross section in Fig. 4), (2) the shift in the SET width distribution towards longer SET widths with temperature is clear for the "P-hit" circuit, and (3) the change in SET width with temperature for the "N-hit" circuit is not quite as obvious. The changes in SET width for the "N-hit" circuit may not be apparent due to the small number of SET events measured.

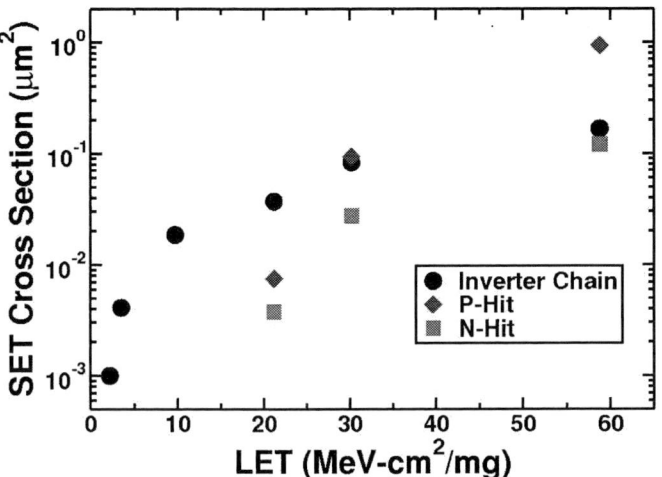

Fig. 4: SET cross section for the different 65-nm test structures. Note that the threshold LET for the "P-hit" and "N-hit" circuits is much larger than the threshold LET for the inverter chain circuit.

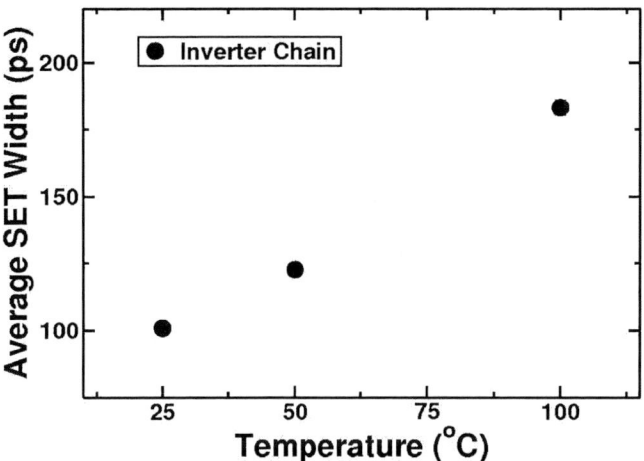

Fig. 6: Average SET width measurements as a function of temperature for the test structure with the 1000-inverter chain target circuit.

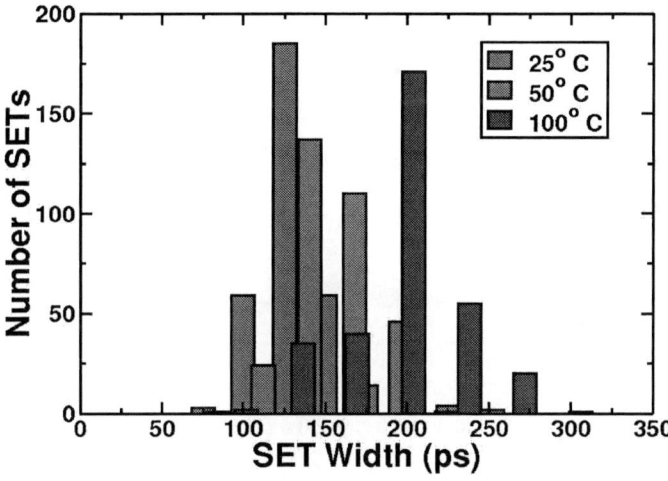

Fig. 7: Measured SET pulse width distribution for the "P-hit" circuit. Note that as the temperature increases the distribution clearly shifts to the longer SET widths

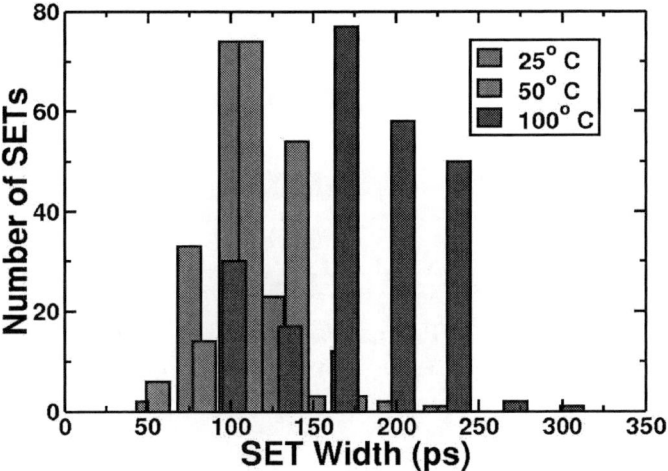

Fig. 5: Measured SET pulse width distribution for the inverter chain circuit. Note that as the temperature increases the distribution shifts to the longer SET widths

In Figs. 7 and 8, the measured SET pulse width distributions for the "N-hit" and "P-hit" circuit are shown. Several important items to note from the histograms are: (1) the number of

The average measured SET widths for the "N-hit" and "P-hit" circuit are plotted in Fig. 9. At room temperature the average SET width in the "P-hit" circuit was only slightly (~10 ps) larger than the average SET width in the "N-hit" circuit. However, the average SET width increased from 128 ps to 202 ps from 25° C to 100° C for the "P-hit" circuit (58% increase), while the average SET width for the "N-hit" circuit increased from 118 ps to 158 ps (34% increase). The error bars for the "N-hit" data represent the standard error in the average measured width. The standard error is found by dividing the standard deviation by the square root of the number of counts.

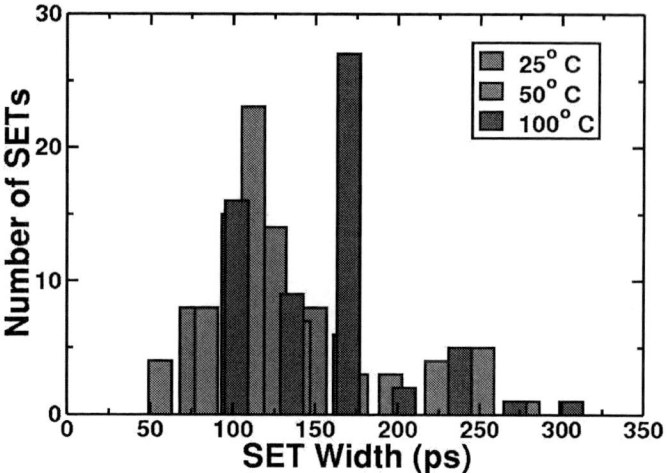

Fig. 8: Measured SET pulse width distribution for the "N-hit" circuit. Due primarily to the small number of SETs measured changes in the SET width distribution are difficult to observe.

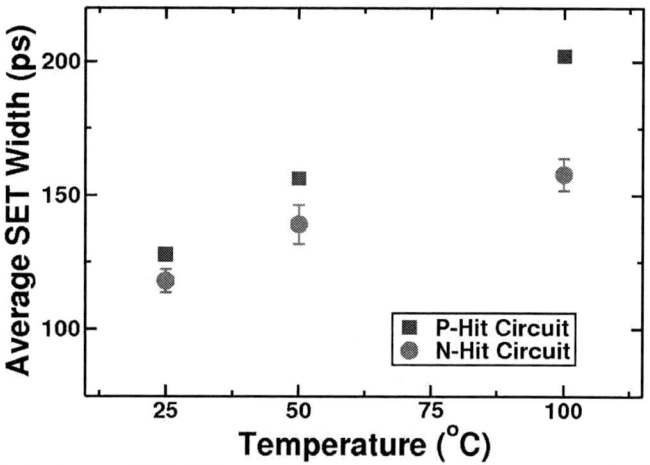

Fig. 9: Average SET width measurements as a function of temperature for the test structures with the "N-hit" and "P-hit" target circuits.

IV. DISCUSSION

The larger increase in SET width with temperature for the "P-hit" circuit suggests that the enhancement of the parasitic-bipolar effect with temperature is greater for the pMOS transistor than for the nMOS transistor. This is mainly caused by the de-biasing of the n-well region surrounding the pMOS transistor [4-5]. The enhancement in the parasitic bipolar effect with temperature and its impact on SET widths in bulk technologies has been explored in detail by Shuming et al. [1] and Gadlage et al. [2-3].

Another potential mechanism that can contribute to the lengthening of SET pulses is the change in individual transistor currents with temperature. After a single event hit on an "OFF" transistor, the corresponding "ON" transistor connected to the circuit node restores the original voltage. In the "N-hit" stage shown in Fig. 1, the restoring transistors are pMOS transistors in the NAND gate for hits on the nMOS transistor. The drive strength of the restoring transistor is proportional to mobility, which in turn is directly related to temperature. In the temperature range of concern here, mobility decreases with increasing temperature. A decrease in the restoring current means that it will take a longer time for the "ON" transistor to restore the same amount of charge to a struck node, resulting in longer SET pulses. To further explore how the restoring current changes with temperature, the maximum drive currents for nMOS and pMOS transistors were simulated at different temperatures at the circuit-level using the PDK for this technology. In Fig. 10, the drain currents for an nMOS transistor with a W/L ratio of 200 nm/50 nm and for a pMOS transistor with a W/L ratio of 400 nm/50 nm are plotted as a function of temperature. These W/L ratios correspond to the W/L ratios of the transistor in the inverter chain circuit. As can be seen, the drain current decreases with temperature, but the change from 20° C to 100° C is less than 6% for both devices. These simulations show that the change in the restoring current with temperature has a small impact on SET widths compared to the enhancement in the parasitic bipolar action with temperature.

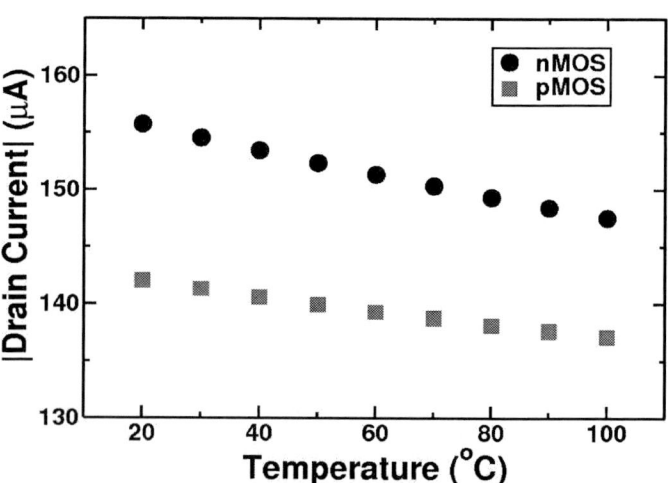

Fig. 10: Simulated drain current as a function of temperature for the nMOS and pMOS transistors used in the inverter chain target circuit.

V. CONCLUSIONS

Soft-error rates are a strong function of SET pulse widths generated by incident neutrons and ions. In extreme environments, electronic devices are often required to operate over wide temperature ranges so any potential change in the soft-error rate with temperature is of great concern. In this work, experimental measurements of heavy-ion induced single-event transient pulse widths are reported for a 65-nm bulk CMOS

978-1-4244-5430-3/10 $26.00 © 2010 IEEE

process over a large temperature range. The average SET pulse width increases by 80% as the temperature increases from 25° C to 100° C for transients induced in an inverter chain. Results from a unique test structure designed to separate transients from strikes on nMOS and pMOS transistors show that SET widths increase more with temperature (58% compared to 34%) for ion strikes on pMOS transistors than for ion strikes on nMOS transistors. The predominant mechanism causing the increase in SET width with temperature is an enhancement in the parasitic bipolar transistor as the temperature is increased. The change in the restoring current with temperature was shown to have little impact on SET widths.

ACKNOWLEDGMENT

The authors would like to thank NAVSEA Crane, Cisco Systems, and the Defense Threat Reduction Agency for their support of this effort.

REFERENCES

[1] C. Shuming, L. Bin, L. Biwei, and L. Zheng, "Temperature dependence of digital SET pulse width in bulk and SOI technologies," *IEEE Trans. Nucl. Sci.*, vol. 55, no. 6, pp. 2914-2920, Dec. 2008.

[2] M. J. Gadlage, J. R. Ahlbin, B. Narasimham, V. Ramachandran, C. A. Dinkins, B. L. Bhuva, R. D. Schrimpf, and R. L. Shuler, "The effect of elevated temperature on digital single event transient pulse widths in a bulk CMOS technology," *Reliability Physics Symposium, 2009 IEEE International*, pp. 170-173, April 2009.

[3] M. J. Gadlage, J. R. Ahlbin, V. Ramachandran, P. Gouker, C. A. Dinkins, B. L. Bhuva, B. Narasimham, R. D. Schrimpf, M. W. McCurdy, M. L. Alles, R. A. Reed, M. H. Mendenhall, L. W. Massengill, R. L. Shuler, and D. McMorrow, "Temperature characterization of digital single event transients in a bulk and fully depleted SOI technology," *IEEE Trans. Nucl. Sci.*, vol. 56, no. 6, pp. 3115-3121, Dec. 2009.

[4] B. D. Olson, O. A. Amusan, S. Dasgupta, L. W. Massengill, A. F. Witulski, B. L. Bhuva, M. L. Alles, K. M. Warren, and D. R. Ball, "Analysis of parasitic PNP bipolar transistor mitigation using well contacts in 130 nm and 90 nm CMOS technology," *IEEE Trans. Nucl. Sci*, vol. 54, no. 4, pp. 894-897, Aug. 2007.

[5] O. A. Amusan, L. W. Massengill, B. L. Bhuva, S. DasGupta, A. F. Witulski, and J. R. Ahlbin, "Design techniques to reduce SET pulse widths in deep-submicron combinational logic," *IEEE Trans. Nucl. Sci.*, vol. 54, no. 6, pp. 2060-2064, Dec. 2007.

[6] B. Narasimham, V. Ramachandran, B. L. Bhuva, R. D. Schrimpf, A. F. Witulski, W. T. Holman, L. W. Massengill, J. D. Black, W. H. Robinson, and D. McMorrow, "On-chip characterization of single event transient pulse widths", *IEEE Trans. on Dev. and Mat. Rel.*, vol. 6, pp. 542-549, 2006.

[7] B. Narasimham, B. L. Bhuva, R. D. Schrimpf, L. W. Massengill, M. J. Gadlage, O. A. Amusan, W. T. Holman, A. F. Witulski, W. H. Robinson, J. D. Black, J. M. Benedetto, and P. H. Eaton, "Characterization of digital single event transient pulse-widths in 130-nm and 90-nm CMOS technologies,", *IEEE Trans. on Nucl. Sci.*, vol. 54, no. 6, pp. 2506-2511, 2007.

[8] P. Gouker, J. Brandt, P. Wyatt, B. Tyrrell, A. Soares, J. Knecht, C. Keast, D. McMorrow, B. Narasimham, M. Gadlage, and B. Bhuva, "Generation and propagation of single event transients in 0.18-um fully depleted SOI," *IEEE Trans. on Nucl. Sci.*. 55, no. 6, pp. 2854-2860, Dec. 2008

[9] T. Makino, D. Kobayashi, K. Hirose, Y. Yanagawa, H. Saito, H. Ikeda, D. Takahashi, S. Ishii, M. Kusano, S. Onoda, T. Hirao, and T. Ohshima, "LET dependence of single event transient pulse-widths in SOI logic cell," *IEEE Trans. Nucl. Sci.*, vol. 56, pp. 202-207, Feb. 2009.

Practicality of Evaluating Soft Errors in Commercial sub-90 nm CMOS for Space Applications

Jonathan A. Pellish and Kenneth A. LaBel
Flight Data Systems and Radiation Effects Branch
NASA/GSFC Code 561.4
Greenbelt, MD 20771 USA
Phone: +001 301-286-6523, Fax: +001 301-286-6523, Email: jonathan.a.pellish@nasa.gov

Abstract—Inclusion of commercial technologies in civil spaceflight applications is reality. These technologies enable higher performance, reduce power consumption, and ultimately yield better science. However, the benefits do not come without cost, and radiation-induced soft errors in advanced, sub-90 nm CMOS technologies present new challenges. These challenges include sensitivity to proton direct ionization, memory technology evaluation, as well as testing and evaluation complexity.

Keywords-space environment, soft errors, CMOS, proton, heavy ion, memory

I. INTRODUCTION

The National Aeronautics and Space Administration (NASA) faces many radiation hardness assurance challenges as microelectronic components used in spacecraft scale below the 90 nm process node. This is particularly true for commercial off the shelf (COTS) complementary metal oxide semiconductor (CMOS) parts. While these parts enable improved scientific investigations, evaluating the semiconductor technologies required for the missions creates unique testing challenges like the examination of low-energy proton-induced soft errors [1-4].

Another key area of ongoing investigation in scaled commercial spacecraft electronics concerns volatile and non-volatile memory applications. These applications include processor program storage, temporary data buffers, mass data storage in solid-state recorders, and configuration storage for static random access memory (SRAM)-based field programmable gate arrays (FPGAs) [5, 6]. Each of these memory applications carries with it different levels of soft error criticality risk – some soft errors may result in scientific or housekeeping data loss, while others may require ground-based intervention for spacecraft safe-hold conditions. Engineers can determine this risk a number of different ways, one of which is a radiation- specific form of failure mode, effects, and criticality analysis called single-event effects criticality analysis (SEECA) [7], another is a Bayesian analysis approach [8].

Testing and evaluation challenges include matching the space environment using ground-based accelerator and pulsed laser facilities, experimental coverage of operational modes, budgetary concerns over non-recurring engineering, the limited lifetime of commercial product generation manufacturing relative to typical spacecraft mission development lifetime, and confronting things like controlled collapse chip connection, or flip-chip, device packaging styles. The use of advanced COTS CMOS in space-based applications has given rise to new soft error experimental and modeling evaluation techniques [5, 9-14] to overcome these challenges.

II. LOW-ENERGY PROTON SOFT ERRORS

Traditional proton soft errors are caused by inelastic nuclear reactions, much the same as high-energy neutron soft errors. Since the inception of space-based radiation effects [15] until very recently, indirect ionization soft errors were the only proton-based concern aside from ionizing dose and displacement damage. For scaled, sensitive COTS parts, protons are able to generate enough charge through electronic stopping, called direct ionization, to cause soft errors. K. P. Rodbell *et al.* [1] and D. F. Heidel *et al.* [2] published the first demonstration of low-energy proton direct ionization soft errors in 2007 and 2008 for a commercial 65 nm silicon-on-insulator (SOI) CMOS process; the results from D. F. Heidel *et al.* [3] are shown in Fig. 1. Scientists and engineers within the radiation effects community predicted the onset of low-energy proton direct ionization soft errors when heavy ion linear energy transfer (LET) thresholds dropped below 1 (MeV·cm^2)/mg while maintaining a sufficient sensitive volume structure; *cf.* [16]. LET is often referred to as mass stopping power. It is the electronic stopping power, dE/dx, normalized by the density of the target material, which is either given as g/cm^3 or mg/cm^3. LET is, by definition, a measure of direct ionization.

Proton direct ionization soft errors represent a significant threat to spacecraft electronics. They cannot be effectively shielded due to the fact that proton energies in space exceed several hundred megaelectron volts for solar, trapped, and galactic cosmic ray environments [17-19]. The external high-energy protons will lose energy and become low-energy protons as they transit the mass between outer space and the electronics boxes within the spacecraft. The spacecraft shielding distribution will determine which portion of the external proton energy spectrum becomes the low-energy spectrum that impacts sensitive microelectronic devices [2]. Low-energy protons have thus far been defined as protons with a kinetic energy less than 10 MeV, though energies that result in soft errors are typically below 2 MeV for the 65 and 45 nm process technologies documented thus far [1-4].

This work was supported in part by the NASA Electronic Parts and Packaging program, NASA flight projects, and the DTRA Radiation Hardened Microelectronics program under IACRO #09-4587I to NASA.

978-1-4244-5430-3/10 $26.00 © 2010 IEEE

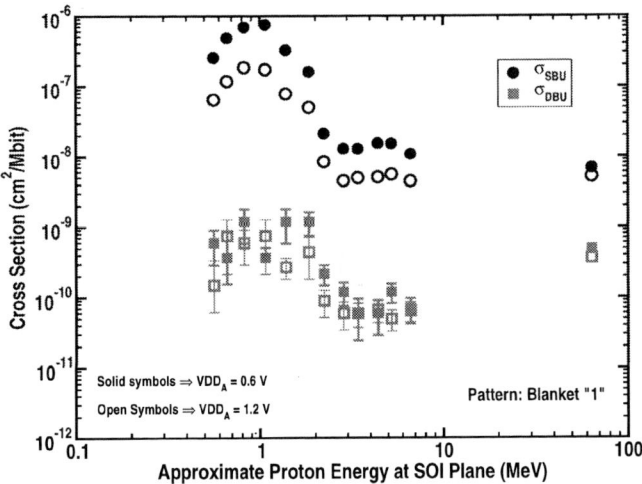

Fig. 1: Single- and double-bit proton upsets (SBU and DBU) in an IBM 45 nm SOI CMOS SRAM, after [3]. The cross sections for proton energies below 2 MeV are the points dominated by direct ionization, resulting in a 100x increase for SBU and a 10x increase for DBU. These irradiations were carried out at the UC Davis Crocker Nuclear Laboratory, which is pictured in Fig. 2.

Fig. 2: Test setup for experimental low-energy proton testing at the University of California at Davis Crocker Nuclear Laboratory. The cyclotron at the Crocker Lab can provide low-energy proton tunes of a few megaelectron volts that can be degraded further by using micrometer-thick aluminum and Mylar foils in air, downstream of the beam collimator. The daughter card is attached to the NASA/GSFC Xilinx Spartan-II-based low cost digital tester [20].

Recent results published by B. D. Sierawski *et al.* [4] show that for space environments with large proton populations – low Earth orbit, highly-elliptical orbit, and solar particle events – direct ionization soft errors from low-energy protons either dominate the overall soft error rate or constitute a significant fraction of it. The problem facing radiation engineers then becomes one of hardness assurance. However, guaranteeing component performance in the space environment by conducting ground-based low-energy proton tests, like the one shown in Fig. 2 using the NASA Goddard

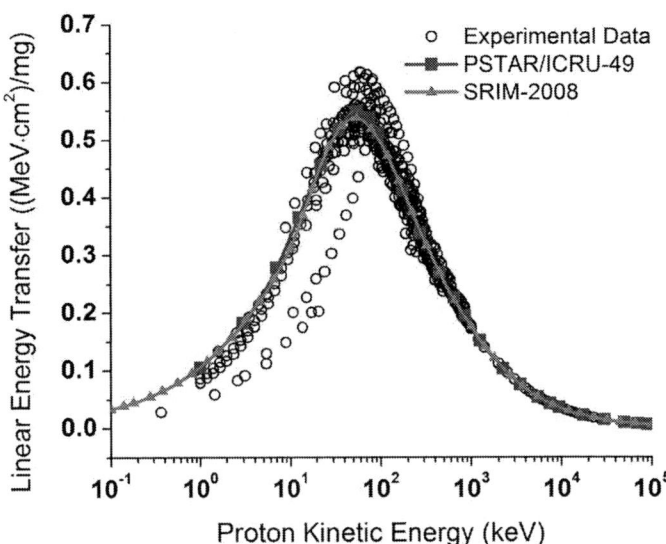

Fig. 3: Experimental and simulated proton linear energy transfer (mass stopping power) as a function of energy in silicon. The experimental values, shown as open circles, are from Helmut Paul's database [22, 23]. The simulated linear energy transfer curves were calculated using SRIM-2008 [24, 25] and NIST PSTAR [26, 27]. The points on the simulation curves are sparse to aid viewing. Note that the PSTAR calculations do not go below 1 keV.

Space Flight Center's low-cost digital tester [20], is fraught with physics-imposed difficulties.

The issues with accelerated low-energy proton testing can be summarized as limited range, energy straggling, and uncertainty in electronic stopping power. A 2 MeV proton has a range of approximately 50 μm in silicon and 74 mm in air, which means that testing either has to be carried out in a vacuum or tested in air using foil degraders. The inconvenience of testing in vacuum aside, the difficulty is exacerbated by the fact that at 2 MeV the LET of the proton is too low to generate enough charge to cause a soft error. Facilities can lower the proton energy below the beam tune energy using a combination of aluminum and Mylar degraders from hundreds of nanometers to several micrometers thick along with air columns and the semiconductor die itself. Particle range limitations become severe with flip-chip ball grid arrays where irradiation has to be done through the substrate. In-situ device thinning is often necessary, which is problematic because the ball grid array is under stress and will crack the die without sufficient mechanical support [21].

Fig. 3 shows experimental measurements of proton LET in silicon, compiled from Helmut Paul's database [22, 23], as well as two theoretical calculations of proton LET in silicon using SRIM-2010 [24, 25] and the National Institute of Standards and Technology's PSTAR tool [26]; the latter is based on ICRU Report 49 [27]. As the figure shows, at high energy there is good agreement between experiment and theory. However, below 1 MeV, moving up towards the Bragg peak, the spread in experimental data becomes large. These low-energy transmission measurements require thin foils, making the presence of pin holes and other material variations critical. The critical angle for ion channeling also increases at low energy along with the importance of multiple scattering [27]. These experimental facts translate to uncertainty in

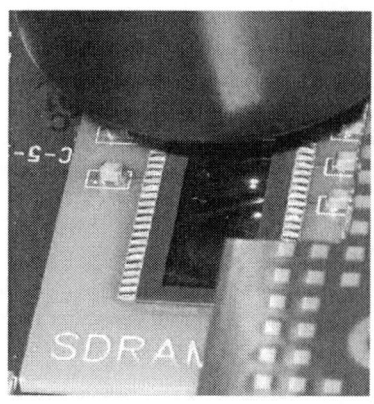
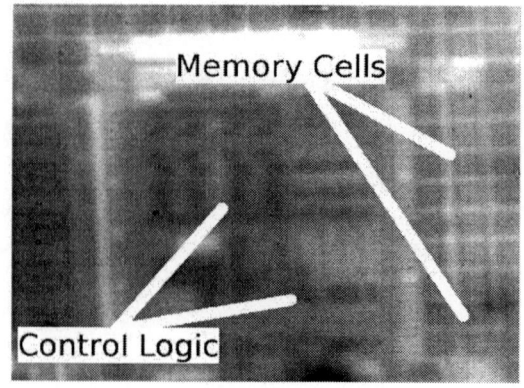

(a)   (b)

Fig. 4: (a) A close-up of the setup for two-photon absorption pulsed laser carrier generation on an Elpida 512 Mbit SDRAM, after [5]. The surface has to be polished to a near-specular finish, evident from the reflection of the objective lens, in order to keep optical losses to a minimum. (b) Infrared image of the SDRAM control logic and memory cells as viewed through the microscope optics.

empirical stopping power formulations that rely on these data, such as SRIM, PSTAR, and GEANT4 [14]. The same difficulties present in measuring the stopping power are also present in soft error testing.

There are generally two options for accelerating proton beams: Van de Graaff accelerators and cyclotrons. Van de Graaff accelerators have much tighter energy spectrums than cyclotrons – a few kiloelectron volts wide versus several hundred kiloelectron volts. However, cyclotrons offer the benefits of in-air irradiation and higher energies. The spread in beam tune energy matters since the protons at the Bragg peak, around 50 keV in silicon, generate the most charge and are therefore the most likely to cause soft errors. Since protons at the Bragg peak have a range of approximately 0.5 μm in silicon, the soft error cross section effect is sharp and dramatic. A beam that has a large energy spread will smear out the dramatic increase in soft errors at low proton energies for susceptible technologies. This makes interpreting results and mechanisms difficult if not impossible. As a general rule, more mass between the tuned beam and the device under test will result in poor energy resolution and less conclusive data. The best scenario is to tune the beam to the exact energy desired and to avoid external degraders. B. D. Sierawski *et al.* [4] has a nice example of this effect shown in their Fig. 13.

The radiation effects community is moving in several directions with regard to low-energy proton testing and soft error rate evaluation. B. D. Sierawski *et al.* [4] advocate characterizing the device under test with high-energy, low-LET, light ions like helium and nitrogen that have LETs close to low-energy protons. Using these ions provides a well-defined electronic stopping power as the ion traverses the device and makes model calibration easier; *cf.* [12, 28, 29]. Other groups, like D. F. Heidel *et al.* [2, 3], are continuing to pursue improved low-energy proton irradiation techniques that reduce systematic errors and unlock underlying soft error mechanisms.

III. EVALUATING SPACECRAFT MEMORY TECHNOLOGIES

Spacecraft memory has gone through several evolutions, from magnetic core, also known as Forrester core, memory in the 1960s and 1970s, to magnetic tape memory in the 1970s and 1980s, and finally to silicon solid-state recorders and other applications in the 1990s and beyond. Current technologies include both volatile and non-volatile random access memory (RAM). For space use, volatile memories consist of dynamic random access memory (DRAM) and SRAM. Non-volatile memory currently in use is limited to NAND flash, but there are many varieties of non-volatile memory currently under investigation for space applications.

From a radiation effects perspective, the only current radiation-hardened memory solutions are SRAMs. Of these, the largest amount of memory per die is 16 Mbit. Radiation-hardened computer offerings still use this type of memory extensively; however some designs have transitioned to DRAM. Due to memory size and power limitations, flight projects transitioned their solid-state recorders from SRAM to DRAM in the mid-1990s. Synchronous DRAMs (SDRAMs) are currently in-flight, with many designs using double data rate (DDR) and DDR2 interfaces. SDRAM is dense and low-power, making it ideal for mass storage applications that need to accommodate fast access times.

However, soft error evaluations of SDRAM indicate that they suffer from cross-contamination from multiple error modes, testability issues related to packaging, a large number of functional modes, soft error latency, and test data repeatability [5, 21, 30-36]. Despite all these drawbacks, commercial SDRAMs are indispensable for mass data storage in solid-state recorders. Though, due to soft error-induced data loss, they are not usually used to store mission-critical information.

Soft error data loss in SDRAMs can occur through single- and multiple-bit upset, control logic errors, block errors, and single-event functional interrupts (SEFIs). The classification categories are not standardized and vary among test groups [5, 32-34]. Block errors and SEFIs can circumvent error detection

978-1-4244-5430-3/10 $26.00 © 2010 IEEE   770

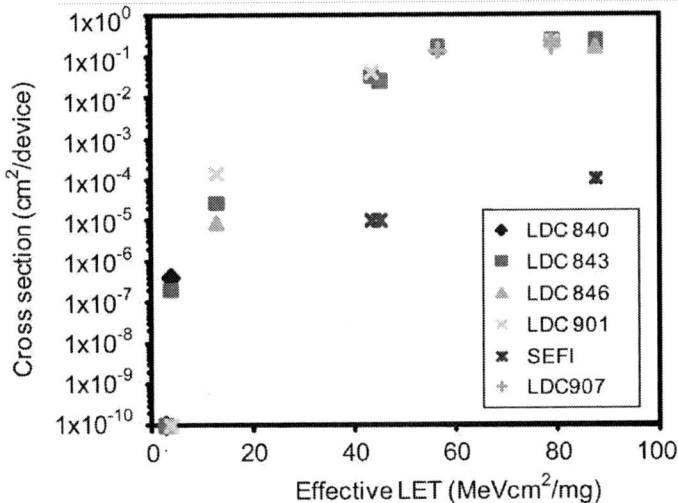

Fig. 5: Static bit-error and SEFI cross sections for five Samsung K9F4G08U0A 4 Gbit single-level cell NAND flash lots, after [6]. T. R. Oldham *et al.* [6] conducted these irradiations at the Texas A&M Cyclotron Facility.

and correction schemes without periodic scrubbing, and in some cases require power cycling causing complete data loss. Since mass memory is assembled into 3-dimensional stacks, having to recycle one die in the stack means losing the data in the entire stack, which can affect other data words if they are split across multiple stacks.

The complex nature of the possible errors in SDRAMs necessitates careful soft error evaluation. The standard technique employs ground-based broadbeam heavy ion testing, but because of latency and the time it takes to read out an entire memory, it could be many seconds before the errors get registered. Broadbeam testing lacks spatial correlation, so there is no definitive connection between the ion strike location and the observed error signature. This limitation has led to testing of SDRAMs with pulsed laser sources [5, 35, 36], similar to the test shown in Fig. 4.

Two-photon absorption [37, 38] is ideally suited for testing SDRAMs since it injects photons through the backside of the die and most SDRAMs are flip-chip mounted – shown in Fig. 4(a). For two-photon absorption, the laser spot size is approximately 2 μm in diameter. While this is large compared to the size of a single memory cell, it is easy to differentiate between stimulating control logic and memory cells, as shown in Fig. 4(b). This type of testing provides a relative measure of soft error sensitivity for different portions of the device by changing the laser pulse energy.

The approaching challenge for SDRAMs in space systems is two-fold. The cost and time required to qualify a SDRAM technology means that by the time a vendor's product is approved for flight use it is nearly obsolete in the marketplace with limited availability. Furthermore, SDRAM scaling beyond the 40 nm process node faces many challenges related to the present ArF lithography process, capacitor dielectric equivalent dielectric thickness, and equivalent electric field of the capacitor dielectric [39]. The International Technology Roadmap for Semiconductors predicts these challenges on its current roadmap as soon as 2012 with no known

manufacturing solutions. This raises concerns about the current qualification efforts focused on DDR3 SDRAMs and what might replace SDRAM.

Along with SDRAMs, NAND flash is the other major component in spacecraft memory applications [6, 40-42]. Like other space memory technologies, non-volatile memories have evolved from early one-time programmable PROMs, to EPROMs, to EEPROMs, to the current generation of single- and multi-level cell NAND flash technologies. The high density of NAND flash makes it attractive for mass storage in solid-state recorders. There are obvious power consumption benefits as well.

However, these benefits come at the cost of access time and requirements for high voltage to complete the read/erase/write cycle. The high voltage, generated with a charge pump, increases the threat of hard errors. The presence of mode registers, as with SDRAMs, means that flash devices are susceptible to SEFIs. However, flash storage density per die is much larger than SDRAM. Static (no read/write modes) heavy ion soft error cross sections for a Samsung 4 Gbit NAND flash are shown in Fig. 5. The onset LET for SEFIs is a critical parameter that will affect how the memory has to be protected on orbit.

Other types of non-volatile memory under investigation for space applications include phase-change RAM, ferroelectric RAM, resistive RAM, spin-torque transfer RAM – a type of magnetoresistive memory, and carbon nanotube RAM. A clear leader has not yet emerged from this group, but access time and thermal stability are key issues for a technology if it does replace SDRAMs in the coming years.

IV. DISCUSSION AND SUMMARY

For commercial technology soft errors, low-energy protons represent one of the greatest mitigation threats due to sheer abundance in the space environment. The radiation effects community must devise a clear and uniform path forward to test, evaluate, and predict the consequences of low-energy proton direct ionization soft errors. Ground testing with low-energy protons will be an inextricable part of the process, but modeling and simulation will also play an important role. The MRED [14] and NOVICE [43-46] codes are two examples that might hold a solution given proper experimental data constraints. The debate at this point revolves around how to calibrate the simulation models – with high-energy light heavy ions or directly with low-energy protons. The answer may depend on differences in interaction mechanisms and track structure between protons and alpha particles.

Along the same path, we need a soft error simulation solution that's not dependent on technology intellectual property or destructive physical analysis. Most of the detailed soft error calculations that have been done to date – *i.e.*, [10, 28, 47-49] – rely on detailed technology information. This is often not available or affordable to obtain for standard commercial products. The radiation effects community needs to work towards a general-purpose modeling technique that can be applied to a variety of problems while maintaining predictive power based on limited data.

978-1-4244-5430-3/10 $26.00 © 2010 IEEE

Constraining soft error analysis parameter spaces leads to the inevitable problem of incomplete state space coverage when doing ground-based heavy ion or proton testing. With the possible exception of SRAMs, most modern commercial memory technologies like SDRAMs and NAND flash have too many operational modes to have full test coverage when ion species, tilt and roll angles, biasing, data patterns, and temperature are incorporated. K. A. LaBel *et al.* [50] calculated the costs of a scaled-down 1 Gbit SDRAM heavy ion test at the Texas A&M University Cyclotron facility and arrived at approximately $80,000 for 16 hours of beam time. Full state space coverage would require several years of continuous testing. This means that any testing has to be application specific without omitting important variables that could affect on-orbit operation. Effective modeling and simulation approaches could alleviate some of this burden by prescribing the data needed to optimally constrain subsequent simulations used to extrapolate coverage to more of the operational state space.

In lieu of heavy ion testing, pulsed laser irradiation has grown as an evaluation technique, largely due to the work of several groups [37, 38, 51-53]. While there have been several direct comparisons of pulsed laser data to heavy ion data, *cf.* [54-56], the pulsed laser technique's speed, spatial correlation, and ease of energy adjustment are the most valuable features. However, as technologies scale, the once relatively small laser spot size of 1-2 μm is now large compared to single transistors meaning that it is impossible to probe single devices in 65 and 45 nm process technologies with current pulsed laser techniques. There is movement in the radiation effects community to overcome this difficulty, perhaps with the use of solid immersion lenses [57], though a practical solution has not yet been demonstrated.

The challenge of evaluating soft errors in commercial technologies for space applications has been and will continue to be centered on memory technologies. Aside from FPGAs and microprocessors, memory technology represents one of the fastest moving semiconductor development sectors and an ideal point for space systems to leverage commercial non-recurring engineering and technological advances. However, this rapid development cycle puts the space electronics community at a disadvantage due to the inherent dichotomy in development cycle time constants. Spacecraft memory choices in the next several years will likely continue their transition to more non-volatile technologies as the search for one or more SDRAM successors continues.

ACKNOWLEDGMENT

The authors wish to thank the members of the NASA/GSFC Radiation Effects and Analysis Group and their many collaborators who have provided great opportunities for research and furthered spacecraft electronics' radiation reliability. The authors are grateful for the ongoing support of the Defense Threat Reduction Agency, Lt. Col. Warren Nuibe USAF, Maj. Eric Heigel USAF, and Lewis Cohn.

REFERENCES

[1] K. P. Rodbell, D. F. Heidel, H. H. K. Tang, M. S. Gordon, P. Oldiges, and C. E. Murray, "Low-Energy Proton-Induced Single-Event-Upsets in 65 nm node, Silicon-on-Insulator, Latches and Memory Cells," vol. 54, no. 6, pp. 2474-2479, Dec. 2007.

[2] D. F. Heidel, P. W. Marshall, K. A. LaBel, J. R. Schwank, K. P. Rodbell, M. C. Hakey, M. D. Berg, P. E. Dodd, M. R. Friendlich, A. D. Phan, C. M. Seidleck, M. R. Shaneyfelt, and M. A. Xapsos, "Low energy proton single-event upset test results on 65 nm SOI SRAM," *IEEE Trans. Nucl. Sci.*, vol. 55, no. 6, pp. 3394-3400, Dec. 2008.

[3] D. F. Heidel, P. W. Marshall, J. A. Pellish, K. P. Rodbell, K. A. LaBel, J. R. Schwank, S. E. Rauch, M. C. Hakey, M. D. Berg, C. M. Castaneda, P. E. Dodd, M. R. Friendlich, A. D. Phan, C. M. Seidleck, M. R. Shaneyfelt, and M. A. Xapsos, "Single-event upsets and multiple-bit upsets on a 45 nm SOI SRAM," *IEEE Trans. Nucl. Sci.*, vol. 56, no. 6, pp. 3499-3504, Dec. 2009.

[4] B. D. Sierawski, J. A. Pellish, R. A. Reed, R. D. Schrimpf, K. M. Warren, R. A. Weller, M. H. Mendenhall, J. D. Black, A. D. Tipton, M. A. Xapsos, R. C. Baumann, D. Xiaowei, M. J. Campola, M. R. Friendlich, H. S. Kim, A. M. Phan, and C. M. Seidleck, "Impact of low-energy proton induced upsets on test methods and rate predictions," *IEEE Trans. Nucl. Sci.*, vol. 56, no. 6, pp. 3085-3092, Dec. 2009.

[5] R. L. Ladbury, J. Benedetto, D. McMorrow, S. P. Buchner, K. A. LaBel, M. D. Berg, H. S. Kim, A. B. Sanders, M. R. Friendlich, and A. Phan, "TPA laser and heavy-ion SEE testing: complementary techniques for SDRAM single-event evaluation," *IEEE Trans. Nucl. Sci.*, vol. 56, no. 6, pp. 3334-3340, Dec. 2009.

[6] T. R. Oldham, M. R. Friendlich, A. B. Sanders, C. M. Seidleck, H. S. Kim, M. D. Berg, and K. A. LaBel, "TID and SEE response of advanced Samsung and Micron 4G NAND flash memories for the NASA MMS mission," in *IEEE Radiation Effects Data Workshop*, Quebec City, Quebec Canada, 2009, pp. 114-122.

[7] M. Gates. *Single Event Effect Criticality Analysis*. Available: http://radhome.gsfc.nasa.gov/radhome/papers/seecai.htm

[8] R. Ladbury, J. L. Gorelick, M. A. Xapsos, T. O'Connor, and S. Demosthenes, "A Bayesian treatment of risk for radiation hardness assurance," in *8th European Conference on Radiation and Its Effects on Components and Systems*, Cap d'Agde, France, 2005, pp. PB1-1-PB1-8.

[9] H. H. K. Tang and E. H. Cannon, "SEMM-2: a modeling system for single event upset analysis," *IEEE Trans. Nucl. Sci.*, vol. 51, no. 6, pp. 3342-3348, Dec. 2004.

[10] R. A. Reed, R. A. Weller, M. H. Mendenhall, J.-M. Lauenstein, K. M. Warren, J. A. Pellish, R. D. Schrimpf, B. D. Sierawski, L. W. Massengill, P. E. Dodd, N. F. Haddad, R. K. Lawrence, J. H. Bowman, and R. Conde, "Impact of ion energy and species on single event effects analysis," *IEEE Trans. Nucl. Sci.*, vol. 54, no. 6, Dec. 2007.

[11] V. Ferlet-Cavrois, D. McMorrow, D. Kobayashi, N. Fel, J. S. Melinger, J. R. Schwank, M. Gaillardin, V. Pouget, F. Essely, J. Baggio, S. Girard, O. Flament, P. Paillet, R. S. Flores, P. E. Dodd, M. R. Shaneyfelt, K. Hirose, and H. Saito, "A new technique for SET pulse width measurement in chains of inverters using pulsed laser irradiation," *IEEE Trans. Nucl. Sci.*, vol. 56, no. 4, pp. 2014-2020, Aug. 2009.

[12] G. Hubert, S. Duzellier, C. Inguimbert, C. Boatella-Polo, F. Bezerra, and R. Ecoffet, "Operational SER calculations on the SAC-C orbit using the multi-scales single event phenomena predictive platform (MUSCA SEP^3)," *IEEE Trans. Nucl. Sci.*, vol. 56, no. 6, pp. 3032-3042, Dec. 2009.

[13] H. Park, D. J. Cummings, R. Arora, J. A. Pellish, R. A. Reed, R. D. Schrimpf, D. McMorrow, S. E. Armstrong, U. Roh, T. Nishida, M. E. Law, and S. E. Thompson, "Laser-induced current transients in strained-Si diodes," *IEEE Trans. Nucl. Sci.*, vol. 56, no. 6, pp. 3203-3209, Dec. 2009.

[14] R. A. Weller, R. A. Reed, K. M. Warren, M. H. Mendenhall, B. D. Sierawski, R. D. Schrimpf, and L. W. Massengill, "General framework for single event effects rate prediction in microelectronics," *IEEE Trans. Nucl. Sci.*, vol. 56, no. 6, pp. 3098-3108, Dec. 2009.

[15] D. Binder, E. C. Smith, and A. B. Holman, "Satellite anomalies from galactic cosmic rays," *IEEE Trans. Nucl. Sci.*, vol. 22, no. 6, pp. 2675-2680, Dec. 1975.

[16] R. A. Reed, P. W. Marshall, H. S. Kim, P. J. McNulty, B. Fodness, T. M. Jordan, R. Reedy, C. Tabbert, M. S. T. Liu, W. Heikkila, S. Buchner, R. Ladbury, and K. A. LaBel, "Evidence for angular effects in proton-induced single-event upsets," *IEEE Trans. Nucl. Sci.*, vol. 49, no. 6, pp. 3038-3044, Dec. 2002.

[17] R. A. Nymmik, M. I. Panasyuk, T. I. Pervaja, and A. A. Suslov, "A model of galactic cosmic ray fluxes," *Nucl. Tracks Radiat. Meas.*, vol. 20, no. 3, pp. 427-429, 1992.

[18] M. A. Xapsos, G. P. Summers, J. L. Barth, E. G. Stassinopoulos, and E. A. Burke, "Probability model for cumulative solar proton event fluences," *IEEE Trans. Nucl. Sci.*, vol. 47, no. 3, pp. 486-490, Jun. 2000.

[19] M. A. Xapsos, C. Stauffer, G. B. Gee, J. L. Barth, E. G. Stassinopoulos, and R. E. McGuire, "Model for solar proton risk assessment," *IEEE Trans. Nucl. Sci.*, vol. 51, no. 6, pp. 3394-3398, Dec. 2004.

[20] J. W. Howard, H. Kim, M. Berg, K. A. LaBel, S. Stansberry, M. Friendlich, and T. Irwin, "Development of a low-cost and high-speed single event effects testers based on reconfigurable field programmable gate arrays (FPGA)," presented at the SEE Symposium, Long Beach, CA, 2006. Available: http://ntrs.nasa.gov/archive/nasa/casi.ntrs.nasa.gov/20060028135_2006228899.pdf

[21] R. Ladbury, M. D. Berg, K. A. LaBel, M. Friendlich, A. Phan, and H. Kim, "Radiation performance of 1 Gbit DDR2 SDRAMs fabricated with 80-90 nm CMOS," in *IEEE Radiation Effects Data Workshop*, Tucson, AZ, 2008, pp. 42-46.

[22] H. Paul and A. Schinner, "Judging the reliability of stopping power tables and programs for protons and alpha particles using statistical methods," *Nucl. Instr. and Meth. B*, vol. 227, no. 4, pp. 461-470, 2005.

[23] H. Paul. (2010). *Stopping Power for Light Ions: Graphs, Data, Comments and Programs*. Available: http://www.exphys.jku.at/stopping/

[24] J. F. Zeigler, "SRIM-2003," *Nucl. Instr. and Meth. B*, vol. 219-220, pp. 1027-1036, Jun. 2004.

[25] J. F. Zeigler and J. P. Biersack. (2010). *Stopping and Range of Ions in Matter*. Available: http://www.srim.org/

[26] M. J. Berger, J. S. Coursey, M. A. Zucker, and J. Chang. (2005). *ESTAR, PSTAR, and ASTAR: computer programs for calculating stopping-power and range tables for electrons, protons, and helium Ions (version 1.2.3)*. Available: http://physics.nist.gov/Star

[27] "Stopping Powers and Ranges for Protons and Alpha Particles," International Commission on Radiation Units and Measurements Report 49, May 1993.

[28] K. M. Warren, A. L. Sternberg, R. A. Weller, M. P. Baze, L. W. Massengill, R. A. Reed, M. H. Mendenhall, and R. D. Schrimpf, "Integrating Circuit Level Simulation and Monte-Carlo Radiation Transport Code for Single Event Upset Analysis in SEU Hardened Circuitry," *IEEE Trans. Nucl. Sci.*, vol. 55, no. 6, pp. 2886-2894, Dec. 2008.

[29] K. M. Warren, R. A. Weller, B. D. Sierawski, R. A. Reed, M. H. Mendenhall, R. D. Schrimpf, L. W. Massengill, M. Porter, J. Wilkinson, K. A. LaBel, and J. H. Adams Jr, "Application of RADSAFE to Model Single Event Upset Response of 0.25 μm CMOS SRAM," *IEEE Trans. Nucl. Sci.*, vol. 54, no. 4, pp. 898-903, Aug. 2007.

[30] K. A. LaBel, P. W. Marshall, J. L. Barth, R. B. Katz, R. A. Reed, H. W. Leidecker, H. S. Kim, and C. J. Marshall, "Anatomy of an in-flight anomaly: investigation of proton-induced SEE test results for stacked IBM DRAMs," *IEEE Trans. Nucl. Sci.*, vol. 45, no. 6, pp. 2898-2903, Dec. 1998.

[31] T. Langley, R. Koga, and T. Morris, "Single-event effects test results of 512MB SDRAMs," in *IEEE Radiation Effects Data Workshop*, Monterey, CA, 2003, pp. 98-101.

[32] J. Benedetto, "Examination of single event functional interrupts (SEFIs) in COTS SDRAMs," presented at the Single-Event Effects Symp., Long Beach, CA, 2006.

[33] R. K. Lawrence, "Radiation characterization of 512 Mb SDRAMs," in *IEEE Radiation Effects Data Workshop*, Honolulu, HI, 2007, pp. 204-207.

[34] J. Benedetto, J. D. Black, and G. Ott, "Soft error case study: Single event functional interrupts (SEFIs) in COTS SDRAMs," presented at the IEEE Nuclear and Space Radiation Effects Conf. Short Course, Tucson, AZ, 2008.

[35] A. Bougerol, F. Miller, and N. Buard, "SDRAM architecture & single event effects revealed with laser," in *IEEE Int. On-Line Testing Symp.*, Rhodes, Greece, 2008, pp. 283-288.

[36] A. Bougerol, F. Miller, N. Guibbaud, R. Gaillard, F. Moliere, and N. Buard, "Use of laser to explain heavy ion induced SEFIs in SDRAMs," *IEEE Trans. Nucl. Sci.*, vol. 57, no. 1, pp. 272-278, Jan. 2010.

[37] D. Lewis, V. Pouget, F. Beaudoin, P. Perdu, H. Lapuyade, P. Fouillat, and A. Touboul, "Backside laser testing of ICs for SET sensitivity evaluation," *IEEE Trans. Nucl. Sci.*, vol. 48, no. 6, pp. 2193-2201, Dec. 2001.

[38] D. McMorrow, W. T. Lotshaw, J. S. Melinger, S. Buchner, and R. L. Pease, "Subbandgap laser-induced single event effects: carrier generation via two-photon absorption," *IEEE Trans. Nucl. Sci.*, vol. 49, no. 6, pp. 3002-3008, Dec. 2002.

[39] *International Technology Roadmap for Semiconductors*. Available: http://www.itrs.net/Links/2009ITRS/Home2009.htm

[40] F. Irom and D. N. Nguyen, "Single event effect characterization of high density commercial NAND and NOR nonvolatile flash memories," *IEEE Trans. Nucl. Sci.*, vol. 54, no. 6, pp. 2547-2553, Dec. 2007.

[41] T. R. Oldham, R. L. Ladbury, M. Friendlich, H. S. Kim, M. D. Berg, T. L. Irwin, C. Seidleck, and K. A. LaBel, "SEE and TID characterization of an advanced commercial 2 Gbit NAND flash nonvolatile memory," *IEEE Trans. Nucl. Sci.*, vol. 53, no. 6, pp. 3217-3222, Dec. 2006.

[42] H. R. Schwartz, D. K. Nichols, and A. H. Johnston, "Single-event upset in flash memories," *IEEE Trans. Nucl. Sci.*, vol. 44, no. 6, pp. 2315-2324, Dec. 1997.

[43] T. M. Jordan, "An adjoint charged particle transport method," *IEEE Trans. Nucl. Sci.*, vol. 23, no. 6, pp. 1857-1861, Dec. 1976.

[44] T. M. Jordan, "The accuracy of NOVICE electron shielding calculations," in *1st European Conf. on Radiation and Its Effects on Components and Systems*, La Grande-Motte, France, 1991, pp. 320-324.

[45] T. M. Jordan, "Space system analysis using the NOVICE code system," in *Radiation and its Effects on Devices and Systems, 1991. RADECS 91., First European Conference on*, 1991, pp. 312-316.

[46] P. Calvel, C. Barillot, A. Porte, G. Auriel, C. Chatry, P. F. Peyrard, G. Santin, R. Ecoffet, and T. M. Jordan, "Review of Deposited Dose Calculation Methods Using Ray Tracing Approximations," *IEEE Trans. Nucl. Sci.*, vol. 55, no. 6, pp. 3106-3113, Dec. 2008.

[47] K. M. Warren, R. A. Weller, M. H. Mendenhall, R. A. Reed, D. R. Ball, C. L. Howe, B. D. Olson, M. L. Alles, L. W. Massengill, R. D. Schrimpf, N. F. Haddad, S. E. Doyle, D. McMorrow, J. S. Melinger, and W. T. Lotshaw, "The contribution of nuclear reactions to heavy ion single event upset cross-section measurements in a high-density SEU hardened SRAM," *IEEE Trans. Nucl. Sci.*, vol. 52, no. 6, pp. 2125-2131, Dec. 2005.

[48] K. M. Warren, B. D. Sierawski, R. A. Reed, R. A. Weller, C. Carmichael, A. Lesea, M. H. Mendenhall, P. E. Dodd, R. D. Schrimpf, L. W. Massengill, H. Tan, W. Hsing, J. L. De Jong, R. Padovani, and J. J. Fabula, "Monte-Carlo based on-orbit single event upset rate prediction for a radiation hardened by design latch," *IEEE Trans. Nucl. Sci.*, vol. 54, no. 6, pp. 2419-2425, Dec. 2007.

[49] A. D. Tipton, J. A. Pellish, J. M. Hutson, R. Baumann, X. Deng, A. Marshall, M. A. Xapsos, H. S. Kim, M. R. Friendlich, M. J. Campola, C. M. Seidleck, K. A. LaBel, M. H. Mendenhall, R. A. Reed, R. D. Schrimpf, R. A. Weller, and J. D. Black, "Device-Orientation Effects on Multiple-Bit Upset in 65 nm SRAMs," vol. 55, no. 6, pp. 2880-2885, Dec. 2008.

[50] K. A. LaBel, R. L. Ladbury, L. M. Cohn, and T. R. Oldham, "Radiation test challenges for scaled commercial memories," *IEEE Trans. Nucl. Sci.*, vol. 55, no. 4, pp. 2174-2180, Aug. 2008.

[51] S. P. Buchner, D. Wilson, K. Kang, D. Gill, J. A. Mazer, W. D. Raburn, A. B. Campbell, and A. R. Knudson, "Laser simulation of

single event upsets," *IEEE Trans. Nucl. Sci.,* vol. 34, no. 6, pp. 1227-1233, Dec. 1987.

[52] J. S. Melinger, S. Buchner, D. McMorrow, W. J. Stapor, T. R. Weatherford, A. B. Campbell, and H. Eisen, "Critical evaluation of pulsed laser method for single-event effects testing and fundamental studies," *IEEE Trans. Nucl. Sci.,* vol. 41, no. 6, pp. 2574-2584, Dec. 1994.

[53] A. Douin, V. Pouget, F. Darracq, D. Lewis, P. Fouillat, and P. Perdu, "Influence of Laser Pulse Duration in Single Event Upset Testing," in *8th European Conference on Radiation and Its Effects on Components and Systems,* Cap d'Agde, France, 2005, pp. C13-1-C13-7.

[54] S. Buchner, K. Kang, W. J. Stapor, A. B. Campbell, A. R. Knudson, P. McDonald, and S. Rivet, "Pulsed laser-induced SEU in integrated circuits: a practical method for hardness assurance testing," *IEEE Trans. Nucl. Sci.,* vol. 37, no. 6, pp. 1825-1831, Dec. 1990.

[55] D. McMorrow, J. S. Melinger, S. Buchner, T. Scott, R. D. Brown, and N. F. Haddad, "Application of a pulsed laser for evaluation and optimization of SEU-hard designs," *IEEE Trans. Nucl. Sci.,* vol. 47, no. 3, pp. 559-565, Jun. 2000.

[56] S. Buchner, D. McMorrow, C. Poivey, J. Howard, Y. Boulghassoul, L. W. Massengill, R. Pease, and M. Savage, "Comparison of single-event transients induced in an operational amplifier (LM124) by pulsed laser light and a broad beam of heavy ions," *IEEE Trans. Nucl. Sci.,* vol. 51, no. 5, pp. 2776-2781, Oct. 2004.

[57] K. A. Serrels, E. Ramsay, and D. T. Reid, "70 nm resolution in subsurface optical imaging of silicon integrated-circuits using pupil-function engineering," *Appl. Phys. Lett.,* vol. 94, no. 7, p. 073113, Feb. 2009.

Analysis of the Breakdown Spots Spatial Distribution in Large Area MOS Structures

E. Miranda

Departament d'Enginyeria Electrónica
Escola d'Enginyeria, Universitat Autònoma de Barcelona
Barcelona, Spain
phone: 34-93-5813183, email: enrique.miranda@uab.es

E. O'Connor and P.K. Hurley
Tyndall National Institute, University College Cork
Cork, Ireland

Abstract—The spatial distribution of multiple breakdown (BD) spots in large area MOS structures was investigated. By means of applying image processing and point pattern analysis techniques we provide for the first time direct evidence that the BD spots' locations are spatially uncorrelated as expected for a Poisson process. For completeness, we show how the available mathematical tools might be utilized to detect interactions (repulsion or attraction) between the spots as well as weak regions in the dielectric layer. In this way, the methods considered here can complement standard reliability techniques based on a large set of samples.

Keywords: breakdown, reliability, high-κ, MOS

I. INTRODUCTION

In a recent paper Alam *et al* [1] raised the question whether the BD spot locations in MOS devices are spatially correlated or they truly follow a *complete spatial randomness* (CSR) model, as a Poisson process is frequently termed in spatial statistics [2]. The issue has major reliability implications since the transistor lifetime is sensitive to the degree of spatial and temporal correlations among BD spots. In spite of the fact that the temporal correlation has received extensive attention in the last years [3,4], the problem of the 2-D spatial distribution is still an open question that needs to be solved. The interaction between spots may arise because of local variations of the potential distribution in the semiconductor substrate in the vicinity of the leakage site [1,5], or by local enhancements of the trap generation rate [3]. In order to assess whether the BD spots follow a Poisson process, two approaches are possible: first, one can infer the spatial distribution from the first BD event statistics in many devices or alternatively, one can analyze the generation of a large number of BD spots in a single sample. Regarding the first method, the fact that the Weibull slopes (β) corresponding to different gate areas are independent of this parameter is well known to be a consequence of the Poisson area scaling [6]. However, this approach relies on the certainty that there is only a single BD event per device, which precludes the possibility of exploring the interaction between spots. In connection with the second

method, in [1], the correlation between spots was evaluated using simulated data compatible with the current distributions in a four-terminal device fabricated to that aim. Again the reported results led to the conclusion that the BD spots do not exhibit spatial correlation and therefore that they are well represented by a CSR process.

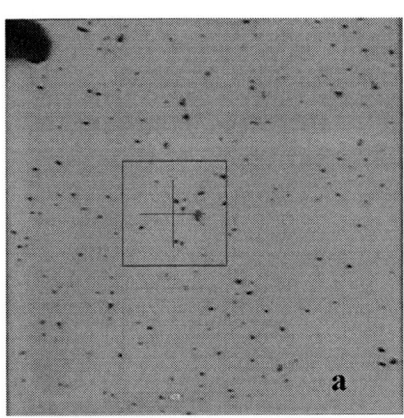

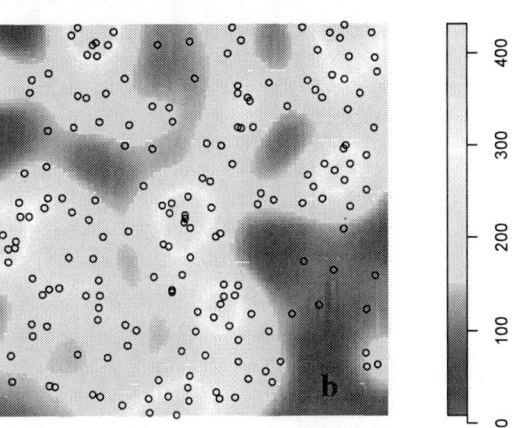

Figure 1. a) Photograph of the gate electrode showing multiple BD events. b) Digitalized image with an intensity plot. The dark shadow in the upper left corner is the voltage probe needle. The size of the sample is 400x400 μm².

In this work, we have investigated the 2-D distribution of BD spots again but contrary to previous analysis our conclusions are strictly drawn from data obtained by direct observation. Furthermore, we show that, beyond the distance distribution, a number of techniques frequently used in spatial point pattern analysis can be applied to characterize the multiple dielectric BD phenomenon. This kind of study has been made possible thanks to the recent development of specialized software packages designed for this purpose [7]. It is also worth mentioning that the investigation has been carried out in a high-κ dielectric material deposited on a high mobility compound semiconductor with metal gate, which is the material combination expected for the next generation of MOS transistors [8].

II. THE SAMPLES

The samples used in this study are MgO films with nominal thickness t_{ox}=20 nm. The films were deposited by e-beam evaporation from 99.9% MgO pellets at a rate of 0.2 Å/s at 180 °C on p- and n-type InP substrates. The MgO/InP samples were capped in-situ with 100 nm of amorphous silicon (α-Si) using a second e-beam source. For the NiSi gate process, nickel was deposited by e-beam evaporation (~80 nm) through a patterned resist mask followed by a lift-off process. The rapid thermal annealing is a one step process at 500 °C for 30 s in N_2. For this study we utilized the largest area available of $1.6x10^{-3}cm^2$. Further details about the sample fabrication and conduction characteristics can be found in [9]. It is worth pointing out that the spots can also be observed with Si substrate but in this case the marks in the metal electrode are less visible and require image treatment.

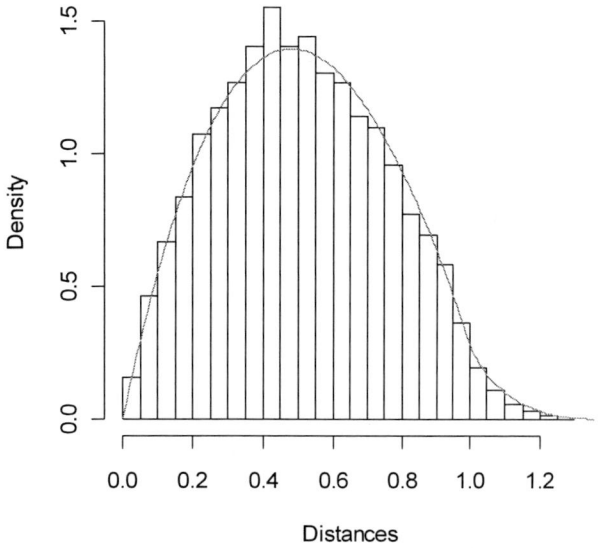

Figure 2. Experimental (histogram) and theoretical (red solid line) probability distribution for the distances between the points shown in Fig.1b. Notice that the distances ranges from 0 to 1.41, the latter being the length of the diagonal of the unit square.

III. EXPERIMENTAL RESULTS AND DISCUSSION

In order to generate a large number of BD spots the device was subject to a voltage ramp (up to -13 V). Fig. 1a shows a typical photograph of a stressed device without any image treatment. Although the spot sizes are dissimilar and therefore their conducting properties are different as well, for the analysis that follows all the BD spots are treated on equal grounds as a point pattern dataset. Additionally, the lateral dimensions of the device were normalized to unity. Fig. 1b is the digitalized image of Fig. 1a with 182 detected spots. The color map corresponds to the local intensity estimator (weighted by an isotropic Gaussian smoothing kernel), which is the average density of points (expected number of points per unit area). Notice that for this particular realization the intensity varies from location to location, but this is consistent with a homogeneous Poisson process in which the points are not uniformly spread but there are empty gaps and clusters of points. The observed point pattern does not seem to be affected by the location of the voltage probe needle.

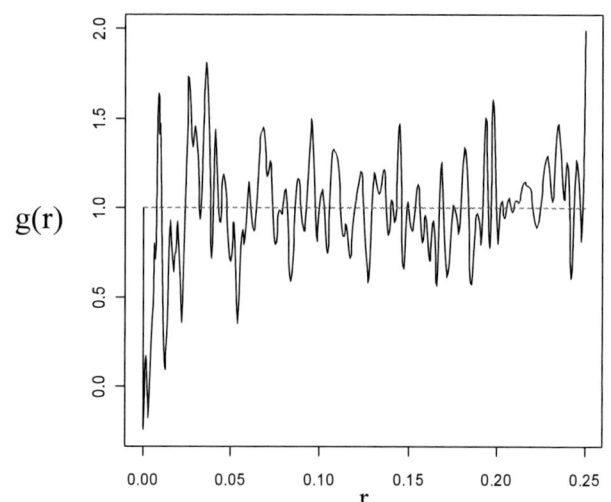

Figure 3. Pair correlation function as a function of r.

The histogram in Fig. 2 shows the distribution of the distances d between all experimental points ($(182^2-182)/2$=16471 values) and the theoretical probability density (red line) given by the expression [10]:

$$f(d) = \begin{cases} 4d\left[\dfrac{\pi}{2} - 2d + \dfrac{d^2}{2}\right] & 0 \le d \le 1 \\[2mm] 4d\left[\arcsin\left(\dfrac{1}{d}\right) - \arccos\left(\dfrac{1}{d}\right) - 1 - \dfrac{d^2}{2} + 2\sqrt{d^2-1}\right] & 1 < d \le \sqrt{2} \end{cases}$$

(1)

This latter function is reached under the hypothesis that the x- and y-coordinates of each data point are uniformly distributed in the range [0,1]. It must be noted that the agreement is very good, which is indicative of a CSR process.

In order to confirm that this is definitely the case, we have performed an alternative statistical test that involves analyzing the interpoint interaction using the pair correlation function $g(r)$, with r a generic radius. This second-order summary characteristic is defined as:

$$g(r) = \frac{1}{2\pi r}\frac{dK(r)}{dr}\begin{cases} =1 & r \geq 0 & \textit{CRS case} \\ \geq 1 & \textit{small r} & \textit{cluster process} \\ \leq 1 & \textit{small r} & \textit{regular process} \\ = 0 & r \leq r_0 & \textit{hard core potential} \end{cases} \quad (2)$$

where $K(r)$ is Ripley's function [2], which is a measure of the expected number of points at many distance scales (for an homogeneous Poisson process $K(r)=\pi r^2$). Fig. 3 illustrates the estimated $g(r)$ for the image shown in Fig. 1b. Notice that the data distribution is consistent with the CRS case (red dashed line) except in a very small range close to the origin (distance of about 4μm), but this is in the range of our digitalization system resolution.

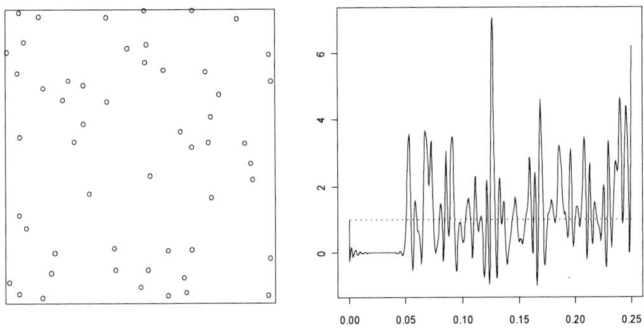

Figure 4. g-plot corresponding to a repulsion process.

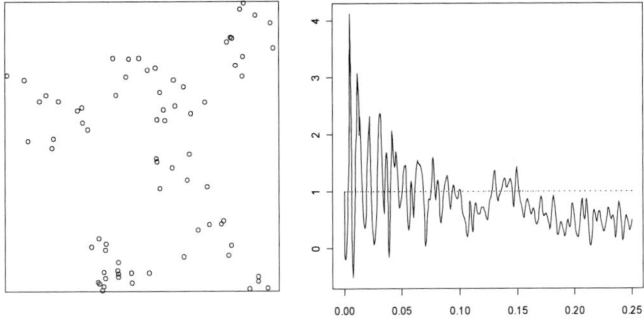

Figure 5. g-plot corresponding to a cluster process.

To demonstrate the power of these techniques, we illustrate two examples that could help to identify situations in which the correlation between BD spots [5] or departures from the Poisson area scaling because of the layout particular features [11,12] cannot be neglected. Figure 4 shows the realization of a Matérn's Model I process with $r_0=0.05$ [7]. The g-plot clearly reveals the existence of a hard core repulsion potential, *i.e.* the events cannot occur within a minimum distance of each

other. On the contrary, Fig. 5 corresponds to a cluster process with a characteristic length of $r=0.1$. Note that without using the g estimator it would be almost impossible to detect the underlying point pattern structure.

IV. CONCLUSIONS

By means of applying spatial point pattern analysis techniques we were able to demonstrate that the locations of the BD spots in a MOS device are uncorrelated. If we associate the generation of a leakage site with the creation of defects, we can affirm that this creation is also uniform, which in turn is an indicator of the quality of the dielectric layer. There are no weak regions. Ultimately, the appearance of the spots is a consequence of the 2-D weakest link character of the BD statistics.

ACKNOWLEDGMENT

E.M. acknowledges the support of the Spanish Ministry of Science and Technology under contract number TEC2009-09350 and the Departament d'Universitats, Recerca i Societat de la Informació de la Generalitat de Catalunya under contract number 2009SGR783. P.K.H. acknowledges the SFI grant (05/IN/1751). The authors acknowledge D. O'Connell from Tyndall National Institute for samples processing.

REFERENCES

[1] M. Alam, D. Varghese, and B. Kaczer, "Theory of breakdown position determination by voltage- and current-ratio methods", IEEE Trans Elect Dev 55, 3150 (2008)

[2] J. Illian, A. Penttinen, H. Stoyan, and D. Stoyanet, in Statistical analysis and modelling of spatial point patterns, Wiley, 2008

[3] M. Alam and R. Smith, "A phenomenological theory of correlated multiple soft-breakdown events in ultra-thin gate dielectrics", Proc. IRPS 2003, p. 406

[4] J. Suñé and E. Wu, "Statistics of successive breakdown events in gate oxides", IEEE Elect Dev Lett 24, 272 (2003)

[5] S. Lombardo, F. Crupi, A. La Magna, C. Spinella, A. Terrasi, and A. La Mantia, "Electrical and thermal transient during dielectric breakdown of thin oxides in metal-SiO2-silicon capacitors", J Appl Phys 84, 472 (1998)

[6] E. Wu, J. Stathis, and L. Hanet, "Ultra-thin oxide reliability for ULSI applications", Semicond Sci Technol 15, 425 (2000)

[7] A. Baddeley and R. Turner, "Spatstat: An R Package for Analyzing Spatial Point Patterns", J Stat Software 12, 1 (2005)

[8] A. Dimoulas, E. Gusev, P. McIntyre, and M. Heyns (Eds.), in Advanced gate stacks for high-mobility semiconductors, Springer, 2007

[9] E. Miranda, E. O'Connor, G. Hughes, P. Casey, K. Cherkaoui, S. Monaghan, R. Long, D. O'Connell, and P.K. Hurley, "Effects of the electrical stress on the conduction characteristics of metal gate/MgO/InP stacks", Mic Rel 49, 1052 (2009)

[10] A. Mathai, R. Moschopoulos, and G. Pederzoli, "Random points associated with rectangles", Rendiconti del Circolo Matematica di Palermo XLVIII, 163 (1999)

[11] H. Uchida, I. Aikawa, N. Hirashita, T. Ajioka, "Enhanced degradation of oxide breakdown in the peripheral region by metallic contamination", Proc. IEDM 1990, p. 405

[12] Y. Li, Zz. Tokei, Ph. Roussel, G. Groeseneken, and K Maex, "Layout dependency induced deviation from Poisson area scaling in BEOL dielectric reliability", Mic Rel 45, 1299 (2005)

New Statistical Model to Decode the Reliability and Weibull Slope of High-κ and Interfacial Layer in a Dual Layer Dielectric Stack

N. Raghavan[1, *], K.L. Pey[1, ▲], W.H. Liu[1] and X. Li[1,2]

[1]Division of Microelectronics, School of Electrical & Electronic Engineering,
Nanyang Technological University (NTU), Singapore – 639798.
[2]Institute of Microelectronics (IME), Singapore Science Park II, Singapore – 117685.
[*]Ph: (+65) 9862-1185, [▲]Fax: (+65) 6792-0415, [*]E-mail: – naga0009@ntu.edu.sg

Abstract — **Reliability study of high-κ (HK) gate dielectric based transistors has become imperative for the current and future CMOS technology nodes as the industry shifts towards replacement of conventional silicon oxynitride (SiON) with hafnium-based oxides. One of the key requirements of any oxide reliability study is a quantitative assessment of the time dependent dielectric breakdown (TDDB) lifetime using suitable statistical models. Direct extension of the simple statistical model used for SiON to the HK is complicated by the presence of the interfacial sub-oxide layer (IL, SiO_x) which is sandwiched between the HK and Si substrate. Given the dual-layer HK-IL dielectric stack, it is necessary to develop new statistical models and electrical test algorithms that can enable us to decode the reliability and Weibull slope of the individual HK and IL layers so that the relative reliability of these two layers can be studied to identify the layer which serves as a "savior" in prolonging the front end reliability of current HK based logic devices. In this study, we propose a new *cumulative damage* statistical model in conjunction with a two step voltage stress electrical test algorithm for sequential HK-IL breakdown which enables us to analyze the TDDB reliability of HK and IL separately.**

Keywords – Cumulative damage model, Grain boundary, Interfacial layer, Poole-Frenkel emission, Time dependent dielectric breakdown (TDDB), Weibull slope.

I. INTRODUCTION

In the sub 45nm technology node that we currently operate in, high-κ (HK) dielectric materials are becoming an integral part of the CMOS transistor process flow replacing the conventional SiON gate oxides in order to reduce gate leakage significantly and enable downscaling of the equivalent oxide thickness (EOT) at the same time. Some of the materials considered include HfO_2, $HfSiON$, ZrO_2, Al_2O_3, Y_2O_3 and La_2O_3 [1]. Process compatibility and reliability are the two key issues to consider during HK implementation. From a reliability point of view, it is necessary to study the time-dependent-dielectric breakdown (TDDB) phenomenon in these novel HK stacks and perform a quantitative lifetime study making use of statistical accelerated life tests and extrapolation models. As for the SiON, various models have been proposed and well established in the recent past [2] – [5]. However, direct application of the previous SiON case statistical models to the HK material is complicated by the presence of the sub-

oxide SiO_x (x < 2) interfacial layer (IL) sandwiched between the HK and silicon substrate. The presence of this IL layer (which is detrimental to the aggressive EOT scaling that we set to achieve) is unavoidable as it tends to form (by partial oxidation of the Si substrate) during the high temperature annealing and HK thin film deposition processes. Therefore, the overall gate stack comprises a dual-layer HK-IL dielectric with the HK and IL having their respective breakdown voltages and different TDDB robustness. A good reliability study for a multi-layer dielectric system should be able to decipher the role and impact of each dielectric on the overall reliability and identify the weakest link in the stack. This study aims to address this issue by making use of the cumulative damage model (CDM) [6, 7] in conjunction with the stepwise sequential "arrested" breakdown of the HK and IL layers [8]. The combination of this electrical and statistical characterization approach is instrumental in fully understanding the nature of TDDB failures in multi-layer dielectric stacks.

The novelty of the work lies in the ability of the proposed methodology to decode the reliability of the HK and IL layers from an electrical and statistical perspective. Analysis of the failure data suggest that both the HK and IL show bimodal failure distributions. The physics behind these observations will also be discussed, supported by failure analysis results.

The layout of the paper is as follows. Section II provides the details of the samples tested and the electrical stress conditions used. This is followed by a detailed insight into the electrical characterization results in Section III and verification of high-κ breakdown using the Poole-Frenkel emission mechanism. Section IV focuses on developing the new statistical model and its application to HK-IL failure data analysis based on the TDDB tests in the previous section. The physical analysis results along with a discussion of the effect of grain boundary microstructural defects on HK reliability are presented in Section V. Finally, we conclude the study with potential suggestions for further investigations aimed at gaining in-depth understanding of the physics of failure.

II. EXPERIMENTAL DETAILS

Devices tested are N-MOSFET with N^+ poly-Si gate having dimensions of W × L = 0.5μm × 0.25μm at room temperature. As shown by the high resolution transmission electron

---

This work is sponsored by the Ministry of Education (MOE), Singapore Grant No. T206B1205 and NTU RGM 33/03.

978-1-4244-5430-3/10 $26.00 © 2010 IEEE

microscope (HRTEM) image of a fresh device in Fig.1, the thickness of the HK and IL layers are 44Å and 8Å respectively. The material used for HK is polycrystalline HfO_2, which is the most common dielectric currently under study. Since the breakdown voltage (V_{BD}) for the gate stack was found to range from 4-5V, considering a safety margin of 0.5-0.7V for intrinsic reliability testing, the gate voltage stress (V_g) is confined to a maximum of 3.5V.

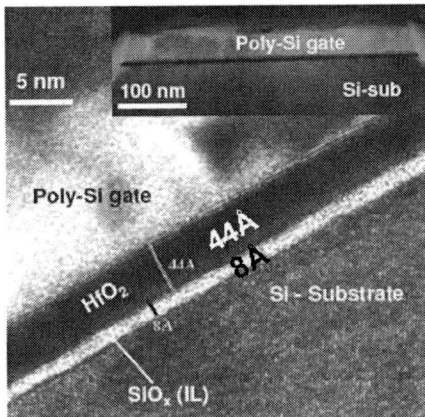

Figure 1. HRTEM image of the poly-Si HfO_2-SiO_x gate stack showing the HK thickness, t_{HK} = 44Å and IL layer thickness, t_{IL} = 8Å.

III. ELECTRICAL CHARACTERIZATION

In order to decode the reliability of the HK and IL layers, it is necessary to be able to initiate the TDDB breakdown for the HK and IL separately. However, in previous studies [9, 10], this could not be accomplished because the voltage stress applied was too high such that after one of the dielectric layers broke down, the electric field across the surviving dielectric was way above its breakdown field thereby causing uncontrolled catastrophic breakdown of the complete stack almost instantly. Another reason was the use of high compliance current settings [11] that cannot arrest the BD after a single layer TDDB. Use of a single stress voltage and compliance level to observe a sequential two-step HK-IL BD is therefore unsuitable. We develop here a two-step constant voltage stress (CVS) algorithm with carefully tuned compliance levels to clearly observe stepwise breakdown.

A. Sequential Breakdown of HK and IL Layers

Fig.2 shows the flowchart describing the new TDDB algorithm. It involves two stages of stressing. Stage I has a higher V_{g1} ~ 3.2-3.5V with a very low current compliance I_{g1} = 0.35-0.5μA (which is typically 5-6 times of the initial stress induced leakage current (SILC)). Stage II stress conditions involve a lower stress of V_{g2} ~ 3 - 3.2V and higher compliance of I_{g2} = 5μA. Although the value of I_{g2} can be arbitrarily chosen, the success of this approach lies in the precise control of $\{I_{g1}, V_{g1}, V_{g2}\}$. Observations of the clear two-step BD are shown in Fig.3 for some of the samples tested. Fig.4 presents the I_g-V_g trends for some of the tested devices at four different stages – (A) fresh device, (b) after 1-layer BD, (c) after 2-layer BD and (d) progressive BD corresponding to a compliance of I_{g1} ~ 100μA.

B. Poole-Frenkel Emission Characteristics

After one layer breakdown, in order to identify the layer which has broken down, we perform I_g-V_g measurement and check for the existence of the Poole-Frenkel (P-F) conduction mechanism which is unique to thick high-κ dielectric materials (typically 4nm and above) [12]. If P-F emission is valid after a one-layer BD, then we can confirm that the HK remains intact while the IL has broken down. However, if the IL is the intact layer, then we should expect to see a deviation from the P-F mechanism as conduction through SiO_x is by inelastic trap assisted tunneling (ITAT) which has a different I_g-V_g relationship [13].

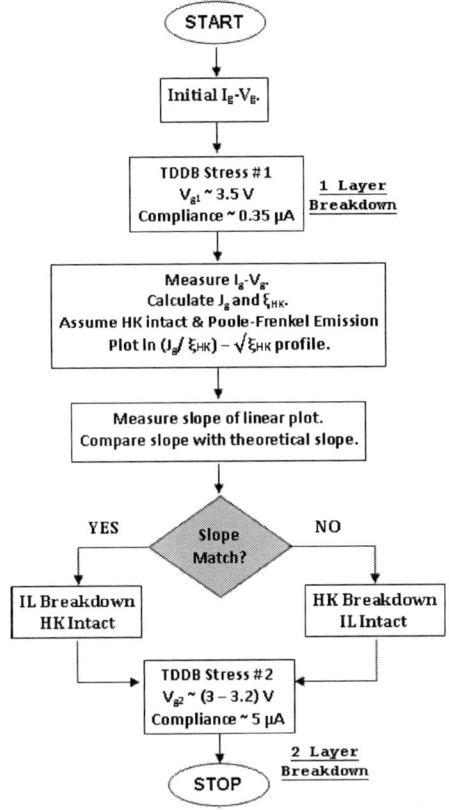

Figure 2. Flowchart of the proposed two-step sequential HK-IL breakdown methodology by precise tuning of the stress voltage and compliance settings.

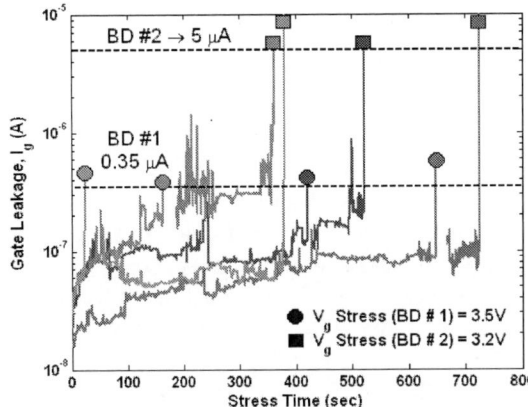

Figure 3. Two-step SILC-TDDB breakdown observed in the bi-layer stacks using the proposed TDDB methodology in Fig.2.

978-1-4244-5430-3/10 $26.00 © 2010 IEEE

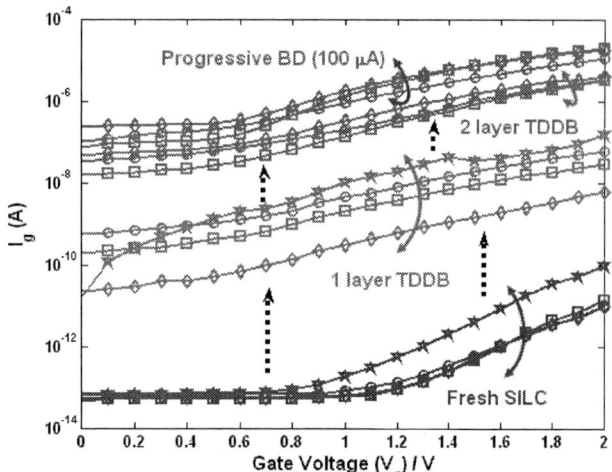

Figure 4. Typical I_g-V_g characteristics for the bi-layer stack for fresh device, after 1-layer TDDB, after 2-layer TDDB and progressive BD (100μA) stages.

The analytical relationship of current density ($J_g = I_g/A$) and electric field across the HK layer ($\xi_{HK} = V_g/t_{HK}$) for P-F emission is given by Eqns.(1) and (2) where C is a proportionality constant which depends on the carrier mobility in the HK as well as the intrinsic process induced trap concentration, ϕ_B is the barrier height for the traps below the conduction band (typically 0.5 – 1 eV), q is the electronic charge, T is the Kelvin temperature and $\{\varepsilon_0, \kappa\}$ represent the free-space and relative permittivity respectively. Poole-Frenkel is a highly temperature sensitive conduction process involving thermally activated electron hopping across potential barriers that are reduced upon application of electric field [14]. These potential barriers correspond to the shallow trap potential wells in the bulk of the HK. The measured I_g-V_g data after one-layer BD is converted into $\ln(J_g/\xi_{HK})$ and $\sqrt{\xi_{HK}}$, to be plotted on a linearized P-F scale plot, as illustrated by Fig.5(a).

$$J_g = C \cdot \xi_{HK} \cdot \exp\left[\frac{-q \cdot \left(\phi_B - \sqrt{q\xi_{HK}/\pi\varepsilon_0\kappa}\right)}{kT}\right] \quad (1)$$

$$\ln\left(\frac{J_g}{\xi_{HK}}\right) = \left[\ln(C) - \frac{q\phi_B}{kT}\right] + \left(\sqrt{\frac{q^3}{\pi\varepsilon_0\kappa}} \cdot \frac{1}{kT}\right) \cdot \sqrt{\xi_{HK}} \quad (2)$$

The P-F analysis is carried out for $V_g > 1V$ only as this is the regime where this conduction mechanism is dominant as illustrated by the band diagram in Fig.5(b). The experimental value of the slope in Fig.5(a) is compared with its theoretical value, estimated to range between $(5.38 – 6.58) \times 10^{-3}$, considering the κ for HfO$_2$ which is expected to vary between $20 – 30$ [15]. Note that we arrived at the condition of $V_g > 1V$ for the P-F emission from the band diagram simulations [16], with a prior knowledge of the trap depth in the HK, ϕ_B, determined from the temperature dependence behavior of the P-F mechanism, to be 0.48eV as shown in Fig.6 for T = 25°C – 125°C. This value closely matches with the trap depth of 0.5–1.0eV reported in the literature [17, 18] for oxygen vacancy defects (V_0^+, V_0^{2+}) in HfO$_2$. For $V_g < 1V$, the conduction mechanism is expected to be direct tunneling.

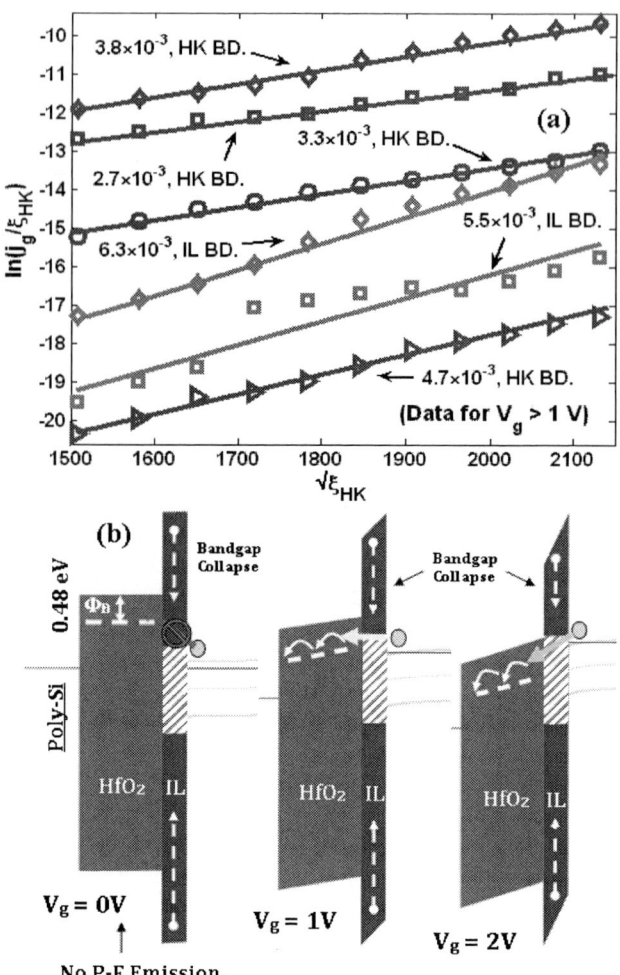

Figure 5. (a) Poole-Frenkel plot of I_g-V_g data after one-layer BD with the experimental linear fit slope values indicated. (b) Band diagram illustrating the existence of P-F emission for $V_g > 1V$, when the electrons tunneling from the Si conduction band can access the energy levels of the shallow traps in the bulk HK (assuming HK is still intact). The hatched regions in the IL represent the bandgap collapse of SiO$_x$ → Si due to oxygen vacancy percolation.

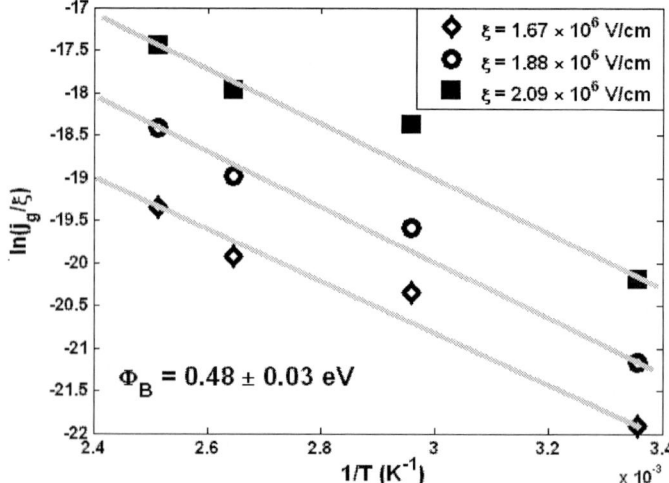

Figure 6. Arrehenius plot of temperature dependence tests for the Poole-Frenkel mechanism aimed at determining the "effective" trap depth for a fresh HK-IL device.

Based on the above approach of using the P-F slope as a criteria to distinguish HK and IL failures, we found that of the 36 devices tested, 31 show HK to be the first layer breaking down (experimental slope falls outside the theoretical range) and only 5 devices suggest IL to suffer a first BD. Therefore, we may conclude that for NMOS under substrate injection stress, HK is almost always the first layer to break down, followed later by the IL BD at the same location. Having obtained the two-step TDDB trends and determined the sequence of breakdown, the time instants for the two BD events (T_{BD1} and T_{BD2}) are recorded and in the next section, we develop the statistical model that makes use of these values to quantitatively estimate the reliability lifetime for the HK and IL layers.

IV. STATISTICAL MODELING AND ANALYSIS

As mentioned previously, the simplest model of accelerated life tests (ALT) involving a single CVS for the complete TDDB duration in a HK-IL stack, which has been adopted by most previous research studies [9, 10], does not provide a true estimate of the intrinsic reliability of the whole gate stack, as the second surviving layer dielectric tends to fail "abruptly" when subjected to a stress more than its breakdown field. Moreover, the stress profile across the surviving dielectric layer is "non-constant" as it bears the additional voltage load after the first dielectric fails. For such non-constant time variant stress conditions, standard reliability models are no longer applicable. In line with the two-step CVS algorithm proposed in the previous section, we present here a new statistical model that accounts for the time-varying voltage stress in the two-step TDDB BD and $\{T_{BD1}, T_{BD2}\}$ values as well as the P-F emission data (which we used to conclude that HK is predominantly the first layer to breakdown). Although the mathematical formulation of this model has been developed and established previously, we focus on applying it to our HK-IL bi-layer stack here. In reliability literature, the model is called the *cumulative damage model* (CDM) [6, 7]. As the name suggests, it is a model that accounts for the "cumulative effect" of the time-varying voltage stress profile that the dielectric is subjected to during ALT.

A. Cumulative Distribution Function

Consider the HK layer to fail first at time $t = t_0$. For $0 < t < t_0$, the voltage across the stack (V_{ox1}) is shared by the HK and IL layer according to Eqns.(3) and (4), which are based on the Gauss's Law [14]. Given this situation, the voltage stress profile as a function of time across the HK and IL layers is shown in Fig.7(a). At $t > t_0$, $V_{IL} = V_{ox2}$ i.e. the voltage drops completely across the IL layer. Note that $V_{ox2} < V_{ox1}$, since the second stage gate stress is kept low in our algorithm. Here, we assume the voltage drop across the broken down HK (very low resistance) to be negligible. Taking κ (HfO$_2$) ~ 25 and κ (SiO$_x$) ~ 3.9, $V_{HK} = 0.46V_{ox1}$ and $V_{IL} = V_{ox1} - V_{HK} = 0.54V_{ox1}$.

$$\varepsilon_{HK}\xi_{HK} = \varepsilon_{IL}\xi_{IL} \qquad (3)$$

$$V_{HK} = V_{ox1} \cdot \left(\frac{\kappa_{HK}}{t_{HK}} \cdot \frac{t_{IL}}{\kappa_{IL}} + 1\right)^{-1} \qquad (4)$$

Considering the failure of both the HK and IL layer to obey Weibull statistics based on the percolation theory [19], the cumulative distribution functions (CDF) for HK and IL may be expressed by Eqns.(5) – (7) where $\{K_{HK}, K_{IL}\}$ and $\{n_{HK}, n_{IL}\}$ are the proportionality constant and power law exponent for the inverse power law (IPL) life-stress model of the HK and IL layers respectively. The quantities β_{HK} and β_{IL} refer to the Weibull slope of the corresponding HK and IL layers.

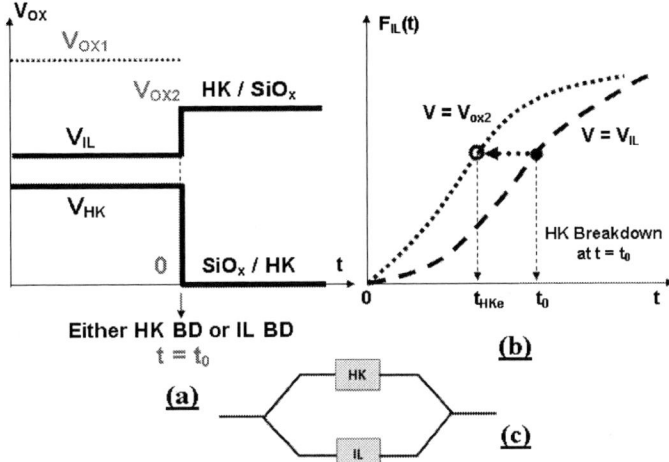

Figure 7. (a) Illustration of the time varying voltage step stress profile across each layer of the dielectric stack. (b) Cumulative failure plot of the surviving IL layer showing the scaling of the first layer TDDB failure time to an equivalent time corresponding to a higher level stress of $V_{IL} \sim V_{ox2}$. (c) Reliability block diagram for the HK-IL system.

$$F_{HK}(t) = 1 - e^{-\left[t \cdot K_{HK} \cdot (0.46V_{ox1})^{n_{HK}}\right]^{\beta_{HK}}} \quad ; \; 0 < t < t_0 \quad (5)$$

$$F_{IL}(t) = 1 - e^{-\left[t \cdot K_{IL} \cdot (0.54V_{ox1})^{n_{IL}}\right]^{\beta_{IL}}} \quad ; \; 0 < t < t_0 \quad (6)$$

$$F_{IL}(t) = 1 - e^{-\left[(t-t_0+t_{HKe}) \cdot K_{IL} \cdot (V_{ox2})^{n_{IL}}\right]^{\beta_{IL}}} \quad ; \; t > t_0 \quad (7)$$

An important quantity to take note of in Eqn.(7) is t_{HKe}, which is the "equivalent time" of survival for the IL layer at the higher stress of $V_{IL} \sim V_{ox2}$ corresponding to the same fraction of failures at $V_{IL} = 0.54V_{ox1}$ at first breakdown time $t = t_0$ (refer to Fig.7(b)). This may be mathematically expressed by Eqn.(8). This is the key link that associates the first and second stages of BD and accounts for the degradation in the IL layer (in addition to the HK layer which degrades and fails) in the first stage. Note here that the CDM model assumes that the remaining life of the surviving dielectric layer depends only on the current cumulative fraction failed and the current stress level, regardless of how the fraction accumulated, which is the typical Markovian property [7].

$$F_{IL}(V = V_{ox2}, t = t_{HKe}) \equiv F_{IL}(V = 0.54V_{ox1}, t = t_0) \quad (8)$$

The same analogy above applies to the case when the IL breaks down first and HK survives the increasing load. Having formulated the separate probability distributions for the HK and IL layers, we may find the optimum values for the statistical parameters $\{\beta_{HK}, \beta_{IL}, \eta_{HK}, \eta_{IL}, n_{HK}, n_{IL}\}$ by optimizing the likelihood function using the maximum likelihood estimation (MLE) approach [7]. Note that η here denotes to the mean-time-to-failure for the Weibull distribution (63.2% percentile). Fig.7(b) clearly shows the leftward shift in the distribution function for the IL layer due to an increased stress after the HK breaks down, as expressed by Eqns.(6) and (7). In all the analysis to be carried out, the actual voltage drop across the stack during inversion ($V_{ox} \neq V_g$) has been precisely calculated, accounting for the flat band voltage (V_{FB}) and surface potential ($2\phi_F$), given by Eqn.(9).

$$V_{ox} = V_g - V_{FB} - 2\phi_F \qquad (9)$$

B. Load Sharing System Reliability

The above CDF formulation helps to analyze the HK and IL data separately. As a further extension, we can make use of the HK and IL reliability expressions to determine the overall HK-IL stack reliability, which is also a quantity of interest. We do this by looking at the reliability block diagram (RBD) for the gate stack. Although the HK and IL layers are serially connected from the capacitance point of view, they are parallel connected from reliability point of view. This is because upon failure of the HK layer, the entire voltage load is borne by the surviving IL. This implies that the failure distribution of surviving IL is "*dependent*" on the reliability and breakdown distribution of the HK layer. Hence, we call this a *load sharing system* [20]. Fig.7(c) illustrates the parallel connectivity of the HK and IL in the RBD.

The overall system reliability for the dual layer stack may be expressed by Eqn.(10) which describes that the dielectric stack is functional under three cases – (A) when both HK+IL layers are intact, represented by $R_{HK\&IL}(t)$, (B) when the HK breaks down and IL is still surviving ($R_{HK\,BD}(t)$) and (C) when the IL breaks down first and HK is still surviving ($R_{IL\,BD}(t)$). Here, $R_{sys}(t)$ is the overall system reliability.

The expressions for $R_{HK\&IL}(t)$ and $R_{HK\,BD}(t)$ are given by Eqns.(11) and (12). The probability expression in Eqn.(12) comprises three product terms. The first term represents the probability of the HK layer failing at time $t = t_0$, the second term is the probability that the IL survived up to time $t = t_0$ (during the lower voltage stress load) and the last term is the conditional probability that the IL still survives under the increased voltage load given that it has already survived for an equivalent time, t_{HKe}, which is the effective time the IL would have functioned had it been operating at the higher load stress since the beginning ($t = 0$). Similar expression as in Eqn.(12) holds for $R_{IL\,BD}(t)$ as well. Having determined the set of parameters $\{\beta_{HK}, \beta_{IL}, \eta_{HK}, \eta_{IL}, n_{HK}, n_{IL}\}$ in the previous section, $R_{sys}(t)$ can be evaluated for any voltage stress condition (V_g). Using the CDM model above, we shall now present the statistical results of our analysis on the HfO_2-SiO_x stack obtained based on the two-step BD electrical tests performed on various samples.

$$R_{sys}(t, V_{ox}) = R_{HK\&IL}(t, V_{ox}) + R_{HK\,BD}(t, V_{ox}) + R_{IL\,BD}(t, V_{ox}) \qquad (10)$$

$$R_{HK\&IL}(t, V_{ox}) = R_{HK}(t, V_{HK}) \cdot R_{IL}(t, V_{IL}) \qquad (11)$$

$$R_{HK\,BD}(t, V_{ox}) = \int_0^t f_{HK}(t_0, V_{HK}) \cdot R_{IL}(t_0, V_{IL}) \cdot \frac{R_{IL}(t_{HKe} + (t - t_0), V_{ox})}{R_{IL}(t_{HKe}, V_{ox})} dt_0 \qquad (12)$$

C. Statistical Failure Data Analysis

Fig.8 shows the respective Weibull plots for the HK and IL layers, extrapolated to $V_g = 1V$, which is the operating voltage condition. It can be seen from the time scale that the IL layer has a lifetime that is almost 9 orders of magnitude more than the HfO_2. Moreover, both the HK and IL data show a certain degree of curvature implying bimodal Weibull distribution suggestive of the existence of two failure mechanisms. The general trend is that the low Weibit region has a steeper slope and this slope become more shallow for high percentile cases. Table I lists out the values for all the statistical parameters of the bimodal distributions for both the HK and IL. The symbols $\{p_1, p_2\}$ are the proportion of failures for bimodal distribution.

TABLE I. STATISTICAL PARAMETERS OF THE HIGH-K (HfO_2) AND INTERFACIAL SiO_x LAYERS

Weibull Parameters	High-K (HfO_2)	IL Layer (SiO_x)
β_1	1.376	1.295
η_1	1.33×10^{16}	2.93×10^{24}
p_1	62.6%	24.3%
β_2	1.268	0.505
η_2	8.46×10^{16}	5.05×10^{25}
p_2	37.4%	75.7%

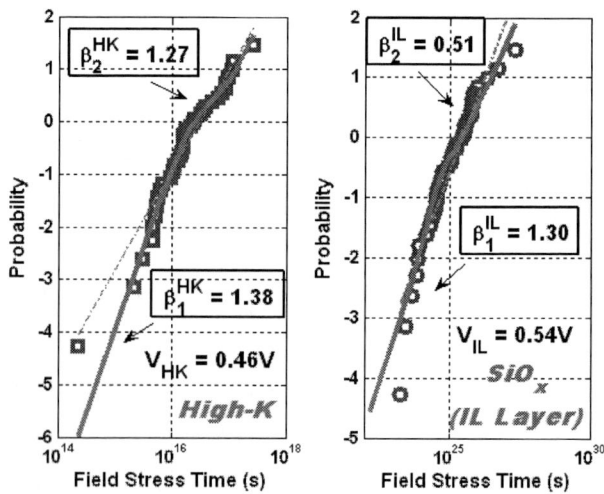

Figure 8. Statistical bimodal Weibull plot for HK and IL layers extrapolated to operating voltage condition of V_g=1V using the inverse power law acceleration model.

Using the Eqns.(10)-(12), the system reliability at $V_g = 1V$ was calculated and is shown in Fig.9. When calculating the system reliability, it was assumed that both the HK and IL failure distributions are monomodal. In spite of assuming this monomodal distribution, the system reliability plot has a "convexity" on the Weibull scale at low Weibit values. Such observations were made previously in [21] through Monte Carlo simulations and hard breakdown (HBD) TDDB data analysis. Our load sharing system model here further verifies the convexity observed experimentally in [21].

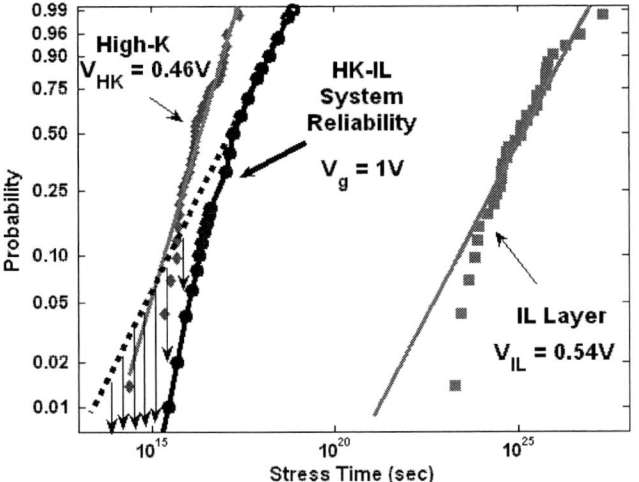

Figure 9. System reliability plot for the "load sharing" HK-IL dual layer stack obtained using the model proposed in Eqns. (10)-(12). The convexity of the line at low Weibit clearly suggests that overall HK-IL stack BD is "non-Weibull".

Furthermore, to confirm the validity of our model, we performed some HBD TDDB tests for a single CVS (V_g = 3.5V) with compliance of I_{gl} = 100µA (refer to inset of Fig.10). The data plotted, in Fig.10, also shows a convexity at low Weibit. This clearly suggests that although the individual HK and IL layers obey Weibull statistics, the overall HK-IL stack is "non-Weibull" in nature and has no specific closed form statistical distribution, as can be confirmed by the complex reliability expressions in Eqns.(10)-(12). Therefore, use of Weibull distribution to fit overall HK-IL stack BD data is statistically inappropriate and could lead to erroneous reliability projections.

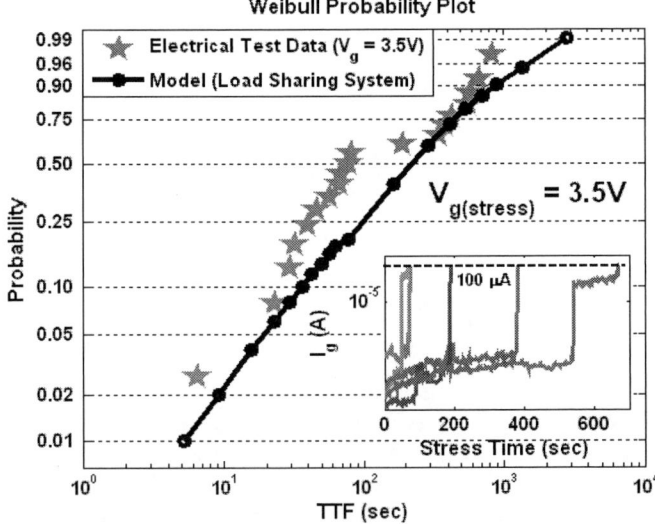

Figure 10. Comparison of the load-sharing HK-IL dependent system model with the HBD data after bi-layer BD at V_g = 3.5V. The close match of the test data and model imply that the model well describes HK-IL failure statistics. Inset shows some of the HBD leakage evolution trends in the bi-layer stack.

The system reliability is again calculated in Fig.10 at V_g = 3.5V to compare the model estimates (●) with the HBD data (✱). The model and the data fit relatively well suggesting that the proposed load sharing system model correctly describes the degradation trends of the HK-IL stack. Some deviations of the model from the data are expected because as mentioned previously use of a single stage CVS at V_g = 3.5V may not fully represent the "intrinsic" nature of failure of the gate stack.

From Table I and Figs.8-9, it can be seen that the mean lifetime for both the HK and IL layers is very long, many orders more than the required standard reliability target of 10 years at operating conditions. All the analysis above is at the "device level" only. It is necessary to extend these results to the "circuit level" in order to assess whether we are able to meet the minimum reliability specifications.

D. Weibull Slope Analysis

Weibull slope (β) is an important parameter, indicative of the number of traps needed to create a percolation. It is expressed as in Eqn.(13) where α is the SILC degradation rate, a_0 is the trap radius and $N = (t_{ox}/a_0)$ represents the number of traps in the percolation path [19]. Our new statistical model has enabled us to extract the β values separately for the HK and IL and additionally the β values for the sub distributions in the bimodal distribution. The early failures in both HK and IL have a larger β compared to the wear-out failures. We speculate this to be a combined effect of a high α and low (t_{ox}/a_0) since these early failures are usually "extrinsic" in nature or occur in highly defective samples which have a high process-induced trap (PIT) concentration, thus requiring very few stress induced traps (SIT) to cause percolation. The term (t_{ox}/a_0) in Eqn.(13) only corresponds to the SIT.

$$\beta = \alpha \cdot \left(\frac{t_{ox}}{a_0}\right) = \alpha \bullet N \qquad (13)$$

The wear-out failures which generally show a low β, correspond to a combined effect of low α and high (t_{ox}/a_0) since wear-out failures are "intrinsic" and occur in defect-free (or low defect density) materials which have a lower degradation rate. This qualitative trend is summarized in Table II.

TABLE II. MAGNITUDE OF THE VARIOUS FACTORS AFFECTING THE WEIBULL SLOPE FOR THE EARLY AND WEAR OUT FAILURE MECHANISMS (FM) IN THE HK AND IL LAYERS

Failure Mechanism (FM)	β	α	(t_{ox} / a_0)	PIT
HK / IL – FM 1 (Early)	↑	↑	↓	↑
HK / IL – FM 2 (Wear-out)	↓	↓	↑	↓

Comparing the β for the wear-out failures in the HK and IL, β_{HK} = 1.27 >> β_{IL} = 0.51. This trend is expected since the HfO$_2$ layer is very thick (44Å) compared to the IL (8Å) and hence requires much more traps (N) to breakdown (higher β). The observation of $\beta_{HK} \gg \beta_{IL}$ confirms our Poole-Frenkel based analysis results that HK is almost always the first layer to breakdown.

E. Area Scaling and Circuit Reliability Implications

In order to compare our statistical lifetime predictions with the reliability specifications, it is necessary to consider "circuit level" reliability based on the device level studies conducted through the use of the "area scaling" law which has been frequently used for both SiO₂/SiON [22] as well as high-κ [23, 24]. It is to be noted that only the first layer to breakdown in the HK-IL stack can obey the area scaling law. The breakdown of the second layer is always "localized" and confined to the region of percolation of the first layer.

Fig.11 shows the life-stress relationship for HK and IL, accounting for the area scaling effect in HK (assuming a circuit comprising 10^9 transistors). Although the power law exponents are very similar ($n_{HK} \sim n_{IL}$), the 8Å IL layer is far more reliable than the 44Å HfO₂ by many orders of magnitude. Taking into account the area scaling further reduces the HK reliability very significantly. As shown by the dotted line in Fig.11, for $V_g =$ 1V which corresponds to $V_{HK} \sim 0.46V$ based on Eqn.(4), the mean-time-to-failure, η, for the first HK BD spot is as low as 10^4 seconds. Subsequent IL failure at the percolated HK region is improbable as the corresponding η for IL is as high as 10^{25} seconds. This large difference in η clearly suggests that circuit level failure is likely to occur by the presence of multiple HK BD events rather than a single sequential HK→IL breakdown.

From the I_g-V_g trends in Fig.4, at $V_g = 1V$, the leakage current after first layer HK BD is around 1nA. Considering a 10 year period, for $\eta = 10^4$ sec, even the most optimistic estimate will suggest as many as 10000 BD spots to occur within the 10-year period with a leakage current of about 10µA which is quite high. The number of BD spots may seem to be too high, but similar results on the expected number of BD spots (~15000) have been recently shown by the IMEC group as well [25]. Since these HK BD events are "soft", their occurrence might be difficult to detect electrically and calls for the need for failure analysis tools.

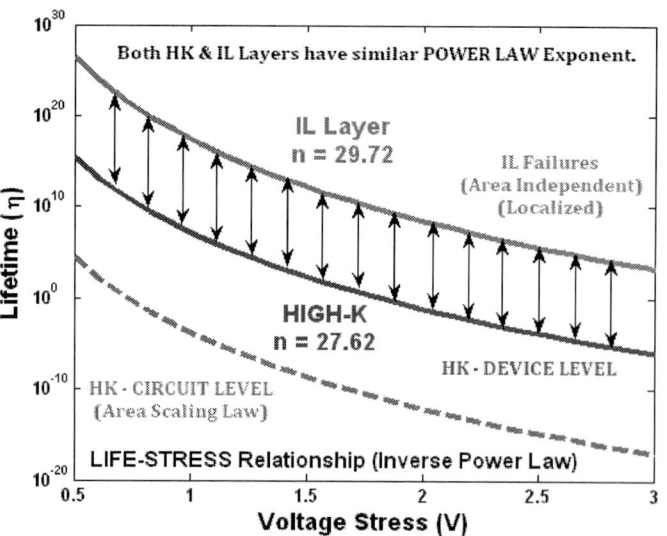

Figure 11. Inverse power law model of lifetime for the HK and IL layers shows that IL and HK layer have similar power law exponents, but IL is always far more reliable than the HK.

The circuit level analysis above assumed the validity of area scaling for the HK layer. However, it is critical to note that polycrystalline HK films may not follow the area scaling rule as defect generation is no longer fully random and does not obey Poisson statistics. Defect generation in polycrystalline HK materials happens preferentially along the grain boundary (GB) dislocation lines which serve as defective weak links for breakdown. It is only for amorphous HK materials that the area scaling law can be expected to hold true. Therefore, the results in Fig.11 only serve as a rough guide and might be an over-optimistic estimate. The presence of weak link GBs might result in a much lower circuit lifetime for this HfO₂ gate stack, which is hard to predict using analytical models. In the following section, we look at the physical failure analysis result for one of the samples using HRTEM and discuss the role played by grain boundaries.

V. PHYSICAL FAILURE ANALYSIS

To uncover the physical nature of failure, one of the samples subjected to HRTEM imaging is shown in Fig.12 [26]. Bright contrast lines along the GB region is clearly observed. Some silicon epitaxy defects (dielectric breakdown induced epitaxy – DBIE) at the vicinity of the GB is also seen. These suggest the preferential occurrence of failures at or close to the GB. This is further confirmed by the bimodal distribution results in Table I and Fig.8, where 62.6% of samples show early HK failure which is to be attributed to percolation in the GB region.

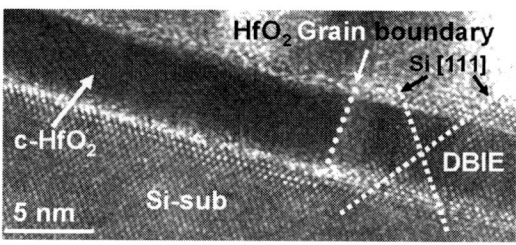

Figure 12. TEM cross-section of a failed device clearly shows the grain boundary site failure in the HK layer along with other microstructural DBIE transformations [26].

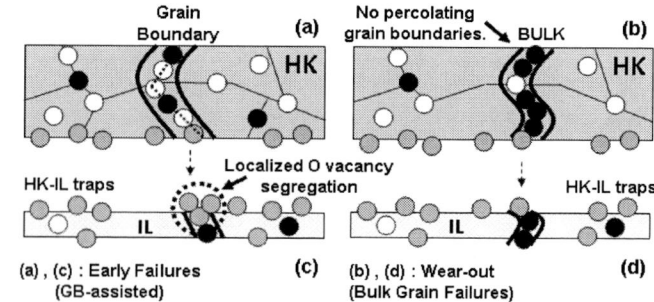

Figure 13. Illustration showing trap formation and percolation path evolution for the different failure mechanisms in the HK and IL layers. The *black*, *white* and *grey* circles represent stress-induced traps (SIT), process induced traps (PIT) and interface traps respectively.

Fig.13 summarizes the two failure mechanisms responsible for failure in the HK and IL layers. The early and wear-out failures in the HK correspond to GB and bulk percolation respectively and in all cases oxygen vacancies are the precursors for trap generation and breakdown. These oxygen vacancies which thermodynamically prefer to segregate along the GB [27] at the HK-IL interface affect the reliability of IL

978-1-4244-5430-3/10 $26.00 © 2010 IEEE

as well. The early failures in the IL could be due to extrinsic factors (for example, highly non-stoichiometric sub-oxide of 8Å – only 2-3 monolayers of atoms).

VI. CONCLUSION

A new statistical cumulative damage model was proposed in this work to model and analyze separately the reliability of the HK and IL layers in a bi-layer HfO_2-SiO_x gate dielectric stack and account for the time-variant stress levels during the TDDB test. The proposed model helped us to compare the lifetime and Weibull slope of the HK and IL layers and infer the existence of bimodal failure distributions. The successful demonstration of this approach lies in the ability to cause a two-stage TDDB breakdown of the gate stack using the new electrical test algorithm in Section III which involves careful tuning of the compliance levels and stressing conditions. Finally, we make use of physical failure analysis results to confirm the role played by grain boundaries in making HK highly susceptible to early failures. From a reliability point of view, therefore, amorphous HK materials are expected to be more robust and prolong the lifetime of the gate stack. A critical aspect of design for reliability (DFR) in front end CMOS technology should therefore involve efforts to optimize the process flow and conditions such that the deposited HK thin films remain amorphous, unaffected by any of the subsequent annealing steps.

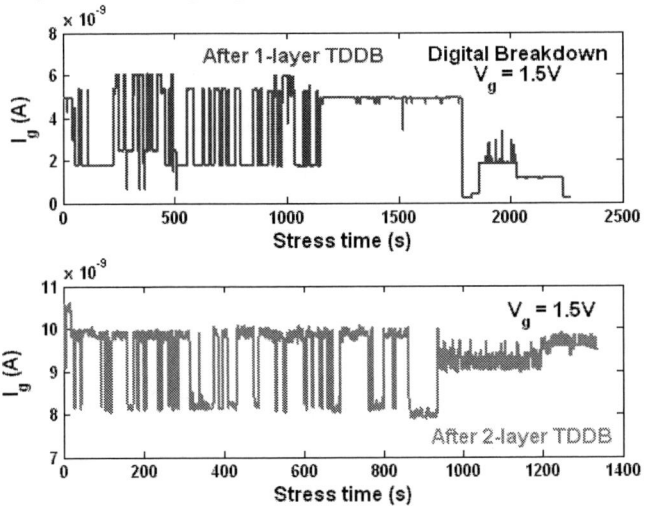

Figure 14. Post TDDB random telegraph noise signatures for the poly-Si-HfO_2-SiO_x gate stack at V_g = 1.5V after 1-layer and 2-layer TDDB BD respectively.

We summarize below the key inferences from this study and their implications on future high-κ gate stack technology.

- The ultra-thin IL layer has a lifetime many orders of magnitude more than the thick polycrystalline HK film. The presence of the IL, although detrimental to EOT scaling, is a reliability "savior" for the overall gate stack. This raises concern regarding the implementation of a zero-IL (ZIL) solution recently proposed in [28], from a reliability perspective.

- Both the HK and IL layers show bimodal statistics in general. In HK, this is attributed to GB early failures and bulk wear-out failures. Amorphous HK materials are likely to exhibit a monomodal behavior.

- The Weibull slope of HK is much more than that of IL for the wear-out (intrinsic) failures, as expected.

- Overall gate stack does not obey Weibull statistics. It does not have any closed form statistical distribution. It is a complex function of many Weibull CDFs.

- It is predicted that circuit level failure occurs by multiple HK BD events rather than a sequential HK→IL BD.

The approach presented here has been successfully applied to FUSI gated stack devices as well [29]. Further work is currently under way to extend the analysis to state-of-the-art metal gate (MG-HK) stacks. A more in-depth understanding on HK BD mechanism can only be achieved through the use of physical failure analysis tools and techniques such as scanning tunneling microscopy (STM) [30], conductive atomic force microscopy (CAFM) [31, 32] and electron energy loss spectroscopy (EELS) [33].

There is further scope for study of post breakdown reliability in HK gate stacks [34]. The analysis in this study was confined only to the TDDB stage. Accounting for the post-BD stage when random telegraph noise fluctuations are observed, as illustrated by Fig.14, further enhances the reliability margin.

ACKNOWLEDGMENT

The authors would like to thank the Interuniversity Microelectronics Centre (IMEC), Belgium for provision of the samples used in this study and Reliasoft® Inc. for licensed access to reliability software tools. This work is sponsored by the Ministry of Education (MOE), Singapore Grant No. T206B1205 and NTU RGM 33/03.

REFERENCES

[1] S. Guha and V. Narayanan, "High-κ / Metal Gate Science and Technology", *Annual Review of Materials Research*, Vol. 39, pp.181-202, (2009).

[2] J. Sune, E.Y. Wu, D. Jimenez and W.L. Lai, "Statistics of soft and hard breakdown in thin SiO_2 gate oxides", *Microelectronics Reliability*, Vol. 43, pp.1185-1192, (2003).

[3] M.A. Alam, R.K. Smith, B.E. Weir and P.J. Silverman, "Uncorrelated breakdown of integrated circuits", *Nature*, Vol. 420, pp.378, (2002).

[4] T. Pompl and M. Rohner, "Voltage acceleration of time-dependent breakdown of ultra-thin gate dielectrics", *Microelectronics Reliability*, Vol. 45, pp.1835-1841, (2005).

[5] A. Kerber, T. Pompl, M. Rohner, K. Mosig and M. Kerber, "Impact of failure criteria on the reliability prediction of CMOS devices with ultrathin gate oxides based on voltage ramp stress", *IEEE Electron Device Letters*, Vol. 27, No. 7, pp.609-611, (2006).

[6] A. Mettas and P. Vassiliou, "Modeling and analysis of time-dependent stress accelerated life data", *IEEE Annual Reliability and Maintainability Symposium (RAMS)*, pp.343-348, (2002).

[7] W. Nelson, "Accelerated Testing: Statistical Models, Test Plans and Data Analyses", *John Wiley & Sons*, (1990).

[8] N. Raghavan, K.L. Pey and X. Li, "Detection of high-κ and interfacial layer breakdown using the tunneling mechanism in a dual layer dielectric stack", *Applied Physics Letters*, Vol. 95, 222903, (2009).

[9] R. Degraeve, T. Kauerauf, M. Cho, M. Zahid, L-Å. Ragnarsson, D.P. Brunco, B. Kaczer, Ph. Roussel, S. De Gendt and G. Groeseneken, "Degradation and breakdown of 0.9 nm EOT SiO₂ / ALD HfO₂ / metal gate stacks under positive constant voltage stress", *IEEE International Electron Device Meeting (IEDM)*, (2005).

[10] N.A. Chowdhury, G. Bersuker, C. Young, R. Choi, S. Krishnan and D. Misra, "Breakdown characteristics of nFETs in inversion with metal / HfO₂ gate stacks", *Microelectronic Engineering*, Vol. 85, pp.27-35, (2008).

[11] S.J. Lee, C.H. Lee, C.H. Choi and D.L. Kwong, "Time-dependent dielectric breakdown in poly-Si CVD HfO₂ gate stack", *IEEE International Reliability Physics Symposium (IRPS)*, pp.409-414, (2002).

[12] D.S. Jeong and C.S. Hwang, "Tunneling-assisted Poole-Frenkel conduction mechanism in HfO₂ thin films", *Journal of Applied Physics*, Vol. 98, 113701, (2005).

[13] J. Lee and G. Bosman, "Model and analysis of gate leakage current in ultrathin nitrided oxide MOSFETs", *IEEE Transactions on Electron Devices*, Vol. 49, No. 7, pp.1232-1241, (2002).

[14] M. Houssa, "High-κ gate dielectrics", *Institute of Physics*, (2004).

[15] H.R. Huff and D.C. Gilmer, "High dielectric constant materials – VLSI MOSFET applications", *Springer Publications*, Berlin, (2005).

[16] R.G. Southwick III and W.B. Knowlton, "Stacked dual-oxide MOS energy band diagram visual representation program", *IEEE Transactions on Device and Materials Reliability*, Vol. 6, No. 2, pp.136-145, (2006).

[17] J. Robertson, "High dielectric constant gate oxides for metal oxide Si transistors", *Reports on Progress in Physics*, Vol. 69, pp.327-296, (2006).

[18] A. Kerber and E.A. Cartier, "Reliability challenges for CMOS technology qualification with hafnium oxide / titanium nitride gate stacks", *IEEE Transactions on Device and Materials Reliability*, Vol. 9, No. 2, pp.147-162, (2009).

[19] A.T. Krishnan and P.E. Nicollian, Analytic extension of the cell-based oxide breakdown model to full percolation and its implications", *IEEE International Reliability Physics Symposium (IRPS)*, pp.232-239, (2007).

[20] A. Mettas and P. Vassiliou, "Application of quantitative accelerated life models on load sharing redundancy", *IEEE Annual Reliability and Maintainability Symposium (RAMS)*, pp.293-296, (2004).

[21] T. Nigam, A. Kerber and P. Peumans, "Accurate model for time-dependent dielectric breakdown of high-κ metal gate stacks", *IEEE International Reliability Physics Symposium (IRPS)*, pp.523-530, (2009).

[22] E.Y. Wu, J.H. Stathis and L.K. Han, "Ultra-thin oxide reliability for ULSI applications", *Semiconductor Science and Technology*, Vol. 15, pp.425-435, (2000).

[23] G. Ribes, S. Bruyere, M. Denais, F. Monsieur, D. Roy, E. Vincent and G. Ghibaudo, "High-κ dielectrics breakdown accurate lifetime assessment methodology", *IEEE International Reliability Physics Symposium (IRPS)*, pp.61-66, (2005).

[24] Y.H. Kim, K. Onishi, C.S. Kang, H.J. Cho, R. Nieh, S. Gopalan, R. Choi, J. Han, S. Krishnan and J.C. Lee, "Area dependence of TDDB characteristics for HfO₂ gate dielectrics", *IEEE Electron Device Letters*, Vol. 23, No. 10, pp.594-596, (2002).

[25] S. Sahhaf, R. Degraeve, P.J. Roussel, B. Kaczer, T. Kauerauf and G. Groeseneken, "A new TDDB reliability prediction methodology accounting for multiple SBD and wear out", *IEEE Transactions on Electron Devices*, Vol. 56, No. 7, pp.1424-1432, (2009).

[26] R. Ranjan, K.L. Pey, C.H. Tung, L.J. Tang, G. Groeseneken, L.K. Bera and S. De Gendt, "A comprehensive model for breakdown mechanism in HfO₂ high-κ gate stacks", *IEEE International Electron Device Meeting (IEDM)*, pp.725-728, (2004).

[27] K. McKenna and A. Shluger, "The interaction of oxygen vacancies with grain boundaries in monoclinic HfO₂", *Applied Physics Letters*, Vol. 95, 222111, (2009).

[28] J. Huang, D. Heh, P. Sivasubramani, P.D. Kirsch, G. Bersuker, D.C. Gilmer, M.A. Quevedo-Lopez, M.M. Hussain, P. Majhi, P. Lysaght, H. Park, N. Goel, C. Young, C.S. Park, C. Park, M. Cruz, V. Diaz, P.Y. Hung, J. Price, H.H. Tseng and R. Jammy, *IEEE VLSI Technology Symposium*, Kyoto, pp.34, (2009).

[29] N. Raghavan, K.L. Pey, W.H. Liu, X. Wu and X. Li, "Unipolar recovery of dielectric breakdown in fully silicided high-κ gate stack devices and its reliability implications", *Applied Physics Letters*, Submitted, (2010).

[30] D.S. Ang, Y.C. Ong, S.J. O'Shea, K.L. Pey, C.H. Tung, T. Kawanago, K. Kakushima and H. Iwai, "Polarity dependent breakdown of the high-κ/SiOₓ gate stack: a phenomenological explanation by scanning tunneling microscopy", *Applied Physics Letters*, Vol. 92, 192904, (2008).

[31] W. Polspoel, P. Favia, J. Mody, H. Bender and W. Vandervorst, "Physical degradation of gate dielectrics induced by local electrical stress using conductive atomic force microscopy", *Journal of Applied Physics*, Vol. 106, 024101, (2009).

[32] X. Blasco, M. Nafria, X. Aymerich, J. Petry and W. Vandervorst, "Nanoscale post-breakdown conduction of HfO₂/SiO₂ MOS gate stacks studied by enhanced-CAFM", *IEEE Transactions on Electron Devices*, Vol. 52, No. 12, pp.2817-2819, (2005).

[33] X. Li, K.L. Pey, M. Bosman, W.H. Liu and T. Kauerauf, "Direct visualization and in-depth physical study of metal filament formation in percolated high-κ dielectrics", *Applied Physics Letters*, Vol. 96, 022903, (2010).

[34] R. Pagano, S. Lombardo, F. Palumbo, P. Kirsch, S.A. Krishnan, C. Young, R. Choi, G. Bersuker and J.H. Stathis, "A novel approach to characterization of progressive breakdown in high-κ / metal gate stacks", *Microelectronics Reliability*, Vol. 48, pp.1759-1764, (2008).

978-1-4244-5430-3/10 $26.00 © 2010 IEEE

ROLE OF INTERFACE LAYER IN STRESS-INDUCED LEAKAGE CURRENT IN HIGH-K/METAL-GATE DIELECTRIC STACKS

W.L. Chang
IBM Microelectronics
Hopewell Junction, NY 12533
845-892-2028, winglai@us.ibm.com

J.H. Stathis and E. Cartier
T. J. Watson Research Center, IBM
Yorktown Heights, NY 10598

Abstract— **The impact of the Silica-based interface layer (IL) thickness on stress induced leakage current (SILC) on high-k/metal-gate transistors is studied at various constant voltage stresses (CVS) and at various temperatures. It is shown that high-k/metal-gate transistors reliability can be greatly improved with interface layer optimization.**

Keywords-component; high-k dielectrics, SILC, TDDB

I. INTRODUCTION

Stress induced leakage current (SILC) is a major consideration for reliability of high-k/metal-gate transistors, in particular nFETs [1-2]. The origin of SILC is still debated. One view is that the interface layer (IL) is the 'weak link' in the gate stack, and defects generated in the IL are responsible for SILC and breakdown [3-5]. This is still controversial, e.g. others claim that defects are more easily generated in the high-k layer and that under some conditions this layer controls the reliability [2, 6].

In this paper we distinguish two SILC components: a component with a shallow power-law time dependence which decreases as the IL thickness increases, and a quasi-linear component which is independent of IL thickness. The results demonstrate the importance of interface layer optimization to improve SILC behavior in high-k/metal-gate devices.

II. EXPERIMENTAL

The samples are nFETS fabricated in a 32 nm bulk technology using a Hafnium-based high-k dielectric (HK) with metal gate. Processing included a narrow range of Silica-based interface layer (IL) thickness with identical HK layers and gate metal. Fig. 1 shows the gate IV characteristics. A 3× range of gate current at 1.0 V corresponds to approximately 2 Å thickness range [7]. The gate area is ~3 um² (comprised of an array of smaller FETs). We stressed nFETs in inversion.

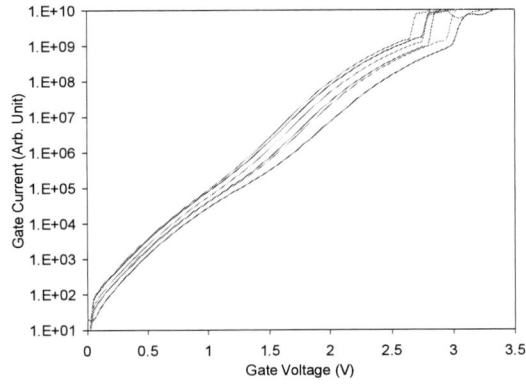

FIGURE 1: *Ig-Vg* of samples under study. Gate leakage is modulated by IL thickness. All samples have identical high-k and metal gate.

III. RESULTS & DISCUSSION

The evolution of the gate current during constant voltage stress (CVS) at V_{stress} = 2.2 V at 125 °C is shown in Fig. 2. Also shown is SILC measured at V_{sense} = 1.0 V. Breakdown (BD) appears as a strong gate current increase at V_{sense}.

A significant increase in gate current, approximately linear in time and accompanied by fluctuations occurs near the breakdown time. This current increase and associated fluctuations, which we refer to as 'linear SILC', were shown to be associated with trapping in generated (stress-induced) electron traps [2]. The linear SILC component is suppressed at lower stress temperature. Figs. 3(a) and (b) compare *I-t* traces at 125 °C and 30 °C with the stress voltages chosen to keep the breakdown times the same. For equal stress times, the linear SILC is strongly suppressed at the lower temperature. However, this may be partly a result of the higher stress voltage employed at 30 °C to keep the same breakdown time as at 125 °C. The higher voltage may cause the breakdown to become harder, masking the linear SILC component. *I-t* traces at 125 °C and 30 °C at equal stress voltages shown in Figs. 3(c) and 3(d) show that linear SILC does occur even at 30 °C if the breakdown time is long enough.

978-1-4244-5430-3/10 $26.00 © 2010 IEEE   787

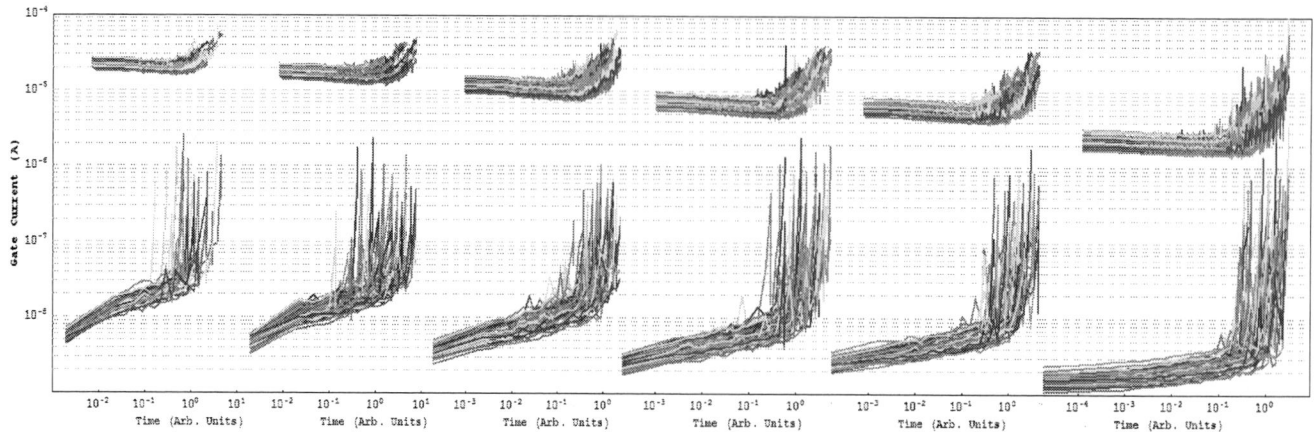

FIGURE 2: *I-t* traces for constant voltage stress V_{stress} = 2.2 V at 125 °C. Samples have different IL thickness (thinnest to thickness arranged left-to-right) corresponding to the gate IVs in Fig. 1.

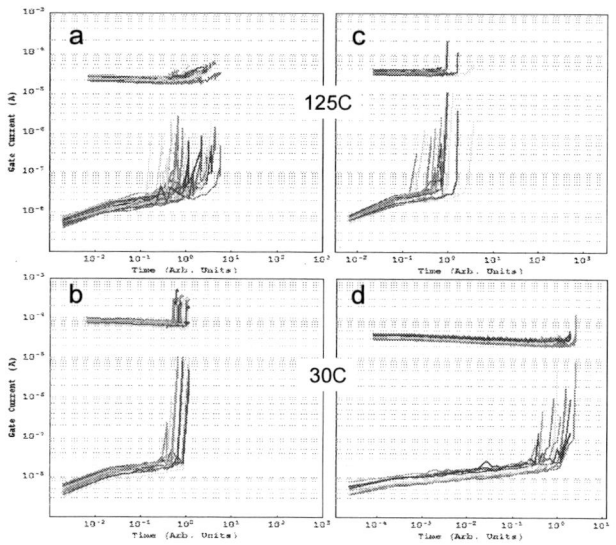

FIGURE 3: Temperature dependence of *I-t* traces, for identical IL thickness stressed at 125 °C and 30 °C. (a) and (b): voltage adjusted to have the same breakdown time at high and low temperature; (c) and (d): stressed at equal voltage at high and low temperature.

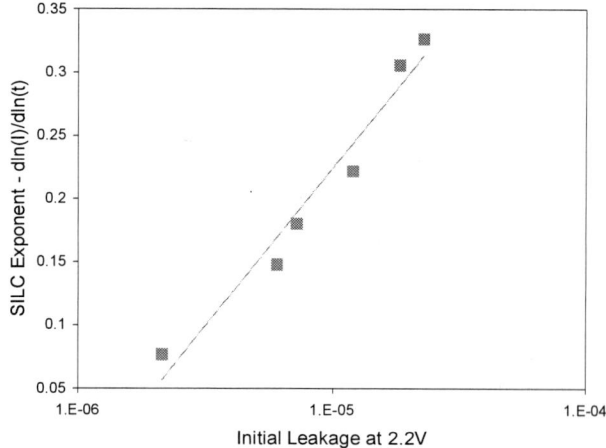

FIGURE 4: Power law time exponent of the pre-BD SILC, d$ln(I)$/d$ln(t)$, vs. the initial leakage.

Another SILC process with a shallow power-law time dependence occurs during earlier stress times. Fig 4 shows the power law exponent of the pre-BD SILC component, d$ln(I)$/d$ln(t)$, for constant V_{stress}, plotted vs. the initial leakage at stress voltage, which is a measure of IL thickness. Each data point is the average slope taken from several *I-t* traces, plotted vs. the median gate leakage. This figure shows that the pre-BD power-law SILC is suppressed for thicker IL. This can be explained by less effective trap-assisted tunneling through defects in the HK layer or at the HK/IL interface, when the HK layer is farther from the injecting interface. [4] In contrast, the linear SILC does not show a strong dependence on IL thickness at constant V_{stress} (see Fig. 2).

Fig 5 shows Weibull plots of time to breakdown for the same samples as in Fig. 2. The breakdown time was determined from the sense current at 1.0 V, using a fixed current level (different for each IL thickness) corresponding to <50% instantaneous current increase above the SILC background. The average Weibull slope for these data is 1.7±0.1. According to Linder [7] the Weibull slope should increase for thinner IL, however we are unable to discern any clear trend among these data because the range of thickness is small. The breakdown times have a super-linear dependence on the gate leakage, Fig. 6.

The shallow power law dependence of the SILC component is much reduced at lower stress voltage. Fig. 7 shows the SILC *I-t* traces measured at 1.0 V for stress voltages ranging from 2.3 – 1.5 V at 125 °C for the thinnest IL samples in Fig. 2. A power-law fit to the early pre-BD SILC is shown. All samples were stressed to breakdown except those stressed at 1.5 V and 1.2 V, since at these low voltages BD cannot be achieved at a reasonable time frame. Also, the sub-linear SILC exists at the earlier part of the stress, there is no need to stress to BD to extract the power law exponent. 1.2 V data was not shown in this plot since the SILC increase is very low, and it overlaps with the 1.5 V data. Fig. 8 shows the power law exponent of the pre-BD SILC exponent vs. stress voltage. The SILC exponent decreases drastically at low voltages. Fig. 7 inset shows the close-up of the end of 1.5 V *I-t* traces. With the sub-linear part of the SILC much suppressed at low voltage the linear SILC can be seen As these data are

978-1-4244-5430-3/10 $26.00 © 2010 IEEE

clearly far from BD at 1.5 V, this means this part of SILC starts earlier but increases much faster, and at high stress voltage it is normally masked by the much larger sub-linear SILC until it catches up near BD, or BD may occur before it becomes visible. Therefore, although at low voltage the SILC contribution due to thinner IL may become insignificant, the SILC due to electron trapping remains a concern

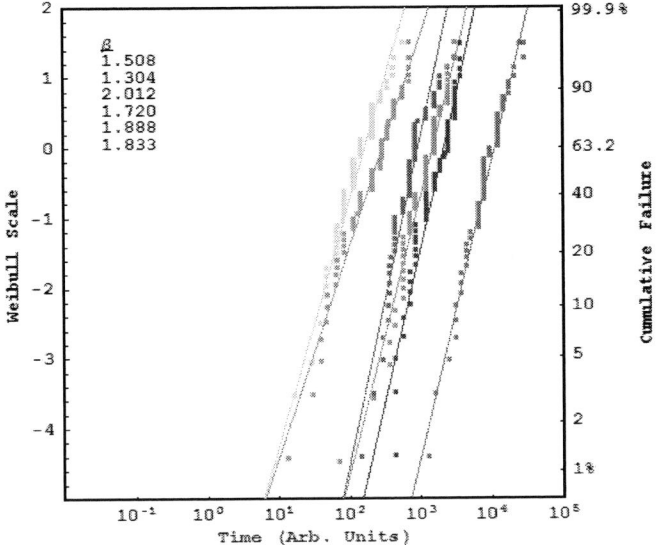

FIGURE 5: Weibull distributions for constant V_{stress} = 2.2 V at 125 °C. Samples have different IL thickness (thinnest to thickness arranged left-to-right). No clear trend for Weibull slope is evident in these data.

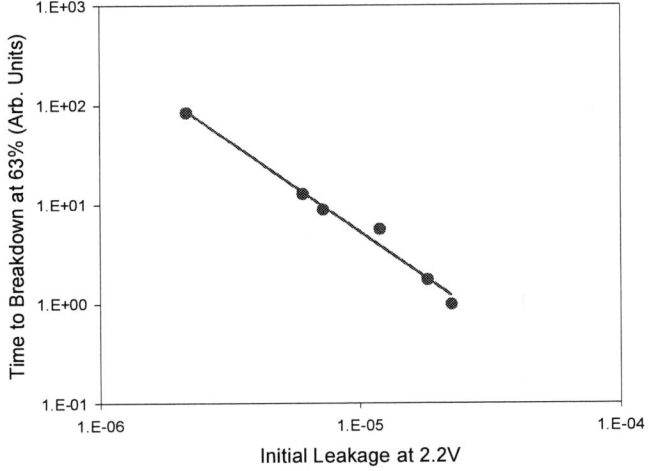

FIGURE 6: Dependence of breakdown time at constant V_{stress} = 2.2 V on initial leakage current.

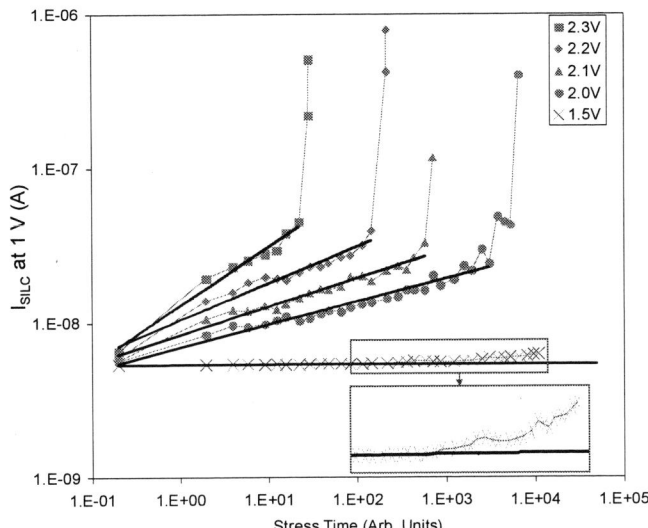

FIGURE 7: *I-t* traces of SILC measured at 1.0 V for CVS at 2.3, 2.2, 2.1, 2.0 and 1.5 V applied to the thinnest IL thickness sample in Fig. 2 at 125 °C. 1.2V data was also collected but due to very little increase, it overlaps with 1.5V data and is not shown in here.

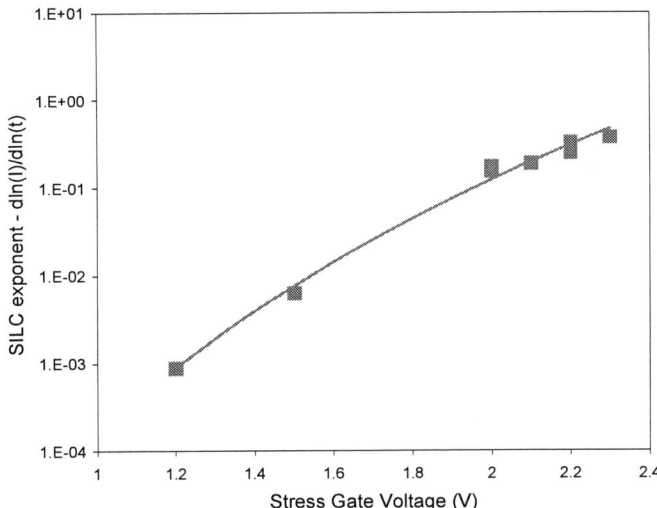

FIGURE 8: Power law time exponent of the early pre-BD SILC, d*ln(I)/*d*ln(t)*, vs. stress voltage for the thinnest IL thickness samples in Fig. 2 at 125 °C.

Similar to the quasi-linear SILC component, this shallow power-law time dependent SILC component is also suppressed at low temperatures when stressed at same voltage. Fig. 9 shows the SILC exponent vs. temperature. SILC exponent was extracted from a range of temperatures from 140 to 30 °C at a constant voltage stress of 2.3 V for same IL thickness.

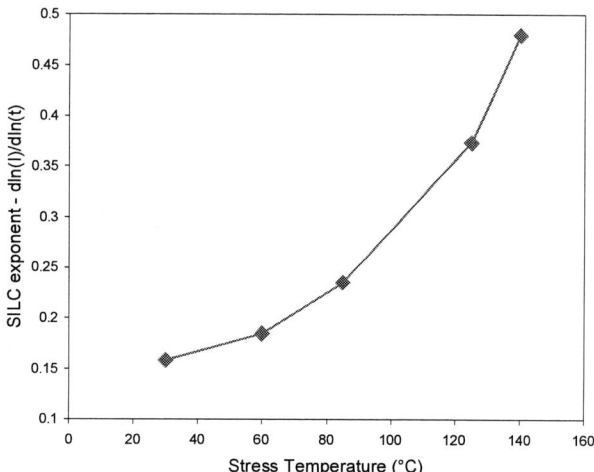

FIGURE 9: Power law time exponent of the pre-BD SILC, $d ln(I)/d ln(t)$, vs. stress voltage the thinnest IL thickness sample in Fig. 2 at 125 °C.

The increase in SILC may have a significant impact on chip reliability in terms of power consumption. A reliability assessment and projection methodology based on SILC increase was previously developed and published [8] and is used here to assess the impact of time to SILC increase for the range of IL thicknesses investigated in this study. Fig. 10 shows that the time evolution of SILC increase follows a power-law as $\Delta J/J_0 = at^b$. The time to SILC (T_{SILC}) increase can be expressed by inverting the power-law expression, as in Eq. (1), for any value of SILC increase failure criterion. (eg. $\Delta J/J_0 = 1$ for 2x SILC increase).

$$T_{SILC} = \left(\frac{1}{a} \frac{\Delta J}{J_0} \right)^{\frac{1}{b}} \qquad (1)$$

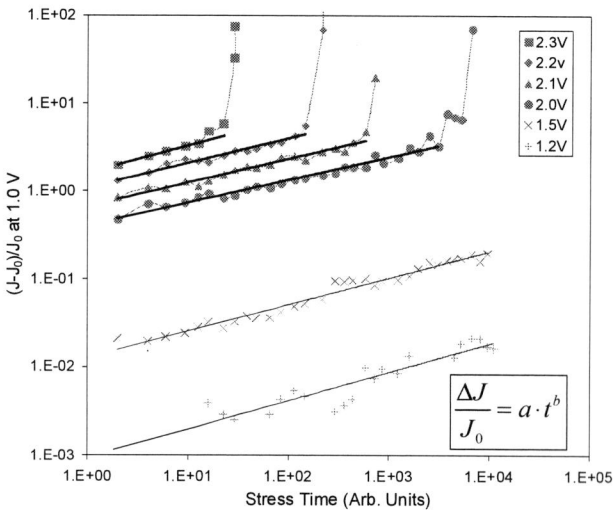

FIGURE 10: Time-evolution of SILC increase measured at 1.0V for CVS at 2.3, 2.2, 2.1, 2.0, 1.5 and 1.2 V applied to the thinnest IL thickness sample in Fig. 2 at 125 °C.

The experimental $\Delta J/J_0$ of each sample was fitted to derive the corresponding values of a and b to calculate T_{SILC}. Fig. 11 shows the mean-time to 2x SILC increase vs. voltage for the thinnest and thickest IL shown in Fig. 2. T_{SILC} is calculated for a number of samples at each stress voltage condition. At lower voltages (1.5 and 1.2 V) the SILC is much smaller as seen in Fig. 7 for 1.5 V, so the data become quite scattered. For the thicker IL in particular, the low voltage stress data was too noisy for fitting due to very low level of SILC increase. These data are not shown in Fig. 11. For the thinner IL, the mean T_{SILC} vs. voltage fits well to a power-law voltage dependent trend for the whole measured voltage range. For the thicker IL, only a very narrow high voltage range was available, so the same power-law voltage acceleration from the thinner IL is used for its projection to low voltage. Under this assumption, the thicker IL projects to ~100x longer time.

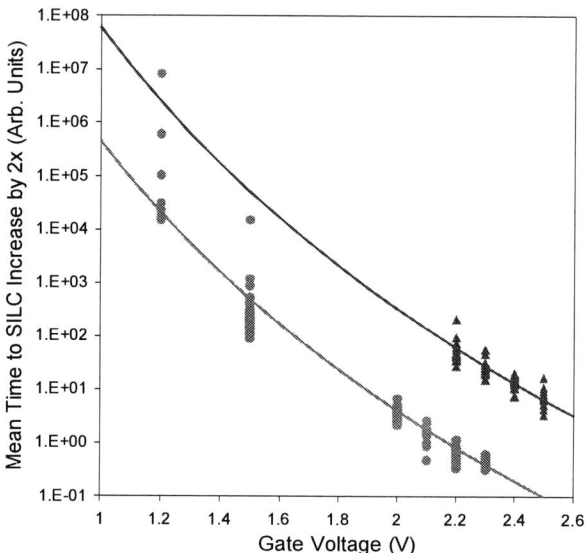

FIGURE 11: Mean time to SILC Increase by 2x vs. stress voltage at 125 °C for the thinnest (red circle symbols) and the thickest (blue triangle symbols) IL samples from Fig. 2.

IV. CONCLUSION

SILC in high-k/metal-gate nFETs is shown to have two components, which differ in their dependence on interface layer thickness ILs. A shallow power-law SILC is strongly suppressed for thicker. Both components seem to be thermally activated. But the "linear" defect generation is independent of IL thickness over the range investigated in this study. These results demonstrate the critical importance of interface layer optimization to improve reliability in high-k/metal-gate devices.

This work was performed at the IBM Microelectronics Div. Semiconductor Research & Development Center, Hopewell Junction, NY 12533.

REFERENCES

[1] C. Prasad et al., IRPS 2008, p. 667; S. Pae et al., IRPS 2009, p.499.
[2] E. Cartier and A. Kerber, IRPS 2009, p. 486.
[3] G. Ribes et al., IRPS 2005, p. 61.

[4] C. Young et al., IPRS 2006, p. 169.
[5] G. Bersuker et al., IRPS 2007, p. 49 ; IEDM 2008, pp. 791.
[6] T. Nigam, A. Kerber, and P. Peumans, IRPS 2009, p. 523.
[7] B. Linder et al., IRPS 2009, p. 510.
[8] W. Lai et al., IRPS 2004, p. 102.

A compact analytic model for the breakdown distribution of gate stack dielectrics

Santi Tous[1], Ernest Y. Wu[2] and Jordi Suñé[1],

(1) Departament d'Enginyeria Electrònica, Universitat Autònoma de Barcelona (SPAIN)
(2) IBM Microelectronics Division, Essex Junction, VT (USA)

Abstract

Recently, the cell-based model of breakdown percolation has been generalized to deal with multilayer gate dielectric stacks. In this work, we depart from the cell-based model to derive a simple analytic compact model for the cumulative failure distribution of dual-layer stack dielectrics in terms of five parameters. The model is validated by large sample size experiments on PFET and NFET transistors with different areas. The model is also applied to single-layer oxides showing progressive breakdown, thus showing that this is a general tool for the reliability assessment of advanced gate dielectrics.

I. Introduction

The exponential growth of the tunneling current leakage associated to the reduction of oxide thickness has fueled the substitution of the thermally grown SiO_2 gate oxide by deposited high-K (HK) dielectrics. However, to preserve the quality of the oxide-semiconductor interface, an interfacial layer (IL) of SiO_2-based material is required, thus leading to the formation of a stack with at least two layers of different dielectric materials. Recently, the percolation model of oxide breakdown (BD) was applied to Hf-based gate dielectric stacks with excellent results [1]. Nigam and coworkers demonstrated that the main features of the breakdown statistics are captured when the percolation model is applied under the assumption that the defect generation dynamics is different in each dielectric layer of the stack [1]. Subsequently, the cell-based model of BD [2,3] was generalized to multilayer stack dielectrics [4] and was demonstrated to be fully equivalent to the percolation picture by comparing with kinetic Monte Carlo (kMC) simulations. Both models are equivalent physics-based pictures that capture the scaling of the BD distribution with device area and with the thickness of each dielectric layer in terms of a number of parameters [1]. However, for practical reliability assessment, we require a model for the BD cumulative failure distribution function, F_{BD}, in terms of a minimum set of convenient parameters. Here, we depart from the cell-based model to develop an analytic model for F_{BD} with five independent parameters, which is validated by large sample size experiments on PFET and NFET transistors with different areas. This compact model is shown to be applicable to situations in which the cumulative failure distribution converges to different Weibull limits at high and low percentiles of failure. In this regard, we show how the model provides a good approximation to statistics of successive breakdowns [5] and we demonstrate that it can also be applied to model the final failure distribution of single-layer oxides showing progressive breakdown (PBD). Thus, our five-parameter compact model is shown to be a very general tool for the reliability assessment of advanced gate dielectrics.

II. Cell-based percolation model applied to gate stack dielectrics

In the cell-based approach to BD percolation, the insulator volume is divided in an array of cells with area σ and thickness a_0. Since the nature of the defects involved in the BD of the HK and the IL is probably different, we consider different cell thickness in each layer, i.e. a_{0HK} and a_{0IL} in the HK and IL, respectively. However, for simplicity, we keep the same cell area in both dielectrics.

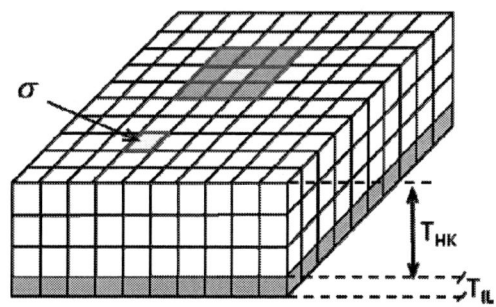

Fig. 1. Schematic representation of the division of the insulator volume in a 3D array of cells of area σ and thickness a_{0IL} in the IL and a_{0HK} in the HK layer. The nine nearest neighbors of a cell in the overlying layer represent the allowed connections in the formation of non-columnar percolation paths.

Defects are assumed to be generated at random in the cells and a cell is assumed to be defective if it contains one or more defects. The BD occurs when a percolation path of defective cells is formed across the whole stack. Defect generation is uniform throughout the dielectric area. This uniformity and the weakest-link character of the BD ensure that the so-called Poisson area scaling property applies. Only defect paths with minimum number of cells are considered, including non-columnar paths connecting each cell to n_C possible nearest neighbor cells in the overlaying layer [3]. The probability to find a BD path departing from an arbitrary cell of the bottom layer and reaching the top electrode is:

$$P_{PATH} = \frac{1}{n_C} \lambda_{HK}^{n_{HK}} \lambda_{IL}^{n_{IL}} \qquad [1]$$

where n_{HK} and n_{IL} are the number of cell layers in the HK and IL, and λ_{HK} and λ_{IL} are defined as the probabilities of finding at least one defective cell among n_C cells in the HK and IL, respectively. In a device with an oxide area A_{OX}, the number of cells in the each layer is $N_{CELL}=A_{OX}/\sigma$ and, applying the weakest-link property, F_{BD} is obtained as:

$$(1 - F_{BD}) = (1 - P_{PATH})^{N_{CELL}} \qquad [2]$$

978-1-4244-5430-3/10 $26.00 © 2010 IEEE

For large enough devices, the number of cells is large, i.e. $N_{CELL} \gg 1$, and the value of P_{PATH} at BD is much smaller than unity for all failure percentiles of interest. Thus, it is possible to approximate the previous equation as:

$$(1 - F_{BD}) = \exp(-N_{CELL}P_{PATH}) \qquad [3]$$

so that:

$$F_{BD} = 1 - \exp\left(-\frac{A_{OX}}{n_C \sigma} \lambda_{HK}^{n_{HK}} \lambda_{I\tilde{N}}^{n_{IL}}\right) \qquad [4]$$

where the Poisson area scaling property of the BD is explicit. Relating this model to the experimental data requires a model for the time dependence of λ_{HK} and λ_{IL}. The density of defects is known to evolve as a power law of time, $N_{OT} = \xi t^{\alpha}$. Thus, the average number of defects in n_C cells is

$$n_{DEF} = n_C \sigma a_0 \xi t^{\alpha} \equiv A t^{\alpha} \qquad [5]$$

and, according to the Poisson distribution, it follows that $\lambda = 1 - \exp\left(-A t^{\alpha}\right)$. Notice that this ensures that $\lambda \to 1$ for $t \to \infty$, as required for a quantity defined as a cumulative damage probability. Considering different defect generation parameters, (ξ_{IL}, α_{IL}) and (ξ_{HK}, α_{HK}), for the IL and HK layers, it follows that:

$$\lambda_{IL} = 1 - \exp\left(-A_{IL} t^{\alpha_{IL}}\right) \qquad [6a]$$

$$\lambda_{HK} = 1 - \exp\left(-A_{HK} t^{\alpha_{HK}}\right) \qquad [6b]$$

The combination of equations 4 and 5 provides a simple analytical model for the cumulative BD distribution function which is completely equivalent to the percolation picture, as it is shown by comparison with Nigam's kMC simulations [1]. An example of this comparison is shown in Figures 2 and 3. For this comparison, we have considered n_C=9 to approach the kind of connections allowed in [1]. However, considering non-columnar percolation paths with connections to n_C neighbours is equivalent to allowing only columnar paths with cells of area $\sigma' = n_C \sigma$. This is evident from eq. 4 and in the definition of A which is implicit in eq. 5. Since σ is a model parameter which is difficult to relate to the microscopic physics, a cell-based model considering only columnar paths as in [2] is perfectly adequate. In Fig. 2, the BD cumulative distribution is depicted for an insulator stack formed by one layer of IL cells and two layers of HK cells. The ratio between defect generation rates $A_{HK}:A_{IL}$ is varied from 1 to 3000 showing how the asymmetry in the defect generation rate modifies the shape of the BD distribution. For equal degradation rates, the BD distribution is a Weibull distribution while for larger ratios, it is evident that the distribution becomes non-Weibull. In fact, as shown in the case of the 3000:1 ratio, the distribution shows a much smaller Weibull slope at high than at low percentiles of failure. At low percentiles (LP) of failure, i.e. short stress times, the cell failure probabilities of eq. 6 can be approximated by power laws of time, $\lambda_{IL} \approx A_{IL} t^{\alpha_{IL}}$ and $\lambda_{HK} \approx A_{HK} t^{\alpha_{HK}}$, and the

distribution converges towards a Weibull distribution with shape factor (Weibull slope)

$$\beta_{LP} = \alpha_{HK} n_{HK} + \alpha_{IL} n_{IL} \qquad [7]$$

and scale factor (characteristic time for % 63.2 failure):

$$T_{LP} = \left(\frac{n_C}{N_{CELL} \, A_{IL}^{n_{IL}} \, A_{HK}^{n_{HK}}}\right)^{\frac{1}{\beta_{LP}}} \qquad [8]$$

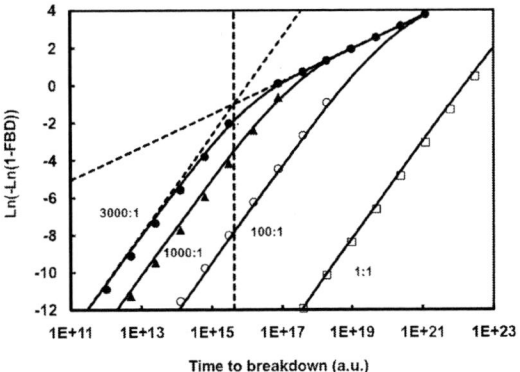

Fig. 2. Weibull plot of the BD distributions given by the cell-based model of [4] (lines) compared to Nigam's kMC simulations (symbols) of 3D percolation of [1]. A gate stack with 1 IL and 2 HK layers and $N_{CELL}=10^4$ is considered. Defect generation rate in the HK layer is varied from being equal to that of the IL (1:1) to being 3000 times higher (3000:1). As explicitly shown in the latter case with the aid of two dashed lines, the distribution is non-Weibull but converges to two different Weibull distributions at the high and low percentile limits. The vertical dashed line represents the transition time T_T, between these two limits. Figure reprinted from [4].

On the other side, at high percentiles (HP), if the defect generation rate is much larger in the HK layer than in the IL (this is a likely assumption in accordance with literature data [1]), the cell failure probability in the HK layer saturates to one, i.e. $\lambda_{HK} \to 1$, for long stress times, while the power law approximation is still valid for λ_{IL}. Hence, at HP, the BD distribution converges to another Weibull distribution with shape factor:

$$\beta_{HP} = \alpha_{IL} n_{IL} \qquad [9]$$

and scale factor:

$$T_{HP} = \left(\frac{n_C}{N A_{IL}^{n_{IL}}}\right)^{\frac{1}{\beta_{HP}}} \qquad [10]$$

Note that this Weibull distribution is the BD distribution of the IL, so that we can conclude that the HK layer plays no relevant role at HP. In fact, as shown by Nigam's kMC simulations, the HK layer is fully degraded at long stress times and the BD times are fully determined by the formation of the percolation path in the IL. The LP and HP Weibull limits are represented by dashed lines in the case of the 3000:1 defect generation ratio in Fig. 2. The weibit of the distribution being defines as $W \equiv Ln(-Ln(1-F))$, a transition time T_T between these Weibull limits can be obtained from the condition $W_{LP}(T_T)=W_{HP}(T_T)$:

$$T_T = (A_{HK})^{-1/\alpha_{HK}} = \left(\frac{T_{LP}^{\beta_{LP}}}{T_{HP}^{\beta_{HP}}} \right)^{\frac{1}{\beta_{LP}-\beta_{HP}}} \qquad [11]$$

Notice that, being related to the saturation of λ_{HK} to unity, this transition time only depends on the defect generation rate in the HK layer.

The most notable merit of the percolation model is the explanation of the scaling of the BD distribution with oxide thickness [5]. In the case of gate stacks, the percolation picture and also the cell-based model explain the scaling with the thickness of each layer. Fig. 3 shows that our cell-based picture fully reproduces the kMC simulation results reported in [1] for different values of HK and IL layer thickness.

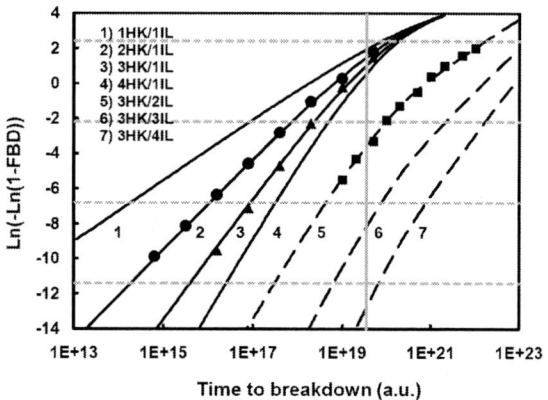

Fig. 3. Weibull plot of the BD distributions obtained from the cell-based model of (lines) compared to Nigam's kMC simulations (symbols) of 3D percolation as reported in [1]. Legend indicates the number of cells in each dielectric of the stack and is proportional to the corresponding thickness. Figure reprinted from [4].

Although the cell-based picture can be generalized straightforwardly to stacks with an arbitrary number of layers of different dielectric materials [4], in this work we will only consider the case of dual-layer stack dielectrics.

III. A five-parameter analytic compact model for the cumulative distribution of dual-layer dielectric stacks.

The cell-based model reviewed in the previous section is a physics-based picture which is an analytic formulation of the percolation model. As such, it relates defect generation and dielectric BD and it depends on a number parameters which are related to the BD physics. In fact, some parameters of the model are related to the properties of the defects involved in the BD and others to the defect generation dynamics. In our model, defects are represented by cells, which are described in terms of geometrical parameters such as the cell area, σ, and the cell thickness

in each layer, a_{0HK} and a_{0IL}. Defect generation dynamics are described in terms of four parameters, namely A_{HK}, A_{IL}, α_{HK} and α_{IL}. These are 7 parameters required to model the BD distribution and its scaling properties with device area and with the thickness of each dielectric layer. However, if our goal is to deal with a single set of experimental data (i.e. the experimental estimation of F_{BD}), it would be convenient to have a compact model in terms of only the minimum number of required parameters that can be extracted from these data. In the case of single-layer dielectrics, the cell-based percolation model is implemented in terms of four parameters, σ, a_0, A and α, while F_{BD} is a Weibull distribution that is fitted with only two parameters, namely the scale and shape factors, T_{BD} and β, respectively. In the case of a dual-layer dielectric stack, we can use equations 7 to 11 to write the F_{BD} of eq. 4 in terms of only 5 independent parameters:

$$F_{BD} = 1 - \exp\left\{ -\left(\frac{t}{T_{HP}} \right)^{\beta_{HP}} \left(1 - \exp\left\{ -\left(\frac{t}{T_T} \right)^{\alpha_{HK}} \right\} \right)^{\frac{\beta_{LP}-\beta_{HP}}{\alpha_{HK}}} \right\} \qquad [12]$$

Here, T_{HP}, T_T, β_{LP}, β_{HP} and α_{HK} have been chosen as independent parameters because the expression of F_{BD} is more compact than using T_{LP} instead of T_T. However, using eq. [11] we can change T_T by T_{LP} so as give a more symmetric role to the LP and HP parameters.

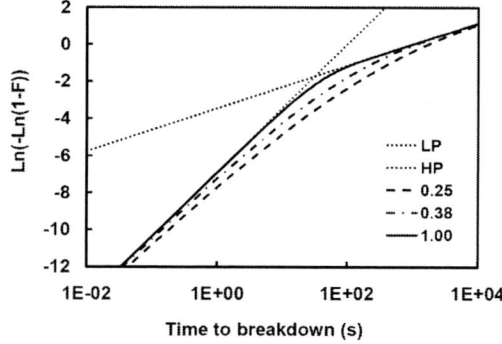

Fig.4. Breakdown distribution function represented in the Weibull plot for $T_{HP}=1000s$, $T_{LP}=100s$, $\beta_{HP}=0.5$, $\beta_{LP}=1.5$ and the three values of α_{HK} shown in the legend.

This is a compact analytic model which includes two parameters to describe the HP limit (T_{HP},β_{HP}), two parameters for the LP limit (T_{LP},β_{LP}) and one parameter, α_{HK}, to model the transition between them. As shown in Fig. 4, eq. 12 perfectly captures the HP and LP asymptotic limits and α_{HK} determines the transition smoothness. The model has five independent (non-correlated) parameters. This number might seem very large in comparison with the standard Weibull model which only has two parametes. However, none of them is arbitrary (all appear naturally in the percolation model) nor redundant, and all of them are required to describe the complexity of F_{BD} (two Weibull distributions and the transition region). In dealing with experimental data, we treat them as fitting parameters, although we know that, within the percolation model, they are directly related to the thickness of the dielectric layers, the effective defect sizes and to the defect generation kinetics.

In any reliability assessment methodology there are three key issues: percentile scaling, area scaling and stress condition (voltage and temperature) scaling. The latter is related to the physics of defect generation and it is not explicitly considered in the percolation model (it is implicit in the stress voltage and temperature dependences of the model parameters). Area and percentile scaling, however, are perfectly captured by our compact model. Percentile scaling is directly provided by the F_{BD} equation, and area scaling is given by the weakest-link property. Area scaling of the time at which the weibit is zero (scale factor if it were a Weibull distribution) and of the local Weibull slope (at a fixed failure percentile) are shown in Fig. 5 and 6, respectively.

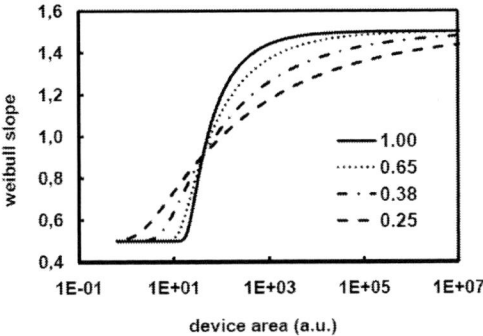

Fig. 5. Change of the local Weibull slope (at a fixed failure percentile) with device area for the distributions considered in Fig. 4. Several values of α_{HK} (shown in the legend) are considered, showing that the smaller is this parameter, the wider is the area range required to capture the full transition.

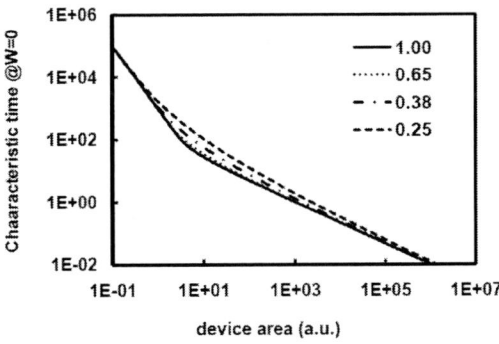

Fig. 6. Change of the scale time @ W=0 with device area. While in single layer BD this double-log plot is a straight line with slope $(-1/\beta)$, in dual-layer dielectric stacks there is a transition between two lines with slopes $(-1/\beta_{LP})$ to $(-1/\beta_{HP})$. Four values of α_{HK} are considered as shown in the legend.

The effects of the transition between Weibull limits are evident in both cases. Moreover, these figures reveal that a good extraction of α_{HK} is crucial for the interpretation of statistical data with limited sample size.

IV. Application to experiment

Large sample size BD experiments were performed on metal-gate PFETs and NFETs (W=152nm/L=40nm) with two-layer stack dielectric formed by a SiO₂-based interfacial layer and a Hf-based HK dielectric layer. Fig. 7 shows the BD distribution for ~2500 PFET devices fitted to the model for two fixed values of α_{HK} (1 and 0.38,

respectively). The model provides an excellent fit in both cases but with different LP parameters (T_{LP}=12.2s β_{LP}=2.7 for α_{HK}=1 and T_{LP}=1.1s ; β_{LP}=4.1 for α_{HK}=0.38).

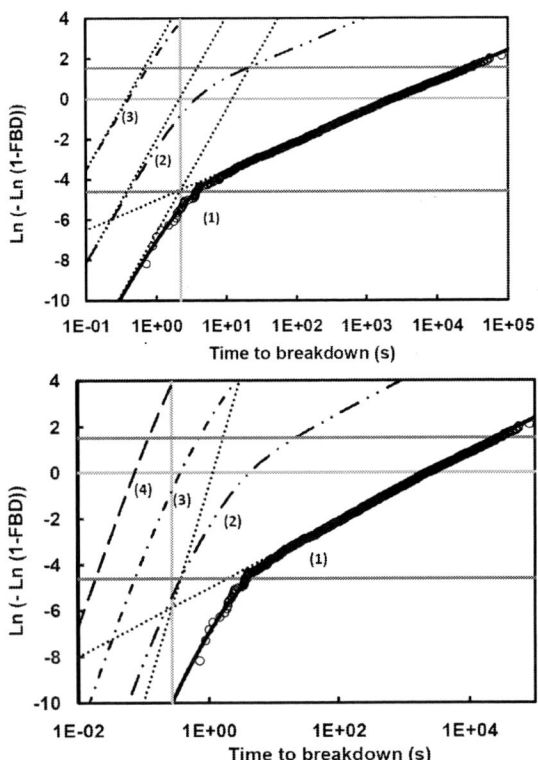

Fig. 7. Experimental BD distribution (dots) measured in ~2500 PFETs with SiO₂/HK stack dielectric stressed to BD @125°C and V_G=-2.8V. The data is fitted fixing two values of α_{HK} (α_{HK}=1 in top figure and α_{HK}=0.38 in bottom one). The fit is excellent and the values of T_{HP} and β_{HP} are the same in both cases. However, the values of T_{LP} and β_{LP} strongly differ from each other, as it is evident from the different LP Weibull asymptotic limits (black dashed lines). As a consequence, the projection to LP for reliability extrapolation is different in both cases. This illustrates the large uncertainty in the extraction of the LP parameters, even with extremely large sample size. The numbered lines correspond to area-scaled distributions, (1)A_{OX}; (2)$100A_{OX}$; (3)10^4A_{OX} and (4)10^6A_{OX}, respectively. The change of the *apparent* slope in the experimental percentile window is evident.

This illustrates that i) the model perfectly captures the shape of the experimental distribution, ii) it is dangerous to fix α_{HK} a-priori and iii) it is difficult to extract LP parameter even with large sample size experiments. Fig. 8 illustrates how the model captures the area scaling property by considering the same type of PFETs but different equivalent areas constructed by connecting in parallel 2, 5, 20 or 100 unit transistors. The uncertainty in the experimental shape of the distribution at low percentiles is again clearly revealed by comparing the scaled distributions. Fig. 9 shows the fitting of the experimental BD distribution for ~3000 NFETs for the same values of α_{HK}. The fitting is also excellent in this case but, again, the accuracy of the extracted LP parameters is rather limited. This uncertainty is neither due to a limitation of the model nor of the extraction methods but it is inherent to the limited percentile window.

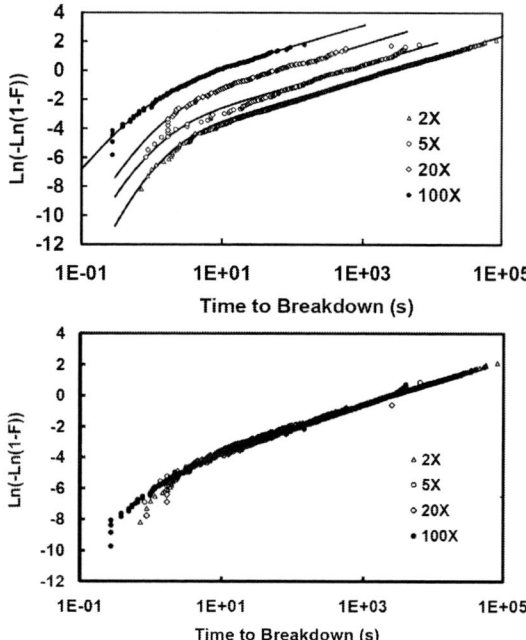

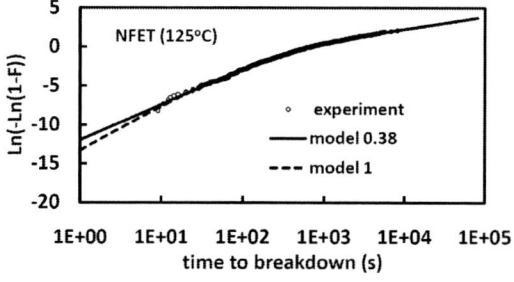

Fig. 8 Experimental BD distributions obtained by stressing PFETs (same devices and stress conditions as in Fig. 7) with four device areas constructed by connecting in parallel 2, 5, 20 or 100 unit transistors. In the top figure, the data are fitted to model using an MLE algorithm keeping α_{HK}=0.38. The extracted T_{HP} nicely scales with area, β_{LP} remains practically constant but T_{LP} and β_{LP} show much larger variations. These variations are due to the large statistical uncertainty of the data at low percentiles. The area-scaled experimental data of the bottom figure further confirm the LP uncertainty.

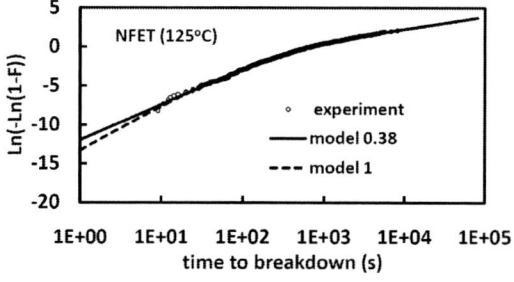

Fig. 9. BD distribution measured in ~3000 NFET devices with Hf-based dielectric stack fitted to the model for two fixed values of α_{HK}. The LP extracted parameters differ (T_{LP}=400s β_{LP}=2 for α_{HK}=1 and T_{LP}=95s ; β_{LP}=2.8 for α_{HK}=0.38). While both sets of parameters provide an excellent fit in the observation window, projection to low percentiles yields significantly different results.

Although the previous experiments are performed in a large number of devices (large statistical sample size), we have seen that there is still a large uncertainty at LP. This issue has been further explored by means of Monte Carlo (MC) simulations of F_{BD}. To this purpose, we have generated random BD times using the model equation with fixed parameters and then we have extracted the model parameters by fitting the resulting *experimental* cumulative distribution. Some examples of simulated F_{BD}

distributions are shown in Fig. 10. The nominal parameters of the MC simulations are very similar to those extracted from the experiments of Fig. 8 so as to allow a direct comparison. Fig. 10 reveals a very similar behavior in all respects, and in particular as far as the uncertainty at LP is concerned. This procedure has been repeated a large number of times, i.e. 2500 random BD distributions (3000 samples each distribution) have been MC simulated, so as to study the statistics of extracted parameters in comparison with the nominal values (Fig. 11).

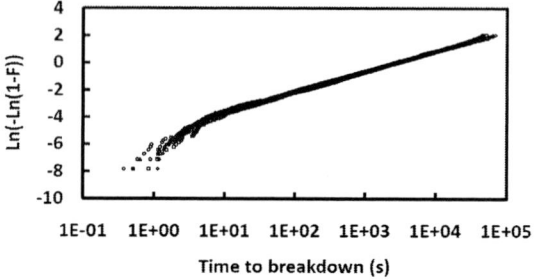

Fig. 10. MC simulation of the BD distribution with fixed model parameters. Comparison with the experimental data of Fig. 8 reveals that the variability at LP is perfectly captured.

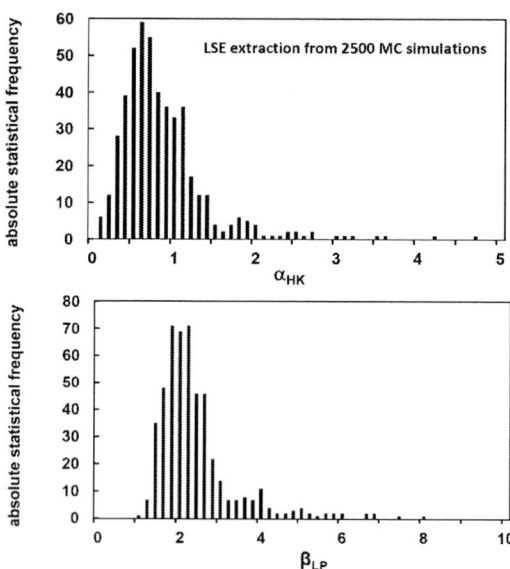

Fig. 11 Histograms showing the extracted values of α_{HK} and β_{LP} from 2500 MC simulations. The MC data were fitted using a LSE algorithm with 3 free parameters (α_{HK}, β_{LP} and T_{LP}). The input values of these parameters for the MC simulation (α_{HK}=0.8, β_{LP}=2.8 and T_{LP}=9s) were obtained by fitting the experimental data of Fig. 7.

Since the HP parameters are extracted with little uncertainty from the considered data (because the HP Weibull limit occupies most of the observation window) we have followed the same procedure as with the analysis of the experimental data. We have kept T_{HP} and β_{HP} fixed at their nominal values and we have extracted only T_{LP}, β_{LP} and α_{HK} by fitting the MC data using either Least Square Estimation (LSE) or Maximum Likelihood

978-1-4244-5430-3/10 $26.00 © 2010 IEEE

Estimation (MLE) methods. Some preliminary results indicate that LSE provides better results than MLE for the extraction of LP parameters. However, this is a preliminary conclusion that needs further systematic work to be confirmed. Fig. 11 shows the histograms of the values of α_{HK} and β_{LP} extracted with an LSE algorithm to be compared with the nominal values $\alpha_{HK}=0.8$ and $\beta_{LP}=2.8$. The distributions of both parameters show a large standard deviation. The distribution of β_{LP}, moreover, shows a significant bias, the extracted values being substantially smaller than the nominal value in most cases, the main peak of the distribution being well below 2.8.

Although the extraction of parameters might be difficult and show some uncertainty due to limited statistical sample size, the previous analysis of experimental results indicate that the 5-parameter model is an excellent compact tool to model the distribution of BD both for PFETs and NFETs with HK dielectric gate stacks. Moreover, this compact model might find extended application in other cases such as single-layer progressive breakdown [6] and even to model the distribution of successive BD events [7,8]. The shape of the BD distribution of two-layer stack dielectrics is very similar to that of the failure distribution in single-layer oxides affected by PBD and also to that of successive BD distributions. In all these cases, the distribution converges to HP and LP Weibull limits with the low percentile Weibull slope being larger than the HP one.

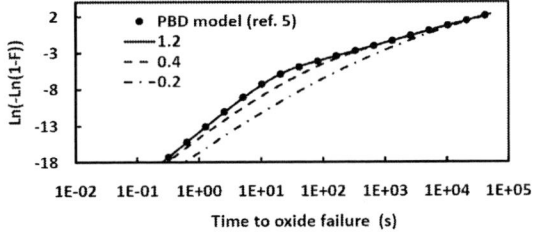

Fig. 12 Demonstration that our model allows to perfectly fit the PBD model of ref. 5. In this particular case, the PBD model is perfectly captured with $\alpha_{HK}=1.2$. Other values of α_{HK} (0.4 and 0.2) are also shown for comparison.

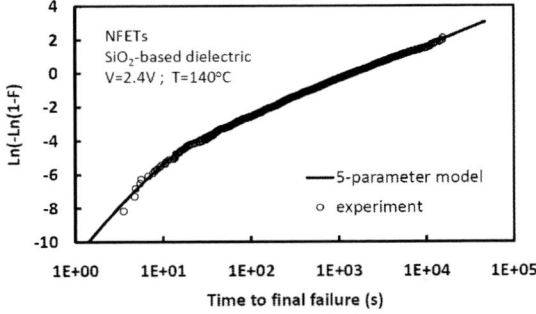

Fig. 13. Experimental failure distribution for ~2500 small NFETs with SiO_2-based single layer dielectric stressed to BD in inversion at 2.4V and 140°C. These oxides show PBD effects. The current failure limit was fixed here to 10μA. Fitting to the 5-parameter model of eq. 8 is excellent with $\alpha_{HK}=1$, $T_{HP}=1600s$, $\beta_{HP}=0.9$; $T_t=10s$, and $\beta_{LP}=2.7$.

This has motivated us to apply our model to fit PBD failure distributions. To this purpose we have compared the results of our stack BD model with those obtained with

the PBD failure distribution model of [6]. Fig. 12 shows this comparison in one particular case and for different values of α_{HK}. It is demonstrated that with an appropriate value of this parameter, our analytical model can perfectly be applied to single-layer PBD failure distributions. The 5-parameter model has been also successfully applied to experimental BD data showing PBD, as shown in Fig. 13. The successive BD theory provides analytic exact equations for the distribution of the 1st, 2nd, ...K^{th} breakdown events. Thus, there is no need of a compact model to deal with this issue. However, we have also applied the 5-parameter model to the statistics of successive BDs as an ulterior demonstration of the general applicability of this model to model failure distributions showing two Weibull limits at LP and HP. This is shown in Fig. 14 for the distribution of the 2nd and 3rd events.

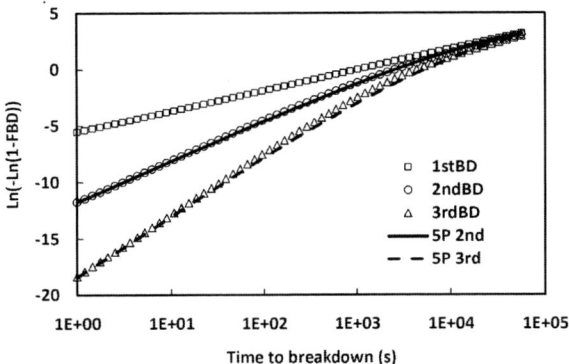

Fig. 14. Application of the 5-parameter compact model developed in this work to the case of successive BD statistics. Symbols correspond to the exact successive BD theory [8] and lines to the the fitting with the 5-parameter model. The values of α_{HK} are 0.45 and 0.4 to fit the 2nd and 3rd BD distribution, respectively.

V. Conclusions.

A compact model for the BD distribution of stack dielectrics has been presented and validated by MC simulation of the percolation theory and by large sample size experiments performed on metal-gate PFETs and NFETs with small EOT Hf-based dielectric stacks. This is a 5-parameter model which is expected to become the kernel of reliability assessment methodology for HK gate-stack dual-layer dielectrics. Our results also reveal the inherent difficulty of exploring the BD distribution at low failure percentiles even with large statistical sample size experiments. Due to its flexibility, the 5-parameter model has also been shown to be applicable to fit progressive BD failure distributions of single layer SiO_2-based dielectrics.

Acknowledgements

J. Suñé and S. Tous acknowledge funding by the MICINN under contract TEC2009-09350, the EU FEDER program, and the DURSI of the Generalitat de Catalunya under project no. 2009SGR783. J.S. also acknowledges the ICREA ACADEMIA award. The experimental work and hardware fabrication was performed by the Research Alliance Teams at various IBM Research and Development facilities.

References

[1]. T. Nigam, A. Kerber and P.Peumans, *Accurate model for time-dependent dielectric breakdown of high-K metal gate stacks*, Proc. of the International Reliability Physics Symposium 2009, p. 523-530.

[2] J. Suñé, *New Physics-Based Analytic Approach to the Thin-Oxide Breakdown Statistics*, IEEE Electron Device Letters, vol. 22, pp. 296-298 (2001).

[3] A. T. Krishnan and P. Nicollian, *Analytic extension of the cell-based oxide breakdown model to full percolation and its implications*, Proc. of the International Reliability Physics Symposium 2007, p. 232-239.

[4] J. Suñé, S. Tous and E.Y. Wu, *Analytical cell-based model for the breakdown statistics of multilayer insulator stacks,* IEEE Electron Device Letters, 30 (12), 1359 (2009)

[5] R. Degraeve, G. Groeseneken, R. Bellens, J. L. Ogier, M. Depas, P. J. Roussel, H. E. Maes, *A consistent model for the thickness dependence of intrinsic breakdown in ultra-thin oxides*, Int. Electron Device Meeting Tech. Dig, pp. 866-869, 1995

[6] S.Tous, E.Y. Wu and J. Suñé, *A compact model for oxide breakdown failure distribution in ultrathin oxides showing progressive breakdown,* IEEE Electron Device Letters, vol. 29, 949-951 (2008).

[7] M.A.Alam, R.K. Smith, B.E. Weir and P.J. Silverman, *Uncorrelated breakdown of integrated circuits,* Nature, vol. 420, pp. 378-379 (2002).

[8] J. Suñé and E. Y. Wu, *Statistics of Successive Breakdown Events in Gate Oxides,* IEEE Electron Device Letters, vol. 24, pp. 272-274 (2003).

Time Dependent Dielectric Breakdown and Stress Induced Leakage Current Characteristics of 8Å EOT HfO₂ N-MOSFETS

Robert O'Connor, Greg Hughes
School of the Physical Sciences, Dublin City University,
Dublin, Ireland.
+353 –1-700 5732, roc@physics.dcu.ie

Thomas Kauerauf, Lars-Åke Ragnarsson
CMOS Device Research Dept, IMEC,
Leuven, Belgium.

Abstract—**In this work we present the time dependent dielectric breakdown (TDDB) characteristics of LaO capped HfO₂ layers with an equivalent oxide thickness of 8Å. The layers show maximum operating voltages in excess of 1V. Such high reliability can be attributed to very high Weibull slopes. We examine the origin of the high slopes by a detailed study of the evolution of the stress induced leakage current with time, temperature and stress voltage.**

Keywords- Hafnium oxide; High-k dielectric; stress induced leakage current, CMOS reliability, time dependent dielectric breakdown.

I. INTRODUCTION

High-k dielectrics to replace SiO₂ have been successfully incorporated at the 45-nm complementary metal-oxide-semiconductor (CMOS) technology node. However, continued research into the scaling of the high-k layer below 1nm equivalent oxide thickness (EOT) is necessary in order to adhere to the technology roadmap. At present there appear to be several barriers to further scaling including film uniformity, significant negative bias temperature instability (NBTI) , and uncertain time dependent dielectric breakdown (TDDB) and stress induced leakage current (SILC) characteristics [1,2,3].

This work focuses on the time dependent dielectric breakdown (TDDB) of LaO capped HfO₂ layers with an equivalent oxide thickness of 8Å. The layers show exceptionally high operating voltages, greater than 1V. In section III we show that this high reliability can be attributed to very high Weibull slopes. We examine the origin of the high slopes in section IV by a detailed study of the evolution of the stress induced leakage current behaviour, and evidence suggests that a defect level deep in the silicon bandgap is responsible for the breakdown in these layers.

II. EXPERIMENT

Two identical gate stacks were considered in this work with one having Si as the substrate material and the other SiGe. The dielectrics were grown on 300mm (100) Si-wafers using a metal inserted poly-Si process (MIPS). The gate stack was formed with an initial IMEC clean followed by 36 cycles of atomic layers deposition (ALD) HfO₂. A LaO capping layer was used to control the work function. The gate was formed by 5nm physical vapour deposition (PVD) TiN followed by 2nm PVD Si.

All experiments were performed on square gated transistors with areas of 1×10^{-8} cm² and 4×10^{-8} cm². This processing resulted in an EOT of 7.8Å and 8.1Å for the Si and SiGe substrates respectively. These values were extracted from Hauser fitting of C-V curves at 100 kHz to 100MHz [4]. The transmission electron microscope (TEM) image from a sample with almost identical processing in figure 1 shows no detectable interfacial layer and a physical thickness of ~2nm.

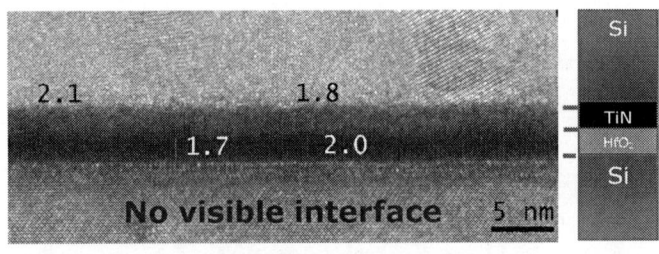

Figure 1. TEM image of a layer from the same lot as the wafers in this work with 2nm TiN gate. No interfacial SiO₂-like layer is observed.

978-1-4244-5430-3/10 $26.00 © 2010 IEEE

III. TIME DEPENDENT DIELECTRIC BREAKDOWN

The initial TDDB evaluation was carried out at 125°C and voltages from 2.2-2.5V. The devices measured were 1μm x 1μm transistors. The time-to-breakdown (t_{BD}) Weibull distributions were fitted with the maximum likelihood method, to ensure accurate extraction of the Weibull slope (β) and the 63%-value (η). Figure 2a shows the resulting TDDB distributions. Using a soft breakdown criterion of a current jump of 10μA, a Weibull slope of 1.54 is extracted for the 7.8Å layer and 1.82 for the 8.1Å case. Such slopes are well in excess of the ~0.8-1 that is commonly reported in the literature for layers of similar thickness [5].

Due to the positive impact of a high Weibull slope when scaling the TDDB distribution to smaller device areas and to the low percentile failure region, the power-law reliability extrapolation in figure 2b shows 10-year maximum operating voltages of 1.24 and 1.3V for the Si and SiGe substrates respectively.

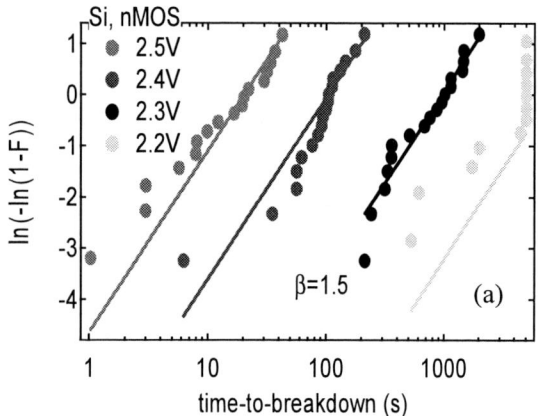

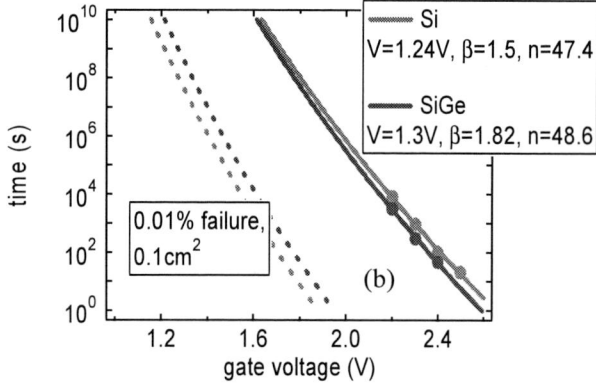

Figure 2. (a) Weibull plot of the tddb distribution for the Si substrate devices showing a Weibull slope of 1.5. (b) Power law lifetime extrapolation scaled to 0.1% failure rate and 0.1cm2. Operating voltages of >1V are forecasted.

The percolation model of oxide degradation [6] shows that the t_{BD}-distribution is Weibull distributed with slope $\beta = m.\beta_{ot}$ where 'm' is the logarithmic trap generation rate and β_{ot} is equal to the number of traps in the path

Thus, a high Weibull slope is caused either by more traps being required to trigger the breakdown or a higher trap generation rate. The model suggests that for SiO$_2$ layers of equivalent physical thickness to those under consideration here, 3 traps should be required to cause breakdown [7]. If the percolation distance is taken to be the same for HfO$_2$ it would mean that for a 3-trap path (with $\beta=1.82$) the 8.1Å layer would have a defect generation rate of 0.61, much higher than the 0.35-0.4 commonly seen in the literature [8]. However, it is commonly accepted that due the reduced barrier height of the high-k compared with SiO$_2$ the effective capture cross section of a high-k defect is larger than that of an SiO$_2$ defect, meaning that the percolation distance is more likely to be reduced in the high-k [9]. If anything then, we should expect that 2 well placed traps or perhaps 3 poorly aligned traps is the maximum needed to cause breakdown in these layers.

IV. STRESS INDUCED LEAKAGE CURRENT ANALYSIS

The I-t traces used to evaluate the TDDB reliability in the SiGe devices are shown in figure 3 and display a significant SILC component which emerges during the device lifetime. To further investigate the possibility of a high defect generation rate in these devices we examined the I-t traces during stress.

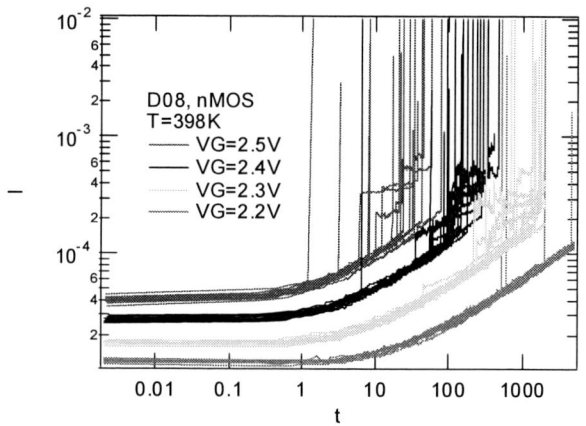

Figure 3. I-t traces measured during constant voltage time dependent dielectric breakdown show significant SILC of up to 5x initial leakage at breakdown for the stress voltage considered.

It has been shown that subtracting the initial charging component from the I-t trace, the pure leakage during the stress can be extracted [10]. This is shown in figure 4a for the SiGe devices. These measurements are performed on larger 2μm x 2μm FETs to maximise the SILC at low stress voltages.

The extracted SILC generation rate from the slope of these curves is 0.63 in line with that suggested by a 3 trap TDDB model with high trap generation rate. Note however that the current in the region of the fit is in the micro-amp range, suggesting that 2-trap conduction is already dominant. If this is the case then, as shown by Degraeve et al [10], the SILC generation rate would be twice the TDDB trap generation rate.

Probing of lower SILC levels with this method is not possible as illustrated schematically in figure 4b. Directly fitting the I-t trace to factor out the charging component is not

978-1-4244-5430-3/10 $26.00 © 2010 IEEE

correct because the I-t trace in this region is partly made up of SILC generated early in the stress from single traps; the region where the logarithmic trap generation rate and the SILC generation rate are the same. This is the reason why the power law fit in figure 4a fails in the low SILC regime.

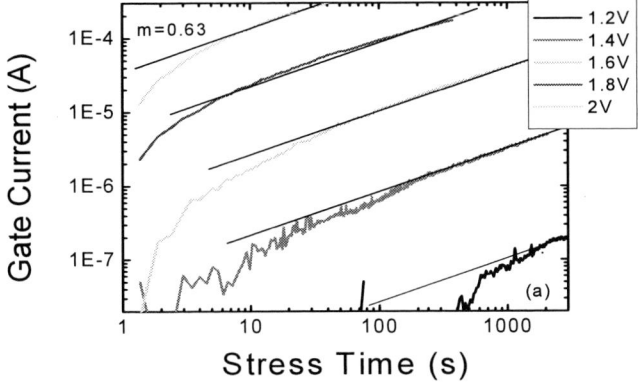

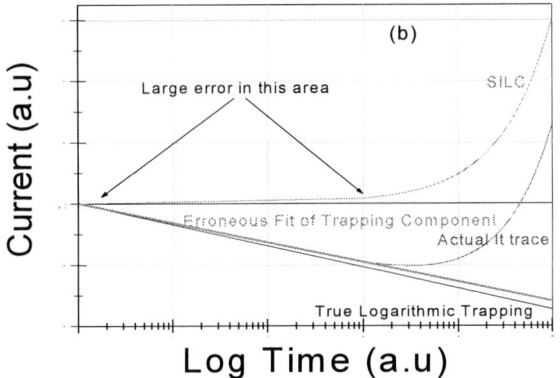

Figure 4. (a) Pure leakage current component extracted from the I-t trace by fitting and subtracting the charging component. (b) Schematic diagram of the error introduced in the single trap regime by directly fitting the I-t trace. The blue curve represents the real I-t trace. Fitting the trace with the pink line does not take into account the presence of single trap SILC early in the stress.

Commonly, SILC measurements involve stressing the dielectric at high field and measuring the increase in leakage current either 'on the fly' as detailed above, or by interrupting the stress and sensing at a fixed voltage close to the operation condition, which may or may not be favourable for tunneling through traps created during the stress. For an nMOSFET, the chosen voltage is typically ~0.5V, which represents a band alignment where resonant tunneling through defects at, or slightly below the silicon conduction band is the dominant trap assisted conduction mechanism. In such a case, the Fermi level is aligned with these defects whereas defects with other energy positions will not contribute significantly to the SILC at this gate voltage.

Measuring the SILC using a stress-and-sense methodology where the stress is interrupted and a full I_gV_g curve is taken

allows sensing defects across the bandgap by direct tunneling where the on-the-fly method will only sense deeper defects by valence band tunneling and is insensitive to resonance effects [11], which is far removed from the operation condition as illustrated in fig. 5.

Furthermore the possibility exists to perform a discharge between the stress and sense conditions to minimise the effect of oxide charging. The stress and sense method was employed for a range of gate voltages and temperatures up to 175°C. Fig. 6 shows the SILC in these thin layers is relatively unaffected by discharging steps indicating that it is largely an irreversible process. Nonetheless, in all stress-and-sense measurements detailed in this work, the discharge step is performed.

There still exists in the literature some confusion over whether SILC is mediated through empty states or if a site needs to trap a carrier to enable it to become involved in the SILC process [12]. In this work the thickness of the layers is such that one would not expect significant trapping, though the SILC component still remains after the discharge indicating that it is mediated through neutral electron traps. However, further work needs to be done to understand the link (if any) between the traps which cause SILC and those which contribute to BTI effects.

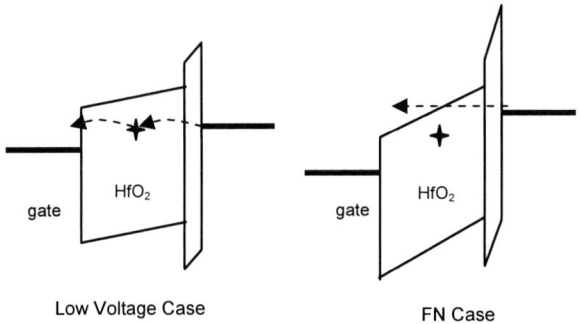

Figure 5. When sensing in the direct tunneling regime the Si Fermi level is resonant with defects in the HfO2 bandgap, whereas the band alignment when directly measuring from the I-t trace may not sense the created defects efficiently.

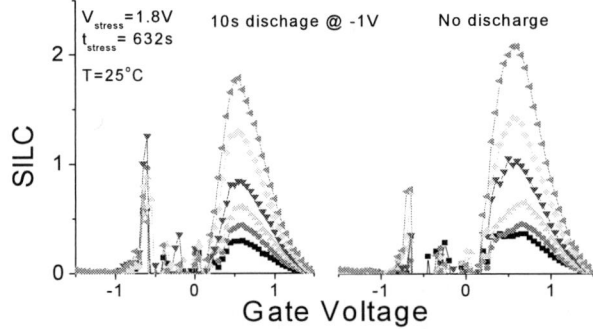

Figure 6. SILC profile for devices stressed with and without a 10s discharge at -1V prior to each stress phase. The discharge step removes less than 10% of the total 'bulk' SILC indicating that the SILC is mediated through empty defect states which remain in the oxide after the discharging step.

978-1-4244-5430-3/10 $26.00 © 2010 IEEE   801

Figure 5 shows the evolution of the SILC as a function of time for several stress conditions at a sense voltage of 0.5V. In this case we see a strong dependence of the SILC generation rate on voltage (fig 5a) and indeed temperature (fig. 5b) which disagrees with the model proposed in [12]. However, even the gentlest stress in the single trap regime only yields a SILC generation rate of 0.32 which cannot explain the TDDB results and so an increased bulk HfO_2 trap generation rate seems unlikely.

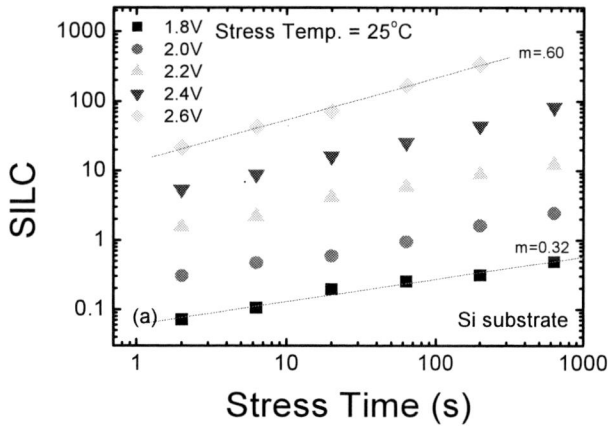

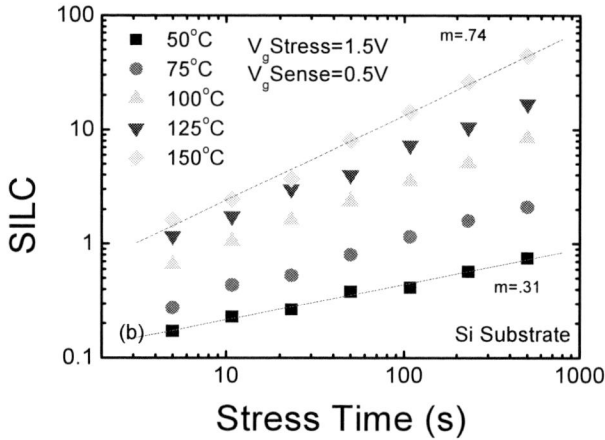

Figure 7. SILC generation as a function of (a) voltage for a series of stresses at 25°C and (b) temperature for a series of stresses at V_g=1.5V. A sense voltage of 0.5V is used in both cases. The SILC generation rate is dependent on both temperature and voltage.

Previous work [11] has shown that a sense voltage just below V_{FB} is resonant with a defect band close to the Si/SiO_2 interface. In order to examine the degradation rate close to the interface, we took the same stresses as above but changed the sense voltage to -0.7V. The kinetics are compared in figure 8 for sense voltages of (a) 0.5V and (b) -0.7V. The SILC generation rate is found to be 0.59 even at the lowest stress condition in the -0.7V case where the leakage increase after 1000s is 2nA and single trap conduction is dominant, rising to 1.4 at very high stress conditions. As for the case of 0.5V sense voltage, the sensitivity to stressing gate voltage is evident

showing the transition from one-trap to two-trap SILC during the stress.

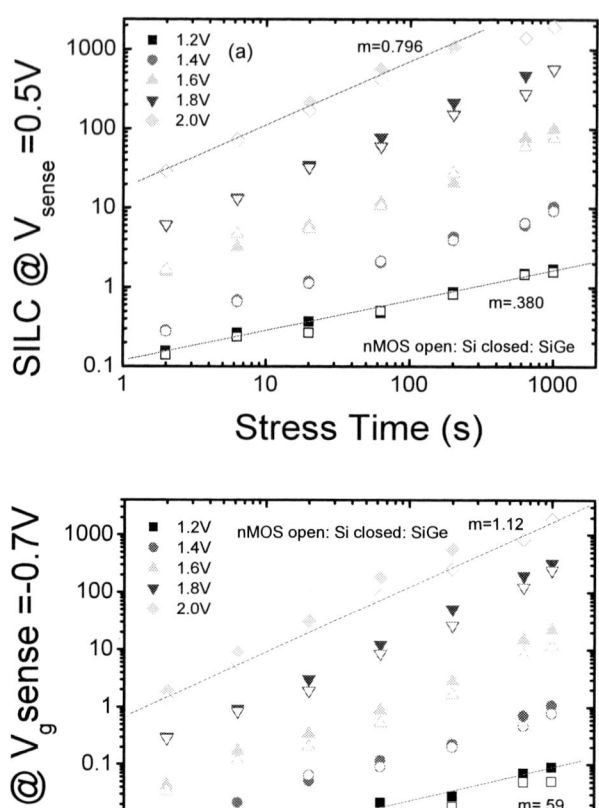

Figure 8. Comparison of the SILC generation behvaiour for a sense voltage of (a) 0.5V and (b) -0.7V. The measured SILC generations rates at -0.7V are much higher than those at 0.5V indicating a higher defect generation rate deep in the bandgap close to the interface.

As mentioned earlier, previous work has shown that SILC sensed at ~ -0.7V in high-k stacks with an SiO_2 interfacial region is linked to defects in the SiO_2. However the TEM image in figure 1 displays no visible interfacial region indicating that in this case, the defect level must result from the HfO_2 layer. Substrate hot electron stresses showed that the defect generation at -0.7V is not increased any more than elsewhere in the SILC spectrum for intensive interface stressing, indicating that the defect level is indeed present through the bulk HfO_2.

The TDDB statistics we have obtained can be explained if these traps are the ones responsible for the breakdown behavior. As shown in figure 8b, a trap generation rate of 0.6 is measured at the lowest stress condition. Thus a 3 trap breakdown path with this generation rate would result in a Weibull slope of 1.8, in line with the value we have measured.

Further investigation of this defect level shows an increased sensitivity to temperature. Measurements show an activation energy of 0.6eV for a sense voltage of -0.7V compared with 0.4eV at 0.5V sense. Figure 9 shows the SILC generation rate as a function of temperature for the two sense voltage during a 100s stress at a number of stress conditions. At room temperature and the lowest stress condition, both sense voltages show a SILC generation rate of ~0.4. This consistent with the Weibull slope of a TDDB distribution at room temperature of 1.22 and again indicates a 3-trap breakdown.

For the 125°C case, the single trap SILC generation rate at the lowest stress condition is again consistent with the defect at -0.7V resulting in the breakdown of the oxide, and at 175°C we can see single trap SILC dominant for both sense conditions at low stress voltage, and the dominance of 2-trap SILC at V_{stress}=1.4V where the SILC generation rate in each case is 2x the trap generation rate. The emergence of new defect generation mechanisms at elevated temperature in high-k materials has previously been reported in [13] and the current results show the effect is still apparent in thinner high-k layers without an interfacial SiO_2.

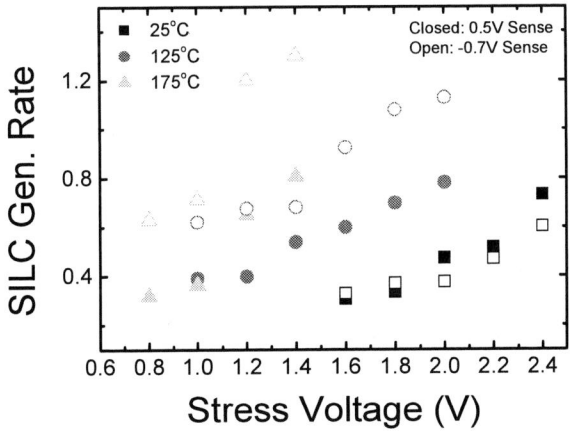

Figure 9. SILC GENERATION RATES EXTRACTED FROM A 100s STRESS AS A FUNCTION OF TEMPERATURE FOR TWO DIFFERENT SENSE VOLTAGES. THE DEFECT SENSED AT -0.7V IS MORE TEMPERATURE SENSITIVE THAN THAT AT 0.5V.

V. CONCLUSION

TDDB studies of high-performance HfO_2 based nMOSFETs show maximum operating voltages in excess of 1V. Within error-bar, the SiGe substrate devices show identical reliability to those with Si substrates. SILC measurements to analyze the defect generation behavior show a voltage and temperature dependent SILC generation rate and suggest that the defect which causes the breakdown is located deep in the silicon bandgap and is highly temperature sensitive. The high trap generation rate associated with this defect leads to an increased Weibull slope during TDDB experiments which in turn drives the high maximum operating voltage.

ACKNOWLEDGEMENTS

The authors wish to acknowledge the IMEC pilot-line for sample processing and Robin Degraeve and Ben Kaczer for helpful discussions. R O'C acknowledges funding from the Irish Research Council for Science, Engineering and Technology under the Embark Initiative and from Science Foundation Ireland.

REFERENCES

[1] Harris, H.R. Choi, R. Lee, B.H. Young, C.D. Sim, J.H. Mathews, K. Zeitzoff, P. Majhi, P. Bersuker, G., "Recovery of NBTI degradation in HfSiON/metal gate transistors", Inegrated Reliability Workshop, pp.132–135, Oct 2004.

[2] Okada, K. Ota, H. Nabatame, T. Toriumi, A., "TDDB and BTI reliabilities of high-k stacked gate dielectrics - Impact of initial traps in high-k layer", ICICDT, pp.87-90, June 2008.

[3] A. Kerber, E. Cartier, B.P. Linder, S.A. Krishnan, T. Nigam, "TDDB failure distribution of metal gate/high-k CMOS devices on SOI substrates", IRPS, pp.505-509, 2009.

[4] J.R. Hauser, A. Ahmed, "Characterization of ultrathin oxides using C-V and I-V measurements", AIP Int. Conf. Characterization Metrology for ULSI Technology, pp. 235-239, 1998.

[5] In-Shik Han; Won-Ho Choi; Hyuk-Min Kwon; Min-Ki Na; Ying-Ying Zhang; Yong-Goo Kim; Jin-Suk Wang; Chang Yong Kang; Bersuker, G.; Byoung Hun Lee; Yoon Ha Jeong; Hi-Deok Lee; Jammy, R.; "Time-dependent dielectric breakdown of La_2O_3 -doped high-k/metal gate stacked NMOSFETs", Electron Device Letters, Vol. 30, Issue 3, pp. 298-301, 2009.

[6] R. Degraeve, G. Groeseneken, R. Bellens, J.L. Ogier, M. Depas, P.J. Roussel and H.E. Maes, "New insights in the relation between electron trap generation and the statistical properties of oxide breakdown", IEEE Trans. Electron Devices 45, pp. 904–911, 1998.

[7] R. Degraeve, B. Kaczer, F. Schuler, M. Lorenzini, D. Wellekens, P. Hendrickx, J. Van Houdt, L. Haspeslagh, G. Tempel, G. Groeseneken, "Statistical model for stress-induced leakage current and pre-breakdown current jumps in ultra-thin oxide layers", IEDM, pp. 6.2.1-6.2.4, 2001.

[8] F. Crupi, R. Degraeve, A. Kerber, D. H. Kwak, G. Groeseneken, "Correlation between Stress-Induced Leakage Current (SILC) and the HfO_2 bulk trap density in a SiO_2/HfO_2 stack", IRPS, pp. 181-187, April 2004.

[9] J. Sune, E.Y. Wu, S. Tous, "A physics-based deconstruction of the percolation model of oxide breakdown", Microelec. Eng. Volume 84, Issues 9-10, pp.1917-1920, 2007.

[10] R. Degraeve, T. Kauerauf, M. Cho, M. Zahid, L.A. Ragnarsson, D.P. Brunco, B. Kaczer, Ph. Roussel, S. De Gendt, G. Groeseneken, "Degradation and breakdown of 0.9 nm EOT SiO_2 ALD HfO_2 metal gate stacks under positive constant voltage stress", IEDM, pp.408-411, 2005.

[11] R. O'Connor, L. Pantisano, R. Degraeve, T. Kauerauf, B. Kaczer, P.J.Roussel, G. Groeseneken, "SILC defect generation spectroscopy in HfSiON using constant voltage stress and substrate hot electron injection" IRPS, pp.324-329, 2008.

[12] E. Cartier, A. Kerber, "Stress-induced leakage current and defect generation in nFETs with HfO2/TiN gate stacks during positive-bias temperature stress", IRPS, pp. 486-492, 2009.

[13] S. Sahhaf, R. Degraeve, R. O'Connor, B. Kaczer,, M.B, Zahid, P.J Roussel, L.Pantisano, G. Groeseneken, "Evidence of a new degradation mechanism in high-k dielectrics at elevated temperatures", IRPS, pp.494-498, 2009.

Frequency-Dependent Charge-Pumping: The Depth Question Revisited

F. Zhang, K. P. Cheung*, J.P. Campbell, J. Suehle
National Institute of Standards and Technology
Semiconductor Electronics Division
Gaithersburg, MD, USA
*kpckpc@ieee.org

Abstract— A popular defect depth-profiling technique, frequency-dependent charge-pumping is carefully re-examined. Without complicated math of modeling, the physics behind the technique is examined clearly. It is shown that there is no unique relationship between the measurement frequency and the probed depth. The conclusion is that frequency-dependent charge-pumping is not a defect depth-profiling technique.

Keywords-frequency dependent charge pumping

I. INTRODUCTION

Frequency-dependent (FD) charge-pumping (CP) is considered a "power technique" to profile defect depths within the gate dielectric. It has been used to illustrate that electrical stress can increase defect density in the high-k layer [1]. However, this claim is disputed [2] by data obtained using the same technique. This dispute is at the heart of a raging controversy on how to correlate depth with frequency in the FD-CP experiments. The disagreement is, however, only on quantitative level. Everyone involved expects that a relationship between depth and frequency exists and is sufficiently unique. **It is the purpose of this paper to show that a simple correlation does not exist and that the technique cannot be used for the stated purpose.**

II. "BASIC PRINCIPLE" OF FD-CP

The basic CP process consists of a two steps. The principle is as shown in figure 1. The basis of FD-CP is that if the time spent in inversion/accumulation is sufficiently long, "near interface" traps can also contribute the CP current as illustrated in figure 2. This formalism has been around for more than 30 years [3], and has been treated by several groups [2-7]. The relationship between the maximum depth at which oxide traps contribute to CP current and the time spent (τ) in inversion/accumulation (half of a CP period) is usually expressed using the tunneling-front formalism [8]:

$$\tau = t_0 e^{2\beta x} \tag{1}$$

where β is the characteristic length, x is the maximum depth, and t_0 is a pre-factor, or the time required when the depth is zero. β has to do with the barrier height and effective mass of the tunneling process – something that can be well-defined if the model is clear. The controversy surrounds the proper choice of the value of t_0. Every group seems to champion its own

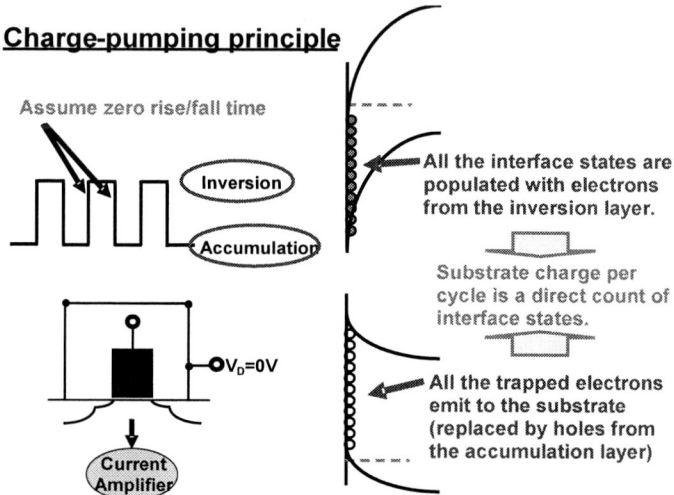

Figure 1. The basic charge-pumping principle involves switching the MOSFET from strong accumulation to strong inversion (and back) rapidly while keeping the source/drain grounded. The DC component of the substrate current is measured at zero volts and is taken as a direct measure of the recombination events happening at the SiO$_2$/Si interface.

value with very little agreement from group to group. Within the groups involved, it is generally believed that if t_0 can be pinned down, then a unique relationship between frequency and probing depth within the dielectric can be attained.

The standard way to obtain t_0 is via its relationship with the mean capture cross section (σ):

$$t_0 = \frac{1}{\sigma v_{th} n} \tag{2}$$

where v_{th} is the thermal velocity of the carriers and n is the carrier density. The value of n is often taken as the inversion charge density (on the order of 10^{13} cm^{-2}). Problems arise when one realizes that a survey of the literature yields reported σ values which vary widely (from 10^{-14} cm^2 to 10^{-19} cm^2) [9]. While most groups choose σ values between 10^{-15} cm^2 to 10^{-17} cm^2, the utilized value of t_0 varies from 6.6×10^{-14} to 10^{-8} s (depending the details of the model). This leads to a significant difference in the depth associated with a given CP frequency and hence the surrounding controversy.

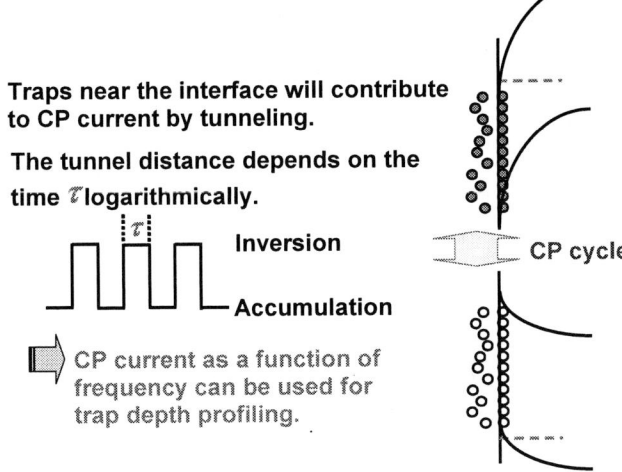

Traps near the interface will contribute to CP current by tunneling.

The tunnel distance depends on the time τ logarithmically.

CP current as a function of frequency can be used for trap depth profiling.

Figure 2. Frequency-dependent charge-pumping (FD-CP) assumes that when given enough time at inversion/accumulation, near interface oxide traps can also participate in producing CP current through tunneling. Lower frequency means longer time available for tunneling, and deeper oxide traps can be accessed.

Inversion:
Interface traps capture electrons + electrons tunnel into oxide traps

Accumulation:
Trapped electrons capture holes + holes tunnel into oxide traps

Figure 3. 4-steps sequence of the frequency dependent charge-pumping process: (1) Interface traps capture electrons, (2) electrons tunnel into oxide traps, (3) trapped electrons capture holes, and (4) holes tunnel into the oxide traps.

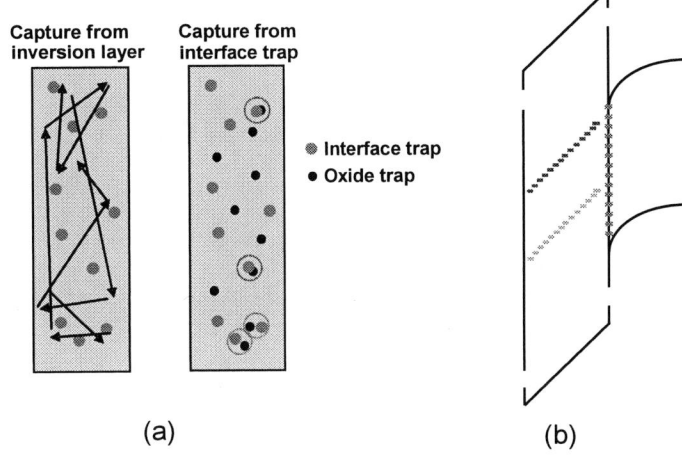

Figure 4. (a) spatial picture of carrier-capture by interface states (left), and carrier-tunneling from interface states to oxide trap states (right). (b) Energy band diagram showing a uniform energy distribution of interface states and two oxide trap states that are uniformly distributed (in depth) within the oxide layer.

While the disagreement on the correct choice of t_0 is well noted [10], the majority of the groups agree that there is a simple logarithmic relationship between CP frequency and the maximum probed depth. In this study, we will show that (1) a simple logarithmic relationship between CP frequency and the maximum depth probed does not exist, and (2) that the surrounding controversy is indeed unjustified because FD-CP cannot be used for the stated purpose.

III. PHYSICAL PROCESS DETAILS

Many groups have published models on the FD-CP depth relationship [2-7], but the complex mathematics utilized renders these works somewhat difficult for readers to gauge the appropriateness of the physics. Here, we focus on the physics of the phenomenon involved and present a clear argument, without complex modeling, that there is flaw in the basic concept itself.

In FDCP, the sequence involves four steps (fig. 3). To simplify matters, let us assume that (1) hole capture has the same rate as electron capture, and (2) that hole-tunneling has the same probability of electron-tunneling. Immediately, we come to the conclusion that the proper expression for τ is not equation (1). Instead, it should be:

$$\tau = t_{if} + t_0 e^{2\beta x} \qquad (3)$$

where t_{if} is the interface trap-fill time. If t_{if} is very small compared to τ, the equation reduces back to (1). Some reports take t_{if} as t_0 with values from 10 ns [6] to a fraction of ns [2]. They are obviously in error because t_{if} and t_0 are referring to different physical processes.

The physical pictures responsible for the two terms in (3) are quite different and are shown in fig.4. Fig 4a schematically illustrates the spatial picture of the two charge capturing processes of eqn. (3). The first term, filling of interface traps, involves inversion layer charge capture. These charges can

move within the inversion layer and every charge has a finite chance of being captured by any one of the interface traps. In this case, equation (2) can be used, burying all the physics of the capture process into a capture cross section σ. (As mention above, the reported cross section varies wildly.)

The right hand side of fig. 4a is a more correct picture for the second term in eqn. (3), and by extension FD-CP. Interface traps are fixed in space and interface state captured charge can only tunnel into oxide traps that have a spatial overlap the interface states (in this 2-D view). This is shown as those traps in the broken blue circles. If an oxide trap and interface trap are sufficiently close (like those in the broken red circles), tunneling is still possible. The probability of tunneling reduces rapidly with the separation of these two traps. We can immediately see that not only does oxide trap depth relate to τ, the lateral position also plays just as big a role. This point alone already eliminates any possibility of a simple relationship between CP frequency and depth, like many have hoped.

The FD-CP depth profiling picture gets even more complicated when we account for the interface and oxide trap energies (schematic band diagram is shown in fig. 4b). In fig. 4b, we depict uniformly distributed (in energy) interface traps and two different oxide traps with uniform distribution in depth. These are both commonly used simplifications. The key point here is that even if the interface trap overlaps with an

978-1-4244-5430-3/10 $26.00 © 2010 IEEE

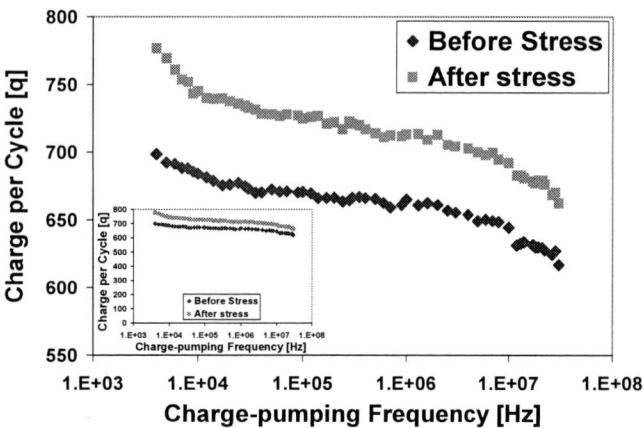

Figure 5.   CP charge per as a function of frequency for a production quality transistor where little bulk traps exists. (V_{acc} = -1.5V, V_{inv} = 1V, t_{rise}=t_{fall} = 2ns)

oxide trap spatially in the 2-D view, it may not be at the same energy level. Since tunneling is only possible when the energy is identical, a deeper oxide trap with better energy alignment may have a higher probability of capturing an interface charge than a shallower oxide trap with poorer energy alignment.

Note that even with all the constraints discussed above, it is still possible to have CP current coming from oxide traps. However, the combined constraints of spatial overlap, spatial separation and energy matching completely destroy the relationship between depth and time (frequency). Thus, any hope of depth profiling using FD-CP is dashed. Furthermore, if the interface-trap energy distribution and/or oxide trap depth distribution are not uniform, the picture becomes even murkier.

It should be noted that while the frequency/depth relationship is seemingly impossible to determine, all the factors discussed thus far still support the common observation of increased CP charge per cycle with decreasing frequency due the contributions of near interface oxide traps. If the oxide-trap density increases, CP charge per cycle will increase as well. We just cannot quantitatively relate the CP frequency to any oxide depth or trap density.

On the other hand, an increase in CP charge per cycle with decreasing frequency does not necessary denote participation by near interface oxide traps. If we take the CP process as filling the available traps with electrons during the inversion cycle and then emptying these electrons during the accumulation cycle, then the filling and emptying processes are given by the first order kinetic equations:

$$N_e = N_0 \left(1 - e^{-Kt}\right) \tag{4}$$

$$N_c = N_e e^{-Kt} \tag{5}$$

where N_0, N_e, and N_c are the number of traps, the number of trapped electrons, and the number of charges contributing to the CP current, respectively, K is the capture time constant, and t is the time spent in either accumulation or inversion. Here we maintain the assumption that hole- and electron-capture has the same time constant (emptying an electron from a trap is the

same as capturing a hole). As the CP frequency decreases, t increases for both electron capture and emptying, N_e increases, and N_c will increase even more. Thus, even in the absence of slower (oxide) traps, one should still expect a CP charge per cycle increase with decreasing CP frequency. When $t \gg K$, the effect is relatively small; otherwise, the effect can be large. Fig. 5 shows the FD-CP for a production-quality pure thermal oxide (2.4 nm physical thickness) 15.6 x 0.16 μm^2 nMOSFET. Oxide traps densities in these devices are expected to be very low (if not absent). Yet, the charge per cycle increases as the frequency decreases. This increase is mild but unmistakable.

This device was subject to a gate injection mode stress to increase the interface state density. We note that stress does increase the CP charge per cycle but also increases the absolute slope of the FD-CP curve. The stress is relatively mild (interface state density increase by ~10%), yet the increase in FD CP slope is evident. While we cannot rule out that oxide traps are also created, it is well-known that interface states are much more readily generated than bulk traps. Thus, this indicates that the FD-CP slope can possibly be due purely to interface defect participation.

IV.   CONCLUSIONS

In summary, we showed that the FD-CP technique is fundamentally unsuitable for defect depth profiling. Here we forgo the complicated mathematics and fuzzy concepts employed by other groups for a clear physical argument. We clearly show that a relationship between the frequency of FD-CP and the probed oxide depth cannot exist. In addition, we show that while stress does induce an increase in the absolute FD-CP slope, it cannot necessarily be explained solely by an increase in near interface oxide trap density. The interpretation is therefore not unique.

V.   ACKNOWLEDGMENTS

The authors wish to acknowledge financial support from the NIST Office of Microelectronic Programs.

VI.   REFERENCES

[1]   R. Degraeve, A. Kerber, P. Roussel, et al. IEDM 2003, pp935-938.

[2]   D. Heh, C. D. Young, G. A. Brown, *et al. Applied Physics Letters* **88**(15) 152907-3(2006).

[3]   W.V. Backensto and C. R. Viswanathan, IEDM 1976, pp287-291.

[4]   R. Paulsen, R. E. and M. H. White, *IEEE Transactions on Electron Devices,* **41**(7) 1213-1216(1994).

[5]   Y. Maneglia, D. Bauza and T. Ouisse, Semiconductor Conference, 1995. CAS'95 Proceedings., pp155-158.

[6]   Y. Maneglia and D. Bauza, *J. Appl. Phys.* **79**(8) 4187 4192(1996).

[7]   M. Masuduzzaman, A. E. Islam and M. A. Alam, *IEEE Transactions on Electron Dev.* **55**(12) 3421-3431(2008).

[8]   F.P. Heiman and G. Warfield, *IEEE Transactions on Electron Dev,* **12**(4) 167-178(1965).

[9]   Y. Wang and K. P. Cheung, *Appl. Phys. Lett.*, **91**, 113509(2007).

[10] Y. Wang, V. Lee, and K. P. Cheung, IEDM 2006, p763.

Progressive Schottky Junction Reaction Induced Degradation in Pt-Sunken Gate InP HEMT MMICs for High Reliability Applications

Y. C. Chou, D. L. Leung, M. Biedenbender, D. Buttari, D. C. Eng, R. S. Tsai, C. H. Lin, L. S. Lee, X. B. Mei,
M. Wojtowicz, M. E. Barsky, R. Lai, A. K. Oki, and T. R. Block

Microelectronics Processes and Products Center
Northrop Grumman Corporation
Redondo Beach, CA90278, USA
Phone: (01) –(310)-812-3550, e-mail: yeong-chang.chou@ngc.com

Abstract—**Reliability performance of 0.1-μm Pt-sunken gate InP HEMT MMICs on 4-inch InP substrates was evaluated under elevated temperature lifetesting. The primary degradation mechanism was observed to be the progressive Schottky junction reaction with the Schottky barrier InAlAs and the InGaAs channel. Despite the progressive Schottky junction reaction with the InAlAs and InGaAs materials, the lifetest at $T_{ambient}$ of 280 °C projects the median-time-to-failure exceeding 1×10^6 hours at $T_{channel}$ of 125 °C. This result indicates that the promising initial reliability performance was achieved on Pt-sunken gate InP HEMT MMICs on 4-inch InP substrates.**

Keywords-Schottky junction degradation; lifetest; reliability; Pt-sunken gate; InP; HEMT; MMICs

I. INTRODUCTION

Superior microwave and millimeter wave performance of InGaAs/InAlAs/InP high electron mobility transistor (HEMT) microwave monolithic integrated circuits (MMICs) has been demonstrated for space and military applications over the frequency ranges 44 GHz [1-2], 94 GHz [3-5], 118 GHz [6-7], 155 GHz [8-9], 183-220 GHz [8, 10-12], and beyond 250 GHz [13-15]. To ensure the reliability of InAlAs/InGaAs/InP HEMT MMICs during their lifetime operation, it is important to demonstrate the high reliability performance of InAlAs/InGaAs/InP HEMT MMICs subjected to elevated temperature lifetest. Since 1993, the reliability performance under elevated temperature lifetest on either InAlAs/InGaAs/InP HEMT discrete transistors or MMICs has been extensively investigated [16-18]. As a result, the high reliability performance of InAlAs/InGaAs/InP HEMTs was demonstrated on 0.07-μm [19], and 0.1-μm [20] InAlAs/InGaAs/InP HEMTs in addition to metamorphic HEMT (MHEMT) [21-22] technologies. This demonstrates the maturity of InAlAs/InGaAs/InP HEMT and MHEMT technologies and its readiness for space/military and commercial applications. These achievements of superior microwave performance and high reliability lead to the first insertion of InP HEMT low noise amplifiers operating at Q-band for phased-array applications at Northrop Grumman Corporation (NGC) [2].

Although the reliability performance of InAlAs/InGaAs/InP HEMTs was investigated in the industry, the reliability performance varies from company to company [18]. The variations of reliability performance were mainly attributed to different degradation mechanisms, consisting of fluorine contamination [23], ohmic contact degradation [16], gate metal stacks [24-26], layer structure [27], bias dependence [28], the variation of gate recess depth [29-30], and gate metal sinking [21-22]. It was observed that the degradation mechanisms strongly depend on the process techniques employed in a particular InAlAs/InGaAs/InP HEMT technology, therefore causing the distinct reliability performance from company to company. Consequently, it is essential to identify the degradation mechanisms of InAlAs/InGaAs/InP HEMTs on the established process techniques in order to further improve the reliability performance.

Several degradation causes in InP HEMTs were reported in literature [16, 21-30], however, most of them were eliminated due to the maturity of manufacturing processes and HEMT epitaxial materials. Nevertheless, gate metal reaction with the InAlAs material in InP HEMTs subjected to elevated temperature lifetest was still observed to be the primary degradation mechanism. A recent investigation of InP HEMTs degradation using a scanning-transmission- electron-microscope (STEM) technique [26] reveals the physical evidence of Ti-InAlAs reaction in InP HEMTs under the elevated temperature lifetest. Although the high activation energy (Ea) was reported in InP HEMTs with the gate metal stacks of Ti/Pt/Au [20, 30], the degree of reaction between the gate metal stacks of Ti/Pt/Au and the InAlAs Schottky barrier layer strongly depends on the InAlAs surface conditions prior to the Ti/Pt/Au gate metal deposition. The variations of the InAlAs surface conditions could be induced by either gate recess wet etching and/or process-related dry etching. As a result, the variations of reliability performance were observed from wafer-lot to wafer-lot despite similar Ea performance. Therefore, it is essential to mitigate the Ti-InAlAs reaction of InP HEMTs under elevated temperature lifetest to assure repeatable high reliability performance.

978-1-4244-5430-3/10 $26.00 © 2010 IEEE

To improve the reliability performance of InP HEMTs, new gate metal stacks were explored recently to reduce the Ti-InAlAs reaction [24-26]. The Pt-sunken gate was initially introduced to fabricate high performance and enhancement-mode InP HEMTs [31-32]. Although recent investigation shows that InP HEMTs with Pt-sunken gate have demonstrated superior reliability performance to InP HEMTs with Ti-gate [12, 25-26, 33], the reliability evaluation of 0.1-µm Pt-sunken gate InP HEMT MMICs on 4-inch InP substrates has been lacking. The fabrication of 0.1-µm InP HEMT low noise amplifiers (LNAs) on 4-inch InP substrates is essential for next-generation high performance phased-array applications requiring a high volume of LNA elements [2]. In this study, the reliability performance of 0.1-µm InP HEMT LNAs on 4-inch InP substrates subjected to elevated temperature lifetesting was evaluated for the first time. The primary degradation mechanism was identified to be due to the progressive Schottky junction reaction with the Schottky barrier InAlAs and the InGaAs channel. The progressive Schottky junction reaction was substantiated by the STEM experiment. Despite the progressive Schottky junction reaction with the InAlAs and InGaAs materials, the promising initial reliability performance was achieved on NGC's Pt-sunken gate InP HEMT MMICs fabricated on 4-inch InP substrates

II. STANDARD EVALUATION CIRCUITRY FOR LIFETESTING

To evaluate the reliability performance of 0.1-µm Pt-sunken InP HEMT MMICs on 4-inch InP substrates, a K–band balanced LNAs operating over 27- 40 GHz was designed for the standard evaluation circuitry (SEC). The SEC is a two-staged balanced amplifier with a total gate periphery of 160-µm on the 1st stage's devices Q1 & Q2 and 400-µm on the 2nd stage's devices Q3 & Q4 as shown in Fig. 1. The gate and drain resistors are also added into the gate and drain electrodes for the SEC stability under RF operation.

The MMIC lifetest allows the reliability assessment on both InP HEMT devices and passive elements, such as via-hole integrity, thin film resistors (TFRs), metal-insulator-metal capacitors (MIMCAPs), and metal interconnects. As a result, at NGC, we performed the MMIC lifetesting instead of discrete devices' lifetesting. As shown in Fig. 1, the InP HEMT SECs with devices Q1, Q2, Q3, and Q4 were fabricated on 4-inch InP substrates for the elevated temperature lifetesting to evaluate the reliability performance

of 0.1-µm Pt-sunken gate InP HEMT MMICs on 4-inch InP substrates.

Figure 2 shows the representative characteristics of drain current, I_{DS}, and transconductance, g_m, versus gate voltage, V_{GS}, of 0.1-µm Pt-sunken gate versus conventional Ti/Pt/Au gate InP HEMTs. A positive shift of V_{GS} of approximately 0.15 to 0.2 volts was observed on Pt-sunken gate InP HEMTs, accompanied by the increase of g_m and the decrease of I_{DS}. The V_{GS} shift of Pt-sunken gate InP HEMTs is induced by a slight Pt sinking into the InAlAs barrier layer during the process. Nevertheless, the effect on unity-gain-cut-off frequency (f_T) and maximum frequency of oscillation frequency (f_{MAX}) is negligible. As shown in Fig. 3, the f_T of Pt-sunken gate and conventional Ti/Pt/Au gate InP HEMTs is 170 - 180 and 180 - 190 GHz; the f_{max} of Pt-sunken gate and conventional Ti/Pt/Au gate InP HEMTs is 180 - 190 and 190 - 200 GHz. The slight difference of f_T and f_{MAX} could be compromised by the parasitic capacitances, such as gate-source (C_{GS}) and gate-drain (C_{GD}) capacitances.

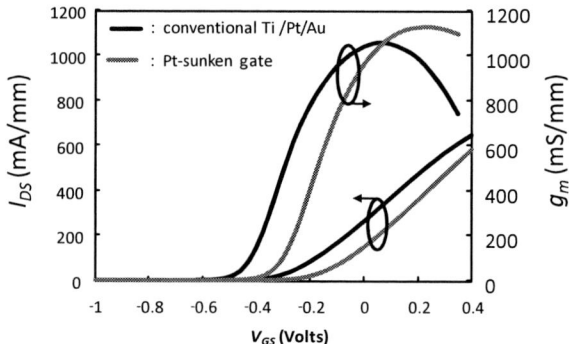

Figure 2. Characteristics of I_{DS} / g_m versus V_{GS} of Pt-sunken gate versus conventional Ti/Pt/Au gate InP HEMTs at V_{DS}= 1 V.

III. ELEVATED TEMPERATURE LIFETESTING

Lifetesting at $T_{ambient}$ of 280 °C was performed on the SECs stressed at V_{DS} = 1.6 V and I_{DS}= 150 mA/mm in N_2 environment. During lifetesting, comprehensive characterization of d. c. and radio frequency (r.f.) parameter from 27 to 40 GHz was performed to achieve insights into device degradation. The failure criterion was based on the ΔS_{21} of -1 dB at 35 GHz. Additionally, some SECs were removed during intermediate intervals to perform the STEM

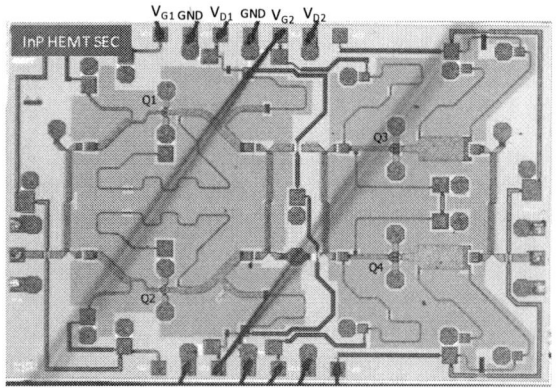

Figure 1. Photograph of an InP HEMT SEC for reliability evaluation.

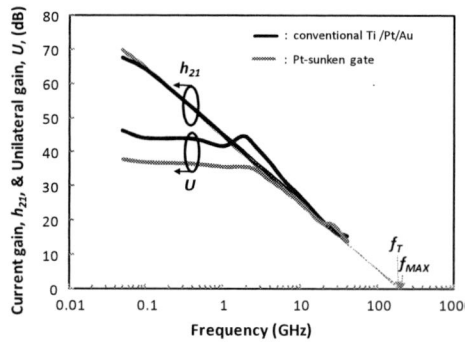

Figure 3. Comparison of f_T and f_{MAX} of Pt-sunken gate versus conventional Ti/Pt/Au gate InP HEMTs at V_{DS}= 1 V and peak g_m.

experiment. The STEM was found to be a powerful tool to substantiate the physical reaction of the gate metals and the AlGaAs Schottky barrier layer in GaAs pHEMTs [34]. Moreover, during the STEM experiment, the high resolution energy-dispersive-analysis with X-ray (EDS) was also introduced to characterize the Schottky junction reaction with the InAlAs Schottky barrier layer and the InGaAs channel.

Although the reliability performance of Pt-sunken InP HEMT was reported based on the transconductance degradation, Δg_m, of -10 % on discrete InP MHEMTs [12] and drain current degradation, $\Delta Idss$, of -20 % on InP HEMT MMICs on 3-inch InP substrates [33], the reliability analysis using the ΔS_{21} of -1 dB at 35 GHz for Pt-sunken InP HEMT MMICs on 4-inch InP substrates is still lacking. Also, the reliability analysis of *r. f.* performance represents the real scenario of InP HEMT LNAs during lifetime operation. As a result, it is essential to achieve the reliability performance using ΔS_{21} as failure criterion in order to have a better lifetime prediction of InP HEMT LNAs during the mission. Therefore, in our study, the SECs were lifetested until reaching ΔS_{21} of -1 dB at 35 GHz. For our SEC lifetest, $\Delta Idss$ of -20 % occurs prior to the ΔS_{21} of -1 dB. Empirically, the Δg_m of -10 % on both the 1st and 2nd stages causes equivalent ΔS_{21} of -1 dB at 35 GHz.

Figure 4 depicts the gradual S_{21} degradation of a Pt-sunken InP HEMT SEC subjected to lifetest at $T_{ambient}$ of 280 °C. The initial S_{21} increase was found to be caused by the initial nominal Pt gate sinking effect [33]. The continuing lifetesting of Pt-sunken InP HEMT SECs after reaching ΔS_{21}= -1 dB could cause more drastic S_{21} degradation. The subsequent STEM analysis reveals that the drastic S_{21} degradation is caused by the progressive Pt Schottky junction into the InAlAs Schottky barrier and the InGaAs channel layers.

IV. PROGRESSIVE SCHOTTKY JUNCTION REACTION

Figure 5 shows the cross section of a Pt-sunken gate InP HEMT on 4-inch InP substrate. The STEM analysis reveals a slight Pt-sinking into the InAlAs Schottky barrier layer. The initial Pt-sinking was introduced by the front-side processes of NGC's Pt-sunken InP HEMT LNAs on 4-inch InP substrates. Accordingly, a thin layer of Pt-InAlAs with a thickness of less than 20 Å was observed on the virgin devices.

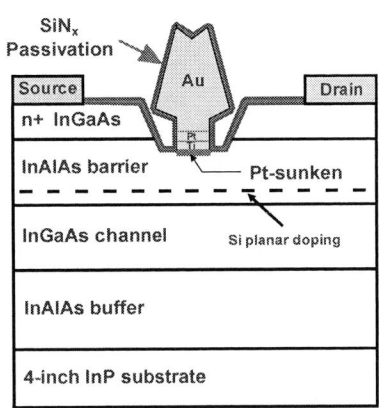

Figure 5. The cross section of a Pt-sunken gate InP HEMT.

Figure 6 shows a cartoon illustration of a Pt-sunken gate InP HEM in a SEC subjected to lifetest at $T_{ambient}$ of 280 °C for 336 hours with ΔS_{21} of -5 dB at 35 GHz. As shown in Fig. 6, the degree of Pt encroachment on the gate edges could be gradually aggravated during the continuing lifetest at $T_{ambient}$ of 280 °C. The progressive Pt encroachment effect could therefore be observed on both the InAlAs Schottky barrier and the InGaAs channel materials. Additionally, the increasing Ti diffusing into the region of Pt sunken metal could be observed. Moreover, the amount of Pt diffusing into the central Ti metal region can also be increased on a Pt-sunken InP HEMT lifetested at $T_{ambient}$ of 280 °C for 336 hours (> 20 Å).

To gain more insights into the Schottky junction reaction of a Pt-sunken gate InP HEMT, a STEM experiment was performed along the gate finger instead of the STEMs were obtained on the region perpendicular to the gate fingers. As shown in Fig. 7, it was observed that the intermetallic Ti-Pt-In-Al-As was formed on the region above the Pt-In-Al-As region. The subsequent EDS analysis substantiates that this is caused by the diffusion of the Ti metal into the Pt-In-Al-As region. Moreover, the Pt-In-Al-As cluster was observed on the InAlAs Schottky barrier layer, which could be related to the Pt encroachment as described in Fig. 6. The continuing lifetest could progressively aggravate the clusters into both the InAlAs Schottky barrier and InGaAs channel layers. This could cause severe Schottky junction degradation as observed on Pt-sunken InP HEMT SECs with drastic S_{21} degradation during lifetest at $T_{ambient}$ of 280 °C for 336 hours.

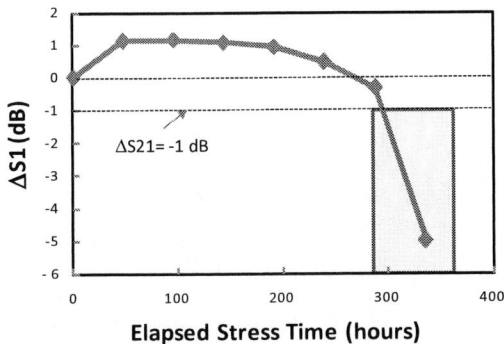

Figure 4. Gradual S_{21} degradation at 35 GHz of a Pt-sunken gate InP HEMT SEC lifetested at $T_{ambient}$ of 280 °C.

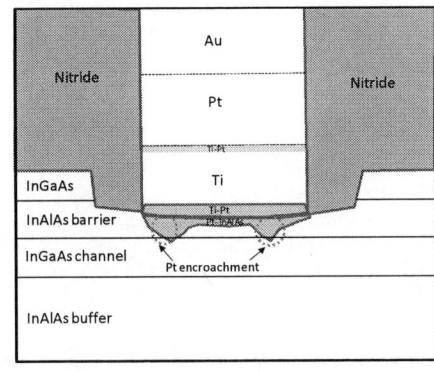

Figure 6. The cartoon illustration of a Pt-sunken gate InP HEMT subjected to lifetest at $T_{ambient}$ of 280 ° C for 336 hours.

EDS analysis was performed on both the InAlAs and InGaAs materials in a degraded Schottky junction to validate the intermetallic formation of Pt-In-Al-As and Pt-In-Ga-As. As shown in Figs. 8 and 9, the intermetallics of Pt-In-Al-As and Pt-In-Ga-As were detected on the InAlAs and InGaAs materials. The intermetallic formation also leads to a decrease of the Schottky barrier height and an increase of the gate leakage current, as shown in Fig. 10. The increase of gate leakage current in an InP HEMT SEC with gate resistors could severely de-bias the devices Q1, Q2, Q3, and Q4 (shown in Fig. 1) during lifetest [35]. As a result, the distinct gate voltage evolution in the oven of the 1st and the 2nd stages of an InP HEMT SEC was observed. The de-bias effect strongly depends on the values of gate resistors and the gate leakage current. In our Pt-sunken gate InP HEMT SEC, the de-bias effect of the 2nd stage is higher than that of the 1st stage, which is due to the higher gate leakage current and higher gate resistor values of Q3 and Q4 devices that those of Q1 and Q2 devices.

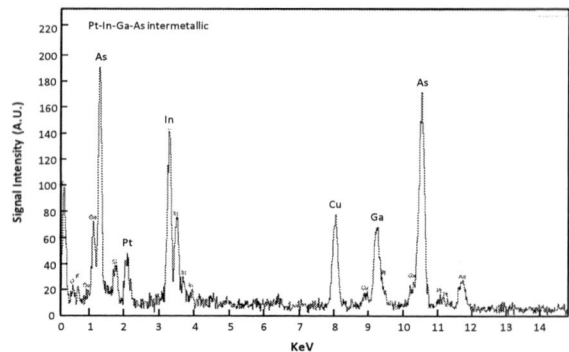

Figure 9. The EDS spectrum of the Pt-In-Ga-As intermetallic observed on the InGaAs channel layer in a degraded Pt-sunken gate InP HEMT subjected to lifetest at T$_{ambient}$ of 280 ° C for 336 hours. The signal of Cu is from the grid target in the STEM equipment.

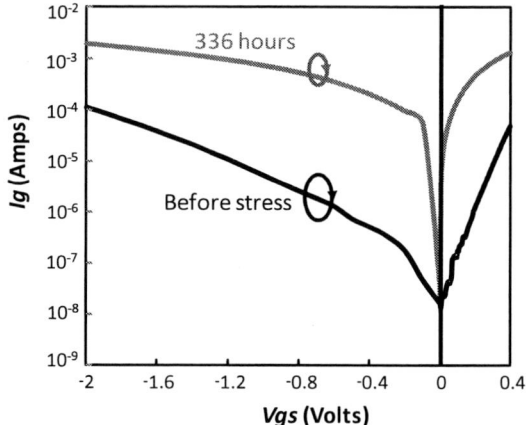

Figure 10. The representative gate current evolution of forward and reverse diodes in a degraded Pt-sunken gate InP HEMT subjected to lifetest at T$_{ambient}$ of 280 ° C for 336 hours.

V. RELIABILITY PERFRORMANCE

Initial reliability performance was evaluated based on the exit criterion of delta $S_{21} \leq -1$ dB at 35 GHz with the assumption of the Ea of 1.5 eV based on the lifetesting results at T$_{ambient}$ of 280 °C. The Ea was validated to be greater than 1.5 eV for Pt-sunken gate InP HEMTs [12, 33]. Moreover, our recent 3-temperature lifetest demonstrates that the Ea of Pt-sunken gate InP HEMT MMICs on 4-inch InP substrates is greater than 1.5 eV [36]. The Arrhenius plot from the reliability analysis shows that the MTF exceeds the typical MTF benchmark of 1×10^{6} hours at T$_{channel}$ of 125 °C. This result indicates that the promising initial reliability performance was achieved, which is essential for the future full qualification of NGC's Pt-sunken gate InP HEMT MMICs on 4-inch InP substrates for space and military applications.

CONCLUSIONS

Reliability performance of 0.1-μm Pt-sunken gate InP HEMT MMICs on 4-inch InP substrates was evaluated under elevated temperature lifetesting. The primary degradation mechanism was observed to be the progressive Schottky junction reaction with the Schottky barrier InAlAs and the InGaAs channel. The progressive Schottky junction reaction was substantiated by the STEM and EDS analyses. Despite the

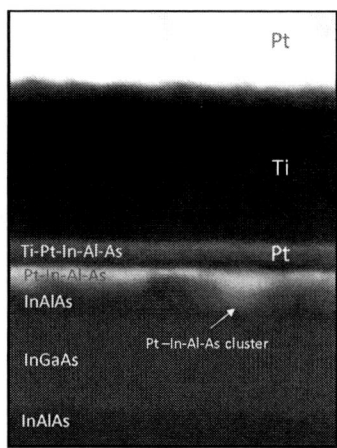

Figure 7. The STEM micrograph of a Pt-sunken gate InP HEMT subjected to lifetest at T$_{ambient}$ of 280 ° C for 336 hours. The STEM was performed along the gate finger.

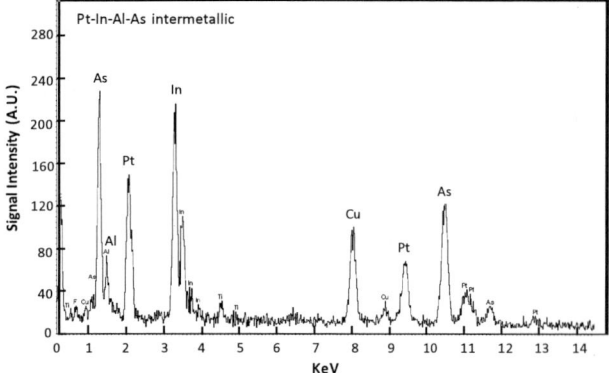

Figure 8. The EDS spectrum of the Pt-In-Al-As intermetallic observed on the InAlAs Schottky barrier layer in a degraded Pt-sunken gate InP HEMT subjected to lifetest at T$_{ambient}$ of 280 ° C for 336 hours. The signal of Cu is from the grid target in the STEM equipment.

progressive Schottky junction reaction with the InAlAs and InGaAs materials, the promising initial reliability performance was achieved on NGC's Pt-sunken gate InP HEMT MMICs fabricated on 4-inch InP substrates.

REFERENCES

[1] R. Lai, R. Grundbacher, Y. C. Chou, M. Barsky, M. Nishimoto, R. Raja, M. Sholley, R. Tsai, D. Leung, Q. Kan, P. H. Liu, M. Wojtowicz, L. Tran, and A. Oki, "InP HEMT MMIC low noise smplifiers for space/military applications," in the Technical Digest of GOMAC conference, pp.358-361, 2003.

[2] R. Grundbacher, Y. C. Chou, R. Lai, K. Ip, S. Kam, M. Barsky, G. Hayashibara, D. Leung, D. Eng, R. Tsai, M. Nishimoto, T. Block, P. H. Liu and A. Oki, "High performance and high reliability InP HEMT low noise amplifiers for phased-array applications," in the Technical Digest of IEEE International Microwave Symposium, pp. 157-160, 2004.

[3] R. Grundbacher, R. Lai, M. Barsky, R. Tsai, T. Gaier, S. Weinreb, D. Dawson, J. J. Bautista, J. F. Davis, N. Erickson, T. Block, and A. Oki "0.1 µm InP HEMT devices and MMICs for cryogenic low noise amplifiers from X-Band to W-band," in the Technical Digest of IEEE International Conference on InP and Related Materials, pp. 455-458, 2002.

[4] A. Leuther, A. Tessmana, M. Dammann, W. Reinert, M. Schlechtweg, and M. Mikulla, "70 nm low-noise metamorphic HEMT technology on 4 inch GaAs wafers," in the Technical Digest of IEEE International Conference on InP and Related Materials, pp. 215-218, 2003.

[5] P. M. Smith, S. M. Liu, M. Y. Kao, P. Ho, S. C. Wang, K. H. G. Duh, S. T. Fu, and P. C. Chao, "W-band high-efficiency InP-based power HEMT with 600 GHz $fmax$," IEEE Microwave Guided Wave Lett., vol. 5, no. 7, pp. 230-232, 1995.

[6] M. Nishimoto, M. Sholley, H. Wang, R. Lai, M. Barsky, D. Streit, Y. Chung, M. Aust, B. Osgood, R. Raja, C. Gage, T. Gaier, and K. Lee, "High performance D-Band (118 GHz) monolithic low noise amplifier," in the Technical Digest of IEEE Radio Frequency Integrated Circuits Symposium, pp. 99-102, 1999.

[7] M. Sholley, G. Barber, R. Lai, M. Barsky, R. Tsai, D. Streit, B. Osgood, J. Dowsing, M. Nishimoto, Y. Yok, P. Huang, C. Gage, T. Huang, H. Nang, C. Jackson, Y. Chung, S. Chan, and J. Mitchell, "118 GHz MMIC radiometer for the (IMAS) integrated multi-spectral atmospheric sounder," in the Technical Digest of IEEE International Microwave Symposium, pp. 479-483, 1999.

[8] R. Lai, M. Barsky, R. Grundbacher, P. H. Liu, T. P. Chin, M. Nishimoto, R. Elmadjian, R. Rodriguez, L. Tran, A. Gutierrez, A. Oki, and D. Streit, "InP HEMT amplifier development for G-band (140-220 GHz) applications," in the Technical Digest of IEEE International Electron Devices Meeting, pp. 175-177, 2000.

[9] R. Grundbacher, R. Raja, Y. C. Chou, M. Nishimoto, T. Gaier, D. Dawson, P. H. Liu, M. Barsky, and A. Oki, "A 150-215 GHz InP HEMT low noise amplifier with 12 dB gain," in the Technical Digest of IEEE International Conference on InP and Related Materials, pp. 613-616, 2005.

[10] R. Raja, M. Nishimoto, M. Barsky, M. Sholley, Q. Quon, G. Barber, P. Liu, R. Lai, and F. Hinte, "A 183 GHz low noise amplifier module for the conical-scanning microwave imager sounder (CMIS) program," in the Technical Digest of IEEE International Microwave Symposium, pp. 987-990, 2000.

[11] R. Raja, M. Nishimoto, B. Osgood, M. Barsky, M. Sholley, R. Quon, G. Barber, P. Liu, R. Lai, F. Hinte, G. Haviland, and B. Vacek, "A 183 GHz low noise amplifier module with 5.5 dB noise figure for the conical-scanning microwave imager sounder (CMIS) program," in the Technical Digest of IEEE International Microwave Symposium, pp. 1851-1854, 2001.

[12] A. Tessmann, A. Leuther, H. Massler, M. Kuri, C. Schwoerer, M. Schlechtweg, and G. Weimann, "A 220 GHz metamorphic HEMT amplifier MMIC," in the Technical Digest of IEEE Compound Semiconductor IC (CSIC) Symposim, pp. 297-200, 2004.

[13] W. R. Deal, S. Din, V. Radisic, J. Padilla, X. B. Mei, W. Yoshida, P. H. Liu, and R. Lai, "Demonstration of sub-millimeter wave integrtaed cicrcuit (S-MMIC) using InP HEMT with a 35 nm gate," in the Technical Digest of IEEE Compound Semiconductor IC (CSIC) Symposim, pp. 33-36, 2006.

[14] X. B. Mei, W. Yoshida, W. R. Deal P. H. Liu, J. Lee, J. Uyeda, L. Dang, J. Wang, W. Liu, D. Li, M. Barsky, Y. K. Kim, M. Lange, T. P. Chin, V. Radisic, T. Gaier, A. Fung, L. Samoska, and R. Lai, "35 nm InP HEMT SMMIC amplifier with 4.5 dB gain at 308 GHz," IEEE Electron Device Lett., vol. 28, no. 6, pp. 470-472, 2007.

[15] R. Lai, X. B. Mei, W. R. Deal, W. Yoshida, K. M. Kim, P. H. Liu, J. Lee, J. Uyeda, V. Radisic, T. Gaier, L. Samoska, and A. Fung, "Sub 50 nm InP HEMT devices with $fmax$ greater than 1 THz," in the Technical Digest of IEEE International Electron Devices Meeting, pp. 609-611, 2007.

[16] D. J. LaCombe, W. W. Hu, and F. R. Bardsley, "Reliability of 0.1 µm InP HEMTs," in the Technical Digest of IEEE Reliability Physics Symposium, pp. 364-371, 2003.

[17] Y. C. Chou, D. Leung, R. Lai, R. Grundbacher, J. Scarpulla, M. Barsky, M. Nishimoto, D. Eng, P. H. Liu, A. Oki, and D. Streit, "High Reliability of 0.1 µm InGaAs/InAlAs/InP high electron mobility transistors microwave monolithic integrated circuit on 3-inch InP substrate," Jpn. J. App. Phys., vol. 41, pp. 1099-1103, 2002.

[18] Y. C. Chou, R. Lai, R. Grundbacher, M. Barsky, Q. Kan, R. Tsai, M. Wojtowicz, D. Eng, L. Tran, T. Block, P. H. Liu, M. Nishimoto, and A. Oki, "Reliability investigation of 0.07-µm InGaAs-InAlAs-InP HEMT MMICs with pseudomorphic $In_{0.75} Ga_{0.25}As$ channel," IEEE Electron Device Lett., vol. 24, no. 6, pp. 378-380, 2003.

[19] G. Meneghesso, and E. Zanoni, "Failure modes and mechanisms of InP-based and metamorphic high electron mobility transistors," Microelectronics Reliability, vol. 42, pp. 685-708, 2002.

[20] Y. C. Chou, D. Leung, R. Grundbacher, R. Lai, M. Barsky, Q. Kan, D. Eng, M. Wojtowicz, T. Block, S. Olson, P. H. Liu, A. Oki, and D. Streit, "0.1 µm InGaAs/InAlAs/InP HEMT MMICs – a flight qualified technology," in the Technical Digest of IEEE GaAs IC Symposium, pp. 77-80, 2002.

[21] M. Dammann, A. Leuther, F. Benkhelifa, T. Feltgan, and W. Jantz, "Reliability and degradation mechanism of AlGaAs/InGaAs and InAlAs/InGaAs HEMTs," Phys. Stat. Sol. (a), vol. 195, pp. 81-86, 2003.

[22] M. Chertouk, M. Dammann, and A. Leuther, "The first 0.15 µm MHEMT 6'' GaAs foundry service: highly reliable process for 3V drain bias operations," in the Technical Digest of European GaAs IC Symposium, pp. 9-12, 2003.

[23] N. Hayafuji, Y. Yamamoto, N. Yoshida, T. Sonoda, S. Takamiya, and S. Mitsui, "Thermal stability of AlInAs/GaInAs/InP heterostructures," Appl. Phys. Lett., vol. 66, no. 7, pp. 863-865, 1995.

[24] T. Ishida, Y. Yamamoto, N. Hayafuji, S. Miyakuni, R. Hattori, T. Ishikawa, and Y. Mitsui, "Fluorine limited reliability of AlInAs/InGaAs high electron mobility transistor with molybdenum gates," in the Technical Digest of IEEE International Conference of InP and Related Materials, pp. 201-204, 1997.

[25] M. Dammann, A. Leuther, H. Konstanzer, and W. Jantz, "Effect of gate metal on reliability of metamorphic HEMTs," in the Technical Digest of IEEE GaAs REL Workshop, pp. 87-88, 2001.

[26] Y. C. Chou, R. Grundbacher, D. Leung, R. Lai, Q. Kan, D. Eng, P. Chin, and A. Oki, "Degradation mechanism and reliability improvement of InGaAs/InAlAs/InP using new gate metal electrode technology," in the Technical Digest of IEEE International Conference of InP and Related Materials, pp. 223-226, 2005.

[27] N. Hayafuji, Y. Yamamoto, and K. Sato, "Fluorine passivation effects in AlInAs/InGaAs HEMT material," in the Technical Digest of Compound Semiconductor Manufacturing Technology, pp. 143-146, 1998.

[28] M. Cleryouk, M. Massler, M. Dammann, K. Kohler, and G. Weimann, "Manufacturable 0.15 µm InP-based HEMT MMICS process with high yield and reliability on 2-inch InP Substrate," in the Technical Digest of Compound Semiconductor Manufacturing Technology, pp. 25-28, 1999.

[29] B. M. Paine, R. Wong, A. Schmitz, R. Walden, L. Nguyen, M. Delaney, and K. Hum, "Ka-band InP HEMT MMIC reliability," in the Technical Digest of IEEE GaAs REL Workshop, pp.21-44, 2000.

[30] Y. C. Chou, R. Grundbacher, D. Leung, R. Lai, M. Barsky, Q. Kan, M. Wojtowicz, R. Tsai, D. Eng, P. H. Liu, and A. Oki, "Tradeoff of DC/RF performance versus reliability in 0.1 μm InP HEMTs," in the Technical Digest of 16th International Conference of Indium Phosphide and Related Material, pp. 389-392, 2004.

[31] K. J . Chen, T. Enoki, K. Maezawa, K. Arai, and M. Yamamoto, "High-performance InP-based enhancement-mode HEMT's using non-alloyed ohmic contacts and Pt-based buried-gate technolgies," IEEE Trans. Electron Devices, vol. 45, no. 2, pp. 252-257, 1996.

[32] A. Mahajan, M. Arafa, P. Fay, C. Caneau, and H. Adesida, "Enhancement-mode high electron mobility transistors (E-HEMT's) lattice-matched to InP," IEEE Trans. Electron Devices, vol. 45, no. 12, pp. 2422-2429, 1998.

[33] Y. C. Chou, R. Lai, D. Leung, Q. Kan, D. Farkas, D. Eng, M. Wojtowicz, P. Chin, T. Block and A. Oki, "Gate sinking effect of 0.1-μm InP HEMT MMICs using Pt/Ti/Pt/Au," in the Technical Digest of International Conference of Indium Phosphide and Related Materials, pp. 188-191, 2006.

[34] Y. C. Chou, R. Grundbacher, D. Leung, R. Lai, P. H. Liu, Q. Kan, M. Biedenbender, M. Wojtowicz, D. Eng, and A. Oki, "Physical identification of gate metal interdiffusion in GaAs pHEMTs,"IEEE Electron Device Lett., pp. 64-66, 2003.

[35] Y.C. Chou, M. Truong, D. Leung, R Grundbacher, R. Lai, Q. Kan, D. Eng and A. Oki, "The de-bias effects of gate current on lifetesting in InP HEMT MMICs", in the Technical Digest of International Conference of Indium Phosphide and Related Materials, pp. 393-396, 2004.

[36] Y. C. Chou, D. L. Leung, M. Biedenbender, D. C. Eng, D. Buttari, L. J. Lee, X. B. Mei, C. H. Lin, R. S. Tsai, R. Lai, M. E. Barsky, A. Cavus, M. Wojtowicz, A. K. Oki and T. R. Block, "High reliability performance of Pt-Sunken gate InP HEMT low noise amplifiers on 100 mm InP substrates", to be appeared in the Technical Digest of International Conference of Indium Phosphide and Related Materials, 2010.

ELECTRICAL STRESS INDUCED DEGRADATION IN INAS – ALSB HEMTS

S. DasGupta, R. A. Reed, R. D. Schrimpf, D.M. Fleetwood

EECS Dept., Vanderbilt University, VU Station B 351825, Nashville, TN 37235-1825, USA, phone: 615-343-6704; fax: 615-343-6614; e-mail: sandeepan.dasgupta@vanderbilt.edu

X. Shen, S. T. Pantelides

Department of Physics and Astronomy, Vanderbilt University, Nashville, TN 37235-1825, USA

J. Bergman and B. Brar

Teledyne Scientific and Imaging, Thousand Oaks, CA 91360, USA

Abstract- InAs - AlSb HEMTs stressed with hot electrons may exhibit shifts in the peak transconductance towards more negative gate voltages. The devices are most degradation prone in operating conditions with high vertical gate field. Annealing trends and theoretical calculations indicate the possible role of an oxygen-induced metastable defect.

Keywords- **HEMT, degradation, metastable defect, DFT.**

I. INTRODUCTION

InAs has generated interest as a candidate for very high speed, low power electronic devices. The electronic band structure of InAs allows for faster electron transport on account of its lower effective mass $(0.023m_o)$ in the Γ-valley relative to almost all other common III-V semiconductors, except InSb, and a large Γ - L valley separation (relative to band gap) of 0.72 eV. The decrease in effective mass directly impacts the low-field mobility of each semiconductor material, as evidenced by a very high 300 K electron mobility of 33,000 cm²/V-s in nominally undoped InAs. The InAs - AlSb HEMT derives its high-speed performance from the inherently fast electron transport properties of the channel semiconductor. In addition to this, a huge conduction band offset of 1.35 eV with the nearly lattice matched AlSb (a = 6.2 Å) allows for high electron confinement in the quantum well, large subband spacing and large 2 DEG densities ~ 2 × 10¹² cm⁻² [1]. For these reasons, the InAs/AlSb HEMT has the capability of being an ideal device for high speed, low power RF circuits [2-4]. While some studies of high temperature life testing have been performed on this device [5], fundamental degradation mechanisms for the device are yet to be adequately understood. The degraded devices show signs of oxygen contamination, but no explicit connection of oxygen-induced defects to the observed shifts in electrical characteristics was established. In this paper, based on electrical stress performed on devices with varied starting characteristics, we show that devices are most degradation prone in operating conditions where vertical gate field is high. Experimental results, coupled with device simulations and Density Functional Theory (DFT) calculations, indicate the possible role of a metastable defect dominating the device degradation.

II. THE DEVICES & STRESS CONDITIONS

Teledyne Scientific fabricated the tested devices. The devices employ a composite top barrier using AlSb and $In_{0.5}Al_{0.5}As$ with a total thickness of 14 nm [6]. The vertical structure is shown in Fig. 1.

Most devices had threshold voltages more negative than -0.7 V. For devices that pinched off at more negative gate voltages, or in some cases did not pinch off at all with the specified usable range of gate voltages ($V_{gs} > -0.8$ V), the gate leakage current was the smallest of all devices. Most devices

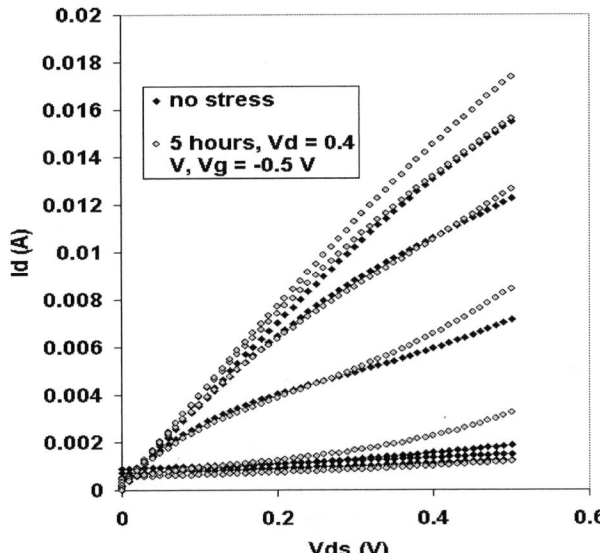

Fig. 2. Degradation in I_d and shift in threshold voltage under 5 hours stress at $V_g = -0.5$ V, $V_d = 0.4$ V. $I_d – V_d$ plot of a 2 × 10 μm HEMT with 100 nm gate. $V_{gs} = 0$ to –1 V in steps of –0.2 V.

pinched off around –0.6 V, but also had a high gate leakage current.

50 devices were tested on wafer at room temperature. The devices showed no signs of degradation when biased at the

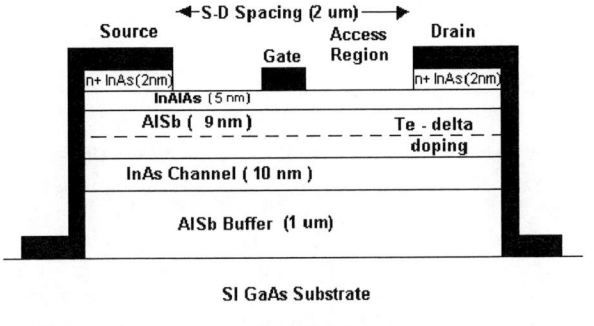

Fig. 1. Cross-section of the tested devices. $In_{0.5}Al_{0.5}As$ forms a stable, low-leakage cap. Remote Te Δ - doping is used to achieve high 2DEG density with minimal scattering.

978-1-4244-5430-3/10 $26.00 © 2010 IEEE

maximum current condition ($V_{gs} = 0$, $V_{ds} = 0.5$ V). At large negative gate voltages ($V_{gs} \sim -0.5$ V, $V_{ds} = 0.5$ V) there were perceptible changes in the *I-V* characteristics.

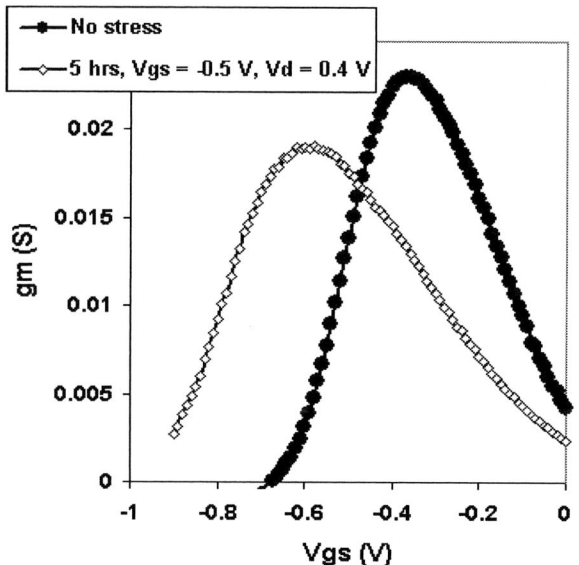

Fig. 3. Degradation in peak g_m and shift in threshold voltage ($V_d = 0.4$ V) under 5 hours of stress at $V_g = -0.5$ V, $V_d = 0.4$ V. g_m plots of a 2×10 μm wide HEMT with 100 nm gate.

III. DEGRADATION – THRESHOLD VOLTAGE AND TRANSCONDUCTANCE PEAK SHIFT

More than 60 per cent of these devices showed no visible signs of degradation under the stress conditions considered here. The degradation observed in the other 40% of the devices was a shift of the threshold voltage and transconductance peak towards more negative gate voltage (Figs. 2 and 3). This trend has been observed earlier for InAs – AlSb HEMTs under thermal stress [5]. When the devices were left at room temperature with no bias, they slowly recovered almost completely back to their initial characteristics. The evolution of gate current showed a much more erratic behavior, increasing in some cases and decreasing in others. Since the gate current is relatively high in these devices (~ 10 mA/mm) at high nega-

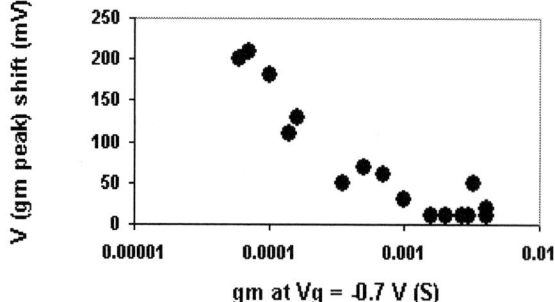

Fig. 4. g_m ($V_d = 0.4$ V) peak shift in 16 2×20 μm HEMTs as a function of g_m at $V_{gs} = -0.7$ V. There is a clear indication of increased degradation at high vertical fields in the channel.

tive gate voltages, it contributes to a significant increase in the drain current by lowering the source gate potential barrier. Devices with gate lengths greater than 250 nm showed no degradation.

IV. G$_M$ PEAK SHIFT AS A FUNCTION OF VERTICAL FIELD, BIASING CURRENT AND STARTING GATE CURRENT

Because of the fairly high variation in threshold voltage, it was not possible simply to change the gate voltage and determine the relationship of the vertical field with the shift in the threshold voltage or g_m peak. For this reason, the stressing gate voltage was kept the same, and the transconductance at $V_{gs} = -0.7$ V, (where many of the "good" devices were found to pinch off) was taken as a measure of the vertical field in the

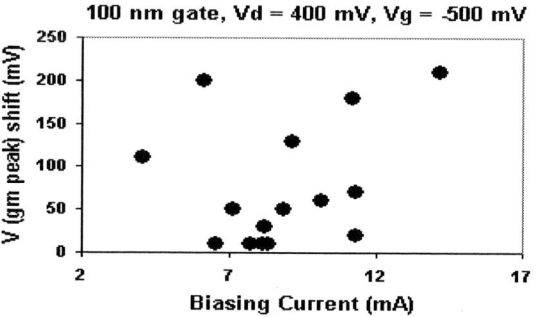

Fig. 5. g_m peak shift as a function of biasing current ($V_d = 0.4$ V). Here, there is no clear trend of peak shift vs. biasing current.

channel. As we can see in Fig. 4, there is a fairly clear tionship between the g_m-peak shift and the pinch-off voltage. In addition to the 16 devices shown in Fig. 4, there were other devices, which were biased differently. All devices that did not pinch off in the allowed region of operation were extremely resistant to g_m-peak shifts. The ones that showed degradation in spite of a high transconductance at $V_g = -0.7$ V were the ones showing high kink-effect signatures.

A. Biasing Current

There was no simple relationship between the biasing current and degradation (Fig. 5), at least in the short time range for which these devices were biased (5 hours). However, it is possible that the differences in the vertical field and the hot carrier-producing gradient between gate and drain (derived from V_{gd}) dominate over the thermal degradation mechanism.

B. Starting Gate Current

The very high gate leakage devices were more prone to g_m shifts. Apart from this there were very few other definite trends to be concluded from the gate current evolution. The gate current being quite high, it caused a significant amount of injection induced source gate barrier lowering, which tended to restore any reduction in the magnitude of peak g_m or drive current at 0 gate voltage. However, the nature of degradation in the gate current itself is unpredictable. The mechanism of gate leakage is not fully clear at this stage – given that the leakage levels at depletion biases are orders of magnitude higher than what would be predicted by Schottky theory, and

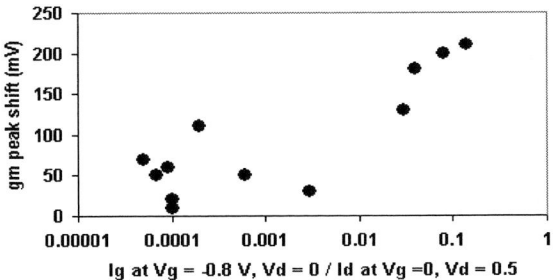

100 nm gate, Vd = 400 mV, Vg = -500 mV

Fig. 6. g_m peak shift ($V_d = 0.4$ V) as a function of starting gate leakage relative to drive current. Very high gate leakage devices show more degradation. At the low gate leakage region, there is no definite trend of degradation as a function of starting gate leakage.

that high leakage and complete channel pinch off can be achieved simultaneously.

V. ROOM TEMPERATURE ANNEALING OF DEGRADED DEVICES

Fig. 7 shows the degradation and recovery trends of a 2×10 μm HEMT with 100 nm gate length and 2 μm source-drain spacing. Devices were stressed at $V_{gs} = -0.5$ V and $V_{ds} = 0.4$ V for 5 hours. The complete annealing to the pre-stress values in ~2 days suggests that a metastable defect is responsible for the threshold-voltage shift.

VI. PHYSICAL MECHANISM OF DEGRADATION – METASTABLE DEFECTS

Since the stress-induced degradation results in a negative shift of the threshold voltage, it is likely that donor traps are being activated, or existing acceptors are being deactivated. Hole trapping in a very nonconductive region is unlikely due to the high gate current (about $1 - 5$ % of the drain current at the bias condition). In these devices, this trap could be in the InAs channel or the AlSb buffer. However, if the defects were in the channel, it would be hard to explain that we see no indication of significant mobility degradation, even for a large significant negative shift of the g_m peak. So the more likely

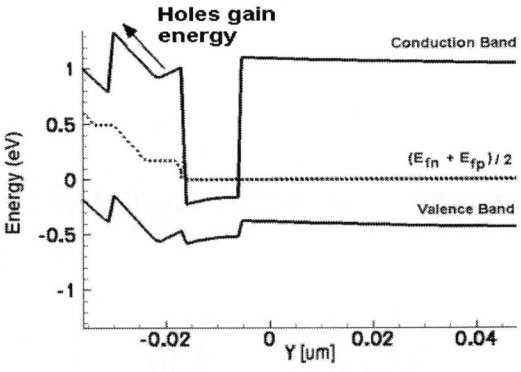

Fig. 8. Vertical band structure of the HEMT under the gate for $V_{gs} = -0.5$ V. Avalanche generated holes gain energy due to type –II alignment with InAs and applied gate bias.

candidate is a trap in the top or the bottom AlSb barrier. The other consideration here is the metastable nature of degradation and gradual recovery of the device to its initial state in a few days. It is unlikely that deep traps with long emission times are responsible for this behavior, since the high gate currents would prevent traps from stabilizing.

The greater degradation at high vertical fields in the channel indicates the role of hot carriers in the degradation process. While hot electrons are unlikely to retain sufficient energy in the AlSb away from the interface (given the high $\Delta E_c = 1.3$ eV), a sufficiently large number of energetic holes (derived from avalanching in the channel) exists at the stressing biases (Fig. 8).

Recent DFT calculations [7] show that the formation energy

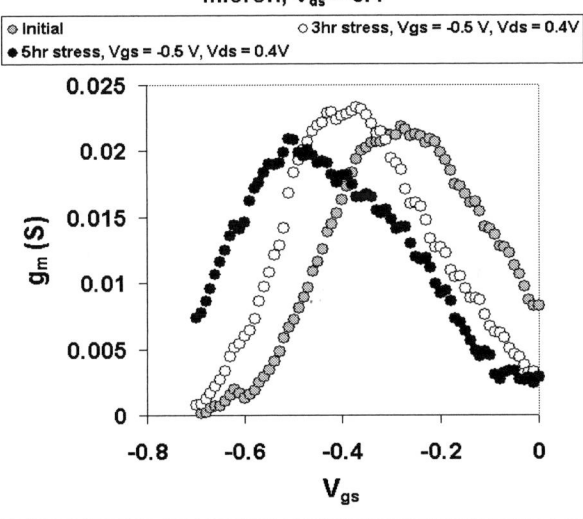

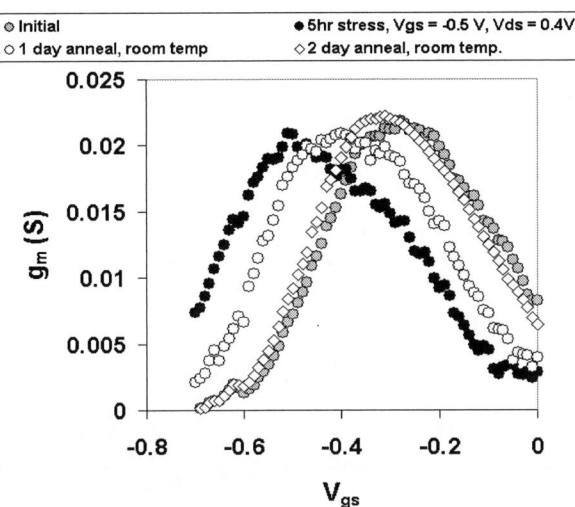

Fig. 7. Degradation (threshold and peak g_m shift) of a 2×10 μm HEMT with 100 nm gate length and 2 μm source-drain spacing. Devices were stressed at $V_{gs} = -0.5$ V and $V_{ds} = 0.4$ V for 5 hours. Annealing results at room temperature are shown. Device recovers to almost unbiased state in 2 days. $V_d = 0.4$ V.

978-1-4244-5430-3/10 $26.00 © 2010 IEEE

of the Sb_{Al} antisite at the growth conditions of AlSb is the lowest among all native defects. Calculations have also demonstrated the possibility of such a defect changing its configuration to a metastable state with a more positive charge state than its lower energy configuration. However, our calculations (Fig. 9) on the formation energies of the ground state and metastable configurations of the Sb_{Al} antisite clearly show the (0/+1) transition energy to be much shallower for T_d defects than for the metastable C_{3v} structure – which is the higher energy configuration. This rules out the antimony antisite as

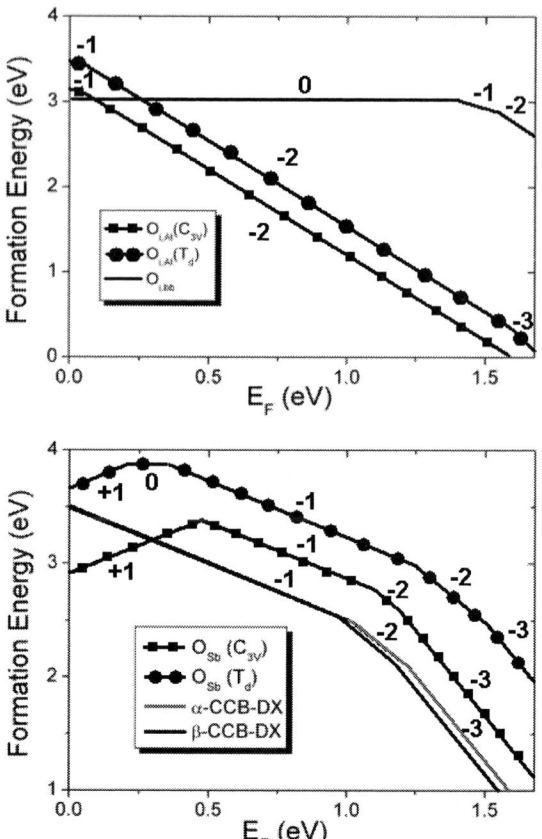

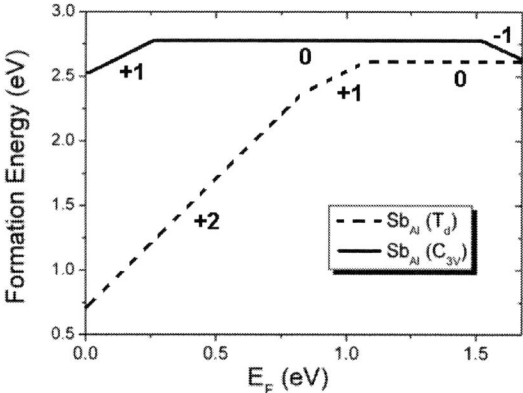

Fig. 9. Thermodynamic transition levels for the ground-state (T_d) and the metastable (C_{3v}) Sb_{Al} defect in Al-rich conditions. The donor like transition level (0/+1) is shallower for T_d than for the metastable C_{3v} structure. This precludes the possibility of threshold voltage left shifts due to transition of some antisites from T_d to C_{3v} under applied stress.

the defect responsible for the observed degradation.

A. Metastable Defects Associated with Oxygen

The basic criterion for the negative shift in threshold voltage under electrical stress is satisfied by several impurity-related defects that exhibit metastability. Because of the experimental evidence of oxygen as a key contaminant in AlSb [5], we concentrate on the energetics of two oxygen-related metastable defects. Fig. 10 gives a schematic picture of the metastable states of substitutional oxygen. Transition from β-CCBDX to C_{3v} configuration at $E_f \sim 0.4$ eV or lower changes the defect charge state from −1 to +1, causing a negative shift in threshold voltage. Also, the difference between the formation energies is sufficiently small near $E_f \sim 0.4$ eV that enough holes have the requisite energy to overcome the formation energy barrier. The values of the defect energy levels and barriers are only approximate due to limitations of the calculation technique. The measured degradation only occurs if the Fermi level is at a position where the ground state and metastable excited state have different charge states. For the part of the energy gap close to the conduction band, there is no difference in the charge states of the different metastable configurations (−1 for all). This could be the reason for the resistance of the devices with large negative threshold voltages to degradation. Since $E_f \sim 0.4$ eV (or lower values) corresponds to small electron densities, it is even possible for complete pinch-off devices to have E_f in the top AlSb layer above 0.4 eV for the

Fig. 10. Transition levels for (bottom) substitutional and (top) interstitial oxygen shown. β-CCBDX is the lowest energy configuration for O_{sb} followed by a ground (T_d) and excited (C_{3v}) state. Transition from β-CCBDX to either of the 2 defects at $E_f \sim 0.4$ eV will change the defect charge state from −1 to +1 or 0, causing a left shift in threshold voltage. A transition from $O_{i, tet, Al}$ to $O_{i,bb}$ for the interstitial oxygen will give the same effect.

entire range of operation. This could be a possible explanation for almost 70 % of the devices not degrading at all.

The metastable configurations $O_{i,Al}$ (C_{3v}) and $O_{i,bb}$ of interstitial oxygen can also explain the degradation trends. For the range of E_f in AlSb, the higher energy configuration $O_{i,bb}$ is neutral, and hence is more positive than the −2 state of the most stable $O_{i,Al}(C_{3v})$. Towards the conduction band, the states are separated by fairly large energies. Near $E_f \sim 0.15$ eV, the energy difference is reduced to almost 0. So, if this defect is responsible for the V_t and g_m peak shifts, its effects would be more pronounced at high gate fields. Further calculations, annealing experiments are necessary to determine whether the energies and relaxation barriers of the mentioned defects are consistent with the recovery rates of the degraded devices. Stress experiments performed at high temperatures can further indicate or disprove the role of each one of these defects in device degradation.

VII. CONCLUSIONS

A significant fraction of InAs - AlSb HEMTs exhibit degradation under stress. Degradation involves shifts in the peak transconductance toward more negative gate voltages, with no specific indications of mobility degradation, indicating the activation of new donor defects or deactivation of existing acceptors in the AlSb layers flanking the channels. Devices with large negative threshold voltages and/or long gate length are very resistant to degradation – indicating the role of hot carrier induced degradation. The defects anneal within a few days, suggesting that a defect with a metastable deep acceptor state is modified by hot carriers. Formation energies and ionization levels of two oxygen-based defects in AlSb are consistent with degradation trends.

REFERENCES

[1] C. Nguyen, B. Brar, C. R. Bolognesi, J. J. Pekarik, H. Kroemer, and J. H. English, "Growth of InAs-AlSb quantum wells having both high mobilities and high concentrations," J. Electronic Materials, vol. 22, n. 2, pp. 255-258,1993.

[2] J. B. Boos, W. Kruppa, B. R. Bennett, D. Park, S. W. Kirchoefer, R. Bass, and H. B. Dietrich, "AlSb/InAs HEMT's for low-voltage, high-speed applications," IEEE Trans. Electron Devices, vol. 45, pp. 1869-1875, Sept. 1998.

[3] B. Brar, G. Nagy, J. Bergman, G. Sullivan, P. Rowell, H. K. Lin, M. Dahlstrom, C. Kadow, and M. Rodwell, "RF and DC characteristics of low-leakage InAs/AlSb HFETs," 2002 Proc. Lester Eastman Conference, pp. 409-413.

[4] C. R. Bolognesi, E. J. Caine, and H. Kroemer, "Improved charge control and frequency performance in InAs/AlSb based InAs/AlSb based heterostructure field-effect transistors," IEEE Electron Device Lett., vol. 15, pp. 16-18, Jan. 1994.

[5] Y. C. Chou, J.M. Yang, M.D. Lange, S.S. Tsui, D.L. Leung, C.H. Lin, M.Wojtowicz, A.K. Oki, "Degradation mechanisms of 0.1 µm AlSb/InAs HEMTS for ultralow-power applications," Proceedings of IRPS, 2008, pp. 236 – 240.

[6] Bergman, J., Nagy, G., Sullivan, G., Ikhlassi, A. and Brar, B., "Low voltage, high-performance InAs/AlSb HEMTs with power gain above 100 GHz at 100 mV drain bias", DRC 2004, pp. 243-244.

[7] Mao-Hua Du, "Defects in AlSb: A density functional study," PHYSICAL REVIEW B 79, 045207 (2009).

InAlAs/InGaAs MHEMT Degradation during DC and Thermal Stressing

E. A. Douglas [1] K.H. Chen [2], C.Y. Chang [1], L. C. Leu [1], C. F. Lo [2], B. H. Chu [2], F. Ren [2], and S. J. Pearton [1]
[1] Department of Materials Science and Engineering, University of Florida [2] Department of Chemical Engineering, University of Florida
E. A. Douglas, Department of Materials Science and Engineering, University of Florida, Gainesville, Florida 32611, 352-316-1309
Rede0001@ufl.edu. S. J. Pearton, Department of Materials Science and Engineering, University of Florida, Gainesville, Florida 32611, 352-846-1086, spear@mse.ufl.edu

ABSTRACT

Reliability studies of InAlAs/InGaAs metamorphic high electron mobility transistors (MHEMTs) grown on GaAs substrates for high frequency/power applications are reported. The MHEMTs were stressed at a drain voltage of 3 V for 36hrs, as well as undergoing a thermal storage test at 250°C for 48hrs. Under both stress conditions, the drain current density decreased about 12.5%. The gate current, however, increased more after the thermal storage as opposed to DC bias. The main degradation mechanism during thermal storage was reaction of the Ohmic contact with the underlying semiconductor. Transmission electron microscopy verified that gate sinking occurred in devices that underwent DC bias stressing.

INTRODUCTION

An assortment of degradation mechanisms have been identified in both GaAs and InP-based based high electron mobility transistors (HEMTs). These mechanisms range from contact degradation (especially gate sinking), hot carrier-induced impact ionization at the gate edge, surface states that contribute to gate lag, avalanche breakdown in the semiconductor, fluorine contamination, mechanical stress due to hydrogen absorption into Ti metallization, and corrosion (mainly related to Al oxidation). [1-23] The use of metamorphic buffer layers has been utilized in order to grow InAlAs/InGaAs on larger diameter GaAs substrates. There is significant interest to develop GaAs MHEMTs due to the fact that it is possible to grow on larger, less brittle wafers at a lower cost as compared to InP. Due to the commercial applications of metamorphic HEMTs (MHEMTs), such as low noise mm-wave amplifiers for radio communications, automotive collision avoidance radar and high bit-rate fiber systems, device reliability is of great concern. [16] The choice of whether they can be used in place of InP-based HEMTs depends upon their DC/RF performance and the chip cost requirements.

A number of studies have shown that MHEMTs can exhibit similar reliability to InP HEMTs, with over 10^6 hours mean-time –to-failure at 125°C. [19, 20] However, a burn-in step is required in order to improve device stability and remove devices that suffer from infant mortality. This burn-in step is carried out for 24-60 hours at certain gate and drain voltages. A decrease in drain current and an increase in contact resistance occurs over time and levels off around 36 hours. Minimizing or eliminating this burn-in step is crucial to decrease fabrication costs. In order to effectively identify the failure mechanisms, both a high temperature storage test and DC stress were used in this study. Additionally, transmission line method (TLM) patterns were used to isolate the effect of the gate on device degradation.

EXPERIMENTAL

The MHEMTs were obtained from a commercial vendor and employed a two finger design with a Ti/Pt-based Schottky gate width of

75µm, as shown in the optical micrograph in Figure 1 (left). The gate length was 150 nm, with 1.2µm spacing between both the gate/drain and gate/source. The TLM patterns also present on the device chip, Figure 2 (right), employed 45 × 70 µm pads with gaps of 3, 6, 9, 12 and 15 µm. A schematic of the device structure is shown in the cross-section of Figure 2.

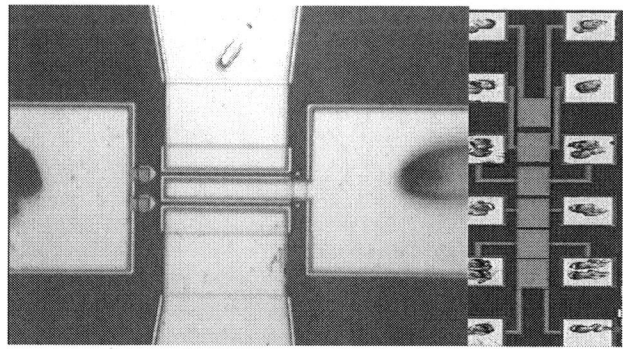

FIGURE 1. TOP VIEW OM IMAGE OF 2 FINGER InAlAs/InGaAs MHEMT DEVICE (LEFT), TLM PATTERN (RIGHT).

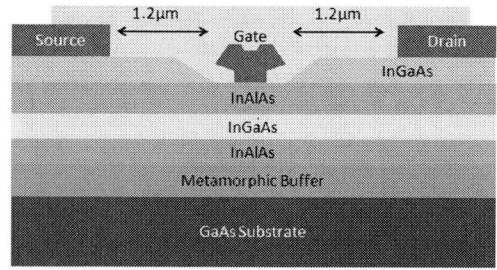

FIGURE 2. CROSS-SECTION SCHEMATIC OF MHEMT

Both DC and RF characteristics of the MHEMTs were measured prior to stressing and are shown in Figure 3. The maximum drain-source current density was 0.25 A.mm, and the gate current was in the hundreds of nA range. The unity current gain, f_t, was 94GHz with a maximum frequency of oscillation, f_{max}, of 124GHz.

978-1-4244-5430-3/10 $26.00 © 2010 IEEE

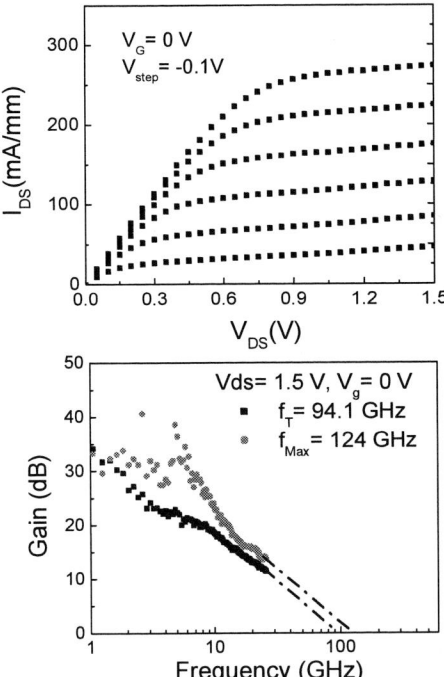

FIGURE 3. DRAIN CURRENT DENSITY (I_{DS}) OF VIRGIN MHMET AS A FUNCTION OF DRAIN VOLTAGE (V_{DS}) (TOP) AND MICROWAVE CHARACTERISTICS AT ROOM TEMPERATURE (BOTTOM).

The first lot of InAlAs/InGaAs MHEMTs was stressed with a source-drain bias of 3V for 36 hours at 165°C. The second lot of devices was subjected to a thermal storage test in an oven at 250°C for 36 hours. DC characteristics of the devices were measured before and after both stress tests with an Agilent 4145 parameter analyzer. A select few devices were also examined by cross-sectional Transmission Electron Microscopy (TEM) in order to examine the reaction of the contacts with the underlying semiconductor. Energy-dispersive x-ray spectroscopy (EDS) elemental analysis was also performed to acquire the elemental profiles near the reacted contacts.

Drain-current density decreased by ~12.5% after both a DC stress at 3V for 36 hours and high temperature stress at 250°C for 36 hours as compared to the unstressed devices (Figure 3 top). Figure 4 shows the drain current (I_{DS}) and gate current (I_{GS}) as a function of drain voltage (V_{DS}) after the thermal stress (left) and the I_{DS} after the DC stress (right). This indicates that after the burn-in process the devices suffer from higher parasitic resistance.

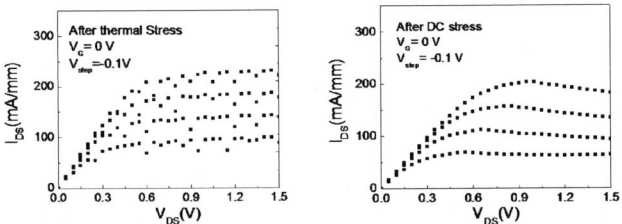

FIGURE 4. DRAIN CURRENT (I_{DS}) OF MHEMT AFTER THERMAL STRESS AS A FUNCTION OF DRAIN VOLTAGE (V_{DS}) AT 250°C FOR 36HRS (LEFT) AND DRAIN CURRENT (I_{DS}) OF MHEMT AFTER DC STRESS AT 3V FOR 36HRS (RIGHT).

In order to further investigate the origin of the degradation in drain-source current, the sheet resistance and specific contact resistance of the devices were obtained from TLM measurements as a

function of thermal storage time and as a function of constant bias voltage stress time. The TLM patterns eliminated any effect on the degradation of the drain-source current due to gate sinking. This allowed for examination of Ohmic metal contact degradation. The results of the DC stressed TLM patterns are shown in Figure 5. Over the 48 hours, the sheet resistance increased around 18%, while specific contact resistance was reduced by 40%. The device resistance increase was dominated by changes in the sheet resistance instead of contact resistance. As shown in Figure 6, the total resistance of TLMs increases significantly within the first 12 hrs of thermal storage at 250 °C, while the specific contact resistivity increases much more than sheet resistance. This indicates that the contact between the Ohmic metal and semiconductor dominated the degradation during the thermal storage.

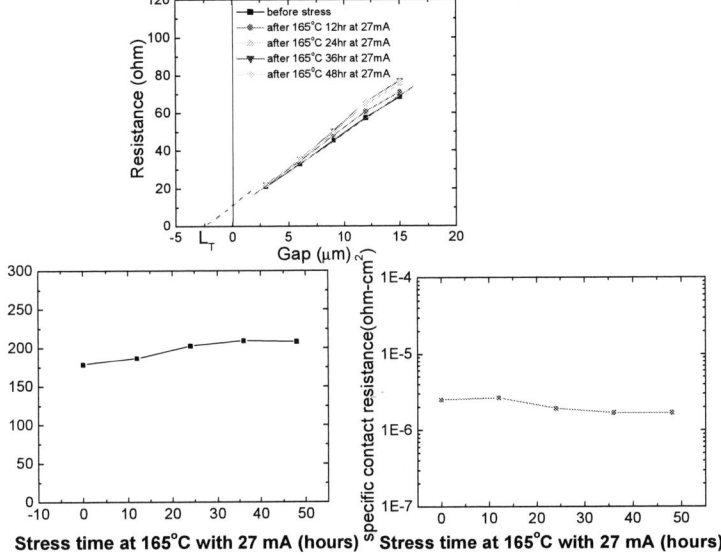

FIGURE 5. DC STRESS OF TLM OF 27 mA AT 165°C FOR 48HRS: RESISTANCE VERSUS GAP DISTANCE (TOP), SHEET RESISTANCE (BOTTOM LEFT), AND SPECIFIC CONTACT RESISTANCE (BOTTOM RIGHT).

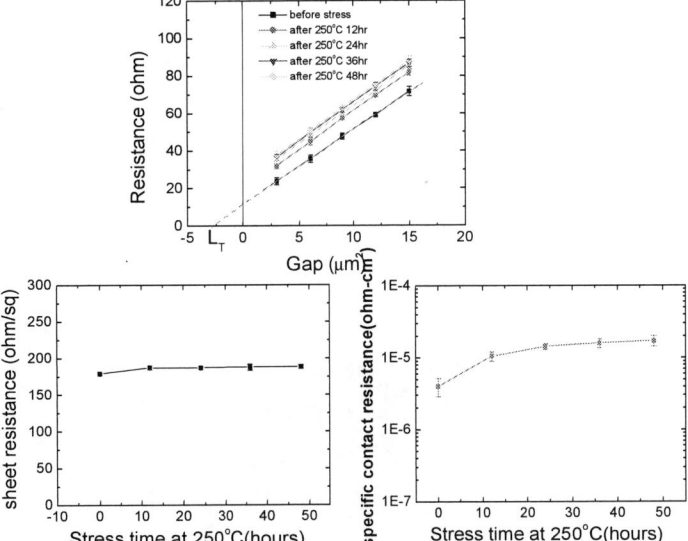

FIGURE 6. THERMAL STRESS OF TLM AT 250°C FOR 48HRS: RESISTANCE VERSUS GAP DISTANCE (TOP), SHEET RESISTANCE (BOTTOM LEFT), AND SPECIFIC CONTACT RESISTANCE (BOTTOM RIGHT).

TEM cross-sections of a degraded thermal storage HEMT and a constant current stressed HEMT are illustrated in Figure 7 (top) and Figure 7 (bottom), respectively. Both samples showed metal spikes of Ohmic metal diffusing into epitaxial layer. These spikes were formed during the high temperature Ohmic annealing. The constant current stressed samples displayed a greater number of spikes around the edge of the source and drain Ohmic contact pads, as illustrated in the Figure 8 (top). Interestingly enough, the region of the high density spikes in the TEM picture matched the estimated transfer length of the TLM measurement as shown in top plot in Figure 5. Thus, the formation of the high density spikes could result from current-induced electromigration.

FIGURE 7. LOW MAGNIFICATION CROSS-SECTION TEM IMAGE OF InAlAs/InGaAs MHEMT AFTER THERMAL STRESS AT 250°C FOR 36 HOURS (TOP) AND AFTER DC STRESS FOR 36HRS (BOTTOM)

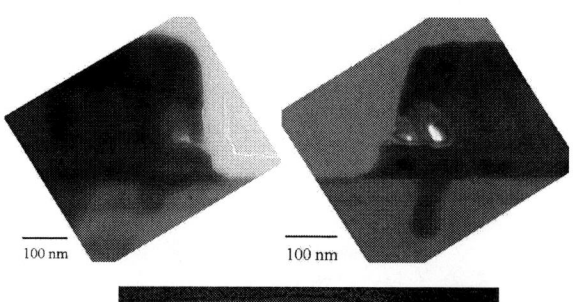

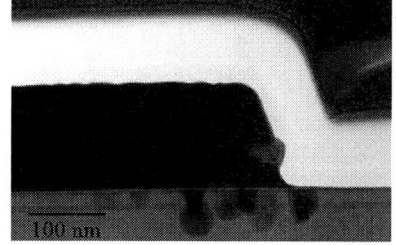

FIGURE 8. HIGHER MAGNIFICATION CROSS –SECTION TEM IMAGE SOURCE (TOP LEFT), DRAIN (TOP RIGHT), AND EDGE OF OHMIC CONTACT (BOTTOM)

The drain current density of the commercial power MHEMTS fabricated with the structure described here is around 0.5 A/mm to 1A/mm. With this structure, the final metal often has a thickness of 4-6 μm. The Ohmic metal contact, on the other hand, has a thickness of less than 0.3 μm. In these devices, the Ohmic metal is alloyed with the semiconductor, which results in a larger resistance of the alloyed metal stack compared to that of the unalloyed metal stack. The 4-6 μm thick final metal does not exhibit any issues with a current density of 0.5 A/mm to 1A/mm. However, as shown by the TEM image of Figure 8, the metal thickness of the region at the edge of the Ohmic contact pad is too thin to sustain the current density to avoid electromigration. For example, the Ohmic metal thickness for the device shown in Figures 8 is roughly 250 nm. During the burn-in process, 20 mA was used to stress a device with a gate width of 75 μm (20 mA/75 μm = 266 mA/mm). Therefore, the current density is given as 20mA/(75 μm × 0.25 μm) = 1×10^5 A/cm², which is the current density through the Ohmic pad. Such a high current density flowing through the thin Ohmic metal, across the metal semiconductor interface, and into the semiconductor causes the Ohmic metal to diffuse during the burn-in process. This resulted in electromigration-induced voids and the formation of additional metal spikes at the edge of the Ohmic metal contact pads, as shown in Figure 8, of the source contact pad (left) and drain contact pad (right) after performing the constant current stressing.

The gate ideality factor extracted from the gate current-voltage (I-V) characteristics of the unstressed HEMT was ~1.6, indicative of both recombination and thermionic emission electron transport mechanisms (Figure 9). The gate characteristics of the thermal and DC stressed HEMTs showed significant degradation. The gate current increased several orders of magnitude in both the forward and reverse bias conditions. Figure 10 (top) shows the low magnification cross-section view TEM image of a Pt/Ti/Pt/Au mushroom gate after 165°C, V_{DS} 3V, J_{DS} 300 mA/mm for 36 hr. From TEM and energy-dispersive x-ray spectroscopy (EDS) elemental analysis, it was determined that the bottom Pt layer of that Pt/Ti/Pt/Au mushroom gate diffused into the InAlAs gate contact layer. This is observed in Figure 10 (bottom) of a high magnification TEM image as a white area in the gate (Ti) and a dark area (Pt). As shown in Figure 11, 5-10 nm of Pt diffused into the InAlAs layer.

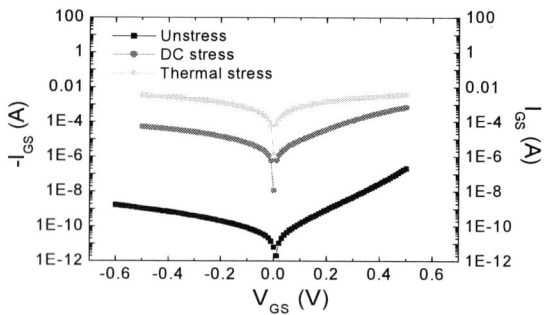

FIGURE 9. GATE CURRENT OF MHEMT DEVICE FOR UNSTRESSED, THERMAL STRESS (250°C, 36hrs) AND DC STRESS (V_{DS} 2.7 V, I_{DS} 167 mA/mm, 36 hrs).

978-1-4244-5430-3/10 $26.00 © 2010 IEEE   820

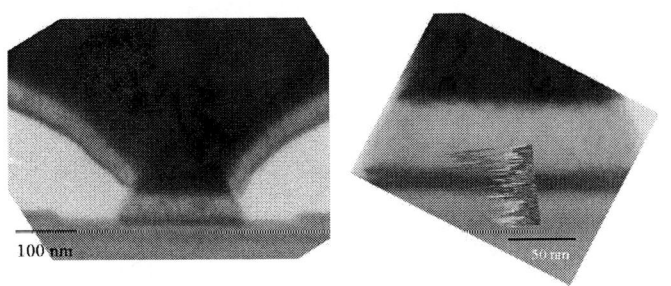

FIGURE 10. LOW MAGNIFICATION TEM CROSS-SECTION (TOP) AND HIGHER MAGNIFICATION TEM IMAGE (BOTTOM) OF MUSHROOM GATE AFTER DC STRESS FOR 36HR.

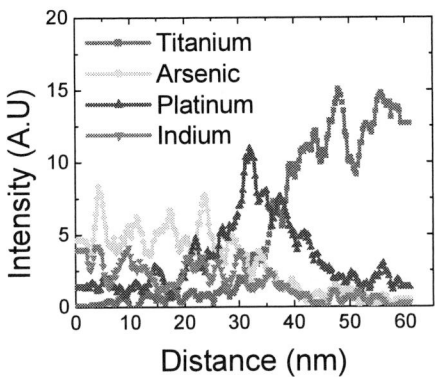

FIGURE 11. ENERGY-DISPERSIVE X-RAY SPECTROSCOPY (EDS) ELEMENTAL ANALYSIS OF MUSHROOM GATE AFTER 36HR DC STRESS TEST.

CONCLUSION

In order to address the needs for elimination of burn-in process and better reliability, we studied the mechanisms of InAlAs/InGaAs MHEMT degradation. DC stressing at a drain voltage of 3V and a current density of 167 mA/mm for 36 hours resulted in Pt from the gate diffusing into the underlying InAlAs layer, which was confirmed with TEM and EDS analysis. The high voltage bias also led to drain current degradation and gate sinking. Gate current increased more significantly after thermal stress (oven storage at 250°C for 36hrs) than after DC stress. TLM measurements showed that the deterioration of R_C was dominant in the thermal storage test and the increase in R_S was observed in the DC stress. Metal spike formation and Ohmic metal diffusion are observed after thermal stress.

ACKNOWLEDGMENTS

This work is supported by a DOD MURI monitored by AFOSR (Dr. Kitt Reinhardt)

REFERENCES

[1] W. J. Roesch, Microelectronics Rel.46, 1218 (2006).

[2]. J. A. del Alamo and A. A. Villanueva, IEDM Tech.Dig., 2004, pp. 1019–1022.

[3] Y.K. Kukai, S.Sugitani, T.Enoki and Y. Yamane, IEEE Trans. Device and Materials Reliability, 8,289 (2008).

[4] G. Meneghesso, C. Canali, P. Cova, E. De Bortoli, and E. Zanoni, IEEE Electron Device Lett., 17, 232 (1996).

[5]Y. C. Chou, R. Grundbacher, D. Leung, R. Lai, P. H. Liu, Q. Kan,M. Beidenbender, M. Wojtowicz, D. Eng, and A. Oki, IEEE Electron Device Lett., 25, 64 (2004).

[6] W. Roesch, GaAs REL Workshop, pp.119-126, 1999.

[7] A.A.Villanueva, J.A.del Alamo, F.Hisaka, K.Hayashi and M. Somerville, IEEE Trans. Device and Materials Reliability, 8,283 (2008)

[8] M. Borgarino, R. Menozzi, Y. Baeyens, P. Cova, and F. Fantini, IEEE Trans. Electron Devices, 45, 366 (1998).

[9] R. E. Leoni and J. C. M. Hwang, IEEE Trans. Electron Devices, 46, 1608 (1999).

[10] K. J. Choi and J. L. Lee, J. Electron. Mater., 30, 885 (2001).

[11] T. Hisaka, Y. Nogami, H. Sasaki, A. A. Villanueva, A. Hasuike, N. Yoshida, and J. A. del Alamo, Proc. Gallium Arsenide Integr. Circuits Symp., 2003, pp. 67–70.

[12] A. Villanueva, J. A. del Alamo, T. Hisaka, and T. Ishida, IEDM Tech. Dig., 2007, pp. 393–396.

[13]S. Mertens and J. del Alamo, IEEE Trans. Eletron Devices, 49,1849 (2002).

[14] P. Ersland, H. Jen, and X. Yang, GaAs REL Workshop, pp.3-6, (2003).

[15] T. Hisaka, Reliability On Compound Semiconductors (ROCS) Workshop, pp. 81-88, (2004).

[16] I. Thayne, K. Elgaid, D. Moran, X. Cao, E. Boyd, H. McLelland, M. Holland, S. Thoms and C. Stanley, Thin Solid Films, 515, 4373 (2007).

[17] S. Bollaert, Y. Cordier, M. Zaknoune, H. Happy, V. Hoel, S. Lepilliet, D. Théron, A. Cappy, Solid-State Electron, 44, 1021 (2000).

[18] M. Dammann, M. Chertouk, W. Jantz, K. Köhler, G. Weimann, Microelectronics Reliability, 40, 1709 (2000)

[19] P. F. Marsh, C. S. Whelan, W. E. Hoke, R. E. Leoni and T. E. Kazior, Microelectron. Reliability, 42, 997 (2002).

[20] M. Dammann, A. Leuther, R. Quay, M. Meng, H. Konstanzer, W. Jantz, and M. Mikulla, Microelectronics Reliability, 44, 939 (2004).

[21] M. Dammann, F. Benkhelifa, M. Meng, W. Jantz, Microelectronics Reliability, 42, 1569, (2002).

[22] G. Meneghesso and E. Zanoni, Microelectronics Reliability, 42, 685 (2002).

[23] Tetsuya Suemitsu, Thin Solid Films, 515, 4378 (2007).

A Built-in Aging Detection and Compensation Technique for Improving Reliability of Nanoscale CMOS Designs

Hamed F. Dadgour and Kaustav Banerjee
Department of Electrical and Computer Engineering
University of California, Santa Barbara
Santa Barbara, CA 93106

Abstract—The time-dependent degradation (aging) of device characteristics caused by Bias Temperature Instability (BTI) and Hot-Carrier Injection (HCI) are one of the major threats to the reliability of nanoscale digital CMOS designs. To address this challenge, a novel built-in aging "detection" and "compensation" technique is proposed. Performance degradation is detected using a novel area- and power-efficient sensor. Then, to improve the reliability, an adaptive Time-Borrowing (TB)-based compensation technique is employed, which decreases the timing failure probabilities in spite of aged transistors. It is shown via simulations that by employing these techniques, the reliability of circuits can be improved by approximately 10X.

Keywords- Aging, Bias-Temperature Instability, Diagnostics, Fault-Tolerance, Hot-carrier Effect, NBTI, PBTI, Process Variation, Reliability, Robustness, Timing Analysis.

I. INTRODUCTION

Transistor aging (I_{ON} degradation) due to BTI and HCI is one of the most challenging reliability concerns in the design of nanoscale VLSI circuits [1]. Over the past few years, several works have attempted to address the BTI challenge at fabrication [2], circuit [3] and system-level [4]. Other research papers have focused on statistical modeling and design optimization [5]-[6]. While such studies are valuable, design and implementation of built-in mechanisms to improve the reliability of circuits is indispensible and as important as modeling and characterization efforts. There have been a number of proposals on the design of "aging sensors", which attempt to detect the degradation of device characteristics (such as I_{ON} current) [7]-[8]. I_{ON} monitoring techniques would be effective if one could assume that the degradation of circuits can be projected based on the data obtained from monitoring of I_{ON} of a single transistor (because each circuit degrades at a different rate under the same aging scenario). Therefore, in this work, a built-in sensor that can measure "the performance degradation of circuits" is proposed. After the aging is detected, proper actions must be taken to reduce the probability of timing failures. In this paper, for the first time, a novel Time-Borrowing (TB)-based method is proposed, which employs reconfigurable flip-flops (FFs) to achieve this goal. Therefore, this work has two major contributions: **(1)** a built-in area and power efficient sensor is proposed for early detection of aging and **(2)** a novel adaptive TB-based compensation technique is introduced that is capable of improving the reliability of aged circuits.

II. BUILT-IN AGING DETECTION SENSOR

The early detection of aging is vital for built-in aging-resilient design techniques. To achieve this goal, this paper proposes a sensor to monitor the arrival times of different data signals (the outputs of Combinational Logic Blocks (CLBs)) at the input of FFs. The proposed sensor is connected to the FF of the pipeline stage and its CLB as shown in **Fig. 2 (a)**. When CLB is fresh, D arrives well before clock ($t_{DCLK} > t_{DX}$), the output (OUT) remains low (**Fig. 2 (b)**). However, when CLB ages ($t_{DCLK} < t_{DX}$), the output (OUT) becomes high (**Fig. 2 (c)**). The advantage of the proposed sensor over other existing designs is its area and power efficiency as will be shown later. The gate-level implementation of the proposed built-in sensor circuit is presented in **Fig. 3**. This sensor is composed of two stages; Stage #1 generates a "voltage glitch" on its output node, F, only if the circuit is aged. Since a glitch is not a valid logic value, Stage #2 is designed to convert the glitches at the node F to stable logic levels at OUT. The area overhead of the proposed circuit is esti-mated to be less than *42%* compared to a standard flip-flop without any sensor circuit. The area overhead for the pre-sample flip-flop and the one presented in [9], are over *145%* and *95%*, respectively.

A. Basic Operation

A schematic diagram presenting the signal waveforms at different nodes of **Fig. 3** is shown in **Fig. 4** to demonstrate its basic operation. The output of Stage #1 (*F*) becomes high, only if all three inputs of the NAND gate (X, Y and Q) are '1'. When the CLB is aged, the signal D arrives late and there will be a time interval when all three inputs of the NAND gate are '1' generating a "voltage glitch" at the output node *F* (**Fig. 4 (b)**). As shown in **Fig. 5 (a)**, when the CLB is fresh, the width of this voltage glitch is not large enough to trigger the "voltage glitch detector" circuit. Therefore, the output of the voltage glitch detector, *OUT*, remains low. However, if the voltage glitch is "wide enough" (**Fig. 5 (b)**), the dynamic node of the inverter will be discharged to ground, forcing *OUT* to become high.

B. Timing Analysis

More detailed signal waveforms of the proposed circuit (**Fig. 3**) are shown in **Fig. 4 (c)**. Note that for accurate operation of this circuit, high-to-low transition of X (T_X) and low-to-high transition of Q (T_Q) must occur during $T_{CLK} < t < T_{CLK} + t_{CLKY}$ interval, which leads to the following inequalities (1)-(2) (**Fig. 6**). Furthermore, the voltage glitch on F must be at least t_{Min} (a design parameter) wide to ensure the proper switching of Stage #2 which results in (3). Calculating the arrival time of X and Q in terms of other variables shown in **Fig. 4 (c)**, one gets (4) and (5). By plugging (4)-(5) in (1)-(3), one can obtain five design constraints as shown in (6)-(10).

C. Impact of Process Variations

Monte Carlo simulations are performed to investigate the impact of parameter fluctuations, suggesting that even under severe process variations ($\sigma_{Vth} = 20\% \mu_{Vth}$), the proposed sensor can accurately detect the aging of CLBs in *80%* of cases by taking only one sample (**Fig. 7 (a)**). By capturing more samples (>5), the accuracy is increased to $\approx 99.7\%$ when $\sigma_{Vth} = 20\% \mu_{Vth}$ (**Fig. 7 (b)**).

D. Simulation Results

The proposed circuit is simulated using Predictive Technology Models (PTM) [10] at 45 nm technology node (**Fig. 8**) where various arrival times (*470*, *480* and *490 ps*) for *D* is considered. It can be observed that nodes *F* and *OUT* remain low when *D* arrives well ahead of the clock (i.e. at *470* or *480 ps*). However, when D arrives at $\approx 490\ ps$ (indicating severe circuit aging), *OUT* becomes high. To examine the efficiency of the proposed circuit, two existing aging sensors in the literature are also considered as shown in **Fig. 9**. In the approach presented in **Fig. 9 (a)**, the inputs of CLBs are constantly screened using a delay element and an extra "pre-sample flip-flop (FF)". The pre-sampled data is then compared by a XOR gate to the actual value captured by the original FF. If CLB is fresh, the pre-sample FF captures the same value as the other FF does and hence, node *F* remains '0'. Otherwise, the two FFs sample different logic values triggering '1' at node *F*. The second aging circuit, which is proposed in [9] is depicted in **Fig. 9 (b)**. The area and power overheads associated with the proposed sensor and the two existing approaches are compared (**Fig. 10 (a)**). As it can be observed, the proposed circuit has clear advantages in terms of having minimal design overheads. Furthermore, the impact of various levels of process variations on

978-1-4244-5430-3/10 $26.00 © 2010 IEEE

both sensors is investigated in **Fig. 10 (b)** where one can observe that the detection error rates (the number of faulty detections to total detections) are almost identical for both designs.

III. AN ADAPTIVE TIME-BORROWING TECHNIQUE

In this work, a novel adaptive TB-based aging-resilient pipelining technique is proposed to improve the reliability once the aging is detected. The TB technique allows CLBs to automatically use slack time from the next cycle and hence, enables them to use more than one clock cycle (in conventional pipelines, only one cycle is available for CLBs). While the TB approach can improve the reliability in the aged circuits, it can result in excessive timing failures in the fresh circuits. Therefore, a modified version of TB method is used here that can overcome such shortcomings.

A. Timing Constraints in Pipelined Architectures

Timing constraints in pipelined circuits are imposed by *setup* and *hold* time requirements of FFs which are shown in **Fig. 11 (a)** and **Fig. 11 (b)**, respectively. According to this figure, the setup time requirement determines the maximum acceptable delay of CLBs and the hold time constraint establishes the minimum tolerable delay of CLBs. Therefore, the setup and hold time constraints together determine the available design space in terms of the delay of CLBs.

B. Time-Borrowing in Pipelined Architectures

In the pipelined architectures, time borrowing technique allows CLBs to borrow some slack time from the successive clock cycle. Time borrowing is usually implemented by replacing the conventional flip-flops with their time-borrowing counterparts. The operations of time-borrowing and conventional flip-flops are shown in **Fig. 12**. In the conventional flip-flop (**Fig. 12 (b)**), the master and slave latches are never transparent simultaneously. On the contrary, in the time-borrowing design, the master and slave latches remain transparent simultaneously for a period of time because the clock of the master latch is delayed with respect to that of the slave latch (**Fig. 12 (c)**). This "transparency window" increases the maximum time available to signals to propagate through the pipeline stages as shown in **Fig. 13**.

Timing requirements for a time-borrowing pipelined architecture (**Fig. 11**) are summarized in (11)-(15) as shown in **Fig. 14**. A graphical representation of the constraints imposed by (11)-(15) is provided in **Fig. 15** where the X-Y plane represents the available design space in terms of the delays of CLB1 and CLB2 (CLBs of two consecutive pipeline stages). While the square-shaped hashed area indicates the design space available using a conventional pipelining strategy, the other hashed area represents the design space offered by the TB method where numbers in parentheses next to each line indicate the corresponding equation from **Fig. 14** that imposes that particular constraint. It can be observed that the design space corresponding to the TB method can be shifted in the X-Y plane by changing the value of t_B.

C. Conventional vs. Time-Borrowing Pipelining

The pros and cons of the conventional and TB approaches are demonstrated using the design space sketches of **Fig. 16** where dark solid circles represent the delay of different critical paths through the pipeline. In an ideal case, all dark circles must reside inside the design space in order to prevent any timing failure. When the CLBs are fresh (**Fig. 16 (a)**), the conventional technique offers a design space that can contain all circles while the TB method fails to do so due to hold time constraint violations. When the CLBs age (**Fig. 16 (b)**), the distribution of dark circles move in the top-right direction (since both CLBs age). As a result, all circles fit in the design space offered by the TB approach while some of them reside outside of the design space of the conventional approach, indicating setup time failures. Using **Fig. 16**, one can quantitatively summarize setup and hold time failure probability of different pipelined architectures as shown in **Fig. 17**. This figure suggests that the conventional pipelining method is more desirable when CLBs are fresh and the TB-based approach is preferable when CLBs age. Therefore, in this work, an adaptive TB-based approach is proposed that can combine the advantages of both techniques while avoiding their drawbacks.

D. Adaptive TB Approach using Reconfigurable FFs

The proposed adaptive TB approach employs "reconfigurable FFs", which operate as conventional FFs when CLBs are fresh and perform as TB FFs when CLBs are aged. The gate-level representations of the conventional, TB and reconfigurable TB FFs are shown in **Fig. 18**. These FFs employ identical master/slave latch architecture (**Fig. 18 (a)**). However, they differ in the clocking schemes that have been used to drive the master and slave latches (**Fig. 18 (b)-(d)**). The design of the reconfigurable FF employs a multiplexer, which allows the selection of the clock signal for the master latch and hence, provides a mechanism for switching between the conventional and TB operational modes.

Using Monte Carlo simulations of 10,000,000 two-stage pipeline architectures, the timing failure probabilities due to setup and hold time violations of the three pipelining techniques (conventional, TB and adaptive TB) are calculated (**Fig. 19**). It can be observed that the conventional technique results in a large rate of setup time violations (**Fig. 19 (a)**) when CLBs are aged while the TB approach suffers from high hold time failures (**Fig. 19 (b)**) when CLBs are fresh. The proposed adaptive TB technique, however, effectively eliminates such timing failures in all scenarios. Note that the results obtained from **Fig. 19** are in agreement with the conclusions of **Fig. 17**.

Moreover, to investigate the impact of process variations, the effectiveness of three pipelining approaches is evaluated in terms of their failure probabilities (which are defined as the highest failure probability considering both setup and hold time failures over the life-time of the circuit). The failure probabilities are shown in **Fig. 20 (a)** for various process variation levels (σ_{Vth}=10%, 15% and 20% μ_{Vth}). Note that the likelihoods of timing failures for all pipelining techniques increase approximately at the same rate by escalating the parameter fluctuations. It can be observed that by using the proposed method, the failure probability is decreased by approximately 10X. Furthermore, the timing failure probabilities are evaluated for deeper pipelined architectures with various numbers of pipeline stages as shown in **Fig. 20 (b)**. Note that although the timing violations increase dramatically for the conventional and TB techniques, the failure probability for the proposed approach increases at a lower rate. Hence, the proposed technique can be more effective for deeper pipelines that are preferred in all high-performance designs including microprocessors.

IV. CONCLUSIONS

Novel built-in aging "detection" and "compensation" techniques are proposed, which enable nanoscale VLSI circuits to sustain their reliable operation in spite of their degraded transistors. In this work, the performance degradation of circuits is detected using an area and power efficient built-in sensor and an adaptive TB-based compensation technique is introduced that can improve the reliability of digital circuits by approximately 10X. The proposed techniques are independent of the aging mechanism and the degree of aging, and can be widely employed for increasing the robustness of deeply pipelined digital circuits.

REFERENCES

[1] S. Borkar, "Designing Reliable Systems from Unreliable Components: The Challenges of Transistor Variability and Degradation," *Micro*, pp. 10-16, 2005.
[2] S. J. Doh et al., "Improvement of NBTI and electrical characteristics by ozone pre-treatment and PBTI issues in HfAlO(N) high-k gate dielectrics," *IEDM*, 2003, pp. 38.7.1-38.7.4.
[3] X. Liang, D. Brooks, and G.-Y. Wei, "A Process-Variation-Tolerant Floating-Point Unit with Voltage Interpolation and Variable Latency," *ISSCC*, 2008, pp. 404-405.
[4] D. Blaauw, et al., "Razor II: In Situ Error Detection and Correction for PVT and SER Tolerance," *ISSCC*, 2008, pp. 400-401.
[5] R. Vattikonda, W. Wang and Y. Cao ,"Modeling and minimization of PMOS NBTI effect for robust nanometer design," *DAC*, 2006, pp. 1047-1052.
[6] S. V. Kumar, C. H. Kim, and S. Sapatnekar, "NBTI-Aware Synthesis of Digital Circuits," *DAC*, 2007, pp. 370-375.
[7] M. Denais, et al., "On-the-fly Characterization of NBTI in Ultra-thin Gate Oxide PMOS-FET's," *IEDM*, 2004, pp. 109-112.
[8] K. Tae-Hyoung, R. Persaud, C.H. Kim, "Silicon Odometer: An On-Chip Reliability Monitor for Measuring Frequency Degradation of Digital Circuits," *JSSC*, Vol. 43, pp. 874-880, 2008.
[9] M. Agarwal, B. Paul, M. Zhang and S. Mitra, "Circuit Failure Prediction and its Application to Transistor Aging," *Proc. IEEE VLSI Test Symp.*, pp. 277-284, 2007.
[10] PTM: http://www.eas.asu.edu/~ptm/.
[11] J. M. Rabaey, A. P. Chandrakasan, and B. Nikolic, *Digital Integrated Circuits: A Design Perspective*, 2 ed., Prentice Hall Electronics and VLSI Series, Upper Saddle River, NJ: Prentice Hall/Pearson Education, 2003.

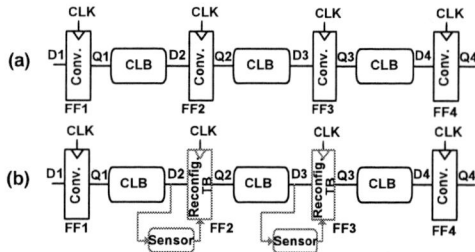

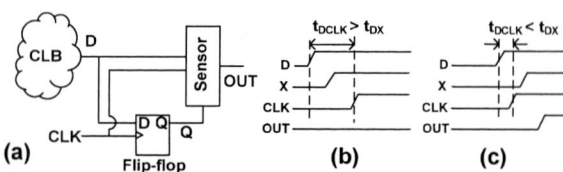

Fig. 1. (a) A conventional pipelined architecture where Combinational Logic Blocks (CLBs) are separated by Flip-Flops (FFs), and (b) the proposed aging-resilient pipelined architecture This architecture employs novel aging monitoring circuits (denoted as "sensor") and reconfigurable TB FFs (referred to as "Reconfig. TB"). Note that unlike the middle stages, the first and last stages do not use TB FFs; hence the proposed scheme does not affect any prior or subsequent pipelined stages.

Fig. 2. The aging detection circuit proposed in this work: (a) circuit-level block diagram. This figure show how the sensor circuit is wired to the other circuit components (the CLB and the FF of the pipeline stage). (b) Signal waveforms when input D arrives well before clock ($t_{DCLK} > t_{DX}$) when CLB is fresh, and (c) signal waveforms after CLB ages ($t_{DCLK} < t_{DX}$).

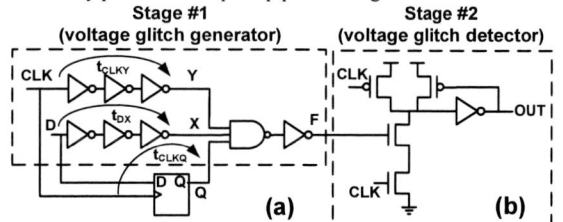

Fig. 3. The schematic of the proposed aging detection circuit: (a) voltage glitch generator and (b) voltage glitch detector. Here, it is assumed that the inputs of each FF (D) are the outputs of the CLBs of the pipelined architecture (as shown in **Fig. 1 (b)**).

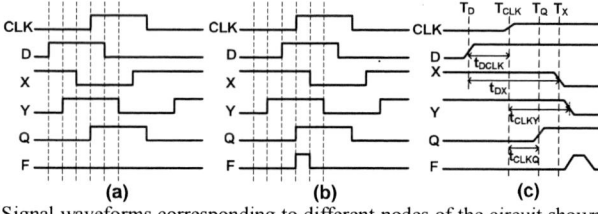

Fig. 4. Signal waveforms corresponding to different nodes of the circuit shown in **Fig. 3 (a)**: (a) fresh circuit, (b) aged circuit and (c) different variables used in the timing analysis.

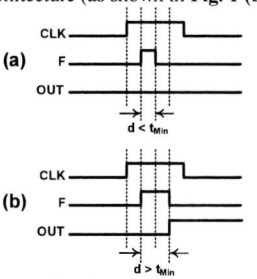

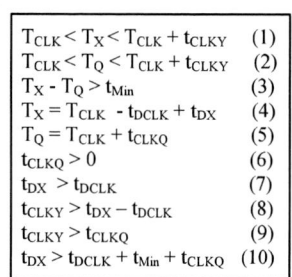

$$T_{CLK} < T_X < T_{CLK} + t_{CLKY} \quad (1)$$
$$T_{CLK} < T_Q < T_{CLK} + t_{CLKY} \quad (2)$$
$$T_X - T_Q > t_{Min} \quad (3)$$
$$T_X = T_{CLK} - t_{DCLK} + t_{DX} \quad (4)$$
$$T_Q = T_{CLK} + t_{CLKQ} \quad (5)$$
$$t_{CLKQ} > 0 \quad (6)$$
$$t_{DX} > t_{DCLK} \quad (7)$$
$$t_{CLKY} > t_{DX} - t_{DCLK} \quad (8)$$
$$t_{CLKY} > t_{CLKQ} \quad (9)$$
$$t_{DX} > t_{DCLK} + t_{Min} + t_{CLKQ} \quad (10)$$

Fig. 5. Signal waveforms corresponding to different nodes of the circuit shown in **Fig. 3 (b)**: (a) fresh circuit, (b) aged circuit. Here, t_{Min} is a parameter that depends on the design of the dynamic gate in **Fig. 3 (b)**.

Fig. 6. Timing analysis of the proposed sensor (**Fig. 3**).

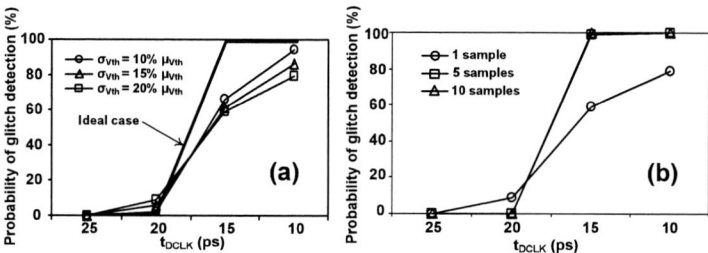

Fig. 7. (a) Probability of accurate detection of aging with one test sample under different process variation scenarios and (b) the accuracy of prediction with the proposed sensor for various numbers of samples. The accuracy of detection can be significantly improved by obtaining more samples.

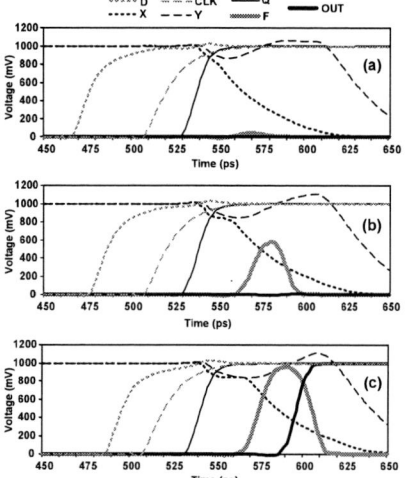

Fig. 8. Simulation results for the proposed sensor where signal D arrives at flip-flop's input at (a) 470 ps, (b) 480 ps and (c) 490 ps emulating the aging of circuit. Only in the last case, OUT becomes high indicating the aging of the combinational logic. Note that in the second case, although a voltage glitch is generated on F by stage #1 (**Fig. 3**), the width of the glitch is not large enough to trigger the output of stage #2.

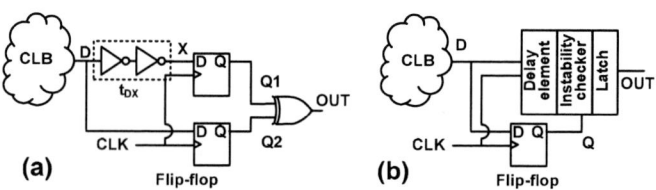

Fig. 9. Two existing aging sensors in the literature: (a) pre-sample flip-flop-based approach and (b) the aging sensor proposed in [9]. The basic operation of both sensors is based on monitoring the arrival time of D signal. If the arrival time of D is sufficiently close to the edge of the clock (due to aging), the output of both sensors become high.

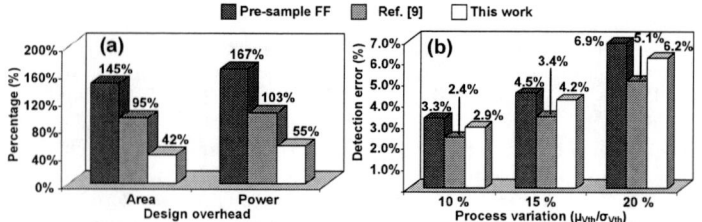

Fig. 10. (a) Design overheads for the proposed sensor in this work, the pre-sample flip-flop technique and the one introduced in [9] and (b) comparison between the detection error rates of the three designs for various degrees of process variations. It can be observed that the proposed sensor circuit is more area and power efficient. The impact of process variation is almost identical on all the designs.

978-1-4244-5430-3/10 $26.00 © 2010 IEEE

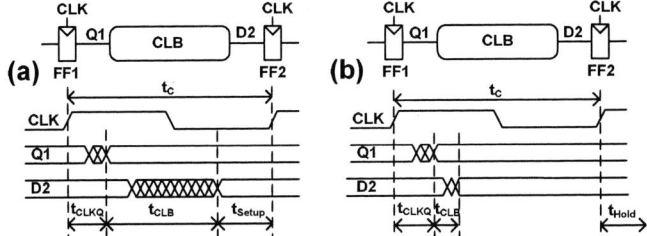

Fig. 11. Schematic diagram to illustrate the timing constraints in pipelined architectures: (a) setup time requirement, which determines the maximum acceptable delay of the CLBs and (b) hold time requirement, which establishes the minimum tolerable delay of CLBs.

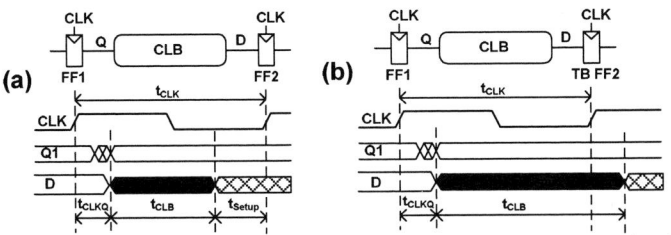

Fig. 13. The time-borrowing concept: (a) timing diagram of a conventional pipelined system. The signal must propagate through the FF1 and CLB in one clock cycle. Therefore, the dark area represents the maximum acceptable delay of CLB. (b) Timing diagram for a time-borrowing architecture where the time-borrowing flip-flop (TB FF2) remains transparent after the clock edge allowing more time for propagating signals across FF1 and CLB. Thus, in this figure, the dark area (the maximum tolerable delay of CLB) is extended.

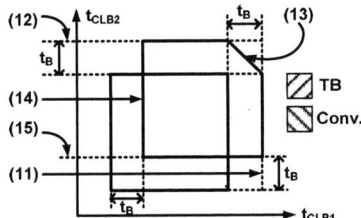

Fig. 15. Different constraints imposed by setup and hold time requirements: each number in the parenthesis refers to the equation from **Fig. 14** that imposes that particular timing constraint. The square-shaped area depicts all the possible combinations for CLB1 and CLB2 delays for which the conventional pipelining approach can operate accurately while the other hashed region highlights part of the design space that is made available by using the TB technique.

Design	setup time failure Prob.		hold time failure Prob.	
	Fresh circuit	Aged circuit	Fresh circuit	Aged circuit
Conv.	Low	High	Low	Very low
TB	Very low	Low	High	Low
Adaptive TB	Low	Low	Low	Low

Fig. 17. Setup and hold time failure probability for aged and fresh circuits using conventional (Conv.), time-borrowing (TB) and adaptive time-borrowing (Adaptive TB) pipelined approaches. While the conventional method results in high setup timing violation when CLBs age, the TB method suffers from high hold time failures when CLBs are fresh. The adaptive TB approach combines the advantages of both the conventional and the TB approach.

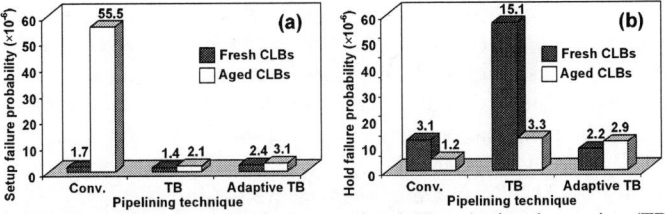

Fig. 19. Comparison between the conventional (Conv.), time-borrowing (TB) and adaptive time-borrowing (Adaptive TB) techniques when CLBs are fresh and aged, in terms of: (a) setup failure and (b) hold failure probabilities.

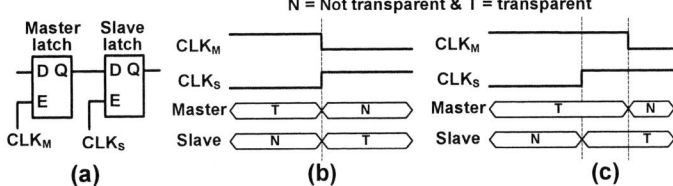

Fig. 12. The operation of time-borrowing and conventional flip-flops: (a) a master-slave flip-flop, which is created by connecting two gated latches in series, (b) clocking scheme used for a conventional flip-flop and state of its latches (transparent (T) or not transparent (N)) and, (c) clocking scheme used for a time-borrowing flip-flop. Here, CLK_M and CLK_S denote the clock signal of the master and slave latches, respectively. In the time-borrowing design, the master and slave latches remain transparent simultaneously for a short time because the clock of the master latch is delayed with respect to that of the slave latch.

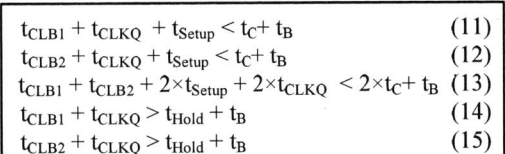

$$t_{CLB1} + t_{CLKQ} + t_{Setup} < t_C + t_B \qquad (11)$$
$$t_{CLB2} + t_{CLKQ} + t_{Setup} < t_C + t_B \qquad (12)$$
$$t_{CLB1} + t_{CLB2} + 2 \times t_{Setup} + 2 \times t_{CLKQ} < 2 \times t_C + t_B \qquad (13)$$
$$t_{CLB1} + t_{CLKQ} > t_{Hold} + t_B \qquad (14)$$
$$t_{CLB2} + t_{CLKQ} > t_{Hold} + t_B \qquad (15)$$

Fig. 14. The timing analysis of a two-stage pipelined architecture (based on **Fig. 11**). Timing constraints are imposed by the setup and hold time requirements of FFs. Equations (11)-(12) show the setup time requirements while (14)-(15) represent hold time conditions. Finally, (13) indicates the fact that the total amount of time borrowing (t_B) is shared among two consecutive pipeline stages and can not exceed an upper limit. Note that setup and hold time constraints determine the upper and lower limits of the delay of CLBs, respectively. In the above equations, by setting $t_B = 0$ and $t_B \neq 0$, one can obtain the timing constraints for the conventional and TB pipelining methods, respectively.

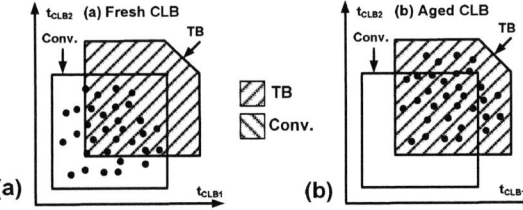

Fig. 16. Design spaces available for a two-stage pipeline using the conventional and TB techniques: when combinational circuits (CLBs) are (a) fresh and (b) aged. The dark solid circles represent the delay of various critical paths. Here, each dark circle represents a pair of t_{CLB1} (the delay of CLB1) and t_{CLB2} (the delay of CLB2).

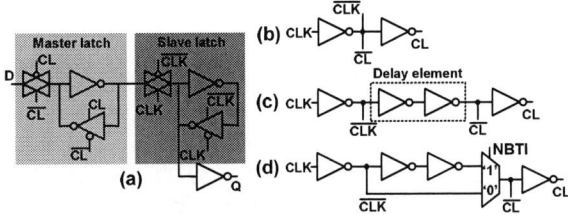

Fig. 18. Gate-level implementation of different flip-flops: (a) master/slave latches (identical for all flip-flops) and clocking scheme for (b) conventional, (c) time-borrowing (note that time-borrowing is performed by delaying the clock of the master latch with respect to that of the slave latch) and (d) reconfigurable flip-flops, which act as conventional flip-flops when NBTI='0' and perform as time-borrowing flip-flops when NBTI = '1'. The NBTI signal is the OUT signal in **Fig. 3(b)**.

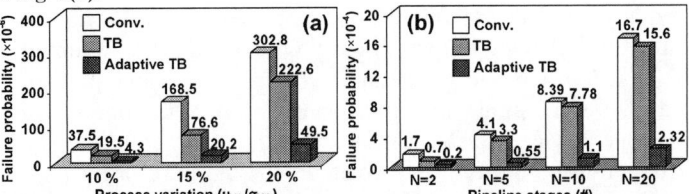

Fig. 20. Timing failure probability for different pipelining approaches considering: (a) various levels of process variations and (b) different number of pipeline stages (in this simulation, the process variation is assumed to be $\sigma_{Vth}=15\% \mu_{Vth}$.

978-1-4244-5430-3/10 $26.00 © 2010 IEEE

A Test Concept For Circuit Level Aging Demonstrated By A Differential Amplifier

Florian R. Chouard, Christoph Werner, Doris Schmitt-Landsiedel
Lehrstuhl für Technische Elektronik
Technische Universität München
Germany
+49–89–289–22930, florian.chouard@tum.de

Michael Fulde
Infineon Technologies AG
Villach, Austria

Abstract— **In this work a general concept for accelerated aging of linear analog circuit blocks is proposed. Due to the interaction of diverse aging mechanisms, circuit behavior in an arbitrary effect accelerated stress setup may show large deviation to the aging under nominal circuit conditions. The proposed analytical small signal analysis proves to be a fast and easy way to obtain the contributions of all degrading transistors with respect to the output monitor of the circuit. Circuit aging simulations over temperature rise as well as supply and input voltage scaling show that single effect acceleration varies significantly between the involved mechanisms. This causes deviations in the aging output monitor compared to the aging under nominal circuit conditions. Based on these findings an accelerated circuit level aging test concept – applicable to linear circuits - is developed and evaluated for the example of a two-stage differential amplifier.**

Keywords: circuit reliability, degradation, NBTI, PBTI, HCI, aging acceleration, testing

I. INTRODUCTION

A. Circuit Level Aging

The state-of-the-art procedure to accelerate aging of integrated circuits is the measurement under raised temperature and voltages. This acceleration technique – adopted from single device measurement, where solely one degradation effect is monitored – may cause large deviations in the circuit aging monitor compared to the aging under realistic nominal conditions. The problem in the circuit level aging arises, as typically diverse degradation mechanisms - each with its own temperature, voltage and time dependence – appear and interact. So, emerging aging mechanisms in the circuit each meet individual effect acceleration that attributes to the overall aging output monitor. Obviously the assignment of an equivalent age to the stressed circuit or a mapping of end-of-lifetime aging under nominal circuit conditions based in single device effect acceleration factors may induce large errors. Furthermore, a temperature rise itself influences sensitive transistor parameters and leads to variations of internal operating voltages. However, for linear analog circuits

the equal increase of all external voltages should approximately lead to an equivalent rise of voltages at internal nodes. In the further analyses we apply a close interaction of circuit design and device physics to obtain realistic accelerated aging characterization of complex analog circuits.

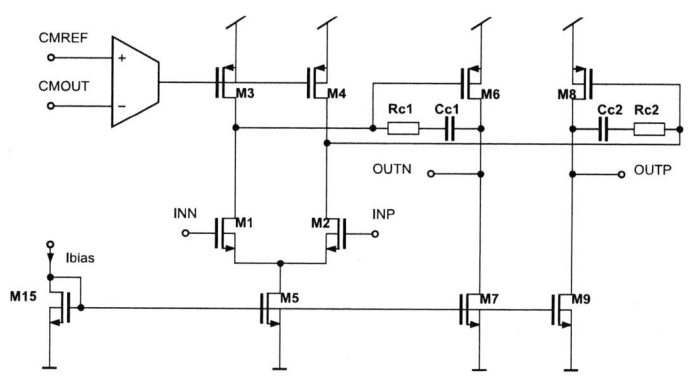

Figure 1.   nMOS input Miller compensated 2-stage OTA

B. Device Under Test

In the following evaluation a fully differential 2-stage Miller compensated operational transconductance amplifier (OTA), as depicted in figure 1, is analyzed as an example for a linear analog block. Previous investigations on differential amplifiers showed that degradation induced device mismatch mainly leads to offset [1]. In [2] we already determined two degradation related critical states for differential amplifiers: the closed-loop worst case, where a small differential input voltage is applied, but large enough that the output sees the full voltage swing, and the open-loop or comparator worst case with full swing at the input. To account for all degradation mechanisms in current technologies the circuit is designed and fabricated in a 32nm high-k, metal gate technology [3]. Simulated and measured aging induced offsets of the output voltage $V_{outp} - V_{outn}$ are depicted in table 1. For

---

This study was supported by Infineon Technologies AG

978-1-4244-5430-3/10 $26.00 © 2010 IEEE

both use cases an arbitrary accelerated aging test setup featuring a voltage rise of 150 % nominal VDD @ 125°C as well as an end-of-lifetime simulation of the circuit aging under nominal conditions are performed. To provide a basis of comparison, all degradation induced offset drifts, manifesting in an output referred offset, are evaluated under nominal circuit conditions (VDD, 25°C).

TABLE I.   DEGRADATION INDUCED OUTPUT REFERRED OFFSET IN MV

	Closed-loop		Open-loop	
	Sim.	*Msm.*	*Sim.*	*Msm.*
1000s, 150% V_{scale}, 125°C	+9	+19	-666	-562
10y, nom. supply, 25°C	+0.8	n.a.	-54.8	n.a.

Comparing the aging induced offsets of the arbitrary accelerated setup and the aging behavior under nominal conditions in table 1, a large deviation between both circuit environments can be observed. Regarding the mapping quality, the acceleration highly exceeds the nominal aging case and provides a poor mapping.

II.   AGING MECHANISMS

Degradation estimation in this work uses semi-empirical models, based on [4] - [6], that are adapted to measurement data. We account for permanent drifts of Negative Bias Temperature Instability (NBTI) for pMOSFETs, Positive BTI (PBTI) for nMOSFETs and Hot-Carrier Injection (HCI) for both transistor types in conducting (HCI) and non-conducting (NCHCI) state. Both types of BTI degradation are modeled via a drift of the device threshold voltage in the form:

$$\Delta V_{th} = A \cdot (V_{gs}/t_{inv})^m \cdot \exp(\Delta E/kT) \cdot L^\alpha \cdot W^\beta \cdot t^n \quad (1)$$

Hot-Carrier effects are modeled as a direct degradation of transistor drain current.

$$\Delta I(\%) = B \cdot V_{ds}^p \cdot \exp(\Delta H/kT) \cdot L^\chi \cdot t^q \quad (2)$$

Evaluated drift values are considered via sub-circuit models in circuit simulations. Based on the formulas in (1) and (2), effect acceleration in time can be derived, by evaluating the time quotient for two stress conditions providing equal drift values. To give an example, the acceleration factor for BTI acceleration is evaluated in (3).

$$ACC_{BTI} = t_{nom}/t_{stress}|_{\Delta Vth,stress=\Delta Vth,nom} \quad (3)$$

III.   REALISTIC AGING CONCEPT

A.   Aging Monitor Partitioning

The typical field of application of differential amplifiers is the closed-loop configuration. Here the amplifier mostly handles only small amplitude signals. In this range of operation the amplifier exhibits a linear amplification behavior and is typically approximated by its linear small signal equivalent. As the aging induced offset can be expected as the induced parameter drift of a device in consideration of its sensitivity towards the output and as parameter drifts exhibit only small values, it is suggestive to apply the small signal equivalent to estimate the resulting offset. The total generated offset can be evaluated by accumulation of all contributors. A main advantage of this approach is the fast and easy way of obtaining the percentage of each degradation mechanism to the total offset in dependence of various parameters.

To prove the small signal approach, temperature and voltage rise is applied to the OTA in figure 1 for both use cases. The induced device degradation is evaluated and generated offsets are evaluated via the small signal approach as well as via circuit simulation.

TABLE II.   INPUT REFERRED OFFSET AFTER 10Y IN MV EVALUATED VIA SIMULATION AND SMALL-SIGNAL ESTIMATION FOR CLOSED-LOOP USE CASE

V	VDD		125%VDD		150%VDD	
T	*Sim.*	*Est.*	*Sim.*	*Est.*	*Sim.*	*Est.*
25°C	-0.01	-0.01	-0.22	-0.22	-1.74	-1.54
85°C	-0.012	-0.011	-0.39	-0.38	-3.62	-2.86
125°C	-0.011	-0.011	-0.51	-0.49	-5.38	-3.81

TABLE III.   INPUT REFERRED OFFSET AFTER 10Y IN MV EVALUATED VIA SIMULATION AND SMALL-SIGNAL ESTIMATION FOR OPEN-LOOP USE CASE

V	VDD		125%VDD		150%VDD	
T	*Sim.*	*Est.*	*Sim.*	*Est.*	*Sim.*	*Est.*
25°C	0.63	0.63	3.89	3.89	18.84	18.92
85°C	1.71	1.71	10.78	10.89	53.90	54.21
125°C	2.78	2.78	17.99	18.00	89.15	89.86

Table 2 and 3 shows good agreement between the simulation and the small signal estimation for the 10y end-of-lifetime aging case. Incidentally, for both test cases offset generating fractions show little variations over a 1y-10y range, which can be attributed to the similar time dependencies of the interacting mechanisms. Hence, Figure 2 and 3 depict mean values over time of the offset composition due to a rise in temperature as well as a rise in voltage.

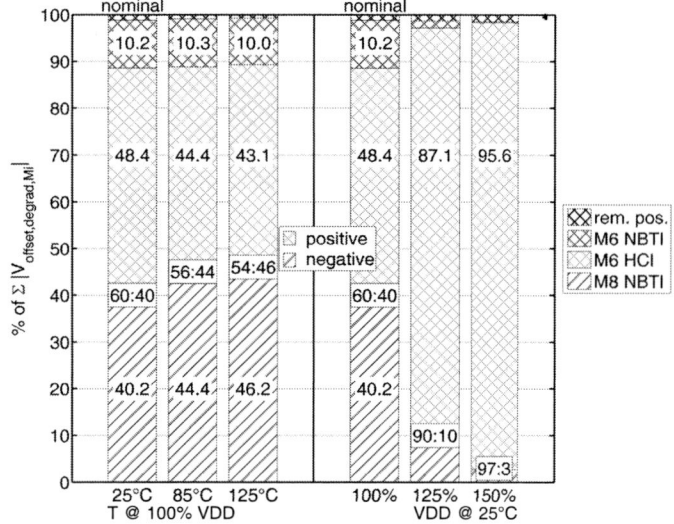
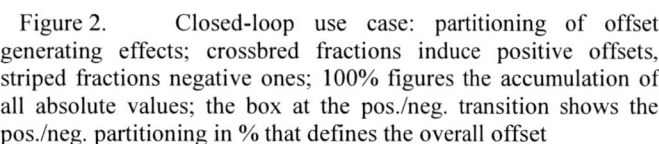
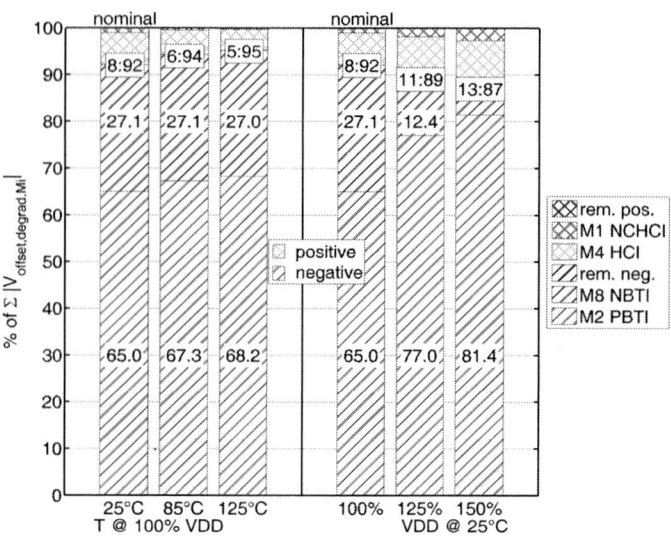

Figure 2. Closed-loop use case: partitioning of offset generating effects; crossbred fractions induce positive offsets, striped fractions negative ones; 100% figures the accumulation of all absolute values; the box at the pos./neg. transition shows the pos./neg. partitioning in % that defines the overall offset

Figure 3. Open-loop use case: partitioning of offset generating effects; crossbred fractions induce positive offsets, striped fractions negative ones; 100% figures the accumulation of all absolute values; the box at the pos./neg. transition shows the pos./neg. partitioning in % that defines the overall offset

Fig. 2 shows that for the closed-loop use case under nominal conditions, offset is not dominated by one single degradation mechanism. Here, positive and negative shifting parts appear in a similar order of magnitude, so fractions almost compensate each other. This is also true for accelerated stress by temperature increase. Rise in voltage, however, leads to large variations of the partitioning, inducing large deviations of the output monitor offset in relation to the nominal case.

Generally, in the open-loop use case in figure 3, aging induced offset is dominated by the negative shifting parts. Here, the main contributor is PBTI degradation at the input transistor M2. For a rise in temperature offset composition again remains nearly constant. Applying elevated voltages leads to a variation of the contribution of PBTI at M2 and of NBTI at M8, but the cumulative negative shifting parts exhibit a quite constant behavior.

B. Simple Circuit Level Aging Acceleration

By means of the above investigations, a general concept for a realistic circuit level aging can be derived. To map circuit aging, e.g. for nominal conditions, in an accelerated test setup, ideally requires the equal composition of the contributing aging mechanisms. A suitable way to obtain this composition for linear analog circuits is the application of the small signal equivalent. As stated in figure 2 and 3, for the investigated differential amplifier, the composition of the output monitor offset shows little deviation if only a rise in temperature is applied.

TABLE IV. OPEN-LOOP USE CASE: ACCELERATION IN TIME OF MAIN OFFSET GENERATING EFFECTS DUE TO RISE OF TEMPERATURE

T	25°C			85°C			125°C		
	M2 PBTI	*M8* NBTI	*M4* HCI	*M2* PBTI	*M8* NBTI	*M4* HCI	*M2* PBTI	*M8* NBTI	*M4* HCI
ACC	1	1	1	161	110	10	1985	1124	32

In table 4 effect acceleration factors of the main offset contributors for the open-loop use case are depicted. It can be seen that HCI is much less accelerated, but in open-loop, stress of M4 is not dominant. Unfortunately, dominant BTI effect acceleration due to temperature rise towards the aging case under nominal conditions is small and end of lifetime aging tests would last long times. Particularly, if worst case circuit aging for high temperatures should be mapped, a simple acceleration via a rise of temperature does not involve satisfactory effect acceleration.

C. Advanced Circuit Level Aging Acceleration

To achieve practicable acceleration we expand the evaluated acceleration concept. Our approach to get acceptable aging effect acceleration in combination with best achievable mapping of the realistic aging case bases on the following consideration. A good mapping is achieved if the output monitor partitioning remains constant. This can be obtained in a quite accurate manner if the necessary rise in voltage is derived from the main contributing mechanisms by exhibiting the same degradation value in the stress and in the nominal case. As we already determined for the investigated circuit that a rise in temperature does not largely affect the aging output monitor composition, but a rise in voltage excessively changes the composition, we ensure that with this approach temperature is maximum, and only the necessary voltage rise to meet acceptable acceleration is applied. This approach can also be considered as deriving the best fit and simultaneously compromise between acceleration and mapping accuracy. This means that accepting lower acceleration – thus longer stress times - implies decreased voltage rise values and makes the mapping more accurate.

To prove the concept, an accelerated test setup to map aging in both use cases under nominal end-of-lifetime circuit conditions (25°C, nominal voltages, 10y) for the differential amplifier in figure 1 should be derived. By means of an error value, expressing the deviation of the output monitor in the stress setup with respect to the nominal case, a rating of the setup can be performed. To give an example, table 5 compares aging simulations of the simple 125°C/150% VDD setup and adapted ones.

TABLE V. RESULTS OF SIMPLE SCALING APPROACH AND ADAPTED SETUP FOR ACCELERATED TESTING; THE ERROR VALUE REPRESENTS THE DEVIATION OF THE OFFSET GENERATED IN THE ACCELERATED STRESS CASE IN RELATION TO THE NOMINAL (25°C, NOM. VOLTAGES, 10Y) CASE

T=125°C		Closed-loop		Open-loop	
t_{stress} [s]	ACC	V_{scale} [%]	Error [%]	V_{scale} [%]	Error [%]
1000	n.a.	150	-950.0	150	1233.3
1000	315360	129.5	80.0	110.1	-12.7
100000	3154	113.1	20.0	100.1	-4.8

Results show that the error may be very large if arbitrary rise in temperature and voltage is applied, as can be seen in the 1st row. In contrast, aligning the setup to the dominating degradation mechanism, errors can be kept small. The mapping quality depends on the degree of domination by one type of effects. For the open-loop use case the stress setup is derived from PBTI at M2, as this effect dominates induced offset (see fig. 3) and resulting errors are rather low. For the closed-loop use case the largest contributing effect is HCI at M6, but counteractive NBTI at M8 is of similar magnitude and shows different acceleration behavior. Here, the degree of domination for HCI at M6 is marginal, leading to a poor mapping of the nominal aging case. But the derived setup is still superior towards the simple scaling approach. Via the application of longer stress durations, stress voltages can be kept smaller which leads to a further reduction of the error in both use cases.

D. Prove Of Concept: Measurement Results

In Table 6 simulation and measurement data for the arbitrary stress setup, the nominal aging case and the derived stress setup for the open-loop use case are quoted. The comparison of the simulated and measured induced offsets for the derived open-loop 1000s test setup shows good agreement, which proves the validity of this approach.

TABLE VI. DEGRADATION INDUCED OUTPUT REFERRED OFFSET IN MV

	Closed-loop		Open-loop	
	Sim.	Msm.	Sim.	Msm.
1000s, 150% V_{scale}, 125°C	+9	+19	-666	-562
10y, nom. supply, 25°C	+0.8	n.a.	-54.8	n.a.
1000s, 110% V_{scale}, 125°C			-48.3	-47

E. Outlook

Further extensions of the methodology are possible. When the measured though unrealistic stress case can be reproduced in the circuit model, the realistic circuit aging can be extrapolated by simulation. In addition to the error reduction, knowledge of the output monitor offset partitioning gives the circuit designer the opportunity to develop in-circuit countermeasures or even benefit from self cancelling effects.

IV. CONCLUSION

State-of-the-art accelerated aging technique by simple temperature and voltage rise applied to complete circuits may lead to significant deviation in the degradation output monitor and may not map to the realistic end-of-lifetime aging case under nominal circuit conditions. For the investigated differential amplifier, temperature rise would be the most suitable approach, as partitioning of the degradation induced offset varies only slightly - but acceleration factors are impractically small. The evaluated concept of deriving an accelerated test setup from the dominating degradation effect shows acceptable aging acceleration, while errors can be kept assessable. The more the output monitor is dominated by one effect, the smaller is the error implied by the stress setup. We further showed that spending longer stress times errors can be reduced. The proposed methodology to combine circuit analysis with physical testing allows controlled and realistic end-of-lifetime aging assessment of complex linear analog circuits.

ACKNOWLEDGMENT

The authors would like to thank the entire reliability team at Infineon for their contributions to these studies.

REFERENCES

[1] Agostinelli M., et al., PMOS NBTI-induced Circuit Mismatch in Advanced Technologies, 2004, IRPS 2004

[2] F. Chouard et al., Impact of Degradation Mechanisms on Analog Differential Amplifiers, 2009, ESSCIRC 09 Fringe

[3] X.Chen, et al., in IEEE Symp. On VLSI Tech. 2008, p.88

[4] V. Huard, et al., CMOS Device Design-In Reliability Approach in Advanced Nodes, 2009, IRPS 2009

[5] A. Bravaix, et al., Hot-Carrier Acceleration Factors for Low Power Management in DC-AC stressed 40nm NMOS node at High Temperature, 2009, IRPS 2009

[6] S. Chakravarthi, et al., A Comprehensive Framework For Predictive Modeling of Negative Bias Temperature Instability, 2004, IRPS 2004

The Hot Carrier Degradation Rate Under AC Stress

Guido T. Sasse, Jaap Bisschop

NXP Semiconductors

Gerstweg 2, 6534 AE, Nijmegen, The Netherlands

phone: +31 243536987, e-mail: guido.sasse@nxp.com

Abstract— **In this work the methodology used for predicting hot carrier device degradation under AC stressing conditions is critically re-examined. Having an accurate method is a key prerequisite of developing useful tools for the reliability simulation of any circuit. It will be shown that existing methods are not generally applicable. A new, better applicable method is presented and verified with experimental data.**

Keywords: reliability simulation, hot carrier degradation, reaction-diffusion, RF reliability.

I. INTRODUCTION

Reliability simulation can be a very powerful tool during circuit design as it allows the design of circuits that are both reliable and optimized for circuit performance. The value of such a tool is largely dependent on the accuracy of the models that underlie the simulator. Many models have been presented in the past years describing device lifetime under DC hot carrier stress, such as the Lucky Electron Model (LEM) [1], or more recent models that are better applicable for stress conditions with $V_{DS} < 3$ V [2-4].

Whichever the most appropriate model for DC stress signals, however, for accurate reliability simulation an important question remains what approach is most suitable for determining the hot carrier degradation rate under circuit conditions, which are typically of AC nature. In this respect an important observation is the fact that hot carrier degradation under periodic stress signals can be considered a frequency independent mechanism, which was recently shown to be valid for frequencies into the GHz range [5]. This indicates that for a proper translation from DC hot carrier models to AC stress signals, only information on the stress signal over one period of the signal is required, the frequency does not have any impact.

Currently reliability simulators typically make use of an empirically found time dependency function, validated for the DC case, and then apply it to AC stress signals using a quasi-static approximation. The main justification of this approach is, however, purely empirical [6] and only verified if the DC device degradation can be written as a power law function of time. From literature it is known that this power law may not always be applicable [7-10], and other time dependency functions for the device degradation are more appropriate. Justification for the quasi-static approach is lacking for such time dependency functions.

In this paper we will show that this conventional approach for modeling device degradation under AC hot carrier stress is not generally applicable, even if the used time dependency

function for the degradation has been experimentally verified for the DC case; a more generally applicable approach will be presented and verified with experimental data.

II. DEGRADATION RATE UNDER DC STRESS

A. Empirical models

The most commonly used time dependency function describing device degradation under hot carrier stress is the power law, empirically verified in [11]:

$$\Delta P(t) = A \cdot t^n \qquad (1)$$

In this expression t is the stress time and $\Delta P(t)$ is the degradation of any device parameter P at time t. It can, e.g., represent a relative increase in the saturated drain current or an absolute shift of the threshold voltage. Parameter n is considered a technology dependent parameter and parameter A is dependent on technology, temperature, device geometry and stress conditions. An appropriate DC hot carrier model, such as the LEM, gives an explicit expression for the dependency of this parameter A on temperature, geometry and stress condition.

Although this power law time dependency function is widely applied for describing device degradation under hot carrier stress, it is known from literature that it yields overly pessimistic results for high stress times, as a saturation effect can be observed [7]. Various models that take this effect into account have been proposed and typically it involves a two-mechanism process [8]. The time dependency functions for the device degradation, that follow from these models, consist of a power law for short stress times, while for long stress times different expressions for the time dependency of device degradation are used. In [9] it was proposed to use a logarithmic expression for the time dependency function of degradation for high stress times. In [10] an easily implementable, empirical expression was presented that combines these two stress regions. In a slightly adjusted form (we made it independent on the units used for the stress time by adding τ_c and valid for $0 \leq t/\tau_c \leq 1$ by adding +1 and) this expression for the time dependency function of degradation looks like:

$$\Delta P(t) = B \cdot \left[\ln\left(\frac{t}{\tau_c} + 1 \right) \right]^m \qquad (2)$$

Parameter B is considered dependent on technology, device geometry, temperature and stress condition; parameter m is

---

This work is supported by the Dutch ministries of economic affairs and education, culture and science under the knowledge workers agreement KWR 0917.

assumed to depend only on the technology used. Parameter τ_c is a constant that we set to 1 s.

B. Reaction-Diffusion model

The two models discussed above are examples of purely empirical models. A qualitative physical explanation of the observed degradation rate under hot carrier stress conditions is often found in a reaction-diffusion (R-D) framework. In this model it is assumed that under the influence of the hot carrier stress Si-H bonds at the silicon-oxide interface are broken, resulting in a generated interface state and a hydrogen atom; the latter can diffuse into the oxide or it can recombine with an interface state. A general expression describing this R-D mechanism is given by [1,12]:

$$\frac{dN_{it}}{dt} = k_F \cdot \left(N_0 - N_{it}\right) - k_R \cdot N_{it} \cdot N_H^{(0)} \qquad (3)$$

Here N_{it} represents the number of interface states, N_0 the number of available Si-H bonds that can be broken under the given hot carrier stress conditions and $N_H^{(0)}$ is the concentration of hydrogen atoms at the interface. Parameters k_F and k_R represent the forward and reverse reaction rate respectively. k_F is considered stress dependent, while k_R is considered only dependent on technology and temperature. According to [1] the interfacial hydrogen concentration can be assumed to be linearly proportional to dN_{it}/dt and hence, after some rewriting we get:

$$\frac{dN_{it}}{dt} = \frac{k_F \cdot \left(N_0 - N_{it}\right)}{1 + k_R \cdot \alpha \cdot N_{it}} \qquad (4)$$

In this expression parameter α is considered a constant, independent of stress conditions. Besides a qualitative explanation of the empirically observed power law (for degradation levels where $N_{it} \ll N_0$), the R-D model can also be used directly to describe device parameter degradation. For this purpose equation (4) should be solved; no explicit function describing N_{it} as a function of t can be found by analytically solving (4), therefore we make use of a numerical solution. Furthermore the link between N_{it} and ΔP must be known. In this paper we will only consider the threshold voltage shift, which we assume to be linearly proportional to N_{it}. This results in the following differential equation:

$$\frac{d\Delta V_T}{dt} = \frac{k_F \cdot \left(\Delta_{VT,S} - \Delta_{VT}\right)}{1 + \beta \cdot \Delta V_T} \qquad (5)$$

In this expression ΔV_T represents the threshold voltage shift and $\Delta V_{T,s}$ is the maximum threshold voltage shift, associated with the situation in which all N_0 bonds are broken. β is considered a technology, temperature and geometry dependent parameter.

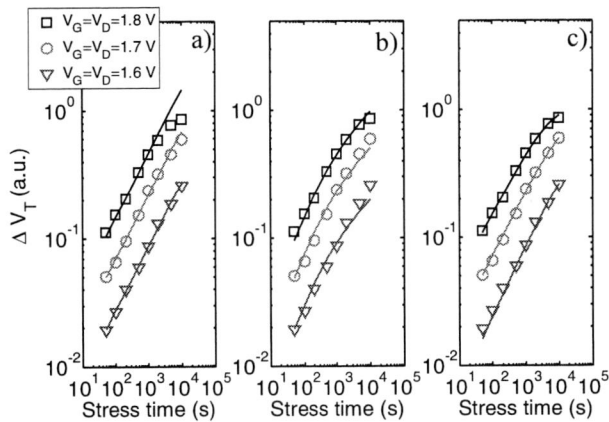

Figure I: V_T shift versus stress time for three different DC stress conditions. The squares, circles and triangles represent data points and the solid lines are fits to the power law, logarithmic function and the R-D equation in figure a), b), respectively c).

III. EXPERIMENTAL VERIFICATION DC MODELS

A. Time dependency functions

The applicability of the power law, the logarithmic expression and the expressions that follow from the R-D model for describing the time dependency of device degradation under DC hot carrier stress is tested by fitting expressions (1), (2) and (5) to a measured threshold voltage shift (ΔV_T). Measurements were performed on nMOS devices with $W=10$ μm and $L= 0.04$ μm. All measurements were performed at room temperature. DC hot carrier experiments were performed at three different values of V_D; gate voltages were chosen such that $V_G=V_D$. In figure I the results of this experiment are plotted. The figure shows that all three time dependency functions give accurate fits to the data for low t. For large t, the power law fails to describe the data accurately, as saturation effects become clearly visible. The solid lines in this figure represent the best fit of the three different models to the data; it was assumed that all three models have only one voltage dependent parameter (stress factor), i.e. A for the power law expression, B for the logarithmic expression and k_F for the R-D expression.

B. Voltage dependency stress factors

The three stress factors A, B, k_F in the three different time dependency functions are dependent on the drain voltage of the applied stress signal. For making a translation from the DC time dependency functions to AC stress signals, an expression describing the relation between drain voltage and the stress factor must be known. We make use of the Takeda law [11], in which this drain voltage dependency can be written as:

$$A, B, k_F \propto e^{\frac{C}{V_{G,D}}} \qquad (6)$$

In this expression C is considered a technology dependent parameter. C can be different for every stress factor considered. $V_{G,D}$ is the applied drain and gate voltage; in this paper we consider only the $V_G = V_D$ stress condition. We evaluated the applicability of (6) to the three time dependency functions described above, at $V_G=V_D$. In figure II the stress factor dependency, as extracted from figure I, on the applied drain

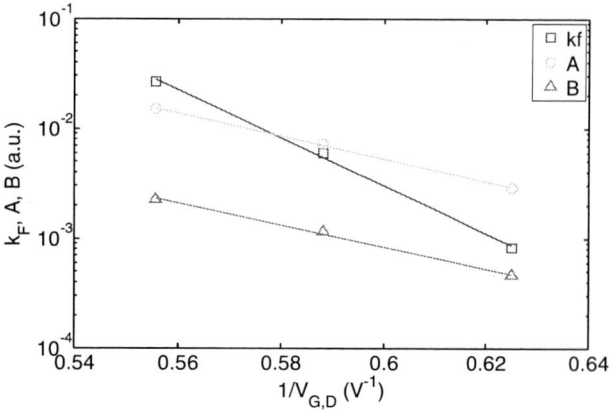

Figure II: Stress factors A, B and k_F, as extracted from figure I, plotted against drain voltage at $V_G=V_D$ condition. The solid lines represent fits to the Takeda law.

and gate voltage is plotted. The solid lines represent the best fits of the Takeda law to the data. The figure clearly reveals that all three different stress factors can be described as a function of the drain and gate voltage using the Takeda law.

IV. QUASI-STATIC AC DEGRADATION MODELING

The conventional approach of determining hot carrier degradation under AC stress conditions is to make use of an expression for the time dependency function describing device degradation, empirically verified for the DC case and using a quasi-static approximation [6] to determine the degradation rate under AC stress conditions. The assumption made in this approximation is that the device degradation rate - the time derivative of the device degradation - for all stress times can be derived from the DC time dependency function. Given a device degradation function f_{DC}:

$$\Delta P_{DC}(t) = f_{DC}(t, S_{DC}) \qquad (9)$$

In this expression ΔP_{DC} represents the degradation of device parameter P at time t for a DC stress signal. f_{DC} denotes the time dependency function, which can be, e.g., a power law or the logarithmic expression discussed above. S_{DC} represents the stress condition, i.e. the stress the device sees in terms of voltages and currents. A function g_{DC}, describing the device degradation rate under DC stress conditions can be defined as:

$$g_{DC}(t) = \frac{d\Delta P_{DC}(t)}{dt} = \frac{df_{DC}(t, S_{DC})}{dt} \qquad (10)$$

If we replace S_{DC} with an AC stress condition $S_{AC}(t)$, we can define a function g_{AC} using the quasi-static assumption:

$$g_{AC}(t) = \frac{df_{DC}(t, S_{AC}(t))}{dt} \qquad (11)$$

From this we can derive the quasi-static approximation for the degradation function:

$$\Delta P_{AC}(t) = \int_0^t g_{AC}(t')dt' = \int_0^t \frac{df_{DC}(t', S_{AC}(t'))}{dt'} \cdot dt' \qquad (12)$$

Here ΔP_{AC} is the expected degradation of device parameter P under an AC stress signal, using the quasi-static approximation. If we follow the above approach for the threshold voltage shift using the power law time dependency function we can derive:

$$\frac{d\Delta V_T}{dt}(t) = A \cdot n \cdot t^{n-1} = A^{\frac{1}{n}} \cdot n \cdot \Delta V_T^{1-\frac{1}{n}}(t) \qquad (13)$$

which leads to:

$$\Delta V_T(t) = \int_0^t \frac{d\Delta V_T(dt')}{dt'} \cdot dt' = \left[\int_0^t A^{\frac{1}{n}}(t')dt' \right]^n \qquad (14)$$

For the logarithmic time dependency function we can derive:

$$\frac{d\Delta V_T}{dt}(t) = \frac{1}{\tau_c} \cdot e^{-\left[\left(\frac{\Delta V_T(t)}{B} \right)^{1/m} \right]} \cdot B(t)^{1/m} \cdot m \cdot \Delta V_T(t)^{1-1/m} \qquad (15)$$

This differential equation does not have a general analytic solution but can be solved numerically.

It is important to note here that the assumptions that lead to (14) and (15) can only be considered valid if the time dependency function used is applicable for all stress times. In practice this is not the case for any empirical degradation function, as it is typically confirmed only for relatively long stress times (i.e. > 1 s). This is a very long time compared to the periodic time of periodic stress signals encountered in real circuits (which can be in sub-ns range).

To exemplify the risk in the above approach we calculated the results of the quasi-static solutions for AC stress signals using the power law and the logarithmic function. For this purpose we made use of the parameters extracted from the data presented in figures I and II. In figure III the results are plotted for a sinusoidal voltage signal ($V_G=V_D$) with a frequency of 1 kHz, ranging between 1.6 V and 1.8 V. In figure III a) the calculated V_T degradations using quasi-static approximations for the power law function and the logarithmic function are plotted. We can see that the results align only for a relatively short time period (roughly between 100 s and 5000 s).

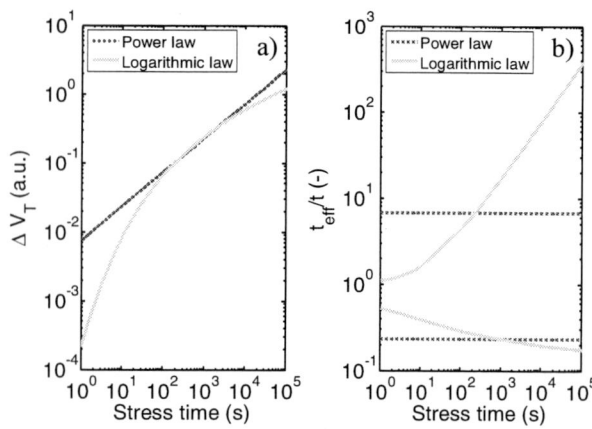

Figure III: a) Calculated V_T degradation using quasi-static approximations for the power law and logarithmic function. The degradation was calculated for a sinusoidal stress voltage with a frequency of 1 kHz ranging between 1.6 V and 1.8 V and using the parameter values as extracted from figures I and II. b) The limits between which the effective stress time t_{eff}, associated with the results in a), ranges; t is the stress time.

Even more remarkable is the result shown in figure III b): here we plotted t_{eff} against the stress time for the two models. We defined t_{eff} as the stress time under DC stress required to achieve the same level of degradation as given at time t. The upper lines in figure III b) represent t_{eff} for a DC stress of 1.6 V and the lower line for a DC stress of 1.8 V. For this AC stress signal t_{eff} continuously ranges between these two lines. For the power law the upper- and lower t_{eff} limits progresses linearly with time, as can be seen as a constant value in the t_{eff}/t vs. t plot. The logarithmic function yields a dramatically different result, as t_{eff}/t is strongly dependent on the actual stress time.

This difference in result is remarkable, as one would expect t_{eff} to be independent on the time dependency function used, if the quasi-static approach can be considered a generally applicable approach. In this example the DC time dependency functions are empirically verified for the entire voltage range considered in the calculation.

The results in figure III make us conclude that at least one of the two presented time dependency function is not suitable for a quasi-static approximation for AC stress signals. Given the successful application of the quasi-static approximation for the power law in the past years, we consider it most probable that only the logarithmic function yields incorrect results when applied for AC stress signals with a quasi-static approximation. It is important to note here, that this means that the method of deriving quasi-static approximation cannot be considered a generally applicable approach for modeling device degradation under AC stress signals using time dependency functions describing device degradation that have only been verified under DC stress conditions.

V. R-D BASED AC DEGRADATION MODELLING

Having shown that the quasi-static approximation is not a generally applicable approach for modeling device degradation under AC stress conditions, the need arises for finding an approach that is applicable for conditions for situations in which saturation effects prohibit the use of the power law. In contrast to the purely empirical power law and logarithmic function, the R-D model in itself describes the degradation rate rather than a device degradation function. This means that no assumption is required to describe the device degradation rate under AC stress conditions. Hence, besides giving a qualitative physical explanation for the observed time dependency of device degradation, the R-D model also provides a convenient approach for determining the device degradation rate under AC hot carrier stress. The device degradation rate as it follows from the R-D model under AC stress conditions can be written as:

$$\frac{d\Delta V_T}{dt}(t) = \frac{k_F(t) \cdot \left(\Delta V_{T,S} - \Delta V_T(t)\right)}{1 + \beta \cdot \Delta V_T(t)} \quad (16)$$

In this expression $k_F(t)$ represents the time-dependent forward reaction rate. It is a function of the stress condition at time t and is related to the applied voltage level as in (6). The result of this expression was compared with quasi-static approximation of the power law and logarithmic function for a sinusoidal voltage stress ($V_G = V_D$) ranging between 0 and 2 V, making use of the

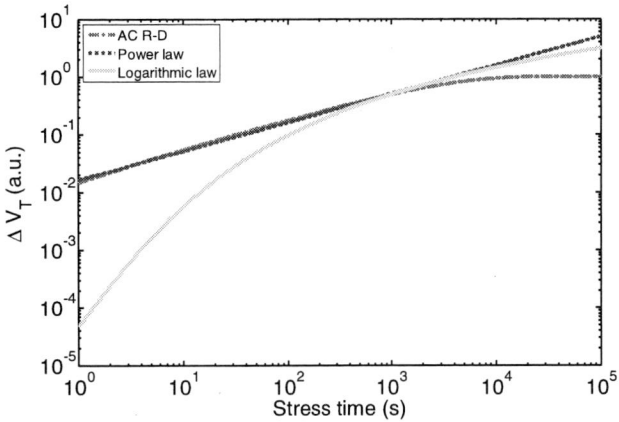

Figure IV: Calculated V_T degradation for a sinusoidal voltage signal with a frequency of 1 kHz ranging between 0 and 2 V. Use was made of the parameter values as extracted from figure I and II.

parameter values, extracted from figures I and II. The results are shown in figure IV. From this figure we can clearly see that the expressions that follow from the R-D model and the power law function align very well, up to the point where the R-D model starts to reveal a saturation effect. The results of the logarithmic time dependency function deviate strongly from the results using the power law and the R-D model.

These results indicate that the R-D model is indeed a suitable model for describing the device degradation rate under AC stress conditions. In order to simplify the implementation of this model in a reliability simulator we propose to make use of the following approximation, with T the periodic time of the stress signal:

$$\frac{d\Delta V_T}{dt}(t) \approx \frac{\Delta(\Delta V_T)}{\Delta t} \approx$$
$$\frac{\left[\frac{1}{T}\int_0^T k_F(t')dt'\right] \cdot \left[\Delta V_{T,0} - \Delta V_T(t)\right]}{1 + \beta \cdot \Delta V_T(t)} \quad (17)$$

The first approximation is valid for sufficiently small Δt, so that $\Delta(\Delta V_T) \ll \Delta V_T$ and $\Delta t \ll t$. The second approximation can be assumed to be valid under the condition that $T \ll \Delta t$. In this expression we replaced $k_F(t)$ by the time-average of k_F, allowing us to integrate over only one period of time of the stress signal and then finding a solution in the same way as can be done for the DC case. In figure V the results of expressions (16) and (17) are compared for a sinusoidal voltage signal with a frequency of 1 kHz, ranging between 0 and 2 V. The results clearly show that the approximation made in (17) is valid, already after a very small number of periods (~ 10) of the stress signal. We have also verified this for other stress signals, including triangular voltage signals and block waves with various amplitudes, all yielding similar results.

VI. EXPERIMENTAL VERIFICATION

Measurements were performed on the same type of devices (nMOSFETs with $W/L = 10/0.04$ μm) as used in the DC experiments illustrated in figure I. Devices were stressed, at room temperature, using a triangular voltage signal with $V_G = V_D$

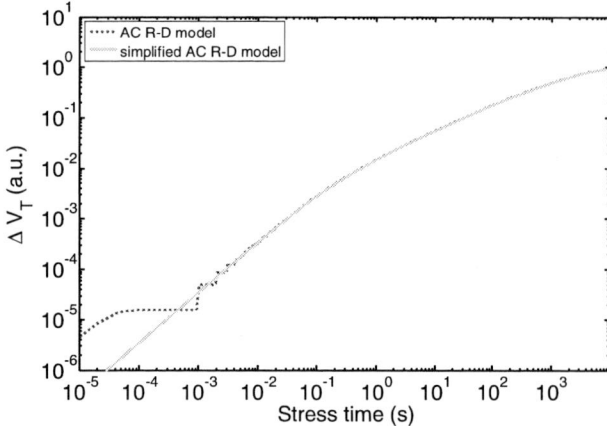

Figure V: Calculated V_T degradation for a sinusoidal voltage signal with a frequency of 1 kHz, ranging between 0 and 2 V. Use was made of the parameter values as extracted from figures I and II. The R-D model (16) was compared to the simplified R-D model (17).

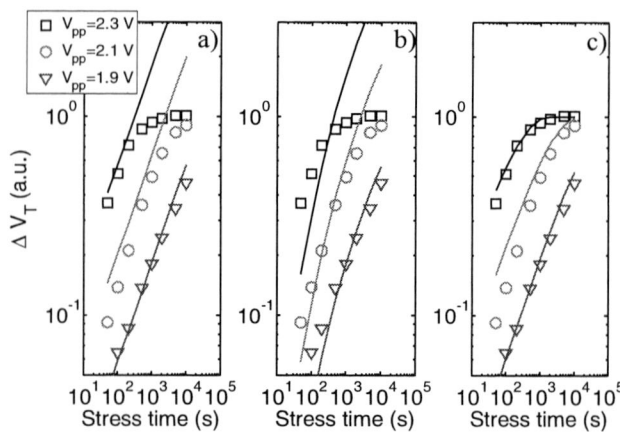

Figure VI: Measured V_T degradation under an AC hot carrier stress plotted agains stress time. The stress signal consisted of a triangular voltage signal of 1 kHz with $V_G = V_D$, ranging between 0 and V_{pp}. The solid lines represent the V_T degradation as calculated using the parameter values extraced from figure I and II for the power law (14), logarithmic function (15) and the R-D model (17) in a), b) respectively c).

for all t. The voltage signal ranges from 0 to V_{pp} where we investigated three different values of V_{pp}. The frequency of the stress signal was set to 1 kHz. The measurement setup consisted of an Agilent 4156 C parameter analyzer, Agilent 41501B pulse generator expander and an Agilent E5250A switch mainframe. All equipment was externally controlled using Labview software. The results of these experiments are shown in figure VI. In this figure we also plotted the expected V_T shift using expressions for the power law (14), logarithmic function (15) and the simplified AC R-D model (17), and using the parameter values as extracted from figures I and II.

The figure clearly reveals that expression (17) has an accurate prediction for all stress times and voltages shown; the power law gives accurate results for low stress time, before any saturation effect can be observed. The calculation using the logarithmic model does not give accurate results; the error between the model and the measured data is considerable for both low and high stress times.

VII. CONCLUSIONS

The use of quasi-static approximations in combination with time dependency functions describing device degradation, only verified under DC stress, is not a generally applicable approach for estimating the device degradation rate under AC hot carrier stress conditions. While for the power law this approach will give fairly accurate results over a broad range of stress times and stress conditions, considerable inaccuracies may arise for other time dependency functions, even if the expressions are verified for the DC case over the entire voltage range of interest A better approach is to make use of the R-D framework. For periodic signals it suffices to solve the R-D equation using the average reaction rate. This approach not only provides an accurate estimation of the AC degradation rate, it also provides a way to accurately model saturation effects in hot carrier degradation. Finally, in addition, this paper shows that AC hot carrier experiments can be used to discriminate between correct and wrong physics-based models for hot carrier degradation.

VIII. REFERENCES

[1] C. Hu, S.C. Tam, F-C. Hsu, P-K. Ko, T-Y. Chan, K.W. Terrill, "Hot-electron-induced MOSFET degradation – model, monitor and improvement," IEEE Trans. El. Dev., vol. 32, pp. 375-385, February 1985.

[2] S.E. Rauch III, F.J. Guarin, G. LaRosa, "Impact of e-e scattering to the hot carrier degradation of deep submicron nMOSFET's," IEEE El. Dev. Lett., vol. 19, pp. 463-465, December 1998.

[3] D.S. Ang, T.W.H. Phua, H. Liao, C.H. Ling, "High-energy tail electrons as the meachansim for the worst-case hot-carrier stress degradation of the deep submicrometer n-MOSFET," IEEE El. Dev. Lett., vol. 24, pp. 469-471, July, 2003.

[4] C. Guerin, V. Huard, A. Bravaix, "The energy-driven hot carrier degradation modes," in Proc. IEEE Int. Reliab. Phys. Symp., 2007, pp. 692-693.

[5] G.T. Sasse, J. Schmitz, F.G. Kuper, "MOSFET degradation under RF stress," IEEE Trans. El. Dev., vol. 55, pp. 3167-3174, November 2008.

[6] M.M. Kuo, K. Seki, P.M. Lee, J.Y. Choi, P.K. Ko, C. Hu, "Simulation of MOSFET lifetime under AC stress hot-electron stress," IEEE Trans. El. Dev., vol. 35, pp. 1004-1010, July 1988.

[7] K.M. Cham, J. Hui, P. Vande Voorde, H.S. Fu, "Self-limiting behavior of hot carrier degradation and its implication on the validity of lifetime extraction by accelerated stress," in Proc. IEEE Int. Reliab. Phys. Symp., 1987, pp. 191-194.

[8] V-H. Chan, J.E. Chung, "Two-stage hot-carrier degradation and its impact on submicrometer LDD nMOSFET lifetime prediction," IEEE Trans. El. Dev., vol. 42, pp. 957-962, May 1995.

[9] C. Liang, H. Gaw, P. Cheng, "An analytical model for self-limiting behavior of hot-carrier degradation in 0.25-µm n-MOSFET's," IEEE El. Dev. Lett., vol. 13, pp. 569-571, November 1992.

[10] B. Szelag, S. Kubicek, K. De Meyer, F. Balestra, "Time-dependent degradation law for reliable lifetime prediction in sub-0.25 µm bulk silicon n-MOSFETs," Electronics Letters, vol. 35, pp. 1385-1386, August 1999.

[11] E. takeda, N. Suzuki, "An empirical model for device degradation due to hot-carrier injection," IEEE El. Dev. Lett., vol.. 4, pp. 111-113, April 1983.

[12] H. Kufluogu, M.A. Alam, "A geometrical unification of the theories of NBTI and HCI time-exponents and its implications for ultra-scaled planar and surround-gate MOSFETs," in Proc. Int. Electron Devices Meeting, 2004, pp-113-116.

ESD Protection for High-Speed Receiver Circuits

Nathan Jack and Elyse Rosenbaum

Dept. of Electrical and Computer Engineering
University of Illinois at Urbana-Champaign
1308 W. Main St., Urbana, IL 61801
ndjack2@illinois.edu

Abstract—**ESD-induced gate oxide breakdown is studied in high-speed receiver circuits. A novel biasing circuit increases the breakdown voltage by modulating the potential of the input transistor's source during ESD. The effectiveness of dual-diode and DTSCR protection of high-speed receiver circuits is examined under various bias conditions.**

I. INTRODUCTION

Dual-diode and SCR-based devices are commonly used for ESD protection of I/O pads because of the high current density that these devices can handle [1]. In this work, dual-diodes and diode-triggered SCRs (DTSCR) [2] were used to protect an input circuit that resembles the differential amplifier shown in Fig. 1; a differential amplifier is the input stage for any high-speed receiver. For the case of positive stress from an input to V_{SS}, SCR-based protection provides lower voltage clamping in steady-state than does dual-diode protection [3][4]. However, in response to a pulse with sub-nanosecond risetime, the voltage across the SCR grossly overshoots, greatly exceeding the clamping voltage and even the DC trigger voltage V_{t1}. This voltage overshoot can break down the oxide of the transistor being protected [5].

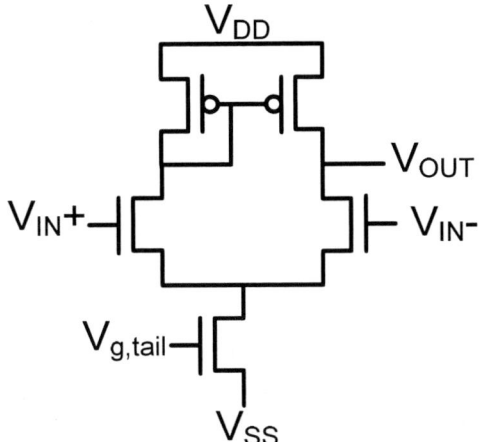

Fig. 1: CMOS differential amplifier. The input circuits used in this work resemble one branch of this amplifier.

Secondary protection is often used to protect the input transistor from the voltage overshoot at the pad [6]. This secondary protection is placed close to the input transistor and is decoupled from the primary protection by a series resistance. Although secondary protection enhances ESD reliability, the added capacitance and the series resistance can degrade high-frequency performance. When performance specifications only allow for primary protection, optimization of the ESD protection scheme is crucial for reliability.

The gate oxide breakdown voltage V_{BD} of a MOS transistor is a function of the potential V_S at its source node [7]. This work examines whether controlling the gate bias of the tail transistor $V_{g,tail}$ during ESD can adjust V_S of the input transistor in a way that improves reliability. A bias circuit is presented that controls the value of $V_{g,tail}$ during ESD. The effectiveness of dual-diode and DTSCR input protection is also examined as a function of $V_{g,tail}$.

II. DESCRIPTION OF TEST STRUCTURES

To test the effectiveness of an ESD protection device in protecting gate oxides, an oxide monitor may be placed in parallel with the protection device [5]. The drain, source, and bulk terminals of the oxide monitor are shorted together [5], resulting in $V_{SB} = 0$ V. However, in real I/O designs, the source and drain of the input transistor are not hard wired to the body, source degeneration may be used, and V_{SB} of the input transistor is non-zero. Because V_{OX} is a function of V_{SB} [7], an I/O circuit has a different V_{BD} than does an oxide monitor.

Test circuits were fabricated in a 130 nm CMOS process. The input circuit used in this work is a differential amplifier half-circuit—one branch of the amplifier of Fig. 1. The test circuit topology is illustrated in Fig. 2. 1.2 V transistors with a physical dielectric thickness of 2.2 nm were used in the input circuit. Each test circuit has input ESD protection, consisting of either STI-bound dual-diodes or a DTSCR with a 2-diode string trigger. The protection devices were sized for equal I_{fail} as opposed to equal capacitance. The test circuits also contain a rail clamp and decoupling capacitance between V_{DD} and V_{SS}. The worst-case stress for the gate oxide of the input NMOS is from the *Input* pad to V_{SS}; this stress case will be the focus of this paper. Under this stress condition, the input NMOS is biased in inversion.

In the benchmark circuit, shown in Fig. 2, a current mirror is used to bias the PMOS and the NMOS tail transistor, as in a real input circuit. V_{bias} was designed to be 0.4 V, under normal operating conditions.

978-1-4244-5430-3/10 $26.00 © 2010 IEEE

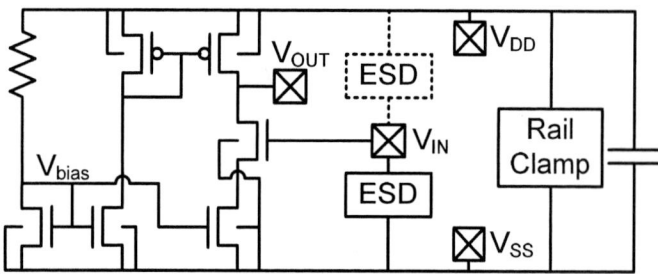

Fig. 2: Schematic of the input circuit test structures used in this work. The current mirror on the left biases the PMOS and the tail NMOS, as is typical in a real high speed input circuit. This is the benchmark case for this work.

In [7], it was shown that increasing V_S of the input transistor will increase its V_{BD} since $V_{ox} \approx V_{GS} = V_G - V_S$ in inversion. V_G of the input transistor is, of course, the pad voltage. The V_S of the input transistor is controlled by the tail transistor. For a given $I_{DS,tail}$ during ESD, $V_{DS,tail}$ must increase if $V_{g,tail}$ is decreased. Note that $V_{DS,tail}$ is equal to V_S. Therefore, reducing $V_{g,tail}$ will reduce the stress on the input oxide: lower $V_{g,tail}$ produces a higher V_S which reduces V_{ox} and thus the magnitude of the gate tunneling current. Hence, it is expected that V_S, and therefore V_{BD}, can be maximized by tying $V_{g,tail}$ to V_{SS} during ESD.

To study the effect of $V_{g,tail}$ on ESD reliability, test structures with a pad at the gate of the tail transistor were designed. Using this pad, $V_{g,tail}$ was biased from an external source. These structures are depicted in Fig. 3 with (a) DTSCR protection and (b) dual-diode protection.

(a)

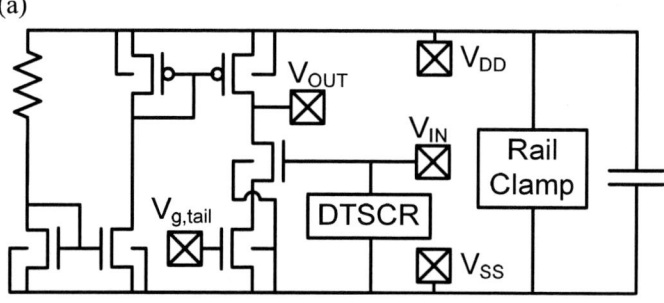

(b)

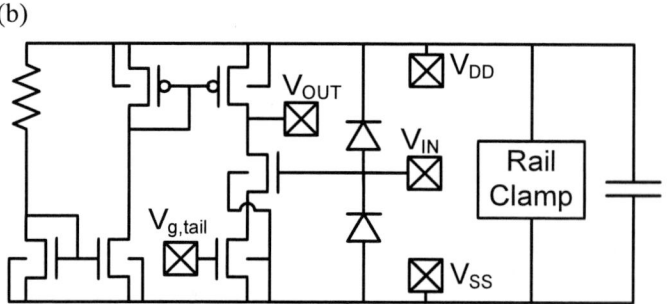

Fig. 3: Input circuits with (a) DTSCR protection and (b) dual-diode protection. $V_{g,tail}$ was biased externally to modulate V_S of the input transistor during stress.

III. TLP MEASUREMENT RESULTS

The test circuits were subjected to TLP stress using 100 ns wide pulses with a 10 ns risetime. Kelvin probes were used to measure the voltage V_{DUT} across the device under test (DUT). To confirm measurement repeatability, at least three test structures were used in each experiment; averaged results are reported.

The TLP I-V curve for the DTSCR protected circuit of Fig. 3a is shown in Fig. 4. The failure current was roughly 1.55 A, regardless of the applied $V_{g,tail}$. The failure current of a standalone DTSCR under TLP stress was also roughly 1.55 A. These identical failure currents, and the fact that $V_{g,tail}$ had no influence on the failure level, indicate that the DTSCR fails before the oxide is damaged. Therefore, the DTSCR protected the oxide within its current handling capability.

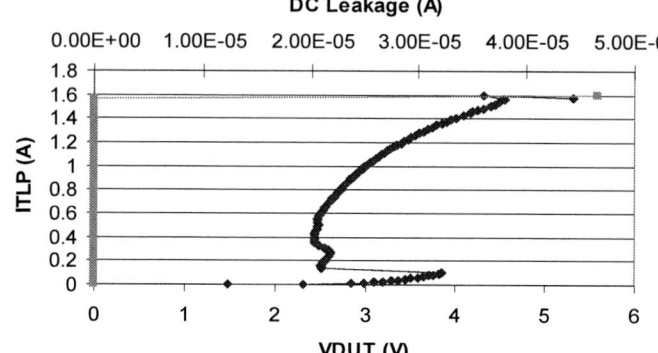

Fig. 4: DTSCR protected circuit (Fig. 3a) under 100 ns TLP stress. The DTSCR failed at 1.55 A, before oxide failure. VDUT is the voltage measured between pads V_{IN} and V_{SS}.

The TLP I-V curve for the dual-diode protected circuit of Fig. 3b is shown in Fig. 5. As with the DTSCR protected circuit, the failure current was roughly 1.55 A. This was true regardless of the applied $V_{g,tail}$. In principle, the failure may have occurred in the input transistor, the top diode, the rail clamp circuit, or the bottom diode. However, a standalone rail clamp passed TLP stress beyond 2.5 A, as shown in Fig. 6, excluding it as the cause of the 1.55 A failure level. During the circuit TLP stress, the bottom diode is reverse biased. The TLP reverse I-V of a standalone bottom diode is shown in Fig. 7; the device enters breakdown at about 15 V and fails above 25 V. When the dual-diode protected circuit failed, the reverse bias across the bottom diode was only approximately 5.6 V, excluding the bottom diode as the failed device. The TLP I-V curve of a standalone top diode is shown in Fig. 8. The failure current is 1.59 A—almost identical to the failure current found during TLP stress of the input circuit. From these data, we conclude that, under TLP stress, the top diode failed in the dual-diode protected circuit, not the oxide. Therefore, the dual-diodes protected the input transistor within the current handling capability of the top diode.

TLP stress was applied to the gate terminal of a standalone transistor with $V_{SB} = 0$ V; the measured V_{BD} was roughly 6.1 V. From Fig. 4 and Fig. 5, it is seen that both protection circuits fail before the breakdown voltage of the oxide is

978-1-4244-5430-3/10 $26.00 © 2010 IEEE

reached. Therefore, both the DTSCR and the dual-diodes can protect gate oxide against TLP stress, with its slow risetime (10 ns) and pulse width much larger than the protection device turn-on time. It is reasonable to assume that during HBM testing, failure would occur in the protection devices, not in the oxide.

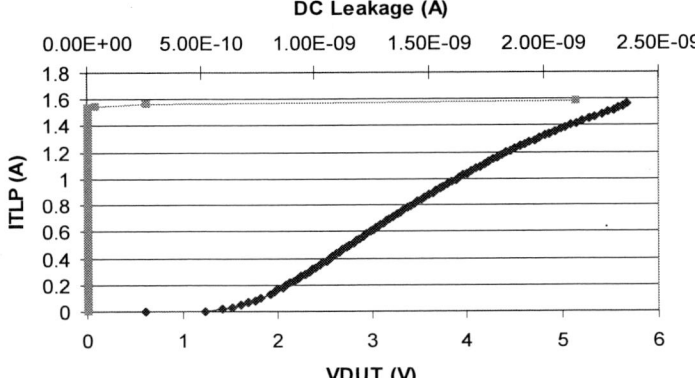

Fig. 5: Dual-diode protected circuit (Fig. 3b) under 100 ns TLP stress. The circuit failed at 1.55 A—roughly the same failure current of the standalone top diode, as shown in Fig. 8. The top diode failed before the oxide was damaged.

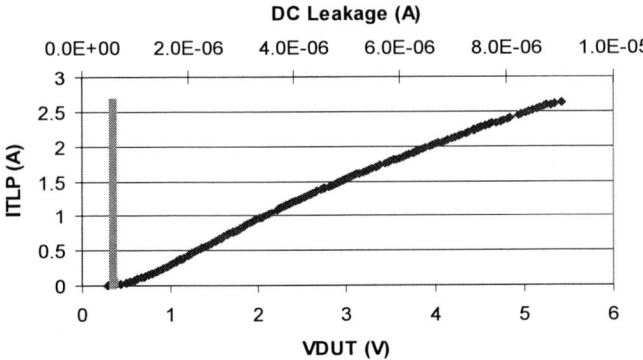

Fig. 6: Standalone rail clamp under 100 ns TLP stress. The failure point (not shown) is greater than 2.5 A; it was not the rail clamp that failed in the stress shown in Fig. 5.

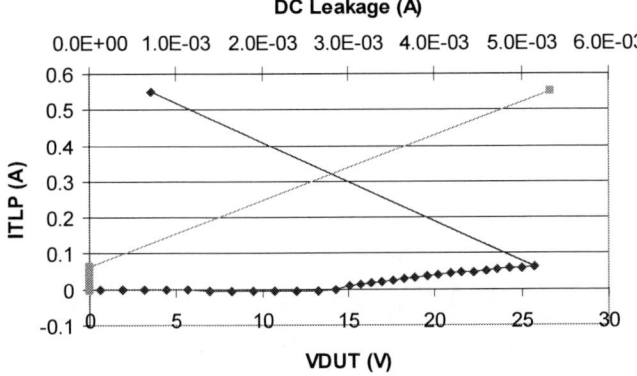

Fig. 7: Standalone bottom diode reverse biased under 100 ns TLP stress. The diode is damaged at a reverse bias of approximately 26 V, indicating that this diode does not fail for the input-to-V_{SS} stress of less than 6 V (see Fig. 5).

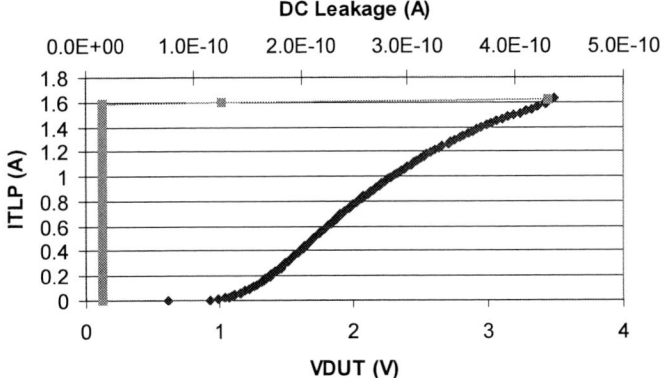

Fig. 8: Standalone top diode under 100 ns TLP stress. The diode failed at 1.59 A and was the source of failure in the dual-diode protected circuit (see Fig. 3b).

IV. VF-TLP MEASUREMENT RESULTS

VF-TLP pulses 9 ns wide with a 100 ps risetime were applied to the structures. Voltage overshoot in ESD protection devices is known to occur for risetimes on this short time scale, with the SCR overshoot being the largest [4][5][8]. Shown in Fig. 9 are the waveforms measured between the input and V_{SS} of the dual-diode and DTSCR protected input circuits when a 50 V VF-TLP stress was applied. To verify that the observed transient voltage overshoot was a characteristic of the protection devices rather than of the VF-TLP pulse delivery system, pulses were also applied to a diffusion resistor; no overshoot occurred, as shown in Fig. 9.

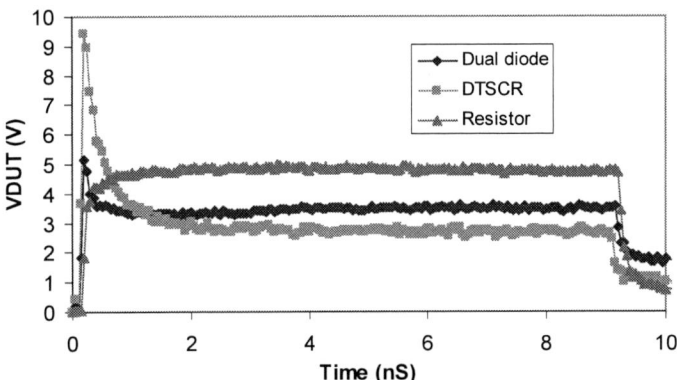

Fig. 9: $V_{DUT}(t)$ at a current of 0.82 A for the dual-diode and DTSCR protected circuits. The VF-TLP input pulse had t_{rise}=100 ps. The DTSCR has larger voltage overshoot than the diode; a pulsed diffusion resistor has none, indicating that the observed overshoots are not a characteristic of the test system.

Failure voltage may be defined as the pad voltage during the current pulse right before the one which causes breakdown. Since the pad voltage is not constant during a single VF-TLP stress, two values of V_{fail} may be defined: the first is the peak voltage; the second is the voltage sampled toward the end of the pulse, when the voltage has reached its quasi-steady-state value. Both values were recorded in this work.

The measurement results for the DTSCR protected circuit are summarized in Table I. When $V_{g,tail}$ was 1.2 V, I_{fail} was 0.72 A and peak V_{fail} was 8.2 V. This is the expected worst-case stress condition since the tail transistor provides a low impedance path to V_{SS}, resulting in V_S near 0 V. When $V_{g,tail}$ was lowered to 0 V, I_{fail} increased to 0.9 A—a 25% improvement—and peak V_{fail} increased by 2 V to 10.2 V.

TABLE I

Input circuit failure conditions under VF-TLP stress using DTSCR protection. Decreasing $V_{g,tail}$ from 1.2 V to 0 V increased peak V_{fail} by 2 V and I_{fail} by 0.18 A.

DTSCR Protection			
$V_{g,tail}$ during VF-TLP	I_{fail}	V_{fail}, steady state	V_{fail}, peak
0 V	0.9 A	2.84 V	10.2 V
1.2 V	0.72 A	2.69 V	8.2 V

The results for the dual-diode protected circuits are summarized in Table II. For the worst-case condition of $V_{g,tail}$ = 1.2 V, I_{fail} was 1.31 A—significantly higher than the best-case I_{fail} of the DTSCR circuit. For the best-case condition of $V_{g,tail}$ = 0 V, I_{fail} increased to 1.93 A and peak V_{fail} increased from 8.2 V to 10.3 V.

Standalone DTSCR, diode, and rail clamp protection devices all passed VFTLP stress beyond 2.3 A. None of the input circuits had this high of a failure current, suggesting that all the failures reported in this section resulted from gate oxide failure in the input transistor. This was confirmed by post-stress electrical characterization of the ESD-protected input circuits that had been damaged under VFTLP stress; all showed damage to the input transistor's gate oxide. Therefore, all V_{fail} values reported in this section refer to oxide failure conditions, not the failure of the ESD protection devices.

TABLE II

Input circuit failure conditions under VF-TLP stress using dual-diode protection. Decreasing $V_{g,tail}$ from 1.2 V to 0 V increased peak V_{fail} by 2.1 V and I_{fail} by 0.62 A. V_{fail} in the benchmark circuit was close to that when $V_{g,tail}$ was biased at 0.5 V.

Dual-diode Protection			
$V_{g,tail}$ during VF-TLP	I_{fail}	V_{fail}, steady state	V_{fail}, peak
0 V	1.93 A	5.86 V	10.3 V
0.5 V	1.75 A	5.4 V	9.7 V
1.2 V	1.31 A	4.41 V	8.2 V
Biased at V_{bias} (benchmark)	1.83 A	5.6 V	9.8 V

Fig. 10 shows VF-TLP I-V characteristics of a dual-diode protected circuit and a DTSCR protected circuit. Current is plotted both as a function of steady-state voltage and peak voltage. The larger voltage overshoot of the DTSCR is

evident, which further highlights the reason why the DTSCR protected circuits fail at lower stress currents.

For a given $V_{g,tail}$, peak V_{fail} was constant, regardless of whether dual-diode or DTSCR protection was used. There was no such equivalence between the steady-state V_{fail} values. When $V_{g,tail}$ =1.2 V, peak V_{fail} was roughly 8.2 V for both the dual-diode and DTSCR protected circuits. When $V_{g,tail}$ =0 V, peak V_{fail} for both circuits was roughly 10.3 V. By decreasing $V_{g,tail}$ from 1.2 V to 0 V (and thereby increasing V_S), V_{BD} was increased by approximately 2 V for both circuits.

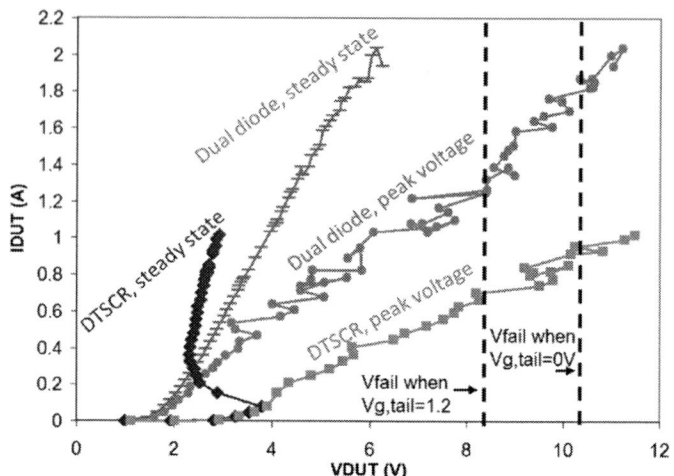

Fig. 10: Steady state and peak voltage VF-TLP I-V characteristics of dual-diode and DTSCR protected circuits. V_{fail} was increased in both circuits by 2 V when $V_{g,tail}$ was decreased from 1.2 V to 0 V. Oxide failure occurred at lower stress current with DTSCR protection due to larger voltage overshoot.

The lowest measured peak V_{fail} of 8.2 V for the ESD protected input circuits was larger than the V_{BD} of 6.5 V obtained from subjecting the gate of standalone transistors to VF-TLP pulses. This is not surprising, as the voltage overshoot duration is far less than the 9 ns VF-TLP pulse width, and breakdown voltage is a decreasing function of stress duration. Additionally, V_{SB} was equal to 0 V for the standalone transistors; V_{SB} was nonzero for the input circuit, even with $V_{g,tail}$ = 1.2 V.

In the benchmark case (Fig. 2), peak V_{fail} was comparable to that obtained when $V_{g,tail}$ is externally biased at 0.5 V, suggesting that V_{bias} during the stress was close to the designed-for normal operating bias of 0.4 V. Peak V_{fail} in the benchmark case was 1.6 V greater than that obtained when $V_{g,tail}$ = 1.2 V and V_{SB} is near zero. This demonstrates that traditional oxide monitors, in which the source is tied to the body [5], yield overly conservative estimates of the failure voltage for real input circuits, which have nonzero V_{SB}.

V. BIAS CONTROL CIRCUIT

It was established in the previous section that modulating $V_{g,tail}$ during CDM-like stress will modulate V_{fail} of the input oxide and therefore I_{fail}. It was shown that the best-case bias was $V_{g,tail}$ = 0 V. In the test circuit shown in Fig. 11, control circuitry was used to ground the gate of the tail transistor during ESD to increase V_{BD} at the input. Control signals ESD and ESDB are generated within the rail clamp when it is activated by an ESD event; the active rail clamp schematic is shown in Fig. 12. During normal operating conditions, the tail transistor is biased by the NMOS current mirror, as in the benchmark case (see Fig. 2). For a positive stress from the input to V_{SS}, the rail clamp is not activated in the DTSCR protected circuit; thus the control circuit was tested only with dual-diode protection.

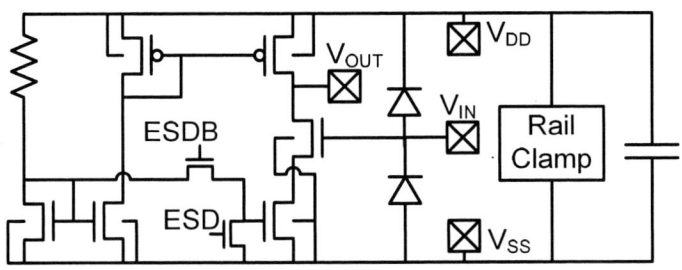

Fig. 11: Dual-diode protected input circuit with a control circuit to ground the gate of the tail transistor, thereby increasing V_{BD} by increasing V_s.

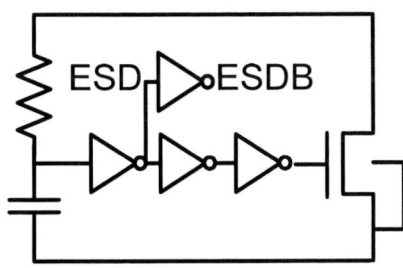

Fig. 12: Active rail clamp. During an ESD event, signals ESD and ESDB enable the tail transistor control circuit which grounds the gate.

In full chip designs, noise-mitigating, i.e. decoupling, capacitors are placed between V_{DD} and V_{SS}. This capacitance reduces the dV/dt on the VDD rail and could affect the control circuit of Fig. 11 by slowing down the control signals ESD and ESDB. On the other hand, very large decoupling capacitance can suppress transient voltage overshoots, thereby reducing the stress at the input and relaxing the clamping requirements [9]. In order to design test structures which emulate a worst-case scenario, simulation was used to select a decoupling capacitance value that would slow the control circuitry response while still allowing for appreciable voltage overshoot. A 75 pF capacitor placed between V_{DD} and V_{SS} was found to be a challenging test case and was included in all the test circuits.

The VF-TLP results for the dual-diode protected input circuit with the bias control circuit are displayed in Table III. V_{fail} and I_{fail} were equal to those measured when $V_{g,tail} = 0$ V. *This result confirms that the control circuit succeeded in clamping $V_{g,tail}$ to 0 V, thereby increasing I_{fail} and V_{BD} relative to the benchmark case.*

In certain input circuit topologies, $V_{g,tail}$ may be pulled closer to V_{DD} during the stress than it was for the benchmark case in this work. A large $V_{g,tail}$ during ESD would increase the channel conductivity of the tail transistor and its V_{DS} would

be closer to 0 V. In such a situation, the ESD bias circuit would have an even more pronounced effect than was observed in this work.

TABLE III

Input circuit failure conditions under VF-TLP stress with dual-diode protection and the bias control circuit in place. The failure levels are equal to those obtained when $V_{g,tail}$ was 0 V, indicating that the control circuit clamped $V_{g,tail}$ to 0 V, thereby increasing V_{fail}.

Dual-diode Protection			
$V_{g,tail}$ during VF-TLP	I_{fail}	V_{fail}, steady state	V_{fail}, peak
Grounded by control circuit	1.92 A	5.84 V	10.3 V

VI. CONCLUSIONS

This work demonstrates that oxide breakdown V_{BD} in input circuits is greatly influenced by the input transistor's source potential V_S. A tail transistor connected between the input transistor's source and V_{SS} will modulate V_S. V_{BD}, and thus I_{fail}, can be increased by decreasing $V_{g,tail}$; this increases the source-to-V_{SS} impedance, thereby increasing V_S. Traditional oxide monitor devices have $V_S = V_D = V_B$ and thus provide overly conservative estimates of oxide V_{BD} for input circuits. A novel biasing circuit was presented which increases V_{BD} by grounding the gate of the tail transistor during an ESD event. This circuit would be useful for input circuits in which $V_{g,tail}$ was biased unfavorably during the stress.

For pulse stress with $t_{rise} = 100$ ps, transient voltage overshoot in the STI-bound diodes and the DTSCR were the cause of oxide failure at the input. The failure current at which this peak voltage stress damaged the oxide could be increased by decreasing $V_{g,tail}$. The larger overshoot of the DTSCR resulted in a lower failure current than when using dual-diode protection.

For slow-rising ESD stress ($t_{rise} = 10$ns), both dual-diode and DTSCR devices protect oxides within their current handling capabilities.

ACKNOWLEDGEMENTS

The authors gratefully acknowledge UMC for fabricating the test structures through the UMC University Program.

REFERENCES

[1] C. Richier, P. Salome, G. Mabboux, I. Zaza, A. Juge, and P. Mortini, "Investigation on different ESD protection strategies devoted to 3.3 V RF applications (2 Ghz) in a 0.18um CMOS process," *Proc. EOS/ESD Symp.*, pp. 251-259, 2000.

[2] M. Mergens et al., "Diode-Triggered SCR (DTSCR) for RF-ESD protection of BiCMOS SiGe HBTs and CMOS ultra-thin gate oxides," Intl. Electron Devices Meeting Technical Digest, pp. 515-518, 2003.

[3] K. Chatty et al., "Study of factors limiting ESD diode performance in 90 nm CMOS technologies and beyond," *Proc. Intl. Reliability Physics Symp.*, pp. 98-105, 2005.

[4] Y. Morishita et al., "An investigation of input protection for CDM robustness in 40nm CMOS technology," *Proc. EOS/ESD Symp.*, pp. 2A.8-1 - 2A.8-6, 2009.

[5] T. Smedes and N. Guitard, "Harmful voltage overshoots due to turn-on behaviour of ESD protections during fast transients," *Proc. EOS/ESD Symp.*, pp. 357-365, 2007.

[6] C. Duvvury, R. Rountree, and R. McPhee, "ESD Protection: Design and layout issues for VLSI circuits," *IEEE Transactions on Industry Applications*, pp. 41-47, 1989.

[7] J. Lee and E. Rosenbaum, "Voltage clamping requirements for ESD protection of inputs in 90nm CMOS technology," *Proc. EOS/ESD Symp.*, pp. 50-58, 2008.

[8] R. Gauthier, M. Abou-Khalil, K. Chatty, S. Mitra, and J. Li, "Investigation of voltage overshoots in diode triggered silicon controlled rectifiers (DTSCRs) under very fast transmission line pulsing (VFTLP)," *Proc. EOS/ESD Symp.*, pp. 5A.4-1 - 5A.4-10, 2009.

[9] V. Shukla, N. Jack, and E. Rosenbaum, "Predictive simulation of CDM events to study effects of package, substrate resistivity and placement of ESD protection circuits on reliability of integrated circuits," *Intl. Reliability Physics Symp.*, 2010.

On the Failure Mechanism and Current Instabilities in RESURF Type DeNMOS Device Under ESD Conditions

Mayank Shrivastava[1], Jens Schneider[2], Maryam Shojaei Baghini[1], Harald Gossner[2], V. Ramgopal Rao[1]

[1]Center for Excellence in Nanoelectronics, Department of Electrical Engineering, Indian Institute of Technology-Bombay Mumbai-400076, India, mailto:mayank@ee.iitb.ac.in; mailto:rrao@ee.iitb.ac.in
[2]Infineon Technologies AG, P.O. Box 80 09 49, D-81609 Munich, Germany, mailto:Harald.Gossner@infineon.com

Abstract- **We present 3D device modeling of RESURF or non-STI type DeNMOS device under ESD conditions. The impact of base push-out, pulse-to-pulse instability and electrical imbalance on the various phases of filamentation is discussed. A new phenomenon called "week NPN action" and the cause of early and fast failure is identified. A modification of the device is proposed which achieved an improvement of ~5X in failure threshold (I_{T2}) and ~2X in ESD window without degrading its I/O performance.**

Index Terms- **DEMOS, ESD Failure, space charge build-up, Filamentation, pulse-to-pulse instability.**

I. INTRODUCTION

Shallow Trench Isolation (STI) and non-STI type *Drain extended MOS* (DeMOS) devices, used for high voltage I/O applications, have been found to be extremely vulnerable to ESD events. In order to improve the behavior of these devices under ESD conditions, there have been several attempts from various groups which provide an insight into their failure mechanism [1]-[4]. Recently it has been proposed that carrier heating in the high-field region influences the saturation velocity and the impact ionization, triggering a regenerative NPN action which causes the second breakdown [5]. Space charge limited current leads to filamentation and device failure [6-7]. Regenerative turn on of parasitic NPN causes a short circuit power dissipation leading to a failure [8]. Further, a complete understanding of various phases of filamentation and the final thermal runaway has been discussed for STI type DeNMOS devices [9]. However, a complete picture of the transient device behavior during base push-out with an understanding of the various phases of filamentation based on 3D simulation studies is still missing for RESURF (or non-STI) type DeNMOS, which is needed for improving its ESD behavior in order to design robust I/O devices.

II. DEVICE DESCRIPTION AND SIMULATION RESULTS

Before we discuss the simulation results, it is worth mentioning that in order to capture the current filamentation, we created physical non-uniformity along the device width in the following way: (i) Non-Uniform contact strip i.e. Multiple contacts with unequal spacing between different contact holes or Single contact strip having width less than the total device width and (ii) Single sided termination (along the width) of device with a dummy STI. A triangular meshing scheme was

used with a grid size resolution less than 200nm along the Z-axis. This simulation study was performed for silicide blocked RESURF type DeNMOS devices with thin gate oxide, as shown in Fig. 1, in a typical state-of-art 65 nm node CMOS technology. A well-calibrated process/device simulation deck was used as discussed in our previous work [9].The advantage of RESURF device over STI-DeMOS is its better R_{ON} v/s V_{BD} performance and smaller gate length, which provides a higher current driving capability and less gate-oxide capacitance. In order to model filamentation and device failure mechanism, this work extensively utilizes systematic 3D device simulations. This device can be used in two ways (i) I/O driver or (ii) high-voltage ESD clamp.

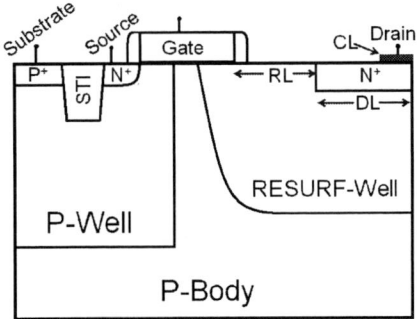

Fig. 1: RESURF DeNMOS device structure (*Type 1*).

The RESURF device has a good electrical performance in order to meet the I/O driving capability, however the device has shown only a low ESD robustness (Fig. 2 and 3). These reliability issues make the RESURF device less attractive as an output driver or power clamp. Fig. 3 shows that the devices with a larger width are prone to early failures, i.e. just at the onset of base push-out driven snapback.

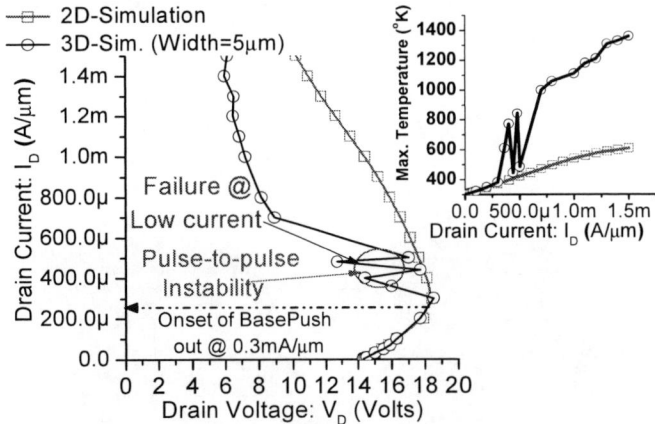

Fig. 2: Simulated (2D and 3D) TLP characteristics of RESURF DeNMOS under HBM like ESD condition. Width in 3D simulation is chosen to 5 μm. 3D simulations are done with single contact strip.

978-1-4244-5430-3/10 $26.00 © 2010 IEEE

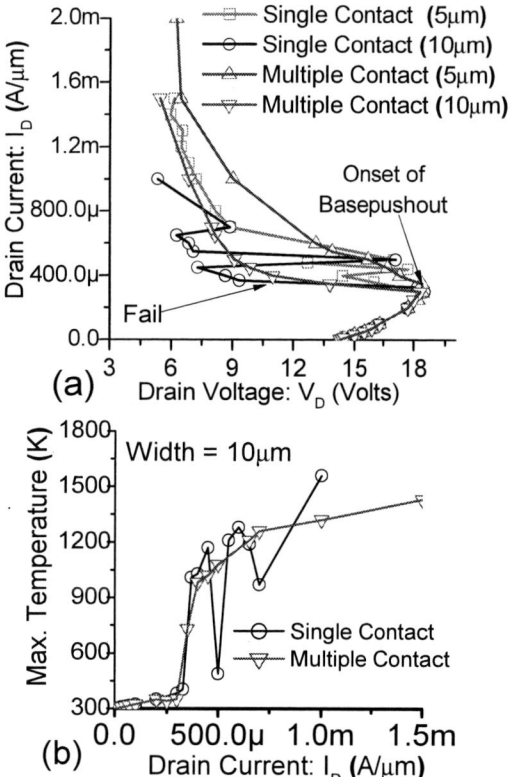

Fig 3: (a) Simulated TLP characteristics (HBM) of single contact strip and multiple contact strip devices with different widths. Devices with higher width fail at the onset of base push out driven snapback. (b) Shows Temperature rise with increasing TLP current for device with width = 10μm.

Two different states (i.e. pulse-to-pulse instability) were observed in TLP characteristics (Fig. 2 and 3), which is attributed to different locations of current filament along the device width. After the formation of current filamentation, i.e. after base push-out, we found that only very few bipolar keeps turned on along the device width. For most of the pulses (Fig. 4b & 4d), device leads to a filament formation at any one of the corners and causes deep snapback (low resistance state).

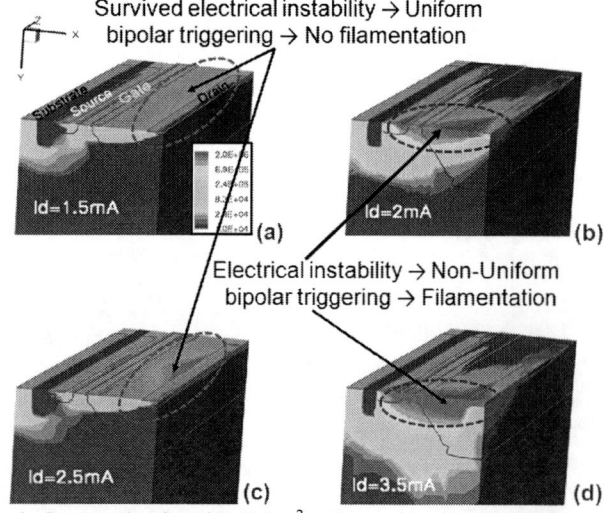

Fig. 4: Current density plot (A/cm²). Figure describes two different conditions- (i) uniform bipolar triggering along the width and (ii) non uniform triggering or current filamentation (Width=5μm).

However, for a few pulses, NPN triggers uniformly across the device width (Fig 4a & 4c) and causes no significant filamentation or snapback (high resistance state). Fig. 5 shows that pulse-to-pulse instability was observed for slow HBM pulses, whereas it was absent for fast pulses. It is worth mentioning that we have observed these pulse-to-pulse instabilities for various cases, i.e. device with different widths, device with different contact scheme (Fig. 2 & 3) and for different rise time of the TLP pulse. This validates that the observed pulse-to-pulse instabilities have some physical significance and are not due to some numerical inaccuracies. Figures 3,6 & 7 summarize the device behavior: (i) The junction breakdown occurs at a very low current followed by a parasitic bipolar triggering at 0.1mA/μm. The device fails at even higher currents, which proves that the parasitic bipolar turn-on is not the dominant cause of device failure unlike to the proposed model in [2][4][8] (ii) The onset of base push-out at 0.3mA/μm which leads to a deep snap back (iii) Fig. 6 (b & d) shows that the current leading to filamentation after the base push out, which causes a localized (along the width) hot spot under the drain diffusion. This is leading to significant temperature rise, which can eventually cause the device failure for higher width devices. Whereas in smaller width device it is visible in the form of lower resistance state. In contrast, when the device survives filamentation (a & c) or when the filament forms in the center of the device, the temperature is quite relaxed along the width and leads to high resistance state. (iv) Fig. 7 shows that when the filamentation occurs (b & d), the device shows an even higher carrier modulation in the localized region along the device width, which is named as "localized base push-out" in the later sections.

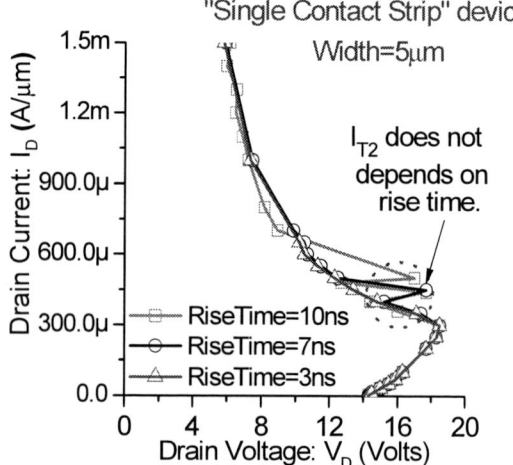

Figure 5: Dependence of simulated TLP characteristics on rise time (or pulse shape). Pulse-to-pulse instability depends on rise time of pulse, eventually on pulse shape.

III. PHYSICS OF ESD FAILURE

Before we discuss the ESD failure mechanism, it is worth pointing out the failure criterion used in this work. Fig. 2 and 3 shows a significant rise in the temperature just after the base push out driven snap-back. This is due to the current filamentation and hot spot formation. Furthermore, it also shows that temperature rise for device with larger width (Fig.

3, 10μm device) is significantly higher compared to device with smaller widths (Fig. 2, 5μm). For ex: @ I_{TLP} = 0.4mA/μm, when width was increased from 5μm to 10 μm, maximum temperature inside the hot spot increases from 800K to 1200K respectively. Another simulation (data not shown here) for device with 15μm width @ I_{TLP}~0.35mA/μm shows temperature greater than 1600K at 100ns. These observations are due to the following fact- We found that the cross-sectional diameter of current filament does not changes with the device width. Due to this, device with higher widths leads to increased current density inside the filament, which eventually attributes to larger temperature rise inside the hot spot. Due to the very huge TLP simulation time required for device with width ≥ 15μm, we have restricted our studies to device with width = 10μm. In this way, the onset of current filamentation, i.e. the onset of base push-out is defined here as the failure point.

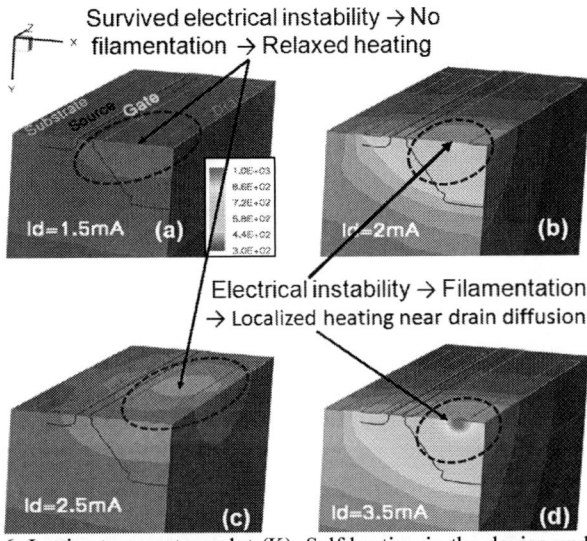

Fig. 6: Lattice temperature plot (K). Self heating in the device under two different conditions. (i) Relaxed temperature when bipolar triggers uniformly and device survived electrical instability, (ii) localized heating after current filamentation. (Width=5μm).

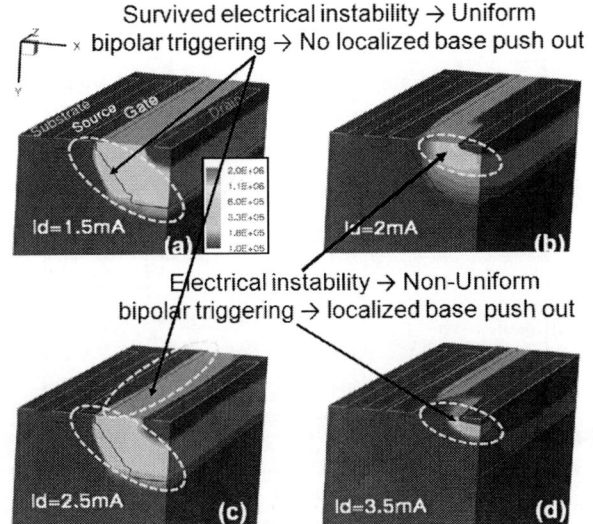

Figure 7: Electric field (V/cm) plot (Width=5μm) at different TLP level. Figure shows uniform bipolar triggering after base push out in 2D plane for few pulses (Fig. 7a and 7c), whereas filament, shrink at the corner, (Fig. 7b and 7d) leads to higher carrier modulation (causes deep snapback) for other pulses.

Figures 4c-4d, 6c-6d and 7c-7d differentiate the device behavior under two different conditions- (i) almost uniform bipolar triggering along the width and filament formation in the center of the device and (ii) non-uniform triggering or current filamentation at any one of the corners of the device. Furthermore, Fig. 8 illustrates the failure model of the device.

(i) *Onset of base push out*, which causes snapback. This happens at ~ 0.3mA/mm, at which the carrier concentration in the drain extension region exceeds the background doping.

(ii) Base push-out in the 2D plane causes snapback, which may lead to an *electrical instability*. If the device survives this instability, current flows almost uniformly along the width without any significant temperature rise.

(iii) Deep snapback because of the base push out-driven electrical instability which leads to a lower Impact Ionization (II) generated carriers, making it less efficient in triggering the distributed parasitic bipolar along the width. Such a *non-uniform bipolar triggering* causes the current to shrink (filamentation) at any one of the corner along the width and leads to significant temperature rise inside the localized hot spot. We refer to this phenomenon as *"Week NPN action"*.

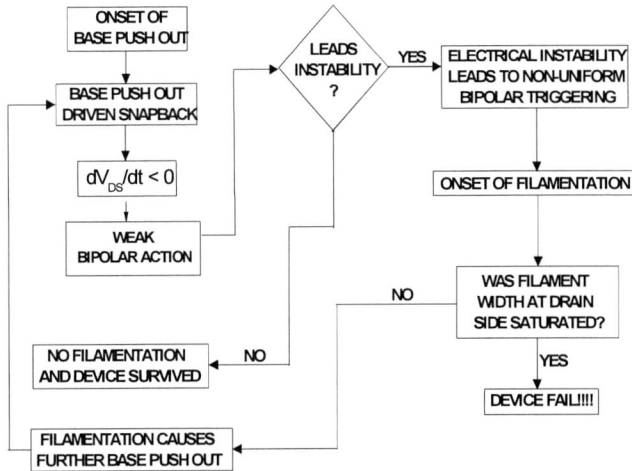

Fig. 8: Flowchart representation of current filamentation and device failure model.

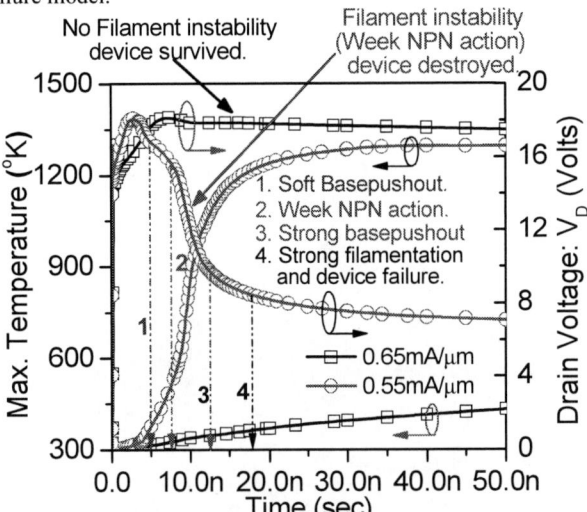

Fig. 9: Figure shows voltage drop because of base push out and early heating. Also describes various phases of filamentation.

978-1-4244-5430-3/10 $26.00 © 2010 IEEE

(iv) Increased carrier/current density in a localized region along the device width due to the onset of current filamentation leads to further carrier modulation in that localized region (i.e. *localized base push out*) and causes further shrink of the current filament in a way similar to (iii).

(v) Week NPN action, localized base push out and current filamentation act as positive feedbacks to each other, eventually leading to a very fast and dense filament formation (within 2-3ns) causing early device failures (Fig. 9).

IV. VALIDATION OF FAILURE MODEL

It was claimed in the past that these devices gets fail due to the regenerative triggering of parasitic bipolar [5][8]. As discussed in [9], drain hole current is the direct measure of excess carrier generation, which leads to regenerative NPN triggering. Fig. 10 shows the absence of regenerative NPN (i.e. significantly small amount of drain hole current) action after the base push out, different from previously reported works [5][8]. Fig. 9 also shows that when the device survives an electrical instability, there is no significant voltage drop and temperature rise, whereas in the presence of electrical instability temperature rises significantly with a simultaneous drop in the potential within a very short time.

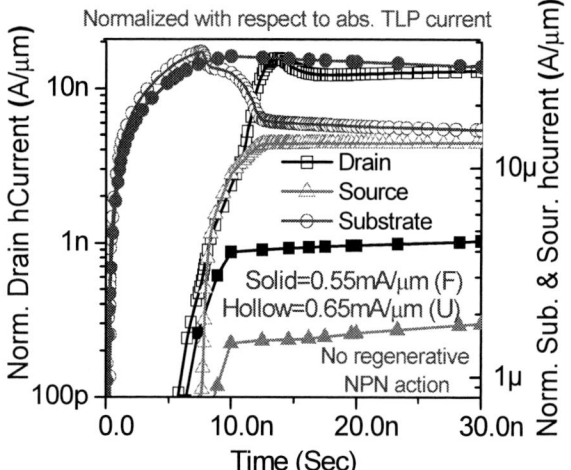

Fig. 10: Figure shows hole current at various terminals, proves absence of regenerative NPN action. (Markers: F→Filamentation and U→Uniform)

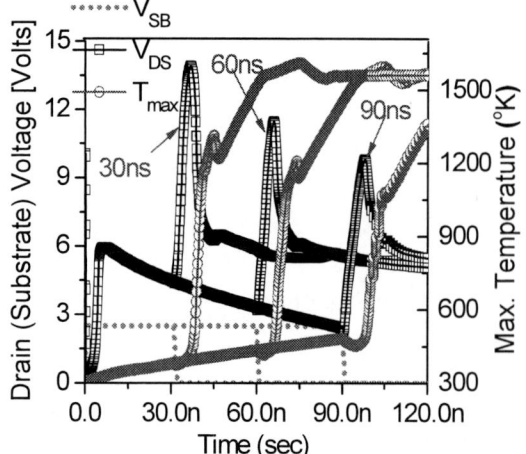

Figure 11: Validation of filament behavior and failure model.

Fig. 11 validates the failure model and the time evolution. It shows that when parasitic bipolar is triggered using an external substrate bias, there is no electrical instability and temperature rise. On the other hand, when the substrate bias is removed, the device goes into a filamentation and elevated temperature because of the electrical instability. Fig. 12 shows that the device failure occurs because of filamentation even under CDM like conditions.

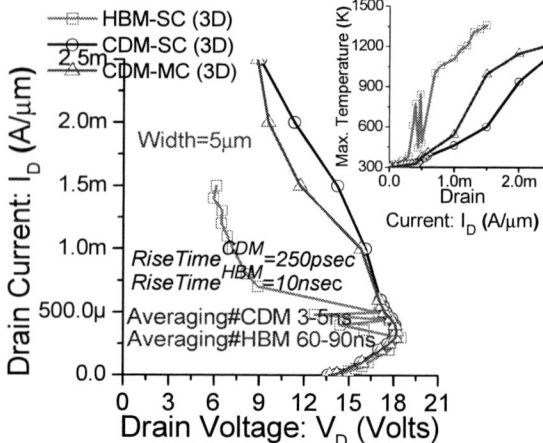

Fig. 12: Simulated HBM and CDM characteristics (TLP). Figure shows early failure even under CDM conditions. Inset shows significant heating under CDM time domain. (Markers:- SC→Single contact strip and MC→Multiple contact strip).

V. DEVICE MODIFICATION AND PERFORMANCE IMPROVEMENT

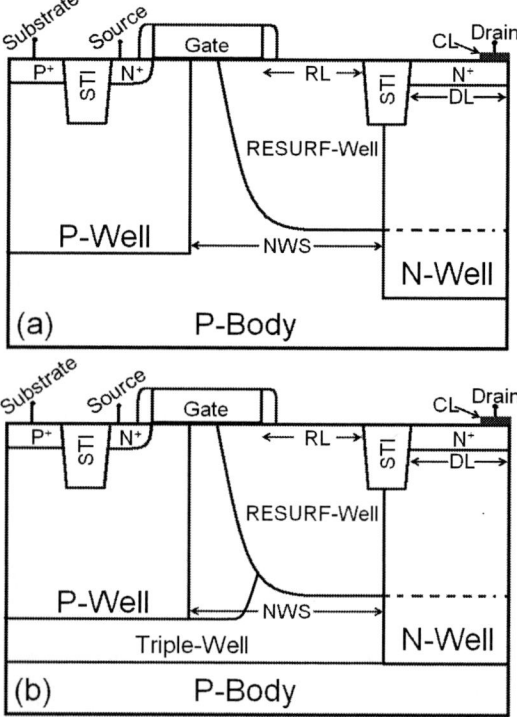

Fig. 13: Various modified RESURF DeMOS devices. (a) Device with extra STI and N-Well -"*Type 2*" (b) Further modified device with added triple well- "*Type 3*".

978-1-4244-5430-3/10 $26.00 © 2010 IEEE

In order to improve the device ESD robustness, modifications are required which should ideally push the onset of base push out to higher current levels. Furthermore, the mixed signal performance of the device should not be sacrificed while incorporating any changes. Fig. 13 shows the modified RESURF device where an extra n-well implant far from the p-well to RESURF-well junction is added, which increases the onset of base push out by increasing the background doping. A STI region is also added which pushes the current deep into the n-well and RESURF-well in order to overcome the electrical instability. Since the STI and n-well implant exist in the standard sub 100nm node CMOS technologies, the modified device has no extra processing cost.

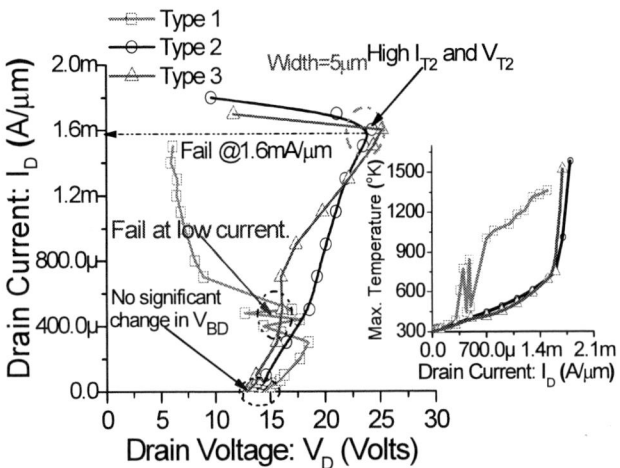

Fig. 14: Comparison of simulated TLP characteristics of standard and modified RESURF devices. Modified RESURF device show higher I_{T2} and ESD window.

Figure 14 shows that the modified device has a significantly improved ESD performance and from TABLE-I one can see that the R_{ON} v/s V_{BD} performance is degraded only by about 10%, which is still in the optimum range [10]. This performance improvement is attributed to the absence of base push out.

TABLE I

DIMENSIONS AND SIMULATED ELECTRICAL PERFORMANCE PARAMETERS OF VARIOUS INVESTIGATED DEVICE TYPES.

Device types & Dimensional Parameters							Electrical Performance			
Device Number	Device Type	RL (nm)	DL (nm)	CL (nm)	NWS (nm)	X (nm)	R_{ON} (KΩ μm)	V_{BD} (V)	V_{T2} (V)	I_{T2} (mA/μm)
1	1	500	375	200	NA	NA	3.8	14.3	6.1	0.4
2	1	500	775	200	NA	NA	3.9	14.4	6.3	0.4
3	1	900	375	200	NA	NA	4.9	14.4	7.6	0.4
4	2	500	500	200	900	200	5.2	14	19.3	1.8
5	2	800	500	200	900	200	4.8	14	27	1.7
6	2	500	500	200	800	200	4.8	13.8	21	1.7
7	2	500	500	200	700	200	4.5	13.5	24.3	1.6
8	3	500	500	200	900	200	4.7	13	24.4	1.5
9	3	500	500	200	800	200	4.5	13	25.2	1.6

V. CONCLUSION

We found that the filamentation is caused by the electrical instability at the onset of base push out. The degree of confinement and the rate of filament formation strongly depends on the existence of week NPN action. The regenerative NPN triggering is shown to be absent even at higher currents (and temperature) and the bipolar turn-on is therefore not the dominant cause of device failure. Further a modified-RESURF DeNMOS device is proposed with an extra N-Well and STI region near the drain diffusion in order to avoid the base push out, which improved the ESD window by a factor of ~2X and the failure threshold by a factor ~5X. The proposed modification has been shown to not degrade the R_{ON} v/s V_{BD} performance and has no extra processing cost.

REFRENCES

[1] P. L. Hower, J. Lin, S. Haynie, S. Paiva, R. Shaw and N. Hepfinger, "Proceedings of International Symposium on Power Semiconductor Devices and ICs, ISPSD '99, pp. 55-58.

[2] P. L. Hower, J. Lin and Steve Merchant, "Snapback and Safe operating area of LDMOS Transistors", Proceedings of International Electron Device Meeting, pp. 193-196, 99.

[3] P. Moens, S. Bychikhin, K. Reynders, D. Pogony and M. Zubeidat, " Effects of hot spot hopping and drain ballasting in Integrated Vertical DMOS devices under TLP stress", Proceedings of International Reliability Physics Symposium, pp. 393-398, 2004.

[4] Gianluca Boselli, Vesselin Vassilev and Charvaka Duvvury, " Drain Extended NMOS High Current behavior and ESD protection strategy for HV application in sub 100nm CMOS technologies", Proceedings of International Reliability Physics Symposium, pp. 342-347, 2007.

[5] A. Chatterjee, C. Duvvury and K. Banerjee, "New Physical Insight and Modeling of Second Breakdown (It2) Phenomenon in Advanced ESD Protection Devices", Proceedings of International Electron Device Meeting 2005, pp 195-198.

[6] A. Chatterjee, S. Pendharkar, Y-Y Lin, C. Duvvury and K. Banerjee , "An Insight into the High Current ESD Behavior of Drain Extended NMOS (DENMOS) Devices in Nanometer Scale CMOS Technologies", IEEE International Reliability Physics Symposium , 2007, pp 608-609.

[7] A. Chatterjee, S. Pendharkar, Y-Y. Lin, C. Duvvury and K. Banerjee, "A Microscopic Understanding of DENMOS Device Failure Mechanism Under ESD Conditions", IEEE International Electron Devices Meeting (IEDM), 2007, pp 181-184.

[8] A. Chatterjee, S. Pendharkar, H. Gossner, C. Duvvury and K. Banerjee, "3D Device Modeling of Damage due to Filamentation under an ESD Event in Nanometer Scale Drain Extended NMOS (DE-NMOS)", IEEE International Reliability Physics Symposium (IRPS), 2008, pp 639-640.

[9] Mayank Shrivastava, Jens Schneider, M. S. Baghini, Harald Gossner and V. Ramgopal Rao, "A new physical insight and 3D device modeling of STI type DeNMOS device failure under ESD conditions", IEEE International Reliability Physics Symposium, 2009, pp. 669-675.

[10] R. A. Bianchi, F. Monsieur, F. Blanchet, C. Raynaud, O. Noblanc, "High voltage devices integration into advanced CMOS technologies", Proceedings of International Electron Device Meeting, IEDM 2008, pp 137-140.

978-1-4244-5430-3/10 $26.00 © 2010 IEEE

Characterization of High-k/Metal Gate Stack Breakdown in the Time Scale of ESD Events

Yang Yang[1], James Di Sarro[2], Robert J. Gauthier[2], Kiran Chatty[2], Junjun Li[2], Rahul Mishra[2], Souvick Mitra[2], Dimitris E. Ioannou[1]

[1]ECE Dept., George Mason University, Fairfax, VA 22030, USA
phone: (+1) (703)-993-1580, e-mail: yyange@gmu.edu
[2]IBM Semiconductor Research and Development Center, System and Technology Group, Essex Junction, VT 05452, USA

Abstract—Catastrophic gate oxide breakdown of MOSFETs with high-k gate was characterized under ESD-like pulsed stress. It was found that the excessive gate current after gate oxide failure may result in a loss of gate contact and form a resistive path between the drain and source. Using constant voltage stress (CVS) method, the gate oxide breakdown voltages (V_{BD}) of NMSOFETs and PMOSFETs were extracted. NMOSFETs under positive stress were found to have the smallest V_{BD}, while the V_{BD} of the PMOSFETs under positive stress were significantly increased due to the well resistance. Compared to that measured using the CVS method, the V_{BD} from the transmission line pulse method (TLP) was smaller by only less than 10%. Despite the cumulative damages caused by the TLP method, the result is a conservative estimation of the breakdown voltage. The V_{BD} corresponding to the failure time of 1-ns measured using TLP method agrees well with the extrapolation result from the CVS measurements on the time scale ranging from ∼100 ns to ∼20 μs, suggesting that the failure mechanism remains the same as in the longer time scale.

Index Terms—high-k dielectrics; metal gate; electrostatic discharge (ESD); transmission line pulse (TLP); gate oxide breakdown.

I. INTRODUCTION

MOSFET gate oxide thickness has been reduced for many years to improve device performance. As a result, the transistor snapback voltage and the gate-oxide breakdown voltage are converging. This draws concern on the impact of electrostatic discharge (ESD) events on gate oxide integrity and long-term reliability. Especially for devices in the input/output circuits, the transistors may be connected to external pins, exposing the thin oxide directly to the high-current ESD events. Quite a few studies have been made on the breakdown characteristics of conventional SiO_2/SiON gate dielectric under ESD stress [1]–[6]. However, the excessive off-state gate leakage current caused by direct tunneling through the gate dielectric has stopped the dielectric thickness scaling in MOSFETs. Substantial efforts have been put into the development of high-k dielectric to replace the conventional SiO_2/SiON gate. This opens new questions about gate oxide integrity and reliability. Although there were numerous reports on the breakdown of high-k gate under low electric field stress related to normal device operation [7]–[10], the knowledge about the effect of high-field stress in the ESD time domain is still limited [11], [12].

In this paper, catastrophic gate oxide breakdown of MOS-FETs with high-k gate was characterized under ESD-like pulsed stress. It was found that the excessive gate current after oxide failure may result in a loss of gate contact and short the drain/source the transistor. Extending our previous work [11], the measurements were made on NMOSFETs with two different high-k gate thicknesses and on PMOSFETs as well. The effect of well resistance was found to have a significant impact on the breakdown voltage (V_{BD}) of PMOSFETs under positive stress. NMOSFET under positive stress (inversion) was identified to have the smallest breakdown voltage. Compared to those with SiON gate, NMOSFETs with high-k gate have a larger breakdown voltage for a given effective oxide thickness. The results obtained from the Constant Voltage Stress (CVS) and the conventional Transmission Line Pulse (TLP) methods were compared and showed that the TLP method gives a conservative estimation compared to the CVS method. The breakdown voltage corresponding to time-to-breakdown of 1-ns was measured on both NMOSFETs and PMOSFETs using the TLP method. It correlates well with the extrapolated data from CVS measurements.

II. DEVICES AND EXPERIMENTS

Measurements were conducted on NMOSFETs in 32 nm bulk CMOS technology. Devices with two dielectric thicknesses, $t_{inv} = 1.46$ nm (thin, SG) and $t_{inv} = 3.2$ nm (thick, EG), were tested. SG devices are used in the core circuits and EG devices are used mainly in input/output circuits. All reported dielectric thicknesses are effective oxide thicknesses t_{inv} measured in inversion ($t_{inv} = \epsilon_0 \epsilon_{SiO2}/C_{inv}$).

The data acquisition system used for pulsed breakdown measurements is shown in Fig. 1. This system is a conventional TLP system with a π resistance network for impedance matching and waveform shaping. The voltage pulse is applied at the gate of the device-under-test (DUT) with all other terminals grounded. The current of the DUT was measured by the current probe. To determine the time-to-breakdown (t_{BD}) of the gate oxide, voltage and current waveforms were captured by the oscilloscope and saved for analysis. The semiconductor parameter analyzer was used to measure device DC characteristics before and after the pulsed stress.

The CVS method is used to determine the breakdown voltage V_{BD} for different ESD events on different time scales (100 ns, 30 ns and 1 ns). The DUT is stressed by only one pulse. The pulse width is chosen by trial and error so as to ensure that it is long enough for the device to fail within

978-1-4244-5430-3/10 $26.00 © 2010 IEEE

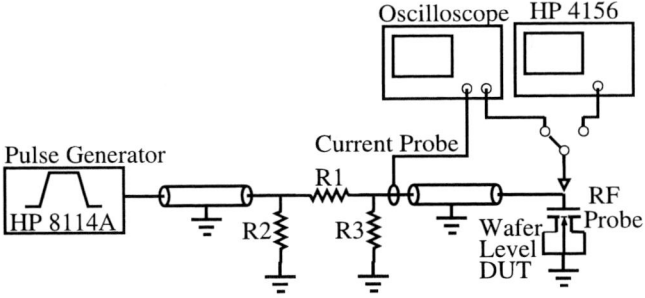

Fig. 1. Experiment setup for pulsed gate oxide breakdown measurements.

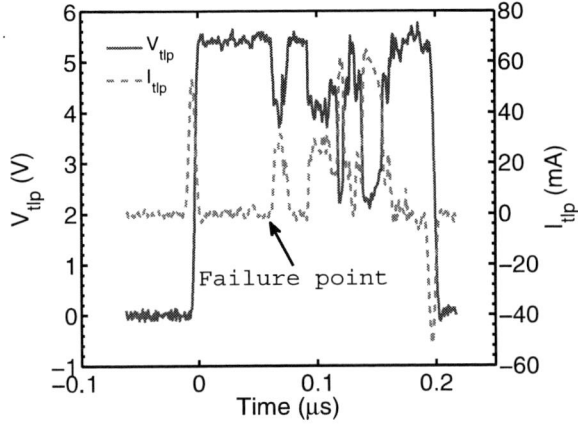

Fig. 2. Typical gate voltage waveform for an NMOSFET with high-k gate dielectric illustrating the feature of gate dielectric breakdown.

one pulse. The same set of experiment is repeated at several stress voltages, V_G, to determine the mathematical relationship between t_{BD} and V_G. Then V_{BD} for a specific t_{BD} can be calculated by either interpolation or extrapolation. Fig. 2 illustrates the representative voltage and current waveforms of a high-k gate NMOSFET stressed in accumulation. A significant impedance drop after oxide breakdown is reflected by the voltage drop (and current increase) on the waveform. This point is taken as the failure point. The current flowing through the device after oxide failure can be larger than 20 mA. The high current density will cause a loss of contact with the damaged gate finger, which leads to the "recovery" of voltage in the waveform. The DUT has multiple fingers and consequently a voltage drop and "recovery" can be observed several times.

The CVS results were compared with the results obtained by the conventional TLP method which is widely used in the characterization of devices under conditions similar to those of an ESD event. Unlike the CVS method, the TLP method applies a series of pulses with successively increasing amplitudes until a failure is detected, and the breakdown voltage V_{BD} is measured directly. The corresponding time-to-breakdown can be taken as the applied pulse width t_{pw}. Although the multiple pulses applied during a TLP test introduce error caused by the

cumulative damage, the TLP method is more convenient and efficient than the CVS method.

Since gate oxide breakdown is caused by the generation of defects at random positions inside the gate stack, time-to-breakdown t_{BD} and breakdown voltage V_{BD} are statistical quantities which follow the Weibull distribution

$$F(t_{BD}) = 1 - \exp\left(\frac{t_{BD}}{t_{63\%}}\right)^{\beta}. \tag{1}$$

In Eq. (1), F is the cumulative failure probability, t_{BD} (V_{BD}) is the time-to-breakdown (breakdown voltage) of a specific sample and $t_{63\%}$ ($V_{63\%}$) is the time-to-breakdown (breakdown voltage) at which 63.2% of the samples have failed. In this paper, $t_{63\%}$ and $V_{63\%}$ are used as the characteristic time-to-breakdown and the characteristic breakdown voltage of the DUT, respectively.

III. RESULTS AND DISCUSSION

A. Gate oxide breakdown of devices with high-k/metal gate

In this section, we will discuss gate oxide breakdown of MOSFETs with high-k/metal gate. As indicated previously, gate oxide breakdown is due to random generation of defects in the gate oxide. Its failure probability can be described by randomly distributed defects following the Poisson model [13] expressed by

$$F = 1 - \exp(-DA), \tag{2}$$

where A is the gate oxide area and D is the defect density. From Eq. (2) and Eq. (1) can be seen that the characteristic breakdown time $t_{63\%,1}$ and $t_{63\%,2}$ of two distributions follow the area scaling law A is the gate oxide area and D is the defect density. From Eq. (2) and Eq. (1) can be seen that the two breakdown distributions F_1 and F_2 with two different gate oxide areas, A_1 and A_2, follow the area scaling law

$$\ln\left[-\ln\left(1 - F_1\right)\right] - \ln\left[-\ln\left(1 - F_2\right)\right] = \ln(A_1/A_2). \tag{3}$$

Fig. 3 shows that the thin-oxide NMOSFETs with high-k gate follow this area scaling law very well. This reveals that the breakdown is related to the intrinsic breakdown process which is caused by defects homogeneously distributed across the gate oxide area as assumed by the Poisson random statistics [13]. The pre-existing defects does not cause early oxide failure. It is interesting to notice that the slope of the Weibull distribution increases from the high failure percentile (corresponding to smaller gate oxide area) to the low failure percentile (corresponding to larger gate oxide area). Similar observations were also made on SiON and high-k gate dielectrics under low stress voltages [10], [14], [15]. For the high-k dielectric, this is attributed to much higher defect generation rate in the high-k layer compared to that in the interfacial layer [10], [15], while the progressiveness of the breakdown path wearout is believed to cause the slope change in SiON dielectrics [14].

Due to the limited resolution of the current probe, the minimum current that can be measured during a TLP pulse is 1 mA. Consequently, detection of soft breakdown, which

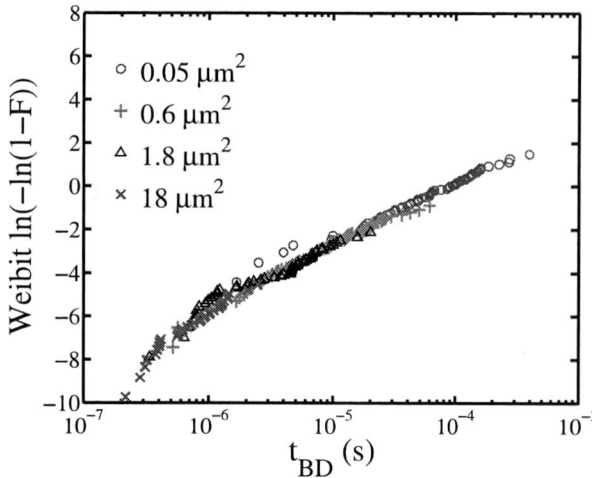

Fig. 3. Gate oxide failure follows the area scaling law, indicating that the breakdown follows the intrinsic breakdown process.

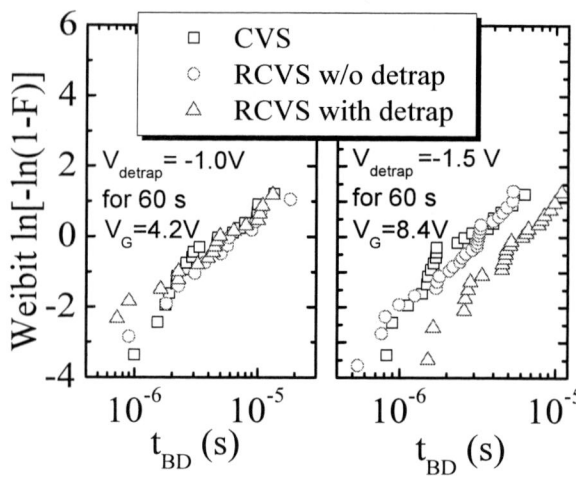

Fig. 5. Effect of stress interruption and charge detrapping on the t_{BD} of thin (left) and thick (right) oxide devices.

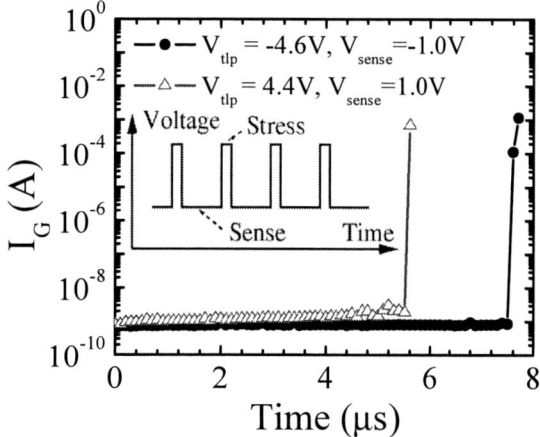

Fig. 4. RCVS shows that gate leakage current changes slightly before the hard failure. The breakdown features an abrupt increase in I_G.

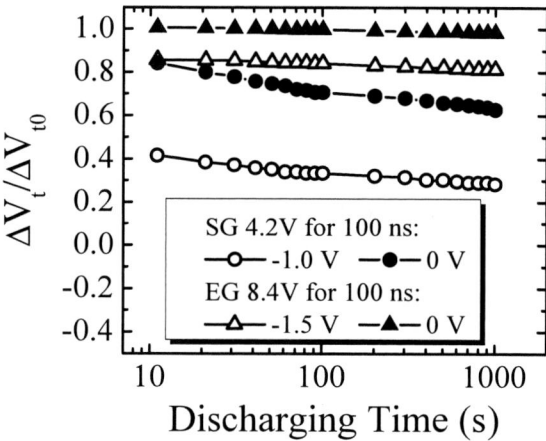

Fig. 6. Recovery of threshold voltage at different gate voltages indicates that the trapped electrons inside the gate stack were detrapped during the period when the stress was interrupted.

features a small increase in gate leakage current (I_G) magnitude or noise, is not possible. Additionally, recent studies [16] showed that the breakdown transient for stress voltages larger than 4 V or for devices with metal gate can be extremely fast. This also indicates that soft breakdown is generally not relevant to this study, which used stress voltages larger than 4 V in most cases and DUTs with metal gate electrodes. The gate oxide breakdown reported in this study is catastrophic. The failure features an abrupt increase of I_G in both stress polarities as shown by Fig. 4. In order to measure the gate leakage current immediately after oxide breakdown, the gate oxide was stressed by a series of pulses with a duration of 100 ns and constant amplitude, i.e. Repetitive Constant Voltage Stress (RCVS), as shown by the inset of Fig. 4. The gate leakage current only changes slightly before hard failure. In PMOSFETs, the breakdown results showed similar characteristics. Moreover, the drain to source current after gate

oxide breakdown under either positive or negative stresses showed a resistor-like characteristics for devices with different channel lengths (SG: 50 nm and EG: 150 nm). This may be caused by electromigration due to the high electric field and the large temperature gradient during the breakdown transient.

High-k dielectric contains a large amount of electron traps, presumably oxygen vacancies [8]. Electrons trapped during the positive stress may affect the time-to-breakdown distribution as they change the electric field distribution inside the gate stack. To study the effects of trapped electrons on t_{BD}, the time-to-breakdown distributions measured from CVS and RCVS are compared as shown in Fig. 5. For one group of devices under RCVS, the stress was interrupted periodically only for DC leakage measurements; for the other group of devices under RCVS, a small negative bias was applied on the gate for 60 s between the stress pulses. Interruption of the stress resulted in electron detrapping from the gate stack;

978-1-4244-5430-3/10 $26.00 © 2010 IEEE

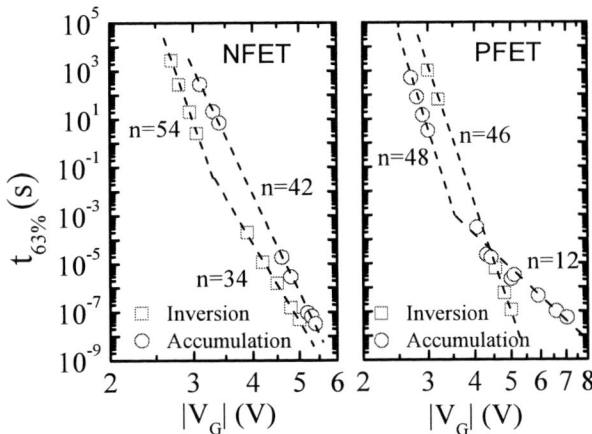

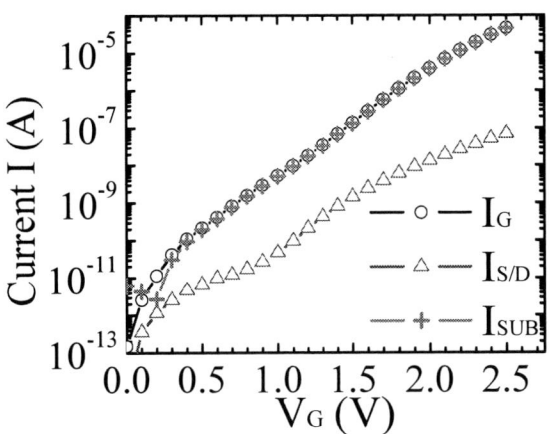

Fig. 7. Time-to-breakdown of thin-oxide MOSFETs with high-k gate dielectrics, obtained using the CVS method.

Fig. 8. Terminal currents measured on a PMOSFET in accumulation (positive stress).

a negative bias on the gate helped accelerate the detrapping process. To avoid any damage caused by the negative bias, -1 V was used for SG devices and -1.5 V for EG devices. Fig. 5 shows that interrupting the stress ($\sim$ 5 s) for DC leakage measurement did not change the gate oxide time-to-breakdown distribution of both SG and EG devices compared to CVS measurement. However, the negative bias between stress pulses increased the t_{BD} of EG devices by a factor of two. The change of the device threshold voltage reflects the amount of trapped electrons. Fig. 6 shows the influence of interrupting the stress ($V_G = 0$ V) and the influence of the explicit negative discharging bias on the device threshold voltage. The explicit discharging process helped remove a considerable portion of the trapped electrons. A much bigger portion of the trapped charges was removed by the negative bias in SG devices compared to EG devices, but surprisingly, the discharging process did not change the SG t_{BD} distribution, whereas it increased the EG t_{BD}. This indicates that the time-to-breakdown of the thick high-k gate is more sensitive to the trapped electrons in the gate stack.

Fig. 7 shows $t_{63\%}$ obtained using the CVS method as a function of gate voltage for NMOSFETs and PMOSFETs with thin gate oxide (SG) in both inversion and accumulation. For $t_{63\%} > 1$ s, the data was extracted using the semiconductor parameter analyzer in the sampling mode. Time-to-breakdown in this case is defined as the time at which the gate current at the stress voltage increased abruptly by at least 10 μA. In order to predict V_{BD} for different ESD events (with different durations), V_{BD} should be expressed as a function of $t_{63\%}$. Fig. 7 shows that the data $t_{63\%}$ can be fitted well by the power law [17]:

$$t_{63\%} = c \cdot V_G^{-n} \qquad (4)$$

in a limited time range. For NMOSFETs and PMOSFETs stressed negatively, the power law exponent n stays constant in the entire time range evaluated in this study. However, the power law exponent for PMOSFETs under positive stress

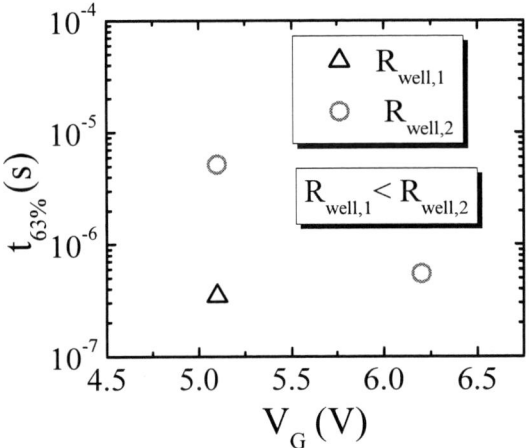

Fig. 9. The effect of well resistance on $t_{63\%}$ of PMOSFETs in accumulation.

(accumulation) decreases considerably as the stress voltage increases. This significant change of n in PMOSFETs is due to the effect of well resistance. For devices under positive stress, the majority of the gate current is coming from the tunneling of conduction band electrons. In PMOSFETs, the electron is provided by the N-well through the body contact. Fig. 8 shows that the majority of I_G flows through the N-well to ground. As the leakage current at the large stress voltages is orders of magnitude higher than for small stress voltages, a significant portion of the stress voltage drops across the well resistance and the effective stress on the gate oxide is reduced. This bends the t_{BD} vs. V_G curve and decreases the power law exponent. To verify the above argument, devices with two different distances between the gate edge and the body contact (i.e. different well resistance) were measured. As illustrated by Fig. 9, for a given V_G, the $t_{63\%}$ of the transistors with a larger well resistance is about 10 times longer. Because the dielectric constant of the high-k layer is much larger than that of the

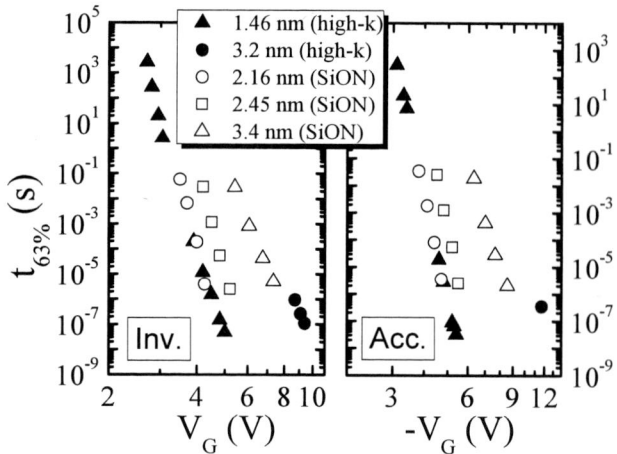

Fig. 10. Comparison of $t_{63\%}$ between high-k and SiON gate dielectrics.

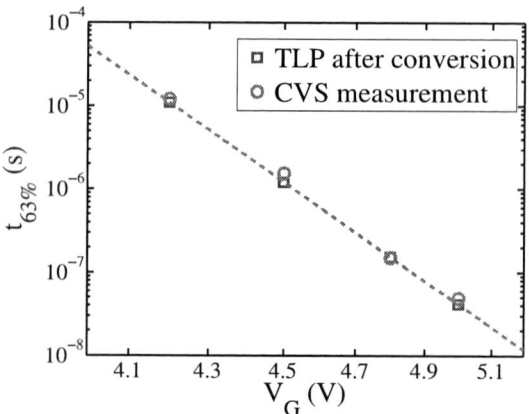

Fig. 11. Data collected using the TLP method can be converted to $t_{63\%}$, which is comparable to $t_{63\%}$ measured using the CVS method.

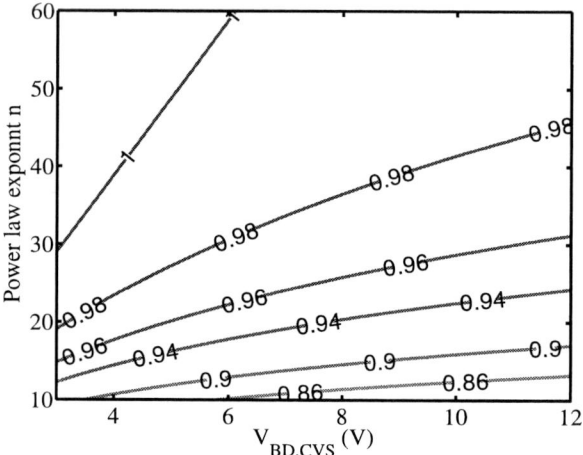

Fig. 12. Contour plot of $V_{BD,TLP}/V_{BD,CVS}$ as a function of power law exponent (n) and the breakdown voltage from CVS ($V_{BD,CVS}$) with TLP voltage increment step $\Delta V = 0.1$ V.

interfacial layer (IL), most of the voltage drops across the IL. Consequently, the high-k layer under large positive stress is almost transparent to electron tunneling from the substrate. On the contrary, for negative stress, the barrier for electrons tunneling from gate to body consists of both the high-k layer and the interfacial layer, which results in a much smaller gate leakage current compared to the case of positive stress. Therefore, the voltage drop over the well resistance is not large enough to affect the $t_{63\%}$ of NMOSFETs in accumulation (under negative stress). A small reduction of the power law exponent was also observed for NMOSFETs under positive stress (inversion). However, this is not a general trend as a constant power law exponent was observed for $t_{63\%}$ in the range from 100 ns to 100 s on some of the wafers .

In SiO$_2$ gate, the charge-to-breakdown ($Q_{BD} = J_G \cdot t_{BD}$) measured in gate injection mode (negative stress) is smaller than that measured in substrate injection mode (positive stress) [17]. The DC measurement on high-k gate shows the same trend. However, as discussed above, the gate leakage current for positive bias is much larger than that for negative bias. Considering this, the $t_{63\%}$ is smaller when stressed by positive voltages than by negative stresses for a given voltage magnitude, as shown by the $t_{63\%}$ of the NMOSFETs and that of the PMOSFETs under small stress voltages in Fig. 7. Because of the well resistance, the trend is reversed for PMOSFETs at large stress voltages. As a result, NMOSFET in inversion was found to have the smallest breakdown voltage.

Fig. 10 compares the $t_{63\%}$ of NMOSFETs with high-k gate and SiON gate in inversion and accumulation. Devices with SiON gate are from the 45 nm technology. Although t_{inv} of the thin-oxide high-k NMOSFETs is much smaller than that of the thin-oxide SiON device (1.46 nm compared to 2.16 nm), $t_{63\%}$ of these two devices are comparable. The result suggests that for a given effective oxide thickness, high-k gate have a larger breakdown voltage and hence increased ESD robustness compared to SiON gate, despite its smaller ultimate breakdown strength E_{BD} (electric field at breakdown) [18]. E_{BD} of

a dielectric material is found to decrease as the dielectric constant ϵ increases; it is proportional to $\epsilon^{-0.65}$ [18]. However, for the same effective oxide thickness, the physical thickness of the high-k gate is $\epsilon_{HK}/\epsilon_{SiON}$ times larger than that of the SiON gate. Therefore, for the same effective oxide thickness, the breakdown voltage of the high-k gate is still larger.

B. Comparison between CVS and TLP method

It is desirable to use TLP method since it is more convenient and efficient than the CVS method. During TLP, the magnitude of the voltage pulse with a fixed width (t_{pw}) is ramped up with an increment, ΔV, in each step until the gate oxide breakdown is detected. The breakdown voltage corresponding to the breakdown time $t_{BD} = t_{pw}$ can be acquired directly from the measurement. The cumulative damage from the multiple stresses has to be taken into account to get the exact t_{BD} or V_{BD}. Following the procedure proposed in [19], V_{BD} data acquired during a TLP test can be converted to $t_{63\%}$ using Eq. (5) in Appendix. The converted $t_{63\%}$ data agree

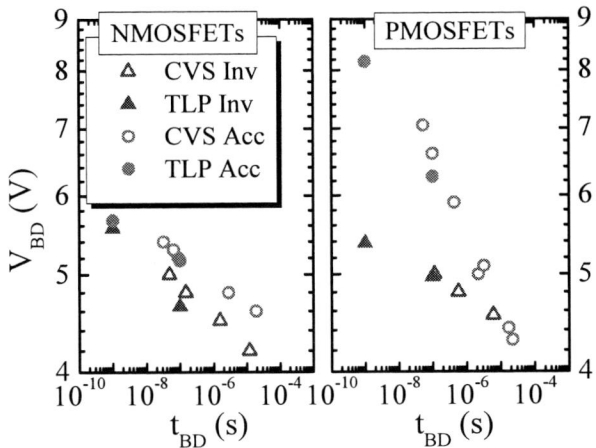

Fig. 13. Comparison between $V_{BD,CVS}$ and $V_{BD,TLP}$ experiment data. $V_{BD,CVS}$ is directly from Fig. 7. t_{BD} from the TLP method is the pulse width t_{pw}.

with CVS measurements very well, as shown by Fig. 11. However, this conversion requires the power law exponent, which is extracted using the CVS method. Time-to-breakdown is very sensitive to the stress voltage, since the power law exponent is typically in the range from 30 to 50. Therefore, only the last few voltage steps in TLP method make significant contributions to the final gate oxide failure. This indicates that the V_{BD} directly from the TLP method should not deviate from the exact breakdown voltage (from CVS method) too much. Fig. 12 compares the breakdown voltage from the TLP method, $V_{BD,TLP}$, and from the CVS method, $V_{BD,CVS}$, by a contour plot of $V_{BD,TLP}/V_{BD,CVS}$ within the typical range of two parameters n and $V_{BD,CVS}$. $V_{BD,TLP}/V_{BD,CVS}$ is calculated following the procedure described in Appendix. The voltage increment ΔV in the TLP method is assumed to be 0.1 V. It is shown that $V_{BD,TLP}$ correlates well with $V_{BD,CVS}$ in the typical parameter range, and in most cases the difference is less than 10%. Fig. 13 compares the experimental $V_{BD,TLP}$ and $V_{BD,CVS}$ measured on thin-oxide MOSFETs. They correspond to the failure percentile $F = 63.2\%$. $V_{BD,CVS}$ is taken directly from the data in Fig. 7. As expected, V_{BD} data from both methods are fairly close, with the difference less than 10%. This suggests that V_{BD} from the TLP method can be used as a conservative estimation of the breakdown voltage.

TLP method was also used to measure V_{BD} down to $t_{BD} = 1$ ns on a commercially available very fast TLP system. The results are shown in Fig. 13. The 1-ns data stays close to the line extrapolated from the data measured using the CVS method. This suggests that the breakdown mechanism remains the same down to 1 ns range.

IV. CONCLUSIONS

Catastrophic gate oxide breakdown of MOSFETs with high-k gate was studied under ESD-like pulsed stress. The high voltage stress changes the gate leakage current only slightly until the final hard breakdown. After gate oxide failure, the large gate current may result in a loss of gate contact, and the drain-to-source current showed a resistor-like behavior. Interrupting the stress briefly does not change the t_{BD} statistical distribution. The time-to-breakdown of the transistors with thick high-k dielectric is more sensitive to the trapped charge inside the gate stack compared to those with thin high-k dielectric. It is found that NMOSFETs under positive stress have the smallest V_{BD}. For PMOSFETs under positive stress the voltage drop on the well resistance increases the breakdown voltage significantly. The comparison between MOSFETs with high-k gate and SiON gate showed that for a given effective gate oxide thickness, the $t_{63\%}$ of the high-k gate stack is longer. Comparison between the results obtained by CVS and TLP methods showed that the difference between $V_{BD,TLP}$ and $V_{BD,CVS}$ is less than 10%. The TLP method can be used as an efficient method to make a conservative estimation of the breakdown voltage of the gate oxide under certain ESD events. V_{BD} data down to $t_{BD} = 1$ ns was extracted using the TLP method. The data suggests that the failure mechanism of the high-k dielectric remains the same in the 1 ns range as in the longer time scale.

APPENDIX

Based on the power law relationship between $t_{63\%}$ and V_{BD}, stressing the device at V_{G1} for time t_1 is equivalent to stressing the device at V_{G2} for time $t_2 = t_1 \cdot (V_{G1}/V_{G2})^n$. Assuming the duration of the ESD event is t_{pw}, the breakdown voltage, $V_{BD,CVS}$, corresponding to t_{pw} (obtained from CVS method) can be related to the result from the TLP measurement with the pulse width equal to t_{pw} by Eq. (5) [19]:

$$t_{pw} = \sum_i^m t_{pw} \cdot \left(\frac{V_i}{V_{BD,CVS}} \right)^n. \tag{5}$$

V_i is the TLP voltage in the ith step, m is the number of voltage pulses until the gate oxide failure and n is the power law exponent. The breakdown voltage from the TLP method is $V_{BD,TLP}$, corresponding to the magnitude of the last voltage pulse. $V_{BD,TLP}$ is equal to $m \cdot \Delta V$. Approximating the summation in Eq. (5) by an integral yields

$$t_{pw} = \int_0^{m \cdot t_{pw}} \left(\frac{RR \cdot t}{V_{BD,CVS}} \right)^n dt. \tag{6}$$

$RR = \Delta V / t_{pw}$ and ΔV is the voltage increment in each TLP step. Calculating this integration gives

$$t_{pw} = \frac{V_{BD,CVS}}{RR \cdot (n+1)} \left(\frac{V_{BD,TLP}}{V_{BD,CVS}} \right)^{n+1}. \tag{7}$$

Reformulating Eq. (7) gives the relationship between $V_{BD,TLP}/V_{BD,CVS}$ and $V_{BD,CVS}$:

$$\frac{V_{BD,TLP}}{V_{BD,CVS}} = \left(\frac{\Delta V \cdot (n+1)}{V_{BD,CVS}} \right)^{1/(n+1)}. \tag{8}$$

Fig. 12 is obtained by plotting $V_{BD,TLP}/V_{BD,CVS}$ against n and $V_{BD,CVS}$.

978-1-4244-5430-3/10 $26.00 © 2010 IEEE

ACKNOWLEDGMENTS

This work was performed at the IBM Microelectronics Div. Semiconductor Research & Development Center, Essex Junction, VT-05452. The authors appreciate Ralph Halbach and Christopher Seguin in support of experiment setup. GMU acknowledges funding by NSF grant # ECCS 0901236.

REFERENCES

[1] J. Wu and E. Rosenbaum, "Gate oxide reliability under ESD-like pulse stress," *Electron Devices, IEEE Transactions on*, vol. 51, no. 9, pp. 1528–1532, Sept. 2004.

[2] B. E. Weir, C.-C. Leung, P. J. Silverman, and M. A. Alam, "Gate dielectric breakdown in the time-scale of ESD events," *Microelectronics Reliability*, vol. 45, no. 3-4, pp. 427–436, 2005.

[3] A. Ille, W. Stadler, A. Kerber, T. Pompl, T. Brodbeck, K. Esmark, and A. Bravaix, "Ultra-thin gate oxide reliability in the ESD time domain," in *Proc. EOS/ESD Symp.*, 2006, pp. 185–194.

[4] A. Ille, W. Stadler, T. Pompl, H. Gossner, T. Brodbeck, K. Esmark, P. Riess, D. Alvarez, K. Chatty, R. Gauthier, and A. Bravaix, "Reliability aspects of gate oxide under ESD pulse stress," in *Proc. EOS/ESD Symp.*, 2007, pp. 328–337.

[5] J.-C. Tseng and J.-G. Hwu, "Oxide-trapped charges induced by electrostatic discharge impulse stress," *Electron Devices, IEEE Transactions on*, vol. 54, no. 7, pp. 1666–1671, July 2007.

[6] ——, "Effects of electrostatic discharge high-field current impulse on oxide breakdown," *Journal of Applied Physics*, vol. 101, no. 1, p. 014103, 2007.

[7] G. Ribes, J. Mitard, M. Denais, S. Bruyere, F. Monsieur, C. Parthasarathy, E. Vincent, and G. Ghibaudo, "Review on high-k dielectrics reliability issues," *Device and Materials Reliability, IEEE Transactions on*, vol. 5, no. 1, pp. 5–19, March 2005.

[8] A. Kerber and E. Cartier, "Reliability challenges for CMOS technology qualifications with Hafnium Oxide/Titanium Nitride gate stacks," *Device and Materials Reliability, IEEE Transactions on*, vol. 9, no. 2, pp. 147–162, June 2009.

[9] A. Kerber, E. Cartier, B. Linder, S. Krishnan, and T. Nigam, "Tddb failure distribution of metal gate/high-k cmos devices on soi substrates," in *Reliability Physics Symposium, 2009 IEEE International*, April 2009, pp. 505–509.

[10] T. Nigam, A. Kerber, and P. Peumans, "Accurate model for time-dependent dielectric breakdown of high-k metal gate stacks," in *Reliability Physics Symposium, 2009 IEEE International*, April 2009, pp. 523–530.

[11] J. Di Sarro, Y. Yang, K. Chatty, R. Gauthier, A. Ille, S. Mitra, J. Li, C. Russ, E. Rosenbaum, and D. Ioannou, "ESD time-domain characterization of High-k gate dielectric in a 32 nm CMOS technology," in *Proc. ESD/EOS Symp.*, 2009.

[12] C.-H. Chen, H.-L. Hwang, and F.-C. Chiu, "Reliability characterization of stress-induced charge trapping in HfO_2 by electrostatic discharge impulse stresses," *Journal of Applied Physics*, vol. 105, no. 10, p. 103910, 2009.

[13] E. Wu and R.-P. Vollertsen, "On the Weibull shape factor of intrinsic breakdown of dielectric films and its accurate experimental determination. Part I: theory, methodology, experimental techniques," *Electron Devices, IEEE Transactions on*, vol. 49, no. 12, pp. 2131–2140, Dec 2002.

[14] J. Sune, E. Wu, and S. Tous, "Failure-current based oxide reliability assessment methodology," in *Reliability Physics Symposium, 2008. IRPS 2008. IEEE International*, 27 2008-May 1 2008, pp. 230–239.

[15] N. Raghavan, K. L. Pey, and X. Li, "Detection of high-kappa and interfacial layer breakdown using the tunneling mechanism in a dual layer dielectric stack," *Applied Physics Letters*, vol. 95, no. 22, p. 222903, 2009.

[16] S. Lombardo, J. H. Stathis, B. P. Linder, K. L. Pey, F. Palumbo, and C. H. Tung, "Dielectric breakdown mechanisms in gate oxides," *Journal of Applied Physics*, vol. 98, no. 12, p. 121301, 2005.

[17] E. Y. Wu and J. Su, "Power-law voltage acceleration: A key element for ultra-thin gate oxide reliability," *Microelectronics and Reliability*, vol. 45, no. 12, pp. 1809 – 1834, 2005.

[18] J. McPherson, J. Kim, A. Shanware, H. Mogul, and J. Rodriguez, "Trends in the ultimate breakdown strength of high dielectric-constant materials," *Electron Devices, IEEE Transactions on*, vol. 50, no. 8, pp. 1771–1778, Aug. 2003.

[19] A. Kerber, L. Pantisano, A. Veloso, G. Groeseneken, and M. Kerber, "Reliability screening of high-k dielectrics based on voltage ramp stress," *Microelectronics Reliability*, vol. 47, no. 4-5, pp. 513 – 517, 2007.

978-1-4244-5430-3/10 $26.00 © 2010 IEEE

Robust High Current ESD Performance of Nano-Meter Scale DeNMOS By Source Ballasting

Amitabh Chatterjee and Forrest Brewer

Department of Electrical and Computer Engineering, University of California, Santa Barbara, CA - 93106, USA

Harald Gossner

Infineon Technologies, Am Campeon 1, D-85579 Neubiberg, Germany

Sameer Pendharkar and Charvaka Duvvury

Silicon Technology Development, Texas Instruments Inc Dallas, TX -75243, USA.

Abstract—"Strong Snapback" in DeNMOS transistors leads to weak ESD performance which is often represented by low It2 and strong die to die dependence. We report here the first experimental evidence that this can be controlled with introduction of source-resistance Rs. A new microscopic model has been analyzed to understand the physics of strong snapback and explain the experimental observations. Impact of current crowding phenomenon and role of adding a resistor across the source and ground has been broadly addressed in this paper. Also the current crowding phenomenon has been macroscopically modeled and a circuit model has been established.

Keywords- StrongSnapback, Current Crowding; Kirk Effect, 2D & 3D localization, Bipolar Turn-on

I. INTRODUCTION

Vulnerability of Drain Extended NMOS under ESD events is a critical reliability issue in I/O circuits for high voltage and other mixed signal applications [1-8]. Extremely low failure current in Gate-Grounded DeNMOS has been attributed to damage triggered by *Strong snapback, i.e.*(essentially $It_2=It_1$). Strong snapback in DeNMOS is explained due to 2D turn-on of the distributed parasitic bipolar [1-3], where subsequent localization along the width (i.e Z-axis-3D plane) triggers filamentation due to localized J.E heating during the strong snapback, which can lead to thermal runaway [1]. However, so far the highly complex electro-thermal runaway is not very well understood. The bipolar turn-on triggers localization first in the 2D plane and subsequently the 3D localization occurs along the device width [10]. The localization leads to self-heating and current crowding mechanisms across the ballast region. Dynamical turn-on process of the parasitic bipolar has a profound impact on the ballasting behavior triggered by crowding due to electrons and holes. In-fact, the current crowding mechanisms also lead to self-heating effects. Now under a very short scale power dissipation localized self heating leads to permanent damage in the device. Strong and deep snapback prevents electro-thermal relaxation and exacerbates the thermal runaway. Lack of detailed knowledge of the coupled electro-thermal nature of strong snapback has prevented robust design of De-NMOS. In this abstract, we first present an analysis where we try to control strong-snapback through circuit topology. Next, a microanalysis based on the parasitic bipolar activity in DeNMOS is presented where we investigate the critical role of ambipolar (i.e *holes and electrons*) current crowding phenomenon.

II. OPTIMALLY WEAKENING STRONG SNAPBACK

The drain extended device often fails as it undergoes snapback (It1) also exhibits extremely high die-to-die variations (i.e a

minimum of 0.16mA/μm & a maximum of 0.76mA/μm observed for $R_s=0$). The dynamic electro-thermal process which involves self-heating triggers thermal runaway. Eventually it leads to highly localized silicon meltdown. The *nonlinear electro-thermal processes* which involve coupled interaction between circuit and the device are the principal reason why devices fail due to *strong snapback*. Moreover, thermal runaway involving the electro-thermal process under an event of *strong snapback* is an extremely sensitive phenomenon which explains such a high die to die variations (*fig.1*).

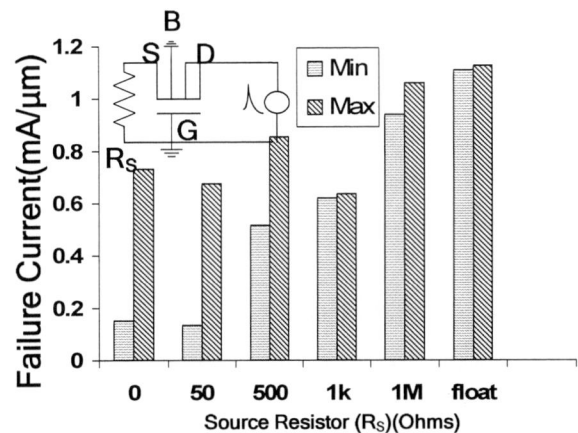

FIG. 1. LESS VARIATIONS IN FAILURE CURRENT FOR HIGHER SOURCE RESISTOR Rs

Now, a *weaker snapback* leads to less localized heating and the thermal runaway is mitigated during the complex electro-thermal process. The filamentation and eventually the self-heating triggers highly localized heat dissipation mechanisms. An intuitive design strategy for the protection element can be weakening the strong snapback, which alleviates highly localized thermal dissipation and eventual thermal runaway.

A. Adding Ballast Resistor (R_D) Across the Drain

The turn-on bipolar can be limited by adding a ballast resistor [2] across the drain (*fig. 2*). A resistor R_D introduced on the drain side prevents the external current from building after the snapback has been triggered. Thus *resistive slope* (*fig. 2*) leads to buildup of high voltage under an ESD event which can have deleterious effect on the gate oxide of the devices across the protection element. Since the failure is not due to the diode failure across the drain/substrate junction, thus employing a ballast resistor through a silicide block in the drain region is not very advantageous. In-fact, the design window [7] for the ESD

978-1-4244-5430-3/10 $26.00 © 2010 IEEE   853

protection is determined by the maximum voltage which the core devices can withstand in lieu of their gate-oxide reliability, in the process it limits the performance of this configuration. The drain resistor R_D plays a critical role in *restricting 3D localization*.

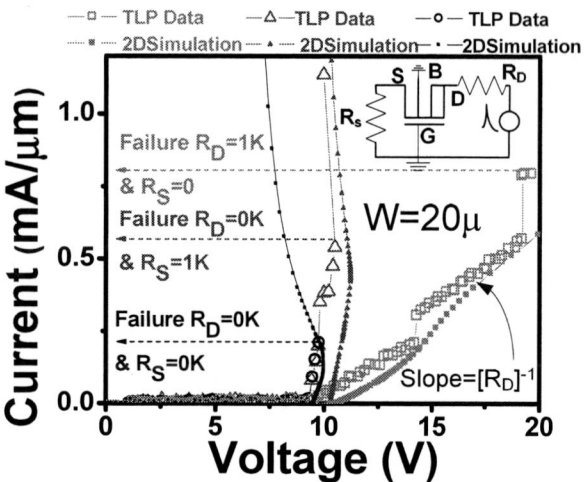

FIG. 2 TLP DATA & 2D SIMULATION SHOW WEAK FIRST SNAPBACK THROUGH INTRODUCTION OF (SOURCE -RESISTOR) R_S & (DRAIN-RESISTOR) R_D

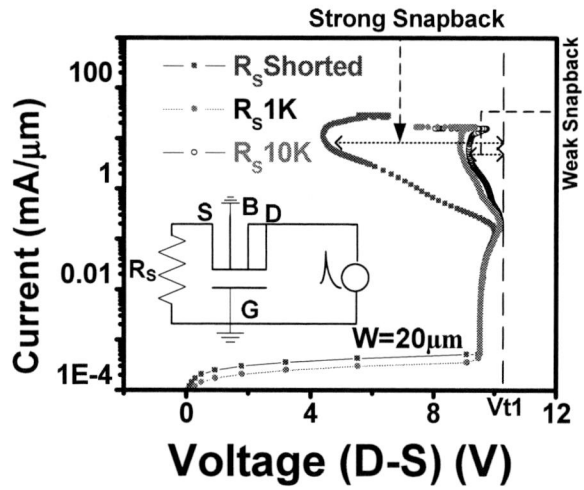

FIG. 3. I(DRAIN CURRENT) VS VOLTAGE [TCAD SIMULATION]ACROSS (DRAIN & SOURCE) SHOWING WEAKER FIRST SNAPBACK DUE TO SOURCE RESISTOR R_S

B.   *Improved Performance adding Source Resistor(R_S)*

Failure performance also improves 2X by limiting the source injection through introduction of R_S (*fig. 2*). Now, adding a resistor (R_S) across the source and the ground weakens the first snapback (*fig. 3*). The source resistor R_S plays a subtle role in restricting the 2D bipolar turn-on and in the process its role is fundamentally different from adding a resistor at the drain end.

However, in a DeNMOS impact of R_S is more critical for alleviating the die to die variations of I_{t2} (*fig. 1*). Also, the device failure at second snapback is relatively less impacted by source resistor R_S. As the value of source resistor, R_S increases both

maximum and minimum value of the failure current increases, while their variation is reduced (*fig. 1*).

In summary, resistive slope across De-NMOS under high current ESD event, can be efficiently controlled through introduction of a resistance across the source i.e R_S. However, the detailed understanding of the slope requires a microscopic analysis of the current crowding phenomenon in the device.

III.   IMPACT OF R_S ON FAILURE CURRENT VARIATION & RESISTIVE SLOPE ($R_{BALLAST}$)

A.   *Physics of Ambipolar Current & Highly sensitive electrostatics below the oxide*

In nano-meter scale De-NMOS (which is a thin gate oxide device), the flow of avalanche-generated carriers is critically related to the highly distributed nature of the E-field lines near the drain region. The 2D-Electrostatics is given by Poisson's Equation.(eqn. 1) The net charge density in the silicon is determined by the ionized dopants and space charge of the mobile carriers.

Poisson Equation

Silicon

$$\nabla.\vec{E} = \frac{1}{\varepsilon_{Si}}\left[N_D - N_A + n - p\right] \approx \frac{1}{\varepsilon_{Si}}\left[N_D - \frac{J}{Vs}\right]$$

$$\left(J \geq N_D V_S\right)\left[Kirk - Effect\right] \tag{1}$$

The E-field is determined by the space charge limited transport of avalanche generated carriers, when the current density becomes equal to product of background doping concentration (N_D) & velocity saturation V_{sat}, which effectively leads to modulation of the 2D-E-field. Macroscopically, the phenomenon is similar to high current phenomenon the *kirk effect*.

Moreover, under higher current injection of holes and electrons, the charge density term in the poisson equation is given by the difference of electrons (*n*) and holes (*p*).

$$\nabla^2[\psi] = \left[\frac{n-p}{\varepsilon_{si}}\right] \left\{n = \frac{J_n}{V_{nSat}} \& p = \frac{J_p}{V_{pSat}}\right\} \tag{2}$$

Contours showing the potential distribution across the device (at I=0.05mA/µm) is shown (*fig.4*) & (at I=0.1mA/µm) is shown (*fig.5*). Flow of holes causes build-up of voltage in the device. The potential in the device is electro-statically coupled not only with the source and the body but also coupled to the gate. Note that the series of electric lines of force originating from the source end-up (E_1, E_2, E_3, E_4) (*fig.4*) which terminate on the gate and serves as a boundary between the diode formed by the source and the body (*source of minority carrier injection*) and a diode formed between the drain and substrate (*source of avalanche generated holes*).

Also one can note that the vertical component of the E-Field vector (i.e E_Y) undergoes a directional change (along the gate), which also causes the accumulated holes to flow down into the substrate in a conical angle Θ i.e *holes to funnel down into the substrate.*

978-1-4244-5430-3/10 $26.00 © 2010 IEEE   854

Using, the radial symmetry in the problem, Poisson's Equation can be solved in the cylindrical co-ordinates. And incremental potential $\Delta\Psi$ can be approximated

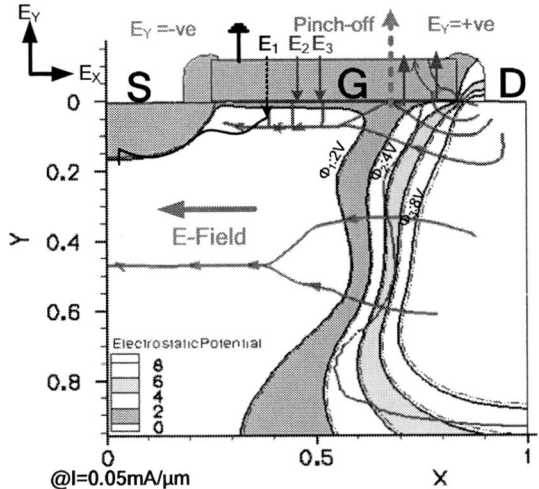

FIG. 4 TCAD SIMULATIONS SHOWING PROFILE OF ELECTROSTATIC POTENTIAL & E-FIELD LINES

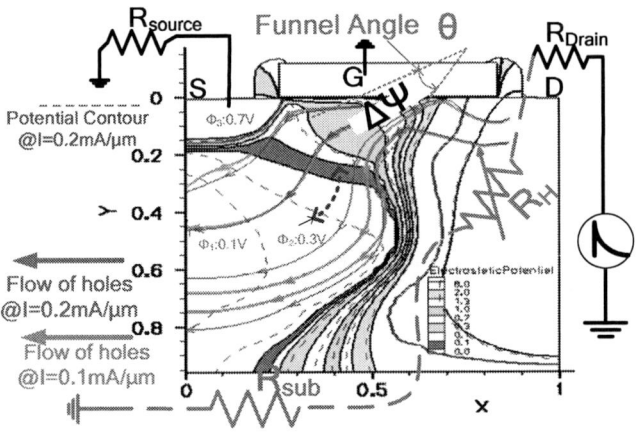

FIG. 5 TCAD SIMULATIONS SHOWING POTENTIAL DISTRIBUTION CURRENT CROWDING DUE TO ATTRACTED HOLES LEADS TO POTENTIAL BUILDUP ($\Delta\Psi$). MODELING HOLE BALLAST RESISTOR (R_H) [RED COMPONENT INTRINSIC CKT PARAMETERS & **BLACK COMPONENT** EXTRINSIC CKT PARAMETERS]

$$\Delta\psi \propto \frac{I_p - I_n}{\varepsilon_{si} \times \theta \times V_{sat}} \qquad (3)$$

Θ is the funnel angle, V_{sat} is the saturated drift velocity. Impact of current crowding due to accumulated holes can be observed, as the funnel angle Θ reduces, it causes incremental potential $\Delta\Psi$ needed to establish a given hole current to increases. In the limit, $\Theta\to 0$, $\Delta\Psi\to\infty$

Therefore, $$\Delta\psi = R_H\left(I_p - I_n\right) \qquad (4)$$

Where R_H is the intrinsic ballast resistance and is inversely proportional to the funnel angle Θ.

B. Current Crowding & Snapback

Avalanche generated holes which accumulate below the gate near the N-well junction causes a build-up of potential as shown by the change in potential contour (fig.5). Thus, a potential buildup ($\Delta\Psi$) is required to push the avalanche-generated holes, leads to *self ballast or intrinsic ballast action* across the drain region (fig. 6).

The potential distribution can be determined using the radial symmetry which involves the flow of holes into the substrate. The ballasting phenomenon due to current crowding of holes can be modeled through R_H (Ballast action due to current crowding of holes). Therefore, the potential established due to space charge of holes can be modeled through constant co-efficient defined as R_H [(fig. 6 & eqn (3&4)]

In the process, R_H limits the drain current during an ESD event and thus prevents early snapback. Also the potential buildup ($\Delta\Psi$) across the drain region is electro-statically coupled in the device which also leads to barrier lowering at the source end. This results in electron injection into the depleted channel below the oxide, which is then swept into the drain region.

Injected minority carriers (i.e electrons) compensate the net space charge as they modulate the breakdown electric field at the drain end. In the process, modulation of electric field decouples the impact of the gate on the flow of holes (fig. 6). The snapback is triggered; whenever the derivative of the potential drop is zero (i.e. *w.r.t incremental increase in the net current*). And the snapback continues till an incremental drop in potential due to electron injection (from source) equals an incremental rise in potential required to establish both the flow of avalanche generated holes into the substrate (i.e $\Delta\Psi$) and also maintain the flow of electrons across the drain contact. Thus for each incremental holes which are pushed into the substrate, injected electrons should be able to compensate the buildup of potential needed to push the accumulated holes. Hence, the deeper snapback in absence of resistor across the source (R_s) can be explained due to higher incremental injection of electrons for each injected avalanche holes (fig.3 i.e R_S=shorted) in the substrate.

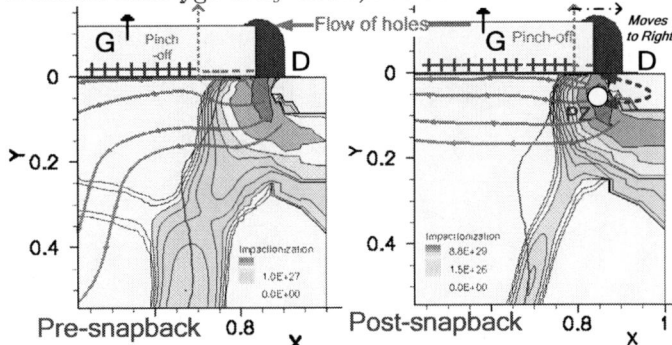

FIG. 6 TCAD SIMULATIONS SHOWING FLOW OF HOLES PRE- SNAPBACK AND AFTER SNAPBACK

Now, as the snapback is triggered the E-field modulation helps to decouple the impact of gate (i.e attraction of holes by the gate) on the flow of avalanche generated carriers (fig. 6). Moreover, the current crowding due to electron pushes the peak electric field to the edge of drain contact (i.e PZ), while the holes are attracted by electric field established due to injected electrons. Once the bipolar turns on, the potential buildup shifts from the

978-1-4244-5430-3/10 $26.00 © 2010 IEEE

region of accumulated holes to the region of accumulated electrons (i.e PZ). Hence, for higher current, the ballast resistor consists of two distinct components: 1) due to current crowding of holes (R_H), and 2) due to current crowding of electrons. And as a consequence the current component comprises of electron current through R_{Source} and hole current through R_{Sub}. In effect, the current crowding phenomenon in the De-nMOS determines the microscopic features of snapback.

C. Impact of Source Resistor & Gate Coupling

Role of source resistor R_S (*fig. 7 & 8*) is to primarily restrict the injection of electrons from the source. Injected electrons help to alleviate the hole current crowding (i.e. compensate the net space charge (*p-n*)) as the *Funnel angle* theta (Θ), expands during the snapback (*fig. 7*). Physically injected minority carriers modulate the current crowding as the holes are pushed into the substrate. Moreover, the potential build up due to holes is electro-statically coupled with the gate.

In the process, the current crowding behavior is profoundly impacted by the source injection which modulates the ballast resistor, RH (*fig. 8*).

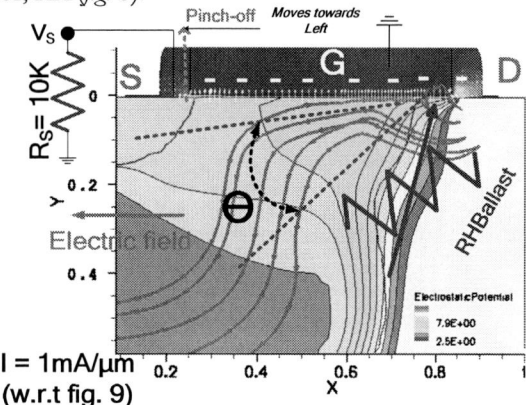

FIG. 7. ALLEVIATION OF HOLE CROWDING BEHAVIOR DUE TO INCREASED FUNNEL ANGLE Θ - (CRITICALLY RELATED TO GATE COUPLING)

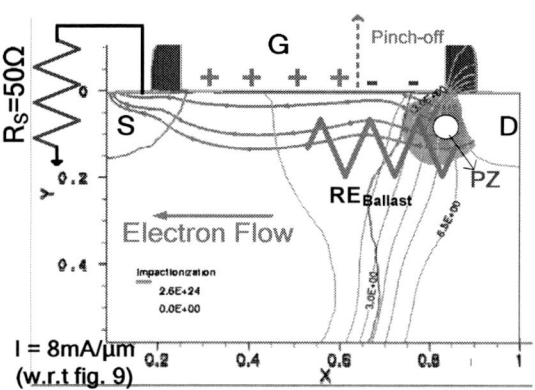

FIG. 8. HIGH ELECTRON INJECTION (DUE TO LOW R_S) DETERMINES THE BALLAST RESISTOR RE$_{BALLAST}$

Now, key to understanding the phenomenon is electro-statics behind the gate coupling which shifts the pinch off to the left (*towards the source*) (*fig.8*). Therefore, as the electron current in the branch containing R_{source} increases, the potential across the source (i.e R_s) begins to increase, while it also increase the gate coupling between the source and the gate pushes the pinch-off

region to the left. Thus shifting the pinch-off helps to ease the current crowding of holes. As the pinch-off region is pushed to the left, the funnel angle expands and in the process eases the current crowding (*fig. 8*)

While large values of R_S show only a marginal improvement of die to die variation it has a profound impact on the slope of I-V curve and leads to buildup of excess voltage under an ESD event. Therefore the implementation of R_S needs to be properly adjusted.

CONCLUSION:

We presented the first experimental evidence that low It2 and its wide die to die variations in DeNMOS can be controlled with introduction of source resistance. This new insight offers a better control of the DeNMOS which is very sensitive for ESD applications.

ACKNOWLEDGEMENT

Authors would like thank Dr Junjun Li, IBM for mentoring this work. A.C will like to thank Dr P. Hower, TI for useful technical discussion. We would also like to thank Dr. Y. Lin, formerly of TI and now with Diode Inc., for his TLP data.

[1] M. Srivastava et. al., "Highly resistive body STI: An Optimized DEMOS to achieve moving current filament for Robust ESD protection", " *in Proc. Int. Reliability Physics Symp. (IRPS)*, 2009,pp. 754

[2] G. Boselli,, V.Vassilev,, C. Duvvury, ".Drain Extended NMOS High Current Behavior and ESD Protection Strategy for HV Applications in Sub-100nm CMOS Technologies", " *in Proc. Int. Reliability Physics Symp. (IRPS)*, 2007 Page(s):342 - 347

[3] A. Chatterjee, S. Pendharkar, Y-Y Lin, C. Duvvury and K. Banerjee "An Insight into the High Current ESD Behavior of Drain Extended NMOS (DENMOS) Devices in Nanometer Scale CMOS Technologies" *in Proc. Int. Reliability Physics Symp. (IRPS)* , 2007.

[4] V. Vassilev, V. Vashchenko, Ph. Jansen, G. Groeseneken, M. Terbeek "ESD circuit model based protection network optimisation for extended-voltage NMOS drivers," *Microelectronics and Reliability*, Volume 45, Issues 9-11, 2005, Pages 1430-1435

[5] M. Mergens et al., "Analysis and Compact Modeling of Lateral DMOS Power Devices Under ESD Stress Conditions," *ESD. Symp. Proc*, 1999, pp. 1-10.

[6] Bart Keppens, Markus P.J. Mergens, Cong Son Trinh, Christian C. Russ¹, Benjamin Van Camp and Koen G. Verhaege, "**ESD** protection solutions for high voltage technologies,"*Microelectronics Reliability*2005 46(5-11) pp 677-688.

[7] James W. Miller, Michael G. Khazhinsky, James C. Weldon: Layout and bias options for maximizing V_{t1} in cascoded NMOS output buffers. *Microelectronics Reliability*,2001,41(11): 1751-1760.

[8] Markus Mergens, Wolfgang Wilkening, Stephan Mettler, Heinrich Wolf and Wolfgang Fichtner "Modular approach of a high current MOS compact model for circuit-level ESD simulation including transient gate-coupling behaviour," *Microelectronics Reliability*, 40(1), 2000 pp.99-115

[9] Kai Esmark, Harald Gossner and Wolfgang Strandler *Simulation Methods for ESD Protection Development*, Elsevier Publication, 2003

[10] A. Chatterjee, S. Pendharkar, H. Gossner, C. Duvvury and K. Banerjee "3D Device Modeling of Damage due to Filamentation under an ESD Event in Nanometer Scale Drain Extended NMOS (DE-NMOS) " *in Proc. Int. Reliability Physics Symp. (IRPS)*, 2008, pp.

[11] A. Chatterjee, S. Pendharkar, Y-Y. Lin, C. Duvvury and K. Banerjee "A Microscopic Understanding of DENMOS Device Failure Mechanism Under ESD Conditions" *Int. Electron Devices Meeting Tech. Dig(IEDM)*., 2007, pp. 677–680.

978-1-4244-5430-3/10 $26.00 © 2010 IEEE

A BENDING N-WELL BALLAST LAYOUT TO IMPROVE ESD ROBUSTNESS IN FULLY-SILICIDED CMOS TECHNOLOGY

Yong-Ru Wen[1], Ming-Dou Ker[1,2], and Wen-Yi Chen[1]

[1] Institute of Electronics, National Chiao-Tung University, Hsinchu, Taiwan
[2] Department of Electronic Engineering, I-Shou University, Kaohsiung, Taiwan

ABSTRACT

Ballast technique has been reported as a cost effective method to improve ESD robustness of fully-silicided devices without using silicide block. In this work, a new ballast technique, the bending N-Well (BNW) ballast structure, is proposed to enhance ESD robustness of fully-silicided NMOS. With a deep N-Well to cover the fully-silicided NMOS with BNW ballast structure, ESD robustness of the NMOS can be further improved by enhancing the turn-on uniformity among the multi-fingers of the NMOS.

INTRODUCTION

In order to enhance the operating speed, silicidation of CMOS devices has become a standard process step in advanced CMOS technologies. Unfortunately, silicidation also reduces ballast resistance of ESD protection devices, which causes ESD current to concentrate within the shallow silicided surface of ESD devices. This results in low ESD robustness of fully-silicided CMOS devices [1], [2]. To restore the lowered ballast resistance, silicide blocking (SB) is an essential and useful technique to improve ESD robustness of fully-silicided NMOS. As the layout top-view and device cross-sectional view of an ESD protection NMOS shown in Fig. 1, the SB can prevent silicidation from being formed on the selected area. By using the SB on ESD protection devices, the ESD robustness of CMOS ICs can be restored without affecting the operating speed of internal circuits [3], [4]. However, additional process steps and mask layer for the SB increase the cost for production. To maintain enough ballast resistance for ESD protection NMOS without using the SB, several ballasting designs have been proposed [5]-[9]. In addition to the lowered ballast resistance of fully-silicided NMOS, asymmetry of parasitic substrate resistances (Rsub) among fingers of ESD protection NMOS has been recognized as another factor to limit ESD robustness of ESD protection NMOS [10], [11]. As shown in the device cross-sectional view of an NMOS in Fig. 1, different distances from the base regions of parasitic n-p-n bipolar junction transistors (BJTs) to the grounded P+ guard ring result in different parasitic Rsub values of parasitic BJTs. The asymmetry of Rsub leads to non-uniform triggering of ESD protection NMOS during ESD stresses. Accordingly, reducing the asymmetry of Rsub among fingers of fully-silicided NMOS is another design target of this work.

BALLAST RESISTANCE TO ESD PROTECTION NMOS

Fig. 2 shows the layout top view and cross-sectional view of NMOS with the traditional N-Well ballast structure [2]. The shallow trench isolation (STI) structure in the N-Well ballast structure blocks the silicidation in the drain side. To improve the ESD robustness of fully-silicided NMOS, an N-Well covers the drain side to increase the ballast resistance and to preserve the driving capability of the NMOS. With the high sheet resistance of N-Well, the N-Well ballast structure has been reported to effectively increase ESD robustness of CMOS ICs [5]-[6], [12].

Besides the N-Well ballast structure to increase ballast resistance,

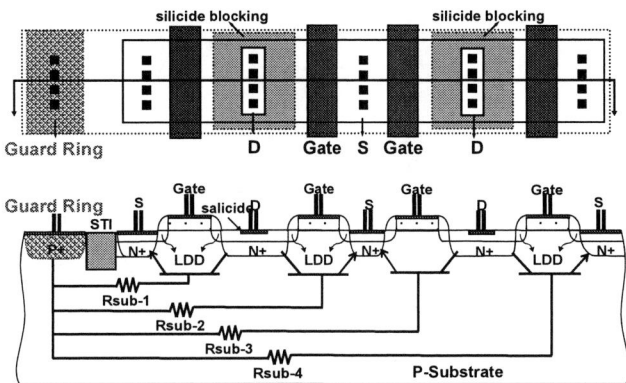

FIGURE 1. THE LAYOUT TOP-VIEW AND THE DEVICE CROSS-SECTIONAL VIEW OF NMOS WITH SILICIDE BLOCKING.

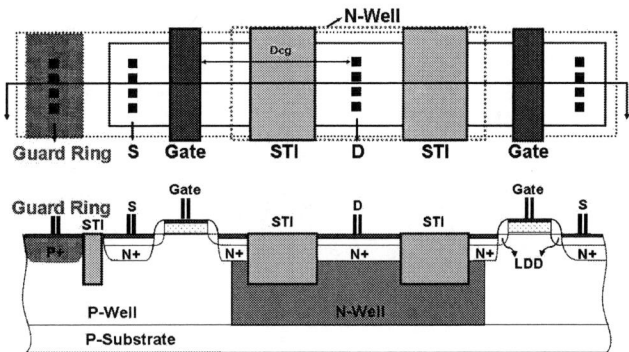

FIGURE 2. THE LAYOUT TOP-VIEW AND THE DEVICE CROSS-SECTIONAL VIEW OF FULLY-SILICIDED NMOS WITH THE N-WELL BALLAST STRUCTURE.

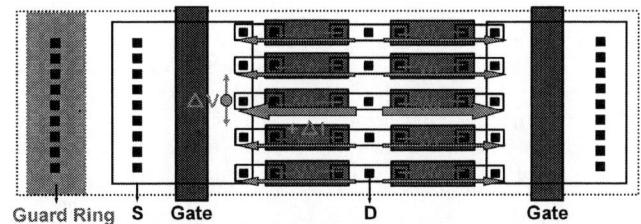

FIGURE 3. MECHANISM OF CURRENT DEFOCUS TO IMPROVE ESD CURRENT UNIFORMITY IN FULLY-SILICIDED ESD PROTECTION NMOS (REDRAWN AFTER [7]).

several techniques exploiting the current defocus have been proposed [7]-[9]. The idea of current defocus is illustrated in Fig. 3 [7]. ESD current is first divided into segments by using techniques such as the poly back-end ballast (BEB) or active-area segmentation (AAS) method [7], [8]. During ESD stresses, local current crowding (ΔI)

leads to locally increased voltage (ΔV) due to the ballast resistance along the crowded current path. This voltage difference (ΔV) forces ESD current to redistribute, which improves ESD robustness by enhancing the uniformity of current distribution.

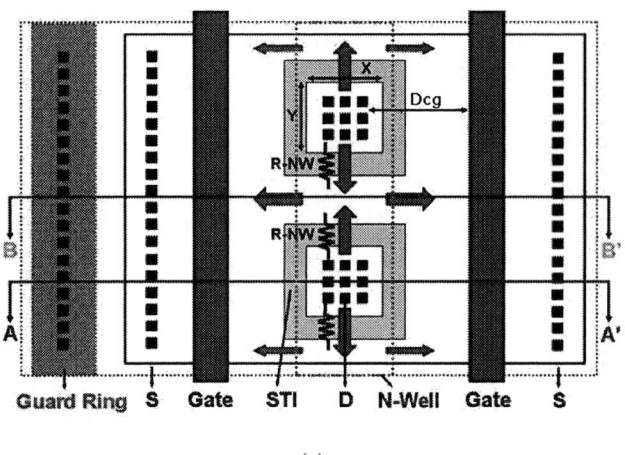

(a)

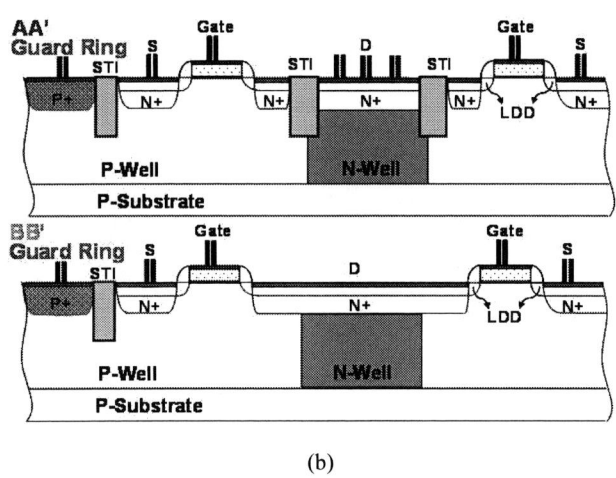

(b)

FIGURE 4. (A) THE LAYOUT TOP-VIEW AND (B) THE DEVICE CROSS-SECTIONAL VIEW OF NMOS WITH THE BENDING N-WELL BALLAST RESISTOR STRUCTURE.

NEW PROPOSED DEVICE STRUCTURES

With the purpose of increasing ballast resistance without using the SB, the bending N-Well ballast resistor (BNW) structure is proposed in this work. Figs. 4(a) and 4(b) respectively show the layout top view and the device cross-sectional view of an ESD protection NMOS with the BNW structure. Along the AA' cross-section, the drain (D) diffusion region is split by a STI ring and covered with the N-Well. N-Well is intentionally drawn in halfway of the STI region, leaving the reverse biased N-Well/P-Well junction along AA' line. Along the BB' cross-section, the N-Well connects to a diffusion region which is adjacent to the drain-side STI rings. During positive ESD stresses at the drain side with source relatively grounded, the ESD current flowline is shown as the blue arrows in Fig. 4(a). The direction of the ballast resistor (R-NW) in the BNW structure is lengthwise as that in the traditional N-Well ballast structure is widthwise. Because the lengthwise N-Well ballast design does not enlarge overall layout length or width of the device but the widthwise N-Well ballast design does, fully-silicided NMOS with

BNW design can be more compact than that with the traditional N-Well ballast structure. For example, the distance from drain contact to poly gate edge (Dcg) of a traditional N-Well ballast structure is 2.38μm, and it can be as small as 0.89μm in the BNW structure.

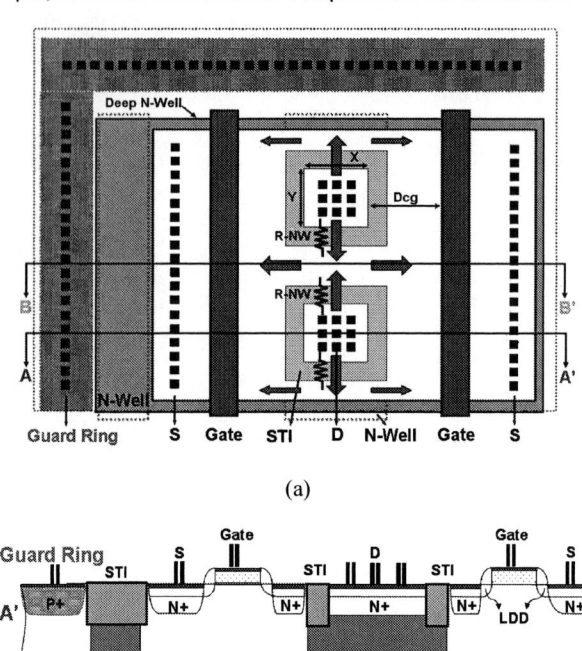

(a)

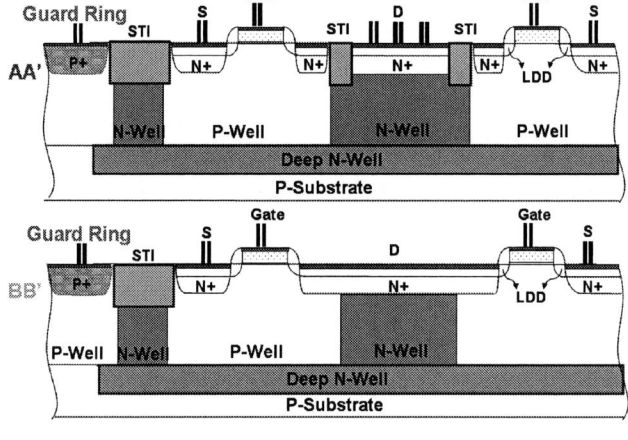

(b)

FIGURE 5. (A) THE LAYOUT TOP-VIEW AND (B) THE CROSS-SECTIONAL VIEW OF NMOS WITH THE BENDING N-WELL BALLAST RESISTOR STRUCTURE COVERED WITH DEEP N-WELL.

Besides the ballast resistance from N-Well, the small silicided N+ active area between two drain-side STI rings provides the function of current segmentation. As a result, when ESD current starts to localize, current defocus can help improve ESD current uniformity and enhace ESD robusntess of the fully-silicided NMOS with the BNW structure at the same time.

To further improve the turn-on uniformity among fingers of the ESD protection NMOS, Figs. 5(a) and 5(b) respectively show the layout top view and the device cross-sectional view of the BNW structure covered with deep N-Well. With the deep N-Well, the electrical short circuit from body regions (P-Well) to P+ guard rings at left and right sides through the P-Substrate is blocked. Base regions of parasitic n-p-n BJTs are grounded through P+ guard rings at up and down sides. This mitigates asymmetry of parasitic base resistors (Rsub) between fingers, which can enhance turn-on uniformity among multi-finger ESD protection NMOS during ESD stresses.

978-1-4244-5430-3/10 $26.00 © 2010 IEEE   858

EXPERIMENTAL RESULTS

The TLP-measured secondary breakdown currents (It2) and Human-Body-Model (HBM) ESD levels were measured to investigate the ESD robustness of fully-silicided NMOS devices. I-V curves of NMOS devices were measured using a transmission line pulse (TLP) system with a pulse width of 100ns. The ESD failure criterion of devices is that the leakage current is greater than 1mA under the drain bias of 1.0V with gate grounded (GGNMOS). Fig. 6 shows the TLP-measured I-V curves of fully-silicided NMOS with the traditional N-Well ballast structure and the BNW structure with or without deep N-Well. Devices studied in this work were fabricated in a 55-nm CMOS process.

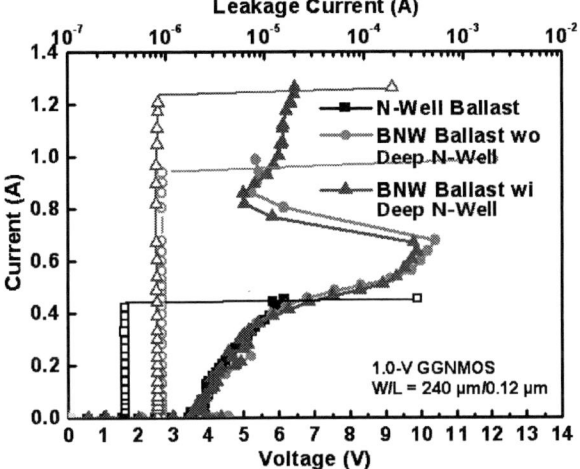

FIGURE 6. THE TLP-MEASURED I-V CURVES OF THE TRADITIONAL N-WELL BALLAST STRUCTURE AND THE BENDING N-WELL BALLAST RESISTOR STRUCTURE IN A 55-NM CMOS PROCESS. THE LEAKAGE TEST VOLTAGE FOR ALL DEVICES IS 1.0V.

From the measurement data shown in Fig. 6, It2 of fully-silicided NMOS with the traditional N-Well ballast structure is 0.44A. With the current defocus to alleviate current localization during ESD stresses, the It2 of the fully-silicided NMOS with the BNW structure was increased to 0.93A when Dcg was drawn with 1.61μm. Further enlarging the Dcg spacing to 2.33μm can improve It2 to 1.15A, which may come from the higher voltage difference (ΔV in Fig. 3) to redistribute the localized ESD current. With the deep N-Well to alleviate the asymmetry of parasitic Rsub between fingers, the measured It2 for fully-silicided NMOS with the BNW structure and

1.61-μm Dcg is 1.24A. The It2 per area of a GGNMOS with N-Well ballast, the BNW ballast without deep N-Well, and the BNW ballast with deep N-Well are 0.32, 0.52, and 0.69 mA/μm², respectively.

From the TLP measurement results, a large turn-on resistance is observed and increases with increasing TLP current level. Moreover, there is a second snapback around the current level of 0.6A. The observed second snapback is related to the second snapback inside the N-Well [5]. While the TLP pulse energy increases step by step, the N+/P-Well junction breaks down and triggers on the parasitic n-p-n BJT. At high current region, velocity saturation of electrons in the ballast N-Well results in the substantially increased turn-on resistance. As the build-up electric field is large enough, the second snapback happens. This phenomenon also happens in NMOS with the N-Well ballast structure, but the measured NMOS failed before the onset of second snapback. With the current defocus, fully-silicided NMOS with the BNW structures can sustain higher It2 values so that the second snapback phenomenon can be observed.

Scanning electron microscopy (SEM) image of the ESD-stressed fully-silicided NMOS with the BNW strucutre is shown in Fig. 7. Metamorphosed silicide was uniformly found over the drain of the device.

FIGURE 7. SEM IMAGE OF THE ESD-STRESSED FULLY-SILICIDED NMOS WITH BNW STRUCTURE.

Comparisons between the traditional slicide-blocked multi-finger GGNMOS, the fully-silicided GGNMOS with the traditional N-Well ballast structure, or with the proposed BNW structure (without deep N-Well) are summarized in Table I. Different drain-side contact

TABLE I
DIFFERENT ARRANGEMENTS OF DRAIN-SIDE CONTACT ARRAY TO IT2, HBM ESD LEVEL, AND IT2 PER AREA.

	Silicide Blocking	Contact Array	X (μm)	Y (μm)	Dcg (μm)	W/L (μm/μm)	It2 (A)	HBM (kV)	Area (μm²)	It2/Area (mA/μm²)
Traditional Multi-Finger NMOS	Yes	--	--	--	2.15	120/0.12	0.860	2.0	839	1.03
		--	--	--	2.15	240/0.12	1.677	3.0	1512	1.11
N-Well Ballast	No	--	--	--	2.38	240/0.12	0.44	0.8	1404	0.32
BNW Ballast without Deep N-Well	No	6x3	1.32	0.6	1.91	240/0.12	0.67	1.8	1540	0.43
		6x5	1.32	1.08	1.91	240/0.12	0.73	2.4	1540	0.47
		12x3	2.28	0.6	0.89	240/0.12	0.70	2.2	1540	0.46
		12x3	2.28	0.6	1.61	240/0.12	0.93	2.8	1791	0.52
		12x3	2.28	0.6	2.33	240/0.12	1.15	3.2	2027	0.57

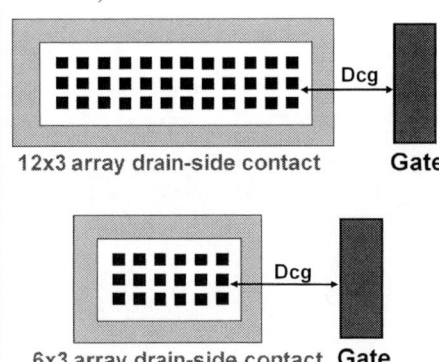

array arrangements were used to investigate the corresponding ESD robustness. Measurement results showed that ESD robustness of the GGNMOS with the BNW structure increases as the contact array expands. Measurement results also showed that the Dcg parameter is crucial to ESD robustness of fully-silicided NMOS with the BNW structure. Though fully-silicided NMOS devices with the BNW structure showed the lower It2/Area ratios compared with traditional silicide-blocked NMOS devices, fully-silicided NMOS devices with the BNW structure are still attractive because process steps and mask layers for the silicide blocking are not needed in these devices.

CONCLUSION

To improve the ESD robustness of fully-silicided NMOS, the bending N-Well ballast resistor structure has been proposed and verified in a 55-nm CMOS process. Experimental results have confirmed that fully-silicided NMOS with the new proposed BNW structure has better ESD robustness than that with the traditional N-Well ballast structure. The insertion of deep N-Well can mitigate the asymmetry of parasitic base resistance between fingers of the fully-silicided NMOS with the BNW structure, which further improves the ESD robustness of the NMOS. Without using the silicide blocking, the new proposed BNW structure in this work provides a cost effective ESD design solution to nano-scale CMOS technologies.

ACKNOWLEDGEMENT

This work was supported in part by Ministry of Economic Affairs, Taiwan, R.O.C., under Grant 98-EC-17-A-01-S1-104, and in part by the "Aim for the Top University Plan" of National Chiao-Tung University and Ministry of Education, Taiwan, R.O.C. The authors would like to thank United Microelectronics Corporation for the chip fabrication.

REFERENCES

[1] G. Notermans, A. Heringa, M. Dort, S. Jansen, and F. Kuper, "The effect of silicide on ESD performance," in Proc. *IEEE Int. Reliab. Phys. Symp.*, 1999, pp. 154–158.

[2] A. Amerasekera and C. Duvvury, *ESD in Silicon Integrated Circuits.*, 2nd ed. New York: Wiley, 2002.

[3] A. Amerasekera and C. Duvvury, "The impact of technology scaling on ESD robustness and protection circuit," *IEEE Trans. Compon., Packag. Manuf. Technol. A*, vol. 18, no. 2, pp. 314–320, Jun. 1995.

[4] S. G. Beebe, "Methodology for layout design and optimization of ESD protection transistors," in *Proc. EOS/ESD Symp.*, 1996, pp. 265–275.

[5] G. Notermans, "On the use of N-well resistors for uniform triggering of ESD protection elements," in *Proc. EOS/ESD Symp.*, 1997, pp. 221–229.

[6] H.-C. Hsu and M.-D. Ker, "Dummy-gate structure to improve ESD robustness in a fully-salicided 130-nm CMOS technology without using extra salicide-blocking mask," in *Proc. IEEE Int. Symp. Quality Electronic Design*, 2006, pp. 503–506.

[7] B. Keppens, M. Mergens, J. Armer, P. Jozwiak, G. Taylor, R. Mohn, C. Trinh, C. Russ, K. Verhaege, and F. Ranter, "Active-area-segmentation (AAS) technique for compact, ESD robust, fully silicided NMOS design," in *Proc. EOS/ESD Symp.*, 2003, pp. 250–258.

[8] K. Verhaege and C. Russ, "Wafer cost reduction through design of high performance fully silicided ESD devices," in *Proc. EOS/ESD Symp.*, 2000, pp. 18–28.

[9] E.Worley, "New ballasting method for MOS output drivers and power bus clamps," in *Proc. IEEE Int. Reliab. Phys. Symp.*, 2005, pp. 458–461.

[10] M. Mergens, K. Verhaege, C. Russ, J. Armer, P. Jozwiak, G. Kolluri, and L. Avery, "Multi-finger turn-on circuits and design techniques for enhanced ESD performance and width-scaling," in *Proc. EOS/ESD Symp.*, 2001,pp. 1–11.

[11] T.-Y. Chen and M.-D. Ker, "Investigation of the gate-driven effect and substrate-triggered effect on ESD robustness of CMOS devices," *IEEE Trans. Device Mater. Rel.*, vol. 1, no. 4, pp. 190–203, Dec. 2001.

[12] M.-D. Ker, W.-Y. Chen, W.-T. Shieh, and I.-J. Wei, "New ballasting layout schemes to improve ESD robustness of I/O buffers in fully silicided CMOS process," *IEEE Trans. Electron Devices*, vol. 56, no. 12, pp. 3149–3159, Dec. 2009.

Isolating Marginally Defective Gate Using Photoperturbation Induced via a C-AFM Laser Beam

Hung Sung Lin, Mong Sheng Wu

United Microelectronics Corporation, Ltd.
No. 3, Li-Hsin Rd. II, Hsinchu Science Park, Taiwan 300, R.O.C.
Tel: 886-3-5782258 ext. 33231; Fax: 886-3-563-6722; Email: giant_lin@umc.com

Abstract—The photoperturbation effects induced via a Conductive Atomic Force Microscope (C-AFM) laser beam, in which the surface photovoltaic effect and the carrier injection effect included in the photoperturbation can cause significant deterioration in the characterization accuracy have been widely investigated [1-5]. This study, however, successfully demonstrates how to take advantage of these non-negligible photoelectric effects to isolate marginally defective gates, which are usually difficult to uncover using the traditional approach of passive voltage contrast (PVC) carried out using a scanning electron microscope (SEM) or focused ion beam (FIB), or even when using an advanced SEM-based nanoprobing technique. Using this technique, such failures, which pose potential reliability issues on devices as the affected circuit degrades over time or under stress, can be easily screened before any quality assurance test.

Keywords-C-AFM, AFM, photoperturbation, photovoltaic

I. BACKGROUND

The continuous scaling of feature dimensions makes degradation mechanisms more pronounced. These mechanisms exhibit a gradual degradation of electrical parameters instead of an abrupt breakdown phenomenon, and have become increasingly relevant when considering the different causes of process yield loss and product reliability failures. In particular, marginally defective gates have become one of the most important aspects of these degradation mechanisms. Unlike defective gates caused by open-failure or short-failure, marginally defective gates cannot be easily addressed or screened out via conventional failure analysis techniques such as passive voltage contrast (PVC) under a scanning electron microscope (SEM) or a focused ion beam (FIB), or even with an advanced SEM-based nanoprobing technique. In order to overcome this issue, a novel approach using Conductive Atomic Force Microscope (C-AFM) to characterize and isolate marginally defective gates by taking advantage of the photoperturbation effects induced by laser beam is proposed in this paper.

When a sample is exposed to light, directly absorbed photons with an energy of $h\nu > E_g$, the bandgap of silicon, excite electron-hole pairs and these excess carriers move either toward or away from the silicon surface depending on the polarity of the DC gate bias. Some of the minority carriers diffuse toward the illuminated surface, establishing a surface potential or surface photovoltage V_{SPV}, which is proportional to the excess minority-carrier concentration Δn at the silicon surface following the relationship described in equation (1) [6].

$$V_{SPV} = C_1 \frac{(1-R)\Phi L}{(s_1 + D/L)(L + 1/\alpha)} \quad (1)$$

Where R is the reflection, Φ is the photon flux density, L is the minority carrier diffusion length, D is the minority carrier diffusion coefficient, s_1 is the surface recombination velocity at the front surface, C_1 is a constant of proportionality, and α is the absorption coefficient.

When a constant voltage is applied to the gate, this voltage is divided between the oxide and the semiconductor, the voltage across the oxide and the voltage across the silicon depletion region. The photovoltaic effect induced by the photoperturbation can lower the potential drop across the silicon depletion region. As the voltage applied on the gate is constant, the voltage across the oxide will be higher than that which is free from the photoperturbation. As a result, when the sample is exposed to light, a greater gate leakage current signal can be detected. Fig. 1 shows the IV curve for the poly gate in a normal SRAM cell extracted using both an SEM based nanoprober and the C-AFM. Compared with the result for -1pA@-1V obtained using the nanoprober, a higher gate leakage current signal of about -1.6nA@-1V can be detected using C-AFM. This study will demonstrate how to take advantage of these non-negligible photoelectric effects to isolate marginally defective gates.

Two case studies are described in this paper. Case study I demonstrates how an example of a resistive gate caused by partial M1 Cu loss was isolated. Case study II presents an example of how a marginally defective gate caused by a tiny amount of silicon residue buried under the spacer was located. PVC carried out using SEM and FIB, and nanoprobing analyses did not reveal any abnormalities in either of these two cases. As a result, C-AFM analysis using a CP-II system was introduced to allow internal probing to be performed in order to understand the electrical behavior and to accurately pinpoint the location of the defect. The DC voltage was supplied from

978-1-4244-5430-3/10 $26.00 © 2010 IEEE

the bulk, which formed a closed path in the circuit with a grounded probe tip, as shown in Fig. 2. The laser diode was powered by a low voltage supply with a maximum output of 0.2mW in a wavelength range 600 to 700nm.Cross-sectional transmission electron microscope (TEM) samples for nanoscopic defect identification were made via FIB using the TEM lift-off method.

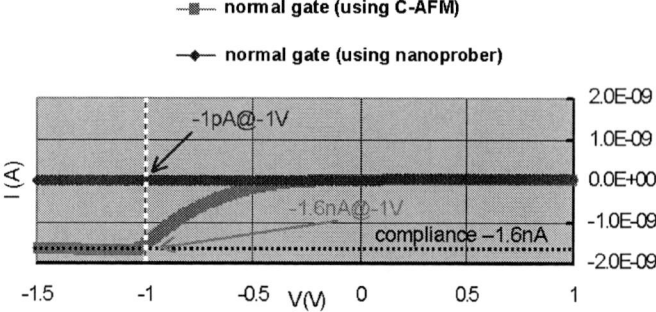

Figure 1. IV curves measured on the Mvia2 using both an SEM based nanoprober and a C-AFM.

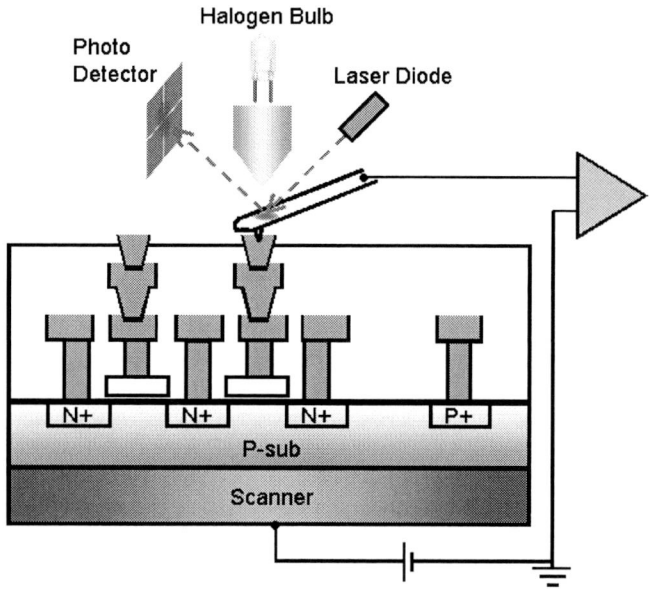

Figure 2. Schematic diagram of a C-AFM apparatus.

II. RESULTS AND DISCUSSION

A. Case Study I

In this study, the failure mechanism that degrades SRAM access time in high performance CMOS devices was analyzed. Fig. 3 shows that PVC carried out using SEM and FIB on the Mvia2 did not reveal any abnormalities in terms of contrast. In order to understand the electrical behavior and accurately pinpoint the location of the defect, electrical characterization using C-AFM was performed on the Mvia2 of a marginal SRAM cell, as shown in Fig. 2. A negative DC bias was

applied to the sample stage during scanning, and the topography and current mapping were recorded simultaneously, as shown in Fig. 4. Current mapping uses different colors to indicate the different current levels in the Mvia2, the darker the Mvia2, the higher the current. NG denotes a normal gate, and DG denotes a defective gate. Fig. 5 shows the IV curves for both the DG and the NG extracted using the C-AFM, and the IV curve for the NG extracted using the nanoprober that was used as a reference. The NG shows a lower leakage current of about −1pA@-1V when extracted using the nanoprober than that of about −1.6nA@-1V when extracted using the C-AFM. It can be observed that when the sample is exposed to the light, a greater gate leakage current signal can be detected, which is induced by the photoperturbation. Under the C-AFM, the NG shows a higher leakage current of about -1.6nA@-1V, which was the maximum pre-determined compliance value, compared to the DG, which was of about -0.1nA@-1V. This indicates that the DG has a higher resistance than the NG. As the failure location was accurately pinpointed after electrical isolation, the sample was prepared for TEM analysis to highlight the physical difference between the DG and the NG. The arrow shown in Fig. 6 indicates the defect, identified as partial M1 Cu loss causing the resistive gate. In this case study, the application of C-AFM was demonstrated in the electrical charaterization and fault isolation of a resistive gate by taking advantage of the photoperturbation effects induced by the laser beam.

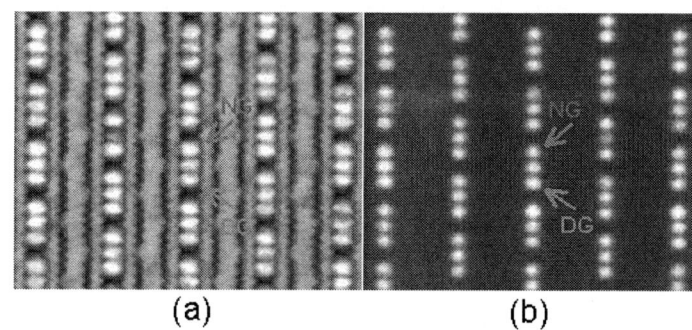

Figure 3. PVC carried out using SEM (a) and FIB (b) did not reveal any abnormalities in terms of contrast.

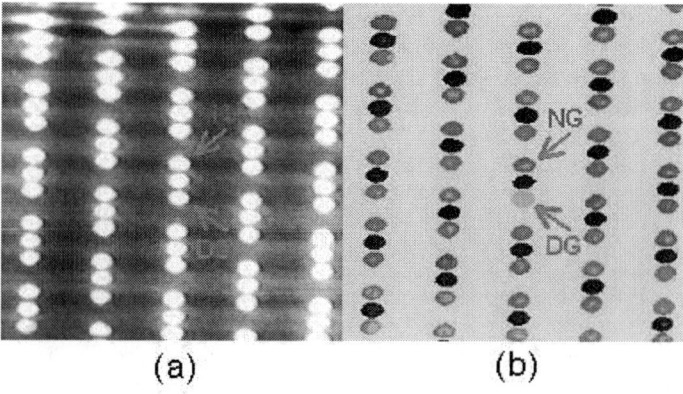

Figure 4. Corresponding topography image (a) and current mapping (b) of C-AFM on the Mvia2.

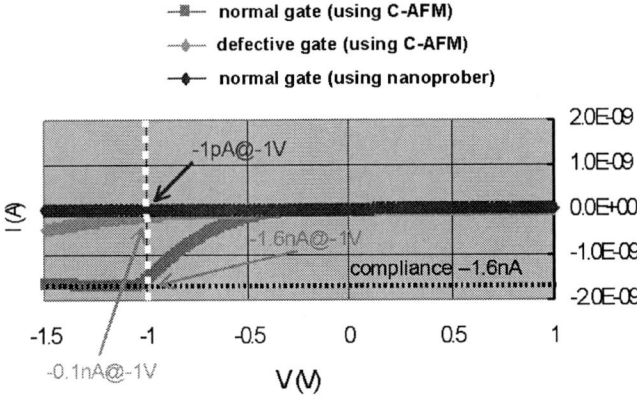

Figure 5. IV curves measured on the Mvia2 using both an SEM based nanoprober and a C-AFM.

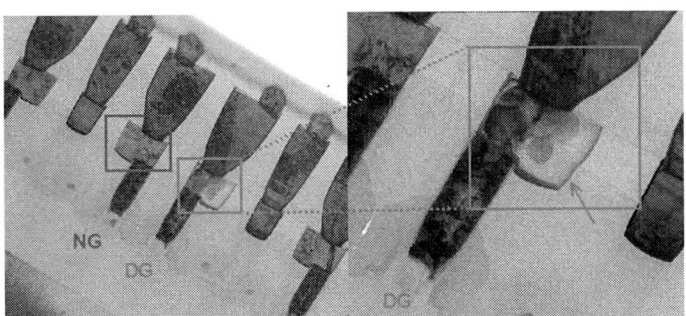

Figure 6. Cross-sectional TEM images.

B. Case Study II

In this study, the failure mechanism that causes a higher sleeping mode current in CMOS HV technology was analyzed. Electrical characterization using C-AFM was performed on the contact layer, as shown in Fig. 7. The current mapping, as shown in Fig. 8, indicates the defective poly gate in which a leakage current occurs when a negative DC bias is applied to the sample stage during scanning. Fig. 9 shows the IV curve for the defective poly gate extracted using both the C-AFM and the nanoprober that was used as a reference. The defective poly gate shows a higher leakage current of about –3.6nA@-0.7V when extracted using the C-AFM than that of about –5pA@-1V when extracted using the nanoprober. The electrical results indicate that the defective poly gate has a capacitance leakage. As the failure location was accurately pinpointed after electrical isolation, the sample was prepared for TEM analysis to understand and identify the physical root cause. The arrow shown in Fig. 10 indicates the defect buried under the spacer. To determine the composition of this buried defect, electron energy loss spectroscopy (EELS) mapping was performed. As shown in Fig. 11, the silicon-rich defect was identified as being buried under the spacer. This is another example using C-AFM to characterize and isolate a marginally defective gate by taking advantage of the photoperturbation effects induced by the laser beam.

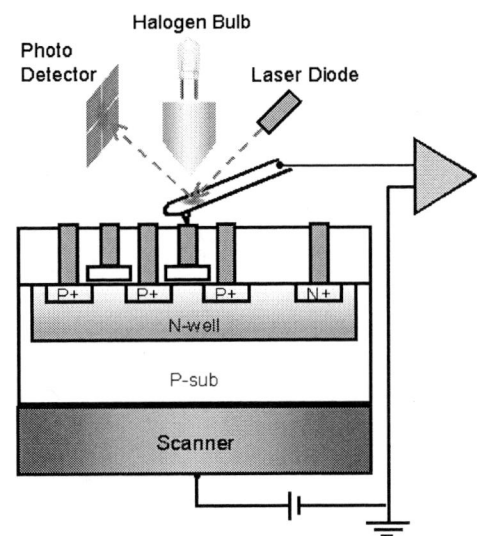

Figure 7. Schematic diagram of a C-AFM apparatus.

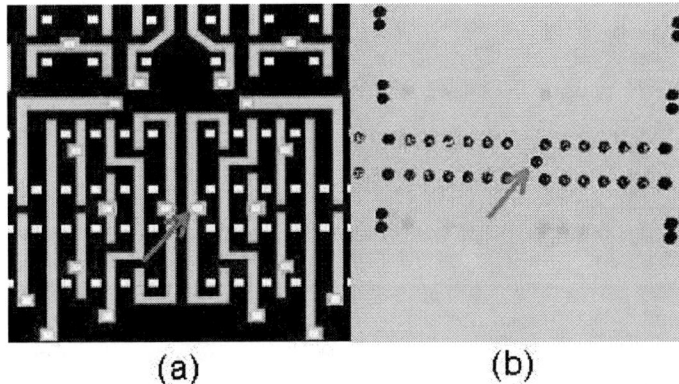

Figure 8. Corresponding layout image (a), and current mapping of C-AFM on the contact level (b).

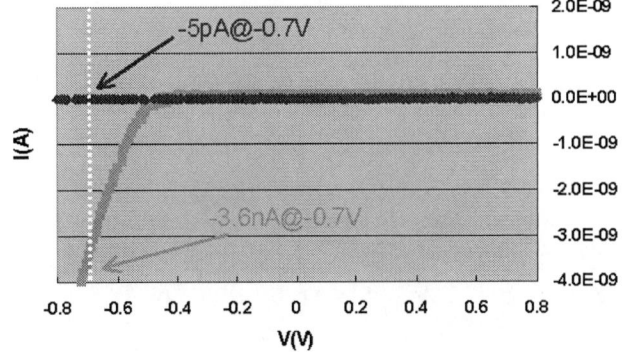

Figure 9. IV curves measured on the contact level using both an SEM based nanoprober and a C-AFM.

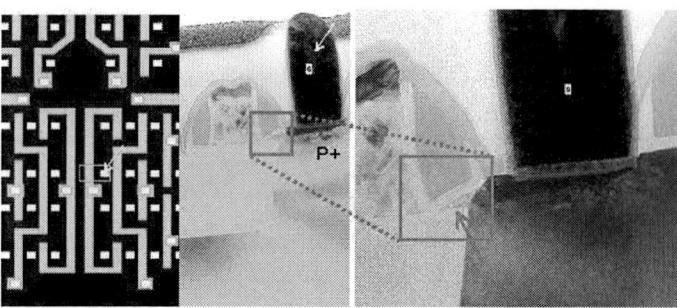

Figure 10. Cross-sectional TEM images.

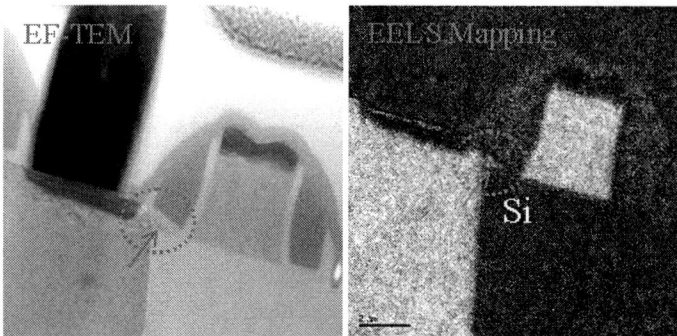

Figure 11. EELS mapping.

III. CONCLUSION

In this study, the use of the C-AFM technique by taking advantage of the photoperturbation effects induced by laser beam was successfully demonstrated when analyzing the characterization of marginally defective gates due to nanoscale defects, such as partial Cu loss and tiny silicon residue buried under spacer which are difficult to uncover using the traditional PVC approach, or even when using an advanced SEM-based nanoprobing technique. After identifying the physical root cause of the electrical failure, corrective actions were then taken to quickly improve yield. Additional applications using this method to perform failure analysis of soft failures are currently under investigation.

ACKNOWLEDGMENT

The author would like to acknowledge the contribution of the SRAM&Logic Product Engineering group in making device level IV measurements. We would also like to thank the TEM group for obtaining cross-sectional TEM images.

REFERENCES

[1] S. Shin, J.-I. Kye, U. H. Pi, and Z. G. Khim, "Effect of photoenhanced minority carriers in metal-oxide-semiconductor capacitor studied by scanning capacitance microscopy," J. Vac. Sci. Technol. B, Vol. 18, No. 6, pp. 2664-2668, 2000.

[2] M. N. Chang, C. Y. Chen, F. M. Pan, J. H. Lai, W. W. Wan, and J. H. Liang, "Photovoltaic effect on differential capacitance profiles of lowuenergy-BF2$^+$-implanted silicon wafers," Appl. Phys. Lett., Vol. 82, No. 22, pp. 3955-3957, 2003.

[3] M. N. Chang, C. Y. Chen, W. W. Wan and J. H. Liang, "Influence of Photoperturbation on the Characterization Accuracy of Scanning Capacitance Microscopy," 11th IPFA Proceedings, Jul. 2004, pp. 295-298.

[4] Hung Sung Lin, Mong Sheng Wu, " A Study of Bipolar Phototransistor Action Existing in CMOS Process Triggered by a Laser Beam Used in a C-AFM System," 47th IRPS Proceedings, Apr. 2009, pp. 801-803.

[5] Hung Sung Lin, Mong Sheng Wu, "A Study of the Photoelectric Effect Caused by a Laser Beam Used in a Beam Bounce Technique in a C-AFM System," 34th ISTFA Proceedings, Nov. 2008, pp.256-259.

[6] Dieter K. Schroder, Semiconductor Material and Device Characterization, John Wiley & Sons, New York, 1990.

A Novel Sample Preparation Technique for Visualizing Invisible Defects Embedded in Poly Gate

Hung Sung Lin

United Microelectronics Corporation, Ltd.

No. 3, Li-Hsin Rd. II, Hsinchu Science Park, Taiwan 300, R.O.C.

Tel: 886-3-5782258 ext. 33231; Fax: 886-3-563-6722; Email: giant_lin@umc.com

Abstract—**The use of nanoprobing techniques to accomplish transistor parametric data extraction has been widely reported as a method of failure analysis in nanometer scale science and technology. Certain failure mechanisms causing parametric transistor fails are, however, not always successfully identified, even using advanced imaging tools, such as transmission electron microscopy (TEM). Therefore, additional techniques are needed that can reveal the physical root cause of the electrical defects. In this paper, an approach using dopant selective etching of samples for TEM is adopted to enable visualization of invisible defects embedded in the poly gate.**

Keywords-nanoprobing, TEM, invisible, selective, etching

I. BACKGROUND

Devices and structures whose dimensions are below 100nm are defined as being in the nanoscale realm. Nanoscale fault isolation and failure analysis is extremely difficult. The ability to isolate physical defects in sub-100nm technologies has become more problematic where no better alternative characterization technique exists. Direct measurement utilizing nanoprobing techniques is commonly used to collect the family of IV curves for the transistors. This availability of accurate electrical measurements at the device level can accurately reveal parametric transistor failures. The more electrical data that can be collected, the better the hypothesis that can help identify the physical root cause will be. An educated guess at the failure scenario without better alternative physical characterization techniques very seldom resolves these failures [1-2]. In this paper, two case studies will be presented in which invisible defects embedded in the poly gate were visualized using a dopant selective etching technique. With this capability, these soft mechanisms and invisible defects can be visualized. This is helpful for understanding the root cause, not merely what can be hypothesized based on the electrical data.

II. RESULTS AND DISCUSSION

A. Case Study I

In this study, the failure mechanism that causes an increase in the threshold voltage and asymmetrical behavior of a failed transistor was analyzed. Fig. 1 shows a typical example of the failure bitmapping used in this study, and indicates that the major failure mode is a random single bit (RSB) in the SRAM.

The failed bits are distributed randomly in both the die and the wafer. Fig. 2 shows the transfer characteristics of a failed pass gate (PG), compared to a normal PG. The on-state current of the failed PG =1.6E-05A@1.2V is about one order of magnitude smaller than that of the normal PG =1.1E-04A@1.2V. The threshold voltage of the failed PG is about 0.25V higher than that of the normal PG. Fig. 3 shows the transfer characteristics of the failed PG, compared with itself, using measurements obtained when an interchange of the sweep direction was performed. The on-state current of the failed PG with a normal S/D =1.6E-05A@1.2V is lower than that when the reversed S/D=1.9E-05A@1.2V, but the off-state current of the failed PG with a normal S/D =5.9E-11A@0V is higher than that when the reversed S/D=2.1E-11A@0V. As there is a barrier at the source-channel interface and another barrier at the drain-channel interface, and these two barriers were treated the same in a symmetrical device, the off-state current of the failed PG should almost invariably be symmetrical. The concentration of electrons in the channel at the surface can be expressed using equation (1). The electron concentration or the conductance of the channel varies exponentially with the barrier height E_B.

$$n_s = N_C e^{-E_B/kT} \qquad (1)$$

Where n_s is the electron concentration at the surface, N_C is the effective density of states function in the conduction band, E_B is the energy barrier height, k is Boltzmann's constant, and T is the temperature.

A proposed hypothesis is that the barrier height E_B of the failed PG is asymmetrical, or the barrier height at the source-channel interface is lower than that at the drain-channel interface. The reduced E_B at the source-channel interface will cause a higher concentration of electrons in the channel at the surface, and then produce a higher off-state current. As the barrier height E_B is mainly bound to the built-in potential of the junction when the PG is operated in the off-state, the doping concentration of the lightly doped drain (LDD) at the source-channel interface was suspected to be lower than that at the drain-channel interface. If this was the case, this shadowing LDD would form a resistance on the source side, and further

978-1-4244-5430-3/10 $26.00 © 2010 IEEE

produce a reverse-bias voltage between the source and the body when the failed PG was turned on using a normal S/D, which, in turn, results in the reduction of the on-state current, and can be used to account for the behavior where the on-state current of the normal S/D is lower than that when the S/D is reversed. As there is only a small degree of asymmetry, the doping concentration of the LDD should be deviated slightly. Although these characteristics suggest that slight LDD shadowing at the source site is one possible explanation for the asymmetrical behavior, it is still hard to explain why the threshold voltage of the failed PG is about 0.25V higher than that of the normal PG. In order to obtain physical evidence, an approach using wet chemical etching of Transmission Electron Microscope (TEM) specimens was implemented. The TEM lamella was wet chemically treated to reveal the various dopant regions. Fig. 4 shows the etched TEM lamella for both the failed PG and the normal PG. It can be observed that etching was homogenous in the complete highly doped area, as indicated by the bright contrast. In comparison to the normal PG, the poly pimple caused by the embedded defect, indicated by the arrow marker denoted label 1 at the bottom of the poly gate, can be observed in the failed PG. Using this sample preparation technique, not only can the embedded defect be seen, but the blocked LDD implant shadowed by this embedded defect, highlighted by the arrow marker denoted label 2, can also be clearly observed in the area on the source side close to the gate of the failed PG. The physical evidence of the LDD shadowing at the source side is consistent with the previous inferences that were drawn from the IV characteristics indicated in Fig. 3. This remaining defect after the wet etching embedded in the poly gate should be able to be used to explain why the threshold voltage of the failed PG is about 0.25V higher than that of the normal PG, as indicated in Fig. 2. In order to identify which elements present in the defect, Electron Energy Loss Spectrum (EELS) was employed to enable the identification of the material. Fig. 5 shows the Energy Filter EF-TEM image, and the results of the EELS analysis of the defect. Aluminum, oxygen, and silicon signals were detected. This indicates that this material is composed of Al, O and Si, which means the defect can be treated as a film with a high dielectric constant placed on the oxide layer. The results explain why the threshold voltage of the failed PG is about 0.25V higher than that of the normal PG as this higher-κ defect increases the effective oxide thickness (EOT) of the failed PG. As the atomic number of Al(13) is close to Si(14), the Z contrast of the TEM image can not easily distinguish Al from Si [2]. In this study, an approach using dopant selective etching of samples for TEM was successfully demonstrated to enable visualization of the higher-κ defect embedded in the poly gate.

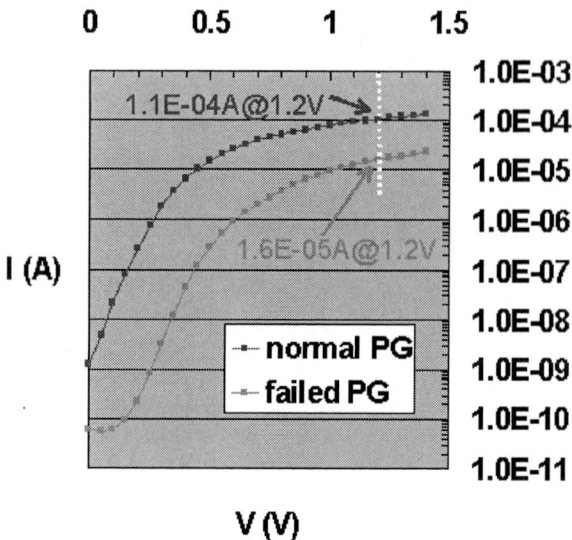

Figure 2. Transfer characteristics (Id-Vg) of the pass gate NMOS in a failed bit and a normal bit.

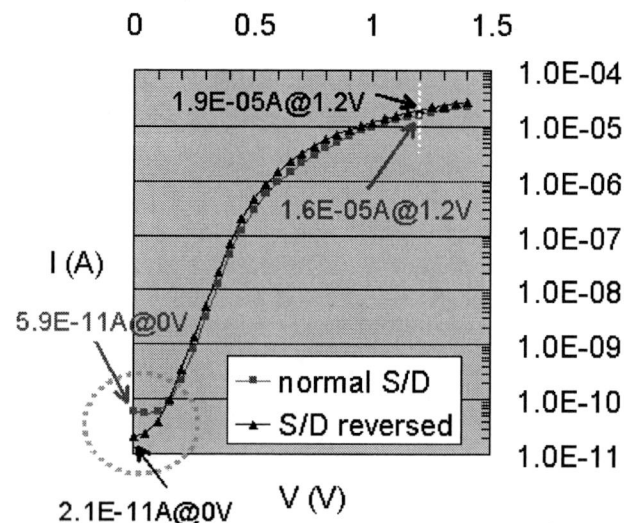

Figure 3. Transfer characteristics (Id-Vg) of the pass gate NMOS in a failed bit.

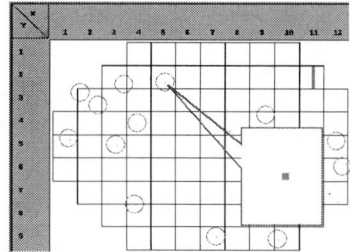

Figure 1. Illustration of bitmapping analysis.

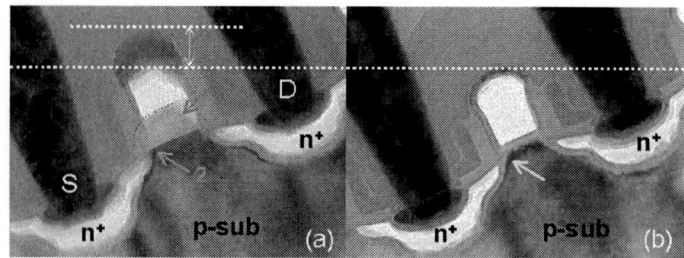

Figure 4. TEM image of wet chemically treated cross-sectional lamella of the failed pass gate NMOS (a), and normal pass gate NMOS (b).

978-1-4244-5430-3/10 $26.00 © 2010 IEEE

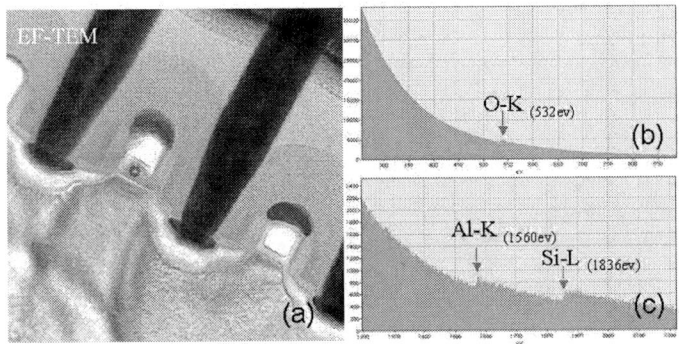

Figure 5. Energy Filter TEM image (a), and EELS spectrum (b)(c).

B. Case Study II

In this study, another failure mechanism that also causes an increase in the threshold voltage and asymmetrical behavior of a failed transistor was analyzed. Fig. 6 shows a typical example of the failure bitmapping used in this study. It shows that the major failure mode is the random single bit in the SRAM (RSB). A large number of failed bits are distributed randomly in both the die and the wafer. Fig. 7 shows the transfer characteristics of a failed pass gate (PG), compared to a normal PG. The on-state current of the failed PG =6.3E-06A@1.2V is about one order of magnitude smaller than that of the normal PG =3.9E-05A@1.2V. The threshold voltage of the failed PG is about 0.28V higher than that of the normal PG. Fig. 8 shows the transfer characteristics of the failed PG, compared with itself, however, using measurements obtained when an interchange of the sweep direction was performed. The on-state current of the failed PG with a normal S/D =6.3E-06A@1.2V is lower than that when the reversed S/D=8.9E-06A@1.2V, but the off-state current of the failed PG is, however, almost the same as that when the S/D is reversed. It is worth noticing that the off-state IV characteristics are different from those observed in case study I. Based on equation (1), a proposed hypothesis is that the barrier height E_B of the failed PG is asymmetrical, or the barrier height at the source-channel interface is higher than that at the drain-channel interface when a positive voltage is applied to the gate of the failed PG with the source and drain connected to 0V. The barrier height E_B of the failed PG becomes symmetrical when 0V is applied to the gate of the failed PG. This hypothesis excludes the possibility that the asymmetrical behavior is caused by LDD shadowing, as illustrated in case study I. As the gate-source voltage Vgs can affect the barrier height E_B, a higher Vgs will cause a lower energy barrier between the source and the channel at the Si surface, and then more electrons are able to enter the channel. So the effective gate voltage used to lower the channel potential close to the source-channel interface was suspected to be lower than that close to the drain-channel interface when a positive voltage is applied to the gate of the failed PG with the source and drain connected to 0V. Poly depletion can form an electrically thicker gate oxide and cause a reduction in the on-state current, and poly depletion effect associated with the grain size distribution in the poly-Si gate will create both an additional and a non-uniform potential drop in the poly layer, and then reduce the effective gate voltage used to lower the channel potential according to the relationship shown in equation (2).

$$Q_{inv} = \frac{\varepsilon_{ox}}{t_{ox} + \dfrac{W_{dpoly}}{3}} \left(V_g - V_{tn}\right) \tag{2}$$

Where Q_{inv} is inversion channel charge density per unit area, ε_{ox} is permittivity of an oxide, t_{ox} is gate oxide thickness, W_{dpoly} is poly depletion width, and V_g is gate voltage, and V_{tn} is threshold voltage of NMOS.

If this was the case, gate depletion caused by a large poly-Si grain close to the source-channel interface can be used to account for the behavior where the on-state current with a normal S/D is lower than that when the S/D is reversed, while the off-state current with a normal S/D is almost the same as that when the S/D is reversed. In order to obtain physical evidence, the approach using wet chemical etching of TEM specimens was implemented. The TEM lamella was wet chemically treated to reveal the various dopant regions. Fig. 9 shows the etched TEM lamella of the failed PG. It can be observed that etching was homogenous in the complete highly doped area, as indicated by the bright contrast. A large low-doped poly-Si grain, indicated by the arrow marker, close to the source-channel interface can be observed in the failed PG. The physical evidence of the large low-doped poly-Si grain on the source side is consistent with the previous inferences that were drawn from the IV characteristics, and this remaining defect after wet etching embedded in the poly gate should be able to be used to explain why the threshold voltage of the failed PG is about 0.28V higher than that of the normal PG, as indicated in Fig. 7. In order to observe the grain structure, both X-S and P-V Dark Field (DF) TEM techniques were adopted to enable identification of the grain boundaries on other samples using the distinguishing feature of different diffraction contrasts of respective grains in different orientations in DF. Fig. 10 shows that a single poly-Si grain, indicated by the bright contrast, blocks the channel using an X-S DF TEM technique, and Fig. 11 shows that a large poly-Si grain, indicated by the bright contrast, blocks the channel using a P-V DF TEM technique. In the large grain, dopant diffusion is suppressed and the gate is depleted. The results of the failure analyses suggest that minimizing the polysilicon grain size is effective in suppressing this failure mode. To make the grains smaller, a furnace-based thermal treatment process instead of Rapid Thermal Oxidation (RTO) after amorphous Si deposition was applied. Fig. 12 shows the P-V TEM images for the gate polysilicon grain using the furnace treatment, and using RTO after amorphous Si deposition. It can be observed that an amorphous Si using a furnace treatment has a smaller and more uniform grain than that using RTO over the whole area. In this study, the approaches using dopant selective etching of samples for TEM and the DF TEM technique were successfully demonstrated to enable visualization of the large and non-uniform poly grain defects embedded in poly gate.

978-1-4244-5430-3/10 $26.00 © 2010 IEEE

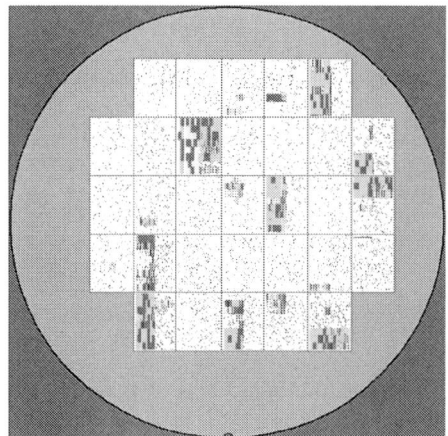

Figure 6. Illustration of bitmapping analysis.

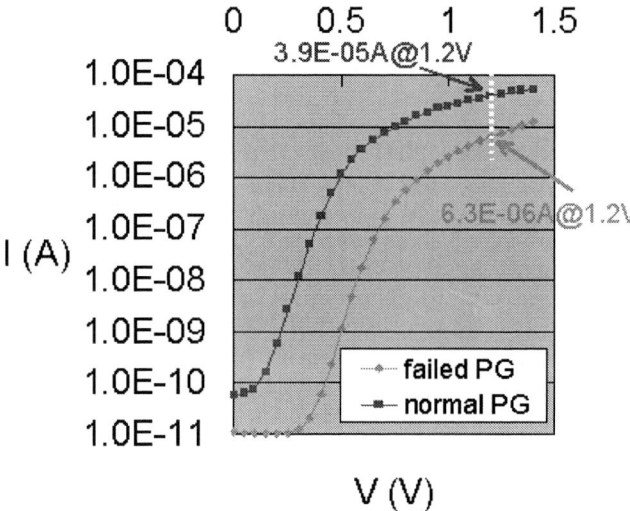

Figure 7. Transfer characteristics (Id-Vg) of the pass gate NMOS in a failed bit and a normal bit.

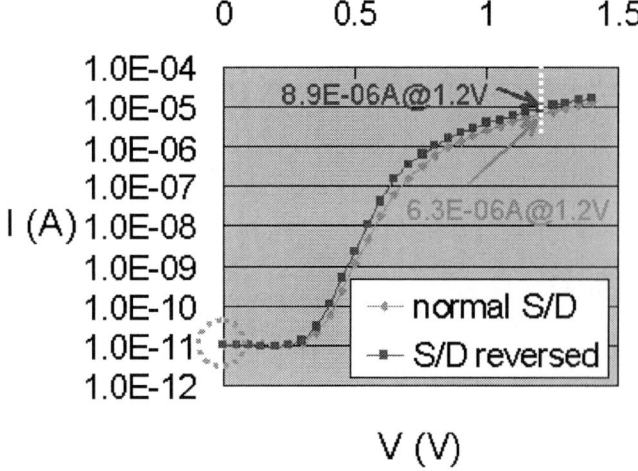

Figure 8. Transfer characteristics (Id-Vg) of the pass gate NMOS in a failed bit.

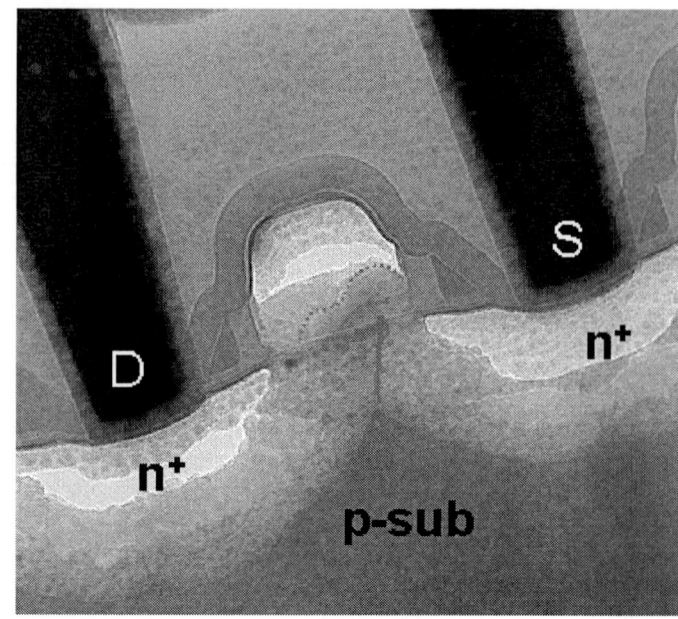

Figure 9. TEM image of wet chemically treated cross-sectional lamella of a failed pass gate NMOS.

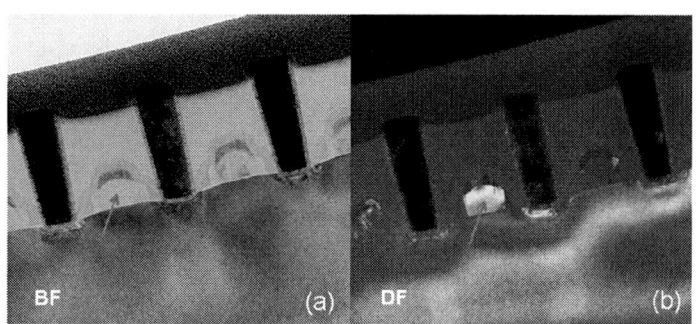

Figure 10. X-S BF TEM image (a), and X-S DF TEM image (b).

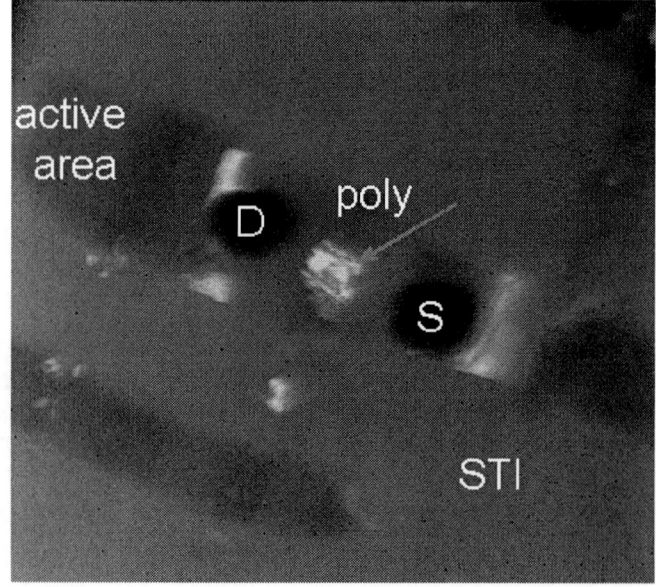

Figure 11. P-V DF TEM image.

978-1-4244-5430-3/10 $26.00 © 2010 IEEE

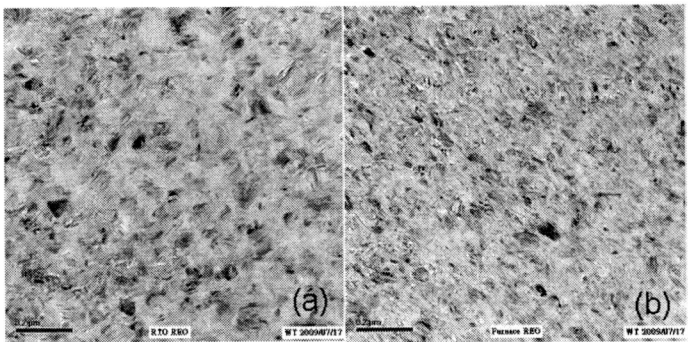

Figure 12. P-V TEM images of polysilicon using RTO (a), and using a furnace treatment (b).

III. CONCLUSION

In this paper, the use of wet chemical etching of samples for TEM was successfully demonstrated in the study of the characterization of the invisible defects embedded in the poly gate. This was demonstration for the higher-κ defect as illustrated in case study I, and a large low-doped poly-Si grain as illustrated in case study II. After identifying the physical root cause of the electrical failure, corrective actions were taken to quickly improve yield. Additional applications using this method to perform failure analysis of soft failures are currently under investigation.

ACKNOWLEDGMENT

The author would like to acknowledge the contribution of the SRAM&Logic Product Engineering group in making device level IV measurements. We would also like to thank the TEM group for obtaining TEM images of several decorated cross-sectional lamellas.

REFERENCES

[1] Larry Liu, Yuguo Wang, Hal Edwards, David Sekel, Dan Corum, "Combination of SCM/SSRM Analysis and Nanoprobing Technique for Soft Single Bit Failure Analysis," 30th ISTFA Proceedings, Nov. 2004, pp.38-41.

[2] Tom X Tong, A N Erickson, "Current Image Atomic Force Microscopy (CI-AFM) Combined with Atomic Force Probing (AFP) for Location and Characterization of Advanced Technology Node," 30th ISTFA Proceedings, Nov. 2004, pp.42-46.

[3] Kun Lin, Chia Hung Chao, Tsui Hua Huang, Hsiu Mei Fan, " Electron Energy Loss Spectrum Application for Failure Mechanism Investigation in Semiconductor Analysis," 46th IRPS Proceedings, Apr. 2008, pp. 589-592.

A Case Study of High Temperature Pass Analysis Using Thermal Laser Stimulation Technique

Hung Sung Lin, Mong Sheng Wu

United Microelectronics Corporation, Ltd.
No. 3, Li-Hsin Rd. II, Hsinchu Science Park, Taiwan 300, R.O.C.
Tel: 886-3-5782258 ext. 33231; Fax: 886-3-563-6722; Email: giant_lin@umc.com

Abstract—The intrinsic carrier concentration n_i is highly temperature dependent in non-degenerate semiconductors. At high temperatures, thermally generated electron-hole pairs contribute to the carrier concentrations. As a result, the generation of free carriers can be achieved through light-to-heat conversion using near-infrared (NIR) laser. This paper describes the use of the thermal laser stimulation (TLS) technique to modify the electrical properties of a heated passive device so that the direct failure location can be revealed in order to clarify the cause of the iddq leakage that will disappear during high temperature tests.

Keywords-temperature, TLS, NIR, thermal, laser

I. BACKGROUND

Iddq leakage is a common failure mode in integrated circuits [1]. Temperature dependent Iddq failure is, however, definitely one of the most difficult areas of fault isolation work [2]. For this type of failure, traditional fault isolation techniques, such as liquid crystal analysis (LCA) and photon emission microscope (PEM) are less successful in exactly isolating the failing area in order to allow physical failure analysis. The hot spots detected using these techniques may be a secondary effect rather than the exact physical defect location. The high temperature test also frequently results in stronger background thermal noise making the failure locations less distinguishable.

Laser based techniques, such as thermally induced voltage alternation (TIVA) and infrared optical beam induced resistance change (IR-OBIRCH) have become increasingly important for failure localization in integrated circuits [3]. The scanning laser beam can be applied during the stimulation and changes the properties of the passive and active devices from both the front-side and the reverse-side of the chip. The effect of laser stimulation can be thermal or photoelectric depending on the energy of the scanning laser beam and the illuminated device and material. When the energy of the scanning beam is higher than the silicon bandgap energy, the photonic stimulation is much stronger than the thermal stimulation. When the energy of the scanning laser beam is lower than the bandgap energy of silicon, then the thermal stimulation will dominate [2]. Thermal stimulation is due to the absorption of light in a material. Heating occurs when the energy of the photon is transferred to the material lattice without losing the energy for the generation of free carriers. The intrinsic carrier concentration n_i can be expressed as equation (1), and so is highly temperature dependent. For an undoped intrinsic

semiconductor, an effect known as impurity-induced bandgap narrowing, which occurs in the high doping levels and changes the electronic properties of the materials enough to significantly affect certain device characteristics, can be neglected. With increasing temperature, the intrinsic carrier concentration n_i increases.

$$n_i = \sqrt{N_C N_V}\, e^{-E_g/2kT} \tag{1}$$

n_i: intrinsic concentration of electrons

N_C: effective density of states in conduction band

N_V: effective density of states in valence band

E_g: bandgap energy

k: Boltzmann's constant

T: Temperature in Kelvin

The mobility is also sensitive to temperature. With increasing temperature, the carrier mobility due to lattice scattering decreases because the increased phonon concentration causes increased scattering. For low doping levels, such as in intrinsic semiconductors, the thermal-equilibrium conductivity becomes better with increasing temperature. The effect of increasing intrinsic carrier concentrations is usually much stronger than that of the mobility degradation due to increased scattering, and the increasing intrinsic carrier concentrations dominate the thermal-equilibrium conductivity. When excess carriers are generated thermally in the intrinsic semiconductor, there is an increase in the conductivity of the material. The change in conductivity can be expressed as equation (2).

$$\Delta\sigma = e(\delta n)(\mu_n + \mu_p) \tag{2}$$

$\Delta\sigma$: the change in conductivity

e: electronic charge

δn: excess carriers concentration

μ_n: electron mobility

978-1-4244-5430-3/10 $26.00 © 2010 IEEE

μ_p: hole mobility

This paper describes the use of the thermal laser stimulation (TLS) technique to modify the passive device properties of the carrier concentrations which will further cause a change in conductivity based on equation (2) in order to indicate the direct failure location. In combination with studying the circuit schematic and layout, and performing a nanoprobing analysis, the cause of the iddq leakage was identified. In this case, the iddq leakage will disappear during high temperature testing.

II. RESULTS AND DISCUSSION

In this study, the failure mechanism that causes room temperature iddq failure was analyzed. The effect of temperature on the IV characteristics is shown in Fig. 1.

Temperature (K)	Iddq reading(mA) at Vdd=1.2V
298 (25C)	17.9
328 (55C)	0.9
358 (85C)	0.5

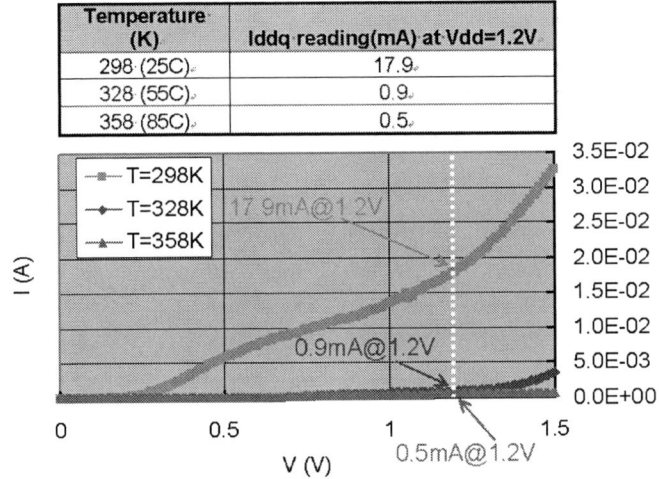

Figure 1. IV curves for the failed sample at three temperatures, 298K, 328K and 358K.

The results indicate that the leakage current becomes lower at high temperatures. As the temperature can alter the leakage current, a global fault isolation technique using IR-OBIRCH was performed from the reverse side of the chip in an attempt to isolate any device which can cause the chip to produce significant thermally induced current changes when the device is locally heated using an IR laser. The effect of thermal laser power on the IV characteristics is shown in Fig. 2. The results indicate that the thermal laser can result in a change in the leakage current, and the higher the laser power the lower the leakage current. The results are similar to those shown in Fig. 1. Fig. 3 indicates that a dark spot was found in the ESD protection circuit using IR-OBIRCH. The layout image shows that the spot was located in the poly resistor. As the failure location was pinpointed accurately, the sample was physically de-processed in order to perform a physical failure analysis (PFA). However, no visible defect could be physically observed in the spot using plane-view scanning electron microscope (SEM), as shown in Fig. 3(d).

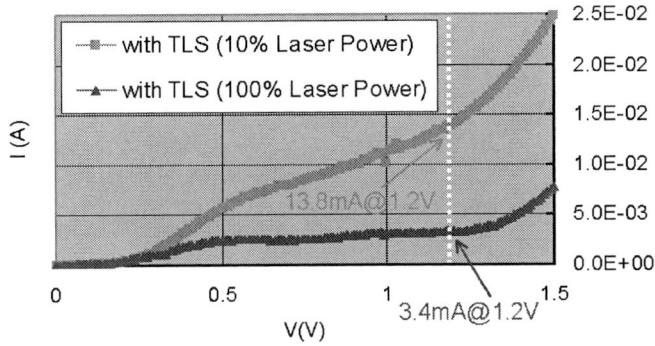

Figure 2. IV curves for the failed sample at two laser power levels, 10% and 100%. (100% laser power correlates to 120mW@5X, 90mW@20X, or 5mW@100X)

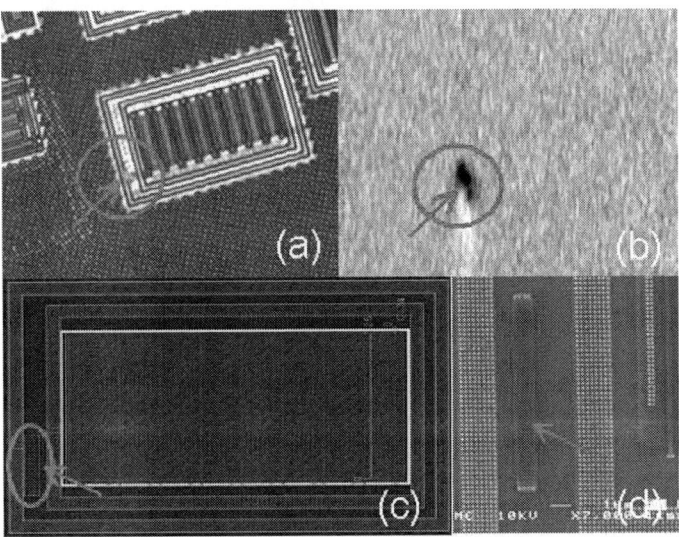

Figure 3. Reverse-side reflected light image overlaid with a TLS image (a), reverse-side TLS image (b), layout image (c), and plane-view SEM image(d).

Fig. 4 shows the schematic of the NFET transistor in the ESD protection circuit. The poly resistor is used to connect the gate and the source terminal of the NFET transistor.

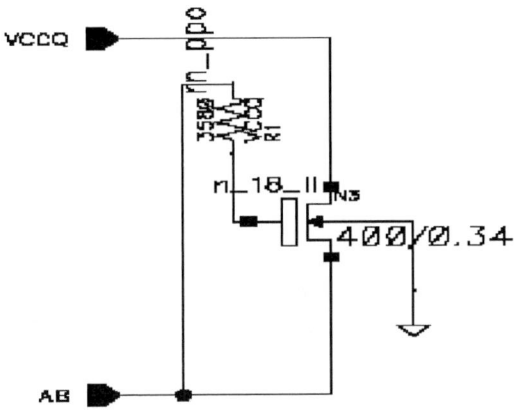

Figure 4. Schematic of the ESD protection circuits

The contrasts in the images in Fig. 3(b) represent current changes. The dark and bright contrasts signify the current decrease and increase, respectively. The results obtained using IR-OBIRCH indicate that current decrease occurred when the laser illuminated the poly resistor, or when the poly resistor was locally heated, providing some reliable hypotheses regarding the physical failure. Another sample was subjected PEM analysis from the reverse side of the chip to help clarify the cause of the electrical failure as shown in Fig 5.

Figure 5.   Illustration of the reverse-side PEM analysis.

Photons were from the NFET transistor in the ESD protection circuit. As most photon emission is generated by the scattering of field accelerated carriers, there is a high possibility that the NFET transistor is in a saturated state rather than the off state. Based on the results obtained using IB-ORIRCH and PEM, it is hypothesized that the cause of the failure could be due to insufficient N doping at the poly resistor, or, even worse, an undoped poly resistor. The photons emitted from the NFET could be explained as being caused by the floating gate due to the undoped poly resistor, and the dark contrast of the TLS signal explains that the gate would be altered from a tri-state to the off state when the undoped poly resistor was locally heated, as there was an increase in the conductivity of the poly resistor when excess carriers were generated thermally.

SEM-based nanoprobing was performed to measure the resistance of poly resistors. Compared with a normal poly resistor whose resistance is about 7.4k ohm, as shown in Fig. 6, the failing poly resistor shows an intensively high resistance, meaning that the resistor is nearly open, as shown in Fig. 7. These electrical results give proof of the previous hypotheses and provide direct evidence of the electrical failure.

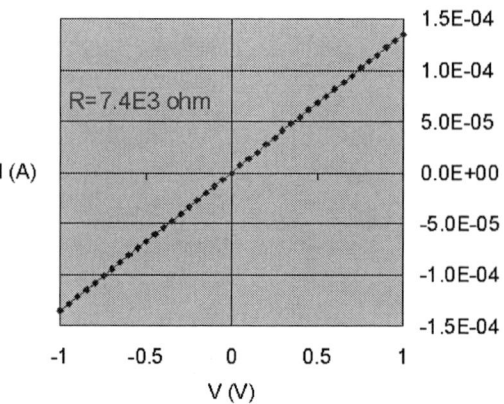

Figure 6.   IV curve for a normal poly resistor.

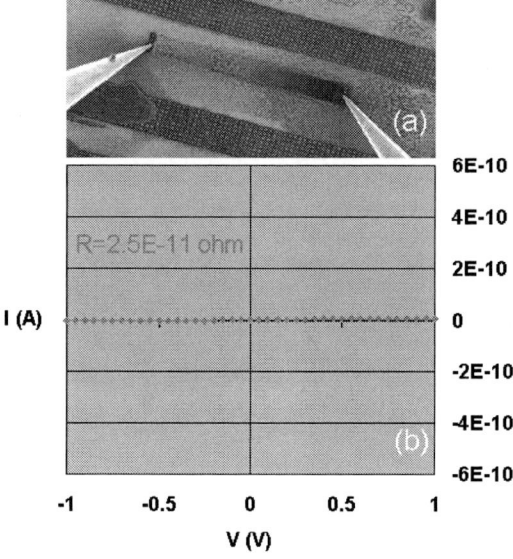

Figure 7.   SEM photograph of nanoprober (a), and IV curve for the failing poly resistor (b).

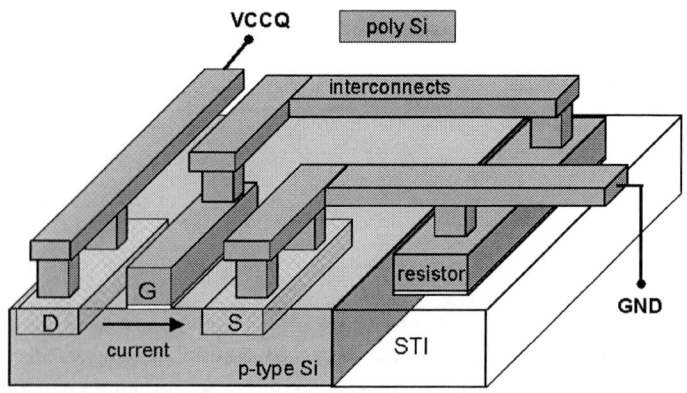

Figure 8.   Schematic for an NFET and poly resistor.

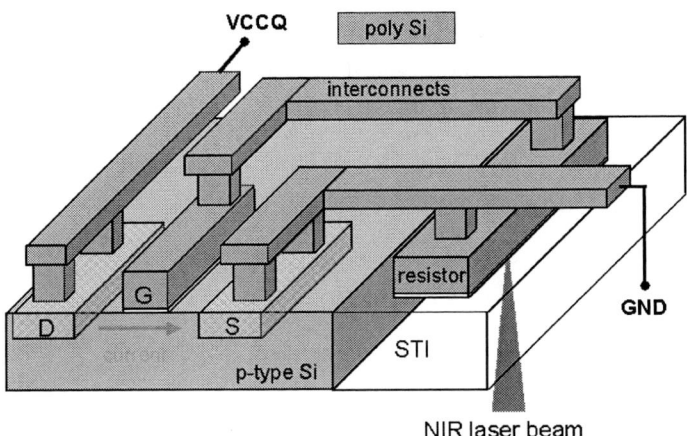

Figure 9. Reverse-side TLS schematic for an NFET and poly resistor.

III. CONCLUSION

This study successfully demonstrates the use of the thermal laser stimulation (TLS) technique in increasing the conductivity of the poly resistor to show the direct failure location. The undoped poly resistor causing a lose of control of the ESD protection circuit NFET gate leads to a high iddq current flowing in the NFET transistor, as shown in Fig. 8. When the undoped poly resistor was locally heated, as shown in Fig. 9, the NFET gate was switched from the tri-state to the off state because there was an increase in the conductivity of the poly resistor, occurring when excess carriers were generated thermally, which explains why the iddq leakage in this case is more serious at low temperatures than at high temperatures. The root cause of the physical failure of the undoped poly resistor was identified as being caused by an unacceptable Boolean operation. After identifying the physical root cause of the electrical failure, corrective actions were then taken to quickly improve yield.

ACKNOWLEDGMENT

The author would like to thank the SRAM&Logic Product Engineering group for IR-OBIRCH analysis and helpful discussions.

REFERENCES

[1] Z. G. Song, S. B. Ippolito, P. J. McGinnis, A. Shore, B. Paulucci, T. Kane, M. P. Tenney and F. G. Trudeau, "Vdd Leakage Analysis by a Combination of Various Failure Analysis Techniques," 34th ISTFA Proceedings, Nov. 2008, pp.75-78.

[2] Fubin Zhang, Corey Lewis, Tim Duryea "Case Study of High Temperature Failure Analyses Using an On-Chip Heater," 34th ISTFA Proceedings, Nov. 2008, pp.273-276.

[3] Arkadiusz M. Glowacki, Sanjib Kumar Brahma, Hiroyoshi Suzuki, and Christian Boit, "Systematic Characterization of Integrated Circuit Standard Components as Stimulated by Scanning Laser Beam," *IEEE Transactions on Device and Materials Reliability*, vol. 7, no.1, pp. 31-49, March, 2007.

HIGH SPATIAL AND TEMPORAL RESOLUTION THERMAL IMAGING FOR LSI CIRCUITS WITH PHASE MICROSCOPY

Tomonori Nakamura, Hidenao Iwai, Toyohiko Yamauchi, Hirotoshi Terada and Hithoshi Iida
Hamamatsu Photonics K.K.
812 JOKO-CHO, HIGASHI-KU, HAMAMATSU-CITY, 431-3196 JAPAN, phone: +81-53-431-0159 fax: +81-53-431-0121 e-mail: export@sys.hpk.co.jp

ABSTRACT

A new thermal imaging method that senses the change in Si index of refraction as a function of temperature to visualize temperature distributions of LSI circuits is described. The resolution is not limited by the heat radiation wavelength but by the sensing light source, which is usually around 1μm. The method further extends the application into the areas that require both high spatial and temporal resolution by utilizing a pulsed light source. [*Keywords: Phase microscopy, Thermal imaging, Heat, NIR,.*]

INTRODUCTION

Thermal measurement has gained popularity in the field of failure analysis as well as in LSI circuit reliability. Present thermal measurement is based on detecting heat radiation with a wavelength that varies as a function of the temperature (T). The blackbody radiation approximation predicts a 3000/T μm wavelength at the radiation maximum. Therefore the wavelength of the heat radiation primarily limits the spatial resolution of thermal radiation images. Alternatively, interferometric measurement of the refractive index change caused by the temperature or free-carrier density has been developed. The time-resolved interferometric mapping method measurement can yield not only localization of the hot spot but also thermal and current flow dynamics. Scanning heterodyne interferometry can easily retrieve the optical phase delay although scanning is troublesome[4-6]. Holographic interferometry can retrieve the optical phase in a single-shot using a 2D image sensor[7]. In this time-domain interferometric method a short stress pulse is applied to the device and then the phase delay or refractive index change is recoded with time.

In this paper we show frequency domain interferometric methods for localization of hot spots in the device to which the stress pulse is applied sinusoidally. The optical path length of the LSI substrate is modulated with the temperature due to the Si expansion and largely the refractive index change. The Si thermal expansion coefficient α is 3.5E-6 K^{-1} and the refractive index coefficient β is 5.0E-5 K^{-1}. Optical path change is calculated as in Eq. (1), with n the refractive index of Si at the illumination wavelength, D the Si thickness, L the optical thickness and T_s the temperature shift. If there is thermal distribution in the DUT the optical path change will be the average. For example, a 400 μm Si substrate has a 1400 μm optical thickness with a refraction parameter n = 3.5, yielding an optical path length change of 75 μm/K. The actively stabilized quantitative phase microscopy (QPM) detesct optical thickness changes less than 1 nm.[1-3].

$$\Delta L = L' - L$$
$$= n(1 + \beta Ts)D(1 + \alpha Ts) - nD \qquad (1)$$
$$\approx (\alpha + \beta)nTs$$

SETUP

Figure 1 shows the interference optical system. The light source is a 15 mW super luminescence diode (SLD) with a wavelength λ of 1.31 μm with 0.02 μm FWHM. At this wavelength the DUT's Si substrate is transparent, hence the system can sense the circuit's temperature from the backside of the chip.

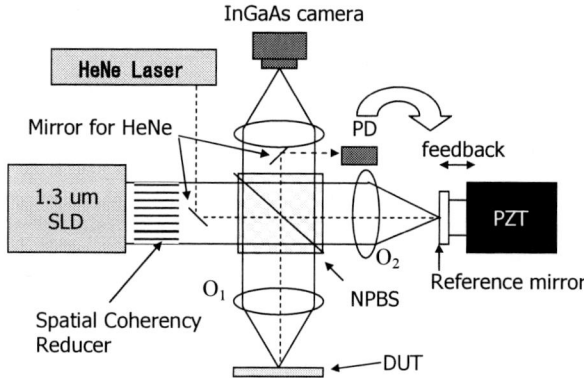

Figure 1. Michelson type low coherency Interference Optical System

The light beam is passed through a multi fiber light guide in order to reduce the spatial coherency. The light beam is split into two at the no polarized beam splitter (NPBS) and through the objective lenses O_1 and O_2. One beam illuminates a DUT from the backside and the other is reflected on the reference mirror having a PZT actuator. Finally, the light beams are combined at the NPBS and an interference optical image is brought into focus on the InGaAs camera. The other side of DUT is attached to a thermoelectric (TE) device for the temperature control and stabilizing. A HeNe laser beam is overlapped with the light beam from a SLD. Reflected laser beams from the DUT backside surface and reference mirror are combined and the interference signal is captured by a photo diode (PD). The signal is used for feedback to the PZT actuator to control the difference of the optical path length of two arms in the interferometer. The interference of the optical image is changed by the optical length in the DUT.

978-1-4244-5430-3/10 $26.00 © 2010 IEEE

EXPERIMENTAL I

STATIC MEASUREMENT

Static phase images are calculated by a quarter-wavelength phase-shifting algorithm [1]. The algorithm needs 4 or more interference images changing the mirror side optical path length (OPL) by λ/4. The OPL is controlled by a PZT with PD feedback. Fig. 2 shows a phase image of a DUT hotspot and the optical thickness change as a function of power dissipations.

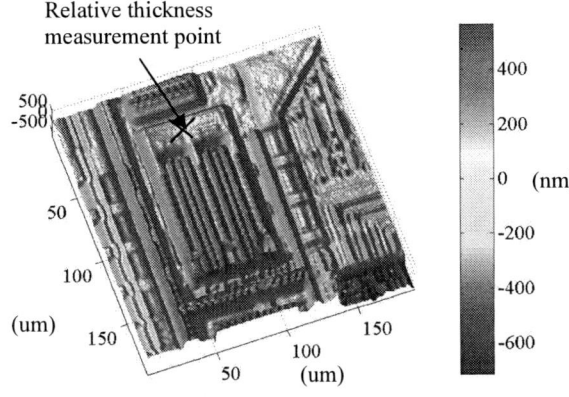

(a) Phase image of hot spot

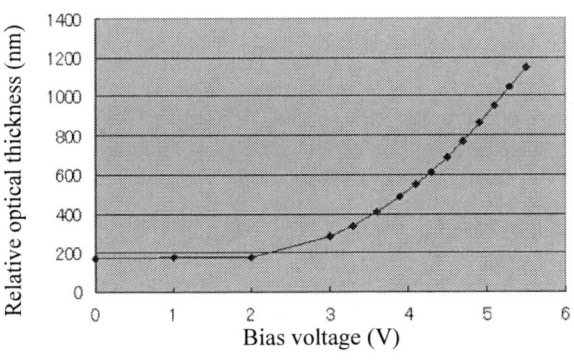

(b) OPL change in DUT substrate

Figure 2.　Interference Phase Image and the OPL variation with power dissipation..

Fig.3 shows the optical path length change dependence with the TE device temperature for a 400 μm Si substrate. The optical path was measured by the λ / 4 shift method. 1 Phase rotation means λ / 2 (655 nm) optical thickness and 8.8 K temperature changing. This QPM method can detect less than 1 nm shift. Therefore, around a 10 mK temperature change can be detected with this method.

However, the thermal spread speed in Si substrate is too fast to be measured with this static measurement method.

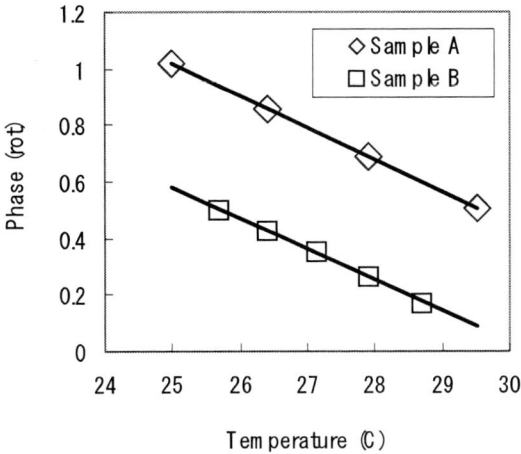

Figure 3.　Interference Phase Rotation depends on the Si substrate temperature.

EXPERIMENTAL II

DYNAMIC MEASUREMENT

Dynamic phase image measurement is a lock-in imaging approach and the data set is calculated by a Fourier transfer method. This method uses power source modulation of the DUT. The power modulation generates thermal waves in the DUT and the thermal waves are observed as optical length changes. The amplitude is observed as a temperature distribution and the phase delay images can be used as thermal spreading images. Fig.4 shows pixel intensity modulation of a DUT interferometer image with 1 and 10 Hz power modulation taken by a 60 Hz InGaAs camera.

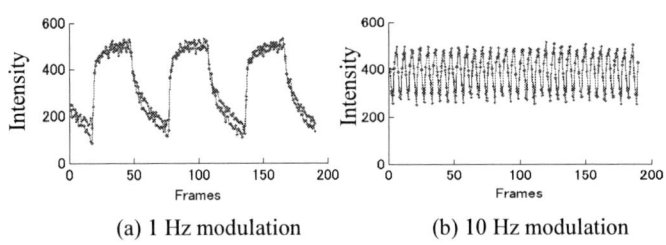

(a) 1 Hz modulation　　　　(b) 10 Hz modulation

Figure 4.　OPL variation with power source modulation. The horizontal axis shows picture frames and the vertical axis shows pixel intensity.

In the lock-in measurement, the InGaAs camera's 60 Hz frame-signal is used for system synchronization where a 1ms SLD pulse signal is generated at each frame. The DUT is powered on for a 4 frame period and off for the next 4 frames (7.5Hz). In this measurement 2000 images are taken in 33 seconds and 8 timing images are generated from each 125 same timing images as shown in Fig.5. The reference optical path length is shifted by λ/4 and the measurement is repeated until sufficient signal-to-noise is achieved. Finally, amplitude and phase components are calculated for each pixel as in Eq. (2-7).

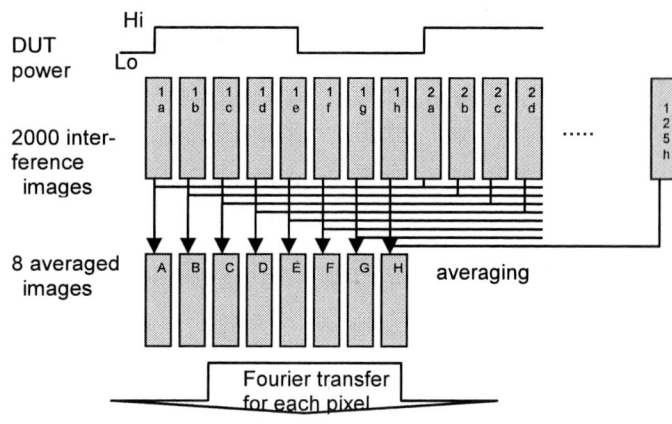

Figure 5.   Calculation process

$$a_2(x,y) = \frac{2}{T}\int_{-T/2}^{T/2} I(t,x,y)\cos(2\Omega t)dt$$

$$= AJ_2(m)\cos\phi\cos\Phi_2 \qquad (8)$$

$$b_2(x,y) = \frac{2}{T}\int_{-T/2}^{T/2} I(t,x,y)\sin(2\Omega t)dt$$

$$= AJ_2(m)\cos\phi\sin\Phi_2 \qquad (9)$$

$$\Phi_1 = \frac{1}{2}\Phi_2 = \frac{1}{2}\tan^{-1}\frac{b_2}{a_2} \qquad (10)$$

$$AJ_1 = \sqrt{a_1^2 + b_1^2 + (J_1/J_2)^2\,(a_2^2 + b_2^2)} \qquad (11)$$

$$\frac{J_2}{J_1} \approx \sqrt{\frac{\sum(a_2^2 + b_2^2)}{\sum(a_1^2 + b_1^2)}} \qquad (12)$$

In most cases this calculation method reduces measurement time in half. But in actuality , J2/J1 is almost 0.1 and it is the cause of image quality degradation. And about Eq.(6), b_1/a_1 and b_1'/a_1' are the same mathematically, but they are different in actual cases depending on camera noise. In this case, compare $|a_1|$ and $|a_1'|$ or $|a_1|$ and $|a_2|$, and chose the larger one: Φ_1 or Φ_1', or Φ_1 or $\Phi_2/2$, then the phase image noise level can be reduced.

RESULT AND DISCUSSION

Fig.6 shows the reflectance image of the DUT with the InGaAs camera. The substrate thickness is 400 μm. The objective lens magnification is x5 and the FOV is 2.4 by 1.8 mm. Phase images are shown in Fig.7 and amplitude images are shown in Fig. 8. The applied voltages and currents are shown in table 1.

$$I(t,x,y)$$
$$= I_r + I_s + 2\sqrt{I_r I_s}\,\cos(\phi + m\sin(\Omega t + \Phi)) \qquad (2)$$

$$I'(t,x,y)$$
$$= I_r + I_s + 2\sqrt{I_r I_s}\,\sin(\phi + m\sin(\Omega t + \Phi)) \qquad (3)$$

$$a_1(x,y) = \frac{2}{T}\int_{-T/2}^{T/2} I(t,x,y)\cos(\Omega t)dt$$
$$= AJ_1(m)\sin\phi\cos\Phi_1 \qquad (4)$$

$$b_1(x,y) = \frac{2}{T}\int_{-T/2}^{T/2} I(t,x,y)\sin(\Omega t)dt$$
$$= AJ_1(m)\sin\phi\sin\Phi_1 \qquad (5)$$

$$\Phi_1 = \tan^{-1}\frac{b_1}{a_1} = \Phi_1' = \tan^{-1}\frac{b_1'}{a_1'} \qquad (6)$$

$$AJ_1 = \sqrt{\left(a_1^2 + b_1^2\right) + \left(a_1'^2 + b_1'^2\right)} \qquad (7)$$

In Eq.(2), I(t,x,y) shows 8 images generated from 2000 InGaAs images. Eq.(3) and ' shows a second set of images. I_s shows the light intensity from a DUT and I_r from a reference mirror. I(t,x,y) shows the pixel intensity on the InGaAs camera. Ω shows the DUT power source frequency, m is the modulation depth, Φ the phase depends on power and ϕ is the phase of each interference pixel in the image. Using a Fourier transfer as in (4-5), the thermal wave phase Φ is calculated as in (6) and the amplitude is shown in (7). J_1 is a Bessel function depending on m.

A twice wave calculation step can further reduce the second measurement. The amplitude A_J is calculation shown in Eqs.(8-12).

Table 1..Bias voltages, power dissipations and OPL amplitude.

Bias Voltage (V)	Current (mA)	Power (duty 50%) (mW)	Maximum Amplitude (A.U.)
4.00	28.1	56.2	316
3.50	22.3	39.0	244
3.00	16.9	25.4	156
2.50	11.4	14.2	83.1
2.00	3.26	3.26	11.2

Noise level: $\sigma = 0.38$ (A.U.)

Figure 6. InGaAs pattern image

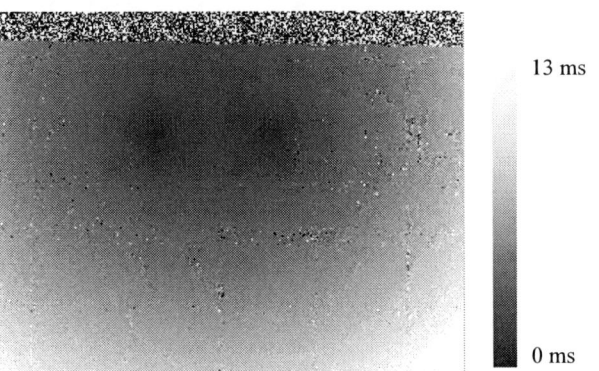

(a) Timing Image (4 V)

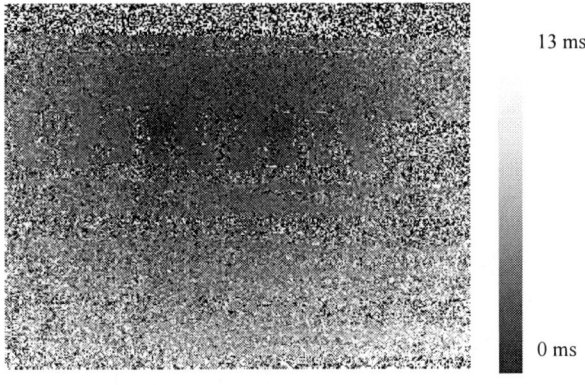

(b) Timing Image (2 V)

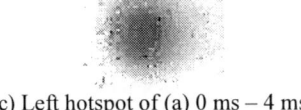

(c) Left hotspot of (a) 0 ms – 4 ms

Figure 7. Timing images calculated by Eq.(6).

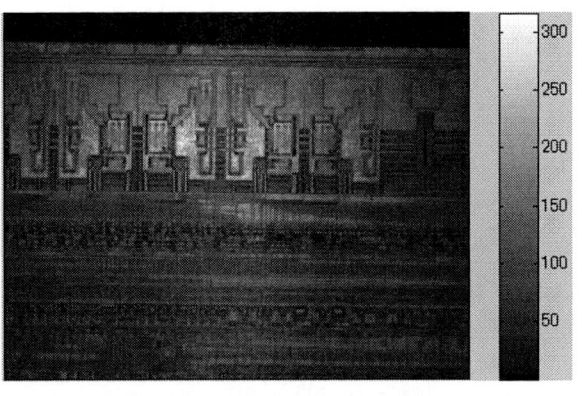

(a) Amplitude Image (4 V)

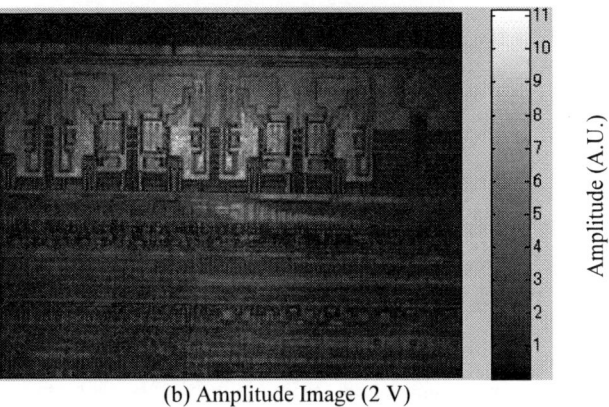

(b) Amplitude Image (2 V)

Figure 8. Amplitude images calculated by Eq.(7).

Fig. 7 and Fig.8, show where the heat originates and where the maximum temperature is reached. The phase image shows the heat wave spreading in the Si substrate.

Fig.7 (a) and (b) show heat phase images with 4 and 2 V bias voltages. The heat source can be detected by phase imaging. And the heat source shape, as shown in Fig.2(a), can be observed in Fig.7(c). In Fig. 7(c), 30 μm pitch 3 lines can be detected. The time resolution and special resolution is limited by light power of SLD.

Fig.8 amplitude images show the amplitude of image intensity modulation. The noise level only depends on the InGaAs camera noise and can be calculated as random noise. The noise level σ is calculated as 0.38 (A.U.) and the signal-to-noise ratio is about 30 with a 3.26 mW heat source modulation. The result shows that less than a 1 mW heat source modulation can be detected by the amplitude imaging.

The amplitude images show the product of temperature modulation and amplitude phase images. Therefore the Si substrate temperature distribution can be calculated with amplitude image, phase image taken by quarter-wavelength phase-shifting algorithm and Eq.(1).

CONCLUSION

In this paper, we have demonstrated the new thermal imaging method using NIR actively stabilized quantitative phase microscopy. The static measurement results show that this measurement can be used for measure the DUT static temperature. The dynamic measurement results demonstrate acquisition of the phase and amplitude of heat wave propagation images of LSI circuits with high spatial and temporal resolution.

REFERENCES

[1] H. Iwai, "Quantitative phase imaging using actively stabilized phase-shifting low-coherence interferometry," Opt. Lett., 29, 2399, (2004).

[2] X. Li, "Full-field quantitative phase imaging by white-light interferometry with active phase stabilization and its application to biological samples," Opt. Lett, 31, 1830, (2006).

[3] T. Yamauchi, "Low-coherent quantitative phase microscope for nanometer-scale measurement of living cells morphology," Opt. Express. 16, 12227, (2008).

[4] M.Goldstein, "Heterodyne interferometer for the detection of electric and thermal signals in integrated circuits through the substarate," Rev. Sci. Instrum. 64, 3009, (1993).

[5] C.Furbock, "Thermal and free carrier concentration mapping during ESD event in smart power ESD protection devices using an improved laser interferometric technique," Microelectronics reliab., 40, 1365, (2000).

[6] M.Litzenberger, "Scanning heterodyne interferometer setup for the time-resolved thermal and free-carrier mapping in semiconductor devices," IEEE Trans. On instrumentation and measurement, 54, 2438, (2005).

[7] D.Pogany, "Single-shot thermal energy mapping of semiconductor devices with nanosecond resolution using holographic interferometry," IEEE Electron device letters, 23, 606, (2002).

RELIABILITY OF ELECTRONIC EQUIPMENT EXPOSED TO CHLORINE DIOXIDE USED FOR BIOLOGICAL DECONTAMINATION

G. E. Derkits [1], M. L. Mandich [1], W. D. Reents[1], J. P. Franey[1], C. Xu[1], D. Fleming[1], R.Kopf [2], and S. Ryan[3],
[1] Alcatel-Lucent Carrier Product Group, [2] Alcatel-Lucent Bell Laboratories, [3] U.S. Environmental Protection Agency National Homel- and Security Research Center.
Address of Primary Contact: G. Derkits, Alcatel-Lucent, Room 1E-241, 600 Mountain Ave., Murray Hill, NJ, 07974, phone: 1-908-582-7050; fax: 1-908-582-7112; e-mail: Gus.Derkits@alcatel-lucent.com
Address of Second Contact: M. Mandich, , Alcatel-Lucent, Room 1E-347, phone: 1-908-582-3396; fax: 1-908-582-6228; e-mail: Mary.Mandich@alcatel-lucent.com

SUMMARY

We have studied the effects of chlorine dioxide fumigation on the reliability of electronic equipment using personal computers as examples of current commercial systems. Unit and subunit failure were objectively defined by standard commercial software. After the initial one-day exposures to the fumigation conditions, the systems were tested to assess impacts and retested monthly for six months. Cumulative failures of decontaminated systems were many times higher than unexposed systems and increased progressively for the harshest fumigation conditions. Failures occurred in electronic, mechanical, optoelectronic, and thermal subsystems. Failure mode and root-cause analyses were performed on a blind sample of systems. Corrosion of metals and degradation of organic materials were predominant causes of failure. Metal corrosion continued to progress well after the initial exposure. [*Keywords:* Decontamination, reliability, personal computer, chlorine dioxide, diagnostic software.]

INTRODUCTION

Chlorine dioxide gas was used to successfully decontaminate several major buildings in the U.S. following the anthrax letter attacks in 2001. The Department of Homeland Security and the Environmental Protection Agency are interested in generating specific data on the impact of chlorine dioxide (ClO_2) decontamination on electronic equipment, to better understand the usefulness and limitations of the decontamination technology. The current study was designed to provide information on electronic systems through the use of commercial personal computers as test vehicles. Personal computers have a variety of advantages in this application. Due to intense competition among suppliers, they are relatively low in cost and the technology they use is very current. They contain subsystems with a variety of technologies – electronic, electromechanical, optical, magneto-optical. They are highly standardized, with the result that commercial diagnostic software capable of giving precise diagnostic information of failure is available at low cost. They are small enough that several identical systems can be exposed to identical conditions simultaneously in a single chamber. Finally, they are easily disassembled for analysis. Pure metal coupons were placed inside the computers as precision corrosion monitors. The progression of failure was followed using diagnostic software over six months after the initial exposure.

METHOD

Test Vehicles and Standard Materials

The system test vehicles in this study were Dell OptiPlex® 745 mini-tower personal computers with Intel Core 2 Duo® E6400 Processors at 2.13 GHz with a single DIMM RAM card, running Windows XP. A 16X DVD-RW optical drive and a 3.5" floppy drive were included in the system configuration. Diagnosis of hardware faults was obtained using PC Doctor® Service Center 6, a commercial kit used by a number of large PC manufacturers for their final test and embedded diagnostics. This product has over 300 diagnostic tests, most of which can be run without operator intervention. Metal coupons of 99.999% copper, 99.99% aluminum, and 99.998% silver purchased to specification from Alfa Aesar were used as corrosion monitors. These coupons were weighed using a Mettler UMT2 microbalance accurate to 0.25ug. ClO_2 gas was generated by a ClorDysis® Solutions, Inc. gas generation system.

Exposure Conditions

All exposures were performed at the EPA National Homeland Security Center (NHSRC) in Research Triangle Park, NC. Three computers were exposed to the same conditions in each test cell (see Table 1). Immediately following the exposures, the systems were tested at NHSRC with PC Doctor®. Of the three computers in each cell, two were deployed for diagnostic testing. The remaining systems from each cell were submitted for in-depth, destructive, analysis. Computers to be destructively analyzed were then packed into shipping bags made of Static Intercept® material which preserved the state of the system during shipping and prevented toxic gases from entering the environment. Metal coupons from all systems were similarly packed and sent for analysis. The range of test conditions was designed to exercise the standard decontamination conditions, minimal exposure, and exposure to high humidity alone. Principle test variables included the power state of the computer, the concentration of ClO_2, the relative humidity, and the time duration of the fumigation. The matrix of test cell conditions is given in Table 1.

TABLE 1. MATRIX OF DECONTAMINATION TEST CONDITIONS

Cell	Pwr	Treatment	ClO_2 [ppmv]	RH [%]	Temp [°F]	Time [hrs]
1	On	High RH	3000	90	75	3
2	Off	Standard	3000	75	75	3
3	On	Standard	3000	75	75	3
4	On	Low ClO_2	75	75	75	12
5	On	Low ClO_2 & RH	75	40	75	12
6	On	No ClO_2 High RH	0	90	75	3
7	On	Ambient (Control)	0	40	75	--

978-1-4244-5430-3/10 $26.00 © 2010 IEEE

Analysis

Analysis of a blind sample of test vehicles, one from each test cell, was performed at the Alcatel-Lucent Reliability Physics facility at Murray Hill, NJ. After the computers were received, they were unpackaged, placed in chemical hoods, and photographed with a high-resolution camera at a set of pre-determined sites to establish an immediate visual record. Additional photographs were taken at the discretion of the analysts. Photographs of some sites were taken multiple times in the course of the investigation to establish a visual record of progressive corrosion. An example of this is shown in Figure 1, in which progressive cut-edge corrosion of the steel frame is demonstrated.

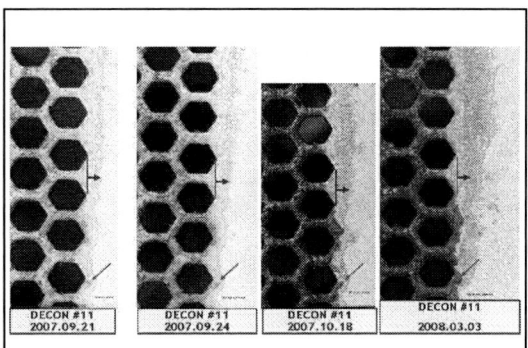

FIGURE 1. PROGRESSIVE CUT-EDGE CORROSION OF STEEL CASE OF PC.

PC Doctor was run on each system and the results recorded. Resistance readings were taken across predetermined test points on the motherboard and the systems were then subjected to destructive physical analysis.

RESULTS

Failure Progression

The relation between the harshness of decontamination conditions and system failure is illustrated in Figure 2. The red line with diamond symbols is the plot of the harshest conditions. The black line at the bottom is the control computer, which saw only ambient conditions. Note that the harshest conditions (3000 ppm, 90%RH) are outside of normal, target, ClO_2 use conditions, chosen to bracket decontamination application conditions.

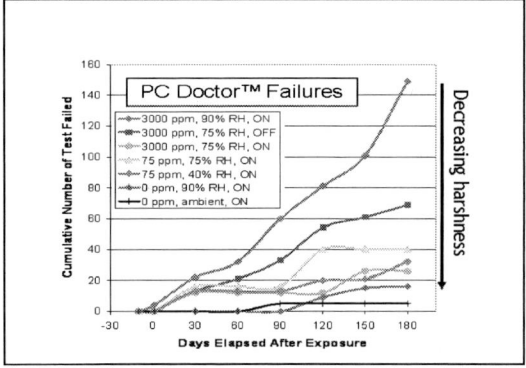

FIGURE 2. PROGRESSIVE FAILURES OF PCS THROUGH TIME AS GAUGED BY PC DOCTOR SOFTWARE, FOR VARIOUS DECONTAMINATION EXPOSURES.

Corrosion of Pure Metal Coupons

Figure 3 provides the weight gain of the pure metal coupons in the form of a bar chart overlaid with a line showing the average of all three metals. A figure of merit (FOM) representing the harshness of the decontamination process as the product of the concentration of ClO_2 times the duration of the exposure times the relative humidity (RH) is shown on the right-hand scale.

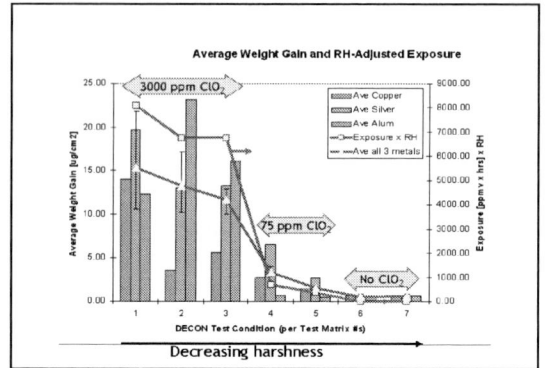

FIGURE 3. COMPARISON OF COUPON WEIGHT GAIN WITH HARSHNESS OF DECONTAMINATION PROCESS.

$$FOM = [ClO_2] \times time \times RH \qquad (1)$$

CONCLUSIONS

ClO_2 fumigation causes permanent and unrecoverable damage to a variety of material and subsystems in the test computers that progresses in time following the exposure. The extent of damage and its impact was a strong function of fumigation conditions. Material degradation included corrosion to metals including aluminum, steel, silver, nickel, and plated copper, and bleaching and discoloration of cable coatings. Subsystem damage included corrosion of gold plated connectors and catastrophic destruction of optical components in the CD/DVD drives.

Failures were detected by PC Doctor® in all computers, including the control (ambient), but the number of failures was significantly higher for the fumigated computers and they showed an increasing failure rate with time.

Novel aspects of this study include the use of personal computers as standard test vehicles, objective definition of failure by the use of independent standard commercial software, long-term (6 months) follow-through after initial fumigation, root-cause analysis of multiple subsystems down to the material level, application of pure metal coupons as standards in decontamination studies, and application of a standardized, repeatable set of target areas for high resolution photography in lieu of visual inspection.

ACKNOWLEDGEMENTS

This work was supported by the U.S. Environmental Protection Agency and the U.S. Department of Homeland Security under contract number USMMM235W9 to LGS Innovations, LLC. We wish to thank Lance Brooks of USDHS for his help and support.

978-1-4244-5430-3/10 $26.00 © 2010 IEEE

Analysis of HCS in STI-based LDMOS transistors

Susanna Reggiani, Stefano Poli, Elena Gnani, Antonio Gnudi, Giorgio Baccarani,
ARCES and Dept. of Electronics
University of Bologna, Bologna, Italy.
phone: +39–051-209-3557, e-mail sreggiani@arces.unibo.it

Marie Denison, Sameer Pendharkar, Rick Wise, Sridhar Seetharaman
Texas Instruments, Dallas, Texas.

Abstract— **A numerical investigation of the hot-carrier behavior of a lateral DMOS transistor with shallow trench isolation (STI) is carried out. The measured drain-current degradation induced by hot-carrier stress (HCS) is nicely reproduced by TCAD results revealing that interface traps are mainly formed at the STI corner close to the channel. The effect of typical device design variations on hot-carrier degradation is analyzed.**

TCAD analysis;LDMOS;Hot-carrier stress

I. INTRODUCTION

Different lateral DMOS (LDMOS) transistor designs were recently proposed to improve the tradeoff between safe operating area (SOA) and specific on-resistance (R_{sp}) versus breakdown voltage (V_{BD}): among them, a new LDMOS transistor with a buried-body implant was recently presented [1-3]. For such devices very little information is available on the HCS reliability [4-5]. In addition, no TCAD predictions of the HCS drift curves have been shown in previous analyses. In this work, the mechanisms of HCS degradation are investigated in a 30V STI-based LDMOS device: TCAD predictions of the linear drain current ($I_{d,lin}$) and maximum transconductance ($g_{m,max}$) shift with stress time are calibrated on experiments over a range of drain biases. A quantitative understanding of carrier transport under HCS conditions is achieved. Furthermore, a number of device design variations have been simulated, illustrating how such modeling can be used to predict design influences on HCS drift and to understand their physical origin.

II. HCS CHARACTERIZATION AND SIMULATION

Fig. 1 shows the schematic view of the transistor. The simulation set-up has been calibrated to experiments in DC and pulsed regimes [6]. Simulations have been performed using the drift-diffusion transport model available in the Sentaurus-Device tool by Synopsys [7]. All measured devices have widths which are significantly larger than the extension of the finger ends: in such a way a 2D simulation approach is fully justified. Self-heating effects have been addressed as well by using the electro-thermal model. The body current

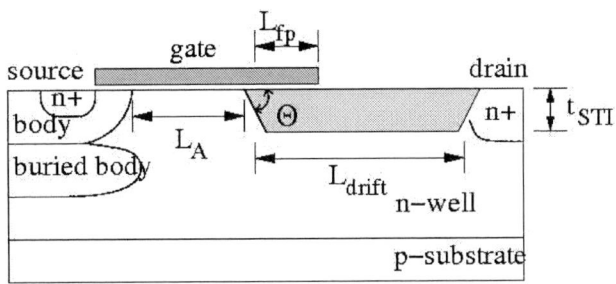

Figure 1. Schematic view of the STI-based LDMOS under study. The geometrical parameters used to define the device variations are indicated.

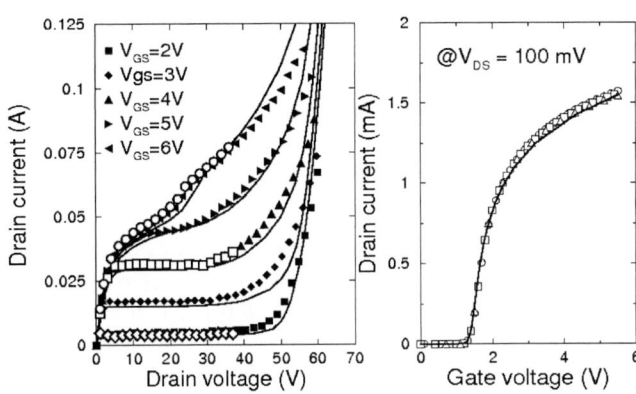

Figure 2. Drain characteristics of the LDMOS device. Left: I_D vs. V_{DS}. Closed symbols: TLP experiments on a single device. Open symbols: Data extracted from devices with different widths by using a differential extraction procedure [6]. Solid lines: simulation data. Right: I_D vs. V_{GS}. Symbols: experiments on different devices. Solid lines: simulation data.

characteristics are nicely predicted on an extended range of biases and temperatures.

Fig. 2 shows the comparison of drift-diffusion simulations with *I-V* characteristics measured under 100 ns transmission-line pulses (TLP) up to the boundary of the SOA [2] (left) and with the DC turn-on curve (right). Fig. 3 shows numerical simulations carried out with the electro-thermal model and

---

Work supported by the SRC Research Contract No. 2007-VJ-1667

978-1-4244-5430-3/10 $26.00 © 2010 IEEE

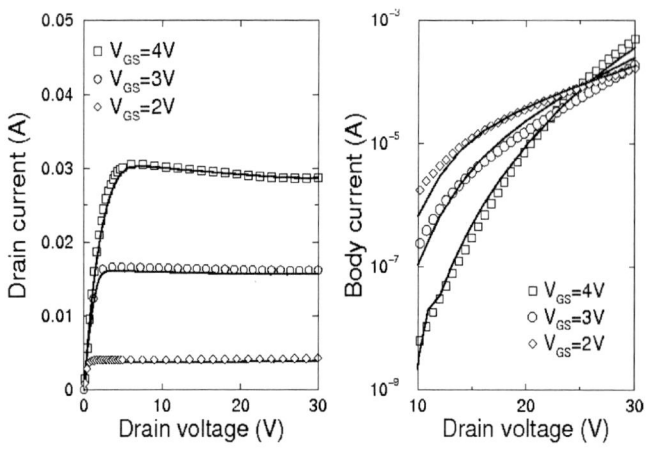

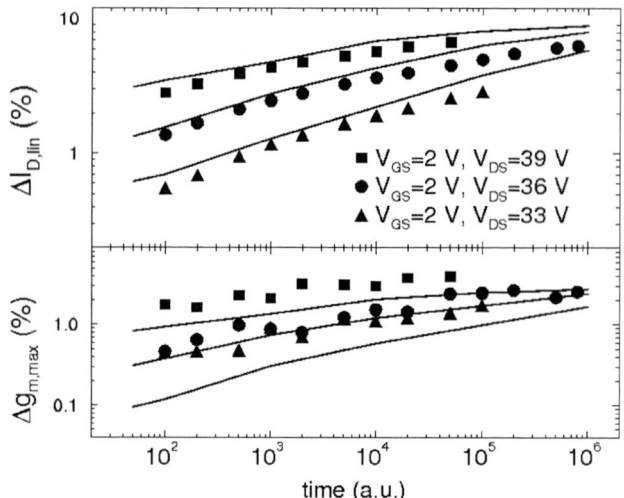

Figure 3. DC characteristics of the LDMOS device. Left: I_D vs. V_{DS}. Right: I_{BODY} vs. V_{GS}. Symbols: experiments. Solid lines: simulation data.

Figure 4. Relative drift of the linear drain current $\Delta I_{d,lin}$ (top) and maximum transconductance $\Delta g_{m,max}$ (bottom) of the device as a function of the stress time at $V_{GS} = 2$ V and different V_{DS} biases. Symbols: experiments. Solid lines: simulation data.

compared with measured DC *I-V* characteristics (left) and body-current curves (right). The excellent agreement indicates that the simulation set-up provides a good description of the device behavior in any operation regime up to its electrical failure.

HCS measurements have been performed by applying high V_{DS} biases for increasing time intervals. A fixed $V_{GS} = 2$V is considered, corresponding to a worst-case stress condition (maximum $I_{D,lin}$ shift). Transient simulations were carried out by using the kinetic equation of trap formation at the Si/SiO$_2$ interface [8]. The HCS model is based on the solution of the Si-H defect kinetics equation, and takes into account the interface disorder and the Si-H bond activation energy evolution as the bonds are broken. It is a reaction-diffusion model based on the Arrhenius approximation with additional empirical dependencies on the local components of the electric field vector and on the hot-carrier current density at the interface. Although the model was calibrated against experiments on conventional CMOS devices [9], no validation was carried out on STI-based LDMOS devices, which are characterized by Si/SiO$_2$ interfaces all along the STI edge and under the STI, where HCS is expected to play its major role. In addition, the shape of the STI angle itself would play some role in determining local trap formation. Thus the HCS model parameters have been calibrated on the experimental $I_{d,lin}$ and $g_{m,max}$ shifts (fig. 4). In particular, the default parameters are used in all the interface regions with the exclusion of the STI edge close to the channel, the STI angle and the STI under the field plate, where hot carriers are mainly present. In the latter regions, an "ad-hoc" parameter set has been used to overcome the limited validity range of the empirical parameters in the kinetic equation [8]. More specifically, the trap generation rate has been enhanced so to obtain a hot-carrier-induced interface state formation in the gate overlapped STI which is dominant with respect to the channel and accumulation regions in the analyzed stress conditions. The low gate voltage limits the normal electric field within the channel well below critical values, and negligible shifts of the threshold voltage have been measured confirming that the gate oxide interface in the

channel region is not degraded by the applied HCS. Moreover, we observed a strict correlation between the trap transient formation at the STI angle and the $I_{D,lin}$ shift curves. This is confirmed also by the experimental and numerical analysis shown in [10] on a high-voltage DEMOS. The driving forces responsible for the trap formation at the STI corner are the components of the electric field normal to the interface and parallel to the current density, as they are strictly related to the carrier injection probability and to the hot-carrier distribution. The effect of the parallel electric field on the carrier energy distribution can be easily monitored by calculating the electron temperature: the "hot" spots at the STI angle and under the field plate are key quantities for the HCS analysis. The calibration has been carried out at a fixed lattice temperature, T = 300 K, as negligible thermal effects are found by using the thermodynamic model in the whole range of analyzed biases at $V_{GS} = 2$ V (see also fig. 3). To further verify the device degradation resulting from the trap formation at the STI angle, the calibration work will be extended to various designs, geometries and temperature variations in the next future.

III. HCS DEPENDENCE ON DEVICE VARIATIONS

The variability of the device performance and HCS degradation with some key design parameters have been investigated by simulation. More specifically, four different geometrical variations have been addressed, namely, i) the length L_A of the accumulation region; ii) the angle θ of the STI trench edge; iii) the STI thickness t_{STI}; iv) the length of the field plate contact L_{fp}. The geometrical parameters are defined as indicated in fig.1. The relative variations of R_{sp} vs. V_{BD} are reported in fig. 5. As far as the R_{sp} figure of merit is concerned, it is worth noting that the pitch of the device is fixed in all the proposed variations except for the L_A case, which induces both a change in the resistive path within the device and a variation of the overall length L of the device. In table I, the relative variations of R_{ON} and L are reported for the different L_A cases

978-1-4244-5430-3/10 $26.00 © 2010 IEEE

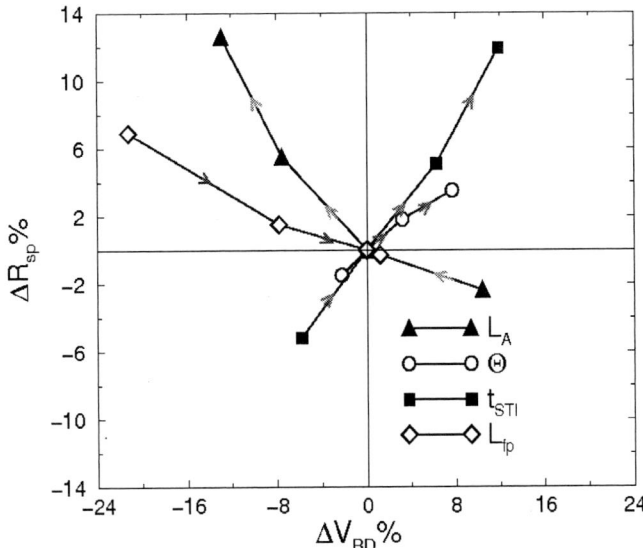

Figure 5. Relative variations of R_{sp} and V_{BD} for the simulated device with different L_A, θ, t_{STI} and L_{fp}. Arrows show the trend of each curve with respect to the parameter variations. $\Delta L_A = -0.25$, $+0.25$ and $+0.5$ μm; $\Delta\theta = \pm 5°$ and $\theta = 90°$; $\Delta t_{STI} = -50$, $+50$, $+100$ nm; $L_{fp} = 0$ and $\Delta L_{fp} = \pm 0.43$ μm.

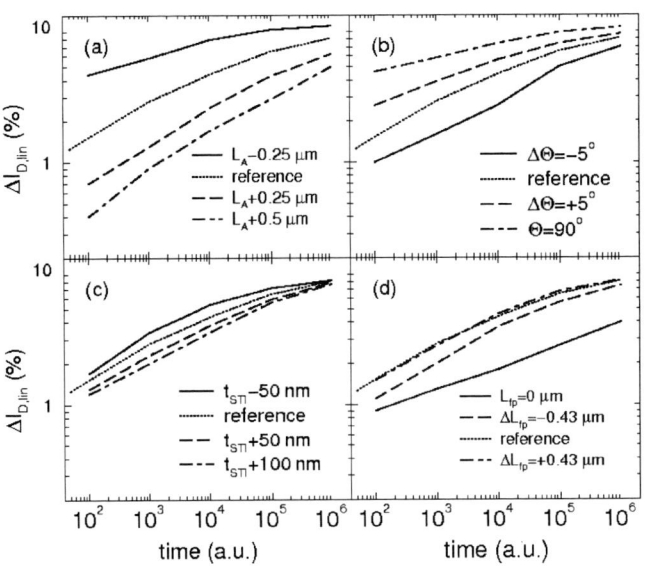

Figure 6. $I_{d,lin}$ drift vs. stress time for the different variations of the device by changing (a) L_A, (b) θ, (c) t_{STI} and (d) L_{fp}. The design parameter values/relative changes are reported in the labels. The stress bias condition is $V_{GS} = 2$ V and $V_{DS} = 0.88$ V_{BD}.

along with the relative R_{sp} and V_{BD} variations, showing that, as opposed to L, R_{ON} slightly decreases with increasing L_A in the range of $\Delta L_A = \pm 0.25$ μm, and tends to increase with L_A in a longer range ($\Delta L_A = 0.5$ μm). This can be ascribed to the relative variations of the accumulation resistance: for short L_A the resistive path in the accumulation region is mainly vertical and L_A is correlated to the section of the resistive path, whereas in devices with longer L_A the most relevant contribution to the resistive path is no more vertical but lateral and the increase of the accumulation resistance is consistent with the increase of L_A, and with the consequent decrease of V_{BD}. The overall R_{sp} resulting from the two concurrent effects increases with L_A due to the dominant effect of the pitch variation.

By observing the trends of the relative variations of R_{sp} vs. V_{BD} in fig.5, the L_A and L_{fp} cases show an overall R_{sp} vs. V_{BD} enhancement with decreasing L_A and increasing L_{fp}, respectively. The reference device is shown to be very close to the optimum choice for the L_{fp} design parameter: the data corresponding to $\Delta L_{fp} = 0.43$ μm are very close to the reference case, and a further increase of L_{fp} would give a reduction of V_{BD} due to the gate-to-drain interaction. The other trends show the expected tradeoff between R_{sp} and V_{BD} enhancements. Significant variations of R_{sp} and V_{BD} are obtained for the considered L_A and t_{STI} parameter ranges. Differently, smaller variations are observed for changes in θ and L_{fp}, with the exclusion of the $L_{fp} = 0$ case ($\Delta V_{BD} = -22\%$), which is reported as a worst-case condition.

The values of V_{BD} obtained for each variation are taken into account in the definition of the stress bias condition for the HCS analysis, which is $V_{DS} = 0.88$ V_{BD}. In fig. 6, the $I_{D,lin}$ shifts are reported as a function of stress time for each parameter variation. A strong dependence of the drift curves is observed with the L_A and θ variations (fig. 6, top).

TABLE I.   RELATIVE VARIATIONS OF R_{ON}, L, R_{sp} AND V_{BD} VS. L_A

Relative variations	ΔL_A variations		
	−0.25 um	*+0.25 um*	*+0.5 um*
ΔR_{ON}	+4.3%	-0.8%	-0.1%
ΔL	-6.4%	+6.4%	+12.8%
ΔR_{sp}	-2.4%	+5.5%	+12.6%
ΔV_{BD}	+10.4%	-7.5%	-12.9%

On the contrary, the HCS drift curves show negligible dependence on the t_{STI} and L_{fp} variations (fig. 6, bottom), with the exclusion of the $L_{fp} = 0$ case again, which shows a significant reduction of the HCS degradation (while the breakdown voltage of the device is degraded, see fig. 5). In addition, the drift curves in fig. 6 tend to saturate towards longer stress times. At short stress times, relatively high differences in the N_{it} densities may be expected, whereas at long stress times, N_{it} tends to saturate to its maximum value at the STI corner for all the analyzed cases.

To explain the obtained results, physical quantities which are correlated with the generation of the interface trap density N_{it} along the STI boundary have been monitored, along with the current density distribution at the bias used to define $I_{D,lin}$. HCS-induced drifts at short and long stress times exhibit the highest and lowest range of variability for the L_A and t_{STI} variations, respectively. Therefore, in the following, we concentrate on a comparative analysis of these two parameters under short and long stress durations.

A. Short stress duration

As shown in fig. 7, the shortest and longest L_A cases exhibit significant differences in the electron temperature distribution and in the current density flow. A high and extended hot-electron spot under the STI corner is visible in both cases, with a higher peak and distribution in the shorter L_A case, which is partly due to the different stress biases, as the hot-spot at the STI angle is mainly due to the longitudinal electric field. Moreover, an equivalently significant normal electric field is found at the STI angle in both cases. The different distribution of the current flowing through the accumulation and drift region observed in the short and long L_A cases is the main cause of their divergent HCS behavior. The shorter L_A case clearly shows a current flowing very close to the STI interface: under such conditions, a fast trap formation is expected due to both high current density and high electric field at the interface, leading to high N_{it} densities at short stress times. On the contrary, when the accumulation region is long enough to more effectively deplete the surface region and allow electrons to flow into the substrate under the STI, the current density at the oxide interface is strongly reduced preventing a fast trap formation, thus inducing limited N_{it} densities at short stress times.

Similar considerations can be applied also to the $I_{d,lin}$ variations with θ. In particular, the $I_{d,lin}$ shifts obtained with $\theta = 90°$ are very close to the shortest L_A case. This is partly due to the high stress bias used in both cases and, in addition, to the slight reduction of the effective length of the accumulation region due to the angle variation which induces the current to flow closer to the STI interface.

The shallowest and deepest t_{STI} cases exhibit less variations in the electron temperature distribution and values, with a higher temperature peak in the thinner STI case (see fig. 8), which can be ascribed to the effects of a smaller resistive path and of a higher gate-to-drain interaction. Differently from the L_A cases, the current flowing through the accumulation and drift region in the shallowest and deepest t_{STI} cases are at a quite similar depth in the substrate (fig. 8). As a consequence the thinner STI case shows a lower surface current density. The concurrent effects of a higher temperature (larger electric field) and a lower current density at the interface give rise to N_{it} densities similar to the deepest t_{STI} case, thus limiting the variability at short stress times.

In fig. 9, the cutlines of the electron concentrations at $t = 0$ along the STI boundary for the two different L_A and t_{STI} cases are reported confirming that the geometrical variations under investigation give rise to significant changes in the electron densities at the interface, with differences for L_A ranging over more than four orders of magnitude at the STI corner, which strongly influence the probability of trap generation and the consequent interface degradation.

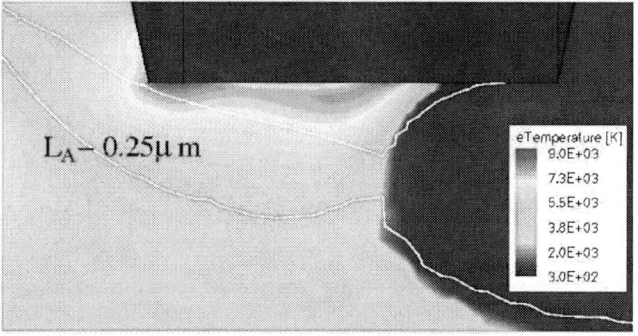

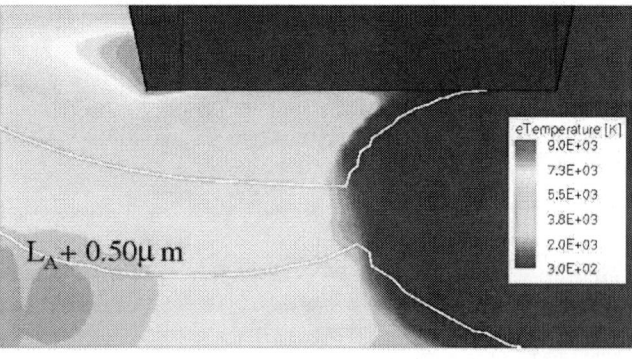

Figure 7. 2D plot of the electron temperature at t = 0 (fresh device) in the LDMOS device with two L_A values corresponding to the shortest and longest cases under study, respectively. A higher peak of electron temperature is shown in the shortest L_A case (top). Simulations were carried out at $V_{DS} = 0.88\ V_{BD}$ and $V_{GS} = 2$ V. White lines: depletion region contours.

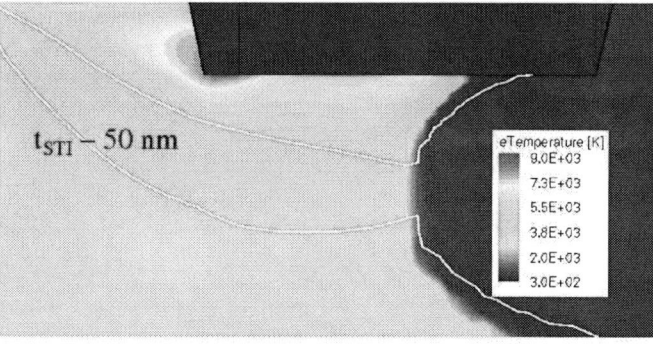

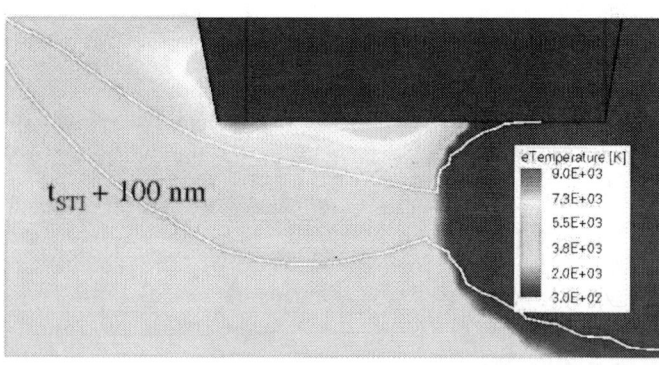

Figure 8. 2D plot of the electron temperature at t = 0 (fresh device) in the LDMOS device with two t_{STI} values corresponding to the shallowest and deepest cases, respectively. A slightly higher peak of electron temperature is shown in the thinner STI case (top). Simulations were carried out at $V_{DS} = 0.88\ V_{BD}$ and $V_{GS} = 2$ V. White lines: depletion region contours.

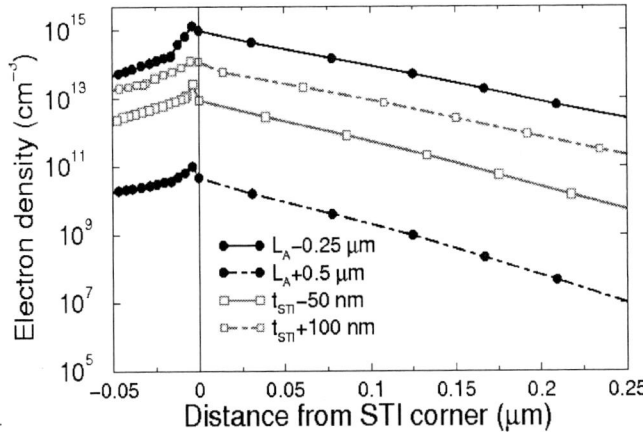

Figure 9. Cutlines along the Si/SiO₂ interface of the electron density at t = 0 for the cases corresponding to the max./min. variations of L_A and t_{STI}. The peak of the electron density occurs at the STI corner in all cases.

B. Long stress duration

Under the maximum tested stress duration, the devices with different L_A show more widely spread $\Delta I_{D,lin}$ values (ranging from 5% to 10%) than the t_{STI} variations, which show very close $\Delta I_{D,lin}$ values reaching a saturation level at about 8%. To examine the mechanisms responsible for such a different dependence on variations, the N_{it} distribution at the maximum stress time has been analyzed along with the current density at the interface at the bias corresponding to the linear regime.

At long stress times, N_{it} tends to saturate to a fixed value at the STI corner for all the analyzed cases (see fig. 10, top): in particular, the worst cases (shortest L_A and thinnest t_{STI}) show an equivalently distributed N_{it} along the interface, while reporting different $\Delta I_{D,lin}$ values (see fig. 6). Thus, the different variations in the $I_{D,lin}$ shifts with L_A and t_{STI} can be mainly ascribed to the variations of the current density at the STI interface when operating in linear regime and to the significant electrostatic effect induced by the trapped charges on the current density when localized at the STI angle.

In order to explain such effect, the linear-regime current density at the STI interface, normalized with respect to the total current $I_{D,lin}$, has been reported in fig. 10, center, for the max./min. variations of L_A and t_{STI}. The shortest L_A case shows an increase of the normalized current density at the STI corner of about 25% with respect to the longest L_A one, which may partly explain, along with the partial saturation of N_{it} (fig. 10, top), the difference (of a factor 2) in the maximum $\Delta I_{D,lin}$ drifts at long stress times. The shallowest and deepest t_{STI} cases show very similar normalized current densities at the STI corner, which, along with a similar saturation of N_{it} (fig. 10, top), confirms the limited variability of the $\Delta I_{D,lin}$ drift curves with t_{STI}.

As far as the effect of the trapped charges on transport is concerned, a very strong variation of the current density at the interface is observed at the STI angle and under the STI, while a limited change is shown at the source-side STI edge, as shown by the cutlines of the linear-regime current densities at

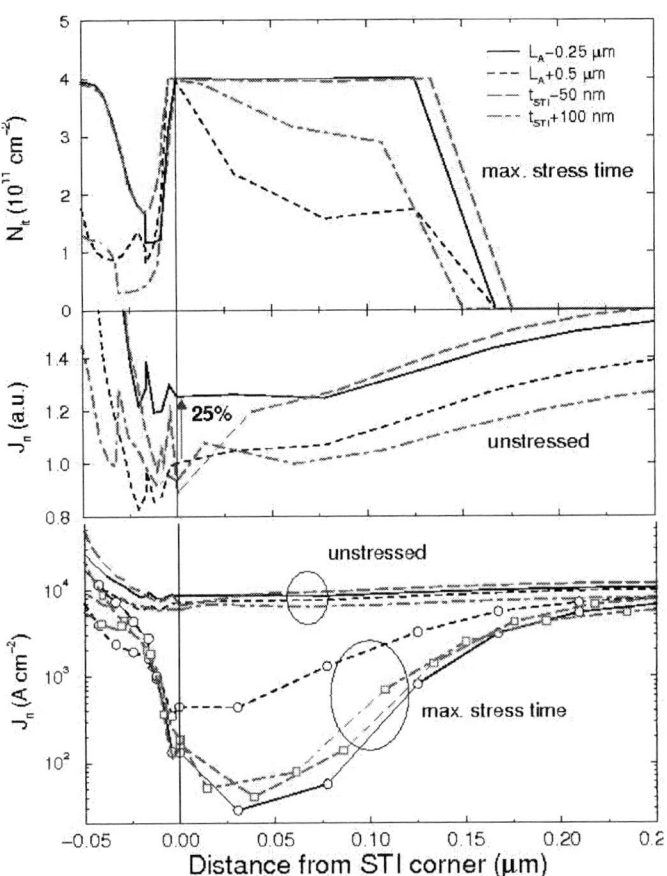

Figure 10. Cutlines along the Si/SiO₂ interface for (top:) the interface trap concentration at the maximum stress time, which is almost equal to ΔN_{it} because initial N_{it} is 10^8 cm⁻², (center:) the electron current density normalized with respect to the total current $I_{D,lin}$ at t = 0, and (bottom:) the current densities (lines) at t=0 and (lines with open symbols) in stressed conditions for the cases corresponding to the max./min. variations of L_A and t_{STI}. The critical position where interface traps play a role in changing the current density and thus inducing HCS drift is close to the STI corner in all cases.

t = 0 and after stress reported in fig. 10, bottom. Such result confirms that the drain current degradation in the analyzed LDMOS device is mainly caused by N_{it} localized at the STI angle.

IV. CONCLUSIONS

A numerical analysis of the HCS degradation has been carried out on a STI-based LDMOS device. The HCS degradation model has been calibrated on the main electrical parameter shifts at different V_{DS}. The 2D device simulation nicely predicts the relative drift curves of the linear drain current and maximum transconductance at different stress biases. As HCS in STI-based LDMOS devices is a critical aspect, its variability with a number of geometrical parameters has been analyzed to gain indications on reliability and performance trends.

REFERENCES

[1] P. Hower, J. Lin, S. Pendharkar, B. Hu, J. Arch, J. Smith, and T. Efland, "A rugged LDMOS for LBC5 technology," Proc. of the Int. Symp. On Power Semiconductor Devices and ICs, ISPSD'05, 23-26 May 2005, pp. 327–330.

[2] J. Lin, and P. Hower, "Two-carrier current saturation in a lateral DMOS," Proc. of the Int. Symp. On Power Semiconductor Devices and ICs, ISPSD'06, 4-8 June 2006, pp. 1–4.

[3] S. Reggiani, E. Gnani, A. Gnudi, G. Baccarani, M. Denison, S. Pendharkar, R. Wise, and S. Seetharaman, "Investigation on saturation effects in the rugged LDMOS transistor," Proc. of the Int. Symp. On Power Semiconductor Devices and ICs, ISPSD'09, 14-18 June 2009, pp. 208–211.

[4] J.F. Chen, K.-S. Tian, S.-Y. Chen, K.-M. Wu, and C. M. Liu, "On-Resistance Degradation Induced by Hot-Carrier Injection in LDMOS Transistors With STI in the Drift Region," IEEE Electron Device Letters, Vol. 29, pp.1071–1073, 2008.

[5] Y. Rey-Tauriac, J. Badoc, B. Reynard, R.A. Bianchi, D. Lachenal, and A. Bravaix, "Hot-carrier reliability of 20V MOS transistors in 0.13 μm CMOS technology," Microelectronics Reliability, Vol. 45, pp. 1349–1354, 2005.

[6] S. Reggiani, E. Gnani, A. Gnudi, G. Baccarani, M. Denison, S. Pendharkar, R. Wise, and S. Seetharaman, "Explanation of the Rugged LDMOS Behavior by Means of Numerical Analysis," IEEE Trans. On Electron Devices, Vol. 56, pp. 2811–2818, 2009.

[7] Synopsys Inc., "Sentaurus device simulator (release z-2007.03)," 2007.

[8] O. Penzin, A. Haggag, W. McMahon, E. Lyumkis, and K. Hess, "MOSFET degradation kinetics and its simulation," IEEE Trans. On Electron Devices, Vol. 50, pp. 1445–1450, 2003.

[9] Z. Chen, K. Hess, J. Lee, J.W. Lyding, E. Rosenbaum, I. Kizilyalli, S. Chetlur, and R. Huang, "On the Mechanism for Interface Trap Generation in MOS Transistors Due to Chanel Hot Carrier Stressing," IEEE Electron Device Letters, Vol. 21, pp.24–26, 2000.

[10] J.F. Chen, S.-Y. Chen, K.-M. Wu, J.R. Shih, and K. Wu, "Convergence of Hot-Carrier-Induced Saturation Region Drain Current and On-Resistance Degradation in Drain Extended MOS Transistors," IEEE Trans. On Electron Devices, Vol. 56, pp. 2843–2847, 2009.

LIFETIME EXTRAPOLATION FOR ELECTROMIGRATION TESTS AT WAFER LEVEL WITH A DEDICATED DEVICE

Chappaz, C., Nakkala P.
STMicroelectronics
850 rue Jean Monnet, 38926 CROLLES CEDEX, +33 492382108; fax: + 33 492382956;

e-mail: cedrick.chappaz@st.com

ABSTRACT

A dedicated new device is proposed to address the electromigration on BEOL (Back-End of Line) at wafer level. A local heat is directly generated around the tested Electromigration via ended Metal line (EVEM) based on a NIST (National Institute of Standard Technology) [1]. A coil plays the role of heater and allows separating the temperature stress from the current stress. Black's parameters obtained and extrapolated lifetimes are closer to package level results. Thermal gradient is minimized and so extrapolated lifetimes are valid. [*Keywords:* electromigration, self heating, coil, lifetime extrapolation, wafer level]

INTRODUCTION

It's mostly admitted that electromigration parameters can only be precisely determined at package level. Indeed, the well known Black's equation [2] implicates to separate current and temperature effects. Usual electromigration at wafer level on standard structures are done using the self-heating brought by a high current in the line. This direct correlation between current and temperature of test doesn't allow extracting Ea and n properly [3].

Previous studies proposed a polysilicon resistor below the metal line to heat the structure [4]. They emphasize good correlation with package level tests. Limitation in this case is that efficiency is lower for high metal level because of dissipation in bulk silicon.

To avoid this problem, we designed a new structure that locally heats the tested metal line separately with the current effect.

In a previous paper, we demonstrated that the heating coil was able to heat a standard mono level metal line [5]. We extended this results on a via ended device (V2M2) with the heating coil surrounding the current leads. Indeed, to be as close as possible to package level tests, heater must heat current leads to reduce the thermal gradient.

A HIGH EFFICIENCY HEATER

To ensure a Mean Time to failure compatible with wafer levels tests constraints, the heating coil has to be able to reach temperatures above 350°C. We evaluated the target temperature on the EVEM versus the dissipated power (Pd) in the heating coil (Fig. 2).

Technologies tested are 120nm & 65nm. A temperature above 400°C can easily be reached. Same measurements were carried out in C65nm. Results clearly demonstrate a drastically reduction of the dissipated power for the same temperature increase. Indeed, the thicknesses and material properties impact the thermal resistance.

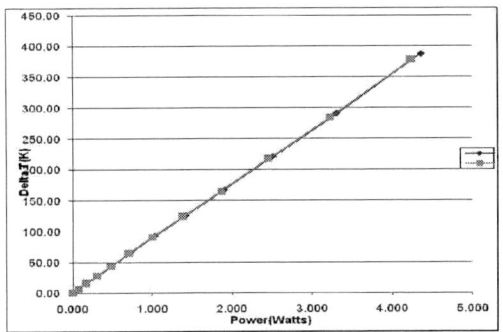

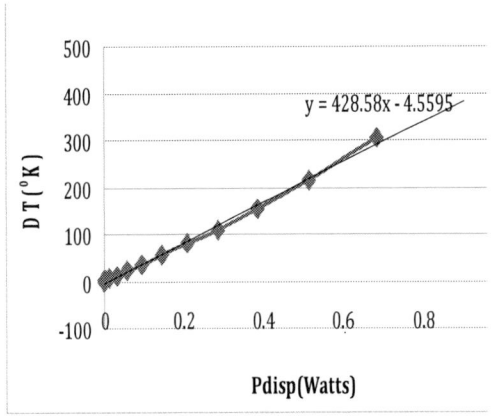

FIG. 2: ΔT VS PD FOR M2 (BLUE CURVE) AND V2M2 (PURPLE) DEVICES SURROUNDED BY THE HEATING COIL IN 120NM TECHNOLOGY. REQUIRED PD IS DIVIDED BY AN ORDER OF 5 FOR THE SAME TEMPERATURE IN 65NM TECHNOLOGY (BOTTOM FIGURE).

For both technologies and configuration, the efficiency, defined as the ratio of $\Delta T_{EVEM} / \Delta T_{HCOIL}$, is above 85% whatever the dissipated power value (see Fig. 3).

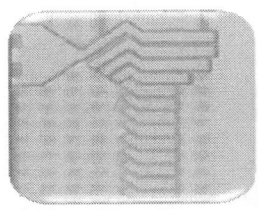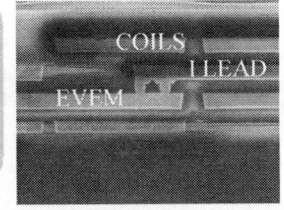

FIG. 1: SCHEMATIC VIEW OF THE HEATING COIL SURROUNDING THE METAL LINE & CURRENT LEADS TESTED (LEFT). SEM SIDE VIEW OF THE HEATING COIL WITH V2M2.

978-1-4244-5430-3/10 $26.00 © 2010 IEEE

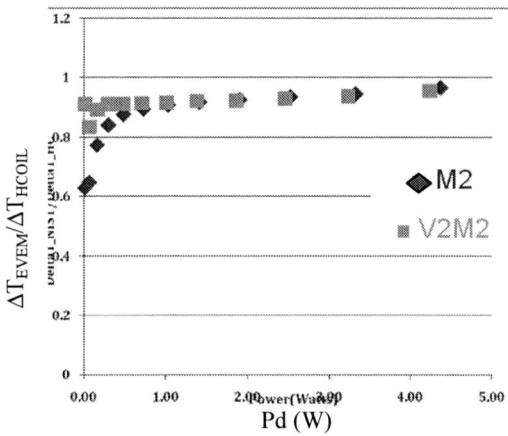

FIG. 3: $\Delta T_{EVEM}/\Delta T_{HCOIL}$ vs Pd for M2 (blue curve) and V2M2 (purple) devices surrounded in 120nm technology node.

RESULTS

Black's parameters extraction

In order to extrapolate the lifetime, different couple of stress conditions (T°_{target}, J_{EVEM}) with a statistical population of 30 dice were used. Table 1 summarizes MTF values obtained for different experiments.

TABLE I:
MTF vs. TARGET TEMPERATURE AND J_{EML} ON METAL 2

T_{Target} [°C]	J_{EVEM} [mA/μm²]	MTF [s]	SD
330	333	1332	0.24
350	288	1010	0.19
370	333	967	0.21
370	291	591	0.26

It is important to note that joule heating in the EVEM was necessary for 2 reasons: Firstly to reduce the Time to Failure for each die. Secondly, to ensure that the failure occurs in the EVEM and not in the heating coil.

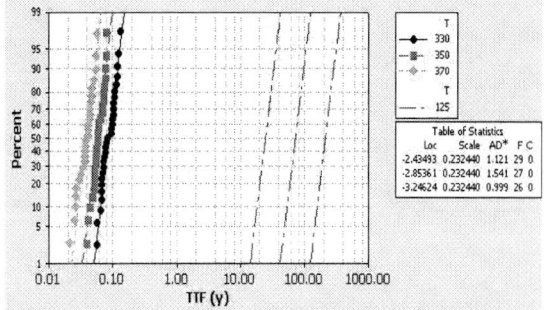

FIG. 4: TTF DISTRIBUTIONS (@T=330, 350 & 370°C) EXTRAPOLATED @ T=125°C, J_{OP} WITH BLACK'S PARAMETERS.

From Fig. 4, we can deduce that the same failure mechanism was emphasized for all experiments. Indeed, the standard deviation is almost constant around 0.2, value which is inline with package level tests.

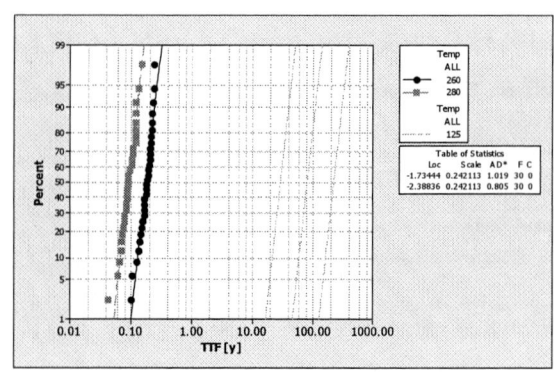

FIG. 5: TTF DISTRIBUTIONS (@T=260 & 280°C) FOR PACKAGE LEVEL TESTS EXTRAPOLATED @ T=125°C, J_{OP}.

Moreover, we extracted the activation energy (Ea=0.67eV) and the acceleration factor was found close to 2. These values, which are slightly different from package level ones, are the sign of the Joule heating impact in the EVEM.

Concerning extrapolated lifetimes, we found close values between wafer level tests (TTF=47y) and package level tests (TTF=32y) (Fig. 5).

Thermal gradient along EVEM

To ensure that extrapolated lifetime is in agreement with standard package level tests, we evaluated the thermal gradient along the device. Indeed, X. Federspiel [6] demonstrated that the mismatch between wafer level and package level tests mainly came from a high thermal gradient on the EVEM edge.

We used the following heat diffusion equation:

$$-kcu.S.\frac{\partial^2 T}{\partial x^2} + \frac{\rho_{cu}(T).I^2}{s} - \frac{T-T_{ext}}{R_{th}} = 0 \quad (1)$$

With kcu: copper thermal conductivity, ρ_{cu}: thermal resistivity of copper, R_{th} thermal resistance of the device and T_{ext}: external temperature. In the case of the EVEM with a heating coil, T_{ext} is defined as the temperature fixed by the heating coil.

Solving this equation, we plotted the temperature along the EVEM.

For the same target temperature (T_{target}=350°C), we observe a huge thermal gradient in the case of a standard EVEM (ΔT=330°C) which is not the case with the heating coil with a ΔT<20°C between the current lead and the middle of the NIST (see Fig. 6).

978-1-4244-5430-3/10 $26.00 © 2010 IEEE

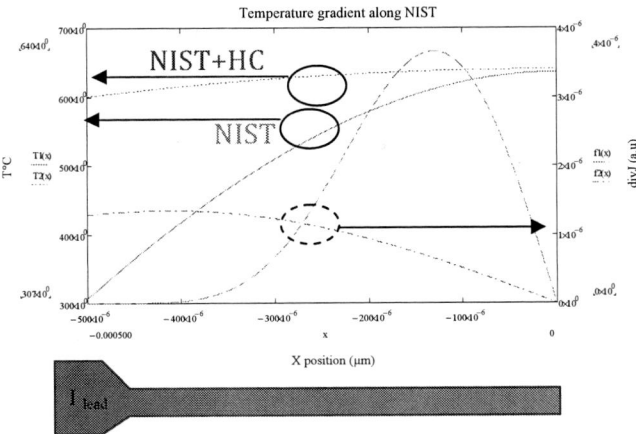

FIG. 6: THERMAL GRADIENT FROM THE CURRENT LEAD TO THE MIDDLE OF THE EVEM IN CASE OF A STANDARD NIST (RED) AND A EVEM WITH AN HEATING COIL (BLUE).

REFERENCES

[1] Schafft, H.A., Staton, T.C., Mandel, J., Jschott,D., "Reproducibility of électromigration measurements", IEEE Transactions on Electron Devices, 1987, Vol. ED-34, N°3,p. 673-681.

[2] J.R Black, "Electromigration failure modes in Aluminum metallization for semiconductor devices" Proc IEEE, 1969, Vol. 57, n°9, pp. 1587-1594.

[3] M. Sakimoto, T. Itoo, T. Fujii, H. Yamaguchi, K. Eguchi, "Temperature measurement of Al metallization and the study of Black's model in high current density", Reliability Physics Symposium, 1995. 33rd Annual Proceedings., IEEE International
4-6 April 1995, pp.333 – 341

[4] Hin-Kiong Yap; Kin-Leong Yap; Yew-Chee Tan; Keng-Foo Lo; "Wafer level electromigration testing on via/line structure with a poly-heated method in comparison to standard package level tests",
Physical and Failure Analysis of Integrated Circuits, 2003. IPFA 2003. Proceedings of the 10th International Symposium on the
7-11 July 2003 pp. 75 – 79

[5] Chappaz C., Electromigration: from package level to wafer level thanks to a heating coil structure, ICMTS 2008.

[6] Federspiel X., Effect of Joule heating on the determination of Electromigration parameters, IEEE/IRW 2003.

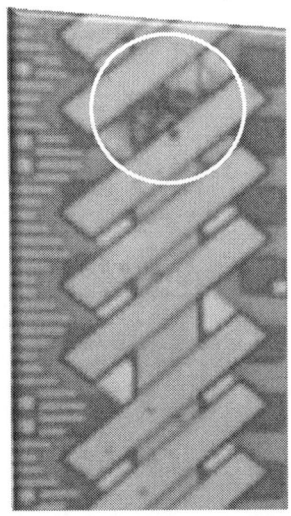

FIG. 7: FAILURE OBSERVATION ON NIST AFTER ELECTROMIGRATION TEST @350°C WITH THE HEATING COIL.

The failure location on M2 NIST confirms the thermal gradient position and impacts on the electromigration test (Fig. 7).

CONCLUSIONS

We developed and tested a new heating coil dedicated to electromigration tests at wafer level. This free device can address all metal levels of BEOL up to M_{N-1}. Efficiency above 85% gives the capability to reach temperature above 400°C.

Comparison with package levels clearly validates the strategy to use this device for lifetime extrapolation.

Whatever joule heating is used to reduce the time to failure for each tested die, almost no thermal gradient is generated on the EVEM edge.

Finally, this device is a very powerful architecture to check the electromigration behavior on all kind of technologies.

978-1-4244-5430-3/10 $26.00 © 2010 IEEE

Ultra-Low-k Dielectric Degradation before Breakdown

T. Breuer*, U. Kerst, C. Boit

Semiconductor Devices Division
TU Berlin -- Berlin Institute of Technology
Einsteinufer 19, Sekr. E2
D-10587 Berlin, Germany
*corresponding author: phone (+49) (0)30 314 78543, taro.breuer@tu-berlin.de

E. Langer, H. Ruelke

GLOBALFOUNDRIES Dresden Module One Limited Liability Company & Co. KG
Wilschdorfer Landstr. 101
D-01109 Dresden, Germany

Abstract— **This paper presents a basic investigation of Ultra Low K (ULK) SiCOH dielectrics degradation before breakdown. For the first time very early stages of degradation before breakdown have been revealed and a theory of the basic process of ULK alteration under electrical stress has been proposed. Tip electrode test structures have been specifically designed for this investigation in order to determine the location of degradation and breakdown. A stepwise increased voltage stress test with a meticulously observed current in fA range was developed and successfully applied.**

Keywords: dielectric, degradation, ultra-low-k, ULK, SiCOH, breakdown, pores

I. INTRODUCTION

Recently signal delays due to interconnect construction have become a limiting factor for the performance of high-speed integrated circuits. An option to lower signal delays ("RC-delay") is to lower the capacitance of the dielectrics between the interconnect lines [1]. As capacitance is dependent on the spacing, it becomes worse when moving to technologies with smaller dimensions. The main solution is to use a different material ('low-k' material) with a lower dielectric constant than k=3.9 of the standard dielectric SiO_2.

Improvement can be achieved by using organosilicate glass (OSG) with a suitable barrier layer on top [2] [3]. This kind of materials are also known as SiCOH, SiOC:H, carbon doped oxide (CDO) or several industry brand names.

The introduction of pores into the dielectric allows further reduction of the effective k-value ('ultra-low-k' material). Ideally the k value approaches pure air properties (k=1). These materials, especially the porous ones, are considerable less rugged than the standard glasses and affect reliability robustness and therefore have to be investigated.

Degradation behavior is assumed to be dependent on the mechanical construction of the interconnect stack consisting of different layers and is also impacted by the density of defects.

II. TIP-TO-TIP TEST STRUCTURES

Time- Dependent- Dielectric- Breakdown (TDDB) measurements or voltage ramp tests on comb-structures are a frequently applied method to determine the degradation and lifetime of the dielectric of long adjacent lines [4]. These structures are very sensitive to defect density. However these tests are not specific to the intrinsic electrical properties of dielectrics and the degradation process itself cannot be investigated by failure analysis because the area of breakdown is unknown in advance.

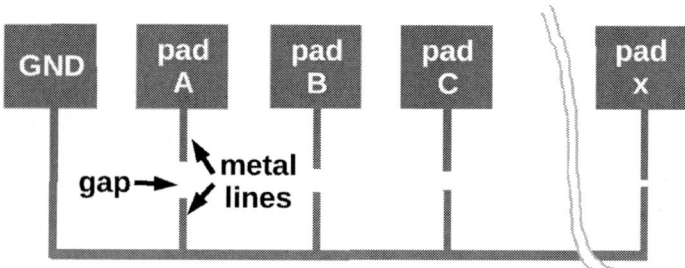

Figure 1: Layout of the test structures. Ground pad GND ('cathode') and several tip-to-tip structures with different gap sizes. Each of them is connected to a corresponding pad A-X ('anode') where a voltage ramp can be applied.

To characterize the fundamental electrical properties of low-k dielectrics, measurements were performed on test structures with a gap between copper lines acting as two opposed electrodes embedded in ULK dielectric (k=2,4) and covered with a thin layer of BLOk™ barrier [3]. Therefore the damage region is in contrast to comb-structures located to a very limited area (Fig. 1) and the forced degradation can be directly investigated focusing on dielectric intrinsic incidents.

978-1-4244-5430-3/10 $26.00 © 2010 IEEE

III. INDUCING STRESS BY VOLTAGE RAMPING

In order to inspect the degradation process before the structure will be destroyed by a breakdown, a sensitive current monitoring setup in the fA range has been developed for measurement during voltage ramp application. Previously unstressed samples have been stressed with a voltage ramp stopped at selected voltages in order to investigate the respective degrees of dielectric degradation.

Figure 2 depicts the typical current characteristic of a voltage ramp, operating at voltages well below breakdown. It shows two phases. Phase 1 shows a steep rise above the noise level. In the second phase the current increase is less pronounced indicated by a smaller gradient of current. This is shown in Fig. 2 by the region between upper and lower line in phase 2. In a few cases the current can even slightly decrease, drawn as the lower line in phase 2. These individual differences depend on individual samples comprising statistically distributed material properties. The current is believed to degrade the dielectric properties further and impacts the overall degradation behavior.

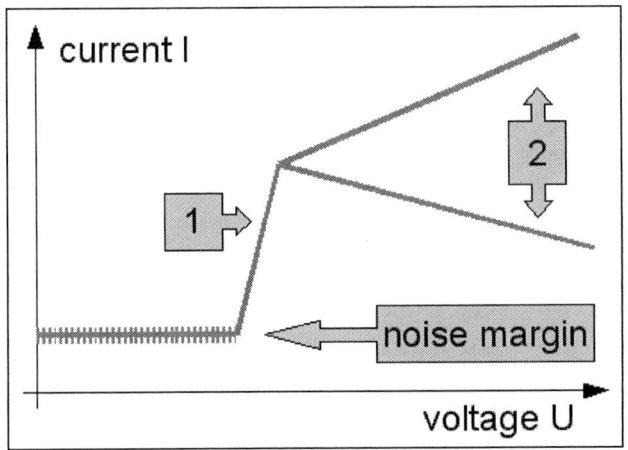

Figure 2: Typical characteristics of leakage current when voltage ramp is applied. Two phases. In the second phase (2) the current does not increase as much as in phase one and is found between the indicated upper and lover line.

An example is shown in Fig. 3. The sample was stressed using a 0.22V/s voltage ramp. The corresponding degradation level is shown in Fig. 7 and Fig. 8.

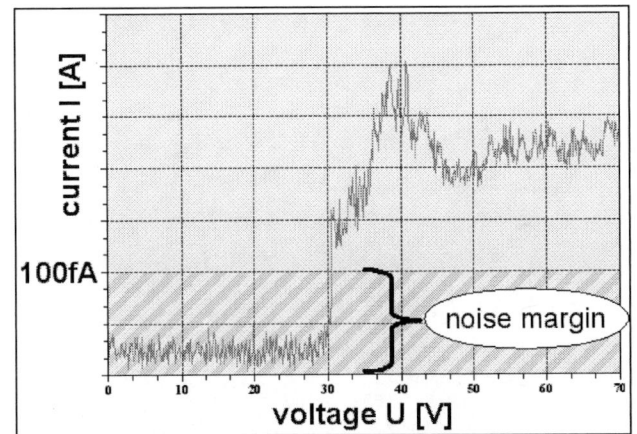

Figure 3: Leakage current (linear axis) vs. voltage ramp. Example Sample F.

IV. SEM CROSS SECTIONS OF STRESSED SAMPLES

The subsequent Scanning Electron Microscopy (SEM) pictures of cross sections made by Focused Ion Beam (FIB) show three examples of degradation of 100nm-gap test structures.

The voltage ramp applied on sample A was stopped at 50V. The current characteristic showed beginning of phase 2 (Fig. 2), but no degradation effect in the SEM picture is visible.

Sample B, voltage ramp stopped at 60V and medium phase 2 current, shows a decomposition of the left electrode (anode) extending into the dielectric filled gap.

Sample C, voltage ramp stopped just before estimated breakdown at 70V, shows an expanded alteration of dielectric into bubble-shaped holes at the anode. The current showed a pronounced phase 2.

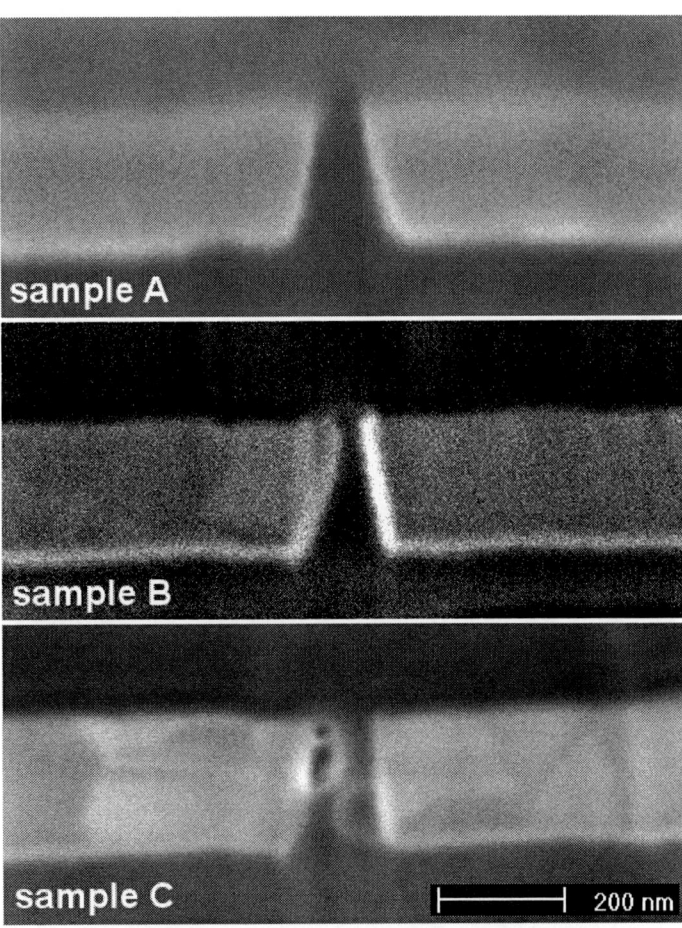

Figure 4: Samples showing three different degradation levels.

V. HIGHEST STRESS LEVEL AT THE ANODE

The distribution of the electric field was simulated (Fig. 5). The highest field at the top interface between dielectric and BLOk barrier layer. The close up (different color scaling) shows the highest field at the anode causing highest stress. This is an explanation, why degradation usually starts at the anode. See Fig. 4, 6, 7, 8.

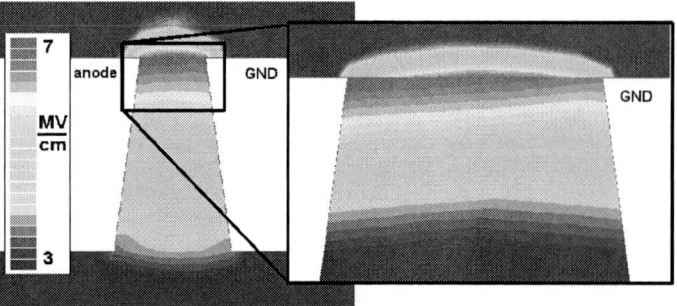

Figure 5: Simulation of electrical field distribution at the test structure when applying 70V at 100nm gap. Close up (different color scaling) shows highest field at the top corner of the anode.

VI. TEM INVESTIGATION AND MATERIAL ANALYSIS OF THE DEGRADATION PROCESS

Transmission Electron Microscopy (TEM) images of the stressed structures (Fig. 6-8) show the significant alteration in the area between the tips in more detail. In sample D (left sample in Fig. 6), at the edge of the positive electrode, a small blurry area is visible. More degradation forming a small void is visible in sample E (right sample in Fig. 6). This is an example of the beginning void formation at the location of highest stress in the electrical field (Fig. 5). An example of highly developed voiding is shown in Fig. 7. A big bubble shaped hole is visible at the anode.

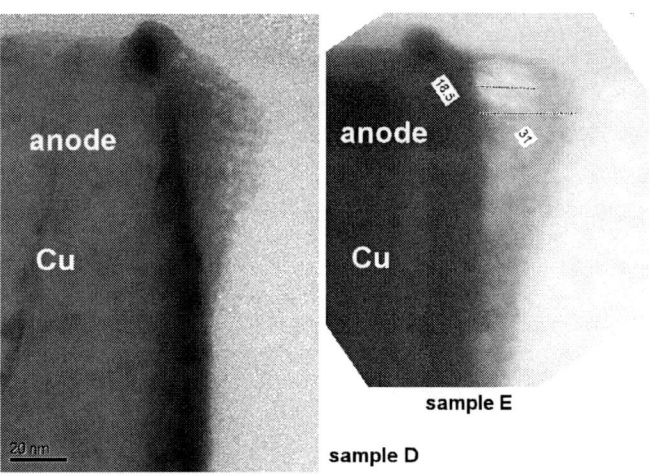

Figure 6: TEM image showing blurry area of tantalum decomposition in detail. Sample showing visible degradation. Dielectric looks bright.

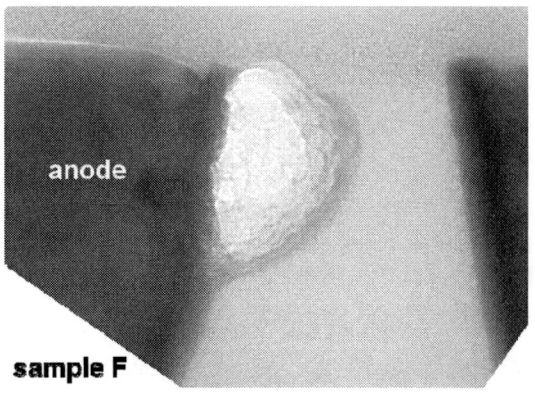

Figure 7: TEM image showing bubble-shaped hole of sample F in detail.

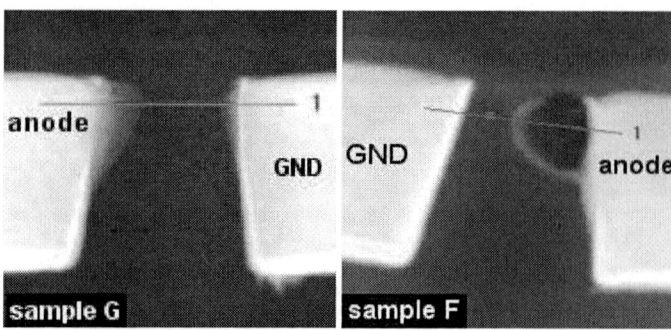

Figure 8: TEM-images showing different degradation levels. Sample G shows a deconstruction of the tantalum layer from the anode (left). Sample F shows formation of a bubble-shaped hole at the anode (right, sample flipped).

Two TEM samples have undergone an Energy Dispersive X-ray material analysis (EDX). EDX plots of the samples G and F reveal a decomposition of the tantalum layer between copper and dielectric (Fig. 9 and Fig. 10). The EDX plots in Fig. 9-12 show the results of the investigation in the samples along the lines from left to right in Fig. 8. Sample G shows a field assisted transport of tantalum into the dielectric and the blurry area can be identified as strongly enriched with tantalum (Fig. 9). Sample F shows that the surface of the stress-formed bubble is covered with tantalum as well (Fig. 10).

The tantalum transport is accompanied with a small amount of copper atoms shown in Fig. 11 and Fig. 12.

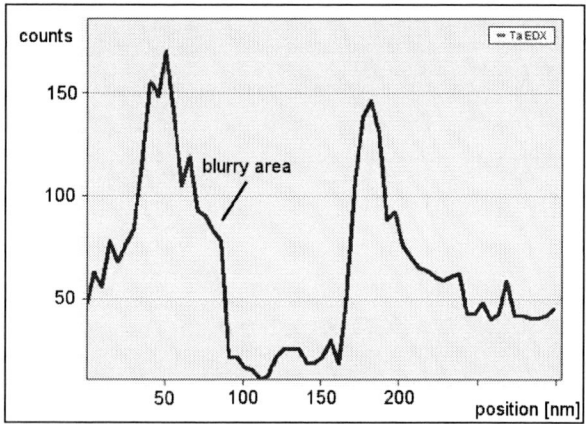

Figure 9: EDX analysis of tantalum. Sample G (anode left).

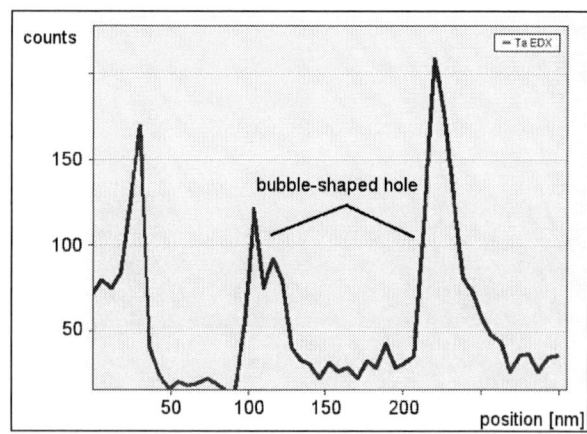

Figure 10: EDX analysis of tantalum. Sample F (anode right).

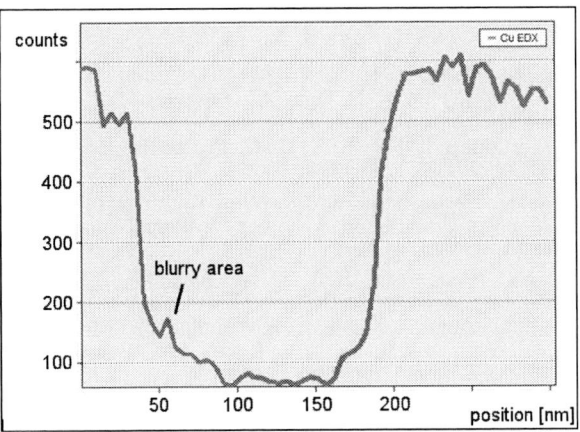

Figure 11: EDX analysis of copper. Sample G (anode left).

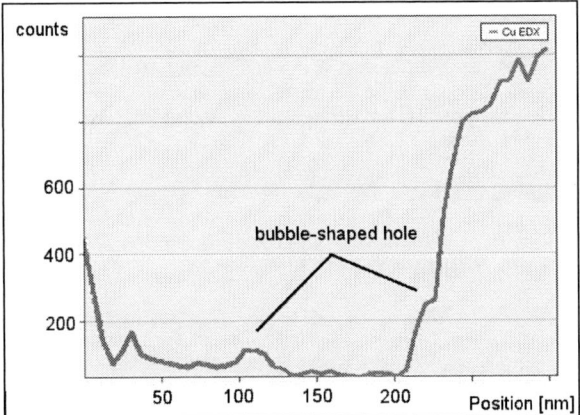

Figure 12: EDX analysis of copper. Sample F (anode right).

VII. DELAMINATION MODELS

A theory to explain the observations is that the observed tantalum transport and the formation of a void starting from the edge of the anode could be caused by delamination (Fig.13). This delamination might be due to field assisted tantalum transport from the liner in the direction of the dielectric. The liner between copper and dielectric is a Ta/TaN bilayer and is used as a barrier layer to avoid diffusion of copper into the dielectric. Interface reliability is an important topic in interconnect technology and has been previously investigated at a SiCN/SiCOH interface [5].

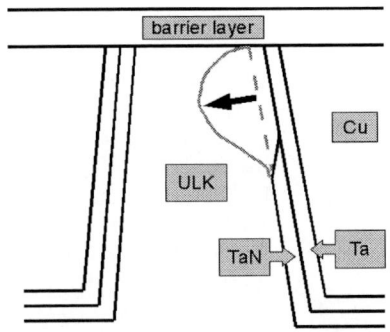

Figure 13: Delamination of tantalum in the test structure

Two mechanisms for tantalum transport are proposed:

1. delamination of dielectric at the dielectric/TaN interface. The tantalum diffuses out along the surface of the void.
2. delamination in the liner itself, leaving behind the Ta layer and taking away part of the TaN layer covering the void.

VIII. MOLUCULAR STRUCTURE OF POROUS SiCOH DIELECTRIC

Ultra-low-k SiCOH dielectric is based on an amorphous SiO_2 base structure. The introduction of pores is essentially made by adding hardly linking methyl groups (Fig. 14) connected to the basic SiO_2 structure and an additional UV curing [6] [7].

Spots of missing linkage in the molecular structure after curing leave small pores in the dielectric causing a lower overall density and lower mechanical stability of the dielectrics.

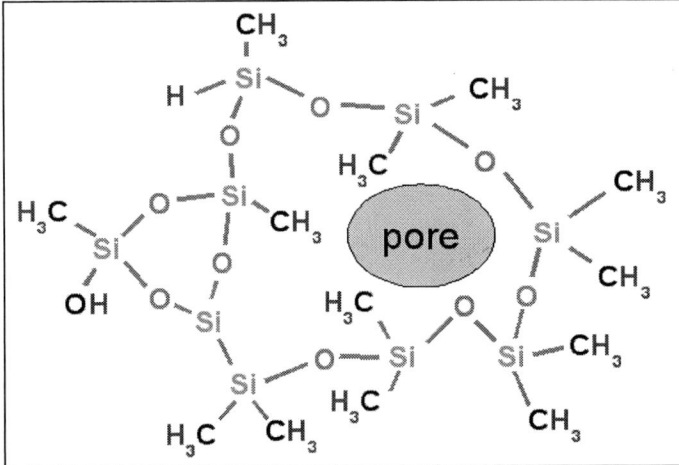

Figure 14: Molecular structure of porous SiCOH with CH_3 methyl groups [7]

IX. THEORY OF DIELECTRIC DEGRADATION AND CONCLUSIONS

In Fig. 4, 7 and 8 a formation of bubble-shaped holes at the dielectric-liner interface after inducing electrical stress was found. The formation of large holes inside the dielectric without migration of metal deeply into the dielectric suggest a dielectric-inherent origin of voiding. The proposed way of tantalum transport (along the interface) does not indicate electromigration induced transportation but rather a clearly delimited distribution of metal with a pretty big surface while forming the void in the porous dielectric.

This may be explained as a dynamic process of shrinkage of dielectric. A decomposition of methyl groups is assumed to enable a densification and creating a modified material comprising different stronger linkage and higher density in the absence of pores. It is also assumed to contain a larger amount of carbon as in dense SiCOH and presumably having different properties.

The high stability of dielectric while applying the voltage ramp and forming a conductive tantalum bridge may indicate a dense, very hard dielectric of high breakdown robustness. Therefore the k value of the modified material is assumed be higher than before [8], e.g. k>>2.4. The lower gradient of current in phase 2 (Fig. 2) also implies increased robustness. It is also assumed that because of extinction of pores in the proposed process after some time under normal IC operation conditions the modified dielectric may reach an equilibrium between a denser dielectric and voids.

ACKNOWLEDGMENT

Part of this investigation was carried out as a part of the Project "PULSAR" (Nr.: 01M3173) of the German Federal Ministry of Education and Research (BMBF). TEM and EDX investigations were kindly provided by Globalfoundries, Dresden, Germany (Fig. 6-12). Fig. 1-5 and 13 are property of TU Berlin -- Berlin Institute of Technology.

REFERENCES

[1] Schindler, G.; Steinhögl, W.; Steinlesberger, G.; Traving, M.; Engelhardt, M.: "Scaling of parasitics and delay times in the backend-of-line." Microelectronic Engineering, Vol. 70, Issue 1, 2003, pp 7-12. doi:10.1016/S0167-9317(03)00285-5

[2] Grill, A.; Gates, S.; Dimitrakopoulos, C.; Patel, V.; Cohen, S.; Ostrovski, Y.; Liniger, E.; Simonyi, E.; Restaino, D.; Sankaran, S.; Reiter, S.; Demos, A.; Yim, K. S.; Nguyen, V.; Rocha, J.; Ho, D.: "Development and optimization of porous pSiCOH interconnect dielectrics for 45nm and beyond." 2008 IEEE IITC Proceedings 2008, pp. 28-30 . doi:10.1109/IITC.2008.4546915

[3] Ping Xu; Kegang Huang; Patel, A.; Rathi, S.; Tang, B.; Ferguson, J.; Huang, J.; Ngai, C.; Loboda, M.: "BLOk^{TM} – A Low-k Dielectric Barrier/Etch Stop Film for Copper Damascene Applications". 1999 IEEE IITC Proceedings, pp. 109-111. doi:10.1109/IITC.1999.787093

[4] Aubel, O., Kiene, M., Yao, W.: "New Approach of 90nm Low-k Interconnect Evaluation Using a Voltage Ramp Dielectric Breakdown (VRDB) Test", 2005 IEEE IRPS Proc., pp. 483-489

[5] Guedj, C.; Claret, N.; Arnal, V.; Aimadeddine, M.; Barnes, J.P.; Barbe, J.C.; Arnaud, L.; Reimbold, G.; Torres, J.; Passemard, G.; Boulanger, F.: "Dielectric Conduction Mechanisms of Advanced Interconnects: Evidence for Thermally-induced 3D/2D Transition." 2006 IEEE IRPS Proceedings, p. 502-506. doi:10.1109/RELPHY.2006.251269

[6] Grill, A.; Neumayer, D. A.: "Structure of low dielectric constant to extreme low dielectric constant SiCOH films: Fourier transform infrared spectroscopy characterisation" Journal of Applied Physics, Vol. 94 (2003) p. 6697-6707. doi:10.1063/1.1618358

[7] Yuan, C.A.; van der Slus, O.; Zhang, G.Q., Ernst, L.J.; van Driel, W. D.; Flower, A.E.; van Silfhout, R.B.R.: "Molecular simulation strategy for mechanical modeling of amorphous/porous low-dielectric constant materials." Applied Physics Letters 92 (2008). doi:10.1063/1.2832639

[8] Ogawa, E.T.; Jinyoung Kim; Haase, G.S.; Mogul, H.C.; McPherson, J.W.: "Leakage, Breakdown, and TDDB Characteristics of Porous Low-k Silica Based Interconnect Dielectrics". 2003 IEEE IRPS Proceedings, pp. 166-172. doi:10.1109/RELPHY.2003.1197739

Analysis of the Impact of Linewidth Variation on Low-K Dielectric Breakdown

Muhammad M. Bashir and Linda Milor

Microelectronics Research Center
Georgia Institute of Technology
Atlanta, GA 30332 USA

Abstract— **Low-k time-dependent dielectric breakdown (TDDB) has been found to vary as a function of metal linewidth, when the distance between the lines is constant. Modeling requires determining the relationship between TDDB and layout geometries. Therefore, comb test structures have been designed and implemented that vary pattern density and linewidth independently in 45nm technology. Models are computed to estimate TDDB as a function of linewidth, and the cause of variation in TDDB behavior is investigated.**

I. INTRODUCTION

Low-k time-dependent dielectric breakdown is considered to be one of the most important reliability issues during Copper/Low-k (Cu/low-k) technology development and its qualification. Lower breakdown field strengths of porous low-k materials, the susceptibility of low-k materials to mechanical damage by chemical mechanical polishing (CMP), contamination due to photoresist poisoning, and copper drift are some of the reasons Cu/low-k interconnects systems are vulnerable to breakdown. Reduced supply voltage scaling with respect to feature size scaling compounds the problem and results in exponentially increasing electric fields among interconnects every technology generation. Time-dependent dielectric breakdown (TDDB) of damascene structures has to be assessed as a system of a dielectric, diffusion barrier, cap layer, and copper interconnect. An example of this system is shown in Fig. 1(a).

Backend dielectric reliability testing relies on comb structures, shown in Fig. 1(b). The comb structures create a lateral stress across the dielectric between the fingers of the comb, which are separated by the minimum distance design rule. A voltage difference, V, is applied to the comb, which creates a lateral electric field through the intra-layer dielectric.

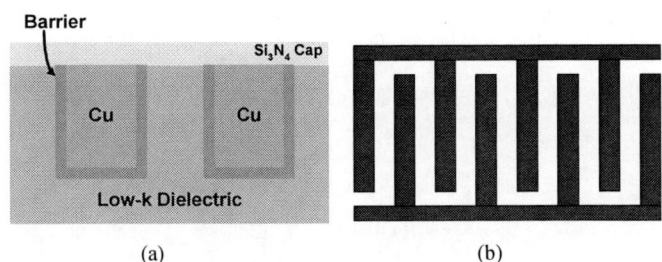

(a)   (b)

Figure 1.   (a) Cross-section of an example copper/low-k interconnect system (b) Comb test structure.

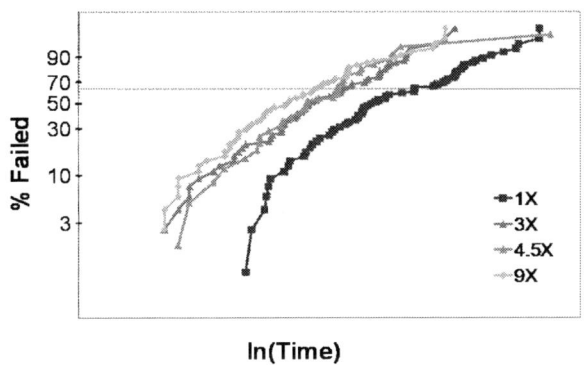

ln(Time)

Figure 2.   Example Weibull plot for comb test structures with four areas: 1X, 3X, 4.5X, and 9X.

Data are collected for a sample of comb structures. The data are ordered from shortest to longest breakdown time. Each time point is assigned a probability point, P_i, by partitioning the probability scale equally. An example is shown in Fig. 2.

The data is fit by a distribution, either the Weibull distribution or the Log-Normal distribution [1], in order to enable extrapolations to lifetimes at low percentiles [2]. The resulting data is then scaled to use conditions [3]-[5] and to the vulnerable area corresponding to the chip [6].

This paper looks at the variation in low-k dielectric breakdown times as a function of metal width, a parameter that is not supposed to impact failure rates. Metal width is the width of the Cu interconnect lines drawn on the mask. Any change in the actual linewidth on chip, from the linewidth drawn on the mask, will also cause a change in actual linespace between the interconnects. Note that in this work we use the term metal width and linewidth interchangeably. Similarly, we use the terms space and linespace interchangeably. We refer to the fraction of the area of interconnect lines on the mask to the fraction of the area of the mask as pattern density.

This paper begins with a summary of prior work on modeling variation in failure rates as a function of linewidth in Section 2. Section 3 describes the test structures that were used in this work, which were implemented on a 45nm test chip, and summarizes the TDDB measurement results. In Section 4 we consider several possible explanations of the observed variation in characteristic lifetime as a function of linewidth. Section 5 uses the results in Section 4 to create a model of characteristic

978-1-4244-5430-3/10 $26.00 © 2010 IEEE

lifetime as a function of linewidth. Section 6 concludes the paper with a summary.

II. PRIOR WORK ON THE IMPACT OF LINEWIDTH ON LOW-K DIELECTRIC BREAKDOWN

The two dominant models of dielectric lifetime, the E Model [3] and the $\sqrt{E}$ Model [4], [5], relate time-to-failure (TF), to electric field. In both models, the only factor that determines TF for structures manufactured using the same low-k dielectric is the electric field (E). Electric field in backend structures is a function of the distance between the interconnect lines, termed linespace (S),

$$ E = \frac{V}{S}. \qquad (1) $$

In our test structures and in prior work on this topic, S is constant, and only the linewidth is varied. In prior work, with 180nm technology, experimental data indicated that time-to-breakdown was a function of linewidth [7]. Analysis found that the difference in time-to-breakdown was due to a physical difference in the distance. The data was analyzed to determine an explanation for the difference in distance. The explanation that best matched the data was microloading in etch [7]. The microloading effect was explained as a sensitivity of etch rate to pattern density [8], [9] .

The SEM results were then used to model variation in distance between the lines as a function of etchable area. Using the data, a model of lifetime as a function of linewidth that matched the data well was found [7]. However, the test structures used to analyze the impact of metal linewidth confounded the impact of linewidth with pattern density, as can be seen in Fig. 3. Fig. 3 shows that whenever linewidth is increased, while keeping linespace constant, the pattern density also increases.

Hence, although the theory associates the time-to-breakdown difference with pattern density, we could not conclusively verify that pattern density, rather than linewidth, produced the time-to-breakdown difference. This paper aims to distinguish between these two factors.

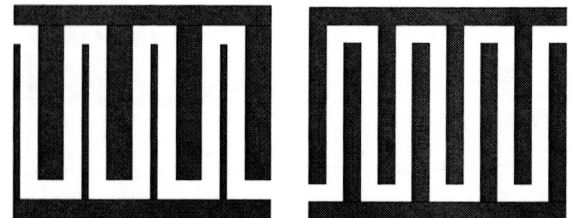

Figure 4. A test structure pair that can distinguish between the impact of density and linewidth (top view). Both the test structures have different linewidths but the same pattern density.

III. TEST STRUCTURE DESIGN AND RESULTS

We have designed test structures that vary metal linewidth and density separately, with the aim to distinguish the impact of linewidth and density. Two of the test structures are shown in Fig.3. Two other test structures have the same density and are shown in Fig. 4. One has non-uniform linewidth, with thin and wide lines that match those in Fig. 3.

We use the following terminology. The structure with minimum linewidth is referred as 1X, since the drawn width of the lines is 1X. The structure with linewidths that are N times the minimum linewidth is referred as NX. We have 3X and 5X test structures. The test structure with non-uniform linewidth is referred as 1X/5X, since one of the combs has 1X linewidth and other has 5X linewidth. Note that 1X, 3X and 5X test structures have different linewidths, but the same linespace on the mask. Test structures 3X and 1X/5X have the same pattern density.

If TF is a function of density, then TF should be the same for the two test structures in Figure 4. On the other hand, if linewidth is the cause of the TF difference, then there will be a difference in TF distributions for the two structures. Furthermore, TF of the non-uniform structure in Fig. 4 should be predictable using TF distributions of the two test structures in Fig. 3 that match the linewidths of the fingers of the non-uniform structure, using area scaling, combined with the method we have proposed to predict the failure rate when there are two independent failure mechanisms [10], [11].

Fig. 5 shows the failure rate distribution for the test structures with 1X, 3X, and 5X linewidths. These test structures vary both linewidth and density.

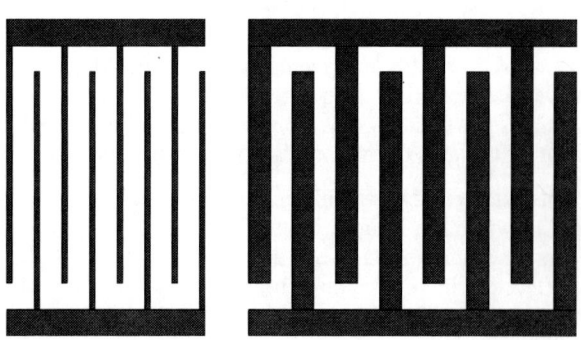

Figure 3. Test structures that vary both linewidth and density concurrently (top view). Both the test structures have the same line space.

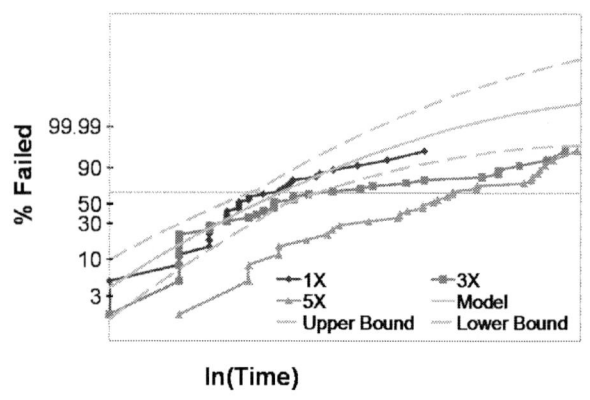

Figure 5. Time-to-failure distributions for test structures with 1X, 3X, and 5X linewidths. 90% confidence bounds are added for the 1X test structure.

Note that for the test structures in Figure 5, only the linewidth and the pattern density are varied and the distance between the lines has remained constant. The characteristic lifetime (x-intercept in Figure 5) increases with an increase in linewidth, with the structure with 5X linewidth showing the largest characteristic lifetime.

Fig. 5 also shows a model and confidence bounds for the 1X linewidth data set, indicating that the changes in the characteristic lifetime cannot be attributed to random variation. The model was computed based on data from 1X, 3X, 4.5X, and 9X area test structures, using area scaling to find slope of Weibull curve, β, and an analysis of the Weibull curves to extract random die-to-die variation [10],[11]. The improvement in characteristic lifetime for wide lines is statistically significant.

Fig. 6 compares data from the non-uniform 1X/5X structure with the 3X structure, which matches its density. Their failure rate curves do not match. Consequently, density does not appear to be a major factor causing a difference in lifetime. We also checked if the data from the 1X and 5X models can be combined to predict the results for the non-uniform 1X/5X structure. As can be seen from Fig. 7, the model and the data do not match. For more information on Fig. 7, see the Appendix and [10],[11].

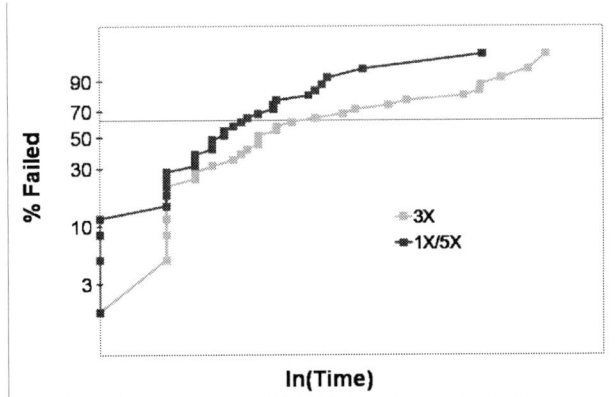

Figure 6.   Failure rate distribution comparison for structures with equal density.

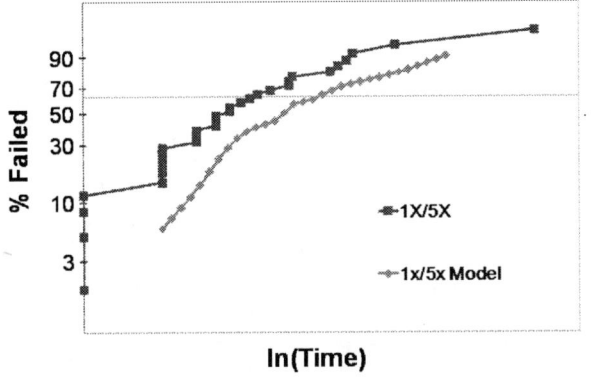

Figure 7.   Predicted failure rate for the 1X/5X structure, based on data from the 1X and 5X structures.

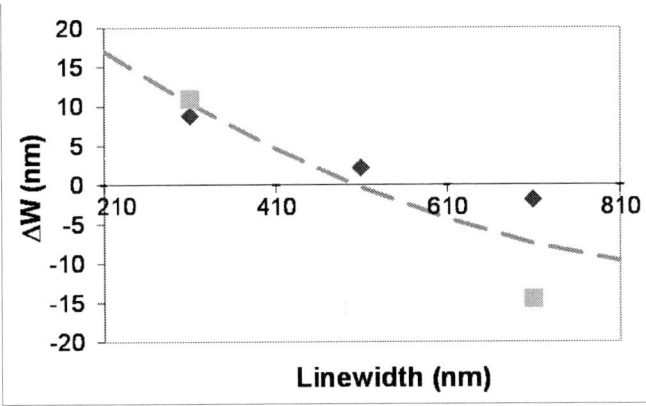

Figure 8.   The manufactured shift in linewidth as a function of linewidth on the mask. The light grey dots correspond to the non-uniform test structure and the black dost correspond to the uniform test structures. The model is computed with regression.

IV.   ANALYSIS OF POTENTIAL CAUSES OF VARIATION

In this section, we consider some possible explanations for variation in characteristic lifetime as a function of either linewidth or density.

A.   Variation as a Function of Printed Geometry

Manufactured geometries were collected for our structures using scanning electron microscopy. The data is shown in Fig. 8.

In this graph, we define the linewidth difference as

$$\Delta W = W_{ACTUAL} - W_{DRAWN} . \qquad (2)$$

In Fig. 8 grey dots correspond to the non-uniform test structure and give the ΔW for the 1X comb and the 5X comb on the 1X/5X test structure. The black dots correspond to the uniform test structure with 1X, 3X, and 5X linewidths. The graph indicates that the narrow lines are wider than drawn, and the wide lines are narrower than drawn. We now turn to two potential explanations for variation in the printed linewidth: lithography and etch.

B.   Lithography

The aerial image of a test structure varies with pattern density because of the optical proximity effect. The optical proximity effect is a function of focal depth and pitch. The radius of influence is around 400nm for an illumination system with a wavelength of 193nm. Hence, the optical proximity effect can influence the narrow lines, but is less likely to influence the wider ones. It tends to increase the linewidths of dense structures, depending on exposure dose. The narrow linewidths are the least dense structures, and therefore the manufactured linewidth difference (in Fig. 8) should be the most negative. Our data is the opposite, and inconsistent with the optical proximity effect.

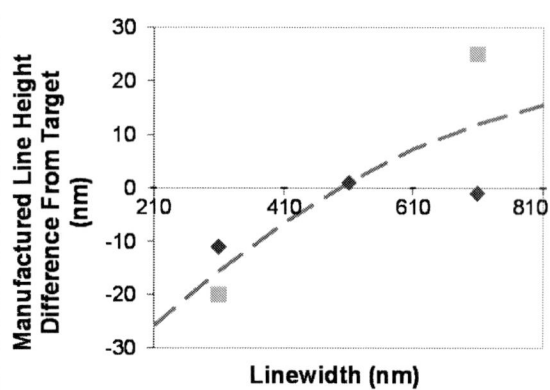

Figure 9. The manufactured line height as a function of linewidth. The black and grey dots correspond to the uniform and non-uniform test structures, respectively. The model is computed with regression.

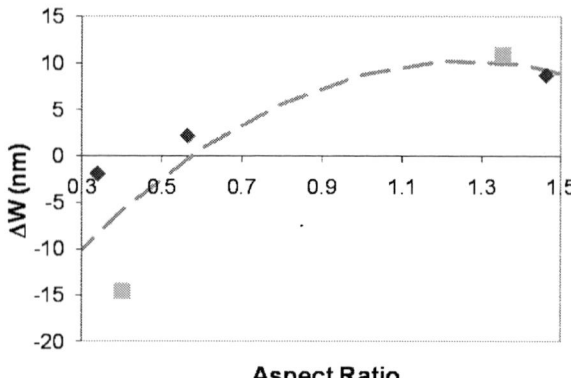

Figure 11. The variation in linewidth in the fabricated chip as a function of aspect ratio. The black dots and lines correspond to the uniform test structures, and the grey dots correspond to the non-uniform test structure. The model is fit by regression.

C. Etching

Etch rate and etch selectivity have been shown to be strongly dependent on pattern density. Previous models have established a link between the line space, linewidth, etch rate, and mean-time-to-failure (MTTF) [7]. These models use Mogab's model [8] that predicts a decrease in etch rate with an increase in linewidth. However our dataset shows variation in the height of the structures. This is summarized in Figure 9. This is because the process uses a timed etch, rather than an etch stop layer. The data in Fig. 9 shows a correlation between linewidth and line height. Based on this data we can assume that the line height is proportional to the etch rate. A model was computed for etch rate, and is shown in Fig. 10.

Pattern density causes spatial variation in etch rate by changing the concentration of reactants in areas with different pattern densities, as different features compete for reactants over short distances [12]. Taylor *et al.* [13] and Abrokwah *et al.* [12] report a decrease in etch rate with increasing pattern density. Our test structures show an increase in etch rate with increasing pattern density, as in [14], [15], if there is any relationship at all, opposite to the trend reported in [12], [13].

Aspect ratio dependant etching (ARDE) manifests itself in submicron feature sizes having high aspect ratios (feature height/feature width). In the presence of ARDE, higher aspect

ratio trenches etch slower [16]. Fig. 10 shows the etch rates for our test structures, along with their aspect ratios.

When the etch rate increases with trench size, this indicates that the process is chemically-controlled. Ion bombardment is not controlling the etch, but rather the concentration of etchant species entering the trench increases with increasing trench width. Therefore, as the trench width increases, more etchant can enter the trench (since etchant arrives at random angles), thereby increasing the etch rate [12], [16].

Hence, it appears that the etch rate is composed of two different etch rate components, the lateral etch rate and the vertical etch rate, both of which depend on aspect ratio. The impact of the lateral component on line width as a function of aspect ratio is as illustrated in Fig. 11.

The trend observed in the actual linewidths can be attributed to the lateral component of etch rate. If we take the line heights as an indicator of vertical etch rate, then the line heights indicate that the vertical etch rate decreases with increasing aspect ratio, while the lateral etch rate increases. The difference in linewidths, attributed to ARDE, partly explains the data in Figure 8 and the increase in TF with linewidth.

D. Variation as a Function of Electric Field Enhancement

A potential cause of variation in characteristic lifetime is electric field enhancement due to fringing effects.

Finite element simulations were carried out using ANSYS to determine the effect of geometry on electric field. Finite element simulations show that high electric field intensities are observed at bends and tips. This is consistent with results in the literature [17].

The locations of high electric field in finite element simulations coincide with vulnerable locations in the Cu/Low-k dielectric structure. Specifically, after formation of the trench, barrier metals are blanket deposited, followed by the deposition of a Cu seed layer and Cu deposition via electroplating. After Cu deposition via electroplating, CMP is carried out to remove the excess Cu covering the dielectric. The difference in hardness between the barrier layer, the soft Cu layer, and the

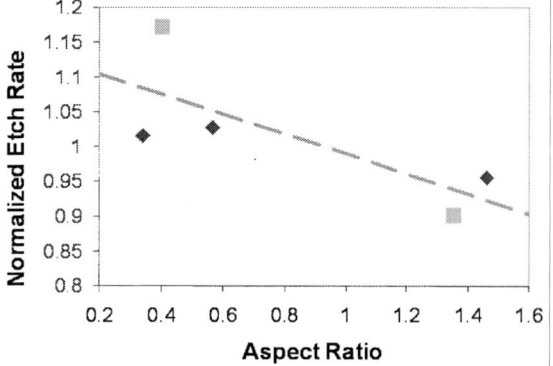

Figure 10. The etch rates for test structures were found to vary as a function of aspect ratio, showing ARDE. The aspect ratio is computed using measured data. The black dots and grey dots correspond to the uniform and non-uniform structures, respectively. The model is computed with regression.

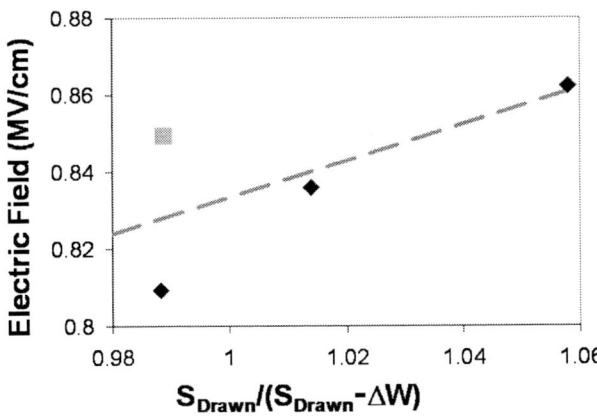

Figure 12. Maximum electric field at the midpoint between the lines vs. the scale factor that accounts for the difference in distance line space between actual and ideal structures. The black dots and grey dot correspond to the uniform and non-uniform structures, respectively. The model is computed with regression.

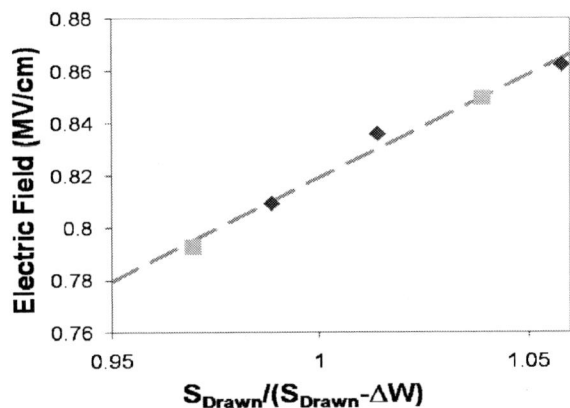

Figure 13. Maximum electric field at the midpoint between the lines vs. the scale factor for the shift in distance between the lines, accounting for the fact that the non-uniform structure has non-uniform line space. The black dots and grey dots correspond to the uniform and non-uniform structures, respectively. The model is computed with regression.

softer low-k dielectric layer, can lead to an uneven profile along the top edge of the trench. This uneven profile, along with high electric fields at the corners of the trench, can trigger Cu diffusion and lead to breakdown. In the literature, breakdown sites have been observed around the corners of the trenches [18], [19].

The electric field distribution should indicate the potential defect sites in the dielectric. However, the exact value of the maximum electric field is determined by the corner rounding at the corners, as noted in [19]. Our simulation results show that the maximum electric field at the corners does not follow any particular trend as linewidth increases. Hence, the role of the maximum electric field at the corners is excluded in this analysis. We compute the peak electric field intensities in the bulk of the dielectric (along a line centered between the interconnect lines) as a function of linewidth.

In these simulations, the physical dimensions of the lines were used. If the distance between the lines were to determine the maximum electric field for a fixed applied voltage, then the maximum electric field, E_{FEM}, would relate to the change in width, ΔW, as follows

$$E_{FEM} = \frac{S_{DRAWN}}{S_{DRAWN} - \Delta W}. \tag{3}$$

S_{DRAWN} is the drawn linespace and $S_{DRAWN} - \Delta W$ is the actual linespace. This relationship is shown in Figure 12. Figure 12 shows that the maximum electric field in FEM simulations is primarily a function of distance between the lines for the uniform structures. It is not a function of distance for the non-uniform structure, because the wide lines are not equally spaced between the narrow lines. Therefore, the maximum electric field is a function of the minimum spacing between the lines. If we take into account the non-uniform spacing between the lines, there is a direct relationship between electric field and (3), as illustrated in Figure 13. This shows that there is no unexpected field enhancement.

V. MODELING CHARACTERISTIC LIFETIME

Variation in width can potentially explain the observed difference in lifetime caused by ARDE.

Characteristic lifetime is assumed to be a function of electric field, E, in the dielectric. The electric field is a function of the distance between the lines. For a pitch, P, then the space, S_{ACTUAL}, is

$$S_{ACTUAL} = P - W_{ACTUAL}. \tag{4}$$

Since,

$$P = W_{DRAWN} + S_{DRAWN}, \tag{5}$$

we have that

$$S_{ACTUAL} = S_{DRAWN} - \Delta W. \tag{6}$$

The electric field is proportional to $1/S_{ACTUAL}$, as noted in (1). Using (1) we have,

$$\ln \eta = A + \frac{B}{S_{DRAWN} - \Delta W} \tag{7}$$

for the E model, and

$$\ln \eta = A + \frac{B}{\sqrt{S_{DRAWN} - \Delta W}} \tag{8}$$

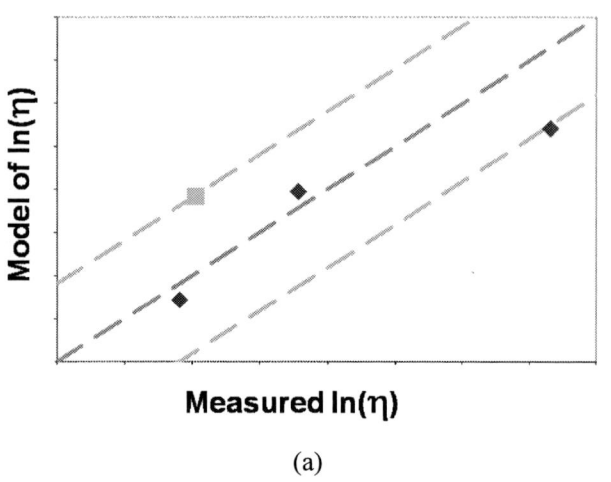
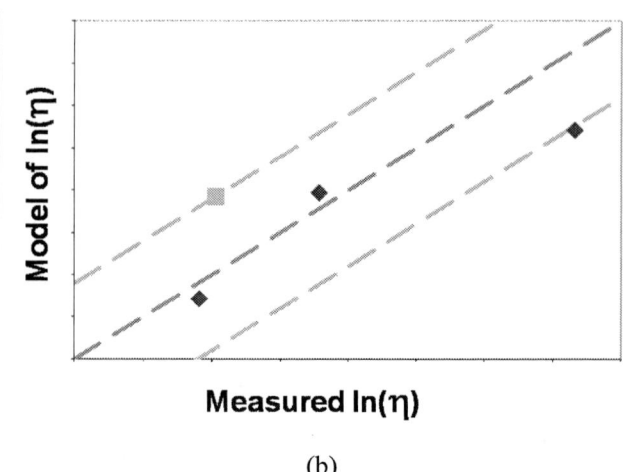

(a)   (b)

Figure 14. Model of the characteristic lifetime as a function of the measured value of charateristic lifetime using regression based on (a) $1/(S_{DRAWN} - \Delta W) = 1/S_{ACTUAL}$ for the E Model and (b) $1/\sqrt{S_{DRAWN} - \Delta W} = 1/\sqrt{S_{ACTUAL}}$ for the $\sqrt{E}$ Model. The black dots correspond to the uniform test structure, from which the model is constructed. Based on this model, the 2σ confidence bounds are computed. The grey dot corrsponds to the non-uniform test structure.

for the $\sqrt{E}$ model.

We use the characteristic lifetime data to find the best fit for A and B using only the data from the uniform structures. The results are shown in Fig. 14.

We now use the models, and computed constants, A and B, to determine a predicted lifetime for the non-uniform structure, which has non-uniform actual linespace. This structure is equivalent to two structures in parallel, with the measured linespace, and half of the area.

To do this we know that for the Weibull distribution, the cumulative probability density function is

$$P(t) = 1 - \exp\left(-\left(\frac{t}{\eta}\right)^{\beta}\right). \tag{8}$$

From the Poisson distribution, the cumulative probability density function is related to defect generation as follows

$$P(t) = 1 - \exp\left(-\lambda(t)A\right). \tag{9}$$

Combining (8) and (9), we have a computed defect generation function:

$$\lambda(t) = \frac{1}{A}\left(\frac{t}{\eta}\right)^{\beta} \tag{10}$$

which produces $d(t) = \lambda(t)A$ as a function of time.

For each of the measured linespaces, the model can be used to compute characteristic lifetimes η_1 and η_2 for the test structures with area A. The two test structures have the same shape parameter, β, which was computed in [10],[11]. The combined defect generation function, $d(t) = \lambda_1(t) A/2 + \lambda_2(t) A/2$, is

$$d(t) = \frac{1}{2}\left(\left(\frac{t}{\eta_1}\right)^{\beta} + \left(\frac{t}{\eta_2}\right)^{\beta}\right). \tag{11}$$

Substituting into (9), we have the following joint cumulative probability density function

$$P(t) = 1 - \exp\left(\frac{-1}{2}\left(\left(\frac{t}{\eta_1}\right)^{\beta} + \left(\frac{t}{\eta_2}\right)^{\beta}\right)\right). \tag{12}$$

The characteristic lifetime, η, of the joint structure is

$$P(\eta) = 62.5\%. \tag{13}$$

Combining (12) and (13), we have

$$\left(\frac{\eta}{\eta_1}\right)^{\beta} + \left(\frac{\eta}{\eta_2}\right)^{\beta} = 2. \tag{14}$$

which results in

$$\eta = \left(\frac{2}{\frac{1}{\eta_1^{\beta}} + \frac{1}{\eta_2^{\beta}}}\right)^{1/\beta}. \tag{15}$$

Figure 14 shows the predicted values of the characteristic lifetime based on (15) for both the E and $\sqrt{E}$ models. It can be seen that the model matches the data reasonably well.

VI. CONCLUSIONS

Test structures have been designed to model the impact of metal linewidth and pattern density on the behavior of TDDB. The test structures vary both pattern density and linewidth independently to enable separation of the impact of these factors. TDDB behavior is found to be dependent on aspect ratio dependent etching, which modulates the linewidths in printed structures and causes the distance between fingers of a comb structure to be non-uniform. Characteristic lifetime has been modeled as a function of the line space, and for the case where the line space is non-uniform. A good fit was obtained for the experimental data.

APPENDIX

In order to generate a Weibull plot for the combination of the two structures, we need to compute a defect generation function, $\lambda(t)$. We use the cumulative Weibull distribution in (8) and the cumulative Poisson model in (9) to obtain $\lambda(t)$ in (10).

Let's suppose that the test structure area is A. Then the defect generation functions for the 1X and 5X structures are

$$\lambda_{1X}(t) = \frac{1}{A}\left(\frac{t}{\eta_{1X}}\right)^{\beta_{1X}} \tag{16}$$

and

$$\lambda_{5X}(t) = \frac{1}{A}\left(\frac{t}{\eta_{5X}}\right)^{\beta_{5X}}. \tag{17}$$

The total number of defects at any time is $\lambda(t)A$. Therefore, the number of defect for the combined structure with $A/2$ of 1X and $A/2$ of 5X is

$$\lambda(t)A = \lambda_{1X}(t)\frac{A}{2} + \lambda_{5X}(t)\frac{A}{2}, \tag{18}$$

and the cumulative probability density function is

$$P(t) = 1 - \exp\left(\frac{-1}{2}\left(\left(\frac{t}{\eta_{1X}}\right)^{\beta_{1X}} + \left(\frac{t}{\eta_{5X}}\right)^{\beta_{5X}}\right)\right). \tag{19}$$

This equation is converted to Weibull format, as follows

$$\ln(-\ln(1-P(t))) = \ln\left(\frac{1}{2}\left(\left(\frac{t}{\eta_{1X}}\right)^{\beta_{1X}} + \left(\frac{t}{\eta_{5X}}\right)^{\beta_{5X}}\right)\right). \tag{20}$$

Note that we have Weibull plots for 1X and 5X structures and for these plots

$$\ln\left(-\ln\left(1-P_{1X}(t)\right)\right) = \ln\left(\left(\frac{t}{\eta_{1X}}\right)^{\beta_{1X}}\right) \tag{21}$$

and

$$\ln\left(-\ln\left(1-P_{5X}(t)\right)\right) = \ln\left(\left(\frac{t}{\eta_{5X}}\right)^{\beta_{5X}}\right). \tag{22}$$

Hence

$$\ln(-\ln(1-P(t))) =$$
$$\ln\left(\frac{1}{2}\exp\left(\ln(-\ln(1-P_{1X}(t)))\right) + \frac{1}{2}\exp\left(\ln(-\ln(1-P_{5X}(t)))\right)\right). \tag{23}$$

Figure 7 is computed by interpolation at each point, t, in the Weibull plots for the 1X and 5X structures. Let

$$y_{1X} = \ln(-\ln(1-P_{1X}(t))) \tag{24}$$

be y-axis values from the Weibull plot for the 1X structure. Similarly, let

$$y_{5X} = \ln(-\ln(1-P_{5X}(t))). \tag{25}$$

Then, at any time point we compute

$$\ln(-\ln(1-P(t))) = \ln\left(\frac{\exp(y_{1X}(t))}{2} + \frac{\exp(y_{5X}(t))}{2}\right). \tag{26}$$

to generate Figure 7.

ACKNOWLEDGEMENT

The authors would like to thank the Semiconductor Research Corporation for financial support, under task 1376.001, and AMD for providing the funding for this custom funded project, together with the wafers used to collect the data in this study. The authors would also like to thank Changsoo Hong and Sohrab Aftabjahani for designing and laying out the test structures.

978-1-4244-5430-3/10 $26.00 © 2010 IEEE

REFERENCES

[1] W.R. Hunter, "The Analysis of Oxide Reliability Data," *Int. Integrated Reliability Workshop Final Report*, 1998, pp. 114-134.

[2] E.Y. Wu and R.P. Vollertsen, "On the Weibull Shape Factor of Intrinsic Breakdown of Dielectric Films and its Accurate Experimental Determination – Part 1: Theory, Methodology, Experimental Techniques," *IEEE Trans. Electron Devices,* vol. 49, no. 12, pp. 2131-2140, Dec. 2002.

[3] J. W. McPherson and D. A. Baglee, "Acceleration Factors for Thin Gate Oxide Stressing," in *Proc. International Reliability Physics Symposium (IRPS)*, 1985, pp. 1-5.

[4] F. Chen, *et al.*, "A Comprehensive Study of Low-k SiCOH TDDB Phenomena and Its Reliability Lifetime Model Development," in *Proc. International Reliability Physics Symposium (IRPS)*, 2006, pp. 46-53.

[5] N. Suzumura, *et al.*, "A New TDDB Degradation Model Based on Cu Ion Drift in Cu Interconnect Dielectrics," in *Proc. International Reliability Physics Symposium (IRPS)*, 2006, pp. 484-489.

[6] C. Hong, L. Milor, M. Choi, and T. Lin, "Study of Area Scaling Effect on Integrated Circuit Reliability Based on Yield Models," *Microelectronics and Reliability,* vol. 45, pp. 1305-1310, 2005.

[7] L. Milor and C. Hong, "Backend Dielectric Breakdown Dependence on Linewidth and Pattern Density," *Microelectronics Reliability,* vol. 47, pp. 1473-1477, 2007.

[8] C. J. Mogab, "The Loading Effect in Plasma Etching," *Journal of the Electrochemical Society,* vol. 124, pp. 1262-1268, 1977.

[9] A. Misaka, *et al.*, "A Simulation of Micro-Loading Phenomena in Dry-Etching Process Using a New Adsorption Model," in *Technical Digest, International Electron Devices Meeting (IEDM)*, 1993, pp. 857-860.

[10] M. Bashir and L. Milor, "A Methodology to Extract Failure Rates for Low-k Dielectric Breakdown with Multiple Geometries and in the Presence of Die-to-Die Linewidth Variation," *Microelectronics Reliability,* vol. 49, pp. 1096-1102, 2009.

[11] M. Bashir and L. Milor, "Modeling Low-k Dielectric Breakdown to Determine Lifetime Requirements," *IEEE Design & Test of Computers,* vol. 26, pp. 18-27, 2009.

[12] K. O. Abrokwah, P. R. Chidambaram, and D. S. Boning, "Pattern Based Prediction for Plasma Etch," *IEEE Transactions on Semiconductor Manufacturing,* vol. 20, pp. 77-86, 2007.

[13] H. K. Taylor, *et al.*, "Characterizing and Predicting Spatial Nonuniformity in the Deep Reactive Ion Etching of Silicon," *Journal of the Electrochemical Society,* vol. 153, pp. 575-585, 2006.

[14] S. Jensen, O. Hansen, and M. C. Mic, "Inverse Microloading Effect in Reactive Ion Etching of Silicon," in *Proceedings of the International Symposium on Plasma Processing XIV, Electrochemical Society,* 2002, pp. 218-226.

[15] V. Bliznetsov, *et al.*, "Challenges of Pattern Transfer for Ultra-Low-k OSG Film Aurora(TM)ULK," *Thin Solid Films,* vol. 462-463, pp. 235-239, 2004.

[16] R. A. Gottscho, C. W. Jurgensen, and D. J. Vitkavage, "Microscopic Uniformity in Plasma Etching," *Journal of Vacuum Science and Technology* vol. 10, pp. 2133-2147, 1992.

[17] C. Hong, L. Milor, and M. Z. Lin, "Analysis of the Layout Impact on Electric Fields in Interconnect Structures Using Finite Element Method," *Microelectronics Reliability,* vol. 44, pp. 1867-1871, 2004.

[18] F. Chen, *et al.*, "Reliability Characterization of BEOL Vertical Natural Capacitor Using Copper and Low-k SiCOH Dielectric for 65nm RF and Mixed-Signal Applications," in *Proc. International Reliability Physics Symposium (IRPS)*, 2006, pp. 490-495.

[19] F. Chen, *et al.*, "Investigation of CVD SiCOH Low-k Time-Dependent Dielectric Breakdown at 65nm Node Technology," in *Proc. International Reliability Physics Symposium (IRPS)*, 2005, pp. 501-507.

EFFECT OF PRE-EXISTING VOID IN SUB-30NM CU INTERCONNECT RELIABILITY

Zungsun Choi, Matsuda Tsukasa, Jong Myeong Lee, Gil-Heyun Choi, Siyoung Choi, Joo-Tae Moon
Samsung Electronics Semiconductor Bussiness Memory Division Process Development Team
San #16 Banwol-Dong, Hwasung-City, Gyeonggi-Do, Korea, 445-701
phone: 82-31-208-2410; fax: 82-31-208-0699; e-mail: jmlee97@samsung.com

Abstract — Pre-existing void effect during electromigration in a sub-30nm wide Cu interconnect was observed. Two types of void are intentionally produced in a single damascene interconnect: 1) A void between Cu and capping dielectric layer (center void) is mainly produced from an excessive overhang by depositing a thick seed layer. 2) A void between Cu and barrier metal (side void) is produced from depositing a thin, discontinuous seed layer. Bi-modality was observed in center voided samples. 44% of lines with center voids show stiff resistance rises at high current density and most of them failed shortly after the resistance rise. No stiff resistance rise was observed at lower current density up to 3000 A.U. In side voided samples, no early failures was observed and the failure show no bi-modal trend. Change in local current density around the void is expected to be the major factor for the electromigration performance difference between lines with center and side voids. We were able to show that shape and location of the pre-existing void have a significant effect on the reliability of Cu interconnect, and also the void behavior is highly sensitive to current density. [*Keywords:* electromigration, copper, void, reliability]

I. INTRODUCTION

Continuing trend of increase in transistor density lead to miniaturization of transistors as well as metal interconnects in semiconductor devices. Miniaturization is especially evident in Flash memory devices where minimum width of less than 40nm metal interconnect is under production as of year 2009 and sub-30nm width interconnect is expected to be implemented by year 2010. On the other hand, the total length of interconnect per chip is more than a kilometer and continuously increasing [1]. Therefore, in order to effectively benefit from the high density devices, the number of defects per length in interconnect must be lowered [1]. One of the fatal defects in the interconnect is a pre-existing void, which can cause open failure by electromigration while device is in operation, however, more works on electromigration assumes a defect-free metal line rather than pre-existing defects. In a void-free interconnect, stress accumulation during electromigration causes void nucleation at the *cathode end*, and the void grows to a fatal size to cause a failure. Void growth is predicted by calculating net atomic flux around the void with an assumption that there is no flux into the void located at the cathode end [2]. Time to failure at operating condition is extrapolated from accelerated condition using a model developed by Black [3]. Pre-existing voids that form during a series of fabrication processes make reliability predictions significantly more difficult compared to that of void-free interconnect because depending on the specific process, the void can form in any location of the interconnect with wide ranges of sizes and various shapes. Also, these voids can grow, shrink, move and change their shapes depending on the initial condition of the voids and their surroundings [4,5]. Thus, it is important to optimize the processes to minimize pre-existing voids in the interconnect. However, since obtaining a void-free interconnect especially in a narrower line is becoming more difficult, it is also critical to identify voids that are the most fatal to reliability and focus on minimizing the fatal voids.

One of the critical processes in Cu interconnect fabrication that leads to void formation is in-situ Ta barrier metal and Cu seed layer deposition before electroplating. It has been reported that an excessive seed layer thickness causes overhang at the top of the trench which eventually leads to a void between Cu and capping layer (center void) [6]. However, if the seed layer is too thin, seed Cu layer with discontinuous region exposing barrier metal occurs more frequently. Exposed barrier metal (Ta) easily oxidizes and makes Cu unable to grow during electroplating, which eventually leads to voiding between Cu and barrier metal (side void) [7]. Therefore, there is a range of seed layer thickness for minimizing both center and side voids, and this range is expected to decrease with decreasing line width. In case if minimizing both types of voids is unachievable and must have one type of the void more than the other, then it is important to find out which type of the void is more prone to reliability degradation. In this paper, two types of pre-existing voids were intentionally created and observed their effect on interconnect reliability during electromigration.

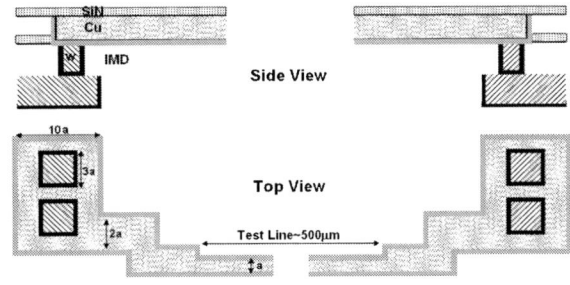

FIGURE 1. SCHEMATIC OF THE INTERCONNECT STRUCTURE.

II. EXPERIMENTAL PROCEDURE

500μm long Cu interconnect was used for the experiment as shown in Figure 1. Test line is connected to W plugs below each ends of the test line and wide W lines below W plugs connect to a current source. Sub-30nm wide M1-Cu line was fabricated by using Double Pattern Technique (DPT) and single damascene process. Width to height ratio is 1 to 2.85. There are 100 dummy lines with identical dimensions at both sides of the test line that are intended for uniform planarization during chemical-mechanical polishing (CMP). Spacing between Cu lines is similar to the line width. By manipulating Cu seed layer deposition process, two types of voids were intentionally produced. In the first set of samples, dominant portion of voids resides between Cu and barrier metal (BM) at the side of the line as shown in Figure 2a (side void) which was obtained by depositing relatively a thin Cu seed layer. In the second set of samples, most voids are between Cu and capping layer at the top of the line as shown in Figure 2b (center void); the center void was obtained by depositing a seed layer that is about 50% thicker than that of the first set of samples. Cross-sectional areas of W plugs and regions at both ends of the test line that connect to the plugs are fabricated to imitate a structure in the actual operating device. Silicon Nitride passivation layer is deposited after the CMP. Individual test structures in a 12 inch wafer were dissected and mounted on a ceramic package and wire-bonded. Aetrium 1164 Reliability Test System was used for the accelerated electromigration test. 16

978-1-4244-5430-3/10 $26.00 © 2010 IEEE
903

samples per split were tested at a temperature of 250°C with a constant current density of 2.2MA/cm² and 4.4MA/cm². Resistance of each sample was monitored throughout the test. Failure was assumed when the resistance increase exceeded 10% of the initial resistance, and the current is no longer applied to the test line.

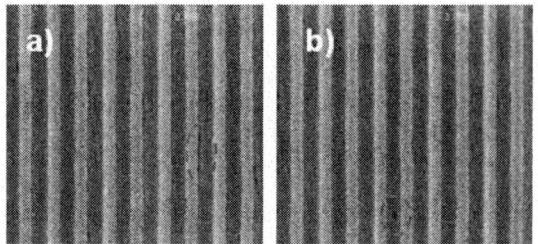

FIGURE 2. TOP VIEWS OF A) SAMPLES WITH MOSTLY SIDE VOIDS AND B) SAMPLES WITH MOSTLY CENTER VOIDS. IMAGES ARE TAKEN BEFORE PASSIVATION DIELECTRIC LAYER.

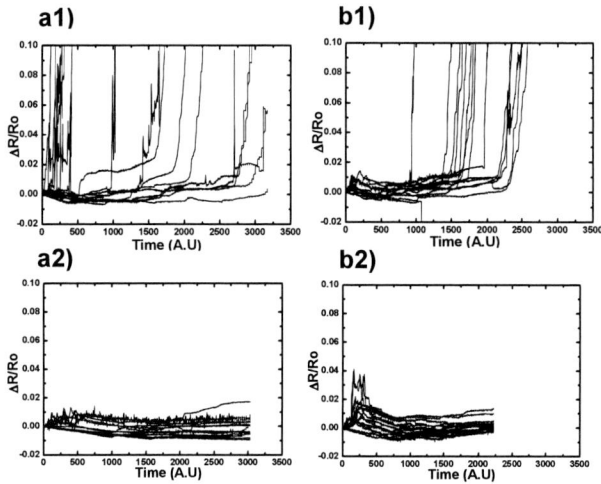

FIGURE 3-1. ΔR VS. TIME PLOTS OF SAMPLES WITH CENTER VOIDS WITH A1) J=4.4MA/CM², AND A2) J=2.2MA/CM2. SAMPLES WITH SIDE VOIDS WITH B1) J=4.4MA/CM², AND B2) J=2.2MA/CM².

III. RESULTS AND DISCUSSION

Resistance Change vs. Time plots were obtained for each split. As shown in Figure 3-1, there are distinct differences in resistance trends with time between samples with center and side voids. Out of 16 samples with center voids (Figure 3-1 a1), 7 of them (44%) showed stiff resistance rise before 500A.U of time. Out of 7 samples with stiff resistance rise, 6 of them lead to failure shortly after initial resistance rise (Figure 3-2 a). Resistance of one sample returned to the value that is similar to the resistance before the jump (Figure 3-2 b). Most of these samples have strong resistance fluctuations after the initial stiff resistance rise. Failure times of the other 8 samples vary between 500 A.U to 3000 A.U and one showed no resistance degradation after 3200A.U. Earlier 2 failures (988A.U, 1580A.U) show severe resistance fluctuation similar to that of failures before 500A.U. Later 6 failures show exponential resistance rise rather than abrupt resistance jump. Lognormal plot shows a clear bi-modality of center voided samples at 4.4MA/cm² (Figure 4). No stiff resistance rise or failures were observed up to 3000 A.U when the current density is lowered to 2.2MA/cm², but 6 out of 16 (38%) show resistance perturbation within 2% of the initial resistance in the first

500A.U (Figure 3-3 a), but the fluctuations disappear after about 500A.U. For the side voided samples with 4.4MA/cm², all samples failed between 900A.U and 2500A.U. Most samples tend to fail with exponential resistance rise and only 2 samples had severe resistance fluctuation before leading to failures, which is similar to that of center voided samples at 4.4MA/cm². 12 out of 16 showed weak resistance fluctuations within 2% of the initial resistance (Figure 3-4 a) before 500A.U. For the side voided samples with reduced current density of 2.2MA/cm2, there are relatively large resistance fluctuations from onset of the test to about 500 A.U. 14 of 16 samples showed resistance fluctuations (Figure 3-4 b). Out of 14 samples with resistance fluctuations, 11 of them had the fluctuation within 2% of the initial resistance, but 3 samples showed resistance rise close to 5%, which is more than two times larger than the resistance fluctuation observed in side voided samples tested at 4.4MA/cm2 or center voided samples tested at 2.2MA/cm2 within the first 500A.U. However, the resistance fluctuation is no longer observed after about 500 A.U., and no failures were observed up to 2500 A.U.

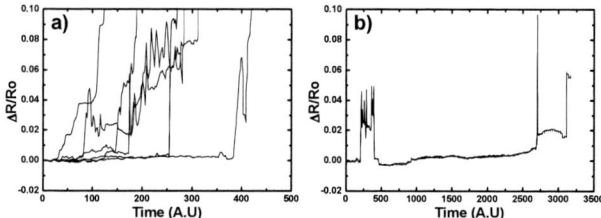

FIGURE 3-2. ΔR VS. TIME PLOTS OF SAMPLES WITH CENTER VOIDS WITH J=4.4MA/CM². MOST EARLY FAILS ARE INITIATED BY STIFF RESISTANCE RISE AS SHOWN IN A), BUT RESISTANCE OF ONE SAMPLE DOES NOT FAIL WITHIN 500A.U AFTER A STIFF RESISTANCE RISE AS SHOWN IN B).

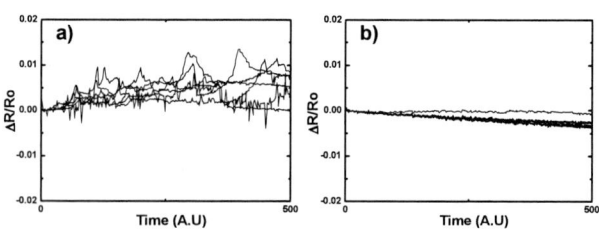

FIGURE 3-3. ΔR VS. TIME PLOTS OF SAMPLES WITH CENTER VOIDS WITH J=2.2MA/CM². A) 6 OUT OF 16 SHOW ~2% RESISTANCE FLUCTUATION IN THE FIRST 500A.U, WHILE B) 10 SAMPLES HAVE NO PERTURBATION.

Electromigration performances of center voids show that failures highly depend on current density. Center voided samples tested with high current density show bimodality in lognormal plot (Figure 4), suggesting that there are two causes for the failures. Early failures tend to have a stiff resistance rise and severe resistance fluctuations while the later failures tend to have an exponentially increasing resistance. We suspect that the observed stiff resistance rise is due to pre-existing void, especially the ones near the cathode end, which grows to span width and thickness of the line so that the current starts to shunt through a thin barrier metal layer [5]. Unlike the void nucleation and growth at the cathode end, growth of the pre-existing void in the middle of the line is expected to be more complex, which

978-1-4244-5430-3/10 $26.00 © 2010 IEEE   904

may be the reason for the wide range of times to failures. As mentioned earlier, there may be a number of reasons for the added complexity, but we expect that the following two reasons have the dominant impact. First, flux into the void in the case of the cathode end is assumed to be zero, however, in the case of the void in the middle of the line, void growth depends on the net amount of atoms diffusing into and out of the void, which highly depends on surrounding conditions such as Cu texture and line width [4,5]. Second, in void-free lines, all voids are expected to *nucleate* and grow at the cathode end, which means that their initial condition is similar, leading to relatively narrow distribution of times to failures with mono-modal trend. On the other hand, size of the pre-existing voids can vary significantly, and depending on their initial sizes, times to failure can vary significantly as observed. At lower current density however, no failure was observed up to 3500 A.U. This suggests that current density needs to be over a certain critical limit to activate a state alteration of a pre-existing void in the line. With a sufficient current, a void may de-pin from a grain boundary [8], change shape for energy minimization [5], and grow. These changes can lead to a void to span the cross-sectional area of the line to cause the failure [5]. Disappearance of early failures at low current density suggests that time to failure extrapolation from accelerated current density to operating current density must be re-evaluated.

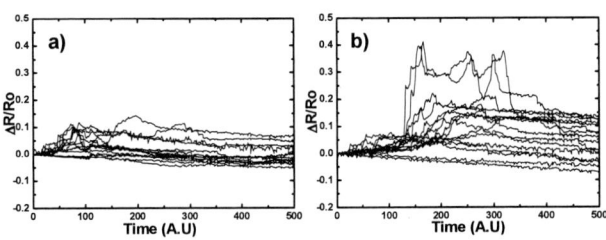

FIGURE 3-4. ΔR vs. TIME PLOTS OF SAMPLES WITH SIDE VOIDS WITH A) J=4.4MA/CM² AND B) J=2.2MA/CM² IN THE FIRST 500 A.U. RESISTANCE PERTURBATION IS LARGER IN LOWER CURRENT DENSITY.

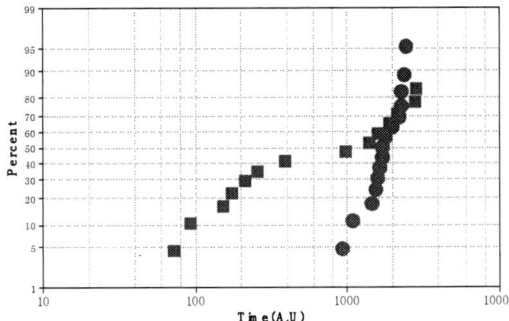

FIGURE 4. LOGNORMAL PLOT OF CENTER VOID (SQUARE DOT) AND SIDE VOID (CIRCULAR DOT) WITH J=4.4MA/CM².

Side voided samples with high (4.4MA/cm2) current density show no early (<500A.U) stiff resistance jump. The cause of the difference in resistance behavior between center and side voids is likely due to the difference in cross-sectional area of the voids perpendicular to the direction of the current flow. Cross-sectional area distribution of center voids is obtained to be wider than that of side voids, and some of the center voids are found to be significantly

larger than the others, with cross-sectional areas occupying more than half of the total area. Side voids tend to span the length of the line, but do not occupy cross-sectional area as much as the center voids. Thus, the local current density at the region with large center void should be more than twice that of the rest of the region, probably causing to accelerate the void growth [9,10]. Failures in side voided samples have narrower distribution compared to that of center voided samples (Figure 4). Narrow distribution and mono-modality suggest that all failures had similar failure mechanisms. No failures are observed at low current density (2.2MA/cm2) up to 2250A.U, however, resistance fluctuation in the first 500A.U is higher compared to that of high current density with side voided samples and that of low current density with center voided samples. It is not clear to what causes the higher resistance fluctuation at low current density with side voided samples.

IV. CONCLUSION

Electromigration performance in sub-30nm wide Cu interconnect lines was observed. Two splits were made with 1) center voids, and 2) side voids which are distributed along the length of the lines. Results at high current density (4.4MA/cm²) showed that center voids lead to bi-modal times to failures with about half of the samples fail by the stiff resistance rise whereas most sided voided samples showed failures after exponentially increasing resistance. The main reason for the difference is expected to be center voids occupying larger cross-sectional area in the line compared to that of side voids, which lead to local current density rise for short times to failures in center voided samples. Results at low current density (2.2MA/cm²) in center voided samples show no early failures, which suggests that there is a critical current density limit to activate a change in state of a void to cause a fatal failure. Low current density in side voided samples show large resistance fluctuation in the first 500A.U compared to other samples, and it is unclear what the exact cause of the high resistance fluctuation.

The Experiment showed that the location and shape of the void is critical to Cu interconnect reliability. Also, we were able to show that a behavior of the void highly depends on the current density, suggesting that the behavior of the pre-existing void should be re-evaluated in order to correctly extrapolate and predict the reliability of the Cu interconnect.

REFERENCES

[1] International Technology Roadmap for Semiconductors, 2007 edition, Interconnect Chapter; *http://www.itrs.net/Links/2007 ITRS/ 2007_Chapters/2007_Interconnect.pdf*

[2] M.A. Korhonen, P. Borgensen, K.N. Tu, and C.-Y. Li, *J. Appl. Phys.,* 73, p3790 (1993)

[3] J.R Black, *In: Proc 6th Annual Int Rel Phy Symp*, p148 (1967)

[4] E. Zschech, et al, *In: Proc. Of 12th IPFA*, p85 (2005)

[5] Z.-S. Choi, R. Mönig, and C.V. Thompson, *J. Mater. Res,* 23, p383 (2008)

[6] L. Bonou, M. Eyraud, R. Denoyel, and Y. Massiani, Electrochim. Acta, 47, p4139 (2002)

[7] Internal experiments on Cu seed layer thickness effect on filling performance by M. Tsukasa.

[8] Børgesen et al, AIP Conf. Proc. 263, p219 (1992)

[9] E. Arzt, O. Kraft, W.D. Nix, and J.E. Sanchez, Jr., *J. Appl. Phys.,* 76, p1563 (1994)

[10] J. Choy and K.L. Kavanagh, *Appl. Phys. Lett.,* 84, p5201 (2004)

STUDY OF UPSTREAM ELECTROMIGRATION BIMODALITY AND ITS IMPROVEMENT IN CU LOW-K INTERCONNECTS

W. Liu, Y.K. Lim, F. Zhang, H. Liu, Y.H. Zhao, A.Y. Du, B.C. Zhang, J.B. Tan, D.K. Sohn and L.C. Hsia
GLOBALFOUNDRIES Singapore, 60 Woodlands Industrial Park D, Street 2, Singapore 738406
Phone: 65-6413-7621; Fax: 65-6413-7506; Email: liuwei@globalfoundries.com

Abstract – **The bimodality of upstream electromigration (EM) failures in the dual damascene structure of 45nm Cu interconnection process with low-k material is investigated, and the improvement is demonstrated. Two major early failure modes with voids forming in via or at the chamfer of via-trench transition area are revealed and attributed to liner process weakness at the respective locations, which can be reduced and eliminated with optimized via and chamfer aspect ratios (ARs) defined by the dual damascene profile. Dielectric thickness, trench etch depth, via critical dimension (CD), chamfer of via-trench transition profile etc. can serve as the tuning factors for the upstream EM bimodality improvement. However, a balance between EM, time-dependent dielectric breakdown (TDDB) and resistance-capacitance (RC) performance need to be achieved.**

Keywords: electromigration, copper metallization, interconnect, dual damascene

INTRODUCTION

With continuous technology scaling, the realization of homogenous and defect-free barrier metal and Cu seed coverage in smaller feature sizes becomes more and more difficult. Consequently, EM reliability of Cu interconnects becomes increasingly challenging. The bimodal phenomena of downstream and upstream EM failure distributions are commonly observed in 45nm node and below and considered to be an intrinsic behavior caused by small feature size and grain boundary etc [1]. In recent years, the influence of sputtering liner deposition parameters on the EM failure time was studied [2]. Some significant theoretical work has also been done to analyze the EM bimodal distribution data in terms of threshold failure time [3] and critical current density [4].

In this paper, our work shows that the upstream EM bimodality in 45nm technology can be strongly affected by the liner and Cu seed coverage with the AR defined by the dual-damascene profile. Consequently, two major early failure modes due to weakness forming inside via or at the via-trench chamfer area were revealed and identified. Such EM bimodality behavior is hard to be improved only by liner process tuning itself, like Tantalum nitride (TaN) and Tantalum (Ta) thickness, sputtering energy of deposit species, Cu seed bias power and thickness etc. However, it can be minimized or even eliminated by the optimization of dual-damascene structure profile with proper tuning of dielectric thickness, trench etch depth, via CD and chamfer profile.

EXPERIMENTAL

The fabrication process used in this study is based on a 45nm technology with dual damascene Cu thin wire levels in carbon-doped oxide (SiCOH) as the low-k dielectric material (k≈2.7). A typical V1-M2 test structure that allows electrons flow upward from a via below the metal line with minimum design rule width was used for upstream EM reliability test, as schematically shown in Fig. 1. The nominal V1 diameter is 0.07 μm, and the M2 line width is 0.07 μm with a length of 200 μm. M1 beneath the via is 0.4 μm wide (only for connection

purpose). The metal lines are passivated with a SiCN cap layer. Sputtering TaN/Ta liners were used as the barrier layer for the Cu interconnects. The tests were done at package-level with stress temperature and current density of 295 ˚C and 25 mA/μm², respectively. A 10% resistance increase was used as failure criteria.

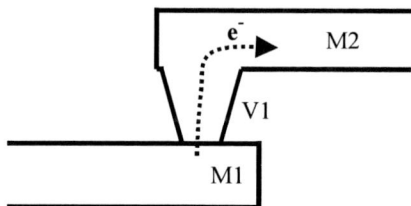

Figure 1. Schematic cross-section of the upstream EM structure. The "e⁻" with arrow indicates the electron flow direction.

RESULTS AND DISCUSSION

A.. Failure modes

The bimodality of upstream EM failures observed in the state-of-the-art technologies is usually correlated to early and late failure populations with voids generating in via and trench, respectively. In our study, it is found that two early failure modes exist actually: failing inside via or at the chamfer of via-trench transition area, as shown in Fig. 2 (a) and (b) as an example. Sometimes both of them show up on the same wafer, and normally the chamfer failure mode is later than the via failure mode. For instance, the group of time-to-failure (TTF) below ~9 hours is dominated by the via failure mode (Sample A), while the group of TTF between ~9 to 15 hours is dominated by the chamfer failure mode (Sample B) in Fig.2. In contrast, the late failure mode with TTF>15 hours is usually due to voids nucleation and migration in the upper metal line (Samples C and D) and the mechanism has been explicitly clarified and understood in literature [5]. Sometimes only one early mode occurs while the other is absent, and the reason will be understood in the following section. It has also been found that most early failures happen at wafer edge area, which is a clear shading effect signature with liner process marginality. For instance, the chamfer failure sample at wafer edge usually has the chamfer open direction against to wafer center and hence the weakest chamfer Cu seed coverage compared to other wafer locations.

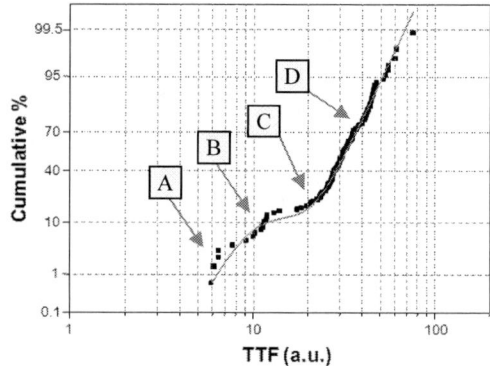

(a) Upstream EM failure distribution

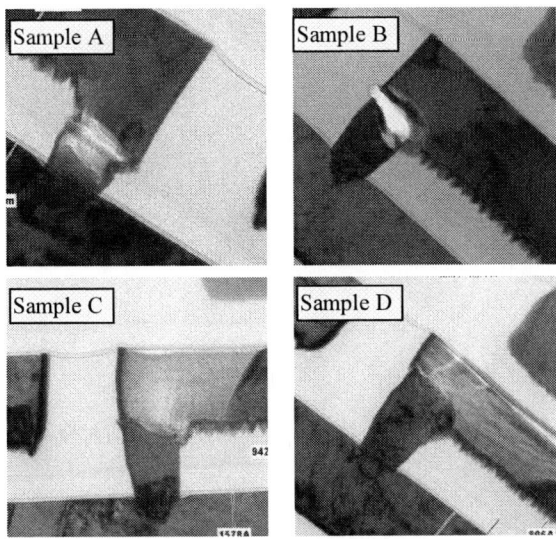

(b) Transmission electron microscope (TEM) cross-sectional micrographs of failure samples A-D

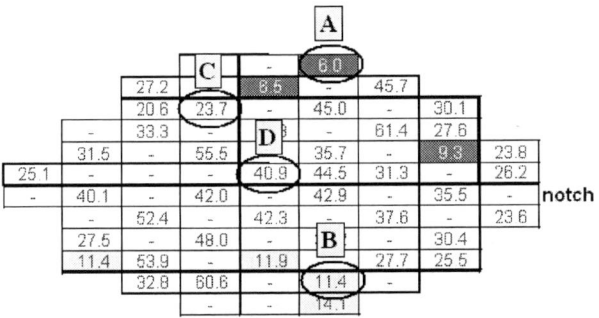

(c) TTF wafer map showing die locations of the failure samples A-D. The early failure dies marked with red and yellow colors are due to via and chamfer failure modes, respectively.

Figure 2. Observed V1-M2 EM failure distribution with three major modes with void forming in via (sample A), chamfer (sample B), and trench (samples C and D), respectively. The die locations on wafer show a clear shading effect signature

B. Correlation of failure mode with AR

It is found that the early failure mode is hard to be minimized only by optimizing the liner process, indicating that other process portions should be evaluated to solve the problem. In Fig. 3 We have defined some key factors for the dual damascene profile which is determined by a combination effect of Litho CD, via etch, trench etch, interlayer dielectric (ILD) thickness and so on. Two key ARs are defined as via AR = VH/D2 (post-etch via height divided by via bottom CD) and chamfer AR = MH/D1 (post-etch trench depth divided by via open CD at chamfer), in order to account for the two different early failure modes. Here D2 (via bottom CD) was used for via AR calculation instead of the conventional way by mid or top CD, because it is easily visible and identified in such structure with inline measurement or cross-section check and it is consistent to the top CD under a definite sidewall angle.

Some experimental data are summarized in Table 1 in terms of EM early failure modes and the two defined ARs. We can clearly see

a strong correlation between them. The four wafers (W1-W4) were processed with different ILD thicknesses and litho or etch recipes, resulting in different via and chamfer ARs as listed in the table. When both ARs are high (Via AR=5.6 and chamfer AR=2.11), both via and chamfer failures are observed (W1 and W2). When only chamfer AR is relatively high, the chamfer failure mode is dominant (W3). When both ARs are low (Via AR=4.57 and chamfer AR=1.75), both the two early failure modes are almost absent (W4).

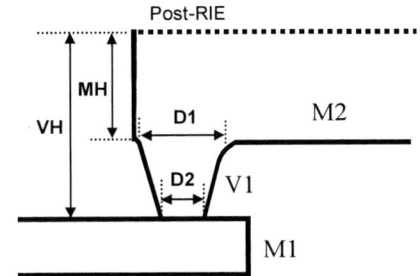

Figure 3. Key factors defined for the dual damascene structure: MH is post-etch trench depth, VH is post-etch via height, D1 is via open CD at the chamfer area and D2 is via bottom CD.

Table 1. EM failure mode vs. AR of processes A-D.

Wafer	Via AR		Chamfer AR		Early fail mode	
	Value	Level	Value	Level	Via	Chamfer
W1	5.6	High	2.11	High	Yes	Yes
W2	4.84	Medium	2.16	High	Yes	Yes
W3	4.57	Low	1.96	Medium	No	Yes
W4	4.57	Low	1.75	Low	No	No

To reveal the root cause of via failure mode, we tried to stop the EM test on an early failure sample at only 0.5% resistance increase, and then sent it for TEM cross-section check (Fig.4). Compared to the normal stress stopped at 10% resistance increase, this experiment can allow us to capture a small void generated in early stage under the stress condition. It can be seen that the void originated from the bottom part of via. The integrity and continuity of TaN/Ta barrier metal in this location is quite good as shown in Fig. 4. This is understandable since the resputter liner process assures a robust coverage around via bottom and sidewall. Since the via early failure mode is not due to barrier metal, it is most likely due to Cu seed coverage weakness in the bottom part of via associated with the high via AR.

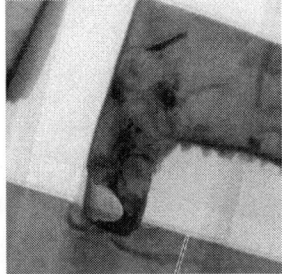

Figure 4. TEM cross-sectional micrograph of a upstream EM sample with test stopped at only 0.5% resistance increase.

Another interesting phenomenon observed is that the via early failure samples don't distribute only at wafer edge, but also spread to

978-1-4244-5430-3/10 $26.00 © 2010 IEEE   907

the wafer center area gradually at significantly high via AR (Fig. 5). It indicates that the Cu seed coverage inside via deteriorates seriously and throughout the whole wafer when via AR goes up too much.

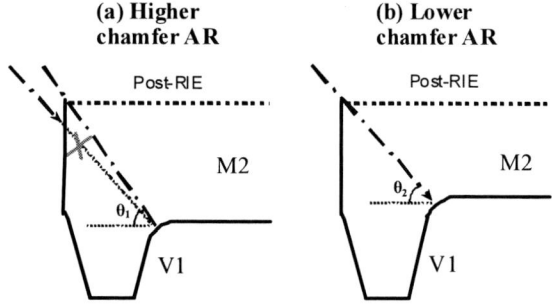

Figure 5. TTF wafer map showing early failure die locations marked with red color in the case of significantly high via AR ~5.6.

Compared to the via failure mode, the chamfer failure mode has a different mechanism. Unlike via bottom and sidewall, the chamfer geography is difficult to be covered by the resputter liner. Therefore, we can only expect it to be covered by incident liner sputtering directly. However, when the chamfer AR is high resulted from deep trench or small via open CD, the incident angle is too high to allow a good coverage of sputtered liner (TaN/Ta and Cu seed) at the chamfer area as schematically shown in Fig. 6. This weak area will easily wear out under the EM stress and results in the early failure mode at chamfer (Fig.1 (b) Sample B). Therefore, it was observed that the chamfer early failure mode usually occurs at wafer edge area only.

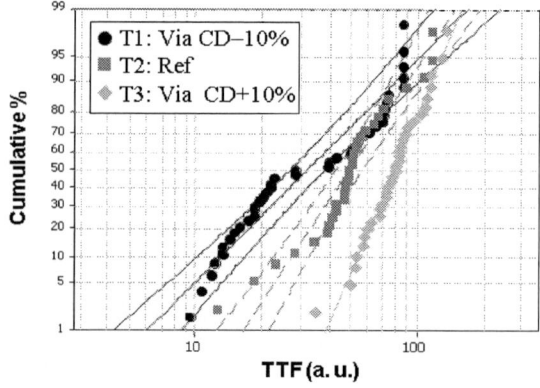

Figure 6. Lower Chamfer AR with shallower trench can gain margin for liner coverage at the chamfer area with decreased incidence angle ($\theta 2 < \theta 1$).

C. Minimize bimodality by ILD thickness and trench etch optimization

Since the two EM upstream early failure modes are already well understood, they can be minimized effectively by reducing ILD thickness and trench etch depth to lower down the AR. The results are shown in Fig. 7 and Table 2. S1 shows the highest percentage (29%) of early failures. S2 has lower via AR because the via height is decreased by ILD-4% (ILD thickness reduced by 4%), and consequently the early failure population decreases to 8%. S3 is also better than S1 because the chamfer AR is lowered down by trench etch depth-6% (trench etch depth reduced by 6%). This can be easily realized by reducing SiCOH main etch time. S4 has the least early failures (<5%) and the best distribution shape factor (sigma=0.32), as both via and chamfer ARs are lowered down with ILD-4% and trench depth-6%. The projected lifetime is larger by more than two times compared to other splits (>2×10^5 hours). In other words, we cannot expect the

upstream EM bimodality to disappear completely if either via AR or chamfer AR is not low enough.

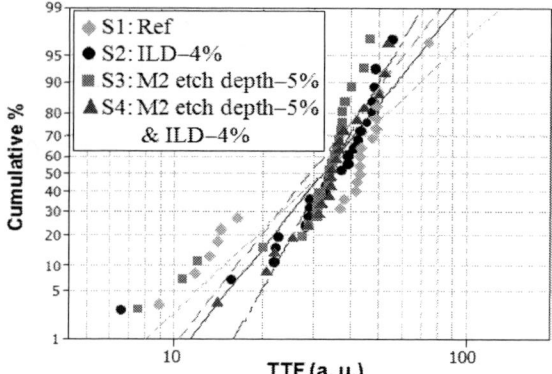

Figure 7. Upstream EM failure distributions of S1-S4 processed with different trench depths and ILD thicknesses for comparison (Table 2).

Table 2. EM performance vs. AR of four process splits S1-S4

Splits/wafers		S1	S2	S3	S4
Process differences		Reference	ILD-4%	M2 etch depth-5%	M2 etch depth-5% &ILD-4%
Aspect ratio (AR)	Chamfer AR (MH/D1)	1.96	1.96	1.81	1.81
	Via AR (VH/D2)	4.71	4.57	4.71	4.57
Sigma		0.60	0.45	0.44	0.32
Percentage of Early failures		29%	8%	17%	<5%

D. Minimize bimodality by via CD tuning

Another effective approach to modify EM upstream bimodality is to increase via CD, and the results are shown in Fig. 8 and Table 3. T1 (CD-10%) has the highest ARs and hence the most early mode population (~50%). T3 (CD+10%) has a much better failure distribution (perfectly single model) and the early failure is absent. Its via AR of 1.75 and chamfer AR of 4.57 define a confidence level to eliminate the early mode. As a result, the T50 (mid TTF of the curve) of T3 also increases by about one time relative to T1 and T2.

Figure 8. Upstream EM failure distributions of three wafers (T1-T3) processed with different via CD for comparison (Table 3).

Table 3. EM performance vs. AR of three process splits T1-T3

Splits		T1	T2	T3
Process differences		Via CD-10%	Reference	Via CD +10%
Aspect ratio (AR)	Chamfer AR (MH/D1)	2.10	2.00	1.75
	Via AR (VH/D2)	5.52	4.92	4.57
Percentage of early failures		50%	10%	~0

E. Balance between upstream EM, via chain TDDB reliability and RC performance

Note that although the presented approaches can be applied to solve the EM bimodality problem, a balance between upstream EM, via chain TDDB and RC performance should be met when tuning the processes. For instance, the reduced ILD thickness will cause the increase of interlevel capacitance since the interlayer metals are more closing. Figure 9 shows the measured electric data from the four process splits S1-S4. We can see that S2 with ILD-4% shows average capacitance higher by ~10% compared to S1. Fortunately, the approach of trench etch reduction can not only benefit the upstream EM performance, but also pull back the interlevel capacitance as it compensates the interlayer metal spacing reduction with increased ILD thickness. With both approaches used, S4 shows a comparable capacitance level to S1, while the upstream EM performance is much better (Fig. 7).

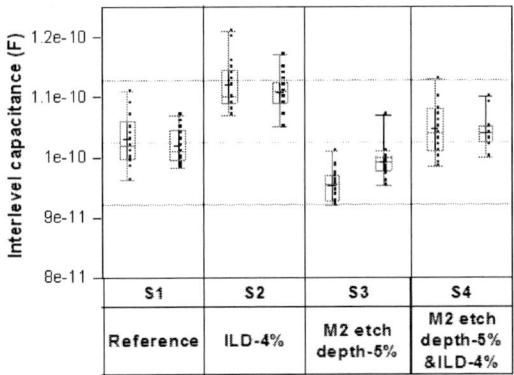

Figure 9. Boxplots of interlevel capacitance measured from the four process splits: S1-S4 for comparison.

Metal resistance is another concern when doing the process tuning. With reduced trench etch depth, the metal resistance will run at high side as shown in Fig. 10 (S4 vs. S1). This problem can be solved by reducing SiCOH loss in the chemical mechanical polishing (CMP) process accordingly to maintain the same trench height. Another compensation way is to increase metal CD a bit to pull back the resistance.

From Section D we can see that via CD blowing-up is an effective way to minimize the upstream EM early model. However, a penalty of degraded via-chain TDDB performance will come out. This is understandable since the via chain top spacing will be narrowed as shown in Fig. 11. A set of TDDB test data is shown in Fig. 12 for comparison. With the via CD + 15%, the mid TTF drops from about 3000 to 1000 (a. u.) and the shape factor also becomes worse significantly. Therefore, a best way for us is to increase via CD in a limited

range and use other approaches (ILD, trench etch etc) together to solve the upstream EM bimodality issue.

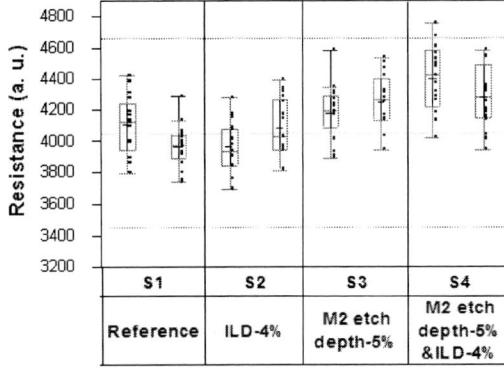

Figure 10. Boxplots of Metal resistance measured from the four process splits: S1-S4 for comparison.

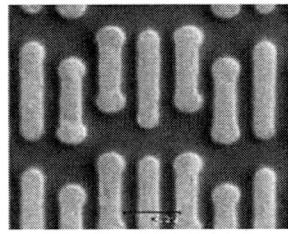

Figure 11. Via-chain top-down image showing reduced top spacing due to big via CD. Consequently TDDB reliability is degraded.

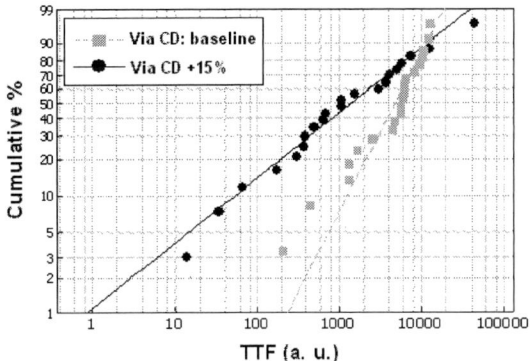

Figure 12. Via-chain TDDB failure distributions tested at 22 V of two process splits: via CD baseline vs. via CD+15%.

CONCLUSION

The results and analyses show that the upstream EM failure early modes of the 45nm Cu low-*k* process can be minimized significantly by process tuning like ILD thickness, trench etch depth, via CD to optimize via and chamfer ARs. At the same time a balance between EM reliability and RC electrical performance (may impacted by interlayer metal spacing and metal height) and TDDB reliability (may impacted by via CD) should be achieved. Looking beyond 45nm, it will be even more challenging to achieve robust EM reliability based on the current process scheme due to more stringent liner step coverage requirement. Therefore, an advanced liner process is very critical for robust reliability and competitive RC performance.

ACKNOWLEDGMENT

The authors would like to express their gratitude to San Leong Liew, Wuping Liu and Meng Guan Ong for useful discussion, Zhehui Wang for etch process, and Wenyi Zhang and Chai Wah Ng for TDDB work and data analysis. Also, thanks to all in GLOBAL-FOUNDRIES Singapore who had contributed, in one way or another, towards the work.

REFERENCES

[1] Jim Lloyd, "Reliability issues with low-k interlevel dielectrics," Tutorial 5, IPFA 2008, July 7-11, Singapore.

[2] A. H. Fischer, O. Aubel, J.Gill, T. C. Lee, B. Li, C. Christiansen, F. Chen, M. Angyal, T. Bolom, E. Kaltalioglu, "Reliability Challenges in Copper Metallizations arising with the PVD Resputter Liner Engineering for 65nm and Beyond.," IEEE-IRPS 2007, pp. 511-515.

[3] R. G. Filippi, P.-C. Wang, A. Brendler, P. S. McLaughlin, J. Poulin, B. Redder, and J. R. Lloyd, "The effect of a threshold failure time and bimodal behavior on the electromigration lifetime of copper interconnects", IEEE-IRPS 2009, pp. 444-451.

[4] A. S. Oates and M.H. Lin, "Void nucleation and growth contributions to the critical current density for failure of Cu vias," IEEE-IRPS 2009, pp. 452-456.

[5] A. V. Vairagar, S. G. Mhaisalkar, M. A. Meyer, E. Zschech, and Ahila Krishnamoorthy, "Direct evidence of electromigration failure mechanism in dual-damascene Cu interconnect tree structures," in Appl. Phys. Lett. Vol. 87, 2005, pp. 081909 1-3.

978-1-4244-5430-3/10 $26.00 © 2010 IEEE

Modeling of Stress Evolution of Electroplated Cu Films during Self-annealing

Rui Huang

Kompetenzzentrum Automobil-und Industrieelektronik (KAI) GmbH
Europastrasse 8, 9524 Villach, Austria
0043-4242-34890-16, rui.huang@k-ai.at

Werner Robl

Infineon Technologies AG
Wernerwerkstrasse 2, 93049 Regensburg, Germany

Thomas Detzel

Infineon Technologies Austria AG
Siemensstrasse 2, 9500 Villach, Austria

Hajdin Ceric

Institute for Microelectronics, TU Wien
Gusshausstrasse 27-29, 1040 Vienna, Austria

Abstract—**Electroplated Cu films are known to change their microstructure at room temperature due to the self-annealing effect. This recrystallization process results in a film-thickness-dependent stress evolution. Films with the thickness of 5μm and below decrease in stress with time, while thicker films reveal initially an increase in film stress followed by a stress relaxation at a later stage. This behavior is explained by the superposition of grain growth and grain size dependent yielding. Existing models have been used and improved to describe the mechanisms related to stress evolution. In general, the models proposed in this study provide a satisfactory description of the stress evolution of electroplated Cu films and the simulated results show good agreement with the experimental data. This gives the possibility to evaluate and predict mechanical behavior of electroplated Cu films at room temperature.**

Keywords-modeling; stress; Cu films; self-annealing

I. INTRODUCTION

Copper has replaced aluminum for interconnect applications in semiconductor devices due to its higher electrical conductivity, increased electromigration resistance and better thermal conductivity [1]. Electro-chemical deposition (ECD) of copper has been demonstrated as one of the best methods to be adopted for high performance logic devices using dual damascene technology and for power devices using pattern plating technology [2-4]. The recrystallization of ECD Cu at room temperature, which is termed self-annealing, is a very distinct phenomenon [5-8]. During self-annealing, stress change is observed [7-11], which is associated with a transition from an as-deposited ultrafine-

This work was jointly funded by the Federal Ministry of Economics and Labor of the Republic of Austria (contract 98.362/0112-C1/10/2005 and the Carinthian Economic Promotion Fund (KWF) (contract 18911 | 13628 | 19100).

grained microstructure to a coarse-grained microstructure. It may require expensive post deposition treatments to overcome the self-annealing which might cause reliability impairment. Thus, the control of stress development is of great importance to ensure the lifetime and reliability performance of Cu metallization.

So far, most of the research has been devoted to analyze mechanisms and kinetics of self-annealing of ECD Cu films [5, 9, 12-18]. Only very few publications can be found which present the modeling results regarding the self-annealing effect. Still, most of them only focus on the modeling in terms of sheet resistivity evolution [19, 20]. The purpose of the present work is to develop a model which is able to describe the mechanisms of stress evolution of ECD Cu films based on experimental studies published elsewhere [21] so that the prediction of the stress evolution at room temperature becomes feasible.

II. EXPERIMENTAL OBSERVATIONS

The film stress of electroplated Cu films with the thickness between 1.5 and 20μm was measured at room temperature by the time elapsed. As illustrated in Fig. 1, all films show a tensile stress. Thin Cu films (1.5-5μm) have relatively high initial stress compared to thick Cu films (8-20μm). The stress evolution shows a disparate tendency between these two groups: i) thin films with the thickness of 5μm and below, ii) thick films with the thickness of 8-20μm. The as-deposited film stress shows also an inverse relationship between film stress and film thickness. The tensile stress of thin films continues to decrease with time, while for thick Cu films the stress first

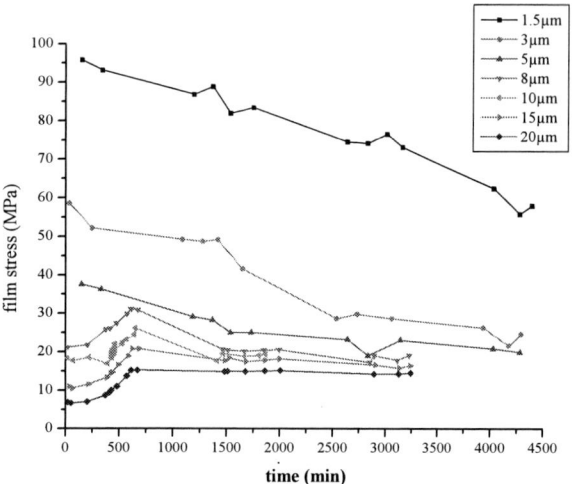

Figure 1. Stress evolution of 1.5-20μm thick Cu films at room temperature. Disparate tendencies of stress are observed between thin Cu films (1.5-5μm) and thick Cu films (8-20μm).

increases and subsequently starts to decrease or stagnate depending on the maximum stress value. Finally, the tensile stress values of all Cu films stagnate at a certain value, which is determined by the film thickness. Thicker films reveal a lower final stress than thinner films.

Meanwhile, the images in Fig. 2 made by focused ion beam (FIB) technique show a remarkable grain growth of 8μm thick Cu film during self-annealing. The first FIB image taken 2h after deposition reveals a fine globular grain structure. After 20h, a significant change in grain size has occurred. The grains which frequently contain twins have a bimodal grain size distribution. It should be noted, that at the Cu/substrate interface, a ~300nm thick region of fine grains still exists. After 44h, all fine grains have been consumed by large grains. Some of the grains (columnar grains) extend through the complete thickness of the film and contain twins. The coexistence of columnar grains and twins leads again to a bimodal grain size distribution. With the help of the linear intercept method, average grain size can be calculated based on each FIB image. Twins are counted as grains. The 45° tilt angle of the FIB images has been taken into account for the grain size analysis. The average grain sizes of 8μm thick Cu film are approximately 135nm, 210nm and 290nm according to the FIB images taken after 2h, 20h and 44h at room temperature, respectively.

The observed stress evolution of ECD Cu films can be explained by the superposition of grain growth and grain size

dependent yielding. On the one hand, grain growth leads to annihilation of excess film volume by reducing the amount of grain boundaries. This gives rise to the stress increase in tension due to the shrinkage of the film volume if the film remains bonded to the substrate. On the other hand, dislocation plasticity leads to the stress decrease by relaxing the film. When initial stress is high, dislocation glide acts as the dominant mechanism and overtakes the effect of grain growth. Thereby it exhibits the continuous stress decrease in Cu films with the thickness of 1.5–5μm. When initial stress is low such as in the case of Cu films with the thickness of 8-20μm, dislocation glide cannot be activated immediately. Instead, grain growth plays a dominant role in the early stage of stress evolution. Until film stress reaches a threshold value, dislocation plasticity is then triggered. Afterwards, the stress decreases like in the 8 and 10μm thick Cu films when the effect of dislocation plasticity becomes more prominent than that of grain growth. For the 15 and 20μm thick films the peak stresses are still so low that dislocation glide and grain growth counterbalance each other. As a consequence, stress evolution of ECD Cu films ascribes to the competing mechanisms of grain growth and dislocation plasticity.

III. MODEL

A. Stress increase due to the loss of grain boundary volume

The loss of grain boundary volume due to grain growth induces the shrinkage of the film volume. If the film remains constrained to the substrate, it causes stress increase in tension. Chaudhari has presented a model which explains the stress development due to grain growth [8, 22]. It calculates the strain in the film by estimating the change in volume associated with the coalescence of grain boundaries. The density of a region containing a grain boundary is usually lower than that of a region containing no boundary. Assuming a grain boundary parameter α $(0< \alpha <1)$, the boundary region has the same density as the grain if $\alpha = 0$, and a monolayer of atoms is missing in the boundary if $\alpha = 1$. Hereby the grain boundary parameter α can be defined as $\alpha = \dfrac{w-a}{a}$ in terms of the grain boundary width w and the atomic diameter a. When grain growth starts, two boundaries coalesce to generate a single boundary. Thus, the elastic distortion in the film due to grain growth can be given by $2w-(w+a)=w-a=\alpha a$.

Considering a thin film that has an average grain size L_0 in the as-deposited condition, and assuming the grains grow to final grain size L, the total elastic strain associated with grain

Figure 2. Microstructural evolution of 8μm thick Cu film at room temperature. a). 2 hours after deposition only fine grains are existing in the film; b). 20 hours after deposition coarse grains are formed while fine grains are still existing at the bottom of the film; c). 44 hours after deposition, coarse grains occupy the film, where twins are embedded.

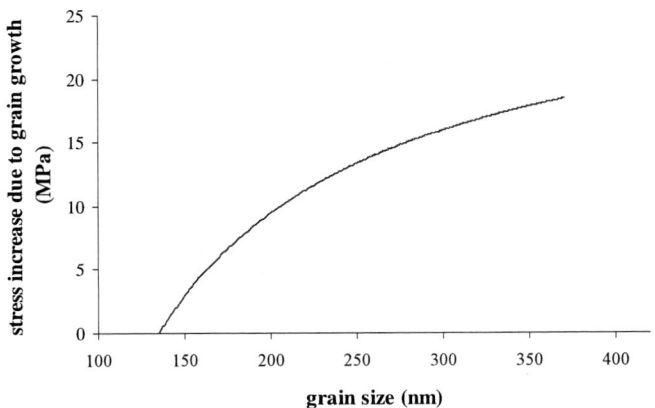

Figure 3. Stress increase due to grain growth, calculated by Chaudhari's model [22]. Original grain size is 135nm. (E=121GPa, v=0.33)

growth is [22]

$$\Delta\varepsilon_{gg} = \frac{\alpha a}{2}\left(\frac{1}{L_0} - \frac{1}{L}\right) \quad (1)$$

Then the total change of biaxial stress associated with grain growth is given by

$$\Delta\sigma_{gg} = \frac{E}{1-v}\Delta\varepsilon_{gg} = \frac{E}{(1-v)}\frac{\alpha a}{2}\left(\frac{1}{L_0} - \frac{1}{L}\right) \quad (2)$$

where $\Delta\sigma_{gg}$ is the stress increase induced by grain growth, E is the elastic modulus of the film, v is the Poisson's ratio of the film, and a is the bulk atom layer spacing (0.36148nm) [8].

If it is assumed that original and final grain size are 135 and 370nm, respectively, the total stress increase calculated is about 25MPa which is plotted in Fig. 3. It must be pointed out that the value of the grain boundary parameter α can vary with the deposition conditions, such as electrolyte composition or current density [23]. Moreover, grain growth and coalescence of grain boundaries are not homogeneous over the film, as seen in Fig. 2. It is very difficult to set the global α value accurately based on the actual local values over the film. In our model, it is assumed that α is the average over transformed and untransformed grains and it is estimated to be 0.12 based on the best fitting, which is fairly close to the value of 0.15 found by Cabral [24].

B. Grain growth rate

As Chaudhari's model describes stress evolution as function of grain size, it additionally needs an intermediate model which introduces the grain size development as function of time elapsed so that the stress evolution can be correlated with time. Doerner and Nix have successfully developed such a grain growth model to describe the rate of grain size development [25].

There are two major contributions to the total energy E_{total} of a film during grain growth. One is the grain boundary energy E_{gb}, and another is the elastic strain energy W_{el}, Therefore, E_{total} can be written as follows:

$$E_{total} = E_{gb} + W_{el} \quad (3)$$

As grain growth proceeds, the grain boundary energy reduces due to the decrease of the grain boundary area, whereas elastic strain increases because of accumulated stress and strain in the film.

The grain boundary energy is given by [22, 25]:

$$E_{gb} = \frac{\beta}{L}\gamma \quad (4)$$

where γ is the grain boundary energy per unit area, L is the average grain size, and β is the geometrical factor determined by the shape of the grains. β has a value of 2 if grains have a square cross section, a value of 3 for grains with a circular cross section and amounts to $4/\sqrt{3}$ for grains with a hexagonal cross section [22, 25, 26].

The strain energy, which is associated with stress, is give by [25]

$$W_{el} = \frac{1}{2}\sigma_{xx}\varepsilon_{xx} + \frac{1}{2}\sigma_{yy}\varepsilon_{yy} \quad (5)$$

Using Using Hook's law, this becomes

$$W_{el} = \left(\frac{E}{1-v}\right)\varepsilon^2 = \left(\frac{E}{1-v}\right)\frac{(\alpha a)^2}{4}\left(\frac{1}{L_0} - \frac{1}{L}\right)^2 \quad (6)$$

Therefore,

$$E_{total} = \frac{\beta}{L}\gamma + \left(\frac{E}{1-v}\right)\frac{(\alpha a)^2}{4}\left(\frac{1}{L_0} - \frac{1}{L}\right)^2 \quad (7)$$

An infinitesimal amount of grain growth dL produces an energy change given by

$$dE = -\frac{\beta\gamma}{L^2}dL - \left(\frac{E}{1-v}\right)\frac{(\alpha a)^2}{2L^2}\left(\frac{1}{L} - \frac{1}{L_0}\right)dL \quad (8)$$

This energy variation is equal to that generated by the migration of grain boundaries in response to a driving pressure difference ΔP.

$$dE = -\Delta P\frac{\beta}{L}\frac{dL}{2} \quad (9)$$

Combining (8) and (9) results in,

$$\Delta P = \frac{2\gamma}{L} + \left(\frac{E}{1-v}\right)\frac{(\alpha a)^2}{\beta L}\left(\frac{1}{L} - \frac{1}{L_0}\right) \quad (10)$$

Following expressions given by Shewmon [27] and Smith [28], the rate of grain growth as a velocity is given by

$$\frac{dL}{dt} = \frac{D^*\Omega}{kT\delta}\Delta P \quad (11)$$

where D^* is the diffusivity given by $D_0 \exp(- E_a / kT)$, D_0 is the pre-exponential factor for grain boundary diffusion, E_a is the activation energy of grain boundary diffusion, Ω is the atomic volume, δ is the average jump distance in the grain boundary, k is the Bolzmann constant, and T is the absolute temperature.

Finally, the kinetics of grain growth can be derived by combining (10) and (11).

$$\frac{dL}{dt} = \frac{D^* \Omega}{kT\delta} \frac{2\gamma}{L} \left(1 - \left(\frac{E}{1-\nu} \right) \frac{(\alpha a)^2}{2\beta\gamma} \left(\frac{1}{L_0} - \frac{1}{L} \right) \right) \quad (12)$$

As mentioned in section II, grain size can be determined by using linear intercept method. Based on the FIB images of 3 and 8µm thick Cu films which were taken at 2h, 20h and 44h, the grain sizes have been extracted and plotted as the dots in Fig. 4. It exhibits that both films undergo a continuous increase of grain size during the whole measurement period. Meanwhile, the simulated curves of 3 and 8µm thick Cu films are also plotted in Fig. 4 by the deduced differential equation (12). The grain sizes of 105 and 135nm accordingly obtained from FIB images of 3 and 8µm thick Cu films are applied as original grain size L_0 in the simulation. As a consequence, the simulated curves converge with the experimental data very well.

By the incorporation of the grain growth rate model (12) into the grain growth model (2), we obtain the direct relationship between stress increased due to grain growth and time. Thus, stress increase as a function of time can be plotted as in Fig. 5. The modeled stress curves display for both films the stagnation after ~1500min, though the grain size still continuously dramatically increases as shown in Fig. 4. This indicates that the grain growth from fine grains to moderate grains contributes the major part to the stress increase. Later on, the grain growth from moderate grains to coarse grains only plays a minor role in stress increase. As a result, it shows clearly that stress increase due to grain growth is film thickness/grain size dependent. 3µm thick Cu films have a

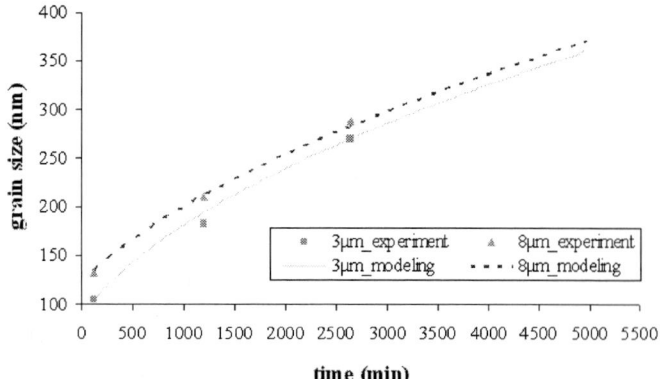

Figure 4. Grain size development of 3 and 8µm thick ECD Cu films at room temperature. The modeled curves are in good agreement with the experimental data (markers) which is determined from FIB images. The following material properties are used in the simulation: $E_a = 1.46 \times 10^{-19}$ J, $D_0 = 3.04 \times 10^{-5}$ m^2/s, $\Omega = 1.182 \times 10^{-29}$ m^3, $\gamma = 0.625$ J/m^2, $\delta = 0.40486$ nm, $\beta = 2$.

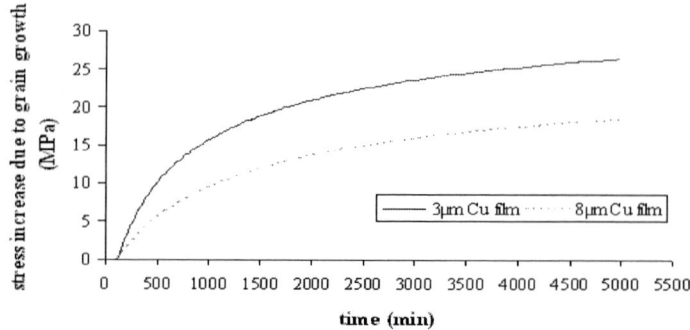

Figure 5. Stress increase of 3 and 8µm thick ECD Cu films as function of time induced by grain boundary loss due to grain growth.

stress increase of ~27MPa compared to ~18MPa in 8µm thick films. This is due to the fact that original grain size is smaller in thinner films. As the amount of stress increase is mainly determined by the fine grains, a smaller original grain size will result in a larger stress increase.

C. Grain boundary strengthening to dislocation plasticity

At room temperature, grain boundaries as well as dislocations act as obstacles to dislocation motion. Impeding the dislocation motion will hinder the onset of plasticity and hence increase the yield strength of the film. The dislocation pileup which determines the stress concentration in the grains, relates to the grain diameter and dislocation density. If the grain size is large and dislocation density is high, a greater stress concentration is developed in the grains, and thus the applied stress needed to activate plastic flow in the grains is relatively low, and vice versa. Hall-Petch law well represents the grain boundary barrier to dislocation motion as function of grain size [8, 29-31]. By the modified Hall-Petch equation the stress decrease due to dislocation motion can be calculated:

$$\Delta\sigma_{gb} = \sigma_L - \sigma_{L_0} = K \left(\frac{1}{\sqrt{L}} - \frac{1}{\sqrt{L_0}} \right) \quad (13)$$

where $\Delta\sigma_{gb}$ is the stress change due to dislocation glide, σ_L and σ_{L_0} are the flow stress of the film in the case of the grain size of L and L_0, respectively, K is the Hall-Petch coefficient.

If the grain growth rate model (12) is incorporated into Hall-Petch equation (13), the direct relationship between stress decrease by dislocation glide and time is obtained and shown in Fig. 6. As discussed in section II, in thick Cu films (8-20µm) dislocation plasticity can only be activated if a certain value of the threshold stress is reached. Therefore, according to the experimental data in Fig. 1, it is assumed that dislocation plasticity in 8µm thick Cu films is activated when grain growth reaches the grain size of ~190nm after 820min. Similar to the stress increase by grain growth, the stress decrease due to dislocation plasticity is film thickness/grain size dependent as well. It indicates that thicker films end up with less stress relaxation, which explains why stress decrease becomes less prominent with the increasing film thickness. The stress decrease tendency is even not visible in 20µm thick Cu film, because dislocation plasticity is so marginal that it is

978-1-4244-5430-3/10 $26.00 © 2010 IEEE

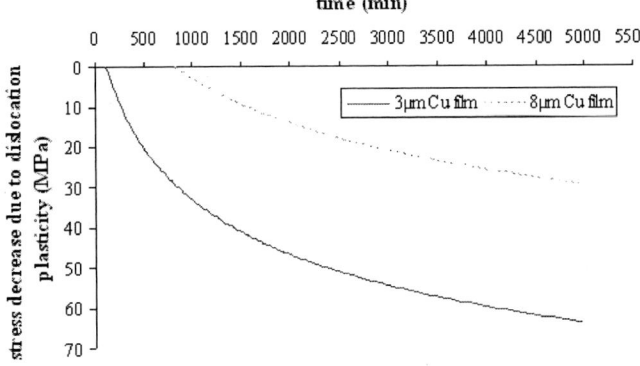

Figure 6. Stress decrease of 3μm and 8μm thick ECD Cu films induced by grain boundary strengthened dislocation plasticity. Note that for thick Cu films, e.g. 8μm, dislocation motion doesn't start immediately after deposition, but only at 820min after deposition.

compensated by grain growth easily.

The Hall-Petch coefficient K reflects the resistance of grain boundaries to dislocation motion. Typically, a lower value of K refers to less resistance of dislocation motion, and vice versa. In Fig. 6, the modeled curves are calculated with the K value of 0.0448 MN/m$^{3/2}$, which is ~40% of the well-known Hall-Petch constant 0.112 MN/m$^{3/2}$ for face-centered cubic (FCC) copper [31]. During the abnormal grain growth, subgrains with low angle grain boundaries (<15°) may be strongly involved in the coarse-grained structure, which weakens the blocking of dislocations compared to high angle grain boundaries (>15°C)[32-34]. Consequently, Hall-Petch coefficient related to subgrains is typically 1/2 to 1/5 of the well-known value [31]. In order to capture the process of subgrain coarsening, in-situ electron backscatter diffraction (EBSD) or transmission electron microscopy (TEM) during grain growth would be required.

D. Model with competing mechanisms

As aforementioned, the individual mechanism of grain growth or dislocation plasticity does not enable us to interpret the complex stress evolution of ECD Cu films. Instead, a competing model combining the above two mechanisms, as shown in (14), needs to be applied in combination with grain growth rate model (12).

$$\sigma = \sigma_0 + \Delta\sigma_{gg} + \Delta\sigma_{gb} \qquad (14)$$

where σ is the film stress, σ_0 is the as-deposited stress, $\Delta\sigma_{gg}$ and $\Delta\sigma_{gb}$ is the stress change due to grain growth and dislocation glide, respectively.

Fig. 7 shows the results of two representative numerical calculations in comparison with the corresponding experimental data labeled as dots. The as-deposited stress σ_0 is 60MPa and 23MPa for 3μm and 8μm thick Cu films, respectively. In Fig. 7, both calculated curves are qualitatively similar to the measured stress plots. As dislocation glide is prominent in the Cu films (1.5-5μm), continuous

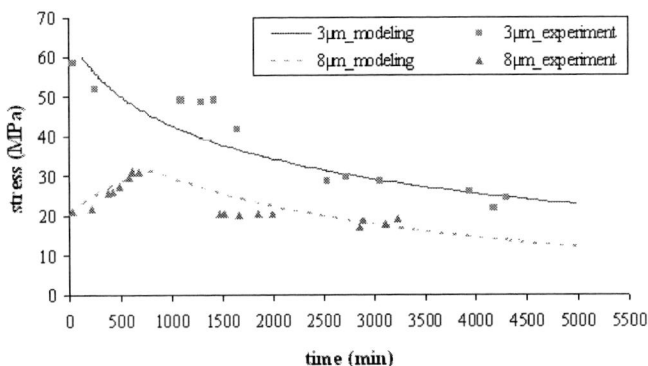

Figure 7. Experimental and modeled stress evolution of 3 and 8μm thick ECD Cu films

decrease of stress, which is fully represented by the calculated curve of 3μm thick Cu films, can be obtained by the described model with competing mechanisms. Meanwhile, the curve of 8μm thick Cu films is well on the behalf of thicker films (8-20μm), which conforms to grain growth and successive dislocation glide mechanisms.

However, there is a notable limitation of the model, which is associated with incubation of grain growth. As observed in Fig. 1, particularly in the stress evolution curves of thick Cu films (8-20μm), stress increase does not start immediately after deposition, but at ~200min after the deposition. This implies that the onset of grain growth is retarded and all fine grains maintain as-deposited undergoing an incubation phase. The incubation of grain growth is not considered in the present model. Thus, the simulated results only emphasize on the stress evolution after the commencement of grain growth, but neglect the stress stabilization during incubation. Nevertheless, compared to the total evolution time, the incubation time is relatively short.

IV. CONCLUSION

A model with competing mechanisms has been developed to simulate the stress evolution of ECD Cu films. The model is based on the mechanisms of grain growth and dislocation plasticity. On the one hand, the stress increase is modeled by the loss of grain boundary volume due to grain growth. On the other hand, dislocation glide is responsible for the stress decrease. The incorporation of the model of grain size development makes it possible to include the time as a parameter into the competing model. The results show an accurate correlation between experimental data and simulated values. To extend the model, incubation phase of grain growth may be considered. Further, EBSD or TEM for the investigation of subgrains is required in order to determine important model parameters, such as the Hall-Petch coefficient.

ACKNOWLEDGMENT

The authors would like to thank M. Roemet for FIB measurements. Furthermore, Prof. G. Dehm is acknowledged with gratitude for many helpful discussions.

REFERENCES

[1] M. D. Thouless, J. Gupta, and J. M. E. Harper, "Stress Development and Relaxation in Copper-Films During Thermal Cycling," Journal of Materials Research, vol. 8, pp. 1845-1852, Aug 1993.

[2] L. T. Romankiw, "A path: from electroplating through lithographic masks in electronics to LIGA in MEMS," Electrochimica Acta, vol. 42, pp. 2985-3005, 1997.

[3] P. C. Andricacos, C. Uzoh, J. O. Dukovic, J. Horkans, and H. Deligianni, "Damascene copper electroplating for chip interconnections," IBM JOURNAL OF RESEARCH AND DEVELOPMENT, vol. 42, pp. 567-574, 1998.

[4] W. Robl, M. Melzl, B. Weidgans, R. Hofmann, and M. Stecher, "Last Metal Copper Metallization for Power Devices," in IEEE Transactions on Semiconductor Manufacturing, 2008, pp. 358-362.

[5] K. Pantleon and M. A. J. Somers, "In situ investigation of the microstructure evolution in nanocrystalline copper electrodeposits at room temperature," Journal of Applied Physics, vol. 100, Dec 2006.

[6] C. Lingk and M. E. Gross, "Recrystallization kinetics of electroplated Cu in damascene trenches at room temperature," Journal of Applied Physics, vol. 84, pp. 5547-5553, Nov 1998.

[7] S. H. Brongersma, E. Richard, I. Vervoort, H. Bender, W. Vandervorst, S. Lagrange, G. Beyer, and K. Maex, "Two-step room temperature grain growth in electroplated copper," Journal of Applied Physics, vol. 86, p. 3642, 1999.

[8] J. M. E. Harper, C. Cabral Jr, P. C. Andricacos, L. Gignac, I. C. Noyan, K. P. Rodbell, and C. K. Hu, "Mechanisms for microstructure evolution in electroplated copper thin films near room temperature," Journal of Applied Physics, vol. 86, pp. 2516-2525, 1999.

[9] C. H. Seah, G. Z. You, C. Y. Li, and R. Kumar, "Characterization of electroplated copper films for three-dimensional advanced packaging," Journal of Vacuum Science & Technology B, vol. 22, pp. 1108-1113, May-Jun 2004.

[10] W. H. Teh, L. T. Koh, S. M. Chen, J. Xie, C. Y. Li, and P. D. Foo, "Study of microstructure and resistivity evolution for electroplated copper films at near-room temperature," Microelectronics Journal, vol. 32, pp. 579-585, 2001.

[11] V. A. Vas'ko, I. Tabakovic, S. C. Riemer, and M. T. Kief, "Effect of organic additives on structure, resistivity, and room-temperature recrystallization of electrodeposited copper," Microelectronic engineering, vol. 75, pp. 71-77, 2004.

[12] V. Weihnacht and W. Brückner, "Abnormal grain growth in {111} textured Cu thin films," Journal of Applied Physics, vol. 418, pp. 136-144, 2002.

[13] S. P. Hau-Riege and C. V. Thompson, "In situ transmission electron microscope studies of the kinetics of abnormal grain growth in electroplated copper films," Applied Physics Letters, vol. 76, pp. 309-311, Jan 2000.

[14] M. T. Pérez-Prado and J. J. Vlassak, "Microstructural evolution in electroplated Cu thin films," Journal of Applied Physics, vol. 47, pp. 817-823, 2002.

[15] Q. T. Jiang and M. E. Thomas, "Recrystallization effects in Cu electrodeposits used in fine line damascene structures," Journal of

Vacuum Science & Technology B: Microelectronics and Nanometer Structures, vol. 19, pp. 762-766, 2001.

[16] K. B. Yin, Y. D. Xia, C. Y. Chan, W. Q. Zhang, Q. J. Wang, X. N. Zhao, A. D. Li, Z. G. Liu, M. W. Bayes, and K. W. Yee, "The kinetics and mechanism of room-temperature microstructural evolution in electroplated copper foils," Scripta Materialia, vol. 58, pp. 65-68, 2008.

[17] K. B. Yin, Y. D. Xia, W. Q. Zhang, Q. J. Wang, X. N. Zhao, A. D. Li, Z. G. Liu, X. P. Hao, L. Wei, C. Y. Chan, K. L. Cheung, M. W. Bayes, and K. W. Yee, "Room-temperature microstructural evolution of electroplated Cu studied by focused ion beam and positron annihilation lifetime spectroscopy," Journal of Applied Physics, vol. 103, pp. 066103-3, 2008.

[18] H. Lee, W. D. Nix, and S. S. Wong, "Studies of the driving force for room-temperature microstructure evolution in electroplated copper films," Journal of Vacuum Science & Technology B, vol. 22, pp. 2369-2374, Sep-Oct 2004.

[19] W. Zhang, S. H. Brongersma, Z. Li, D. Li, O. Richard, and K. Maex, "Analysis of the size effect in electroplated fine copper wires and a realistic assessment to model copper resistivity." vol. 101: Journal of applied physics, 2007, p. 063703.

[20] M. Stangl and M. Militzer, "Modeling self-annealing kinetics in electroplated Cu thin films," in JOURNAL OF APPLIED PHYSICS vol. 103: AIP, 2008, p. 113521.

[21] R. Huang, "Stress, sheet resistance and microstructure evolution of electroplated Cu films during self-annealing," IEEE transactions on device and materials reliability, accepted, 2009.

[22] P. Chaudhari, "Grain Growth and Stress Relief in Thin Films," in Journal of Vacuum Science and Technology. vol. 9, 1972, pp. 520-522.

[23] H. Lee, S. S. Wong, and S. D. Lopatin, "Correlation of stress and texture evolution during self- and thermal annealing of electroplated Cu films," Journal of Applied Physics, vol. 93, pp. 3796-3804, Apr 2003.

[24] C. Cabral, P. C. Andricacos, L. Gignac, I. C. Noyan, K. P. Rodbell, T. M. Shaw, R. Rosenberg, J. M. E. Harper, P. W. DeHaven, and P. S. Locke, "Room temperature evolution of microstructure and resistivity in electroplated copper films," in Advanced Metallization Conference, Colorado, 1998, p. 81.

[25] M. F. Doerner and W. D. Nix, "Stresses and deformation processes in thin films on substrates," Critical Reviews in Solid State and Materials Sciences, vol. 14, pp. 225-268, 1988.

[26] Q. Z. Hong, J. M. E. Harper, and S. Q. Hong, "Stresses and morphological instabilities in silicide/polycrystalline Si layered structures." vol. 62, 1993, p. 2637.

[27] P. G. Shwemon, Transformations in Metals. New York: McGraw-Hill, 1969.

[28] C. M. F. Rae and D. A. Smith, "On the Mechanisms of Grain-Boundary Migration," Philosophical Magazine a-Physics of Condensed Matter Structure Defects and Mechanical Properties, vol. 41, pp. 477-492, 1980.

[29] R. Venkatraman and J. C. Bravman, "Separation of film thickness and grain boundary strengthening effects in Al thin films on Si," in Journal of Materials Research. vol. 7, 1992, pp. 2040–2048.

[30] R. M. Keller, S. P. Baker, and E. Arzt, "Quantitative analysis of strengthening mechanisms in thin Cu films: Effects of film thickness, grain size, and passivation," in Journal of Materials Research. vol. 13: Pittsburgh, PA: Published for the Materials Research Society by the American Institute of Physics, c1986-, 1998, pp. 1307-1317.

[31] T. H. Courtney and T. Hugh, Mechanical behavior of materials. New York McGraw-Hill New York, 1990.

[32] F. J. Humphreys, "A unified theory of recovery, recrystallization and grain growth, based on the stability and growth of cellular microstructures--I. The basic model," Acta Materialia, vol. 45, pp. 4231-4240, 1997.

[33] E. A. Holm, M. A. Miodownik, and A. D. Rollett, "On abnormal subgrain growth and the origin of recrystallization nuclei," Acta Materialia, vol. 51, pp. 2701-2716, 2003.

[34] W. Blum and X. H. Zeng, "A simple dislocation model of deformation resistance of ultrafine-grained materials explaining Hall-Petch strengthening and enhanced strain rate sensitivity," Acta Materialia, vol. 57, pp. 1966-1974, 2009.

THE TDDB FAILURE MODE AND ITS ENGINEERING STUDY FOR 45NM AND BEYOND IN POROUS LOW κ DIELECTRICS DIRECT POLISH SCHEME

Chia-Lin Hsu[1,2], Kuan-Ting Lu[2], Wen-Chin Lin[2], Jeh-Chieh Lin[2], Chih-Hsien Chen[2], Teng-Chun Tsai[2], Climbing Huang[2], J Y Wu[2], and Dung-Ching Perng[1]

1. Institute of Microelectronics, Department of Electrical Engineering, National Cheng Kung University, No. 1 University Road, Tainan 70101, Taiwan, R.O.C.

2. United Microelectronics Corp., No 18, Nanke 2nd Rd. Tainan Science Park, Sinshih, Tainan County 741, Taiwan, R.O.C.

Tel: 886-6-505-4888 Ext.87-12513, fax: 886-6-505-0960, e-mail: chia_lin_hsu@umc.com

ABSTRACT

To keep pursuing the chip resistance capacitance (RC) delay improvement, it is necessary to further reduce k value. Accordingly, direct polished porous type ultra-low-k (ULK) film instead of non-porous low-k materials is integrated into Cu interconnects from 45 nm. However, because of the ULK characteristics and the minimized feature size, the time-to-break-down (TDDB) failure mode behaves different from silica glass or non-porous low-k film. And it is not only sensitive to geometries but also very sensitive to the engineering in the fabrication process. In this paper, we identified three TDDB failure modes, Cu protrusion from trench top interface, sidewall, and bottom corner, in the direct polished ULK scheme. In addition, on the basis of those failure modes, the related mechanisms in conjunction with the sensitivity to the processes are reported as well. [*Keywords: time-dependent dielectric breakdown, TDDB, Cu interconnect, Failure mode, low k, Direct Polish.*]

INTRODUCTION

In order to benefit to chip resistance capacitance delay improvement, low-*k* dielectrics are indispensable for the continuous scaling of advanced VLSI circuits. And in order to further pursuing more low effective k value, porous type ultra-low-*k* dielectric materials with direct chemical mechanical polish (CMP) are applied from the 45nm technology node and beyond. However, because of its porosity, ultra-low-*k* materials generally have lower intrinsic breakdown strengths than non-porous low-*k* dielectrics. Moreover, direct polish process exposes this fragile film in the various wet chemicals such as slurries and post-CMP clean solutions under certain pressure. Accordingly, the long-term reliability, especially time dependent dielectric breakdown (TDDB), of such materials is becoming one of the most critical challenges for technology qualification. This problem is further exacerbated due to continuously shrinkage of the interconnect pitch size in the advance technology nodes.

Lots of articles have discussed the sensitivities and acceleration models of time-dependent-dielectric-breakdown characteristics to the geometries such spacing, area and line edge roughness for low-*k* and ultra-low-*k* dielectrics integrated scheme [1.2.3]. However, as for the minimized feature size in 45 nm and beyond, the time-to-break-down behaves not only sensitive to geometries but also highly sensitive to the engineering in the fabrication process. In addition, the high porosi-

ty characteristics of ULK dielectrics film lead the ULK TDDB failure mode behaving diverse from silica glass or non-porous low-k film. In this work, we identified three TDDB failure modes in direct ULK polish scheme. And on the basis of these failure modes, the related mechanisms in conjunction with the sensitivity to the process are reported as well.

EXPERIMENTAL

45 nm and its sub-node patterned wafers were utilized in all experiments. The structure wafers were prepared with the following steps. The nano-porogen pre-mixed low-*k* dielectric film was deposited on the wafers by plasma-enhanced chemical vapor deposition (PECVD). The porogens were then removed by UV lamp curing to form porous low-*k* film (*k* value 2.4~2.6). The formation of interconnects were based on dual damascene process with the trench 1st metal hard mask scheme. The patterned wafers were then processing PVD type barrier and seed, Cu electroplating and Cu chemical mechanical polish (Cu CMP) sequentially to finish the Cu interconnects.

Break down behavior tests were probed on a comb-comb testkey on the structured wafers comprised of 3 Cu interconnect and an Al layers. The failure location was detected by optical beam induced resistance change (OBIRCH) and then applied transmission electro microscope (TEM) to do the failure mode analysis. Dynamic Secondary ion mass spectrometry (SIMS) was applied for the element component analysis in depth profile.

RESULTS AND DISCUSSIONS

1. Top Interface Failure

The 1st type failure occurred at the dielectric barrier above the Cu and barrier interface located on the metal line top corner (Fig. 1a). Both M1 and 1XDD tests have observed similar failure mode. The non-stressed TEM image (Fig. 1b) showed that a seam existed within the following dielectric barrier cap layer if the surface of metal was recess at the Cu and barrier interface. The existed seam in the dielectric barrier became the weak point to suppress the Cu depletion when applied voltage and formed the 1st type of failure mode. The failure is easy to be seen as an early failure mode especially when voids at the metal trench sidewall presented after Cu CMP, as shown in Fig 2a. Fig 2b illustrated that this sidewall void formation was due to incomplete Cu gap-fill at trench sidewall. In the etching step, because of the etching rate difference between ULK film and dielectric hard mask, the trench profile was re-entrant at the border of these two films. It leaded the followed PVD barrier/ seed coverage to be discontinuous. The void formed in the following plating process and exposed on the surface after Cu CMP process.

Except sidewall void, there are some other factors which generated Cu recess on the surface also possibly conducting to the similar surface breakdown mode, such as the non-equivalent removal rate of Cu to barrier during polishing and the Cu surface roughness. How the post-CMP Cu roughness effecting on EM at this generation has been addressed in a prior article [4]. Moreover, the correlation between TDDB and Cu roughness would be shown here. Cu roughness could be induced at Cu CMP itself and deteriorate in the following thermal process after CMP. As can be seen in Fig. 3a, with an amine type clean solution, different Cu roughness was generated with different post-CMP clean time. The longer clean time, the rougher Cu surface. Fig. 3b showed that if plotting the mean time to failure (MTTF) versus average roughness (Ra), a very well linear correlation was observed in this roughness region. It is because that the probability of dielectric seam formation increased as the Cu roughness is larger and results in the degradation of MTTF.

The other factor to change the Cu surface roughness is the thermal after CMP. The thermal after CMP provides Cu enough energy for the grain re-growth and thus results in the change of Cu roughness. As shown in Fig. 4, it indicated that Cu roughness increased in the post-CMP following steps. As similar to the roughness effect in the previous post-CMP clean time experiment, this roughness increment of Cu surface resulted in the degradation of TDDB. Fig. 5 presented this effect. Anneal 2 condition is with less thermal budget than anneal 1 condition. The wafer with anneal 2 was almost with 2-order MTTF improvement rather than the one with anneal 1.

From the process point of view, this type of failure could be improved by etching profile smoothness, line edge roughness improvement, gap-filling capability enhancement, post-CMP cleaning optimization and the following thermal optimization.

2. Sidewall Failure

The second failure mode took place at the middle of trench sidewall. Fig. 6 demonstrated the typical TEM picture of this failure mode that Cu extruded though bulk dielectrics film. It seemed that there was a tunnel underneath the polished surface when breakdown occurred. As the comparison represented in the table 1, this specific TDDB failure mode did not observed in non-porous type carbon-doped silica dielectric or F-doped silica glass (FSG) but only in the porous type ULK dielectrics integrated wafers no matter M1 single damascene and 1X dual damascene (1XDD) layers. It indicated that comparing with non-porous dielectric film, the existence of pores in the ULK film deteriorated the barrier capability at the interface of porous dielectric film to barrier. The Cu protruded from this interface when under electrical field stress. On the other hand, if comparing the M1 and 1XDD failure pictures, even through the M1 ULK dielectric film thickness much thinner than 1XDD one, the failure occurrence were at similar depth from polished surface in both interconnect layers. It implied this failure weak point might not correlate to the ULK bulk film deposition and possibly formed after CMP process.

The way to elongate the mean time to failure of this sidewall failure mode is to enhance the barrier capability at the metal barrier and ULK interface. Both ultra low *k* film property and barrier step coverage optimization could achieve it.

3. Bottom Corner Failure

The third failure mode happened at the corner of trench bottom as shown in Fig 7a. It was just observed at 1XDD layers but not at M1. In PVD barrier deposition process, re-sputter technology was widely applied to improve the barrier sidewall step coverage. In addition, in punch-thru approach, it was used to punch through via bottom barrier in order to reduce via resistance and enhance the electro-migration as well. However, for the porosity of ultra low-*k* silica film, too aggressive re-sputter amount would harm the trench bottom ULK dielectric film and caused the following barrier deposition anchoring into ULK film. Fig 7b presented the whole trench bottom was serious damaged with tremendous re-sputter amount intentionally. This property was not seen in non-porous low-k or FSG dielectric film. Besides, the PVD barrier deposition was with thinner thickness at bottom corner than other place. Barrier would be penetrated starting from this location and formed two teeth anchoring into ULK film if with mild re-sputter amount. The unwilling teeth were with higher local electric-field enhancement and tended to breakdown when under voltage stressing. Since M1 trench bottom is traditional silica glass film, therefore, no such failure mode was seen at M1 layers.

To extend the failure time of this type, the barrier anchoring should be avoided. By optimizing the barrier thickness in conjunction with its re-sputter process, the trench bottom damage could be eliminated and the time to break down was extended successfully.

CONCLUSIONS

Because of the porosity in ultra low-*k* dielectric film, the TDDB failure mode performs different from non-porous low-k dielectrics. The TDDB failure modes for direct polished ULK scheme in L45 and sub-node generation has been reported. Three TDDB failure types, Cu protrusion from trench top interface, sidewall, and bottom corner were identified. Top interface failure is related to the seam formation in the dielectric barrier capping layer due to sidewall void or high Cu roughness induced Cu recess at the Cu and barrier interface. Sidewall void is specific failure mode for porous ultra-low-*k* film. The existence of pores in the ULK film deteriorated the barrier capability at the interface of porous dielectric film to barrier. Bottom corner failure is due to the trench bottom damage in the barrier re-sputter process. High local electric-field enhancement at barrier anchoring area leads to breakdown.

REFERENCES

[1] Jinyoung Kim, et al., "Time Dependent Dielectric Breakdown Characteristics of Low-k Dielectric (SiOC) Over a Wide Range of Test Areas and Electric Fields" in *Proceedings of the 2007 IEEE International Reliability Physics Symposium*, 2007, pp. 399-404.

[2] F. Chen, et al., "Line Edge Roughness and Spacing Effect on Low-k TDDB Characteristics," in *Proceedings of the 2008 IEEE International Reliability Physics Symposium*, 2008, pp. 132-137.

[3] Gaddi S. Haase, "An Alternative Model For Interconnect Low-K Dielectric Lifetime Dependence On Voltage" in *Proceedings of the 2008 IEEE International Reliability Physics Symposium*, 2008, pp. 556-565.

[4] Chia Lin Hsu, et al., "Defect Study of Manufacturing Feasible Porous Low k Dielectrics Direct Polish for 45nm Technology and beyond," in *Proceedings of the 2009 IEEE International Interconnect Technology Conference*, 2009, pp. 140-142.

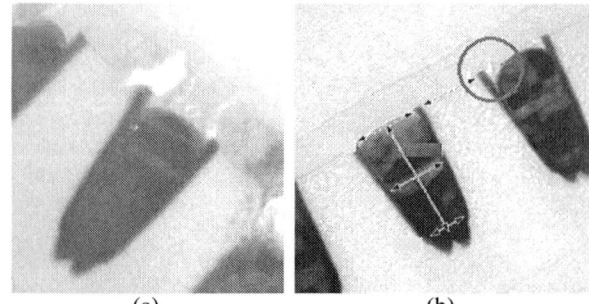

FIGURE 1. THE 1ST TDDB FAILURE TYPE: CU PROTRUSION FROM TOP INTERFACE (a) AND UN-STRESSED TEM (b)

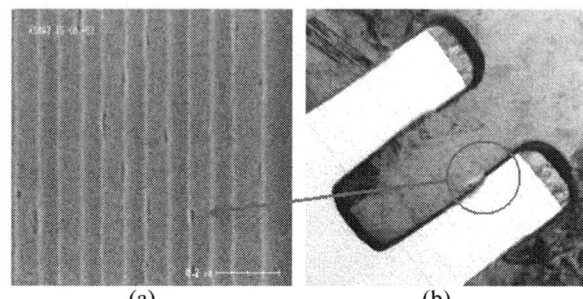

FIGURE 2. TRENCH SIDEWALL VOID OBSERVED BY TOP-VIEW SEM (AFTER POLISH) (a) AND CROSS SECTIONAL TEM (BEFORE POLISH) (b)

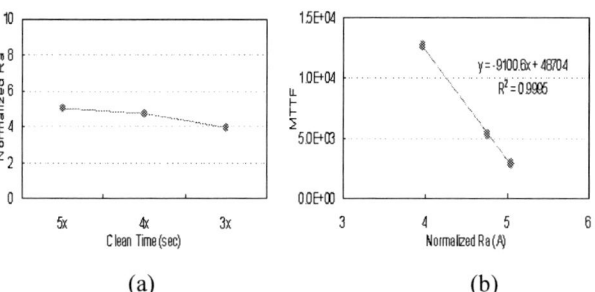

FIGURE 3. HIGHER CU ROUGHNEES WITH LONGER CLEAN TIME (a) MTTF PROPORTIONAL TO CU ROUGHNESS (b)

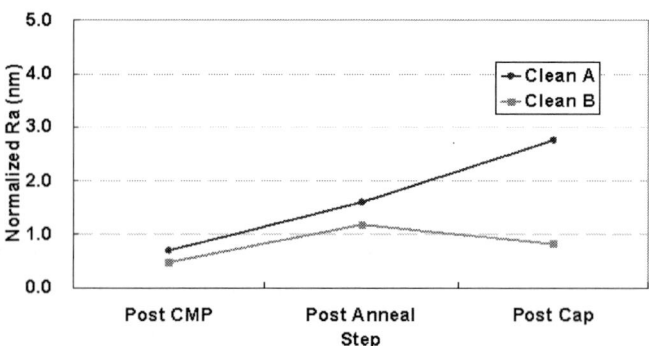

FIGURE 4. THE ROUGHNESS INCREMENT IN THE FOLLOWING PROCESS AFTER CMP

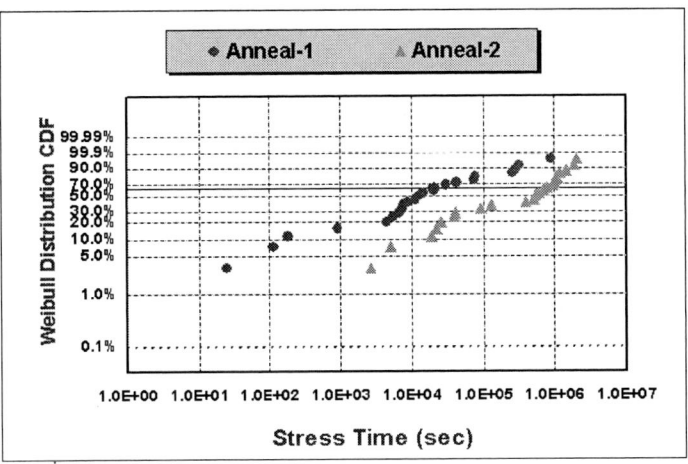

FIGURE 5. TDDB COMPARISON FOR DIFFERENCE ANNEALING
CONDITIONS WHERE ANNEAL 2 IS WITH LESS THERMAL BUDG-
ET THAN ANNEAL 1.

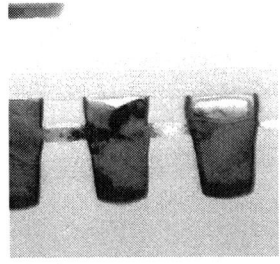

FIGURE 6. THE 2ND TDDB FAILURE TYPE: CU PROTRUSION
AT SIDEWALL AND ACROSS THE ULK FILMS.

	Porous LK		Non-Porous LK	
	M1	1XDD	M1	1XDD
Top Interface Failure	X	X	X	X
Sidewall Failure	X	X	-	-
Bottom Corner Failure	-	X	-	X

TABLE 1. FAILURE MODE COMPARISON BETWEEN POROUS LOW-K
AND NON-POROUS LOW-K DIELECTRIC FILMS

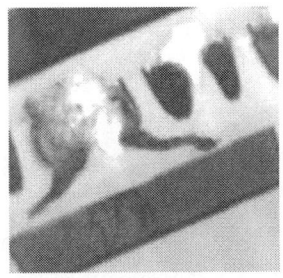

(a)   (b)

FIGURE 7. THE 3RD TDDB FAILURE TYPE: CU PROTRUSION
FROM BOTTOM CORNER (a) AND RE-SPUTTER DAMAGED
TRENCH BOTTOM (b)

Resistance Trace Modeling and Electromigration Immortality Criterion Based on Void Growth Saturation

P. Lamontagne[1,2], D. Ney[1], L. Doyen[1], E. Petitprez[1]

[1]STMicroelectronics
850, rue Jean Monnet, 38926 Crolles, France
line 4: phone: (33) –438-922-346, patrick.lamontagne@st.com

Y. Wouters[2]

[2]SIMaP, CNRS-UJF-Grenoble-INP
1130, rue de la piscine, 38402 Saint-Martin-d'Hères Cedex, France

Abstract—**In this paper, we present our investigations on time evolution of the resistance during EM tests for various j and L conditions at three different temperatures. These resistance traces have been modeled calculating the void volume kinetic. A good agreement with experimental data was found; in particular the resistance saturation regime at low jL was simulated. From this modeling, we propose an immortality criterion depicted by the product jL^2, based on the limitation of the void growth prior to an electrical detection. In the temperature range investigated, no significant variation was found for this criterion.**

Keywords-electromigration; resistance modeling; immortality criterion

I. INTRODUCTION

The downscaling of interconnect cross-sections induces a strong limitation of the current level permitted to designers in order to prevent failures by electromigration (EM) during a minimum operating time-period. A simple way to get around the limitation of the process capability is to take advantage of the Blech effect for short lines[1], in which a stress-induced backflow balances the EM flux at stationary state. As a consequence, interconnects are assumed immortal as the product current density j x interconnect length L remains below the well-known EM threshold product jL_c. However, even if the above condition is satisfied, interconnects are not necessarily safe from degradations[2], [3]. Our previous study reports saturated resistance traces for $jL < jL_c$ which can trigger the electrical failure [4], [5]. Before including this length effect in design rules, a deep and rigorous understanding of the backflow impact on the void growth kinetic and thus with the EM failure time is essential. We propose to use the EM resistance traces with respect to j , L and the temperature to study the interconnect immortality to EM.

II. EXPERIMENTAL

This study was performed on dual damascene copper interconnects embedded in low-k SiOCH dielectric (k = 2.8) from 65 nm process technology. 0.09 µm wide and 0.18 µm thick copper lines are surrounded by TaN/Ta sidewall barriers and SiCN liner. All samples are tested in downstream configuration. The studied lengths ranged from 25 to 250 µm.

EM experiments were carried out at package level at 260, 300 and 330°C and at current density ranging from 10 to 30 mA/µm².

III. RESULTS AND RESITANCE MODELING

The resistance evolution during EM tests for advanced technologies with copper and TaN/Ta sidewall is composed by three steps. First, the resistance remains constant which corresponds to the void nucleation and growth until its electrical detection. The second stage consists of a resistance step that pictures the line opening as the void spans over the whole section of the line. In most cases, this resistance step is associated with the interconnect time to failure (TTF). The third stage is the consequence of the longitudinal void growth and exhibits a resistance increase dR/dt. Figure 1 shows examples of experimental resistance traces for three jL conditions where $jL_1 > jL_2 > jL_3$. As the jL product decreases the TTF increases and dR/dt is different due to the Blech effect. For $jL_1 \gg jL_c$, resistance slope is constant with time contrary to jL_2 where the resistance increase tends to saturate as the EM degradation occurs. The EM-induced stress gradient along the line leads to a backflow flux becoming more dominant on the total net flux. jL_3 trace presents a resistance saturation for hundreds of hours. It pictures an EM flux entirely compensated by the backflow flux: stationary state is reached and the EM-induced void no longer grows.

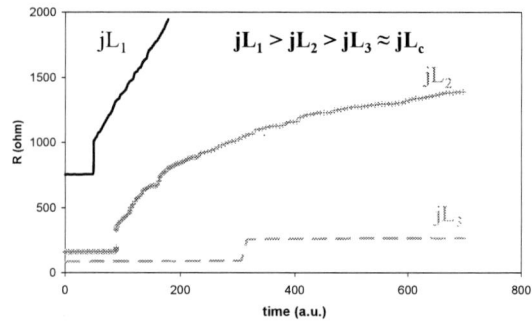

Figure 1. Resistance traces for three jL conditions

The latter example shows that even if a test condition can lead to the stationary state in the line, failure may occur. Interestingly, the resistance step at the line opening is very close to the saturation level. Since the resistance increase is related to the void growth, it is relevant to consider the evolution of the void volume at the cathode side for short lines. The former is given by the integral calculus of σ/B over L. B is the effective elastic modulus of the interconnect. Here σ corresponds to the solution of Korhonen's equation for the case of a finite line, in which the flux is blocked at one end while the stress is annihilated at the other extremity [6]. This case is representative of failure mechanism in copper lines where the void is "initially" present.

$$V(t) = V_{sat}\left[1 + \frac{32}{\pi^3}\sum_{n=1}^{\infty}\frac{(-1)^n}{(2n-1)^3}\exp\left[-\left(\frac{2n-1}{2}\pi\right)^2\frac{t}{\tau}\right]\right] \quad (1)$$

$$\tau = \frac{L^2 kT}{DB\Omega} \quad (2) \qquad V_{sat} = \frac{e\rho Z^* A_{cu}}{2\Omega B}\cdot jL^2 \quad (3)$$

Where D, ρ, Ω and Z^* are respectively the copper diffusion coefficient, resistivity, atomic volume and effective charge. A_{cu} is the line section. e is the elementary charge. For $t \to \infty$, $V(t) \to V_{sat}$ which represents the void volume needed to induce a stress gradient that stops void growth.

From the void growth kinetic $V(t)$ given by (1) [7], the resistance traces $R(t)$ can be reproduced whatever j, L and T. With this aim in view, we assume the presence of a horizontal slit void at the upper interface at $t = 0$, presenting a length l_{voidi}. As depicted by the figure 2, this void will evolve in two unidirectional growth stages: the first stage is the transversal growth from the upper interface to the bottom of the line. The void growth spreads through the line thickness and is defined by $h_{void}(t)$; the void length is fixed to a given l_{voidi}. The second stage starts when $h_{void}(t) = (h - t_b)$, where h and t_b are respectively the line and the barrier thickness. Thus the entire thickness of the line is opened which corresponds to the failure. Then the void extends longitudinally and is depicted by $l_{voidi} + l_{void}(t)$. The equations related to these two phases as function of the line characteristics are (4) and (5).

Phase 1: vertical growth   **Phase 2: longitudinal growth**

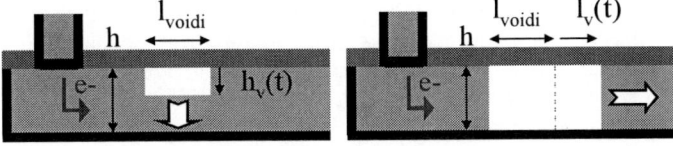

Figure 2. Schematization of the void growth procedure for the resistance trace modeling

$$h_{void}(t) = \frac{V(t)}{l_{voidi}(w-2t_b)} \quad (4) \qquad l_{void}(t) = \frac{V(t)}{(h-t_b)(w-2t_b)} \quad (5)$$

Where w is the line width and t_b corresponds to the sidewall thickness.

Global resistance $R(t)$ during EM test is the sum of two resistances in series R_{cu} and $R_{void}(t)$. The first term corresponds to the copper resistance not associated with the void. Its variation during void growth is neglected in our test time frame

and R_{cu} is set to the initial value of the line resistance. The second resistance $R_{void}(t)$ is related to the resistance variation consequently to void growth: its expression varies hence with the growth phase. In the phase 1, $R_{void}(t)$ depicts the resistance variation of the copper under the void and is dependent of $h_{void}(t)$. During the phase 2, the void opens the line thickness and the current is forced through the sidewall: $R_{void}(t)$ pictures the resistance increase induced by the shunt layer and depends on $l_{void}(t)$.

$$R(t) = R_{cu} + R_{voidj}(t)$$

Phase 1:   **Phase 2:**

$$R_{void1}(t) = \rho\frac{l_{voidi}}{(w-2t_b).h_{void}(t)} \qquad R_{void2}(t) = \rho\frac{l_{voidi}+l_{void}(t)}{(w-2t_b)(h-t_b)}$$

Figure 3 shows simulations achieved from this procedure for different jL conditions. We can notice the simulation of the TTF increase and the saturation of dR/dt when jL decreases.

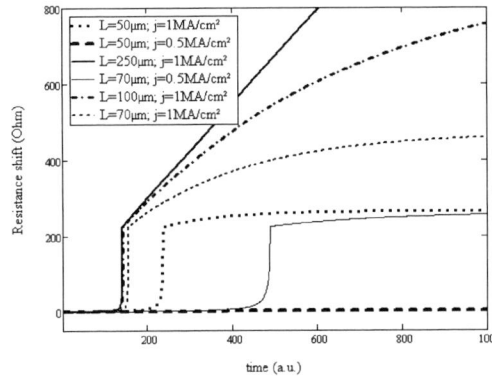

Figure 3.   Resistance trace simulations for various j / L conditions

Furthermore the smallest jL condition does not present resistance variation. For 250 and 25 μm-long lines, comparison with experimental resistance curves is illustrated on the figure 4. A good agreement is obtained after optimization of B, D and Z^* in accordance with values already published. l_{voidi} is fixed to 150 nm which corresponds to a fixed resistance step value in the simulation [5]. The times to failure are in the same order of magnitude and the resistance kinetics are reasonably similar. Experimental behaviors are well reproduced which proves a good confidence in this simulation.

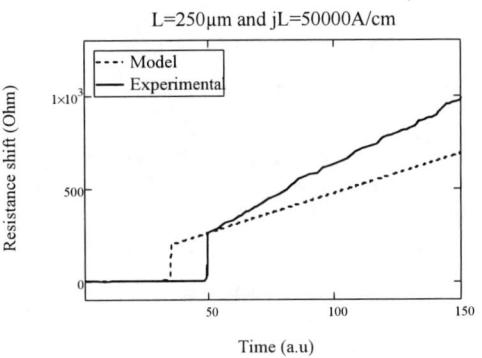

978-1-4244-5430-3/10 $26.00 © 2010 IEEE

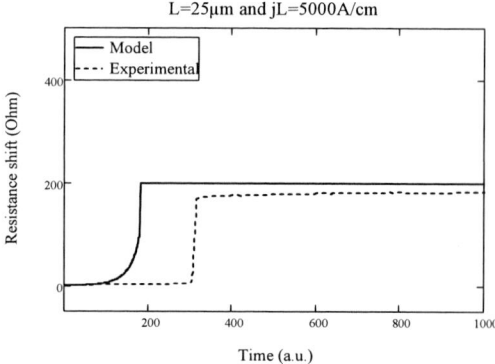

Figure 4.   Resistance traces: simulations vs experiments

It is worth noting that for L = 25 μm lines, experimental failure was not achieved for all samples unlike L = 250 μm lines that exhibit 100% of failure. Increasing the initial void length l_{voidi}, no resistance variation is observed by simulation for L = 25 μm and the interconnect can be considered safe from EM degradation. This could explain the censored TTF distributions near jL_c [3][4] considering a dispersion on the void length at the early stage of EM degradation. In this particular case, the resistance saturations occurred short after the steps: the void volume needed to induce failure was close to the one needed to stop the flux. For the same void volume the void configuration in length and in height does seem critical for failure. In addition, scanning electronic microscopy (SEM) observations on samples without failure confirm a larger upper base length of the void compared to the one observed on failed samples as illustrated by the figure 5.

Considering dR/dt, a linear trend with time is modeled for L = 250 μm whereas an asymptotic value is reached quickly after line opening for L = 25 μm. Interestingly, the simulation also provides saturation for longer lines but in a time frame and a resistance level not experimentally observable.

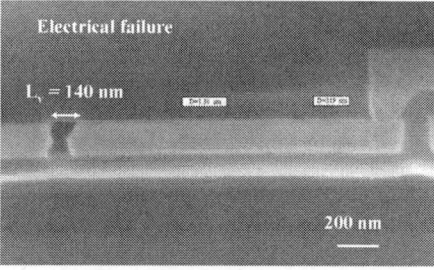

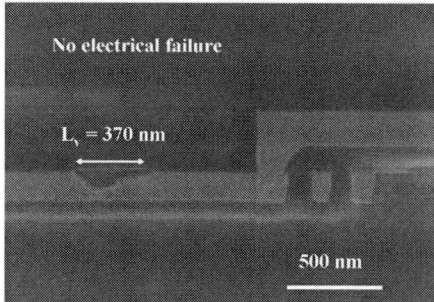

Figure 5.   Void morphologies leading to failure or not for the same j and L test condition

The asymptotic resistance value is given by (1) for an infinite time: $V(t \to \infty) = V_{sat}$. Physically, it corresponds to the EM induced void volume needed to cancel the net flux in the interconnect. The removed matter from the void builds up a sufficient backflow flux that can counter the EM flux. When this asymptotic volume is reached the stationary state is obtained and the resistance does not increase anymore. The resistance saturation seems to be a suitable experimental observation to define a stationary state condition during EM tests. From (3), this direct observation of steady state is described proportional to the product jL^2 contrary to the classical jL_c condition. Experimental V_{sat} extracted from resistance traces assuming a void with a parallelepiped shape or measured by SEM observations assuming the length measured in cross-section is the same along the line width, for the lowest test conditions of currents or line lengths verify a linear trend with jL^2 as depicted in figure 6. This correlation was investigated also at three different temperatures, 260, 300 and 330°C. In this temperature range, no significant variation was found on the saturation volume with respect to jL^2.

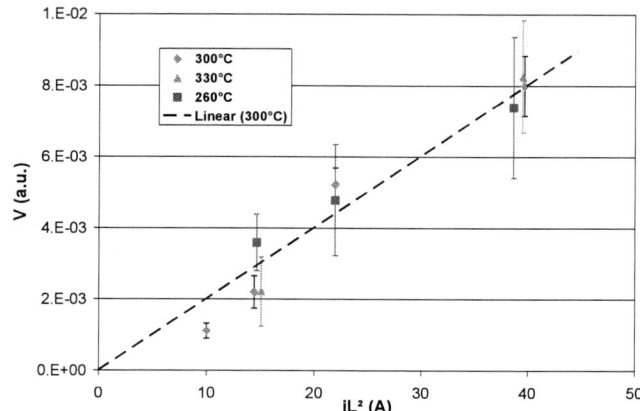

Figure 6.   Confirmation of the linear trend of V_{sat} with jL^2 at 260, 300 and 330°C

Unfortunately, the observation of the resistance saturation proves that the stationary state setting up does not immune interconnects from electrical degradations and some condition can lead to a fraction of failed samples. Assuming that we know the minimal void volume V_c below which no failure will be detected, any interconnect will be immortal provided the EM-induced void volume at stationary state V_{sat} remains lower than V_c: $V_{sat}(j,L) < V_c$. The immortality condition will be hence given by the product jL^2. As the classical jL immortality condition, the criterion as function of jL^2 takes into account the condition of stationary state but prevents also from potential failure in the transient state.

As an example to estimate jL^2_c, we can define a range on the V_c value from SEM observations of samples tested at the lowest jL^2 condition. From this consideration on V_c, we can estimate a range on the threshold below which interconnect immortality would be observed. jL^2_c is estimated between 4 and 7 A. For the time being, jL^2_c value cannot be precisely determined due to the assumption of a constant void dimension along the line width which can induce large error on V_c. Moreover, a deeper study on the void configuration physically achievable is necessary to accurate a jL^2_c value.

We verified that the linear trend followed by V_{sat} as function of jL^2 was not affected by the temperature between 260°C and 330°C (figure 6) and no significant variation was found in this temperature range. Tests at operation temperature are not experimentally achievable but this result seems to show jL^2 derived at typical experimental temperature around 300°C could be used at operation temperature without any modification.

IV. CONCLUSION

Using the expression of the stress during EM tests determined by Korhonen et al., the void growth kinetic $V(t)$ is available. The latter enables the modeling of the resistance evolution with time, from standard linear trends to saturation behavior with a good agreement compared to experiments. The asymptotic behavior of the resistance value is defined by $V(t\rightarrow\infty) = V_{sat}$. It evidences the stationary state condition in the interconnect and is experimentally observable for the lowest test couple j and L. We propose an immortality criterion based on the limitation of the void at stationary state V_{sat} below an electrical detection threshold volume. Unlike the classical one, this immortality criterion is proportional to the product jL^2. Investigations show no temperature dependence of the jL^2 criterion between 260 and 330°C. The extrapolation of the immortality criterion at operation conditions from accelerated tests seems to be possible without any modification.

ACKNOWLEDGMENT

The authors would like to thank the Failure Analysis Team of STMicroelectronics, Crolles for performing FIB/SEM analysis of the EM-induced void morphologies and locations.

REFERENCES

[1] I.A. Blech, "Electromigration on Thin Aluminium Films on Titanium Nitride", J. Appl. Phys., 47, p1203 (1976)

[2] S.P. Hau-Riege, "Probabilistic Immortality of Cu damascene Interconnects", J. Appl. Phys., 91, p2014 (2002)

[3] A.S. Oates and M.H. Lin, "Analysis and Modeling of Critical Current Density Effects on Electromigration Failure Distribution of Cu Dual-Damascene Vias", IRPS proceedings, p385 (2008)

[4] D. Ney, X. Federspiel, V. Girault, O. Thomas, P. Gergaud, "Electromigration Threshold in Copper Interconnects and Consequences on Lifetime Extrapolations", IITC proceedings, p105 (2005)

[5] L. Doyen et al., "Extensive analysis of Resistance Evolution due to Electromigration Induced Degradation", J. Appl. Phys., 104, p123521 (2008)

[6] M.A. Korhonen et al., "Stress evolution due to Electromigration in Confined Metal Lines", J. Appl. Phys., p3790 (1993)

[7] J. He et al., "Electromigration Lifetime and Critical Void Volume", Appl. Phys. Lett., 85, p4639 (2004)

PRACTICAL CONSIDERATIONS OF PROCESS CORNER EVALUATION FOR DEEP-SUB MICRON TECHNOLOGY NODES USING THE EXAMPLE OF ITS IMPACT ON ELECTROMIGRATION

Oliver Aubel
Quality and Reliability Engineering
GLOBALFOUNDRIES Dresden Module One LLC & Co. KG,
Wilschdorfer Landstrasse 101, 01109 Dresden, Germany
phone: +49-351-277-4652; fax: +49-351-277-94652; e-mail: oliver.aubel@globalfoundries.com

Thomas Hoffmann
Quality and Reliability Engineering
GLOBALFOUNDRIES Dresden Module One LLC & Co. KG,
Wilschdorfer Landstrasse 101, 01109 Dresden, Germany

Abstract — **In this paper a practical approach of investigating reliability performance over a specific process window is described. Based on an example of electromigration robustness evaluation, we present the calculation methods which need to be applied to correctly take reliability performance variation into account. This method also provides the possibility to investigate extreme corner reliability even failing the overall requirements so that the complete process window can be budgeted to fulfill the reliability requirement on the process.**

[Keywords: electromigration, process window, reliability budgeting]

I. INTRODUCTION

Back End of Line (BEoL) reliability has been a major concern in recent technology nodes and will most likely become even more important in future nodes. In particular electromigration (EM) and dielectric breakdown (BEOL-TDDB) are in focus for reliability investigations and process improvements. During the process qualification the target process will be qualified with respect to reliability performance. JEDEC [1] proposes to run the qualification based on three different lots processed separately in a timely manner to take process variations into account. Nevertheless high volume production of state-of-the-art semiconductor devices suffers from unavoidable process variations. These variations are caused by all kinds of processes and will not be caught completely by the three lot strategy.

For example one group of very common issues for the BEoL are patterning inhomogeneities distributed over the wafer resulting in dimensional variations. These within wafer variations (WWV) yield increased spread in lifetime distributions or even bimodal or multi-modal distributions [2]. Most of the time this kind of issues can be attributed to wafer patterns such as radial center edge signals (Fig 1).

Another group of variations is the equipment based variation resulting in wafer to wafer (WTW) variation. Typically the root cause is the processing on different tools of the same kind or even different tool sets for the same process step. Tool or even tool-chamber matching cannot always be guaranteed exactly. Additionally the aging of certain consumables or the tool itself can generate slightly different conditions during processing even on the same tool. Usually, these aspects are very well controlled by statistical process control (SPC) techniques and by very intense process monitoring. Nevertheless in most recent technology nodes (32nm and beyond) the relation between process window and variation becomes more and more challenging.

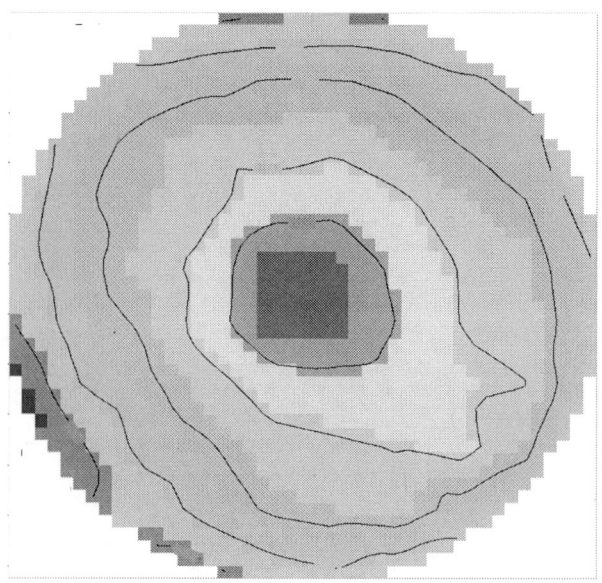

Figure 1. Example of a wafer map representing a severe within wafer variation (WWV) (e.g. red=high and blue=low resistance).

$$V_{max} = \frac{1}{V_{acc}^2} \cdot \left(\ln \frac{EOL}{A} - \frac{E_A}{k_B T} - \sigma \cdot u_{pEOL} \right)^2 \tag{3}$$

To use performance data for the populations it is necessary to have a reference level for the maximum allowed cumulative fail percentage p_{EOL} at the end-of-life time EOL expressed by the failure rate (F(EOL)):

$$p_{EOL} = P(t < EOL) = F(EOL) \tag{4}$$

This value can be expressed in terms of the normalized average failure rate (AFR) in FITs:

$$\text{AFR(EOL)/FITs} = F(t)/t \cdot 10^9 h = p_{EOL} / EOL \cdot 10^9 h \tag{5}$$

A mixture of N populations is characterized by the relative amount h_i of every population i with

$$\sum_{i=1}^{N} h_i = 1 \tag{6}$$

Combining the above ideas yields the performance measure for the mixture which can be computed with an implicit formula. A general format which is independent of the acceleration and extrapolation model is:

$$p_{EOL} = \sum_i h_i \, p_i (EOL) \tag{7}$$

The cumulative fail percentage at EOL is the weighed sum of the cumulative failure percentages of the corner populations given in equation 7. For lognormal fail time, at end-of-life the following amount will have failed:

$$p_i (EOL) = \Phi \left(\frac{\ln(EOL) - \ln t_{50i}}{\sigma_i} \right) \tag{8}$$

The corner's median lifetime is a function of temperature and current density ("Black's equation" [3]):

$$t_{50i} = A_i \, j^{-n_i} \exp \left(\frac{Ea_i}{kT} \right) \tag{9}$$

J_{max} of a corner is defined as the current density for which the fail percentage at end-of-life (EOL) exactly matches the allowed maximum fail percentage p_{EOL}:

$$EOL = A_i \, J_{max_i}^{-n_i} \exp \left(\frac{Ea_i}{kT} \right) \exp \left(\sigma_i u_{pEOL} \right) \tag{10}$$

Combining equation 9 and 10 with equation 8 gives a straight forward expression of the individual failure percentage $p_i(EOL)$.

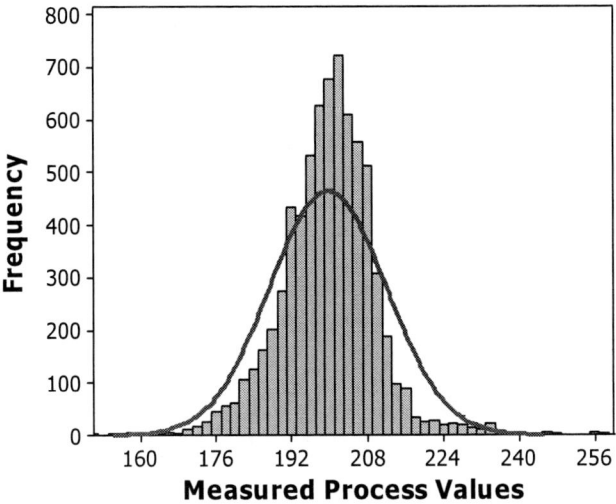

Figure 2. Histogram and Gaussian Distribution Fit for measured value of 45nm 1x module process showing the frequency (occurrence).

An example of a 45nm process variation is given in fig. 2. Small variations in processing can result in significantly degraded reliability lifetime in some aspects.

Therefore in this paper we are presenting a model of weighted reliability performance to characterize the complete process window (process budgeting) and are proposing a simple way to define process corners for the reliability characterization of the process window.

II. MODEL CALCULATION

A. Derivation of the necessary statistics

The underlying reliability data for this investigation is a set of electromigration data for relevant process corners (examples are given in fig. 7, 8, 9). For the quantification of the electromigration reliability robustness, we are using the J_{max} value calculated by eg.2. This is a modified "Black's equation" [3] (eq. 1) where u_{pEOL} stands for the quantile of the desired failure rate [4]. The parameters t_{50} and σ represent lifetime distributions measured at stress conditions of j_{stress} and T_{stress}. End of Life (EOL) is the maximum required lifetime at use conditions of T_{op}. The parameters n (current density exponent), Ea (activation energy) and A (geometry factor) are taken from "Black's equation". k_B is the "Boltzmann" constant.

$$\frac{t_{50}}{EOL} = \left(\frac{J_{max}}{j_{stress}} \right)^n \exp \left(\frac{E_A}{k_B} \left(\frac{1}{T_{Stress}} - \frac{1}{T_{op}} \right) \right) \cdot \exp \left(\sigma \cdot u_{pEOL} \right) \tag{1}$$

$$J_{max} = \sqrt[n]{A \cdot \exp \left(\frac{E_A}{k_B T_{op}} \right) \exp \left(\sigma \cdot u_{pEOL} \right) / EOL} \tag{2}$$

A similar approach can be taken using the square root E relation used for BEOL-TDDB. Based on this relation one can generate an equivalent performance measure V_{max} which describes the maximum voltage which could be applied and still ensure EOL reliability.

978-1-4244-5430-3/10 $26.00 © 2010 IEEE

$$p_i(EOL) = \Phi\left(\frac{\ln\left(A_i\, J_{maxi}^{-ni}\, \exp\left(\frac{Ea_i}{kT}\right)\exp(\sigma_i\, u_{p_{EOL}})\right) - \ln\left(A_i\, j^{-ni}\exp\left(\frac{Ea_i}{kT}\right)\right)}{\sigma_i}\right)$$

$$= \Phi\left(\frac{\ln A_i - n_i \ln J_{maxi} + \frac{Ea_i}{kT} + \sigma_i\, u_{p_{EOL}} - \ln A_i + n_i \ln j - \frac{Ea_i}{kT}}{\sigma_i}\right)$$

$$= \Phi\left(u_{p_{EOL}} - \frac{n_i}{\sigma_i}\ln\left(\frac{J_{maxi}}{j}\right)\right) \qquad (11)$$

Therefore the final equation in terms of the individual performance measures depends on the acceleration and extrapolation model. For an Arrhenius/power law model of lognormal distributed lifetimes (as typically used for electromigration) the equation is (j here is the common performance parameter of the mixture of the populations):

$$p_{EOL} = \sum_{i=1}^{N} h_i \Phi\left(u_{p_{EOL}} - \frac{n_i}{\sigma_i}\ln\frac{J_{maxi}}{j}\right) \qquad (12)$$

Another example is the Arrhenius/square root law with lognormal life time, which gives for v as performance parameter:

$$p_{EOL} = \sum_{i=1}^{N} h_i \Phi\left(u_{p_{EOL}} + \frac{\gamma_i}{\sigma_i}\sqrt{v - V_{maxi}}\right) \qquad (13)$$

To compute the joint performance measure, an iterative algorithm is used:

1. Get h_i and population data (e.g., J_{maxi}, n_i, σ_i)
2. Get reliability target p_{EOL}
3. Vary initial mixture performance guess (e.g. j or v) until the sum equals p_{EOL}, the resulting value of the performance measure is the mixed-population performance (e.g. J_{max}, V_{max})

As all cumulative failure probabilities are monotonically non-decreasing functions, a simple bisection or Newton-Raphson iteration can quickly find the solution [6]. It has to be taken care to not leave the range of physical validity of the used models during the iteration. Advanced models (e.g. bimodal distribution models and models with a threshold parameter) need refined iteration algorithms, but follow the same logic as explained above.

B. Simplified evaluation

It is possible to evaluate quickly the influence of decreased reliability performance of one corner on the entire population mixture.

For this an idealized mixture is considered: All populations but one contribute virtually nothing to the failure budget of the mixture. Given this, what is the maximum allowed percentage of one "bad" population that fails the reliability target for the mixture? As no corner is allowed to contribute more than p_{EOL} to the system failure budget, the requirement is:

$$p_{EOL} \geq h_i\, p_i \rightarrow h_{iMAX} = p_{EOL}\, /\, p_i \qquad (14)$$

As an example for the Arrhenius/power law model of lognormal distributed life time follows:

$$h_{iMAX} = p_{EOL}\, /\, \Phi\left(u_{p_{EOL}} - \frac{n_i}{\sigma_i}\ln\frac{J_{maxi}}{J_{max}}\right) \qquad (15)$$

Using a graph with normalized technology and population data, a quick pre-check can be made to decide if it is necessary to start a full mixed-populations calculation (fig. 3). This plot shows the relative amount (frequency or occurrence) of a bad corner in percent dependent on the measured σ/n of the respective corner.

As an example: Measuring a "bad" corner with an EM performance J_i (with a σ of e.g. 0.2) which is only 70% of the required J_{max} of the whole population ($J_i/J_{max}=0.7$). To fulfill the overall reliability the occurrence (h_i) of this corner must not be over 0.01% (fig. 3) for n=1.4.

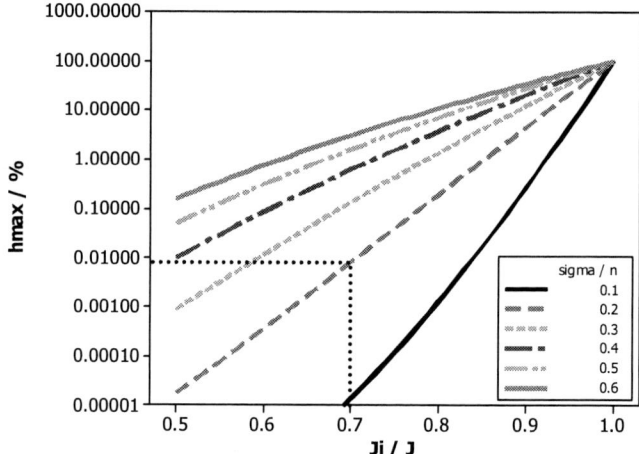

Figure 3. Allowed occurrence in Percent (h_{max}) of a corner with bad electromigration performance (based on a current density of n~1.4 and p=1e-7%).

Another way of interpretation is how much failure percent (p_i) the investigated corner contributes. Here, an approximately linear reduction in J_i is leading to an approximately exponential increase in failure percent (fig. 4).

Overall this method gives a very good indication on how much J_{max} reduction (below target) can be tolerated (dependent on the expected occurrence of the respective corner), so that the complete data set (complete process window) is still meeting the reliability targets.

On the first look the influence of the process variation appears paradoxical: Processes with large σ benefit most from this calculation. Fig. 5 gives a pictorial explanation of this fact. Here the same t_{50} reduction for samples with higher σ would yield a smaller CDF value increase compared to the smaller σ samples.

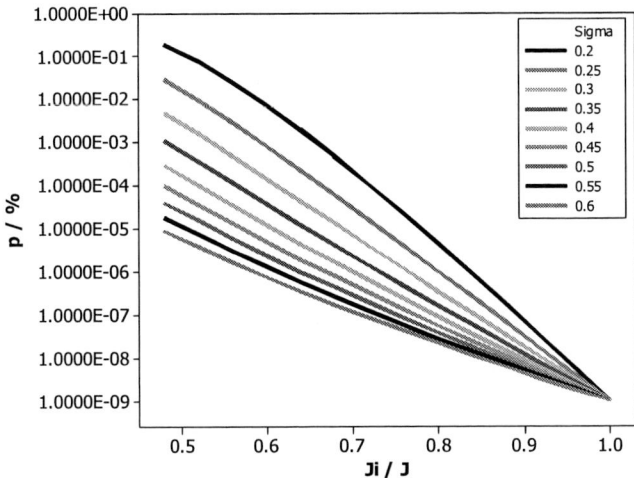

Figure 4. Fail percentage p_i of corner with a given J_i for required J_{max}. The failure percentage increases almost exponentially.

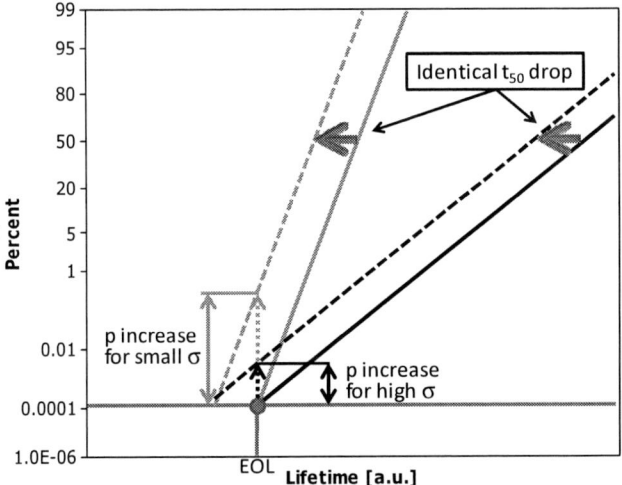

Figure 5. Explanation of the apparently paradoxical situation, where high sigma value results benefit more from the process-budgeting idea.

C. Evaluation of confidence intervals

Upper and lower confidence limits for the cumulative system failure probability at EOL can be computed from variance data for the individual $p_i(EOL)$ and h_i using the delta method [6]. In a first step the system variance for independent corners will be estimated:

$$Var(\hat{p}_{EOL}) = \sum_i \hat{h}_i^2 \cdot Var(\hat{p}_{iEOL}) \qquad (16)$$

From the system variance follows the system standard error:

$$\hat{se}_{pEOL} = \sqrt{Var(\hat{p}_{EOL})} \qquad (17)$$

As p_{EOL} is a positive parameter, a logit transformation will be used to get the confidence interval:

$$\left[\underline{p_{EOL}}, \overline{p_{EOL}}\right] = \left[\frac{p_{EOL}}{\hat{p}_{EOL} + (1 - \hat{p}_{EOL}) \times w}, \frac{p_{EOL}}{\hat{p}_{EOL} + (1 - \hat{p}_{EOL})/w}\right] \qquad (18)$$

$$w = \exp\left(z_{1-\frac{a}{2}} \frac{\hat{se}_{pEOL}}{\hat{p}_{EOL} \cdot (1 - \hat{p}_{EOL})}\right) \qquad (19)$$

The approximate calculation of confidence limits relies on assumptions, e.g. on statistical independent failure probabilities at the different process corners that would have to be checked for practical applications.

III. PROCESS CORNER DEFINITION

Having established the proposed model we suggest to run process corner splits to fully characterize the high volume process performance. The process corners are defined by shifting the process distribution (fig. 2) of the "on-target" process in a way, that 1% of the distribution is out of specification (fig. 6). Due to the processing complexity of cross corner splits and the very unlikely occurrence of these splits, we did not evaluate these types of corners in this study. Therefore we focus on the most critical "single corner" splits for electromigration and BEOL-TDDB performance.

1. Cu seed layer thickness variation enhances the early failure electromigration mode related to the via fill weaknesses [7]. The seed layer is essential to ensure copper fill (fig. 7).

2. Variation in critical (horizontal) dimensions e.g. M2 width reduction decreases Via 1 upstream performance (fig. 8) whereas it improves Via 1 BEOL-TDDB performance at the same time.

3. Increase of aspect ratio for PVD (barrier & seed processes) and copper fill. Figure 9 indicates a slight tendency to early failures.

Corner splits generated this way were tested for their electromigration performance (fig. 7-9). Even though in the discussed example most of the splits do not fail the absolute reliability target, the electromigration performance of the "complete" process (incl. window) is significantly smaller then the performance of the "on-target" process alone. The split "M2-20%" represents an out of specification scenario.

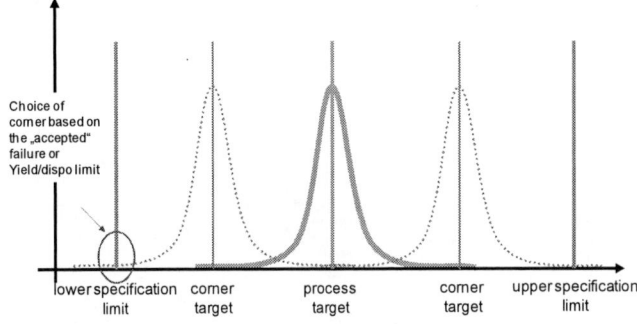

Figure 6. Suggested shift of the "on-target" process to define the process corners for reliability testing.

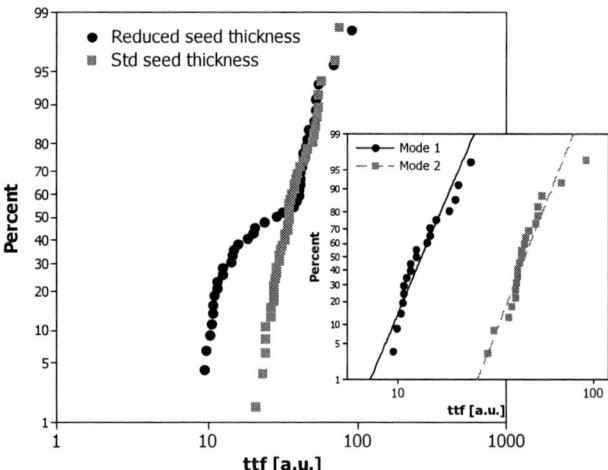

Figure 7. Electromigration test results for the standard and reduced seed thickness corner evaluation. The corner shows a bimodal behavior and the modes have been separated for evaluation (see insert).

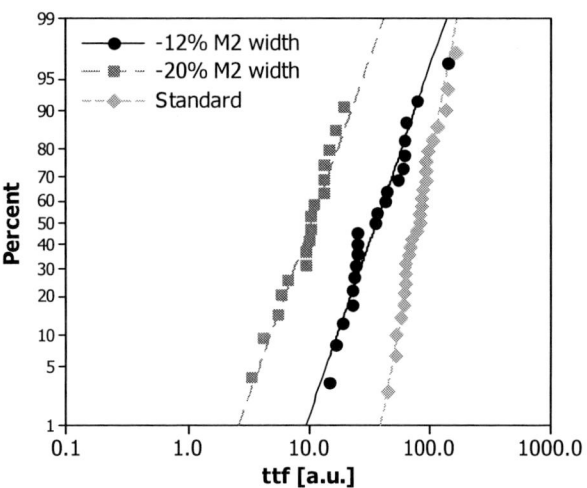

Figure 8. Electromigration test results for the metal 2 CD splits, where "-20%" represents an out-of-SPEC scenario.

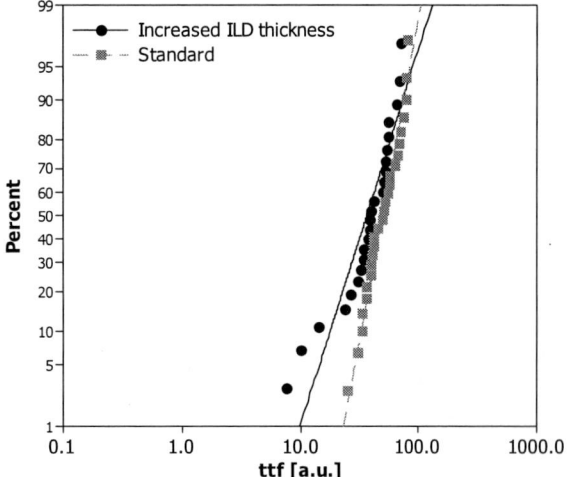

Figure 9. Electromigration test results generated on process corner splits with ILD thickness variations (variations in via height).

Using equation 11, we can run a quick calculation applying the following parameters: $p_{EOL}=10^{-7}\%$, $u_{pEOL}=-6$, $n=1.4$ and preset parameter for h_i occurrence which need to be generated out of the Fab performance (process stability) (Table 1). The measure J_{maxi} values for each split, the σ_i values as well as the occurrence values h_i are given in table 1. Please note that the M2 width splits have been measured at a lower temperature than for the other two split investigations.

One can see that the standard performance for the split is very comparable and is yielding a very solid electromigration performance whereas the corner splits sometimes deliver a significantly reduced performance. The reduced seed thickness corner split showed a bimodal behavior and the modes where separated accordingly. For the further calculation we used the individual modes instead of the total data set (fig. 7 inlet). The 1% occurrence for this corner was then divided by half so that each of both modes occur 0.5%.

The basic idea of this corner methodology is that the "bad" corner most significantly impacts the overall performance. Investigating each corner set individually one notices that the "bad" corner performance can be compensated to a limited amount. E.g. for the seed thickness split the standard (target) process delivers a J_{maxi} of ~4.5MA/cm^2 whereas the combined evaluation reduces this number by ~2x. The effect is caused by a corner which is assumed to happen only at 0.5% of all processes (and therefore all products).

Combining all investigated corners deliver a $J_{max(tot)}$ of 1.14MA/cm^2, this is ~4x lower than all standard (target) electromigration tests. This total electromigration performance is by far dominated by the "-20% M2 width corner "-split. Removing this corner from the process e.g. by adjusting resistance targets for M2 one would significantly improve $J_{max(tot)}$.

TABLE 1. MEASURED ELECTROMIGRATION PERFROMANCE FOR CORNER SPLITS.

Split		J_{Maxi} [MA/cm^2]	σ_i	h_i [%]	J_{max} (tot)
Seed thickness (Fig. 7)	Std.	4.49	0.31	99	2.70
	Red. (tot)	0.5	0.7	---	
	Red. (1)	2.25	0.27	0.5	
	Red. (2)	6	0.24	0.5	
M2 width (Fig. 8)	Std.	4.7	0.32	99.98	1.14
	-12%	0.86	0.58	0.0019	
	-20%	0.327	0.6	0.0001	
ILD thickness (Fig. 9)	Std.	4.55	0.32	98	1.67
	Thick	1.3	0.56	2	

Total sum of the process window: Corner h_i stays the same, Std h_i reduced to 96.998% accordingly → $J_{max(tot)}$=1.14MA/cm^2

IV. CONCLUSION

In this paper we are presenting a method to fully characterize the reliability performance of a high volume process taking the unavoidable process variations into account. This method can be used either to define the process limitations or to characterize a process with a given window. One can even think of the situation where some corners are not meeting the reliability target but can be compensated by the "on-target" process or benign corners ending up with a process budgeting concept.

A key message is that a simple sum of weighted J_{max} values (eq. 20) is significantly underestimating the impact of the "bad" process corner reliability:

$$J_{max(tot)} = \sum_{i=1}^{N} h_i * J_{Maxi} \qquad (20)$$

In the example (table 1) the result from eq. 20 would be: $J_{Max(tot)}$=4.42MA/cm^2. Since this $J_{max(tot)}$ value is mainly influenced by the standard electromigration performance, low performing corners do not have an impact on the overall reliability performance.

As a conclusion: for all reliability investigations of different splits, process corners or even bimodal or multimodal distribution one can sum the FIT or failure rate value but not the regular performance parameters. If one particular split or corner is resulting in a significant FIT contribution, this performance can be compensated only to a very limited amount.

On the other hand reliability results passing the EOL requirements may not result in significant failure rates and would allow a process budgeting for corners with occurrences (hi) in the very low percentile. These specific corners can even fail the reliability requirement but using this method the overall process can pass requirements.

This approach may also be used to budget different layer performances in the BEOL stack.

ACKNOWLEDGMENT

The authors like to acknowledge the reliability team in GLOBALFOUNDRIES Fab1 for generating the data, here in particular Christian Hennesthal, Jens Poppe and Jens Paul. Additionally we like to thank the integration team for providing the samples and for helpful discussion; in particular we thank Frank Feustel, Juergen Boemmels and Matthias Lehr.

REFERENCES

[1] JEDEC JP-001, "Foundry Process Qualification Guideline", March 2002

[2] A.H. Fischer, et al, "Experimental Data and Statistical Models for Bimodal EM Failures", Conf. Proc. IEEE International Reliability Physics Symposium 2000, p. 359-363

[3] J.R. Black, "Electromigration - A brief survey and some recent results", IEEE Transactions on Electron Devices, Vol. 16, No. 4, p. 338-347, April 1969

[4] B. Li, et al, "Minimum void size and 3-parameter lognormal distributions for electromigration failure in Cu interconnect", Conf. Proc. IEEE International Reliability Physics Symposium 2006, p. 115-122

[5] S. Nassif, et al, "Modeling and analysis of manufacturing variations", IEEE Custom Integrated Circuits Conference 2001

[6] W.Q. Meeker, L.A. Escobar, Statistical Methods for Reliability Data, Wiley 1998

[7] A.H. Fischer, O. Aubel, et al., "Challenges in Copper Metallizations arising with the PVD Resputter Liner Engineering for 65nm and beyond", IEEE International Reliability Physics Symposium 2007

Investigating The Electro-Thermal Origin Of Breakdown In Low-K/Cu Dielectrics Under Short Duration Over Stressed Pulsed Regime

Amitabh Chatterjee and Forrest Brewer

Department of Electrical and Computer Engineering, University of California, Santa Barbara, CA - 93106, USA

S.C. Lee and A. S. Oates

R&D, Taiwan Semiconductor Manufacturing Company, Hsin-Chu, Taiwan 300-77, R.O.C.

Abstract—We present a detailed study of low-K/Cu structures under short duration pulse regime and establish a microscopic understanding of breakdown behavior under high current stressing. Random and anomalous behavior of the breakdown characteristics observed under very fast pulsing conditions are explained through electro-thermal instability. A model based on extensive experimental study has been developed to show switching behavior of a conducting path (short failure) due to meltdown of copper. The dumping of critical energy can lead to permanent damage and leads to an open failure. Established breakdown model has been critically linked to the material behavior and extrapolated to understand the TDDB behavior of the low K dielectric.

Keywords-component LowK/Cu, ESD, Electrothermal Runaway,

I. INTRODUCTION

A. Background : Low K Reliability & Scaling Technology

Miniaturization in adherence to Moore's law has increased both the density of interconnects and the level of power dissipation - a critical figure of merit determining its performance. The ITRS projection is that the IC thickness will decrease to <50 μm, wire diameter to <15 μm, interconnect pitch to <20 μm, copper film thickness to <10 μm, and via diameter in substrates to <20 μm. This has necessitated the use of novel ultra low-k dielectric material; critical for next generation VLSI technologies e.g., interconnect for 32nm SoC for future mobile, digital home appliance and automotive products [1, 3-9]. However, these novel materials have serious consequences from the perspective of structural integrity of interconnects due to substantially different thermo-mechanical properties, as their elastic modulus is two orders of magnitude lower. Thus one needs to monitor the adhesion between low-k and the adjacent materials, which is very poor due to high porosity of these materials [10-14]. Therefore the dedicated damascene process for copper requires needs a better characterization to study the effect of interconnects scaling and low-k dielectric material on the thermal behavior of the IC metal is desirable to provide thermal design guidelines in the near future. Importantly porosity or material softness has a deleterious effect on the electrical breakdown behavior, which can make it vulnerable under harsh environments [1].

B. Correlating Backend Reliability with Gate Oxide Reliability under TDDB stressing

Time dependent dielectric breakdown in gate oxide is extensively studied phenomenon [15-16]. There is a consensus that oxide breakdown results from defect generation (Si/SiO$_2$ interface traps, neutral electron traps, anomalous positive charge) in the Si/SiO$_2$ system. The defects accumulate with time and eventually reach a critical density, at which point a conductive path is created in the gate oxide layer (*fig.1*). A surge of current produces a large localized rise in temperature which leads to permanent thermal damage at the local breakdown path and possible lateral propagation of the breakdown spot.

However, the backend dielectrics are different than thin gate oxides in several ways. The backend dielectric undergoes many process steps, such as chemical mechanical polishing (CMP) and photoresist strip/ashing. These steps can damage the interfaces, which can become trap sites and assist in conduction. Moreover, the quality of the backend dielectric, which is deposited, rather than thermally grown, is much poorer. As a result, defect densities are likely to be much higher.

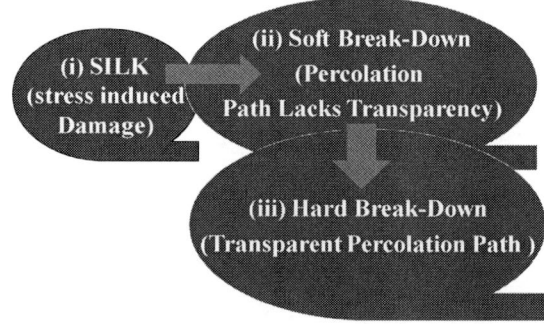

FIG1. STAGES OF DEGRADATION DURING THE WEAROUT UNDER TDDB STRESSING

Accordingly, several theories have been proposed for backend dielectric breakdown. One approach has proposed a multi- step model and explains breakdown as induced by copper diffusion in the dielectric. In the first stage, the applied bias causes the injection of positively charged copper ions into the dielectric. The lack of a neutralizing electron current results in space-charge build up which sets up an opposing field and reduces the ionic current. Next, the copper ions and neutral atoms thermally diffuse through the insulator, which leads to dielectric degradation and an increase in dielectric leakage. Finally, in the final stage, enhanced electric fields in the dielectric lead to breakdown. An alternative theory associates

978-1-4244-5430-3/10 $26.00 © 2010 IEEE

breakdown of low-k materials, not with the presence of copper in the dielectric, but with the presence of pores that act as defective sites in the dielectric. Lower-k materials start with a large numbers of pores (defective sites). During electrical stressing, more defective sites are generated with time due to bond breakage. These defective sites serve to create a percolation path, and failure results when such a path connects previously isolated copper lines.

The purpose of this paper is to comprehend the implications of interconnect scaling on low-k dielectric structures. The purpose of this paper is to comprehend and build insight into electro-thermal instabilities under short duration electrical pulsing. We establish a model to comprehend the impact of novel low-k dielectric material under ultra short duration over-stressing and examine their thermal behavior and degradation under an electrical breakdown, so that it can provide critical electrical and thermal design guidelines for the scaling technologies.

II. EXPERIMENTAL STUDY: DIFFERENT PULSE REGIMES EASE OF USE

A. Anamolous Snapback Behavior Under ESD (HBM) Pulse

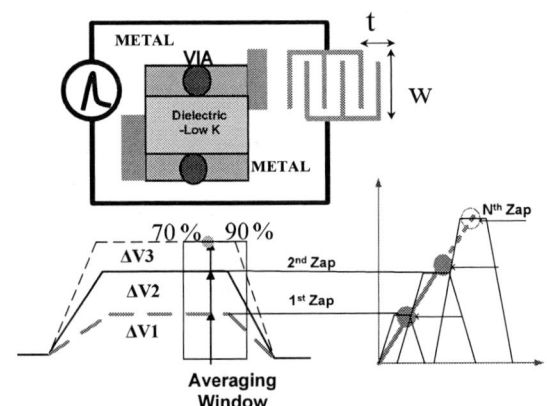

FIG2. INTERCONNECT STRUCTURE & QUASI-STEADY MEASUREMENT UNDER TLP SETUP

In our experiments (*fig. 2*) different Low-K/Cu dielectric structures were stressed using high voltage TLP pulses (t_{rise} < 10ns i.e ESD HBM model) (*fig. 2*). The breakdown I-V curve (*fig. 3*) generated by averaging time domain waveforms (i.e voltage and current) obtained from reflections across the transmission line show the electro-thermal origin of the breakdown phenomenon (~time scale few 10ns >> thermal diffusion time constant). TLP setup involves quasi-steadily incrementing the pulse amplitude. Thus an I-V relationship generated quasi-steadily through TLP measurement comprises of series of zaps with each pulse comprising of increasing amplitude. Under short-duration pulse the dielectric structure can exhibit electro-thermal instability under short duration pulse and leads to high and low conducting breakdown state.

While high temperature annealing (*fig 4 & 5*) can make the structures leakier, interestingly, it can also show recovery, which demonstrates the duality in the electro-thermal generation of defects. Sudden switching to a high current state (ON/OFF switching of conducting paths) and an open failure

higher voltages marks the *anomalous behavior* (*fig. 4*)[2]. Sudden switching to higher conducting state and then jump to low current state which makes it look like almost a reversible phenomenon. To under-stand these random jumps in current the stressed samples were analyzed through FA pictures (fig. 5) and TEM analysis of the cross-section (*fig. 7*).

B. Failure Analysis of Anomalous Behavior: Open and short failure

The anomalous behavior, wherein the breakdown current suddenly shoots up is marked by varying degree of delamination in the stressed samples (*fig. 6*). This can be explained due to self-heating electro-thermal during the transient process of filamentation. This results in intense and highly localized heating. For certain samples de-lamination is very prominent, whereby dark spots appears due to highly localized heating, while few samples showed lighter delamination. However, few samples show lighter delamination an short damaged samples showed.

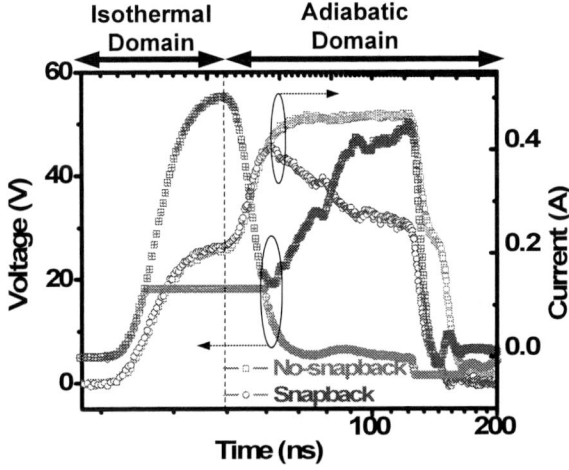

FIG3. SAMPLED CURRENT (I) & VOLTAGE (V) FROM TIME DOMAIN WAVEFORMS SHOWING *ELECTRO-THERMAL DOMAINS*

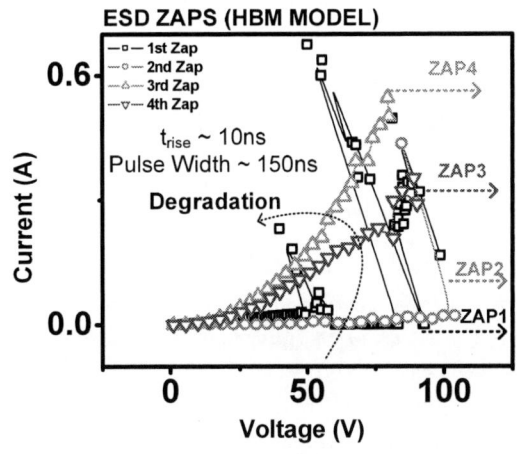

FIG4. ANAMOLOUS SNAPBACK & ANNEALING EFFECT SHOWING BOTH DEGRADATION & RECOVERY

978-1-4244-5430-3/10 $26.00 © 2010 IEEE

Preliminary FA analysis show that a metal short formed in the dummy layer under an ESD pulse condition leads to *short* mode of failure. Interestingly, the FA pictures show that return to lower conduction state (beyond high current conduction state) was marked by extensive damage to the comb structure, which explains the open nature of the failure [17]. Shorter failure can be explained due to formation of conducting filament. Subsequently, very high conduction filamentary state (i.e short failure) leads to open failure.

Thus the anomalous behavior is characterized by the dielectric structures going into either into an open failure, or a short failure mode. Interestingly, the sample exhibited short and subsequent open failure over wide range of voltages. Observed instability during the process of filamentation corresponds to certain voltages as *high conducting filaments burns out*. It can be noted that abrupt filament formation and highly localized heating; can be controlled by shorter duration pulse, TLP like pulse.

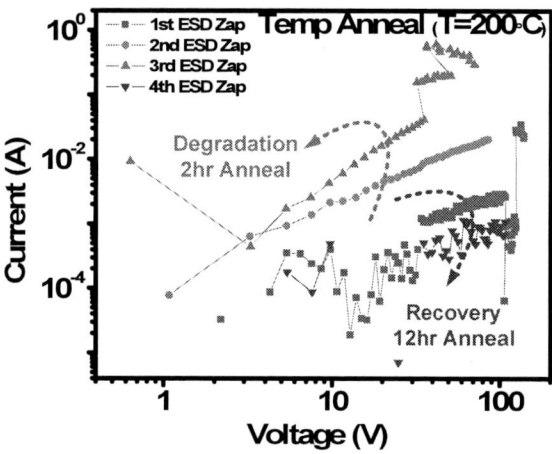

FIG.5. TEMPERATURE DEPENDENCE SHOWING *ANNEALING EFFECT & DEGRADATION*

Thus one observes a series of random jumps (i.e 1st, 2nd,3rd,4th) (*sample A*) as the voltage across the sample is increased and the open failure is always preceded by short failure (*fig. 8*). These sample exhibits *short failure* at different voltages which is followed by burn-out of the shorted path. Few of the samples (from different wafer lot) (*inset fig. 8*) showed lighter de-lamination as it entered a high conduction state (zap 1- *inset fig. 8*) (*sample B*). However, when the pulse sequence was restarted (zap2- *inset fig. 8*) it was relatively unaffected (indicating *a soft breakdown*). In-fact, it was during the 3rd Zap that the sample showed a trend towards shorter failure. Thus a *short failure* can remain dormant and need not be followed by a *hard or open failure*. Now to further analyze whether these shorted conditions were typical of an ESD pulse condition, the impact was studied by slowing the pulse.

C. Impact of slowing the ramp

When the speed of the ramp was slowed, (*fig.9*) (rise time ~ 10's μs and interleaved between two ESD zaps); anomalous

behavior was significantly alleviated (*fig. 6*), while soft breakdown eventually lead to permanent damage (after multiple zap).

The dielectric breakdown mechanism can be summarized as occurring through three different stages: in the first stage stress induced breakdown current goes into reversible soft breakdown (high current conducting state); secondly, as electric field and temperature increases with carrier flow the rate of damage increases; third, irreversible hard breakdown is observed when sufficiently large energy is dumped. Surprisingly, some of the samples, which were permanently damaged by slower ramps, show recovery when zapped with

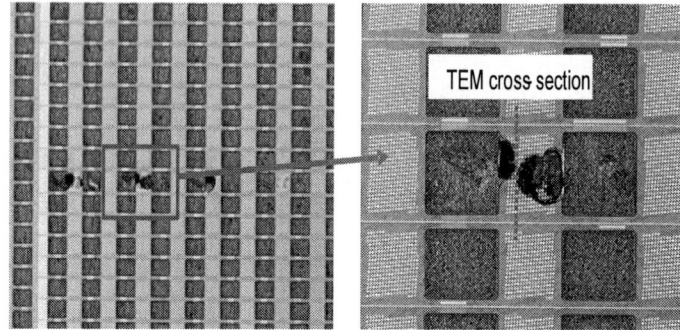

FIG 6 . FA PICTURES OF SAMPLE SHOWING VERY HARD OPEN FAILURE WHICH RESULTS

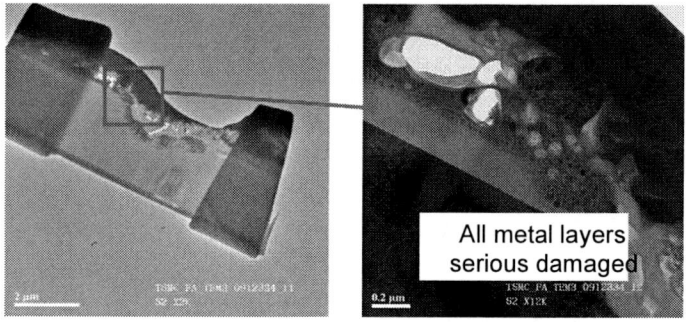

FIG 7 EXTENSIVE DAMAGE WHEREBY ALL THE METAL LAYERS GET DAMAGED OBSERVED THROUGH TEM CROSS-SECTION

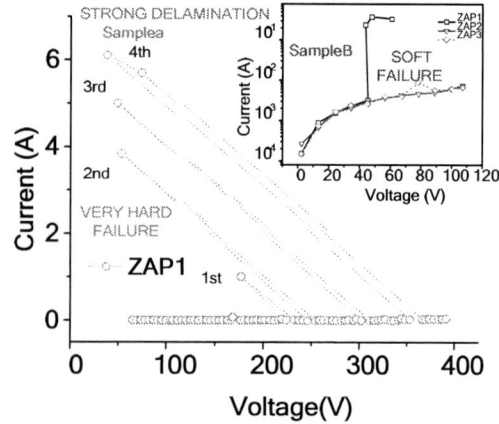

FIG 8. TLP DATA OF THE SAMPLES SHOWING DARK AND WEAK DELAMINATION

978-1-4244-5430-3/10 $26.00 © 2010 IEEE

ESD pulses, indicating open failure mechanisms under ESD pulse condition due to short time scale heating. The difference in slope between two ramping condition (*fig. 8*) can be explained due to higher localized temperature due to adiabatic behavior under an ESD cvent, where localized heating leads to thermal generation of defects(*fig.10a*). Lower K materials [K=2.0] compared to [K=2.5] show a gradual degradation with multiple short ramps (*fig. 9*).

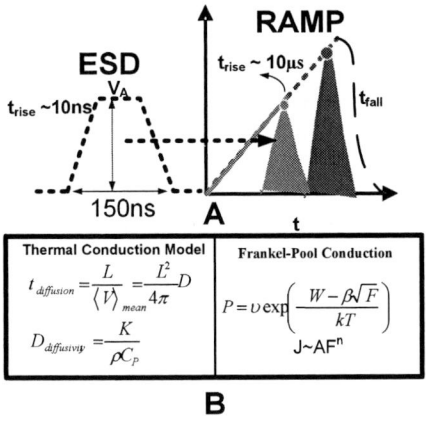

FIG.9 DUMPED ENERGY PROPORTIONAL TO AREA UNDER THE CURVE ENERGY OF THE PULSE (B) E-T MODEL

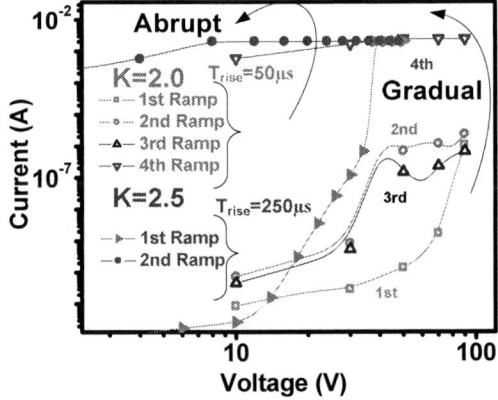

FIG.10. PERMANENT DAMAGE UNDER RAMP CONDITION GRADUAL FOR [K=2.0] & ABRUPT FOR [K=2.5]

	Anomalous Behavior	
Higher Level Metal	↓	Higher Thermal Capacitance (C_T)
	Less Prominent	
Via Structures	↑	Higher Thermal Resistance (R_T)
	More Prominent	
Wider W	↓	Higher Thermal Capacitance (C_T)
	Less Prominent	

FIG11: TRENDS IN ANAMOLOUS BEHAVIOR INDICATING *ELECTRO-THERMAL ORIGIN* OF *DEGRADATION*

D. Electro-thermal Origin & Summary of Experiments

Better thermal performance under short duration pulsing makes the higher metal level more resilient towards reversible anamolous snapback behavior i.e *soft breakdown and subsequent hard breakdown (fig. 11)*. However when the pulse amplitude is very high, structures show permanent damage without entering soft breakdown stage

Summary:

1. Self-heating causes localized heat dissipation which leads to de-lamination of the cap layer (*fig. 6 & fig. 7*).

2. Localized J.E and elector-thermal instability during the filament formation is related to defects. This is corroborated by the fact that under higher breakdown voltages, the low- K structures are not prone to abrupt jumps (*fig. 8*)

3. The delamination is also observed during the *short failure* in the samples which is subsequently followed by an *open failure* (*fig. 8*).

4. Burnout of filaments is triggered at certain voltages. This can be related to instability during filamentation, which are again related to the defects in the cap layer (*fig. 8*)

5. Low-K structure below the cap layer is relatively unaffected even when zapped with very voltage TLP pulses.

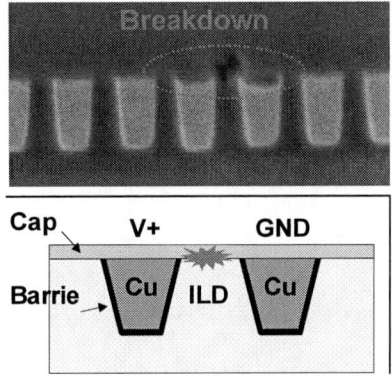

FIG:12 DAMAGE IN THE CAP LAYER LEADING TO FAILURE

Thus all these experimental data gives very strong evidence that the burnout is strongly related to electro-thermal instability and formation of conducting channel or filaments (*fig. 12*). The transient unstable process of filamentation is strongly related to the defect sites.

III. MODELING ELECTRO-THERMAL INSTABILITY

A. Understanding non-linear breakdown behavior: Anomalous Effect in shorter time scale

The slope of the I-V curve, which obeys the power-law, results from series expansion of F-P emission model (high field conduction mechanism in a dielectric) (*fig.13*)[3]. Multiple zaps clearly show a stressing effect (making the dielectric more leaky), which has been attributed to an increased number of defects triggered by a de-lamination of the cap layer [3]. Dielectric breakdown phenomenon, often modeled through defect generation (e.g. interface traps, neutral electron traps, anomalous positive charge) can be exacerbated by copper diffusion, which can be highly temperature sensitive [3-9].

978-1-4244-5430-3/10 $26.00 © 2010 IEEE   935

Sudden short failure can be related to formation of conduction can be related to defects where the copper ions can thermally diffuse into these defect sites (*fig. 15 & fig. 16*).

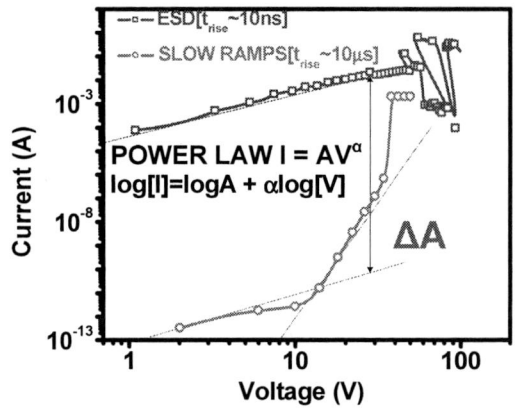

FIG 13. BREAKDOWN SLOPE IN THE I-V CHARACTERISTICS COMPARING ESD ZAPS VS SLOWER RAMPS

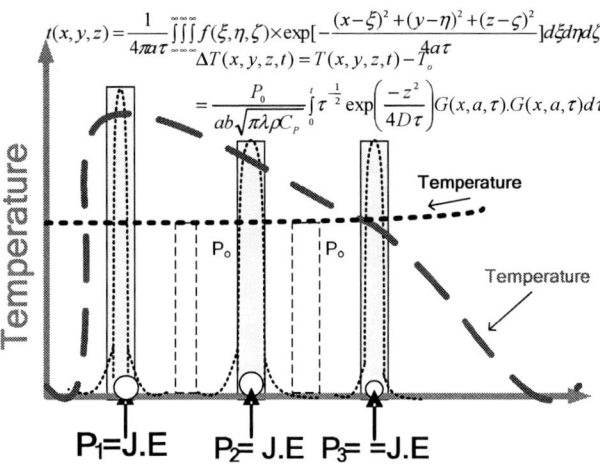

FIG 14 LOCALIZED HEAT DISSIPATION & ITS IMPACT ON TEMPERATURE DISTRIBUTION

More prominently, very fast ESD pulse trigger electro-thermal instability, due to initial isothermal build up of breakdown current and subsequent adiabatic self heating process, leads to hot and cold domains as it becomes unstable, giving rise to the bifurcation triggered by the external circuit and the stressed sample. This leads to a highly localized rise in temperature, at which point a conductive path is created in the dielectric layer (*fig. 15 & fig. 16*). More prominently higher E-field due to space-charge of mobile ions, can lead to *crack of the whip* mechanisms – characterizing instability in very short duration time scale event.

Variations were observed between the wafer lots (having process split) - as some of the samples showed an early short and subsequent open failure. Now, the variation in the heat dissipation centers is central to understanding the sensitivity of the process during ultra short time scale ESD event (*fig. 14*). The coupled electro-thermal process is highly nonlinear and

randomness can be explained due to variability of heat dissipation mechanism due to localized heating.

The proposed randomness in the breakdown behavior can be explained through the electro-thermal model as the copper move into the defect centers, which leads to redistribution of J.E heat dissipation mechanisms as the electric field is modulated. Thus certain areas see a fall in temperature while it intensifies in certain region. Net heat dissipation is thus sum of heat dissipated from all regions and depending upon the *anisotropic diffusion time constant* (along the x-axis→T_a, along the y-axis→ T_b, along the z-axis→T_c) – (*fig. 15*), which can trigger thermal runaway.

The variations of defect centers have a bearing on how the copper can diffuse into the structure. Moreover, these variations can significantly impact the maximum temperature attained under an ESD pulse condition. However, critically under certain conditions, it can lead to electro-thermal instability which triggers thermal runaway. The localized J.E event and unstable self heating process leads to meltdown or catastrophic burn-out of filament. Now, shorter failures results when the damage is not extensive under these shorter duration pulse and while only a conduction path is established. However, extensive damage is observed after a short has been established and the dielectric structure when again stressed under ESD pulse leads to open failure. This also confirms the stage wise degradation as open failure is preceded by short failure which implies role of diffusing copper in forming bridges which often lead to metal short.

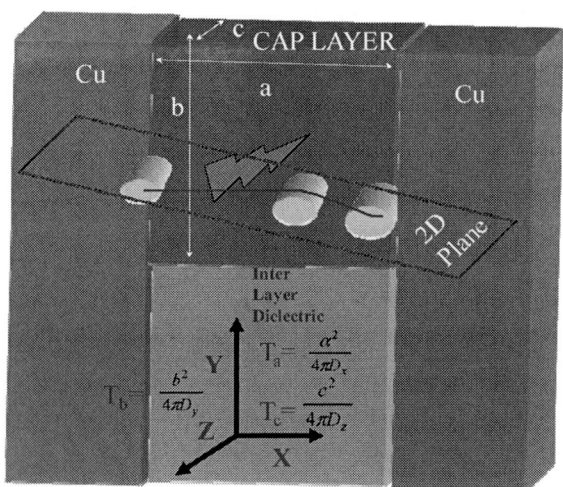

FIG15 HEAT DISSIPATION MECHANISMS IN THE CAP LAYER

However, under slower ramps (i.e similar to V_{BD} experiments) when the structure was ramped under longer duration, heat dissipation under the pulsed regime can-not be controlled and hence the stage wise process cannot be probed. Under the V_{BD} device exhibits either an open or short failure

B. *Extrapolating the model to longer time scale: Time Dependent Dilelectric Breakdown TDDB*

At longer time scales, the degradation (i.e defect generation) under competing mechanisms is due to a cumulative effect when a current channel has been established (i.e once the structure is in the state of breakdown) and is less significantly impacted by early stressing history. A model based on thermally diffusing copper ions and neutral atoms or the presence of pores (acting as defective sites) predicts that the dielectric degradation accumulates over time and eventually reaches a critical density - a threshold triggering permanent damage [3-9]. However, these models have overlooked precursors and latent damage, which can significantly impact the TDDB behavior in novel porous low-K/Cu dielectrics.

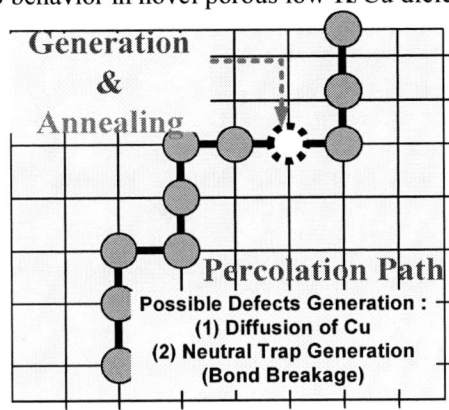

FIG 15. 2D *PERCOLATION MODEL* SHOWING FORMATION OF CONDUCTION PATH UNDER SOFT BREAKDOWN

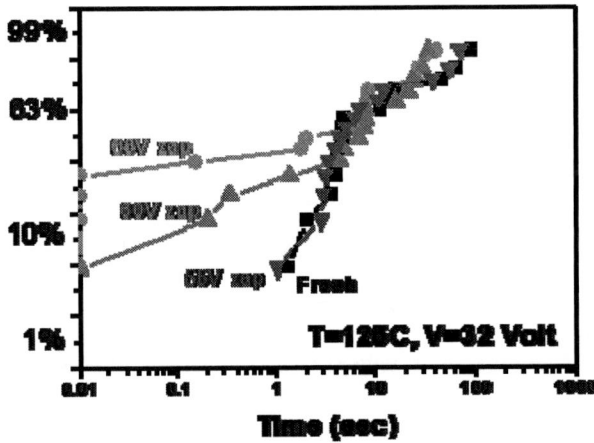

FIG.16 LONGER TAIL IN THE TDDB CURVE AFTER THE SAMPLES WERE ZAPPED WITH ESD PULSES (HBM MODEL)

Such materials start with very larger numbers of pores (defective sites), and accrued defect sites under stressing serve to create a percolation path. In this scenario, ESD events can leave precursor defects impacting the TDDB behavior by reducing the time required for the cumulative critical damage. (*fig 16*).

CONCLUSION

In this work, we have presented experiments and models to understand temporal domain breakdown behavior of Low-K/Cu dielectric and established a model to understand the implication of latent damage due to ESD events impacting long term reliability.

ACKNOWLEDGMENT

Authors would like to acknowledge Dr Young-Joon Park from TI in mentoring this work.

REFERENCE

[1] S-C Lee, Shou-Chung Lee; A.S Oates,.; Kow-Ming Chang "Fundamental understanding of porous low-k dielectric breakdown," *IRPS*,2009, pp. 481-485.

[2] T. P. Chen, Man Siu Tse, Chang Qing Sun and Steve Fung, "Post-Breakdown Conduction Instability of Ultrathin SiO₂ Films Observed in Ramped-Current and Ramped-Voltage Current–Voltage Measurements," *JJAP*, Vol. 41, 2002, pp. 3047-3051.

[3] Nam Hwang T.L. Tan, Cheng Kuo Cheng, · A. Du, C.L. Gan, K.L. Pey, , "Investigation of intrinsic dielectric breakdown mechanism in Cu/low-κ interconnect system," *IEEE Elect. Dev. Letters*, Vol. 27, 2006, pp. 234- 236.

[4] Junji Noguchi, "Impact of Low-κ Dielectrics and Barrier Metals on TDDB Lifetime of Cu Interconnects," *IRPS*, 2001, pp. 355-359.

[5] J. Kim, E. T. Ogawa and J. W. McPherson, "A Statistical Evaluation of the Field Acceleration Parameter Observed During Time Dependent Dielectric Breakdown Testing of Silica-Based Low-k Interconnect Dielectrics," *IRPS*, 2006, pp.478-483.

[6] E.T. Ogawa, Kim Haase Jinyoung, G.S. Mogul, J.W. McPherson, J.W Leakage, "Breakdown, and TDDB characteristics of porous low-k silica-based interconnect dielectrics," *IRPS*, 2003, pp. 166-172.

[7] N.Suzumura, S. Yamamoto, Kodama, D. Makabe, K. Komori, J. Murakami, E. Maegawa, S. Kubogta, K. "A New TDDB Degradation Model Based on Cu Ion Drift in Cu Interconnect Dielectrics,"*IRPS*, 2006, pp. 484-48.

[8] Zs. T kei, Y.-L. Li, G. P. Beyer," Reliability challenges for copper low-k dielectrics and copper diffusion barriers," *Microelectronics and Reliability*, Vol. 45(9-11), 2005, pp. 1436-442.

[9] J. Noguchi., "Dominant factors in TDDB degradation of Cuinterconnects," *IEEE Trans. Electron. Devices*, Vol. 52(8), pp.1743-1750, 2005.

[10] FenChen, M.Shinosky, "Addressing Cu/Low- Dielectric TDDB-Reliability Challenges for Advanced CMOS Technologies," *IEEE Trans.Electron Devices*, Vol. 56 (1), pp. 2 – 12, 2009.

[11] J. R. Lloyd, E. Liniger, and T. M. Shaw, "Simple model for time dependent dielectric breakdown in inter- and intralevel low-$k\psi$ dielectrics,"*J. Appl. Phys.*, 98(8), pp. 084 109, Oct. 2005.

[12] F. Chen, K. Chanda, J. Gill, M. Angyal, J. Demarest, T. Sullivan.,"Investigation of CVD SiCOH low-$k\psi$ time-dependent dielectric breakdown at 65 nm node technology," *IRPS*, 2005, pp. 501–507.

[13] Shou-Chung Lee, A. S. Oates, and Kow-Ming Chang, "Limitation of Low-k Reliability due to Dielectric Breakdown at Vias," *IITC*, 2008, pp.177-179.

[14] F. Chen, P. McLaughlin, J. Gambino, E. Wu, J. Demarest, D. Meatyard, and M. Shinosky, " The Effect of Metal Area and Line Spacing on TDDB Characteristics of 45nm Low-k SiCOH Dielectrics," *IRPS*, 2007, pp. 382-389.

[15] F Crupi, R Degraeve, G Groeseneken, Nigam T and Maes H E "On the properties of the gate and substrate current after soft breakdown in ultra-thin oxide layers," *IEEE Trans. Electron. Devices*, Vol. **45**, pp. 2329-2334, 1998.

[16] Ernest Y Wu , James H Stathis and Liang-Kai Han "Ultra-thin oxide reliability for ULSI applications," *Semicond. Sci. Technol.* Vol. 15, **pp.** 425-435, 2000.

[17] Tam Lyn Tan, Nam Hwang, and Chee Lip Gan, "Bimodal Dielectric Breakdown Failure Mechanisms in Cu–SiOC Low-k Interconnect System," *IEEE Trans Dev. and Mat. Reliability*, Vol. 7, pp. 373-378, 2007.

STUDY OF ELECTRIC FIELD–BASED LIFETIME PROJECTION METHOD IN IMD TDDB

W. Zhang, X. Zeng , W. Liu, Y. K. Lim , J. F. Liu, and E. C. Chua

GLOBALFOUNDRIES Singapore Ltd., 60 Woodlands Industrial Park D, Street 2, Singapore 738406

Phone: 65-6360-4751; Fax: 65-6362-2935; Email: galorzhang@@globalfoundries.com

Abstract—**Effect of backend interconnect critical dimension variation on IMD TDDB is studied. Statistical data shows that low-*k* dielectric TDDB time to failure correlates well with leakage current, which reflects actual trench-to-trench or trench-to-via spacing. So a lifetime projection method, based on equal electric field, is reported. A more realistic lifetime is achieved while predicting whole lot TDDB life-time. Moreover, monitoring leakage current could be adopted into process monitoring strategy.**

Keywords-: time-dependent dielectric breakdown, low-k, Schottky emission, equal electric field

I. INTRODUCTION

Since the introduction of low-*k* and ultro low-*k* dielectric materials into 90nm below technology node, constant voltage based TDDB has been widely used for back end of line dielectric characterization. It is used to qualify process and make lifetime projection [1-3]. Essentially, IMD TDDB is adopted from gate oxide TDDB. Three stress voltages are applied on three groups of samples in order to get voltage acceleration factor γ. Time to failure (TTF) follows Weibull or lognormal distribution. One point to be noted here is that during data analysis there is an assumption that equal voltage is equivalent to equal electric field within same group of samples. In other words, this method assumes equal spacing cross whole wafers. However, this is not true. With device dimension scaled down, wafer uniformity becomes worse and worse. Critical dimension of metal trench could vary >10% from wafer center to edge. The shape factor of final TTF Weibull distribution would be a convolution of "weakest link distribution", which describes intrinsic dielectric breakdown and follows Weibul or lognormal distribution, and spacing variation, which would follows normal distribution. As a consequence, it is very often to see that shape factor of Weibull distribution in IMD TDDB is less than 1. It is possible that a very pessimistic lifetime is derived. Worse still, the exact root cause for the shorter lifetime, which is related with trench-to-trench (or trench-to-via), spacing uniformity or dielectric materials, could be mixed up.

Obviously, electric field-based lifetime projection would be a way to explore. In order to get electric field, trench to trench (or trench-to-Via) spacing is needed. Spacing could be extracted by physical or electrical methods. As physical methods, SEM or TEM could provide direct measurement but the measured value could be challenged on its representativeness since electron microscope gives only very

local snapshot. A dual ramp rate method was reported to derive trench-to-trench spacing [3]. But it is a destructive test. Samples that are used for spacing derivation could not be used anymore for subsequent TDDB test. A non-destructive spacing determination method with combined I-V and C-V was proposed [2,4]. In this paper, a simpler lifetime projection method, equal electric field (EEF), is reported. In this method, relative spacing is derived by comparing dielectric leakage current. In addition, leakage current was used as a preliminary indicator of TDDB performance in process control.

II. EXPERIMENTAL

Wafers investigated in this work were fabricated by GLOUBLYFOUNDRIES' 65 and 45nm Cu dual damascece process. The low-*k* material is SiCOH with an effective dielectric constant of *k*=2.7. Two commonly used test structures, comb-serpentine structure and intertwined via chain structure are used for trench and via chain study, respectively. Fig. 1 shows the test structure schematics.

For spacing effect study on metal trench and Via chain splits with different trench-to-trench and M2 to V1 spacing in 45nm (pitch 140nm) and 65nm (pitch 200nm) technology node were used. Spacing varied from 55nm to 100nm.

Test pattern definition which is done during testing setup is important in back end of line. Each test pattern should have good wafer coverage. And more important, different test patterns should have same wafer center-to-edge coverage. Otherwise, voltage acceleration factor γ would vary a lot while swapping test patterns. To meet these requirements, firstly we use checkerboard pattern method for individual test pattern. Secondly, wafer coverage is quantitatively evaluated by calculating each sample's distance to central reference site. Adjustment such as swapping of dice in different test pattern is needed if there is big difference of summary (SUM), average (Ave) and standard deviation (STDEV) between patterns. Data in Table 1 is about one wafer which is divided into four patterns. By following above rules and after adjustment, four patterns show quite similar wafer coverage.

978-1-4244-5430-3/10 $26.00 © 2010 IEEE

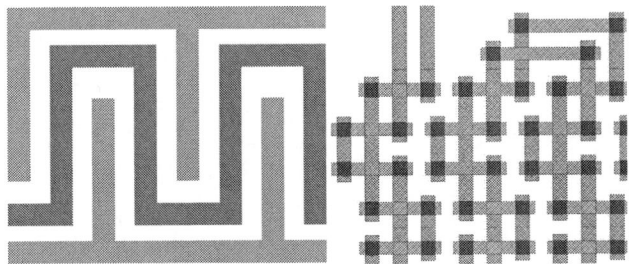

Figure 1. Schematics of comb-serpentine structure and intertwined via chain structure

TABLE 1. CALCULATING DISTANCE FROM INDIVIDUAL DUT TO REFERENCE SITE AT WAFER CENTER. COMPARABLE WAFER COVERAGE OF 4 PATTERN IS ACHIEVED.

DUT S/N	pattern 1 SQRT(x^2+y^2)	pattern 2 SQRT(x^2+y^2)	pattern 3 SQRT(x^2+y^2)	pattern 4 SQRT(x^2+y^2)
1	5.10	5.00	5.10	4.47
2	4.12	4.00	4.12	4.47
≈≈≈	≈≈≈	≈≈≈	≈≈≈	≈≈≈
≈≈≈	≈≈≈	≈≈≈	≈≈≈	≈≈≈
19	5.39	5.10	5.00	5.10
20	5.39	6.08	6.00	6.08
SUM	68.02	69.45	69.54	68.11
Ave	3.40	3.47	3.48	3.41
STDEV	1.27	1.33	1.36	1.41

III. RESULTS AND DISCUSSION

A. Correlation Between Leakage Current And Dielectric Spacing

During TDDB test, two leakage mechanisms would dominate at low and high voltage regimes: Schottky emission (SE) which is thermal driven dominating at low voltages, and Pool-Frenkel (PF) emission which is trap-assisted dominating at high voltage [2][5]. Schottky emission equation is given by

$$J = A * T^2 \exp\left(\frac{\beta_S \sqrt{E} - \Phi_S}{k_B T}\right), \quad \beta_S = \sqrt{\frac{q^3}{4\pi\varepsilon_o \varepsilon_r}} \quad (1)$$

where J denotes current density $A*$ Richardson constant, T temperature in Kelvin, E electric field, k_B Boltzmann constant, q electron charge, ε_o permittivity of free space, ε_r relative permittivity, Φ_S barrier height. From the equation, leakage current density is a function of electric field, which is determined by spacing.

In this study, leakage current was measured at 10V (called Iuse in this paper). Measurement at 10V instead of lower voltage like 3.3V which is normally used because, on one hand, it is high enough to tell spacing difference; and on the other hand, PF emission is negligible at 10V so we can use single model, SE, to model the trench-to-trench spacing. Moreover, we have data demonstrating that fast measurement at 10V would not affect on subsequent TDDB lifetime, shown in Fig. 2. TTF at same stress voltage shows correlation based on confidence interval and t-test result.

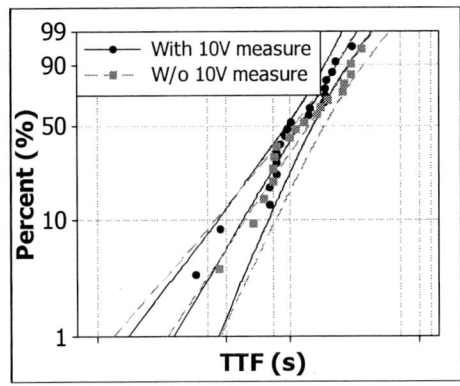

Figure 2. Weibull distribution of samples at same stress voltage, with and without 10V measurement before TDDB test. Performance is comparable

Fig. 3a is box plot of leakage current in trench-to-trench spacing split. The median value of leakage current is shown in Fig. 3b. It fits well by using Schottky emission. Similarly, splits with via overlay are shown in Fig. 4a and 4b. The minimum spacing in via chain is determined by V1 to neighboring M2. Bigger via overlay gives in smaller spacing between via to metal line and, as a result, higher leakage current is expected. Fig. 4b shows that, in general trend, leakage current correlates with V1 to M2 overlay. So it is possible to use Iuse to predict spacing.

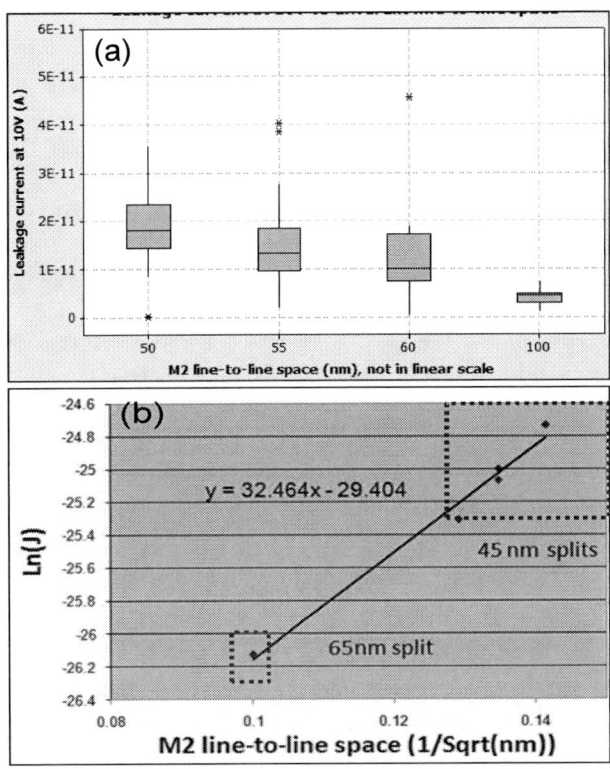

Figure 3. Leakage current as a function of line-to-line (trench-to-trench) splits, shown by (a) box plot and (b) fitting by Schottky emission.

978-1-4244-5430-3/10 $26.00 © 2010 IEEE

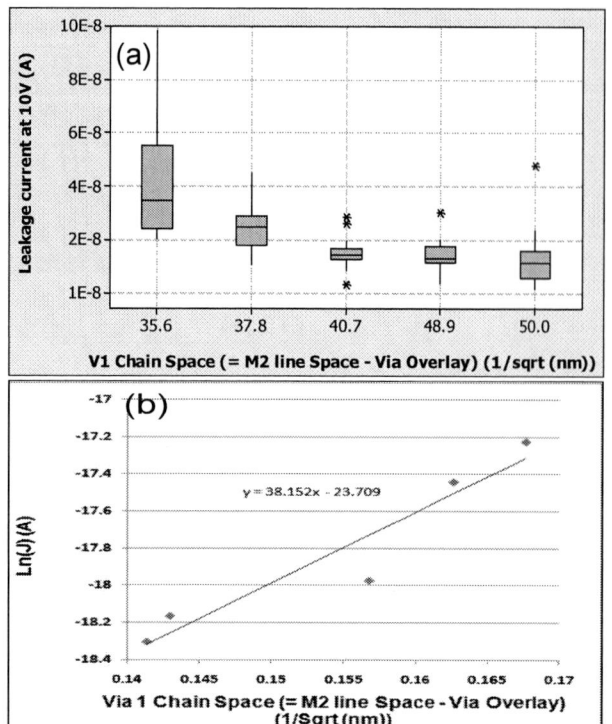

Figure 4. Leakage current as a function of M2-to-V1 overlay splits, shown by (a) box plot and (b) fitting by Schottky emission.

Samples from Fig. 4 are put into TDDB stress. TDDB TTF at same stress voltage but from different splits is plot against Iuse. As we know, physical top spacing between via and the adjacent metal line is decisive parameter for intrinsic TDDB [6]. Fig. 5 shows a very good correlation.

As we know, TDDB is an electric field-driven dielectric breakdown process. TTF strongly depends on spacing. And spacing could be further reflected by Iuse. Since results above show good correlation between spacing and leakage, it is reasonable to use leakage to normalize DUT at equal voltage into equal electric field.

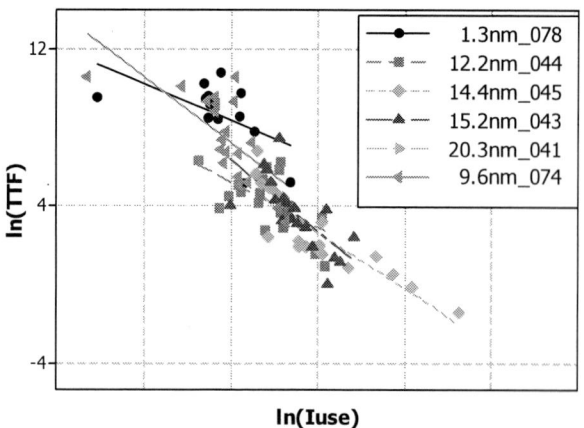

Figure 5. Correlation between leakage current and TTF at same stress voltage but in different splits

B. Equal Electric Field in IMD TDDB Lifetime Projection

TDDB is an electric field-driven dielectric breakdown. TTF is governed by

$$TTF_2 = \exp[(-\gamma * (\sqrt{V} / \sqrt{S})] \tag{2}$$

where S is the spacing. Based on equations (1) and (2), TTF could be normalized into equal electric field.

First case study by EEF is about metal trench. Results are shown in Fig. 6. By normalizing into equal electric field (EEF), shape factors at lower stress voltages change from 0.68 to 0.83 and 0.76 to 0.84, respectively. The increase of shape factor is not significant. It indicates that Weibull β at each stress voltage is mainly determined by "weakest link distribution". Global spacing uniformity does not contribute much to small beta value.

Second case study is about via chain. Fig. 7 shows Weibull distribution of 3 wafers from same lot. The effect of both wafer-to-wafer and within wafer center-to-edge variation gives a very pessimistic product lifetime (under 3.63V). It is only 1.8E-7 years, or 5.6 seconds, due to extremely small shape factor and voltage acceleration factor. Applying EEF, shape factor is corrected to 0.78, which is very close to the value of single wafer, and a more reasonable lifetime 0.01 years is achieved. This value is close to the value of single wafer. Important message from this method is uniformity is the root cause for IMD TDDB failure. Both wafer to wafer variation and within wafer uniformity need to be addressed. Subsequent process changes in litho, etch and CMP improve the uniformity.

One interesting observation in the statistical distribution is the curve bending at early TTF portion either with Weilbull or lognormal (see inset) fitting. Such a curve bending is only obvious with large sample size, ~70 DUT at one stress voltage. The minimum TTF is more than 1.5s (high resolution SMU used in test is demonstrated to be able to capture TTF<1s). It seems low-*k* dielectric has a minimum lifetime to sustain high stress voltage. In other words, there is a threshold volume of trap accumulation and a threshold time for Cu to drift to reach dielectric breakdown. The mechanism can be understood through reviewing the low-*k* dielectric breakdown mechanism. Chen and Suzumara et al reported that low-*k* dielectric is an electron-fluence driven process, and Cu clusters [7] or Cu ion concentration [8] will determines dielectric breakdown. Since Cu is a heavy atom with big atom size, Cu or Cu ions need time to drift and diffuse in dielectric. During data analysis, without considering the curve bending, extrapolating Weibull or lognormal distribution to low PPM level will give a very pessimistic lifetime. So a modified data analysis model, maybe referring the idea of 3-paramer model in EM in which critical void size is considered to cause EM failure [9], is strongly needed.

In summary, deriving actual physical spacing is the key for electric field-based TDDB lifetime projection. EEF method derives the relative spacing ratio by a simply pre-test leakage

978-1-4244-5430-3/10 $26.00 © 2010 IEEE

current measurement. After that, TTF of each DUT is normalized into a value at equal electric field. This method shows more value while applying on cross-wafer or cross-lot lifetime projection. The one point leakage measurement (at 10V) is easy to implement during test. But the dependence of measurement voltages, which could be lower and higher than 10V will be further studied. Moreover, in order to implement EEF, finding more accurate and easy way to drive actual physics spacing, especially in via chain, is the area to explore.

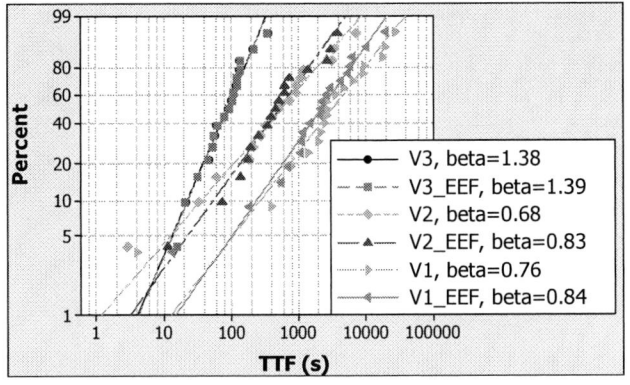

Figure 6. TTF Weibull distribution of metal lines with EEF method. Weibull shape factors increase at lower stress voltages.

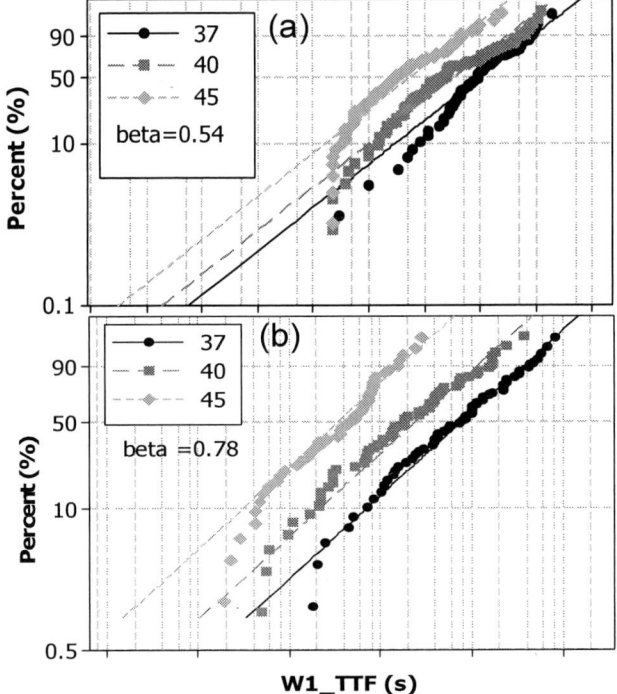

Figure 7. TTF Weibull distribution of via chain with (a) raw data and (b) EEF. Weibull shape factor increases from 0.54 to 0.78

C. Leakage Current Control

Knowing the correlation between leakage current and spacing, leakage current could be used as one important parameter to monitor TDDB performance.

One study on via chain V1 overlay is shown. M2 to V1 overlay is intentionally misaligned in 6 splits. As expected, TDDB lifetime degrades as a function of overlay misalignment, shown in table 2. But from this experiment, it is observed that metal-to-via overlay is just one of the parameters determining IMD TDDB, because the final physical dielectric spacing is also determined by other process modules, especially CMP oxide loss. So leakage current would be a more accurate parameter to monitor process drift. Fig. 8 shows the leakage current at different via-to-metal overlay. For process control purpose, leakage current of 1.5E-8A serves as process control spec of IMD TDDB performance.

TABLE 2. VIA CHAIN LIFETIME AS A FUNCTION OF M1-TO-VIA1 OVERLAY SPLITS. BETA IS COMPARABLE BECAUSE PROCESS HAS SIMILAR SIGMA. BUT LIFETIME DEGRADES· WITH WORSE MISALIGNMENT (LIFETIME REQUIREMENT IS 10years.

	Split 1	Split 2	Split 3	Split 4	Split 5	Split 6
OVL-X mean	-1.3	9.6	14.4	12.2	15.2	20.3
beta	0.64	0.58	0.64	0.72	0.65	0.53
gamma	11.58	12.85	11.55	12.01	10.48	10
lifetime (years)	40.49	33.60	0.08	4.53	0.01	0.0008

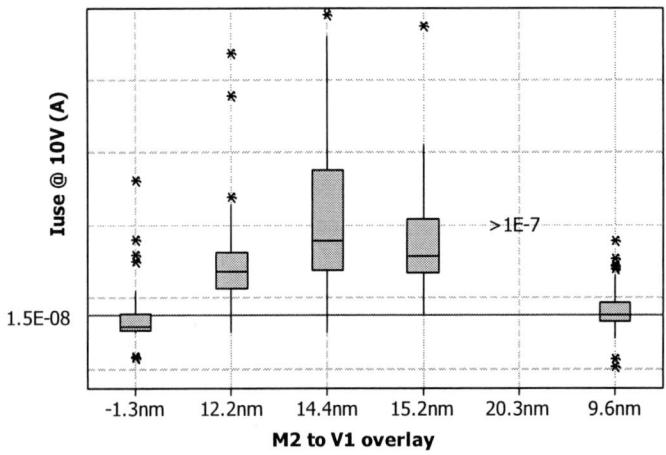

Figure 8. Leakage current of via chain as a function of M2-V1 overlay. Leakage current of 1.5E08Acould be set as spec for process control.

IV. CONCLUSION

In this work, the correlation between leakage current and dielectric spacing, as well as TTF, is demonstrated. A lifetime prediction method, EEF which is based on equal electric field, is reported. Relative physical spacing is derived by comparing pre-test leakage current. TTF of each DUT is normalized into a value at equal electric field. This method shows more value while applying on cross-wafer or cross-lot lifetime projection. In addition, leakage current is proposed as a monitoring parameter for inline production.

ACKNOWLEDGMENT

The authors would like to thank colleagues in GLOBALFOUNDRIES Fab 7 integration, modules and TD Singapore site who planned process splits and run hardware. Also thanks go to TE groups in QRA who provide test support.

REFERENCES

[1] E. T. Ogawa, J. Kim, G. Haase, H. Mogul and J. McPherson "Leakage, breakdown, and TDDB characteristics of porous low-*k* silica-based interconnect dielectrics," IRPS 2003, pp. 166-172.

[2] F. Chen, "Addressing Cu/Low-*k* Dielectric TDDB Reliability Challenges," IRPS Tutorial 2008.

[3] G. S. Haase, Ennis T. Ogawa and Joe W. McPherson "Breakdown Characterization of Inter-connect Dielectrics," IRPS 2005, pp. 466-473.

[4] F. Chen, P. S. McLaughlin, J. Gambino, J. Gill, "A Comparison of Voltage Ramp and Time Dependent Dielectric Breakdown Tests for Evaluation of 45nm Low-*k* SiCOH Reliability," IITC 2007, pp. 120-122.

[5] K. H. Cheng, "Effect of ramp rate on dielectric breakdown in CU-SiOC interconnect," Thin Solid Films vol. 462, p. 316, 2004.

[6] F. chen, J. R. Lloyd, K. Chanda, R. Achanta, O. Bravo, A. Strong, P. S. McLaughlin, M. Shinosky, S. Sankaran, E. Gebreselasie, A. K. Stamper, and Z.X. He "Line Edge Roughness and Spacing Effect on Low-*k* TDDB Characteristics"," IRPS 2008, pp. 132-137.

[7] F. Chen, O. Bravo, K. Chanda, P. McLaughlin, T. Sullivan, J. Gill, J. Lloyd, R. Kontra, and J. Aitken, "A Comprehensive Study of Low-*k* SiCOH TDDB Phenomena and Its Reliability Lifetime Model Development," IRPS 2006, pp. 46-52.

[8] N. Suzumara, S. Yamamoto1, D. Kodama2, K. Makabe3, J. Komori1, E. Murakami3, S. Maegawa4 and K. Kubota, "A new TDDB degradation model based on Cu ion drift in Cu interconnect dielectrics," IRPS 2006, pp. 484-489.

[9] B. Li, Cathryn Christiansen, Jason Gill "Minimum void size and 3-parameter lognormal distribution for EM failures in Cu interconnects" IRPS 2006, pp. 115-122.

On the Physical Interpretation of the Impact Damage model in TDDB of low-k dielectrics

J.R. Lloyd,

SUNY Albany, College of Nanoscale Science and Engineering

Albany NY 12203 USA

(01)-518-956-7062   jlloyd@uamail.albany.edu

Abstract— In one of the new "root-E" models for TDDB, there is a second term in the exponent representing the probability that there is enough energy to create damage that contributes to breakdown failure. This provides an exponential "1/E" character to the extrapolation to use conditions that predicts lifetimes many orders of magnitude longer than the simple root-E extrapolation. It also predicts a threshold voltage for damage independent of the applied field. It is argued here that the requirement for damage may not be the energy, but the momentum transfer to cause physical damage. The consequences are discussed.

Keywords-component; TDDB, Low-k Dielelctrics, Impact Damage

I. ROOT-E MODELS

There are at least three distinct "root-E" models for TDDB in low-k dielectrics [1-4] and another that mimics it but from a different perspective [7].   In all of these models the exponential dependence of the lifetime on the square root of the applied field originates in the conduction mechanism, being either Schottky or Poole Frenkel conduction.

$$t_f = AE^n \exp\left(\frac{-\beta\sqrt{E} + \phi}{kT}\right) \qquad [1]$$

where E is the applied field, β, n and A are is a quantities that depends on the conduction mechanism and φ is the barrier height. In two of the models, the diffusion of Cu is necessary for breakdown to occur. Suzumura et. al. [4] postulated that the copper is transported through the dielectric according to Poole-Frenkel or Schottky kinetics, but why a heavy Cu atom should act this way is unexplained. In the model of Chen et. al. [3] it is postulated that Cu ions are injected into the dielectric, then neutralized and diffuse through the dielectric as neutral atoms according to Fick's Law. Since the ionization of the Cu would depend on the current density, this then follows root-E behavior. However, a similar model was treated subsequently with the difference that Cu was treated to diffuse as an ion and the kinetics were seen not to follow root-E type behavior. [5,6]

In one of the models [1,2], known as the "Impact Damage" model,   there is an additional term in the exponent that represents the probability that there is enough energy to create damage from the collision of a ballistic electron accelerated by the electric field with the dielectric material. This term has been suggested experimentally by deviations from the root-E model at low applied fields, however, the data are not so strong as to unambiguously determine exactly what the relationship is. [8]

It is of some importance to understand the precise failure mechanism, since the difference in extrapolation between the other root-E models and the Impact Damage [ID] model to use conditions can be tens of orders of magnitude. It could be said that if the ID model were correct, TDDB in interlevel dielectrics may be impossible under most operational conditions and further engineering work in that direction need not be undertaken.

The ID model is similar in concept to the 1/E model [9,10] best described as the "Lucky Electron" model where it is assumed that an electron must have escaped collision until enough energy has been gained from the field such that in a subsequent collision damage to the dielectric will result. In the ID model, as written, the particulars of the damage mechanism were ignored, with a stipulation that enough energy must be gained by the electron to impart some unspecified type of damage during the eventual collision that would lead to breakdown. This produced a model where the equation describing failure time [t_f] at constant temperature is

$$t_f = A \exp\left(-\alpha\sqrt{E} + \frac{\beta}{E}\right) \qquad [2]$$

where E is the applied electric field, α is a parameter related to the conduction mechanism and β is given by $\beta = \dfrac{\varepsilon_t}{e\mu}$ where ε_t is the threshold energy for damage formation, e is the electronic charge and μ is the mean free path of an electron. β can be treated as a property of the material. The only major assumption here was that the distance an electron would travel between collisions was exponentially distributed.

In a criticism of the lucky electron concept, McPherson and Khamamkar [11] argued that the creation of a trap would

978-1-4244-5430-3/10 $26.00 © 2010 IEEE

require that an atom be physically moved from its normal position and this could not occur because enough energy would not be transferred in a collision to an atom by an electron due to the huge disparity in mass. As will be seen, the argument of ref 11 may be valid for silica based dielectrics, but for the low-k dielectrics currently in use, it fails. As in ref [11] the argument here is basically classical.

II. IMPACT DAMAGE MODEL REVISITED

Let us consider an electron initially at rest that is then accelerated by the electric field. It is recognized that this electron is really never at rest, in fact possessing considerable energy as a consequence of the Pauli Exclusion Principle. However, we are not interested in the average electron, but in the "averaged" electron and we must be most concerned with the momentum possessed by the electron [a vector] and not the energy [a scalar]. Thus, even though the energy may be on the order of a few eV, in the absence of an applied field, the momentum is averaged to naught. Therefore, with many collisions taking place, the average energy, in the absence of an electric field, could be considerable whereas the average momentum would vanish. It is also recognized that the energy available for damage creation is actually the difference between the total energy and the Fermi Energy, the energy below this being unavailable for interaction as the states are filled. Although this is strictly true only at 0K, it is still a good approximation when the energy gained from the field is large compared with kT. Therefore, the energy that can be lost from the electron is essentially that gained from the field in between collisions. Once the electric field is applied, the momentum of the averaged electron takes on a non-zero value with the kinetic energy increased by λeE, where λ is the length of the path traveled before the electron is scattered. The momentum gained between collisions, p, is;

$$p = \sqrt{2eE\lambda m}_e \qquad [3]$$

where m_e is the mass of the electron.

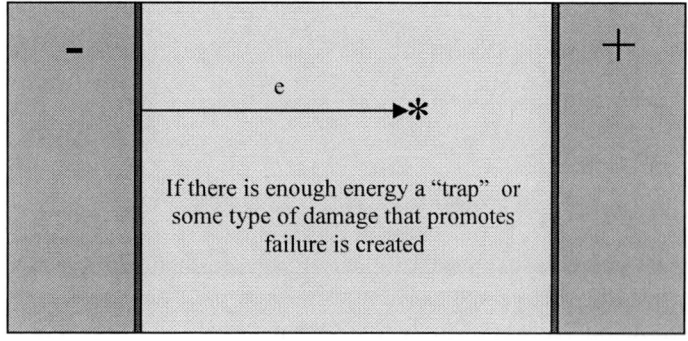

Fig 2 Schematic of the ID model of ref (2)

After McPherson and Khamamkar [11] it is proposed here that damage is most likely the creation of a "dangling bond" originating from the displacement of an atom from it's normal position in the structure of the dielectric. In order to remove this atom and create the charge trap, the bond must be broken and then the atom must be moved.

The collision of an electron with a dielectric molecule will be in some part elastic and some part inelastic. At low energies [~10 eV], the inelastic component of a collision with a complex low-k dielectric dominates, probably due to the many vibrational modes available to absorb energy in the rather complex molecular structure. [12] Thus the energy could be absorbed in a manner very closely to that of a classical particle and quantum effects can be ignored. If, for the sake of argument, an atom in the molecule were to absorb all the energy in a collision [100% inelastic collision], the energy of that atom would be increased by the energy that the electron had gained by the field before the collision. Complete thermalization occurs in less than 1 ps in dielectrics. [13] Therefore the stricken atom would be in a very high energy state for a short time that could be described as a very high effective temperature, T_e:

$$T_e = \frac{\lambda eE}{k} \qquad [4]$$

Taking the analogy further, we can expect that the probability that the chemical bond of that atom in the molecule will be broken by the collision after the electron had traveled λ, P'_B, is expressed by an Arrhenius like relation

$$P'_B(\lambda) = e^{-\frac{\varepsilon_B}{\lambda eE}} \qquad [5]$$

where ε_B is the bond energy. It can be argued that λ would be exponentially distributed with a mean free path in the dielectric, $\mu = \sim 3$ nm. [14,15] Therefore, the probability that a bond will break, P_B, for any given electron/atom collision is the product of P'_B and the probability that the electron would have been able to travel λ before the collision occurred.

$$P_B = e^{-\frac{\varepsilon_B}{\lambda eE} - \frac{\lambda}{\mu}} \qquad [6]$$

For a given electric field, as λ increases the probability that the bond will break at the collision increases, but the probability that λ can be achieved decreases. The product of these two probabilities will then have a maximum when

$$\lambda = \sqrt{\frac{\varepsilon_B \mu}{eE}} \qquad [7]$$

as illustrated below. Plotted below is the probability of breaking a 4.2 eV bond with an electric field of 1 x 10^6 V/cm and a mean free path of 3 nm.

Substituting into eqn [5], the highest probability for bond breakage due to a collision with an electron becomes

$$P'_B = e^{-2\sqrt{\frac{\varepsilon_B}{\mu e E}}} \qquad [8]$$

Eqn [7] is a remarkably simple expression. All that is important is the ratio of the bond strength and the mean free path of the electron in the dielectric. This can be considered a property of the material.

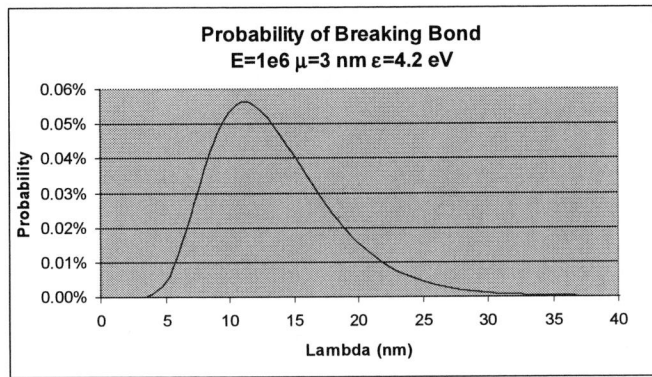

Figure 1 Probability that a 4.2 eV bond will break in a collision of an atom with an electron in an applied field of 1 MV/cm with a mean free path of 3 nm

For a perfectly inelastic collision, P'_B for a SiCOH based ULK can be easily estimated. The possible bond energies would vary between 1.5 eV [O-O] to 4.7 eV [Si-O] and μ will be on the order of 3 nm. [14,15] Therefore, the highest probability of an electron breaking a bond would between .04% and 1% for a field of 1 MV/cm. Breaking the chemical bond, however, is not sufficient to create a trap. The atom struck by the electron must be displaced sufficiently such that the bond will not spontaneously reform when the atom cools from the collision. The momentum transfer from the electron to the atom will displace the atom a mean distance, l, calculated with the use of eqn [2]

$$l = \frac{\tau\sqrt{2eE\lambda m_e}}{m_a} \qquad [9]$$

where τ is the time it takes for the hot atom to thermalize and m_a is the mass of the stricken atom. To calculate the probability of and the time to failure, we need to solve for λ_C, which is the critical distance through which the electron must be accelerated in order to have enough momentum to displace a hydrogen atom. If we assume that the bond can be permanently broken when the atom is displaced twice the bond length, l_B, we arrive at

$$\lambda_C = \frac{2\left(\frac{l_B m_a}{\tau}\right)^2}{eEm_e} \qquad [10]$$

The probability that a hydrogen atom would be displaced and a trap generated, P_t, would be

$$P_t = \exp\left(-\frac{\lambda_C}{\mu}\right) = \exp\left[\frac{-2\left(\frac{l_B m_a}{\tau}\right)^2}{\mu e E m_e}\right] \qquad [11]$$

Clearly the easiest atom to move is hydrogen [H]. The next lightest atom in the structure would be carbon [C] with a mass 12 times that of H. It is instructive to compare what the relative probability of creating a trap with a H atom as compared to a C would be. Using available figures for the bond length and assuming τ is 100 fs, for an applied field of 1 MV/cm, the ratio is a ridiculous exp [-5.6 x 10^{10}]. Obviously, only H need be considered and this model would only be viable for traps generated by the displacement of a hydrogen atom. Therefore, for the case of TDDB in SiO_2 based dielectrics, the model would not be appropriate and the McPherson "E" model [16] should be used instead.

It is interesting to note that the criterion for damage in the form of a dangling bond due to hydrogen ejection is not that there be enough energy to break a bond, but that there be enough momentum to move the hydrogen atom. If we have enough momentum we easily have enough energy to break the bond. First estimates are that if we have the momentum, the probability that a bond is broken [substituting λ_C for λ in eqn [8]] exceeds 50%.

The rate of trap formation by this mechanism then becomes the product of the current density and the probability that there is enough momentum to move the hydrogen and the probability that the bond is broken. We replace eqn. [2] with

$$t_f = A\exp\left(-\alpha\sqrt{E} + \frac{\beta}{E} - \gamma\right) \qquad [12]$$

where A and α retain their identities,

$$\beta = \frac{2\left(\frac{l_B m_a}{\tau}\right)^2}{\mu m_e} \qquad [12a]$$

and

$$\gamma = \frac{\varepsilon_B m_e}{2} \left(\frac{l_B m_a}{\tau} \right)^{-2} \qquad [12b]$$

All the quantities in 12a are well established except τ, which has not been measured for the materials of interest. If we compare eq. [12a] with that estimated from data in the literature [2] we arrive at an upper limit of about 500 fs, which compares favorably with measurements made on other dielectrics. [18,19]

The question arises as to what effect the assumption of a perfectly inelastic collision makes on the argument, or would the model hold if the electron/atom collision were not perfectly elastic. It is seen that since the momentum exchange is the critical phenomena for trap generation and when there is enough momentum to remove the hydrogen atom from its place, there is more than sufficient energy available to break the bond. If, in fact, the collision were mostly elastic, there would be enough energy to break the bond with a probability reasonably high so as to cause failure in reasonable times. It must be remembered that the momentum exchange is the same regardless of whether the collision is elastic or inelastic. Of course a 100% elastic collision would not be energetic enough to break a bond, but it is seen that at the low energies we are interested in here, the collisions are usually significantly inelastic in character.

III. SUMMARY

We have investigated the contention of McPherson and Khamamkar that the traps leading to failure in TDDB are dangling bonds and that an atom has to be displaced in the structure of the dielectric to generate the trap. Their further contention that a "lucky electron" model can't account for TDDB because of insufficient energy transfer is correct for the materials they considered, but does not hold for low-k and ultra-low-k dielectrics with hydrogen in the structure. The Impact Damage model was extended to account for the "1/E" character by posing a specific defect generation, that of displacement of a hydrogen atom and was found to be reasonable. It was seen that the criterion for trap generation was not the attainment of enough energy, but that there be enough momentum exchanged, leading to the conclusion that Impact Damage is a viable model whether or not the collisions are purely inelastic. Furthermore, from first principles, we see that the calculated effect is consistent with the data.

Thus, the impact damage model for TDDB in low and ultra-low k dielectrics is viable and accounts for the poorer TDDB performance of these materials as compared to silica based dielectrics and also the observed root-E kinetics. However, there exists a 1/E term that makes the extrapolation of the reliability to use conditions much more robust than the other root-E models.

ACKNOWLEDGMENTS

The author would like to thank Dr. Joanna Atkin of Columbia University, Dr. Robert Rosenberg of IBM, and Dr. Joe McPherson formerly of Texas Instruments for stimulating discussions [not necessarily full agreement] in the generation of this model.

REFERENCES

[1] J.R. Lloyd, E. Liniger and S.T. Chen, Microelectronics. Reliability, **44**, 1861 [2004].

[2] J.R. Lloyd, E. Liniger and T.M. Shaw, J. Appl. Phys., **98**, 084109 [2005]

[3] F. Chen, O. Bravo, K. Chanda, P. McLaughlin, T. Sullivan, J. Gill, J. Lloyd, R. Kontra, J. Aitken, Proc. 44th Int. Reliab. Phys. Symp. [2006]

[4] N. Suzumura, et. al. Proc. 44th Ann. IRPS San Jose CA, 484 [2006]

[5] J. R. Lloyd, C. E. Murray, S. Ponoth, S. Cohen, and E. Liniger: Microelectronics Reliability 46, 1643 [2006]

[6] R.S. Achanta, W.N. Gill and J.L. Plawsky, J. Appl. Phys. 103, 014907 [2008]

[7] G.S. Haase, Proc. 47th Ann. Reliab. Phys. Symp. 556 [2008]

[8] K. Croes, G. Cannata, L. Zhao and Zs. Tokei, Microelectronics Reliability, 48, 1384 [2008]

[9] S. Tam, P.K. Ko and C. Hu, Lucky-electron model of channel hot-electron injection in MOSFET's, *IEEE Tran. Electron Dev.* **31**, 1116 (1984)

[10] I.C. Chen, S. Holland and C. Hu, Proc. 23rd Ann. Int'l. Reliab. Phys. Symp., 24 [1984]

[11] J.W. McPherson and R.B. Khamankar, Semicond. Sci. Technol., 15, 462 [2000]

[12] J.B. Sokoloff, Phys. Rev. B, 61, 9380 [2000]

[13] H.-J. Fitting, V.S. Kortov and G. Petite, Journal of I Luminescence 122-123, 542 (2007)

[14] B. Lesiak, A. Jablonski, J. Zemek, M. Trchova and J. Stejskal, Langmuir, 16, 1415 (2000)

[15] K.-B. Chua and U. Osterberg, J. Appl. Phys., 95, 6204 (2004)

[16] J.W. McPherson and H.C. Mogul, J. Appl. Phys. **84**, 1513 (1998);

[17] R.C.J. Wang, K.S.Chang-Liao, T.K. Wang, M.N. Chang, C.H. Lin, C.C. Lee, C.C. Chiu and K.Wu, Thin Solid Films, 517, 1230 (2008)

[18] R. Tommasi, P. Langot and F. Vallee, Appl. Phys. Lett. 66, 1461 (1995)

[19] S.-Z. Sun, Y.-C. Wen, S.-H. Guol, H.-M. Lee, S. Gwo and C-.K. Sun, J. Appl. Phys., 103, 123513 (2008)

Reliability and Performance Limiting Defects in Low-κ Dielectrics for use as Interlayer Dielectrics

B.C. Bittel [1*], P.M. Lenahan [1], S. King [2]

[1] The Pennsylvania State University, [2] Intel Corporation

*212 EES Building, University Park PA 16802 phone: 814-863-4630, fax: 814-863-7967, email: bcb183@psu.edu

ABSTRACT

Reliability issues of low-κ dielectric thin films are important problems in present day ULSI development.[1-6] Leakage currents in general as well as reliability issues such as time dependent dielectric breakdown (TDDM) and stress induced leakage currents (SILC) are critical problems that are not yet well understood. A topic of current interest is ultraviolet light curing (UV curing) of low-k materials.[5,6] An atomic scale understanding of the defects involved in reliability problems of these films is virtually non-existent. We have initiated a study utilizing electron spin resonance (ESR) and electrical measurements which provides some fundamental understanding of the deep level defects likely involved in these reliability problems.

INTRODUCTION

There is interest in finding new materials for use as interlayer dielectrics (ILDs) and etch stop layers (ESLs) for use in ULSI. The reliability of these novel ILDs and ESLs are of particular concern.[1-6] TDDM and SILC are of particular interest due to the relatively low breakdown strength of these films.[1-6]

There is a vast literature dealing with ESR in Si based dielectric thin films as well as large volume samples of silicon based amorphous insulators.[7-14] The wealth of knowledge provided in the literature on previous ESR studies of Si based dielectrics, bulk Si, and glass offers a foundation for understanding these reliability related defects in ILDs and ESLs. In this work, we compare ESR and electronic measurements to provide insight into performance limiting defects associated with reliability concerns.

We have made ESR and current density versus voltage measurements on a set of dielectric/silicon structures involving materials of importance to low-k interlayer systems. We have compared ESR and current density versus voltage measurements both before and after exposing the dielectrics to UV light (hc/λ ≤ 5 eV). We observe extremely gross differences in the ESR spectra and leakage current versus voltage response of these low-k films. In many cases we observe that UV exposure increases both the density of paramagnetic defects and the leakage current density at a given field. This result suggests a cause / effect relationship. Paramagnetic point defects observed in these films include, E' centers, silicon dangling bond defects in which the central silicon is back bonded to oxygens ,hydrogen complexed E' centers, K-centers, silicon dangling bond defects in which the central silicon is back-bonded to nitrogen's, other possibly carbon or silicon dangling bond centers. Our preliminary results suggest the UV curing process creates paramagnetic centers which take part in trap assisted tunneling and also indicate that processing parameters have extremely gross effects upon defect densities within these films.

EXPERIMENTAL

A variety of film compositions were investigated including SiO_2, SiN:H, SiCN:H, and SiOC:H. All films were deposited by plasma enhanced chemical vapor deposition (PECVD) using various combinations of SiH_4, Methylsilane, NH_3, N_2, H_2, He, and Oxidizers. EPR measurements were carried out on a Bruker ER200D X-band spectrometer with a 300 series bridge utilizing a TE104 double cavity. Bias was applied across the dielectric structures with a corona discharge apparatus. Surface potential was measured with a Kelvin probe electrostatic voltmeter. Most films were irradiated with a low intensity UV pen lamp (hc/λ ≤ 5 eV).

RESULTS

Figure 1 shows the ESR results taken on a k = 6.5 ESL SiN:H film pre and post UV irradiation. We observe a large increase in defect density from 0.3 x 10^{14}/cm^{-2} to 4 x 10^{14}/cm^{-2} with UV irradiation. The ESR response is dominated by a spectrum with a zero crossing g = 2.0027 and a peak to peak line width of 15 Gauss (1.5 mill-Tesla.) (In the simplest cases, the ESR magnetic resonance condition is given by hv=gβH, where h is Planck's constant, β is the Bohr magneton and v is the frequency of the microwave radiation and H is the magnetic field at resonance. The g typically depends on the relationship between magnetic field vector and the orientation of the defect under observation.) We assign this spectrum to that of the well known K-center since the zero crossing g-value and line width match precisely with literature values.[9-10]

The leakage current of the k = 6.5 SiN:H film is also greatly increased by exposure to UV irradiation as figure 2 demonstrates. This result suggests that leakage current likely involves the K-centers via trap assisted tunneling.

Figure 3 shows ESR results from a k = 4.4 ILD SiO_2 film pre and post UV irradiation. We observe a large increase in the amplitude of the zero crossing 2 Gauss wide g = 2.0007 spectrum from 0.3 x 10^{14}/cm^{-2} to 1.3 x 10^{14}/cm^{-2}. We assign this spectrum to E' centers since the observed spectrum is very similar in zero crossing g-value and line width to that of the E' centers reported in the literature.[8] The most prominent E' variant has a zero crossing g= 2.0005+/- 0.0003 and a 2.5 Gauss line width. Figure 4 shows a pre- UV irradiated sample with gain increased to illustrate additional side structure in this SiO_2 film. This additional structure is likely due to the presence of organic contaminants and probably nonbridging

978-1-4244-5430-3/10 $26.00 © 2010 IEEE

oxygen hole or peroxy centers.[13] (Note: This side structure vanishes after the UV exposure.) Figure 5 shows a post-UV SiO_2 film with the gain increased compared to figure 3 to illustrate a very different post UV side structure. (The gain in figure 5 is lower than the gain in figure 4.) Note the pair of sidepeaks symmetric with respect to the E' center with a separation of 74 Gauss. This structure is almost certainly due to a hydrogen complexed E' center called the 74 Gauss doublet as the separation (74 Gauss) and the width of the two sidepeaks (~3 Gauss) are all consistent with the literature for this center.[14-15]

The leakage current of the k = 4.4 SiO_2 film is also greatly increased by exposure to UV irradiation as figure 6 demonstrates. This result suggests that leakage current is likely correlated to the existence of the paramagnetic centers detected in the film. Since E' centers dominate the ESR spectrum and have energy levels in the SiO_2 band-gap appropriate for trap assisted tunneling our results suggest that they are largely responsible for the increase leakage in these films. [14-15]

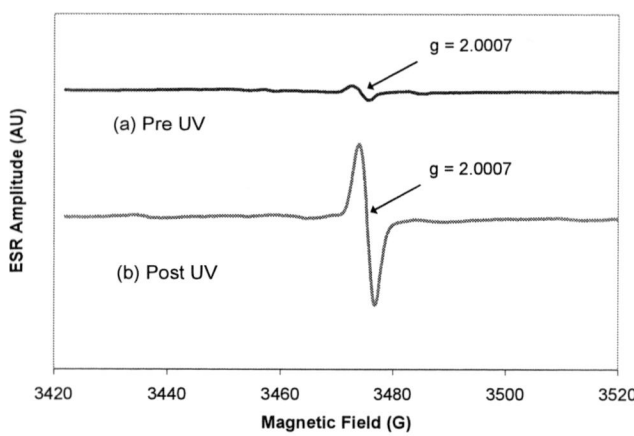

Figure 3. ESR results on a k = 4.4 SiO_2 film (a) pre and (b) post UV irradiation. Note that post UV ESR trace is significantly larger and both spectra have a g-value of 2.0007 and narrow line-width which is consistent with an E' spectra.

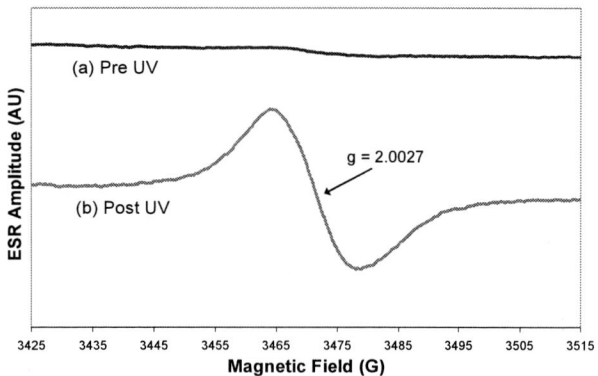

Figure 1. ESR results on a k = 6.5 SiN:H film (a) pre and (b) post UV irradiation. Note that post UV ESR trace is much larger and has a g-value of 2.0027 and a peak to peak line width of 15 Gauss, consistent with a K-center spectrum.

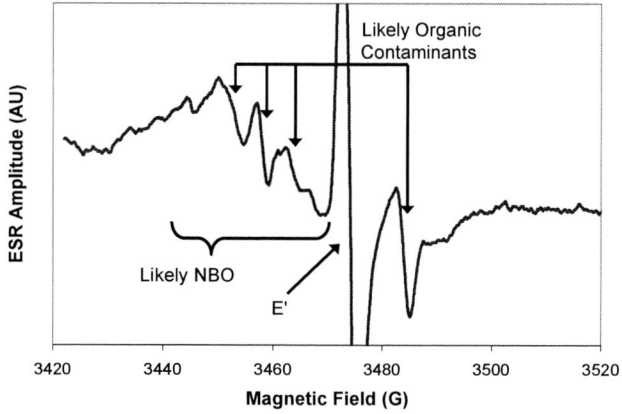

Figure 4. ESR results on a k = 4.4 SiO_2 pre UV film with gain increased to showcase additional structure. This structure is likely due to organic contaminants and possibly due to the nonbridging oxygen hole center (NBO) or peroxy centers.

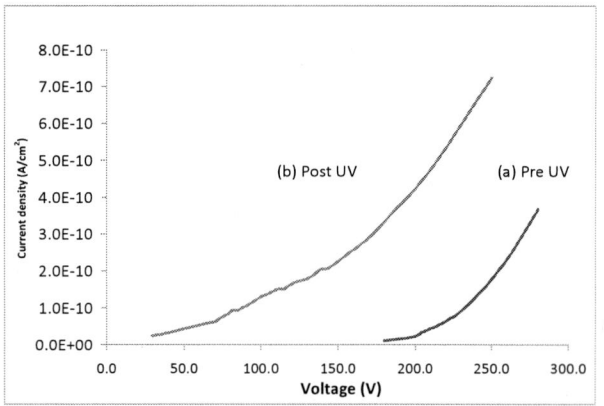

Figure 2. Leakage current results on a k = 6.5 SiN:H film (a) pre and (b) post UV irradiation. Note that UV irradiation greatly increase the amount of leakage.

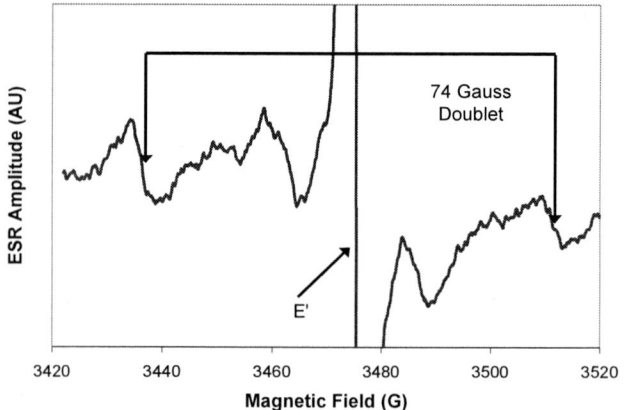

Figure 5. ESR results on a k = 4.4 SiO_2 post UV film with gain increased to showcase additional structure. Distant side peaks are almost certainly due to the 74 Gauss doublet.

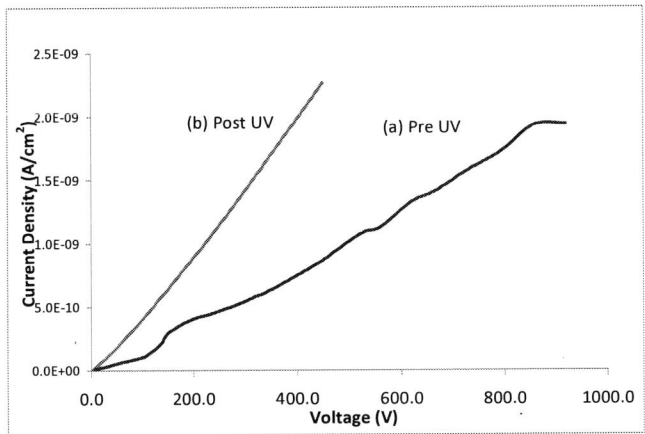

Figure 6. Leakage current results on an k = 4.4 SiO₂ film (a) pre and (b) post UV irradiation. Note that UV irradiation greatly increase the amount of leakage.

We examined three low-k SiOC:H films with ESR and electrical measurements. The three SiOC:H films differ significantly in composition. The k=3.0 film is 33% Si, 29% C and 39% O. The k=2.8 film is 32% Si, 31% C and 37% O. The k = 2.5 film is 33% Si, 46% C and 21% O. These apparently modest changes in composition result in large changes in defect density, which corresponds to a large change in leakage currents between the three films. The three films (without UV irradiation or UV curing) have defect densities that differ by a factor of about 15. The UV cured sample results also suggests that our UV irradiation has a similar effect on the films as the post deposition UV curing method.

Figure 7 shows ESR results from a k = 3.0 ILD SiOC:H film pre and post UV irradiation. We observe a large increase in defect density of the g = 2.0026 spectrum from 0.05 x 10¹⁴/cm⁻² to 0.15 x 10¹⁴/cm⁻². We tentatively assign this spectrum to a silicon or carbon dangling bond defect. The leakage current of the k = 3.0 SiOC:H film is also greatly increased by exposure to UV irradiation as figure 8 demonstrates.

Figure 9 shows ESR results from a k = 2.8 ILD SiOC:H film pre and post UV irradiation. We observe a large increase in defect density of the g = 2.003 spectrum from 0.006 x 10¹⁴/cm⁻² to 0.02 x 10¹⁴/cm⁻². We tentatively assign this spectrum to that of silicon or carbon dangling bond. The leakage current of the k = 2.8 SiOC:H film is also greatly increased by exposure to UV irradiation as figure 10 demonstrates.

Figure 11 shows ESR results from a k = 2.5 ILD SiOC:H film prepared with and without a post deposition UV cure. We observe a large increase in defect density of the g = 2.003 spectrum from 0.1 x 10¹⁴/cm⁻² to 0.25 x 10¹⁴/cm⁻² in the film that has received the cure. We tentatively assign this spectrum to a silicon or carbon dangling bond defect. The leakage current of the k = 2.5 SiOC:H film is also greatly increased by exposure to UV irradiation as figure 12 demonstrates.

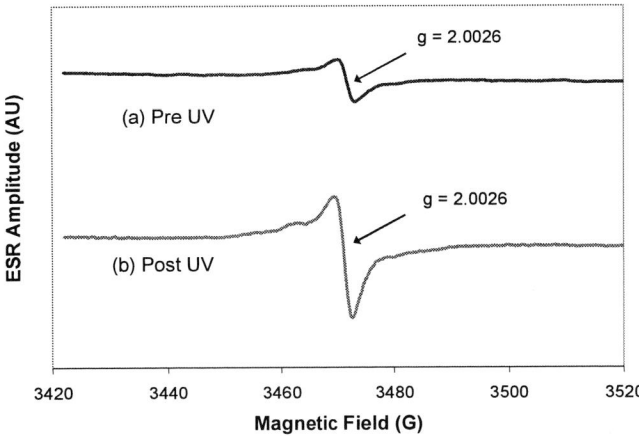

Figure 7. ESR results on a k = 3.0 SiOC:H film (a) pre and (b) post UV irradiation. Note that post UV ESR trace is significantly larger and both spectra have a g-value of 2.0026 which is likely some type of silicon dangling bond.

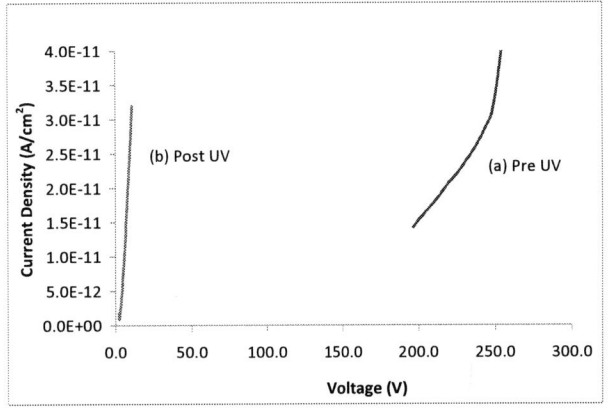

Figure 8. Leakage current results on an k = 3.0 SiOC:H film (a) pre and (b) post UV irradiation. Note that UV irradiation greatly increase the amount of leakage.

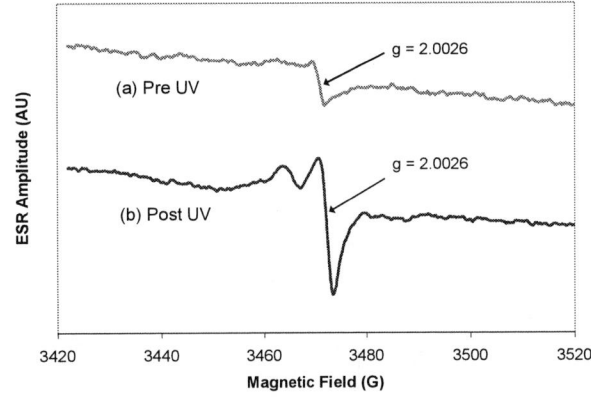

Figure 9. ESR results on a k = 2.8 SiOC:H film (a) pre and (b) post UV irradiation. Note that post UV ESR trace is significantly larger and both dominating spectra have a g-value of 2.0026 which is likely some type of silicon or carbon dangling bond.

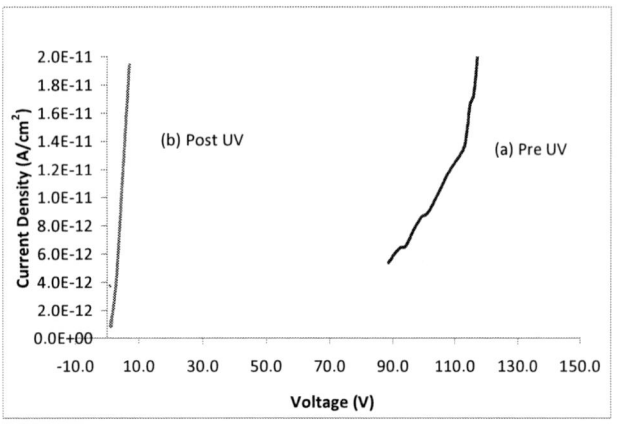

Figure 10. Leakage current results on an k = 2.8 SiOC:H film (a) pre and (b) post UV irradiation. Note that UV irradiation greatly increase the amount of leakage.

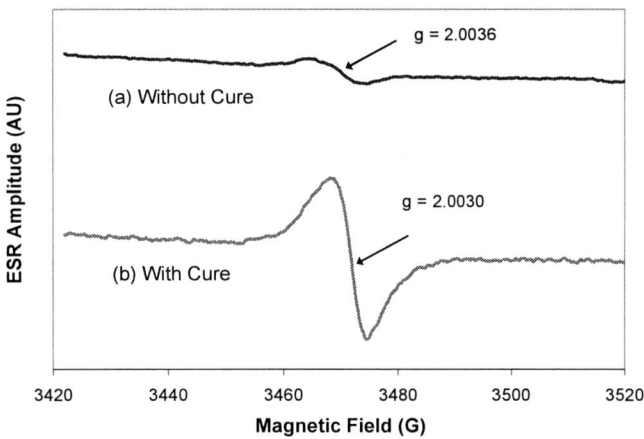

Figure 11. ESR results on a k = 2.5 SiOC:H film (a) with and (b) without a UV cure. Note that post UV ESR trace is significantly larger and both spectra have a g-value of 2.003 which is likely some type of silicon or carbon dangling bond.

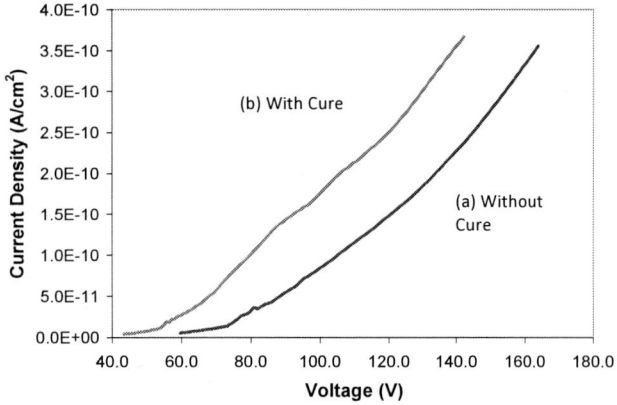

Figure 12. Leakage current results on an k = 2.5 SiOC:H film (a) without (b) with a post deposition UV cure. Note that UV irradiation greatly increase the amount of leakage.

CONCLUSIONS

Low-k dielectrics show great promise for use in ULSI, but their performance limiting defects are not well understood. This work identifies several performance limiting defects and the effect of UV irradiation on various films with applications as ESLs and low - k ILDs. Reliability phenomenon are likely closely correlated to the existence of these performance limiting defects.

Acknowledgements: Work at Penn State was supported by Intel Corporation. We wish to thank Thomas Pomorski for assistance with the electrical measurements.

REFERENCES

[1] F. Chen, K. Chanda, J. Grill, et al., "Investigation of CVD SiCOH low-k Time-dependent Dielectric Breakdown at 65nm node Technology," Proceedings of the Forty Third International Reliability Physics Symposium, 2008, pp. 501-507.

[2] Y. Ou, P. Wang, M. He, et al., "Conduction Mechanisms of Ta/Porous SiCOH Films under Electrical Bias," J. Electrochem Soc., 2008, pp. 283-286.

[3] J. Michelon and R. J. O.M. Hoofman, "Moisture Influenece on Porous Low-k Reliability," IEEE Trans on Dev. and Mcr. Rel., 2006, pp. 169-174.

[4] C. Y. Kim, R. Navamathan, H. J. Lee, C.K. Chio, "Electrical characterisization of low-k films with nano-pore structure prepared with DMDMOS/O2 precursors," Surface and Coatings Technology, 2008, pp. 5688-5692.

[5] S. Eslava, G. Eymery, P. Marsik, et al., "Optical Property Changes in Low-k Films upon Untraviolet-Assisted Curing," Electrochem Soc. , 2008, pp. 115-120.

[6] E. Marhrez, N. Rochet, C. Guedj, et al., " Influence of electron-beam and ultraviolet treatments on low-k porous dielectrics," J. Applied Phys. , 2006, pp. 124106-1-124106-5.

[7] Y. Nishi, K. Tanaka, A. Ohwada, "Study of Silicon-Silicon Dioxide Structures by Electron Spin Resonance II," Japanese Journal of Applied Physics, 11, 1972, pp. 85-91.

[8] P. M. Lenahan and P. V. Dressendorfer, "Hole traps and trivalent silicon centers in metal/oxide/silicon devices," Journal of Applied Physics, 55, 1984, pp. 3495-3499.

[9] D. T. Krick, P. M. Lenahan, and J. Kanicki, "Electrically Active Point-Defects in Amorphous-Silicon Nitride - An Illumination and Charge Injection Study," J. Appl. Phys., 64, 1988, pp. 3558-3563.

[10] P. M. Lenahan and S. E. Curry, "First observation of the ^{29}Si hyperfine spectra of silicon dangling bond centers in silicon nitride," Appl. Phys. Lett., 56, 1990, pp. 157-159.

[11] P. J. Caplan, E. H. Poindexter, B. E. Deal, and R. R. Razouk, "ESR centers, interface states, and oxide fixed charge in thermally oxidized silicon wafers," Journal of Applied Physics, 50, 1979, pp. 5847-5854.

[13] D.L. Griscom and E.J. Friebele, "Fundamental defect centers in glass: 29 Si hyperfine structure of the nonbridging oxygen hole center and the peroxy radical in a-SiO2," Phys. Rev. B., 1981, pp. 4896-4898.

[14] J. Vitko, "ESR Studies of hydrogen spectra in irradiated vitreous silica,: J. Applied Phys., 1978, pp. 5530-5535.

[15] J.F. Conley and P.M. Lenahan, "Room temperature reactions involving silicon dangling bond centers and molecular hydrogen in amorphous SiO2 thin films on silicon," Appl. Phys. Lett., 1992, pp. 40-42

978-1-4244-5430-3/10 $26.00 © 2010 IEEE

A HIGH-ENDURANCE (>100K) BE-SONOS NAND FLASH WITH A ROBUST NITRIDED TUNNEL OXIDE/SI INTERFACE

Szu-Yu Wang, *Hang-Ting Lue, *Tzu-Hsuan Hsu, *Pei-Ying Du, *Sheng-Chih Lai, *Yi-Hsuan Hsiao, Shih-Ping Hong, Ming-Tsung Wu, Fang-Hao Hsu, Nan-Tzu Lian, Chi-Pin Lu, Jung-Yu Hsieh, Ling-Wu Yang, Tahone Yang, Kuang-Chao Chen, *Kuang-Yeu Hsieh, and Chih-Yuan Lu

Macronix International Co., Ltd., Technology Development Center
*Macronix International Co., Ltd., Emerging Central Lab.
16 Li-Hsin Road, Science Park, Hsin-chu 300, Taiwan
Corresponding author: Szu-Yu Wang; tel: +886-3-5786688 ext. 78177; fax: +886-3-5789087; e-mail: kirinwang@mxic.com.tw

ABSTRACT

For Solid-State Drive (SSD) applications cycling endurance of NAND flash is a critical challenge. In this work the endurance reliability of BE-SONOS NAND is thoroughly examined. Using dual *CV/IV* tests the impact of interface state (Dit) generation/annealing and real charge trapping (Q) on the endurance degradation has been clearly identified. For BE-SONOS with pure thermal oxide O1, the endurance degradation mainly comes from Dit generation at Si/O1 interface, while charge trapping in the thin ONO barrier is negligible even after 100K cycles of stressing. Meanwhile, the high-temperature V_T loss mainly comes from interface state annealing, while the real charge loss due to electron de-trapping is much smaller. This indicates that our nitride-trapping device has "deep" traps that well retain charges even after the tunnel barrier is damaged. Based on this understanding, we have introduced nitrided O1 to strengthen the Si/O1 interface, and both the endurance and retention are greatly improved. We demonstrate high-endurance BE-SONOS NAND devices of P/E > 5K for MLC and P/E > 100K for SLC operations with excellent retention, promising for solid-state drive (SSD) applications.

I. Introduction

Charge-trapping (CT) devices promise to continue the NAND Flash scaling and open the door for 3D NAND Flash. BE-SONOS [1, 2] is one of the promising solutions because it uses mature and mass-production proven materials of oxide, nitride and poly gate. These well understood materials help further improve the reliability.

Real charge loss (Q) and interface state (Dit) generation are two major causes for Flash memory device degradation. V_T will be impacted by both factors and the independent effects are difficult to be separated. Unlike the floating gate device, a large planar CT device can be operated in FN mode because there is no gate coupling ratio restrictions. This allows the measuring of *CV* and $I_D Vg$ on the same large-area device, where the V_FB is revealed from *CV* curve, and $I_D Vg$ provides the Vt. In this way both Q and Dit can be extracted separately by comparing V_FB and V_T, thus provides further understanding of endurance degradation.

II. Dual *CV/IV* Testing - Differentiating Charge Trapping (Q) from Interface State (Dit)

The schematic structure of BE-SONOS device is shown in Fig. 1. The O1 layer in this work was either a pure thermal oxide or thermally nitrided oxide with very well controlled thickness around 13Å. The dual *CV* and $I_D Vg$ testing is designed to differentiate the real charge trapping Q from interface state Dit. *CV* provides V_FB shift, which is related to the Q and is independent of Dit. On the other hand, $I_D Vg$ provides V_T shift, which is related to both Q and Dit. Thus by analyzing V_T and V_FB we can extract both Q and Dit. Moreover, the subthreshold swing (S.S.) change from $I_D Vg$ is directly related to Dit.

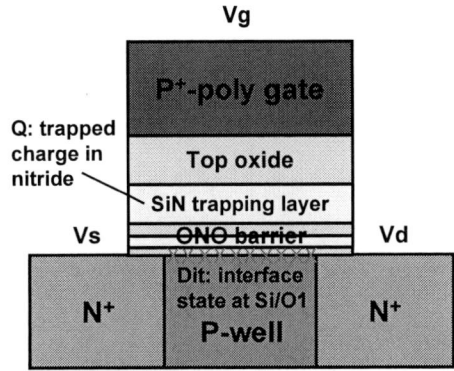

Fig. 1 Schematic structure of BE-SONOS device. Both the pure O1 and the nitrided O1 are applied to evaluate the post-cycled real charge loss (Q) and interface state (Dit) generation.

The basic equations for the V_FB and V_T shifts are illustrated as followed:

V_FB equation:

$$V_{FB} = (\phi_{MS} - \frac{Q}{C_{OX}}) \tag{1}$$

V_T equation:

$$V_T = V_{FB} + 2\Psi_B + \frac{\sqrt{4\varepsilon_S q N_A (2\Psi_B + V_{BS})}}{C_{OX}} + \Delta V_T(Dit) \tag{2}$$

$$\Delta V_T(Dit) = S.S. \times [Log_{10}(I_D(@V_T)) - Log_{10}(I_D(@V_{MG}))]$$
$$\sim \Delta S.S. \times 10 \tag{3}$$

According to the interface state model [3-6], interface state is neutral and inactive at mid-bandgap voltage (when surface band bending is at mid-bandgap, both acceptor and donor interface traps are neutral). The corresponding mid-bandgap current (I_MG) is unchanged after Dit generation. This is confirmed by simulation - the TCAD simulated I_MG is around 10^{-17}A for our device. Since the typical V_T is defined at 10^{-7}A, the V_T shift caused by S.S.

978-1-4244-5430-3/10 $26.00 © 2010 IEEE   951

degradation can be simply approximated by $\Delta V_T \sim 10 \times \Delta S.S.$. This provides a way to estimate the contribution of S.S. changes to reliability degradation.

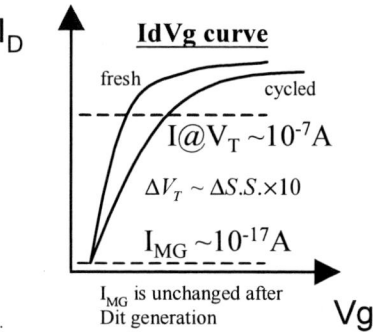

Fig. 2 Schematic diagram of $I_D Vg$ curves of fresh and P/E cycled devices. Midgap current (I_{MG}) is unchanged with Dit generation. However, S.S. is increased with Dit generation. Thus the IdVg curve rotates with respect to I_{MG}. Assuming $I_{MG}=10^{-17}A$, and Vt is defined at $10^{-7}A$, a simple approximation gives $\Delta V_T - \Delta V_{FB} \sim 10 \times \Delta S.S.$

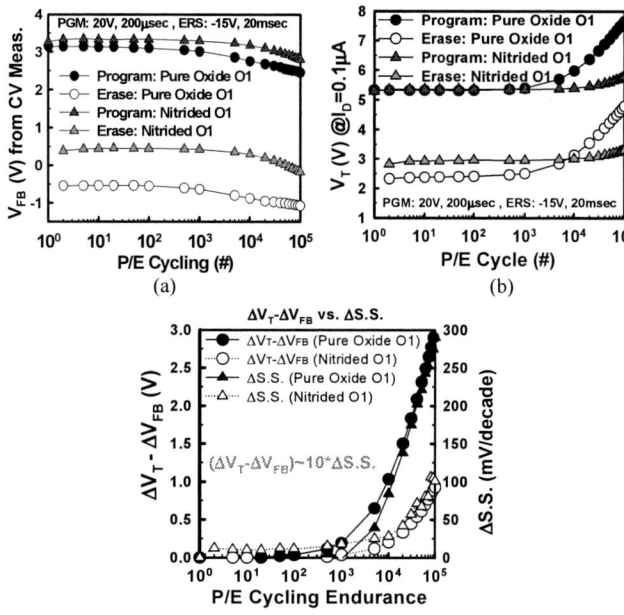

Fig. 3 Endurance comparison of two BE-SONOS devices with pure oxide O1 and nitrided O1. (a) V_{FB}'s during P/E cycling are relatively stable even after 100K cycling. The small V_{FB} roll-down is due to the interface state generation and CV distortion (in Fig. 5(a)). (b) The measured V_T during cycling. V_T roll-up happens severely for the pure oxide O1. On the other hand, the memory window is kept even V_T roll-up happens. Nitrided O1 shows significant improvement in endurance. (c) The subthreshold slope (S.S.) degradation ($\Delta S.S.$, right Y axis) is highly correlative with the V_T roll-up. $\Delta V_T - \Delta V_{FB} \sim 10 \times \Delta S.S.$ is observed, consistent with the model in Fig. 2.

III. Reliability Study of BE-SONOS by Dual *CV/IV* Method

Figure 3 compares the endurance of BE-SONOS with pure oxide O1 and nitrided O1. In Fig. 3(a), the V_{FB} is very stable even after 100K cycling stressing. On the other hand, V_T in Fig. 3(b) shows

very significant roll-up for pure oxide O1. Note that the cycling window is kept constant even with V_T roll up. This suggests that the FN tunneling speed is not degraded and the amount of stored charge remained the same even after 100K P/E cycles. On the other hand, Dit generation caused severe degradation of S.S. (Fig. 3(c)), leading to the V_T roll-up. It is interesting that the measured $\Delta V_T - \Delta V_{FB} \sim 10 \times \Delta S.S.$, very consistent with our model.

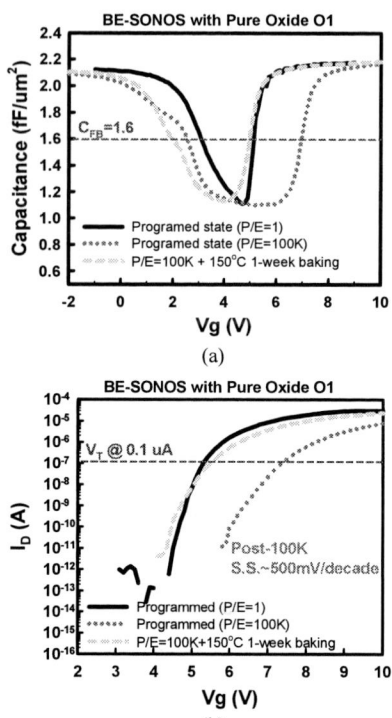

Fig. 4 (a) CV curves of post 100K P/E cycling and 150°C baking for BE-SONOS with **pure oxide O1**. CV curve has a large distortion and the inversion part is broadened after 100K PE cycling. At accumulation region, the CV curve is also distorted and V_{FB} shifts leftward. (b) $I_D Vg$ curves for the same device. IV curve shifts rightward, corresponding to the CV curve at the inversion region. Meanwhile, S.S. is strongly degraded. After 150°C 1-week baking, both CV and IV curves almost completely recover to fresh state. It should be noted that V_T and V_{FB} loss are different, which provides understanding of real charge loss (Q) and interface state generation (Dit).

Figure 4 compares the details of CV and $I_D Vg$ during P/E cycling for pure oxide O1. Dit generation is the major reason to the endurance degradation, which leads to CV stretch-out as well as V_T increase.

Note that the endurance degradation of BE-SONOS is quite different from the conventional floating gate device, where not only Dit is generated, but also memory window is gradually closed due to trapped electrons in tunnel oxide [5]. In BE-SONOS, both electron and hole tunneling happens during cycling, which neutralizes any charge trapped in the tunneling barrier. Thus the major endurance degradation of BE-SONOS comes from Dit generation instead of FN tunneling speed degradation.

The nitrided tunnel oxide is widely used in floating gate technology [5] to produce a stronger interface against Si dangling bond generation. For this work we have developed a very thin nitrided O1 to reduce Dit generation in BE-SONOS device. By comparing to pure oxide O1, great improvements of endurance to >100K are clearly shown with nitrided O1 (Fig. 3). Both CV and IV degradation are also much smaller, as shown in Fig. 5.

978-1-4244-5430-3/10 $26.00 © 2010 IEEE

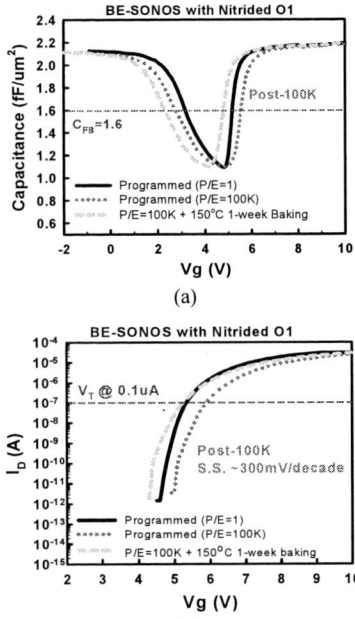

(a)

(b)

Fig. 5 (a) Detail *CV* curves during 100K P/E cycling and 150°C baking for BE-SONOS with **nitrided O1**. (b) Detail *IV* curve for the same device. Both *CV* and *IV* degradation are much smaller for nitrided O1. In *CV* curve, the *CV* stretching after cycling is suppressed. During baking, *CV* curves shift in parallel, thus V_T and V_{FB} loss are similar.

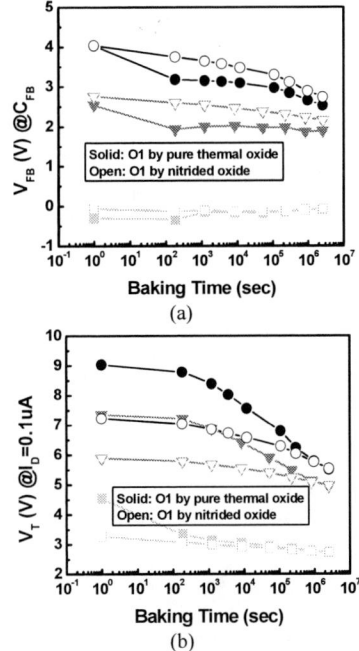

(a)

(b)

Fig. 6 (a) V_{FB} loss during 150°C baking after 100K P/E cycling. V_{FB} loss is small (<500mV) even after 100K cycling, indicating the real charge loss is very small even after 100K cycling. (b) V_T loss for the same device. V_T loss is much larger than V_{FB} loss for **pure oxide O1**, while the two values are quite similar for **nitrided O1**. Moreover, pure oxide O1 has much more V_T loss than nitrided O1 and less dependent on initial V_{FB} state, suggesting that the V_T loss is not dominated by real charge loss but Dit annealing. After 1-week baking, V_T loss of pure oxide O1 is about 2 V, while V_{FB} loss is ~500mV. Therefore, the estimated interface annealing effect contributes to ~1.5 V, which is approximately 10 times of S.S. recovery in Fig. 7.

The retention results are shown in Figs. 6-7. For pure oxide O1, the V_{FB} loss (real charge loss) is much smaller than V_T loss, indicating that the V_T loss mainly comes from interface annealing [5, 6], while real charge loss is much smaller. The annealing of interface state (Fig. 7(a)) is also observed from the S.S. recovery during baking (Fig. 7(b)). The total V_T loss can be approximated by real charge loss (V_{FB} loss) + interface anneal ($\sim 10 \times \Delta S.S.$), or $\Delta V_T - \Delta V_{FB} \sim 10 \times \Delta S.S.$ The experimental results show very good agreement with this model.

For nitrided O1, the suppressed S.S. degradation and V_T roll-up are shown (Fig. 3). Furthermore, V_T loss is comparable to V_{FB} loss during baking (Fig. 6), indicating the suppressed Dit generation after P/E cycling.

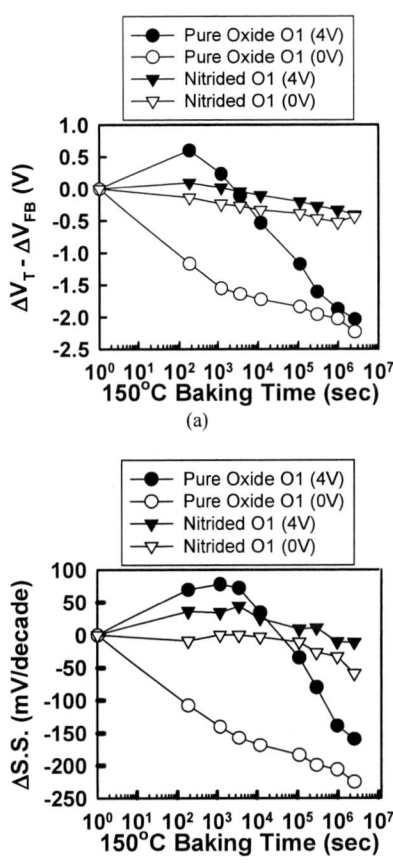

(a)

(b)

Fig. 7 (a) Interface-state annealing ($\Delta V_T - \Delta V_{FB}$) and (b) S.S. recovery (Δ S.S.) during 150°C baking for pure oxide O1 and nitrided O1 samples after 100K cycling. Significant suppression of interface-state annealing and S.S. recovery are revealed in the devices with nitrided O1. In addition, $\Delta V_T - \Delta V_{FB} \sim 10 \times \Delta S.S.$ is also observed.

IV. Performances of BE-SONOS NAND Flash Using the Nitrided O1

(1) Cycling Endurance: The cycling endurance of BE-SONOS NAND cell is shown in Fig. 8. S.S. starts to degrade only after 10K cycling. After 100K cycling, S.S. only degrades by 100mV/decade, and the read current degradation is less than 20%. With suitable P/E conditions, more than 1M cycling endurance is achieved, as shown in Fig. 9.

978-1-4244-5430-3/10 $26.00 © 2010 IEEE

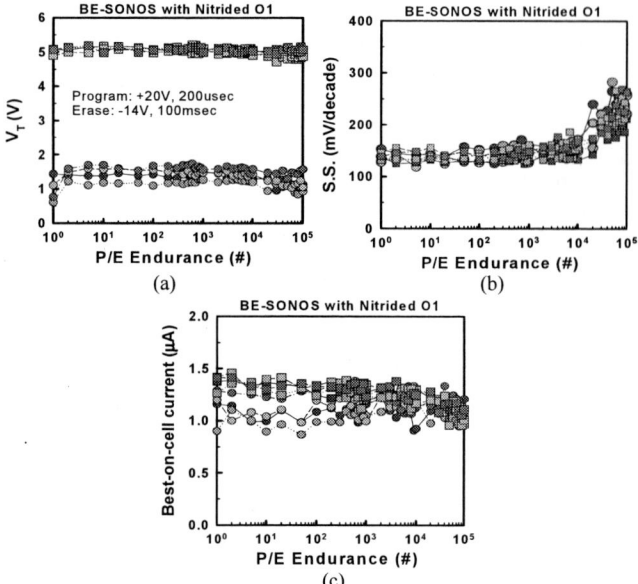

(a)

(b)

(c)

Fig. 8 (a) Endurance characteristics of BE-SONOS NAND device with **nitrided O1**. Several identical devices are collected. Dumb-mode P/E cycling (without any P/E verify) shows no degradation even after 100K cycling. (b) S.S. during 100K P/E cycling. S.S. shows no degradation below 10K cycling. After 100K, S.S. degradation is only about 100mV/decade. (c) the best-on-cell current under V_{pass}=7V, and read gate-overdrive (Vg-V_T) =1V.

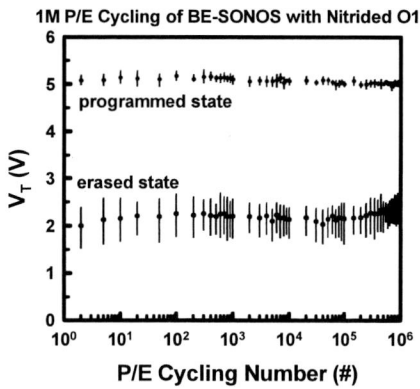

Fig. 9 1M PE cycling endurance distribution collected from many cells. With suitable P/E algorithms, 1M cycling can be achieved.

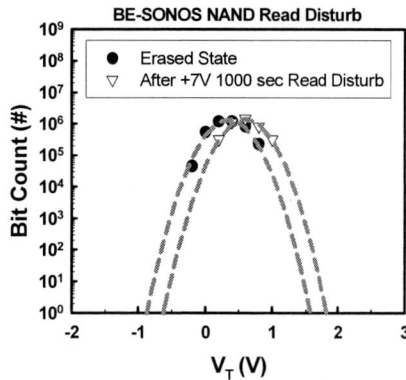

Fig. 10 Typical read disturb behavior at $V_{pass,read}$=7V and 1000 sec. The corresponding read cycle time is more than 10M.

(2) Read Disturb: Typical read disturb distribution analysis is shown in Fig. 10. More than 10M read is guaranteed.

(3) Self-Boosting Characteristics: The typical global self-boosting characteristics are shown in Fig. 11. More than 4V disturb-free window is achieved with V_{pass} = 10V. After 100K P/E cycling, self-boosting is only slightly degraded, allowing enough memory window design.

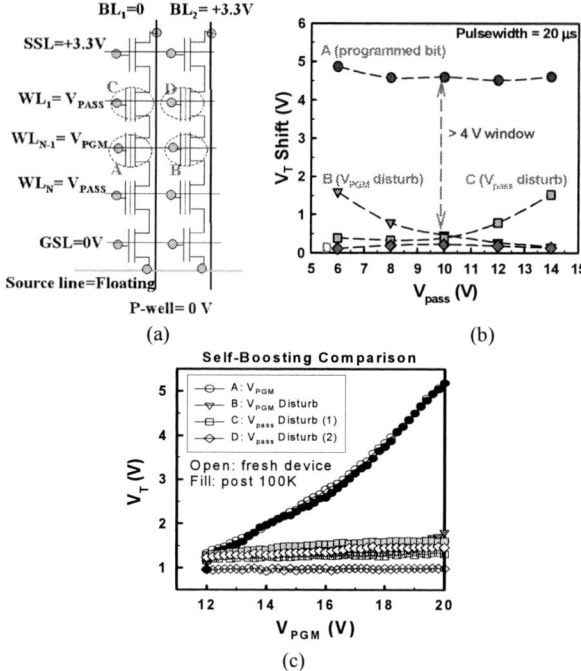

(a)   (b)

(c)

Fig. 11 (a) Self-boosting operation method. (b) Self-boosting V_{pass} disturb window. Increased V_{pass} shows suppressed V_{PGM} disturb, but increased V_{pass} disturb. V_{pass} ~ 10V is the optimized condition. The disturb-free memory window is more than 4V. (c) Self-boosting operation before and after 100K cycling. Post 100K cycling still shows successful self-boosting disturb window.

(4) Data Retention: The memory window and post-cycling retention properties are shown in Figs. 12-13. Figure 12 shows that below 5K cycling the retention properties are quite enough for MLC design. After a strong P/E cycling (>100K) as shown in Fig. 13, the retention is still reasonable for SLC design.

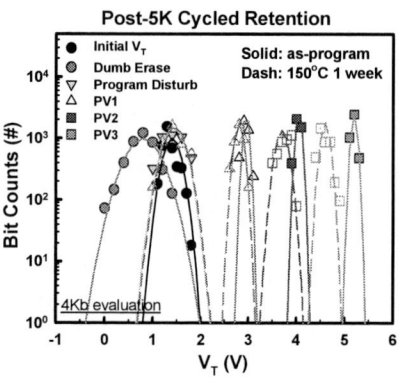

Fig. 12 The MLC distribution and post-5K cycled 150°C retention result for BE-SONOS with nitrided O1. Post 5K cycling and 150°C baking shows reasonable window for MLC operation.

978-1-4244-5430-3/10 $26.00 © 2010 IEEE   954

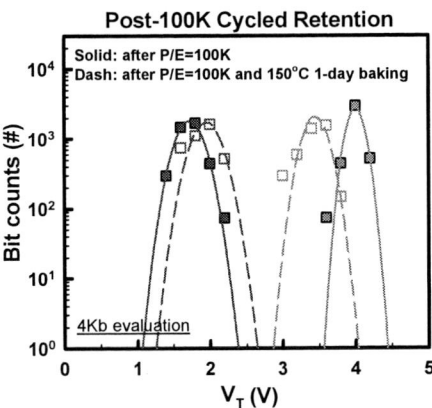

Fig. 13 The SLC distribution and post-100K cycled 150°C retention result for BE-SONOS with nitrided O1. Post 100K cycling and 150°C baking shows enough window for SLC operation.

V. Conclusions

We have clarified that the major endurance degradation mechanism of BE-SONOS NAND using pure oxide O1 is the interface state (Dit) generation/annealing. By developing a nitrided O1 that strengthens the Si/O1 interface, the endurance and retention are greatly improved. A high-performance BE-SONOS NAND Flash with P/E > 5K for MLC and P/E > 100K for SLC operation is developed.

References

[1] H. T. Lue, *et al, IEDM* 2006, pp. 547-550.
[2] H. T. Lue et al, *VLSI Symposia,* 2008, pp. 116-117.
[3] D. K. Shorder, "Semiconductor material and device characterization", John Wiley, 1998.
[4] C. H. Tsai et al, *IRPS* 2009, pp. 294-300.
[5] J. D. Lee et al, *IEEE TDMR*, 2004, pp. 110-117.
[6] Y. H. Shih et al, *IEDM* 2006, pp.503-506.

Transition of Erase Mechanism for MONOS memory depending on SiN Composition and its Impact on Cycling Degradation

Shosuke Fujii, Jun Fujiki, and Naoki Yasuda

Advanced LSI Technology Laboratory, Corporate R&D Center,
Toshiba Corporation
8, Shinsugita-cho, Isogo-ku, Yokohama 235-8522, Japan
phone: (+81) -(45)-776-5926, shosuke.fujii@toshiba.co.jp

Ryota Fujitsuka and Katsuyuki Sekine
Process & Manufacturing Engineering Center, Toshiba Corporation
8, Shinsugita-cho, Isogo-ku, Yokohama 235-8522, Japan

Abstract— We clarify the origin of erase improvement in MONOS memories with Si-rich SiN layer, and investigate the impact of erase mechanism on cycling degradation. It is demonstrated that cycling degradation is uniquely determined by charges injected during erase operations irrespective of program/erase condition, number of program/erase cycling, or MONOS structure.

Keywords; MONOS, TANOS, BE-MONOS, Cycling degradation

I. INTRODUCTION

MONOS-type devices are candidates to replace conventional floating-gate non-volatile memory devices because of their low program/erase (P/E) voltage and reduced cell-to-cell interference effects. Recently, band-engineered (ONO) MONOS and MONOS with silicon-rich SiN (SRN) layer have been proposed to improve P/E characteristics and reliability of MONOS devices [1-4]. Whereas the improvement of the erase performance by ONO tunnel structures clearly originates from the enhancement of hole injection from a Si substrate as already reported in [1], the mechanism of the erase improvement by SRN layer remains to be clarified [2-4]. Thus, it is of great importance to elucidate the impact of the SRN layer in detail for the performance optimization of MONOS devices.

In addition, cycling degradation is also a critical issue in the memory operation [5], as well as P/E performance and data retention. Therefore, the purpose of this study is to clarify the origin of the erase improvement in MONOS with SRN layer by using our new evaluation technique [6,7], and to investigate the impact of the erase mechanism on cycling degradation.

We clearly demonstrate that the cycling degradation is uniquely determined by the total amount of the injected charges during erase operations irrespective of P/E conditions, number of P/E cycling, or MONOS structures, and also show that the cycling degradation is induced mainly by holes injected from a Si-substrate.

II. EXPERIMENTAL

MONOS capacitors with n^+diffusion layer on p-type Si substrates were fabricated. The 5-nm-thick tunnel oxide was thermally grown, and the 5-nm-thick SRN layer was deposited by ALD method using dichlorosilane (DCS) and NH_3. The composition (N/Si ratio) of SRN was modulated by controlling the DCS/NH_3 gas supply ratio. Refractive indices (R.I.) of the SRN layers are described in Table I. Since R.I. increases with decreasing the N/Si ratio, higher R.I. indicates more Si-rich composition. After the SRN deposition, the Al_2O_3 block layer (13nm) and TaN gate electrode were formed. Areas of the capacitors are all $100 \times 100 \mu m^2$. EOTs of the MONOS devices are estimated by C-V measurements.

III. DEPENDENCE OF ERASE MECHANISM ON SiN COMPOSITION

First, we investigate the effects of the SiN composition on P/E characteristics. Fig. 1 shows the P/E characteristics of the MONOS devices. As shown in Fig. 1(a), the program speed slightly decreases with increasing R.I., whereas the erase speed drastically increases. Next, by using our developed technique [6,7], we directly measured the amount of charges injected during P/E operations, and extracted the injected current as a function of electric field across the tunnel oxide, as plotted in Fig. 2.

TABLE I.   MONOS DEVICES USED IN THIS STUDY.

Sample name	Tunnel ox	Charge SiN	Block Al₂O₃	EOT
2.07	SiO₂ 5nm	R.I.=2.07 5nm	13nm	12.9 nm
2.13		R.I.=2.13 5nm		13.1 nm
2.23		R.I.=2.23 5nm		12.9 nm
2.30		R.I.=2.30 5nm		13.2 nm
ONO+2.07	ONO	R.I.=2.07 5nm		13.0 nm

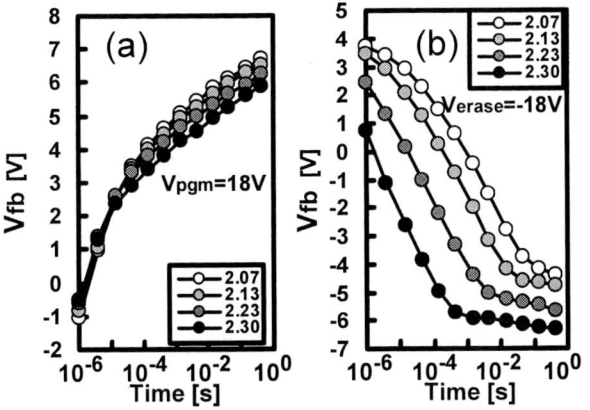

Figure 1. (a) Program and (b) erase characteristics for MONOS devices of "2.07", "2.13", "2.23", and "2.30".

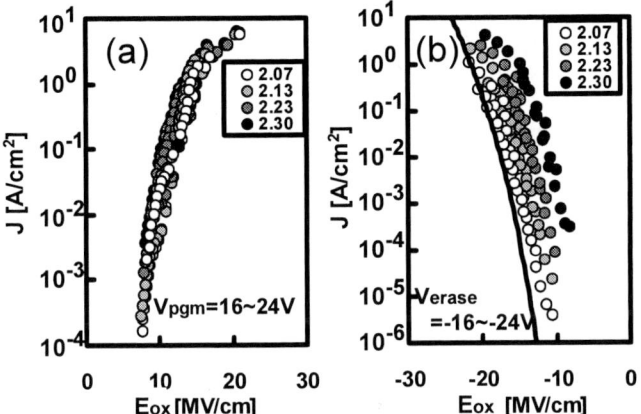

Figure 2. J-E_{ox} characteristics during (a) program and (b) erase operations extracted by our new method[6,7].

Although the program speed slightly decreases as R.I. increases (Fig. 1(a)), the amount of electrons injected during the program pulse is independent of the SiN composition (Fig. 2(a)), indicating that the trapping efficiency, i.e., $Q_{trap}/(Q_{trap}+Q_{leak})$, weakly depends on the SiN composition. In contrast, J-E_{ox} characteristics during the erase operation are strongly dependent on the SiN composition (Fig. 2(b)). We also plot the theoretical hole current from the substrate, which is calculated by the WKB approximation, as a solid line in Fig. 2(b). According to the WKB approximation, the tunneling current J is:

$$J = \frac{m^* e}{2\pi^2 \hbar^3} \int k_B T \log\{1 + \exp[\frac{(E_F - E_x)}{k_B T}]\} T_{tunneling} dE_x$$

$$T_{tunneling} = \exp[-2\int \sqrt{\frac{2m^* e(V(x) - E_x)}{\hbar^2}} dx]$$

where $V(x)$ is the potential barrier height at position x. E_x is the longitudinal energy of hole. The other symbols m^*, e, k_B, T, E_F, and $\hbar$ have the usual meanings. In the present calculation, potential barrier height for hole was assumed to be 3.9eV and effective mass of hole was assumed to be $0.6m$, where m is electron mass in vacuum. The theoretical curve is in fairly good agreement with the extracted J-E_{ox} curve of "2.07" MONOS,

suggesting that the erase operation of "2.07" MONOS proceeds by the holes injected from the substrate. However, the extracted erase current increases and deviates from the theoretical hole current with increasing R.I. This result reveals that the erase mechanism of the MONOS with higher R.I. cannot be explained only by the hole injection. Since the erase operation can proceed by hole injection and/or electron de-trapping from the SRN layer, we can conclude that the electron de-trapping component becomes more dominant in the erase operation as R.I. increases. Program and erase mechanisms as a function of the SiN composition are summarized in Fig. 3. With increasing Si content in the SiN layer, the program speed slightly decreases due to reduced trapping efficiency, whereas the erase speed increases due to the transition of erase mechanism from hole injection to electron de-trapping.

IV. CYCLING DEGRADATION OF MONOS DEVICES

In order to evaluate the degradation of tunnel oxide, we measure the amount of "C-V stretch" during P/E cycling. The definition of "C-V stretch" is described in Fig. 4. From C-V characteristics, threshold voltage and flat-band voltage have been evaluated considering $0.85 \times C_{max}$ in inversion and depletion configuration, respectively. Therefore, evaluation of interface state generation via "C-V stretch" is important not only for understanding the degradation mechanism of MONOS devices, but also for predicting endurance characteristics, since it is reported that the instability of P/E window during endurance is primarily caused by interface traps generated by electrical stress [8].

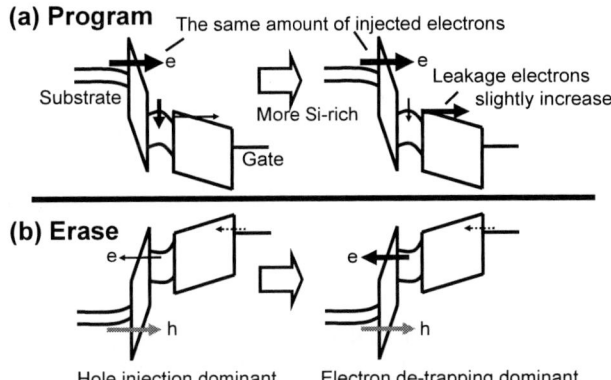

Figure 3. Schematic diagrams for explaining the transition of (a) program and (b) erase mechanisms depending on SiN composition.

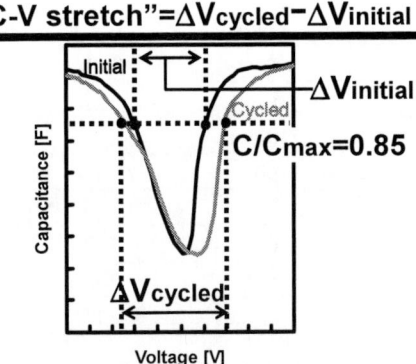

Figure 4. Definition of "C-V stretch" evaluated in this study.

Table II summarizes P/E conditions to MONOS devices. Under each condition, P/E cycling was performed and "C-V stretches" after 5~10000 cycles were evaluated.

Fig. 5 shows the amount of "C-V stretch" of "2.07" MONOS as functions of (a)Q_{pgm} and (b)Q_{era}. Here, Q_{pgm} and Q_{era} are defined as the total amount of injected charges during program operations and erase operations in the P/E cycling, respectively, which are measured by our developed technique [6,7]. It is found that the amount of "C-V stretch" is strongly correlated with Q_{era}, while it has no correlation with Q_{pgm}. In other words, the trend with Q_{era} in Fig. 5(b) can fit with a single line. Since each point in Fig. 5 is obtained under different P/E conditions and different numbers of P/E cycling, these results clearly reveal that the cycling degradation of tunnel oxide is uniquely determined by Q_{era} irrespective of P/E condition or number of P/E cycling. Therefore, cycling degradation is induced by charges flowing in the erase operation. The dependence of "C-V stretch" on the SiN composition is depicted in Fig. 6. The cycling degradation in all the MONOS devices are determined by Q_{era}, and the degradation is suppressed with increasing R.I. Considering the above-discussed fact that the erase mechanism shifts toward electron de-trapping as Si content in the SiN layer increases, it is concluded that holes injected in erase operations during P/E cycling mainly degrade the tunnel oxide.

To confirm the above findings, we performed the P/E cycling to MONOS with ONO tunnel structure listed in Table I. By employing ONO structure as a tunnel layer, the erase characteristic is improved as seen in Fig. 7. This is because the enhancement of hole injection by band engineering of the tunnel layer as already reported in [1]. However, even though the improvement of erase performance can be achieved by the ONO structure, our findings predict that the cycling degradation cannot be suppressed as long as the erase operation is caused by hole injection. Fig. 8 shows the amount of "C-V stretch" of MONOS with the ONO structure as a function of Q_{era}, along with the result of "2.07" MONOS. Although the erase performance is improved, as shown in Fig. 7, we find that the amount of cycling degradation is almost the same as that of "2.07" MONOS. Thus, this result reconfirms the fact that the holes injected in erase operations determine the cycling degradation, which is independent of MONOS structure.

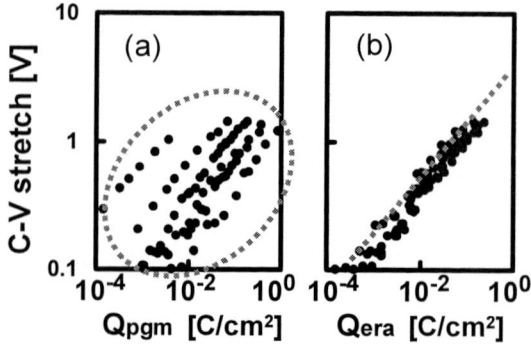

Figure 5. "C-Vstretch" as a function of Q_{pgm} or Q_{era}. "C-Vstretch" strongly correlates with the sum of injected charges in the erase operations. The measured device is "2.07".

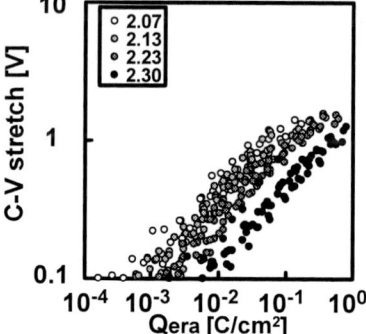

Figure 6. "C-Vstretch" vs. Q_{era} for MONOS devices of "2.07", "2.13", "2.23", and "2.30".

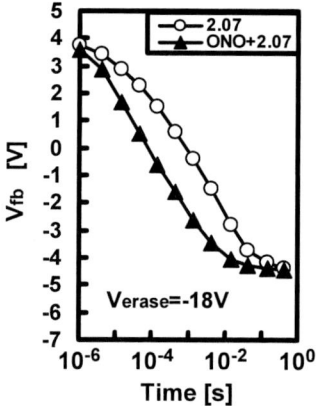

Figure 7. Erase characteristics for MONOS devices of "ONO+2.07" and "2.07".

It should be noted that there is no improvement of cycling degradation by introducing the MONOS with ONO tunnel structure which enhances hole injection.

Finally, it is pointed out that data retention property remains as a critical issue. Although the MONOS with the SRN layer can suppress the cycling degradation, it is well known that the Si-rich SiN severely degrades data retention characteristics [2-4]. Therefore, alternative structures and/or materials are demanded as a charge trapping layer, which improves data retention property while maintaining the erase mechanism of electron de-trapping [9].

TABLE II. P/E CONDITIONS FOR ENDURANCE

Endurance conditions	Program		Erase	
	V_{pgm}	T_{pgm}	Verase	Terase
1	20 V	1 ms	-18 V	1 or 10 ms
2	18 V	1 ms	-20 V	1 or 10 ms
3	20 V	100 μs	-20 V	1 or 10 ms
4	20 V	100 μs	-18 V	10 or 100 ms
5	18 V	1 ms	-18 V	10 or 100 ms
6	18 V	100 μs	-18 V	1 or 10 ms
7	20 V	1 ms	-20 V	10 or 100 ms
8	18 V	100 μs	-20 V	10 or 100 ms

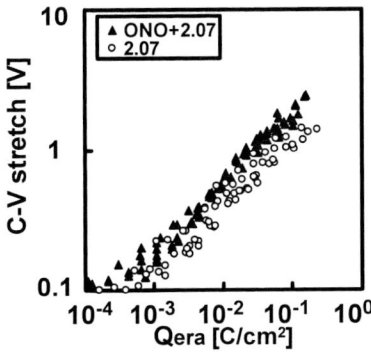

Figure 8. "C-V stretch" as a function of Q_{era}. The devices are "ONO+2.07" and "2.07".

V. CONCLUSIONS

In this work, we investigated the dependence of P/E mechanisms on the SiN composition by using our charge evaluation technique. The erase mechanism shifts from hole injection to electron de-trapping as Si content in the SiN layer increases, which is the origin of the improvement of the erase performance in the MONOS with SRN. Furthermore, the cycling degradation is induced mainly by holes injected during

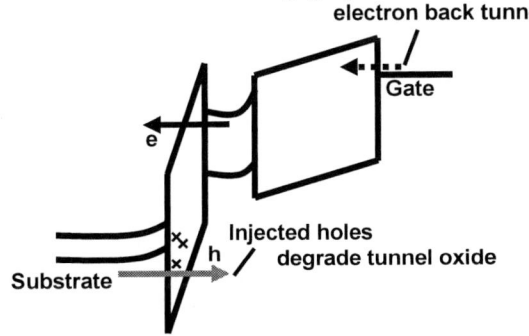

Figure 9. Schematic diagram for explaining the origin of MONOS degradation.

erase operations, as summarized in Fig. 9, and the degradation can be suppressed in the MONOS with SRN as a result of transition of the erase mechanism. These findings will be helpful to design reliable MONOS structures based on the accurate understanding of operation mechanisms.

REFERENCES

[1] H. T. Lue, S. Y. Wang, Y. H. Hsiao, E. K. Lai, L. W. Yang, T. Yang, K. C. Chen, K. Y. Hsieh, R. Liu, and C. Y. Lu, "Reliability model of bandgap engineered SONOS (BE-SONOS)," IEDM Tech. Dig., (2006) p.495.

[2] C. Sandhya, U. Ganguly, N. Chattar, C. Olsen, S. M. Seutter, L. Date, R. Hung, J. M. Vasi, and S. Mahapatra, "Effect of SiN on performance and reliability of charge trap flash (CTF) under Fowler–Nordheim tunneling program/erase operation," IEEE Electron Dev. Lett. 30 (2009) 171.

[3] N. Goel, D. C. Gilmer, H. Park, V. Diaz, J. Sun, J. Price, C. Park, P. Pianetta, P. D. Kirsh, and R. Jammy, "Erase and retention improvements in charge trap flash through engineered charge storage layer," IEEE Elecron Dev. Lett. 30 (2009) 216.

[4] G. Van den bosch, A. Funemont, M. B. Zahid, R. Degraeve, L. Breuil, A. Cacciato, A. Rothschild, C. Olsen, U. Ganguly, and J. Van Houdt, "Nitride engineering for improved erase performance and retention of TANOS NAND Flash memory," NVSMW (2008) p.128.

[5] G. Ghidini, C. Scozzari, N. Galbiati, A. Modelli, E. Camerlenghi, M. Alessandri, A. Del Vitto, G. Albini, A. Grossi, T. Ghilardi, and P. Tessariol, "Cycling degradation in TANOS stack" Microelectron. Eng. 86 (2009) 1822.

[6] S. Fujii, N. Yasuda, J. Fujiki, and K. Muraoka, "A new method to extract the charge centroid in the program operation of MONOS memories," SSDM (2009) p.158.

[7] J. Fujiki, S. Fujii, N. Yasuda, and K. Muraoka, "Direct measurement of back-tunneling current during program/erase operation of MONOS memories and its dependence on gate work function," SSDM (2009) p.857.

[8] C. H. Lee, W. H. Tu, S. H. Gu, C.W. Wu, S. W. Lin, T. H. Yeh, K.F. Chen, Y. J. Chen, J. Y. Hsieh, I. J. Huang, N. K. Zous, T. T. Han, M. S. Chen, W. P. Lu, K. C. Chen, T. Wang, and C. Y. Lu, "Cell endurance prediction from a large-area SONOS capacitor," IRPS (2009) p.891.

[9] R. Fujitsuka, K. Sekine, A. Sekihara, A. Fukumoto, J. Fujita, F. Aiso, and Y. Ozawa, "Engineering of Si-rich nitride charge-trapping layer for highly reliable MONOS type NAND flash memory with MLC operation," SSDM (2009) p.861.

Use of Random Telegraph Signal as Internal Probe to Study Program/Erase Charge Lateral Spread in a SONOS Flash Memory

Y.L. Chou[†], J.P. Chiu[†], H.C. Ma[†], Tahui Wang[†,*], Y.P. Chao[*], K.C. Chen[*], and Chih-Yuan Lu[*]

[†]Dept. of Electronics Engineering, National Chiao-Tung University, Hsin-Chu, Taiwan

[*]Macronix International Co., No. 16, Li-Hsin Road, Science Based Industrial Park, Hsin-Chu, Taiwan

email:twang@cc.nctu.edu.tw

Abstract—A novel random telegraph signal (RTS) method is proposed to study the lateral spread of injected charges in program/erase of a NOR-type SONOS flash memory. The concept is to use RTS to extract an interface trap position and to detect a local potential variation near the trap due to injection of program/erase charges. By using this method, we find that CHISEL program has a broader charge distribution than CHE program. A mismatch of CHE program electrons and band-to-band erase holes is observed directly from this method.

I. INTRODUCTION

Two-bit/cell SONOS flash memory has been realized by storing bit charges in two sides of the channel by CHE program and band-to-band hot hole erase [1]. The control of program and erase charge lateral distributions of each bit is a major research thrust to improve cell endurance and scalability [2,3]. Attempts have been made in the past to characterize a trapped charge distribution in a SONOS cell. An inverse modeling approach is used to extract a program charge distribution from measured I-V characteristics [4]. Besides, a modified charge pumping (CP) technique [5] is employed to probe the lateral profile of programmed charges at the source and drain junctions separately. However, the inverse I-V modeling suffers from some limitations, for example, knowledge of precise device doping profile and lack of a unique solution. On the other side, the CP method is based on an assumption that interface traps have a uniform distribution along the channel [6], which is not correct in a buried diffusion bit-line SONOS cell. In addition, a charge pumping current is hardly sensed in a small area SONOS device.

In this work, we will use RTS arising from charge emission and capture at an oxide trap to investigate program/erase charge lateral spread. First, we determine the trap position from RTS without the need to know doping profile. Second, because RTS is very sensitive to a local potential change near the trap, we can use oxide traps as internal probes to detect a potential change due to program/erase charges. Finally, by using this technique, we find that CHISEL program has a broader charge distribution than CHE program and a mismatch of CHE program electrons and band-to-band erase holes is observed.

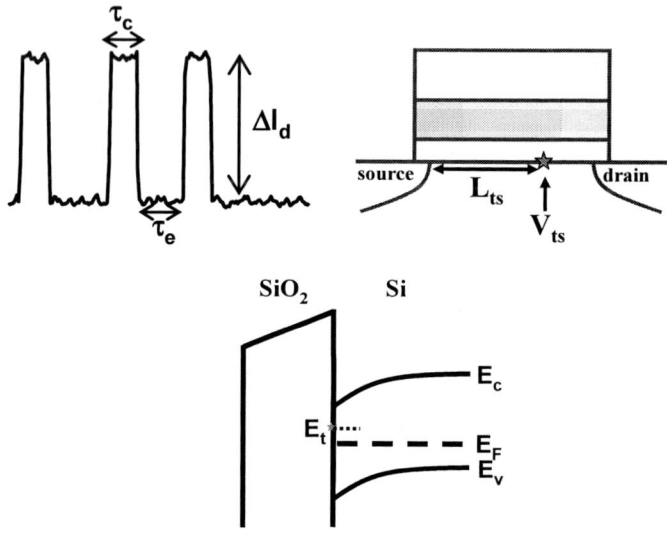

Fig. 1 Illustration of an interface trap induced RTS and the energy band diagram.

II. MEASUREMENT RESULT

Experiments were performed on SONOS flash cells with an ONO thickness of 8.5nm (top oxide), 7nm and 5.8nm, respectively. The cell size is W/L=0.11μm/0.1μm. The CHE program condition is V_{gs}=8V and V_{ds}=3.7V. The band-to-band hot hole (BTBH) erase is performed at V_{gs}= -4V and V_{ds}=5V.

A. Extraction of a Trap Position

An oxide trap position in the channel can be extracted in a way similar to the method in [7]. The RTS capture time τ_c, as illustrated in Fig. 1, can be expressed below,

$$<\tau_c>=1/n_e\sigma v_{th} \qquad (1)$$

where σ is the trap cross-section and v_{th} is the thermal velocity. The channel electron concentration n_e is a function of gate overdrive, i.e.V_{gs}-V_{ts}, where V_{ts} is the channel potential at the trap position and is equal to V_{ts}=(L_{ts}/L)V_{ds}. L_{ts} is the distance of the trap from the source and L is the channel length, as shown in Fig. 1.

Two different V_{ds} (=0.05V and 0.3V) are used in RTS and capture time (τ_c) measurement. Note that the device is in the linear region at the measurement biases. Since the capture

978-1-4244-5430-3/10 $26.00 © 2010 IEEE

time is dependent on an electron concentration near the trap, or in other words, a voltage drop between the gate (V_{gs}) and the channel right below the trap (V_{ts}), the amount of the lateral shift of these two curves (ΔV_{ts}) in Fig. 2 is equal to the difference of the voltages, raised by the two drain voltages, at the point (L_{ts}) of the trap. Therefore, the trap position in the channel can be extracted from $\Delta V_{ts}/\Delta V_{ds}=L_{ts}/L$. In this work, the RTS extraction is conducted in more than 30 un-cycled devices. For simplicity, we only record devices with two-level RTS (i.e., a single trap). The extracted trap position distribution is shown in Fig. 3. With this information, we can choose devices with appropriate trap positions as internal probes to investigate program/erase charge lateral spread.

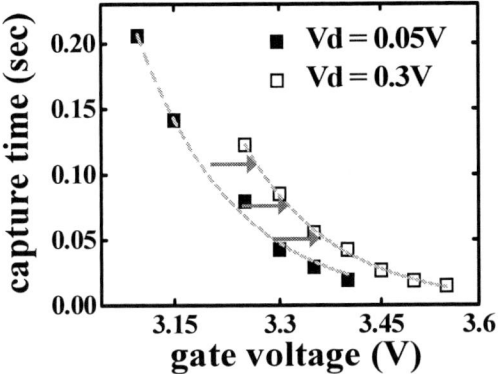

Fig. 2 The gate voltage dependence of average capture time in RTS at two drain voltages, $V_{ds}=0.05V$ and $0.3V$. The lateral shift of these two curves corresponds to ΔV_{ts}.

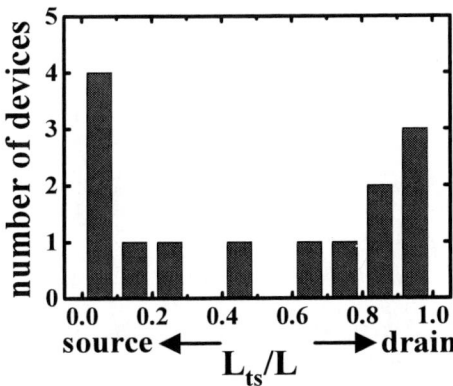

Fig. 3 Cumulative trap position distribution along the channel. L is the channel length and L_{ts} is the distance of a trap to the source.

B. Detection of a local potential change

The ratio of τ_c/τ_e is dependent on the local potential at the trap position, i.e.

$$<\tau_c>/<\tau_e>=g\exp[(E_t-E_F)/kT]\sim\exp(-q\Delta\varphi_s/kT) \qquad (2)$$

where g is a pre-factor, E_t is the trap energy and $\Delta\varphi_s$ is a local potential change at the trap position.

C. CHE and CHISEL Programming

In order to detect a surface potential change near the drain during CHE program, a SONOS cell having a trap at $x_t=0.2L$ from the drain edge is used. Fig. 4 shows the average emission

time (τ_e) and capture time (τ_c) versus program ΔV_T [8,9]. Fig. 5(a) illustrates the band diagrams during CHE program. The trap energy level increases with respect to the Fermi level as program charge increases. The ratio of (τ_c/τ_e) and the corresponding surface potential change ($\Delta\varphi_s$) from Eq. (2) are plotted in Fig. 5(b). As more electrons are injected into the nitride layer, the conduction band at x_t and the trap level move upward away from the Fermi level. Thus, the τ_c/τ_e ratio becomes larger and larger.

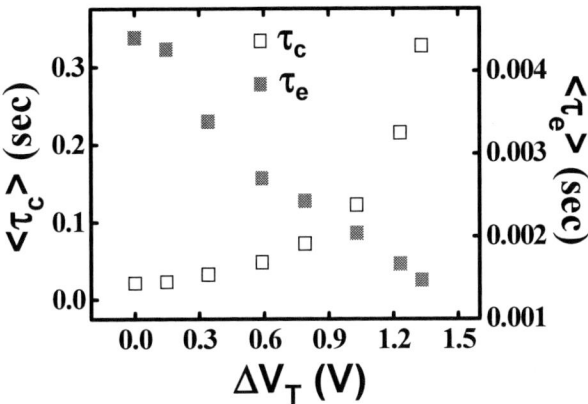

Fig. 4 Average capture time (τ_c) and emission time (τ_e) versus program ΔV_T.

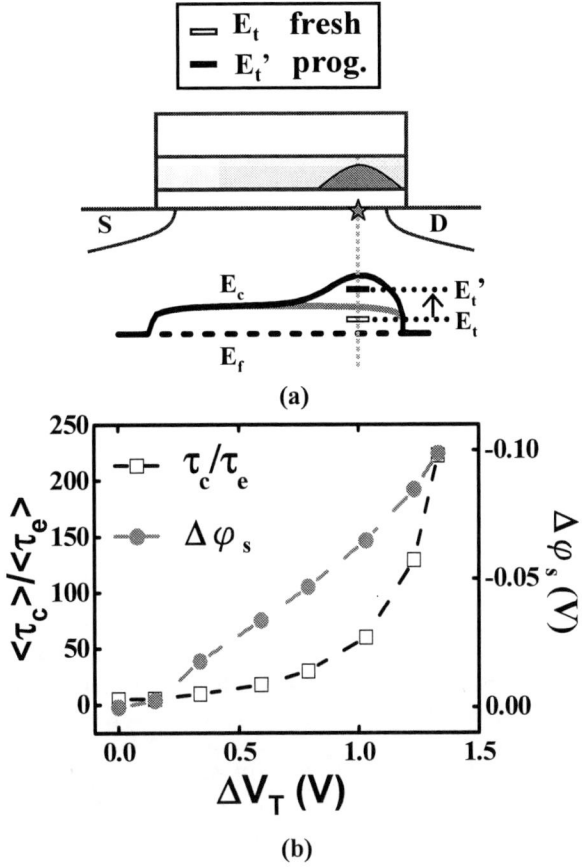

Fig. 5 (a)Band diagrams before and after CHE program. (b) $<\tau_c>/<\tau_e>$ and surface potential change versus program ΔV_T. The local potential change ($\Delta\varphi_s$) at the trap position is calculated from Eq. (2).

Fig. 6 compares the τ_c/τ_e in source side programming and drain side programming. The τ_c/τ_e remains unchanged during source-side programming, as expected, since program charges are near the source junction while the trap is near the drain edge. We also compare the program charge spread by CHE and CHISEL [10] in Fig. 7. The substrate bias is -2V in CHISEL operation. During CHISEL program, holes generated by channel electron impact ionization flow to the substrate and result in secondary impact ionization. The secondary electrons, accordingly, would be accelerated by the drain voltage and inject into the nitride layer. Fig. 7 shows that CHISEL has a broader injected charge distribution than CHE because the τ_c/τ_e and the potential change are larger at the same program ΔV_T.

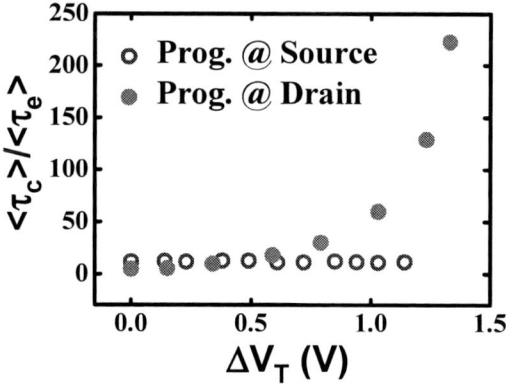

Fig. 6 Comparison of $<\tau_c>/<\tau_e>$ evolutions in source-side CHE program and in drain-side CHE program. Since the trap is near the drain, $<\tau_c>/<\tau_e>$ remains unchanged in source-side programming.

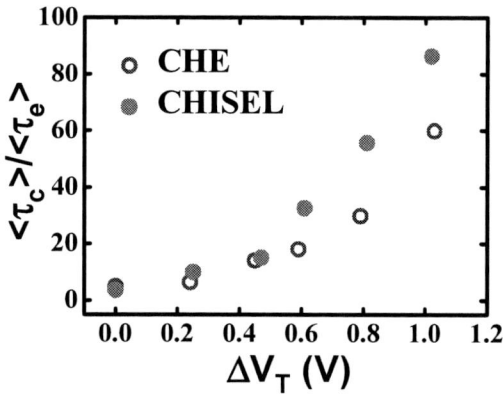

Fig. 7 Comparison of $<\tau_c>/<\tau_e>$ evolutions by CHE program and by CHISEL program. A substrate bias of -2V is applied in CHISEL injection.

D. BTBH Erase and Program/Erase Charge Mismatch

For comparison, we choose two devices with a respective trap position at 0.05L and 0.3L from the drain. CHE program and BTBH erase are performed. The τ_c/τ_e evolutions during program/erase are shown in Fig. 8 (in the x_t=0.05L cell) and in Fig. 9 (x_t=0.3L cell). The τ_c/τ_e increases as program V_T increases by CHE program and decreases by hot hole erase. For the point near the drain (i.e., x_t=0.05L), the τ_c/τ_e curves in program and in erase match very well, suggesting that program electrons at 0.05L are totally neutralized by erase holes. In contrast, at the point of x_t=0.3L from the drain, the τ_c/τ_e does not return to its original value after a P/E cycle. The larger τ_c

/τ_e value during erase implies that program electrons (at 0.3L) are not completely compensated although the cell has been erased to its original V_T.

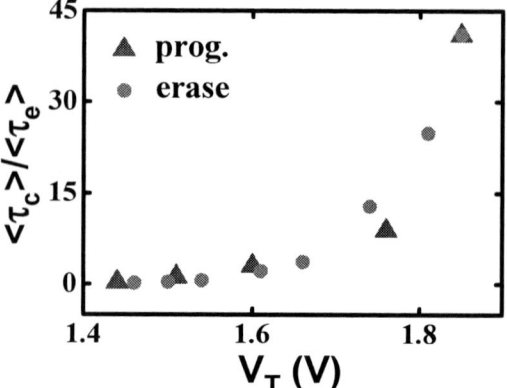

Fig. 8 The $<\tau_c>/<\tau_e>$ evolutions during CHE program and BTBH erase. The device has a trap at 0.05L from the drain. The $<\tau_c>/<\tau_e>$ has the same path in P/E.

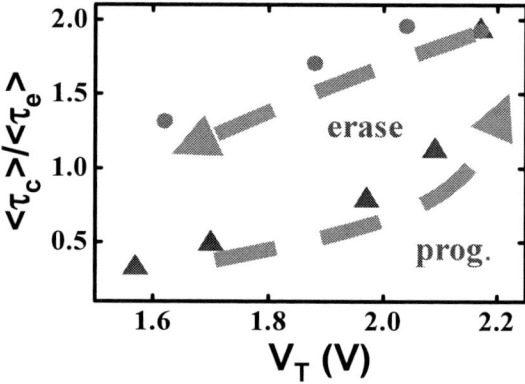

Fig. 9 The $<\tau_c>/<\tau_e>$ evolutions during CHE program and BTBH erase. The device has a trap at 0.3L from the drain. The erase path of $<\tau_c>/<\tau_e>$ is different from the program path.

Fig. 10 illustrates the program/erase charge distributions and the band diagrams after program and erase. The trap position is at x_t=0.05L from the drain in Fig. 10(a), and 0.3L from the drain in Fig. 10(b). In Fig. 10(a), the trap level increases with respect to the Fermi level due to program electron injection and decreases due to erase hole injection. In Fig. 10(b), the trap level increases due to program electrons, but is not affected by injected erase holes. The result in Fig. 9 implies that injected holes have a narrower distribution than program electrons.

III. CONCLUSION

We propose a novel RTS method to characterize program and erase charge lateral spread in a SONOS flash memory without the need to know a doping profile. Since the RTS method is very sensitive to a local potential change due to program/erase charges, it can provide a better resolution than a charge pumping method or an inverse I-V modeling approach. An evidence of a mismatch between program electrons and erase holes is shown by this method.

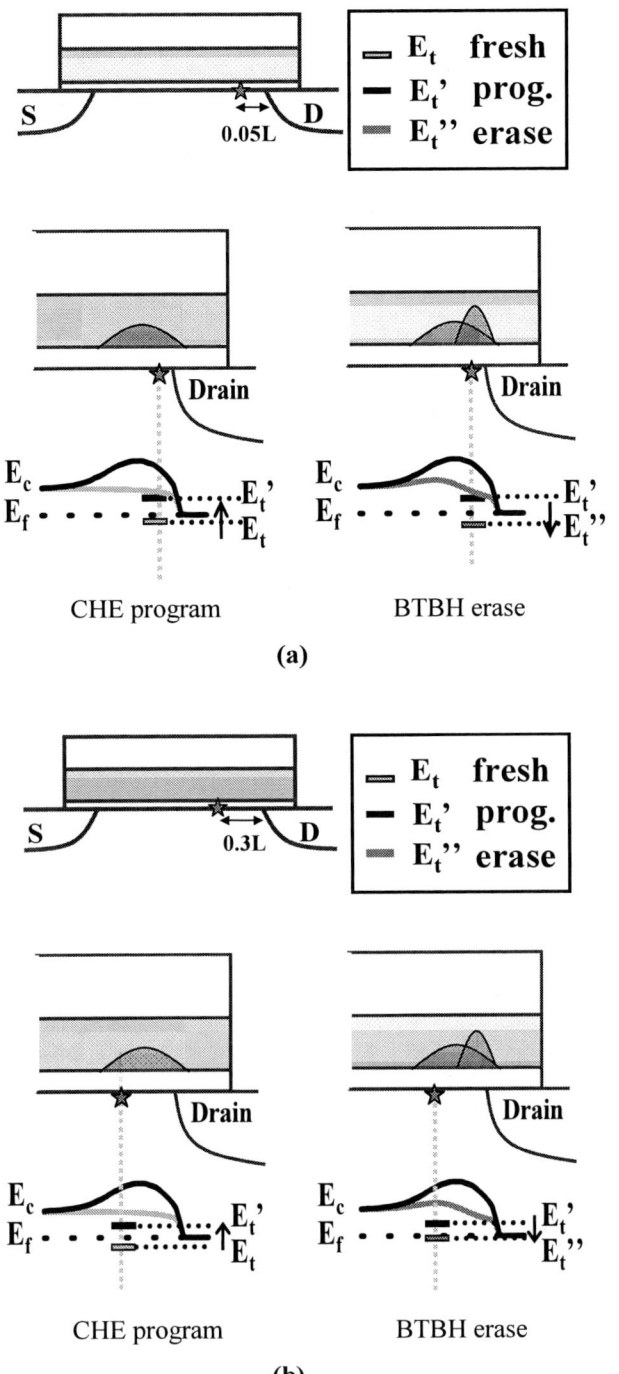

Fig. 10 Illustration of program/erase charge distributions and the band-diagrams after program and erase. The star represents interface trap position at (a) x_t=0.05L and (b) x_t=0.3L from the drain edge. The program electrons at x_t=0.05L are completely compensated, but some far electrons at x_t=0.3L are not compensated after erase.

ACKNOWLEDGMENT

The authors (Y.L. Chou, J.P. Chiu, H.C. Ma and Tahui Wang) would like to acknowledge financial support from National Science Council under contact #NSC96-2628-E009-165

REFERENCES

[1] Boaz Eitan, Paolo Pavan, Ilan Bloom, Efraim Aloni, Aviv Frommer, and David Finzi, "NROM: A Novel Localized Trapping, 2-Bit Nonvolatile Memory Cell," *IEEE Trans. Electron Devices*, vol. 21, pp. 543–545, Nov. 2000.

[2] Arnaud Furnemont, Maarten Rosmeulen, Koen van der Zanden, Jan Van Houdt, Kristin De Meyer, and Herman Maes, "Physical Modeling of Retention in Localized Trapping Nitride Memory Devices" *IEDM Tech. Dig.*, pp.397–400, 2006

[3] Yao-Wen Chang, Tao-Cheng Lu, Sam Pan, and Chih-Yuan Lu, "Modeling for the 2nd-Bit Effect of a Nitride-Based Trapping Storage Flash EEPROM Cell Under Two-Bit Operation," *IEEE Electron Device Lett.*, vol. 25, pp. 95–97, Feb. 2004.

[4] Luca Larcher, Giovanni Verzellesi, Paolo Pavan, E. Lusky, Ilan Bloom, and Boaz Eitan, "Impact of Programming Charge Distribution on Threshold Voltage and Subthreshold Slope of NROM Memory Cells," *IEEE Tran. Electron Devices*, vol. 49, pp. 1939–1946, Nov. 2002

[5] Shaw-Hung Gu, Tahui Wang, Wen-Pin Lu, Wenchi Ting, Yen-Hui Joseph Ku, and Chih-Yuan Lu, "Characterization of Programmed Charge Lateral Distribution in a Two-Bit Storage Nitride Flash Memory Cell by Using a Charge-Pumping Technique," *IEEE Tran. Electron Devices*, vol. 53, pp. 103–108, Jan. 2006

[6] Chun Chen, and Tso-Ping Ma, "Direct Lateral Profiling of Hot-Carrier-Induced Oxide Charge and Interface Traps in Thin Gate MOSFET's," *IEEE Tran. Electron Devices*, vol. 45, pp. 512–520, Feb. 1998

[7] Phillip Restle, "Individual oxide traps as probes into submicron devices," *Appl. Phys. Lett.*, 53, pp. 1862–1864, Nov. 1988

[8] K. Kandiah, M. O. Deighton, and F. B. Whiting, "A Physical Model for Random Telegraph Signal Currents in Semiconductor devices," *J. Appl. Phys.*, vol.66, pp. 937–948, July 1989

[9] K. Kandiah, "Random Telegraph Signal Currents and Low-Frequency Noise in Junction Field Effect Transistors," *IEEE Tran. Electron Devices*, vol. 41, pp. 2006–2015, Nov. 1994

[10] Jeff D. Bude, Mark R. Pinto, and R. Kent Smith, "Monte Carlo Simulation of the CHISEL Flash Memory Cell," *IEEE Tran. Electron Devices*, vol. 47, pp. 1873–1881, Oct. 2000

BIAS TEMPERATURE INSTABILITY OF BINARY OXIDE BASED RERAM

Z. Fang[1], #H.Y. Yu[1], W. J. Liu[1], K.L. Pey[1], X. Li[1], L. Wu[1], Z.R. Wang[1], Patrick G.Q. Lo[2], B.Gao[3], J. F. Kang[3]

[1]School of EEE, Nanyang Technological University, Singapore 639798; #E-Mail: hyyu@ntu.edu.sg

[2]Institute of Microelectronics /A*STAR, Singapore; [3]Institute of Microelectronics, Peking University, China.

ABSTRACT

Bias temperature instability of TiN/HfO$_x$/Pt resistive random access memory (ReRAM) device is investigated in this work for the first time. As temperature increases (up to 100°C in this work), it is observed that: 1) leakage current at high resistance state (HRS) increases, which can be explained by the higher density of traps inside dielectrics (related to trap-assistant tunneling), leading to a lower On/Off ratio; 2) set and reset voltages decrease, which may be attributed to the higher oxygen ion mobility, in addition to the reduced potential barrier to create / recover oxygen ions (or oxygen vacancies); 3) electrical stress plays a negligible role as compared to the temperature impact on the ReRAM degradation; and 4) multi-level switching behavior exhibited at room temperature might not be retained.

Indexed terms: Bias temperature instability, Resistance random access memory

INTRODUCTION

As conventional flash memory scaling is expected to face technical and physical limitations beyond sub-30 nm node, it is necessary to identify replacement emerging memory devices with good scalability, high speed, high density, low power as well as CMOS compatibility. Resistive random access memory (ReRAM) is considered as one of the most promising candidates in this regard [1]. Transition metal oxide (TMO) has attracted great attention as ReRAM candidate materials largely owing to their simple composition. Various TMO, including NiO$_x$ [2], TiO$_x$ [3], CuO$_x$ [4] have been reported for resistive switching behavior, and HfO$_x$ has shown exceptional electrical performance [5]. In this work, bias temperature instability of the resistive switching behavior is investigated for the first time on Ti/HfO$_x$/Pt ReRAM, and it was found temperature effect is more important on the degradation of ReRAM as compared to the electrical stress. All the observations at high temperature can be correlated with the oxygen vacancies (V_o) inside binary oxide (HfO$_x$ in this work) dielectrics, which further reinforce the role of V_o assisted conduction filament formation and rupture in the ReRAM switching behavior.

EXPERIMENT

Hafnium oxide based ReRAM devices are fabricated by high temperature oxidation of Hf thin films. First, Hf film is prepared by DC sputtering of Hf target on Pt/Ti/SiO$_2$/Si substrates, follows by 450°C oxygen furnace annealing for 10 minutes. After that, TiN top electrode is deposited by reactive sputtering and then devices are patterned with lithography and dry etched with final devices area ranging from 5625 to 99225 μm^2. The morphologies of the MIM capacitor are examined by high-resolution transmission electron microscopy (HRTEM) and chemical composition of the blank HfO$_x$ film is analyzed by X-ray photoelectron spectroscope (XPS).

RESULTS AND DISCUSSION

Fig.1(a) shows the XPS Hf 4f binding energy (BE) peak of 17.1eV and 18.6eV, indicating the HfO$_x$ film prepared in this work is fully oxidized prior to TiN metal deposition, consistent with the XPS composition study. Fig.1(b) shows the XTEM of the TiN/HfO$_x$/Pt MIM capacitor with HfO$_x$ thickness of ~8nm, revealing an interfacial layer (IL) of TiO$_x$N$_y$ between HfO$_x$ and top TiN electrode, which is further confirmed by energy dispersive X-ray spectroscopy (EDX) profile (Fig.1c). The formation of TiO$_x$N$_y$ layer can be attributed to that Ti acts as an oxygen gathering layer, indicating that HfO$_x$ layer in the device possibly becomes oxygen deficient. The IL might serve as oxygen reservoir during the resistive switching.

Fig.2 shows the well-behaved 100 repetitive DC sweeps current-voltage (I-V) characteristics of the HfO$_x$ MIM capacitor after a forming process at 25°C. The current increases suddenly at the set voltage (V_{set}) of ~ +1.4V during positive sweep, which shows it turns from high resistance state (HRS) to low resistance state (LRS). The LRS reserves till with a negative voltage sweep to the reset voltage (V_{reset}) of ~ -2.5V, and HRS will resume. Fig.3 demonstrates good endurance of the fabricated HfO$_x$ ReRAM devices (DC sweeps). The resistive switching phenomenon can be well explained by formation and rupture of oxygen vacancies related conduction filaments, as schematically depicted in Fig.4 [6].

The TiN/HfO$_x$/Pt device resistive switching at room temperature and high temperature (100°C) is compared in Fig.5. It is clearly observed the HRS current increases under both polarities with temperature, which can be explained by the schematics as inset of this figure. When device is reset to HRS, the conduction filament is believed to be broken near metal interface, cutting off the current. With the increase of temperature, V_o–related traps inside the dielectrics would be increased according to crystal defect theory due to effectively lowered energy barrier at high temperature, as shown in Fig.6. The increased V_o-related trap density would thus increase the trap-assist tunneling (TAT) leakage. As a result, a smaller On/Off ratio would also be observed, as illustrated in Fig.7. Note that LRS leakage doesn't vary with temperature due to the conduction filament formation. On the other hand, from Fig.7, it is worthy mentioning the retention at high temperature is still acceptable despite of higher leakage. In Fig.8, V_{set} / V_{reset} decrease with increasing temperature, which is correlated with the increased oxygen ion mobility [7], in addition to reduced potential barrier to create/recover oxygen ions (or V_o, see Fig.6).

Electrical stress of 0.1V to 0.8V at both room temperature and high temperature on the reliability of ReRAM devices is studied in Fig. 9. The HRS resistance variation after stress of different period ($|\Delta R/R_{t=0}|$, where $R_{t=0}$ stands for the resistance prior to stressing) is used to evaluate the degradation of ReRAM. It is seen that the variation at room temperature is much smaller compared to high temperature case for all the stressing conditions. The resistance variation at high temperature is less than 50% and is almost independent of stressing condition. This variation will not degrade the reliability (retention) of ReRAM, as shown in Fig.7. The possible reason is due to the small stressing voltage (which must be smaller than V_{set}).

Fig.10 shows that HfO$_x$ based ReRAM exhibits multi-level switching capability at room temperature. However, at high temperature, due to high leakage of HRS, the multi-level switching can not be retained, as shown in Fig.11. The temperature dependant switching properties might pose a challenge on commercialization of the binary oxide based ReRAM devices. It is believed that materials solution should be actively sought to address this concern by controlling the filament formation/rupture, and by improving the materials thermal stability.

CONCLUSIONS

In summary, bias temperature instability of TiN/HfO$_x$/Pt ReRAM is studied for the first time. It was found temperature effect plays a more significant role on the degradation of ReRAM as compared to the electrical stress. As temperature increases, HRS leakage current increases, and set and reset voltages decrease, which are correlated to the higher V_o related traps concentration and/or higher oxygen ion mobility. In addition, multi-level switching behavior exhibited at room temperature might not be reserved at high temperature, which poses a challenge in order to bring the binary oxide-based ReRAM devices into market.

REFERENCES

[1] I. G. Baek et al., IEDM, p.750, 2005. [2] D. C. Kim et al., APL, vol. 88, p.202102, 2008. [3] D. B. Strukov et al., Nature, vol. 453,p.80-83, 2008. [4] R. Dong et al., APL, vol. 90, p.042107, 2007.[5] H. Y. Lee et al., IEDM, p.297-300, 2008 [6] N. Xu et al., VLSI, p.100-101, 2008.[7] R. Meyer et al. NVMTS, p.1-5, 2008. [8] B. Gao et al., IEDM, p.563-566, 2008.

978-1-4244-5430-3/10 $26.00 © 2010 IEEE

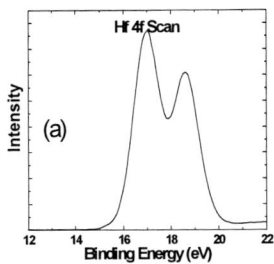 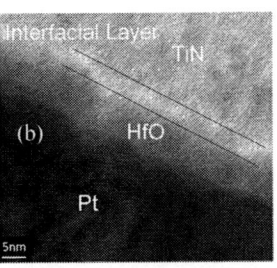 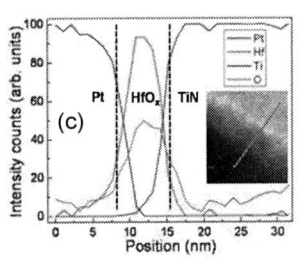 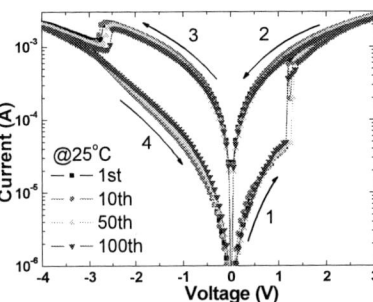

Fig.1 (a) XPS Hf 4*f* scan of HfO$_x$ blank film, binding energy (BE) peaks at 17.1eV and 18.6 eV show the as prepared Hf thin film has been fully oxidized. The BE was calibrated with C 1*s*. Physical characterization of HfO$_x$ MIM capacitor (b) XTEM image of the TiN/HfO$_x$/Pt memory cell, an interfacial layer of TiO$_x$N$_y$ is identified; (c) EDX study of elemental spatial profile (inset shows the direction of profile): it is observed that the oxygen profile is attracted to the TiN electrode side.

Fig. 2 The switching characteristics of 100 repetitive DC sweeping current-voltage of the HfO$_x$ based ReRAM after a forming process at 25°C. During positive sweep (1), device is set to LRS (2) and with negative sweep (3), device is reset to HRS (4).

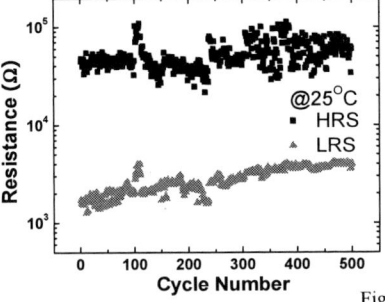 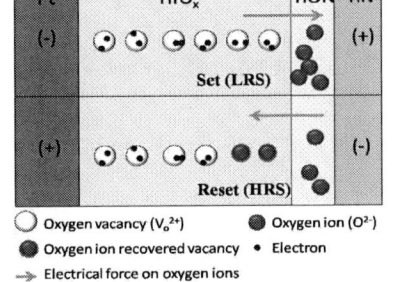 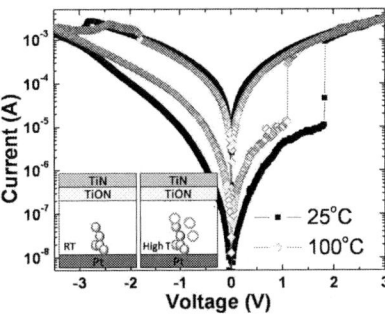

Fig. 3 Switching endurance of 500 cycles in DC sweeping mode at room temperature.

Fig. 4 The schematics of set and reset processes. During set, oxygen ions move from its lattice by applied electric field and form the conduction path with oxygen vacancies. During reset, oxygen ions move back to recover with its lattice vacancies and then conduction path disappears.

Fig. 5 Comparison of resistive switching at both 25°C and 100°C. (inset shows schematic high temperature HRS leakage current (TAT) increases). At high temperature, trap concentration inside dielectric increases, which gives rise to TAT current. Therefore, the On/Off ratio decreases at high temperature.

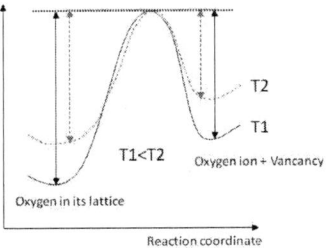 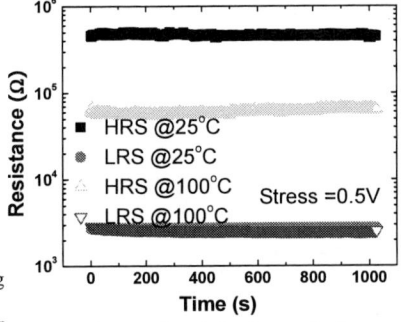 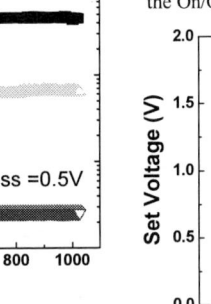

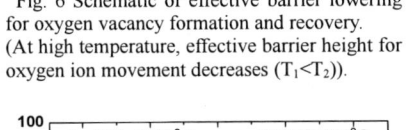

Fig. 6 Schematic of effective barrier lowering for oxygen vacancy formation and recovery. (At high temperature, effective barrier height for oxygen ion movement decreases (T$_1$<T$_2$)).

Fig. 7 Retention behavior at both room temperature and high temperature with stress of 0.5V.

Fig. 8 Set (V$_{set}$) and reset (V$_{reset}$) voltages decrease with increasing temperature. Data are averaged from 10 cycles of DC sweeps at each temperature.

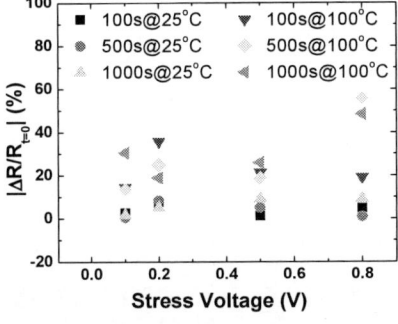 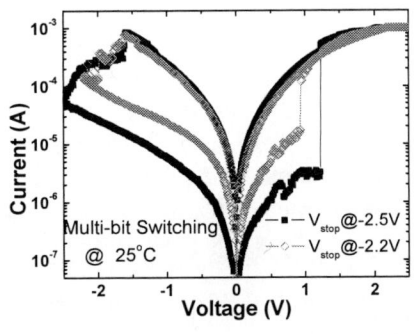 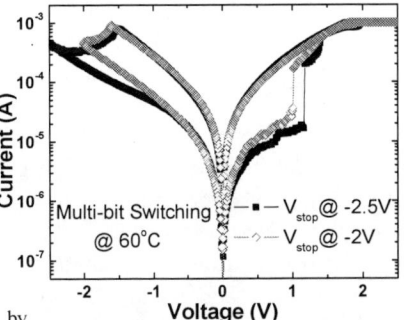

Fig. 9 Resistance variation ($|\Delta R/R_{t=0}|$) of HRS at different stress conditions and temperatures. It shows $|\Delta R/R_{t=0}|$ (original resistance is defined as the resistance prior to stressing) does not vary much, which is negligible as compared with memory window of ~100x.

Fig.10 Multi-bit switching at room temperature by controlling V$_{stop}$. Different reset stop voltages can reset the device to different HRS resistance values, indicating that some oxygen ions need higher energy to recover with vacancies.

Fig.11 Multi-bit switching disappears at high temperature. This may be due to higher leakage current at high temperature, which is equivalent to a lowered HRS resistance.

978-1-4244-5430-3/10 $26.00 © 2010 IEEE

Reliability Constraints for TANOS Memories due to Alumina Trapping and Leakage

Salvatore M. Amoroso*[†], Aurelio Mauri[†], Nadia Galbiati[†], Claudia Scozzari[†], Evelyne Mascellino*, Elisa Camozzi[†], Armando Rangoni[†], Tecla Ghilardi[†], Alessandro Grossi[†], Paolo Tessariol[†], Christian Monzio Compagnoni*, Alessandro Maconi*, Andrea L. Lacaita*[‡], Alessandro S. Spinelli*[‡], Gabriella Ghidini[†]

* Dipartimento di Elettronica e Informazione, Politecnico di Milano–IU.NET,
piazza L. da Vinci 32, 20133 Milano, Italy, e-mail: amoroso@elet.polimi.it
[†] Numonyx, R&D Technology Development via C. Olivetti 2, 20041 Agrate Brianza (MI), Italy,
e-mail:salvatore.amoroso@numonyx.com
[‡] IFN-CNR, Milano, Italy

Abstract—**In this work we present a detailed investigation of TANOS memory reliability, focusing on issues raised by Al_2O_3 trapping/detrapping and leakage. These effects are investigated as a function of alumina thickness, electric field and temperature, comparing experimental and modeling results for trap parameters extraction. For TANOS devices, Al_2O_3 charge storage modifies program and erase saturation level particularly when higher Al_2O_3 thikness are considered. Threshold instability in early steps for endurance and retarded behavior for retention can be also ascribed to the Al_2O_3 trapping. Moreover, Al_2O_3 layer has been shown to provide the main leakage path for bottom oxides thickness in the 4.5 nm or above range.**

Index Terms—**Flash memories, charge-trapping memories, high-k dielectrics, semiconductor device modeling.**

I. INTRODUCTION

The TANOS stack is today considered a viable solution to overcome the erase–data retention trade-off of SONOS devices [1], and has been successfully integrated within non-volatile NAND arrays [2], [3]. One of the key concepts is the use of a high-k dielectric (usually Al_2O_3) on top of the nitride trapping layer, to both increase the coupling efficiency between the trapping layer and the metal gate and reduce the electron injection from the gate during erase. Notwithstanding the clear benefits in terms of program/erase (P/E) speed and threshold voltage window brought by the introduction of the alumina layer, its properties have seldom been studied, with most of the research activity being focused on the nitride layer. In particular, the alumina constraints on TANOS reliability only received very few attention, though the works in [4], [5] had already highlighted the degradation of the Al_2O_3 layer during electrical stress and its effects on floating-gate memory when employed as interpoly dielectric.

Aim of this work is to correlate the Al_2O_3 trapping and leakage properties with the transient behavior of the TANOS stack, namely program/erase and retention. The paper is organized as follows: Sec. II presents the experimental characterization of TAOS devices, whose results are then compared with those coming from a model for alumina trapping and detrapping developed in Sec. III. Finally, Sec. IV discusses

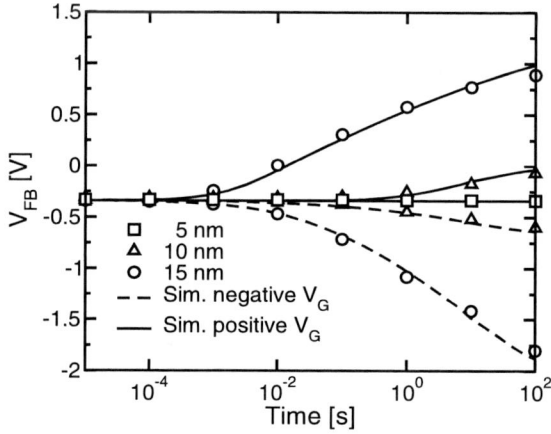

Fig. 1. V_{FB} transients for different TAOS capacitors during positive and negative V_G stress at the same field and $T = 25^oC$.

the implications of the Al_2O_3 trapping dynamics on TANOS performances.

II. EXPERIMENTAL RESULTS

Alumina has been characterized on large-area capacitors with TaN metal gate and Al_2O_3 directly grown by ALD on a 1 nm chemical SiO_2 layer on a p-type substrate. An n-type ring provides the minority carriers for inversion. A 1100^oC inert anneal was used for alumina crystallization. Fig. 1 shows the V_{FB} evolution during constant-voltage stress, applying to all samples the same initial electric field $F_{Al_2O_3} = 4.6$ and 4.9 MV cm^{-1} for positive and negative stress, respectively. A large shift in V_{FB} (more than 1 V) can be observed in Fig. 1 for the 15 nm stack, as reported also in [6], while the V_{FB} shift decreases when reducing $t_{Al_2O_3}$ until no variation appears in the 5 nm case up to 100 s stress time. This reveals an increase of the bulk trap density as the Al_2O_3 thickness increases beyond 5 nm, which could be ascribed to oxygen-vacancy creation in the alumina layer.

The behavior of the 15 nm sample is further investigated in

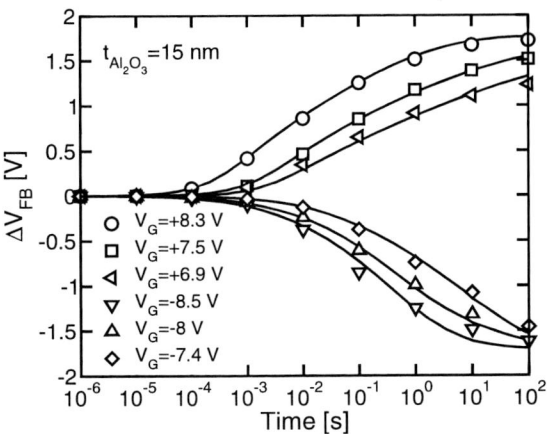

Fig. 2. ΔV_{FB} transients on a TAOS capacitor subjected to positive and negative stresses at different V_G at $T = 25^{o}$C.

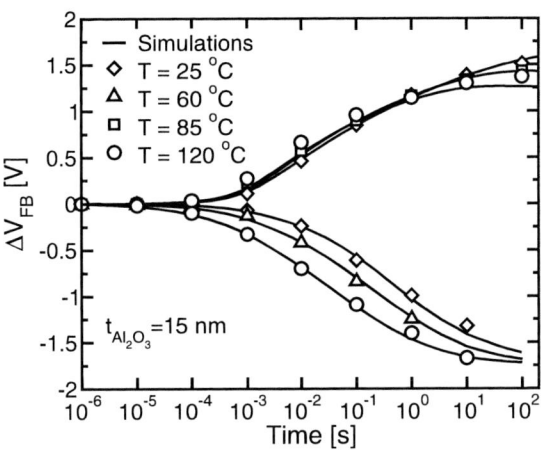

Fig. 3. Effect of temperature on the ΔV_{FB} transients of TAOS capacitors for positive and negative V_G stress.

Fig. 2, showing the room temperature V_{FB} variation (ΔV_{FB}) during positive and negative stress of 100 s at different gate voltages V_G, starting from the same $V_{FB} \approx -0.1$ V. Note that different behaviors can be observed for the two polarities: for positive stress, a saturation level is neatly reached only at the highest electric field, whereas for lower values a saturation is not reached within the observation time of 100 s. In the case of negative stress, instead, the ΔV_{FB} curves seem to converge toward the same level, independently of the electric field strength. A different behavior for program and erase can also be seen in Fig. 3, where the ΔV_{FB} transients are shown as a function of temperature. A negligible variation with temperature can be seen for positive V_G, where only the saturation value changes slightly, while results for the negative stress polarity feature a much stronger dependence on temperature. These results will be discussed in the next Section, where a numerical model for the charge trapping in the alumina layer is developed.

III. MODELING RESULTS

Figs. 1 and 2 make clear that both positive and negative charges can be stored in the Al_2O_3 layer. While negative charges are usually interpreted as electrons being trapped into acceptorlike traps, positive charges could arise from both hole injection from the substrate and trapping into acceptorlike traps or from electron detrapping from donorlike trap sites. To discriminate between these phenomena, we have resorted to numerical simulations, modifying our numerical tool [8] for studying the transient charge trapping and detrapping in nitride layers to also account for alumina. The final model includes Poisson and electron current continuity equations, dealt within a drift-diffusion framework. Capture/emission processes of carriers from traps in the Al_2O_3 layer are accounted for, including both thermal and tunneling-assisted phenomena.

Numerical simulations first confirmed that hole injection plays a negligible role during erase due to the high injection barrier and long injection distance, and has thus been neglected. To reproduce both program and erase transients,

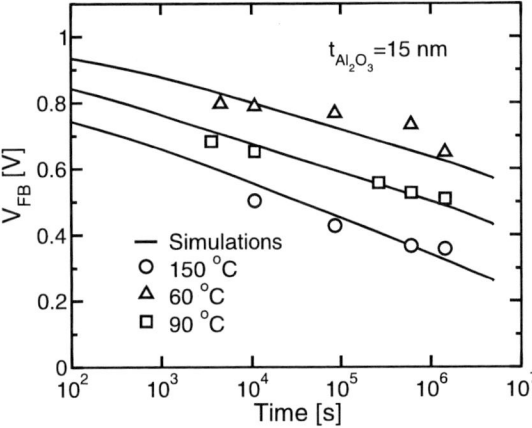

Fig. 4. V_{FB} recovery during retention for TAOS capacitors after program.

we considered both acceptorlike (N_{TA}) and donorlike (N_{TD}) traps in the alumina, having densities $N_{TA} = 1.5 \times 10^{19}$ and 5×10^{18} cm^{-3} for the 15 and 10 nm sample, respectively, with a negligible density in the 5 nm case [7]. The donorlike traps were assumed to be located in correspondence of the interface between the alumina and the 1 nm SiO_2, with density $N_{TD} = 4$ and 6×10^{12} cm^{-2} for the 10 and 15 nm respectively, energy $E_T = 1.8$ eV below the alumina conduction band and electron capture cross-section $\sigma = 1.5 \times 10^{-14}$ cm^2.

Simulation results are shown in Figs. 1-3 (solid lines), featuring a good agreement with data. In particular, the program and erase variations with temperature shown in Fig. 3 are correctly accounted for. The difference in the behavior is due to the strong dependence on temperature of the electron emission from traps, as opposed to a program transient which is ruled by electron capture. The latter is not dependent on temperature except for the final part of the transient, where a balance between capture and emission is reached. This explains the slightly-different program saturation levels found with increasing temperature as well as the stronger dependence of erase, and provides further support to the existence of

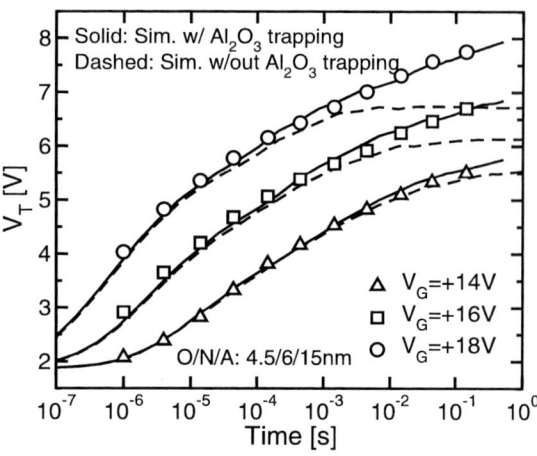

Fig. 5. Programming transients for TANOS devices. Simulation results with and without alumina trapping are reported.

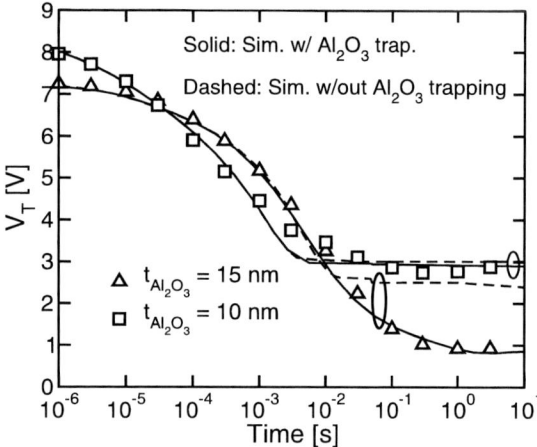

Fig. 6. Effect of alumina trapping on erase for TANOS devices with different $t_{Al_2O_3}$ and O/N. Simulation results with and without alumina trapping are also reported.

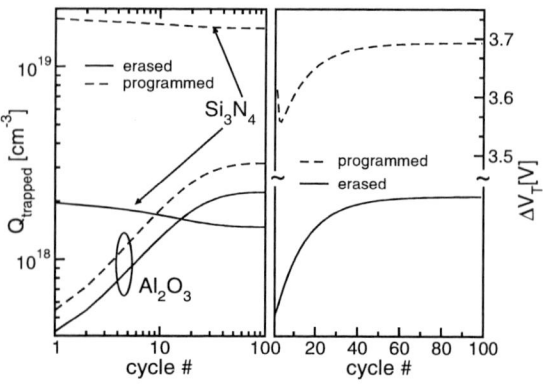

Fig. 7. Simulated endurance of TANOS devices including Al_2O_3 trapping: trapped charge (left) and V_T (right) vs. cycles.

donorlike traps, as opposed to hole injection and capture.

Fig. 4 shows results of a retention experiment on the 15 nm sample, starting from the programmed level. Note the temperature acceleration of the charge loss, which is consistent with electron detrapping from bulk traps with the assumed energy level of 1.8 eV. A reasonable agreement between simulations and data can be seen for the long-term loss. The behavior for shorter times may be instead ruled by fast detrapping of interface charge, which requires a more accurate investigation.

IV. IMPACT ON TANOS RELIABILITY

To directly evaluate the impact of the previous effects on TANOS device performances, we have characterized TANOS cells with 4.5 nm bottom oxide, 6 nm nitride, and different Al_2O_3 layer thicknesses ($t_{Al_2O_3}$). To account for the trapping phenomena in alumina, the numerical model for TANOS simulations presented in [8] was extended to include the trapping/detrapping dynamics for this layer, too.

Fig. 5 shows a comparison between experimental data and simulations for the programming transients of the sample with $t_{Al_2O_3} = 15$ nm. Simulation results are shown with and without alumina trapping. Note that trapping in the Al_2O_3 allows to explain the non-saturating behavior of the programming transients, which is typically observed in TANOS stacks (as opposed to SONOS devices): when alumina trapping is included, V_T continues to increase for large programming times, while simulations predict a saturation in V_T, in particular when high V_G is applied, when alumina trapping is neglected. The previously-extracted parameters were used for the alumina, with the exception of the donorlike traps, which have been here neglected as a consequence of the different interface (alumina on nitride rather than over oxide). For the nitride, we have used the following parameters: $N_{TA} = 6 \times 10^{19}$ cm^{-3}, $E_T = -1.5$ eV and $\sigma = 5.5 \times 10^{-15}$ cm^2. Fig. 6 reveals that Al_2O_3 trapping should also be invoked in the erase operation, in particular for explaining the position of the saturation level. Given the trapping characteristics shown in Fig. 1, this effect is stronger for thicker Al_2O_3 layers.

We have also investigated the impact of the Al_2O_3 trapping properties on endurance and retention of TANOS devices. Simulation results for the endurance have been obtained by allowing the final state of a program or erase operation to become the initial one of the following, and are shown in Fig. 7 for the trapped charge and V_T. Note that alumina and nitride seems to show different behaviors as a function of the cycle number, with a significant charge buildup within the alumina which saturates after a few P/E cycles as a result of the equilibrium between trapping and emission. However, due to the computational time required, this procedure can only be used for a few tens of cycles.

Fig. 8 shows that misleading results can be obtained for data retention if Al_2O_3 trapping is neglected: deep traps in the alumina lead to a slower charge-loss transient, in better agreement with data, which cannot be reproduced by the model if alumina is considered an ideal dielectric and all the V_T shift is ascribed to charge trapped in the nitride. Note that the effect is stronger for large initial ΔV_T (see also Fig. 5). Data retention of TANOS devices at 60°C is also investigated in Fig. 9 for different $t_{Al_2O_3}$. Devices were first programmed to

978-1-4244-5430-3/10 $26.00 © 2010 IEEE

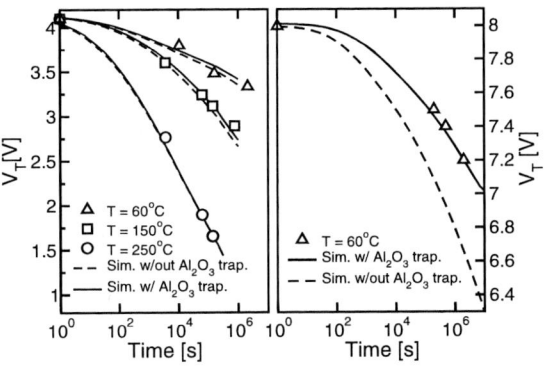

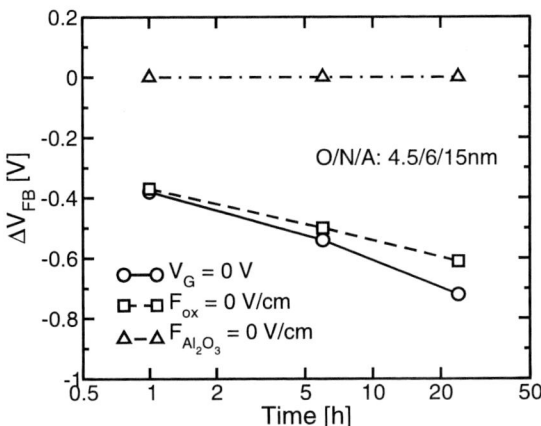

Fig. 8. Retention transients for TANOS devices at different temperatures (left) and program level (right). Simulation results with or without alumina trapping are also included.

Fig. 10. Experimental results for data retention of TANOS devices under different applied V_G.

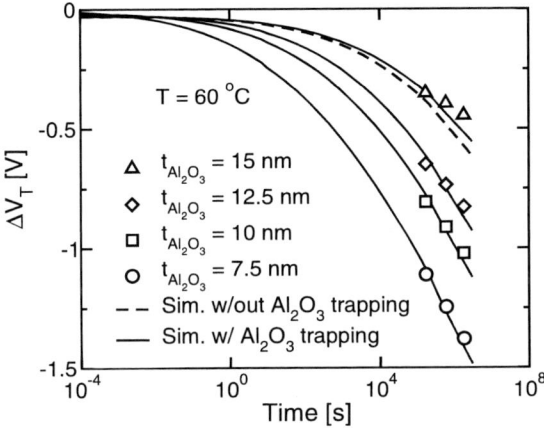

Fig. 9. Data retention transients at $T = 60°C$ for TANOS devices with different alumina thickness. Simulation results with and without alumina trapping are also reported.

the same moderate ΔV_T, to avoid significant charge injection in the alumina (see Fig. 8, left), and retention was studied to investigate the effect of the alumina on the charge stored in the nitride. Note that samples with thicker Al_2O_3 feature smaller charge losses, suggesting a non-negligible role of Al_2O_3 on data retention. However, as changing the Al_2O_3 thickness also affects the fields and potentials during retention, further analysis is needed, which is shown in Fig. 10, comparing retention data at different conditions. Experimental data show that the retention charge loss (at $V_G = 0$) is barely affected by the reduction of the oxide field, while a neutralization of the Al_2O_3 field drastically improves the data retention. This shows that, for samples with 4.5 nm bottom oxide, the main charge leakage path is through the top alumina layer. Obviously, a stronger leakage through the oxide is expected in samples where its thickness is reduced.

V. CONCLUSIONS

We thoroughly investigated the impact of Al_2O_3 leakage and trapping on TANOS reliability as a function of the alumina thickness. Program/erase as well as retention are strongly affected by the alumina, which determines the saturation

behavior for long-term program or erase and provides the main leakage path for bottom oxides thickness in the 4.5 nm or above range. This work has shown that Al_2O_3 non-negligibly modifies the TANOS behavior and should be considered an important ingredient to optimize charge-trap devices.

VI. ACKNOWLEDGEMENTS

Authors would like to thank P. Cappelletti, E. Camerlenghi and R. Bez from Numonyx for discussions and support. This work has been partially supported by the European Commission under the FP7 research contract 214431 "GOSSAMER" and by MIUR under the FIRB Project No. RBIP06YSJJ.

REFERENCES

[1] C. H. Lee, K. I. Choi, M. K. Cho, Y. H. Song, K. C. Park, and K. Kim, "A novel SONOS structure of SiO2/SiN/Al2O3 with TaN metal gate for multi-giga bit flash memories", *IEDM Tech. Dig.*, pp. 613-616, 2003.
[2] Y. Shin, J. Choi, C. Kang, C. Lee, K.-T. Park, J.-S. Lee, J. Sel, V. Kim, B. Choi, J. Sim, D. Kim, H.-J. Cho, and K. Kim, "A novel NAND-type MONOS memory using 63 nm process technology for multi-gigabit Flash EEPROM", *IEDM Tech. Dig.*, pp. 337-340, 2005.
[3] Y. Park, J. Choi, C. Kang, C. Lee, Y. Shin, B. Choi, J. Kim, S. Jeon, J. Sel, J. Park, K. Choi, T. Yoo, J. Sim, and K. Kim, "Highly manufacturable 32Gb Multi-Level NAND Flash memory with 0.0098 μm^2 cell size using TANOS (Si-Oxide-Al2O3-TaN) cell technology", in *IEDM Tech. Dig.*, pp. 29-32, 2006.
[4] R. Degraeve, T. Kauerauf, A. Kerber, E. Cartier, B. Governeanu, P. Roussel, L. Pantisano, P. Blomme, B. Kaczer, and G. Groeseneken, "Stress polarity dependence of degradation and breakdown of SiO/high-k stacks", *Proc. IRPS*, pp. 23-28, 2003.
[5] A. H. Miranda , R. Van Schaijk, M. Van Duuren, N.Akil, D. S. Golubovi, "Reliability comparison of Al2O3 and HfSiON for use as interpoly dielectric in flash arrays", *Proc. ESSDERC*, pp. 234-237, 2006
[6] M. Specht, H. Reisinger, F. Hofmann, T. Schulz, E. Landgraf, R. J. Luyken, W. Rosner, M. Grieb, L. Risch, "Charge trapping memory structures with Al2O3 trapping dielectric for high-temperature applications", *Solid State Electron.*, pp. 716-720, 2005.
[7] A. Kerber, E. Cartier, R. Degraeve, P. J. Roussel, L. Pantisano, T. Kaureauf, G. Groeseneken, H. E. Maes, U. Schwalke, "Charge trapping and dielectric reliability of SiO2-Al2O3 gate stacks with TiN electrodes", *IEEE Trans. Electron Devices*, pp. 1261-1269, 2003.
[8] A. Mauri, C. Monzio Compagnoni, S. Amoroso, A. Maconi, F. Cattaneo, A. Benvenuti, A. S. Spinelli, A. L. Lacaita "A new physics-based model for TANOS memories program/erase", *IEDM Tech. Dig.*, pp. 555-558, 2008.

978-1-4244-5430-3/10 $26.00 © 2010 IEEE

Variability Effects on the V_T Distribution of Nanoscale NAND Flash Memories

Alessio Spessot*, Alessandro Calderoni*, Paolo Fantini*, Alessandro S. Spinelli[†‡],
Christian Monzio Compagnoni[†], Fabrizio Farina[†], Andrea L. Lacaita[†‡], and Andrea Marmiroli*

* Numonyx, R&D - Technology Development, via C. Olivetti 2, 20041 Agrate Brianza (MI), Italy,
e-mail: alessio.spessot@numonyx.com

[†] Dipartimento di Elettronica e Informazione, Politecnico di Milano–IU.NET, piazza L. da Vinci 32, 20133 Milano, Italy

[‡] IFN-CNR, Milano, Italy

Abstract—This work investigates the variability effects on the threshold voltage distribution of deca-nanometer NAND Flash memories. Different sources of variability have been considered, evaluating their impact on the neutral, programmed and erased distributions. A compact model that is able to account for the variability effects on the array performance and reliability is presented and used. Monte Carlo simulations have been employed to analyze the contributions of variability when technology nodes scale down and to compare the intrinsic variability with the electron injection statistical fluctuations. A good agreement with experimental data is reached, opening the application of the proposed methodology to investigate the reliability impact of variability on future technology nodes.

Index Terms—Flash memories, variability effects, threshold voltage distribution, semiconductor device modeling.

I. INTRODUCTION

The aggressive scaling trend of the NAND Flash technology recently led to the development of storage cells with dimensions of few tens of nanometers. One of the consequences of the reduced device dimensions is the increasing role played by variability effects at the single-device level [1], [2], that strongly influence the threshold voltage (V_T) distribution of nanoscaled NAND arrays, affecting their performance and reliability. The standard way of assessing the impact of these phenomena is by 3D numerical simulations [3], but the computational load required makes this approach unpractical when extensive statistical analyses under different operating conditions are needed. In this work, we present a compact yet detailed model for studying the impact of variability effects on the V_T distribution of nanoscale NAND Flash memories not only when cells are in the neutral state but also after program/erase and retention. Both intrinsic and technological sources of fluctuation have been investigated and implemented in the model, assessing their importance under real operating conditions of the device.

II. THE MODEL

The starting point for our analysis is the compact model for the NAND Flash memory array presented in [4]. The model includes 3 strings of 32 cells each plus the select transistors (Fig. 1, left), and accounts for interference effects [2] between adjacent cells via cell-to-cell parasitic capacitors. Only the

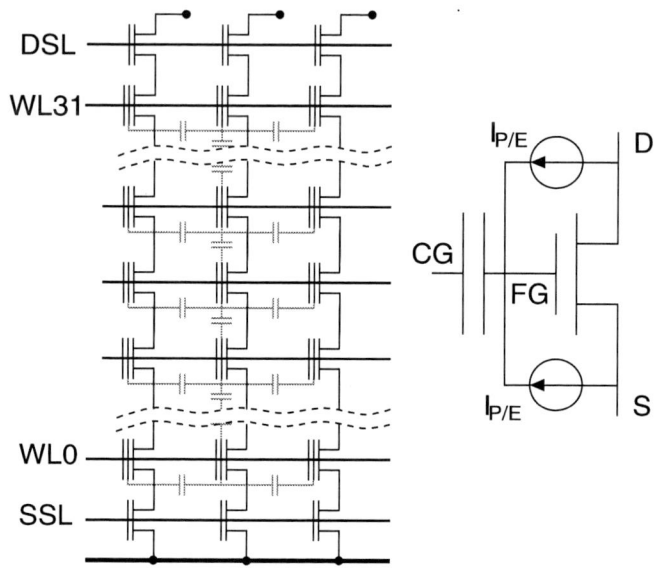

Fig. 1. Schematic of the simulated structure, comprising 3 strings of 32 cells plus 2 select transistors each (left). The central string is actually simulated, while the lateral two account for interference effects. On the right, the model adopted for the floating-gate cell.

central string is simulated, while the two lateral strings set the boundary conditions for the electrostatic couplings among the cells. The floating-gate devices are described via capacitors in series to MOS transistors (Fig. 1, right), whose parameters were extracted as detailed in [4] for any technology node we investigated.

The SPICE compact model was used in a Monte Carlo framework, running simulations to obtain the V_T distribution from the calculated string current under read conditions. In each simulation, the device parameters were randomly changed, to account for the different variability effects. In particular, we considered both process-induced fluctuations in the cell geometry and more fundamental ones, due for example to the discrete nature of the charge. The former include W, L, tunnel and interpoly dielectric thickness fluctuations (indicated as WF, LF, TOXF, IPDF, respectively) as well as fluctuations in the control-to-floating gate coupling coefficient; the latter

978-1-4244-5430-3/10 $26.00 © 2010 IEEE

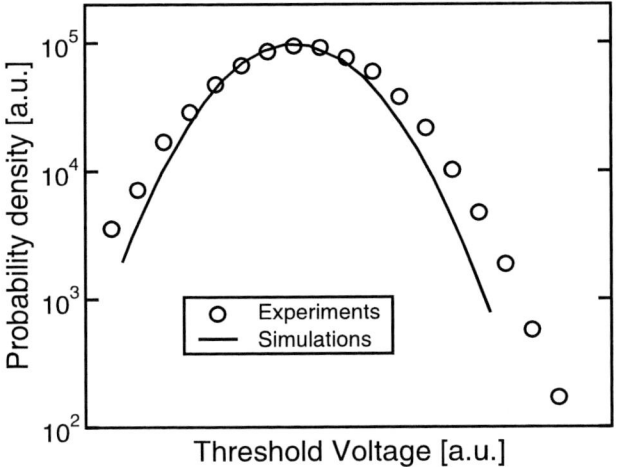

Fig. 2. V_T distribution for a page of a 41 nm NAND Flash array and the corresponding simulation results.

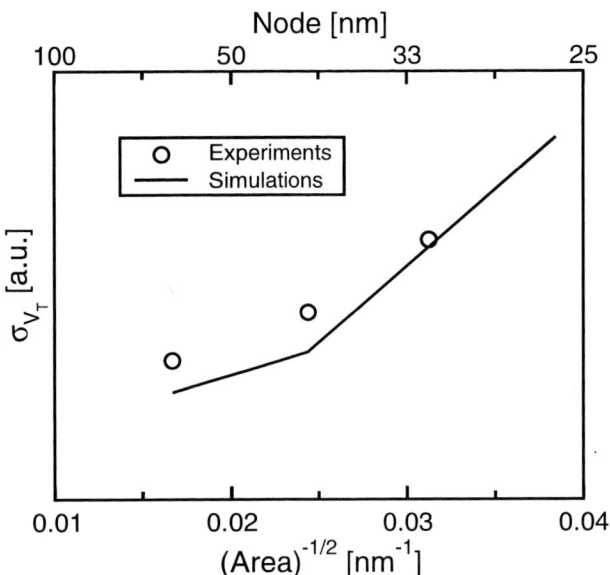

Fig. 3. Comparison between modeling results and experimental data for σ_{V_T} as a function of technology node.

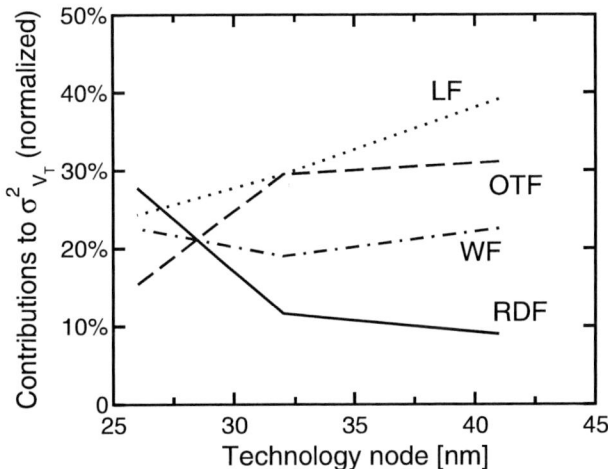

Fig. 4. Most important contributions to $\sigma^2_{V_T}$ of neutral cells (normalized) for different technology nodes.

account for random dopant (RDF) and oxide trap fluctuations (OTF). Process-induced fluctuations are directly inserted in the compact model by changing the device parameters (W, L, etc.) in each Monte Carlo run, according to Gaussian distributions whose spreads are extracted from process data. The implementation of the so-called intrinsic contributions is instead carried out as follows: RDF effect on V_T was accounted for by the analytical formula reported in [1], while the V_T variability due to OTF was implemented as $\sigma_{OTF} = K_{ox}Q_{ox}^\alpha/\sqrt{WL}$ with $\alpha \approx 0.5$ and K_{ox} and Q_{ox} fitted on cycled distribution data.

III. VARIABILITY EFFECTS ON NEUTRAL V_T

Fig. 2 shows the experimental V_T distribution measured on a page of a 41 nm NAND Flash technology, together with our simulated results including the spread of neutral cell V_T. Note that a good agreement is reached, supporting the correctness of the included variability models. The slight underestimation of the spread was indeed expected and is ascribed to the fact that the experimental V_T distribution was obtained by a soft erase operation from the programmed state, resulting in a V_T distribution that is slightly above 0. Note, in fact, that only positive V_T values can be experimentally read on the NAND test-chips, limiting the possibility to observe the V_T distribution when cells are completely neutral, as this condition corresponds to a V_T level around 0 V. The need for soft program (or soft erase) operations to reach a V_T distribution that approximates the neutral case is due to the ineffectiveness of UV erase to bring deca-nanometer NAND devices to their neutral state, due to the small pitch size and the number of metal layers above the cells. In order to further improve the agreement between modeling and experimental results in Fig. 2, the effect of the soft erase operation on the spread should be accounted for in the simulations, as will be discussed in detail in the next section.

Once results have been calibrated on a technology node, the model can be used to assess the scaling projections for

variability. Fig. 3 shows the computed standard deviation of the neutral-cell V_T distribution (σ_{V_T}) as a function of the technology node, compared against available experimental data. The model correctly captures the increase in the V_T variability down to the 32 nm technology, and can be used to draw predictions for future nodes. It is interesting to note that a simple straight line proportional to $1/\sqrt{WL}$ does not represent the real behavior over different technologies.

A breakdown of the most important contributions to $\sigma^2_{V_T}$ is shown in Fig. 4 for three different nodes. The contributions are shown in a normalized scale, allowing for their direct comparison. The results highlight that balanced scaling of the cell has resulted in a condition where there is not a single "killer" phenomenon and, instead, variability is dominated by the interplay of several effects, the most important being

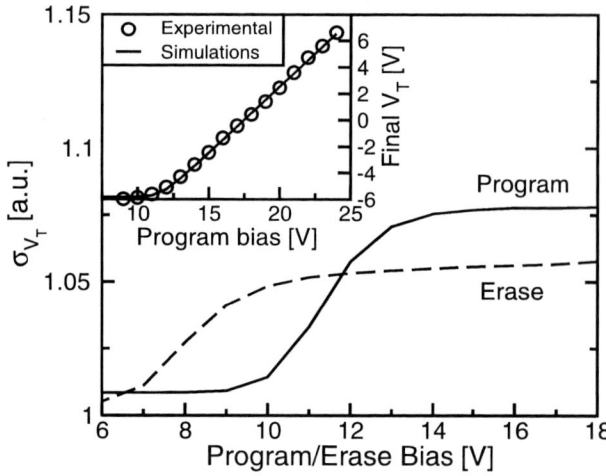

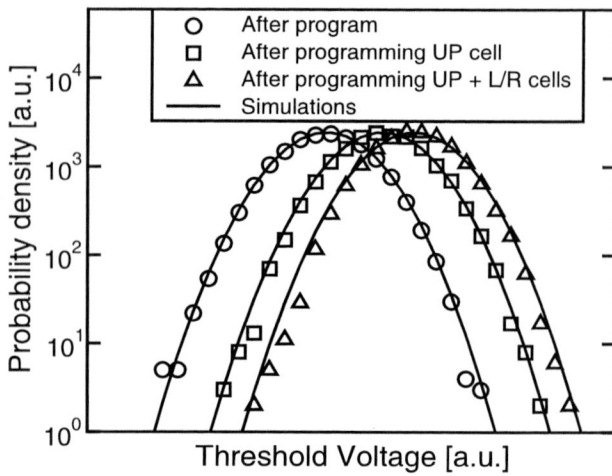

Fig. 5. Statistical spread of V_T after P/E as a function of the P/E bias. The inset shows the average experimental and simulated V_T for a 41 nm cell after constant-voltage programming at different bias.

Fig. 6. V_T distribution for a page of the 41 nm array right after program and after programming the cells on its top (UP) and the ones in the adjacent strings (L and R). Lines are simulation results.

random dopants, trapped charge, and fluctuations in L and W. Moreover, it is interesting to note the increasing contribution played by RDF in scaled cells, which eventually dominates over WF and LF.

IV. VARIABILITY EFFECTS ON PROGRAM

The model for the memory cell presented in [4] and used in the previous section to investigate the behavior of the neutral cell was then expanded to simulate the program and erase (P/E) operations, adding two voltage-controlled current sources between the floating gate and the source and drain electrodes (Fig. 1, right). The Fowler-Nordheim (FN) and direct tunnel equations were used to model the tunneling current over the active area, with the spread in the number of injected electrons [5] added to the variability sources. The inset of Fig. 5 shows that the model correctly reproduces the average experimental V_T shift after a constant-voltage program operation on a single 41 nm Flash cell.

Fig. 5 shows the Monte Carlo simulation results for σ_{V_T} after a constant-voltage program or erase operations. When comparing these results with those shown in the inset, it is clear that σ_{V_T} increases as soon as V_T moves from the starting condition. This additional spread is related to the fluctuation in the number of electrons injected/emitted to/from the floating gate and to the spread in device parameters affecting the FN current. In particular, the different values of σ_{V_T} after P and E are related to the different values of cell parameters in the two bias conditions. The increased spread consequence of the program operation allows the simulated V_T distribution to achieve a better agreement with experimental data for a single page of the array (Fig. 6) when compared with previous results of Fig. 2. Moreover, note that the model correctly reproduces the V_T shift determined by the electrostatic interference when cells in the page above (UP in the figure) and those adjacent to (left/right, LR in the figure) the ones under investigation are programmed.

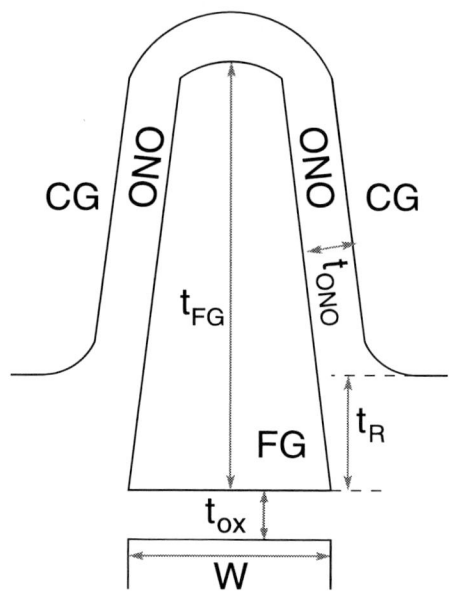

Fig. 7. Schematic view of the cross-section of a memory cell along the W direction, showing the floating-gate geometry.

One of the most important parameters that plays a fundamental role in the σ_{V_T} during P/E operations is the control-gate to floating-gate coupling coefficient (α_G). This parameter is related to the geometry adopted for the floating-gate definition, and its fluctuations depend on the spread in the geometrical parameters, which are schematically shown in Fig. 7. The contributions of such factors to the spread in α_G are shown in Fig. 8, on a normalized scale. In this case, fluctuations in t_{FG} and t_R (see Fig. 7) play the major role for all technology nodes.

V. IMPACT ON RELIABILITY

Despite the effect of neutral V_T spread on the program placement can be largely reduced by adopting the incremental

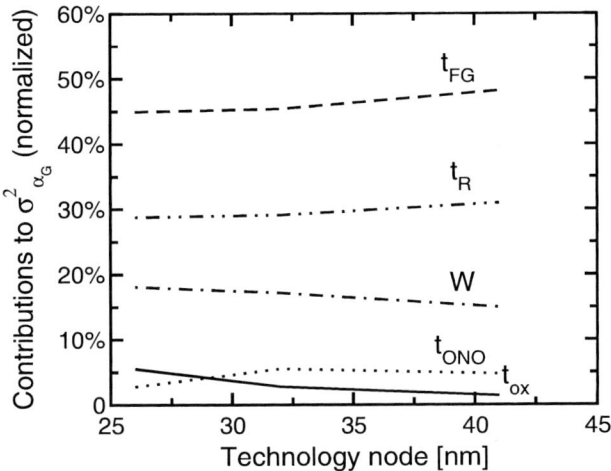

Fig. 8. Individual contributions to the spread in the coupling coefficient α_G for the different technology nodes.

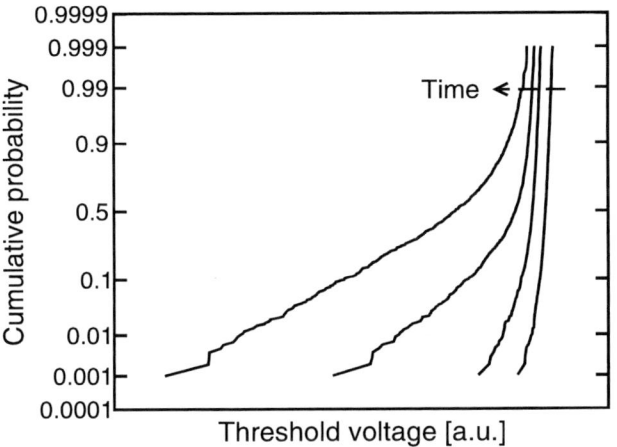

Fig. 9. Monte Carlo simulation results for an accelerated data retention experiment on the 41 nm technology, starting from an ISPP programmed V_T distribution and neglecting EES.

step pulse programming (ISPP) algorithm [6], the neutral V_T dispersion still impacts the array reliability during retention. In order to investigate this point, we referred only to data retention on fresh, *i.e.* uncycled, devices, where floating gate charging/discharging rules the V_T transient as a function of elapsing time. In so doing, we did not consider the effect of SILC and detrapping on the V_T distribution which are very important physical effects when studying data retention after cycling. We conducted Monte Carlo simulations of a field-accelerated retention test on 1000 cells programmed to a high-V_T state, evaluating their V_T evolution as a function of time. Fig. 9 shows the results obtained assuming a very tight initial V_T distribution, considering increasing elapsed time during data retention. A gaussian tail of fast-moving cells emerges from the rest of the distribution and grows with time until the entire V_T distribution matches the spread of the intrinsic one (even if the cells are still programmed). This effect is a clear signature of variability phenomena: in fact,

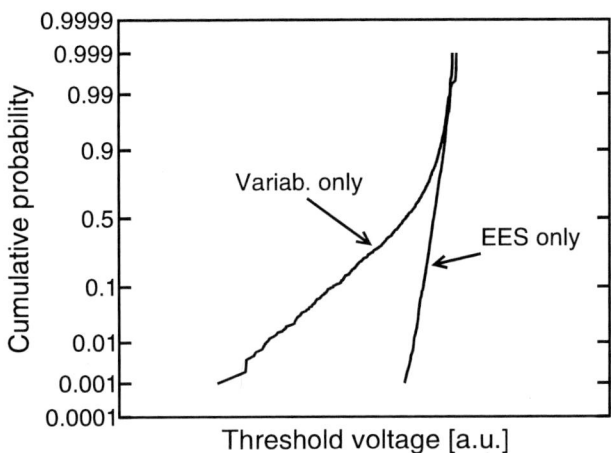

Fig. 10. Comparison between results of Fig. 9 (including only the spread in the neutral V_Ts) and data retention simulations obtained including only the EES effect.

even if the ISPP algorithm allows a tight programmed V_T distribution to be obtained, cells having different neutral V_Ts experience different electric fields on their tunnel oxide during data retention, which in turn means different rates of charge loss from the floating gate.

In the above simulations, the tunneling current itself was not subjected to variability. In reality, however, even fluctuations in the number of electrons emitted from the floating gate could also give rise to a widening of the distribution [7], and must be accounted for. To compare the two effects, we have included the electron emission statistics (EES) in the Monte Carlo code, adopting as a worst case a Poissonian distribution for the number of emitted electrons. A comparison between the impact of EES and the intrinsic variability on the resulting V_T after retention is shown in Fig. 10: it is clear that the increased spread due to EES is overwhelmed by the intrinsic variability, which remains the main issue to be addressed in the development of future NAND devices from the data retention standpoint.

VI. CONCLUSIONS

We presented a scalable compact model accounting for variability effects in nanoscale NAND Flash memories. The model captures the impact of different variability sources with a scaling perspective and can predict the V_T distribution under different operating conditions. We also investigate the impact of variability on retention, and show that EES is not a major issue for Flash reliability, but rather intrinsic variability plays a dominant role.

VII. ACKNOWLEDGMENTS

Authors would like to thank P. Cappelletti, E. Camerlenghi and R. Bez from Numonyx for helpful discussions and support. This work has been also partially supported by ENIAC under the MODERN project 120003.

REFERENCES

[1] A. Asenov, A. R. Brown, J. H. Davies, S. Kaya, and G. Slavcheva, "Simulation of intrinsic parameter fluctuations in decananometer and nanometer-scale MOSFETs," *IEEE Trans. Electron Devices*, vol. 50, pp. 1837–1852, Sept. 2003.

[2] J.-D. Lee, S.-H. Hur, and J.-D. Choi, "Effects of floating-gate interference on NAND Flash memory cell operation," *IEEE Electron Device Lett.*, vol. 23, pp. 264–266, May 2002.

[3] A. Cathignol, B. Cheng, D. Chanemougame, A. R. Brown, K. Rochereau, G. Ghibaudo, and A. Asenov, "Quantitative evaluation of statistical variability sources in a 45-nm technological node LP N-MOSFET," *IEEE Electron Dev. Lett.*, vol. 29, pp. 609–611, June 2008.

[4] L. Larcher, A. Padovani, P. Pavan, P. Fantini, A. Calderoni, A. Mauri, and A. Benvenuti, "Modeling NAND Flash memories for IC design," *IEEE Electron Dev. Lett.*, vol. 29, pp. 1152–1154, Oct. 2008.

[5] C. Monzio Compagnoni, A. S. Spinelli, R. Gusmeroli, S. Beltrami, A. Ghetti, and A. Visconti, "Ultimate accuracy for the NAND Flash program algorithm due to the electron injection statistics," *IEEE Trans. Electron Devices*, vol. 55, pp. 2695–2702, Oct. 2008.

[6] G. J. Hemink, T. Tanaka, T. Endoh, S. Aritome, and R. Shirota, "Fast and accurate programming method for multi-level NAND EEPROMs," in *1995 Symp. VLSI Tech. Dig.*, pp. 129–130, 1995.

[7] G. Molas, D. Deleruyelle, B. De Salvo, G. Ghibaudo, M. Gely, S. Jacob, D. Lafond, and S. Deleonibus, "Impact of few electron phenomena on floating-gate memory reliability," in *IEDM Tech. Dig.*, pp. 877–880, 2004.

NAND Flash Reliability Degradation Induced by HCI in Boosted Channel Potential

Milim Park, Sukkwang Park, Seokwon Cho, Dong-Kyu Lee, YeonJoo Jeong, Chonga Hong,
Ho Seok Lee, Myoung Kwan Cho, Kun-Ok Ahn and Yohwan Koh

**Flash Memory Division, Hynix Semiconductor Inc., 55 Hyangjeong-dong Hungduk-gu Cheongju-si 361-725, Korea,
Phone : +82-43-280-6677, E-mail : myoungkwan.cho@hynix.com**

Abstract— In this paper, we present the impact of hot carrier injection (HCI) during programming operation in NAND Flash, and describe how HCI degrades reliability characteristics. In order to understand reliability degradation induced by HCI, we evaluated the reliability characteristics under various stress conditions including the number of disturbance pulses, pulse shapes and temperatures. We have concluded that the programming pulse and boosting bias should be carefully optimized to reduce the impact of HCI.

Keywords; NAND Flash; Hot carrier; Channel potential

I.   INTRODUCTION

NAND flash memories are widely used for data storage because of large capacities, low cost and low power consumption [1]. While capacities are being rapidly increased with the help of fine lithography. reliability is being degraded due to thinning of the tunnel and interpoly dielectrics. In this paper we present how HCI impacts NAND flash reliability, especially in erase/write (E/W) cycles and retention standpoint. The channel potential is steeply bent near edge wordlines (W/Ls) in conventional boosting and near local boosted (LB) biased W/Ls during local boosting, therefore the generated hot carrier is injected into floating gate or Si/SiO$_2$ interface degrading the reliability. The hot carrier induced degradation mechanism during programming operation is investigated and optimal operating conditions are suggested.

II.   EXPERIMENT AND RESULT

The NAND cell array consists of string select transistors and series connected cell transistor shown in Fig.1.   To program the selected cell, the unselected cells are under program disturbance situation from local or conventional self boosting [1]. If the unselected W/L is located adjacent to the select transistor in this disturbance situation, edge W/L cells are affected by HCI. The Vth shift due to HCI degradation of the edge W/Ls is shown in Fig.2. This phenomenon is well explained by Medici simulation as shown in Fig.3. The energy band bends steeply near the select transistor and the generated hot carriers in this area are injected into the floating gate of the edge W/Ls causing larger Vth shifts compared to center W/L cells which are not impacted by HCI. In addition to Vth shifts, the E/W cycles of edge W/L is also worse than center W/Ls in Fig.4. It is evident that the Vth shift is increased as E/W cycles and which causes E/W cycle degradation. The impact of the increased Vth of the edge W/Ls can be seen in Fig. 4 where

the edge wordlines are shown to degrade faster than inner W/Ls leading to eventual erase failure.   Fig.5 shows the comparison of retention characteristics as a function of W/L position. During HCI, some of hot electrons are trapped in Si/SiO$_2$ interface which is supported by subthreshold swing degradation.   The trapped hot electrons make degrade the retention characteristics since interface trapped charges are easily de-trapped [2]. In order to understand the HCI induced degradation mechanism, the disturbance characteristics are studied from various stress conditions. In Fig.6, as the number of program pulses increases, the HCI induced Vth shift is increases in same stress time, which means that the hot carriers are generated and injected into the F/G when the energy bands are being bent rather than already band bent state. This implies that the number of program pulse should be balanced with step bias of ISPP for narrow Vth distribution. Another characteristic is the temperature dependence of the HCI disturbance shown in Fig.7. As temperature increases the Vth shift increases unlike conventional CHEI NOR flash cell. This means that the HCI disturbances are not a carrier energy limit process but carrier generation limit process. Fig.8 shows that the edge W/L has a larger Vth shift with longer pulse rise time. This also means that the HCI induced Vth shift is more affected when the energy band is being bent. In local boosting schemes, there is also channel potential bending in the mid of W/Ls as shown in Fig. 9. There are hot electrons injected in erased cells and hot holes injected in programmed cells, causing charge gain and loss, respectively.

III.   CONCLUSION

This paper shows that hot carrier injection significantly affects the reliability degradation in edge W/Ls of NAND flash. The reliability characteristics are evaluated in terms of the number of disturbance pulse, temperature, and pulse condition. Through these experiments, the HCI induced degradations are a hot carrier generation limited process and it suggest that double verification and ISPP should be considered in terms of HCI. In local boosting schemes, the LB bias should be optimized since there are both hot holes and electrons injected according to the state of F/G.

References

[1] B. Dipert, L. Hebert, "Flash memory goes mainstream," IEEE Spectrum, vol. 30, pp. 48-52, October 1993.

[2] J. -D. Lee, J. -H. Choi, D. Park, and K. Kim, "Data retention characteristics of sub-100 nm NAND flash memory cells," IEEE Electron Device Lett., vol. 24, pp. 748-750, December 2003.

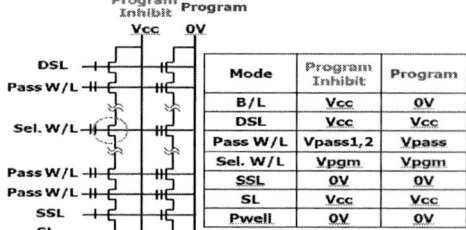

Fig.1 NAND flash cell array and operating conditions.

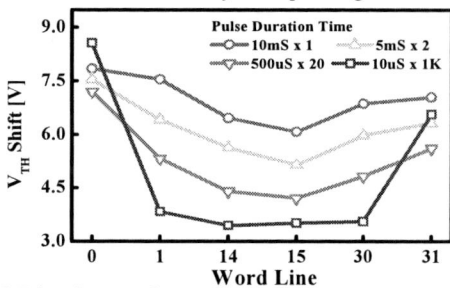

Fig.2 Disturbance of erased cell during program operation. A Vth increase is observed in edge W/Ls(0,31) comparing to center W/Ls.

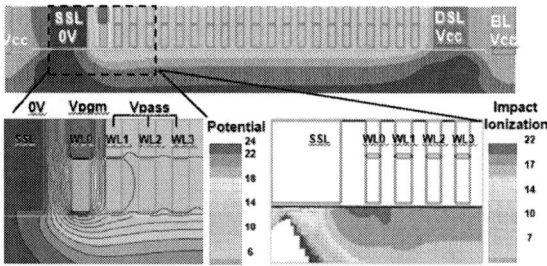

Fig.3 Steep potential contour changes are noticed near select transistors. The hot carriers are generated due to this drastic energy band and injected into the floating gate of edge W/L.

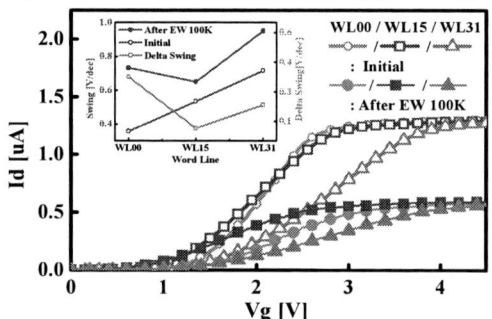

Fig.4 The subthreshold swing change versus W/L position after E/W 100K cycles. The larger swing shift is observed in W/L 0 and 31, causing E/W cycle failure.

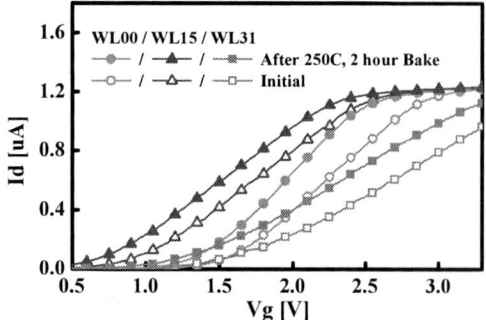

Fig.5 The edge W/L shows more severe degradation than center W/L. The trapped hot electrons near the silicon-oxide interface are easily de-trapped during retention test.

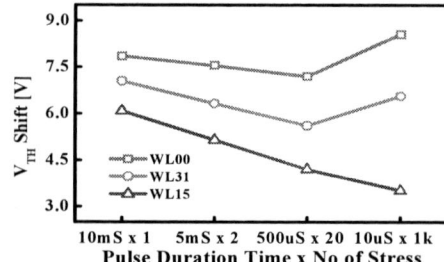

Fig.6 The Vth shift as a function of the number of pulses among W/Ls. The edge W/Ls(0,31) show different characteristics from center W/L15.

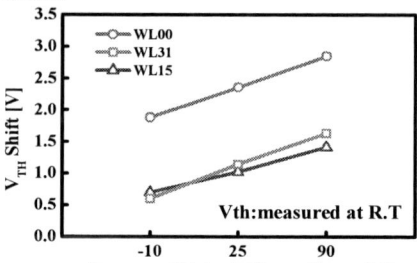

Fig.7 The Vth shift as a function of stress temperature among W/Ls. One of edge W/L shows larger Vth shift than center W/Ls.

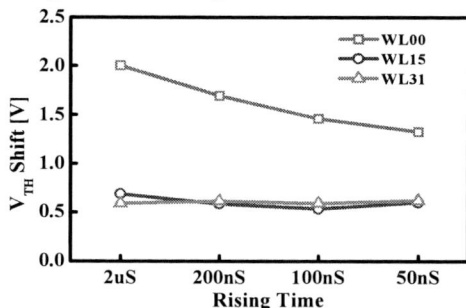

Fig.8 The Vth shift as a function of the rising time of stress pulses among W/Ls. One of edge W/L shows different characteristics from center W/L.

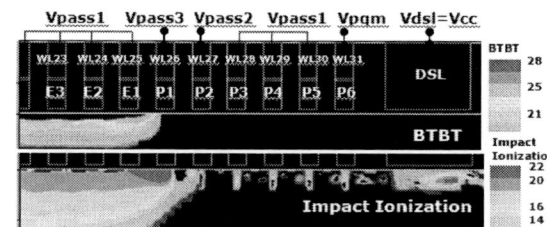

Fig.9 Near Vpass3 biased W/L in local boosting scheme, Electron-Hole pairs are generated and hot carriers are injected into cells, E1, E2, P1 and P2 according to F/G potentials and cell position..

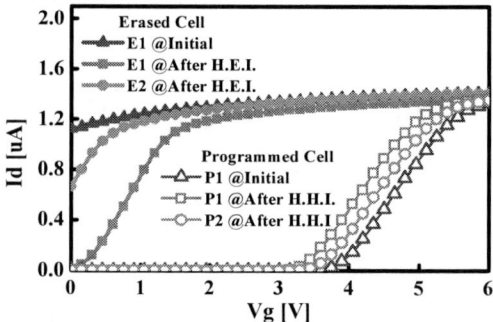

Fig.10 Hot electrons are injected into E1 and E2 cells, hot holes are injected into P1 and P2, respectively, which causes charge gain and loss.

978-1-4244-5430-3/10 $26.00 © 2010 IEEE

Interface-Trap Modeling for Silicon-Nanowire MOSFETs

Zuhui Chen, Xing Zhou, Guojun Zhu, and Shihuan Lin

School of Electrical & Electronic Engineering
Nanyang Technological University
Singapore
Phone: (0065)-6790-4523, zhchen@ntu.edu.sg, exzhou@ntu.edu.sg

Abstract—**Interface traps generated during device operation or stress is directly related to transistor electrical characteristics and reliability as well as critical to device performance. In this paper, an interface-trap model is included in the unified compact model (Xsim) in order to physically and accurately characterize the interface-trap behavior in silicon-nanowire (SiNW) MOSFETs. The interface-trap model is verified by TCAD simulation data. Very good agreement is achieved and the effect of interface traps is accurately captured in the drain-source characteristics of SiNW MOSFETs. The physical interface-trap model is readily applicable for circuit and reliability modeling with SiNW transistors as building blocks.**

Keywords: interface traps, drain-source current, generation current, reliability, unified compact model, SiNW MOSFETs.

I. INTRODUCTION

The scaling of conventional metal-oxide-semiconductor (MOS) transistors will reach its physical limit in 2020 as projected by the 2005 ITRS (International Technology Roadmap for Semiconductors) [1]. Because of the superior electrostatic control along the surface channel region, the silicon-nanowire (SiNW) MOSFET with surrounding gates, as shown in Fig. 1, is generally believed to be one of the most promising devices for extending transistor scaling in the future generations of integrated circuit chips [2]. The generation of interface traps along the channel region shifts the threshold voltage, reduces the drive current, shortens the holding time of dynamic memory transistors, increases leakage current, and consequently increases device power consumption [3], [4]. Degradation of device reliability may be exacerbated under the conditions of strong electric field and high temperature. In this paper, we will extend the explicit analytic approach (Xsim) [5], [6] to explore the interface traps on device electrical characteristics in SiNW MOSFETs. A paired-linear distribution of neutral interface traps with linear energy distributions of neutral electron and hole interface traps will be theoretically studied. The behavior of interface traps will be captured in the I_{DS}–V_{GS} characteristics of SiNW MOSFETs, and the interface-trap model is verified by TCAD simulations.

II. MODEL FORMULATIONS

As Slater's perturbation theory [7] is applied to the SiO₂/Si interface [8], the neutral interface trap, which is due to the random variations of Si-O and Si-Si bond angles and lengths

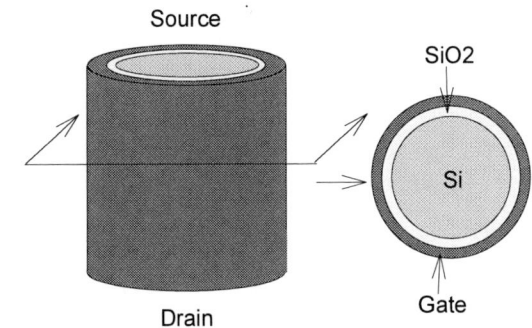

FIGURE 1. Schematic of an ideal silicon nanowire MOSFET.

along the SiO₂/Si interface, is defined as an electrically neutral trapping potential well that can bind only one electron or hole. The neutral interface traps become negative when capturing an electron or positive when capturing a hole. Thus, the neutral electron traps have the charge states 0 or −1, and neutral hole traps have the charge states 0 or +1. The charge formulae of the neutral electron traps Q_{ETi} and neutral hole traps Q_{HTi} are, respectively, given by [9]

$$Q_{ETi} = -qN_{ETi} \times \frac{c_{ns}n_s + e_{ps}}{c_{ns}n_s + e_{ns} + c_{ps}p_s + e_{ps}}, \quad (1a)$$

$$Q_{HTi} = +qN_{HTi} \times \frac{c_{ps}p_s + e_{ns}}{c_{ns}n_s + e_{ns} + c_{ps}p_s + e_{ps}}, \quad (1b)$$

where N_{ETi} and N_{HTi} are the neutral electron-trap and hole-trap densities (cm⁻²eV) at the ith energy level, respectively. c_{ns}, c_{ps}, e_{ns}, and e_{ps} are the electron and hole capture and emission coefficients. n_s and p_s are the electron and hole surface concentrations, given respectively by

$$n_s = n_i e^{(\phi_s - \phi_{Fn})/v_{th}}, \quad (2a)$$

$$p_s = n_i e^{(\phi_{Fp} - \phi_s)/v_{th}}. \quad (2b)$$

From detailed balance near thermal equilibrium and assuming Boltzmann distribution of electron and hole concentrations, e_{ns} and e_{ps} are given by

$$e_{ns} = c_{ns}n_i e^{+E_{TI}/k_B T}, \quad (3a)$$

$$e_{ps} = c_{ps}n_i e^{-E_{TI}/k_B T}, \quad (3b)$$

978-1-4244-5430-3/10 $26.00 © 2010 IEEE

where E_{TI} is the interface-trap energy level measured from the intrinsic Fermi level E_I. The total interface-trap charge is the sum of the charges at the neutral electron and hole interface traps, given as

$$Q_{IT} = \int_{E_V}^{E_c} (Q_{ETi} + Q_{HTi}) dE_{TI} . \qquad (4)$$

Since the applied gate voltage modulates the surface carrier density along the channel region, the total interface-trap charge not only depends on the trap density but also the gate voltage because charge trapping probability is directly related to the surface carrier concentration.

For a three-dimensional bulk model, the ratio of neutral electron traps to neutral hole traps is theoretically given by [10]

$$R_{eh} = \frac{D_{ET0}}{D_{HT0}} = \frac{1.09412}{0.5228205} = 2.095526 , \qquad (5)$$

where D_{ET0} and D_{HT0} are the densities of state (cm^{-2}) at conduction- and valence-band edges, respectively. The R_{eh} value is a fitting parameter in the model computation since R_{eh} varies under different stress conditions. For a linear distribution of interface traps, which is an approximation of the U-shaped energy distribution of interface trap in the silicon energy gap [8], the densities of the neutral electron traps and neutral hole traps are, respectively, given by

$$N_{ET} = \int_0^1 (D_{ET0} \times E) dE = \frac{D_{ET0}}{2} , \qquad (6a)$$

$$N_{HT} = \int_0^1 D_{HT0} \times (1 - E) dE = \frac{D_{HT0}}{2} , \qquad (6b)$$

where $E = E/E_G$ is normalized to the silicon energy gap E_G.

In SiNW MOSFETs as shown in Fig. 1, the electrostatic potential is governed by the Poisson equation. After integrating Poisson's equation and applying the boundary conditions, the following coupled equation with interface traps can be obtained [5], [6]

$$V_g - \Phi_{MS} - \phi_s + \frac{Q_{IT}}{C_{ox}} = \Upsilon_i \sqrt{v_{th}} e^{(\phi_s - V_c)/2v_{th}} \sqrt{1 - e^{-(\phi_s - \phi_0)/2v_{th}}} , \qquad (7)$$

where Φ_{MS} is the work-function difference between the (metal) gate and semiconductor. V_c is the channel voltage, and $\Upsilon_i = (2q\varepsilon_{Si}n_i)^{1/2}/C_{ox}$ is the body factor. $C_{ox} = \varepsilon_{ox}/[R\ln(1 + T_{ox}/R)]$ is the oxide capacitance in cylindrical coordinate. Other symbols have their usual meanings.

The net charge is dominated by electrons for nMOSFETs in the strong-inversion region. The surface potential without including interface-trap charges can be expressed as [5], [6]

$$\phi_s [V_c(y)] = V_g - \Phi_{MS} - 2v_{th} \mathcal{L} \left\{ \frac{\Upsilon_i}{2\sqrt{v_{th}}} e^{(V_g - \Phi_{MS} - V_c)/2v_{th}} \right\} , \qquad (8)$$

where $\mathcal{L}\{w\}$ is the *Lambert W* function. Once the surface potential ϕ_s is solved in (8), then the Q_{IT} can be obtained from (4). In order to obtain the final surface potential for devices with interface traps, ϕ_s is computed by several iterations. For each iteration, the surface potential is computed by replacing Φ_{MS} with $\Phi_{MS} - Q_{IT}/C_{ox}$ in (8). With the surface potential, zero-field potential for SiNW MOSFETs can be explicitly expressed in terms of surface potential as

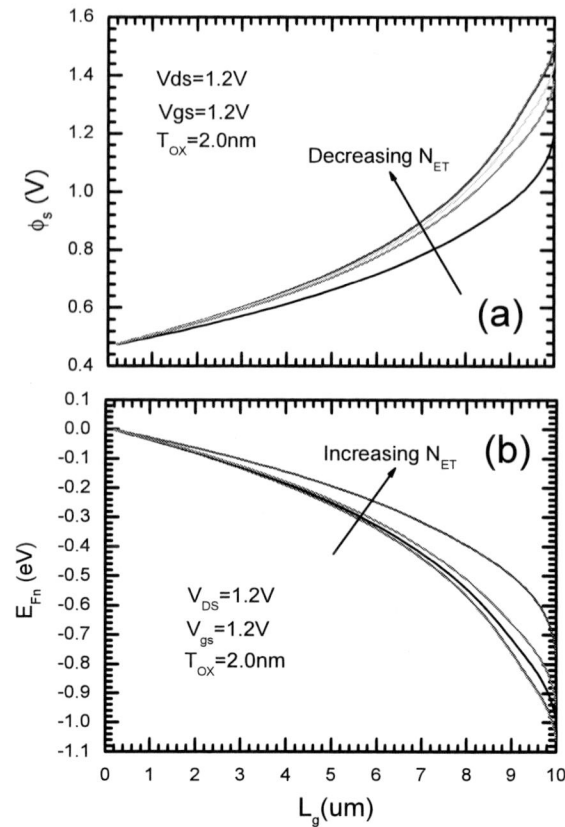

Fig. 2. Extracted (a) surface potential and (b) electron quasi-Fermi level with various neutral electron interface-trap densities along the channel region from Medici simulation for silicon nanowire MOSFETs. $N_{ET} = 0$, 1.0×10^{10}, 1.0×10^{11}, 1.0×10^{12}, 2.0×10^{12}, and 5.0×10^{12} cm^{-2}.

$$\phi_0 [V_c(y)] = \phi_s [V_c(y)] + 2v_{th} \ln \left\{ \frac{-1 + \sqrt{1 + 4\omega}}{2\omega} \right\} , \qquad (9)$$

$$\omega = \delta e^{\frac{\phi_s[V_c(y)] - V_c(y)}{v_{th}}} , \qquad (9a)$$

$$\delta = \frac{R^2}{8\varepsilon_{si}L_i^2} , \qquad (9b)$$

$$L_i = \sqrt{v_{th}\varepsilon_{si}/qn_i} . \qquad (9c)$$

Then, (7) with interface traps can be further transformed into:

$$V_g - \Phi_{MS} - \phi_s + \frac{Q_{IT}}{C_{ox}} = \frac{R\varepsilon_{si}v_{th}}{2L_i^2 C_{ox}} e^{(\phi_s + \phi_0 - 2V_c)/2v_{th}} . \qquad (10)$$

Therefore, the final expression of I_{ds} for SiNW MOSFETs with interface traps can be obtained as [5], [6]

$$I_{dd} = 2\mu_{eff0}C_{ox}\frac{\pi R}{L}\left(V_g - \Phi_{MS} + \frac{Q_{IT}}{C_{ox}} - \frac{\phi_{s,eff} + \phi_{d,eff}}{2} + 2v_{th}\right)V_{ds,eff} . \qquad (11)$$

where $V_{ds,eff}$ is the effective drain-source voltage, and $\phi_{s,eff}$ and $\phi_{d,eff}$ are the effective surface potentials at the source and drain sides, respectively .

During regular operations in nMOSFETs with p-type nanowire and n-type source and drain contacts, the source-body p/n junction is zero biased while the drain-body p/n junction is reversed biased. Thus, there are electron-hole generations due to the emission of trapped electrons and holes from interface traps. The electron-hole recombination-

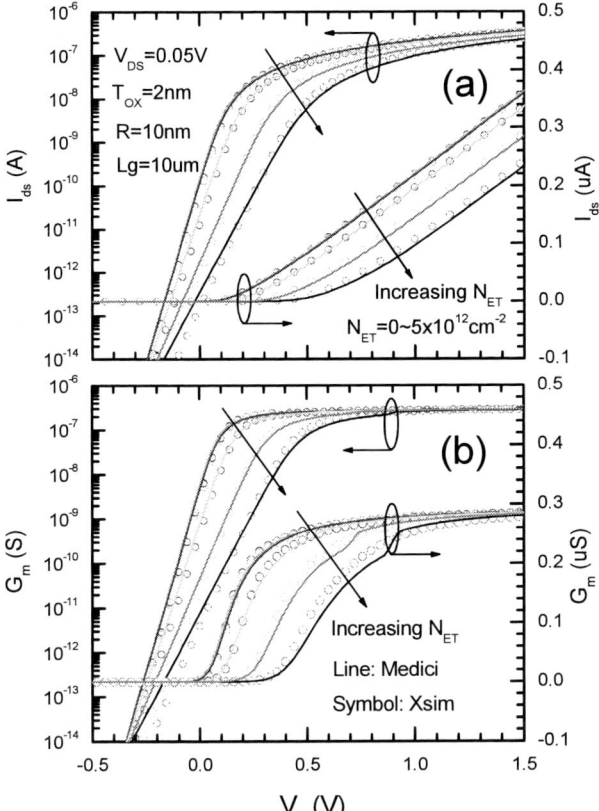

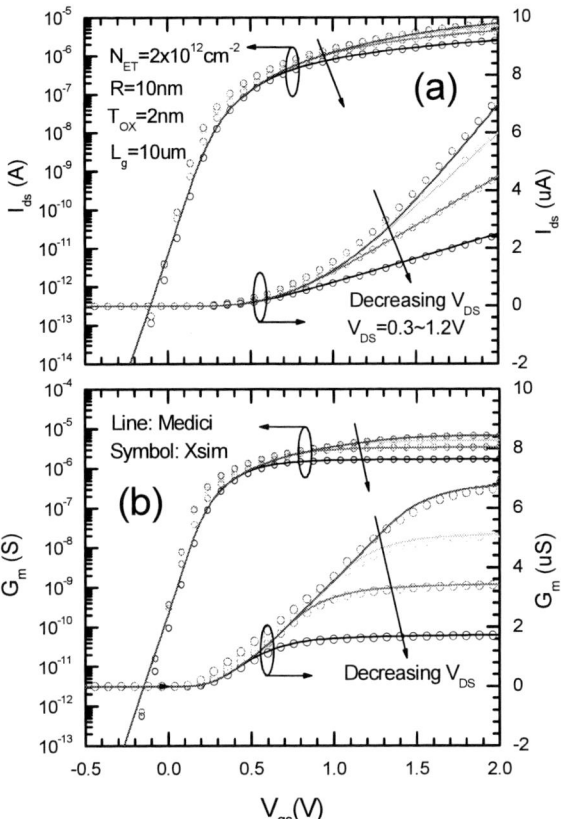

Figure 3. (a) Interface-trap effect on drain-source current, (b) transconductance at various neutral electron trap density $N_{ET} = 0$, 1.0×10^{10}, 1.0×10^{11}, 1.0×10^{12}, 3.0×10^{12}, and 5.0×10^{12} cm^{-2}.

FIGURE 4. (a) drain-source current and (b) transconductance at various drain-source voltage $V_{DS} = 0.3$, 0.6, 0.9, and 1.2 V, and $N_{ET} = 2.0 \times 10^{12}$ cm^{-2}.

generation rate R_{SS}, at a discrete energy-level of interface traps and with an interface-trap density of N_{IT}, is given by the Shockley-Read-Hall (SRH) formula:

$$R_{SS} = \frac{c_{ns}c_{ps}n_s n_s - e_{ns}e_{ps}}{c_{ns}n_s + e_{ns} + c_{ps}n_s + e_{ps}} N_{IT}, \qquad (12)$$

Assuming small deviation from thermal equilibrium and using Boltzmann statistics, the R_{SS} can be expressed by a more convenient form as follows [11]

$$R_{SS} = \frac{(c_{ns}c_{ps})^{1/2}n_i}{2} \frac{\left[\exp\left(\frac{-V_c(y)}{v_{th}}\right) - 1\right]}{\exp\left(\frac{-V_c(y)}{2v_{th}}\right)\cosh\left(\frac{\phi_s^*(Q_{IT}, y)}{v_{th}}\right) + \cosh\left(\frac{E_{TI}^*}{k_B T}\right)} N_{IT}, \qquad (13)$$

$$\phi_s^* = \phi_s + \frac{1}{2}v_{th}\ln\left(\frac{c_{ns}}{c_{ps}}\right) - \frac{\phi_{Fp} + \phi_{Fn}}{2}, \qquad (13a)$$

$$E_{TI}^* = (E_T - E_I) + \frac{1}{2}k_B T \ln\left(\frac{c_{ns}}{c_{ps}}\right), \qquad (13b)$$

where ϕ_s^* and E_{TI}^* are the effective surface potential and interface-trap energy level, respectively. The generation current I_G is obtained by integrating the SRH recombination-generation rate at the interface traps over the channel area [11]:

$$I_G(V_{GS}) = \frac{q(c_{ns}c_{ps})^{1/2}n_i W}{2} \iint \frac{\left[\exp\left(\frac{-V_c(y)}{v_{th}}\right) - 1\right] N_{IT}(E_{TI})dydE_{TI}}{\exp\left(\frac{-V_c(y)}{2v_{th}}\right)\cosh\left(\frac{\phi_s^*(Q_{IT}, y)}{v_{th}}\right) + \cosh\left(\frac{E_{TI}^*}{k_B T}\right)} \qquad (14)$$

Therefore, the drain-terminal current is sum of the drift, diffusion currents I_{dd} and the generation current I_G given as

$$I_{ds} = I_{dd} + I_G. \qquad (15)$$

III. RESULTS AND DISCUSSION

Figure 2 shows the extracted surface potential ϕ_s and electron quasi-Fermi level E_{Fn} due to the influence of interface traps along the surface channel region in SiNW MOSFETs. The neutral electron-trap density N_{ET} is varied from zero to 5.0×10^{12} cm^{-2} along the surface channel region. Because of the interface traps, the ϕ_s at the SiO$_2$/Si interface adjusts itself to a value in order to keep the net interface-trap charge nearly zero. Thus, the gate voltage is not zero at zero interface-trap charge which is affected by the interface-trap distribution, crystallographic orientation, and bias applied to the source and drain terminals. The higher interface-trap density, the more prominent effect of interface traps is on device surface electric characteristics. Since the ϕ_s and E_{Fn} are key parameters for device performance, the generation of interface traps can dramatically impact the interface electric property and drain-source current characteristics.

Figure 3 shows the effect of the interface-trap density on drain-source current I_{ds} and transconductance G. Because interface traps shift the threshold voltage, which reduce the useful gate voltage at a given condition, the lineshapes of I_{ds} and G_m are distorted increasingly as the neutral interface-trap

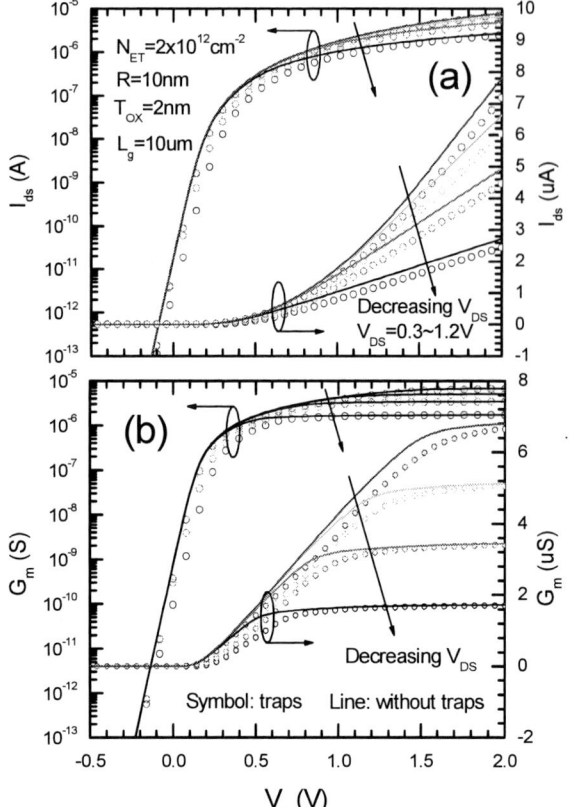

FIGURE 5. Comparisons between devices with and without interface traps: (a) drain-source current and (b) transconductance at various drain-source voltage V_{DS} = 0.3, 0.6, 0.9, and 1.2 V, and N_{ET} = 2.0×10^{12} cm^{-2}.

density is increased. Not all interface traps are trapped with a charge (electron or hole) and the charge trapping rate depends on the surface carrier concentration, which is modulated by the gate voltage as anticipated in (1). Thus, the interface-trap charge not only depends on the interface-trap density but also the applied gate voltage. For a SiNW transistor with N_{IT} = 1.0×10^{12} cm^{-2} and T_{OX} = 2.0 nm, the threshold-voltage variation could be up to $\Delta V_T \approx -qN_{IT}/C_{OX}$ = −0.0846 V. Therefore, interface traps could significantly reduce the drain-source current and lower the transconductance as shown in Fig. 3(a) and 3(b), respectively.

Figure 4 shows the dependence of drain-source current and transconductance on the gate voltage as a function of drain-source voltage. V_{ds} is varied from 0.3 V to 1.2 V for SiNW transistors with N_{ET} = 2.0×10^{12} cm^{-2}. Such a high interface-trap concentration can be generated during repeated program-erase cycling in nonvolatile floating-gate and SNONS memory transistors [12]. The interface-trap effect is verified by TCAD simulation data. Very good agreement is achieved between the model computation and TCAD simulation data. Fig. 5 shows the comparison of drain-source current between devices with and without interface traps. These curves show that the slope of the drain-source current with interface traps is lower than the counterpart without traps because the existence of interface-trap charges lowers the surface carrier density and gate-voltage modulation as anticipated by (7). Thus, interface traps along the channel region can impact the behavior of drain-source current in SiNW. It is important to include interface traps when investigating device properties,

particularly for a mid-life device with high interface-trap density.

IV. CONCLUSIONS

In conclusion, an interface-trap model with energy distribution in the silicon energy gap is included into the unified compact model to investigate the effect of interface traps on the device characteristics. Devices with and without interface traps are also studied to show the difference in the current-voltage curves. The trapped charges at the SiO$_2$/Si interface along the surface channel region can significantly impact the surface potential, which affects the drain-source current and terminal charges. This interface-trap effect is usually neglected in conventional MOSFET compact models for designing logic circuits. The presented model is based on the device physics of bias dependence of interface-trap charges, which can well characterize the effect on transistor performance. The modeling approach has been verified by the TCAD simulation data with excellent agreement in the silicon-nanowire MOSFETs.

ACKNOWLEDGMENT

This work was supported by Lee Kuan Yew Postdoctoral Fellowship from Nanyang Technological University. The original work on interface traps was supported by CTSAH Associates founded by late Linda Chang Sah.

REFERENCES

[1] ITRS International technology roadmap for semiconductors 2005. http://www.itrs.net/Links/2005ITRS/Home2005.htm

[2] S. L. Jang and S. S. Liu, "An analytical surrounding gate MOSFET model," Solid State Electron., vol. 42, pp. 721–726, 1998.

[3] A. Kumar, M.V. Fischetti, T. H. Ning, E. Gusev, "Hot-carrier charge trapping and trap generation in HfO2 and Al2O3 field-effect transistors," J. Appl. Phys., vol. 94, pp.1728, 2003.

[4] M. A. Alam, H. Kufluoglu, D. Varghese, S. Mahapatra, "A comprehensive model for PMOS NBTI degradation: Recent progress," Microelectronics Reliability, vol. 47, pp. 583–862, 2007.

[5] X. Zhou, G. J. Zhu, G. H. See, J. B. Zhang, S. H. Lin, C. Q. Wei, Z. Chen, M. Srikanth, Y. F. Yan, R. Selvakumar, and W. Chandra, "Unified compact modeling for Bulk/SOI/FinFET/SiNW MOSFETs," in Proc. IEDST, pp. 1–6, 2009.

[6] G. J. Zhu, X. Zhou, G. H. See, S. H. Lin, C. Q. Wei and J. B. Zhang, "A unified compact model for FinFET and silicon nanowwire MOSFETs" in Proc. NSTI Nanotech 2009, pp. 588–591, 2009.

[7] J. C. Slater, Insulators Semiconductors and Metals Quantum Theory of Molecules and Solids, McGraw, 1967, vol. 3, pp. 292–307.

[8] Z. Chen, B. B. Jie, and C.-T. Sah, "Effects of energy distribution of interface traps on recombination DC current-voltage lineshape," J. Appl. Phys., vol. 100, 114511, 2006.

[9] C.-T. Sah, "Equivalent circuit models in semiconductor transport for thermal, optical, auger-impact, and tunneling recombination-generation-trapping processes," Phys. Status. Solidi., vol. 7, pp. 541–549, 1971.

[10] Z. Chen, B. B. Jie, and C.-T. Sah, "High concentration effects of neutral-potential-well interface traps on recombination DC current-voltage lineshape in metal-oxide-silicon transistors," J. Appl. Phys., vol. 104, 094512, 2008.

[11] C.-T. Sah, "DCIV Diagnosis for Submicron MOS Transistor Design, Process, Reliability and Manufacturing," in Proc. ICSICT, pp. 1-15, 2001.

[12] J. D. Lee, J. H Choi, D. Park, and K. Kim, "Effects of interface trap generation and annihilation on the data retention characteristics of flash memory cells," IEEE Trans. Device and Materials Reliability, vol. 4, pp. 110–117, March 2004.

978-1-4244-5430-3/10 $26.00 © 2010 IEEE

Applicability of Dual Layer Metal Nanocrystal Flash Memory for NAND 2 or 3-bit/cell Operation: Understanding the Anomalous Breakdown and Optimization of P/E Conditions

Pawan Singh, Sandhya C, Kshitij Auluck, Gaurav Bisht, Sivatheja M, Ralf Hofmann[*]
Gautam Mukhopadhyay, and Souvik Mahapatra

Pawan K Singh, Sandhya C, Kshitij Auluck Gaurav Bisht, Sivatheja M, Gautam Mukhopadhyay, Souvik Mahapatra
Indian Institute of Technology Bombay
Mumbai-400076. India
Ph: +91-22-25767412, e-mail: souvik@ee.iitb.ac.in

Ralf Hofmann,
Applied Materials
Santa Clara, CA, 95054 USA

Abstract— Large memory window (6-9V) program/erase (P/E) cycling endurance is studied for evaluating their suitability for MLC operation. Effect of NC area coverage and device size is evaluated using statistical method. Constant voltage stress (CVS) measurements and 2-D simulations are extensively used to evaluate the impact of carrier; type, fluence, and energy on the defect generation process in the gate stack. Degradation during P and E are isolated to allow individual optimization for improving the cycling reliability. P/E cycling endurance >10^4 at 8V MW and >2.5×10^3 at 9V MW are shown for first time in metal NC memory devices using the proposed distributed cycling scheme.

Keywords-component; Metal nanocrystal, Flash memory, MLC, reliability

I. INTRODUCTION

Physical scaling of floating gate (FG) Flash [1] cell size is unlikely below the 30nm node , 2bit/cell multi-level-cell (MLC) techniques and recently 3-bits/cell super-MLC (SMLC) are being investigated to continue density scaling. In addition, localized charge storage structures like metal nanocrystal (NC) and charge trap (CTF) devices are considered as possible alternatives for FG devices. Discretization of charge storage increases reliability of the memory device by providing immunity to complete charge loss through a single defect, reduction in TO and CD thickness and better channel control. So far, reported CTF devices have shown good memory window (MW) but poor retention , while standard single layer NC devices poor memory window . Poor retention of CTF is linked to shallow trap depth of the nitride storage layer which is an inherent material property and is difficult to control. Choice of tunable workfunction (WF) of the metal-NC storage layer is advantageous for achieving good retention, provided

optimal NC area coverage (AC) is obtained for good memory window .

Use of dual layer (DL) NC structure is reported in literature to increase the memory window over single layer (SL) NC structure. Author's recent work has shown a significant improvement in P/E cycling reliability of DL device over SL devices without adversely affecting retention reliability, making DL devices suitable for NAND applications.

In this work, optimized dual layer (DL) metal nanocrystal (NC) Flash memory with large workfunction storage node is proposed as suitable FG replacement. Large memory window (6-9V) program/erase (P/E) cycling endurance is studied for evaluating their suitability for MLC and SMLC operation. Effect of NC area coverage and device size is evaluated using statistical method. Constant voltage stress (CVS) measurements and 2-D simulations are extensively used to evaluate the impact of carrier type, fluence, and energy on the defect generation process in the gate stack. Degradation during P and E are isolated to allow individual optimization for improving the cycling reliability. P/E cycling endurance >10^4 at 8V and >2.5×10^3 at 9V memory window are shown for first time in metal NC memory devices using the proposed distributed cycling scheme. Overall this work shows the excellent reliability of the DL devices, making them suitable choices for FG replacement at sub 30nm technology nodes, provided some technological integration issues are solved.

II. FABRICATION DEAILS

Device fabrication is performed on low resistivity p-type wafers with doping of ~1×10^{15} cm^{-3}. After surface cleaning, 40Å SiO$_2$ was thermally grown using in-situ steam generation

(ISSG) process as tunnel oxide (TO) in an Applied Materials Centura RTP tool. Subsequently, a thin film of Pt is deposited and annealed to form metal NCs. 60Å Al_2O_3 film is deposited as the inter layer film (ILF). This is followed by another thin film of Pt and NC-anneal to form the second NC layer. 120Å Al_2O_3 is then deposited and annealed to form the control dielectric (CD). 1000Å Pt metal is deposited as control gate (CG) to complete the device formation process. Schematic and process detail of the DL devices is shown in Fig.1.

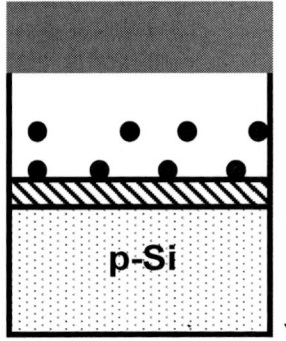

Wafer Clean
Tunnel Oxide: ISSG (40Å)
Surface Preparation
Pt Deposition: 5-10Å
NC Anneal
ILF Deposition: 40–60Å Al_2O_3
Pt Deposition: 5-10Å
NC Anneal
Control Dielectric: 120Å Al_2O_3
Post Deposition Anneal
Metallization: 1000Å Pt

Figure 1. Schematic of SL and DL NC devices showing the random position of NC in the gate stack.

TABLE I.   DL DEVICE SPLITS WITH THE RESPECTIVE LAYER AREA COVERAGE AND NUMBER DENSITY

Device ID	$Pt/Al_2O_3/Pt$ (Å)	Area Coverage	Density (#/cm^2)
DL3	5Å/ 60Å / 5Å	26%	3.40×10^{12}
DL9	10Å/ 60Å /10 Å	30%	2.50×10^{12}

III.   ANALYTICAL RESULTS

Cross-section and plan-view TEM images of DL devices in Fig. 2(a) and Fig. 2(b) respectively, show formation of discrete NC embedded in the gate dielectric. NC layers of the DL device are clearly separated from each other by Al_2O_3 ILF.

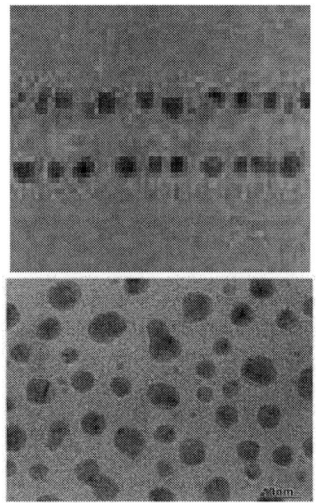

Figure 2. Cross sectional and plan-view TEM images

Plan-view TEM image is used to extract information regarding the number density, area coverage and size distribution of the NCs listed along with the device splits in table I. Area coverage is observed to be in the range of 26 – 30%, while number density is >2.5×10^{12} cm^{-2}. The average size of NC for 5Å Pt is ~3nm, while it is ~4nm for 10Å Pt film. Area coverage and number density were estimated to be between ~26% - 30% and 2.5×10^{12} – 3.4×10^{12} cm^{-2} respectively. Average NC size is ~4nm. More information on NC formation and statistics generation can be found in [18]. Obtained area coverage is found to be very close to the optimal area coverage required for good performance .

IV.   RESULTS AND DISCUSSION

A.   Memory Window and Retention

Capacitors with 100µm diameter are used for electrical testing. Fig. 3 shows the memory window and retention characteristics of the DL devices. Fig. 3 (a) shows the memory window of DL3 and DL9 with P/E performed at ±19V - ±21V measured after 10ms. Memory window is obtained by measuring V_{FB} shifts using HFCV measurements performed at 1MHz. Large memory window and overerase (due to low-leakage CD film) in both devices allowing split window operation. High quality low leakage CD film prevents tunneling of charge from the NC to control gate during programming and (along with high WF metal gate) prevents electron injection from CG during erase.

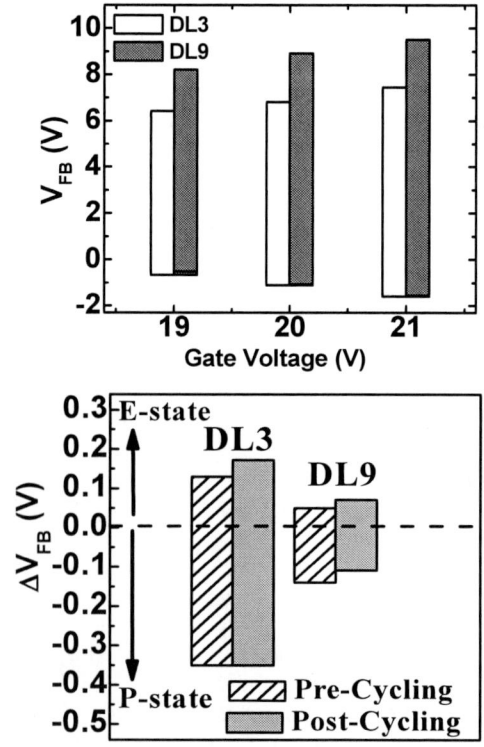

Figure 3. Memory window at ±20V V_G (b) pre and post-cycling

Retention loss in the devices is observed to be small as shown in Fig. 3(b). The measured retention loss at 25°C is −0.3V and −0.1V respectively for DL3 and DL9 (Fig. 3(b)). E-state loss is even smaller and shows insignificant change after P/E cycling (measured after 10^4 cycles with 6V memory window). The retention loss shows small increase with temperature from both the P- and E-states (not shown).

B. Endurance

P/E cycling endurance measurements are performed with a constant pulse program and constant pulse erase, with no correction used to control device V_{FB} during the measurement. All P/E cycling is performed at room temperature. The measurements are stopped after 10^4 cycles or breakdown, whichever occurs earlier. Fig. 4 (a) and (b) show the P/E cycling transients of DL9 at 7V, and 8V memory window from different staring P and E levels. No degradation in the cycling memory window is noted, but both P and E levels shift negative due to permanent hole trapping [22]. Fig. 4(c) lists the maximum achievable cycles with 6 – 9V memory window.

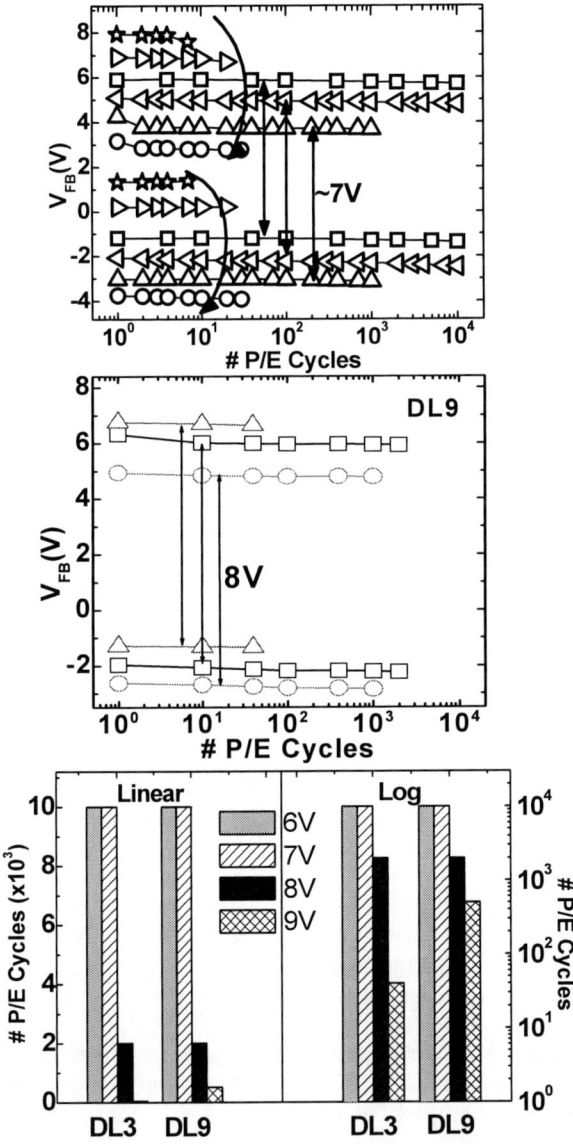

Figure 4. (a) P /E cycling transients at 7V (DL9), (b) P /E cycling transients at 8V (DL9)and (c) maximum endurance possible with 6 – 9 V memory window for MLC and SMLC operation

DL3 and DL9 show 10^4 P/E cycles endurance at 6 – 7V memory window, which is sufficient for MLC, but is insufficient for SMLC, which will require 8 – 9V MW for maintaining good separation between levels. DL9 could be cycled to ~7.5×10^4 cycles at 6V memory window before breakdown, when the measurement is not interrupted after 104 cycles. At 8V memory window, both DL3 and DL9 show 2×10^3 P/E cycle endurance. Increasing the memory window to 9V causes breakdown in DL9 after ~500 cycles, while 9V memory window could not be sustained in DL3 for more than a few cycles. DL9 therefore shows better cycling endurance than DL3, suggesting NC area coverage dependence of P/E cycling reliability. Understanding the defect generation mechanism in the gate stack is therefore most crucial for improving the overall endurance reliability.

Statistical distribution in the breakdown point of devices with P/E cycling must be established for quantitative assessment. Fig. 5 shows the results P/E cycling in DL devices at 9V memory window. The measurement is performed on 10 devices for each case. Impact of NC area coverage and device area is studied. Devices with diameter of 100µm and 200µm are studied in DL3 and 200µm in DL9. From table 1, DL9 has larger NC area coverage of ~30%, while DL3 has area coverage of ~26%. P/E cycling endurance is found to be dependent on both device size and NC AC. DL3-200µm devices show breakdown at lower cycles compared to DL3-100µm devices, and DL9-200µm (higher NC AC) shows better reliability than DL3-200µm under identical memory window of 9V. If the device area scaling trend for breakdown cycles holds for smaller sizes, the practical NC cycling will be significantly larger in the sub 30nm dimension devices,

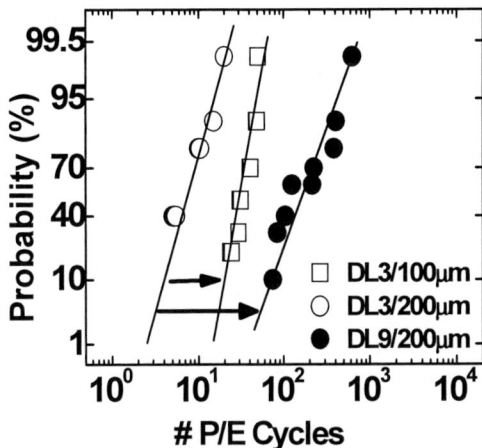

Figure 5. Impact of device size scaling and NC area on the breakdown of DL NC devices at 9V memory window.

Impact of P/E voltage is measured by varying the P and E V_G separately, while keeping the total memory window constant. Fig. 6 shows the impact of independently varying P/E V_G on the maximum number of P/E cycles obtained in DL3 measured at P- and E-states fixed constant at +6V and -2V respectively (8V memory window). Reducing E V_G causes endurance to increase to 5000 cycles, while changing P V_G does not have any impact, and the breakdown is roughly at 3000 cycles.

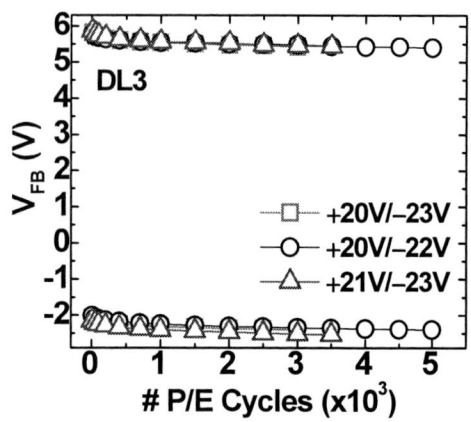

Figure 6. Impact of P and E voltage on the P/E cycling endurance reliability at equal MW

Measurements performed by keeping P(E)-states constant while varying the E(P)-state. Fig. 7(a) shows early breakdown on increasing (negative) E-state by increasing E-V_G from -20 to $-24V$, keeping P-state and P-V_G constant in DL3. Fig. 7(b), in contrast, shows no impact of P V_G on device BD when E-state is kept constant ($-24V$) and P-state is varied by varying V_G from $+18$ to $+21V$. BD occurs within measured range of error as shown in Fig. 5. Cycling also improves if breaks are included in between cycle steps, possibly due to annealing of defects. BD is found to be abrupt even with 10 cycle measurement steps.

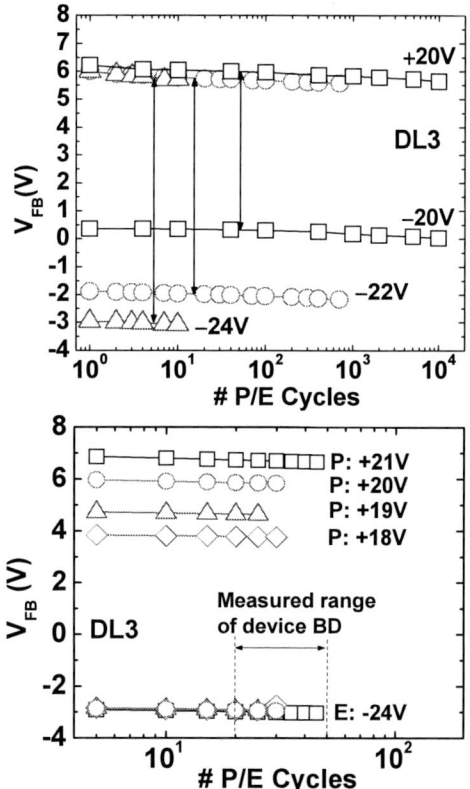

Figure 7. P /E cycling transients (DL3) with (a) constant P and variable E, and (b) constant E, variable P. P and E levels are varied by changing applied V_G.

C. Constant Voltage Stress

Constant voltage stress (CVS) measurements are performed with negative V_G ($-V_G$ CVS) and positive V_G ($+V_G$ CVS) to isolate and study defect generation process during P and E V_G. This is essential to isolate the impact of carrier type, fluence and energy on the breakdown behavior of the NC gate stacks. During the measurement, constant V_G is applied to the control gate and the corresponding gate current is measured. Current compliance of 1mA is used to detect breakdown. Fig. 8 (a) and (b) show current transients during $-V_G$ and $+V_G$ CVS in DL3. Under $-V_G$ CVS, devices show hard BD (HBD) (gate current reaches compliance) after a few soft BD (SBD) events. The HBD is prominent and is noted in every device. In contrast, $+V_G$ CVS shows SBD, but no real HBD (I_G never reaches compliance). Similar trends are also noted in DL9, $-V_G$ CVS causes HBD, while $+V_G$ CVS causes only SBD.

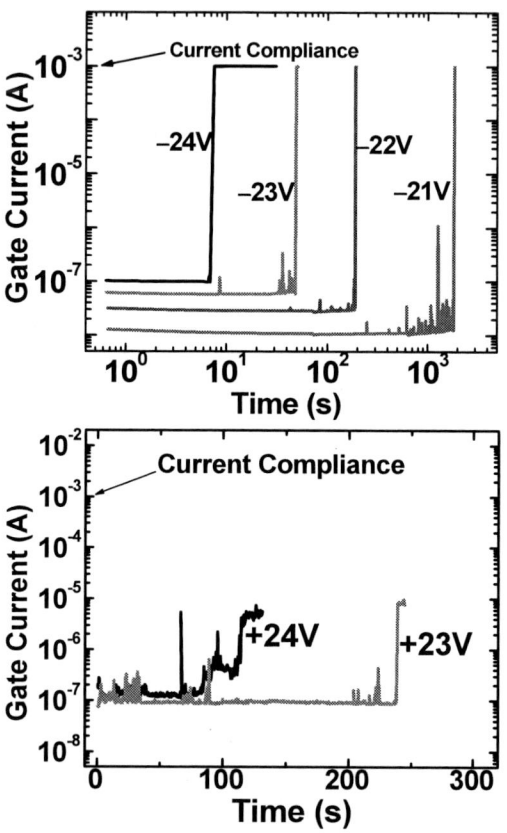

Figure 8. Current transients during $-V_G$ CVS and $+V_G$ CVS till breakdown.

Fig. 9 shows the time to BD in DL3 after $-V_G$ CVS. Time to breakdown reduces significantly with increase in (negative) V_G during stress. Time to BD (TBD) after $-V_G$ CVS follows similar trend as variable E cycling. The breakdown time follow a linear trend on the log(time)-linear(V_G) plot, indicating exponential dependence of the breakdown time on V_G. Since the time to BD could not be determined unambiguously in the $+V_G$ CVS, similar plot could not be drawn.

To prove that $+V_G$ CVS does not cause complete HBD, P/E cycling is performed on one device after $+V_G$ CVS. Fig. 10 shows the comparison between the P/E cycling endurance of a

978-1-4244-5430-3/10 $26.00 © 2010 IEEE

fresh device vs. stressed device (+23V CVS). The inset shows the gate current transient during the +23V V_G CVS. P/E cycling is performed immediately after the end of stress. 1500 cycles can be achieved after +V_G CVS compared to 3000 cycles in an unstressed device under identical P/E conditions, proving the lack of complete device failure after +V_G stress.

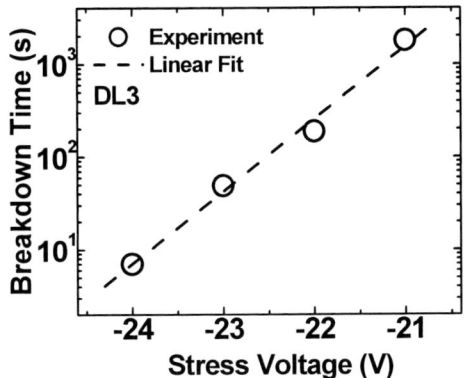

Figure 9. Time to BD vs. stress V_G for –V_G CVS.

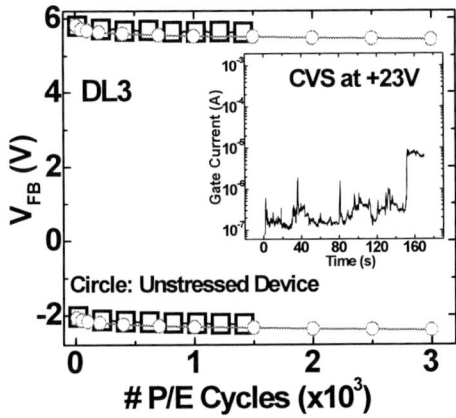

Figure 10. Comparison of P/E cycling of stress (+V_G CVS) and unstressed device (inset) current transients at +23 V_G CVS

Fig. 11 shows the comparison of fresh device CV, CV after +V_G CVS, after –V_G CVS and CV after P/E cycling BD.

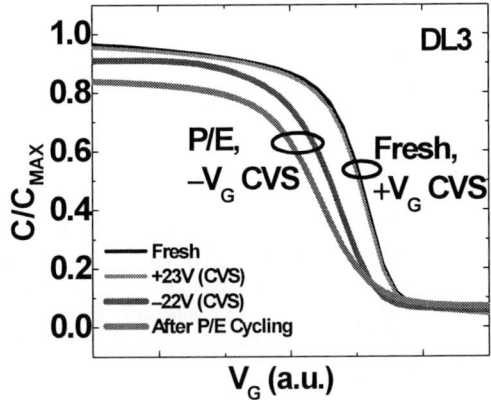

Figure 11. HFCV Comparison of Fresh CV, CV after +V_G CVS, CV after –V_G CVS stress, and CV after P/E BD.

HFCV measurements after +V_G CVS shows excellent similarity to fresh device HFCV, while HFCV after –V_G CVS and P/E cycled devices after BD are similar. It should be noted that HFCV measurements are possible even after P/E cycling induced breakdown and –VG CVS breakdown, the measured conductivity increases significantly causing the decrease in C_{MAX}. The conductivity of the +VG CVS device is similar to the fresh device CV.

The HFCV measurements also confirm the similarity between the device breakdown caused by –V_G CVS and P/E cycling; both result in HBD of the gate stack preventing further programming of the device. +V_G CVS however, only causes a reduction in the sustainable number of cycles, but does not cause HBD by itself. The defects generated during +V_G CVS may cause rapid degradation during further P/E cycling, thus causing an earlier BD. Therefore no definite trend in P/E cycling could be noted by varying the +VG during P/E cycling in Fig. 7 (b).

D. Physical Mechanism

In author's previous paper , 2-region model for NC gate stack degradation was proposed to explain the observed endurance reliability, post-cycling MW and retention of single layer and DL devices. NC device gate area was divided in two regions; region 1 (R1) was defined as the area under NC coverage and region 2 (R2) is the area between NC with no NC coverage. Degradation was proposed to occur in R2 (no NC coverage) due to holes injected during erase. Holes injected in R2 cause impact ionization in the CD/ILF causing defect generation and permanent trapping. Holes injected in R1, reach the NC and get thermalized, thus becoming unavailable for defect generation in R1. Since the P/E and retention characteristics are primarily governed by quality of the dielectric directly above and below the NC (R1), the model explains the negligible degradation in post-cycling P/E, retention and P/E cycling endurance and the eventual breakdown behavior by dielectric degradation only in R2. Breakdown of NC devices is explained by the leakage path formed in R2 which shorts CG and substrate preventing charging of NC. Difference in DL3 and DL9 endurance can be explained by the difference in area coverage of the NC layers, where DL9 with larger area coverage shows better endurance.

We performed 2-D simulations on DL NC devices to understand the electrostatics and the current flow in the devices, during erase and program operation. Fig. 12 (a) and (b) show the simulated electric field lines for 25% – 35% area coverage (for one NC layer) and hole current density (25% area coverage) respectively. Electric field lines (using method in) at –20V V_G for 25% - 35% NC AC show considerable reduction in R2 for 30% area coverage over 25% area coverage, explaining the better reliability of DL9 compared to DL3. Simulations also predict elimination of R2 at ~35% area coverage (simulation procedure details in). By incorporating current flow in the gate stack, the model is expanded to include the degradation due to both +V_G and –V_G observed during P/E. Simulations suggest that during erase, majority of charge transport is by hole tunneling from the substrate, which increases considerably with (negative) E V_G as shown in Fig. 12(b). Increase of ~200 times is noted in hole current density as the V_G increases from -18V to -24V. The CD electric field increases from 5.5MV/cm to 7.5MV/cm for the same corresponding increase in erase V_G. Electron injection from the CG during erase is found to be small, as expected due to the observed overerase, while electrons are the only carriers involved during program. Since electric field in R1 and R2 are slightly different due to localized effects, there is a small difference in the value of simulated hole current density.

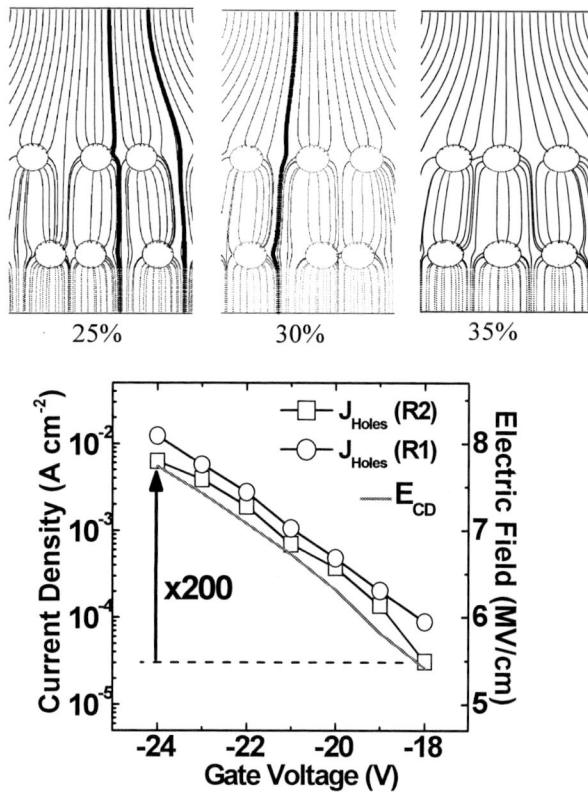

Figure 12. (a) Electric field lines showing the path of current flow, dark lines indicate R2 with AC from 25% - 35% and (b) Simulated current Increase in hole current density and CD electric field with V_G

Fig 13 (a) and (b) show the simulated band diagram in R2 at ±20V at start of P/E. During P, injected electrons (sole carriers) travel through the gate stack gaining energy, some of them cause defect generation in the only in the CD (energy >9eV), and is monitored as SBD events during +V_G CVS. During E, holes injected from the substrate cause impact ionization in the ILF/CD, while electrons injected from the CG cause defect generation in the TO/ILF, leading to formation of short between the substrate and gate, similar to –V_G CVS measurements which show HBD and complete device failure. Defects generated during P do not cause catastrophic failure, but will aid defect generation during E, causing early failure.

E. Distributed P/E Cycling

An intelligent distributed P/E cycling scheme utilizing P to intermediate levels is proposed to improve the cycling reliability. Step size of 10 cycles is chosen to more closely resemble practical device usage where the data can be randomly written to even the intermediate levels. Fig. 14 shows the results of the distributed cycling scheme on DL9 device. By

use of this scheme, 3×10^4 P/E cycles at P/E: +6V/-2V, and 10^4 P/E cycles at P/E: +7V/-1 could be achieved without breakdown with minimal P/E degradation. In contrast, only 2000 and 50 P/E cycles could be obtained in the normal scheme at the P/E: +6V/-2V and P/E: +7V/-1V respectively. 2.5×10^3 cycles at P/E: +7V/-2V (9V MW) are possible using the new scheme in comparison to ~40 cycles using the normal method. Similar trends are also seen in DL3 with considerable improvement in P/E cycling endurance using this scheme. It is therefore argued that P/E cycling to maximum memory window severely underestimates the endurance reliability of the devices. This scheme also uses optimized P/E V_G reducing defect generation in the gate stack for improved reliability,

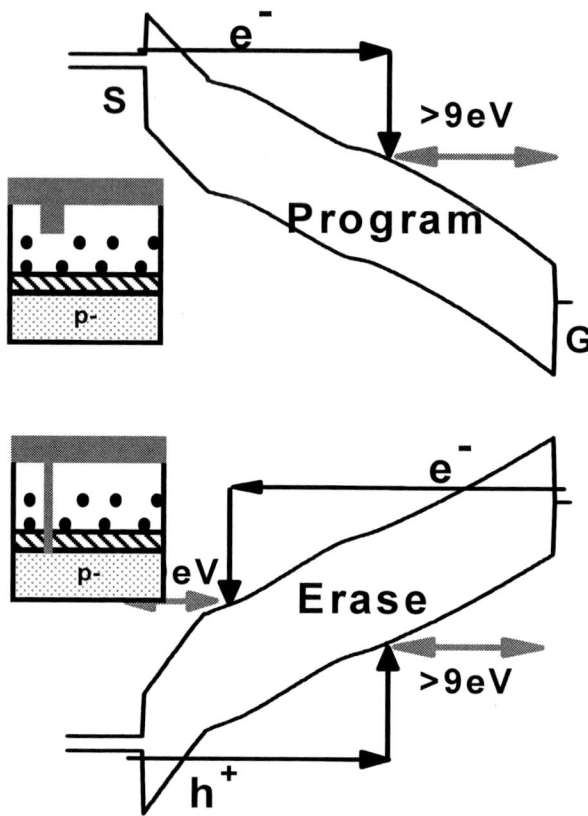

Figure 13. Simulated band diagrams in R2 showing electron and hole current during P/E at ±20V. Double sided arrows mark the region of defect generation where carrier energy >9eV.

V. CONCLUSIONS

In conclusion, optimization of P/E conditions and gate stack optimization are identified as the most crucial parameters for the successful operation of the NC memory devices. Lowest possible E V_G must be used to prevent defect generation in the gate stack. This work clearly demonstrates the unique breakdown behavior of NC devices, and shows impact of different parameters like device area, NC AC, V_G, and carrier fluence/energy. Large area coverage of the NC is required for reliable operation. Finally, a more realistic approach to testing P/E cycling endurance reliability at large MW (8-9V) is demonstrated with >10^4 P/E cycle endurance, sufficient to meet NAND Flash requirements .

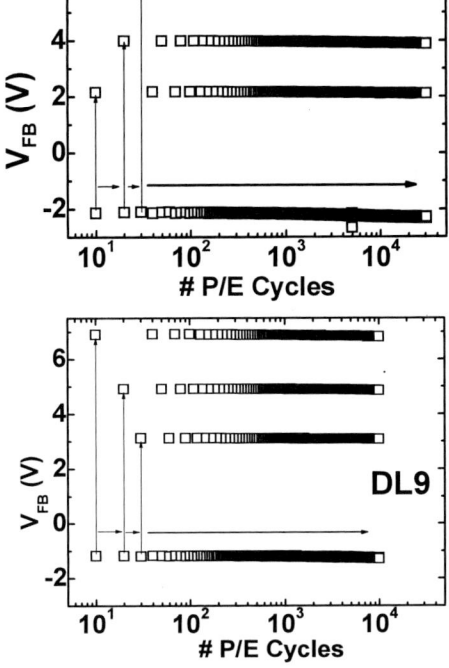

Figure 14. Optimized intelligent cycling scheme showing enhanced P/E cycling endurance at 8V memory window with (a) +6V/-2 V and (b) +7V/-1 V V_{FB} states.

ACKNOWLEDGMENT

The authors acknowledge partial funding received from the Department of Information Technology, Government of India, through the Centre of Excellence in Nanoelectronics. Authors also acknowledge Dr. Omkaram Nalamasu from Applied Materials, Santa Clara for his support to this project.

REFERENCES

[1] P. Cappelletti, C. Golla, P. Olivio, and E. Zanoni, Flash Memories. Boston, MA: Kluwer Academic, 1999.

[2] International Technology Roadmap for Semiconductors, ITRS 2008, available at http://www.itrs.net.

[3] J. H. Park, *et al.*, "8Gb MLC (multi-level cell) NAND flash memory using 63nm process technology," in *IEEE International Electron Devices Meeting*, IEDM, 2004, pp. 873-876.

[4] T. Kamigaichi, *et al.*, "Floating gate super multilevel NAND flash memory technology for 30 nm and beyond," in *International Electron Devices Meeting* IEDM 2008., p. 4.

[5] C. Lee, J. Meteer, V. Narayanan, and E. C. Kan, "Self-assembly of metal nanocrystals on ultrathin oxide for nonvolatile memory applications," *Journal of Electronic Materials*, vol. 34, no. 1, pp. 1-11, Jan.2005.

[6] H. Tuo-Hung, L. Chungho, V. Narayanan, U. Ganguly, and E. C. Kan, "Design optimization of metal nanocrystal memory - Part II: gate-stack engineering," *IEEE Transactions on Electron Devices*, vol. 53, no. 12, p. 7, Dec.2006.

[7] H. Tuo-Hung, L. Chungho, V. Narayanan, U. Ganguly, and E. C. Kan, "Design optimization of metal nanocrystal memory - Part I: nanocrystal array engineering," *IEEE Transactions on Electron Devices*, vol. 53, no. 12, p. 8, Dec.2006.

[8] C. Sandhya, *et al.*, "The effect of band gap engineering of the nitride storage node on performance and reliability of charge trap flash," in *International Symposium on the Physical and Failure Analysis of Integrated Circuits, IPFA* 2008.

[9] C. Sandhya, *et al.*, "Nitride engineering and the effect of interfaces on Charge Trap Flash performance and reliability," in *IEEE International Reliability Physics Symposium Proceedings* 2008, pp. 406-411.

[10] C. Sandhya, *et al.*, "Effect of SiN on performance and reliability of charge trap flash (CTF) under Fowler-Nordheim tunneling program/erase operation," *IEEE Electron Device Letters*, vol. 30, no. 2, pp. 171-173, 2009.

[11] M. H. White, D. A. Adams, and J. Bu, "On the go with SONOS," *IEEE Circuits and Devices Magazine*, vol. 16, no. 4, pp. 22-31, July2000.

[12] Y. Park, *et al.*, "Highly manufacturable 32Gb multi-level NAND flash memory with 0.0098 m² cell size using TANOS (Si-oxide-Al2O3-TaN) cell technology," in *IEEE International Electron Devices Meeting* 2006, p. 4.

[13] B. De Salvo, *et al.*, "Performance and reliability features of advanced nonvolatile memories based on discrete traps (silicon nanocrystals, SONOS)," *IEEE Transactions on Device and Materials Reliability*, vol. 4, no. 3, pp. 377-389, Sept.2004.

[14] S. Choi, *et al.*, "Al2O3 with Metal-Nitride nanocrystals as a charge trapping layer of MONOS-type nonvolatile memory devices," *Microelectronic Engineering*, vol. 80, pp. 264-267, June2005.

[15] C. Lee, T. H. Hou, and E. C. Kan, "Metal nanocrystal/nitride heterogeneous-stack floating gate memory," in *Device Research Conference*, 2005, pp. 97-98.

[16] S. K. Samanta, *et al.*, "Tungsten nanocrystals embedded in high-k materials for memory application," *Applied Physics Letters*, vol. 87, no. 11, pp. 113110-113111, Sept.2005.

[17] S. K. Samanta, et al., "Enhancement of memory window in short channel non-volatile memory devices using double layer tungsten nanocrystals," in *IEEE International Electron Devices Meeting* 2005, p. 4.

[18] R. Hofmann and N. Krishna, "Self-assembled metallic nanocrystal structures for advanced non-volatile memory applications," *Microelectronic Engineering*, vol. 85, no. 10, pp. 1975-1978, 2008.

[19] L. Chungho, A. Gorur-Seetharam, and E. C. Kan, "Operational and reliability comparison of discrete-storage nonvolatile memories: advantages of single- and double-layer metal nanocrystals," in *IEEE International Electron Devices Meeting* 2003, pp. 22-26.

[20] R. Seong-Wan, *et al.*, "A thickness modulation effect of HfO2 interfacial layer between double-stacked Ag nanocrystals for nonvolatile memory device applications," *Journal of Applied Physics*, vol. 101, no. 2, pp. 26109-1, Jan.2007.

[21] P. K. Singh, *et al.*, "Metal Nanocrystal Memory with Pt Single- and Dual-Layer NC with Low-Leakage Al2O3 Blocking Dielectric," *Electron Device Letters, IEEE*, vol. 29, no. 12, pp. 1389-1391, 2008.

[22] P. K. Singh, *et al.*, "Reliability of Single and Dual Layer Pt Nanocrystal Devices for NAND Flash Applications: A 2-Region Model for Endurance Defect Generation," in *IEEE International Reliability Physics Symposium Proceedings* 2009, pp. 301-306.

[23] A. Nainani, *et al.*, "Development of a 3D simulator for metal nanocrystal (NC) flash memories under NAND operation," in *Technical Digest - International Electron Devices Meeting, IEDM*, 2007, pp. 947-950.

Pattern-independent, Fine-morphology Ni-Pt Silicide Formation by Partial Conversion with Low Metal-consumption Ratio

Takuya Futase[1,4,*], Takeshi Kamino[2], Naoto Hashikawa[3], Yutaka Inaba[1],
Tetsuo Fujiwara[1], Hirohiko Yamamoto[1], and Hisanori Tanimoto[4]

[1]Wafer Process Manufacturing Technology Dept. 2, Renesas Electronics Corporation, Hitachinaka, Ibaraki 312-8504, Japan
[2]SOC Device Technology Dept., Renesas Electronics Corporation, Hitachinaka, Ibaraki 312-8504, Japan
[3]Process & Device Analysis Engineering Development Dept., Renesas Electronics Corporation, Kodaira, Tokyo 187-8588, Japan
[4]Graduate School of Pure and Applied Sciences, University of Tsukuba, Tsukuba, Ibaraki 305-8573, Japan
*Phone: +81-29-354-0748, e-mail address: takuya.futase.vt@renesas.com

Abstract—We applied partial conversion as initial silicidation to control the morphologies of Ni-Pt silicide, viz., the thickness, crystal grain, and Pt concentration of the Ni-Pt silicide. This partial conversion kept the thickness of Ni-Pt silicide constant regardless of the device pattern, i.e., by controlling silicidation with thermal diffusion. The key to partially converting Ni-Pt silicide was leaving a thick Ni-Pt alloy on the silicide, viz., a low metal-consumption ratio, at the narrow active line. This process made the crystal grain finer and enriched the Pt of Ni-Pt silicide, thereby suppressing the increase in resistivity in Ni-Pt silicide.

Keywords—crystal size; diffusion; metal-consumption ratio; Ni-Pt; rapid thermal annealing; resistivity; partial conversion; silicide

I. INTRODUCTION

Nickel-platinum mono-silicide (Ni-Pt silicide) [1]–[3] is a key material for 28-nm-node logic devices and beyond. It is important to keep the silicide thickness constant after the silicide process, regardless of the pattern, even between narrow and wide active lines. However, the silicide thickness on narrow active lines at such small dimensions is actually thinner than that on wide active lines, even if collimator sputtering is used. If the entire thickness is increased to compensate for a narrow active line, a wide active line becomes too thick, deteriorating its electrical properties. In contrast, if the entire thickness is decreased to fit the wide active line, the narrow active line becomes too thin, also deteriorating its electrical properties [4]. This trade-off in thickness is caused by the conventional two-step silicidation scheme, i.e., sputtered metal is fully transformed into silicide during the initial silicidation (full conversion) [5].

We propose the application of partial conversion instead of full conversion to enhance the electrical properties of Ni-Pt silicide with 28-nm-node dimensions. Partial conversion has been reported for pure nickel silicide; it effectively suppresses the formation of nickel di-silicide (NiSi$_2$) at the edge of the gate and the active line [5]. In contrast, we applied partial conversion to Ni-Pt silicide, solving the thickness trade-off between narrow and wide active lines, as shown in the silicidation scheme in Fig. 1. In partial conversion, a stack of

Ni-Pt alloy, di-metal silicide, and silicon is formed after the initial silicidation. We found that the morphology of the silicide after the second silicidation depends on the ratio between the sputtered Ni-Pt alloy thickness and that consumed during the initial silicidation (metal-consumption ratio). Thus, using the partial conversion with a low metal-consumption ratio improved the electrical properties of the silicide on the narrow active-line between the gates. We further found that a finer crystal-grain size and Pt enrichment in the Ni-Pt silicide formed by partial conversion.

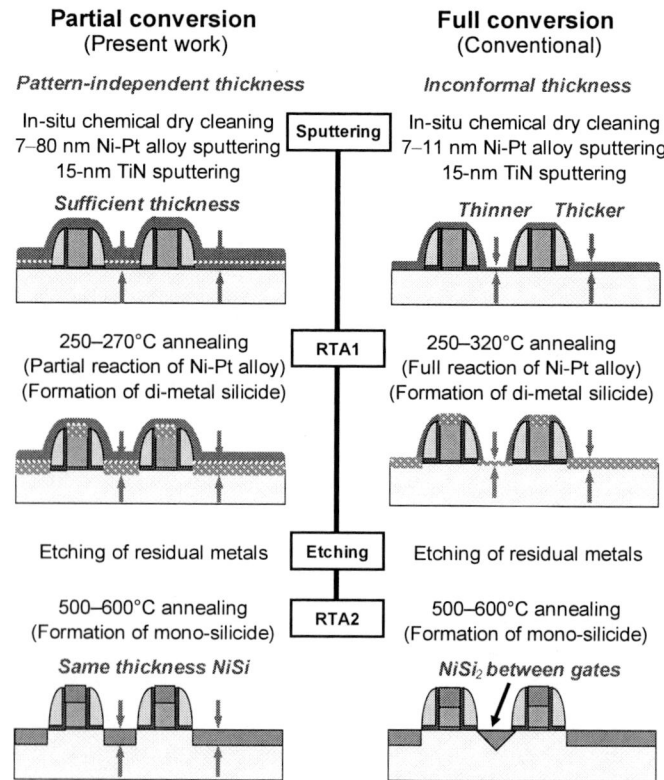

Fig. 1. Comparison of two-step silicidation scheme.

II. EXPERIMENTAL

A. Sample preparation

Logic devices with 28-nm-node dimensions were prepared on a 300-mm-diameter (001) silicon wafer to observe the cross section of n- and p-channel metal-oxide-semiconductor field-effect transistors (MOSFETs) with transmission electron microscopy (TEM) and to evaluate the electrical properties of p-channel MOSFETs. The doping concentrations of arsenic, phosphorus, and boron on source and drain region corresponded to 4×10^{15} cm^{-2}, 5×10^{14} cm^{-2}, and 4×10^{15} cm^{-2}. The gate length L_g and the active-line width W of the patterns used in the sheet-resistance measurements were minimally 38 nm for the former and 35 nm for the later. As seen in the process flow in Fig. 1, silicide was fabricated by using the two-step rapid thermal annealing (RTA) scheme as follows. *In situ* chemical-dry cleaning using a remote plasma of ammonia and nitrogen trifluoride [6] was carried out before sputtering Ni-Pt (5 atomic %) alloy with a collimator. A heat-conductor furnace with excellent uniformity in temperature was used for the initial RTA (RTA1) to transform Ni-Pt alloy into di-metal silicide. The Ni-Pt alloy was partially consumed by a diffusion-controlled reaction during the RTA1 at 250–270°C for 15–240 sec in a nitrogen ambient. The Ni-Pt alloy remaining on the silicide was etched and followed by spike annealing at 500–600°C in a nitrogen ambient as the second RTA (RTA2) to transform the di-metal silicide into mono-silicide [7]. The Ni-Pt alloys were sputtered with several thicknesses and annealed under several RTA conditions for comparison.

Blanket wafers were also prepared to investigate the silicide properties, viz., the crystal-grain size and Pt concentration of the silicide, as functions of the thicknesses of Ni-Pt alloys and RTA conditions. The doping conditions for the blanket wafers were the same as those for the above logic devices. The crystal-grain size of silicide was analyzed with TEM and the Pt concentration of silicide was quantified with ionized coupled plasma atomic emission spectroscopy (ICP-AES) using 5-cm-square specimens.

B. Definition of metal-consumption ratio (MCR)

We defined the metal-consumption ratio (MCR) to characterize the stack structure of residual metal, di-metal silicide, and silicon formed after the initial silicidation. The formation of the di-metal silicides is controlled by the mutual diffusion of metal and silicon. Therefore, the diffusion length is an indication of the amount of metal consumed by silicide formation,

$$d_\mathrm{MC} \propto \sqrt{Dt}, \tag{1}$$

where d_MC, D, and t correspond to the thickness of metal consumption, the diffusivity of the metal, and the reaction time during RTA1. The consumed metal thickness was estimated from the x-ray fluorescence intensity for nickel in the di-metal silicide after unreacted Ni-Pt alloy was removed from the silicide.

In partial conversion, the stack structure of residual metal, di-metal silicide, and silicon was formed by RTA1, as shown in Fig. 2. We defined the metal-consumption ratio R_MC in percent as the fraction of the consumed metal thickness b to the sputtered metal thickness a,

$$R_\mathrm{MC} = \frac{b}{a} \times 100, \tag{2}$$

where b is always smaller than a by definition because the metal-consumption ratio is less than a hundred percent. Furthermore, we took into consideration the metal-consumption ratio not only in partial conversion but also in full conversion in the following way. Obviously, a is equal to b in full conversion. However, even after all the sputtered metal has been consumed, annealing continues and can partially transform the di-metal silicide into mono-silicide. We denote this excessive thermal-budget as the thickness of possibly diffused metal c, as shown in Fig. 3, extending the definition of the metal-consumption ratio in full conversion by,

$$R_\mathrm{MC} = \frac{a+c}{a} \times 100. \tag{3}$$

The method by which c was obtained is graphically shown in Fig. 4 as the dependence of metal-consumption on the square root of the reaction time. The c is estimated as the exploded thickness from the metal consumed during partial conversion when there is sufficient sputtered metal from a, even if the diffusivity and reaction time have shifted because of changing

In partial conversion

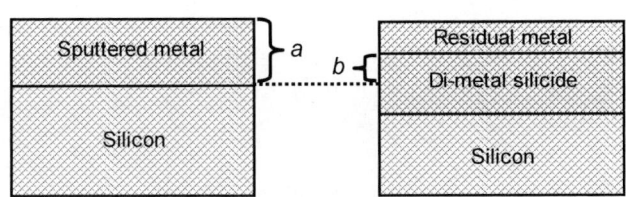

a : Sputtered metal thickness
b : Consumed metal thickness during RTA1

Fig. 2. Schematic model of stacked film before and after partial conversion.

In full conversion

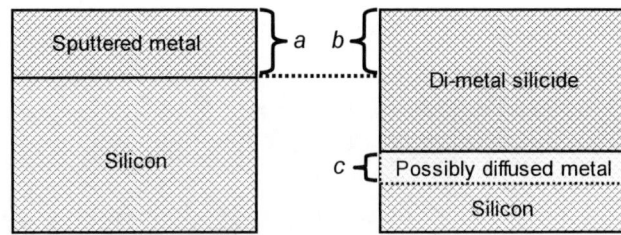

a : Sputtered metal thickness
b : Consumed metal thickness during RTA1
c : Possibly diffused metal during RTA1

Fig. 3. Schematic model of stacked film before and after partial conversion.

reaction temperature.

As previously mentioned, the metal-consumption ratio indicates, regardless of the temperature or the reaction time of RTA1, not only the extent of the thickness ratio between consumed and sputtered metal in partial conversion but also the amount of excessive thermal budget in full conversion. Consequently, by using the metal-consumption ratio, we can generally discuss the dependence of the physical quantities of silicide on one parameter regardless of partial or full conversion.

III. RESULTS AND DISCUSSION

A. Reducing thickness trade-off and making crystal grain finer

There are cross-sectional images of a transistor consisting of nominal 15-nm-thick silicide in Fig. 5. Both silicides on the gate and the narrow active line are nearly of equal thickness obtained by partial conversion. The p-channel MOSFET

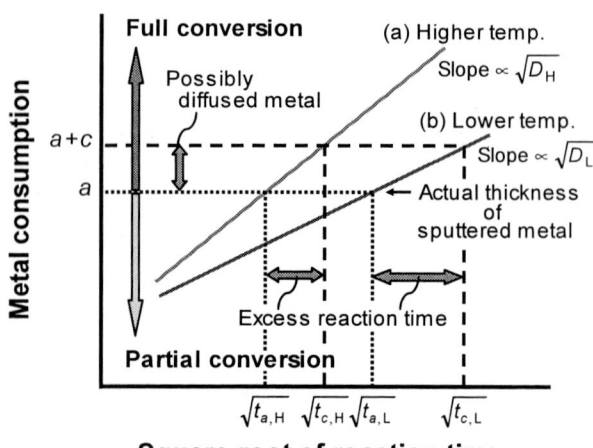

Fig. 4. Dependence of metal consumption on reaction-time at (a) higher temperature and (b) lower temperatures. The possibly diffusing metal is exploded as the square root of the product of diffusivity and excess reaction time, viz., $[D_H \times (t_{c,H} - t_{a,H})]^{1/2}$ and $[D_L \times (t_{c,L} - t_{a,L})]^{1/2}$.

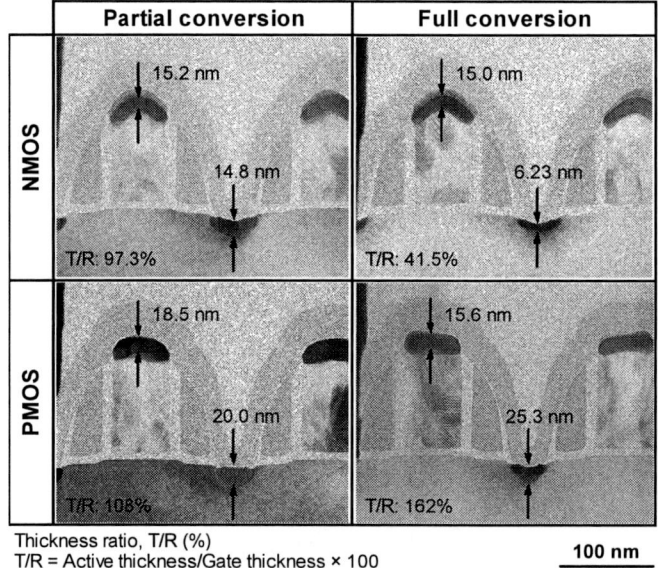

Thickness ratio, T/R (%)
T/R = Active thickness/Gate thickness × 100

Fig. 5. Reducing thickness trade-off between gate and narrow active line by partial conversion (gate length = 60 nm and active line width = 35 nm).

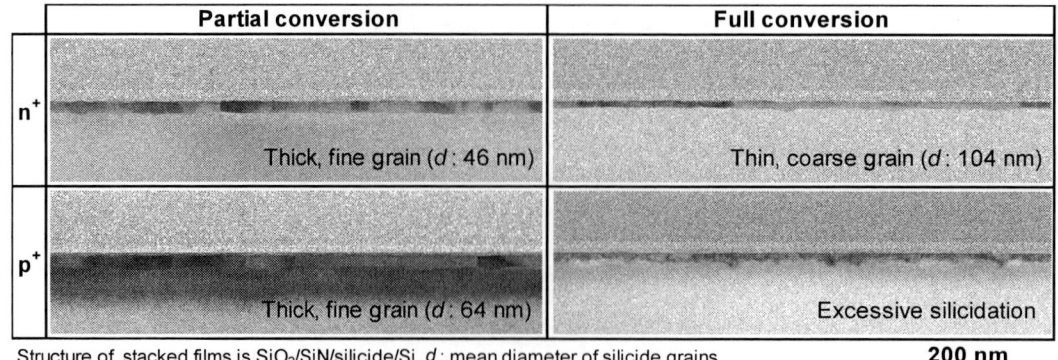

Structure of stacked films is SiO₂/SiN/silicide/Si, d: mean diameter of silicide grains

Fig. 6. Decreasing grain size of silicide on narrow active line between gates by partial conversion (active line width = 35 nm).

978-1-4244-5430-3/10 $26.00 © 2010 IEEE

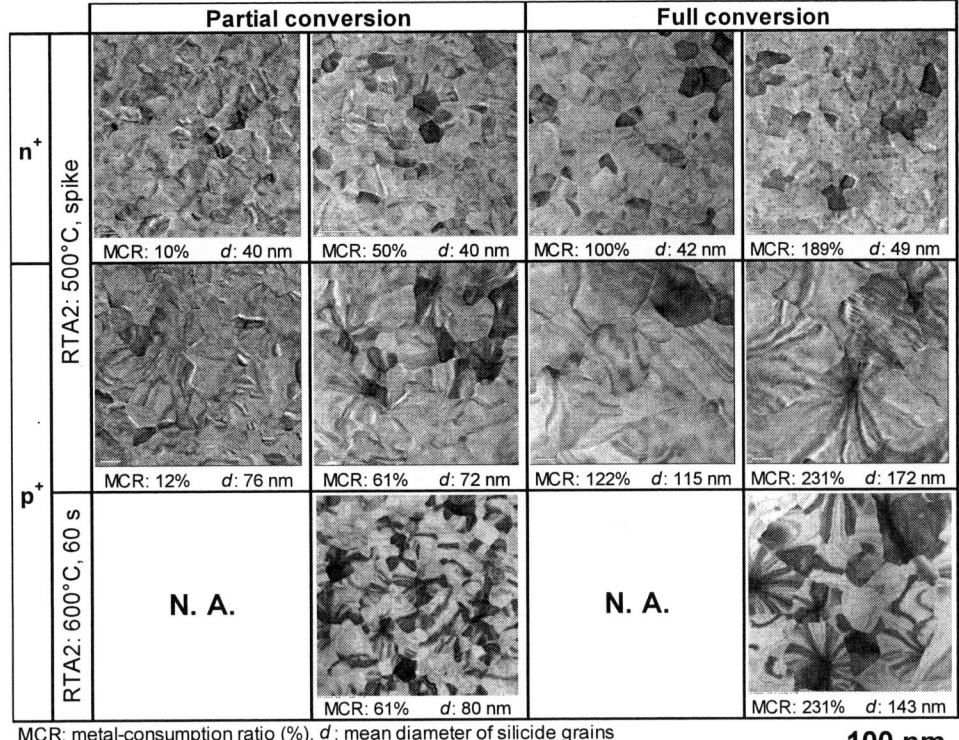

MCR: metal-consumption ratio (%), d: mean diameter of silicide grains

100 nm

Fig. 7. Comparison of silicide textures between fine grains by partial conversion and course grains by full conversion captured with TEM.

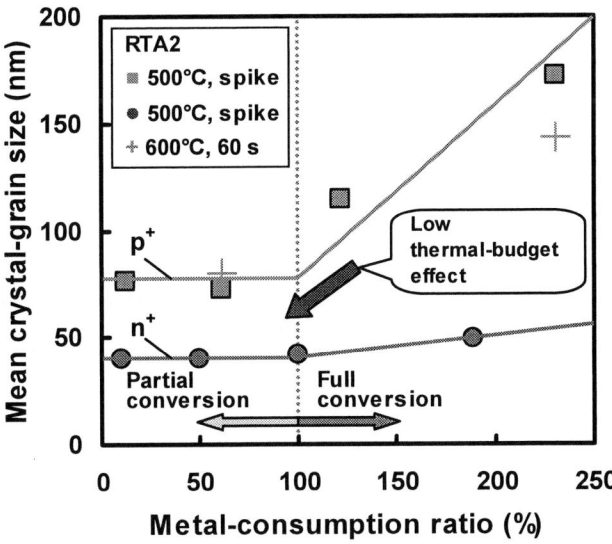

Fig. 8 Making crystal-grain size of blanket silicide finer after RTA2 by MCR <100%.

(PMOS) is thicker than n-channel MOSFET (NMOS), because the reaction rate of metal consumption on p^+-silicon is higher than that on n^+-silicon. In contrast, using full conversion, the silicide on the narrow active line of the NMOS has been thinned to about 40% of the thickness of the gate, and that of the PMOS has been thickened to about 160% of the thickness of the gate, forming a pyramidal shape. This indicates that the

silicide on the narrow active line of PMOS by using full conversion has been transformed into $NiSi_2$.

There are also cross-sectional images of a long active line in Fig. 6. In the n^+-doped active line, the grains of silicide obtained by partial conversion are fine, but those obtained by full conversion are coarse. In the p^+-doped active line, the grains of silicide are also fine, but those obtained by full conversion form a rough interface, viz., the $NiSi_2$.

To find the mechanisms responsible for decreasing the crystal-grain size obtained by partial conversion, the textures of the blanket wafers were also analyzed. The TEM images of silicides formed at several values of the metal-consumption ratio are compared in Fig. 7. As shown in Fig. 8, the crystal-grain size of silicide obtained by partial conversion is fine when the metal-consumption ratio is less than 100%. In contrast, that of silicide obtained by full conversion is coarse with a high metal-consumption ratio. As can be seen by comparing the temperature and annealing time with RTA2 in Fig. 8, the grain size of silicide depends on the RTA1 conditions and not on those of RTA2. Therefore, these results indicate an excessive thermal budget in metal-consumption ratios of more than 100% that coarsen the grains of silicide.

B. Enhancing resistivity on narrow active line

We analyzed the electrical properties of PMOS in detail, since its silicide morphology is significantly different to that of NMOS, i.e., the thicker active line, the smoother silicide-silicon interface, and the finer crystal grains of POMS than

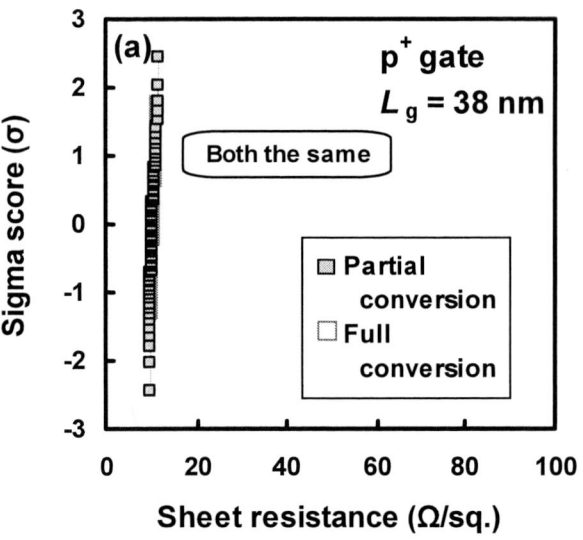 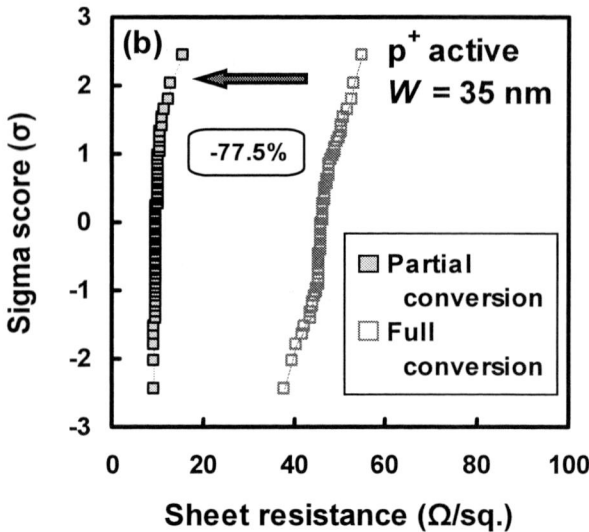

Fig. 9. Enhancing sheet resistance of nominal 15-nm-thick silicide on narrow p$^+$ active line (active line width = 35 nm) by partial conversion.

those of NOMS. Figure 9 shows the sheet-resistance distribution of 15-nm-thick silicide. The vertical axis plots the sigma score [8], which is determined by the number of sample, denoting the inverse function of a normal distribution function. The horizontal axis plots the sheet resistance of the silicide. Therefore, the reciprocal of the line slope coincides exactly with the standard deviation of the sheet resistance when the data follow the normal distribution. The results indicate that there is no difference in the sheet resistance of the gate between partial and full conversion. In contrast, there is a difference in the sheet resistance of the narrow active line between partial and full conversion, i.e., the sheet resistance of NiSi obtained by partial conversion was the same as that of the gate; however, the sheet resistance of NiSi obtained by full conversion was a 77.5 percent increase compared with that of partial conversion.

The enhancement obtained in the sheet resistance of the narrow active line by partial conversion was not attributed to the formation of NiSi$_2$, indicated by the active-line-width (approximately 395–35 nm) dependence on resistivity of nominal 15-nm-thick silicide. The resistivity between partial and full conversion is compared in Fig. 10. The resistivity may be determined by the following equation using the original resistance-measurement results,

$$\frac{R \cdot t}{L} = \frac{\rho}{\alpha} \cdot \frac{1}{W}, \tag{4}$$

where R, t, L, ρ, α, and W correspond to the measured resistance, the thickness of the consumed Ni-Pt alloy, the length of the active line, the resistivity of the silicide, the coefficient of the silicide phase, and the width of the active line. The α for the NiSi and NiSi$_2$ phases are 2.1 for the former and 3.3 for the later. The incline in the plot in Fig. 10 indicates that the silicide produced by partial conversion (MCR 50%) has a resistivity of 14.6×10^{-8} $\Omega \cdot$m, which is consistent with the value of NiSi and not that of NiSi$_2$ [5], [9]–[12]. However, the

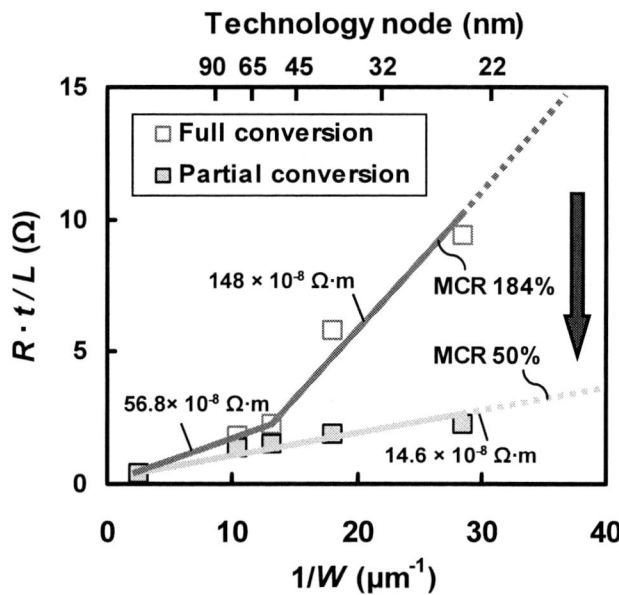

Fig. 10. Enhancing resistivity of nominal 15-nm-thick silicide on less than 55-nm-narrow p$^+$ active lines by partial conversion.

silicide produced by full conversion (MCR 184%) has higher resistivity than that by partial conversion, particularly, with active lines that are less than 45-nm wide.

The dependence of the resistivity of Ni-Pt silicide on the metal-consumption ratio under several conditions is also compared in Fig. 11. The resistivity of silicide at metal-consumption ratios of more than 150% is significantly higher than that of NiSi$_2$. This could indicate that the silicide is disconnected or agglomerated [4], [7]. In full conversion at metal-consumption ratios of less than 150%, the resistivity of the silicide is almost the same as that of NiSi$_2$ [13]. Even if partial conversion was used, the resistivity of the silicide was

978-1-4244-5430-3/10 $26.00 © 2010 IEEE

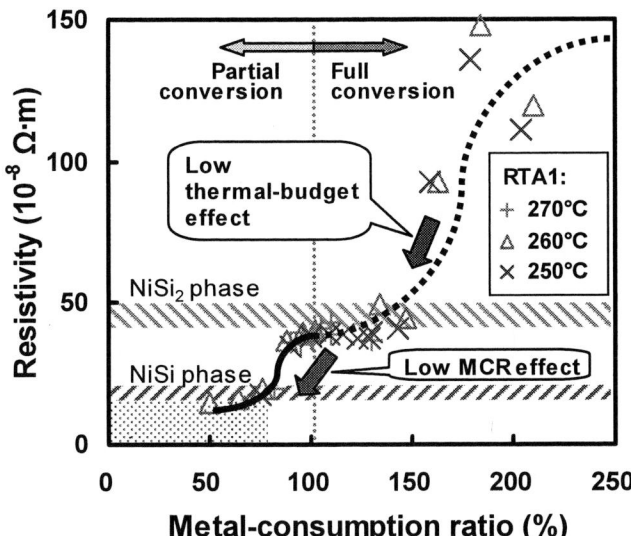

Fig. 11. Achieving resistivity of NiSi phase by using MCR <80%.

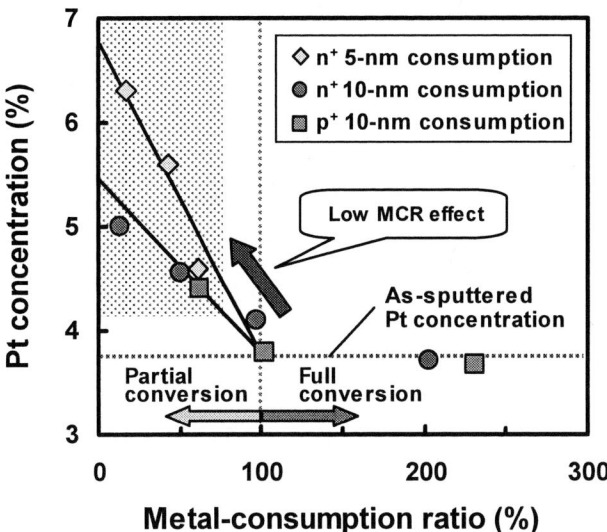

Fig. 12. Enriching Pt by using MCR <100%. (ICP-AES after RTA2 of blanket silicide).

close to that of $NiSi_2$ at metal-consumption ratios from 80% to 100%. In contrast, the resistivity of silicide decreased to as low as that of NiSi at metal-consumption ratios of less than 80%.

C. Enriching Pt using low metal-consumption ratio

The Pt concentration in blanket samples was analyzed to find the mechanisms responsible for the decrease in silicide resistivity at the lower metal-consumption ratios. The dependence of the Pt concentration of silicide on the metal-consumption ratio under several conditions is shown in Fig. 12. This Pt concentration indicates the Pt atomic ratio between Pt and Ni except for Si in the silicide, i.e., when the composition of the mono-silicide is $(Ni_{1-x}Pt_x)Si$, this Pt concentration is x-percent. The horizontal dashed line in Fig. 12 denotes the Pt concentration of as-sputtered Ni-Pt alloy film, i.e., 3.7%, being lower than that of the Ni-Pt target, i.e., 5%. The Pt could have been trapped in the collimator during Ni-Pt sputtering. This is because the sputter angularity of Pt is higher than that of Ni [14], [15]. The Pt concentration of the silicide obtained by full conversion is the same as that of Ni-Pt alloy film. In contrast, the Pt was enriched by partial conversion with the lower metal-consumption ratios and saturated at 5.3% with a thickness of 10 nm, and 6.8% with a thickness of 5 nm from 3.7%, i.e., the original Pt concentration of as-sputtered Ni-Pt alloy film. The Pt concentration of the thinner silicide is higher than that of the thicker silicide even if the metal-consumption ratio is the same. This indicates the Pt could have accumulated in the bottom of silicide in the vicinity of the silicide-silicon interface. Consequently, enriching the Pt of Ni-Pt silicide by using partial conversion at metal-consumption ratios of less than 80% more effectively suppressed the formation of $NiSi_2$ than thickening the silicide on the narrow active lines between gates, as shown in Figs. 11 and 12.

We think that, for these Pt enrichments of silicide, it is important to have a difference in diffusivity between Pt and Ni during RTA1. Figure 13 plots the dependence of diffusivity for Pt and Ni in bulk silicon on the reciprocal of the absolute

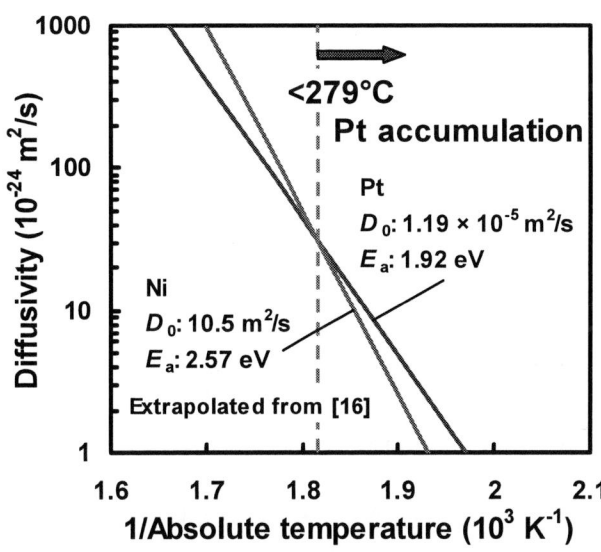

Fig. 13. Comparison of diffusivity of platinum and nickel in silicon.

temperature near the RTA1 temperature, viz., an Arrhenius plot, which had the diffusivity extrapolated from a previous study [16]. The lines cross at 279°C, and the activation energy of Pt is smaller than that of Ni. Therefore, the diffusivity of Pt is larger than that of Ni below 279°C. Consequently, we think that a larger diffusivity of Pt compared with that of Ni made the Pt accumulate in the silicide during RTA1.

IV. CONCLUSION

We applied partial conversion with a low metal-consumption ratio as the initial silicidation, thus improving the electrical properties of Ni-Pt silicide due to control of the morphologies of the silicide on narrow active lines between gates. Partial conversion not only resulted in the formation of

Ni-Pt silicide with pattern-independent thickness, finer crystal-grain and enriched Pt but it also minimized the resistivity. The crystal grain of silicide was made finer due to minimizing the thermal budget of initial silicidation and not that of second silicidation. The Pt was enriched due to the use of partial conversion with a low metal-consumption ratio. The metal-consumption ratio is a key parameter in Ni-Pt silicidation. It is important to leave the Ni-Pt alloy on the silicide after the initial silicidation, making the structure of the narrow active line become the stack of the Ni-Pt alloy, di-metal silicide, and silicon substrate. The resistivity of Ni-Pt silicide is as low as that of NiSi at metal-consumption ratios of less than 80%. We successfully attained Ni-Pt mono-silicide with low resistivity, i.e., 14.6×10^{-8} $\Omega \cdot m$, consisting of nominal 15-nm-thick silicide on 35-nm-width p^+-doped active lines between gates by using partial conversion at a metal-consumption ratio of 50%. This partial conversion can thus be applied to 28-nm-node logic devices and beyond.

ACKNOWLEDGMENT

The authors would like to thank Messrs. M. Nakamura and K. Kihara of Renesas Semiconductor Engineering for performing the sample preparation, and Dr. T. Furusawa, Messrs. Y. Takada, K. Maekawa, K. Funayama, T. Yamaguchi, A. Hiraiwa, M. Takamori, and H. Kozawa of Renesas Electronics Corporation. for their useful comments on our silicidation scheme. The authors also wish to thank Messrs. N. Kobayashi and H. Onji of Renesas Semiconductor Engineering for providing the TEM observations.

REFERENCES

[1] C. Ortolland, E. Rosseel, N. Horiguchi, C. Kerner, S. Mertens, J. Kittl, E Verleysen, H. Bender, W. Vadervost, A. Lauwers, P. P. Absil, S Biesemans, S. Muthurishnan, S. Srinivasan, A. J. Mayur, R. Schreutelkamp, and T. Hoffmann, "Silicide yield improvement with NiPtSi formation by laser anneal for advanced low power platform CMOS technology," in *Tech. Dig. Int. Electron Devices meeting (IEDM)*, Baltimore, MD, Dec. 6–9, 2009, pp. 23–26.

[2] B. S. Haran, A. Kumar, L. Adam, J. Chang, V. Basker, S. Kanakasabapathy et al., "22 nm technology compatible fully function $0.1\mu m^2$ 6T-SRAM cell," in *Tech. Dig. Int. Electron Devices meeting (IEDM)*, San Francisco, CA, Dec. 15–17, 2008, pp. 625–628.

[3] Y. Y. Chiang, Y. L. Chang, T. Y. Hung, Y. W. Chen, C. C. Shieh, C. C. Huang, and S. F. Tzou, "Study of Pt addition to solve NiSi integration issues on CMOS devices," in *Proc. Advanced Metallization Conference 2005 (AMC 2005)*, Colorad Springs, CO, Oct. 13–14, 2005, Warrendale, PA: Materials Research Society, 2006, pp. 193–198.

[4] F. F. Zhao, J. Z. Zheng, Z. X. Shen, T. Osipowicz, W. Z. Gao, and L. H. Chan, "Thermal stability study of NiSi and NiSi$_2$ thin films," *Microelectronic Engineering*, vol. 71, no. 1, pp. 104–111, Jan. 2004.

[5] K. Funk, X. Pages, V. I. Kuznetsov, and E. H. A. Granneman, "NiSi contact formation - process integration advantages with partial Ni conversion," in *Proc. 12th IEEE Int. Conf. Advanced Thermal Processing of Semiconductors (RTP)*, Portland, OR, Sep. 28–30, 2004, pp. 94–98.

[6] T. Kuratomi, K. Tanaka, D. L. Diehl, S.E. Phan, X. Lu, D. Or, J. Lei, G. Lai, K. Lavu, C. Jiang, K. Moraes, C.T. Kao, T. Futase, and K. Maekawa, "Native oxide removal application using NF$_3$/NH$_3$ remote plasma for Ni silicide process," in *Proc. Advanced Metallization Conf. 2006 (AMC 2006)*, San Diego, CA, Oct. 17–19, 2006, Warrendale, PA: Materials Research Society, 2007 pp. 611–616.

[7] T. Futase, N. Hashikawa, T. Kamino, T. Fujiwara, Y. Inaba, T. Suzuki, and H. Yamamoto, "Spike annealing as second rapid thermal annealing to prevent pure nickel silicide from decomposing on a gate," *IEEE Trans. Semicond. Manuf.*, vol. 22, pp. 475–481, Nov. 2009.

[8] D. Freedman, R. Pisani, and R. Purves, *Statistics 4th ed.*, New York, NY: W.W. Norton & Co., 2007.

[9] H. Iwai, T. Ohguro, and S. Ohmi, "NiSi salisaid technology for scaled CMOS," *Microelectronic Engineering*, vol. 60, pp. 157–169, 2004.

[10] J. Eberhardt and E. Kasper, "Ni/Ag metallization for SiGe HBTs using a Ni silicide contact," *Semicond. Sci. Technol*, vol. 16, pp. 47–49, 2001.

[11] M. C. Poon, F. Deng, M. Chana, W. Y. Chana, and S. S. Lau, "*Resistivity and thermal stability of nickel mono-silicide*," Applied Surface Science, vol. 157, pp. 29–34, Mar. 2000.

[12] L. W. Cheng, S. L. Cheng, J. Y. Chen, L. J. Chen, and B. Y. Tsui, "Effect of nitrogen ion implantation on the formation of nickel silicide contacts on shallow junction," *Thin Solid Films*, vol. 355, pp. 412–416, 1999.

[13] Y. N. Erokhin, F. Hong, S. Pramanick, and G. A. Rozgonyi, "Spatially confined nickel disilicide formation at 400°C on ion implantation preamorphized silicon," *Appl. Phys. Lett.* vol.63, pp. 3173–3175, 1993.

[14] R. Behrisch ed., *Sputtering by particle bombardment II*, Berlin: Springer-Verlag, 1983.

[15] R. Behrisch, and K. Wittmaack eds., *Sputtering by particle bombardment III*, Berlin: Springer-Verlag, 1991.

[16] O. Madelung, M. Schulz, and H. Weiss eds., *Landolt-Börnstein Zahlenwerte und Funktionen aus Naturwissenshaften und Technik*, p. 494, Berlin: Springer-Verlag, 1984.

978-1-4244-5430-3/10 $26.00 © 2010 IEEE

Disconnection of NiSi Shared Contact and Its Correction Using NH_3 Soak Treatment in Ti/TiN Barrier Metallization

Takuya Futase[1,5,*], Kota Funayama[2], Naoto Hashikawa[3], Hiroshi Tobimatsu[4],
Hirohiko Yamamoto[1], and Hisanori Tanimoto[5]

[1]Wafer Process Manufacturing Technology Dept. 2, Renesas Electronics Corporation, Hitachinaka, Ibaraki 312-8504, Japan
[2]MCU Device Technology Dept., Renesas Electronics Corporation, Hitachinaka, Ibaraki 312-8504, Japan
[3]Process & Device Analysis Engineering Development Dept., Renesas Electronics Corporation, Kodaira, Tokyo 187-8588, Japan
[4]Value Engineering Planning Dept., Renesas Electronics Corporation, Chiyoda, Tokyo 100-0004, Japan
[5]Graduate School of Pure and Applied Sciences, University of Tsukuba, Tsukuba, Ibaraki 305-8573, Japan
[*]Phone: +81-29-354-0748, e-mail address: takuya.futase.vt@renesas.com

Abstract—During Ti/TiN barrier metallization of a shared contact in SRAM, an NH_3 soak treatment selectively deoxidized silicon oxide on NiSi at the gate shoulder, improving the resistance of the contact. This deoxidizing NH_3 soak treatment drastically reduced the drawbacks of conventional NH_3 plasma treatment: plasma-induced damage of gate oxide and excessive nitridation of Ti/TiN. Although NH_3 gas does not kinetically deoxidize silicon oxide, it does selectively deoxidize silicon oxide on the NiSi. We think that this is because the NiSi surface promotes the deoxidization of silicon oxide by NH_3.

Keywords—ammonia; barrier metal; contact; deoxidation; nickel silicide; PECVD; plasma-induced damage; SRAM; $TiCl_4$; titanium

I. INTRODUCTION

Titanium and titanium nitride (Ti/TiN) films are used as contact barriers between nickel silicide (NiSi) and tungsten (W) plugs [1]–[4]. Barriers prepared by plasma-enhanced chemical vapor deposition (PECVD) using titanium chloride ($TiCl_4$) as a precursor reduce the contact resistance by half compared with others [1], and therefore they are used for 65-nm-node-and-beyond static random access memory (SRAM). In the SRAM circuit shown in Fig. 1, p-type transistors have three types of contacts: shared contacts, active-line contacts, and gate contacts, as shown in Fig. 2. The gate shoulder of a shared contact is often damaged by fluorine during contact etching. The fluorine promotes oxidation of the NiSi [5]. Thus, the gate shoulder of the shared contact can be easily disconnected because of the oxide on the NiSi at a poor contact area.

We selectively removed the oxide on the NiSi of the gate shoulder during Ti/TiN barrier metallization by NH_3 soak or NH_3 plasma treatment; while the conventional NH_3 plasma treatment caused plasma-induced damage caused by the electron shading effect [6], [7] or excessive nitridation of the Ti/TiN, the NH_3 soak treatment caused none of the drawbacks of the NH_3 plasma.

II. EXPERIMENTAL

As shown in the process flow in Fig. 3, the contact was fabricated as follows. NiSi formation on the active-lines and gates [8] was followed by deposition of silicon nitride and silicon oxide as pre-metal dielectric layers. Dry etching to form the contact hole was followed by wet cleaning using a mixture of sulfuric acid and peroxide hydrogen. Prior to Ti deposition,

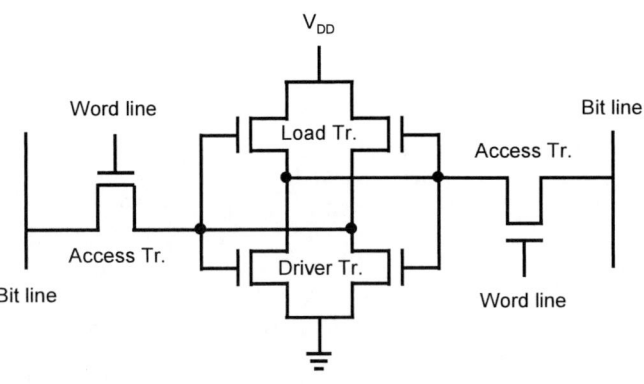

Fig. 1. Configuration of SRAM circuit.

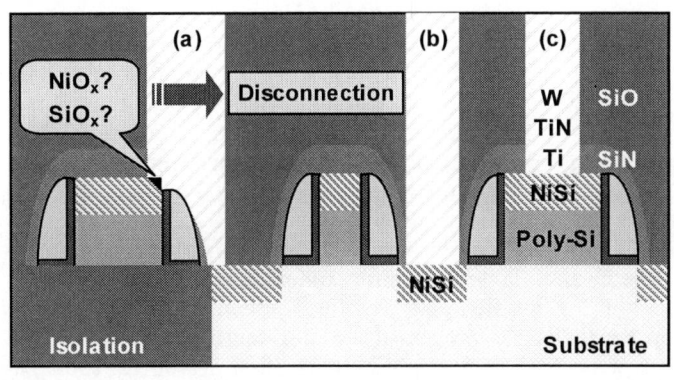

Fig. 2. Schematic cross-section of contacts: (a) shared contact, (b) active-line contact, and (c) gate contact.

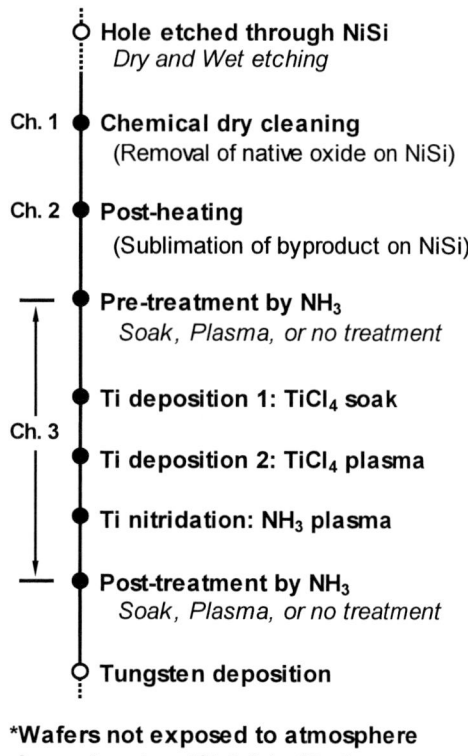

○ **Hole etched through NiSi**
  Dry and Wet etching

Ch. 1 ● **Chemical dry cleaning**
  (Removal of native oxide on NiSi)

Ch. 2 ● **Post-heating**
  (Sublimation of byproduct on NiSi)

● **Pre-treatment by NH₃**
  Soak, Plasma, or no treatment

● **Ti deposition 1: TiCl₄ soak**

Ch. 3

● **Ti deposition 2: TiCl₄ plasma**

● **Ti nitridation: NH₃ plasma**

● **Post-treatment by NH₃**
  Soak, Plasma, or no treatment

○ **Tungsten deposition**

***Wafers not exposed to atmosphere
from chamber (Ch.) 1 to Ch. 3**

Fig. 1. Process flow of contact barrier.

chemical dry cleaning to remove native oxide on the NiSi using the reaction shown in (1) was followed by post-heating to sublimate byproduct on the NiSi at 200°C using the reaction shown in (2).

$$SiO_2(g) + 6HF(g) + 2NH_3(g)$$
$$\longrightarrow 2H_2O(g) + (NH_4)_2SiF_6(s) \tag{1}$$

$$(NH_4)_2SiF_6(s) \xrightarrow[\geq 100°C]{} (NH_4)_2SiF_6(g) \tag{2}$$

Then NH₃ pre-treatment was performed in the Ti deposition chamber. The chemical dry cleaning, the post-heating, and the NH₃ pre-treatment were performed without breaking vacuum. Ti was formed by a two-step deposition process [1], namely, TiCl₄ soak and TiCl₄ PECVD at 450°C, 666.6 Pa using hydrogen as reductant, and was followed by nitridation of the Ti surface by NH₃ plasma treatment at 450°C, 666.6 Pa. The total thickness of the Ti/TiN barrier is approximately 5 nm. Then NH₃ post-treatment was performed. The effect of NH₃ pre- or post-treatment, an NH₃ soak—NH₃ gas flow without radio frequency power—or an NH₃ plasma were examined before or after Ti deposition for comparison, as listed in Table 1. Finally, a tungsten plug was filled in contact hole by a two-step thermal CVD process using tungsten hexafluoride (WF₆) as a precursor [1]: 7-nm-thick nucleation at 350°C, 1000 Pa using di-borane as reductant and 200-nm-thick main deposition at 390°C, 10666 Pa using hydrogen as reductant.

TABLE I.   COMPARISON OF NH₃ TREATMENT CONDITIONS

ID	Pre-treat	Ti depo.	Ti nitridation	Post-treat
Pre-soak	Soak 15 s	5 nm	20 s	None
Pre-plasma	Plasma 15 s	↑	↑	None
Baseline	None	↑	↑	None
Post-soak	None.	↑	↑	Soak 60 s
Post-plasma	None	↑	↑	Plasma 40 s

Logic devices with 0.54-μm²-cell SRAMs were prepared on a 300-mm-diameter (001) silicon wafer to measure the contact resistances on the p⁺-doped active-line and gate. The contact hole sizes of the shared, active-line, and gate contacts are approximately 80 × 200, 80-, and 80-nm-diameter, respectively, and maximally 225-nm-high. Cross-sections of the logic devices were analyzed with a transmission electron microscope with an energy dispersive x-ray spectrometer (TEM-EDX). Logic devices were also used to evaluate the plasma-induced damage to the gate oxide. The antenna ratios of the contact antenna and gate antenna were about 40 and 900 times, respectively, on a 0.1-μm² p⁺-doped active area.

Blanket wafers were also prepared to measure sheet resistance of the Ti/TiN film under several NH₃ post-plasma exposure times with the four-probe method and to analyze the NiSi films with x-ray photoelectron spectroscopy (XPS) before and after TiCl₄ soak. The XPS measurements used monochromatized Al Kα₁,₂ radiation. The binding energy scale was calibrated to the Si 2p₃/₂ peak at 99.2 eV.

III. RESULTS AND DISCUSSION

A. Disconnection in the NiSi shared contact

Resistances of shared, active-line, and gate contacts are compared in Fig. 4. The vertical axis plots the sigma score [9], which is determined by the sample number, denoting the inverse function of a normal distribution function. The horizontal axis plots the contact resistance. Therefore, the reciprocal of the line slope coincides exactly with the standard deviation of the contact resistance when the data follow the normal distribution. Results for the shared contact shown in Fig. 4(a) show that the disconnection was observed only for the baseline conditions, and not for the NH₃ treatment. Results for the active-line and gate contacts shown in Figs. 4(b) and (c) show that the resistance only increased for the NH₃ post-plasma treatment condition. We discuss this increase caused by NH₃ post-plasma later.

Figure 5 shows the TEM cross-sectional images of (a) connected and (b) disconnected shared contacts fabricated under the baseline conditions. As shown in Fig. 5(a), no intercept film exists on the NiSi of the gate shoulder; in contrast, as shown in Fig. 5(b), there is an approximately 10-nm-thick dielectric film on the NiSi of the gate shoulder. Figure 6 shows the TEM-EDX maps for oxygen and carbon analyses around the disconnected shared contact area shown in Fig. 5(b), indicating that the dielectric film is oxide and not

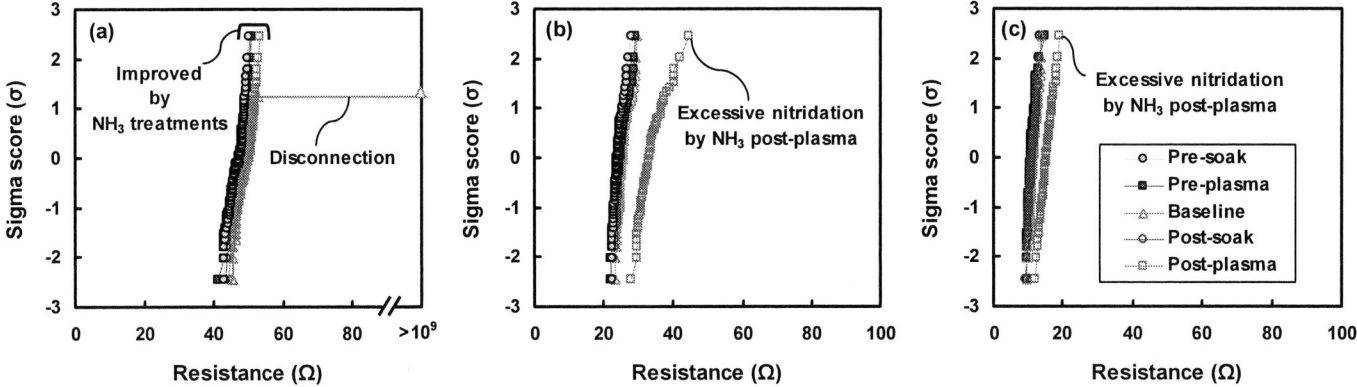

Fig. 4. Disconnection improvement by NH₃ treatment (a) in shared contact and drawbacks of NH₃ post-plasma treatment (b) in active-line contact and (c) in gate contact.

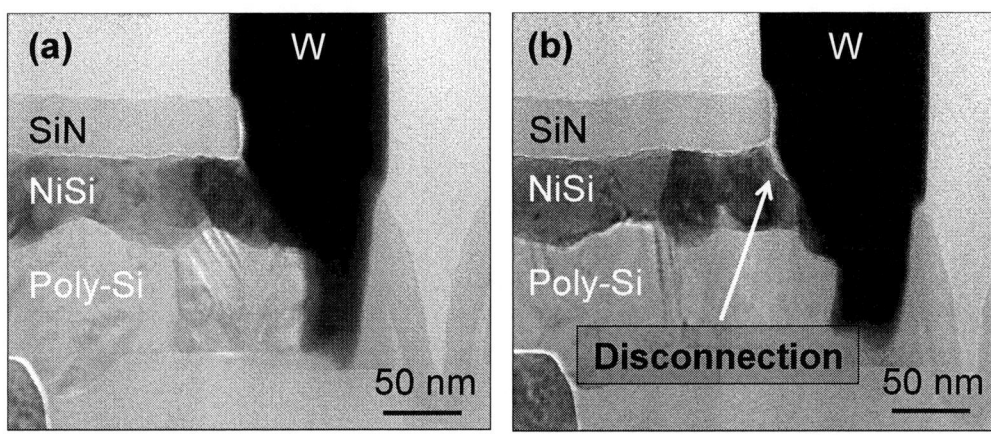

Fig. 5. Cross-sectional TEM images: (a) connected and (b) disconnected shared contact of baseline sample.

carbide derived from a dry-etching polymer. Such a thick oxide could not be removed by chemical dry cleaning without expanding the hole size. As shown in Fig. 4, however, no decrease in the sheet resistance due to expansion of the hole was observed. Therefore, NH₃ pre- and post-treatment, ether soak or plasma, selectively deoxidize the oxide on the NiSi; in the NH₃ pre-soak, NH₃ gas deoxidizes the oxide directly, and in the NH₃ post-soak, NH₃ gas deoxidizes the oxide through the Ti/TiN film.

B. Drawbacks of NH₃ plasma treatment

NH₃ plasma treatment is widely used to deoxidize native oxides and to nitridize the surface on copper interconnects [10], [11]. However, we show that NH₃ plasma treatment, ether before or after titanium deposition, is not appropriate for contact metallization due to two drawbacks: plasma-induced damage of the gate oxide and excessive nitridation of Ti/TiN.

1) Plasma-induced damage: The voltage-threshold (V_{th}) shift of the p-channel metal-oxide-semiconductor field-effect transistor (PMOS) is compared with the NH₃ pre-plasma and the baseline condition. The V_{th} of the PMOS fabricated with

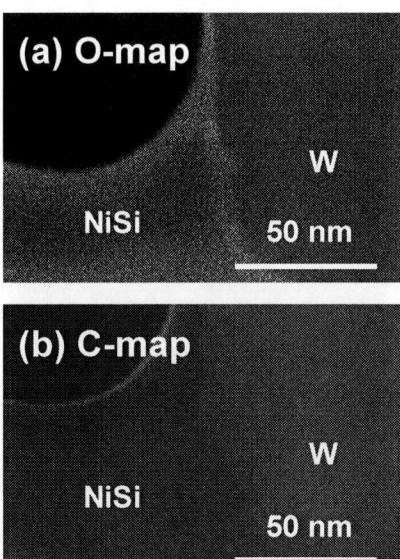

Fig. 6. Cross-sectional TEM-EDX maps of disconnected shared contact: (a) oxygen map and (b) carbon map of baseline sample.

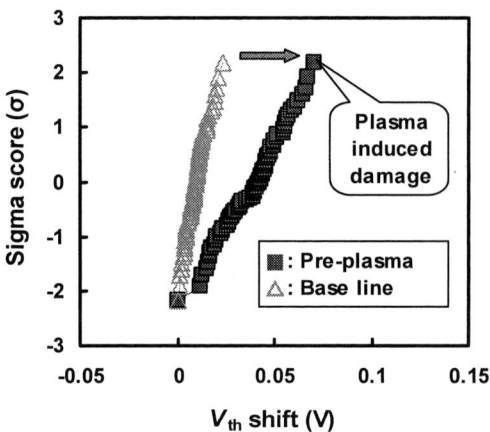

Fig. 7. First drawback of NH₃ plasma: plasma-induced damage before Ti deposition.

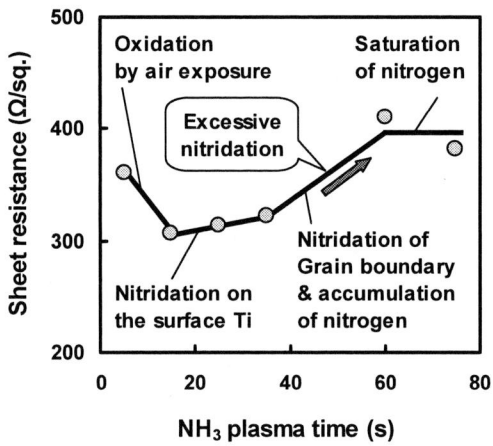

Fig. 8. Second drawback of NH₃ plasma: excessive nitridation after Ti deposition.

the NH₃ pre-plasma treatment is larger than that with the baseline condition. The larger V_{th} shifts indicate the deterioration of gate oxide due to charge accumulation through the gate electrode by NH₃ plasma. We think that this deterioration is caused by the electron shading effect of the contact hole [6], [7].

2) Excessive Nitridation: NH₃ post-plasma treatment increased the sheet resistances of the active-line and the gate contact, as shown in Fig. 4. We think that the increase could be due to excessive nitridation of the Ti/TiN caused by long exposure to NH₃ plasma. This nitridation is shown in sheet-resistance measurements of the blanket Ti/TiN film, as shown in Fig. 8. The Ti/TiN film was prepared by depositing 5-nm-thick Ti on the thermal silicon oxide film and then treating with NH₃ plasma. We think that the sheet-resistance measurements reflect four states corresponding to the exposure time to NH₃ plasma: decreasing up to 15-second NH₃ plasma exposure due to oxidation of the surface Ti by air exposure, slightly increasing between 15- and 35-second NH₃ plasma exposure due to nitridation of the surface Ti, drastically increasing

TABLE II. RESULTS OF NH₃ TREATMENT CONDITIONS

ID	Shared contact disconnection	Plasma-induced damage	Excessive nitridation
Pre-soak	✓	✓	✓
Pre-plasma	✓	NG	✓
Baseline	NG	✓	✓
Post-soak	✓	✓	✓
Post-plasma	✓	✓	NG

between 35- and 60-second NH₃ plasma exposure due to nitridation of the Ti grain boundary and accumulation of nitrogen in the Ti crystals, and saturating at more than a 60-second NH₃ plasma exposure. It is shown that an approximately 20-second NH₃ plasma exposure is enough to form a Ti/TiN barrier against WF₆, because its sheet resistance is not deteriorated by the WF₆, as shown in Fig. 4 and in a former study [1]. Therefore, more than 35-second NH₃ plasma exposure is excessive for the Ti/TiN and increases the sheet resistance of the contacts.

Therefore, the NH₃ pre- and post-soak treatments are appropriate for barrier Ti/TiN metallization because they improve the disconnection of the shared contact and cause none of the drawbacks of using the NH₃ plasma treatment—the plasma-induced damage or the excessive nitridation—as listed in Table 2. In an NH₃ post-soak, the deoxidation proceeds when NH₃ passes through Ti/TiN. As listed in Table 1, the NH₃ pre-soak is preferable to the NH₃ post-soak because of the throughput required.

C. Deoxidization mechanism by NH₃ soak treatment

During NH₃ soak treatment, NH₃ gas selectively deoxidizes the oxide on the NiSi of the contact gate shoulder, as mentioned in Section A. To analyze the mechanism, we investigated the composition of the oxide on the NiSi by XPS and estimated the reaction energy for the deoxidization of the oxide by NH₃ gas.

1) Composition of NiSi surface: Figure 9 shows the XPS spectrum of F 1s, Ni 2p, and Si 2p for the NiSi surface under several conditions: dry etching, wet etching, and TiCl₄ soak. The contact window of the samples was 1-mm-square after dry etching of the blanket silicon nitride on the NiSi film. The sample soaked in TiCl₄ was not pretreated with NH₃. The peak position of nickel fluoride (NiF₂) is 685.3 eV for F 1s [12] and 858 eV for Ni 2p [13]. The peak positions of nickel oxides—Ni₂O₃ and NiO—are 855.8 and 854.5 eV for Ni 2p [14]. The peaks of 684.8 eV, as shown in Fig. 9(a), and the high-energy-side broad peak of 856.7 eV, as shown in Fig. 9(b) for the dry etching sample, indicate that there is NiF₂ on the surface of the NiSi. The fluoride peaks of F 1s are inconsequential for the wet etching and TiCl₄-soaked samples, indicating that wet etching cleaned the NiF₂. The peak position of adsorbed fluoride is 686.4–685.5 eV [15]. The small peak of 686.2 eV for the TiCl₄-soaked sample, as shown in Fig. 9(a), indicates that a small amount of fluorine adsorbs on the NiSi surface. We think

978-1-4244-5430-3/10 $26.00 © 2010 IEEE

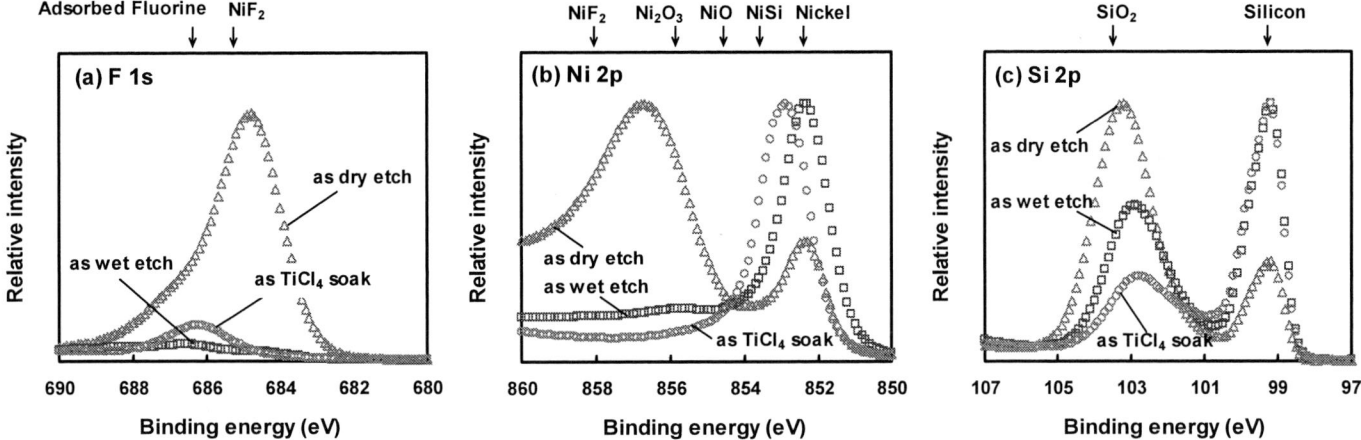

Fig. 9. XPS results of blanket NiSi film: (a) fluoride, (b) nickel oxide, and (c) silicon oxide peaks exist on dry etched NiSi; all the peaks decrease after wet etching; of the three, only the silicon oxide peak remains after TiCl$_4$ soak.

that the adsorbed fluorine is caused by the chemical dry cleaning using HF and NH$_3$, as shown in (1). The peak position of NiSi, di-nickel silicide (Ni$_2$Si), and Ni are 853.5, 853.0, and 852.4 eV, respectively [16]. The sharp peak of 852.9 eV for the TiCl$_4$-soaked sample, as shown in Fig. 9(b), indicates that there is no nickel oxide on the nickel-rich silicide. The peak position of silicon oxide (SiO$_2$) and Si are 103.4 and 99.2 eV, respectively [17]. The low-energy-side broad peak of 102.7 eV for the TiCl$_4$-soaked sample, as shown in Fig. 9(c), indicates that there is non-stoichiometric SiO$_2$ on the NiSi. These results show that the oxide disconnection on the gate shoulder NiSi as shown in Fig. 5(b) is not nickel oxide, but silicon oxide, even after the chemical dry cleaning prior to Ti deposition.

2) Kinetics of NiSi deoxidization: We showed that the oxide on the NiSi is silicon oxide and that a small amount of fluorine adsorbs on the NiSi. We do not think that the adsorbed fluorine is attributable to deoxidization by the similar reaction shown in (1). This is because the NH$_3$ soak treatment did not expand the contact hole compared with the baseline treatment, as mentioned in Section A, even if the fluorine derived from the chemical dry cleaning adsorbed on the silicon oxide that composes the sidewall of the contact hole. Therefore, we discuss the deoxidization of the silicon oxide by NH$_3$ gas.

The Gibbs' free energy change ΔG for the reaction shown in (3) at 450°C and 6.208×10^{-4} Pa of NH$_3$ partial pressure, which are the conditions for our NH$_3$ soak, may be calculated as 215.8 kJ/mol [18]. This shows that silicon oxide is not kinetically deoxidized by NH$_3$ gas because the ΔG is positive. We think that non-stoichiometric silicon oxide cannot be deoxidized by NH$_3$ gas either.

$$SiO_2 (s) + 2NH_3 (g) = Si(s) + N_2 (g) + H_2 (g) + 2H_2O(g) \quad (3)$$

The ΔG for the reaction shown in (4) at 450°C and 9.172×10^{-6} Pa of TiCl$_4$ partial pressure, which are the conditions for our TiCl$_4$ soak, may be calculated as 207.1 kJ/mol [18]. This also shows that TiCl$_4$ gas does not kinetically decompose because the ΔG is positive.

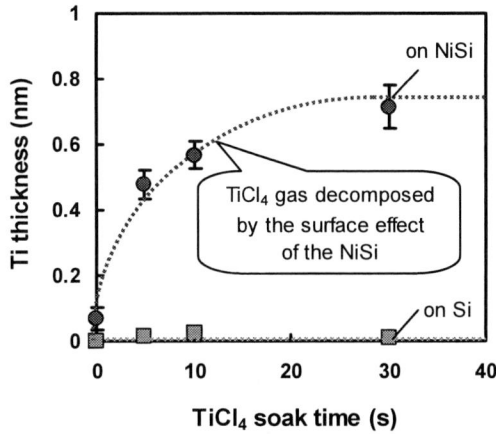

Fig. 10. Ti selective growth on blanket NiSi film by TiCl$_4$ soak.

$$TiCl_4 (g) + 2H_2 (g) = Ti(s) + 4HCl(g) \quad (4)$$

Nevertheless, TiCl$_4$ did decompose on the surface of NiSi, although it did not decompose on the surface of Si, as shown in Fig. 10, though it did in our previous study [1]. The thickness of Ti caused by the TiCl$_4$ decomposition asymptotically approaches 0.7 nm, which is extremely thin. We thus think that TiCl$_4$ gas selectively decomposes due to the surface effect of the NiSi and not the catalysis effect of the NiSi. Like the decomposition of TiCl$_4$ on the NiSi, we think that NH$_3$ gas also selectively deoxidizes silicon oxide on the NiSi due to the surface/interface effects of the NiSi.

IV. CONCLUSION

Disconnection of the nickel silicide shared contact in SRAM was efficiently prevented by an NH$_3$ soak treatment during Ti/TiN barrier metallization. The NH$_3$ soak treatment caused none of the drawbacks of conventional NH$_3$ plasma treatment, such as plasma-induced damage of gate oxide or excessive nitridation of Ti/TiN. TEM-EDX analysis showed

978-1-4244-5430-3/10 $26.00 © 2010 IEEE

that the disconnection was caused by oxide on the NiSi at the gate shoulder of the shared contact. XPS analysis showed that the oxide is silicon oxide and not nickel oxide. Although NH_3 gas does not kinetically deoxidize silicon oxide, it does selectively deoxidize silicon oxide on the NiSi. We think that this selective deoxidization of the silicon oxide by NH_3 gas is promoted by the NiSi surface. This novel NH_3 soak treatment for the Ni silicide contact metallization enhances the reliability of contact resistance for 65-nm-node-and-beyond logic devices.

ACKNOWLEDGMENT

The authors thank K. Uesugi, K. Sasahara, T. Hayashi, and Y. Koide of Renesas Electronics Corporation for performing the sample preparation, and Dr. T. Furusawa, K. Makabe, N. Abe, T. Fujiwara, and H. Kozawa of Renesas Electronics Corporation for their useful comments on our metallization scheme. The authors also thank S. Ogawa of Toray Research Center for performing the XPS analysis of the nickel silicide.

REFERENCES

[1] T. Futase, N. Hashikawa, T. Hayashi, H. Tobimatsu, H. Yamamoto, and H. Kozawa, "Low contact-resistance metallization process for a nickel self-aligned contact of beyond 65 nm node CMOS," in *Proc. 16th IEEE Int. Symp. Semiconductor Manufacturing (ISSM)*, Santa Clara, CA, Oct. 15–17, 2007, pp. 471–474.

[2] H. Ai, M. Jackson, and S. H. Yu, "Low temperature MOCVD TiN process for 45nm contact metallization," in *Proc. 16th IEEE Int. Symp. Semiconductor Manufacturing (ISSM)*, Santa Clara, CA, Oct. 15–17, 2007, pp. 445–448.

[3] E. Gerritsen, "Material and integration scaling in the contact module beyond the 65nm node," in *Proc. 24th Advanced Metallization Conf.*, Albany, NY, Oct. 9–11, 2007, Warrendale, PA: Materials Research Society, 2008, pp. 597–603.

[4] K. Ichinose, A. Yutani, K. Maekawa, K. Asai, and M. Yoneda, "New barrier metal and ALD-W process for low stable resistance in contact metallization beyond 45-nm CMOS," in *Proc. 24th Advanced Metallization Conf.*, Albany, NY, Oct. 9–11, 2007, Warrendale, PA: Materials Research Society, 2008, pp. 604–610.

[5] S. Sakamori, K. Yonekura, N. Fujiwara, T. Kosaka, M. Ohkuni, and K. Tateiwa, "Control of oxidation on $NiSi_x$ during etching and ashing processes," *Thin Solid Films*, vol. 515, pp. 4933–4936, 2007.

[6] K. Hashimoto, "New phenomena of charge damage in plasama etching: Heavy damage only through dense-line antenna," *Jpn. J. Apple Phys.*, vol. 32, pp. 6109–6113, 1993.

[7] K. Hashimoto, "Chage damage caused by electron shading effect," *Jpn. J. Apple Phys.*, vol. 33, pp. 6013–6018, 1994.

[8] T. Futase, N. Hashikawa, T. Kamino, T. Fujiwara, Y. Inaba, T. Suzuki, and H. Yamamoto, "Spike annealing as second rapid thermal annealing to prevent pure nickel silicide from decomposing on a gate," *IEEE Trans. Semicond. Manuf.*, vol. 22, pp. 475–481, Nov. 2009.

[9] D. Freedman, R. Pisani, and R. Purves, *Statistics 4th ed.*, New York, NY: W.W. Norton & Co., 2007.

[10] C. C. Huang, J. L. Huang, Y. L. Wang, and K. Y. Lo, "Study of pretreatment prior to silicon-oxycarbide deposition on Cu interconnect," *J. Vac. Sci. Technol. B*, vol. 26, pp. 96–101, 2009.

[11] J. Noguchi, N. Ohashi, J. Yasuda, T. Jimbo, H. Yamaguchi, N. Owada, K. Takeda, and K. Hinode, "TDDB improvement in Cu metallization under bias stress," in *Proc. 38th Annual IEEE Int. Reliability Physics Symp. (IRPS)*, San Jose, CA, Apr. 10–13, 2000, pp. 339–343.

[12] K. Murai, Y. Suzuki, T. Moriga, and A. Yoshiasa, "EXAFS and XPS study of rutile-type difluorides of first-row transition metals," in *Proc. 13th AIP Int.Conf. X-ray Adsorption Fine Structure*, 2007, pp. 463–465.

[13] G. G. Totir, G. S. Chottiner, C. L. Gross, W. V. Childs, and D. A. Scherson, "X-ray photolectron spectroscopy and morphological studies of plycrystalline nickel surface exposed to anhydrous HF," *J. electrochem. Soc.*, vol. 147, pp. 4212–4216, 2000.

[14] K. S. Kim, W. E. Baitinger, J. W. Amy, and N. Winograd, S. A. Shabalovskaya and J. W. Anderegg, "ESCA studies of metal-oxygen surfaces using argon and oxygen ion-bombardmen," *J. Electron Spectrosc. Relat. Phenom.*, vol. 5, pp. 351–367, 1974.

[15] E. T. Paul Benny and J. Majhi, "An x-ray photoelectron study of the dependence of HF concentration on an etched silicon surface," *J. Electron Spectrosc. Relat. Phenom.*vol. 58, no. 4, pp. 261–270, Jun. 1992.

[16] N. W. Cheung, P. J. Grunthaner, F. J. Grunthaner, J. W. Mayer, and B. M. Ullrich, "Metal-Semiconductor interfacial reactions: Ni/Si system," *J. Vac. Sci. Technol.*, vol. 18, pp. 917–923, 1980.

[17] C. D. Wagner, W. M. Riggs, L. E. Davis, J. F. Moulder, G. E. Mullenberg, *Handbook of X-ray Photoelectron Spectroscopy: A reference book of standard data for use in x-ray photoelectron spectroscopy*, Eden Prairie, MN: Perkin-Elmer Corporation, 1979.

[18] D. R. Lide, Jr. ed, "The NBS tables of chemical thermodynamic properties," *J. Phys. Chem. Ref. Data*, vol. 11, suppl. 2, 1982.

Analysis of Statistical Variation in NBTI Degradation of HfO$_2$/SiO$_2$ FETs

H. Yoshimoto, D. Hisamoto, Y. Shimamoto, R. Tsuchiya, I. Yanagi, T. Arigane and K. Torii

Nano-process Research Department
Central Research Laboratory, Hitachi, Ltd.
1-280 Higashi-Koigakubo, Kokubunji, Tokyo 185-8601, Japan
Phone: +81-42-323-1111, E-mail: hiroyuki.yoshimoto.ev@hitachi.com

K. Funayama, T. Hashimoto, H. Makiyama, K. Horita, T. Iwamatsu, K. Shiga, M. Mizutani, M. Inoue, and

T. Kaneoka

RENESAS Technology Corp.
751 Horiguchi, Hitachinaka, Ibaraki 321-8504, Japan
Phone: +81-29-272-3111

Abstract— **The variations in negative bias temperature instability (NBTI) degradation in transistors with HfO$_2$/SiO$_2$ gate dielectric in which thin HfO$_2$ was deposited on SiO$_2$ by atomic layer deposition (ALD) were investigated. The median value of NBTI degradation in the FET with HfO$_2$/SiO$_2$ is lower than that with SiON, but the NBTI variation is large and increases as HfO$_2$ becomes thicker. We propose a model in which the NBTI variation is caused by the variations of the initial hydrogen density in the gate oxide and the small variation of the initial interface trap density. By using this model, the observed large NBTI variation in the high-k device can be well reproduced.**

[Keywords- NBTI; HfO$_2$; variation; R&D mode; hydrogen]

I. INTRODUCTION

Due to the aggressive scaling of MOSFET, both reliability and threshold voltage (V_{th}) variability have become important. V_{th} variation in high-k gate stack has particularly been attracting increasing attention [1-2]. Since NBTI degradation ex-

acerbates V_{th} variation [3], understanding statistical variation of NBTI is indispensable. However, to our knowledge, NBTI in high-k gate stack has mainly been studied for process de-

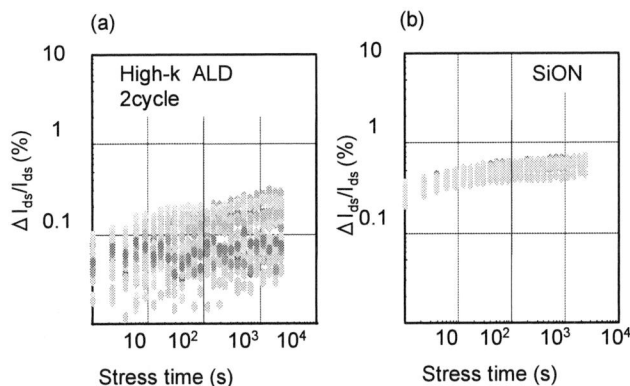

FIGURE. 1 NBTI DEGRADATIONS MEASURED FROM 28 CHIPS ON A WAFER. (A) MEASURED FROM HFO2 FETS. (B) MEASURED FROM SION FETS.

TABLE I. PROCESS FLOW OF THE HFO2 ALD GATE STACK

Process flow	
In-site steam generation	2.4 nm
HfO$_2$ ALD	0 , 1, 2 cycle
Post deposition nitridation	Normal, Strong

TABLE. II NBTI MEASURED CONDITION.

Temperature	125C°
L/W	1 μm/ 10 μm
Stress voltage	-1.7 V
Drain voltage	-0.625 V
Evaluated gate voltage of $\Delta I_{ds}/I_{ds}$	(Initial Vth) -0.5 V

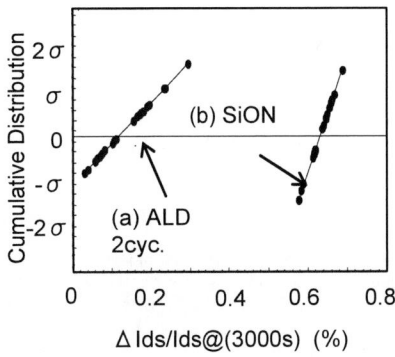

FIGURE. 2 CUMULATIVE DISTRIBUTION OF NBTI DEGRADATIONS SHOWN IN FIG. 1 (A) AND FIG. 1 (B).

pendence [4-5] and short time degradation [6-7], but little work has been reported on NBTI variation itself [8].

In this study, we evaluated NBTI variation in high-k gate stack. From the measurements, we observed anomalous large NBTI variation in PMOS transistors with HfO_2 thin film deposited by ALD. We discuss the cause of this NBTI variation below.

II. EXPERIMENTAL DETAILS

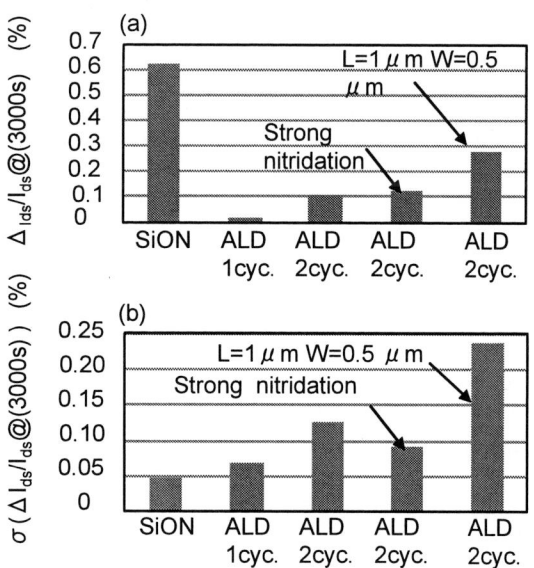

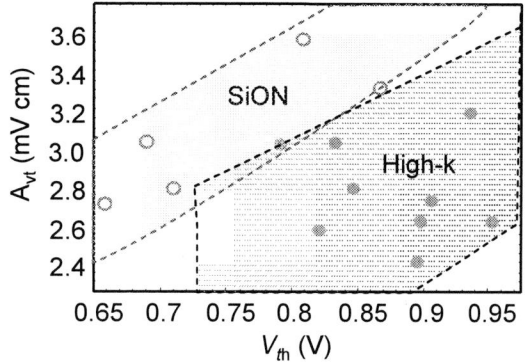

FIGURE. 3 THE MEDIAN VALUE (A) AND THE STANDARD DEVIATION (B) OF NBTI DEGRADATIONS FOR VARIOUS DEVICES.

FIGURE. 4 VTH DEPENDENCE OF PELGROM COEFFICIENT (AVT) MEASURED FROM SEVERAL SION AND HFO2 FETS.

After well and isolation formation, a 2.4-nm-thick gate oxide was formed by in-situ steam generation (ISSG). Then HfO_2 film was deposited by ALD. After HfO_2-ALD, the gate dielectric was nitrided (Table 1). To evaluate the variation of

the degradation, 28 chips on a wafer were measured for each sample. Typical sample size and the measurement conditions are given in Table 2.

III. RESULTS AND DISCUSSION

Figures 1-2 show NBTI degradation measured from HfO_2 and SiON devices. The median value of the degradation rate ($\Delta I_{ds}/I_{ds}$) and the standard deviation of the degradation ($\sigma(\Delta I_{ds}/I_{ds})$) of various HfO_2 devices are compared with those of a SiON device in Fig. 3. The median values of $\Delta I_{ds}/I_{ds}$ of all HfO_2 devices are smaller than that of the SiON device (Fig. 3 (a)) [6] [9-10], while the $\sigma(\Delta I_{ds}/I_{ds})$ of all HfO_2 devices are larger than that of the SiON device (Fig. 3 (b)). The degradation varies in proportion to the ALD cycle number, suggesting that the amount of the HfO_2 on the gate oxide correlates with the variation. For the same ALD cycle number, a strongly nitridated device varies less. Moreover, this figure shows the variation measured from a small gate dimension (L/W = 1μm/0.5μm) was larger than that measured from a larger gate dimension (L/W = 1μm/10μm), which is similar to V_{th} variation caused by the statistical variation of the number of impurities in the channel. Thus, these results suggest the cause of the variation exists in the SiO_2/HfO_2 gate.

Since high-k gate stack shows a significant V_{th} shift due to Fermi-level pinning, we investigated the effect of local and global V_{th} variation of virgin devices on NBTI variation. Fig. 4 shows that the local V_{th} variations in HfO_2 devices were rather smaller than that of SiON devices for the same V_{th}. Moreover, correlation between global V_{th} variation and $\Delta I_{ds}/I_d$ was not observed (Fig. 5).

On the other hand, since hole trapping under NBT condition was observed in high-k gate stack [6-7], the variation of hole trapping may cause NBTI variation. However, it is reported that hole trapping phenomena were significant only in the short stress time less than 1000s [6, 11]. Thus, they also cannot be the predominant factor in the variation either because this variation increases even if the stress time is long (~3000s) (Fig. 1 (a)). These results show the charge trapping in the SiO_2/HfO_2 gate stack cannot be the predominant factor.

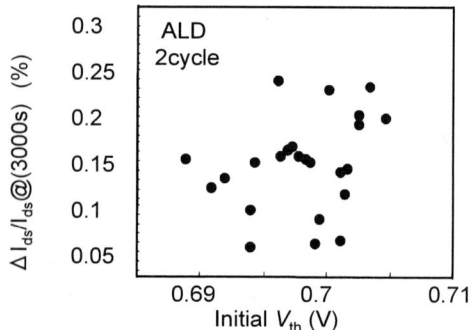

FIGURE. 5 INITIAL VTH VS NBTI DEGRADATION MEASURED FROM 28 DEVICES.

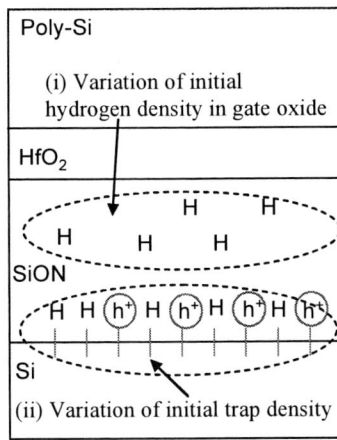

FIGURE. 6 ORIGIN OF THE NBTI VARIATION ADOPTED IN THIS STUDY.

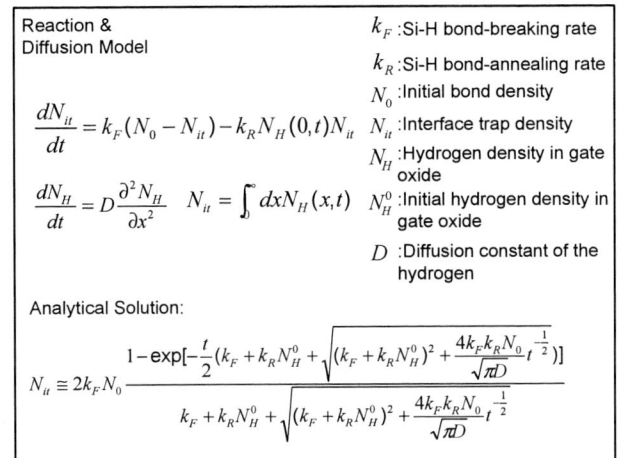

FIGURE. 7 REACTION & DIFFUSION MODEL AND THE ANALYTICAL SOLUTION OF THE INTERFACE TRAP DENSITY.

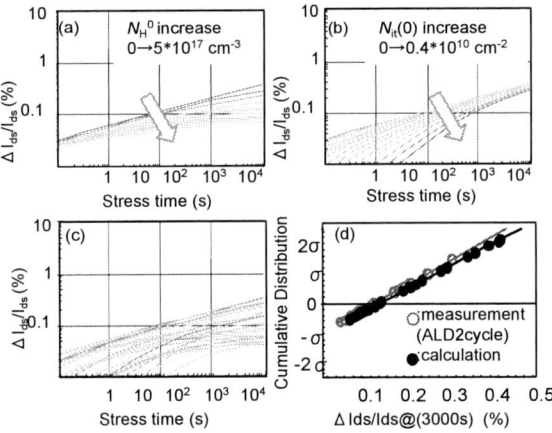

FIGURE. 8 SIMULATION OF THE NBTI DEGRADATION. (A) SHOWS THE VARIATION SHOWN IN FIG. 7 (I). (B) SHOWS THAT SHOWN IN FIG. 7 (II). BOTH (I) AND (II) ARE INCLUDED IN (C).(D) COMPARES DISTRIBUTION DERIVED FROM THE CALCULATION AND THE MEASUREMENT.

Accordingly, in this study, we went back to the conventional reaction and diffusion (R&D) model [12] where NBTI was modulated by diffusing neutral hydrogen atoms or molecules. We attributed the NBTI variation to the variation of atomic hydrogen density in the gate oxide and small variation of the Si-H bond, which barely affects the initial V_{th} variation (Fig. 6). The variation of the hydrogen in the gate oxide is assumed to come from the ALD process, where organic hafnium compounds are used as a precursor. Note that even if we choose the diffusing element to be molecular hydrogen [13], our conclusion basically does not change. Using an approximating technique (Fig. 7), we analytically solved the R&D model, where we assumed the initial hydrogen density N_H^0 and the initial bond trap density $N_{it}(0)$ are not zero. The calculation results on the ΔI_{ds} variations are shows in Fig. 8. Parameters in the R&D model were determined from Krishnan et al. [8] and Islam et al. [13]. Relationships between V_{th} and I_{ds} used in the calculations were determined from the measured results. Figure 8 (a) shows the increase in N_H^0 (0 cm^{-3} < N_H^0 < 5×10^{17} cm^{-3}) enhances the hydrogen termination leading to lower the $N_{it}(t)$ when stress time is long. Figure 8 (b) shows the ΔI_{ds} variation caused by the variation of $N_{it}(0)$ (0 cm^{-2} < $N_{it}(0)$ < 0.4 × 10^{10} cm^{-2}). If $N_{it}(0)$ is large, the initial ΔI_{ds} is small, but the degradation rate becomes large. Unlike the N_H^0 variation (Fig. 8 (a)), the $N_{it}(0)$ variation disappears when stress time is long. If variations both in N_H^0 and in $N_{it}(0)$ exist, the resultant ΔI_{ds} variation agrees well with the measured NBTI variation (Fig. 8 (c)-(d)), where Gaussian distribution in N_H^0 and in $N_{it}(0)$ are assumed. These results indicate that the large NBTI variation in the high-k transistor is caused by the variations in the hydrogen density in gate oxide and in the interface trap density

CONCLUSION

We found that the variation of NBTI degradation in FET with HfO$_2$/SiO$_2$ gate dielectric is larger than that with SiON. Using a simple model based on the R&D model, we showed that the initial part of NBTI variation is mainly caused by the variation in the hydrogen density in the gate oxide while the latter part is due to the variation in pre-existing interface trap density. In conclusion, to predict the precise NBTI lifetime in HfO$_2$/SiO$_2$ gate stack, the variation of degradation should be included in the NBTI model.

REFERENCES

[1] K. J. Kuhn, *Tech. Digest. of IEDM*, pp471-474, 2007

[2] G. Tsutsui et al., *Symp. VLSI Tech.*, pp158-159, 2008.

[3] K. Kang, S. P. Park, K. Roy, and M. A. Alam., *Proc. of ICCAD*, pp.730-734, 2007.

[4] H.-H. Tseng et al., *Tech. Digest. of IEDM*, pp83-86, 2003.

[5] M. Sato et al., *Symp. VLSI Tech*, pp66-67, 2008.

[6] S. Pae et al., *Proc. IRPS.* pp. 352-357, 2008.

[7] A. Shimizu, H. Ota, A. Toriumi., *Proc. IRPS.* pp. 669-670, 2008.

[8] A. T. Krishnan et al., *Tech. Dig. of IEDM*, pp705-708, 2005.

[9] K. Onishi et al., *IEEE Trans. Elec. Dev.*, 54, pp. 1517-1524, 2003.

[10] K. Onishi et al., *Tech. Digest. of IEDM*, pp659-662, 2001.

[11] H. Reisinger, U. Brunner, W, Heinrigs, W. Gustin, C. Schlunder., *IEEE Trans. Elec. Dev.*, 7, pp. 119-129, 2007.

[12] K. O. Jeppson et al., *JAP*, 48, pp 2004-2014, 1977.

[13] A. E. Islam, H. Kufluoglu, D. Varghese, S. Mahapatra, M. A. Alam., *IEEE Trans. Elec. Dev.*, 54, pp. 2143-2154, 2007

Method of Deciding Burn-in Stress Voltage in Conceptual Design Phase

Jae Yong Seo, Noh Seok Park, Hyung-Jin Park, Hong Sik Park, Woo Sup Kim, Se Young Lim, Hyun Kim, Nam Hyun Cha, Ju Seong Kang and Byung Se So

Product Quality Team, Memory Division, Samsung Electronics
Banwol-Dong, Hwasung-City, Gyeonggi-Do, Korea, 445-701
82-31-208-6369; fax: 82-31-208-6699; e-mail: jy_seo@samsung.com

INTRODUCTION

Randomly unintended process defects or process variations are the major contributor to semiconductor component reliability failures. DRAM manufacturers rely on burn-in (BI) to achieve required field failure rates as shown in fig.1. [1] For the voltage acceleration an exponential relation between time and voltage is used [2] (In addition, there is acceleration factors for applying stress pattern[3] & frequency that is not considered here.) Typically, voltage acceleration parameters with multi defects are experimentally determined for etch product and technology. [4] However, The decision of BI voltage in conceptual design level (such as a planning phase to develop the new product using new process as shown in fig 2) without evaluating product has not been discussed yet. In this paper, we proposes new ratio method that could be used in the conceptual design, because designer has not known defect characterizations, when the BI voltage should be decided without acceleration factors to develop the next generation's DRAM in fig.3. First, to establish this method, the history of BI stress voltage and early life failure rate (ELFR) data was reviewed over past ten years. Second, based on the existing trend data, we will find the individual increasing ratio between internal stress voltage (V_{stress}) and nominal use-voltage (V_{use})[1] compared to etch design rule product. Finally, we will show that both bit line (storage node cell cap.) and word line stress voltage (Vpp) ratio with single BI condition for screening multi defect are investigated and discuss the design guide by considering BI strategy in planning phase.

BURN-IN MODELING AND V_{STRESS}

In the case of BI, wherein it uses both stress voltage and temperature (85~140`c) stresses, the effects of both should be taken into consideration. For BI modeling, similar equations as for gate oxide are used as show in fig.4. For temperature acceleration, the well-known Arrheius equation is applied, and for the voltage acceleration an exponential relation between time and voltage is used. The acceleration parameters are experimentally determined for each product and technology. The weibull distribution (fig.7) is accepted as probability distribution. Typical BI voltage is usually 1.1 to 1.7 times higher than the nominal voltage. This also applies to chip internal voltages which are generated on the product. Traditionally, the BI stress voltage could be calculated from equation (1) with T_{user} and V_{user} (use-conditions). (Where: E_a=Activation energy (eV), k= Boltzmann's constant, β= Voltage acceleration rate factor.(V^{-1}))

$$V_{stress} = V_{User} + \left\{ \left(\frac{1}{\beta} \right) \ln AF_{Total} - \frac{Ea}{\kappa} \left(\frac{1}{T_{User}} - \frac{1}{T_{Stress}} \right) \right\} \quad (1)$$

To predict lifetime of multiple defects in fig.5 and fig.6, the equation to describe distribution after BI is shown in equation (2) and fig.7 [4]

$$\sum_{defect\,A-D} F_{pBI}(t) = \frac{F(t + t_{BI\,equiv}) - F(t_{BI\,equiv})}{1 - F(t_{BI\,equiv})} \quad (2)$$

These equations apply to extrinsic and intrinsic distribution [5].By using these equations, ELFR could also be obtained by estimating the product defective part per million (DPPM) from user condition.

RESULTS AND DISCUSSION

Burn-in effectiveness relies on the capability to accelerate any latent defects to failure. However the margin between operating voltages and any additional voltage that may be applied during burn-in is shrinking because the intrinsic reliability margins to voltage are shrinking. Consequently, elevated voltage are not only uncovering latent defects, but are impacting the expected lifetime of the intrinsic material properties of semiconductor device. Conventionally, when designers decide the stress voltage, limitations such as acceleration parameters should be taken into consideration. In this study, in order to help circuit and process designers who work early design stage, we propose method of determining the Burn-in stress voltage without acceleration parameters. For obtaining the optimized stress voltage, we investigated the history of Burn-in stress voltage over past ten years as shown in fig.8. We found the increase ratio of stress voltage which is to meet field ELFR (ppm) target as shown in fig 9.

Methodology

Various histories of products and process were reviewed to confirm the increasing ratio of BI stress voltage in fig 10, 11. The ratio involves a two node approach. The first node which determines charge level of DRAM cell capacitor is Bit line (BL) voltage. Second power node is word line (WL) voltage level with operating gate node of cell transistor. [6] In fig.10 (a), constant increasing ratio of BL voltage is observed 64% ± 14%. Furthermore, as shown in Fig. 10(b), increasing ratio of monitoring voltage of Product Reliability Test (PRT) which is evaluated ELFR (~1year) is observed 33.1% ±9%. Fig.11 shows the average increasing ratio of WL voltage is observed 24.3% ±8%. Also, as shown in Fig. 11(b), increasing ratio of monitoring voltage of PRT is observed 11.7% ±3%, even though materials and processes, architectures, was changed during past ten year. It is indicated that circuit designer can obtain the screen voltage without product defect information. The above descried concept was verified from ELFR which were returned field data to meet reliability target as shown in fig.9. Utilizing this BI stress concept is effective and shortens the required development time substantially.

CONCLUSION

Design for next generation DRAM on early development stage should be concerned about the intrinsic reliability margins and maximum Burn-in stress voltage. This paper proposes that BI stress voltage, when latent defects could not be evaluated by burn-in to obtain the acceleration parameters, is decided by the increasing ratio (as shown in Table.1.) of historical burn-in stress voltage trend.

REFERENCES

[1] "International Technology Roadmap for Semiconductors" http://public.itrs.net/
[2] C.Glenn Shirley., "Infant Mortality Control" *IRPS tutorial*, 2002.
[3] Klaus Nierle., et al., *IRW*, 2000, pp.183-184.
[4] Rolf-P. Vollerten., et al., *IRW*, 1999, pp. 167-173
[5] Thomas J.Anderson., et al., *IRPS*, 2006, pp. 545-551
[6] J.Y.Seo et al., *Microelectronics Relibility*, 2005, pp. 1317-1320

978-1-4244-5430-3/10 $26.00 © 2010 IEEE

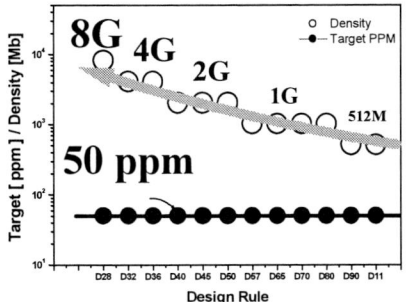

Fig.1. DRAM Product Generation vs Early failures target (Failures during the first 4000 hours of operation (~1 year's use at 50% duty cycle)) [1]

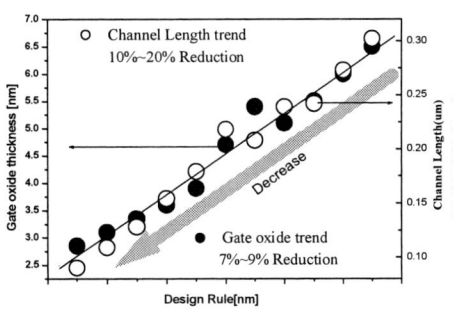

Fig.2.Design rule versus Gate oxide thickness / channel length reduction

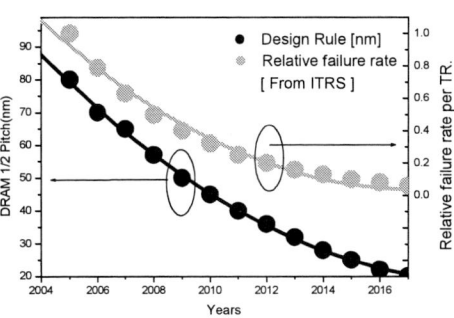

Fig.3. Scaling of DRAM Half Pitch and Relative failure rate per TR. [1] from ITRS

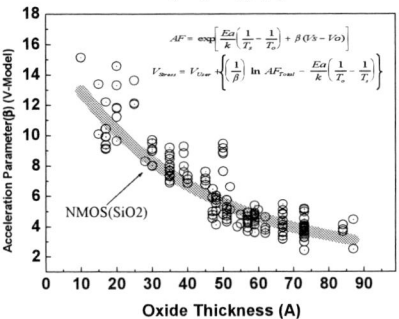

Fig.4 Voltage acceleration parameter values (β) as function of Gate oxide thickness (Å).

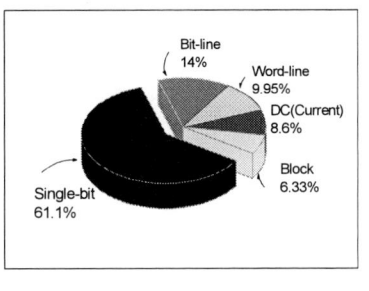

Fig.5 Distribution of the analyzed early failure case by stressing burn-in.

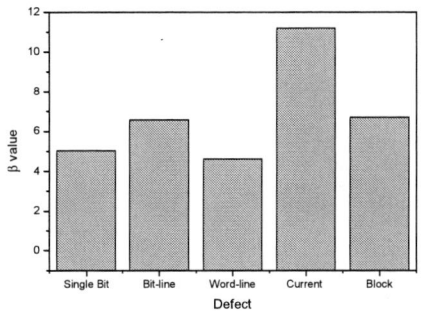

Fig.6 Acceleration parameters (β) with multiple defects are experimentally determined for etch defect.

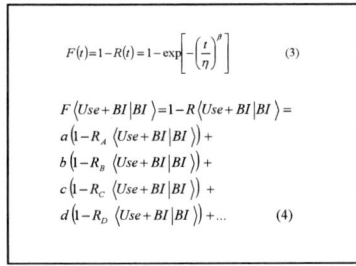

Fig.7.Equation (3): Weibull distribution and Equation (4): Cumulative failure of components post Burn-in stress [5]

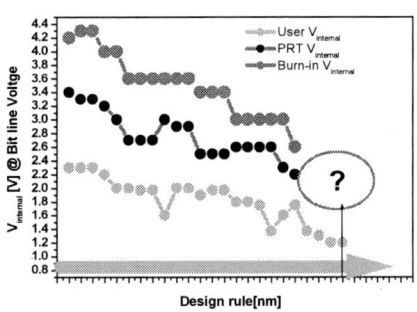

Fig.8. Voltage reduction trend (Use, Monitoring, Burn-in) versus DRAM design rule

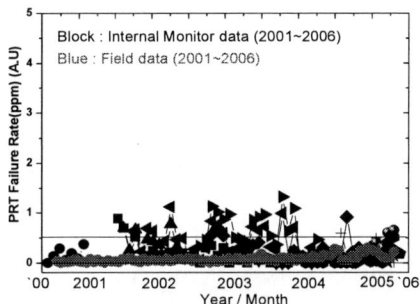

Fig.9. Early life failure (ppm) of internal monitoring (Block color) trend and filed monitoring trend (Blue color)

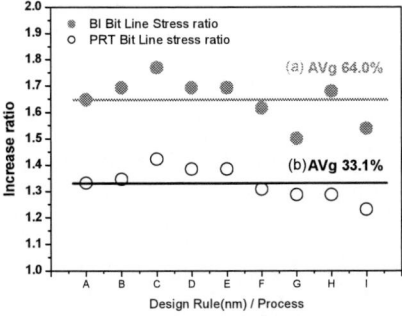

Fig.10. Increase ratio of Bit line (BL): (a) Burn-in stress, (b) ELFR monitor stress (~1 year's equivalent condition)

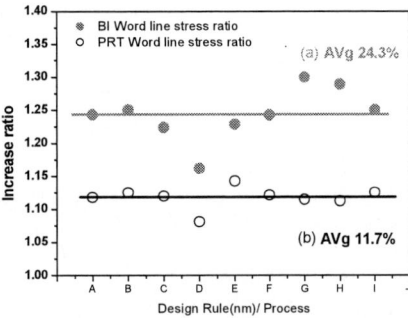

Fig.11 Voltage increase ratio of word line (WL): (a) Burn-in stress, (b) ELFR monitor stress (~1 year's equivalent condition)

Stress Voltage	Bit Line voltage	Word Line voltage
*User (Reference)	* x1	* x1
PRT(Use-like monitor)	x1.33	x1.17
Burn-in	x1.64	x1.24

Table.1. Summary of Increase ratio (Comparison between word line and bit line level)

Device-Level Reliability Simulation for High Temperature Applications of a Modular CMOS Foundry Process

Markus Ackermann

X-FAB Semiconductor Foundries AG
D-99085 Erfurt, Germany
+49 -361 -427 8353, markus.ackermann@xfab.com

Abstract— **Introduction of CMOS processes with extended temperature operating conditions up to 448K had posed great challenges for device reliability. This work will detail a simulator that enables support of robust IC design by application specific device-level lifetime calculation based on reliability models for the most severe intrinsic failure mechanisms of MOSFETs.**

I. INTRODUCTION

Reliability tests are part of foundry process qualification [1]. Typically pass/fail tests are performed in order to ensure reliable device performance according to the process specification. Shrinked device dimensions as well as extended fields of application make an sufficient reliability prediction difficult due to discrepancies between the lifetime values extracted from static stress tests and the actual behaviour during operation, where the stress often applies rather partially regarding time and extent and further mechanisms as relaxation may occur. An example is a high temperature CMOS foundry process which is specified for junction temperatures up to 448K as required by automotive applications. The lifetime evaluation at this stress temperature will result in a too pessimistic assessment in many cases since temperature and bias will either not occur in combination or the maximum temperature will only occur for short times. The impact of the operating conditions on the device lifetime needs to be modelled to prevent such an underestimation. The presented approach enables an application specific lifetime prediction which is based on lifetime models that are already determined within normal reliability tests. The absence of any additionally required investigations to the standard qualification procedures makes the method very attractive for foundry companies.

II. MODEL EXTRACTION

During wafer-level qualification of the CMOS process, reliability investigations at different elevated operating conditions were performed. The reliability analysis included tests on MOS transistors as well as on further components of the process offering as capacitances and metallization. The lifetime at stress conditions was determined with respect to a defined maximum shift of a device characteristic. By using a lifetime model the lifetime finally was extrapolated to the target operating conditions.

The lifetime models including the measured model parameters were taken as a base for the reliability simulation. An overview of the implemented lifetime models [2] is given by table 1. The model parameters represent the worst case which has been found with respect to the variation of the CMOS process as well as to the design and to the bias conditions. For example, in terms of Hot Carrier Injection (HCI) analysis of MOS transistors, devices of minimum channel length have been stressed at maximum drain voltage and at gate voltage according to the maximum of degradation.

III. SIMULATION METHODOLOGY

Based on a device lifetime at maximum operating conditions (respectively worst case within the specified range of operating conditions) and a lifetime model, the simulator recalculates the device lifetime (effective lifetime) according to a given set of operating conditions (mission profile). Therefor a statical approach is used; the transition phases between the different operating points are not considered.

The mission profile is broken down to different operating phases. The adjustable parameters of each phase are the maximum junction temperature T, the percentage X of the operating time with respect to the total operating time as well as the operating voltages V_{ds} (drain-source) and V_{gs} (gate-source) in terms of MOS transistors, the operating voltage V in terms of capacitances or the operating current density J in terms of metallization, including a duty factor for each operating condition.

Dependent on the selected device category, the lifetime model is applied in order to calculate the acceleration factor AF which is defined as the ratio of lifetime at use conditions and lifetime at worst case conditions. For each operating phase, AF is determined by the relation of the operating conditions at use case and at worst case. The equations for the acceleration factors of the different considered degradation mechanisms are given by table 1. The effective lifetime factor of an operating phase i is then simply described by

$$EF_i = AF_i * X_i. \qquad (1)$$

A sum approach as a rough method of integration is used to calculate the total effective lifetime factor and the effective

---

This work was supported by the German Federal Ministry of Education and Research (BMBF) within the project 01M3184B.

978-1-4244-5430-3/10 $26.00 © 2010 IEEE

lifetime, $EF = \Sigma_i(EF_i)$ and $t_{life} = t_0/EF$. Since the calculation of the acceleration factors is realized separately for different degradation mechanisms which all affect one device, the respective calculated effective lifetimes need to be combined. If all mechanisms affect the same device parameter (an example is the degradation of an interconnect due to Electro Migration (EM) and Stress Migration (SM) which both influence the resistance), the Sum-Of-Failure-Rate (SOFR) approach is used [3]. Otherwise, if different device parameters are affected by different mechanisms, the effective lifetimes are calculated separately. The effective lifetime factor relates to an average junction temperature T_{av}, which can be calculated either from $T_{av} = \Sigma_i(X_i * T_i)$ (not thermally activated mechanism) or by solving the equation $AF(T_{av}) = EF$.

In addition to pure device-level reliability estimation, the tool also includes a failure rate calculation which is based on circuit-level tests. According to a defined average junction temperature the estimated failure rate is calculated. A simple Arrhenius model is used to consider the temperature dependence of the failure rate.

TABLE I.   ACCELERATION FACTOR MODELS FOR THE DEGRADATION MECHANISMS HOT CARRIER INJECTION (HCI), NEGATIVE BIAS TEMPERATURE INSTABILITY (NBTI), TIME-DEPENDENT DIELECTRIC BREAKDOWN (TDDB), ELECTRO MIGRATION (EM) AND STRESS MIGRATION (SM).

Model	Acceleration factor	
	Equation	*Parameter definition*
HCI	$AF = \exp[r*(1/V_{ds} - 1/V_{ds0})]$	r - voltage acceleration coefficient V_{ds} - drain-source voltage
NBTI	$AF = \exp[-r*(V_{gs} - V_{gs0})] *$ $\exp[E_a * (T_0 - T) / (k*T*T_0)]$	r - voltage acceleration coefficient V_{gs} - gate-source voltage E_a - activation energy k - Boltzmann's const. T - temperature
TDDB	$AF = \exp[-r(T) * (E - E_0)] *$ $\exp[E_a * (T_0 - T) / (k*T*T_0)]$	r - voltage acceleration coefficient E - electrical field E_a - activation energy k - Boltzmann's const. T - temperature
EM	$AF = (J/J_0)^{-n} *$ $\exp[E_a * (T_0 - T) / (k*T*T_0)]$	J - current density n - current density acceleration coefficient E_a - activation energy k - Boltzmann's const. T - temperature
SM	$AF = [(T_d - T) / (T_d - T_0)]^{-N} *$ $\exp[E_a * (T_0 - T) / (k*T*T_0)]$	N - stress acceleration coefficient E_a - activation energy k - Boltzmann's const. T - temperature T_d - deposition temp.

IV.   RESULTS

The simulation approach is demonstrated using an inverter circuit as an example. Table 2 includes the relevant device characteristics of the NMOS and the PMOS which are part of the inverter.

The devices are offered by a high-temperature CMOS foundry process and specified for junction temperatures up to 448K and operating voltages $|V_{dsmax}| = |V_{gsmax}| = 5.5$V. A typical automotive mission profile (motor control) contains the following operating phases:

1) $T_1 = 233$K, $X_1 = 5\%$, non working mode

2) $T_2 = 298$K, $X_2 = 20\%$, working mode

3) $T_3 = 378$K, $X_3 = 65\%$, working mode

4) $T_4 = 448$K, $X_4 = 10\%$, working mode

It is assumed that during the non working mode the bias is set to zero and during the working mode V_{ds} and V_{gs} are applied with a duty cycle of 2% respectively 50%. If the peak voltages are equal to V_{dsmax} and V_{gsmax}, the effective operating lifetime calculated for the temperature mission profile is 1.0E+05h for the NMOS and the PMOS, what corresponds to an increase by a factor 50 respectively 10 compared to the worst case approximation. The calculated average junction temperatures are 362K (NMOS) and 414K (PMOS). The difference shows that the PMOS degradation is significantly more temperature accelerated than the NMOS degradation according to the implemented models. The simulated lifetime meets automotive requirements and could be further increased if reducing the peak values of V_{ds} respectively V_{gs} during operation.

TABLE II.   CHARACTERISTICS OF NMOS AND PMOS IN THE EXAMPLE INVERTER CIRCUIT.

	Device in the example circuit	
	NMOS	*PMOS*
T_{max}	448K	448K
V_{dsmax}	+/- 5.5V	+/- 5.5V
V_{gsmax}	+/- 5.5V	+/- 5.5V
t_{life}	2.0E+03h [a]	1.0E+04h [b]

a. DC lifetime at max. HCI stress.

b. DC lifetime at max. NBTI stress.

V.   CONCLUSION

The presented approach enables a lifetime simulation which is based on conventional device-level reliability tests. The benefit is a more realistic assessment of device reliability during circuit operation than represented by the lifetime measured during statical stress tests. Especially the aspect of different temperature environments can be considered accurately. Further the reliability simulation enables design optimization regarding the bias conditions (voltage, current).

REFERENCES

[1] JEDEC Solid State Technology Association, "Foundry process qualification guidelines", JEDEC/FSA Joint Publication JP001.01, 2004.

[2] JEDEC Solid State Technology Association, "Failure mechanisms and models for semiconductor devices", JEDEC Publication JEP122E, 2009.

[3] J. B. Bernstein, M. Gurfinkel, X. Li, J. Walters, Y. Shapira, M. Talmor, "Electronic circuit reliability modeling", Microelectronics Reliability, vol. 46, pp. 1957-1979, 2006.

978-1-4244-5430-3/10 $26.00 © 2010 IEEE

Accurate Projection of V_{ccmin} by Modeling "Dual Slope" in FinFET based SRAM, and impact of Long Term Reliability on End of Life V_{ccmin}

H. Park[*+], S. C. Song[+], S. H. Woo[*+], M. H. Abu-Rahma[+], L. Ge[+], M. G. Kang[*+], B. M. Han[+], J. Wang[+], R. Choi[^],
J.W.Yang[#], S. O. Jung[*+], and G. Yeap[+]

[+] Qualcomm Incorporated, San Diego, CA 92121, United States
[^] Inha University, Incheon, Korea
[#] Korea Univeristy, Seochang, Korea
[*] Yonsei University, Seoul, Korea, email: sjung@yonsei.ac.kr

Abstract— Supply voltage (Vcc) scaling is mostly used method to achieve low power consumption. However, a high Vccmin is required to meet the high target yield because the SRAM yield according to Vcc scaling shows "dual slope". In this paper, the root causes of "dual slope" are analyzed. Both side effect of SRAM bitcell on the yield is also considered to accurately project Vccmin, which results in 40mV increase of Vccmin to meet 99% target yield for 32nm HK/MG planar 1M SRAM. The "dual slope" effect on the yield is compared for 32nm HK/MG planar and FinFET 32M SRAMs with high (HD) and low doping (LD). Under the "dual slope" effect, the channel length adjustment method for pass gate transistor is proposed to reduce Vccmin of FinFET SRAM. When the number of finis is 1:2:2(=PU:PG:PD), HD and LD 32M FinFET SRAMs improve Vccmin by 370mV and 500mV, respectively, compared to 32M planar counterparts using the proposed the channel length adjustment method. Effect of NBTI and PBTI on Vccmin is also investigated. BTI degradation is greatly dependent on HK thickness and surface plane orientation of FinFET. End of Life (EOL) Vccmin optimization therefore requires careful selection of HK thickness and surface orientation.

Keywords-dual slope; read stability; write ability; SRAM; FinFET; NBTI ; PBTI;

I. INTRODUCTION

As Power limited applications increase, low power implementation has become important. Because SRAMs occupy large portion of a chip, the supply voltage scaling for SRAMs is widely used to reduce power consumption [1]. However, because the process variations increase due to technology scaling, a tradeoff between Vcc scaling and yield becomes severe. Vccmin projection of SRAM array is known to be very difficult especially at high yield target, as the SRAM array yield shows two different slopes (Fig. 1) with respect to operation Vcc scaling [2] and the mechanism of this behavior is not well understood. As lowering Vcc is important for low power design especially for scaled SRAM bitcell, it's utmost important to understand the dual slope issue in SRAM yield and project Vccmin accurately. In this paper, we report

- For the first time, detailed mechanism of dual slope issue in SRAM yield behavior with respect to Vcc of

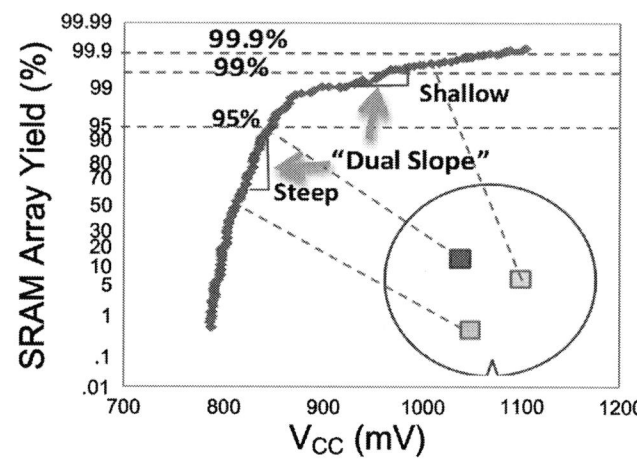

Figure 1. Measured SRAM array yield with respect to Vcc across wafer. Depending on the target yield, minimum operation Vcc (Vccmin) differs significantly, especially in high yield region due to significantly different slope in yield curve.

the bitcell was explained, which allows accurate projection of Vccmin in SRAM design.

- FinFET bitcell is shown to be more beneficial in reducing Vccmin at high yield at 22nm and below.

- Based on the model, further scaling of FinFET SRAM Vccmin is demonstrated through adjusting physical and electrical design of FinFET bitcell.

- Vt shift from Bias Temperaure Instabilty (BTI) affects End of Life (EOL) Vccmin significantly.

- As NBTI and PBTI are greatly dependent on High-K (HK) thickness and surface orientation, selecting appropriate HK thickness and Fin surface is improtant to improve EOL Vccmin.

II. SIMULATION METHOD

We used 32nm HK/MG planar and FinFET models to extract SRAM yield and perform comparison each other. SRAM yield for read and write operations was measured using

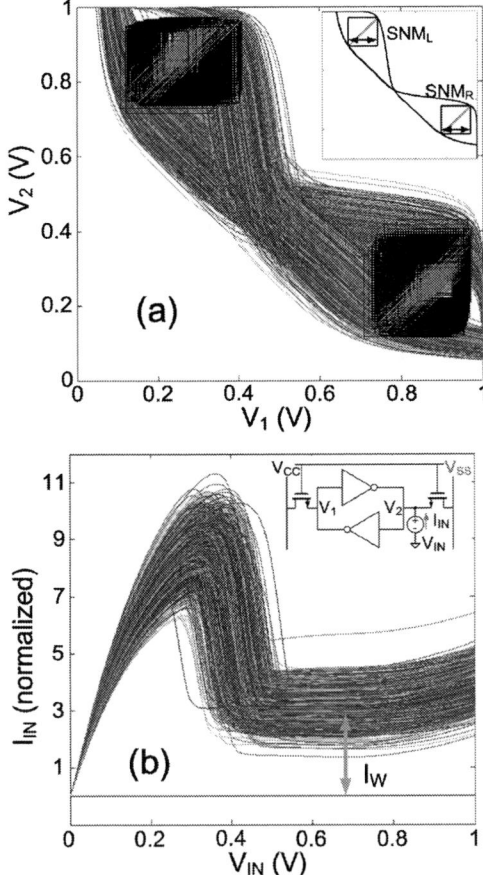

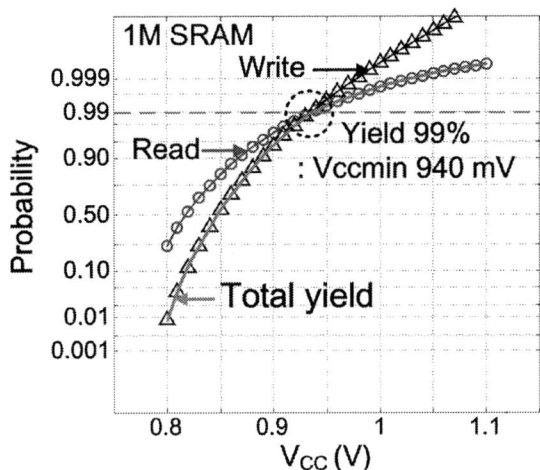

Figure 3. Both read and write yield show yield drop as Vcc scales, but with different sensitivity, making crossing point at certain point.

slope in the plot. When overall yield (i.e., limiting yield either from read or write operation) is plotted, two different slopes ("dual slope") are revealed. Detailed explanation of the origin of dual slope is following.

- When V_t increases, SNM improves as butterfly curve becomes sharper. Reducing V_{gs} has similar effect as V_t increase in terms of V_{gs}-V_t. Therefore, degradation of read margin due to Vcc scaling is mitigated by decreasing (V_{gs}-V_t), making rate of read margin drop with Vcc scaling slowed down.

- Write margin is primarily dependent on alpha ratio which is determined by relative ratio of V_t of PG (Pass-Gate) and PU (Pull-Up) transistors $[(V_{gs}-V_{t\_PG})/(V_{gs}-V_{t\_PU})]^\gamma$ with layout geometry [6]. When V_{t_PU} is lower than V_{t_PG}, therefore, the ratio decreases with Vcc scaling. Due to degraded ratio, write yield drops more

Figure 2. (a) SNM from square box method in conventional butterfly curve was used for read operation evaluation. Length of squares was defined as SNM (inset). (b) I_W was used for write operation. I_W is defined as minimum I_{IN} with respect to V_{IN} in the bias configuration of bitcell shown in the inset.

static noise margin (SNM) and write current (I_W) methods respectively (Fig. 2) [3,4]. Process variation has been modeled into V_t variation [5], from which distributions of SNM and I_W can be fitting with Gaussian distribution [4, 7]. The tail region of SNM distribution seriously varies according to the number of MC simulation. This tail region plays an important role when determining mean (μ) and standard deviation (σ) of read and write metric (i.e., SNM and I_W). It's important, therefore, to determine appropriate the number of MC simulation to obtain accurate distribution curve. In an attempt to find out the effect of the number of MC simulation, μ/σ, where μ is normalized by σ, was used, which allows us to study the relation between the number of MC simulation and change in distribution curve statistically. μ/σ oscillates as the number of MC simulation increases, but finally converges after 10k simulation, which indicates that failure probability is settled down after 10k MC simulation. 10k MC simulation for all the data in this paper was used.

III. DUAL SLOPE EFFECT ON SRAM YIEDL

1M SRAM yield using 32nm HK/MG planar model was simulated. As shown in Fig. 3, read and write yield drop with different rate with respect to Vcc scaling, which forms different

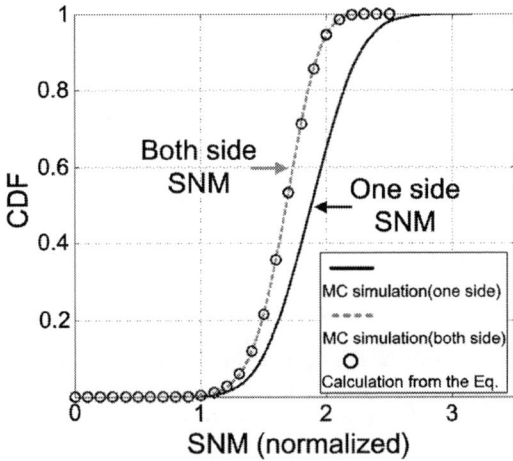

Figure 4. CDF calculation from the both side equation predicts SNM distribution of both side accurately, coinciding with Monte Carlo simulation line.

978-1-4244-5430-3/10 $26.00 © 2010 IEEE

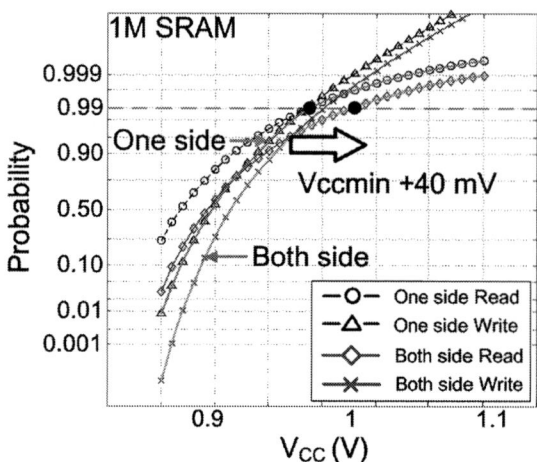

Figure 5. When both side of SNM is taken into account, more refined Vccmin can be projected. For 1M SRAM array using 32nm HK/MG planar device, >40mV Vccmin difference is observed between one side and both side yield curve.

TABLE I. SPILIT TABLE OF FINFET CONFIGURATION IN THE BITCELL

Conventional size of HD & LD [4]	Case 1			Case 2		
	PU	PG	PD	PU	PG	PD
H_{fin} (nm)	35					
T_{fin} (nm)	15					
Length (nm)	32					
N_f	1	1	2	1	2	2

Balanced size	LD			HD		
	PU	PG	PD	PU	PG	PD
H_{fin} (nm)	35					
T_{fin} (nm)	15					
Length (nm)	32	**45**	32	32	**35**	32
N_f	1	2	2	1	2	2

sharply at lower Vcc region.

- Because of mechanisms explained in 1 and 2 above, read and write yield making dual slope in the yield vs. Vcc plot, leading to cross over point at certain Vcc.

Shallow slope in high yield portions is especially important, as small change in the target yield could make significant difference in target Vccmin. It's important, therefore, to project accurate slope especially at high yield portion. More accurate yield curve can be obtained from both side evaluation of SNM, as SNM is decided by lower value of either side in butterfly curve [7]. This can be achieved by using Cumulative distribution function (CDF) of bivariable Gaussian distribution [8].The model from the equation fits the simulation results well,

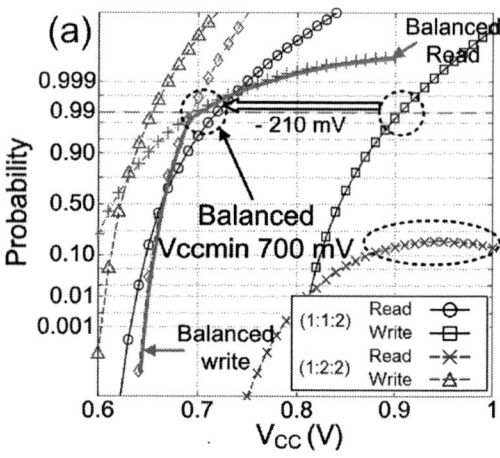

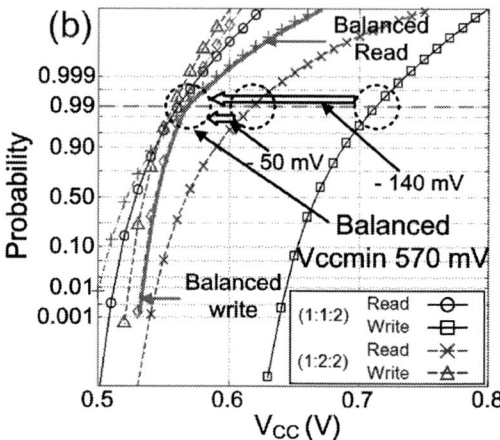

Figure 6. Conventionally sized vs. balanced (a) HD 1:2:2 and (b) LD 1:2:2 ; Slight adjustment in PG transistor Lg improves Vccmin scaling significantly both for HD and LD split.

indicating the mathematical model is accurate (Fig.4). From this improved yield projection method taking both side of SNM, more accurate Vccmin projection is possible. From Fig. 5, one can see that 99% yield target Vccmin is modified more than 40mV when both side of SNM is taken into account.

IV. FINFET MODEL AND COMPARISON OF PLANAR AND FINFET SRAMS

In order to evaluate the effect of dual slope in 32nm FinFET, SRAM yield of 32M array was simulated using bulk Si-substrate based HK/MG FinFET model [9]. Several different FinFET bitcell configurations were used. Detailed FinFET geometry of each transistor in the bitcell is described in table.1 [4]. In conventional FinFET bitcell, only number of Fin varies within bitcell while fixing other FinFET geometry including Lg. Channel doping level was also varied to High Doping (HD) and Low Doping (LD) splits to evaluate random dopant fluctuation effect. Number of Fin varies among PU, PG and PD (Pull-Down) transistors. FinFET bitcell split with number of Fin ratio 1:1:2 (=PU:PG:PD) shows high read yield but low write yield due to larger beta ratio (=PD/PG) and smaller alpha ratio (=PG/PU). In order to improve Vccmin

978-1-4244-5430-3/10 $26.00 © 2010 IEEE

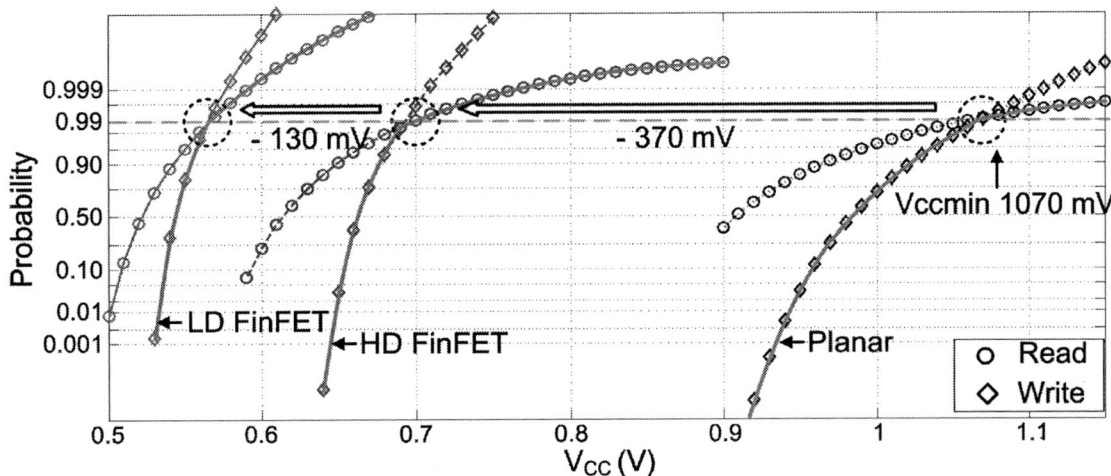

Figure 7. SRAM yield curve between 32nm HK/MG FinFET SRAM and planar counterpart. At the same doping level (HD), FinFET has 370mV lower Vccmin, while lower doping (LD) further reduces Vccmin by additional 130mV. Owing to strong electrostatic control, FinFET allows much lower channel doping with better short channel control, of which benefit is manifested in significantly lower Vccmin. Larger effective width of FinFET also enhances Vccmin scaling. This improvement of Vccmin from FinFET would become critically important for low power design as bitcell size scales <0.1 μm2 in 22nm and below.

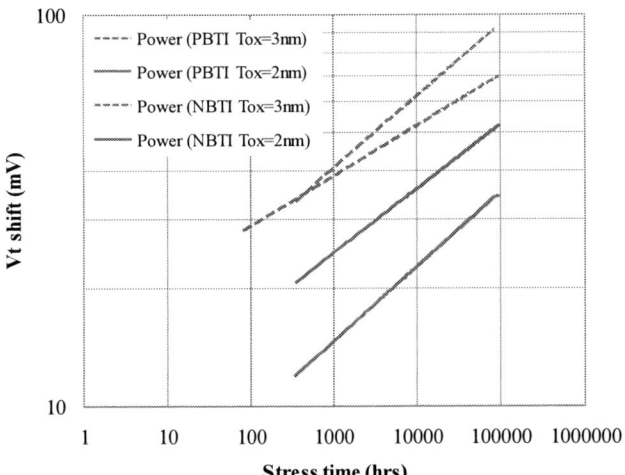

Figure 8. NBTI and PBTI V_t shift with two different HK thicknesses. PBTI improves significantly as HK thickness (electrical) scales from 3 to 2nm.

scaling, read and write stabilities have to be balanced. However, due to quantization of effective width from discrete nature of Fin, fine adjustment of beta and alpha ratio through width control is not possible. Instead, L_g of PG transistor was adjusted as shown in table.1, optimized for both HD and LD splits with 1:2:2 Fin configuration. Fig. 6 compares SRAM yield curves with and without PG L_g adjustment. Despite fixed width ratio, slight adjustment of L_g in HD split with 1:2:2 Fin enables Vccmin scaling by 210mV compared to HD split with 1:1:2 Fin. A balanced LD split with 1:2:2 Fin improves Vccmin by 140mV and 50 mV compared to LD split with 1:1:2 and 1:2:2 Fins respectively. Fig. 7 shows SRAM yield curve between 32nm HK/MG FinFET SRAM and planar counterpart. At the same doping level (HD), FinFET has 370mV lower Vccmin, and lower doping (LD) further reduces Vccmin by additional 130mV. Owing to strong electrostatic control, FinFET allows much lower channel doping with better short

channel control, of which benefit is manifested in significantly lower Vccmin. Larger effective width of FinFET also enhances Vccmin scaling. This improvement of Vccmin from FinFET would become critically important for low power design as bitcell size scales <0.1 μm2 in 22nm and below.

V. NBTI AND PBTI EFFET ON DUAL SLOPE

Fig. 8 shows NBTI and PBTI V_t shift with two different HK thicknesses. PBTI improves significantly as HK thickness scales from Tinv=3nm to 2nm, while PBTI improvement is not as significant as NBTI. The significant PBTI improvement with HK scaling is mainly attributed to less amount of trapping sites in the bulk of HK as the HK volume reduces. NBTI is, on the other hand, mostly interface related degradation, which is less susceptible to HK thickness scaling.

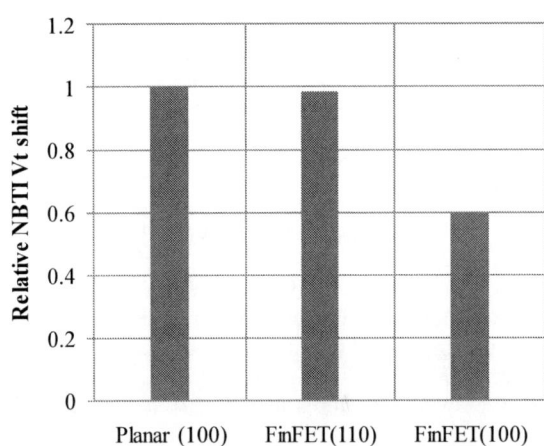

Figure 9. NBTI effect from Planar vs FinFET with different surface orienttions. Due to lower surface potential, FinFET has ~40% less V_t shift compared to planar at the same (100) surface. At (110) surface, V_t shift from FinFET becomes similar to planar (100) due to higher surface atomic density.

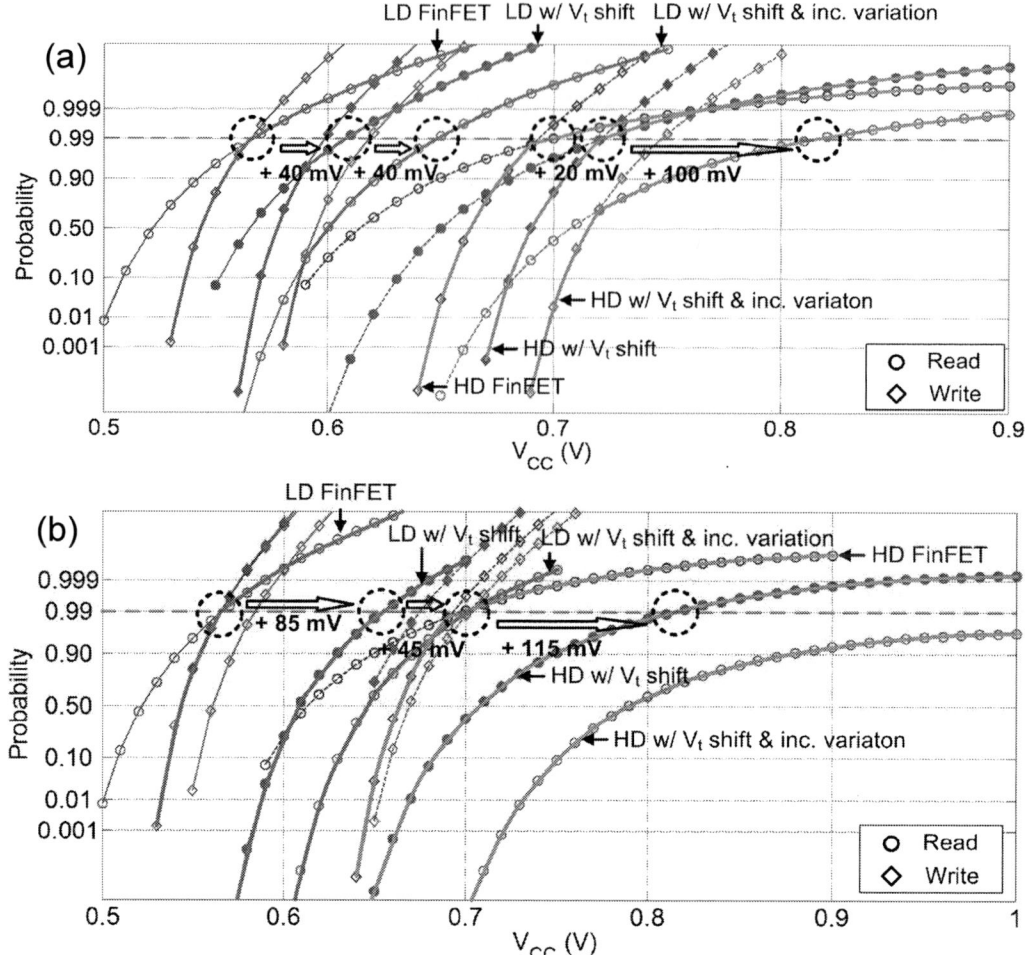

Figure 10. Impact on Vccmin from NBTI and PBTI V_t shift. (a) PU suffers NBTI and both PD & PG suffere PBTI. (b) PU suffers NBTI and only PD sufferes PBTI. Due to significantly degraded read SNM in the case of (b), Vccmin degradation is much more significant than the case (a). This indicates that intentially weaker PG might need to be used to prevent significantly Vccmin degradation along the aging process.

In addition to HK thickness, surface plane is also play a role in BTI degradation. Fig. 9 shows NBTI effect from Planar vs. FinFET with different surface orientations. Lower surface potential due to lower body doping allows FinFET to have ~40% less V_t shift compared to planar at the same (100) surface. At (110) surface, V_t shift from FinFET becomes similar to planar (100) due to higher surface atomic density.

Vccmin projection applying BTI V_t shift is shown in Fig. 10. HK thickness 2nm is used at 10 years (1×10^5 hrs) lifetme from Fig. 8. Conventional (110) FinFET surface is assumed as Fig. 8 is from planar (100) surface. As expected, BTI V_t shift degrades Vccmin. Fig.10 (a) assumes that PU suffers NBTI and both PD & PG suffer PBTI, while (b) assumes PU suffers NBTI and only PD sufferes PBTI. (b) is considered more realistic as PG is stressed only during WL is selected. Due to significantly degraded PD/PG ratio (thus read SNM) in the case of (b), Vccmin degradation is much more significant than the case of (b). This indicates that intentially weaker PG might need to be used to prevent significantly Vccmin degradation along the aging process.

VI. CONCLUSION

- Physical modeling of dual slope in SRAM yield with respect to different operation voltage (Vcc) was made, which allow accurate projection of Vccmin .

- Different yield behavior between read and write operations with Vcc scaling is the root cause of the dual slope, and the effect is quantitatively modeled with simulation verification.

- Vccmin projection can be further refined taking both sides of bitcell stability into account during dual slope modeling,

- FinFET bitcell shows much lower Vccmin (i.e., lower Vccmin) once the FinFET bitcell is balanced in read and write operation with considering dual slope issue.

- V_t shift from Bias Temperaure Instabilty (BTI) affects End of Life (EOL) Vccmin significantly.

- As NBTI and PBTI are greatly dependent on High-K (HK) thickness and surface orientation, selecting appropriate HK thickness and Fin surface is important to improve EOL Vccmin.

REFERENCES

[1] International Technology Roadmap for Semiconductors (ITRS), Albuquerque, NM, 2008.

[2] S. Hasegawa et al., "A cost-conscious 32nm CMOS platform technology with advanced single exposure lithography and gate-first metal gate/high-k process," Proc. IEDM, pp.1-3, 2008.

[3] E. Seevinck, F. List, and J. Lohstroh, "Static-noise margin analysis of MOS SRAM cells," IEEE J. Solid-State Circuits, vol. sc-22, no. 5, pp.748-754, October 1987.

[4] H. Kawasaki et al., "Demonstration of highly scaled FinFET SRAM cells with high-k/metal gate and investigation of charactersitic variability for the 32 nm node and beyond," Proc. IEDM, p.1-4, 2008.

[5] K. Kuhn, "Reducing variation in advanced logic technologies: approaches to process and design for manufacturability of nanoscale CMOS," Proc. IEDM, p. 471-474, 2007.

[6] K. A. Bowman et al., "A physical alpha-power law MOSFET model," IEEE J. Solid-State Circuits, vol. 34, no. 10, pp.1410-1414, October 1999.

[7] A. Bhavnagarwala et al., "A sub-600-mV, fluctuation tolerant 65-nm CMOS SRAM array with dynamic cell biasing," IEEE J. Solid-State Circuits, vol. 36, no. 4, pp.946-955, April 2001.

[8] R. D. Yates and D. J. Goodman, Probability and stochastic process, 2nd ed., Rosewood Drive, Danvers: John Wiley & Sons, Inc., 2005.

[9] M. Dunga et al., "BSIM-MG: a versatile multi-gate FET model for mixed-signal design," Symp. on VLSI Techn., pp. 60-61, 2007.

System-level analysis of soft error rates and mitigation trade-off explorations

Zhe Ma*, Francky Catthoor*, Frank Vermunt[†] and Teun Hendriks[‡]

* IMEC, Kapeldreef 75, 3000 Leuven, Belgium

[†] NXP Semiconductors, Nijmegen, The Netherlands

[‡] Embedded System Institute, Eindhoven, The Netherlands

Abstract—**This paper presents a novel system-level analysis of soft error rates (SER) based on the Transaction Level Model (TLM) of a targeted System-On-a-Chip (SoC). This analysis runs 1000x faster than the conventional SoC analysis using a gate-level model. Moreover, it allows accurate prediction in the early design phase of a SoC, when only limited application details are available. Preliminary validation results from accelerated SER tests on the physical system have shown that the analysis can predict the SER with a reasonable accuracy (within 5x of the results from tests on physical systems). This system-level analysis is particularly suitable to handle the black-box models for industrial semiconductor IP libraries. Based on this system-level analysis, we also propose a SE mitigation solution using selective protection of SRAM of a SoC. This solution provides a series of trade-offs between the system dependability and cost (in terms of silicon area).**

I. ANALYSIS TECHNIQUE

Our analysis provides estimation of SER at the system-level for embedded systems. Such embedded systems usually include several DSP processors, hardware accelerator IPs and a relatively large on-chip SRAM. SE can take place in both on-chip memories (SRAM) and flip-flops. When data-intensive applications are running on such systems, the SEs originating in an unprotected SRAM have a dominant influence on the system-level failures (as reported for 65-nm or larger CMOS technology by [1]). Therefore, our analysis presented in this paper focuses on the SE from SRAM.

A. System-level SE derating ratio and its relevance to dependability cost trade-offs

The SER at system-level is typically much lower than the SER at the memory level, this effect is referred to as SE derating. This derating is similar to the Timing Derating and the Logic Derating [7] at the circuit level (also known as the TVF (Timing Vulnerability Factor) and AVF (Architecture Vulnerability Factor), see [5]), but at a higher level of abstraction. That is, our system-level SE derating is concerned with the data transfers and consumptions in memories. At the circuit level, the TVF is caused by the fact that a circuit element is not susceptible to SE in its entire clock cycle; while the AVF is caused by the fact that all circuit elements in a chip are not working simultaneously. At the transaction level modeling, the primary reasons for our system-level SE derating are that (1) only corrupted data that have yet to be consumed could cause a system failure and (2) algorithms/data-structures used in the

applications may have inherent fault tolerance. It is important to have an accurate estimation of the SER at the system-level for a product with high dependability requirements. Such an estimation must be able to handle different working modes of the target system (e.g. AM mode vs. FM mode for a radio system) as very different SERs at the system-level could be exhibited when a system works with different modes [11] (this phenomenon is opposite to the cases of scientific applications on general purpose CPUs as reported by [9]). In practice, the SER at the system-level can be expressed as the product of the raw SER in memories and the derating ratio of a specific system. For example, if the raw SER of the SRAM used in a target embedded system is 100 FIT (Failure In Time), and the derating ratio of this system is 5%; then the SER at the system-level is 5 FIT. The essential purpose of our SE analysis is to give the SE derating ratio of the system running with a real-life workload; then a system architect is able to use this ratio and the raw SER at the memory level to calculate the user-perceived SER at the system-level.

For products under well defined levels of safety regulations (e.g. automotive electronics), such system-level SER estimation is crucial at an early stage of system design to allow for a trade-off between the final product's dependability and its cost. For the SER of a system running its real-life workload, NXP has developed its estimation model based on empirical data as well as accelerated tests on physical systems [11]. Nevertheless, as systems become more complex and workloads become more dynamic, the accuracy of estimations based on empirical data becomes less reliable. In the meantime, to use SE counter measures in a cost-effective manner requires an early identification of a system's SE vulnerability. Therefore, our high-level analysis based on a Transaction Level Model (TLM) becomes increasingly important.

B. Related analysis techniques

Many techniques exist at different abstraction levels to estimate the SER. For memories, SER is usually derived by testing the memory alone without running the functionalities on processor or other hardware IP blocks. Many raw SE at the memory system level are eventually filtered out [3]. When the user-experienced SER at the system level should be identified for a specific system, the complete system must be analyzed. If the target system has already been fabricated, then a physical system running the real-life workload can be tested

978-1-4244-5430-3/10 $26.00 © 2010 IEEE

with SER acceleration by applying radioactive sources [3]. If SER acceleration by a radioactive source is unavailable or inconvenient, a random fault injection can be used to accelerate the test [8], [10]. This fault injection is a cheaper alternative than a real test. But it requires knowledge of the raw SER in the memories. It also requires the instrumentation of the memory of a system such that memory faults can be injected at run-time when the system is executing. When the size of the memories of the system under test is large, such random fault injections at a low abstraction level require a long process. When the physical system is unavailable, then the analysis can be performed by either a hardware architecture-level emulation [6] or a gate-level emulation [2]. The system models from both levels are available later than TLM which is a higher-level abstraction model.

C. A fast and accurate high-level SER analysis

Our high-level analysis provides a fast TLM simulation of a system. A system is modeled as a set of connected components. It treats all components as black-boxes and hence no detailed specifications are required inside each component. This is important from an industrial point of view as some IP blocks are only available as black-boxes. We are mostly interested in the I/O and (real-)time behaviors of each component when memory faults are injected, and we will simulate the propagations of errors in a system. This high-level analysis provides a balance between modeling a system entirely as a black-box and modeling a system as a white-box with all intra-component details.

D. Required input for the high-level SER analysis

We work with a customized transaction level model [4] called Thread Node Graph (TNG). A TNG is a directed acyclic graph where each vertex is a Thread Node (TN) and each edge between two TNs represents a data transfer. A TN is the basic unit for computation in our model and hence cannot be divided further. Each TN represents a self-contained function that can be executed without calling functions in other TNs. A TN can start after all data associated with the incoming edges are ready; and it finishes with all data associated with its outgoing edges become ready. Therefore, an edge between two TNs represents an ordering relation that the succeeding TN can only start when the preceding TN has finished. Each schedule has the timing information on the executions of TNs, including start/end times and memory accesses schedules. Such schedules present a deterministic view on the reads and writes to the memories. A valid schedule of a TNG is defined as one in which each TN is executed once and all orders on TNs are satisfied. A target embedded system can be specified with one or many different TNGs; and a TNG can have different schedules.

To evaluate the SER at system-level, we first need to have the raw SER at the memory level to calculate the memory faults number. Each memory fault is a single bit error taking place in a SRAM. The number of SE-induced memory faults in a particular memory can be calculated as:

$$MemFaults = SER \times MemSize \times ElapsedTime \quad (1)$$

In most cases, the raw SER is determined by whether ECC (Error Correction Code) protection is used for a SRAM with a specific process technology.

The vulnerability of each TN (as a black-box) to the SE-induced data corruptions is measured by both its failure probability to bit errors and its silent data corruption probability to bit errors.

The TN failures caused by memory faults from a particular memory is:

$$TNFailures = TNFailProbability \times MemFaults \quad (2)$$

where $MemFaults$ is the number of memory faults from formula (1). This $TNFailProbability$ can be identified by fault injection simulations in Instruction-Set Simulator (ISS) for software IPs or in VHDL simulator for hardware IPs. And a TN can have different failure probabilities to bit errors from different data objects (e.g. input buffer, internal data or control parameters).

In addition to the failure probabilities, we also need the probabilities of silent data corruption at the output data. These probabilities can be obtained in the similar manner as the failure probabilities. The only difference is that for silent data corruptions, we need to calculate them for each output data object. For a TN with a set of output data objects $D = \{d_0, d_1, ...\}$, we define $\forall d_i \in D$,

$$\begin{aligned} DataCorruptions_i = CorruptionProbability_i \\ \times MemFaults \end{aligned} \quad (3)$$

II. SELECTIVELY PROTECTING MEMORY/DATA

Once a baseline analysis (without any counter measures against SE) is complete, the designers can find a derating ratio of their system and calculate the SE-induced system failure rate. If the failure rate is higher than an acceptable number, designers can use this high-level analysis to have a quick evaluation of trade-offs between the failure rate and the size of protected memories. Our high-level analysis allows a user to specify the SER at the granularity of a TN's memories. The designers can specify a set of memory address ranges for a TN and give either a raw SER or a SER with memory-protecting for each range. Due to the combinatorial nature of selective memory protection, further optimizations are still required to have a scalable exploration with a large number of TNs. Our experiments have shown that when the number of TNs is 10 or less, and the number of memory address ranges of each TN is 3 or less, our high-level analysis can complete the exploration of selective protection in several minutes on an ordinary desktop PC.

In Section III, we will show a simple case where SRAMs can be protected by ECC. And our analysis can help to find a number of optimal selectively protecting configurations.

III. EXPERIMENTAL RESULTS

A. Baseline analysis for NXP car audio processing system

We show a simple case study of applying our high-level analysis on a audio processing system. The TNG of this audio processing system is depicted in Fig.1. The bubbles are TNs. Each TN is annotated with numbers of memory accesses and execution times (not shown in the figure). This system is running on a single DSP processor with two SRAMs: a data memory and a coefficient memory (the program memory is immune to SE and hence is excluded in this case study). The numbers of read and write accesses of each individual TN for a single invocation are measured. Also, the execution times (in terms of clock cycles at a clock frequency of 130MHz) of TNs are measured on the physical systems.

Our analysis calculated a SE derating ratio of 5% for TN failures at system-level. This means that for every 100 SEs in the memory used by this audio processing system, only 5 TN failures would take place. In addition to this derating ratio of TN failures, our analysis also provided the probabilities of (SE-induced) bit errors in the memory address ranges of TN output buffers. The preliminary validations based on accelerated tests of the physical system have shown that our 5% derating ratio is relatively close to test results.

B. Selectively protecting memories

The on-chip SRAM can be partly protected by an Error Detection And Correction (EDAC) method based on the Hamming code with a minimum distance of 4. This EDAC method is also known as the Single Error Correction-Double Error Detection (SEC-DED) Error Correction Code (ECC). In general, increasing the percentage of ECC protected on-chip SRAMS can reduce the probability of SE-induced chip failures. Since ECC protection incurs a SRAM area overhead, a chip architect may need to have SRAMs partly protected to obtain an optimal trade-off between the silicon area cost and the reliability of a target chip. Our analysis of system-level failure rate can identify such trade-offs. The workflow of this identification is actually an iteration of many system-level failure rate analyses. Each analysis gives a failure rate for an unique configuration of partial ECC protection. If an architect can list all relevant configurations of partial ECC protection, then this iteration of our workflow can automatically go through the list and reports a failure rate for each configuration. At the end, the workflow can prune all non-Pareto optimal trade-offs and reports a Pareto curve which contains the optimal trade-off points of all explored configurations. The overall workflow for this exploration is depicted in Fig. 2.

We have explored the trade-offs between the dependability cost (in terms of protected memory) and the system failure rate for the target audio processing system. The pruned trade-offs from the exploration are shown in Fig.3. Each point in the figure represents a specific allocation of ECC-protection and the corresponding failure rate. This result has identified that when the protected SRAM area grows from 0 to 15kbit (extra silicon area due to the protection is about 13% of the protected area), this audio processing system's failure rate

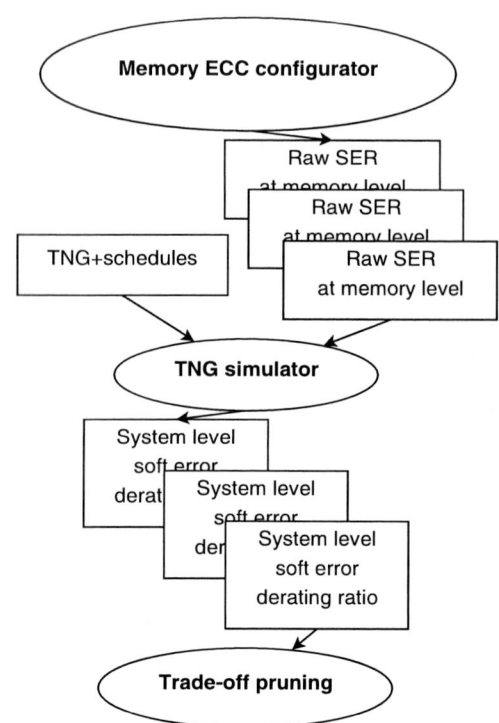

Figure 2. Workflow of the selectively memory protection trade-offs exploration

drops from 5% to 0.5% (the remaining 0.5% is due to the fact that communicating buffers between TNs are not protected).

IV. CONCLUSIONS

We have presented a system-level SER analysis technique for predicting the user-perceived SER of a complex SoC. Running with a TLM model of a targeted system, our technique can predict the system's vulnerability to SE in a few minutes. Based on this analysis technique, we have shown a selective SRAM protecting that can give a fine-granularity trade-offs between the system dependability and the memory area cost.

REFERENCES

[1] Robert Baumann. Soft errors in advanced computer systems. *IEEE Design & Test of Computers*, 22(3):258–266, 2005.

[2] Jean-Marc Daveau, Alexandre Blampey, Gilles Gasiot, Joseph Bulone, and Philippe Roche. An industrial fault injection platform for soft-error dependability analysis and hardening of complex system-on-a-chip. In *IRPS*, pages 212–220. IEEE, 2009.

[3] P. Kudva, J. Kellington, P. Sanda, R. McBeth, J. Schumann, and R. Kalla. Fault injection verification of IBM POWER6 soft error resilience. In *Workshop on Architectural Support for gigascale Integration (ASGI)*, 2007, 2007.

[4] Zhe Ma, Paul Marchal, Daniele Scarpazza, Peng Yang, Chun Wong, José Ignacio Gómez, Stefaan Himpe, Chantal Ykman-Couvreur, and Francky Catthoor. *Systematic methodology for real-time cost-effective mapping of dynamic concurrent task-based systems on heterogeneous platforms*. Springer, 2007.

[5] Subhasish Mitra, Norbert Seifert, Ming Zhang, Quan Shi, and Kee Sup Kim. Robust system design with built-in soft-error resilience. *IEEE Computer*, 38(2):43–52, 2005.

[6] Shubhendu S. Mukherjee, Christopher T. Weaver, Joel S. Emer, Steven K. Reinhardt, and Todd M. Austin. A systematic methodology to compute the architectural vulnerability factors for a high-performance microprocessor. In *MICRO*, pages 29–42. ACM/IEEE, 2003.

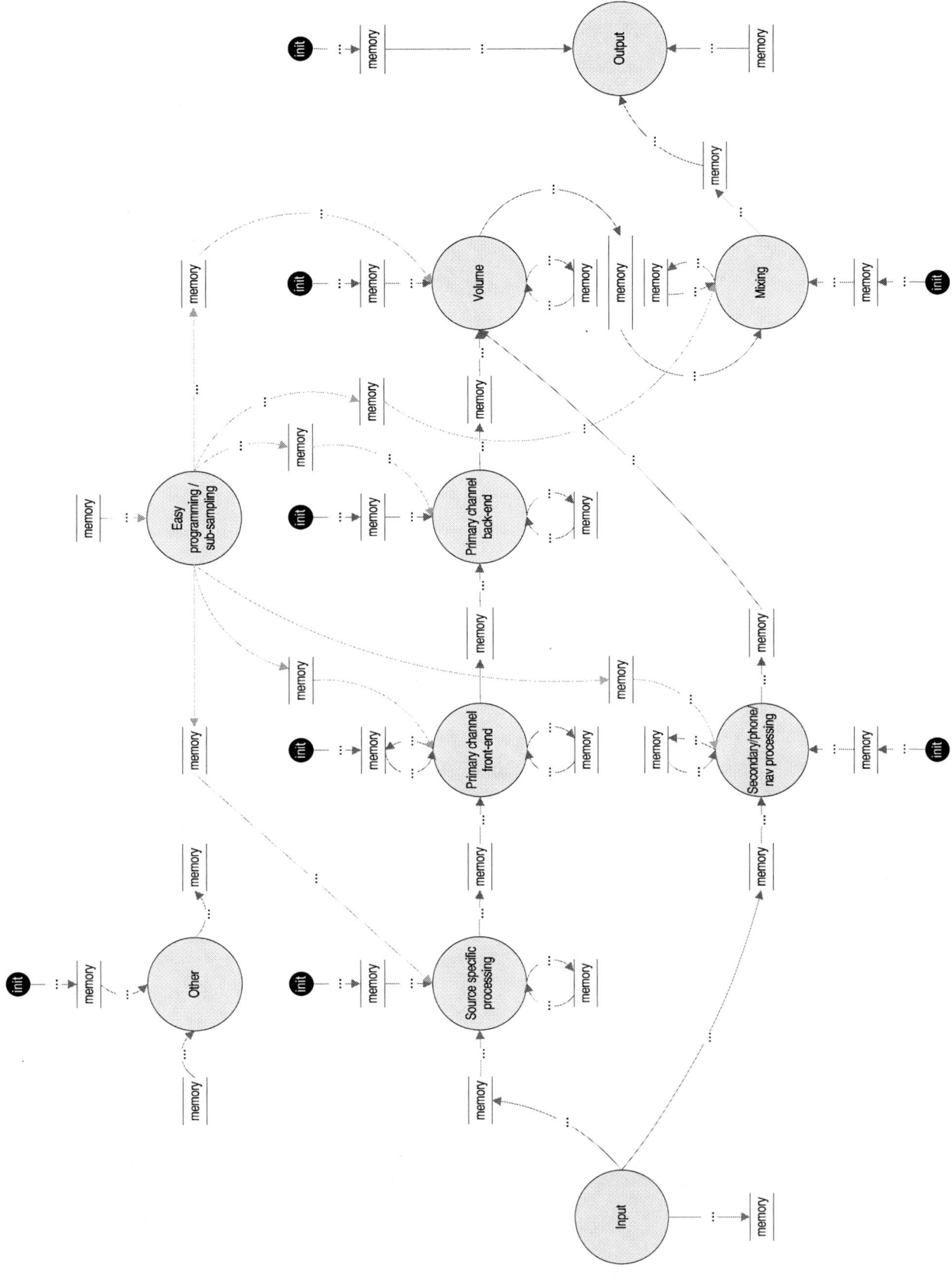

Figure 1. Audio process system TNG

978-1-4244-5430-3/10 $26.00 © 2010 IEEE

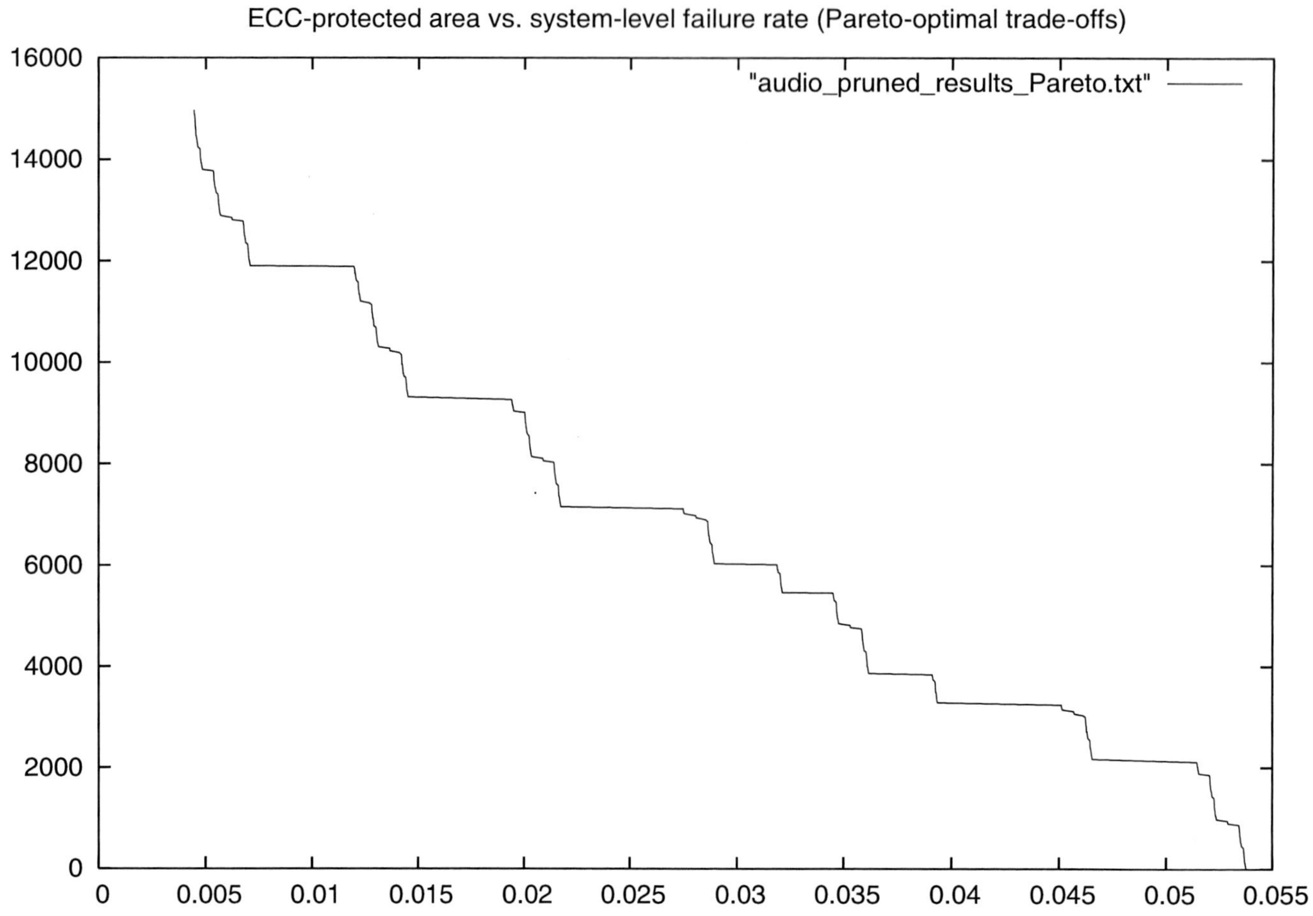

ECC-protected area vs. system-level failure rate (Pareto-optimal trade-offs)

Figure 3. Selectively memory protection trade-offs

[7] Hang T. Nguyen and Yoad Yagil. A systematic approach to SER estimation and solutions. In *IRPS*, pages 60–70. IEEE, 2003.

[8] Pradeep Ramachandran, Prabhakar Kudva, Jeffrey W. Kellington, John Schumann, and Pia Sanda. Statistical fault injection. In *DSN*, pages 122–127. IEEE Computer Society, 2008.

[9] Sonny Rao, Pia Sanda, Jerry Ackaret, Adrian Barrera, Jorge Yanez, and Subhasish Mitra. Examing workload dependence of soft error rates. In *SELSE*, 2008.

[10] Daniel Skarin, Martin Sanfridson, and Johan Karlsson. Impact of soft errors in a brake-by-wire system. In *SELSE*, 2007.

[11] Frank Vermunt. SER experiment 1 (doc. rev. 1.0). Technical Report ExperimentSERonXXXXXX9141.doc, NXP, March 2009.

Soft Errors from Neutron and Proton-induced Multiple-node Events

Ethan H. Cannon
The Boeing Company
Seattle, WA, USA
253-657-5104, ethan.cannon@boeing.com

Abstract—We present geometric models to calculate the rate of multiple node events that can defeat soft error mitigation based on spatial redundancy. One aspect of our model determines the probability of an ion crossing two nodes, while the second part considers two different daughter particles from a neutron or proton collision affecting two nodes.

Keywords-soft errors, single event upsets, cosmic ray neutrons, protons

I. INTRODUCTION

Many mitigation techniques for radiation-induced soft errors rely on spatial redundancy, where information is stored on multiple nodes; examples include triple module redundancy (TMR), the dual interlocked storage cell (DICE) [1], and error detection and correction codes (EDAC) in arrays. Spatial redundancy protects against charge collection at a single node, but is still susceptible to charge collection at multiple nodes.

Multiple cell upsets (MCUs) in SRAM arrays occur when a single radiation event flips multiple SRAM cells. Charge collection by more than one node is known as charge sharing and can be caused by multiple physical mechanisms, including: particle trajectories that directly strike multiple nodes, charge diffusing to multiple nodes, parasitic bipolar current due to voltage droop in the Nwell or Pwell potential, and multiple daughter particles from a single nuclear reaction striking different nodes, see, e.g., [2]. The same physical mechanisms engender charge sharing in MCU and in spatially redundant circuits, such as a DICE latch [3].

Recent experiments on 90-nm bulk CMOS DICE flip-flops have shown strong orientational effects for the soft error cross section: Upsets are likely when the ion beam is parallel to the line connecting pairs of jointly-sensitive devices, since a single ion can directly impact both devices. On the other hand, upsets are much less likely when the beam is perpendicular to the line connecting sensitive pairs [4,5].

Experiments on SOI SRAMs yielded a lower rate of MCU than bulk SRAMs [2,6]. Diffusion and well potential droop cannot cause charge sharing in SOI technologies because there is full dielectric isolation between devices in different silicon islands. Hence charge sharing in an SOI technology will be dominated by particle trajectories that directly strike multiple nodes.

We present a two-part geometric model for events with ions crossing multiple nodes. The first aspect of the geometric model identifies ion trajectories that cross both sensitive nodes;

some results of this model have been presented previously [7]. Next, we extend the model to consider the case of two daughter particles from a neutron or proton collision striking two different sensitive nodes. This model does not include charge sharing by diffusion or by a bipolar current induced by well-charging effects. However, the case of ions directly crossing each node is the dominant source of multiple-node events for large node spacing in a bulk CMOS technology, such as EDAC in an SRAM array with sufficient bit interleaving, and for an SOI CMOS technology. For example, the DICE flip-flop in [8] separates NMOS and PMOS pairs by at least 5 μm to reduce charge sharing by diffusion or a parasitic bipolar current; heavy ion testing showed peaks in the upset cross section for angles where ions traverse two sensitive nodes, as expected.

II. MODEL

A. Single-ion Multiple-node Events

We first develop a geometric model of a radiation particle crossing two sensitive nodes with equal width, W, whose top and bottom are aligned. The charge collection depth is t, and the spacing between nearest edges of the two gates is D. Fig. 1a shows a side-view of devices in an SOI technology where the device body forms the sensitive volume; the figure shows the trajectory of an ion that crosses the inside top corner of the sensitive volume of the device on the left and the inside bottom corner of the sensitive volume of the device on the right. This trajectory has the maximum tilt angle that can hit both devices, Φ_m. From Fig. 1a, it is clear that

$$\tan(\Phi_M) = \frac{t}{D} \qquad . \qquad (1)$$

Fig. 1b shows a top-view of the devices, with the trajectory of an ion that has the maximum in-plane rotation angle able to hit both devices, Θ_m; from geometric considerations

$$\tan(\Theta_M) = \frac{W}{D} \qquad . \qquad (2)$$

In Fig. 1a we assume the particle is orthogonal to the two gates, and in Fig.1b we assume the particle moves parallel to the active silicon surface. While Fig. 1a and Fig. 1b show the sensitive volume of an SOI technology, extension to a bulk technology is straightforward. Note that in Fig. 1 angles are measured from the plane of the active silicon, while many papers measure angles from the normal to the plane of the active silicon.

This work was supported in part by DARPA and Defense Threat Reduction Agency contract HDTRAI-05-D-0001.

978-1-4244-5430-3/10 $26.00 © 2010 IEEE

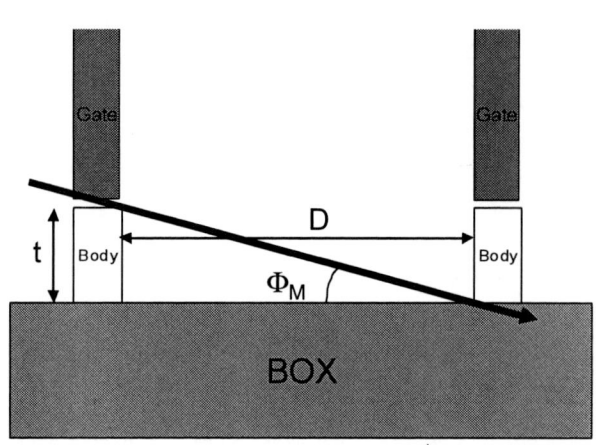

Figure 1a. Side view of two sensitive volumes. The trajectory of an ion that hits both sensitive volumes with maximum tilt angle, Φ_M, is indicated.

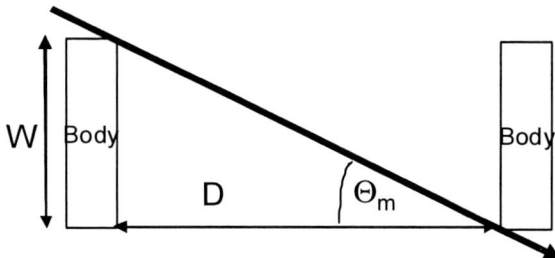

Figure 1b. Top view of two sensitive volumes. The trajectory of an ion that hits both sensitive volumes with maximum rotation angle, Θ_M, is indicated.

In Fig. 2a and Fig. 2b we take into account the trapezoidal shape of the body and the finite diameter of the ion track, D_C. For bulk silicon devices, the ion track radius is often defined as the distance at which the generated charge density is 10^{-16} cm^{-3}, the typical substrate doping level. For an SOI transistor, the ion track radius is better defined as the distance at which the generated charge density is equal to the body doping. In Fig. 2a, the ion track effectively increases the silicon layer thickness to $t+D_C/cos(\Phi_m)$, and the separation, D, is decreased by the difference between the real and drawn body length, D_b. Equation (1) becomes

$$\tan(\Phi_M) = \frac{t+\dfrac{D_C}{\cos(\Phi_M)}}{D-D_b}, \quad (3)$$

which can be rearranged to

$$\cos(\Phi_M) = \frac{-2D_C t + \sqrt{4D_C^2 t^2 - 4[t^2+(D-D_b)^2][D_C^2-(D-D_b)^2]}}{2[t^2+(D-D_b)^2]}$$

(4)

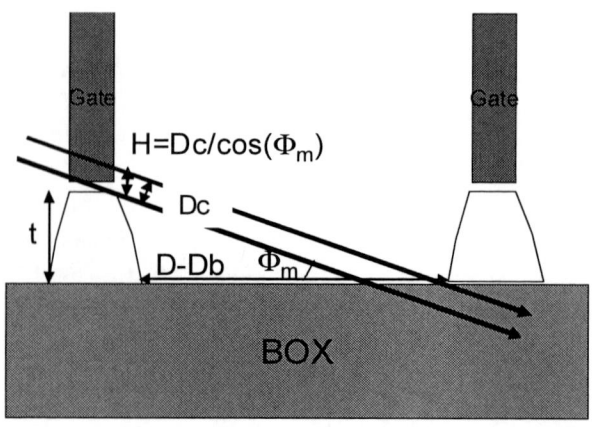

Figure 2a. Improved side view of two sensitive volumes, including size of the ion track and shape of the device body. The trajectory of an ion that hits both sensitive volumes with maximum tilt angle, Φ_M, is indicated

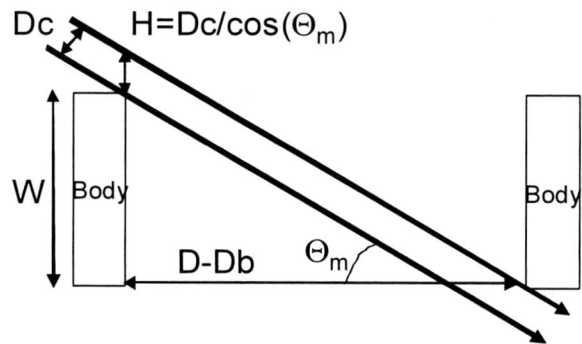

Figure 2b. Improved top view of two sensitive volumes, including the size of the ion track. The trajectory of an ion that hits both sensitive volumes with maximum rotation angle, Θ_M, is indicated.

The expression for Θ_m is similar, with t replaced by the gate width W.

Fig. 3 presents the situation where the two sensitive devices are offset by a distance W_{offset}. The offset angle satisfies the expression

$$\tan(\beta) = \frac{W_{offset}}{D}. \quad (5)$$

The maximum rotation angle in the opposite direction as the offset angle, Θ_1, can be determined from the geometry

$$\tan(\Theta_1) = \frac{W\cos(\beta)}{\dfrac{D}{\cos(\beta)}-W\sin(\beta)}$$

$$\tan(\Theta_1) = \frac{W\cos^2(\beta)}{D-W\sin(\beta)\cos(\beta)} \quad (6)$$

The maximum rotation angle in same direction as the offset angle, Θ_2, has a similar form

978-1-4244-5430-3/10 $26.00 © 2010 IEEE

$$\tan(\Theta_2) = \frac{W\cos^2(\beta)}{D + W\sin(\beta)\cos(\beta)}. \quad (7)$$

The offset breaks the symmetry, thus Θ_1 and Θ_2 are not equal.

The model can be extended to include the ion track structure and body shape. In order to hit both sensitive nodes, the rotation and tilt angles must meet the conditions

$$\beta - \Theta_1 < \Theta < \beta + \Theta_2$$
$$-\Phi_M < \Phi < \Phi_M \quad (8)$$

where

$$\cos(\Theta_1) = \frac{-2D_C'W' + \sqrt{4D_C'^2W'^2 - 4(W'^2 + D_1'^2)(D_C'^2 - D_1'^2)}}{2(W'^2 + D_1'^2)}$$

$$\cos(\Theta_2) = \frac{-2D_C'W' + \sqrt{4D_C'^2W'^2 - 4(W'^2 + D_2'^2)(D_C'^2 - D_2'^2)}}{2(W'^2 + D_2'^2)}$$

$$D_C' = D_C\cos(\beta)$$

$$W' = W\cos^2(\beta)$$

$$D_1' = D - D_B - W\sin(\beta)\cos(\beta)$$

$$D_2' = D - D_B + W\sin(\beta)\cos(\beta)$$

$$\cos(\Phi_M) = \frac{-2D_Ct + \sqrt{4D_C^2t^2 - 4[t^2 + \frac{(D-D_b)^2}{\cos^2(\Theta)}][D_C^2 - \frac{(D-D_b)^2}{\cos^2(\Theta)}]}}{2[t^2 + \frac{(D-D_b)^2}{\cos^2(\Theta)}]}$$

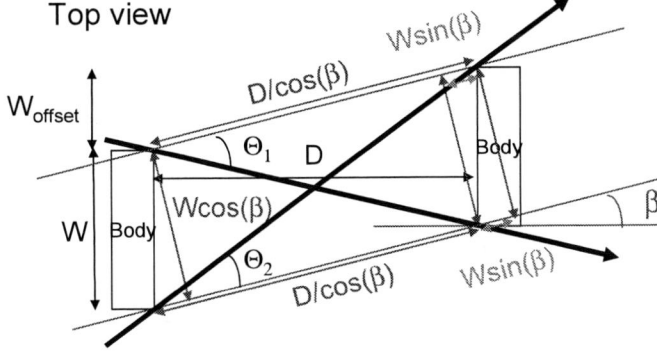

Top view

Figure 3 Top view of two sensitive volumes which are offset from each other, including size of the ion track. The trajectories of ions that hit both sensitive volumes with largest possible negative and positive tilt angles, Θ_1 and Θ_2, are indicated.

. (9)

Fig. 4a and Fig. 4b plot the maximum tilt angle that can hit two sensitive nodes, Φ_m, as a function of node-to-node separation, D, for both the simple model in (1) and the improved model in (4). For the improved model, an ion track diameter of 0.1 μm is used. Fig. 4a represents an SOI technology with t=80 nm and D_b=40 nm. Fig. 4b represents a bulk technology with t=1 μm; note that this model is a rough

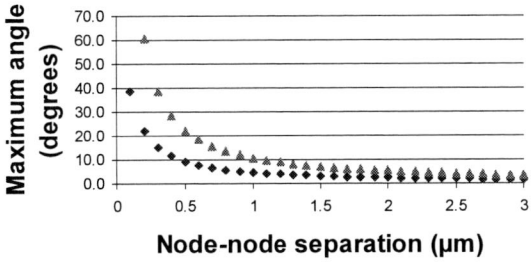

Figure 4a. Maximum tilt angle as a function of sensitive node spacing for parameters representative of an SOI technology, from both the simple and improved models. Charge collection depth is 80nm.

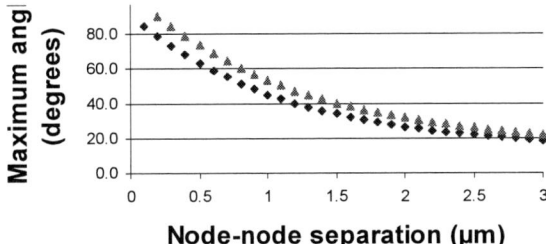

Figure 4b. Maximum tilt angle as a function of sensitive node spacing for parameters representative of a bulk technology, from both the simple and improved models. Charge collection depth is 1μm.

approximation to bulk technologies, as it neglects charge sharing by diffusion and well potential droop which can be significant factors. At the same node separation in Fig. 4a and 4b, an SOI technology is sensitive to a much smaller range of angles because of its shallow sensitive volume.

For the bulk technology in Fig. 4b, the improved model leads to a 10% increase in maximum angle for a charge radius of 100 nm, and a 20% increase for 200 nm radius. For the SOI technology in Fig. 4a, the maximum angle in the improved model is more than twice as large as in the simple model. However, for node separation more than 1.2 μm, the difference is less than 5°. Consequently, the simple model provides useful guidance on which trajectories affect two nodes.

Using the simple model in (1) and (2), the solid angle for a particle exiting the center of one sensitive node to hit a 2nd sensitive node is

$$SolidAngle = 4\arcsin\left(\frac{tw}{\sqrt{4D^2 + t^2}\sqrt{4D^2 + w^2}}\right); (10)$$

[9] has the *arctan* form of this expression. For large sensitive node separation, $D \to \infty$, this reduces to the approximate form of solid angle

$$SolidAngle = \frac{tw}{D^2}. (11)$$

For small separation, $D \to 0$, the solid angle (10) is 2π, since the particle will hit the second sensitive volume for any angle towards the right in Fig. 1a, which covers half the unit sphere.

We use (10) to estimate the solid angle for a trajectory to strike two nodes for representative cases of interest. For the 90-nm DICE FF in [8], we assume t=1 µm and w=0.5 µm. For L=5 µm (NMOS-NMOS/PMOS-PMOS spacing), the solid angle is 0.020 sr; for L=2.5 µm (minimally space NMOS-PMOS pair), the solid angle is 0.078 sr. For comparison, we consider a DICE FF in a 45-nm SOI technology assuming t=0.08 µm and w=0.25 µm; to obtain the same solid angle to hit sensitive node pairs, the sensitive node spacing can be reduced by a factor of 5. Thus the sensitive node spacing necessary to maintain DICE FF soft error rate is less strict in a 45-nm SOI technology compared to a 90-nm bulk technology. Assuming an isotropic flux, 0.16% of incident ions will hit both sensitive nodes when they subtend a solid angle of 0.020 sr.

Multiple cell upsets in a 45-nm SOI and bulk SRAMs were studied in [10,16] and [11], respectively. The SRAM bit cell in [10] is 0.4 µm along the bitline direction and 1.0µm along the wordline direction. Estimating the separation of NMOS pairs for a blanket pattern as 0.125 µm, and assuming a 0.2 µm gate width, for an isotropic flux the probability an ion hits both transistors calculated from (10) is 6.1%. This compares well with 63 MeV proton beam tests, where the daughter particles approach an isotropic distribution, and the measured probability of MCU is 6%. Assuming 4, 8, and 16-bit interleaving schemes, the probability of an ion hitting two cells in the same word is 8.0E-5, 2.0E-5 and 5.0E-6, respectively. A single error correction, double error detection (SECDED) EDAC scheme detects two soft errors in a single word, but cannot correct them. The rate of such events is expected to be at least four orders of magnitude lower than the total soft error rate for typical bit interleaving schemes. For a qualitative comparison with a 45-nm bulk SRAM, we assume a charge collection depth of 1 µm and use the cell and transistor dimensions for the SOI SRAM. At a separation of L=1 µm, (10) gives a 1.4% probability for an ion to cross two sensitive nodes, which agrees with neutron data in [11]. For an 8-bit interleaving scheme, with a separation of 8 µm, the probability of hitting sensitive nodes of two bits in the same word decreases to 2.5E-4, which is an order of magnitude larger than calculated for an SOI SRAM. The only difference between the 45-nm SOI and bulk SRAM in these calculations is the charge

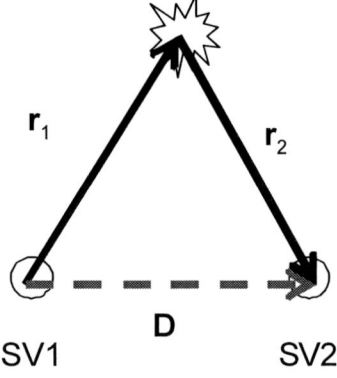

Figure 5. Positions of neutron collision and sensitive volumes for Eq. (4), with sensitive volume 1 centered about the origin.

collection depth, which directly impacts the probability to strike two nodes.

B. Two-ion Multiple-node Events

Soft errors from cosmic ray neutrons are a significant concern in terrestrial environments. In space applications, protons in trapped belts and from solar events are a source of soft errors. Proton and neutron collisions generate similar daughter particles for incident energy greater than 50 MeV, where the incident energy greatly exceeds the energy of the proton-nucleus interaction, thus this geometric model is applicable for both types of interactions.

We extend the model for the soft error cross section from neutron collisions in [12] to consider two daughter particles hitting distinct sensitive nodes. In [12], the soft error cross section, σ, is

$$\sigma = N\sigma_{col} \int_0^\infty P_{geom}(r) P_{ion}(E_n, Q_{crit}, R_S, r) 4\pi r^2 dr \quad (12)$$

where N is the atomic density of silicon, σ_{col} is the cross section for a neutron-silicon (or proton-silicon) interaction, P_{geom} is the probability of a daughter particle hitting the sensitive node for a collision a distance r from the center of the sensitive node, and P_{ion} is the probability that a daughter particle from a neutron (or proton) of energy E_n deposits at least the critical charge Q_{crit} in the sensitive volume of radius R_S. In this approach, the center of the sensitive volume is at the origin, and the cross section is obtained by integrating the number of nuclear collisions over all space. The factor of $N\sigma_{col} 4\pi r^2 dr$ gives the number of nuclear collisions in a spherical shell of thickness dr a distance r from the center of the sensitive node.

The geometric function, Pgeom, gives the probability for a daughter particle to hit the sensitive node when the collision is a distance r from the center of the sensitive node. It has the form [12]

$$P_{geom}(r) = \frac{1}{2}[1 - \sqrt{1 - (\frac{R_S}{r})^2}] \quad r > R_S$$
$$P_{geom}(r) \approx \frac{R_S^2}{4r^2} \quad r \gg R_S \quad (13)$$

Ref. [12] calculates $P_{ion}(E_n, Q_{crit}, R_S, r)$, the probability a daughter particle deposits more than the critical charge. Roughly speaking, the probability has two plateaus: A higher plateau at short radius due to heavy ion recoils, and a lower plateau due to light recoils such as alpha particles that extends to much larger radius. For circuits with significant alpha particle sensitivity, long range daughter particles are responsible for a majority of soft errors.

The probability a daughter particle deposits enough charge to upset a node, P_{ion}, can be approximated by distinct constant values in two different regions,

$$P_{ion}(r) = P_1 \quad r < r_a$$
$$P_{ion}(r) = P_2 \quad r_a < r < r_b \quad (14)$$

giving an analytical expression for the soft error cross section (10)

$$\sigma = N\sigma_{col}\pi R_S^2[P_1 r_a + P_2(r_b - r_a)]. \qquad (15)$$

Following [12], the cross section for two daughter particles to hit two identical sensitive nodes is

$$\sigma_{2-nodes} = N\sigma_{col}P_{2-ions}\int_0^\infty r_1^2 dr_1 \int_0^\pi \sin(\theta)d\theta \int_{-\pi}^\pi d\phi * \qquad (16)$$
$$* P_{geom}(r_1)P_{ion}(r_1)P_{geom}(r_2(r_1,\theta,\phi))P_{ion}(r_2)$$

where P_{2-ions} is the probability a neutron (or proton) collision generates two daughter particles [13]. Fig 5 shows the set of coordinates used in (16): r_1 is the position vector of the neutron collision, with the first sensitive volume at the origin, as in (12); and r_2 is the position vector from the collision to the second sensitive volume, at position **D**. With two sensitive volumes, the integration is not rotationally symmetric, so integration over the angles θ and ϕ must be explicitly performed.

An upper bound solution to (16) is obtained in the appendix,

$$\sigma_{2-nodes} = N\sigma_{col}P_{2-ions}\frac{\pi P_1 R_S^2}{4D}\frac{R_S^2}{4}I_3 \qquad (17).$$

The appendix also provides an approximate analytical expression for the term I_3 in (17); alternatively, the integral I_3 can be solved numerically. The following simplifying assumptions are made to obtain (17)

- The large distance form for P_{geom} in (13) is used
- The probability for an ion crossing the 2nd sensitive volume to deposit the critical charge is assumed to be constant, $P_{ion}(r_2)=P_1$
- Strikes in the 2nd sensitive volume are not treated exactly, due to mathematical singularities

The ratio of the cross section for two daughter particles striking separate sensitive nodes compared to the soft error cross section follows from (17) and (12)

$$\frac{\sigma_{2-nodes}}{\sigma} = \frac{P_{2-ions}P_1 R_S^2}{16D}\frac{I_3}{P_1 r_a + P_2(r_b - r_a)} \qquad (18).$$

The integral I_3 decreases weakly with node separation, D, for node separation less than the range of daughter particles, so the probability for two daughter particles to strike two sensitive nodes decreases approximately linearly with node separation, D. In contrast, the probability for a single ion to cross two sensitive nodes decreases more steeply, as the square of the node separation in (11).

To estimate the rate of multiple-node events from two daughter particles, we need to understand the probability of a nuclear collision generating multiple daughter particles. Fig. 3 in [13] shows that more than 85% of neutron-silicon collisions generate two or more daughter particles for energies above 50

MeV. However, Fig. 2 in [13] shows that protons comprise the majority of the daughter particles. The low level of proton ionization can directly upset deep submicron SRAMs, when protons arrive at the active silicon with energy in a very narrow range around the Bragg peak. Consequently, protons have very low probability to upset a circuit, and we do not include them in P_{2-ions}. The majority of events with two daughter particles upsetting two nodes create an alpha particle, a heavier recoil, and possibly some protons. Fig. 2 in [13] shows that alpha particles make up less than 10% of the daughter particles; considering that the most common event generates two daughter particles, we assume P_{2-ions}=20%. Table 1 presents

TABLE I. REPRESENTATIVE CIRCUITS WITH DIFFERENT SENSITVE NODE DIMENSIONS AND SPACING, INCLUDING PROBABILITY OF A SINGLE PARTICLE STRIKING TWO NODES (FOR AN ISOTROPIC FLUX), AND THE PROBABILITY TWO DAUGHTER PARTICLES FROM A NEUTRON OR PROTON COLLISION STRIKE TWO NODES.

	R_S	t	W	D	Probability 1 particle strikes 2 nodes	Probability 2 particles strike 2 nodes
90-nm DICE	1	1	0.5	2.5	6.2E-3	1.2E-3
90-nm DICE	1	1	0.5	5	1.6E-3	5.3E-4
SOI DICE	0.14	0.08	0.25	0.5	6.2E-3	1.3E-4
SOI DICE	0.14	0.08	0.25	1	1.6E-3	6.3E-5
45-nm SRAM	0.44	1	0.2	1	1.4E-2	6.3E-4
45-nm SRAM	0.44	1	0.2	8	2.5E-4	5.4E-5
45-nm SOI SRAM	0.13	0.08	0.2	0.25	1.9E-2	2.1E-4
45-nm SOI SRAM	0.13	0.08	0.2	8	2.0E-5	5.5E-6

the probability of multiple-node events for the DICE FF and SRAM cells discussed in section IIA. The probability of two daughter particles striking two different nodes is at most 30% the probability of a single particle crossing two sensitive nodes, assuming an isotropic flux. For the 2 daughter particle model, the sensitive volume is represented as a sphere with radius equal to the geometric mean of the charge collection depth, t, and device width, W. The situation of two daughter particles striking different nodes becomes more important at larger separation distance, D, because it decreases less rapidly with separation than the solid angle for a single particle to hit two nodes.

III. CONCLUSION

We developed geometric models to predict the rate of soft errors from multiple-node events, which are able to defeat soft error mitigation techniques employing spatial redundancy, such as the DICE flip-flop and EDAC in arrays. We first derived the probability for a single ion to strike two sensitive nodes; we then obtained the probability for two distinct daughter particles from a neutron or proton collision to strike two different sensitive nodes. These models can be used to design spatially redundant systems with adequate node spacing to meet soft error requirements. DICE flip-flops and SRAM cells in 90-nm and 45-nm SOI technologies were studied as examples. The probability of a single ion crossing two nodes agrees well with experimental data on SRAM multiple-cell upsets. For neutron or proton collisions, multiple-node events are dominated by single ions crossing two sensitive nodes for all examples we evaluated. The probability for two different daughter particles to affect two nodes decreases less rapidly with node spacing, and causes a larger fraction of multiple-node events at large node separation.

IV. APPENDIX

In this appendix we obtain an upper bound solution to the cross section for two daughter particles to strike two sensitive nodes (16). The position vectors in Fig. 5 can be expressed in terms of the polar angle, θ, and azimuthal angle, ϕ,

$$
\vec{r}_1 = r_1 \sin(\theta)\cos(\phi)\hat{x} + r_1 \sin(\theta)\sin(\phi)\hat{y} + r_1 \cos(\theta)\hat{z}
$$
$$
\vec{r}_1 + \vec{r}_2 = \vec{D} = D\hat{x}
$$
$$
\vec{r}_2 = [D - r_1 \sin(\theta)\cos(\phi)]\hat{x} - r_1 \sin(\theta)\sin(\phi)\hat{y} - r_1 \cos(\theta)\hat{z} \qquad \text{. (A1)}
$$
$$
r_2^2 = D^2 + r_1^2 - 2Dr_1 \sin(\theta)\cos(\phi)
$$

We use two simplifying assumptions that give an upper bound to $\sigma_{2\text{-nodes}}$. First, the probability an ion hitting the second node deposits at least the critical charge is held constant, i.e. $P_{ion}(r_2) = P_1$. Second, we use the large distance form of the geometric factor, given in (13). With these assumptions, we first perform the integration over azimuthal angle, ϕ,

$$
I_1 = \int_{-\pi}^{\pi} d\phi \, P_{geom}(r_2) P_{ion}(r_2)
$$

$$
I_1 \approx \int_{-\pi}^{\pi} d\phi \frac{P_1 R_S^2}{4[D^2 + r_1^2 - 2Dr_1 \sin(\theta)\cos(\phi)]}
$$

$$
I_1 \approx \frac{P_1 R_S^2}{2\sqrt{(D^2 + r_1^2)^2 - [2Dr_1 \sin(\theta)]^2}}
$$
$$
\tan^{-1}\left(\frac{\sqrt{(D^2 + r_1^2)^2 - [2Dr_1 \sin(\theta)]^2}}{D^2 + r_1^2 - 2Dr_1 \sin(\theta)} \tan(\frac{\phi}{2})\right)\Big|_{-\pi}^{\pi}
$$

$$
I_1 \approx \frac{P_1 R_S^2}{2\sqrt{(D^2 + r_1^2)^2 - [2Dr_1 \sin(\theta)]^2}}\left[\frac{\pi}{2} - \frac{-\pi}{2}\right]
$$

$$
I_1 \approx \frac{\pi P_1 R_S^2}{2\sqrt{(D^2 + r_1^2)^2 - [2Dr_1 \sin(\theta)]^2}}, \text{ (A2)}
$$

using [14] p. 264, integral 341. The integral I_1 diverges when the collision occurs near the 2nd sensitive volume. Using the exact form of the geometric factor, (13), would eliminate the divergence. Integration over the polar angle yields

$$
I_2 = \int_0^\pi \sin(\theta)d\theta \int_{-\pi}^{\pi} d\phi P_{geom}(r_2(r_1, \theta, \phi)) P_{ion}(r_2)
$$

$$
I_2 = \frac{\pi P_1 R_S^2}{2} \int_0^\pi \frac{\sin(\theta)d\theta}{\sqrt{1 - \frac{4D^2 r_1^2}{(D^2 + r_1^2)^2}\sin^2(\theta)}\sqrt{(D^2 + r_1^2)^2}}
$$

$$
I_2 = \frac{\pi P_1 R_S^2}{2(D^2 + r_1^2)} \int_0^\pi \frac{\sin(\theta)d\theta}{\sqrt{1 - b^2 \sin^2(\theta)}}, \quad b^2 = \frac{4D^2 r_1^2}{(D^2 + r_1^2)^2}
$$

$$
I_2 = \frac{\pi P_1 R_S^2}{2(D^2 + r_1^2)}(\frac{-1}{b})\log[b\cos(\theta) + \sqrt{1 - b^2 \sin^2(\theta)}]\Big|_0^\pi
$$

$$
I_2 = \frac{\pi P_1 R_S^2}{2(D^2 + r_1^2)}(\frac{-1}{b})[\log(1 - b) - \log(1 + b)]
$$

$$
I_2 = \frac{\pi P_1 R_S^2}{2(D^2 + r_1^2)}\frac{D^2 + r_1^2}{2Dr_1}\log(\frac{1 + \frac{2Dr_1}{D^2 + r_1^2}}{1 - \frac{2Dr_1}{D^2 + r_1^2}})
$$

$$
I_2 = \frac{\pi P_1 R_S^2}{4Dr_1}\log[(\frac{D + r_1}{D - r_1})^2], \text{ (A3)}
$$

see [14] p. 271, integral 433. The divergence for $r_1 = D$ follows from using the large distance form of the geometric factor for small values of r_2.

In the integration over collision radius, r_1, we use the large distance form of the geometric factor $P_{geom}(r_1)$ (13). We change variables in order to divide the logarithm into multiple pieces. To obtain the final analytical form, in (A5) we break I_3 into three pieces, assuming $r_a/D > 1$. The first piece includes nearby collisions where heavy recoils can reach the 1st sensitive volume, $r_1 < r_a$. The second and third pieces describe the region where only light ions can reach the sensitive volume, $r_a < r_1 < r_b$. The integral is broken at distance $x=1$, i.e. $r_1 = D$, in order to use the positive square root of the argument of the logarithm. To avoid divergences in the logarithm at x=1 (r1=D), we adjust the limits of integration from $x=1$ to $x=1 \pm R_S/D$ ($r_1 = D \pm R_S$). When $D < r_a$, the integral I_3 must be evaluated in three regions: $0 < r_1 < D$, $D < r_1 < r_a$, and $r_a < r_1 < r_b$, giving a slightly different analytical expression. For the cases in Table 1, the analytical expression for I_3 is 10-50% lower than an exact result from numerically performing the integration. The analytical solution better approximates the exact integral for large node separation, D.

978-1-4244-5430-3/10 $26.00 © 2010 IEEE

$$\sigma_{2-nodes} = N\sigma_{col}P_{2-ions}\int_0^\infty r_1^2 dr_1 I_2 P_{geom}(r_1)P_{ion}(r_1)$$

$$\sigma_{2-nodes} = N\sigma_{col}P_{2-ions}\frac{\pi P_1 R_S^2}{4D}\frac{R_S^2}{4}\int_0^\infty dr_1 P_{ion}(r_1)\frac{1}{r_1}\log[(\frac{D+r_1}{D-r_1})^2]$$

$$x=\frac{r_1}{D}$$

$$I_3 = \int_0^\infty \frac{Ddx}{Dx}P_{ion}(Dx)\log[(\frac{1+x}{1-x})^2]$$

$$P_{ion}(Dx)=P_1 \quad (Dx)<r_a$$

$$P_{ion}(Dx)=P_2 \quad r_a<(Dx)<r_b$$

(A4)

$$I_3 = 2P_1\int_0^1 dx\frac{\log(1+x)-\log(x-1)}{x}+$$

$$+2P_1\int_1^{r_a/D}dx\frac{\log(x+1)-\log(1-x)}{x}+$$

$$+2P_2\int_{r_a/D}^{r_b/D}dx\frac{\log(x+1)-\log(x-1)}{x}$$

, (A5)

$$\int dx\frac{\log(1+x)}{x}=-Li_2(-x)$$

$$\int dx\frac{\log(1-x)}{x}=-Li_2(x)$$

$$\int dx\frac{\log(x-1)}{x}=\log(x)\log(x-1)+Li_2(1-x)$$

see, e.g., [15],

$$I_3 =2P_1[Li_2(1-\frac{R_S}{D})-Li_2(\frac{R_S}{D}-1)]+$$

$$+2P_1[Li_2(-1-\frac{R_S}{D})+Li_2(-\frac{R_S}{D})-Li_2(-\frac{r_a}{D})-Li_2(1-\frac{r_a}{D})+$$

$$+\log(1+\frac{R_S}{D})\log(\frac{R_S}{D})-\log(\frac{r_a}{D})\log(\frac{r_a}{D}-1)]+$$

(A6)

$$+2P_2[Li_2(-\frac{r_a}{D})+Li_2(1-\frac{r_a}{D})-Li_2(-\frac{r_b}{D})-Li_2(1-\frac{r_b}{D})+$$

$$+\log(\frac{r_a}{D})\log(\frac{r_a}{D}-1)-\log(\frac{r_b}{D})\log(\frac{r_b}{D}-1)]$$

ACKNOWLEDGMENT

I would like to thank my colleagues Manuel Cabanas-Holmen, Jeremy Popp and Tony Amort for fruitful discussions.

REFERENCES

[1] T. Calin, M. Nicolaidis, and R. Valazco, "Upset hardened memory design for submicron CMOS technology," IEEE Trans. Nuc. Sci., vol. 43, pp. 2874-2878, Dec. 1996.

[2] E. H. Cannon, et al., "Multi-bit upsets in 65nm SOI SRAMs ," Proc. 46th Ann. Intl Rel. Phys. Symp., pp. 195-201, 2008.

[3] N. Seifert, B. Gill, V. Zia, M. Zhang, and V. Ambrose, "On the scalability of redundancy based SER mitigation schemes," Proc. 2007 IEEE Intl Conf on IC Design and Tech., pp. 263-271, 2007.

[4] O. A. Amusan, et al., "Directional sensitivity of single event upsets in 90 nm CMOS due to charge sharing," IEEE Trans. Nuc. Sci., vol. 54, pp. 2584-2589, Dec, 2007.

[5] O. A. Amusan, et al., "Mitigation techniques for single event induced charge sharing in a 90 nm bulk CMOS process," Proc. 46th Ann. Intl Rel. Phy. Symp., pp. 468-472, 2008.

[6] G. Gasiot, P. Roche, and P. Flatresse, "Comparison of multiple cell upset response of BULK and SOI 130nm technologies in the terrestrial environment," Proc. 46th Ann. Intl Rel. Phys. Symp., pp. 192-194, 2008.

[7] E. H. Cannon, S. Rabaa, M. Cabanas-Holmen, and J. Popp, "Multiple-node strikes in 45 nm SOI technology," presented at 2009 HEART Conf.

[8] M. P. Baze, et al., "Angular dependence of single event sensitivity in hardened flip/flop designs," IEEE Trans. Nuc. Sci., vol. 55, pp. 3295-3301, Dec. 2008.

[9] L. Wielopolski, "Monte Carlo calculation of the average solid angle subtended by a parallelepiped detector from a distributed source," Nucl. Instr. and Meth. in Phys. Res., vol. 226, pp. 436-448, 1984.

[10] D. Heidel, et al., "Proton and heavy ion testing of 45nm and 65nm SOI SRAMs," presented at 2009 Single Event Effects Symp.

[11] N. Seifert, B. Gill, K. Foley, and P. Relangi, "Multi-cell upset probabilities of 45nm high-k + metal gate SRAM devices in terrestrial and space environments," Proc. 46th Ann. Intl. Rel. Phys. Symp., pp. 181-186, 2008.

[12] F. Wrobel, et al., "Methodology to compute neutron-induced alphas contribution on the SEU cross section in sensitive RAMs," IEEE Trans. Nuc. Sci., vol. 51, pp. 3291-3297, Dec. 2004.

[13] F. Wrobel, J.-M. Palau, M. C. Calvet, O. Bersillon, and H. Duarte, "Incidence of multi-particle events on soft error rates caused by n-Si nuclear reactions," IEEE Trans. Nuc. Sci., vol. 47, pp. 2580-2585, Dec. 2000.

[14] W. H. Beyer, ed., CRC Standard Mathematical Tables, 27th ed.. Boca Raton, FL: CRC Press, 1984.

[15] Wolfram Research website, see http://integrals.wolfram.com/index.jsp and http://mathworld.wolfram.com/Polylogarithm.html.

[16] D. F. Heidel, et al., "Single-Event Upsets and Multiple-Bit Upsets on a 45 nm SOI SRAM," IEEE Trans. Nuc. Sci., vol. 56, pp. 3499-3504, Dec. 2009.

Effects of multi-node charge collection in flip-flop designs at advanced technology nodes

Vijay B Sheshadri, Bharat L Bhuva, Robert A Reed, Robert A Weller, Marcus H Mendenhall, Ron D Schrimpf
Department of Electrical Engineering,
Vanderbilt University, Nashville, USA
Phone: +1 615-343-6704, e-mail address: vijay.b.sheshadri@vanderbilt.edu

Kevin M Warren, Brian D Sierawski
Institute for Space and Defense Electronics,
Vanderbilt University, Nashville, USA

Shi-Jie Wen, Rick Wong
Cisco Systems Inc.
San Jose, USA

Abstract—Circuit-level simulations predict increased vulnerability of flip-flop designs and increased occurrence of single-event upsets in advanced technologies due to multi-node charge collection from single-ion strikes. This trend is examined by simulating 3D models of the flip-flops in a terrestrial neutron environment with Monte-Carlo simulations of charge generation in several technology generations.

Keywords-component; Flip-flops, DICE, DFF,charge sharing, multi-node upsets, MRED

I. INTRODUCTION

Advances in fabrication technologies for semiconductor integrated circuits (ICs) have resulted in sub-100 nm feature sizes. Along with this desired reduction in dimension has come an undesired increase in vulnerability of flip-flops to soft errors (figure 1) [1]. Soft errors are caused by energetic particles that either directly or indirectly deposit energy and create electron-hole pairs in the semiconductor material. These charges either recombine or are collected at circuit nodes. The charge collected at a circuit node perturbs the associated node voltage, creating a transient pulse. If the voltage transient is created on a node within a flip-flop, it may alter the data stored in the flip-flop; this is known as a single- event upset (SEU). There are many factors that control this error mechanism, namely physical structure, nodal capacitances, transistor currents, etc. From a circuit point of view, if the voltage perturbation (also referred to as a single-event transient or SET) is longer than the feedback-loop delay of the flip-flop, the data stored in the flip-flop will change. As the voltage perturbation characteristics are directly related to the total charge collected at a node, the semiconductor industry has used critical charge (threshold charge at which the flip-flop will upset) as the standard metric for indicating the vulnerability of a flip-flop design to SEUs. In addition to SET

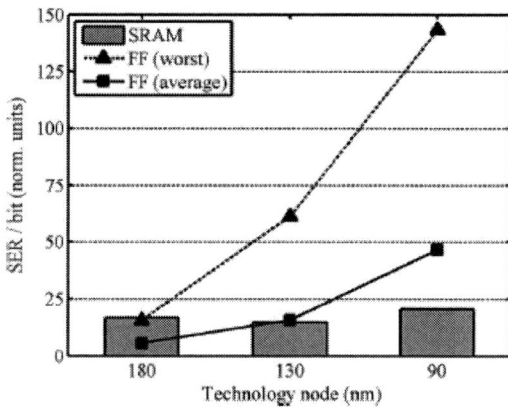

Fig. 1. Soft error rate per flip-flop as a function of technology. [1]

pulse width, individual transistor currents and nodal capacitances also play major roles in determining the critical charge of a flip-flop [2].

Most previous studies for flip-flop upsets have assumed that single-event related charge is collected at only one circuit-node (usually the node that is hit by the incident ion). This assumption was valid for older technologies where individual transistors are relatively far apart, but not for advanced deep-submicron technologies where distance between transistors is measured in nanometers. Decreasing technology feature size has resulted in higher packing densities, or close proximity of transistors to each other. As a result of this, single-event related charge may be collected at multiple nodes [3, 4]) in a circuit. Many traditional design approaches to mitigate soft errors - such as Triple Mode Redundancy (TMR) [5] or Dual Interlocked Cell (DICE) [6] - are based on the assumption that

978-1-4244-5430-3/10 $26.00 © 2010 IEEE

an incident particle affects only one circuit node. This assumption is valid for technologies in which an ion strike results in charge collection on a single node within the circuit, referred to as single-node upset. However, in deep-submicron technologies, this isolation assumption is not valid, and charge collection by multiple nodes will lead to increased susceptibility of hardened flip-flop designs [7]. In this paper, we explore the trends for soft-error rates in flip-flop designs at 90 nm, 65 nm, and 45 nm CMOS bulk technologies as a result of multi-node charge collection. Simulation results show that upsets caused by multi-node charge collection in these technologies are significantly higher than those for older technologies.

II. FLIP-FLOP DESIGNS

Three different flip-flop designs with varying levels of area, power, speed, and soft-error hardness were studied. The flip-flops designs considered are a conventional master-slave D flip-flop design (MS DFF), a SEU-tolerant master-slave Dual-Interlocked Cell flip-flop (MS DICE FF) [6], as shown in figure 2, and a low-power version of the D flip-flop design (LPFF). The power and delay of the flip-flops, normalized with respect to the DFF for each technology node considered, are given in Table 1.

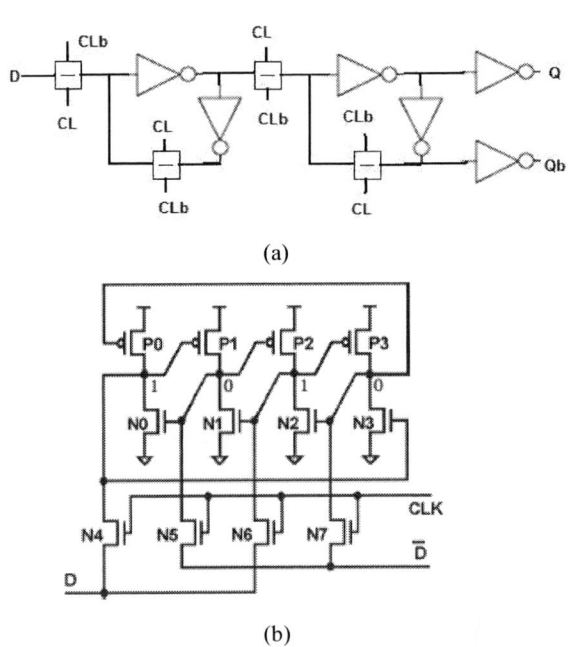

(a)

(b)

Fig 2: Flip-flop designs (a) MS DFF (b) MS DICE FF

III. CRITICAL CHARGE

Circuit-level simulations were carried out to estimate critical charge requirements for single-node and multiple-node upsets for each flip-flop design. Results were used to identify sensitive nodes and node pairs. Sensitive nodes are the nodes, which, upon sufficient charge collection, will cause an upset. Sensitive node pairs are defined as nodes that, upon simultaneous charge collection, cause the latch to upset. It has been shown previously that the charge required to cause an upset when a single node collects charge is usually more than the charge required on either node of a sensitive node pair for an upset [7]. Ion strikes were simulated by connecting a current source based on 3D TCAD device simulations to each of the nodes under consideration. The critical charge of each node was determined by varying the charge deposited at the node until an upset was observed. The DFF and the LPFF were susceptible to single-node strikes. Figure 3 shows Q_{crit} for the master stage of the DFF and LPFF. As expected, the DICE flip-flop did not show any upsets due to single-node hits for a large range of deposited charge.

However, the DICE flip-flop was susceptible to strikes on combinations of two nodes (referred to as node pairs). To simulate charge collection by two nodes, charge was deposited

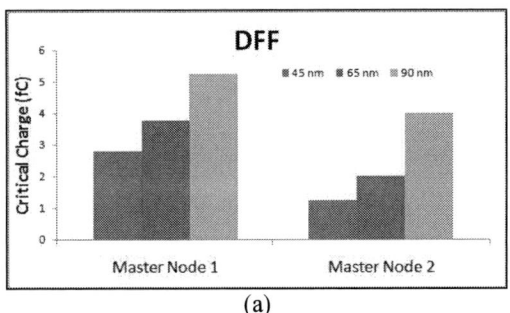

(a)

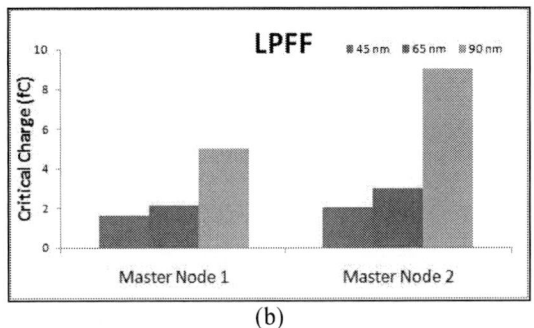

(b)

Fig. 3. Critical charge for the flip-flops decreases as the feature size decreases. (a) Qcrit for the master stage of the LPFF (b) Qcrit for the master stage of the DFF

TABLE 1. POWER AND DELAY OF THE FLIP-FLOPS NORMALIZED W.R.T DFF

Flip-flop	No. of transistors	45 nm		65 nm		90 nm	
		max. C-Q delay	Power	max. C-Q delay	Power	max. C-Q delay	Power
LPFF	28	0.64	0.90	0.68	0.88	0.92	0.83
DFF	20	1	1	1	1	1	1
DICE	40	1.38	2.23	1.26	1.87	1.11	2.65

simultaneously on the node pairs using multiple current sources. As multi-node charge collection is a strong function of layout, and the layout may contain any of these nodes in physical proximity, all possible combinations of node pairs were simulated. From these simulations, the most vulnerable node-pair for each flip-flop design was identified. Fig. 4 illustrates the vulnerable pair of transistors on a layout and the

associated schematic design for the DICE flip-flop. For this design, transistors (P1, N0) and (P3, N2) are the most vulnerable node pairs, as seen from Fig. 5 while transistor pairs (P1, N2) and (P3, N0) are not vulnerable at all. Fig. 6 shows the effects of multi-node charge collection in these flip-flops for the three technologies considered. The data points on the curves indicate the charge required to be deposited simultaneously on each node to cause an upset. For example, for the 65 nm DFF design (fig. 6(a)), if ~ 1 fC is deposited on both nodes of the vulnerable pair simultaneously, an upset occurs. Similarly, in the 90 nm DFF design, if 1 fC is deposited on node 1, ~2.5 fC must be deposited on node 2 to produce an upset. However, for a single-node upset, at least 2 fC and 4 fC are required for these 65 nm and 90 nm DFF designs, respectively. Any combination of charge deposition that falls in the region above the curve will cause an error whereas any

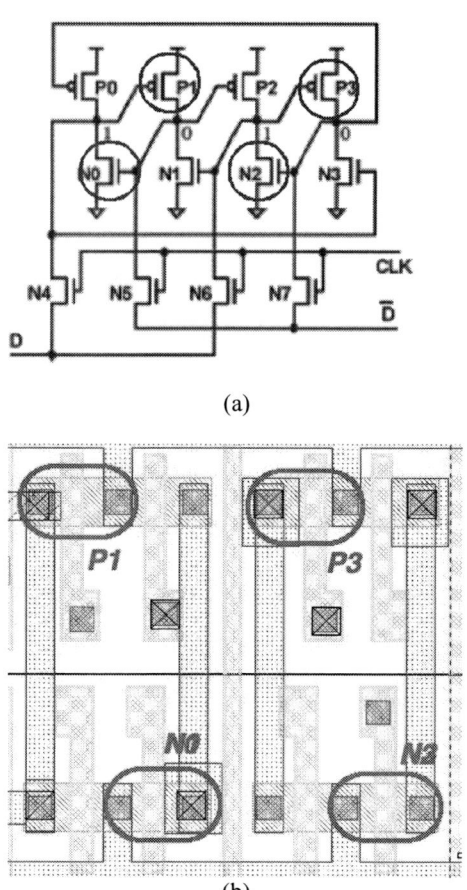

(a)

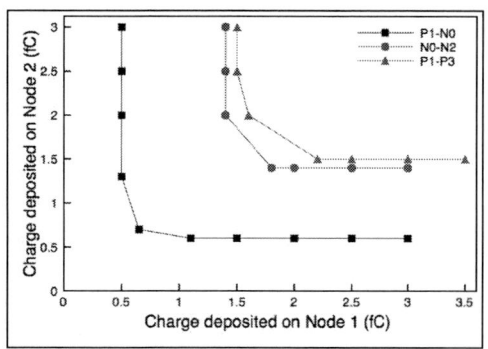

(b)

Fig.4: Sensitive nodes of a DICE latch are marked on the schematic (a) and their position in the layout is shown in (b)

Fig. 5. Critical charge combinations for all the vulnerable node pairs of DICE. P1-N0 and P3-N2 exhibit same charge combination requirements

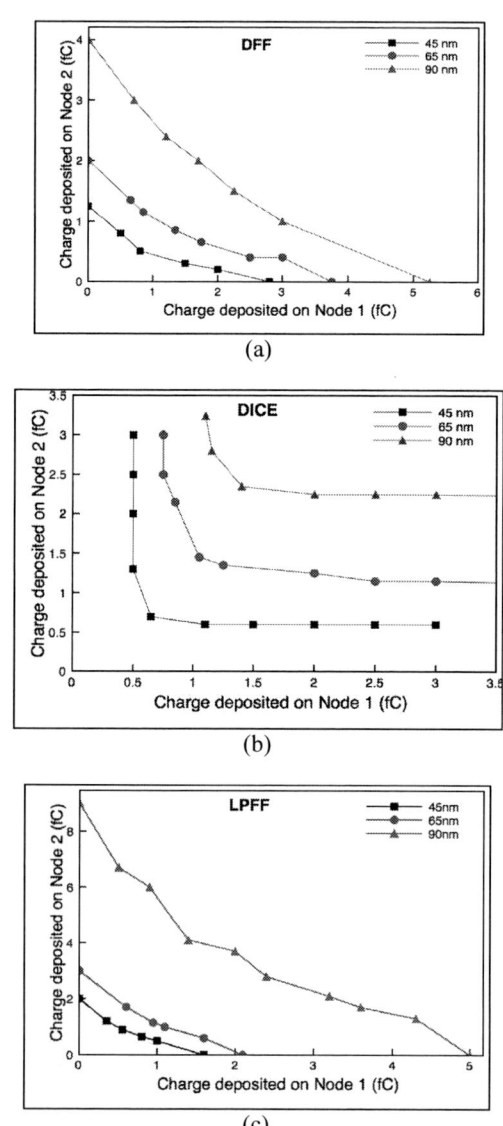

Fig. 6. Critical charge combinations for most vulnerable node pairs of (a) DFF (b) DICE (c) LPFF

combination that falls in the region below the curve does not cause an upset. The point of intersection of the curve and the axis gives the amount of single-node charge collection required to cause an upset. Since the DICE FF did not upset for single-node strikes, the curves in fig. 6(b) do not intersect the axes. These results clearly support the assertion that the charge required to cause multi-node upsets is less than that for a single-node upset.

IV.   MONTE CARLO SIMULATIONS

Monte-Carlo Radiative Energy Deposition (MRED) is a simulation tool for quantifying the energy deposited by radiation in microelectronic devices and is based upon the Geant4 class libraries, which comprise computational physics models for the transport of radiation through matter [8]. In this paper, we make use of the MRED tool to study charge generation in the sensitive nodes due to a single incident particle.

A 3D geometrical model of a DFF, including the sensitive nodes and the metallization layers, was developed. The following assumptions were used to develop the model: (i) energy deposition calorimetry was computed using a single rectangular parallelepiped volume, (ii) the placement of these volumes was based on the position of the sensitive nodes in the flip-flop's layout, (iii) the size of the volume was based on the well dimensions in the layout, i.e., the surface dimensions of the volume was equal to the width of the well, (figure 7), and (iv) the depth of the volume was 1 μm (corresponding to the well depth).

Scaled models were built based on the above assumptions to represent the flip-flop design in 90 nm, 65 nm, and 45 nm technologies. Since these models are based on the layout, the volumes differ in both size and relative distance between them. The sizes of the volumes representing the sensitive nodes and the distance between them are given in Table 2.

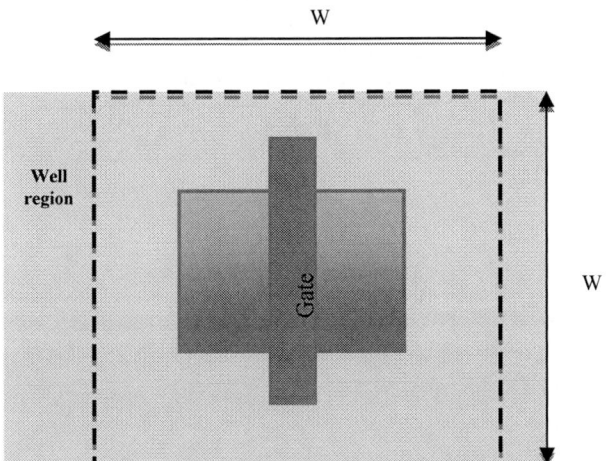

Fig.7: The sensitive volume is indicated by the dashed lines. 'W' is the width of the well.

Simulation Results

The simulations were carried out using the New-York-City terrestrial neutron spectrum. For every particle striking the 3D structure, the amount of energy deposited in each calorimetry volume is recorded; energy deposition was converted to charge generation using 3.6 eV per electron-hole pair for silicon. For every particle incident on the 3D model, there are three possibilities: (1) the incident particle does not affect either of the volumes and hence there is no charge generation in either of the volumes (2) the incident particle affects only one of the volumes, causing charge generation in that volume (3) the particle affects both volumes simultaneously, causing charge generation in those affected volumes. The number of instances where such a particle strike resulted in charge generation of at least 1 fC in both the volumes is calculated.

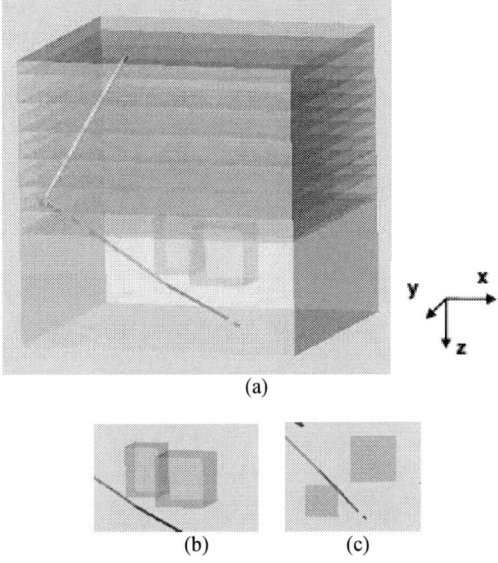

Fig.8: (a) MRED model of the DFF (b) 3D view of the sensitive volumes (c) top-view, showing the placement of the sensitive volumes

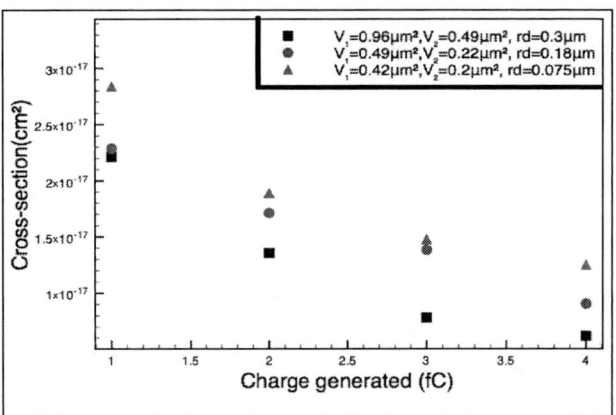

Fig. 9. Cross-section for a fixed amount of charge generated in two sensitive volumes simultaneously.

TABLE 2. THE SIZES OF THE VOLUMES REPRESENTING THE SENSITIVE NODES AND THE DISTANCE BETWEEN THEM

	Sizes of the volumes (μm^2)		Relative distance (μm)
	V1	*V2*	*rd*
3D model based on 90nm layout	0.96	0.49	0.3
3D model based on 65nm layout	0.49	0.22	0.18
3D model based on 45nm layout	0.42	0.2	0.075

This number, divided by the total fluence, gives the cross-section. Similarly, the number of instances where a single incident particle resulted in charge generation of 2, 3 and 4 fC in both the volumes and the corresponding cross-section values is calculated. The cross-section values are plotted vs. generated charge in Fig. 9. The plot shows that a higher occurrence of multiple node charge generation is observed in the 3D model based on the 45nm DFF layout as compared to the model based on the corresponding 90nm layout. The increase in the cross-section can be attributed to shrinking the distance between the volumes, indicating that the occurrence of multiple node charge generation is a function of the proximity of sensitive nodes.

These results, however, are layout dependant. Layout isolation techniques such as interleaving master and slave stages may reduce the probability of multi-node charge generation. These results may also differ from one layout to another, allowing designers to optimize layout for soft errors with minimum performance penalty.

V. CONCLUSION

Trends observed in critical charge for dual-node strikes in all three technology nodes indicate that as technology scales, the charge required to cause an upset decreases, increasing the flip-flop vulnerability. However, it is not sufficient to determine the critical charge for single-node and multi-node upsets. The probability of occurrence for multi-node charge collection must also be evaluated. If multi-node charge-collection increases as technology scales, it will strongly affect the flip-flop soft-error rate. Multiple-node charge collection is a complex phenomenon that can occur due to the following mechanisms: (1) carrier motion in the semiconductor, which causes two nodes to collect charge and (2) single incident particle that affects multiple nodes simultaneously resulting in multi-node charge generation. Simulations for volumes representing the sensitive nodes of the flip-flops showed increased probability of multi-node charge-generation as the distance between the transistors decreased. This may result in higher soft-error vulnerability as technology scales for a given flip-flop design and layout

REFERENCES

[1] T. Heijmen, P. Roche, G. Gasiot, K. R. Forbes and D. Giot, "A Comprehensive Study on the Soft-Error Rate of Flip-flops from 90-nm Production Libraries," *IEEE Trans.Device and Materials Reliability*, vol. 7, no. 1, March 2007

[2] S.M. Jahinuzzaman, M. Sharifkhani and M. Sachdev, "An Analytical Model for Soft Error Critical Charge of Nanometric SRAMs," *IEEE Trans. VLSI Systems*, vol. 17, no. 9, September 2009

[3] B. D. Olson, D. R. Ball, K. M. Warren, L. W. Massengill, N. F. Haddad, S. E. Doyle and D. McMorrow, "Simultaneous single event charge sharing and parasitic bipolar conduction in a highly-scaled SRAM design," *IEEE Trans. Nucl. Sci.*, vol. 52, no. 6, pp. 2135-2136, December 2005

[4] O. A. Amusan, A. F. Witulski, L. W. Massengill, B. L. Bhuva, P. R. Fleming, M. L. Alles, A. L. Sternberg, J. D. Black and R. D. Schrimpf, "Charge collection and charge sharing in a 130 nm CMOS technology," *IEEE Trans. Nucl. Sci.*, vol. 53, no. 6, pp. 3253-3258, December 2006

[5] W. Peterson, *Error-Correcting Codes*,2nd ed. Cambridge,MA: MIT Press, 1980, 560p.

[6] T. Calin, M. Nicolaidis, and R. Velazco, "Upset hardened memory design for submicron CMOS technology," *IEEE Trans. Nucl. Sci.*, vol. 43, no. 6,pp. 2874–2878, Dec. 1996. B

[7] O. A. Amusan, L. W. Massengill, M. P. Baze, A. L. Sternberg, A. F. Witulski, B. L. Bhuva and J. D. Black, "Single event upsets in deep-submicrometer technologies due to charge sharing," *IEEE Trans. Device and Materials Reliability*, vol. 8, no. 3, pp. 582-589, September 2008

[8] R. A. Weller, M. H. Mendenhall, R. A. Reed, R. D. Schrimpf, K. M. Warren, B. D. Sierawski, and L. W. Massengill, "Monte Carlo Simulation of Single Event Effects," Accepted for publication in IEEE Trans. Nuc. Sci. 2010

978-1-4244-5430-3/10 $26.00 © 2010 IEEE

ANALYSIS OF SOFT ERROR RATES IN COMBINATIONAL AND SEQUENTIAL LOGIC AND IMPLICATIONS OF HARDENING FOR ADVANCED TECHNOLOGIES

N. N. Mahatme[1], I. Chatterjee[1], B. L. Bhuva[1], J. Ahlbin[1], L. W Massengill[1] and R. Shuler[2]

[1]Vanderbilt University, Nashville, TN 37235, USA; [2]NASA Johnson Space Center, USA

N. N. Mahatme, Vanderbilt University, Nashville, TN 37235 Phone: 615-473-2671; e-mail: nihaar.n.mahatme@vanderbilt.edu

Abstract — Previous results and models have predicted that combinational logic errors would dominate over flip-flop errors for the past few technology nodes. However, recent experimental results show very little contribution from combinational-logic soft errors to overall soft-error rates. A model that explains the soft error rates as a function of frequency is developed to account for the inconsistency in observed data. Implications for hardening against soft errors for advanced technologies are discussed.

Keywords--Soft error rates, single event effects, single event transient, single event upset, transient propagation, transient pulse-width.

INTRODUCTION

With the decrease in feature sizes on an Integrated Circuit (IC), the amount of collected charge required to cause a Soft Error (SE) has decreased due to smaller nodal capacitances and operating voltages [1]. This, coupled with higher operating frequencies, has increased the vulnerability of circuits to soft errors to the point where they are now a major concern for semiconductor manufacturers and designers. Soft errors are generally classified into two categories: a) Flip-flop upsets and b) Combinational-logic upsets. Flip-flop (FF) upsets result when an ion is incident on a node contained within a flip-flop cell. Combinational-logic (CL) upsets are errors latched by storage elements due to transients generated in combinational- logic that propagate to a storage cell. Combinational-logic upsets increase linearly with frequency while flip-flop upsets are independent of frequency [2].

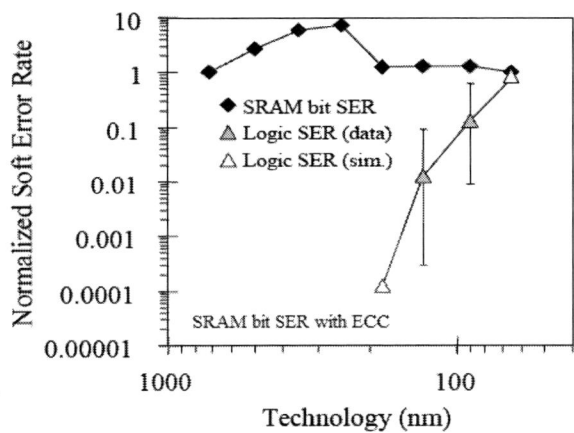

Fig. 2. Data showing increase in SER for Combinational and Sequential Logic [4]

All previous predictive models show flip-flop upsets to dominate in older technologies where operating frequencies are lower than those for advanced technologies. However, some of the recent results do not show the expected increase in combinational-logic upsets. Figures 1 and 2 show that with technology scaling, combinational-logic upsets increase at a rapid rate and dominate beyond 65 nm technology node. On the other hand, Figure 3 shows only a slight contribution from combinational-logic errors to the overall SER for a 32nm technology node [5]. The main focus of this work is to investigate the effects of operating frequency and technology scaling on soft-error rates. A model is presented to explain possible reasons for the differences in the sets of published data.

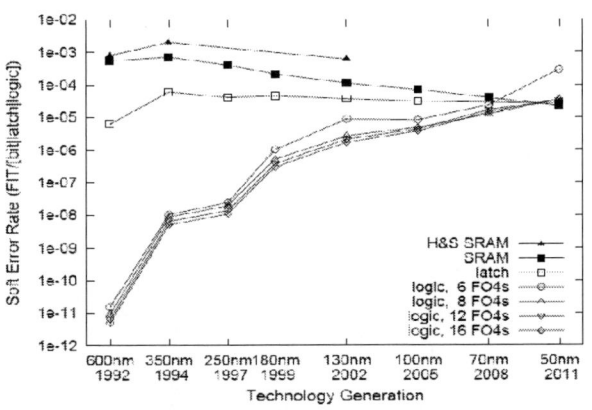

Fig.1. Predicted SER trend for flip-flops and combinational-logic upsets due to incident ions on ICs [3].

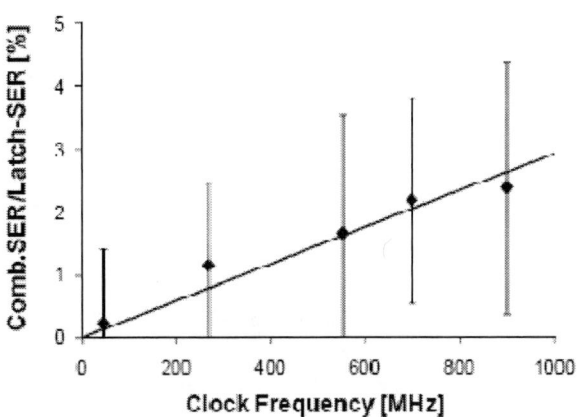

Fig. 3. The contribution of combinational-logic upsets is very small even at 32 nm technology node [5].

FACTORS AFFECTING SOFT ERROR RATES

As technology scales, the soft error rates are affected strongly by the sensitive area per logic gate (which in turn is controlled by the charge collection processes and drain area). As technology scales, the sensitive area of logic gates increases because of the lower critical charge (Q_{crit}) requirements. Previously, researchers have found that the Q_{crit} value reduces by a factor of 2 from one technology node to another [3]. One implication of this reduction in critical charge is that an ion hit away from a transistor may result in sufficient charge collection to cause an upset. In physical terms, this implies that the sensitive area of a transistor will increase as technology scales, even though drain region dimensions (or the area of the logic gate itself) decreases.

Figure 4 shows the increase in sensitive area as an increase in upset cross-section of a logic gate. These simulations for 90 nm and 65 nm technologies were performed using neutrons with energies corresponding to the LANL spectrum with Monte-Carlo Radiative Deposition software based on GeantIV libraries [6, 7]. These curves were obtained using experimental data for 90 nm technology and calibrated models for 65 nm technology. The physical structure used for simulations was large enough (50μm X 50μm X 16.25 μm) to allow accurate simulation of particle strikes over a wide range of incident angles. The overlayers for the processes were obtained from PDK available from the manufacturers. The increase in cross-section results in higher number of soft-error events (measurable voltage perturbations) as technology scales. The increase in cross-section (the cross-section includes sensitive area as well as the probability for sufficient charge collection) has been experimentally verified for multiple technology nodes [8-10]. The increase in cross-section as technology scaled from 90 nm to 65 nm is approximately 40% as seen in Figure 4.

CROSSOVER-FREQUENCY ESTIMATION

Due to the linear frequency dependence of combinational-logic upsets and near frequency independence of flip-flop upsets, frequency may be used to understand the relative contribution of these upsets. In combinational logic, there is linear frequency dependence for upsets because of a greater number of latching edges presented to the SETs by the receiving flip-flops with increasing frequency [11]. However, there is little frequency dependence on FF upsets because a latch upset results due to a direct hit by an ion on a transistor within the latch. Previously, researchers have presented a model that predicts a cross-over frequency at which combinational-logic upsets would exceed flip-flop upsets [2]. Beyond the crossover frequency, combinational-logic errors will dominate flip-flop errors as shown in Figure 5.

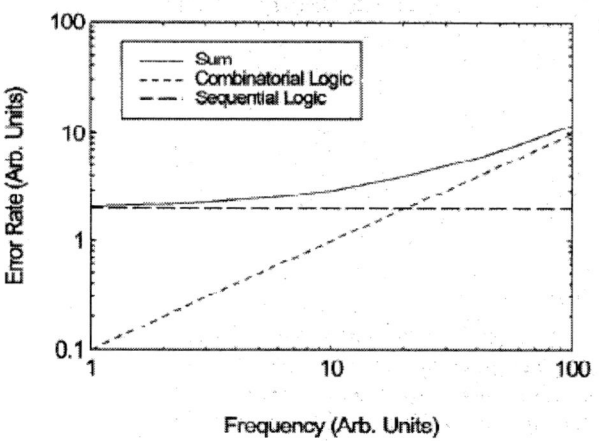

Fig. 5. Error rate as a function of frequency for combinational and sequential logic elements as well as their sum[2].

The frequency dependence for a combinational-logic error results from the fact that the voltage perturbation due to an ion-hit must get latched. For a combinational-logic error to occur, a particle must hit a sensitive region, the associated circuit node must collect sufficient charge to generate a voltage perturbation (known as single-event transient (SET)), this SET must propagate through the combinational logic to reach a storage cell, and the storage cell must latch incorrect data. Each node in the circuit must be evaluated for these events to determine overall circuit SER. The following equation is conceptually similar to that developed by [12]. The combinational-logic error probability (worst case) for a circuit may be given by:

$$(Flux)\left\{\sum_{i=1}^{n}(A_i)(Q_i)(P_{prop})\right\}T_{mask} \quad (1)$$

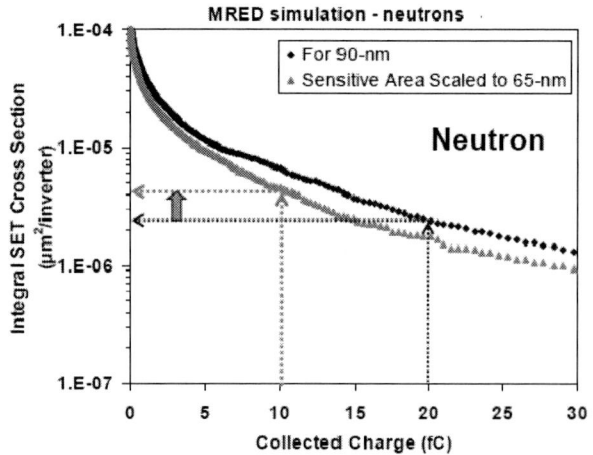

Fig. 4. Estimation of scaling trends in neutron SET cross sections Integral SET cross section is a measure of the number of SETs that are latched [6].

Where,

$Flux$ = Atmospheric Flux of particles. This factor is independent of scaling.

A_i = Probability that the sensitive node i being considered is hit. This is a scaling dependent factor for soft errors. This is the sensitive area of a transistor. As technology scales the physical area of a transistor decreases, but due to lower critical charge requirements, the sensitive area of a transistor increases.

Q_i = Probability that sufficient amount of charge is collected at the node i, to generate a transient pulse of sufficient amplitude and duration. This is also related to electrical masking. As technology scales, the lower nodal capacitances and lower drive currents increase this factor[1].

P_{prop} = Probability that the SET propagates to a latch input. This is also known as logical masking. This is independent of scaling.

T_{mask} = Probability that an SET is latched by a receiving flip-flop. This is also known as timing-window masking. This factor is directly related to the operating frequency.

The factor $(A_i \bullet Q_i)$ is evaluated in MRED as the integral cross-section per logic gate as shown in Figure 4. Factor P_{prop} is independent of scaling for this discussion as identical circuits and bias conditions are assumed for each technology generation. Factor $T_{mask,}$ is linearly related to the operating frequency and will increase as technology scales. Taken together, all of these factors will increase the probability of the event that will generate an SET and result in a combinational-logic soft error as technology scales (assuming same circuit design is used for all technology nodes). These factors also imply that, for the same operating frequency, the soft error rates for combinational-logic circuit will increase as technology scales (assuming identical combinational logic circuit at each node). The rate of increase in soft-errors will be a strong function of the circuit topology as the factor $(A_i \bullet Q_i)$ is dependent on topology and layout.

For a flip-flop error probability, the same equation may be used with the condition that the generated SET pulse must be longer than the feedback of the storage cell. As this requirement replaces the T_{mask} factor in Eqn. 1, the frequency dependence is absent. With scaling, the transistors switch faster, making it easier for shorter transients to upset the latches. Previously published FF SER data shows a varied response to technology scaling [6, 13]. The increases in error rates for these flip-flops varied from an order-of-magnitude to very little as technology scales. As the critical charge, charge collection efficiency of sensitive volumes and sensitive area of a flip-flop is a strong function of layout and individual transistor sizes, care must be taken to compare such a diverse set of data without any knowledge of exact designs. However, based on the discussion in the previous sections, it may be postulated that for similar layouts for a latch design across technology nodes, error-rate per bit should increase.

The effects of frequency and technology scaling on combinational-logic and flip-flop errors were investigated using MRED and circuit-level simulations. The MRED curves for 90 nm and 65 nm technologies from Fig. 4 were used for inverter cross-sections. Next a circuit consisting of a chain of inverters (20 stages) feeding into a flip-flop was used for each technology to estimate critical charge for each inverter. The critical charge here is defined as the charge required to generate a SET pulse at least as wide as the setup-and-

hold time of the flip-flop. Based on these critical charges, the cross-section for each inverter was obtained for each technology node based on MRED data. This represents the $(A_i \bullet Q_i)$ factor in Eqn.1. The T_{mask} factor in Eqn. 1 was obtained by $T_{setup-\&-hold} / T_{clock}$. For the given chain of inverters, P_{prop} factor is always 1. Using these parameters, error probabilities for combinational-logic (inverter chain) and flip-flop designs were evaluated for each technology node as a function of frequency and are shown in Figure 6. Flip-flop soft-error probabilities were determined based on the Q_{crit} for the flip-flop. As only the comparative values are of importance, the relative error probabilities are shown as a function of frequency. For these simulation results, the crossover frequency at which combinational-logic errors will dominate increases as technology scales. This effect was observed by other researchers through experiments on similarly designed circuits at different technology nodes [10, 11]. At the same time, any increase in flip-flop vulnerability increases the crossover frequency. These two factors will determine, the overall soft error vulnerability for a given design at a given technology node. For different circuits at different technology nodes, this will result in varying amount of contribution from combinational-logic errors, as seen in previously published data.

As increases in FF and CL errors are design topology and layout dependent, the curves for any other circuit will be different and each design should be calibrated along this model to estimate the crossover frequency. For example, if a logic design is used instead of an inverter chain, P_{prop} will be less than 1 resulting in lower rate of increase for combinational-logic errors as technology scales. Different designs fabricated at the same technology node may yield different crossover frequency.

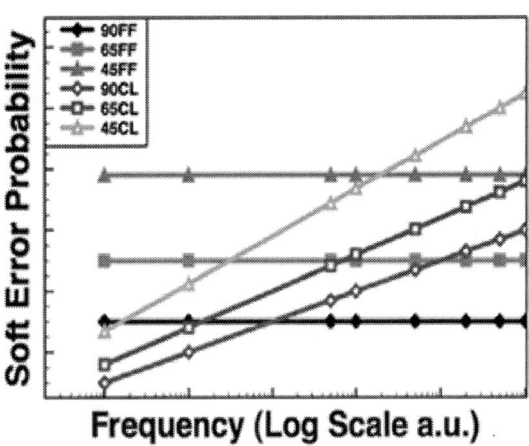

Fig. 6. Simulated FF and CL upset rates as a function of frequency for 90, 65, and 45 nm technology nodes.

Same design from different technology nodes will show an increase in crossover frequency as individual logic gate cross-section will increase. Inconsistencies in published results showing varying levels of contribution from combinational-logic errors may be caused by different crossover frequencies for the circuits used.

HARDENING IMPLICATIONS

The implications of this model are enormous for the hardening strategies employed by the designers. Presently, the conventional approaches regarding soft-error hardening is aimed towards hardening flip-flops only. In earlier technology generations, operating frequencies were lower and the associated crossover frequencies were higher than operating frequencies (relative contribution of combinational logic upsets to overall SER was low) of the circuits. As a result, hardening flip-flops resulted in improved overall SER for the circuit. However, as the crossover frequencies get closer to operating frequencies due to scaling, any hardening of flip-flops will result in a lower crossover frequency. If the crossover frequency decreases below operating frequency, combinational-logic errors will dominate, negating any improvement in soft-error rates due to flip-flop hardening. Thus, designers will need to identify the crossover frequency before determining the hardening approach for a given circuit. If the crossover frequency is very close to operating frequency, hardening flip-flop designs will yield very little improvement in overall SER of the circuit. In such cases, hardening flip-flops alone will not decrease soft-error rates significantly as combinational-logic errors will dominate. Additional resources or higher performance penalty to address combinational-logic errors will be necessary to bring soft-error rates to desired levels.

A test-chip to demonstrate the effects of flip-flop hardening when crossover frequency is close to operating frequency was designed and tested [11]. The IC was designed and fabricated using 90 nm IBM 9SF CMOS technology. It was then tested at the Lawrence Berkeley National Laboratory using the 10 MeV/u cocktail of Ne, Ar, Cu, Kr, and Xe at normal incidence. The test IC contained conventional D-Flip Flop (DFF) and Dual-Interlocked Storage Cell (DICE) FF shift registers to measure the upset rates of flip-flops only [14]. DICE FF is considered harder to upset compared to a conventional DFF design. DFF shift register contained 2000 FF stages. DICE shift register contained 576 FF stages. Another structure consisted of DICE latches and combinational logic. For this test structure, each DICE latch was fed by a string of 74 current matched inverters as shown in Figure 7 and as described in [11].

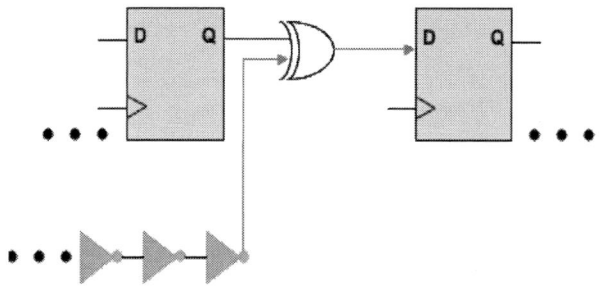

Fig: 7. Schematic representing test structure using an inverter chain [11].

The heavy-ion exposure for these test circuits show (a) vulnerability of conventional DFF designs, (b) vulnerability of hardened DICE FF, and (c) dependence of CL upsets as a function of frequency. Figure 8 shows the test results for Krypton ion exposure with Linear Energy Transfer (LET) value of 30.85 MeV-cm^2/mg. As expected, the FF errors are mostly independent of operating frequency and DFF show higher error rates then DICE FF. The main point of interest here is the relative contribution of combinational-logic errors compared to DFF and DICE FF. If designers had hardened this circuit using DICE FF instead of DFF, the overall error rate, at 200 MHz operating frequency, would not have changed significantly due to dominance of CL errors as crossover frequency increases above 50 MHz for the hardened circuit, implementing DICE FFs.

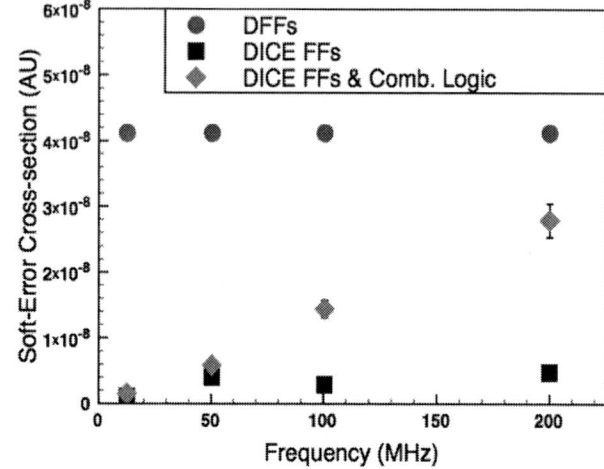

Fig. 8. Experimental results for cross-section for DFF, DICE FF and DICE FF with inverters as a function of frequency per FF stage[11].

CONCLUSION

As technology scales, the critical charge for flip-flop and combinational-logic upsets decrease while the sensitive area for logic gates increase. These factors should increase the overall soft error rates for flip-flop and combinational circuits as technology scales. Also for a given circuit, the crossover frequency at which combinational-logic errors dominate flip-flop errors may increase as technology scales.

For older technologies, the operating frequency was well below crossover frequency resulting in dominance of flip-flop errors over combinational-logic errors. As technology scales, the operating frequency will get closer to crossover frequency, and may eventually cause combinational-logic errors to dominate. The main implication of this model is the hardening approaches taken by designers for advanced technologies. If only flip-flop hardening is considered, as it is the most conventional approach, combinational-logic errors may dominate and the overall error rate may not change significantly. Overall circuit design topology and layout must be considered together for determining the most efficient hardening approach for future designs.

ACKNOWLEDGEMENT

The authors wish to thank DTRA and CISCO Systems, Inc. for their support. The authors also thank Dr. Jeff Black and Vanderbilt Radiation Effects and Reliability Group (RER) students for valuable discussions.

REFERENCES

[1] Heijmen T. et al., "A Comprehensive Study on the Soft-Error Rate of Flip-Flops From 90-nm Production Libraries" IEEE Transactions On Device And Materials Reliability, Vol. 7, No. 1, pp 84-96, 2007.

[2] Buchner S. et al., "Comparison of Error Rates in Combinational and Sequential Logic", IEEE Transactions On Nuclear Science, Vol. 44, No. 6, pp 2209-2216, 1997.

[3] Shivakumar P. et al., "Modeling the Effect of Technology Trends on the Soft Error Rate of Combinational Logic", Proceedings of the International Conference on Dependable Systems and Networks, pp 389- 398, 2002.

[4] Baumann R, "Radiation Induced Soft Errors in Advanced Semiconductor Technologies", IEEE Transactions on Device Materials and Reliability, Vol 5, No3, pp 305-316, 2005

[5] Gill B. et al., "Gill. B. et al., "Comparison of Alpha-particle and Neutron-induced Combinational and Sequential Logic Error Rates at the 32nm Technology Node" Proceedings of 47th Annual International Reliability Physics Symposium, pp 199-205, 2009.

[6] Narasimham B. "Characterization Of Heavy-Ion, Neutron And Alpha Particle Induced Single-Event Transient Pulse Widths In Advanced CMOS Technologies", PhD Thesis, Vanderbilt University, 2008

[7] Agostinelli S. et. al, " GEANT-4 a simulation Toolkit," Nucl. Instrum. Meth. Phys Res. Vol A, 506, pp 250-306, 2003

[8] Hass K..J. et. al, "Single Event Transients in Deep Submicron CMOS", Proceeding of 42nd Midwest Symposium on Circuits and Systems, vol 1, pp 12-125, 1999.

[9] Karnik T. et.al, "Characterization of Soft Errors Caused by Single Event Upsets in CMOS Processes", IEEE Transactions on Dependable and Secure Computing, Vol. 1, pp 128-143, April-June 2004

[10] Gadlage M. J. et al., "Digital Device Error Rate Trends in Advanced CMOS Technologies", IEEE Transactions On Nuclear Science, Vol. 53, No. 6, pp3462-3464, 2006

[11] Ahlbin J. et al., "C-CREST Technique for Combinational Logic SET Testing" IEEE Transactions on Nuclear Science, Vol. 55, No. 6, pp 3347-33512008.

[12] Seifert N. et. al, "Frequency dependence of Errors for Deep Sub-Micron CMOS Technologies", IEDM technical digest, pp14.4.1-14.4.4, 2001

[13] Seifert N. et al., "Radiation Induced Soft Error Rates of Advanced CMOS Bulk Devices", Proc of International Reliability Physics Symposium,, pp 217-225, 2006.

[14] Calin T. et. al, "Upset hardened memory design for submicron CMOS technology," IEEE Transactions on Nuclear Science, vol. 43, no. 6, pp. 2874–2878, 1996.

978-1-4244-5430-3/10 $26.00 © 2010 IEEE

THERMAL NEUTRON SOFT ERROR RATE FOR SRAMS IN THE 90NM-45NM TECHNOLOGY RANGE

ShiJie Wen [1], Richard Wong [1], Michael Romain [1], Nelson Tam [2]

[1] Cisco Systems Inc, [2] Marvell Semiconductor
ShiJie Wen, 170 W Tasman drive, San Jose, CA 95134. Phone: 408-525-5171; e-mail: shwen@cisco.com

Richard Wong, 170 W Tasman drive, San Jose, CA 95134. Phone: 408-525-5171; e-mail: rickwon@cisco.com

ABSTRACT

The thermal neutron soft error rate (SER) was measured systematically on SRAM cells in the technology range of 90nm to 45nm. We report here a substantial SER sensitivity with neutron energies below 0.4eV for many SRAM cells.

Keywords – Thermal neutron, soft error rate, SRAM

INTRODUCTION

SER has remained as a reliability challenge for IC chips as Si technology scales. The key contributors to the SER are: (a) high energy neutrons originating from cosmic rays in the natural environment; (b) alpha particles originating from trace radioactive material contamination in the IC package or wafer fab process; and (c) thermal neutrons originating from B_{10} in the BPSG layer used during the IC fab process. [1-5]

For Si technology smaller than 0.18/0.15um, BPSG is no longer used in the IC fab process. It is the current industry belief that thermal neutron SER is not a concern in modern Si. Although some work has been done in attempt to invalidate thermal neutron SER sensitivity even without BPSG, the result was inconclusive due to the lack of systematic tests. [6] For doping level in the range of 2E13 – 3E15/cm² for B or/and BF_2, the consensus is that the small traces of B_{10}, if any, shouldn't be sensitive to thermal neutrons due to the lower concentration found compared to that found in BPSG.

In this paper, we systematically examine the SRAM cell thermal neutron SER sensitivity with a variety of Si technologies and technology derivatives from different fabs. The result shows that some sub micron SRAM cells have surprisingly high thermal neutron SER rate. This is the first report of high thermal neutron sensitivity with sub-micron ICs.

MEASUREMENT PROCEDURE

The thermal neutron tests were conducted in MNRC (McClellan Nuclear Research Center) at UC-Davis, located in California. It is a nuclear reactor-based facility. The reactor core is surrounded by a graphite reflector, which was modified to accept the source ends of four tangential neutron beam tubes that deliver thermal neutrons to four neutron radiography bays. The neutron beam is thermalized by a 11"-thick of sapphire crystal with relatively low contamination, reducing the fast neutrons and gamma rays. This neutron beam is normally used for high quality neutron radiography. The beam aperture is 1.25" x 1.25" and built with a 1"-thick B_4C piece. The beam exit size is about 9" x 9" at the shutter location. A 2 mm-thick cadmium sheet with a 1" x 1" hole cut into it is mounted at the shutter location to reduce the irradiation field down to the chip size. The thermal neutron flux at the IC DUT location is about 8.3 x 10⁵ n/cm²/sec at 1 MW reactor operating power. The thermal component (< 0.4 eV) of

the neutron beam is about 90%. Theoretically, there are few high energy neutrons (even >10 MeV) in this beam facility, the average fission or fast neutron energy is about 2 MeV. Figure 1 shows the reactor and beam set up.

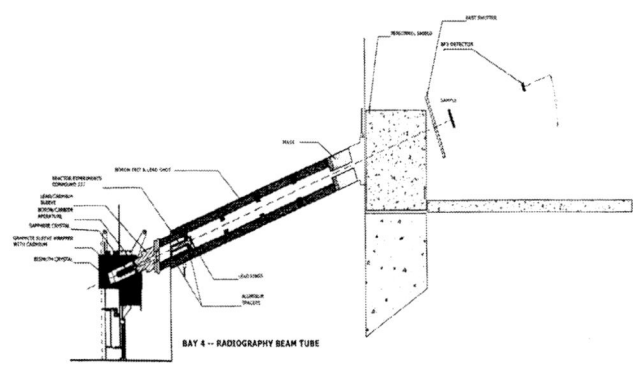

Figure 1. Beam setup at McClellan Nuclear Radiation Center

Since <10% of the neutron flux are fast or epithermal neutrons, some bit errors can be attributed to these neutrons. To determine the error rate due to the fast neutrons only, a 2mm Cd shield was placed at the fast shutter covering the 1" x 1" beam exit (about 6-10" away from tested chips), which attenuates the thermal neutrons below 0.4eV by 1e-8 (effectively removing them). Figure 2 shows the normalized neutron flux at the fast shutter with the reactor at 1MW.

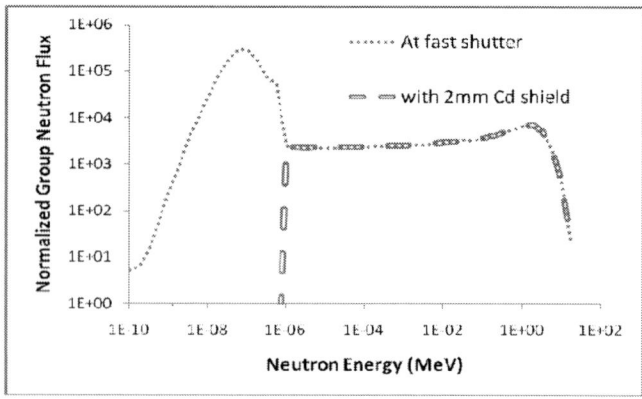

Figure 2. Neutron Flux with and without Cd shielding

Table 1 below lists the IC chips tested in this paper. For chip ID, the 1st numeric digit defines the Si technology process size and the 2nd alphabetic letter defines the technology derivative/fab/vendor. Some devices are SRAM test chips, while others are ASIC with embedded SRAM. One chip with 0.15um technology (containing a BPSG layer) was used for reference purposes. The test was conducted at room temperature. The test algorithm is static in nature:

978-1-4244-5430-3/10 $26.00 © 2010 IEEE

write a date pattern, read-back multiple times to validate the write, turn on the beam for a certain amount time duration / fluence delivered and then perform multiple reads to compute the bit error count and validate the stability of bit error count. Multiple runs were done for each device, varying the data pattern, neutron fluence and with and without the 2mm Cd shield. The following parameters were also varied for some of the devices tested: voltage, with and without package mold compound.

TABLE 1. CHIPS USED IN THERMAL NEUTRON MEASUREMENT

Chip ID in test & SRAM size (Mbit)	Technology (nm)
(1-a, 1-b, 1Mb)	45
(2-a, 2-b, 2-c, 2-d, 1Mb), (2-e, 15.4Mb), (2-f 11Mb)	65
(3-a, 3-b, 3-c, 3-d 1Mb), (3-e 22.6Mb), (3-f 40Mb), (3-g 72Mb)	90
(4-a 1.7Mb)	150

RESULTS

TABLE 2. THERMAL NEUTRON RAW TEST DATA W & W/O CD SHIELD

Chip ID	Total Neutron Fluence	Cd shield	Bit Flip count	Event Count	Bit FIT/Mbit	Thermal neutron sensitivity
1-a	1.36E9	No	134	100	108	Yes
1-a	1.36E9	Yes	13	12	NA	
2-a	2.03E9	No	106	51	313	Yes
2-a	2.03E9	Yes	8	5	NA	
2-b	1.36E9	No	75	50	332	Yes
2-b	1.36E9	Yes	2	1	NA	
2-c	1.36E9	No	72	43	319	Yes
2-c	1.36E9	Yes	6	5	NA	
2-e	8.31E8	No	16	NA	7.5	No
2-e	8.31E8	Yes	12	NA	NA	
3-a	2.38E9	No	5	5	12	No
3-b	3.54E9	No	7	7	11	
3-e	6.24E8	No	36	NA	5	No
3-e	8.03E8	Yes	56	NA	NA	
3-e	4.16E8	No	47	NA	30	Yes
3-e	4.16E8	Yes	9	NA	NA	
4-a	6.48E8	No	447	NA	2450	Yes
4-a	6.48E8	Yes	14	NA	NA	

Table 2 details the test run bit error counts, event counts and thermal neutron fluence delivered for each device. Due to the space constraints in for this paper, not all of the test data is listed. The estima-

tion of FIT/Mbit based on bit flip count or event count is done under the following assumptions: (a) Total neutron fluence is for neutrons with energy < 0.4 eV; (b) Sea level flux of 6 n/cm^2/hr is used as a reference. The FIT estimation for fast neutrons (done with a Cd shield) is not listed here and the bit flip count is provided for judgment. The purpose of this table is to provide a relative comparison for thermal neutron (< 0.4eV) and fast neutron (>0.4eV) sensitivity.

If the bit error or event count as tested with the high energy neutron beam only (w/ Cd shield) is 5-20% of those with the thermal neutron dominant beam (without Cd shield), we concluded that the thermal neutron effect is dominant and that the device is sensitive to thermal neutrons. Some of the tested devices showed substantial thermal neutron sensitivity, as high as rates of up to 10% of the device with BPSG. Furthermore, we compared FIT/Mbit data from high energy (>10MeV) neutron tests with this data; for example, chip 4-a has ~800 FIT/Mbit rate for >10MeV neutrons. The thermal neutron FIT/Mbit rate in some devices is comparable to that of high energy neutrons reported for these technologies ranges. Finally, analysis based on the physical bit map of the bit errors observed showed all the flipped bits were distributed randomly in the SRAM cells. This randomness of the physical bit map is detailed in Figure 3 for chip 1-a. In this chip, there are many 6T-SRAM instances. Each uses the same SRAM bit cell. The errors occur in every memory and the error count was roughly proportional to the size of the memory. The memories with lower bit error counts have larger statistical error.

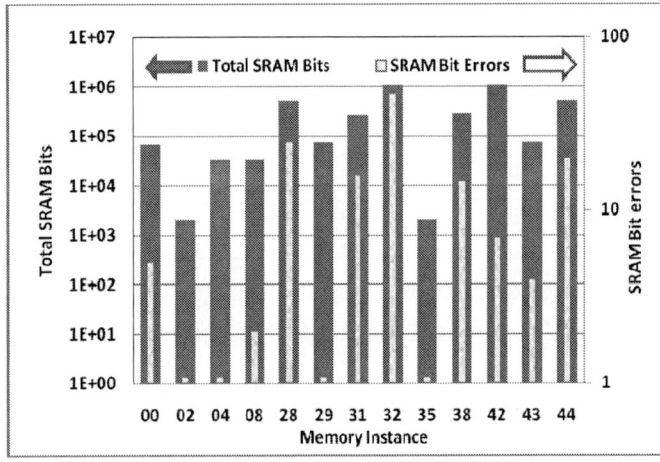

Figure 3. Bit error distribution in test chip 1-a memories.

Table 2 also lists several devices that do not show thermal neutron sensitivity, even at the same manufacturing process size, which concludes that this sensitivity is specific to particular fabs, and not to a whole process size. As the data shows, devices from the same process size but from different fabs are either moderately sensitive to thermal neutrons or not at all. This implies that thermal neutron test is needed for all IC suppliers in the future to consider its overall SER rate and that testing with a high energy (>10MeV) beam only is no longer sufficient for a full device test.

One additional phenomena noticed during testing is that for the chips showing no thermal neutron sensitivity, the fast/epithermal neutron (ie, with the Cd shield) SER rate is also significant. However, this paper focuses solely on the results of the devices sensitive to thermal neutrons, we plan to study this phenomena in further work.

Figure 4 is a plot of thermal neutron (as beam tested without Cd shield) FIT_event/Mbit vs. technology. All devices tested in this plot are from the same IC vendor, the only variable is process size. Is it clear from the data that there is no clear technology trend in thermal

978-1-4244-5430-3/10 $26.00 © 2010 IEEE   1037

neutron sensitivity, and that different devices from the same vendor and the same process size do exhibit the same sensitivity (see 65nm, 3 different devices were tested, 2-a, 2-b and 2-c).

Fig 5 is a plot of thermal neutron (without Cd shield) FIT_bit/Mbit vs SRAM test voltage for device 1-a. In this case, 1V is the nominal voltage for this 45nm technology device, but was beam tested scaled down across three different levels. The data does not show a voltage impact on FIT that is normally observed with high energy neutron (>10MeV) and alpha SER test, which is surprising.

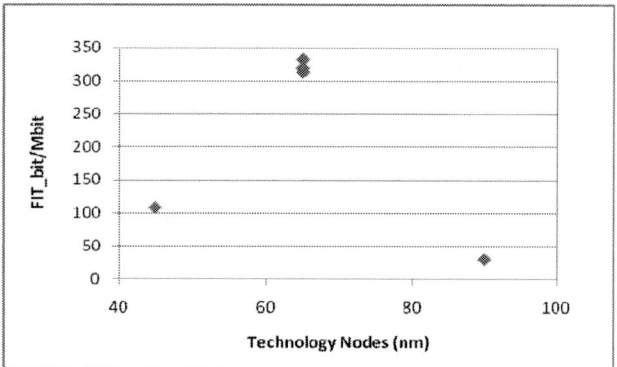

Figure 4: Thermal neutron bit error rate vs. technology

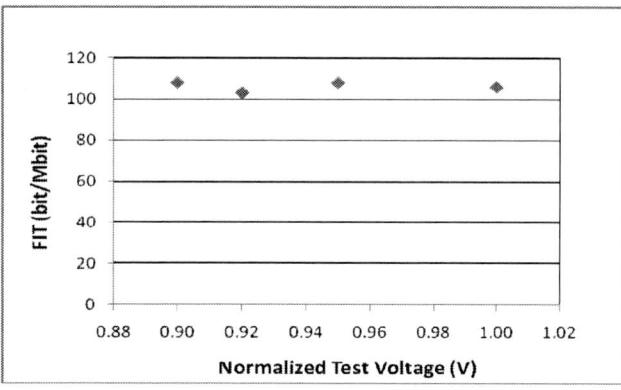

Figure 5: Thermal neutron bit error vs. test voltage for device 1-a

Figure 6 plots bit error count linearity vs. neutron fluence. This confirms that the failure rate is proportional to the total amount of neutron put onto the chip (a straight line is expected here).

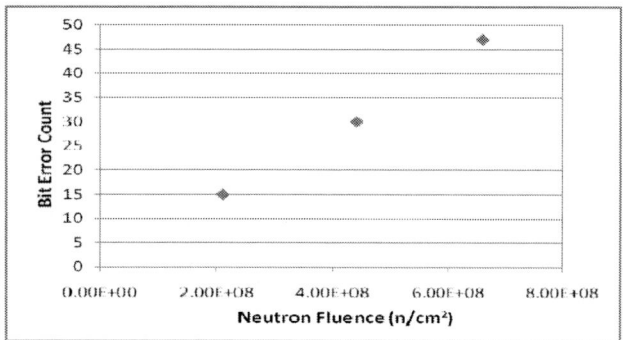

Figure 6: Bit error count vs. neutron fluence for device 1-a

A small data pattern dependency is often observed in high energy neutron as well as alpha SER tests, and it is confirmed here as well with thermal neutrons. Device 1-a was tested with different data patterns and some variability of bit error counts was observed for different data patterns using the same neutron fluence for each run (see Figure 7).

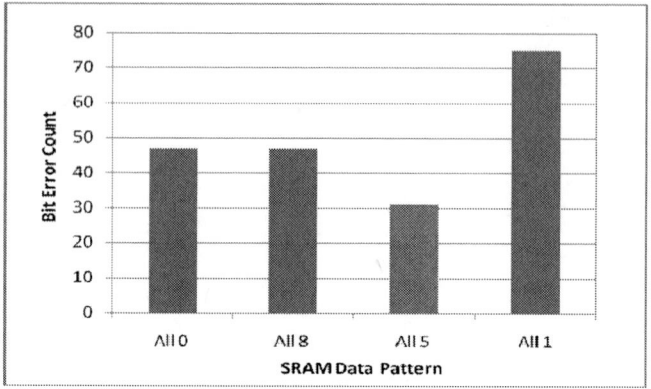

Figure 7: The bit error counter vs. SRAM data pattern.

A surprisingly high ratio of MCU's was found during the testing with thermal neutrons, Table 3 lists some results of this. This may suggest that the thermal neutron impact in silicon is very close to the transistor's sensitive area. If we closely look at one chip example (45nm), the MCU shape is lised in table 4.

TABLE 3. THERMAL NEUTRON SENSITIVITY MCU STATISTICS

	45nm	65nm-a	65nm-b	65nm-c
Total bit flip	135	107	75	73
Total Events	100	51	50	43
2-bit flip #	15	14	14	10
>2-bit flip #	6	10	6	7
Max # of bit flip	6	16	4	4
Overall MCU%	21%	48%	26%	40%

The MCU patterns were analyzed in detail with the 45nm chip. Although the total number of events is small, there is a large number of 4 bit errors with all the affected bits in the same row in the memory. Although this preliminary points to substantial MCU sensitivity with thermal neutrons, more data needs to be taken and analyzed in conjunction with physical layout to determine the significance and the cause. The MCU shape distribution is shown in Table 4.

The fact that there is a large number of MCU events observed with thermal neutrons and no voltage dependence may imply that the material is responsible for the thermal neutron reaction (local or adjacent to transistor sensitive areas) causing SER sensitivity. This should be further validated with simulation work.

Tested chips with the molding compound removed versus molding compound intact (the whole package) have the same FIT rates. This concludes that the package mold compound for this wire package chip is not a contributor to the thermal neutron failure rate observed here. 10 different chips with and without mold compound (only one example is listed in table 5).

TABLE 4. THERMAL NEUTRON MCU SHAPE FOR 45NM CHIPS.

Pattern (Row x Column)	2-bit	3-bit	4-bit	6-bit	8-bit
1 x 2	3				
2 x 1	4				
2 x 2 for 2 bit	1				
2 x 4					2
3 x 1		1			
3 x 2				2	
4 x 1			14		

TABLE 5. Molding compound sensitivity to thermal neutrons

Chip ID	Total Neutron Fluence	Molding compound on die	Bit Flip count	Event Count	Bit FIT/Mbit	Molding compound sensitivity
2-a	1.36E9	No	82	39	363	No
2-a	2.10E8	Yes	12	9	345	

DISCUSSION

With the beam on during our testing, we performed continuous read operations and recorded the cumulative bit flip count and associated logical address. After the run with the beam off, we continued the read operations for an additional 30 mins and did not observe any additional bit flips. This suggests that the failures are not associated to thermal neutron activation of a nuclide with a half life longer than several seconds. Otherwise, failures should have continued to occur after the beam was turned off.

Many elements in the semiconductor fab processing such as F, In, Cu, Ti, W, Si, O, etc. can be activated with thermal neutrons. However, these elements have relatively small thermal small cross sections and decay with beta or gamma emission, which are not likely to cause a bit flip and extremely unlikely to cause the large ratio of MCU events that we see in the test data.

Boron has two naturally occurring isotopes, 20% B_{10} and 80% B_{11}.[7] Previous work has shown that B_{10} in the BPSG layer is responsisble for the thermal neutron contribution to device SER in semiconductors with BPSG layers. The B_{10} isotope has a large thermal neutron cross section [8] that captures the thermal neutron to form a B_{11} nuclide in an excited state. This B_{11} nuclide undergoes a prompt fission into a Li ion and an alpha particle. When these charged ions enter the silicon, they can generate charges which cause the bit errors.

In this paper, all devices tested except chip 4-a have no BPSG layer, but many of them are susceptible to thermal neutrons (not to the same rate as with BPSG, but a substantial contribution to overall device FIT nonetheless). The root cause of the thermal neutron interaction for devices without BPSG is not understood and needs further investigation.

One possible approach to potentially determine the root cause is to look at the materials in the different chips and focus on the materials that differ for the thermal neutron sensitive group and that of the non-sensitive group. Since most fabs consider this information to be highly proprietary, we may be able to analyze the materials in the chips using Secondary Ion Mass Spectrometry. Another method, Prompt Gamma Neutron Activation Analysis [9], can be used to look directly at the effect of thermal neutrons on the various samples.

CONCLUSION

We have found that many SRAM cells with sub-micron technology are sensitive to thermal neutrons, and some SRAM cells which were completely insensitive; these results are fab specific and not process specific. As the previously held industry belief was that all sub-micron SRAM cells should be insensitive to thermal neutrons, we encourage the industry to study the root cause so that this sensitivity can be mitigated. We are going to continue to perform this testing on DRAM, eDRAM and Flip-Flops, these results may be presented in the final paper.

Acknowledgements

We would like to acknowledge the MNRC beam facility's support, the technical discussions with Hungyuan B. Lui of MNRC; Bharat Bhuva, Robert Reed, Robert Weller, Brian Sierawski, and Arthur Witulski of Vanderbilt University.

REFERENCES

[1] R. Baumann, et al, "Neutron-induced boron fission as a major source of soft errors in deep submicron SRAM devices". IRPS 2000 Proceeding, pp. 152-157.

[2] H. Kobayashi et al., "Soft errors in SRAM devices induced by high energy neutrons, thermal neutrons and alpha particles", in *Proceedings of International Electron Devices Meeting*, 2002, pp. 337-340.

[3] J. F. Ziegler, "Terrestrial cosmic rays", in *IBM J. Res. Develop.*, 1996, vol.40 , pp. 19–40.

[4] J. M. Armani, G. Simon, P. Poirot, "Low-energy neutron sensitivity of recent generation SRAMs", in *IEEE Transactions on Nuclear Science*, 2004, vol. 51, pp. 2811-2816.

[5] E Normand, K Vranish, A Sheets, M Stitt, R Kim, "Quantifying the Double-Sided Neutron SEU Threat, From Low Energy (Thermal) and High Energy (>10 MeV) Neutrons," IEEE Transactions on Nuclear Science, Vol 53, No. 6, Dec 2006, pp. 3587-3594.

[6] M Olmos et al. "Investigation of Thermal Neutron Induced Soft Error Rates in Commercial SRAMs with 0.35 μm to 90 nm Technologies" IRPS 2006 Proceeding.

[7] "Atomic Weights and Isotopic Compositions for All Elements". National Institute of Standards and Technology. Retrieved 2009-12-08

[8] Hughes, D.J., Schwartz, R.D, *Neutron Cross Sections,*. BNL-325, 2nd ed., BNL, 1958

[9] Alfassi, Zeev B., Chung, Chien, *Prompt Gamma Neutron Activation Analysis*. CRC Press 1995

978-1-4244-5430-3/10 $26.00 © 2010 IEEE

Evaluation of Self-Heating and Hot Carrier Degradation of Poly-Si Thin-Film Transistors using Charge Pumping technique

Xiaowei LU, *Mingxiang WANG, Kai SUN, and Lei LU

Dept. of Microelectronics, Soochow University, No.1 Shizi Street, Suzhou, 215006, China
*Email: Mingxiang_wang@suda.edu.cn

Abstract—Self-heating (SH) and hot carrier (HC) degradation of n-type poly-Si thin-film transistors (TFTs) is evaluated by using charge pumping (CP) technique. By extracting trap state energy distribution, it is demonstrated that SH degradation is mainly attributed to the generation of deep states. For HC stressed TFTs, an anomalous I_{CP} decrease with the stress time is observed in a low V_g stress condition controlled by hole trapping; while in a mid V_g condition, CP signal clearly indicates the trap states generation controlled by electron trapping.

Keywords: charge pumping, poly-Si TFTs, self-heating, hot carrier

I. INTRODUCTION

Charge pumping (CP) is a technique providing direct information on interface trap properties [1], and has been extensively employed to investigate device degradation in MOSFETs [2]. While in poly-Si thin-film transistors (TFTs), there are very limited applications of CP for hot carrier (HC) or NBTI degradation [3, 4]. Recently, we showed that measurement optimization was critical to achieve reliable CP characterization in poly-Si TFTs [5]. In this work, such optimized CP technique is employed to evaluate their self-heating (SH) and HC degradation. For SH degradation, trap states generation mainly on the deep states is observed. While for HC degradation, different CP characteristics are observed in low and mid V_g stress conditions. An abnormal I_{CP} decrease with the stress time is observed in low V_g stressed TFTs, which is different from previous observations [3], whereas in mid V_g stressed TFTs, CP characterization clearly indicates the electron trapping dominated trap states generation.

II. EXPERIMENTAL

N-type TFT with a p$^+$ side contact connecting the poly-Si body is used in this study. As shown in Fig.1a is a planar view of the device under test. The 100 nm poly-Si active layer was formed by solution based metal-induced crystallization of a-Si. Wafers were first dipped in nickel nitrate solution, and then crystallized at 630 °C in N_2 ambient. Subsequently the poly-Si film was recrystallized at 900 °C. Gate oxide was formed by polyoxidation at 950 °C in dry O_2 for 48 mins. Phosphorous implantation was introduced to form the source and drain with a dose of 4×10^{15} cm^{-2}, then followed by a dopant activation at 900 °C for 1.5 h. Device W/L for SH stress test is 15/4 μm while for HC stress is 10/4 or 20/4 μm. For SH stress,

V_g/V_d=12/14 V is applied with a stress power about 47 mW. For HC stress, low V_g and mid V_g stress are applied with bias stress of V_g/V_d =2/10 V and 6/12 V, respectively.

Fig.1b is a cross sectional view of CP measurement. All stress test and CP measurement are performed at room temperature. Before and after a HC or SH stress, the TFT is pulsed from accumulation to inversion using a square V_g pulse with the source (S) and drain (D) grounded. By adjusting the pulse base voltage (V_{gb}) while keeping a constant pulse height (V_{ph}), an Elliot curve is measured from the p$^+$ side contact [1, 5] for CP analysis. In the CP measurement, the pulse period (T_p) is 100 μs and V_{ph} is fixed at 5 V. The pulse rising and falling time (T_r, T_f) are chosen as 5 μs, long enough to minimize the geometric effect [5]. Thus reliable and well shaped CP curves can be obtained. Besides, transfer curves are also measured before and after the degradation by using Agilent 4156C analyzer and Vector MX-1100B prober.

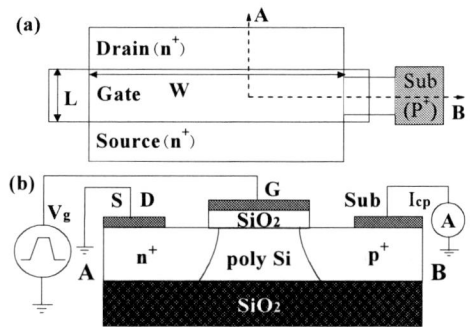

Figure 1. (a). Planar view of the high temperature processed TFT under test. (b). Cross sectional view of CP measurement.

III. RESULTS AND DISCUSSIONS

SH Degradation

Shown in Fig.2 is typical transfer curve degradation of a SH stressed TFT [6]. The subthreshold slope degrades slightly, resulting in a positive shift of device V_{th}. On current (I_{on}) decreases while leakage current (I_{off}) increases. In Fig.3 the Elliot curves measured at different stress times are shown. Clearly, CP current (I_{CP}) significantly increases with stress time, indicating the generation of interface traps. However, almost no shift in the Elliot curves is observed at either transition edge. Thus there should be no net charge generated in the gate oxide during the stress.

978-1-4244-5430-3/10 $26.00 © 2010 IEEE

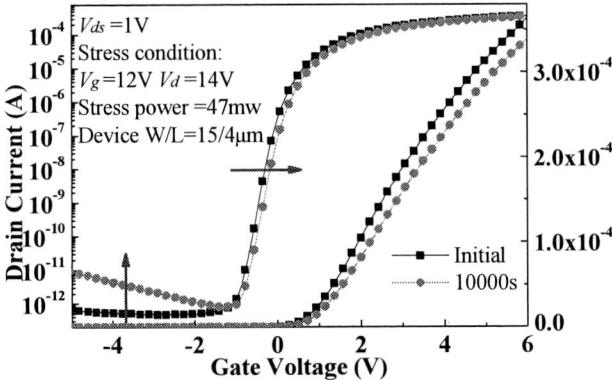

Figure 2. Linear and logarithmic plots of the typical transfer curves under SH stress, V_g/V_d=12/14 V, with a stress power of about 47 mw.

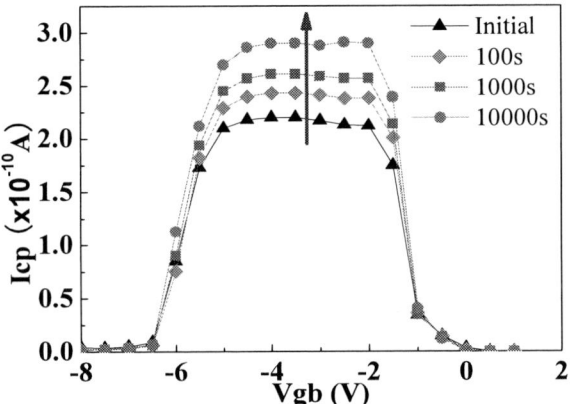

Figure 3. Charge pumping characteristics as a fucntion of pulse base voltage at various stress times. Distinct increase of I_{CP} is observed.

In Fig.4 inset, the maximum I_{CP} linearly increases with the measurement pulse frequency f, from which average trap densities (D_t) are extracted for all stress times using the equation [1] as below:

$$I_{cp} = 2D_t[\ln(V_{th}n_i\sqrt{\sigma_n\sigma_p}) + \ln(\frac{|V_{fb} - V_t|}{V_{ph}}\sqrt{T_rT_f})]kTA_gqf \quad (1)$$

The capture cross section $(\sigma_n\sigma_p)^{1/2}$ is set as 2.4×10^{-16} cm^2 [5]. Shown in Fig.4 is the extracted D_t as a function of stress time. In Fig.5 normalized degradation curves for key device parameters as well as D_t are compared. All curves follow a power law dependence on the stress time with about the same time exponent of 0.21, showing the correlation between the device degradation and trap states generation.

Furthermore, by measuring I_{CP} with fixed T_r and variable T_f, trap energy distribution $D_t(E)$ within the upper half of the band gap for each stress time can be extracted using the equation [1]:

$$D_t(E) = -\frac{T_f}{qfA_gkT}\frac{dI_{cp}}{dT_f} \quad (2)$$

In Fig.6 both deep states and tail states are seen to steadily increase with stress time, however, larger increase is observed near the mid-gap than that near the band edge. Deep states are normally attributed to dangling bonds due to bond breaking,

while tail states to strained bonds [6]. Thus the observation here is a direct evidence to the SH trap generation mechanism related to Si-H or Si-Si breaking at the Si-SiO$_2$ interface or grain boundaries. In previous studies [7, 8], the same mechanism was proposed based on simulation or indirect observations, to which such direct CP evidence can be a better support. On the other hand, it also demonstrates the feasibility of the CP technique in probing the trap states generation in SH degraded poly-Si TFTs.

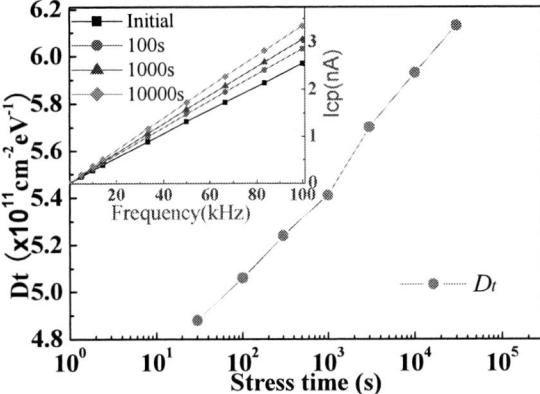

Figure 4. Stress time dependence of trap state density (D_t). The inset is frequency dependence of CP current, where average D_t can be extracted by the given equation (1).

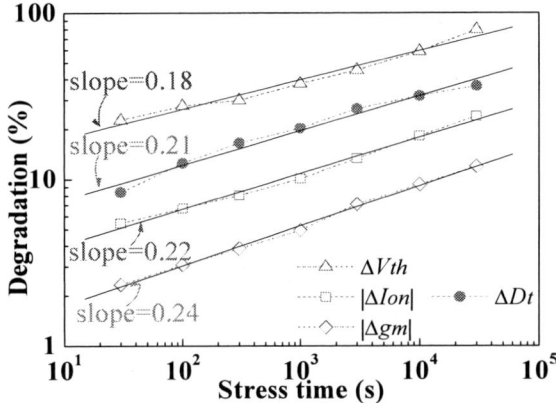

Figure 5. Stress time dependence of key device parameters degradation. D_t and device parameters follow a power law with about the same time exponent of 0.21.

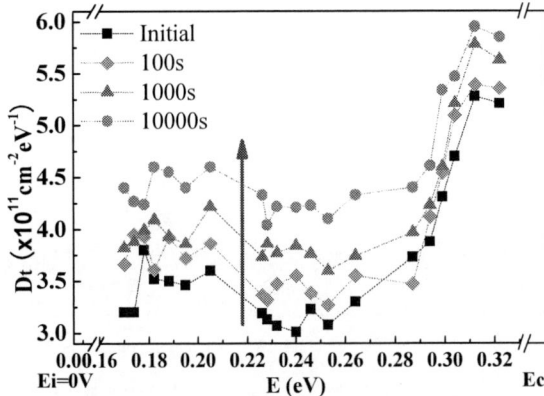

Figure 6. Trap state energy distribution $D_t(E)$ is extracted by keeping pulse T_r constant and varying T_f. SH degradation is mainly attributed to the generation of deep states.

HC Degradation

Shown in Fig.7 is transfer curve degradation of a low V_g HC stressed TFT. The subthreshold region continuously shifts to the negative and degrades severely. I_{off} increases distinctly, whereas negligible change occurs in I_{on}. Similar transfer curve degradation was also seen in other high temperature processed TFTs under similar low V_g HC stress condition [9]. Shown in Fig.8 are CP curves similarly measured by optimized CP technique as in SH degradation. The right edge of the Elliot curve clearly shifts to the negative, in accordance with the negative shift of the subthreshold region in Fig.7, indicating the generation of positive charges in the gate oxide [2]. Indeed, hot hole injection into the gate oxide is favored under a low V_g HC stress condition [2]. Surprisingly, I_{CP} clearly *decreases* with the stress times. It's a different observation from previous CP result also in low V_g HC stressed TFTs [3]. In order to verify such observation, pulse T_r and T_f is varied from 100 ns to 500 µs, all measured I_{CP} similarly decreases, eliminating the possibility of CP measurement error.

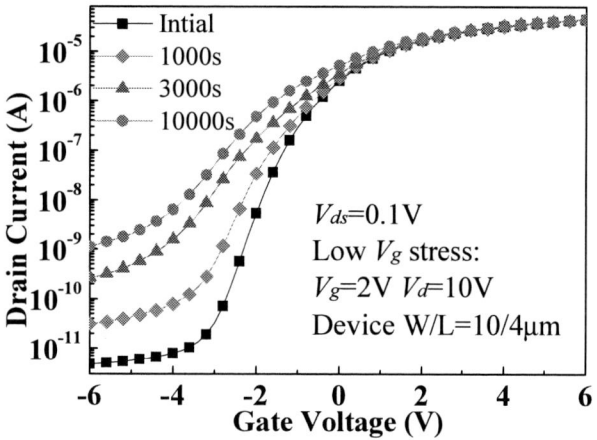

Figure 7. Stress time dependence of transfer curves under low V_g HC stress, V_g/V_d=2/10 V. Device W/L=10/4 µm.

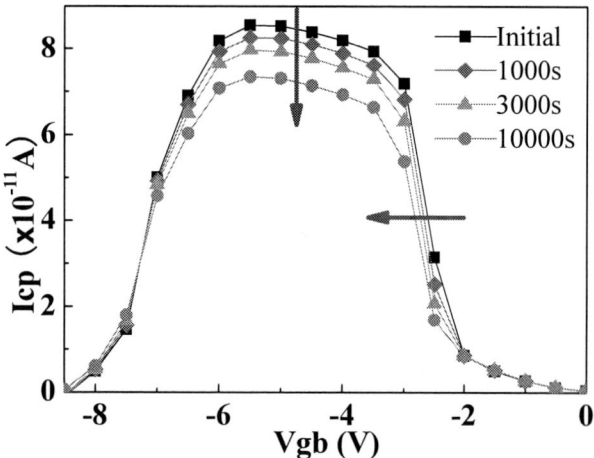

Figure 8. Charge pumping current versus pulse base voltage of the device under a low V_g HC stress.

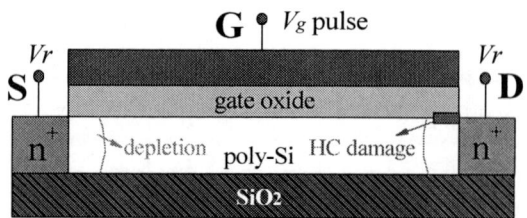

Figure 9. Schematic cross-sectional view of device in CP measurement. HC induced damage region can be masked by drain depletion region due to applied reverse bias V_r.

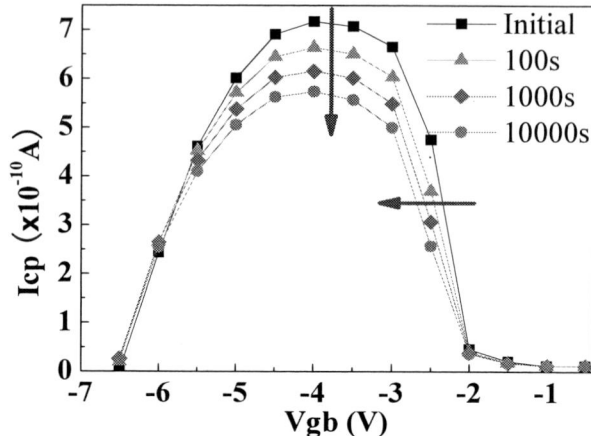

Figure 10. Charge pumping characteristics of device under a low V_g HC stress measured with the source and drain reversely biased at V_r=1 V.

Obviously, a conclusion of interface trap reduction based on the observed I_{CP} decrease can not be true. Thus one may consider that the effective channel length shortening due to HC injection into the gate oxide [2] may cause an I_{CP} reduction. To clarify this point, CP curve is measured with the source and drain reversely biased at V_r=1V, as shown in Fig.9. In such condition, HC induced damaged region [6] or HC injection induced local inversion region [2] at the drain end can be masked by a wider drain depletion region. Therefore, only the central channel region is probed by CP. As expected, such measured I_{CP} is reduced compared to that with S/D grounded, which is attributed to the reduction of effective gate area (A_g) as given in Eq.(1). However, the continuous I_{CP} decrease with the stress time still remains as shown in Fig.10. Thus we tend to believe that conventional I_{CP} measurement is not sensitive to the HC trap states generation dominated by hole trapping mechanism [10].

In a mid V_g HC stress condition, interface traps are generated through both electron and hole trapping [2]. Fig.11 is transfer curve degradation of a TFT stressed at V_g/V_d=6/12 V. In the inset, a positive shift of the right edge of the Elliot curve indicates the generation of net negative charges due to more electron trapping. Interestingly, here I_{CP} clearly *increases* with stress time, in agreement with the HC induced trap generation. Therefore, in both cases of HC degradation of poly-TFTs, shift of CP Elliot curve is a good indication of the respective trap generation mechanism, while one should be tentative when quantitatively evaluating trap generation using the I_{CP} variation. However, further investigation is needed to clarify the origin of the anomalous CP decrease under a low V_g HC stress.

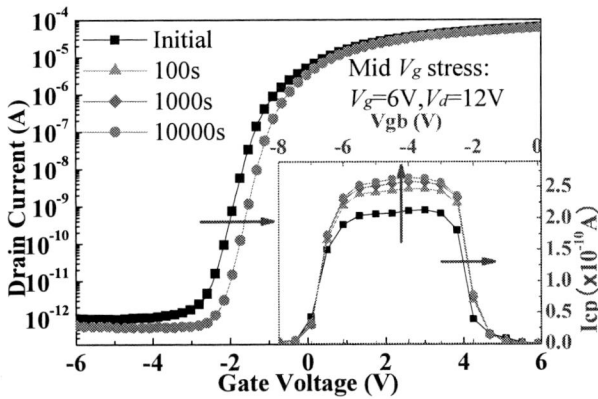

Figure 11. Transfer curve degradation of device under a mid V_g HC stress, V_g/V_d=6/12 V. Inset is degradation of CP characteristics, indicating the electron trapping dominated trap states generation. Device W/L=20/4 μm.

IV. CONCLUSION

The optimized CP technique was employed to evaluate the self-heating and hot-carrier degradation in poly-Si TFTs. SH degradation is mainly attributed to the generation of deep states. For HC degradation, CP exhibits different behaviors associated with different trap generation mechanisms. In the low V_g stress condition, the anomalous I_{CP} decrease is in contradiction to the HC trap states generation. While for the mid V_g HC stress induced degradation, CP characterization clearly indicates the electron trapping dominated trap states generation.

REFERENCES

[1] G. Groeseneken, H. E. Maes, N. Beltran, and R. F. De Keersmaecker, "A reliable approach to charge-pumping measurements in MOS transistors," *IEEE Trans. Electron Devices*, vol ED-31, NO. 1, pp. 42-53, Jan. 1984.

[2] Paul Heremans, Rudi Bellens, Guido Groeseneken, and Herman E. Maes, "Consistent Model for the Hot-Carrier Degradation in n-Channel and p-Channel MOSFET's," *IEEE Trans. Electron Devices*, vol. 35, NO. 12, December, 1988.

[3] T. Yoshida, K. Yoshino, M. Takei, A. Hara, N. Sasaki, and T. Tsuchiya "Experimental Evidence of Grain-Boundary Related Hot-Carrier Degradation Mechanism in Low-Temperature Poly-Si Thin-Film-Transistors," *IEEE IEDM*, 2003.

[4] Chih-Yang Chen, Ming-Wen Ma, Wei-Cheng Chen, Hsiao-Yi Lin, Kuan-Lin Yeh, Shen-De Wang, and Tan-Fu Lei, "Analysis of Negative Bias Temperature Instability in Body-Tied Low-Tempreature Polycrystalline Silicon Thin-Film Transistors," *IEEE Electron Device Letters*, vol. 29, NO. 2, February 2008.

[5] Lei Lu, Mingxiang Wang, and Man Wong, "Geometric Effect Elimination and Reliable Trap State Density Extraction in Charge Pumping of Polysilicon Thin-Film Transistors," *IEEE Electron Device Letters*, vol. 30, NO. 5, May 2009.

[6] N.A. Hatas, A. Archontas, C.A. Dimitriadis, G. Kamarinos, T. Nikolaidis, N. Georgoulas, and A. Thanailakis, "Substrate current and degradation of n-channel polycrystalline silicon thin film transistors," *Microelectronics Reliability*, vol 45, 2005, pp. 341-348.

[7] Huaisheng Wang, Mingxiang Wang, Zhenyu Yang, Han Hao, and Man Wong, "Stress Power Dependent Self-Heating Degradation of Metal-Induced Laterally Crystallized n-Type Polycrystalline Silicon Thin-Film Transistors," *IEEE Trans. Electron Devices*, vol.54, NO. 12, December 2007.

[8] Satoshi Inoue, Hiroyuki Ohshima, and Tatsuya Shimoda, "Analysis of Degradation Phenomenom Caused by Self-Heating in Low-Temperature-Processed Polycrystalline Silicon Thin Film Transistors," *Jpn. J. Appl. Phys.* Vol. 41, 2002, pp. 6313-6319.

[9] F.V. Farmakis, C.A. Dimitriadis, J. Brini, G. Kamarinos, V. K. Gueorguiev and Tz. E. Ivanov, "Hot-carrier phenomena in high temperature processed undoped-hydrogenated n-channel polysilicon thin film transistors," *Solid-State Electronics* (1999) 1259-1266.

[10] D. J. DiMaria, D. A. Buchanan, J. H. Stathis, and R. E. Stahlbush, "Interface states induced by the presence of trapped holes near the silicon-silicon-dioxide interface," *Journal of Applied Physics*, vol 77, 1995, pp. 2032-2040.

PBTI response to Interfacial Layer Thickness variation in Hf-based HKMG nFETs

D.P. Ioannou, E. Cartier[2], Y. Wang[3], and S. Mittl

IBM Microelectronics, Semiconductor R&D Center, Essex Junction, VT, USA [2] IBM T.J. Watson Research Center, Yorktown Heights, NY, USA, [3] IBM Microelectronics, Semiconductor R&D Center, Hopewell Junction, NY USA

ABSTRACT

The impact of SiO$_2$ interfacial layer (IL) thickness on the Positive Bias Temperature Instability (PBTI) is investigated for nMOSFETs with an IL/High-K/metal/poly-Si gate stack architecture. Results from extensive PBTI measurements using three different measurement methodologies consistently demonstrate that thickening the IL results in threshold voltage (V_T) instability reduction and thus significantly enhances PBTI device lifetime. The voltage acceleration is found to increase with thicker IL, while the PBTI fractional recovery is independent of the IL thickness, providing new insights into the PBTI buildup and recovery mechanisms.

Keywords- High-k dielectrics, HfO$_2$, metal gate, PBTI, interface layer, PBTI recovery

INTRODUCTION

High-k gate stacks must have an SiO$_2$ based dielectric interface (IL) between the high-k and the channel to maintain high channel mobility [1]. This IL significantly decreases the total effective capacitance which increases inversion thickness (T_{inv}) and impedes gate length scaling. Performance optimization generally forces the IL to be as thin as possible. However, the IL does not only impact device performance but it has been reported to also impact threshold voltage instability in Metal Gate/High-k (MGHK) nMOSFETs [2-4]. Young *et al.* [2] reported enhanced electron trapping for gate stacks with thinner IL and they attributed this effect to the corresponding reduction of the critical time for electron capture (assuming electron tunneling from the inversion layer through the IL into the trapping sites in the high-k layer). According to Pantisano *et al.* [3], the effectiveness of the IL to modulate the V_T instability depends on the mechanism of electron trapping at work which in turn depends on the actual IL thickness; for a thickness of ~ 2 nm capture of HfO$_2$ conduction band electrons is responsible for electron trapping (the instability is not expected to be sensitive to IL thickness variation) whereas for thinner IL (~1 nm), a direct trap filling by electron tunneling appears to control the instability.

In this paper, we present a systematic study on the PBTI response to IL thickness scaling by various interlayer processes. A wide range of stress bias conditions including those closer to the transistor operational voltage were studied using a variety of measurement techniques. We consistently find that the V_T instability is improved when thickening the IL and that this improvement is more significant for the lower stress bias conditions. Furthermore, the

equally important PBTI recovery dependence on IL thickness is investigated in this work. It is found that PBTI fractional recovery [5], is independent of the IL thickness providing new insights into the net PBTI (including both charge build up and recovery) dependence on IL thickness.

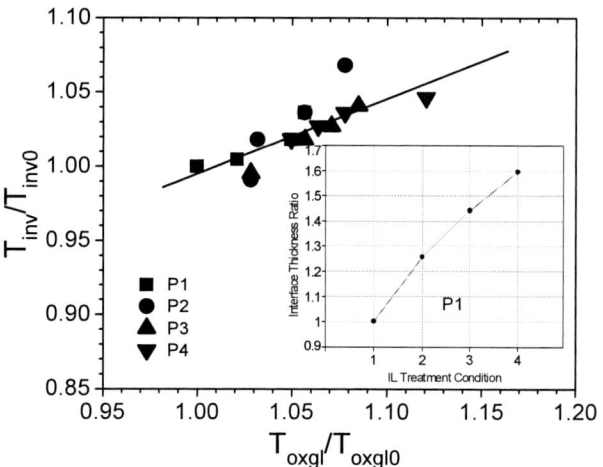

Fig. 1 Comparison of the normalized inversion thickness (Tinv), versus the normalized gate leakage (Toxgl), for the 4 base processes. The inset shows the relative IL thickness increase with four IL treatment conditions for process P1.

EXPERIMENTAL DETAILS

A. Devices

Planar CMOS devices were fabricated with identical metal gates and a single high-k dielectric thickness. Four different base processes (P1-P4) were used to create IL of varying thickness. As shown in Fig. 1, the gate leakage can be substantially reduced by IL processing with only a small T_{inv} penalty. The change of the physical IL thickness with IL treatment conditions is illustrated by the inset in Fig. 1, demonstrating that the gate leakage reduction is caused by a physical increase in the IL thickness.

B. Measurements

To study the impact of the IL thickening on PBTI, three different measurement techniques were used; 1) a recently proposed Voltage Ramp Stress (VRS) methodology [6], 2) a fast Constant Voltage Stress (CVS) [7] and 3) an uninterrupted CVS method (proposed here) featuring uninterrupted stresses at various stress times (t_{stress}) followed by a

978-1-4244-5430-3/10 $26.00 © 2010 IEEE

prolonged recovery monitor phase of fixed duration. Fig. 2 is a schematic illustration of the voltage time traces for the

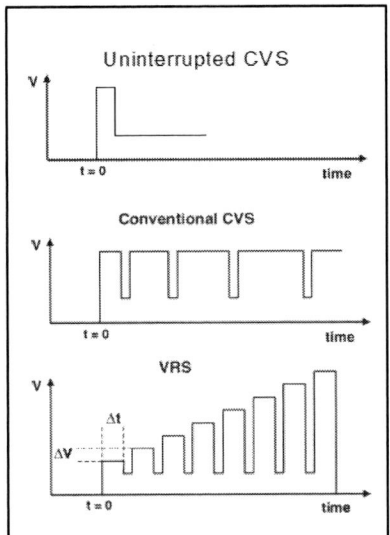

Fig. 2 Schematics, describing the gate voltage – time traces for VRS (for process screening and voltage acceleration, CVS (for time and voltage acceleration, and uninterrupted CVS (for PBTI relaxation) used in this study.

VRS (lower panel), the fast CVS (middle panel) and the uninterrupted CVS procedures (upper panel). For the VRS, the voltage is stepped up in small voltage intervals ΔV, and each voltage condition is held for a time Δt. Before each step, the drain current is measured at a lower voltage, called the sense voltage. The threshold voltage shift can then be calculated from the drain current degradation. An important parameter for this technique is the ramp rate RR which is defined as $\Delta V/\Delta t$. In [6] it was shown that the threshold voltage shift, (ΔV_T), corresponding to a fixed gate voltage (Vg) and as measured by the VRS method is defined by Eq. (1). The parameters m and n are the voltage and time power exponents extracted from a CVS experiment with ΔV_T parameterized according to Eq. (2).

$$\Delta V_T (Vg) = C(RR) \times Vg^{m+n} \qquad (1)$$

$$\Delta V_T (CVS) = A \times Vg^m \times t^n \qquad (2)$$

For the fast CVS method, a constant stress voltage is applied to the gate and the stress is periodically interrupted for a fast measurement at the sense voltage similarly to the VRS technique. Finally, for the uninterrupted CVS method, the measurement at the sense voltage occurs only upon stress completion. In this case, the V_T shift time dependency is obtained by subjecting different groups of devices to uninterrupted bias stresses of different time durations. The minimum delay between the end of the stress phase and the beginning of the sense measurement for all CVS measurements is 280 µs. All measurements are performed at T=125 ^{0}C.

RESULTS & DISCUSSION

A. Voltage Ramp Stress Results

In Fig. 3, the results of the VRS tests are summarized. Fig. 3a) shows VRS data for IL process, P1. As, can be seen, the power exponent, n+m (slope) [6], increases with increasing IL thickness. This trend applies to all IL processes, as summarized in Fig. 3b).

The voltage acceleration, m, systematically increases with IL thickness (the value of n is small) and a substantial PBTI reduction at low voltages is observed for thicker IL, as shown in Fig. 3c).

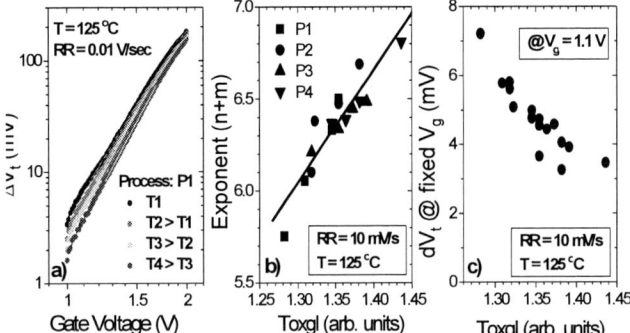

Fig. 3 VRS stress data: a) VRS versus V_g traces at a ramp rate, RR = 0.01 V/sec for four IL with different thickness, T_i, for process P1. b) Comparison of the slope in the VRS data versus gate leakage for 16 ILs. c) Reduction of PBTI instability at low stress voltages versus IL thickness.

B. Fast CVS vs. uninterrupted CVS methods comparison

To further investigate the PBTI scaling trends observed with the VRS method, extensive CVS PBTI measurements were performed. Prior to applying the new uninterrupted CVS method for the IL thickness study, we first performed a series of experiments to carefully compare various CVS implementations to gain insight into the impact of subtle differences in measurement procedures on the PBTI results. Fig. 4 is a schematic illustration of the voltage time traces and the sense conditions used. The upper panel shows the fast CVS sequence corresponding to a sense time durations between stress intervals of 1 ms, while the bottom panel shows a sense interval of 100 ms. The three different cases labeled "case a", "case b", and "case c" in Fig. 4 have the following characteristics: In case a, the sense time duration is 1 ms and the threshold voltage is measured at 280 µs. In cases b and c, the sense time duration is 100 ms, however, in case b, V_T is measured at the end of the sense phase whereas in case c it is measured in the beginning, at 280 µs. The V_T shift vs. stress time characteristics obtained by these three variants of the CVS method are shown in Fig. 5 where they are also compared to the characteristics corresponding to the uninterrupted CVS method (measurement schematics are shown in the upper panel of Fig. 2). There are a number of interesting observations. First, it is seen that the fast CVS (case a) and the uninterrupted CVS methods yield equal PBTI (the measured V_T shifts overlay on top of each other). This result suggests that the multiple stress interruptions occurring in the fast CVS method do not trigger any additional V_T relaxation as long as the V_T is measured at equal delay time (280 μs in our case). A typical time-power exponent value (for fast V_T measurements) of ~0.16 is obtained from both methods.

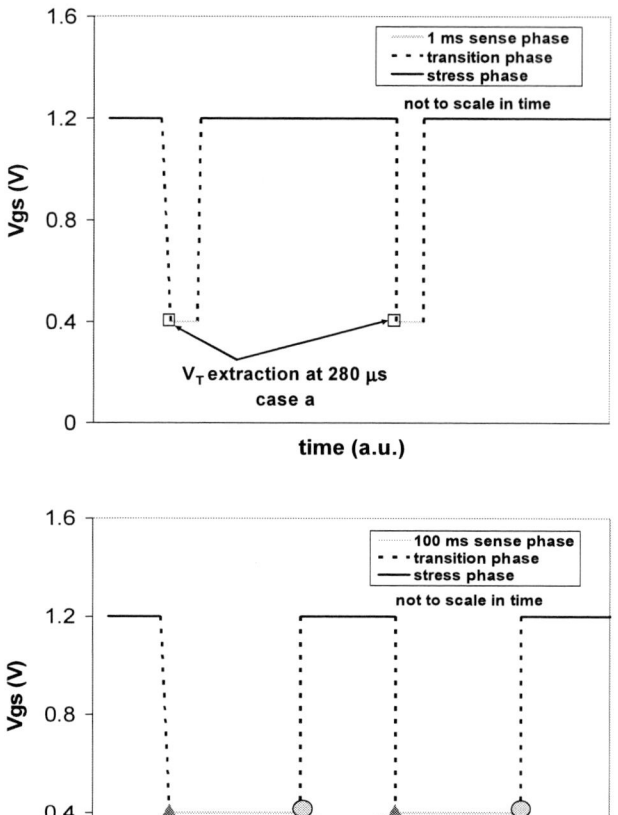

Fig. 4 Schematics, describing the gate voltage – time traces for three variants of the fast CVS method: utilizing a short (1 ms) sense phase between stress time intervals (upper panel) and a longer one (100 ms) (lower panel). For the implementation with the long sense phase, (lower panel), two distinct cases are illustrated: in case b, the V_T is measured at the end of the sense phase whereas in case c it is measured in the beginning at 280 μs.

By contrast, significantly enhanced relaxation is observed with the fast CVS (circular symbols in Fig. 5) for which the V_T is measured at the end of a 100 ms sense phase (case b) and as expected, it also manifests itself in a higher time-power exponent value close to 0.19. The most intriguing observation though from Fig. 5 is that the ΔV_T vs. stress time characteristics corresponding to the fast CVS measurement (case c) are very close to the characteristics measured with the uninterrupted method as well as to the fast CVS method utilizing the short sense phase (case a). This result suggests that most of the electron traps are being repopulated during the subsequent stress phase despite the significant V_T relaxation during the longer sense phase.

Additional work is required to better understand the nature of these trapping-detrapping dynamics and assess its possible implications on the AC stress induced device instabilities more relevant to circuit operation. Finally, although we find that the fast CVS and the uninterrupted CVS methods are equivalent in terms of the ΔV_T vs. stress time measurement results, there is a significant advantage for the uninterrupted method because it also provides a large database associated with the PBTI recovery behavior in addition to the V_T drifts during stress. This important aspect of the uninterrupted CVS method is utilized to study the PBTI recovery as a function of the IL thickness in section D.

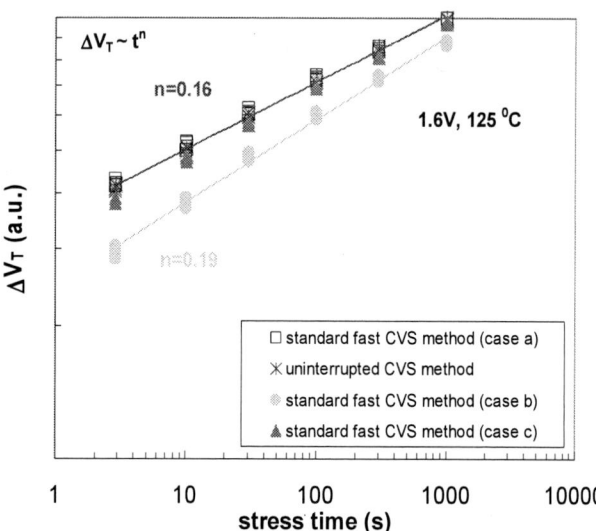

Fig. 5 ΔV_T vs. stress time characteristics measured with the fast CVS and the uninterrupted CVS methods. Measurement details for the variants of fast CVS (cases a, b and c) and the uninterrupted CVS methods are shown in Fig. 4 and Fig. 2 respectively. It is seen that the two methods, fast CVS (case a) and uninterrupted CVS yield the same results when V_T is measured at minimum delay time (280 μs for our set up).

C. V_T instability vs. IL thickness (fast CVS method)

Fig. 6 shows the threshold voltage shift vs. stress time characteristics for the four IL splits from process P1 as measured by the fast CVS method under gate bias stress of 1.2V. It is seen that for all splits the time evolution of ΔV_T follows a time-power law with a time-power exponent value of ~ 0.16. It can also be seen that for increasing IL thickness the magnitude of the threshold voltage shifts reduce, consistent with VRS (Fig. 3a). This result is in agreement with the predictions from an electron trapping model which assumes filling of trap sites in the high-k layer via electron tunneling from the inversion layer through the interfacial layer [2]; a thinner interface results in higher tunneling current and hence larger electron trapping.

Fig. 7 illustrates the bias dependence of ΔV_T at a fixed stress time for the same splits. It should be noted that the rate of ΔV_T increase with stress time was found to be independent of the stress voltage (not shown here.) A power-law dependence for voltage acceleration ($\Delta V_T \sim V_g^m$) which

978-1-4244-5430-3/10 $26.00 © 2010 IEEE

is often used for lifetime projections seems to fit the data very well in all cases. The interesting observation though is that with increasing IL thickness there is a corresponding increase in the voltage acceleration factor, again consistent with the VRS data.

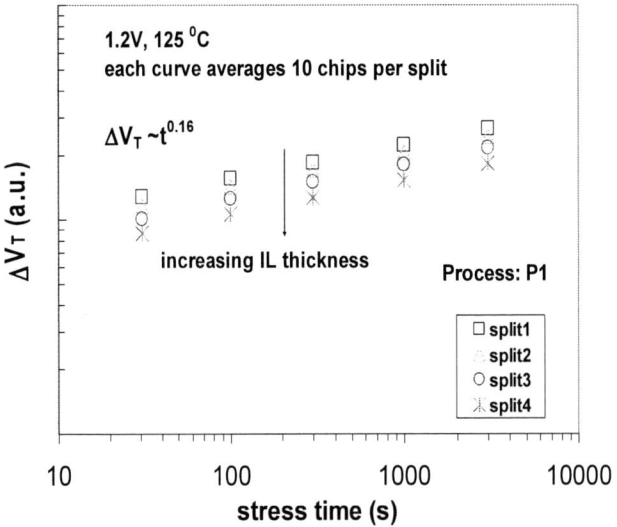

Fig. 6 ΔV_T vs. stress time characteristics corresponding to the four IL splits from process P1. For increasing IL thickness the magnitude of the threshold voltage shifts reduce, consistent with VRS results (Fig. 3a).

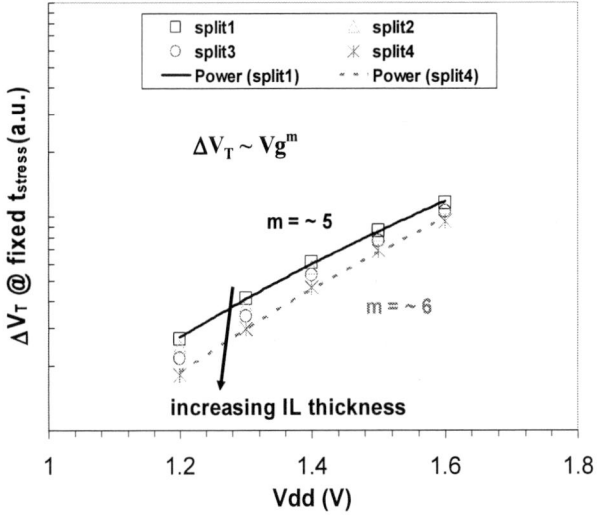

Fig. 7: Voltage acceleration factor increases with IL thickness.

This indicates that PBTI reduction through IL thickness scaling becomes progressively more effective with decreasing bias, resulting in a significant PBTI lifetime improvement at use condition. This observation is consistent with the assumption of a change of the predominant trapping mechanism from electron trap filling by the capture of HfO_2 conduction band electrons (under high bias levels / injected charge densities) to electron trap filling by tunneling from the channel to the traps (under lower bias levels / injected charge densities) [4].

D. PBTI fractional recovery vs. IL thickness (uninterrupted CVS method)

For a more comprehensive assessment of the IL thickness scaling effect on PBTI degradation, in addition to the V_T shift during stress, the ΔV_T recovery needs to be characterized. Fig. 8 illustrates the ΔV_T recovery vs. relaxation time characteristics (for one of the splits) following uninterrupted stresses at 1.2V of various durations. It is seen that the ΔV_T recovery for all stress times follows a log(t) dependence. This behavior has been previously observed and has pointed to the suggestion that PBTI recovery is associated with electron detrapping via back tunneling through the IL into the inversion channel [8]. However, we are presenting below some experimental data that can not easily be rationalized by this assumption.

To capture the net impact of stress and recovery on the PBTI combined, the recovery data shown in Fig. 8 is plotted in Fig. 9 as the Fraction Remaining (FR) vs. the normalized time, $t_{sress}/(t_{stress}+t_{relax})$. This time has the same mathematical form as the definition of the duty cycle fraction in an AC experiment. The fraction remaining, FR, has previously been used for the modeling of the NBTI recovery [5,9] and more recently it has been found to show good scalability for PBTI recovery as well [7,10]. This is also demonstrated by our data in Fig. 9, where the recovery data for all stress durations are compared. As can be seen, all data sets follow a universal curve irrespective of the stress time. It is seen that with decreasing duty cycle fraction, an initial sharp decrease in the fraction remaining is followed by a much weaker duty cycle dependence. A significant PBTI improvement is observed for lower duty cycle fraction values in agreement with the findings reported in [10].

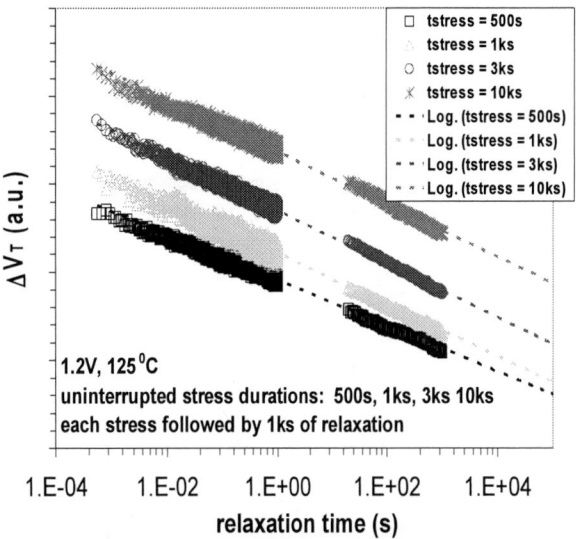

Fig. 8 ΔV_T recovery vs. relaxation time traces (for one of the splits) measured immediately after uninterrupted stresses at 1.2V of various durations. ΔV_T recovery for all stress times follows a log(t) dependence.

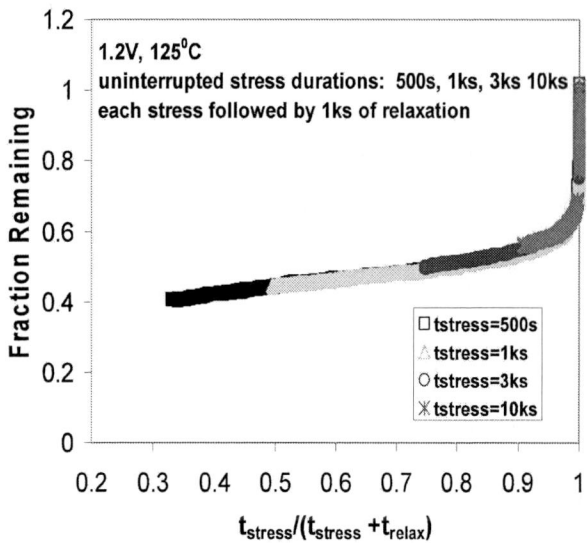

Fig. 9 Fraction remaining scaling with $t_{sress}/(t_{stress}+t_{relax})$ for data shown in Fig. 8. Very good scalability is observed with the recovery data from all stress durations forming a single trend.

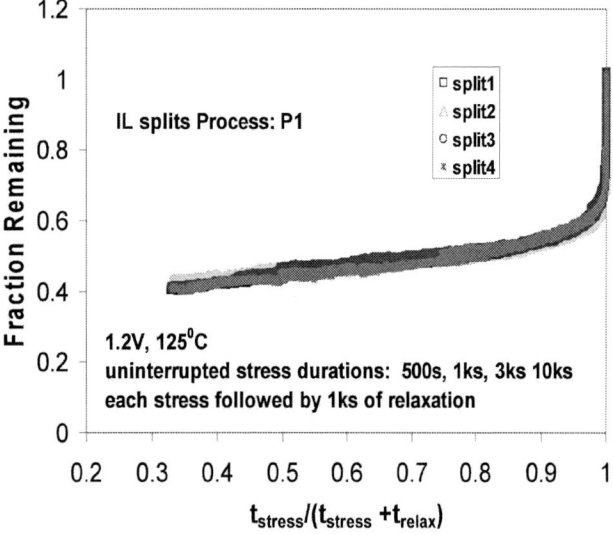

Fig. 10 Similar fraction remaining vs. duty cycle fraction characteristics are obtained for all P1 IL splits suggesting that PBTI recovery behavior is IL thickness independent.

Finally, in Fig. 10, the FR vs. $t_{sress}/(t_{stress}+t_{relax})$ characteristics corresponding to the devices from all P1 IL splits are compared. These data were obtained by uninterrupted CVS measurements under the same stress bias of 1.2V and for the same stress durations. Interestingly, it can be observed that the FR values are very similar for all splits. This IL thickness independent recovery behavior may suggest that PBTI recovery could originate from a mechanism other than electron detrapping via tunneling back to the channel

through the IL. If this was indeed the mechanism at work, one would expect that the recovery rate decreases exponentially with IL thickness. This is not supported by the data shown in Fig. 10. It should be noticed, however, that the process variations used to modulate the IL thickness may also impact the band gap and the band offsets of the IL with the substrate. Irrespective of these details, the practical significance of this result lays in the fact that it demonstrates that thickening the IL will result in reduced PBTI shifts during stress (Fig. 6) without degrading the recovery rate and hence resulting overall in an improved net PBTI.

ACKNOWLEDGMENT

This work was supported by the research alliance teams at various IBM research and development facilities. Dimitris Ioannou would like to thank D. Brochu for his electrical measurement work and D. Badami for management support.

REFERENCES

[1] K. Maitra, M. M. Frank, V. Narayanan, V Misra, and E. A. Cartier, "Impact of metal gates on remote phonon scattering in titanium nitride/hafnium dioxide n-channel metal–oxide–semiconductor field effect transistors–low temperature electron mobility study", Journal of Applied Physics, 102, December 2007.

[2] C. D. Young, R. Choi, J.H. Sim, B.H. Lee, P. Zeitzoff, Y. Zhao, K. Matthews, G.A. Brown, and G. Bersuker, "Interfacial layer dependence of $H_FSI_xO_Y$ Gate stacks on V_T Instability and charge trapping using Ultra-Short Pulse I-V Characterization, " International Reliability Physics Symposium, pp. 75-79, 2005.

[3] L. Pantisano, E. Cartier, A. Kerber, R. Degraeve, M. Lorenzini, M. Rosmeulen, G. Groeseneken, and H.E. Maes, "Dynamics of Threshold Voltage Instability in Stacked High-k Dielectrics : Rrole of Interfacial Oxide," Symp. VLSI Technol, pp. 163-164, 2003.

[4] C. Leroux, J. Mitard, G. Ghilbaudo, X. Garros, G. Reimbold, B. Guillaumot, and F. Martin, "Characteristics and modeling of hysteresis phenomena in high K dielectrics," IEDM Tech. Dig., pp. 737-740, 2004.

[5] T. Grasser, W. Gös, V. Sverdlov, and B. Kaczer, "The Universality of NBTI Relaxation And Its Implications For Modeling And Characterization," International Reliability Physics Symposium, pp. 268-280, 2007.

[6] A. Kerber, and E. Cartier, "Reliability Challenges for CMOS technology qualifications with Hafnium Oxide/Titanium Nitride Gate Stacks," IEEE Trans. Dev. Mat., vol. 9, no. 2, pp. 147-162, Jun. 2009.

[7] D.P. Ioannou, S. Mittl, and G. La Rosa, " Positive Bias Temperature Instability Effects in nMOSFETs with HfO_x/TiN Gate Stacks," IEEE Trans. Dev. Mat., vol. 9, no. 2, pp. 128-134, Jun. 2009.

[8] A. Kerber, K. Maitra, A. Majumdar, M. Hargrove, R.J. Carter, and E. Cartier, "Characterization of Fast Relaxation during BTI stress in Conventional and Advanced CMOS Devices with HfO2/TiN Gate Stacks, " IEEE Trans. Electron Devices, vol. 55, no. 11, pp. 3175-3183, Nov. 2008.

[9] D.P. Ioannou, D. Harmon, and W. Abadeer, "Investigation of Plasma Charging Damage Impact on Device and Gate dielctric Reliability in 180 nm SOI CMOS RF Switch Technology," International Reliability Physics Symposium, pp. 1011-1013, 2009.

[10] S. Ramey, C. Prasad, M. Agostinelli, S. Pae, S. Walstra, S. Gupta, and J. Hicks, , "Frequency and Recovery Effects in High-k BTI degradation, " International Reliability Physics Symposium, pp. 1023-1027, 2009.

HCI and NBTI including the Effect of Back-Biasing in Thin-BOX FD-SOI CMOSFETs

T. Ishigaki, R. Tsuchiya, Y. Morita, H. Yoshimoto, N. Sugii, and S. Kimura

Central Research Laboratory, Hitachi, Ltd.
1-280 Higashi-Koigakubo, Kokubunji, Tokyo 185-8601, Japan
phone: +81-42-323-1111, e-mail: takashi.ishigaki.ug@hitachi.com

Abstract—Hot carrier injection (HCI) and negative bias temperature instability (NBTI) of fully depleted silicon-on-insulator (FD-SOI) CMOSFETs with thin-buried oxide (BOX) were investigated for the first time. A comparison with conventional bulk devices showed that no halo implant in this structure produces better reliability. The impact of back-biasing in thin-BOX FD-SOI devices on reliability is also reported.

Keywords-HCI, NBTI, FD-SOI, thin-BOX, back-bias

I. INTRODUCTION

Increasing channel dopant concentration of conventional bulk CMOSFETs against a short-channel effect raises the threshold voltage (V_{th}) variation, which significantly increases power consumption in LSIs. We have proposed and developed FD-SOI CMOSFETs with an ultrathin BOX, called silicon on thin BOX (SOTB) as a solution (Fig. 1) [1-4]. In addition to its superior scalability, another significant feature is the back-bias controllability through the thin BOX ($t_{BOX} \sim 10$ nm), whose flexibly optimizes the performance without increasing standby leakage currents.

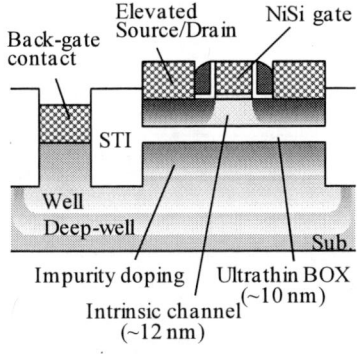

1. Strong SCE immunity due to FD-SOI structure
2. Small V_{th} variation due to intrinsic channel
3. Multiple V_{th} by adjusting substrate doping
4. Variation reduction and power optimization by V_b
5. Integrated I/O bulk trs. by removing BOX

Figure 1. Schematic cross section of SOTB CMOSFET and its features.

While some intrinsic problems of an FD-SOI structure such as low breakdown voltage are well known [5], the reliability of this structure is not fully understood. In this study, the HCI and NBTI of SOTB CMOSFETs, including its back-biasing influences, were examined.

II. DEVICE FABRICATION

SOTB CMOSFETs were fabricated based on a 65-nm low standby-power technology combined with additional raised source/drain (S/D) and fully silicided (FUSI) metal-gate processes [3]. The thin SOI channel ($t_{SOI} \sim 12$ nm) was very lightly ($\sim 10^{17}$ cm^{-3}) doped without a halo implant. Conventional Poly-Si gate bulk devices with both a doped channel and halo implant were also fabricated for comparison. The gate dielectrics in all devices were SiON with 1.9-nm EOT.

III. RESULTS AND DISCUSSION

A. Hot carrier injection in SOTB NMOSFETs

The device characteristics of $L_g/W_g = 55$ nm/8 μm, such as V_{th}, I_{on} and I_{off}, were nearly identical at $V_{dd} = 1.2$ V across the samples in this study. Figure 2 compares I_d - V_d characteristics of the SOTB and the bulk NMOSFETs. The SOTB device has comparable driving performance because the raised S/D structure makes the S/D resistances sufficiently low [2]. Neither kink nor self-heating effects were observed because of the FD operation due to thin SOI and thin BOX, respectively. The drain breakdown voltage of the SOTB device is about 1-V lower than that of the bulk device because non-destructive breakdown occurs due to the parasitic bipolar action at the source region.

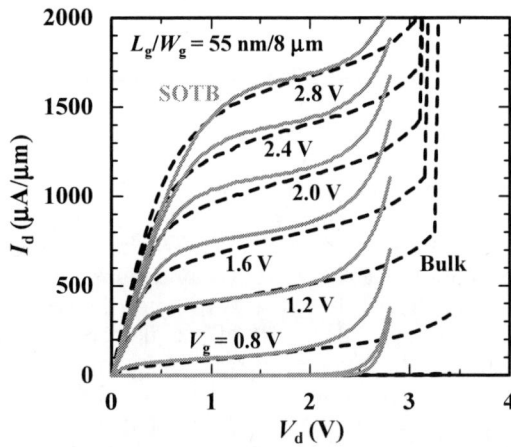

Figure 2. I_d-V_d characteristics of SOTB and bulk NMOSFETs.

978-1-4244-5430-3/10 $26.00 © 2010 IEEE

The dependence of the V_{th} shift and I_{dsat} degradation on V_g under HCI stress and the dependence of the V_{th} shift on stress time are compared in Fig. 3 and 4. The worst case degradation at $V_g = V_d$ (Fig. 3) and the power-law slopes of about 0.5 (Fig. 4) indicate that the interface trap generation due to the channel hot electrons is the dominant mechanism in both SOTB and bulk devices.

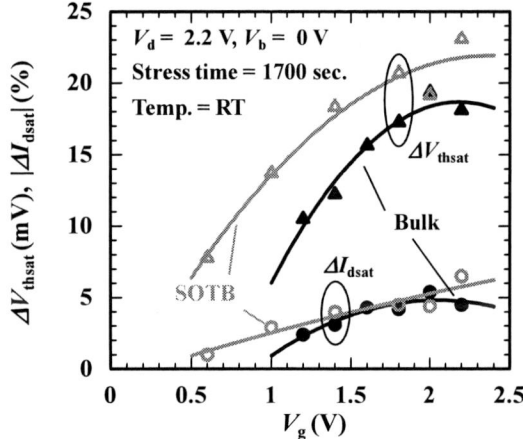

Figure 3.   Comparison of dependencies of HCI degradation on V_g.

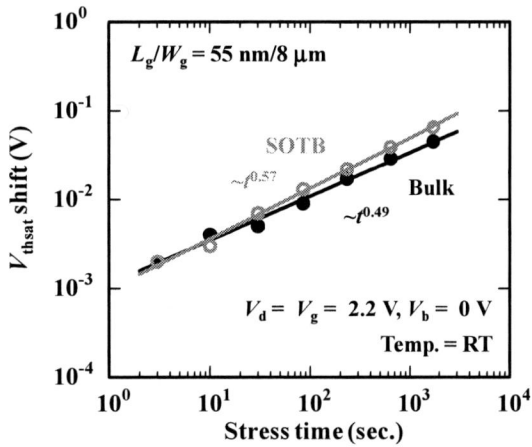

Figure 4.   HCI characteristics of SOTB and bulk NMOSFETs.

The SOTB lifetime dependence on V_{stress} produces a longer lifetime at the operating voltage (Fig. 5). Figure 6 shows the simulated lateral electric field strength E_x along the channel at $V_{stress} = 1.2$ and 2.4 V by using a 2D device simulator. The E_x at the drain edge of the devices with and without the halo implant are almost the same at $V_{stress} = 2.4$ V due to high drain voltage. Meanwhile, that of the device without the halo implant becomes 13% lower than that of the device with the halo implant at 1.2 V, which is due to the abrupt p-n junction between the channel and drain. The advantage of no halo implant is also experimentally-observed in the Poly-Si gate SOTB devices (Fig. 7). These results also indicate that the gate electrode material has only a modest influence on the degradation due to HCI.

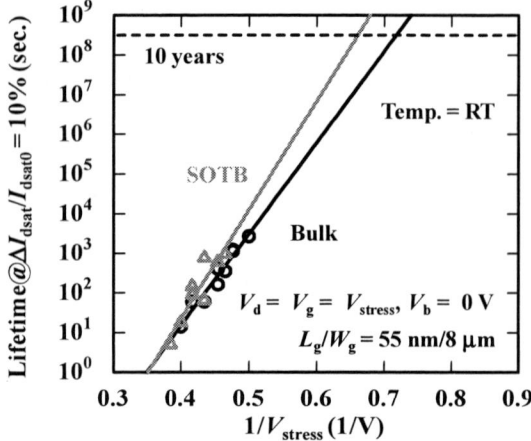

Figure 5.   HCI lifetime of SOTB and bulk NMOSFETs.

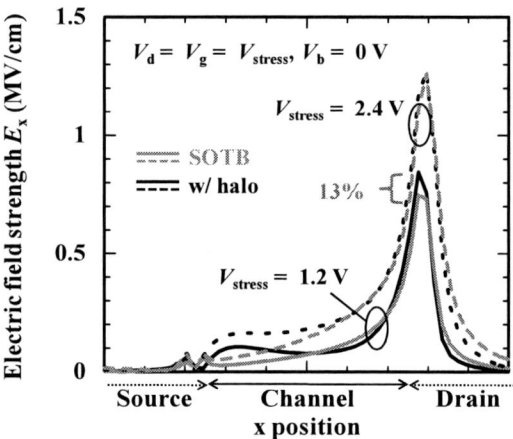

Figure 6.   Simulated electric field strength along the channel.

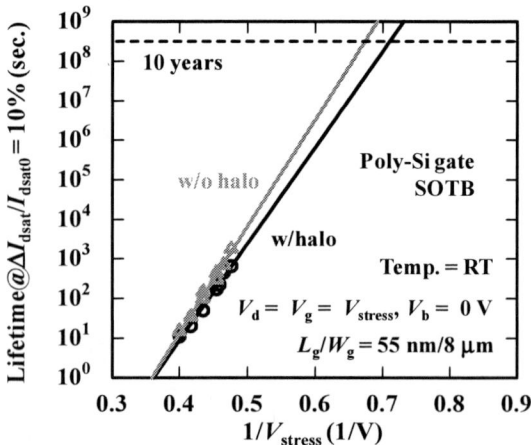

Figure 7.   Impact of halo implant on HCI lifetime in Poly-Si gate SOTB NMOSFETs.

In the SOTB devices, wide-range back-bias V_b controllability significantly changes not only the drive currents but also the strength of the vertical electric field E_y. Figure 8 shows the dependencies of the HCI characteristics on V_b at $V_{stress} = 2.4$ V. The slope of the power-law increases as V_b decreases because the enhanced E_y, especially at the drain edge, increases electron trapping. However, the worst condition is $V_b = V_{stress}$ because of increased drain current. Even under a forward bias condition, the HCI lifetime is sufficiently longer than 10 years (Fig. 9).

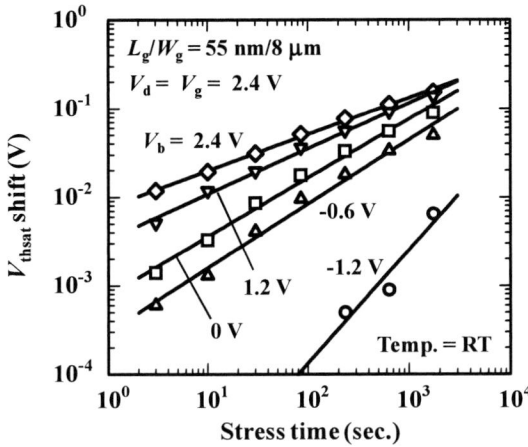

Figure 8. Dependencies of HCI characteristics on V_b at $V_{stress} = 2.4$ V.

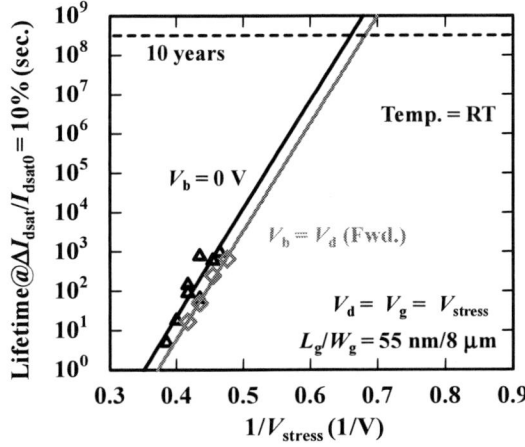

Figure 9. The worst case of HCI lifetime of SOTB NMOSFET is in forward back-biasing.

B. Negative bias temperature instability in SOTB PMOSFETs

The NBTI characteristics of the SOTB PMOSFETs with and without the halo implant are compared in Fig. 10. The power-law slopes of about 0.3 suggest the degradation mechanism in the SOTB device is the same as that in the bulk device: the interaction of holes with hydrogen breaks the Si-H bonds, creating interface traps. The NBTI of the device with the halo implant is severer than that without the halo implant. The degradation in the bulk device is further enhanced. These results indicate the NBTI degradation is enhanced due to

additional exposed Si-H bonds caused by the halo and/or a high energy implant [6].

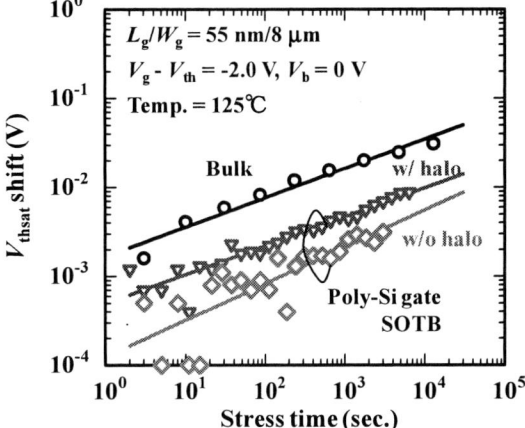

Figure 10. Impact of halo implant in NBTI characteristics.

We observed a small impact of back-bias V_b on NBTI degradation (Fig. 11). It is presumably because the influence of back-biasing on the electric field of the channel surface is mostly blocked by an inversion layer (Fig. 12). In addition, the substrate hot hole injection in reverse back-biasing [7] does not happen due to the BOX layer. The NBTI lifetime for SOTB PMOSFETs was estimated to exceed greatly the 10-year specification (not shown).

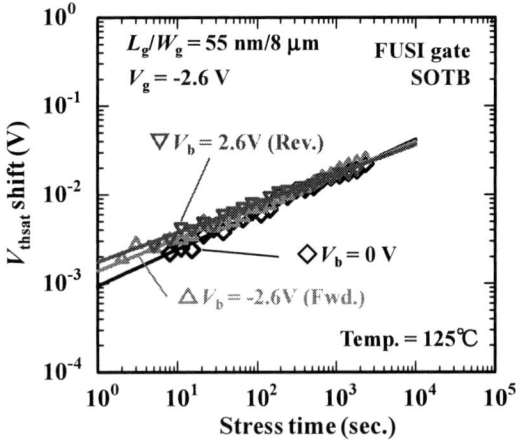

Figure 11. Dependencies of NBTI characteristics on V_b in SOTB PMOSFETs.

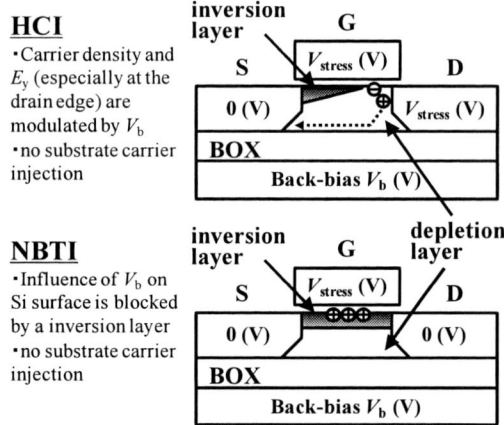

HCI

· Carrier density and E_y (especially at the drain edge) are modulated by V_b

· no substrate carrier injection

NBTI

· Influence of V_b on Si surface is blocked by a inversion layer

· no substrate carrier injection

Figure 12. Schematic back-bias influence in SOTB CMOSFETs.

IV. CONCLUSION

We showed that the HCI degradation in SOTB NMOSFETs is mainly caused by channel hot electron and that longer lifetime is possible in such devices than in conventional bulk devices due to suppressed lateral electric field E_x at the drain edge because of no halo implant. Forward back-biasing slightly shortens the HCI lifetime due to enhanced drive current. However, the back-biasing has a small impact on the NBTI lifetime of SOTB PMOSFETs due to its structure. These results indicate the superior features of SOTB technology in reliability even under a high level of performance boosting.

REFERENCES

[1] R. Tsuchiya et al., "Silicon on thin BOX : a new paradigm of the CMOSFET for low-power and high-performance application featuring wide-range back-bias control," IEDM, pp. 631-634, 2004.

[2] T. Ishigaki et al., "Wide-Range Threshold Voltage Controllable Silicon on Thin Buried Oxide Integrated with Bulk Complementary Metal Oxide Semiconductor Featuring Fully Silicided NiSi Gate Electrode," JJAP Vol. 47, pp. 2585-2588, 2008.

[3] Y. Morita et al., "Smallest Vth Variability Achieved by Intrinsic Silicon on Thin BOX (SOTB) CMOS with Single Metal Gate," VLSI tech., pp. 166-167, 2008.

[4] R. Tsuchiya et al., "Low Voltage (Vdd~0.6V) SRAM operation achieved by reduced threshold voltage variability in SOTB (silicon on thin BOX)," VLSI tech., pp. 150-151, 2009.

[5] M. Yoshimi et al., "Analysis of the Drain Breakdown Mechanism in Ultra-Thin-Film SOI MOSFET's," IEEE TED. Vol. 37, pp. 2015-2021, 1990.

[6] D. Brisbin et al., "Enhanced PMOS NBTI Degradation Due to Halo Implant Channeling," IRPS, pp. 61-66, 2008.

[7] M. Togo et al., "Power-aware 65 nm Node CMOS Technology Using Variable Vdd and Back-bias Control with Reliability Consideration for Back-bias Mode," VLSI tech., pp. 88-89, 2004.

978-1-4244-5430-3/10 $26.00 © 2010 IEEE

The Understanding of Strain-Induced Device Degradation in Advanced MOSFETs with Process-Induced Strain Technology of 65nm Node and Beyond

M. H. Lin[1], E. R. Hsieh[1], Steve S. Chung[1,*], C. H. Tsai[2], P. W. Liu[2], Y. H. Lin[2], C. T. Tsai[2], G. H. Ma[2]

[1]Department of Electronics Engineering, National Chiao Tung University, Hsinchu, Taiwan * Email: schung@cc.nctu.edu.tw

[2]United Microelectronics Corporation (UMC), Hsinchu, Taiwan

Abstract- In this paper, the origin of the strained-induced degradation in the MOSFETs with process-induced strain has been investigated by the I_D-RTN (Drain Current Random Telegraph Noise)technique. The process-induced strain on devices will make worse the device reliability, as reported in [1-2]. First, the I_D-RTN has been employed to study the reliability of two different types of strain devices, i.e., the CESL strain and SiC S/D strain on nMOSFETs. Both CESL and SiC S/D nMOSFETs exhibit poorer reliability compared to bulk devices. However, their impacts to the much worse degradation are different. Results demonstrated that, for the strain in CESL device, it introduced extra mobility scattering in the vertical direction, while in SiC S/D device, the tensile strain along the channel causes an increase of trap generation via the horizontal field only. The CESL process introduces an additional compressive strain vertical to the channel such that it shows much worse reliability than the SiC S/D ones.

Indexed terms: Random Telegraph Noise, Strained-silicon, MOSFET

Introduction- Recent developments [2] in CMOS have a consensus that uniaxial strain is easier to fabricate and also provide flexibility in designing n- and p-MOSFETs individually. Whatever the strain technology we used, the device exhibits much worse reliability comparing to conventional ones [3]. In seeking for solutions to enhance the device driving current, we need to consider a trade-off between reliability and current enhancement capability. Therefore, to understand the process-induced reliability becomes essential toward a better design of the devices. As noted in [2], the I_D-RTN technique provides a way to understand the device trapping behavior, which can be used to understand the physics mechanism underlying the origins of reliability.

In this paper, by applying I_D-RTN method on the strained n-MOSFET with different process-induced strains, we will be able to systematically explore the physics underlying the origins of the strain induced degradation. Then, the correlation between those different strains (physical strain) and the electrical reliability can be found.

1. Device Preparation

The devices were fabricated by the advanced 65 nm CMOS technology. Two sets of the devices have the same dimension (W/L=0.2/0.12μm). The first set is bulk-Si and strained n-MOSFET with nitride-cap layer with <100> channel on (100) substrate (CESL). The second set is bulk-Si and strained n-MOSFET with Si:C S/D with <100> channel on (100) substrate in Fig. 1. All the devices have 14 Å (physical thickness) gate oxide with SION process.

2. Results and Discussion
A. Basic Measurements of RTN

Because of the slow oxide trap, the carrier from source to the drain could be captured and emitted as shown in Fig. 2(a). Fig. 2 (b) shows the I_D-RTN waveform with three major parameters (capture time, τ_c, emission time, τ_e, and current amplitude ΔI_D[4]). First, measurement has been done and no process induced traps existing in the fresh devices. Then, the hot carrier stress was applied to the devices and produced the oxide traps which made two-level fluctuation of drain current in our stress condition. Fig. 3 shows the I_D-RTN spectra for the bulk nMOSFET, from which the RTN fluctuation was increased as gate bias increases. Also, for both bulk and CESL devices, τ_e and τ_c vary with different gate biases, which can be evaluated in a certain time period as shown in Fig. 4. The capture time, τ_c of the slow trap in the CESL is larger than the bulk, which implies that the trap is deeper in CESL than that in the bulk. While the magnitudes of emission time τ_e do not show much difference. Also, the trapping and de-trapping events happen more frequently in CESL so that the capture time over the emission

time decreases more quickly in CESL than bulk Fig. 5(a). This also assures that the HC stress produces more damage in the Si/SiO2 for the CESL and the trap's location is deeper in the CESL than the bulk, Fig. 5(b). The drain current fluctuation $\Delta I_D/I_D$ rolls-off because of the increasing of carrier concentration. Further, the variation of the RTN amplitude $\Delta I_D/I_D$ is proportional to the normalized conductance change g_m/I_D (i.e., $\Delta I_D/I_D \alpha g_m/I_D$, Eq.(3) in Table 1) in the bulk device [6]; while the variation in CESL changes rapidly (Fig.6). The RTN is neither influenced by the change of carrier's number fluctuation ΔN_s nor by the mobility $\Delta \mu$ [4]. Here, the $\Delta I_D/I_D$ roll-off quickly in CESL reveals that there is carrier scattering induced. This becomes an additional mobility degradation factor of the CSEL device after the HC-stress.

B. RTN in the Strained CSEL and SiC S/D nMOSFETs

Similarly, experiments have been demonstrated on the SiC S/D nMOSFETs and its comparison with bulk ones, Fig. 7. The drain current degradation and the two-level I_D fluctuations are also observed in Figs. 8 and 9 respectively. The SiC S/D has larger impact ionization rate near the drain region and induces more current degradation than the bulk ones [7-8]. The capture and emission of electrons are examined for SiC and the bulk devices. They are about the same order for the electrons τ_c and τ_e. Both emission mechanisms are about the same for the two traps (Fig. 10). Fig. 10 shows the $\Delta I_D/I_D$ of the RTN amplitude which is proportional to the normalized conductance change for both devices. This implies that the slow oxide trap in SiC S/D follows the same mechanism to induce the two-level I_D fluctuation as that of bulk devices. However, in Fig. 11, $\Delta I_D/I_D$ is proportional to the normalized conductance change g_m/I_D for both devices, which means that SiC behaves differently from the CSEL devices.

C. The Origins of the Worse Reliability in Strained Devices

Fig. 12 shows the comparison of I_D degradation for devices before and after the HC-stress. The CESL device *shows larger drain current degradation* than the SiC ones, as a result of *an extra degradation factor*. Fig. 13 illustrates the strain direction for the CESL and SiC S/D respectively. For the CESL nMOSFET, the capping layer in the nMOSFET provides the tensile strain along the channel direction and also the compressive strain along the vertical direction. The SiC on S/D inducing the tensile strain along the channel region only. Under strong inversion, for comparable N_s, $\Delta I_D/I_D$ should be proportional to the mobility [9]. From Fig. 11, the $\Delta I_D/I_D$ roll-off more quickly in the CESL compared to the bulk nMOSFET while the SiC S/D and bulk nMOSFETs show comparable trend. Because the process induced extra vertical compressive strains in the dielectric (Fig. 13), the CESL gives rise to more scattering and degrade the Si/SiO2 interface quality after the HC stress; the SiC S/D device with the tensile strain along the channel induces an increase of trap generation via the horizontal field only. In short: (1) the strain techniques can enhance the device performance while on the contrary they show poor hot carrier reliabilities (2) the vertical strain which would cause more scattering can change the carrier mobility and may make worse the device reliability than the SiC S/D device ones.

In summary, the enhanced hot carrier degradation was found in uniaxial-strained devices. The correlation between physical strain and the electrical reliability has been proposed. It is believed that the SiC S/D provides a one-dimensional strain to the channel, while CSEL provides two-dimensional strains to the channel such that the latter one induces larger I_D degradation. This can be verified that the lateral degradation comes from the generated interface traps caused by the tensile strain, while the more enhanced degradation in CSEL comes from the extra compressive strain in the vertical direction.

978-1-4244-5430-3/10 $26.00 © 2010 IEEE

Acknowledgments This work was supported by the National Science Council under NSC96-2221-E009-185.

[1] S. S. Chung et al., in *Symp. on VLSI Tech.*, p. 86, 2005.

[2] M. H. Lin et al., in *Symp. on VLSI Tech.*, p. 52, 2009.

[3] S. S. Chung et al., in *Symp. on VLSI Tech.*, p. 86, 2005.

[4] K. K. Hung et al., in *IEEE EDL.*, vol. 11, p. 90, 1990.

[5] S. S. Chung et al., in *Tech. Dig. IEDM*, pp. 325–328, 2006.

[6] E. Simoen et al, in *Mater. Sci. Eng.*, vol. B91-92, pp. 136–143, 2002.

[7] S. S. Chung et al., in *Tech. Dig. IEDM*, p. 435, 2008.

[8] S. S. Chung et al., in *Symp. on VLSI Tech.*, p. 158, 2009.

[9] Z. Shi et al., in *IEEE T-ED*, vol. 41, p.1161, July 1994.

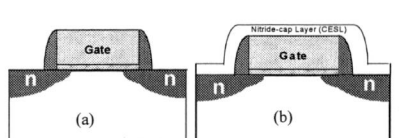

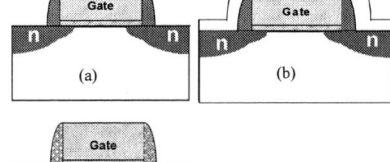

$$\frac{d\ln\left|\frac{\tau_c}{\tau_e}\right|}{dV_G} = -\frac{q Z_{eff}}{kT} \qquad \text{Eq. (1) [4]}$$

$$\frac{\Delta I_D}{I_D} = -\frac{1}{W_{eff} \times L_{eff}}(\frac{1}{N_{eff}} \pm \alpha\mu) \qquad \text{Eq. (2) [4]}$$

$$\frac{d I_D}{I_D} = \frac{g_m}{I_D}\frac{q}{W_{eff}L_{eff}C_{ox}}(1-\frac{x_T}{t_{ox}}) \qquad \text{Eq. (3) [6]}$$

Table 1 Model equations used in this work.

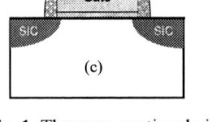

Fig. 1 The cross-sectional views of the experimental strained devices:.(a) Bulk n-MOSFET (b) CESL strained N-MOSFET (c) SiC S/D strained n-MOSFET.

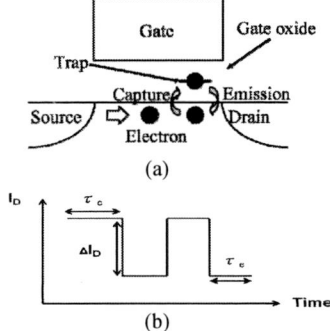

Fig. 2 (a) Carrier trapping and de-trapping by the slow oxide trap near the drain side. (b) Illustration of the three parameters of the RTN measurement: capture time, τ_c, emission time, τ_e, and the amplitude of current fluctuation ΔI_D.

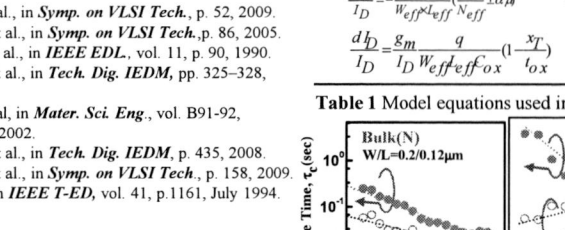

Fig. 4 Comparison of the capture time (filled circles) and emission time (open circles) for devices after the stress. (left) Bulk-Si, (right) CESL n-MOSFET.

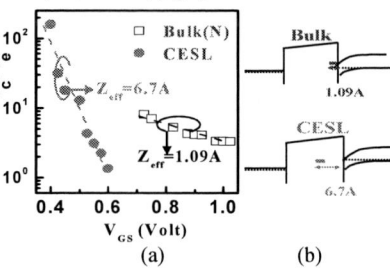

Fig. 5 nMOSFETs. (a) Plot of τ_c/τ_e versus gate voltage. (b) The effective depth location for the two traps.

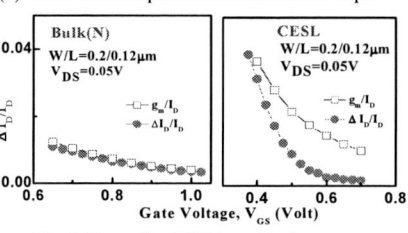

Fig. 6 Normalized RTN amplitude versus gate voltages. (left) Bulk-Si (right) CESL n-MOSFET. The open squares are the normalized conductance change g_m/I_D. Note that CSEL device shows a drastic change of the drain current fluctuation.

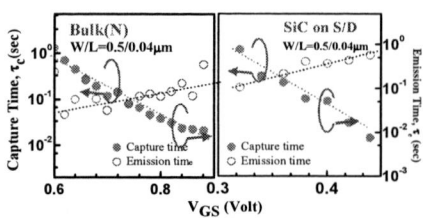

Fig. 8 Comparison of the capture time (filled circles) and emission time (open circles) for devices after the stress. (left) Bulk-Si (right) SiC S/D n-MOSFET.

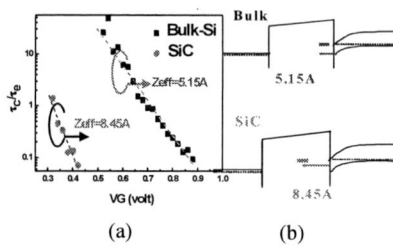

Fig. 9 nMOSFETs. (a) Plot of τ_c/τ_e versus gate voltages. (b) The effective depth location for the two traps.

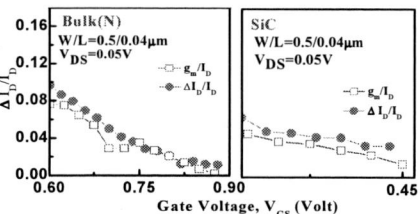

Fig. 10 Normalized RTN amplitude versus gate voltages. (left) Bulk-Si (right) SiC n-MOSFET. The open squares are the normalized conductance change g_m/I_D.

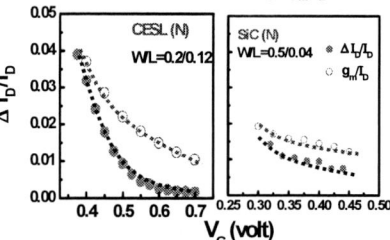

Fig. 11 Normalized RTN amplitude versus gate voltages. (left) Bulk-Si (right) CESL n-MOSFET. Note that these results show that CSEL exhibits a different drain current fluctuation compared to the SiC ones.

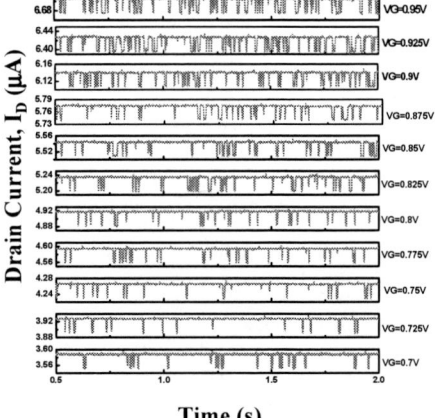

Time (s)

Fig. 3 Drain current waveforms of the Bulk-Si device at various gate biases..

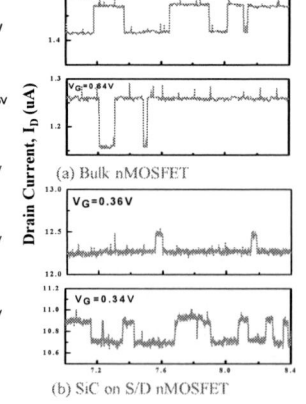

Fig. 7 Drain current waveforms: (a) Bulk-Si (b) SiC S/D n-MOSFET.

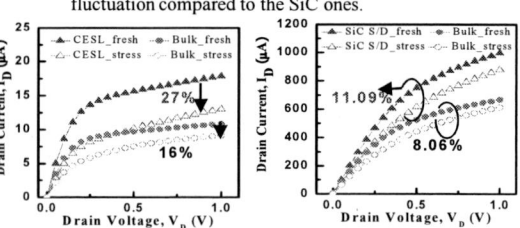

Fig. 12 Comparison of drain current degradation results for CESL (left) SiC on S/D nMOSFETs (right). CSEL device shows much larger degradation than the SiC S/D ones.

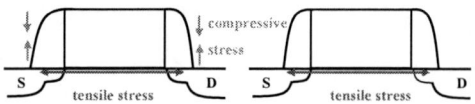

Fig. 13 Illustration of the various strains for CESL nMOSFET (left) and SiC on S/D nMOSFET (right).

978-1-4244-5430-3/10 $26.00 © 2010 IEEE

Effect of Strain on Negative Bias Temperature Instability of Germanium p-Channel Field-Effect Transistor with High-κ Gate Dielectric

Bin Liu, Phyllis Shi Ya Lim, and Yee-Chia Yeo*

Department of Electrical and Computer Engineering, and
NUS Graduate School for Integrative Sciences and Engineering
National University of Singapore (NUS), Singapore 117576
*Phone: +65-6516-2298, E-mail: yeo@ieee.org

Abstract—**We report the first investigation of the effect of strain on NBTI of Germanium (Ge) p-channel Field Effect Transistors (p-FETs) with high-κ gate dielectric. In this study, a mechanical wafer bending tool was used to alter strain in the Ge channel. It is found that higher longitudinal tensile strain in the channel of Ge p-FETs leads to worse NBTI performance. By reducing the tensile strain in the longitudinal direction by wafer bending, improvement in drive current and reduction of NBTI degradation are achieved. Gate width W_G dependence of NBTI in Ge p-MOSFET is also reported.**

Keywords: Ge, p-FET, high-κ, NBTI, strain, wafer bending

I. INTRODUCTION

Strain engineering provides significant boost in carrier mobility and drive current I_{Dsat} in CMOS technology [1]-[3]. The impact of strain on various aspects of reliability in Si channel MOSFET is important, e.g. Negative Bias Temperature Instability (NBTI) in p-channel MOSFETs or p-FETs with SiO$_2$ and/or high-κ gate dielectrics [4]-[7]. It was reported that both compressive and tensile strain could lead to further NBTI degradation in comparison with unstrained control [4]-[7]. For Ge-channel p-FETs, in which there is interest due to the high hole mobility in Ge [8]-[11], the effect of strain on NBTI has not been investigated before. In addition, most work on the effect of strain on NBTI actually used process-induced strain, for example from a liner stressor. Hydrogen from SiN liner [5] may also affect NBTI, and this complicates the isolation of strain-induced effect on NBTI.

In this work, we perform the first investigation of the effect of strain on NBTI of Ge p-FETs with HfO$_2$ dielectric. By altering strain using a mechanical wafer bending tool, the effect of varying the channel strain on NBTI characteristics is studied. The effect of strain on NBTI can thus be isolated. Gate width W_G dependence of NBTI will also be reported.

II. DEVICE FABRICATION

(001) 8-inch Si bulk wafers were used as starting substrates for epitaxial growth of a Si-capped (4 nm) Ge layer (~340 nm).

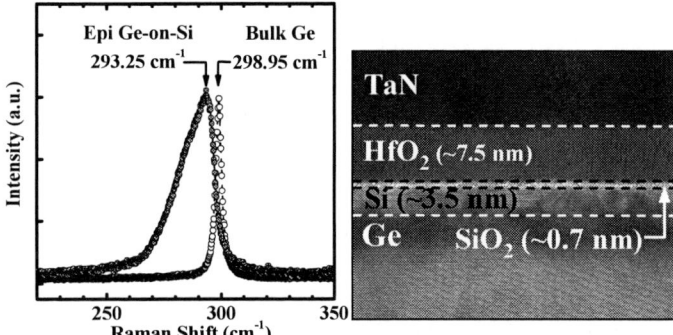

Fig. 1. (a) Visible Raman (514 nm wavelength) spectra of relaxed bulk Ge and Ge-on-Si samples, indicating the latter has a tensile strain of ~1.4 %. (b) High resolution TEM image of gate stack consisting of TaN/HfO$_2$/SiO$_2$/Si-cap/Ge.

The final Ge layer has tensile strain of ~1.4 %, as shown by Raman spectra in Fig. 1(a), and this is due to difference in thermal coefficient of expansion between Si and Ge [8]. P-FET device fabrication started with gate stack (70 nm TaN on 7.5 nm HfO$_2$) formation. Fig. 1(b) shows a high resolution cross-sectional TEM image of the gate stack consisting of TaN/HfO$_2$/SiO$_2$/Si/Ge. Source/Drain Extension (SDE) implant and SiN spacer formation were then performed. This was followed by deep Source/Drain (S/D) implant and NiGe formation in the S/D regions. This completed the device fabrication. Device width W_G ranges from 50 μm to 200 μm.

The device wafers were then cut into long strips with a longitudinal axis parallel to the source-to-drain direction (<110> direction) of the Ge p-FETs. Some strips were bent so that an additional compressive stress of ~300 MPa was introduced in the longitudinal direction. Due to the large intrinsic tensile strain in Ge layer, the additional compressive stress due to wafer bending will only reduce the magnitude of tensile strain, and does not change the polarity of the strain along the longitudinal direction. Devices measured under additional compressive stress will be referred to as "p-FET with compressive stress (of 300 MPa) applied".

978-1-4244-5430-3/10 $26.00 © 2010 IEEE

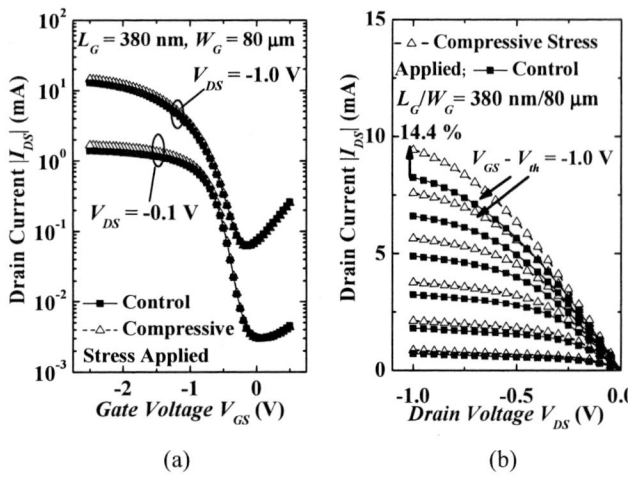

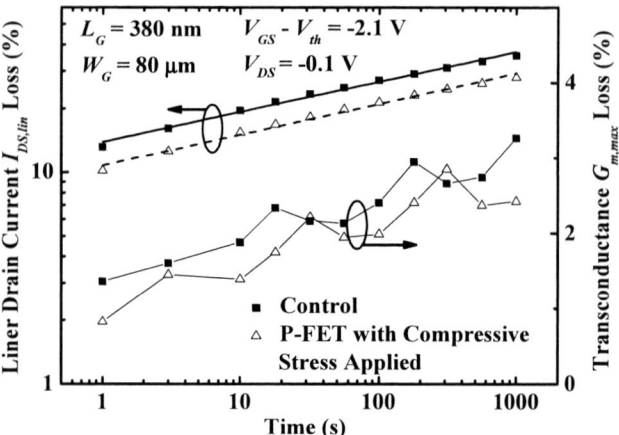

Fig. 2. (a) $|I_{DS}|$-V_{GS} and (b) $|I_{DS}|$-V_{DS} characteristics of a p-FET before and after wafer bending. The wafer bending provides 300 MPa compressive stress which effectively reduces the longitudinal tensile stress in the Ge channel and leads to an increase in hole mobility.

Fig. 4. $I_{DS,lin}$ loss (left axis) and $G_{m,max}$ degradation (right axis) at the NBT stress of $V_{GS} - V_{th} = -2.1$ V and $V_{DS} = -0.1$ V for p-FET with compressive stress applied and control (with intrinsic tensile stress in the Ge channel). The control p-FET has more severe NBTI degradation.

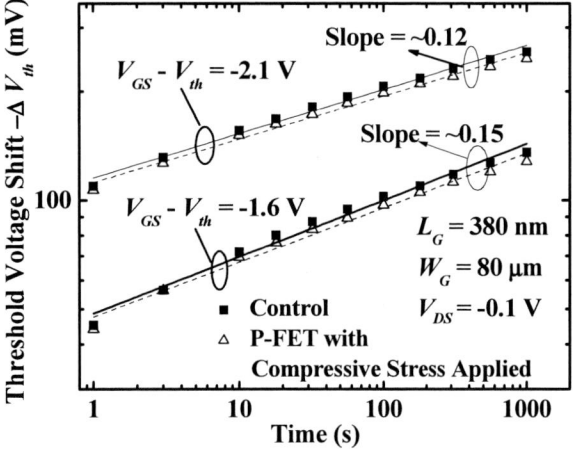

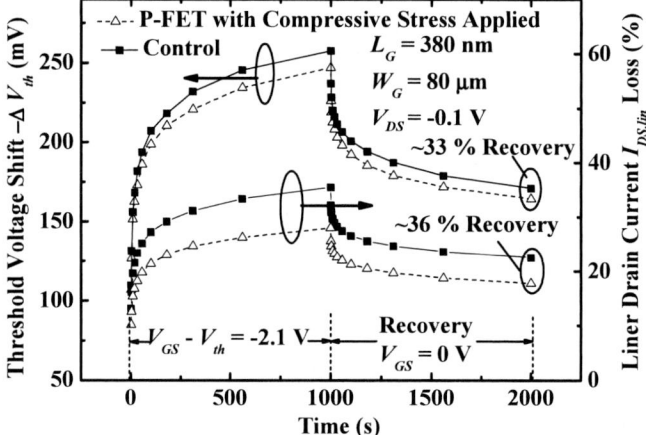

Fig. 3. V_{th} shifts for p-FETs with compressive stress applied and control p-FETs under NBT stress of $V_{GS} - V_{th} = -1.6$ V and -2.1 V, respectively. $V_{DS} = -0.1$ V. P-FETs with compressive stress (of 300 MPa) applied show slightly smaller V_{th} degradation, as compared with control p-FETs.

Fig. 5. Time evolution of V_{th} shift (left axis) and $I_{DS,lin}$ loss (right axis) during NBT stress phase and recovery phase, showing similar recovery rates for both splits.

III. RESULTS AND DISCUSSION

Fig. 2 (a) shows the $|I_{DS}|$-V_{GS} characteristics of a p-FET with a gate length of 380 nm before and after being loaded compressively along the longitudinal direction. Similar subthreshold swing (SS) and drain induced barrier lowering (DIBL) are observed. Fig. 2 (b) demonstrates that compressive wafer bending of 300 MPa in longitudinal direction leads to a drive current enhancement of 14.4 % at a gate over-drive of -1 V, suggesting that hole mobility is enhanced by bending the wafer to increase the compressive strain or to reducing the tensile strain in the channel.

Fig. 3 shows the NBTI degradation in terms of V_{th} shift for control p-FETs and p-FETs with compressive stress of 300 MPa applied. V_{th} was measured using maximum G_M method and NBTI characterization was done by DC method in this work. All devices in Fig. 3 have a gate length of 380 nm and were stressed at the fixed gate over-drive, V_{GS} - V_{th}. Slightly larger V_{th} shifts are observed on control p-FETs at both NBT

stress conditions of V_{GS} - V_{th} = -1.6 V and -2.1 V, respectively, as compared with the compressively stressed p-FETs. Fig. 4 shows the $I_{DS,lin}$ loss under NBT stress with $I_{DS,lin}$ taken at V_{GS} = -1 V and V_{DS} = -0.1 V. It is observed that p-FETs with compressive stress applied have smaller $I_{DS,lin}$ loss at the same NBT stress of $V_{GS} - V_{th} = -2.1$ V, as compared with control p-FETs. It is also found (Fig. 4) that $G_{m,max}$ loss is larger for the control p-FETs, which is consistent with the results of $I_{DS,lin}$ loss and V_{th} degradation (Fig. 3). Fig. 3 and 4 indicate that a higher tensile strain in the channel of control Ge p-FETs leads to slightly worse NBTI performance, and that reducing the tensile strain can lead to less NBTI degradation for a given gate over-drive. The degradation of NBTI caused by higher tensile strain can be possibly attributed to the weakened Si-H bonds in presence of strain. Weaker Si-H bonds could in turn lead to larger interface trap generation under NBT stress. Besides, bonding between Hf and hydrogen related species at the edge of HfO$_2$ layer may also be changed by strain, contributing to change in NBTI characteristics. The results of this work are consistent with previous studies on NBTI of

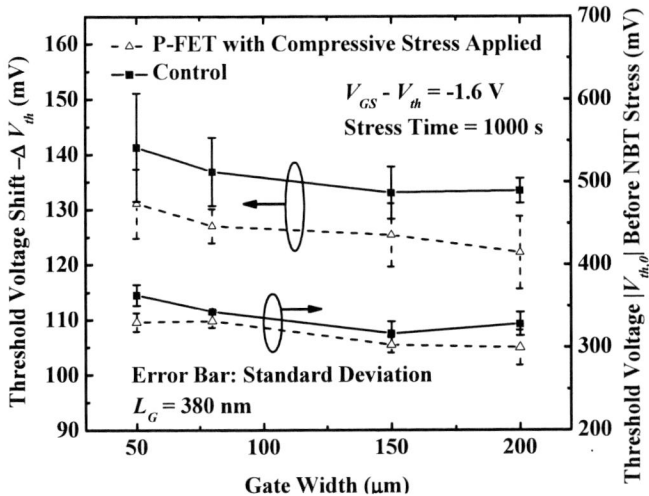

Fig. 6. Gate width dependence of NBTI degradation (left axis) and initial threshold voltage $|V_{th,0}|$ (right axis) for p-FETs with compressive stress applied and control p-FETs.

strained Si channel devices with high-κ dielectric [7]. Fig. 5 compares the recovery behavior of compressively stressed p-FET and the control p-FET. V_{th} and $I_{DS,lin}$ recovery behaviors are shown in Fig. 5, demonstrating similar V_{th} and $I_{DS,lin}$ recovery rate for p-FETs with and without compressive stress of 300 MPa applied. $I_{DS,lin}$ recoveries for both splits are found to be slightly higher than those of V_{th}.

Fig. 6 shows the gate width W_G dependence of NBTI degradation, as well as initial threshold voltage (before NBT stress) $|V_{th,0}|$, for p-FETs with and without compressive stress applied in the longitudinal direction. It is found that at the NBT stress condition of $V_{GS} - V_{th} = -1.6$ V, a larger gate width generally leads to a slightly smaller V_{th} shift. For all gate widths ranging from 50 μm up to 200 μm, p-FETs with compressive stress applied in longitudinal direction always show slightly better NBTI performance, as compared with control.

IV. CONCLUSION

We investigated the effect of strain on NBTI of p-FETs with HfO_2 dielectric fabricated on Ge-on-Si substrate. Wafer bending was used to introduce 300 MPa of compressive stress in longitudinal direction in Ge p-FETs, effectively reducing the tensile strain in the channel and leading to better device performance and less NBTI degradation for a given gate over-drive. Gate width dependence of NBTI of Ge p-FETs was also reported.

REFERENCES

[1] A. Oishi, O. Fujii, T. Yokoyama, K. Ota, T. Sanuki, H. Inokuma, K. Eda, T. Idaka, H. Miyajima, S. Iwasa, H. Yamasaki, K. Oouchi, K. Matsuo, H. Nagano, T. Komoda, Y. Okayama, T. Matsumoto, K. Fukasaku, T. Shimizu, K. Miyano, T. Suzuki, K. Yahashi, A. Horiuchi, Y. Takegawa, K. Saki, S. Mori, K. Ohno, L. Mizushima, M. Saito, M. Iwai, S. Yamada, N. Nagashima, and F. Matsuoka, "High performance CMOSFET technology for 45nm generation and scalability of stress-induced mobility enhancement technique," in *IEDM Technical Digest.*, 2005, pp. 229-232.

[2] P. Bai, C. Auth, S. Balakrishnan, M. Bost, R. Brain, V. Chikarmane, R. Heussner, M. Hussein, J. Hwang, D. Ingerly, R. James, J. Jeong, C. Kenyon, E. Lee, S. H. Lee, N. Lindert, M. Liu, Z. Ma, T. Marieb, A. Murthy, R. Nagisetty, S. Natarajan, J. Neirynck, A. Ott, C. Parker, J. Sebastian, R. Shaheed, S. Sivakumar, J. Steigerwald, S. Tyagi, C. Weber, B. Woolery, A. Yeoh, K. Zhang, M. Bohr, and ieee, "A 65nm logic technology featuring 35nm gate lengths, enhanced channel strain, 8 Cu interconnect layers, low-k ILD and 0.57 μm² SRAM cell," in *IEEE International Electron Devices Meeting*, 2004, pp. 657-660.

[3] W. H. Lee, A. Waite, H. Nii, H. M. Nayfeh, V. McGahay, H. Nakayama, D. Fried, H. Chen, L. Black, R. Bolam, J. Cheng, D. Chidambarrao, C. Christiansen, M. Cullinan-Scholl, D. R. Davies, A. Domenicucci, P. Fisher, J. Fitzsimmons, J. Gill, M. Gribelyuk, D. Harmon, J. Holt, K. Ida, M. Kiene, J. Kluth, C. Labelle, A. Madan, K. Malone, P. V. McLaughlin, M. Minami, D. Mocuta, R. Murphy, C. Muzzy, M. Newport, S. Panda, I. Peidous, A. Sakamoto, T. Sato, G. Sudo, H. VanMeer, T. Yamashita, H. Zhu, P. Agnello, G. Bronner, G. Freeman, S. F. Huang, T. Ivers, S. Luning, K. Miyamoto, H. Nye, J. Pellerin, K. Rim, D. Schepis, T. Spooner, X. Chen, M. Khare, and Ieee, "High performance 65 nm SOI technology with enhanced transistor strain and advanced-low-K BEOL," in *IEEE International Electron Devices Meeting*, 2005, pp. 61-64.

[4] B. Liu, K. M. Tan, M. C. Yang, and Y. C. Yeo, "NBTI Reliability of P-Channel Transistors With Diamond-Like Carbon Liner Having Ultrahigh Compressive Stress," *IEEE Electron Device Letters*, vol. 30, pp. 867-869, Aug 2009.

[5] H. S. Rhee, H. Lee, T. Ueno, D. S. Shin, S. H. Lee, Y. Kim, A. Samoilov, P. O. Hansson, M. Kim, H. S. Kim, N. I. Lee, and Ieee, "Negative bias temperature instability of carrier-transport enhanced pMOSFET with performance boosters," in *IEEE International Electron Devices Meeting*, 2005, pp. 709-712.

[6] T. Irisawa, T. Numata, E. Toyoda, N. Hirashita, T. Tezuka, N. Sugiyama, and S. Takagi, "Physical Understanding of Strain-Induced Modulation of Gate Oxide Reliability in MOSFETs," *IEEE Transactions on Electron Devices*, vol. 55, pp. 3159-3166, Nov 2008.

[7] J. C. Liao, Y. K. Fang, Y. T. Hou, C. L. Hung, P. F. Hsu, K. C. Lin, K. T. Huang, T. L. Lee, and M. S. Liang, "Strain effect and channel length dependence of bias temperature instability on complementary metal-oxide-semiconductor field effect transistors with high-k/SiO₂ gate stacks," *Applied Physics Letters*, vol. 93, pp. 092101, Sep 2008.

[8] H. Zang, W. Y. Loh, J. D. Ye, G. Q. Lo, and B. J. Cho, "Tensile-strained germanium CMOS integration on silicon," *IEEE Electron Device Letters*, vol. 28, pp. 1117-1119, Dec 2007.

[9] T. Yamamoto, Y. Yamashita, M. Harada, N. Taoka, K. Ikeda, K. Suzuki, O. Kiso, N. Sugiyama, S. Takagi, and Ieee, "High performance 60 nm gate length germanium p-MOSFETs with Ni germanide metal source/drain," in *IEEE International Electron Devices Meeting*, 2007, pp. 1041-1043.

[10] R. L. Xie, T. H. Phung, W. He, Z. Q. Sun, M. B. Yu, Z. Y. Cheng, C. X. Zhu, and Ieee, "High Mobility High-k/Ge pMOSFETs with 1 nm EOT -New Concept on Interface Engineering and Interface Characterization," in *IEEE International Electron Devices Meeting*, 2008, pp. 393-396.

[11] J. Mitard, B. De Jaeger, F. E. Leys, G. Hellings, K. Martens, G. Eneman, D. P. Brunco, R. Loo, J. C. Lin, D. Shamiryan, T. Vandeweyer, G. Winderickx, E. Vrancken, C. H. Yu, K. De Meyer, M. Caymax, L. Pantisano, M. Meuris, M. M. Heyns, and Ieee, "Record I-ON/I-OFF performance for 65nm Ge pMOSFET and novel Si passivation scheme for improved EOT scalability," in *IEEE International Electron Devices Meeting*, 2008, pp. 873-878.

A ROBUST ULTRAFAST SWITCHING METHODOLOGY FOR DEVICE PARAMETER CHARACTERIZATION OF BIAS-TEMPERATURE INSTABILITY

Y. Z. Hu, D. S. Ang[+], Z. Q. Teo, and G. A. Du

Nanyang Technological University, School of Electrical and Electronic Engineering,
Nanyang Avenue, Singapore 639798 (E-mail[+]: edsang@ntu.edu.sg)

ABSTRACT

A robust methodology for extracting MOSFET parameters (threshold voltage, V_t and effective mobility, $\bar{\mu}_{eff}$) affected by the bias-temperature instability phenomenon is presented. The method is based on customary pulsed current-voltage measurement hardware and hence may be readily implemented using commercially available set-ups. Application of this method for V_t and $\bar{\mu}_{eff}$ extraction of a 1.8- nm thick SiON gate p-MOSFET under negative-bias temperature stress is demonstrated.

I. INTRODUCTION

The fast recovery of bias-temperature instability (BTI) poses a difficult challenge to device parameter extraction. Quasi-static (or dc) drain current versus gate voltage (I_d-V_g) method, which takes several seconds to complete, could no longer fulfil its role. Fast measurement methods, though suppress recovery, yield very limited information. In particular, I_d-V_g measurement (a pre-requisite for precise extraction of device parameters) is generally not possible with existing fast measurement methods (e.g. on-the-fly (OTF) [1], ramped-V_g [2], etc.) due to cumulative recovery [3] which "distorts" subsequent measurement points. To-date, there is no known fast and yet relatively *accurate* methodology for the extraction of parameters such as threshold voltage (V_t), low-field mobility ($\bar{\mu}_0$), etc. in scaled MOSFETs subjected to BTI test.

In this work, we describe a systematic approach for extracting V_t and $\bar{\mu}_0$, which enables the "aging" of these parameters to be examined in relation to the total drain current degradation (ΔI_d). The approach is based on the pulsed I_d-V_g technique [4], whose very short delay (60 ns) allows the I_d-V_g curve to be measured under the same state of device degradation and a constant recovery determined only by the short delay and not the aggregate measurement time.

II. EXPERIMENTAL

The test devices were p[+] polysilicon gate p-MOSFETs with drawn channel width and length of 2 μm and 60 nm, respectively. The gate dielectric was prepared by exposing a 16 Å base SiO_2 to decoupled plasma nitridation. The nitrogen concentration at the gate/SiON interface was estimated to be ~13 atomic percent. The final physical thickness is 18 Å, determined from high resolution transmission electron microscopy. Negative-bias temperature stress was carried out at 100 °C, with the oxide field capped at ~10 MV/cm.

III. RESULTS AND DISCUSSION

The OTF method promises recovery-free measurement of BTI but is not suitable for precise V_t and $\bar{\mu}_0$ extraction because of the following reasons, as illustrated in Fig. 1: (i) Since the gate stress voltage $V_g{}^s$ is relatively large, the effect of a given $|\Delta V_t|$ on ΔI_d is diminished (as is well known to a first-order that $\Delta I_d/I_{d0} \sim \Delta\bar{\mu}_{eff}/\bar{\mu}_{eff} - \Delta V_t/(V_g - V_t)$). This is evident from a ΔI_d of 28 μA at $V_g{}^s = -2.5$ V, as compared to that of 51 μA at $V_g = -1$ V, for a given state of degradation. The former is further reduced by an underestimation of I_d at $V_g = -2.5$ V due to spontaneous degradation effect (open circle). (ii) Onset of substantial I_d degradation for $|V_g| > 1$ V, due to generation of interface trapped charge – this is apparent

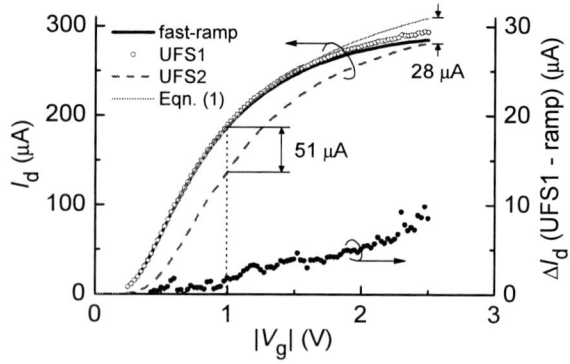

Fig. 1. Left-axis: For the normal pulsed I_d-V_g method (UFS1), the gate was pulsed from 0 V to the measurement voltage $V_g{}^m$ for 100 ns (I_d measurement taken after 60 ns). The nominal delay at each $V_g{}^m$ for the "ramp" method was 40 ms. For the inverse gate pulsing method (UFS2) [4], the gate was held at −2.5 V for 10^4 s before measurement was taken by pulsing the gate down to each $V_g{}^m$ for 100 ns. Right-axis: Id difference between UFS1 and the ramp methods, showing onset of I_d degradation due to high-field effect for $|V_g{}^m| > 1$ V. Also shown in thin line is the I_d-V_g curve as predicted by Eqn. (1) (parameters are extracted using low-field data as illustrated in Fig. 3 and 7).

978-1-4244-5430-3/10 $26.00 © 2010 IEEE

from the non-negligible "recovery" of I_d degradation (open circle) for much shorter measurement interval (60 ns). Further I_d recovery towards the degradation-free curve (thin line), whereby the curvature of the I_d-V_g curve is determined by surface scattering only (Eqn. (1)), may be possible at even shorter measurement interval.

$$I_d = \frac{\beta \cdot V_d \cdot |V_g - V_t|}{1 + \theta_1 \cdot |V_g - V_t|}; \quad \beta = \overline{\mu_o} \cdot C_{ox} \cdot \frac{W}{L}; \quad \theta_1 = \theta_0 + \beta R_{sd} \quad (1)$$

(V_d is applied drain voltage; C_{ox} is oxide capacitance per unit area; W and L are channel width and length, respectively; θ_0 is a parameter characterizing vertical field induced mobility reduction due to surface scattering and R_{sd} is total source/drain series resistance.) This observation implies extremely fast generation of interface trapped charge. This fast degradation effect renders accurate determination of stress-free $I_d(@V_g^s)$ or I_d^s almost impossible for present instrumentation (even with delay down to microseconds). An underestimated I_d^s further reduces its sensitivity to subsequent stress evolution (I_d^s further decreases by only 4 µA after the gate was held at $V_g^s = -2.5$ V for another 10^4 s upon completion of the fast-ramp measurement, which took ~2 seconds to complete). In conjunction with the relatively poor sensitivity of ΔI_d to a given $|\Delta V_t|$ at a relatively high V_g, (cf. (i)), a small error in ΔI_d may lead to a large error in $|\Delta V_t|$. (iii) Despite several orders of magnitude reduction in delay when one moves from the fast-ramp method (bold line) to the pulsed I_d-V_g method (open circle), the increase in I_d (285 → 293 µA) is small. Thus, if one does not vary the delay substantially, the small difference in I_d would wrongly suggest that the given delay is fast enough to achieve negligible I_d degradation at V_g^s. In view of the above problems, a first-order I_d-V_g model (Eqn. (1)) cannot be used for parameter extraction in the high V_g regime (cf. Fig. 1 – thin line). This is apparent in the non-linear dependence of the Y-function (Eqn. 1(a)) [5] for $|V_g| > 1$ V (Fig. 2) [6]. To account for the deviation, some authors [6] have adopted a second-order mobility function (Eqn. (2)). The complexity of this equation renders precise parameter extraction from limited I_d measurement data (cf. OTF) in the high V_g regime almost impossible.

$$Y = \frac{I_d}{\sqrt{g_m}} = \sqrt{\beta \cdot V_d} \cdot |V_g - V_t| \quad (1a)$$

$$I_d = \frac{\beta \cdot V_d \cdot |V_g - V_t|}{1 + \theta_1 \cdot |V_g - V_t| + \theta_2 |V_g - V_t|^2} \quad (2)$$

(g_m is transconductance and θ_2 is a constant taking into account secondary high-field effects.)

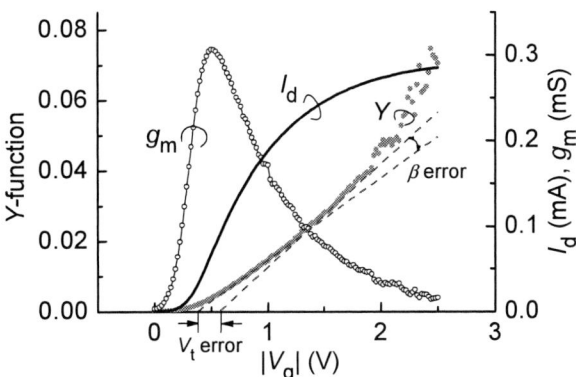

Fig. 2. High-field effect results in deviation of the Y-function (= $I_d/g_m^{0.5}$; left axis) [5] from linearity for $|V_g^m| > 1$ V, rendering parameter extraction using first-order model invalid in this voltage regime. Onset of I_d degradation for $|V_g^m| > 1$ V (Fig. 1) introduces a gradual "stretch-out" effect in the I_d-V_g curve (cf. Fig 1), resulting in piecewise curvilinear behaviour of the Y-function [6]. This can lead to large uncertainties in threshold voltage V_t (x-intercept) and current factor β (= $\overline{\mu}_0 C_{ox} W/L$) extraction for $|V_g^m| > 1$V.

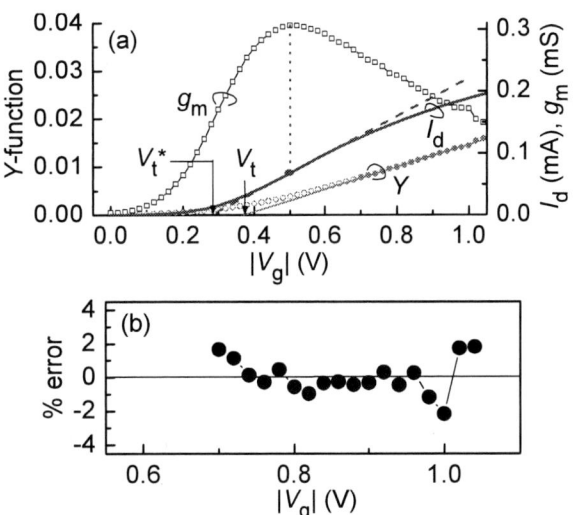

Fig. 3. (a) Solid line shows a good linear fit to the Y-function (filled circle) for relatively low $|V_g^m|$. For $|V_g^m| \sim |V_t|$ (x-intercept of Y-function), deviation from linearity occur as Eqn. (1) is not valid in this voltage regime [6]. Also shown for comparison is an approximate V_t^* extracted by the maximum g_m method (dashed line). V_t^* and V_t are equal only if mobility is constant. (b) Error for linear fitting of the Y-function is within ±2-3 %.

On the other hand, the Y-function remains linear for $|V_g|$ ranging from ~0.65 to 1 V (Fig. 3(a)), which is directly relevant to circuit operation and where high-field effects are negligible (cf. Fig. 1 – excellent agreement of the fast-ramp and pulsed I_d-Vg curves). Linear regression of the Y-function gives error less than 2 % in this V_g range (Fig. 3(b)) and thus the data can be used for reliable parameter extraction based on Eqn. (1) provided the I_d-V_g curve can be measured precisely (as shown later). The excellent linear fit in this V_g regime further validates the implicit assumption of a V_g-independent R_{sd} or scattering parameter θ_1 in the derivation of the Y-function. The x-intercept of the Y-function yields V_t while its slope gives the current factor β (which corresponds to $\bar{\mu}_0$) (Fig. 3(a)) [5].

Pulsed and ramped V_g methods with delay in the range of milliseconds and microseconds, respectively, have been demonstrated [2], [3], but they suffer from significant *cumulative* recovery which distorts the I_d-V_g curve [3]. This is, however, not a problem for the inverse pulsed I_d-V_g technique (the offset is held at V_g^s instead of 0 V) [4]. Fig. 4 shows that the power-law exponents of V_g shifts extracted at different I_d criteria (which correspond to the $|V_g|$ range of interest, i.e. $0.65 < |V_g| < 1$ V) are identical, implying that the amount of recovery is the same at each gate measurement voltage. As the pulse width is narrow (100 ns) and the duty cycle of the pulse is small (1 %), the state of device degradation (as denoted by I_d^s) is approximately constant (largest deviation is only 0.3 %) before each pulsed measurement (Fig. 5). These advantages imply that a precise and almost distortion-free I_d-V_g curve can be measured.

Fig. 6 and 7 further verify that the Y-function (calculated based on the pulsed I_d-V_g) is linear over the $|V_g|$ range of interest for both the pre- and post-stress device, thus enabling straightforward yet precise extraction of V_t, $\bar{\mu}_0$ and θ_1 (Eqn. (1)). Fig. 8 compares experimentally

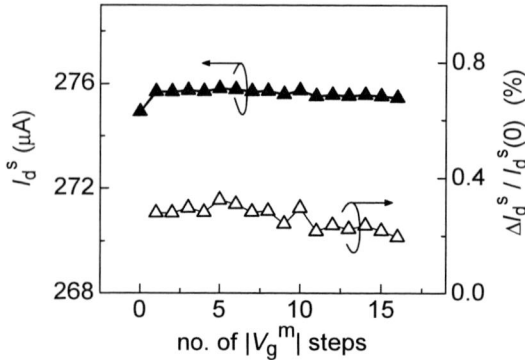

Fig. 5. Drain current at gate stress voltage (−2 V), I_d^s, measured before the gate was pulsed to a given V_g^m for I_d measurement (UFS2), as a function of the number of V_g^m steps. I_d^s is increased at the first V_g^m pulse due to recovery. But this deviation is only 0.3 % (maximum) of I_d^s before measurement was executed. $|V_g^m|$ was varied from 0.6 to 1 V in steps of 25 mV.

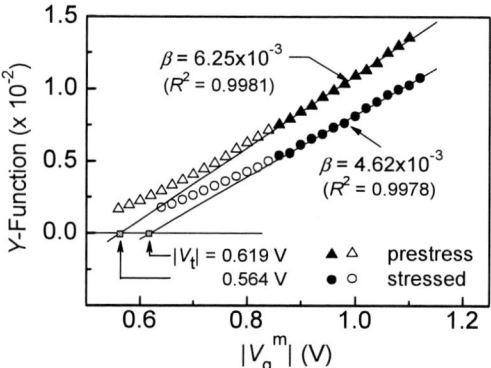

Fig. 6: $|V_t|$ is increased by 55 mV and β (which corresponds to low-field mobility) is degraded by ~25% after 10^4 s stress at −2 V. This is a high-V_t device, which explains the higher V_t compared to that in Fig. 3. The $|V_g^m|$ range over which linear fitting of the Y-function is carried out is adjusted correspondingly. Excellent linear fit is apparent.

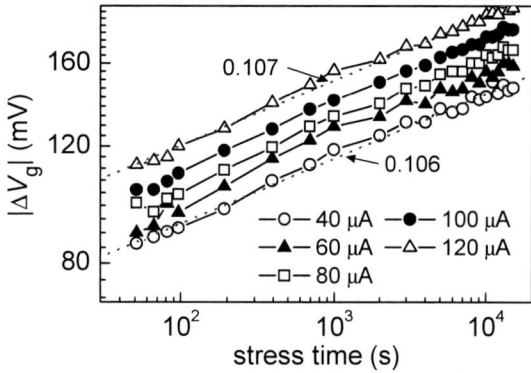

Fig. 4. Power-law time dependence of gate voltage shift $|\Delta V_g|$, showing constant exponent for different I_d criteria (for $|V_g^m|$ ranging from 0.6 to 1 V). The I_d-V_g curve at a given stress interval was measured by the UFS2 method (cf. Fig. 1). An exponent of 0.1 implies that the UFS2 method is able to probe a fraction of hole traps which normally exhibit very fast relaxation and is not measured by slower techniques [3]. Stress condition: −2 V; 100 °C.

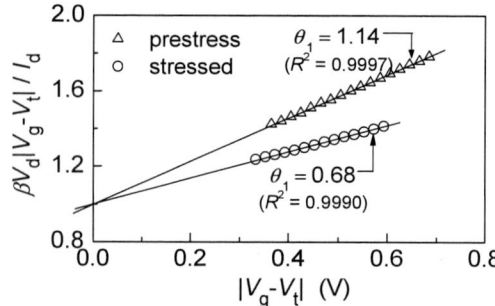

Fig. 7: After obtaining V_t and β, θ_1 may be determined by plotting $\beta V_d(V_g - V_t)/I_d$ versus $|V_g - V_t|$ (see Eqn. (1)) over the range of V_g where the Y-function is linear. A straight line is obtained whose y-intercept is always unity and slope gives the fitting parameter θ_1.

978-1-4244-5430-3/10 $26.00 © 2010 IEEE

extracted $|\Delta V_g|$ to that calculated from Eqn. (1), based on the $|\Delta V_t|$ and $\Delta\bar{\mu}_0$ extracted by the Y-function plot. Good agreement is observed, which indicates that difference between $|\Delta V_g|$ and $|\Delta V_t|$ can be consistently described in terms of $\bar{\mu}_0$ and θ_1 degradation. Substantial mobility degradation effect is apparent. Finally, the power-law exponent of $|\Delta V_t|$ is ~0.113 (Fig. 9), which implies some degree of hole trapping. This does not mean that the gate dielectric is of a poor quality. On the other hand, it indicates that the inverse pulsed I_d-V_g method is able to probe trapped holes which generally exhibit very fast relaxation. Our method shares the same advantage as OTF in that it is recovery-free (cf. Fig. 8 and 9) (gate is kept 99 % at V_g^s) but is free from the disadvantages of OTF. Assuming that hole trapping saturates at ~1 s [7] and subtracting this component from $|\Delta V_t|$, an exponent of ~0.171, typically associated with interface state generation [8], is revealed. If this demarcation is valid, the respective contributions of trapped holes and interface states towards $\Delta\bar{\mu}_0$ can be readily clarified from Eqn. (1).

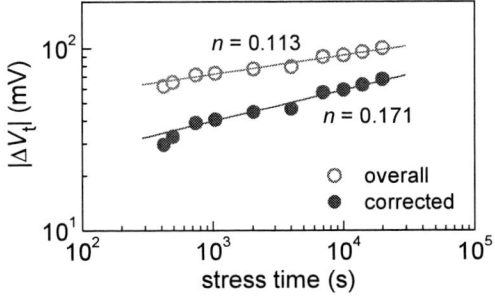

Fig. 9: Open circle denotes $|\Delta V_t|$ extracted via the method depicted in Fig. 6 (cf. open circle in Fig. 8). The relatively small exponent of 0.113 implies that UFS2 also probes a component of hole traps (which relax very rapidly) by virtue of the extremely short delay (60 ns). After subtracting away a component of $|\Delta V_t|$ due to hole traps (estimated at 1 s [7]), the resultant $|\Delta V_t|$ (filled circle) due to NBTI exhibits the familiar exponent of 0.171 [8].

IV. CONCLUSION

A robust ultrafast switching methodology for MOSFET parameter extraction under BTI stress condition is described. The main advantages of this method, as compared to existing methods, are as follows: (i) It employs the established pulsed current-voltage measurement hardware and hence may be readily implemented using commercially set-ups; (ii) it enables complete measurement of the I_d-V_g curve to be made under a constant state of degradation and at a minimal delay of ~60 ns (this ensures distortion-free measurement of the *whole* I_d-V_g curve). Experimental results based on this method reveal substantial mobility degradation effect arising from negative-bias temperature stress.

ACKNOWLEDGEMENT

This work is supported in part by Singapore Ministry of Education under research grant MoE2009-T2-1-050.

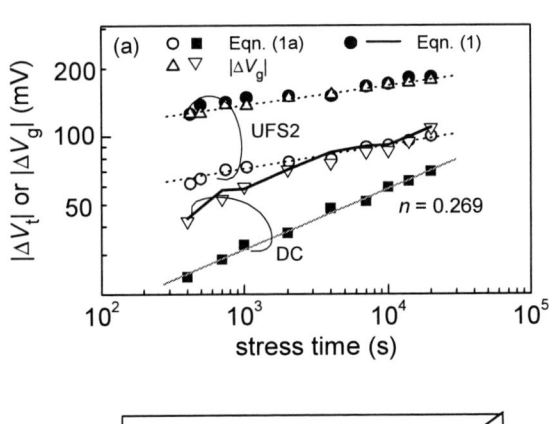

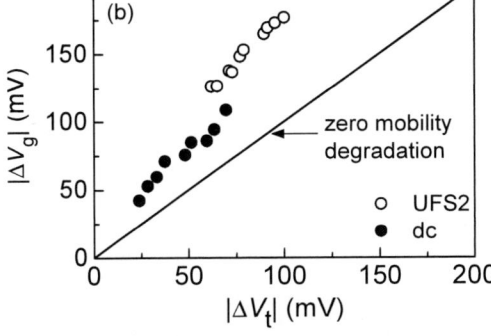

Fig. 8. Comparison of $|\Delta V_g|$ (extracted at $V_g^m = -1$ V) and $|\Delta V_t|$. Difference between these 2 parameters can be consistently explained by taking into account stress induced degradation of $\bar{\mu}_0$ and θ_1, as in Eqn. (1). The data from dc I_d-V_g measurement (delay ~5 s) is also shown for comparison. Compared to the UFS2 data, difference between $|\Delta V_g|$ and $|\Delta V_t|$ for the dc case is clearly smaller, indicating recovery of $\bar{\mu}_0$ and θ_1 after relaxation of positive oxide and interface trapped charge.

REFERENCES

[1] M. Denais, A. Bravaix, V. Huard, C. Parthasarathy, G. Ribes, F. Perrier, Y. Rey-Tauriac, N. Revil, *"On-the-fly* characterization of NBTI in ultra-thin gate oxide PMOSFET's," in *Int. Electron Dev. Mtg. Tech. Dig.*, 2004, pp. 109-112.

[2] A. Kerber, E. Cartier, L. Pantisano, M. Rosmeulen, R. Degraeve, T. Kauerauf, G. Groeseneken, H. E. Maes, and U. Schwalke, "Characterization of the V_T-instability in SiO$_2$/HfO$_2$ gate dielectric," in Proc. *Int. Reliab. Phys. Symp.*, 2003, pp. 41-44.

[3] C. Schlünder, M. Hoffmann, R.-P. Vollertsen, G. Schindler, W. Heinrigs, W. Gustin and H. Reisinger, "A novel multi-

point NBTI characterization methodology using smart intermediate stress," in *Proc. Int. Reliab Phys. Symp.*, 2008, pp. 79-86.

[4] G. A. Du, D. S. Ang, Z. Q. Teo, and Y. Z. Hu, "Ultrafast measurement on NBTI," *IEEE Electron Dev. Lett.*, vol. 30, no. 3, pp. 275-277, Mar. 2009.

[5] G. Ghibaudo, "New method for the extraction of MOSFET parameters," *Electron. Lett.*, vol. 24, no. 9, pp. 543-545, Apr. 1988.

[6] D. Fleury, A. Cros, H. Brut, and G. Ghibaudo, "New *Y*-function-based methodology for accurate extraction of electrical parameters on nano-scaled MOSFETs," in *Proc. Conf. Microelectron. Test Struct.*, 2008, pp. 160-165.

[7] H. Reisinger, O. Blank, W. Heinrigs, A. Mühlhoff, W. Gustin, and C. Schlünder, "Anaslysis of NBTI degradation- and recovery-behavior based on ultra fast V_T-measurements," in *Proc. Int. Reliab. Phys. Symp.*, 2006, pp. 448-453.

[8] S. Chakravarthi, A. T. Krishnan, V. Reddy, C. F. Machala, and S. Krishnan, "A comprehensive framework for predictive modeling of negative bias temperature instability," in *Proc. Int. Reliab Phys. Symp.*, 2004, pp. 273-282..

Impact of Hydrogen on Recoverable and Permanent Damage following Negative Bias Temperature Stress

T. Aichinger[1], S. Puchner[1], M. Nelhiebel[2], T. Grasser[3], and H. Hutter[4]

[1]Kompetenzzentrum für Automobil- und Industrieelektronik (KAI), Europastrasse 8, 9524 Villach, Austria
phone: 0043-4242-34890-24, email: thomas.aichinger@k-ai.at
[2]Infineon Technologies Austria, Siemensstrasse 2, 9500 Villach, Austria
[3]Christian Doppler Laboratory for TCAD, Institute for Microelectronics TU Wien, Gusshausstrasse 27-29, 1040 Wien, Austria
[4]Institute for Chemical Technologies and Analytics TU Wien, Getreidemarkt 9/E164, 1060 Wien, Austria

Abstract—**By subjecting selected split wafers to a specifically adapted measure-stress-measure (MSM) procedure, we analyze negative bias temperature stress (NBTS) and recovery characteristics of PMOS devices with respect to the impact of hydrogen. We control the hydrogen incorporation within the gate oxide during Back End of Line (BEOL) fabrication by varying the titanium barrier thickness below the routing metallization. Differences in the initial passivation degree of the gate oxide are verified electrically by Charge Pumping (CP) measurements and physically by time-of-flight secondary ion mass spectrometry (TOFSIMS). Our results indicate that the total V_{TH} shift is the sum of quasi permanent and recoverable damage which are of comparable scale but have completely different physical and electrical characteristics. While the permanent component seems to be strongly linked to hydrogen release from the interface (increase in CP current), the recoverable component is widely independent of hydrogen and its recovery can be controlled via carrier exchange with the silicon substrate. Hence, our results suggest different trap precursors for the individual components which challenge some predictions of the classical reaction-diffusion (RD) model and support the concepts of an alternative model based on permanent interface state creation via hydrogen transfer to recoverable E' centers which have their origin in oxygen vacancies, whose density is roughly independent of the hydrogen concentration.**

NBTI, hydrogen, Pb center, E' center, oxygen vacancy

I. INTRODUCTION

Hydrogen has often been reported to play a crucial role in the negative bias temperature instability (NBTI) of MOS devices [1-4]. This has been linked to its ability to passivate/de-passivate dangling bonds at the gate-oxide silicon-substrate interface [5]. In particular, the performance and defect densities of *virgin* silicon devices improve considerably by incorporating hydrogen into the gate oxide during the Back End of Line (BEOL) process [6]. This can be done either directly by exposing the wafers to pure hydrogen or forming gas anneals [7-8] or indirectly as a consequence of plasma-enhanced

chemical vapor deposition (PECVD) of silicon nitride (SNIT) layers [9]. Such layers are considered to be efficient hydrogen sources since they contain a large concentration of hydrogen [10-11] which may be released immediately after deposition to diffuse toward the gate oxide. Provided there is no diffusing barrier below the SNIT, some hydrogen may reach the gate oxide where it can passivate dangling bonds at the interface, thereby improving the virgin performance of the MOS device. However, once passivated, previously captured hydrogen may be released from the interface during NBTS leaving behind donor-like P_b centers that are reported to cause a negative shift in the threshold voltage of a p-doped metal oxide semiconductor (PMOS) transistor. Thus, one may expect that the initial passivation degree of the interface, namely the total number of Si-H bonds present at the interface before stress, crucially determines the NBTI sensitivity of the technology. This is a generally accepted fact often reported in literature. However, concerning the underlying micro-structural physics behind degradation and recovery, two competing models (namely the reaction diffusion (RD) model [2] and the alternative Grasser model [12]) come to different conclusions and hence make different predictions.

In order to check the fundamental statements of those models with respect to the impact of hydrogen, we have fabricated a wafer split, where we modified the hydrogen budget within the gate oxide (and hence the number of Si-H bonds at the interface). The hydrogen incorporation is measured physically by TOFSIMS analysis (counting secondary H ions) and electrically by CP measurements (counting dangling bonds at the interface). Having fabricated the hardware, we characterize and compare stress and recovery dynamics of two selected split wafers. Finally, we check our results against the predictions of the classical RD model [2] and against the alternative model introduced by Grasser et al. [12].

II. SAMPLE PREPARATION AND HARDWARE

Our basic idea for modifying the hydrogen content within the gate oxide is to control hydrogen diffusion from the upper SNIT layer toward the gate oxide. This is basically achieved by introducing titanium (Ti) liners below the routing metallization. Titanium is known to be an effective hydrogen barrier [13-15] which can absorb a certain amount of diffusing hydrogen.

This work was jointly funded by the Federal Ministry of Economics and Labour of the Republic of Austria (contract 98.362/0112-C1/10/2005) and the Carinthian Economic Promotion Fund (KWF) (contract 98.362/0112-C1/10/2005).

(a)

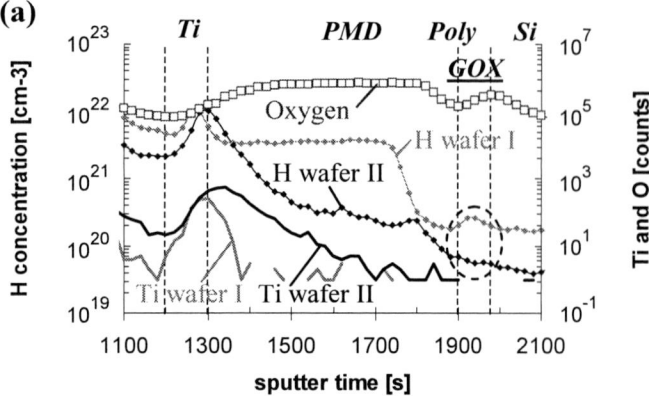

(b)

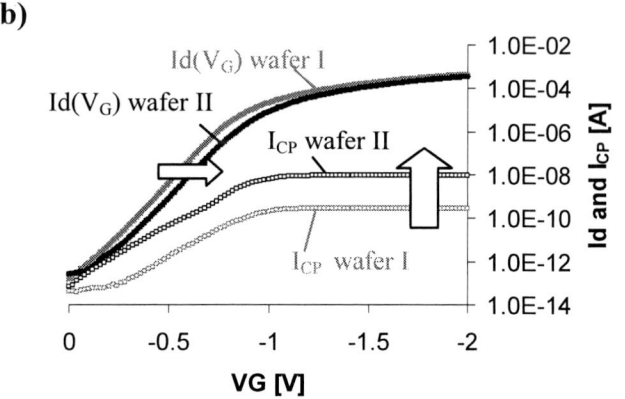

Figure 1. (a) TOFSIMS image of the BEOL process split wafers I and II. By modifying the titanium layer thickness, the hydrogen budget within the post metal dielectric (PMD), the gate-poly and within the gate oxide (GOX) can be controlled. Wafer I has a thinner Ti layer than wafer II and therefore more hydrogen within the GOX. (b) Virgin transfer curves (full squares) and CP currents (open squares) of the split wafers. Due to the higher hydrogen concentration within the GOX, wafer I has a more efficiently passivated interface, a lower CP signal and initially less positively charged defects than wafer II.

To produce split wafers with different hydrogen concentrations within the gate oxide, we simply vary the Ti liner thickness between the SNIT and the gate oxide. The thinner the titanium liner, the higher is its hydrogen permeability and the more hydrogen arrives actually at the gate oxide. In particular, as the titanium layer thickness exceeds several tens of nanometers, the hydrogen concentration within the gate oxide approaches a minimum which is reflected electrically by a maximum initial CP signal (low interface state passivation degree). In fact, the CP signal of such a sample is comparable to the one of a split wafer pulled before SNIT deposition indicating that hardly any hydrogen can pass a titanium barrier provided the barrier is sufficiently thick.

In our measurements discussed in the following, we focus on two representative split wafers (extreme cases), the first (wafer I) having a minimum titanium barrier (high hydrogen concentration in the gate oxide), the second (wafer II) having a very thick titanium barrier (minimum hydrogen concentration in the gate oxide). Except for the Ti-liner thickness, the two wafers were processed identically in the same wafer lot. Our devices under test (DUTs) are isolated PMOS transistors with 30 nm pure SiO_2 gate oxides. The structures are surrounded by

in-situ polyheaters that enable us to perform on-chip fast heating and cooling [16-17]. We use such thick oxide devices in order to improve the hydrogen resolution within the gate oxide during the TOFSIMS measurement and in order to reduce tunneling currents through the gate oxide during stress and during charge pumping. It has recently been shown that the basic mechanisms of NBTI are essentially the same in thin and thick SiO_2 and SiON technologies [18-20]. The use of pure SiO_2 gate oxides guarantees that our general conclusions are not distorted by the strongly process dependent impact of nitridation [21].

III. HYDROGEN DETECTION USING TOFSIMS

The hydrogen concentrations of the different split wafers were investigated by a time of flight secondary ion mass spectrometer (TOF.SIMS[5]) equipped with a liquid metal ion gun (LMIG) producing primary ions and a dual source (DSC-S) column eroding the sample surface. TOFSIMS measurements were performed on large reference capacitors (1 mm^2) located in the vicinity of the electrically measured PMOS devices. Depth profiling was carried out in the dual beam mode [22] using common techniques for insulators [23]. Pulsed Bi_1^+ primary ions (25 keV) were used to analyze the target surface. Alternating, a second beam of Cs^+ ions erodes the sample surface and enhances the yield of negatively charged secondary ions (e.g. H^-, O^-). In order to avoid distortion of the hydrogen signal, we abstained from exposing our samples to a chemical back preparation via etching. Hence, at the expense of depth resolution, we were forced to use a 2 keV sputter beam ablating the thick stack of layers (3.5 µm) on the top of the gate oxide within a reasonable time interval. To quantify the hydrogen signal, a relative sensitivity factor (RSF) was used for converting the secondary ion intensity into absolute concentrations. For determining this hydrogen-related RSF, we used reference samples providing a defined standard implantation dose ($\Phi_H = 10^{15}$ cm^2). The obtained RSF for hydrogen was 5×10^{21}, which is close to literature values taking into account different matrix signals and SIMS techniques [24].

Fig. 1 (a) shows a TOFSIMS image of the sub-metal BEOL layer stack of the two selected split wafers. Displayed are the oxygen and the titanium signals for the orientation within the BEOL stack. Underneath the Ti liners we measure significantly different hydrogen concentrations in the PMD, the poly and the GOX for wafer I and II. In perfect agreement with the TOFSIMS results we obtain in Fig. 1 (b) that (i) the initial CP signal of wafer I (~ 0.3 nA) is about 30 times lower than the one of wafer II (~ 9 nA). This is consistent with the assumption that the interface of wafer I (Dit ~ 2.3×10^9 $eV^{-1}cm^{-2}$) is more efficiently passivated by hydrogen than the interface of wafer II (Dit ~ 6.9×10^{10} $eV^{-1}cm^{-2}$). The virgin CP currents were recorded at a temperature of 50°C using a pulsing frequency of 500 kHz and rising/falling slopes of 10 V/µs scanning roughly 500 meV of the silicon band gap around mid gap; (ii) wafer II has a more negative threshold voltage (~ -160 mV) than wafer I which can be explained by an initially larger number of positively charged P_b centers and oxide defects which both remain unpassivated in wafer II due to the low hydrogen concentration within the gate oxide.

978-1-4244-5430-3/10 $26.00 © 2010 IEEE

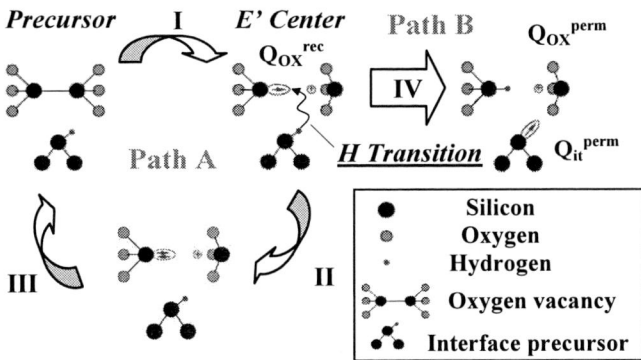

Figure 2. A schematic illustration of the Grasser model transitions. During stress, recoverable oxide traps are created by temperature and field induced bond breakage. In the positive charge state the E' center may become neutralized and anneals permanently (Path A) or gets locked-in by H capture form the interface (Path B).

IV. QUALITATIVE MODEL DISCRIPTIONS

Before starting with the description of the measurement setup and presenting results, we briefly summarize some of the main characteristics of the discussed NBTI models.

A. The RD model:

The classic reaction-diffusion (RD) model as described in [2] explains V_{TH} degradation and recovery by a non-dispersive diffusion process of neutral hydrogen atoms or molecules (H or H_2) which are released during stress from Si-H bonds at the interface via a field-dependent reaction. Once released, the hydrogen species diffuse away from the interface, leaving behind a positively charged interface state (Si^+) that is responsible for the higher threshold voltage and the lower transconductance. Once the stress is removed, previously released hydrogen can diffuse back and recombine with silicon dangling bonds restoring them to their passive Si-H state.

Consequences: (i) the increase in CP current fully explains the total V_{TH} shift; (ii) the recovery rate is proportional to the amount of created interface traps, released hydrogen respectively; (iii) recovery is not influenced by slight variations in gate biasing during read-out.

B. The Grasser model:

The chemical transitions suggested by the reaction controlled model of Grasser et al. [12] are schematically depicted in Fig. 2. **Path A:** During NBTS, oxygen vacancies located close to the interface are assumed to break up and charge positively (transition I) due to the presence of the high electric field and a majority of holes at the gate oxide substrate interface. During recovery, where the field and the carrier situation at the interface is quite different, some of these so-called E' centers (Q_{ox}^{rec}) may become neutralized by hole emission (transition II). Once in the neutral charge state, the E' center can anneal permanently via structural relaxation thereby restoring the initial precursor state again (transition III). **Path B:** Once created during stress, the dangling bond of the E' center can optionally attract a hydrogen atom from the interface which converts the recoverable oxide defect (Q_{ox}^{rec}) and the

passivated interface state (Si-H bonds) into a *locked-in* positive oxide defect (Q_{ox}^{perm}) and an electrically active P_b center (Q_{it}^{perm}) (transition IV). In principle, the reverse reaction of path B, where the H atom is released from the dangling bond of the E' center and travels back to the un-passivated interface state, is feasible as well. However, in a first order approximation, this back transition is neglected assuming that the Si-H bond is stable within the E' center. Consequently, once created, locked-in oxide defects and interface states are considered as permanent.

Consequences: (i) the total V_{TH} degradation consists of permanent interface states and locked-in oxide charges (which emerge in a physical 50:50 relation due to entropy driven hydrogen exchange between Si-H bonds at the interface and positively charged E' centers (cf. **Path B**)) plus a recoverable portion of positively charged E' centers; (ii) recovery (cf. **Path A**) is largely independent of the hydrogen incorporation (at least, when neglecting the loss of recoverable E' centers by transition IV); (iii) recovery (cf. **Path A**) is highly dependent on the gate bias during read-out since the carrier situation at the interface (and hence the Fermi-level position) governs neutralization and relaxation of positively charged E' centers.

V. ELECTRICAL MEASUREMENT SETUP

Our experimental setup for NBTI characterization is illustrated in Fig. 3. The measurement procedure is particularly designed to separate time and bias dependent recoverable damage from apparently permanent degradation. During a basic measure-stress-measure (MSM) cycle we subject PMOS devices to NBTS (200°C; < 7.0 MV/cm) for a defined time t_S. By making use of the in-situ polyheater technique, the following recovery phase can be performed at a much lower temperature of 50°C which decelerates thermo dynamical recovery mechanisms and improves the charge pumping measurement resolution. Right before the end of the stress phase, we quickly (< 10 s) cool down to 50°C while the stress bias is maintained at the gate, thereby quenching the degradation [16-17]. Then we initiate a 1000 s recovery phase (t_{R1}), by switching the gate bias from its stress level to -2.0 V.

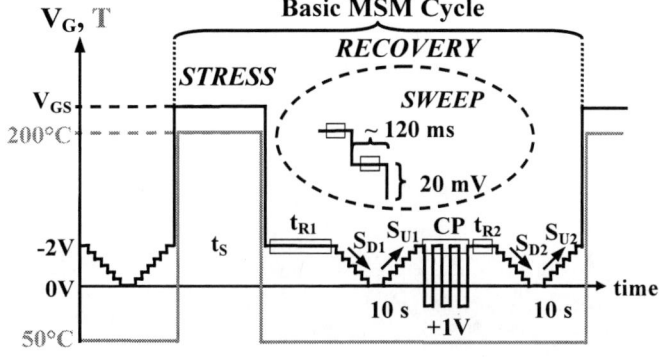

Figure 3. Our basic MSM procedure used for degradation/recovery analysis. During stress, we use the polyheater tool to generate an elevated stress temperature. During recovery, we perform gate bias sweeps and CP measurements in order to monitor V_{TH} recovery and interface state creation.

We denote the relative amount of V_{TH} recovery during t_{R1} between the first measured point after the removal of the stress bias (40 ms post stress) and the last measured point (1000 s post stress) as the *time dependent recovery* contribution (ΔV_{TH}^{time}). Subsequently to t_{R1}, we ramp down (SD1) the gate bias in 20 mV steps from strong inversion (-2.0 V) toward depletion (0.0 V) and monitor in parallel the V_{TH} shift as a function of the current gate bias. One full gate bias ramp takes approximately 10 s. Approaching depletion, the Fermi level moves from the valance band edge toward the conduction band edge thereby gradually changing the ratio of free holes and electrons at the interface. After staying for 10 s at 0.0 V, we ramp the gate bias back (SU1) to -2.0 V. The difference in the V_{TH} shift recorded at -2.0 V at the beginning of SD1 and at the end of SU1 is denoted as the *bias dependent recovery* contribution (ΔV_{TH}^{bias}). After the first ramp down-up cycle, we record the maximum CP current for 10 s by pulsing the gate junction between strong inversion (-2.0 V) and accumulation (+1.0 V) at a frequency of 500 kHz. In the analysis, we convert the maximum CP signal into an interface state dependent threshold voltage shift ($\Delta V_{TH}^{perm,it}$) by assuming an amphoteric nature of interface traps [25] and a flat density of state profile [26]. Subsequent to the CP cycle, we again perform a short 10 s constant gate bias phase at -2.0 V (t_{R2}) followed by a second down-up ramp (SD2; SU2). This basic MSM cycle is repeated six times on both devices of the wafer split with increasing stress times t_S (1/10/100/1,000/10,000/100,000 s).

VI. RESULTS AND DISCUSSION

The V_{TH} shifts measured during the different stages of the experiment (c.f. Fig. 3) are illustrated for wafer I (thin Ti/high H) in Fig. 4 (a) and for wafer II (thick Ti/low H) in Fig. 4 (b). Shown are six curves corresponding to the six subsequent stress-runs. In Fig. 4 (c), we have depicted separately the individual V_{TH} shifts for both devices as a function of the stress time. For both devices of the wafer split we obtain the following characteristics: (i) within the initial 1000 s constant bias phase in strong inversion (-2.0 V), we measure a similar amount of time dependent recovery (ΔV_{TH}^{time}) for both H-levels, cf. Fig. 4 (c1); (ii) the total V_{TH} shift decreases during SD1 and increases during SU1; (iii) a significant bias dependent reduction in the V_{TH} shift is observed after SU1 which is similar for both H-levels, cf. Fig. 4 (c2); (iv) after the intermediate CP cycle, the remaining degradation level is permanent and cannot be reduced further by an additional gate bias ramp toward 0.0 V; (v) The remaining permanent V_{TH} shift and the interface state dependent V_{TH} shift are much larger for wafer I than for wafer II, cf. Fig. 4 (c3); (vi) the interface state dependent V_{TH} shift is smaller than the permanent V_{TH} shift, cf. Fig. 4 (c3); (vii) interface state dependent and permanent V_{TH} shift coincide when multiplying $\Delta V_{TH}^{perm,it}$ by a factor 3, cf. Fig. 4 (c4).

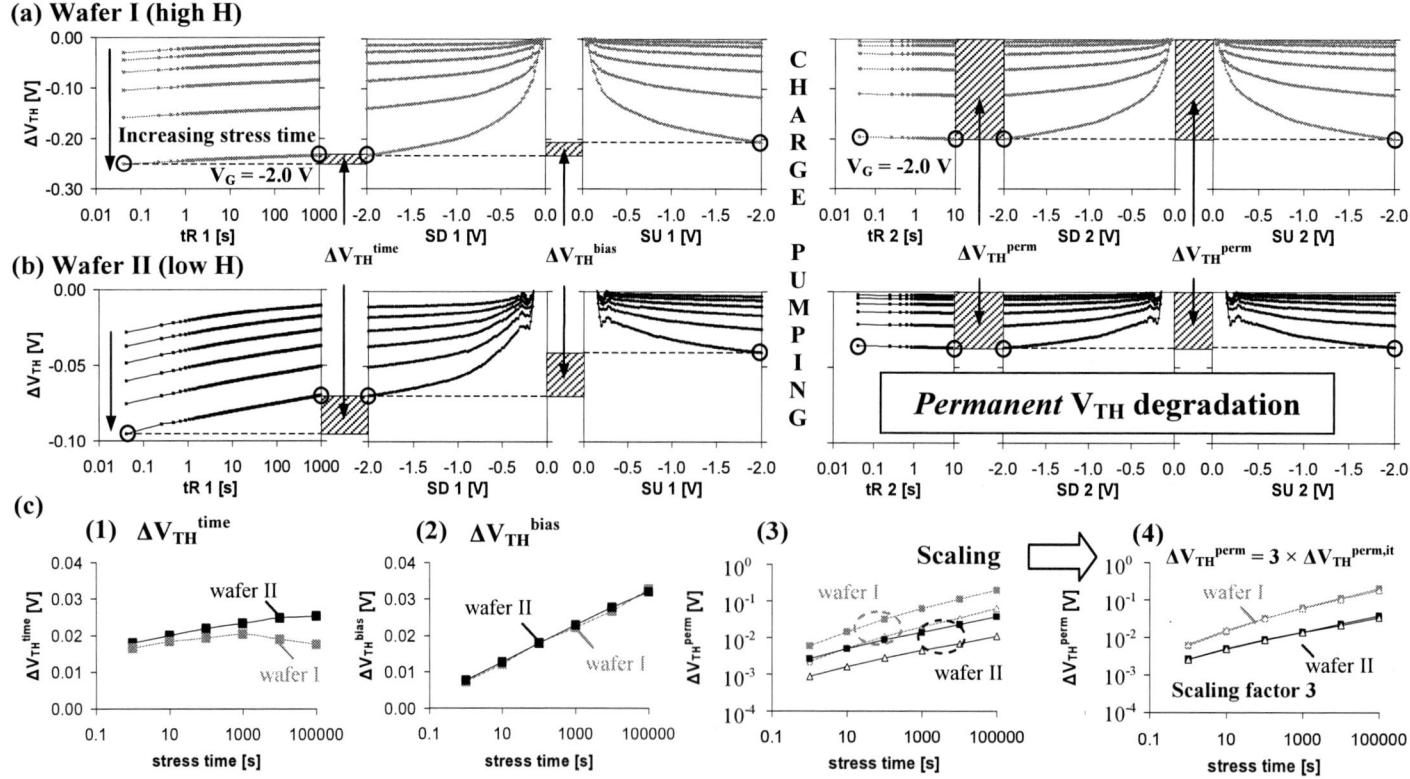

Figure 4. The V_{TH} shifts recorded after six subsequent stress-runs (1/10/100/1,000/10,000/100,000 s) at 50°C at different stages of the experiment (wafer I (a); wafer (II) (b)). The *time dependent V_{TH} recovery* (ΔV_{TH}^{time}) is recored at -2.0 V directly post stress for 1000 s (tR1). The *bias dependent recovery* (ΔV_{TH}^{bias}) is the difference in V_{TH} shift between SD1 and SU1 recorded at -2.0 V. Subsequently to SU1, we measure the maximum CP current and convert changes in I_{CP}^{max} into appropriate *interface state dependent V_{TH} shifts* ($\Delta V_{TH}^{perm,it}$). After gate pulsing, the remaining V_{TH} shift at -2.0 V is *permanent* (ΔV_{TH}^{perm}). In (c) the respective V_{TH} shifts are summarized for both wafers as a function of stress time: (c1) → ΔV_{TH}^{time}; (c2) → ΔV_{TH}^{bias}; (c3) → ΔV_{TH}^{perm} (full symbols) & $\Delta V_{TH}^{perm,it}$ (open symbols). By multiplying $\Delta V_{TH}^{perm,it}$ by a scaling factor 3, ΔV_{TH}^{perm} and $\Delta V_{TH}^{perm,it}$ coincide, cf. (c4).

These seven findings on the bias and time dependence of the recovery, on interface state creation and on permanent damage, as a function of the H content within the gate oxide, represent a significant collection of NBTI characteristics challenging the reliability of suggested models.

Cross check with the proposed NBTI models:

Point (i) – Recovery over time is H independent: Agrees only with the Grasser model which predicts that recovery (path A) is *nearly* independent of the hydrogen budget within the gate oxide except for the small fraction of recoverable damage being converted to locked-in oxide charge by hydrogen capture from the interface (path B), cf. Fig. 4 (c1). The RD model is challenged since it suggests a balance between hydrogen release and re-capture and would therefore predict a much larger amount of recovery for wafer I.

Point (ii) – V_{TH} shift is dependent on the applied gate bias: Agrees with both models which suggest that oxide traps and/or interface states may become neutralized and charge up positively again as the Fermi level crosses the silicon band gap [27-28].

Point (iii) – Ramping the gate toward 0.0 V accelerates V_{TH} recovery, the effect being H independent: Agrees only with the Grasser model which suggests bias dependent neutralization and relaxation of positively charged hydrogen-independent E' centers. The RD model cannot explain bias accelerated trap recovery since it assumes the hydrogen diffusion species to be neutral.

Point (iv) – The degradation level is quasi permanent after the intermediate CP cycle: In the Grasser model, this is explained by the bias-accelerated relaxation of E' centers which is finished after the first gate bias double ramp. The RD model would not expect significant recovery within the following 10-100 s either, since already more than 1000 s elapsed since the actual end of stress.

Point (v) – The remaining permanent V_{TH} shift is larger for the H rich wafer: Agrees with both models considering that the larger permanent damage of the H-rich wafer I is a logical consequence of its initially higher Si-H precursor concentration.

Point (vi): – $\Delta V_{TH}^{perm,it}$ is smaller than ΔV_{TH}^{perm}: Provided that our conversion of the CP signal into an interface state dependent V_{TH} shift ($\Delta V_{TH}^{perm,it}$) is correct, statement (vi) agrees only with the Grasser model which suggests that the permanent V_{TH} shift is the sum of locked-in oxide defects and interface states ($Q_{ox}^{perm} + Q_{it}^{perm}$). The RD model attributes the entire V_{TH} shift solely to interface states.

Point (vii) – $\Delta V_{TH}^{perm,it}$ is proportional to ΔV_{TH}^{perm}: Agrees with the Grasser model which predicts the simultaneous creation of $\Delta V_{TH}^{perm,it}$ and ΔV_{TH}^{perm} via hydrogen exchange between passivated interface states and E' centers and hence suggests a physical 50:50 relation and a strong correlation between $\Delta V_{TH}^{perm,it}$ and ΔV_{TH}^{perm}. The deviation in the measured factor 3 from the proposed factor 2 (50:50) may be due to different energy distributions of interface and oxide charges, leading to a different electrical response of both trap types. It has to be remarked that basically the profiled energy

range during CP and the energy range of defects being positively charged during the drain current measurement do not coincide. We have considered this energy mismatch assuming an amphoteric nature of interface traps [25] and a flat density of state profile [26].

Discussion on the absolute degradation potential

Fig. 5 illustrates the development of the absolute CP current (a) and the effective difference in the absolute V_{TH} between wafer I and II (b). Before stress (stress time = 0 s) wafer II has a considerable higher CP signal and also a much lower V_{TH} than wafer I. Note that this initial difference in the CP current (~ 92 nA) accounts only for ~ 53 mV of the difference in the initial V_{TH} (~ 160 mV) which reflects even on the virgin device the factor 3 measured between interface state creation and V_{TH} shift. The mismatch is increasingly compensated with increasing stress time. Indeed, after 100,000 s of stress, both devices have a similar CP signal and almost the same V_{TH}. However, since neither the CP signal nor the V_{TH} development of wafer I tends to saturate, it is likely to assume that wafer I would even have exceeded wafer II in the CP current and in the V_{TH} at longer stress times. Following Ref. [6], this would indicate that the sum of Si-H bonds plus interface traps (Si•) is initially higher in a thoroughly passivated gate oxide, implying an additional latent damage by excessive hydrogen ingress. This may lead eventually to a higher absolute drift potential for the H-rich wafer I at longer stress times. Additional measurements are required in order to verify this suggestion unambiguously.

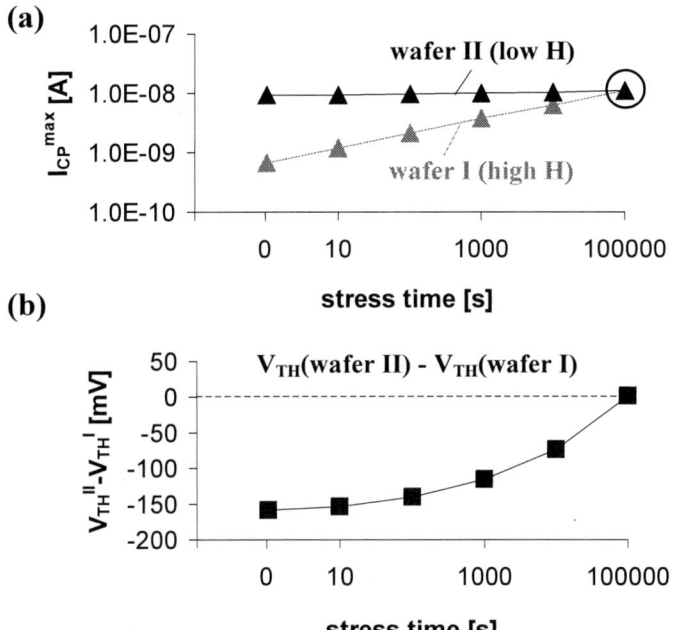

Figure 5. Development of the maximum CP current as a function of the stress time for wafer I and II (a). After 100,000 s of stress, the maximum CP signal of wafer I equals the maximal CP signal of wafer II indicating a similar number of interface traps at the end of the last stress run. Also, the initial difference in V_{TH} between wafer I and II is compensated at the end of the last stress run (b) indicating a similar degradation level of the oxide and the interface.

VII. CONCLUSIONS

By varying the Ti thickness during BEOL processing, we have produced two split wafers with vastly different hydrogen contents within the gate oxide and subjected them to NBTS. In agreement with the Grasser model, we have demonstrated that the recoverable part of the NBT degradation is largely independent of hydrogen while the permanent V_{TH} shift component is strongly linked to the total hydrogen budget within the gate oxide. Also, a strong correlation between the increase in the CP signal and the permanent V_{TH} shift component was found which is also consistent with the Grasser model.

REFERENCES

[1] K. O. Jeppson, and C. M. Svensson, "Negative bias stress of MOS devices at high electric fields and degradation of MNOS devices," J. Appl. Phys.,Vol. 48, Issue 5, pp. 2004-2014, 1977.

[2] M. A. Alam , and S. Mahapatra, "A comprehensive model of PMOS NBTI degradation," Microelectron. Reliab., Vol. 45, Issue 1, pp. 71-81, 2005.

[3] A. T. Krishnan, S. Chakravarthi, P. Nicollian, V. Reddy, and S. Krishnan, "Negative bias temperature instability mechanism: The role of molecular hydrogen," Appl. Phys. Lett., Vol. 88, Issue 15, pp. 153518 1-3, 2006.

[4] T. Grasser, B. Kaczer, and W. Goes, "An energy-level perspective of negative bias temperature instability," In Proc. 46th Annual International International Reliability Physics Symposium (IRPS), pp. 28-38, 2008.

[5] E. Cartier, J. H. Stathis, and D. A. Buchanan, "Passivation and depassivation of silicon dangling bonds at the Si/SiO_2 interface by atomic hydrogen," Appl. Phys. Lett., Vol. 63, Issue 11, pp. 1510-1512, 1993.

[6] M. Nelhiebel, J. Wissenwasser, Th. Detzel, A. Timmerer, and E. Bertagnolli, "Hydrogen-related influence of the metallization stack on characteristics and reliability of a trench gate oxide," Microelectron. Reliab., Vol. 45, Issue 9-11, pp. 1355-1359, 2005.

[7] Y. Nissan-Cohen, "The effect of hydrogen on hot carrier and radiation immuity of MOS devices," Appl. Surf. Sci., Vol. 39, Issue 1-4, pp.511-522, 1989.

[8] L. J. Jin, H. P. Kuan, D. Sim, and M. Mukhopadhyay, "Influence of hydrogen annealing on NBTI performance," In Proc. 15th IEEE International Symposium on the Physical and Failure Analysis of Integrated Cicuits (IPFA), pp. 1-4, 2008.

[9] J. Z. Xie, and S. P. Murarka, "Stability of hydrogen in silicon nitride films deposited by low-pressure and plasma enhanced chemical vapor deposition techniques," J. Vac. Sci. Technol. B, Vol. 7, Issue 2, pp. 150-152, 1989.

[10] C. Y. Chang, and S. M. Sze, "ULSI Technology," McGraw-Hill, Chapt. 5.6.3, Table 5, 1996.

[11] I. Jonak-Auer, R. Meisels, and F. Kuchar, "Determination of the hydrogen concentration of silicon nitride layers by Fourier transform infrared spectroscopy," Infrared Phys. Technol., Vol. 38, Issue 4, pp. 223-226, 1997.

[12] T. Grasser, B. Kaczer, W. Goes, Th. Aichinger, Ph. Hehenberger, and M. Nelhiebel, "A two-stage model for negative bias temperature instability," In Proc. 47th Annual International International Reliability Physics Symposium (IRPS), pp. 33-44, 2009.

[13] T. Pompl, K.-H. Allers, R. Schwab, K. Hofmann, and M. Roehner, "Change of acceleration behavior of time-dependent dielectric breakdown by the BEOL process: indicators for hydrogen induced transition in dominat degradation mechanism," In Proc. 43th Annual International International Reliability Physics Symposium (IRPS), pp. 388-397, 2005.

[14] A. D. Marwick, J. C. Liu, and K. P. Rodbell, "Hydrogen redistribution and gettering in AICu/Ti thin films," J. Appl. Phys.,Vol. 69, Issue 11, pp. 7921-7923, 1991.

[15] D. E. Woon, and D. S. Marynick, "Titanium and Copper in Si: barriers for diffusion and interactions with hydrogen," Phys. Rev. B: Condens. Matter, Vol. 45, Issue 23, pp. 13383-13389, 1992.

[16] T. Aichinger, M. Nelhiebel, S. Einspieler, and T. Grasser, "In-Situ Polyheater – A reliable tool for performing fast and defined temperature switches on chip," IEEE Trans. Device Mater. Reliab., 2009, in press.

[17] T. Aichinger, M. Nelhiebel, and T. Grasser, "On the temperature dependence of NBTI recovery," Microelectron. Reliab., Vol. 48, Issue 8-9, pp. 1178-1184, 2008.

[18] G. Pobegen, T. Aichinger, M. Nelhiebel, T. Grasser, "The Influence of the gate oxide thickness on negative bias temperature instability", In Proc. 48th Annual International International Reliability Physics Symposium (IRPS), 2010, in press.

[19] T. Grasser, B. Kaczer, T. Aichinger, W. Gös, and M. Nelhiebel, "Defect Creation Stimulated by Thermally Activated Hole Trapping as the Driving Force Behind Negative Bias Temperature Instability in SiO_2, SiON, and High-k Gate Stacks," In Proc. IEEE International Integrated Reliablity Workshop (IIRW), pp. 91-95, 2008.

[20] H. Reisinger, R. Vollertsen, P.-J. Wagner, T. Huttner, A. Martin, S. Aresu, W. Gustin, T. Grasser, and C. Schlünder, "The Effect of Recovery on NBTI Characterization of Thick Non-Nitrided Oxides," In Proc. IEEE International Integrated Reliablity Workshop (IIRW), pp. 1-6, 2008.

[21] S. Mahapatra, K. Ahmed, D. Varghese, A. E. Islam, G. Gupta, L. Madhav, D. Saha, and M. A. Alam, "On the Physical Mechanism of NBTI in Silicon Oxynitride p-MOSFETs: Can Differences in Insulator Processing Conditions Resolve the Interface Trap Generation versus Hole Trapping Controversy?," In Proc. 45th Annual International International Reliability Physics Symposium (IRPS), pp. 1-9, 2007.

[22] E. Niehuis, and T. Grehl, "Secondary Ion Mass Spectrometry," In Proc. 12th International Conference on Secondary Ion Mass Spectrometry (SIMS XII), pp. 49-54, 2000.

[23] B. Hagenhoff, D. Van Leyen, E. Niehuis, and A. Benninghoven, "Time-of-flight secondary ion mass spectrometry of insulators with pulsed charge compensation by low-energy electrons," J. Vac. Sci. Technol., A: Vacuum, Surfaces, and Films, Vol. 7, Issue 5, pp. 3056-3064, 1989.

[24] R.G. Wilson, "SIMS quantification in Si, GaAs, and diamond - an update," Int. J. Mass Spectrom. Ion Processes, Vol. 143, pp. 43-49, 1995

[25] P. V. Gray, and D. M. Brown, "Density of $Si-SiO_2$ interface states," Appl. Phys. Lett., Vol. 8, Issue 2, pp. 31-33, 1966.

[26] G. Groeseneken, H. E. Maes, N. Beltran, and R. F. De Keersmaecker, "A reliable approach to charge-pumping measurements in MOS transistors," IEEE Trans. Electron Devices, Vol. 31, No. 1, pp. 42-53, 1984.

[27] V. Huard, M. Denais, and C. Parthasarathy, "NBTI degradation: From physical mechanisms to modelling," Microelectron. Reliab., Vol. 46, Issue 1, pp. 1-23, 2006.

[28] D. S. Ang, S. Wang, G. A. Du, and Y. Z. Hu, "A consistent deep-level hole trapping model for negative bias temperature instability," IEEE Trans. Device Mater. Reliab., Vol. 8, No. 1, pp. 22-34, 2008.

A multi-probe correlated bulk defect characterization scheme for ultra-thin high-κ dielectric

*M. Masuduzzaman, A.E. Islam, and M.A. Alam

Dept. of ECE, Purdue University
West Lafayette, IN-47907, USA
*Phone: (765)-496-6517, Email: mmasuduz@purdue.edu

Abstract – **Various characterization techniques such as charge pumping (CP) and its variants (MFCP, VT²CP), 1/f noise, random telegraph noise (RTN), stress-induced leakage current (SILC) *etc.* are being used to identify the trap location within the dielectric. However, the most defective regions within the SiO₂/HfO₂ gate stacks identified by various methodologies are often not unique, leading to different optimization strategies for the gate stack. To resolve this issue, we develop a single theoretical framework to self-consistently interpret the MFCP, RTN, and SILC. Our analysis not only provides a consistent interpretation of different experimental observations of MFCP, SILC and RTN, but also identifies the capabilities and limitations of these techniques in terms of trap probing region. We show that the transition of the quasi Fermi level plays a vital role for the probing region with CP, SILC and consequently the classical interpretations of trap location from these experiments are not always correct for composite HK transistors. We demonstrate that none of these techniques can (in isolation) unambiguously back-extract the position of the traps and we suggest a correlation method to circumvent the 'uniqueness' problem.**

[Keywords: Bulk trap, high-k dielectric, trap profiling, charge pumping, Stress Induced Leakage Current, Random Telegraph Noise, Correlated probing.]

I. INTRODUCTION

High-k dielectrics (HK) are recently being used in the gate stack to achieve low EOT as required for the ultra-scaled transistors. However, due to the higher coordination number of the usual HK materials (mostly HfO₂) and lower annealing temperature during processing, it is well known that High-k (HK) gate dielectrics in modern CMOS transistors suffer from process-induced defects and are prone to defect formation under electrical stress. In order to optimize the process and thereby make the device reliable, it is thus extremely important to know the locations (both in energy and position) of the initial process induced as well as the dynamic stress induced defects within the HK gate stack.

Various characterization techniques such as charge pumping (CP) and its variants (MFCP [1,2], VT²CP [3]), 1/f noise [4,5], random telegraph noise (RTN) [5], stress-induced leakage current (SILC) [6-8] etc. have recently been used with a comprehensive effort to identify the location and energy levels of bulk defects ($N_T(x,E)$) within the dielectric. The conclusions reached by various techniques regarding $N_T(x,t)$ often appear not only to be at odds with each other (*e.g.*, the recent IL [Group-A: Ref. 6] vs. HK [Group-B: Refs. 2,7] debate, the most favorable SILC location [7], etc.), but also lead to results that appear anomalous/counterintuitive (RTN, [5]). It has thus become very important to take a closer look at the various

interpretation methodologies of different classical characterization techniques.

In this paper, we attribute different controversies and confusions to the theoretical approximations used to map back $N_T(x,E)$ from experimental measures of CP, SILC, and RTN. In an effort to resolve this apparent multi-facet puzzle, we (a) develop a single theoretical framework (Fig. 1) for interpreting MFCP, RTN, and SILC and provide a consistent interpretation of the experimental observations (Fig. 3-6). In particular, we (b) show that the probing regions of MFCP, VT²CP, SILC within composite HK transistors are significantly different than that of SiO₂. This invalidates some of the classical approximations used to interpret experimental data [8]. In addition, our theoretical analysis can (c) explain the experimental trends observed in Campbell's RTN experiment [5]. Moreover, similar to our analysis of MFCP in [9-10], we (d) show that $N_T(x,E)$ back-extraction is not unique (Fig. 4) for SILC (as presumed in [6,7]). Finally, we use the single comprehensive theoretical framework to (e) suggest how the cross-correlation among different characterization techniques (Fig. 7) could be a powerful tool for the unique back-extraction of $N_T(x,E)$ in certain regions within the dielectric.

II. THEORETICAL MODEL

We calculate the response in MFCP and SILC using the transient ($f_T(t)$) and steady state (f_{T0}) occupancy of traps, respectively as a function of gate voltage. Fig. 1 schematically shows different fluxes involved in calculating the occupancy of the traps. For each flux, the associated carrier density has been calculated using a self-consistent Schrodinger-Poisson solver [11]. The value of capture cross section used in this calculation is 10^{-17} cm² which is within the range of 10^{-13} cm²

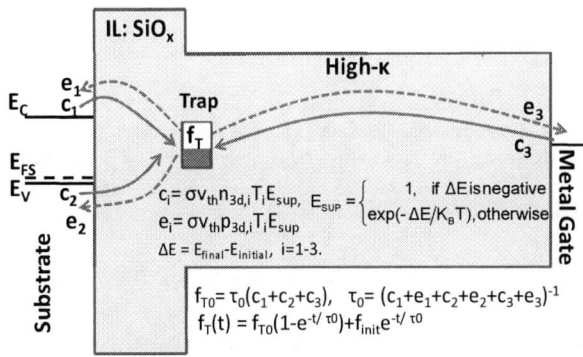

Fig. 1: The common theoretical framework for MFCP, SILC and RTN. Different flaxes are shown in the flatband condition.

978-1-4244-5430-3/10 $26.00 © 2010 IEEE

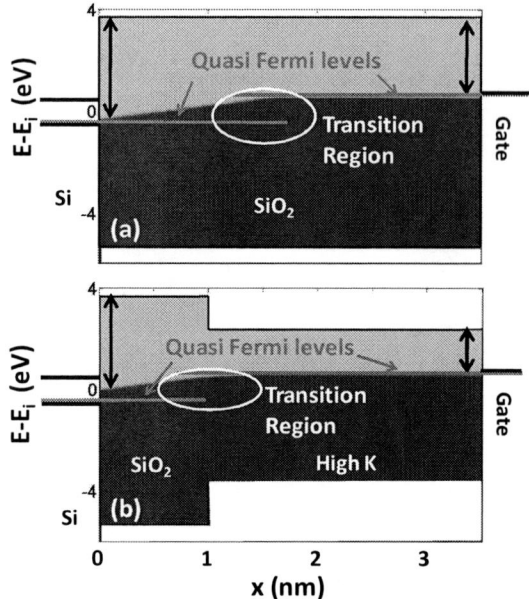

Fig. 2 **Transition Region of Quasi Fermi Level** – Important for CP depth, SILC area: The profile of f_{T0} (dark: filled, light: empty) at flat band for NMOS transistors having (a) 3.5nm SiO_2 and (b) for 1.0nm SiO_2 +2.5nm HfO_2 dielectric. Note that, f_{T0} is dictated by the quasi Fermi level at either side, as expected, and there is a *transition region* in between. For (a), this transition region is near the middle of the oxide. However for (b), this is near the IL/High-K interface, due to the presence of lower tunneling barrier for HK at the gate side.

[12] and 10^{-20} cm^2 [13] reported in the literature.

The calculation of charge pumping current as a function of different trap location can be found in [9,10] in detail. The trap assisted tunneling current in SILC (I_{SILC}) is calculated from the steady state occupancy, f_{T0} using the formula- $I_{SILC} = qA_G[c_1(1-f_{T0})-e_1 f_{T0}]N_T \Delta x \; \Delta E$, where A_G=gate area (in cm^2), Δx ΔE is the area of interest in the position energy space (in cm-eV), N_T=trap density in that area (in cm^{-3}eV^{-1}). Finally, the simulation of RTN involves the calculation of capture/emission times (τ_c/τ_e), drain current fluctuation ($\Delta I_D/I_D$), *etc.* at different gate overdrives (V_G-V_T), considering the contribution from substrate conduction band only. Here, the mean capture (emission) time is calculated from the capture (emission) rate as- $\tau_c = 1/c_1$ ($\tau_e = 1/e_1$). The fractional drain current fluctuation is calculated from the formula: $\Delta Id/Id = \Delta Q_n/Q_n$, where $Q_n= A_G n_{2d,1} =$ total no. of channel carrier, $\Delta Q_n= \Delta Q_T(t_{ox}-d_t)/t_{ox}$, $\Delta Q_T =$ no. of traps responsible for the drain current fluctuation and is taken as unity for RTN, t_{ox} and d_t are the physical oxide thickness and the trap depth from interface respectively.

Our simulations indicate that the profile of f_{T0} (Fig. 2) and the transition region of the Quasi Fermi level *plays significant role* in defining the scannable regions in the (x,E) space for MFCP, VT^{2}CP, SILC *etc.* (sections 3,4) and is significantly different for SiO_2 and HK.

III. DEPTH AND NON-UNIQUENESS OF MFCP

Since its introduction in early 1990s [1], MFCP technique has widely been used for extracting N_T(x,E). Although both Group–A and B use MFCP, they reach at different conclusion on the *weak link* [2,6] because of their disagreement on the probing depth with MFCP. Group-A mainly limits MFCP to

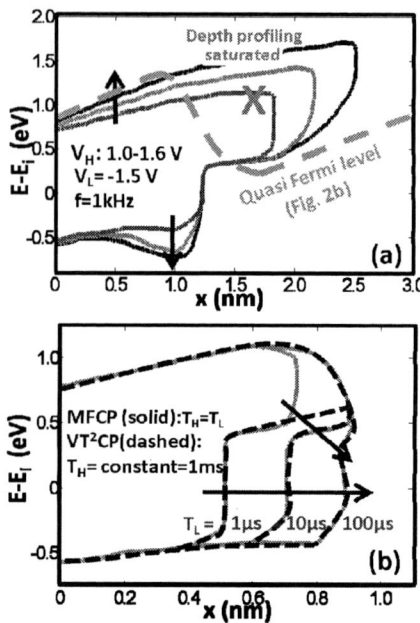

Fig. 3 **Charge Pumping/VT^{2}CP depth and uncertainties**: (a) Contours of (x,E) scanned by MFCP during HI voltage (V_H) sweep for the transistor of Fig. 2b without considering contributions from gate. Influence from gate Fermi level restricts the scanning depth beyond the dashed line [10]. (b) T_L of the gate pulse has been varied for: fixed HI voltage duration T_H (as in VT^{2}CP) and T_H=T_L (as in MFCP). Moreover, note that as V_H(a) or T_L(b) increase, the probing region extends in different (x,E) as marked by the arrows. Hence, N_T back extraction is not unique.

the IL (~1nm), whereas Group-B claims that it can probe beyond IL into the HK. We, for the first time, explore that it is the transition region of Quasi Fermi level (Fig.2,3), that determines the maximum probing depth in CP, and beyond this point, the gate Fermi level takes over and no further probing with CP (both MFCP and VT^{2}CP) is allowed. The transition region primarily depends on the relative barrier height on either side of the oxide (Fig. 2). Unfortunately, for a typical HK structure [2], this can only extend 1-2 Å beyond the IL/HK boundary, and hence it is not possible to confidently comment on the generated defect location (IL or HK) based on CP alone (contradicting both Group-A,B).

Fig. 3b compares the probing depth between MFCP and VT^{2}CP for the variation of LOW voltage duration T_L of the gate pulse. In one case, T_L has been varied with fixed HI voltage duration T_H (as in VT^{2}CP [3]) whereas in the other case T_H is kept equal to T_L (as in MFCP). Since classical CP depth (for short channel length) is limited by T_L only (due to higher hole barrier [9-10]), (x,E) scanned by MFCP and VT^{2}CP are same. Both of them saturates at ~1nm due to transition of Quasi-Fermi level. Moreover, with the variation of different parameters in CP (MFCP/VT^{2}CP), the probing region extends along different (x,E) directions (shown by arrows, Fig. 3), indicating the non-uniqueness in the extracted N_T(x,E) using charge pumping alone.

IV. PROBING LOCATION BY SILC

In addition to MFCP, SILC has been widely used by Groups-A, B to prove their respective claim about the *weak link* in HK [6-8]. It is thus worthwhile to precisely examine the primary SILC location and the uniqueness of the extracted defect location. Our simulation suggests that (Fig. 4a) in HK

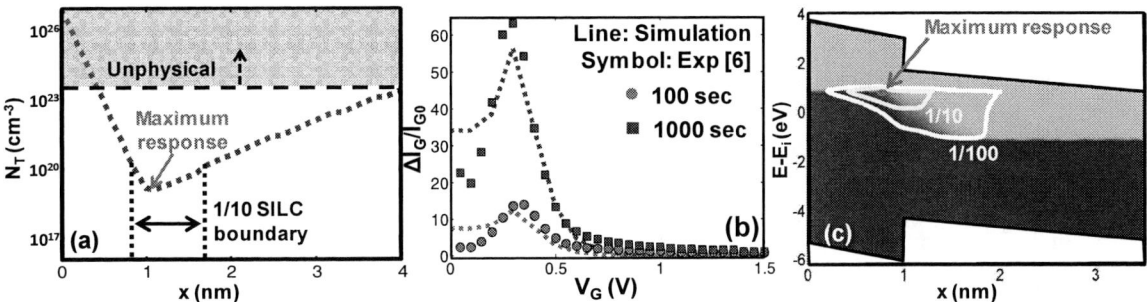

Fig. 4: **SILC region in HK and uncertainties**: (a) The required N_T as a function of x for a given I_{SILC} (or ΔI_G) in a HK transistor. Here, 1/10 boundary is defined as the region, where the required N_T for *same I_{SILC}* is 10 times, compared to the one required at maximum response point. Although this point is located at IL/HK interface, higher (but physical) amount of N_T at different location could produce same ΔI_G. (b) Experimental match of ΔI_G for a particular defect location. (c) Different SILC response area within the (x,E) space of the dielectric. The defect occupancy level f_{T0} is also shown in (c) for convenience with dark for $f_{T0} = 1$, light for $f_{T0} = 0$. This shows that SILC primarily originates around the transition region (see Fig. 2) of the quasi Fermi level.

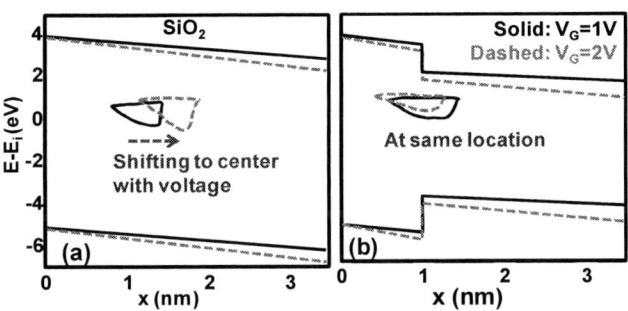

Fig. 5: **V_G dependent SILC: HK is different from SiO_2**. 1/10 contour lines for (a) SiO_2: Fig. 2a and (b) HK: Fig. 2b transistors. Increase in V_G has more influence on the maximum response point in SiO_2 compared to HK. This is because the relative influence of substrate Fermi level increases in SiO_2 with V_G. However, this cannot happen in HK, due to the already greater influence from gate Fermi level (as tunneling barrier is lower near the gate end).

transistors the main SILC contribution (*i.e.*, minimum N_T required for a definite I_{SILC}) results from the defects near the IL/HK interface. However, it is entirely possible to get same SILC from a different location (on either IL/HK sides) with higher (but within physical range) N_T. Thus although for a particular (x,E), it is possible to reproduce the experimental trends (Fig. 4b), the uncertainties in the back-extracted $N_T(x,E)$ using SILC alone must be considered. We also identify that the maximum SILC location in the (x,E) space is primarily determined by the transition of the two Quasi Fermi level and for HK transistor, it is at the IL/HK boundary (Fig. 4c).

Moreover, Fig. 5 signifies the influence of V_G on the maximum response point, which is more pronounced for transistors having SiO_2 dielectric. These observations suggest revision in the SILC spectroscopy technique [7-8], which assumes that SILC originates from the middle of the oxide regardless of the dielectric structure and gate bias.

V. CONTROVERSY ON RTN

Noise has long been used as a characterization tool for bulk defects, and it has most recently been used to locate deep bulk defects in high-κ dielectric [4]. Our theoretical framework provides intuitive explanation (Fig. 6a-c) of the (sometimes counterintuitive) trends of τ_c, τ_e, $\Delta I_D/I_D$ [5] for low V_G-V_T. Although noise/RTN in [5] for a SiON structure, appears to deviate from theoretical prediction at high V_G-V_T, and suggesting the possibility that classical model is incorrect, consider the following simple explanation (Fig. 6c): If the mean of τ_c/τ_e distribution (PDF) lies beyond the measurement window (limited by the bandwidth and sample size), one would therefore over- (under-) estimate mean τ_c (τ_e), if mean τ_c/τ_e are estimated by fitting residual tail of the PDF (measured data) with an exponential function [5]. Calculations based on model PDF systems and given bandwidth limits appear to consistently support this interpretation. Therefore, primarily the classical theory does not appear to be inconsistent with experimental data (contrary to the claim in [5]). Here, we emphasize on the need for accounting the distribution of τ_c/τ_e, it's cut-off by the limited bandwidth of the measurement, and the extraction methodology of the representative τ_c/τ_e while comparing it with the theoretical *mean* value. A further study with all these considerations is required to investigate any non-

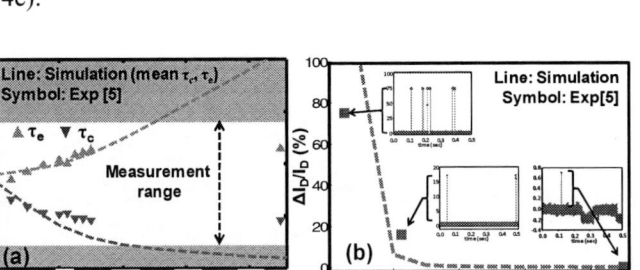

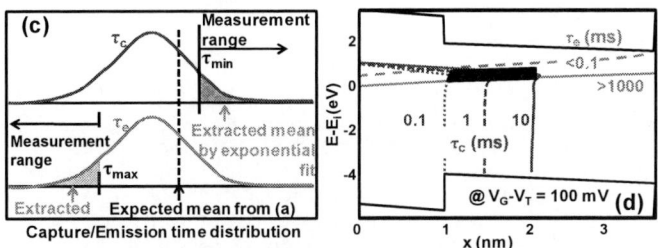

Fig. 6: **RTN- Experimental trends and probing region**: The simulated (a) mean τ_c, τ_e and (b) $\Delta I_D/I_D$ at different V_G-V_T are consistent with the experimental trends of [5] for a defect located at certain (x,E). The shaded regions in (a) indicate the timing limitations ($\tau_{max} \sim 300s$ and $\tau_{min} \sim 16\mu s$ [5]) of the RTN setup. (c) When the expected mean τ_c, τ_e are beyond the measurement window, only the data points from the tail of the τ_c, τ_e distribution (shaded regions) can be recorded. An exponential fitting [5] to such tail of the PDF would greatly over- (under-) estimate the extracted τ_c (τ_e). (d) The contour lines for different τ_c and τ_e for the transistor in Fig. 2b

classical origin of RTN.

Based on the present theory, the contour lines for different τ_c and τ_e for the transistor in Fig. 2b (HK structure) are shown in Fig. 6d. For a particular RTN setup, we obtain the marked regions at the center, which indicates the scannable (x,E) regions by RTN.

VI. CORRELATION

After identifying the probing regions for three different techniques (MFCP, SILC, and RTN) along with their uncertainties, we establish a guideline (Fig. 7) for back-extracting $N_T(x,E)$ using these experiments. For example, since a differential signal (*e.g.,* V_H variation of Fig. 3a) in MFCP can back-extract $N_T(x,E)$ at both edges in a non-unique way, one can use SILC to differentiate between the marked regions 1 and 2. In a similar effort, Group-A [6] uses correlation between MFCP and SILC to show that the main generated defects are in IL. However, in order to establish that this is the sole defect location, it is important to use a third experiment, which probes only within HK (*e.g.,* noise/RTN), to show that such (or stronger) correlation does not exist (see [4] for similar efforts). Using our numerical framework, we have established a correlation table between these experiments, which can be used to probe $N_T(x,E)$ in the marked regions (numbered 1-6), thus complementing the previous studies [4,6].

VII. CONCLUSION

We have studied MFCP, RTN, and SILC techniques under a common theoretical framework. Our simulation identifies that VT^2CP cannot probe deeper within the HK structure than conventional MFCP due to the gate detrapping effect. It can further explain the experimental trends observed in recent RTN experiments on ultra-scaled transistors. The probing regions have been shown to be non-unique and for SILC, this region is completely different form the one presumed in recent reports [6-8]. Our analysis demonstrates the impossibility of determining defect-location uniquely by any of the above techniques alone. However, we suggest the necessary and sufficient conditions to locate trap by different correlation measurements. Therefore, we assert the need for performing a set of experiments, rather than one single type, in order to uniquely determine $N_T(x,E)$ and understand the *weak-link* [6] for HK reliability.

ACKNOWLEDGEMENT

The authors would like to thank the Network for Computational and Nanotechnology (NCN) for providing computational facilities for this work.

REFERENCES

[1] R. E. Paulsen and M. H. White, "Theory and Application of Charge Pumping for the Characterization of Si-SiO₂ Interface and Near-Interface Oxide Traps", *IEEE Trans. Electron Devices*, vol. 41, pp. 1213-1216, 1994.

[2] E. Cartier, B. P. Linder, V. Narayanan, *et al.*, "Fundamental Understanding and Optimization of PBTI in nFETs with SiO₂/HfO₂ Gate Stack," *IEDM Tech. Dig.*, pp. 317-320, 2006.

[3] M. B. Zahid, R. Degraeve, L. Pantisano, *et al.*, " Defects Generation in SiO₂/HfO₂ Studied with Variable T$_{CHARGE}$-T$_{DISCHARGE}$ Charge Pumping (VT²CP)," *Proc IEEE IRPS*, pp. 55-60, 2007.

[4] H. D. Xiong, D. Heh, M. Gurfinkel, et al., "Characterization of electrically active defects in high-k gate dielectrics by using low frequency noise and charge pumping measurements," *Microelectron Eng.*, vol. 84, pp. 2230-2234, 2007.

[5] J. P. Campbell, J. Qin, K. P. Cheung, *et al.*, "Random Telegraph Noise in Highly Scaled nMOSFETs," *Proc IEEE IRPS*, pp. 382-388, 2009.

[6] G. Bersuker, D. Heh, C. Young, *et al.*, "Breakdown in the metal/high-k gate stack: Identifying the "weak link" in the multilayer dielectric," *IEDM Tech. Dig.*, pp. 791, 2008.

[7] E. Cartier, A. Kerber, "Stress-Induced Leakage Current and Defect Generation in nFETs with HfO2/TiN Gate Stacks during Positive-Bias Temperature Stress," *Proc IEEE IRPS*, pp. 486-492, 2009.Aa

[8] R. O'Connor, L. Pantisano, R. Degraeve, *et al.*, "SILC Defect Generation Spectroscopy in HfSiON using Constant Voltage Stress and Substrate Hot Electron Injection," *Proc IEEE IRPS*, pp. 324-329, 2008.

[9] M. Masuduzzaman, A. E. Islam and M. A. Alam, "Exploring the Capability of Multi-Frequency Charge Pumping in Resolving Location and Energy Levels of Traps within Dielectric," *IEEE Trans. Electron Devices*, vol. 55, pp. 3421-3431, 2008.

[10] M. Masuduzzaman, A. E. Islam and M. A. Alam, "Physics and Mechanisms of Dielectric Trap Profiling by Multi-Frequency Charge Pumping (MFCP) Method," *Proc IEEE IRPS*, pp. 13-20, 2009.

[11] A. Ghetti, A. Hamad, P. J. Silverman, et al., "Self-consistent simulation of quantization effects and tunneling current in ultra-thin gate oxide MOS devices," *SISPAD*, pp. 239 - 242, 1999..

[12] T. H. Ning, "Capture Cross-Section and Trap Concentration of Holes in Silicon Dioxide," *J. Appl. Phys.*, vol. 47, pp. 1079-1081, 1976.

[13] S. Zafar, A. Kumar, E. Gusev, et al., "Threshold voltage instabilities in high-k gate dielectric stacks," *IEEE Trans. on Dev. and Mat. Rel.*, vol. 5, pp. 45-64, 2005.

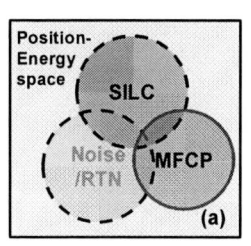
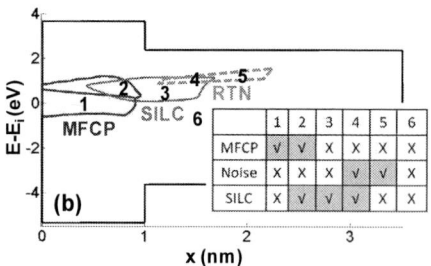

Fig. 7 **Multi-probe Characterization**: The probing regions for three different techniques (MFCP, RTN, and SILC) are shown together in (a) Venn diagram, (b) the position-energy (x-E) space. In order to uniquely determine (x,E) for as-grown or stress-induced defects, it is required to perform different type of experiments. For example, a differential signal (*e.g.,* V_H variation of Fig. 3a) in MFCP can back-extract $N_T(x,E)$ at the edges of region 1,2. However, one can use SILC to differentiate between these two regions. Similarly, defects in region 3 will only create response in the SILC experiment. Fig. b summarizes the correlation table between these experiments by explicitly identifying the responses expected from different (x,E) locations (numbered 1-6).

Dependence of the Negative Bias Temperature Instability on the Gate Oxide Thickness

Gregor Pobegen[*][†], Thomas Aichinger[*], Michael Nelhiebel[‡] and Tibor Grasser[§]

[*]Kompetenzzentrum für Automobil- und Industrieelektronik (KAI) GmbH
Europastraße 8, A-9500 Villach
+43-4242-34890-24, gregor.pobegen@k-ai.at, thomas.aichinger@k-ai.at
[†]Institute for Solid State Physics, Graz University of Technology
Petersgasse 16/II, A-8010 Graz
[‡]Infineon Technologies Austria AG
Siemensstraße 2, A-9500 Villach
[§]Institute for Microelectronics, Vienna University of Technology
Gusshausstraße 27-29, A-1040 Vienna

Abstract—The exact location and type of defects created under negative bias temperature (NBT) stress in pMOS field effect transistors is still a highly debated topic. We present a detailed study on equivalent devices with different oxide thicknesses (5 to 30 nm) where we show experimentally that the basic mechanisms behind the NBT instability are the same in thin and thick oxide technologies. In particular, voltage driven degradation like impact ionization or anode hole injection are not the driving forces for the larger degradation of thick oxide devices. Finally, we show that defects created under NBT stress are not solely located at the interface but extend a few nanometers into the oxide.

I. Introduction

During negative bias temperature stress (NBTS), a negative shift in the threshold voltage of pMOS devices is observed [1–5]. This shift is due to energetically widely spread defects which emerge via bond breaking and subsequent charging of donor-like defects or positively charged oxide defects. However, besides the position in energy, also the exact location and type of the defects within the gate oxide is a widely debated topic [1,6–8] raising the question, if the physical mechanisms behind the negative bias temperature instability (NBTI) are the same when comparing thin and thick oxide technologies. In particular, thick oxide devices are often assumed to have a larger contribution of oxide defects, while NBTI degradation of thin oxide devices is sometimes mainly attributed to the creation of interface states [4,9,10]. This view is often physically argued by degradation mechanisms like impact ionization or anode (hot) hole injection which would increase the observed degradation of thick oxide devices. During impact ionization, carriers, which had tunneled through a part of the oxide layer or had been thermally activated, become accelerated by the electric field. They are then capable of creating electron-hole pairs in the SiO_2, which may cause fixed positively charged oxide traps [11]. In the anode hole injection model, it is assumed that electrons tunnel from the poly gate towards the silicon substrate, where they excite deep valence band electrons. These electrons leave behind (hot) holes which may penetrate back into the oxide causing oxide damage [12]. Both processes are initiated by electrons which

tunnel through a part of the gate oxide barrier and then are accelerated by the electric field in the oxide or the substrate. The maximum energy of the electrons arriving from the poly gate depends highly on the voltage drop across the oxide (V_{ox}). Hence V_{ox} has to exceed the band gap of the SiO_2 insulator of $\approx 9\,eV$ before these mechanisms become efficient [13–16]. This means, when applying conventional oxide fields during NBTS $|E_{ox}| < 7\,MV/cm$, this critical voltage drop is reached as the gate oxide thickness exceeds roughly $13\,nm$.

We chose our oxide thicknesses such that identical stress fields imply voltage drops from far below to well above the band gap of the oxide. This helps to investigate whether the voltage driven degradation effects mentioned above contribute and thus completes the work of earlier studies [17]. We paid particular attention to equivalent stress and measurement conditions for the different devices, namely identical oxide fields during stress and equal Fermi level positions relative to the conduction and valence band edges during recovery. In the analysis of our measurement results we account for the different capacitances of the oxides. Considering all this, we are able to perform an exact comparison of thin and thick oxide technologies.

II. Experimental Setup

For our study we used pMOS devices with n^{++} poly gates with effective oxide thicknesses of $5.6\,nm$, $13.9\,nm$ and $29.1\,nm$, calculated out of capacitance measurements in accumulation, and stressed them at the same temperature (125°C) and under the same negative oxide field with $|E_{ox}| < 7\,MV/cm$. The temperature was kept constant throughout the whole experiment. Note that when applying the same electric fields to different oxide thicknesses, we end up with completely different voltage drops across the oxide (V_{ox}), which are roughly half of the SiO_2 band gap for the thin oxide device and roughly twice the band gap for the thick oxide device. In order to compare NBTI induced V_{TH} shifts of different oxide thicknesses, it is important to assure the following experimental conditions:

978-1-4244-5430-3/10 $26.00 © 2010 IEEE

- All devices must be stressed equivalently. It is generally assumed that the oxide field and/or the hole population at the interface are the driving forces causing NBTI degradation [8]. However, we remark that the same oxide field implies similar carrier concentrations at the interface for different oxide thicknesses.
- During recovery, the Fermi level position and the carrier concentrations at the interface must be identical as well to assure equivalent occupancy levels of the interface states and identical recovery conditions for the defects created during stress.

A. Equivalent stress conditions

To determine the relationship between the oxide field (E_{ox}) and the voltage applied to the gate (V_G), we used the method

$$E_{ox}(V_G) = \frac{\int_{V_1}^{V_G} C(V)\, dV}{\varepsilon_{SiO_2}} + \text{const} \tag{1}$$

proposed in [17]. The constant in Eq. (1) is determined by the condition that the oxide field has to be close to zero at the flat-band voltage. V_1 can be chosen arbitrarily except for the requirement $V_G < V_{FB} < V_1$ for negative stress voltages.

B. Equivalent recovery conditions

Using a drain current criterion, we estimated the correct gate voltages during recovery that guarantee a similar Fermi level position. The drain current is directly proportional to the density of inversion channel carriers and therefore also proportional to the Fermi level position at the interface. We determined the appropriate readout voltages V_R for different oxide thicknesses from virgin transfer curves to be $V_R = -0.76\,V, -0.82\,V$ and $-1.04\,V$ for $T_{ox} = 5.6\,nm, 13.9\,nm$ and $29.1\,nm$, respectively which is around the V_{TH} of all devices.

By using Berglund's method [18], which gives the dependency of the surface potential ψ_S on the applied gate voltage V_G, we experimentally verified that the drain current criterion actually assures similar surface potentials with an accuracy of around $60\,mV$ (c.f. Fig. 1).

C. Experimental details

Fig. 2 shows the evolution of the voltages applied to the gate and to the drain during a basic measurement cycle. During stress, we applied the stress voltage V_S for $t_{str} = 10^0, 10^1, \ldots, 10^5\,s$ to the gate of the device. After stressing the device, the gate voltage was switched to the appropriate readout level V_R and simultaneously the drain bias was switched from $0\,V$ to $V_D = -100\,mV$ for $t_{rec} = 10^0, 10^1, \ldots, 10^5\,s$. The V_{TH} recovery of the device was monitored by measuring the change in the drain current and subsequent conversion to a corresponding V_{TH} shift [19]. Following constant bias recovery, the interface trap density was monitored for $t_{CP} = 10\,s$ by recording the maximum charge pumping (CP) current [20, 21]. We used the same experimental CP setup for all three oxide thicknesses, namely $1\,MHz$ gate pulsing frequency and a pulse amplitude of $2\,V$ ranging from deep accumulation ($+0.5\,V$)

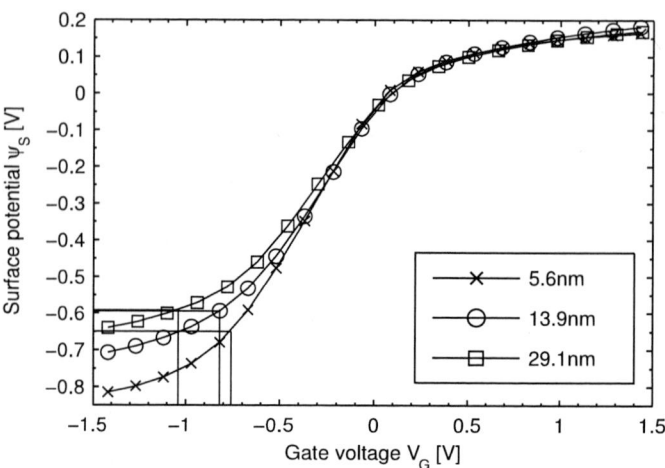

Fig. 1. Measured change of the surface potential ψ_S with the gate voltage V_G calculated according [18]. In inversion different voltages correspond to the same surface potential for different oxide thicknesses.

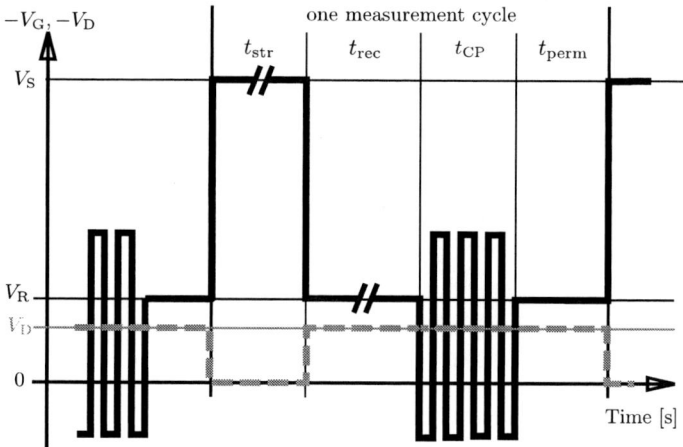

Fig. 2. Evolution of the voltage applied to the gate (V_G) and the drain (V_D) over time. During t_{str}, t_{rec} and t_{perm} a constant bias is applied to the gate and the drain contact. During t_{CP} a charge pumping measurement is performed.

to deep inversion ($-1.5\,V$). Since we observed a vanishing influence of V_{TH} and V_{FB} for different T_{ox} on the energy range scanned by CP [21], the dependence of the maximum CP current on the oxide thickness is negligible. By switching the gate bias between inversion and accumulation during CP, repeatedly majority electrons become attracted to the interface during the accumulation phase, favoring neutralization and annealing of positively charged oxide defects located close to the interface [8, 22]. Recent findings [23] suggest the that remaining V_{TH} shift after CP can be considered to be permanent on the timescale of an experiment. The remaining degradation was recorded for $t_{perm} = 10\,s$.

The experiments described above were performed on three different devices on three different wafers having different oxide thicknesses. The devices had been carefully selected to have similar virgin maximum CP currents, which was found to guarantee similar degradation behavior [24].

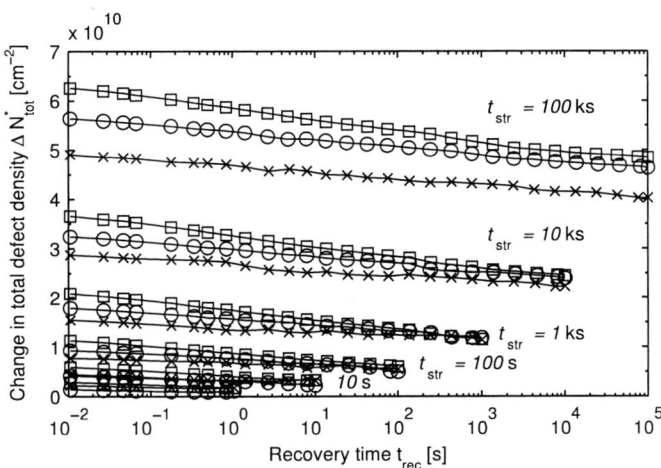

Fig. 3. The change of total effective defect charge density per area ΔN_{tot}^* over recovery time t_{rec} for six different stress times. Crosses $\times$ indicate the 5.6 nm, circles $\odot$ the 13.9 nm and squares $\boxdot$ the 29.1 nm device.

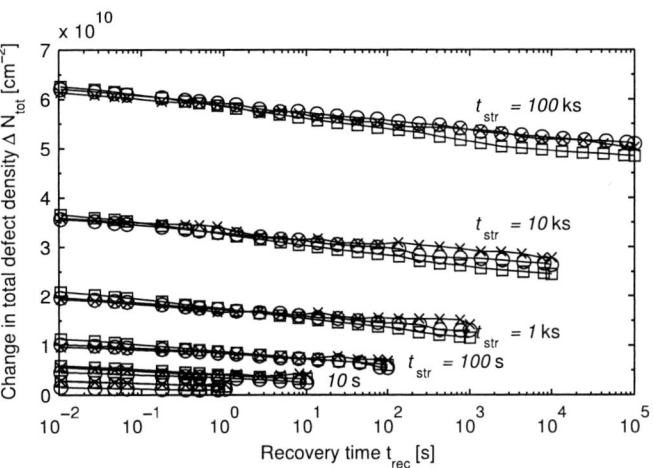

Fig. 4. Scalability holds for different oxide thicknesses. Depicted is the data of Fig. 3 multiplied with 1.25 for the 5.6 nm device and 1.10 for the 13.9 nm device.

III. EXPERIMENTAL RESULTS

As already mentioned in [17], the different oxide capacitances C_{ox} have to be taken into account when comparing the V_{TH} shifts of different oxide thicknesses. This is because a single charge q located at the silicon substrate – silicon dioxide interface causes a threshold voltage shift of $\Delta V_{\text{TH}} = q/C_{\text{ox}}$. Thus, a more general way to compare the V_{TH} shifts of devices with different capacitances is to characterize them by their change in effective defect charge density per area

$$\Delta N_{\text{tot, perm}}^* = \frac{\Delta V_{\text{TH}} C_{\text{ox}}}{q}, \tag{2}$$

with q the elementary charge, assuming that all defects are located close to the interface.

The recovery traces of the devices right after stress are illustrated in Fig. 3. Remarkably, multiplying the data of the 5.6 nm device with 1.25 and the data of the 13.9 nm device with 1.10 leads to a very good agreement of the recovery characteristics, as can be seen in Fig. 4. The very first data points in each trace in Fig. 3, that is to say the actual drifts 10ms after termination of stress, are plotted as a function of stress time t_{str} in Fig. 5. We remark that apart from these small scaling factors, the overall stress and recovery behavior of the devices is similar, if not equivalent.

A. Permanent degradation and interface states

In Fig. 6 the total change in defect density as well as the change in permanent defect density is illustrated as a function of the stress time. The values for ΔN_{tot}^* are taken from the data of Fig. 3 (recorded 10 ms after termination of stress). The permanent degradation ΔN_{perm}^* was measured after constant bias recovery and CP. During the positive bias phase of the CP gate pulse, when majority electrons accumulate at the interface, positive bulk traps may exchange carriers with the silicon substrate by tunneling processes [6]. Once an electron is captured, the trap is neutralized and it may anneal

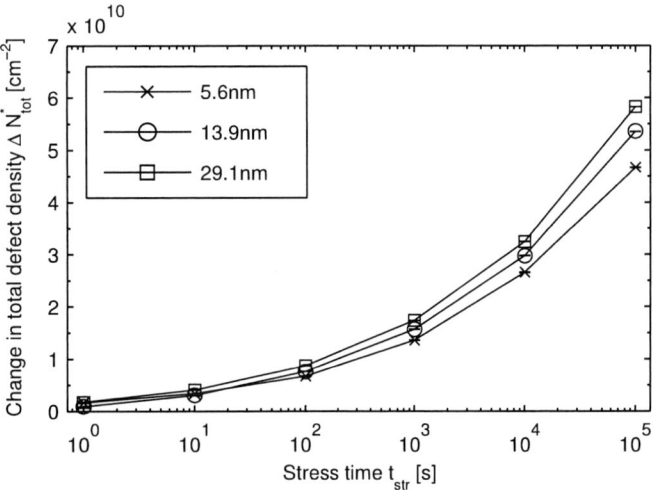

Fig. 5. The total change in effective defect charge density ΔN_{tot}^* per area 10 ms after termination of stress is illustrated over the stress time t_{str}.

subsequently [2]. The remaining defects are assumed to have very long time constants compared to the time scale of our experiment. Hence the remaining shift after CP is labeled as permanent in this study. The change in the maximum CP current ΔI_{CP} is depicted in Fig. 7. Note that the increase in the maximum CP current after stress is similar for all three oxide thicknesses indicating an equivalent increase of interface traps.

Impact ionization or anode hole injection during NBTI are of vanishing probability in our thinnest oxide device (5.6 nm) because of the small voltage drop across the oxide ($V_{\text{ox}} <$ 4 V) [25]. Since all analyzed larger oxide thickness devices have a similar degradation behavior we suggest that impact ionization or anode hole injection do not play a crucial role when subjecting devices to NBTS at commonly used oxide fields. In particular, this argument holds even for devices where V_{ox} may greatly exceed the band gap of the dielectric.

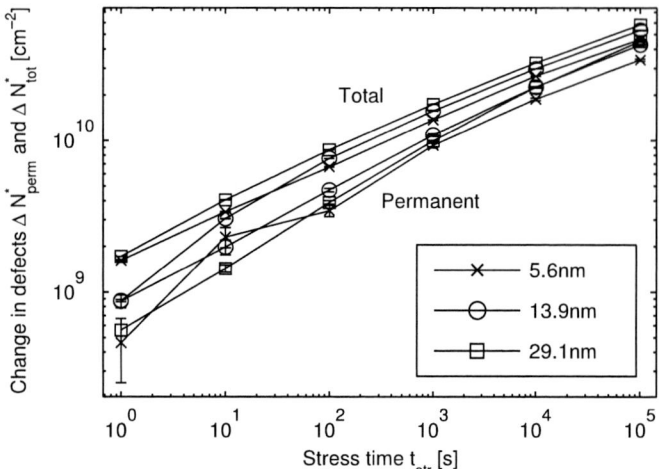

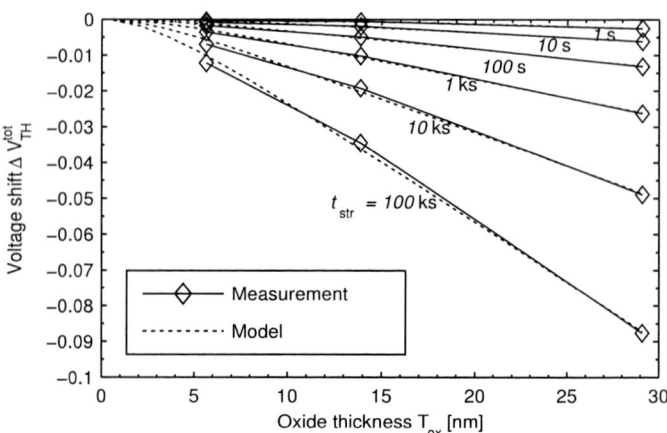

Fig. 6. The total change in effective defect charge density ΔN_{tot}^* per area 10 ms after termination of stress with t_{str} seconds duration and the permanent change in effective defect charge density per area ΔN_{perm}^* is illustrated over the stress time t_{str}. The data for ΔN_{perm}^* are the average values of measurements with $t_{perm} = 10$ s duration. We have drawn error bars according the standard deviation of the measured ΔN_{perm}^* data.

Fig. 8. Threshold voltage shifts 10 ms after termination of stress over oxide thickness. The six different lines indicate six different stress times $t_{str} = 10^0, \ldots, 10^5$ s. The dotted lines show the ideal relationship assuming the exponential defect density distribution Eq. (4).

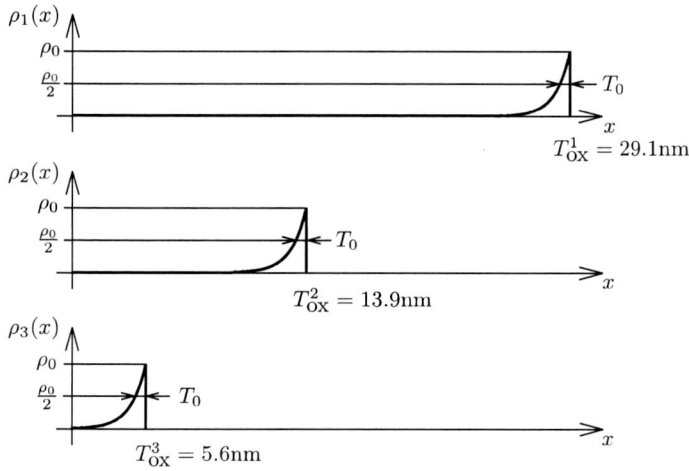

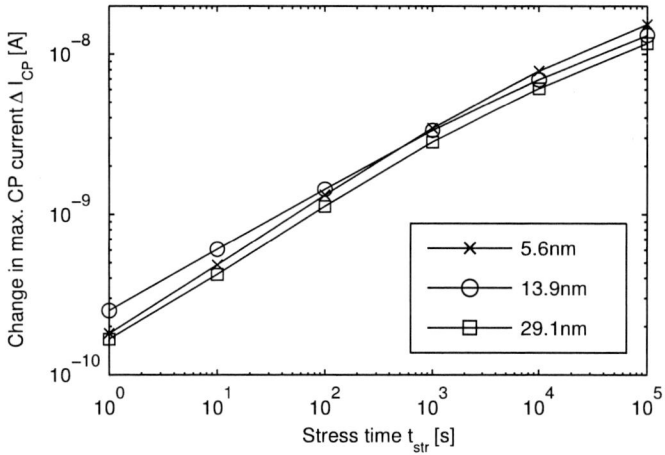

Fig. 7. Change of the maximum charge pumping current after $t_{str} = 10^0, \ldots, 10^5$ s of stress and $t_{rec} = 10^0, \ldots, 10^5$ s of recovery measured for $t_{CP} = 10$ s. Error bars of the standard deviation of the measurement data would be smaller than the size of the symbols and are thus not depicted here.

Fig. 9. Spatial defect density distributions for different oxide thicknesses $T_{ox}^1, T_{ox}^2, T_{ox}^3$ under the assumption that equivalent NBTS causes equivalent defect density distribution.

B. Spatial defect distribution

The remaining discrepancy of the recovery traces in Fig. 3 can be fully explained by assuming that the defects are not located directly at the interface, as assumed by the capacitance scaling approach Eq. (2), but rather extend a few nanometers into the bulk of the oxide. In Fig. 8 the V_{TH} shifts of the devices 10 ms after termination of stress are depicted as a function of oxide thickness. Assuming equal effective defect charge density for the different devices, the capacitance scaling approach Eq. (2) would suggest a linear correlation with an interception in the origin in Fig. 8. It is clearly shown that the capacitance scaling approach is not sufficient to fully

explain the observed behavior and consequently defects at the interface alone cannot be made responsible for the observed degradation.

A spatial defect distribution $\rho(x)$, with x being the distance from the poly gate – oxide interface towards the semiconductor, causes a shift in the threshold voltage according to Gauss's law of

$$\Delta V_{TH} = -\frac{1}{C_{ox}} \left(\frac{1}{T_{ox}} \int_0^{T_{ox}} x \rho(x) \mathrm{d}x \right). \qquad (3)$$

Assuming that equivalent NBTS causes equivalent spatial defect distributions for the different oxide thicknesses (c.f. Fig. 9), we may draw conclusions on the spatial defect distribution within the bulk of the oxide. We observe strikingly good fits (c.f. Fig. 8) using the distribution

$$\rho(x) = \rho_0 \exp\left(\frac{x - T_{ox}}{T_0/\ln(2)} \right), \qquad (4)$$

with a half-value defect depth T_0 and a maximum charge density per volume ρ_0. Our data fitting results give an average value of T_0 of roughly $2\,\mathrm{nm}$, slightly decreasing with increasing stress time t_{str}. However, we remark that our study might be insufficient to make unambiguous statements on the exact distribution of defects in the bulk of the oxide because our data are limited to only three different oxide thicknesses. Nevertheless, it is clearly demonstrated that interface states alone can not be made responsible for all of the NBTS induced V_{TH} shift. We speculate that this statement holds for arbitrary oxide thicknesses, at least in the 5 to $30\,\mathrm{nm}$ range.

IV. CONCLUSIONS

By monitoring NBTI degradation on equivalent devices with different gate oxide thicknesses ranging from $5.6\,\mathrm{nm}$ to $29.1\,\mathrm{nm}$, we draw three main conclusions on the influence of the oxide thickness:

- Oxide bulk degradation originating from impact ionization or anode hole injection, which depends highly on the voltage drop across the oxide V_{ox}, do not play a crucial role in NBTI degradation.
- The increase of the charge pumping current is similar for all tested oxide thicknesses. Thus comparable stress induces a comparable number of broken silicon-hydrogen bonds at the interface, independent of the oxide thickness.
- The total number of NBTI induced defects is the same within a 15% range assuming all defects at the interface. This discrepancy vanishes if we assume the defects to extend a few nanometers into the bulk of the oxide.

Consequently, it has to be concluded that the physical mechanisms leading to NBTI are the same for thin and thick oxide devices. Future studies should involve a larger number of oxide thicknesses. This would help to derive more precise information about the particular distribution of oxide defects in the bulk of the oxide.

ACKNOWLEDGMENT

This work was jointly funded by the Federal Ministry of Economics and Labor of the Republic of Austria (contract 98.362/0112-C1/10/2005) and the Carinthian Economic Promotion Fund (KWF) (contract 98.362/0112-C1/10/2005).

REFERENCES

[1] K. O. Jeppson and C. M. Svensson, "Negative bias stress of MOS devices at high electric fields and degradation of MNOS devices," *Journal of Applied Physics*, vol. 48, no. 5, pp. 2004–2014, 1977.

[2] T. Grasser, B. Kaczer, W. Goes, T. Aichinger, P. Hehenberger, and M. Nelhiebel, "A two-stage model for negative bias temperature instability," in *Reliability Physics Symposium, 2009 IEEE International*, 2009, pp. 33–44.

[3] D. K. Schroder, "Negative bias temperature instability: What do we understand?" *Microelectronics Reliability*, vol. 47, no. 6, pp. 841 – 852, 2007.

[4] M. A. Alam and S. Mahapatra, "A comprehensive model of PMOS NBTI degradation," *Microelectronics Reliability*, vol. 45, no. 1, pp. 71 – 81, 2005.

[5] J. Stathis and S. Zafar, "The negative bias temperature instability in MOS devices: A review," *Microelectronics and Reliability*, vol. 46, no. 2-4, pp. 270 – 286, 2006.

[6] T. L. Tewksbury and H.-S. Lee, "Characterization, modeling, and minimization of transient threshold voltage shifts in MOSFETs," *Solid-State Circuits, IEEE Journal of*, vol. 29, no. 3, pp. 239–252, 1994.

[7] S. Mahapatra, K. Ahmed, D. Varghese, A. Islam, G. Gupta, L. Madhav, D. Saha, and M. Alam, "On the Physical Mechanism of NBTI in Silicon Oxynitride p-MOSFETs: Can Differences in Insulator Processing Conditions Resolve the Interface Trap Generation versus Hole Trapping Controversy?" in *Reliability Physics Symposium, 2007 IEEE International*, 2007, pp. 1–9.

[8] V. Huard, M. Denais, and C. Parthasarathy, "NBTI degradation: From physical mechanisms to modelling," *Microelectronics and Reliability*, vol. 46, no. 1, pp. 1 – 23, 2006.

[9] S. Mahapatra, D. Saha, D. Varghese, and P. Kumar, "On the generation and recovery of interface traps in MOSFETs subjected to NBTI, FN, and HCI stress," *Electron Devices, IEEE Transactions on*, vol. 53, no. 7, pp. 1583–1592, 2006.

[10] Y. Wang, "On the recovery of interface state in pMOSFETs subjected to NBTI and SHI stress," *Solid-State Electronics*, vol. 52, no. 2, pp. 264 – 268, 2008.

[11] D. J. DiMaria, E. Cartier, and D. Arnold, "Impact ionization, trap creation, degradation, and breakdown in silicon dioxide films on silicon," *Journal of Applied Physics*, vol. 73, no. 7, pp. 3367–3384, 1993.

[12] K. Schuegraf and C. Hu, "Hole injection SiO2 breakdown model for very low voltage lifetime extrapolation," *Electron Devices, IEEE Transactions on*, vol. 41, no. 5, pp. 761–767, 1994.

[13] Z. A. Weinberg and M. V. Fischetti, "Investigation of the SiO2-induced substrate current in silicon field-effect transistors," *Journal of Applied Physics*, vol. 57, no. 2, pp. 443–452, 1985.

[14] A. Kinoshita, Y. Mitani, K. Matsuzawa, H. Kawashima, C. Sutoh, J. Kurihara, T. Hiraoka, I. Hirano, M. Muta, M. Takayanagi, and N. Shigyo, "Breakdown Voltage Prediction of Ultra-Thin Gate Insulator in Electrostatic Discharge (ESD) Based on Anode Hole Injection Model," in *Reliability Physics Symposium, 2006 IEEE International*, 2006, pp. 623–624.

[15] M. Alam, J. Bude, and A. Ghetti, "Field acceleration for oxide breakdown-can an accurate anode hole injection model resolve the E vs. 1/E controversy?" in *Reliability Physics Symposium, 2000 IEEE International*, 2000, pp. 21–26.

[16] U. Schwalke, M. Pölzl, T. Sekinger, and M. Kerber, "Ultra-thick gate oxides: charge generation and its impact on reliability," *Microelectronics Reliability*, vol. 41, no. 7, pp. 1007 – 1010, 2001.

[17] H. Reisinger, R. Vollertsen, P. Wagner, T. Huttner, A. Martin, S. Aresu, W. Gustin, T. Grasser, and C. Schlunder, "The effect of recovery on NBTI characterization of thick non-nitrided oxides," in *Integrated Reliability Workshop, 2008 IEEE International*, 2008, pp. 1–6.

[18] C. Berglund, "Surface states at steam-grown silicon-silicon dioxide interfaces," *Electron Devices, IEEE Transactions on*, vol. 13, no. 10, pp. 701–705, 1966.

[19] B. Kaczer, V. Arkbipov, R. Degraeve, N. Collaert, G. Groeseneken, and M. Goodwin, "Disorder-controlled-kinetics model for negative bias temperature instability and its experimental verification," in *Reliability Physics Symposium, 2005 IEEE International*, 17-21, 2005, pp. 381–387.

[20] J. Brugler and P. Jespers, "Charge pumping in MOS devices," *Electron Devices, IEEE Transactions on*, vol. 16, no. 3, pp. 297–302, 1969.

[21] G. Groeseneken, H. Maes, N. Beltran, and R. De Keersmaecker, "A reliable approach to charge-pumping measurements in MOS transistors," *Electron Devices, IEEE Transactions on*, vol. 31, no. 1, pp. 42–53, 1984.

[22] D. Ang, S. Wang, G. Du, and Y. Hu, "A Consistent Deep-Level Hole Trapping Model for Negative Bias Temperature Instability," *Device and Materials Reliability, IEEE Transactions on*, vol. 8, no. 1, pp. 22–34, 2008.

[23] T. Aichinger, M. Nelhiebel, and T. Grasser, "Correlation between threshold voltage shift and charge pumping current in NBTI degradation," *Electron Devices, IEEE Transactions on*, submitted.

[24] ——, "On the temperature dependence of NBTI recovery," *Microelectronics Reliability*, vol. 48, no. 8-9, pp. 1178–1184, 2008.

[25] Y. Mitani, T. Yamaguchi, H. Satake, and A. Toriumi, "Reconsideration of Hydrogen-Related Degradation Mechanism in Gate Oxide," in *Reliability Physics Symposium, 2007 IEEE International*, 2007, pp. 226–231.

978-1-4244-5430-3/10 $26.00 © 2010 IEEE

Interpretation of PBTI/ TDDB predicted lifetime based on trap characterization by TSCIS in V_{th}-adjusted transistors

S. Sahhaf[1], R. Degraeve, V. Srividya[2], M. Cho, T. Kauerauf, G. Groeseneken[1]

IMEC, Kapeldreef 75, B-3001, Leuven ,Belgium, tel +32 16 287669, email: Sahar.Sahhaf@imec.be
[1] KULeuven, ESAT Department , Leuven , Belgium
[2] Micron Technology, USA

Abstract— **We investigate the change in the energy profile of the initially present HfSiO defects in nMOSFETs after V_{th}-adjustment by As and Ar implantation. We demonstrate that after implantation, PBTI lifetime is considerably improved as the present shallow traps in the implanted devices are not accessible at real operating conditions. We also study the TDDB to consider the effect of the generated defects with stress and we conclude that the same lifetime within specs is achieved for both no-implant and implanted devices.**

Keywords; initial V_{th}, As and Ar implantation, trap density, PBTI, TDDB

I. INTRODUCTION

Hafnium-based dielectric/metal gate stacks have been implemented as the replacement of traditional SiO_2/poly-Si gate stacks [1]. However, a major issue that must be addressed is the MOSFET threshold voltage (V_{th}) control, which has been the subject of frequent studies [2-7]. V_{th} tuning by implantation of As [3] or Ar is a viable option for metal gate CMOS integration.

In this paper, we focus on the impact of As and Ar implantation on the reliability of V_{th}-adjusted transistors. i) Applying Trap Spectroscopy by Charge Injection and Sensing (TSCIS) [8], we study the defect profile of devices with and with-out implantation and explain how the energy profile of initially present defects determines PBTI lifetime. ii) We also characterize the generated defects under constant voltage stress (CVS) and show that the stress-induced traps in these devices are different from the initial defects.

II. DEVICES

For this study we used devices with a 3.2 nm HfSiO dielectric (60% Hf), deposited by MOCVD on a 1 nm In-situ steam generated (ISSG) SiO_2 interface layer. 10 nm PVD TiN was used as metal gate. In order to adjust the threshold voltage, As or Ar was implanted through the gate at energies in the range 4keV to 6keV with implantation dose from 4×10^{15} at/cm²

to 6×10^{15} at/cm². Stopping and Range of Ions in Matter (SRIM) simulations place the implant primarily in the gate metal. A 5nm Si-cap was deposited on the TiN before implantation to prevent metal contamination in the implanter. No-implant devices with EOT of 1.4 nm and implanted devices with EOT of 1.35nm are obtained. The transistors showed state of the art drive and off-current performance.

III. TRAP CHARACTERIZATION OF AS OR AR IMPLANTED DEVICES

Fig.1 summarizes the effect of As or Ar implantation lowering the threshold voltage up to 250 mV compared to the reference devices without implantation. Note that as Ar is inert, the possibility of counterdoping due to the implantation is ruled out. We observe that the dependence of Vth shift on the implantation energy is much stronger for As than for Ar implant. This is an indication that channel counterdoping plays a role only in the case of As implant. Both As and Ar implants show less V_{th} shift with higher implant dose.

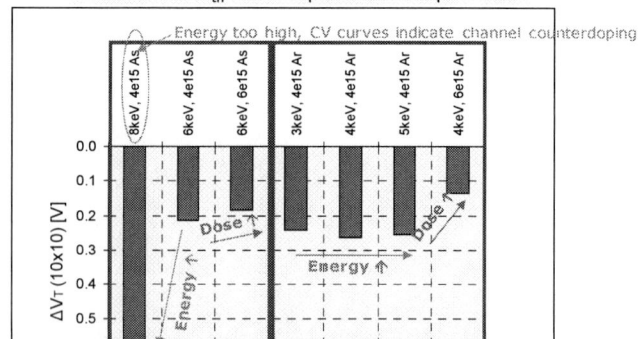

Fig.1: The reduction in V_{th} due to As or Ar implant (different energies and doses) with respect to no-implant reference device. Channel counterdoping plays a role only in the case of As implant when implant energy is high.

In order to investigate the implant-induced V_{th} shift, we studied the spatial and energy profile of the initially present defects by TSCIS [8]. Originally, this technique was developed for memory applications where traps in thick oxide layers (10-20 nm) were scanned. TSCIS relies on a controlled charging of defects by direct tunneling as a function of charging time (t_{charge}) and charging voltage (V_{charge}) [8]. In this work, we have adapted the technique by shortening t_{charge} as V_{charge} increases in order to scan thin layers (1nm SiO_2/3.2 nm HfSiO).

As shown in Fig.2(a) and 3(a), the measured V_{th}-shifts vs t_{charge} for increasing V_{charge} can be transformed to a trap density profile (Fig.2(b) and 3(b)) using the WKB-approximation [9] and a Poisson solver to find the correct trap location spatially and energetically.

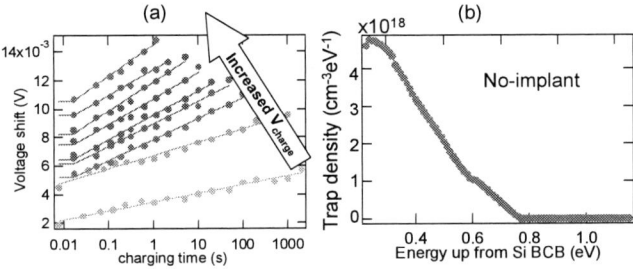

Fig.2 (a) V_{th}-shifts as a function of t_{charge} for increasing V_{charge} on no-implant reference devices are translated into (b) the trap density profile (TSCIS technique [8]). A defect band between 0 and 0.7eV above Si conduction band is observed.

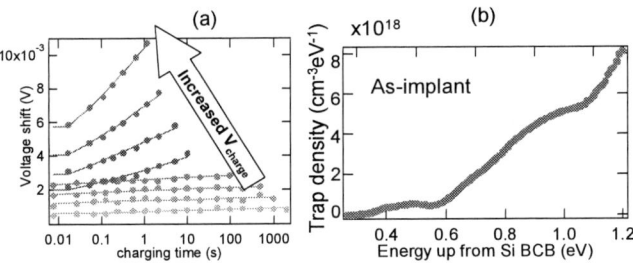

Fig.3: (a) V_{th}-shifts as a function of t_{charge} for increasing V_{charge} on a 6keV, 4×10^{15} at/cm^2 As-implanted device are translated [8] into (b) the trap density profile which differs from the trap profile of the no-implant devices. A defect band > 0.7eV above Si conduction band is observed. The same defect profile is also observed in devices with Ar implant (not shown here).

A clear difference between the energy profiles of the initial traps for the As-implanted (Ar-implanted not shown here) and the reference devices is observed. The no-implant samples contain a typical defect band at deep energy levels (between 0 and 0.7eV above Si conduction band) while in As-implanted samples energetically shallow defects (>0.7eV above Si conduction band) are found. In the next paragraph, we investigate the effect of these different trap profiles on the reliability of the devices.

IV. INTERPRETATION OF PBTI/TDDB PREDICTED LIFETIME BASED ON TRAP CHARACTERIZATION

PBTI was performed on $W \times L = 10 \times 1$ cm^2 MOSFETs at 125°C. The drain current was measured on the fresh device. Then a low V_{G_stress} was applied at the gate of the transistor with source and bulk grounded. At regular intervals, the gate voltage was switched to V_{G_sense} and the drain current was measured in a short time ~1ms to reduce the V_{th} recovery effect. The drain current shift was then translated into a V_{th}-shift and lifetime was determined at 30mV V_{th}-shift.

The ΔV_{th} kinetics during PBTI stress follow a power-law time dependence and as shown in Fig.4, the implanted samples have a larger V_G acceleration factor compared to no-implant samples. Consequently, PBTI lifetime is considerably improved with As or Ar implantation. In order to explain these results, we first make the link between PBTI results and defect profiles characterized by TSCIS. One should keep in mind that the TSCIS methodology (performed at room temperature) is similar to fast V_{th}-evaluation methods developed for minimizing the relaxation during PBTI measurements [10].

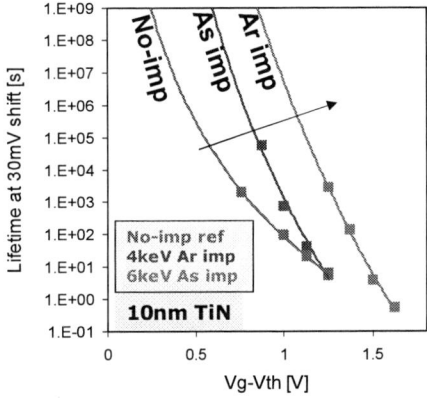

Fig.4: PBTI lifetime as a function of V_g-V_{th} for no-implant and implanted devices. Different stress voltages (Vg) in the range 1.38 to 1.96V and different sense voltages in the range 0.34 to 0.62V were used. PBTI lifetime is considerably increased with implants. This is mainly due to the steeper Vg-acceleration for implanted devices.

In Fig.5 (a) and (b), a cross section of the V_{th} shifts vs t_{charge} data (resulted from TSCIS) at a fixed threshold voltage shift is taken. The corresponding charging times are displayed vs V_g-V_{th} as shown in Fig.5(c). It is clear that voltage shifts due to shallow traps (in devices with implantation) result in a steeper charging time extrapolation curve compared to deep traps (in reference devices), which is consistent with the steeper PBTI lifetime curves shown in Fig. 4. This proves that PBTI is directly correlated to the energy profile of initially present defects.

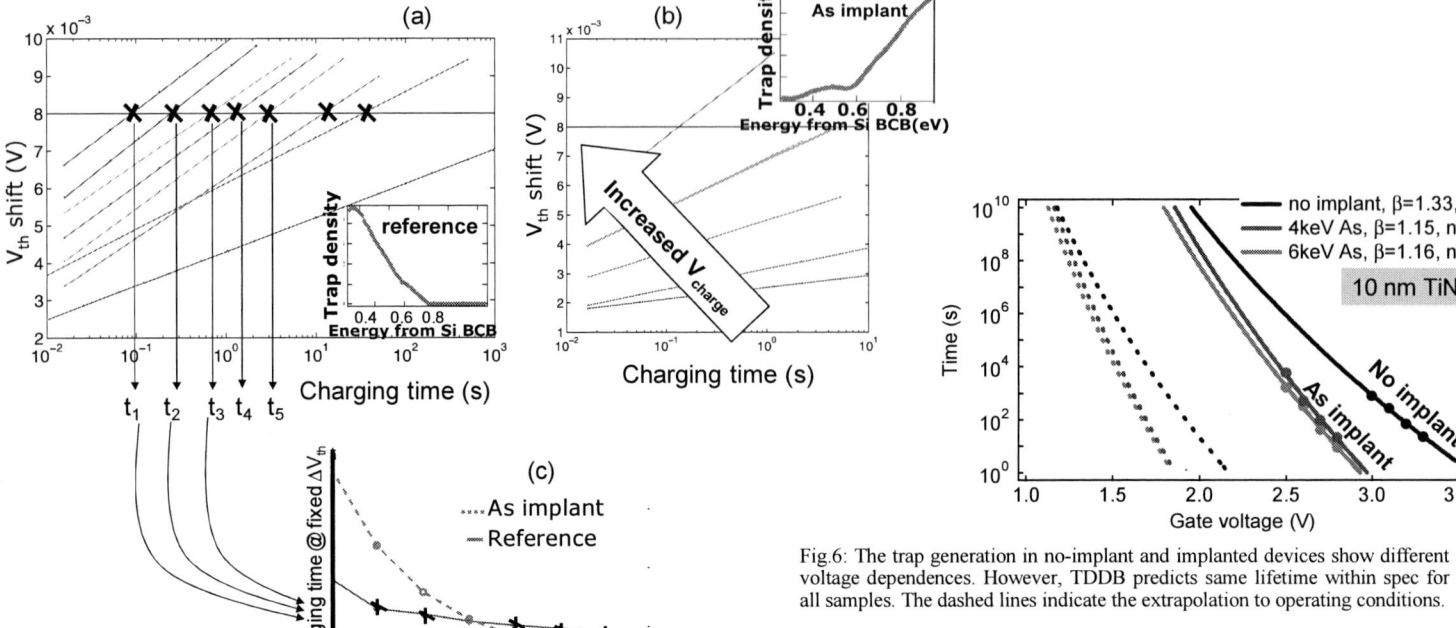

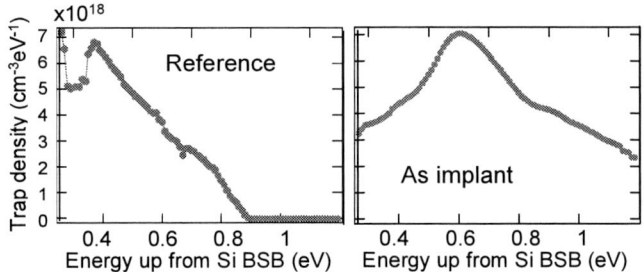

Fig.5: (a) and (b) Cross section of the V_{th} shifts vs t_{charge} for increasing V_{charge} (resulted from TSCIS) at a certain threshold is taken. (c) The corresponding charging times are displayed vs V_g-V_{th}. Energetically shallow defects in devices with implant result in a steeper charging time extrapolation curve. As TSCIS measurements are performed at 25°C, lower threshold (compared to 30mV for PBTI) is chosen.

Fig.6: The trap generation in no-implant and implanted devices show different voltage dependences. However, TDDB predicts same lifetime within spec for all samples. The dashed lines indicate the extrapolation to operating conditions.

Fig.7: Defect profiles of the stressed devices characterized by TSCIS. Generated defects with CVS are different in no- implant and implanted devices. Also these profiles differ from the profiles of initially present defects (Figs 2b and 3b).

We remind that electron trap filling is the main underlying mechanism in PBTI [11]. As or Ar-implanted devices contain shallow traps which are only accessible at high gate voltages. At lower voltages filling of these shallow traps does not occur, hence resulting in higher PBTI lifetimes and steeper voltage acceleration.

In order to study the defect generation, Constant Voltage Stress (CVS) is applied on an L=1 μm and W=1 μm transistors. Note that much higher stress voltages are used compared to PBTI. The breakdown (BD) is triggered with a current step of 500nA and a reliability spec of 0.01% failures after 10 years on 0.1 cm² effective area is used for the lifetime extrapolation.

While PBTI lifetime prediction is based on filling of existing traps, TDDB studies the effect of generated defects. Fig.6 shows different voltage dependences of the trap generation in no-implant and implanted devices. This indicates different kinds of generated defects in these devices.

This conjecture is confirmed by TSCIS characterizing different defect energy profiles in the stressed devices (see Fig.7). Note that the energy profile of the generated defects also differs from the initially present defects shown in Fig.2 (b) and 3(b). However, for lifetime extrapolation to operating conditions, the difference in voltage dependence (different power law exponents in Fig. 6) is compensated by difference in β (=a measure for the number of traps participating in BD-path) resulting in nearly identical lifetime for both devices.

V. CONCLUSIONS

We performed trap characterization by TSCIS and found that energetically deep traps in no-implant devices are passivated after gate implantation. We also detected additionally shallow traps in the V_{th}-adjusted devices.

Further, we investigated the reliability of low V_{th} nMOS transistors with As or Ar implantation. We showed that, as PBTI is determined by filling of the existing traps, the PBTI lifetime is mainly dominated by the energy profile of initially present defects. As the shallow traps in implanted devices are not accessible at low voltages, PBTI lifetime is considerably improved at real operating conditions. On the other hand, TDDB is determined by generated defects which have different energy profile compared to the initially present defects.

ACKNOWLEDGMENTS

This work is part of IMEC's Industrial Affiliation Program, funded by IMEC's core partners: Intel, Texas Instruments, Micron, Infineon, NXP, ST, Matsushita, TSMC, Samsung, and Elpida.

REFERENCES

[1] Y. Kim *et al.*, "Conventional n-Channel MOSFET Devices Using Single Layer HfO2 and ZrO2 as High-k Gate Dielectrics with Polysilicon Gate Electrode", IEDM Techn. Dig., pp. 455-458, 2001.

[2] H.H. Tseng *et al.*, "The progress and challenges of threshold voltage control of high-k/metal-gated devices for advanced technologies", Microelectronic Engineering 86, 1722–1727, (2009).

[3] J. K. Schaeffer *et al.*, "Contributions to the effective work function of platinum on hafnium dioxide", Appl. Phys. Lett. 85, 1826 ,2004.

[4] R. Singanamalla, *et al.*, "The Study of Effective Work Function Modulation by As Ion Implantation in TiN/TaN/HfO2 Stacks", Japanese Journal of Applied Physics, Vol. 46, pp L320-L322, 2007.

[5] N. Mise, *et al.*, "Universal Correlation between Flatband Voltage and Electron Mobility in TiN/HfSiON Devices with MgO or La2O3 Incorporation and Stack Variation ", Japanese Journal of Applied Physics, Vol. 47, No. 10, pp. 7780–7783, 2008.

[6] H.S. Jung *et al.*, "Dual high-k gate dielectric technology using selective AlOx etch (SAE) process with nitrogen and fluorine incorporation", VLSI Technology, pp. 162-163, 2006.

[7] P. Sivasubramani, *et al.*, "Dipole moment model explaining nFET Vt tuning utilizing La, Sc, Er, and Sr doped HfSiON dielectrics", VLSI Technology, pp. 68-69, 2007.

[8] R. Degraeve, *et al.*, "Trap Spectroscopy by Charge Injection and Sensing (TSCIS): a quantitative electrical technique for studying defects in dielectric stacks", Technical Digest International Electron Devices Meeting - IEDM, pp. 775-778, 2008.

[9] M. Cho, *et al.*, "How far can we analyze oxide traps spatially with charge injection techniques?", 39th IEEE Semiconductor Interface Specialists Conference (SISC), December 2008.

[10] B. Kaczer *et al.*," Ubiquitous relaxation in VTI stressing new evaluation and insights", *IRPS,* pp. 20-27, , 2008.

[11] M. Aoulaiche1 *et al.*, "Positive and Negative Bias Temperature Instability in La2O3 and Al2O3 capped high-k MOSFETs", *IRPS,* pp.1014-1018, 2009.

Improvements of NBTI Reliability in SiGe p-FETs

J. Franco[1], B. Kaczer, M. Cho, G. Eneman[1,2], G. Groeseneken[1]

IMEC

Kapeldreef 75, B-3000 Leuven, Belgium

Phone: +32 16 28 10 85; fax: +32 16 28 17 06; e-mail: Jacopo.Franco@imec.be

[1]also at ESAT Dept., K.U. Leuven, Belgium

[2] also FWO-Vlaanderen, Belgium

T. Grasser

Christian Doppler Laboratory for TCAD

Institute for Microelectronics, TU Wien

Gusshausstrasse 27-29, 1040 Wien, Austria

Abstract—**NBTI reliability of buried SiGe channel p-FETs is investigated as a function of Ge concentration, SiGe layer thickness and Si cap thickness. Measurements show that NBTI reliability can be dramatically improved by varying these three parameters, i.e., increasing the Ge fraction, increasing the thickness of the SiGe layer, and reducing the Si cap thickness. Consequently, it is demonstrated that SiGe devices are a promising option for improving NBTI in highly-scaled sub-1nm EOT pFETs.**

Keywords: NBTI, SiGe, Ge, Reliability, pFETs, thin EOT, high-mobility substrates

I. INTRODUCTION

The negative bias temperature instability (NBTI) is a major reliability problem for the semiconductor industry [1]. Reduction of effective oxide thickness (EOT), which is one of the most efficient ways to improve MOSFET performance, enhances NBTI due to increased oxide electric field (E_{ox}). As a consequence, 10 year lifetime can be guaranteed for sub-1nm EOT Si pFETs only at gate overdrive voltages far below the expected operating voltages (Fig. 1).

Another way to improve FET performance is the use of a strained SiGe channel [2-6]. In this paper the NBTI reliability of such SiGe devices is investigated. All SiGe pFETs under test showed better NBTI reliability with respect to their Si channel counterparts. As can be seen from Fig. 1, an *ad hoc* optimization of SiGe device gate-stacks at the moment seems to be the only solution to the NBTI issue for sub-1nm EOT devices, as it allows to significantly increase the operating gate overdrive (V_{op}) while still guaranteeing 10 year device lifetime.

II. EXPERIMENTAL

The buried SiGe channel p-FETs used in this work were fabricated at IMEC on 300 mm (100) Si wafers. A cross-sectional sketch of the final device and a band diagram of this type of stack in inversion are depicted in Fig. 2.

After a pre-epi clean, an HCl etchback of Si was performed to level the surface of the final grown structure to the shallow trench isolation (STI). A 2nm-thick Si buffer was grown at 500°C from SiH$_4$ precursor to guarantee a high quality starting Si surface for the following growth of a compressively strained thin Si$_{1-x}$Ge$_x$ layer with thickness varying between

3nm and 7nm. Ge fractions were x=0.45 or x=0.55. Epitaxial growth was performed from SiH$_4$ and GeH$_4$ precursors at 500°C for x=0.45 and at 450°C for x=0.55. On top of the Si$_{1-x}$Ge$_x$ layer, a thin undoped Si cap was grown from SiH$_4$ at 500°C for x=0.45 and at 475°C for x=0.55. The physical thicknesses of this thin Si cap varied between 0.65nm and 2nm (as estimated from C-V curves and TEM pictures of the final device). A detailed description of the epi-process can be found elsewhere [7-8].

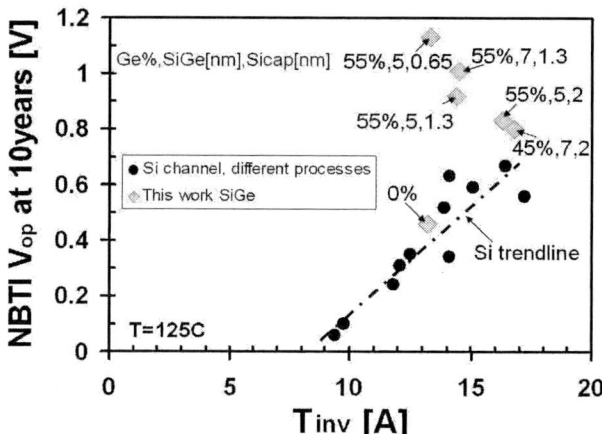

Figure 1. Plot of the operating overdrive voltage (V_{op}) for 10 year lifetime assuming a 30mV threshold voltage shift criterion under NBTI stress condition (T=125°C) vs. the capacitance equivalent thickness evaluated in inversion (T_{inv}), for Si channel devices with different processing used as a reference and for SiGe pFETs used in this work. For low T_{inv}, Si devices V_{op} is below the expected operating voltage. In contrast to that, optimized SiGe devices presented in this study show improved lifetime, and therefore provide a potential solution for the NBTI issue toward sub-1nm EOT devices.

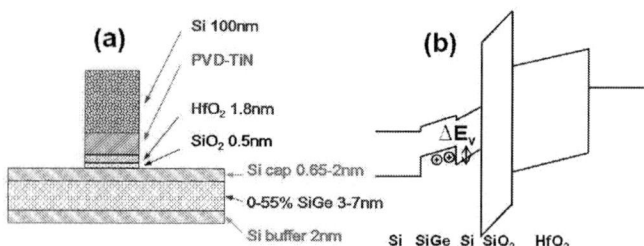

Figure 2. (a) Gate-stack sketch of the SiGe devices under test; (b) Band diagram in inversion. Si cap acts as a barrier (ΔE_v) for holes.

978-1-4244-5430-3/10 $26.00 © 2010 IEEE

Gate stack fabrication started with a very thin (0.8nm) wet chemical oxidation of the Si cap. On top of this SiO_2 interfacial layer (IL), ~1.8nm of HfO_2 were deposited using atomic layer deposition (ALD). Finally, a TiN metal gate was deposited. Channel width and length of the devices used in this work were 10 μm and 1 μm respectively. Effective mobility enhancement factor of our SiGe devices with respect to Si control ranged between 1.5x and 2.2x, depending on the process parameters.

NBTI stress experiments were performed using the extended measure-stress-measure technique [9]. Devices were stressed at $T = 125°C$ with several gate overdrive conditions and up to 2000s of stress time. To monitor the degradation, stress was interrupted several times for sensing source current at $V_G \sim V_{th}$. Measured source current was converted to threshold voltage shift using the IsVg curve of the fresh device as a conversion table. Relaxation traces were recorded for ~12s. However for lifetime prediction ΔV_{th} was evaluated at $t_{relax}=2ms$, i.e., the minimum delay to get a reliable measurement with the used setup (2x Keithley 2602 Source Meter). Such delay was fixed in all the experiments to allow cross-comparison. For each gate voltage the stress time needed to reach a failure criterion, assumed as 30mV threshold voltage shift, is extracted. The 10 year lifetime operating overdrive (V_{op}) is then extrapolated fitting a power law to the lifetime vs. gate overdrive dataset (Fig. 3-5).

The three major process parameters of our SiGe pMOSFETs, i.e. the Ge concentration, the SiGe layer thickness and the Si cap thickness, were varied separately in order to assess their impact on NBTI reliability.

III. RESULTS AND DISCUSSION

We now report the main experimental observations followed by a discussion of possible explanations for the observed NBTI improvement.

A. Experimental Observations

Because of the valence band offset between the SiGe and the Si cap (see Fig. 2b) inversion channel holes are confined in the SiGe layer, which therefore acts as a quantum well (QW) for holes. This causes the Si cap thickness to lower the inversion capacitance as compared to the accumulation capacitance. For these devices it is therefore necessary to report the capacitance-equivalent thickness in inversion (T_{inv}, evaluated at $V_G=V_{th}-0.6V$) which will be affected by the thickness of the Si cap.

A large set of stress experiments was performed on SiGe p-FETs changing only one of the gate-stack parameters under investigation at the time, while fixing the other two. For comparison purposes, a second set of standard Si channel devices with identical dimensions and gate stack were considered. It is worth noting that, although having the same gate stack, Si channel devices show a lower T_{inv} as compared to the SiGe channel devices due to the absence of any additional displacement for holes (i.e. the Si cap acting as a barrier in SiGe devices). The T_{inv} for the Si channel reference devices was estimated as 1.32nm.

From NBTI stress experiments the following observations were made:

1) *Ge content:* As is shown in Fig. 3, the introduction of Ge in the channel dramatically improved the NBTI reliability. The extrapolated operating overdrive voltage for a 10 year lifetime (V_{op}), assuming a 30mV threshold voltage shift criterion, increased from 0.46V for the pure Si reference up to 0.8V for 45% Ge content device with a SiGe layer thickness of 7nm and a Si cap thickness of 2nm. Increasing the Ge concentration up to 55%, while keeping constant the other parameters, boosted the operating overdrive voltage even more, reaching 0.9V.

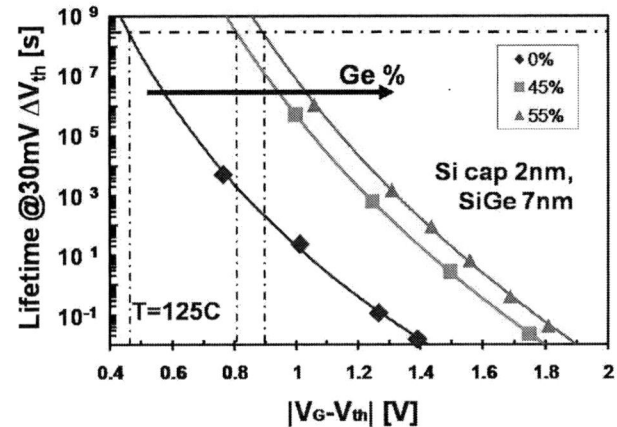

Figure 3. Higher Ge content improves extrapolated operating overdrive voltage for 10 year lifetime. (triangles: rescaled from 5nm SiGe layer).

2) *SiGe QW thickness:* Increasing the thickness of the SiGe QW had also a positive impact on the NBTI reliability (Fig. 4): V_{op} increased from 0.85V up to 1.01V when moving from a 3nm-thick SiGe layer to a 7nm one. This observation was made while fixing the Si cap thickness to 2nm on 55% Ge channel concentration devices.

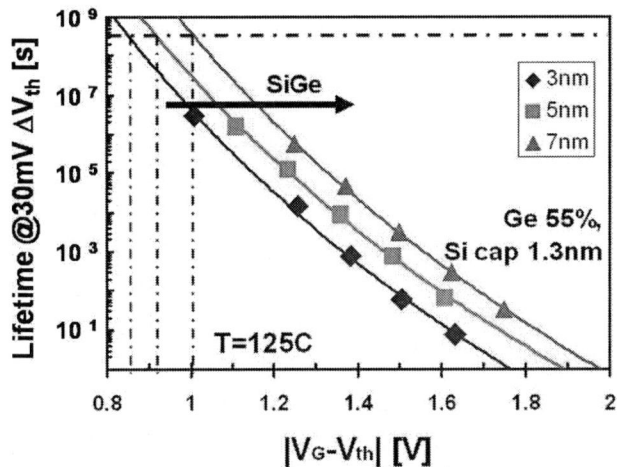

Figure 4. A thicker SiGe channel, i.e. a thicker QW for holes, improves the NBTI reliability.

3) *Si cap thickness:* The Si cap had also a significant impact on the NBTI reliability (Fig. 5). A thicker Si cap clearly degraded the NBTI performance. V_{op} increased from 0.82V to

1.14V while the Si cap thickness decreased from 2nm to 0.65nm. The thickness of the Si cap impacts, as already mentioned, the C_{ox} value in inversion and therefore the T_{inv}; hence a fixed overdrive stress voltage leads to different electric fields (E_{ox}) for devices with different Si cap. Since this may affect the interpretation of NBTI data, in order to have a more fair comparison, the effective degradation ($\Delta N_{eff} = \Delta V_{th} \cdot C_{ox}/q$) is plotted as function of E_{ox} and compared for different Si cap thickness (Fig. 6). E_{ox} extraction is not trivial for such complex structures. Here we used the following method: a Q-V curve was obtained by integrating the measured C-V trace; a line was fitted to the almost linear part of the Q-V curve; the slope of the line represented the estimated C_{ox}; the capacitance equivalent thickness (CET) was calculated from C_{ox}; finally E_{ox} was calculated as $|V_G - V_{th}|/CET$. As one can see from Fig. 6, the impact of the Si cap thickness on NBTI degradation is still clear when correcting for the differences in E_{ox}.

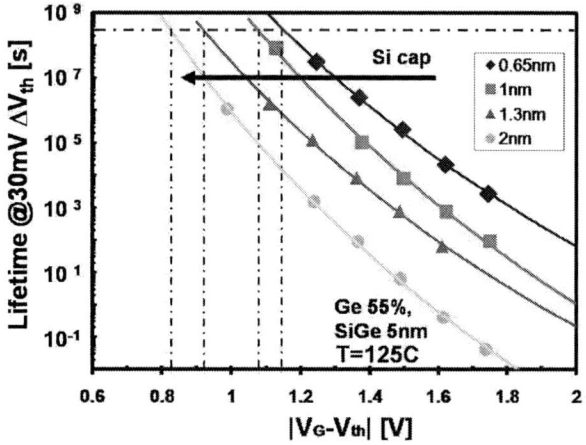

Figure 5. A thicker Si cap reduces NBTI reliability. On the contrary, the use of a thin Si cap improves NBTI reliability while enabling T_{inv} reduction.

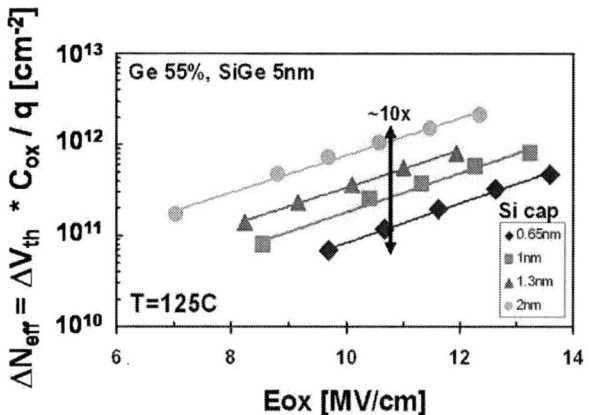

Figure 6. Comparison of degradation as a function of E_{ox} confirms NBTI dependence on Si cap thickness is not an artifact due to different T_{inv}.

B. Discussion

Previous work attributed improvement of NBTI to induced strain at the interface [10]. This explanation does not apply to our devices. The SiGe layer thicknesses considered here were well below the reported critical relaxation thickness for the used epitaxial process, causing the channel to be compressively strained [7]. Avoiding relaxation of the SiGe results in a Si cap that is lattice-matched to the underlying Si substrate. Therefore we conclude that strain effects at the Si/SiO$_2$ interface were not involved here.

As shown in the band diagram of Fig. 2b, channel holes are confined in the SiGe QW due to the valence band offset (ΔE_v) between SiGe and Si. Fig. 7 shows ΔE_v as a function of the Ge concentration as calculated using MEDICI [11]. The higher offset caused by a higher Ge fraction in the channel reduces the hole tunneling probability through the Si cap layer acting as a potential barrier for holes. This can explain the observed NBTI reduction when increasing the Ge content (Fig. 3).

A similar explanation can justify also the observed NBTI trend when changing the SiGe layer thickness (Fig. 4): for a very thin QW, i.e. 3nm, quantum mechanical effects increase the hole energy, and therefore artificially reduce the ΔE_v [12]. This can lead to degraded NBTI performance due to an enhanced hole tunneling probability.

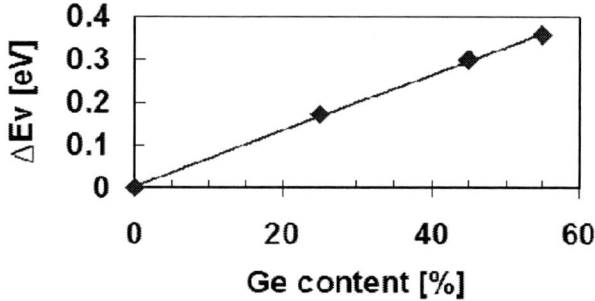

Figure 7. While increasing the Ge content, the SiGe bandgap is reduced and therefore the valence band offset between SiGe and Si increases.

This explanation, however, does not apply to our third experimental observation: a thicker Si cap, i.e. a thicker barrier for holes (Fig. 2), is expected to reduce the tunneling probability [13]. In contrast to this, a thicker Si cap degrades the NBTI reliability, which, on the other hand, is strongly improved when using a thin Si cap (Fig. 5).

Interestingly, the use of a thin Si cap for the passivation of similar structures has been reported to be detrimental for the initial interface quality as compared with thicker Si cap [14]. Charge pumping measurements on our SiGe pFETs confirmed that the Si cap thickness has a considerable impact on the initial interface quality: the use of a 0.65nm Si cap increases the initial N_{it} values by almost two orders of magnitude (Fig. 8) as compared to the 2nm thick Si cap. Therefore NBTI experimental data suggest an anti-correlation between initial N_{it} and NBTI reliability for SiGe devices: a better initial interface quality (i.e., thicker Si cap) leads to degraded NBTI performance. A possible explanation for this trend is that a higher initial N_{it} value, i.e. an already partially degraded interface, could slow down the extra defect creation during NBTI stress [15]. This could explain why devices with thinner Si cap, showing a bad initial interface quality, suffer less NBTI.

The impact of the Si cap thickness on the NBTI reliability is of particular interest because it provides an additional benefit to the advantages of T_{inv} scaling. As one can notice in Fig. 1, a thinner Si cap improves NBTI reliability while reducing T_{inv} due to reduced hole displacement.

978-1-4244-5430-3/10 $26.00 © 2010 IEEE

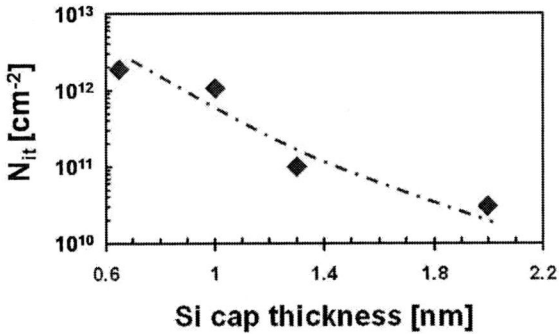

Figure 8. Charge pumping measurements show degraded interface quality for thinner Si cap.

IV. CONCLUSIONS

The NBTI reliability of buried SiGe channel p-FETs was investigated as a function of Ge concentration, SiGe layer thickness and Si cap thickness. Measurements showed that the NBTI reliability can be dramatically improved by increasing the Ge content, increasing the SiGe QW thickness and decreasing the Si cap thickness. By means of optimizing the three gate-stack parameters under investigation, SiGe devices can yield excellent NBTI reliability with respect to their Si counterparts. Buried channel SiGe devices therefore are extremely promising for helping to solve the NBTI issue in sub-1nm EOT pFETs.

ACKNOWLEDGEMENTS

The IMEC sub-32nm program members, in particular Drs. T.Y. Hoffman and S. Takeoka (Panasonic), the IMEC pilot line, and Amsimec are acknowledged for their support. We also gratefully acknowledge Drs. N. Collaert and G. Pourtois (IMEC) and Profs. A. Stesmans and V.V. Afanas'ev (Physics and Astronomy Dept., University of Leuven) for useful discussions.

REFERENCES

[1] V. Huard, M. Denais, C. Parthasarathy, "NBTI degradation: From physical mechanism to modeling", in Microelectronics Reliability, Vol. 46, No. 1, Jan. 2006, pp. 1-23

[2] N. Collaert, P. Verheyen, K. De Meyer, R. Loo and M. Caymax, "High-Performance Strained Si/SiGe pMOS Devices With Multiple Quantum Wells", IEEE Transactions on nanotechnology, Vol. 1, No. 4, Dec. 2002, pp. 190-194

[3] N. Collaert, P. Verheyen, K. De Meyer, R. Loo and M. Caymax, "Influence of the Ge-concentration and RTA on the device performance ofstrained Si/SiGe pMOS devices", in Proc. ESSDERC 2002, pp. 263-266

[4] P. Majhi, P. Kalra, R. Harris, K. J. Choi, D. Heh, J. Oh, D. Kelly, R. Choi, B. J. Cho, S. Banerjee, W. Tsai, H. Tseng, and R. Jammy, "Demonstration of High-Performance PMOSFETs Using Si–Si$_x$Ge$_{1-x}$–Si Quantum Wells With High-κ/Metal-Gate Stacks", IEEE Electron Device Letters, Vol. 29, No. 9, Sept. 2008, pp. 99-101

[5] S. H. Lee, P. Majhi, J. Oh, B. Sassman, C. Young, A. Bowonder, W.-Y. Loh, K.-J. Choi, B.-J. Cho, H.-D. Lee, P. Kirsch, H.R. Harris, W. Tsai, S. Datta, H.-H. Tseng, S.K. Banerjee, and R. Jammy, "Demonstration of Lg=55nm pMOSFETs With Si/Si$_{0.25}$Ge$_{0.75}$/Si Channels, High Ion/Ioff (> 5 × 10^4), and Controlled Short Channel Effects

(SCEs)", IEEE Electron Device Letters, Vol. 29, No. 9, Sept. 2008, pp. 1017-1020

[6] N. Tamura, Y. Shimamune and H. Maekawa, "Embedded Silicon Germanium (eSiGe) technologies for 45nm nodes and beyond", in Proc. IEEE IWJT 2008, pp. 73-77

[7] A. Hikavyy, R. Loo, L.Witters S. Takeoka , J. Geypen, B.Brijs, C. Merckling, M. Caymax and J. Dekoster, "SiGe SEG Growth For Buried Channel p-MOS Devices", ECS Transactions, Vol. 25, No. 7, 2009, pp. 201-210

[8] R. Loo, C. Walczyk, P. Verheyen, R. Rooyackers, F.E. Leys, G. Eneman, D. Shamiryan, P.P. Absil, T. Delande, A. Moussa, H. Bender, C. Drijbooms, L. Geenen, M. Caymax, J.W. Weijtmans, R. Wise, V. Machkaoutsan, P. Tomasini, C. Arena, J. McCormack, S. Passefort, H. Sorada, A. Inoue, B.C. Lee, S. Hyun, S. Jakschik, S. Godny, "Selective Epitaxy of Si/SiGe to Improve pMOS Devices by Recessed Source/Drain and/or Buried SiGe Channels", ECS Transactions, Vol. 3, No. 7, 2006, pp. 453-465

[9] B. Kaczer, T. Grasser, Ph. J. Roussel, J. Martin-Martinez, R. O'Connor, B.J. O'Sullivan, G. Groeseneken, "Ubiquitous Relaxation in BTI Stressing–New Evaluation and Insights", in Proc. IEEE IRPS 2008, pp. 20-27

[10] A.E. Islam, J.H. Lee, W.-H. Wu, A. Oates, and M.A. Alam, "Universality of Interface Trap Generation and Its Impact on I_D Degradation in Strained/Unstrained PMOS Transistors Under NBTI Stress", in Proc. IEEE IEDM 2008, pp. 107-110

[11] S.S. Iyer, G.L. Patton, J.M.C. Stork, B.S. Meyerson, D.L. Harame, "Heterojunction bipolar transistors using Si-Ge alloys", IEEE Transactions on Electron Devices, Vol.36, No. 10, Oct. 1989, pp. 2043-2064

[12] M.V. Fischetti and S.E. Laux, "Band Structure, Deformation Potentials, and Carrier Mobility in Strained Si, Ge, and SiGe Alloys", in Journal of Applied Physics, Vol. 80, No. 4, Aug. 1996, pp. 2234-2252

[13] B. Kaczer, J. Franco, J. Mitard, Ph.J. Roussel, A. Veloso, and G. Groeseneken, "Improvement in NBTI Reliability of Si-passivated Ge/high-k/metal gate pFETs", in Microelectronics Engineering, Vol. 86, No. 7-9, Jul.-Sep. 2009, pp. 1582-1584

[14] J. Mitard, K. Martens, B. De Jaeger, J. Franco, C. Shea, C. Plourde, F.E. Leys, R. Loo, G. Hellings, G. Eneman, W.E. Wang, J.C. Lin, B. Kaczer, K. De Meyer, T. Hoffman, S. De Gendt, M. Caymax, M. Meuris, and M.M. Heyns, "Impact of Epi-Si Growth Temperature on Ge-pFET Performance", in Proc. ESSDERC 2009, pp. 411-414

[15] J. Franco, B. Kaczer, A. Stesmans, V.V. Afanas'ev, K. Martens, M. Aoulaiche, T. Grasser, J. Mitard, G. Groeseneken, "Impact of Si-Passivation Thickness and Processing on NBTI Reliability of Ge and SiGe pMOSFETs", as discussed at IEEE SISC 2009, Washington, DC

[16] T. Grasser, B. Kaczer, W. Goes, Th. Aichinger, Ph. Hehenberger, and M. Nelhiebel, "A Two-Stage Model for Negative Bias Temperature Instability", in Proc. IEEE IRPS 2009, pp. 33-44

A Model for NBTI in Nitrided Oxide MOSFETs Without Hydrogen or Diffusion

P.M. Lenahan
Department of Engineering Science and Mechanics
The Pennsylvania State University
University Park, PA 16802
Tel: 814-863-4630, Fax: 814-863-7967, Email: pmlesm@engr.psu.edu

Abstract -- The negative bias temperature instability (NBTI) is, arguably, the single most important reliability problem in present day metal oxide silicon field effect transistor (MOSFET) technology. This paper presents a model for NBTI which is radically different from the quite widely utilized reaction diffusion models which dominate the current day NBTI literature. The proposed model is relevant to technologically important nitrided oxide pMOSFETs. The model is clearly not, at least in its entirety, relevant to pure silicon dioxide gate pMOSFETs. Reaction diffusion models involve hydrogen/silicon bond breaking events at the silicon/ silicon dioxide interface initiated by the presence of an interface hole, followed by the diffusion of a hydrogenic species from the interface as well as potential rebonding of hydrogen and interface trap defect centers. This model does not invoke hydrogen in any form whatsoever but does simply account for the observed NBTI power law response and provides a reasonably accurate value for this exponent. The model also provides a reasonable explanation for recovery which includes a simple explanation for the extremely rapid rate of recovery at short times. In addition, the model provides a very simple explanation why the introduction of nitrogen greatly enhances NBTI. Finally, the model is consistent with recent electron paramagnetic resonance studies of NBTI defect chemistry.

Key words: negative bias temperature instability, oxide traps, interface traps.

I. Introduction

In NBTI, pMOSFETs subjected to moderately elevated temperature and modest oxide fields resulting from negative gate bias experience negative threshold voltage shifts and a decrease in drive current. The threshold voltage shifts are most commonly modeled in terms of expressions of the form: [1, 2,3]

$$\Delta V_{th} = A(T, E_{ox})t^n \qquad (1)$$

where $A(T, E_{ox})$ is a strong function of both oxide field Eox and absolute temperature T. In the most recent literature, $n \cong 1/6$.

The process is generally viewed as a result of both interface traps and dielectric space charge generation. [1,2] The introduction of significant nitrogen concentration into the gate dielectric causes a substantial enhancement in NBTI. [2,4,5] It has also been reported that both the interface trap generation and oxide charge generation are increased by nitridation. It has been proposed that the enhanced charge trapping and interface trap generation have a common origin. [4]

The vast NBTI literature has long been dominated by various forms of the reaction diffusion model. [1,2] Reaction diffusion models invoke hydrogen/silicon bond breaking events caused by inversion layer holes at the Si/dielectric boundary, the release of a hydrogenic species, diffusion of a hydrogenic species, and reforming of hydrogen/silicon bonds. Although these reaction diffusion models "make sense" in that they explain much of the NBTI phenomena, some aspects of the phenomena are quite difficult to fully explain in a reaction diffusion framework, especially the rapid recovery which occurs when bias stress is removed. [1,6,7] Could it be that, in some important cases, hydrogen actually has nothing to do with this? Could it be that the reaction diffusion model is completely wrong for a technologically important class of devices, nitrided oxide MOSFETs? This paper offers a radically different model for NBTI which does not involve hydrogen diffusion, does not invoke hydrogen bond breaking events, and does not involve reforming of hydrogen/silicon bonds. The model's predictions are generally consistent with both macroscopic and atomic scale experimental observations on NBTI.

978-1-4244-5430-3/10 $26.00 © 2010 IEEE

II.   The Model

The proposed model involves K centers, defects involving silicon atoms bonded to nitrogen atoms. Recent magnetic resonance studies of Campbell et al. [8] and Ryan et al. [9] demonstrate a dominating role for these defects in NBTI of nitrided oxide pMOSFETs. A schematic illustration of the K center appears in figure 1. In the proposed model, the NBTI process is triggered by the tunneling of an electron from a K center to an inversion layer hole. The triggering of NBTI by such a tunneling process from another defect, an E' center, was recently proposed by Campbell et al. [8].

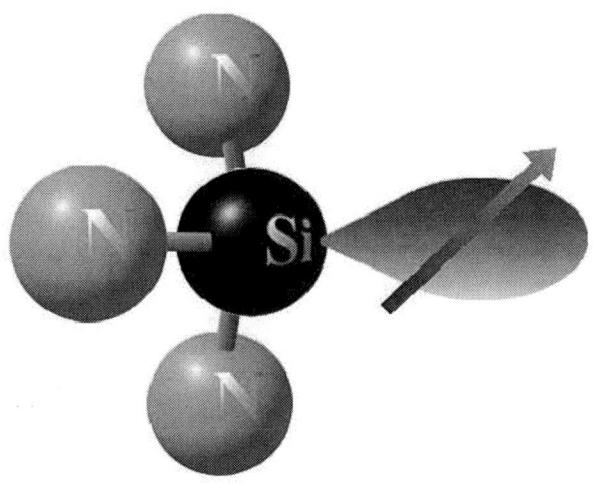

Figure 1: This figure schematically illustrates the part of the K center observed in magnetic resonance: a silicon "dangling bond" defect in which the central silicon is back bonded to nitrogen atoms.

Campbell et al. [8] proposed that, in pure silicon dioxide gate devices, NBTI is triggered by the tunneling of electrons from a neutral E' center precursor (a neutral oxygen vacancy) to unoccupied valence band states (occupied by holes in the silicon valence band). The E' center generation leads to a thermodynamic instability with the E' centers and passivated Si/SiO$_2$ interface Pb center precursors. Quite recently, Grasser et al. [10] have developed a comprehensive physically based predictive model for NBTI in pure SiO$_2$ devices based on these and other assumptions. Recently, Campbell et al. [11] showed that, in at least some nitrided silicon dioxide devices, K centers play a dominating role in NBTI. The K center involves silicon back bonded to three nitrogen atoms (N3≡Si•). Very recently, Ryan et al. [9] have found that the K centers dominate NBTI in a wider variety of oxynitride devices. It is possible to construct an NBTI model for these devices based on K center generation, without any consideration of hydrogen reaction/diffusion, which predicts a great deal of NBTI phenomena. The model described herein contains several ideas similar to those expressed in

the recent work of Campbell et al. [8] and Grasser et al. [10] but does not invoke any direct role for hydrogen. To the best of the author's knowledge, the model proposed herein does not conflict with the ideas expressed by Campbell et al.[8] or the recent (quite detailed) model of Grasser and co-workers dealing with NBTI in pure silicon dioxide based devices. [10] In fact, this model's K center defect plays a role which is, in part, similar to the E' center defect's role in the Grasser et al. model.

Suppose that nitrided oxide NBTI has a single dominating precursor defect type: N3≡Si-Si≡O$_3$. The fine details here don't matter except that the precursor involves a silicon-silicon bond and that one side of the precursor differs from the other. An alternative assumption, discussed later in the paper, is that the process of tunneling and an asymmetric splitting of the precursor results in different energy levels on two sides of the defect. The precursor defect would almost certainly have an energy level slightly below the silicon valence band edge, because K centers have levels in the lower part of the Si band gap [9]. An electron could tunnel out of the K center precursor into the Si valence band only if holes were present at the K center levels. Drawing an analogy with the fairly well understood E' center precursor (O3≡Si-Si≡O3) the defect will undergo a gross structural relaxation upon hole capture [12, 13, 14]. With hole capture, the E' precursor opens up; one silicon is a neutral dangling bond site the other part of a positively charged site. The hybridization of the positively charged silicon becomes sp^2 for the back bonds and thus almost pure p for the empty silicon orbital. The neutral dangling bond orbital retains significant s character and is thus lower in energy than the empty orbital of the other side of the defect. This structural rearrangement brings defect levels into the silicon band gap making the near interface K centers interface traps. This process can only occur at significant negative bias, since it requires Si valence band holes. It should be significantly enhanced at elevated temperatures: the higher the temperature, the deeper the hole distribution in the valence band also the greater the availability of phonons to facilitate the structural relaxation. The process would also exhibit recovery, if the negative bias were to be removed. If the silicon at the higher energy side of the defect (+Si≡O3 in this provisional model) were to accept an electron, the Si-Si bond might reform [10]. A process quite similar to this one is fairly well understood for the E' case in radiation damage. [10, 11] The interface trap formation would occur rapidly at first then much more slowly. A quantitative treatment can be directly adapted from a radiation effects model for E' centers developed by Oldham et al. [12]. The calculations presented in this paper in fact mirror those of Oldham et al. [12] developed for the radiation effects model involving E' centers.

Another plausible, and only slightly different alternative, would be to invoke N$_3$≡Si-Si≡N$_3$ as the precursor. In this slightly different case: after an electron tunnels from the precursor defect into the empty silicon valence band state, the hole, an asymmetry in the two sides of the defect occurs. The cause for this is essentially the same as that for the generation of E' centers discussed by Oldham [12] and others. [13,14] The positively charged side of the defect, with an empty silicon dangling bond orbital, rehybridizes so that the empty orbital becomes nearly pure p like, allowing the back bonding orbitals to become sp^2 like. This lowers the energy of their side of the defect overall, but raises the energy level of the empty orbital. (An s wave function energy is lower than a p wave function energy.) The neutral side of the defect will thus have a lower energy; a partially filled dangling bond energy level will be lower because it, unlike the empty orbital on the other side, retains some s character.

Consider an electron tunneling from a K center precursor into a hole in the silicon VB. The process is schematically illustrated in figure 2. To first order, take the barrier to be rectangular. (The defect will be within a nanometer or two of the Si/dielectric boundary; so this isn't a terrible approximation.) Utilize the WKB approximation, a standard first order approach. The transmission coefficient for a defect a distance X_m away from the interface will be

$$T = exp\left\{-2\sqrt{\frac{2m*}{\hbar^2}}\int_0^{X_m}\sqrt{U(x)-E}\,dx\right\} =$$

$$= exp\left[-2\sqrt{\frac{2m*}{\hbar^2}E_t}X_m\right], \tag{2}$$

where E_t is the barrier experienced by the electron tunneling into a hole near the top of the silicon VB and m* is an appropriate effective mass.

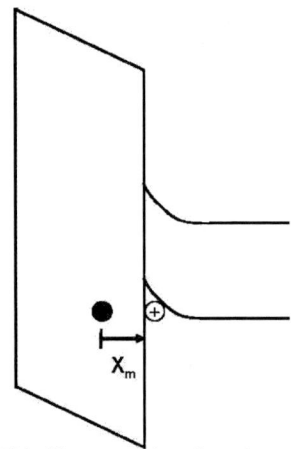

Figure 2: This illustrates the triggering step in NBTI in the proposed model. An electron tunnels from a K center precursor to an inversion layer hole in the silicon.

The transmission coefficient T should be inversely proportional to the time it takes for the electron to tunnel into the VB hole (for a given hole density in the VB). So we could write $T = t_0/t$, where t_0 is the constant of proportionality, which would roughly correspond to the tunneling time to traps right at the boundary. We could rewrite equation (2) as

$$\frac{t_0}{t} = exp\left[-2\sqrt{\frac{2m*}{\hbar^2}E_t}X_m\right] = exp[-2\beta X_m], \tag{3}$$

Where

$$\beta = \sqrt{\frac{2m*}{\hbar^2}E_t}. \tag{4}$$

So, at any given time t after the application of the negative bias, the K centers would have given up their electron to a silicon VB hole if they are within a distance $X_m(t)$ of the Si/dielectric interface where

$$\ln\left(\frac{t}{t_0}\right) = -2\beta X_m(t) \tag{5}$$

$$\frac{1}{2\beta}\ln\left(\frac{t}{t_0}\right) = X_m(t) \tag{6}$$

The V_t shift resulting from this process would be

$$\Delta V_t = \frac{\Delta Q}{C_{ox}} = \frac{q}{C_{ox}}\int_0^{X_m(t)} N(x)\,dx, \tag{7}$$

where ΔQ is the charge in the centers, Cox is the dielectric capacitance, and $N(x)$ is the physical distribution of K center precursors within the dielectric. In a somewhat analogous circumstance involving E' centers, Oldham et al. assume a simple expression for the distribution of defect centers in the dielectric: $N(x) = N_0 e^{\lambda x}$, as illustrated in figure 3.

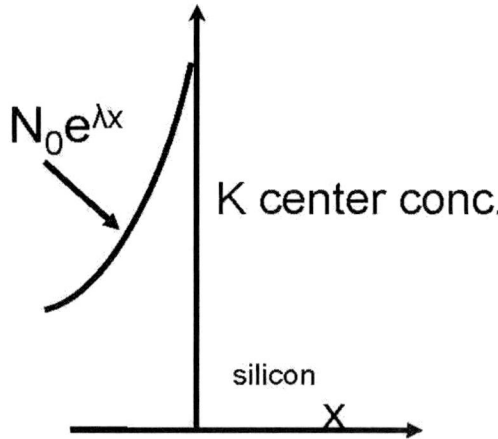

Figure 3: A schematic illustration of K center precursor density plotted versus position in the dielectric.

The positive exponent is a little confusing, but remember we are measuring from the defect in the dielectric over to the Si/dielectric boundary. So, making the approximation that $N(x) = N_0 e^{\lambda x}$,

$$T = \exp\left\{-2\sqrt{\frac{2m*}{\hbar^2}}\int_0^{X_m}\sqrt{U(x) - E}\,dx\right\} =$$

$$= \exp\left[-2\sqrt{\frac{2m*}{\hbar^2}E_t}X_m\right] \tag{8}$$

Remember that $\ln(\gamma)\delta = \delta\ln\gamma$, so

$$\Delta V_t = \left[\left[\frac{q}{C_{ox}}\right]N_0\frac{1}{\lambda}\right]\left\{\left(\frac{t}{t_0}\right)^n - 1\right\} \tag{9}$$

where

$$n = \frac{\lambda}{2\beta}$$

Note that t_0 will be very small, so

$$\Delta V_t = \left\{\left[\frac{q}{C_{ox}}\right]\frac{N_0}{\lambda}\left(\frac{1}{t_0}\right)^n\right\}t^n \tag{10}$$

Notice that this is a power law.

If $m* \cong 0.5m$ and $E_t \cong 4eV$:

$$\beta = \left[\frac{2m*}{\hbar^2}E_t\right]^{\frac{1}{2}} = \frac{7.24}{nm} \tag{11}$$

The exponent in the power law would be a small number, much less than one. If $\lambda = 2nm$, $n = 1/7.24 = 0.14$. Note also that t_0 would depend upon both oxide electric field and temperature; the higher the temperature and the higher the oxide field, the larger the pre factor. A more precise model would also have an enhancement factor from the oxide field in the β term. One could make essentially any reasonable hypothesis for a defect distribution and the outcome would be semi-quantitatively the same, a power law with a small exponent. Note also that the model also predicts a strong negative bias/field dependency, since inversion layer holes are required.

It should be emphasized that the small exponent n is not simply an arbitrary filling factor, but, to zero order at least, the inevitable outcome of the model premise. (It should also be noted that the power law exponent is the result of an approximation which is not valid over all

conceivable time ranges.) The parameter β depends only upon the product of the electron tunneling effective mass m* and the trap depth E_t. Both values are approximately known quantities; since β depends upon the square root of the m* Et product, the values must be close to the value in expression 11 that is about 7.24/nm. The exponent depends upon the ratio of λ and β. The value chosen for the lambda parameter is far less accurately determined. One might also argue that the proposed distribution is not necessarily appropriate. Nevertheless, several literature reports indicate that nitrogen atoms are concentrated within 1 or 2 nanometers of the Si/dielectric interface in plasma nitrided oxide devices, a result generally consistent with the assumptions made with regard to the defect precursor distribution. [15, 16]

Rapid recovery is also an inevitable outcome of the model premise as it would be triggered by quantum mechanical tunneling as well, presumably involving the higher energy side of the dominating defects. The tunneling process would be facilitated by the removal of the negative stressing bias. The process would be extremely rapid at first because it would involve electron tunneling to the defect centers closest to the dielectric /silicon boundary.

One more point should be emphasized: the proposed model is not directly relevant to NBTI in pure SiO$_2$ gate devices. Recent conventional electron paramagnetic resonance measurements by Fujieda et al. [17] and electrically detected magnetic resonance measurements by Campbell et al.[18, 8, 11] clearly demonstrate that Si/SiO$_2$ interface trap defects called P_b centers dominate interface trap generation in pure SiO$_2$ gate devices. This result is inconsistent with the model described herein. Campbell [8] and Lenahan [19] proposed that, in pure SiO$_2$ device structures, NBTI is triggered by the capture of silicon inversion layer holes, which statistical mechanics arguments [19] indicate would lead to subsequent Si/SiO$_2$ P_b defect generation. Recently, Grasser et al.[20] have developed a comprehensive quantitative model for NBTI in pure oxide devices in which NBTI is also triggered by inversion layer hole capture at E' centers, a process leading to P_b center interface trap generation. The qualitative ideas proposed by Campbell et al. and Lenahan as well as the comprehensive model of Grasser et al. are clearly relevant to NBTI in pure SiO$_2$ gate devices. The model proposed in this paper is not so directly relevant.

III. Summary and Conclusions

This simple model predicts an instability which requires a significant negative gate bias, and is enhanced by increasing negative bias and elevated temperature. The model also predicts a power law for threshold voltage shift: $\Delta V \cong t^n(const)$ with $n \cong 0.14$ being a reasonable estimate. The model also predicts recovery when the negative bias is withdrawn and that the recovery would be most rapid at first, because it would involve tunneling first to centers closest to the Si/dielectric boundary. It predicts an enhanced NBTI response due to nitrogen incorporation (since it is based on K centers) and predicts that the NBTI centers will be K centers and it predicts that the D_{it} would behave so that there would be net positive charge when the interface Fermi level is below about mid-gap and likely net zero charge with the Fermi level above mid-gap. The model predicts that the response will depend upon the physical distribution of K center precursors in the oxide and on the total number of K center precursors per unit area. Note that no reaction diffusion or hydrogenic species are involved. Work supported by Texas Instruments through SRC Custom Funding.

REFERENCES

[1] T. Grasser, W. Goes, and B. Kaczer, "Toward engineering modeling of negative bias temperature instability," in Defects in Microelectronic Materials and Devices, D.M. Fleetwood, S.T. Pantelides, and R.D. Schrimpf. Boca Raton, London, New York: CRC Press, 2009, pp. 399-436.

[2] G. LaRosa, "Negative bias temperature instabilities in pMOSFET devices," in IEEE Series on Microelectronic Systems, A.W. Strong, E.Y. Wu, R.P. Vollertsen, J. Sune, G. LaRosa, S.E. Rauch, and R.P. Sullivan. Hoboken: Wiley, 2009, pp. 331-439.

[3] A. T. Krishnan, S. Chanravarti, P. Nicollian, V. Reddy, and S. Krishnan, Appl. Phys. Lett. 88, #153518 (2006)

[4] S.S. Tan, T.P. Chen, C.H. ang, and L. Chan, "Relationship between interfacial nitrogen concentration and activation energies of fixed charge trapping and interface trap generation under bias temperature stress conditions." Appl. Phys Lett. 82, 269 (2003)

[5] S.S. Tan, T.P. Chen, T.M. Soon, K.P. Loh, C.H. Ang, and L. Chan, Nitrogen-enhanced negative bias instability: An insight by experiment and first principle calculations, Appl. Phys. Lett. 82, 1881 (2003)

[6] H. Reisinger, O. Blank, W. Heinrigs, A. Muhlhoff, W. Gustin, and C. Schlunder, "Analysis of NBTI Degradation and Recovery Behavior Based on Ultra-Fast Vt Measurements" Proceedings of the 2006 IRPSS, 448 (2006)

[7] C. Shen, M.P. Li, C.E. Foo, T. Yang, R.M. Huang, A. Yzp, G.S. Samudra, Y.C. Yeo, "Characterization and Physical origin of Fast Vt Transition NBTI of p MOSFETs with SiON dielectric," Proc, IEDM, San Francisco, CA, 333 (2006)

[8] J.P. Campbell, P.M. Lenahan, A.T. Krishnan, and S. Krishnan, "Observations of NBTI-induced atomic-scale defects," IEEE Trans. Dev. Mater. Reliab., 6, pp. 117-122 (2006).

[9] J.T. Ryan, P.M. Lenahan, A.T. Krishnan, and S. Krishnan, "Energy resolved spin dependent tunneling in 1.2 nm dielectrics," Appl. Phys. Lett., 95, pp. 103503 (2009).

[10] T. Grasser, B. Kaczer, W. Goes, T. Aichinger, P. Hehenberger, and M. Nelhiebel, "A two-stage model for negative bias temperature instability," Proc. IEEE Intl. Reliab. Phys. Symp., pp 33-44 (2009).

[11] J.P. Campbell, P.M. Lenahan, A.T. Krishnan, and S. Krishnan, "Identification of the atomic-scale defects involved in the negative bias temperature instability in plasma-nitrided p-channel metal-oxide-silicon field-effect transistors," J. Appl. Phys., 103, pp. 044505 (2008).

[12] T.R. Oldham, A.J. Lelis, and F.B. McLean, "Spatial dependence of trapped holes determined from tunneling analysis and measured annealing," IEEE Trans. Nucl. Sci., 33, pp. 1203-1209 (1986).

[13] J.F. Conley, P.M. Lenahan, A.J. Lelis, and T.R. Oldham, "Electron spin resonance evidence that E'(gamma) centers can behave as switching oxide traps," IEEE Trans. Nuc. Sci., 42, pp. 1744-1749 (1995).

[14] P.M. Lenahan and J.F. Conley; What can electron paramagnetic resonance tell us about the Si/SiO2 system?" J. Vac. Sci. Tech. B. 16, 2134 (1998)

[15] S. W. Lim, S. Parsons, T.Y. Luo, "Effect of repetition of plasma nitride oxide integration," Thin Solid Films, 515, 2673 (2006)

[16] K. Takasuki, K. Irino, T. Aoyama, Y. Momiyama, T. Nauanishi, Y. Tamura, T. Ito, Impact of Nitrogen Profile in Gate Nitrided Oxide on Deep-Submicron CMOS Performance and Reliability, Fujitsu Sci. Tech. J. 39, 40 (2003)

[17] S.Fujleda, Y.Micera, M.Saitoh, E. Hasagawa, S.Koyama, and KAndo, "Interface defects responsible for negative bias temperature instability in plasma nitride SiON/Si (100) systems," Appl. Phys. Lett. 82,3677 (2003)

[18] J.P. Campbell, P.M. Lenahan, A.T. Krishnan, and S. Krishnan, "Direct observation of the structure of defect centers involved in the negative bias temperature instability," Appl. Phys. Lett. 87, art.# 204106 (2005)

[19] P.M. Lenahan, "Deep level defects Involved in MOS Devices Instabilities," Microelectronic. Reliability 41, 890 (2007)

[20] T. Grasser, B. Kaczer, W. Goes, T. Aichinger, P. Hehenbeger, and M. Nelheibel, "A Two Stage Model for the Negative Bias Temperature Instability," Proceedings of the 2009 IEEE International Reliability Physics Symposium 33 (2009)

A Generalized, I_B-independent, Physical HCI Lifetime Projection Methodology based on Universality of Hot-Carrier Degradation

Dhanoop Varghese, Muhammad Ashraful Alam
School of Electrical and Computer Engineering
Purdue University
West Lafayette, IN, 47907, USA
1-765-494-5988, alam@purdue.edu

Bonnie Weir
LSI
Allentown, PA, 18109, USA

Abstract—**We develop a novel approach for hot carrier lifetime prediction based on the 'universality of HCI degradation' that not only generalizes the classical theory by obviating the measurement of I_B, but also allows prediction of HCI lifetime over a broad range of technology nodes, bias conditions, and device geometries. We explain the shape of the degradation vs. time characteristic based on the energy distribution of the Si-O bonds, and we show, based on the bond-dispersion model, that the degradation shows similar features for both ON- and OFF-state bias conditions.**

Keywords- on-state hot-carrier; off-state hot-carrier; universal degradation; bond-dispersion model; voltage acceleration model;

I. INTRODUCTION

Despite significant scaling in voltage even at the 40nm technology node, hot carrier injection (HCI) remains an important reliability issue for CMOS transistors [1, 2]. The degradation is believed to be due to energetic carriers that get injected into the gate dielectric and create damage by breaking bonds (Si-H, Si-O *etc.*) within the dielectric. Classical hot carrier theories (a) assume a constant power law exponent *n* for degradation ($\Delta=At^n$) and (b) use the change in impact ionization current measured at the bulk terminal (I_B) as a measure of voltage acceleration from stress to operating conditions (see Fig. 1). Both of these observations do not apply to modern MOSFETs: HCI degradation shows a saturating behavior with the time exponents gradually reducing at longer stress time and at higher degradation levels. Also the traditional voltage acceleration model [1] (and an even more recent variant of it [3,4]) based on the bulk current (I_B) is no longer a viable predictor of HCI lifetime of ultra-scaled devices because I_B for transistors with thin oxides is irretrievably contaminated by gate leakage (I_G). The alternate empirical model uses linear temporal extrapolation of degradation data and an $e^{B/VD}$ voltage acceleration model, which are not physically justified. In addition, as worst-case HCI lifetimes become shorter with

transistor scaling, and designers increasingly rely on reliability modeling for lifetime projection, one must correctly estimate the HCI lifetimes at all V_G and V_D combinations (not only at $V_G=V_D$ or $V_G=V_D/2$). For example, HCI lifetime at $V_G>V_D$ bias conditions deviates from the empirical $e^{B/VD}$ model (see Fig. 7b) and is several orders of magnitude lower compared to that predicted by classical/empirical hot carrier models [4].

In this paper we demonstrate that the HCI in logic transistors exhibits a universal time-dependent (temporal)

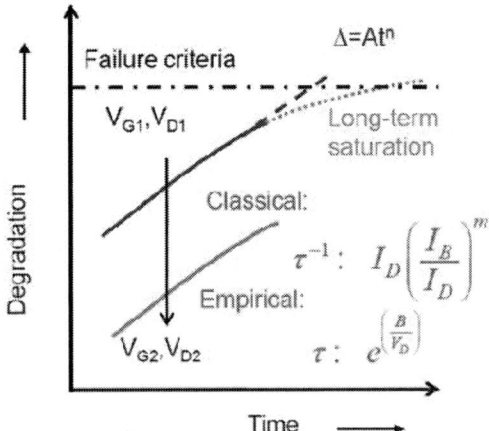

Figure 1. Lifetime extrapolation based on classical ON-state hot carrier theories assumes constant power law time exponent (dashed line) while we show that the degradation has saturating characteristics at longer stress time (dotted line). The bulk current for ultra-scaled 40nm technology node transistor is dominated by gate leakage and therefore cannot be used to compute HCI voltage acceleration factors. The alternate empirical model lacks explicit physical justification and HCI lifetime at $V_G>V_D$ bias conditions is shown to deviate from $e^{B/VD}$ model (see Fig. 7b) [4]. In this paper we demonstrate that hot carrier degradation (both at ON- and OFF-state) exhibit universal behavior, which can be used to perform fast and accurate lifetime extrapolation for a range of technology nodes, device geometries and bias conditions.

TSMC, AMAT, NCN

degradation. The saturating nature of the universal curve is explained based on a bond-dispersion model [5]. Even though, temporal universality of ON state HCI, *i.e.* $N_{IT} \sim f(t/t_0(V_G, V_D, T))$, has been previously suggested in literature [6-8], its application to a wide range of devices and bias conditions and implications for lifetime projection have not been explored. A judicious use of this universality enables fast and efficient (I_B-independent) methodology of lifetime extrapolation that provides more accurate results than simple linear extrapolations of short-term data.

II. UNIVERSALITY OF DEGRADATION

Fig. 2 shows the HCI-induced parameter degradation for transistors from 40nm technology node. The high gate leakage current in these transistors prevents direct monitoring of interface damage through charge pumping (CP) or DCIV [9] techniques. The lack of recovery on removal of the stress (data not shown) [10], however, indicates dominant contribution to interface degradation arises primarily from permanent interface traps. The theory of N_{IT}-generation [5] suggests (see Fig. 5) that the individual curves can be scaled laterally by factors S (or shifted in log scale) to overlap in time. The resulting curve is universal over > 6 orders of magnitude in time. A single long

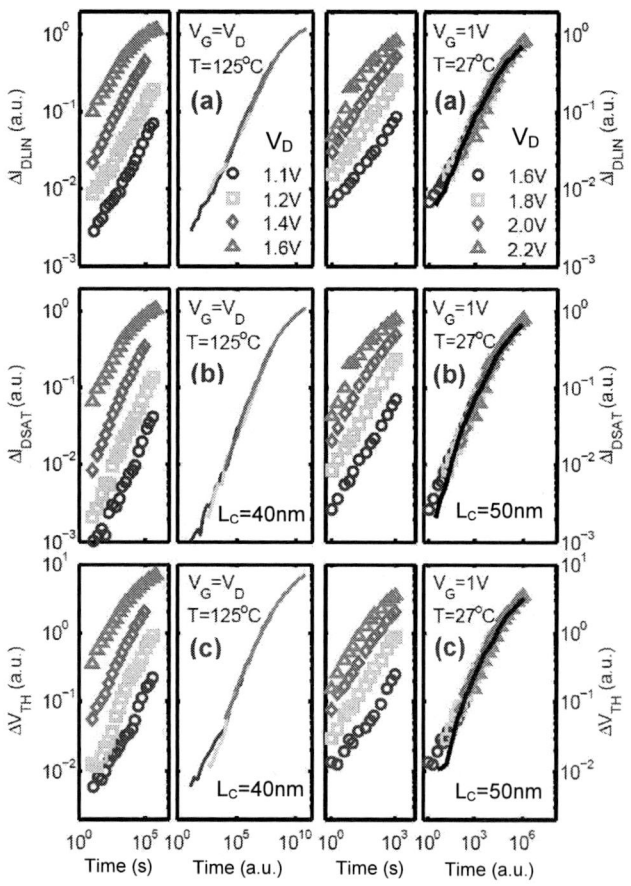

Figure 2. Degradation in (a) linear drain current, (b) saturation drain current and (c) threshold voltage at (LHS) $V_G=V_D$, T=125°C and (RHS) $V_G=1V$, T=25°C ON-state hot carrier stress for ultra-scaled logic transistors. Solid black line (rightmost panel) represents data from single long term (300ks) measurement.

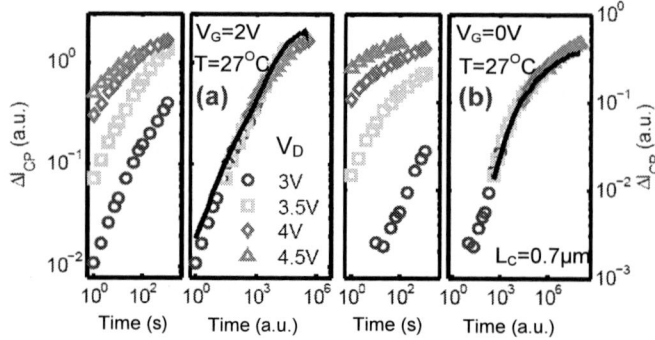

Figure 3. N_{IT} time evolution measured through charge pumping at (a) ON-state and (b) OFF-state hot carrier stress conditions for 130nm technology logic transistors. The solid black line represents data from single long term (300ks) measurement.

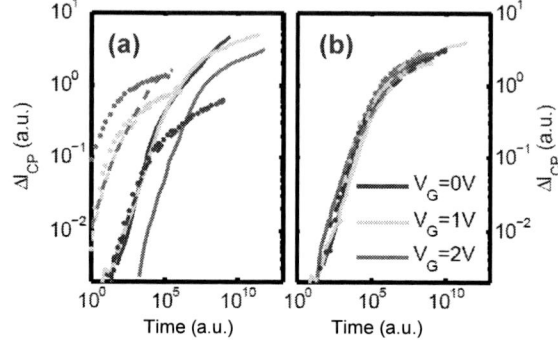

Figure 4. (a) Universal degradation curves for various stress biases (ON/OFF) and device structures (solid: Drain extended NMOS for I/O applications [7], dashed: Logic 0.7μm, dotted: Logic 0.16 μm) formed through the scaling of degradation data (see Fig. 3). (b) Various degradation curves show very similar time evolution over several orders of magnitude in time.

term measurement at a given stress bias (Fig. 2, rightmost panel) verifies the hypothesis that various stress biases trace different parts of a single time-dependent universal curve.

Fig. 3a shows the HCI-induced interface damage measured by CP during ON-state stress ($V_G=2V$, $V_D \geq 3V$) of 130nm logic transistors. Fig. 3b shows that the universality of interface damage remains unaffected as the bias condition is changed from ON ($V_G=2V$) to OFF ($V_G=0V$) in spite of the reduction in drain current over four orders of magnitude.

Fig. 4a summarizes the universal curves from hot carrier measurements carried out on a wide range of device geometries (drain extended NMOS, long and short channel core logic NMOS) and bias conditions ($V_G=0V$, 1V, and 2V). The universal curves are formed by scaling multiple short-term degradation data as shown in Fig. 3. Since the universal curves are for various device geometries and bias conditions, they differ in their relative magnitudes and degradation rates. However, all the universal curves shown in Fig. 4a have very similar shape as given by the equation:

$$\Delta I_{CP}=A\xi(St), \qquad (1)$$

where ξ is the universal function, and the pre-factor A and scaling factor S are determined by the device geometry and

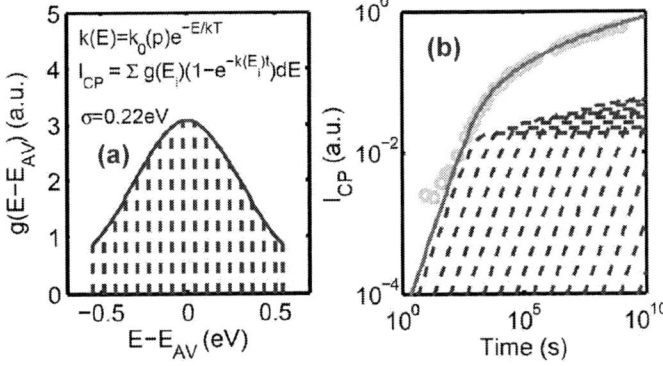

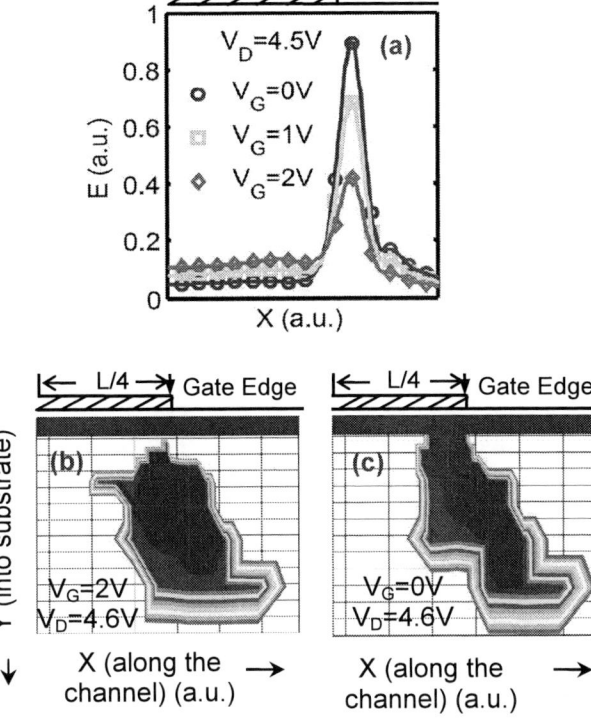

Figure 5. (a) Energy distribution of Si-O bond precursors. E_{AV} denotes the average and σ denotes the standard deviation of Si-O bond energies. The bonds with lower energies break first and are followed by bonds with progressively higher energies. (b) Total damage is obtained by summing up (red solid line) contributions from precursors at each energy grid point (dashed lines). The B-D model explains the CP universal curve over several orders of magnitude in time (symbols: measurement, Lines: simulation).

applied bias. To prove this, the universal curves are scaled along x-axis and y-axis directions (see Fig. 4b), and the resultant curve overlaps over six orders of magnitude in time. The similar shape for all universal curves indicates the robustness of the underlying hot carrier degradation mechanism.

III. THEORY OF UNIVERSAL DEGRADATION

Let us now consider the shape of the universal curve via the framework of Si-O bond-dispersion (B-D) model [5]. The Si-O bond precursors have a finite energy distribution in a disordered medium like SiO_2 (see Fig. 5a) [5]. The precursors at various energies break at different rates, which are then added together to obtain the net interface damage (see Fig. 5b). Bond-dispersion simulation with an energy spread specified by $\sigma=0.22eV$ is found to explain the measurement data over several orders of magnitude in time. The reaction rate constant being dependent on hot hole density (p), a change in V_G and V_D combination changes the hot hole injection, thereby changing the degradation rate and shifting the curves laterally. Indeed, this universality of damage is model-agnostic and could be also arrived at within the R-D formulation of Si-H bond-dissociation [11].

To understand why the extremes of ON- and OFF-state degradation showed identical universal features, we obtained electric field and hot carrier profiles from device- and Monte Carlo simulations (see Fig. 6). The magnitude of the electric field and hot carriers are different at various V_G, but their spatial profile close to interface almost remained the same. The B-D model therefore predicts similar universal degradation at both ON- and OFF-state bias conditions. The difference in absolute hot hole densities at these bias conditions, however, results in different degradation rates for the universal curves (rate constant $k_0 \sim$ hot hole density p) as shown in Fig. 4a.

Figure 6. (a) Electric field profile close to the Si/SiO_2 interface for long channel transistors at ON- and OFF-state bias conditions obtained from device simulations. The electric field profiles are found to be similar at all bias conditions and peak close to the gate edge. The resultant 2D hot hole profile (E>4.7eV) obtained from Monte Carlo simulations at (b) ON-state and (c) OFF-state biases are also similar, resulting in similar universal degradation. Though the spatial profiles are similar, the absolute hot hole densities are different at both bias conditions, resulting in different degradation rates for the universal curves.

IV. VOLTAGE ACCELERATION MODEL

The scaling factors used to construct the universal curves are proportional to the degradation rate at each bias, and hence can be used to obtain the voltage acceleration of device lifetime ($T_F \sim 1/S$) (see Fig. 7). Unlike the actual lifetime, the scaling factors can be obtained using short term measurements. Moreover, the method no longer relies on the problematic measure of I_B for HCI lifetime projection. As shown in Fig. 7a, the linear extrapolations lead to inaccurate values at low stress voltages which in turn lead to inaccurate projections at operating conditions (Fig. 7b). The universality of hot carrier degradation, therefore, proves to be a more accurate alternative for lifetime extrapolation based on short term stress data, especially in the regimes of interest for reliability modeling. Moreover, by adopting this model-agnostic methodology, we no longer rely on the model of I_G-I_B relationship and obviate the debate as to whether hot hole or hot electron injection is responsible for HCI degradation, to make accurate HCI-lifetime projection possible.

978-1-4244-5430-3/10 $26.00 © 2010 IEEE

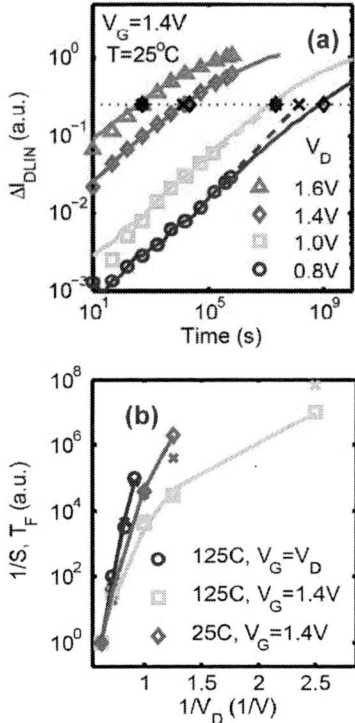

Figure 7. (a) Lifetime estimation using long-term extrapolation (dashed line) and that using scaling factors and universal degradation curve (solid line) for 40nm devices. By knowing the shape of the universal curve (obtained by scaling multiple short-term degradation data) and the respective scaling factors, long-term degradation at each bias point and the corresponding device lifetime can be obtained accurately. (b) Voltage acceleration of normalized hot carrier lifetime thus obtained are plotted using $e^{B/VD}$ model, and clearly shows deviations at $V_G > V_D$ bias condition. Both techniques agree well at lower lifetimes where errors due to extrapolation are smaller, but show differences as lifetime gets longer, resulting in significant differences when curves are extended to low V_D values. The universality of hot carrier degradation allows quick estimation of scaling factors based on short term measurements and facilitates fast and accurate lifetime prediction for the entire range of operating bias conditions.

V. CONCLUSIONS

Theory, modeling and measurements over a wide range of technology nodes and operating conditions are used to demonstrate universality of HCI degradation and to develop a new lifetime prediction methodology that is independent of substrate and gate current measurements or any specific physical models of HCI degradation. We also find that this generalized model anticipates other popular classical models as special cases.

ACKNOWLEDGEMENT

We gratefully acknowledge Texas Instruments for supporting the research, Network of Computational Nanotechnology at Purdue for the computational facilities and Birck Nanotechnology Center at Purdue for the experimental facilities.

REFERENCES

[1] C. Hu, S. C. Tam, F.-C. Hsu, P.-K. Ko, T.-Y. Chan, and K. W. Terril, "Hot-electron-induced MOSFET degradation – model, monitor, and improvement", IEEE Trans. on Electron Devices, vol. 32, pp. 375–385, February, 1985.

[2] A. Bravaix, C. Guerin, V. Huard, D. Roy, J. M. Roux, and E. Vincent, "Hot-carrier acceleration factors for low power management in DC-AC stressed 40nm NMOS node at high temperature", in Proc. Int. Reliability Physics Symposium, pp. 531–548, April, 2009.

[3] S. E. Rauch, F. J. Guarin, and G. LaRosa, "Impact of E-E scattering on the hot-carrier degradation of deepsubmicron NMOSFETs", IEEE Electron Device Letters, vol. 19, pp. 463–465, December, 1998.

[4] C. Guerin, V. Huard, and A. Bravaix, "The energy-driven hot-carrier degradation modes of nMOSFETs", IEEE Trans. on Device and Materials Reliability, vol. 7, pp. 225–235, June, 2007.

[5] K. Hess, A. Haggag, W. McMahon, B. Fischer, K. Cheng, J. Lee and J. Lyding, "Simulation of Si-SiO$_2$ defect generation in CMOS chips: from atomistic structure to chip failure rates", in Proc. Int. Electron Devices Meeting, pp. 93–96, December, 2000.

[6] J.-S. Goo, Y.-G. Kim, H. L'Yee, H.-Y. Kwon, and H. Shin, "An analytical model for hot-carrier-induced degradation of deep-submicron n-channel LDD MOSFETs", Solid-State Electronics, vol. 38, pp. 1191–1196, June, 1995.

[7] D. S. Ang, and C. H. Ling, "On the time-dependent degradation of LDD n-MOSFETs under hot-carrier stress", Microelectronics Reliability, vol. 39, pp. 1311–1322, September, 1999.

[8] D. Varghese, H. Kufluoglu, V. Reddy, H. Shichijo, D. Mosher, S. Krishnan, and M. A. Alam, "Off-state degradation in drain-extended NMOS transistors: Interface damage and correlation to dielectric breakdown", IEEE Trans. on Electron Devices, vol. 54, pp. 2669–2678, October, 2007.

[9] A. Asenov, J. Berger, W. Weber, M. Bollu, and F. Koch, "Hot-carrier degradation monitoring in LDD n-MOSFETs using drain gated-diode measurements", Microelectronic Enginnering, vol. 15, pp. 445-448, October, 1991.

[10] D. Varghese, "Multi-probe experimental and 'bottom-up' computational analysis of correlated defect generation in modern nanoscale transistors", Ph.D dissertation, Purdue University, West Lafayette, IN, USA, December, 2009.

[11] D. Lachenal, F. Monsieur, Y. Rey-Tauriac, and A. Bravaix, "HCI degradation model based on the diffusion equation including the MVHR model", Microelectronic Engineering, vol. 84, pp. 1921–1924, September-October, 2007.

POSITIVE AND NEGATIVE BIAS TEMPERATURE INSTABILITY

ON SUB-NANOMETER EOT HIGH-κ MOSFETs

Moonju Cho, Marc Aoulaiche, Robin Degraeve, Ben Kaczer, Jacopo Franco[2], Thomas Kauerauf,

Philippe Roussel, Lars Å. Ragnarsson, Joshua Tseng[1], Thomas Y. Hoffmann and Guido Groeseneken[2]

IMEC, Kapeldreef 75, B-3001 Leuven, Belgium
[1] TSMC assignee at IMEC, [2] also at K.U. Leuven, ESAT Department
phone: +32 16 28 78 11; fax: +32 16 28 17 06; e-mail: Moon.Ju.Cho@imec.be

Abstract - For the first time, positive and negative bias temperature instability (P/NBTI) mechanisms in sub-nanometer EOT devices are investigated in this study. It is shown that PBTI degradation in sub-nanometer EOT devices occurs by interface degradation, additionally to the oxide bulk trap filling which is the dominant mechanism in over 1nm EOT devices. For NBTI, interface degradation remains as the main mechanism in sub-nano EOT devices, and additional high contribution of the high-κ bulk defects can increase the degradation below 6A EOT.

[Keywords: NBTI, PBTI, thin EOT, high-κ dielectrics.]

INTRODUCTION

The P/NBTI mechanism has been widely studied on high-κ oxide/ metal gate devices [1-3]. However, the degradation mechanism in sub-nano EOT devices has not clearly been reported yet. In this paper, we report P/NBTI mechanisms in very thin devices with reduced interfacial layer.

EXPERIMENTAL

The CMOS devices were fabricated on 300 mm (100) Si-wafers using a metal-inserted poly-Si process (MIPS). The high-κ dielectric is mainly HfO_2. An Al_2O_3 cap in pMOS or a La_2O_3 cap in nMOS are added in some devices for work function control. The thin EOT is obtained by reducing the interfacial layer as in figure 1, with a thin TiN/ Si-cap gate structure [4]. The EOT is extracted on large area devices at 100 kHz-100 MHz.

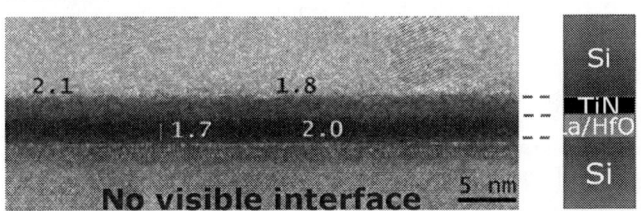

Figure 1. High resolution TEM image of a Si/ SiO₂/ HfO/ LaO/ TiN/ Si device; SiO_2 interfacial layer disappears during processing.

P/NBTI was performed on W×L=10×1cm² MOSFETs at 125°C, unless otherwise specified. An I_D–V_G characteristic was measured on the fresh device, then a pulse-like stress was applied at the gate of the transistor with source and bulk

grounded. The transistor was stressed at V_{G_stress}. The gate voltage was then switched to $V_{G\_sense} \sim V_{TH}$, and the drain current was measured with minimized delay (~1ms) to reduce the V_{TH} recovery effect. The measured drain current is then transformed into a V_{th}-shift.

RESULTS AND DISCUSSION

Positive bias temperature instability mechanism

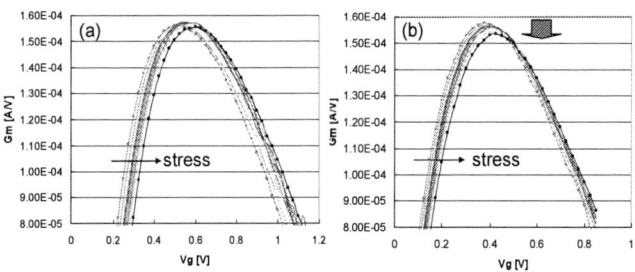

Figure 2. G_m degradation after 1800s stress at 16.3MV/cm on (a) 11.4A, (b) 6.1A EOT devices with HfO_2 dielectric: maximum G_m decreases during PBTI stress in sub-nano EOT devices.

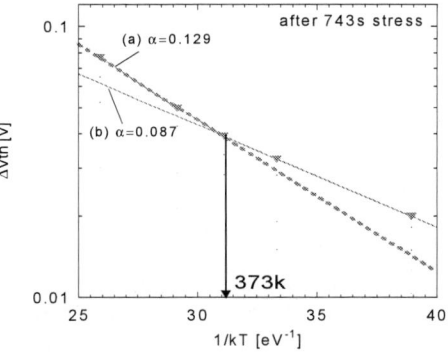

Figure 3. Strong temperature activation is shown by an Arrhenius plot in high temperature (>373K) during PBTI stress on a 6.1A EOT device with HfO_2 dielectric.

978-1-4244-5430-3/10 $26.00 © 2010 IEEE

PBTI degradation in over 1nm EOT devices is known to be caused by bulk trap filling of initially present traps [1]. In this case, G_m degradation is not observed after PBT stress, as is shown in figure 2 (a) for a 11.4A EOT device. However, for a 6.1A EOT device a clear G_m degradation is observed (figure 2 (b)), which indicates trap generation near the interface. Strong temperature activation is also shown in the high temperature (>373K) regime as shown in figure 3 by an Arrhenius plot. The activation energy is ~0.13eV, which is higher than the value found from PBTI in over 1nm EOT (<0.1eV) [1,5,6]. This confirms again that generation of traps contribute to the PBTI degradation, additionally to the direct tunneling of charges into oxide bulk traps, and contrary to the case of over nm EOT devices.

Finally in all the devices measured, the V_{TH} shift follows a power-law dependency with stress time in sub-nm EOT devices, unlike in over 1nm EOT devices which show a different V_{TH} shift trend in each stack (figure 4).

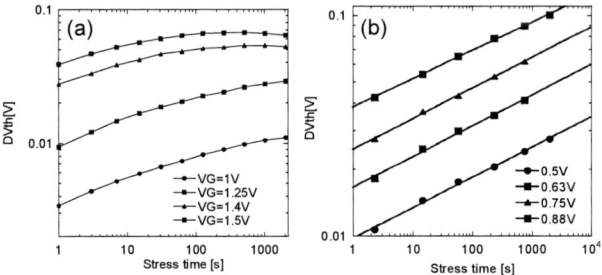

Figure 4. The threshold voltage shift after stress for (a) 13.0Å and (b) 7.6Å EOT devices. Sub-nano EOT devices show a power-law dependency.

Previously in [1], V_{TH} shift kinetics in thick EOT devices is modeled by assuming direct tunneling of electrons from the channel into preexisting defects in the dielectric and using the S-R-H statistics and WKB for the direct tunneling approximation [7]. V_{TH} shift during PBTI is given by,

$$\Delta V_{th}(t) = \frac{q}{\varepsilon_0 \varepsilon_{ox}} \cdot \int_0^{x(t)} (t_{ox} - x).N(x)dx \qquad (1)$$

where q is the elementary charge, ε_o and ε_{ox} are the relative and oxide dielectric constants, t_{ox} is the oxide thickness and $N(x)$ is the defects density at the distance x from the Si/SiO_2 interface. The correlation between the distance (x) from the SiO_2 interface and the time (t) is given by defect filling probability $f(t,x)$.

$$f(t,x) = 1 - \exp(c_n.P(x)t) \qquad (2)$$

where c_n is the electron capturing rate and $P(x)$ is the direct tunneling probability using the WKB approximation.

An advanced PBTI degradation model is derived for sub-nm EOT applications in figure 5. As the interfacial layer with higher conduction band offset decreases in sub-nm EOT devices as in figure 1, electrons in the silicon conduction band can tunnel deeper into the high k oxide according to the WKB approximation. More traps in the high-k layer can be filled and contribute to the PBTI degradation in this case. Additionally, trap generation near the interface occurs as shown in figures 2-4, which makes PBTI in sub-nano EOT devices a

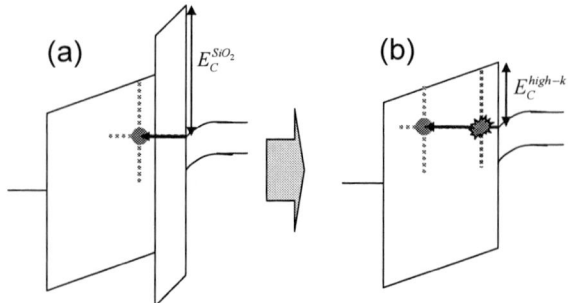

Figure 5. Schematic diagrams showing PBTI degradation mechanism change: in sub-nano EOT devices with reduced interfacial layer, both increased electron tunneling into oxide bulk traps, and trap generation near interface contribute to the PBTI as in (b).

more important reliability concern than in over 1 nm devices. Therefore, Eq. (1) should be updated as,

$$\Delta V_{th}(t) = A \cdot \left[\frac{q}{\varepsilon_0 \varepsilon_{ox}} \cdot \int_0^{x(t)} (t_{ox} - x).N(x)dx \right] + B \cdot \frac{q\Delta N_{it}}{\varepsilon_0 \varepsilon_{ox}} EOT \quad (3)$$

where ΔN_{it} is the interfacial trap density generated.
The newly generated defects at the interface can be originated from non-bridging oxygen, weak Si-Si bond, oxygen vacancy, etc [8]. However Pb-center, which is suspected to be the main reason of interface trap generated during NBTI, cannot contribute to this N_{it} generation since it didn't appear in over 1nm-EOT PBTI.

Negative bias temperature instability mechanism

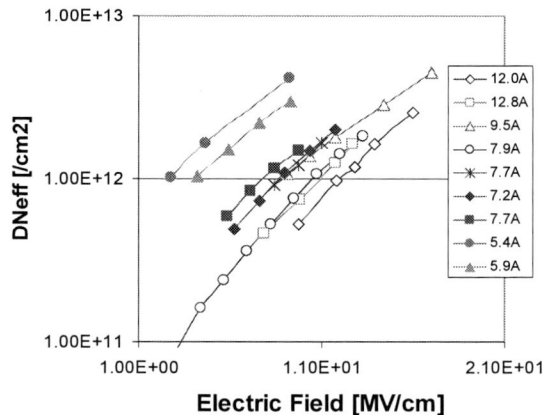

Figure 6. Effective trap density generation calculated from V_{TH} shift, vs electric field: 'zero interfacial layer' below 6A of EOT changes interface characteristic, and generates more effective traps contributing to the NBTI degradation.

Figure 6 shows effective trap density generated (ΔN_{eff}) under NBTI stress vs electric field calculated from EOT. ΔN_{eff} is similar down to ~7A EOT devices, which shows that the main NBTI degradation mechanism remains interfacial layer degradation. For even thinner layers, however, a significantly larger degradation is observed. The HRTEM picture on a sample with EOT thinner then ~7A shows a 'zero-interface layer thickness' as in figure 1, and this implies that the chemical

978-1-4244-5430-3/10 $26.00 © 2010 IEEE

interface quality can be changed. As a result, these devices under 6A EOT show a higher trap density generation (ΔN_{eff}) at the interface, and/or hole trapping into the high-k bulk defects increases due to removal of the high tunneling barrier oxide at the interface. In this case, we need to consider V_{TH} shift during NBTI with both interfacial layer degradation by Pb-center generation from reaction-dispersive proton transport model [9] and bulk trap filling by direct tunneling as,

$$\Delta V_{th}(t) = A\left[\equiv Si_3 - SiH + h^+ \underset{kr}{\overset{kf}{\rightleftharpoons}} \equiv Si_3 - Si^\bullet + H^+ \right] + B\left[\frac{q}{\varepsilon_0 \varepsilon_{ox}} \cdot \int_0^{x(t)} (t_{ox} - x).N(x)dx \right] \quad (4)$$

where $\equiv Si_3 - Si^\bullet$ is so-called Pb-centers, h^+ is hole, H^+ is proton. Most widely accepted NBTI degradation model in over-1nm EOT regime is the 'reaction-diffusion model [10]', but we'd like to use the 'reaction-dispersive proton transport model' here. This is because the transport of protons in SiO_2 is highly dispersive, i.e., protons are randomly hopping from bridging oxygen atoms to the others in the SiO_2 lattice.

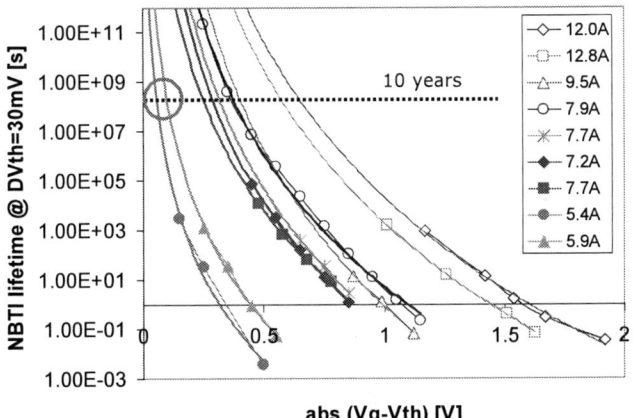

Figure 7. NBTI lifetime vs gate voltage over-drive: 'zero interfacial layer' below 6A EOT changes interface characteristic, and degrades NBTI lifetime severely.

Finally, the increased ΔN_{eff} results in a severe reliability problem, as shown on Figure 7. Lifetime is extracted at 30mV of V_{th} shift criterion, and 10 years lifetime under 6A EOT devices is short as expected.
Note that this lifetime is calculated from V_{th} shift including both of permanent and recoverable components. If we define the BTI as degradation only from permanently generated defects, recoverable component should be subtracted from the total V_{th} shift. Indeed, a simulation separating permanent and recoverable components based on V_{th} shift relaxation measurement [11] is performed on a EOT 5.9 nm device (figure 8). High amount of recoverable component is extracted, which indicate a possible lifetime improvement. However, this definition needs a thorough discussion in the semiconductor society.

P/NBTI lifetime in sub-nano EOT devices

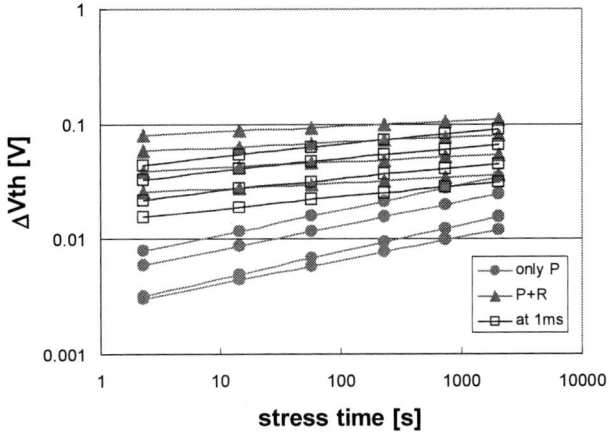

Figure 8. Permanent P and recoverable R components are separated by a simulation based on V_{th} relaxation measurement between stressing: high amount of recoverable component is extracted.

Figure 9 shows that V_G over-drive at 10 years after NBTI on pMOSFETs decreases continuously according with decreasing EOT, as shown in figure 7, and more severely so below 6A EOT. PBTI for over-1nm EOT devices improves with thinner EOT due to lower bulk trap density to fill, but degrades in sub-nano EOT devices due to the lower direct tunneling barrier and additional trap generation near the interface as shown previously. Finally ~5A EOT devices show serious P/NBTI degradation under V_G over-drive at the 0.7V target, calculated from an operating voltage of 1.0V and a V_{TH} of 0.3V. However, considering the BTI target by *iso*-electric field as in figure 10 can decrease the concern: for example, as a target with V_G over-drive 0.7V at 15A EOT corresponds to ~4.67MV/cm, then PBTI is still reliable in the whole sub-nano EOT regime. Note that the electric field at 10 years after NBTI decreases in sub-nano devices, due to the change of the interfacial layer characteristic.

CONCLUSIONS

The P/NBTI mechanisms in sub-1nm EOT devices were reported. PBTI degradation occurs by (1) trap generation near

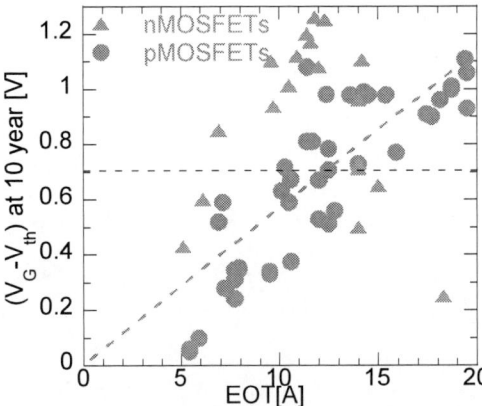

Figure 9. V_G over-drive at 10 years after P/NBTI vs EOT: both N/PBTI are serious concern in sub-nano EOT regime, especially below 7A EOT where interface disappears.

978-1-4244-5430-3/10 $26.00 © 2010 IEEE

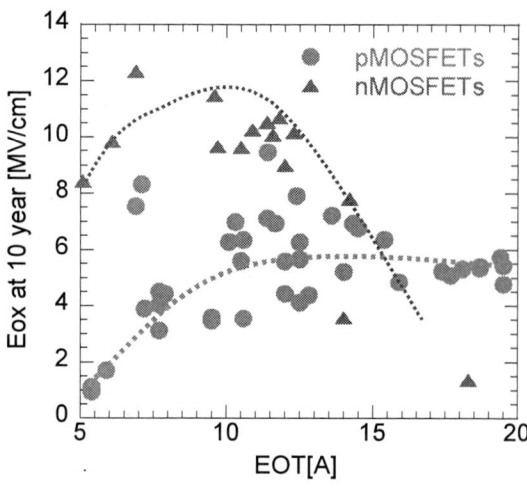

Figure 10. Electric field at 10 years after P/NBTI vs EOT. Considering *iso*-electric field can decrease the concern.

the interface apparent as indicated from G_m degradation and higher energy activation, and (2) increased bulk trap filling due to the lower direct tunneling barrier from the reduced interfacial layer. The main mechanism for NBTI remains the interfacial trap degradation as in over 1nm EOT devices. For sub-7nm EOT devices, NBTI lifetime becomes problematic, caused by the disappearance of the interfacial layer, and the corresponding change in the chemical interface quality.

ACKNOWLEDGEMENTS

The IMEC sub 32 nm program members, the IMEC pilot line and amsimec are acknowledged for their support.

REFERENCES

[1] M. Aoulaiche, B. Kaczer, M. Cho, M. Houssa, R. Degraeve, T. Kauerauf, A. Akheyar, T. Schram, Ph. Roussel, H.E. Maes, T. Hoffmann, S. Biesemans and G. Groeseneken, IRPS, Proc., pp. 1014-1018, 2009.

[2] E. Cartier, B. P. Linder, V. Narayanan, V. K. Paruchuri, IEDM Technical Digest, pp.1-4, 2006.

[3] S. Pae, M. Agostinelli, M. Brazier, R. Chau, G. Dewey, T. Ghani, M. Hattendorf, J. Hicks, J. Kavalieros, K. Kuhn, M. Kuhn, J. Maiz, M. Metz, K. Mistry, C. Prasad, S. Ramey, A. Roskowski, J. Sandford, C. Thomas, J. Thomas, C. Wiegand, and J. Wiedemer, IRPS, Proc., pp. 352-357, 2008.

[4] L. Ragnarsson, Z. Li, J. Tseng, T. Schram, E. Rohr, M. J. Cho, T. Kauerauf, T. Conard, Y. Okuno, B. Parvais, P. Absil, S. Biesemans, and T. Y. Hoffmann, IEDM Technical Digest, pp. 663-666, 2009.

[5] M. Houssa, S. De Gendt, J. L. Autran, G. Groeseneken, and M. M. Heyns, Symp. VLSI Tech. dig, pp. 212-213, 2004.

[6] S. Kalpat, H.-H. Tseng, M. Ramon, M. Moosa, D. Tekleab, P. J. Tobin, D. C. Gilmer, R. I. Hegde, C. Capasso, C. Tracy, and B. E. White, Jr, IEEE TDMR, pp. 26-35, 2005.

[7] S. M. Sze, *Physics of Semiconductor Devices*, John Wiley & Sons Inc., 2nd Ed., pp. 520-531, 1981.

[8] G. Barbottin, and A. Vapaille, *Instabilities in Silicon Devices*, Vol. 1, Elsevier Science Publishers B. V., Chapter 2, 1986.

[9] M. Houssa, M. Aoulaiche, S. De Gendt, G. Groeseneken, M. M. Heyns, and A. Stesmans, Appl. Phys. Lett. 86, pp. 093506, 2005.

[10] K.O. Jeppson, and C. M. Svensson, J. Appl. Phys., Vol. 48, Iss. 5, pp. 2004-2014, 1977.

[11] B. Kaczer, T. Grasser, Ph. J. Roussel, J. Martin-Martinez, R. O'Connor, B. J. O'Sullivan, and G. Groeseneken, IRPS, Proc., pp. 20-27, 2008.

HOT-CARRIER DEGRADATION IN UNDOPED-BODY ETSOI FETs AND SOI FINFETs

Miaomiao Wang, Pranita Kulkarni, Kangguo Cheng, Ali Khakifirooz, V.S Basker, Hemanth Jagannathan, Chun-Chen Yeh, Vamsi Paruchuri, Bruce Doris, Huiming Bu

IBM Research at Albany Nanotech
257 Fuller Road, Albany, NY, 12203
phone: 518-292-7291, e-mail: mwang@us.ibm.com

Chung-Hsun Lin, James H. Stathis
IBM T. J. Watson Research Center
Yorktown Heights, NY, USA

Kingsuk Maitra
GLOBALFOUNDRIES Inc
Albany, NY, USA

Philip J. Oldiges
SRDC, 2070 Rt 52,
Hopewell Junction, NY, USA

Hot-carrier degradation (HCI) in aggressively scaled undoped-body devices is carefully studied and compared for high-k/metal gate FINFETs and extremely thin silicon-on-insulator (ETSOI) transistors. We show that HCI involves different degradation mechanisms for silicon-on-insulator (SOI)-FINFETs and ETSOI devices though both are fabricated on undoped body. For FINFETs, the HC degradation correlated with interface trap generation in the channel region, whereas for ETSOI, trap generation and electron trapping in the spacer-nitride region were observed.

Keywords-ETSOI; FINFET; fully-depleted; undoped body; hot carrier induced degradation; high-k/metal gate

I. INTRODUCTION

Fully-depleted device structures such as undoped-body extremely thin silicon-on-insulator (ETSOI) and silicon-on-insulator (SOI)-FINFET have been recognized as promising candidates for sub-22nm CMOS technology due to better gate control and excellent short-channel effects. [1] In this work, the HCI of these technologically-relevant structures are investigated and compared. HCI reliability for various fabrication processes such as surface orientation, SOI layer thickness (Tsi) and drive-in temperature (Tdr) is explored. The ETSOI and FINFET devices were simulated using FIELDAY device simulator. [2] The physical models in the simulator were calibrated to experimental data and predicted the electrostatics of the devices accurately. Simulations are used to obtain electrical field during different HCI stress conditions

for different device structures and doping profiles to explain the experimental observations.

II. EXPERIMENTAL DETAILS

High-k/metal gate SOI nFINFETs with (110) orientated side surface and ETSOI nFETs with (100) surface orientation (Tinv=14nm) are used for this study. HC stresses are applied at room temperature with source grounded. The linear threshold voltage (Vt), sub-threshold swing (SS), saturation drain current in forward region (Idsat) and reverse region (Issat measured by flipping the source and drain connections) were monitored during the stress interruptions. Recovery at Vg=Vd=0V is also monitored after the HCI stress. For both ETSOI FETs and FINFETs, the most damaging condition occurs at Vg=Vd.

III. EXPERIMENTAL DETAILS

A. Tsi Effect on HCI

Issat degradation (%) of ETSOI nFETs vs stress time for 3 different Tsi are plotted in Figure 1 under the most damaging condition (Vg=Vd). [3] Similar trends were obtained for different Vg under the same Vd: thinner Tsi causes more HCI degradation. For undoped ultra-thin body devices, the availability of inversion charges is quite limited when the surface potential is at $2\Phi f$, whereas Φf is the fermi potential, given by equation (1) [4]:

$$\Phi f = \frac{kT}{q} \ln(Na/Ni) \qquad \dots\dots\dots\dots (1)$$

Additional surface potential to 2Φf is needed to bring enough inversion charge into the channel for the transistor to reach strong inversion (threshold). Thus, Vt is slightly higher for thinner Tsi as is shown in Figure 2. Figure 3 shows the calculated vertical electrical field distribution. The horizontal cut was made 1nm below the Si/dielectric interface. We found that the vertical field is slightly smaller for thinner Tsi. This is in accordance with the measured Vt and drain current under stress for different Tsi (lower stress drain current for thinner Tsi) as is shown in Figure 2. Figure 4 illustrates the calculated lateral electrical field as a function of distance from the Si/dielectric interface. The vertical cut was made 2.5 nm from the gate edge near the drain side. We see that the lateral field increases with decreasing Tsi near the drain region, which results in more HCI degradation.

B. Tdr Effects on HCI

The Issat degradation (%) of ETSOI nFETs vs stress time for 3 different Tdr under the most damaging condition (Vg=Vd). is shown in Figure 5. Two major features are worth noting here: (1) HCI increases with decreasing Tdr. The higher the Tdr, the more the overlap between the source/drain junction and the channel region. We would expect higher HCI due to the lower Vt and higher drive current as illustrated in Figure 6. However, a more graded junction would be formed under higher Tdr, which results in a reduction in the maximum lateral field intensity as is verified by the simulation results shown in Figure 7a for three different Tdr. In addition, the vertical field near the drain region also increases with decreasing Tdr as shown in Figure 7b. The migration of worst stress condition from Vg=Vd/2 to Vg=Vd implies that hot electron injection plays a more important role than hot carrier recombination. Hence higher vertical field also leads to more severe HCI.

C. Different HCI degradation mechanism for ETSOI devices and FINFETs

ΔIssat, ΔIdsat, Vt shift (ΔVt) and normalized SS degradation at Vg=Vd=1.5V for FINFET and ETSOI devices with 3 different Tdr are shown in Figure 5, Figure 8, Figure 9 and Figure 10 respectively. For ETSOI devices Tsi is around 6.1~6.8nm and for SOI FINFETs Wsi=13nm. Little ΔVt (<15mV) and SS change are observed for ETSOI devices while significant ΔVt (>100mV) and SS degradation occur in FINFETs after 800 seconds of HC stress. It is also worth pointing out that the defect generation is more localized in the drain region for ETSOI (ΔIdsat/ΔIssat ~30%) than FINFET (ΔIdsat/ΔIssat ~70%). Figure 11 illustrates the time exponent n for (a) FINFET and (b) ETSOI devices with varies bias conditions for different gate lengths (Lgate) obtained from simple power law fitting:

$$n = \frac{\partial \ln(\Delta Issat)}{\partial \ln(t)} = \frac{t}{\Delta Issat} \times \frac{\partial \Delta Issat}{\partial t} \quad \ldots\ldots\ldots\ldots(2)$$

Good correlation is observed between the time exponent n and the saturation drain current degradation regardless of stress conditions and Lgate for FINFET. The time exponent n increases slightly during the first few seconds and decreases (saturation occurs) after 1000 sec stress, indicating both interface trap generation in the channel region (n varies from 0.3 ~ 0.7, typical of HCI degradation) and degradation in the spacer nitride region (possibly related to small n caused by degradation saturation in spacer nitride). In contrast, n is not correlated with saturation drain current degradation for ETSOI devices. Figure 12 illustrates the remaining fraction for FINFET and ETSOI devices after 1000 seconds at Vg=Vd=1.4V followed by recovery at Vg=Vd=0 Volt for 800 seconds. The remaining fraction is defined as the ratio of ΔIssat after recovery to the maximum ΔIssat after stress. Significant recovery >10% is observed for ETSOI devices after 800 seconds, whereas <1% of recovery is detected for FINFETs. The above observations suggest different HCI degradation mechanisms involved for FINFETs and ETSOI. Hot carrier trapping in defects generated during HC stress accounts for part of the degradation in ETSOI device but it does not contribute significantly in FINFET degradation. For ETSOI, few interface traps are generated in the channel region during stress. The Id degradation may be caused by interface trap generation in the spacer nitride region. [5] FINFET has much higher Si-H bond density than ETSOI at the Si/dielectric interface since the dielectric is grown on (110) side surface. [6] As a result, we observe severe SS degradation for FINFET, a sign of interface trap generation in the channel. Figure 7a shows TCAD simulation of the lateral field distribution of FINFET and ETSOI devices with Tdr=1020°C, 1065°C and 1070°C. We can see the shift of maximum lateral field location toward the spacer nitride region for ETSOI as compared with FINFET. For ETSOI device, the lateral field starts from the gate edge and propagates into the spacer nitride region. For FINFET, the lateral field also occurs at the gate edge but slightly higher field intensity than ETSOI is observed in the channel region. This may also account partly for the difference in the HCI behavior for FINFET and ETSOI devices.

D. Modeling HCI

In this work the model proposed by R. Dreesen et al. [7] for conventional nMOSFETs with LDD structure has been modified to fit the HCI for nFINFETs and ETSOI nFETs:

$$\Delta Issat(\%) = \frac{A * t^m}{1 + B * At^m} + C * t^n \quad \ldots\ldots\ldots\ldots\ldots\ldots\ldots(3)$$

where $A \sim Lgate^{-\beta} \times \exp(Vd/Vo)$. [8] Vo and C are constants. B is a constant only determined by technology and junction profile and predicts the saturation behavior. The term $\frac{A * t^m}{1 + B * At^m}$ forms a good description of the interface trap generation in spacer oxide region. [7] For FINFET, $C*t^n$ represents the interface trap generation in channel region. For ETSOI, $C*t^n$ describes the electron charge trapping. FINFETs

978-1-4244-5430-3/10 $26.00 © 2010 IEEE

and ETSOI devices with various gate lengths (Lgate = 25 nm, 29 nm, 33nm and 37 nm) under various stress voltages were used to obtain the fitted value of *m*, *B* and *n* (as listed in Table 1). Fits from the model are in Figure 13 for FINFETs and Figure 14 for ETSOI devices. We can see that the small time exponent for ETSOI extracted from simple power law fitting might come from the combined effects of the degradation saturation in spacer oxide region and electron charge trapping.

IV. CONCLUSIONS

HCI effects in aggressively scaled high-k/metal gate SOI FINFETs and ETSOI devices fabricated on undoped body are studied and compared. For FINFET, interface trap generation in the channel region is the dominant HCI mechanism whereas both electron charge trapping and interface trap generation in the spacer oxide account for the degradation in ETSOI devices. The dependence of HCI on Tsi and junction profile is also investigated.

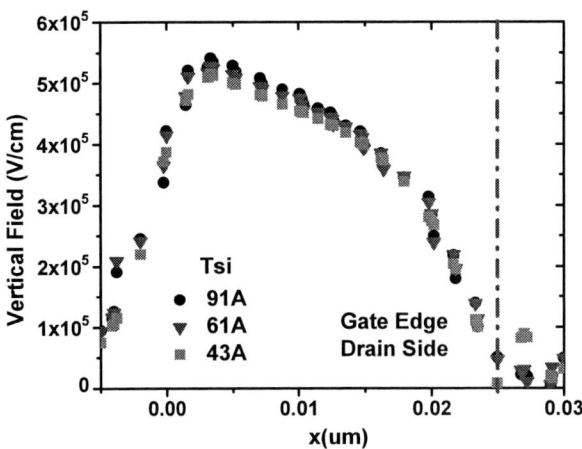

Figure 3. Vertical field distribution of ETSOI nFETs for Tsi=43A, 61A and 91A

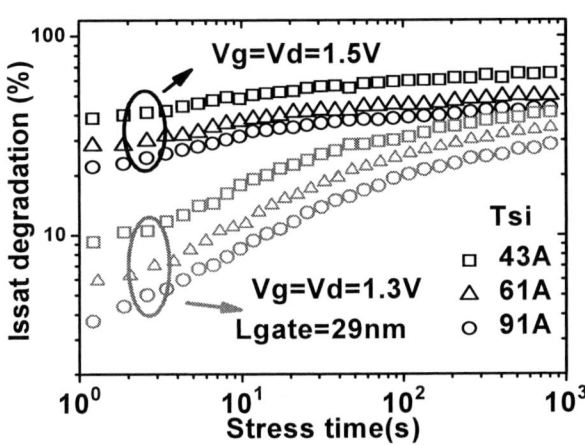

Figure 1. Issat degradation (%) of ETSOI nFETs vs stress time for Tsi=43A, 61A and 91A (Tdr=1020°C)

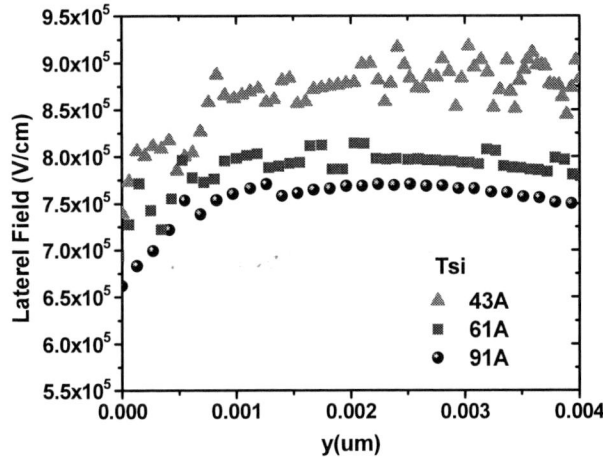

Figure 4. Lateral field distribution of ETSOI nFET for Tsi=43A, 61A and 91A

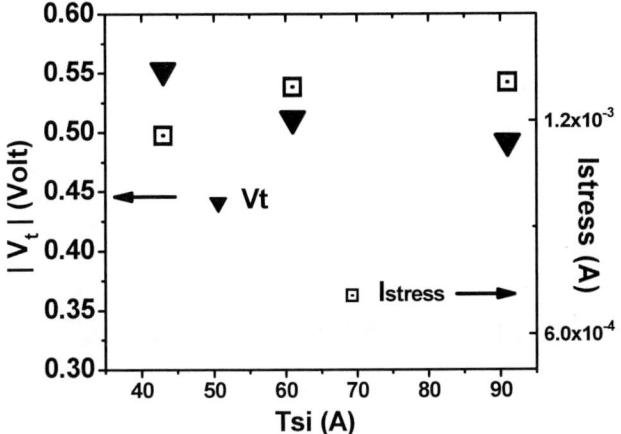

Figure 2. Vt and stress drain current for ETSOI devices with different Tsi

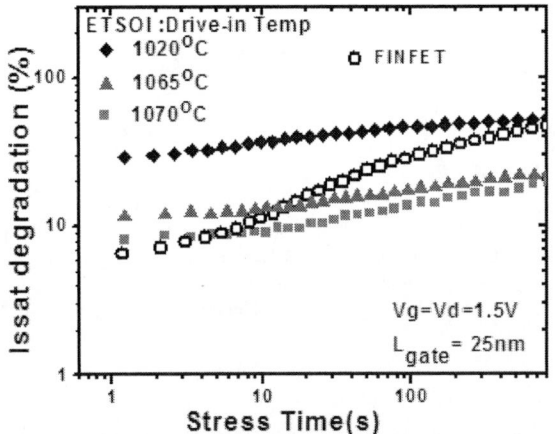

Figure 5. Issat degradation at Vg=Vd =1.5V for FINFET and ETSOI devices

978-1-4244-5430-3/10 $26.00 © 2010 IEEE   1101

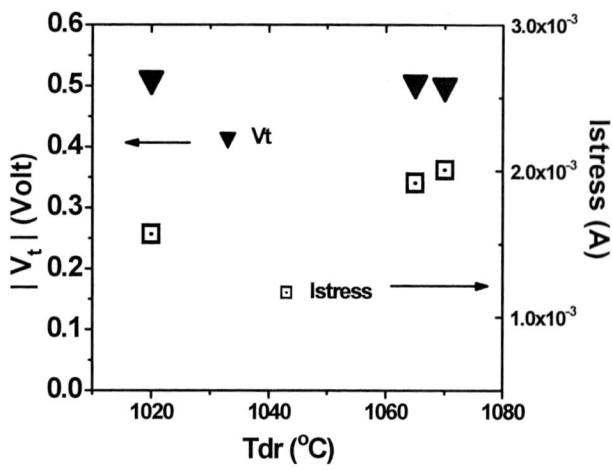

Figure 6. Vt and stress drain current for ETSOI devices with different Tdr

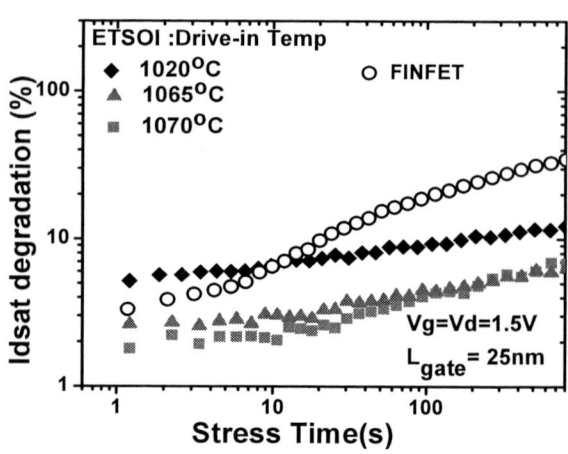

Figure 8. Idsat degradation at Vg=Vd=1.5V for FINFET and ETSOI devices

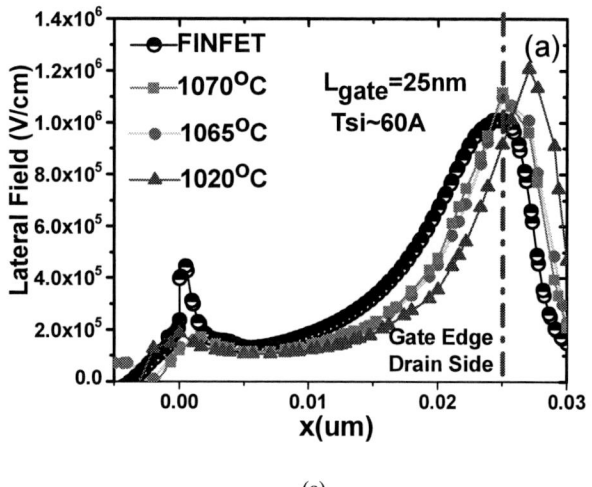

(a)

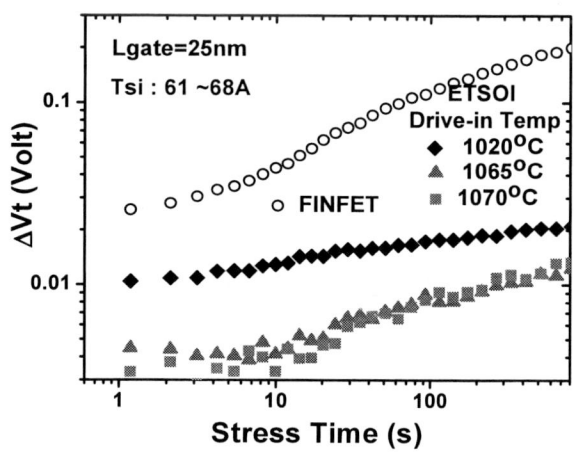

Figure 9. ΔVt at Vg=Vd =1.5V for FINFET and ETSOI devices

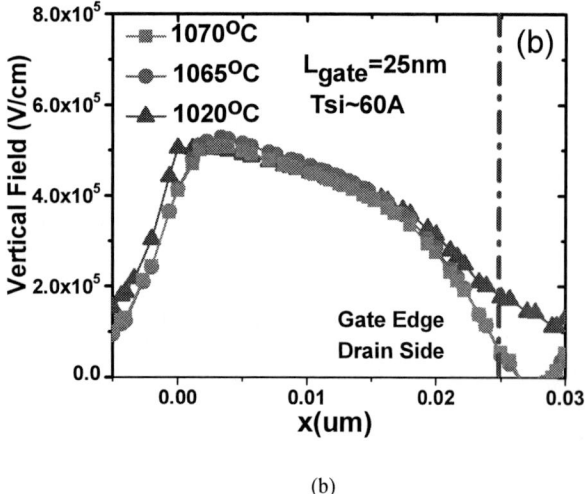

(b)

Figure 7. TCAD simulation of the (a) lateral and (b) vertical field distribution of FINFET and ETSOI devices at Vg=Vd=1.5V

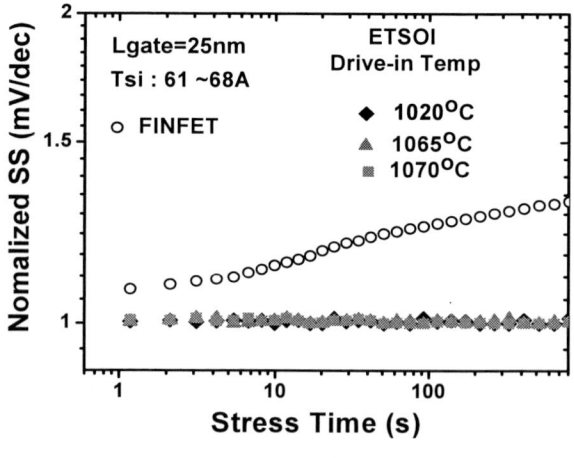

Figure 10. Normalized SS degradation at Vg=Vd =1.5V for FINFET and ETSOI devices

978-1-4244-5430-3/10 $26.00 © 2010 IEEE

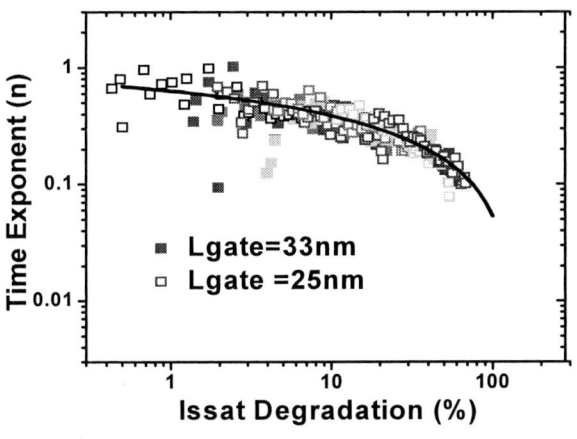

(a)

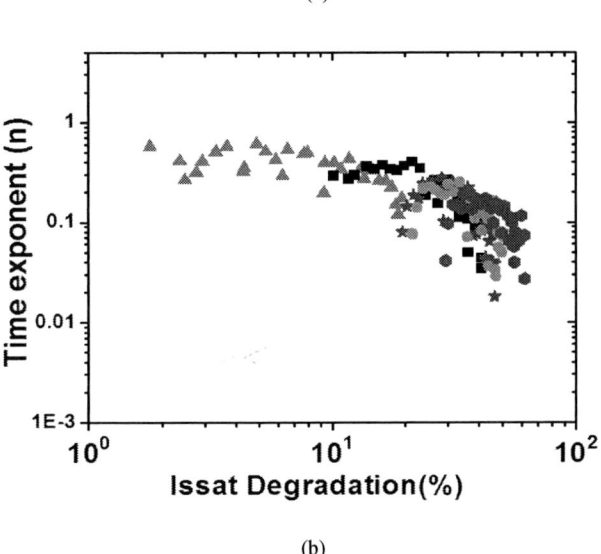

(b)

Figure 11. time exponent n for (a) FINFET and (b) ETSOI devices under varies bias conditions for different Lgate

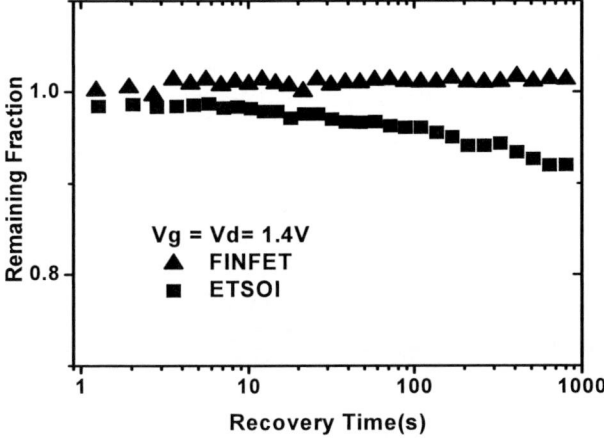

Figure 12. Recovery of FINFET and ETSOI after HCI stress at Vg=Vd=1.4V for 800 sec

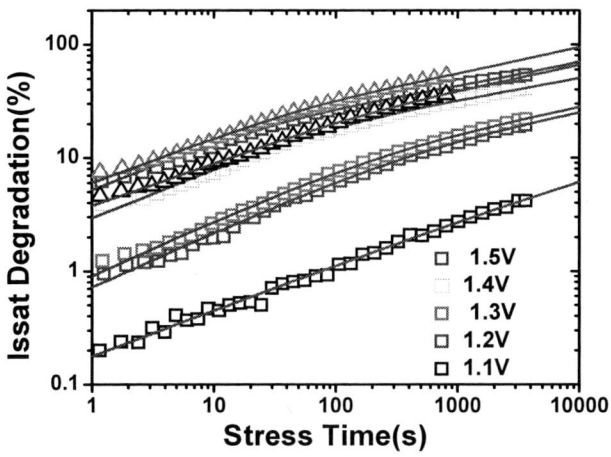

Figure 13. Fits from Model for SOI-FINFETs

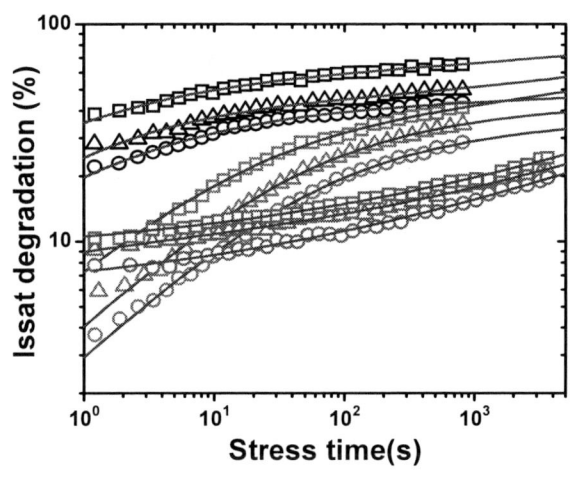

Figure 14. Fits from Model for ETSOI devices

Table I Fitting Parameters for Eq. (3) for FINFET and ETSOI

Parameters	m	B	n
FINFET			
Wsi=13nm	0.4~0.7	0.056	0.3~0.4
ETSOI Tdr=1020°C			
Tsi~43nm	0.4~0.7	0.031	0.1~0.2
Tsi~61nm	0.4~0.7	0.033	0.1~0.2
Tsi~91nm	0.4~0.7	0.035	0.1~0.2
ETSOI Tsi=6.1~6.8nm			
T=1060°C	0.4~0.7	0.123	0.15~0.25
T=1065°C	0.4~0.7	0.141	0.15~0.25
T=1070°C	0.4~0.7	0.200	0.15~0.25

ACKNOWLEDGMENT

This work was performed by the Research Alliance Teams at various IBM Research and Development Facilities.

REFERENCES

[1] Hyunjin Lee, et al. "A Study of Negative-Bias Temperature Instability of SOI and Body-Tied FinFETs", *IEEE Electron Device Letters*. Vol. 26, No. 5, May 2005

[2] E. Buturla *et al.*, "New 3D device simulation formulation", *NASECODE VI*, p. 291, 1989.

[3] Esteve Amat et al. "Channel Hot-Carrier Degradation in Short-Channel Transistors with High-k/Metal Gate Stacks", *IEEE Transactions on Device and Materials reliability*. Vol. 9, No. 3 September 2009.

[4] Weize (Wade)Xiong, "Multigate MOSFET Technology" in FINFETs and other Multigate MOSFET Technology, Jean-Pierre Colinge (Ed.), (Springer Science+Business Media, LLC, New York, 2008), p 71.

[5] J. S. Got, et al., "An Analytical Model for Hot-Carrier-Induced Degradation of Deep-submicron n-Channel LDD MOSFETs", *Japanese Journal of Applied Physics*, Vol. 33, pp. 606-611, 1994.

[6] Sang-Yun Kim et al, "Negative Bias Temperature Instability of Bulk Fin Field Effect Transistor", *J. Appl. Phys.*Vol.45, 2006 pp. 1467-1470.

[7] R. Dreesen, et al., "A new degradation model and lifetime extrapolation technique for lightly doped drain nMOSFETs under hot carrier degradation", *Microelectronics and Reliability*, pp 437-443, 2001.

[8] T. Nigam, et al., "Accurate Product Lifetime Predictions Based on Device Level Measurements", *IEEE International Reliability Physics Symposium*, pp. 634-639, 2009.

978-1-4244-5430-3/10 $26.00 © 2010 IEEE

NBTI Lifetime Prediction in SiON p-MOSFETs by H/H2 Reaction-Diffusion(RD) and Dispersive Hole Trapping Model

S. Deora, V. D. Maheta and S. Mahapatra

Department of Electrical Engineering, Indian Institute of Technology Bombay, Mumbai 400076, India
Phone: +91-222-572-0408, Fax: +91-222-572-3707, Email: souvik@ee.iitb.ac.in

Abstract — I_{DLIN} shift due to NBTI is measured using UF-OTF I_{DLIN} method in PNO, RTNO and RTNO+PN SiON p-MOSFETs having a wide range of EOT and %N. Time evolution of I_{DLIN} shift at different stress E_{OX} and T is modeled from ultra-short to long stress time using non-dispersive H/H_2 RD model governed N_{IT} and dispersive N_h components. N_{IT} and N_h model parameters show consistent E_{OX} and T dependent behavior across all devices. Finally, extrapolated tt_F values are obtained for different E_{OX} from conventional power-law fit and the proposed model, and are compared across different measurement delay. Inconsistencies associated with conventional power-law fit extrapolation method are highlighted, which justifies the use of proposed model.

Keywords - NBTI; SiON; PNO; PNA; RTNO; hole trapping; interface traps; Reaction-Diffusion (RD) model.

I. INTRODUCTION

Negative Bias Temperature Instability (NBTI) is a serious reliability concern in silicon oxynitride (SiON) p-MOSFETs [1]-[6]. It is now universally accepted that NBTI degradation recovers substantially upon the removal of stress [7], [8] when conventional stress-measure-stress (SMS) techniques [9] are used. These measurement methods suffer from recovery related artifacts and result in incorrect degradation magnitude, time and temperature (T) dependence [2], [8], [10] and hence cannot be used for estimating extrapolated degradation at end-of-life. Ultra-Fast SMS [5], [11], [12] can be used to measure delay free degradation and overcome such recovery artifacts. However, these methods cannot provide degradation from short ($t<1s$) stress time (t_{STS}), which is crucial to understand NBTI physical mechanism. Another Ultra-Fast SMS method [13] can provide measured data from very short (~µs) to long stress time, but needs complicated subthreshold slope correction [14] to yield proper threshold voltage (V_T) shift. Finally, the Ultra-Fast On-The-Fly linear drain current (UF-OTF I_{DLIN}) method [15], [16] can be used to determine NBTI degradation from very short (~µs) to long stress time. However to obtain V_T shift if needed, SMS based post-processing is required to convert measured I_{DLIN} degradation as discussed in detail in [17].

It is also universally believed now that both interface-trap generation (N_{IT}) and hole trapping (N_h) contributes to overall NBTI, although their relative dominance, time dependence, T activation, (non-) correlation etc. are debated [1], [4], [6], [11], [13], [15], [18]-[23]. It is important to reliably isolate N_{IT} and N_h contributions for proper modeling of the underlying physical mechanism, such that realistic methodologies can be developed to extrapolate moderately long-time stress data to end-of-life. In the past, several attempts were made to isolate N_{IT} and N_h from overall measured degradation [4], [13], [20], [21], [24]. Note that [4], [13] assumed an unrealistic power-law time exponent (n) value of 0.25 for N_{IT} time dependence; [20]

assumed an ad-hoc 1s first data point subtraction method that cannot be justified; and [24] incorrectly assumed no N_{IT} recovery and only N_h recovery while subtracting degradation measured at different measurement delay to separate N_{IT} and N_h components. In [21], a correct exponent value of n=0.16 [25], [27] for N_{IT} buildup at long t_{STS} has been used to consistently isolate the N_{IT} and N_h components for wide variety of Plasma Nitrided Oxide (PNO) SiON films. However, as n varies with range of t_{STS} over which the fit (log-log degradation versus time plot) is made [16], the isolated N_{IT} and N_h components might show some uncertainty and conclusions made in [21] needs to be re-verified by using alternative methods. Moreover, the robustness of the isolation methodology must be verified for other type of SiON devices, such as Rapid Thermal Nitrided Oxide (RTNO), RTNO+PN, PNO with varying Post Nitridation Anneal (PNA), etc. Finally, though few attempts were made to model NBTI time transients for the entire stress duration (t_{STS} from µs or ms to ~hours) [4], [6], [13], [19], none attempted this on different type of SiON devices having a wide range of N content (%N) and EOT (SiO_2 equivalent oxide thickness), for stress using different stress gate bias (V_G) and T, which is absolutely essential to check the validity and robustness of any model.

Past works by the authors have unequivocally established strong SiON process dependence of NBTI [6], [15], [16], [21], [22]. Similar to conventional SiO_2, it has been shown that low to moderate %N PNO devices that undergo proper, 2-step PNA [28] show N_{IT} dominated degradation. This is generally true for industrial grade devices [3], though non-negligible presence of N_h for such devices cannot be ruled out. However, PNO with proper PNA but high %N, PNO with low %N but improper PNA as used in [5], [13], and RTNO devices show moderate to very high N_h contribution on top of N_{IT}. It has been shown that generated N_{IT} and N_h demonstrate very different time and T dependence but identical oxide field (E_{OX}) dependence [21]. Non-negligible N_h presence in-turn influences the magnitude and rate (power-law slope, n) of time evolution of overall NBTI at longer stress time [6], [21]. N_h is presumed to be faster than N_{IT} and saturates early [4], [6], [13], [19], [21], [34]. Magnitude of N_h captured at the onset of stress depends on measurement speed [5], [13]. This affects time-to-fail (tt_F) values obtained by extrapolating measured degradation using power-law fit (as is conventionally done), since it makes tt_F a strong function of measurement speed. Therefore, accurate prediction of NBTI time evolution beyond measurement window would require proper understanding and modeling of the underlying physical mechanisms that governs NBTI.

In this work, NBTI in PNO (different %N, EOT, PNA), RTNO and RTNO+PN SiON p-MOSFETs is measured from ~µs through longer t_{STS} at different stress V_G and T using the UF-OTF I_{DLIN} technique [15], [16]. Overall I_{DLIN} degradation

978-1-4244-5430-3/10 $26.00 © 2010 IEEE

($\Delta I_{DLIN}/I_{DLIN0}$, I_{DLIN0} being the first point measured at the onset of stress, see [16] for details) is modeled by an N_{IT} *influenced* component (denoted as ΔI_{DLIN}-N_{IT}) governed by non-dispersive H/H_2 Reaction-Diffusion (RD) model [19], together with an N_h *influenced* component (denoted by ΔI_{DLIN}-N_h) governed by dispersive hole trapping in pre-existing N related traps [29] in SiON bulk. The relative contribution of ΔI_{DLIN}-N_{IT} and ΔI_{DLIN}-N_h are obtained for all devices under different stress conditions. The time evolution of individual ΔI_{DLIN}-N_{IT} and ΔI_{DLIN}-N_h components and hence overall $\Delta I_{DLIN}/I_{DLIN0}$ are evaluated from very short to long t_{STS}, using consistent set of T activation and E_{OX} dependent parameters across all devices. This work relies and improves upon the past work by the authors, provides credibility to the assumptions and conclusions made in [21], and unequivocally establishes underlying physical mechanism governing NBTI time evolution from very short to long t_{STS} for wide range of SiON device type and stress conditions. Finally, the tt_F values at different E_{OX} are obtained using conventional power-law fit and model extrapolation, and are compared for measurements with different time-zero delay (t_0). Inconsistencies associated with conventional power-law fit approach are highlighted, justifying the use of physics-based extrapolation to reliably determine end-of-life degradation.

II. DEVICE AND EXPERIMENTAL DETAILS

This study uses a wide variety of SiON p-MOSFET devices fabricated using different gate insulator processes, as shown in Tables-I and II, to have different N distribution profile [22], N density and film thickness. Nitridation condition (PNO, RTNO or RTNO+PN), starting base oxide thickness (T_{BASE}), total N dose (N_{DOSE}) and post nitridation thickness (T_{XPS}) from XPS, EOT from CV, and calculated atomic %N are shown. Type-I devices (SiO_2 and PNO with proper PNA having low to moderately high N_{DOSE}) are listed in Table-I. All other devices (PNO with very high N_{DOSE}, RTNO, RTNO+PN and PNO without proper PNA) are labeled as Type-II and are listed in Table-II. Type-I PNO devices have low to high N density at the Si/SiON interface, which can be estimated by %N/T_{XPS}. Note that for PNO devices, %N/T_{XPS} increases with increase in N_{DOSE} for fixed T_{BASE} or with reduction in T_{BASE} for similar N_{DOSE}. Type-II devices in general have higher Si/SiON N density compared to Type-I devices.

TABLE I. T_{BASE}, N DOSE, T_{XPS}, EOT, %N AND %N/T_{XPS} FOR VARIOUS PNO DEVICES (TYPE-I)

D no.	Type	T_{BASE} (nm)	N dose ($\times 10^{15} cm^{-2}$)	T_{XPS} (nm)	EOT (nm)	%N	%N/ T_{XPS}
D1	RTO	2.0	0.054	2.19	2.19	0.38	0.017
D2	PNO	2.5	0.0+3.1	2.81	2.35	16.6	0.6
D3	PNO	1.5	0.0+2.8	1.85	1.40	22.64	1.23
D4	PNO	2.5	0.0+5.54	2.85	2.14	29.4	1.304
D5	PNO	2.0	0.0+5.28	2.32	1.55	35.0	1.49
D6	PNO	2.0	0.0+6.84	2.35	1.46	42.5	1.8
D7	PNO	1.5	0.0+5.75	2.12	1.23	41.26	1.95

TABLE II. T_{BASE}, N DOSE, T_{XPS}, EOT AND %N FOR DIFFERENT TYPES OF SiON DEVICES (TYPE-II)

D no.	Type	T_{BASE} (nm)	N dose ($\times 10^{15} cm^{-2}$)	T_{XPS} (nm)	EOT (nm)	%N
D8	PNO	1.5	0.0+7.756	2.39	1.18	49.2
D9	RTNO+PN	2.0	0.8+5.1	2.29	1.305	39.1
D10	RTNO	2.0	0.8+0.0	2.11	1.85	5.84
D11	RTNO	2.5	0.8+0.0	2.64	2.2	6
D12	PNO (moderate PNA)	2.0	0.0+2.7	2.41	2.02	16.7

Electrical measurements were performed on fully processed p-MOSFETs having W/L=15/0.16 (μm). UF-OTF I_{DLIN} method [15], [16] was used to measure NBTI degradation from ~μs through longer t_{STS} at different stress V_G and T. The measured I_{DLIN} degradation is expressed as $\Delta I_{DLIN}/I_{DLIN0}$, and the first data point, *i.e.*, I_{DLIN0} was recorded within 1μs (unless specifically mentioned otherwise) after application of the stress V_G.

Before proceeding further, it is important to note that the UF-OTF I_{DLIN} method is chosen as it can provide degradation in sub 1s (especially sub 1ms) t_{STS} range where N_h contribution is important. Note that most fast SMS methods [5], [12] are unsuitable to measure degradation in the sub 1s range. Of course, the fast SMS method as described in [13] can in principle be used, though it is left out as it involves complicated subthreshold slope correction [14]. Finally, conventional OTF methods [2], [3], [8] cannot provide sub 1ms data, while conventional SMS [9] has too much delay to be effectively used in this work.

Furthermore, also note that measured $\Delta I_{DLIN}/I_{DLIN0}$ can be readily converted to ΔV_T using the post-processing algorithm described in [17]. However, the procedure is effective for low %N devices with negligible N_h contribution, and in general it is not applicable in devices showing moderately high to very high N_h contribution. As this work involves a wide variety of SiON devices that show very low to very high N_h contribution (to validate the proposed model), no attempt has been made to convert $\Delta I_{DLIN}/I_{DLIN0}$ to ΔV_T to remain consistent across all type of devices. Therefore, as mentioned above, overall measured $\Delta I_{DLIN}/I_{DLIN0}$ data is modeled by using a sum of N_{IT} and N_h *influenced* components, ΔI_{DLIN}-N_{IT} and ΔI_{DLIN}-N_h respectively. The relation between actual defects (N_{IT} and N_h) and the corresponding *influenced* components (ΔI_{DLIN}-N_{IT} and ΔI_{DLIN}-N_h) can be readily ascertained by noting that generated N_{IT} impacts V_T and effective mobility and hence ΔI_{DLIN}-N_{IT}, while N_h impacts V_T and hence ΔI_{DLIN}-N_h.

III. MODELING OF NBTI DEGRADATION

In spite of several years of active research, NBTI modeling is a hotly debated area. As mentioned above, it is now accepted that NBTI degradation consists of N_{IT} and N_h components. The RD framework [19], [30], [33], [34] has been most successful for modeling N_{IT}, as it is the only model that has been successfully utilized to explain (i) SiON process dependence [21], (ii) time exponent of measured degradation at ultra long-

978-1-4244-5430-3/10 $26.00 © 2010 IEEE

time [25], [27], (iii) AC frequency dependence [31], [32], and also (iv) AC duty cycle dependence, as well as NBTI recovery [34].

Time evolution of $\Delta I_{DLIN}/I_{DLIN0}$ at different stress E_{OX} and T is modeled using non-dispersive H/H_2 RD solution for N_{IT} [19] hence $\Delta I_{DLIN}-N_{IT}$, along with dispersive hole trapping model for N_h hence $\Delta I_{DLIN}-N_h$. The $\Delta I_{DLIN}-N_{IT}$ component is computed by using the analytic expression for H/H_2 RD [19] as follows

$$\frac{N_{IT}}{t} - \frac{\left[\delta k_H \left(k_f N_0 - \frac{N_{it}}{t}\right)^2\right]}{k_r^2 N_{IT}^2} + \frac{\delta k_{H2} N_{IT}}{\sqrt{6 D_{H2} t}} = 0$$

where N_0 ($=10^{12}$ cm^{-2}) is initial Si-H bond density before stress, N_{IT} is generated interface trap density, t is stress time (t_{STS}), and δ is interfacial thickness ($\sim$1-2A^O). Adjustable parameters k_F and k_R are Si-H bond breaking and annealing rates, k_H and k_{H2} are generation and dissociation rates of H to H_2 conversion, and D_{H2} is diffusion constant of H_2. At 125°C, the following values are used for all Type-I and II devices: $k_R = 3\times10^{-9}$ cm^{3}s^{-1}, $k_H = 9\times10^{-2}$ cm^{3}s^{-1}, $k_{H2} = 95.4$ s^{-1} and $D_{H2} = 1.8\times10^{-16}$ cm^{2}s^{-1}. Only k_F has been varied (1.6×10^{-3} s^{-1} to 8×10^{-3} s^{-1}) for different %N to reflect enhanced forward reaction due to more N near Si/SiON interface [35]. Similar T dependence is assumed for k_H and k_{H2} (no impact has been observed by identically changing both k_H and k_{H2}) to maintain detailed balance. Arrhenius T activation is used for other parameters with following activation energy: $E_A(k_R) = 0.2$eV, $E_A(D_{H2}) = 0.55$eV [35], with adjustable $E_A(k_F)$ that is similar to $E_A(k_R)$ for low %N [10], [21] and reduces for high Si/SiON N density, as explicitly mentioned later.

$\Delta I_{DLIN}-N_h$ is calculated using purely analytical formulation, by the equation $A*[1-\exp\{-(t/\tau)^\beta\}]$, with adjustable τ and β to represent the dispersive nature of hole trapping whose evidence is given below. It is interesting to note that all SiON devices, as shown in Tables I and II, stressed under a wide range of stress E_{OX} and T, can be modeled with τ in the range of 1ms to 20ms. The parameter β has been found to be 0.22 for most type of devices, though for a few type of devices β has been found to vary between 0.20 to 0.25. Note that the equation used here to calculate dispersive hole trapping has also been used by others [34], although on a *completely* different context. Also note that detailed derivation of an exact hole trapping model based on physics based trapping-detrapping kinetics is beyond the scope of the present paper. Finally, it is important to realize that the saturated $\Delta I_{DLIN}-N_h$ value ("A" in the model equation) has the most crucial influence on extrapolated long-time degradation and hence needs to be accurately determined.

IV. OXIDE FIELD DEPENDENCE

Figure 1 shows time evolution of measured $\Delta I_{DLIN}/I_{DLIN0}$ (normalized to 1000s data to highlight unique features at short and long t_{STS}) under NBTI stress at T=125°C but different V_G for low N_{DOSE} (D2), moderately high N_{DOSE} (D5) and very high N_{DOSE} (D6) PNO devices. A random spread for short t_{STS} data is seen with no apparent stress V_G dependence, while all long t_{STS} data merge and show unique time evolution, which is true

for all %N devices. The model fit to overall $\Delta I_{DLIN}/I_{DLIN0}$ and its underlying $\Delta I_{DLIN}-N_{IT}$ and $\Delta I_{DLIN}-N_h$ components are also shown. $\Delta I_{DLIN}/I_{DLIN0}$ including the upper and lower bounds of short t_{STS} random spread in data can be modeled by identical $\Delta I_{DLIN}-N_{IT}$ but different $\Delta I_{DLIN}-N_h$ components. Note, $\Delta I_{DLIN}-N_h$ impacts short-time data but saturates quickly, as expected for ultra-thin gate insulators [4], [6], [13], [19], [21], [34]. $\Delta I_{DLIN}-N_h$ calculation to capture the upper and lower bounds of short t_{STS} random spread in data was done by varying hole trapping parameter τ, while β was kept constant. The magnitude of $\Delta I_{DLIN}-N_{IT}$, although is relatively low at short t_{STS}, eventually surpass the saturated $\Delta I_{DLIN}-N_h$ value at longer t_{STS}. Fractional $\Delta I_{DLIN}-N_h$ contribution to overall $\Delta I_{DLIN}/I_{DLIN0}$ increases with increase in %N [6], [21], and therefore, $\Delta I_{DLIN}-N_{IT}$ eventually surpasses $\Delta I_{DLIN}/I_{DLIN0}$ at longer t_{STS} as %N is increased. This has important implication on calculated end-of-life degradation as discussed later.

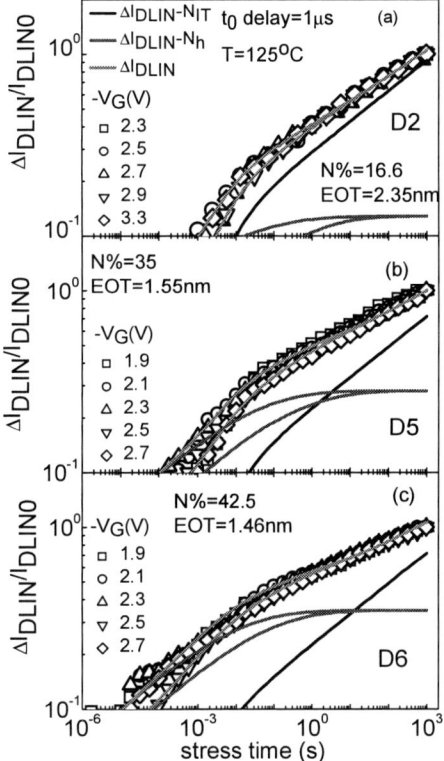

FIGURE.1. TIME EVOLUTION OF MEASURED $\Delta I_{DLIN}/I_{DLIN0}$ TRANSIENTS (NORMALIZED TO 1000s VALUE), OVERALL MODEL FIT AND EXTRACTED $\Delta I_{DLIN}-N_h$ AND $\Delta I_{DLIN}-N_{IT}$ UNDER IDENTICAL STRESS T AND DIFFERENT STRESS V_G FOR (A) D2, (B) D5 AND (C) D6 PNO DEVICES. HOLE TRAP PARAMETERS USED ARE: $\beta = 0.22$ (FOR ALL), τ RANGES FROM 1.0ms TO 10ms.

Stress bias dependent measurements were performed and modeled using the above framework for all devices used in this work. As an example, Figure 2 shows measured $\Delta I_{DLIN}/I_{DLIN0}$ and calculated $\Delta I_{DLIN}-N_{IT}$ and $\Delta I_{DLIN}-N_h$ at t_{STS} of 1000s as a function of stress E_{OX} for D2, D5 and D6 PNO devices having different %N. Note that $\Delta I_{DLIN}/I_{DLIN0}$ increases with increase in %N. $\Delta I_{DLIN}-N_{IT}$ component is dominant for low %N devices

978-1-4244-5430-3/10 $26.00 © 2010 IEEE

and increases with increase in %N. However, the ΔI_{DLIN}-N_h contribution increases dramatically with increase in %N, and to a large extent contributes to the overall increase in $\Delta I_{DLIN}/I_{DLIN0}$ for such devices. For all devices, similar E_{OX} dependent slope (Γ) for ΔI_{DLIN}-N_{IT} and ΔI_{DLIN}-N_h components results in similar Γ for overall $\Delta I_{DLIN}/I_{DLIN0}$ [21]. However, it is important to note that Γ reduces with increase in %N [22], and has important implication when accelerated stress data is extrapolated to low E_{OX} use conditions. For identical degradation under accelerated stress, a device with higher Γ would show lower extrapolated degradation at use condition and vice-versa.

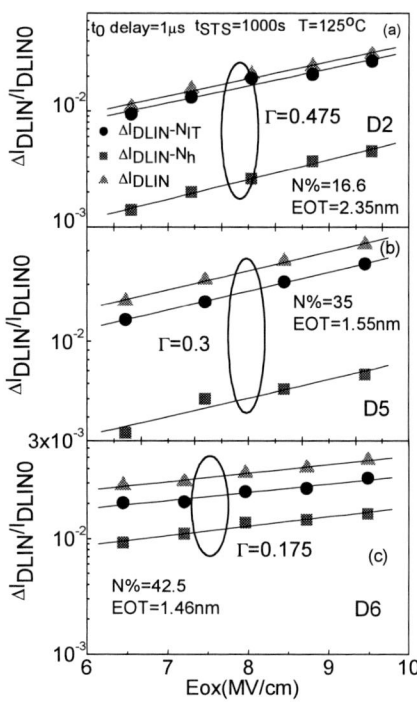

FIGURE.2. E_{OX} DEPENDENCE OF MEASURED $\Delta I_{DLIN}/I_{DLIN0}$, EXTRACTED ΔI_{DLIN}-N_h AND ΔI_{DLIN}-N_{IT} UNDER IDENTICAL STRESS T AT t_{STS}=1000S FOR (a) D2, (b) D5 AND (c) D6 PNO.

Note that NBTI degradation is governed by N density at the Si/SiON interface and not on total %N [22]. For PNO devices with proper PNA, %N/T_{XPS} can be a useful indicator of Si/SiON N density. Figures 3 and 4 respectively show ΔI_{DLIN}-N_{IT} and ΔI_{DLIN}-N_h components as well as overall $\Delta I_{DLIN}/I_{DLIN0}$ at t_{STS} of 1000s for stress using T=125°C and E_{OX}~8.5MV/cm (Figure 3), and E_{OX} dependent slope Γ (Figure 4) as a function of %N/T_{XPS} for Type-I devices having different %N (Table-I). ΔI_{DLIN}-N_{IT} dominates overall $\Delta I_{DLIN}/I_{DLIN0}$ for lower %N/T_{XPS}. Both ΔI_{DLIN}-N_{IT} and ΔI_{DLIN}-N_h components increase (the latter shows more drastic increase compared to the former) which increases overall $\Delta I_{DLIN}/I_{DLIN0}$ for very high %N/T_{XPS}. Γ also shows a consistent behavior, it remains almost constant for low %N/T_{XPS} and reduces with increase in %N/T_{XPS}, as also shown before [21], [22], and can be used as an estimator of N density at the Si/SiON interface. Note that lower Γ generally implies higher N density at the Si/SiON interface, higher hole trapping and higher overall NBTI degradation.

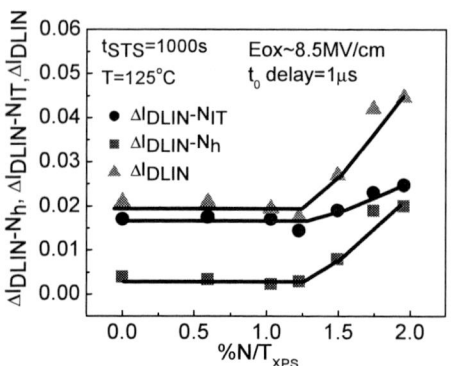

FIG.3. %N/T_{XPS} DEPENDENCE OF $\Delta I_{DLIN}/I_{DLIN0}$, ΔI_{DLIN}-N_{IT} AND ΔI_{DLIN}-N_h EXTRACTED AT SIMILAR t_{STS}(1000S) FOR t_0 DELAY=1μS UNDER IDENTICAL STRESS E_{OX} (8.5MV/cm) AND STRESS T(125°C). LINES ARE GUIDE TO THE EYE.

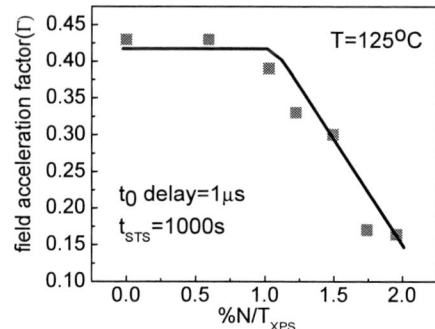

FIG.4. %N/T_{XPS} DEPENDENCE OF Γ EXTRACTED AT SIMILAR t_{STS}(1000S) FOR t_0 DELAY=1μS UNDER IDENTICAL STRESS T(125°C). LINE IS GUIDE TO THE EYE.

TABLE III. $\Delta I_{DLIN}/I_{DLIN0}$, EXTRACTED ΔI_{DLIN}-N_{IT}, ΔI_{DLIN}-N_h AND Γ FOR TYPE-II DEVICES

	%N	$\Delta I_{DLIN}/I_{DLIN0}$	ΔI_{DLIN}-N_{IT}	ΔI_{DLIN}-N_h	Γ
D8	49.2	0.0738	0.041	0.0328	0.145
D9	39	0.0595	0.037	0.0225	0.14
D10	6	0.079	0.047	0.032	0.14
D11	6	0.072	0.037	0.035	0.174
D12	16.7	0.0335	0.023	0.0105	0.28

Finally, $\Delta I_{DLIN}/I_{DLIN0}$, ΔI_{DLIN}-N_{IT} and ΔI_{DLIN}-N_h at T=125°C, t_{STS}=1000s, and E_{OX}~8.5MV/cm and Γ for all Type-II devices are listed in Table-III. Very low value of Γ for all such devices (except for PNO with improper PNA device, D12) is consistent with very high N density at the Si/SiON interface, and as a result very high $\Delta I_{DLIN}/I_{DLIN0}$, ΔI_{DLIN}-N_h and higher ΔI_{DLIN}-N_{IT}. Device D12 shows reasonably high NBTI irrespective of very low N_{DOSE} (note Γ shows intermediate value) as improper PNA could not cure the plasma induced defects that are generated during PNO and failed to perform a very crucial re-oxidation of the Si/SiON interface [28]. It is sometimes assumed that low N_{DOSE} should result in proper PNO SiON gate dielectric and low NBTI, which is not always the case [5], [13] unless proper PNA is employed.

978-1-4244-5430-3/10 $26.00 © 2010 IEEE

V. TIME-ZERO DELAY DEPENDENCE

As discussed above, hole trapping is a very fast process and dominates the early part of NBTI degradation. The magnitude of hole trapping captured by UF-OTF method would depend on SiON process as well as t_0 delay, i.e., the delay in time between application of stress V_G and measurement of first I_{DLIN} point I_{DLIN0}. It is expected that the impact of t_0 delay would be higher for devices with high N density at Si/SiON interface that show larger N_h contribution.

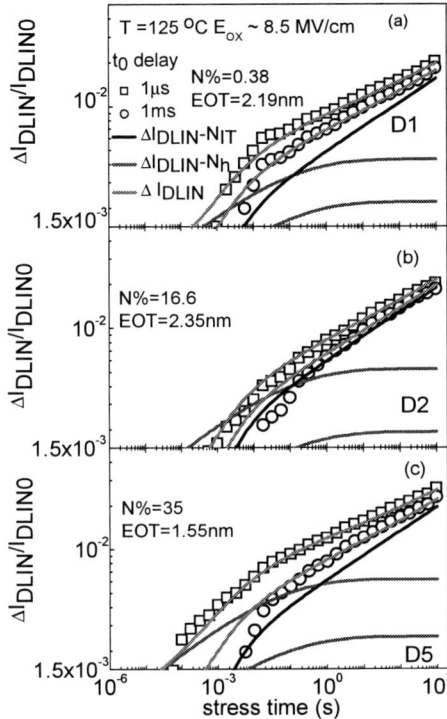

FIGURE.5. TIME EVOLUTION OF MEASURED $\Delta I_{DLIN}/I_{DLIN0}$ TRANSIENT, OVERALL MODEL FIT AND EXTRACTED ΔI_{DLIN}-N_{IT} AND ΔI_{DLIN}-N_h AT DIFFERENT t_0 DELAY FOR (a) D1, (b) D2 AND (c) D5 PNO UNDER CONSTANT EOX AND T. FOR D2, ΔI_{DLIN}-N_h IS MULTIPLIED BY 2.2 TO MAKE IT VISIBLE IN GIVEN Y-AXIS RANGE. HOLE TRAP PARAMETERS: β=0.22(ALL), τ(t_0 DELAY=1µs,1ms) IS (20ms,20ms), (20ms, 20ms) AND (10ms,10ms) FOR D1, D2 AND D5 RESPECTIVELY.

To verify the above hypothesis, Figures 5, 6 and 7 show $\Delta I_{DLIN}/I_{DLIN0}$ measured using t_0 delay of 1µs and 1ms at stress E_{OX}~8.5MV/cm and T=125°C, on SiO$_2$ (D1), low N_{DOSE} (D2) and moderately high N_{DOSE} (D5) PNO devices (Figure 5), very high N_{DOSE} PNO (D6), RTNO (D10) and RTNO+PN (D9) devices (Figure 6) and low N_{DOSE} PNO but with improper PNA (D12) device (Figure 7). Degradation measured at short t_{STS} increases with increase in %N due to larger N_h contribution, and as expected, the impact of t_0 delay, while more prominent at shorter t_{STS}, is also more prominent for devices having high N_h contribution (i.e., Type II devices show more t_0 impact as compared to Type I devices) [22].

The overall model fit to $\Delta I_{DLIN}/I_{DLIN0}$ and underlying ΔI_{DLIN}-N_{IT} and ΔI_{DLIN}-N_h components are also shown in Figures 5, 6 and 7. It can be clearly seen that for all devices, $\Delta I_{DLIN}/I_{DLIN0}$ measured at t_0=1µs differs from that measured at t_0=1ms *only*

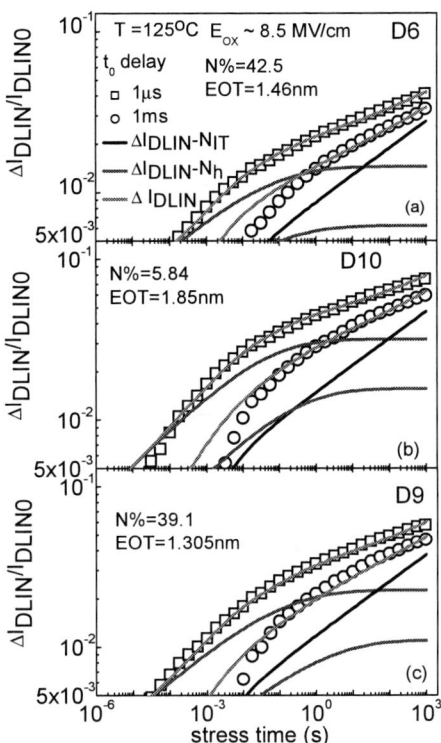

FIGURE.6. TIME EVOLUTION OF MEASURED $\Delta I_{DLIN}/I_{DLIN0}$ TRANSIENT, OVERALL MODEL FIT AND EXTRACTED ΔI_{DLIN}-N_{IT} AND ΔI_{DLIN}-N_h AT DIFFERENT t_0 DELAY FOR (a) D6, (b) D10 AND (c) D9 UNDER CONSTANT E_{OX} AND T. HOLE TRAP PARAMETERS: β=0.22(D6,D9), β=0.25(D10) AND τ(t_0 DELAY=1µs,1ms) IS (10ms,10ms), (20ms,50ms) AND (10ms,70ms) FOR D6, D10 AND D9 RESPECTIVELY.

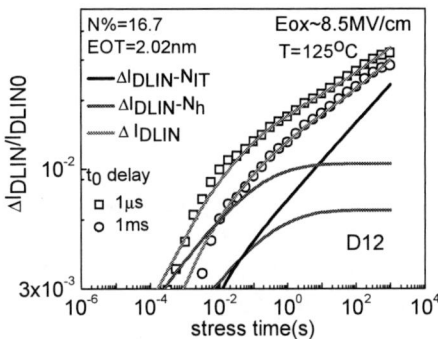

FIG.7. TIME EVOLUTION OF MEASURED $\Delta I_{DLIN}/I_{DLIN0}$ TRANSIENT, OVERALL MODEL FIT AND EXTRACTED ΔI_{DLIN}-N_{IT} AND ΔI_{DLIN}-N_h AT DIFFERENT t_0 DELAY FOR D12 UNDER CONSTANT E_{OX} AND T. HOLE TRAP PARAMETERS: β=0.25, τ(t_0DELAY=1µs,1ms) IS (20ms,50ms).

due to higher, short-time ΔI_{DLIN}-N_h component, consistent with previous reports [15], [16], [21], [22]. Also note that ΔI_{DLIN}-N_h component calculated at 1ms differs from that at 1µs only due to a different value of "A", while τ and β remains identical. As expected, the difference in ΔI_{DLIN}-N_h between t_0 of 1µs and 1ms is higher for devices having high N density at the Si/SiON interface, with device D12 being the only exception (showing large impact irrespective of low N_{DOSE}) as it is an un-optimized device. It is interesting to note that ΔI_{DLIN}-N_{IT} remains identical

978-1-4244-5430-3/10 $26.00 © 2010 IEEE

between $t_0=1\mu s$ and 1ms, and any faster measurement using t_0 less than 1μs would likely increase contribution from $\Delta I_{DLIN}-N_h$ only, especially so for devices having very high N at Si/SiON interface. Larger N_h contribution captured by UF-OTF ($t_0=1\mu s$) when compared to conventional OTF ($t_0=1ms$) influences power-law time exponent n for longer t_{STS} [22], more so for devices having higher %N. This has important implications when measured data is extrapolated to end-of-life and is discussed later.

VI. TEMPERATURE DEPENDENCE

Note that interface trap generation is a strong T activated process while hole trapping is not [5], [6], [19], [21], [22], [34]. It is therefore important to study the T dependence of NBTI for different SiON gate dielectric processes to verify that the early and longer-time degradation are indeed dominated by N_h and N_{IT} respectively, to identify their relative contributions as SiON processes are varied, and to also verify the robustness of the proposed model and underlying parameters.

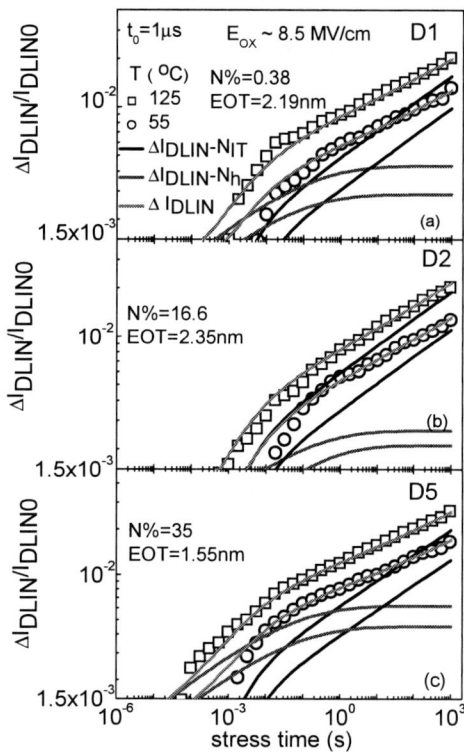

FIGURE.8. TIME EVOLUTION OF MEASURED $\Delta I_{DLIN}/I_{DLIN0}$ TRANSIENT, OVERALL MODEL FIT AND EXTRACTED $\Delta I_{DLIN}-N_{IT}$ AND $\Delta I_{DLIN}-N_h$ AT DIFFERENT T FOR (A) D1, (B) D2 AND (C) D5 UNDER CONSTANT E_{OX} AND t_0 DELAY. HOLE TRAP PARAMETERS: $\beta=0.22$(ALL), τ(125°C,55°C) IS (20ms,10ms), (20ms, 10ms) AND (10ms,10ms) FOR D1, D2 AND D5 RESPECTIVELY.

Figures 8, 9 and 10 show $\Delta I_{DLIN}/I_{DLIN0}$ measured at stress T of 55°C and 125°C and stress $E_{OX}\sim8.5MV/cm$, on SiO_2 (D1), low N_{DOSE} (D2) and moderately high N_{DOSE} (D5) PNO devices (Figure 8), very high N_{DOSE} PNO (D6), RTNO (D10) and RTNO+PN (D9) devices (Figure 9) and low N_{DOSE} PNO but

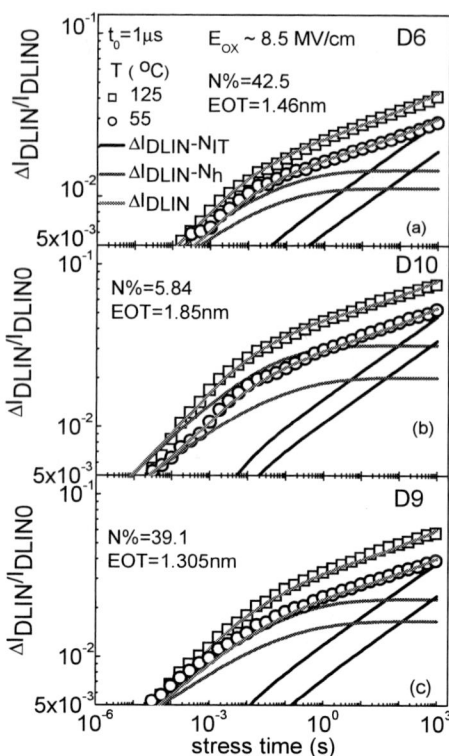

FIGURE.9. TIME EVOLUTION OF MEASURED $\Delta I_{DLIN}/I_{DLIN0}$ TRANSIENT, OVERALL MODEL FIT AND EXTRACTED $\Delta I_{DLIN}-N_{IT}$ AND $\Delta I_{DLIN}-N_h$ AT DIFFERENT T FOR (A) D6, (B) D10 AND (C) D9 UNDER CONSTANT E_{OX} AND t_0 DELAY. HOLE TRAP PARAMETERS: $\beta=0.22$(ALL), τ(125°C,55°C) IS (10ms,6ms), (20ms, 5ms) AND (10ms.10ms) FOR D6. D10 AND D9 RESPECTIVELY.

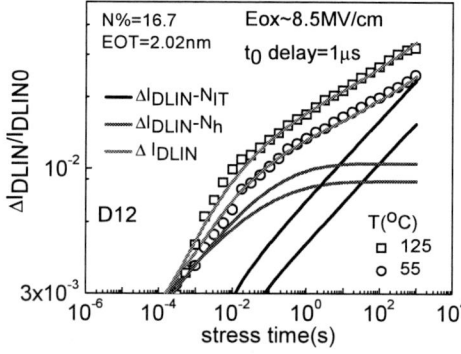

FIGURE.10. TIME EVOLUTION OF MEASURED $\Delta I_{DLIN}/I_{DLIN0}$ TRANSIENT, OVERALL MODEL FIT AND EXTRACTED $\Delta I_{DLIN}-N_{IT}$ AND $\Delta I_{DLIN}-N_h$ AT DIFFERENT T FOR D12 UNDER CONSTANT EOX AND t_0 DELAY. HOLE TRAP PARAMETERS: $\beta=0.25$(ALL), τ(125°C,55°C) IS (20ms,10ms).

with improper PNA (D12) device (Figure 10). Overall model fit to $\Delta I_{DLIN}/I_{DLIN0}$ and the underlying $\Delta I_{DLIN}-N_{IT}$ and $\Delta I_{DLIN}-N_h$ components are also shown. Due to relatively lower $\Delta I_{DLIN}-N_h$ contribution, $\Delta I_{DLIN}/I_{DLIN0}$ measured over the entire range of t_{STS} for low to moderately high %N devices show strong T activation (Figure 8), which validates N_{IT} dominated NBTI for these devices [2], [3], [6], [21], [22]. However, NBTI measured at short t_{STS} increases with increase in Si/SiON interfacial N density (Figure 9), and the increased degradation show weak T dependence. As shown, short t_{STS} degradation for such devices is dominated by weak T activated ΔI_{DLIN}-N_h, while long t_{STS} data is again governed by ΔI_{DLIN}-N_{IT} with stronger T activation. A very large increase in ΔI_{DLIN}-N_h at short t_{STS} increases overall $\Delta I_{DLIN}/I_{DLIN0}$ at short t_{STS} as well as at longer t_{STS} for such high nitrided devices, although increased ΔI_{DLIN}-N_{IT} also contributes to some extent to the overall increase in $\Delta I_{DLIN}/I_{DLIN0}$ at longer t_{STS}. Higher fractional ΔI_{DLIN}-N_h contribution results in negligible T activation at short t_{STS} and reduced T activation at longer t_{STS}, and lowers the overall long-time power-law time exponent n as shown later. Finally, note that low N_{DOSE} PNO with improper PNA device (D12) shows higher, T independent degradation at short t_{STS} and hence larger, weak T dependent degradation at longer t_{STS} due to larger ΔI_{DLIN}-N_h contribution, like RTNO and RTNO+PN devices.

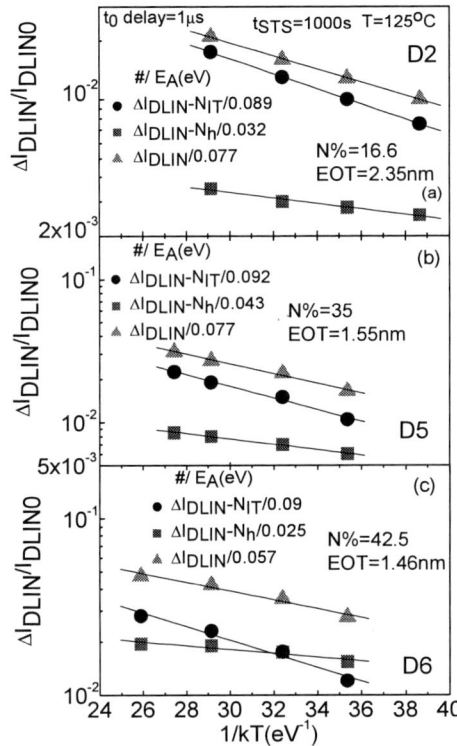

FIG.11. T DEPENDENCE OF MEASURED $\Delta I_{DLIN}/I_{DLIN0}$, EXTRACTED ΔI_{DLIN}-N_h AND ΔI_{DLIN}-N_{IT} UNDER IDENTICAL STRESS E_{OX} AT t_{STS}=1000S FOR (a) D2, (b) D5 AND (c) D6 PNO.

Similar stress T dependent measurements were performed and modeled using the above framework for all devices used in this work. Figure 11 shows $\Delta I_{DLIN}/I_{DLIN0}$, ΔI_{DLIN}-N_{IT} and ΔI_{DLIN}-N_h at a fixed t_{STS} of 1000s as a function of stress T for D2, D5 and D6 devices under identical E_{OX}. ΔI_{DLIN}-N_{IT} component is much higher than ΔI_{DLIN}-N_h for low %N device (D2) at all T. Both ΔI_{DLIN}-N_{IT} and ΔI_{DLIN}-N_h components and therefore the overall $\Delta I_{DLIN}/I_{DLIN0}$ increases with increase in %N, while the relative increase is much more for ΔI_{DLIN}-N_h. Identical E_A is

extracted for ΔI_{DLIN}-N_{IT} and ΔI_{DLIN}-N_h for all devices, $E_A(N_{IT})$ being larger than $E_A(N_h)$ as expected [21]. This verifies the original assumption of strong T dependence for N_{IT} and weak T dependence for N_h components. Irrespective of identical $E_A(N_{IT})$ and $E_A(N_h)$ across all %N, E_A for overall $\Delta I_{DLIN}/I_{DLIN0}$ reduces for higher %N due to larger relative contribution by N_h having lower $E_A(N_h)$, as also explained in detail in [21].

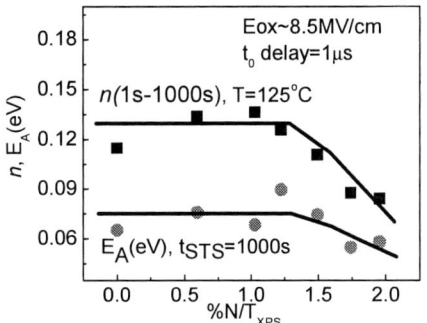

FIGURE.12. %N/T_{XPS} DEPENDENCE OF N AND E_A AT t_0 DELAY=1μS UNDER IDENTICAL STRESS E_{OX} (8.5MV/cm). LINES ARE GUIDE TO THE EYE.

Figure 12 shows measured power-law time exponent n (from log-log fit of $\Delta I_{DLIN}/I_{DLIN0}$ versus time data for t_{STS} range of 1-1000s, observed to be independent of stress E_{OX} and T [22]) and E_A of overall $\Delta I_{DLIN}/I_{DLIN0}$ at t_{STS}=1000s as a function of %N/T_{XPS} for Type-1 devices. Note that both n and E_A remain constant for low Si/SiON N density and decrease with increase in %N/T_{XPS} for N>30% [22]. It is worthwhile to mention that Γ, a measure of Si/SiON interfacial N density, also show similar dependence with %N/T_{XPS} (Figure 4). The reduction in n and E_A is due to higher relative contribution of ΔI_{DLIN}-N_h over ΔI_{DLIN}-N_{IT}, although increase in %N/T_{XPS} increases both ΔI_{DLIN}-N_{IT} and ΔI_{DLIN}-N_h as shown before (Figure 3).

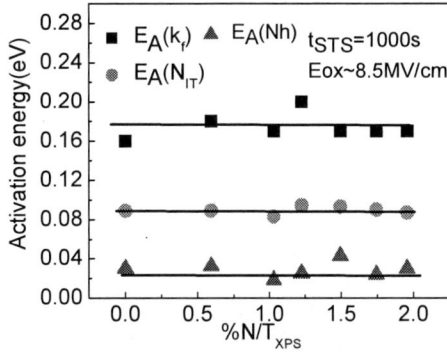

FIGURE.13. %N/T_{XPS} DEPENDENCE OF $E_A(N_{IT})$ AND $E_A(N_h)$ AT t_0 DELAY=1μS UNDER IDENTICAL STRESS E_{OX} (8.5MV/cm). LINES ARE GUIDE TO THE EYE.

For a physically sound model, E_A for individual ΔI_{DLIN}-N_{IT} and ΔI_{DLIN}-N_h components should remain constant across all devices as %N is varied. To verify this, Figure 13 shows extracted $E_A(N_{IT})$ and $E_A(N_h)$ as a function of %N/T_{XPS} for all Type-1 devices. Note that both $E_A(N_{IT})$ and $E_A(N_h)$ remain constant, which validates the robustness of the proposed model. Figure 13 also shows $E_A(k_F)$, used as a fitting parameter, versus $\%N/T_{XPS}$ for Type-1 devices. Note that $E_A(k_F)$ also remains constant with a value very similar to $E_A(k_R)$ for all Type-I devices [10], [21].

TABLE IV: EXTRACTED n AND ACTIVATION ENERGY FOR N_{IT}, N_h AND Si-H FORWARD REACTION FOR TYPE-II DEVICES

	n	$E_A(eV)$	$E_A\text{-}N_{IT}(eV)$	$E_A\text{-}N_h(eV)$	$E_A\text{-}k_F$ (eV)
D8	0.075	0.051	0.047	0.056	0.1
D9	0.085	0.066	0.08	0.05	0.16
D10	0.08	0.074	0.087	0.052	0.13
D11	0.065	0.044	0.05	0.036	0.12
D12	0.09	0.066	0.07	0.03	0.14

Finally, Table IV summaries extracted parameters n, E_A, $E_A(N_{IT})$, $E_A(N_h)$ and $E_A(k_F)$ for Type-II devices. These devices show lower n and E_A due to larger fractional N_h contribution, and lower $E_A(N_{IT})$ due to lower $E_A(k_F)$[1], possibly due to the presence of high N density at the Si/SiON interface that favors forward reaction [35].

VII. LIFETIME EXTRACTION

Figure 14 shows overall $\Delta I_{DLIN}/I_{DLIN0}$ measured at longer t_{STS} for t_0 delay of 1μs and 1ms for low and high %N devices. To determine tt_F, shorter time measured data are extrapolated to 10% degradation using brute power-law fit (t_{STS} range of 10-1000s) and proposed model. Time evolution of $\Delta I_{DLIN}\text{-}N_{IT}$ obtained from RD solution at longer t_{STS} is also shown as a reference. Figure 15 shows E_{OX} dependence of tt_F as obtained using brute fit and model fit of $\Delta I_{DLIN}/I_{DLIN0}$ data measured using t_0 of 1μs and 1ms (discussed above) for low and high %N devices. Obtained tt_F by only RD model extrapolation is also shown for comparison. For both low and high %N, brute fit tt_F is higher than model fit, the difference being larger for higher %N. Moreover, brute fit results in higher tt_F for t_0=1μs when compared to t_0=1ms, which is more prominent for higher %N device. This is due to lower n at lower t_0 [15], [16] and is clearly an artifact, as the use of lower t_0 would result in the capture of higher degradation at the onset of stress [16], which would result in higher overall degradation at longer t_{STS} and earlier tt_F. Though not shown, brute fit tt_F also depends on range of t_{STS} used for fit, as power-law time slope n depends on

---

[1] Long time RD solution (when dominated by H_2 diffusion in the poly-Si) [19] yields $N_{IT} = [k_F*N_0/k_R]^m$. $(Dt)^n$, with m=0.67 and n=0.16. The T activation energies are related by $E_A(N_{IT}) = m*[E_A(k_F)] - E_A(k_R)] + n*E_A(D)$. As $E_A(k_F) \sim E_A(k_R)$ for Type-I devices, $E_A(N_{IT}) \sim n*E_A(D)$ [10][21], see Figure 13.

range of data used for fit [16]. Therefore, brute fit tt_F suffers from severe uncertainty and cannot be used to reliably extrapolate measured degradation to end-of-life.

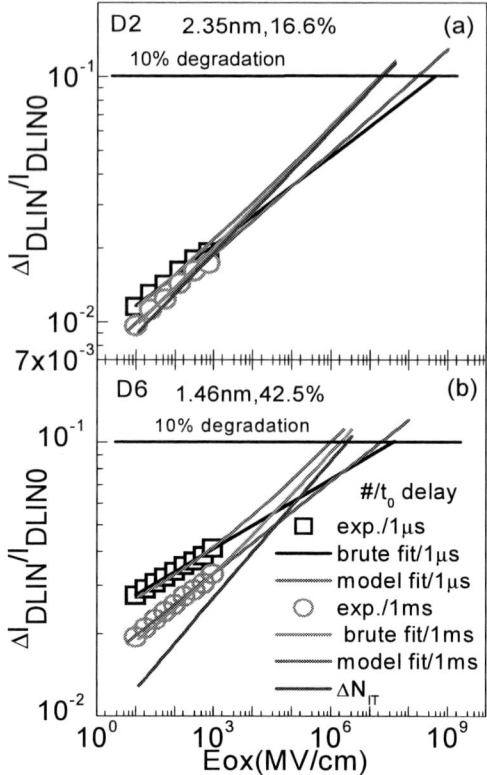

FIGURE.14. $\Delta I_{DLIN}/I_{DLIN0}$ EXTRAPOLATED TO 10% DEGRADATION FOR EXPERIMENTAL (10S-1KS FIT) AND MODEL FIT DATA FOR LOW AND HIGH %N DEVICES UNDER SIMILAR STRESS E_{OX} AND T.

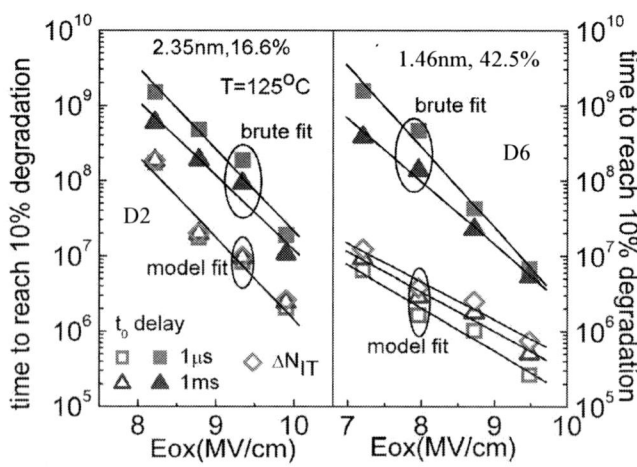

FIGURE.15. TIME TO REACH $\Delta I_{DLIN}/I_{DLIN0}$ 10% DEGRADATION FOR MODEL AND BRUTE FIT AS A FUNCTION OF E_{OX} FOR DIFFERENT t_0 DELAY UNDER CONSTANT STRESS T.

For low %N device, model fit to t_0=1μs and 1ms data show similar tt$_F$, which is also similar to that obtained by pure RD extrapolation. Due to negligible N_h contribution for low %N devices and since N_{IT} component is in any case negligible at short t_{STS} (<1ms), not much difference exists if $\Delta I_{DLIN}/I_{DLIN0}$ is measured using t_0 of 1μs or 1ms, as shown. Note that for low to moderately high %N devices, conventional OTF can be used to predict reliable end-of-life degradation and there is no need to use ultra-fast methods, as short-time hole trapping is negligible.

However for higher %N devices, model fit yields lower tt$_F$ for lower t_0 (correctly so) as overall measured $\Delta I_{DLIN}/I_{DLIN0}$ is larger due to larger N_h contribution at short t_{STS}. This higher ΔI_{DLIN}-N_h contribution also results in lower tt$_F$ for $\Delta I_{DLIN}/I_{DLIN0}$ (both t_0's) when compared to tt$_F$ corresponding to ΔI_{DLIN}-N_{IT} obtained from pure RD extrapolation. However as ΔI_{DLIN}-N_h saturates after some time, its impact diminishes at higher t_{STS}. Therefore, the difference between tt$_F$ at 1μs and 1ms reduces at lower E_{OX}, as measured $\Delta I_{DLIN}/I_{DLIN0}$ is lower, and needs to be extrapolated over larger t_{STS} to end-of-life. For high %N devices, reliable extrapolated end-of-life degradation cannot possibly be obtained by using conventional OTF methods (unless very low E_{OX} is used for accelerated stress), and ultra-fast measurements are necessary to correctly capture the hole trapping dominated early degradation phase that influences long-time data.

VIII. CONCLUSION

To summarize, non-dispersive H/H$_2$ RD model solution for N_{IT} and an analytic dispersive model for N_h are used to predict NBTI time transients from ~μs to long t_{STS} for different stress E_{OX} and T, in PNO (different N dose, EOT, PNA), RTNO and RTNO+PN SiON p-MOSFETs. Two broad categories of SiON devices have been identified. Type-I devices have relatively lower Si/SiON interfacial N density, whose NBTI is dominated by N_{IT} generation. Type-II devices have high interfacial N density and show large, short-time N_h dominated degradation that influences overall NBTI degradation even at longer stress time. The proposed model can predict NBTI transients over such wide range of SiON processes with consistent set of N_{IT} and N_h parameters. It is shown that captured N_h contribution at short-t_{STS} (depends on SiON process and t_0 delay) influences measured n at longer t_{STS}, and hence tt$_F$ obtained from brute-fit extrapolation is unrealistic. Model fit results in correct tt$_F$ for different SiON processes and t_0 delay. Finally for realistic, well optimized devices (low to moderately high %N PNO but with proper PNA, hence low N_h contribution), tt$_F$ is shown to be governed by N_{IT}, and UF-OTF (t_0=1μs) show similar tt$_F$ values as conventional OTF (t_0=1ms), making fast measurements redundant.

ACKNOWLEDGMENT

Authors would like to thank K. Ahmed and C. Olsen (Applied Materials) for providing devices and process details, M. A. Alam and A. E. Islam (Purdue University) for useful discussion, SRC/GRC and Department of IT, Govt. of India for funding and Center for Excellence in Nanoelectronics, IIT Bombay for providing measurement facilities.

REFERENCES

[1] Y. Mitani, "Influence of nitrogen in ultra-thin SiON on negative bias temperature instability under AC stress," in *IEDM Tech.Dig.*, pp. 117-120, 2004.

[2] D. Varghese, D. Saha, S. Mahapatra, K. Ahmed, F. Nouri, and M. Alam, "On the dispersive versus Arrhenius temperature activation of NBTI time evolution in plasma nitrided gate oxides: measurements, theory, and implications," in *IEDM Tech.Dig.*, pp. 684-687, 2005.

[3] A. T. Krishnan, C. Chancellor, S. Chakravarthi, P. E. Nicollian, V. Reddy, A. Varghese, R. B. Khamankar, and S. Krishnan, "Material dependence of hydrogen diffusion: implications for NBTI degradation," in *IEDM Tech.Dig.*, pp. 705-708, 2005.

[4] K. Sakuma, D. Matsushita, K. Muraoka, and Y. Mitani, "Investigation of nitrogen originated NBTI mechanism in SiON with high nitrogen concentration," in *Proc.Int.Rel.Phys.Symp.*, pp. 454-460, 2006.

[5] C. Shen, M. F. Li, C. E. Foo, T. Yang, D. M. Huang, A.Yap, G. S. Samudra, and Y. C. Yeo, "Characterization and physical origin of fast Vth transient in NBTI of pMOSFETs with SiON dielectric," in *IEDM Tech.Dig.*, pp. 333-336, 2006.

[6] S. Mahapatra, K. Ahmed, D. Varghese, A. E. Islam, G. Gupta, L. Madhav, D. Saha, and M. A. Alam, "On the Physical Mechanism of NBTI in Silicon Oxynitride p-MOSFETs: Can Differences in Insulator Processing Conditions Resolve the Interface Trap Generation versus Hole Trapping Controversy?," in *Proc., Int. Rel. Phys. Symp.*, pp. 1-9, 2007.

[7] M. Ershov, S. Saxena, H. Karbasi, S. Winters, S. Minehane, J. Babcock, R. Lindlev, P. Clifton, M. Redford and A. Shibkov, " Dynamic recovery of negative bias temperature instability in p-type metal-oxide-semiconductor field-effect transistors", in *Appl. Phys. Lett.*, vol. 83, p. 1647, Aug. 2003.

[8] S. Rangan, N. Mielke and E. C. C.Yeg, "Universal recovery behavior of negative bias temperature instability," in proc., *Int. Electron Device Meet.*, p331, 2003.

[9] B. Kaczer, V. Arkhipov, R. Degraeve, N. Collaert, G. Groeseneken, and M. Goodwin, "Disorder controlled kinetics model for negative bias temperature instability and its experimental verification," in *Proc.Int.Rel.Phys.Symp.*, pp. 381-387, 2005.

[10] S. Mahapatra and M. A. Alam, "Defect Generation in p-MOSFETs under Negative Bias Stress: An Experimental Perspective," in *IEEE Trans. Device Mater. Rel.*, vol. 8, no. 1, pp. 35-46, Mar. 2008.

[11] T. L. Yang, M. F. Li, C. Shen, C. H. Ang, Z. Chunxiang, Y. C. Yeo, G. Samudra, S. C. Rustagi, and M. B. Yu, "Fast and slow dynamic NBTI components in p-MOSFET with SiON dielectric and their impact on device life-time and circuit application," in *VLSI Symp.Tech.Dig.*, pp. 92-93, 2005.

[12] Y. Z. Hu, D. S. Ang and G. A. Du, "An improved methodology for monitoring NBTI induced threshold voltage shift of scaled p-MOSFETS," in *Proc.Int.Rel.Phys.Symp.*, pp.743-744, 2008.

[13] H. Reisinger, O. Blank, W. Heinrigs, A. Muhlhoff, W. Gustin, and C. Schlunder, "Analysis of NBTI degradation- and recovery- behavior based on ultra fast V$_T$ measurments," in *Proc.Int.Rel.Phys.Symp.*, pp. 448-453, 2006.

[14] C. Schlunder, M. Hoffman, R.-P. Vollertsen, G. Schindler, W. Heinrigs, W. Gustin and H. Reisinger,"A novel multi-point NBTI characterization methodology using Smart Intermediate Stress (SIS)," in *Proc.Int.Rel.Phys.Symp* , pp.79-86, 2008.

[15] E. N. Kumar, V. D. Maheta, S. Purawat, A. E. Islam, C. Olsen, K. Ahmed, M. Alam and S. Mahapatra, "Material dependence of NBTI physical mechanism in silicon oxynitride (SiON) p-MOSFETs: A comprehensive study by ultra-fast On-the-fly (UF-OTF) I_{DLIN} technique," in *IEDM Tech. Dig.*, pp. 809-812, 2007.

[16] V. D. Maheta, E. N. Kumar, S. Purawat, C. Olsen, K. Ahmed and S. Mahapatra, "Development of an Ultra-Fast On-The-Fly I_{DLIN} Technique to Study NBTI in Plasma and Thermal Oxynitride p-MOSFETs," in *IEEE Trans. Electron Devices*, vol. 55, no. 10, pp. 2614 -2622, Oct. 2008.

[17] A. E. Islam, V. D. Maheta, H. Das, S. Mahapatra, and M. A. Alam, "Mobility degradation due to interface traps in plasma oxinitride PMOS devices," in *Proc. Int. Rel. Phys. Symp.*, p. 87, 2008.

[18] V. Huard and M. Denais, "Hole trapping effect on methodology for DC and AC negative bias temperature instability measurement in PMOS transistors," in *Proc.Int.Rel.Phys.Symp.*, pp. 40-45, 2004.

[19] A. E. Islam, H. Kufluoglu, D. Varghese, S. Mahapatra, and M. A. Alam, "Recent issues in negative bias temperature instability: Initial degradation, field dependence of interface trap generation, hole trapping effects and relaxation," *IEEE Trans. Electron Devices*, vol. 54, no. 9, pp. 2143-2154, Sep. 2007

[20] A. Neugroschel, G. Bersuker, R. Choi, C. Cochrane, P. Lenahan, D. Heh, C. Young, C. Y. Kang, B. H. Lee, and R. Jammy, "An Accurate Lifetime Analysis Methodology Incorporating Governing NBTI Mechanisms in High-K/SiO_2 Gate Stacks," in *IEDM Tech.Dig.*, pp.317-329, 2006.

[21] S. Mahapatra, V. D. Maheta, A. E. Islam and M. A. Alam, "Isolation of NBTI Stress Generated Interface Trap and Hole Trapping Components in PNO p-MOSFETs," *IEEE Trans. Electron Devices*, Mahapatra, *IEEE Trans. Electron Devices*, vol. 56, pp. 236-242, 2009.

[22] V. D. Maheta, C. Olsen, K. Ahmed and S. Mahapatra, "The Impact of Nitrogen Engineering in Silicon Oxynitride Gate Dielectric on Negative Bias Temperature Instability of p-MOSFETs: A study by Ultra-Fast On-The-Fly I_{DLIN} Technique," *IEEE Trans. Electron Devices*, vol. 55, no. 7, pp. 1630-1638, Jul. 2008.

[23] T. Grasser, B. Kaczer, W. Goes, Th. Aichinger, Ph. Hehenberger and M. Nelhiebel, "Two stage model for negative bias temperature instability," in *Proc.Int.Rel.Phys.Symp.*, pp. 33-44, 2009.

[24] J. H. Lee, W. H. Wu, A. E. Islam, M. A. Alam, A. S. Oates, "Separation method of hole trapping and interface trap generation and their roles in NBTI reaction-diffusion model," in *Proc., Int. Rel. Phys. Symp.*, pp. 745-746, 2008.

[25] C. L. Chen, Y. M. Lin, C. J. Wang, and K. Wu, "A new finding on NBTI lifetime model and an investigation of NBTI degradation characteristic for 1.2nm ultra thin oxide," in *Proc.Int.Rel.Phys.Symp.*, pp. 705-706, 2005.

[26] A. Haggag, Anderson G., Parihar S., Burnett D., Abeln G., Higman J., and Moosa M., "Understanding SRAM high temperature operating life NBTI: statistics and permanent vs recoverable damage," in *Proc.Int.Rel.Phys.Symp.*, pp. 452-456, 2007.

[27] A. E. Islam, G. Gupta, S. Mahapatra, A. T. Krishnan, K. Ahmed, F. Nouri, A. Oates and M. A. Alam, " Gate leakage Vs NBTI in plasma nitrided oxides: characterization, physical principle, and optimization," in *IEDM Tech. Dig.*, pp. 329-332, 2006.

[28] C. S. Olsen, "Two-step post nitridation annealing for lower EOT plasma nitrided gate dielectrics,"U.S.Patent 017 596 1A1, Sept.9, 2004

[29] G. Kapila, K. Neeraj, V. D. Maheta and S. Mahapatra, "A Comprehensive Study of Flicker Noise in Plasma Nitrided SiON p-MOSFETs: Process Dependence of Pre-existing and NBTI Stress Generated Trap Distribution Profiles", in *IEDM Tech. Dig*, pp.x, 2008.

[30] K. O. Jeppson and C. M. Svensson, "Negative bias stress of MOS devices at high electric fields and degradation of MOS devices," in *J. Appl. Phys.*, vol. 48, no. 5, pp. 2004-2014, May 1977.

[31] M. A. Alam, "A Critical Examination of the Mechanics of Dynamic NBTI for p-MOSFETs," in *IEDM Tech.Dig.*, pp. 345-348, 2003.

[32] M. Alam and S. Mahapatra, "A comprehensive model of PMOS NBTI degradation," *Microelectron. Reliab.*, vol. 45, no. 1, pp. 71-81, Jan. 2005.

[33] S. Chakravarthi, A. T. Krishnan, V. Reddy, C. F. Machala and S. Krishnan,"A comprehensive framework for predictive modeling of negative bias temperature instability," in *Proc., Int. Rel. Phys. Symp*, 2004.

[34] A. E. Islam, S. Mahapatra, S. Deora, V. D. Maheta, and M. A. Alam, "On The Differences Between Ultra-fast NBTI Experiments and Reaction-Diffusion Theory", in *IEDM Tech. Dig.*, 2009, pp. 733, 2009.

[35] S. S. Tan, T. P. Chen, J. M. Soon, K. P. Loh, C. H. Ang, W. Y. Teo, and L. Chan, "Neighboring effect in nitrogen-enhanced negative bias temperature instability," in *Proc.Solid State Devices Mater.*, pp. 70-71, 2003.

[36] M. L. Reed and J. D. Plummer, "Chemistry of Si-SiO_2 interface trap annealing," *J. Appl. Phys.*, vol. 63, no. 12, pp. 5776-5793, Jun. 1988.

[37] S. Zafar, B. H. Lee, J. Stathis, A. Callegari and T. Ning, " A model for negative bias temperature instability(NBTI) in oxide and high-k pFETs", in *VLSI Test Symp.*, pp. 208-209, 2004.

978-1-4244-5430-3/10 $26.00 © 2010 IEEE

Product NBTI Distribution and Voltage Dependence – Impact of Relaxation and Droops

Amr Haggag, Ning Liu, Peter Abramowitz, Mohamed Moosa, Gary Anderson, David Burnett, Sanjay Parihar, Glenn Abeln, Jack Higman
Freescale Semiconductor, Austin, Texas
amr.haggag@freescale.com

Abstract: The impact of relaxation and droops in determining the voltage dependence of NBTI during product stressing is discussed. This has implications on the Fmax and Vmin guardbands extrapolated from product stress to use conditions. Voltage dependence for the shift of the 3sigma tail bits in both new and old technology nodes are compared to illustrate the importance of relaxation and droops.

Keywords: NBTI; distribution; relaxation; product

Introduction

Gate dielectric scaling has played a pivotal role in enabling conventional CMOS to be offered with high performance, low power and higher levels of integration within and across technology generations, without significantly altering process cost or complexity. As a consequence of this gate dielectric scaling, NBTI which increases device V_t has become a dominant mechanism degrading device max operating frequency (F_{max}) and min operating voltage (V_{min}). This paper shares some V_{min} shift measurements on SRAM arrays and attempts to correlate them to device level V_t shifts from NBTI identifying some critical parameters such as time- and voltage- dependence and their dependence on relaxation and droops during product stressing.

Impact of Relaxation and Droops

Under burn-in stress/test, relaxation (transistor bias interruption) and droops (IR drops making actual transistor bias smaller than applied bias) must be taken into account when evaluating the voltage dependence of NBTI. NBTI has recoverable hole trapping and permanent Si-H breaking components:

$$\Delta V_t \sim \Delta V_{t\,recoverable} + \Delta V_{t\,permanent}$$

$$\Delta V_{t,\,permanent} \sim V^{(5m/3)\alpha} t^\alpha$$

$$\Rightarrow TTF \sim V^{-5m/3} \quad (m \sim 10-13 \text{ excitations to break Si-H})$$

$$\textbf{Eq. 2}$$

As evident in Fig 1, the voltage dependence is expected to be different for product-like long-stress/relaxation where permanent NBTI dominates vs short-

stress/relaxation where recoverable NBTI dominates. Indeed this is the case as shown in Fig 2, where the permanent NBTI from relaxed measurements has a TTF power-law 17-22 consistent with Eq. 2 whereas with recoverable NBTI the TTF power-law is much greater approaching 30.

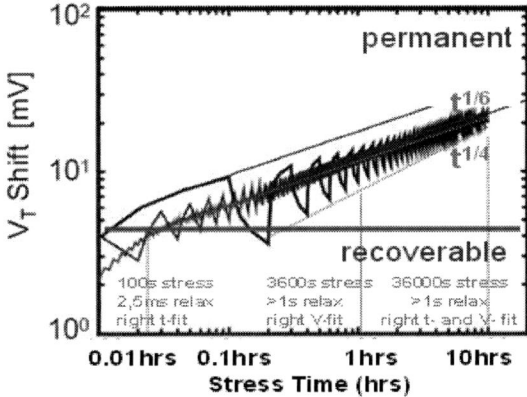

Fig 1 – NBTI consists of recoverable and permanent component. The voltage dependence is expected to be different for product-like long-stress/relaxation where permanent NBTI dominates vs short-stress/relaxation where recoverable NBTI dominates.

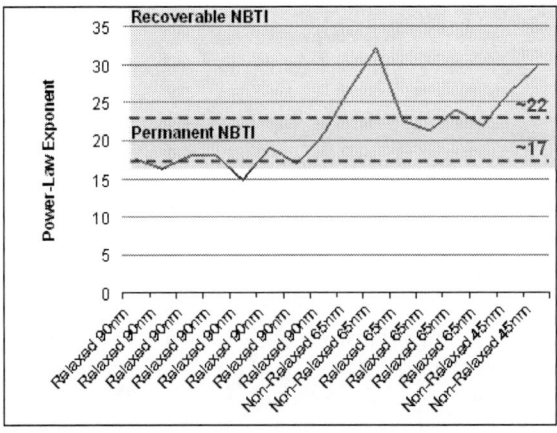

Fig 2 – TTF power-law for NBTI showing the lower value (17-22) for permanent NBTI compared to higher value (approaches 30) with recoverable NBTI.

Also, in the case of new technology, careful droops analysis must be performed as the actual voltage seen by the product may droop vs stress voltage applied due to increased leakage for higher stress voltages (Fig 3):

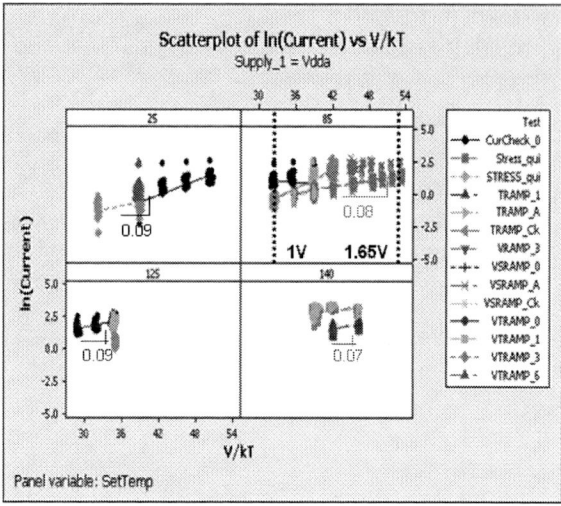

Fig 3 – Droops analysis on array supply during HTOL showing the increased log(Ioff) vs V/kT from 1V to 1.65V. This data showed ~2X more Ioff at 1.65V than 1.4V so ~2X more droop.

Product NBTI (Vmin Shift)

One of the earliest verification of reliability models on circuits is SRAM high temperature operating lifetests (HTOL). The min operating voltage V_{min} at which a bit functions will shift to higher values due to an increase in gate leakage from TDDB or pMOS Vt shift from NBTI. A pMOS Vt shift will cause a shift in static noise margin (SNM) which in turn will cause a shift in minimum operating bitcell voltage (V_{min}). The probability F_{part} that the part Vmin is > V can be related to the exponential probability F_{bit} that the tail bit's Vmin > V:

$$CDF = 1 - Fpart = (1 - Fbit)^N$$
$$= (1 - \frac{e^{-\lambda(V-Vwcsbit)}}{N})^N = exp(-e^{-\lambda(V-Vwcsbit)}) \quad \textbf{Eq 3}$$
$$= LEV(V, scale = 1/\lambda, location = Vwcsbit)$$

where N is the number of bits in the part, λ and Vwcsbit (Vmin of worst case bit in array) can be obtained by plotting the avg failing bitcounts on a log-linear plot. Note this predicts a LEV CDF for part Vmin as expected. The mean Vmin shift of N bits due to NBTI can then be estimated using:

$$\Delta V_{mean} = (1 - \ln N) * (1/\lambda - 1/\lambda 0) + V_{wcsbit} - V0_{wcsbit} \quad \textbf{Eq 4}$$

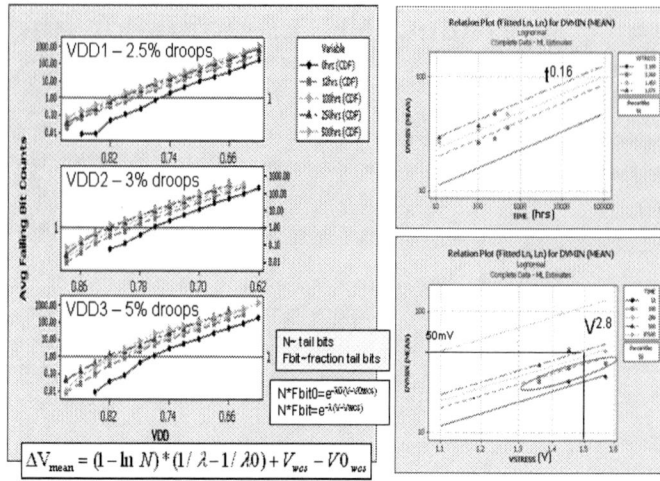

$$\Delta V_{mean} = (1 - \ln N) * (1/\lambda - 1/\lambda 0) + V_{wcs} - V0_{wcs}$$

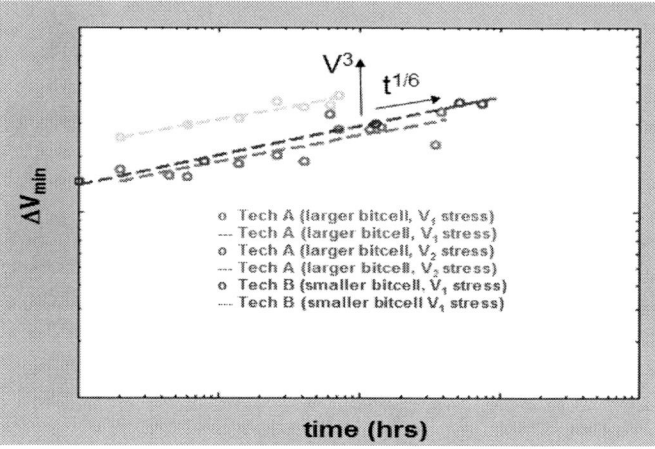

Fig 4 – Shift of 3sigma tail bits $\sim V^{2.8} t^{0.16}$ in new technology node when droops are included similar to the result in old technology node with negligible droops $\Rightarrow$ TTF$\sim V^{-17}$

Conclusion

Fig 4 shows Vmin distributions where with droops included the estimated NBTI TTF power-law ~17 – if droops had not been included the power-law would have been smaller. This is consistent with the NBTI TTF power-law ~17 from old technology where droops are negligible. **For all nodes in Fig 2, this permanent NBTI voltage power-law (17-22) is more realistic when scaling product NBTI stress to use conditions.**

References

[1] H. Reisinger et. al., p. 448-453, IRPS 2006.
[2] D. Schroder et.al., J. Appl. Phys. 94, p. 1, July 2003
[3] V. Huard et al, p. 733-734, IRPS 2006.
[4] A. Haggag et al., p. 665-666, IRPS 2006.
[5] A. Haggag et. al., IRPS 2007

Analysis of the Relationship between Random Telegraph Signal and Negative Bias Temperature Instability

Yasumasa Tsukamoto[1,2], Seng Oon Toh[2], Changhwan Shin[2], Andrew Mairena[2],
Tsu-Jae King Liu[2] and Borivoje Nikolić[2]

[1]Renesas Technology Corp., Japan. [2]University of California at Berkeley, USA
Email address: tsukamot@eecs.berkeley.edu, tsukamoto.yasumasa@renesas.com

Abstract— Random telegraph signal (RTS) is shown to be an intrinsic component of the shift in MOSFET threshold voltage (V_{th}) due to bias temperature instability (BTI). This is done by starting from a well-known model for negative BTI (NBTI), to derive the formula for RTS-induced V_{th} shift. Based on this analysis, RTS simply contributes an offset in NBTI degradation, with an acceleration factor that is dependent on the gate voltage and temperature. This is verified by 3-dimensional (3-D) device simulations and measurements of 45nm-node bulk-Si PMOS transistors. It has an important implication for design of robust SRAM arrays in the future: design margin for RTS should not be simply added, because it is already partially accounted for within the design margin for NBTI degradation.

Keywords-component; random telegraph signal; negative bias temperature instability; hole traps; SRAM

I. INTRODUCTION

Random telegraph signal (RTS) is a temporal variation in $I_{ds}(V_{th})$ caused by the capture and emission of mobile charge carriers, and it has been studied for over two decades [1-6]. Since the amplitude of RTS is proportional to the inverse of the MOSFET channel area [3], memory devices manufactured using the most advanced fabrication technologies are facing severe challenges due to this phenomenon. Being different from flash memory devices which have thick gate-oxide films, SRAM devices until recently were not considered to be seriously affected by RTS because they utilize thin gate-oxide films in accordance with MOS scaling [7]. However, because V_{th} variation due to RTS eventually (with scaling) will become comparable to that due to random dopant fluctuations (RDF) [5] – a major source of degradation in SRAM characteristics [8] –studies of the impact of RTS on highly scaled SRAM have been reported [9-12]. Although it is still controversial how much the SRAM minimum operation voltage (V_{min}) is degraded by RTS, SRAM designers should allocate some design margin for RTS in addition to a (larger) design margin for RDF. Furthermore, degradation in PMOS V_{th}, *i.e.* negative bias temperature instability (NBTI) [13-15], should be considered. With the advent of high-k/metal-gate transistors in SRAM, V_{min} degradation due to positive bias temperature instability (PBTI) will also be significant [16, 17].

There is therefore a need to establish appropriate design margins to guarantee sufficient SRAM yield for operation at V_{min}, at minimum cost (*i.e.* without overdesigning the cells). In this regard, it is necessary to account for inter-dependencies between the physical phenomena that cause V_{th} variation and degradation. Islam *et al.* introduced a model for NBTI comprising two components, associated with hole traps and with interface traps [18, 19]. The effect of hole trapping appears for short stress times and eventually saturates at some constant value. On the other hand, interface trapping (attributed to dangling Si-H bonds at the Si-SiO$_2$ boundary) follows a power-law relationship with stress time and is the dominant component of V_{th} degradation for long stress times. This behavior has been observed experimentally [20], so it is recognized to be a plausible explanation of the NBTI effect. Meanwhile, RTS has been explained to be due to hole trapping/detrapping, and therefore should be related to NBTI. In fact, Kaczer *et al.* pointed out a broad similarity between the relaxation process of NBTI and 1/f noise [21]. Furthermore, Grasser *et al.* succeeded in reproducing the time constants of RTS by using a recovery model of NBTI [22], which indicates that care must be taken to properly account for these two phenomena for worst-case SRAM design margins.

In this paper, a relationship between the V_{th} degradation of RTS and that of NBTI is established, to allow for minimization of design margin. This paper is organized as follows. In section II, the relationship between RTS- and NBTI-induced V_{th} degradation is derived from theory. In section III, the theory is verified by analyzing device simulation results and some measured data for SRAM fabricated using 45nm CMOS technology. Finally, in Section IV, conclusions are drawn.

II. THEORETICAL ANALYSIS

V_{th} degradation due to NBTI is modeled as the sum of an interface-trap effect (ΔV_{IT}) and a hole-trap effect (ΔV_H) [18]:

$$\Delta V_{th} = \Delta V_{IT} + \Delta V_H$$

$$= a\frac{q\Delta N_{IT}(t)}{C_{ox}} + \frac{\int_0^{Tox} \int_E x\rho_H(x,E,t)dEdx}{C_{ox}T_{ox}} \quad (1)$$

where a is the fraction of donor-like traps, ΔN_{IT} is the areal density of interface traps, C_{ox} is the areal gate capacitance, T_{ox} is the gate oxide thickness, and ρ_H is the areal charge density due to trapped holes in the gate oxide, respectively. The first term is dominant at long stress times, and shows a power-law relationship with stress time t; it is proportional to t^n, where n ranges from 1/6 to 1/3. The second term saturates at some

978-1-4244-5430-3/10 $26.00 © 2010 IEEE

value within a relatively short time (milliseconds). According to [18],

$$\rho_H(x, E_i, t) = N_0 \delta(E - E_i) f_T(x, E_i, t) \qquad (2)$$

where N_0 is the areal density of pre-existing traps within the gate oxide and E_i is the energy level of the trap. f_T is the probability that a hole trap is filled, which evolves with time:

$$\frac{df_T}{dt} = \sigma \cdot v_{th} \cdot \left[p_h T_1 (1 - f_T) - n_S T_1 f_T - n_G T_2 f_T \right] \qquad (3)$$

where σ is the capture cross-section, v_{th} is the thermal velocity, p_h is the inversion-layer hole areal density, and n_S and n_G are the areal concentrations of states in the substrate and in the gate, respectively. T_1 and T_2 are the probabilities for tunneling between the substrate and the gate oxide, and for tunneling between the gate oxide and the gate, respectively. This equation was solved in [18] for the initial condition $f_T(0) = 0$, i.e. zero initially trapped holes. Taking RTS into consideration, however, $f_T(0)$ should have a non-zero value. We therefore propose to separate f_T into a shallow-trap term f_{T-S} and a deep-trap term f_{T-D}, and subsequently focus on f_{T-S} since the derivation of f_{T-D} is the same as that given in [18]. From [23], $f_T = \tau_e/(\tau_e + \tau_c)$ under equilibrium conditions, where τ_c is the mean time to hole capture and τ_e is the mean time to hole emission. This is the appropriate initial value for f_{T-S}. Note that $\tau_e/(\tau_e + \tau_c)$ is the average duty cycle for hole trapping: if the hole trap is mostly filled ($\tau_e \to \infty$), it approaches 1; if the hole trap is mostly empty ($\tau_c \to \infty$), it approaches 0. The solution to (3) for the shallow-trap term is therefore

$$f_{T-S}(t) = \frac{T_1}{\left(1 + \dfrac{n_S}{p_h}\right) T_1 + \dfrac{n_G}{p_h} T_2} \left[1 - \exp\left(-\frac{t}{\tau_{NBTI}}\right)\right]$$

$$+ \frac{\tau_e}{\tau_e + \tau_c} \exp\left(-\frac{t}{\tau_{NBTI}}\right) \qquad (4)$$

where $1/\tau_{NBTI} = \sigma v_{th}[(p_h + n_S)T_1 + n_G T_2]$. It should be noted that the last term in (4) does not appear in the expression for f_{T-D} [18]. Based on the Wentzel-Kramers-Brillouin (WKB) approximation, it can be assumed that $T_1 \gg T_2$ if the trap is located closer to the channel than to the gate. Since $n_S = 1/\sigma v_{th} \tau_e$ and $p_h = 1/\sigma v_{th} \tau_c$ [23], the saturated value of f_{T-S} is simply given by

$$f_{T-S}(t) = \frac{\tau_e}{\tau_e + \tau_c}. \qquad (5)$$

Assuming that the hole traps are located at a distance z from the substrate-oxide interface, $N_0 = (q/(L_{eff}W_{eff})) \cdot \delta(x - (T_{ox} - z))$. The integral for ΔV_H can then be evaluated:

$$\Delta V_H \big|_{t \to \infty} = \frac{q}{L_{eff}W_{eff}C_{ox}}\left(1 - \frac{z}{T_{ox}}\right)\frac{\tau_e}{\tau_e + \tau_c} \equiv \Delta V_{RTS}\frac{\tau_e}{\tau_e + \tau_c}. \qquad (6)$$

Note that $\Delta V_{RTS} = (q/(L_{eff}W_{eff}C_{ox})) \cdot (1 - z/T_{ox})$ is the expression for the V_{th} shift due to RTS derived in [3] in a different manner, and that $\tau_e/(\tau_e + \tau_c)$ is the average duty cycle for hole

trapping. Therefore, the expression in (6) represents the time-averaged value of V_{th} shift due to RTS. Substituting (6) into (1), the total V_{th} degradation due to NBTI is expressed by

$$\Delta V_{th}(t) = \Delta V_{RTS}\frac{\tau_e}{\tau_e + \tau_c} + A\left(1 - \exp\left(-\frac{t}{\tau_{NBTI}}\right)\right) + Bt^n \qquad (7)$$

where A and B are constants. In (7), the first term corresponds to shallow traps, the second to deep traps, and the third to interface traps. Considering that actual NBTI measurements of V_{th} vs. t can be fitted to the empirical model $V_{th} = C + Dt^n$ [20], the first term is included in C, i.e. the effect of RTS is included in NBTI degradation.

The relationship between RTS and NBTI is clarified further by examining the factor $\tau_e/(\tau_c + \tau_e) = 1/(1 + \tau_c/\tau_e)$. τ_c/τ_e is dependent on the gate voltage V_g and absolute temperature T, decreasing with V_g and increasing with T [3]:

$$\frac{\tau_c}{\tau_e} = \exp\left(K - \frac{q}{kT}\left[\left(1 - \frac{z}{T_{ox}}\right)\Psi_S + \frac{z}{T_{ox}}V_g\right]\right). \qquad (8)$$

where K is a constant and Ψ_S is the channel surface potential. Generally, RTS is measured under normal operating conditions (V_{g0}, T_0) whereas NBTI is measured under accelerated stress conditions ($V_{g0} + \Delta V$, $T_0 + \Delta T$), so that one must be careful to account for ΔV and ΔT when correlating the first term in (7) with RTS. This can be done by introducing the concept of an acceleration factor $\alpha(\Delta V, \Delta T)$:

$$\frac{\Delta V_{RTS}}{1 + (\tau_c/\tau_e)}\bigg|_{(V_{g0}+\Delta V, T_0+\Delta T)} \equiv \Delta V_{RTS} \cdot \alpha(\Delta V, \Delta T). \qquad (9)$$

Note that (7) deals with a single-trap RTS with depth z where both ΔV_{RTS} and α are the function of z. Therefore we can expand (7) to include the multiple traps with depths z_i (where i represents the trap number) as,

$$\Delta V_{th}(t) = \sum_i \Delta V_{RTS}^i \cdot \alpha_i + A\left(1 - \exp\left(-\frac{t}{\tau_{NBTI}}\right)\right) + Bt^n. \qquad (10)$$

The dependence of τ_c/τ_e on V_g is illustrated in the representative example shown in Fig. 1(a), for which the normal operating conditions are (1.0V, 300K) and the NBTI stress conditions are (2.0V, 400K). It can be seen that $\tau_c/\tau_e = 1$ under normal operating conditions, so that $\alpha = 50\%$ for (ΔV, ΔT) = (0V, 0K) in Fig. 1(b). In contrast, $\tau_c/\tau_e \approx 0.01$ under NBTI stress conditions, so that $\alpha \approx 100\%$ for (ΔV, ΔT) = (1.0V, 100K). Thus, under NBTI stress conditions, 100% of ΔV_{RTS} appears as an offset in the NBTI-induced V_{th} shift. However, when the stress is removed, α is decreased to its normal value. This implies that the traps responsible for RTS are also responsible for the NBTI recovery as reported in [21, 22]. Note that Fig. 1 is a sample using $z/T_{ox} = 0.25$ and $K = 18.35$ condition and thus, α also depends on the values of z/T_{ox} and K; small z and large K result in large τ_c/τ_e (since $V_g > \Psi_S$) and hence small α.

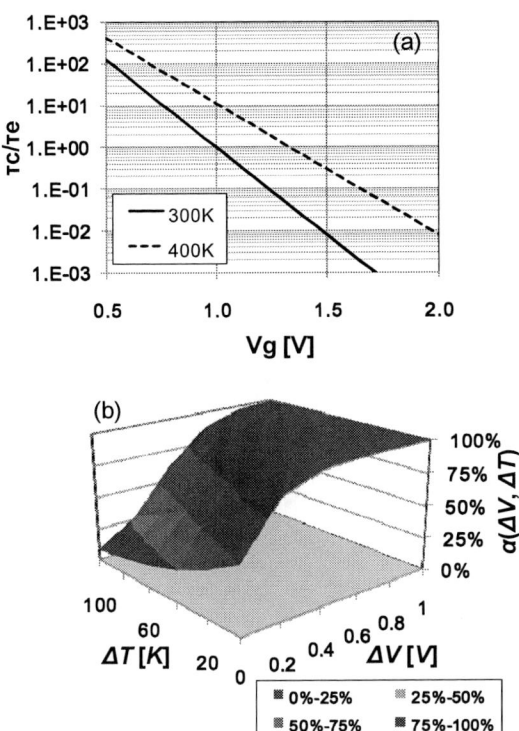

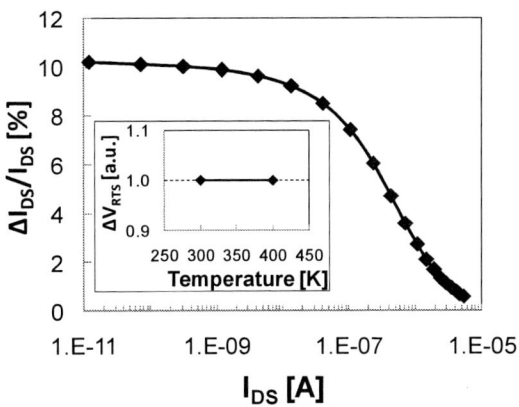

Figure 2: $\Delta I_{DS}/I_{DS}$ dependence on I_D using 3-D device simulation results for a PMOSFET with channel dimensions $L/W = 60\text{nm}/60\text{nm}$ and uniform body doping $N_{SUB}=10^{18}$ cm^{-3}. Inset shows temperature dependence of ΔV_{RTS}.

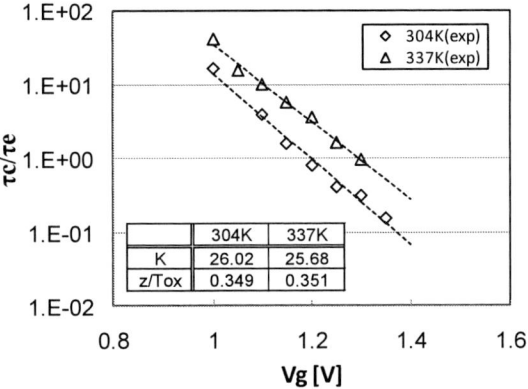

Figure 1: V_g and T dependence of τ_c/τ_e (a) and of the acceleration factor (b). The origin (0,0) in (b) corresponds to the nominal operating condition while (1,100) corresponds to the NBTI stress condition.

III. VALIDATION OT THE THEORY

Both 3-D device simulations using Sentaurus v.2009.06 (Synopsys, Inc.) and measurements of RTS and NBTI-induced V_{th} shifts for pull-up devices in 45nm-node SRAM cells [12] were performed in order to validate the theory as developed above. Fig. 2 shows the 3-D device simulation results for a p-channel MOSFET ($L/W = 60\text{nm}/60\text{nm}$, $T_{ox} = 2\text{nm}$) which has a single trap located at the center of the channel area and a distance 0.1nm away from the substrate-oxide interface ($z = 0.1\text{nm}$). The expected dependence of RTS on drain current (I_D) [4] is reproduced, as shown in Fig. 2. Also, the simulated ΔV_{RTS} shows no T dependence (inset), as expected.

Fig. 3 shows the measured dependence of τ_c/τ_e on V_g and T for a pull-up transistor in a 45nm-node bulk-Si SRAM cell. Each τ_c/τ_e data point is extracted from the RTS time-domain measurement. Inset table indicates z/T_{ox} and K values extracted from the slope and y-intercept for 304K and 337K plots, respectively. Those values are almost equivalent to each other, confirming the validity of using (8) at different V_g and T conditions. For instance, since τ_c/τ_e at $V_g = 1.1\text{V}$ at 304K is 3.997, we observe 20% of RTS-induced V_{th} shift as the mean-time value at nominal condition. On the other hand, τ_c/τ_e at $V_g = 2.0\text{V}$ and $T = 400\text{K}$ is evaluated as 1.124E-2, which projects 0.989 of α, *i.e.* 98.9% of the RTS-induced V_{th} shift is expected to be included in the NBTI degradation.

Figure 3: Measured V_g and T dependence of τ_c/τ_e, for a pull-up transistor within a 45nm-node bulk-Si SRAM cell.

Fig. 4(a) shows measured V_{th} degradation at $|V_g| = 1.5\text{V}$ and $T = 400\text{K}$ using the on-the-fly (OTF) technique [24, 25]. A prior work reveals that the hole trap effect saturates within 1-s after stress, while the interface trap component starts degrading after 1-s [26]. Based on this fact, V_{th} degradation during the first second can be entirely attributed to RTS with no interface trap effect. Fig. 4(b) is a close-up graph of Fig 4(a), and it also shows the RTS observed at $V_g = 1.1\text{V}$, $T = 300\text{K}$. Note that (i) the RTS was measured for about 20 sec using the alternating bias method [12], where the time axis is normalized to 1-s scale duration to compare the amplitudes, (ii) the $\Delta V_{th}=0$ point (bottom line of RTS data) is being shifted upward. From this graph, NBTI behavior can be separated into three phases; first, there is a deep trap effect that appears in the second term of (10) as well as in the fluctuation (Phase 1). Second, there appears the RTS related fluctuation with an amplitude equivalent to that observed at the nominal condition (Phase 2). A rough estimate of the acceleration factor α from the phase 2 is 57%, compared to 23% for the nominal operating condition, confirming that α is larger for higher V_g and T. Third, additional RTS fluctuations appear with amplitudes of ~3mV and ~5mV in arbitrary units (a.u.), equivalent to those observed in Phase 1, by subtracting out the deep trap effect (Phase 3). In summary, for V_{th} degradation due to NBTI, a deep-trap effect appears during the first 1-s of stress, then RTS (with larger α value than observed for the nominal operating condition) appears, followed by some additional RTS induced by the NBTI stress.

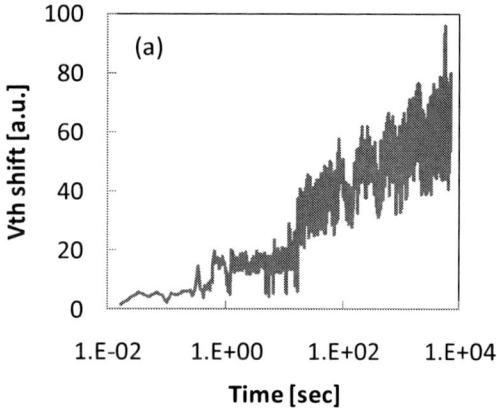

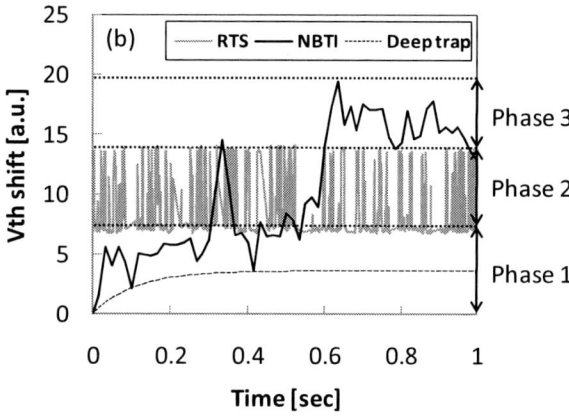

Figure 4: V_{th} degradation under NBTI stress for long time (a), for 1s after stress (b). In (b), the gray line is the RTS observed for nominal bias condition. Black dashed line indicates deep trap effect which is obtained by fitting $A(1-\exp(-t/\tau_{NBTI}))$ to the local minimum points.

Figure 5(a) shows the V_{th} shift induced by NBTI for long stress times (from 5000 sec to 7200 sec in Fig. 4(a)) using a time lag plot (TLP) developed by T. Nagumo *et al.* [27], which was originally used for analyzing RTS. In this method, based on a sequence of V_{th} shifts represented by x_i (i=data number), we generate a two dimensional variable (x_i, x_{i+1}) and plot it on the x-y plane. We use the TLP because it allows us to analyze the RTS amplitude more effectively than using a complex time-domain waveform. Since the NBTI-induced V_{th} shift includes a time-dependent component associated with the interface trap, the data distributes toward the upper right-hand side compared with the case of RTS analysis [27]. From this figure, three diagonal trend lines can be identified. The difference between those lines is found to be 7mV (a.u.) and corresponds to the RTS amplitude observed in the NBTI degradation. To see this more effectively, we extract the

absolute values of all points measured from the diagonal line in Fig. 5(a) (see Fig. 5(b)). A peak at 7mV is observed as well as at 5mV, which is related to the RTS observed in Phases 1 and 3 as discussed with regard to Fig 4(b). Regarding the RTS with 3mV amplitude, it is hard to distinguish a peak in this graph; however, it is noticeable that some large V_{th} shifts with ~15mV are observed, which can be explained by coincident RTS with amplitudes of 7, 5 and 3mV. Therefore, RTS observed during the first 1s of NBTI degradation is retained for long stress times, showing the validity of equation (10).

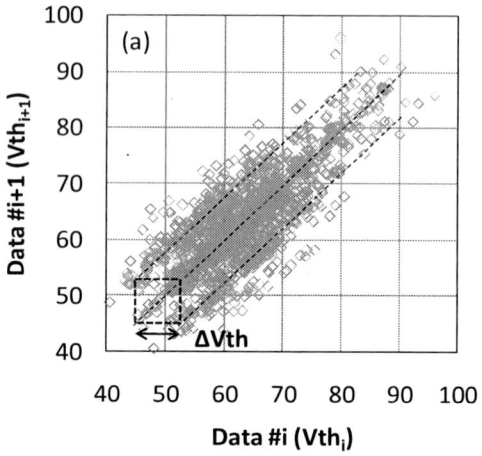

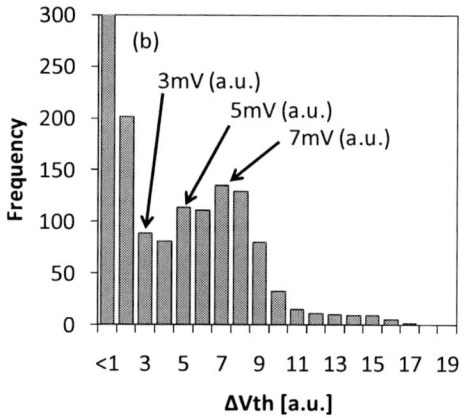

Figure 5: V_{th} degradation of NBTI during 5000 s to 7200 s stress time using TLP metric (a) and the distribution of amplitudes extracted from the diagonal line (b).

To compare the RTS and NBTI data more quantitatively, it is effective to take the time-averaged V_{th} shift after 1 sec stress because the first and the second terms in (10) represent the time-averaged amplitude. Regarding Fig. 4(b), we obtain 10.15mV (a.u.) for NBTI and 2.07mV (a.u.) for RTS, indicating that the time-averaged V_{th} shift of NBTI has a larger value than that of RTS. We performed the same experiment for multiple transistors in our chip and compared the time-averaged V_{th} shift of RTS with that of NBTI (see Fig. 6). All points appear above the diagonal line, confirming that the RTS-induced shift is included in the NBTI-induced shift, validating the theory developed in this work.

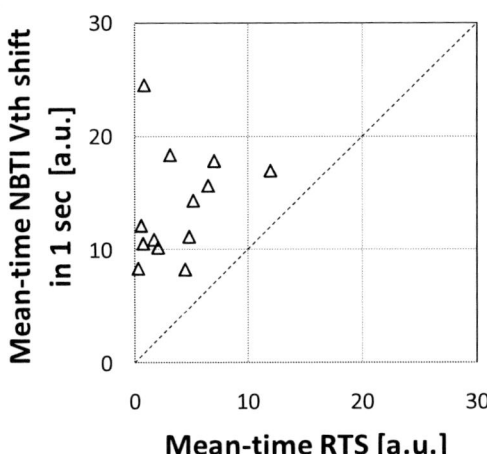

Figure 6: Comparison between mean-time V_{th} shift of RTS and that of NBTI during 1 sec after applying stress.

IV. CONCLUSION

The expression for the V_{th} shift due to RTS is derived starting from a well-known model for V_{th} degradation due to NBTI. This indicates that the hole traps primarily responsible for the time-independent V_{th} shift (modeled by the mean-time V_{th} shift during 1 sec after stress) under NBTI stress are the same ones responsible for RTS. The mean-time V_{th} shift of NBTI during 1 sec that is manifested as RTS, referred to herein as an acceleration factor α, is dependent on the NBTI stress conditions (gate voltage and temperature). 3-D device simulations and SRAM pull-up transistor measurements are consistent with this theory. It can be concluded, therefore, that adding design margin for RTS independently of the design margin for NBTI would result in overly conservative VLSI design.

ACKNOWLEDGMENT

The authors would like to thank Koji Nii, Makoto Yabuuchi in Renesas Technology Corp., and Zheng Guo in University of California, Berkeley for their technical support and valuable discussions. The first author also thank to Hiroshi Makino in Osaka Institute of Technology and Hirofumi Shinohara, Hisashi Matsumoto, Yuji Kihara, Yohinobu Nakagome and Toshifumi Takeda in Renesas Technology Corp. for their support.

REFERENCES

[1] G. Ghibaudo, "On the Theory of Carrier Number Fluctuations in MOS Devices," *Solid-State Electronics* Vol.32, No.7, pp. 563-565, 1989.

[2] K. K. Hung *et. al.*, "Random Telegraph Noise of Deep-Submicrometer MOSFET's," *Electron Device Letters*, Vol. 11, No. 2, pp. 90-92, 1990.

[3] O. Buisson *et. al.*, "Model for Drain Current RTS Amplitude in Small-Area MOS Transistors," *Solid-State Electronics* Vol.35, No.9, pp. 1273-1276, 1992.

[4] A. Asenov, *et. al.*, "RTS Amplitudes in Decananometer MOSFETs: 3-D Simulation Study," *Transactions on Electron Devices*, Vol. 50, No. 3, pp. 839-845, 2003.

[5] K. Abe *et. al.*, "Random Telegraph Signal Statistical Analysis using a Very Large-Scale Array TEG with 1M MOSFETs," *Symposium on VLSI Tech. Dig.*, pp. 210-211, 2007.

[6] J.P. Campbell *et. al.*, "Random Telegraph Noise in Highly Scaled nMOSFETs," *IRPS Tech. Dig.*, pp. 382-388, 2009.

[7] K. Sonoda *et. al.*, "Discrete Dopant Effects on Statistical Variation of Random Telegraph Signal Magnitude," *Transactions on Electron Devices*, Vol. 54, No. 8, pp. 1918-1925, 2007.

[8] Y. Tsukamoto *et al.*, "Worst Case Analysis to Obtain Stable Read/Write DC Margin of High Density 6T-SRAM-Array with Local Vth Variability," *ICCAD Tech. Dig.*, pp.398-405, 2005.

[9] M. Agostinelli *et. al.*, "Erratic Fluctuations of SRAM Cache Vmin at the 90nm Process Technology Node," *IEDM Tech. Dig.*, pp. 655-658, 2005.

[10] N. Tega. *et. al.*, "Impact of Threshold Voltage Fluctuation due to Random Telegraph Noise on Scaled-down SRAM," *IRPS Tech. Dig.*, pp. 541-546, 2008.

[11] K. Takeuchi *et. al.*, "Single-Charge-Based Modeling of Transistor Characteristics Fluctuations Based on Statistical Measurement of RTN Amplitude," *Symposium on VLSI Tech. Dig*, pp. 54-55, 2009.

[12] S. O. Toh *et. al.*, "Impact of Random Telegraph Signals on V_{min} in 45nm SRAM," *IEDM Tech. Dig.*, pp.767-770, 2009.

[13] J.C. Lin *et. al.*, "Prediction and Control of NBTI – Induced SRAM V_{ccmin} Drift," *IEDM Tech. Dig.*, 2006.

[14] A.T. Krishnan *et. al.*, "SRAM Cell Static Noise Margin and VMIN Sensitivity to Transistor Degradation," *IEDM Tech. Dig.*, 2006.

[15] V. Huard *et. al.*, "NBTI Degradation: From Transistor to SRAM Arrays," *IRPS Tech. Dig.*, pp. 289-300, 2008.

[16] J.C. Lin *et. al.*, "Time Dependent V_{ccmin} Degradation of SRAM Fabricated with High-k Gate Dielectrics," *IRPS Tech. Dig.*, pp. 439-444, 2007.

[17] A. Bansal *et. al.*, "Impact of NBTI and PBTI on SRAM static/dynamic noise margins and cell failure probability," *Microelectronics Reliability*, Vol 49, pp. 642-649, 2009.

[18] A.E. Islam *et. al.*, "Recent Issues in Negative-Bias Temperature Instability: Initial Degradation, Field Dependence of Interface Trap Generation, Hole Trapping Effects, and Relaxation," *Transactions on Electron Devices*, Vol. 54, No. 9, pp. 2143-2154, 2007.

[19] S. Mahapatra *et. al.*, "On the Physical Mechanism of NBTI in Silicon Oxynitride p-MOSFETs: Can Differences in Insulator Processing Conditions Resolve the Interface Trap Generation versus Hole Trapping Controversy?" *IRPS Tech. Dig.*, pp. 1-4, 2007.

[20] J.H. Lee *et. al.*, "Separation Method of Hole Trapping and Interface Trap Generation and Their Roles in NBTI Reaction-Diffusion Model," *IRPS Tech. Dig.*, pp. 745-746, 2008.

[21] B. Kaczer *et. al.*, "NBTI from the perspective of defect states with widely distributed time scales," *IRPS Tech. Dig.*, pp. 55-60, 2009.

[22] T. Grasser, *et. al.*, "Switching Oside Traps as the Missing Link between Negative Bias Temperature Instability and Random Telegraph Noise," *IEDM Tech. Dig.*, pp. 729-732, 2009.

[23] E. Hoekstra, "Large Signal Excitation Measurement Techniques for RTS Noise in MOSFETs," *EUROCON*, pp, 1863-1866, 2005.

[24] M. Denais *et. al.*, "On-the-fly characterization of NBTI in ultra-thin gate oxide PMOSFET's," *IEDM Tech. Dig.*, pp. 109-112, 2004.

[25] H. Aono *et. al.*, "A Study of SRAM NBTI by OTF Measurement," *IRPS Tech. Dig.*, pp. 67-71, 2008. *IEDM Tech. Dig.*, pp. 109-112, 2004.

[26] A.E. Islam, *et. al.*, "On the Differences between Ultra-fast NBTI Measurements and Reaction-Diffution Theory," *IEDM Tech. Dig.*, pp. 733-736, 2009.

[27] T. Nagumo, *et. al.*, "New Analysis Methods for Comprehensive Understanding of Random Telegraph Noise," *IEDM Tech. Dig.*, pp. 759-762, 2009.

ENERGY RESOLVED SPIN DEPENDENT TRAP ASSISTED TUNNELING INVESTIGATION OF SILC RELATED DEFECTS

J.T. Ryan [1], P.M. Lenahan [1], A.T. Krishnan [2], and S. Krishnan [2]
[1] The Pennsylvania State University, [2] Texas Instruments
Primary Contact: J.T. Ryan, 212 EES Bldg, University Park, PA 16802, phone: 814-863-6484; fax: 814-863-7967; jtr16@psu.edu
Second Contact: P.M. Lenahan, 212 EES Bldg, University Park, PA 16802, phone: 814-863-4630; fax: 814-863-7967; pmlesm@engr.psu.edu

PURPOSE

We demonstrate energy resolved spin dependent trap assisted tunneling in 1.2nm effective oxide thickness silicon oxynitride film subject to room temperature electric field stressing. Our observations introduce a simple method to link point defect structure and energy levels in a very direct way. We obtain defect energy level resolution of SILC related defects by exploiting the enormous difference between the capacitance of the very thin dielectric and the capacitance of the depletion of the silicon. The simplicity of the technique and the robust character of the response make it, at least potentially, of widespread utility in the understanding of defects important in microelectronics. [*Keywords:* Electron Paramagnetic Resonance, Spin Dependent Tunneling, SILC, K Centers]

INTRODUCTION

Although a vast amount of literature on the mechanisms of stress induced leakage currents (SILC) exists, the atomic scale nature of the defects responsible for SILC in ultra thin MOS devices is not yet fully understood.[1-3] SILC is manifested as a continuous increase in gate leakage current and eventual oxide breakdown during the application of high oxide electric fields.[1-3] Aggressive gate oxide scaling has exacerbated SILC in recent years and poses a serious concern for future gate oxide scaling.[4] SILC is believed to arise from an oxide wear out process in which defect centers are continuously generated in the oxide and/or interface as a result of electric field stress.[1-3] The increased concentration of defect centers leads to an increase in gate leakage current due to trap assisted tunneling. For thicker oxides (greater than about 3.5nm) SILC related trap assisted tunneling is thought to be an inelastic tunneling process [5,6] which proceeds through neutral electron traps in the oxide (likely oxygen vacancies otherwise known as E' centers in SiO_2 dielectrics [7,8]). However, when the gate oxide thickness is scaled down below about 3.5nm, SILC related trap assisted tunneling is thought to be an elastic (or low loss inelastic) process which proceeds through an interface trap to interface trap tunneling mechanism.[9,10] This process is usually referred to as low-voltage SILC (LV-SILC) because the SILC current is usually only observed at low sense (gate) voltages within a volt or two of flat band conditions.[9,10] Additionally, Nicollian et al. suggest that LV-SILC is only observed when the energy states of the interface defects are within the same range of electrostatic potential, supporting the idea that LV-SILC is due to interface trap to interface trap tunneling.[9,10]

We have developed an energy resolved spin dependent tunneling (ER-SDT) technique which allows us to directly observe the atomic-scale defects participating in trap assisted tunneling in 1.2nm equivalent oxide thickness (EOT) SILC stressed MOS devices. The approach provides excellent sensitivity and also provides direct information about defect energy levels involved in SILC.

EXPERIMENTAL

The samples used in this study are large area (10,000 μm^2) 1.2nm EOT silicon oxynitride pMOS capacitors. Defects were generated in the devices with a room temperature stress of $V_G = 2.2V$ for 10,000s. ER-SDT measurements where made on a custom built X-band spectrometer based on a Resonance Instruments 8330 series microwave bridge and TE_{102} resonant cavity. I_G-V_G measurements where made before and after stress to monitor the change in gate leakage current (not shown).

RESULTS AND DISCUSSION

Figure 1 illustrates the difference between the pre- and post- I_G-V_G (not shown) curves by plotting $\Delta J/J_0$ vs. V_G. J_0 is the gate current density pre-stress and ΔJ is the difference between J_0 and the gate current density post-stress (J_t). The peak in the curve (0.35V) is caused by a trap assisted tunneling current in the post-stress I_G-V_G measurement.[10] Values for $\Delta J/J_0$ around $V_G = 0V$ are not included because the amplitude of the currents are below the detection limit of the I_G-V_G measurements.

Figure 2 illustrates a typical ER-SDT spectrum obtained with the Si/dielectric interface normal parallel to the magnetic field and V_G biased to correspond to the peak in figure 1; the point with the maximum fractional contribution of trap assisted tunneling current. The spectrum has a g = 2.0030 and a linewidth of about 15G. (In the simplest cases, the electron paramagnetic resonance (EPR) resonance condition is given by $h\nu = g\beta H$, where h is Planck's constant, ν is the microwave frequency at resonance, β is the Bohr magneton, and H is the magnetic field at resonance. The g varies with the orientation of the defect in the magnetic field for nearly all defects. It is described by a matrix often referred to as the g tensor.[11]) Measurements taken with the Si/dielectric interface normal rotated in the magnetic field (not shown) indicate that the spectrum of figure 2 does not change.

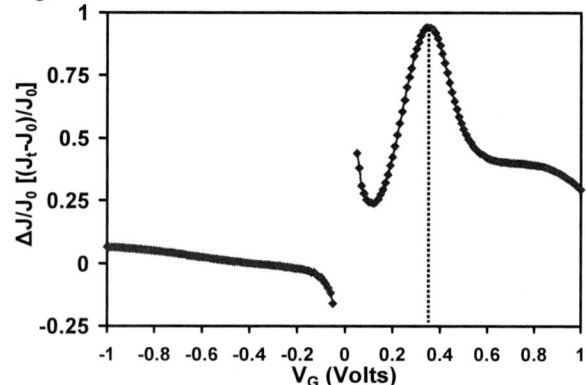

FIGURE 1: $\Delta J/J_0$ VERSUS V_G. THE PEAK IN THE CURVE (0.35V) IS CAUSED BY A TRAP ASSISTED TUNNELING CURRENT IN THE STRESSED I_G-V_G MEASUREMENT (NOT SHOWN).

The magnetic field orientation independence, the zero crossing g value of 2.0030, and the 15G linewidth of the observed defect spectrum are all consistent with the K center found in Si_3N_4 and some SiO_xN_y films.[12-14] K centers are silicon dangling bond defects in which the central silicon atom is back bonded to three nitrogen atoms. When the center is paramagnetic, a single electron occupies a high p-character wave function in a silicon atom back bonded to three nitrogen atoms. The K center is responsible for trapping in conventional Si_3N_4 films [13] and dominates NBTI in at least some nitrided oxide MOS devices.[12]

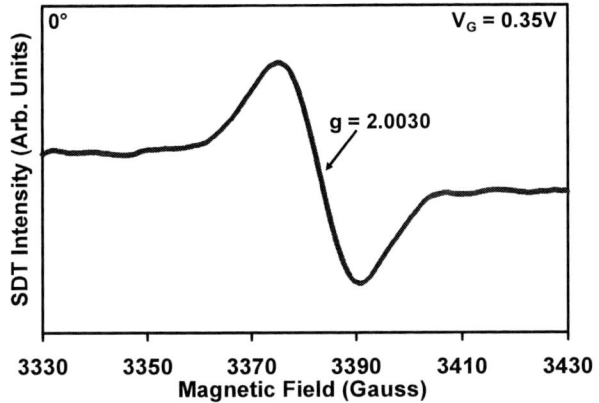

FIGURE 2: REPRESENTATIVE SDT MEASUREMENT TAKEN WITH V_G BIASED TO CORRESPOND TO THE PEAK IN THE $\Delta J/J_0$ CURVE OF FIGURE 1 ($V_G = 0.35$V). THE MEASUREMENT WAS TAKEN WITH THE MAGNETIC FIELD PARALLEL TO THE SI/DIELECTRIC INTERFACE NORMAL. THE ZERO CROSSING G = 2.0030 ± 0.0002.

Figure 3a illustrates the normalized SDT intensities versus V_G achieved by dividing the spin dependent modification to the tunneling current (ΔI_{SDT}) by the total DC current (I) at a particular gate voltage. The $\Delta I_{SDT}/I$ response very closely follows the characteristic trap assisted tunneling peak of figure 1, indicating that we are observing spin dependent trap assisted tunneling through the defects largely responsible for the tunneling current. In an attempt to delineate between spin dependent trap assisted tunneling current and direct tunneling current (not spin dependent), figure 3b shows ΔI_{SDT} versus V_G. It peaks at about $V_G = 0.5$V, indicating that the peak at 0.35V in $\Delta I_{SDT}/I$ (figure 3a) is shifted downward because direct tunneling overwhelms the trap assisted tunneling process at higher V_G. Since the direct tunneling is not spin dependent, the SDT response is not affected by the large direct tunneling current response which overwhelms the "electrically" measured trap assisted tunneling current at higher bias.

Figure 4 illustrates the poly-Si/SiOxNy/crystalline-Si band diagram for the device at three quite different biasing conditions: V_G = 0, 0.55, and 1.0V. For simplicity of presentation, only two levels of a single dielectric trap are included in diagrams. (These band diagrams were calculated using the Boise State band diagram program.[15]) Note first that there is very little band bending in the dielectric at any of the illustrated biasing levels. The dielectric is so thin that the relationship between the crystalline-Si/dielectric Fermi level (E_F) and the defect energy level is nearly independent of the physical position of the defect with respect to the crystalline-Si/dielectric interface. This is so because of the enormous difference between the capacitance of the 1.2nm EOT dielectric and the much thicker Si depletion region. Nearly all the voltage appears across the Si. Figure 3b shows that the SDT response appears at a V_G of about 0.2V, peaks at 0.5V, and has completely disappeared at about 0.65V. At $V_G = 0.2$V, where SDT appears, the crystalline-Si/dielectric E_F is 0.26eV above the valence band edge (VBE). At $V_G = 0.65$V, where the SDT disappears, the E_F is about 0.68eV above the Si VBE. This narrow response must reflect a narrow distribution in K center levels.

An explanation of the response can be gleaned from a brief consideration of the physics of spin. The SDT process, like all EDMR

processes, must involve a pair of spins initially separated physically. One of the spin sites is a K center. K centers, especially those nearest the crystalline-Si/dielectric boundary, can act like interface traps in that, as the E_F is advanced from the VBE toward the conduction band edge (CBE), the empty dangling bond trap levels (+/0) will accept an electron as the E_F crosses the relevant energy. This process is not spin dependent whether or not it was to involve paramagnetism at the K center site, it does not involve paramagnetism from the valence band. However, once the K center is rendered paramagnetic, interactions of the K center site with another paramagnetic site would be spin dependent and thus susceptible to SDT. Should the K center accept an additional electron, it would be rendered diamagnetic again, insensitive to the SDT process.

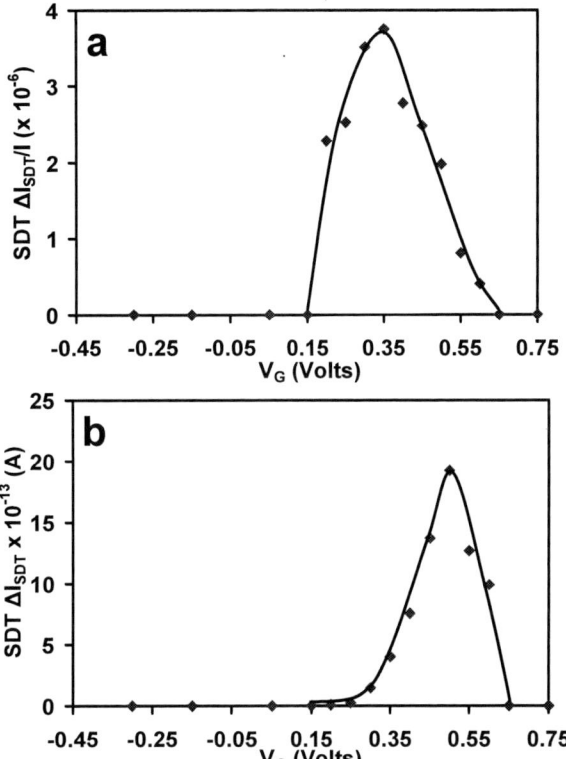

FIGURE 3: COMPARISON BETWEEN NORMALIZED SDT INTENSITIES ($\Delta I_{SDT}/I$) VS. V_G (A) AND THE SPIN DEPENDENT MODIFICATION TO THE TUNNELING CURRENT (ΔI_{SDT}) VS. V_G (B). NOTE THAT THE ΔI_{SDT} OF 3B PEAKS AT ABOUT 0.5V INDICATING THE PEAK AT 0.3V IN THE SDT $\Delta I_{SDT}/I$ OF 3A IS SHIFTED DOWNWARD BECAUSE DIRECT TUNNELING OVERWHELMS THE TRAP ASSISTED TUNNELING PROCESS AT HIGHER V_G.

Consider tunneling of an electron from a paramagnetic K center site to another paramagnetic site in the (highly defective) poly-Si gate. The process would be allowed only if the unpaired electron spins have opposite spin quantum numbers. If the two sites had electron spins with the same spin quantum number, the tunneling process would be forbidden (Pauli Exclusion principle). However, if the K center electron spin were to be "flipped" via ESR the previously forbidden tunneling event would be allowed. Thus, magnetic resonance could modulate such a tunneling process. The SDT process would thus "turn on" when E_F crosses the energy level corresponding to the first K center electron (+/0) transition which places one electron in the defect's dangling bond orbital. Figure 5a, a replotting of the results of figure 6 in which V_G is replaced by E_F, indicates that the SDT response begins to appear with E_F at about 0.26eV above the VBE. The process peaks with E_F at about 0.54eV. Very crudely speaking, the energy range of 0.26eV to 0.54eV would correspond to the range of energy over which the K centers accept the first electron (+/0 transition). The SDT response drops from 0.54eV to below our detection limit at 0.68eV. So, to a rough approximation, the energy range of 0.54eV to 0.68eV corresponds to the range of energy over which the K centers accept the second electron (0/- transition).

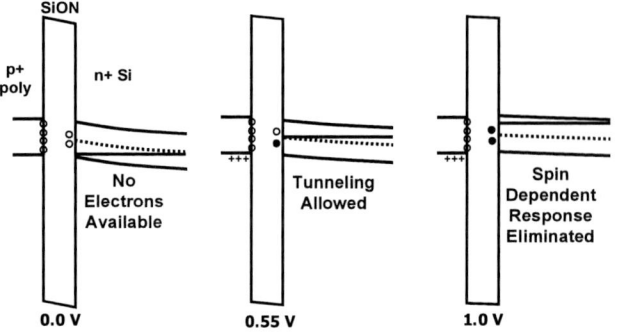

FIGURE 4: ENERGY BAND DIAGRAMS FOR THE SAMPLE AT THREE DIFFERENT V_G VALUES. NOTE THAT THE ONLY PLAUSIBLE EXPLANATION FOR THE TUNNELING CURRENT MUST INVOLVE ELECTRON TUNNELING THROUGH DEFECTS WITH LEVELS CORRESPONDING TO THE RANGE OF THE SILICON BAND GAP. THE SIMPLIFIED SKETCH ILLUSTRATES TWO DIELECTRIC DEFECT LEVELS, CONSISTENT WITH THE EXPERIMENTAL RESULT.

To a very crude approximation, we could approximate the collective K center density of states by the absolute value of the derivative $d\Delta I_{SDT}/dE_F$. This is illustrated in figure 5b. The cartoons of figure 5c illustrate the spin states (and charge) of the K centers vs. E_F. We can understand how this is so by first considering an array of precisely identical defects which have precisely identical energy levels. This array of defects would have a density of states as illustrated in figure 6. Figure 7a illustrates a more physically reasonable density of states in which each of the levels is broadened to take into account disorder. If the E_F is below the (+/0) level, the defect's unoccupied dangling bond orbital does not have an electron to contribute to the tunneling. The defect is also diamagnetic (no unpaired electron) and cannot take part in magnetic resonance. Thus, with E_F below the (+/0) level, no SDT signal can be observed.

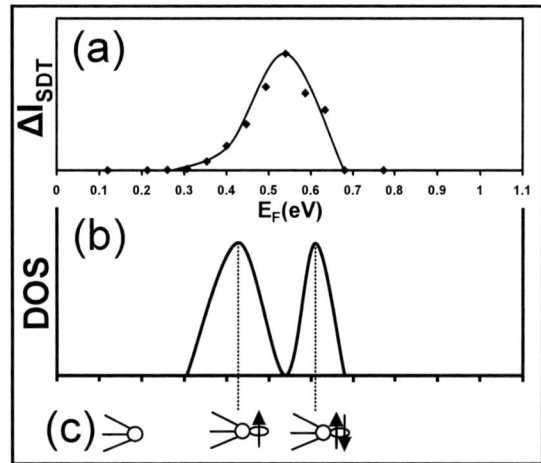

FIGURE 5: (A) THE SDT RESPONSE AS A FUNCTION OF INTERFACE E_F, (B) A CRUDE SCHEMATIC REPRESENTATION OF K CENTER DENSITY OF STATES, AND (C) A CARTOON REPRESENTATION OF THE CHARGE STATES OF THE K CENTERS.

However, if E_F crosses the (+/0) level of some of the K centers, these centers can contribute to the tunneling and are paramagnetic and do take part in magnetic resonance. Therefore, the SDT response begins to turn on as the E_F level crosses the lower (+/0) levels and increases as long as E_F continues to cross these levels. However, as the E_F begins to cross the (0/-) level, the orbitals begin to accept a second electron and become negative. When this happens, the centers lose their paramagnetism and can no longer take part in magnetic resonance, so the SDT response is reduced. The SDT response drops to zero when all of the K centers accept the second electron. This SDT response is illustrated in figure 7b.

Figure 7c illustrates the derivative of the SDT amplitude vs. energy response of figure 7b. Notice that the maximum on the left side of the trace occurs at the same energy as the (+/0) peak in figure 7a. This is so because the increase in SDT amplitude vs. energy will be greatest at the lower peak of the curve in figure 7a. Analogously, since the rate of decrease in SDT amplitude vs. energy will occur at the (0/-) peak, the minimum on the right will occur at that (0/-) energy. Thus, the absolute value of the derivative shown in figure 7d is a fairly good first order representation of the defect density of states illustrated in 7a.

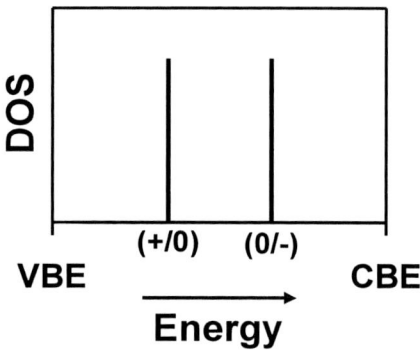

FIGURE 6: SCHEMATIC ILLUSTRATION OF THE DENSITY OF STATES FOR AN ARRAY OF PRECISELY IDENTICAL DEFECTS WITH PRECISELY IDENTICAL ENERGY LEVELS.

It is important to point out that this absolute value of the derivative is only a first order representation of the actual density of states. If the (+/0) and (0/-) transition peaks overlap, the absolute value of the derivative will incorrectly indicate a zero in the density of states between the two peaks. Also, the tunneling transmission probability from the K centers to defects in the poly-Si gate will not be precisely constant throughout the energy range (about 0.4eV) over which the SDT is observed. However, the transmission probability will vary relatively slowly over the energy range.

Note that this very crude representation is correct to the extent that the average energy of the first (+/0) transition is almost certainly higher than 0.26eV and the average energy of the (0/-) transition is almost certainly lower than 0.68eV but above 0.54eV. As mentioned previously, this approximation is illustrated in figure 5b. The cartoons of figure 5c illustrate the charge and spin states of the K centers. Figure 5b indicates that the K center electron-electron correlation energy is quite small, roughly 0.2eV. These results and their interpretation are qualitatively consistent with ideas of Nicollian et al. who developed a model for LV-SILC based on interface trap to interface trap tunneling.[9,10]

978-1-4244-5430-3/10 $26.00 © 2010 IEEE   1124

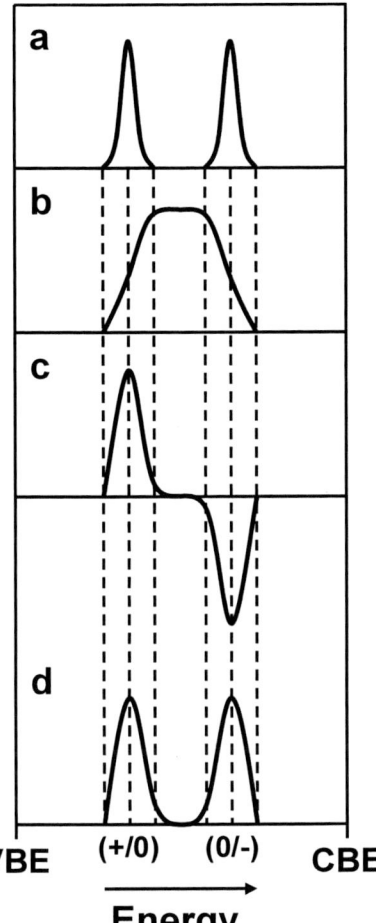

VBE   **(+/0)**   **(0/-)**   **CBE**

Energy

FIGURE 7: (A) A MORE PHYSICALLY REASONABLE DENSITY OF STATES IN WHICH EACH OF THE LEVELS OF FIGURE 6 IS BROADENED TO TAKE INTO ACCOUNT DISORDER. (B) THE SDT RESPONSE FROM THE LEVELS OF (A). (C) SCHEMATIC ILLUSTRATION OF THE DERIVATIVE OF THE SDT AMPLITUDE VS. ENERGY RESPONSE OF (B). (D) THE ABSOLUTE VALUE OF THE DERIVATIVE (C). THE PLOT ILLUSTRATED IN (D) IS, AS DISCUSSED IN THE TEXT, AN APPROXIMATION OF THE DEFECT DENSITY OF STATES.

CONCLUSIONS

This very simple measurement offers the capability to directly link the analytical power of magnetic resonance with defect energy levels. One could envision extending this approach to more complex device structures, different semiconductor substrates (for example SiC), and thicker dielectrics. One could also envision extending this approach to quantum computing as it provides a simple sensitive method for magnetic resonance which could, at least in principle, be utilized at temperatures low enough to assure quite long spin lattice relaxation times.[16] Work at Penn State has been supported by Texas Instruments through Semiconductor Research Corporation Custom Funding.

REFERENCES

[1] E. Rosenbaum and L. F. Register, "Mechanism of stress-induced leakage current in MOS capacitors," *IEEE Trans. Elec. Dev.*, **44**, pp. 317-323, 1997.

[2] L. F. Register, E. Rosenbaum, and K. Yang, "Analytic model for direct tunneling current in polycrystalline silicon-gate metal-oxide-semiconductor devices," *Appl. Phys. Lett.*, **74**, pp. 457-459, 1999.

[3] J. S. Suehle, in *Defects in Microelectronic Materials and Devices*, edited by D. M. Fleetwood, S. T. Pantelides, and R. D. Schrimpf (CRC Press, Boca Raton, London, New York, 2009), pp. 437-463.

[4] M. L. Green, E. P. Gusev, R. Degraeve, and E. L. Garfunkel, "Ultrathin (< 4 nm) SiO2 and Si-O-N gate dielectric layers for silicon microelectronics: Understanding the processing, structure, and physical and electrical limits," *J. Appl. Phys.*, **90**, pp. 2057-2121, 2001.

[5] S. Takagi, N. Yasuda, and A. Toriumi, "Experimental evidence of inelastic tunneling and new I-V model for stress-induced leakage current," *IEEE IEDM Tech. Dig.*, pp. 323-326, 1996.

[6] A. Ghetti, M. A. Alam, J. Bude, D. Monroe, E. Sangiorgi, and H. Vaidya, "Analysis of trap-assisted conduction mechanisms through silicon dioxide films using quantum yield," *IEEE IEDM Tech. Dig.*, pp. 723-726, 1999.

[7] P. M. Lenahan and J. J. Mele, "E ' centers and leakage currents in the gate oxides of metal oxide silicon devices," *J. Vac. Sci. Technol., B*, **18**, pp. 2169-2173, 2000.

[8] P. M. Lenahan, J. J. Mele, J. P. Campbell, A. Y. Kang, R. K. Lowry, D. Woodbury, S. T. Liu, and R. Weimer, "Direct experimental evidence linking silicon dangling bond defects to oxide leakage currents," *IEEE Int. Reliab. Phys. Symp.*, pp. 150-155, 2001.

[9] P. E. Nicollian, A. T. Krishnan, and V. K. Reddy, "Two-trap model for low voltage stress-induced leakage current in ultra-thin SiON dielectrics," *J. Appl. Phys.*, **104**, pp. 053718, 2008.

[10] P. E. Nicollian, M. Rodder, D. T. Grider, P. Chen, R. M. Wallace, and S. V. Hattangady, "Low voltage stress-induced-leakage-current in ultrathin gate oxides," *IEEE Int. Reliab. Phys. Symp.*, pp. 400-404, 1999.

[11] P. M. Lenahan and J. F. Conley, "What can electron paramagnetic resonance tell us about the Si/SiO$_2$ system?" *J. Vac. Sci. Technol., B*, **16**, pp. 2134-2153, 1998.

[12] J. P. Campbell, P. M. Lenahan, A. T. Krishnan, and S. Krishnan, "Identification of the Atomic-Scale Defects Involved in the Negative Bias Temperature Instability in Plasma-Nitrided p-Channel Metal-Oxide-Silicon Field-Effect Transistors," *J. Appl. Phys.*, **103**, pp. 044505, 2008.

[13] D. T. Krick, P. M. Lenahan, and J. Kanicki, "Electrically Active Point-Defects in Amorphous-Silicon Nitride - An Illumination and Charge Injection Study," *J. Appl. Phys.*, **64**, pp. 3558-3563, 1988.

[14] W. L. Warren and P. M. Lenahan, "Electron-Nuclear Double-Resonance and Electron-Spin-Resonance Study of Silicon Dangling-Bond Centers in Silicon-Nitride," *Phys. Rev. B*, **42**, pp. 1773-1780, 1990.

[15] R. G. Southwick and W. B. Knowlton, "Stacked dual-oxide MOS energy band diagram visual representation program," *IEEE Trans. Dev. Mater. Reliab.*, **6**, pp. 136-145, 2006.

[16] M. A. Nielsen and I. L. Chuang, Quantum Computation and Quantum Information, Cambridge University Press, Cambridge, UK, 2000

BIOGRAPHIES

Abe, Kenichi (5F.2)
Kenichi Abe (S'07) was born in Shiogama, Japan in 1983. He received the B.S. degrees in electronic engineering and M.S. degree in management of science and technology from Tohoku University, Sendai, Japan, in 2006 and 2008, respectively. He is currently working toward the Ph.D. degree in Tohoku University from April 2008. His current research topics are low-frequency noise and random telegraph noise in deep submicron devices and variability of MISFET characteristics in ULSI. Mr. Abe received the IEEE Electron Devices Society Japan Chapter Student Award in 2008.

Ackermann, Markus (RM.2)
Markus Ackermann is with X-FAB Semiconductor Foundries, being responsible for process characterisation and primitive device reliability in the field of CMOS technologies. He works on characterisation and modelling of semiconductor device degradation. Before joining X-FAB he studied physics at the University of Jena and received the diploma in 2008.

Acosta, Antonio (5F.3)
Antonio Acosta received his B.S. from Georgia Tech in 2005 and M.S. from the University of Florida in 2009, both in Electrical Engineering. He is currently working towards his Ph.D. in Electrical Engineering at the University of Florida. His main research interests are the development of next-generation semiconductor devices and nanotechnology. He has worked as an intern within the Analog Technology Development group at Texas Instruments since the summer of 2008. His past projects have been the characterization of PZT ferroelectric capacitors for decoupling applications, and scaling reliability and modeling of fully embedded ferroelectric capacitors.

Afanas'ev, Valery (5A.2)
Valery Afanas'ev received the Ph. D. degree from Leningrad State University in 1985 and worked there as a member of research staff till 1992. In 1992 he joined Delft Institute of Microelectronics as research fellow. In 1994, he moved the University of Erlangen to develop the internal photoemission spectroscopy of oxidized SiC. Since 1995 he is with the Laboratory of Semiconductor Physics, University of Leuven, focussing on applications of internal photoemission spectroscopy to wide spectrum of solid interfaces. The main research results were published in more than 250 journal papers, 150 conference contributions, 10 book chapters, and 1 monography.

Ahlbin, Jonathan (3A.2, 6E.2)
Jonathan R. Ahlbin received B.E. and M.S. degrees in electrical engineering from Vanderbilt University in 2005 and 2009 respectively, and is currently pursuing his Ph.D. there. His research interests include microelectronic circuit analysis and design, and the effects of radiation on integrated circuits, specifically the modeling of circuit-level soft errors.

Ahn, Kun-Ok (MY.7)
Kun-ok Ahn received the B.S. degree in electrical engineering for Dankook University, Seoul Korea in 1982. In 1983, he joined Samsung Electronics, Keheung Korea. Since then he has been engaged in development and producction of nonvolatile memory devices such as MaskROM, NOR and NAND flash memory production. He left Samsung Electronics in 2003 and he has worked for Hynix Semiconductor from 2007, Cheongju-si, Korea and then he has been engaged in production for NAND flash memory devices as the VP of Process Integration and Device Engineering now.

Aichinger, Thomas (XT.6, XT.8)
Thomas Aichinger received his M.S. degree in physics from the Karl-Franzens University in Graz in May 2007. During his diploma thesis, concerning the implementation of Charge Pumping in the laboratory soft- and hardware environment, he was integrated in the Quality and Reliability group of Infineon Technologies Villach. Currently he is toward his doctor degree at Infineon Villach and KAI in cooperation with the "Christian Doppler Laboratory" group of the Institute for Microelectronics in Vienna. In 2007, he recieved the Best Paper arward at ESREF.

Akira, Toriumi (3D.4)
Akira Toriumi received the B.S. degree in physics, the M.S. and Ph.D. degrees in applied physics from the University of Tokyo in Japan, 1978, 1980 and 1983, respectively. He joined R&D Center of Toshiba Corporation in 1983. In May 2000, he moved to the University of Tokyo. His research interests have been on silicon device physics and related materials science throughout his professional carrier. He is currently studying the materials science of high-k dielectrics, and physics and technology of Ge as well as organic electron devices. He is a member of the JSAP, JPS, APS, ECS and IEEE.

Alain, Bravaix (2B.2)
Alain Bravaix graduated from the University of Sciences of Paris and joined the R&D research laboratory of BULL S.A. in 1988 (Les Clayes sous Bois-France). He received the Ph.D. degree in microelectronics in 1991. From 1991 to 1993, he worked on gate-oxide nitridation optimization as a postdoctorate fellowship in the Solid-State Physics Research group of the Institut d'Electronique et de Microélectronique du Nord (IEMN). Since 1994 he is developing research activities and teaching for Engineering and Master degrees at the Institut Supérieur d'Electronique et du Numérique (ISEN-Toulon) as a member of the Institut Matériaux Microélectronique Nanosciences de Provence (IM2NP) UMR 6242. He works in collaboration with ST Microelectronics Crolles since 1995 on the reliability and optimization of CMOS and BICMOS technologies. His research interests cover device to circuit reliability developing electrical characterization techniques for novel and ultra-small CMOS nodes.

Alam, Muhammad Ashraful (2B.3, 3E.2, XT.7)
Muhammad Ashraful Alam received the B.S. from Bangladesh University of Engineering and Technology in 1988, the M.S. degree from Clarkson University in 1991, and the Ph.D. degree from Purdue University in 1994, all in electrical engineering. He is a Professor of Electrical and Computer Engineering at Purdue University where his research and teaching focus on physics, simulation, characterization and technology of classical and novel semiconductor devices, including theory of oxide reliability, nanocomposite thin film transistors and nano-bio sensors. From 1995 to 2001, he was with Bell Laboratories as a Member of Technical Staff in the Silicon ULSI Research Department. From 2001 to 2003, he was a Distinguished Member of Technical Staff at Agere Systems. Dr. Alam has published over 80 papers in international journals and has presented many invited and contributed talks at international conferences. He is an IEEE Fellow, Purdue University Faculty Scholar, and also received the 2006 IEEE Kiyo Tomiyasu Award for his contributions to device technology for communication systems.

Albin, David (3E.3)
David Albin is a Senior Scientist at the National Renewable Energy Laboratory (NREL) in Golden, CO. He obtained his Ph.D. in Materials Science and Engineering from the University of Arizona with a Masters Degree in Ceramic Engineering from the University

of Illinois. He has over 20 years experience in both Cu(In,Ga)Se2 and CdTe thin film solar cell development, with over 100 scientific publications, and three patents. Approximately half of his research at NREL is funded through collaborative research with outside companies including PrimeStar Solar and the GE Global Research Center in New York.

Allmon, Randy (3A.1)
Randy Allmon received a B.S.E.E. degree from the University of Cincinnati in 1981. Since that time he has worked at Digital Equipment Corporation, SiByte and Broadcom Corporation as a Circuit Designer and Technical Leader, and has been active in the Circuit and Layout design of numerous PDP-11, VAX, Alpha and MIPS high-performance microprocessors. Since joining Intel in 2003, he has worked in both the Itanium and Xeon design teams, and is currently in a Technical Leader in the High Performance Computing Group. He holds 4 Patents and has 1 Patent Pending

Ambrose, Vinod (3A.1)
Vinod Ambrose works in the Server Microprocessor Group in Intel. Starting in 1996, he has worked in the areas of Device Modeling, Reliability, Power Estimation, and Circuit Design for microprocessors in DEC, Compaq, and HP. He joined Intel in 2003. Vinod holds a BS in Physics/Math from Monmouth College, IL, and a PhD in Physics from Texas Christian University.

Amoroso, Salvatore Maria (MY.5)
Salvatore Maria Amoroso was born in Catania, Italy, in 1983. He received the B.S. and M.S. degrees in physics engineering from the Politecnico di Milano, Milano, Italy, in 2005 and 2008, respectively, where he is currently working toward the Ph.D. degree in information technology in the Dipartimento di Elettronica e Informazione. He is also with the Italian University Nano-Electronics Team, Milano, Italy. His research activities include modeling and numerical simulation of semiconductor devices, with particular interest on innovative nonvolatile memories.

Andrea, Marmiroli (MY.6)
Andrea Marmiroli received the Degree in Physics (cum laude) from Milano University in 1984, with a thesis work on an algebraic approach to quantum field theory. In September 1985 he joined SGS (now STMicroelectronics). He worked at first within the New Technologies Development group, then within the Technology CAD team. Since its foundation (March 2008), he is with Numonyx, in the position of "Technology and Device Modelling Director". He coordinated ST activity within different European funded project. He is reviewer for the IEEE- TED. His major interests are in Non Volatile Memory modeling.

Andreani, Carla (4B.3)
Carla Andreani (CA) received a Laurea Cum Laude in Physics in 1977 at the Università di Roma "La Sapienza". CA is now full professor in Condensed Matter at the University of Rome Tor Vergata. She is an experimental physicist, with a background in condensed matter physics. CA designed several neutron scattering instruments to study the structure and dynamics of fluids and materials. Most recently CA has used neutron spectroscopy to produce quantitative measurements and 3D images of the elemental composition and physical structure of artifacts. CA has authored about 150 publications on international journals and delivered over 80 invited talks.

Ang, Diing Shenp (4A.4, XT.5)
D. S. Ang received the B.Eng. (Hons) and Ph.D. degrees in electrical engineering from the National University of Singapore (NUS) in 1994 and 1998, respectively. He was a teaching staff of the Bachelor of Technology Programme in NUS before joining the School of Electrical and Electronic Engineering, Nanyang Technological University as an assistant professor in 2002. From 2008, he has been an associate professor. His current research interests are bias-temperature instability of advanced CMOS devices, application of scanning probe microscopy for gate stack characterization, and resistance memory devices. He has published more than 90 papers in international journals and conferences and has served on the technical programme committee of several conferences (IRPS, IIRW, and IPFA).

Angyal, Matthew (5A.5)
Matt Angyal received his B. S. in Electrical Engineering from the University of Maryland in 1991 and his Ph. D. in Electrical Engineering from Cornell University in 1996. From 1996 through 2001 he worked at Motorola's Advanced Products Research and Development Lab where he contributed to the development and qualification of copper interconnect technology. He joined the IBM Semiconductor Research and Development Center in 2001, where he has been involved in the development and qualification of copper / low-k interconnect technology for advanced CMOS technologies.

Aoulaiche, Marc (XT.13)
Marc Aoulaiche received the M.S. degree in physics and Microelectronics from the University of Provence (Aix-Marseille I), France, in 2004. He pursued the Ph.D degree in microelectronics on Bias-Temperature-Instability effects in MOSFET's with high-k dielectrics and metal gates, characterization and modeling, at the Interuniversity Microelectronics Center (IMEC) and Department of Electrical Engineering, Katholieke Universiteit of Leuven (K.U.Leuven), Belgium. Since January 2009, he is working at IMEC in the Logic Device Design group. His main research topics are reliability, electrical characterization of floating body RAM devices and compact modeling.

Arreghini, Antonio (6C.1)
Antonio Arreghini was born in Udine, Italy, in 1979. He received the Laurea degree in electronic engineering and the Ph.D. degree in industrial and information engineering from the University of Udine, in 2004 and 2008, respectively. From July 2006 to July 2007, he was a visiting scientist with NXP Semiconductors, Leuven, Belgium, where he worked on the modeling and simulations of long-term retention, program, and erase transients of SONOS non volatile memories. Since 2008, he has been with IMEC, Leuven. His research topics include modeling, simulation and advanced experimental characterization of charge-trapping non volatile memory cells.

Asheghi, Mehdi (2C.5)
Dr. Mehdi Asheghi completed his Ph.D. (1999) and postdoctoral (2000) at the Stanford university conducting research in the area of nanoscale thermal engineering of microelctronic devices. He led a well-known and funded research program (2000-2006) at the Carnegie Mellon University that focused on nanoscale thermal phenomena in semiconductor and data storage devices. He is currently a consulting associate professor at the Stanford University focusing on further development of PCRAM technology. He is the author of more that 100 book chapters, journal publications and fully-reviewed conference papers.

Atkinson, Nick (3A.2)
Nicholas M. Atkinson received the B.E. degree in electrical engineering from Vanderbilt University, Nashville in 2008. He is currently working toward the M.S. degree at Vanderbilt University. His research interests include VLSI circuit design and single-event effects.

Aubel, Oliver (2D.3 , 5A.4, 5B.2, IC.9)
Dr. Oliver Aubel has earned his diploma (M.S.) in electrical engineering with focus on microelectronic engineering from the University of Hannover (Germany) in 2000. Early 2004 he finished his PhD study at the same university with focus on ultra highly accelerated electromigration testing. Immediately after that he joined GLOBALFOUNDRIES (formerly AMD) (Germany) as a reliability engineer for interconnect reliability. In his early years at GLOBALFOUNDRIES he was assigned to the IBM technology development alliance at Burlington site in Vermont, US (2004-2005) responsible for the reliability transfer to GLOBALFOUNDRIES. He is now coordinating all BEOL related reliability activities in GLOBALFOUNDRIES (Germany).

Auluck, Kshitij (NA.2)
Kshitij Auluck is currently working towards B.Tech and M.Tech in Microelectronics (dual-degree) from the Department of Electrical Engineering, Indian Institute of Technology Bombay, Mumbai, India, and will graduate in 2010. From May 2008 to July 2008, he was a student intern at Corporate Technology Group, Intel Corp., Bangalore, India. His research interests are physics, fabrication, modeling and simulation of novel semiconductor devices for logic and memory.

Baccarani, Giorgio (HV.1)
Received the B.S. degree in electrical engineering and in physics from the University of Bologna, Bologna, Italy, in 1967 and 1969. In 1969, he joined the Bell Laboratories, Murray Hill, NJ. In 1970, he became Research Assistant and in 1980 Full Professor with the University of Bologna. In 1981, 1983 and 1989 he was with the IBM T.J. Watson Research Center, Yorktown Heights, NY. He is the Director of the ARCES Research Center at the University of Bologna. His current research interests include device simulation, synthesis of analog circuits, analog and digital architectures for image processing, and integrated circuit design.

Baek, Chang-Ki (2C.4)
Received the B.S. degree in electronic engineering from Chungnam National University, Daejeon, Korea, in 1999, the M.S. degree in electronic and electrical engineering from Pohang University of Science and Technology (POSTECH), Korea, in 2002, and the Ph. D. degree in electrical engineering and computer science from Seoul National University, Seoul, Korea, in 2008. Currently he is a research fellow in the Computational Sciences Division, Korea Institute for Advanced Study (KIAS), Seoul, Korea. His current research interests are in 3-dimensional quantum transport, semiconductor device analysis, modeling and simulation of MOSFET, Silicon Nanowire and Flash memory cells, particularly on oxide reliability issues in nano-scale CMOS devices.

Baek, Rock-Hyun (2C.4)
Received the B.S. (2004) in electronics engineering from the Korea University, Korea. He received the M.S. (2006) in electronics engineering from the Pohang University of Science and Technology (POSTECH), Pohang, Korea. He is currently in Ph.D. course at POSTECH. His main research interests include CMOS devices and nanowires FET.

Bagatin, Marta (4B.3)
Marta Bagatin received the Laurea degree (cum laude) in Electronic Engineering from the University of Padova, Italy, in 2006. Since January, 2007, she has been following the Ph.D. School in Information Science and Technology at the Department of Information Engineering, University of Padova. Her research interests concern radiation and reliability effects on electronic devices, especially on volatile and non-volatile semiconductor memories. The results of her work were recognized with the Outstanding Student Paper Award at NSREC 2008, the Best Student Presentation Award at RADECS 2008, and the NPSS Phelps Award 2009.

Baghini, M. Shojaei (4D.4, EL.2)
Maryam Shojaei Baghini received M.S. and Ph.D. degrees in Electrical Engineering from Sharif University of Technology, Tehran, in 1991 and 1999, respectively. She worked for two years in industry on the design of analog and mixed-signal VLSI ICs. In 2001, she joined IIT-Bombay as a Postdoctoral Fellow, where she is currently a faculty member. She has been a Designer/Co-designer of several analog chips in industry and academia. As a part of her research in IIT Bombay, she has designed one of the most power-efficient CMOS instrumentation amplifiers for biomedical applications in 2004. Dr. Shojaei is author/co-author of 49 Int. journal & conference papers, inventor/co-inventor of seven patent applications, and author of one invited book chapter. She is a co-recipient of the best research award in circuit design at Intel Corporation AAF'08 and the third award on research and development at 15th International Festival of Kharazmi in 2002. Her team of students won the first Cadence Design Systems, Inc. Student Design Contest, among SAARC countries in 2006. Dr. Shojaei is member of Emerging Applications and Technologies sub-committee of IEEE A-SSCC and member of Low-Power Design/Circuits and Technology Track of IEEE Int. Conf. on VLSI Design. Her current research interests include device-circuit interaction in emerging technologies, high-performance low-power analog/mixed-signal/RF VLSI design and test, analog/mixed-signal/RF EDA, power management for SoCs, high-speed interconnects and circuit design with organic thin film components.

Balasubramanian, Anu (3A.1)
Anu Balasubramanian joined the Microprocessor and Graphics development (MGD) Quality and Reliability Team at Intel in Dec 2008. Her focus is in the areas of ESD, High Speed Bus reliability, and SER modeling. Anu Balasubramanian received her Ph.D. in Electrical Engineering in 2008 from Vanderbilt University. Her areas of research interests include biological sensors, microelectronic-circuit analysis and design, and effects of radiation on integrated circuits, specifically designing for radiation hardness, reliability and modeling circuit-level soft errors.

Balk, Ludwig Josef (3C.4)
Ludwig Josef Balk studied physics at the RWTH Aachen and received his diploma in 1971, hereafter he received his PhD in electrical engineering from the same university in 1976. After an intermediate stay at the University of Duisburg as academic director he joined University of Wuppertal in 1991 as full professor for electronics. Additionally he became director of the institute of polymer technology in 2005. His main research is in the area of micro and nano characterization of materials and devices.

Barbato, Marco (3F.2)
Marco Barbato was born in Dolo (Venice), Italy, in 1984. He graduated in Electronics Engineering at University of Padova in 2009 with a thesis on reliability of RF micro electro mechanical systems. He currently holds a scholarship with University of Padova, working on the development of measurement set-up in particular on RF-MEM systems. Moreover he works on development of HV systems and power converters. In his career he has coauthored 5 papers.

Bashir, Muhammad (IC.3)
Muhammad Bashir is a doctoral candidate in electrical and computer engineering at the Georgia Institute of Technology. His research interests include modeling yield and reliability of semiconductors. He has an MS in electrical and computer engineering from the Georgia Institute of Technology.

Baumann, Robert (4B.2)
Dr. Baumann heads the Texas Instruments radiation effects program and is a TI and IEEE Fellow and EDS Distinguished Lecturer with 20 years of experience in semiconductor reliability. He co-led the SIA Panel investigating the impact of the ITAR on export restrictions and was directly responsible for changes that reduced the risk of U.S. commercial electronics becoming inadvertently controlled. He led and was one of the authors of the JEDEC JESD89 test standard for radiation testing of commercial microelectronics and was awarded the JEDEC Chaiman's Award. He has published > 55 papers, two book chapters, and holds eight U.S. patents.

Bauza, Daniel (3C.2)
Daniel Bauza received the M.S. degree in electronics and the Ph. D degree in Energetics and Complex Systems Dynamics, both from the University of Paris XII Créteil-France in 1982 and 1986, respectively. In 1986, he was an engineer with CNRS and the "Laboratoire de Physique des Composants à Semiconducteurs, Grenoble-France", which became the Institute of Microelectronics, Electromagnetism and Photonics (IMEP) in 2001. Since 2003, he has been an Associate Scientist with CNRS. Currently with the IMEP-LAHC, MINATEC, his research interests are on the characterization of defects in silicon devices, on the electrical characterization of silicon-insulator interface traps, on the electrical properties of MOSFET's gate insulator, and on the development of related characterization tools.

Beltrami, Silvia (4B.3, 5C.2)
Silvia Beltrami was born in Bergamo, Italy, in 1978. She received the Laurea degree in Electronic Engineering from Politecnico di Milano, Italy, in 2003. She joined STMicroelectronics in 2003 and now she is working in the R&D Technology Center of Numonyx. Her work is about the reliability improvement of flash memory and, in particular, NAND and multilevel devices. She is a co-author of various scientific papers about flash memories.

Benakanakere Sheshadri, Vijay (SE.3)
Vijay B Sheshadri received the B.E. degree in Electronics & Communication engineering from BMS College of Engineering, Bangalore, India (affiliated to the Vishweshwariah Technological University, Belgaum, India) in 2006 and is currently working toward the M.S. degree in electrical engineering at Vanderbilt University, Nashville, TN,

Bender, Hugo (6A.3)
Hugo Bender is team leader of the Structural Analysis (SA) team in the Materials and Components Analysis (MCA) group at imec. He has over 25 years experience in materials characterization for semiconductor applications, in particular with transmission electron microscopy, scanning electron microscopy, focused ion beam, Auger spectroscopy, spectroscopic ellipsometry, infrared spectroscopy. In these fields he is author/co-author of about 200 journal papers, 280 conference contributions and 3 book chapters. His current interests are focused on the application of transmission electron microscopy for strain analysis, electron tomography, and chemical analysis by STEM/EDS/EELS/ELNES on device structures for next generation technologies.

Benvenuti, Augusto (5C.4)
Augusto Benvenuti received the Ph.D. degree in electronic engineering from Politecnico di Torino in 1993. During 1991-1992 he was a research visitor at Bell Labs, Murray Hill; in 1993 he was at the Centre d'Electronique de Montpellier. During 1994 he worked at ETH Zuerich on the implementation of the hydrodynamic model in the device simulation program Dessis. Since 1995 he works in Agrate, initially with STMicroelectronics and later with Numonyx R&D, where he is currently managing TCAD activities. He has been involved in the support of several EEPROM, NOR, NAND and PCM technology platforms, and has coauthored about fifty papers.

Bersuker, Gennadi (2B.4, 4A.4, 4F.4)
Gennadi Bersuker (M'05) received the M.S. degree in physics from Leningrad State University, St. Petersburg, Russia, and the Ph.D. degree in physics from Kishinev State University, Kishinev, Moldova. He was with the Moldavian Academy of Sciences, Kishinev; Leiden University, Leiden, The Netherlands; and the University of Texas, Austin. Since 1994, he has been with SEMATECH Inc., Austin, TX, where he is working on processinduced charging damage, electrical characterization of Cu/low K interconnect, high-κ gate dielectrics, and advanced CMOS process development

Beyer, Gerald P. (5B.4)
Gerald Beyer is the program manager of Cu/low-k program. His background is sputter deposition. He earned a PhD in Materials Science from Imperial College, London.

Bhatia, Charanjit (4E.4)
Prof Bhatia, Charanjit Singh received his MS and PhD in Electrical Engineering from University of Minnesota, Minneapolis MN USA in 1978 & 1979 respectively. Prof Bhatia joined NUS as a Professor of Electrical & Computer Engineering and also has 25% appointment in the Institute of Materials Research & Engineering (IMRE). Prof Bhatia was a Temasek Professor in NUS from 2001 to 2005. He worked as a Senior Technical Staff Member (STSM) in the Advanced Magnetic Storage Lab (AMRL) of IBM and Hitachi GST. Prof Bhatia was awarded the coveted IBM's Faculty award in 2008 for his project on Fabrication, Characterization and Performance of Thin Film Si Photovoltaic (PV) Cells. He also set up a Joint Study agreement (JSA) with IBM's T J Watson Jr Research lab at Yorktown Heights, NY, to work on the Si based photovoltaic cells.

Bhuva, Bharat (3A.2, 6E.2, SE.3)
Bharat L. Bhuva received the B.S. degree in electrical engineering from Maharaja Sayajirao University of Baroda, Baroda, India, in 1982 and the M.S. and Ph.D. degrees in electrical engineering from North Carolina State University, Raleigh, in 1984 and 1987, respectively. He is currently an Associate Professor of electrical engineering with the Department of Electrical Engineering and Computer Science, Vanderbilt University, Nashville, TN. His research interests include radiation effects on electronics, biosensors, optical-signal transmission using all Si process, and cognitive science.

Bisht, Gaurav (NA.2)
Gaurav Bisht obtained his M.Sc (2007) in physics from Indian Institute of Technology (IIT) Roorkee. Since 2007 he is pursuing M. Tech. in microelectronics from department of Electrical Engineering at IIT Bombay India. His research interests are in the areas of non-volatile memories, device reliability and characterization.

Bisschop, Jaap (CR.3)
Jaap Bisschop received the M.Sc. in physics in 1978 and Ph.D. in electrical engineering in 1983. He worked on oxide reliability at the University of Groningen, The Netherlands. In 1986 he joined Philips Research, where he worked on device physics and reliability of solid state image sensors. In NXP Semiconductors he has been responsible for coordination of company reliability activities. His current focus is on wafer level reliability, building in reliability and automotive reliability. He is convenor of the IEC TC47 Wafer Level Reliability working group. He is a member of the ESREF steering committee.

Bittel, Brad (IC.13)
Brad Bittel received his B.S. degree in Electrical Engineering from the Pennsylvania State University in 2007 and is now pursuing a Ph.D. in Materials Science and Engineering also from the Pennsylvania State University.

Blaauw, David (5F.1)
David Blaauw received his B.S. in Physics and Computer Science from Duke University in 1986, and his Ph.D. in Computer Science from the University of Illinois, Urbana, in 1991. Until August 2001, he worked for Motorola, Inc. in Austin, TX, were he was the manager of the High Performance Design Technology group. Since August 2001, he has been on the faculty at the University of Michigan where he is a Professor. He has published over 300 papers and hold 30 patents. His work has focussed on VLSI design with particular emphasis on ultra low power and high performance design. He was the Technical Program Chair and General Chair for the International Symposium on Low Power Electronic and Design. He was also the Technical Program Co-Chair of the ACM/IEEE Design Automation Conference and a member of the ISSCC Technical Program Committee.

Bluet, Jean-Marie (2E.3)
Jean-Marie BLUET his PhD from University of Montpellier II in 1997, with speciality in condensed matter. He has then worked in LMGP lab in Grenoble and in CEA LETI as a post doc . He then obtained an assistant professor position at INSA de Lyon in 1999. His research activity is focused on wide band gap semiconductors materials and devices characterizations (SiC and GaN). At present, he also performs spectroscopic studies of semiconductors at the nanoscale with investigations of Si and SiC nanoparticules and porous SiC. He is author and co-author 82 publications in international journals and conferences.

Boit, Christian (IC.2)
Prof. Dr. Ing. Chris Boit is dean of the faculty for CS&EE in the TUB Berlin University of Technology (Germany) and head of the semiconductor device department with special focus on functional analysis and debug of Integrated Circuits and semiconductor devices.   From 1986 to 2002 he has been with Siemens AG's Research Laboratories and Semiconductor Group, and Infineon Technologies, respectively. 1990 to 1993 he participated in the joint IBM-Siemens 64M DRAM development team in East Fishkill NY. In the following years, he took responsibility as Manager and then Director of Failure Analysis. He received the Physics Diploma and the PhD in Electrical Engineering from TUB Berlin. In the field of electronics, Chris has been co-founder of several international

societies on failure analysis like EDFAS and EUFANET and has served in their Boards of Directors. He has been supporting the respective international conferences for many years, for ISTFA as Seminar Chair in 2000 and General Chair in 2002. He has also been co-editor of the Electronics Failure Analysis Desk Reference. He is member of the German Academy of Science and Engineering, ACATECH. In the German association of electrical engineers, VDE, he is speaker of the section for reliability and failure analysis of microelectronics.

Boku, Katsushi (6C.4)
Katsushi Boku received a B.Sc. degree in material science from University of Electro Communications (Tokyo, Japan) in 1988. Katsushi joined Texas Instruments Japan on 1989, and has worked for DRAM development. In 2002, Katsushi moved to the US and became involved in FeRAM reliability and yield analysis. He has been issued 10 patents and has coauthored several technical papers. In his spare time, Katsushi loves playing golf and soft ball.

Boselli, Gianluca (4D.3)
Gianluca Boselli (MS '96, Ph.D '01) joined Texas Instruments, Dallas, Texas, in 2001, where was responsible for advanced CMOS ESD/Latch-up development. Recently his responsibilities extended into ESD development of Analog technologies.
He authored several papers in the area of ESD/Latch-up. He presented his work at major conferences, including EOS/ESD Symposium, IEDM, and IRPS. He also presented several invited papers and/or tutorials at the EOS/ESD Symposium, IRPS, IEDM, ESREF and RCJ. Dr. Boselli has been the recipient of the "Best Paper Award" on behalf of Microelectronics Reliability Journal in 2000. He received "The Best Paper Award" at the EOS/ESD Symposium 2002. He also received the "The Best Presentation Award" at the EOS/ESD Symposium in 2002 and in 2006. Dr. Boselli is a member of ESD Association and an IEEE senior member. Dr. Boselli serves on the TPC of the EOS/ESD Symposium, IRPS and ESREF. Dr. Boselli has served as TPC Chair at the EOS/ESD Symposium 2006, Vice-General Chair at the EOS/ESD Symposium 2007 and General Chair at the EOS/ESD Symposium 2008.

Bosman, Michel (4A.1, BD.2)
Michel Bosman received his M.Sc. degree in Materials Science from the Delft University of Technology, Delft, Netherlands, and Ph.D. degree in Electron Microscopy from the University of Sydney, Australia. He is currently with the Institute of Microelectronics, A*STAR, Singapore. His research interests include nanoscale optics and atomic-resolution spectroscopy using transmission electron microscopes.

Bounasser, Mounaim (3A.3)
Mounaim Bounasser received his B.E. and M.S. in Electrical Engineering from University of Montpellier II, Montpellier, France, in 2004 and 2006, respectively. He has previously worked on radiation effects on circuits and radiation hardened circuit design at Vanderbilt University and with Ridgetop Group Inc. He is currently working on radhard circuit design and simulation with Robust Chip Inc.

Breuer, Taro (IC.2)
Taro Breuer received his diploma degree in electrical engineering (equivalent to M.S.) in 2006 from TU Berlin (Berlin University of Technology). In 1998-2003 he studied electrical engineering with focus on telecommunications engineering at TU Hamburg-Harburg. His PhD research at TU Berlin in cooperation with Advanced Micro Devices (AMD) Dresden (now GLOBALFOUNDRIES) focuses on ultra-low-k dielectric reliability .

Brewer, Forrest (IC.10)
Professor Brewer (B.S (Physics), Caltech; PhD (Computer Science) ,UIUC) joined the UCSB faculty in 1988, He was formerly a consulting engineer and a senior engineer at Northrop Corp. Advanced Technology Division. His research interests are in mixed signal VLSI design as well as computer aided design tools and analysis. His research in mixed signal design involves investigating solutions for data transmission for inter-chip and intra-chip communications which includes the application of communications and signal processing theory and techniques to analyze high-speed I/O links and looking into ESD solutions for mixed signal I/Os. Recent work includes development of a family of specialized microprocessors for low-power/ high-performance embedded closed loop control. This work spans mixed signal design at the sensor and actuator interfaces to multi-threaded digital system design in the digital processing parts. He teaching interest includes VLSI design and coupled tools development based on systematic and formal approaches as well as heuristic analysis. Other interest includes low Power Control: to drastically reduce the power consumption of digital feedback controller which have become ubiquitous in embedded systems used in automotive,cars, appliances and portable phones.

Bru-Chevallier, Catherine (2E.3)
Catherine Bru-Chevallier received her PhD from Paris XI University. She is currently senior scientist (CNRS) at Lyon Institute of Nanotechnology (INL), head of "Spectroscopy and Nanomaterials " group and of « Material Department » at INL. She is mainly involved in optical spectroscopy of III-V heterostructures and nanostructures using photoluminescence as well as photoreflectance spectroscopy. She is developing local optical spectroscopy in order to characterize operating devices (transistors) as well as to study radiative emission from single nanostructures. She is author and co-author of more than 120 publications in international journals and conferences.

Butera, Geni (6A.3)
Geni Butera was born in Lamezia Terme, Italy, in 1982. He received the Bachelor's degree in electronics engineering and the Master's degree in electronic engineering cum laude from University of Calabria, Rende, Italy, in 2006 and 2009, respectively. From October 2008 to April 2009, he was with imec, Leuven, Belgium, as Internship student (Erasmus Project) for Katholieke Universiteit Leuven, Belgium, focusing on the topic: electrical characterization of advanced MOS devices with Cu contacts.

Bychikhin, Sergey (4D.4)
Sergey Bychikhin studied physics at Moscow State University and received Ph.D. in 2000. During his study he was specialized on noise phenomena in electronic systems and devices. His Ph.D. thesis was dedicated to fluctuations in scanning tunneling microscope. Since 2000 he is with device characterization group at Vienna Technical University. He is engaged in thermal and electrical characterization of devices with optical methods and in thermal modeling.

Cagli, Carlo (5D.1)
Carlo Cagli was born in Naples, Italy, in 1984. He received the B.S. and M.S. degrees (cum laude) from Università degli Studi di Milano, Italy, in 2005 and 2007, respectively. For his M.S. thesis, he worked on resistive-switching memories (RRAMs) within the Dipartimento di Elettronica e Informazione, Politecnico di Milano, Milano, where he is currently working toward the Ph.D. degree. In 2009 he was a User at the Lawrence Berkeley National Laboratories, Berkeley, California (USA). His research interests include the characterization and modeling of electrical and physical properties in RRAM devices as well as nanowire-based nanostructure for memory applications.

Calderoni, Alessandro (6C.2, MY.6)
Alessandro Calderoni received the second (master) level Laurea degree in Electrical Engineering (cum laude) from Politecnico di Milano, Italy, in 2006, with an experimental thesis on MOSFET's noise under microwave irradiation. In 2006 he joined the Research and Development department of STMicroelectronics (now Numonyx) within the Compact Modeling team. Since that, he has been working on advanced CMOS modeling and Floating Gate-based Memory noise characterization and modeling. Than he worked on Phase Change Memories transport mechanism and low-frequency noise. His current research interests include statistical physical characterization and reliability of PCM over large arrays.

Camozzi, Elisa (MY.5)
Elisa Camozzi was born in Milan, Italy in 1980. In 2004 she received her Master Degree in Physics (cum laude) from the Università degli Studi Milano-Bicocca, Milano, Italy, as a result of one full-time year of research work in the field of biophysics. In 2008 she received a Bachelors degree in Philosophy (cum laude) from the Università degli Studi di Milano, Milano, Italy, with a dissertation about human mind, knowledge and the best way to describe both of them. From 2005 she works in the R&D of STMicroelectronics (now Numonyx) in the field of NOR/NAND memories.

Cannon, Ethan (SE.2)
Ethan Cannon is a Radiation Effects Engineer with the Solid State Electronics Development organization in Boeing Research & Technology in Seattle, where he works on characterization and mitigation of radiation sensitivity of advanced deep sub-micron technologies. Previously, he worked as a Reliability Engineer focusing on radiation-induced soft error simulations, measurements and qualifications in the IBM Systems and Technology Group in Essex Junction, Vermont. He received the B.S. degree in engineering physics from the University of California, Berkeley, and the M.S. and Ph.D. degrees in physics from the University of Illinois at Urbana-Champaign.

Cao, Yiqun (4D.1)
Yiqun Cao studied electrical and electronic engineering at the Rheinisch-Westflische Technische Hochschule (RWTH) Aachen in Germany. He received his Dipl. Ing. degree in 2007. The same year he joined smart power technology R&D at Infineon Technologies in Munich, where he is currently working on his Ph.D. thesis in close collaboration with the Institute for On-Board Systems Lab at the Technische Universität Dortmund. His field of research is ESD protection concepts in high voltage automotive technologies, with focus on chip- and system-level ESD.

Carine, Besset (2B.2)
Carine Besset received the Engineer Degree in physics (2001) from the INPG (Institut National Polytechnique de Grenoble, France). She joined STMicroelectronics (Crolles, France) in 1992, where she worked on back-end dielectric process developments and industrialization during 9 years. Since 2001, she is in the Reliability Group where she is involved in process development-qualification and process support for reliability aspects in CMOS technologies, Analog and derivative applications. Her main interests are BEOL dielectrics reliability and front-end device reliability.

Cartier, Eduard (2B.1, 4A.3, BD.3)
Eduard Albert Cartier was born in Switzerland in 1951. He earned a bachelors degree (1971), a masters degree (ETH, Zurich, Switzerland, 1977) and Ph.D degree (Dr.sc.nat., granted with honors) from ETH in 1982. From 1984 till 1987 he worked as a research staff member at the ASEA Brown Boweri (ABB) Research

Center in Baden-Dattwil, Switzerland. Since 1988, he works as a research staff member of the IBM Research Division at the T.J. Watson Research Center in Yorktown Heigths, NY, USA. Over the last 10 years he has been working on the development of alternative, high-k gate dielectrics for CMOS applications.

Catthoor, Francky (SE.1)
Francky Catthoor received a Ph.D. in EE from the Katholieke Univ. Leuven, Belgium in 1987. Between 1987 and 2000, he has headed several research domains in the area of high-level and system synthesis techniques and architectural methodologies, including related application and deep submicron technology aspects, all at IMEC Leuven, Belgium. Currently he is an IMEC fellow. He is also part-time full professor at the EE department of the K.U.Leuven. He has been elected IEEE fellow in 2005.

Cellere, Giorgio (4B.3)
Giorgio Cellere got his MS and Ph.D. degrees in 1998 and 2002 at Padua University. As a post-doc researcher at Padua University, he worked on ultrathin gate oxides reliability, on radiation effects on advanced nonvolatile devices, and on microelectronic devices for biological applications. He is now with Applied Materials Baccini, Treviso, Italy, where he is working at the development of advanced tools and processes for crystalline silicon solar cells. He holds more than 100 scientific publications, including two book chapters and ten patents. He served as Reviewer and Session Chair in international conferences and his research works won several prices.

Ceric, Hajdin (IC.6)
Hajdin Ceric was born in Sarajevo, Bosnia and Herzegovina, in 1970. He studied electrical engineering with the University of Sarajevo, Sarajevo, and received the Dipl.Ing. degree in 2000 and the Ph.D.degree in technical sciences in 2005 from Technische Universität Wien, Vienna, Austria. In June 2000, he joined the Institute for Microelectronics, Technische Universität Wien, where he is currently a Postdoctoral Researcher. His scientific interests include interconnect and process simulation.

Cester, Andrea (4F.2)
Andrea Cester received the degree (magna cum laude) in electronic engineering and the Ph.D. degree in electronic and telecommunication engineering from the Padova University, Italy, in 1998 and 2002, respectively. He is currently an Assistant Professor with the Department of Information Engineering, Padova University. He is the author of more than 120 papers published in international journals and conference proceedings. His previous research interest included the reliability issues of deep-submicrometer CMOS devices and advanced nonvolatile memories. His current research interests are characterization, reliability, and modeling of organic electronic devices such as OTFT, OLED, and organic and hybrid solar cells.

Chancellor, Cathy (4A.6, 5E.5)
Cathy Chancellor received her Bachelor of Science degree in Engineering Technology from the University of Arkansas at Little Rock in 1983. She is a Member of the Group Technical Staff in the Texas Instruments External Development and Manufacturing Reliability group. She has been with TI for over 25 years and is responsible for reliability testing for TI's leading edge technology development. She is co-author on ten papers covering TDDB, NBTI, and process integration and co-inventor on one patent.

Chang, C. Y. (CD.3)
C. Y. Chang received the Ph.D. degree in 2006 in physics from the National Central University, Taiwan, R.O.C. During 2005–2006, he was a visiting student in Department of Chemical Engineering, University of Florida, Gainesville. He is a postdoctoral associate in the Department of Materials Science & Engineering, University of Florida, Gainesville, FL. His interests are in growth of carbon nanotube and GaN nanowires, and characterization of III-V compounds materials and devices. He has published over 30 journal publications.

Chang, Wing L. (BD.3)
Wing L. Chang received the bachelor of science degree in Mathematics and Chemistry from Lyon College in 1995 and the PhD in Physical Chemistry/Material Science from the University of North Carolina at Chapel Hill in 1999. Currently, she is a technology reliability engineer with the IBM Technology Group in Hopewell Junction, NY.

Chao, Yuan-Peng (MY.3)
Yuan-Peng Chao was born in HsinChu, Taiwan, R.O.C., in 1984. He received the B.S. and M.S. degrees in electrical engineering from National Chiao-Tung University, HsinChu, Taiwan, R.O.C., in 2006 and 2008, respectively. In 2008, he joined Macronix International Co., Ltd. (MXIC), Hsinchu, Taiwan, R.O.C., and has been with the Device Engineering Department, where he has engaged in MOS device characterization and modeling, interconnect capacitance characterization, and high-voltage device modeling.

CHAPPAZ, CEDRICK (IC .1)
Cedrick Chappaz received the M.S. in electrical and optronics engineering from the Ecole Nationale Superieure d'Ingenieur (ENSI CAEN) in 1998. Afterwards, he pursued a Ph.D. in the field of Microsystems. He obtained its Ph.D. in 2003. its expertise in MEMS, integrated optics and characterization lead him to do a post-doctorate year in the Commissariat a l'Energie Atomique (CEA) in LETI laboratories from 2003 to 2004. Then, he joined STMicroelectronics as an expert in Reliability and Characterization fields in 2004. Since this date, he's in charge of the qualification of BEOL of advanced technologies and specific 3D integration.

Chatterjee, Amitabh (EL.5, IC.10)
Amitabh Chatterjee graduated from IIT Powai with (5yrs Int.) M.Tech in electrical engineering and subsequently, he joined as a scientist at the Centre for Advanced Technology, Indore in the Laser Physics Division in 1996. His work involved development of ultra fast avalanche transistor based HV Marx Bank Circuit ands its impact on the stability of ultra-short laser pulses. His research included analysis of bipolar snapback due to low doped resistive drift region of a HV BJT under avalanche break-down and study of avalanche injection mechanisms from the ohmic contacts. Currently he is working on his doctoral dissertation on the reliability of Drain Extended NMOS (DeNMOS) under ESD stressing at the Univ. Of CA, Santa Barbara.

Chatty, Kiran (EL.3)
Kiran Chatty received his B. Tech in Electrochemical Engineering in 1995 from Central Electrochemical Research Institute (CECRI), India, and M.S. in 1996 in Materials Science from New Jersey Institute of Technology (NJIT), Newark, NJ and his Ph.D. in 2001 in Electrical Engineering from Rensselaer Polytechnic Institute (RPI), Troy, NY. He is a Senior Engineer in IBM's Semiconductor Research and Development Center (SRDC) focusing on ESD and Latchup development in digital and analog & mixed signal CMOS and BiCMOS technologies. Kiran served as the chair of the ESD/Latchup session at the 2006 IRPS, co-chair of the 2009 IRPS, chair of the Bipolar, RF and HV ESD session at 2009 EOS/ESD symposium, served as a member of EOS/ESD symposium and IRPS technical committees. He has authored or co-authored over 35 technical papers in journals/conferences, has 10 issued patents and many patent applications.

Chen, An (2C.2)
An Chen is a Member of Technical Staff at GLOBALFOUNDRIES. He is working in the Strategic Technology Group on emerging logic and memory technologies. He is an assignee of the Nanoelectronics Research Initiative (NRI) within SRC, and collaborates with university research groups to develop beyond-CMOS logic solutions. An Chen received his Ph.D. in Electrical Engineering from Yale University in 2004. Before joining AMD (currently GLOBALFOUNDRIES) in 2007, he worked in the Advanced Memory Development Group at Spansion LLC.

Chen, Fen (5A.5)
Fen Chen received his Ph.D. degree in Electrical Engineering in 1998 from University of Delaware. From 1997 to 1998, he was with IBM System Group at Rochester, MN and Intel Component Research at Santa Clara, CA as a graduate intern working on system stress and IC interconnect reliability. He joined IBM microelectronics at Essex Junction, VT in 1998 and has worked on semiconductor technology reliability issues since that time. During the past several years he has focused on low-k ILD TDDB issue for various IBM and IBM Alliance development programs.

Chen, Kangguo (XT.14)
Dr. Kangguo Cheng is currently a Lead Engineer of IBM Research at Albany Nanotech, responsible for leading the exploration of fully depleted devices for 22nm node and beyond. He received a B.Eng. and a M.Eng. from Tsinghua University, China, and a Ph.D. degree from University of Illinois at Urbana-Champaign (UIUC). Since joining IBM in 2001, he has worked on process/device integration in a variety of advanced semiconductor technologies including memory, embedded memory, bulk, SOI, finFET, ETSOI, etc. He is a Master Inventor of IBM with over 200 patents and patent applications. He has over 50 peer-reviewed journal/conference publications.

Chen, Ke-Hung (CD.3)
Ke-Hung Chen is currently working toward the Graduate degree in the Department of Chemical Engineering,University of Florida, Gainesville. He is the author or coauthor of more than 12 papers published in refereed journals. His current research interests include advanced wide bandgap sensors and III-V compound electronics reliability.

Chen, Kuan-Fu (5D.3)
Kuan-Fu Chen was born in Taipei, Taiwan, R.O.C. on September 21, 1978. He received the B.S. degree in physics from National Taiwan University, Taipei, Taiwan, R.O.C. in 2000 and the M.S. degree in electrical engineering from National Taiwan University, Taipei, Taiwan, R.O.C. in 2002. He joined Macronix International Company, Ltd., Hsinchu, in 2002 as a Process Integration Engineer. He worked on process integration and devices characteristics analysis for nonvolatile floating-gate-type and nitride storage flash memory.

Chen, Kuang-Chao (5D.3, MY.1, MY.3)
Kuang-Chao Chen received the M.S. degrees in chemistry from the National Chong-Shan University, Taiwan, in 1987. From 1989 to 1995, he joined Electronic Research and Service Organization (ERSO), Hsinchu, Taiwan, where he has been involved in the development of BEOL planarization process technology. From 1995 to 1998, he was with Mosel-Vitelic International Co., Ltd, Hsinchu, Taiwan. He performed yield improvement in manufacturing line. In 1998, he joined Vanguard International Semiconductor Co., Ltd, Hsinchu, Taiwan, as a department manager. He was responsible for thin film module development. In 2000, he joined Macronix International (MXIC), where he worked on advanced module

development. He is currently Executive Director of Technology Development Center.

Chen, Ming-Shiang (5D.2, 5D.3)
Ming-Shiang Chen was born in Tainan, Taiwan, R.O.C. on September 22, 1970. He received the B.S. and M.S. degrees in electronics engineering from National Chiao-Tung University, Hsinchu, Taiwan, in 1992 and 1994, respectively. He joined Macronix International Company, Ltd., Hsinchu, Taiwan, in 1996 as a Device Engineer. From 1996 to 1998, he has worked on nonvolatile memory devices characteristics analysis. Since 1998, he has been engaged in the development floating-gate Flash memory technology and Nitride-based NBit technology.

Chen, Wen-Yi (EL.6)
Wen-Yi Chen received the B.S. degree and the M.S. degree both from the the Institute of Electronics, National Chiao-Tung University, Hsinchu, Taiwan, in 2003 and 2005, respectively. In 2005, he joined the Circuit Design Department, SoC Technology Center, Industrial Technology Research Institute (ITRI), Hsinchu, as a circuit design engineer. In 2006, he joined the Amazing Microelectronic Corporation and worked with system-level ESD protection design. He is currently working toward the Ph.D. degree in the Institute of Electronics, National Chiao-Tung University, Hsinchu, Taiwan.

Chen, Yi Ning (2C.3)
Yi Ning Chen received the B.Eng degree from Zhejiang University, China, in 2004, and the M.Eng in materials for microelectronics in Catholic University of Leuven, Belgium, in 2005. He joined IMEC (Inter-university MicroElectronic Centre) for student internship in 2005. Since 2007, he is pursuing his Ph.D in Electrical and Electronic Engineering at Nanyang Technological University, Singapore.

Chen, Yin-Jen (5D.3)
Yin-Jen Chen was born in Taipei, Taiwan, R.O.C. on October 30, 1976. He received the B.S. and M.S. degrees in electronics engineering from National Chiao-Tung University, Hsinchu, Taiwan, R.O.C. in 1999 and 2001, respectively. He joined Macronix International Company, Ltd., Hsinchu, in 2001 as a Process Integration Engineer. From 2001 to 2003, he worked on process integration and reliability testing for floating-gate-type Flash memory. Since 2004, he worked on nitride stroage memory and focus on the optimization of operation condition.

Chen, Zuhui (NA. 1)
Zuhui Chen received the B.S. degree in physics from Fujian Normal University and M.S. degree in solid-state physics from Xiamen University, Fujian, China, respectively in 1998 and 2001, and Ph.D. in engineering science from University of Florida, FL, USA, in 2005. In 2006, he was a faculty of the Pen-Tung Sah MEMS Research Center, Xiamen University, China. In 2007, he joined Nanyang Technological University as a research fellow with Lee Kuan Yew Postdoctoral Fellowship. His current interests include NBTI and interface-trap modeling in silicon MOSFETs.

Cheng, Cheng-Hsien (5D.3)
Cheng-Hsien Cheng was born in Yunlin, Taiwan, ROC., on December 11, 1982. He recieved the BS and MS degree in Engineering and System Science from National Tsing Hua University, Hsinchu, Taiwan, in 2005 and 2007. In 2007, he joined the Advanced Device Department of Macronix International Company Ltd., Hsinchu, Taiwan. His research interests include the device characterization of Flash memory devices.

Chi Wen, Soo (3C.5)
Chi Wen was born in Johor, Malaysia. He received his B.Sc (Merit) and B.Sc (2nd upper Hons.) in Materials Science from National

University of Singapore in 2000 and 2001 respectively. He is still pursuing a Master degree in Management Science in National University of Singapore. He is currently working in GLOBALFOUNDRIES Singapore and responsible for Semiconductor chip Failure Analysis work. His strength is Transmission Electron Microscopy (TEM) Imaging Analysis. His area of interests include TEM Imaging techniques and Management Science. He is one of the authors of 1paper in In-situ TEM studies published in Applied Physics Letter.

Chihiro, Uchibori (3A.5)
Dr. Chihiro J. UCHIBORI received his Ph. D. degree from Kyoto University, before he started his carrier in Fujitsu Laboratories LTD in 1996. In 2002, he joined a project in the University of Texas at Austin as a visiting scientist. Since 2004, he is working for Fujitsu Laboratories of America as a Senior Researcher. His interests are in the electrical and mechanical reliability of ULSI devices. Not only evaluating the reliability and the manufacturing yield, he is also working on clarification of the reliability degradation mechanism by experiment and simulation.

Chih-Yuan, Lu (5D.4)
Chih-Yuan Lu received B.S. degree from National Taiwan University in 1972, and Ph.D. degree in physics from Columbia University, NYC, in 1977. In Dr. Lu has been a professor in National Chiao-Tung Univ. and with AT&T Bell Labs from 1984-1989; later joined ERSO/ITRI in 1989 as a Deputy General Director responsible for the MOEA grand Submicron Project. This project successfully developed Taiwan first 8-inch manufacturing technology with high density DRAM/SRAM. He was therefore granted the highest honor prize--National Science & Technology Achievement Award by the Prime Minister of ROC, due to his leadership and achievement in this Submicron Project. In 1994, Dr. Lu becomes the co-founder of Vanguard International Semiconductor Corporation, which is a spin-off memory IC Company from ITRI's Submicron Project. He was the VP of Operation, VP of R&D, and later President from 1994-99. Dr. Lu now is the founding chairman and CEO of Ardentec Corp. a VLSI testing service company; and also serves Macronix International as a Senior VP/CTO, and now the President. Dr. Lu led MXIC's technology development team to successfully achieve the state of the art nonvolatile memory technology and now responsible for MXIC's overall operation. Dr. Lu has published more than 300 papers and has been granted 140 worldwide patents, and was elected a Fellow of IEEE, and a Fellow of APS. He also received IEEE Millennium Medal, and the most prestige semiconductor R&D Award in Taiwan from Pan Wen Yuan Foundation.

Chikhaoui, Walf (2E.3)
Walf Chikhaoui was born in Tunisia in 1983. He obtained his Master degree at the National Institute of Applied Science "INSA Lyon" in 2007. He is currently PhD student at Institute of Nanotechnology of Lyon (INL). His work consists in the analysis of degradation mechanisms of AlGaN/GaN and AlInN/GaN HEMT structures by electrical and optical methods.

Child, Craig (5A.5)
Craig Child received a Ph.D. degree in Chemical Physics in 1996 from The University of Texas at Austin. From 1996 to 2001, he was with 3M Corporation in Austin, Tx working as a laser physicist in the Electronic Products Division. In 2001, he joined Intel Corporation in Hillsboro, OR as a backend of line integration engineer working on 130nm, 90nm, 65nm, and 45nm technology nodes. In 2007, Craig joined the IBM ASTA/ISDA alliance as an AMD employee working on ULK integration in the 45nm and 32nm technology nodes.

Chin, Melida (4C.6, 5A.4)
Melida Chin received her M.S. and Ph.D. in Mechanical Engineering from the University of Michigan in 2001 and 2005, respectively. After graduation, she joined the Technology and Development group in GLOBALFOUNDRIES (formerly AMD), where she currently works as a Senior Technology and Integration Engineer. Her research work focuses on thermo-mechanical analysis and modeling of subjects such as chip-package interaction and reliability of back-end-of-line (BEOL). Before attending graduate school, she worked as a general engineer at the Panama Canal for the design, build, repair and improvement of electrical and mechanical infrastructure systems.

Chioko, Kaneta (4A.5)
Chioko Kaneta received the B.S., M.S., and D.S. degrees in Physics from Tohoku University, Sendai, Japan in 1980 and 1982, and 1985, respectively. She joined Fujitsu Laboratories Ltd., Japan in 1985 and has been engaged in atomic scale simulations of defects in Si crystal, gate dielectrics, and the interfaces between them for advanced silicon devices. She is a member of the Japan Society of Applied Physics and The Physical Society of Japan. She received a fellow award from the Japan Society of Applied Physics for contributions in the atomic scale simulations.

Chiu, Jung-Piao (MY.3)
Jung-Piao Chiu was born in Kaohsiung, Taiwan, R.O.C., in 1985. He received the B.S. degree in electronics engineering from National Chiao-Tung University, Hsinchu, Taiwan, in 2007, where he is currently working toward the Ph.D. degree in electronics engineering. His main research interests are in the field of reliability analysis in high-k devices and Monte Carlo simulation.

Chloé, Guerin (2B.2)
Chloé Guérin obtained the Engineer degree in materials and microtechnologies from the Institut National des Sciences Appliquées (INSA), Toulouse, France, and the M.S. in nanophysics from the Université Paul Sabatier, Toulouse, France in 2005. She worked in the Reliability Group of STMicroelectronics, Crolles, France, in collaboration with the Institut Matériaux Microélectronique Nanoscience de Provence (IM2NP), Toulon, France to receive her Ph.D in microelectronics from the Université de Provence, Aix-Marseille I, France in 2008. The aim was to study Hot Carrier degradation on advanced CMOS technologies under both static and dynamic supply voltage. She is currently working in Photovoltaic for Roth and Rau Switzerland, on high efficiency c-Si cells development.

Cho, Moonju (XT.9, XT.10, XT.13)
Received the M.Sc. and the Ph.D. degrees in materials science and engineering from Seoul National University, Korea, in 2003 and 2007, respectively. Her study was focused on the physical properties of ALD-HfO2 films and reliability problems based on chlorine/carbon residues.In 2005, she stayed at IMEC, Leuven, Belgium, working on TDDB and charge pumping. From 2007 till February 2009, she did postdoctoral study at IMEC in theoretical modeling for oxide trap characterization. She is currently working at the Reliability Group, IMEC. Her current interest is on hot carrier injection and trap characterization study on ultrathin high-k gate oxide.

Cho, Myoung Kwan (MY.7)
Myoung Kwan Cho received the B.S. degree in E.E. from Yonsei University, in 1988, and the M.S. and Ph.D. degrees in E.E. from Pohang University of Science and Technology 1997 and 2000, respectively. In 1988, he joined Samsung Electronics Company, Korea, where he had been engaged in the development of 2 Tr. EEPROM, Mask ROM, OTP-EPROM, NOR flash, SONOS, NAND flash. Since 2004, he had worked in Korea Intellectual Patent Office as a patent examiner for Non volatile semiconductor memory patent. And received the B.S degree from KNOU and M.S. degree in law

from Yonsei University in 2007, 2008, respectively and in pursuit of Ph.D in law. Since joined Hynix Semiconductor in 2009, he has been working on device and process integration of SLC and 3bit/cell NAND in 32, 41nm technology.

Choi, Gil-Bok (2C.4)
Received the B.S. (2004) and Ph.D (2010) in electronics engineering from the Pohang University of Science and Technology (POSTECH), Pohang, Korea. His main research interests include CMOS devices and RF measurement.

Choi, Hyun Ki (5C.1)
Hyun Ki Choi was born in Busan, Korea, on May 3, 1980. He received the B.S. degree in electronic engineering from Inha University, Korea, in 2006. He has been working for Samsung electronics, Gyunggi-Do, Korea since 2006 and has been in a Flash Memory Process Architecture team and worked in 42nm 16Gb NAND flash and 35nm, 32nm 32Gb NAND flash. His recent research interests include 27nm NAND flash memory development and reliability.

Choi, Hyun-Sik (2C.4)
Received the B.S. (2004), M.S. (2006) and Ph.D (2010) in electronics engineering from the Pohang University of Science and Technology (POSTECH), Pohang, Korea. His main research is related to CMOS devices.

Choi, Rino (RM.3)
Rino Choi received his BS and MS in the Department of Inorganic Materials Engineering of the Seoul National University in 1992 and 1994, respectively. After he worked for Daewoo Motors Company from 1994 to 1999, he joined a Ph.D program of Materials Science and Engineering in the University of Texas at Austin. After his completion of his Ph.D in 2004, he worked as a project manager of the electrical characterization and reliability of advanced gate stacks at SEMATECH.
Since September, 2007, he is with School of Materials Science and Engineering, Inha University, Korea

Choi, Jeong-Hyuk (5C.1)
Jeong-Hyuk Choi was born on March 8, 1962, in Seoul, Korea. He received the B.E. degree in chemical engineering from Inha University, Korea in 1985. He joined the Memory Division, Samsung Electronics, Gyunggi-Do, Korea, in 1985, where he has been working on the process integration of EEPROMs, NOR flash memories, and NAND flash memories. Currently, he is an executive director of Flash Product & Technology division, and leader of Flash Process Architecture team. His current work is focus on high-density NAND flash memories sub 30nm technology, and research interests are NVM reliability, cell technology, yield modeling and field.

Chong, Lit-Ho (5D.3)
Was born in Johor Bahru, Malaysia, in 1976. He received the PhD degree in Microelectronics from the University of Southampton in 2006. He joined Macronix International Company, Ltd., Hsinchu, Taiwan, in 2006 as a device engineer. He has been working on the research and development of the nitride storage Flash memory.

Chou, H.L. (2F.3)
H.L. Chou received the M.S. degree in Mechanical Engineering from National Chiao-Tung University, Hsinchu, Taiwan, R.O.C. in 2002. He joined the Taiwan Semiconductor Manufacturing Company (TSMC) , Hsinchu, Taiwan, R.O.C. in 2002. He works in Division of Anolog Power IC and Special Technology. He is currently the Principle Engineer for the development on power IC and BCD technology.

Chou, Tso-Min (4F.3)
Tso-Min Chou graduated in Physics from the Tamkang University, Taipei in 1988, and obtained the M.S. and Ph.D. degrees from Southern Methodist University, Dallas, TX in 1990 and 1996 respectively. Between 1996 and 2000, he worked at Southern Methodist University as a post-doctoral researcher. Currently he works as researcher scientist in the R&D organization at TriQuint Semiconductor, TX.

Chou, Y. C. (CD.1)
Y. C. Chou received the Ph.D. degree in electrical engineering and computer science from the University of California, Irvine, in 1997. Since 1996, he has been a HEMT product engineer at Northrop Grumman Aerospace Systems (NGAS, Redondo Beach, CA). At NGAS, he has been involved in the technology development of high reliability assurance on 4-inch GaAs and InP-based HEMT MMICs for commercial, military, and space applications. In 2009, he was elevated to the IEEE Senior member. He has authored and co-authored more than 100 papers in the journals and conferences in the areas of GaAs, InP, GaN-based , and AlSb/InAs HEMT devices and MMICs.

Chou, You-Liang (MY.3)
You-Liang Chou received the B.S. degree in electrical engineering from National Chiao-Tung University, Hsinchu, Taiwan, in 2006. He is currently pursuing the Ph.D. degree at the same university. His research is devoted to the study of microscopic mechanisms in flash memory devices. His research interests include reliability analysis in semiconductor memory and high-voltage devices.

Chouard, Florian Raoul (CR.2)
Florian Raoul Chouard received the Dipl.-Ing. degree in electrical engineering from the Technische Universität München (TUM), Germany, in 2007. He is currently working towards the Dr.-Ing. degree on reliability aspects of advanced analog circuits at the Institute for Technical Electronics, TUM, in collaboration with Analog Circuit Exploration, Infineon Austria AG, Austria. His research interests are aging mechanisms of advanced CMOS devices and the impacts on analog/RF circuits.

Chowdhury, Uttiya (4F.3)
Uttiya Chowdhury received the B. Sc. Engg. in Electrical and Electronic Engineering from Bangladesh University of Engineering and Technology in 1995 and the Ph. D. (2002) in Electrical Engineering from University of Texas at Austin. He has worked as researcher at Georgia Institute of Technology and Arizona State University and presently works at TriQuint Semiconductor, Texas as a development engineer. His present focus is failure analysis and reliability of GaN HFET devices.

Chu, Fan (6C.4)
Dr. Fan Chu was born on Dec. 26, 1960 in Beijing, China. He received the B.S. and M.S. degree in electrical engineering from Xi'an Jiaotong University of Xi'an, China, in 1982 and 1987 respectively. He received PhD. Degree in electronic material science from Swiss Federal Polytechnique Institute (EPFL) in Lausanne, Switzerland, in 1994. From 1994 to 1996, he was a post-doctoral research fellow in EPFL , Switzerland and in MRL in the Pennsylvania State University, USA. He is presently a principal scientist in Ramtron International Corp. dealing with F-RAM process and product development.

Chu, Byung-Hwan (CD.3)
Byung-Hwan Chu received his B.S. degree in chemical engineering at New Mexico State University in 2006. He joined the Chemical Engineering department at the University of Florida in 2006 and earned his M.S. degree in 2008. He is currently continuing his study at University of Florida as a Ph.D. student. His research interests are in GaN devices.

Chung, Chilhee (5C.1)
Chilhee Chung is an Executive Vice President of Semiconductor R&D Canter in Samsung Electronics. He is in charge of Technology Development for DRAM, Flash Memory, PRAM, and Advanced Logic in Semiconductor R&D Center. He has been with Samsung Electronics, Semiconductor Business since 1979. he was promoted to Senior Vice President in January 2005. He was in charge of LSI Technology Development as wall as product development of Image sensor and Driver IC. Along with promotion to Executive VP in January 2009, he moved to Memory Division as a head of Flash Product and Technology group. Dr. Chung has got Bachelor of Science in Physics from Seoul National University, Master of Science in Physics from KAIST (Korea Advanced Institute of Science and Technology), and PhD in Physics from Michigan State University.

Cirba, Claude (5E.5)
Claude R. Cirba was born in August, 21, 1965 in Dakar, Senegal. He attended the University of Montpellier in France, receiving his Ph.D. in Electrical Engineering in 1996. He was a Senior Research Associate in the field of radiation effects on semiconductors at Vanderbilt University, Nashville TN, from 1997 to 2003. Since 2003 he is a modeling specialist within the RF-CMOS Spice Modeling Lab at Texas Instruments, Dallas TX. For several years he has been an industrial liaison on TI-sponsored university research projects and actively participated on the GEIA Compact Model Council, an international consortium which promotes standardization of models for integrated circuit design. Claude is author or co-author of 18 refereed professional publications.

Croes, Kristof (5A.1, 5A.2, 5B.4, 6A.3)
Kristof Croes received his BSc in physics at the Catholic University of Louvain (Belgium) in 1993 and his MSc in biostatistics at the Limburgs Universitair Centrum (LUC) in 1994. In 1999, he obtained his PhD, concerning the development of statistical techniques for planning reliability experiments. After that, he joined the reliability business unit of XPEQT, first as the software responsible and than as the manager of the R&D. From 2003 till end 2006, he was product and application manager of the package level reliability products of Chiron holdings. Beginning 2007, he went back to research, working as a BEOL reliability engineer in IMEC.

Curutchet, Arnaud (2E.3)
Arnaud Curutchet was born in Dax, France, in 1975. He received the Ph.D. degree in electronics from the University of Bordeaux, Talence, France, in 2005. In 2006, he held a post-doctoral position with the Institut d'Electronique, de microelectronique et de nanotechnnologies (IEMN) laboratory. One topic of his post-doctoral position was non linear characterization of CNFETs. He is currently an Associate Professor with the University of Bordeaux (ENSEIRB-MATMECA), and carries out his research at the Intégration du Matériau au Système (IMS) laboratory. He has authored or coauthored 13 publications in international conferences and journals. His research concerns electric characterization (low-frequency noise, pulsed and microwave measurements) and modeling of GaN HEMTs devices.

De Jaeger, Jean Claude (2E.3)
Jean-Claude DE JAEGER is currently Professor of electronics.He is head of the Microwave Power Devices research group at IEMN. Current research and developed projects concern the simulation, the design, the fabrication and the measurement of microwave power HEMTs based mainly on GaN as well as devices based on wide bandgap semiconductors such as BN, AlN and diamond. From 2002 to 2007, he was also deputy manager of TIGER, a common laboratory between IEMN and ALCATEL - THALES III-V Lab, working on wide bandgap semiconductors. He is the author or a co-author of about 250 publications and communications.

Defrance, Nicolas (2E.3)
Nicolas Defrance received his M.Sc. and Ph.D. degrees in Electrical Engineering in 2004 and 2007, respectively. He worked on characterization, modeling and technology of AlGaN/GaN HEMTs for microwave power applications. After graduation he worked at Thalès Alenia Space - Belgium within Research & Innovation group. Since 2009 he joined the Institute of Electronics, Microelectronics and Nanotechnology (IEMN) as an associate professor. His current research activities rely on the study of advanced wide bandgap-based devices for millimeter wave applications with emphasis on electrical characterization. He carries out his teaching activities at the University of Lille in the field of electrical engineering.

Degraeve, Robin (2A.3, XT.9, XT.13)
Robin Degraeve received the M.Sc. degree in electrical engineering from the University of Gent, Belgium, in 1992, and the Ph.D. degree from the Catholic University of Leuven, Belgium, in 1998. He joined the Interuniversity Microelectronics Center (IMEC), Leuven, in 1992 in the CMOS Reliability and Characterization group, where he is working as a Principal Scientist. His work currently focuses on advanced and novel characterization techniques for studying electrical defects in dielectrics and related reliability aspects. This includes characterization and reliability of high-k material in transistor and memory applications, hot-carrier related issues and breakdown physics.

Della Marca, Vincenzo (6C.1)
Vincenzo Della Marca was born in Modena, Italy, in 1983. He received B.S. and M.S. degrees in electronic engineering from the Università degli Studi di Modena e Reggio Emilia, Modena, Italy, in 2005 and 2008, respectively. During his M.S. degree thesis, he studied and characterized electrical properties, in Phase Change Memory (PCMs) devices. Since 2008 he has been working for Università degli Studi di Modena e Reggio Emilia, on characterization of electrical properties in TANOS devices. His research interests also include the characterization and modeling of resistive-switching memories (RRAMs).

Demirtas, Sefa (2E.2)
Sefa Demirtas graduated from Bogazici University in June 2007 with a Bachelor of Science degree in Electrical and Electronics Engineering. He earned his Master's degree from MIT in June 2009 in the area of device physics and reliability of GaN high electron mobility transistors. He is currently pursuing his Ph.D at MIT.

Demuynck, Steven (6A.3)
Steven Demuynck received a PhD in Physics from the Katholieke Universiteit Leuven in 2000. In 2001 he joined imec where he joined the interconnect integration team. His research focuses on advanced contact module and narrow pitch low-k integration. He has (co-) authored over 70 conference and journal publications in the field of microelectronics.

Denison, Marie (HV.1)
Obtained her M.Sc. degree in Applied Physics from the University of Liege, Belgium in 1997 and the Ph.D. degree in Electrical Engineering from the University of Bremen, Germany in 2004. From 1998 to 2005 she worked on the development of Smart Power devices and technologies at Infineon Technologies in Germany. Since 2005, she has been working at Texas Instruments, Dallas, Texas on the development of BCD (Linear BiCMOS) technologies and on high-voltage component design and process integration.

Deora, Shweta (XT.15)
Shweta Deora received her bacholar of engineering degree in Electronics and communication Engineering from Sardar Patel University, Gujarat, India in 2004. She was at IIT Bombay working as research assistant on radtion sensor project during 2005-2006. Since 2006, she is doing her PhD in electrical engineering department, IIT Bombay. Her reserch interest are in field of semiconductor device physics, reliability and modeling of CMOS devices. She is currently working on NBTI in SiON and high-k gate dielectrics.

Depner, Martin (6C.4)
"Marty" Depner lives in Colorado Springs with his wife, daughter, and three sons. He graduated in 1989 with B.S. in Physics from the University of Colorado at Colorado Springs. He started in the semiconductor industry at Ramtron, working from 1995 to 2001 as an Engineering Technician in the FAB. He worked for Atmel, Corp., in the FAB, for most of 2001 and returned to Ramtron in February 2002 where he worked on the joint TI/RIC project helping to create the world's first high density F-RAM product. He is presently a Product Engineer at Ramtron.

Derkits, Gustav (FA.5)
Gustav Derkits, Ph.D., is a Reliability Physicist in the Alcatel-Lucent Reliability Physics Group in Murray Hill, NJ. He joined Bell Laboratories in 1982 and has worked in a variety of technical areas including design and fabrication of III-V electronic and optoelectronic devices, and physics and chemistry issues affecting yield, quality, and reliability of telecommunications products. Gus has over 30 US patents and has authored over 20 papers in peer-reviewed journals. He has been certified by the American Society for Quality as a Six Sigma Black Belt.

Detcheverry, Celine (2D.2)
Celine Detcheverry received her PhD in Electronics from the University of Montpellier II, France. She joined Philips Research in 1998 to work on Polymer Electronics circuits and models. In 2001 she moved to Philips Resarch Leuven, Belgium, to explore process options in advanced CMOS for RF applications. In 2004 she joined Philips Semiconductors 300mm Foundry in Crolles, France, within the Alliance of Philips (to become NXP in 2007) / ST microelectronics and Freescale Semiconductors. She hold there a position of RFCMOS Project coordinator for the 3 companies. In 2007 she became Process Owner CMOS065 at NXP Semiconductors in Nijmegen.

Detzel, Thomas (IC.6)
Thomas Detzel received the M.S. degree in physics from the University of Konstanz, Konstanz, Germany, and the Ph.D. degree in surface and thin film physics in 1994 from the Max-Planck-Institute, Garching, Germany. In 1995, he was with Rodel Europe GmbH, where he was an Application Manager for chemical–mechanical polishing. In 1999, he joined Infineon Technologies Austria AG, Villach, Austria, where he was responsible for the metallization development of power semiconductors, has been the Project Manager of different power integrated circuit developments since 2004, and has been leading the research project Robust Metallization and Interconnect in the Competence Center for Automotive and Industrial Electronics since 2006.

Di Sarro, James (EL.3)
James Di Sarro received the B.S. degree in electrical engineering and economics from Duke University in 2004. He received the M.S. degree in 2006 and the Ph.D. degree in 2009 from the University of Illinois at Urbana-Champaign, both in electrical engineering. In 2009, he joined IBM as an Advisory Engineer/Scientist in the Semiconductor Research & Development Center (SRDC) focusing on ESD device and model development in SOI technologies.

Didier, Goguenheim (2B.2)
Didier GOGUENHEIM received the Engineer degree in electrical engineering from Institut Supérieur d'Electronique du Nord (ISEN-Lille) in 1987 and the Ph.D degree in Physics from the University of Lille in 1992. In 1992, he joined ISEN-Toulon (Institut Supérieur de l'Electronique et du Numérique) school, where he still works as Director of the Research activities. Since 2000, he has been member of the IM2NP laboratory (Institut Matérieux Microélectronique et Nanosciences de Provence, UMR CNRS 6242). His current research fields concern static and dynamic degradation modes in advanced MOSFETs, reliability of ultra-thin gate insulators (< 2 nm), and defect characterization in insulators and semiconductors.

Digh, Hisamoto (PI.3)
Digh Hisamoto received the B.S., M.S. degrees in reaction chemistry and the Ph. D. degree in electronic engineering from the University of Tokyo, Tokyo, Japan, in 1984, 1986, and 2003, respectively. In 1986, he joined Central Research Laboratory, Hitachi Ltd., Tokyo, where he has been working on ULSI device physics and process technologies. His current research interests include thin-film SOI materials, short-channel MOSFETs, semiconductor memories, Si-RF devices, and Si-Photonics devices. Dr. Hisamoto is a member of IEEE Electron Device Society, the Japan Society of Applied Physics, and the Institute of Electronics and Communication Engineers of Japan.

Djelassi, Christian (4C.5)
Christian Djelassi received his Diploma degree in electronics and equipment engineering from the Carinthia University of Applied Science in Villach in July 2009. Since his diploma thesis he works as a researcher in the Quality and Reliability Group of KAI and Infineon, respectively. His main research fields contain package stress and short circuit measurements as well as FEM simulations.

Domengie, Florian (3C.2)
Florian Domengie was born in France in 1984. He received the engineer degree in physics engineering and the M.S. degree in materials for electronics and plasma engineering in 2007 from the Institut National des Sciences Appliquées (INSA), Toulouse, France. Since 2007, he is working as a Ph.D. student in STMicroelectronics (Crolles, France) in collaboration with the Institute of Microelectronics, Electromagnetism and Photonics (IMEP-LAHC, Grenoble, France). His current research interests are the electrically active defects in CMOS image sensors and the metallic contamination in silicon.

Douvry, Yannick (2E.3)
Yannick Douvry was born in Cambrai, France, in 1985. He obtained both his master degree at the University of Lille1 and his engineer diploma at the "Institut Supérieur d'Electronique et du Numérique (ISEN)" in 2008. He is currently working as a PhD student at the Institute of Electronics, Microelectronics and Nanotechnology (IEMN). The topic of the thesis consists to improve design and manufacture of AlGaN / GaN HEMTs for microwave power application in K-band and Ka-band. His main activities are processing and characterization. He is also following an applied mathematics' master degree.

Doyen, Lise (IC.8)
Lise Doyen received the Engineering degree in materials and microelectronics from the Institut National des Sciences Appliquées, Toulouse, France, in 2005, and the Ph.D. degree in micro- and nanoelectronics from the Université Joseph Fourier, Grenoble, in 2009. Her dissertation focused on electromigration in advanced copper interconnects. in collaboration with the CEA-Leti/Minatec

and the Science et Ingeniére des Matériaux et Procédés Laboratory, Grenoble, France. She is now with Central CAD Design Solutions, STMicroelectronics, Crolles, France, working on silicon integrity.

Du, Guoan (XT.5)
G. A. Du received the B.Eng. (Hons) degree from the School of Electrical and Electronic Engineering, Nanyang Technological University, Singapore. He is currently pursuing the Ph.D. degree in the same school under a NTU Graduate Research Scholarship. His research project is on the characterization of bias-temperature instability in advanced P-MOSFETs.

Du, Pei-Ying (MY.1)
Pei-Ying Du was born in Taipei, Taiwan in 1982. She received her B.S in engineering and system science from National Tsing-Hua University (NTHU) in 2004, and Ph.D. degree in electrical engineering from National Chiao-Tung University (NCTU) in 2009. She joined Emerging Central Lab. (ECL) in Macronix International (MXIC) in 2006, where her current research is engaged in the theoretical modeling and reliability physics of nitride trapping Flash Memories.

Dua, Christian (2E.3)
Christian Dua received the Engineer degree in Physics from the University of Clermont Ferrand (France) in 1980. He has worked in different Units of THOMSON-CSF Group (previous name of THALES) in the field of microwave devices and optoelectronic components. From 1982 to 1996 he gained experience in crystal growth and physical and electrical characterization of semiconductor materials. In 1997 he joined the research unit of Thales, TRT, and is participating in the development of wide band gap semiconductor technologies. His present activities include the study of reliability of GaN HEMTs.

Eliason, Jarrod (6C.4)
Jarrod Eliason received his BSEE in Electrical Engineering from GMI Engineering and Management Institute (now Kettering University) and currently leads the memory macro design group at Ramtron. Between 2001 and 2004, Jarrod participated in the joint development program between Ramtron and Texas Instruments to commercialize F-RAM on TI's 130nm process. Jarrod holds 18 patents related to ferroelectric memory. Along with his father, Jarrod developed a side scan sonar system which has been used to locate 10 Lake Superior Shipwrecks. They are currently planning a trip to Newfoundland to search for U-656, the first U-boat sunk by U.S. forces in WWII.

Emmanuel, Vincent (2B.2)
Emmanuel Vincent received the Engineer degree in electronics, the M.S. degree in microelectronics in 1992 and the Ph.D. degree in microelectronics in 1996 from the Institut National Polytechnique de Grenoble (INPG), Grenoble, France. He received the Ph.D. degree through a collaboration between STMicroelectronics Central R&D Labs, Crolles, France, and the Laboratoire de Physique des Composants à Semiconducteur (now IMEP/ENSERG), Grenoble. Since 1993, he has been with STMicroelectronics, where he held various positions in the reliability area. He is currently an Electrical Characterization and Reliability Manager in the Crolles 2 Alliance.

Eneman, Geert (XT.10)
Geert Eneman received the B.S. and M.S. degrees in electrical engineering and the Ph.D. degree on the topic of "Design, fabrication, and characterization of strained silicon transistors" from the Catholic University of Leuven, Leuven, Belgium, in 1999, 2002, and 2006, respectively. His Ph.D. work was done in the Interuniversity MicroElectronics Center (IMEC), Leuven. He is currently with the CMOS Technology Department, IMEC.

He also holds a postdoctoral position at the Catholic University of Leuven. He is a Postdoctoral Fellow with the Fund for Scientific Research-Flanders, Belgium. His current research interests include the characterization and modeling of Ge MOS devices, Ge junction analysis, and modeling of alternative device structures.

Enichlmair, Hubert (2A.5)
Hubert Enichlmair received the M.S.(1991) from the University of Graz and the Ph.D.(1995) in solid state physics from the Technical University in Linz, Austria. In 1995 he joined austriamicrosystems AG, Unterpremstaetten, Austria, where his present activities are focussing on the reliability of integrated power devices. He is author and co-author of over 30 papers in international journals and proceedings and issued several patents.

Erica, Douglas (CD.3)
Erica Douglas received her B.S. degree in Physics at the University of Floirda in 2008. She is currently a graduate student in the Department of Materials Science & Engineering at the University of Florida. She has published 5 papers in refereed journals, and her interests are in wide bandgap devices and their reliability.

Fang, Zheng (MY.4)
Fang Zheng received his B.Eng in Electrical and Electronics Engineering (EEE) from Nanyang Technological University (NTU) Singapore in 2008. He is currently pursuing his PhD degree in EEE, NTU working on metal oxide based resistive random access memory. His research interest includes memory device fabrication as well as physical and electrical characterization.

Fantini, Paolo (6C.2, MY.6)
Paolo Fantini received the Laurea and the Ph.D. degree in physics from Modena University, Italy in 1999. In 2000 he has been engaged by STMicroelectronics to work in the Compact Modeling team. Since 2004 he is a Team Leader of Compact Modeling of STMicroelectronics. After the Numonyx foundation in 2008 he held the Compact Modeling Manager position in Numonyx. He published more 50 papers covering many fields, from the solid-state physics to device physics, modeling, low-frequency noise.

Faqir, Mustapha (2E.5)
Received the M.S. degree (summa cum laude) in electronics engineering from the University of Modena e Reggio Emilia, Italy. In 2009, he received the Ph.D. degree in electronics engineering, jointly from the University of Modena e Reggio Emilia and the University of Bordeaux 1, France. He was with MD Microdetectors, Modena, where he worked for two years as an R&D Engineer. His research interests include the study and the analysis through experimental measurements and numerical simulations of trapping effects in gallium nitride devices and their reliability, as well as thermo-mechanical modeling and characterization of GaN electronics packaging. Currently, he is a research assistant in the CDTR at the University of Bristol, UK.

Farbiz, Farzan (4D.2)
Farzan Farbiz is a Ph.D. candidate in the Department of Electrical and Computer Engineering at the University of Illinois at Urbana-Champaign. He received his B.S. degree in electrical engineering from the University of Tehran, Iran. Since 2005, he has been at the University of Illinois, studying integrated circuit reliability in the Illinois Center for Integrated Microsystems (iCIMS). He has held summer positions at Freescale and Texas Instruments. He has received the 2008 IRPS best student paper award and the 2010 Gregory Stillman semiconductor research award from University of Illinois for excellence in semiconductor research.

Farina, Fabrizio (MY.6)
Fabrizio Farina was born in Brindisi, Italy, in 1985. He received the B. degree in Electronics Engineering from the Politecnico di Milano, Milan, Italy in 2007. Since 2009 he has been working on characterization and modelling of Advanced non-volatile memories. His research interest is in the areas of application-specific integrated circuit design.

Ferro, Massimo (6C.2)
Massimo Ferro was born in Italy, in 1983. He received the master Laurea in electronic engineering from the Politecnico di Milano, Milan, Italy, in 2009. During his thesis, he worked on phase change memories with the Department of Electronic Engineering, Politecnico di Milano, collaborating with the Modeling & Characterization team of the Advanced R&D, Numonyx, Agrate Brianza, Italy, where he has been working on modeling and characterization of transport properties of amorphous chalcogenide-based devices.

Fleming, Debra (FA.5)
Debra Fleming is a Member of the Technical Staff for Reliability Engineering Group at Alcatel-Lucent in Murray Hill, New Jersey. Debra is a materials engineer with expertise in materials processing, characterization, failure mode analysis, and component prototyping. Her present work is focused on the reliability of lead-free solder and the failure mode analysis of telecommunication components.. She received her B.S. and M.S. in Ceramic Science and Engineering from Rutgers University in New Brunswick, NJ. Debra has 18 U.S. patents and has coauthored more than 50 papers in the area of novel optical, electronic, and magnetic materials and components.

Francis, Rick (2D.3)
Rick Francis is an associate engineer at GLOBALFOUNDRIES in Sunnyvale California where he works on reliability testing for lifetime and design manual characterization of sub-micron devices. He graduated from Heald College in San Jose with a degree in electrical engineering in 2000 when he joined Advanced Micro Devices. He has been with AMD and GLOBALFOUNDRIES for over nine years as an associate engineer and software programmer. He is currently attending University of Phoenix to obtain a B. S. degree in software engineering.

Franco, Jacopo (2A.3, XT.10, XT.13)
Jacopo Franco received the B.Sc. and M.Sc. degrees in Electronic Engineering from the University of Calabria - Italy, in 2005 and 2008 respectively. His M.Sc. thesis was developed during an internship at IMEC, Leuven - Belgium, and is related to reliability issue in advanced Silicon and Germanium MOSFETs. Since Feb. 2009 he is working toward a Ph.D. degree in the reliability group of IMEC and at the Katholieke Universiteit Leuven, on the topic "Interface stability and reliability of Ge and III-V transistors for future CMOS applications". He is the recipient of the Ed Nicollian Best Student Paper Award at the 40th IEEE Semiconductor Interface Specialists Conference (SISC).

Franey, John (FA.5)
John Franey P.E. is a Distinguished Member of Technical Staff in the Alcatel-Lucent Reliability Physics Group in Murray Hill, NJ. He joined Bell Laboratories in 1970. His work has been focused on failure analysis and the interactions of materials with people and atmospheric environments. With degrees in Electronics, Chemistry, and Material science he has been a corrosion consultant on the Restoration of the Statue of Liberty, and, the State Department in Washington, DC. John has 36 US and Foreign Patents in the areas of Corrosion and ESD

Protection, and has authored over 100 papers in those areas of expertise.

Frei, Stephan (4D.1)
Stephan Frei was born in Germany in 1966. He received his diploma in electrical engineering in 1995. From 1995 till 1999 he was a research assistant for EMC at the University of Technology in Berlin. There he investigated the influence of ESD on electronic devices and the occurrence rate of ESD in typical environments. In 1999 he received his Dr.-Ing. degree. From 1999 till 2005 he worked at the car manufacturer AUDI AG in Germany. Here he introduced and developed, among other things new methods for the computation of EMC in automobiles. In 2006 he was appointed as professor for vehicular electronics at the University of Technology in Dortmund, Germany. His recent research interests include automotive EMC, Signal Integrity of automotive bus systems, ESD, and numerical modelling. Professor Frei is active and chairman in several national and international EMC standardization groups. From 2008 till 2009 he was appointed as Distinguished Lecturer from the IEEE EMC Society.

Frost, Christopher (4B.3)
Christopher Frost has been a research scientist at ISIS, the UK's pulsed neutron source sited at the Rutherford Appleton Laboratory, UK, since 1998. He took his BA degree at Trinity College, Cambridge University where he stayed to complete an M.Phil. and Ph.D. in neutron science at the Cavendish Laboratory. He undertook a Post-Doctoral position at Warwick University where he was seconded to ISIS to help develop the world leading MAPS spectrometer. He currently leads the design and development of CHIPIR, a new neutron irradiation facility for electronics. He is a co-author on over sixty scientific and technical papers in the areas of condensed matter and fast neutron science.

Fugazza, Davide (6C.3)
Davide Fugazza was born in 1981. He received the M.S. degree (cum laude) in Electronic Engineering from Politecnico di Milano, Italy, in 2006. In 2006-07 he worked as a digital hardware engineer, developing base band processing algorithms for PtP microwave radio systems. In January 2008 he joined the Dipartimento di Elettronica e Informazione, Politecnico di Milano, as a Ph.D. student in Information Technology. His primary research interests are in the area of microelectronics devices and his research activity is actually mainly focused on the characterization and modeling of switching and reliability characteristics for phase change non volatile memories (PCMs).

Fujii, Shosuke (MY.2)
Shosuke Fujii received the B.S. (2005) and M.S. (2007) in materials science and engineering from Kyoto University, Kyoto, Japan. He joined Advanced LSI Technology Lab, Toshiba Corp., Yokohama, Japan, in 2007, where he has been engaged in the research on reliability physics of non-volatile memories.

Fujiki, Jun (MY.2)
Jun Fujiki received the B.S. (2003) and M.S. (2005) in applied physics from the department of applied physics school of engineering, Tokyo University, Tokyo, Japan. He joined the Advanced LSI Technology Lab, Toshiba Corp., Yokohama, Japan, in 2005, where he has been engaged in the research on reliability physics of non-volatile memory devices.

Fujisawa, Takafumi (5F.2)
Takafumi Fujisawa was born in Nagano, Japan, on March 22, 1985. He received the B.S. degree in electronic engineering from Tohoku University, Sendai, Japan in 2008. He is currently working toward the M.S. degree in the Graduate School of Engineering, Tohoku University (MC2).

His research interests are the development of processes suppressing the variability of MOSFETs and Random Telegraph Signal (RTS) noise.

Fujitsuka, Ryota (MY.2)
Ryota Fujitsuka received the B.S. (2003) and M.S. (2005) degrees, from Nagoya University, Nagoya, Japan. In 2005, He joined Process and Manufacturing Engineering Center, Toshiba Corporation, Semiconductor Company, Yokohama, Japan, where he worked on development of dielectric film for advanced nonvolatile memory device. He is currently engaged in development of nonvolatile memory device in Advanced Memory Development Center, Semiconductor Company, Toshiba Corporation, Yokkaich, Japan

Fujiwara, Tetsuo (PI.1)
Tetsuo Fujiwara received his B.E. and M.E. in materials science from Kyoto University, Kyoto, Japan in 1987 and 1989, respectively. He joined Hitachi Research Laboratory of Hitachi, Ltd., Ibaraki, Japan in 1989. Since 2005, he has been working as a senior engineer for Renesas Technology Corp. (now Renesas Electronics Corporation), Ibaraki, Japan, where he is engaged in developing metallization processes.

Fukatsu, Shigeto (4C.2)
Shigeto Fukatsu, received the B.E. (2000), M.S. (2002), and Ph.D. (2005) in applied physics from Keio University, Japan. In 2005, he joined Advanced LSI Technology Laboratory, Toshiba Corporation, Yokohama, Japan. He has been engaged in the research on the reliability physics of MOSFETs.

Fuketa, Hiroshi (3A.4)
Hiroshi Fuketa received the B.E. degree from Kyoto University, Kyoto, Japan, in 2002 and the M.E. degree in information systems engineering from Osaka University, Osaka, Japan, in 2008. He is currently pursuing the Ph.D. degree from the Graduate School of Information Science and Technology, Osaka University. His research interests include ultra-low-power circuit design and variation modeling. Mr. Fuketa is a student member of IEEE and IEICE.

Fukuda, Toshikazu (5F.4)
Toshikazu Fukuda received B.S. degree and M.S. degree in mathematics from the Osaka University, Osaka, Japan, in 1993 and 1995, respectively. He joined TOSHIBA Corporation in 1995 as a TCAD development engineer. Since 2000, he has worked on static-RAM development: circuit and layout design, library development, yield estimation and soft-error related issues. He joined Environmental Variability Tolerant Device Technology Program of MIRAI project in 2008.

Fukutani, Atsuyuki (4C.1)
Received D.Sc. from the University of Tokyo in 1990. From 1990 to 1995, he worked as a research associate in the Surface Science group at Institute for Solid State Physics of the University of Tokyo. Since 1995, he has been at Institute of Industrial Science of the University of Tokyo where he is a Professor of the Department of Fundamental Engineering. His major is surface and interface physics, and he is currently interested in chemistry and physics of hydrogen at solid surfaces and interfaces.

Fulde, Michael (CR.2)
Michael Fulde received the Dipl.-Ing. degree in electrical engineering from the Technische Universität München (TUM), Germany, in 2005 and the Dr.-Ing. degree in 2009. From 2005 to 2008, he was with the Institute for Technical Electronics, Technische Universität München, where he worked on technology oriented analog and mixed-signal circuit design in emerging multi-gate CMOS technologies. In 2008 he joined the Analog Circuit Exploration group at Infineon Technologies Austria AG. He is currently working on high-speed mixed signal and digital enhanced RF circuits in advanced CMOS nodes.

Funayama, Kota (PI.2)
Kota Funayama received his B.E. and M.E. in materials science from Tohoku University, Miyagi, Japan in 1995 and 1997, respectively. He joined Hitachi, Ltd., Tokyo, Japan in 1997. Since 2003, he has been working as an engineer for Renesas Technology Corp. (now Renesas Electronics Corporation), Ibaraki, Japan, where he is engaged in the research and development of advanced MCU technology.

Futase, Takuya (PI.1, PI.2)
Takuya Futase received his B.E. and M.E. in materials science and engineering from Muroran Institute of Technology, Hokkaido, Japan in 1994 and 1996, respectively. He began pursuing a Ph. D. in materials science from the University of Tsukuba, Ibaraki, Japan in 2010. He joined Hitachi ULSI Engineering Corp. in 1996 and then joined Hitachi, Ltd. and Renesas Technology Corp. in 2001 and 2002, respectively. Since 2010, he has been working for Renesas Electronics Corporation, where he is engaged in front-end metallization for advanced logic devices. Mr. Futase is a member of The Japan Society of Applied Physics and of IEEE.

Gadlage, Matthew (3A.2, 6E.2)
Matthew J. Gadlage received his B.S. in electrical engineering from the University of Evansville in 2002 and his M.S. in electrical engineering from Vanderbilt University in 2004. Since 2002, he has been a member of the Radiation Sciences Branch at Crane Naval Surface Warfare Center. His primary research interests include the effects of radiation on electronic circuits and soft errors. He has authored or co-authored over 30 papers in these research areas. He is planning to finish his Ph.D. in electrical engineering at Vanderbilt University in the spring of 2010.

Galbiati, Nadia (MY.5)
Nadia Galbiati graduated in physics from the University of Milano Italy, (1994) and received the Material Science Diploma in (1997) from the University of Milano Italy. In 1998 she joined the STMicroelectronics in Agrate Brianza (Italy) working in the Non Volatile Memory Process Development group within the Dielectric Reliability group of Central R&D. Her research activities include failure and wear-out mechanism of all active dielectric of non valatile memories, in particular characterization of dielectrics and new materials for charge trap memory applications.

Gasiot, Gilles (4B.4)
Gilles GASIOT received the M.S. (2000) and Ph.D. (2004) in microelectronics from the University of Bordeaux, France. His thesis research (in collaboration with STMicroelectronics and French Atomic Energy Commission, Military applications centre at Bruyères-le-Chatel) focused on the reliability of bulk and SOI devices in the natural radioactive terrestrial environment (neutrons and alpha particles). In 1999, he joined for a trainee period the Radiation Effect Group at Vanderbilt University, USA. Since 2004, he has been with Central R&D, STMicroelectronics, Crolles, France, and has been actively working on soft error rate characterization (experimental and simulated) for ultradeep submicrometer CMOS processes. Dr. Gasiot has coauthored more than 40 articles and holds 2 patents in radiation hardening. He is also member of the IEEE.

Gauthier Jr., Robert (EL.3)
Robert Gauthier has worked in IBM's Microelectronics Division since 1995. He has worked in the areas of ASIC design, CMOS device design (350nm -180nm technology nodes), ESD/Latchup technology and design enablement development (250nm - 15nm technology nodes). He is currently a Senior Technical Staff Member

(STSM) and Manager in IBM's System and Technology Group (Semiconductor Research and Development Center's - SRDC) where his department focuses on ESD/Latchup development for IBM's analog and mixed signal, leading edge bulk CMOS, RF-CMOS, silcon-on-insulator (SOI) and research technologies. His current research interests are in the area of next generation ESD devices/solutions for 15nm and beyond technology nodes. His external involvement outside of IBM consists of serving as the vice-chair of the ESD/Latchup session at the International Reliability Physics Symposium (IRPS) for the 2000 symposium and chair of the ESD/Latchup session in 2001. He was session chair and moderator of the On-chip CMOS session at the EOS/ESD symposium in 2003 and in 2007 the session chair of the On-chip physics session at the EOS/ESD symposium. In addition, from 2002-2009, he has been actively serving on the EOS/ESD symposium TPC sub-committees. In 2007 he was one of the founders of the International ESD Workshop (IEW) where he has served roles as Communication Chair and Technical Program Chair. In 2009 he was the EOS/ESD symposium Technical Program Chair and in 2010 is the EOS/ESD symposium Vice General Chair. In 2010 he will be serving as a member of the ESD Association Board of Directors and will be taking on the mission of Secretary on the ESD Association Executive Committee. He has authored or co-authored over 50 papers and has received the 33rd level invention plateau at IBM (over 128 patents filed).

Geinzer, Thomas (3C.4)
Thomas Geinzer studied electrical engineering at the University of Cooperative Education Ravensburg, Germany, gaining his Bachelor in 2003. He continued his studies at the University of Wuppertal and received his Master degree within the Faculty of Electrical, Information and Media Engineering. Since 2005 he is PhD student at the Department of Electronics of the University of Wuppertal.

Genoe, Jan (4F.2)
Jan Genoe was born in Leuven, Belgium on May 19, 1965. He obtained his Ph.D. degree in electronic engineering on May 31, 1994. Afterwards, he joined the Grenoble High Magnetic Field Laboratory as a Human Capital and Mobility (HCM) Fellow of the European Community. Currently, he's professor at the Katholieke Hogeschool Limburg (KHLim) and head of the Polymer and Molecular Electronics (PME) group of imec.

Gerardin, Simone (3F.2, 4B.3)
Simone Gerardin received the Laurea degree (cum laude) in Electronics Engineering in 2003, and a Ph.D. in Electronics and Telecommunications Engineering in 2007, both from the University of Padova - Italy. He is currently a research assistant at the same university. His research is focused on soft and hard errors induced by ionizing radiation in advanced CMOS technologies, and on their interplay with device aging and ESD. Simone has authored or co-authored more than 40 papers published in international journals and more than 50 conference presentations, three of which won awards at RADECS 2007, NSREC 2008, and RADECS 2008.

Ghetti, Andrea (5C.4)
Andrea Ghetti received the Laurea (summa cum laude) and Ph.D. degrees in electrical engineering both from the University of Bologna, Bologna, Italy. In 1994, he was a Visiting Scientist at the TCAD Department of Intel Corporation. From March 1997 to March 2000, he held a postdoctoral position with Lucent Technologies, Bell Laboratories, Murray Hill, NJ. In May 2000, he joined STMicroelectronics, that later became Numonyx. He has authored or coauthored more than 70 peer-reviewed papers, contributed a chapter to a book on Oxide Reliability and served in the IEDM Modeling and Simulation subcommittee. His research interests are in the field of nonvolatile memory development and device modeling, simulation and characterization. Dr. Ghetti received twice the "Outstanding paper award" at the International Reliability Physics Symposium in 2000 and 2008.

Ghidini, Gabriella (MY.5)
Gabriella Ghidini received the Ph.D. in physics in 1983 from the City College of New York. In 1983 she joined the Central R&D department of STMicroelectronics in Agrate Brianza, Italy. In 1987 she moved to the Non-Volatile Memory division, becoming the leader of the Dielectric Reliability Group. Her research activities include failure and wear-out mechanism of active dielectric of Flash NOR and NAND memory devices and the evaluation of new technologies for the future generations. She is coauthored over 130 papers and 12 patents in the abovementioned topics. From 2008 she joined Numonyx (Agrate Brianza, Italy) with the same role.

Ghidotti, Michele (5C.2)
Michele Ghidotti was born in Cremona, Italy, in 1981. He received the Bachelor's and Master's degrees in Electrical Engineering from the Politecnico di Milano, Milano, Italy, in 2003 and 2006, respectively. He is currently working toward the Ph.D. degree with the Dipartimento di Elettronica e Informazione, Politecnico di Milano, working on experimental characterization and modeling of ultra-scaled Flash memory reliability.

Ghilardi, Tecla (MY.5)
Tecla Ghilardi, graduated in Physics in 1990 at the University of Milan. After degree, she joined the Research and Development group of STMicroelectronics, and, since 1998, of Numonyx. Her field of work is NOR and NAND Flash memories process development.

Giai Gischia, Gianni (5A.2)
Gianni Giai Gischia received the M.Sc. Degree in physic in 2002 from the University of Turin, Italy and the advanced master in material science in 2008 from the University of Pavia, Italy. He is part of the BEOL group at Imec, Belgium and his research activity focuses on dielectric reliability in advanced copper/low-k interconnect.

Giliberto, Valentina (3F.2)
Valentina Giliberto was born in Brindisi, Italy, in 1984.
She graduated in Electronics Engineering at the University of Padova in 2009, working on the reliability of switch RF-MEMS, in particular on the study of the degradation of contact resistance.

Gill, Balkaran (3A.1)
Balkaran Gill received the B.E. degree in electronics and communication engineering from Gulberga University, Gulberga, India in 1998 and the M.S. and Ph.D. degrees in computer engineering from the Case Western Reserve University, Cleveland Ohio in 2002 and 2005 respectively. Dr. Gill worked with IROC Technologies in 2003 and 2004. He is currently a component design engineer in architecture for quality and reliability with Intel Corporation, Hillsboro, Oregon. His interests are modeling and mitigation of transient errors and robust circuit design methodologies for low power systems. He has published more than ten technical papers and holds several pending U.S. patent

Glaser, Ulrich (4D.1)
Ulrich Glaser received the Dipl.-Phys. degree in physics from the University of Bayreuth, Bayreuth, Germany, in 2002 and the Ph.D. degree from the Swiss Federal Institute of Technology (ETH), Zurich, Switzerland, in 2007. He worked in the field of quantum information theory in his Diploma thesis.In 2002, he joined the Integrated Systems Laboratory, ETH, where he was involved in research and development projects about device simulation of electrostatic discharge (ESD) phenomena in close collaboration with

the CMOS ESD group of Infineon Technologies AG, Munich, Germany. In 2006, he joined the automotive power technology research and development group of Infineon Technologies, where, since then, he has been responsible for the development of ESD devices and concepts, as well as the ESD on-chip protection in advanced smart power technologies.

Glavanovics, Michael (4C.5)
Michael Glavanovics studied electrical engineering at the Technical University of Vienna, Austria, where he obtained his PhD in 1994 . In 1997 he went to Villach, Austria to join Infineon technologies as a Senior Staff Engineer for smart power design. Since 2006 he has been involved in the foundation and building up of the KAI competence center for automotive and industrial electronics in Villach. His main focus is on the development of power cycling reliability test methods for smart power semiconductors.

Gnani, Elena (HV.1)
Received the Laurea and Ph.D. degrees in electrical engineering from the University of Bologna, Bologna, Italy, in 1999 and 2003, respectively. Since October 1999, she has been with the Department of Electronics (DEIS), University of Bologna, in the field of investigations on physics of carrier transport and numerical analysis of semiconductor devices, where she is also currently with the Advanced Research Center for Electronic Systems (ARCES).

Gnudi, Antonio (HV.1)
Received the B.S. and the Ph.D. degree in electrical engineering and computer science from the University of Bologna, Bologna, Italy, in 1983 and 1989. From 1989 to 1990, he was a Visiting Scientist with the IBM T. J. Watson Research Center, NY. He became a Research Assistant in 1990 and has been an Associate Professor of electronics since 1998 with the University of Bologna, where he worked on the design of analog CMOS circuits for RF applications and of RF MEMS devices. His current research interests include numerical simulation of nanometric devices and efficient algorithms for their solution.

Goodson, Kenneth (2C.5)
Dr. Kenneth E. Goodson is Professor and Vice Chair of Mechanical Engineering at Stanford University. Goodson was educated at MIT (Ph.D. 1993, MS 1991, BSME 1989, BSH 1989) and spent two post-doctoral years with the Materials Group at Daimler-Benz AG. His Stanford research group includes approximately 20 students, research associates, and affiliated faculty. The group studies thermal transport phenomena in semiconductor nanostructures, energy conversion devices, and microfluidic heat sinks, with a focus on those occurring with very small length and time scales. Goodson is a co-founder and former CTO of Cooligy, Inc., which builds microfluidic cooling systems for computers and was acquired by Emerson, Inc., in 2005. Goodson received the ASME Journal of Heat Transfer Outstanding Reviewer Award, and now serves as an Associate Editor for this Journal. Goodson serves as Editor-in-Chief of Nanoscale and Microscale Thermophysical Engineering. He has been a JSPS Visiting Professor at The Tokyo Institute of Technology and received the ONR Young Investigator Award and the NSF CAREER Award. He and his group have published more than 100 archival journal articles, 150 conference papers, and ten books and book chapters, which have been recognized through best paper awards at SEMI-THERM, the Multilevel interconnect Symposium, SRC TECHCON, and the IEDM.

Gorini, Giuseppe (4B.3)
Giuseppe Gorini. After graduating cum laude in Physics at Pisa University and Scuola Normale Superiore in 1985, Giuseppe

Gorini (GG) received a PhD cum Laude in Physics at Scuola Normale Superiore (Pisa) in 1991. Since 2000 Giuseppe Gorini is associate professor at the Physics Department of Milano-Bicocca University. For the past 20 years GG has been engaged in the development of new experimental methods for neutron measurements in fusion and material science. The scientific production by GG is documented in 130 journal papers.

Gornik, Erich (4D.4)
Prof. Erich Gornik has studied at the Technical University of Vienna, where he finished his Ph.D. in 1972. He was Postdoc from 1975 to 1977 with the Bell Laboratories. From 1979 until 1988 he was full Prof. for Experimental Physics at the University of Innsbruck from where he changed to the Technical University of Munich as director of the Walter Schottky Institute. Since 1993 he is full Prof. for Semiconductor Electronics and since 1995 he is head of the Microstructure Center at the Technical University of Vienna. In 2003-2008 he was a managing director of Austrian Research Center Seibersdorf. Erich Gornik is author and co-author of more than 750 publications in scientific journals mainly in the field of semiconductor physics and technology. His main expertise is current transport and mid-infrared to far-infrared emission spectroscopy of semiconductor nanostructures.

Gossner, Harald (4D.4, EL.2)
Harald Gossner received the degree in physics (Dipl.Phys.) from the Ludwig–Maximilians University, Munich, Germany, in 1990, and the Ph.D. degree in electrical engineering from the Universität der Bundeswehr, Munich, in 1995. Since 1995, he has been with Infineon technologies AG, Munich, working on the development of ESD protection concepts for bipolar, BiCMOS, and CMOS technologies. He is the Head of the team of Infineon's center of competence for ESD and external latchup development and also a Senior Principal who guides the company activities in the field of overvoltage robust design. He has authored and coauthored more than 40 technical papers and one book in the field of ESD. Dr. Gossner is a member of the management board of the International ESD Workshop (IEW) and a Cochair of the Industry Council on ESD Target Values. He is serving in the TPC of EOS/ESD Symposium, IEDM, and IEW.

Goto, Masakazu (4C.2)
Masakazu Goto received the B. E. and M. E. degrees in applied physics from University of Tsukuba, Tsukuba, Japan, in 2003 and 2005, respectively. He joined the Center for Semiconductor Research & Development of Toshiba Semiconductor Company, Yokohama, Japan, in 2005. His current work is the development of metal gate /high-k CMOS devices. He is a member of the Japan Society of Applied Physics.

Grasser, Tibor (2A.1, 2A.2, 2A.3, 2A.5, XT.6, XT.8, XT.10)
Tibor Grasser received his Ph.D. degree in technical sciences from the TU Wien where he is currently employed as an Associate Professor. In 2003 he was appointed head of the Christian Doppler Laboratory for TCAD in Microelectronics. Dr. Grasser is the co-author or author of over 250 scientific articles, editor of a book on advanced device simulation, a senior member of IEEE, has been involved in the program committees of SISPAD, IWCE, ESSDERC, IRPS, IIRW, and ISDRS, and is a recipient of the Best Paper Awards at IRPS and ESREF. He was also a chairman of SISPAD 2007.

Green, Keith (5F.3)
Keith Green received the Ph.D. and M.S. degrees in Electrical Engineering from the University of Florida in 1993 and 1990, respectively, and the B.S. degree in Electrical Engineering from the University of Delaware in 1988. He has been with Texas Instruments in Dallas, TX since 1993, with a career dedicated to the development and extraction of compact models for integrated circuit

design. He has worked on models for CMOS and BiCMOS technologies for analog, digital, and RF design applications. He has modeled almost every major component type: MOSFETs, LDMOS, BJTs, diodes, capacitors, resistors, and many more. He is currently a Distinguished Member of the Technical Staff and responsible for the development of new compact models to support analog technology and circuit development. He is the Chairman of the Semiconductor Research Corporation's Compact Modeling Technical Advisory Board and the Vice-Chair of the Compact Model Council, a standards consortium of approximately 40 semiconductor and EDA companies from Asia, Europe, and the United States.

Groeseneken, Guido (2A.3, 5A.2, 6A.3, XT.9, XT.10, XT.13)
received the M.Sc. degree in electrical and mechanical engineering (1980) and the Ph.D degree in applied sciences (1986), both from the Katholieke Universiteit Leuven, Belgium. In 1987 he joined the R&D Laboratory of IMEC (Interuniversity Microelectronics Center) in Leuven, Belgium, where he is responsible for research in reliability physics for deep submicron CMOS technologies. From October 2005 until April 2007 he was also responsible for the IMEC Post CMOS Nanotechnology program within IMEC's core partner research program. Since 2001 he is also Professor at the KU Leuven,where he is Program Director of the Master in Nanoscience and Nanotechnology, and where he is also coordinating a European Erasmus Mundus Master program in Nanoscience and nanotechnology. He became an IEEE Fellow in 2005 and an IMEC Fellow in 2007.He has made contributions to the fields of non-volatile semiconductor memory devices and technology, reliability physics of VLSI-technology, hot carrier effects in MOSFET's, time-dependent dielectric breakdown of oxides, Negative-Bias-Temperature Instability effects, ESD-protection and –testing, plasma processing induced damage, electrical characterization of semiconductors and characterization and reliability of high k dielectrics. Recently he has also interest in nanotechnology for post-CMOS applications, such as carbon nanotubes for interconnect applications, tunnel FET's for alternative nanowire devices etc.He has served as a technical program committee member of several international scientific conferences, among which the IEEE International Electron Device Meeting (IEDM), the European Solid State Device Research Conference (ESSDERC), the International Reliability Physics Symposium (IRPS), the IEEE Semiconductor Interface Specialists Conference (SISC) and the EOS/ESD Symposium. From 2000 until 2002 he also acted as European Arrangements Chair of IEDM. In 2005 he was the General Chair of the Insulating Films on Semiconductor (INFOS) conference, organized in Leuven, Belgium.He has authored or co-authored more than 500 publications in international scientific journals and in international conference proceedings, 6 book chapters and 10 patents in his fields of expertise.

Grossi, Alessandro (MY.5)
Alessandro Grossi received the Laurea degree (cum laude) in Physics from the University of Milan in 1991. In 1993 he joined the STMicroelectronics R&D Non Volatile Memory Development group. From 1999 he has been Technology Development Project Leader for NOR memory applications and from 2007 he is involved in NAND development. Since 2008, he has been with Numonyx R&D as Flash Memory Manager working on new architectures for Flash devices. He is coauthor of several publications and patents on abovementioned topics. He has been lecturer in Electron Device Physics at the University of Milan, University of Parma and University of Udine.

Gustin, Wolfgang (2A.1)
Wolfgang Gustin received the diploma in physics (1990) from the University of Stuttgart, and the Ph.D. (1994) from the Max-Planck-Institut Stuttgart. From 1994-1998, he had been with Philips & IBM, working on Integration and Unit Process Issues for Logic and DRAM Technologies. In 1998 he joined the DRAM Development Group at Infineon Technolgies. Currently, he is the Manager of the Device Reliability Group at Infineon Technologies.

Haggag, Amr (2D.5, XT.16)
Amr Haggag, from Egypt, was born in Buenos Aires, Argentina, in 1975. He received the B.S. degree in computer engineering and the M.S. and Ph.D. degrees in electrical engineering from the University of Illinois at Urbana-Champaign, USA, in 1996, 1999, and 2002, respectively. Since September 2002, he has been with Freescale (previously Motorola Semiconductors) in Austin, TX. He serves on both management and technical committees for the International Reliability Physics Symposium and the Integrated Reliability Workshop and has given invited talks/tutorials at conferences such as International Reliability Physics Symposium, VLSI Test Symposium, International Test Conference, Design for Variability-Reliability Workshop and also at Middle East Nanotechnology Conference patronage of King Abdullah of Jordan. His primary research interests include transistor and chip-level reliability specifically those which integrate logic, SRAM and NVM memory and has led the certification of multiple technology nodes co-integrating these devices specifically for automotive and networking applications.

Han, T. T. (5D.2, 5D.3)
Tzung-Ting Han was born in I-Lan, Taiwan, ROC., on November 18, 1973. He received the B.S. degree in engineering science from National Cheng Kung University, Tainan, Taiwan, R.O.C. in 1997 and the M.S. degree in electrical engineering from National Cheng Kung University, Tainan, Taiwan, R.O.C. in 1999. In 1999, he joined Macronix International Co., Ltd., Hsinchu, Taiwan, to work on advanced diffusion module process development. Since 2002, he has worked on process integration of advanced non-volatile memory.

Hang-Ting, Lue (5D.4)
Hang-Ting Lue was born in Hsinchu, Taiwan in 1975. He received his B.S and M.S degrees in physics from National Tsing-Hua University (NTHU) in 1997 and 1999, respectively, and Ph.D degree in electrical engineering in National Chiao-Tung University (NCTU) in 2002. He joined Emerging Central Lab. (ECL) in Macronix International (MXIC) in 2003. Currently he is the department manager of nano-technology R&D, and leads a team to develop the advanced Flash Memory devices, and the related theoretical modeling and reliability physics. One of his famous invention is "bandgap engineered SONOS" (BE-SONOS), which solves the fundamental problems of SONOS, providing both efficient hole tunneling erase as well as good data retention. Currently he is the project manager of BE-SONOS NAND Flash R&D, and deep involved in the NAND Flash test chip developments in the company. From 2004 he has published more than 30 papers in the premier semiconductor conferences including 10 IEDM, 5 VLSI and 15 IRPS, and a total of more than 75 technical papers in IEEE journal/letter/conference. He was also invited to give invited papers in ICSICT, MRS, and several workshops. So far he has 11 granted US patents, and more than 30 patents in applications. In 2007, he received the "Outstanding Young Innovator Award of the Industrial Technology Advancement Awards" by Taiwan's government. Currently, he is the chair of IRPS 2010 memory subcommittee, and committee member in VLSI-TSA.

Hashikawa, Naoto (PI.1, PI.2)
Naoto Hashikawa received his B.S. from the National Defense Academy in 1987 and his M.S. and Ph. D. in physics from Tokyo University of Science, Tokyo, Japan in 1992 and 1996, respectively.
Since 1996, he has been working for the Process Technology Development Div., Renesas Technology Corp. (now Renesas Electronics Corporation), on developing ULSI processes and device-analysis engineering. He is engaged in failure analysis of ULSIs using advanced TEM technology.

Hashimoto, Masanori (3A.4)
Masanori Hashimoto received the B.E., M.E., and Ph.D. degrees in communications and computer engineering from Kyoto University, Kyoto, Japan, in 1997, 1999, and 2001, respectively. Since 2004, he has been an Associate Professor with the Department of Information Systems Engineering, Osaka University, Osaka, Japan. His research interests include computer-aided-design for digital integrated circuits, and high-speed circuit design. Dr. Hashimoto was a recipient of the Best Paper Award at ASP-DAC 2004. He is a member of IEEE, IEICE, and IPSJ. He served on the technical program committees for international conferences including DAC, ICCAD, ASP-DAC, ICCD, and ISQED.

Heh, Dawei (4A.4)
Dawei Heh received the B.S. degree in physics from National Taiwan University, Taipei, Taiwan, R.O.C., in 1996 and the M.S. and Ph.D. degrees in electrical engineering from the University of Maryland, College Park, in 2001 and 2005, respectively.He was with the CMOS and Novel Devices Group, NIST, where he studied the reliability and breakdown mechanisms of ultrathin dielectrics. He joined SEMATECH as a research engineer in the electrical characterization and reliability group from 2005 to 2009. He is currently a associate researcher in the National Nano Device Laboratories in Taiwan. His research focuses on simulating and characterizing advanced gate dielectrics on different substrate materials. He also participates in developing new characterization techniques for future devices.

Heiderhoff, Ralf (3C.4)
Dr.-Ing. Ralf Heiderhoff became head of the group: "Optical and Thermal Microscopy Techniques" at the Department of Electronics within the Faculty of Electrical, Information, and Media Engineering at the University of Wuppertal, Germany, in 1997. 1999 he was Visiting Lecturer at the National University of Singapore followed by a Visiting Professorship at the Belarusian State University of Informatics and Radioelectronics 2001 and at the Beijing University of Technology 2008. His major research is focused on failure analyses and reliability investigations in the nanometer range as well as the combination of Scanning Electron Microscopes and Scanning Probe Microscopes to hybrid systems.

Hendriks, Teun (SE.1)
Teun Hendriks received his MSc degree from Delft University of Technology in 1986, with a specialization in aerospace engineering. He subsequently joined Philips Research Laboratories in Eindhoven, as a member of Research Staff, continuing in 1988 at Philips Laboratories in Briarcliff Manor, NY, USA. In 1996, he returned to The Netherlands to work for Philips on car navigation systems with emphasis on the integration of traffic information. Since 2005, he is employed by the Embedded Systems Institute (ESI) in Eindhoven, The Netherlands as a research fellow. His research interests lie in system reliability and interoperability in context of embedded systems.

Heryanto, Anson (5B.3)
Anson Heryanto was born in Indonesia in 1985. He received his B.Eng (2006) and M.Sc (2007) degrees in Electrical and Electronic Engineering from Nanyang Technological University (NTU), Singapore. He is currently working toward a Ph.D degree at the same institution. He is a recipient of the Singapore GLOBALFOUNDRIES-NTU Graduate Research Scholarship award. His current research interests are reliability of copper interconnects and 3D-chipstacks. He is currently a Graduate Student Member of IEEE.

Hideaki, Tsuchiya (6A.4)
Hideaki Tsuchiya received the B.S. and M.S. degrees in physics from Tohoku University, Sendai, Japan, 2000 and 2002, respectively. In 2002, he was with NEC Corporation, Sagamihara, Japan. Since 2002, he has been with the Advanced Device Development Division, NEC Electronics Corporation, Kawasaki, Japan, where he has been working on reliability engineering of Cu interconnects.

Hideki, Makiyama (PI.3)
Hideki Makiyama received the B.S. and M.S.degrees in material engineering from Kyushu University in 2005 and 2007, respectively. In 2008, he joined the Renesas Technology Crop., Tokyo, Japan. Since then, he has been working on research and development of Low Leakage MOSFET devices.

Hideya, Matsuyama (3A.5, 4A.5)
Hideya Matsuyama received his B.S. from Nagoya University (1985), Nagoya, Japan. He joined FUJITSU LIMITED in 1985. He has worked for Product reliability for several years, and has worked for Wafer Level reliability of BEOL and FEOL. Currently he is a director of ULSI reliability. He is a chairman of JEITA (Japan Electronics and Information Technology industries Association) Failure Mechanism Wafer Reliability Project Group and IEC (International Electro technical Commission) TC47/WG5 Japanese International Expert.

Hirano, Izumi (4C.2)
Izumi Hirano, received the B.S. (2000) and M.S.(2002) in physics from Kyoto University, Kyoto, Japan. She joined the Advanced LSI Technology Laboratory, TOSHIBA Corporation, Yokohama, Japan, in 2002, where she has been engaged the study of the reliability of gate dielectrics, especially High-k gate dielectrics, for ULSI technology. She is a member of the Japan Society of Applied Physics.

Hiroko, Mori (4A.5)
Hiroko Mori received the B.S. and M.S. degrees in Physics from Keio University, Kanagawa, Japan, in 1996 and 1998, respectively. She joined Fujitsu Ltd. in 1998. Her currently work is reliability physics for gate dielectrics of ULSI devices.

Hiroshi, Minakata (4A.5)
Hiroshi Minakata was born in 1969 in Aichi Pref., Japan. He graduated from Higashiyama technical highschool, Aichi, Japan in 1988. He joined Fujitsu Lab. Ltd., Atsugi, Japan in 1988. He is currently concerned with development of the gate-dielectric, SiON and High-k, process at Fujitsu Microelectronics Pacific Asia LTD., R.O.C

Hiroshima, Shoichi (4C.1)
Joined NEC Corporation. in 1987. He is working on the physical analysis of LSI process from 2002, in NEC Electronics Corporation.

Hirotoshi, Terada (FA.4)
Hirotoshi Terada received the B.S. degree in Mechanical Engineering from Keio Univ., Japan in 1985. He joined Hamamatsu Photonics K.K., and worked as an optics engineer. He has studied video microscopy and developed confocal microscope at Marine Biological Laboratory, M.A.,U.S.A. in 1991. He is now in charge of developing optical systems in Hamamatsu Photonics K.K..He is a member of Reliability Engineering Association of Japan.

Hiroyuki, Yoshimoto (PI.3)
Hiroyuki Yoshimoto received the B.S., M.S., and Ph.D. degrees in solid state physics from Waseda University in 2000, 2002, and 2006, respectively. In 2006, he joined the Central Research Laboratory, Hitachi, Ltd., Tokyo, Japan. Since then, he has been working on research and development of CMOS devices including SOI and high-k MOSFETs. Since 2009, he has also been working on research and development of power electronics devices.

Ho, Paul (5B.2)
Dr. Paul S. Ho is the Director of the Laboratory for Interconnect and Packaging at The University of Texas at Austin. He received his Ph.D. degree in physics from Rensselaer Polytechnic Institute. In 1972, he joined the IBM T.J. Watson Research Center and became the Senior Manager of the Interface Science Department in 1985. In 1991, he joined the faculty at the University of Texas and was appointed the Cockrell Family Regents Chair in Materials Science and Engineering department. His current research is in the areas of materials and processing science for interconnect and packaging applications.

Hoel, Virginie (2E.3)
Virginie Hoel received the Ph.D. degree from the university of Lille, Lille, France, in 1998. She is currently an Assistant Professor at Institut d'Electronique, de Microelectronique et de Nanotechnologie (IEMN), Villeneuve d'Ascq, France. From 1995 to 2000, she was working on design, fabrication and characterization of InP HEMTs for application in millimeter- and submillimeter-wave ranges. Since 2000, she worked on power devices. His main research interests include design, fabrication and characterization of AlGaN/GaN HEMTs for high power applications in centimeter- and millimeter-wave ranges. She carries out his teaching activities at Lille1 University in the field of electrical engineering.

Hoffmann, Thomas (IC.9, XT.13)
Dr. Thomas Hoffmann has earned his diploma in electrical engineering, specializing in engineering statistics from the Dresden University of Technology (Germany) in 1988. He finished his PhD study at the same university on topics of finite-element modeling (FEM) of plasma CVD reactors in 1991. After post-doc research work in the fields of computer simulation and statistics for semiconductors and microsystems he joined GLOBALFOUNDRIES (formerly AMD) (Germany) as quality engineer in 2001. At GLOBALFOUNDRIES he established and improved methods and processes for manufacturing quality and reliability statistics. He is responsible for the coordination of manufacturing quality support at GLOBALFOUNDRIES (Germany).

Hofmann, Ralf (NA.2)
Ralf Hofmann is a Member of Technical Staff in the Advanced Technology Group under the Office of the CTO at Applied Materials. His work involves the evaluation and development of new processing techniques and equipment for new and emerging applications in the IC space as well as the energy sector. He joined Applied Materials in 1994 in the PVD Technology group,

working on various aspects of metal deposition and process integration. In 2001 he joined ZMD America, Inc., a mixed-signal ASIC design company, where he worked on two successfully productized IC designs. He worked from 2002 to 2003 as an independent consultant for a telecommunications start-up company as well as a small company in the transportation sector before rejoining Applied Materials in 2004. Mr. Hofmann received a Diplom-Ingenieur degree in Electrical Engineering (MSEE) from Chemnitz University of Technology in Germany.

Holman, Tim (3A.2)
W. Timothy Holman received the B.S.E.E. degree from the University of Tennessee, Knoxville, in 1986 and the M.S.E.E. and Ph.D.E.E. degrees from the Georgia Institute of Technology, Atlanta, in 1988 and 1994, respectively.From 1994 to 2000, Dr. Holman was an Assistant Professor at the University of Arizona. In 2000, he joined the Department of Electrical Engineering and Computer Science at the Vanderbilt University School of Engineering, Nashville, Tennessee. His current research interests include the modeling and mitigation of radiation effects in analog and mixed-signal microelectronics. Dr. Holman is a member of the Institute for Space and Defense Electronics at Vanderbilt University.

Hong, Chong-A (MY.7)
Chong-A Hong received the B.S. degree in electrical engineering from Chung-Nam University, Korea, in 2006 She joined the Flash Memory Division of Hynix Semiconductor Inc.,Gyunggi-Do, Korea , where she has been working on the failure analysis of Nand Flash Memories. Currently, she is involved in the development of 3x-nm Nand flash.

Hong, Shih-Ping (MY.1)
Shih-Ping Hong was born in Pingtung, Taiwan in 1975. He received his PhD. in Chemistry from National Tsing-Hua University, Taiwan, R.O.C., in 2002. He is currently a project deputy dept. manager of Etch Process Development Department in Technology Development Center in Macronix International (MXIC).

Hsiao, Yi-Hsuan (MY.1)
Yi-Hsuan Hsiao was born in Chia-Yi, Taiwan in 1980. He received his B.S and M.S degrees in electrical engineering from National Chiao-Tung University (NCTU) in 2002 and 2004, respectively. His M.S thesis focused on TFT devices. He joined Emerging Central Lab. (ECL) in Macronix International. Co. (MXIC) in 2005, where his current research includes developing new nonvolatile memory devices and reliability studies of nitride trapping Flash Memories.

Hsieh, Jung-Yu (MY.1)
Jung-Yu Hsieh was born in Hsinchu, Taiwan, R.O.C., in 1971. He received the B.S. degree in chemical engineering from the National Tsing Hua University, Hsinchu, Taiwan, in 1995, and the M.S. degree in chemical engineering from the National Tsing Hua University, Hsinchu, Taiwan, in 1997. He joined the Advanced Module Process Development Division of Macronix International (MXIC), in 1997 where he has been working on the process development of chemical mechanical planarization (CMP). His current works are engaged in advanced diffusion module process development.

Hsieh, Kuang-Yeu (5D.2, MY.1)
Kuang-Yeu Hsieh was born in Tainan Taiwan in 1958. He received his B.S (Physics) and MS (Materials Science) degrees from National Tsing-Hua University (NTHU) and National Sun Yet-sen University (NSYSU) in 1980 and 1985, respectively. He obtained his Ph.D. at North Carolina State University in Materials Science in 1989. Before he joined Macronix International in 2001, he had been an Associate Professor at Institute of Material Science in National Sun Yet-sen University since 1992. His research interests include MBE thin film growth, characterization of material, solid-state physics, IC

fabrication, and optoelectronic materials. Currently, he is the Director of Nano-technology R & Div./Emerging Central Lab in Macronix and involves in developing new nonvolatile memory devices and exploring new material for the next generation nonvolatile memory. Meanwhile, he has more than 70 journal papers published and 15 patents granted.

Hsieh, Sunnys (2F.3)
Sunnys received a B.S degree in Chemistry major from Chung-Yuan Christian University, Taiwan in 1993, and the M.S. degree in Chemical Engineering from National Tsing-Hua University, Taiwan in 1995. He worked in Mosel-Vitelic in 1995 and joined TSMC since 2000 till now. He involved process integration in R&D for SRAM development, and then joined HV reliability team as a section manager in 2005. Currently, he works on HV PMIC Field Technology Service in Asia-Pacific Business Unit.

Hsu, Fang-Hao (MY.1)
Fang-Hao Hsu received the M.S. degree in mechanical engineering from the National Taiwan University of Science and technology, Taipei, Taiwan, R.O.C., in 2006. He is currently a Principal Engineer in the etch process development department of Technology Development Center, Macronix International (MXIC).

Hsu, Tzu-Hsuan (5D.3, MY.1)
Tzu-Hsuan Hsu received the B.S. and M.S. degrees from the Department of Electrical Engineering and Institute of Microelectronic Engineering, National Cheng-Kung University, Tainan, Taiwan, R.O.C., in 2000 and 2002, respectively, and Ph.D. degree in Institute of Electronics, National Tsing-Hua University (NTHU) in 2009. He joined Emerging Central Lab (ECL) in Macronix International Company, Ltd., Hsinchu, Taiwan, R.O.C., in 2002 working on nanodevice research and development. His current research areas include high-density memory development, nitride-trapping memory devices, and advanced nonvolatile memory technologies.

Hu, Youzhou (XT.5)
Y. Z. Hu, received the B.Sc. and M.Sc. degrees in Applied Physics from TianJin University, China, in 2001 and 2004, respectively. He worked as a Process Integration Engineer in SMIC of China and Chartered Semiconductor Manufacturing for two and half years. He is currently pursuing the Ph.D. degree in the School of Electrical and Electronic Engineering, Nanyang Technological University. His research interest lies in the meaurement methodologies and mechanisms of bias temperature instability.

Huang, I. J. (5D.2)
I-Jen Huang received the B.S. degree in electrical engineering from National Taiwan University, Taiwan, in 1992, and the M.S. degree in electrical engineering from National Taiwan University in 1994. He joined Macronix International Co., Ltd., Hsinchu, Taiwan, in 1996 as a process integration engineer. He has worked on process integration of non-volatile memory and SRAM for several generations. He is presently engaged in array characterization of NVM products.

Huang, Elbert (5A.5)
Elbert Huang received B. S. degrees in Chemical Engineering and Materials Science from the University of Minnesota in 1995 and his Ph. D. in Polymer Science and Engineering from the University of Massachusetts in 1999. From 1999 through 2001 he worked at the IBM Almaden Research Center on the development and characterization of ultralow-k porous dielectrics. He moved the IBM T. J. Watson Resarch Center in Yorktown Heights in 2001, where he has been involved in the research and development of exploratory materials and integration for advanced CMOS technologies.

Huang, I-Jen (5D.3)
I-Jen Huang received the B.S. degree in electrical engineering from National Taiwan University, Taiwan, in 1992, and the M.S. degree in electrical engineering from National Taiwan University in 1994. He joined Macronix International Co., Ltd., Hsinchu, Taiwan, in 1996 as a process integration engineer. He has worked on process integration of non-volatile memory and SRAM for several generations. He is presently engaged in array characterization of NVM products.

Huang, Jyun-Siang (5D.3)
Jyun-Siang Huang was born in Chiayi, Taiwan,R.O.C., in 1981. He received the M.S. degree from Electrophysics Department, National Chiao Tung University, Hsinchu, Taiwan, in 2006. In 2006, he joined the Advanced Device Department of Macronix International Company Ltd., Hsinchu, Taiwan. His research interests include the device characterization of Flash memory devices

Huang, Rui (IC.6)
Rui Huang received the B.S. degree in materials science and engineering from Zhejiang University, Zhejiang, China, and the M.S. degree in materials science and engineering from Kiel University, Kiel, Germany. He is currently working toward the Ph.D. degree with Kompetenzzentrum Automobil-und Industrieelektronik GmbH (KAI), Villach, Austria, in cooperation with the institute of Microelectronics, Technische Universität Wien (TU Vienna), Vienna, Austria.
His research is focused on mechanical and thermo-mechanical properties of copper metallization used in semiconductor devices.

Huang, Yu-Hui (2F.3)
Yu-Hui Huang was born in Taoyuan, Taiwan, R.O.C., on Sep. 19, 1979. She received the B.S. degree in Physics from National Central University, Taoyuan, Taiwan, R.O.C., in 2002, and the M.S. degree in electronics engineering from National Chiao Tung University, Hsinchu, Taiwan, R.O.C., in 2004. She has been working in Taiwan Semiconductor Manufacturing Company (TSMC), Hsinchu, Taiwan, R.O.C., as an High Voltage Process Integration Engineer from 2004 to 2008 and a HV Reliability Engineer now.

Hurkx, G. A. M. (2C.5)
Fred Hurkx was born in Best, The Netherlands, in 1956. He received the MSc degree in physical engineering and the Ph.D. degree from the Technical University of Eindhoven, The Netherlands, in 1985 and 1990, respectively. His thesis involved the modeling of downscaled bipolar transistors. In 1979 he joined Philips Research Laboratories, Eindhoven, where he has been working in the field of semiconductor research since 1983. Currently he is with NXP-TSMC Research Center, Leuven, Belgium.

Hurley, Paul (BD.1)
Paul Hurley received his Ph.D. (1990) and B.Eng.(1985 - First class honors) in Electronic Engineering at the University of Liverpool. He is a Senior Research Scientist at the Tyndall National Institute, University College Cork where his work focuses on high dielectric constant (high-k) materials intended for use as gate level insulators in transistors for future integrated circuits. The emphasis of his current research is the formation and characterisation of high-k films on silicon and III-V semiconductor substrates. Paul is a member of the Technical Committee of the Insulating Films on Semiconductors (INFOS) conference and the International Workshop on Dielectrics in Microelectronics (WoDiM). In addition to research activities, he is a part time lecturer in the Department of Microelectronic Engineering at University College Cork. He has published over sixty papers in the field of microelectronics.

Hutter, Herbert (XT.6)
Herbert Hutter received his Ph.D. degree in technical sciences from the TU Wien where he is currently employed as an Associate Professor. Dr. Hutter is the co-author or author of over 140 scientific articles; all of them were published in the field of Secondary Ion Masspectrometry (SIMS) of materials and measurement data treatment. His main working areas are material analysis, corrosion of nitrides, and diffusion of oxygen in oxides (using isotope enriched oxygen) and oxidation of metals. In the field of semiconductors he is working on diffusion and migration of light mobile ions and high energy implantation

Ielmini, Daniele (5D.1, 6C.2, 6C.3)
Daniele Ielmini received the Laurea (cum laude) and Ph.D. in Nuclear Engineering from Politecnico di Milano in 1995 and 1999, respectively. In 1999, he joined the Dipartimento di Elettronica e Informazione, Politecnico di Milano, where he is an Assistant Professor from 2002. In 2006 he was Visiting Scientist at Intel Corporation and the Center for Integrated Systems (CIS), Stanford University. His most recent research interests include the modeling and the characterization of emerging phase change memory (PCM) and resistive switching memory (RRAM). He authored/coauthored two book chapters, more than 140 papers in international journals and international conferences.

Iida, Hitoshi (FA.4)
Hitoshi Iida Received MSEE from Tokyo Institute of Technology in 1971. Has been involved in technical management for the development of various image acquisition/analysis/processing systems at Hamamatsu Photonics K.K. Currently, General Manager of Systems Division at Hamamatsu Photonics K.K. Japan

Inaba, Yutaka (PI.1)
Yutaka Inaba received his B.E. in electrical engineering from Kinki University, Osaka, Japan in 1988. He joined Mitsubishi Electric Corporation, Hyogo, Japan in 1988. Since 2003, he has been working as a senior engineer for Renesas Technology Corp. (now Renesas Electronics Corporation), Ibaraki, Japan, where he is engaged in developing diffusion processes. Mr. Inaba is a member of The Japan Society of Applied Physics.

Inan, Umran (3A.3)
Umran S. Inan is Professor of Electrical Engineering and Director of the Space, Telecommunications, and Radioscience Laboratory at Stanford University. Prof. Inan has published over 300 refereed journal articles and 2 textbooks in electromagnetism, and actively conducts research on near-Earth low frequency electromagnetic waves and their applications in space telecommunications. Prof. Inan is a Fellow of IEEE, the American Geophysical Union (AGU), and the American Physics Society (APS). He is the recipient of the Appleton Prize of the International Union of Radio Science and Royal Society. He is also currently President of Koç University, Istanbul, Turkey.

Inumiya, Seiji (4C.2)
Seiji Inumiya received B.S. and M.S. degrees in physics from Waseda University, Tokyo, Japan, in 1992 and 1994, respectively. He joined the R & D Center, Toshiba Corporation, in 1994, where hewas engaged in the research on the process and the reliability of thin silicon oxide in Si MOS devices. In 1997, he moved to the Process & Manufacturing Engineering Center, Semiconductor Company, Toshiba Corporation, Yokohama, Japan, where he has been engaged in the research and the development of high-k gate dielectrics and silicon oxynitride for logic LSI devices. In 2004, he joined Semiconductor Leading Edge Technology, Inc., where he had worked on the process development and the reliability of high-k

gate stacks. Currently, he is in charge of the high-k gate stack process development at Process & Manufacturing Engineering Center, Toshiba Corporation.He is a member of the Japan Society of Applied Physics.

Ioannou, Dimitris (EL.3, XT.1)
Dimitris E. Ioannou: BS in Physics (1974), University of Thessaloniki, Greece; MS (1975) and PhD (1978) in Solid-State Electronics, University of Manchester, UK. Prior to joining George Mason University as a professor of electrical engineering, held positions at Manchester and Middlesex Universities (UK), Democritus University of Thrace (Greece), University of Maryland (USA) and the Institute National Polytechnique de Grenoble (France). Authored or coauthored over two hundred fifty research papers and conference presentations, and advised over thirty five research students. Technical program chairman (2001) and general chairman (2002) of the IEEE Inter. SOI Conf. Recipient (twice: 04 and 05) of the IBM Faculty award and Fellow IEEE (2010).

Irene, Tee (3C.5)
Irene TEE was born in Malaysia, she attended Informatics Computer School where she completed diploma & advance diploma in computer science discipline before gained her Bachelor Degree, Information System & Software Engineering from Oxford Brooke University (UK) in 2005.
Irene is currently working as Failure Analysis Engineer for GLOBAL FOUNDRIES SINGAPORE. Her main responsibilities include Transmission Electron Microscopy (TEM) sample preparation & analysis and overall laboratory IT supports. She is the author of 1 paper and co-author of several papers plus 1 patent on failure analysis and TEM sample preparation technique.

Ishigaki, Takashi (XT.2)
Takashi Ishigaki received the B.E. and M.E. degrees in electrical and electronic engineering from Kobe University, Hyogo, Japan, in 1999, 2001, respectively. He joined Compound Semiconductor Division, NEC Corporation in 2001, where he was involved in the research and development of GaAs HJFETs and HBTs for microwave power amplifiers and RF switch ICs. Since 2004, he has been working in Central Research Laboratory, Hitachi, Ltd. on the research and development of high-capacity flash memories and SOI CMOSFETs. Mr. Ishigaki is a member of the IEEE Electron Devices Society.

Islam, Ahmad Ehteshamul (2B.3, XT.7)
Ahmad Ehteshamul Islam received the B.S.E.E from Bangladesh University of Engineering and Technology (BUET) in 2004. He is currently enrolled in the direct Ph.D. program at ECE, Purdue University. During 2004-2005, he worked as Lecturer in the Department of EEE, BUET. His current research focuses on the variation resilience aspects of strained/III-V CMOS transistors. He has (co)-authored more than 20 journals and conference papers. He is a student member of the IEEE EDS and APS and also serves as a reviewer for several IEEE, Elsevier, APS and ECS journals. He is also the recipient of Kintar-Ul-Haque Gold Medal (2005) for his undergraduate result, and IEEE EDS PhD Fellowship (2008), Intel PhD Fellowship (2009-2010) for his work on transistor reliability.

Itaru, Yanagi (PI.3)
Itaru Ynagi received the B.S., M.S. degrees in solid state physics from Waseda University in 2003, 2005. In 2005, he joined the Central Research Laboratory, Hitachi, Ltd., Tokyo, Japan. Since then, he has been working on research and development of non-volatile memory.

Ito, Shuu (4C.1)
Received the B.S. (1990) and M.S. (1992) in mathematical science from Osaka Prefecture University, Osaka, Japan. In 1992, he joined NEC Corp. He is currently engaged in the research and development

of embedded flash memory LSIs. He is a member of the Physical Society of Japan and the Japan Society of Applied Physics.

Iwai, Hidenao (FA.4)
Hidenao Iwai received the B.S. (1995) in Mechanical Engineering and the M.S. (1997) in Biomedical Engineering from Keio Univ., Japan. In 1997, he joined Hamamatsu Photonics K.K., and worked on diffused optical tissue spectroscopy. He was at Massachusetts Institute of Technology from 2001-2002 as a visiting scientist, where he was involved in low-coherence interferometry. Currently he moved back to Hamamatsu Photonics K.K. His research interests include biological application of interferometric quantitative phase microscopy.

Jack, Nathan (4D.5, EL.1)
Nathan Jack received a B.S. from Utah State University in 2007 and a M.S. in 2009 from the University of Illinois at Urbana-Champaign, both in Electrical and Computer Engineering. He is currently pursuing a Ph.D. degree at UIUC in ECE. His research interests include on-chip ESD protection and ESD test methods. He has completed five summer internships at Micron Technology, Inc, in Boise, Idaho, of which two were in the ESD/LUP R&D Reliability Group.

Jagannathan, Hemanth (XT.14)
Hemanth Jagannathan (M'98) received the M.S. and Ph.D. degrees in electrical engineering from Stanford University in 2003 and 2007 respectively. He is currently a Research Staff Member at IBM Research based at IBM @ Albany NanoTech, Albany, NY and IBM T. J. Watson Research Center, Yorktown Heights, NY. His research interests include high-k/metal gate CMOS devices and integration, semiconductor technology, semiconductor nanowire synthesis and device design, nanoscale science, materials, and technology.

Jain, Palkesh (6A.1)
Palkesh Jain graduated from the Indian Institute of Technology (IIT) Bombay in 2004, with Bachelors and Masters in Electrical Engineering. His Masters thesis research was on soft errors and Radiation Reliability. Since then, he has been with the Reliability CAD Group at ASIC, Texas Instruments India. He has contributed to the CAD and methodology development for several technology nodes for reliability assessment. He has over ten international publications and 6 filed patents to date.

Jammy, Raj (2B.4)
Raj Jammy received his doctoral degree in Electrical Engineering from Northwestern University (1996). He joined IBM's Semiconductor Research and Development Center in East Fishkill, NY, where he worked on various aspects of DRAM technology development in engineering and managerial roles. In 2002 he moved to IBM T. J. Watson Research Center in Yorktown Heights, NY, to manage IBM's efforts on high k gate dielectrics and metal gates. From 2005 to 2008 he was an IBM assignee to SEMATECH as the Director of the Front End Processes Division in Austin, TX. Since June 2008 he has been with SEMATECH and currently serves as Vice President of Materials and Emerging Technologies. He holds more than 50 patents and is an author/co-author of over 200 publications/presentations.

Jeng-Hwa, Liao (5D.4)
Jeng-Hwa Liao was born in Taipei, Taiwan in 1979. He received his Ph. D degree in material science and engineering from National Tsing-Hua University (NTHU) in 2007. He joined Technology Development Center in Macronix International (MXIC) since 2007. His current works are focused on advanced diffusion module process development, and technology development of NAND Flash and charge trapping memories.

Jeong, YeonJoo (MY.7)
YeonJoo Jeong received the B.S. degree in Quantum and Electronic Engineering from the University of Tsukuba, Tsukuba, Japan, in 2007, and the M.S. degree in Electrical Engineering from the University of Tokyo, Tokyo, Japan, in 2009. He is currently with the Flash Memory Division, Hynix Semiconductor Inc., Cheongju, Korea, working on device engineering of Hynix's multi-level Flash cell.

Jeong, Yoon-Ha (2C.4)
Received the Ph.D. (1987) in electronics engineering from the University of Tokyo, Japan, where he pioneered in situ vapor phase deposition and the development of photo-chemical vapor deposition (CVD) technology for InP metal–insulator–semiconductor field-effect transistors (MISFETs). In 1987, he joined the Pohang University of Science and Technology (POSTECH), Pohang, Korea, where he is a Professor with the Department of Electronics Engineering and a Director of the National Nano Devices Center for Industry, where he is involved with nano-CMOS devices and circuits for RF applications. His research interests include microwave and millimeter-wave device fabrication, RF circuit design, single electron transistors, and nano-CMOS devices. He was a conference chair in IEEE Nanotechnology Materials and Devices Conference 2006 (IEEE NMDC). He was nominated on Who`s Who in Science and Technology. He is IEEE senior member, director in National Nanodevices Center for Industry, IEE fellow member, director in National Center for Nanomaterials Technology (NCNT) of the MKE (Korea), industrial developement committee member in Ministry of Commerce Industry and Energy (MOCIE), chair in IEEE EDS Yeongnam chapter (Korea), and board member in National Science & Technology Council (Korea).

Jimenez, Jose (4F.3)
Jose Jimenez received the B.A (1992) in Electrical Engineering from the Universidad Politecnica de Madrid and the Ph. D. (1996) in Electrical Engineering from Columbia University in New York. He has worked in both integrated optics and in transport and optoelectronics semiconductor devices for the last 15 years in Telefonica R&D (Spain), T. J. Watson IBM Research Laboratory, Beckman Institute and Nanovation. For the last six years, he has been part of the R&D organization of TriQuint Semiconductor focusing early on in 4" inch optoelectronics devices (DFB lasers and high speed photodetectors) and later in GaN FET technology (physics, test and reliability).

Jo, Das (2E.4)
Jo Das received the M.Sc. and Ph.D. degrees in electrical engineering from the Catholic University of Leuven, Leuven, Belgium, in 1998 and 2003, respectively. His research topic for his Ph.D. dissertation is on magnetic random access memories. Since 2003, he has been Senior Engineer with imec, Leuven, Belgium. His current research interests are the fabrication and circuit integration of GaN-based Devices for Power switching and RF power applications.

John, Aitken (5A.5)
John Aitken joined IBM in 1974 at the T.J. Watson Research Center. He is a currently a Senior Technical Staff Member at IBM Burlington managing the Semiconductor Technology Reliability Engineering Department evaluating reliability of new technologies and materials for advanced semiconductor device technologies . He is a Senior Member of the IEEE and past committee member and General Chairman of the International Electron Devices Meeting . He received a MS and Ph.D. in Physics /Materials Science from Rensselaer Polytechnic Institute 1972 and a BS in Physics from

Fordham University. John is an Adjunct Professor at University of Vermont.

Jung, Seong-Ook (RM.3)
Seong-Ook Jung received the Ph.D. degree from University of Illinois at Urbana Champaign in 2002. From 1989 to 1998, he was with Samsung Electronics where he worked on specialty memories and merged memory logic. He led thyristor based memory design team in T-RAM Inc. from 2001 to 2003. He also worked on embedded memories, process variation tolerant circuit , and low power circuit in Qualcomm Inc. from 2003 to 2006. He has been an associate professor in Yonsei University since 2006. His research interest includes process variation tolerant circuit, low power circuit, mixed-mode circuit, and future generation memory.

Jungemann, Christoph (2F.4)
Christoph Jungemann holds the chair for Microelectronics at the Bundeswehr University. He has more than 15 years of experience in the field of physics-based device simulation. He developed the first full-band Monte Carlo device simulator for SiGe HBTs and pioneered numerical methods for the solution of the Langevin-Boltzmann equation. He is a co-developer of hierarchical device simulation including the first 2D bipolar hydrodynamic model for noise in SiGe devices, where all transport and noise parameters are generated by consistent Monte Carlo simulations. He developed CPU efficient models of impact ionization for the classical device simulators. He has authored one book on hierarchical device simulation and more than 180 journal and conference papers in this field. He is a member of the editorial board of the IEEE Transaction on Electron Devices and a recipient of the IEEE EDS Paul-Rappaport-Award for 2005. He was or is a program committee member of IEDM, ICCAD, and IWCE.

Jung-Yu, Hsieh (5D.4)
Jung-Yu Hsieh was born in Hsinchu, Taiwan, R.O.C., in 1971. He received the B.S. degree in chemical engineering from the National Tsing Hua University, Hsinchu, Taiwan, in 1995, and the M.S. degree in chemical engineering from the National Tsing Hua University, Hsinchu, Taiwan, in 1997. He joined the Advanced Module Process Development Division of Macronix International (MXIC), in 1997 where he has been working on the process development of chemical mechanical planarization (CMP). His current works are engaged in advanced diffusion module process development.

Jurczak, Malgorzata (6C.1)
Malgorzata Jurczak received M.Sc. and Ph.D. in electrical engineering from the Warsaw University of Technology where she worked as teaching assistant and research scientist. She was with NMRC, Ireland and Kyung Hee University, Korea in 1994 and 1997, respectively. In 1998-99 she worked on 0.18 and 0.12μm CMOS and alternative CMOS approaches at CNET, France Telecom. In 2000 she joined IMEC, Belgium. From 2000 to 2003 she was leading the IMEC-Philips JDP on 90nm and 65nm CMOS. From 2003 to 2007 she coordinated FINFET project. From 2008 she has been the manager of NVM and Emerging Memories programs.

K Singh, Pawan (2C.3)
Pawan K Singh received B. Tech. and M. Tech. combined degrees (2006) from Electrical Engineering department of the Indian Institute of Technology (IIT) Bombay, India. Since 2006, he is pursuing Ph.D. in Electrical Engineering department in IIT Bombay. He is an AMAT fellow for PhD in IIT Bombay since 2006. He has been a graduate intern at Applied Materials, Santa Clara, CA from June 2007 - July 2008. His research interests include technology development, physics, and

reliability characterization of non-volatile memories. He is currently working on metal nanocrystal based non-volatile memories for NAND application. He is a student member of the IEEE.

Kaczer, Ben (2A.3, XT.10, XT.13)
Ben Kaczer received the M.S. degree in physical electronics from Charles University, Prague, Czech Republic, in 1992 and the M.S. and Ph.D. degrees in physics from Ohio State University (OSU), Columbus, in 1996 and 1998, respectively. For his Ph.D. research on the ballistic-electron emission microscopy of SiO2 and SiC films, he received the OSU Presidential Fellowship and support from Texas Instruments, Inc. Since 1998, he has been with the reliability group of IMEC, Leuven, Belgium, where his activities have included the research of the degradation phenomena and reliability assessment of SiO2, SiON, high-k, and ferroelectric films, planar and multiple-gate FETs, circuits, and characterization of Ge/III-V and MIM devices. He is the author or a coauthor of more than 100 journal and conference papers and has presented eight invited presentations and two International Reliability Physics Symposium (IRPS) tutorials. Dr. Kaczer is the recipient of Best and the Outstanding Paper Awards at IRPS and the Best Paper Award at IPFA. He has served or is serving at various functions at the International Electron Devices Meeting (IEDM), IRPS, IEEE Semiconductor Interface Specialists Conference (SISC), and Conference on Insulating Films on Semiconductors (INFOS).

Kalya, Shubhakar (4A.1, BD.2)
Shubhakar K received his B.E (Electronics and Communication, 2000) and M.E (Microelectronics, 2007) from NMAMIT Nitte, Mangalore and Indian Institute of Science (I.I.Sc), Bangalore respectively. He is currently pursuing his Ph.D. at the Division of Microelectronics, School of EEE, Nanyang Technological University working on reliability physics and degradation mechanisms in novel high-κ dielectric materials using scanning tunneling microscopy (STM) and conductive atomic force microscopy (CAFM) technique.

Kamino, Takeshi (PI.1)
Takeshi Kamino received his B.E. and M.E. in materials physics and engineering from Osaka University, Osaka, Japan in 1992 and 1994, respectively. He joined Sumitomo Metal Industries, Ltd., Osaka, Japan in 1994 and worked on the development of magnetic thin-film heads for hard disk drives. In 2000, he joined Mitsubishi Electric Corp., Hyogo, Japan and was then transferred to Renesas Technology Corp. (now Renesas Electronics Corporation), Hyogo, Japan in 2003. He has been engaged in the research and development of deep-submicron CMOS devices and process integration. He is working on the research and development of advanced SoC technology.

Kang, Han-Byul (3C.3)
Han-Byul Kang received the Ph. D. degree in advanced material science & engineering from Sungkyunkwan University, Korea, in 2004. His thesis topic was interfacial reaction between environmental friendly solder and Ni-P UBM by using analytical transmission electron microscopy. He is currently with Technology Reliability, Q&R team in SYSTEM LSI division, Samsung Electronics as a senior engineer and is involved in package level reliability. He has authored/coauthored over 30 international papers and patents.

Kang, Jinfeng (MY.4)
Kang Jinfeng received his B.S. degree in physics from Dalian University of Technology in 1984, and M.S. and Ph.D degrees in solid-state electronics from Peking University in 1992 and 1995 respectively. Next, he joined Institute of Microelectronics in Peking University as a post-doctoral fellow. In 1997 he joined the faculty first as an associate professor then professor in 2001. He is the

author of a book and more than 100 conference and journal papers. Currently his research interests are in the areas of novel memory technology, high-k/metal gate technology and MOS device physics, as well as solar cell technology.

Kang, Min-gu (RM.3)
Min-gu Kang was born in Seoul, Korea, in 1981. He received the M. S. degree in the electrical and electronic engineering from Yonsei University, Seoul, Korea, in 2009. Currently, He is working in SAMSUNG Semiconductor. His current research interests is FinFET SRAM circuitary.

Kato, Koichi (4C.2)
Koichi Kato, received the B.S., M.S., and Ph.D. degrees in physics from the University of Tokyo, Japan, in 1977, 1979, and 1982, respectively. He joined Toshiba Research & Development Center from 1982. His early activiy includes SOI device technology of body floating effects and thin body effects. He joined the Atom Technology national project from 1993 to 1997 in Tsukuba. His careers have moved to more fundamental technologies including layer-by-layer oxidation, novel SiON films, high-k insulating films, and dipole comforting Schottky barriers mostly through a theoretical viewpoint, but including an experimental viewpoint. His current research interest is interface engineering with atom manipulation. He is a member of the Japan Society of Applied Physics and a member of the Physical Society of Japan.

Katsuji, Ono (4A.5)
Katsuji Ono received the M.S. degree in electrical engineering from The University of Electro-Communications, Tokyo, Japan in 1987, where his research focused on Liquid-phase selective epitaxial growth of GaAs for Junction FET. He joined Compound Semiconductor Devices Laboratory of Fujitsu laboratories, Kanagawa, Japan in 1987. He is currently working on characterizing CMOS devices for advancing device performance and fabrication technologies in Fujitsu Microelectronics Limited (FML).

Katsuto, Tanahashi (4A.5)
Katsuto Tanahashi was born in Gifu, Japan, in December 1964. He received the B.S. and M.S. degrees in materials science from the Nagoya Institute of Technology, Aichi, Japan, in 1992 and 1994, and Ph.D. degree in Physics from Osaka Prefecture University, Osaka, Japan, in 2001. He was a Research Fellowship for Young Scientist of Japan Society for the Promotion of Science in 2001 and 2001. He joined Fujitsu Laboratories Ltd., Atsugi, Japan, in 2001, where he studied physical analysis of semiconductor devices using FT-IR, photoluminescence, light scattering. Since 2010, he has been studying failure analysis of semiconductor devices.

Katsuya, Shiga (PI.3)
Katsuya Shiga received the B.S. (1994) and M.S. (1996) in electronics engineering from Kansai Univ., Osaka, Japan. He joined Mitsubishi Electric Corporation in 1996 (Renesas Technology Corporation from 2003) and he has been engaged in development of reliability engineering for ULSI devices.

Katsuyuki, Horita (PI.3)
Katsuyuki Horita received the B.S. and M.S. degrees in electronics from Waseda University in 1993 and 1995, respectively.In 1995, he joined the ULSI Laboratory of Mitsubishi Electric Corporation, where he was engaged in the research and development of CMOS device and process technology. He is now with Renesas Technology Corporation, continuing the research and development of CMOS.

Kauerauf, Thomas (6A.3, XT.9, XT.13)
Thomas Kauerauf received his degree in electrical engineering from the Technical University of Ilmenau, Germany, in 2001 and the Ph.D. degree from the Katholieke Universiteit Leuven, Belgium, in 2007. In 2006 he joined imec, Leuven, Belgium, where he is currently working in the Device Reliability and Electrical characterization group focusing on high-k gate stacks and the impact of Cu contacts on the FEOL reliability. From 1999 to 2000 he stayed several months at Bell Laboratories, Murray Hill, USA. Thomas Kauerauf has authored or co-authored more than 50 publications and received the IEEE SISC Ed Nicollian Award for the best student paper in 2001.

Kazuyoshi, Torii (PI.3)
Dr. Torii received B.S and M.S degrees in physics from Keio University,Tokyo, Japan, in 1986 and 1988, and a Ph.D. degree in engineering from Tokyo Institute of Technology, Tokyo, Japan, in 1998. From 1988 to 1999, he was with Central Research Laboratory, Hitachi, Ltd., Tokyo, Japan. He worked on process and integration of capacitor dielectrics for DRAM and FeRAM. From 2000 to 2004, he was working on high-k gate dielectrics for CMOS. In 2002, he joined Semiconductor Leading Edge Technology, Inc (SELETE). Dr. Torii is a member of the Japan Society of Applied physics and IEEE.

Kees, Beenakker (3F.3)
Kees Beenakker studied chemistry and physics at Leiden University, the Netherlands. He received the Ph.D. degree at the FOM-Institute for Atomic and Molecular Physics, Amsterdam, the Netherlands, in 1974. In 1974, he joined Philips' Research Laboratories, Eindhoven, the Netherlands. In 1982, he moved to the Philips Semiconductor Division, Nijmegen, the Netherlands. In 1987, he co-founded Eurasem, a European hi-rel IC Assembly Company. Since 1990, he has been a full Professor, Faculty of Electrical Engineering, Mathematics and Computer Science (EEMCS), Delft University of Technology, Delft, the Netherlands and Chairman of the Department of Microelectronics.

Keiji, Takahisa (3A.5)
Keiji Takahisa received the B.S. degree from the Hokkaido University, in 1986, and the M.Sc. and Ph.D. degrees in physics from Tohoku University, in 1988 and 1991, respectively. In 1991, he began work on nuclear physics in RCNP, Osaka University. He is currently Radiation Handling Supervisor, Division of Radiation Security Control, RCNP, Osaka University.

Keita, Nishigaya (4A.5)
Keita Nishigaya is an engineer in Si processes development department at Fujitsu microelectronics limited and is involved in gate dielectrics Reliability of 45nm technologies. He received his Bachelor's degree and Master's degree in engineering from Tokyo Institute of Technology.

Keller, Robert (2F.4)
Robert Keller received his Ph.D. in Geophysics from the Ludwig-Maximilians University in Munich in 1998. Since that time he is working at Infineon Technologies Munich.

Ken, Shono (3A.5)
Ken Shono received M.S Degree in Applied Physics from Osaka university, Japan in 1979. He joined Fujitsu in 1979 and he has been managed reliability-related activities in technology development of deep submicron MOSFET and multi-layer metallization system. He is also interested in mitigation of soft errors in LSI.

Ker, Ming-Dou (EL.6)
Ming-Dou Ker is now served as Chair Professor and Vice President of I-Shou University, Kaohsiung, Taiwan. On the topic of reliability and quality design for integrated circuits, he has published over 380 technical papers in international journals and conferences, and has been granted with 154 U.S. patents and 146 R.O.C. (Taiwan) patents. He was selected as the Distinguished Lecturer in the IEEE Circuits and Systems Society (2006–2007) and in the IEEE Electron Devices Society (2008-2010). In 2008, he has been elevated as an IEEE Fellow. He was the President of Foundation in Taiwan ESD Association.

Kerber, Andreas (2B.1, 4A.3)
Andreas Kerber was born in Schnann, Austria, in 1973. He received his Diploma in physics from the University of Innsbruck, Austria, in 2001 and a PhD in electrical engineering from the TU-Darmstadt, Germany in 2004 (granted with honor). He worked as intern at Bell Laboratories, Lucent Technologies, Murray Hill, NJ, USA (1999-2000), at IMEC in Leuven, Belgium (2001-03) as Infineon Technologies assignee to International SEMATECH, for the Reliability Methodology Department at Infineon Technologies in Munich, Germany (2004-06), for AMD in Yorktown Heights, NY (2006-09), and for GLOBALFOUNDRIES in Yorktown Heights, NY (since 2009). Much of his work centered around Front-End-Of-Line (FEOL) reliability research with focus on metal gate / high-k CMOS technologies. He has co-authored 65 papers in Journals and Conferences.

Kerst, Uwe (IC.2)
Dr.-Ing. Uwe Kerst is heading the electronic devices failure analysis lab at Berlin University of Technology. Additionally he is lecturing Quality Management and Reliability in the Semiconductor Industry. Before joining the University he worked for Infineon Technologies in the LDA alliance East Fishkill subsequently in the Fiber Optics division. His Ph.D. thesis (2000) covered the integrated series interconnection in thin film solar cells.

Killat, Nicole (4F.3)
Nicole Killat received the Dipl. Ing. degree in technical physics from the Ilmenau University of Technology, Germany, in 2008. She is currently working towards the Ph.D. degree at the Applied Spectroscopy Group, University of Bristol. Her current research interests include studying thermal and electrical properties of GaN-based devices using micro-Raman and electroluminescence spectroscopy.

Kim, Dae Mann (2C.4)
Received his B. S. degree from Seoul National University, Seoul, Korea and M.S. and Ph. D. degrees from Yale University, New Haven, CT, all in physics. He was a research associate at the Massachusetts Institute of Technology, MA, and taught at three levels of professorship with the Department of Electrical and Computer Engineering, Rice University, Houston, TX from 1970 to 1984. After working as a Principal Scientist at Tektronix, Inc., he became a professor with the Department of Electronics and Electrical Engineering, Pohang University of Science and Technology, serving as the Chairman of the department and the Dean of the Graduate School. Currently he is a professor in the Computational Sciences Division, Korea Institute for Advanced Study and a visiting professor in the department of electrical engineering, Seoul National University, Seoul, Korea. His current research interests are in nano-CMOS devices, flash memory technology, and Schottky contacts in molecular and CNT devices. He is a member of American Physical Society and a fellow of Korean Academy of Science and Technology.

Kim, Daesig (6C.4)
Daesig Kim received the B.S. degree in metallurgical engineering from Hanyang University, Seoul, Korea in 1982. He received the M.S. and Ph.D. degrees in materials science and engineering from Stevens Institute of Technology, Hoboken, New Jersey, in 1991. From 1991 to 2001, he was with Samsung Advanced Institute of Technology, Korea, as a principal researcher in Material and Device Division working on the development of new thin film materials and thin film processes. He joined Ramtron International Corporation, Colorado Springs, CO, in 2004 as a senior reliability engineer and has been working on the reliability of high density FeRAM.

Kim, SangBum (2C.5)
SangBum Kim received the B.S. degree from Seoul National University, Seoul, Korea, in 2001, and the M.S. degree from Stanford University, Stanford, CA, in 2005, all in electrical engineering. He is currently working toward the Ph.D. degree in electrical engineering at Stanford University, Stanford, CA, USA. His current research focuses on fabrication and characterization of novel phase-change memory (PCM) structures with reduced programming power and cell size, and measurement of phase-change material properties to understand how they can affect PCM operation.

Kim, Seong Soo (5C.1)
Seong Soo Kim was born in Busan, Korea, on May 22, 1974. He received the M.S. and the Ph. D. degree
in physics from Seoul National University, Seoul, Korea in 2001 and 2006, respectively. He has been working for Samsung electronics since 2006 as a member of Flash Memory Process Architecture team. His main activities have been in TCAD simulation and analysis of device operation.

Kim, Yong Seok (5C.1)
Yong Seok Kim was born in Seoul, Korea, on July 16, 1969. He received Ph.D. degrees in electronic engineering from Seoul National University, Seoul, Korea, in 2004. Since 2005, he has been working in device engineering of the NAND Flash memories at Samsung Electronics Corporation, Korea. His current interests include reliability issues and the scaling of floating type NAND Flash memories.

Kimura, Shin'ichiro (XT.2)
Shin'ichiro Kimura received the B.S. and M.S. degrees in materials science from Tohoku University, Sendai, Japan, in 1978 and 1980, respectively, and the Ph.D. degree from the University of Tokyo, Japan, in 1989. Since 1980, he has been with the Central Research Laboratory, Hitachi, Ltd.. During 1988 to 1989, he was a Visiting Research Associate at the University of Warwick, Coventry, U. K. Currently, he is a Chief Senior Researcher. His current research interests include new submicrometer MOSFET devices and processes. Dr. Kimura is a member of the Japan Society of Applied Physics and the IEEE Electron Devices Society.

King, Sean (IC.13)
Dr. Sean King is a Senior Technical Contributor for Intel Corporation's Portland Technology Development (PTD) Division. Dr. King received a B.S. degree in Materials Engineering from Virginia Tech in 1991, and a Ph.D. in Materials Science and Engineering from North CarolinaStateUniversity in 1997. Since joining Intel in 1997, Dr. King has held a variety of technical positions in the development of Intel's 0.35 mm - 22 nm technologies. Currently, Dr. King is leading development of low-k dielectrics for 16 nm Cu interconnects. Dr. King's research interests include thin film deposition, diffusion barriers, and electrical properties of interfaces.

Köck, Helmut (4C.5)
Helmut Köck studied communication engineering at the Carinthia University of Applied Science, Klagenfurt, Austria. He received the M.S. degree in 2007. He is currently working toward the Ph.D. degree at the Technical University of Vienna, Vienna, Austria. Since 2006 he has been with Infineon Technologies and KAI GmbH, Villach, Austria, respectively. His research involves the development of optical methods for device characterization and reliability studies on smart power semiconductors.

Koh, Yohwan (MY.7)
In 1989, he joined Hyundai Electronics Industries Co., Ltd (changed to "Hynix" after merging with LG semiconductor in 2001), Korea where he has been engaged in Advanced DRAM device and process architecture development for 64M DRAM to 256M DRAM until 2002. During his R&D period, he and his team developed the world's 1st fully functional SOI 64Mb DRAM in 1996, and then the SOI 1Gb DRAM with 0.18um technology in 1997. From 2002, he worked in Manufacturing Fab. in Eugene, OR where he was Vice President of Device Engineering group until 2005. He was project manager for 80nm DRAM development. He also took care of ProMOS foundry business and Graphics/Mobile DRAM development with 80nm & 66nm technologies. From the end of 2007, He moved his position to Flash Development Division. He also developed 41nm 32Gb, 32nm 32Gb and 26nm 64Gb NAND Flash. He is now Senior VP and Head of Flash Development Division.

Koichi, Hashimoto (4A.5)
Koichi Hashimoto received B.S. degree in applied chemistry and M.S. degree in chemical energy engineering from the University of Tokyo, Tokyo, Japan, in 1983 and 1985, respectively. In 1985, he joined Fujitsu Ltd., Kawasaki, Japan, where he engaged in the research and development of dry etching technology and DRAM process technology. Since 2002, he has been engaged in the research and development of high performance CMOS logic devices for CPU applications, such as SPARC64™ series.

Kopf, Rose (FA.5)
Rose F. Kopf is a Member of the Technical Staff at Alcatel-Lucent Technologies, Bell Labs, in the High-Speed Electronics and Optoelectronics Research Dept.. She is also a Member of The Alcatel-Lucent Technical Academy. She received a B.S degree in chemistry from Northeastern University in Boston, MA in 1982, and M.S. and Ph.D. degrees in materials science and engineering from Steven Institute of Technology in Hoboken, NJ in 1987 and 1991, respectively. Since 1992, she has been involved in process development and integration for HBT high-speed circuits and optoelectronic integrated circuits. She has over 200 publications and holds 20 patents.

Kota, Funayama (PI.3)
Kota Funayama received B.E. and M.E. degrees in materials science from Tohoku University, Sendai, Japan, in 1995 and 1997, respectively. He joined Hitachi Ltd., Tokyo, Japan, in 1997. Since 2003, he has been working as an engineer for Renesas Technology Corporation (present Renesas Electronics Corporation), Ibaraki, Japan, where he is currently engaged in the research and development of advanced MCU technology.

Koyama, Shin (4C.1)
Received BS and MS degree of physics from Tohoku University in 1995 and 1997, respectively. He joined NEC in 1997, had worked on the development of gate dielectric process. He is now participating in JDA with IBM from 2009, and working on the gate dielectric and metal electrode for 32nm/28nm node and beyond.

Krause, Jonathan (3A.1)
Jon Krause coordinates the design, verification, and release of Standard Cell Libraries for several Intel Server Development Group Processors. Jon has been involved in the implementation, backend circuit verification, and full chip integration of multiple high-performance microprocessors, including the Digital Alpha 21264 and 21364 processors. Jon obtained a BSEE from the University of Cincinnati in 1998.

Krick, John (5E.5)
John joined Texas Instruments in 1993 and is currently the Director of Advanced CMOS SPICE Models and Design Kits.

Krishnan, Anand (4A.6, 5E.5, XT.18)
Anand T. Krishnan received his Bachelor of Technology degree in Metallurgical Engineering from Institute of Technology, BHU, Varanasi, India in 1994, and M.S. and Ph.D degrees in Materials from The Pennsylvania State University in 1997 and 2000, respectively. In 2000, he joined the Silicon Technology Development Group at Texas Instruments, where he is currently working as a Reliability Engineer. His interests and activities are in the areas of negative bias temperature instability, dielectric breakdown physics and plasma charging damage. He has served in the technical program committee for the International Reliability Physics Symposium (IRPS), Integrated Reliability Workshop (IRW) and for the Plasma Process-Induced Damage Symposium (P2ID) and as a Guest Editor for IEEE Transactions on Device and Materials Reliability. He has authored or co-authored more than 50 papers, including 10 at IEDM/VLSI, and holds 12 patents.

Krishnan, Srikanth (5F.3, 6A.1, XT.18)
Srikanth Krishnan received his B.Tech in Electrical Engineering from the Indian Institute of Technology (IIT) in 1985, M.S. and Ph.D from Pennsylvania State University (PSU) in 1988 and 1992. He is a Distinguished Member of Technical Staff and Manager, responsible for Analog Technology Reliability at Texas Instruments. Srikanth has published or presented 60 papers. He serves on the IRPS Management committee. He has received 3 Outstanding Paper awards and 1 Best Paper award at IRPS, as a co-author and mentor. He is a Centennial Fellow of the Penn State Engineering Science Department.

Ku, Shaw-Hung (5D.3)
Shaw-Hung Ku was born in Taipei, Taiwan, R.O.C. 1977. He received the M.S. and Ph.D. degree in electronics engineering from the National Chiao-Tung University, Hsinchu, Taiwan, R.O.C., in 2001 and in 2006. Then, he joined Macronix International Company, Ltd. and is responsible for the developement of nitride-based and floating-gate storage memories.

Kuang-Chao, Chen (5D.4)
Kuang-Chao Chen received the M.S. degrees in chemistry from the National Chong-Shan University, Taiwan, in 1987. From 1989 to 1995, he joined Electronic Research and Service Organization (ERSO), Hsinchu, Taiwan, where he has been involved in the development of BEOL planarization process technology. From 1995 to 1998, he was with Mosel-Vitelic International Co., Ltd, Hsinchu, Taiwan. He performed yield improvement in manufacturing line. In 1998, he joined Vanguard International Semiconductor Co., Ltd, Hsinchu, Taiwan, as a department manager. He was responsible for thin film module development. In 2000, he joined Macronix International (MXIC), where he worked on advanced module development. He is currently Executive Director of Technology Development Center.

Kuball, Martin (2E.5, 4F.3)
Received his Ph.D. from the Max-Planck Institute for Solid State Physics, Stuttgart, Germany in 1995. After a two year stay at Brown University, Providence, USA as Feodor-Lynen Postdoctoral Fellow,

he joined the faculty of the University of Bristol (UK) in 1997, where he is presently Professor in Physics and Director of the CDTR. His current research interests include the optical, thermal, and electrical study of advanced electronic and optoelectronic materials and devices, especially of III-nitride, GaAs, and boron-based structures, with focus on reliability physics and engineering. He is author or co-author of more than 150 scientific publications.

Kufluoglu, Haldun (5E.5)
He obtained his B.S., M.S., and PhD in 2001, 2003, and 2007, respectively from Purdue University. His research interests include MOSFET reliability, characterization and modeling of semiconductor devices. His PhD research focus was measurements and theoretical modeling of MOSFET degradation mechanisms such as NBTI, HCI and TDDB, and their implications on VLSI design. He also participated in off-state transistor reliability assessment. In 2006, he held a summer internship at Intel Corporation, LTD FE Q&R, Hillsboro, OR, on experimental NBTI reliability and recovery modeling. He is now with Texas Instruments and working on parametric reliability modeling with EDA focus.

Kulkarni, Pranita (XT.14)
Pranita Kulkarni received her B. Engg. in Materials Science and Engineering from University of Pune, India in 1999, and M.S. and Ph.D. degrees in the same discipline from Carnegie Mellon University (CMU), Pittsburgh in 2006 and 2008, respectively. Prior to joining CMU for graduate studies, she worked at General Electric Co., Niskayuna NY (2003) and Cosmo Films Ltd., India (2000) as Polymer Engineer. Since 2008 she is a Research Staff Member at IBM Corp., NY where she performs process and device modeling of electronic devices that would meet the scaling requirements of future generation(s) of semiconductor chips.

Kumashiro, Shigetaka (5F.4)
Shigetaka Kumashiro received his B.E. and M.E. degrees from the University of Tokyo in 1981 and 1983, respectively, and his Ph.D. degree from Carnegie Mellon University in 1992. He joined NEC Corporation in 1983 and has been working in the field of the modeling and simulation of ULSI process and device. Now he is a Chief Engineer of Core Development Division, NEC Electronics Corporation. In 2007, he was appointed as theme leader of Environmental Variability Tolerant Device Technology Program of MIRAI project. Dr. Kumashiro is a senior member of the Institute of Electrical and Electronics Engineers.

Kuper, Fred (2D.2)
Fred G. Kuper starting working in the field of wafer level reliability in 1987 with Philips Semiconductors, working on GOI, EM, ESD etc. In 1996 he received the outstanding paper award of the IRPS for his paper on the relation between yield and reliability. In 2002 Fred joined the automotive division of Philips that is now part of NXP Semiconductors. First as quality manager and since 2008 in an automotive quality and reliability specialist position. Furthermore, since 1998 Fred Kuper is an extraordinary professor at the University of Twente in the field of Integrated Circuit Reliability.

Kuroda, Rihito (5F.2)
Rihito Kuroda (S'05) was born in Tokyo, Japan, on July 23, 1982. He received the B.S. degree in electronic engineering and the M.S. degree and Ph. D. degree in the Graduate School of Engineering, Tohoku University, Sendai, Japan in 2007 and 2010, respectively. He is engaged in researches on advanced semiconductor device and process technologies, such as novel-structure SOI CMOS developments; silicon surface processing,

such as flattening, cleaning, and high-integrity gate insulator film formation; compact MISFET modeling developments for circuit simulation; and reliability characterization, including negative-bias-temperature and hot-carrier instabilities. Dr. Kuroda was the recipient of the IEEE Electron Devices Society Japan Chapter Student Award in 2005.

Labat, Nathalie (2E.3)
Nathalie LABAT joined the University of Bordeaux, France in 1987. She received the Ph.D. degree in Electronics in 1990 and the dissertation "Habilitation à Diriger des Recherches" in 1999, all from the University of Bordeaux, France. She is currently Professor in IMS Laboratory in the Department of Electronics Engineering at the University of Bordeaux and head of the Doctoral School on Physical Sciences, Electronics and Mechanical Engineering. Her research interests include the reliability of microwave technologies using the finite element simulation, electrical and physical analysis and low frequency noise characterization.

Lacaita, Andrea Leonardo (5C.2, 5D.1, 6C.3, MY.5, MY.6)
Prof. Andrea L. Lacaita. Full Professor of Electronics at the Politecnico di Milano. Scientist since 1987, he has been Visiting Professor at the AT&T Bell Laboratories, Murray Hill, NJ (1989-90), IBM T.J. Watson Research Center, Yorktown Heights, NY (1999). 2009 IEEE Fellow. He has contributed to study quantum effects as well as experimental characterization techniques and numerical models of non-volatile memories, both Flash and emerging (PCM,RRAM). He is co-author of more than 200 papers, patents and several educational books in Electronics.

Lai, Sheng-Chih (MY.1)
Sheng-Chih Lai was born in Taichung, Taiwan in 1976. He received his B.S., M.S. and Ph.D. degrees in department of materials science and engineering from National Tsing-Hua University, Taiwan, in 1999, 2001 and 2008, respectively. He joined Emerging Central Laboratory (ECL) in Macronix International (MXIC) in 2001, where his current research is engaged in reliability studies of nitride trapping Flash memories, especially incorporated with high-K and metal gate.

Lamontagne, Patrick (IC.8)
Patrick Lamontagne received the engineering degree in Materials and Nanotechnologies from the Institut National des Sciences Appliquées de Rennes, Rennes, France and M.S. degree of Physic from the Université de Rennes 1, Rennes, France in 2007. He is currently working toward the Ph.D. degree with the Electrical Characterization and Reliability Group at STMicroelectronics, Crolles, France. His research field is the electromigration issues in advanced copper interconnects.

Langer, Eckhard (IC.2)
Eckhard Langer received his PhD in electrical engineering & micro systems technology from the Technical University of Chemnitz, Germany. He started his professional carrier at the Fraunhofer Institute of Mechanics of Materials Halle, where he worked in the field of microelectronics and materials analysis. In 1997 Eckhard Langer joined Advanced Micro Devices (AMD) in Dresden (today Globalfoundries). As a senior member of technical staff he is responsible for the electron microscopy group (SEM, TEM, FIB.) and Failure Localization. He is member of the EUFANET board.

Larcher, Luca (4A.4, 4F.4, 6C.1)
Luca Larcher graduated in Electronics Engineering from University of Padova in 1998. He received the PhD degree in 2001 from the University of Modena and Reggio Emilia, where he is currently Associate Professor of Electronics. His research interests are twofold. He focused on the experimental characterization, reliability and modeling non-volatile memories and logic devices. He worked on the characterization, reliability and design of both RF Integrated

Circuits for telecommunications and circuits for energy harvesting from renewable sources in CMOS technology. He authored and co-authored a book, more than 85 technical papers published on international journals and conferences.

Laurin, Luca (5C.4)
Luca Laurin was born in Monza, Italy on 10/02/1982. He received the master degree in physics from Bicocca university, Milan, Italy in 2008. He studied, in an atomistic approach, Boron diffusion and clustering during his Master thesis work. In 2008, he joined Numonyx Corp., in Agrate, Italy, to work in TCAD simulation team, dealing with the develop of innovative PCM memory technology.

Lavizzari, Simone (6C.3)
Simone Lavizzari was born in 1982. He took his first level Laurea (Bachelor) degree in 2004, and the second level Laurea (Master) degree in 2006, both in electronic engineering with 1st class honors at the Politecnico di Milano, Italy, where he was also awarded with the best student medal within his course. He worked on Phase Change Memories (PCMs) within the Department of Electrical Engineering of Politecnico di Milano, where he achieved his PhD degree in 2009 working on PCMs and design of VLSI analog instrumentation. At the moment he is employed in Numonyx R&D working on PCM products.

Lee, Byoungil (2C.5)
Byoungil Lee received the B.S. degree in Electrical Engineering from the Korea Advanced Institute of Science and Technology (KAIST) in 2005. He received the M.S. degree in Electrical Engineering from Stanford University in 2007. He is currently pursuing the Ph.D. degree in Electrical Engineering at Stanford University. He is a recipient of the Samsung Scholarship since 2005. His research interests are on resistive non-volatile memories including metal-oxide memory and phase change memory.

Lee, Chi Kyoung (5C.1)
Chi Kyoung Lee was born in Seoul, Korea, on March 08, 1971. He received the B.S. degree in electronics engineering from Korea Aerospace University, Gyunggi-Do, Korea, in 1995 and the M.S degree in electronics engineering from Korea University, Seoul, Korea, in 2004. He has been working at Samsung electronics, Gyunggi-Do, Korea since 1995. he has been in a Flash Memory Process Architecture team and his main research is fail analysis for yield & cell characteristic improvement. His current work is for the reliability improvement of high-density NAND flash memories sub 30nm technology, and research interests are NVM reliability, cell technology, cell characteristic modeling.

Lee, Dong Jun (5C.1)
Dong Jun Lee was born in Seoul, Korea, on August 18, 1976. He received the B.S.degree in materials science and engineering from Yonsei University, Seoul, Korea, in 2001 and the M.S degrees in materials science and engineering from Pohang University of Science and Technology, Pohang, Korea, in 2003. He has been working at Samsung electronics, Gyunggi-Do, Korea since 2003. From 2003 to 2010, he was a Flash Process Architecture team and worked on 120nm 1Gb NAND, 90nm 4Gb NAND, 51nm 16Gb NAND, 42nm 16Gb NAND, and 35nm, 32nm, 27nm 32Gb NANDs.

Lee, Dong-Kyu (MY.7)
Dong-Kyu Lee received the B.S. degree and the M.S. degree in electrical engineering from Sungkyunkwan University, Korea in 1994 and in 2002 respectively. Since he joined Hynix semiconductor Inc. in 2008, he has been working in process integration and reliability improvement in advanced NAND Flash technologies. Prior to joining Hynix, he had worked at Leadis technology, Inc. and Samsung Electronics, Co. in Korea, where he was involved in product engineering and process integration of EEPROM embedded display driver IC, and NAND and NOR Flash respectively for 14 years.

Lee, Hsiao-Heng (3A.3)
Hsiao-Heng Kelin Lee received his B.Eng in Computer Engineering from McGill University, Montréal, Canada, in 2000 and his M.S. in Electrical Engineering from Stanford University, Stanford CA, in 2002, where he is currently pursuing his Ph.D. in the Department of Electrical Engineering. His research interests include low power/high performance digital circuits, digital signal processing algorithms and soft error resilient circuits. While at Stanford, Mr. Lee was supported by the National Sciences and Engineering Research Council of Canada (NSERC) Postgraduate Scholarship from 2000 to 2002.

Lee, Jeong-Soo (2C.4)
Received the B.S. (1991), M.S. (1993) and Ph.D (1996) in electronics engineering from the Pohang University of Science and Technology (POSTECH), Pohang, Korea. His main research is related to fabrication and characterization of nano-CMOS devices and its reliability issues.

Lee, Sang-Hyun (2C.4)
Received the B.S. (2009) in electronics engineering from the Pohang University of Science and Technology (POSTECH), Pohang, Korea. He is currently in M.S. and Ph.D. integrative program course at POSTECH. His main research is related to nanowire FET.

Lee, S-C (IC.10)
S.C. Lee received the B.S. (1994) in physics from National Sun Yat-Sen Univ., Kaohsiung, TW, and M.S. (1996) from Institute of Electro-Optical Engineering of National Chiao-Tung Univ., Hsin-chu. In 1998, he joined TSMC where he work on Cu/Low-k interconnect reliability including Cu electromigration and Low-k dielectric reliability breakdown physics.

Lee, Y.H. (2F.3)
Yung-Huei Lee (S'82-M'86-SM'06) received Ph.D. in EE from Ohio State University. He is a 23 years veteran of Intel and has worked in the development of various CPU, RF/analog, and Flash technology generations. He is currently a Technical Director at TSMC. Dr. Lee holds 2 US patents and has published over 60 technical papers.

Lenahan, Patrick (2A.5, IC.13, XT.11, XT.18)
Patrick Lenahan earned a B.S. from Notre Dame and a Ph.D. from Illinois and did a brief post-doc at Princeton. From 1980-1985, he was with Sandia. Since 1985, he's been at Penn State where he is Distinguished Professor of Engineering Science. In 2001, he was Visiting Professor of Electronics and Computer Engineering at Nihon University. He served as Technical and General Program Chairman of the IEEE IIRW in 2008 and 2009 respectively. He's authored about 130 journal articles which have been cited approximately 3500 times. He is a fellow of IEEE.

Leu, Lii-Cherng (CD.3)
Lii-Cherng Leu received received his Ph.D. in Materials Science and Engineering at University of Florida in 2008 with an emphasis on electronic thin film materials processing and characterization. After his postdoctoral work with Dr. Pearton and Dr. Ren on the failure mechanism of InGaAs/InAlAs and AlGaN/InGaN based HEMTs, he joined the Department of Materials Science and Engineering at Boise State University to continue his postdoctoral research in August, 2009. His current research is focusing on structural analysis of functional oxide ceramics and minerals using JEOL2100 TEM.

Li, Xiang (MY.4)
Li Xiang received his B.Eng in Electrical and Electronics Engineering (EEE) from Nanyang Technological University (NTU) Singapore in 2005. He is currently pursuing his PhD degree in EEE, NTU working on failure analysis of advanced gate stacks. His research interest includes device reliability and nano-scale characterization using transmission electron microscopy and electron energy loss spectroscopy. Since 2006, he has been a student member of IEEE.

Lian, Nan-Tzu (MY.1)
Nan-Tzu Lian was born in Kaohsiung, Taiwan in 1969. He received his M.S in material science and engineering from National Taiwan University (NTU), Taiwan, R.O.C., in 1993. He joined Macronix International (MXIC) since 1995. He is currently a project deputy division manager of the advanced module process development division in Technology Development Center.

Liang, James (5A.3)
James Laing received his B.S. (1992) in Ceramic Engineering from Rutgers University and M.S. (1994) in Material Science from New Jersey Institute of Technology. He joined United Microelectronics Corp. in 2002 and has been involved in the quality and reliability of CMOS process. Currently he is a staff engineer of reliability department.

Liang, Zhongning (2D.2)
Zhongning Liang received the MSc (1987) in Solid States Physics from ZhongShan University, Guangzhou, China and the PhD (1994) from Groningen University of the Netherlands. He joined Philips Semiconductors in 1995 as a process reliability engineer and currently is the reliability manager in a business line at NXP semiconductors. He is a member of the Automotive Electronic Council Technical Committee, with broad and deep interests in automotive requirements and quality.

Lilja, Klas (3A.3)
Klas Lilja is CEO and co-founder of Robust Chip Inc. (RCI). Prior to founding RCI, he was CEO and VP Engineering of Integrated Systems Engineering and Head of TCAD at Avant! Corporation. He received his M.S. and Ph.D. in Physics, from Chalmers University of Technology, Gothenburg, Sweden, and the Swiss Federal Institute of Technology, Zurich, Switzerland, respectively. At RCI, Dr. Lilja has lead the development and market introduction of the company's new simulation and design tools. In the area of soft-error hardened design, he is working on layout techniques and on RCI's single event simulation for cross-section and error-rate prediction.

Lim, Phyllis Shi Ya (XT.4)
Phyllis S.Y Lim received the B.Eng. (hons) degree in electrical and electronic engineering from the National University of Singapore (NUS) in 2007. She is now working towards a Ph.D. degree at NUS under the NUS Graduate School for Integrative Sciences and Engineering (NGS) scholarship. Her research interests include nanofabrication and CMOS device physics, high mobility channel materials and advanced junction engineering

Lim, Yeow Kheng (5B.3)
Yeow-Kheng Lim received Bachelor (Hons.), Master and Ph.D. degrees in Electrical and Electronic Engineering from Nanyang Technological University, Singapore in 1999, 2001 and 2008 respectively. He works in Technology Development Department of Global Foundries Singapore (formerly Chartered Semiconductor Manufacturing Ltd) for more than 9 years. Yeow-Kheng is a senior member of IEEE and the Executive Committee Member and Treasurer of the Singapore IEEE

REL/CPMT/ED Chapter in 2009 and 2010 respectively. He is the Organizing Committee Member, Co-chair of Technical Sub-committee and Technical Co-chair of IPFA 2008, 2009 and 2010 respectively; and the Technical Sub-committee Member of IRPS 2009 - 2010.

Lin, Chung-Hsun (CD.1, XT.14)
Chung-Hsun Lin received the B.S. and M.S. degrees in electrical engineering from National Taiwan University, Taipei, Taiwan, R.O.C., in 1999 and 2001, and the Ph.D. degree in electrical engineering from the University of California, Berkeley, in 2007. He joined the IBM Thomas J. Watson Research Center in 2008 as a Research Staff Member in the area of CMOS technology and device modeling for 22nm node and beyond. He has authored or coauthor of more than 50 technical papers. He is currently a Reviewer of the IEEE TRANSACTIONS ON ELECTRON DEVICES and ELETRON DEVICE LETTERS.

Lin, Hung Sung (FA.1, FA.2, FA.3)
Hung-Sung Lin was born in Hsinchu, Taiwan in 1972. He received the M.S. in MEMS from National Tsing Hua University, Hsinchu, Taiwan in 1997. He joined Micro System Laboratory, ITRI, Hsinchu, Taiwan in 1999 where he worked on the research of advanced MEMS devices and process development. In March 2000, he joined Product Engineering Division, UMC, Hsinchu, Taiwan. He is a section manager of Logic&MM group and presently in charge of failure analysis of Logic and MM products. He has published 13 papers at international conferences, and he received UMC ten best composing invention disclosure Award in 2006 and 2008 for the contribution on composing patents in Microelectronic Device.

Lin, Mingte (5A.3)
Mingte Lin received his B.Eng. (1987) in Power Mechanical engineering and M.S. (1992) in Physics from Tsing Hua University Taiwan. He joined United Microelectronics Corp. in 1997 and has been involved in the quality and reliability of CMOS process. Currently he is a staff engineer of reliability department.

Lin, Shang-Wei (5D.3)
Shang-Wei Lin was born in Kaohsiung, Taiwan, ROC., on July 5, 1979. He recieved the M.S. degree in Electronics Engineering from National Central University, Taoyuan County, Taiwan, in 2004. He joined Macronix International Company, Ltd., Hsinchu, in 2005, as a Process Integration Engineer in technology development center. Since 2005, he has been engaged in the development and integration for the nitride-based NBit technology at Macronix.

Lin, Shihuan (NA. 1)
Shihuan Lin recevied B.E. degree in electrical engineering from Beijing Insititute of Technology, Beijing, China, in 2001. Since 2006, has been a research student in Nanyang Technology University, research interests are nano device physics and modeling.

Ling-Wu, Yang (5D.4)
Ling-Wu Yang received the M.S. degrees in material science and engineering from the University of National Sun Yat-Sen, Taiwan, R.O.C., in 1996. From 1996 to 1999, he joined Vanguard International Semiconductor Co., Ltd, Hsinchu, Taiwan, R.O.C., as a process engineer. In 2000, he joined Macronix International (MXIC) for advanced diffusion module development. He is currently a Project Deputy Director of Advanced Module Process Development Division in Technology Development Center.

Linscott, Ivan (3A.3)
Ivan Linscott is a Senior Research Associate in the Space Telecommunications and Radio Science Laboratory, in the Electrical Engineering Department of Stanford University. Dr. Linscott received his Ph.D. in Elementary Particle Physics from the

University of California, Berkeley CA, in 1974, and after a Post-Doctoral residence at Syracuse University, and the Brookhaven National Laboratory, transitioned to radio astronomy at the Arecibo National Observatory before accepting a National Academy of Sciences Fellowship with the SETI Project at the NASA/Ames Research Center, Moffett Field CA, to develop high performance signal processing systems. He has been at Stanford since 1984.

Lipp, Dieter (4A.3)
Dieter Lipp was born in Goerlitz, Germany, in 1972. From the Technical University of Dresden he received his Diploma in physics in 1997 and a PhD (Dr. rer. nat.) in physics in the field of thermal low temperature properties of superconductors in 2002. From 2002 till 2004 he worked as scientific staff memeber at the University of Dresden were he worked on Tantalum based barriers against copper diffusion in CMOS copper damascene technologies. From 2004 until 2007, Dieter worked at AMD as manufacturing engineer in chemical mechanical planarisation (CMP) operations. Since November 2007, Dieter has been working as reliability engineer at AMD from 2007 till 2009 and since 2009 at GLOBALFOUNDRIES with focus on Front-End-of-Line reliability (gate oxide reliability/TDDB) for CMOS technologies. He has co-authored 13 papers in Journals and conferences.

Liu, Bin (XT.4)
Bin Liu received the B.Eng. degree in electrical engineering from the National University of Singapore (NUS) in 2008, where he is currently working toward a Ph.D. degree at the Silicon Nano Device Laboratory (SNDL) with the NUS Graduate School for Integrative Sciences and Engineering (NGS) scholarship. His research interests include strained-silicon MOSFETs and reliability physics of CMOS device

Liu, Chien-Chih (2F.3)
Chien-Chih Liu received the M.S. degree in Materials Science and Engineering from National Chiao-Tung University, Hsinchu, Taiwan, R.O.C.in 1995. He joined the Taiwan Semiconductor Manufacturing Company (TSMC) , Hsinchu, Taiwan, R.O.C. in 2000. He has worked on the development for several logic technologies in R&D, and then moved to FEOL reliability department for HV device reliability qualification. Currently, he is the HV reliability section manager in Technology Quality and Reliability Division.

Liu, Tsu-Jae King (XT.17)
Tsu-Jae King Liu received her Ph.D. degree in Electrical Engineering from Stanford University. She joined the Xerox Palo Alto Research Center as a Member of Research Staff in 1992. In August 1996 she joined the faculty of the University of California at Berkeley, where she is now Conexant Systems Distinguished Professor of Electrical Engineering and Computer Sciences and Associate Dean for Research in the College of Engineering. Her awards include the DARPA Significant Technical Achievement Award (2000) for development of the FinFET and the 2010 IEEE Kiyo Tomiyasu Award for contributions to nanoscale MOS transistors, memory devices, and MEMs devices.

Liu, Ziyuan (4C.1)
Received the B. S. (1982) and the M. S. (1985) in Materials science and engineering from Beijing Institute of Aeronautics, China, and the Dr. Sc. (1994) in Material Science from Tokyo Institute of Technology, Japan. From 1996 to 1997, she worked as a frontier researcher in RIKEN (The Institute of Physical and Chemical Research). Since 1997 she joined NEC Corporation, and was engaged in the development of device analysis technology. She is working on the development of gate oxide

reliability in Logic and Flash devices from 2002 at NEC Electronics Corp. She is currently interested in chemistry and physics of hydrogen at surfaces and interfaces contained in the MOS stacks.

Liu, Wei (5B.3, IC.5)
Wei Liu received his B.E. (2000) and M.E. (2003) in physical electronics and optoelectronics from Huazhong University of Science and Technology, China, and Ph.D. (2009) in microelectronics from Nanyang Technological University, Singapore. He has been working on the advanced back-end-of-line process integration and technology qualification in the Technology Development department of GLOBALFOUNDRIES Singapore since Oct 2007.

Lo, Chien-Fong (CD.3)
Chien-Fong Lo received the B.Sc. and the M.Sc. degrees from the Chemical Engineering Department, National Taiwan University, and then joined the Department of Chemical Engineering, University of Florida, in 2007 for the Ph.D. degree pursuing. He has been working on the research in AlN/GaN high electron mobility transistors (HEMTs) and InGaAsSb based double heterojunction bipolar transistors (DHBTs).

Lofrano, Melina (5B.4, 6A.3)
Melina Lofrano received the BSc degree in Physics in 2001 and the MSc in Mechanical Engineering in 2003 at the University of São Paulo, Brazil. From 2003 until 2006 she worked in research and product development, where she was responsible for the mechanical modeling and analysis. From 2006 she was involved with several CAE research projects at Katholieke Universiteit Leuven, Belgium. In 2008 she joins IMEC modeling and reliability team.

Lu, Chih-Yuan (5D.2, 5D.3, MY.1, MY.3)
Chih-Yuan Lu received B.S. degree from National Taiwan University in 1972, and Ph.D. degree in physics from Columbia University, NYC, in 1977. In Dr. Lu has been a professor in National Chiao-Tung Univ. and with AT&T Bell Labs from 1984-1989; later joined ERSO/ITRI in 1989 as a Deputy General Director responsible for the MOEA grand Submicron Project. This project successfully developed Taiwan first 8-inch manufacturing technology with high density DRAM/SRAM. He was therefore granted the highest honor prize--National Science & Technology Achievement Award by the Prime Minister of ROC, due to his leadership and achievement in this Submicron Project. In 1994, Dr. Lu becomes the co-founder of Vanguard International Semiconductor Corporation, which is a spin-off memory IC Company from ITRI's Submicron Project. He was the VP of Operation, VP of R&D, and later President from 1994-99. Dr. Lu now is the founding chairman and CEO of Ardentec Corp. a VLSI testing service company; and also serves Macronix International as a Senior VP/CTO, and now the President. Dr. Lu led MXIC's technology development team to successfully achieve the state of the art nonvolatile memory technology and now responsible for MXIC's overall operation. Dr. Lu has published more than 300 papers and has been granted 140 worldwide patents, and was elected a Fellow of IEEE, and a Fellow of APS. He also received IEEE Millennium Medal, and the most prestige semiconductor R&D Award in Taiwan from Pan Wen Yuan Foundation.

Lu, Chi-Pin (MY.1)
Chi-Pin Lu was born in Hsinchu, Taiwan, R.O.C., in 1978. He received the B.S. degree in mechanical engineering from the National Central University, Taoyuan, Taiwan, in 2001, and the M.S. degree in material science and engineering from the National Tsing-Hua University, Hsinchu, Taiwan, in 2003. Since 2003, he has been with the Technology Development Center, Macronix International (MXIC), where his current works are engaged in advanced diffusion module process development as well as the

dielectric characteristics studies of the nitride-trapping non-volatile memories.

Lu, Lei (TF.1)
Lei Lu was born in Xuzhou, China, in 1985. He received the B.S. degree in microelectronics in 2007 from Soochow University, Suzhou, China, where he is currently working toward the M.S. degree in the Department of Microelectronics. His current research work is about charge pumping and substrate current of poly-Si TFTs.

Lu, Wen-Pin (5D.3)
Wen-Pin Lu was born in I-Lan, Taiwan, R.O.C. on December 20, 1967. He received the B.S. degree in electronics engineering from National Chiao-Tung University, Hsinchu, Taiwan, R.O.C. in 1990 and the M.S. degree in electrical engineering from National Taiwan University, Taipei, Taiwan, R.O.C. in 1992. He joined Macronix International Co., Ltd., Hsinchu, Taiwan, in 1994 as a device engineer. From 1994 to 1999, he has worked on device analysis of non-volatile memory, especially in floating-gate flash memory. Since 2000, he has been engaged in the development of PACAND Flash memory technology, and has accomplished 0.18mm and 0.15mm of delivering. He is presently responsible for the Nitrite-based NBit technology and also floating gate NOR flash development at Macronix.

Lu, Xiaowei (TF.1)
Xiaowei Lu was born in Suzhou, China, in 1986. He received the B.S. degree in microelectronics in 2009 from Soochow University, Suzhou, China, where he is currently working toward the M.S. degree in the Department of Microelectronics. His main research interests include charge pumping of poly-Si TFTs and device noise models.

Lue, Hang-Ting (5D.25D.3, MY.1)
Hang-Ting Lue was born in Hsinchu, Taiwan in 1975. He received his B.S and M.S degrees in physics from National Tsing-Hua University (NTHU) in 1997 and 1999, respectively, and Ph.D degree in electrical engineering in National Chiao-Tung University (NCTU) in 2002. He joined Emerging Central Lab. (ECL) in Macronix International. Co., Ltd. (MXIC) in 2003. Currently he is the project manager of BE-SONOS NAND Flash and 3D TFT Memory. From 2004 he has published more than 30 papers in the premier semiconductor conferences including IEDM, VLSI and IRPS, and a total of more than 75 technical papers in IEEE journal/letter/conference. In 2007, he received the "Outstanding Young Innovator Award of the Industrial Technology Advancement Awards" by Taiwan's government. Currently, he is the chair of IRPS 2010 memory subcommittee, and committee member in VLSI-TSA.

Lwin, Zin Zar (2C.3)
Zin Zar Lwin received B.Eng degree from Nanyang Technological University, Singapore, in 2009. She is currently pursuing her Ph.D in Electrical and Electronic Engineering at Nanyang Technological University, Singapore. Her research interests include physics and reliability characterization of memory devices. She is a student member of IEEE.

M, Sivatheja (NA.2)
Siva Theja M is pursuing B. Tech. and M. Tech. combined degrees from Electrical Engineering department of the Indian Institute of Technology (IIT) Bombay, since 2004. His research interests include semiconductor device modeling and simulation. He is currently working on metal nanocrystal based non-volatile memories.

Ma, Huan-Chi (MY.3)
Huan-Chi Ma was born in Tainan, Taiwan, R.O.C., in 1981. He received the B.S. and M.S. degrees in electronics engineering from National Chiao-Tung University, Hsinchu, Taiwan, in 2003 and 2005, respectively, where he is currently working toward the Ph.D. degree. His research interests include negative bias temperature instability degradation, nonvolatile memory reliability issues, and flicker noise characterization.

Ma, Zhe (SE.1)
Zhe ma received his Ph.D. in E.E. from K.U.Leuven, Belgium in 2006. Since then he has been working in the Digital Component group at IMEC, Leuven as a senior researcher. His main research activies focus on the optimization of embedded systems design.

Maconi, Alessandro (MY.5)
Alessandro Maconi was born in Carate Brianza, Italy, in 1983. He received the Laurea degree in electronics engineering from the Politecnico di Milano, Milano, Italy, in 2008, where he is currently working toward the Ph.D. degree in the Dipartimento di Elettronica e Informazione. He is also with the Italian University Nano-Electronics Team, Milano, Italy. His research activities mainly involve characterization and modeling of advanced nonvolatile memories, with particular interest to TANOS memories.

Mahapatra, Souvik (2C.3NA.2, XT.15)
Souvik Mahapatra received his PhD in Electrical Engineering from the Indian Institute of Technology (IIT) Bombay, India, in 1999. He was at Bell Laboratories, Murray Hill, NJ, USA during 2000-2001. Since 2002 he is with the Department of Electrical Engineering, IIT Bombay, India, and presently holds the position of Professor. His research interests are in the area of characterization, modeling and simulation of CMOS and Flash memory devices, and device reliability. He has published more than 90 papers in peer reviewed journals and conferences, delivered invited talks at leading international conferences in the USA, Europe and Asia-pacific including at the IEEE IEDM, delivered reliability tutorials at the IEEE IRPS, and acted as a reviewer of several international journals and conferences. He also holds an honorary graduate faculty position at Purdue University, USA, is a distinguished lecturer of IEEE EDS and senior member of the IEEE.

Maheta, Vrajesh (XT.15)
Vrajesh D. Maheta received the B.E. degree in Electronics Engineering from Sardar Patel University, Gujarat, India, in 1993, the M.E. degree in Microelectronics from Birla Institute of Technology and Science, Pilani, India, in 2002, and the Ph.D. degree in electrical engineering from Indian Institute of Technology (IIT) Bombay, Mumbai, India, in 2009. From 1998 to 2009, he was with G. H. Patel College of Engineering & Technology, as a regular faculty. Since 2010, he has been with Middle East College of Information Technology, Muscat, Oman, where he is currently working as an Associate Professor in the department of electronics and communication engineering. His research interests are in the field of semiconductor device physics and simulation, modeling and characterization of CMOS silicon devices

Mairena, Andrew (XT.17)
Andrew Mairena is a fourth year undergraduate in the Electrical Engineering and Computer Sciences Department at the University of California, Berkeley. For the past two years, his research interests have included reliability issues in SRAM such as HCI, NBTI, PBTI, and RTS.

Makabe, Kazuya (2D.4)
Kazuya Makabe received the B.S. degree in physics from Chuo University, Tokyo, Japan, in 1983. He joined Hitachi ULSI Engineering Corp., Japan, in 1983. He engaged in process development of bipolar devices, DRAM devices, and so on. He is

now with Renesas Technology Corp. Currently, he engaged in the process reliability of BEOL. He is a member of the Japan Society of Applied Physics.

Makabe, Mariko (4C.1)
Received the B.S. (1990), and the M.S. (1992) in physics from Tokyo University of Science, Japan. Since 1992 she has been engaged in development effort and manufacturing technology of FEOL at NEC Corporation. From 2002 she was in charge of gate oxide process and reliability in Logic and Flash devices at NEC Electronics Corporation.

Malbert, Nathalie (2E.3)
Nathalie Malbert received the Ph.D. degree in Electronics in 1996; and the professoral dissertation (Habilitation à Diriger des Recherches) in 2004, all from the University of Bordeaux, France. She is currently Full Professor in IMS Laboratory in the Department of Electronics Engineering at the University of Bordeaux. Since 2007, she is head of the team "Characterisation and Reliability of microwave technologies" in the Nanoelectronic group. Her research topics covered electrical characterization, modelling, physical simulation and reliability assessment of compound semiconductor based HEMT such as GaN HEMT.

Mandich, Mary (FA.5)
Mary Mandich is a Distinguished Member of Technical Staff in the Reliability Physics Group in the Chief Technology Office Reliability Engineering organization of Alcatel-Lucent. Her current work is focused on reliability and failure mode analysis of wireline and wireless telecommunication systems. Previous research areas include high speed electrical and optical backplanes, and low cost optical platforms for high capacity optical networks. Mandich is the Alcatel-Lucent project leader for this research effort and contributes technically to the collaborative ALU-EPA-DHS program to assess the impact of biodecontamination agents on electronic equipment. Mandich has a Ph.D. in Physical Chemistry from Columbia University.

Marathe, Amit (2D.3, 4C.6, 5A.4)
Amit P. Marathe earned his M.S. and Ph.D. in Materials Science and Engineering from the University of California, Berkeley in May 1991 and August 1996 respectively. His research work was in the field of high temperature superconductor thin film processes for integrated circuits with emphasis on high-Tc SQUIDs and Josephson Junctions. After graduating from Berkeley, he joined Analog Devices, Inc. in Santa Clara, CA where he was working on the process development and integration of BiCMOS Integrated circuits technology for over a year. Since joining AMD in Sunnyvale CA in October of 1997, he has led the reliability development of interconnect metallization for AMD's memory and microprocessor technologies. Currently, he is the Department Manager of the Technology Reliability Development Group at GLOBALFOUNDRIES. His group is involved with development of reliability methodologies and modeling failure mechanisms of ultra thin gate dielectrics, advanced transistor structures as well as Cu/low-k metallization. He has co-authored over 40 technical research publications. He is also a co-inventor of over 15 patents granted and over 50 pending US patents in the area of technology & reliability development.

Marcon, Denis (2E.4)
Denis Marcon was born in Conegliano, Italy, on March 12, 1981. He received a M.S. degree in computer science with the thesis entitled "Assessment of trap mechanisms in in-situ passivated Si3N4/AlGaN/GaN HEMTs by means of pulse IV measurements: impact of field-plate and passivation technology" from the University of Padova in 2006. Since 2007, he is

working toward a Ph.D. investigating potential reliability issues of GaN-based devices both for RF and switching applications at the Catholic University of Leuven and imec, Leuven, Belgium.

Marshall, Andrew (5E.5)
Andrew Marshall is an analog and digital process verification expert, working on leading edge and future technologies, including sub-45nm processes and SOI. He is a Distinguished Member of Technical Staff at Texas Instruments Incorporated, Dallas, TX. Dr. Marshall has authored/co-authored approximately 50 patents and 60 papers. He is co-author of the book 'SOI Design: Analog, Memory and Digital Techniques' and sole author of "Mismatch and Noise in Modern IC Processes". Dr. Marshall is a Fellow of both the IEEE and Institute of Physics.

Masaharu, Mizutani (PI.3)
Masaharu Mizutani received the B.S. (2000) and the M.S. (2002) in electrical engineering science from the University of Osaka, Japan. He joined the Process Development Dept, MITSUBISHI Ltd., Hyougo, Japan in 2002. In 2003, he moved to Process Development Dept., RENESAS Technology Corp., Hyogo, Japan. Throughout his career at MITSUBISHI and RENESAS, he has been working on research and development of advanced CMOS devices with high-k/metal gate stacks.

Masao, Inoue (PI.3)
Masao Inoue received the B.S., M.S. and Ph.D. degrees in electrical engineering from Osaka University, Osaka, Japan, in 1993, 1995 and 1997, respectively. In 1997, he joined Mitsubishi Electric Co. He has been working on research and development of gate oxide, tunnel oxide and high-k/metal gate CMOS devices. He is currently with Process Development Department in Renesas Technology Co.

Masataka, Kase (4A.5)
Masataka Kase joined Fujitsu Limited, Kawasaki, Japan in 1986, where he has been engaged in development of advanced Si LSI processes. Since 2008, he has been deputy general manager in the device development division of Fujitsu microelectronics limited. His present activities include the advanced process development of LSI device technologies, especially, which are including leading edge technology and science of ion implantation, msec annealing, gate dielectric formation, gate dielectric reliability, and stress control. He has served as committee of several international technology conferences. He has authored or coauthored more than 50 articles.

Mascellino, Evelyne (MY.5)
Evelyne Mascellino was born in Petralia Sottana, Italy, in 1982. She received the Bachelor Degree in Electronics Engineering from the Politecnico di Milano, Italy, in 2005. She's currently working on her Master Degree thesis on Charge-Trap Memories, collaborating with the Dipartimento di Elettronica e Informazione, Politecnico di Milano, Milano, Italy, and the R&D Technology Development of Numonyx, Agrate Brianza, Italy.

Massengill, Lloyd (3A.2, 6E.2)
Lloyd Massengill received the Ph.D. degree in solid state circuits from North Carolina State University, Raleigh, in 1987. He is currently a Professor with the Department of Electrical Engineering and Computer Science, Vanderbilt University, Nashville, TN, where he teaches microelectronic circuit analysis and design, and studies the effects of radiation on the operation of integrated circuits, particularly the modeling of circuit-level soft errors. He also serves as the Director of Engineering for the Vanderbilt Institute for Space and Defense Electronics, Nashville.

Masuduzzaman, Muhammad (XT.7)
Muhammad Masuduzzaman received the B.S. degree in Electrical and Electronic Engineering (EEE) from Bangladesh University of Engineering and Technology (BUET), Dhaka, Bangladesh, in 2004.

During 2005-2006, he worked as a Lecturer in the Department of EEE, BUET. He is currently enrolled in direct Ph.D. program at the Department of Electrical and Computer Engineering, Purdue University, West Lafayette, IN, USA. His research interest includes physics, simulation and characterization of nanoscale devices. Currently he is working on reliability issues in ferro-electric and high-κ devices.

Mauri, Aurelio (MY.5)
Aurelio Mauri was born in 1969. He received the M.S. degree in plasma physics (cum laude) from the University of Milano, Milano, Italy, in 1995. In 1996, he started to work for a semiconductor company focused on the chemistry treatment of silicon surfaces. In 2004, he joined the nonvolatile technology development of STMicroelectronics in the TCAD group working particularly on NOR/NAND memories and then with the same function in the R&D—Technology Development, Numonyx, Agrate Brianza, Italy. He is a coauthor of more than 20 scientific conference papers on different physics topics.

Mendenhall, Marcus (4B.2, SE.3)
Marcus Mendenhall received his PhD from Caltech in 1983. He has been involved with ion-beam analytical techniques and computational methods for ion scattering and transport. He served as the associate director for operations of the Vanderbilt Free Electron Laser, and was the physics lead on a project to develop the first practical tunable Compton Xray source. His current work at Vanderbilt is in computer modeling of radiation effects in solids, numerical computing methods, and development of new xray sources for radiobiological applications.

Meneghesso, Gaudenzio (3F.2, 4F.2)
Gaudenzio Meneghesso graduated in Electronics Engineering at the University of Padova in 1992 working on the failure mechanism induced by hot-electrons in GaAs MESFETs and HEMTs. His research interests include Electrical characterization, modeling and reliability of microwave and optoelectronic devices like compound semiconductors HEMTs and MESFETs, RF-MEMS switches, and organic semiconductors devices. He is also developing ESD protection structures. Within these activities he published over 350 technical papers (of which more than 35 Invited). He is reviewer of several international journals and he is Associate Editor of the IEEE Electron Device Letter for the compound semiconductor devices area since 2007.

Meneghini, Matteo (4F.2)
Matteo Meneghini received the degree in electronics engineering from the University of Padova, Italy. In 2008 he received the PhD in Electronic and Telecommunication Engineering (University of Padova), working on the optimization of GaN-based LED and laser structures. He is now Research Fellow at the Department of Information Engineering of the University of Padova. His main interest is the characterization, reliability and simulation of compound semiconductor devices. On these (and related) subjects, he has coauthored approximately 80 papers published in international journal and conference proceedings, and a number of invited papers

Meng, Lei (4E.4)
Lei Meng received her B.Eng. degree in Electrical Engineering from National University of Singapore in 2009. She worked as an Associate Engineer in GLOBALFOUNDRIES Singapore from 2005 to 2006 and an intern in Seagate Technology, Bloomington MN USA for a period of three months during her B. Eng. program in 2008. She is currently a Master student at the Centre for Integrated Circuit Failure Analysis and Reliability

(CICFAR). Her field of research is in failure analysis techniques and their potential applications in semiconductor industry.

Mertens, Robert (2E.4)
Robert P. Mertens received the electrical engineering and the Ph.D. degree from the Catholic University of Leuven, Belgium, in 1969 and in 1972 respectively. Today he is Senior Vice President of imec, heading the Scientific Leadership Team of imec. He is also professor at the University of Leuven, teaching courses on semiconductor devices and on technology of electronic and optoelectronic systems. In 1995 Robert Mertens was elected Fellow of the IEEE for contributions to heavily doped semiconductors, bipolar transistors and silicon solar cells. He has authored or co-authored more than 450 publications and has received several best paper awards.

Miccoli, Carmine (5C.2)
Carmine Miccoli was born in Cantù, Italy, in 1984. He received the Bachelor (BS) and the Master (MS) degrees with full marks (cum laude) in Electronics Engineering from the Politecnico di Milano, Milan, Italy, in 2006 and 2009, respectively. Since 2009 he has been with the Dipartimento di Elettronica e Informazione, Politecnico di Milano, where he is currently pursuing the Ph.D. degree in Information Technology. His research activities include characterization and modeling of ultra-scaled Flash memories.

Milor, Linda (IC.3)
Linda Milor is an associate professor of electrical and computer engineering at the Georgia Institute of Technology. Her research interests include yield and reliability modeling, testing, and design-for-testability of analog and digital circuits. She has a PhD in electrical engineering from the University of California, Berkeley.

Miranda, Enrique (BD.1)
Enrique Miranda received his Ph.D. degrees in Electronics Engineering and Physics from Universitat Autònoma de Barcelona (UAB), Spain and Universidad de Buenos Aires (UBA), Argentina in 1999 and 2001, respectively. From 1987 to 2003, he was Associated Professor at the Faculty of Engineering-UBA and from 2001 to 2003, Associated Researcher at the National Council of Science and Technology (CONICET), Argentina. Since 2004, he is Professor at the Escola d'Enginyeria-UAB. Dr. Miranda serves as Editorial Advisor of Microelectronics Reliability and is member of the Distinguished Lecturer program of the IEEE-Electron Devices Society. He has served in the technical committees of INFOS'07&09 and IRPS'08,09&10. His research interests include dielectric physics and reliability.

Mishra, Rahul (EL.3)
Rahul Mishra received his M.Sc.(Engg.) degree in Instrumentation in 2004 from Indian Institute of Science , Bangalore, where his research was focused on synthesis and optimization of ZnO thin-films with nano-particles for gas sensors. He received his PhD degree from George Mason University in Electrical Engineering in 2008 with research focus on interaction of ESD, NBTI and HCI in nano-scale bulk and SOI MOSFETs. During the summer and fall 2006 he worked at IBM Microelectronics on a student internship in ESD/Latchup Development Group. In summer 2007 he was again at IBM Microelectronics on a student internship in TCAD Technology Enablement group where he worked on 3D device simulations for substrate noise isolation. In 2008, he joined IBM's Semiconductor Research and Development Center (SRDC) focusing on ESD device development and compact modeling in 32nm, 28nm and 20nm CMOS technologies.

Mishra, Umesh K. (2E.5)
Received the M.S. degree in electrical engineering from Lehigh University, Bethlehem, PA, in 1981 and the Ph.D. degree in electrical engineering from Cornell University, Ithaca, NY, in 1984.

He is a Professor with the Department of Electrical and Computer Engineering, University of California, Santa Barbara. He made major contributions in the area of high-speed field effect transistors at every laboratory and academic institution that he was with, including North Carolina State University, Raleigh, Hughes Research Laboratories, Malibu, CA, University of Michigan, Ann Arbor, and General Electric, Syracuse, NY. His research interests include electronics and photonics: high-speed transistors, semiconductor device physics, quantum electronics, optical control, design and fabrication of millimeter-wave devices, in situ processing, and integration techniques.

Mitani, Yuichiro (4C.2)
Yuichiro Mitani received the B. E. and M. E. in material science and engineering from Tohoku University, Sendai, Japan, in 1990 and 1992, respectively. He received the Ph.D. from the University of Tokyo in 2009. He joined the R&D Center, Toshiba Corporation in 1992. His primary works were concerned in the Si-CVD and the ultra-shallow junction process technology. Since 1999, he has been with the Advanced LSI Technology Laboratory, Corporate R&D Center, Toshiba Corporation, Yokohama, Japan. His present research interests and activities cover the ultra-thin oxide process technology and the study of the reliability of ultra-thin gate dielectrics (SiO2, SiON and High-k) for ULSI technology. He serves (or served) on the technical committees of International Conference on IC Design & Technology (ICICDT) and IEEE International Reliability Physics Symposium (IRPS). He is a member of the JSAP.

Mitra, Subhasish (3A.3)
Subhasish Mitra is an Assistant Professor in the Departments of Electrical Engineering and Computer Science at Stanford University where he leads the Stanford Robust Systems Group. His research interests include: 1. Robust system design; 2. VLSI design, CAD, validation and test; 3. Design for emerging nanotechnologies. Prof. Mitra has co-authored over 125 technical papers, and is the recipient of multiple honors including the Presidential Early Career Award for Scientists and Engineers, National Science Foundation CAREER Award, Terman Fellowship, IEEE CAS/CEDA Donald O. Pederson Award, ACM SIGDA Outstanding New Faculty Award and the Intel Achievement Award, Intel's highest corporate honor.

Mitsuaki, Hori (4A.5)
He received the B.S. degree in Physics from Tokyo University of Science in 1992. He joined Fujitsu limited,Kawasaki,Japan in 1992, where he has been engaged in development of LSI processes. He is now working on development of advanced FEOL such as ultra thin gate dielectrics formation and gate dielectric reliability at Process development dept. of Fujitsumicroelectronics limited, Japan.

Mitsuhiro, Fukuda (3A.5)
Mitsuhiro Fukuda received the B.Sc., the M.Sc. and Ph.D. degrees in physics from Osaka University, Osaka, Japan, in 1983, 1985 and 1988, respectively. From 1988 to 2005, he worked on the acceleration technologies of cyclotrons and ion beam irradiation techniques for the related applications at Japan Atomic Energy Agency. In 2006 he joined Research Center for Nuclear Physics, Osaka University, and worked in the fields of accelerator physics. He is currently responsible for upgrading the RCNP cyclotrons for nuclear physics experiments and various ion beam applications such as SEU analysis.

Mitsuyama, Yukio (3A.4)
Yukio Mitsuyama received the B.E. and M.E. degrees in information systems engineering from Osaka University, Osaka, Japan, in 1998 and 2000, respectively. He is currently an Assistant Professor with Graduate School of Engineering, Osaka University. His research interests include reconfigurable architecture and its VLSI design. Mr. Mitsuyama is a member of IEEE, IEICE, and IPSJ.

Moise, Ted (5F.3, 6C.4)
Ted Moise (M '91) earned the B.S. degree in Physics and Engineering from Trinity College, Hartford, CT, in 1987 and the Ph.D. degree in electrical engineering from Yale University, New Haven, CT, in 1992. He joined Texas Instruments in 1992, where he was responsible for the development of high-performance III-V quantum-effect devices and circuits. In 1997, he initiated work on the development of scaled ferroelectric capacitors leading to the first demonstration of low-voltage, high-density, embedded ferroelectric memory in 2002. In conjunction with Ramtron International Corporation, Ted and his team have also produced the first high-density (4Mb) ferroelectric memory products on an advanced (130nm) silicon technology node. Ted is a distinguished member of TI's technical staff and is currently the non-volatile memory department manager within TI's Analog Technology Development organization. Ted has authored or co-authored over 60 papers, served as conference and session chair for several international technical conferences, presented numerous invited lectures, and holds more than 35 issued patents. Ted was presented with an outstanding achievement award at the 2008 ISIF conference.

Monaco, Gianni (3F.2)
Gianni Monaco received the Laurea degree in Material Science in 2004, and a Ph.D. in Material Science and Engineering in 2009, both from the University of Padova - Italy. He is currently a research assistant at LUXOR Laboratory of the National Institute of Nanophotonics (CNR-INF). His research is focused on deposition of thin films for optical applications (from Soft-X ray to Infrared) by means of Pulsed Laser Deposition (PLD) and e-beam. He is working on thin films analysis by studying the optical constants (in the Soft X-ray), Atomic Force Microscope (AFM) and spectroscopic measurements using synchrotron light.

Monzio Compagnoni, Christian (5C.2, MY.5, MY.6)
Christian Monzio Compagnoni received the Laurea degree (cum laude) in Electronics Engineering and the Ph.D. degree in Information Technology from the Politecnico di Milano, Milan, Italy, in 2001 and 2005, respectively. Since 2002, he has been with the Dipartimento di Elettronica e Informazione, Politecnico di Milano, where he became an Assistant Professor in 2006. His research activities include characterization and modeling of advanced non-volatile memories and MOS devices. Dr. Monzio Compagnoni received the Outstanding Paper Award at the IRPS in 2008 and was a member of the memory committee of the IRPS in 2009 and 2010.

Mora, Pascal (4A.2)
Pascal Mora Ph.D, STMicroelectronics, Hopewell Junction, NY Dr. Mora received his Engineering degree from "L' Ecole Nationale Supérieure de Physique de Grenoble", M.S.E.E. from "L'Institut National Polytechnique de Grenoble", and Ph.D. in Micro and Nano Electronics from "L'Institut National Polytechnique de Grenoble", Grenoble, France. His doctoral research studied the Reliability of Embedded Non-Volatile Memories in Advanced CMOS and Bi-CMOS technologies. He has been a direct contributor for the qualification of several of STMICROELECTRONICS's most advanced technologies and has actively worked in Semiconductor reliability for 7 years.

Morassi, Luca (4A.4, 4F.4)
Morassi Luca, received in 2009 his academic master degree in Electronic Engineering from University of Modena e Reggio Emilia, Italy. During 2009 he collaborates with University of Modena and Reggio Emilia with a research activity focused on electrical characterization of high-k material for NVM devices. Since 2010 he

is a PhD student at ICT Electronics & Telecommunications Doctorate School, Modena, Italy. Currently his research activity is based on III-V compound semiconductor FETs characterization.

Morin, Pierre (3C.2)
Pierre Morin was born in France in 1965. He received the Ph.D. degree in electronics in 1995 from Pierre & Marie Curie University, Paris. From 1995 to 2000 he was with Phillips, involved in electron optics and physical processes developments for cathode ray tubes. He joined ST Microelectronics in 2000 to work on thin film processes development. He was then in charge of the integration of low thermal budget deposition processes in FEOL CMOS flows and of stressor modules in 65nm and 45nm CMOS technology nodes. He is member of the R&D technical staff, in charge of FEOL processes for CMOS and Flash memories and has managed the project dedicated to solve the pattern effects issues in the 45/40nm node. He has authored or co-authored more than 50 publications or conferences presentations in the microelectronics field and owns 3 patents.

Morita, Yusuke (XT.2)
Yusuke Morita received the B.S. and M.S. degrees in materials engineering from Shonan Institute of Technology, Kanagawa, Japan, in 1999 and 2001, respectively, and the Ph.D. degree from the Tokyo Institute of Technology, Tokyo, Japan, in 2004. He joined the Central Research Laboratory, Hitachi, Ltd. in 2005 where he has been working on the research and development of CMOS devices including SOI MOSFETs. Dr. Morita is a member of the Japan Society of Applied Physics.

Mottadelli, Riccardo (5C.2)
Riccardo Mottadelli was born in Seregno, Italy, in 1985. He received the Laurea degree in Physics Engineering from the Politecnico di Milano, Milan, Italy in December 2009. Since 2009 he has been working on the reliability of Flash memories and multilevel products.

Mukherjee, Shubu (3A.1)
Shubu Mukherjee is a Principal Engineer and Director in Intel's Microprocessor and Graphics Architecture Group. His interests include computer architecture, fault tolerance, and innovation "confluencing." He is the winner of the 2009 Maurice Wilkes award, a Fellow of IEEE, and has written a book titled, "Architecture Design for Soft Errors."

Mukhopadhyay, Gautam (NA.2)
Gautam Mukhopadhyay received M.Sc.(Physics), with 1st rank from IIT-Kharagpur, India (1966); BARC Training School (1967) with 2nd rank, Bombay, India; Ph.D.(1973) in Solid State Theory from Theory Group, Tata Institute of Fundamental Reasearch (TIFR), Bombay, India; International Atomic Energy Agency (IAEA) Fellow (Jan-July, 1973) at International Centre for Theoretical Physics (ICTP), Trieste, Italy; Visiting Scientist (March, 1973- September 1978) in Institute of Theoretical Physics, Chalmers University of Technology, Gothenburg, Sweden; Assistant Professor (1978-1987) at Physics Department, IIT-Bombay; Associate of ICTP, Trieste (1980-85); Senior Solid State Fellow (1985-87) at ICTP, Trieste, Italy; Professor of Physics (1987-) at IIT-Bombay. He has worked on various areas of Theoretical Condensed Matter Physics, like electronic energy band calculations for Ce, electron correlations for homogeneous and inhomogeneous electron systems, surface physics, Magnetism in Rare Earth Iron Garnets (RIG), optical properties of dielectric and magnetic nanoparticles, etc. He has more than 100 papers in peer reviewed journals and conference Proceedings.

Murakami, Eiichi (2D.4)
Eiichi Murakami received the B.S./M.S./Ph.D. degree in applied physics from Waseda University, Tokyo, Japan, in 1981/1983/1995, respectively. He joined the Central Research Laboratory, Hitachi, Ltd. Japan, in 1983. He worked on Si-SPE, SiGe, ultrashallow-junction, and MOSFET's design & characterization studies. He is now the manager of Process & Device Analysis Engineering Development Department in Renesas Technology Corp. His current interest is in reliability and failure physics in Si-LSI. Dr. Murakami is a member of the Japan Society of Applied Physics and the IEEE EDS. He served as a sub-committee member of CMOS and Interconnect Reliability in IEDM 2004,05.

Myny, Kris (4F.2)
Kris Myny was born in Hasselt, Belgium on July 26, 1980. He received the master degree at the Katholieke Hogeschool Limburg in Diepenbeek, Belgium in 2002. He joined imec in Leuven in 2004 as a member of the Large Area Electronics group. In 2008, he started a PhD on the design of organic circuits. His main research interests are the design, fabrication and optimization of digital organic circuits for, amongst others, organic RFID tags and AMOLED-backplanes

Nagalingam, Dayanand (4E.4)
Dayanand Nagalingam received his B.E from Anna University, India in 2006 and then M.Sc from National University of Singapore in 2009. He worked as an intern in Advanced Micro Devices for a period of 7 months during his M.Sc program. Currently he is working in National University of Singapore as a Research Engineer since 2009. His field of research is Characterization of Solar Cells.

Nagarajan, Raghavan (4A.1, BD.2)
Nagarajan Raghavan was born in Bangalore, India in 1985. He received his B.Eng, 1st Class Honors, (Electronics Engineering, 2007), S.M. (Advanced Materials for Micro & Nano Systems, 2008) and M.Eng (Materials Science and Engineering, 2008) from Nanyang Technological University (NTU), National University of Singapore (NUS) and Massachusetts Institute of Technology (MIT) respectively. He was the recipient of the prestigious Nanyang Scholarship, NTU President Research Scholar and Singapore-MIT Alliance (SMA) Graduate Fellowship awards. He is also one of the five recipients to be bestowed with the IEEE Reliability Society Graduate Scholarship award in 2008 for his research accomplishments in reliability and its application to nanoelectronics. He is currently pursuing his Ph.D at the Division of Microelectronics, School of EEE, NTU focusing on reliability modeling and statistical characterization of novel high-κ dielectric materials in nanodevices. He serves on the review committee for IEEE Transactions on Device and Materials Reliability (TDMR). He is currently a Graduate Student Member of IEEE (2005-present).

Nakamura, Hideyuki (5F.4)
Hideyuki Nakamura received the B.S. and M.S. degrees in electronic engineering from University of Electro-Communications, Tokyo, Japan, in 1991 and 1993, respectively. He joined NEC Corporation in 1993. He has been working in NEC Electronics Corporation from its establishment in 2002. He has been working on development of soft error reliability technologies of SRAM and logic circuits on SoC. He joined Environmental Variability Tolerant Device Technology Program of MIRAI project in 2007.

Nakamura, Tomonori (FA.4)
Tomonori Nakamura received the Master's and Doctor's degree in Engineering from the Tohoku University in 2001 and 2004. He has been working at system department of Hamamatsu Photonics K.K. Japan. Currently, He is in charge of reserching and developping semiconductor circuits failure analysis systems.

Nakasaki, Yasushi (4C.2)
Yasushi Nakasaki, received the B.E. (1983) and M.E. (1985) in nuclear engineering from Kyoto University, Kyoto, Japan. He joined the ULSI Research Labs, Research & Development Center, Toshiba Corp., Kawasaki in 1985. Since 1996, he joined the Advanced LSI Technology Lab, Research & Development Center, Toshiba Corp., Kawasaki. He has been engaged in the research on metallization and dielectrics in both frontend and back-end process technology using first principles calculations.

Naoyoshi, Tamura (4A.5)
He received the B.S. in Physics from Yokohama City University in 1985. He also joined FUJITSU LIMITED working on Advanced FEOL such as New Rapid Thermal technologies, Reliability of ultra thin gate dielectrics and Advanced process induced strained technologies for 20 years (He joined Fujitsu Laboratories from 2003 to 2008 in order to research embedded SiGe technologies) Now He is working on the classification of 1/f noise and Random Telegraph Signal originated Si/SiO2 interface and improvement of interface on Advanced RF/Mixed Signal device. He is a member of the Japan Society of Applied Physics.

Narasimham, Balaji (3A.2)
Balaji Narasimham received the B.E. degree in electrical engineering from the University of Madras, Chennai, India, in 2003 and the M.S. and Ph.D. degrees in electrical engineering from Vanderbilt University, Nashville, TN, in 2005 and 2008, respectively. He is currently a Staff Reliability Scientist with Broadcom Corporation, Irvine, CA, where his work focuses on device- and circuit-level reliability and characterization of soft errors for memory and logic circuits. He was with Intel Corporation, Hillsboro, OR, and IBM T. J. Watson Research Center, Yorktown Heights, NY, where he held a graduate level cooperative position. Dr. Narasimham's research interests include CMOS circuit design, radiation effects and reliability of semiconductor devices and circuits. He has authored or co-authored over 30 papers related to his research and has authored a book chapter on single-event transients. He has served in the technical committee of IRPS and is the recipient of the Best Paper Award at the 2007 RADECS conference.

Nardi, Federico (5D.1)
Federico Nardi was born in 1984 in Milano, Italy. He received his Bachelor degree (BS) in 2006 and his Master degree (MS) in 2008, both in Electronic Engineering from Politecnico di Milano, Italy. For his first level graduation thesis he worked on organic non volatile memories and for his second level graduation thesis he succeeded in studying resistive-switching effects in oxide-based memories (RRAMs). He is currently pursuing his Ph.D degree in Information Technology Engineering in the Dipartimento di Elettronica ed Informazione, Politecnico di Milano, Italy. He is also with the Italian Universities Nanoelectronics Team, Politecnico di Milano.

Nassif, Nevine (3A.1)
Nevine Nassif is a member of the Massachusetts Microprocessor Design Center at Intel, Massachusetts, and is responsible for full chip physical integration. She also focused on the design of sequential circuits that are resilient to soft errors. Throughout her career at Digital Equipment Corporation, Compaq, Hewlett Packard and Intel, she has worked on several vax, alpha, x86, and itanium microprocessor designs. Her major contributions are in the area of timing, including algorithm development, methodology, and modeling for which she has been awarded 7 patents. She holds a Ph.D. in Electrical Engineering from McGill University

Nelhiebel, Michael (XT.6, XT.8)
Michael Nelhiebel received the M.Sc. degree in physics from the Vienna University of Technology, Austria, and the PhD degree in solid state physics from Ecole Centrale Paris, France, working on interferometry in electron energy loss spectrometry. In 1999, he joined Infineon Technologies Austria as a Reliability Engineer of the silicon wafer production. He is currently a Senior Staff Engineer with the quality department of the Automotive Business Division, responsible during the development phase for technology related product reliability. He has coordinated the qualification of major automotive technology platforms and participates in research activities of Infineon Technologies Austria targeting technology reliability.

Nelson, Tan (SE.5)
Received the B.S. degree in chemical engineering and the M.S. and Ph.D. degrees in EECS from the University of California, Berkeley, in 1984, 1989, and 1991, respectively. At UC Berkeley, his research was on the characterizing and modeling of optical resists under electron-beam lithography. In 1991, he joined Intel Corporation, Santa Clara, CA, where he worked on the development of phase shifting mask (PSM). In 1997, he joined the Enterprise Processor Division as a Quality and Reliability Engineer focusing on microprocessor reliability issues. His research interests include simulation and experimental techniques for determining radiation effects on microprocessors. He joined Marvell Semiconductor, Inc. in 2006 as a Principle Reliability Engineer focusing on soft error reliability issues in various ASIC and SoC devices.

Ney, David (IC.8)
David Ney is a graduate of the engineering school of Physics of Grenoble, France (ENSPG) in 2003. In 2006, he received his PhD in Microelectronics from INPG (Institut National Polytechnique de Grenoble, France). His graduate work focused on electromigration issues in advanced copper interconnects. Since 2006, he is working on interconnect reliability issues at Central R&D labs of STMicroelectronics, Crolles.

Ngan, Paul (2D.2)
Paul Ngan is Reliability Manager of Regional Quality Center at NXP Semiconductors, San Jose. He holds a BS, MS in Physics and MBA in Finance. His previous works include qualifying BiCMOS technologies for RF and championed the implementation of knowledge-based qualification. He is a member of JC14.1 and 14.3 subcommittee and various task groups, including JEP122 working group. His primary research interest is in reliability circuit simulation and ESD robustness design techniques (especially for RF). Occasionally when he is taking a break from reliability, you will see him "playing" with general relativity, quantum field theory and quantitative finance.

Nicollian, Paul (4A.6)
Paul E. Nicollian received the B.S. degree in Physics from The Pennsylvania State University in 1983, the M.S. degree in Physics from The University of Texas at Dallas in 1990, and the Ph.D. degree in Electrical Engineering from The University of Twente (The Netherlands) in 2007. He was employed by Mostek Corporation in 1984. He joined Texas Instruments in 1985 and is currently a Senior Member of the Technical Staff in the Advanced CMOS Technology-Design Integration department. His research interests include the reliability physics of dielectric materials. He has co-authored 32 publications and is a recipient of the 2000 IRPS Best Paper Award. He has served on the IEDM and IRPS Technical Program Committees. Dr. Nicollian is a Senior Member of the IEEE.

Nicolosi, Piergiorgio (3F.2)

Piergiorgio Nicolosi is full professor at University of Padua since 2004. His research activity has been mainly devoted to plasma and atomic spectroscopy, spectroscopic studies of laser generated plasmas, to the development of spectroscopic instrumentation for laboratory, synchrotron and FEL sources and space applications, to the development of nano-structured multilayer coatings for the extreme ultraviolet spectral range.

Nigam, Tanya (4A.3)

Tanya Nigam received her Bachelor's degree in Physics (Hons.) from St. Stephens College, Delhi University. She obtained a M.Sc in Physics from IIT Kanpur and a M.Sc in Electrical Engineering from the Katholieke Universiteit Leuven in 1995. Between 1995 and 1999, she obtained Ph.D in the area of ultra-thin gate oxides at IMEC, Belgium. From 1999 until 2001, Tanya was a Member of Technical Staff at Bell Labs where she worked on novel device geometries to overcome sub-50nm device challenges. From 2001 until 2005, she was with Agere Systems, formerly the Microelectronics Division of Lucent Technology. At Agere, she worked on reliability issues for power LDMOS devices, and HCI/NBTI reliability concerns for CMOS. From October 2005 till 2007 Tanya worked as a Senior Staff at Cypress Semiconductor involved in the optimization of 65nm CMOS. In 2008 she was with AMD and since 2009 she is with GLOBALFOUNDRIES as SMTS working on the correlation between device and product level degradation. She has co-authored 30 papers in Journals and Conferences.

Nikolic, Borivoje (XT.17)

Borivoje Nikolic is a Professor of Electrical Engineering and Computer Sciences at the University of California, Berkeley. He received the Dipl.Ing. and M.Sc. degrees in electrical engineering from the University of Belgrade, Serbia, in 1992 and 1994, respectively, and the Ph.D. degree from the University of California at Davis in 1999. His research activities include digital and analog integrated circuit design in scaled technologies and VLSI implementation of communications and signal processing algorithms.

Nobuyuki, Yoshioka (2D.4)

Nobuyuki Yoshioka received B.S. (1980), M.S. (1982) and Ph.D (1989) degrees in physics from Collage of Science and Technology, Nihon University, Tokyo, Japan. In 1982, he joined LSI research and development laboratory, Mitsubishi Electric Corp. From 1982 to 2000, he worked on development of x-ray lithography and photomask technology. From 2000 to 2004, he worked on infrastructure development of photomask technology in Selete (a consortium for development of semiconductor technologies in Japan). Currently, He is responsible for the DFM (Design For Manufacturability) technology in Renesas Technology Corp..

Oates, Anthony (IC.10)

Tony Oates received his Ph.D. in physics from the University of Reading, U.K. in 1985. He then joined AT&T Bell Laboratories in Allentown, PA, as a post-doctoral member of the technical staff, where his research focused on defects in silicon crystals. In 1987, he joined the VLSI technology development laboratory of AT&T Bell Laboratories and since then he has studied failure mechanisms in CMOS technologies. He is currently a member of the technology development organization of Agere Systems (formerly the Microelectronics Division of Lucent Technologies), where he is a technical manager with responsibility for technology reliability. He has published over 40 papers in the areas of interconnect and circuit reliability. He is a member of the management committee of the International Reliability Physics Symposium, serving as the symposium General Chair in 2001. He is also involved in paper selection activities for the International Electron Devices Meeting. He has edited 2 conference proceedings on microelectronics materials reliability for the Materials Research Society.

Obradovic, Borna (5F.3)

Borna Obradovic was born in Zagreb, Croatia, in 1970. He received a B.S. degree in Physics in 1993, and MSE and Ph.D. degrees in Electrical Engineering in 1996 and 1999, from the University of Texas at Austin. From 1999 to 2006 he was a TCAD engineer at Intel Corp, working on simulator infrastructure, physical models, and applications, in particular regarding stress-mobility effects, SiGe HBTs and Graphene transistors. Since 2006, he has been at Texas Instruments, developing SPICE models for Non-Volatile Memory devices.

O'Connor, Eamon (BD.1)

Eamon O'Connor was born in Cork, Ireland in 1980. He received his B.E. in Electrical and Microelectronic Engineering from University College Cork in 2002. He was awarded an MEngSc in 2005 for research at the Tyndall Institute (University College Cork) on the fabrication and characterization of electroluminescent devices based on organic and inorganic materials. Since 2006 his research has been focused on the electrical characterisation of MOS device structures utilising high-k dielectric materials on high-mobility III-V compound semiconductors. He is currently a PhD student in the research group of Dr. Paul Hurley at the Tyndall Institute.

Ogasawara, Makoto (2D.4)

Makoto Ogasawara received B.S. (1981) and M.S. (1983) degrees in electronics from Toyohashi university of technology , Aichi, Japan. In 1983, he joined Device development center, Hitachi ,Ltd. From 1983 to 2003, he engaged in process development of DRAM and gate oxide reliability in MOS devices. He is now with Renesas Technology Corp. Currently, He is responsible for the process reliability of FEOL and BEOL

Ohmi, Tadahiro (5F.2)

Tadahiro Ohmi received the B.S., M.S., and Ph.D. degrees in electrical engineering from Tokyo Institute of Technology, Tokyo, Japan, in 1961, 1963, and 1966, respectively. Prior to 1972, he served as a Research Associate in the Department of Electronics, Tokyo Institute of Technology, where he worked on Gunn diodes such as velocity overshoot phenomena, multivalley diffusion and frequency limitation of negative differential mobility due to an electron transfer in the multi-valleys, high-field transport in semiconductor such as unified theory of space-charge dynamics in negative differential mobility materials, Bloch-oscillation-induced, negative mobility and Bloch oscillators, and dynamics in injection lasers. In 1972, he moved to Tohoku University, Sendai, Japan, where he is currently a Professor at the New Industry Creation Hatchery Center. He is engaged in researches on high-performance ULSI such as ultrahigh-speed ULSI based on gas-isolated-interconnect metal-substrate SOI technology, base store image sensor (BASIS) and high-speed flat-panel display, and advanced semiconductor process technologies such as low kinetic-energy particle bombardment processes including high-quality oxidation, high-quality metallization, very-low-temperature Si epitaxy, and crystallinity-controlled film growth technologies from single-crystal, grain-size-controlled polysilicon and amorphous highly selective CVD, highly selective RIE, and high-quality ion implantation with low-temperature annealing capability based on ultraclean technology concept supported by newly developed ultraclean gas supply system, ultrahigh vacuum-compatible reaction chamber with self-cleaning function, and ultraclean wafer surface cleaning technology. His research activities are summarized by the publication of over 1300 original papers and the application of 1600 patents. Dr. Ohmi serves as the President of the Institute of Basic Semiconductor Technology-Development (Ultra Clean Society). He is a Fellow of the Institute of Electricity, Information and

Communication Engineers of Japan. He is a member of the Institute of Electronics of Japan, the Japan Society of Applied physics, and the Electrochemical Society. He received the Ichimura Award in 1979, the Inoue Harushige Award in 1989, the Ichimura Prizes in Industry-Meritorious Achievement Prize in 1990, the Okouchi Memorial Technology Prize in 1991, the Minister of State for Science and Technology Award for the Promotion of Invention (the Invention Prize) in 1993, the IEICE Achievement Award in 1997, the Okouchi Memorial Technology Prize in 1999, the Werner Kern Award in 2001, the ECS Electronics Division Award, the Medal with Purple Ribbon from Government of Japan and the Best Collaboration Award (the Prime Minister's Award) in 2003.

Ok, Injo (4F.4)
Injo Ok received the B.S. and M.S. degrees in Electrical Engineering from Changwon National University, Changwon, Korea in 2000 and 2002, respectively and the Ph.D. degree in Electrical and Computer Engineering at the University of Texas, Austin. In April 2008, he joined SEMATECH at Albany, Albany, NY, where he working on fabrication and characterization of III-V and SiGe FinFET transistors for 22-nm node and beyond. He has authored or coauthored more than 47 technical papers.

Olney, Andrew (3B.1)
Andrew Olney is the Director of Reliability, Product Analysis, Calibration & ESD at Analog Devices, Inc. in Wilmington, Massachusetts. He received a BS degree from Lehigh University and an MS degree from Boston University, both in Electrical Engineering. Andrew is responsible for managing ADI's worldwide Reliability, Product Analysis, Calibration, and ESD labs and associated engineering organizations in Ireland, the Philippines, and the United States. Andrew has published several papers and holds several patents related to ESD testing and on-chip protection. He also represents ADI on the Semiconductor Industry Association (SIA) Anti-Counterfeiting Task Force.

Olson, Nicholas (4D.3)
Nicholas Olson received his B.S. degree in electrical engineering from Iowa State University of Science and Technology in 2005. He received his MS degree from the University of Illinois at Urbana-Champaign in 2008 and is working towards his Ph.D at the same university. His research is in the field of ESD and he plans to graduate by 2011.

Ong, Yi Ching (4A.4)
Y. C. Ong received the B.Eng. (Hons) and Ph.D. degrees in electrical and electronics engineering from Nanyang Technological University (NTU) in 2004 and 2010, respectively. She is currently a post doctoral fellow in the University of Tokyo. Her research interest is in the application of scanning probe microscopy technique on surface and reliability physics.

Onoye, Takao (3A.4)
Takao Onoye received the B.E. and M.E. degrees in electronic engineering, and the Dr.Eng. degree in information systems engineering, all from Osaka University, Osaka, Japan, in 1991, 1993, and 1997, respectively. He was an Associate Professor with the Department of Communications and Computer Engineering, Kyoto University, Kyoto, Japan. Since 2003, he has been a Professor in the Department of Information Systems Engineering, Osaka University. He has published more than 200 research papers in the field of VLSI design and multimedia signal processing in reputed journals and proceedings of international conferences. His current research interests include media-centric low-power architecture and its SoC implementation. Dr. Onoye has served as a member of the CAS

Society Board of Governors since 2008. He is a member of IEEE, IEICE, IPSJ, and ITE-J.

Ottogalli, Federica (5C.4)
Federica Ottogalli, since 28-Oct-2007 with Numonyx, Via C.Olivetti 2, 20041, Agrate Brianza (Milan), Italy. She received the doctor degree in Physics from the University of Padova, Italy, in 1998 with a thesis on crystallographic characterization by RBS-Channeling and modeling of III-V compounds. She joined the Non-Volatile Memory Technology Development Group of the Central R&D of STMicroelectronics in Agrate Brianza (Milan) in 1999. Since 2002, she has been working on the process development for phase-change memories based on chalcogenide materials.

Oualli, Mourad (2E.3)
Mourad Oualli was born in France in 1982. He was graduated from the Ecole Polytechnique, Palaiseau, France and the Ecole Supérieure d'Electricité (Supélec), Gif-sur-Yvette, France, in 2007. He then joined the Alcatel-Thales III-V lab as a research engineer in the development of the AlGaN/GaN and AlInN/GaN HEMT technologies. He especially works on the improvement of device reliability.

Ouchi, Tomohiko (2D.4)
Tomohiko Ouchi received B.S.(1988) and M.S.(1991) degrees in physics from Ibaraki University, Ibaraki, Japan. He joined the Hitachi Research Laboratory, Hitachi ,Ltd, in 1991. He worked on molecular CAD development, Flash memory design and ferroelectric memory development. He is now with Renesas Technology Corp. Currently, he is working on the development of DFM method.

Paccagnella, Alessandro (3F.2, 4B.3)
Alessandro Paccagnella is Full Professor of Electronics and Director of the Department of Information Engineering at the University of Padova. He is the author of more than 300 scientific papers, and about 200 of them have been published on international journals. In the past, his research has been directed to the study of different aspects of physics, technology, and reliability of semiconductor devices. At present, he coordinates the activity of a research group focused on the study of ultra-thin gate dielectrics in MOS devices and on Total Ionizing Dose and Single Event Effects induced by ionizing radiation on integrated circuits.

Padovani, Andrea (4A.4, 4F.4, 6C.1)
Andrea Padovani graduated in Electronics Engineering at the University of Modena and Reggio Emilia, Italy, in 2005. He received his Ph.D. in 2009 from the University of Ferrara, Italy. He is currently a post-doc at the University of Modena and Reggio Emilia, Italy.His research activity focuses on the reliability and modeling of logic transistors based on high-k/metal gate technology, and of innovative non-volatile memories, such as resistive rams (RRAM) and charge-trapping devices (NROM, TANOS). He authored and co-authored more that 25 technical papers in international journals and conference proceedings. He serves as reviewer for several international journals.

Pan, Liu (3C.5)
Liu Pan was born in Xi'an, Shaanxi province, China, He received the B.S and M.S degree in material science and technology from the University of Science and Technology Beijing in 2000 and 2003, respectively. He is currently working for GLOBALFOUNDRIES Singapore company and responsible for semiconductor chip failure analysis. His strength is in transmission electron microscopy (TEM) sample preparation and image analysis. His interested areas include chip function failure analysis and reliability assessment. He also is the author of 8 papers and 2 patents on the failure analysis, material characterization and TEM sample preparation technique.

Park, Chan-Hoon (2C.4)
Received the B.S. (2008) in electronics engineering from the Kyungbuk National University, Korea. He received the M.S. (2010) in electronics engineering from the Pohang University of Science and Technology (POSTECH), Pohang, Korea. He is currently in Ph.D. course at POSTECH. His main research interests include nanowires fabrication.

Park, Hokyung (2B.4)
Hokyung Park received a B.S. (2001) in Avionics from Korea Aerospace University and M.S. (2003) and Ph. D (2007) in Material Science and Engineering from Gwangju Institute of Science and Technology, KOREA. He is device characterization and reliability engineer at SEMATECH from 2008. His current research area is characterization of memory devices and high-k gate dielectric for logic application and reliability evaluation.

Park, Hyun-Kook (RM.3)
Hyun-Kook Park was born in Ik-san, Jeollabuk-do, Korea, in 1983. He received the B. S. degree in the electrical and electronic engineering from Yonsei University, Seoul, Korea, in 2008. He is studying for M. S degree in Yonsei University. His current research interests is SRAM stability and FinFET SRAM design.

Park, Jongwoo (2D.1, 3C.3)
Jongwoo Park received the Ph. D. degree from Lehigh University, Bethlehem, PA, in 1998. After post doctoral research at Lehigh University in 1999, he joined Lucent Technologies as a Member of Technical Staff. His research projects were packaging and reliability associated with application and characterization of polymeric materials used for microelectronic devices and interfacial failure mechanism. In 2002, he joined Princeton Optronics as a Manager of Quality/Reliability. He was involved in hermetic/nonhermetic package development and reliability focused on tunable laser module including pump laser, MEMS and VCSEL. Since 2003, he is with Technology Reliability, Q&R team in SYSTEM LSI, Samsung Electronics as a Director responsible for reliability qualification of CMOS process and product and has contributed to the development and reliability of 90, 65, 45, and 32nm logic process technology nodes.

Park, Milim (MY.7)
Milim Park received the B.S. degree in electrical engineering from Chonbuk National Univercity, Chonjo, Korea, in 2005. From 2005 to 2010, she is with the Flash Device Team, Hynix Semiconductor Inc., Korea. She is currently focusing on triple level cell technology development in Nand Flash Memeory.

Park, Suk-Kwang (MY.7)
Suk-Kwang Park received the B.S. degree in Physics from Dong A University, Busan, Korea, in 1999, and the M.S. degrees in Physics from Pusan National University Busan, Korea, in 2002 , Since joining the hynix Semiconductor Inc. Ichon,Korea, in 2002, he has been working on device and process integration of high density memory as well as deep submicron devices. From 2002 to 2005.He and his team developed hynix's first MLC product ,the world's first commercial 16MB 3bits NAND flash in 2008, He is currently working on the development 3bits NAND.

Park, Young-Joon (6A.1)
Young-Joon Park received his BS from Seoul National University in 1986, and MS.D. and Ph.D. in material science and engineering from KAIST in 1988 and 1995. He was a post doctoral associate in MIT during 1995-1996. Young-Joon joined Texas Instruments (TI) in 2002 and is a Senior Member of Technical Staff, working in the BEOL reliability area. Prior to TI, He had researched on various topics in Korean governmental research institutes: 1988-1991 in ETRI and 1996-2002 in KIST. Young-Joon has authored or co-authored 70+ papers/presentations and holds 10+ patents issued or filed.

Pavan, Paolo (4F.4, 6C.2)
Paolo Pavan graduated in Electrical Engineering at the University of Padova, Italy, in 1990. He received his PhD in 1994 from the same University. From 1992 to 1994 he was at the University of California at Berkeley. He is currently Full Professor of Electronics at the University of Modena and Reggio Emilia. He is the President of the IU.net Consortium. His research activity deals with electrical characterization, modeling and reliability of integrated circuits and nonvolatile memory devices. He also works on "By-wire" systems for automotive and wireless embedded systems. He has been involved in the technical committees of international conferences (IEDM, ESREF). He authored and co-authored many papers, one book and two chapters in edited books.

Pei, Yi (2E.5)
Received the B.S. degree in Electrical Engineering from Peking University, Beijing, China, in 2000, the M.S. and Ph.D. degrees in Electrical Engineering from University of Santa Barbara, USA, in 2005 and 2009, respectively. He joined in Dynax Semiconductor, Inc. in 2009. His research interests focus on design, fabrication, and characterization and applications of compound semiconductors, especially in Nitride-based devices. Dr. Pei has authored and co-authored more than 50 papers in technical journals and conferences.

Pendharkar, Sameer (HV.1)
Graduated from The University of Wisconsin-Madison in 1996. In 1996, he joined Texas Instruments Inc. where he worked on developing more than 5 generations of BiCMOS-DMOS (LBCTM) technologies. His primary focus was on developing highly efficient, robust and cost efficient integrated high voltage and high power semiconductor devices. He is currently a TI Fellow and manages the High Voltage Component Development Group at Texas Instruments Inc.

Petitprez, Emmanuel (IC.8)
Emmanuel Petitprez received the Engineering degree in solid state physics from the Institut National des Sciences Appliquées, Toulouse, France, in 1994, and the Ph.D. degree in materials science from the University of São Paulo, Brazil, in 2001. He was with Serma Technologies, Grenoble, France, as a Characterization Engineer. In 2005 he moved to Freescale Semiconductor, Crolles, France working on interconnect reliability issues in SOI technologies. Since 2007, he has been with the STMicroelectronics Central R&D Laboratory, Crolles, France where he works on the reliability of interconnects for advanced CMOS technology.

Pey, Kin Leong (2C.3, 4A.1, 4A.4, 5B.3, BD.2, MY.4)
Kin Leong Pey received his Bachelor of Engineering (1989) and Ph.D. (1994) in Electrical Engineering from the National University of Singapore (NUS). He has held various research positions in the Institute of Microelectronics, Chartered Semiconductor Manufacturing, Agilent Technologies and National University of Singapore. He is currently a Visiting Professor at Nanyang Technological University (NTU) and an Associate Provost of Singapore University of Technology and Design (SUTD), Singapore and also holds a concurrent Fellowship appointment in the Singapore-MIT Alliance (SMA). He has published more than 150 international refereed publications and 160 technical papers at international meetings/conferences and holds 33 US patents. Dr. Pey is a senior member of IEEE and an IEEE EDS Distinguished Lecturer.

Phang, Jacob Chee Hong (3C.4, 4E.4)
Jacob CH Phang received both his BA and PhD degrees from the University of Cambridge in 1975 and 1979 respectively. He joined the National University of Singapore in 1979 where he is now Professor at the Centre for Integrated Circuit Failure Analysis and Reliability (CICFAR), Faculty of Engineering. His field of research is in microelectronic device failure analysis and reliability. He is also Executive Chairman of SEMICAPS Corporation, an NUS spin-off company he co-founded in 1988 to commercialise the technologies developed at CICFAR for world-wide distribution.

Piazza, Michele (2E.3)
Michele Piazza was born in Pordenone, Italy, in 1981. He graduated in Electronic Engineering at the University Of Padova (Italy) in 2008. He's currently a research Engineer at Alcatel-Thales 3-5 Lab and registered as PhD student at X-LIM, France. He works on GaN HEMTs reliability within the Thales Microelectronics Group.

Pickholtz, Jeffrey (3A.1)
Jeffrey Pickholtz joined Intel in 2003 where he contributed to the design of multiple Itanium microprocessors. He is currently an engineering manager working on a next generation microprocessor. Prior to joining Intel, Jeffrey worked for Digital Equipment, Compaq and HP where he contributed to a variety of VAX and Alpha microprocessor designs

Pinato, Alessandro (4F.2)
Alessandro Pinato was born in Piove di Sacco (Padova), Italy, in 1982. In 2007 he received the degree (summa cum laude) in Electronics Engineering at the University of Padova, working on the development of an Ion Sensitive Field Effect Transistor (ISFET) based on organic electronics. He is currently working toward the Ph.D. degree in Information Engineering at the University of Padova. His main interests are the characterization and reliability analysis of organic semiconductor devices, in particular OLEDs and organic photovoltaic solar cells.

Pirovano, Agostino (5C.4)
Agostino Pirovano was born in Italy in 1973. He received the Laurea degree in electrical engineering in 1997, and the Ph.D. degree at Politecnico di Milano, Italy, in 2000. He joined the Department of Electrical Engineering in 2000, working on the modeling and characterization of transport properties in MOSFET devices. In 2001 and 2002 he was a consultant for STMicroelectronics. From 2002 he teaches 'Optoelectronics' at the Politecnico di Milano, where he was a lecturer from 1999. In 2003 he joined the Non-Volatile Memory Technology Development Group of STMicroelectronics, working on the modeling and characterization of phase-change memory devices. From 2008 he is with Numonyx, R&D Technology Development, being in charge for the investigation of emerging NVM technologies.

Pobegen, Gregor (XT.8)
Gregor Pobegen received his BSc degree in technical physics from the Graz University of Technology, Austria, in 2007. He is currently working toward his MSc degree at KAI GmbH in cooperation with Infineon Technologies Austria. His master thesis focuses on the electrical characterization of NBTI induced oxide defects.

Pogany, Dionyz (4D.4)
Dionyz Pogany received his Dipl. - Ing. degree in solid state engineering from the Slovak Technical University in Bratislava in 1987. In 1994 he received a Ph.D. degree at INSA de Lyon, France. In 1994-95 he was a postdoc at France Telecom, CNET-Grenoble. Since 1995 he is with the Institute of Solid State Electronics, TU Vienna, Austria, where he leads a research team. Since 2003 he is Associate Professor at TU Vienna. He published on defect states in semiconductors, low frequency noise and device reliability physics. His current research interest is in ESD-phenomena, self-heating effects, current filamentation and thermal breakdown, power electronics, GaN HEMTs and LEDs, failure analysis, device reliability and development of new optical methods for device characterization. He is author or co-author of more than 250 scientific contributions.

Poli, Stefano (HV.1)
Received the Ph.D. degree in Information Technology from the University of Bologna in 2009. In 2005 he was with the Interuniversity Micro Electronic Center, Leuven, Belgium, as an Internship under-graduate student. Since 2006 he is with the ARCES Research Center, working on the modeling and simulation of ultra-scaled CMOS, post-CMOS and CNT based device. In 2007 he was with CEA-LETI, Grenoble, France, as a graduate student. He is currently involved in the modeling, design and TCAD analysis of low-Rsp power MOSFETs in the frame of a SRC Project in collaboration with Texas Instruments (Dallas, Texas).

Puchner, Stefan (XT.6)
Stefan Puchner finished his diploma thesis, concerning thin layer surface analysis with time of flight secondary ion mass spectrometry in May 2007 and received his M.S. degree in physics from the Vienna University of Technology, Austria. Currently he is working toward his doctor degree at KAI in cooperation with Infineon Villach and the Institute for Chemical Technologies and Analytics.

Rafik, Mustapha (4A.2)
Mustapha Rafik received the engineering degree in physics from the Instiut Nationale Polytechnique de Grenoble, and the M.S.E.E degree from Joseph Fourier University of Grenoble, France, 2005. In joined STMicroelectronics, Crolles, France in 2005 where he obtained a PhD degree in micro and nanoelectcronics in 2008. He is currently a reliability engineer with STMicroelectronics and his research focus on advanced gate stack reliability.

Raghavan, Nagarajan (5B.3)
Nagarajan Raghavan was born in Bangalore, India in 1985. He received his B.Eng, 1st Class Honors, (Electronics Engineering, 2007), S.M. (Advanced Materials for Micro & Nano Systems, 2008) and M.Eng (Materials Science and Engineering, 2008) from Nanyang Technological University (NTU), National University of Singapore (NUS) and Massachusetts Institute of Technology (MIT) respectively.

Ragheb, Tamer (5E.5)
Tamer Ragheb received the Ph.D. degree in electrical and computer engineering from Rice University, Houston, TX, in 2008. Since 2008, he has been with Texas Instruments Incorporated, Dallas, TX, where he currently works with Advanced CMOS Technology-Design Integration Group. His research interests include interconnect modeling, parasitic extraction, substrate noise characterization, and the development of circuit designs to evaluate and verify performance and power metrics for deep submicron technologies. He has published more than 30 papers in peer reviewed journal and conference proceedings.

Ragnarsson, Lars-Ake (2A.3, XT.13)
Lars-Åke Ragnarsson received a M.S. degree in 1993 and a Ph.D. degree in 1999 in Electrical Engineering from Chalmers University of Technology, Goteborg, Sweden. He did post-doctoral studies at the IBM T.J. Watson Research Center in Yorktown Heights, NY, USA between 2000 and 2002, focusing mainly on electrical characterization of high-k dielectrics. He is since 2002 employed by IMEC in Leuven, Belgium as a senior research scientist on high-k dielectrics and metal gates.

Rangoni, Armando (MY.5)
Armando Rangoni was born in 1971 in Piacenza, Italy. In 1996 he graduated in Industrial Chemistry and in 1999 he took the Specialization in Science of Polymers. In 2000 he was hired from STMicrolectronics as R&D Process Engineer and after, from 2004, he worked as R&D SPC Engineer. From 2008, under Numonyx, he works in the field of NAND memories electrical characterization and reliability.

Rao, V. Ramgopal (4D.4, EL.2)
V. Ramgopal Rao received the M.Tech. degree from Indian Institute of Technology (IIT) Bombay, Mumbai, India, in 1991 and Dr. Ingenieur degree from the Faculty of Electrical Engineering, Universitaet der Bundeswehr Munich, Germany, in 1997. During 1997–1998 and again in 2001, he was a Visiting Scholar with the Electrical Engineering Department, University of California, Los Angeles. He is currently a Professor in the Department of Electrical Engineering, IIT Bombay. His areas of interest include physics, technology, and characterization of silicon CMOS devices for logic and mixed-signal application and Nanoelectronics. He has over 250 publications in these areas in refereed international journals and conference proceedings and holds three patents with eight patents currently pending. Prof. Rao is a Fellow of the Indian National Academy of Engineering, a Fellow of the Indian Academy of Sciences and a Fellow of the Institution of Electronics and Telecommunication Engineers (IETE). He received the Shanti Swarup Bhatnagar Prize in Engineering Sciences in 2005 for his work on electron devices. He also received the Swarnajayanti Fellowship Award for 2003–2004, instituted by the Department of Science and Technology, Government of India, 2007 IBM Faculty award, 2008 'The Materials Research Society of India (MRSI) Superconductivity & Materials Science Prize' and the 2009 TechnoMentor award instituted by the Indian Semiconductor Association. . He is an Editor for the IEEE TRANSACTIONS ON ELECTRON DEVICES in the CMOS devices and technology area and is a Distinguished Lecturer (DL), IEEE Electron Devices Society. Prof. Rao was the organizing committee Chair for the 17th International Conference on VLSI Design and the 14 th International Workshop on the Physics of Semiconductor Devices and serves on the program/organizing committees of various international conferences including the International Electron Devices Meeting (IEDM), IEEE Asian Solid-State Circuits Conference, 2006 IEEE Conference on Nano-Networks, ACM/IEEE International Symposium on Low Power Electronics and Design, 11 th IEEE VLSI Design & Test Symposium among others. He was Chairman, IEEE AP/ED Bombay Chapter during 2002-2003 and currently serves on the executive committee of IEEE Bombay Section besides being the vice-chair, IEEE Asia-Pacific Regions/Chapters Subcommittee.

Recchia, Charles (3A.1)
Dr Charles Recchia is currently senior product reliability engineering manager at Intel Corporation, having previously focused on microlithography process development for next-generation silicon technologies with Portland Technology Development. Dr Recchia holds a doctorate in condensed matter physics from Ohio State University, holding 3 patents and author on over twenty technical publications spanning fundamental research, technology development and microprocessor reliability.

Redaelli, Andrea (5C.4)
Andrea Redaelli was born in Italy in 1978. He received the Laurea (cum laude) and Ph.D. degrees in electronic engineering from the Politecnico di Milano, Italy, in 2003 and 2007 respectively. During the Ph.D. thesis he worked on Phase Change Memories in the Department of Electrical and Electronic Engineering, Politecnico di Milano. His research interests include the modeling and characterization of transport properties and phase-change transition of chalcogenide-based devices. From 2007, he joined STMicroelectronics working on advanced technologies for non volatile memories and since 2008 he is employed in the emerging non volatile memory group of Numonyx. From 2005 to 2009, he cooperated with Politecnico di Milano, in holding master's classes on electronics and signal conditioning.

Reddy, Vijay (5E.5)
After receiving the Ph.D. (1994) in Electrical Engineering from the University of Texas at Austin, Vijay Reddy joined Texas Instruments and has worked on several topics concerning transistor, circuit reliability and product qualification methodologies. He is currently a Distinguished Member Technical Staff and is focusing on the digital, analog, and RF spaces. He has served on the IRPS/IEDM program committees and has presented papers at IRPS/IEDM and invited tutorials at IRPS/ICTMS/VLSI Test Symposium. He received the 2002 IRPS Outstanding Paper Award, 2004 IRPS Outstanding Paper Award, and the 2002 ESD/EOS Symposium Best Paper/Best Presentation Awards. He has received eleven patents with several pending along with more than 26 publications.

Reed, Robert (4B.2, SE.3)
Robert A. Reed received his M.S. and Ph.D. degrees in Physics from Clemson University in 1993 and 1994. After completion of his Ph.D. he worked as a post-doctoral fellow at the Naval Research Laboratory and later worked for Hughes Space and Communication. From 1997 to 2004, Robert was a research physicist at NASA Goddard Space Flight Center where he supported NASA space flight and research programs. He is currently a Research Associate Professor at Vanderbilt University. His radiation effects research activities include topics such as single event effect and displacement damage basic mechanisms and on-orbit performance analysis and prediction techniques.

Reents, William (FA.5)
William Reents is a Consulting Member of Technical Staff and director of the Reliability Physics Group in Murray Hill, NJ. He joined Bell Laboratories in 1980, working in several areas of research and root cause analysis. His areas of specialty include package hermeticity, organic contamination and gas and particle contamination issues.

Reggiani, Susanna (HV.1)
Received the B.S. and Ph.D. degrees in electrical engineering from the University of Bologna, Bologna, Italy, in 1997 and 2001. Since April 1997, she has been working with the Department of Electronics (DEIS), University of Bologna, in the field of numerical simulation of semiconductor devices. She became a Research Associate in 2001 and is currently with the ARCES Research Center. Since 1999, she has been working on the simulation of electron transport in nanoscale devices such as silicon nanowires, carbon nanotubes and graphene nanoribbons. She is currently involved in a SRC project in collaboration with Texas Instruments.

Regolini, Jorge Luis (3C.2)
Jorge Luis Regolini received his M.Sc. from Centro Atomico Bariloche (Argentina) and graduated from Strasbourg University (France) with a Thesis in Solid State Physics. He was then with the Atomic Energy Commission in Argentina as a researcher on Semiconductor Physics and Technology. At the EE Dept. (Stanford University - USA) he worked as a post-doctoral fellow in laser processing for silicon recrystalisation and silicide formation. From 1986 he was with France Telecom CNET-CNS involved on RT/RPCVD for the selective deposition of epitaxial Si/SiGeC and silicides. He has presented several invited papers on those fields, concerning material fabrication and characterisation, kinetics and

kinematics aspects of reduced pressure single wafer reactors. From 2000 he was with STMicroelectronics (Crolles-France) as Senior Staff Engineer on Advanced Dielectrics for new generations of DRAMs and gate stacks, and circuit passivation issues. He is now R&D expert and he has participated to the development of new Image Sensors Technologies. He holds several patents and more than 100 publications in these fields.

Reifenberg, John (2C.5)
John P. Reifenberg received the B. S. degree in mechanical engineering from Carnegie Mellon University in 2003, and the M. S. and Ph. D degrees in mechanical engineering from Stanford University in 2006 and 2010, respectively. His graduate research developed models and measurements of nanoscale thermal transport phenomena in phase change memory data storage devices. He is a recipient of the National Defense Science and Engineering Graduate (NDSEG) Fellowship and an Honorary Stanford Graduate Fellowship (SGF).

Reisinger, Hans (2A.1, 2A.3, 2F.4)
Hans Reisinger received his diploma in physics (1979) and his Ph.D. (1982) both from the Technical University of Munich. In 1982/83 he was with the IBM T.J.Watson Research Ctr. in Yorktown Heights working on electronic properties of 2d-systems. In 1986 he joined the Siemens Semiconductor Department (now Infineon). His work was focused on the study of thin dielectrics and interfaces in DRAMs and NVMs. Currently he is with the Infineon Central Reliability Methodology group and mainly works on threshold instabilities of MOSFETs.

Relangi, Prasanthi (3A.3)
Prasanthi Relangi received her B.E. in Electrical and Electronic Engineering with Honors from Birla Institute of Technology and Science (BITS), Pilani, India in 2005 and her M.S. in Electrical Engineering from Stanford University, Stanford CA, in 2007, where she is currently pursuing her Ph.D. in the Department of Electrical Engineering. Her research interests include erratic bit errors, soft errors and digital circuits.

Remack, Keith (6C.4)
Keith Remack (IEEE M'81) was born in Chicago, IL. He received the B.Sc. degree in Electrical Engineering from The Milwaukee School Of Engineering, Milwaukee, WI. He currently works in the Medical Business Unit of Texas Instruments, Inc., Dallas, TX. His current research interests are the development of product and reliability test methods for embedded FRAM memory arrays. Mr. Remack is a registered Professional Engineer in the state of Texas and has authored or coauthored over 20 technical papers.

Ren, Fan (CD.3)
Fan Ren is Charles Stokes Professor of Chemical Engineering at the University of Florida, Gainesville, FL, USA. He joined UF in 1997 after 12 years as a Member of Technical Staff at AT&T Bell Laboratories, where he was responsible for high speed compound semiconductor device development. He is a Fellow of ECS. APS and AVS.

Renard, Sophie (2B.2)
Sophie Renard received the PhD degree in microelectronics from the Université de Provence, Marseille, France in 2003 working on EEPROM memories reliability through the collaboration between STMicroelectronics, Rousset, France and the L2MP laboratory, Marseille, France. Afterwards, she joined the reliability group of STMicroelectronics Central R&D, Crolles, France where she has been working on Front-end Reliability, and more specifically on Hot-Carriers Injection

mechanism. Since 2008, she has been Technical Leader of the Non Volatile Memories Reliability.

Ribes, Guillaume (4A.2)
Guillaume Ribes received the engineer degree in electrical engineering from the ISEN (Institut Supérieur d'Electronique et du Numérique, France), in 2002. He obtained in 2005, the Ph.D degree in microelectronics from INPG (Institut National Polytechnique de Grenoble). He worked as a reliability engineer at STMicroelectronics in the area of advanced devices reliability. He works now for ISDA as ST assignee in the field of High-K and Low K reliability. He has published 40 papers and 20 as first author. He serves the IRPS Technical program committee in 2006 and 2007 for High-K and in the 2010 for BEOL dielectrics & Interconnects.

Riedlberger, Eva (2F.4)
Eva Riedlberger obtained her Diploma in Physics from the Technical University Munich in 2007, where she worked on the growth of III-V based heterostructures by Selective Area Epitaxy at the Walter-Schottky-Institute. She is currently pursuing a Ph.D. degree in Electrical Engineering from the Bundeswehr University in Neubiberg, to graduate by Sept. 2010. She is with Infineon Technologies since 2007 and is working on the modeling of the degradation of Lateral DMOS Transistors due to Hot Carrier Injection.

Robl, Werner (IC.6)
Werner Robl received his PhD in Physics from the University of Regensburg in 1994. After his degree he joined Infineon (former Siemens Semiconductors). Since then he has been working on development of new metallization schemes in Regensburg and Munich, Germany and East Fishkill, NY. Currently, he is working as a Principal for Metallization on new metalliza¬tion schemes for semiconductor devices and chip packaging. He is a member of the German Physical Society.

Roche, Philippe (4B.4)
Philippe Roche received the M.S. (1995) and Ph.D. (1999) in semiconductor physics from the University of Montpellier, France. From 1995 to 1999, he worked consecutively at the University of Eindhoven, the Netherlands, at the French Atomic Energy Commission, Military applications centre at Bruyères-le-Chatel, at the University of Montpellier and in the Radiation Effects Group at Vanderbilt University, USA. Since 1999, he has been with STMicroelectronics, Central CAD and Design Solutions, Crolles France, as senior expert and manager of a group in charge of both SER safety/reliability aspects and subthreshold (~0.3V) IP designs. His primary research activities are Single Event Effects and Total Ionizing Dose, as well as Ultra Low Voltage IPs, on sub-0.25µm commercial technologies down to CMOS 20nm. He has been serving in conferences since 1997, as session chairman and short course instructor, in 10 international conferences, such as IRPS, NSREC, IOLTS, RADECS and SOI conference. Philippe has coauthored 90 papers and has filed 19 patents and 3 trade marks in radiation hardening.

Rodriguez, John (5F.3, 6C.4)
Dr. Rodriguez is a Senior Member of the Technical Staff in the Analog Technology Development Reliability group at Texas Instruments, Dallas, where he has focused on FRAM technology since 2001. He earned his BS, MS and PhD ('99) degrees at Rice University, all in Electrical and Computer Engineering. He joined TI's advanced process development lab in Houston in 1993 as a summer intern, where he contributed TCAD and SPICE modeling for submicron modular merged technology components including high voltage, analog, non-volatile memories and ESD. He has co-authored over 25 conference and journal publications and has been awarded 11 US patents.

Rodriguez Latorre, Jose (6C.4)
Jose A. Rodriguez was born in Moca, Puerto Rico on June 21, 1983. He received the B.S. degree in Electrical Engineering from the University of Puerto Rico at Mayaguez in 2007. Currently, he is a Research Assistant at the University of Puerto Rico at Mayaguez and working toward his M.S. degree. Jose has worked at Texas Instruments as a summer intern since 2007 mostly focusing on product engineering and test development.

Rosenbaum, Elyse (4D.5, EL.1)
Elyse Rosenbaum received the B.S. degree (with distinction) from Cornell University in 1984, the M.S. degree from Stanford University in 1985, and the Ph.D. degree from the University of California, Berkeley in 1992. All of these degrees were in electrical engineering. From 1984 through 1987, she was a Member of Technical Staff at AT&T Bell Laboratories in Holmdel, NJ. She is currently a Professor in the Department of Electrical and Computer Engineering at the University of Illinois at Urbana-Champaign. Dr. Rosenbaum's present research interests include design, testing, modeling and simulation of ESD protection circuits, design of high-speed circuits with ESD protection, and latch-up. She has presented tutorials on reliability physics at the International Reliability Physics Symposium, the EOS/ESD Symposium, and the RFIC Symposium. She has authored or co-authored over 100 technical papers and is an editor for IEEE Transactions on Device and Materials Reliability. Dr. Rosenbaum has been the recipient of a Best Student Paper Award from the IEDM, a Technical Excellence Award from the SRC, an NSF CAREER award, an IBM Faculty Award, and a UIUC Bliss Faculty Scholar Award.

Roussel, Philippe J. (2A.3, 6A.3, XT.13)
Philippe J. Roussel was born in Bruges, Belgium, on May 18, 1955. He recei¬ved the diploma in Electrical Engineering from the Industriële Hogeschool of Ghent, Belgium, in 1983. In 1984, he joined the ESAT labo-ratory of the KULeuven, Belgium as an assistant in a reli¬ability research project on plastic en¬capsulation. From 1987, he assisted in a gov¬ernment funded project on electromigration at imec, Leuven, which also resulted in a maxi¬mum likelihood fitting program for the assessment of statistical distributions and acceleration models from reliability test data. From 1991 till 2000, he worked on analysis techniques like microprobe, TEM and Spectroscopic Ellipsometry (SE) in the PT/MCA group, while continuing his research on Reliability Statistics, SPC and Data Analysis in the PT/DRE group. At present he also works in a multi¬disciplinary team on the statistics required for Technology Aware Design (TAD). He co-authored papers in fields as diverse as scientific handheld programming, virology, SE, ESD, oxide and interconnect reliability, and TAD.

Roy, David (2B.2)
David Roy received the BS (1997) in physics, the M.S.(1998) in Physics (from the Institut National Polytechnique de Grenoble (INPG) and the Magistere (1998) of Physics Research from University Joseph Fourier de Grenoble. He worked for the CEA-Grenoble on the 3D-optical micro-system in the "Laboratoire d Electronique et des Technologies de l'information" (LETI) in 1999. In 1999, he joined STMicroelectronics as a reliability engineer, working on oxide and device reliability. Since 2007, he is in charge of the Front-end Reliability team. His current research interests include transistor reliability as well as low k interconnect reliability.

Ruelke, Hartmut (IC.2)
Dipl. Phys. Hartmut Ruelke is a GLOBALFOUNDRIES Fellow Process Engineer working for Module One LLC&Co KG Dresden, Germany. He is leading the CVD and PVD projects in the Thin Films Module at GLOBALFOUNDRIES, former AMD Fab36 LLC & Co KG, 300mm Micro-Processor Fabrication. During 1996 to 2004 he was the lead CVD Engineer in AMD's 200mm fabrication line, responsible for PE-CVD, HDP and Tungsten CVD processes. Prior joining AMD in 1996 he was working for System Microelectronic Innovation GmbH Frankfurt (Oder) as Senior Process Engineer for CVD - Processes and Epitaxy. During 1991 to 1992 Hartmut Ruelke was Head of the Group Doping Processes / CVD in the Technology Department of Halbleiter Elektronik GmbH Frankfurt (Oder). From 1979 to 1991 he worked at the Halbleiterwerk Frankfurt (Oder), a Bipolar and CMOS Semiconductor Fabrication, as Process Engineer for Diffusion, Epitaxy and Semiconductor Measurement.Hartmut Ruelke received his Diploma degree in Physics in 1979 from Technical University of Magdeburg, Germany and completed postgraduate study of Semiconductor Technology 1982 at Technical University of Karl-Marx-Stadt, Germany.

Rumyantsev, Sergey (2B.4)
Sergey L. Rumyantsev received the M.S.E.E. degree from Leningrad Electrotechnical Institute, Leningrad, USSR, in 1977, the Ph.D. degree in physics from Leningrad Polytechnical Institute in 1986, and the Doctor of Science (Habilitation) degree from A.F. Ioffe Institute of Physics and Technology, Lenningrad, in 1996. From 1999 he is with Rensselaer Polytechnic Institute. He is also a Research Fellow of A.F. Ioffe Institute of Physics and Technology. His current research interests include low frequency noise, wide bandgap semiconductors, terahertz electronics, nanowires and semiconductor nanotubes, graphene. He published a number of papers and he is a coeditor of 5 books.

Ryan, Jason (2A.5, XT.18)
Jason T. Ryan (S'04) received the B.S. degree in Physics from Millersville University, Millersville, PA in 2004. He received the M.S. degree in Engineering Science and the Ph.D in Materials Science and Engineering from The Pennsylvania State University, University Park, PA in 2006 and 2010 respectively. He is currently employed in the Semiconductor Electronics Division at the National Institute of Standards and Technology. His current research involves the atomic-scale mechanisms involved in bias temperature instabilities and stress induced leakage currents in ultra thin gate dielectrics.

Ryan, Shawn (FA.5)
Dr. Shawn Ryan is the decontamination research area lead in EPA's National Homeland Security Research Center's (NHSRC) Decontamination and Consequence Management Division. His research includes determining methods for assessing decontamination efficacy; investigating technologies as a function of agent and operation; decontaminant-material interactions (demand, compatibility, by-products). He received his B.S. in Environmental Engineering, M.S. in Chemical Engineer, and Ph.D. in Chemical Engineering all from Rensselaer Polytechnic Institute, Troy, NY.

Ryoichi, Ishihara (3F.3)
Ryoichi Ishihara received the Ph.D. degree from the Tokyo Institute of Technology, Tokyo, Japan, in 1996. Since 1996, he has been with the Delft Institute of Microsystems and Nanotechnology (DIMES), Delft University of Technology, Delft, The Netherlands, where he has been focusing on location control of grains through a novel excimer-laser crystallization process and fabrication and characterization of high-performance TFTs inside a single grain. He is currently an Associate Professor with the Faculty of Electrical Engineering Mathematics and Computer Science, Delft University of Technology.

Ryuta, Tsuchiya (PI.3)
Ryuta Tsuchiya received the B.S., M.S., and Ph.D. degrees in material science from the Tokyo Institute of Technology, Tokyo, Japan, in 1993, 1995, and 1998, respectively. He joined the Central

Research Laboratory, Hitachi, Ltd., Tokyo, Japan, in 1998, where he has been engaged in research on fabrication and characterization of high-performance and low-power MOSFETs including thin-film SOI and BOX transistors. Dr. Tsuchiya is a member of the Japan Society of Applied Physics.

Sagong, Hyun Chul (2C.4)
Received the B.S. (2008) in electronics engineering from the Pusan National University (PNU), Pusan, Korea. He is currently in M.S. and Ph.D. integrative program cource at the Pohang University of Science and Technology (POSTECH), Pohang, Korea. His main research interests include CMOS devices and RF measurement.

Sahhaf, Sahar (XT.9)
Received the M.Sc. degree in electrical engineering from the Catholic University of Leuven, Belgium, in 2006. Currently, she is working toward the Ph.D. degree at the KULeuven university (Belguim) and IMEC. The main focus of her work is on the reliability of high-κ gate dielectrics and metal gates.

Salman, Akram (4D.3)
He received his B.Sc. from Alexandria University, M.Sc. from AAST, Egypt, and Ph.D. from George Mason University, Fairfax, VA. He held several positions in the semiconductor industry at IBM and AMD. In 2008 he joined Texas Instruments as and ESD specialist working on the ESD development for various analog technologies. He is an author/co-author of more than 38-refereed publications; he holds 5 U.S patents with 4 more pending. Dr. Salman was awarded, the distinguished academic achievement award from GMU, the IBM co-operative fellowship and the best paper award from the 2005 SOI conference.

San, Tamer (5F.3)
Tamer San received the B.S. degrees in electrical engineering and physics from Bogazici University, Istanbul, Turkey. He received the M.S and Ph.D. degrees in electrical engineering in 1991 and 1994, respectively, from Yale University. He joined Texas Instruments in 1994 where he worked on memory technology development. He was involved in development of nonvolatile memory (Flash) technology at various technology nodes and SRAM technology at 90nm node. From 2005 to 2007, he was at IMEC as an assignee from TI and worked on MUGFET (multi-gate) devices. Since 2009, he is leading FRAM development at 180nm node. Since 1996, he has been a member of technical committee of non-volatile semiconductor memory workshop. He is currently serving as the technical program chair of international memory workshop.

Sandhya, C. (NA.2)
C. Sandhya (S'07) received the M.Tech. degree in microelectronics in 2006 and Ph.D. degree in electrical engineering in 2010 from the Indian Institute of Technology Bombay, Mumbai, India. During her graduation, she was a recipient of the SRC (GRC) fellowship. Since November 2009, she is working with Samsung Semiconductor R&D Center, South Korea. Her research interests are in the field of Flash memory optimization through device characterization and understanding device physics through modeling, design, and simulation.

Sasse, Guido (CR.3)
Guido Sasse received the M.Sc and Ph.D degree, both in electrical engineering, from the University of Twente, The Netherlands, in 2003 respectively 2008. He performed his Ph.D. research at the MESA+ Institute for Nanotechnology, where he studied RF CMOS reliability, reliability simulation, and advanced CMOS characterization techniques. In 2008 he joined

NXP Semiconductors, where he works on wafer level reliability. His current activities have a focus on RF reliability and reliability simulation. He received the IRPS 2006 best poster award.

Sato, Motoyuki (4C.2)
Motoyuki Sato received the B.S. and M.S. degrees in applied physics of engineering from the University of Tokyo, Tokyo, Japan, in 1996 and 1998, respectively, and the Ph.D. degree in engineering from University of Tsukuba in 2008. He joined Toshiba Corporation, Yokohama, Japan in 1998, and has been engaged in the research for semiconductor device fabrication and its physics. He joined the Semiconductor Leading Edge Technologies Inc.(Selete) since 2006, and moved to Toshiba in 2010. He received the best paper award in International Workshop on Dielectric Thin Film (IWDTF) 2008. His current interests and activities include the study of the reliability physics in high-k gate dielectrics. Dr. Sato is a member of technical program committee member in IRPS since 2008.

Schlünder, Christian (2A.1)
Dr. Christian Schlünder has received his Dipl.-Ing. (1999) in electrical engineering and his doctoral degree in engineering science (2006) accompanying his regular work (both from the Technical University of Dortmund, Germany). From 1998-1999 he worked in a cooperative program between Siemens Corporate Research Labs in Munich and the Technical University of Dortmund. 1999 he joined Infineon as a member of the Corporate Research Department, where he was active in research on hot carrier stress in analog and mixed signal applications. Since 2000 he works in the Corporate Reliability Methodology Group. There he manages technology qualification and quality assurance for various state-of-the-art CMOS-Technologies and technology transfers to silicon foundries. Furthermore he evaluates the device reliability of innovative technologies. He leads the NBTI-research for Infineon. His current work is focussed on recovery phenomena. Christian Schlünder has published some 30 papers in various conference proceedings and microelectronic journals. Additionally, he has presented invited talks and tutorials at many conferences such as 'IRPS', 'ESSDERC' or "ZuE". He is frequently a member of the Technical Program Committee of the IEEE-conferences 'IRPS', 'IRW' and referee of several microelectronic journals. Moreover he is involved in the JEDEC NBTI standard development.

Schmitt-Landsiedel, Doris (CR.2)
Doris Schmitt-Landsiedel received the Dipl. Ing. degree in electrical engineering from the Technical University of Karlsruhe, the diploma in physics from the University of Freiburg and the Dr. rer. nat. degree from the Technical University of Munich. She joined the Corporate Research and Development Department of Siemens AG, Munich, Germany, in 1981. There she worked on scaling problems in MOS devices and on the design of high speed logic and SRAM circuits. Since 1989 she has been manager of a research section with projects in future generation memory design, analog and digital CMOS and BICMOS circuits and design-based yield analysis. Since 1996 she is a professor of electrical engineering and director of the Institute for Technical Electronics at the Technical University of Munich. Her research interests are in robust CMOS circuit design and reliability as well as in circuits with novel devices and nanomagnetic computing.

Schneider, Jens (4D.4, EL.2)
Jens Schneider received his diploma in physics (Dipl. Phys.) in 1995 from the University of Kaiserlautern working on the condensation of quasi-particles in field theories. He then moved to Munich where he he received his Ph.D. from Ludwig-Maximilians-University in 2001 in the filed of atom lasers and Bose-Einstein condensation. Since then he is working with Infineon Technologies, first in the field of modelling of process effects in photomask lithography. Since 2003 he is working in the Infineon ESD/LU group where is involved with

process and device simulation of ESD phenomena and with the development of HV CMOS ESD protection concepts.

Schrimpf, Ron (SE.3)

Ron Schrimpf received his BEE, MSEE, and Ph.D. degrees from the University of Minnesota in 1981, 1984, and 1986, respectively. He is currently a Professor of Electrical Engineering at Vanderbilt and he previously was a Professor at the University of Arizona. Ron's research deals with the effects of radiation on semiconductor devices and integrated circuits. Current projects include use of high performance parallel computing to simulate single-event effects and soft errors in integrated circuits, atomic-scale modeling of radiation-induced defects, and design techniques for reliable integrated circuits. Ron is the director of the Institute for Space and Defense Electronics.

Schrimpf, Ronald (4B.2, 6E.2)

Ron Schrimpf is the Orrin Henry Ingram Professor of Electrical Engineering at Vanderbilt University and the Director of Vanderbilt's Institute for Space and Defense Electronics (ISDE). He received his B.E.E., M.S.E.E., and Ph.D. degrees from the University of Minnesota in 1981, 1984, and 1986, respectively, and served as a Professor of Electrical and Computer Engineering at the University of Arizona prior to joining Vanderbilt in 1996. Ron's research activities focus on radiation effects and reliability in microelectronics and semiconductor devices. He has served as General Chairman and Technical Chairman of the IEEE Nuclear and Space Radiation Effects Conference and Chairman of the IEEE Radiation Effects Steering Group.

Scozzari, Claudia (MY.5)

Claudia Scozzari graduated in physics from the University of Trieste Italy (2005). In 2006 she joined the STMicroelectronics in Agrate Brianza (Italy) working in the Non-Volatile Memory Process Development group within the Dielectric Reliability group of Central R&D. She is coauthored 10 papers published in journals or presented in international conferences. Recently, the memory section of STMicroelectronics has been separated forming a new company, Numonyx, and she joined the new company.

Seetharaman, Sridhar (HV.1)

Received the B.Tech degree from the Indian Institute of Technology in Madras, India, and MS and PhD degrees in Electrical engineering from North CarolinaStateUniversity in Raleigh. He joined Texas Instruments Inc. in 1996 where he worked as a process integration engineer in developing 4 generations of deep submicron CMOS. Since 2007 he has been working in the areas of power component design and BCD process integration. He is currently a Distinguished Member of Technical staff at TI and a BCD development manager.

Seifert, Norbert (3A.1)

Norbert Seifert currently heads all radiation effects efforts in the Manufacturing Technology Group at Intel Corporation. Prior to joining Intel in 2003, he worked in the fields of compact modeling, device reliability and digital design in the Alpha Development Group (DEC, CPQ, HP). He received Diplom Ingenieur and Ph.D. degrees in physics from the Technical University of Vienna, Vienna, Austria, in 1990 and 1993, respectively. He also holds an M.S. degree in physics from Vanderbilt University (Nashville, TN, May 1994).

Sekine, Katsuyuki (4C.2, MY.2)

Katsuyuki Sekine was born in Saitama, Japan, in 1970. He received the B.S., M.S., and Ph.D. degrees in electronic engineering from Tohoku University, Sendai, Japan in 1995,

1997, and 2000, respectively. In 2000, He joined Process and Manufacturing Engineering Center, Toshiba Corporation, Semiconductor Company, Yokohama, Japan, where he worked on development of advanced gate dielectric film formation process for CMOS device. He is currently engaged in development of dielectric film for nonvolatile memory device in Advanced Memory Development Center, Semiconductor Company, Toshiba Corporation, Yokkaich, Japan

Shao, Ingrid (4E.5)

Ingrid Shao is a research staff member at IBM T.J. Watson Research Center. She got her Ph.D. in materials science and engineering department at Johns Hopkins University in 2002. She joined IBM Research soon after graduation. She has worked in research related to hard disk drive, CMOS interconnect, Phase change memory, and currently working on Si photovoltaics. Ingrid has won a graduate student gold medal from Materials Research Society in 2001.

Sheldon, Douglas (6E.1)

Doug Sheldon is currently the VLSI Reliability Center Lead at the Jet Propulsion Laboratory. Dr. Sheldon has been at JPL for 7 years and involved in a wide variety of electronic part reliability and radiation issues. He is also a principal investigator on several NASA and JPL R&D projects regarding new FPGA and memory technology applications to spacecraft. Prior to coming JPL Dr. Sheldon had 20 years experience in various engineering and management positions in the commercial integrated circuit industry including Lattice Semiconductor, Ramtron, and Inmos. Dr. Sheldon has a bachelor and master's degrees in physics from the University of Colorado and the University of Oregon respectively and a doctorate degree in management from Colorado Technical University.

Shi, Quan (3A.1)

Quan Shi received M.S. from Beijing Normal University in Radio Frequency Electronics in 1986 and Ph.D. degree from University of New Mexico in Electrical Engineering in 2000. From 1998 to 2002, he worked as a design engineer and Research Assistant Professor with NASA Institute of Advanced MicroElectronics to design Radiation Tolerant special purpose processors for NASA applications. He joined Intel in 2002. He is currently working in standard cell library team focusing on library reliability and cell architecture related issues. His research interests include radiation hardened sequential design, circuit electromigration and standard cell layout architecture.

Shih, J.R. (2F.3)

Jiaw-Ren Shih received his M.S. and Ph.D. from National Tsing-Hua University, Taiwan, in 1992 and 2000, respectively. All are with Electrical Engineering. Dr. Shih is the Academician in TSMC Academy now, and also serves as the manager of Advanced Reliability Development Program and Mainstream Technology Quality and Reliability Department at Technology Quality and Reliability Division. Prior to joining reliability division in 2000, he served as the transistor designer at device department of R&D since 1992. Dr. Shih has served on the IRPS Technical Committee in 2002, 2004 ~ 2009 for Transistor, Process Integration and Dielectric Sections, respectively. He holds 53 U.S. patents and 40 Taiwan patents, and also published more than 41 technical papers.

Shin, Changhwan (XT.17)

Changhwan Shin received the B.S. degree with top honors in electrical engineering from Korea University, Seoul, Korea, in 2006. Since 2004, he has received a fellowship from the Korea Foundation for Advanced Studies (KFAS). He is currently working toward the Ph.D. degree in electrical engineering in the Department of Electrical Engineering and Computer Sciences, University of California, Berkeley, CA. He won the Best Paper Award and the Best Student Paper Award at the 2009 IEEE International SOI

Conference. His research interests include non-conventional advanced CMOS device designs and their applications to variation-robust SRAM cell.

Shinosky, Mike (5A.5)
Mike Shinosky received an Associate of Engineering degree in Electronic Engineering from A.T.E.S. Technical Institute of Niles, Ohio in 1982. Mike joined IBM in 1982, where he was responsible for building and running Soft Error Rate testers for DRAM Memory products. In 1985 he began T2 Functionality Qualification testing of DRAM memory cards for large scale server applications. In 1999 Mike transferred to an ASICS department doing book level Physical Design Mask Layout work. In 2003 Mike joined Technology Reliability Engineering, where he is currently and primarily responsible for TDDB testing of BEOL Dielectric technologies.

Shrivastava, Mayank (4D.4, EL.2)
Mayank Shrivastava was born in Lucknow, India, in 1984. He received the B.S. degree in engineering from Rajiv Gandhi Technical University, Bhopal, India, in 2006. He is currently working toward the Ph.D. degree with the Center for Excellence in Nanoelectronics, Department of Electrical Engineering, Indian Institute of Technology (IIT) Bombay, Mumbai, India In July 2006, he joined IIT Bombay as a Research Fellow. He was a Visiting Research Scholar with Infineon Technology AG, Munich, Germany, from April 2008 to October 2008. His current research interest includes ESD- and HCD-aware I/O device design, ESD-aware technology development, FinFET and UTB-Planar SOI devices, nonvolatile analog memories, and electrothermal modeling and simulation. He has five U.S. patents and one Indian patent pending in the fields of ESD, I/O devices, FinFETs, and nonvolatile analog memory. Mr. Shrivastava was a recipient of the Intel's 2008 Asia Academic Forum (AAF'08) Best Research Paper Award in the circuit design category. He also served as a Reviewer for the IEEE TRANSACTIONS ON ELECTRON DEVICES, the 2009 IEEE International Electron Devices Meeting (IEDM'09), and the 2010 International Conference on VLSI Design (VLSI'10).

Shukla, Vrashank (4D.5)
Vrashank Shukla received his B.E. degree from the National Institute of Technology, Karnataka in 2001. He is currently working towards a PhD degree in Electrical Engineering at University of Illinois Urbana Champaign. He worked at Texas Instruments, Bangalore, India from 2001 to 2007 in the ASIC CAD group. He worked on the development of EDA tools for reliability analysis of circuits.
His current research focus is on full-chip and package modeling for electro-static discharge (ESD) simulations, in particular, simulation of charged device model of ESD. He works in the ESD group of Co-ordinated Science Laboratory at University of Illinois Urbana Champaign.

Shur, Michael (2B.4)
Michael Shur received MSEE degree (with honors) from LETI, PhD and Dr. Sc. degrees from Ioffe Institute. He is Patricia W. and C. Sheldon Roberts Professor, Acting Director of Center for Integrated Electronics and Director of the NSF I/UCRC at RPI. He is also co-founder and Vice-President of Sensor Electronic Technology, Inc. His area of expertise is physics of semiconductor devices. Dr. Shur is Foreign Member of the Lithuanian Academy of Sciences and Fellow of IEEE, APS, IET, ECS, MRS, and AAAS. He received IEEE and other awards and holds Honorary Doctorate from St. Petersburg Technical State University.

Sierawski, Brian (4B.2, SE.3)
Brian Sierawski is a Staff Engineer at the Institute for Space and Defense Electronics. He joined ISDE in January 2005 where his efforts include modeling of semiconductor devices using Technology Computer Aided Design (TCAD) simulations and developing tools for the prediction of single event error rates. He earned a BSE in Computer Engineering and MSE in Computer Science and Engineering from the University of Michigan in 2002 and 2004 respectively and is currently pursuing his Ph.D. in Electrical Engineering at Vanderbilt University.

Simms, Richard J.T. (2E.5)
Received the M.S. degree in physics in 2006 from the University of Bristol, Bristol, U.K., where he is currently working toward the Ph.D. degree at the CDTR group. His current research interests include studying electrical effects in AlGaN/GaN HFET devices using dc and time-resolved micro-Raman spectroscopy, electroluminescence and electrical characterisation techniques.

Simoen, Eddy (2A.3)
Eddy Simoen obtained his Masters ('80) and Doctoral Degree ('85) in Engineering from the University of Gent (Belgium). His doctoral thesis was devoted to the study of trap levels in high-purity germanium by deep-level transient spectroscopy. In 1986, he joined imec to work in the field of low temperature electronics. His current interests cover the field of device physics and defect engineering in general, with particular emphasis on the study of low-frequency noise, low-temperature behavior and of radiation defects in semiconductor components and materials. He is an imec Researcher currently involved in the study of defect and strain engineering in high-mobility and epitaxial substrates and defect studies in germanium. In these fields, he has (co-) authored over 1000 Journal and Conference papers.

Singh, Pawan (NA.2)
Pawan Singh received B. Tech. and M. Tech. combined degrees and PhD from Electrical Engineering department of the Indian Institute of Technology (IIT) Bombay, India in 2006 and 2010 (January) respectively. He was the recipient of Applied Materials fellowship during his PhD in IIT Bombay.. He worked as a graduate intern at Applied Materials, Santa Clara, CA from June 2007 - July 2008. Since February 2010 he is with CEA-LETI-MINATEC at Grenoble, France. His research interests include technology development, physics, and reliability characterization of non-volatile memories and metal/high-k integration.

Smout, Steve (4F.2)
Steve Smout was born in Leuven, Belgium on March 17, 1984. He received the bachelor degree at the Katholieke Hogeschool Leuven in Leuven, Belgium in 2007. He joined imec in Leuven in 2007 as a member of the Large Area Electronics group. His main focus is the development of new transistor of foil technologies, both for organic and oxide semiconductor device concepts.

Song, Du Heon (5C.1)
Du Heon Song received the Ph.D. degree in electronics engineering from Seoul National University. He joined Samsung Electronics in 2000. Since 2009, he has been working on the process integration of NAND flash memories. His current interests include sub-30nm NAND flash memory development and reliability.

Song, Jai Hyuk (5C.1)
Jai Hyuk Song was born in Seoul, Korea, on August 3, 1967. He received Ph.D. degrees in electronic engineering from Seoul National University, Seoul, Korea, in 1996. Since 1996, he has been working in device engineering of sub-micron DRAM & sub-60nm NAND Flash memories at Samsung Electronics Corporation, Korea. His current interests include reliability issues and operation optimization and scaling of floating type NAND Flash memories.

Song, Seung-Chul (RM.3)
Seung-Chul Song received Ph.D in Solid State Electronics from the Univeristy of Texas at Austin in 2000. Since then, he had been in engineering and managemental positions in various organizations including Motorola, Samsung and Sematech. Since 2007, he has been with Qualcomm, leading projects of 28nm HK/MG technology enablement and advanced transistor technology development for future technology nodes.

Song, Seung-Hyun (2C.4)
Received the B.S. (2004) and Ph.D (2010) in electronics engineering from the Pohang University of Science and Technology (POSTECH), Pohang, Korea. His main research is related to CMOS devices.

Soo Sien, Seah (3C.5)
Seah Soo Sien received his bachelors in Materials Engineering from Nanyang Technological University, and is a Engineer in Quality Reliability Assurance Failure Analysis at Global Foundries of Singapore.

Spessot, Alessio (MY.6)
Alessio Spessot received the Laurea degree (cum laude) in Physics from the University of Triest (Italy) and the Ph.D. degree in Solid State Physics from the University of Modena (Italy), in 2003 and 2006, respectively. In 2006 he joined the Non-Volatile Memory Process Development Group, Central Research and Development department of STMicroelectronics (now Numonyx), Agrate Brianza, Milano, Italy. His research activities include characterization and modeling of MOS devices and advanced non-volatile memories. He published more than 30 papers in device physics and solid state physics. Dr. Alessio Spessot received the Best Paper Award at the EMRS in 2006.

Spinelli Sottocornola, Alessandro (5C.2, MY.5, MY.6)
Alessandro S. Spinelli is a Full Professor of electronics with the Politecnico di Milano. He has conducted experimental and theoretical research in electronics instrumentation and microelectronics, co-authoring more than 130 papers published in international journals or presented at international conferences, and serving in the technical committees of the IEDM and IRPS conferences. His current research interests include experimental characterization and modeling of non-volatile memory cells performance and reliability, development of innovative non-volatile memory technologies and circuit design for biological signal readout.

Spitzer, Andreas (2F.4)
Andreas Spitzer received his Diploma in Physics in 1979 and the Ph.D. in 1982 from the Technical University Aachen, Germany. From 1982 to 1984, he specialized there on adsorption on single crystal surfaces and on GaAs Schottky contacts. In 1984, he joined Siemens AG Corporate Research and Development in Munich, where his research activities have focused on the physics of dielectrics with a special emphasis on multilayer films. From 1999 to 2006 he was with the Memory Products Division of Infineon Technologies AG as head of the process and device simulation department. In 2006, he joined the Automotive Division of Infineon. He is now responsible for the device development and simulation of the technology development in Munich.

Srividya, Vidya (XT.9)
Earned a B.S. in Chemical Engineering from Anna University in Chennai, India and a Ph.D. in Chemical Engineering in 1997 from Clarkson University in Potsdam, New York. After graduating, she worked at CVC as a Process engineer developing PVD and CVD metal films. She joined Micron Technology, Boise in 2001 as a CVD Process engineer. She has worked on CVD/ALD deposition of various metal and high-k films for DRAM capacitor module application. For the past 3 years, she has been working on DRAM- and flash-related programs at IMEC, Belgium where she is on an international assignment, representing Micron Technology.

Stathis, James H. (2B.1, BD.3, XT.14)
Jim Stathis received the bachelor's in physics from Washington University in St. Louis (1980), and the Ph.D. in physics from the Massachusetts Institute of Technology (1986), joining the IBM Research Division the same year. At IBM the focus of his work has been the electrical properties of point defects in SiO2, including basic studies of defect structure using magnetic resonance and electrical measurement techniques, and the role of defects in wearout and breakdown. He is the author or coauthor of more than 100 research papers and over 60 invited talks. From November 2005 to February 2007 he served as Technical Assistant to the Vice President for Science and Technology, IBM Research Division. In February 2007 he became manager of High-k/Metal-Gate Characterization and Reliability, IBM Research. Jim has served on technical program committees for SISC, INFOS, IRPS, ESREF, IPFA, MIEL and other conferences, served as Chair of the dielectrics sub-committee for the 2003 IRPS in Dallas, and is on the IRPS management committee. He was the Technical Program Chair for IRPS 2009 and is Vice-General Chair for IRPS 2010. He has presented tutorials on CMOS reliability at IRPS, ESREF, MRS, and IPFA and is an Associate Editor of the journal Microelectronics Reliability. He is a Senior Member of IEEE and a Fellow of the American Physical Society.

Stecher, Matthias (2F.4, 4D.1)
Matthias Stecher studied Electrical and Electronic Engineering at the VaTech (USA) and at the Rheinisch-Westfälische Technische Hochschule Aachen (Germany) where he received his PhD in 1995. Between 1998 and 1994 he was involved in the development of device and circuit simulation tools. He joined Infineon Technologies in 1994 and has been project manager for several Smart Power Technologies of Infineon. Since 2003 he has been involved in the thermo-mechanical optimization of chip-package systems. Currently, he holds the position of a technical advisor in the fields of power technology and package development. He holds more than 20 patents and has published more than 70 papers in technical journals and conferences.

Stellari, Franco (4E.5)
Franco Stellari received M.S. and Ph.D. degrees in electronics engineering from the Politecnico di Milano, Italy, in 1998 and 2002 respectively. He subsequently joined IBM Watson Research Center as a PostDoc becoming a Research Staff Member in 2004. His major interests are the development and use of new optical methodologies for testing VLSI circuits. He has more than 55 international publications, 9 granted patents and several more pending. He has won the IEEE EDS Paul Rappaport Award for the best Trans. on Electron Devices of 2004, and two Best Paper Awards at the ESREF conference in 2002 and 2004.

Steve, Pearton (CD.3)
Steve Pearton is Distinguished Professor and Alumni Chair of Materials Science and Engineering at the University of Florida, Gainesville, FL, USA. He has a Ph.D in Physics from the University of Tasmania and was a postdoc at UC Berkeley prior to working at AT&T Bell Laboratories from 1994-2004 . His interests are in the electronic and optical properties of semiconductors. He is a Fellow of the IEEE, AVS, ECS, MRS, TMS and APS.

Su, K.C. (5A.3)
K C Su received the B.S. degree (1986) and M.S. degree (1988) in Electrical Engineering from National Cheng Kung university, Taiwan, ROC. Then, he joined United Microelectronic Corp.

(UMC) and worked with process technology development division for over 10 years. He is currently focusing on reliability technology & methodology development. His main interests are reliability engineering, device engineering and logic technology development.

Sugawa, Shigetoshi (5F.2)
Shigetoshi Sugawa received his M.S. in 1982 in Physics from Tokyo Institute of Technology and his Ph.D. in 1996 in Electrical Engineering from Tohoku University. In 1982-1999 he worked in Canon Inc., where he researched high S/N ratio solid-state imaging devices, high performance amorphous silicon devices, high-speed low power SOI devices and high-density liquid crystal display devices. In 1999 he moved to Tohoku University and he is presently a professor at Graduate School of engineering, Tohoku University. He is currently engaged in researches on CMOS image sensors, high-performance ULSI's and advanced displays such as high-performance, high-speed and low-power circuits/devices, and advanced semiconductor process technologies related to high-quality low-temperature oxidation, nitridation, CVD and etching process using microwave-exited high-density plasma. Dr. Sugawa is a member of the IEEE, the Institute of Image Information and Television Engineering of Japan.

Sugii, Nobuyuki (XT.2)
Nobuyuki Sugii received the B.S., M.S., and Ph.D. degrees in applied chemistry from the University of Tokyo in 1986, 1988, and 1995, respectively. He joined the Central Research Laboratory, Hitachi, Ltd. in 1988 where he had engaged in R&D of oxide superconducting materials and devices until 1996. Since 1996, he has been working in CRL, Hitachi on R&D of CMOS devices including SOI and strained-silicon MOSFETs. Since 2004, he also serves as a visiting professor for the Tokyo Institute of Technology. Dr. Sugii is a member of the Japan Society of Applied Physics and the IEEE Electron Devices Society.

Suh, Kang-Deog (5C.1)
Kang-Deog Suh was born in Gyunggi-Do, Korea, on October 2, 1956. He received the B.S. degree in electrical engineering from Seoul National University, Seoul, Korea, and the M.S. and Ph.D. degrees from the Korea Advanced Institute of Science and Technology (KAIST), Daejeon, Korea, in 1979, 1981, and 1991, respectively. He joined Samsung Electronics Company, Gyunggi-Do, Korea, where he was engaged in the CMOS logic IC design. Since 1991, he has been developing EEPROMs, flash memories, and MaskROMs. From 2009, he was a professor of information and communication engineering in SungKyunKwan University, Gyunggi-Do, Korea.

Summerfelt, Scott (5F.3, 6C.4)
Dr. Summerfelt is DMTS in Analog Technology Development at Texas Instruments. He has two BS degrees in 1984 from Iowa State University (Metallurgical Engineering, double major Physics and Mathematics) and MS and Ph.D. degree (1990) from Cornell University in MS&E. He joined Texas Instruments in 1990 to work on (Ba,Sr)TiO3 for DRAM and later Pb(Zr,Ti)O3 for FRAM. His research interests are ferroelectric materials for electronics including electrical properties, electrode materials, processing, design rules, cell design, ferroelectric spice models and integration issues. He is an author of over 34 refereed, 40 proceeding publications, two book chapters and 130 US patents

Sun, Kai (TF.1)
Kai Sun was born in Taizhou, China, in 1987. He received the B.S. degree in microelectronics in 2009 from Soochow University, Suzhou, China. He is currently working toward the M.S. degree in the Department of Microelectronics. His current research work is about Hall effect devices and sensors.

Suñé, Jordi (BD.4)
Jordi Suñé is Professor of Electronics at the Universitat Autònoma de Barcelona (UAB) and IEEE Fellow. He has (co)authored more than 280 papers in international journals and relevant conferences, among which and 5 tutorials on oxide reliability at IRPS. He has served in technical subcommittees of IRPS, IEDM, and INFOS. He has received the the IBM Faculty award (2008) and the ICREA ACADEMIA award (2010). His main interests are oxide reliability physics and statistics, and the modeling of novel nanoelectronic devices. He coordinates NANOCOMP research group at the UAB.

Sury, Charlotte (2E.3)
Charlotte Sury was born in Bordeaux, France, in 1981. She obtained both her master degree at the University of Bordeaux and her engineer diploma at the "Ecole Nationale Supérieure d'Electronique, d'Informatique, et de Radiocommunication de Bordeaux" (ENSEIRB)" in 2006. She is currently working as a PhD student in the team "Characterisation and Reliability of microwave technologies" in the Nanoelectronic group of the IMS Laboratory of Bordeaux. Her thesis was supported by the DGA and takes part in the ANR project CARDYNAL, which is centered on the parasitic effects and the degradation mechanisms of AlGaN/GaN HEMTs designed for microwave power applications.

Suzuki, Hiroyoshi (5F.2)
Hiroyoshi Suzuki was born in Sendai, Japan on May 16, 1986. He received the B.S. degree in electronic engineering from Tohoku University, Sendai, Japan in 2010. He is currently working toward the M.S. degree in the Graduate School of Engineering, Tohoku University.
His reserch interest is fabrication technologies of ultra low-noise MOSFETs which have structures to suppress low-frequency noise and Random Teregraph Signal (RTS) noise.

Tahone, Yang (5D.4)
Tahone Yang received the B.S. degree in chemistry from Fu Jan Catholic University, Taipei, Taiwan, in 1989, and the M.S. degree in chemistry from National Tsing-Hua University, Hsinchu, Taiwan, in 1991. In 1991, he was with FAB1, Macronix International (MXIC), working on the start-up and manufacturing of Etching /Lithography Department. In 1999, he transferred to Technology Development Center. Currently, he is leading the division of advanced module process development for nonvolatile memory devices. He has publications and holds patents in the areas of lithography, OPC, etching, thin film, diffusion, yield improvement and nonvolatile memory processing.

Taiki, Uemura (3A.5)
Taiki Uemura received M. S. degrees in chemical engineering from Yamagata University, Japan, in 2004. He joined Fujitsu Laboratories in 2004. He is currently a staff of LSI Quality Assurance in Fujitsu Microelectronics. He has worked on developing SER mitigation technologies and SER evaluation methodology on LSI, CPU for HPC and ASIC. He joined Environmental Variability Tolerant Device Technology Program of MIRAI project in 2008.

Takashi, Hashimoto (PI.3)
Takashi Hashimoto received B.E. and M.E. degrees in applied chemistry from Tohoku University, Sendai, Japan, in 1983 and 1985, respectively. He joined Hitachi Ltd., Tokyo, Japan, in 1985. Since 2003, he has been working as an engineer for Renesas Technology Corporation (present Renesas Electronics Corporation), Ibaraki, Japan, where he is currently engaged in the research and development of advanced MCU technology.

Takeuchi, Kan (5F.4)
Kan Takeuchi is a group manager of Renesas Technology Corp., after some experiences as a researcher at the Central Research Laboratory, Hitachi, Ltd. Renesas was established in 2003, merging two semiconductor department of Hitachi and Mitsubishi. He is now working on the SoC reliability design methodology at Renesas Technology Corp. He joined Environmental Variability Tolerant Device Technology Program of MIRAI project in 2008. He was a visiting researcher at the Cavendish Laboratory, University of Cambridge, U. K., from April 1990 to March 1991.

Tan, Juan Boon (5B.3)
Juan-Boon Tan received his B.Eng. with Honours in Engineering Science (Electrical) from the University of Exeter, UK, and Ph.D. in Opto-Electronics from the University of Oxford, UK. He is currently the BEOL Integration Deputy Director of Technology Development of GlobalFoundries, Singapore.

Tanaka, Katsuhiko (5F.4)
Katsuhiko Tanaka received the B.S. and M.S. degrees in applied mathematics from University of Tokyo, Japan, in 1984 and 1986, respectively. In 1986, he joined NEC Corp., Sagamihara, Japan, and transferred to NEC Electronics Corp., Sagamihara, Japan, in 2008. He has been engaged in the research of silicon MOS devices and development of three dimensional mesh generation system. In 2007, he joined Environmental Variability Tolerant Device Technology Program of MIRAI project. His research interest is in the analysis of advanced CMOS devices through 2D/3D process and device simulations.

Tang, Zhao (3F.4)
Dr. Zhao Tang was born in Xi'an, Shaanxi, P.R. China. He received B.S. degree in Electronics from Peking University in 2001, M.S. degree and Ph.D. degree in Electrical Engineering from the State University of New York, University at Buffalo in 2003 and 2010, respectively. He is a student member of IEEE, Society of Information Display (SID) and American Association for the Advancement of Science (AAAS). His current research interests included the thin film transistors based on amorphous Si, polycrystalline Si, organic and amorphous oxide, thin film solar cells and reliability issues.

Tanimoto, Hisanori (PI.1, PI.2)
Hisanori Tanimto received his Ph. D. in engineering from Osaka University, Osaka, Japan in 1990. He gained a position at the University of Tsukuba, Ibaraki, Japan in 1990 and is now an Associate Professor. His main scientific interest is the properties of nanostructured materials such as nanocrystalline metals, thin metal films and ultrafine metal particles. Dr. Tanimoto is a member of The Physical Society of Japan and of The Japan Institute of Metals.

Tao, Chen (3F.3)
Tao Chen received the Bachelor of Science degree in microelectronics from Fudan University, Shanghai in 2005. And he got Master of Science degree (cum laude) in electrical engineering from Delft University of Technology, Netherlands, in April 2007. After completing his master thesis with the Faculty of Electrical Engineering Mathematics and Computer Science, Delft University of Technology, he continued to work as a Phd student with Delft Institute of Microsystems and Nanotechnology, Delft University of Technology. His research interests include low temperature (<100oC) high performance TFTs on flexible substrate, laser crystallization of IGZO and location and orientation controlled single grain TFT.

Ťapajna, Milan (2E.5)
Received the M.S. and Ph.D. degrees in 2003 and 2007, respectively, in electrical engineering form Slovak University of Technology, Bratislava (Slovakia). In 2004, he joined Slovak Academy of Sciences, Bratislava (Slovakia). Currently he is a post-doctoral research assistant in the Center for Device Thermography and Reliability (CDTR) at University of Bristol (UK). His research interest includes electrical and optical study of advanced electronic structures and devices, especially III-nitride and Si-based structures with focus on defect characterization and reliability.

Tatsunori, Kaneoka (PI.3)
Tatsunori Kaneoka received the B.S. and M.S. degrees Electrical and Electronic Engineering in from Toyohashi University of Technology in 1985 and 1987, respectively. He joined Mitsubishi Electric Corp. where he was engaged in the research and development of the gate dielectrics . He is presently with RENESAS Technology Corp. where he works on the development of the gate dielectric and process integration for the CMOS and embedded Flash devices.

Taylor, Bill (2B.4, 4A.4)
Bill Taylor earned a PhD in Mechanical Engineering and Materials Science at Duke University in 1992. He then joined Motorola's Advanced Products R&D Labs, spending 4 years in the process and device simulation group, and 11 years on front-end materials development. He joined Sematech in 2008, with responsibilities in front-end integration, and electrical & physical characterization. He has authored or co-authored over 60 papers, and holds 12 patents in these areas

Tazzoli, Augusto (3F.2, 4F.2)
Augusto Tazzoli (IEEE Member) was born in Padova, Italy, in 1978. He graduated in Electronics Engineering at the University of Padova in 2003, and he received the Ph.D. degree in Electronics and Telecommunications Engineering from the same University in 2006 working on the reliability of silicon, compound and MEMS devices. He currently holds a post-doc position, working on the develop of measurement systems in genre, ESD phenomena, HV systems, compound and MEMS devices. In his career he has coauthored about 80 papers, awarded with 3 best paper awards. He serves as reviewer of several International Journals and in Symposia TPCs.

Teo, Zhiqiang (XT.5)
Z. Q.Teo received the B.Eng. (Hons.) degree from the School of Electrical and Electronic Engineering, Nanyang Technological University, Singapore, where he is currently working toward the Ph.D. degree under a GLOBALFOUNDRIES Singapore-NTU Graduate Research Scholarship. His research project is on the characterization of bias-temperature instability in state-of-the-art P-MOSFETs.

Teramoto, Akinobu (5F.2)
Akinobu Teramoto (M'02) received the B.S. and M.S. degrees in electronic engineering from Tohoku University in 1990 and 1992, respectively and Ph.D. in 2001 in Electrical Engineering from Tohoku University. In 1992-2002, he worked Mitsubishi Electric Corporation, Hyogo, Japan, where he has been engaged in the research and development of thin silicon dioxide films. In 2002, he moved to Tohoku University and he is presently an associate professor at New Industry Creation Hatchery Center, Tohoku University. He is currently engaged in an advanced semiconductor device technologies and process technologies, such as SOI MOS transistors, accumulation-mode transistors, variation and noise of transistors, high-quality low-temperature oxidation, nitridation, and CVD process using microwave-exited high-density plasma. Dr. Teramoto is a member of the IEEE, the Electrochemical Society, the Japan Institute of Electronics Packaging, Information and

Communication Engineers of Japan and the Japan Society of Applied Physics.

Tessariol, Paolo (MY.5)
Paolo Tessariol received his PhD in Physics from University of Padova, Padova, Italy in 1997. In 1998 he joined the INFM institute, Meran, Italy. From 1999 to 2008 he has been with ST Microelectronics Non Volatile Memory Group of Central R&D (Agrate Brianza, Italy) working on NOR and NAND Flash memory process development down to 45nm node. Since 2008 he is with Numonyx appointed NAND Flash memory Manager, working in the development of high density NAND memories with particular emphasis in Charge Trap technologies. He is Technical Coordinator and Member of Strategic Board of EC FP7 "GOSSAMER" project.

Tobimatsu, Hiroshi (PI.2)
Hiroshi Tobimatsu received his B.E. and M.E. in industrial chemistry from Kyushu Institute of Technology, Fukuoka, Japan in 1985 and 1987, respectively.
He joined Mitsubishi Electric Co., Ltd. in 1987, and then joined Renesas Technology Corp., engaged mainly in developing insulation and metal film deposition technology for DRAM and logic devices. Since 2010 he has been working for Renesas Electronics Corporation, where he is engaged in the planning of value engineering.

Toh, Seng Oon (XT.17)
Seng Oon Toh received the B.S. degree (highest honors) in computer engineering from the Georgia Institute of Technology, Atlanta, in 2002. He received the M.S. degree in electrical engineering from the University of California at Berkeley in 2008, where he continues to work towards the Ph.D. degree. His research emphasis is on power-performance optimization as well as robust design of nanoscale SRAM, with emphasis on dynamic stability, RTS, and BTI. Mr. Toh was awarded an IBM Ph.D. fellowship in 2010.

Tőkei, Zsolt (5A.1, 5A.2, 5B.4, 6A.3)
Zsolt Tőkei is program director for the advanced interconnect program. He joined IMEC in 1999 and since then he is working in the field of copper low-k interconnects. He obtained his M.S. (1994) in physics from the University Kossuth in Debrecen, Hungary. In the framework of a co-directed thesis between the Hungarian University Kossuth and the French University Aix Marseille-III, he earned his PhD (1997) in materials science. From 1998 he worked at the Max-Planck Institute of Düsseldorf, Germany, as a post-doctorate researcher. In 1999, he joined the Interuniversity Microelectronics Center (IMEC), Belgium, where he has worked on a range of interconnect issues such as metallization, process development, electrical performance, interconnect scaling and dielectric reliability. He has authored or co-authored more than 100 publications in international scientific journals and in international scientific proceedings.

Tomohiro, Kubo (4A.5)
Tomohiro Kubo received the B.S. and M.S. degree in physics from Nagoya University, Nagoya, Japan in 1991 and 1993, respectively. In 1993, he joined Fujitsu Ltd., Kawasaki, Japan, where he was engaged in research and development of ultra shallow junction technology. He is now mainly workings on development of millisecond anneal technology.

Tortorelli, Innocenzo (5C.4)
Innocenzo Tortorelli, since 28-Oct-2007 with Numonyx, Via C.Olivetti 2, 20041, Agrate Brianza (Milan), Italy. He received the Laurea degree in Electronic Engineering in 2002 from the Politecnico di Torino, Italy, with a thesis on smart antennas and relative beamforming algorithms for UMTS. In 2004 he joined

Central R&D department of STMicroelectronics in Agrate Brianza (Italy) and since then he has been working in process development and electrical characterization of phase-change memories based on chalcogenide materials. He has authored papers and patents on phase-change memories.

Toshiaki, Iwamatsu (PI.3)
Toshiaki Iwamatsu received the M.S. degree from Kyushu University, in 1989, and the Ph.D. degree from Osaka University, in 1998. In 1989, he joined the ULSI Development Center, Mitsubishi Electric Corporation, Itami, Japan, where he has been engaged in SOI process and device technology research. Since 2003 he has been with the Advanced Technology Development Div., Renesas Technology Corp., Itami, Japan. Presently, he is responsible for the research and development of the advanced MOSFETs. He has also been a Visiting Professor at Osaka University since 2007. Dr. Iwamatsu is a Member of the Japan Society of Applied Physics.

Toshifumi, Mori (4A.5)
Mori Toshifumi joined Fujitsu Limited in 1989. He has been engaged in development of advanced Si LSI processes, such as CVD (poly-Si and WSix for gate electrode, dielectric films for side-wall spacer, and stress control) of FEOL. His present activities are the process development of High Performance LSI device technologies, which are including gate dielectric formation and reliability, Si and SiGe epitaxial films.

Tous, Santi (BD.4)
Santi Tous is a Ph. D. student in the NANOCOMP research group of the Universitat Autònoma de Barcelona. He graduated in Physics in 1996 and in Electronic Engineering in 2008, both at the UAB. Master in Micro and Nanoelectronics Engineering from the UAB in 2008. He's been working towards Ph.D. on oxide reliability for MOS gate applications since 2005. He has published several papers on the modelling of progressive breakdown and high-K stack insulator breakdown.

Truong, Connie (5A.5)
Connie Truong received her B. S. in Chemical Engineering from the University of Massachusetts in 1988, her M.S. in Materials Science from Columbia University in 1994, and her M.B.A. degree from Marist College in 2001. She is also a PMP® certified by PMI and IBM since 2001. Connie started working for IBM since 1988, where she held various process engineering positions in IBM East Fishkill Ceramic packaging manufacturing between 1988 and 1999. In 1999, Connie took on a management assignment and subsequently managed various manufacturing process engineering departments in IBM System & Technology Group. In 2006, Connie joined the IBM Semiconductor Research and Development Center as manager of Development Manufacturing Engineering department enabling the technology process transfer from Development to Manufacturing FAB. Since 2007, Connie has been managing the Unit Process Development Engineering department of IBM and Alliance engineers focussing on developing Surface Prep, Plating, and CMP processes for advanced CMOS technologies.

Tseng, Joshua (XT.13)
Joshua Tseng received the Ph.D. degree in materials science and engineering from National Taiwan University, Taipei, Taiwan, in 1998. His study was focused on materials, processes and electrical properties of BaTiO3 based dielectrics with Ni electrode. In 2000, he joined Taiwan Semiconductor Manufacturing Company (TSMC), Hsinchu, Taiwan as a process engineer for CVD, PVD and CMP processes. From 2002, he moved to TSMC's R&D group for advanced technologies related to clean, control, low k film and then HK/MG as a technical manager. In 2007, he began his work as a TSMC assignee to the Interuniversity Microelectronics Center (IMEC), Leuven, Belgium, focusing on process and integration of high-k and metal gates with Si or high mobility channel.

Tsuchiya, Ryuta (XT.2)
Ryuta Tsuchiya received the B.S., M.S., and Ph.D. degrees in material science from the Tokyo Institute of Technology, Tokyo, Japan, in 1993, 1995, and 1998, respectively. He joined the Central Research Laboratory, Hitachi, Ltd., Tokyo, Japan, in 1998, where he has been engaged in research on fabrication and characterization of high-performance and low-power MOSFETs including thin-film SOI and BOX transistors. Dr. Tsuchiya is a member of the Japan Society of Applied Physics.

Tsukamoto, Yasumasa (XT.17)
Yasumasa Tsukamoto received his Ph. D. degree in Applied Physics from Osaka University, Japan. After his graduation in 2001, he entered Mitsubishi Electric Corporation and was transferred to Renesas Technology Corp in 2003. He has been engaged in the research and development of the embedded SRAM design using advanced CMOS processes. Since 2008, he has been a visiting industrial fellow at the University of California at Berkeley, where he is conducting research on the variability and reliability issues on the advanced SRAM design as well as the new device application to the SRAM memory cells.

Tsunehisa, Sakoda (4A.5)
Tsunehisa Sakoda received the B.S. (1999) and M.S. (2001) degrees in physical electronics from Tokyo Institute of Technology, Tokyo, Japan. He joined Fujitsu Laboratories Ltd. in 2001 and has been engaged in the research and development of semiconductor technologies. His present activities are the process development of High Performance LSI device technologies, which are including reliability of gate dielectric in Fujitsu Microelectronics Ltd.He is a member of the Japan Society of Applied Physics.

Tuyoshi, Arigane (PI.3)
Tsuyoshi Arigane was born in Ibaraki, Japan, on July 30, 1971. He received the B.E, M.E., and D.E. degrees in 1995, 1997, and 2000, respectively, from the University of Tohoku, Miyagi, Japan. In 2000, he joined Hitachi Ltd., Tokyo, Japan, where he has been engaged in the research and development of non-volatile memory and high-k MOSFETs.

Udayakumar, K.R. (6C.4)
K. R. Udayakumar earned his Ph.D. in Solid State Science from Pennsylvania State University. He joined TI in 1995 where he was involved in the development of monolithic un-cooled infrared detectors for TI's Defense Systems & Electronics Group. From 2000, he has been working on process integration of embedded ferroelectric memories. Udayakumar has published extensively in the area of energetics of point defects in alkaline earth titanates through atomistic simulation, thermodynamic phenomenology of perovskites, piezoelectrics for MEMS, pyroelectric detectors, and FRAMs. His other interests are in comparative literature and political science.

Uemura, Taiki (5F.4)
Taiki Uemura received M. S. degrees in chemical engineering from Yamagata University, Japan, in 2004. He joined Fujitsu Laboratories in 2004. He is currently a staff of LSI Quality Assurance in Fujitsu Microelectronics. He has worked on developing SER mitigation technologies and SER evaluation methodology on LSI, CPU for HPC and ASIC. He joined Environmental Variability Tolerant Device Technology Program of MIRAI project in 2008.

Uznanski, Slawosz (4B.4)
Slawosz Uznanski received the Engineer Degree and the M.Sc. degree in microelectronics from Technical University of Lodz, Poland, and from University of Nantes, France, both in 2008. He is currently working towards the Ph.D. degree at STMicroelectronics Crolles, France, and the Institute of Materials, Microelectronics and Nanosciences of Provence, Marseille, France. His research interests include modeling of Single Event Effects in commercial ultra-deep sub-micrometer CMOS technologies in terrestrial and space environments.

Van den bosch, Geert (6C.1)
Geert Van den bosch received the M. Sc. (1987) and Ph. D. (1993) in applied sciences from the Katholieke Universiteit Leuven, Belgium. He is with imec in Leuven since 1987. As a Senior Researcher, he has made contributions to the fields of hot-carrier degradation, plasma and process induced damage in ultrathin oxides, back end - front end of line interaction, and smart power devices. Since 2007 his activity moved to charge trap Flash memory. He has (co-)authored more than 80 international journal publications and conference contributions, and served as program committee member of P2ID, ICICDT, and IRPS.

van Dijk, Kitty (2D.2)
Kitty van Dijk received her MSc (1992) and PhD (1997) degree in Solid State Physics from the Technical University of Delft, and from the University of Nijmegen, respectively, both in The Netherlands. In 1998 she joined the Research and Development group of ASMI at IMEC in Leuven, Belgium, working on gate-oxide quality. In 2000 she moved to NXP (former Philips) Semiconductors in Nijmegen, where she started in non-volatile memory reliability, eventually broadening to process and product reliability in general. She now works on the interaction between process and product reliability, with a focus to Automotive and extreme mission profile qualifications.

Van Houdt, Jan (6C.2)
Jan Van Houdt received a Ph.D. from the Katholieke Universiteit Leuven in 1994. During his PhD work, he invented the HIMOS™ Flash memory, which he transferred to several industrial production lines. In 1999 he became responsible for Flash memory at IMEC and as such was the driving force behind the expansion of IMEC's memory program. Today he is IMEC's Memory Device Design Group Manager and Flash Memory Program Manager. He has published more than 160 papers in international journals (incl. 2 book chapters) and accumulated more than 140 conference contributions (incl. about 30 invited papers and panel contributions). He has filed about 50 patent applications and served on the program committees of 10 major conferences.

Van Hove, Marleen (2E.4)
Marleen Van Hove obtained her degree in Physics from the Catholic University of Louvain in 1980. She continued research at the Institute of Nuclear and Radiation Physics group at the Physics Department of the same university and obtained her PhD in 1985. After this, she joined imec, Leuven, Belgium. She specialised in GaAs and InP III-V processing in the period between 1986 and 1997. Afterwards she was responsible for CMOS back-end integration in imec. At the end 2006 she returned to compound semiconductor research, being responsible for the development of high-power GaN switching devices.

Vandelli, Luca (6C.1)
Vandelli Luca was born in Reggio Emilia, Italy, in 1985. He received the Master degree in electronic engineering from the University of Modena and Reggio Emilia Italy, in 2009. Since 2010, he has been pursuing the Ph.D. degree at the Department of Engineering Sciences and Methods of the University of Modena and Reggio Emilia. His work focuses on modeling and characterization of thin gate oxides in nonvolatile memory devices and MOS transistors.

Vandevelde, Bart (6A.3)
Bart Vandevelde received his Masters degree in mechanical engineering from the Catholic University of Leuven (Belgium) in June 1994. In March 2002, he received a PhD degree at imec in the field of thermo-mechanical modeling for electronic systems. Currently, he is team leader for the packaging reliability activities at imec and internal project coordinator for several Flemish and European projects. He is co-founder and member of the organisation committee for the Eurosime conference. Dr. Vandevelde is author and co-author of about 100 publications in international conferences and journals.

Vayshenker, Alex (4A.3)
Alex Vayshenker received masters degree in opto-electronic engineering from Moscow Institute of Geodesy (1972 - 1978). He worked as an engineer in Moscow State Optical Enterprise qualifying hybrid opto-electronic components (1978 - 1981). He joined IBM Technology group in 1984 and held several engineering positions in the fields of Quality and Reliability of semiconductor components/technology. Most recently he worked on reliability of ultra-thin gate dielectrics. He co-authored several papers on this subject.

Veksler, Dmitry (2B.4, 4F.4)
Dmitry Veksler received his B.S. in 1996 and M.S. in 1998 from Nizhniy Novgorod State University and Ph.D. in Physics in 2007 from Rensselaer Polytechnic Institute, Troy, NY. From 1998 till 2003 he was with Institute for Physics of Microstructures, Nizhniy Novgorod (Russian Academy of Sciences) investigating optical and transport phenomena in low-dimensional semiconductor hetero structures. During 2003-2008 he was with Rensselaer Polytechnic Institute working on plasma wave electronics for THz spectroscopy and imaging applications. In 2009 he joined SEMATECH Inc., Albany, NY. His current research focus is on characterization/reliability of high-k and alternative substrate MOSFETs.

Velamala, Jyothi Bhaskarr (5E.2)
Jyothi Bhaskarr Velamala received his Bachelor of Technology degree in Electronics and Communications Engineering from Indian Institute of Technology (IIT), Guwahati, India in 2008. He is currently working towards his Masters degree in Arizona State University. His research interests include reliability effects in scaled CMOS technology and circuits; design and test solutions for resilience.

Ventrice, Domenico (6C.2)
Domenico Ventrice was born in Italy in 1979. He received the Laurea degree in electrical engineering from the Politecnico di Milano, Italy, in 2006. He is currently working as a process development engineer at Numonyx R&D. His research interests include the modeling of electrical and phase-change properties in PCM devices.

Vereecke, Bart (5B.4)
Bart Vereecke received a Ph. D. degree in nuclear particle physics in 2001 at the Catholic University of Leuven, Belgium. In 2002, he started working in a development aid project as computer advisor for the Zambian ministry of education. In 2007, he joined IMEC, Leuven, where he works on the development of Cu/low-k integration projects.

Vermunt, Frank (SE.1)
Frank Vermunt received his B.Sc. degree in computer technology in 1987 at the 'Hogere Technische School' in Venlo. In 1989 he joined AT&T in Hilversum, and in 1996 Philips in Eindhoven, both in The Netherlands. Since November 1999 he is with NXP Semiconductors, which is a former division of Philips. Frank is currently working as a system architect at Car Entertainment's development department for Car DSPs, where his focus is on system infrastructure for next generation software defined analogue and digital radios. He has a special interest in assessing and addressing the perceived impact of SER for (end-)users.

Verzellesi, Giovanni (4F.4)
Giovanni Verzellesi, "Laurea" degree from the Univ. of Bologna, Italy, in 1989, PhD in Electrical Engineering from the Univ. of Padova, Italy, in 1994, Full Professor of Electronics with the Univ. of Modena and Reggio Emilia, Italy, since 2006. His research activity has dealt with: (i) impact-ionization effects in silicon BJTs; (ii) silicon optical, chemical, and ionizing radiation sensors; (iii) compound-semiconductor FETs. He has coauthored more than 130 papers in international journals and conference proceedings. He is, or has been, member of the technical program committee of IEDM, IRPS, ESREF. He is member of the steering committee of HETECH. He is Senior Member of the IEEE.

Vincent, Huard (2A.4, 2B.2, 5E.3)
Vincent Huard received the B.S. (1996) in physics and the M.S. (1997) in electrical engineering from the Institut National Polytechnique de Grenoble (INPG). He worked for the CEA-Grenoble on the MBE growth of II-VI based doped heterostructures and their magneto-optical and electrical characterizations. He received his Ph.D. (2000) in physics from the university of Grenoble. In 2000 and 2001, he was a Visiting Scholar at the University of California, where he worked on devices made of ferromagnetic materials on top of semiconductors. In 2002, he joined Philips Semiconductors as a reliability engineer, working on oxide and device reliability. Since 2007, he is at STMicroelectronics as Design-in Reliability project leader, working on device and circuit reliability modeling and product qualification tests. His current research interests include NBTI and hot carrier degradation both at wafer and product levels as well as Design for reliability. He authored and co-authored more than 70 regular papers, several invited papers and is IRPS 2009 Circuit Reliability Chairman.

Visalli, Domenica (2E.4)
Domenica Visalli obtained her degree in Electronic Engineering from the University of Messina in Italy in 2006. She joined imec in 2005 where she worked on the development of a new methodology for the characterization of advanced Cu/low k interconnects. Afterwards, in 2007, she started the PhD program at the Catholic University of Leuven and she joined the group of GaN in imec. She is currently working on the simulation and characterization of GaN-based devices for power switching applications.

Visconti, Angelo (4B.3, 5C.2)
Angelo Visconti (M'07) received the Laurea degree in physics (cum laude) from the University of Milano, in 1997. The same year he joined the Non-Volatile Memory Process Development Group, R&D department of STMicroelectronics (now Numonyx), Agrate Brianza, Italy. Since that, he has been involved in the developments of ten-generations of Flash Memory Process. His current research interests include reliability, radiation effects, and multilevel applications of Flash cells. He is co-author of more than 100 scientific publications and 14 patents. He is a Lecturer on NVM Reliability and Radiation Effects on Flash Memory with the University of Padova and Politecnico di Milano. Since 2005, he has been the ST-rapresentative, and chair from 2009, within the JEDEC 14.3 task-group on NVM reliability. Finally, he was member of the IEDM-CIR committee, IRPS-PIR committee, IRPS-Memory committee and co-chair in 2010. Mr. Visconti won the IRPS 2008 "Outstanding Paper Award" and the JEDEC 2010 Team Award.

Volf, Paul (2D.2)
Paul A. J. Volf (M'93) received his MSc degree in 1994, and his PhD degree in 2002 for his research work on data compression, both

in electrical engineering from the Eindhoven University of Technology, The Netherlands. In 2000 he joined Philips Semiconductors, now NXP Semiconductors, where he started as DfM-engineer, followed by product engineering positions in operations and in the automotive business line. Currently he holds a product engineering position in the central department for process development and foundry interfacing. His activities cover the process and product qualifications and product introductions in internal specialty processes and advanced CMOS foundry processes.

Wang, Li (6C.4)
Li Wang received her Ph. D. in physics from the Colorado School of Mines. She joined Texas Instruments in 1997 and is currently working as a product engineer in the Analog Technology Development group. Her interests are in the area of FeRAM functional and reliability testing and product yield improvement.

Wang, Miaomiao (XT.14)
Miaomiao Wang Received her B.S. in Electrical Enginnering from Peking University, China in 2003, and M.S. and Ph.D. degrees in the same discipline from Yale University in 2005 and 2008 respectively. She is Currently a Research Staff Member at IBM Research @Albany Nanotech, Albany, NY. She is working on Electrical Characterization and Reliability for a variety of high-k metal gate MOS devices including SOI, FINFET, ETSOI and etc.

Wang, Mingxiang (TF.1)
Mingxiang Wang (M'07) received the B.S. degree in Physics and Ph.D. degree in Condensed Matter Physics from Nanjing University, Nanjing, China, in 1993 and 1998, respectively. From 1998 to 2001, he was a Postdoctoral Research Associate with the Department of Electrical and Electronic Engineering, the Hong Kong University of Science and Technology, Hong Kong. Afterwards, he joined Semiconductor Manufacturing International Corporation (Shanghai), as a member of technical staff in process reliability engineering. Currently he is with the Department of Microelectronics, Soochow University, Suzhou, China. His research interests include thin-film materials and devices, semiconductor device physics and device reliability.

Wang, Szu-Yu (5D.2, MY.1)
Szu-Yu Wang was born in Kauhsiung, Taiwan in 1974. He received his B.S and M.S degrees in material science and engineering from National Tsing-Hua University (NTHU) in 1996 and 1998, respectively. Now he is pursuing his Ph.D degree in electronic engineering in National Tsing-Hua University (NTHU). In 2000, he joined Technology Development Center in Macronix International. Co. Ltd., where his current work is focused on dielectric characterizations in nitride trapping Flash Memories.

Wang, Tahui (MY.3)
Tahui Wang (S'84-M'85-SM'94) was born in Taoyuan, Taiwan, R.O.C., on May 3, 1958. He received the B.S.E.E. degree from National Taiwan University, Taipei, Taiwan, in 1980, and the Ph.D. degree in electrical engineering from the University of Illinois at Urbana–Champaign, Urbana, in 1985. From 1985 to 1987, he was with Hewlett-Packard Laboratories, Palo Alto, CA, where he was engaged in the development of GaAs HEMT devices and circuits. Since 1987, he has been with the Department of Electronics Engineering, National Chiao-Tung University, Hsinchu, Taiwan, where he is currently a Professor. His research interests include hot-carrier phenomena characterization and reliability physics in very large scale integration (VLSI) devices, RF CMOS devices, and nonvolatile semiconductor devices. Dr. Wang was given the Best Teacher Award by Taiwan's Ministry of Education. He has served as technical committee member of many international conferences, among them International Electron Devices Meeting, International Reliability Physics Symposium, and VLSI Technology, Systems, and Applications.

Wang, Zhongrui (MY.4)
Wang Zhongrui received his B.Eng in Electrical and Electronics Engineering (EEE) from Nanyang Technological University (NTU) Singapore in 2009. He is currently pursuing his PhD degree in EEE, NTU working on metal oxide based resistive random access memory device simulation and characterization.

Warren, Kevin (4B.2, SE.3)
Kevin Warren received the B.S. degree in Chemistry from Tennessee Technological University in 1994 and his M.S. in Chemistry at Vanderbilt University in 1997. He received his M.S. in Electrical Engineering from Vanderbilt University in 1999. From 1998-2000 he worked for Raytheon ITSS as a contractor at the Marshall Space Flight Center where he supported the component group as a radiation effects engineer for the International Space Station and the Space Shuttle Programs. From 2000-2003 he worked at the Johns Hopkins University Applied Physics Laboratories as a specialist in the testing and qualification of electronic parts. In 2003, he joined the Institute for Space and Defense Electronics where he has focused on modeling the soft error response of microelectronics for terrestrial and space applications.

Watabe, Shunichi (5F.2)
Shunichi Watabe was born in Japan, on September 9, 1982. He received the M.S. degree and Ph. D. degree in the Graduate School of Engineering, Tohoku University, Sendai, Japan in 2007 and 2010, respectively.
His research interests are the variability of characteristics in MOSFETs and the development of device and fabrication processes to suppress the variability.

Wei, Jun (5B.3)
Jun Wei received the Ph.D. degree from Tsinghua University in Materials Science and Engineering. He is currently a Group Manager and Senior Scientist at Singapore Institute of Manufacturing Technology. Currently, he is working on microsystems and nanosystems fabrication and packaging, low temperature wafer/substrate bonding, advanced interconnection materials, packaging for electronics and optoelectronics, nanofabrication, nanocomposites as well as advanced characterization techniques. He is serving several international conference committees. He has published more than 250 technical papers in the areas of micro/nano-interconnection, electronics, nanotechnology, microsystems and nanosystems and solid oxide fuel oxide, and currently holds 30 patents and technology disclosures.

Weller, Robert (4B.2, SE.3)
Robert A. Weller received his Ph.D. in physics from Caltech in 1978, and is now a Professor of Electrical Engineering at Vanderbilt University. His research interests include the computational analysis of radiation effects in electronics, device physics, radiation effects in materials, and applications of the techniques of nuclear physics.

Wen, Shi-Jie (SE.3, SE.5)
Shi-Jie Wen received his Ph.D in Material Engineering from University of Bordeaux I in 1993. He joined Cisco Systems Inc., San Jose, CA in 2004, where he has been engaged in IC component technology reliability assurance. His main interest is in silicon technology reliability, such as SEU, WLR, and complex failure analysis, etc. He is a member of DFR, SEU core teams in Cisco. Before Cisco, he worked in Cypress Semiconductor where he was

involved in the area of product reliability qualification with technology in 0.35u, 0.25u, 0.18u, 0.13u and 90nm.

Wen, Yong-Ru (EL.6)
Yong-Ru Wen received the B.S. degree from the Department of Mechenical Engineering, National Chiao-Tung University, Hsinchu, Taiwan, R.O.C., in 2007, and the M.S. degree from the Institute of Electronics, National Chiao-Tung University, Hsinchu, Taiwan, R.O.C., in 2009. He is currently with military service in the Chinese Army.

Wen Hu, Liu (4A.1, BD.2)
LIU WENHU was born in Shandong, China in 1986. He received his B.Eng, 1st Class Honors (Electrical and Electronics Engineering) from Nanyang Technological University (NTU) in 2008. He is currently pursuing his Ph.D. at the Division of Microelectronics, School of Electrical and Electronic Engineering in NTU. His project is focusing on electrical characterization and reliability of novel high-κ gate dielectrics in nano-scale MOSFETs. He is currently a Graduate Student Member of IEEE (2008-present).

Werner, Christoph (CR.2)
Christoph Werner received the Ph.D. degree from Technical Unversity Munich in 1980. From 1980 until 2007 he was with Siemens/Infineon working in research and development of advanced CMOS devices and processes and circuits. Since 2007 he has been with Technical University Munich as a senior scientific consultant. He has published more than 50 scientific papers and has filed 20 patent applications. His current research include Low Power digital CMOS systems, CMOS long term degradation effects as well as future device concepts beyond CMOS.

Wie, Chu-Ryang (3F.4)
Chu Ryang Wie received Ph.D. degree in Applied Physics from Caltech in 1985. Ever since, he was Professor of Electrical Engineering at State University of New York at Buffalo. His research was on developing analysis methods of III-V heterostructures such as strained buried quantum wells and barriers and lattice-mismatched structures and the strain relaxation, by high resolution x-ray diffraction techniques; and on the effects of lattice strain and relaxation on the device characteristics such as strained resonant tunneling diodes, laser diodes and p-n junctions. Recently, he is interested in thin film device and reliability physics He also developed "The Semiconductor Applet Service" website during 1996-2000 period for semiconductor education.

Wilde, Markus (4C.1)
Obtained master and doctoral degrees in Physical Chemistry from the University of Bochum, Germany, and is currently Associate Professor at the Institute of Industrial Science, University of Tokyo (Japan). He investigates the reactive behavior of hydrogen (H) on surfaces, in shallow surface-bulk transition regions and at thin film interfaces, combining surface science techniques with quantitative high-resolution H depth profiling by nuclear reaction analysis (NRA). Current research focuses on H-absorption and surface/subsurface H exchange at transition metal single crystals and nanoparticles. Other interests include H impurities at MOS electrode/dielectric interfaces, surface hydroxylation and hydrophilicity of oxides, and semiconductor H-passivation.

Wilson, Christopher J. (5B.4, 6A.3)
Christopher J. Wilson (S'07) received the M.Eng. degree (first class honors) in Electronics and Ph.D. degree in interconnect reliability and mechanical stress in thin films from Newcastle University, Newcastle upon Tyne, U.K. in 2006 and 2009

respectively. Between 2007 and 2008 he spent 16 months as Marie Curie APROTHIN Fellow and Visiting Scholar in the Advanced Interconnect Program at the Interuniversitair Micro-Elektronica Centrum (IMEC). He was an Institution of Electrical Engineers Robinson Research Scholar and held an Institution of Electrical and Electronic Engineers Reliability Society Scholarship. Dr. Wilson is now the International Mobility Post-Doctoral Fellow in IMEC's Interconnect Process Technology unit, where he is working on advanced metallization integration and reliability.

Wise, Rick (HV.1)
Received the B.S. degree in chemical engineering from the University of Arkansas, Fayetteville and the M.S. and Ph.D. degrees in engineering and applied science from Southern Methodist University, Dallas, Texas. He joined Texas Instruments (TI) Incorporated, Dallas, as a process engineer in 1983. He has worked on the development of semiconductor device fabrication front-end processes utilizing his expertise in CVD and silicon materials. He currently works on external research & consortia projects for the Analog Technology Development group. He was elected Texas Instruments Fellow in 1998.

Witulski, Arthur (3A.2)
Arthur F. Witulski received the Ph.D. degree in electrical engineering from the University of Colorado, Boulder in 1988. He has worked in analog electronics, power electronics, and microelectronics in industry and academia. He has held positions as design engineer at Storage Technology in Colorado, Associate Professor at the University of Arizona, and currently as Senior Research Engineer at the Institute for Space and Defense Electronics at Vanderbilt University, where he also holds an appointment as Research Associate Professor. His technical interests include single event radiation effects in sub-100 nm CMOS technologies, as well as power devices and power electronics

Wong, H.-S. Philip (2C.5)
H.-S. Philip Wong joined Stanford University as Professor of Electrical Engineering in September, 2004. From 1988 to 2004, he was with the IBM T.J. Watson Research Center. He held various positions from Research Staff Member to Manager, and Senior Manager. While he was Senior Manager, he had the responsibility of shaping and executing IBM's strategy on nanoscale science and technology as well as exploratory silicon devices and semiconductor technology. His present research covers a broad range of topics including carbon nanotubes, semiconductor nanowires, self-assembly, exploratory logic devices, nanoelectromechanical relays, device modeling, and novel memory devices such as phase change memory and metal oxide resistance change memory. He is a Fellow of the IEEE and served on the Electron Devices Society AdCom as elected member (2001 – 2006). He served as the Editor-in-Chief of the IEEE Transactions on Nanotechnology in 2005 – 2006, sub-committee Chair of the ISSCC (2003 – 2004), General Chair of the IEDM (2007), and is currently a member of the Executive Committee of the Symposia of VLSI Technology and Circuits (2007 – 2009). He received the B.Sc. (Hons.), M.S., and Ph.D. from the University of Hong Kong, Stony Brook University, and Lehigh University, respectively.

Wong, Richard (SE.3, SE.5)
Richard Wong received his M.S. degree in electrical engineering from Santa Clara University in 1988 and his B.S. degree in chemical engineering from UC Berkeley in 1982. He joined Cisco Systems Inc., San Jose, CA in 2006. He is engaged in IC component technology reliability assurance in issues such as SEU, ESD, WLR, failure analysis and reliability modeling. Prior to Cisco, he had worked on ASICs, FPGAs, TCAMs and memories.

Woo, Seung-Han (RM.3)
Seung-Han Woo was born in Seoul, Korea, in 1983. He received the B. S. degree in the electrical and electronic engineering from Yonsei University, Seoul, Korea, in 2009. He is studying for M. S degree in Yonsei University. His current research interests is process variation tolerant circuit design.

Wouters, Yves (IC.8)
Yves Wouters is Professor of chemistry at the University of Grenoble (France). After a PhD from the Grenoble Institute of Technology, he joined the technical centre of Juelich (Germany) in 1996 in the team of Dr W.J. Quaddakers. In 2000 Yves Wouters joined the SIMaP (the Materials and Processes Science and Engineering Laboratory). His main research topic concerns the preparation, forming, assembly and properties of materials for structural and functional applications and in particular the durability of metallic materials (energy, microelectronics, etc.).

Wrachien, Nicola (3F.2, 4F.5)
Nicola Wrachien was born in Treviso, Italy, in 1982. He received the degree (magna cum laude) in electronic engineering in 2006 from the University of Padova, Italy, and the PhD in Information Engineering in 2010, working on advanced non-volatile memories. He currently holds a post-doc position working on organic semiconductor devices and on III-V MOSFETs.

Wu, Ernest (BD.4)
Ernest Y. Wu is a senior technical staff member in Technology Reliability Department at Semiconductor Research and Development Center (SRDC) in IBM Microelectronics Division in IBM System and Technology Group. He received M.S. and Ph.D. degrees in physics from University of Kansas in 1986 and 1989, respectively. Dr. Wu joined IBM Microelectronics Division in 1994 at Essex Junction, Vermont after transferring from IBM Rochester, Minnesota.

Wu, Kenneth (2F.3)
Dr. Kenneth Wu is the Director of Technology Quality & Reliability at Taiwan Semiconductor Manufacturing Company (TSMC), responsible for technology reliability characterization and qualification. Prior to join TSMC in 2002, Dr. Wu worked at Intel Corporation since 1982 in technology development and reliability qualification. Dr. Wu has published more than 50 papers on a variety of technology and reliability subjects; and received IRPS The Best Paper Award in 1990. Dr. Wu received his B.S from National Taiwan University in 1975, an M.S. from Northwestern University in 1978 and his Ph.D. from Princeton University in 1982; all are in Electrical Engineering.

Wu, Ling (MY.4)
Wu Ling received his B.Eng in Electrical and Electronics Engineering (EEE) from Nanyang Technological University (NTU) Singapore in 2008. He is currently pursuing his PhD degree in EEE, NTU working on High-k/Metal gate of advanced gate stacks.

Wu, Ming-Tsung (MY.1)
Ming-Tsung Wu was born in Hualien, Taiwan in 1977. He received the M.S. degree in Institute of Environmental Engineering & Science from Feng Chia University (FCU), Taiwan, R.O.C., in 2002. He is currently a project manager of Etch Process Development Department in Technology Development Center in Macronix International (MXIC).

Wu, Mong Sheng (FA.1, FA.3)
M.S. Wu, born in Chia-yi in 1982, is a engineer in UMC Corporation. Wu received a bachelor degree from National Tsing Hua University in Taiwan. Now, he has been a FA engineer for 4 years.

Wu, Xing (4A.1, BD.2)
Xing Wu was born in Xi'an, China in 1984. She received the B.E. degree in Electronic Science and Technology in 2008 from Xi'an Jiaotong University, People's Republic of China. She was an international exchange student at Osaka University from 2006 to 2007. She is currently pursuing her Ph.D. degree at the Division of Microelectronics, School of Electrical and Electronic Engineering, Nanyang Technological University (NTU), Singapore. Her current research interests include reliability study of novel high-κ dielectric materials in nanodevices using atomistic simulations and transmission electron microscopy (TEM) analysis. She is currently a graduate student member of the IEEE (2008-present).

Xiang, Li (4A.1, BD.2)
Xiang Li received his B.Eng in Electrical and Electronics Engineering (EEE) from Nanyang Technological University (NTU), Singapore in 2005. He is currently pursuing his PhD degree in EEE, NTU working on failure analysis of advanced gate stacks. His research interests include front-end device reliability and nano-scale characterization using transmission electron microscopy (TEM) and electron energy loss spectroscopy (EELS). Since 2006, he has been a student member of IEEE.

Xu, Chen (FA.5)
Dr. Chen Xu is a reliability engineer with Alcatel-Lucent Reliability Enginerring Department. He has extensive experience in the fields of semiconductor and electronic packaging, electronic manufacturing, surface finishes and coatings, failure mode analysis and reliability assessment. Chen Xu received his BS Degree in Chemistry from Tongji University in Shanghai, China and his Ph.D. in Physical Chemistry from Ruhr-University Bochum, Germany. Prior to joining Alcatel-Lucent, he was a Senior Scientist with Enthone, Cookson Electroncis. Dr. Xu's career also includes the position of project leader/MTS with Lucent Technologies and R&D positions at Hoechst Research and Technology Center.

Yamamoto, Hirohiko (PI.1, PI.2)
Hirohiko Yamamoto received his B.E. in industrial chemistry from Kumamoto University, Kumamoto, Japan in 1981. He joined Hitachi Microcomputer Engineering, Ltd. (now Hitachi ULSI systems, Co., Ltd.), Tokyo, Japan in 1981, where he worked on developing and improving CVD technology for mass production. He then worked at Trecenti Technologies, Inc., Ibaraki, Japan from 2000 to 2004. Since 2005, he has been working at the Naka Sector of Renesas Technology Corp. (now Renesas Electronics Corporation), Ibaraki, Japan. He is a chief engineer of the Wafer Process Manufacturing Technology Dept. 2.

Yamauchi, Toyohiko (FA.4)
Mr. Yamauchi graduated with Bachelor's and Master's degrees in Engineering from the University of Tokyo in 2002 and 2004, respectively. He has been working at Hamamatsu Photonics K.K. (Japan) since 2004 as a researcher in the central research group. He has also been a visiting researcher for a collaborative study at the MIT Spectroscopy Laboratory since 2008. He is engaged in research on biomedical sensing with interference optics.

Yang, Ling-Wu (MY.1)
Ling-Wu Yang received the M.S. degrees in material science and engineering from the University of National Sun Yat-Sen, Taiwan, R.O.C., in 1996. From 1996 to 1999, he joined Vanguard International Semiconductor Co., Ltd, Hsinchu, Taiwan, R.O.C., as a process engineer. In 2000, he joined Macronix International (MXIC) for advanced diffusion module development. He is currently a Project Deputy Director of Advanced Module Process Development Division in Technology Development Center.

Yang, Tahone (MY.1)
Tahone Yang received the B.S. degree in chemistry from Fu Jan Catholic University, Taipei, Taiwan, in 1989, and the M.S. degree in chemistry from National Tsing-Hua University, Hsinchu, Taiwan, in 1991. In 1991, he was with FAB1, Macronix International (MXIC), working on the start-up and manufacturing of Etching /Lithography Department. In 1999, he transferred to Technology Development Center. Currently, he is leading the division of advanced module process development for nonvolatile memory devices. He has publications and holds patents in the areas of lithography, OPC, etching, thin film, diffusion, yield improvement and nonvolatile memory processing.

Yang, Yang (EL.3)
Yang Yang received the B.Sc. (2004) in electrical engineering from Beijing University of Aeronautics and Astronautics, Beijing, China and the M.Sc. (2005) in electronics from Queen's University Belfast, Belfast, UK. He is currently a Ph.D. student at George Mason University, VA, USA. His research focuses on the design & optimization of field-effect-diode (FED) and ESD-like stress on the gate dielectric reliability of nano CMOS. From Jan. 2009 to Dec. 2009, he was at IBM Microelectronics-System and Technology Group, SRDC, on a student internship in ESD/Latchup Development Group.

Yasuda, Naoki (MY.2)
Naoki Yasuda received the B.S., M.S., and Ph.D. degrees in electronic engineering from Osaka University, Japan. He joined Toshiba Corporation in 1992, where he was engaged in the research on SiO2 in Si MOSFETs. From 1998 to 1999, he was a visiting scientist at Rutgers University, NJ, where he studied the stabilities of high-k dielectrics. When he returned to Toshiba, he was engaged in silicon oxynitride in CMOS devices. From 2001 to 2004, he worked for MIRAI Project, Tsukuba, Japan, as a researcher in High-k Gate Stack Group. Since 2005, he has been working in the area of non-volatile memories.

Yasuhiro, Shimamoto (PI.3)
Yasuhiro Shimamoto received the Ph.D. degrees in physical engineering from The University of Tokyo, Tokyo, Japan, in 1996. He joined the Central Research Laboratory, Hitachi Ltd., Tokyo, in 1996, where he was engaged in the development of FeRAM, DRAM capacitors and high-k gate dielectrics. From 2002 to 2004, he was an Industrial resident in the Interuniversity Microelectronics Center (IMEC), Leuven, Belgium, where he studied on high-k gate dielectrics and metal gate. He is currently working on reliability of gate dielectrics for non-volatile memory.

Yasuo, Nara (4A.5)
Yasuo Nara received the B.S. (1980), M.S. (1982), and Ph.D (1985) degrees in physical electronics from Tokyo Institute of Technology, Tokyo, Japan. He joined Fujitsu Laboratories Ltd. in 1985 and has been engaged in the research and development of Si-based semiconductor technologies. He is currently in charge of advanced Si LSI technology development in Fujitsu Microelectronics Ltd. He is the author or coauthor of more than 270 journals and international conference papers in the field of semiconductor research. Dr. Nara is a member of the Japan Society of Applied Physics, the Institute of Electronics, Information and Communication Engineers, and IEEE Electron Devices Society.

Yau, Anthony (2D.2)
Anthony Yau received the B.E. in Material from Cheng Kung University, Tainan, Taiwan. He has joined Taiwan Semiconductors Manufacturing Company on product reliability engineering since 2000.

Ye, Chen (3C.5)
Chen Ye is an Engineer in Quality Reliability Assurance Failure Analysis at Global Foundries of Singapore. She received his bachelors in Materials Engineering from Nanyang Technological University,Singapore.

Yeh, Teng-Hao (5D.3)
Teng-Hao Yeh was born in I-Lan, Taiwan, ROC., on January 8, 1981. He recieved the BS and MS degree in Material Science and Engineering from National Tsing Hau University, Hsinchu, Taiwan, in 2003 and 2005. He joined Macronix International Company, Ltd., Hsinchu, in 2006, as a Process Integration Engineer in technology development center. Since 2006, he has been engaged in the development and integration for the nitride-based NBit technology at Macronix.

Yeo, Yee-Chia (XT.4)
Y.-C. Yeo received the B. Eng and M. Eng degrees from National University of Singapore (NUS), and the M.S. and Ph.D degrees from UC Berkeley, all in Electrical Engineering. He was with TSMC in 2001-2003. He is now an Assistant Professor of Electrical and Computer Engineering at NUS and a Program Manager for a nanoelectronics research program at the Agency for Science, Technology, and Research, Singapore. His research interests include strained-silicon MOSFETs, compound semiconductor transistors, and transistors with steep subthreshold swing. He co-authored ~330 papers, and has 82 US Patents.

Yew, Kwang Sing (4A.4)
K. S. Yew received the B.Eng. (Hons) degree in electrical and electronics engineering from the University of Technology, Malaysia (UTM) in 2007. From 2007 to 2009, he was with Altera Corporation, Penang, Malaysia as a Backend Device Modeling Engineer. He is currently working toward the Ph.D. degree in high-κ gate stacks electrical characterization by scanning probe microscopy at Nanyang Technological University (NTU) at Singapore.

Yiang, Kok-Yong (2D.3, 5A.4)
Kok-Yong Yiang received his Ph.D in Electrical Engineering from the National University of Singapore. He joined Advanced Micro Devices (Sunnyvale CA) in 2005 as a BEOL reliability engineer, and is currently with GLOBALFOUNDRIES (Sunnyvale CA) as a Member of Technical Staff. His main interest is in BEOL and middle-of-line (MOL) TDDB reliability. Before joining the semiconductor industry, Kok was a military engineer in the procurement of extreme-damage missile systems, and a software engineer involved in high-security smartcard encryption and authentication technologies.

Yoann, Mamy Randriamihaja (2B.2)
Yoann Mamy Randriamihaja was born in Grenoble, France, in 1985. He received the M.S. degree in micro- nano- electronic from University Joseph Fourier and the Engineering degree in 2009 from the Ecole Nationale Supérieure d'Electronique et de Radioélectricité de Grenoble (ENSERG) of the Institut National Polytechnique de Grenoble (INPG). Since 2009, he is pursuing the Ph.D. degree in micro- and nanoelectronic through a collaboration between the Institut Matériaux Microélectronique Nanosciences de Provence (IM2NP), Toulon, and the Electrical Characterization and Reliability Department of STMicroelectronics Crolles. His Ph.D. work focuses on the improvement of electrical characterization techniques of oxide defects, and the understanding of physical phenomena observed in advanced MOSFETs.

Yokogawa, Shinji (6A.4)
Shinji Yokogawa recieved the B.S.(1992), M.S.(1994), and Ph.D(2008) in engineering from The University of Electro-Communications, Tokyo, Japan. In 1994 he joined NEC Corp., Japan. He is currently with the Advanced Device Development

Division, NEC Electronics Corporation, Kawasaki, Japan. Currently, he serves as a leading engineer of advanced technology qualification in NEC Electronics. He is a member of the Japan Society of Applied Physics and Reliability Engineering Association of Japan.

Yoshiharu, Tosaka (3A.5, 4A.5)
Yoshiharu Tosaka was born in Akita, Japan, in April 1962. He received B. S., M. S. and Ph. D. degrees in physics from Niigata University, Japan, in 1985, 1987, and 1990, respectively. He joined Fujitsu Laboratories Ltd in 1990. His current research interests are in the reliability physics of VLSI devices, especially for soft-error reliability. Dr. Tosaka is a member of the Japan Society of Applied Physics and the Physical society of Japan.

Yoshimoto, Hiroyuki (XT.2)
Hiroyuki Yoshimoto received the B.S., M.S., and Ph.D. degrees in solid state physics from Waseda University in 2000, 2002, and 2006, respectively. In 2006, he joined the Central Research Laboratory, Hitachi, Ltd., Tokyo, Japan. Since then, he has been working on research and development of CMOS devices including SOI and high-k MOSFETs. Since 2009, he has also been working on research and development of power electronics devices.

Young, Chadwin (2B.4, 4A.4)
Chadwin D. Young received his B.S. degree in Electrical Engineering from the University of Texas at Austin in 1996. He then went on to receive a M.S. and Ph.D. from the North Carolina State University in 1998 and 2004, respectively. He has held several internships during this time until 2001 when he joined SEMATECH. Here, he completed his dissertation research on high-k gate stacks. He now continues this research at SEMATECH as a Member of the Technical Staff working on electrical characterization and reliability methodologies for the evaluation of high-k gate stacks, and he has authored or co-authored 150+ journal and conference papers.

Yu, Hongyu (MY.4)
Yu Hongyu obtained his BEng degree from Tsinghua University, MASc degree from University of Toronto and PhD from National University of Singapore respectively. From June 2004 to January 2008, he worked as a Senior Researcher in IMEC, Belgium. He joined the School of EEE starting from January 2008, receiving the inaugural Nanyang Assistant Professorship. His research is on sustainable Si-based Nano electronic device, e.g. emerging memories / sub-22nm CMOS devices for green IC; advanced solar cell / nano-photonic devices. He has authored / co-authored more than 150 peer-reviewed international papers. He also has published / been granted with > 20 USA/EU patents. His research achievements have gained international recognition. He has received many awards including NUS president graduate fellowship, IEEE EDS PhD fellowship, Nanyang Assistant Professorship, Tan Chin Tuan Academic Exchange Fellowship, and one high-light paper in Tech. Sym. VLSI (2007).

Yuichiro, Mitani (3D.4)
Yuichiro Mitani received the B. E. and M. E. in material science and engineering from Tohoku University, Sendai, Japan, in 1990 and 1992, respectively. He received the Ph.D. from the University of Tokyo in 2009. He joined the R&D Center, Toshiba Corporation in 1992. His primary works were concerned in the Si-CVD and the ultra-shallow junction process technology. Since 1999, he has been with the Advanced LSI Technology Laboratory, Corporate R&D Center, Toshiba Corporation, Yokohama, Japan. His present research interests and activities cover the ultra-thin oxide process technology and the study of the reliability of ultra-thin gate dielectrics (SiO_2,

SiON and High-k) for ULSI technology. He serves (or served) on the technical committees of International Conference on IC Design & Technology (ICICDT) and IEEE International Reliability Physics Symposium (IRPS). He is a member of the JSAP.

Yuko, Kobayashi (4A.5)
Yuko Kobayashi received the B.S. degree in Materials Property, and the M.S. degree in Materials Processing and Characterization from Tohoku University, Miyagi, Japan, in 2003 and 2005, respectively. She joined Fujitsu Ltd. in 2005. Her currently work is reliability technology for ULSI devices.

Yumi, Kakuhara (6A.4)
Yumi Kakuhara received the B.E. degree from Aoyama Gakuin University, Tokyo, Japan, the M.E. degree from Shinshu University, Nagano, Japan, and Ph.D (2010) in engineering from Shibaura Institute of Technology, Tokyo, Japan. She was with the Semiconductor Group, NEC Corporation, Sagamihara, Japan. She is currently focused on reliability issues in advanced Cu./Low-k interconnect. She is currently with the Advanced Device Development Division, NEC Electronics Corporation, Kawasaki, Japan.

Zhang, Lijuan (5B.2)
Lijuan Zhang is a graduate student in the Department of Electrical Engineering at The University of Texas at Austin. She received her M.S. degree in solid state physics from the Institute of Physics, Chinese Academy of Sciences (CAS) in 2004 and her B.S. degree in physics from Peking University in China in 2001. She joined The University of Texas at Austin in fall 2004 and pursued her PhD in the Interconnect and Packaging group under the supervision of Prof. Paul Ho. Zhang's research focuses on the BEOL reliability issues such as electromigration and stress migration reliability of Cu damascene interconnects.

Zhao, Kai (2B.1)
Kai Zhao received the bachelor's and master 's in physics from University of Science and Technology of China in 2000 and 2002 respectively and the Ph.D. from Physics department at University of California, San Diego in 2008. In the same year, he joined IBM research division as a postdoctroal researcher, working in the reliability and characterization group. At IBM research, he is focused on understanding the fundamental reliability aspects of advanced Field Effect Transistors, including HKMG FETs and Carbon Nanotube Transistors.

Zhao, Larry (5A.2)
Larry Zhao joined Intel Corporation in 2002 and is currently working at IMEC as an assignee from Intel. He has more than 10 years of experience in the research and development of Cu interconnects and holds 6 U.S. patents. He has also co-authored more than 40 scientific publications. Larry Zhao received his B.S. degree in Materials Science from Shanghai Jiaotong University in 1984 and M.S. degree in Engineering from Dartmouth College, Hanover, NH in 1993. He studied for Ph.D. at Columbia University, NYC from 1993 to 1997.

Zheng, Rui (5E.2)
Rui Zheng received Bachelor of Science from School of EECS, Peking University, China in 2008.He is a master student in Arizona State University since 2008. Currently researching in circuit reliability modeling and prediction.

Zhi Qiang, Mo (3C.5)
Mo Zhi Qiang is Manager of failure analysis operations group in Global Foundries Singapore Pte Ltd. He has been working in microelectronics failure analysis area for more than 18 years. His areas of interest includ surface analysis, materials/chemical analysis and physical analysis. Prior to join Chartered, he was Assistant Vice

President in TUV-SUD-PSB Corp Singapore and worked as Senior Research Engineer in Institute of Microelectronics (IME) Singapore. He received his Bachelor and Master's degree in electronic engineering from Tsinghua University, Beijing in 1987 and 1993 respectively

Zhou, Carl (6C.4)
Carl Zhou has worked at Texas Instruments since 1999. He has developed yield enhancement tools such as critical area analysis and logic mapping. He also works on Non-Volatile Memory, flash and FRAM, testing, characterization, and reliability study. Prior to joining TI, Carl Zhou has worked in several companies, including Integrated Device Technology, Inc. and SunPower. He received his Ph.D in Physics at Stanford University.

Zhou, Jiping (5B.2)
J. P. Zhou is a Research Scientist and Facility Manager in Texas Materials Institute in the University of Texas at Austin. His research career is covered to high Tc superconductors, semiconductors, nano-materials and their syntheses, fabrications and characterizations. Recently, he is engaged in the area for nano-materials structures, defects and properties, corresponding to crystal orientations by using Diffraction-Scanning Transmission Electron Microscope (D-STEM) and Precession Microscope.

Zhou, Xing (NA. 1)
Xing Zhou received the B.E. degree from Tsinghua University in 1983 and the M.S. and Ph.D. degrees in electrical engineering from the University of Rochester in 1987 and 1990, respectively. Since 1992 he has been a faculty of the School of Electrical and Electronic Engineering, Nanyang Technological University, Singapore. He was a visiting professor with Stanford University in 1997 and 2001 and with Hiroshima University in 2003. He is a distinguished lecturer of the IEEE EDS and an editor for the IEEE Electron Device Letters. His research focuses on development of compact models for circuit simulation for nanoscale MOS devices.

Zhu, Goujun (NA. 1)
Guojun Zhu was born in China in 1984. He received the B.E. degree (Hons.) in electrical and electronic engineering in 2007 from Nanyang Technological University, Singapore, where he is currently working toward the Ph.D. degree at the School of Electrical and Electronic Engineering. His current research interests include unified compact modeling of bulk/SOI/DG/GAA and Schottky-barrier MOSFETs. From July to December 2005, he was an Intern with Chartered Semiconductor Manufacturing Ltd., Singapore, where he worked on electrical testing.

Zhu, Vivian (4B.2)
Xiaowei (Vivian) Zhu received her Ph.D. in Electrical Engineering from Vanderbilt University. She joined Texas Instruments, Inc. in 2002 as a Reliability Engineer in the Silicon Technology Development group. Dr. Zhu has published several papers in the field of radiation induced single event upset, served as technical session chair for IEEE Nuclear and Space Radiation Effects Conference, and International Conference on the Application of Accelerators in Research and Industry. She also served as a committee member and section author of JEDEC JESD89A standard, and is a frequent reviewer for IEEE TNS, European Conference on Radiation and Its Effects on Components and Systems, and Microelectronics Reliability.

Zous, Nian-Kai (5D.3)
Nian-Kai Zous received the Ph.D. degree in electronics engineering from National Chiao-Tung University (NCTU), Hsin-chu, Taiwan, R.O.C., in 2002. He joined the Macronix International Co., Ltd. (MXIC) in Hsinchu, Taiwan, in 2003, working in the advanced device group of the Device Technology Division and his research activities include the study of reliability and scaling issues in nitride storage devices. From 2005 to 2007, he is working in the NBIT Flash Tech. Dept. II of MXIC and is responsible for the reliability issue on NBIT product. Since 2007, he has been engaged in the methodologies for technology development and yield enhancement .

Zschech, Ehrenfried (5B.1, 5B.2)
Ehrenfried Zschech is Division Director at Fraunhofer Institute for Nondestructive Testing in Dresden, which he joint in 2009. His responsibilities include micro- and nanoanalysis. He received his Dr. degree from Dresden University of Technology. Ehrenfried Zschech gathered experience in industry, during 17 years in several technical and management positions at Airbus and AMD. He has published three books and more than 100 papers in scientific journals. He is an honorary professor for nanomaterials at the Brandenburg University of Technology in Cottbus, Germany. In 2009, Ehrenfried Zschech was elected as Vice President of the Federation of European Materials Societies (FEMS).

AUTHOR INDEX

Abe, Kenichi 683
Abeln, Glenn 1115
Abramowitz, Peter 125, 1115
Abu-Rahma, M. H.......................................1008
Ackermann, Markus...................................1006
Acosta, Antonio G. 689
Afanas'ev, Valery 549
Ahlbin, J. R..198, 763, 1031
Ahn, Kun-Ok...975
Aichinger, Thomas.........................1063, 1073
Aitken, J..566
Alam, Muhammad Ashraful 65, 312, 1069, 1091
Albin, David S. 318
Alers, G. B. 323, 499
Allee, David R..644
Allmon, R. 188
Ambrose, V. 188
Amoroso, Salvatore M. 966
Anderson, Gary 125, 1115
Andreani, C..400
Ang, D. S...373, 1058
Angyal, M. 566
Aoulaiche, Marc.......................................1095
Arigane, T..1001
Arreghini, Antonio...................................731
Asheghi, Mehdi...99
Ashok, Ashwin...287
Atkinson, N. M...198
Aubel, Oliver 117, 562, 574, 581, 926
Auluck, Kshitij...981
Baburske, Roman......................................162
Baccarani, Giorgio...................................881
Badami, D. 566
Bae, Kidan 104
Baek, Chang-Ki...94
Baek, Rock-Hyun.......................................94
Bagatin, M. 400
Baghini, Maryam Shojaei.....................480, 841
Bai, P. 293
Balasubramanian, A. 188
Balk, L. J. 271
Banerjee, Kaustav.....................................822
Barbato, M. 246
Barsky, M. E..807
Bashir, Muhammad M. 895
Basker, V. S...1099
Baumann, Robert C....................................395
Bauza, D. 259
Beenakker, C. I. M....................................342
Bellaio, N...334
Beltrami, Silvia...................................400, 604
Bender, Hugo...712
Benvenuti, A..615
Bergman, J..813
Bersuker, G. 73, 373, 532

Besset, C. 55
Beyer, G. P. 591
Bhatia, C. S. 503
Bhuva, Bharat L. 198, 763, 1026, 1031
Biedenbender, M. 807
Bisht, Gaurav 981
Bisschop, Jaap 830
Bittel, B. C. 947
Blaauw, D. 676
Bloch, Didier 231
Block, T. R. 807
Bluet, J.-M. 139
Boit, C. 890
Boku, K. 750
Bonilla, G. 566
Borghs, Gustaaf 146
Boselli, Gianluca 474
Bosman, M. 354
Bounasser, Mounaim 203
Bourgeois, F. 129
Brar, B. 813
Bravaix, A. 55
Breuer, T. 890
Brewer, Forrest 853, 932
Briggs, Benjamin 80
Bronner, W. 129
Bru-Chevallier, C. 139
Bu, Huiming..1099
Burnett, David 125, 1115
Butendeich, Rainer 522
Butera, Geni 712
Buttari, D. 807
Bychikhin, S. 480
Cacho, Florian 655
Cagli, C. 620
Cai, Xiao Xiao 411
Calderoni, Alessandro.........................738, 970
Camozzi, Elisa 966
Campbell, J. P. 804
Cannon, Ethan H. 1019
Cao, Yiqun 458
Cao, Yu 650
Carter, S. A. 323
Cartier, E. 50, 369, 787, 1044
Cäsar, M. 129
Catthoor, Francky 1014
Cavelaars, Jan 724
Cellere, G. 400
Ceric, Hajdin 911
Cester, A..334, 536
Cha, Nam Hyun 1004
Chancellor, Cathy A............................385, 670
Chang, C. S. 627
Chang, C. Y. 818
Chang, W. L. 364, 787

AUTHOR INDEX

Chang, Y. F. 627
Chao, Y. P. 960
Chappaz, C. 887
Chatterjee, Amitabh 853, 932
Chatterjee, I. 1031
Chatty, Kiran 846
Chen, An 84
Chen, Chih-Hsien 918
Chen, F. 566
Chen, K. C. 627, 960
Chen, K. F. 627
Chen, K. H. 818
Chen, Kuan-Fu 634
Chen, Kuang-Chao 634, 639, 951
Chen, M. J. 665
Chen, M. S. 627
Chen, Ming-Shiang 634
Chen, Tao...342
Chen, Wen-Yi 857
Chen, X. 566
Chen, Y. J. 627
Chen, Y. N. 89
Chen, Ye 277
Chen, Yin-Jen 634
Chen, Zuhui 977
Cheng, Cheng-Hsien 634
Cheng, Kai 146
Cheng, Kangguo 1099
Cheung, K. P. 804
Chevallier, Remy 655
Chikhaoui, W. 139
Child, C. 566
Chimeno, Alejandro 655
Chin, Melida 453, 562
Chiu, J. P. 960
Cho, Eun Suk 611
Cho, M. 1078, 1082
Cho, Moonju 1095
Cho, Myoung Kwan 975
Cho, Seokwon 975
Choi, Gil-Bok 94
Choi, Gil-Heyun 903
Choi, Hyun Ki 599
Choi, Hyun-Sik 94
Choi, Jeong-Hyuk 599, 611
Choi, Jingyoo 287
Choi, R. 1008
Choi, Siyoung 282, 903
Choi, Zungsun 903
Chong, Lit Ho 634
Chopra, Sanjeev 655
Chou, H. L. 170
Chou, T.-M. 528
Chou, Y. C. 807
Chou, Y. L. 960

Chouard, Florian R. 826
Chowdhury, U. 528
Christiansen, C. 566
Chu, B. H. 818
Chu, F. 750
Chua, E. C. 938
Chung, Chilhee 599, 611
Chung, Steve S. 1053
Cirba, C. 670
Coccetti, F. 237
Compagnoni, Christian Monzio 604, 966, 970
Croes, Kristof...................543, 549, 591, 712
Curutchet, A. 139
Dadgour, Hamed F. 822
Dammann, M. 129
Das, Jo 146
DasGupta, S. 813
De Jaeger, J.-C. 139
Decoutere, Stefaan 146
Defrance, N. 139
Degraeve, Robin 26, 1078, 1095
Degroote, Stefan 146
Del Alamo, Jesús A. 134
Del Cueto, Joseph A. 318
Demirtas, Sefa 134
Demuynck, Steven 712
Denison, Marie 881
Deora, S. 1105
Depner, M. 750
Derkits, G. E. 879
Derluyn, Joff 146
Detcheverry, C. 111
Detzel, Thomas 911
Dhere, Neelkanth G. 306
Di Sarro, James 846
Djelassi, Christian 446
Domengie, F. 259
Dongaonkar, S. 312
Donnet, David 724
Doris, Bruce 1099
Douglas, E. A. 818
Douvry, Y. 139
Doyen, L. 922
Drijbooms, Chris 712
Du, A. Y. 906
Du, G. A. 1058
Du, Pei-Ying 951
Dua, C. 139
Duvvury, Charvaka 853
Eaton, P. H. 198
Eliason, J. 750
Eneman, G. 1082
Eng, D. C. 807
Enichlmair, H. 43
Fang, Z. 964

AUTHOR INDEX

Fantini, Paolo 738, 970
Faqir, M. 152
Farbiz, Farzan 466
Farina, Fabrizio 970
Ferro, M. 738
Fisher, Kathryn C. 508
Fleetwood, D. M. 813
Fleming, D. 879
Forbes, Keith 125
Forsythe, Eric 644
France, C. E. 323
Francis, Rick 117
Franco, Jacopo 26, 1082, 1095
Franey, J. P. 879
Frei, M. 312
Frei, Stephan 458
Frost, C. D. 400
Fugazza, Davide 743
Fujii, Shosuke 956
Fujiki, Jun 956
Fujisawa, Takafumi 683
Fujitsuka, Ryota 956
Fujiwara, Tetsuo 988
Fukatsu, Shigeto 424
Fuketa, Hiroshi 213
Fukuda, Mitsuhiro 218
Fukuda, Toshikazu 694
Fukutani, Katsuyuki 417
Fulde, Michael 826
Funayama, Kota 995, 1001
Futase, Takuya 988, 995
Gadlage, M. J. 198, 763
Galbiati, Nadia 966
Gannavaram, S. 293
Gao, B. 964
Gasiot, Gilles 407
Gauthier, Robert J. 846
Ge, L. 1008
Geinzer, T. 271
Genoe, J. 334
Gerardin, S. 246, 400
Germain, Marianne 146
Gertas, J. 750
Ghani, Tahir 287
Ghetti, A. 615
Ghidini, Gabriella 966
Ghidotti, Michele 604
Ghilardi, Tecla 966
Gilbert, N. 566
Giliberto, V. 246
Gill, B. 188
Gischia, Gianni Giai 549
Glaser, Ulrich 458
Glavanovics, Michael 446
Gnade, Bruce 644

Gnani, Elena 881
Gnudi, Antonio 881
Goes, W. 16
Goguenheim, D. 55
Goodnick, Stephen M. 516
Goodson, Kenneth 99
Gorini, G. 400
Gornik, Erich 480
Gossner, Harald 480, 841, 853
Goto, Masakazu 424
Goyal, Deepak 252
Goyal, S. 430
Graham, R. L. 323
Grasser, Tibor 7, 16, 26, 43, 1063, 1073, 1082
Green, Keith 689
Groat, J. 750
Groeseneken, Guido 26, 549, 712, 1078, 1082, 1095
Grossi, Alessandro 966
Grzegorczyk, Andrzej 724
Guarin, F. 364
Guérin, C. 55
Guerra, Diego 516
Gustin, Wolfgang 7, 175
Ha, Sungmok 104
Hafez, W. 293
Haggag, Amr 125, 1115
Hahn, Berthold 522
Han, B. M. 1008
Han, Jeong Hee 282
Han, Tzung-Ting 627, 634
Hashikawa, Naoto 988, 995
Hashimoto, Koichi 379
Hashimoto, Masanori 213
Hashimoto, T. 1001
Hatanaka, Kichiji 218
He, Jun 287
Heh, D. 373
Heiderhoff, R. 271
Heinze, Birk 162
Hendriks, Teun 1014
Hennesthal, C. 581
Heryanto, A. 586
Hicks, J. 293
Higman, Jack 1115
Hinze, Peter 327
Hirano, Izumi 424
Hiroshima, Shoichi 417
Hisamoto, D. 1001
Hitoshi IIda 874
Ho, Paul S. 574, 581
Hoel, V. 139
Hoffmann, Thomas Y. 926, 1095
Hofmann, Ralf 981
Holman, W. T. 198
Hong, Chonga 975

AUTHOR INDEX

Hong, Shih-Ping 951
Hori, Mitsuaki 379
Horita, K. 1001
Hsia, L. C. 906
Hsiao, Yi-Hsuan 951
Hsieh, E. R. 1053
Hsieh, Jung-Yu 639, 951
Hsieh, Kuang-Yeu 627, 951
Hsieh, Sunnys 170
Hsu, Chia-Lin 918
Hsu, Fang-Hao 951
Hsu, Tzu-Hsuan 634, 951
Hu, Y. Z. 1058
Huang, Climbing 918
Huang, E. 566
Huang, I-Jen 627, 634
Huang, Jyun-Siang...............................634
Huang, Rui 911
Huang, Yu-Hui 170
Huard, Vincent 33, 55, 655
Hübner, René 574
Hughes, Greg 799
Hurkx, G. A. M. 99
Hurley, P. K. 775
Hutter, H. 1063
Hwang, Heedon 282
Hyun, Sangjin 282
Ielmini, Daniele 620, 738, 743
Im, J. 581
Inaba, Yutaka 988
Inan, Umran S. 203
Inoue, M. 1001
Inumiya, Seiji 424
Ioannou, D. P. 1044
Ioannou, Dimitris E. 846
Ishigaki, T. 1049
Ishihara, Ryoichi 342
Islam, Ahmad Ehteshamul 65, 1069
Ito, Shuu 417
Iwai, Hidenao 874
Iwamatsu, T. 1001
Jack, Nathan...............................485, 835
Jagannathan, Hemanth 1099
Jain, Palkesh 698
Jammy, R. 73
Jan, C.-H. 293
Jeong, YeonJoo 975
Jeong, Yoon-Ha 94
Jha, N. K. 665
Jimenez, J. 528
Johannes, Hans-Hermann 327
Jones, M. 293
Jung, S. O. 1008
Jungemann, C. 175
Jurczak, Malgorzata 731

Kaapor, Neeraj 655
Kaczer, Ben...................16, 26, 1082, 1095
Kakuhara, Y. 717
Kamino, Takeshi 988
Kane, T. 566
Kaneoka, T. 1001
Kaneta, Chioko 379
Kang, Han-Byul 265
Kang, J. F. 964
Kang, Ju Seong 1004
Kang, M. G. 1008
Karthik, Y. 312
Kase, Masataka 379
Kato, Koichi 424
Kauerauf, Thomas...............712, 1078, 1095
Kaul, Ashwani 306
Kaureauf, Thomas 799
Keller, R. 175
Kendig, D. 499
Ker, Ming-Dou 857
Kerber, A. 50, 369
Kerst, U. 890
Khakifirooz, Ali 1099
Kiefer, R. 129
Killat, N. 528
Kim, Byoung Taek 611
Kim, D. 750
Kim, Dae Mann 94
Kim, Dong-Won 94
Kim, Gun-Rae 104, 265
Kim, Hyun 1004
Kim, Hyun Jung 611
Kim, Kinam 94
Kim, Min 104
Kim, SangBum 99
Kim, Seong Soo 599
Kim, Woo Sup 1004
Kim, Yong Seok 599
Kim, Yongshik 104
Kimura, S. 1049
King, S. 947
Ko, T.-M. 566
Kobayashi, Yuko 379
Köck, Helmut 446
Koh, Yohwan 975
Kolics, A. 566
Komeyli, K. 293
Konstanzer, H. 129
Kopf, R. 879
Koszewski, Adam 231
Kotlyar, R. 293
Kowalsky, Wolfgang 327
Koyama, Shin 417
Krause, J. 188
Krick, J. 670

AUTHOR INDEX

Krishnan, Anand T. 385, 670, 1122
Krishnan, Srikanth 650, 689, 698, 1122
Ku, Shaw-Hung 634
Kuball, M. 152, 528
Kubo, Tomohiro 379
Kufluoglu, Haldun 670
Kulkarni, Pranita 1099
Kulshrestra, Vishal 655
Kumashiro, Shigetaka 694
Kuper, Fred 111, 724
Kuroda, Rihito 683
Labat, N. 139
LaBel, Kenneth A. 768
Lacaita, Andrea L. 604, 620, 743, 966, 970
Lai, R. 807
Lai, Sheng-Chih 951
Lamontagne, P. 922
Langer, E. 890
Larcher, Luca 373, 532, 731
Laurin, L. 615
Lavizzari, Simone 743
Law, S. B. 566
Lee, Byoungil 99
Lee, Chi Kyoung 599
Lee, Dong Jun 599
Lee, Dong-Kyu 975
Lee, Eun-Kyu 80
Lee, Ho Seok 975
Lee, Hsiao-Heng Kelin 203
Lee, J. H. 665
Lee, Jeong-Soo 94
Lee, Jian-Hsing 182
Lee, Jong Myeong 903
Lee, L. S. 807
Lee, S. C. 705, 932
Lee, Sang-Hyun 94
Lee, Seok-hee 287
Lee, T. 566
Lee, Woon-Hak 265
Lee, Y. C. 627
Lee, Y.-H. 170, 665
Lemay, Karen 287
Lenahan, P. M. 43, 947, 1086, 1122
Leu, L. C. 818
Leung, D. L. 807
Leys, Maarten 146
Li, B. 566
Li, Junjun 846
Li, Li 440
Li, X. 354, 778, 964
Li, Yuan 724
Lian, Nan-Tzu 951
Liang, James W. 556
Liang, Z. 111
Liao, Jeng-Hwa 639

Liaw, M. H. 627
Lilja, Klas 203
Lim, K. Y. 73
Lim, Phyllis Shi Ya 1055
Lim, Se Young 1004
Lim, Sunme 104
Lim, Y. K. 586, 906, 938
Lin, Chung-Hsun 807, 1099
Lin, Hung Sung 861, 865, 870
Lin, J. 293
Lin, Jeh-Chieh 918
Lin, M. H. 705, 1053
Lin, Mingte 556
Lin, Shang-Wei 634
Lin, Shihuan 977
Lin, Wen-Chin 918
Lin, Y. H. 1053
Linscott, Ivan R. 203
Lipp, D. 369
Liu, Bin 1055
Liu, C. C. 170
Liu, H. 906
Liu, J. F. 938
Liu, Mark 287
Liu, Ning 1115
Liu, P. W. 1053
Liu, Pan 277
Liu, Tsu-Jae King 1117
Liu, W. 586, 906, 938
Liu, W. H. 354, 778
Liu, W. J. 964
Liu, Ziyuan 417
Lloyd, J. R. 943
Lo, Chester 627, 818
Lo, Patrick G. Q. 964
Lofrano, Melina 591, 712
Lu, Chih-Yuan 627, 634, 639, 951, 960
Lu, Kuan-Ting 918
Lu, Lei 1040
Lu, Ryan 287
Lu, Wen-Pin 627, 634
Lu, Xiaowei 1040
Lue, Hang-Ting 627, 634, 639, 951
Lutz, Josef 162
Lwin, Z. Z. 89
Ma, G. H. 1053
Ma, H. C. 960
Ma, Zhe 1014
Maconi, Alessandro 966
Mahapatra, Souvik 89, 312, 981, 1105
Mahatme, N. N. 1031
Maheta, V. D. 1105
Mairena, Andrew 1117
Maitra, Kingsuk 1099
Makabe, K. 120

AUTHOR INDEX

Makabe, Mariko 417
Makiyama, H. 1001
Malbert, N. 139
Mandich, M. L. 879
Manfredi, Manfredo 522
Marathe, Amit 117, 453, 562
Marca, Vincenzo Della 731
Marcon, Denis 146
Marino, Fabio Alessio 516
Marmiroli, Andrea 970
Marshall, A. 670
Mascellino, Evelyne 966
Massengill, L. W. 198, 763, 1031
Masuduzzaman, M. 1069
Matsuyama, Hideya 218, 379
Mauri, Aurelio 966
Medjdoub, Farid 146
Mei, X. B. 807
Mendenhall, Marcus H. 1026
Mendenhall, Marcus M. 395
Meneghesso, Gaudenzio 1, 246, 334, 522, 536
Meneghini, Matteo 1, 334, 522
Meng, L. 503
Mertens, Robert 146
Meyer, Jens 327
Miccoli, Carmine 604
Mikulla, M. 129
Milor, Linda 895
Minakata, Hiroshi 379
Miranda, E. 775
Mishra, Anand 655
Mishra, Rahul 846
Mishra, U. K. 152
Mistry, K. 293
Mitani, Yuichiro 299, 424
Mitra, Souvick 846
Mitra, Subhasish 203
Mitsuyama, Yukio 213
Mittl, S. 1044
Mizutani, M. 1001
Mo, Zhi Qiang 277
Moise, Ted 689, 750
Monaco, G. 246
Monsieur, F. 364
Moon, Joo-Tae 282, 903
Moosa, Mohamed 125, 1115
Mora, P. 364
Morassi, L. 373, 532
Mori, Hiroko 379
Mori, Toshifumi 379
Morin, P. 259
Morita, Y. 1049
Morton, David 644
Mottadelli, Riccardo 604
Mukherjee, S. 188

Mukhopadhyay, Gautam 981
Muller, K. Paul 391
Müller, S. 129
Murakami, E. 120
Myny, K. 334
Nagabhirava, Bhaskar 80
Nagalingam, D. 503
Nakamura, Hideyuki 694
Nakamura, Tomonori 874
Nakasaki, Yasushi 424
Nakkala, P. 887
Nam, Kab-Jin 282
Nara, Yasuo 379
Narasimham, B. 198
Nardi, F. 620
Nassif, N. 188
Nelhiebel, Michael 1063, 1073
Ney, David 655, 922
Ngan, P. 111
Nicholson, L. 566
Nicollian, Paul E. 385
Nicolosi, P. 246
Nigam, T. 369
Nikolic, Borivoje 1117
Nishigaya, Keita 379
Nowodzinski, Antoine 231
Oates, A. S. 705, 932
Obradovic, Borna 689
O'Connor, E. 775
O'Connor, Robert 799
Ogasawara, M. 120
Oh, M. 566
Ohmi, Tadahiro 683
Ok, Injo 532
Oki, A. K. 807
Oldiges, Philip J. 1099
Olney, Andrew 224
Olson, Nicholas 474
Ong, Tong-Chern 182
Ong, Y. C. 373
Ono, Katsuji 379
Onoye, Takao 213
Orita, Kenji 1
Ottogalli, F. 615
Oualli, M. 139
Ouchi, T. 120
Paccagnella, A. 246, 400
Pacheco, Mario 252
Packan, Paul 287
Padovani, Andrea 373, 532, 731
Pae, Sangwoo 287
Pan, JiFong 627
Pantelides, S. T. 813
Papaioannou, G. J. 237
Parihar, Sanjay 1115

AUTHOR INDEX

Park, Chan-Hoon 94
Park, H. 73, 1008
Park, Hong Sik 1004
Park, Hyung-Jin 1004
Park, Hyun-Woo 265
Park, Jongwoo 104, 265
Park, Junkyun 104
Park, Milim 975
Park, Noh Seok 1004
Park, Sukkwang 975
Park, Young-Joon 698
Parker, Chris 287
Parthasarathy, Chittoor 655
Paruchuri, Vamsi 1099
Pavan, Paolo 532, 731
Pavesi, Maura 522
Pearton, S. J. 818
Pei, Y. 152
Pellish, Jonathan A. 768
Pendharkar, Sameer 853, 881
Perng, Dung-Ching 918
Persin, Flore 655
Pethe, Shirish A. 306
Petitdidier, S. 566
Petitprez, E. 922
Pey, K. L. 89, 354, 373, 586, 778, 964
Phang, J. C. H. 271, 503
Piazza, M. 139
Pickholtz, J. 188
Pinato, A. 334
Pion, Emmanuel 655
Pirovano, A. 615
Plana, R. 237
Planes, Nicolas 655
Platt, S. P. 411
Pobegen, Gregor 1073
Pogany, D. 480
Poli, Stefano 881
Pons, P. 237
Post, I. 293
Prasad, C. 293
Prokofiev, A. V. 411
Puchner, S. 1063
Purser, Richard 287
Quay, R. 129
Quevedo-Lopez, Manuel 644
Rafik, M. 364
Raghavan, N. 354, 586, 778
Ragheb, T. 670
Ragnarsson, Lars-Åke 26, 799, 1095
Randriamihaja, Y. Mamy 55
Rangoni, Armando 966
Ranjan, R. 665
Ranzato, Enrico 522
Rao, V. Ramgopal 480, 841

Recchia, C. 188
Redaelli, An. 615
Reddy, Vijay 650, 670
Reed, Robert A. 395, 813, 1026
Reents, W. D. 879
Reggiani, Susanna 881
Regolini, J. L. 259
Reifenberg, John 99
Reisinger, Hans 7, 16, 26, 175
Relangi, Prasanthi 203
Remack, K. 750
Ren, F. 818
Renard, S. 55
Ribes, G. 364, 566
Riedl, Thomas 327
Riedlberger, E. 175
Riepe, K. 129
Rigoutat, O. 566
Robert, Vincent 655
Robl, Werner 911
Roche, Philippe 407
Rödle, T. 129
Rodriguez, John 689, 750
Rodriguez-Latorre, J. 750
Romain, Michael 1036
Rosenbaum, Elyse 466, 474, 485, 835
Roussel, J. 26
Roussel, Philippe 712, 1095
Roy, D. 55, 364
Ruelke, H. 890
Ruiz-Amador, Natalia 655
Rumyantsev, S. 73
Rupp, Roland 156
Ryan, J. T. 43, 1122
Ryan, S. 879
Sagong, Hyun Chul 94
Sahhaf, S. 1078
Sakoda, Tsunehisa 379
Salman, Akram 474
San, Tamer 689
Sanda, Pia N. 391
Sandhya, C. 981
Saraniti, Marco 516
Sasse, Guido T. 830
Sato, Motoyuki 424
Sawada, H. 566
Schanovsky, F. 16
Schlünder, Christian 7
Schmidt, Hans 327
Schmitt-Landsiedel, Doris 826
Schneidenbach, Daniel 327
Schneider, Jens 480, 841
Schrimpf, Ronald D 395, 763, 813, 1026
Scozzari, Claudia 966
Seah, Soo Sien 277

AUTHOR INDEX

Seetharaman, Sridhar 881
Seifert, N. 188
Seirawski, Brian D. 1026
Sekine, Katsuyuki 424, 956
Seo, Jae Yong 1004
Shakouri, A. 499
Shao, Xiaoyan 508
Sheldon, Douglas J. 759
Shen, X. 813
Sheshadri, Vijay B. 1026
Shi, Q. 188
Shiga, K. 1001
Shih, J. R. 170, 182, 665
Shimamoto, Y. 1001
Shin, Changhwan 1117
Shin, Yu Gyun 282
Shinosky, M. 566
Shono, Ken 218
Shrivastava, Mayank 480, 841
Shubhakar, K. 354
Shukla, Vrashank 485
Shuler, R. 1031
Shur, M. 73
Siegmund, T. 430
Sierawski, Brian D. 395
Simms, R. J. T. 152
Simoen, E. 26
Singh, P. 676, 981
Singh, P. K. 89
Sites, James R. 494
Sivatheja, M. 981
Smout, S. 334
So, Byung Se 1004
Sohn, D. K. 586, 906
Sölkner, Gerald 156
Song, Du Heon 599, 611
Song, Jai Hyuk 599, 611
Song, S. C. 1008
Song, Seung Hyun 94
Soo, Chi Wen 277
Spessot, Alessio 970
Spinelli, Alessandro S. 604, 966, 970
Spitzer, A. 175
Srinivasan, K. 430
Srividya, V. 1078
St Amour, Anthony 287
Stathis, J. 364
Stathis, James H. 50, 787, 1099
Stecher, Matthias 175, 458
Steen, Steven E. 508
Stellari, Franco 508
Street, A. G. 503
Su, K. C. 556
Subbarayan, G. 430
Suehle, J. 804

Sugawa, Shigetoshi 683
Sugii, N. 1049
Suh, Kang-Deog 599, 611
Summerfelt, Scott 689, 750
Sun, Kai 1040
Suñe, Jordi 792
Sury, C. 139
Suzuki, Hiroyoshi 683
Sylvester, D. 676
Takahisa, Keiji 218
Takeuchi, Kan 694
Tam, Nelson 1036
Tamura, Naoyoshi 379
Tan, J. B. 586, 906
Tanahashi, Katsuto 379
Tanaka, Katsuhiko 694
Tang, T. J. 566
Tang, Z. 347
Tanimoto, Hisanori 988, 995
Tapajna, M. 152
Taylor, W. 73, 373
Tazzoli, Augusto 246, 334, 522
Tee, Irene 277
Teo, Z. Q. 1058
Terada, Hirotoshi 874
Teramoto, Akinobu 683
Tessariol, Paolo 966
Tobimatsu, Hiroshi 995
Toh, Seng Oon 1117
Tokei, Zsolt 543, 549, 591, 712
Torii, K. 1001
Toriumi, Akira 299
Tortorelli, I. 615
Tosaka, Yoshiharu 218, 379
Tous, Santi 792
Toussaint, Thibaut 231
Treu, Michael 156
Trivellin, Nicola 1, 522
Truong, C. 566
Tsai, C. 293
Tsai, C. H. 1053
Tsai, C. T. 1053
Tsai, R. S. 807
Tsai, Teng-Chun 918
Tsai, Y. S. 665
Tseng, Joshua 1095
Tsuchiya, H. 717
Tsuchiya, R. 1001, 1049
Tsukamoto, Yasumasa 1117
Tsukasa, Matsuda 903
Uchibori, Chihiro J. 218
Udayakumar, K. R. 750
Uemura, Taiki 218, 694
Uznanski, Slawosz 407
Van Den Bosch, Geert 731

AUTHOR INDEX

Van Der Wel, P. J. 129
Van Dijk, K. 111
Van Houdt, Jan 731
Van Hove, Marleen 146
Vandelli, Luca 731
Vandevelde, Bart.................................712
Varghese, Dhanoop 1091
Vayshenker, A. 369
Veksler, D. 73, 532
Velamala, Jyothi B. 650
Ventrice, D..738
Venugopal, Sameer M. 644
Vereecke, B. 591
Vermunt, Frank 1014
Verzellesi, G. 532
Vialle, Nicolas 655
Vincent, E. 55
Visalli, Domenica 146
Visconti, Angelo 400, 604
Volf, P. A. J. 111
Wagner, P.-J. 16
Waltereit, P. 129
Wang, D. 312
Wang, J. 1008
Wang, L. 750
Wang, Miaomiao 1099
Wang, Mingxiang 1040
Wang, Szu-Yu 627, 951
Wang, Tahui 960
Wang, Wayne 665
Wang, Y. 566, 1044
Wang, Z. R. 964
Warren, Kevin M. 395, 1026
Watabe, Shunichi 683
Wei, J. 586
Weimann, Thomas 327
Weir, Bonnie......................................1091
Weller, Robert A. 395, 1026
Wen, Shi-Jie 1026, 1036
Wen, Yong-Ru 857
Werner, Christoph 826
Wie, C. R. 347
Wilde, Markus 417
Wilson, Christopher J.....................591, 712
Winkler, Thomas 327
Wise, Rick 881
Witulski, A. F. 198
Wojtowicz, M. 807
Wong, H.-S. Philip 99
Wong, Richard 1026, 1036
Woo, S. H. 1008
Woolery, Bruce....................................287
Wouters, Y. 922
Wrachien, N......................................334, 536
Wu, Ernest Y. 792

Wu, J. Y. 918
Wu, Kenneth 170, 182, 665
Wu, L. 964
Wu, Ming-Tsung 951
Wu, Mong Sheng 861, 870
Wu, X. 354
Wu, Y. Q. 536
Xu, C. 879
Xue, Jie 440
Yamamoto, Hirohiko 988, 995
Yamauchi, Toyohiko 874
Yanagi, I. 1001
Yang, G. 364, 566
Yang, J. W. 1008
Yang, Ling-Wu 639, 951
Yang, Tahone 627, 639, 951
Yang, Yang 846
Yasuda, Naoki 956
Yau, A. 111
Ye, P. D. 536
Yeap, G. 1008
Yeh, Chun-Chen 1099
Yeh, Teng-Hao 634
Yeo, Kyoung Hwan 94
Yeo, Yee-Chia 1055
Yeoh, Yun Young 94
Yew, K. S. 373
Yiang, Kok-Yong 117, 562
Yokogawa, S. 717
Yoo, Jae-Yoon 104
Yoon, Joo-Byoung 265
Yoshimoto, H. 1001, 1049
Yoshioka, N. 120
Young, C. D. 73, 373
Yu, Bin 80
Yu, H. Y. 964
Yu, Tianhua 80
Yuri, Masaaki 1
Zaghloul, U. 237
Zaitz, M. 566
Zanoni, Enrico 1, 522, 536
Zehnder, Ulrich 522
Zeng, X. 938
Zhang, B. C. 906
Zhang, F. 804, 906
Zhang, L. 581
Zhang, W. 938
Zhao, K. 50
Zhao, Larry..549
Zhao, Y. H. 906
Zheng, Rui 650
Zhou, C. 750
Zhou, J. P. 581
Zhou, Xing 977
Zhu, Guojun 977

AUTHOR INDEX

Zhu, Vivian 395
Zous, Nian-Kai 634
Zschech, Ehrenfried 574, 581

CURRAN ASSOCIATES INC.
proceedings
.com

9781424454303